RSMeans® Building Construction Cost Data

2015 73rd annual edition

Stephen C. Plotner, Senior Editor

Engineering Director
Bob Mewis, CCC (*1, 14, 41*)

Contributing Editors
Christopher Babbitt
Adrian C. Charest, PE (*26, 27, 28, 48*)
Gary W. Christensen
Cheryl Elsmore
David Fiske
Robert Fortier, PE (*2, 31, 32, 33, 34, 35, 44, 46*)
Robert J. Kuchta (*8*)
Thomas Lane (*6, 7*)
Robert C. McNichols
Genevieve Medeiros
Melville J. Mossman, PE (*21, 22, 23*)
Jeannene D. Murphy
Marilyn Phelan, AIA (*9, 10, 11, 12*)
Stephen C. Plotner (*3, 5*)
Siobhan Soucie (*4, 13*)

**Numbers in italics are the divisional responsibilities for each editor. Please contact the designated editor directly with any questions.*

Vice President Data & Engineering
Chris Anderson

Cover Design
Sara Rutan

Product Manager
Andrea Sillah

Production Manager
Debbie Panarelli

Production
Jill Goodman
Jonathan Forgit
Sara Rutan
Mary Lou Geary

Technical Support
Gary L. Hoitt
Kathryn S. Rodriguez

RSMeans
Construction Publishers & Consultants
700 Longwater Drive
Norwell, MA 02061
United States of America
1-877-784-5289
www.rsmeans.com

Cover photo ©
29mokara/iStock/Thinkstock

Printed in the United States of America
ISSN 0068-3531
ISBN 978-1-940238-50-0

0015 $206.95 per copy (in United States)
Price is subject to change without prior notice.

This book is printed on 50% recycled paper using soy-based ink and is entirely recyclable.

Related RSMeans Products and Services

This book is aimed primarily at commercial and industrial projects or large multi-family housing projects costing $3,500,000 and up. For civil engineering structures such as bridges, dams, highways, or the like, please refer to *RSMeans Heavy Construction Cost Data*.

The engineers at RSMeans suggest the following products and services as companion information resources to *RSMeans Building Construction Cost Data*:

Construction Cost Data Books

Assemblies Cost Data 2015
Square Foot Costs 2015

Reference Books

Estimating Building Costs
RSMeans Estimating Handbook
Green Building: Project Planning & Estimating
How to Estimate with RSMeans Data
Plan Reading & Material Takeoff
Project Scheduling & Management for Construction

Seminars and In-House Training

Unit Price Estimating
RSMeans Online® Training
Practical Project Management for Construction Professionals
Scheduling with MSProject for Construction Professionals
Mechanical & Electrical Estimating

RSMeans Online Bookstore

Visit RSMeans at **www.rsmeans.com** for a one-stop portal for the most reliable and current resources available on the market. Here you can learn about our more than 20 annual cost data books and eBooks, also available on CD and online versions, as well as our libary of reference books and RSMeans seminars.

RSMeans Electronic Data

Get the information found in RSMeans cost books electronically at **RSMeansOnline.com**.

RSMeans Business Solutions

Cost engineers and analysts conduct facility life cycle and benchmark research studies and predictive cost modeling, as well as offer consultation services for conceptual estimating and real property management. Research studies are designed with the application of proprietary cost and project data from RSMeans' extensive North American databases. Analysts offer building product manufacturers qualitative and quantitative market research, as well as time/motion studies for new products, and Web-based dashboards for market opportunity analysis. Clients are from both the public and private sectors, including federal, state, and municipal agencies; corporations; institutions; construction management firms; hospitals; and associations.

Construction Costs for Software Applications

More than 25 unit price and assemblies cost databases are available through a number of leading estimating and facilities management software partners. For more information see the "Other RSMeans Products and Services" pages at the back of this publication.

RSMeans data is also available to federal, state, and local government agencies as multi-year, multi-seat licenses.

For information on our current partners, visit: **www.rsmeans.com/partners** or call 877.784.5289.

Table of Contents

Foreword

Our Mission

Since 1942, RSMeans has been actively engaged in construction cost publishing and consulting throughout North America.

Today, more than 70 years after RSMeans began, our primary objective remains the same: to provide you, the construction and facilities professional, with the most current and comprehensive construction cost data possible.

Whether you are a contractor, owner, architect, engineer, facilities manager, or anyone else who needs a reliable construction cost estimate, you'll find this publication to be a highly useful and necessary tool.

With the constant flow of new construction methods and materials today, it's difficult to find the time to look at and evaluate all the different construction cost possibilities. In addition, because labor and material costs keep changing, last year's cost information is not a reliable basis for today's estimate or budget.

That's why so many construction professionals turn to RSMeans. We keep track of the costs for you, along with a wide range of other key information, from city cost indexes . . . to productivity rates . . . to crew composition . . . to contractor's overhead and profit rates.

RSMeans performs these functions by collecting data from all facets of the industry and organizing it in a format that is instantly accessible to you. From the preliminary budget to the detailed unit price estimate, you'll find the data in this book useful for all phases of construction cost determination.

The Staff, the Organization, and Our Services

When you purchase one of RSMeans' publications, you are, in effect, hiring the services of a full-time staff of construction and engineering professionals.

Our thoroughly experienced and highly qualified staff works daily at collecting, analyzing, and disseminating comprehensive cost information for your needs. These staff members have years of practical construction experience and engineering training prior to joining the firm. As a result, you can count on them not only for accurate cost figures, but also for additional background reference information that will help you create a realistic estimate.

The RSMeans organization is always prepared to help you solve construction problems through its variety of data solutions, including online, CD, and print book formats, as well as cost estimating expertise available via our business solutions, training, and seminars.

Besides a full array of construction cost estimating books, RSMeans also publishes a number of other reference works for the construction industry. Subjects include construction estimating and project and business management, green building, and a library of facility management references.

In addition, you can access all of our construction cost data electronically in convenient CD format or on the web. Visit **RSMeansOnline.com** for more information on our 24/7 online cost data.

What's more, you can increase your knowledge and improve your construction estimating and management performance with an RSMeans construction seminar or in-house training program. These two-day seminar programs offer unparalleled opportunities for everyone in your organization to become updated on a wide variety of construction-related issues.

RSMeans is also a worldwide provider of construction cost management and analysis services for commercial and government owners.

In short, RSMeans can provide you with the tools and expertise for constructing accurate and dependable construction estimates and budgets in a variety of ways.

Robert Snow Means Established a Tradition of Quality That Continues Today

Robert Snow Means spent years building RSMeans, making certain he always delivered a quality product.

Today, at RSMeans, we do more than talk about the quality of our data and the usefulness of our books. We stand behind all of our data, from historical cost indexes to construction materials and techniques to current costs.

If you have any questions about our products or services, please call us toll-free at 1-877-784-5289. Our customer service representatives will be happy to assist you. You can also visit our website at **www.rsmeans.com.**

How the Book Is Built: An Overview

The Construction Specifications Institute (CSI) and Construction Specifications Canada (CSC) have produced the 2014 edition of MasterFormat®, a system of titles and numbers used extensively to organize construction information.

All unit price data in the RSMeans cost data books is now arranged in the 50-division MasterFormat® 2014 system.

A Powerful Construction Tool

You have in your hands one of the most powerful construction tools available today. A successful project is built on the foundation of an accurate and dependable estimate. This book will enable you to construct just such an estimate.

For the casual user the book is designed to be:

- quickly and easily understood so you can get right to your estimate.
- filled with valuable information so you can understand the necessary factors that go into the cost estimate.

For the regular user, the book is designed to be:

- a handy desk reference that can be quickly referred to for key costs.
- a comprehensive, fully reliable source of current construction costs and productivity rates so you'll be prepared to estimate any project.
- a source book for preliminary project cost, product selections, and alternate materials and methods.

To meet all of these requirements, we have organized the book into the following clearly defined sections.

Quick Start

See our "Quick Start" instructions on the following page to get started right away.

Estimating with RSMeans Unit Price Cost Data

Please refer to these steps for guidance on completing an estimate using RSMeans unit price cost data.

How to Use the Book: The Details

This section contains an in-depth explanation of how the book is arranged . . . and how you can use it to determine a reliable construction cost estimate. It includes information about how we develop our cost figures and how to completely prepare your estimate.

Unit Price Section

All cost data has been divided into the 50 divisions according to the MasterFormat system of classification and numbering. For a listing of these divisions and an outline of their subdivisions, see the Unit Price Section Table of Contents.

Estimating tips are included at the beginning of each division.

Reference Section

This section includes information on Equipment Rental Costs, Crew Listings, Historical Cost Indexes, City Cost Indexes, Location Factors, Reference Tables, Change Orders, Square Foot Costs, and a listing of Abbreviations.

Equipment Rental Costs: This section contains the average costs to rent and operate hundreds of pieces of construction equipment.

Crew Listings: This section lists all of the crews referenced in the book. For the purposes of this book, a crew is composed of more than one trade classification and/or the addition of power equipment to any trade classification. Power equipment is included in the cost of the crew. Costs are shown both with bare labor rates and with the installing contractor's overhead and profit added. For each, the total crew cost per eight-hour day and the composite cost per labor-hour are listed.

Historical Cost Indexes: These indexes provide you with data to adjust construction costs over time.

City Cost Indexes: All costs in this book are U.S. national averages. Costs vary because of the regional economy. You can adjust costs by CSI Division to over 700 locations throughout the U.S. and Canada by using the data in this section.

Location Factors: You can adjust total project costs to over 900 locations throughout the U.S. and Canada by using the data in this section.

Reference Tables: At the beginning of selected major classifications in the Unit Price Section are reference numbers shown in a shaded box. These numbers refer you to related information in the Reference Section. In this section, you'll find reference tables, explanations, estimating information that support how we develop the unit price data, technical data, and estimating procedures.

Change Orders: This section includes information on the factors that influence the pricing of change orders.

Square Foot Costs: This section contains costs for 59 different building types that allow you to make a rough estimate for the overall cost of a project or its major components.

Abbreviations: A listing of abbreviations used throughout this book, along with the terms they represent, is included in this section.

Index

A comprehensive listing of all terms and subjects in this book will help you quickly find what you need when you are not sure where it occurs in MasterFormat.

The Scope of This Book

This book is designed to be as comprehensive and as easy to use as possible. To that end we have made certain assumptions and limited its scope in two key ways:

1. We have established material prices based on a national average.
2. We have computed labor costs based on a 30-city national average of union wage rates.

For a more detailed explanation of how the cost data is developed, see "How To Use the Book: The Details."

Project Size/Type

The material prices in RSMeans cost data books are "contractor's prices." They are the prices that contractors can expect to pay at the lumberyards, suppliers/distributers warehouses, etc. Small orders of speciality items would be higher than the costs shown, while very large orders, such as truckload lots, would be less. The variation would depend on the size, timing, and negotiating power of the contractor. The labor costs are primarily for new construction or major renovation rather than repairs or minor alterations.
With reasonable exercise of judgment, the figures can be used for any building work.

Absolute Essentials for a Quick Start

If you feel you are ready to use this book and don't think you will need the detailed instructions that begin on the following page, this Absolute Essentials for a Quick Start page is for you. These steps will allow you to get started estimating in a matter of minutes.

1 Scope
Think through the project that you will be estimating, and identify the many individual work tasks that will need to be covered in your estimate.

2 Quantify
Determine the number of units that will be required for each work task that you identified.

3 Pricing
Locate individual Unit Price line items that match the work tasks you identified. The Unit Price Section Table of Contents that begins on page 1 and the Index in the back of the book will help you find these line items.

4 Multiply
Multiply the Total Incl O&P cost for a Unit Price line item in the book by your quantity for that item. The price you calculate will be an estimate for a completed item of work performed by a subcontractor. Keep adding line items in this manner to build your estimate.

5 Project Overhead
Include project overhead items in your estimate. These items are needed to make the job run and are typically, but not always, provided by the General Contractor. They can be found in Division 1. An alternate method of estimating project overhead costs is to apply a percentage of the total project cost.

Include rented tools not included in crews, waste, rubbish handling, and cleanup.

6 Estimate Summary
Include General Contractor's markup on subcontractors, General Contractor's office overhead and profit, and sales tax on materials and equipment.

Adjust your estimate to the project's location by using the City Cost Indexes or Location Factors found in the Reference Section.

Editors' Note: We urge you to spend time reading and understanding the supporting material in the front of this book. An accurate estimate requires experience, knowledge, and careful calculation. The more you know about how we at RSMeans developed the data, the more accurate your estimate will be. In addition, it is important to take into consideration the reference material in the back of the book such as Equipment Listings, Crew Listings, City Cost Indexes, Location Factors, and Reference Tables.

Estimating with RSMeans Unit Price Cost Data

Following these steps will allow you to complete an accurate estimate using RSMeans Unit Price cost data.

1 Scope Out the Project

- Think through the project and identify those CSI Divisions needed in your estimate.
- Identify the individual work tasks that will need to be covered in your estimate.
- The Unit Price data in this book has been divided into 50 Divisions according to the CSI MasterFormat® 2014—their titles are listed on the back cover of your book.
- The Unit Price Section Table of Contents on page 1 may also be helpful when scoping out your project.
- Experienced estimators find it helpful to begin with Division 2 and continue through completion. Division 1 can be estimated after the full project scope is known.

2 Quantify

- Determine the number of units required for each work task that you identified.
- Experienced estimators include an allowance for waste in their quantities. (Waste is not included in RSMeans Unit Price line items unless so stated.)

3 Price the Quantities

- Use the Unit Price Table of Contents, and the Index, to locate individual Unit Price line items for your estimate.
- Reference Numbers indicated within a Unit Price section refer to additional information that you may find useful.
- The crew indicates who is performing the work for that task. Crew codes are expanded in the Crew Listings in the Reference Section to include all trades and equipment that comprise the crew.
- The Daily Output is the amount of work the crew is expected to do in one day.
- The Labor-Hours value is the amount of time it will take for the crew to install one unit of work.
- The abbreviated Unit designation indicates the unit of measure upon which the crew, productivity, and prices are based.
- Bare Costs are shown for materials, labor, and equipment needed to complete the Unit Price line item. Bare costs do not include waste, project overhead, payroll insurance, payroll taxes, main office overhead, or profit.
- The Total Incl O&P cost is the billing rate or invoice amount of the installing contractor or subcontractor who performs the work for the Unit Price line item.

4 Multiply

- Multiply the total number of units needed for your project by the Total Incl O&P cost for each Unit Price line item.
- Be careful that your take off unit of measure matches the unit of measure in the Unit column.
- The price you calculate is an estimate for a completed item of work.
- Keep scoping individual tasks, determining the number of units required for those tasks, matching each task with individual Unit Price line items in the book, and multiply quantities by Total Incl O&P costs.
- An estimate completed in this manner is priced as if a subcontractor, or set of subcontractors, is performing the work. The estimate does not yet include Project Overhead or Estimate Summary components such as General Contractor markups on subcontracted work, General Contractor office overhead and profit, contingency, and location factor.

5 Project Overhead

- Include project overhead items from Division 1 – General Requirements.
- These items are needed to make the job run. They are typically, but not always, provided by the General Contractor. Items include, but are not limited to, field personnel, insurance, performance bond, permits, testing, temporary utilities, field office and storage facilities, temporary scaffolding and platforms, equipment mobilization and demobilization, temporary roads and sidewalks, winter protection, temporary barricades and fencing, temporary security, temporary signs, field engineering and layout, final cleaning and commissioning.
- Each item should be quantified, and matched to individual Unit Price line items in Division 1, and then priced and added to your estimate.
- An alternate method of estimating project overhead costs is to apply a percentage of the total project cost, usually 5% to 15% with an average of 10% (see General Conditions, page ix).
- Include other project related expenses in your estimate such as:
  - Rented equipment not itemized in the Crew Listings
  - Rubbish handling throughout the project (see 02 41 19.19)

6 Estimate Summary

- Include sales tax as required by laws of your state or county.
- Include the General Contractor's markup on self-performed work, usually 5% to 15% with an average of 10%.
- Include the General Contractor's markup on subcontracted work, usually 5% to 15% with an average of 10%.
- Include General Contractor's main office overhead and profit:
  - RSMeans gives general guidelines on the General Contractor's main office overhead (see section 01 31 13.60 and Reference Number R013113-50).
  - RSMeans gives no guidance on the General Contractor's profit.
  - Markups will depend on the size of the General Contractor's operations, his projected annual revenue, the level of risk he is taking on, and on the level of competition in the local area and for this project in particular.
- Include a contingency, usually 3% to 5%, if appropriate.
- Adjust your estimate to the project's location by using the City Cost Indexes or the Location Factors in the Reference Section:
  - Look at the rules on the pages for "How to Use the City Cost Indexes" to see how to apply the Indexes for your location.
  - When the proper Index or Factor has been identified for the project's location, convert it to a multiplier by dividing it by 100, and then multiply that multiplier by your estimated total cost. The original estimated total cost will now be adjusted up or down from the national average to a total that is appropriate for your location.

Editors' Notes:
We urge you to spend time reading and understanding the supporting material in the front of this book. An accurate estimate requires experience, knowledge, and careful calculation. The more you know about how we at RSMeans developed the data, the more accurate your estimate will be. In addition, it is important to take into consideration the reference material in the back of the book such as Equipment Listings, Crew Listings, City Cost Indexes, Location Factors, and Reference Tables.

How to Use the Book: The Details

What's Behind the Numbers? The Development of Cost Data

The staff at RSMeans continually monitors developments in the construction industry in order to ensure reliable, thorough, and up-to-date cost information. While overall construction costs may vary relative to general economic conditions, price fluctuations within the industry are dependent upon many factors. Individual price variations may, in fact, be opposite to overall economic trends. Therefore, costs are constantly tracked and complete updates are published yearly. Also, new items are frequently added in response to changes in materials and methods.

Costs—$ (U.S.)

All costs represent U.S. national averages and are given in U.S. dollars. The RSMeans City Cost Indexes can be used to adjust costs to a particular location. The City Cost Indexes for Canada can be used to adjust U.S. national averages to local costs in Canadian dollars. No exchange rate conversion is necessary.

G The processes or products identified by the green symbol in our publications have been determined to be environmentally responsible and/or resource-efficient solely by the RSMeans engineering staff. The inclusion of the green symbol does not represent compliance with any specific industry association or standard.

Material Costs

The RSMeans staff contacts manufacturers, dealers, distributors, and contractors all across the U.S. and Canada to determine national average material costs. If you have access to current material costs for your specific location, you may wish to make adjustments to reflect differences from the national average. Included within material costs are fasteners for a normal installation. RSMeans engineers use manufacturers' recommendations, written specifications, and/or standard construction practice for size and spacing of fasteners. Adjustments to material costs may be required for your specific application or location. The manufacturer's warranty is assumed. Extended warranties are not included in the material costs. Material costs do not include sales tax.

Labor Costs

Labor costs are based on the average of wage rates from 30 major U.S. cities. Rates are determined from labor union agreements or prevailing wages for construction trades for the current year. Rates, along with overhead and profit markups, are listed on the inside back cover of this book.

- If wage rates in your area vary from those used in this book, or if rate increases are expected within a given year, labor costs should be adjusted accordingly.

Labor costs reflect productivity based on actual working conditions. In addition to actual installation, these figures include time spent during a normal weekday on tasks such as, material receiving and handling, mobilization at site, site movement, breaks, and cleanup.

Productivity data is developed over an extended period so as not to be influenced by abnormal variations and reflects a typical average.

Equipment Costs

Equipment costs include not only rental, but also operating costs for equipment under normal use. The operating costs include parts and labor for routine servicing such as repair and replacement of pumps, filters, and worn lines. Normal operating expendables, such as fuel, lubricants, tires, and electricity (where applicable), are also included. Extraordinary operating expendables with highly variable wear patterns, such as diamond bits and blades, are excluded. These costs are included under materials. Equipment rental rates are obtained from industry sources throughout North America—contractors, suppliers, dealers, manufacturers, and distributors.

Rental rates can also be treated as reimbursement costs for contractor-owned equipment. Owned equipment costs include depreciation, loan payments, interest, taxes, insurance, storage and major repairs.

Equipment costs do not include operators' wages; nor do they include the cost to move equipment to a job site (mobilization) or from a job site (demobilization).

Equipment Cost/Day—The cost of power equipment required for each crew is included in the Crew Listings in the Reference Section (small tools that are considered as essential everyday tools are not listed out separately). The Crew Listings itemize specialized tools and heavy equipment along with labor trades. The daily cost of itemized equipment included in a crew is based on dividing the weekly bare rental rate by 5 (number of working days per week) and then adding the hourly operating cost times 8 (the number of hours per day). This Equipment Cost/Day is shown in the last column of the Equipment Rental Cost pages in the Reference Section.

Mobilization/Demobilization—The cost to move construction equipment from an equipment yard or rental company to the job site and back again is not included in equipment costs. Mobilization (to the site) and demobilization (from the site) costs can be found in the Unit Price Section. If a piece of equipment is already at the job site, it is not appropriate to utilize mob./demob. costs again in an estimate.

Overhead and Profit

Total Cost including O&P for the *Installing Contractor* is shown in the last column on the Unit Price and/or the Assemblies pages of this book. This figure is the sum of the bare material cost plus 10% for profit, the bare labor cost plus total overhead and profit, and the bare equipment cost plus 10% for profit. Details for the calculation of Overhead and Profit on labor are shown on the inside back cover and in the Reference Section of this book. (See "How RSMeans Unit Price Data Works" for an example of this calculation.)

General Conditions

Cost data in this book is presented in two ways: Bare Costs and Total Cost including O&P (Overhead and Profit). General Conditions, or General Requirements, of the contract should also be added to the Total Cost including O&P when applicable. Costs for General Conditions are listed in Division 1 of the Unit Price Section and the Reference

Section of this book. General Conditions for the *Installing Contractor* may range from 0% to 10% of the Total Cost including O&P. For the *General* or *Prime Contractor*, costs for General Conditions may range from 5% to 15% of the Total Cost including O&P, with a figure of 10% as the most typical allowance. If applicable, the Assemblies and Models sections of this book use costs that include the installing contractor's overhead and profit (O&P).

Factors Affecting Costs

Costs can vary depending upon a number of variables. Here's how we have handled the main factors affecting costs.

Quality—The prices for materials and the workmanship upon which productivity is based represent sound construction work. They are also in line with U.S. government specifications.

Overtime—We have made no allowance for overtime. If you anticipate premium time or work beyond normal working hours, be sure to make an appropriate adjustment to your labor costs.

Productivity—The productivity, daily output, and labor-hour figures for each line item are based on working an eight-hour day in daylight hours in moderate temperatures. For work that extends beyond normal work hours or is performed under adverse conditions, productivity may decrease. (See "How RSMeans Unit Price Data Works" for more on productivity.)

Size of Project—The size, scope of work, and type of construction project will have a significant impact on cost. Economies of scale can reduce costs for large projects. Unit costs can often run higher for small projects.

Location—Material prices in this book are for metropolitan areas. However, in dense urban areas, traffic and site storage limitations may increase costs. Beyond a 20-mile radius of large cities, extra trucking or transportation charges may also increase the material costs slightly. On the other hand, lower wage rates may be in effect. Be sure to consider both of these factors when preparing an estimate, particularly if the job site is located in a central city or remote rural location. In addition, highly specialized subcontract items may require travel and per-diem expenses for mechanics.

Other Factors——

- season of year
- contractor management
- weather conditions
- local union restrictions
- building code requirements
- availability of:
  - adequate energy
  - skilled labor
  - building materials
- owner's special requirements/restrictions
- safety requirements
- environmental considerations

Unpredictable Factors—General business conditions influence "in-place" costs of all items. Substitute materials and construction methods may have to be employed. These may affect the installed cost and/or life cycle costs. Such factors may be difficult to evaluate and cannot necessarily be predicted on the basis of the job's location in a particular section of the country. Thus, where these factors apply, you may find significant but unavoidable cost variations for which you will have to apply a measure of judgment to your estimate.

Rounding of Costs

In general, all unit prices in excess of $5.00 have been rounded to make them easier to use and still maintain adequate precision of the results. The rounding rules we have chosen are in the following table.

Prices from...	Rounded to the nearest...
$.01 to $5.00	$.01
$5.01 to $20.00	$.05
$20.01 to $100.00	$.50
$100.01 to $300.00	$1.00
$300.01 to $1,000.00	$5.00
$1,000.01 to $10,000.00	$25.00
$10,000.01 to $50,000.00	$100.00
$50,000.01 and above	$500.00

How Subcontracted Items Affect Costs

A considerable portion of all large construction jobs is usually subcontracted. In fact, the percentage done by subcontractors is constantly increasing and may run over 90%. Since the workers employed by these companies do nothing else but install their particular product, they soon become expert in that line. The result is, installation by these firms is accomplished so efficiently that the total in-place cost, even adding the general contractor's overhead and profit, is no more, and often less, than if the principal contractor had handled the installation himself/herself. Companies that deal with construction specialties are anxious to have their product perform well and, consequently, the installation will be the best possible.

Contingencies

The allowance for contingencies generally provides for unforeseen construction difficulties. On alterations or repair jobs, 20% is not too much. If drawings are final and only field contingencies are being considered, 2% or 3% is probably sufficient, and often nothing need be added. Contractually, changes in plans will be covered by extras. The contractor should consider inflationary price trends and possible material shortages during the course of the job. These escalation factors are dependent upon both economic conditions and the anticipated time between the estimate and actual construction. If drawings are not complete or approved, or a budget cost is wanted, it is wise to add 5% to 10%. Contingencies, then, are a matter of judgment. Additional allowances for contingencies are shown in Division 1.

Important Estimating Considerations

The productivity or daily output of each craftsman or crew assumes a well-managed job where tradesmen with the proper tools and equipment, along with the appropriate construction materials, are present. Included are daily mobilization and cleanup time, break time and plan layout time. Unless otherwise indicated, time for material movement on site (for items that can be transported by hand) of up to 200' into the building and to the first or second floor is also included. If material has to be transported by other means, over greater distances, or to higher floors, an additional allowance should be considered by the estimator.

While horizontal movement is typically a sole function of distances, vertical transport introduces other variables that can significantly impact productivity. In an occupied building, the use of elevators (assuming access, size, and required protective measures are acceptable) must be understood at the time of the estimate. For new construction, hoist wait and cycle times can easily be 15 minutes and may result in scheduled access extending beyond the normal work day. Finally, all vertical transport will impose strict weight limits likely to preclude the use of any motorized material handling.

The productivity, or daily output, also assumes installation that meets manufacturer/designer/ standard specifications. A time allowance for quality control checks, minor adjustments and any task required to ensure the proper function or operation is also included. For items that require connections to services, time is included for positioning, leveling, securing the unit and for making all the necessary connections (and start up where applicable) ensuring a complete installation. Estimating of the services themselves (electrical, plumbing, water, steam, hydraulics, dust collection, etc.) is separate.

In some cases, the estimator must consider the use of a crane and an appropriate crew for the installation of large or heavy items. For those situations where a crane is not included in the assigned crew and as part of the line item cost, equipment rental costs, mobilization and demobilization costs, and operator and support personnel costs must be considered.

Labor-Hours

The labor-hours expressed in this publication are derived by dividing the total daily labor hours for the crew by the daily output. Based on average installation time and the assumptions listed above, the labor-hours include: direct labor, indirect labor, and nonproductive time. A typical day for a craftsman might include, but not be limited to:

- Direct Work
  - Measuring and layout
  - Preparing materials
  - Actual installation
  - Quality assurance/quality control
- Indirect Work
  - Reading plans or specifications
  - Preparing space
  - Receiving materials
  - Material movement
  - Giving or receiving instruction
  - Miscellaneous
- Non-Work
  - Chatting
  - Personal issues
  - Breaks
  - Interruptions (i.e., sickness, weather, material or equipment shortages, etc.)

If any of the items for a typical day do not apply to the particular work or project situation, the estimator should make any necessary adjustments.

Final Checklist

Estimating can be a straightforward process provided you remember the basics. Here's a checklist of some of the steps you should remember to complete before finalizing your estimate.

Did you remember to . . .

- factor in the City Cost Index for your locale?
- take into consideration which items have been marked up and by how much?
- mark up the entire estimate sufficiently for your purposes?
- read the background information on techniques and technical matters that could impact your project time span and cost?
- include all components of your project in the final estimate?
- double check your figures for accuracy?
- call RSMeans if you have any questions about your estimate or the data you've found in our publications? Remember, RSMeans stands behind its publications. If you have any questions about your estimate . . . about the costs you've used from our books . . . or even about the technical aspects of the job that may affect your estimate, feel free to call the RSMeans editors at 1-877-784-5289.

Free Quarterly Updates

Stay up-to-date throughout 2015 with RSMeans' free cost data updates four times a year. Sign up online to make sure you have access to the newest data. Every quarter we provide the city cost adjustment factors for hundreds of cities and key materials. Register at: **http://info.thegordiangroup.com/RSMeans.html**.

Unit Price Section

Table of Contents

Table of Contents (cont.)

Table of Contents (cont.)

How RSMeans Unit Price Data Works

All RSMeans unit price data is organized in the same way.

It is important to understand the structure, so that you can find information easily and use it correctly.

RSMeans **Line Numbers** consist of 12 characters, which identify a unique location in the database for each task. The first 6 or 8 digits conform to the Construction Specifications Institute MasterFormat® 2014. The remainder of the digits are a further breakdown by RSMeans in order to arrange items in understandable groups of similar tasks. Line numbers are consistent across all RSMeans publications, so a line number in any RSMeans product will always refer to the same unit of work.

RSMeans engineers have created **reference** information to assist you in your estimate. If there is information that applies to a section, it will be indicated at the start of the section. In this case, R033105-10 provides information on the proportionate quantities of formwork, reinforcing, and concrete used in cast-in-place concrete items such as footings, slabs, beams, and columns. The Reference Section is located in the back of the book on the pages with a gray edge.

RSMeans **Descriptions** are shown in a hierarchical structure to make them readable. In order to read a complete description, read up through the indents to the top of the section. Include everything that is above and to the left that is not contradicted by information below. For instance, the complete description for line 03 30 53.40 3550 is "Concrete in place, including forms (4 uses), Grade 60 rebar, concrete (Portland cement Type 1), placement and finishing unless otherwise indicated; Equipment pad (3000 psi), 4' x 4' x 6" thick."

When using **RSMeans data**, it is important to read through an entire section to ensure that you use the data that most closely matches your work. Note that sometimes there is additional information shown in the section that may improve your price. There are frequently lines that further describe, add to, or adjust data for specific situations.

03 30 Cast-In-Place Concrete

03 30 53 – Miscellaneous Cast-In-Place Concrete

03 30 53.40 Concrete In Place

0010	**CONCRETE IN PLACE**	R033105-10
0020	Including forms (4 uses), Grade 60 rebar, concrete (Portland cement	R033105-20
0050	Type I), placement and finishing unless otherwise indicated	R033105-50
0300	Beams (3500 psi), 5 kip per L.F., 10′ span	R033105-65
0350	25′ span	R033105-70
0500	Chimney foundations (5000 psi), over 5 C.Y.	R033105-85
0510	(3500 psi), under 5 C.Y.	
0700	Columns, square (4000 psi), 12″ x 12″, less than 2% reinforcing	
3450	Over 10,000 S.F.	
3500	Add per floor for 3 to 6 stories high	
3520	For 7 to 20 stories high	
3540	Equipment pad (3000 psi), 3′ x 3′ x 6″ thick	
3550	4′ x 4′ x 6″ thick	
3560	5′ x 5′ x 8″ thick	
3570	6′ x 6′ x 8″ thick	

The data published in RSMeans print books represents a "national average" cost. This data should be modified to the project location using the **City Cost Indexes** or **Location Factors** tables found in the Reference Section (see pages 748–796). Use the location factors to adjust estimate totals if the project covers multiple trades. Use the city cost indexes (CCI) for single trade projects or projects where a more detailed analysis is required. All figures in the two tables are derived from the same research. The last row of data in the CCI, the weighted average, is the same as the numbers reported for each location in the location factor table.

Crews include labor or labor and equipment necessary to accomplish each task. In this case, Crew C-14H is used. RSMeans selects a crew to represent the workers and equipment that are typically used for that task. In this case, Crew C-14H consists of one carpenter foreman (outside), two carpenters, one rodman, one laborer, one cement finisher, and one gas engine vibrator. Details of all crews can be found in the reference section.

Crews

Crew No.	Bare Costs		Incl. Subs O&P		Cost Per Labor-Hour	
Crew C-14H	**Hr.**	**Daily**	**Hr.**	**Daily**	**Bare Costs**	**Incl. O&P**
1 Carpenter Foreman (outside)	$48.95	$391.60	$75.35	$602.80	$46.33	$71.08
2 Carpenters	46.95	751.20	72.25	1156.00		
1 Rodman (reinf.)	52.55	420.40	82.10	656.80		
1 Laborer	37.60	300.80	57.85	462.80		
1 Cement Finisher	45.00	360.00	66.65	533.20		
1 Gas Engine Vibrator		31.20		34.32	.65	.71
48 L.H., Daily Totals		$2255.20		$3445.92	$46.98	$71.79

The **Daily Output** is the amount of work that the crew can do in a normal 8-hour workday, including mobilization, layout, movement of materials, and cleanup. In this case, crew C-14H can install thirty 4' x 4' x 6" thick concrete pads in a day. Daily output is variable, based on many factors, including the size of the job, location, and environmental conditions. RSMeans data represents work done in daylight (or adequate lighting) and temperate conditions.

Bare Costs are the costs of materials, labor, and equipment that the installing contractor pays. They represent the cost, in U.S. dollars, for one unit of work. They do not include any markups for profit or labor burden.

Crew	Daily Output	Labor-Hours	Unit	Material	2015 Bare Costs Labor	Equipment	Total	Total Incl O&P
C-14A	15.62	12.804	C.Y.	320	605	48	973	1,325
"	18.55	10.782		335	510	40.50	885.50	1,200
C-14C	32.22	3.476		150	157	.97	307.97	405
"	23.71	4.724		177	213	1.32	391.32	525
C-14A	11.96	16.722		365	790	63	1,218	1,700
	2200	.025		.94	1.07	.33	2.34	3.02
	31800	.002			.07	.02	.09	.13
	21200	.003			.11	.03	.14	.21
C-14H	45	1.067	Ea.	47	49.50	.69	97.19	128
	30	1.600		69.50	74	1.04	144.54	191
	18	2.667		122	124	1.73	247.73	325
	14	3.429		164	159	2.23	325.23	425

The **Total Incl O&P column** is the total cost, including overhead and profit, that the installing contractor will charge the customer. This represents the cost of materials plus 10% profit, the cost of labor plus labor burden and 10% profit, and the cost of equipment plus 10% profit. It does not include the general contractor's overhead and profit. Note: See the inside back cover for details of how RSMeans calculates labor burden.

The **Total column** represents the total bare cost for the installing contractor, in U.S. dollars. In this case, the sum of $69.50 for material + $74.00 for labor + $1.04 for equipment is $144.54.

The figure in the **Labor Hours** column is the amount of labor required to perform one unit of work—in this case the amount of labor required to construct one 4' x 4' equipment pad. This figure is calculated by dividing the number of hours of labor in the crew by the daily output (48 labor hours divided by 30 pads = 1.6 hours of labor per pad). Multiply 1.600 times 60 to see the value in minutes: 60 x 1.6 = 96 minutes. Note: the labor hour figure is not dependent on the crew size. A change in crew size will result in a corresponding change in daily output, but the labor hours per unit of work will not change.

All RSMeans unit cost data includes the typical **Unit of Measure** used for estimating that item. For concrete-in-place the typical unit is cubic yards (C.Y.) or each (Ea.). For installing broadloom carpet it is square yard, and for gypsum board it is square foot. The estimator needs to take special care that the unit in the data matches the unit in the take-off. Unit conversions may be found in the Reference Section.

How RSMeans Unit Price Data Works (Continued)

Sample Estimate

This sample demonstrates the elements of an estimate, including a tally of the RSMeans data lines, and a summary of the markups on a contractor's work to arrive at a total cost to the owner. The RSMeans Location Factor is added at the bottom of the estimate to adjust the cost of the work to a specific location.

Work Performed: The body of the estimate shows the RSMeans data selected, including line number, a brief description of each item, its take-off unit and quantity, and the bare costs of materials, labor, and equipment. This estimate also includes a column titled "SubContract." This data is taken from the RSMeans column "Total Incl O&P," and represents the total that a subcontractor would charge a general contractor for the work, including the sub's markup for overhead and profit.

Division 1, General Requirements: This is the first division numerically, but the last division estimated. Division 1 includes project-wide needs provided by the general contractor. These requirements vary by project, but may include temporary facilities and utilities, security, testing, project cleanup, etc. For small projects a percentage can be used, typically between 5% and 15% of project cost. For large projects the costs may be itemized and priced individually.

Bonds: Bond costs should be added to the estimate. The figures here represent a typical performance bond, ensuring the owner that if the general contractor does not complete the obligations in the construction contract the bonding company will pay the cost for completion of the work.

Location Adjustment: RSMeans published data is based on national average costs. If necessary, adjust the total cost of the project using a location factor from the "Location Factor" table or the "City Cost Index" table. Use location factors if the work is general, covering multiple trades. If the work is by a single trade (e.g., masonry) use the more specific data found in the "City Cost Indexes."

Project Name: Pre-Engineered Steel Building		Architect: As Shown		
Location:	**Anywhere, USA**			
Line Number	**Description**	**Qty**	**Unit**	**Material**
03 30 53.40 3940	Strip footing, 12" x 24", reinforced	34	C.Y.	$4,726.00
03 30 53.40 3950	Strip footing, 12" x 36", reinforced	15	C.Y.	$1,995.00
03 11 13.65 3000	Concrete slab edge forms	500	L.F.	$175.00
03 22 11.10 0200	Welded wire fabric reinforcing	150	C.S.F.	$2,580.00
03 31 13.35 0300	Ready mix concrete, 4000 psi for slab on grade	278	C.Y.	$29,746.00
03 31 13.70 4300	Place, strike off & consolidate concrete slab	278	C.Y.	$0.00
03 35 13.30 0250	Machine float & trowel concrete slab	15,000	S.F.	$0.00
03 15 16.20 0140	Cut control joints in concrete slab	950	L.F.	$57.00
03 39 23.13 0300	Sprayed concrete curing membrane	150	C.S.F.	$1,657.50
Division 03	**Subtotal**			**$40,936.50**
08 36 13.10 2650	Manual 10' x 10' steel sectional overhead door	8	Ea.	$8,800.00
08 36 13.10 2860	Insulation and steel back panel for OH door	800	S.F.	$3,800.00
Division 08	**Subtotal**			**$12,600.00**
13 34 19.50 1100	Pre-Engineered Steel Building, 100' x 150' x 24'	15,000	SF Flr.	$0.00
13 34 19.50 6050	Framing for PESB door opening, 3' x 7'	4	Opng.	$0.00
13 34 19.50 6100	Framing for PESB door opening, 10' x 10'	8	Opng.	$0.00
13 34 19.50 6200	Framing for PESB window opening, 4' x 3'	6	Opng.	$0.00
13 34 19.50 5750	PESB door, 3' x 7', single leaf	4	Opng.	$2,340.00
13 34 19.50 7750	PESB sliding window, 4' x 3' with screen	6	Opng.	$2,280.00
13 34 19.50 6550	PESB gutter, eave type, 26 ga., painted	300	L.F.	$2,085.00
13 34 19.50 8650	PESB roof vent, 12" wide x 10' long	15	Ea.	$540.00
13 34 19.50 6900	PESB insulation, vinyl faced, 4" thick	27,400	S.F.	$11,508.00
Division 13	**Subtotal**			**$18,753.00**
		Subtotal		**$72,289.50**
Division 01	**General Requirements @ 7%**			**5,060.27**
		Estimate Subtotal		**$77,349.77**
		Sales Tax @ 5%		**3,867.49**
		Subtotal A		**81,217.25**
		GC O & P		**8,121.73**
		Subtotal B		**89,338.98**
		Contingency @ 5%		
		Subtotal C		
		Bond @ $12/1000 +10% O&P		
		Subtotal D		
		Location Adjustment Factor		
		Grand Total		

This estimate is based on an interactive spreadsheet. A copy of this spreadsheet is located on the RSMeans website at **http:/www.reedconstructiondata.com/rsmeans/extras/546011.** You are free to download it and adjust it to your methodology.

This example shows the cost to construct a pre-engineered steel building. The foundation, doors, windows, and insulation will be installed by the general contractor. A subcontractor will install the structural steel, roofing, and siding.

		01/01/15	STD	
Labor	Equipment	SubContract	Estimate Total	
$3,570.00	$22.10	$0.00		
$1,260.00	$7.80	$0.00		
$1,190.00	$0.00	$0.00		
$4,050.00	$0.00	$0.00		
$0.00	$0.00	$0.00		
$4,753.80	$158.46	$0.00		
$9,000.00	$450.00	$0.00		
$380.00	$85.50	$0.00		
$952.50	$0.00	$0.00		
$25,156.30	**$723.86**	**$0.00**	**$66,816.66**	**Division 03**
$3,320.00	$0.00	$0.00		
$0.00	$0.00	$0.00		
$3,320.00	**$0.00**	**$0.00**	**$15,920.00**	**Division 08**
$0.00	$0.00	$352,500.00		
$0.00	$0.00	$2,280.00		
$0.00	$0.00	$9,200.00		
$0.00	$0.00	$3,330.00		
$672.00	$0.00	$0.00		
$576.00	$67.20	$0.00		
$789.00	$0.00	$0.00		
$3,165.00	$0.00	$0.00		
$9,042.00	$0.00	$0.00		
14,244.00	**$67.20**	**$367,310.00**	**$400,374.20**	**Division 13**
$42,720.30	**$791.06**	**$367,310.00**	**$483,110.86**	**Subtotal**
2,990.42	**55.37**	**25,711.70**		**Gen. Requirements**
$45,710.72	**$846.43**	**$393,021.70**	**$483,110.86**	**Estimate Subtotal**
	42.32	**9,825.54**		**Sales tax**
45,710.72	**888.76**	**402,847.24**		**Subtotal**
24,912.34	**88.88**	**40,284.72**		**GC O & P**
70,623.06	**977.63**	**443,131.97**	**$604,071.64**	**Subtotal**
			30,203.58	**Contingency**
			$634,275.22	**Subtotal**
			8,372.43	**Bond**
			$642,647.66	**Subtotal**
102.30			**14,780.90**	**Location Adjustment**
			$657,428.55	**Grand Total**

Sales Tax: If the work is subject to state or local sales taxes, the amount must be added to the estimate. Sales tax may be added to material costs, equipment costs, and subcontracted work. In this case, sales tax was added in all three categories. It was assumed that approximately half the subcontracted work would be material cost, so the tax was applied to 50% of the subcontract total.

GC O&P: This entry represents the general contractor's markup on material, labor, equipment, and subcontractor costs. RSMeans' standard markup on materials, equipment, and subcontracted work is 10%. In this estimate, the markup on the labor performed by the GC's workers uses "Skilled Workers Average" shown in Column F on the table "Installing Contractor's Overhead & Profit," which can be found on the inside-back cover of the book.

Contingency: A factor for contingency may be added to any estimate to represent the cost of unknowns that may occur between the time that the estimate is performed and the time the project is constructed. The amount of the allowance will depend on the stage of design at which the estimate is done, and the contractor's assessment of the risk involved. Refer to section 01 21 16.50 for contigency allowances.

Estimating Tips

01 20 00 Price and Payment Procedures

- Allowances that should be added to estimates to cover contingencies and job conditions that are not included in the national average material and labor costs are shown in section 01 21.
- When estimating historic preservation projects (depending on the condition of the existing structure and the owner's requirements), a 15%–20% contingency or allowance is recommended, regardless of the stage of the drawings.

01 30 00 Administrative Requirements

- Before determining a final cost estimate, it is a good practice to review all the items listed in Subdivisions 01 31 and 01 32 to make final adjustments for items that may need customizing to specific job conditions.
- Requirements for initial and periodic submittals can represent a significant cost to the General Requirements of a job. Thoroughly check the submittal specifications when estimating a project to determine any costs that should be included.

01 40 00 Quality Requirements

- All projects will require some degree of Quality Control. This cost is not included in the unit cost of construction listed in each division. Depending upon the terms of the contract, the various costs of inspection and testing can be the responsibility of either the owner or the contractor. Be sure to include the required costs in your estimate.

01 50 00 Temporary Facilities and Controls

- Barricades, access roads, safety nets, scaffolding, security, and many more requirements for the execution of a safe project are elements of direct cost. These costs can easily be overlooked when preparing an estimate. When looking through the major classifications of this subdivision, determine which items apply to each division in your estimate.
- Construction Equipment Rental Costs can be found in the Reference Section in section 01 54 33. Operators' wages are not included in equipment rental costs.
- Equipment mobilization and demobilization costs are not included in equipment rental costs and must be considered separately in section 01 54 36.50.
- The cost of small tools provided by the installing contractor for his workers is covered in the "Overhead" column on the "Installing Contractor's Overhead and Profit" table that lists labor trades, base rates and markups and, therefore, is included in the "Total Incl. O&P" cost of any Unit Price line item.

01 70 00 Execution and Closeout Requirements

- When preparing an estimate, thoroughly read the specifications to determine the requirements for Contract Closeout. Final cleaning, record documentation, operation and maintenance data, warranties and bonds, and spare parts and maintenance materials can all be elements of cost for the completion of a contract. Do not overlook these in your estimate.

Reference Numbers

Reference numbers are shown in shaded boxes at the beginning of some major classifications. These numbers refer to related items in the Reference Section. The reference information may be an estimating procedure, an alternate pricing method, or technical information.

Note: Not all subdivisions listed here necessarily appear in this publication. ■

01 11 Summary of Work

01 11 31 – Professional Consultants

01 11 31.10 Architectural Fees

	01 11 31.10 Architectural Fees	Crew	Daily Output	Labor-Hours	Unit	2015 Bare Costs Material	Labor	Equipment	Total	Total Incl O&P
0010	ARCHITECTURAL FEES R011110-10									
0020	For new construction									
0060	Minimum				Project				4.90%	4.90%
0090	Maximum								16%	16%
0100	For alteration work, to $500,000, add to new construction fee								50%	50%
0150	Over $500,000, add to new construction fee								25%	25%
2000	For "Greening" of building G								3%	3%

01 11 31.20 Construction Management Fees

	01 11 31.20 Construction Management Fees	Crew	Daily Output	Labor-Hours	Unit	Material	Labor	Equipment	Total	Total Incl O&P
0010	CONSTRUCTION MANAGEMENT FEES									
0020	$1,000,000 job, minimum				Project				4.50%	4.50%
0050	Maximum								7.50%	7.50%
0060	For work to $100,000								10%	10%
0070	To $250,000								9%	9%
0090	To $1,000,000								6%	6%
0100	To $5,000,000								5%	5%
0110	To $10,000,000								4%	4%
0300	$50,000,000 job, minimum								2.50%	2.50%
0350	Maximum								4%	4%

01 11 31.30 Engineering Fees

	01 11 31.30 Engineering Fees	Crew	Daily Output	Labor-Hours	Unit	Material	Labor	Equipment	Total	Total Incl O&P
0010	ENGINEERING FEES R011110-30									
0020	Educational planning consultant, minimum				Project				.50%	.50%
0100	Maximum				"				2.50%	2.50%
0200	Electrical, minimum				Contrct				4.10%	4.10%
0300	Maximum								10.10%	10.10%
0400	Elevator & conveying systems, minimum								2.50%	2.50%
0500	Maximum								5%	5%
0600	Food service & kitchen equipment, minimum								8%	8%
0700	Maximum								12%	12%
0800	Landscaping & site development, minimum								2.50%	2.50%
0900	Maximum								6%	6%
1000	Mechanical (plumbing & HVAC), minimum								4.10%	4.10%
1100	Maximum								10.10%	10.10%
1200	Structural, minimum				Project				1%	1%
1300	Maximum				"				2.50%	2.50%

01 11 31.50 Models

	01 11 31.50 Models	Crew	Daily Output	Labor-Hours	Unit	Material	Labor	Equipment	Total	Total Incl O&P
0010	MODELS									
0500	2 story building, scaled 100′ x 200′, simple materials and details				Ea.	5,000			5,000	5,500
0510	Elaborate materials and details				"	30,000			30,000	33,000

01 11 31.75 Renderings

	01 11 31.75 Renderings	Crew	Daily Output	Labor-Hours	Unit	Material	Labor	Equipment	Total	Total Incl O&P
0010	RENDERINGS Color, matted, 20″ x 30″, eye level,									
0020	1 building, minimum				Ea.	2,050			2,050	2,275
0050	Average					2,875			2,875	3,175
0100	Maximum					4,625			4,625	5,100
1000	5 buildings, minimum					4,125			4,125	4,525
1100	Maximum					8,250			8,250	9,075
2000	Aerial perspective, color, 1 building, minimum					3,000			3,000	3,275
2100	Maximum					8,150			8,150	8,950
3000	5 buildings, minimum					5,875			5,875	6,450
3100	Maximum					11,700			11,700	12,900

01 21 Allowances

01 21 16 – Contingency Allowances

01 21 16.50 Contingencies

	01 21 16.50 Contingencies	Crew	Daily Output	Labor-Hours	Unit	2015 Bare Costs Material	Labor	Equipment	Total	Total Incl O&P
0010	**CONTINGENCIES**, Add to estimate									
0020	Conceptual stage				Project				20%	20%
0050	Schematic stage								15%	15%
0100	Preliminary working drawing stage (Design Dev.)								10%	10%
0150	Final working drawing stage				↓				3%	3%

01 21 55 – Job Conditions Allowance

01 21 55.50 Job Conditions

	Description	Crew	Daily Output	Labor-Hours	Unit	Material	Labor	Equipment	Total	Total Incl O&P
0010	**JOB CONDITIONS** Modifications to applicable									
0020	cost summaries									
0100	Economic conditions, favorable, deduct				Project				2%	2%
0200	Unfavorable, add								5%	5%
0300	Hoisting conditions, favorable, deduct								2%	2%
0400	Unfavorable, add								5%	5%
0500	General Contractor management, experienced, deduct								2%	2%
0600	Inexperienced, add								10%	10%
0700	Labor availability, surplus, deduct								1%	1%
0800	Shortage, add								10%	10%
0900	Material storage area, available, deduct								1%	1%
1000	Not available, add								2%	2%
1100	Subcontractor availability, surplus, deduct								5%	5%
1200	Shortage, add								12%	12%
1300	Work space, available, deduct								2%	2%
1400	Not available, add				↓				5%	5%

01 21 57 – Overtime Allowance

01 21 57.50 Overtime

	Description	Crew	Daily Output	Labor-Hours	Unit	Material	Labor	Equipment	Total	Total Incl O&P
0010	**OVERTIME** for early completion of projects or where R012909-90									
0020	labor shortages exist, add to usual labor, up to				Costs		100%			

01 21 61 – Cost Indexes

01 21 61.10 Construction Cost Index

	Description	Crew	Daily Output	Labor-Hours	Unit	Material	Labor	Equipment	Total	Total Incl O&P
0010	**CONSTRUCTION COST INDEX** (Reference) over 930 zip code locations in									
0020	the U.S. and Canada, total bldg. cost, min. (Fayetteville, AR)				%				75.40%	75.40%
0050	Average								100%	100%
0100	Maximum (New York, NY)				↓				131.80%	131.80%

01 21 61.20 Historical Cost Indexes

	Description	Crew	Daily Output	Labor-Hours	Unit	Material	Labor	Equipment	Total	Total Incl O&P
0010	**HISTORICAL COST INDEXES** (See Reference Section)									

01 21 61.30 Labor Index

	Description	Crew	Daily Output	Labor-Hours	Unit	Material	Labor	Equipment	Total	Total Incl O&P
0010	**LABOR INDEX** (Reference) For over 930 zip code locations in									
0020	the U.S. and Canada, minimum (Beaufort, SC)				%		44.90%			
0050	Average						100%			
0100	Maximum (New York, NY)				↓		168.90%			

01 21 61.50 Material Index

	Description	Crew	Daily Output	Labor-Hours	Unit	Material	Labor	Equipment	Total	Total Incl O&P
0010	**MATERIAL INDEX** (Reference) For over 930 zip code locations in									
0020	the U.S. and Canada, minimum (Oil City, PA)				%	92.40%				
0040	Average					100%				
0060	Maximum (Ketchikan, AK)				↓	132%				

01 21 63 – Taxes

01 21 63.10 Taxes

	Description	Crew	Daily Output	Labor-Hours	Unit	Material	Labor	Equipment	Total	Total Incl O&P
0010	**TAXES** R012909-80									
0020	Sales tax, State, average				%	5.04%				
0050	Maximum				↓	7.50%				

01 21 Allowances

01 21 63 – Taxes

01 21 63.10 Taxes		Crew	Daily Output	Labor-Hours	Unit	Material	2015 Bare Costs Labor	Equipment	Total	Total Incl O&P
0200	Social Security, on first $117,000 of wages				%		7.65%			
0300	Unemployment, combined Federal and State, minimum R012909-85						.60%			
0350	Average						7.80%			
0400	Maximum				↓		12.87%			

01 31 Project Management and Coordination

01 31 13 – Project Coordination

01 31 13.20 Field Personnel

		Crew	Daily Output	Labor-Hours	Unit	Material	Labor	Equipment	Total	Total Incl O&P
0010	**FIELD PERSONNEL**									
0020	Clerk, average				Week		450		450	695
0100	Field engineer, minimum						1,075		1,075	1,650
0120	Average						1,400		1,400	2,150
0140	Maximum						1,575		1,575	2,450
0160	General purpose laborer, average						1,500		1,500	2,325
0180	Project manager, minimum						1,975		1,975	3,075
0200	Average						2,275		2,275	3,525
0220	Maximum						2,600		2,600	4,025
0240	Superintendent, minimum						1,925		1,925	2,975
0260	Average						2,125		2,125	3,275
0280	Maximum						2,400		2,400	3,725
0290	Timekeeper, average				↓		1,225		1,225	1,900

01 31 13.30 Insurance

		Crew	Daily Output	Labor-Hours	Unit	Material	Labor	Equipment	Total	Total Incl O&P
0010	**INSURANCE** R013113-40									
0020	Builders risk, standard, minimum				Job				.24%	.24%
0050	Maximum								.64%	.64%
0200	All-risk type, minimum								.25%	.25%
0250	Maximum				↓				.62%	.62%
0400	Contractor's equipment floater, minimum R013113-50				Value				.50%	.50%
0450	Maximum				"				1.50%	1.50%
0600	Public liability, average				Job				2.02%	2.02%
0800	Workers' compensation & employer's liability, average R013113-60									
0850	by trade, carpentry, general				Payroll		14.93%			
0900	Clerical						.50%			
0950	Concrete						12.93%			
1000	Electrical						5.76%			
1050	Excavation						9.70%			
1100	Glazing						13.29%			
1150	Insulation						11.66%			
1200	Lathing						8.38%			
1250	Masonry						13.68%			
1300	Painting & decorating						11.72%			
1350	Pile driving						14.31%			
1400	Plastering						10.90%			
1450	Plumbing						6.98%			
1500	Roofing						31.69%			
1550	Sheet metal work (HVAC)						8.82%			
1600	Steel erection, structural						31.68%			
1650	Tile work, interior ceramic						8.86%			
1700	Waterproofing, brush or hand caulking						6.56%			
1800	Wrecking						26.14%			
2000	Range of 35 trades in 50 states, excl. wrecking, min.				↓		1.31%			

01 31 13 – Project Coordination

01 31 13.30 Insurance		Crew	Daily Output	Labor-Hours	Unit	Material	2015 Bare Costs Labor	Equipment	Total	Total Incl O&P
2100	Average				Payroll		13.50%			
2200	Maximum				↓		132%			

01 31 13.40 Main Office Expense		Crew	Daily Output	Labor-Hours	Unit	Material	Labor	Equipment	Total	Total Incl O&P
0010	**MAIN OFFICE EXPENSE** Average for General Contractors R013113-50									
0020	As a percentage of their annual volume									
0125	Annual volume under 1 million dollars				% Vol.				17.50%	
0145	Up to 2.5 million dollars								8%	
0150	Up to 4.0 million dollars								6.80%	
0200	Up to 7.0 million dollars								5.60%	
0250	Up to 10 million dollars								5.10%	
0300	Over 10 million dollars				↓				3.90%	

01 31 13.50 General Contractor's Mark-Up		Crew	Daily Output	Labor-Hours	Unit	Material	Labor	Equipment	Total	Total Incl O&P
0010	**GENERAL CONTRACTOR'S MARK-UP** on Change Orders									
0200	Extra work, by subcontractors, add				%				10%	10%
0250	By General Contractor, add								15%	15%
0400	Omitted work, by subcontractors, deduct all but								5%	5%
0450	By General Contractor, deduct all but								7.50%	7.50%
0600	Overtime work, by subcontractors, add								15%	15%
0650	By General Contractor, add				↓				10%	10%

01 31 13.60 Installing Contractor's Main Office Overhead		Crew	Daily Output	Labor-Hours	Unit	Material	Labor	Equipment	Total	Total Incl O&P
0010	**INSTALLING CONTRACTOR'S MAIN OFFICE OVERHEAD** R013113-50									
0020	As percent of direct costs, minimum				%				5%	
0050	Average								13%	
0100	Maximum				↓				30%	

01 31 13.80 Overhead and Profit		Crew	Daily Output	Labor-Hours	Unit	Material	Labor	Equipment	Total	Total Incl O&P
0010	**OVERHEAD & PROFIT** Allowance to add to items in this									
0020	book that do not include Subs O&P, average				%				25%	
0100	Allowance to add to items in this book that									
0110	do include Subs O&P, minimum				%				5%	5%
0150	Average								10%	10%
0200	Maximum								15%	15%
0300	Typical, by size of project, under $100,000								30%	
0350	$500,000 project								25%	
0400	$2,000,000 project								20%	
0450	Over $10,000,000 project				↓				15%	

01 31 13.90 Performance Bond		Crew	Daily Output	Labor-Hours	Unit	Material	Labor	Equipment	Total	Total Incl O&P
0010	**PERFORMANCE BOND** R013113-80									
0020	For buildings, minimum				Job				.60%	.60%
0100	Maximum				"				2.50%	2.50%

01 32 Construction Progress Documentation

01 32 13 – Scheduling of work

01 32 13.50 Scheduling		Crew	Daily Output	Labor-Hours	Unit	Material	2015 Bare Costs Labor	Equipment	Total	Total Incl O&P
0010	SCHEDULING									
0020	Critical path, as % of architectural fee, minimum				%				.50%	.50%
0100	Maximum				"				1%	1%
0300	Computer-update, micro, no plots, minimum				Ea.				455	500
0400	Including plots, maximum				"				1,450	1,600
0600	Rule of thumb, CPM scheduling, small job ($10 Million)				Job				.05%	.05%
0650	Large job ($50 Million +)								.03%	.03%
0700	Including cost control, small job								.08%	.08%
0750	Large job				↓				.04%	.04%

01 32 33 – Photographic Documentation

01 32 33.50 Photographs		Crew	Daily Output	Labor-Hours	Unit	Material	Labor	Equipment	Total	Total Incl O&P
0010	PHOTOGRAPHS									
0020	8" x 10", 4 shots, 2 prints ea., std. mounting				Set	475			475	525
0100	Hinged linen mounts					530			530	580
0200	8" x 10", 4 shots, 2 prints each, in color					425			425	470
0300	For I.D. slugs, add to all above					5.30			5.30	5.85
0500	Aerial photos, initial fly-over, 5 shots, digital images					400			400	440
0550	10 shots, digital images, 1 print					450			450	495
0600	For each addtional print from fly-over					200			200	220
0700	For full color prints, add				↓	40%				
0750	Add for traffic control area				Ea.	305			305	335
0900	For over 30 miles from airport, add per				Mile	6			6	6.60
1500	Time lapse equipment, camera and projector, buy				Ea.	2,650			2,650	2,925
1550	Rent per month				"	1,600			1,600	1,750
1700	Cameraman and film, including processing, B.&W.				Day	1,225			1,225	1,350
1720	Color				"	1,425			1,425	1,550

01 41 Regulatory Requirements

01 41 26 – Permit Requirements

01 41 26.50 Permits		Crew	Daily Output	Labor-Hours	Unit	Material	Labor	Equipment	Total	Total Incl O&P
0010	PERMITS									
0020	Rule of thumb, most cities, minimum				Job				.50%	.50%
0100	Maximum				"				2%	2%

01 45 Quality Control

01 45 23 – Testing and Inspecting Services

01 45 23.50 Testing		Crew	Daily Output	Labor-Hours	Unit	Material	Labor	Equipment	Total	Total Incl O&P
0010	TESTING and Inspecting Services									
0015	For concrete building costing $1,000,000, minimum				Project				4,725	5,200
0020	Maximum								38,000	41,800
0050	Steel building, minimum								4,725	5,200
0070	Maximum								14,800	16,300
0100	For building costing, $10,000,000, minimum								30,100	33,100
0150	Maximum				↓				48,200	53,000
0200	Asphalt testing, compressive strength Marshall stability, set of 3				Ea.				145	165
0220	Density, set of 3								86	95
0250	Extraction, individual tests on sample								136	150
0300	Penetration								41	45
0350	Mix design, 5 specimens				↓				182	200

01 45 23.50 Testing		Crew	Daily Output	Labor-Hours	Unit	Material	2015 Bare Costs Labor	Equipment	Total	Total Incl O&P
0360	Additional specimen				Ea.				36	40
0400	Specific gravity								41	45
0420	Swell test								64	70
0450	Water effect and cohesion, set of 6								182	200
0470	Water effect and plastic flow								64	70
0600	Concrete testing, aggregates, abrasion, ASTM C 131								136	150
0650	Absorption, ASTM C 127								42	46
0800	Petrographic analysis, ASTM C 295								775	850
0900	Specific gravity, ASTM C 127								50	55
1000	Sieve analysis, washed, ASTM C 136								59	65
1050	Unwashed								59	65
1200	Sulfate soundness								114	125
1300	Weight per cubic foot								36	40
1500	Cement, physical tests, ASTM C 150								320	350
1600	Chemical tests, ASTM C 150								245	270
1800	Compressive test, cylinder, delivered to lab, ASTM C 39								12	13
1900	Picked up by lab, minimum								14	15
1950	Average								18	20
2000	Maximum								27	30
2200	Compressive strength, cores (not incl. drilling), ASTM C 42				↓				36	40
2250	Core drilling, 4" diameter (plus technician)				Inch				23	25
2260	Technician for core drilling				Hr.				45	50
2300	Patching core holes				Ea.				22	24
2400	Drying shrinkage at 28 days								236	260
2500	Flexural test beams, ASTM C 78								59	65
2600	Mix design, one batch mix								259	285
2650	Added trial batches								120	132
2800	Modulus of elasticity, ASTM C 469								164	180
2900	Tensile test, cylinders, ASTM C 496								45	50
3000	Water-Cement ratio curve, 3 batches								141	155
3100	4 batches								186	205
3300	Masonry testing, absorption, per 5 brick, ASTM C 67								45	50
3350	Chemical resistance, per 2 brick								50	55
3400	Compressive strength, per 5 brick, ASTM C 67								68	75
3420	Efflorescence, per 5 brick, ASTM C 67								68	75
3440	Imperviousness, per 5 brick								87	96
3470	Modulus of rupture, per 5 brick								86	95
3500	Moisture, block only								32	35
3550	Mortar, compressive strength, set of 3								23	25
4100	Reinforcing steel, bend test								55	61
4200	Tensile test, up to #8 bar								36	40
4220	#9 to #11 bar								41	45
4240	#14 bar and larger								64	70
4400	Soil testing, Atterberg limits, liquid and plastic limits								59	65
4510	Hydrometer analysis								109	120
4530	Specific gravity, ASTM D 354								44	48
4600	Sieve analysis, washed, ASTM D 422								55	60
4700	Unwashed, ASTM D 422								59	65
4710	Consolidation test (ASTM D2435), minimum								250	275
4715	Maximum								430	475
4720	Density and classification of undisturbed sample								73	80
4735	Soil density, nuclear method, ASTM D2922								35	38.50
4740	Sand cone method ASTM D1556				↓				27	30

01 45 Quality Control

01 45 23 – Testing and Inspecting Services

01 45 23.50 Testing

	01 45 23.50 Testing	Crew	Daily Output	Labor-Hours	Unit	2015 Bare Costs Material	Labor	Equipment	Total	Total Incl O&P
4750	Moisture content, ASTM D 2216				Ea.				9	10
4780	Permeability test, double ring infiltrometer								500	550
4800	Permeability, var. or constant head, undist., ASTM D 2434								227	250
4850	Recompacted								250	275
4900	Proctor compaction, 4" standard mold, ASTM D 698								123	135
4950	6" modified mold								68	75
5100	Shear tests, triaxial, minimum								410	450
5150	Maximum								545	600
5300	Direct shear, minimum, ASTM D 3080								320	350
5350	Maximum								410	450
5550	Technician for inspection, per day, earthwork								320	350
5650	Bolting								400	440
5750	Roofing								480	530
5790	Welding								480	530
5820	Non-destructive metal testing, dye penetrant				Day				310	340
5840	Magnetic particle								310	340
5860	Radiography								450	495
5880	Ultrasonic								310	340
6000	Welding certification, minimum				Ea.				91	100
6100	Maximum				"				250	275
7000	Underground storage tank									
7500	Volumetric tightness test ,<=12,000 gal.				Ea.				435	480
7510	<=30,000 gal.				"				615	675
7600	Vadose zone (soil gas) sampling, 10-40 samples, min.				Day				1,375	1,500
7610	Maximum				"				2,275	2,500
7700	Ground water monitoring incl. drilling 3 wells, min.				Total				4,550	5,000
7710	Maximum				"				6,375	7,000
8000	X-ray concrete slabs				Ea.				182	200
9000	Thermographic testing, for bldg envelope heat loss, average 2,000 S.F. G				"				500	500

01 51 Temporary Utilities

01 51 13 – Temporary Electricity

01 51 13.80 Temporary Utilities

	01 51 13.80 Temporary Utilities	Crew	Daily Output	Labor-Hours	Unit	Material	Labor	Equipment	Total	Total Incl O&P
0010	**TEMPORARY UTILITIES**									
0350	Lighting, lamps, wiring, outlets, 40000 SF building, 8 strings	1 Elec	34	.235	CSF Flr	4.91	12.85		17.76	24.50
0360	16 strings	"	17	.471		9.80	25.50		35.30	49.50
0400	Power for temp lighting only, 6.6 KWH, per month								.92	1.01
0430	11.8 KWH, per month								1.65	1.82
0450	23.6 KWH, per month								3.30	3.63
0600	Power for job duration incl. elevator, etc., minimum								47	51.50
0650	Maximum								110	121
1000	Toilet, portable, see Equip. Rental 01 54 33 in Reference Section									

01 52 Construction Facilities

01 52 13 – Field Offices and Sheds

01 52 13.20 Office and Storage Space		Crew	Daily Output	Labor-Hours	Unit	Material	2015 Bare Costs Labor	Equipment	Total	Total Incl O&P
0010	**OFFICE AND STORAGE SPACE**									
0020	Office trailer, furnished, no hookups, 20′ x 8′, buy	2 Skwk	1	16	Ea.	10,400	780		11,180	12,600
0250	Rent per month					188			188	206
0300	32′ x 8′, buy	2 Skwk	.70	22.857		15,500	1,100		16,600	18,800
0350	Rent per month					239			239	262
0400	50′ x 10′, buy	2 Skwk	.60	26.667		24,100	1,300		25,400	28,500
0450	Rent per month					340			340	375
0500	50′ x 12′, buy	2 Skwk	.50	32		29,400	1,550		30,950	34,700
0550	Rent per month					410			410	455
0700	For air conditioning, rent per month, add					48.50			48.50	53.50
0800	For delivery, add per mile				Mile	11			11	12.10
0890	Delivery each way				Ea.	200			200	220
0900	Bunk house trailer, 8′ x 40′ duplex dorm with kitchen, no hookups, buy	2 Carp	1	16		37,300	750		38,050	42,200
0910	9 man with kitchen and bath, no hookups, buy		1	16		38,000	750		38,750	43,000
0920	18 man sleeper with bath, no hookups, buy		1	16		49,000	750		49,750	55,000
1000	Portable buildings, prefab, on skids, economy, 8′ x 8′		265	.060	S.F.	25	2.83		27.83	32
1100	Deluxe, 8′ x 12′		150	.107	"	20	5		25	29.50
1200	Storage boxes, 20′ x 8′, buy	2 Skwk	1.80	8.889	Ea.	2,775	430		3,205	3,725
1250	Rent per month					82.50			82.50	91
1300	40′ x 8′, buy	2 Skwk	1.40	11.429		3,850	555		4,405	5,075
1350	Rent per month					101			101	111
5000	Air supported structures, see Section 13 31 13.13									

01 52 13.40 Field Office Expense		Crew	Daily Output	Labor-Hours	Unit	Material	Labor	Equipment	Total	Total Incl O&P
0010	**FIELD OFFICE EXPENSE**									
0100	Office equipment rental average				Month	200			200	220
0120	Office supplies, average				"	80			80	88
0125	Office trailer rental, see Section 01 52 13.20									
0140	Telephone bill; avg. bill/month incl. long dist.				Month	85			85	93.50
0160	Lights & HVAC				"	160			160	176

01 54 Construction Aids

01 54 09 – Protection Equipment

01 54 09.50 Personnel Protective Equipment		Crew	Daily Output	Labor-Hours	Unit	Material	Labor	Equipment	Total	Total Incl O&P
0010	**PERSONNEL PROTECTIVE EQUIPMENT**									
0015	Hazardous waste protection									
0020	Respirator mask only, full face, silicone				Ea.	335			335	370
0030	Half face, silicone					62.50			62.50	68.50
0040	Respirator cartridges, 2 req'd/mask, dust or asbestos					4.03			4.03	4.43
0050	Chemical vapor					2.88			2.88	3.16
0060	Combination vapor and dust					8.60			8.60	9.45
0100	Emergency escape breathing apparatus, 5 minutes					720			720	790
0110	10 minutes					830			830	915
0150	Self contained breathing apparatus with full face piece, 30 minutes					2,375			2,375	2,600
0160	60 minutes					4,050			4,050	4,450
0200	Encapsulating suits, limited use, level A					1,900			1,900	2,100
0210	Level B					435			435	480
0300	Over boots, latex				Pr.	7.80			7.80	8.60
0310	PVC					29.50			29.50	32.50
0320	Neoprene					40.50			40.50	44.50
0400	Gloves, nitrile/PVC					38			38	41.50
0410	Neoprene coated					34			34	37.50

01 54 Construction Aids

01 54 09 – Protection Equipment

01 54 09.60 Safety Nets		Crew	Daily Output	Labor-Hours	Unit	Material	Labor	Equipment	Total	Total Incl O&P
						2015 Bare Costs				
0010	**SAFETY NETS**									
0020	No supports, stock sizes, nylon, 3 1/2" mesh				S.F.	2.91			2.91	3.20
0100	Polypropylene, 6" mesh					1.59			1.59	1.75
0200	Small mesh debris nets, 1/4" mesh, stock sizes					.67			.67	.74
0220	Combined 3 1/2" mesh and 1/4" mesh, stock sizes					4.65			4.65	5.10
0300	Monthly rental, 4" mesh, stock sizes, 1st month					.50			.50	.55
0320	2nd month rental					.25			.25	.28
0340	Maximum rental/year				↓	1.15			1.15	1.27

01 54 16 – Temporary Hoists

01 54 16.50 Weekly Forklift Crew		Crew	Daily Output	Labor-Hours	Unit	Material	Labor	Equipment	Total	Total Incl O&P
0010	**WEEKLY FORKLIFT CREW**									
0100	All-terrain forklift, 45' lift, 35' reach, 9000 lb. capacity	A-3P	.20	40	Week		1,950	2,625	4,575	5,850

01 54 19 – Temporary Cranes

01 54 19.50 Daily Crane Crews		Crew	Daily Output	Labor-Hours	Unit	Material	Labor	Equipment	Total	Total Incl O&P
0010	**DAILY CRANE CREWS** for small jobs, portal to portal									
0100	12-ton truck-mounted hydraulic crane	A-3H	1	8	Day		415	860	1,275	1,575
0200	25-ton	A-3I	1	8			415	990	1,405	1,725
0300	40-ton	A-3J	1	8			415	1,225	1,640	1,975
0400	55-ton	A-3K	1	16			775	1,625	2,400	2,975
0500	80-ton	A-3L	1	16			775	2,350	3,125	3,750
0600	100-ton	A-3M	1	16	↓		775	2,325	3,100	3,750
0900	If crane is needed on a Saturday, Sunday or Holiday									
0910	At time-and-a-half, add				Day		50%			
0920	At double time, add				"		100%			

01 54 19.60 Monthly Tower Crane Crew		Crew	Daily Output	Labor-Hours	Unit	Material	Labor	Equipment	Total	Total Incl O&P
0010	**MONTHLY TOWER CRANE CREW**, excludes concrete footing									
0100	Static tower crane, 130' high, 106' jib, 6200 lb. capacity	A-3N	.05	176	Month		9,100	24,800	33,900	41,100

01 54 23 – Temporary Scaffolding and Platforms

01 54 23.60 Pump Staging		Crew	Daily Output	Labor-Hours	Unit	Material	Labor	Equipment	Total	Total Incl O&P
0010	**PUMP STAGING**, Aluminum R015423-20									
0200	24' long pole section, buy				Ea.	410			410	450
0300	18' long pole section, buy					320			320	350
0400	12' long pole section, buy					215			215	236
0500	6' long pole section, buy					113			113	124
0600	6' long splice joint section, buy					84			84	92.50
0700	Pump jack, buy					172			172	189
0900	Foldable brace, buy					67.50			67.50	74
1000	Workbench/back safety rail support, buy					91			91	100
1100	Scaffolding planks/workbench, 14" wide x 24' long, buy					775			775	850
1200	Plank end safety rail, buy					345			345	380
1250	Safety net, 22' long, buy				↓	410			410	450
1300	System in place, 50' working height, per use based on 50 uses	2 Carp	84.80	.189	C.S.F.	6.95	8.85		15.80	21.50
1400	100 uses		84.80	.189		3.46	8.85		12.31	17.45
1500	150 uses	↓	84.80	.189	↓	2.32	8.85		11.17	16.20

01 54 23.70 Scaffolding		Crew	Daily Output	Labor-Hours	Unit	Material	Labor	Equipment	Total	Total Incl O&P
0010	**SCAFFOLDING** R015423-10									
0015	Steel tube, regular, no plank, labor only to erect & dismantle									
0090	Building exterior, wall face, 1 to 5 stories, 6'-4" x 5' frames	3 Carp	8	3	C.S.F.		141		141	217
0200	6 to 12 stories	4 Carp	8	4			188		188	289
0301	13 to 20 stories	5 Clab	8	5	↓		188		188	289

01 54 23 – Temporary Scaffolding and Platforms

01 54 23.70 Scaffolding		Crew	Daily Output	Labor-Hours	Unit	2015 Bare Costs Material	Labor	Equipment	Total	Total Incl O&P
0460	Building interior, wall face area, up to 16′ high	3 Carp	12	2	C.S.F.		94		94	145
0560	16′ to 40′ high		10	2.400			113		113	173
0800	Building interior floor area, up to 30′ high		150	.160	C.C.F.		7.50		7.50	11.55
0900	Over 30′ high	4 Carp	160	.200	"		9.40		9.40	14.45
0906	Complete system for face of walls, no plank, material only rent/mo				C.S.F.	35.50			35.50	39.50
0908	Interior spaces, no plank, material only rent/mo				C.C.F.	4.33			4.33	4.76
0910	Steel tubular, heavy duty shoring, buy									
0920	Frames 5′ high 2′ wide				Ea.	81.50			81.50	90
0925	5′ high 4′ wide					93			93	102
0930	6′ high 2′ wide					93.50			93.50	103
0935	6′ high 4′ wide					109			109	120
0940	Accessories									
0945	Cross braces				Ea.	15.50			15.50	17.05
0950	U-head, 8″ x 8″					19.10			19.10	21
0955	J-head, 4″ x 8″					13.90			13.90	15.30
0960	Base plate, 8″ x 8″					15.50			15.50	17.05
0965	Leveling jack					33.50			33.50	36.50
1000	Steel tubular, regular, buy									
1100	Frames 3′ high 5′ wide				Ea.	75			75	82.50
1150	5′ high 5′ wide					86.50			86.50	95
1200	6′-4″ high 5′ wide					118			118	129
1350	7′-6″ high 6′ wide					158			158	173
1500	Accessories cross braces					15.50			15.50	17.05
1550	Guardrail post					16.40			16.40	18.05
1600	Guardrail 7′ section					8.70			8.70	9.60
1650	Screw jacks & plates					21.50			21.50	24
1700	Sidearm brackets					30.50			30.50	33.50
1750	8″ casters					31			31	34
1800	Plank 2″ x 10″ x 16′-0″					57.50			57.50	63.50
1900	Stairway section					286			286	315
1910	Stairway starter bar					32			32	35
1920	Stairway inside handrail					58.50			58.50	64
1930	Stairway outside handrail					86.50			86.50	95
1940	Walk-thru frame guardrail					41.50			41.50	46
2000	Steel tubular, regular, rent/mo.									
2100	Frames 3′ high 5′ wide				Ea.	5			5	5.50
2150	5′ high 5′ wide					5			5	5.50
2200	6′-4″ high 5′ wide					5.50			5.50	6.05
2250	7′-6″ high 6′ wide					10			10	11
2500	Accessories, cross braces					1			1	1.10
2550	Guardrail post					1			1	1.10
2600	Guardrail 7′ section					1			1	1.10
2650	Screw jacks & plates					2			2	2.20
2700	Sidearm brackets					2			2	2.20
2750	8″ casters					8			8	8.80
2800	Outrigger for rolling tower					3			3	3.30
2850	Plank 2″ x 10″ x 16′-0″					10			10	11
2900	Stairway section					35			35	38.50
2940	Walk-thru frame guardrail					2.50			2.50	2.75
3000	Steel tubular, heavy duty shoring, rent/mo.									
3250	5′ high 2′ & 4′ wide				Ea.	8.50			8.50	9.35
3300	6′ high 2′ & 4′ wide					8.50			8.50	9.35
3500	Accessories, cross braces					1			1	1.10

01 54 Construction Aids

01 54 23 – Temporary Scaffolding and Platforms

01 54 23.70 Scaffolding		Crew	Daily Output	Labor-Hours	Unit	Material	2015 Bare Costs Labor	Equipment	Total	Total Incl O&P
3600	U-head, 8" x 8"				Ea.	2.50			2.50	2.75
3650	J-head, 4" x 8"					2.50			2.50	2.75
3700	Base plate, 8" x 8"					1			1	1.10
3750	Leveling jack					2.50			2.50	2.75
5700	Planks, 2" x 10" x 16'-0", labor only to erect & remove to 50' H	3 Carp	72	.333			15.65		15.65	24
5800	Over 50' high	4 Carp	80	.400			18.80		18.80	29
6000	Heavy duty shoring for elevated slab forms to 8'-2" high, floor area									
6100	Labor only to erect & dismantle	4 Carp	16	2	C.S.F.		94		94	145
6110	Materials only, rent.mo				"	44.50			44.50	49
6500	To 14'-8" high									
6600	Labor only to erect & dismantle	4 Carp	10	3.200	C.S.F.		150		150	231
6610	Materials only, rent/mo				"	65			65	71.50

01 54 23.75 Scaffolding Specialties		Crew	Daily Output	Labor-Hours	Unit	Material	Labor	Equipment	Total	Total Incl O&P
0010	**SCAFFOLDING SPECIALTIES**									
1200	Sidewalk bridge, heavy duty steel posts & beams, including									
1210	parapet protection & waterproofing (material cost is rent/month)									
1220	8' to 10' wide, 2 posts	3 Carp	15	1.600	L.F.	45	75		120	166
1230	3 posts	"	10	2.400	"	69.50	113		182.50	250
1500	Sidewalk bridge using tubular steel scaffold frames including									
1510	planking (material cost is rent/month)	3 Carp	45	.533	L.F.	8.55	25		33.55	48
1600	For 2 uses per month, deduct from all above					50%				
1700	For 1 use every 2 months, add to all above					100%				
1900	Catwalks, 20" wide, no guardrails, 7' span, buy				Ea.	166			166	183
2000	10' span, buy					233			233	256
3720	Putlog, standard, 8' span, with hangers, buy					162			162	178
3730	Rent per month					20			20	22
3750	12' span, buy					203			203	223
3755	Rent per month					25			25	27.50
3760	Trussed type, 16' span, buy					365			365	400
3770	Rent per month					30			30	33
3790	22' span, buy					420			420	465
3795	Rent per month					40			40	44
3800	Rolling ladders with handrails, 30" wide, buy, 2 step					267			267	294
4000	7 step					760			760	835
4050	10 step					1,075			1,075	1,175
4100	Rolling towers, buy, 5' wide, 7' long, 10' high					1,200			1,200	1,325
4200	For 5' high added sections, to buy, add					204			204	224
4300	Complete incl. wheels, railings, outriggers,									
4350	21' high, to buy				Ea.	2,025			2,025	2,225
4400	Rent/month = 5% of purchase cost				"	200			200	220
5000	Motorized work platform, mast climber									
5050	Base unit, 50' W, less than 100' tall, rent/mo				Ea.	2,900			2,900	3,175
5100	Less than 200' tall, rent/mo					3,700			3,700	4,075
5150	Less than 300' tall, rent/mo					4,375			4,375	4,825
5200	Less than 400' tall, rent/mo					5,000			5,000	5,500
5250	Set up and demob, per unit, less than 100' tall	B-68F	16.60	1.446	C.S.F.		71.50	20	91.50	132
5300	Less than 200' tall		25	.960			47.50	13.30	60.80	87.50
5350	Less than 300' tall		30	.800			39.50	11.05	50.55	73
5400	Less than 400' tall		33	.727			36	10.05	46.05	66.50
5500	Mobilization (price includes freight in and out) per unit				Ea.				1,100	1,225

01 54 Construction Aids

01 54 23 – Temporary Scaffolding and Platforms

01 54 23.80 Staging Aids		Crew	Daily Output	Labor-Hours	Unit	Material	2015 Bare Costs Labor	Equipment	Total	Total Incl O&P
0010	**STAGING AIDS** and fall protection equipment									
0100	Sidewall staging bracket, tubular, buy				Ea.	55			55	60.50
0110	Cost each per day, based on 250 days use				Day	.22			.22	.24
0200	Guard post, buy				Ea.	50.50			50.50	55.50
0210	Cost each per day, based on 250 days use				Day	.20			.20	.22
0300	End guard chains, buy per pair				Pair	36			36	39.50
0310	Cost per set per day, based on 250 days use				Day	.22			.22	.24
1000	Roof shingling bracket, steel, buy				Ea.	9.80			9.80	10.80
1010	Cost each per day, based on 250 days use				Day	.04			.04	.04
1100	Wood bracket, buy				Ea.	16.80			16.80	18.50
1110	Cost each per day, based on 250 days use				Day	.07			.07	.07
2000	Ladder jack, aluminum, buy per pair				Pair	118			118	130
2010	Cost per pair per day, based on 250 days use				Day	.47			.47	.52
2100	Steel siderail jack, buy per pair				Pair	97.50			97.50	107
2110	Cost per pair per day, based on 250 days use				Day	.39			.39	.43
3000	Laminated wood plank, 2″ x 10″ x 16′, buy				Ea.	57.50			57.50	63.50
3010	Cost each per day, based on 250 days use				Day	.23			.23	.25
3100	Aluminum scaffolding plank, 20″ wide x 24′ long, buy				Ea.	755			755	830
3110	Cost each per day, based on 250 days use				Day	3.01			3.01	3.31
4000	Nylon full body harness, lanyard and rope grab				Ea.	172			172	189
4010	Cost each per day, based on 250 days use				Day	.69			.69	.76
4100	Rope for safety line, 5/8″ x 100′ nylon, buy				Ea.	56			56	61.50
4110	Cost each per day, based on 250 days use				Day	.22			.22	.25
4200	Permanent U-Bolt roof anchor, buy				Ea.	35.50			35.50	39.50
4300	Temporary (one use) roof ridge anchor, buy				″	32			32	35.50
5000	Installation (setup and removal) of staging aids									
5010	Sidewall staging bracket	2 Carp	64	.250	Ea.		11.75		11.75	18.05
5020	Guard post with 2 wood rails	″	64	.250			11.75		11.75	18.05
5030	End guard chains, set	1 Carp	64	.125			5.85		5.85	9.05
5100	Roof shingling bracket		96	.083			3.91		3.91	6
5200	Ladder jack		64	.125			5.85		5.85	9.05
5300	Wood plank, 2″ x 10″ x 16′	2 Carp	80	.200			9.40		9.40	14.45
5310	Aluminum scaffold plank, 20″ x 24′	″	40	.400			18.80		18.80	29
5410	Safety rope	1 Carp	40	.200			9.40		9.40	14.45
5420	Permanent U-Bolt roof anchor (install only)	2 Carp	40	.400			18.80		18.80	29
5430	Temporary roof ridge anchor (install only)	1 Carp	64	.125			5.85		5.85	9.05

01 54 26 – Temporary Swing Staging

01 54 26.50 Swing Staging		Crew	Daily Output	Labor-Hours	Unit	Material	Labor	Equipment	Total	Total Incl O&P
0010	**SWING STAGING**, 500 lb. cap., 2′ wide to 24′ long, hand operated									
0020	steel cable type, with 60′ cables, buy				Ea.	4,900			4,900	5,375
0030	Rent per month				″	490			490	540
0600	Lightweight (not for masons) 24′ long for 150′ height,									
0610	manual type, buy				Ea.	10,200			10,200	11,200
0620	Rent per month					1,025			1,025	1,125
0700	Powered, electric or air, to 150′ high, buy					25,600			25,600	28,200
0710	Rent per month					1,800			1,800	1,975
0780	To 300′ high, buy					26,000			26,000	28,600
0800	Rent per month					1,825			1,825	2,000
1000	Bosun's chair or work basket 3′ x 3.5′, to 300′ high, electric, buy					10,400			10,400	11,400
1010	Rent per month					730			730	800
2200	Move swing staging (setup and remove)	E-4	2	16	Move		850	73	923	1,550

01 54 Construction Aids

01 54 36 – Equipment Mobilization

01 54 36.50 Mobilization		Crew	Daily Output	Labor-Hours	Unit	Material	2015 Bare Costs Labor	Equipment	Total	Total Incl O&P
0010	**MOBILIZATION** (Use line item again for demobilization) R015436-50									
0015	Up to 25 mi. haul dist. (50 mi. RT for mob/demob crew)									
1200	Small equipment, placed in rear of, or towed by pickup truck	A-3A	4	2	Ea.		97	39	136	191
1300	Equipment hauled on 3-ton capacity towed trailer	A-3Q	2.67	3			146	67	213	295
1400	20-ton capacity	B-34U	2	8			355	237	592	795
1500	40-ton capacity	B-34N	2	8			365	375	740	965
1600	50-ton capacity	B-34V	1	24			1,125	1,050	2,175	2,850
1700	Crane, truck-mounted, up to 75 ton (driver only)	1 Eqhv	4	2			103		103	157
1800	Over 75 ton (with chase vehicle)	A-3E	2.50	6.400			294	62.50	356.50	515
2400	Crane, large lattice boom, requiring assembly	B-34W	.50	144	↓		6,275	7,500	13,775	17,800
2500	For each additional 5 miles haul distance, add						10%	10%		
3000	For large pieces of equipment, allow for assembly/knockdown									
3001	For mob/demob of vibrofloatation equip, see Section 31 45 13.10									
3100	For mob/demob of micro-tunneling equip, see Section 33 05 23.19									
3200	For mob/demob of pile driving equip, see Section 31 62 19.10									
3300	For mob/demob of caisson drilling equip, see Section 31 63 26.13									

01 55 Vehicular Access and Parking

01 55 23 – Temporary Roads

01 55 23.50 Roads and Sidewalks		Crew	Daily Output	Labor-Hours	Unit	Material	Labor	Equipment	Total	Total Incl O&P
0010	**ROADS AND SIDEWALKS** Temporary									
0050	Roads, gravel fill, no surfacing, 4" gravel depth	B-14	715	.067	S.Y.	4.04	2.67	.51	7.22	9.10
0100	8" gravel depth	"	615	.078	"	8.10	3.10	.59	11.79	14.30
1000	Ramp, 3/4" plywood on 2" x 6" joists, 16" O.C.	2 Carp	300	.053	S.F.	1.58	2.50		4.08	5.60
1100	On 2" x 10" joists, 16" O.C.	"	275	.058	"	2.24	2.73		4.97	6.65

01 56 Temporary Barriers and Enclosures

01 56 13 – Temporary Air Barriers

01 56 13.60 Tarpaulins		Crew	Daily Output	Labor-Hours	Unit	Material	Labor	Equipment	Total	Total Incl O&P
0010	**TARPAULINS**									
0020	Cotton duck, 10 oz. to 13.13 oz. per S.Y., 6'x8'				S.F.	.83			.83	.91
0050	30'x30'					.59			.59	.65
0100	Polyvinyl coated nylon, 14 oz. to 18 oz., minimum					1.36			1.36	1.50
0150	Maximum					1.36			1.36	1.50
0200	Reinforced polyethylene 3 mils thick, white					.04			.04	.04
0300	4 mils thick, white, clear or black					.09			.09	.10
0400	5.5 mils thick, clear					.18			.18	.20
0500	White, fire retardant					.44			.44	.48
0600	12 mils, oil resistant, fire retardant					.39			.39	.43
0700	8.5 mils, black					.57			.57	.63
0710	Woven polyethylene, 6 mils thick					.18			.18	.20
0730	Polyester reinforced w/integral fastening system 11 mils thick					.21			.21	.23
0740	Mylar polyester, non-reinforced, 7 mils thick				↓	1.17			1.17	1.29

01 56 13.90 Winter Protection		Crew	Daily Output	Labor-Hours	Unit	Material	Labor	Equipment	Total	Total Incl O&P
0010	**WINTER PROTECTION**									
0100	Framing to close openings	2 Clab	500	.032	S.F.	.46	1.20		1.66	2.35
0200	Tarpaulins hung over scaffolding, 8 uses, not incl. scaffolding		1500	.011		.24	.40		.64	.88
0250	Tarpaulin polyester reinf. w/integral fastening system 11 mils thick	↓	1600	.010	↓	.21	.38		.59	.81

01 56 Temporary Barriers and Enclosures

01 56 13 – Temporary Air Barriers

01 56 13.90 Winter Protection		Crew	Daily Output	Labor-Hours	Unit	2015 Bare Costs Material	Labor	Equipment	Total	Total Incl O&P
0300	Prefab fiberglass panels, steel frame, 8 uses	2 Clab	1200	.013	S.F.	2.40	.50		2.90	3.41

01 56 16 – Temporary Dust Barriers

01 56 16.10 Dust Barriers, Temporary

		Crew	Daily Output	Labor-Hours	Unit	Material	Labor	Equipment	Total	Total Incl O&P
0010	**DUST BARRIERS, TEMPORARY**									
0020	Spring loaded telescoping pole & head, to 12', erect and dismantle	1 Clab	240	.033	Ea.		1.25		1.25	1.93
0025	Cost per day (based upon 250 days)				Day	.26			.26	.29
0030	To 21', erect and dismantle	1 Clab	240	.033	Ea.		1.25		1.25	1.93
0035	Cost per day (based upon 250 days)				Day	.44			.44	.48
0040	Accessories, caution tape reel, erect and dismantle	1 Clab	480	.017	Ea.		.63		.63	.96
0045	Cost per day (based upon 250 days)				Day	.36			.36	.40
0060	Foam rail and connector, erect and dismantle	1 Clab	240	.033	Ea.		1.25		1.25	1.93
0065	Cost per day (based upon 250 days)				Day	.10			.10	.11
0070	Caution tape	1 Clab	384	.021	C.L.F.	2.70	.78		3.48	4.18
0080	Zipper, standard duty		60	.133	Ea.	8	5		13	16.50
0090	Heavy duty		48	.167	"	9.50	6.25		15.75	20
0100	Polyethylene sheet, 4 mil		37	.216	Sq.	2.90	8.15		11.05	15.70
0110	6 mil		37	.216	"	4.02	8.15		12.17	16.90
1000	Dust partition, 6 mil polyethylene, 1" x 3" frame	2 Carp	2000	.008	S.F.	.30	.38		.68	.90
1080	2" x 4" frame	"	2000	.008	"	.35	.38		.73	.97

01 56 23 – Temporary Barricades

01 56 23.10 Barricades

		Crew	Daily Output	Labor-Hours	Unit	Material	Labor	Equipment	Total	Total Incl O&P
0010	**BARRICADES**									
0020	5' high, 3 rail @ 2" x 8", fixed	2 Carp	20	.800	L.F.	6	37.50		43.50	64.50
0150	Movable	"	30	.533	"	5	25		30	44
0300	Stock units, 6' high, 8' wide, plain, buy				Ea.	390			390	430
0350	With reflective tape, buy				"	405			405	445
0400	Break-a-way 3" PVC pipe barricade									
0410	with 3 ea. 1' x 4' reflectorized panels, buy				Ea.	106			106	117
0500	Plywood with steel legs, 24" wide					59			59	65
0600	Warning signal flag tree, 11' high, 2 flags, buy					257			257	283
0800	Traffic cones, PVC, 18" high					12.50			12.50	13.75
0850	28" high					19.50			19.50	21.50
1000	Guardrail, wooden, 3' high, 1" x 6", on 2" x 4" posts	2 Carp	200	.080	L.F.	1.27	3.76		5.03	7.20
1100	2" x 6", on 4" x 4" posts	"	165	.097		2.46	4.55		7.01	9.70
1200	Portable metal with base pads, buy					15.55			15.55	17.10
1250	Typical installation, assume 10 reuses	2 Carp	600	.027		2.55	1.25		3.80	4.74
1300	Barricade tape, polyethylene, 7 mil, 3" wide x 500' long roll				Ea.	25			25	27.50
3000	Detour signs, set up and remove									
3010	Reflective aluminum, MUTCD, 24" x 24", post mounted	1 Clab	20	.400	Ea.	2.46	15.05		17.51	25.50
4000	Roof edge portable barrier stands and warning flags, 50 uses	1 Rohe	9100	.001	L.F.	.07	.03		.10	.12
4010	100 uses	"	9100	.001	"	.03	.03		.06	.09
5000	Barricades, see Section 01 54 33.40									

01 56 26 – Temporary Fencing

01 56 26.50 Temporary Fencing

		Crew	Daily Output	Labor-Hours	Unit	Material	Labor	Equipment	Total	Total Incl O&P
0010	**TEMPORARY FENCING**									
0020	Chain link, 11 ga., 4' high	2 Clab	400	.040	L.F.	2.95	1.50		4.45	5.55
0100	6' high		300	.053		2.95	2.01		4.96	6.35
0200	Rented chain link, 6' high, to 1000' (up to 12 mo.)		400	.040		4.29	1.50		5.79	7.05
0250	Over 1000' (up to 12 mo.)		300	.053		4.29	2.01		6.30	7.80
0350	Plywood, painted, 2" x 4" frame, 4' high	A-4	135	.178		6.20	7.95		14.15	19
0400	4" x 4" frame, 8' high	"	110	.218		12	9.75		21.75	28

01 56 Temporary Barriers and Enclosures

01 56 26 – Temporary Fencing

01 56 26.50 Temporary Fencing		Crew	Daily Output	Labor-Hours	Unit	Material	2015 Bare Costs Labor	Equipment	Total	Total Incl O&P
0500	Wire mesh on 4" x 4" posts, 4' high	2 Carp	100	.160	L.F.	9.85	7.50		17.35	22.50
0550	8' high	"	80	.200	↓	14.90	9.40		24.30	31

01 56 29 – Temporary Protective Walkways

01 56 29.50 Protection		Crew	Daily Output	Labor-Hours	Unit	Material	Labor	Equipment	Total	Total Incl O&P
0010	**PROTECTION**									
0020	Stair tread, 2" x 12" planks, 1 use	1 Carp	75	.107	Tread	5.10	5		10.10	13.30
0100	Exterior plywood, 1/2" thick, 1 use		65	.123		1.97	5.80		7.77	11.05
0200	3/4" thick, 1 use		60	.133	↓	2.78	6.25		9.03	12.70
2200	Sidewalks, 2" x 12" planks, 2 uses		350	.023	S.F.	.85	1.07		1.92	2.59
2300	Exterior plywood, 2 uses, 1/2" thick		750	.011		.33	.50		.83	1.13
2400	5/8" thick		650	.012		.39	.58		.97	1.32
2500	3/4" thick	↓	600	.013	↓	.46	.63		1.09	1.47

01 56 32 – Temporary Security

01 56 32.50 Watchman		Crew	Daily Output	Labor-Hours	Unit	Material	Labor	Equipment	Total	Total Incl O&P
0010	**WATCHMAN**									
0020	Service, monthly basis, uniformed person, minimum				Hr.				25	27.50
0100	Maximum								45.50	50
0200	Person and command dog, minimum								31	34
0300	Maximum				↓				54.50	60
0500	Sentry dog, leased, with job patrol (yard dog), 1 dog				Week				290	320
0600	2 dogs				"				390	430
0800	Purchase, trained sentry dog, minimum				Ea.				1,375	1,500
0900	Maximum				"				2,725	3,000

01 58 Project Identification

01 58 13 – Temporary Project Signage

01 58 13.50 Signs		Crew	Daily Output	Labor-Hours	Unit	Material	Labor	Equipment	Total	Total Incl O&P
0010	**SIGNS**									
0020	High intensity reflectorized, no posts, buy				Ea.	25			25	27.50

01 71 Examination and Preparation

01 71 23 – Field Engineering

01 71 23.13 Construction Layout		Crew	Daily Output	Labor-Hours	Unit	Material	Labor	Equipment	Total	Total Incl O&P
0010	**CONSTRUCTION LAYOUT**									
1100	Crew for layout of building, trenching or pipe laying, 2 person crew	A-6	1	16	Day		750	55	805	1,200
1200	3 person crew	A-7	1	24			1,225	54.50	1,279.50	1,950
1400	Crew for roadway layout, 4 person crew	A-8	1	32	↓		1,600	54.50	1,654.50	2,500
01 71 23.19 Surveyor Stakes										
0010	**SURVEYOR STAKES**									
0020	Hardwood, 1" x 1" x 48" long				C	70			70	77
0100	2" x 2" x 18" long					78			78	86
0150	2" x 2" x 24" long				↓	140			140	154

01 74 Cleaning and Waste Management

01 74 13 – Progress Cleaning

01 74 13.20 Cleaning Up		Crew	Daily Output	Labor-Hours	Unit	Material	2015 Bare Costs Labor	Equipment	Total	Total Incl O&P
0010	**CLEANING UP**									
0020	After job completion, allow, minimum				Job				.30%	.30%
0040	Maximum				"				1%	1%
0050	Cleanup of floor area, continuous, per day, during const.	A-5	24	.750	M.S.F.	2.18	28.50	2.56	33.24	48.50
0100	Final by GC at end of job	"	11.50	1.565	"	2.31	59	5.35	66.66	99.50
0200	Rubbish removal, see Section 02 41 19.19									

01 76 Protecting Installed Construction

01 76 13 – Temporary Protection of Installed Construction

01 76 13.20 Temporary Protection		Crew	Daily Output	Labor-Hours	Unit	Material	Labor	Equipment	Total	Total Incl O&P
0010	**TEMPORARY PROTECTION**									
0020	Flooring, 1/8" tempered hardboard, taped seams	2 Carp	1500	.011	S.F.	.41	.50		.91	1.22
0030	Peel away carpet protection	1 Clab	3200	.003	"	.11	.09		.20	.26

01 91 Commissioning

01 91 13 – General Commissioning Requirements

01 91 13.50 Building Commissioning		Crew	Daily Output	Labor-Hours	Unit	Material	Labor	Equipment	Total	Total Incl O&P
0010	**BUILDING COMMISSIONING**									
0100	Basic building commissioning, minimum				%				.25%	.25%
0150	Maximum								.50%	.50%
0200	Enhanced building commissioning, minimum								.50%	.50%
0250	Maximum				↓				1%	1%

Division Notes

		CREW	DAILY OUTPUT	LABOR-HOURS	UNIT	BARE COSTS MAT.	LABOR	EQUIP.	TOTAL	TOTAL INCL O&P

Estimating Tips

02 30 00 Subsurface Investigation

In preparing estimates on structures involving earthwork or foundations, all information concerning soil characteristics should be obtained. Look particularly for hazardous waste, evidence of prior dumping of debris, and previous stream beds.

02 40 00 Demolition and Structure Moving

The costs shown for selective demolition do not include rubbish handling or disposal. These items should be estimated separately using RSMeans data or other sources.

- Historic preservation often requires that the contractor remove materials from the existing structure, rehab them, and replace them. The estimator must be aware of any related measures and precautions that must be taken when doing selective demolition and cutting and patching. Requirements may include special handling and storage, as well as security.
- In addition to Subdivision 02 41 00, you can find selective demolition items in each division. Example: Roofing demolition is in Division 7.

02 40 00 Building Deconstruction

This section provides costs for the careful dismantling and recycling of most of low-rise building materials.

02 50 00 Containment of Hazardous Waste

This section addresses on-site hazardous waste disposal costs.

02 80 00 Hazardous Material Disposal/ Remediation

This subdivision includes information on hazardous waste handling, asbestos remediation, lead remediation, and mold remediation. See reference R028213-20 and R028319-60 for further guidance in using these unit price lines.

02 90 00 Monitoring Chemical Sampling, Testing Analysis

This section provides costs for on-site sampling and testing hazardous waste.

Reference Numbers

Reference numbers are shown in shaded boxes at the beginning of some major classifications. These numbers refer to related items in the Reference Section. The reference information may be an estimating procedure, an alternate pricing method, or technical information.

Note: Not all subdivisions listed here necessarily appear in this publication. ■

02 21 Surveys

02 21 13 – Site Surveys

02 21 13.09 Topographical Surveys		Crew	Daily Output	Labor-Hours	Unit	Material	2015 Bare Costs Labor	Equipment	Total	Total Incl O&P
0010	TOPOGRAPHICAL SURVEYS									
0020	Topographical surveying, conventional, minimum	A-7	3.30	7.273	Acre	20	375	16.60	411.60	615
0100	Maximum	A-8	.60	53.333	"	60	2,675	91	2,826	4,250

02 21 13.13 Boundary and Survey Markers		Crew	Daily Output	Labor-Hours	Unit	Material	Labor	Equipment	Total	Total Incl O&P
0010	BOUNDARY AND SURVEY MARKERS									
0300	Lot location and lines, large quantities, minimum	A-7	2	12	Acre	35	620	27.50	682.50	1,025
0320	Average	"	1.25	19.200		55	990	44	1,089	1,625
0400	Small quantities, maximum	A-8	1	32	↓	75	1,600	54.50	1,729.50	2,600
0600	Monuments, 3′ long	A-7	10	2.400	Ea.	40	124	5.45	169.45	240
0800	Property lines, perimeter, cleared land	"	1000	.024	L.F.	.05	1.24	.05	1.34	2.02
0900	Wooded land	A-8	875	.037	"	.07	1.83	.06	1.96	2.95

02 21 13.16 Aerial Surveys		Crew	Daily Output	Labor-Hours	Unit	Material	Labor	Equipment	Total	Total Incl O&P
0010	AERIAL SURVEYS									
1500	Aerial surveying, including ground control, minimum fee, 10 acres				Total				4,700	4,700
1510	100 acres								9,400	9,400
1550	From existing photography, deduct				↓				1,625	1,625
1600	2′ contours, 10 acres				Acre				470	470
1850	100 acres								94	94
2000	1000 acres								90	90
2050	10,000 acres				↓				85	85

02 32 Geotechnical Investigations

02 32 13 – Subsurface Drilling and Sampling

02 32 13.10 Boring and Exploratory Drilling		Crew	Daily Output	Labor-Hours	Unit	Material	Labor	Equipment	Total	Total Incl O&P
0010	BORING AND EXPLORATORY DRILLING									
0020	Borings, initial field stake out & determination of elevations	A-6	1	16	Day		750	55	805	1,200
0100	Drawings showing boring details				Total		335		335	425
0200	Report and recommendations from P.E.						775		775	970
0300	Mobilization and demobilization	B-55	4	6	↓		229	271	500	650
0350	For over 100 miles, per added mile		450	.053	Mile		2.03	2.41	4.44	5.75
0600	Auger holes in earth, no samples, 2-1/2″ diameter		78.60	.305	L.F.		11.65	13.75	25.40	33
0650	4″ diameter		67.50	.356			13.55	16.05	29.60	38
0800	Cased borings in earth, with samples, 2-1/2″ diameter		55.50	.432		14	16.50	19.50	50	62
0850	4″ diameter	↓	32.60	.736		18	28	33	79	99.50
1000	Drilling in rock, "BX" core, no sampling	B-56	34.90	.458			19.75	45.50	65.25	80
1050	With casing & sampling		31.70	.505		14	22	50.50	86.50	104
1200	"NX" core, no sampling		25.92	.617			26.50	61.50	88	108
1250	With casing and sampling	↓	25	.640	↓	15	27.50	63.50	106	129
1400	Borings, earth, drill rig and crew with truck mounted auger	B-55	1	24	Day		915	1,075	1,990	2,600
1450	Rock using crawler type drill	B-56	1	16	"		690	1,600	2,290	2,800
1500	For inner city borings add, minimum								10%	10%
1510	Maximum								20%	20%

02 32 19 – Exploratory Excavations

02 32 19.10 Test Pits		Crew	Daily Output	Labor-Hours	Unit	Material	Labor	Equipment	Total	Total Incl O&P
0010	TEST PITS									
0020	Hand digging, light soil	1 Clab	4.50	1.778	C.Y.		67		67	103
0100	Heavy soil	"	2.50	3.200			120		120	185
0120	Loader-backhoe, light soil	B-11M	28	.571			25	14	39	54
0130	Heavy soil	"	20	.800	↓		35.50	19.60	55.10	75.50
1000	Subsurface exploration, mobilization				Mile				6.75	8.40

02 32 Geotechnical Investigations

02 32 19 – Exploratory Excavations

02 32 19.10 Test Pits		Crew	Daily Output	Labor-Hours	Unit	Material	2015 Bare Costs Labor	Equipment	Total	Total Incl O&P
1010	Difficult access for rig, add				Hr.				260	320
1020	Auger borings, drill rig, incl. samples				L.F.				26.50	33
1030	Hand auger								31.50	40
1050	Drill and sample every 5′, split spoon				↓				31.50	40
1060	Extra samples				Ea.				36	45.50

02 41 Demolition

02 41 13 – Selective Site Demolition

02 41 13.15 Hydrodemolition

		Crew	Daily Output	Labor-Hours	Unit	Material	Labor	Equipment	Total	Total Incl O&P
0010	**HYDRODEMOLITION** R024119-10									
0015	Hydrodemolition, concrete pavement									
0120	20,000 PSI, Crew to include loader/vacuum truck as required									
0130	2″ depth	B-5	1000	.056	S.F.		2.33	1.42	3.75	5.15
0410	4″ depth		800	.070			2.91	1.78	4.69	6.40
0420	6″ depth	↓	600	.093	↓		3.88	2.37	6.25	8.55

02 41 13.17 Demolish, Remove Pavement and Curb

		Crew	Daily Output	Labor-Hours	Unit	Material	Labor	Equipment	Total	Total Incl O&P
0010	**DEMOLISH, REMOVE PAVEMENT AND CURB** R024119-10									
5010	Pavement removal, bituminous roads, up to 3″ thick	B-38	690	.058	S.Y.		2.48	1.85	4.33	5.80
5050	4″ to 6″ thick		420	.095			4.08	3.04	7.12	9.60
5100	Bituminous driveways		640	.063			2.68	1.99	4.67	6.30
5200	Concrete to 6″ thick, hydraulic hammer, mesh reinforced		255	.157			6.70	5	11.70	15.75
5300	Rod reinforced		200	.200	↓		8.55	6.40	14.95	20
5400	Concrete, 7″ to 24″ thick, plain		33	1.212	C.Y.		52	38.50	90.50	122
5500	Reinforced	↓	24	1.667	″		71.50	53	124.50	168
5600	With hand held air equipment, bituminous, to 6″ thick	B-39	1900	.025	S.F.		1	.12	1.12	1.67
5700	Concrete to 6″ thick, no reinforcing		1600	.030			1.19	.15	1.34	1.99
5800	Mesh reinforced		1400	.034			1.36	.17	1.53	2.27
5900	Rod reinforced	↓	765	.063	↓		2.50	.30	2.80	4.16
6000	Curbs, concrete, plain	B-6	360	.067	L.F.		2.75	1.01	3.76	5.30
6100	Reinforced		275	.087			3.60	1.32	4.92	6.95
6200	Granite		360	.067			2.75	1.01	3.76	5.30
6300	Bituminous	↓	528	.045	↓		1.88	.69	2.57	3.63

02 41 13.23 Utility Line Removal

		Crew	Daily Output	Labor-Hours	Unit	Material	Labor	Equipment	Total	Total Incl O&P
0010	**UTILITY LINE REMOVAL**									
0015	No hauling, abandon catch basin or manhole	B-6	7	3.429	Ea.		142	52	194	274
0020	Remove existing catch basin or manhole, masonry		4	6			248	91	339	480
0030	Catch basin or manhole frames and covers, stored		13	1.846			76	28	104	148
0040	Remove and reset	↓	7	3.429			142	52	194	274
0900	Hydrants, fire, remove only	B-21A	5	8			375	94.50	469.50	675
0950	Remove and reset	″	2	20	↓		940	237	1,177	1,675
2900	Pipe removal, sewer/water, no excavation, 12″ diameter	B-6	175	.137	L.F.		5.65	2.08	7.73	10.95
2930	15″-18″ diameter	B-12Z	150	.160			6.75	10.70	17.45	22
2960	21″-24″ diameter		120	.200			8.45	13.40	21.85	27.50
3000	27″-36″ diameter	↓	90	.267			11.30	17.85	29.15	37
3200	Steel, welded connections, 4″ diameter	B-6	160	.150			6.20	2.28	8.48	11.95
3300	10″ diameter	″	80	.300	↓		12.40	4.55	16.95	24

02 41 13.30 Minor Site Demolition

		Crew	Daily Output	Labor-Hours	Unit	Material	Labor	Equipment	Total	Total Incl O&P
0010	**MINOR SITE DEMOLITION** R024119-10									
0100	Roadside delineators, remove only	B-80	175	.183	Ea.		7.55	4.12	11.67	16.05
0110	Remove and reset	″	100	.320	″		13.20	7.20	20.40	28

02 41 Demolition

02 41 13 – Selective Site Demolition

02 41 13.30 Minor Site Demolition		Crew	Daily Output	Labor-Hours	Unit	Material	2015 Bare Costs Labor	Equipment	Total	Total Incl O&P
0800	Guiderail, corrugated steel, remove only	B-80A	100	.240	L.F.		9	3.03	12.03	17.25
0850	Remove and reset	"	40	.600	"		22.50	7.60	30.10	43
0860	Guide posts, remove only	B-80B	120	.267	Ea.		10.75	2.03	12.78	18.75
0870	Remove and reset	B-55	50	.480	"		18.30	21.50	39.80	52
1000	Masonry walls, block, solid	B-5	1800	.031	C.F.		1.29	.79	2.08	2.85
1200	Brick, solid		900	.062			2.59	1.58	4.17	5.70
1400	Stone, with mortar		900	.062			2.59	1.58	4.17	5.70
1500	Dry set		1500	.037			1.55	.95	2.50	3.42
1600	Median barrier, precast concrete, remove and store	B-3	430	.112	L.F.		4.57	6	10.57	13.55
1610	Remove and reset	"	390	.123	"		5.05	6.60	11.65	14.95
4000	Sidewalk removal, bituminous, 2" thick	B-6	350	.069	S.Y.		2.83	1.04	3.87	5.45
4010	2-1/2" thick		325	.074			3.05	1.12	4.17	5.90
4050	Brick, set in mortar		185	.130			5.35	1.97	7.32	10.35
4100	Concrete, plain, 4"		160	.150			6.20	2.28	8.48	11.95
4110	Plain, 5"		140	.171			7.05	2.60	9.65	13.70
4120	Plain, 6"		120	.200			8.25	3.04	11.29	16
4200	Mesh reinforced, concrete, 4 "		150	.160			6.60	2.43	9.03	12.75
4210	5 " thick		131	.183			7.55	2.78	10.33	14.60
4220	6 " thick		112	.214			8.85	3.25	12.10	17.15
4300	Slab on grade removal, plain	B-5	45	1.244	C.Y.		52	31.50	83.50	114
4310	Mesh reinforced		33	1.697			70.50	43	113.50	156
4320	Rod reinforced		25	2.240			93	57	150	206
4400	For congested sites or small quantities, add up to								200%	200%
4450	For disposal on site, add	B-11A	232	.069			3.04	6	9.04	11.25
4500	To 5 miles, add	B-34D	76	.105			4.22	9.75	13.97	17.05

02 41 13.33 Railtrack Removal

		Crew	Daily Output	Labor-Hours	Unit	Material	Labor	Equipment	Total	Total Incl O&P
0010	**RAILTRACK REMOVAL**									
3500	Railroad track removal, ties and track	B-13	330	.170	L.F.		6.95	2.23	9.18	13.10
3600	Ballast	B-14	500	.096	C.Y.		3.82	.73	4.55	6.65
3700	Remove and re-install, ties & track using new bolts & spikes		50	.960	L.F.		38	7.30	45.30	66.50
3800	Turnouts using new bolts and spikes		1	48	Ea.		1,900	365	2,265	3,325

02 41 13.60 Selective Demolition Fencing

		Crew	Daily Output	Labor-Hours	Unit	Material	Labor	Equipment	Total	Total Incl O&P
0010	**SELECTIVE DEMOLITION FENCING** R024119-10									
1600	Fencing, barbed wire, 3 strand	2 Clab	430	.037	L.F.		1.40		1.40	2.15
1650	5 strand	"	280	.057			2.15		2.15	3.31
1700	Chain link, posts & fabric, 8' to 10' high, remove only	B-6	445	.054			2.23	.82	3.05	4.31
1750	Remove and reset	"	70	.343			14.15	5.20	19.35	27

02 41 16 – Structure Demolition

02 41 16.13 Building Demolition

		Crew	Daily Output	Labor-Hours	Unit	Material	Labor	Equipment	Total	Total Incl O&P
0010	**BUILDING DEMOLITION** Large urban projects, incl. 20 mi. haul R024119-10									
0011	No foundation or dump fees, C.F. is vol. of building standing									
0020	Steel	B-8	21500	.003	C.F.		.13	.15	.28	.36
0050	Concrete		15300	.004			.18	.22	.40	.51
0080	Masonry		20100	.003			.14	.16	.30	.39
0100	Mixture of types		20100	.003			.14	.16	.30	.39
0500	Small bldgs, or single bldgs, no salvage included, steel	B-3	14800	.003			.13	.17	.30	.39
0600	Concrete		11300	.004			.17	.23	.40	.52
0650	Masonry		14800	.003			.13	.17	.30	.39
0700	Wood		14800	.003			.13	.17	.30	.39
0750	For buildings with no interior walls, deduct								50%	50%
1000	Demoliton single family house, one story, wood 1600 S.F.	B-3	1	48	Ea.		1,975	2,575	4,550	5,825
1020	3200 S.F.		.50	96			3,925	5,150	9,075	11,700

02 41 Demolition

02 41 16 – Structure Demolition

02 41 16.13 Building Demolition		Crew	Daily Output	Labor-Hours	Unit	Material	2015 Bare Costs Labor	Equipment	Total	Total Incl O&P
1200	Demoliton two family house, two story, wood 2400 S.F.	B-3	.67	71.964	Ea.		2,950	3,850	6,800	8,750
1220	4200 S.F.		.38	128			5,250	6,850	12,100	15,600
1300	Demoliton three family house, three story, wood 3200 S.F.		.50	96			3,925	5,150	9,075	11,700
1320	5400 S.F.		.30	160			6,550	8,575	15,125	19,400
5000	For buildings with no interior walls, deduct								50%	50%

02 41 16.15 Explosive/Implosive Demolition

		Crew	Daily Output	Labor-Hours	Unit	Material	Labor	Equipment	Total	Total Incl O&P
0010	**EXPLOSIVE/IMPLOSIVE DEMOLITION** R024119-10									
0011	Large projects,									
0020	No disposal fee based on building volume, steel building	B-5B	16900	.003	C.F.		.13	.17	.30	.38
0100	Concrete building		16900	.003			.13	.17	.30	.38
0200	Masonry building		16900	.003			.13	.17	.30	.38
0400	Disposal of material, minimum	B-3	445	.108	C.Y.		4.41	5.80	10.21	13.10
0500	Maximum	"	365	.132	"		5.40	7.05	12.45	15.95

02 41 16.17 Building Demolition Footings and Foundations

		Crew	Daily Output	Labor-Hours	Unit	Material	Labor	Equipment	Total	Total Incl O&P
0010	**BUILDING DEMOLITION FOOTINGS AND FOUNDATIONS** R024119-10									
0200	Floors, concrete slab on grade,									
0240	4" thick, plain concrete	B-13L	5000	.003	S.F.		.17	.41	.58	.70
0280	Reinforced, wire mesh		4000	.004			.21	.51	.72	.88
0300	Rods		4500	.004			.18	.46	.64	.78
0400	6" thick, plain concrete		4000	.004			.21	.51	.72	.88
0420	Reinforced, wire mesh		3200	.005			.26	.64	.90	1.10
0440	Rods		3600	.004			.23	.57	.80	.98
1000	Footings, concrete, 1' thick, 2' wide	B-5	300	.187	L.F.		7.75	4.74	12.49	17.10
1080	1'-6" thick, 2' wide		250	.224			9.30	5.70	15	20.50
1120	3' wide		200	.280			11.65	7.10	18.75	25.50
1140	2' thick, 3' wide		175	.320			13.30	8.10	21.40	29.50
1200	Average reinforcing, add								10%	10%
1220	Heavy reinforcing, add								20%	20%
2000	Walls, block, 4" thick	B-13L	8000	.002	S.F.		.10	.26	.36	.44
2040	6" thick		6000	.003			.14	.34	.48	.59
2080	8" thick		4000	.004			.21	.51	.72	.88
2100	12" thick		3000	.005			.28	.68	.96	1.17
2200	For horizontal reinforcing, add								10%	10%
2220	For vertical reinforcing, add								20%	20%
2400	Concrete, plain concrete, 6" thick	B-13L	4000	.004			.21	.51	.72	.88
2420	8" thick		3500	.005			.24	.59	.83	1.01
2440	10" thick		3000	.005			.28	.68	.96	1.17
2500	12" thick		2500	.006			.33	.82	1.15	1.40
2600	For average reinforcing, add								10%	10%
2620	For heavy reinforcing, add								20%	20%
4000	For congested sites or small quantities, add up to								200%	200%
4200	Add for disposal, on site	B-11A	232	.069	C.Y.		3.04	6	9.04	11.25
4250	To five miles	B-30	220	.109	"		4.75	10.95	15.70	19.25

02 41 19 – Selective Demolition

02 41 19.13 Selective Building Demolition

		Crew	Daily Output	Labor-Hours	Unit	Material	Labor	Equipment	Total	Total Incl O&P
0010	**SELECTIVE BUILDING DEMOLITION**									
0020	Costs related to selective demolition of specific building components									
0025	are included under Common Work Results (XX 05)									
0030	in the component's appropriate division.									

02 41 Demolition

02 41 19 – Selective Demolition

02 41 19.16 Selective Demolition, Cutout		Crew	Daily Output	Labor-Hours	Unit	Material	2015 Bare Costs Labor	Equipment	Total	Total Incl O&P
0010	SELECTIVE DEMOLITION, CUTOUT R024119-10									
0020	Concrete, elev. slab, light reinforcement, under 6 C.F.	B-9	65	.615	C.F.		23.50	3.58	27.08	40
0050	Light reinforcing, over 6 C.F.		75	.533	"		20.50	3.10	23.60	34.50
0200	Slab on grade to 6" thick, not reinforced, under 8 S.F.		85	.471	S.F.		17.90	2.74	20.64	30.50
0250	8 – 16 S.F.		175	.229	"		8.70	1.33	10.03	14.80
0255	For over 16 S.F. see Line 02 41 16.17 0400									
0600	Walls, not reinforced, under 6 C.F.	B-9	60	.667	C.F.		25.50	3.88	29.38	43.50
0650	6 – 12 C.F.	"	80	.500	"		19	2.91	21.91	32
0655	For over 12 C.F. see Line 02 41 16.17 2500									
1000	Concrete, elevated slab, bar reinforced, under 6 C.F.	B-9	45	.889	C.F.		34	5.15	39.15	57.50
1050	Bar reinforced, over 6 C.F.		50	.800	"		30.50	4.66	35.16	52
1200	Slab on grade to 6" thick, bar reinforced, under 8 S.F.		75	.533	S.F.		20.50	3.10	23.60	34.50
1250	8 – 16 S.F.		150	.267	"		10.15	1.55	11.70	17.30
1255	For over 16 S.F. see Line 02 41 16.17 0440									
1400	Walls, bar reinforced, under 6 C.F.	B-9	50	.800	C.F.		30.50	4.66	35.16	52
1450	6 – 12 C.F.	"	70	.571	"		21.50	3.33	24.83	37
1455	For over 12 C.F. see Lines 02 41 16.17 2500 and 2600									
2000	Brick, to 4 S.F. opening, not including toothing									
2040	4" thick	B-9	30	1.333	Ea.		50.50	7.75	58.25	86.50
2060	8" thick		18	2.222			84.50	12.95	97.45	144
2080	12" thick		10	4			152	23.50	175.50	260
2400	Concrete block, to 4 S.F. opening, 2" thick		35	1.143			43.50	6.65	50.15	74.50
2420	4" thick		30	1.333			50.50	7.75	58.25	86.50
2440	8" thick		27	1.481			56.50	8.60	65.10	96
2460	12" thick		24	1.667			63.50	9.70	73.20	108
2600	Gypsum block, to 4 S.F. opening, 2" thick		80	.500			19	2.91	21.91	32
2620	4" thick		70	.571			21.50	3.33	24.83	37
2640	8" thick		55	.727			27.50	4.23	31.73	47
2800	Terra cotta, to 4 S.F. opening, 4" thick		70	.571			21.50	3.33	24.83	37
2840	8" thick		65	.615			23.50	3.58	27.08	40
2880	12" thick		50	.800			30.50	4.66	35.16	52
3000	Toothing masonry cutouts, brick, soft old mortar	1 Brhe	40	.200	V.L.F.		7.60		7.60	11.60
3100	Hard mortar		30	.267			10.15		10.15	15.50
3200	Block, soft old mortar		70	.114			4.35		4.35	6.65
3400	Hard mortar		50	.160			6.10		6.10	9.30
6000	Walls, interior, not including re-framing,									
6010	openings to 5 S.F.									
6100	Drywall to 5/8" thick	1 Clab	24	.333	Ea.		12.55		12.55	19.30
6200	Paneling to 3/4" thick		20	.400			15.05		15.05	23
6300	Plaster, on gypsum lath		20	.400			15.05		15.05	23
6340	On wire lath		14	.571			21.50		21.50	33
7000	Wood frame, not including re-framing, openings to 5 S.F.									
7200	Floors, sheathing and flooring to 2" thick	1 Clab	5	1.600	Ea.		60		60	92.50
7310	Roofs, sheathing to 1" thick, not including roofing		6	1.333			50		50	77
7410	Walls, sheathing to 1" thick, not including siding		7	1.143			43		43	66

02 41 19.18 Selective Demolition, Disposal Only		Crew	Daily Output	Labor-Hours	Unit	Material	Labor	Equipment	Total	Total Incl O&P
0010	SELECTIVE DEMOLITION, DISPOSAL ONLY R024119-10									
0015	Urban bldg w/salvage value allowed									
0020	Including loading and 5 mile haul to dump									
0200	Steel frame	B-3	430	.112	C.Y.		4.57	6	10.57	13.55
0300	Concrete frame		365	.132			5.40	7.05	12.45	15.95
0400	Masonry construction		445	.108			4.41	5.80	10.21	13.10

02 41 19 – Selective Demolition

02 41 19.18 Selective Demolition, Disposal Only		Crew	Daily Output	Labor-Hours	Unit	Material	2015 Bare Costs Labor	Equipment	Total	Total Incl O&P
0500	Wood frame	B-3	247	.194	C.Y.		7.95	10.40	18.35	23.50

02 41 19.19 Selective Demolition

Line	Description	Crew	Daily Output	Labor-Hours	Unit	Material	Labor	Equipment	Total	Total Incl O&P
0010	**SELECTIVE DEMOLITION,** Rubbish Handling R024119-10									
0020	The following are to be added to the demolition prices									
0050	The following are components for a complete chute system									
0100	Top chute circular steel, 4′ long, 18″ diameter R024119-30	B-1C	15	1.600	Ea.	266	61	29	356	420
0102	23″ diameter		15	1.600		288	61	29	378	440
0104	27″ diameter		15	1.600		310	61	29	400	465
0106	30″ diameter		15	1.600		330	61	29	420	490
0108	33″ diameter		15	1.600		355	61	29	445	515
0110	36″ diameter		15	1.600		375	61	29	465	540
0112	Regular chute, 18″ diameter		15	1.600		199	61	29	289	345
0114	23″ diameter		15	1.600		222	61	29	312	370
0116	27″ diameter		15	1.600		243	61	29	333	395
0118	30″ diameter		15	1.600		266	61	29	356	420
0120	33″ diameter		15	1.600		288	61	29	378	440
0122	36″ diameter		15	1.600		310	61	29	400	465
0124	Control door chute, 18″ diameter		15	1.600		375	61	29	465	540
0126	23″ diameter		15	1.600		400	61	29	490	565
0128	27″ diameter		15	1.600		420	61	29	510	590
0130	30″ diameter		15	1.600		445	61	29	535	615
0132	33″ diameter		15	1.600		465	61	29	555	635
0134	36″ diameter		15	1.600		490	61	29	580	660
0136	Chute liners, 14 ga., 18-30″ diameter		15	1.600		223	61	29	313	370
0138	33-36″ diameter		15	1.600		280	61	29	370	435
0140	17% thinner chute, 30″ diameter		15	1.600		222	61	29	312	370
0142	33% thinner chute, 30″ diameter		15	1.600		167	61	29	257	310
0144	Top chute cover	1 Clab	24	.333		153	12.55		165.55	187
0146	Door chute cover	″	24	.333		153	12.55		165.55	187
0148	Top chute trough	2 Clab	12	1.333		510	50		560	635
0150	Bolt down frame & counter weights, 250 lb.	B-1	4	6		4,250	230		4,480	5,000
0152	500 lb.		4	6		6,200	230		6,430	7,175
0154	750 lb.		4	6		9,100	230		9,330	10,400
0156	1000 lb.		2.67	8.989		10,200	345		10,545	11,700
0158	1500 lb.		2.67	8.989		13,000	345		13,345	14,800
0160	Chute warning light system, 5 stories	B-1C	4	6		8,975	230	109	9,314	10,400
0162	10 stories	″	2	12		14,300	460	218	14,978	16,600
0164	Dust control device for dumpsters	1 Clab	8	1		102	37.50		139.50	170
0166	Install or replace breakaway cord		8	1		27	37.50		64.50	87.50
0168	Install or replace warning sign		16	.500		11.45	18.80		30.25	41.50
0600	Dumpster, weekly rental, 1 dump/week, 6 C.Y. capacity (2 Tons) R024119-20				Week	415			415	455
0700	10 C.Y. capacity (3 Tons)					480			480	530
0725	20 C.Y. capacity (5 Tons)					565			565	625
0800	30 C.Y. capacity (7 Tons)					730			730	800
0840	40 C.Y. capacity (10 Tons)					775			775	850
2000	Load, haul, dump and return, 0 – 50′ haul, hand carried	2 Clab	24	.667	C.Y.		25		25	38.50
2005	Wheeled		37	.432			16.25		16.25	25
2040	0 – 100′ haul, hand carried		16.50	.970			36.50		36.50	56
2045	Wheeled		25	.640			24		24	37
2080	Haul and return, add per each extra 100′ haul, hand carried		35.50	.451			16.95		16.95	26
2085	Wheeled		54	.296			11.15		11.15	17.15
2120	For travel in elevators, up to 10 floors, add		140	.114			4.30		4.30	6.60

02 41 19 – Selective Demolition

02 41 19.19 Selective Demolition		Crew	Daily Output	Labor-Hours	Unit	Material	2015 Bare Costs Labor	Equipment	Total	Total Incl O&P
2130	0 – 50′ haul, incl. up to 5 riser stair, hand carried	2 Clab	23	.696	C.Y.		26		26	40
2135	Wheeled		35	.457			17.20		17.20	26.50
2140	6 – 10 riser stairs, hand carried		22	.727			27.50		27.50	42
2145	Wheeled		34	.471			17.70		17.70	27
2150	11 – 20 riser stairs, hand carried		20	.800			30		30	46.50
2155	Wheeled		31	.516			19.40		19.40	30
2160	21 – 40 riser stairs, hand carried		16	1			37.50		37.50	58
2165	Wheeled		24	.667			25		25	38.50
2170	0 – 100′ haul, incl. 5 riser stair, hand carried		15	1.067			40		40	61.50
2175	Wheeled		23	.696			26		26	40
2180	6 – 10 riser stair, hand carried		14	1.143			43		43	66
2185	Wheeled		21	.762			28.50		28.50	44
2190	11 – 20 riser stair, hand carried		12	1.333			50		50	77
2195	Wheeled		18	.889			33.50		33.50	51.50
2200	21 – 40 riser stair, hand carried		8	2			75		75	116
2205	Wheeled		12	1.333			50		50	77
2210	Haul and return, add per each extra 100′ haul, hand carried		35.50	.451			16.95		16.95	26
2215	Wheeled		54	.296			11.15		11.15	17.15
2220	For each additional flight of stairs, up to 5 risers, add		550	.029	Flight		1.09		1.09	1.68
2225	6 – 10 risers, add		275	.058			2.19		2.19	3.37
2230	11 – 20 risers, add		138	.116			4.36		4.36	6.70
2235	21 – 40 risers, add		69	.232			8.70		8.70	13.40
3000	Loading & trucking, including 2 mile haul, chute loaded	B-16	45	.711	C.Y.		27.50	15.35	42.85	59
3040	Hand loading truck, 50′ haul	"	48	.667			26	14.40	40.40	55.50
3080	Machine loading truck	B-17	120	.267			10.90	6.45	17.35	24
5000	Haul, per mile, up to 8 C.Y. truck	B-34B	1165	.007			.28	.59	.87	1.07
5100	Over 8 C.Y. truck	"	1550	.005			.21	.45	.66	.80

02 41 19.20 Selective Demolition, Dump Charges		Crew	Daily Output	Labor-Hours	Unit	Material	Labor	Equipment	Total	Total Incl O&P
0010	SELECTIVE DEMOLITION, DUMP CHARGES R024119-10									
0020	Dump charges, typical urban city, tipping fees only									
0100	Building construction materials				Ton	74			74	81
0200	Trees, brush, lumber					63			63	69.50
0300	Rubbish only					63			63	69.50
0500	Reclamation station, usual charge					74			74	81

02 41 19.21 Selective Demolition, Gutting		Crew	Daily Output	Labor-Hours	Unit	Material	Labor	Equipment	Total	Total Incl O&P
0010	SELECTIVE DEMOLITION, GUTTING R024119-10									
0020	Building interior, including disposal, dumpster fees not included									
0500	Residential building									
0560	Minimum	B-16	400	.080	SF Flr.		3.10	1.73	4.83	6.65
0580	Maximum	"	360	.089	"		3.44	1.92	5.36	7.35
0900	Commercial building									
1000	Minimum	B-16	350	.091	SF Flr.		3.54	1.97	5.51	7.55
1020	Maximum	"	250	.128	"		4.95	2.76	7.71	10.65

02 41 19.25 Selective Demolition, Saw Cutting		Crew	Daily Output	Labor-Hours	Unit	Material	Labor	Equipment	Total	Total Incl O&P
0010	SELECTIVE DEMOLITION, SAW CUTTING R024119-10									
0015	Asphalt, up to 3″ deep	B-89	1050	.015	L.F.	.12	.67	.46	1.25	1.65
0020	Each additional inch of depth	"	1800	.009		.04	.39	.27	.70	.93
1200	Masonry walls, hydraulic saw, brick, per inch of depth	B-89B	300	.053		.04	2.34	2.86	5.24	6.75
1220	Block walls, solid, per inch of depth	"	250	.064		.04	2.81	3.43	6.28	8.05
2000	Brick or masonry w/hand held saw, per inch of depth	A-1	125	.064		.05	2.41	.65	3.11	4.46
5000	Wood sheathing to 1″ thick, on walls	1 Carp	200	.040			1.88		1.88	2.89
5020	On roof	"	250	.032			1.50		1.50	2.31

02 41 Demolition

02 41 19 – Selective Demolition

	02 41 19.27 Selective Demolition, Torch Cutting	Crew	Daily Output	Labor-Hours	Unit	Material	2015 Bare Costs Labor	Equipment	Total	Total Incl O&P
0010	SELECTIVE DEMOLITION, TORCH CUTTING R024119-10									
0020	Steel, 1" thick plate	E-25	333	.024	L.F.	.84	1.31	.03	2.18	3.24
0040	1" diameter bar	"	600	.013	Ea.	.14	.73	.02	.89	1.44
1000	Oxygen lance cutting, reinforced concrete walls									
1040	12" to 16" thick walls	1 Clab	10	.800	L.F.		30		30	46.50
1080	24" thick walls	"	6	1.333	"		50		50	77

02 42 Removal and Salvage of Construction Materials

02 42 10 – Building Deconstruction

	02 42 10.10 Estimated Salvage Value or Savings	Crew	Daily Output	Labor-Hours	Unit	Material	Labor	Equipment	Total	Total Incl O&P
0010	ESTIMATED SALVAGE VALUE OR SAVINGS									
0015	Excludes material handling, packaging, container costs and									
0020	transportation for salvage or disposal									
0050	All Items in Section 02 42 10.10 are credit deducts and not costs									
0100	Copper Wire Salvage Value G				Lb.				1.60	1.60
0110	Disposal Savings G								.04	.04
0200	Copper Pipe Salvage Value G								2.50	2.50
0210	Disposal Savings G								.05	.05
0300	Steel Pipe Salvage Value G								.06	.06
0310	Disposal Savings G								.03	.03
0400	Cast Iron Pipe Salvage Value G								.03	.03
0410	Disposal Savings G								.01	.01
0500	Steel Doors or Windows Salvage Value G								.06	.06
0510	Aluminum G								.55	.55
0520	Disposal Savings G								.03	.03
0600	Aluminum Siding Salvage Value G								.49	.49
0630	Disposal Savings G								.03	.03
0640	Wood Siding (no lead or asbestos) G				C.Y.				12	12
0800	Clean Concrete Disposal Savings G				Ton				62	62
0850	Asphalt Shingles Disposal Savings G				"				60	60
1000	Wood wall framing clean salvage value G				M.B.F.				55	55
1010	Painted G								44	44
1020	Floor framing G								55	55
1030	Painted G								44	44
1050	Roof framing G								55	55
1060	Painted G								44	44
1100	Wood beams salvage value G								55	55
1200	Wood framing and beams disposal savings G				Ton				66	66
1220	Wood sheating and sub-base flooring G								72.50	72.50
1230	Wood wall paneling (1/4 inch thick) G								66	66
1300	Wood panel 3/4-1 inch thick low salvage value G				S.F.				.55	.55
1350	high salvage value G				"				2.20	2.20
1400	Disposal savings G				Ton				66	66
1500	Flooring tongue and groove 25/32 inch thick low salvage value G				S.F.				.55	.55
1530	High salvage value G				"				1.10	1.10
1560	Disposal savings G				Ton				66	66
1600	Drywall or sheet rock salvage value G								22	22
1650	Disposal savings G								66	66

02 42 10.20 Deconstruction of Building Components			Crew	Daily Output	Labor-Hours	Unit	Material	2015 Bare Costs Labor	Equipment	Total	Total Incl O&P
0010	**DECONSTRUCTION OF BUILDING COMPONENTS**										
0012	Buildings one or two stories only										
0015	Excludes material handling, packaging, container costs and										
0020	transportation for salvage or disposal										
0050	Deconstruction of Plumbing Fixtures										
0100	Wall hung or countertop lavatory	G	2 Clab	16	1	Ea.		37.50		37.50	58
0110	Single or double compartment kitchen sink	G		14	1.143			43		43	66
0120	Wall hung urinal	G		14	1.143			43		43	66
0130	Floor mounted	G		8	2			75		75	116
0140	Floor mounted water closet	G		16	1			37.50		37.50	58
0150	Wall hung	G		14	1.143			43		43	66
0160	Water fountain, free standing	G		16	1			37.50		37.50	58
0170	Wall hung or deck mounted	G		12	1.333			50		50	77
0180	Bathtub, steel or fiberglass	G		10	1.600			60		60	92.50
0190	Cast iron	G		8	2			75		75	116
0200	Shower, single	G		6	2.667			100		100	154
0210	Group	G		7	2.286			86		86	132
0300	Deconstruction of Electrical Fixtures										
0310	Surface mount incandescent fixtures	G	2 Clab	48	.333	Ea.		12.55		12.55	19.30
0320	Fluorescent, 2 lamp	G		32	.500			18.80		18.80	29
0330	4 lamp	G		24	.667			25		25	38.50
0340	Strip Fluorescent, 1 lamp	G		40	.400			15.05		15.05	23
0350	2 lamp	G		32	.500			18.80		18.80	29
0400	Recessed drop-in fluorescent fixture, 2 lamp	G		27	.593			22.50		22.50	34.50
0410	4 lamp	G		18	.889			33.50		33.50	51.50
0500	Deconstruction of appliances										
0510	Cooking stoves	G	2 Clab	26	.615	Ea.		23		23	35.50
0520	Dishwashers	G	"	26	.615	"		23		23	35.50
0600	Deconstruction of millwork and trim										
0610	Cabinets, wood	G	2 Carp	40	.400	L.F.		18.80		18.80	29
0620	Countertops	G		100	.160	"		7.50		7.50	11.55
0630	Wall paneling, 1 inch thick	G		500	.032	S.F.		1.50		1.50	2.31
0640	Ceiling trim	G		500	.032	L.F.		1.50		1.50	2.31
0650	Wainscoting	G		500	.032	S.F.		1.50		1.50	2.31
0660	Base, 3/4" to 1" thick	G		600	.027	L.F.		1.25		1.25	1.93
0700	Deconstruction of doors and windows										
0710	Doors, wrap, interior, wood, single, no closers	G	2 Carp	21	.762	Ea.	4.50	36		40.50	60
0720	Double	G		13	1.231		9	58		67	99
0730	Solid core, single, exterior or interior	G		10	1.600		4.50	75		79.50	121
0740	Double	G		8	2		9	94		103	155
0810	Windows, wrap, wood, single										
0812	with no casement or cladding	G	2 Carp	21	.762	Ea.	4.50	36		40.50	60
0820	with casement and/or cladding	G	"	18	.889	"	4.50	41.50		46	69
0900	Deconstruction of interior finishes										
0910	Drywall for recycling	G	2 Clab	1775	.009	S.F.		.34		.34	.52
0920	Plaster wall, first floor	G		1775	.009			.34		.34	.52
0930	Second floor	G		1330	.012			.45		.45	.70
1000	Deconstruction of roofing and accessories										
1010	Built-up roofs	G	2 Clab	570	.028	S.F.		1.06		1.06	1.62
1020	Gutters, facia and rakes	G	"	1140	.014	L.F.		.53		.53	.81
2000	Deconstruction of wood components										
2010	Roof sheeting	G	2 Clab	570	.028	S.F.		1.06		1.06	1.62

02 42 10 – Building Deconstruction

02 42 10.20 Deconstruction of Building Components

	02 42 10.20 Deconstruction of Building Components		Crew	Daily Output	Labor-Hours	Unit	Material (2015 Bare Costs)	Labor (2015 Bare Costs)	Equipment (2015 Bare Costs)	Total (2015 Bare Costs)	Total Incl O&P
2020	Main roof framing	G	2 Clab	760	.021	L.F.		.79		.79	1.22
2030	Porch roof framing	G		445	.036			1.35		1.35	2.08
2040	Beams 4" x 8"	G	B-1	375	.064			2.45		2.45	3.77
2050	4" x 10"	G		300	.080			3.06		3.06	4.71
2055	4" x 12"	G		250	.096			3.67		3.67	5.65
2060	6" x 8"	G		250	.096			3.67		3.67	5.65
2065	6" x 10"	G		200	.120			4.59		4.59	7.05
2070	6" x 12"	G		170	.141			5.40		5.40	8.30
2075	8" x 12"	G		126	.190			7.30		7.30	11.20
2080	10" x 12"	G		100	.240			9.20		9.20	14.15
2100	Ceiling joists	G	2 Clab	800	.020			.75		.75	1.16
2150	Wall framing, interior	G		1230	.013			.49		.49	.75
2160	Sub-floor	G		2000	.008	S.F.		.30		.30	.46
2170	Floor joists	G		2000	.008	L.F.		.30		.30	.46
2200	Wood siding (no lead or asbestos)	G		1300	.012	S.F.		.46		.46	.71
2300	Wall framing, exterior	G		1600	.010	L.F.		.38		.38	.58
2400	Stair risers	G		53	.302	Ea.		11.35		11.35	17.45
2500	Posts	G		800	.020	L.F.		.75		.75	1.16
3000	Deconstruction of exterior brick walls										
3010	Exterior brick walls, first floor	G	2 Clab	200	.080	S.F.		3.01		3.01	4.63
3020	Second floor	G		64	.250	"		9.40		9.40	14.45
3030	Brick chimney	G		100	.160	C.F.		6		6	9.25
4000	Deconstruction of concrete										
4010	Slab on grade, 4" thick, plain concrete	G	B-9	500	.080	S.F.		3.04	.47	3.51	5.20
4020	Wire mesh reinforced	G		470	.085			3.23	.50	3.73	5.50
4030	Rod reinforced	G		400	.100			3.80	.58	4.38	6.50
4110	Foundation wall, 6" thick, plain concrete	G		160	.250			9.50	1.46	10.96	16.20
4120	8" thick	G		140	.286			10.85	1.66	12.51	18.55
4130	10" thick	G		120	.333			12.65	1.94	14.59	21.50
9000	Deconstruction process, support equipment as needed										
9010	Daily use, portal to portal, 12-ton truck-mounted hydraulic crane crew	G	A-3H	1	8	Day		415	860	1,275	1,575
9020	Daily use, skid steer and operator	G	A-3C	1	8			390	320	710	940
9030	Daily use, backhoe 48 H.P., operator and labor	G	"	1	8			390	320	710	940

02 42 10.30 Deconstruction Material Handling

	02 42 10.30 Deconstruction Material Handling		Crew	Daily Output	Labor-Hours	Unit	Material	Labor	Equipment	Total	Total Incl O&P
0010	**DECONSTRUCTION MATERIAL HANDLING**										
0012	Buildings one or two stories only										
0100	Clean and stack brick on pallet	G	2 Clab	1200	.013	Ea.		.50		.50	.77
0200	Haul 50' and load rough lumber up to 2" x 8" size	G		2000	.008	"		.30		.30	.46
0210	Lumber larger than 2" x 8"	G		3200	.005	B.F.		.19		.19	.29
0300	Finish wood for recycling stack and wrap per pallet	G		8	2	Ea.	36	75		111	156
0350	Light fixtures			6	2.667		65	100		165	226
0375	Windows			6	2.667		61	100		161	222
0400	Miscellaneous materials			8	2		18	75		93	136
1000	See Section 02 41 19.19 for bulk material handling										

02 43 Structure Moving

02 43 13 – Structure Relocation

02 43 13.13 Building Relocation		Crew	Daily Output	Labor-Hours	Unit	Material	2015 Bare Costs Labor	Equipment	Total	Total Incl O&P
0010	**BUILDING RELOCATION**									
0011	One day move, up to 24′ wide									
0020	Reset on existing foundation				Total				11,500	11,500
0040	Wood or steel frame bldg., based on ground floor area G	B-4	185	.259	S.F.		9.95	2.78	12.73	18.30
0060	Masonry bldg., based on ground floor area G	"	137	.350			13.45	3.75	17.20	24.50
0200	For 24′ to 42′ wide, add				↓				15%	15%

02 56 Site containment

02 56 13 – Waste containment

02 56 13.10 Containment of Hazardous Waste		Crew	Daily Output	Labor-Hours	Unit	Material	Labor	Equipment	Total	Total Incl O&P
0010	**CONTAINMENT OF HAZARDOUS WASTE**									
0020	OSHA hazard level C									
0030	OSHA Hazard level D decrease labor and equipment, deduct						-45%	-45%		
0035	OSHA Hazard level B increase labor and equipment, add						22%	22%		
0040	OSHA Hazard level A increase labor and equipment, add						71%	71%		
0100	Excavation of contaminated soil & waste									
0105	Includes one respirator filter and two disposable suits per work day									
0110	3/4 C.Y. excavator to 10 feet deep	B-12F	51	.314	B.C.Y.	1.59	14	12.85	28.44	37.50
0120	Labor crew to 6′ deep	B-2	19	2.105		10.65	80		90.65	135
0130	6′ - 12′ deep	"	12	3.333	↓	16.85	127		143.85	214
0200	Move contaminated soil/waste upto 150′ on-site with 2.5 C.Y. loader	B-10T	300	.040	L.C.Y.	.27	1.85	1.73	3.85	5.05
0210	300′	"	186	.065	"	.43	2.99	2.80	6.22	8.10
0300	Secure burial cell construction									
0310	Various liner and cover materials									
0400	Very low density polyethylene (VLDPE)									
0410	50 mil top cover	B-47H	4000	.008	S.F.	.42	.39	.08	.89	1.15
0420	80 mil liner	"	4000	.008	"	.52	.39	.08	.99	1.26
0500	Chlorosulfunated polyethylene									
0510	36 mil hypalon top cover	B-47H	4000	.008	S.F.	1.57	.39	.08	2.04	2.42
0520	45 mil hypalon liner	"	4000	.008	"	1.71	.39	.08	2.18	2.57
0600	Polyvinyl chloride (PVC)									
0610	60 mil top cover	B-47H	4000	.008	S.F.	.81	.39	.08	1.28	1.58
0620	80 mil liner	"	4000	.008	"	.97	.39	.08	1.44	1.76
0700	Rough textured H.D. polyethylene (HDPE)									
0710	40 mil top cover	B-47H	4000	.008	S.F.	.40	.39	.08	.87	1.13
0720	60 mil top cover		4000	.008		.46	.39	.08	.93	1.20
0722	60 mil liner		4000	.008		.42	.39	.08	.89	1.15
0730	80 mil liner	↓	3800	.008	↓	.53	.41	.08	1.02	1.31
1000	3/4″ crushed stone, 6″ deep ballast around liner	B-6	30	.800	L.C.Y.	23.50	33	12.15	68.65	90
1100	Hazardous waste, ballast cover with common borrow material	B-63	56	.714		12.40	28.50	3.10	44	60.50
1110	Mixture of common borrow & topsoil		56	.714		18.45	28.50	3.10	50.05	67.50
1120	Bank sand		56	.714		17.85	28.50	3.10	49.45	66.50
1130	Medium priced clay		44	.909		24	36	3.95	63.95	86.50
1140	Mixture of common borrow & medium priced clay	↓	56	.714	↓	18.20	28.50	3.10	49.80	67

02 58 Snow Control

02 58 13 – Snow Fencing

02 58 13.10 Snow Fencing System		Crew	Daily Output	Labor-Hours	Unit	Material	2015 Bare Costs Labor	Equipment	Total	Total Incl O&P
0010	**SNOW FENCING SYSTEM**									
7001	Snow fence on steel posts 10′ O.C., 4′ high	B-1	500	.048	L.F.	.93	1.84		2.77	3.85

02 65 Underground Storage Tank Removal

02 65 10 – Underground Tank and Contaminated Soil Removal

02 65 10.30 Removal of Underground Storage Tanks

			Crew	Daily Output	Labor-Hours	Unit	Material	Labor	Equipment	Total	Total Incl O&P
0010	**REMOVAL OF UNDERGROUND STORAGE TANKS** R026510-20										
0011	Petroleum storage tanks, non-leaking										
0100	Excavate & load onto trailer										
0110	3000 gal. to 5000 gal. tank	G	B-14	4	12	Ea.		475	91	566	830
0120	6000 gal. to 8000 gal. tank	G	B-3A	3	13.333			535	345	880	1,200
0130	9000 gal. to 12000 gal. tank	G	″	2	20			805	515	1,320	1,800
0190	Known leaking tank, add					%				100%	100%
0200	Remove sludge, water and remaining product from tank bottom										
0201	of tank with vacuum truck										
0300	3000 gal. to 5000 gal. tank	G	A-13	5	1.600	Ea.		78	153	231	286
0310	6000 gal. to 8000 gal. tank	G		4	2			97	191	288	360
0320	9000 gal. to 12000 gal. tank	G		3	2.667			130	254	384	475
0390	Dispose of sludge off-site, average					Gal.				6.25	6.80
0400	Insert inert solid CO_2 "dry ice" into tank										
0401	For cleaning/transporting tanks (1.5 lb./100 gal. cap)	G	1 Clab	500	.016	Lb.	1.17	.60		1.77	2.22
0403	Insert solid carbon dioxide, 1.5 lb./100 gal.	G	″	400	.020	″	1.17	.75		1.92	2.45
0503	Disconnect and remove piping	G	1 Plum	160	.050	L.F.		2.94		2.94	4.43
0603	Transfer liquids, 10% of volume	G	″	1600	.005	Gal.		.29		.29	.44
0703	Cut accessway into underground storage tank	G	1 Clab	5.33	1.501	Ea.		56.50		56.50	87
0813	Remove sludge, wash and wipe tank, 500 gal.	G	1 Plum	8	1			58.50		58.50	88.50
0823	3,000 gal.	G		6.67	1.199			70.50		70.50	106
0833	5,000 gal.	G		6.15	1.301			76.50		76.50	115
0843	8,000 gal.	G		5.33	1.501			88		88	133
0853	10,000 gal.	G		4.57	1.751			103		103	155
0863	12,000 gal.	G		4.21	1.900			112		112	168
1020	Haul tank to certified salvage dump, 100 miles round trip										
1023	3000 gal. to 5000 gal. tank					Ea.				760	830
1026	6000 gal. to 8000 gal. tank									880	960
1029	9,000 gal. to 12,000 gal. tank									1,050	1,150
1100	Disposal of contaminated soil to landfill										
1110	Minimum					C.Y.				145	160
1111	Maximum					″				400	440
1120	Disposal of contaminated soil to										
1121	bituminous concrete batch plant										
1130	Minimum					C.Y.				80	88
1131	Maximum					″				115	125
1203	Excavate, pull, & load tank, backfill hole, 8,000 gal. +	G	B-12C	.50	32	Ea.		1,425	2,350	3,775	4,750
1213	Haul tank to certified dump, 100 miles rt, 8,000 gal. +	G	B-34K	1	8			320	950	1,270	1,525
1223	Excavate, pull, & load tank, backfill hole, 500 gal.	G	B-11C	1	16			705	365	1,070	1,475
1233	Excavate, pull, & load tank, backfill hole, 3,000 – 5,000 gal.	G	B-11M	.50	32			1,400	785	2,185	3,000
1243	Haul tank to certified dump, 100 miles rt, 500 gal.	G	B-34L	1	8			390	245	635	860
1253	Haul tank to certified dump, 100 miles rt, 3,000 – 5,000 gal.	G	B-34M	1	8			390	305	695	925
2010	Decontamination of soil on site incl poly tarp on top/bottom										
2011	Soil containment berm, and chemical treatment										
2020	Minimum	G	B-11C	100	.160	C.Y.	7.80	7.05	3.64	18.49	23.50

02 65 Underground Storage Tank Removal

02 65 10 – Underground Tank and Contaminated Soil Removal

02 65 10.30 Removal of Underground Storage Tanks		Crew	Daily Output	Labor-Hours	Unit	Material	2015 Bare Costs Labor	Equipment	Total	Total Incl O&P
2021	Maximum [G]	B-11C	100	.160	C.Y.	10.10	7.05	3.64	20.79	26
2050	Disposal of decontaminated soil, minimum								135	150
2055	Maximum				↓				400	440

02 81 Transportation and Disposal of Hazardous Materials

02 81 20 – Hazardous Waste Handling

02 81 20.10 Hazardous Waste Cleanup/Pickup/Disposal		Crew	Daily Output	Labor-Hours	Unit	Material	Labor	Equipment	Total	Total Incl O&P
0010	**HAZARDOUS WASTE CLEANUP/PICKUP/DISPOSAL**									
0100	For contractor rental equipment, i.e., Dozer,									
0110	Front end loader, Dump truck, etc., see 01 54 33 Reference Section									
1000	Solid pickup									
1100	55 gal. drums				Ea.				240	265
1120	Bulk material, minimum				Ton				190	210
1130	Maximum				"				595	655
1200	Transportation to disposal site									
1220	Truckload = 80 drums or 25 C.Y. or 18 tons									
1260	Minimum				Mile				3.95	4.45
1270	Maximum				"				7.25	7.35
3000	Liquid pickup, vacuum truck, stainless steel tank									
3100	Minimum charge, 4 hours									
3110	1 compartment, 2200 gallon				Hr.				140	155
3120	2 compartment, 5000 gallon				"				200	225
3400	Transportation in 6900 gallon bulk truck				Mile				7.95	8.75
3410	In teflon lined truck				"				10.20	11.25
5000	Heavy sludge or dry vacuumable material				Hr.				140	160
6000	Dumpsite disposal charge, minimum				Ton				140	155
6020	Maximum				"				415	455

02 82 Asbestos Remediation

02 82 13 – Asbestos Abatement

02 82 13.39 Asbestos Remediation Plans and Methods		Crew	Daily Output	Labor-Hours	Unit	Material	Labor	Equipment	Total	Total Incl O&P
0010	**ASBESTOS REMEDIATION PLANS AND METHODS**									
0100	Building Survey-Commercial Building				Ea.				2,200	2,400
0200	Asbestos Abatement Remediation Plan				"				1,350	1,475

02 82 13.41 Asbestos Abatement Equipment		Crew	Daily Output	Labor-Hours	Unit	Material	Labor	Equipment	Total	Total Incl O&P
0010	**ASBESTOS ABATEMENT EQUIPMENT** R028213-20									
0011	Equipment and supplies, buy									
0200	Air filtration device, 2000 CFM				Ea.	960			960	1,050
0250	Large volume air sampling pump, minimum					345			345	380
0260	Maximum					335			335	365
0300	Airless sprayer unit, 2 gun					4,500			4,500	4,950
0350	Light stand, 500 watt				↓	49			49	54
0400	Personal respirators									
0410	Negative pressure, 1/2 face, dual operation, min.				Ea.	26.50			26.50	29
0420	Maximum					29.50			29.50	32.50
0450	P.A.P.R., full face, minimum					124			124	137
0460	Maximum					165			165	182
0470	Supplied air, full face, incl. air line, minimum					168			168	185
0480	Maximum				↓	405			405	445

02 82 Asbestos Remediation

02 82 13 – Asbestos Abatement

02 82 13.41 Asbestos Abatement Equipment		Crew	Daily Output	Labor-Hours	Unit	Material	2015 Bare Costs Labor	Equipment	Total	Total Incl O&P
0500	Personnel sampling pump				Ea.	226			226	249
1500	Power panel, 20 unit, incl. GFI					600			600	660
1600	Shower unit, including pump and filters					1,275			1,275	1,425
1700	Supplied air system (type C)					3,425			3,425	3,750
1750	Vacuum cleaner, HEPA, 16 gal., stainless steel, wet/dry					435			435	475
1760	55 gallon					560			560	615
1800	Vacuum loader, 9 – 18 ton/hr.					94,500			94,500	104,000
1900	Water atomizer unit, including 55 gal. drum					281			281	310
2000	Worker protection, whole body, foot, head cover & gloves, plastic					8.65			8.65	9.55
2500	Respirator, single use					25			25	27.50
2550	Cartridge for respirator					5.75			5.75	6.30
2570	Glove bag, 7 mil, 50″ x 64″					9.20			9.20	10.10
2580	10 mil, 44″ x 60″					5.75			5.75	6.35
2590	6 mil, 44″ x 60″					5.75			5.75	6.35
3000	HEPA vacuum for work area, minimum					305			305	335
3050	Maximum					815			815	900
6000	Disposable polyethylene bags, 6 mil, 3 C.F.					.82			.82	.90
6300	Disposable fiber drums, 3 C.F.					18.20			18.20	20
6400	Pressure sensitive caution labels, 3″ x 5″					3.47			3.47	3.82
6450	11″ x 17″					7.35			7.35	8.05
6500	Negative air machine, 1800 CFM					845			845	930

02 82 13.42 Preparation of Asbestos Containment Area		Crew	Daily Output	Labor-Hours	Unit	Material	Labor	Equipment	Total	Total Incl O&P
0010	**PREPARATION OF ASBESTOS CONTAINMENT AREA**									
0100	Pre-cleaning, HEPA vacuum and wet wipe, flat surfaces	A-9	12000	.005	S.F.	.02	.28		.30	.45
0200	Protect carpeted area, 2 layers 6 mil poly on 3/4″ plywood	″	1000	.064		2.04	3.35		5.39	7.45
0300	Separation barrier, 2″ x 4″ @ 16″, 1/2″ plywood ea. side, 8′ high	2 Carp	400	.040		3.32	1.88		5.20	6.55
0310	12′ high		320	.050		3.32	2.35		5.67	7.25
0320	16′ high		200	.080		2.30	3.76		6.06	8.35
0400	Personnel decontam. chamber, 2″ x 4″ @ 16″, 3/4″ ply ea. side		280	.057		4.34	2.68		7.02	8.90
0450	Waste decontam. chamber, 2″ x 4″ studs @ 16″, 3/4″ ply ea. side		360	.044		4.34	2.09		6.43	8
0500	Cover surfaces with polyethylene sheeting									
0501	Including glue and tape									
0550	Floors, each layer, 6 mil	A-9	8000	.008	S.F.	.04	.42		.46	.70
0551	4 mil		9000	.007		.03	.37		.40	.61
0560	Walls, each layer, 6 mil		6000	.011		.04	.56		.60	.92
0561	4 mil		7000	.009		.03	.48		.51	.78
0570	For heights above 12′, add						20%			
0575	For heights above 20′, add						30%			
0580	For fire retardant poly, add					100%				
0590	For large open areas, deduct					10%	20%			
0600	Seal floor penetrations with foam firestop to 36 sq. in.	2 Carp	200	.080	Ea.	7.75	3.76		11.51	14.30
0610	36 sq. in. to 72 sq. in.		125	.128		15.50	6		21.50	26.50
0615	72 sq. in. to 144 sq. in.		80	.200		31	9.40		40.40	48.50
0620	Wall penetrations, to 36 square inches		180	.089		7.75	4.17		11.92	14.90
0630	36 sq. in. to 72 sq. in.		100	.160		15.50	7.50		23	28.50
0640	72 sq. in. to 144 sq. in.		60	.267		31	12.50		43.50	53.50
0800	Caulk seams with latex	1 Carp	230	.035	L.F.	.17	1.63		1.80	2.70
0900	Set up neg. air machine, 1-2k CFM/25 M.C.F. volume	1 Asbe	4.30	1.860	Ea.		97.50		97.50	152
0950	Set up and remove portable shower unit	2 Asbe	4	4	″		209		209	325

02 82 13 – Asbestos Abatement

02 82 13.43 Bulk Asbestos Removal		Crew	Daily Output	Labor-Hours	Unit	Material	2015 Bare Costs Labor	Equipment	Total	Total Incl O&P
0010	**BULK ASBESTOS REMOVAL**									
0020	Includes disposable tools and 2 suits and 1 respirator filter/day/worker									
0100	Beams, W 10 x 19	A-9	235	.272	L.F.	.79	14.25		15.04	23
0110	W 12 x 22		210	.305		.88	15.95		16.83	26
0120	W 14 x 26		180	.356		1.03	18.65		19.68	30
0130	W 16 x 31		160	.400		1.15	21		22.15	34
0140	W 18 x 40		140	.457		1.32	24		25.32	39
0150	W 24 x 55		110	.582		1.68	30.50		32.18	49.50
0160	W 30 x 108		85	.753		2.17	39.50		41.67	64
0170	W 36 x 150		72	.889		2.56	46.50		49.06	75.50
0200	Boiler insulation		480	.133	S.F.	.45	7		7.45	11.40
0210	With metal lath, add				%				50%	50%
0300	Boiler breeching or flue insulation	A-9	520	.123	S.F.	.36	6.45		6.81	10.45
0310	For active boiler, add				%				100%	100%
0400	Duct or AHU insulation	A-10B	440	.073	S.F.	.21	3.82		4.03	6.20
0500	Duct vibration isolation joints, up to 24 sq. in. duct	A-9	56	1.143	Ea.	3.30	60		63.30	97
0520	25 sq. in. to 48 sq. in. duct		48	1.333		3.85	70		73.85	113
0530	49 sq. in. to 76 sq. in. duct		40	1.600		4.62	84		88.62	136
0600	Pipe insulation, air cell type, up to 4" diameter pipe		900	.071	L.F.	.21	3.73		3.94	6.05
0610	4" to 8" diameter pipe		800	.080		.23	4.19		4.42	6.80
0620	10" to 12" diameter pipe		700	.091		.26	4.79		5.05	7.75
0630	14" to 16" diameter pipe		550	.116		.34	6.10		6.44	9.85
0650	Over 16" diameter pipe		650	.098	S.F.	.28	5.15		5.43	8.35
0700	With glove bag up to 3" diameter pipe		200	.320	L.F.	9.15	16.75		25.90	36
1000	Pipe fitting insulation up to 4" diameter pipe		320	.200	Ea.	.58	10.50		11.08	16.95
1100	6" to 8" diameter pipe		304	.211		.61	11.05		11.66	17.85
1110	10" to 12" diameter pipe		192	.333		.96	17.45		18.41	28
1120	14" to 16" diameter pipe		128	.500		1.44	26		27.44	42.50
1130	Over 16" diameter pipe		176	.364	S.F.	1.05	19.05		20.10	30.50
1200	With glove bag, up to 8" diameter pipe		75	.853	L.F.	6.25	44.50		50.75	76.50
2000	Scrape foam fireproofing from flat surface		2400	.027	S.F.	.08	1.40		1.48	2.26
2100	Irregular surfaces		1200	.053		.15	2.80		2.95	4.52
3000	Remove cementitious material from flat surface		1800	.036		.10	1.86		1.96	3.01
3100	Irregular surface		1000	.064		.13	3.35		3.48	5.35
4000	Scrape acoustical coating/fireproofing, from ceiling		3200	.020		.06	1.05		1.11	1.69
5000	Remove VAT and mastic from floor by hand		2400	.027		.08	1.40		1.48	2.26
5100	By machine	A-11	4800	.013		.04	.70	.01	.75	1.14
5150	For 2 layers, add				%				50%	50%
6000	Remove contaminated soil from crawl space by hand	A-9	400	.160	C.F.	.46	8.40		8.86	13.55
6100	With large production vacuum loader	A-12	700	.091	"	.26	4.79	1.09	6.14	8.95
7000	Radiator backing, not including radiator removal	A-9	1200	.053	S.F.	.15	2.80		2.95	4.52
8000	Cement-asbestos transite board and cement wall board	2 Asbe	1000	.016		.15	.84		.99	1.46
8100	Transite shingle siding	A-10B	750	.043		.21	2.24		2.45	3.72
8200	Shingle roofing	"	2000	.016		.08	.84		.92	1.40
8250	Built-up, no gravel, non-friable	B-2	1400	.029		.08	1.09		1.17	1.76
8260	Bituminous flashing	1 Rofc	300	.027		.08	1.07		1.15	1.92
8300	Asbestos millboard, flat board and VAT contaminated plywood	2 Asbe	1000	.016		.08	.84		.92	1.39
9000	For type B (supplied air) respirator equipment, add				%				10%	10%

02 82 13.44 Demolition In Asbestos Contaminated Area		Crew	Daily Output	Labor-Hours	Unit	Material	Labor	Equipment	Total	Total Incl O&P
0010	**DEMOLITION IN ASBESTOS CONTAMINATED AREA**									
0200	Ceiling, including suspension system, plaster and lath	A-9	2100	.030	S.F.	.09	1.60		1.69	2.59
0210	Finished plaster, leaving wire lath		585	.109		.32	5.75		6.07	9.30

02 82 13.44 Demolition In Asbestos Contaminated Area

	02 82 13.44 Demolition In Asbestos Contaminated Area	Crew	Daily Output	Labor-Hours	Unit	Material	Labor (2015 Bare Costs)	Equipment	Total	Total Incl O&P
0220	Suspended acoustical tile	A-9	3500	.018	S.F.	.05	.96		1.01	1.55
0230	Concealed tile grid system		3000	.021		.06	1.12		1.18	1.81
0240	Metal pan grid system		1500	.043		.12	2.24		2.36	3.62
0250	Gypsum board		2500	.026		.07	1.34		1.41	2.17
0260	Lighting fixtures up to 2′ x 4′		72	.889	Ea.	2.56	46.50		49.06	75.50
0400	Partitions, non load bearing									
0410	Plaster, lath, and studs	A-9	690	.093	S.F.	.88	4.86		5.74	8.50
0450	Gypsum board and studs	"	1390	.046	"	.13	2.41		2.54	3.91
9000	For type B (supplied air) respirator equipment, add				%				10%	10%

02 82 13.45 OSHA Testing

	02 82 13.45 OSHA Testing	Crew	Daily Output	Labor-Hours	Unit	Material	Labor	Equipment	Total	Total Incl O&P
0010	**OSHA TESTING**									
0100	Certified technician, minimum				Day				200	220
0110	Maximum								300	330
0120	Industrial hygienist, minimum								250	250
0130	Maximum								400	440
0200	Asbestos sampling and PCM analysis, NIOSH 7400, minimum	1 Asbe	8	1	Ea.	2.96	52.50		55.46	85
0210	Maximum		4	2		3.26	105		108.26	167
1000	Cleaned area samples		8	1		2.81	52.50		55.31	84.50
1100	PCM air sample analysis, NIOSH 7400, minimum		8	1		31.50	52.50		84	117
1110	Maximum		4	2		3.42	105		108.42	167
1200	TEM air sample analysis, NIOSH 7402, minimum								80	106
1210	Maximum								360	450

02 82 13.46 Decontamination of Asbestos Containment Area

	02 82 13.46 Decontamination of Asbestos Containment Area	Crew	Daily Output	Labor-Hours	Unit	Material	Labor	Equipment	Total	Total Incl O&P
0010	**DECONTAMINATION OF ASBESTOS CONTAINMENT AREA**									
0100	Spray exposed substrate with surfactant (bridging)									
0200	Flat surfaces	A-9	6000	.011	S.F.	.36	.56		.92	1.27
0250	Irregular surfaces		4000	.016	"	.31	.84		1.15	1.65
0300	Pipes, beams, and columns		2000	.032	L.F.	.56	1.68		2.24	3.23
1000	Spray encapsulate polyethylene sheeting		8000	.008	S.F.	.34	.42		.76	1.02
1100	Roll down polyethylene sheeting		8000	.008	"		.42		.42	.65
1500	Bag polyethylene sheeting		400	.160	Ea.	.81	8.40		9.21	13.95
2000	Fine clean exposed substrate, with nylon brush		2400	.027	S.F.		1.40		1.40	2.18
2500	Wet wipe substrate		4800	.013			.70		.70	1.09
2600	Vacuum surfaces, fine brush		6400	.010			.52		.52	.82
3000	Structural demolition									
3100	Wood stud walls	A-9	2800	.023	S.F.		1.20		1.20	1.87
3500	Window manifolds, not incl. window replacement		4200	.015			.80		.80	1.24
3600	Plywood carpet protection		2000	.032			1.68		1.68	2.61
4000	Remove custom decontamination facility	A-10A	8	3	Ea.	15.40	158		173.40	262
4100	Remove portable decontamination facility	3 Asbe	12	2	"	13.05	105		118.05	177
5000	HEPA vacuum, shampoo carpeting	A-9	4800	.013	S.F.	.07	.70		.77	1.17
9000	Final cleaning of protected surfaces	A-10A	8000	.003	"		.16		.16	.25

02 82 13.47 Asbestos Waste Pkg., Handling, and Disp.

	02 82 13.47 Asbestos Waste Pkg., Handling, and Disp.	Crew	Daily Output	Labor-Hours	Unit	Material	Labor	Equipment	Total	Total Incl O&P
0010	**ASBESTOS WASTE PACKAGING, HANDLING, AND DISPOSAL**									
0100	Collect and bag bulk material, 3 C.F. bags, by hand	A-9	400	.160	Ea.	.82	8.40		9.22	13.95
0200	Large production vacuum loader	A-12	880	.073		.86	3.81	.87	5.54	7.85
1000	Double bag and decontaminate	A-9	960	.067		.82	3.49		4.31	6.35
2000	Containerize bagged material in drums, per 3 C.F. drum	"	800	.080		18.20	4.19		22.39	26.50
3000	Cart bags 50′ to dumpster	2 Asbe	400	.040			2.09		2.09	3.26
5000	Disposal charges, not including haul, minimum				C.Y.				61	67
5020	Maximum				"				355	395
9000	For type B (supplied air) respirator equipment, add				%				10%	10%

02 82 Asbestos Remediation

02 82 13 – Asbestos Abatement

02 82 13.48 Asbestos Encapsulation With Sealants		Crew	Daily Output	Labor-Hours	Unit	2015 Bare Costs Material	Labor	Equipment	Total	Total Incl O&P
0010	**ASBESTOS ENCAPSULATION WITH SEALANTS**									
0100	Ceilings and walls, minimum	A-9	21000	.003	S.F.	.31	.16		.47	.59
0110	Maximum		10600	.006		.45	.32		.77	.99
0200	Columns and beams, minimum		13300	.005		.31	.25		.56	.73
0210	Maximum		5325	.012		.51	.63		1.14	1.54
0300	Pipes to 12" diameter including minor repairs, minimum		800	.080	L.F.	.43	4.19		4.62	7
0310	Maximum		400	.160	"	1.12	8.40		9.52	14.30

02 83 Lead Remediation

02 83 19 – Lead-Based Paint Remediation

02 83 19.21 Lead Paint Remediation Plans and Methods		Crew	Daily Output	Labor-Hours	Unit	Material	Labor	Equipment	Total	Total Incl O&P
0010	**LEAD PAINT REMEDIATION PLANS AND METHODS**									
0100	Building Survey-Commercial Building				Ea.				2,050	2,250
0200	Lead Abatement Remediation Plan								1,225	1,350
0300	Lead Paint Testing, AAS Analysis								51	56
0400	Lead Paint Testing, X-Ray Fluorescence								51	56

02 83 19.23 Encapsulation of Lead-Based Paint		Crew	Daily Output	Labor-Hours	Unit	Material	Labor	Equipment	Total	Total Incl O&P
0010	**ENCAPSULATION OF LEAD-BASED PAINT**									
0020	Interior, brushwork, trim, under 6"	1 Pord	240	.033	L.F.	2.30	1.34		3.64	4.56
0030	6" to 12" wide		180	.044		3.06	1.79		4.85	6.05
0040	Balustrades		300	.027		1.84	1.08		2.92	3.64
0050	Pipe to 4" diameter		500	.016		1.12	.65		1.77	2.20
0060	To 8" diameter		375	.021		1.48	.86		2.34	2.93
0070	To 12" diameter		250	.032		2.19	1.29		3.48	4.36
0080	To 16" diameter		170	.047		3.26	1.90		5.16	6.45
0090	Cabinets, ornate design		200	.040	S.F.	2.81	1.61		4.42	5.50
0100	Simple design		250	.032	"	2.24	1.29		3.53	4.41
0110	Doors, 3' x 7', both sides, incl. frame & trim									
0120	Flush	1 Pord	6	1.333	Ea.	28	54		82	112
0130	French, 10 – 15 lite		3	2.667		5.65	108		113.65	168
0140	Panel		4	2		34	80.50		114.50	160
0150	Louvered		2.75	2.909		31	117		148	212
0160	Windows, per interior side, per 15 S.F.									
0170	1 to 6 lite	1 Pord	14	.571	Ea.	19.40	23		42.40	56
0180	7 to 10 lite		7.50	1.067		21.50	43		64.50	89
0190	12 lite		5.75	1.391		29	56		85	117
0200	Radiators		8	1		69	40.50		109.50	137
0210	Grilles, vents		275	.029	S.F.	2.04	1.17		3.21	4.01
0220	Walls, roller, drywall or plaster		1000	.008		.56	.32		.88	1.11
0230	With spunbonded reinforcing fabric		720	.011		.63	.45		1.08	1.37
0240	Wood		800	.010		.69	.40		1.09	1.37
0250	Ceilings, roller, drywall or plaster		900	.009		.63	.36		.99	1.23
0260	Wood		700	.011		.78	.46		1.24	1.55
0270	Exterior, brushwork, gutters and downspouts		300	.027	L.F.	1.84	1.08		2.92	3.64
0280	Columns		400	.020	S.F.	1.38	.81		2.19	2.74
0290	Spray, siding		600	.013	"	.93	.54		1.47	1.83
0300	Miscellaneous									
0310	Electrical conduit, brushwork, to 2" diameter	1 Pord	500	.016	L.F.	1.12	.65		1.77	2.20
0320	Brick, block or concrete, spray		500	.016	S.F.	1.12	.65		1.77	2.20
0330	Steel, flat surfaces and tanks to 12"		500	.016		1.12	.65		1.77	2.20

02 83 Lead Remediation

02 83 19 – Lead-Based Paint Remediation

02 83 19.23 Encapsulation of Lead-Based Paint

	02 83 19.23 Encapsulation of Lead-Based Paint	Crew	Daily Output	Labor-Hours	Unit	Material	2015 Bare Costs Labor	Equipment	Total	Total Incl O&P
0340	Beams, brushwork	1 Pord	400	.020	S.F.	1.38	.81		2.19	2.74
0350	Trusses		400	.020		1.38	.81		2.19	2.74

02 83 19.26 Removal of Lead-Based Paint

	02 83 19.26 Removal of Lead-Based Paint	Crew	Daily Output	Labor-Hours	Unit	Material	2015 Bare Costs Labor	Equipment	Total	Total Incl O&P
0010	**REMOVAL OF LEAD-BASED PAINT** R028319-60									
0011	By chemicals, per application									
0050	Baseboard, to 6″ wide	1 Pord	64	.125	L.F.	.72	5.05		5.77	8.40
0070	To 12″ wide		32	.250	″	1.40	10.10		11.50	16.75
0200	Balustrades, one side		28	.286	S.F.	1.45	11.55		13	18.95
1400	Cabinets, simple design		32	.250		1.31	10.10		11.41	16.65
1420	Ornate design		25	.320		1.58	12.90		14.48	21
1600	Cornice, simple design		60	.133		1.50	5.40		6.90	9.75
1620	Ornate design		20	.400		5.35	16.15		21.50	30.50
2800	Doors, one side, flush		84	.095		1.88	3.84		5.72	7.85
2820	Two panel		80	.100		1.31	4.04		5.35	7.55
2840	Four panel		45	.178		1.40	7.15		8.55	12.35
2880	For trim, one side, add		64	.125	L.F.	.71	5.05		5.76	8.40
3000	Fence, picket, one side		30	.267	S.F.	1.31	10.75		12.06	17.65
3200	Grilles, one side, simple design		30	.267		1.33	10.75		12.08	17.65
3220	Ornate design		25	.320		1.43	12.90		14.33	21
3240	Handrails		90	.089	L.F.	1.33	3.59		4.92	6.85
4400	Pipes, to 4″ diameter		90	.089		1.88	3.59		5.47	7.45
4420	To 8″ diameter		50	.160		3.75	6.45		10.20	13.90
4440	To 12″ diameter		36	.222		5.65	8.95		14.60	19.70
4460	To 16″ diameter		20	.400		7.50	16.15		23.65	33
4500	For hangers, add		40	.200	Ea.	2.49	8.05		10.54	14.90
4800	Siding		90	.089	S.F.	1.24	3.59		4.83	6.75
5000	Trusses, open		55	.145	SF Face	1.96	5.85		7.81	11
6200	Windows, one side only, double hung, 1/1 light, 24″ x 48″ high		4	2	Ea.	23	80.50		103.50	148
6220	30″ x 60″ high		3	2.667		31	108		139	196
6240	36″ x 72″ high		2.50	3.200		37	129		166	236
6280	40″ x 80″ high		2	4		46.50	161		207.50	294
6400	Colonial window, 6/6 light, 24″ x 48″ high		2	4		46.50	161		207.50	294
6420	30″ x 60″ high		1.50	5.333		62	215		277	395
6440	36″ x 72″ high		1	8		93	325		418	585
6480	40″ x 80″ high		1	8		93	325		418	585
6600	8/8 light, 24″ x 48″ high		2	4		46.50	161		207.50	294
6620	40″ x 80″ high		1	8		93	325		418	585
6800	12/12 light, 24″ x 48″ high		1	8		93	325		418	585
6820	40″ x 80″ high		.75	10.667		124	430		554	785
6840	Window frame & trim items, included in pricing above									
7000	Hand scraping and HEPA vacuum, less than 4 S.F.	1 Pord	8	1	Ea.	.82	40.50		41.32	62
8000	Collect and bag bulk material, 3 C.F. bags, by hand	″	30	.267	″	.82	10.75		11.57	17.10

02 85 Mold Remediation

02 85 16 – Mold Remediation Preparation and Containment

02 85 16.40 Mold Remediation Plans and Methods		Crew	Daily Output	Labor-Hours	Unit	Material	2015 Bare Costs Labor	Equipment	Total	Total Incl O&P
0010	**MOLD REMEDIATION PLANS AND METHODS**									
0020	Initial inspection, areas to 2500 S.F.				Total				268	295
0030	Areas to 5000 S.F.								440	485
0032	Areas to 10000 S.F.								440	485
0040	Testing, air sample each								126	139
0050	Swab sample								115	127
0060	Tape sample								125	140
0070	Post remediation air test								126	139
0080	Mold abatement plan, area to 2500 S.F.								1,225	1,325
0090	Areas to 5000 S.F.								1,625	1,775
0095	Areas to 10000 S.F.								2,550	2,800
0100	Packup & removal of contents, average 3 bedroom home, excl storage								8,150	8,975
0110	Average 5 bedroom home, excl storage				↓				15,300	16,800
0600	For demolition in mold contaminated areas, see Section 02 85 33.50									
0610	For personal protection equipment, see Section 02 82 13.41									

02 85 16.50 Preparation of Mold Containment Area		Crew	Daily Output	Labor-Hours	Unit	Material	Labor	Equipment	Total	Total Incl O&P
0010	**PREPARATION OF MOLD CONTAINMENT AREA**									
0100	Pre-cleaning, HEPA vacuum and wet wipe, flat surfaces	A-9	12000	.005	S.F.	.02	.28		.30	.45
0300	Separation barrier, 2" x 4" @ 16", 1/2" plywood ea. side, 8' high	2 Carp	400	.040		3.32	1.88		5.20	6.55
0310	12' high		320	.050		3.32	2.35		5.67	7.25
0320	16' high		200	.080		2.30	3.76		6.06	8.35
0400	Personnel decontam. chamber, 2" x 4" @ 16", 3/4" ply ea. side		280	.057		4.34	2.68		7.02	8.90
0450	Waste decontam. chamber, 2" x 4" studs @ 16", 3/4" ply each side	↓	360	.044	↓	4.34	2.09		6.43	8
0500	Cover surfaces with polyethylene sheeting									
0501	Including glue and tape									
0550	Floors, each layer, 6 mil	A-9	8000	.008	S.F.	.04	.42		.46	.70
0551	4 mil		9000	.007		.03	.37		.40	.61
0560	Walls, each layer, 6 mil		6000	.011		.04	.56		.60	.92
0561	4 mil	↓	7000	.009	↓	.03	.48		.51	.78
0570	For heights above 12', add						20%			
0575	For heights above 20', add						30%			
0580	For fire retardant poly, add					100%				
0590	For large open areas, deduct					10%	20%			
0600	Seal floor penetrations with foam firestop to 36 sq. in.	2 Carp	200	.080	Ea.	7.75	3.76		11.51	14.30
0610	36 sq. in. to 72 sq. in.		125	.128		15.50	6		21.50	26.50
0615	72 sq. in. to 144 sq. in.		80	.200		31	9.40		40.40	48.50
0620	Wall penetrations, to 36 square inches		180	.089		7.75	4.17		11.92	14.90
0630	36 sq. in. to 72 sq. in.		100	.160		15.50	7.50		23	28.50
0640	72 sq. in. to 144 sq. in.	↓	60	.267	↓	31	12.50		43.50	53.50
0800	Caulk seams with latex caulk	1 Carp	230	.035	L.F.	.17	1.63		1.80	2.70
0900	Set up neg. air machine, 1-2k CFM/25 M.C.F. volume	1 Asbe	4.30	1.860	Ea.		97.50		97.50	152

02 85 33 – Removal and Disposal of Materials with Mold

02 85 33.50 Demolition in Mold Contaminated Area		Crew	Daily Output	Labor-Hours	Unit	Material	Labor	Equipment	Total	Total Incl O&P
0010	**DEMOLITION IN MOLD CONTAMINATED AREA**									
0200	Ceiling, including suspension system, plaster and lath	A-9	2100	.030	S.F.	.09	1.60		1.69	2.59
0210	Finished plaster, leaving wire lath		585	.109		.32	5.75		6.07	9.30
0220	Suspended acoustical tile		3500	.018		.05	.96		1.01	1.55
0230	Concealed tile grid system		3000	.021		.06	1.12		1.18	1.81
0240	Metal pan grid system		1500	.043		.12	2.24		2.36	3.62
0250	Gypsum board		2500	.026		.07	1.34		1.41	2.17
0255	Plywood		2500	.026	↓	.07	1.34		1.41	2.17
0260	Lighting fixtures up to 2' x 4'	↓	72	.889	Ea.	2.56	46.50		49.06	75.50

02 85 Mold Remediation

02 85 33 – Removal and Disposal of Materials with Mold

02 85 33.50 Demolition in Mold Contaminated Area		Crew	Daily Output	Labor-Hours	Unit	2015 Bare Costs Material	Labor	Equipment	Total	Total Incl O&P
0400	Partitions, non load bearing									
0410	Plaster, lath, and studs	A-9	690	.093	S.F.	.88	4.86		5.74	8.50
0450	Gypsum board and studs		1390	.046		.13	2.41		2.54	3.91
0465	Carpet & pad		1390	.046		.13	2.41		2.54	3.91
0600	Pipe insulation, air cell type, up to 4″ diameter pipe		900	.071	L.F.	.21	3.73		3.94	6.05
0610	4″ to 8″ diameter pipe		800	.080		.23	4.19		4.42	6.80
0620	10″ to 12″ diameter pipe		700	.091		.26	4.79		5.05	7.75
0630	14″ to 16″ diameter pipe		550	.116		.34	6.10		6.44	9.85
0650	Over 16″ diameter pipe		650	.098	S.F.	.28	5.15		5.43	8.35
9000	For type B (supplied air) respirator equipment, add				%				10%	10%

02 91 Chemical Sampling, Testing and Analysis

02 91 10 – Monitoring, Sampling, Testing and Analysis

02 91 10.10 Monitoring, Chemical Sampling, Testing and Analysis		Crew	Daily Output	Labor-Hours	Unit	Material	Labor	Equipment	Total	Total Incl O&P
0010	**MONITORING, CHEMICAL SAMPLING, TESTING AND ANALYSIS**									
0015	Field Sampling of waste									
0100	Field samples, sample collection, sludge	1 Skwk	32	.250	Ea.		12.15		12.15	18.80
0110	Contaminated soils	″	32	.250	″		12.15		12.15	18.80
0200	Vials and bottles									
0210	32 oz. clear wide mouth jar (case of 12)				Ea.	43.50			43.50	48
0220	32 oz. Boston round bottle (case of 12)					38			38	42
0230	32 oz. HDPE bottle (case of 12)					36			36	39.50
0300	Laboratory analytical services									
0310	Laboratory testing 13 metals				Ea.	187			187	205
0312	13 metals + mercury					220			220	242
0314	8 metals					160			160	176
0316	Mercury only					42			42	46
0318	Single metal (only Cs, Li, Sr, Ta)					43			43	47.50
0320	Single metal (excludes Hg, Cs, Li, Sr, Ta)					43			43	47.50
0400	Hydrocarbons standard					89			89	98
0410	Hydrocarbons fingerprint					161			161	177
0500	Radioactivity gross alpha					154			154	169
0510	Gross alpha & beta					154			154	169
0520	Radium 226					87			87	96
0530	Radium 228					128			128	141
0540	Radon					148			148	162
0550	Uranium					90			90	99
0600	Volatile organics without GC/MS					156			156	172
0610	Volatile organics including GC/MS					179			179	197
0630	Synthetic organic compounds					1,025			1,025	1,125
0640	Herbicides					215			215	237
0650	Pesticides					151			151	166
0660	PCB's					138			138	152

Division Notes

		CREW	DAILY OUTPUT	LABOR-HOURS	UNIT	BARE COSTS MAT.	BARE COSTS LABOR	BARE COSTS EQUIP.	BARE COSTS TOTAL	TOTAL INCL O&P

Estimating Tips

General

- Carefully check all the plans and specifications. Concrete often appears on drawings other than structural drawings, including mechanical and electrical drawings for equipment pads. The cost of cutting and patching is often difficult to estimate. See Subdivision 03 81 for Concrete Cutting, Subdivision 02 41 19.16 for Cutout Demolition, Subdivision 03 05 05.10 for Concrete Demolition, and Subdivision 02 41 19.19 for Rubbish Handling (handling, loading and hauling of debris).
- Always obtain concrete prices from suppliers near the job site. A volume discount can often be negotiated, depending upon competition in the area. Remember to add for waste, particularly for slabs and footings on grade.

03 10 00 Concrete Forming and Accessories

- A primary cost for concrete construction is forming. Most jobs today are constructed with prefabricated forms. The selection of the forms best suited for the job and the total square feet of forms required for efficient concrete forming and placing are key elements in estimating concrete construction. Enough forms must be available for erection to make efficient use of the concrete placing equipment and crew.
- Concrete accessories for forming and placing depend upon the systems used. Study the plans and specifications to ensure that all special accessory requirements have been included in the cost estimate, such as anchor bolts, inserts, and hangers.
- Included within costs for forms-in-place are all necessary bracing and shoring.

03 20 00 Concrete Reinforcing

- Ascertain that the reinforcing steel supplier has included all accessories, cutting, bending, and an allowance for lapping, splicing, and waste. A good rule of thumb is 10% for lapping, splicing, and waste. Also, 10% waste should be allowed for welded wire fabric.
- The unit price items in the subdivisions for Reinforcing In Place, Glass Fiber Reinforcing, and Welded Wire Fabric include the labor to install accessories such as beam and slab bolsters, high chairs, and bar ties and tie wire. The material cost for these accessories is not included; they may be obtained from the Accessories Division.

03 30 00 Cast-In-Place Concrete

- When estimating structural concrete, pay particular attention to requirements for concrete additives, curing methods, and surface treatments. Special consideration for climate, hot or cold, must be included in your estimate. Be sure to include requirements for concrete placing equipment, and concrete finishing.
- For accurate concrete estimating, the estimator must consider each of the following major components individually: forms, reinforcing steel, ready-mix concrete, placement of the concrete, and finishing of the top surface. For faster estimating, Subdivision 03 30 53.40 for Concrete-In-Place can be used; here, various items of concrete work are presented that include the costs of all five major components (unless specifically stated otherwise).

03 40 00 Precast Concrete
03 50 00 Cast Decks and Underlayment

- The cost of hauling precast concrete structural members is often an important factor. For this reason, it is important to get a quote from the nearest supplier. It may become economically feasible to set up precasting beds on the site if the hauling costs are prohibitive.

Reference Numbers

Reference numbers are shown in shaded boxes at the beginning of some major classifications. These numbers refer to related items in the Reference Section. The reference information may be an estimating procedure, an alternate pricing method, or technical information.

Note: Not all subdivisions listed here necessarily appear in this publication. ■

03 01 Maintenance of Concrete

03 01 30 – Maintenance of Cast-In-Place Concrete

03 01 30.62 Concrete Patching		Crew	Daily Output	Labor-Hours	Unit	2015 Bare Costs Material	Labor	Equipment	Total	Total Incl O&P
0010	**CONCRETE PATCHING**									
0100	Floors, 1/4" thick, small areas, regular grout	1 Cefi	170	.047	S.F.	1.46	2.12		3.58	4.74
0150	Epoxy grout	"	100	.080	"	8.05	3.60		11.65	14.25
2000	Walls, including chipping, cleaning and epoxy grout									
2100	1/4" deep	1 Cefi	65	.123	S.F.	7.65	5.55		13.20	16.65
2150	1/2" deep		50	.160		15.35	7.20		22.55	27.50
2200	3/4" deep		40	.200		23	9		32	39

03 05 Common Work Results for Concrete

03 05 05 – Selective Demolition for Concrete

03 05 05.10 Selective Demolition, Concrete

Code	Description	Crew	Daily Output	Labor-Hours	Unit	Material	Labor	Equipment	Total	Total Incl O&P
0010	**SELECTIVE DEMOLITION, CONCRETE** R024119-10									
0012	Excludes saw cutting, torch cutting, loading or hauling									
0050	Break into small pieces, reinf. less than 1% of cross-sectional area	B-9	24	1.667	C.Y.		63.50	9.70	73.20	108
0060	Reinforcing 1% to 2% of cross-sectional area		16	2.500			95	14.55	109.55	162
0070	Reinforcing more than 2% of cross-sectional area		8	5			190	29	219	325
0150	Remove whole pieces, up to 2 tons per piece	E-18	36	1.111	Ea.		58.50	26.50	85	128
0160	2 – 5 tons per piece		30	1.333			70	31.50	101.50	154
0170	5 – 10 tons per piece		24	1.667			87.50	39.50	127	193
0180	10 – 15 tons per piece		18	2.222			117	52.50	169.50	256
0250	Precast unit embedded in masonry, up to 1 C.F.	D-1	16	1			42		42	64.50
0260	1 – 2 C.F.		12	1.333			56		56	85.50
0270	2 – 5 C.F.		10	1.600			67.50		67.50	103
0280	5 – 10 C.F.		8	2			84		84	129
0990	For hydrodemolition see Section 02 41 13.15									

03 05 13 – Basic Concrete Materials

03 05 13.20 Concrete Admixtures and Surface Treatments

Code	Description	Crew	Daily Output	Labor-Hours	Unit	Material	Labor	Equipment	Total	Total Incl O&P
0010	**CONCRETE ADMIXTURES AND SURFACE TREATMENTS**									
0040	Abrasives, aluminum oxide, over 20 tons				Lb.	1.88			1.88	2.07
0050	1 to 20 tons					2.02			2.02	2.22
0070	Under 1 ton					2.10			2.10	2.31
0100	Silicon carbide, black, over 20 tons					2.87			2.87	3.16
0110	1 to 20 tons					3.04			3.04	3.35
0120	Under 1 ton					3.17			3.17	3.49
0200	Air entraining agent, .7 to 1.5 oz. per bag, 55 gallon drum				Gal.	14			14	15.40
0220	5 gallon pail					19.25			19.25	21
0300	Bonding agent, acrylic latex, 250 S.F. per gallon, 5 gallon pail					22.50			22.50	25
0320	Epoxy resin, 80 S.F. per gallon, 4 gallon case					61			61	67
0400	Calcium chloride, 50 lb. bags, TL lots				Ton	810			810	890
0420	Less than truckload lots				Bag	24			24	26.50
0500	Carbon black, liquid, 2 to 8 lb. per bag of cement				Lb.	9.10			9.10	10
0600	Colored pigments, integral, 2 to 10 lb. per bag of cement, subtle colors					2.40			2.40	2.64
0610	Standard colors					3.26			3.26	3.59
0620	Premium colors					5.40			5.40	5.95
0920	Dustproofing compound, 250 S.F./gal., 5 gallon pail				Gal.	6.65			6.65	7.30
1010	Epoxy based, 125 S.F./gal., 5 gallon pail				"	55.50			55.50	61
1100	Hardeners, metallic, 55 lb. bags, natural (grey)				Lb.	.70			.70	.76
1200	Colors					2.27			2.27	2.50
1300	Non-metallic, 55 lb. bags, natural grey					.43			.43	.47
1320	Colors					.94			.94	1.04

03 05 13 – Basic Concrete Materials

03 05 13.20 Concrete Admixtures and Surface Treatments		Crew	Daily Output	Labor-Hours	Unit	Material	Labor	Equipment	Total	Total Incl O&P
						2015 Bare Costs				
1550	Release agent, for tilt slabs, 5 gallon pail				Gal.	17.60			17.60	19.40
1570	For forms, 5 gallon pail					12.50			12.50	13.75
1590	Concrete release agent for forms, 100% biodegradeable, zero VOC, 5 gal pail G					19.95			19.95	22
1595	55 gallon drum G					16.80			16.80	18.50
1600	Sealer, hardener and dustproofer, epoxy-based, 125 S.F./gal., 5 gallon unit					55.50			55.50	61
1620	3 gallon unit					61			61	67.50
1630	Sealer, solvent-based, 250 S.F./gal., 55 gallon drum					24.50			24.50	27
1640	5 gallon pail					30.50			30.50	34
1650	Sealer, water based, 350 S.F., 55 gallon drum					22.50			22.50	25
1660	5 gallon pail					25.50			25.50	28.50
1900	Set retarder, 100 S.F./gal., 1 gallon pail				↓	22.50			22.50	24.50
2000	Waterproofing, integral 1 lb. per bag of cement				Lb.	2.12			2.12	2.33
2100	Powdered metallic, 40 lb. per 100 S.F., standard colors					2.52			2.52	2.77
2120	Premium colors				↓	3.53			3.53	3.88
3000	For integral colored pigments, 2500 psi (5 bag mix)									
3100	Standard colors, 1.8 lb. per bag, add				C.Y.	21.50			21.50	24
3200	9.4 lb. per bag, add					113			113	124
3400	Premium colors, 1.8 lb. per bag, add					29.50			29.50	32.50
3500	7.5 lb. per bag, add					122			122	134
3700	Ultra premium colors, 1.8 lb. per bag, add					49			49	53.50
3800	7.5 lb. per bag, add				↓	203			203	224
6000	Concrete ready mix additives, recycled coal fly ash, mixed at plant G				Ton	58.50			58.50	64
6010	Recycled blast furnace slag, mixed at plant G				"	91			91	100

03 05 13.25 Aggregate

03 05 13.25 Aggregate		Crew	Daily Output	Labor-Hours	Unit	Material	Labor	Equipment	Total	Total Incl O&P
0010	**AGGREGATE** R033105-20									
0100	Lightweight vermiculite or perlite, 4 C.F. bag, C.L. lots G				Bag	23.50			23.50	26
0150	L.C.L. lots R033105-40 G				"	26.50			26.50	29
0250	Sand & stone, loaded at pit, crushed bank gravel				Ton	21.50			21.50	23.50
0350	Sand, washed, for concrete R033105-50					19.35			19.35	21.50
0400	For plaster or brick					19.35			19.35	21.50
0450	Stone, 3/4" to 1-1/2"					17.80			17.80	19.60
0470	Round, river stone					26.50			26.50	29
0500	3/8" roofing stone & 1/2" pea stone					26			26	28.50
0550	For trucking 10-mile round trip, add to the above	B-34B	117	.068			2.74	5.90	8.64	10.65
0600	For trucking 30-mile round trip, add to the above	"	72	.111	↓		4.45	9.60	14.05	17.30
0850	Sand & stone, loaded at pit, crushed bank gravel				C.Y.	30			30	33
0950	Sand, washed, for concrete					27			27	29.50
1000	For plaster or brick					27			27	29.50
1050	Stone, 3/4" to 1-1/2"					34			34	37.50
1055	Round, river stone					36			36	39.50
1100	3/8" roofing stone & 1/2" pea stone					25.50			25.50	28
1150	For trucking 10-mile round trip, add to the above	B-34B	78	.103			4.11	8.85	12.96	15.95
1200	For trucking 30-mile round trip, add to the above	"	48	.167	↓		6.70	14.40	21.10	26
1310	Onyx chips, 50 lb. bags				Cwt.	34			34	37
1330	Quartz chips, 50 lb. bags					33.50			33.50	37
1410	White marble, 3/8" to 1/2", 50 lb. bags				↓	7.50			7.50	8.25
1430	3/4", bulk				Ton	117			117	129

03 05 13.30 Cement

03 05 13.30 Cement		Crew	Daily Output	Labor-Hours	Unit	Material	Labor	Equipment	Total	Total Incl O&P
0010	**CEMENT** R033105-20									
0240	Portland, Type I/II, TL lots, 94 lb. bags				Bag	9.90			9.90	10.90
0250	LTL/LCL lots R033105-30				"	11			11	12.10
0300	Trucked in bulk, per Cwt.				Cwt.	7.10			7.10	7.85

03 05 Common Work Results for Concrete

03 05 13 – Basic Concrete Materials

03 05 13.30 Cement		Crew	Daily Output	Labor-Hours	Unit	Material	2015 Bare Costs Labor	Equipment	Total	Total Incl O&P
0400	Type III, high early strength, TL lots, 94 lb. bags R033105-40				Bag	11.80			11.80	13
0420	L.T.L. or L.C.L. lots					12.10			12.10	13.30
0500	White, type III, high early strength, T.L. or C.L. lots, bags R033105-50					23.50			23.50	26
0520	L.T.L. or L.C.L. lots					24			24	26.50
0600	White, type I, T.L. or C.L. lots, bags					25			25	27.50
0620	L.T.L. or L.C.L. lots					26.50			26.50	29

03 05 13.80 Waterproofing and Dampproofing

		Crew	Daily Output	Labor-Hours	Unit	Material	Labor	Equipment	Total	Total Incl O&P
0010	**WATERPROOFING AND DAMPPROOFING**									
0050	Integral waterproofing, add to cost of regular concrete				C.Y.	12.70			12.70	14

03 05 13.85 Winter Protection

		Crew	Daily Output	Labor-Hours	Unit	Material	Labor	Equipment	Total	Total Incl O&P
0010	**WINTER PROTECTION**									
0012	For heated ready mix, add				C.Y.	4.05			4.05	4.46
0100	Temporary heat to protect concrete, 24 hours	2 Clab	50	.320	M.S.F.	535	12.05		547.05	605
0200	Temporary shelter for slab on grade, wood frame/polyethylene sheeting									
0201	Build or remove, light framing for short spans	2 Carp	10	1.600	M.S.F.	310	75		385	460
0210	Large framing for long spans	"	3	5.333	"	415	250		665	840
0500	Electrically heated pads, 110 volts, 15 watts per S.F., buy				S.F.	10.35			10.35	11.40
0600	20 watts per S.F., buy					13.80			13.80	15.15
0710	Electrically, heated pads, 15 watts/S.F., 20 uses					.52			.52	.57

03 11 Concrete Forming

03 11 13 – Structural Cast-In-Place Concrete Forming

03 11 13.20 Forms In Place, Beams and Girders		Crew	Daily Output	Labor-Hours	Unit	Material	Labor	Equipment	Total	Total Incl O&P
0010	**FORMS IN PLACE, BEAMS AND GIRDERS** R031113-40									
0500	Exterior spandrel, job-built plywood, 12" wide, 1 use	C-2	225	.213	SFCA	3.12	9.75		12.87	18.45
0550	2 use R031113-60		275	.175		1.64	8		9.64	14.10
0600	3 use		295	.163		1.25	7.45		8.70	12.80
0650	4 use		310	.155		1.01	7.10		8.11	12
1000	18" wide, 1 use		250	.192		2.78	8.80		11.58	16.55
1050	2 use		275	.175		1.53	8		9.53	14
1100	3 use		305	.157		1.11	7.20		8.31	12.25
1150	4 use		315	.152		.91	6.95		7.86	11.70
1500	24" wide, 1 use		265	.181		2.54	8.30		10.84	15.55
1550	2 use		290	.166		1.43	7.55		8.98	13.25
1600	3 use		315	.152		1.02	6.95		7.97	11.80
1650	4 use		325	.148		.83	6.75		7.58	11.30
2000	Interior beam, job-built plywood, 12" wide, 1 use		300	.160		3.67	7.30		10.97	15.30
2050	2 use		340	.141		1.77	6.45		8.22	11.90
2100	3 use		364	.132		1.47	6.05		7.52	10.90
2150	4 use		377	.127		1.19	5.80		6.99	10.25
2500	24" wide, 1 use		320	.150		2.60	6.85		9.45	13.40
2550	2 use		365	.132		1.47	6		7.47	10.85
2600	3 use		385	.125		1.03	5.70		6.73	9.90
2650	4 use		395	.122		.84	5.55		6.39	9.45
3000	Encasing steel beam, hung, job-built plywood, 1 use		325	.148		3.20	6.75		9.95	13.90
3050	2 use		390	.123		1.76	5.65		7.41	10.60
3100	3 use		415	.116		1.28	5.30		6.58	9.55
3150	4 use		430	.112		1.04	5.10		6.14	9
3500	Bottoms only, to 30" wide, job-built plywood, 1 use		230	.209		4.27	9.55		13.82	19.40
3550	2 use		265	.181		2.39	8.30		10.69	15.40

03 11 13.20 Forms In Place, Beams and Girders		Crew	Daily Output	Labor-Hours	Unit	Material	2015 Bare Costs Labor	Equipment	Total	Total Incl O&P
3600	3 use	C-2	280	.171	SFCA	1.71	7.85		9.56	13.95
3650	4 use		290	.166		1.39	7.55		8.94	13.20
4000	Sides only, vertical, 36" high, job-built plywood, 1 use		335	.143		5.20	6.55		11.75	15.80
4050	2 use		405	.119		2.86	5.40		8.26	11.50
4100	3 use		430	.112		2.08	5.10		7.18	10.15
4150	4 use		445	.108		1.69	4.93		6.62	9.45
4500	Sloped sides, 36" high, 1 use		305	.157		5	7.20		12.20	16.55
4550	2 use		370	.130		2.80	5.95		8.75	12.25
4600	3 use		405	.119		2.01	5.40		7.41	10.55
4650	4 use		425	.113		1.63	5.15		6.78	9.75
5000	Upstanding beams, 36" high, 1 use		225	.213		6.35	9.75		16.10	22
5050	2 use		255	.188		3.53	8.60		12.13	17.15
5100	3 use		275	.175		2.56	8		10.56	15.10
5150	4 use	↓	280	.171	↓	2.08	7.85		9.93	14.35

03 11 13.25 Forms In Place, Columns		Crew	Daily Output	Labor-Hours	Unit	Material	Labor	Equipment	Total	Total Incl O&P
0010	**FORMS IN PLACE, COLUMNS** R031113-40									
0500	Round fiberglass, 4 use per mo., rent, 12" diameter	C-1	160	.200	L.F.	8.95	8.90		17.85	23.50
0550	16" diameter R031113-60		150	.213		10.65	9.50		20.15	26.50
0600	18" diameter		140	.229		11.95	10.20		22.15	29
0650	24" diameter		135	.237		14.85	10.55		25.40	32.50
0700	28" diameter		130	.246		16.60	11		27.60	35
0800	30" diameter		125	.256		17.35	11.40		28.75	36.50
0850	36" diameter		120	.267		23	11.90		34.90	44
1500	Round fiber tube, recycled paper, 1 use, 8" diameter G		155	.206		1.77	9.20		10.97	16.10
1550	10" diameter G		155	.206		2.40	9.20		11.60	16.80
1600	12" diameter G		150	.213		2.83	9.50		12.33	17.75
1650	14" diameter G		145	.221		3.91	9.85		13.76	19.45
1700	16" diameter G		140	.229		4.88	10.20		15.08	21
1720	18" diameter G		140	.229		5.70	10.20		15.90	22
1750	20" diameter G		135	.237		7.60	10.55		18.15	24.50
1800	24" diameter G		130	.246		9.70	11		20.70	27.50
1850	30" diameter G		125	.256		14.50	11.40		25.90	33.50
1900	36" diameter G		115	.278		17.10	12.40		29.50	38
1950	42" diameter G		100	.320		44.50	14.30		58.80	71
2000	48" diameter G	↓	85	.376	↓	51.50	16.80		68.30	83
2200	For seamless type, add					15%				
3000	Round, steel, 4 use per mo., rent, regular duty, 14" diameter G	C-1	145	.221	L.F.	12.30	9.85		22.15	28.50
3050	16" diameter G		125	.256		12.50	11.40		23.90	31.50
3100	Heavy duty, 20" diameter G		105	.305		13.80	13.60		27.40	36
3150	24" diameter G		85	.376		15.05	16.80		31.85	42.50
3200	30" diameter G		70	.457		17.30	20.50		37.80	50.50
3250	36" diameter G		60	.533		18.60	24		42.60	57
3300	48" diameter G		50	.640		27.50	28.50		56	74.50
3350	60" diameter G	↓	45	.711	↓	34	31.50		65.50	86.50
4500	For second and succeeding months, deduct					50%				
5000	Job-built plywood, 8" x 8" columns, 1 use	C-1	165	.194	SFCA	2.73	8.65		11.38	16.30
5050	2 use		195	.164		1.56	7.30		8.86	12.95
5100	3 use		210	.152		1.09	6.80		7.89	11.65
5150	4 use		215	.149		.90	6.65		7.55	11.20
5500	12" x 12" columns, 1 use		180	.178		2.62	7.95		10.57	15.10
5550	2 use		210	.152		1.44	6.80		8.24	12.05
5600	3 use	↓	220	.145	↓	1.05	6.50		7.55	11.15

03 11 13 – Structural Cast-In-Place Concrete Forming

03 11 13.25 Forms In Place, Columns

	03 11 13.25 Forms In Place, Columns	Crew	Daily Output	Labor-Hours	Unit	Material	2015 Bare Costs Labor	Equipment	Total	Total Incl O&P
5650	4 use	C-1	225	.142	SFCA	.85	6.35		7.20	10.70
6000	16" x 16" columns, 1 use		185	.173		2.62	7.70		10.32	14.75
6050	2 use		215	.149		1.41	6.65		8.06	11.75
6100	3 use		230	.139		1.05	6.20		7.25	10.70
6150	4 use		235	.136		.86	6.05		6.91	10.30
6500	24" x 24" columns, 1 use		190	.168		2.97	7.50		10.47	14.80
6550	2 use		216	.148		1.63	6.60		8.23	11.95
6600	3 use		230	.139		1.19	6.20		7.39	10.85
6650	4 use		238	.134		.96	6		6.96	10.30
7000	36" x 36" columns, 1 use		200	.160		2.31	7.15		9.46	13.55
7050	2 use		230	.139		1.31	6.20		7.51	11
7100	3 use		245	.131		.92	5.85		6.77	9.95
7150	4 use		250	.128		.75	5.70		6.45	9.65
7400	Steel framed plywood, based on 50 uses of purchased									
7420	forms, and 4 uses of bracing lumber									
7500	8" x 8" column	C-1	340	.094	SFCA	2.14	4.20		6.34	8.80
7550	10" x 10"		350	.091		1.87	4.08		5.95	8.35
7600	12" x 12"		370	.086		1.59	3.86		5.45	7.70
7650	16" x 16"		400	.080		1.23	3.57		4.80	6.85
7700	20" x 20"		420	.076		1.09	3.40		4.49	6.45
7750	24" x 24"		440	.073		.79	3.24		4.03	5.85
7755	30" x 30"		440	.073		1	3.24		4.24	6.10
7760	36" x 36"		460	.070		.88	3.10		3.98	5.75

03 11 13.30 Forms In Place, Culvert

	03 11 13.30 Forms In Place, Culvert	Crew	Daily Output	Labor-Hours	Unit	Material	Labor	Equipment	Total	Total Incl O&P
0010	**FORMS IN PLACE, CULVERT** R031113-40									
0015	5' to 8' square or rectangular, 1 use	C-1	170	.188	SFCA	4.03	8.40		12.43	17.35
0050	2 use R031113-60		180	.178		2.42	7.95		10.37	14.85
0100	3 use		190	.168		1.89	7.50		9.39	13.65
0150	4 use		200	.160		1.62	7.15		8.77	12.80

03 11 13.35 Forms In Place, Elevated Slabs

	03 11 13.35 Forms In Place, Elevated Slabs	Crew	Daily Output	Labor-Hours	Unit	Material	Labor	Equipment	Total	Total Incl O&P
0010	**FORMS IN PLACE, ELEVATED SLABS** R031113-40									
1000	Flat plate, job-built plywood, to 15' high, 1 use R031113-60	C-2	470	.102	S.F.	3.86	4.67		8.53	11.45
1050	2 use		520	.092		2.13	4.22		6.35	8.85
1100	3 use		545	.088		1.55	4.03		5.58	7.90
1150	4 use		560	.086		1.26	3.92		5.18	7.45
1500	15' to 20' high ceilings, 4 use		495	.097		1.30	4.43		5.73	8.25
1600	21' to 35' high ceilings, 4 use		450	.107		1.60	4.88		6.48	9.25
2000	Flat slab, drop panels, job-built plywood, to 15' high, 1 use		449	.107		4.42	4.89		9.31	12.35
2050	2 use		509	.094		2.43	4.31		6.74	9.30
2100	3 use		532	.090		1.77	4.13		5.90	8.30
2150	4 use		544	.088		1.43	4.04		5.47	7.80
2250	15' to 20' high ceilings, 4 use		480	.100		2.47	4.57		7.04	9.75
2350	20' to 35' high ceilings, 4 use		435	.110		2.77	5.05		7.82	10.80
3000	Floor slab hung from steel beams, 1 use		485	.099		2.73	4.53		7.26	9.95
3050	2 use		535	.090		2.12	4.10		6.22	8.65
3100	3 use		550	.087		1.92	3.99		5.91	8.25
3150	4 use		565	.085		1.81	3.89		5.70	8
3500	Floor slab, with 1-way joist pans, 1 use		415	.116		6.35	5.30		11.65	15.15
3550	2 use		445	.108		4.36	4.93		9.29	12.40
3600	3 use		475	.101		3.70	4.62		8.32	11.15
3650	4 use		500	.096		3.37	4.39		7.76	10.45
4500	With 2-way waffle domes, 1 use		405	.119		6.50	5.40		11.90	15.50

03 11 13 – Structural Cast-In-Place Concrete Forming

03 11 13.35 Forms In Place, Elevated Slabs		Crew	Daily Output	Labor-Hours	Unit	Material	2015 Bare Costs Labor	Equipment	Total	Total Incl O&P
4520	2 use	C-2	450	.107	S.F.	4.51	4.88		9.39	12.45
4530	3 use		460	.104		3.84	4.77		8.61	11.55
4550	4 use		470	.102		3.51	4.67		8.18	11.05
5000	Box out for slab openings, over 16" deep, 1 use		190	.253	SFCA	4.57	11.55		16.12	23
5050	2 use		240	.200	"	2.51	9.15		11.66	16.80
5500	Shallow slab box outs, to 10 S.F.		42	1.143	Ea.	12.10	52.50		64.60	94
5550	Over 10 S.F. (use perimeter)		600	.080	L.F.	1.61	3.66		5.27	7.45
6000	Bulkhead forms for slab, with keyway, 1 use, 2 piece		500	.096		2.14	4.39		6.53	9.10
6100	3 piece (see also edge forms)		460	.104		2.31	4.77		7.08	9.90
6200	Slab bulkhead form, 4-1/2" high, exp metal, w/keyway & stakes G	C-1	1200	.027		.92	1.19		2.11	2.84
6210	5-1/2" high G		1100	.029		1.12	1.30		2.42	3.23
6215	7-1/2" high G		960	.033		1.31	1.49		2.80	3.73
6220	9-1/2" high G		840	.038		1.43	1.70		3.13	4.19
6500	Curb forms, wood, 6" to 12" high, on elevated slabs, 1 use		180	.178	SFCA	1.68	7.95		9.63	14.05
6550	2 use		205	.156		.93	6.95		7.88	11.70
6600	3 use		220	.145		.67	6.50		7.17	10.75
6650	4 use		225	.142		.55	6.35		6.90	10.35
7000	Edge forms to 6" high, on elevated slab, 4 use		500	.064	L.F.	.21	2.86		3.07	4.62
7070	7" to 12" high, 1 use		162	.198	SFCA	1.27	8.80		10.07	14.95
7080	2 use		198	.162		.70	7.20		7.90	11.85
7090	3 use		222	.144		.51	6.45		6.96	10.45
7101	4 use		350	.091		.21	4.08		4.29	6.55
7500	Depressed area forms to 12" high, 4 use		300	.107	L.F.	1.02	4.76		5.78	8.40
7550	12" to 24" high, 4 use		175	.183		1.38	8.15		9.53	14.05
8000	Perimeter deck and rail for elevated slabs, straight		90	.356		12.45	15.85		28.30	38
8050	Curved		65	.492		17.05	22		39.05	53
8500	Void forms, round plastic, 8" high x 3" diameter G		450	.071	Ea.	.71	3.17		3.88	5.65
8550	4" diameter G		425	.075		1.06	3.36		4.42	6.30
8600	6" diameter G		400	.080		1.75	3.57		5.32	7.45
8650	8" diameter G		375	.085		3.13	3.81		6.94	9.30

03 11 13.40 Forms In Place, Equipment Foundations

		Crew	Daily Output	Labor-Hours	Unit	Material	Labor	Equipment	Total	Total Incl O&P
0010	**FORMS IN PLACE, EQUIPMENT FOUNDATIONS** R031113-40									
0020	1 use	C-2	160	.300	SFCA	3.45	13.70		17.15	25
0050	2 use R031113-60		190	.253		1.90	11.55		13.45	19.90
0100	3 use		200	.240		1.38	11		12.38	18.40
0150	4 use		205	.234		1.12	10.70		11.82	17.75

03 11 13.45 Forms In Place, Footings

		Crew	Daily Output	Labor-Hours	Unit	Material	Labor	Equipment	Total	Total Incl O&P
0010	**FORMS IN PLACE, FOOTINGS** R031113-40									
0020	Continuous wall, plywood, 1 use	C-1	375	.085	SFCA	6.80	3.81		10.61	13.35
0050	2 use R031113-60		440	.073		3.74	3.24		6.98	9.10
0100	3 use		470	.068		2.72	3.04		5.76	7.65
0150	4 use		485	.066		2.22	2.94		5.16	6.95
0500	Dowel supports for footings or beams, 1 use		500	.064	L.F.	.94	2.86		3.80	5.40
1000	Integral starter wall, to 4" high, 1 use		400	.080		.98	3.57		4.55	6.60
1500	Keyway, 4 use, tapered wood, 2" x 4"	1 Carp	530	.015		.22	.71		.93	1.34
1550	2" x 6"		500	.016		.33	.75		1.08	1.52
2000	Tapered plastic		530	.015		1.31	.71		2.02	2.53
2250	For keyway hung from supports, add		150	.053		.94	2.50		3.44	4.88
3000	Pile cap, square or rectangular, job-built plywood, 1 use	C-1	290	.110	SFCA	2.91	4.92		7.83	10.75
3050	2 use		346	.092		1.60	4.13		5.73	8.10
3100	3 use		371	.086		1.16	3.85		5.01	7.20
3150	4 use		383	.084		.95	3.73		4.68	6.80

03 11 13 – Structural Cast-In-Place Concrete Forming

03 11 13.45 Forms In Place, Footings

	03 11 13.45 Forms In Place, Footings	Crew	Daily Output	Labor-Hours	Unit	Material	Labor (2015 Bare Costs)	Equipment	Total	Total Incl O&P
4000	Triangular or hexagonal, 1 use	C-1	225	.142	SFCA	3.40	6.35		9.75	13.50
4050	2 use		280	.114		1.87	5.10		6.97	9.90
4100	3 use		305	.105		1.36	4.68		6.04	8.70
4150	4 use		315	.102		1.11	4.53		5.64	8.15
5000	Spread footings, job-built lumber, 1 use		305	.105		2.28	4.68		6.96	9.70
5050	2 use		371	.086		1.27	3.85		5.12	7.30
5100	3 use		401	.080		.91	3.56		4.47	6.50
5150	4 use		414	.077		.74	3.45		4.19	6.10
6000	Supports for dowels, plinths or templates, 2′ x 2′ footing		25	1.280	Ea.	6.05	57		63.05	94.50
6050	4′ x 4′ footing		22	1.455		12.10	65		77.10	113
6100	8′ x 8′ footing		20	1.600		24	71.50		95.50	137
6150	12′ x 12′ footing		17	1.882		32.50	84		116.50	165
7000	Plinths, job-built plywood, 1 use		250	.128	SFCA	3.38	5.70		9.08	12.50
7100	4 use		270	.119	″	1.11	5.30		6.41	9.35

03 11 13.47 Forms In Place, Gas Station Forms

	03 11 13.47 Forms In Place, Gas Station Forms		Crew	Daily Output	Labor-Hours	Unit	Material	Labor	Equipment	Total	Total Incl O&P
0010	**FORMS IN PLACE, GAS STATION FORMS**										
0050	Curb fascia, with template, 12 ga. steel, left in place, 9″ high	G	1 Carp	50	.160	L.F.	13.90	7.50		21.40	27
1000	Sign or light bases, 18″ diameter, 9″ high	G		9	.889	Ea.	87.50	41.50		129	160
1050	30″ diameter, 13″ high	G		8	1		139	47		186	226
2000	Island forms, 10′ long, 9″ high, 3′- 6″ wide	G	C-1	10	3.200		390	143		533	645
2050	4′ wide	G		9	3.556		400	159		559	685
2500	20′ long, 9″ high, 4′ wide	G		6	5.333		645	238		883	1,075
2550	5′ wide	G		5	6.400		670	286		956	1,175

03 11 13.50 Forms In Place, Grade Beam

	03 11 13.50 Forms In Place, Grade Beam		Crew	Daily Output	Labor-Hours	Unit	Material	Labor	Equipment	Total	Total Incl O&P
0010	**FORMS IN PLACE, GRADE BEAM**	R031113-40									
0020	Job-built plywood, 1 use		C-2	530	.091	SFCA	3.14	4.14		7.28	9.80
0050	2 use	R031113-60		580	.083		1.73	3.78		5.51	7.70
0100	3 use			600	.080		1.25	3.66		4.91	7.05
0150	4 use			605	.079		1.02	3.63		4.65	6.70

03 11 13.55 Forms In Place, Mat Foundation

	03 11 13.55 Forms In Place, Mat Foundation		Crew	Daily Output	Labor-Hours	Unit	Material	Labor	Equipment	Total	Total Incl O&P
0010	**FORMS IN PLACE, MAT FOUNDATION**	R031113-40									
0020	Job-built plywood, 1 use		C-2	290	.166	SFCA	3.08	7.55		10.63	15.05
0050	2 use	R031113-60		310	.155		1.26	7.10		8.36	12.30
0100	3 use			330	.145		.81	6.65		7.46	11.15
0120	4 use			350	.137		.74	6.25		6.99	10.45

03 11 13.65 Forms In Place, Slab On Grade

	03 11 13.65 Forms In Place, Slab On Grade		Crew	Daily Output	Labor-Hours	Unit	Material	Labor	Equipment	Total	Total Incl O&P
0010	**FORMS IN PLACE, SLAB ON GRADE**	R031113-40									
1000	Bulkhead forms w/keyway, wood, 6″ high, 1 use		C-1	510	.063	L.F.	1.04	2.80		3.84	5.45
1050	2 uses	R031113-60		400	.080		.57	3.57		4.14	6.15
1100	4 uses			350	.091		.34	4.08		4.42	6.65
1400	Bulkhead form for slab, 4-1/2″ high, exp metal, incl keyway & stakes	G		1200	.027		.92	1.19		2.11	2.84
1410	5-1/2″ high	G		1100	.029		1.12	1.30		2.42	3.23
1420	7-1/2″ high	G		960	.033		1.31	1.49		2.80	3.73
1430	9-1/2″ high	G		840	.038		1.43	1.70		3.13	4.19
2000	Curb forms, wood, 6″ to 12″ high, on grade, 1 use			215	.149	SFCA	2.44	6.65		9.09	12.90
2050	2 use			250	.128		1.35	5.70		7.05	10.30
2100	3 use			265	.121		.98	5.40		6.38	9.35
2150	4 use			275	.116		.79	5.20		5.99	8.85
3000	Edge forms, wood, 4 use, on grade, to 6″ high			600	.053	L.F.	.35	2.38		2.73	4.04
3050	7″ to 12″ high			435	.074	SFCA	.79	3.28		4.07	5.90
3060	Over 12″			350	.091	″	.91	4.08		4.99	7.30
3500	For depressed slabs, 4 use, to 12″ high			300	.107	L.F.	.75	4.76		5.51	8.15

	03 11 13.65 Forms In Place, Slab On Grade	Crew	Daily Output	Labor-Hours	Unit	Material	Labor (2015 Bare Costs)	Equipment	Total	Total Incl O&P
3550	To 24" high	C-1	175	.183	L.F.	.99	8.15		9.14	13.65
4000	For slab blockouts, to 12" high, 1 use		200	.160		.80	7.15		7.95	11.85
4050	To 24" high, 1 use		120	.267		1.01	11.90		12.91	19.40
4100	Plastic (extruded), to 6" high, multiple use, on grade		800	.040		8.20	1.78		9.98	11.75
5000	Screed, 24 ga. metal key joint, see Section 03 15 16.30									
5020	Wood, incl. wood stakes, 1" x 3"	C-1	900	.036	L.F.	.77	1.59		2.36	3.28
5050	2" x 4"		900	.036	"	.80	1.59		2.39	3.32
6000	Trench forms in floor, wood, 1 use		160	.200	SFCA	1.94	8.90		10.84	15.90
6050	2 use		175	.183		1.06	8.15		9.21	13.70
6100	3 use		180	.178		.77	7.95		8.72	13.05
6150	4 use		185	.173		.63	7.70		8.33	12.55
8760	Void form, corrugated fiberboard, 4" x 12", 4' long G		3000	.011	S.F.	2.87	.48		3.35	3.88
8770	6" x 12", 4' long		3000	.011		3.42	.48		3.90	4.49
8780	1/4" thick hardboard protective cover for void form	2 Carp	1500	.011		.58	.50		1.08	1.41

	03 11 13.85 Forms In Place, Walls	Crew	Daily Output	Labor-Hours	Unit	Material	Labor	Equipment	Total	Total Incl O&P
0010	**FORMS IN PLACE, WALLS** R031113-10									
0100	Box out for wall openings, to 16" thick, to 10 S.F.	C-2	24	2	Ea.	27	91.50		118.50	171
0150	Over 10 S.F. (use perimeter) R031113-40	"	280	.171	L.F.	2.31	7.85		10.16	14.60
0250	Brick shelf, 4" w, add to wall forms, use wall area above shelf									
0260	1 use R031113-60	C-2	240	.200	SFCA	2.50	9.15		11.65	16.80
0300	2 use		275	.175		1.37	8		9.37	13.80
0350	4 use		300	.160		1	7.30		8.30	12.35
0500	Bulkhead, wood with keyway, 1 use, 2 piece		265	.181	L.F.	2.12	8.30		10.42	15.10
0600	Bulkhead forms with keyway, 1 piece expanded metal, 8" wall G	C-1	1000	.032		1.31	1.43		2.74	3.64
0610	10" wall G		800	.040		1.43	1.78		3.21	4.32
0620	12" wall G		525	.061		1.72	2.72		4.44	6.05
0700	Buttress, to 8' high, 1 use	C-2	350	.137	SFCA	4.28	6.25		10.53	14.35
0750	2 use		430	.112		2.35	5.10		7.45	10.45
0800	3 use		460	.104		1.72	4.77		6.49	9.25
0850	4 use		480	.100		1.41	4.57		5.98	8.60
1000	Corbel or haunch, to 12" wide, add to wall forms, 1 use		150	.320	L.F.	2.37	14.65		17.02	25
1050	2 use		170	.282		1.30	12.90		14.20	21.50
1100	3 use		175	.274		.95	12.55		13.50	20.50
1150	4 use		180	.267		.77	12.20		12.97	19.60
2000	Wall, job-built plywood, to 8' high, 1 use		370	.130	SFCA	2.74	5.95		8.69	12.15
2050	2 use		435	.110		1.75	5.05		6.80	9.65
2100	3 use		495	.097		1.27	4.43		5.70	8.20
2150	4 use		505	.095		1.03	4.35		5.38	7.85
2400	Over 8' to 16' high, 1 use		280	.171		3.03	7.85		10.88	15.40
2450	2 use		345	.139		1.34	6.35		7.69	11.25
2500	3 use		375	.128		.95	5.85		6.80	10.05
2550	4 use		395	.122		.78	5.55		6.33	9.40
2700	Over 16' high, 1 use		235	.204		2.71	9.35		12.06	17.35
2750	2 use		290	.166		1.49	7.55		9.04	13.30
2800	3 use		315	.152		1.09	6.95		8.04	11.90
2850	4 use		330	.145		.88	6.65		7.53	11.20
4000	Radial, smooth curved, job-built plywood, 1 use		245	.196		2.55	8.95		11.50	16.60
4050	2 use		300	.160		1.40	7.30		8.70	12.80
4100	3 use		325	.148		1.02	6.75		7.77	11.50
4150	4 use		335	.143		.83	6.55		7.38	11
4200	Below grade, job-built plywood, 1 use		225	.213		2.77	9.75		12.52	18.05
4210	2 use		225	.213		1.53	9.75		11.28	16.70

03 11 Concrete Forming

03 11 13 – Structural Cast-In-Place Concrete Forming

03 11 13.85 Forms In Place, Walls		Crew	Daily Output	Labor-Hours	Unit	Material (2015 Bare Costs)	Labor (2015 Bare Costs)	Equipment (2015 Bare Costs)	Total (2015 Bare Costs)	Total Incl O&P
4220	3 use	C-2	225	.213	SFCA	1.27	9.75		11.02	16.40
4230	4 use		225	.213		.90	9.75		10.65	16
4300	Curved, 2′ chords, job-built plywood, to 8′ high, 1 use		290	.166		2.16	7.55		9.71	14
4350	2 use		355	.135		1.19	6.20		7.39	10.80
4400	3 use		385	.125		.86	5.70		6.56	9.70
4450	4 use		400	.120		.70	5.50		6.20	9.20
4500	Over 8′ to 16′ high, 1 use		290	.166		.93	7.55		8.48	12.65
4525	2 use		355	.135		.51	6.20		6.71	10.05
4550	3 use		385	.125		.37	5.70		6.07	9.15
4575	4 use		400	.120		.31	5.50		5.81	8.80
4600	Retaining wall, battered, job-built plyw'd, to 8′ high, 1 use		300	.160		2.03	7.30		9.33	13.50
4650	2 use		355	.135		1.12	6.20		7.32	10.75
4700	3 use		375	.128		.81	5.85		6.66	9.90
4750	4 use		390	.123		.66	5.65		6.31	9.40
4900	Over 8′ to 16′ high, 1 use		240	.200		2.22	9.15		11.37	16.50
4950	2 use		295	.163		1.22	7.45		8.67	12.80
5000	3 use		305	.157		.89	7.20		8.09	12.05
5050	4 use		320	.150		.72	6.85		7.57	11.35
5500	For gang wall forming, 192 S.F. sections, deduct					10%	10%			
5550	384 S.F. sections, deduct					20%	20%			
7500	Lintel or sill forms, 1 use	1 Carp	30	.267		3.26	12.50		15.76	23
7520	2 use		34	.235		1.79	11.05		12.84	18.95
7540	3 use		36	.222		1.30	10.45		11.75	17.50
7560	4 use		37	.216		1.06	10.15		11.21	16.75
7800	Modular prefabricated plywood, based on 20 uses of purchased									
7820	forms, and 4 uses of bracing lumber									
7860	To 8′ high	C-2	800	.060	SFCA	.94	2.74		3.68	5.25
8060	Over 8′ to 16′ high		600	.080		.99	3.66		4.65	6.75
8600	Pilasters, 1 use		270	.178		3.19	8.15		11.34	16
8620	2 use		330	.145		1.75	6.65		8.40	12.20
8640	3 use		370	.130		1.27	5.95		7.22	10.55
8660	4 use		385	.125		1.04	5.70		6.74	9.90
9010	Steel framed plywood, based on 50 uses of purchased									
9020	forms, and 4 uses of bracing lumber									
9060	To 8′ high	C-2	600	.080	SFCA	.74	3.66		4.40	6.45
9260	Over 8′ to 16′ high		450	.107		.74	4.88		5.62	8.30
9460	Over 16′ to 20′ high		400	.120		.74	5.50		6.24	9.25
9475	For elevated walls, add						10%			
9480	For battered walls, 1 side battered, add					10%	10%			
9485	For battered walls, 2 sides battered, add					15%	15%			

03 11 16 – Architectural Cast-in Place Concrete Forming

03 11 16.13 Concrete Form Liners		Crew	Daily Output	Labor-Hours	Unit	Material	Labor	Equipment	Total	Total Incl O&P
0010	**CONCRETE FORM LINERS**									
5750	Liners for forms (add to wall forms), ABS plastic									
5800	Aged wood, 4″ wide, 1 use	1 Carp	256	.031	SFCA	3.30	1.47		4.77	5.90
5820	2 use		256	.031		1.82	1.47		3.29	4.26
5830	3 use		256	.031		1.32	1.47		2.79	3.71
5840	4 use		256	.031		1.07	1.47		2.54	3.44
5900	Fractured rope rib, 1 use		192	.042		4.79	1.96		6.75	8.25
5925	2 use		192	.042		2.63	1.96		4.59	5.90
5950	3 use		192	.042		1.92	1.96		3.88	5.10
6000	4 use		192	.042		1.56	1.96		3.52	4.72

03 11 16 – Architectural Cast-in Place Concrete Forming

03 11 16.13 Concrete Form Liners		Crew	Daily Output	Labor-Hours	Unit	Material	2015 Bare Costs Labor	Equipment	Total	Total Incl O&P
6100	Ribbed, 3/4" deep x 1-1/2" O.C., 1 use	1 Carp	224	.036	SFCA	4.79	1.68		6.47	7.85
6125	2 use		224	.036		2.63	1.68		4.31	5.50
6150	3 use		224	.036		1.92	1.68		3.60	4.69
6200	4 use		224	.036		1.56	1.68		3.24	4.29
6300	Rustic brick pattern, 1 use		224	.036		3.30	1.68		4.98	6.20
6325	2 use		224	.036		1.82	1.68		3.50	4.58
6350	3 use		224	.036		1.32	1.68		3	4.03
6400	4 use		224	.036		1.07	1.68		2.75	3.76
6500	3/8" striated, random, 1 use		224	.036		3.30	1.68		4.98	6.20
6525	2 use		224	.036		1.82	1.68		3.50	4.58
6550	3 use		224	.036		1.32	1.68		3	4.03
6600	4 use		224	.036		1.07	1.68		2.75	3.76
6850	Random vertical rustication, 1 use		384	.021		6.25	.98		7.23	8.40
6900	2 use		384	.021		3.45	.98		4.43	5.30
6925	3 use		384	.021		2.51	.98		3.49	4.27
6950	4 use		384	.021		2.04	.98		3.02	3.75
7050	Wood, beveled edge, 3/4" deep, 1 use		384	.021	L.F.	.16	.98		1.14	1.69
7100	1" deep, 1 use		384	.021	"	.33	.98		1.31	1.87
7200	4" wide aged cedar, 1 use		256	.031	SFCA	3.30	1.47		4.77	5.90
7300	4" variable depth rough cedar		224	.036	"	4.79	1.68		6.47	7.85

03 11 19 – Insulating Concrete Forming

03 11 19.10 Insulating Forms, Left In Place		Crew	Daily Output	Labor-Hours	Unit	Material	Labor	Equipment	Total	Total Incl O&P
0010	**INSULATING FORMS, LEFT IN PLACE**									
0020	S.F. is for exterior face, but includes forms for both faces (total R22)									
2000	4" wall, straight block, 16" x 48" (5.33 S.F.) G	2 Carp	90	.178	Ea.	19.60	8.35		27.95	34.50
2010	90 corner block, exterior 16" x 38" x 22" (6.67 S.F.) G		75	.213		23	10		33	41
2020	45 corner block, exterior 16" x 34" x 18" (5.78 S.F.) G		75	.213		23	10		33	41
2100	6" wall, straight block, 16" x 48" (5.33 S.F.) G		90	.178		20	8.35		28.35	35
2110	90 corner block, exterior 16" x 32" x 24" (6.22 S.F.) G		75	.213		25	10		35	43
2120	45 corner block, exterior 16" x 26" x 18" (4.89 S.F.) G		75	.213		22.50	10		32.50	40
2130	Brick ledge block, 16" x 48" (5.33 S.F.) G		80	.200		25	9.40		34.40	42
2140	Taper top block, 16" x 48" (5.33 S.F.) G		80	.200		23.50	9.40		32.90	40.50
2200	8" wall, straight block, 16" x 48" (5.33 S.F.) G		90	.178		21.50	8.35		29.85	36.50
2210	90 corner block, exterior 16" x 34" x 26" (6.67 S.F.) G		75	.213		27.50	10		37.50	46
2220	45 corner block, exterior 16" x 28" x 20" (5.33 S.F.) G		75	.213		23.50	10		33.50	41.50
2230	Brick ledge block, 16" x 48" (5.33 S.F.) G		80	.200		26.50	9.40		35.90	43.50
2240	Taper top block, 16" x 48" (5.33 S.F.) G		80	.200		24.50	9.40		33.90	41.50

03 11 19.60 Roof Deck Form Boards		Crew	Daily Output	Labor-Hours	Unit	Material	Labor	Equipment	Total	Total Incl O&P
0010	**ROOF DECK FORM BOARDS** R051223-50									
0050	Includes bulb tee sub-purlins @ 32-5/8" O.C.									
0070	Non-asbestos fiber cement, 5/16" thick	C-13	2950	.008	S.F.	2.73	.41	.05	3.19	3.74
0100	Fiberglass, 1" thick		2700	.009		3.46	.45	.05	3.96	4.62
0500	Wood fiber, 1" thick G		2700	.009		2.24	.45	.05	2.74	3.28

03 11 23 – Permanent Stair Forming

03 11 23.75 Forms In Place, Stairs		Crew	Daily Output	Labor-Hours	Unit	Material	Labor	Equipment	Total	Total Incl O&P
0010	**FORMS IN PLACE, STAIRS** R031113-40									
0015	(Slant length x width), 1 use	C-2	165	.291	S.F.	5.90	13.30		19.20	27
0050	2 use R031113-60		170	.282		3.35	12.90		16.25	23.50
0100	3 use		180	.267		2.50	12.20		14.70	21.50
0150	4 use		190	.253		2.08	11.55		13.63	20
1000	Alternate pricing method (1.0 L.F./S.F.), 1 use		100	.480	LF Rsr	5.90	22		27.90	40.50

03 11 Concrete Forming

03 11 23 – Permanent Stair Forming

03 11 23.75 Forms In Place, Stairs		Crew	Daily Output	Labor-Hours	Unit	Material	2015 Bare Costs Labor	Equipment	Total	Total Incl O&P
1050	2 use	C-2	105	.457	LF Rsr	3.35	21		24.35	35.50
1100	3 use		110	.436		2.50	19.95		22.45	33.50
1150	4 use		115	.417		2.08	19.10		21.18	32
2000	Stairs, cast on sloping ground (length x width), 1 use		220	.218	S.F.	2.36	10		12.36	17.95
2025	2 use		232	.207		1.30	9.45		10.75	15.95
2050	3 use		244	.197		.94	9		9.94	14.90
2100	4 use		256	.188		.77	8.55		9.32	14.05

03 15 Concrete Accessories

03 15 05 – Concrete Forming Accessories

03 15 05.12 Chamfer Strips

			Crew	Daily Output	Labor-Hours	Unit	Material	Labor	Equipment	Total	Total Incl O&P
0010	**CHAMFER STRIPS**										
2000	Polyvinyl chloride, 1/2" wide with leg		1 Carp	535	.015	L.F.	.57	.70		1.27	1.71
2200	3/4" wide with leg			525	.015		.68	.72		1.40	1.85
2400	1" radius with leg			515	.016		.64	.73		1.37	1.82
2800	2" radius with leg			500	.016		1.22	.75		1.97	2.50
5000	Wood, 1/2" wide			535	.015		.15	.70		.85	1.25
5200	3/4" wide			525	.015		.16	.72		.88	1.28
5400	1" wide			515	.016		.33	.73		1.06	1.48

03 15 05.15 Column Form Accessories

			Crew	Daily Output	Labor-Hours	Unit	Material	Labor	Equipment	Total	Total Incl O&P
0010	**COLUMN FORM ACCESSORIES**										
1000	Column clamps, adjustable to 24" x 24", buy	G				Set	145			145	160
1100	Rent per month	G					14.55			14.55	16
1300	For sizes to 30" x 30", buy	G					170			170	187
1400	Rent per month	G					17			17	18.70
1600	For sizes to 36" x 36", buy	G					212			212	233
1700	Rent per month	G					21			21	23.50
2000	Bar type with wedges, 36" x 36", buy	G					112			112	123
2100	Rent per month	G					7.85			7.85	8.60
2300	48" x 48", buy	G					152			152	167
2400	Rent per month	G					10.65			10.65	11.70
3000	Scissor type with wedges, 36" x 36", buy	G					95			95	105
3100	Rent per month	G					9.50			9.50	10.45
3300	60" x 60", buy	G					135			135	149
3400	Rent per month	G					13.50			13.50	14.85
4000	Friction collars 2'-6" diam., buy	G					2,575			2,575	2,825
4100	Rent per month	G					180			180	198
4300	4'-0" diam., buy	G					3,150			3,150	3,450
4400	Rent per month	G					220			220	242

03 15 05.30 Hangers

			Crew	Daily Output	Labor-Hours	Unit	Material	Labor	Equipment	Total	Total Incl O&P
0010	**HANGERS**										
0020	Slab and beam form										
0500	Banding iron										
0550	3/4" x 22 ga., 14 L.F. per lb. or 1/2" x 14 ga., 7 L.F. per lb.	G				Lb.	1.34			1.34	1.47
1000	Fascia ties, coil type, to 24" long	G				C	425			425	465
1500	Frame ties to 8-1/8"	G					525			525	575
1550	8-1/8" to 10-1/8"	G					550			550	605
5000	Snap tie hanger, to 30" overall length, 3000#	G					425			425	465
5050	To 36" overall length	G					480			480	525
5100	To 48" overall length	G					585			585	645

03 15 05 – Concrete Forming Accessories

03 15 05.30 Hangers			Crew	Daily Output	Labor-Hours	Unit	Material	2015 Bare Costs Labor	Equipment	Total	Total Incl O&P
5500	Steel beam hanger										
5600	Flange to 8-1/8″	G				C	525			525	575
5650	8-1/8″ to 10-1/8″	G				″	550			550	605
5900	Coil threaded rods, continuous, 1/2″ diameter	G				L.F.	1.36			1.36	1.50
6000	Tie hangers to 30″ overall length, 4000#	G				C	480			480	530
6100	To 36″ overall length	G					535			535	585
6500	Tie back hanger, up to 12-1/8″ flange	G					1,400			1,400	1,550
8500	Wire, black annealed, 15 gage	G				Cwt.	151			151	167
8600	16 ga					″	203			203	223

03 15 05.70 Shores			Crew	Daily Output	Labor-Hours	Unit	Material	Labor	Equipment	Total	Total Incl O&P
0010	**SHORES**										
0020	Erect and strip, by hand, horizontal members										
0500	Aluminum joists and stringers	G	2 Carp	60	.267	Ea.		12.50		12.50	19.25
0600	Steel, adjustable beams	G		45	.356			16.70		16.70	25.50
0700	Wood joists			50	.320			15		15	23
0800	Wood stringers			30	.533			25		25	38.50
1000	Vertical members to 10′ high	G		55	.291			13.65		13.65	21
1050	To 13′ high	G		50	.320			15		15	23
1100	To 16′ high	G		45	.356			16.70		16.70	25.50
1500	Reshoring	G		1400	.011	S.F.	.58	.54		1.12	1.47
1600	Flying truss system	G	C-17D	9600	.009	SFCA		.43	.08	.51	.75
1760	Horizontal, aluminum joists, 6-1/4″ high x 5′ to 21′ span, buy	G				L.F.	15.45			15.45	17
1770	Beams, 7-1/4″ high x 4′ to 30′ span	G				″	18.10			18.10	19.90
1810	Horizontal, steel beam, W8x10, 7′ span, buy	G				Ea.	57			57	63
1830	10′ span	G					67.50			67.50	74.50
1920	15′ span	G					117			117	128
1940	20′ span	G					164			164	180
1970	Steel stringer, W8x10, 4′ to 16′ span, buy	G				L.F.	6.80			6.80	7.45
3000	Rent for job duration, aluminum joist @ 2′ O.C., per mo.	G				SF Flr.	.39			.39	.43
3050	Steel W8x10	G					.17			.17	.19
3060	Steel adjustable	G					.17			.17	.19
3500	#1 post shore, steel, 5′-7″ to 9′-6″ high, 10000# cap., buy	G				Ea.	148			148	163
3550	#2 post shore, 7′-3″ to 12′-10″ high, 7800# capacity	G					171			171	188
3600	#3 post shore, 8′-10″ to 16′-1″ high, 3800# capacity	G					187			187	206
5010	Frame shoring systems, steel, 12000#/leg, buy										
5040	Frame, 2′ wide x 6′ high	G				Ea.	93.50			93.50	103
5250	X-brace	G					15.50			15.50	17.05
5550	Base plate	G					15.50			15.50	17.05
5600	Screw jack	G					33.50			33.50	36.50
5650	U-head, 8″ x 8″	G					19.10			19.10	21

03 15 05.75 Sleeves and Chases			Crew	Daily Output	Labor-Hours	Unit	Material	Labor	Equipment	Total	Total Incl O&P
0010	**SLEEVES AND CHASES**										
0100	Plastic, 1 use, 12″ long, 2″ diameter		1 Carp	100	.080	Ea.	2.15	3.76		5.91	8.15
0150	4″ diameter			90	.089		6.05	4.17		10.22	13.05
0200	6″ diameter			75	.107		10.65	5		15.65	19.40
0250	12″ diameter			60	.133		31.50	6.25		37.75	44

03 15 05.80 Snap Ties			Crew	Daily Output	Labor-Hours	Unit	Material	Labor	Equipment	Total	Total Incl O&P
0010	**SNAP TIES**, 8-1/4″ L&W (Lumber and wedge)										
0100	2250 lb., w/flat washer, 8″ wall	G				C	90			90	99
0150	10″ wall	G					131			131	144
0200	12″ wall	G					136			136	150
0250	16″ wall	G					150			150	165

03 15 Concrete Accessories

03 15 05 – Concrete Forming Accessories

03 15 05.80 Snap Ties		Crew	Daily Output	Labor-Hours	Unit	Material	2015 Bare Costs Labor	Equipment	Total	Total Incl O&P
0300	18″ wall [G]				C	156			156	172
0500	With plastic cone, 8″ wall [G]					80			80	88
0550	10″ wall [G]					83			83	91.50
0600	12″ wall [G]					89			89	98
0650	16″ wall [G]					98			98	108
0700	18″ wall [G]					101			101	111
1000	3350 lb., w/flat washer, 8″ wall [G]					163			163	179
1150	12″ wall [G]					183			183	201
1200	16″ wall [G]					210			210	231
1250	18″ wall [G]					218			218	240
1500	With plastic cone, 8″ wall [G]					132			132	145
1600	12″ wall [G]					149			149	164
1650	16″ wall [G]					170			170	187
1700	18″ wall [G]				↓	176			176	194

03 15 05.85 Stair Tread Inserts		Crew	Daily Output	Labor-Hours	Unit	Material	Labor	Equipment	Total	Total Incl O&P
0010	**STAIR TREAD INSERTS**									
0105	Cast nosing insert, abrasive surface, pre-drilled, includes screws									
0110	Aluminum, 3″ wide x 3′ long	1 Cefi	32	.250	Ea.	55	11.25		66.25	77
0120	4′ long		31	.258		70	11.60		81.60	94
0130	5′ long	↓	30	.267	↓	87	12		99	113
0135	Extruded nosing insert, black abrasive strips, continuous anchor									
0140	Aluminum, 3″ wide x 3′ long	1 Cefi	64	.125	Ea.	35.50	5.65		41.15	47.50
0150	4′ long		60	.133		55	6		61	69.50
0160	5′ long	↓	56	.143	↓	62	6.45		68.45	77.50
0165	Extruded nosing insert, black abrasive strips, pre-drilled, incl. screws									
0170	Aluminum, 3″ wide x 3′ long	1 Cefi	32	.250	Ea.	49	11.25		60.25	70.50
0180	4′ long		31	.258		63.50	11.60		75.10	87
0190	5′ long	↓	30	.267	↓	74.50	12		86.50	100

03 15 05.95 Wall and Foundation Form Accessories		Crew	Daily Output	Labor-Hours	Unit	Material	Labor	Equipment	Total	Total Incl O&P
0010	**WALL AND FOUNDATION FORM ACCESSORIES**									
2000	Footings, turnbuckle form aligner [G]				Ea.	16.20			16.20	17.85
2050	Spreaders for footer, adjustable [G]				″	25			25	27.50
3000	Form oil, up to 1200 S.F. per gallon coverage				Gal.	13.55			13.55	14.90
3050	Up to 800 S.F. per gallon				″	20.50			20.50	22.50
3500	Form patches, 1-3/4″ diameter				C	27			27	29.50
3550	2-3/4″ diameter				″	47			47	51.50
4000	Nail stakes, 3/4″ diameter, 18″ long [G]				Ea.	3.30			3.30	3.63
4050	24″ long [G]					4.26			4.26	4.69
4200	30″ long [G]					5.45			5.45	6
4250	36″ long [G]				↓	6.60			6.60	7.25

03 15 13 – Waterstops

03 15 13.50 Waterstops		Crew	Daily Output	Labor-Hours	Unit	Material	Labor	Equipment	Total	Total Incl O&P
0010	**WATERSTOPS**, PVC and Rubber									
0020	PVC, ribbed 3/16″ thick, 4″ wide	1 Carp	155	.052	L.F.	1.35	2.42		3.77	5.20
0050	6″ wide		145	.055		2.27	2.59		4.86	6.50
0500	With center bulb, 6″ wide, 3/16″ thick		135	.059		1.99	2.78		4.77	6.45
0550	3/8″ thick		130	.062		3.67	2.89		6.56	8.50
0600	9″ wide x 3/8″ thick		125	.064		6.05	3		9.05	11.25
0800	Dumbbell type, 6″ wide, 3/16″ thick		150	.053		2.02	2.50		4.52	6.05
0850	3/8″ thick		145	.055		3.79	2.59		6.38	8.15
1000	9″ wide, 3/8″ thick, plain		130	.062		5.65	2.89		8.54	10.65
1050	Center bulb	↓	130	.062	↓	9.15	2.89		12.04	14.55

03 15 Concrete Accessories

03 15 13 – Waterstops

03 15 13.50 Waterstops		Crew	Daily Output	Labor-Hours	Unit	2015 Bare Costs Material	Labor	Equipment	Total	Total Incl O&P
1250	Ribbed type, split, 3/16" thick, 6" wide	1 Carp	145	.055	L.F.	1.92	2.59		4.51	6.10
1300	3/8" thick		130	.062		4.38	2.89		7.27	9.25
2000	Rubber, flat dumbbell, 3/8" thick, 6" wide		145	.055		10.60	2.59		13.19	15.65
2050	9" wide		135	.059		16.15	2.78		18.93	22
2500	Flat dumbbell split, 3/8" thick, 6" wide		145	.055		1.92	2.59		4.51	6.10
2550	9" wide		135	.059		4.38	2.78		7.16	9.10
3000	Center bulb, 1/4" thick, 6" wide		145	.055		10.45	2.59		13.04	15.50
3050	9" wide		135	.059		23	2.78		25.78	30
3500	Center bulb split, 3/8" thick, 6" wide		145	.055		14.80	2.59		17.39	20.50
3550	9" wide		135	.059		25.50	2.78		28.28	32.50
5000	Waterstop fittings, rubber, flat									
5010	Dumbbell or center bulb, 3/8" thick,									
5200	Field union, 6" wide	1 Carp	50	.160	Ea.	35.50	7.50		43	51
5250	9" wide		50	.160		47.50	7.50		55	64
5500	Flat cross, 6" wide		30	.267		45	12.50		57.50	69
5550	9" wide		30	.267		64.50	12.50		77	90.50
6000	Flat tee, 6" wide		30	.267		43.50	12.50		56	67.50
6050	9" wide		30	.267		59	12.50		71.50	84.50
6500	Flat ell, 6" wide		40	.200		42.50	9.40		51.90	61
6550	9" wide		40	.200		54.50	9.40		63.90	74.50
7000	Vertical tee, 6" wide		25	.320		31.50	15		46.50	58
7050	9" wide		25	.320		45	15		60	72.50
7500	Vertical ell, 6" wide		35	.229		29.50	10.75		40.25	49
7550	9" wide		35	.229		38	10.75		48.75	58

03 15 16 – Concrete Construction Joints

03 15 16.20 Control Joints, Saw Cut

Line	Description	Crew	Daily Output	Labor-Hours	Unit	Material	Labor	Equipment	Total	Total Incl O&P
0010	**CONTROL JOINTS, SAW CUT**									
0100	Sawcut control joints in green concrete									
0120	1" depth	C-27	2000	.008	L.F.	.04	.36	.08	.48	.66
0140	1-1/2" depth		1800	.009		.06	.40	.09	.55	.75
0160	2" depth		1600	.010		.08	.45	.10	.63	.88
0180	Sawcut joint reservoir in cured concrete									
0182	3/8" wide x 3/4" deep, with single saw blade	C-27	1000	.016	L.F.	.06	.72	.17	.95	1.31
0184	1/2" wide x 1" deep, with double saw blades		900	.018		.11	.80	.19	1.10	1.52
0186	3/4" wide x 1-1/2" deep, with double saw blades		800	.020		.23	.90	.21	1.34	1.82
0190	Water blast joint to wash away laitance, 2 passes	C-29	2500	.003			.12	.03	.15	.22
0200	Air blast joint to blow out debris and air dry, 2 passes	C-28	2000	.004			.18	.01	.19	.28
0300	For backer rod, see Section 07 91 23.10									
0340	For joint sealant, see Sections 03 15 16.30 or 07 92 13.20									
0900	For replacement of joint sealant, see Section 07 01 90.81									

03 15 16.30 Expansion Joints

Line	Description		Crew	Daily Output	Labor-Hours	Unit	Material	Labor	Equipment	Total	Total Incl O&P
0010	**EXPANSION JOINTS**										
0020	Keyed, cold, 24 ga., incl. stakes, 3-1/2" high	G	1 Carp	200	.040	L.F.	.81	1.88		2.69	3.78
0050	4-1/2" high	G		200	.040		.92	1.88		2.80	3.90
0100	5-1/2" high	G		195	.041		1.12	1.93		3.05	4.19
0150	7-1/2" high	G		190	.042		1.31	1.98		3.29	4.48
0160	9-1/2" high	G		185	.043		1.43	2.03		3.46	4.69
0300	Poured asphalt, plain, 1/2" x 1"		1 Clab	450	.018		.69	.67		1.36	1.79
0350	1" x 2"			400	.020		2.76	.75		3.51	4.20
0500	Neoprene, liquid, cold applied, 1/2" x 1"			450	.018		2.88	.67		3.55	4.20
0550	1" x 2"			400	.020		11.50	.75		12.25	13.80
0700	Polyurethane, poured, 2 part, 1/2" x 1"			400	.020		1.28	.75		2.03	2.57

03 15 Concrete Accessories

03 15 16 – Concrete Construction Joints

03 15 16.30 Expansion Joints			Crew	Daily Output	Labor-Hours	Unit	Material	2015 Bare Costs Labor	Equipment	Total	Total Incl O&P
0750	1″ x 2″		1 Clab	350	.023	L.F.	5.10	.86		5.96	6.95
0900	Rubberized asphalt, hot or cold applied, 1/2″ x 1″			450	.018		.40	.67		1.07	1.47
0950	1″ x 2″			400	.020		1.60	.75		2.35	2.92
1100	Hot applied, fuel resistant, 1/2″ x 1″			450	.018		.60	.67		1.27	1.69
1150	1″ x 2″			400	.020		2.40	.75		3.15	3.80
2000	Premolded, bituminous fiber, 1/2″ x 6″		1 Carp	375	.021		.44	1		1.44	2.02
2050	1″ x 12″			300	.027		1.99	1.25		3.24	4.12
2140	Concrete expansion joint, recycled paper and fiber, 1/2″ x 6″	G		390	.021		.42	.96		1.38	1.94
2150	1/2″ x 12″	G		360	.022		.83	1.04		1.87	2.53
2250	Cork with resin binder, 1/2″ x 6″			375	.021		1.11	1		2.11	2.76
2300	1″ x 12″			300	.027		3.11	1.25		4.36	5.35
2500	Neoprene sponge, closed cell, 1/2″ x 6″			375	.021		2.37	1		3.37	4.15
2550	1″ x 12″			300	.027		8.25	1.25		9.50	11.05
2750	Polyethylene foam, 1/2″ x 6″			375	.021		.68	1		1.68	2.29
2800	1″ x 12″			300	.027		2.19	1.25		3.44	4.34
3000	Polyethylene backer rod, 3/8″ diameter			460	.017		.03	.82		.85	1.29
3050	3/4″ diameter			460	.017		.06	.82		.88	1.32
3100	1″ diameter			460	.017		.09	.82		.91	1.36
3500	Polyurethane foam, with polybutylene, 1/2″ x 1/2″			475	.017		1.14	.79		1.93	2.47
3550	1″ x 1″			450	.018		2.88	.83		3.71	4.45
3750	Polyurethane foam, regular, closed cell, 1/2″ x 6″			375	.021		.84	1		1.84	2.46
3800	1″ x 12″			300	.027		3	1.25		4.25	5.25
4000	Polyvinyl chloride foam, closed cell, 1/2″ x 6″			375	.021		2.28	1		3.28	4.05
4050	1″ x 12″			300	.027		7.85	1.25		9.10	10.60
4250	Rubber, gray sponge, 1/2″ x 6″			375	.021		1.92	1		2.92	3.65
4300	1″ x 12″			300	.027		6.90	1.25		8.15	9.55
4400	Redwood heartwood, 1″ x 4″			400	.020		1.16	.94		2.10	2.73
4450	1″ x 6″			375	.021		1.75	1		2.75	3.47
5000	For installation in walls, add							75%			
5250	For installation in boxouts, add							25%			

03 15 19 – Cast-In Concrete Anchors

03 15 19.05 Anchor Bolt Accessories			Crew	Daily Output	Labor-Hours	Unit	Material	Labor	Equipment	Total	Total Incl O&P
0010	**ANCHOR BOLT ACCESSORIES**										
0015	For anchor bolts set in fresh concrete, see Section 03 15 19.10										
8150	Anchor bolt sleeve, plastic, 1″ diam. bolts		1 Carp	60	.133	Ea.	8.20	6.25		14.45	18.70
8500	1-1/2″ diameter			28	.286		14.60	13.40		28	36.50
8600	2″ diameter			24	.333		18.60	15.65		34.25	44.50
8650	3″ diameter			20	.400		34	18.80		52.80	66.50

03 15 19.10 Anchor Bolts			Crew	Daily Output	Labor-Hours	Unit	Material	Labor	Equipment	Total	Total Incl O&P
0010	**ANCHOR BOLTS**										
0015	Made from recycled materials										
0025	Single bolts installed in fresh concrete, no templates										
0030	Hooked w/nut and washer, 1/2″ diameter, 8″ long	G	1 Carp	132	.061	Ea.	1.33	2.85		4.18	5.85
0040	12″ long	G		131	.061		1.48	2.87		4.35	6.05
0050	5/8″ diameter, 8″ long	G		129	.062		2.98	2.91		5.89	7.75
0060	12″ long	G		127	.063		3.67	2.96		6.63	8.60
0070	3/4″ diameter, 8″ long	G		127	.063		3.67	2.96		6.63	8.60
0080	12″ long	G		125	.064		4.59	3		7.59	9.65
0090	2-bolt pattern, including job-built 2-hole template, per set										
0100	J-type, incl. hex nut & washer, 1/2″ diameter x 6″ long	G	1 Carp	21	.381	Set	5.20	17.90		23.10	33.50
0110	12″ long	G		21	.381		5.80	17.90		23.70	34
0120	18″ long	G		21	.381		6.70	17.90		24.60	35

03 15 19 – Cast-In Concrete Anchors

03 15 19.10 Anchor Bolts			Crew	Daily Output	Labor-Hours	Unit	2015 Bare Costs Material	Labor	Equipment	Total	Total Incl O&P
0130	3/4" diameter x 8" long	G	1 Carp	20	.400	Set	10.20	18.80		29	40
0140	12" long	G		20	.400		12	18.80		30.80	42
0150	18" long	G		20	.400		14.75	18.80		33.55	45.50
0160	1" diameter x 12" long	G		19	.421		20.50	19.75		40.25	53
0170	18" long	G		19	.421		24	19.75		43.75	57
0180	24" long	G		19	.421		28.50	19.75		48.25	62
0190	36" long	G		18	.444		38	21		59	74
0200	1-1/2" diameter x 18" long	G		17	.471		37.50	22		59.50	75.50
0210	24" long	G		16	.500		44	23.50		67.50	84.50
0300	L-type, incl. hex nut & washer, 3/4" diameter x 12" long	G		20	.400		12.90	18.80		31.70	43
0310	18" long	G		20	.400		15.75	18.80		34.55	46.50
0320	24" long	G		20	.400		18.60	18.80		37.40	49.50
0330	30" long	G		20	.400		23	18.80		41.80	54
0340	36" long	G		20	.400		25.50	18.80		44.30	57.50
0350	1" diameter x 12" long	G		19	.421		19.55	19.75		39.30	52
0360	18" long	G		19	.421		23.50	19.75		43.25	56.50
0370	24" long	G		19	.421		28.50	19.75		48.25	62
0380	30" long	G		19	.421		33	19.75		52.75	67
0390	36" long	G		18	.444		37.50	21		58.50	73.50
0400	42" long	G		18	.444		45	21		66	81.50
0410	48" long	G		18	.444		50	21		71	87
0420	1-1/4" diameter x 18" long	G		18	.444		30	21		51	65
0430	24" long	G		18	.444		35	21		56	70.50
0440	30" long	G		17	.471		40	22		62	78
0450	36" long	G		17	.471		45	22		67	83.50
0460	42" long	G	2 Carp	32	.500		50.50	23.50		74	91.50
0470	48" long	G		32	.500		57	23.50		80.50	99
0480	54" long	G		31	.516		67	24		91	111
0490	60" long	G		31	.516		73	24		97	118
0500	1-1/2" diameter x 18" long	G		33	.485		51	23		74	91
0510	24" long	G		32	.500		59	23.50		82.50	101
0520	30" long	G		31	.516		66	24		90	110
0530	36" long	G		30	.533		75.50	25		100.50	122
0540	42" long	G		30	.533		85.50	25		110.50	133
0550	48" long	G		29	.552		96	26		122	145
0560	54" long	G		28	.571		116	27		143	170
0570	60" long	G		28	.571		127	27		154	182
0580	1-3/4" diameter x 18" long	G		31	.516		74.50	24		98.50	120
0590	24" long	G		30	.533		87	25		112	134
0600	30" long	G		29	.552		101	26		127	151
0610	36" long	G		28	.571		114	27		141	168
0620	42" long	G		27	.593		128	28		156	184
0630	48" long	G		26	.615		141	29		170	200
0640	54" long	G		26	.615		174	29		203	236
0650	60" long	G		25	.640		187	30		217	252
0660	2" diameter x 24" long	G		27	.593		111	28		139	166
0670	30" long	G		27	.593		125	28		153	181
0680	36" long	G		26	.615		137	29		166	196
0690	42" long	G		25	.640		153	30		183	214
0700	48" long	G		24	.667		175	31.50		206.50	241
0710	54" long	G		23	.696		208	32.50		240.50	280
0720	60" long	G		23	.696		223	32.50		255.50	297
0730	66" long	G		22	.727		239	34		273	315

03 15 19 – Cast-In Concrete Anchors

03 15 19.10 Anchor Bolts			Crew	Daily Output	Labor-Hours	Unit	2015 Bare Costs Material	Labor	Equipment	Total	Total Incl O&P
0740	72″ long	G	2 Carp	21	.762	Set	261	36		297	340
1000	4-bolt pattern, including job-built 4-hole template, per set										
1100	J-type, incl. hex nut & washer, 1/2″ diameter x 6″ long	G	1 Carp	19	.421	Set	7.55	19.75		27.30	39
1110	12″ long	G		19	.421		8.75	19.75		28.50	40
1120	18″ long	G		18	.444		10.55	21		31.55	43.50
1130	3/4″ diameter x 8″ long	G		17	.471		17.55	22		39.55	53.50
1140	12″ long	G		17	.471		21	22		43	57.50
1150	18″ long	G		17	.471		26.50	22		48.50	63.50
1160	1″ diameter x 12″ long	G		16	.500		38	23.50		61.50	77.50
1170	18″ long	G		15	.533		45	25		70	88
1180	24″ long	G		15	.533		54	25		79	98
1190	36″ long	G		15	.533		73	25		98	119
1200	1-1/2″ diameter x 18″ long	G		13	.615		72	29		101	124
1210	24″ long	G		12	.667		85.50	31.50		117	142
1300	L-type, incl. hex nut & washer, 3/4″ diameter x 12″ long	G		17	.471		23	22		45	59.50
1310	18″ long	G		17	.471		28.50	22		50.50	65.50
1320	24″ long	G		17	.471		34.50	22		56.50	72
1330	30″ long	G		16	.500		43	23.50		66.50	83
1340	36″ long	G		16	.500		48.50	23.50		72	89.50
1350	1″ diameter x 12″ long	G		16	.500		36	23.50		59.50	76
1360	18″ long	G		15	.533		44.50	25		69.50	87.50
1370	24″ long	G		15	.533		54	25		79	98
1380	30″ long	G		15	.533		63.50	25		88.50	109
1390	36″ long	G		15	.533		72	25		97	118
1400	42″ long	G		14	.571		87	27		114	138
1410	48″ long	G		14	.571		97.50	27		124.50	149
1420	1-1/4″ diameter x 18″ long	G		14	.571		57	27		84	104
1430	24″ long	G		14	.571		67	27		94	115
1440	30″ long	G		13	.615		77	29		106	129
1450	36″ long	G		13	.615		87	29		116	141
1460	42″ long	G	2 Carp	25	.640		98	30		128	154
1470	48″ long	G		24	.667		112	31.50		143.50	171
1480	54″ long	G		23	.696		131	32.50		163.50	195
1490	60″ long	G		23	.696		144	32.50		176.50	209
1500	1-1/2″ diameter x 18″ long	G		25	.640		99	30		129	155
1510	24″ long	G		24	.667		115	31.50		146.50	174
1520	30″ long	G		23	.696		129	32.50		161.50	193
1530	36″ long	G		22	.727		148	34		182	216
1540	42″ long	G		22	.727		169	34		203	238
1550	48″ long	G		21	.762		189	36		225	263
1560	54″ long	G		20	.800		230	37.50		267.50	310
1570	60″ long	G		20	.800		251	37.50		288.50	335
1580	1-3/4″ diameter x 18″ long	G		22	.727		146	34		180	214
1590	24″ long	G		21	.762		171	36		207	243
1600	30″ long	G		21	.762		199	36		235	273
1610	36″ long	G		20	.800		226	37.50		263.50	305
1620	42″ long	G		19	.842		254	39.50		293.50	340
1630	48″ long	G		18	.889		278	41.50		319.50	370
1640	54″ long	G		18	.889		345	41.50		386.50	445
1650	60″ long	G		17	.941		370	44		414	480
1660	2″ diameter x 24″ long	G		19	.842		220	39.50		259.50	305
1670	30″ long	G		18	.889		247	41.50		288.50	335
1680	36″ long	G		18	.889		272	41.50		313.50	365

03 15 Concrete Accessories

03 15 19 – Cast-In Concrete Anchors

03 15 19.10 Anchor Bolts

	03 15 19.10 Anchor Bolts		Crew	Daily Output	Labor-Hours	Unit	Material	2015 Bare Costs Labor	Equipment	Total	Total Incl O&P
1690	42″ long	G	2 Carp	17	.941	Set	305	44		349	405
1700	48″ long	G		16	1		345	47		392	455
1710	54″ long	G		15	1.067		415	50		465	530
1720	60″ long	G		15	1.067		445	50		495	565
1730	66″ long	G		14	1.143		475	53.50		528.50	605
1740	72″ long	G		14	1.143		520	53.50		573.50	655
1990	For galvanized, add					Ea.	75%				

03 15 19.20 Dovetail Anchor System

	Description		Crew	Daily Output	Labor-Hours	Unit	Material	Labor	Equipment	Total	Total Incl O&P
0010	**DOVETAIL ANCHOR SYSTEM**										
0500	Dovetail anchor slot, galvanized, foam-filled, 26 ga.	G	1 Carp	425	.019	L.F.	.88	.88		1.76	2.33
0600	24 ga.	G		400	.020		1.64	.94		2.58	3.25
0625	22 ga.	G		400	.020		1.92	.94		2.86	3.56
0900	Stainless steel, foam-filled, 26 ga.	G		375	.021		1.41	1		2.41	3.09
1200	Dovetail brick anchor, corrugated, galvanized, 3-1/2″ long, 16 ga.	G	1 Bric	10.50	.762	C	30.50	35		65.50	87
1300	12 ga.	G		10.50	.762		43.50	35		78.50	101
1500	Seismic, galvanized, 3-1/2″ long, 16 ga.	G		10.50	.762		44.50	35		79.50	102
1600	12 ga.	G		10.50	.762		41.50	35		76.50	99
2000	Dovetail cavity wall, corrugated, galvanized, 5-1/2″ long, 16 ga.	G		10.50	.762		38	35		73	95.50
2100	12 ga.	G		10.50	.762		55.50	35		90.50	115
3000	Dovetail furring anchors, corrugated, galvanized, 1-1/2″ long, 16 ga.	G		10.50	.762		17.20	35		52.20	72.50
3100	12 ga.	G		10.50	.762		25.50	35		60.50	81.50
6000	Dovetail stone panel anchors, galvanized, 1/8″ x 1″ wide, 3-1/2″ long	G		10.50	.762		99	35		134	163
6100	1/4″ x 1″ wide	G		10.50	.762		112	35		147	177

03 15 19.30 Inserts

	Description		Crew	Daily Output	Labor-Hours	Unit	Material	Labor	Equipment	Total	Total Incl O&P
0010	**INSERTS**										
1000	Inserts, slotted nut type for 3/4″ bolts, 4″ long	G	1 Carp	84	.095	Ea.	19.65	4.47		24.12	28.50
2100	6″ long	G		84	.095		22	4.47		26.47	31.50
2150	8″ long	G		84	.095		29.50	4.47		33.97	39.50
2200	Slotted, strap type, 4″ long	G		84	.095		21	4.47		25.47	30.50
2250	6″ long	G		84	.095		24	4.47		28.47	33.50
2300	8″ long	G		84	.095		31.50	4.47		35.97	42
2350	Strap for slotted insert, 4″ long	G		84	.095		11.55	4.47		16.02	19.60
4100	6″ long	G		84	.095		12.55	4.47		17.02	20.50
4150	8″ long	G		84	.095		15.65	4.47		20.12	24
4200	10″ long	G		84	.095		19.05	4.47		23.52	28
7000	Loop ferrule type										
7100	1/4″ diameter bolt	G	1 Carp	84	.095	Ea.	2.23	4.47		6.70	9.35
7350	7/8″ diameter bolt	G	″	84	.095	″	10.35	4.47		14.82	18.30
9000	Wedge type										
9100	For 3/4″ diameter bolt	G	1 Carp	60	.133	Ea.	9.20	6.25		15.45	19.80
9800	Cut washers, black										
9900	3/4″ bolt	G				Ea.	1.17			1.17	1.29
9950	For galvanized inserts, add						30%				

03 15 19.45 Machinery Anchors

	Description		Crew	Daily Output	Labor-Hours	Unit	Material	Labor	Equipment	Total	Total Incl O&P
0010	**MACHINERY ANCHORS**, heavy duty, incl. sleeve, floating base nut,										
0020	lower stud & coupling nut, fiber plug, connecting stud, washer & nut.										
0030	For flush mounted embedment in poured concrete heavy equip. pads.										
0200	Stud & bolt, 1/2″ diameter	G	E-16	40	.400	Ea.	72.50	21.50	3.65	97.65	122
0300	5/8″ diameter	G		35	.457		80.50	24.50	4.17	109.17	136
0500	3/4″ diameter	G		30	.533		93	28.50	4.86	126.36	157
0600	7/8″ diameter	G		25	.640		101	34.50	5.85	141.35	177
0800	1″ diameter	G		20	.800		117	43	7.30	167.30	212

03 15 Concrete Accessories

03 15 19 – Cast-In Concrete Anchors

03 15 19.45 Machinery Anchors			Crew	Daily Output	Labor-Hours	Unit	Material	2015 Bare Costs Labor	Equipment	Total	Total Incl O&P
0900	1-1/4" diameter	G	E-16	15	1.067	Ea.	141	57	9.75	207.75	265

03 21 Reinforcement Bars

03 21 05 – Reinforcing Steel Accessories

03 21 05.10 Rebar Accessories			Crew	Daily Output	Labor-Hours	Unit	Material	2015 Bare Costs Labor	Equipment	Total	Total Incl O&P
0010	**REBAR ACCESSORIES**										
0030	Steel & plastic made from recycled materials										
0100	Beam bolsters (BB), lower, 1-1/2" high, plain steel	G				C.L.F.	33			33	36.50
0102	Galvanized	G					39.50			39.50	43.50
0104	Stainless tipped legs	G					460			460	505
0106	Plastic tipped legs	G					49			49	54
0108	Epoxy dipped	G					84			84	92.50
0110	2" high, plain	G					42			42	46
0120	Galvanized	G					50.50			50.50	55.50
0140	Stainless tipped legs	G					465			465	515
0160	Plastic tipped legs	G					57			57	62.50
0162	Epoxy dipped	G					95			95	105
0200	Upper (BBU), 1-1/2" high, plain steel	G					81			81	89
0210	3" high	G					91			91	100
0500	Slab bolsters, continuous (SB), 1" high, plain steel	G					29			29	32
0502	Galvanized	G					35			35	38.50
0504	Stainless tipped legs	G					455			455	500
0506	Plastic tipped legs	G					42			42	46
0510	2" high, plain steel	G					36			36	39.50
0515	Galvanized	G					43			43	47.50
0520	Stainless tipped legs	G					495			495	545
0525	Plastic tipped legs	G					49			49	54
0530	For bolsters with wire runners (SBR), add	G					39			39	43
0540	For bolsters with plates (SBP), add	G					92			92	101
0700	Bag ties, 16 ga., plain, 4" long	G				C	4			4	4.40
0710	5" long	G					5			5	5.50
0720	6" long	G					4			4	4.40
0730	7" long	G					5			5	5.50
1200	High chairs, individual (HC), 3" high, plain steel	G					59			59	65
1202	Galvanized	G					71			71	78
1204	Stainless tipped legs	G					485			485	530
1206	Plastic tipped legs	G					64			64	70.50
1210	5" high, plain	G					86			86	94.50
1212	Galvanized	G					103			103	114
1214	Stainless tipped legs	G					510			510	560
1216	Plastic tipped legs	G					94			94	103
1220	8" high, plain	G					126			126	139
1222	Galvanized	G					151			151	166
1224	Stainless tipped legs	G					550			550	605
1226	Plastic tipped legs	G					139			139	153
1230	12" high, plain	G					300			300	330
1232	Galvanized	G					360			360	395
1234	Stainless tipped legs	G					725			725	800
1236	Plastic tipped legs	G					330			330	360
1400	Individual high chairs, with plate (HCP), 5" high	G					179			179	197
1410	8" high	G					248			248	273

03 21 05 – Reinforcing Steel Accessories

03 21 05.10 Rebar Accessories		Crew	Daily Output	Labor-Hours	Unit	Material	2015 Bare Costs Labor	Equipment	Total	Total Incl O&P
1500	Bar chair (BC), 1-1/2" high, plain steel G				C	38			38	42
1520	Galvanized G					43			43	47.50
1530	Stainless tipped legs G					470			470	515
1540	Plastic tipped legs G					41			41	45
1700	Continuous high chairs (CHC), legs 8" O.C., 4" high, plain steel G				C.L.F.	51			51	56
1705	Galvanized G					61			61	67.50
1710	Stainless tipped legs G					475			475	525
1715	Plastic tipped legs G					69			69	76
1718	Epoxy dipped G					89			89	98
1720	6" high, plain G					70			70	77
1725	Galvanized G					84			84	92.50
1730	Stainless tipped legs G					495			495	545
1735	Plastic tipped legs G					94			94	103
1738	Epoxy dipped G					121			121	133
1740	8" high, plain G					100			100	110
1745	Galvanized G					120			120	132
1750	Stainless tipped legs G					525			525	580
1755	Plastic tipped legs G					119			119	131
1758	Epoxy dipped G					153			153	169
1900	For continuous bottom wire runners, add G					52			52	57
1940	For continuous bottom plate, add G					214			214	235
2200	Screed chair base, 1/2" coil thread diam., 2-1/2" high, plain steel G				C	325			325	360
2210	Galvanized G					390			390	430
2220	5-1/2" high, plain G					395			395	430
2250	Galvanized G					470			470	520
2300	3/4" coil thread diam., 2-1/2" high, plain steel G					410			410	450
2310	Galvanized G					490			490	540
2320	5-1/2" high, plain steel G					505			505	555
2350	Galvanized G					605			605	665
2400	Screed holder, 1/2" coil thread diam. for pipe screed, plain steel, 6" long G					395			395	435
2420	12" long G					605			605	665
2500	3/4" coil thread diam. for pipe screed, plain steel, 6" long G					565			565	625
2520	12" long G					905			905	995
2700	Screw anchor for bolts, plain steel, 3/4" diameter x 4" long G					540			540	595
2720	1" diameter x 6" long G					890			890	975
2740	1-1/2" diameter x 8" long G					1,125			1,125	1,250
2800	Screw anchor eye bolts, 3/4" x 3" long G					2,825			2,825	3,125
2820	1" x 3-1/2" long G					3,800			3,800	4,175
2840	1-1/2" x 6" long G					11,600			11,600	12,800
2900	Screw anchor bolts, 3/4" x 9" long G					1,625			1,625	1,775
2920	1" x 12" long G					3,125			3,125	3,425
3001	Slab lifting inserts, single pickup, galv, 3/4" diam., 5" high G					1,725			1,725	1,900
3010	6" high G					1,750			1,750	1,925
3030	7" high G					1,775			1,775	1,950
3100	1" diameter, 5-1/2" high G					1,800			1,800	2,000
3120	7" high G					1,875			1,875	2,050
3200	Double pickup lifting inserts, 1" diameter, 5-1/2" high G					3,500			3,500	3,850
3220	7" high G					3,875			3,875	4,250
3330	1-1/2" diameter, 8" high G					4,875			4,875	5,375
3800	Subgrade chairs, #4 bar head, 3-1/2" high G					38			38	42
3850	12" high G					45			45	49.50
3900	#6 bar head, 3-1/2" high G					38			38	42
3950	12" high G					45			45	49.50

03 21 Reinforcement Bars

03 21 05 – Reinforcing Steel Accessories

03 21 05.10 Rebar Accessories			Crew	Daily Output	Labor-Hours	Unit	Material	Labor	Equipment	Total	Total Incl O&P
							2015 Bare Costs				
4200	Subgrade stakes, no nail holes, 3/4" diameter, 12" long	G				C	294			294	325
4250	24" long	G					360			360	395
4300	7/8" diameter, 12" long	G					375			375	410
4350	24" long	G				↓	625			625	690
4500	Tie wire, 16 ga. annealed steel	G				Cwt.	203			203	223

03 21 05.75 Splicing Reinforcing Bars			Crew	Daily Output	Labor-Hours	Unit	Material	Labor	Equipment	Total	Total Incl O&P
0010	**SPLICING REINFORCING BARS**	R032110-70									
0020	Including holding bars in place while splicing										
0100	Standard, self-aligning type, taper threaded, #4 bars	G	C-25	190	.168	Ea.	5.95	6.95		12.90	17.80
0105	#5 bars	G		170	.188		7.30	7.75		15.05	20.50
0110	#6 bars	G		150	.213		8.40	8.80		17.20	23.50
0120	#7 bars	G		130	.246		9.80	10.15		19.95	27
0300	#8 bars	G	↓	115	.278		16.60	11.50		28.10	37
0305	#9 bars	G	C-5	105	.533		18.15	27.50	7	52.65	70
0310	#10 bars	G		95	.589		20	30.50	7.75	58.25	77.50
0320	#11 bars	G		85	.659		21.50	34	8.65	64.15	86
0330	#14 bars	G		65	.862		32	44.50	11.35	87.85	116
0340	#18 bars	G		45	1.244		49	64.50	16.35	129.85	171
0500	Transition self-aligning, taper threaded, #18-14	G		45	1.244		51	64.50	16.35	131.85	174
0510	#18-11	G		45	1.244		52	64.50	16.35	132.85	175
0520	#14-11	G		65	.862		34	44.50	11.35	89.85	119
0540	#11-10	G		85	.659		23.50	34	8.65	66.15	88.50
0550	#10-9	G	↓	95	.589		22	30.50	7.75	60.25	80
0560	#9-8	G	C-25	105	.305		20	12.60		32.60	42.50
0580	#8-7	G		115	.278		18.60	11.50		30.10	39
0590	#7-6	G		130	.246		11.75	10.15		21.90	29.50
0600	Position coupler for curved bars, taper threaded, #4 bars	G		160	.200		27.50	8.25		35.75	43.50
0610	#5 bars	G		145	.221		28.50	9.10		37.60	46
0620	#6 bars	G		130	.246		34.50	10.15		44.65	54.50
0630	#7 bars	G		110	.291		36.50	12		48.50	59.50
0640	#8 bars	G	↓	100	.320		38	13.20		51.20	63
0650	#9 bars	G	C-5	90	.622		41.50	32	8.20	81.70	105
0660	#10 bars	G		80	.700		44.50	36	9.20	89.70	115
0670	#11 bars	G		70	.800		46.50	41.50	10.50	98.50	127
0680	#14 bars	G		55	1.018		58	52.50	13.40	123.90	160
0690	#18 bars	G		40	1.400		83.50	72.50	18.40	174.40	224
0700	Transition position coupler for curved bars, taper threaded, #18-14	G		40	1.400		85.50	72.50	18.40	176.40	227
0710	#18-11	G		40	1.400		86.50	72.50	18.40	177.40	228
0720	#14-11	G		55	1.018		60	52.50	13.40	125.90	162
0730	#11-10	G		70	.800		48.50	41.50	10.50	100.50	129
0740	#10-9	G	↓	80	.700		46.50	36	9.20	91.70	117
0750	#9-8	G	C-25	90	.356		43.50	14.70		58.20	71.50
0760	#8-7	G		100	.320		40	13.20		53.20	65.50
0770	#7-6	G		110	.291		38.50	12		50.50	62
0800	Sleeve type w/grout filler, for precast concrete, #6 bars	G		72	.444		21.50	18.35		39.85	53
0802	#7 bars	G		64	.500		25.50	20.50		46	62
0805	#8 bars	G		56	.571		30.50	23.50		54	71.50
0807	#9 bars	G	↓	48	.667		36.50	27.50		64	84.50
0810	#10 bars	G	C-5	40	1.400		43	72.50	18.40	133.90	180
0900	#11 bars	G		32	1.750		47.50	90.50	23	161	218
0920	#14 bars	G	↓	24	2.333		74	121	30.50	225.50	305
1000	Sleeve type w/ferrous filler, for critical structures, #6 bars	G	C-25	72	.444	↓	59	18.35		77.35	94.50

03 21 05 – Reinforcing Steel Accessories

03 21 05.75 Splicing Reinforcing Bars			Crew	Daily Output	Labor-Hours	Unit	2015 Bare Costs Material	Labor	Equipment	Total	Total Incl O&P
1210	#7 bars	G	C-25	64	.500	Ea.	59.50	20.50		80	99
1220	#8 bars	G		56	.571		63	23.50		86.50	107
1230	#9 bars	G	C-5	48	1.167		64.50	60.50	15.35	140.35	181
1240	#10 bars	G		40	1.400		68.50	72.50	18.40	159.40	208
1250	#11 bars	G		32	1.750		83	90.50	23	196.50	257
1260	#14 bars	G		24	2.333		104	121	30.50	255.50	335
1270	#18 bars	G		16	3.500		106	181	46	333	445
2000	Weldable half coupler, taper threaded, #4 bars	G	E-16	120	.133		8.90	7.15	1.22	17.27	23.50
2100	#5 bars	G		112	.143		10.50	7.65	1.30	19.45	26.50
2200	#6 bars	G		104	.154		16.60	8.25	1.40	26.25	34
2300	#7 bars	G		96	.167		19.30	8.95	1.52	29.77	38
2400	#8 bars	G		88	.182		20	9.75	1.66	31.41	41
2500	#9 bars	G		80	.200		22	10.75	1.82	34.57	45
2600	#10 bars	G		72	.222		22.50	11.90	2.03	36.43	47.50
2700	#11 bars	G		64	.250		24.50	13.40	2.28	40.18	52.50
2800	#14 bars	G		56	.286		28	15.35	2.61	45.96	60.50
2900	#18 bars	G		48	.333		45.50	17.90	3.04	66.44	84.50

03 21 11 – Plain Steel Reinforcement Bars

03 21 11.60 Reinforcing In Place			Crew	Daily Output	Labor-Hours	Unit	Material	Labor	Equipment	Total	Total Incl O&P
0010	**REINFORCING IN PLACE**, 50-60 ton lots, A615 Grade 60 R032110-10										
0020	Includes labor, but not material cost, to install accessories										
0030	Made from recycled materials										
0100	Beams & Girders, #3 to #7	G	4 Rodm	1.60	20	Ton	970	1,050		2,020	2,725
0150	#8 to #18 R032110-20	G		2.70	11.852		970	625		1,595	2,050
0200	Columns, #3 to #7	G		1.50	21.333		970	1,125		2,095	2,825
0250	#8 to #18	G		2.30	13.913		970	730		1,700	2,225
0300	Spirals, hot rolled, 8" to 15" diameter	G		2.20	14.545		1,575	765		2,340	2,925
0320	15" to 24" diameter R032110-40	G		2.20	14.545		1,500	765		2,265	2,850
0330	24" to 36" diameter	G		2.30	13.913		1,425	730		2,155	2,725
0340	36" to 48" diameter R032110-50	G		2.40	13.333		1,350	700		2,050	2,600
0360	48" to 64" diameter	G		2.50	12.800		1,500	675		2,175	2,700
0380	64" to 84" diameter R032110-70	G		2.60	12.308		1,575	645		2,220	2,725
0390	84" to 96" diameter	G		2.70	11.852		1,650	625		2,275	2,800
0400	Elevated slabs, #4 to #7 R032110-80	G		2.90	11.034		970	580		1,550	1,975
0500	Footings, #4 to #7	G		2.10	15.238		970	800		1,770	2,325
0550	#8 to #18	G		3.60	8.889		970	465		1,435	1,800
0600	Slab on grade, #3 to #7	G		2.30	13.913		970	730		1,700	2,225
0700	Walls, #3 to #7	G		3	10.667		970	560		1,530	1,950
0750	#8 to #18	G		4	8		970	420		1,390	1,725
0900	For other than 50 – 60 ton lots										
1000	Under 10 ton job, #3 to #7, add						25%	10%			
1010	#8 to #18, add						20%	10%			
1050	10 – 50 ton job, #3 to #7, add						10%				
1060	#8 to #18, add						5%				
1100	60 – 100 ton job, #3 to #7, deduct						5%				
1110	#8 to #18, deduct						10%				
1150	Over 100 ton job, #3 to #7, deduct						10%				
1160	#8 to #18, deduct						15%				
1200	Reinforcing in place, A615 Grade 75, add	G				Ton	92.50			92.50	102
1220	Grade 90, add						125			125	138
2000	Unloading & sorting, add to above		C-5	100	.560			29	7.35	36.35	53
2200	Crane cost for handling, 90 picks/day, up to 1.5 Tons/bundle, add to above			135	.415			21.50	5.45	26.95	39

03 21 Reinforcement Bars

03 21 11 – Plain Steel Reinforcement Bars

03 21 11.60 Reinforcing In Place		Crew	Daily Output	Labor-Hours	Unit	Material	2015 Bare Costs Labor	Equipment	Total	Total Incl O&P
2210	1.0 Ton/bundle	C-5	92	.609	Ton		31.50	8	39.50	58
2220	0.5 Ton/bundle	↓	35	1.600	↓		82.50	21	103.50	151
2400	Dowels, 2 feet long, deformed, #3 G	2 Rodm	520	.031	Ea.	.40	1.62		2.02	2.97
2410	#4 G		480	.033		.71	1.75		2.46	3.52
2420	#5 G		435	.037		1.11	1.93		3.04	4.24
2430	#6 G		360	.044	↓	1.60	2.34		3.94	5.40
2450	Longer and heavier dowels, add G		725	.022	Lb.	.53	1.16		1.69	2.40
2500	Smooth dowels, 12" long, 1/4" or 3/8" diameter G		140	.114	Ea.	.72	6		6.72	10.20
2520	5/8" diameter G		125	.128		1.26	6.75		8.01	11.90
2530	3/4" diameter G	↓	110	.145	↓	1.57	7.65		9.22	13.65
2600	Dowel sleeves for CIP concrete, 2-part system									
2610	Sleeve base, plastic, for 5/8" smooth dowel sleeve, fasten to edge form	1 Rodm	200	.040	Ea.	.53	2.10		2.63	3.86
2615	Sleeve, plastic, 12" long, for 5/8" smooth dowel, snap onto base		400	.020		1.18	1.05		2.23	2.94
2620	Sleeve base, for 3/4" smooth dowel sleeve		175	.046		.53	2.40		2.93	4.33
2625	Sleeve, 12" long, for 3/4" smooth dowel		350	.023		1.33	1.20		2.53	3.34
2630	Sleeve base, for 1" smooth dowel sleeve		150	.053		.67	2.80		3.47	5.10
2635	Sleeve, 12" long, for 1" smooth dowel	↓	300	.027		1.40	1.40		2.80	3.73
2700	Dowel caps, visual warning only, plastic, #3 to #8	2 Rodm	800	.020		.26	1.05		1.31	1.93
2720	#8 to #18		750	.021		.63	1.12		1.75	2.44
2750	Impalement protective, plastic, #4 to #9	↓	800	.020	↓	1.22	1.05		2.27	2.98

03 21 13 – Galvanized Reinforcement Steel Bars

03 21 13.10 Galvanized Reinforcing		Crew	Daily Output	Labor-Hours	Unit	Material	Labor	Equipment	Total	Total Incl O&P
0010	**GALVANIZED REINFORCING**									
0150	Add to plain steel rebar pricing for galvanized rebar				Ton	460			460	505

03 21 16 – Epoxy-Coated Reinforcement Steel Bars

03 21 16.10 Epoxy-Coated Reinforcing		Crew	Daily Output	Labor-Hours	Unit	Material	Labor	Equipment	Total	Total Incl O&P
0010	**EPOXY-COATED REINFORCING**									
0100	Add to plain steel rebar pricing for epoxy-coated rebar				Ton	420			420	465

03 21 21 – Composite Reinforcement Bars

03 21 21.11 Glass Fiber-Reinforced Polymer Reinforcement Bars		Crew	Daily Output	Labor-Hours	Unit	Material	Labor	Equipment	Total	Total Incl O&P
0010	**GLASS FIBER-REINFORCED POLYMER REINFORCEMENT BARS**									
0020	Includes labor, but not material cost, to install accessories									
0050	#2 bar, .043 lb./L.F.	4 Rodm	9500	.003	L.F.	.40	.18		.58	.72
0100	#3 bar, .092 lb./L.F.		9300	.003		.49	.18		.67	.82
0150	#4 bar, .160 lb./L.F.		9100	.004		.71	.19		.90	1.07
0200	#5 bar, .258 lb./L.F.		8700	.004		1.09	.19		1.28	1.50
0250	#6 bar, .372 lb./L.F.		8300	.004		1.41	.20		1.61	1.87
0300	#7 bar, .497 lb./L.F.		7900	.004		1.84	.21		2.05	2.35
0350	#8 bar, .620 lb./L.F.		7400	.004		2.38	.23		2.61	2.97
0400	#9 bar, .800 lb./L.F.		6800	.005		3.10	.25		3.35	3.80
0450	#10 bar, 1.08 lb./L.F.	↓	5800	.006	↓	3.75	.29		4.04	4.58
0500	For Bends, add per bend				Ea.	1.48			1.48	1.63

03 22 Fabric and Grid Reinforcing

03 22 11 – Plain Welded Wire Fabric Reinforcing

03 22 11.10 Plain Welded Wire Fabric		Crew	Daily Output	Labor-Hours	Unit	Material	2015 Bare Costs Labor	Equipment	Total	Total Incl O&P
0010	**PLAIN WELDED WIRE FABRIC** ASTM A185 R032205-30									
0020	Includes labor, but not material cost, to install accessories									
0030	Made from recycled materials									
0050	Sheets									
0100	6 x 6 - W1.4 x W1.4 (10 x 10) 21 lb. per C.S.F. G	2 Rodm	35	.457	C.S.F.	14.50	24		38.50	53.50
0200	6 x 6 - W2.1 x W2.1 (8 x 8) 30 lb. per C.S.F. G		31	.516		17.20	27		44.20	61.50
0300	6 x 6 - W2.9 x W2.9 (6 x 6) 42 lb. per C.S.F. G		29	.552		22.50	29		51.50	70
0400	6 x 6 - W4 x W4 (4 x 4) 58 lb. per C.S.F. G		27	.593		31.50	31		62.50	83
0500	4 x 4 - W1.4 x W1.4 (10 x 10) 31 lb. per C.S.F. G		31	.516		20	27		47	64.50
0600	4 x 4 - W2.1 x W2.1 (8 x 8) 44 lb. per C.S.F. G		29	.552		25	29		54	73
0650	4 x 4 - W2.9 x W2.9 (6 x 6) 61 lb. per C.S.F. G		27	.593		40.50	31		71.50	93
0700	4 x 4 - W4 x W4 (4 x 4) 85 lb. per C.S.F. G	↓	25	.640	↓	50.50	33.50		84	108
0750	Rolls									
0800	2 x 2 - #14 galv., 21 lb./C.S.F., beam & column wrap G	2 Rodm	6.50	2.462	C.S.F.	41.50	129		170.50	248
0900	2 x 2 - #12 galv. for gunite reinforcing G	"	6.50	2.462	"	62.50	129		191.50	271

03 22 13 – Galvanized Welded Wire Fabric Reinforcing

03 22 13.10 Galvanized Welded Wire Fabric		Crew	Daily Output	Labor-Hours	Unit	Material	Labor	Equipment	Total	Total Incl O&P
0010	**GALVANIZED WELDED WIRE FABRIC**									
0100	Add to plain welded wire pricing for galvanized welded wire				Lb.	.23			.23	.25

03 22 16 – Epoxy-Coated Welded Wire Fabric Reinforcing

03 22 16.10 Epoxy-Coated Welded Wire Fabric		Crew	Daily Output	Labor-Hours	Unit	Material	Labor	Equipment	Total	Total Incl O&P
0010	**EPOXY-COATED WELDED WIRE FABRIC**									
0100	Add to plain welded wire pricing for epoxy-coated welded wire				Lb.	.21			.21	.23

03 23 Stressed Tendon Reinforcing

03 23 05 – Prestressing Tendons

03 23 05.50 Prestressing Steel		Crew	Daily Output	Labor-Hours	Unit	Material	Labor	Equipment	Total	Total Incl O&P
0010	**PRESTRESSING STEEL** R034136-90									
0100	Grouted strand, in beams, post-tensioned in field, 50' span, 100 kip G	C-3	1200	.053	Lb.	2.64	2.59	.09	5.32	7
0150	300 kip G		2700	.024		1.14	1.15	.04	2.33	3.09
0300	100' span, 100 kip G		1700	.038		2.64	1.83	.07	4.54	5.80
0350	300 kip G		3200	.020		2.27	.97	.04	3.28	4.05
0500	200' span, 100 kip G		2700	.024		2.64	1.15	.04	3.83	4.74
0550	300 kip G		3500	.018		2.27	.89	.03	3.19	3.92
0800	Grouted bars, in beams, 50' span, 42 kip G		2600	.025		1.04	1.20	.04	2.28	3.06
0850	143 kip G		3200	.020		1	.97	.04	2.01	2.65
1000	75' span, 42 kip G		3200	.020		1.06	.97	.04	2.07	2.72
1050	143 kip G	↓	4200	.015		.89	.74	.03	1.66	2.16
1200	Ungrouted strand, in beams, 50' span, 100 kip G	C-4	1275	.025		.63	1.33	.02	1.98	2.80
1250	300 kip G		1475	.022		.63	1.15	.02	1.80	2.51
1400	100' span, 100 kip G		1500	.021		.63	1.13	.02	1.78	2.48
1450	300 kip G		1650	.019		.63	1.03	.02	1.68	2.32
1600	200' span, 100 kip G		1500	.021		.63	1.13	.02	1.78	2.48
1650	300 kip G		1700	.019		.63	1	.02	1.65	2.27
1800	Ungrouted bars, in beams, 50' span, 42 kip G		1400	.023		.48	1.21	.02	1.71	2.44
1850	143 kip G		1700	.019		.48	1	.02	1.50	2.11
2000	75' span, 42 kip G		1800	.018		.48	.94	.02	1.44	2.02
2050	143 kip G		2200	.015		.48	.77	.01	1.26	1.76
2220	Ungrouted single strand, 100' elevated slab, 25 kip G		1200	.027		.63	1.41	.03	2.07	2.93
2250	35 kip G	↓	1475	.022	↓	.63	1.15	.02	1.80	2.51

03 23 Stressed Tendon Reinforcing

03 23 05 – Prestressing Tendons

03 23 05.50 Prestressing Steel		Crew	Daily Output	Labor-Hours	Unit	Material	2015 Bare Costs Labor	Equipment	Total	Total Incl O&P
3000	Slabs on grade, 0.5-inch diam. non-bonded strands, HDPE sheathed,									
3050	attached dead-end anchors, loose stressing-end anchors									
3100	25′ x 30′ slab, strands @ 36″ O.C., placing	2 Rodm	2940	.005	S.F.	.60	.29		.89	1.11
3105	Stressing	C-4A	3750	.004			.22	.01	.23	.36
3110	42″ O.C., placing	2 Rodm	3200	.005		.53	.26		.79	1
3115	Stressing	C-4A	4040	.004			.21	.01	.22	.34
3120	48″ O.C., placing	2 Rodm	3510	.005		.47	.24		.71	.88
3125	Stressing	C-4A	4390	.004			.19	.01	.20	.31
3150	25′ x 40′ slab, strands @ 36″ O.C., placing	2 Rodm	3370	.005		.58	.25		.83	1.03
3155	Stressing	C-4A	4360	.004			.19	.01	.20	.31
3160	42″ O.C., placing	2 Rodm	3760	.004		.50	.22		.72	.90
3165	Stressing	C-4A	4820	.003			.17	.01	.18	.28
3170	48″ O.C., placing	2 Rodm	4090	.004		.45	.21		.66	.81
3175	Stressing	C-4A	5190	.003			.16	.01	.17	.26
3200	30′ x 30′ slab, strands @ 36″ O.C., placing	2 Rodm	3260	.005		.58	.26		.84	1.04
3205	Stressing	C-4A	4190	.004			.20	.01	.21	.32
3210	42″ O.C., placing	2 Rodm	3530	.005		.52	.24		.76	.95
3215	Stressing	C-4A	4500	.004			.19	.01	.20	.30
3220	48″ O.C., placing	2 Rodm	3840	.004		.47	.22		.69	.85
3225	Stressing	C-4A	4850	.003			.17	.01	.18	.28
3230	30′ x 40′ slab, strands @ 36″ O.C., placing	2 Rodm	3780	.004		.56	.22		.78	.97
3235	Stressing	C-4A	4920	.003			.17	.01	.18	.28
3240	42″ O.C., placing	2 Rodm	4190	.004		.49	.20		.69	.85
3245	Stressing	C-4A	5410	.003			.16	.01	.17	.25
3250	48″ O.C., placing	2 Rodm	4520	.004		.45	.19		.64	.78
3255	Stressing	C-4A	5790	.003			.15	.01	.16	.24
3260	30′ x 50′ slab, strands @ 36″ O.C., placing	2 Rodm	4300	.004		.53	.20		.73	.90
3265	Stressing	C-4A	5650	.003			.15	.01	.16	.24
3270	42″ O.C., placing	2 Rodm	4720	.003		.47	.18		.65	.80
3275	Stressing	C-4A	6150	.003			.14	.01	.15	.22
3280	48″ O.C., placing	2 Rodm	5240	.003		.42	.16		.58	.71
3285	Stressing	C-4A	6760	.002	▼		.12	.01	.13	.20

03 24 Fibrous Reinforcing

03 24 05 – Reinforcing Fibers

03 24 05.30 Synthetic Fibers		Crew	Daily Output	Labor-Hours	Unit	Material	Labor	Equipment	Total	Total Incl O&P
0010	**SYNTHETIC FIBERS**									
0100	Synthetic fibers, add to concrete				Lb.	4.40			4.40	4.84
0110	1-1/2 lb. per C.Y.				C.Y.	6.80			6.80	7.50

03 24 05.70 Steel Fibers		Crew	Daily Output	Labor-Hours	Unit	Material	Labor	Equipment	Total	Total Incl O&P
0010	**STEEL FIBERS**									
0140	ASTM A850, Type V, continuously deformed, 1-1/2″ long x 0.045″ diam.									
0150	Add to price of ready mix concrete G				Lb.	1.13			1.13	1.24
0205	Alternate pricing, dosing at 5 lb. per C.Y., add to price of RMC G				C.Y.	5.65			5.65	6.20
0210	10 lb. per C.Y. G					11.30			11.30	12.45
0215	15 lb. per C.Y. G					16.95			16.95	18.65
0220	20 lb. per C.Y. G					22.50			22.50	25
0225	25 lb. per C.Y. G					28.50			28.50	31
0230	30 lb. per C.Y. G					34			34	37.50
0235	35 lb. per C.Y. G					39.50			39.50	43.50
0240	40 lb. per C.Y. G				▼	45			45	49.50

03 24 Fibrous Reinforcing

03 24 05 – Reinforcing Fibers

03 24 05.70 Steel Fibers			Crew	Daily Output	Labor-Hours	Unit	Material	2015 Bare Costs Labor	Equipment	Total	Total Incl O&P
0250	50 lb. per C.Y.	G				C.Y.	56.50			56.50	62
0275	75 lb. per C.Y.	G					85			85	93
0300	100 lb. per C.Y.	G					113			113	124

03 30 Cast-In-Place Concrete

03 30 53 – Miscellaneous Cast-In-Place Concrete

03 30 53.40 Concrete In Place

			Crew	Daily Output	Labor-Hours	Unit	Material	Labor	Equipment	Total	Total Incl O&P
0010	**CONCRETE IN PLACE**	R033105-10									
0020	Including forms (4 uses), Grade 60 rebar, concrete (Portland cement	R033105-20									
0050	Type I), placement and finishing unless otherwise indicated	R033105-50									
0300	Beams (3500 psi), 5 kip per L.F., 10′ span	R033105-65	C-14A	15.62	12.804	C.Y.	320	605	48	973	1,325
0350	25′ span	R033105-70	"	18.55	10.782		335	510	40.50	885.50	1,200
0500	Chimney foundations (5000 psi), over 5 C.Y.	R033105-85	C-14C	32.22	3.476		150	157	.97	307.97	405
0510	(3500 psi), under 5 C.Y.		"	23.71	4.724		177	213	1.32	391.32	525
0700	Columns, square (4000 psi), 12″ x 12″, less than 2% reinforcing		C-14A	11.96	16.722		365	790	63	1,218	1,700
0720	2% to 3% reinforcing			10.13	19.743		565	935	74	1,574	2,125
0740	Over 3% reinforcing			9.03	22.148		840	1,050	83.50	1,973.50	2,600
0800	16″ x 16″, less than 2% reinforcing			16.22	12.330		286	585	46.50	917.50	1,250
0820	2% to 3% reinforcing			12.57	15.911		480	750	60	1,290	1,750
0840	Over 3% reinforcing			10.25	19.512		735	920	73.50	1,728.50	2,300
0900	24″ x 24″, less than 2% reinforcing			23.66	8.453		241	400	32	673	915
0920	2% to 3% reinforcing			17.71	11.293		425	535	42.50	1,002.50	1,325
0940	Over 3% reinforcing			14.15	14.134		670	670	53	1,393	1,825
1000	36″ x 36″, less than 2% reinforcing			33.69	5.936		212	281	22.50	515.50	690
1020	2% to 3% reinforcing			23.32	8.576		370	405	32.50	807.50	1,075
1040	Over 3% reinforcing			17.82	11.223		625	530	42	1,197	1,550
1100	Columns, round (4000 psi), tied, 12″ diameter, less than 2% reinforcing			20.97	9.537		315	450	36	801	1,075
1120	2% to 3% reinforcing			15.27	13.098		515	620	49.50	1,184.50	1,575
1140	Over 3% reinforcing			12.11	16.515		780	780	62	1,622	2,125
1200	16″ diameter, less than 2% reinforcing			31.49	6.351		289	300	24	613	800
1220	2% to 3% reinforcing			19.12	10.460		490	495	39.50	1,024.50	1,350
1240	Over 3% reinforcing			13.77	14.524		735	685	54.50	1,474.50	1,925
1300	20″ diameter, less than 2% reinforcing			41.04	4.873		289	230	18.30	537.30	690
1320	2% to 3% reinforcing			24.05	8.316		475	395	31.50	901.50	1,150
1340	Over 3% reinforcing			17.01	11.758		735	555	44	1,334	1,700
1400	24″ diameter, less than 2% reinforcing			51.85	3.857		269	182	14.50	465.50	595
1420	2% to 3% reinforcing			27.06	7.391		470	350	28	848	1,100
1440	Over 3% reinforcing			18.29	10.935		715	515	41	1,271	1,625
1500	36″ diameter, less than 2% reinforcing			75.04	2.665		266	126	10	402	500
1520	2% to 3% reinforcing			37.49	5.335		445	252	20	717	900
1540	Over 3% reinforcing			22.84	8.757		695	415	33	1,143	1,425
1900	Elevated slab (4000 psi), flat slab with drops, 125 psf Sup. Load, 20′ span		C-14B	38.45	5.410		261	255	19.55	535.55	700
1950	30′ span			50.99	4.079		276	192	14.75	482.75	615
2100	Flat plate, 125 psf Sup. Load, 15′ span			30.24	6.878		240	325	25	590	790
2150	25′ span			49.60	4.194		249	198	15.15	462.15	595
2300	Waffle const., 30″ domes, 125 psf Sup. Load, 20′ span			37.07	5.611		259	265	20.50	544.50	715
2350	30′ span			44.07	4.720		241	223	17.05	481.05	625
2500	One way joists, 30″ pans, 125 psf Sup. Load, 15′ span			27.38	7.597		310	360	27.50	697.50	920
2550	25′ span			31.15	6.677		292	315	24	631	830
2700	One way beam & slab, 125 psf Sup. Load, 15′ span			20.59	10.102		259	475	36.50	770.50	1,050
2750	25′ span			28.36	7.334		245	345	26.50	616.50	830

03 30 Cast-In-Place Concrete

03 30 53 – Miscellaneous Cast-In-Place Concrete

	03 30 53.40 Concrete In Place	Crew	Daily Output	Labor-Hours	Unit	Material	Labor	Equipment	Total	Total Incl O&P
						2015 Bare Costs				
2900	Two way beam & slab, 125 psf Sup. Load, 15′ span	C-14B	24.04	8.652	C.Y.	250	410	31	691	935
2950	25′ span		35.87	5.799		216	273	21	510	680
3100	Elevated slabs, flat plate, including finish, not									
3110	including forms or reinforcing									
3150	Regular concrete (4000 psi), 4″ slab	C-8	2613	.021	S.F.	1.43	.90	.28	2.61	3.24
3200	6″ slab		2585	.022		2.09	.91	.28	3.28	3.99
3250	2-1/2″ thick floor fill		2685	.021		.94	.87	.27	2.08	2.65
3300	Lightweight, 110# per C.F., 2-1/2″ thick floor fill		2585	.022		1.46	.91	.28	2.65	3.30
3400	Cellular concrete, 1-5/8″ fill, under 5000 S.F.		2000	.028		.99	1.17	.36	2.52	3.27
3450	Over 10,000 S.F.		2200	.025		.94	1.07	.33	2.34	3.02
3500	Add per floor for 3 to 6 stories high		31800	.002			.07	.02	.09	.13
3520	For 7 to 20 stories high		21200	.003			.11	.03	.14	.21
3540	Equipment pad (3000 psi), 3′ x 3′ x 6″ thick	C-14H	45	1.067	Ea.	47	49.50	.69	97.19	128
3550	4′ x 4′ x 6″ thick		30	1.600		69.50	74	1.04	144.54	191
3560	5′ x 5′ x 8″ thick		18	2.667		122	124	1.73	247.73	325
3570	6′ x 6′ x 8″ thick		14	3.429		164	159	2.23	325.23	425
3580	8′ x 8′ x 10″ thick		8	6		350	278	3.90	631.90	815
3590	10′ x 10′ x 12″ thick		5	9.600		595	445	6.25	1,046.25	1,350
3800	Footings (3000 psi), spread under 1 C.Y.	C-14C	28	4	C.Y.	166	180	1.12	347.12	460
3825	1 C.Y. to 5 C.Y.		43	2.605		201	117	.73	318.73	405
3850	Over 5 C.Y.		75	1.493		185	67.50	.42	252.92	310
3900	Footings, strip (3000 psi), 18″ x 9″, unreinforced	C-14L	40	2.400		125	105	.79	230.79	300
3920	18″ x 9″, reinforced	C-14C	35	3.200		148	144	.90	292.90	385
3925	20″ x 10″, unreinforced	C-14L	45	2.133		122	93.50	.70	216.20	278
3930	20″ x 10″, reinforced	C-14C	40	2.800		140	126	.78	266.78	350
3935	24″ x 12″, unreinforced	C-14L	55	1.745		120	76.50	.58	197.08	250
3940	24″ x 12″, reinforced	C-14C	48	2.333		139	105	.65	244.65	315
3945	36″ x 12″, unreinforced	C-14L	70	1.371		116	60	.45	176.45	219
3950	36″ x 12″, reinforced	C-14C	60	1.867		133	84	.52	217.52	276
4000	Foundation mat (3000 psi), under 10 C.Y.		38.67	2.896		204	131	.81	335.81	425
4050	Over 20 C.Y.		56.40	1.986		178	89.50	.56	268.06	335
4200	Wall, free-standing (3000 psi), 8″ thick, 8′ high	C-14D	45.83	4.364		160	204	16.40	380.40	510
4250	14′ high		27.26	7.337		190	345	27.50	562.50	770
4260	12″ thick, 8′ high		64.32	3.109		145	146	11.70	302.70	395
4270	14′ high		40.01	4.999		154	234	18.80	406.80	550
4300	15″ thick, 8′ high		80.02	2.499		139	117	9.40	265.40	345
4350	12′ high		51.26	3.902		139	183	14.65	336.65	450
4500	18′ high		48.85	4.094		157	192	15.40	364.40	485
4520	Handicap access ramp (4000 psi), railing both sides, 3′ wide	C-14H	14.58	3.292	L.F.	320	153	2.14	475.14	585
4525	5′ wide		12.22	3.928		330	182	2.55	514.55	645
4530	With 6″ curb and rails both sides, 3′ wide		8.55	5.614		330	260	3.65	593.65	765
4535	5′ wide		7.31	6.566		335	305	4.27	644.27	840
4650	Slab on grade (3500 psi), not including finish, 4″ thick	C-14E	60.75	1.449	C.Y.	124	67.50	.51	192.01	242
4700	6″ thick	″	92	.957	″	119	44.50	.33	163.83	200
4701	Thickened slab edge (3500 psi), for slab on grade poured									
4702	monolithically with slab; depth is in addition to slab thickness;									
4703	formed vertical outside edge, earthen bottom and inside slope									
4705	8″ deep x 8″ wide bottom, unreinforced	C-14L	2190	.044	L.F.	3.47	1.92	.01	5.40	6.80
4710	8″ x 8″, reinforced	C-14C	1670	.067		5.75	3.02	.02	8.79	10.95
4715	12″ deep x 12″ wide bottom, unreinforced	C-14L	1800	.053		7.05	2.34	.02	9.41	11.35
4720	12″ x 12″, reinforced	C-14C	1310	.086		11.20	3.85	.02	15.07	18.30
4725	16″ deep x 16″ wide bottom, unreinforced	C-14L	1440	.067		11.85	2.92	.02	14.79	17.50
4730	16″ x 16″, reinforced	C-14C	1120	.100		16.80	4.51	.03	21.34	25.50

03 30 Cast-In-Place Concrete

03 30 53 – Miscellaneous Cast-In-Place Concrete

03 30 53.40 Concrete In Place		Crew	Daily Output	Labor-Hours	Unit	2015 Bare Costs Material	Labor	Equipment	Total	Total Incl O&P
4735	20″ deep x 20″ wide bottom, unreinforced	C-14L	1150	.083	L.F.	17.90	3.66	.03	21.59	25.50
4740	20″ x 20″, reinforced	C-14C	920	.122		24	5.50	.03	29.53	35
4745	24″ deep x 24″ wide bottom, unreinforced	C-14L	930	.103		25	4.53	.03	29.56	35
4750	24″ x 24″, reinforced	C-14C	740	.151		33.50	6.80	.04	40.34	47
4751	Slab on grade (3500 psi), incl. troweled finish, not incl. forms									
4760	or reinforcing, over 10,000 S.F., 4″ thick	C-14F	3425	.021	S.F.	1.35	.90	.01	2.26	2.85
4820	6″ thick		3350	.021		1.98	.92	.01	2.91	3.57
4840	8″ thick		3184	.023		2.71	.97	.01	3.69	4.44
4900	12″ thick		2734	.026		4.06	1.13	.01	5.20	6.15
4950	15″ thick		2505	.029		5.10	1.23	.01	6.34	7.45
5000	Slab on grade (3000 psi), incl. broom finish, not incl. forms									
5001	or reinforcing, 4″ thick	C-14G	2873	.019	S.F.	1.32	.82	.01	2.15	2.69
5010	6″ thick		2590	.022		2.07	.91	.01	2.99	3.65
5020	8″ thick		2320	.024		2.70	1.02	.01	3.73	4.50
5200	Lift slab in place above the foundation, incl. forms, reinforcing,									
5210	concrete (4000 psi) and columns, over 20,000 S.F. per floor	C-14B	2113	.098	S.F.	6.90	4.64	.36	11.90	15.15
5250	10,000 S.F. to 20,000 S.F. per floor		1650	.126		7.55	5.95	.46	13.96	17.95
5300	Under 10,000 S.F. per floor		1500	.139		8.20	6.55	.50	15.25	19.65
5500	Lightweight, ready mix, including screed finish only,									
5510	not including forms or reinforcing									
5550	1:4 (2500 psi) for structural roof decks	C-14B	260	.800	C.Y.	166	37.50	2.89	206.39	244
5600	1:6 (3000 psi) for ground slab with radiant heat	C-14F	92	.783		168	33.50	.34	201.84	235
5650	1:3:2 (2000 psi) with sand aggregate, roof deck	C-14B	260	.800		164	37.50	2.89	204.39	242
5700	Ground slab (2000 psi)	C-14F	107	.673		164	29	.29	193.29	224
5900	Pile caps (3000 psi), incl. forms and reinf., sq. or rect., under 10 C.Y.	C-14C	54.14	2.069		168	93.50	.58	262.08	330
5950	Over 10 C.Y.		75	1.493		157	67.50	.42	224.92	277
6000	Triangular or hexagonal, under 10 C.Y.		53	2.113		123	95.50	.59	219.09	283
6050	Over 10 C.Y.		85	1.318		138	59.50	.37	197.87	244
6200	Retaining walls (3000 psi), gravity, 4′ high see Section 32 32	C-14D	66.20	3.021		140	141	11.35	292.35	385
6250	10′ high		125	1.600		134	75	6	215	269
6300	Cantilever, level backfill loading, 8′ high		70	2.857		150	134	10.75	294.75	385
6350	16′ high		91	2.198		145	103	8.25	256.25	325
6800	Stairs (3500 psi), not including safety treads, free standing, 3′-6″ wide	C-14H	83	.578	LF Nose	5.60	27	.38	32.98	47.50
6850	Cast on ground		125	.384	″	4.63	17.80	.25	22.68	33
7000	Stair landings, free standing		200	.240	S.F.	4.52	11.10	.16	15.78	22
7050	Cast on ground		475	.101	″	3.52	4.68	.07	8.27	11.15

03 31 Structural Concrete

03 31 13 – Heavyweight Structural Concrete

03 31 13.25 Concrete, Hand Mix		Crew	Daily Output	Labor-Hours	Unit	2015 Bare Costs Material	Labor	Equipment	Total	Total Incl O&P
0010	**CONCRETE, HAND MIX** for small quantities or remote areas									
0050	Includes bulk local aggregate, bulk sand, bagged Portland									
0060	cement (Type I) and water, using gas powered cement mixer									
0125	2500 psi	C-30	135	.059	C.F.	3.29	2.23	1.28	6.80	8.45
0130	3000 psi		135	.059		3.52	2.23	1.28	7.03	8.70
0135	3500 psi		135	.059		3.66	2.23	1.28	7.17	8.85
0140	4000 psi		135	.059		3.82	2.23	1.28	7.33	9.05
0145	4500 psi		135	.059		4	2.23	1.28	7.51	9.25
0150	5000 psi		135	.059		4.26	2.23	1.28	7.77	9.50
0300	Using pre-bagged dry mix and wheelbarrow (80-lb. bag = 0.6 C.F.)									
0340	4000 psi	1 Clab	48	.167	C.F.	6.25	6.25		12.50	16.55

03 31 Structural Concrete

03 31 13 – Heavyweight Structural Concrete

03 31 13.30 Concrete, Volumetric Site-Mixed		Crew	Daily Output	Labor-Hours	Unit	Material	2015 Bare Costs Labor	Equipment	Total	Total Incl O&P
0010	CONCRETE, VOLUMETRIC SITE-MIXED									
0015	Mixed on-site in volumetric truck									
0020	Includes local aggregate, sand, Portland cement (Type I) and water									
0025	Excludes all additives and treatments									
0100	3000 psi, 1 C.Y. mixed and discharged				C.Y.	177			177	194
0110	2 C.Y.					138			138	151
0120	3 C.Y.					124			124	136
0130	4 C.Y.					113			113	124
0140	5 C.Y.					106			106	117
0200	For truck holding/waiting time past first 2 on-site hours, add				Hr.	78			78	86
0210	For trip charge beyond first 20 miles, each way, add				Mile	3.50			3.50	3.85
0220	For each additional increase of 500 psi, add				Ea.	4.17			4.17	4.59

03 31 13.35 Heavyweight Concrete, Ready Mix

		Crew	Daily Output	Labor-Hours	Unit	Material	Labor	Equipment	Total	Total Incl O&P
0010	HEAVYWEIGHT CONCRETE, READY MIX, delivered R033105-10									
0012	Includes local aggregate, sand, Portland cement (Type I) and water									
0015	Excludes all additives and treatments R033105-20									
0020	2000 psi				C.Y.	97			97	107
0100	2500 psi R033105-30					99.50			99.50	109
0150	3000 psi					102			102	112
0200	3500 psi R033105-40					104			104	115
0300	4000 psi					107			107	118
0350	4500 psi R033105-50					110			110	121
0400	5000 psi					113			113	125
0411	6000 psi					116			116	128
0412	8000 psi					123			123	135
0413	10,000 psi					129			129	142
0414	12,000 psi					135			135	149
1000	For high early strength (Portland cement Type III), add					10%				
1010	For structural lightweight with regular sand, add					25%				
1300	For winter concrete (hot water), add					4.05			4.05	4.46
1410	For mid-range water reducer, add					3.31			3.31	3.64
1420	For high-range water reducer/superplasticizer, add					5.65			5.65	6.20
1430	For retarder, add					3.23			3.23	3.55
1440	For non-Chloride accelerator, add					5.50			5.50	6.05
1450	For Chloride accelerator, per 1%, add					2.90			2.90	3.19
1460	For fiber reinforcing, synthetic (1 lb./C.Y.), add					6.65			6.65	7.30
1500	For Saturday delivery, add					10.80			10.80	11.90
1510	For truck holding/waiting time past 1st hour per load, add				Hr.	94			94	103
1520	For short load (less than 4 C.Y.), add per load				Ea.	60.50			60.50	66.50
2000	For all lightweight aggregate, add				C.Y.	45%				

03 31 13.70 Placing Concrete

		Crew	Daily Output	Labor-Hours	Unit	Material	Labor	Equipment	Total	Total Incl O&P
0010	PLACING CONCRETE R033105-70									
0020	Includes labor and equipment to place, level (strike off) and consolidate									
0050	Beams, elevated, small beams, pumped	C-20	60	1.067	C.Y.		43	13.05	56.05	80.50
0100	With crane and bucket	C-7	45	1.600			65.50	27	92.50	130
0200	Large beams, pumped	C-20	90	.711			28.50	8.70	37.20	53.50
0250	With crane and bucket	C-7	65	1.108			45.50	18.70	64.20	89.50
0400	Columns, square or round, 12" thick, pumped	C-20	60	1.067			43	13.05	56.05	80.50
0450	With crane and bucket	C-7	40	1.800			73.50	30.50	104	146
0600	18" thick, pumped	C-20	90	.711			28.50	8.70	37.20	53.50
0650	With crane and bucket	C-7	55	1.309			53.50	22	75.50	107
0800	24" thick, pumped	C-20	92	.696			28	8.50	36.50	52.50

03 31 13.70 Placing Concrete		Crew	Daily Output	Labor-Hours	Unit	Material	2015 Bare Costs Labor	Equipment	Total	Total Incl O&P
0850	With crane and bucket	C-7	70	1.029	C.Y.		42	17.35	59.35	83
1000	36" thick, pumped	C-20	140	.457			18.45	5.60	24.05	34
1050	With crane and bucket	C-7	100	.720			29.50	12.15	41.65	58.50
1400	Elevated slabs, less than 6" thick, pumped	C-20	140	.457			18.45	5.60	24.05	34
1450	With crane and bucket	C-7	95	.758			31	12.80	43.80	61.50
1500	6" to 10" thick, pumped	C-20	160	.400			16.15	4.89	21.04	30
1550	With crane and bucket	C-7	110	.655			27	11.05	38.05	53
1600	Slabs over 10" thick, pumped	C-20	180	.356			14.35	4.35	18.70	27
1650	With crane and bucket	C-7	130	.554			22.50	9.35	31.85	45
1900	Footings, continuous, shallow, direct chute	C-6	120	.400			15.65	.52	16.17	24.50
1950	Pumped	C-20	150	.427			17.25	5.20	22.45	32.50
2000	With crane and bucket	C-7	90	.800			32.50	13.50	46	65
2100	Footings, continuous, deep, direct chute	C-6	140	.343			13.45	.45	13.90	21
2150	Pumped	C-20	160	.400			16.15	4.89	21.04	30
2200	With crane and bucket	C-7	110	.655			27	11.05	38.05	53
2400	Footings, spread, under 1 C.Y., direct chute	C-6	55	.873			34	1.13	35.13	53.50
2450	Pumped	C-20	65	.985			40	12.05	52.05	74.50
2500	With crane and bucket	C-7	45	1.600			65.50	27	92.50	130
2600	Over 5 C.Y., direct chute	C-6	120	.400			15.65	.52	16.17	24.50
2650	Pumped	C-20	150	.427			17.25	5.20	22.45	32.50
2700	With crane and bucket	C-7	100	.720			29.50	12.15	41.65	58.50
2900	Foundation mats, over 20 C.Y., direct chute	C-6	350	.137			5.35	.18	5.53	8.40
2950	Pumped	C-20	400	.160			6.45	1.96	8.41	12
3000	With crane and bucket	C-7	300	.240			9.80	4.05	13.85	19.45
3200	Grade beams, direct chute	C-6	150	.320			12.55	.42	12.97	19.60
3250	Pumped	C-20	180	.356			14.35	4.35	18.70	27
3300	With crane and bucket	C-7	120	.600			24.50	10.10	34.60	48.50
3500	High rise, for more than 5 stories, pumped, add per story	C-20	2100	.030			1.23	.37	1.60	2.29
3510	With crane and bucket, add per story	C-7	2100	.034			1.40	.58	1.98	2.78
3700	Pile caps, under 5 C.Y., direct chute	C-6	90	.533			21	.69	21.69	33
3750	Pumped	C-20	110	.582			23.50	7.10	30.60	44
3800	With crane and bucket	C-7	80	.900			37	15.15	52.15	72.50
3850	Pile cap, 5 C.Y. to 10 C.Y., direct chute	C-6	175	.274			10.75	.36	11.11	16.80
3900	Pumped	C-20	200	.320			12.95	3.91	16.86	24
3950	With crane and bucket	C-7	150	.480			19.65	8.10	27.75	39
4000	Over 10 C.Y., direct chute	C-6	215	.223			8.75	.29	9.04	13.65
4050	Pumped	C-20	240	.267			10.75	3.26	14.01	20
4100	With crane and bucket	C-7	185	.389			15.95	6.55	22.50	31.50
4300	Slab on grade, up to 6" thick, direct chute	C-6	110	.436			17.10	.57	17.67	26.50
4350	Pumped	C-20	130	.492			19.90	6	25.90	37
4400	With crane and bucket	C-7	110	.655			27	11.05	38.05	53
4600	Over 6" thick, direct chute	C-6	165	.291			11.40	.38	11.78	17.80
4650	Pumped	C-20	185	.346			14	4.23	18.23	26
4700	With crane and bucket	C-7	145	.497			20.50	8.35	28.85	40
4900	Walls, 8" thick, direct chute	C-6	90	.533			21	.69	21.69	33
4950	Pumped	C-20	100	.640			26	7.85	33.85	48
5000	With crane and bucket	C-7	80	.900			37	15.15	52.15	72.50
5050	12" thick, direct chute	C-6	100	.480			18.80	.62	19.42	29
5100	Pumped	C-20	110	.582			23.50	7.10	30.60	44
5200	With crane and bucket	C-7	90	.800			32.50	13.50	46	65
5300	15" thick, direct chute	C-6	105	.457			17.90	.59	18.49	28
5350	Pumped	C-20	120	.533			21.50	6.50	28	40
5400	With crane and bucket	C-7	95	.758			31	12.80	43.80	61.50

03 31 Structural Concrete

03 31 13 – Heavyweight Structural Concrete

03 31 13.70 Placing Concrete		Crew	Daily Output	Labor-Hours	Unit	Material	2015 Bare Costs Labor	Equipment	Total	Total Incl O&P
5600	Wheeled concrete dumping, add to placing costs above									
5610	Walking cart, 50′ haul, add	C-18	32	.281	C.Y.		10.65	1.88	12.53	18.40
5620	150′ haul, add		24	.375			14.20	2.50	16.70	25
5700	250′ haul, add		18	.500			18.90	3.34	22.24	32.50
5800	Riding cart, 50′ haul, add	C-19	80	.113			4.25	1.25	5.50	7.90
5810	150′ haul, add		60	.150			5.65	1.66	7.31	10.60
5900	250′ haul, add		45	.200			7.55	2.22	9.77	14.10
6000	Concrete in-fill for pan-type metal stairs and landings. Manual placement									
6010	includes up to 50′ horizontal haul from point of concrete discharge.									
6100	Stair pan treads, 2″ deep									
6110	Flights in 1st floor level up/down from discharge point	C-8A	3200	.015	S.F.		.61		.61	.92
6120	2nd floor level		2500	.019			.78		.78	1.18
6130	3rd floor level		2000	.024			.97		.97	1.47
6140	4th floor level		1800	.027			1.08		1.08	1.63
6200	Intermediate stair landings, pan-type 4″ deep									
6210	Flights in 1st floor level up/down from discharge point	C-8A	2000	.024	S.F.		.97		.97	1.47
6220	2nd floor level		1500	.032			1.29		1.29	1.96
6230	3rd floor level		1200	.040			1.62		1.62	2.45
6240	4th floor level		1000	.048			1.94		1.94	2.94

03 35 Concrete Finishing

03 35 13 – High-Tolerance Concrete Floor Finishing

03 35 13.30 Finishing Floors, High Tolerance

Code	Description	Crew	Daily Output	Labor-Hours	Unit	Material	Labor	Equipment	Total	Total Incl O&P
0010	**FINISHING FLOORS, HIGH TOLERANCE**									
0012	Finishing of fresh concrete flatwork requires that concrete									
0013	first be placed, struck off & consolidated									
0015	Basic finishing for various unspecified flatwork									
0100	Bull float only	C-10	4000	.006	S.F.		.26		.26	.38
0125	Bull float & manual float		2000	.012			.51		.51	.76
0150	Bull float, manual float, & broom finish, w/edging & joints		1850	.013			.55		.55	.83
0200	Bull float, manual float & manual steel trowel		1265	.019			.81		.81	1.21
0210	For specified Random Access Floors in ACI Classes 1, 2, 3 and 4 to achieve									
0215	Composite Overall Floor Flatness and Levelness values up to FF35/FL25									
0250	Bull float, machine float & machine trowel (walk-behind)	C-10C	1715	.014	S.F.		.60	.03	.63	.92
0300	Power screed, bull float, machine float & trowel (walk-behind)	C-10D	2400	.010			.43	.05	.48	.69
0350	Power screed, bull float, machine float & trowel (ride-on)	C-10E	4000	.006			.26	.06	.32	.45
0352	For specified Random Access Floors in ACI Classes 5, 6, 7 and 8 to achieve									
0354	Composite Overall Floor Flatness and Levelness values up to FF50/FL50									
0356	Add for two-dimensional restraightening after power float	C-10	6000	.004	S.F.		.17		.17	.25
0358	For specified Random or Defined Access Floors in ACI Class 9 to achieve									
0360	Composite Overall Floor Flatness and Levelness values up to FF100/FL100									
0362	Add for two-dimensional restraightening after bull float & power float	C-10	3000	.008	S.F.		.34		.34	.51
0364	For specified Superflat Defined Access Floors in ACI Class 9 to achieve									
0366	Minimum Floor Flatness and Levelness values of FF100/FL100									
0368	Add for 2-dim'l restraightening after bull float, power float, power trowel	C-10	2000	.012	S.F.		.51		.51	.76

03 35 16 – Heavy-Duty Concrete Floor Finishing

03 35 16.30 Finishing Floors, Heavy-Duty

Code	Description	Crew	Daily Output	Labor-Hours	Unit	Material	Labor	Equipment	Total	Total Incl O&P
0010	**FINISHING FLOORS, HEAVY-DUTY**									
1800	Floor abrasives, dry shake on fresh concrete, .25 psf, aluminum oxide	1 Cefi	850	.009	S.F.	.53	.42		.95	1.21
1850	Silicon carbide		850	.009		.79	.42		1.21	1.50

03 35 Concrete Finishing

03 35 16 – Heavy-Duty Concrete Floor Finishing

03 35 16.30 Finishing Floors, Heavy-Duty		Crew	Daily Output	Labor-Hours	Unit	Material	Labor	Equipment	Total	Total Incl O&P
						2015 Bare Costs				
2000	Floor hardeners, dry shake, metallic, light service, .50 psf	1 Cefi	850	.009	S.F.	.54	.42		.96	1.22
2050	Medium service, .75 psf		750	.011		.80	.48		1.28	1.59
2100	Heavy service, 1.0 psf		650	.012		1.07	.55		1.62	2
2150	Extra heavy, 1.5 psf		575	.014		1.61	.63		2.24	2.70
2300	Non-metallic, light service, .50 psf		850	.009		.21	.42		.63	.87
2350	Medium service, .75 psf		750	.011		.32	.48		.80	1.06
2400	Heavy service, 1.00 psf		650	.012		.43	.55		.98	1.29
2450	Extra heavy, 1.50 psf		575	.014		.64	.63		1.27	1.64
2800	Trap rock wearing surface, dry shake, for monolithic floors									
2810	2.0 psf	C-10B	1250	.032	S.F.	.02	1.30	.22	1.54	2.22
3800	Dustproofing, liquid, for cured concrete, solvent-based, 1 coat	1 Cefi	1900	.004		.18	.19		.37	.47
3850	2 coats		1300	.006		.63	.28		.91	1.11
4000	Epoxy-based, 1 coat		1500	.005		.14	.24		.38	.51
4050	2 coats		1500	.005		.28	.24		.52	.67

03 35 19 – Colored Concrete Finishing

03 35 19.30 Finishing Floors, Colored		Crew	Daily Output	Labor-Hours	Unit	Material	Labor	Equipment	Total	Total Incl O&P
0010	**FINISHING FLOORS, COLORED**									
3000	Floor coloring, dry shake on fresh concrete (0.6 psf)	1 Cefi	1300	.006	S.F.	.44	.28		.72	.89
3050	(1.0 psf)	"	625	.013	"	.73	.58		1.31	1.66
3100	Colored dry shake powder only				Lb.	.73			.73	.81
3600	1/2" topping using 0.6 psf dry shake powdered color	C-10B	590	.068	S.F.	4.96	2.75	.46	8.17	10.10
3650	1.0 psf dry shake powdered color	"	590	.068	"	5.25	2.75	.46	8.46	10.45

03 35 23 – Exposed Aggregate Concrete Finishing

03 35 23.30 Finishing Floors, Exposed Aggregate		Crew	Daily Output	Labor-Hours	Unit	Material	Labor	Equipment	Total	Total Incl O&P
0010	**FINISHING FLOORS, EXPOSED AGGREGATE**									
1600	Exposed local aggregate finish, seeded on fresh concrete, 3 lb. per S.F.	1 Cefi	625	.013	S.F.	.47	.58		1.05	1.36
1650	4 lb. per S.F.	"	465	.017	"	.87	.77		1.64	2.11

03 35 29 – Tooled Concrete Finishing

03 35 29.30 Finishing Floors, Tooled		Crew	Daily Output	Labor-Hours	Unit	Material	Labor	Equipment	Total	Total Incl O&P
0010	**FINISHING FLOORS, TOOLED**									
4400	Stair finish, fresh concrete, float finish	1 Cefi	275	.029	S.F.		1.31		1.31	1.94
4500	Steel trowel finish		200	.040			1.80		1.80	2.67
4600	Silicon carbide finish, dry shake on fresh concrete, .25 psf		150	.053		.53	2.40		2.93	4.13

03 35 29.60 Finishing Walls		Crew	Daily Output	Labor-Hours	Unit	Material	Labor	Equipment	Total	Total Incl O&P
0010	**FINISHING WALLS**									
0020	Break ties and patch voids	1 Cefi	540	.015	S.F.	.04	.67		.71	1.03
0050	Burlap rub with grout		450	.018		.04	.80		.84	1.23
0100	Carborundum rub, dry		270	.030			1.33		1.33	1.97
0150	Wet rub		175	.046			2.06		2.06	3.05
0300	Bush hammer, green concrete	B-39	1000	.048			1.91	.23	2.14	3.19
0350	Cured concrete	"	650	.074			2.94	.36	3.30	4.90
0500	Acid etch	1 Cefi	575	.014		.14	.63		.77	1.08
0600	Float finish, 1/16" thick	"	300	.027		.36	1.20		1.56	2.18
0700	Sandblast, light penetration	E-11	1100	.029		.46	1.23	.21	1.90	2.77
0750	Heavy penetration	"	375	.085		.92	3.61	.63	5.16	7.60
0850	Grind form fins flush	1 Clab	700	.011	L.F.		.43		.43	.66

03 35 33 – Stamped Concrete Finishing

	03 35 33.50 Slab Texture Stamping	Crew	Daily Output	Labor-Hours	Unit	2015 Bare Costs Material	Labor	Equipment	Total	Total Incl O&P
0010	**SLAB TEXTURE STAMPING**									
0050	Stamping requires that concrete first be placed, struck off, consolidated,									
0060	bull floated and free of bleed water. Decorative stamping tasks include:									
0100	Step 1 - first application of dry shake colored hardener	1 Cefi	6400	.001	S.F.	.43	.06		.49	.55
0110	Step 2 - bull float		6400	.001			.06		.06	.08
0130	Step 3 - second application of dry shake colored hardener		6400	.001		.21	.06		.27	.31
0140	Step 4 - bull float, manual float & steel trowel	3 Cefi	1280	.019			.84		.84	1.25
0150	Step 5 - application of dry shake colored release agent	1 Cefi	6400	.001		.10	.06		.16	.19
0160	Step 6 - place, tamp & remove mats	3 Cefi	2400	.010		1.43	.45		1.88	2.25
0170	Step 7 - touch up edges, mat joints & simulated grout lines	1 Cefi	1280	.006			.28		.28	.42
0300	Alternate stamping estimating method includes all tasks above	4 Cefi	800	.040		2.17	1.80		3.97	5.05
0400	Step 8 - pressure wash @ 3000 psi after 24 hours	1 Cefi	1600	.005			.23		.23	.33
0500	Step 9 - roll 2 coats cure/seal compound when dry	"	800	.010		.62	.45		1.07	1.35

03 35 43 – Polished Concrete Finishing

03 35 43.10 Polished Concrete Floors

	03 35 43.10 Polished Concrete Floors	Crew	Daily Output	Labor-Hours	Unit	Material	Labor	Equipment	Total	Total Incl O&P
0010	**POLISHED CONCRETE FLOORS** R033543-10									
0015	Processing of cured concrete to include grinding, honing,									
0020	and polishing of interior floors with 22" segmented diamond									
0025	planetary floor grinder (2 passes in different directions per grit)									
0100	Removal of pre-existing coatings, dry, with carbide discs using									
0105	dry vacuum pick-up system, final hand sweeping									
0110	Glue, adhesive or tar	J-4	1.60	15	M.S.F.	19.75	640	135	794.75	1,125
0120	Paint, epoxy, 1 coat		3.60	6.667		19.75	284	60	363.75	515
0130	2 coats		1.80	13.333		19.75	565	120	704.75	1,000
0200	Grinding and edging, wet, including wet vac pick-up and auto									
0205	scrubbing between grit changes									
0210	40-grit diamond/metal matrix	J-4A	1.60	20	M.S.F.	31.50	825	281	1,137.50	1,600
0220	80-grit diamond/metal matrix		2	16		31.50	660	225	916.50	1,275
0230	120-grit diamond/metal matrix		2.40	13.333		31.50	550	187	768.50	1,075
0240	200-grit diamond/metal matrix		2.80	11.429		31.50	470	161	662.50	920
0300	Spray on dye or stain (1 coat)	1 Cefi	16	.500		216	22.50		238.50	272
0400	Spray on densifier/hardener (2 coats)	"	8	1		350	45		395	450
0410	Auto scrubbing after 2nd coat, when dry	J-4B	16	.500			18.80	14.60	33.40	45
0500	Honing and edging, wet, including wet vac pick-up and auto									
0505	scrubbing between grit changes									
0510	100-grit diamond/resin matrix	J-4A	2.80	11.429	M.S.F.	31.50	470	161	662.50	920
0520	200-grit diamond/resin matrix	"	2.80	11.429	"	31.50	470	161	662.50	920
0530	Dry, including dry vacuum pick-up system, final hand sweeping									
0540	400-grit diamond/resin matrix	J-4A	2.80	11.429	M.S.F.	31.50	470	161	662.50	920
0600	Polishing and edging, dry, including dry vac pick-up and hand									
0605	sweeping between grit changes									
0610	800-grit diamond/resin matrix	J-4A	2.80	11.429	M.S.F.	31.50	470	161	662.50	920
0620	1500-grit diamond/resin matrix		2.80	11.429		31.50	470	161	662.50	920
0630	3000-grit diamond/resin matrix		2.80	11.429		31.50	470	161	662.50	920
0700	Auto scrubbing after final polishing step	J-4B	16	.500			18.80	14.60	33.40	45

03 37 Specialty Placed Concrete

03 37 13 – Shotcrete

03 37 13.30 Gunite (Dry-Mix)		Crew	Daily Output	Labor-Hours	Unit	Material	Labor	Equipment	Total	Total Incl O&P
0010	**GUNITE (DRY-MIX)**									
0020	Typical in place, 1" layers, no mesh included	C-16	2000	.028	S.F.	.35	1.17	.20	1.72	2.38
0100	Mesh for gunite 2 x 2, #12	2 Rodm	800	.020		.63	1.05		1.68	2.33
0150	#4 reinforcing bars @ 6" each way	"	500	.032		1.61	1.68		3.29	4.41
0300	Typical in place, including mesh, 2" thick, flat surfaces	C-16	1000	.056		1.32	2.34	.39	4.05	5.45
0350	Curved surfaces		500	.112		1.32	4.69	.79	6.80	9.40
0500	4" thick, flat surfaces		750	.075		2.02	3.13	.52	5.67	7.55
0550	Curved surfaces		350	.160		2.02	6.70	1.12	9.84	13.60
0900	Prepare old walls, no scaffolding, good condition	C-10	1000	.024			1.02		1.02	1.53
0950	Poor condition	"	275	.087			3.71		3.71	5.55
1100	For high finish requirement or close tolerance, add						50%			
1150	Very high						110%			

03 37 13.60 Shotcrete (Wet-Mix)		Crew	Daily Output	Labor-Hours	Unit	Material	Labor	Equipment	Total	Total Incl O&P
0010	**SHOTCRETE (WET-MIX)**									
0020	Wet mix, placed @ up to 12 C.Y. per hour, 3000 psi	C-8C	80	.600	C.Y.	113	25	5.70	143.70	168
0100	Up to 35 C.Y. per hour	C-8E	240	.200	"	101	8.20	2.21	111.41	127
1010	Fiber reinforced, 1" thick	C-8C	1740	.028	S.F.	.83	1.14	.26	2.23	2.94
1020	2" thick		900	.053		1.66	2.20	.51	4.37	5.75
1030	3" thick		825	.058		2.48	2.40	.55	5.43	7
1040	4" thick		750	.064		3.31	2.65	.61	6.57	8.35

03 39 Concrete Curing

03 39 13 – Water Concrete Curing

03 39 13.50 Water Curing		Crew	Daily Output	Labor-Hours	Unit	Material	Labor	Equipment	Total	Total Incl O&P
0010	**WATER CURING**									
0015	With burlap, 4 uses assumed, 7.5 oz.	2 Clab	55	.291	C.S.F.	13.05	10.95		24	31
0100	10 oz.	"	55	.291	"	23.50	10.95		34.45	43
0400	Curing blankets, 1" to 2" thick, buy				S.F.	.19			.19	.21

03 39 23 – Membrane Concrete Curing

03 39 23.13 Chemical Compound Membrane Concrete Curing		Crew	Daily Output	Labor-Hours	Unit	Material	Labor	Equipment	Total	Total Incl O&P
0010	**CHEMICAL COMPOUND MEMBRANE CONCRETE CURING**									
0300	Sprayed membrane curing compound	2 Clab	95	.168	C.S.F.	11.05	6.35		17.40	22
0700	Curing compound, solvent based, 400 S.F./gal., 55 gallon lots				Gal.	24.50			24.50	27
0720	5 gallon lots					30.50			30.50	34
0800	Curing compound, water based, 250 S.F./gal., 55 gallon lots					22.50			22.50	25
0820	5 gallon lots					25.50			25.50	28.50

03 39 23.23 Sheet Membrane Concrete Curing		Crew	Daily Output	Labor-Hours	Unit	Material	Labor	Equipment	Total	Total Incl O&P
0010	**SHEET MEMBRANE CONCRETE CURING**									
0200	Curing blanket, burlap/poly, 2-ply	2 Clab	70	.229	C.S.F.	23	8.60		31.60	38.50

Note: Material, Labor, Equipment and Total are 2015 Bare Costs.

03 41 Precast Structural Concrete

03 41 13 – Precast Concrete Hollow Core Planks

03 41 13.50 Precast Slab Planks		Crew	Daily Output	Labor-Hours	Unit	2015 Bare Costs Material	Labor	Equipment	Total	Total Incl O&P
0010	**PRECAST SLAB PLANKS** R034105-30									
0020	Prestressed roof/floor members, grouted, solid, 4″ thick	C-11	2400	.030	S.F.	5.85	1.56	.76	8.17	9.90
0050	6″ thick		2800	.026		6.65	1.34	.65	8.64	10.30
0100	Hollow, 8″ thick		3200	.023		7.10	1.17	.57	8.84	10.40
0150	10″ thick		3600	.020		7.75	1.04	.50	9.29	10.80
0200	12″ thick		4000	.018		7.90	.93	.45	9.28	10.75

03 41 16 – Precast Concrete Slabs

03 41 16.20 Precast Concrete Channel Slabs		Crew	Daily Output	Labor-Hours	Unit	Material	Labor	Equipment	Total	Total Incl O&P
0010	**PRECAST CONCRETE CHANNEL SLABS**									
0335	Lightweight concrete channel slab, long runs, 2-3/4″ thick	C-12	1575	.030	S.F.	8.50	1.42	.42	10.34	12
0375	3-3/4″ thick		1550	.031		8.75	1.44	.42	10.61	12.30
0475	4-3/4″ thick		1525	.031		9.75	1.46	.43	11.64	13.45
1275	Short pieces, 2-3/4″ thick		785	.061		12.75	2.84	.83	16.42	19.35
1375	3-3/4″ thick		770	.062		13.15	2.90	.85	16.90	19.85
1475	4-3/4″ thick		762	.063		14.65	2.93	.86	18.44	21.50

03 41 16.50 Precast Lightweight Concrete Plank		Crew	Daily Output	Labor-Hours	Unit	Material	Labor	Equipment	Total	Total Incl O&P
0010	**PRECAST LIGHTWEIGHT CONCRETE PLANK**									
0015	Lightweight plank, nailable, T&G, 2″ thick	C-12	1800	.027	S.F.	7.65	1.24	.36	9.25	10.70
0150	For premium ceiling finish, add				″	10%				
0200	For sloping roofs, slope over 4 in 12, add						25%			
0250	Slope over 6 in 12, add						150%			

03 41 23 – Precast Concrete Stairs

03 41 23.50 Precast Stairs		Crew	Daily Output	Labor-Hours	Unit	Material	Labor	Equipment	Total	Total Incl O&P
0010	**PRECAST STAIRS**									
0020	Precast concrete treads on steel stringers, 3′ wide	C-12	75	.640	Riser	135	30	8.70	173.70	203
0300	Front entrance, 5′ wide with 48″ platform, 2 risers		16	3	Flight	485	140	41	666	790
0350	5 risers		12	4		765	186	54.50	1,005.50	1,175
0500	6′ wide, 2 risers		15	3.200		535	149	43.50	727.50	865
0550	5 risers		11	4.364		845	203	59.50	1,107.50	1,300
0700	7′ wide, 2 risers		14	3.429		685	160	46.50	891.50	1,050
0750	5 risers		10	4.800		1,125	223	65.50	1,413.50	1,675
1200	Basement entrance stairwell, 6 steps, incl. steel bulkhead door	B-51	22	2.182		1,450	83.50	11.15	1,544.65	1,725
1250	14 steps	″	11	4.364		2,400	167	22.50	2,589.50	2,900

03 41 33 – Precast Structural Pretensioned Concrete

03 41 33.10 Precast Beams		Crew	Daily Output	Labor-Hours	Unit	Material	Labor	Equipment	Total	Total Incl O&P
0010	**PRECAST BEAMS** R034105-30									
0011	L-shaped, 20′ span, 12″ x 20″	C-11	32	2.250	Ea.	3,000	117	56.50	3,173.50	3,550
0060	18″ x 36″		24	3		4,125	156	75.50	4,356.50	4,875
0100	24″ x 44″		22	3.273		4,925	170	82.50	5,177.50	5,800
0150	30′ span, 12″ x 36″		24	3		5,550	156	75.50	5,781.50	6,475
0200	18″ x 44″		20	3.600		6,775	187	90.50	7,052.50	7,875
0250	24″ x 52″		16	4.500		8,125	234	113	8,472	9,450
0400	40′ span, 12″ x 52″		20	3.600		8,625	187	90.50	8,902.50	9,925
0450	18″ x 52″		16	4.500		9,625	234	113	9,972	11,100
0500	24″ x 52″		12	6		10,800	310	151	11,261	12,600
1200	Rectangular, 20′ span, 12″ x 20″		32	2.250		2,925	117	56.50	3,098.50	3,450
1250	18″ x 36″		24	3		3,600	156	75.50	3,831.50	4,325
1300	24″ x 44″		22	3.273		4,325	170	82.50	4,577.50	5,125
1400	30′ span, 12″ x 36″		24	3		4,875	156	75.50	5,106.50	5,700
1450	18″ x 44″		20	3.600		5,850	187	90.50	6,127.50	6,875
1500	24″ x 52″		16	4.500		7,075	234	113	7,422	8,300

03 41 33 – Precast Structural Pretensioned Concrete

03 41 33.10 Precast Beams		Crew	Daily Output	Labor-Hours	Unit	Material	Labor	Equipment	Total	Total Incl O&P
						2015 Bare Costs				
1600	40′ span, 12″ x 52″	C-11	20	3.600	Ea.	7,225	187	90.50	7,502.50	8,375
1650	18″ x 52″		16	4.500		8,225	234	113	8,572	9,575
1700	24″ x 52″		12	6		9,425	310	151	9,886	11,100
2000	"T" shaped, 20′ span, 12″ x 20″		32	2.250		3,425	117	56.50	3,598.50	4,025
2050	18″ x 36″		24	3		4,600	156	75.50	4,831.50	5,425
2100	24″ x 44″		22	3.273		5,525	170	82.50	5,777.50	6,450
2200	30′ span, 12″ x 36″		24	3		6,325	156	75.50	6,556.50	7,300
2250	18″ x 44″		20	3.600		7,675	187	90.50	7,952.50	8,850
2300	24″ x 52″		16	4.500		9,175	234	113	9,522	10,600
2500	40′ span, 12″ x 52″		20	3.600		10,400	187	90.50	10,677.50	11,900
2550	18″ x 52″		16	4.500		11,000	234	113	11,347	12,700
2600	24″ x 52″	↓	12	6	↓	12,200	310	151	12,661	14,200

03 41 33.15 Precast Columns		Crew	Daily Output	Labor-Hours	Unit	Material	Labor	Equipment	Total	Total Incl O&P
0010	**PRECAST COLUMNS** R034105-30									
0020	Rectangular to 12′ high, 16″ x 16″	C-11	120	.600	L.F.	172	31	15.10	218.10	258
0050	24″ x 24″		96	.750		234	39	18.90	291.90	345
0300	24′ high, 28″ x 28″		192	.375		271	19.50	9.45	299.95	340
0350	36″ x 36″	↓	144	.500	↓	365	26	12.60	403.60	460

03 41 33.25 Precast Joists		Crew	Daily Output	Labor-Hours	Unit	Material	Labor	Equipment	Total	Total Incl O&P
0010	**PRECAST JOISTS** R034105-30									
0015	40 psf L.L., 6″ deep for 12′ spans	C-12	600	.080	L.F.	24	3.72	1.09	28.81	33.50
0050	8″ deep for 16′ spans		575	.083		40	3.88	1.14	45.02	51
0100	10″ deep for 20′ spans		550	.087		70	4.06	1.19	75.25	84.50
0150	12″ deep for 24′ spans	↓	525	.091	↓	96	4.25	1.25	101.50	114

03 41 33.60 Precast Tees		Crew	Daily Output	Labor-Hours	Unit	Material	Labor	Equipment	Total	Total Incl O&P
0010	**PRECAST TEES** R034105-30									
0020	Quad tee, short spans, roof	C-11	7200	.010	S.F.	7.65	.52	.25	8.42	9.55
0050	Floor		7200	.010		7.65	.52	.25	8.42	9.55
0200	Double tee, floor members, 60′ span		8400	.009		9.10	.45	.22	9.77	11
0250	80′ span		8000	.009		11.80	.47	.23	12.50	14.05
0300	Roof members, 30′ span		4800	.015		7.65	.78	.38	8.81	10.15
0350	50′ span		6400	.011		8.40	.58	.28	9.26	10.55
0400	Wall members, up to 55′ high		3600	.020		11.35	1.04	.50	12.89	14.80
0500	Single tee roof members, 40′ span		3200	.023		11.30	1.17	.57	13.04	15
0550	80′ span		5120	.014		11.70	.73	.35	12.78	14.50
0600	100′ span		6000	.012		17.15	.62	.30	18.07	20
0650	120′ span	↓	6000	.012	↓	18.80	.62	.30	19.72	22
1000	Double tees, floor members									
1100	Lightweight, 20″ x 8′ wide, 45′ span	C-11	20	3.600	Ea.	3,025	187	90.50	3,302.50	3,750
1150	24″ x 8′ wide, 50′ span		18	4		3,350	208	101	3,659	4,150
1200	32″ x 10′ wide, 60′ span		16	4.500		5,050	234	113	5,397	6,075
1250	Standard weight, 12″ x 8′ wide, 20′ span		22	3.273		1,225	170	82.50	1,477.50	1,725
1300	16″ x 8′ wide, 25′ span		20	3.600		1,525	187	90.50	1,802.50	2,100
1350	18″ x 8′ wide, 30′ span		20	3.600		1,825	187	90.50	2,102.50	2,450
1400	20″ x 8′ wide, 45′ span		18	4		2,750	208	101	3,059	3,475
1450	24″ x 8′ wide, 50′ span		16	4.500		3,050	234	113	3,397	3,875
1500	32″ x 10′ wide, 60′ span	↓	14	5.143	↓	4,575	267	130	4,972	5,650
2000	Roof members									
2050	Lightweight, 20″ x 8′ wide, 40′ span	C-11	20	3.600	Ea.	2,700	187	90.50	2,977.50	3,375
2100	24″ x 8′ wide, 50′ span		18	4		3,350	208	101	3,659	4,150
2150	32″ x 10′ wide, 60′ span		16	4.500		5,050	234	113	5,397	6,075
2200	Standard weight, 12″ x 8′ wide, 30′ span	↓	22	3.273	↓	1,825	170	82.50	2,077.50	2,400

03 41 Precast Structural Concrete

03 41 33 – Precast Structural Pretensioned Concrete

03 41 33.60 Precast Tees		Crew	Daily Output	Labor-Hours	Unit	Material	2015 Bare Costs Labor	Equipment	Total	Total Incl O&P
2250	16" x 8' wide, 30' span	C-11	20	3.600	Ea.	1,925	187	90.50	2,202.50	2,550
2300	18" x 8' wide, 30' span		20	3.600		2,025	187	90.50	2,302.50	2,650
2350	20" x 8' wide, 40' span		18	4		2,450	208	101	2,759	3,150
2400	24" x 8' wide, 50' span		16	4.500		3,050	234	113	3,397	3,875
2450	32" x 10' wide, 60' span	↓	14	5.143	↓	4,575	267	130	4,972	5,650

03 45 Precast Architectural Concrete

03 45 13 – Faced Architectural Precast Concrete

03 45 13.50 Precast Wall Panels

		Crew	Daily Output	Labor-Hours	Unit	Material	Labor	Equipment	Total	Total Incl O&P
0010	**PRECAST WALL PANELS** R034513-10									
0050	Uninsulated, smooth gray									
0150	Low rise, 4' x 8' x 4" thick	C-11	320	.225	S.F.	20.50	11.70	5.65	37.85	49
0210	8' x 8', 4" thick		576	.125		20.50	6.50	3.15	30.15	37
0250	8' x 16' x 4" thick		1024	.070		20.50	3.65	1.77	25.92	30.50
0600	High rise, 4' x 8' x 4" thick		288	.250		20.50	13	6.30	39.80	52
0650	8' x 8' x 4" thick		512	.141		20.50	7.30	3.54	31.34	39
0700	8' x 16' x 4" thick		768	.094		20.50	4.87	2.36	27.73	33.50
0750	10' x 20', 6" thick	↓	1400	.051		34.50	2.67	1.30	38.47	44
0800	Insulated panel, 2" polystyrene, add					1.04			1.04	1.14
0850	2" urethane, add					.80			.80	.88
1200	Finishes, white, add					2.93			2.93	3.22
1250	Exposed aggregate, add					2.17			2.17	2.39
1300	Granite faced, domestic, add					29.50			29.50	32.50
1350	Brick faced, modular, red, add				↓	9			9	9.90
2200	Fiberglass reinforced cement with urethane core									
2210	R20, 8' x 8', 5" plain finish	E-2	750	.075	S.F.	21	3.86	2.01	26.87	32
2220	Exposed aggregate or brick finish	"	600	.093	"	32	4.83	2.52	39.35	46

03 47 Site-Cast Concrete

03 47 13 – Tilt-Up Concrete

03 47 13.50 Tilt-Up Wall Panels

		Crew	Daily Output	Labor-Hours	Unit	Material	Labor	Equipment	Total	Total Incl O&P
0010	**TILT-UP WALL PANELS** R034713-20									
0015	Wall panel construction, walls only, 5-1/2" thick	C-14	1600	.090	S.F.	5.55	4.16	1.02	10.73	13.60
0100	7-1/2" thick		1550	.093		6.90	4.29	1.05	12.24	15.30
0500	Walls and columns, 5-1/2" thick walls, 12" x 12" columns		1565	.092		8.30	4.25	1.04	13.59	16.80
0550	7-1/2" thick wall, 12" x 12" columns		1370	.105	↓	10.10	4.85	1.19	16.14	19.85
0800	Columns only, site precast, 12" x 12"		200	.720	L.F.	20	33.50	8.10	61.60	82
0850	16" x 16"	↓	105	1.371	"	29	63.50	15.45	107.95	146

03 48 Precast Concrete Specialties

03 48 43 – Precast Concrete Trim

03 48 43.40 Precast Lintels

	03 48 43.40 Precast Lintels	Crew	Daily Output	Labor-Hours	Unit	2015 Bare Costs Material	Labor	Equipment	Total	Total Incl O&P
0010	**PRECAST LINTELS**, smooth gray, prestressed, stock units only									
0800	4″ wide, 8″ high, x 4′ long	D-10	28	1.143	Ea.	25	52.50	16.90	94.40	126
0850	8′ long		24	1.333		60.50	61.50	19.70	141.70	182
1000	6″ wide, 8″ high, x 4′ long		26	1.231		35.50	56.50	18.20	110.20	146
1050	10′ long		22	1.455		91	67	21.50	179.50	226
1200	8″ wide, 8″ high, x 4′ long		24	1.333		45	61.50	19.70	126.20	165
1250	12′ long		20	1.600		146	73.50	23.50	243	299
1275	For custom sizes, types, colors, or finishes of precast lintels, add					150%				

03 48 43.90 Precast Window Sills

	03 48 43.90 Precast Window Sills	Crew	Daily Output	Labor-Hours	Unit	Material	Labor	Equipment	Total	Total Incl O&P
0010	**PRECAST WINDOW SILLS**									
0600	Precast concrete, 4″ tapers to 3″, 9″ wide	D-1	70	.229	L.F.	11.90	9.60		21.50	28
0650	11″ wide		60	.267		15.75	11.25		27	34.50
0700	13″ wide, 3 1/2″ tapers to 2 1/2″, 12″ wall		50	.320		15.50	13.45		28.95	37.50

03 51 Cast Roof Decks

03 51 13 – Cementitious Wood Fiber Decks

03 51 13.50 Cementitious/Wood Fiber Planks

	03 51 13.50 Cementitious/Wood Fiber Planks	Crew	Daily Output	Labor-Hours	Unit	Material	Labor	Equipment	Total	Total Incl O&P
0010	**CEMENTITIOUS/WOOD FIBER PLANKS** R051223-50									
0050	Plank, beveled edge, 1″ thick	2 Carp	1000	.016	S.F.	2.50	.75		3.25	3.91
0100	1-1/2″ thick		975	.016		3.25	.77		4.02	4.77
0150	T & G, 2″ thick		950	.017		2.75	.79		3.54	4.25
0200	2-1/2″ thick		925	.017		3	.81		3.81	4.55
0250	3″ thick		900	.018		3.40	.83		4.23	5
1000	Bulb tee, sub-purlin and grout, 6′ span, add	E-1	5000	.005		2.16	.25	.03	2.44	2.82
1100	8′ span	"	4200	.006		2.16	.30	.03	2.49	2.91

03 51 16 – Gypsum Concrete Roof Decks

03 51 16.50 Gypsum Roof Deck

	03 51 16.50 Gypsum Roof Deck	Crew	Daily Output	Labor-Hours	Unit	Material	Labor	Equipment	Total	Total Incl O&P
0010	**GYPSUM ROOF DECK**									
1000	Poured gypsum, 2″ thick	C-8	6000	.009	S.F.	1.45	.39	.12	1.96	2.31
1100	3″ thick	"	4800	.012	"	2.17	.49	.15	2.81	3.30

03 52 Lightweight Concrete Roof Insulation

03 52 16 – Lightweight Insulating Concrete

03 52 16.13 Lightweight Cellular Insulating Concrete

	03 52 16.13 Lightweight Cellular Insulating Concrete	Crew	Daily Output	Labor-Hours	Unit	Material	Labor	Equipment	Total	Total Incl O&P
0010	**LIGHTWEIGHT CELLULAR INSULATING CONCRETE** R035216-10									
0020	Portland cement and foaming agent G	C-8	50	1.120	C.Y.	123	47	14.40	184.40	222

03 52 16.16 Lightweight Aggregate Insulating Concrete

	03 52 16.16 Lightweight Aggregate Insulating Concrete	Crew	Daily Output	Labor-Hours	Unit	Material	Labor	Equipment	Total	Total Incl O&P
0010	**LIGHTWEIGHT AGGREGATE INSULATING CONCRETE** R035216-10									
0100	Poured vermiculite or perlite, field mix,									
0110	1:6 field mix G	C-8	50	1.120	C.Y.	252	47	14.40	313.40	365
0200	Ready mix, 1:6 mix, roof fill, 2″ thick G		10000	.006	S.F.	1.40	.23	.07	1.70	1.98
0250	3″ thick G		7700	.007		2.10	.30	.09	2.49	2.87
0400	Expanded volcanic glass rock, 1″ thick G	2 Carp	1500	.011		.52	.50		1.02	1.34
0450	3″ thick G	"	1200	.013		1.56	.63		2.19	2.68

03 53 Concrete Topping

03 53 16 – Iron-Aggregate Concrete Topping

03 53 16.50 Floor Topping		Crew	Daily Output	Labor-Hours	Unit	Material	2015 Bare Costs Labor	Equipment	Total	Total Incl O&P
0010	**FLOOR TOPPING**									
0400	Integral topping/finish, on fresh concrete, using 1:1:2 mix, 3/16" thick	C-10B	1000	.040	S.F.	.10	1.62	.27	1.99	2.86
0450	1/2" thick		950	.042		.28	1.71	.29	2.28	3.21
0500	3/4" thick		850	.047		.42	1.91	.32	2.65	3.70
0600	1" thick		750	.053		.56	2.16	.36	3.08	4.28
0800	Granolithic topping, on fresh or cured concrete, 1:1:1-1/2 mix, 1/2" thick		590	.068		.31	2.75	.46	3.52	5
0820	3/4" thick		580	.069		.46	2.80	.47	3.73	5.25
0850	1" thick		575	.070		.62	2.82	.47	3.91	5.45
0950	2" thick		500	.080		1.23	3.24	.55	5.02	6.85
1200	Heavy duty, 1:1:2, 3/4" thick, preshrunk, gray, 20 M.S.F.		320	.125		.80	5.05	.85	6.70	9.45
1300	100 M.S.F.	↓	380	.105	↓	.42	4.27	.72	5.41	7.70

03 54 Cast Underlayment

03 54 13 – Gypsum Cement Underlayment

03 54 13.50 Poured Gypsum Underlayment		Crew	Daily Output	Labor-Hours	Unit	Material	Labor	Equipment	Total	Total Incl O&P
0010	**POURED GYPSUM UNDERLAYMENT**									
0400	Underlayment, gypsum based, self-leveling 2500 psi, pumped, 1/2" thick	C-8	24000	.002	S.F.	.36	.10	.03	.49	.58
0500	3/4" thick		20000	.003		.54	.12	.04	.70	.82
0600	1" thick	↓	16000	.004		.72	.15	.05	.92	1.07
1400	Hand placed, 1/2" thick	C-18	450	.020		.36	.76	.13	1.25	1.71
1500	3/4" thick	"	300	.030	↓	.54	1.13	.20	1.87	2.57

03 54 16 – Hydraulic Cement Underlayment

03 54 16.50 Cement Underlayment		Crew	Daily Output	Labor-Hours	Unit	Material	Labor	Equipment	Total	Total Incl O&P
0010	**CEMENT UNDERLAYMENT**									
2510	Underlayment, P.C based, self-leveling, 4100 psi, pumped, 1/4" thick	C-8	20000	.003	S.F.	1.55	.12	.04	1.71	1.92
2520	1/2" thick		19000	.003		3.09	.12	.04	3.25	3.63
2530	3/4" thick		18000	.003		4.64	.13	.04	4.81	5.35
2540	1" thick		17000	.003		6.20	.14	.04	6.38	7.05
2550	1-1/2" thick	↓	15000	.004		9.30	.16	.05	9.51	10.50
2560	Hand placed, 1/2" thick	C-18	450	.020		3.09	.76	.13	3.98	4.71
2610	Topping, P.C based, self-leveling, 6100 psi, pumped, 1/4" thick	C-8	20000	.003		2.26	.12	.04	2.42	2.71
2620	1/2" thick		19000	.003		4.52	.12	.04	4.68	5.20
2630	3/4" thick		18000	.003		6.80	.13	.04	6.97	7.70
2660	1" thick		17000	.003		9.05	.14	.04	9.23	10.20
2670	1-1/2" thick	↓	15000	.004		13.55	.16	.05	13.76	15.20
2680	Hand placed, 1/2" thick	C-18	450	.020	↓	4.52	.76	.13	5.41	6.30

03 62 Non-Shrink Grouting

03 62 13 – Non-Metallic Non-Shrink Grouting

03 62 13.50 Grout, Non-Metallic Non-shrink		Crew	Daily Output	Labor-Hours	Unit	Material	Labor	Equipment	Total	Total Incl O&P
0010	**GROUT, NON-METALLIC NON-SHRINK**									
0300	Non-shrink, non-metallic, 1" deep	1 Cefi	35	.229	S.F.	6.50	10.30		16.80	22.50
0350	2" deep	"	25	.320	"	13	14.40		27.40	36

03 62 16 – Metallic Non-Shrink Grouting

03 62 16.50 Grout, Metallic Non-Shrink		Crew	Daily Output	Labor-Hours	Unit	Material	Labor	Equipment	Total	Total Incl O&P
0010	**GROUT, METALLIC NON-SHRINK**									
0020	Column & machine bases, non-shrink, metallic, 1" deep	1 Cefi	35	.229	S.F.	10.25	10.30		20.55	26.50
0050	2" deep	"	25	.320	"	20.50	14.40		34.90	44

03 63 Epoxy Grouting

03 63 05 – Grouting of Dowels and Fasteners

03 63 05.10 Epoxy Only		Crew	Daily Output	Labor-Hours	Unit	Material	2015 Bare Costs Labor	Equipment	Total	Total Incl O&P
0010	**EPOXY ONLY**									
1500	Chemical anchoring, epoxy cartridge, excludes layout, drilling, fastener									
1530	For fastener 3/4" diam. x 6" embedment	2 Skwk	72	.222	Ea.	5.65	10.80		16.45	23
1535	1" diam. x 8" embedment		66	.242		8.45	11.80		20.25	27.50
1540	1-1/4" diam. x 10" embedment		60	.267		16.90	12.95		29.85	38.50
1545	1-3/4" diam. x 12" embedment		54	.296		28	14.40		42.40	53.50
1550	14" embedment		48	.333		34	16.20		50.20	62
1555	2" diam. x 12" embedment		42	.381		45	18.55		63.55	78
1560	18" embedment		32	.500		56.50	24.50		81	99.50

03 81 Concrete Cutting

03 81 13 – Flat Concrete Sawing

03 81 13.50 Concrete Floor/Slab Cutting		Crew	Daily Output	Labor-Hours	Unit	Material	Labor	Equipment	Total	Total Incl O&P
0010	**CONCRETE FLOOR/SLAB CUTTING**									
0050	Includes blade cost, layout and set-up time									
0300	Saw cut concrete slabs, plain, up to 3" deep	B-89	1060	.015	L.F.	.14	.66	.46	1.26	1.66
0320	Each additional inch of depth		3180	.005		.05	.22	.15	.42	.55
0400	Mesh reinforced, up to 3" deep		980	.016		.16	.72	.50	1.38	1.80
0420	Each additional inch of depth		2940	.005		.05	.24	.17	.46	.60
0500	Rod reinforced, up to 3" deep		800	.020		.19	.88	.61	1.68	2.21
0520	Each additional inch of depth		2400	.007		.06	.29	.20	.55	.73

03 81 13.75 Concrete Saw Blades		Crew	Daily Output	Labor-Hours	Unit	Material	Labor	Equipment	Total	Total Incl O&P
0010	**CONCRETE SAW BLADES**									
3000	Blades for saw cutting, included in cutting line items									
3020	Diamond, 12" diameter				Ea.	241			241	265
3040	18" diameter					465			465	510
3080	24" diameter					770			770	845
3120	30" diameter					1,100			1,100	1,200
3160	36" diameter					1,425			1,425	1,550
3200	42" diameter					2,625			2,625	2,875

03 81 16 – Track Mounted Concrete Wall Sawing

03 81 16.50 Concrete Wall Cutting		Crew	Daily Output	Labor-Hours	Unit	Material	Labor	Equipment	Total	Total Incl O&P
0010	**CONCRETE WALL CUTTING**									
0750	Includes blade cost, layout and set-up time									
0800	Concrete walls, hydraulic saw, plain, per inch of depth	B-89B	250	.064	L.F.	.05	2.81	3.43	6.29	8.10
0820	Rod reinforcing, per inch of depth	"	150	.107	"	.06	4.68	5.70	10.44	13.45

03 82 Concrete Boring

03 82 13 – Concrete Core Drilling

03 82 13.10 Core Drilling		Crew	Daily Output	Labor-Hours	Unit	Material	Labor	Equipment	Total	Total Incl O&P
0010	**CORE DRILLING**									
0015	Includes bit cost, layout and set-up time									
0020	Reinforced concrete slab, up to 6" thick									
0100	1" diameter core	B-89A	17	.941	Ea.	.18	40.50	6.80	47.48	70
0150	For each additional inch of slab thickness in same hole, add		1440	.011		.03	.48	.08	.59	.86
0200	2" diameter core		16.50	.970		.28	42	7	49.28	72.50
0250	For each additional inch of slab thickness in same hole, add		1080	.015		.05	.64	.11	.80	1.15
0300	3" diameter core		16	1		.38	43	7.25	50.63	75

03 82 13.10 Core Drilling		Crew	Daily Output	Labor-Hours	Unit	Material	2015 Bare Costs Labor	Equipment	Total	Total Incl O&P
0350	For each additional inch of slab thickness in same hole, add	B-89A	720	.022	Ea.	.06	.96	.16	1.18	1.73
0500	4" diameter core		15	1.067		.49	46	7.70	54.19	80
0550	For each additional inch of slab thickness in same hole, add		480	.033		.08	1.44	.24	1.76	2.58
0700	6" diameter core		14	1.143		.76	49.50	8.25	58.51	86
0750	For each additional inch of slab thickness in same hole, add		360	.044		.13	1.92	.32	2.37	3.45
0900	8" diameter core		13	1.231		1.05	53	8.90	62.95	93
0950	For each additional inch of slab thickness in same hole, add		288	.056		.17	2.40	.40	2.97	4.32
1100	10" diameter core		12	1.333		1.48	57.50	9.65	68.63	101
1150	For each additional inch of slab thickness in same hole, add		240	.067		.25	2.88	.48	3.61	5.25
1300	12" diameter core		11	1.455		1.78	62.50	10.50	74.78	110
1350	For each additional inch of slab thickness in same hole, add		206	.078		.30	3.35	.56	4.21	6.10
1500	14" diameter core		10	1.600		2.08	69	11.55	82.63	121
1550	For each additional inch of slab thickness in same hole, add		180	.089		.35	3.83	.64	4.82	7
1700	18" diameter core		9	1.778		2.82	76.50	12.85	92.17	135
1750	For each additional inch of slab thickness in same hole, add		144	.111		.47	4.79	.80	6.06	8.80
1754	24" diameter core		8	2		4.02	86.50	14.45	104.97	153
1756	For each additional inch of slab thickness in same hole, add		120	.133		.67	5.75	.96	7.38	10.65
1760	For horizontal holes, add to above						20%	20%		
1770	Prestressed hollow core plank, 8" thick									
1780	1" diameter core	B-89A	17.50	.914	Ea.	.24	39.50	6.60	46.34	68.50
1790	For each additional inch of plank thickness in same hole, add		3840	.004		.03	.18	.03	.24	.34
1794	2" diameter core		17.25	.928		.38	40	6.70	47.08	69.50
1796	For each additional inch of plank thickness in same hole, add		2880	.006		.05	.24	.04	.33	.46
1800	3" diameter core		17	.941		.51	40.50	6.80	47.81	70.50
1810	For each additional inch of plank thickness in same hole, add		1920	.008		.06	.36	.06	.48	.69
1820	4" diameter core		16.50	.970		.66	42	7	49.66	73
1830	For each additional inch of plank thickness in same hole, add		1280	.013		.08	.54	.09	.71	1.02
1840	6" diameter core		15.50	1.032		1.01	44.50	7.45	52.96	78
1850	For each additional inch of plank thickness in same hole, add		960	.017		.13	.72	.12	.97	1.38
1860	8" diameter core		15	1.067		1.40	46	7.70	55.10	81
1870	For each additional inch of plank thickness in same hole, add		768	.021		.17	.90	.15	1.22	1.75
1880	10" diameter core		14	1.143		1.98	49.50	8.25	59.73	87.50
1890	For each additional inch of plank thickness in same hole, add		640	.025		.25	1.08	.18	1.51	2.13
1900	12" diameter core		13.50	1.185		2.37	51	8.55	61.92	91
1910	For each additional inch of plank thickness in same hole, add		548	.029		.30	1.26	.21	1.77	2.50
3000	Bits for core drilling, included in drilling line items									
3010	Diamond, premium, 1" diameter				Ea.	70.50			70.50	78
3020	2" diameter					113			113	124
3030	3" diameter					152			152	168
3040	4" diameter					197			197	217
3060	6" diameter					305			305	335
3080	8" diameter					420			420	460
3110	10" diameter					595			595	650
3120	12" diameter					710			710	780
3140	14" diameter					835			835	915
3180	18" diameter					1,125			1,125	1,250
3240	24" diameter					1,600			1,600	1,775

03 82 16 – Concrete Drilling

03 82 16.10 Concrete Impact Drilling		Crew	Daily Output	Labor-Hours	Unit	Material	2015 Bare Costs Labor	Equipment	Total	Total Incl O&P
0010	**CONCRETE IMPACT DRILLING**									
0020	Includes bit cost, layout and set-up time, no anchors									
0050	Up to 4" deep in concrete/brick floors/walls									
0100	Holes, 1/4" diameter	1 Carp	75	.107	Ea.	.07	5		5.07	7.75
0150	For each additional inch of depth in same hole, add		430	.019		.02	.87		.89	1.36
0200	3/8" diameter		63	.127		.06	5.95		6.01	9.20
0250	For each additional inch of depth in same hole, add		340	.024		.01	1.10		1.11	1.72
0300	1/2" diameter		50	.160		.06	7.50		7.56	11.60
0350	For each additional inch of depth in same hole, add		250	.032		.01	1.50		1.51	2.33
0400	5/8" diameter		48	.167		.09	7.85		7.94	12.15
0450	For each additional inch of depth in same hole, add		240	.033		.02	1.56		1.58	2.43
0500	3/4" diameter		45	.178		.12	8.35		8.47	13
0550	For each additional inch of depth in same hole, add		220	.036		.03	1.71		1.74	2.66
0600	7/8" diameter		43	.186		.16	8.75		8.91	13.60
0650	For each additional inch of depth in same hole, add		210	.038		.04	1.79		1.83	2.79
0700	1" diameter		40	.200		.17	9.40		9.57	14.65
0750	For each additional inch of depth in same hole, add		190	.042		.04	1.98		2.02	3.09
0800	1-1/4" diameter		38	.211		.26	9.90		10.16	15.50
0850	For each additional inch of depth in same hole, add		180	.044		.06	2.09		2.15	3.28
0900	1-1/2" diameter		35	.229		.39	10.75		11.14	16.95
0950	For each additional inch of depth in same hole, add	▼	165	.048	▼	.10	2.28		2.38	3.61
1000	For ceiling installations, add						40%			

Division Notes

		CREW	DAILY OUTPUT	LABOR-HOURS	UNIT	BARE COSTS MAT.	LABOR	EQUIP.	TOTAL	TOTAL INCL O&P

Estimating Tips

04 05 00 Common Work Results for Masonry

- The terms mortar and grout are often used interchangeably, and incorrectly. Mortar is used to bed masonry units, seal the entry of air and moisture, provide architectural appearance, and allow for size variations in the units. Grout is used primarily in reinforced masonry construction and is used to bond the masonry to the reinforcing steel. Common mortar types are M(2500 psi), S(1800 psi), N(750 psi), and O(350 psi), and conform to ASTM C270. Grout is either fine or coarse and conforms to ASTM C476, and in-place strengths generally exceed 2500 psi. Mortar and grout are different components of masonry construction and are placed by entirely different methods. An estimator should be aware of their unique uses and costs.
- Mortar is included in all assembled masonry line items. The mortar cost, part of the assembled masonry material cost, includes all ingredients, all labor, and all equipment required. Please see reference number R040513-10.
- Waste, specifically the loss/droppings of mortar and the breakage of brick and block, is included in all masonry assemblies in this division. A factor of 25% is added for mortar and 3% for brick and concrete masonry units.
- Scaffolding or staging is not included in any of the Division 4 costs. Refer to Subdivision 01 54 23 for scaffolding and staging costs.

04 20 00 Unit Masonry

- The most common types of unit masonry are brick and concrete masonry. The major classifications of brick are building brick (ASTM C62), facing brick (ASTM C216), glazed brick, fire brick, and pavers. Many varieties of texture and appearance can exist within these classifications, and the estimator would be wise to check local custom and availability within the project area. For repair and remodeling jobs, matching the existing brick may be the most important criteria.
- Brick and concrete block are priced by the piece and then converted into a price per square foot of wall. Openings less than two square feet are generally ignored by the estimator because any savings in units used is offset by the cutting and trimming required.
- It is often difficult and expensive to find and purchase small lots of historic brick. Costs can vary widely. Many design issues affect costs, selection of mortar mix, and repairs or replacement of masonry materials. Cleaning techniques must be reflected in the estimate.
- All masonry walls, whether interior or exterior, require bracing. The cost of bracing walls during construction should be included by the estimator, and this bracing must remain in place until permanent bracing is complete. Permanent bracing of masonry walls is accomplished by masonry itself, in the form of pilasters or abutting wall corners, or by anchoring the walls to the structural frame. Accessories in the form of anchors, anchor slots, and ties are used, but their supply and installation can be by different trades. For instance, anchor slots on spandrel beams and columns are supplied and welded in place by the steel fabricator, but the ties from the slots into the masonry are installed by the bricklayer. Regardless of the installation method, the estimator must be certain that these accessories are accounted for in pricing.

Reference Numbers

Reference numbers are shown in shaded boxes at the beginning of some major classifications. These numbers refer to related items in the Reference Section. The reference information may be an estimating procedure, an alternate pricing method, or technical information.

Note: Not all subdivisions listed here necessarily appear in this publication. ■

04 01 Maintenance of Masonry

04 01 20 – Maintenance of Unit Masonry

04 01 20.20 Pointing Masonry

	04 01 20.20 Pointing Masonry	Crew	Daily Output	Labor-Hours	Unit	Material	Labor (2015 Bare Costs)	Equipment (2015 Bare Costs)	Total (2015 Bare Costs)	Total Incl O&P
0010	**POINTING MASONRY**									
0300	Cut and repoint brick, hard mortar, running bond	1 Bric	80	.100	S.F.	.56	4.62		5.18	7.65
0320	Common bond		77	.104		.56	4.80		5.36	7.90
0360	Flemish bond		70	.114		.59	5.25		5.84	8.70
0400	English bond		65	.123		.59	5.70		6.29	9.30
0600	Soft old mortar, running bond		100	.080		.56	3.69		4.25	6.25
0620	Common bond		96	.083		.56	3.85		4.41	6.45
0640	Flemish bond		90	.089		.59	4.10		4.69	6.90
0680	English bond		82	.098		.59	4.50		5.09	7.50
0700	Stonework, hard mortar		140	.057	L.F.	.74	2.64		3.38	4.85
0720	Soft old mortar		160	.050	"	.74	2.31		3.05	4.34
1000	Repoint, mask and grout method, running bond		95	.084	S.F.	.74	3.89		4.63	6.75
1020	Common bond		90	.089		.74	4.10		4.84	7.05
1040	Flemish bond		86	.093		.78	4.29		5.07	7.40
1060	English bond		77	.104		.78	4.80		5.58	8.15
2000	Scrub coat, sand grout on walls, thin mix, brushed		120	.067		2.96	3.08		6.04	7.95
2020	Troweled		98	.082		4.12	3.77		7.89	10.30

04 01 20.30 Pointing CMU

	04 01 20.30 Pointing CMU	Crew	Daily Output	Labor-Hours	Unit	Material	Labor	Equipment	Total	Total Incl O&P
0010	**POINTING CMU**									
0300	Cut and repoint block, hard mortar, running bond	1 Bric	190	.042	S.F.	.23	1.94		2.17	3.22
0310	Stacked bond		200	.040		.23	1.85		2.08	3.07
0600	Soft old mortar, running bond		230	.035		.23	1.61		1.84	2.70
0610	Stacked bond		245	.033		.23	1.51		1.74	2.55

04 01 20.40 Sawing Masonry

	04 01 20.40 Sawing Masonry	Crew	Daily Output	Labor-Hours	Unit	Material	Labor	Equipment	Total	Total Incl O&P
0010	**SAWING MASONRY**									
0050	Brick or block by hand, per inch depth	A-1	125	.064	L.F.	.05	2.41	.65	3.11	4.46

04 01 20.41 Unit Masonry Stabilization

	04 01 20.41 Unit Masonry Stabilization	Crew	Daily Output	Labor-Hours	Unit	Material	Labor	Equipment	Total	Total Incl O&P
0010	**UNIT MASONRY STABILIZATION**									
0100	Structural repointing method									
0110	Cut / grind mortar joint	1 Bric	240	.033	L.F.		1.54		1.54	2.35
0120	Clean and mask joint		2500	.003		.11	.15		.26	.35
0130	Epoxy paste and 1/4" FRP rod		240	.033		1.67	1.54		3.21	4.19
0132	3/8" FRP rod		160	.050		2.48	2.31		4.79	6.25
0134	1/4" CFRP rod		240	.033		14.50	1.54		16.04	18.30
0136	3/8" CFRP rod		160	.050		18.35	2.31		20.66	23.50
0140	Remove masking		14400	.001			.03		.03	.04
0300	Structural fabric method									
0310	Primer	1 Bric	600	.013	S.F.	.90	.62		1.52	1.93
0320	Apply filling/leveling paste		720	.011		.72	.51		1.23	1.57
0330	Epoxy, glass fiber fabric		720	.011		8.10	.51		8.61	9.70
0340	Carbon fiber fabric		720	.011		18.90	.51		19.41	22

04 01 30 – Unit Masonry Cleaning

04 01 30.20 Cleaning Masonry

	04 01 30.20 Cleaning Masonry	Crew	Daily Output	Labor-Hours	Unit	Material	Labor	Equipment	Total	Total Incl O&P
0010	**CLEANING MASONRY**									
0200	By chemical, brush and rinse, new work, light construction dust	D-1	1000	.016	S.F.	.06	.67		.73	1.09
0220	Medium construction dust		800	.020		.09	.84		.93	1.38
0240	Heavy construction dust, drips or stains		600	.027		.11	1.12		1.23	1.84
0260	Low pressure wash and rinse, light restoration, light soil		800	.020		.13	.84		.97	1.43
0270	Average soil, biological staining		400	.040		.19	1.68		1.87	2.78
0280	Heavy soil, biological and mineral staining, paint		330	.048		.26	2.04		2.30	3.41
0300	High pressure wash and rinse, heavy restoration, light soil		600	.027		.12	1.12		1.24	1.84

04 01 Maintenance of Masonry

04 01 30 – Unit Masonry Cleaning

04 01 30.20 Cleaning Masonry		Crew	Daily Output	Labor-Hours	Unit	2015 Bare Costs Material	Labor	Equipment	Total	Total Incl O&P
0310	Average soil, biological staining	D-1	400	.040	S.F.	.17	1.68		1.85	2.76
0320	Heavy soil, biological and mineral staining, paint		250	.064		.23	2.69		2.92	4.36
0400	High pressure wash, water only, light soil	C-29	500	.016			.60	.14	.74	1.08
0420	Average soil, biological staining		375	.021			.80	.19	.99	1.44
0440	Heavy soil, biological and mineral staining, paint		250	.032			1.20	.28	1.48	2.16
0800	High pressure water and chemical, light soil		450	.018		.16	.67	.16	.99	1.37
0820	Average soil, biological staining		300	.027		.23	1	.23	1.46	2.06
0840	Heavy soil, biological and mineral staining, paint		200	.040		.31	1.50	.35	2.16	3.04
1200	Sandblast, wet system, light soil	J-6	1750	.018		.31	.76	.13	1.20	1.65
1220	Average soil, biological staining		1100	.029		.46	1.21	.21	1.88	2.59
1240	Heavy soil, biological and mineral staining, paint		700	.046		.62	1.91	.34	2.87	3.94
1400	Dry system, light soil		2500	.013		.31	.53	.09	.93	1.25
1420	Average soil, biological staining		1750	.018		.46	.76	.13	1.35	1.82
1440	Heavy soil, biological and mineral staining, paint		1000	.032		.62	1.34	.24	2.20	2.97
1800	For walnut shells, add					.69			.69	.76
1820	For corn chips, add					.69			.69	.76
2000	Steam cleaning, light soil	A-1H	750	.011			.40	.10	.50	.73
2020	Average soil, biological staining		625	.013			.48	.12	.60	.87
2040	Heavy soil, biological and mineral staining		375	.021			.80	.20	1	1.45
4000	Add for masking doors and windows	1 Clab	800	.010		.07	.38		.45	.66
4200	Add for pedestrian protection				Job				10%	10%

04 01 30.60 Brick Washing		Crew	Daily Output	Labor-Hours	Unit	Material	Labor	Equipment	Total	Total Incl O&P
0010	**BRICK WASHING** R040130-10									
0012	Acid cleanser, smooth brick surface	1 Bric	560	.014	S.F.	.04	.66		.70	1.06
0050	Rough brick		400	.020		.06	.92		.98	1.47
0060	Stone, acid wash		600	.013		.07	.62		.69	1.02
1000	Muriatic acid, price per gallon in 5 gallon lots				Gal.	8.70			8.70	9.55

04 05 Common Work Results for Masonry

04 05 05 – Selective Demolition for Masonry

04 05 05.10 Selective Demolition		Crew	Daily Output	Labor-Hours	Unit	Material	Labor	Equipment	Total	Total Incl O&P
0010	**SELECTIVE DEMOLITION** R024119-10									
0200	Bond beams, 8" block with #4 bar	2 Clab	32	.500	L.F.		18.80		18.80	29
0300	Concrete block walls, unreinforced, 2" thick		1200	.013	S.F.		.50		.50	.77
0310	4" thick		1150	.014			.52		.52	.80
0320	6" thick		1100	.015			.55		.55	.84
0330	8" thick		1050	.015			.57		.57	.88
0340	10" thick		1000	.016			.60		.60	.93
0360	12" thick		950	.017			.63		.63	.97
0380	Reinforced alternate courses, 2" thick		1130	.014			.53		.53	.82
0390	4" thick		1080	.015			.56		.56	.86
0400	6" thick		1035	.015			.58		.58	.89
0410	8" thick		990	.016			.61		.61	.93
0420	10" thick		940	.017			.64		.64	.98
0430	12" thick		890	.018			.68		.68	1.04
0440	Reinforced alternate courses & vertically 48" OC, 4" thick		900	.018			.67		.67	1.03
0450	6" thick		850	.019			.71		.71	1.09
0460	8" thick		800	.020			.75		.75	1.16
0480	10" thick		750	.021			.80		.80	1.23
0490	12" thick		700	.023			.86		.86	1.32
1000	Chimney, 16" x 16", soft old mortar	1 Clab	55	.145	C.F.		5.45		5.45	8.40

04 05 Common Work Results for Masonry

04 05 05 – Selective Demolition for Masonry

04 05 05.10 Selective Demolition		Crew	Daily Output	Labor-Hours	Unit	Material	2015 Bare Costs Labor	Equipment	Total	Total Incl O&P
1020	Hard mortar	1 Clab	40	.200	C.F.		7.50		7.50	11.55
1030	16" x 20", soft old mortar		55	.145			5.45		5.45	8.40
1040	Hard mortar		40	.200			7.50		7.50	11.55
1050	16" x 24", soft old mortar		55	.145			5.45		5.45	8.40
1060	Hard mortar		40	.200			7.50		7.50	11.55
1080	20" x 20", soft old mortar		55	.145			5.45		5.45	8.40
1100	Hard mortar		40	.200			7.50		7.50	11.55
1110	20" x 24", soft old mortar		55	.145			5.45		5.45	8.40
1120	Hard mortar		40	.200			7.50		7.50	11.55
1140	20" x 32", soft old mortar		55	.145			5.45		5.45	8.40
1160	Hard mortar		40	.200			7.50		7.50	11.55
1200	48" x 48", soft old mortar		55	.145			5.45		5.45	8.40
1220	Hard mortar		40	.200			7.50		7.50	11.55
1250	Metal, high temp steel jacket, 24" diameter	E-2	130	.431	V.L.F.		22.50	11.60	34.10	50.50
1260	60" diameter	"	60	.933			48.50	25	73.50	109
1280	Flue lining, up to 12" x 12"	1 Clab	200	.040			1.50		1.50	2.31
1282	Up to 24" x 24"		150	.053			2.01		2.01	3.09
2000	Columns, 8" x 8", soft old mortar		48	.167			6.25		6.25	9.65
2020	Hard mortar		40	.200			7.50		7.50	11.55
2060	16" x 16", soft old mortar		16	.500			18.80		18.80	29
2100	Hard mortar		14	.571			21.50		21.50	33
2140	24" x 24", soft old mortar		8	1			37.50		37.50	58
2160	Hard mortar		6	1.333			50		50	77
2200	36" x 36", soft old mortar		4	2			75		75	116
2220	Hard mortar		3	2.667			100		100	154
2230	Alternate pricing method, soft old mortar		30	.267	C.F.		10.05		10.05	15.45
2240	Hard mortar		23	.348	"		13.10		13.10	20
3000	Copings, precast or masonry, to 8" wide									
3020	Soft old mortar	1 Clab	180	.044	L.F.		1.67		1.67	2.57
3040	Hard mortar	"	160	.050	"		1.88		1.88	2.89
3100	To 12" wide									
3120	Soft old mortar	1 Clab	160	.050	L.F.		1.88		1.88	2.89
3140	Hard mortar	"	140	.057	"		2.15		2.15	3.31
4000	Fireplace, brick, 30" x 24" opening									
4020	Soft old mortar	1 Clab	2	4	Ea.		150		150	231
4040	Hard mortar		1.25	6.400			241		241	370
4100	Stone, soft old mortar		1.50	5.333			201		201	310
4120	Hard mortar		1	8			300		300	465
5000	Veneers, brick, soft old mortar		140	.057	S.F.		2.15		2.15	3.31
5020	Hard mortar		125	.064			2.41		2.41	3.70
5050	Glass block, up to 4" thick		500	.016			.60		.60	.93
5100	Granite and marble, 2" thick		180	.044			1.67		1.67	2.57
5120	4" thick		170	.047			1.77		1.77	2.72
5140	Stone, 4" thick		180	.044			1.67		1.67	2.57
5160	8" thick		175	.046			1.72		1.72	2.64
5400	Alternate pricing method, stone, 4" thick		60	.133	C.F.		5		5	7.70
5420	8" thick		85	.094	"		3.54		3.54	5.45

04 05 13 – Masonry Mortaring

04 05 13.10 Cement		Crew	Daily Output	Labor-Hours	Unit	Material	Labor	Equipment	Total	Total Incl O&P
0010	**CEMENT**									
0100	Masonry, 70 lb. bag, T.L. lots				Bag	12.15			12.15	13.35
0150	L.T.L. lots					12.85			12.85	14.15

04 05 Common Work Results for Masonry

04 05 13 – Masonry Mortaring

04 05 13.10 Cement		Crew	Daily Output	Labor-Hours	Unit	Material	Labor (2015 Bare Costs)	Equipment	Total	Total Incl O&P
0200	White, 70 lb. bag, T.L. lots				Bag	15.85			15.85	17.45
0250	L.T.L. lots					16.70			16.70	18.40

04 05 13.20 Lime

		Crew	Daily Output	Labor-Hours	Unit	Material	Labor	Equipment	Total	Total Incl O&P
0010	**LIME**									
0020	Masons, hydrated, 50 lb. bag, T.L. lots				Bag	10.10			10.10	11.10
0050	L.T.L. lots					11.10			11.10	12.20
0200	Finish, double hydrated, 50 lb. bag, T.L. lots					8.65			8.65	9.55
0250	L.T.L. lots					9.55			9.55	10.50

04 05 13.23 Surface Bonding Masonry Mortaring

		Crew	Daily Output	Labor-Hours	Unit	Material	Labor	Equipment	Total	Total Incl O&P
0010	**SURFACE BONDING MASONRY MORTARING**									
0020	Gray or white colors, not incl. block work	1 Bric	540	.015	S.F.	.13	.68		.81	1.19

04 05 13.30 Mortar

		Crew	Daily Output	Labor-Hours	Unit	Material	Labor	Equipment	Total	Total Incl O&P
0010	**MORTAR** R040513-10									
0020	With masonry cement									
0100	Type M, 1:1:6 mix	1 Brhe	143	.056	C.F.	5.25	2.13		7.38	9
0200	Type N, 1:3 mix		143	.056		5.40	2.13		7.53	9.20
0300	Type O, 1:3 mix		143	.056		4.18	2.13		6.31	7.85
0400	Type PM, 1:1:6 mix, 2500 psi		143	.056		6	2.13		8.13	9.85
0500	Type S, 1/2:1:4 mix		143	.056		5.25	2.13		7.38	9.05
2000	With portland cement and lime									
2100	Type M, 1:1/4:3 mix	1 Brhe	143	.056	C.F.	8.95	2.13		11.08	13.05
2200	Type N, 1:1:6 mix, 750 psi		143	.056		7.15	2.13		9.28	11.15
2300	Type O, 1:2:9 mix (Pointing Mortar)		143	.056		8.35	2.13		10.48	12.40
2400	Type PL, 1:1/2:4 mix, 2500 psi		143	.056		6	2.13		8.13	9.85
2600	Type S, 1:1/2:4 mix, 1800 psi		143	.056		8.20	2.13		10.33	12.30
2650	Pre-mixed, type S or N					5.05			5.05	5.55
2700	Mortar for glass block	1 Brhe	143	.056		11.05	2.13		13.18	15.40
2900	Mortar for fire brick, dry mix, 10 lb. pail				Ea.	27			27	29.50

04 05 13.91 Masonry Restoration Mortaring

		Crew	Daily Output	Labor-Hours	Unit	Material	Labor	Equipment	Total	Total Incl O&P
0010	**MASONRY RESTORATION MORTARING**									
0020	Masonry restoration mix				Lb.	.29			.29	.32
0050	White				"	.29			.29	.32

04 05 13.93 Mortar Pigments

		Crew	Daily Output	Labor-Hours	Unit	Material	Labor	Equipment	Total	Total Incl O&P
0010	**MORTAR PIGMENTS**, 50 lb. bags (2 bags per M bricks) R040513-10									
0020	Color admixture, range 2 to 10 lb. per bag of cement, light colors				Lb.	4.37			4.37	4.81
0050	Medium colors					6.40			6.40	7
0100	Dark colors					13.25			13.25	14.60

04 05 13.95 Sand

		Crew	Daily Output	Labor-Hours	Unit	Material	Labor	Equipment	Total	Total Incl O&P
0010	**SAND**, screened and washed at pit									
0020	For mortar, per ton				Ton	19.35			19.35	21.50
0050	With 10 mile haul					33.50			33.50	37
0100	With 30 mile haul					65.50			65.50	72
0200	Screened and washed, at the pit				C.Y.	27			27	29.50
0250	With 10 mile haul					47			47	51.50
0300	With 30 mile haul					91			91	100

04 05 13.98 Mortar Admixtures

		Crew	Daily Output	Labor-Hours	Unit	Material	Labor	Equipment	Total	Total Incl O&P
0010	**MORTAR ADMIXTURES**									
0020	Waterproofing admixture, per quart (1 qt. to 2 bags of masonry cement)				Qt.	3.22			3.22	3.54

04 05 Common Work Results for Masonry

04 05 16 – Masonry Grouting

	04 05 16.30 Grouting	Crew	Daily Output	Labor-Hours	Unit	Material	2015 Bare Costs Labor	Equipment	Total	Total Incl O&P
0010	**GROUTING** R040513-10									
0011	Bond beams & lintels, 8" deep, 6" thick, 0.15 C.F. per L.F.	D-4	1480	.022	L.F.	.66	.92	.09	1.67	2.24
0020	8" thick, 0.2 C.F. per L.F.		1400	.023		1.06	.98	.10	2.14	2.75
0050	10" thick, 0.25 C.F. per L.F.		1200	.027		1.11	1.14	.11	2.36	3.08
0060	12" thick, 0.3 C.F. per L.F.		1040	.031		1.33	1.31	.13	2.77	3.60
0200	Concrete block cores, solid, 4" thk., by hand, 0.067 C.F./S.F. of wall	D-8	1100	.036	S.F.	.30	1.56		1.86	2.71
0210	6" thick, pumped, 0.175 C.F. per S.F.	D-4	720	.044		.77	1.90	.19	2.86	3.94
0250	8" thick, pumped, 0.258 C.F. per S.F.		680	.047		1.14	2.01	.20	3.35	4.53
0300	10" thick, pumped, 0.340 C.F. per S.F.		660	.048		1.50	2.07	.20	3.77	5.05
0350	12" thick, pumped, 0.422 C.F. per S.F.		640	.050		1.87	2.14	.21	4.22	5.55
0500	Cavity walls, 2" space, pumped, 0.167 C.F./S.F. of wall		1700	.019		.74	.80	.08	1.62	2.13
0550	3" space, 0.250 C.F./S.F.		1200	.027		1.11	1.14	.11	2.36	3.08
0600	4" space, 0.333 C.F. per S.F.		1150	.028		1.47	1.19	.12	2.78	3.56
0700	6" space, 0.500 C.F. per S.F.		800	.040		2.21	1.71	.17	4.09	5.20
0800	Door frames, 3' x 7' opening, 2.5 C.F. per opening		60	.533	Opng.	11.05	23	2.22	36.27	49
0850	6' x 7' opening, 3.5 C.F. per opening		45	.711	"	15.45	30.50	2.97	48.92	67
2000	Grout, C476, for bond beams, lintels and CMU cores		350	.091	C.F.	4.42	3.91	.38	8.71	11.25

04 05 19 – Masonry Anchorage and Reinforcing

	04 05 19.05 Anchor Bolts	Crew	Daily Output	Labor-Hours	Unit	Material	Labor	Equipment	Total	Total Incl O&P
0010	**ANCHOR BOLTS**									
0015	Installed in fresh grout in CMU bond beams or filled cores, no templates									
0020	Hooked, with nut and washer, 1/2" diam., 8" long	1 Bric	132	.061	Ea.	1.33	2.80		4.13	5.75
0030	12" long		131	.061		1.48	2.82		4.30	5.95
0040	5/8" diameter, 8" long		129	.062		2.98	2.86		5.84	7.65
0050	12" long		127	.063		3.67	2.91		6.58	8.50
0060	3/4" diameter, 8" long		127	.063		3.67	2.91		6.58	8.50
0070	12" long		125	.064		4.59	2.95		7.54	9.55

	04 05 19.16 Masonry Anchors	Crew	Daily Output	Labor-Hours	Unit	Material	Labor	Equipment	Total	Total Incl O&P
0010	**MASONRY ANCHORS**									
0020	For brick veneer, galv., corrugated, 7/8" x 7", 22 Ga.	1 Bric	10.50	.762	C	14.35	35		49.35	69.50
0100	24 Ga.		10.50	.762		9.85	35		44.85	64.50
0150	16 Ga.		10.50	.762		28	35		63	84
0200	Buck anchors, galv., corrugated, 16 ga., 2" bend, 8" x 2"		10.50	.762		56.50	35		91.50	116
0250	8" x 3"		10.50	.762		58.50	35		93.50	118
0660	Cavity wall, Z-type, galvanized, 6" long, 1/8" diam.		10.50	.762		25	35		60	81
0670	3/16" diameter		10.50	.762		29	35		64	85.50
0680	1/4" diameter		10.50	.762		40.50	35		75.50	98
0850	8" long, 3/16" diameter		10.50	.762		25	35		60	81
0855	1/4" diameter		10.50	.762		49	35		84	108
1000	Rectangular type, galvanized, 1/4" diameter, 2" x 6"		10.50	.762		72.50	35		107.50	134
1050	4" x 6"		10.50	.762		87.50	35		122.50	150
1100	3/16" diameter, 2" x 6"		10.50	.762		45	35		80	103
1150	4" x 6"		10.50	.762		51.50	35		86.50	110
1500	Rigid partition anchors, plain, 8" long, 1" x 1/8"		10.50	.762		235	35		270	315
1550	1" x 1/4"		10.50	.762		277	35		312	360
1580	1-1/2" x 1/8"		10.50	.762		259	35		294	340
1600	1-1/2" x 1/4"		10.50	.762		325	35		360	410
1650	2" x 1/8"		10.50	.762		305	35		340	390
1700	2" x 1/4"		10.50	.762		405	35		440	500

04 05 19 – Masonry Anchorage and Reinforcing

04 05 19.26 Masonry Reinforcing Bars

	04 05 19.26 Masonry Reinforcing Bars	Crew	Daily Output	Labor-Hours	Unit	2015 Bare Costs Material	Labor	Equipment	Total	Total Incl O&P
0010	**MASONRY REINFORCING BARS** R040519-50									
0015	Steel bars A615, placed horiz., #3 & #4 bars	1 Bric	450	.018	Lb.	.48	.82		1.30	1.78
0020	#5 & #6 bars		800	.010		.48	.46		.94	1.23
0050	Placed vertical, #3 & #4 bars		350	.023		.48	1.06		1.54	2.14
0060	#5 & #6 bars		650	.012		.48	.57		1.05	1.40
0200	Joint reinforcing, regular truss, to 6" wide, mill std galvanized		30	.267	C.L.F.	22.50	12.30		34.80	43.50
0250	12" wide		20	.400		26	18.45		44.45	56.50
0400	Cavity truss with drip section, to 6" wide		30	.267		22.50	12.30		34.80	43.50
0450	12" wide		20	.400		26	18.45		44.45	56.50

04 05 23 – Masonry Accessories

04 05 23.13 Masonry Control and Expansion Joints

	Description	Crew	Daily Output	Labor-Hours	Unit	Material	Labor	Equipment	Total	Total Incl O&P
0010	**MASONRY CONTROL AND EXPANSION JOINTS**									
0020	Rubber, for double wythe 8" minimum wall (Brick/CMU)	1 Bric	400	.020	L.F.	2.07	.92		2.99	3.69
0025	"T" shaped		320	.025		1.23	1.15		2.38	3.11
0030	Cross-shaped for CMU units		280	.029		1.45	1.32		2.77	3.61
0050	PVC, for double wythe 8" minimum wall (Brick/CMU)		400	.020		1.47	.92		2.39	3.03
0120	"T" shaped		320	.025		.75	1.15		1.90	2.59
0160	Cross-shaped for CMU units		280	.029		.92	1.32		2.24	3.02

04 05 23.19 Masonry Cavity Drainage, Weepholes, and Vents

	Description	Crew	Daily Output	Labor-Hours	Unit	Material	Labor	Equipment	Total	Total Incl O&P
0010	**MASONRY CAVITY DRAINAGE, WEEPHOLES, AND VENTS**									
0020	Extruded aluminum, 4" deep, 2-3/8" x 8-1/8"	1 Bric	30	.267	Ea.	34	12.30		46.30	56.50
0050	5" x 8-1/8"		25	.320		45	14.75		59.75	72
0100	2-1/4" x 25"		25	.320		78	14.75		92.75	108
0150	5" x 16-1/2"		22	.364		62.50	16.80		79.30	94
0175	5" x 24"		22	.364		84	16.80		100.80	118
0200	6" x 16-1/2"		22	.364		90.50	16.80		107.30	125
0250	7-3/4" x 16-1/2"		20	.400		77	18.45		95.45	113
0400	For baked enamel finish, add					35%				
0500	For cast aluminum, painted, add					60%				
1000	Stainless steel ventilators, 6" x 6"	1 Bric	25	.320		210	14.75		224.75	255
1050	8" x 8"		24	.333		232	15.40		247.40	280
1100	12" x 12"		23	.348		266	16.05		282.05	320
1150	12" x 6"		24	.333		254	15.40		269.40	305
1200	Foundation block vent, galv., 1-1/4" thk, 8" high, 16" long, no damper		30	.267		16.20	12.30		28.50	36.50
1250	For damper, add					3.11			3.11	3.42

04 05 23.95 Wall Plugs

	Description	Crew	Daily Output	Labor-Hours	Unit	Material	Labor	Equipment	Total	Total Incl O&P
0010	**WALL PLUGS** (for nailing to brickwork)									
0020	25 ga., galvanized, plain	1 Bric	10.50	.762	C	26.50	35		61.50	82.50
0050	Wood filled	"	10.50	.762	"	68	35		103	128

04 21 13.13 Brick Veneer Masonry		Crew	Daily Output	Labor-Hours	Unit	Material	Labor (2015 Bare Costs)	Equipment	Total	Total Incl O&P
0010	**BRICK VENEER MASONRY**, T.L. lots, excl. scaff., grout & reinforcing R042110-20									
0015	Material costs incl. 3% brick and 25% mortar waste									
0020	Standard, select common, 4" x 2-2/3" x 8" (6.75/S.F.)	D-8	1.50	26.667	M	615	1,150		1,765	2,425
0050	Red, 4" x 2-2/3" x 8", running bond		1.50	26.667		550	1,150		1,700	2,350
0100	Full header every 6th course (7.88/S.F.) R042110-50		1.45	27.586		550	1,175		1,725	2,400
0150	English, full header every 2nd course (10.13/S.F.)		1.40	28.571		550	1,225		1,775	2,475
0200	Flemish, alternate header every course (9.00/S.F.)		1.40	28.571		550	1,225		1,775	2,475
0250	Flemish, alt. header every 6th course (7.13/S.F.)		1.45	27.586		550	1,175		1,725	2,400
0300	Full headers throughout (13.50/S.F.)		1.40	28.571		545	1,225		1,770	2,475
0350	Rowlock course (13.50/S.F.)		1.35	29.630		545	1,275		1,820	2,550
0400	Rowlock stretcher (4.50/S.F.)		1.40	28.571		555	1,225		1,780	2,500
0450	Soldier course (6.75/S.F.)		1.40	28.571		550	1,225		1,775	2,475
0500	Sailor course (4.50/S.F.)		1.30	30.769		555	1,325		1,880	2,650
0601	Buff or gray face, running bond, (6.75/S.F.)		1.50	26.667		550	1,150		1,700	2,350
0700	Glazed face, 4" x 2-2/3" x 8", running bond		1.40	28.571		1,825	1,225		3,050	3,900
0750	Full header every 6th course (7.88/S.F.)		1.35	29.630		1,750	1,275		3,025	3,875
1000	Jumbo, 6" x 4" x 12",(3.00/S.F.)		1.30	30.769		1,725	1,325		3,050	3,925
1051	Norman, 4" x 2-2/3" x 12" (4.50/S.F.)		1.45	27.586		1,175	1,175		2,350	3,100
1100	Norwegian, 4" x 3-1/5" x 12" (3.75/S.F.)		1.40	28.571		1,400	1,225		2,625	3,425
1150	Economy, 4" x 4" x 8" (4.50 per S.F.)		1.40	28.571		915	1,225		2,140	2,875
1201	Engineer, 4" x 3-1/5" x 8", (5.63/S.F.)		1.45	27.586		645	1,175		1,820	2,500
1251	Roman, 4" x 2" x 12", (6.00/S.F.)		1.50	26.667		1,175	1,150		2,325	3,050
1300	S.C.R. 6" x 2-2/3" x 12" (4.50/S.F.)		1.40	28.571		1,350	1,225		2,575	3,375
1350	Utility, 4" x 4" x 12" (3.00/S.F.)		1.08	37.037		1,600	1,600		3,200	4,175
1360	For less than truck load lots, add					15%				
1400	For battered walls, add						30%			
1450	For corbels, add						75%			
1500	For curved walls, add						30%			
1550	For pits and trenches, deduct						20%			
1999	Alternate method of figuring by square foot									
2000	Standard, sel. common, 4" x 2-2/3" x 8", (6.75/S.F.)	D-8	230	.174	S.F.	4.16	7.45		11.61	16
2020	Red, 4" x 2-2/3" x 8", running bond		220	.182		3.72	7.80		11.52	16
2050	Full header every 6th course (7.88/S.F.)		185	.216		4.34	9.30		13.64	18.90
2100	English, full header every 2nd course (10.13/S.F.)		140	.286		5.55	12.25		17.80	25
2150	Flemish, alternate header every course (9.00/S.F.)		150	.267		4.94	11.45		16.39	23
2200	Flemish, alt. header every 6th course (7.13/S.F.)		205	.195		3.93	8.35		12.28	17.10
2250	Full headers throughout (13.50/S.F.)		105	.381		7.40	16.35		23.75	33
2300	Rowlock course (13.50/S.F.)		100	.400		7.40	17.15		24.55	34
2350	Rowlock stretcher (4.50/S.F.)		310	.129		2.51	5.55		8.06	11.20
2400	Soldier course (6.75/S.F.)		200	.200		3.72	8.60		12.32	17.20
2450	Sailor course (4.50/S.F.)		290	.138		2.51	5.90		8.41	11.80
2600	Buff or gray face, running bond, (6.75/S.F.)		220	.182		3.93	7.80		11.73	16.20
2700	Glazed face brick, running bond		210	.190		11.75	8.15		19.90	25.50
2750	Full header every 6th course (7.88/S.F.)		170	.235		13.70	10.10		23.80	30.50
3000	Jumbo, 6" x 4" x 12" running bond (3.00/S.F.)		435	.092		4.67	3.95		8.62	11.15
3050	Norman, 4" x 2-2/3" x 12" running bond, (4.5/S.F.)		320	.125		5.95	5.35		11.30	14.70
3100	Norwegian, 4" x 3-1/5" x 12" (3.75/S.F.)		375	.107		5.15	4.58		9.73	12.65
3150	Economy, 4" x 4" x 8" (4.50/S.F.)		310	.129		4.08	5.55		9.63	12.95
3200	Engineer, 4" x 3-1/5" x 8" (5.63/S.F.)		260	.154		3.61	6.60		10.21	14.05
3250	Roman, 4" x 2" x 12" (6.00/S.F.)		250	.160		6.95	6.85		13.80	18.10
3300	SCR, 6" x 2-2/3" x 12" (4.50/S.F.)		310	.129		6	5.55		11.55	15.05
3350	Utility, 4" x 4" x 12" (3.00/S.F.)		360	.111		4.65	4.77		9.42	12.40
3360	For less than truck load lots, add				M	15%				

04 21 13.13 Brick Veneer Masonry		Crew	Daily Output	Labor-Hours	Unit	2015 Bare Costs Material	Labor	Equipment	Total	Total Incl O&P
3370	For battered walls, add						30%			
3380	For corbels, add						75%			
3400	For cavity wall construction, add						15%			
3450	For stacked bond, add						10%			
3500	For interior veneer construction, add						15%			
3510	For pits and trenches, deduct						20%			
3550	For curved walls, add						30%			

04 21 13.14 Thin Brick Veneer

		Crew	Daily Output	Labor-Hours	Unit	Material	Labor	Equipment	Total	Total Incl O&P
0010	**THIN BRICK VENEER**									
0015	Material costs incl. 3% brick and 25% mortar waste									
0020	On & incl. metal panel support sys, modular, 2-2/3" x 5/8" x 8", red	D-7	92	.174	S.F.	9	6.65		15.65	19.70
0100	Closure, 4" x 5/8" x 8"		110	.145		8.80	5.55		14.35	17.85
0110	Norman, 2-2/3" x 5/8" x 12"		110	.145		8.85	5.55		14.40	17.90
0120	Utility, 4" x 5/8" x 12"		125	.128		8.55	4.89		13.44	16.65
0130	Emperor, 4" x 3/4" x 16"		175	.091		9.65	3.49		13.14	15.80
0140	Super emperor, 8" x 3/4" x 16"		195	.082		9.95	3.13		13.08	15.60
0150	For L shaped corners with 4" return, add				L.F.	9.25			9.25	10.20
0200	On masonry/plaster back-up, modular, 2-2/3" x 5/8" x 8", red	D-7	137	.117	S.F.	4.21	4.46		8.67	11.25
0210	Closure, 4" x 5/8" x 8"		165	.097		3.98	3.70		7.68	9.85
0220	Norman, 2-2/3" x 5/8" x 12"		165	.097		4.02	3.70		7.72	9.85
0230	Utility, 4" x 5/8" x 12"		185	.086		3.75	3.30		7.05	9
0240	Emperor, 4" x 3/4" x 16"		260	.062		4.85	2.35		7.20	8.80
0250	Super emperor, 8" x 3/4" x 16"		285	.056		5.15	2.14		7.29	8.80
0260	For L shaped corners with 4" return, add				L.F.	9.25			9.25	10.20
0270	For embedment into pre-cast concrete panels, add				S.F.	14.40			14.40	15.85

04 21 13.15 Chimney

		Crew	Daily Output	Labor-Hours	Unit	Material	Labor	Equipment	Total	Total Incl O&P
0010	**CHIMNEY**, excludes foundation, scaffolding, grout and reinforcing									
0100	Brick, 16" x 16", 8" flue	D-1	18.20	.879	V.L.F.	24	37		61	83
0150	16" x 20" with one 8" x 12" flue		16	1		37.50	42		79.50	106
0200	16" x 24" with two 8" x 8" flues		14	1.143		55	48		103	134
0250	20" x 20" with one 12" x 12" flue		13.70	1.168		45.50	49		94.50	125
0300	20" x 24" with two 8" x 12" flues		12	1.333		62.50	56		118.50	154
0350	20" x 32" with two 12" x 12" flues		10	1.600		80	67.50		147.50	191

04 21 13.18 Columns

		Crew	Daily Output	Labor-Hours	Unit	Material	Labor	Equipment	Total	Total Incl O&P
0010	**COLUMNS**, solid, excludes scaffolding, grout and reinforcing R042110-10									
0050	Brick, 8" x 8", 9 brick per V.L.F.	D-1	56	.286	V.L.F.	4.76	12.05		16.81	23.50
0100	12" x 8", 13.5 brick per V.L.F.		37	.432		7.15	18.20		25.35	36
0200	12" x 12", 20 brick per V.L.F.		25	.640		10.60	27		37.60	52.50
0300	16" x 12", 27 brick per V.L.F.		19	.842		14.30	35.50		49.80	69.50
0400	16" x 16", 36 brick per V.L.F.		14	1.143		19.05	48		67.05	94.50
0500	20" x 16", 45 brick per V.L.F.		11	1.455		24	61		85	120
0600	20" x 20", 56 brick per V.L.F.		9	1.778		29.50	75		104.50	147
0700	24" x 20", 68 brick per V.L.F.		7	2.286		36	96		132	187
0800	24" x 24", 81 brick per V.L.F.		6	2.667		43	112		155	218
1000	36" x 36", 182 brick per V.L.F.		3	5.333		96.50	225		321.50	450

04 21 13.30 Oversized Brick

		Crew	Daily Output	Labor-Hours	Unit	Material	Labor	Equipment	Total	Total Incl O&P
0010	**OVERSIZED BRICK**, excludes scaffolding, grout and reinforcing									
0100	Veneer, 4" x 2.25" x 16"	D-8	387	.103	S.F.	5.10	4.44		9.54	12.40
0102	8" x 2.25" x 16", multicell		265	.151		16.10	6.50		22.60	27.50
0105	4" x 2.75" x 16"		412	.097		5.30	4.17		9.47	12.20
0107	8" x 2.75" x 16", multicell		295	.136		16.10	5.80		21.90	26.50
0110	4" x 4" x 16"		460	.087		3.48	3.73		7.21	9.55

04 21 13 – Brick Masonry

04 21 13.30 Oversized Brick		Crew	Daily Output	Labor-Hours	Unit	Material	2015 Bare Costs Labor	Equipment	Total	Total Incl O&P
0120	4" x 8" x 16"	D-8	533	.075	S.F.	4.08	3.22		7.30	9.40
0122	4" x 8" x 16" multi cell		327	.122		15.10	5.25		20.35	24.50
0125	Loadbearing, 6" x 4" x 16", grouted and reinforced		387	.103		10.30	4.44		14.74	18.10
0130	8" x 4" x 16", grouted and reinforced		327	.122		11.30	5.25		16.55	20.50
0132	10" x 4" x 16", grouted and reinforced		327	.122		23	5.25		28.25	33
0135	6" x 8" x 16", grouted and reinforced		440	.091		13.40	3.90		17.30	20.50
0140	8" x 8" x 16", grouted and reinforced		400	.100		14.30	4.29		18.59	22.50
0145	Curtainwall/reinforced veneer, 6" x 4" x 16"		387	.103		14.65	4.44		19.09	23
0150	8" x 4" x 16"		327	.122		17.80	5.25		23.05	27.50
0152	10" x 4" x 16"		327	.122		25	5.25		30.25	35.50
0155	6" x 8" x 16"		440	.091		18.20	3.90		22.10	26
0160	8" x 8" x 16"	↓	400	.100		25.50	4.29		29.79	34.50
0200	For 1 to 3 slots in face, add					15%				
0210	For 4 to 7 slots in face, add					25%				
0220	For bond beams, add					20%				
0230	For bullnose shapes, add					20%				
0240	For open end knockout, add					10%				
0250	For white or gray color group, add					10%				
0260	For 135 degree corner, add				↓	250%				

04 21 13.35 Common Building Brick		Crew	Daily Output	Labor-Hours	Unit	Material	Labor	Equipment	Total	Total Incl O&P
0010	**COMMON BUILDING BRICK**, C62, TL lots, material only R042110-20									
0020	Standard				M	490			490	540
0050	Select				"	500			500	550

04 21 13.40 Structural Brick		Crew	Daily Output	Labor-Hours	Unit	Material	Labor	Equipment	Total	Total Incl O&P
0010	**STRUCTURAL BRICK** C652, Grade SW, incl. mortar, scaffolding not incl.									
0100	Standard unit, 4-5/8" x 2-3/4" x 9-5/8"	D-8	245	.163	S.F.	4.08	7		11.08	15.20
0120	Bond beam		225	.178		4.08	7.65		11.73	16.15
0140	V cut bond beam		225	.178		4.08	7.65		11.73	16.15
0160	Stretcher quoin, 5-5/8" x 2-3/4" x 9-5/8"		245	.163		7.55	7		14.55	19
0180	Corner quoin		245	.163		7.55	7		14.55	19
0200	Corner, 45 deg, 4-5/8" x 2-3/4" x 10-7/16"	↓	235	.170	↓	7.55	7.30		14.85	19.45

04 21 13.45 Face Brick		Crew	Daily Output	Labor-Hours	Unit	Material	Labor	Equipment	Total	Total Incl O&P
0010	**FACE BRICK** Material Only, C216, TL lots R042110-20									
0300	Standard modular, 4" x 2-2/3" x 8"				M	435			435	480
0450	Economy, 4" x 4" x 8"					775			775	855
0510	Economy, 4" x 4" x 12"					1,225			1,225	1,350
0550	Jumbo, 6" x 4" x 12"					1,400			1,400	1,550
0610	Jumbo, 8" x 4" x 12"					1,400			1,400	1,550
0650	Norwegian, 4" x 3-1/5" x 12"					1,225			1,225	1,350
0710	Norwegian, 6" x 3-1/5" x 12"					1,525			1,525	1,675
0850	Standard glazed, plain colors, 4" x 2-2/3" x 8"					1,600			1,600	1,750
1000	Deep trim shades, 4" x 2-2/3" x 8"					2,025			2,025	2,225
1080	Jumbo utility, 4" x 4" x 12"					1,400			1,400	1,525
1120	4" x 8" x 8"					1,775			1,775	1,950
1140	4" x 8" x 16"					5,225			5,225	5,750
1260	Engineer, 4" x 3-1/5" x 8"					525			525	575
1350	King, 4" x 2-3/4" x 10"					485			485	530
1770	Standard modular, double glazed, 4" x 2-2/3" x 8"					2,400			2,400	2,650
1850	Jumbo, colored glazed ceramic, 6" x 4" x 12"					2,525			2,525	2,775
2050	Jumbo utility, glazed, 4" x 4" x 12"					4,750			4,750	5,225
2100	4" x 8" x 8"					5,600			5,600	6,150
2150	4" x 16" x 8"				↓	6,550			6,550	7,200

04 21 Clay Unit Masonry

04 21 13 – Brick Masonry

04 21 13.45 Face Brick		Crew	Daily Output	Labor-Hours	Unit	Material	2015 Bare Costs Labor	Equipment	Total	Total Incl O&P
2170	For less than truck load lots, add				M	15			15	16.50
2180	For buff or gray brick, add				↓	16			16	17.60

04 21 26 – Glazed Structural Clay Tile Masonry

04 21 26.10 Structural Facing Tile		Crew	Daily Output	Labor-Hours	Unit	Material	Labor	Equipment	Total	Total Incl O&P
0010	**STRUCTURAL FACING TILE**, std. colors, excl. scaffolding, grout, reinforcing									
0020	6T series, 5-1/3" x 12", 2.3 pieces per S.F., glazed 1 side, 2" thick	D-8	225	.178	S.F.	8.95	7.65		16.60	21.50
0100	4" thick		220	.182		12.20	7.80		20	25.50
0150	Glazed 2 sides		195	.205		16	8.80		24.80	31
0250	6" thick		210	.190		18.20	8.15		26.35	32.50
0300	Glazed 2 sides		185	.216		21.50	9.30		30.80	38
0400	8" thick		180	.222	↓	24	9.55		33.55	41
0500	Special shapes, group 1		400	.100	Ea.	7.70	4.29		11.99	15
0550	Group 2		375	.107		12.50	4.58		17.08	21
0600	Group 3		350	.114		16.05	4.90		20.95	25
0650	Group 4		325	.123		33.50	5.30		38.80	44.50
0700	Group 5		300	.133		39.50	5.70		45.20	52.50
0750	Group 6		275	.145	↓	54	6.25		60.25	69
1000	Fire rated, 4" thick, 1 hr. rating		210	.190	S.F.	17.45	8.15		25.60	31.50
1300	Acoustic, 4" thick	↓	210	.190	"	34	8.15		42.15	50
2000	8W series, 8" x 16", 1.125 pieces per S.F.									
2050	2" thick, glazed 1 side	D-8	360	.111	S.F.	10.45	4.77		15.22	18.80
2100	4" thick, glazed 1 side		345	.116		14.35	4.98		19.33	23.50
2150	Glazed 2 sides		325	.123		17.15	5.30		22.45	27
2200	6" thick, glazed 1 side		330	.121		24.50	5.20		29.70	35
2250	8" thick, glazed 1 side		310	.129	↓	25	5.55		30.55	35.50
2500	Special shapes, group 1		300	.133	Ea.	14.85	5.70		20.55	25
2550	Group 2		280	.143		19.85	6.15		26	31.50
2600	Group 3		260	.154		21	6.60		27.60	33
2650	Group 4		250	.160		43.50	6.85		50.35	58.50
2700	Group 5		240	.167		39.50	7.15		46.65	54.50
2750	Group 6		230	.174	↓	85.50	7.45		92.95	105
3000	4" thick, glazed 1 side		345	.116	S.F.	13.95	4.98		18.93	23
3100	Acoustic, 4" thick	↓	345	.116	"	18.55	4.98		23.53	28
3120	4W series, 8" x 8", 2.25 pieces per S.F.									
3125	2" thick, glazed 1 side	D-8	360	.111	S.F.	9.60	4.77		14.37	17.85
3130	4" thick, glazed 1 side		345	.116		11.30	4.98		16.28	20
3135	Glazed 2 sides		325	.123		15.55	5.30		20.85	25
3140	6" thick, glazed 1 side		330	.121		16.05	5.20		21.25	25.50
3150	8" thick, glazed 1 side		310	.129	↓	23.50	5.55		29.05	34
3155	Special shapes, group I		300	.133	Ea.	7.50	5.70		13.20	17
3160	Group II	↓	280	.143	"	8.65	6.15		14.80	18.85
3200	For designer colors, add					25%				
3300	For epoxy mortar joints, add				S.F.	1.74			1.74	1.91

04 21 29 – Terra Cotta Masonry

04 21 29.10 Terra Cotta Masonry Components		Crew	Daily Output	Labor-Hours	Unit	Material	Labor	Equipment	Total	Total Incl O&P
0010	**TERRA COTTA MASONRY COMPONENTS**									
0020	Coping, split type, not glazed, 9" wide	D-1	90	.178	L.F.	13.15	7.50		20.65	26
0100	13" wide		80	.200		18.35	8.40		26.75	33
0200	Coping, split type, glazed, 9" wide		90	.178		11.95	7.50		19.45	24.50
0250	13" wide	↓	80	.200	↓	15.75	8.40		24.15	30
0500	Partition or back-up blocks, scored, in C.L. lots									
0700	Non-load bearing 12" x 12", 3" thick, special order	D-8	550	.073	S.F.	16.85	3.12		19.97	23.50

04 21 Clay Unit Masonry

04 21 29 – Terra Cotta Masonry

04 21 29.10 Terra Cotta Masonry Components		Crew	Daily Output	Labor-Hours	Unit	Material	2015 Bare Costs Labor	Equipment	Total	Total Incl O&P
0750	4″ thick, standard	D-8	500	.080	S.F.	5.35	3.43		8.78	11.10
0800	6″ thick		450	.089		7.20	3.81		11.01	13.70
0850	8″ thick		400	.100		9.05	4.29		13.34	16.50
1000	Load bearing, 12″ x 12″, 4″ thick, in walls		500	.080		5.45	3.43		8.88	11.25
1050	In floors		750	.053		5.45	2.29		7.74	9.50
1200	6″ thick, in walls		450	.089		7.35	3.81		11.16	13.90
1250	In floors		675	.059		7.35	2.54		9.89	12
1400	8″ thick, in walls		400	.100		9.15	4.29		13.44	16.60
1450	In floors		575	.070		9.15	2.99		12.14	14.60
1600	10″ thick, in walls, special order		350	.114		24.50	4.90		29.40	34.50
1650	In floors, special order		500	.080		24.50	3.43		27.93	32.50
1800	12″ thick, in walls, special order		300	.133		23	5.70		28.70	34.50
1850	In floors, special order		450	.089		23	3.81		26.81	31.50
2000	For reinforcing with steel rods, add to above					15%	5%			
2100	For smooth tile instead of scored, add					3.66			3.66	4.03
2200	For L.C.L. quantities, add					10%	10%			

04 21 29.20 Terra Cotta Tile		Crew	Daily Output	Labor-Hours	Unit	Material	Labor	Equipment	Total	Total Incl O&P
0010	**TERRA COTTA TILE**, on walls, dry set, 1/2″ thick									
0100	Square, hexagonal or lattice shapes, unglazed	1 Tilf	135	.059	S.F.	4.62	2.54		7.16	8.85
0300	Glazed, plain colors		130	.062		7.10	2.63		9.73	11.70
0400	Intense colors		125	.064		8.30	2.74		11.04	13.20

04 22 Concrete Unit Masonry

04 22 10 – Concrete Masonry Units

04 22 10.11 Autoclave Aerated Concrete Block		Crew	Daily Output	Labor-Hours	Unit	Material	Labor	Equipment	Total	Total Incl O&P
0010	**AUTOCLAVE AERATED CONCRETE BLOCK**, excl. scaffolding, grout & reinforcing									
0050	Solid, 4″ x 8″ x 24″, incl. mortar G	D-8	600	.067	S.F.	1.48	2.86		4.34	6
0060	6″ x 8″ x 24″ G		600	.067		2.24	2.86		5.10	6.85
0070	8″ x 8″ x 24″ G		575	.070		2.98	2.99		5.97	7.85
0080	10″ x 8″ x 24″ G		575	.070		3.64	2.99		6.63	8.55
0090	12″ x 8″ x 24″ G		550	.073		4.47	3.12		7.59	9.70

04 22 10.12 Chimney Block		Crew	Daily Output	Labor-Hours	Unit	Material	Labor	Equipment	Total	Total Incl O&P
0010	**CHIMNEY BLOCK**, excludes scaffolding, grout and reinforcing									
0220	1 piece, with 8″ x 8″ flue, 16″ x 16″	D-1	28	.571	V.L.F.	18.50	24		42.50	57
0230	2 piece, 16″ x 16″		26	.615		23	26		49	64.50
0240	2 piece, with 8″ x 12″ flue, 16″ x 20″		24	.667		31.50	28		59.50	78

04 22 10.14 Concrete Block, Back-Up		Crew	Daily Output	Labor-Hours	Unit	Material	Labor	Equipment	Total	Total Incl O&P
0010	**CONCRETE BLOCK, BACK-UP**, C90, 2000 psi R042210-20									
0020	Normal weight, 8″ x 16″ units, tooled joint 1 side									
0050	Not-reinforced, 2000 psi, 2″ thick	D-8	475	.084	S.F.	1.50	3.61		5.11	7.15
0200	4″ thick		460	.087		1.79	3.73		5.52	7.65
0300	6″ thick		440	.091		2.31	3.90		6.21	8.50
0350	8″ thick		400	.100		2.84	4.29		7.13	9.65
0400	10″ thick		330	.121		2.96	5.20		8.16	11.20
0450	12″ thick	D-9	310	.155		4.05	6.50		10.55	14.40
1000	Reinforced, alternate courses, 4″ thick	D-8	450	.089		1.95	3.81		5.76	7.95
1100	6″ thick		430	.093		2.48	3.99		6.47	8.85
1150	8″ thick		395	.101		3.03	4.35		7.38	10
1200	10″ thick		320	.125		3.12	5.35		8.47	11.65
1250	12″ thick	D-9	300	.160		4.21	6.75		10.96	14.95

04 22 10.16 Concrete Block, Bond Beam

	04 22 10.16 Concrete Block, Bond Beam	Crew	Daily Output	Labor-Hours	Unit	2015 Bare Costs Material	Labor	Equipment	Total	Total Incl O&P
0010	**CONCRETE BLOCK, BOND BEAM**, C90, 2000 psi									
0020	Not including grout or reinforcing									
0125	Regular block, 6" thick	D-8	584	.068	L.F.	2.65	2.94		5.59	7.40
0130	8" high, 8" thick	"	565	.071		2.71	3.04		5.75	7.60
0150	12" thick	D-9	510	.094		4.02	3.96		7.98	10.45
0525	Lightweight, 6" thick	D-8	592	.068		2.73	2.90		5.63	7.45
0530	8" high, 8" thick	"	575	.070		3.29	2.99		6.28	8.20
0550	12" thick	D-9	520	.092		4.43	3.89		8.32	10.80
2000	Including grout and 2 #5 bars									
2100	Regular block, 8" high, 8" thick	D-8	300	.133	L.F.	4.78	5.70		10.48	14
2150	12" thick	D-9	250	.192		6.65	8.10		14.75	19.70
2500	Lightweight, 8" high, 8" thick	D-8	305	.131		5.35	5.65		11	14.50
2550	12" thick	D-9	255	.188		7.05	7.90		14.95	19.90

04 22 10.18 Concrete Block, Column

	04 22 10.18 Concrete Block, Column	Crew	Daily Output	Labor-Hours	Unit	Material	Labor	Equipment	Total	Total Incl O&P
0010	**CONCRETE BLOCK, COLUMN** or pilaster									
0050	Including vertical reinforcing (4-#4 bars) and grout									
0160	1 piece unit, 16" x 16"	D-1	26	.615	V.L.F.	15.90	26		41.90	57
0170	2 piece units, 16" x 20"		24	.667		20.50	28		48.50	65.50
0180	20" x 20"		22	.727		26	30.50		56.50	75.50
0190	22" x 24"		18	.889		39	37.50		76.50	100
0200	20" x 32"		14	1.143		45	48		93	123

04 22 10.19 Concrete Block, Insulation Inserts

	04 22 10.19 Concrete Block, Insulation Inserts	Crew	Daily Output	Labor-Hours	Unit	Material	Labor	Equipment	Total	Total Incl O&P
0010	**CONCRETE BLOCK, INSULATION INSERTS**									
0100	Styrofoam, plant installed, add to block prices									
0200	8" x 16" units, 6" thick				S.F.	1.20			1.20	1.32
0250	8" thick					1.35			1.35	1.49
0300	10" thick					1.40			1.40	1.54
0350	12" thick					1.55			1.55	1.71
0500	8" x 8" units, 8" thick					1.20			1.20	1.32
0550	12" thick					1.40			1.40	1.54

04 22 10.23 Concrete Block, Decorative

	04 22 10.23 Concrete Block, Decorative	Crew	Daily Output	Labor-Hours	Unit	Material	Labor	Equipment	Total	Total Incl O&P
0010	**CONCRETE BLOCK, DECORATIVE**, C90, 2000 psi									
0020	Embossed, simulated brick face									
0100	8" x 16" units, 4" thick	D-8	400	.100	S.F.	2.70	4.29		6.99	9.50
0200	8" thick		340	.118		2.95	5.05		8	10.95
0250	12" thick		300	.133		4.98	5.70		10.68	14.20
0400	Embossed both sides									
0500	8" thick	D-8	300	.133	S.F.	3.97	5.70		9.67	13.10
0550	12" thick	"	275	.145	"	5.20	6.25		11.45	15.30
1000	Fluted high strength									
1100	8" x 16" x 4" thick, flutes 1 side,	D-8	345	.116	S.F.	3.83	4.98		8.81	11.80
1150	Flutes 2 sides		335	.119		4.40	5.10		9.50	12.65
1200	8" thick		300	.133		5.15	5.70		10.85	14.45
1250	For special colors, add					.62			.62	.68
1400	Deep grooved, smooth face									
1450	8" x 16" x 4" thick	D-8	345	.116	S.F.	2.46	4.98		7.44	10.30
1500	8" thick	"	300	.133	"	3.89	5.70		9.59	13.05
2000	Formblock, incl. inserts & reinforcing									
2100	8" x 16" x 8" thick	D-8	345	.116	S.F.	3.44	4.98		8.42	11.40
2150	12" thick	"	310	.129	"	4.55	5.55		10.10	13.45
2500	Ground face									

04 22 10 – Concrete Masonry Units

	04 22 10.23 Concrete Block, Decorative	Crew	Daily Output	Labor-Hours	Unit	2015 Bare Costs Material	Labor	Equipment	Total	Total Incl O&P
2600	8" x 16" x 4" thick	D-8	345	.116	S.F.	3.58	4.98		8.56	11.55
2650	6" thick		325	.123		3.98	5.30		9.28	12.45
2700	8" thick		300	.133		4.39	5.70		10.09	13.55
2750	12" thick	D-9	265	.181		5.60	7.65		13.25	17.80
2900	For special colors, add, minimum					15%				
2950	For special colors, add, maximum					45%				
4000	Slump block									
4100	4" face height x 16" x 4" thick	D-1	165	.097	S.F.	3.89	4.08		7.97	10.55
4150	6" thick		160	.100		5.65	4.21		9.86	12.70
4200	8" thick		155	.103		5.60	4.35		9.95	12.80
4250	10" thick		140	.114		10.95	4.81		15.76	19.40
4300	12" thick		130	.123		11.85	5.20		17.05	21
4400	6" face height x 16" x 6" thick		155	.103		5.25	4.35		9.60	12.45
4450	8" thick		150	.107		7.95	4.49		12.44	15.60
4500	10" thick		130	.123		12.80	5.20		18	22
4550	12" thick		120	.133		13.15	5.60		18.75	23
5000	Split rib profile units, 1" deep ribs, 8 ribs									
5100	8" x 16" x 4" thick	D-8	345	.116	S.F.	3.84	4.98		8.82	11.85
5150	6" thick		325	.123		4.37	5.30		9.67	12.85
5200	8" thick		300	.133		4.94	5.70		10.64	14.20
5250	12" thick	D-9	275	.175		5.80	7.35		13.15	17.55
5400	For special deeper colors, 4" thick, add					1.24			1.24	1.37
5450	12" thick, add					1.28			1.28	1.40
5600	For white, 4" thick, add					1.24			1.24	1.37
5650	6" thick, add					1.27			1.27	1.39
5700	8" thick, add					1.29			1.29	1.42
5750	12" thick, add					1.33			1.33	1.47
6000	Split face									
6100	8" x 16" x 4" thick	D-8	350	.114	S.F.	3.48	4.90		8.38	11.35
6150	6" thick		325	.123		3.98	5.30		9.28	12.40
6200	8" thick		300	.133		4.47	5.70		10.17	13.65
6250	12" thick	D-9	270	.178		5.40	7.50		12.90	17.40
6300	For scored, add					.37			.37	.41
6400	For special deeper colors, 4" thick, add					.60			.60	.66
6450	6" thick, add					.71			.71	.78
6500	8" thick, add					.73			.73	.81
6550	12" thick, add					.76			.76	.83
6650	For white, 4" thick, add					1.23			1.23	1.35
6700	6" thick, add					1.24			1.24	1.37
6750	8" thick, add					1.25			1.25	1.38
6800	12" thick, add					1.28			1.28	1.40
7000	Scored ground face, 2 to 5 scores									
7100	8" x 16" x 4" thick	D-8	340	.118	S.F.	7.55	5.05		12.60	16
7150	6" thick		310	.129		8.40	5.55		13.95	17.65
7200	8" thick		290	.138		9.45	5.90		15.35	19.45
7250	12" thick	D-9	265	.181		12.70	7.65		20.35	25.50
8000	Hexagonal face profile units, 8" x 16" units									
8100	4" thick, hollow	D-8	340	.118	S.F.	3.63	5.05		8.68	11.70
8200	Solid		340	.118		4.65	5.05		9.70	12.80
8300	6" thick, hollow		310	.129		3.82	5.55		9.37	12.65
8350	8" thick, hollow		290	.138		4.36	5.90		10.26	13.85
8500	For stacked bond, add						26%			
8550	For high rise construction, add per story	D-8	67.80	.590	M.S.F.		25.50		25.50	38.50

04 22 Concrete Unit Masonry

04 22 10 – Concrete Masonry Units

04 22 10.23 Concrete Block, Decorative		Crew	Daily Output	Labor-Hours	Unit	Material (2015 Bare Costs)	Labor (2015 Bare Costs)	Equipment (2015 Bare Costs)	Total (2015 Bare Costs)	Total Incl O&P
8600	For scored block, add					10%				
8650	For honed or ground face, per face, add				Ea.	1.13			1.13	1.24
8700	For honed or ground end, per end, add				"	1.13			1.13	1.24
8750	For bullnose block, add					10%				
8800	For special color, add					13%				
04 22 10.24 Concrete Block, Exterior										
0010	**CONCRETE BLOCK, EXTERIOR**, C90, 2000 psi									
0020	Reinforced alt courses, tooled joints 2 sides									
0100	Normal weight, 8" x 16" x 6" thick	D-8	395	.101	S.F.	2.32	4.35		6.67	9.20
0200	8" thick		360	.111		4	4.77		8.77	11.70
0250	10" thick	↓	290	.138		4.24	5.90		10.14	13.70
0300	12" thick	D-9	250	.192		4.84	8.10		12.94	17.65
0500	Lightweight, 8" x 16" x 6" thick	D-8	450	.089		3.10	3.81		6.91	9.20
0600	8" thick		430	.093		3.94	3.99		7.93	10.45
0650	10" thick	↓	395	.101		4.08	4.35		8.43	11.15
0700	12" thick	D-9	350	.137	↓	4.12	5.75		9.87	13.35
04 22 10.26 Concrete Block Foundation Wall										
0010	**CONCRETE BLOCK FOUNDATION WALL**, C90/C145									
0050	Normal-weight, cut joints, horiz joint reinf, no vert reinf.									
0200	Hollow, 8" x 16" x 6" thick	D-8	455	.088	S.F.	2.89	3.77		6.66	8.90
0250	8" thick		425	.094		3.45	4.04		7.49	9.95
0300	10" thick	↓	350	.114		3.57	4.90		8.47	11.40
0350	12" thick	D-9	300	.160		4.68	6.75		11.43	15.45
0500	Solid, 8" x 16" block, 6" thick	D-8	440	.091		3.05	3.90		6.95	9.30
0550	8" thick	"	415	.096		4.15	4.14		8.29	10.85
0600	12" thick	D-9	350	.137	↓	5.95	5.75		11.70	15.35
04 22 10.28 Concrete Block, High Strength										
0010	**CONCRETE BLOCK, HIGH STRENGTH**									
0050	Hollow, reinforced alternate courses, 8" x 16" units									
0200	3500 psi, 4" thick	D-8	440	.091	S.F.	2.24	3.90		6.14	8.40
0250	6" thick		395	.101		2.23	4.35		6.58	9.10
0300	8" thick	↓	360	.111		3.92	4.77		8.69	11.60
0350	12" thick	D-9	250	.192		4.72	8.10		12.82	17.55
0500	5000 psi, 4" thick	D-8	440	.091		2.02	3.90		5.92	8.15
0550	6" thick		395	.101		2.88	4.35		7.23	9.80
0600	8" thick	↓	360	.111		4.09	4.77		8.86	11.80
0650	12" thick	D-9	300	.160	↓	4.73	6.75		11.48	15.50
1000	For 75% solid block, add					30%				
1050	For 100% solid block, add					50%				
04 22 10.30 Concrete Block, Interlocking										
0010	**CONCRETE BLOCK, INTERLOCKING**									
0100	Not including grout or reinforcing									
0200	8" x 16" units, 2,000 psi, 8" thick	D-1	245	.065	S.F.	2.90	2.75		5.65	7.40
0300	12" thick		220	.073		4.30	3.06		7.36	9.40
0350	16" thick	↓	185	.086		6.45	3.64		10.09	12.65
0400	Including grout & reinforcing, 8" thick	D-4	245	.131		7.75	5.60	.54	13.89	17.60
0450	12" thick		220	.145		9.35	6.20	.61	16.16	20.50
0500	16" thick	↓	185	.173	↓	11.70	7.40	.72	19.82	25

04 22 Concrete Unit Masonry

04 22 10 – Concrete Masonry Units

04 22 10.32 Concrete Block, Lintels

	04 22 10.32 Concrete Block, Lintels	Crew	Daily Output	Labor-Hours	Unit	Material	Labor	Equipment	Total	Total Incl O&P
						2015 Bare Costs				
0010	**CONCRETE BLOCK, LINTELS**, C90, normal weight									
0100	Including grout and horizontal reinforcing									
0200	8" x 8" x 8", 1 #4 bar	D-4	300	.107	L.F.	3.74	4.56	.44	8.74	11.55
0250	2 #4 bars		295	.108		3.96	4.63	.45	9.04	11.90
0400	8" x 16" x 8", 1 #4 bar		275	.116		3.83	4.97	.49	9.29	12.35
0450	2 #4 bars		270	.119		4.05	5.05	.49	9.59	12.70
1000	12" x 8" x 8", 1 #4 bar		275	.116		5.20	4.97	.49	10.66	13.85
1100	2 #4 bars		270	.119		5.40	5.05	.49	10.94	14.20
1150	2 #5 bars		270	.119		5.65	5.05	.49	11.19	14.45
1200	2 #6 bars		265	.121		5.95	5.15	.50	11.60	14.95
1500	12" x 16" x 8", 1 #4 bar		250	.128		5.80	5.45	.53	11.78	15.35
1600	2 #3 bars		245	.131		5.80	5.60	.54	11.94	15.50
1650	2 #4 bars		245	.131		6	5.60	.54	12.14	15.70
1700	2 #5 bars		240	.133		6.25	5.70	.56	12.51	16.20

04 22 10.33 Lintel Block

	04 22 10.33 Lintel Block	Crew	Daily Output	Labor-Hours	Unit	Material	Labor	Equipment	Total	Total Incl O&P
0010	**LINTEL BLOCK**									
3481	Lintel block 6" x 8" x 8"	D-1	300	.053	Ea.	1.28	2.25		3.53	4.84
3501	6" x 16" x 8"		275	.058		2	2.45		4.45	5.95
3521	8" x 8" x 8"		275	.058		1.15	2.45		3.60	5
3561	8" x 16" x 8"		250	.064		2.02	2.69		4.71	6.35

04 22 10.34 Concrete Block, Partitions

	04 22 10.34 Concrete Block, Partitions	Crew	Daily Output	Labor-Hours	Unit	Material	Labor	Equipment	Total	Total Incl O&P
0010	**CONCRETE BLOCK, PARTITIONS**, excludes scaffolding R042210-20									
0100	Acoustical slotted block									
0200	4" thick, type A-1	D-8	315	.127	S.F.	4.06	5.45		9.51	12.75
0210	8" thick		275	.145		5.20	6.25		11.45	15.25
0250	8" thick, type Q		275	.145		9.05	6.25		15.30	19.50
0260	4" thick, type RSC		315	.127		6.30	5.45		11.75	15.25
0270	6" thick		295	.136		6.30	5.80		12.10	15.85
0280	8" thick		275	.145		6.30	6.25		12.55	16.50
0290	12" thick		250	.160		6.30	6.85		13.15	17.45
0300	8" thick, type RSR		275	.145		6.30	6.25		12.55	16.50
0400	8" thick, type RSC/RF		275	.145		7.10	6.25		13.35	17.40
0410	10" thick		260	.154		7.65	6.60		14.25	18.50
0420	12" thick		250	.160		8.25	6.85		15.10	19.55
0430	12" thick, type RSC/RF-4		250	.160		9.80	6.85		16.65	21.50
0500	NRC .60 type R, 8" thick		265	.151		8.45	6.50		14.95	19.20
0600	NRC .65 type RR, 8" thick		265	.151		7.90	6.50		14.40	18.55
0700	NRC .65 type 4R-RF, 8" thick		265	.151		8.70	6.50		15.20	19.45
0710	NRC .70 type R, 12" thick		245	.163		9.15	7		16.15	21
1000	Lightweight block, tooled joints, 2 sides, hollow									
1100	Not reinforced, 8" x 16" x 4" thick	D-8	440	.091	S.F.	1.80	3.90		5.70	7.95
1150	6" thick		410	.098		2.56	4.19		6.75	9.20
1200	8" thick		385	.104		3.13	4.46		7.59	10.25
1250	10" thick		370	.108		3.79	4.64		8.43	11.25
1300	12" thick	D-9	350	.137		4.01	5.75		9.76	13.20
2000	Not reinforced, 8" x 24" x 4" thick, hollow		460	.104		1.31	4.39		5.70	8.15
2100	6" thick		440	.109		1.84	4.59		6.43	9
2150	8" thick		415	.116		2.29	4.87		7.16	9.95
2200	10" thick		385	.125		2.77	5.25		8.02	11.05
2250	12" thick		365	.132		2.96	5.55		8.51	11.70
2800	Solid, not reinforced, 8" x 16" x 2" thick	D-8	440	.091		1.61	3.90		5.51	7.70
2900	4" thick		420	.095		1.63	4.09		5.72	8.05

04 22 10.34 Concrete Block, Partitions

	04 22 10.34 Concrete Block, Partitions	Crew	Daily Output	Labor-Hours	Unit	Material	2015 Bare Costs Labor	Equipment	Total	Total Incl O&P
2950	6" thick	D-8	390	.103	S.F.	2.96	4.40		7.36	9.95
3000	8" thick		365	.110		3.80	4.70		8.50	11.40
3050	10" thick		350	.114		3.91	4.90		8.81	11.80
3100	12" thick	D-9	330	.145		3.94	6.10		10.04	13.70
4000	Regular block, tooled joints, 2 sides, hollow									
4100	Not reinforced, 8" x 16" x 4" thick	D-8	430	.093	S.F.	1.70	3.99		5.69	7.95
4150	6" thick		400	.100		2.22	4.29		6.51	9
4200	8" thick		375	.107		2.74	4.58		7.32	10
4250	10" thick		360	.111		2.86	4.77		7.63	10.45
4300	12" thick	D-9	340	.141		3.95	5.95		9.90	13.45
4500	Reinforced alternate courses, 8" x 16" x 4" thick	D-8	425	.094		1.86	4.04		5.90	8.20
4550	6" thick		395	.101		2.39	4.35		6.74	9.30
4600	8" thick		370	.108		2.94	4.64		7.58	10.35
4650	10" thick		355	.113		3.98	4.84		8.82	11.80
4700	12" thick	D-9	335	.143		4.12	6.05		10.17	13.75
4900	Solid, not reinforced, 2" thick	D-8	435	.092		1.44	3.95		5.39	7.60
5000	3" thick		430	.093		1.35	3.99		5.34	7.60
5050	4" thick		415	.096		1.96	4.14		6.10	8.45
5100	6" thick		385	.104		2.39	4.46		6.85	9.45
5150	8" thick		360	.111		3.45	4.77		8.22	11.10
5200	12" thick	D-9	325	.148		5.25	6.20		11.45	15.25
5500	Solid, reinforced alternate courses, 4" thick	D-8	420	.095		2.11	4.09		6.20	8.55
5550	6" thick		380	.105		2.53	4.52		7.05	9.70
5600	8" thick		355	.113		3.59	4.84		8.43	11.35
5650	12" thick	D-9	320	.150		4.38	6.30		10.68	14.45

04 22 10.38 Concrete Brick

	04 22 10.38 Concrete Brick	Crew	Daily Output	Labor-Hours	Unit	Material	Labor	Equipment	Total	Total Incl O&P
0010	**CONCRETE BRICK**, C55, grade N, type 1									
0100	Regular, 4 x 2-1/4 x 8	D-8	660	.061	Ea.	.54	2.60		3.14	4.56
0125	Rusticated, 4 x 2-1/4 x 8		660	.061		.60	2.60		3.20	4.64
0150	Frog, 4 x 2-1/4 x 8		660	.061		.58	2.60		3.18	4.61
0200	Double, 4 x 4-7/8 x 8		535	.075		.95	3.21		4.16	5.95

04 22 10.42 Concrete Block, Screen Block

	04 22 10.42 Concrete Block, Screen Block	Crew	Daily Output	Labor-Hours	Unit	Material	Labor	Equipment	Total	Total Incl O&P
0010	**CONCRETE BLOCK, SCREEN BLOCK**									
0200	8" x 16", 4" thick	D-8	330	.121	S.F.	2.34	5.20		7.54	10.50
0300	8" thick		270	.148		3.71	6.35		10.06	13.80
0350	12" x 12", 4" thick		290	.138		4.27	5.90		10.17	13.75
0500	8" thick		250	.160		7.15	6.85		14	18.35

04 22 10.44 Glazed Concrete Block

	04 22 10.44 Glazed Concrete Block	Crew	Daily Output	Labor-Hours	Unit	Material	Labor	Equipment	Total	Total Incl O&P
0010	**GLAZED CONCRETE BLOCK** C744									
0100	Single face, 8" x 16" units, 2" thick	D-8	360	.111	S.F.	9.60	4.77		14.37	17.90
0200	4" thick		345	.116		9.90	4.98		14.88	18.50
0250	6" thick		330	.121		10.70	5.20		15.90	19.70
0300	8" thick		310	.129		11.45	5.55		17	21
0350	10" thick		295	.136		13.05	5.80		18.85	23.50
0400	12" thick	D-9	280	.171		13.95	7.20		21.15	26.50
0700	Double face, 8" x 16" units, 4" thick	D-8	340	.118		13.85	5.05		18.90	23
0750	6" thick		320	.125		17	5.35		22.35	27
0800	8" thick		300	.133		17.80	5.70		23.50	28.50
1000	Jambs, bullnose or square, single face, 8" x 16", 2" thick		315	.127	Ea.	18.35	5.45		23.80	28.50
1050	4" thick		285	.140	"	19.30	6		25.30	30
1200	Caps, bullnose or square, 8" x 16", 2" thick		420	.095	L.F.	17.05	4.09		21.14	25
1250	4" thick		380	.105	"	18.50	4.52		23.02	27.50

04 22 Concrete Unit Masonry

04 22 10 – Concrete Masonry Units

04 22 10.44 Glazed Concrete Block		Crew	Daily Output	Labor-Hours	Unit	Material	2015 Bare Costs Labor	Equipment	Total	Total Incl O&P
1256	Corner, bullnose or square, 2" thick	D-8	280	.143	Ea.	19.95	6.15		26.10	31.50
1258	4" thick		270	.148		21.50	6.35		27.85	33
1260	6" thick		260	.154		27	6.60		33.60	39.50
1270	8" thick		250	.160		32.50	6.85		39.35	46.50
1280	10" thick		240	.167		33	7.15		40.15	47.50
1290	12" thick		230	.174		35	7.45		42.45	50
1500	Cove base, 8" x 16", 2" thick		315	.127	L.F.	8.55	5.45		14	17.75
1550	4" thick		285	.140		8.65	6		14.65	18.75
1600	6" thick		265	.151		9.30	6.50		15.80	20
1650	8" thick		245	.163		9.75	7		16.75	21.50

04 23 Glass Unit Masonry

04 23 13 – Vertical Glass Unit Masonry

04 23 13.10 Glass Block		Crew	Daily Output	Labor-Hours	Unit	Material	2015 Bare Costs Labor	Equipment	Total	Total Incl O&P
0010	**GLASS BLOCK**									
0100	Plain, 4" thick, under 1,000 S.F., 6" x 6"	D-8	115	.348	S.F.	26.50	14.95		41.45	52
0150	8" x 8"		160	.250		15.35	10.75		26.10	33.50
0160	end block		160	.250		59	10.75		69.75	81.50
0170	90 deg corner		160	.250		53.50	10.75		64.25	75.50
0180	45 deg corner		160	.250		44	10.75		54.75	65
0200	12" x 12"		175	.229		22	9.80		31.80	39.50
0210	4" x 8"		160	.250		28	10.75		38.75	47.50
0220	6" x 8"		160	.250		19.90	10.75		30.65	38.50
0300	1,000 to 5,000 S.F., 6" x 6"		135	.296		26	12.70		38.70	48
0350	8" x 8"		190	.211		15.05	9.05		24.10	30.50
0400	12" x 12"		215	.186		21.50	8		29.50	36
0410	4" x 8"		215	.186		27.50	8		35.50	42
0420	6" x 8"		215	.186		19.50	8		27.50	33.50
0500	Over 5,000 S.F., 6" x 6"		145	.276		25	11.85		36.85	45.50
0550	8" x 8"		215	.186		14.60	8		22.60	28.50
0600	12" x 12"		240	.167		21	7.15		28.15	34
0610	4" x 8"		240	.167		26.50	7.15		33.65	40.50
0620	6" x 8"		240	.167		18.90	7.15		26.05	32
0700	For solar reflective blocks, add					100%				
1000	Thinline, plain, 3-1/8" thick, under 1,000 S.F., 6" x 6"	D-8	115	.348	S.F.	21	14.95		35.95	46
1050	8" x 8"		160	.250		11.60	10.75		22.35	29
1200	Over 5,000 S.F., 6" x 6"		145	.276		19.80	11.85		31.65	40
1250	8" x 8"		215	.186		11.05	8		19.05	24.50
1400	For cleaning block after installation (both sides), add		1000	.040		.16	1.72		1.88	2.80
4000	Accessories									
4100	Anchors, 20 ga. galv., 1-3/4" wide x 24" long				Ea.	4.96			4.96	5.45
4200	Emulsion asphalt				Gal.	10.65			10.65	11.70
4300	Expansion joint, fiberglass				L.F.	.61			.61	.67
4400	Steel mesh, double galvanized				"	.96			.96	1.06

04 24 Adobe Unit Masonry

04 24 16 – Manufactured Adobe Unit Masonry

04 24 16.06 Adobe Brick		Crew	Daily Output	Labor-Hours	Unit	Material	Labor	Equipment	Total	Total Incl O&P
						2015 Bare Costs				
0010	**ADOBE BRICK**, Semi-stabilized, with cement mortar									
0060	Brick, 10" x 4" x 14", 2.6/S.F. G	D-8	560	.071	S.F.	4.50	3.07		7.57	9.60
0080	12" x 4" x 16", 2.3/S.F. G		580	.069		6.85	2.96		9.81	12
0100	10" x 4" x 16", 2.3/S.F. G		590	.068		6.40	2.91		9.31	11.50
0120	8" x 4" x 16", 2.3/S.F. G		560	.071		4.89	3.07		7.96	10.10
0140	4" x 4" x 16", 2.3/S.F. G		540	.074		4.79	3.18		7.97	10.10
0160	6" x 4" x 16", 2.3/S.F. G		540	.074		4.44	3.18		7.62	9.75
0180	4" x 4" x 12", 3.0/S.F. G		520	.077		5.15	3.30		8.45	10.75
0200	8" x 4" x 12", 3.0/S.F. G		520	.077		4.16	3.30		7.46	9.65

04 25 Unit Masonry Panels

04 25 20 – Pre-Fabricated Masonry Panels

04 25 20.10 Brick and Epoxy Mortar Panels

		Crew	Daily Output	Labor-Hours	Unit	Material	Labor	Equipment	Total	Total Incl O&P
0010	**BRICK AND EPOXY MORTAR PANELS**									
0020	Prefabricated brick & epoxy mortar, 4" thick, minimum	C-11	775	.093	S.F.	8	4.83	2.34	15.17	19.50
0100	Maximum	"	500	.144		10	7.50	3.63	21.13	27.50
0200	For 2" concrete back-up, add					50%				
0300	For 1" urethane & 3" concrete back-up, add					70%				

04 27 Multiple-Wythe Unit Masonry

04 27 10 – Multiple-Wythe Masonry

04 27 10.20 Cavity Walls

		Crew	Daily Output	Labor-Hours	Unit	Material	Labor	Equipment	Total	Total Incl O&P
0010	**CAVITY WALLS**, brick and CMU, includes joint reinforcing and ties									
0200	4" face brick, 4" block	D-8	165	.242	S.F.	5.65	10.40		16.05	22
0400	6" block		145	.276		6	11.85		17.85	24.50
0600	8" block		125	.320		6.40	13.75		20.15	28

04 27 10.30 Brick Walls

		Crew	Daily Output	Labor-Hours	Unit	Material	Labor	Equipment	Total	Total Incl O&P
0010	**BRICK WALLS**, including mortar, excludes scaffolding R042110-20									
0020	Estimating by number of brick									
0140	Face brick, 4" thick wall, 6.75 brick/S.F.	D-8	1.45	27.586	M	540	1,175		1,715	2,400
0150	Common brick, 4" thick wall, 6.75 brick/S.F.		1.60	25		595	1,075		1,670	2,300
0204	8" thick, 13.50 bricks per S.F.		1.80	22.222		615	955		1,570	2,125
0250	12" thick, 20.25 bricks per S.F.		1.90	21.053		620	905		1,525	2,050
0304	16" thick, 27.00 bricks per S.F.		2	20		630	860		1,490	2,000
0500	Reinforced, face brick, 4" thick wall, 6.75 brick/S.F.		1.40	28.571		565	1,225		1,790	2,500
0520	Common brick, 4" thick wall, 6.75 brick/S.F.		1.55	25.806		620	1,100		1,720	2,375
0550	8" thick, 13.50 bricks per S.F.		1.75	22.857		640	980		1,620	2,200
0600	12" thick, 20.25 bricks per S.F.		1.85	21.622		645	930		1,575	2,125
0650	16" thick, 27.00 bricks per S.F.		1.95	20.513		655	880		1,535	2,075
0790	Alternate method of figuring by square foot									
0800	Face brick, 4" thick wall, 6.75 brick/S.F.	D-8	215	.186	S.F.	3.65	8		11.65	16.20
0850	Common brick, 4" thick wall, 6.75 brick/S.F.		240	.167		4.03	7.15		11.18	15.35
0900	8" thick, 13.50 bricks per S.F.		135	.296		8.30	12.70		21	28.50
1000	12" thick, 20.25 bricks per S.F.		95	.421		12.55	18.05		30.60	41.50
1050	16" thick, 27.00 bricks per S.F.		75	.533		17	23		40	53.50
1200	Reinforced, face brick, 4" thick wall, 6.75 brick/S.F.		210	.190		3.82	8.15		11.97	16.70
1220	Common brick, 4" thick wall, 6.75 brick/S.F.		235	.170		4.20	7.30		11.50	15.75
1250	8" thick, 13.50 bricks per S.F.		130	.308		8.65	13.20		21.85	29.50
1300	12" thick, 20.25 bricks per S.F.		90	.444		13.05	19.05		32.10	43.50

04 27 Multiple-Wythe Unit Masonry

04 27 10 – Multiple-Wythe Masonry

04 27 10.30 Brick Walls			Crew	Daily Output	Labor-Hours	Unit	Material	2015 Bare Costs Labor	Equipment	Total	Total Incl O&P
1350	16″ thick, 27.00 bricks per S.F.		D-8	70	.571	S.F.	17.65	24.50		42.15	57

04 27 10.40 Steps			Crew	Daily Output	Labor-Hours	Unit	Material	Labor	Equipment	Total	Total Incl O&P
0010	**STEPS**										
0012	Entry steps, select common brick		D-1	.30	53.333	M	500	2,250		2,750	3,975

04 41 Dry-Placed Stone

04 41 10 – Dry Placed Stone

04 41 10.10 Rough Stone Wall			Crew	Daily Output	Labor-Hours	Unit	Material	Labor	Equipment	Total	Total Incl O&P
0011	**ROUGH STONE WALL**, Dry										
0012	Dry laid (no mortar), under 18″ thick	G	D-1	60	.267	C.F.	12.90	11.25		24.15	31.50
0100	Random fieldstone, under 18″ thick	G	D-12	60	.533		12.90	22.50		35.40	48.50
0150	Over 18″ thick	G	″	63	.508		15.50	21.50		37	50
0500	Field stone veneer	G	D-8	120	.333	S.F.	12.20	14.30		26.50	35.50
0510	Valley stone veneer	G		120	.333		12.20	14.30		26.50	35.50
0520	River stone veneer	G		120	.333		12.20	14.30		26.50	35.50
0600	Rubble stone walls, in mortar bed, up to 18″ thick	G	D-11	75	.320	C.F.	15.55	14.10		29.65	38.50

04 43 Stone Masonry

04 43 10 – Masonry with Natural and Processed Stone

04 43 10.05 Ashlar Veneer		Crew	Daily Output	Labor-Hours	Unit	Material	Labor	Equipment	Total	Total Incl O&P
0011	**ASHLAR VENEER** 4″ + or - thk, random or random rectangular									
0150	Sawn face, split joints, low priced stone	D-8	140	.286	S.F.	11.60	12.25		23.85	31.50
0200	Medium priced stone		130	.308		13.35	13.20		26.55	34.50
0300	High priced stone		120	.333		17.90	14.30		32.20	41.50
0600	Seam face, split joints, medium price stone		125	.320		19	13.75		32.75	42
0700	High price stone		120	.333		18.75	14.30		33.05	42.50
1000	Split or rock face, split joints, medium price stone		125	.320		11.40	13.75		25.15	33.50
1100	High price stone		120	.333		17.25	14.30		31.55	41

04 43 10.10 Bluestone		Crew	Daily Output	Labor-Hours	Unit	Material	Labor	Equipment	Total	Total Incl O&P
0010	**BLUESTONE**, cut to size									
0500	Sills, natural cleft, 10″ wide to 6′ long, 1-1/2″ thick	D-11	70	.343	L.F.	13.35	15.15		28.50	37.50
0550	2″ thick	″	63	.381		14.60	16.80		31.40	41.50
1000	Stair treads, natural cleft, 12″ wide, 6′ long, 1-1/2″ thick	D-10	115	.278		12	12.80	4.12	28.92	37
1050	2″ thick		105	.305		13	14	4.51	31.51	41
1100	Smooth finish, 1-1/2″ thick		115	.278		12	12.80	4.12	28.92	37
1150	2″ thick		105	.305		13	14	4.51	31.51	41
1300	Thermal finish, 1-1/2″ thick		115	.278		12	12.80	4.12	28.92	37
1350	2″ thick		105	.305		13	14	4.51	31.51	41

04 43 10.45 Granite		Crew	Daily Output	Labor-Hours	Unit	Material	Labor	Equipment	Total	Total Incl O&P
0010	**GRANITE**, cut to size									
0050	Veneer, polished face, 3/4″ to 1-1/2″ thick									
0150	Low price, gray, light gray, etc.	D-10	130	.246	S.F.	26.50	11.35	3.64	41.49	51
0180	Medium price, pink, brown, etc.		130	.246		29.50	11.35	3.64	44.49	54
0220	High price, red, black, etc.		130	.246		42	11.35	3.64	56.99	68
0300	1-1/2″ to 2-1/2″ thick, veneer									
0350	Low price, gray, light gray, etc.	D-10	130	.246	S.F.	28.50	11.35	3.64	43.49	53
0500	Medium price, pink, brown, etc.		130	.246		33.50	11.35	3.64	48.49	58.50
0550	High price, red, black, etc.		130	.246		52.50	11.35	3.64	67.49	79

04 43 Stone Masonry

04 43 10 – Masonry with Natural and Processed Stone

04 43 10.45 Granite

04 43 10.45 Granite		Crew	Daily Output	Labor-Hours	Unit	Material	2015 Bare Costs Labor	Equipment	Total	Total Incl O&P
0700	2-1/2" to 4" thick, veneer									
0750	Low price, gray, light gray, etc.	D-10	110	.291	S.F.	38.50	13.40	4.30	56.20	67.50
0850	Medium price, pink, brown, etc.		110	.291		44	13.40	4.30	61.70	73.50
0950	High price, red, black, etc.		110	.291		63	13.40	4.30	80.70	94.50
1000	For bush hammered finish, deduct					5%				
1050	Coarse rubbed finish, deduct					10%				
1100	Honed finish, deduct					5%				
1150	Thermal finish, deduct					18%				
2450	For radius under 5', add				L.F.	100%				
2500	Steps, copings, etc., finished on more than one surface									
2550	Low price, gray, light gray, etc.	D-10	50	.640	C.F.	91	29.50	9.45	129.95	155
2575	Medium price, pink, brown, etc.		50	.640		118	29.50	9.45	156.95	185
2600	High price, red, black, etc.		50	.640		146	29.50	9.45	184.95	215
2800	Pavers, 4" x 4" x 4" blocks, split face and joints									
2850	Low price, gray, light gray, etc.	D-11	80	.300	S.F.	13.10	13.25		26.35	34.50
2875	Medium price, pinks, browns, etc.		80	.300		21	13.25		34.25	43
2900	High price, red, black, etc.		80	.300		29	13.25		42.25	52
4000	Soffits, 2" thick, low price, gray, light gray	D-13	35	1.371		37.50	61.50	13.50	112.50	150
4050	Medium price, pink, brown, etc.		35	1.371		64.50	61.50	13.50	139.50	180
4100	High price, red, black, etc.		35	1.371		91.50	61.50	13.50	166.50	209
4200	Low price, gray, light gray, etc.		35	1.371		63	61.50	13.50	138	178
4250	Medium price, pink, brown, etc.		35	1.371		91	61.50	13.50	166	209
4300	High price, red, black, etc.		35	1.371		119	61.50	13.50	194	240

04 43 10.50 Lightweight Natural Stone

04 43 10.50 Lightweight Natural Stone		Crew	Daily Output	Labor-Hours	Unit	Material	Labor	Equipment	Total	Total Incl O&P
0011	**LIGHTWEIGHT NATURAL STONE** Lava type									
0100	Veneer, rubble face, sawed back, irregular shapes G	D-10	130	.246	S.F.	8.50	11.35	3.64	23.49	30.50
0200	Sawed face and back, irregular shapes G	"	130	.246	"	8.50	11.35	3.64	23.49	30.50

04 43 10.55 Limestone

04 43 10.55 Limestone		Crew	Daily Output	Labor-Hours	Unit	Material	Labor	Equipment	Total	Total Incl O&P
0010	**LIMESTONE**, cut to size									
0020	Veneer facing panels									
0500	Texture finish, light stick, 4-1/2" thick, 5' x 12'	D-4	300	.107	S.F.	19.50	4.56	.44	24.50	29
0750	5" thick, 5' x 14' panels	D-10	275	.116		20.50	5.35	1.72	27.57	32.50
1000	Sugarcube finish, 2" Thick, 3' x 5' panels		275	.116		28.50	5.35	1.72	35.57	41.50
1050	3" Thick, 4' x 9' panels		275	.116		22.50	5.35	1.72	29.57	35
1200	4" Thick, 5' x 11' panels		275	.116		30	5.35	1.72	37.07	43
1400	Sugarcube, textured finish, 4-1/2" thick, 5' x 12'		275	.116		31	5.35	1.72	38.07	44
1450	5" thick, 5' x 14' panels		275	.116		32	5.35	1.72	39.07	45.50
2000	Coping, sugarcube finish, top & 2 sides		30	1.067	C.F.	67.50	49	15.80	132.30	166
2100	Sills, lintels, jambs, trim, stops, sugarcube finish, simple		20	1.600		67.50	73.50	23.50	164.50	212
2150	Detailed		20	1.600		67.50	73.50	23.50	164.50	212
2300	Steps, extra hard, 14" wide, 6" rise		50	.640	L.F.	25	29.50	9.45	63.95	83
3000	Quoins, plain finish, 6" x 12" x 12"	D-12	25	1.280	Ea.	37.50	54.50		92	125
3050	6" x 16" x 24"	"	25	1.280	"	50	54.50		104.50	139

04 43 10.60 Marble

04 43 10.60 Marble		Crew	Daily Output	Labor-Hours	Unit	Material	Labor	Equipment	Total	Total Incl O&P
0011	**MARBLE**, ashlar, split face, 4" + or - thick, random									
0040	Lengths 1' to 4' & heights 2" to 7-1/2", average	D-8	175	.229	S.F.	17.45	9.80		27.25	34
0100	Base, polished, 3/4" or 7/8" thick, polished, 6" high	D-10	65	.492	L.F.	12	22.50	7.30	41.80	55.50
0300	Carvings or bas relief, from templates, simple design		80	.400	S.F.	144	18.40	5.90	168.30	193
0350	Intricate design		80	.400	"	335	18.40	5.90	359.30	405
0600	Columns, cornices, mouldings, etc.									
0650	Hand or special machine cut, simple design	D-10	35	.914	C.F.	52.50	42	13.50	108	136
0700	Intricate design	"	35	.914	"	278	42	13.50	333.50	385

04 43 Stone Masonry

04 43 10 – Masonry with Natural and Processed Stone

04 43 10.60 Marble		Crew	Daily Output	Labor-Hours	Unit	2015 Bare Costs Material	Labor	Equipment	Total	Total Incl O&P
1000	Facing, polished finish, cut to size, 3/4″ to 7/8″ thick									
1050	Carrara or equal	D-10	130	.246	S.F.	19	11.35	3.64	33.99	42.50
1100	Arabescato or equal		130	.246		39	11.35	3.64	53.99	64.50
1300	1-1/4″ thick, Botticino Classico or equal		125	.256		24	11.80	3.79	39.59	48.50
1350	Statuarietto or equal		125	.256		41.50	11.80	3.79	57.09	67.50
1500	2″ thick, Crema Marfil or equal		120	.267		42.50	12.25	3.94	58.69	70
1550	Cafe Pinta or equal		120	.267		70	12.25	3.94	86.19	100
1700	Rubbed finish, cut to size, 4″ thick									
1740	Average	D-10	100	.320	S.F.	40	14.70	4.73	59.43	71.50
1780	Maximum	″	100	.320	″	69	14.70	4.73	88.43	104
2200	Window sills, 6″ x 3/4″ thick	D-1	85	.188	L.F.	10.90	7.90		18.80	24
2500	Flooring, polished tiles, 12″ x 12″ x 3/8″ thick									
2510	Thin set, Giallo Solare or equal	D-11	90	.267	S.F.	14.75	11.75		26.50	34
2600	Sky Blue or equal		90	.267		16	11.75		27.75	35.50
2700	Mortar bed, Giallo Solare or equal		65	.369		14.75	16.30		31.05	41.50
2740	Sky Blue or equal		65	.369		16	16.30		32.30	42.50
2780	Travertine, 3/8″ thick, Sierra or equal	D-10	130	.246		9.25	11.35	3.64	24.24	31.50
2790	Silver or equal	″	130	.246		25.50	11.35	3.64	40.49	49.50
2800	Patio tile, non-slip, 1/2″ thick, flame finish	D-11	75	.320		11.90	14.10		26	34.50
2900	Shower or toilet partitions, 7/8″ thick partitions									
3050	3/4″ or 1-1/4″ thick stiles, polished 2 sides, average	D-11	75	.320	S.F.	45.50	14.10		59.60	71.50
3201	Soffits, add to above prices				″	20%	100%			
3210	Stairs, risers, 7/8″ thick x 6″ high	D-10	115	.278	L.F.	15.25	12.80	4.12	32.17	41
3360	Treads, 12″ wide x 1-1/4″ thick	″	115	.278	″	43.50	12.80	4.12	60.42	72
3500	Thresholds, 3′ long, 7/8″ thick, 4″ to 5″ wide, plain	D-12	24	1.333	Ea.	35.50	57		92.50	126
3550	Beveled		24	1.333	″	70.50	57		127.50	164
3700	Window stools, polished, 7/8″ thick, 5″ wide		85	.376	L.F.	21.50	16.05		37.55	48

04 43 10.75 Sandstone or Brownstone

		Crew	Daily Output	Labor-Hours	Unit	Material	Labor	Equipment	Total	Total Incl O&P
0011	**SANDSTONE OR BROWNSTONE**									
0100	Sawed face veneer, 2-1/2″ thick, to 2′ x 4′ panels	D-10	130	.246	S.F.	19	11.35	3.64	33.99	42.50
0150	4″ thick, to 3′-6″ x 8′panels		100	.320		19	14.70	4.73	38.43	48.50
0300	Split face, random sizes		100	.320		13.65	14.70	4.73	33.08	42.50
0350	Cut stone trim (limestone)									
0360	Ribbon stone, 4″ thick, 5′ pieces	D-8	120	.333	Ea.	155	14.30		169.30	192
0370	Cove stone, 4″ thick, 5′ pieces		105	.381		156	16.35		172.35	196
0380	Cornice stone, 10″ to 12″ wide		90	.444		192	19.05		211.05	240
0390	Band stone, 4″ thick, 5′ pieces		145	.276		99.50	11.85		111.35	127
0410	Window and door trim, 3″ to 4″ wide		160	.250		84.50	10.75		95.25	109
0420	Key stone, 18″ long		60	.667		88.50	28.50		117	141

04 43 10.80 Slate

		Crew	Daily Output	Labor-Hours	Unit	Material	Labor	Equipment	Total	Total Incl O&P
0010	**SLATE**									
0040	Pennsylvania - blue gray to black									
0050	Vermont - unfading green, mottled green & purple, gray & purple									
0100	Virginia - blue black									
0200	Exterior paving, natural cleft, 1″ thick									
0250	6″ x 6″ Pennsylvania	D-12	100	.320	S.F.	7.15	13.65		20.80	29
0300	Vermont		100	.320		11.25	13.65		24.90	33.50
0350	Virginia		100	.320		14.95	13.65		28.60	37.50
0500	24″ x 24″, Pennsylvania		120	.267		13.80	11.35		25.15	32.50
0550	Vermont		120	.267		28	11.35		39.35	48
0600	Virginia		120	.267		21.50	11.35		32.85	41.50
0700	18″ x 30″ Pennsylvania		120	.267		15.65	11.35		27	34.50

04 43 Stone Masonry

04 43 10 – Masonry with Natural and Processed Stone

04 43 10.80 Slate		Crew	Daily Output	Labor-Hours	Unit	Material	2015 Bare Costs Labor	Equipment	Total	Total Incl O&P
0750	Vermont	D-12	120	.267	S.F.	28	11.35		39.35	48
0800	Virginia	↓	120	.267	↓	19.30	11.35		30.65	38.50
1000	Interior flooring, natural cleft, 1/2" thick									
1100	6" x 6" Pennsylvania	D-12	100	.320	S.F.	4.24	13.65		17.89	25.50
1150	Vermont		100	.320		9.90	13.65		23.55	32
1200	Virginia		100	.320		11.80	13.65		25.45	34
1300	24" x 24" Pennsylvania		120	.267		8.20	11.35		19.55	26.50
1350	Vermont		120	.267		22.50	11.35		33.85	42.50
1400	Virginia		120	.267		15.60	11.35		26.95	34.50
1500	18" x 24" Pennsylvania		120	.267		8.20	11.35		19.55	26.50
1550	Vermont		120	.267		18	11.35		29.35	37
1600	Virginia	↓	120	.267	↓	15.85	11.35		27.20	35
2000	Facing panels, 1-1/4" thick, to 4' x 4' panels									
2100	Natural cleft finish, Pennsylvania	D-10	180	.178	S.F.	36	8.20	2.63	46.83	55
2110	Vermont		180	.178		29	8.20	2.63	39.83	47
2120	Virginia	↓	180	.178		35	8.20	2.63	45.83	54.50
2150	Sand rubbed finish, surface, add					10.80			10.80	11.85
2200	Honed finish, add					7.80			7.80	8.55
2500	Ribbon, natural cleft finish, 1" thick, to 9 S.F.	D-10	80	.400		13.90	18.40	5.90	38.20	50
2550	Sand rubbed finish		80	.400		18.80	18.40	5.90	43.10	55
2600	Honed finish		80	.400		17.50	18.40	5.90	41.80	54
2700	1-1/2" thick		78	.410		18.05	18.90	6.05	43	55.50
2750	Sand rubbed finish		78	.410		24	18.90	6.05	48.95	61.50
2800	Honed finish		78	.410		22.50	18.90	6.05	47.45	60.50
2850	2" thick		76	.421		21.50	19.35	6.25	47.10	60.50
2900	Sand rubbed finish		76	.421		30	19.35	6.25	55.60	69.50
2950	Honed finish	↓	76	.421		27.50	19.35	6.25	53.10	67
3100	Stair landings, 1" thick, black, clear	D-1	65	.246		21	10.35		31.35	39.50
3200	Ribbon	"	65	.246	↓	23.50	10.35		33.85	41.50
3500	Stair treads, sand finish, 1" thick x 12" wide									
3550	Under 3 L.F.	D-10	85	.376	L.F.	23.50	17.30	5.55	46.35	58
3600	3 L.F. to 6 L.F.	"	120	.267	"	25	12.25	3.94	41.19	50.50
3700	Ribbon, sand finish, 1" thick x 12" wide									
3750	To 6 L.F.	D-10	120	.267	L.F.	21	12.25	3.94	37.19	46.50
4000	Stools or sills, sand finish, 1" thick, 6" wide	D-12	160	.200		12.20	8.50		20.70	26.50
4100	Honed finish		160	.200		11.65	8.50		20.15	26
4200	10" wide		90	.356		18.80	15.15		33.95	43.50
4250	Honed finish		90	.356		17.50	15.15		32.65	42.50
4400	2" thick, 6" wide		140	.229		19.60	9.75		29.35	36.50
4450	Honed finish		140	.229		18.65	9.75		28.40	35.50
4600	10" wide		90	.356		30.50	15.15		45.65	57
4650	Honed finish	↓	90	.356		29	15.15		44.15	55
4800	For lengths over 3', add				↓	25%				

04 43 10.85 Window Sill		Crew	Daily Output	Labor-Hours	Unit	Material	Labor	Equipment	Total	Total Incl O&P
0010	**WINDOW SILL**									
0020	Bluestone, thermal top, 10" wide, 1-1/2" thick	D-1	85	.188	S.F.	10	7.90		17.90	23
0050	2" thick		75	.213	"	10	9		19	24.50
0100	Cut stone, 5" x 8" plain		48	.333	L.F.	12.10	14.05		26.15	35
0200	Face brick on edge, brick, 8" wide		80	.200		5.05	8.40		13.45	18.40
0400	Marble, 9" wide, 1" thick		85	.188		8.65	7.90		16.55	21.50
0900	Slate, colored, unfading, honed, 12" wide, 1" thick		85	.188		8.50	7.90		16.40	21.50
0950	2" thick	↓	70	.229	↓	8.50	9.60		18.10	24

04 51 Flue Liner Masonry

04 51 10 – Clay Flue Lining

04 51 10.10 Flue Lining		Crew	Daily Output	Labor-Hours	Unit	2015 Bare Costs Material	Labor	Equipment	Total	Total Incl O&P
0010	**FLUE LINING**, including mortar									
0020	Clay, 8″ x 8″	D-1	125	.128	V.L.F.	5.50	5.40		10.90	14.30
0100	8″ x 12″		103	.155		8.15	6.55		14.70	18.95
0200	12″ x 12″		93	.172		10.75	7.25		18	23
0300	12″ x 18″		84	.190		21.50	8		29.50	36
0400	18″ x 18″		75	.213		28	9		37	44
0500	20″ x 20″		66	.242		40	10.20		50.20	59.50
0600	24″ x 24″		56	.286		51.50	12.05		63.55	75.50
1000	Round, 18″ diameter		66	.242		37	10.20		47.20	56.50
1100	24″ diameter		47	.340		72.50	14.35		86.85	102

04 54 Refractory Brick Masonry

04 54 10 – Refractory Brick Work

04 54 10.10 Fire Brick		Crew	Daily Output	Labor-Hours	Unit	Material	Labor	Equipment	Total	Total Incl O&P
0010	**FIRE BRICK**									
0012	Low duty, 2000°F, 9″ x 2-1/2″ x 4-1/2″	D-1	.60	26.667	M	1,425	1,125		2,550	3,300
0050	High duty, 3000°F	″	.60	26.667	″	2,600	1,125		3,725	4,575

04 54 10.20 Fire Clay		Crew	Daily Output	Labor-Hours	Unit	Material	Labor	Equipment	Total	Total Incl O&P
0010	**FIRE CLAY**									
0020	Gray, high duty, 100 lb. bag				Bag	33			33	36.50
0050	100 lb. drum, premixed (400 brick per drum)				Drum	41			41	45

04 57 Masonry Fireplaces

04 57 10 – Brick or Stone Fireplaces

04 57 10.10 Fireplace		Crew	Daily Output	Labor-Hours	Unit	Material	Labor	Equipment	Total	Total Incl O&P
0010	**FIREPLACE**									
0100	Brick fireplace, not incl. foundations or chimneys									
0110	30″ x 29″ opening, incl. chamber, plain brickwork	D-1	.40	40	Ea.	550	1,675		2,225	3,175
0200	Fireplace box only (110 brick)	″	2	8	″	157	335		492	685
0300	For elaborate brickwork and details, add					35%	35%			
0400	For hearth, brick & stone, add	D-1	2	8	Ea.	205	335		540	740
0410	For steel, damper, cleanouts, add		4	4		19.70	168		187.70	279
0600	Plain brickwork, incl. metal circulator		.50	32		995	1,350		2,345	3,150
0800	Face brick only, standard size, 8″ x 2-2/3″ x 4″		.30	53.333	M	550	2,250		2,800	4,025

04 71 Manufactured Brick Masonry

04 71 10 – Simulated or Manufactured Brick

04 71 10.10 Simulated Brick		Crew	Daily Output	Labor-Hours	Unit	Material	Labor	Equipment	Total	Total Incl O&P
0010	**SIMULATED BRICK**									
0020	Aluminum, baked on colors	1 Carp	200	.040	S.F.	4.31	1.88		6.19	7.65
0050	Fiberglass panels		200	.040		8.05	1.88		9.93	11.75
0100	Urethane pieces cemented in mastic		150	.053		8.25	2.50		10.75	12.90
0150	Vinyl siding panels		200	.040		10	1.88		11.88	13.90
0160	Cement base, brick, incl. mastic	D-1	100	.160		9.15	6.75		15.90	20.50
0170	Corner		50	.320	V.L.F.	23	13.45		36.45	46
0180	Stone face, incl. mastic		100	.160	S.F.	9.85	6.75		16.60	21
0190	Corner		50	.320	V.L.F.	28	13.45		41.45	51.50

04 72 Cast Stone Masonry

04 72 10 – Cast Stone Masonry Features

04 72 10.10 Coping		Crew	Daily Output	Labor-Hours	Unit	Material	2015 Bare Costs Labor	Equipment	Total	Total Incl O&P
0010	**COPING**, stock units									
0050	Precast concrete, 10" wide, 4" tapers to 3-1/2", 8" wall	D-1	75	.213	L.F.	17.10	9		26.10	32.50
0100	12" wide, 3-1/2" tapers to 3", 10" wall		70	.229		18.45	9.60		28.05	35
0110	14" wide, 4" tapers to 3-1/2", 12" wall		65	.246		20.50	10.35		30.85	38.50
0150	16" wide, 4" tapers to 3-1/2", 14" wall		60	.267		22	11.25		33.25	41.50
0250	Precast concrete corners		40	.400	Ea.	29	16.85		45.85	57
0300	Limestone for 12" wall, 4" thick		90	.178	L.F.	14.70	7.50		22.20	27.50
0350	6" thick		80	.200		22	8.40		30.40	37.50
0500	Marble, to 4" thick, no wash, 9" wide		90	.178		12.05	7.50		19.55	24.50
0550	12" wide		80	.200		16.15	8.40		24.55	30.50
0700	Terra cotta, 9" wide		90	.178		6.35	7.50		13.85	18.40
0750	12" wide		80	.200		8.65	8.40		17.05	22.50
0800	Aluminum, for 12" wall		80	.200		8.90	8.40		17.30	22.50

04 72 20 – Cultured Stone Veneer

04 72 20.10 Cultured Stone Veneer Components		Crew	Daily Output	Labor-Hours	Unit	Material	Labor	Equipment	Total	Total Incl O&P
0010	**CULTURED STONE VENEER COMPONENTS**									
0110	On wood frame and sheathing substrate, random sized cobbles, corner stones	D-8	70	.571	V.L.F.	10.15	24.50		34.65	48.50
0120	Field stones		140	.286	S.F.	7.40	12.25		19.65	27
0130	Random sized flats, corner stones		70	.571	V.L.F.	10.05	24.50		34.55	48.50
0140	Field stones		140	.286	S.F.	8.70	12.25		20.95	28.50
0150	Horizontal lined ledgestones, corner stones		75	.533	V.L.F.	10.15	23		33.15	46
0160	Field stones		150	.267	S.F.	7.40	11.45		18.85	25.50
0170	Random shaped flats, corner stones		65	.615	V.L.F.	10.15	26.50		36.65	51.50
0180	Field stones		150	.267	S.F.	7.40	11.45		18.85	25.50
0190	Random shaped/textured face, corner stones		65	.615	V.L.F.	10.15	26.50		36.65	51.50
0200	Field stones		130	.308	S.F.	7.40	13.20		20.60	28
0210	Random shaped river rock, corner stones		65	.615	V.L.F.	10.15	26.50		36.65	51.50
0220	Field stones		130	.308	S.F.	7.40	13.20		20.60	28
0240	On concrete or CMU substrate, random sized cobbles, corner stones		70	.571	V.L.F.	9.55	24.50		34.05	48
0250	Field stones		140	.286	S.F.	7.10	12.25		19.35	26.50
0260	Random sized flats, corner stones		70	.571	V.L.F.	9.45	24.50		33.95	48
0270	Field stones		140	.286	S.F.	8.40	12.25		20.65	28
0280	Horizontal lined ledgestones, corner stones		75	.533	V.L.F.	9.55	23		32.55	45.50
0290	Field stones		150	.267	S.F.	7.10	11.45		18.55	25.50
0300	Random shaped flats, corner stones		70	.571	V.L.F.	9.55	24.50		34.05	48
0310	Field stones		140	.286	S.F.	7.10	12.25		19.35	26.50
0320	Random shaped/textured face, corner stones		65	.615	V.L.F.	9.55	26.50		36.05	51
0330	Field stones		130	.308	S.F.	7.10	13.20		20.30	28
0340	Random shaped river rock, corner stones		65	.615	V.L.F.	9.55	26.50		36.05	51
0350	Field stones		130	.308	S.F.	7.10	13.20		20.30	28
0360	Cultured stone veneer, #15 felt weather resistant barrier	1 Clab	3700	.002	Sq.	5.40	.08		5.48	6.10
0370	Expanded metal lath, diamond, 2.5 lb./S.Y., galvanized	1 Lath	85	.094	S.Y.	2.72	4.05		6.77	8.95
0390	Water table or window sill, 18" long	1 Bric	80	.100	Ea.	10.30	4.62		14.92	18.40

04 73 Manufactured Stone Masonry

04 73 20 – Simulated or Manufactured Stone

04 73 20.10 Simulated Stone		Crew	Daily Output	Labor-Hours	Unit	2015 Bare Costs Material	Labor	Equipment	Total	Total Incl O&P
0010	**SIMULATED STONE**									
0100	Insulated fiberglass panels, 5/8" ply backer	L-4	200	.120	S.F.	10.80	5.30		16.10	20

Estimating Tips

05 05 00 Common Work Results for Metals

- Nuts, bolts, washers, connection angles, and plates can add a significant amount to both the tonnage of a structural steel job and the estimated cost. As a rule of thumb, add 10% to the total weight to account for these accessories.
- Type 2 steel construction, commonly referred to as "simple construction," consists generally of field-bolted connections with lateral bracing supplied by other elements of the building, such as masonry walls or x-bracing. The estimator should be aware, however, that shop connections may be accomplished by welding or bolting. The method may be particular to the fabrication shop and may have an impact on the estimated cost.

05 10 00 Structural Steel

- Steel items can be obtained from two sources: a fabrication shop or a metals service center. Fabrication shops can fabricate items under more controlled conditions than can crews in the field. They are also more efficient and can produce items more economically. Metal service centers serve as a source of long mill shapes to both fabrication shops and contractors.
- Most line items in this structural steel subdivision, and most items in 05 50 00 Metal Fabrications, are indicated as being shop fabricated. The bare material cost for these shop fabricated items is the "Invoice Cost" from the shop and includes the mill base price of steel plus mill extras, transportation to the shop, shop drawings and detailing where warranted, shop fabrication and handling, sandblasting and a shop coat of primer paint, all necessary structural bolts, and delivery to the job site. The bare labor cost and bare equipment cost for these shop fabricated items is for field installation or erection.
- Line items in Subdivision 05 12 23.40 Lightweight Framing, and other items scattered in Division 5, are indicated as being field fabricated. The bare material cost for these field fabricated items is the "Invoice Cost" from the metals service center and includes the mill base price of steel plus mill extras, transportation to the metals service center, material handling, and delivery of long lengths of mill shapes to the job site. Material costs for structural bolts and welding rods should be added to the estimate. The bare labor cost and bare equipment cost for these items is for both field fabrication and field installation or erection, and include time for cutting, welding and drilling in the fabricated metal items. Drilling into concrete and fasteners to fasten field fabricated items to other work are not included and should be added to the estimate.

05 20 00 Steel Joist Framing

- In any given project the total weight of open web steel joists is determined by the loads to be supported and the design. However, economies can be realized in minimizing the amount of labor used to place the joists. This is done by maximizing the joist spacing, and therefore minimizing the number of joists required to be installed on the job. Certain spacings and locations may be required by the design, but in other cases maximizing the spacing and keeping it as uniform as possible will keep the costs down.

05 30 00 Steel Decking

- The takeoff and estimating of metal deck involves more than simply the area of the floor or roof and the type of deck specified or shown on the drawings. Many different sizes and types of openings may exist. Small openings for individual pipes or conduits may be drilled after the floor/roof is installed, but larger openings may require special deck lengths as well as reinforcing or structural support. The estimator should determine who will be supplying this reinforcing. Additionally, some deck terminations are part of the deck package, such as screed angles and pour stops, and others will be part of the steel contract, such as angles attached to structural members and cast-in-place angles and plates. The estimator must ensure that all pieces are accounted for in the complete estimate.

05 50 00 Metal Fabrications

- The most economical steel stairs are those that use common materials, standard details, and most importantly, a uniform and relatively simple method of field assembly. Commonly available A36 channels and plates are very good choices for the main stringers of the stairs, as are angles and tees for the carrier members. Risers and treads are usually made by specialty shops, and it is most economical to use a typical detail in as many places as possible. The stairs should be pre-assembled and shipped directly to the site. The field connections should be simple and straightforward to be accomplished efficiently, and with minimum equipment and labor.

Reference Numbers

Reference numbers are shown in shaded boxes at the beginning of some major classifications. These numbers refer to related items in the Reference Section. The reference information may be an estimating procedure, an alternate pricing method, or technical information.

Note: Not all subdivisions listed here necessarily appear in this publication. ■

05 01 Maintenance of Metals

05 01 10 – Maintenance of Structural Metal Framing

05 01 10.51 Cleaning of Structural Metal Framing		Crew	Daily Output	Labor-Hours	Unit	Material	2015 Bare Costs Labor	Equipment	Total	Total Incl O&P
0010	**CLEANING OF STRUCTURAL METAL FRAMING**									
6125	Steel surface treatments, PDCA guidelines									
6170	Wire brush, hand (SSPC-SP2)	1 Psst	400	.020	S.F.	.02	.83		.85	1.49
6180	Power tool (SSPC-SP3)	"	700	.011		.09	.47		.56	.92
6215	Pressure washing, up to 5000 psi, 5000-15,000 S.F./day	1 Pord	10000	.001			.03		.03	.05
6220	Steam cleaning, 600 psi @ 300 F, 1250 – 2500 S.F./day		2000	.004			.16		.16	.24
6225	Water blasting, up to 25,000 psi, 1750 – 3500 S.F./day		2500	.003			.13		.13	.19
6230	Brush-off blast (SSPC-SP7)	E-11	1750	.018		.15	.77	.13	1.05	1.59
6235	Com'l blast (SSPC-SP6), loose scale, fine pwder rust, 2.0#/S.F. sand		1200	.027		.31	1.13	.20	1.64	2.41
6240	Tight mill scale, little/no rust, 3.0#/S.F. sand		1000	.032		.46	1.35	.24	2.05	2.99
6245	Exist coat blistered/pitted, 4.0#/S.F. sand		875	.037		.62	1.55	.27	2.44	3.52
6250	Exist coat badly pitted/nodules, 6.7#/S.F. sand		825	.039		1.03	1.64	.29	2.96	4.14
6255	Near white blast (SSPC-SP10), loose scale, fine rust, 5.6#/S.F. sand		450	.071		.86	3.01	.52	4.39	6.45
6260	Tight mill scale, little/no rust, 6.9#/S.F. sand		325	.098		1.06	4.16	.73	5.95	8.80
6265	Exist coat blistered/pitted, 9.0#/S.F. sand		225	.142		1.39	6	1.05	8.44	12.55
6270	Exist coat badly pitted/nodules, 11.3#/S.F. sand		150	.213		1.74	9	1.57	12.31	18.45

05 05 Common Work Results for Metals

05 05 05 – Selective Demolition for Metals

05 05 05.10 Selective Demolition, Metals

Line	Description	Crew	Daily Output	Labor-Hours	Unit	Material	Labor	Equipment	Total	Total Incl O&P
0010	**SELECTIVE DEMOLITION, METALS** R024119-10									
0015	Excludes shores, bracing, cutting, loading, hauling, dumping									
0020	Remove nuts only up to 3/4" diameter	1 Sswk	480	.017	Ea.		.88		.88	1.52
0030	7/8" to 1-1/4" diameter		240	.033			1.75		1.75	3.05
0040	1-3/8" to 2" diameter		160	.050			2.63		2.63	4.57
0060	Unbolt and remove structural bolts up to 3/4" diameter		240	.033			1.75		1.75	3.05
0070	7/8" to 2" diameter		160	.050			2.63		2.63	4.57
0140	Light weight framing members, remove whole or cut up, up to 20 lb.		240	.033			1.75		1.75	3.05
0150	21 – 40 lb.	2 Sswk	210	.076			4.01		4.01	6.95
0160	41 – 80 lb.	3 Sswk	180	.133			7		7	12.20
0170	81 – 120 lb.	4 Sswk	150	.213			11.25		11.25	19.50
0230	Structural members, remove whole or cut up, up to 500 lb.	E-19	48	.500			26	19.75	45.75	65.50
0240	1/4 – 2 tons	E-18	36	1.111			58.50	26.50	85	128
0250	2 – 5 tons	E-24	30	1.067			55.50	24.50	80	121
0260	5 – 10 tons	E-20	24	2.667			138	48.50	186.50	287
0270	10 – 15 tons	E-2	18	3.111			161	84	245	365
0340	Fabricated item, remove whole or cut up, up to 20 lb.	1 Sswk	96	.083			4.39		4.39	7.60
0350	21 – 40 lb.	2 Sswk	84	.190			10.05		10.05	17.40
0360	41 – 80 lb.	3 Sswk	72	.333			17.55		17.55	30.50
0370	81 – 120 lb.	4 Sswk	60	.533			28		28	49
0380	121 – 500 lb.	E-19	48	.500			26	19.75	45.75	65.50
0390	501 – 1000 lb.	"	36	.667			34.50	26.50	61	87
0500	Steel roof decking, uncovered, bare	B-2	5000	.008	S.F.		.30		.30	.47

05 05 13 – Shop-Applied Coatings for Metal

05 05 13.50 Paints and Protective Coatings

Line	Description	Crew	Daily Output	Labor-Hours	Unit	Material	Labor	Equipment	Total	Total Incl O&P
0010	**PAINTS AND PROTECTIVE COATINGS**									
5900	Galvanizing structural steel in shop, under 1 ton R050516-30				Ton	550			550	605
5950	1 ton to 20 tons					505			505	555
6000	Over 20 tons					460			460	505

05 05 19.10 Chemical Anchors		Crew	Daily Output	Labor-Hours	Unit	Material	2015 Bare Costs Labor	Equipment	Total	Total Incl O&P
0010	**CHEMICAL ANCHORS**									
0020	Includes layout & drilling									
1430	Chemical anchor, w/rod & epoxy cartridge, 3/4" diam. x 9-1/2" long	B-89A	27	.593	Ea.	9.80	25.50	4.28	39.58	55
1435	1" diameter x 11-3/4" long		24	.667		17.55	29	4.82	51.37	69
1440	1-1/4" diameter x 14" long		21	.762		36.50	33	5.50	75	96.50
1445	1-3/4" diameter x 15" long		20	.800		65	34.50	5.80	105.30	131
1450	18" long		17	.941		78	40.50	6.80	125.30	156
1455	2" diameter x 18" long		16	1		103	43	7.25	153.25	187
1460	24" long		15	1.067		134	46	7.70	187.70	227

05 05 19.20 Expansion Anchors			Crew	Daily Output	Labor-Hours	Unit	Material	Labor	Equipment	Total	Total Incl O&P
0010	**EXPANSION ANCHORS**										
0100	Anchors for concrete, brick or stone, no layout and drilling										
0200	Expansion shields, zinc, 1/4" diameter, 1-5/16" long, single	G	1 Carp	90	.089	Ea.	.43	4.17		4.60	6.85
0300	1-3/8" long, double	G		85	.094		.54	4.42		4.96	7.40
0400	3/8" diameter, 1-1/2" long, single	G		85	.094		.64	4.42		5.06	7.50
0500	2" long, double	G		80	.100		1.18	4.70		5.88	8.55
0600	1/2" diameter, 2-1/16" long, single	G		80	.100		1.18	4.70		5.88	8.55
0700	2-1/2" long, double	G		75	.107		1.91	5		6.91	9.80
0800	5/8" diameter, 2-5/8" long, single	G		75	.107		2.05	5		7.05	9.95
0900	2-3/4" long, double	G		70	.114		2.70	5.35		8.05	11.20
1000	3/4" diameter, 2-3/4" long, single	G		70	.114		3.09	5.35		8.44	11.65
1100	3-15/16" long, double	G		65	.123		5	5.80		10.80	14.40
2100	Hollow wall anchors for gypsum wall board, plaster or tile										
2300	1/8" diameter, short	G	1 Carp	160	.050	Ea.	.23	2.35		2.58	3.86
2400	Long	G		150	.053		.23	2.50		2.73	4.10
2500	3/16" diameter, short	G		150	.053		.40	2.50		2.90	4.29
2600	Long	G		140	.057		.59	2.68		3.27	4.78
2700	1/4" diameter, short	G		140	.057		.59	2.68		3.27	4.78
2800	Long	G		130	.062		.65	2.89		3.54	5.15
3000	Toggle bolts, bright steel, 1/8" diameter, 2" long	G		85	.094		.19	4.42		4.61	7
3100	4" long	G		80	.100		.25	4.70		4.95	7.55
3200	3/16" diameter, 3" long	G		80	.100		.26	4.70		4.96	7.55
3300	6" long	G		75	.107		.36	5		5.36	8.10
3400	1/4" diameter, 3" long	G		75	.107		.34	5		5.34	8.05
3500	6" long	G		70	.114		.51	5.35		5.86	8.80
3600	3/8" diameter, 3" long	G		70	.114		.84	5.35		6.19	9.15
3700	6" long	G		60	.133		1.36	6.25		7.61	11.15
3800	1/2" diameter, 4" long	G		60	.133		1.88	6.25		8.13	11.70
3900	6" long	G		50	.160		2.45	7.50		9.95	14.25
4000	Nailing anchors										
4100	Nylon nailing anchor, 1/4" diameter, 1" long		1 Carp	3.20	2.500	C	21	117		138	204
4200	1-1/2" long			2.80	2.857		26	134		160	235
4300	2" long			2.40	3.333		30.50	157		187.50	275
4400	Metal nailing anchor, 1/4" diameter, 1" long	G		3.20	2.500		17.85	117		134.85	201
4500	1-1/2" long	G		2.80	2.857		23.50	134		157.50	232
4600	2" long	G		2.40	3.333		27.50	157		184.50	272
5000	Screw anchors for concrete, masonry,										
5100	stone & tile, no layout or drilling included										
5700	Lag screw shields, 1/4" diameter, short	G	1 Carp	90	.089	Ea.	.34	4.17		4.51	6.75
5800	Long	G		85	.094		.38	4.42		4.80	7.20
5900	3/8" diameter, short	G		85	.094		.67	4.42		5.09	7.55
6000	Long	G		80	.100		.87	4.70		5.57	8.20

05 05 Common Work Results for Metals

05 05 19 – Post-Installed Concrete Anchors

05 05 19.20 Expansion Anchors			Crew	Daily Output	Labor-Hours	Unit	Material	2015 Bare Costs Labor	Equipment	Total	Total Incl O&P
6100	1/2" diameter, short	G	1 Carp	80	.100	Ea.	.95	4.70		5.65	8.30
6200	Long	G		75	.107		1.22	5		6.22	9.05
6300	5/8" diameter, short	G		70	.114		1.45	5.35		6.80	9.85
6400	Long	G		65	.123		1.86	5.80		7.66	10.95
6600	Lead, #6 & #8, 3/4" long	G		260	.031		.18	1.44		1.62	2.42
6700	#10 - #14, 1-1/2" long	G		200	.040		.37	1.88		2.25	3.30
6800	#16 & #18, 1-1/2" long	G		160	.050		.41	2.35		2.76	4.06
6900	Plastic, #6 & #8, 3/4" long			260	.031		.04	1.44		1.48	2.26
7000	#8 & #10, 7/8" long			240	.033		.04	1.56		1.60	2.45
7100	#10 & #12, 1" long			220	.036		.05	1.71		1.76	2.69
7200	#14 & #16, 1-1/2" long			160	.050		.07	2.35		2.42	3.69
8000	Wedge anchors, not including layout or drilling										
8050	Carbon steel, 1/4" diameter, 1-3/4" long	G	1 Carp	150	.053	Ea.	.41	2.50		2.91	4.30
8100	3-1/4" long	G		140	.057		.54	2.68		3.22	4.72
8150	3/8" diameter, 2-1/4" long	G		145	.055		.50	2.59		3.09	4.54
8200	5" long	G		140	.057		.88	2.68		3.56	5.10
8250	1/2" diameter, 2-3/4" long	G		140	.057		.99	2.68		3.67	5.20
8300	7" long	G		125	.064		1.70	3		4.70	6.50
8350	5/8" diameter, 3-1/2" long	G		130	.062		1.86	2.89		4.75	6.50
8400	8-1/2" long	G		115	.070		3.95	3.27		7.22	9.40
8450	3/4" diameter, 4-1/4" long	G		115	.070		2.90	3.27		6.17	8.25
8500	10" long	G		95	.084		6.60	3.95		10.55	13.35
8550	1" diameter, 6" long	G		100	.080		9.30	3.76		13.06	16.05
8575	9" long	G		85	.094		12.10	4.42		16.52	20
8600	12" long	G		75	.107		13.10	5		18.10	22
8650	1-1/4" diameter, 9" long	G		70	.114		24.50	5.35		29.85	35.50
8700	12" long	G		60	.133		31.50	6.25		37.75	44
8750	For type 303 stainless steel, add						350%				
8800	For type 316 stainless steel, add						450%				
8950	Self-drilling concrete screw, hex washer head, 3/16" diam. x 1-3/4" long	G	1 Carp	300	.027	Ea.	.20	1.25		1.45	2.15
8960	2-1/4" long	G		250	.032		.24	1.50		1.74	2.57
8970	Phillips flat head, 3/16" diam. x 1-3/4" long	G		300	.027		.21	1.25		1.46	2.16
8980	2-1/4" long	G		250	.032		.23	1.50		1.73	2.56

05 05 21 – Fastening Methods for Metal

05 05 21.10 Cutting Steel		Crew	Daily Output	Labor-Hours	Unit	Material	Labor	Equipment	Total	Total Incl O&P
0010	**CUTTING STEEL**									
0020	Hand burning, incl. preparation, torch cutting & grinding, no staging									
0050	Steel to 1/4" thick	E-25	400	.020	L.F.	.19	1.09	.03	1.31	2.13
0100	1/2" thick		320	.025		.35	1.37	.04	1.76	2.80
0150	3/4" thick		260	.031		.58	1.68	.04	2.30	3.61
0200	1" thick		200	.040		.84	2.19	.06	3.09	4.78

05 05 21.15 Drilling Steel		Crew	Daily Output	Labor-Hours	Unit	Material	Labor	Equipment	Total	Total Incl O&P
0010	**DRILLING STEEL**									
1910	Drilling & layout for steel, up to 1/4" deep, no anchor									
1920	Holes, 1/4" diameter	1 Sswk	112	.071	Ea.	.09	3.76		3.85	6.65
1925	For each additional 1/4" depth, add		336	.024		.09	1.25		1.34	2.28
1930	3/8" diameter		104	.077		.09	4.05		4.14	7.15
1935	For each additional 1/4" depth, add		312	.026		.09	1.35		1.44	2.44
1940	1/2" diameter		96	.083		.10	4.39		4.49	7.70
1945	For each additional 1/4" depth, add		288	.028		.10	1.46		1.56	2.65
1950	5/8" diameter		88	.091		.14	4.79		4.93	8.45
1955	For each additional 1/4" depth, add		264	.030		.14	1.60		1.74	2.93

05 05 21 – Fastening Methods for Metal

05 05 21.15 Drilling Steel		Crew	Daily Output	Labor-Hours	Unit	Material	2015 Bare Costs Labor	Equipment	Total	Total Incl O&P
1960	3/4" diameter	1 Sswk	80	.100	Ea.	.18	5.25		5.43	9.35
1965	For each additional 1/4" depth, add		240	.033		.18	1.75		1.93	3.25
1970	7/8" diameter		72	.111		.23	5.85		6.08	10.40
1975	For each additional 1/4" depth, add		216	.037		.23	1.95		2.18	3.65
1980	1" diameter		64	.125		.24	6.60		6.84	11.70
1985	For each additional 1/4" depth, add		192	.042		.24	2.19		2.43	4.07
1990	For drilling up, add						40%			

05 05 21.90 Welding Steel		Crew	Daily Output	Labor-Hours	Unit	Material	Labor	Equipment	Total	Total Incl O&P
0010	**WELDING STEEL**, Structural R050521-20									
0020	Field welding, 1/8" E6011, cost per welder, no operating engineer	E-14	8	1	Hr.	4.33	54.50	18.25	77.08	120
0200	With 1/2 operating engineer	E-13	8	1.500		4.33	79	18.25	101.58	157
0300	With 1 operating engineer	E-12	8	2		4.33	103	18.25	125.58	194
0500	With no operating engineer, 2# weld rod per ton	E-14	8	1	Ton	4.33	54.50	18.25	77.08	120
0600	8# E6011 per ton	"	2	4		17.30	219	73	309.30	480
0800	With one operating engineer per welder, 2# E6011 per ton	E-12	8	2		4.33	103	18.25	125.58	194
0900	8# E6011 per ton	"	2	8		17.30	415	73	505.30	775
1200	Continuous fillet, down welding									
1300	Single pass, 1/8" thick, 0.1#/L.F.	E-14	150	.053	L.F.	.22	2.91	.97	4.10	6.35
1400	3/16" thick, 0.2#/L.F.		75	.107		.43	5.85	1.94	8.22	12.75
1500	1/4" thick, 0.3#/L.F.		50	.160		.65	8.75	2.92	12.32	19.10
1610	5/16" thick, 0.4#/L.F.		38	.211		.87	11.50	3.84	16.21	25
1800	3 passes, 3/8" thick, 0.5#/L.F.		30	.267		1.08	14.55	4.86	20.49	32
2010	4 passes, 1/2" thick, 0.7#/L.F.		22	.364		1.52	19.85	6.65	28.02	43.50
2200	5 to 6 passes, 3/4" thick, 1.3#/L.F.		12	.667		2.81	36.50	12.15	51.46	80
2400	8 to 11 passes, 1" thick, 2.4#/L.F.		6	1.333		5.20	73	24.50	102.70	159
2600	For vertical joint welding, add						20%			
2700	Overhead joint welding, add						300%			
2900	For semi-automatic welding, obstructed joints, deduct						5%			
3000	Exposed joints, deduct						15%			
4000	Cleaning and welding plates, bars, or rods									
4010	to existing beams, columns, or trusses	E-14	12	.667	L.F.	1.08	36.50	12.15	49.73	78

05 05 23 – Metal Fastenings

05 05 23.10 Bolts and Hex Nuts			Crew	Daily Output	Labor-Hours	Unit	Material	Labor	Equipment	Total	Total Incl O&P
0010	**BOLTS & HEX NUTS**, Steel, A307										
0100	1/4" diameter, 1/2" long	G	1 Sswk	140	.057	Ea.	.06	3.01		3.07	5.30
0200	1" long	G		140	.057		.07	3.01		3.08	5.35
0300	2" long	G		130	.062		.10	3.24		3.34	5.75
0400	3" long	G		130	.062		.15	3.24		3.39	5.80
0500	4" long	G		120	.067		.17	3.51		3.68	6.30
0600	3/8" diameter, 1" long	G		130	.062		.14	3.24		3.38	5.80
0700	2" long	G		130	.062		.18	3.24		3.42	5.85
0800	3" long	G		120	.067		.24	3.51		3.75	6.35
0900	4" long	G		120	.067		.30	3.51		3.81	6.45
1000	5" long	G		115	.070		.38	3.66		4.04	6.75
1100	1/2" diameter, 1-1/2" long	G		120	.067		.40	3.51		3.91	6.55
1200	2" long	G		120	.067		.46	3.51		3.97	6.60
1300	4" long	G		115	.070		.75	3.66		4.41	7.20
1400	6" long	G		110	.073		1.05	3.83		4.88	7.80
1500	8" long	G		105	.076		1.38	4.01		5.39	8.45
1600	5/8" diameter, 1-1/2" long	G		120	.067		.98	3.51		4.49	7.20
1700	2" long	G		120	.067		1.09	3.51		4.60	7.30
1800	4" long	G		115	.070		1.59	3.66		5.25	8.10

05 05 23 – Metal Fastenings

05 05 23.10 Bolts and Hex Nuts			Crew	Daily Output	Labor-Hours	Unit	Material	Labor	Equipment	Total	Total Incl O&P
							2015 Bare Costs				
1900	6″ long	G	1 Sswk	110	.073	Ea.	2.05	3.83		5.88	8.90
2000	8″ long	G		105	.076		3.06	4.01		7.07	10.30
2100	10″ long	G		100	.080		3.87	4.21		8.08	11.55
2200	3/4″ diameter, 2″ long	G		120	.067		1.15	3.51		4.66	7.35
2300	4″ long	G		110	.073		1.65	3.83		5.48	8.45
2400	6″ long	G		105	.076		2.12	4.01		6.13	9.30
2500	8″ long	G		95	.084		3.20	4.43		7.63	11.20
2600	10″ long	G		85	.094		4.20	4.96		9.16	13.20
2700	12″ long	G		80	.100		4.92	5.25		10.17	14.55
2800	1″ diameter, 3″ long	G		105	.076		2.69	4.01		6.70	9.90
2900	6″ long	G		90	.089		3.94	4.68		8.62	12.50
3000	12″ long	G		75	.107		7.10	5.60		12.70	17.55
3100	For galvanized, add						75%				
3200	For stainless, add						350%				

05 05 23.25 High Strength Bolts			Crew	Daily Output	Labor-Hours	Unit	Material	Labor	Equipment	Total	Total Incl O&P
0010	**HIGH STRENGTH BOLTS** R050523-10										
0020	A325 Type 1, structural steel, bolt-nut-washer set										
0100	1/2″ diameter x 1-1/2″ long	G	1 Sswk	130	.062	Ea.	.97	3.24		4.21	6.70
0120	2″ long	G		125	.064		1.05	3.37		4.42	7
0150	3″ long	G		120	.067		1.46	3.51		4.97	7.70
0170	5/8″ diameter x 1-1/2″ long	G		125	.064		1.64	3.37		5.01	7.65
0180	2″ long	G		120	.067		1.77	3.51		5.28	8.05
0190	3″ long	G		115	.070		2.19	3.66		5.85	8.75
0200	3/4″ diameter x 2″ long	G		120	.067		2.65	3.51		6.16	9
0220	3″ long	G		115	.070		3.19	3.66		6.85	9.85
0250	4″ long	G		110	.073		3.92	3.83		7.75	10.95
0300	6″ long	G		105	.076		5.10	4.01		9.11	12.55
0350	8″ long	G		95	.084		10.15	4.43		14.58	18.85
0360	7/8″ diameter x 2″ long	G		115	.070		3.66	3.66		7.32	10.40
0365	3″ long	G		110	.073		4.33	3.83		8.16	11.40
0370	4″ long	G		105	.076		5.25	4.01		9.26	12.70
0380	6″ long	G		100	.080		6.65	4.21		10.86	14.65
0390	8″ long	G		90	.089		10.60	4.68		15.28	19.80
0400	1″ diameter x 2″ long	G		105	.076		4.58	4.01		8.59	12
0420	3″ long	G		100	.080		5.20	4.21		9.41	13
0450	4″ long	G		95	.084		5.85	4.43		10.28	14.15
0500	6″ long	G		90	.089		7.80	4.68		12.48	16.75
0550	8″ long	G		85	.094		13.55	4.96		18.51	23.50
0600	1-1/4″ diameter x 3″ long	G		85	.094		10.70	4.96		15.66	20.50
0650	4″ long	G		80	.100		11.65	5.25		16.90	22
0700	6″ long	G		75	.107		15.20	5.60		20.80	26.50
0750	8″ long	G		70	.114		19.35	6		25.35	32
1020	A490, bolt-nut-washer set										
1170	5/8″ diameter x 1-1/2″ long	G	1 Sswk	125	.064	Ea.	4.48	3.37		7.85	10.80
1180	2″ long	G		120	.067		5.35	3.51		8.86	11.95
1190	3″ long	G		115	.070		6.55	3.66		10.21	13.55
1200	3/4″ diameter x 2″ long	G		120	.067		3.92	3.51		7.43	10.40
1220	3″ long	G		115	.070		4.64	3.66		8.30	11.45
1250	4″ long	G		110	.073		5.40	3.83		9.23	12.60
1300	6″ long	G		105	.076		7.95	4.01		11.96	15.70
1350	8″ long	G		95	.084		13.50	4.43		17.93	22.50
1360	7/8″ diameter x 2″ long	G		115	.070		5.90	3.66		9.56	12.85

05 05 23 – Metal Fastenings

05 05 23.25 High Strength Bolts			Crew	Daily Output	Labor-Hours	Unit	Material	2015 Bare Costs Labor	Equipment	Total	Total Incl O&P
1365	3" long	G	1 Sswk	110	.073	Ea.	6.95	3.83		10.78	14.30
1370	4" long	G		105	.076		8.65	4.01		12.66	16.45
1380	6" long	G		100	.080		12.15	4.21		16.36	20.50
1390	8" long	G		90	.089		17.70	4.68		22.38	27.50
1400	1" diameter x 2" long	G		105	.076		7.80	4.01		11.81	15.55
1420	3" long	G		100	.080		9.45	4.21		13.66	17.70
1450	4" long	G		95	.084		10.95	4.43		15.38	19.70
1500	6" long	G		90	.089		14.65	4.68		19.33	24.50
1550	8" long	G		85	.094		23.50	4.96		28.46	34
1600	1-1/4" diameter x 3" long	G		85	.094		37.50	4.96		42.46	49.50
1650	4" long	G		80	.100		43	5.25		48.25	56.50
1700	6" long	G		75	.107		60.50	5.60		66.10	76.50
1750	8" long	G		70	.114		79	6		85	97.50
05 05 23.30 Lag Screws											
0010	**LAG SCREWS**										
0020	Steel, 1/4" diameter, 2" long	G	1 Carp	200	.040	Ea.	.09	1.88		1.97	2.99
0100	3/8" diameter, 3" long	G		150	.053		.30	2.50		2.80	4.18
0200	1/2" diameter, 3" long	G		130	.062		.64	2.89		3.53	5.15
0300	5/8" diameter, 3" long	G		120	.067		1.15	3.13		4.28	6.10
05 05 23.35 Machine Screws											
0010	**MACHINE SCREWS**										
0020	Steel, round head, #8 x 1" long	G	1 Carp	4.80	1.667	C	3.93	78.50		82.43	124
0110	#8 x 2" long	G		2.40	3.333		6.20	157		163.20	248
0200	#10 x 1" long	G		4	2		5.20	94		99.20	151
0300	#10 x 2" long	G		2	4		8.65	188		196.65	299
05 05 23.50 Powder Actuated Tools and Fasteners											
0010	**POWDER ACTUATED TOOLS & FASTENERS**										
0020	Stud driver, .22 caliber, single shot					Ea.	148			148	163
0100	.27 caliber, semi automatic, strip					"	440			440	485
0300	Powder load, single shot, .22 cal, power level 2, brown					C	5.35			5.35	5.90
0400	Strip, .27 cal, power level 4, red						7.70			7.70	8.50
0600	Drive pin, .300 x 3/4" long	G	1 Carp	4.80	1.667		4.11	78.50		82.61	125
0700	.300 x 3" long with washer	G	"	4	2		12.65	94		106.65	159
05 05 23.55 Rivets											
0010	**RIVETS**										
0100	Aluminum rivet & mandrel, 1/2" grip length x 1/8" diameter	G	1 Carp	4.80	1.667	C	7.65	78.50		86.15	128
0200	3/16" diameter	G		4	2		11.40	94		105.40	158
0300	Aluminum rivet, steel mandrel, 1/8" diameter	G		4.80	1.667		10.25	78.50		88.75	131
0400	3/16" diameter	G		4	2		16.45	94		110.45	163
0500	Copper rivet, steel mandrel, 1/8" diameter	G		4.80	1.667		9.50	78.50		88	130
0800	Stainless rivet & mandrel, 1/8" diameter	G		4.80	1.667		24.50	78.50		103	147
0900	3/16" diameter	G		4	2		39	94		133	188
1000	Stainless rivet, steel mandrel, 1/8" diameter	G		4.80	1.667		15.70	78.50		94.20	137
1100	3/16" diameter	G		4	2		25.50	94		119.50	173
1200	Steel rivet and mandrel, 1/8" diameter	G		4.80	1.667		7.40	78.50		85.90	128
1300	3/16" diameter	G		4	2		12	94		106	158
1400	Hand riveting tool, standard					Ea.	70.50			70.50	77.50
1500	Deluxe						365			365	400
1600	Power riveting tool, standard						500			500	555
1700	Deluxe						1,650			1,650	1,825

05 05 23 – Metal Fastenings

05 05 23.70 Structural Blind Bolts		Crew	Daily Output	Labor-Hours	Unit	Material	Labor	Equipment	Total	Total Incl O&P
						2015 Bare Costs				
0010	**STRUCTURAL BLIND BOLTS**									
0100	1/4" diameter x 1/4" grip G	1 Sswk	240	.033	Ea.	1.24	1.75		2.99	4.41
0150	1/2" grip G		216	.037		1.33	1.95		3.28	4.85
0200	3/8" diameter x 1/2" grip G		232	.034		1.75	1.82		3.57	5.10
0250	3/4" grip G		208	.038		1.84	2.02		3.86	5.55
0300	1/2" diameter x 1/2" grip G		224	.036		3.99	1.88		5.87	7.65
0350	3/4" grip G		200	.040		5.60	2.11		7.71	9.80
0400	5/8" diameter x 3/4" grip G		216	.037		8.25	1.95		10.20	12.50
0450	1" grip G		192	.042		9.50	2.19		11.69	14.25
05 05 23.80 Vibration and Bearing Pads										
0010	**VIBRATION & BEARING PADS**									
0300	Laminated synthetic rubber impregnated cotton duck, 1/2" thick	2 Sswk	24	.667	S.F.	69	35		104	137
0400	1" thick		20	.800		135	42		177	222
0600	Neoprene bearing pads, 1/2" thick		24	.667		26.50	35		61.50	90
0700	1" thick		20	.800		52.50	42		94.50	131
0900	Fabric reinforced neoprene, 5000 psi, 1/2" thick		24	.667		11.50	35		46.50	73.50
1000	1" thick		20	.800		23	42		65	98.50
1200	Felt surfaced vinyl pads, cork and sisal, 5/8" thick		24	.667		29	35		64	93
1300	1" thick		20	.800		52.50	42		94.50	131
1500	Teflon bonded to 10 ga. carbon steel, 1/32" layer		24	.667		51.50	35		86.50	118
1600	3/32" layer		24	.667		77	35		112	146
1800	Bonded to 10 ga. stainless steel, 1/32" layer		24	.667		91	35		126	161
1900	3/32" layer		24	.667		120	35		155	193
2100	Circular machine leveling pad & stud				Kip	6.45			6.45	7.10
05 05 23.85 Weld Shear Connectors										
0010	**WELD SHEAR CONNECTORS**									
0020	3/4" diameter, 3-3/16" long G	E-10	960	.017	Ea.	.53	.89	.47	1.89	2.65
0030	3-3/8" long G		950	.017		.56	.90	.47	1.93	2.70
0200	3-7/8" long G		945	.017		.60	.91	.48	1.99	2.76
0300	4-3/16" long G		935	.017		.63	.92	.48	2.03	2.81
0500	4-7/8" long G		930	.017		.70	.92	.48	2.10	2.90
0600	5-3/16" long G		920	.017		.73	.93	.49	2.15	2.96
0800	5-3/8" long G		910	.018		.74	.94	.49	2.17	2.99
0900	6-3/16" long G		905	.018		.81	.95	.50	2.26	3.09
1000	7-3/16" long G		895	.018		1	.96	.50	2.46	3.32
1100	8-3/16" long G		890	.018		1.10	.96	.50	2.56	3.45
1500	7/8" diameter, 3-11/16" long G		920	.017		.86	.93	.49	2.28	3.11
1600	4-3/16" long G		910	.018		.93	.94	.49	2.36	3.20
1700	5-3/16" long G		905	.018		1.05	.95	.50	2.50	3.36
1800	6-3/16" long G		895	.018		1.17	.96	.50	2.63	3.51
1900	7-3/16" long G		890	.018		1.30	.96	.50	2.76	3.67
2000	8-3/16" long G		880	.018		1.42	.98	.51	2.91	3.82
05 05 23.87 Weld Studs										
0010	**WELD STUDS**									
0020	1/4" diameter, 2-11/16" long G	E-10	1120	.014	Ea.	.35	.77	.40	1.52	2.16
0100	4-1/8" long G		1080	.015		.33	.79	.42	1.54	2.20
0200	3/8" diameter, 4-1/8" long G		1080	.015		.38	.79	.42	1.59	2.26
0300	6-1/8" long G		1040	.015		.49	.83	.43	1.75	2.44
0400	1/2" diameter, 2-1/8" long G		1040	.015		.35	.83	.43	1.61	2.29
0500	3-1/8" long G		1025	.016		.43	.84	.44	1.71	2.40
0600	4-1/8" long G		1010	.016		.50	.85	.44	1.79	2.52

05 05 Common Work Results for Metals

05 05 23 – Metal Fastenings

05 05 23.87 Weld Studs			Crew	Daily Output	Labor-Hours	Unit	Material	2015 Bare Costs Labor	Equipment	Total	Total Incl O&P
0700	5-5/16" long	G	E-10	990	.016	Ea.	.62	.87	.45	1.94	2.69
0800	6-1/8" long	G		975	.016		.67	.88	.46	2.01	2.78
0900	8-1/8" long	G		960	.017		.95	.89	.47	2.31	3.10
1000	5/8" diameter, 2-11/16" long	G		1000	.016		.61	.86	.45	1.92	2.66
1010	4-3/16" long	G		990	.016		.76	.87	.45	2.08	2.85
1100	6-9/16" long	G		975	.016		.99	.88	.46	2.33	3.13
1200	8-3/16" long	G		960	.017		1.33	.89	.47	2.69	3.52

05 05 23.90 Welding Rod			Crew	Daily Output	Labor-Hours	Unit	Material	Labor	Equipment	Total	Total Incl O&P
0010	**WELDING ROD**										
0020	Steel, type 6011, 1/8" diam., less than 500#					Lb.	2.16			2.16	2.38
0100	500# to 2,000#						1.95			1.95	2.15
0200	2,000# to 5,000#						1.83			1.83	2.02
0300	5/32" diameter, less than 500#						2.20			2.20	2.42
0310	500# to 2,000#						1.98			1.98	2.18
0320	2,000# to 5,000#						1.86			1.86	2.05
0400	3/16" diam., less than 500#						2.46			2.46	2.71
0500	500# to 2,000#						2.22			2.22	2.44
0600	2,000# to 5,000#						2.09			2.09	2.30
0620	Steel, type 6010, 1/8" diam., less than 500#						2.23			2.23	2.45
0630	500# to 2,000#						2.01			2.01	2.21
0640	2,000# to 5,000#						1.89			1.89	2.08
0650	Steel, type 7018 Low Hydrogen, 1/8" diam., less than 500#						2.16			2.16	2.38
0660	500# to 2,000#						1.95			1.95	2.15
0670	2,000# to 5,000#						1.83			1.83	2.02
0700	Steel, type 7024 Jet Weld, 1/8" diam., less than 500#						2.50			2.50	2.75
0710	500# to 2,000#						2.25			2.25	2.48
0720	2,000# to 5,000#						2.12			2.12	2.33
1550	Aluminum, type 4043 TIG, 1/8" diam., less than 10#						5.10			5.10	5.65
1560	10# to 60#						4.61			4.61	5.05
1570	Over 60#						4.33			4.33	4.77
1600	Aluminum, type 5356 TIG, 1/8" diam., less than 10#						5.45			5.45	6
1610	10# to 60#						4.90			4.90	5.40
1620	Over 60#						4.61			4.61	5.05
1900	Cast iron, type 8 Nickel, 1/8" diam., less than 500#						22			22	24
1910	500# to 1,000#						19.75			19.75	21.50
1920	Over 1,000#						18.55			18.55	20.50
2000	Stainless steel, type 316/316L, 1/8" diam., less than 500#						7.10			7.10	7.85
2100	500# to 1000#						6.40			6.40	7.05
2220	Over 1000#						6.05			6.05	6.65

05 12 Structural Steel Framing

05 12 23 – Structural Steel for Buildings

05 12 23.05 Canopy Framing			Crew	Daily Output	Labor-Hours	Unit	Material	Labor	Equipment	Total	Total Incl O&P
0010	**CANOPY FRAMING**										
0020	6" and 8" members, shop fabricated	G	E-4	3000	.011	Lb.	1.59	.57	.05	2.21	2.79

05 12 23.10 Ceiling Supports			Crew	Daily Output	Labor-Hours	Unit	Material	Labor	Equipment	Total	Total Incl O&P
0010	**CEILING SUPPORTS**										
1000	Entrance door/folding partition supports, shop fabricated	G	E-4	60	.533	L.F.	26.50	28.50	2.43	57.43	80.50
1100	Linear accelerator door supports	G		14	2.286		121	121	10.40	252.40	355
1200	Lintels or shelf angles, hung, exterior hot dipped galv.	G		267	.120		18.10	6.35	.55	25	31.50

05 12 Structural Steel Framing

05 12 23 – Structural Steel for Buildings

05 12 23.10 Ceiling Supports			Crew	Daily Output	Labor-Hours	Unit	Material	Labor	Equipment	Total	Total Incl O&P
							2015 Bare Costs				
1250	Two coats primer paint instead of galv.	G	E-4	267	.120	L.F.	15.65	6.35	.55	22.55	29
1400	Monitor support, ceiling hung, expansion bolted	G		4	8	Ea.	420	425	36.50	881.50	1,250
1450	Hung from pre-set inserts	G		6	5.333		450	283	24.50	757.50	1,000
1600	Motor supports for overhead doors	G		4	8		214	425	36.50	675.50	1,025
1700	Partition support for heavy folding partitions, without pocket	G		24	1.333	L.F.	60.50	71	6.10	137.60	196
1750	Supports at pocket only	G		12	2.667		121	142	12.15	275.15	390
2000	Rolling grilles & fire door supports	G		34	.941		51.50	50	4.29	105.79	149
2100	Spider-leg light supports, expansion bolted to ceiling slab	G		8	4	Ea.	172	213	18.25	403.25	580
2150	Hung from pre-set inserts	G		12	2.667	"	186	142	12.15	340.15	465
2400	Toilet partition support	G		36	.889	L.F.	60.50	47	4.05	111.55	153
2500	X-ray travel gantry support	G		12	2.667	"	207	142	12.15	361.15	485

05 12 23.15 Columns, Lightweight			Crew	Daily Output	Labor-Hours	Unit	Material	Labor	Equipment	Total	Total Incl O&P
0010	**COLUMNS, LIGHTWEIGHT**										
1000	Lightweight units (lally), 3-1/2" diameter		E-2	780	.072	L.F.	8.25	3.71	1.94	13.90	17.50
1050	4" diameter		"	900	.062	"	10.15	3.22	1.68	15.05	18.40
5800	Adjustable jack post, 8' maximum height, 2-3/4" diameter	G				Ea.	52.50			52.50	58
5850	4" diameter	G				"	84			84	92.50

05 12 23.17 Columns, Structural			Crew	Daily Output	Labor-Hours	Unit	Material	Labor	Equipment	Total	Total Incl O&P
0010	**COLUMNS, STRUCTURAL** R051223-10										
0015	Made from recycled materials										
0020	Shop fab'd for 100-ton, 1-2 story project, bolted connections										
0800	Steel, concrete filled, extra strong pipe, 3-1/2" diameter		E-2	660	.085	L.F.	43.50	4.39	2.29	50.18	58
0830	4" diameter			780	.072		48.50	3.71	1.94	54.15	62
0890	5" diameter			1020	.055		58	2.84	1.48	62.32	70.50
0930	6" diameter			1200	.047		77	2.41	1.26	80.67	90
0940	8" diameter			1100	.051		77	2.63	1.37	81	90.50
1100	For galvanizing, add					Lb.	.25			.25	.28
1300	For web ties, angles, etc., add per added lb.		1 Sswk	945	.008		1.33	.45		1.78	2.23
1500	Steel pipe, extra strong, no concrete, 3" to 5" diameter	G	E-2	16000	.004		1.33	.18	.09	1.60	1.86
1600	6" to 12" diameter	G		14000	.004		1.33	.21	.11	1.65	1.93
1700	Steel pipe, extra strong, no concrete, 3" diameter x 12'-0"	G		60	.933	Ea.	163	48.50	25	236.50	288
1750	4" diameter x 12'-0"	G		58	.966		238	50	26	314	375
1800	6" diameter x 12'-0"	G		54	1.037		455	53.50	28	536.50	620
1850	8" diameter x 14'-0"	G		50	1.120		805	58	30	893	1,025
1900	10" diameter x 16'-0"	G		48	1.167		1,150	60.50	31.50	1,242	1,400
1950	12" diameter x 18'-0"	G		45	1.244		1,550	64.50	33.50	1,648	1,875
3300	Structural tubing, square, A500GrB, 4" to 6" square, light section	G		11270	.005	Lb.	1.33	.26	.13	1.72	2.04
3600	Heavy section	G		32000	.002	"	1.33	.09	.05	1.47	1.66
4000	Concrete filled, add					L.F.	4.20			4.20	4.63
4500	Structural tubing, square, 4" x 4" x 1/4" x 12'-0"	G	E-2	58	.966	Ea.	219	50	26	295	355
4550	6" x 6" x 1/4" x 12'-0"	G		54	1.037		360	53.50	28	441.50	515
4600	8" x 8" x 3/8" x 14'-0"	G		50	1.120		775	58	30	863	985
4650	10" x 10" x 1/2" x 16'-0"	G		48	1.167		1,450	60.50	31.50	1,542	1,700
5100	Structural tubing, rect., 5" to 6" wide, light section	G		8000	.007	Lb.	1.33	.36	.19	1.88	2.28
5200	Heavy section	G		12000	.005		1.33	.24	.13	1.70	2.01
5300	7" to 10" wide, light section	G		15000	.004		1.33	.19	.10	1.62	1.89
5400	Heavy section	G		18000	.003		1.33	.16	.08	1.57	1.82
5500	Structural tubing, rect., 5" x 3" x 1/4" x 12'-0"	G		58	.966	Ea.	212	50	26	288	345
5550	6" x 4" x 5/16" x 12'-0"	G		54	1.037		330	53.50	28	411.50	485
5600	8" x 4" x 3/8" x 12'-0"	G		54	1.037		485	53.50	28	566.50	650
5650	10" x 6" x 3/8" x 14'-0"	G		50	1.120		775	58	30	863	985
5700	12" x 8" x 1/2" x 16'-0"	G		48	1.167		1,425	60.50	31.50	1,517	1,700

05 12 Structural Steel Framing

05 12 23 – Structural Steel for Buildings

05 12 23.17 Columns, Structural		Crew	Daily Output	Labor-Hours	Unit	Material	2015 Bare Costs Labor	Equipment	Total	Total Incl O&P
6800 W Shape, A992 steel, 2 tier, W8 x 24	G	E-2	1080	.052	L.F.	35	2.68	1.40	39.08	44.50
6850 W8 x 31	G		1080	.052		45	2.68	1.40	49.08	55.50
6900 W8 x 48	G		1032	.054		70	2.81	1.46	74.27	83.50
6950 W8 x 67	G		984	.057		97.50	2.94	1.54	101.98	114
7000 W10 x 45	G		1032	.054		65.50	2.81	1.46	69.77	78.50
7050 W10 x 68	G		984	.057		99	2.94	1.54	103.48	116
7100 W10 x 112	G		960	.058		163	3.02	1.57	167.59	187
7150 W12 x 50	G		1032	.054		73	2.81	1.46	77.27	86.50
7200 W12 x 87	G		984	.057		127	2.94	1.54	131.48	146
7250 W12 x 120	G		960	.058		175	3.02	1.57	179.59	199
7300 W12 x 190	G		912	.061		277	3.18	1.66	281.84	310
7350 W14 x 74	G		984	.057		108	2.94	1.54	112.48	126
7400 W14 x 120	G		960	.058		175	3.02	1.57	179.59	199
7450 W14 x 176	G		912	.061		257	3.18	1.66	261.84	289
8090 For projects 75 to 99 tons, add					All	10%				
8092 50 to 74 tons, add						20%				
8094 25 to 49 tons, add						30%	10%			
8096 10 to 24 tons, add						50%	25%			
8098 2 to 9 tons, add						75%	50%			
8099 Less than 2 tons, add						100%	100%			

05 12 23.20 Curb Edging		Crew	Daily Output	Labor-Hours	Unit	Material	Labor	Equipment	Total	Total Incl O&P
0010 **CURB EDGING**										
0020 Steel angle w/anchors, shop fabricated, on forms, 1" x 1", 0.8#/L.F.	G	E-4	350	.091	L.F.	1.67	4.86	.42	6.95	10.75
0100 2" x 2" angles, 3.92#/L.F.	G		330	.097		6.65	5.15	.44	12.24	16.75
0200 3" x 3" angles, 6.1#/L.F.	G		300	.107		10.50	5.65	.49	16.64	22
0300 4" x 4" angles, 8.2#/L.F.	G		275	.116		13.85	6.20	.53	20.58	26.50
1000 6" x 4" angles, 12.3#/L.F.	G		250	.128		20.50	6.80	.58	27.88	35
1050 Steel channels with anchors, on forms, 3" channel, 5#/L.F.	G		290	.110		8.35	5.85	.50	14.70	19.95
1100 4" channel, 5.4#/L.F.	G		270	.119		9	6.30	.54	15.84	21.50
1200 6" channel, 8.2#/L.F.	G		255	.125		13.85	6.65	.57	21.07	27.50
1300 8" channel, 11.5#/L.F.	G		225	.142		19.10	7.55	.65	27.30	35
1400 10" channel, 15.3#/L.F.	G		180	.178		25	9.45	.81	35.26	45
1500 12" channel, 20.7#/L.F.	G		140	.229		33.50	12.15	1.04	46.69	59
2000 For curved edging, add						35%	10%			

05 12 23.40 Lightweight Framing		Crew	Daily Output	Labor-Hours	Unit	Material	Labor	Equipment	Total	Total Incl O&P
0010 **LIGHTWEIGHT FRAMING** R051223-35										
0015 Made from recycled materials										
0200 For load-bearing steel studs see Section 05 41 13.30										
0400 Angle framing, field fabricated, 4" and larger R051223-45	G	E-3	440	.055	Lb.	.77	2.91	.33	4.01	6.25
0450 Less than 4" angles	G		265	.091	"	.80	4.83	.55	6.18	9.90
0460 1/2" x 1/2" x 1/8"	G		200	.120	L.F.	.16	6.40	.73	7.29	12.05
0462 3/4" x 3/4" x 1/8"	G		160	.150		.45	8	.91	9.36	15.40
0464 1" x 1" x 1/8"	G		135	.178		.64	9.50	1.08	11.22	18.35
0466 1-1/4" x 1-1/4" x 3/16"	G		115	.209		1.18	11.15	1.27	13.60	22
0468 1-1/2" x 1-1/2" x 3/16"	G		100	.240		1.43	12.80	1.46	15.69	25
0470 2" x 2" x 1/4"	G		90	.267		2.54	14.20	1.62	18.36	29
0472 2-1/2" x 2-1/2" x 1/4"	G		72	.333		3.26	17.75	2.03	23.04	37
0474 3" x 2" x 3/8"	G		65	.369		4.69	19.70	2.24	26.63	41.50
0476 3" x 3" x 3/8"	G		57	.421		5.70	22.50	2.56	30.76	48
0600 Channel framing, field fabricated, 8" and larger	G		500	.048	Lb.	.80	2.56	.29	3.65	5.65
0650 Less than 8" channels	G		335	.072	"	.80	3.82	.44	5.06	8
0660 C2 x 1.78	G		115	.209	L.F.	1.42	11.15	1.27	13.84	22.50

05 12 Structural Steel Framing

05 12 23 – Structural Steel for Buildings

05 12 23.40 Lightweight Framing			Crew	Daily Output	Labor-Hours	Unit	2015 Bare Costs Material	Labor	Equipment	Total	Total Incl O&P
0662	C3 x 4.1	G	E-3	80	.300	L.F.	3.26	16	1.82	21.08	33.50
0664	C4 x 5.4	G		66	.364		4.29	19.40	2.21	25.90	40.50
0666	C5 x 6.7	G		57	.421		5.35	22.50	2.56	30.41	47.50
0668	C6 x 8.2	G		55	.436		6.30	23.50	2.65	32.45	50.50
0670	C7 x 9.8	G		40	.600		7.80	32	3.65	43.45	68
0672	C8 x 11.5	G		36	.667		9.15	35.50	4.05	48.70	76.50
0710	Structural bar tee, field fabricated, 3/4" x 3/4" x 1/8"	G		160	.150		.45	8	.91	9.36	15.40
0712	1" x 1" x 1/8"	G		135	.178		.64	9.50	1.08	11.22	18.35
0714	1-1/2" x 1-1/2" x 1/4"	G		114	.211		1.86	11.25	1.28	14.39	23
0716	2" x 2" x 1/4"	G		89	.270		2.54	14.40	1.64	18.58	29.50
0718	2-1/2" x 2-1/2" x 3/8"	G		72	.333		4.69	17.75	2.03	24.47	38.50
0720	3" x 3" x 3/8"	G		57	.421		5.70	22.50	2.56	30.76	48
0730	Structural zee, field fabricated, 1-1/4" x 1-3/4" x 1-3/4"	G		114	.211		.60	11.25	1.28	13.13	21.50
0732	2-11/16" x 3" x 2-11/16"	G		114	.211		1.42	11.25	1.28	13.95	22.50
0734	3-1/16" x 4" x 3-1/16"	G		133	.180		2.14	9.60	1.10	12.84	20.50
0736	3-1/4" x 5" x 3-1/4"	G		133	.180		2.92	9.60	1.10	13.62	21
0738	3-1/2" x 6" x 3-1/2"	G		160	.150		4.40	8	.91	13.31	19.75
0740	Junior beam, field fabricated, 3"	G		80	.300		4.53	16	1.82	22.35	35
0742	4"	G		72	.333		6.10	17.75	2.03	25.88	40
0744	5"	G		67	.358		7.95	19.10	2.18	29.23	44
0746	6"	G		62	.387		9.95	20.50	2.35	32.80	49.50
0748	7"	G		57	.421		12.15	22.50	2.56	37.21	55
0750	8"	G		53	.453		14.65	24	2.75	41.40	61
1000	Continuous slotted channel framing system, shop fab, simple framing	G	2 Sswk	2400	.007	Lb.	4.11	.35		4.46	5.15
1200	Complex framing	G	"	1600	.010		4.64	.53		5.17	6
1300	Cross bracing, rods, shop fabricated, 3/4" diameter	G	E-3	700	.034		1.59	1.83	.21	3.63	5.15
1310	7/8" diameter	G		850	.028		1.59	1.51	.17	3.27	4.56
1320	1" diameter	G		1000	.024		1.59	1.28	.15	3.02	4.13
1330	Angle, 5" x 5" x 3/8"	G		2800	.009		1.59	.46	.05	2.10	2.60
1350	Hanging lintels, shop fabricated	G		850	.028		1.59	1.51	.17	3.27	4.56
1380	Roof frames, shop fabricated, 3'-0" square, 5' span	G	E-2	4200	.013		1.59	.69	.36	2.64	3.31
1400	Tie rod, not upset, 1-1/2" to 4" diameter, with turnbuckle	G	2 Sswk	800	.020		1.72	1.05		2.77	3.72
1420	No turnbuckle	G		700	.023		1.66	1.20		2.86	3.91
1500	Upset, 1-3/4" to 4" diameter, with turnbuckle	G		800	.020		1.72	1.05		2.77	3.72
1520	No turnbuckle	G		700	.023		1.66	1.20		2.86	3.91

05 12 23.45 Lintels			Crew	Daily Output	Labor-Hours	Unit	Material	Labor	Equipment	Total	Total Incl O&P
0010	**LINTELS**										
0015	Made from recycled materials										
0020	Plain steel angles, shop fabricated, under 500 lb.	G	1 Bric	550	.015	Lb.	1.02	.67		1.69	2.15
0100	500 to 1000 lb.	G		640	.013		.99	.58		1.57	1.97
0200	1,000 to 2,000 lb.	G		640	.013		.97	.58		1.55	1.94
0300	2,000 to 4,000 lb.	G		640	.013		.94	.58		1.52	1.91
0500	For built-up angles and plates, add to above	G					1.33			1.33	1.46
0700	For engineering, add to above						.13			.13	.15
0900	For galvanizing, add to above, under 500 lb.						.30			.30	.33
0950	500 to 2,000 lb.						.28			.28	.30
1000	Over 2,000 lb.						.25			.25	.28
2000	Steel angles, 3-1/2" x 3", 1/4" thick, 2'-6" long	G	1 Bric	47	.170	Ea.	14.30	7.85		22.15	28
2100	4'-6" long	G		26	.308		26	14.20		40.20	50
2600	4" x 3-1/2", 1/4" thick, 5'-0" long	G		21	.381		33	17.60		50.60	63
2700	9'-0" long	G		12	.667		59	31		90	112

05 12 23 – Structural Steel for Buildings

05 12 23.60 Pipe Support Framing		Crew	Daily Output	Labor-Hours	Unit	Material	2015 Bare Costs Labor	Equipment	Total	Total Incl O&P
0010	PIPE SUPPORT FRAMING									
0020	Under 10#/L.F., shop fabricated G	E-4	3900	.008	Lb.	1.78	.44	.04	2.26	2.75
0200	10.1 to 15#/L.F. G		4300	.007		1.75	.40	.03	2.18	2.65
0400	15.1 to 20#/L.F. G		4800	.007		1.72	.35	.03	2.10	2.54
0600	Over 20#/L.F. G	↓	5400	.006	↓	1.70	.32	.03	2.05	2.45

05 12 23.65 Plates		Crew	Daily Output	Labor-Hours	Unit	Material	Labor	Equipment	Total	Total Incl O&P
0010	PLATES R051223-80									
0015	Made from recycled materials									
0020	For connections & stiffener plates, shop fabricated									
0050	1/8" thick (5.1 lb./S.F.) G				S.F.	6.75			6.75	7.45
0100	1/4" thick (10.2 lb./S.F.) G					13.50			13.50	14.85
0300	3/8" thick (15.3 lb./S.F.) G					20.50			20.50	22.50
0400	1/2" thick (20.4 lb./S.F.) G					27			27	29.50
0450	3/4" thick (30.6 lb./S.F.) G					40.50			40.50	44.50
0500	1" thick (40.8 lb./S.F.) G				↓	54			54	59.50
2000	Steel plate, warehouse prices, no shop fabrication									
2100	1/4" thick (10.2 lb./S.F.) G				S.F.	7.30			7.30	8

05 12 23.70 Stressed Skin Steel Roof and Ceiling System		Crew	Daily Output	Labor-Hours	Unit	Material	Labor	Equipment	Total	Total Incl O&P
0010	STRESSED SKIN STEEL ROOF & CEILING SYSTEM									
0020	Double panel flat roof, spans to 100' G	E-2	1150	.049	S.F.	10.60	2.52	1.31	14.43	17.35
0100	Double panel convex roof, spans to 200' G		960	.058		17.25	3.02	1.57	21.84	25.50
0200	Double panel arched roof, spans to 300' G	↓	760	.074	↓	26.50	3.81	1.99	32.30	37.50

05 12 23.75 Structural Steel Members		Crew	Daily Output	Labor-Hours	Unit	Material	Labor	Equipment	Total	Total Incl O&P
0010	STRUCTURAL STEEL MEMBERS R051223-10									
0015	Made from recycled materials									
0020	Shop fab'd for 100-ton, 1-2 story project, bolted connections									
0100	Beam or girder, W 6 x 9 G	E-2	600	.093	L.F.	13.10	4.83	2.52	20.45	25.50
0120	x 15 G		600	.093		22	4.83	2.52	29.35	35
0140	x 20 G		600	.093		29	4.83	2.52	36.35	43
0300	W 8 x 10 G		600	.093		14.60	4.83	2.52	21.95	27
0320	x 15 G		600	.093		22	4.83	2.52	29.35	35
0350	x 21 G		600	.093		30.50	4.83	2.52	37.85	44.50
0360	x 24 G		550	.102		35	5.25	2.75	43	50.50
0370	x 28 G		550	.102		41	5.25	2.75	49	57
0500	x 31 G		550	.102		45	5.25	2.75	53	61.50
0520	x 35 G		550	.102		51	5.25	2.75	59	68
0540	x 48 G		550	.102		70	5.25	2.75	78	89
0600	W 10 x 12 G		600	.093		17.50	4.83	2.52	24.85	30
0620	x 15 G		600	.093		22	4.83	2.52	29.35	35
0700	x 22 G		600	.093		32	4.83	2.52	39.35	46.50
0720	x 26 G		600	.093		38	4.83	2.52	45.35	52.50
0740	x 33 G		550	.102		48	5.25	2.75	56	65
0900	x 49 G		550	.102		71.50	5.25	2.75	79.50	90.50
1100	W 12 x 16 G		880	.064		23.50	3.29	1.72	28.51	33
1300	x 22 G		880	.064		32	3.29	1.72	37.01	43
1500	x 26 G		880	.064		38	3.29	1.72	43.01	49
1520	x 35 G		810	.069		51	3.58	1.87	56.45	64
1560	x 50 G		750	.075		73	3.86	2.01	78.87	88.50
1580	x 58 G		750	.075		84.50	3.86	2.01	90.37	102
1700	x 72 G		640	.088		105	4.53	2.36	111.89	125
1740	x 87 G	↓	640	.088	↓	127	4.53	2.36	133.89	149

05 12 23 – Structural Steel for Buildings

05 12 23.75 Structural Steel Members			Crew	Daily Output	Labor-Hours	Unit	Material	Labor	Equipment	Total	Total Incl O&P
							2015 Bare Costs				
1900	W 14 x 26	G	E-2	990	.057	L.F.	38	2.93	1.53	42.46	48
2100	x 30	G		900	.062		43.50	3.22	1.68	48.40	55.50
2300	x 34	G		810	.069		49.50	3.58	1.87	54.95	62.50
2320	x 43	G		810	.069		62.50	3.58	1.87	67.95	77
2340	x 53	G		800	.070		77.50	3.62	1.89	83.01	93
2360	x 74	G		760	.074		108	3.81	1.99	113.80	128
2380	x 90	G		740	.076		131	3.92	2.04	136.96	153
2500	x 120	G		720	.078		175	4.02	2.10	181.12	201
2700	W 16 x 26	G		1000	.056		38	2.90	1.51	42.41	48
2900	x 31	G		900	.062		45	3.22	1.68	49.90	57
3100	x 40	G		800	.070		58.50	3.62	1.89	64.01	72
3120	x 50	G		800	.070		73	3.62	1.89	78.51	88
3140	x 67	G		760	.074		97.50	3.81	1.99	103.30	116
3300	W 18 x 35	G	E-5	960	.083		51	4.35	1.73	57.08	65.50
3500	x 40	G		960	.083		58.50	4.35	1.73	64.58	73.50
3520	x 46	G		960	.083		67	4.35	1.73	73.08	83.50
3700	x 50	G		912	.088		73	4.58	1.82	79.40	90
3900	x 55	G		912	.088		80	4.58	1.82	86.40	98
3920	x 65	G		900	.089		94.50	4.64	1.84	100.98	114
3940	x 76	G		900	.089		111	4.64	1.84	117.48	132
3960	x 86	G		900	.089		125	4.64	1.84	131.48	148
3980	x 106	G		900	.089		155	4.64	1.84	161.48	180
4100	W 21 x 44	G		1064	.075		64	3.93	1.56	69.49	79
4300	x 50	G		1064	.075		73	3.93	1.56	78.49	88.50
4500	x 62	G		1036	.077		90.50	4.03	1.60	96.13	108
4700	x 68	G		1036	.077		99	4.03	1.60	104.63	118
4720	x 83	G		1000	.080		121	4.18	1.66	126.84	142
4740	x 93	G		1000	.080		136	4.18	1.66	141.84	158
4760	x 101	G		1000	.080		147	4.18	1.66	152.84	171
4780	x 122	G		1000	.080		178	4.18	1.66	183.84	205
4900	W 24 x 55	G		1110	.072		80	3.76	1.49	85.25	96
5100	x 62	G		1110	.072		90.50	3.76	1.49	95.75	108
5300	x 68	G		1110	.072		99	3.76	1.49	104.25	117
5500	x 76	G		1110	.072		111	3.76	1.49	116.25	130
5700	x 84	G		1080	.074		122	3.87	1.53	127.40	143
5720	x 94	G		1080	.074		137	3.87	1.53	142.40	159
5740	x 104	G		1050	.076		152	3.98	1.58	157.56	175
5760	x 117	G		1050	.076		171	3.98	1.58	176.56	196
5780	x 146	G		1050	.076		213	3.98	1.58	218.56	242
5800	W 27 x 84	G		1190	.067		122	3.51	1.39	126.90	142
5900	x 94	G		1190	.067		137	3.51	1.39	141.90	158
5920	x 114	G		1150	.070		166	3.63	1.44	171.07	191
5940	x 146	G		1150	.070		213	3.63	1.44	218.07	242
5960	x 161	G		1150	.070		235	3.63	1.44	240.07	266
6100	W 30 x 99	G		1200	.067		144	3.48	1.38	148.86	166
6300	x 108	G		1200	.067		157	3.48	1.38	161.86	180
6500	x 116	G		1160	.069		169	3.60	1.43	174.03	194
6520	x 132	G		1160	.069		192	3.60	1.43	197.03	220
6540	x 148	G		1160	.069		216	3.60	1.43	221.03	245
6560	x 173	G		1120	.071		252	3.73	1.48	257.21	285
6580	x 191	G		1120	.071		278	3.73	1.48	283.21	315
6700	W 33 x 118	G		1176	.068		172	3.55	1.41	176.96	197
6900	x 130	G		1134	.071		189	3.68	1.46	194.14	216

05 12 23.75 Structural Steel Members		Crew	Daily Output	Labor-Hours	Unit	Material	2015 Bare Costs Labor	Equipment	Total	Total Incl O&P
7100 x 141	G	E-5	1134	.071	L.F.	206	3.68	1.46	211.14	234
7120 x 169	G		1100	.073		246	3.80	1.51	251.31	279
7140 x 201	G		1100	.073		293	3.80	1.51	298.31	330
7300 W 36 x 135	G		1170	.068		197	3.57	1.42	201.99	224
7500 x 150	G		1170	.068		219	3.57	1.42	223.99	248
7600 x 170	G		1150	.070		248	3.63	1.44	253.07	281
7700 x 194	G		1125	.071		283	3.71	1.47	288.18	320
7900 x 231	G		1125	.071		335	3.71	1.47	340.18	380
7920 x 262	G		1035	.077		380	4.04	1.60	385.64	430
8100 x 302	G		1035	.077		440	4.04	1.60	445.64	495
8490 For projects 75 to 99 tons, add						10%				
8492 50 to 74 tons, add						20%				
8494 25 to 49 tons, add						30%	10%			
8496 10 to 24 tons, add						50%	25%			
8498 2 to 9 tons, add						75%	50%			
8499 Less than 2 tons, add						100%	100%			

05 12 23.77 Structural Steel Projects

05 12 23.77 Structural Steel Projects		Crew	Daily Output	Labor-Hours	Unit	Material	Labor	Equipment	Total	Total Incl O&P
0010 **STRUCTURAL STEEL PROJECTS** R050516-30										
0015 Made from recycled materials										
0020 Shop fab'd for 100-ton, 1-2 story project, bolted connections										
0200 Apartments, nursing homes, etc., 1 to 2 stories R050523-10	G	E-5	10.30	7.767	Ton	2,650	405	161	3,216	3,800
0300 3 to 6 stories	G	"	10.10	7.921		2,700	415	164	3,279	3,850
0400 7 to 15 stories R051223-10	G	E-6	14.20	9.014		2,750	470	130	3,350	3,975
0500 Over 15 stories	G	"	13.90	9.209		2,850	480	133	3,463	4,100
0700 Offices, hospitals, etc., steel bearing, 1 to 2 stories R051223-20	G	E-5	10.30	7.767		2,650	405	161	3,216	3,800
0800 3 to 6 stories	G	E-6	14.40	8.889		2,700	465	128	3,293	3,900
0900 7 to 15 stories R051223-25	G		14.20	9.014		2,750	470	130	3,350	3,975
1000 Over 15 stories	G		13.90	9.209		2,850	480	133	3,463	4,100
1100 For multi-story masonry wall bearing construction, add R051223-30							30%			
1300 Industrial bldgs., 1 story, beams & girders, steel bearing	G	E-5	12.90	6.202		2,650	325	128	3,103	3,625
1400 Masonry bearing	G	"	10	8		2,650	420	166	3,236	3,825
1500 Industrial bldgs., 1 story, under 10 tons,										
1510 steel from warehouse, trucked	G	E-2	7.50	7.467	Ton	3,175	385	201	3,761	4,375
1600 1 story with roof trusses, steel bearing	G	E-5	10.60	7.547		3,125	395	156	3,676	4,300
1700 Masonry bearing	G	"	8.30	9.639		3,125	505	200	3,830	4,525
1900 Monumental structures, banks, stores, etc., simple connections	G	E-6	13	9.846		2,650	515	142	3,307	3,950
2000 Moment/composite connections	G	"	9	14.222		4,400	745	206	5,351	6,325
2200 Churches, simple connections	G	E-5	11.60	6.897		2,475	360	143	2,978	3,475
2300 Moment/composite connections	G	"	5.20	15.385		3,275	805	320	4,400	5,325
2800 Power stations, fossil fuels, simple connections	G	E-6	11	11.636		2,650	610	168	3,428	4,125
2900 Moment/composite connections	G		5.70	22.456		3,975	1,175	325	5,475	6,725
2950 Nuclear fuels, non-safety steel, simple connections	G		7	18.286		2,650	955	264	3,869	4,850
3000 Moment/composite connections	G		5.50	23.273		3,975	1,225	335	5,535	6,825
3040 Safety steel, simple connections	G		2.50	51.200		3,875	2,675	740	7,290	9,625
3070 Moment/composite connections	G		1.50	85.333		5,100	4,450	1,225	10,775	14,500
3100 Roof trusses, simple connections	G	E-5	13	6.154		3,700	320	127	4,147	4,750
3200 Moment/composite connections	G		8.30	9.639		4,500	505	200	5,205	6,025
3210 Schools, simple connections	G		14.50	5.517		2,650	288	114	3,052	3,550
3220 Moment/composite connections	G		8.30	9.639		3,875	505	200	4,580	5,325
3400 Welded construction, simple commercial bldgs., 1 to 2 stories	G	E-7	7.60	10.526		2,700	550	237	3,487	4,175
3500 7 to 15 stories	G	E-9	8.30	15.422		3,125	805	270	4,200	5,125
3700 Welded rigid frame, 1 story, simple connections	G	E-7	15.80	5.063		2,750	264	114	3,128	3,600

05 12 Structural Steel Framing

05 12 23 – Structural Steel for Buildings

05 12 23.77 Structural Steel Projects			Crew	Daily Output	Labor-Hours	Unit	2015 Bare Costs Material	Labor	Equipment	Total	Total Incl O&P
3800	Moment/composite connections	G	E-7	5.50	14.545	Ton	3,575	760	330	4,665	5,575
3810	Fabrication shop costs (incl in project bare material cost, above)										
3820	Mini mill base price, Grade A992	G				Ton	800			800	880
3830	Mill extras plus delivery to warehouse						275			275	305
3835	Delivery from warehouse to fabrication shop						85			85	93.50
3840	Shop extra for shop drawings and detailing						295			295	325
3850	Shop fabricating and handling						920			920	1,000
3860	Shop sandblasting and primer coat of paint						155			155	171
3870	Shop delivery to the job site						120			120	132
3880	Total material cost, shop fabricated, primed, delivered						2,650			2,650	2,925
3900	High strength steel mill spec extras:										
3950	A529, A572 (50 ksi) and A36: same as A992 steel (no extra)										
4000	Add to A992 price for A572 (60, 65 ksi)	G				Ton	80			80	88
4100	A242 and A588 Weathering	G				"	80			80	88
4200	Mill size extras for W-Shapes: 0 to 30 plf: no extra charge										
4210	Member sizes 31 to 65 plf, deduct	G				Ton	.01			.01	.01
4220	Member sizes 66 to 100 plf, deduct	G					5			5	5.50
4230	Member sizes 101 to 387 plf, add	G					55.50			55.50	61
4300	Column base plates, light, up to 150 lb.	G	2 Sswk	2000	.008	Lb.	1.46	.42		1.88	2.33
4400	Heavy, over 150 lb.	G	E-2	7500	.007	"	1.52	.39	.20	2.11	2.55
4600	Castellated beams, light sections, to 50#/L.F., simple connections	G		10.70	5.234	Ton	2,775	271	141	3,187	3,650
4700	Moment/composite connections	G		7	8		3,050	415	216	3,681	4,275
4900	Heavy sections, over 50 plf, simple connections	G		11.70	4.786		2,925	248	129	3,302	3,750
5000	Moment/composite connections	G		7.80	7.179		3,175	370	194	3,739	4,350
5390	For projects 75 to 99 tons, add						10%				
5392	50 to 74 tons, add						20%				
5394	25 to 49 tons, add						30%	10%			
5396	10 to 24 tons, add						50%	25%			
5398	2 to 9 tons, add						75%	50%			
5399	Less than 2 tons, add						100%	100%			
05 12 23.78 Structural Steel Secondary Members											
0010	**STRUCTURAL STEEL SECONDARY MEMBERS**										
0015	Made from recycled materials										
0020	Shop fabricated for 20-ton girt/purlin framing package, materials only										
0100	Girts/purlins, C/Z-shapes, includes clips and bolts										
0110	6" x 2-1/2" x 2-1/2", 16 ga., 3.0 lb./L.F.					L.F.	3.58			3.58	3.94
0115	14 ga., 3.5 lb./L.F.						4.17			4.17	4.59
0120	8" x 2-3/4" x 2-3/4", 16 ga., 3.4 lb./L.F.						4.05			4.05	4.46
0125	14 ga., 4.1 lb./L.F.						4.89			4.89	5.40
0130	12 ga., 5.6 lb./L.F.						6.70			6.70	7.35
0135	10" x 3-1/2" x 3-1/2", 14 ga., 4.7 lb./L.F.						5.60			5.60	6.15
0140	12 ga., 6.7 lb./L.F.						8			8	8.80
0145	12" x 3-1/2" x 3-1/2", 14 ga., 5.3 lb./L.F.						6.30			6.30	6.95
0150	12 ga., 7.4 lb./L.F.						8.80			8.80	9.70
0200	Eave struts, C-shape, includes clips and bolts										
0210	6" x 4" x 3", 16 ga., 3.1 lb./L.F.					L.F.	3.70			3.70	4.07
0215	14 ga., 3.9 lb./L.F.						4.65			4.65	5.10
0220	8" x 4" x 3", 16 ga., 3.5 lb./L.F.						4.17			4.17	4.59
0225	14 ga., 4.4 lb./L.F.						5.25			5.25	5.75
0230	12 ga., 6.2 lb./L.F.						7.40			7.40	8.15
0235	10" x 5" x 3", 14 ga., 5.2 lb./L.F.						6.20			6.20	6.80
0240	12 ga., 7.3 lb./L.F.						8.70			8.70	9.60

05 12 Structural Steel Framing

05 12 23 – Structural Steel for Buildings

05 12 23.78 Structural Steel Secondary Members		Crew	Daily Output	Labor-Hours	Unit	2015 Bare Costs Material	Labor	Equipment	Total	Total Incl O&P
0245	12" x 5" x 4", 14 ga., 6.0 lb./L.F.				L.F.	7.15			7.15	7.85
0250	12 ga., 8.4 lb./L.F.				↓	10			10	11
0300	Rake/base angle, excludes concrete drilling and expansion anchors									
0310	2" x 2", 14 ga., 1.0 lb./L.F.	2 Sswk	640	.025	L.F.	1.19	1.32		2.51	3.60
0315	3" x 2", 14 ga., 1.3 lb./L.F.		535	.030		1.55	1.57		3.12	4.45
0320	3" x 3", 14 ga., 1.6 lb./L.F.		500	.032		1.91	1.68		3.59	5.05
0325	4" x 3", 14 ga., 1.8 lb./L.F.	↓	480	.033	↓	2.15	1.75		3.90	5.40
0600	Installation of secondary members, erection only									
0610	Girts, purlins, eave struts, 16 ga., 6" deep	E-18	100	.400	Ea.		21	9.50	30.50	46
0615	8" deep		80	.500			26.50	11.85	38.35	57.50
0620	14 ga., 6" deep		80	.500			26.50	11.85	38.35	57.50
0625	8" deep		65	.615			32.50	14.60	47.10	71
0630	10" deep		55	.727			38.50	17.25	55.75	84
0635	12" deep		50	.800			42	19	61	92.50
0640	12 ga., 8" deep		50	.800			42	19	61	92.50
0645	10" deep		45	.889			47	21	68	103
0650	12" deep	↓	40	1	↓		52.50	23.50	76	115
0900	For less than 20-ton job lots									
0905	For 15 to 19 tons, add					10%				
0910	For 10 to 14 tons, add					25%				
0915	For 5 to 9 tons, add					50%	50%	50%		
0920	For 1 to 4 tons, add					75%	75%	75%		
0925	For less than 1 ton, add					100%	100%	100%		

05 12 23.80 Subpurlins		Crew	Daily Output	Labor-Hours	Unit	Material	Labor	Equipment	Total	Total Incl O&P
0010	**SUBPURLINS** R051223-50									
0015	Made from recycled materials									
0020	Bulb tees, shop fabricated, painted, 32-5/8" O.C., 40 psf L.L.									
0200	Type 218, max 10'-2" span, 3.19 plf, 2-1/8" high x 2-1/8" wide G	E-1	3100	.008	S.F.	1.66	.40	.05	2.11	2.54
1420	For 24-5/8" spacing, add					33%	33%			
1430	For 48-5/8" spacing, deduct				↓	33%	33%			

05 14 Structural Aluminum Framing

05 14 23 – Non-Exposed Structural Aluminum Framing

05 14 23.05 Aluminum Shapes		Crew	Daily Output	Labor-Hours	Unit	Material	Labor	Equipment	Total	Total Incl O&P
0010	**ALUMINUM SHAPES**									
0015	Made from recycled materials									
0020	Structural shapes, 1" to 10" members, under 1 ton G	E-2	4000	.014	Lb.	3.85	.72	.38	4.95	5.90
0050	1 to 5 tons G		4300	.013		3.52	.67	.35	4.54	5.40
0100	Over 5 tons G		4600	.012		3.34	.63	.33	4.30	5.10
0300	Extrusions, over 5 tons, stock shapes G		1330	.042		3.28	2.18	1.14	6.60	8.50
0400	Custom shapes G	↓	1330	.042	↓	3.41	2.18	1.14	6.73	8.65

	05 15 16.05 Accessories for Steel Wire Rope		Crew	Daily Output	Labor-Hours	Unit	2015 Bare Costs Material	Labor	Equipment	Total	Total Incl O&P
0010	**ACCESSORIES FOR STEEL WIRE ROPE**										
0015	Made from recycled materials										
1500	Thimbles, heavy duty, 1/4"	G	E-17	160	.100	Ea.	.52	5.35		5.87	9.85
1510	1/2"	G		160	.100		2.29	5.35		7.64	11.80
1520	3/4"	G		105	.152		5.20	8.20		13.40	19.90
1530	1"	G		52	.308		10.40	16.50		26.90	40
1540	1-1/4"	G		38	.421		16	22.50		38.50	56.50
1550	1-1/2"	G		13	1.231		45	66		111	165
1560	1-3/4"	G		8	2		93	107		200	288
1570	2"	G		6	2.667		135	143		278	400
1580	2-1/4"	G		4	4		183	215		398	575
1600	Clips, 1/4" diameter	G		160	.100		2.74	5.35		8.09	12.30
1610	3/8" diameter	G		160	.100		3.01	5.35		8.36	12.60
1620	1/2" diameter	G		160	.100		4.84	5.35		10.19	14.60
1630	3/4" diameter	G		102	.157		7.85	8.40		16.25	23.50
1640	1" diameter	G		64	.250		13.05	13.40		26.45	38
1650	1-1/4" diameter	G		35	.457		21.50	24.50		46	66
1670	1-1/2" diameter	G		26	.615		29	33		62	89.50
1680	1-3/4" diameter	G		16	1		67.50	53.50		121	167
1690	2" diameter	G		12	1.333		75	71.50		146.50	207
1700	2-1/4" diameter	G		10	1.600		110	86		196	270
1800	Sockets, open swage, 1/4" diameter	G		160	.100		47	5.35		52.35	61.50
1810	1/2" diameter	G		77	.208		68	11.15		79.15	94
1820	3/4" diameter	G		19	.842		105	45		150	195
1830	1" diameter	G		9	1.778		188	95.50		283.50	375
1840	1-1/4" diameter	G		5	3.200		262	172		434	585
1850	1-1/2" diameter	G		3	5.333		575	286		861	1,125
1860	1-3/4" diameter	G		3	5.333		1,025	286		1,311	1,625
1870	2" diameter	G		1.50	10.667		1,550	570		2,120	2,700
1900	Closed swage, 1/4" diameter	G		160	.100		28	5.35		33.35	40
1910	1/2" diameter	G		104	.154		48	8.25		56.25	67.50
1920	3/4" diameter	G		32	.500		72	27		99	126
1930	1" diameter	G		15	1.067		126	57		183	239
1940	1-1/4" diameter	G		7	2.286		189	123		312	420
1950	1-1/2" diameter	G		4	4		345	215		560	755
1960	1-3/4" diameter	G		3	5.333		505	286		791	1,050
1970	2" diameter	G		2	8		985	430		1,415	1,825
2000	Open spelter, galv., 1/4" diameter	G		160	.100		59.50	5.35		64.85	75
2010	1/2" diameter	G		70	.229		62	12.25		74.25	89.50
2020	3/4" diameter	G		26	.615		93	33		126	160
2030	1" diameter	G		10	1.600		258	86		344	435
2040	1-1/4" diameter	G		5	3.200		370	172		542	705
2050	1-1/2" diameter	G		4	4		785	215		1,000	1,225
2060	1-3/4" diameter	G		2	8		1,375	430		1,805	2,250
2070	2" diameter	G		1.20	13.333		1,575	715		2,290	2,975
2080	2-1/2" diameter	G		1	16		2,900	860		3,760	4,675
2100	Closed spelter, galv., 1/4" diameter	G		160	.100		49.50	5.35		54.85	64
2110	1/2" diameter	G		88	.182		53	9.75		62.75	75.50
2120	3/4" diameter	G		30	.533		80.50	28.50		109	138
2130	1" diameter	G		13	1.231		171	66		237	305
2140	1-1/4" diameter	G		7	2.286		274	123		397	515
2150	1-1/2" diameter	G		6	2.667		590	143		733	900
2160	1-3/4" diameter	G		2.80	5.714		785	305		1,090	1,400

05 15 Wire Rope Assemblies

05 15 16 – Steel Wire Rope Assemblies

05 15 16.05 Accessories for Steel Wire Rope

	05 15 16.05 Accessories for Steel Wire Rope		Crew	Daily Output	Labor-Hours	Unit	2015 Bare Costs Material	Labor	Equipment	Total	Total Incl O&P
2170	2″ diameter	G	E-17	2	8	Ea.	970	430		1,400	1,825
2200	Jaw & jaw turnbuckles, 1/4″ x 4″	G		160	.100		16.25	5.35		21.60	27
2250	1/2″ x 6″	G		96	.167		20.50	8.95		29.45	38
2260	1/2″ x 9″	G		77	.208		27.50	11.15		38.65	49.50
2270	1/2″ x 12″	G		66	.242		31	13		44	56.50
2300	3/4″ x 6″	G		38	.421		40	22.50		62.50	83.50
2310	3/4″ x 9″	G		30	.533		44.50	28.50		73	98.50
2320	3/4″ x 12″	G		28	.571		57.50	30.50		88	117
2330	3/4″ x 18″	G		23	.696		68.50	37.50		106	141
2350	1″ x 6″	G		17	.941		78	50.50		128.50	173
2360	1″ x 12″	G		13	1.231		85.50	66		151.50	209
2370	1″ x 18″	G		10	1.600		128	86		214	290
2380	1″ x 24″	G		9	1.778		141	95.50		236.50	320
2400	1-1/4″ x 12″	G		7	2.286		144	123		267	370
2410	1-1/4″ x 18″	G		6.50	2.462		178	132		310	425
2420	1-1/4″ x 24″	G		5.60	2.857		240	153		393	530
2450	1-1/2″ x 12″	G		5.20	3.077		465	165		630	795
2460	1-1/2″ x 18″	G		4	4		495	215		710	920
2470	1-1/2″ x 24″	G		3.20	5		665	268		933	1,200
2500	1-3/4″ x 18″	G		3.20	5		1,000	268		1,268	1,575
2510	1-3/4″ x 24″	G		2.80	5.714		1,150	305		1,455	1,775
2550	2″ x 24″	G		1.60	10		1,550	535		2,085	2,625

05 15 16.50 Steel Wire Rope

	05 15 16.50 Steel Wire Rope		Crew	Daily Output	Labor-Hours	Unit	Material	Labor	Equipment	Total	Total Incl O&P
0010	**STEEL WIRE ROPE**										
0015	Made from recycled materials										
0020	6 x 19, bright, fiber core, 5000′ rolls, 1/2″ diameter	G				L.F.	.85			.85	.93
0050	Steel core	G					1.12			1.12	1.23
0100	Fiber core, 1″ diameter	G					2.86			2.86	3.15
0150	Steel core	G					3.26			3.26	3.59
0300	6 x 19, galvanized, fiber core, 1/2″ diameter	G					1.25			1.25	1.38
0350	Steel core	G					1.43			1.43	1.57
0400	Fiber core, 1″ diameter	G					3.67			3.67	4.03
0450	Steel core	G					3.84			3.84	4.23
0500	6 x 7, bright, IPS, fiber core, <500 L.F. w/acc., 1/4″ diameter	G	E-17	6400	.003		1.51	.13		1.64	1.89
0510	1/2″ diameter	G		2100	.008		3.68	.41		4.09	4.76
0520	3/4″ diameter	G		960	.017		6.65	.89		7.54	8.90
0550	6 x 19, bright, IPS, IWRC, <500 L.F. w/acc., 1/4″ diameter	G		5760	.003		.94	.15		1.09	1.29
0560	1/2″ diameter	G		1730	.009		1.52	.50		2.02	2.53
0570	3/4″ diameter	G		770	.021		2.64	1.11		3.75	4.84
0580	1″ diameter	G		420	.038		4.47	2.04		6.51	8.45
0590	1-1/4″ diameter	G		290	.055		7.40	2.96		10.36	13.30
0600	1-1/2″ diameter	G		192	.083		9.10	4.47		13.57	17.80
0610	1-3/4″ diameter	G	E-18	240	.167		14.55	8.75	3.95	27.25	35
0620	2″ diameter	G		160	.250		18.65	13.15	5.95	37.75	49.50
0630	2-1/4″ diameter	G		160	.250		25	13.15	5.95	44.10	56.50
0650	6 x 37, bright, IPS, IWRC, <500 L.F. w/acc., 1/4″ diameter	G	E-17	6400	.003		1.11	.13		1.24	1.45
0660	1/2″ diameter	G		1730	.009		1.88	.50		2.38	2.93
0670	3/4″ diameter	G		770	.021		3.04	1.11		4.15	5.30
0680	1″ diameter	G		430	.037		4.83	2		6.83	8.75
0690	1-1/4″ diameter	G		290	.055		7.30	2.96		10.26	13.15
0700	1-1/2″ diameter	G		190	.084		10.45	4.52		14.97	19.35
0710	1-3/4″ diameter	G	E-18	260	.154		16.55	8.10	3.65	28.30	36

05 15 Wire Rope Assemblies

05 15 16 – Steel Wire Rope Assemblies

05 15 16.50 Steel Wire Rope			Crew	Daily Output	Labor-Hours	Unit	Material	2015 Bare Costs Labor	Equipment	Total	Total Incl O&P
0720	2″ diameter	G	E-18	200	.200	L.F.	21.50	10.55	4.74	36.79	46.50
0730	2-1/4″ diameter	G		160	.250		28.50	13.15	5.95	47.60	60
0800	6 x 19 & 6 x 37, swaged, 1/2″ diameter	G	E-17	1220	.013		2.47	.70		3.17	3.94
0810	9/16″ diameter	G		1120	.014		2.87	.77		3.64	4.49
0820	5/8″ diameter	G		930	.017		3.41	.92		4.33	5.35
0830	3/4″ diameter	G		640	.025		4.34	1.34		5.68	7.10
0840	7/8″ diameter	G		480	.033		5.50	1.79		7.29	9.15
0850	1″ diameter	G		350	.046		6.70	2.45		9.15	11.60
0860	1-1/8″ diameter	G		288	.056		8.20	2.98		11.18	14.25
0870	1-1/4″ diameter	G		230	.070		9.95	3.73		13.68	17.45
0880	1-3/8″ diameter	G		192	.083		11.50	4.47		15.97	20.50
0890	1-1/2″ diameter	G	E-18	300	.133		13.95	7	3.16	24.11	30.50

05 15 16.60 Galvanized Steel Wire Rope and Accessories			Crew	Daily Output	Labor-Hours	Unit	Material	Labor	Equipment	Total	Total Incl O&P
0010	**GALVANIZED STEEL WIRE ROPE & ACCESSORIES**										
0015	Made from recycled materials										
3000	Aircraft cable, galvanized, 7 x 7 x 1/8″	G	E-17	5000	.003	L.F.	.20	.17		.37	.52
3100	Clamps, 1/8″	G	″	125	.128	Ea.	1.96	6.85		8.81	14.10

05 15 16.70 Temporary Cable Safety Railing			Crew	Daily Output	Labor-Hours	Unit	Material	Labor	Equipment	Total	Total Incl O&P
0010	**TEMPORARY CABLE SAFETY RAILING**, Each 100′ strand incl.										
0020	2 eyebolts, 1 turnbuckle, 100′ cable, 2 thimbles, 6 clips										
0025	Made from recycled materials										
0100	One strand using 1/4″ cable & accessories	G	2 Sswk	4	4	C.L.F.	209	211		420	595
0200	1/2″ cable & accessories	G	″	2	8	″	440	420		860	1,225

05 21 Steel Joist Framing

05 21 13 – Deep Longspan Steel Joist Framing

05 21 13.50 Deep Longspan Joists			Crew	Daily Output	Labor-Hours	Unit	Material	Labor	Equipment	Total	Total Incl O&P
0010	**DEEP LONGSPAN JOISTS**										
3010	DLH series, 40-ton job lots, bolted cross bridging, shop primer										
3015	Made from recycled materials										
3040	Spans to 144′ (shipped in 2 pieces)	G	E-7	13	6.154	Ton	1,925	320	139	2,384	2,825
3200	52DLH11, 26 lb./L.F.	G		2000	.040	L.F.	24	2.09	.90	26.99	31
3220	52DLH16, 45 lb./L.F.	G		2000	.040		43.50	2.09	.90	46.49	52
3240	56DLH11, 26 lb./L.F.	G		2000	.040		25	2.09	.90	27.99	32
3260	56DLH16, 46 lb./L.F.	G		2000	.040		44.50	2.09	.90	47.49	53.50
3280	60DLH12, 29 lb./L.F.	G		2000	.040		28	2.09	.90	30.99	35.50
3300	60DLH17, 52 lb./L.F.	G		2000	.040		50	2.09	.90	52.99	59.50
3320	64DLH12, 31 lb./L.F.	G		2200	.036		30	1.90	.82	32.72	37
3340	64DLH17, 52 lb./L.F.	G		2200	.036		50	1.90	.82	52.72	59
3360	68DLH13, 37 lb./L.F.	G		2200	.036		35.50	1.90	.82	38.22	43.50
3380	68DLH18, 61 lb./L.F.	G		2200	.036		59	1.90	.82	61.72	68.50
3400	72DLH14, 41 lb./L.F.	G		2200	.036		39.50	1.90	.82	42.22	47.50
3420	72DLH19, 70 lb./L.F.	G		2200	.036		67.50	1.90	.82	70.22	78.50
3500	For less than 40-ton job lots										
3502	For 30 to 39 tons, add						10%				
3504	20 to 29 tons, add						20%				
3506	10 to 19 tons, add						30%				
3507	5 to 9 tons, add						50%	25%			
3508	1 to 4 tons, add						75%	50%			
3509	Less than 1 ton, add						100%	100%			

05 21 13 – Deep Longspan Steel Joist Framing

05 21 13.50 Deep Longspan Joists			Crew	Daily Output	Labor-Hours	Unit	Material	2015 Bare Costs Labor	Equipment	Total	Total Incl O&P
4010	SLH series, 40-ton job lots, bolted cross bridging, shop primer										
4040	Spans to 200′ (shipped in 3 pieces)	G	E-7	13	6.154	Ton	1,975	320	139	2,434	2,875
4200	80SLH15, 40 lb./L.F.	G		1500	.053	L.F.	39.50	2.78	1.20	43.48	49.50
4220	80SLH20, 75 lb./L.F.	G		1500	.053		74.50	2.78	1.20	78.48	88
4240	88SLH16, 46 lb./L.F.	G		1500	.053		45.50	2.78	1.20	49.48	56
4260	88SLH21, 89 lb./L.F.	G		1500	.053		88.50	2.78	1.20	92.48	103
4280	96SLH17, 52 lb./L.F.	G		1500	.053		51.50	2.78	1.20	55.48	62.50
4300	96SLH22, 102 lb./L.F.	G		1500	.053		101	2.78	1.20	104.98	117
4320	104SLH18, 59 lb./L.F.	G		1800	.044		58.50	2.32	1	61.82	69.50
4340	104SLH23, 109 lb./L.F.	G		1800	.044		108	2.32	1	111.32	124
4360	112SLH19, 67 lb./L.F.	G		1800	.044		66.50	2.32	1	69.82	78
4380	112SLH24, 131 lb./L.F.	G		1800	.044		130	2.32	1	133.32	148
4400	120SLH20, 77 lb./L.F.	G		1800	.044		76.50	2.32	1	79.82	89
4420	120SLH25, 152 lb./L.F.	G	▼	1800	.044	▼	151	2.32	1	154.32	171
6100	For less than 40-ton job lots										
6102	For 30 to 39 tons, add						10%				
6104	20 to 29 tons, add						20%				
6106	10 to 19 tons, add						30%				
6107	5 to 9 tons, add						50%	25%			
6108	1 to 4 tons, add						75%	50%			
6109	Less than 1 ton, add						100%	100%			

05 21 16 – Longspan Steel Joist Framing

05 21 16.50 Longspan Joists			Crew	Daily Output	Labor-Hours	Unit	Material	2015 Bare Costs Labor	Equipment	Total	Total Incl O&P
0010	**LONGSPAN JOISTS**										
2000	LH series, 40-ton job lots, bolted cross bridging, shop primer										
2015	Made from recycled materials										
2040	Longspan joists, LH series, up to 96′	G	E-7	13	6.154	Ton	1,825	320	139	2,284	2,725
2200	18LH04, 12 lb./L.F.	G		1400	.057	L.F.	11	2.98	1.29	15.27	18.55
2220	18LH08, 19 lb./L.F.	G		1400	.057		17.40	2.98	1.29	21.67	25.50
2240	20LH04, 12 lb./L.F.	G		1400	.057		11	2.98	1.29	15.27	18.55
2260	20LH08, 19 lb./L.F.	G		1400	.057		17.40	2.98	1.29	21.67	25.50
2280	24LH05, 13 lb./L.F.	G		1400	.057		11.90	2.98	1.29	16.17	19.55
2300	24LH10, 23 lb./L.F.	G		1400	.057		21	2.98	1.29	25.27	29.50
2320	28LH06, 16 lb./L.F.	G		1800	.044		14.65	2.32	1	17.97	21
2340	28LH11, 25 lb./L.F.	G		1800	.044		23	2.32	1	26.32	30
2360	32LH08, 17 lb./L.F.	G		1800	.044		15.60	2.32	1	18.92	22
2380	32LH13, 30 lb./L.F.	G		1800	.044		27.50	2.32	1	30.82	35.50
2400	36LH09, 21 lb./L.F.	G		1800	.044		19.25	2.32	1	22.57	26
2420	36LH14, 36 lb./L.F.	G		1800	.044		33	2.32	1	36.32	41.50
2440	40LH10, 21 lb./L.F.	G		2200	.036		19.25	1.90	.82	21.97	25
2460	40LH15, 36 lb./L.F.	G		2200	.036		33	1.90	.82	35.72	40.50
2480	44LH11, 22 lb./L.F.	G		2200	.036		20	1.90	.82	22.72	26
2500	44LH16, 42 lb./L.F.	G		2200	.036		38.50	1.90	.82	41.22	46.50
2520	48LH11, 22 lb./L.F.	G		2200	.036		20	1.90	.82	22.72	26
2540	48LH16, 42 lb./L.F.	G	▼	2200	.036	▼	38.50	1.90	.82	41.22	46.50
2600	For less than 40-ton job lots										
2602	For 30 to 39 tons, add						10%				
2604	20 to 29 tons, add						20%				
2606	10 to 19 tons, add						30%				
2607	5 to 9 tons, add						50%	25%			
2608	1 to 4 tons, add						75%	50%			
2609	Less than 1 ton, add						100%	100%			

05 21 Steel Joist Framing

05 21 16 – Longspan Steel Joist Framing

05 21 16.50 Longspan Joists			Crew	Daily Output	Labor-Hours	Unit	Material	2015 Bare Costs Labor	Equipment	Total	Total Incl O&P
6000	For welded cross bridging, add							30%			

05 21 19 – Open Web Steel Joist Framing

05 21 19.10 Open Web Joists

Line	Description		Crew	Daily Output	Labor-Hours	Unit	Material	2015 Bare Costs Labor	Equipment	Total	Total Incl O&P
0010	**OPEN WEB JOISTS**										
0015	Made from recycled materials										
0050	K series, 40-ton lots, horiz. bridging, spans to 30′, shop primer	G	E-7	12	6.667	Ton	1,650	350	150	2,150	2,575
0130	8K1, 5.1 lb./L.F.	G		1200	.067	L.F.	4.22	3.48	1.50	9.20	12.20
0140	10K1, 5.0 lb./L.F.	G		1200	.067		4.14	3.48	1.50	9.12	12.10
0160	12K3, 5.7 lb./L.F.	G		1500	.053		4.72	2.78	1.20	8.70	11.25
0180	14K3, 6.0 lb./L.F.	G		1500	.053		4.97	2.78	1.20	8.95	11.50
0200	16K3, 6.3 lb./L.F.	G		1800	.044		5.20	2.32	1	8.52	10.80
0220	16K6, 8.1 lb./L.F.	G		1800	.044		6.70	2.32	1	10.02	12.45
0240	18K5, 7.7 lb./L.F.	G		2000	.040		6.40	2.09	.90	9.39	11.55
0260	18K9, 10.2 lb./L.F.	G		2000	.040		8.45	2.09	.90	11.44	13.85
0440	K series, 30′ to 50′ spans	G		17	4.706	Ton	1,625	246	106	1,977	2,325
0500	20K5, 8.2 lb./L.F.	G		2000	.040	L.F.	6.65	2.09	.90	9.64	11.90
0520	20K9, 10.8 lb./L.F.	G		2000	.040		8.80	2.09	.90	11.79	14.20
0540	22K5, 8.8 lb./L.F.	G		2000	.040		7.15	2.09	.90	10.14	12.40
0560	22K9, 11.3 lb./L.F.	G		2000	.040		9.20	2.09	.90	12.19	14.65
0580	24K6, 9.7 lb./L.F.	G		2200	.036		7.90	1.90	.82	10.62	12.75
0600	24K10, 13.1 lb./L.F.	G		2200	.036		10.65	1.90	.82	13.37	15.80
0620	26K6, 10.6 lb./L.F.	G		2200	.036		8.60	1.90	.82	11.32	13.55
0640	26K10, 13.8 lb./L.F.	G		2200	.036		11.20	1.90	.82	13.92	16.45
0660	28K8, 12.7 lb./L.F.	G		2400	.033		10.30	1.74	.75	12.79	15.15
0680	28K12, 17.1 lb./L.F.	G		2400	.033		13.90	1.74	.75	16.39	19.10
0700	30K8, 13.2 lb./L.F.	G		2400	.033		10.75	1.74	.75	13.24	15.60
0720	30K12, 17.6 lb./L.F.	G		2400	.033		14.30	1.74	.75	16.79	19.55
0800	For less than 40-ton job lots										
0802	For 30 to 39 tons, add						10%				
0804	20 to 29 tons, add						20%				
0806	10 to 19 tons, add						30%				
0807	5 to 9 tons, add						50%	25%			
0808	1 to 4 tons, add						75%	50%			
0809	Less than 1 ton, add						100%	100%			
1010	CS series, 40-ton job lots, horizontal bridging, shop primer										
1040	Spans to 30′	G	E-7	12	6.667	Ton	1,700	350	150	2,200	2,625
1100	10CS2, 7.5 lb./L.F.	G		1200	.067	L.F.	6.40	3.48	1.50	11.38	14.55
1120	12CS2, 8.0 lb./L.F.	G		1500	.053		6.80	2.78	1.20	10.78	13.55
1140	14CS2, 8.0 lb./L.F.	G		1500	.053		6.80	2.78	1.20	10.78	13.55
1160	16CS2, 8.5 lb./L.F.	G		1800	.044		7.25	2.32	1	10.57	13
1180	16CS4, 14.5 lb./L.F.	G		1800	.044		12.35	2.32	1	15.67	18.60
1200	18CS2, 9.0 lb./L.F.	G		2000	.040		7.65	2.09	.90	10.64	12.95
1220	18CS4, 15.0 lb./L.F.	G		2000	.040		12.75	2.09	.90	15.74	18.60
1240	20CS2, 9.5 lb./L.F.	G		2000	.040		8.10	2.09	.90	11.09	13.45
1260	20CS4, 16.5 lb./L.F.	G		2000	.040		14.05	2.09	.90	17.04	20
1280	22CS2, 10.0 lb./L.F.	G		2000	.040		8.50	2.09	.90	11.49	13.90
1300	22CS4, 16.5 lb./L.F.	G		2000	.040		14.05	2.09	.90	17.04	20
1320	24CS2, 10.0 lb./L.F.	G		2200	.036		8.50	1.90	.82	11.22	13.45
1340	24CS4, 16.5 lb./L.F.	G		2200	.036		14.05	1.90	.82	16.77	19.55
1360	26CS2, 10.0 lb./L.F.	G		2200	.036		8.50	1.90	.82	11.22	13.45
1380	26CS4, 16.5 lb./L.F.	G		2200	.036		14.05	1.90	.82	16.77	19.55
1400	28CS2, 10.5 lb./L.F.	G		2400	.033		8.95	1.74	.75	11.44	13.60

05 21 Steel Joist Framing

05 21 19 – Open Web Steel Joist Framing

05 21 19.10 Open Web Joists			Crew	Daily Output	Labor-Hours	Unit	Material	2015 Bare Costs Labor	Equipment	Total	Total Incl O&P
1420	28CS4, 16.5 lb./L.F.	G	E-7	2400	.033	L.F.	14.05	1.74	.75	16.54	19.25
1440	30CS2, 11.0 lb./L.F.	G		2400	.033		9.35	1.74	.75	11.84	14.10
1460	30CS4, 16.5 lb./L.F.	G	↓	2400	.033	↓	14.05	1.74	.75	16.54	19.25
1500	For less than 40-ton job lots										
1502	For 30 to 39 tons, add						10%				
1504	20 to 29 tons, add						20%				
1506	10 to 19 tons, add						30%				
1507	5 to 9 tons, add						50%	25%			
1508	1 to 4 tons, add						75%	50%			
1509	Less than 1 ton, add						100%	100%			
6200	For shop prime paint other than mfrs. standard, add						20%				
6300	For bottom chord extensions, add per chord	G				Ea.	36			36	39.50
6400	Individual steel bearing plate, 6" x 6" x 1/4" with J-hook	G	1 Bric	160	.050	"	7.95	2.31		10.26	12.25

05 21 23 – Steel Joist Girder Framing

05 21 23.50 Joist Girders			Crew	Daily Output	Labor-Hours	Unit	Material	Labor	Equipment	Total	Total Incl O&P
0010	**JOIST GIRDERS**										
0015	Made from recycled materials										
7020	Joist girders, 40-ton job lots, shop primer	G	E-5	13	6.154	Ton	1,650	320	127	2,097	2,500
7100	For less than 40-ton job lots										
7102	For 30 to 39 tons, add						10%				
7104	20 to 29 tons, add						20%				
7106	10 to 19 tons, add						30%				
7107	5 to 9 tons, add						50%	25%			
7108	1 to 4 tons, add						75%	50%			
7109	Less than 1 ton, add						100%	100%			
8000	Trusses, 40-ton job lots, shop fabricated WT chords, shop primer	G	E-5	11	7.273	Ton	5,425	380	151	5,956	6,750
8100	For less than 40-ton job lots										
8102	For 30 to 39 tons, add						10%				
8104	20 to 29 tons, add						20%				
8106	10 to 19 tons, add						30%				
8107	5 to 9 tons, add						50%	25%			
8108	1 to 4 tons, add						75%	50%			
8109	Less than 1 ton, add						100%	100%			

05 31 Steel Decking

05 31 13 – Steel Floor Decking

05 31 13.50 Floor Decking			Crew	Daily Output	Labor-Hours	Unit	Material	Labor	Equipment	Total	Total Incl O&P
0010	**FLOOR DECKING** R053100-10										
0015	Made from recycled materials										
5100	Non-cellular composite decking, galvanized, 1-1/2" deep, 16 ga.	G	E-4	3500	.009	S.F.	3.44	.49	.04	3.97	4.68
5120	18 ga.	G		3650	.009		2.78	.47	.04	3.29	3.91
5140	20 ga.	G		3800	.008		2.22	.45	.04	2.71	3.26
5200	2" deep, 22 ga.	G		3860	.008		1.93	.44	.04	2.41	2.93
5300	20 ga.	G		3600	.009		2.13	.47	.04	2.64	3.20
5400	18 ga.	G		3380	.009		2.73	.50	.04	3.27	3.92
5500	16 ga.	G		3200	.010		3.41	.53	.05	3.99	4.72
5700	3" deep, 22 ga.	G		3200	.010		2.10	.53	.05	2.68	3.28
5800	20 ga.	G		3000	.011		2.34	.57	.05	2.96	3.62
5900	18 ga.	G		2850	.011		2.90	.60	.05	3.55	4.29
6000	16 ga.	G	↓	2700	.012	↓	3.87	.63	.05	4.55	5.40

05 31 Steel Decking

05 31 23 – Steel Roof Decking

05 31 23.50 Roof Decking			Crew	Daily Output	Labor-Hours	Unit	Material	2015 Bare Costs Labor	Equipment	Total	Total Incl O&P
0010	**ROOF DECKING**										
0015	Made from recycled materials										
2100	Open type, 1-1/2" deep, Type B, wide rib, galv., 22 ga., under 50 sq.	G	E-4	4500	.007	S.F.	2.05	.38	.03	2.46	2.95
2200	50-500 squares	G		4900	.007		1.59	.35	.03	1.97	2.38
2400	Over 500 squares	G		5100	.006		1.47	.33	.03	1.83	2.23
2600	20 ga., under 50 squares	G		3865	.008		2.39	.44	.04	2.87	3.43
2650	50-500 squares	G		4170	.008		1.92	.41	.04	2.37	2.86
2700	Over 500 squares	G		4300	.007		1.72	.40	.03	2.15	2.63
2900	18 ga., under 50 squares	G		3800	.008		3.09	.45	.04	3.58	4.22
2950	50-500 squares	G		4100	.008		2.47	.41	.04	2.92	3.48
3000	Over 500 squares	G		4300	.007		2.22	.40	.03	2.65	3.18
3050	16 ga., under 50 squares	G		3700	.009		4.17	.46	.04	4.67	5.45
3060	50-500 squares	G		4000	.008		3.34	.43	.04	3.81	4.45
3100	Over 500 squares	G		4200	.008		3	.41	.03	3.44	4.04
3200	3" deep, Type N, 22 ga., under 50 squares	G		3600	.009		3.01	.47	.04	3.52	4.17
3250	50-500 squares	G		3800	.008		2.41	.45	.04	2.90	3.47
3260	over 500 squares	G		4000	.008		2.17	.43	.04	2.64	3.17
3300	20 ga., under 50 squares	G		3400	.009		3.26	.50	.04	3.80	4.50
3350	50-500 squares	G		3600	.009		2.61	.47	.04	3.12	3.73
3360	over 500 squares	G		3800	.008		2.35	.45	.04	2.84	3.40
3400	18 ga., under 50 squares	G		3200	.010		4.21	.53	.05	4.79	5.60
3450	50-500 squares	G		3400	.009		3.37	.50	.04	3.91	4.62
3460	over 500 squares	G		3600	.009		3.03	.47	.04	3.54	4.19
3500	16 ga., under 50 squares	G		3000	.011		5.55	.57	.05	6.17	7.15
3550	50-500 squares	G		3200	.010		4.45	.53	.05	5.03	5.85
3560	over 500 squares	G		3400	.009		4	.50	.04	4.54	5.35
3700	4-1/2" deep, Type J, 20 ga., over 50 squares	G		2700	.012		3.64	.63	.05	4.32	5.15
3800	18 ga.	G		2460	.013		4.80	.69	.06	5.55	6.55
3900	16 ga.	G		2350	.014		6.25	.72	.06	7.03	8.25
4100	6" deep, Type H, 18 ga., over 50 squares	G		2000	.016		5.75	.85	.07	6.67	7.90
4200	16 ga.	G		1930	.017		7.20	.88	.08	8.16	9.50
4300	14 ga.	G		1860	.017		9.25	.91	.08	10.24	11.90
4500	7-1/2" deep, Type H, 18 ga., over 50 squares	G		1690	.019		6.85	1.01	.09	7.95	9.35
4600	16 ga.	G		1590	.020		8.50	1.07	.09	9.66	11.30
4700	14 ga.	G		1490	.021		10.60	1.14	.10	11.84	13.80
4800	For painted instead of galvanized, deduct						5%				
5000	For acoustical perforated with fiberglass insulation, add					S.F.	25%				
5100	For type F intermediate rib instead of type B wide rib, add	G					25%				
5150	For type A narrow rib instead of type B wide rib, add	G					25%				

05 31 33 – Steel Form Decking

05 31 33.50 Form Decking			Crew	Daily Output	Labor-Hours	Unit	Material	Labor	Equipment	Total	Total Incl O&P
0010	**FORM DECKING**										
0015	Made from recycled materials										
6100	Slab form, steel, 28 ga., 9/16" deep, Type UFS, uncoated	G	E-4	4000	.008	S.F.	1.55	.43	.04	2.02	2.48
6200	Galvanized	G		4000	.008		1.37	.43	.04	1.84	2.29
6220	24 ga., 1" deep, Type UF1X, uncoated	G		3900	.008		1.49	.44	.04	1.97	2.44
6240	Galvanized	G		3900	.008		1.75	.44	.04	2.23	2.73
6300	24 ga., 1-5/16" deep, Type UFX, uncoated	G		3800	.008		1.58	.45	.04	2.07	2.56
6400	Galvanized	G		3800	.008		1.86	.45	.04	2.35	2.87
6500	22 ga., 1-5/16" deep, uncoated	G		3700	.009		2	.46	.04	2.50	3.04
6600	Galvanized	G		3700	.009		2.04	.46	.04	2.54	3.08
6700	22 ga., 2" deep, uncoated	G		3600	.009		2.60	.47	.04	3.11	3.72

05 31 Steel Decking

05 31 33 – Steel Form Decking

05 31 33.50 Form Decking		Crew	Daily Output	Labor-Hours	Unit	Material	2015 Bare Costs Labor	Equipment	Total	Total Incl O&P
6800	Galvanized	G	E-4	3600	.009	S.F.	2.55	.47	.04	3.06	3.67
7000	Sheet metal edge closure form, 12" wide with 2 bends, galvanized										
7100	18 ga.	G	E-14	360	.022	L.F.	4.20	1.21	.41	5.82	7.20
7200	16 ga.	G	"	360	.022	"	5.70	1.21	.41	7.32	8.80

05 35 Raceway Decking Assemblies

05 35 13 – Steel Cellular Decking

05 35 13.50 Cellular Decking

			Crew	Daily Output	Labor-Hours	Unit	Material	Labor	Equipment	Total	Total Incl O&P
0010	**CELLULAR DECKING**										
0015	Made from recycled materials										
0200	Cellular units, galv, 1-1/2" deep, Type BC, 20-20 ga., over 15 squares	G	E-4	1460	.022	S.F.	8.10	1.17	.10	9.37	11.05
0250	18-20 ga.	G		1420	.023		9.20	1.20	.10	10.50	12.30
0300	18-18 ga.	G		1390	.023		9.40	1.22	.11	10.73	12.60
0320	16-18 ga.	G		1360	.024		11.25	1.25	.11	12.61	14.65
0340	16-16 ga.	G		1330	.024		12.50	1.28	.11	13.89	16.10
0400	3" deep, Type NC, galvanized, 20-20 ga.	G		1375	.023		8.90	1.24	.11	10.25	12.05
0500	18-20 ga.	G		1350	.024		10.75	1.26	.11	12.12	14.10
0600	18-18 ga.	G		1290	.025		10.70	1.32	.11	12.13	14.15
0700	16-18 ga.	G		1230	.026		12.05	1.38	.12	13.55	15.85
0800	16-16 ga.	G		1150	.028		13.15	1.48	.13	14.76	17.15
1000	4-1/2" deep, Type JC, galvanized, 18-20 ga.	G		1100	.029		12.40	1.55	.13	14.08	16.50
1100	18-18 ga.	G		1040	.031		12.30	1.64	.14	14.08	16.55
1200	16-18 ga.	G		980	.033		13.90	1.74	.15	15.79	18.40
1300	16-16 ga.	G	↓	935	.034	↓	15.10	1.82	.16	17.08	20
1500	For acoustical deck, add						15%				
1700	For cells used for ventilation, add						15%				
1900	For multi-story or congested site, add							50%			
8000	Metal deck and trench, 2" thick, 20 ga., combination										
8010	60% cellular, 40% non-cellular, inserts and trench	G	R-4	1100	.036	S.F.	16.20	1.94	.13	18.27	21

05 41 Structural Metal Stud Framing

05 41 13 – Load-Bearing Metal Stud Framing

05 41 13.05 Bracing

			Crew	Daily Output	Labor-Hours	Unit	Material	Labor	Equipment	Total	Total Incl O&P
0010	**BRACING**, shear wall X-bracing, per 10' x 10' bay, one face										
0015	Made of recycled materials										
0120	Metal strap, 20 ga. x 4" wide	G	2 Carp	18	.889	Ea.	17.55	41.50		59.05	83.50
0130	6" wide	G		18	.889		29.50	41.50		71	96
0160	18 ga. x 4" wide	G		16	1		30.50	47		77.50	106
0170	6" wide	G	↓	16	1	↓	45	47		92	122
0410	Continuous strap bracing, per horizontal row on both faces										
0420	Metal strap, 20 ga. x 2" wide, studs 12" O.C.	G	1 Carp	7	1.143	C.L.F.	53	53.50		106.50	141
0430	16" O.C.	G		8	1		53	47		100	131
0440	24" O.C.	G		10	.800		53	37.50		90.50	116
0450	18 ga. x 2" wide, studs 12" O.C.	G		6	1.333		75	62.50		137.50	179
0460	16" O.C.	G		7	1.143		75	53.50		128.50	165
0470	24" O.C.	G	↓	8	1	↓	75	47		122	155

05 41 Structural Metal Stud Framing

05 41 13 – Load-Bearing Metal Stud Framing

05 41 13.10 Bridging

05 41 13.10 Bridging		Crew	Daily Output	Labor-Hours	Unit	Material	Labor	Equipment	Total	Total Incl O&P
						2015 Bare Costs				
0010	**BRIDGING**, solid between studs w/1-1/4" leg track, per stud bay									
0015	Made from recycled materials									
0200	Studs 12" O.C., 18 ga. x 2-1/2" wide G	1 Carp	125	.064	Ea.	.89	3		3.89	5.60
0210	3-5/8" wide G		120	.067		1.07	3.13		4.20	6
0220	4" wide G		120	.067		1.13	3.13		4.26	6.05
0230	6" wide G		115	.070		1.48	3.27		4.75	6.70
0240	8" wide G		110	.073		1.84	3.41		5.25	7.25
0300	16 ga. x 2-1/2" wide G		115	.070		1.13	3.27		4.40	6.30
0310	3-5/8" wide G		110	.073		1.38	3.41		4.79	6.75
0320	4" wide G		110	.073		1.47	3.41		4.88	6.85
0330	6" wide G		105	.076		1.87	3.58		5.45	7.55
0340	8" wide G		100	.080		2.34	3.76		6.10	8.40
1200	Studs 16" O.C., 18 ga. x 2-1/2" wide G		125	.064		1.14	3		4.14	5.85
1210	3-5/8" wide G		120	.067		1.37	3.13		4.50	6.35
1220	4" wide G		120	.067		1.45	3.13		4.58	6.40
1230	6" wide G		115	.070		1.90	3.27		5.17	7.15
1240	8" wide G		110	.073		2.36	3.41		5.77	7.85
1300	16 ga. x 2-1/2" wide G		115	.070		1.45	3.27		4.72	6.65
1310	3-5/8" wide G		110	.073		1.77	3.41		5.18	7.20
1320	4" wide G		110	.073		1.88	3.41		5.29	7.30
1330	6" wide G		105	.076		2.39	3.58		5.97	8.15
1340	8" wide G		100	.080		3	3.76		6.76	9.10
2200	Studs 24" O.C., 18 ga. x 2-1/2" wide G		125	.064		1.65	3		4.65	6.45
2210	3-5/8" wide G		120	.067		1.98	3.13		5.11	7
2220	4" wide G		120	.067		2.10	3.13		5.23	7.15
2230	6" wide G		115	.070		2.75	3.27		6.02	8.05
2240	8" wide G		110	.073		3.41	3.41		6.82	9
2300	16 ga. x 2-1/2" wide G		115	.070		2.10	3.27		5.37	7.35
2310	3-5/8" wide G		110	.073		2.55	3.41		5.96	8.05
2320	4" wide G		110	.073		2.72	3.41		6.13	8.25
2330	6" wide G		105	.076		3.46	3.58		7.04	9.30
2340	8" wide G		100	.080		4.34	3.76		8.10	10.60
3000	Continuous bridging, per row									
3100	16 ga. x 1-1/2" channel thru studs 12" O.C. G	1 Carp	6	1.333	C.L.F.	48	62.50		110.50	150
3110	16" O.C. G		7	1.143		48	53.50		101.50	136
3120	24" O.C. G		8.80	.909		48	42.50		90.50	119
4100	2" x 2" angle x 18 ga., studs 12" O.C. G		7	1.143		75	53.50		128.50	165
4110	16" O.C. G		9	.889		75	41.50		116.50	147
4120	24" O.C. G		12	.667		75	31.50		106.50	131
4200	16 ga., studs 12" O.C. G		5	1.600		94.50	75		169.50	220
4210	16" O.C. G		7	1.143		94.50	53.50		148	187
4220	24" O.C. G		10	.800		94.50	37.50		132	162

05 41 13.25 Framing, Boxed Headers/Beams

05 41 13.25 Framing, Boxed Headers/Beams		Crew	Daily Output	Labor-Hours	Unit	Material	Labor	Equipment	Total	Total Incl O&P
0010	**FRAMING, BOXED HEADERS/BEAMS**									
0015	Made from recycled materials									
0200	Double, 18 ga. x 6" deep G	2 Carp	220	.073	L.F.	5.10	3.41		8.51	10.85
0210	8" deep G		210	.076		5.65	3.58		9.23	11.70
0220	10" deep G		200	.080		6.90	3.76		10.66	13.40
0230	12" deep G		190	.084		7.55	3.95		11.50	14.40
0300	16 ga. x 8" deep G		180	.089		6.50	4.17		10.67	13.55
0310	10" deep G		170	.094		7.90	4.42		12.32	15.45
0320	12" deep G		160	.100		8.60	4.70		13.30	16.70

05 41 13 – Load-Bearing Metal Stud Framing

05 41 13.25 Framing, Boxed Headers/Beams		Crew	Daily Output	Labor-Hours	Unit	Material	Labor	Equipment	Total	Total Incl O&P
0400	14 ga. x 10" deep G	2 Carp	140	.114	L.F.	9.10	5.35		14.45	18.25
0410	12" deep G		130	.123		10	5.80		15.80	19.85
1210	Triple, 18 ga. x 8" deep G		170	.094		8.20	4.42		12.62	15.80
1220	10" deep G		165	.097		9.85	4.55		14.40	17.85
1230	12" deep G		160	.100		10.85	4.70		15.55	19.15
1300	16 ga. x 8" deep G		145	.110		9.50	5.20		14.70	18.35
1310	10" deep G		140	.114		11.35	5.35		16.70	20.50
1320	12" deep G		135	.119		12.40	5.55		17.95	22
1400	14 ga. x 10" deep G		115	.139		12.40	6.55		18.95	23.50
1410	12" deep G		110	.145		13.70	6.85		20.55	25.50

(2015 Bare Costs: Material, Labor, Equipment, Total)

05 41 13.30 Framing, Stud Walls		Crew	Daily Output	Labor-Hours	Unit	Material	Labor	Equipment	Total	Total Incl O&P
0010	**FRAMING, STUD WALLS** w/top & bottom track, no openings,									
0020	Headers, beams, bridging or bracing									
0025	Made from recycled materials									
4100	8' high walls, 18 ga. x 2-1/2" wide, studs 12" O.C. G	2 Carp	54	.296	L.F.	8.50	13.90		22.40	31
4110	16" O.C. G		77	.208		6.80	9.75		16.55	22.50
4120	24" O.C. G		107	.150		5.10	7		12.10	16.40
4130	3-5/8" wide, studs 12" O.C. G		53	.302		10.05	14.15		24.20	33
4140	16" O.C. G		76	.211		8.05	9.90		17.95	24
4150	24" O.C. G		105	.152		6.05	7.15		13.20	17.65
4160	4" wide, studs 12" O.C. G		52	.308		10.55	14.45		25	33.50
4170	16" O.C. G		74	.216		8.45	10.15		18.60	25
4180	24" O.C. G		103	.155		6.35	7.30		13.65	18.20
4190	6" wide, studs 12" O.C. G		51	.314		13.40	14.75		28.15	37
4200	16" O.C. G		73	.219		10.75	10.30		21.05	27.50
4210	24" O.C. G		101	.158		8.10	7.45		15.55	20.50
4220	8" wide, studs 12" O.C. G		50	.320		16.30	15		31.30	41
4230	16" O.C. G		72	.222		13.10	10.45		23.55	30.50
4240	24" O.C. G		100	.160		9.90	7.50		17.40	22.50
4300	16 ga. x 2-1/2" wide, studs 12" O.C. G		47	.340		10.10	16		26.10	35.50
4310	16" O.C. G		68	.235		8	11.05		19.05	26
4320	24" O.C. G		94	.170		5.90	8		13.90	18.80
4330	3-5/8" wide, studs 12" O.C. G		46	.348		12.05	16.35		28.40	38.50
4340	16" O.C. G		66	.242		9.55	11.40		20.95	28
4350	24" O.C. G		92	.174		7.05	8.15		15.20	20.50
4360	4" wide, studs 12" O.C. G		45	.356		12.65	16.70		29.35	39.50
4370	16" O.C. G		65	.246		10	11.55		21.55	29
4380	24" O.C. G		90	.178		7.40	8.35		15.75	21
4390	6" wide, studs 12" O.C. G		44	.364		15.80	17.05		32.85	44
4400	16" O.C. G		64	.250		12.55	11.75		24.30	32
4410	24" O.C. G		88	.182		9.30	8.55		17.85	23.50
4420	8" wide, studs 12" O.C. G		43	.372		19.50	17.45		36.95	48.50
4430	16" O.C. G		63	.254		15.50	11.90		27.40	35.50
4440	24" O.C. G		86	.186		11.50	8.75		20.25	26
5100	10' high walls, 18 ga. x 2-1/2" wide, studs 12" O.C. G		54	.296		10.20	13.90		24.10	32.50
5110	16" O.C. G		77	.208		8.05	9.75		17.80	24
5120	24" O.C. G		107	.150		5.95	7		12.95	17.35
5130	3-5/8" wide, studs 12" O.C. G		53	.302		12.05	14.15		26.20	35.50
5140	16" O.C. G		76	.211		9.55	9.90		19.45	25.50
5150	24" O.C. G		105	.152		7.05	7.15		14.20	18.75
5160	4" wide, studs 12" O.C. G		52	.308		12.65	14.45		27.10	36
5170	16" O.C. G		74	.216		10.05	10.15		20.20	26.50

05 41 13 – Load-Bearing Metal Stud Framing

05 41 13.30 Framing, Stud Walls			Crew	Daily Output	Labor-Hours	Unit	Material	2015 Bare Costs Labor	Equipment	Total	Total Incl O&P
5180	24" O.C.	G	2 Carp	103	.155	L.F.	7.40	7.30		14.70	19.35
5190	6" wide, studs 12" O.C.	G		51	.314		16	14.75		30.75	40
5200	16" O.C.	G		73	.219		12.70	10.30		23	30
5210	24" O.C.	G		101	.158		9.40	7.45		16.85	22
5220	8" wide, studs 12" O.C.	G		50	.320		19.50	15		34.50	44.50
5230	16" O.C.	G		72	.222		15.50	10.45		25.95	33
5240	24" O.C.	G		100	.160		11.50	7.50		19	24
5300	16 ga. x 2-1/2" wide, studs 12" O.C.	G		47	.340		12.20	16		28.20	38
5310	16" O.C.	G		68	.235		9.55	11.05		20.60	27.50
5320	24" O.C.	G		94	.170		6.95	8		14.95	19.95
5330	3-5/8" wide, studs 12" O.C.	G		46	.348		14.55	16.35		30.90	41
5340	16" O.C.	G		66	.242		11.40	11.40		22.80	30
5350	24" O.C.	G		92	.174		8.30	8.15		16.45	21.50
5360	4" wide, studs 12" O.C.	G		45	.356		15.25	16.70		31.95	42.50
5370	16" O.C.	G		65	.246		12	11.55		23.55	31
5380	24" O.C.	G		90	.178		8.70	8.35		17.05	22.50
5390	6" wide, studs 12" O.C.	G		44	.364		19	17.05		36.05	47.50
5400	16" O.C.	G		64	.250		14.95	11.75		26.70	34.50
5410	24" O.C.	G		88	.182		10.90	8.55		19.45	25
5420	8" wide, studs 12" O.C.	G		43	.372		23.50	17.45		40.95	53
5430	16" O.C.	G		63	.254		18.50	11.90		30.40	39
5440	24" O.C.	G		86	.186		13.50	8.75		22.25	28.50
6190	12' high walls, 18 ga. x 6" wide, studs 12" O.C.	G		41	.390		18.65	18.30		36.95	48.50
6200	16" O.C.	G		58	.276		14.70	12.95		27.65	36
6210	24" O.C.	G		81	.198		10.75	9.25		20	26
6220	8" wide, studs 12" O.C.	G		40	.400		22.50	18.80		41.30	54
6230	16" O.C.	G		57	.281		17.90	13.20		31.10	40
6240	24" O.C.	G		80	.200		13.10	9.40		22.50	29
6390	16 ga. x 6" wide, studs 12" O.C.	G		35	.457		22.50	21.50		44	57.50
6400	16" O.C.	G		51	.314		17.40	14.75		32.15	41.50
6410	24" O.C.	G		70	.229		12.55	10.75		23.30	30.50
6420	8" wide, studs 12" O.C.	G		34	.471		27.50	22		49.50	64.50
6430	16" O.C.	G		50	.320		21.50	15		36.50	46.50
6440	24" O.C.	G		69	.232		15.50	10.90		26.40	34
6530	14 ga. x 3-5/8" wide, studs 12" O.C.	G		34	.471		21	22		43	57.50
6540	16" O.C.	G		48	.333		16.55	15.65		32.20	42.50
6550	24" O.C.	G		65	.246		11.90	11.55		23.45	31
6560	4" wide, studs 12" O.C.	G		33	.485		22.50	23		45.50	59.50
6570	16" O.C.	G		47	.340		17.55	16		33.55	44
6580	24" O.C.	G		64	.250		12.65	11.75		24.40	32
6730	12 ga. x 3-5/8" wide, studs 12" O.C.	G		31	.516		29.50	24		53.50	70
6740	16" O.C.	G		43	.372		23	17.45		40.45	52
6750	24" O.C.	G		59	.271		16.05	12.75		28.80	37.50
6760	4" wide, studs 12" O.C.	G		30	.533		31.50	25		56.50	73
6770	16" O.C.	G		42	.381		24.50	17.90		42.40	54
6780	24" O.C.	G		58	.276		17.15	12.95		30.10	39
7390	16' high walls, 16 ga. x 6" wide, studs 12" O.C.	G		33	.485		28.50	23		51.50	66.50
7400	16" O.C.	G		48	.333		22.50	15.65		38.15	48.50
7410	24" O.C.	G		67	.239		15.80	11.20		27	34.50
7420	8" wide, studs 12" O.C.	G		32	.500		35.50	23.50		59	75
7430	16" O.C.	G		47	.340		27.50	16		43.50	55
7440	24" O.C.	G		66	.242		19.50	11.40		30.90	39
7560	14 ga. x 4" wide, studs 12" O.C.	G		31	.516		29	24		53	69.50

05 41 Structural Metal Stud Framing

05 41 13 – Load-Bearing Metal Stud Framing

05 41 13.30 Framing, Stud Walls			Crew	Daily Output	Labor-Hours	Unit	Material	2015 Bare Costs Labor	Equipment	Total	Total Incl O&P
7570	16" O.C.	G	2 Carp	45	.356	L.F.	22.50	16.70		39.20	50
7580	24" O.C.	G		61	.262		15.90	12.30		28.20	36.50
7590	6" wide, studs 12" O.C.	G		30	.533		36.50	25		61.50	78.50
7600	16" O.C.	G		44	.364		28.50	17.05		45.55	57.50
7610	24" O.C.	G		60	.267		20	12.50		32.50	41.50
7760	12 ga. x 4" wide, studs 12" O.C.	G		29	.552		41	26		67	85
7770	16" O.C.	G		40	.400		31.50	18.80		50.30	63.50
7780	24" O.C.	G		55	.291		22	13.65		35.65	45
7790	6" wide, studs 12" O.C.	G		28	.571		51.50	27		78.50	98.50
7800	16" O.C.	G		39	.410		39.50	19.25		58.75	73
7810	24" O.C.	G		54	.296		27.50	13.90		41.40	52
8590	20' high walls, 14 ga. x 6" wide, studs 12" O.C.	G		29	.552		45	26		71	89
8600	16" O.C.	G		42	.381		34.50	17.90		52.40	65.50
8610	24" O.C.	G		57	.281		24	13.20		37.20	47
8620	8" wide, studs 12" O.C.	G		28	.571		48.50	27		75.50	95
8630	16" O.C.	G		41	.390		37.50	18.30		55.80	69
8640	24" O.C.	G		56	.286		26.50	13.40		39.90	49.50
8790	12 ga. x 6" wide, studs 12" O.C.	G		27	.593		64	28		92	113
8800	16" O.C.	G		37	.432		48.50	20.50		69	84.50
8810	24" O.C.	G		51	.314		33.50	14.75		48.25	59.50
8820	8" wide, studs 12" O.C.	G		26	.615		77.50	29		106.50	130
8830	16" O.C.	G		36	.444		59	21		80	97
8840	24" O.C.	G	↓	50	.320	↓	41	15		56	68

05 42 Cold-Formed Metal Joist Framing

05 42 13 – Cold-Formed Metal Floor Joist Framing

05 42 13.05 Bracing			Crew	Daily Output	Labor-Hours	Unit	Material	Labor	Equipment	Total	Total Incl O&P
0010	**BRACING**, continuous, per row, top & bottom										
0015	Made from recycled materials										
0120	Flat strap, 20 ga. x 2" wide, joists at 12" O.C.	G	1 Carp	4.67	1.713	C.L.F.	55	80.50		135.50	185
0130	16" O.C.	G		5.33	1.501		53.50	70.50		124	167
0140	24" O.C.	G		6.66	1.201		51.50	56.50		108	144
0150	18 ga. x 2" wide, joists at 12" O.C.	G		4	2		74	94		168	227
0160	16" O.C.	G		4.67	1.713		73	80.50		153.50	204
0170	24" O.C.	G	↓	5.33	1.501	↓	71.50	70.50		142	187

05 42 13.10 Bridging			Crew	Daily Output	Labor-Hours	Unit	Material	Labor	Equipment	Total	Total Incl O&P
0010	**BRIDGING**, solid between joists w/1-1/4" leg track, per joist bay										
0015	Made from recycled materials										
0230	Joists 12" O.C., 18 ga. track x 6" wide	G	1 Carp	80	.100	Ea.	1.48	4.70		6.18	8.90
0240	8" wide	G		75	.107		1.84	5		6.84	9.70
0250	10" wide	G		70	.114		2.29	5.35		7.64	10.75
0260	12" wide	G		65	.123		2.60	5.80		8.40	11.75
0330	16 ga. track x 6" wide	G		70	.114		1.87	5.35		7.22	10.30
0340	8" wide	G		65	.123		2.34	5.80		8.14	11.50
0350	10" wide	G		60	.133		2.91	6.25		9.16	12.85
0360	12" wide	G		55	.145		3.35	6.85		10.20	14.20
0440	14 ga. track x 8" wide	G		60	.133		2.93	6.25		9.18	12.90
0450	10" wide	G		55	.145		3.64	6.85		10.49	14.50
0460	12" wide	G		50	.160		4.20	7.50		11.70	16.15
0550	12 ga. track x 10" wide	G		45	.178		5.35	8.35		13.70	18.70
0560	12" wide	G	↓	40	.200	↓	5.45	9.40		14.85	20.50

05 42 Cold-Formed Metal Joist Framing

05 42 13 – Cold-Formed Metal Floor Joist Framing

05 42 13.10 Bridging			Crew	Daily Output	Labor-Hours	Unit	Material	2015 Bare Costs Labor	Equipment	Total	Total Incl O&P
1230	16" O.C., 18 ga. track x 6" wide	G	1 Carp	80	.100	Ea.	1.90	4.70		6.60	9.35
1240	8" wide	G		75	.107		2.36	5		7.36	10.30
1250	10" wide	G		70	.114		2.94	5.35		8.29	11.50
1260	12" wide	G		65	.123		3.33	5.80		9.13	12.55
1330	16 ga. track x 6" wide	G		70	.114		2.39	5.35		7.74	10.90
1340	8" wide	G		65	.123		3	5.80		8.80	12.20
1350	10" wide	G		60	.133		3.73	6.25		9.98	13.75
1360	12" wide	G		55	.145		4.29	6.85		11.14	15.20
1440	14 ga. track x 8" wide	G		60	.133		3.76	6.25		10.01	13.80
1450	10" wide	G		55	.145		4.67	6.85		11.52	15.65
1460	12" wide	G		50	.160		5.40	7.50		12.90	17.45
1550	12 ga. track x 10" wide	G		45	.178		6.85	8.35		15.20	20.50
1560	12" wide	G		40	.200		7	9.40		16.40	22
2230	24" O.C., 18 ga. track x 6" wide	G		80	.100		2.75	4.70		7.45	10.25
2240	8" wide	G		75	.107		3.41	5		8.41	11.45
2250	10" wide	G		70	.114		4.25	5.35		9.60	12.90
2260	12" wide	G		65	.123		4.82	5.80		10.62	14.20
2330	16 ga. track x 6" wide	G		70	.114		3.46	5.35		8.81	12.05
2340	8" wide	G		65	.123		4.34	5.80		10.14	13.70
2350	10" wide	G		60	.133		5.40	6.25		11.65	15.60
2360	12" wide	G		55	.145		6.20	6.85		13.05	17.35
2440	14 ga. track x 8" wide	G		60	.133		5.45	6.25		11.70	15.65
2450	10" wide	G		55	.145		6.75	6.85		13.60	17.95
2460	12" wide	G		50	.160		7.80	7.50		15.30	20
2550	12 ga. track x 10" wide	G		45	.178		9.90	8.35		18.25	23.50
2560	12" wide	G	↓	40	.200	↓	10.10	9.40		19.50	25.50

05 42 13.25 Framing, Band Joist

			Crew	Daily Output	Labor-Hours	Unit	Material	Labor	Equipment	Total	Total Incl O&P
0010	**FRAMING, BAND JOIST** (track) fastened to bearing wall										
0015	Made from recycled materials										
0220	18 ga. track x 6" deep	G	2 Carp	1000	.016	L.F.	1.21	.75		1.96	2.49
0230	8" deep	G		920	.017		1.50	.82		2.32	2.91
0240	10" deep	G		860	.019		1.87	.87		2.74	3.40
0320	16 ga. track x 6" deep	G		900	.018		1.52	.83		2.35	2.95
0330	8" deep	G		840	.019		1.91	.89		2.80	3.48
0340	10" deep	G		780	.021		2.37	.96		3.33	4.09
0350	12" deep	G		740	.022		2.73	1.02		3.75	4.56
0430	14 ga. track x 8" deep	G		750	.021		2.39	1		3.39	4.17
0440	10" deep	G		720	.022		2.97	1.04		4.01	4.88
0450	12" deep	G		700	.023		3.42	1.07		4.49	5.40
0540	12 ga. track x 10" deep	G		670	.024		4.35	1.12		5.47	6.50
0550	12" deep	G	↓	650	.025	↓	4.45	1.16		5.61	6.70

05 42 13.30 Framing, Boxed Headers/Beams

			Crew	Daily Output	Labor-Hours	Unit	Material	Labor	Equipment	Total	Total Incl O&P
0010	**FRAMING, BOXED HEADERS/BEAMS**										
0015	Made from recycled materials										
0200	Double, 18 ga. x 6" deep	G	2 Carp	220	.073	L.F.	5.10	3.41		8.51	10.85
0210	8" deep	G		210	.076		5.65	3.58		9.23	11.70
0220	10" deep	G		200	.080		6.90	3.76		10.66	13.40
0230	12" deep	G		190	.084		7.55	3.95		11.50	14.40
0300	16 ga. x 8" deep	G		180	.089		6.50	4.17		10.67	13.55
0310	10" deep	G		170	.094		7.90	4.42		12.32	15.45
0320	12" deep	G		160	.100		8.60	4.70		13.30	16.70
0400	14 ga. x 10" deep	G	↓	140	.114	↓	9.10	5.35		14.45	18.25

05 42 13 – Cold-Formed Metal Floor Joist Framing

05 42 13.30 Framing, Boxed Headers/Beams		Crew	Daily Output	Labor-Hours	Unit	Material	Labor	Equipment	Total	Total Incl O&P
						2015 Bare Costs				
0410	12" deep	G	2 Carp	130	.123	L.F.	10	5.80		15.80	19.85
0500	12 ga. x 10" deep	G		110	.145		12	6.85		18.85	23.50
0510	12" deep	G		100	.160		13.25	7.50		20.75	26
1210	Triple, 18 ga. x 8" deep	G		170	.094		8.20	4.42		12.62	15.80
1220	10" deep	G		165	.097		9.85	4.55		14.40	17.85
1230	12" deep	G		160	.100		10.85	4.70		15.55	19.15
1300	16 ga. x 8" deep	G		145	.110		9.50	5.20		14.70	18.35
1310	10" deep	G		140	.114		11.35	5.35		16.70	20.50
1320	12" deep	G		135	.119		12.40	5.55		17.95	22
1400	14 ga. x 10" deep	G		115	.139		13.15	6.55		19.70	24.50
1410	12" deep	G		110	.145		14.50	6.85		21.35	26.50
1500	12 ga. x 10" deep	G		90	.178		17.50	8.35		25.85	32
1510	12" deep	G		85	.188		19.40	8.85		28.25	35

05 42 13.40 Framing, Joists			Crew	Daily Output	Labor-Hours	Unit	Material	Labor	Equipment	Total	Total Incl O&P
0010	**FRAMING, JOISTS**, no band joists (track), web stiffeners, headers,										
0020	Beams, bridging or bracing										
0025	Made from recycled materials										
0030	Joists (2" flange) and fasteners, materials only										
0220	18 ga. x 6" deep	G				L.F.	1.58			1.58	1.73
0230	8" deep	G					1.86			1.86	2.04
0240	10" deep	G					2.18			2.18	2.40
0320	16 ga. x 6" deep	G					1.93			1.93	2.13
0330	8" deep	G					2.31			2.31	2.54
0340	10" deep	G					2.70			2.70	2.97
0350	12" deep	G					3.07			3.07	3.37
0430	14 ga. x 8" deep	G					2.90			2.90	3.19
0440	10" deep	G					3.34			3.34	3.67
0450	12" deep	G					3.80			3.80	4.18
0540	12 ga. x 10" deep	G					4.86			4.86	5.35
0550	12" deep	G					5.50			5.50	6.10
1010	Installation of joists to band joists, beams & headers, labor only										
1220	18 ga. x 6" deep		2 Carp	110	.145	Ea.		6.85		6.85	10.50
1230	8" deep			90	.178			8.35		8.35	12.85
1240	10" deep			80	.200			9.40		9.40	14.45
1320	16 ga. x 6" deep			95	.168			7.90		7.90	12.15
1330	8" deep			70	.229			10.75		10.75	16.50
1340	10" deep			60	.267			12.50		12.50	19.25
1350	12" deep			55	.291			13.65		13.65	21
1430	14 ga. x 8" deep			65	.246			11.55		11.55	17.80
1440	10" deep			45	.356			16.70		16.70	25.50
1450	12" deep			35	.457			21.50		21.50	33
1540	12 ga. x 10" deep			40	.400			18.80		18.80	29
1550	12" deep			30	.533			25		25	38.50

05 42 13.45 Framing, Web Stiffeners			Crew	Daily Output	Labor-Hours	Unit	Material	Labor	Equipment	Total	Total Incl O&P
0010	**FRAMING, WEB STIFFENERS** at joist bearing, fabricated from										
0020	Stud piece (1-5/8" flange) to stiffen joist (2" flange)										
0025	Made from recycled materials										
2120	For 6" deep joist, with 18 ga. x 2-1/2" stud	G	1 Carp	120	.067	Ea.	.85	3.13		3.98	5.75
2130	3-5/8" stud	G		110	.073		1	3.41		4.41	6.35
2140	4" stud	G		105	.076		1.05	3.58		4.63	6.65
2150	6" stud	G		100	.080		1.32	3.76		5.08	7.25
2160	8" stud	G		95	.084		1.60	3.95		5.55	7.85

05 42 13.45 Framing, Web Stiffeners		Crew	Daily Output	Labor-Hours	Unit	Material	2015 Bare Costs Labor	Equipment	Total	Total Incl O&P
2220	8" deep joist, with 2-1/2" stud	G	1 Carp	120	.067	Ea.	1.14	3.13		4.27	6.05
2230	3-5/8" stud	G		110	.073		1.34	3.41		4.75	6.70
2240	4" stud	G		105	.076		1.41	3.58		4.99	7.05
2250	6" stud	G		100	.080		1.77	3.76		5.53	7.75
2260	8" stud	G		95	.084		2.14	3.95		6.09	8.45
2320	10" deep joist, with 2-1/2" stud	G		110	.073		1.41	3.41		4.82	6.80
2330	3-5/8" stud	G		100	.080		1.66	3.76		5.42	7.65
2340	4" stud	G		95	.084		1.74	3.95		5.69	8
2350	6" stud	G		90	.089		2.19	4.17		6.36	8.80
2360	8" stud	G		85	.094		2.66	4.42		7.08	9.70
2420	12" deep joist, with 2-1/2" stud	G		110	.073		1.70	3.41		5.11	7.10
2430	3-5/8" stud	G		100	.080		2	3.76		5.76	8
2440	4" stud	G		95	.084		2.10	3.95		6.05	8.40
2450	6" stud	G		90	.089		2.64	4.17		6.81	9.30
2460	8" stud	G		85	.094		3.20	4.42		7.62	10.30
3130	For 6" deep joist, with 16 ga. x 3-5/8" stud	G		100	.080		1.25	3.76		5.01	7.20
3140	4" stud	G		95	.084		1.31	3.95		5.26	7.55
3150	6" stud	G		90	.089		1.62	4.17		5.79	8.20
3160	8" stud	G		85	.094		2	4.42		6.42	9
3230	8" deep joist, with 3-5/8" stud	G		100	.080		1.68	3.76		5.44	7.65
3240	4" stud	G		95	.084		1.76	3.95		5.71	8.05
3250	6" stud	G		90	.089		2.17	4.17		6.34	8.80
3260	8" stud	G		85	.094		2.68	4.42		7.10	9.75
3330	10" deep joist, with 3-5/8" stud	G		85	.094		2.08	4.42		6.50	9.10
3340	4" stud	G		80	.100		2.17	4.70		6.87	9.65
3350	6" stud	G		75	.107		2.69	5		7.69	10.65
3360	8" stud	G		70	.114		3.32	5.35		8.67	11.90
3430	12" deep joist, with 3-5/8" stud	G		85	.094		2.50	4.42		6.92	9.55
3440	4" stud	G		80	.100		2.62	4.70		7.32	10.15
3450	6" stud	G		75	.107		3.24	5		8.24	11.25
3460	8" stud	G		70	.114		4	5.35		9.35	12.65
4230	For 8" deep joist, with 14 ga. x 3-5/8" stud	G		90	.089		2.08	4.17		6.25	8.70
4240	4" stud	G		85	.094		2.20	4.42		6.62	9.20
4250	6" stud	G		80	.100		2.76	4.70		7.46	10.30
4260	8" stud	G		75	.107		2.95	5		7.95	10.95
4330	10" deep joist, with 3-5/8" stud	G		75	.107		2.57	5		7.57	10.55
4340	4" stud	G		70	.114		2.72	5.35		8.07	11.25
4350	6" stud	G		65	.123		3.42	5.80		9.22	12.65
4360	8" stud	G		60	.133		3.65	6.25		9.90	13.65
4430	12" deep joist, with 3-5/8" stud	G		75	.107		3.10	5		8.10	11.10
4440	4" stud	G		70	.114		3.28	5.35		8.63	11.85
4450	6" stud	G		65	.123		4.12	5.80		9.92	13.45
4460	8" stud	G		60	.133		4.40	6.25		10.65	14.50
5330	For 10" deep joist, with 12 ga. x 3-5/8" stud	G		65	.123		3.72	5.80		9.52	13
5340	4" stud	G		60	.133		3.97	6.25		10.22	14
5350	6" stud	G		55	.145		5	6.85		11.85	16
5360	8" stud	G		50	.160		6.05	7.50		13.55	18.20
5430	12" deep joist, with 3-5/8" stud	G		65	.123		4.48	5.80		10.28	13.85
5440	4" stud	G		60	.133		4.78	6.25		11.03	14.90
5450	6" stud	G		55	.145		6	6.85		12.85	17.10
5460	8" stud	G	↓	50	.160	↓	7.30	7.50		14.80	19.55

05 42 23 – Cold-Formed Metal Roof Joist Framing

05 42 23.05 Framing, Bracing

	05 42 23.05 Framing, Bracing		Crew	Daily Output	Labor-Hours	Unit	Material	Labor	Equipment	Total	Total Incl O&P
0010	**FRAMING, BRACING**										
0015	Made from recycled materials										
0020	Continuous bracing, per row										
0100	16 ga. x 1-1/2" channel thru rafters/trusses @ 16" O.C.	G	1 Carp	4.50	1.778	C.L.F.	48	83.50		131.50	181
0120	24" O.C.	G		6	1.333		48	62.50		110.50	150
0300	2" x 2" angle x 18 ga., rafters/trusses @ 16" O.C.	G		6	1.333		75	62.50		137.50	179
0320	24" O.C.	G		8	1		75	47		122	155
0400	16 ga., rafters/trusses @ 16" O.C.	G		4.50	1.778		94.50	83.50		178	232
0420	24" O.C.	G	↓	6.50	1.231	↓	94.50	58		152.50	193

05 42 23.10 Framing, Bridging

	05 42 23.10 Framing, Bridging		Crew	Daily Output	Labor-Hours	Unit	Material	Labor	Equipment	Total	Total Incl O&P
0010	**FRAMING, BRIDGING**										
0015	Made from recycled materials										
0020	Solid, between rafters w/1-1/4" leg track, per rafter bay										
1200	Rafters 16" O.C., 18 ga. x 4" deep	G	1 Carp	60	.133	Ea.	1.45	6.25		7.70	11.25
1210	6" deep	G		57	.140		1.90	6.60		8.50	12.25
1220	8" deep	G		55	.145		2.36	6.85		9.21	13.10
1230	10" deep	G		52	.154		2.94	7.20		10.14	14.35
1240	12" deep	G		50	.160		3.33	7.50		10.83	15.20
2200	24" O.C., 18 ga. x 4" deep	G		60	.133		2.10	6.25		8.35	11.95
2210	6" deep	G		57	.140		2.75	6.60		9.35	13.15
2220	8" deep	G		55	.145		3.41	6.85		10.26	14.25
2230	10" deep	G		52	.154		4.25	7.20		11.45	15.75
2240	12" deep	G	↓	50	.160	↓	4.82	7.50		12.32	16.85

05 42 23.50 Framing, Parapets

	05 42 23.50 Framing, Parapets		Crew	Daily Output	Labor-Hours	Unit	Material	Labor	Equipment	Total	Total Incl O&P
0010	**FRAMING, PARAPETS**										
0015	Made from recycled materials										
0100	3' high installed on 1st story, 18 ga. x 4" wide studs, 12" O.C.	G	2 Carp	100	.160	L.F.	5.30	7.50		12.80	17.40
0110	16" O.C.	G		150	.107		4.52	5		9.52	12.65
0120	24" O.C.	G		200	.080		3.73	3.76		7.49	9.90
0200	6" wide studs, 12" O.C.	G		100	.160		6.80	7.50		14.30	19
0210	16" O.C.	G		150	.107		5.80	5		10.80	14.05
0220	24" O.C.	G		200	.080		4.80	3.76		8.56	11.10
1100	Installed on 2nd story, 18 ga. x 4" wide studs, 12" O.C.	G		95	.168		5.30	7.90		13.20	18
1110	16" O.C.	G		145	.110		4.52	5.20		9.72	12.90
1120	24" O.C.	G		190	.084		3.73	3.95		7.68	10.20
1200	6" wide studs, 12" O.C.	G		95	.168		6.80	7.90		14.70	19.60
1210	16" O.C.	G		145	.110		5.80	5.20		11	14.30
1220	24" O.C.	G		190	.084		4.80	3.95		8.75	11.40
2100	Installed on gable, 18 ga. x 4" wide studs, 12" O.C.	G		85	.188		5.30	8.85		14.15	19.45
2110	16" O.C.	G		130	.123		4.52	5.80		10.32	13.85
2120	24" O.C.	G		170	.094		3.73	4.42		8.15	10.90
2200	6" wide studs, 12" O.C.	G		85	.188		6.80	8.85		15.65	21
2210	16" O.C.	G		130	.123		5.80	5.80		11.60	15.25
2220	24" O.C.	G	↓	170	.094	↓	4.80	4.42		9.22	12.10

05 42 23.60 Framing, Roof Rafters

	05 42 23.60 Framing, Roof Rafters		Crew	Daily Output	Labor-Hours	Unit	Material	Labor	Equipment	Total	Total Incl O&P
0010	**FRAMING, ROOF RAFTERS**										
0015	Made from recycled materials										
0100	Boxed ridge beam, double, 18 ga. x 6" deep	G	2 Carp	160	.100	L.F.	5.10	4.70		9.80	12.85
0110	8" deep	G		150	.107		5.65	5		10.65	13.90
0120	10" deep	G		140	.114		6.90	5.35		12.25	15.85
0130	12" deep	G	↓	130	.123	↓	7.55	5.80		13.35	17.20

(Material, Labor, Equipment and Total columns are 2015 Bare Costs.)

05 42 23 – Cold-Formed Metal Roof Joist Framing

05 42 23.60 Framing, Roof Rafters			Crew	Daily Output	Labor-Hours	Unit	2015 Bare Costs Material	Labor	Equipment	Total	Total Incl O&P
0200	16 ga. x 6″ deep	G	2 Carp	150	.107	L.F.	5.80	5		10.80	14.05
0210	8″ deep	G		140	.114		6.50	5.35		11.85	15.40
0220	10″ deep	G		130	.123		7.90	5.80		13.70	17.55
0230	12″ deep	G		120	.133		8.60	6.25		14.85	19.10
1100	Rafters, 2″ flange, material only, 18 ga. x 6″ deep	G					1.58			1.58	1.73
1110	8″ deep	G					1.86			1.86	2.04
1120	10″ deep	G					2.18			2.18	2.40
1130	12″ deep	G					2.52			2.52	2.77
1200	16 ga. x 6″ deep	G					1.93			1.93	2.13
1210	8″ deep	G					2.31			2.31	2.54
1220	10″ deep	G					2.70			2.70	2.97
1230	12″ deep	G					3.07			3.07	3.37
2100	Installation only, ordinary rafter to 4:12 pitch, 18 ga. x 6″ deep		2 Carp	35	.457	Ea.		21.50		21.50	33
2110	8″ deep			30	.533			25		25	38.50
2120	10″ deep			25	.640			30		30	46
2130	12″ deep			20	.800			37.50		37.50	58
2200	16 ga. x 6″ deep			30	.533			25		25	38.50
2210	8″ deep			25	.640			30		30	46
2220	10″ deep			20	.800			37.50		37.50	58
2230	12″ deep			15	1.067			50		50	77
8100	Add to labor, ordinary rafters on steep roofs							25%			
8110	Dormers & complex roofs							50%			
8200	Hip & valley rafters to 4:12 pitch							25%			
8210	Steep roofs							50%			
8220	Dormers & complex roofs							75%			
8300	Hip & valley jack rafters to 4:12 pitch							50%			
8310	Steep roofs							75%			
8320	Dormers & complex roofs							100%			

05 42 23.70 Framing, Soffits and Canopies			Crew	Daily Output	Labor-Hours	Unit	Material	Labor	Equipment	Total	Total Incl O&P
0010	**FRAMING, SOFFITS & CANOPIES**										
0015	Made from recycled materials										
0130	Continuous ledger track @ wall, studs @ 16″ O.C., 18 ga. x 4″ wide	G	2 Carp	535	.030	L.F.	.97	1.40		2.37	3.22
0140	6″ wide	G		500	.032		1.27	1.50		2.77	3.70
0150	8″ wide	G		465	.034		1.57	1.62		3.19	4.22
0160	10″ wide	G		430	.037		1.96	1.75		3.71	4.84
0230	Studs @ 24″ O.C., 18 ga. x 4″ wide	G		800	.020		.92	.94		1.86	2.47
0240	6″ wide	G		750	.021		1.21	1		2.21	2.87
0250	8″ wide	G		700	.023		1.50	1.07		2.57	3.30
0260	10″ wide	G		650	.025		1.87	1.16		3.03	3.84
1000	Horizontal soffit and canopy members, material only										
1030	1-5/8″ flange studs, 18 ga. x 4″ deep	G				L.F.	1.26			1.26	1.39
1040	6″ deep	G					1.58			1.58	1.74
1050	8″ deep	G					1.92			1.92	2.11
1140	2″ flange joists, 18 ga. x 6″ deep	G					1.80			1.80	1.98
1150	8″ deep	G					2.12			2.12	2.34
1160	10″ deep	G					2.50			2.50	2.75
4030	Installation only, 18 ga., 1-5/8″ flange x 4″ deep		2 Carp	130	.123	Ea.		5.80		5.80	8.90
4040	6″ deep			110	.145			6.85		6.85	10.50
4050	8″ deep			90	.178			8.35		8.35	12.85
4140	2″ flange, 18 ga. x 6″ deep			110	.145			6.85		6.85	10.50
4150	8″ deep			90	.178			8.35		8.35	12.85
4160	10″ deep			80	.200			9.40		9.40	14.45

05 42 Cold-Formed Metal Joist Framing

05 42 23 – Cold-Formed Metal Roof Joist Framing

05 42 23.70 Framing, Soffits and Canopies		Crew	Daily Output	Labor-Hours	Unit	Material	2015 Bare Costs Labor	Equipment	Total	Total Incl O&P
6010	Clips to attach facia to rafter tails, 2" x 2" x 18 ga. angle G	1 Carp	120	.067	Ea.	.88	3.13		4.01	5.80
6020	16 ga. angle G	"	100	.080	↓	1.12	3.76		4.88	7.05

05 44 Cold-Formed Metal Trusses

05 44 13 – Cold-Formed Metal Roof Trusses

05 44 13.60 Framing, Roof Trusses		Crew	Daily Output	Labor-Hours	Unit	Material	Labor	Equipment	Total	Total Incl O&P
0010	**FRAMING, ROOF TRUSSES**									
0015	Made from recycled materials									
0020	Fabrication of trusses on ground, Fink (W) or King Post, to 4:12 pitch									
0120	18 ga. x 4" chords, 16' span G	2 Carp	12	1.333	Ea.	59	62.50		121.50	161
0130	20' span G		11	1.455		73.50	68.50		142	186
0140	24' span G		11	1.455		88	68.50		156.50	202
0150	28' span G		10	1.600		103	75		178	229
0160	32' span G		10	1.600		118	75		193	245
0250	6" chords, 28' span G		9	1.778		129	83.50		212.50	270
0260	32' span G		9	1.778		148	83.50		231.50	291
0270	36' span G		8	2		166	94		260	330
0280	40' span G		8	2		185	94		279	350
1120	5:12 to 8:12 pitch, 18 ga. x 4" chords, 16' span G		10	1.600		67	75		142	190
1130	20' span G		9	1.778		84	83.50		167.50	221
1140	24' span G		9	1.778		101	83.50		184.50	239
1150	28' span G		8	2		118	94		212	274
1160	32' span G		8	2		134	94		228	293
1250	6" chords, 28' span G		7	2.286		148	107		255	330
1260	32' span G		7	2.286		169	107		276	350
1270	36' span G		6	2.667		190	125		315	400
1280	40' span G		6	2.667		211	125		336	425
2120	9:12 to 12:12 pitch, 18 ga. x 4" chords, 16' span G		8	2		84	94		178	238
2130	20' span G		7	2.286		105	107		212	281
2140	24' span G		7	2.286		126	107		233	305
2150	28' span G		6	2.667		147	125		272	355
2160	32' span G		6	2.667		168	125		293	380
2250	6" chords, 28' span G		5	3.200		185	150		335	435
2260	32' span G		5	3.200		211	150		361	465
2270	36' span G		4	4		238	188		426	550
2280	40' span G	↓	4	4		264	188		452	580
5120	Erection only of roof trusses, to 4:12 pitch, 16' span	F-6	48	.833			37	13.60	50.60	71.50
5130	20' span		46	.870			38.50	14.20	52.70	74.50
5140	24' span		44	.909			40	14.85	54.85	78
5150	28' span		42	.952			42	15.55	57.55	81.50
5160	32' span		40	1			44	16.35	60.35	85.50
5170	36' span		38	1.053			46.50	17.20	63.70	90.50
5180	40' span		36	1.111			49	18.15	67.15	95.50
5220	5:12 to 8:12 pitch, 16' span		42	.952			42	15.55	57.55	81.50
5230	20' span		40	1			44	16.35	60.35	85.50
5240	24' span		38	1.053			46.50	17.20	63.70	90.50
5250	28' span		36	1.111			49	18.15	67.15	95.50
5260	32' span		34	1.176			52	19.25	71.25	101
5270	36' span		32	1.250			55	20.50	75.50	107
5280	40' span		30	1.333			59	22	81	115
5320	9:12 to 12:12 pitch, 16' span	↓	36	1.111	↓		49	18.15	67.15	95.50

05 44 Cold-Formed Metal Trusses

05 44 13 – Cold-Formed Metal Roof Trusses

05 44 13.60 Framing, Roof Trusses		Crew	Daily Output	Labor-Hours	Unit	Material	2015 Bare Costs Labor	Equipment	Total	Total Incl O&P
5330	20′ span	F-6	34	1.176	Ea.		52	19.25	71.25	101
5340	24′ span		32	1.250			55	20.50	75.50	107
5350	28′ span		30	1.333			59	22	81	115
5360	32′ span		28	1.429			63	23.50	86.50	123
5370	36′ span		26	1.538			68	25	93	132
5380	40′ span		24	1.667			73.50	27.50	101	143

05 51 Metal Stairs

05 51 13 – Metal Pan Stairs

05 51 13.50 Pan Stairs

			Crew	Daily Output	Labor-Hours	Unit	Material	Labor	Equipment	Total	Total Incl O&P
0010	**PAN STAIRS**, shop fabricated, steel stringers										
0015	Made from recycled materials										
0200	Cement fill metal pan, picket rail, 3′-6″ wide	G	E-4	35	.914	Riser	500	48.50	4.17	552.67	640
0300	4′-0″ wide	G		30	1.067		560	56.50	4.86	621.36	720
0350	Wall rail, both sides, 3′-6″ wide	G		53	.604		380	32	2.75	414.75	480
1500	Landing, steel pan, conventional	G		160	.200	S.F.	66	10.65	.91	77.56	92.50
1600	Pre-erected	G		255	.125	"	118	6.65	.57	125.22	142
1700	Pre-erected, steel pan tread, 3′-6″ wide, 2 line pipe rail	G	E-2	87	.644	Riser	550	33.50	17.35	600.85	680

05 51 16 – Metal Floor Plate Stairs

05 51 16.50 Floor Plate Stairs

			Crew	Daily Output	Labor-Hours	Unit	Material	Labor	Equipment	Total	Total Incl O&P
0010	**FLOOR PLATE STAIRS**, shop fabricated, steel stringers										
0015	Made from recycled materials										
0400	Cast iron tread and pipe rail, 3′-6″ wide	G	E-4	35	.914	Riser	535	48.50	4.17	587.67	675
0500	Checkered plate tread, industrial, 3′-6″ wide	G		28	1.143		330	60.50	5.20	395.70	470
0550	Circular, for tanks, 3′-0″ wide	G		33	.970		370	51.50	4.42	425.92	500
0600	For isolated stairs, add							100%			
0800	Custom steel stairs, 3′-6″ wide, economy	G	E-4	35	.914		500	48.50	4.17	552.67	640
0810	Medium priced	G		30	1.067		660	56.50	4.86	721.36	835
0900	Deluxe	G		20	1.600		825	85	7.30	917.30	1,075
1100	For 4′ wide stairs, add						5%	5%			
1300	For 5′ wide stairs, add						10%	10%			

05 51 19 – Metal Grating Stairs

05 51 19.50 Grating Stairs

			Crew	Daily Output	Labor-Hours	Unit	Material	Labor	Equipment	Total	Total Incl O&P
0010	**GRATING STAIRS**, shop fabricated, steel stringers, safety nosing on treads										
0015	Made from recycled materials										
0020	Grating tread and pipe railing, 3′-6″ wide	G	E-4	35	.914	Riser	330	48.50	4.17	382.67	450
0100	4′-0″ wide	G	"	30	1.067	"	430	56.50	4.86	491.36	580

05 51 23 – Metal Fire Escapes

05 51 23.25 Fire Escapes

			Crew	Daily Output	Labor-Hours	Unit	Material	Labor	Equipment	Total	Total Incl O&P
0010	**FIRE ESCAPES**, shop fabricated										
0200	2′ wide balcony, 1″ x 1/4″ bars 1-1/2″ O.C., with railing	G	2 Sswk	10	1.600	L.F.	58.50	84		142.50	211
0400	1st story cantilevered stair, standard, with railing	G		.50	32	Ea.	2,425	1,675		4,100	5,600
0500	Cable counterweighted, with railing	G		.40	40	"	2,250	2,100		4,350	6,125
0700	36″ x 40″ platform & fixed stair, with railing	G		.40	40	Flight	1,075	2,100		3,175	4,825
0900	For 3′-6″ wide escapes, add to above						100%	150%			

05 51 23.50 Fire Escape Stairs

			Crew	Daily Output	Labor-Hours	Unit	Material	Labor	Equipment	Total	Total Incl O&P
0010	**FIRE ESCAPE STAIRS**, portable										
0100	Portable ladder					Ea.	112			112	123

05 51 Metal Stairs

05 51 33 – Metal Ladders

05 51 33.13 Vertical Metal Ladders

	05 51 33.13 Vertical Metal Ladders		Crew	Daily Output	Labor-Hours	Unit	Material	2015 Bare Costs Labor	Equipment	Total	Total Incl O&P
0010	**VERTICAL METAL LADDERS**, shop fabricated										
0015	Made from recycled materials										
0020	Steel, 20" wide, bolted to concrete, with cage	G	E-4	50	.640	V.L.F.	64	34	2.92	100.92	133
0100	Without cage	G		85	.376		38	20	1.72	59.72	79
0300	Aluminum, bolted to concrete, with cage	G		50	.640		120	34	2.92	156.92	194
0400	Without cage	G		85	.376		51	20	1.72	72.72	93

05 51 33.16 Inclined Metal Ladders

	05 51 33.16 Inclined Metal Ladders		Crew	Daily Output	Labor-Hours	Unit	Material	Labor	Equipment	Total	Total Incl O&P
0010	**INCLINED METAL LADDERS**, shop fabricated										
0015	Made from recycled materials										
3900	Industrial ships ladder, steel, 24" W, grating treads, 2 line pipe rail	G	E-4	30	1.067	Riser	193	56.50	4.86	254.36	315
4000	Aluminum	G	"	30	1.067	"	270	56.50	4.86	331.36	400

05 51 33.23 Alternating Tread Ladders

	05 51 33.23 Alternating Tread Ladders	Crew	Daily Output	Labor-Hours	Unit	Material	Labor	Equipment	Total	Total Incl O&P
0010	**ALTERNATING TREAD LADDERS**, shop fabricated									
0015	Made from recycled materials									
0800	Alternating tread ladders, 68-degree angle of incline									
0810	8 foot vertical rise, steel, 149 lb., standard paint color	B-68G	3	5.333	Ea.	2,275	281	111	2,667	3,100
0820	Non-standard paint color		3	5.333		2,625	281	111	3,017	3,500
0830	Galvanized		3	5.333		2,650	281	111	3,042	3,500
0840	Stainless		3	5.333		3,850	281	111	4,242	4,825
0850	Aluminum, 87 lb.		3	5.333		2,800	281	111	3,192	3,675
1010	10 foot vertical rise, steel, 181 lb., standard paint color		2.75	5.818		2,750	305	121	3,176	3,700
1020	Non-standard paint color		2.75	5.818		3,150	305	121	3,576	4,150
1030	Galvanized		2.75	5.818		3,200	305	121	3,626	4,200
1040	Stainless		2.75	5.818		4,625	305	121	5,051	5,750
1050	Aluminum, 103 lb.		2.75	5.818		3,375	305	121	3,801	4,400
1210	12 foot vertical rise, steel, 245 lb., standard paint color		2.50	6.400		3,225	335	133	3,693	4,275
1220	Non-standard paint color		2.50	6.400		3,650	335	133	4,118	4,750
1230	Galvanized		2.50	6.400		3,750	335	133	4,218	4,850
1240	Stainless		2.50	6.400		5,400	335	133	5,868	6,675
1250	Aluminum, 103 lb.		2.50	6.400		5,200	335	133	5,668	6,450
1410	14 foot vertical rise, steel, 281 lb., standard paint color		2.25	7.111		3,700	375	147	4,222	4,875
1420	Non-standard paint color		2.25	7.111		4,175	375	147	4,697	5,400
1430	Galvanized		2.25	7.111		4,325	375	147	4,847	5,550
1440	Stainless		2.25	7.111		6,200	375	147	6,722	7,600
1450	Aluminum, 136 lb.		2.25	7.111		4,525	375	147	5,047	5,775
1610	16 foot vertical rise, steel, 317 lb., standard paint color		2	8		4,175	420	166	4,761	5,500
1620	Non-standard paint color		2	8		4,700	420	166	5,286	6,050
1630	Galvanized		2	8		4,875	420	166	5,461	6,250
1640	Stainless		2	8		6,975	420	166	7,561	8,575
1650	Aluminum, 153 lb.		2	8		5,100	420	166	5,686	6,525

05 52 Metal Railings

05 52 13 – Pipe and Tube Railings

05 52 13.50 Railings, Pipe			Crew	Daily Output	Labor-Hours	Unit	Material	Labor (2015 Bare Costs)	Equipment	Total	Total Incl O&P
0010	**RAILINGS, PIPE**, shop fab'd, 3'-6" high, posts @ 5' O.C.										
0015	Made from recycled materials										
0020	Aluminum, 2 rail, satin finish, 1-1/4" diameter	G	E-4	160	.200	L.F.	36	10.65	.91	47.56	59
0030	Clear anodized	G		160	.200		44	10.65	.91	55.56	68
0040	Dark anodized	G		160	.200		49	10.65	.91	60.56	73.50
0080	1-1/2" diameter, satin finish	G		160	.200		42.50	10.65	.91	54.06	66
0090	Clear anodized	G		160	.200		47.50	10.65	.91	59.06	71.50
0100	Dark anodized	G		160	.200		52.50	10.65	.91	64.06	77
0140	Aluminum, 3 rail, 1-1/4" diam., satin finish	G		137	.234		54.50	12.40	1.07	67.97	82
0150	Clear anodized	G		137	.234		67.50	12.40	1.07	80.97	97
0160	Dark anodized	G		137	.234		75	12.40	1.07	88.47	105
0200	1-1/2" diameter, satin finish	G		137	.234		64.50	12.40	1.07	77.97	93.50
0210	Clear anodized	G		137	.234		74	12.40	1.07	87.47	104
0220	Dark anodized	G		137	.234		80.50	12.40	1.07	93.97	111
0500	Steel, 2 rail, on stairs, primed, 1-1/4" diameter	G		160	.200		25	10.65	.91	36.56	47
0520	1-1/2" diameter	G		160	.200		27	10.65	.91	38.56	49.50
0540	Galvanized, 1-1/4" diameter	G		160	.200		34	10.65	.91	45.56	57
0560	1-1/2" diameter	G		160	.200		38.50	10.65	.91	50.06	61.50
0580	Steel, 3 rail, primed, 1-1/4" diameter	G		137	.234		37.50	12.40	1.07	50.97	63.50
0600	1-1/2" diameter	G		137	.234		39	12.40	1.07	52.47	65.50
0620	Galvanized, 1-1/4" diameter	G		137	.234		52.50	12.40	1.07	65.97	80
0640	1-1/2" diameter	G		137	.234		60.50	12.40	1.07	73.97	89
0700	Stainless steel, 2 rail, 1-1/4" diam. #4 finish	G		137	.234		110	12.40	1.07	123.47	144
0720	High polish	G		137	.234		178	12.40	1.07	191.47	219
0740	Mirror polish	G		137	.234		223	12.40	1.07	236.47	268
0760	Stainless steel, 3 rail, 1-1/2" diam., #4 finish	G		120	.267		166	14.15	1.22	181.37	209
0770	High polish	G		120	.267		275	14.15	1.22	290.37	330
0780	Mirror finish	G		120	.267		335	14.15	1.22	350.37	395
0900	Wall rail, alum. pipe, 1-1/4" diam., satin finish	G		213	.150		20	8	.69	28.69	36.50
0905	Clear anodized	G		213	.150		25	8	.69	33.69	42
0910	Dark anodized	G		213	.150		29.50	8	.69	38.19	47
0915	1-1/2" diameter, satin finish	G		213	.150		22.50	8	.69	31.19	39.50
0920	Clear anodized	G		213	.150		28	8	.69	36.69	45.50
0925	Dark anodized	G		213	.150		35	8	.69	43.69	53
0930	Steel pipe, 1-1/4" diameter, primed	G		213	.150		15	8	.69	23.69	31
0935	Galvanized	G		213	.150		22	8	.69	30.69	38.50
0940	1-1/2" diameter	G		176	.182		15.50	9.65	.83	25.98	35
0945	Galvanized	G		213	.150		22	8	.69	30.69	38.50
0955	Stainless steel pipe, 1-1/2" diam., #4 finish	G		107	.299		88	15.90	1.36	105.26	126
0960	High polish	G		107	.299		179	15.90	1.36	196.26	226
0965	Mirror polish	G		107	.299		212	15.90	1.36	229.26	262
2000	2-line pipe rail (1-1/2" T&B) with 1/2" pickets @ 4-1/2" O.C.,										
2005	attached handrail on brackets										
2010	42" high aluminum, satin finish, straight & level	G	E-4	120	.267	L.F.	198	14.15	1.22	213.37	244
2050	42" high steel, primed, straight & level	G	"	120	.267		127	14.15	1.22	142.37	166
4000	For curved and level rails, add						10%	10%			
4100	For sloped rails for stairs, add						30%	30%			

05 52 16 – Industrial Railings

05 52 16.50 Railings, Industrial			Crew	Daily Output	Labor-Hours	Unit	Material	Labor	Equipment	Total	Total Incl O&P
0010	**RAILINGS, INDUSTRIAL**, shop fab'd, 3'-6" high, posts @ 5' O.C.										
0020	2 rail, 3'-6" high, 1-1/2" pipe	G	E-4	255	.125	L.F.	27	6.65	.57	34.22	42
0100	2" angle rail	G	"	255	.125		25	6.65	.57	32.22	39.50

05 52 Metal Railings

05 52 16 – Industrial Railings

05 52 16.50 Railings, Industrial			Crew	Daily Output	Labor-Hours	Unit	2015 Bare Costs Material	Labor	Equipment	Total	Total Incl O&P
0200	For 4" high kick plate, 10 ga., add	G				L.F.	5.65			5.65	6.20
0300	1/4" thick, add	G					7.40			7.40	8.15
0500	For curved level rails, add						10%	10%			
0550	For sloped rails for stairs, add						30%	30%			

05 53 Metal Gratings

05 53 13 – Bar Gratings

05 53 13.10 Floor Grating, Aluminum			Crew	Daily Output	Labor-Hours	Unit	Material	Labor	Equipment	Total	Total Incl O&P
0010	**FLOOR GRATING, ALUMINUM**, field fabricated from panels										
0015	Made from recycled materials										
0110	Bearing bars @ 1-3/16" O.C., cross bars @ 4" O.C.,										
0111	Up to 300 S.F., 1" x 1/8" bar	G	E-4	900	.036	S.F.	17.35	1.89	.16	19.40	22.50
0112	Over 300 S.F.	G		850	.038		15.80	2	.17	17.97	21
0113	1-1/4" x 1/8" bar, up to 300 S.F.	G		800	.040		18.05	2.13	.18	20.36	23.50
0114	Over 300 S.F.	G		1000	.032		16.40	1.70	.15	18.25	21
0122	1-1/4" x 3/16" bar, up to 300 S.F.	G		750	.043		27.50	2.27	.19	29.96	34.50
0124	Over 300 S.F.	G		1000	.032		25	1.70	.15	26.85	30.50
0132	1-1/2" x 3/16" bar, up to 300 S.F.	G		700	.046		32	2.43	.21	34.64	39.50
0134	Over 300 S.F.	G		1000	.032		29	1.70	.15	30.85	35
0136	1-3/4" x 3/16" bar, up to 300 S.F.	G		500	.064		35.50	3.40	.29	39.19	45
0138	Over 300 S.F.	G		1000	.032		32.50	1.70	.15	34.35	38.50
0146	2-1/4" x 3/16" bar, up to 300 S.F.	G		600	.053		45	2.83	.24	48.07	54.50
0148	Over 300 S.F.	G		1000	.032		41	1.70	.15	42.85	48
0162	Cross bars @ 2" O.C., 1" x 1/8", up to 300 S.F.	G		600	.053		30.50	2.83	.24	33.57	38.50
0164	Over 300 S.F.	G		1000	.032		27.50	1.70	.15	29.35	33.50
0172	1-1/4" x 3/16" bar, up to 300 S.F.	G		600	.053		49.50	2.83	.24	52.57	59.50
0174	Over 300 S.F.	G		1000	.032		45	1.70	.15	46.85	52.50
0182	1-1/2" x 3/16" bar, up to 300 S.F.	G		600	.053		57	2.83	.24	60.07	68
0184	Over 300 S.F.	G		1000	.032		52	1.70	.15	53.85	60
0186	1-3/4" x 3/16" bar, up to 300 S.F.	G		600	.053		62	2.83	.24	65.07	73
0188	Over 300 S.F.	G		1000	.032		56	1.70	.15	57.85	65
0200	For straight cuts, add					L.F.	4.30			4.30	4.73
0300	For curved cuts, add						5.30			5.30	5.85
0400	For straight banding, add	G					5.50			5.50	6.05
0500	For curved banding, add	G					6.60			6.60	7.25
0600	For aluminum checkered plate nosings, add	G					7.15			7.15	7.85
0700	For straight toe plate, add	G					10.80			10.80	11.90
0800	For curved toe plate, add	G					12.60			12.60	13.85
1000	For cast aluminum abrasive nosings, add	G					10.50			10.50	11.55
1200	Expanded aluminum, .65# per S.F.	G	E-4	1050	.030	S.F.	12.45	1.62	.14	14.21	16.65
1400	Extruded I bars are 10% less than 3/16" bars										
1600	Heavy duty, all extruded plank, 3/4" deep, 1.8# per S.F.	G	E-4	1100	.029	S.F.	26.50	1.55	.13	28.18	32
1700	1-1/4" deep, 2.9# per S.F.	G		1000	.032		29	1.70	.15	30.85	35
1800	1-3/4" deep, 4.2# per S.F.	G		925	.035		40.50	1.84	.16	42.50	48
1900	2-1/4" deep, 5.0# per S.F.	G		875	.037		60.50	1.94	.17	62.61	70
2100	For safety serrated surface, add						15%				

05 53 13.70 Floor Grating, Steel			Crew	Daily Output	Labor-Hours	Unit	Material	Labor	Equipment	Total	Total Incl O&P
0010	**FLOOR GRATING, STEEL**, field fabricated from panels										
0015	Made from recycled materials										
0300	Platforms, to 12' high, rectangular	G	E-4	3150	.010	Lb.	3.21	.54	.05	3.80	4.52
0400	Circular	G	"	2300	.014	"	4.01	.74	.06	4.81	5.75

05 53 Metal Gratings

05 53 13 – Bar Gratings

05 53 13.70 Floor Grating, Steel

Line	05 53 13.70 Floor Grating, Steel		Crew	Daily Output	Labor-Hours	Unit	Material	Labor	Equipment	Total	Total Incl O&P
							2015 Bare Costs				
0410	Painted bearing bars @ 1-3/16"										
0412	Cross bars @ 4" O.C., 3/4" x 1/8" bar, up to 300 S.F.	G	E-2	500	.112	S.F.	8.25	5.80	3.02	17.07	22
0414	Over 300 S.F.	G		750	.075		7.50	3.86	2.01	13.37	16.95
0422	1-1/4" x 3/16", up to 300 S.F.	G		400	.140		12.75	7.25	3.78	23.78	30.50
0424	Over 300 S.F.	G		600	.093		11.60	4.83	2.52	18.95	23.50
0432	1-1/2" x 3/16", up to 300 S.F.	G		400	.140		14.85	7.25	3.78	25.88	32.50
0434	Over 300 S.F.	G		600	.093		13.50	4.83	2.52	20.85	25.50
0436	1-3/4" x 3/16", up to 300 S.F.	G		400	.140		18.50	7.25	3.78	29.53	37
0438	Over 300 S.F.	G		600	.093		16.80	4.83	2.52	24.15	29.50
0452	2-1/4" x 3/16", up to 300 S.F.	G		300	.187		23	9.65	5.05	37.70	47.50
0454	Over 300 S.F.	G		450	.124		21	6.45	3.36	30.81	37.50
0462	Cross bars @ 2" O.C., 3/4" x 1/8", up to 300 S.F.	G		500	.112		15.15	5.80	3.02	23.97	29.50
0464	Over 300 S.F.	G		750	.075		12.60	3.86	2.01	18.47	22.50
0472	1-1/4" x 3/16", up to 300 S.F.	G		400	.140		19.80	7.25	3.78	30.83	38.50
0474	Over 300 S.F.	G		600	.093		16.50	4.83	2.52	23.85	29
0482	1-1/2" x 3/16", up to 300 S.F.	G		400	.140		22.50	7.25	3.78	33.53	41
0484	Over 300 S.F.	G		600	.093		18.60	4.83	2.52	25.95	31.50
0486	1-3/4" x 3/16", up to 300 S.F.	G		400	.140		33.50	7.25	3.78	44.53	53.50
0488	Over 300 S.F.	G		600	.093		28	4.83	2.52	35.35	42
0502	2-1/4" x 3/16", up to 300 S.F.	G		300	.187		33	9.65	5.05	47.70	58
0504	Over 300 S.F.	G		450	.124		27.50	6.45	3.36	37.31	44.50
0690	For galvanized grating, add						25%				
0800	For straight cuts, add					L.F.	6.15			6.15	6.75
0900	For curved cuts, add						7.80			7.80	8.60
1000	For straight banding, add	G					6.60			6.60	7.25
1100	For curved banding, add	G					8.70			8.70	9.55
1200	For checkered plate nosings, add	G					7.65			7.65	8.40
1300	For straight toe or kick plate, add	G					13.70			13.70	15.05
1400	For curved toe or kick plate, add	G					15.50			15.50	17.05
1500	For abrasive nosings, add	G					10.15			10.15	11.15
1510	For stair treads, see Section 05 55 13.50										
1600	For safety serrated surface, bearing bars @ 1-3/16" O.C., add						15%				
1700	Bearing bars @ 15/16" O.C., add						25%				
2000	Stainless steel gratings, close spaced, 1" x 1/8" bars, up to 300 S.F.	G	E-4	450	.071	S.F.	73	3.78	.32	77.10	87.50
2100	Standard spacing, 3/4" x 1/8" bars	G		500	.064		84	3.40	.29	87.69	98
2200	1-1/4" x 3/16" bars	G		400	.080		80	4.25	.36	84.61	96
2400	Expanded steel grating, at ground, 3.0# per S.F.	G		900	.036		8.05	1.89	.16	10.10	12.35
2500	3.14# per S.F.	G		900	.036		7.15	1.89	.16	9.20	11.35
2600	4.0# per S.F.	G		850	.038		8.60	2	.17	10.77	13.15
2650	4.27# per S.F.	G		850	.038		9.30	2	.17	11.47	13.90
2700	5.0# per S.F.	G		800	.040		13.80	2.13	.18	16.11	19.05
2800	6.25# per S.F.	G		750	.043		17.80	2.27	.19	20.26	24
2900	7.0# per S.F.	G		700	.046		19.65	2.43	.21	22.29	26
3100	For flattened expanded steel grating, add						8%				
3300	For elevated installation above 15′, add							15%			

05 53 16 – Plank Gratings

05 53 16.50 Grating Planks

Line	05 53 16.50 Grating Planks		Crew	Daily Output	Labor-Hours	Unit	Material	Labor	Equipment	Total	Total Incl O&P
0010	**GRATING PLANKS**, field fabricated from planks										
0020	Aluminum, 9-1/2" wide, 14 ga., 2" rib	G	E-4	950	.034	L.F.	25.50	1.79	.15	27.44	31.50
0200	Galvanized steel, 9-1/2" wide, 14 ga., 2-1/2" rib	G		950	.034		16.80	1.79	.15	18.74	22
0300	4" rib	G		950	.034		18.50	1.79	.15	20.44	24
0500	12 ga., 2-1/2" rib	G		950	.034		16.90	1.79	.15	18.84	22

05 53 Metal Gratings

05 53 16 – Plank Gratings

05 53 16.50 Grating Planks			Crew	Daily Output	Labor-Hours	Unit	2015 Bare Costs Material	Labor	Equipment	Total	Total Incl O&P
0600	3" rib	G	E-4	950	.034	L.F.	22.50	1.79	.15	24.44	28
0800	Stainless steel, type 304, 16 ga., 2" rib	G		950	.034		37	1.79	.15	38.94	44
0900	Type 316	G		950	.034		57	1.79	.15	58.94	66.50

05 53 19 – Floor Grating Frame

05 53 19.30 Grating Frame			Crew	Daily Output	Labor-Hours	Unit	Material	Labor	Equipment	Total	Total Incl O&P
0010	**GRATING FRAME**, field fabricated										
0020	Aluminum, for gratings 1" to 1-1/2" deep	G	1 Sswk	70	.114	L.F.	3.78	6		9.78	14.60
0100	For each corner, add	G				Ea.	5.65			5.65	6.25

05 54 Metal Floor Plates

05 54 13 – Floor Plates

05 54 13.20 Checkered Plates			Crew	Daily Output	Labor-Hours	Unit	Material	Labor	Equipment	Total	Total Incl O&P
0010	**CHECKERED PLATES**, steel, field fabricated										
0015	Made from recycled materials										
0020	1/4" & 3/8", 2000 to 5000 S.F., bolted	G	E-4	2900	.011	Lb.	.88	.59	.05	1.52	2.05
0100	Welded	G		4400	.007	"	.84	.39	.03	1.26	1.63
0300	Pit or trench cover and frame, 1/4" plate, 2' to 3' wide	G		100	.320	S.F.	11.35	17	1.46	29.81	43.50
0400	For galvanizing, add	G				Lb.	.29			.29	.32
0500	Platforms, 1/4" plate, no handrails included, rectangular	G	E-4	4200	.008		3.21	.41	.03	3.65	4.27
0600	Circular	G	"	2500	.013		4.01	.68	.06	4.75	5.65

05 54 13.70 Trench Covers			Crew	Daily Output	Labor-Hours	Unit	Material	Labor	Equipment	Total	Total Incl O&P
0010	**TRENCH COVERS**, field fabricated										
0020	Cast iron grating with bar stops and angle frame, to 18" wide	G	1 Sswk	20	.400	L.F.	213	21		234	271
0100	Frame only (both sides of trench), 1" grating	G		45	.178		1.37	9.35		10.72	17.75
0150	2" grating	G		35	.229		3.23	12.05		15.28	24.50
0200	Aluminum, stock units, including frames and										
0210	3/8" plain cover plate, 4" opening	G	E-4	205	.156	L.F.	21	8.30	.71	30.01	38
0300	6" opening	G		185	.173		25.50	9.20	.79	35.49	45
0400	10" opening	G		170	.188		34.50	10	.86	45.36	56.50
0500	16" opening	G		155	.206		48	10.95	.94	59.89	73
0700	Add per inch for additional widths to 24"	G					2			2	2.20
0900	For custom fabrication, add						50%				
1100	For 1/4" plain cover plate, deduct						12%				
1500	For cover recessed for tile, 1/4" thick, deduct						12%				
1600	3/8" thick, add						5%				
1800	For checkered plate cover, 1/4" thick, deduct						12%				
1900	3/8" thick, add						2%				
2100	For slotted or round holes in cover, 1/4" thick, add						3%				
2200	3/8" thick, add						4%				
2300	For abrasive cover, add						12%				

05 55 Metal Stair Treads and Nosings

05 55 13 – Metal Stair Treads

05 55 13.50 Stair Treads			Crew	Daily Output	Labor-Hours	Unit	Material	2015 Bare Costs Labor	Equipment	Total	Total Incl O&P
0010	**STAIR TREADS**, stringers and bolts not included										
3000	Diamond plate treads, steel, 1/8″ thick										
3005	Open riser, black enamel										
3010	9″ deep x 36″ long	G	2 Sswk	48	.333	Ea.	91.50	17.55		109.05	131
3020	42″ long	G		48	.333		96.50	17.55		114.05	137
3030	48″ long	G		48	.333		101	17.55		118.55	142
3040	11″ deep x 36″ long	G		44	.364		96.50	19.15		115.65	140
3050	42″ long	G		44	.364		102	19.15		121.15	146
3060	48″ long	G		44	.364		108	19.15		127.15	153
3110	Galvanized, 9″ deep x 36″ long	G		48	.333		144	17.55		161.55	190
3120	42″ long	G		48	.333		154	17.55		171.55	200
3130	48″ long	G		48	.333		163	17.55		180.55	210
3140	11″ deep x 36″ long	G		44	.364		149	19.15		168.15	198
3150	42″ long	G		44	.364		163	19.15		182.15	213
3160	48″ long	G		44	.364		169	19.15		188.15	220
3200	Closed riser, black enamel										
3210	12″ deep x 36″ long	G	2 Sswk	40	.400	Ea.	110	21		131	158
3220	42″ long	G		40	.400		119	21		140	168
3230	48″ long	G		40	.400		126	21		147	175
3240	Galvanized, 12″ deep x 36″ long	G		40	.400		173	21		194	228
3250	42″ long	G		40	.400		189	21		210	245
3260	48″ long	G		40	.400		198	21		219	255
4000	Bar grating treads										
4005	Steel, 1-1/4″ x 3/16″ bars, anti-skid nosing, black enamel										
4010	8-5/8″ deep x 30″ long	G	2 Sswk	48	.333	Ea.	54	17.55		71.55	90
4020	36″ long	G		48	.333		63.50	17.55		81.05	100
4030	48″ long	G		48	.333		97.50	17.55		115.05	138
4040	10-15/16″ deep x 36″ long	G		44	.364		70	19.15		89.15	111
4050	48″ long	G		44	.364		100	19.15		119.15	144
4060	Galvanized, 8-5/8″ deep x 30″ long	G		48	.333		62	17.55		79.55	99
4070	36″ long	G		48	.333		73.50	17.55		91.05	112
4080	48″ long	G		48	.333		108	17.55		125.55	150
4090	10-15/16″ deep x 36″ long	G		44	.364		86.50	19.15		105.65	129
4100	48″ long	G		44	.364		112	19.15		131.15	157
4200	Aluminum, 1-1/4″ x 3/16″ bars, serrated, with nosing										
4210	7-5/8″ deep x 18″ long	G	2 Sswk	52	.308	Ea.	50	16.20		66.20	83
4220	24″ long	G		52	.308		59	16.20		75.20	93
4230	30″ long	G		52	.308		68.50	16.20		84.70	103
4240	36″ long	G		52	.308		177	16.20		193.20	223
4250	8-13/16″ deep x 18″ long	G		48	.333		67	17.55		84.55	105
4260	24″ long	G		48	.333		99.50	17.55		117.05	140
4270	30″ long	G		48	.333		115	17.55		132.55	157
4280	36″ long	G		48	.333		194	17.55		211.55	245
4290	10″ deep x 18″ long	G		44	.364		130	19.15		149.15	177
4300	30″ long	G		44	.364		173	19.15		192.15	224
4310	36″ long	G		44	.364		210	19.15		229.15	265
5000	Channel grating treads										
5005	Steel, 14 ga., 2-12″ thick, galvanized										
5010	9″ deep x 36″ long	G	2 Sswk	48	.333	Ea.	103	17.55		120.55	144
5020	48″ long	G	″	48	.333	″	139	17.55		156.55	183

05 55 Metal Stair Treads and Nosings

05 55 19 – Metal Stair Tread Covers

05 55 19.50 Stair Tread Covers for Renovation		Crew	Daily Output	Labor-Hours	Unit	Material	Labor	Equipment	Total	Total Incl O&P
						2015 Bare Costs				
0010	**STAIR TREAD COVERS FOR RENOVATION**									
0205	Extruded tread cover with nosing, pre-drilled, includes screws									
0210	Aluminum with black abrasive strips, 9″ wide x 3′ long	1 Carp	24	.333	Ea.	107	15.65		122.65	142
0220	4′ long		22	.364		142	17.05		159.05	183
0230	5′ long		20	.400		177	18.80		195.80	223
0240	11″ wide x 3′ long		24	.333		139	15.65		154.65	177
0250	4′ long		22	.364		179	17.05		196.05	223
0260	5′ long		20	.400		230	18.80		248.80	281
0305	Black abrasive strips with yellow front strips									
0310	Aluminum, 9″ wide x 3′ long	1 Carp	24	.333	Ea.	116	15.65		131.65	151
0320	4′ long		22	.364		154	17.05		171.05	197
0330	5′ long		20	.400		196	18.80		214.80	245
0340	11″ wide x 3′ long		24	.333		149	15.65		164.65	188
0350	4′ long		22	.364		192	17.05		209.05	238
0360	5′ long		20	.400		247	18.80		265.80	300
0405	Black abrasive strips with photoluminescent front strips									
0410	Aluminum, 9″ wide x 3′ long	1 Carp	24	.333	Ea.	156	15.65		171.65	195
0420	4′ long		22	.364		172	17.05		189.05	216
0430	5′ long		20	.400		215	18.80		233.80	265
0440	11″ wide x 3′ long		24	.333		153	15.65		168.65	192
0450	4′ long		22	.364		204	17.05		221.05	251
0460	5′ long		20	.400		255	18.80		273.80	310

05 56 Metal Castings

05 56 13 – Metal Construction Castings

05 56 13.50 Construction Castings

Line	Description		Crew	Daily Output	Labor-Hours	Unit	Material	Labor	Equipment	Total	Total Incl O&P
0010	**CONSTRUCTION CASTINGS**										
0020	Manhole covers and frames, see Section 33 44 13.13										
0100	Column bases, cast iron, 16″ x 16″, approx. 65 lb.	G	E-4	46	.696	Ea.	142	37	3.17	182.17	223
0200	32″ x 32″, approx. 256 lb.	G	″	23	1.391		520	74	6.35	600.35	710
0400	Cast aluminum for wood columns, 8″ x 8″	G	1 Carp	32	.250		41	11.75		52.75	63.50
0500	12″ x 12″	G	″	32	.250		68.50	11.75		80.25	93.50
0600	Miscellaneous C.I. castings, light sections, less than 150 lb.	G	E-4	3200	.010	Lb.	8.95	.53	.05	9.53	10.80
1100	Heavy sections, more than 150 lb.	G		4200	.008		4.66	.41	.03	5.10	5.90
1300	Special low volume items	G		3200	.010		11.25	.53	.05	11.83	13.30
1500	For ductile iron, add						100%				

05 58 Formed Metal Fabrications

05 58 13 – Column Covers

05 58 13.05 Column Covers

Line	Description		Crew	Daily Output	Labor-Hours	Unit	Material	Labor	Equipment	Total	Total Incl O&P
0010	**COLUMN COVERS**										
0015	Made from recycled materials										
0020	Excludes structural steel, light ga. metal framing, misc. metals, sealants										
0100	Round covers, 2 halves with 2 vertical joints for backer rod and sealant										
0110	Up to 12′ high, no horizontal joints										
0120	12″ diameter, 0.125″ aluminum, anodized/painted finish	G	2 Sswk	32	.500	V.L.F.	30.50	26.50		57	79
0130	Type 304 stainless steel, 16 gauge, #4 brushed finish	G		32	.500		46.50	26.50		73	96.50
0140	Type 316 stainless steel, 16 gauge, #4 brushed finish	G		32	.500		54	26.50		80.50	105
0150	18″ diameter, aluminum	G		32	.500		46	26.50		72.50	96

05 58 Formed Metal Fabrications

05 58 13 – Column Covers

05 58 13.05 Column Covers			Crew	Daily Output	Labor-Hours	Unit	2015 Bare Costs Material	Labor	Equipment	Total	Total Incl O&P
0160	Type 304 stainless steel	G	2 Sswk	32	.500	V.L.F.	69.50	26.50		96	122
0170	Type 316 stainless steel	G		32	.500		81	26.50		107.50	135
0180	24″ diameter, aluminum	G		32	.500		61	26.50		87.50	113
0190	Type 304 stainless steel	G		32	.500		93	26.50		119.50	148
0200	Type 316 stainless steel	G		32	.500		108	26.50		134.50	165
0210	30″ diameter, aluminum	G		30	.533		76.50	28		104.50	133
0220	Type 304 stainless steel	G		30	.533		116	28		144	177
0230	Type 316 stainless steel	G		30	.533		135	28		163	197
0240	36″ diameter, aluminum	G		30	.533		91.50	28		119.50	150
0250	Type 304 stainless steel	G		30	.533		139	28		167	202
0260	Type 316 stainless steel	G		30	.533		162	28		190	227
0400	Up to 24′ high, 2 stacked sections with 1 horizontal joint										
0410	18″ diameter, aluminum	G	2 Sswk	28	.571	V.L.F.	48	30		78	106
0450	Type 304 stainless steel	G		28	.571		73	30		103	133
0460	Type 316 stainless steel	G		28	.571		85	30		115	146
0470	24″ diameter, aluminum	G		28	.571		64	30		94	123
0480	Type 304 stainless steel	G		28	.571		97.50	30		127.50	160
0490	Type 316 stainless steel	G		28	.571		113	30		143	178
0500	30″ diameter, aluminum	G		24	.667		80	35		115	150
0510	Type 304 stainless steel	G		24	.667		122	35		157	195
0520	Type 316 stainless steel	G		24	.667		142	35		177	217
0530	36″ diameter, aluminum	G		24	.667		96.50	35		131.50	167
0540	Type 304 stainless steel	G		24	.667		146	35		181	222
0550	Type 316 stainless steel	G		24	.667		170	35		205	248

05 58 21 – Formed Chain

05 58 21.05 Alloy Steel Chain

			Crew	Daily Output	Labor-Hours	Unit	Material	Labor	Equipment	Total	Total Incl O&P
0010	**ALLOY STEEL CHAIN**, Grade 80, for lifting										
0015	Self-colored, cut lengths, 1/4″	G	E-17	4	4	C.L.F.	765	215		980	1,225
0020	3/8″	G		2	8		980	430		1,410	1,825
0030	1/2″	G		1.20	13.333		1,550	715		2,265	2,975
0040	5/8″	G		.72	22.222		2,450	1,200		3,650	4,750
0050	3/4″	G	E-18	.48	83.333		3,300	4,375	1,975	9,650	13,300
0060	7/8″	G		.40	100		5,725	5,275	2,375	13,375	17,800
0070	1″	G		.35	114		7,975	6,025	2,700	16,700	22,000
0080	1-1/4″	G		.24	166		13,400	8,775	3,950	26,125	34,000
0110	Hook, Grade 80, Clevis slip, 1/4″	G				Ea.	26			26	28.50
0120	3/8″	G					30.50			30.50	33.50
0130	1/2″	G					49.50			49.50	54.50
0140	5/8″	G					73			73	80
0150	3/4″	G					101			101	111
0160	Hook, Grade 80, eye/sling w/hammerlock coupling, 15 Ton	G					340			340	375
0170	22 Ton	G					925			925	1,025
0180	37 Ton	G					2,825			2,825	3,100

05 58 23 – Formed Metal Guards

05 58 23.90 Window Guards

			Crew	Daily Output	Labor-Hours	Unit	Material	Labor	Equipment	Total	Total Incl O&P
0010	**WINDOW GUARDS**, shop fabricated										
0015	Expanded metal, steel angle frame, permanent	G	E-4	350	.091	S.F.	23	4.86	.42	28.28	34
0025	Steel bars, 1/2″ x 1/2″, spaced 5″ O.C.	G	″	290	.110	″	15.90	5.85	.50	22.25	28.50
0030	Hinge mounted, add	G				Opng.	46			46	50.50
0040	Removable type, add	G				″	29			29	32
0050	For galvanized guards, add					S.F.	35%				
0070	For pivoted or projected type, add						105%	40%			

05 58 Formed Metal Fabrications

05 58 23 – Formed Metal Guards

05 58 23.90 Window Guards			Crew	Daily Output	Labor-Hours	Unit	2015 Bare Costs Material	Labor	Equipment	Total	Total Incl O&P
0100	Mild steel, stock units, economy	G	E-4	405	.079	S.F.	6.25	4.20	.36	10.81	14.60
0200	Deluxe	G		405	.079		12.70	4.20	.36	17.26	21.50
0400	Woven wire, stock units, 3/8" channel frame, 3' x 5' opening	G		40	.800	Opng.	169	42.50	3.65	215.15	264
0500	4' x 6' opening	G		38	.842		270	45	3.84	318.84	380
0800	Basket guards for above, add	G					233			233	256
1000	Swinging guards for above, add	G					79.50			79.50	87.50

05 58 25 – Formed Lamp Posts

05 58 25.40 Lamp Posts			Crew	Daily Output	Labor-Hours	Unit	Material	Labor	Equipment	Total	Total Incl O&P
0010	**LAMP POSTS**										
0020	Aluminum, 7' high, stock units, post only	G	1 Carp	16	.500	Ea.	82	23.50		105.50	126
0100	Mild steel, plain	G	"	16	.500	"	73	23.50		96.50	116

05 71 Decorative Metal Stairs

05 71 13 – Fabricated Metal Spiral Stairs

05 71 13.50 Spiral Stairs			Crew	Daily Output	Labor-Hours	Unit	Material	Labor	Equipment	Total	Total Incl O&P
0010	**SPIRAL STAIRS**										
1805	Shop fabricated, custom ordered										
1810	Aluminum, 5'-0" diameter, plain units	G	E-4	45	.711	Riser	575	38	3.24	616.24	705
1820	Fancy units	G		45	.711		1,100	38	3.24	1,141.24	1,275
1900	Cast iron, 4'-0" diameter, plain units	G		45	.711		540	38	3.24	581.24	665
1920	Fancy Units	G		25	1.280		735	68	5.85	808.85	935
2000	Steel, industrial checkered plate, 4' diameter	G		45	.711		540	38	3.24	581.24	665
2200	6' diameter	G		40	.800		650	42.50	3.65	696.15	795
3100	Spiral stair kits, 12 stacking risers to fit exact floor height										
3110	Steel, flat metal treads, primed, 3'-6" diameter	G	2 Carp	1.60	10	Flight	1,275	470		1,745	2,125
3120	4'-0" diameter	G		1.45	11.034		1,450	520		1,970	2,400
3130	4'-6" diameter	G		1.35	11.852		1,600	555		2,155	2,625
3140	5'-0" diameter	G		1.25	12.800		1,750	600		2,350	2,850
3210	Galvanized, 3'-6" diameter	G		1.60	10		1,525	470		1,995	2,400
3220	4'-0" diameter	G		1.45	11.034		1,750	520		2,270	2,725
3230	4'-6" diameter	G		1.35	11.852		1,925	555		2,480	2,975
3240	5'-0" diameter	G		1.25	12.800		2,100	600		2,700	3,225
3310	Checkered plate tread, primed, 3'-6" diameter	G		1.45	11.034		1,500	520		2,020	2,450
3320	4'-0" diameter	G		1.35	11.852		1,700	555		2,255	2,725
3330	4'-6" diameter	G		1.25	12.800		1,850	600		2,450	2,975
3340	5'-0" diameter	G		1.15	13.913		2,025	655		2,680	3,225
3410	Galvanized, 3'-6" diameter	G		1.45	11.034		1,800	520		2,320	2,800
3420	4'-0" diameter	G		1.35	11.852		2,050	555		2,605	3,100
3430	4'-6" diameter	G		1.25	12.800		2,225	600		2,825	3,375
3440	5'-0" diameter	G		1.15	13.913		2,425	655		3,080	3,675
3510	Red oak covers on flat metal treads, 3'-6" diameter			1.35	11.852		2,600	555		3,155	3,725
3520	4'-0" diameter			1.25	12.800		2,875	600		3,475	4,100
3530	4'-6" diameter			1.15	13.913		3,100	655		3,755	4,400
3540	5'-0" diameter			1.05	15.238		3,375	715		4,090	4,825

05 73 Decorative Metal Railings

05 73 16 – Wire Rope Decorative Metal Railings

05 73 16.10 Cable Railings		Crew	Daily Output	Labor-Hours	Unit	Material	2015 Bare Costs Labor	Equipment	Total	Total Incl O&P
0010	**CABLE RAILINGS**, with 316 stainless steel 1 x 19 cable, 3/16" diameter									
0015	Made from recycled materials									
0100	1-3/4" diameter stainless steel posts x 42" high, cables 4" OC G	2 Sswk	25	.640	L.F.	33	33.50		66.50	95

05 73 23 – Ornamental Railings

05 73 23.50 Railings, Ornamental		Crew	Daily Output	Labor-Hours	Unit	Material	Labor	Equipment	Total	Total Incl O&P
0010	**RAILINGS, ORNAMENTAL**, 3'-6" high, posts @ 6' O.C.									
0020	Bronze or stainless, hand forged, plain G	2 Sswk	24	.667	L.F.	122	35		157	195
0100	Fancy G		18	.889		223	47		270	325
0200	Aluminum, panelized, plain G		24	.667		13.60	35		48.60	76
0300	Fancy G		18	.889		30	47		77	115
0400	Wrought iron, hand forged, plain G		24	.667		81	35		116	150
0500	Fancy G		18	.889		159	47		206	257
0550	Steel, panelized, plain G		24	.667		20.50	35		55.50	83.50
0560	Fancy G		18	.889		32	47		79	117
0600	Composite metal/wood/glass, plain		18	.889		119	47		166	213
0700	Fancy	↓	12	1.333	↓	238	70		308	385

05 75 Decorative Formed Metal

05 75 13 – Columns

05 75 13.10 Aluminum Columns		Crew	Daily Output	Labor-Hours	Unit	Material	Labor	Equipment	Total	Total Incl O&P
0010	**ALUMINUM COLUMNS**									
0015	Made from recycled materials									
0020	Aluminum, extruded, stock units, no cap or base, 6" diameter G	E-4	240	.133	L.F.	10.50	7.10	.61	18.21	24.50
0100	8" diameter G		170	.188		13.85	10	.86	24.71	33.50
0200	10" diameter G		150	.213		18.95	11.35	.97	31.27	42
0300	12" diameter G		140	.229		32.50	12.15	1.04	45.69	58
0400	15" diameter G	↓	120	.267	↓	48	14.15	1.22	63.37	78.50
0410	Caps and bases, plain, 6" diameter G				Set	23.50			23.50	26
0420	8" diameter G					30			30	33
0430	10" diameter G					42			42	46.50
0440	12" diameter G					68.50			68.50	75
0450	15" diameter G					110			110	121
0460	Caps, ornamental, plain G					340			340	370
0470	Fancy G				↓	1,675			1,675	1,850
0500	For square columns, add to column prices above				L.F.	50%				
0700	Residential, flat, 8' high, plain G	E-4	20	1.600	Ea.	97.50	85	7.30	189.80	263
0720	Fancy G		20	1.600		190	85	7.30	282.30	365
0740	Corner type, plain G		20	1.600		168	85	7.30	260.30	340
0760	Fancy G	↓	20	1.600	↓	330	85	7.30	422.30	520

05 75 13.20 Columns, Ornamental		Crew	Daily Output	Labor-Hours	Unit	Material	Labor	Equipment	Total	Total Incl O&P
0010	**COLUMNS, ORNAMENTAL**, shop fabricated R051223-10									
6400	Mild steel, flat, 9" wide, stock units, painted, plain G	E-4	160	.200	V.L.F.	9.15	10.65	.91	20.71	29.50
6450	Fancy G		160	.200		17.75	10.65	.91	29.31	39
6500	Corner columns, painted, plain G		160	.200		15.75	10.65	.91	27.31	37
6550	Fancy G	↓	160	.200	↓	31	10.65	.91	42.56	54

Estimating Tips

06 05 00 Common Work Results for Wood, Plastics, and Composites

- Common to any wood-framed structure are the accessory connector items such as screws, nails, adhesives, hangers, connector plates, straps, angles, and hold-downs. For typical wood-framed buildings, such as residential projects, the aggregate total for these items can be significant, especially in areas where seismic loading is a concern. For floor and wall framing, the material cost is based on 10 to 25 lbs. per MBF. Hold-downs, hangers, and other connectors should be taken off by the piece.

  Included with material costs are fasteners for a normal installation. RSMeans engineers use manufacturer's recommendations, written specifications, and/or standard construction practice for size and spacing of fasteners. Prices for various fasteners are shown for informational purposes only. Adjustments should be made if unusual fastening conditions exist.

06 10 00 Carpentry

- Lumber is a traded commodity and therefore sensitive to supply and demand in the marketplace. Even in "budgetary" estimating of wood-framed projects, it is advisable to call local suppliers for the latest market pricing.
- Common quantity units for wood-framed projects are "thousand board feet" (MBF). A board foot is a volume of wood, 1" x 1' x 1', or 144 cubic inches. Board-foot quantities are generally calculated using nominal material dimensions— dressed sizes are ignored. Board foot per lineal foot of any stick of lumber can be calculated by dividing the nominal cross-sectional area by 12. As an example, 2,000 lineal feet of 2 x 12 equates to 4 MBF by dividing the nominal area, 2 x 12, by 12, which equals 2, and multiplying by 2,000 to give 4,000 board feet. This simple rule applies to all nominal dimensioned lumber.
- Waste is an issue of concern at the quantity takeoff for any area of construction. Framing lumber is sold in even foot lengths, i.e., 10', 12', 14', 16' and, depending on spans, wall heights, and the grade of lumber, waste is inevitable. A rule of thumb for lumber waste is 5%–10% depending on material quality and the complexity of the framing.
- Wood in various forms and shapes is used in many projects, even where the main structural framing is steel, concrete, or masonry. Plywood as a back-up partition material and 2x boards used as blocking and cant strips around roof edges are two common examples. The estimator should ensure that the costs of all wood materials are included in the final estimate.

06 20 00 Finish Carpentry

- It is necessary to consider the grade of workmanship when estimating labor costs for erecting millwork and interior finish. In practice, there are three grades: premium, custom, and economy. The RSMeans daily output for base and case moldings is in the range of 200 to 250 L.F. per carpenter per day. This is appropriate for most average custom-grade projects. For premium projects, an adjustment to productivity of 25%–50% should be made, depending on the complexity of the job.

Reference Numbers

Reference numbers are shown in shaded boxes at the beginning of some major classifications. These numbers refer to related items in the Reference Section. The reference information may be an estimating procedure, an alternate pricing method, or technical information.

Note: Not all subdivisions listed here necessarily appear in this publication. ■

06 05 05 – Selective Demolition for Wood, Plastics, and Composites

06 05 05.10 Selective Demolition Wood Framing		Crew	Daily Output	Labor-Hours	Unit	Material	2015 Bare Costs Labor	Equipment	Total	Total Incl O&P
0010	**SELECTIVE DEMOLITION WOOD FRAMING** R024119-10									
0100	Timber connector, nailed, small	1 Clab	96	.083	Ea.		3.13		3.13	4.82
0110	Medium		60	.133			5		5	7.70
0120	Large		48	.167			6.25		6.25	9.65
0130	Bolted, small		48	.167			6.25		6.25	9.65
0140	Medium		32	.250			9.40		9.40	14.45
0150	Large		24	.333			12.55		12.55	19.30
2958	Beams, 2″ x 6″	2 Clab	1100	.015	L.F.		.55		.55	.84
2960	2″ x 8″		825	.019			.73		.73	1.12
2965	2″ x 10″		665	.024			.90		.90	1.39
2970	2″ x 12″		550	.029			1.09		1.09	1.68
2972	2″ x 14″		470	.034			1.28		1.28	1.97
2975	4″ x 8″	B-1	413	.058			2.22		2.22	3.42
2980	4″ x 10″		330	.073			2.78		2.78	4.28
2985	4″ x 12″		275	.087			3.34		3.34	5.15
3000	6″ x 8″		275	.087			3.34		3.34	5.15
3040	6″ x 10″		220	.109			4.17		4.17	6.40
3080	6″ x 12″		185	.130			4.96		4.96	7.65
3120	8″ x 12″		140	.171			6.55		6.55	10.10
3160	10″ x 12″		110	.218			8.35		8.35	12.85
3162	Alternate pricing method		1.10	21.818	M.B.F.		835		835	1,275
3170	Blocking, in 16″ OC wall framing, 2″ x 4″	1 Clab	600	.013	L.F.		.50		.50	.77
3172	2″ x 6″		400	.020			.75		.75	1.16
3174	In 24″ OC wall framing, 2″ x 4″		600	.013			.50		.50	.77
3176	2″ x 6″		400	.020			.75		.75	1.16
3178	Alt method, wood blocking removal from wood framing		.40	20	M.B.F.		750		750	1,150
3179	Wood blocking removal from steel framing		.36	22.222	″		835		835	1,275
3180	Bracing, let in, 1″ x 3″, studs 16″ OC		1050	.008	L.F.		.29		.29	.44
3181	Studs 24″ OC		1080	.007			.28		.28	.43
3182	1″ x 4″, studs 16″ OC		1050	.008			.29		.29	.44
3183	Studs 24″ OC		1080	.007			.28		.28	.43
3184	1″ x 6″, studs 16″ OC		1050	.008			.29		.29	.44
3185	Studs 24″ OC		1080	.007			.28		.28	.43
3186	2″ x 3″, studs 16″ OC		800	.010			.38		.38	.58
3187	Studs 24″ OC		830	.010			.36		.36	.56
3188	2″ x 4″, studs 16″ OC		800	.010			.38		.38	.58
3189	Studs 24″ OC		830	.010			.36		.36	.56
3190	2″ x 6″, studs 16″ OC		800	.010			.38		.38	.58
3191	Studs 24″ OC		830	.010			.36		.36	.56
3192	2″ x 8″, studs 16″ OC		800	.010			.38		.38	.58
3193	Studs 24″ OC		830	.010			.36		.36	.56
3194	"T" shaped metal bracing, studs at 16″ OC		1060	.008			.28		.28	.44
3195	Studs at 24″ OC		1200	.007			.25		.25	.39
3196	Metal straps, studs at 16″ OC		1200	.007			.25		.25	.39
3197	Studs at 24″ OC		1240	.006			.24		.24	.37
3200	Columns, round, 8′ to 14′ tall		40	.200	Ea.		7.50		7.50	11.55
3202	Dimensional lumber sizes	2 Clab	1.10	14.545	M.B.F.		545		545	840
3250	Blocking, between joists	1 Clab	320	.025	Ea.		.94		.94	1.45
3252	Bridging, metal strap, between joists		320	.025	Pr.		.94		.94	1.45
3254	Wood, between joists		320	.025	″		.94		.94	1.45
3260	Door buck, studs, header & access., 8′ high 2″ x 4″ wall, 3′ wide		32	.250	Ea.		9.40		9.40	14.45
3261	4′ wide		32	.250			9.40		9.40	14.45
3262	5′ wide		32	.250			9.40		9.40	14.45

06 05 05 – Selective Demolition for Wood, Plastics, and Composites

06 05 05.10 Selective Demolition Wood Framing		Crew	Daily Output	Labor-Hours	Unit	Material	2015 Bare Costs Labor	Equipment	Total	Total Incl O&P
3263	6′ wide	1 Clab	32	.250	Ea.		9.40		9.40	14.45
3264	8′ wide		30	.267			10.05		10.05	15.45
3265	10′ wide		30	.267			10.05		10.05	15.45
3266	12′ wide		30	.267			10.05		10.05	15.45
3267	2″ x 6″ wall, 3′ wide		32	.250			9.40		9.40	14.45
3268	4′ wide		32	.250			9.40		9.40	14.45
3269	5′ wide		32	.250			9.40		9.40	14.45
3270	6′ wide		32	.250			9.40		9.40	14.45
3271	8′ wide		30	.267			10.05		10.05	15.45
3272	10′ wide		30	.267			10.05		10.05	15.45
3273	12′ wide		30	.267			10.05		10.05	15.45
3274	Window buck, studs, header & access, 8′ high 2″ x 4″ wall, 2′ wide		24	.333			12.55		12.55	19.30
3275	3′ wide		24	.333			12.55		12.55	19.30
3276	4′ wide		24	.333			12.55		12.55	19.30
3277	5′ wide		24	.333			12.55		12.55	19.30
3278	6′ wide		24	.333			12.55		12.55	19.30
3279	7′ wide		24	.333			12.55		12.55	19.30
3280	8′ wide		22	.364			13.65		13.65	21
3281	10′ wide		22	.364			13.65		13.65	21
3282	12′ wide		22	.364			13.65		13.65	21
3283	2″ x 6″ wall, 2′ wide		24	.333			12.55		12.55	19.30
3284	3′ wide		24	.333			12.55		12.55	19.30
3285	4′ wide		24	.333			12.55		12.55	19.30
3286	5′ wide		24	.333			12.55		12.55	19.30
3287	6′ wide		24	.333			12.55		12.55	19.30
3288	7′ wide		24	.333			12.55		12.55	19.30
3289	8′ wide		22	.364			13.65		13.65	21
3290	10′ wide		22	.364			13.65		13.65	21
3291	12′ wide		22	.364			13.65		13.65	21
3360	Deck or porch decking		825	.010	L.F.		.36		.36	.56
3400	Fascia boards, 1″ x 6″		500	.016			.60		.60	.93
3440	1″ x 8″		450	.018			.67		.67	1.03
3480	1″ x 10″		400	.020			.75		.75	1.16
3490	2″ x 6″		450	.018			.67		.67	1.03
3500	2″ x 8″		400	.020			.75		.75	1.16
3510	2″ x 10″		350	.023			.86		.86	1.32
3610	Furring, on wood walls or ceiling		4000	.002	S.F.		.08		.08	.12
3620	On masonry or concrete walls or ceiling		1200	.007	″		.25		.25	.39
3800	Headers over openings, 2 @ 2″ x 6″		110	.073	L.F.		2.73		2.73	4.21
3840	2 @ 2″ x 8″		100	.080			3.01		3.01	4.63
3880	2 @ 2″ x 10″		90	.089			3.34		3.34	5.15
3885	Alternate pricing method		.26	30.651	M.B.F.		1,150		1,150	1,775
3920	Joists, 1″ x 4″		1250	.006	L.F.		.24		.24	.37
3930	1″ x 6″		1135	.007			.27		.27	.41
3940	1″ x 8″		1000	.008			.30		.30	.46
3950	1″ x 10″		895	.009			.34		.34	.52
3960	1″ x 12″		765	.010			.39		.39	.61
4200	2″ x 4″	2 Clab	1000	.016			.60		.60	.93
4230	2″ x 6″		970	.016			.62		.62	.95
4240	2″ x 8″		940	.017			.64		.64	.98
4250	2″ x 10″		910	.018			.66		.66	1.02
4280	2″ x 12″		880	.018			.68		.68	1.05
4281	2″ x 14″		850	.019			.71		.71	1.09

06 05 05 – Selective Demolition for Wood, Plastics, and Composites

06 05 05.10 Selective Demolition Wood Framing		Crew	Daily Output	Labor-Hours	Unit	Material	2015 Bare Costs Labor	Equipment	Total	Total Incl O&P
4282	Composite joists, 9-1/2"	2 Clab	960	.017	L.F.		.63		.63	.96
4283	11-7/8"		930	.017			.65		.65	1
4284	14"		897	.018			.67		.67	1.03
4285	16"		865	.019			.70		.70	1.07
4290	Wood joists, alternate pricing method		1.50	10.667	M.B.F.		400		400	615
4500	Open web joist, 12" deep		500	.032	L.F.		1.20		1.20	1.85
4505	14" deep		475	.034			1.27		1.27	1.95
4510	16" deep		450	.036			1.34		1.34	2.06
4520	18" deep		425	.038			1.42		1.42	2.18
4530	24" deep		400	.040			1.50		1.50	2.31
4550	Ledger strips, 1" x 2"	1 Clab	1200	.007			.25		.25	.39
4560	1" x 3"		1200	.007			.25		.25	.39
4570	1" x 4"		1200	.007			.25		.25	.39
4580	2" x 2"		1100	.007			.27		.27	.42
4590	2" x 4"		1000	.008			.30		.30	.46
4600	2" x 6"		1000	.008			.30		.30	.46
4601	2" x 8 or 2" x 10"		800	.010			.38		.38	.58
4602	4" x 6"		600	.013			.50		.50	.77
4604	4" x 8"		450	.018			.67		.67	1.03
5400	Posts, 4" x 4"	2 Clab	800	.020			.75		.75	1.16
5405	4" x 6"		550	.029			1.09		1.09	1.68
5410	4" x 8"		440	.036			1.37		1.37	2.10
5425	4" x 10"		390	.041			1.54		1.54	2.37
5430	4" x 12"		350	.046			1.72		1.72	2.64
5440	6" x 6"		400	.040			1.50		1.50	2.31
5445	6" x 8"		350	.046			1.72		1.72	2.64
5450	6" x 10"		320	.050			1.88		1.88	2.89
5455	6" x 12"		290	.055			2.07		2.07	3.19
5480	8" x 8"		300	.053			2.01		2.01	3.09
5500	10" x 10"		240	.067			2.51		2.51	3.86
5660	Tongue and groove floor planks		2	8	M.B.F.		300		300	465
5682	Rafters, ordinary, 16" OC, 2" x 4"		880	.018	S.F.		.68		.68	1.05
5683	2" x 6"		840	.019			.72		.72	1.10
5684	2" x 8"		820	.020			.73		.73	1.13
5685	2" x 10"		820	.020			.73		.73	1.13
5686	2" x 12"		810	.020			.74		.74	1.14
5687	24" OC, 2" x 4"		1170	.014			.51		.51	.79
5688	2" x 6"		1117	.014			.54		.54	.83
5689	2" x 8"		1091	.015			.55		.55	.85
5690	2" x 10"		1091	.015			.55		.55	.85
5691	2" x 12"		1077	.015			.56		.56	.86
5795	Rafters, ordinary, 2" x 4" (alternate method)		862	.019	L.F.		.70		.70	1.07
5800	2" x 6" (alternate method)		850	.019			.71		.71	1.09
5840	2" x 8" (alternate method)		837	.019			.72		.72	1.11
5855	2" x 10" (alternate method)		825	.019			.73		.73	1.12
5865	2" x 12" (alternate method)		812	.020			.74		.74	1.14
5870	Sill plate, 2" x 4"	1 Clab	1170	.007			.26		.26	.40
5871	2" x 6"		780	.010			.39		.39	.59
5872	2" x 8"		586	.014			.51		.51	.79
5873	Alternate pricing method		.78	10.256	M.B.F.		385		385	595
5885	Ridge board, 1" x 4"	2 Clab	900	.018	L.F.		.67		.67	1.03
5886	1" x 6"		875	.018			.69		.69	1.06
5887	1" x 8"		850	.019			.71		.71	1.09

06 05 05 – Selective Demolition for Wood, Plastics, and Composites

06 05 05.10 Selective Demolition Wood Framing		Crew	Daily Output	Labor-Hours	Unit	Material	Labor	Equipment	Total	Total Incl O&P
						2015 Bare Costs				
5888	1″ x 10″	2 Clab	825	.019	L.F.		.73		.73	1.12
5889	1″ x 12″		800	.020			.75		.75	1.16
5890	2″ x 4″		900	.018			.67		.67	1.03
5892	2″ x 6″		875	.018			.69		.69	1.06
5894	2″ x 8″		850	.019			.71		.71	1.09
5896	2″ x 10″		825	.019			.73		.73	1.12
5898	2″ x 12″		800	.020			.75		.75	1.16
6050	Rafter tie, 1″ x 4″		1250	.013			.48		.48	.74
6052	1″ x 6″		1135	.014			.53		.53	.82
6054	2″ x 4″		1000	.016			.60		.60	.93
6056	2″ x 6″		970	.016			.62		.62	.95
6070	Sleepers, on concrete, 1″ x 2″	1 Clab	4700	.002			.06		.06	.10
6075	1″ x 3″		4000	.002			.08		.08	.12
6080	2″ x 4″		3000	.003			.10		.10	.15
6085	2″ x 6″		2600	.003			.12		.12	.18
6086	Sheathing from roof, 5/16″	2 Clab	1600	.010	S.F.		.38		.38	.58
6088	3/8″		1525	.010			.39		.39	.61
6090	1/2″		1400	.011			.43		.43	.66
6092	5/8″		1300	.012			.46		.46	.71
6094	3/4″		1200	.013			.50		.50	.77
6096	Board sheathing from roof		1400	.011			.43		.43	.66
6100	Sheathing, from walls, 1/4″		1200	.013			.50		.50	.77
6110	5/16″		1175	.014			.51		.51	.79
6120	3/8″		1150	.014			.52		.52	.80
6130	1/2″		1125	.014			.53		.53	.82
6140	5/8″		1100	.015			.55		.55	.84
6150	3/4″		1075	.015			.56		.56	.86
6152	Board sheathing from walls		1500	.011			.40		.40	.62
6158	Subfloor/roof deck, with boards		2200	.007			.27		.27	.42
6159	Subfloor/roof deck, with tongue & groove boards		2000	.008			.30		.30	.46
6160	Plywood, 1/2″ thick		768	.021			.78		.78	1.21
6162	5/8″ thick		760	.021			.79		.79	1.22
6164	3/4″ thick		750	.021			.80		.80	1.23
6165	1-1/8″ thick		720	.022			.84		.84	1.29
6166	Underlayment, particle board, 3/8″ thick	1 Clab	780	.010			.39		.39	.59
6168	1/2″ thick		768	.010			.39		.39	.60
6170	5/8″ thick		760	.011			.40		.40	.61
6172	3/4″ thick		750	.011			.40		.40	.62
6200	Stairs and stringers, straight run	2 Clab	40	.400	Riser		15.05		15.05	23
6240	With platforms, winders or curves	″	26	.615	″		23		23	35.50
6300	Components, tread	1 Clab	110	.073	Ea.		2.73		2.73	4.21
6320	Riser		80	.100	″		3.76		3.76	5.80
6390	Stringer, 2″ x 10″		260	.031	L.F.		1.16		1.16	1.78
6400	2″ x 12″		260	.031			1.16		1.16	1.78
6410	3″ x 10″		250	.032			1.20		1.20	1.85
6420	3″ x 12″		250	.032			1.20		1.20	1.85
6590	Wood studs, 2″ x 3″	2 Clab	3076	.005			.20		.20	.30
6600	2″ x 4″		2000	.008			.30		.30	.46
6640	2″ x 6″		1600	.010			.38		.38	.58
6720	Wall framing, including studs, plates and blocking, 2″ x 4″	1 Clab	600	.013	S.F.		.50		.50	.77
6740	2″ x 6″		480	.017	″		.63		.63	.96
6750	Headers, 2″ x 4″		1125	.007	L.F.		.27		.27	.41
6755	2″ x 6″		1125	.007			.27		.27	.41

06 05 05 – Selective Demolition for Wood, Plastics, and Composites

06 05 05.10 Selective Demolition Wood Framing		Crew	Daily Output	Labor-Hours	Unit	Material	2015 Bare Costs Labor	Equipment	Total	Total Incl O&P
6760	2" x 8"	1 Clab	1050	.008	L.F.		.29		.29	.44
6765	2" x 10"		1050	.008			.29		.29	.44
6770	2" x 12"		1000	.008			.30		.30	.46
6780	4" x 10"		525	.015			.57		.57	.88
6785	4" x 12"		500	.016			.60		.60	.93
6790	6" x 8"		560	.014			.54		.54	.83
6795	6" x 10"		525	.015			.57		.57	.88
6797	6" x 12"		500	.016			.60		.60	.93
7000	Trusses									
7050	12' span	2 Clab	74	.216	Ea.		8.15		8.15	12.50
7150	24' span	F-3	66	.606			29	9.90	38.90	55.50
7200	26' span		64	.625			30	10.20	40.20	57.50
7250	28' span		62	.645			31	10.55	41.55	59
7300	30' span		58	.690			33	11.30	44.30	63
7350	32' span		56	.714			34	11.70	45.70	65.50
7400	34' span		54	.741			35.50	12.10	47.60	68
7450	36' span		52	.769			37	12.60	49.60	70.50
8000	Soffit, T & G wood	1 Clab	520	.015	S.F.		.58		.58	.89
8010	Hardboard, vinyl or aluminum	"	640	.013			.47		.47	.72
8030	Plywood	2 Carp	315	.051			2.38		2.38	3.67
9500	See Section 02 41 19.19 for rubbish handling									

06 05 05.20 Selective Demolition Millwork and Trim		Crew	Daily Output	Labor-Hours	Unit	Material	Labor	Equipment	Total	Total Incl O&P
0010	**SELECTIVE DEMOLITION MILLWORK AND TRIM** R024119-10									
1000	Cabinets, wood, base cabinets, per L.F.	2 Clab	80	.200	L.F.		7.50		7.50	11.55
1020	Wall cabinets, per L.F.	"	80	.200	"		7.50		7.50	11.55
1060	Remove and reset, base cabinets	2 Carp	18	.889	Ea.		41.50		41.50	64
1070	Wall cabinets		20	.800			37.50		37.50	58
1072	Oven cabinet, 7' high		11	1.455			68.50		68.50	105
1074	Cabinet door, up to 2' high	1 Clab	66	.121			4.56		4.56	7
1076	2' - 4' high	"	46	.174			6.55		6.55	10.05
1100	Steel, painted, base cabinets	2 Clab	60	.267	L.F.		10.05		10.05	15.45
1120	Wall cabinets		60	.267	"		10.05		10.05	15.45
1200	Casework, large area		320	.050	S.F.		1.88		1.88	2.89
1220	Selective		200	.080	"		3.01		3.01	4.63
1500	Counter top, straight runs		200	.080	L.F.		3.01		3.01	4.63
1510	L, U or C shapes		120	.133			5		5	7.70
1550	Remove and reset, straight runs	2 Carp	50	.320			15		15	23
1560	L, U or C shape	"	40	.400			18.80		18.80	29
2000	Paneling, 4' x 8' sheets	2 Clab	2000	.008	S.F.		.30		.30	.46
2100	Boards, 1" x 4"		700	.023			.86		.86	1.32
2120	1" x 6"		750	.021			.80		.80	1.23
2140	1" x 8"		800	.020			.75		.75	1.16
3000	Trim, baseboard, to 6" wide		1200	.013	L.F.		.50		.50	.77
3040	Greater than 6" and up to 12" wide		1000	.016			.60		.60	.93
3080	Remove and reset, minimum	2 Carp	400	.040			1.88		1.88	2.89
3090	Maximum	"	300	.053			2.50		2.50	3.85
3100	Ceiling trim	2 Clab	1000	.016			.60		.60	.93
3120	Chair rail		1200	.013			.50		.50	.77
3140	Railings with balusters		240	.067			2.51		2.51	3.86
3160	Wainscoting		700	.023	S.F.		.86		.86	1.32

06 05 23 – Wood, Plastic, and Composite Fastenings

06 05 23.10 Nails		Crew	Daily Output	Labor-Hours	Unit	Material	2015 Bare Costs Labor	Equipment	Total	Total Incl O&P
0010	**NAILS**, material only, based upon 50# box purchase									
0020	Copper nails, plain				Lb.	9.85			9.85	10.80
0400	Stainless steel, plain					9.20			9.20	10.15
0500	Box, 3d to 20d, bright					1.37			1.37	1.51
0520	Galvanized					1.68			1.68	1.85
0600	Common, 3d to 60d, plain					1.30			1.30	1.43
0700	Galvanized					1.97			1.97	2.17
0800	Aluminum					10.80			10.80	11.90
1000	Annular or spiral thread, 4d to 60d, plain					2.63			2.63	2.89
1200	Galvanized					3.15			3.15	3.47
1400	Drywall nails, plain					1.50			1.50	1.65
1600	Galvanized					2.20			2.20	2.42
1800	Finish nails, 4d to 10d, plain					1.25			1.25	1.38
2000	Galvanized					1.86			1.86	2.05
2100	Aluminum					5.85			5.85	6.45
2300	Flooring nails, hardened steel, 2d to 10d, plain					3.38			3.38	3.72
2400	Galvanized					4.04			4.04	4.44
2500	Gypsum lath nails, 1-1/8", 13 ga. flathead, blued					2			2	2.20
2600	Masonry nails, hardened steel, 3/4" to 3" long, plain					1.71			1.71	1.88
2700	Galvanized					3.27			3.27	3.60
2900	Roofing nails, threaded, galvanized					1.76			1.76	1.94
3100	Aluminum					4.85			4.85	5.35
3300	Compressed lead head, threaded, galvanized					2.49			2.49	2.74
3600	Siding nails, plain shank, galvanized					2.02			2.02	2.22
3800	Aluminum					5.85			5.85	6.45
5000	Add to prices above for cement coating					.11			.11	.12
5200	Zinc or tin plating					.14			.14	.15
5500	Vinyl coated sinkers, 8d to 16d				↓	2.50			2.50	2.75

06 05 23.40 Sheet Metal Screws		Crew	Daily Output	Labor-Hours	Unit	Material	Labor	Equipment	Total	Total Incl O&P
0010	**SHEET METAL SCREWS**									
0020	Steel, standard, #8 x 3/4", plain				C	3.20			3.20	3.52
0100	Galvanized					3.20			3.20	3.52
0300	#10 x 1", plain					4.40			4.40	4.84
0400	Galvanized					4.40			4.40	4.84
0600	With washers, #14 x 1", plain					9.65			9.65	10.60
0700	Galvanized					9.65			9.65	10.60
0900	#14 x 2", plain					15			15	16.50
1000	Galvanized					15			15	16.50
1500	Self-drilling, with washers, (pinch point) #8 x 3/4", plain					7.30			7.30	8
1600	Galvanized					7.30			7.30	8
1800	#10 x 3/4", plain					7.80			7.80	8.55
1900	Galvanized					7.80			7.80	8.55
3000	Stainless steel w/aluminum or neoprene washers, #14 x 1", plain					8.10			8.10	8.95
3100	#14 x 2", plain				↓	10			10	11

06 05 23.50 Wood Screws		Crew	Daily Output	Labor-Hours	Unit	Material	Labor	Equipment	Total	Total Incl O&P
0010	**WOOD SCREWS**									
0020	#8 x 1" long, steel				C	2.70			2.70	2.97
0100	Brass					12.20			12.20	13.40
0200	#8, 2" long, steel					4.16			4.16	4.58
0300	Brass					22			22	24
0400	#10, 1" long, steel					3.30			3.30	3.63
0500	Brass				↓	15.60			15.60	17.15

06 05 23 – Wood, Plastic, and Composite Fastenings

06 05 23.50 Wood Screws		Crew	Daily Output	Labor-Hours	Unit	Material	2015 Bare Costs Labor	Equipment	Total	Total Incl O&P
0600	#10, 2" long, steel				C	5.30			5.30	5.85
0700	Brass					27			27	29.50
0800	#10, 3" long, steel					8.85			8.85	9.75
1000	#12, 2" long, steel					7.30			7.30	8.05
1100	Brass					34.50			34.50	38
1500	#12, 3" long, steel					11.15			11.15	12.30
2000	#12, 4" long, steel				↓	18.75			18.75	20.50

06 05 23.60 Timber Connectors		Crew	Daily Output	Labor-Hours	Unit	Material	Labor	Equipment	Total	Total Incl O&P
0010	**TIMBER CONNECTORS**									
0020	Add up cost of each part for total cost of connection									
0100	Connector plates, steel, with bolts, straight	2 Carp	75	.213	Ea.	30.50	10		40.50	49
0110	Tee, 7 ga.		50	.320		35	15		50	61.50
0120	T- Strap, 14 ga., 12" x 8" x 2"		50	.320		35	15		50	61.50
0150	Anchor plates, 7 ga., 9" x 7"	↓	75	.213		30.50	10		40.50	49
0200	Bolts, machine, sq. hd. with nut & washer, 1/2" diameter, 4" long	1 Carp	140	.057		.75	2.68		3.43	4.96
0300	7-1/2" long		130	.062		1.37	2.89		4.26	5.95
0500	3/4" diameter, 7-1/2" long		130	.062		3.20	2.89		6.09	7.95
0610	Machine bolts, w/nut, washer, 3/4" diam., 15" L, HD's & beam hangers		95	.084	↓	5.95	3.95		9.90	12.65
0800	Drilling bolt holes in timber, 1/2" diameter		450	.018	Inch		.83		.83	1.28
0900	1" diameter		350	.023	"		1.07		1.07	1.65
1100	Framing anchor, angle, 3" x 3" x 1-1/2", 12 ga		175	.046	Ea.	2.61	2.15		4.76	6.15
1150	Framing anchors, 18 ga., 4-1/2" x 2-3/4"		175	.046		2.61	2.15		4.76	6.15
1160	Framing anchors, 18 ga., 4-1/2" x 3"		175	.046		2.61	2.15		4.76	6.15
1170	Clip anchors plates, 18 ga., 12" x 1-1/8"		175	.046		2.61	2.15		4.76	6.15
1250	Holdowns, 3 ga. base, 10 ga. body		8	1		25	47		72	100
1260	Holdowns, 7 ga. 11-1/16" x 3-1/4"		8	1		25	47		72	100
1270	Holdowns, 7 ga. 14-3/8" x 3-1/8"		8	1		25	47		72	100
1275	Holdowns, 12 ga. 8" x 2-1/2"		8	1		25	47		72	100
1300	Joist and beam hangers, 18 ga. galv., for 2" x 4" joist		175	.046		.73	2.15		2.88	4.10
1400	2" x 6" to 2" x 10" joist		165	.048		1.39	2.28		3.67	5.05
1600	16 ga. galv., 3" x 6" to 3" x 10" joist		160	.050		2.96	2.35		5.31	6.85
1700	3" x 10" to 3" x 14" joist		160	.050		4.92	2.35		7.27	9
1800	4" x 6" to 4" x 10" joist		155	.052		3.07	2.42		5.49	7.10
1900	4" x 10" to 4" x 14" joist		155	.052		5	2.42		7.42	9.25
2000	Two-2" x 6" to two-2" x 10" joists		150	.053		4.19	2.50		6.69	8.45
2100	Two-2" x 10" to two-2" x 14" joists		150	.053		4.68	2.50		7.18	9
2300	3/16" thick, 6" x 8" joist		145	.055		65	2.59		67.59	75.50
2400	6" x 10" joist		140	.057		67.50	2.68		70.18	78.50
2500	6" x 12" joist		135	.059		70.50	2.78		73.28	82
2700	1/4" thick, 6" x 14" joist	↓	130	.062		73	2.89		75.89	85
2900	Plywood clips, extruded aluminum H clip, for 3/4" panels					.24			.24	.26
3000	Galvanized 18 ga. back-up clip					.18			.18	.20
3200	Post framing, 16 ga. galv. for 4" x 4" base, 2 piece	1 Carp	130	.062		16.40	2.89		19.29	22.50
3300	Cap		130	.062		23	2.89		25.89	29.50
3500	Rafter anchors, 18 ga. galv., 1-1/2" wide, 5-1/4" long		145	.055		.49	2.59		3.08	4.53
3600	10-3/4" long		145	.055		1.46	2.59		4.05	5.60
3800	Shear plates, 2-5/8" diameter		120	.067		2.38	3.13		5.51	7.45
3900	4" diameter		115	.070		5.65	3.27		8.92	11.25
4000	Sill anchors, embedded in concrete or block, 25-1/2" long		115	.070		12.70	3.27		15.97	19
4100	Spike grids, 3" x 6"		120	.067		.96	3.13		4.09	5.90
4400	Split rings, 2-1/2" diameter		120	.067		1.96	3.13		5.09	7
4500	4" diameter	↓	110	.073	↓	2.97	3.41		6.38	8.50

06 05 23.60 Timber Connectors		Crew	Daily Output	Labor-Hours	Unit	Material	Labor	Equipment	Total	Total Incl O&P
						2015 Bare Costs				
4550	Tie plate, 20 ga., 7" x 3 1/8"	1 Carp	110	.073	Ea.	2.97	3.41		6.38	8.50
4560	Tie plate, 20 ga., 5" x 4 1/8"		110	.073		2.97	3.41		6.38	8.50
4575	Twist straps, 18 ga., 12" x 1 1/4"		110	.073		2.97	3.41		6.38	8.50
4580	Twist straps, 18 ga., 16" x 1 1/4"		110	.073		2.97	3.41		6.38	8.50
4600	Strap ties, 20 ga., 2 -1/16" wide, 12 13/16" long		180	.044		.94	2.09		3.03	4.24
4700	Strap ties, 16 ga., 1-3/8" wide, 12" long		180	.044		.94	2.09		3.03	4.24
4800	21-5/8" x 1-1/4"		160	.050		2.96	2.35		5.31	6.85
5000	Toothed rings, 2-5/8" or 4" diameter		90	.089		1.78	4.17		5.95	8.35
5200	Truss plates, nailed, 20 ga., up to 32' span		17	.471	Truss	12.90	22		34.90	48
5400	Washers, 2" x 2" x 1/8"				Ea.	.40			.40	.44
5500	3" x 3" x 3/16"				"	1.06			1.06	1.17
6000	Angles and gussets, painted									
6012	7 ga., 3-1/4" x 3-1/4" x 2-1/2" long	1 Carp	1.90	4.211	C	1,100	198		1,298	1,525
6014	3-1/4" x 3-1/4" x 5" long		1.90	4.211		2,150	198		2,348	2,675
6016	3-1/4" x 3-1/4" x 7-1/2" long		1.85	4.324		4,075	203		4,278	4,775
6018	5-3/4" x 5-3/4" x 2-1/2" long		1.85	4.324		2,625	203		2,828	3,200
6020	5-3/4" x 5-3/4" x 5" long		1.85	4.324		4,175	203		4,378	4,900
6022	5-3/4" x 5-3/4" x 7-1/2" long		1.80	4.444		6,175	209		6,384	7,125
6024	3 ga., 4-1/4" x 4-1/4" x 3" long		1.85	4.324		2,800	203		3,003	3,375
6026	4-1/4" x 4-1/4" x 6" long		1.85	4.324		6,000	203		6,203	6,900
6028	4-1/4" x 4-1/4" x 9" long		1.80	4.444		6,750	209		6,959	7,750
6030	7-1/4" x 7-1/4" x 3" long		1.80	4.444		4,825	209		5,034	5,650
6032	7-1/4" x 7-1/4" x 6" long		1.80	4.444		6,525	209		6,734	7,500
6034	7-1/4" x 7-1/4" x 9" long		1.75	4.571		14,600	215		14,815	16,400
6036	Gussets									
6038	7 ga., 8-1/8" x 8-1/8" x 2-3/4" long	1 Carp	1.80	4.444	C	4,625	209		4,834	5,400
6040	3 ga., 9-3/4" x 9-3/4" x 3-1/4" long	"	1.80	4.444	"	6,400	209		6,609	7,375
6101	Beam hangers, polymer painted									
6102	Bolted, 3 ga., (W x H x L)									
6104	3-1/4" x 9" x 12" top flange	1 Carp	1	8	C	19,600	375		19,975	22,200
6106	5-1/4" x 9" x 12" top flange		1	8		20,500	375		20,875	23,100
6108	5-1/4" x 11" x 11-3/4" top flange		1	8		23,200	375		23,575	26,200
6110	6-7/8" x 9" x 12" top flange		1	8		21,200	375		21,575	23,900
6112	6-7/8" x 11" x 13-1/2" top flange		1	8		24,400	375		24,775	27,500
6114	8-7/8" x 11" x 15-1/2" top flange		1	8		26,100	375		26,475	29,300
6116	Nailed, 3 ga., (W x H x L)									
6118	3-1/4" x 10-1/2" x 10" top flange	1 Carp	1.80	4.444	C	20,500	209		20,709	22,800
6120	3-1/4" x 10-1/2" x 12" top flange		1.80	4.444		20,500	209		20,709	22,800
6122	5-1/4" x 9-1/2" x 10" top flange		1.80	4.444		20,900	209		21,109	23,300
6124	5-1/4" x 9-1/2" x 12" top flange		1.80	4.444		20,900	209		21,109	23,300
6128	6-7/8" x 8-1/2" x 12" top flange		1.80	4.444		21,400	209		21,609	23,800
6134	Saddle hangers, glu-lam (W x H x L)									
6136	3-1/4" x 10-1/2" x 5-1/4" x 6" saddle	1 Carp	.50	16	C	15,300	750		16,050	18,000
6138	3-1/4" x 10-1/2" x 6-7/8" x 6" saddle		.50	16		16,100	750		16,850	18,900
6140	3-1/4" x 10-1/2" x 8-7/8" x 6" saddle		.50	16		16,900	750		17,650	19,800
6142	3-1/4" x 19-1/2" x 5-1/4" x 10-1/8" saddle		.40	20		15,300	940		16,240	18,300
6144	3-1/4" x 19-1/2" x 6-7/8" x 10-1/8" saddle		.40	20		16,100	940		17,040	19,200
6146	3-1/4" x 19-1/2" x 8-7/8" x 10-1/8" saddle		.40	20		16,900	940		17,840	20,100
6148	5-1/4" x 9-1/2" x 5-1/4" x 12" saddle		.50	16		18,300	750		19,050	21,300
6150	5-1/4" x 9-1/2" x 6-7/8" x 9" saddle		.50	16		20,000	750		20,750	23,200
6152	5-1/4" x 10-1/2" x spec x 12" saddle		.50	16		21,800	750		22,550	25,100
6154	5-1/4" x 18" x 5-1/4" x 12-1/8" saddle		.40	20		18,300	940		19,240	21,600
6156	5-1/4" x 18" x 6-7/8" x 12-1/8" saddle		.40	20		20,000	940		20,940	23,500

06 05 23 – Wood, Plastic, and Composite Fastenings

06 05 23.60 Timber Connectors		Crew	Daily Output	Labor-Hours	Unit	Material	2015 Bare Costs Labor	Equipment	Total	Total Incl O&P
6158	5-1/4" x 18" x spec x 12-1/8" saddle	1 Carp	.40	20	C	21,800	940		22,740	25,400
6160	6-7/8" x 8-1/2" x 6-7/8" x 12" saddle		.50	16		21,800	750		22,550	25,200
6162	6-7/8" x 8-1/2" x 8-7/8" x 12" saddle		.50	16		22,600	750		23,350	26,000
6164	6-7/8" x 10-1/2" x spec x 12" saddle		.50	16		21,800	750		22,550	25,200
6166	6-7/8" x 18" x 6-7/8" x 13-3/4" saddle		.40	20		21,800	940		22,740	25,500
6168	6-7/8" x 18" x 8-7/8" x 13-3/4" saddle		.40	20		22,600	940		23,540	26,300
6170	6-7/8" x 18" x spec x 13-3/4" saddle		.40	20		24,400	940		25,340	28,300
6172	8-7/8" x 18" x spec x 15-3/4" saddle		.40	20		38,000	940		38,940	43,300
6201	Beam and purlin hangers, galvanized, 12 ga.									
6202	Purlin or joist size, 3" x 8"	1 Carp	1.70	4.706	C	2,025	221		2,246	2,575
6204	3" x 10"		1.70	4.706		2,175	221		2,396	2,750
6206	3" x 12"		1.65	4.848		2,500	228		2,728	3,100
6208	3" x 14"		1.65	4.848		2,675	228		2,903	3,275
6210	3" x 16"		1.65	4.848		2,825	228		3,053	3,450
6212	4" x 8"		1.65	4.848		2,025	228		2,253	2,600
6214	4" x 10"		1.65	4.848		2,200	228		2,428	2,775
6216	4" x 12"		1.60	5		2,600	235		2,835	3,200
6218	4" x 14"		1.60	5		2,750	235		2,985	3,375
6220	4" x 16"		1.60	5		2,925	235		3,160	3,575
6222	6" x 8"		1.60	5		2,625	235		2,860	3,225
6224	6" x 10"		1.55	5.161		2,675	242		2,917	3,325
6226	6" x 12"		1.55	5.161		4,575	242		4,817	5,425
6228	6" x 14"		1.50	5.333		4,850	250		5,100	5,725
6230	6" x 16"		1.50	5.333		5,125	250		5,375	6,025
6250	Beam seats									
6252	Beam size, 5-1/4" wide									
6254	5" x 7" x 1/4"	1 Carp	1.80	4.444	C	7,725	209		7,934	8,800
6256	6" x 7" x 3/8"		1.80	4.444		8,650	209		8,859	9,850
6258	7" x 7" x 3/8"		1.80	4.444		9,250	209		9,459	10,500
6260	8" x 7" x 3/8"		1.80	4.444		10,900	209		11,109	12,300
6262	Beam size, 6-7/8" wide									
6264	5" x 9" x 1/4"	1 Carp	1.80	4.444	C	9,225	209		9,434	10,400
6266	6" x 9" x 3/8"		1.80	4.444		12,000	209		12,209	13,500
6268	7" x 9" x 3/8"		1.80	4.444		12,200	209		12,409	13,700
6270	8" x 9" x 3/8"		1.80	4.444		14,400	209		14,609	16,200
6272	Special beams, over 6-7/8" wide									
6274	5" x 10" x 3/8"	1 Carp	1.80	4.444	C	12,500	209		12,709	14,000
6276	6" x 10" x 3/8"		1.80	4.444		14,600	209		14,809	16,300
6278	7" x 10" x 3/8"		1.80	4.444		15,200	209		15,409	17,000
6280	8" x 10" x 3/8"		1.75	4.571		16,300	215		16,515	18,200
6282	5-1/4" x 12" x 5/16"		1.75	4.571		12,600	215		12,815	14,200
6284	6-1/2" x 12" x 3/8"		1.75	4.571		20,900	215		21,115	23,300
6286	5-1/4" x 16" x 5/16"		1.70	4.706		18,600	221		18,821	20,700
6288	6-1/2" x 16" x 3/8"		1.70	4.706		24,200	221		24,421	27,000
6290	5-1/4" x 20" x 5/16"		1.70	4.706		21,800	221		22,021	24,300
6292	6-1/2" x 20" x 3/8"		1.65	4.848		28,500	228		28,728	31,800
6300	Column bases									
6302	4 x 4, 16 ga.	1 Carp	1.80	4.444	C	780	209		989	1,175
6306	7 ga.		1.80	4.444		2,850	209		3,059	3,475
6308	4 x 6, 16 ga.		1.80	4.444		1,825	209		2,034	2,325
6312	7 ga.		1.80	4.444		2,975	209		3,184	3,600
6314	6 x 6, 16 ga.		1.75	4.571		2,050	215		2,265	2,600
6318	7 ga.		1.75	4.571		4,075	215		4,290	4,800

06 05 23 – Wood, Plastic, and Composite Fastenings

06 05 23.60 Timber Connectors		Crew	Daily Output	Labor-Hours	Unit	Material	2015 Bare Costs Labor	Equipment	Total	Total Incl O&P
6320	6 x 8, 7 ga.	1 Carp	1.70	4.706	C	3,175	221		3,396	3,825
6322	6 x 10, 7 ga.		1.70	4.706		3,400	221		3,621	4,075
6324	6 x 12, 7 ga.		1.70	4.706		3,675	221		3,896	4,400
6326	8 x 8, 7 ga.		1.65	4.848		6,200	228		6,428	7,175
6330	8 x 10, 7 ga.		1.65	4.848		7,425	228		7,653	8,525
6332	8 x 12, 7 ga.		1.60	5		8,075	235		8,310	9,225
6334	10 x 10, 3 ga.		1.60	5		8,225	235		8,460	9,400
6336	10 x 12, 3 ga.		1.60	5		9,475	235		9,710	10,800
6338	12 x 12, 3 ga.		1.55	5.161		10,300	242		10,542	11,700
6350	Column caps, painted, 3 ga.									
6352	3-1/4" x 3-5/8"	1 Carp	1.80	4.444	C	10,400	209		10,609	11,800
6354	3-1/4" x 5-1/2"		1.80	4.444		10,400	209		10,609	11,800
6356	3-5/8" x 3-5/8"		1.80	4.444		8,550	209		8,759	9,725
6358	3-5/8" x 5-1/2"		1.80	4.444		8,550	209		8,759	9,725
6360	5-1/4" x 5-1/2"		1.75	4.571		11,200	215		11,415	12,600
6362	5-1/4" x 7-1/2"		1.75	4.571		11,200	215		11,415	12,600
6364	5-1/2" x 3-5/8"		1.75	4.571		12,100	215		12,315	13,600
6366	5-1/2" x 5-1/2"		1.75	4.571		12,100	215		12,315	13,600
6368	5-1/2" x 7-1/2"		1.70	4.706		12,100	221		12,321	13,600
6370	6-7/8" x 5-1/2"		1.70	4.706		12,600	221		12,821	14,100
6372	6-7/8" x 6-7/8"		1.70	4.706		12,600	221		12,821	14,100
6374	6-7/8" x 7-1/2"		1.70	4.706		12,600	221		12,821	14,100
6376	7-1/2" x 5-1/2"		1.65	4.848		13,100	228		13,328	14,900
6378	7-1/2" x 7-1/2"		1.65	4.848		13,100	228		13,328	14,900
6380	8-7/8" x 5-1/2"		1.60	5		13,900	235		14,135	15,700
6382	8-7/8" x 7-1/2"		1.60	5		13,900	235		14,135	15,700
6384	9-1/2" x 5-1/2"		1.60	5		18,800	235		19,035	21,000
6400	Floor tie anchors, polymer paint									
6402	10 ga., 3" x 37-1/2"	1 Carp	1.80	4.444	C	5,000	209		5,209	5,825
6404	3-1/2" x 45-1/2"		1.75	4.571		5,250	215		5,465	6,100
6406	3 ga., 3-1/2" x 56"		1.70	4.706		9,250	221		9,471	10,500
6410	Girder hangers									
6412	6" wall thickness, 4" x 6"	1 Carp	1.80	4.444	C	2,800	209		3,009	3,425
6414	4" x 8"		1.80	4.444		3,150	209		3,359	3,775
6416	8" wall thickness, 4" x 6"		1.80	4.444		3,250	209		3,459	3,900
6418	4" x 8"		1.80	4.444		3,250	209		3,459	3,900
6420	Hinge connections, polymer painted									
6422	3/4" thick top plate									
6424	5-1/4" x 12" w/5" x 5" top	1 Carp	1	8	C	35,900	375		36,275	40,100
6426	5-1/4" x 15" w/6" x 6" top		.80	10		38,100	470		38,570	42,600
6428	5-1/4" x 18" w/7" x 7" top		.70	11.429		40,100	535		40,635	44,900
6430	5-1/4" x 26" w/9" x 9" top		.60	13.333		42,700	625		43,325	47,900
6432	1" thick top plate									
6434	6-7/8" x 14" w/5" x 5" top	1 Carp	.80	10	C	43,700	470		44,170	48,800
6436	6-7/8" x 17" w/6" x 6" top		.80	10		48,600	470		49,070	54,000
6438	6-7/8" x 21" w/7" x 7" top		.70	11.429		53,000	535		53,535	59,500
6440	6-7/8" x 31" w/9" x 9" top		.60	13.333		58,000	625		58,625	65,000
6442	1-1/4" thick top plate									
6444	8-7/8" x 16" w/5" x 5" top	1 Carp	.60	13.333	C	54,500	625		55,125	61,000
6446	8-7/8" x 21" w/6" x 6" top		.50	16		60,000	750		60,750	67,000
6448	8-7/8" x 26" w/7" x 7" top		.40	20		68,000	940		68,940	76,000
6450	8-7/8" x 39" w/9" x 9" top		.30	26.667		84,500	1,250		85,750	95,000
6460	Holddowns									

06 05 23.60 Timber Connectors		Crew	Daily Output	Labor-Hours	Unit	Material	Labor	Equipment	Total	Total Incl O&P
						2015 Bare Costs				
6462	Embedded along edge									
6464	26" long, 12 ga.	1 Carp	.90	8.889	C	1,375	415		1,790	2,150
6466	35" long, 12 ga.		.85	9.412		1,800	440		2,240	2,675
6468	35" long, 10 ga.		.85	9.412		1,450	440		1,890	2,250
6470	Embedded away from edge									
6472	Medium duty, 12 ga.									
6474	18-1/2" long	1 Carp	.95	8.421	C	825	395		1,220	1,525
6476	23-3/4" long		.90	8.889		965	415		1,380	1,700
6478	28" long		.85	9.412		985	440		1,425	1,750
6480	35" long		.85	9.412		1,350	440		1,790	2,150
6482	Heavy duty, 10 ga.									
6484	28" long	1 Carp	.85	9.412	C	1,750	440		2,190	2,625
6486	35" long	"	.85	9.412	"	1,925	440		2,365	2,775
6490	Surface mounted (W x H)									
6492	2-1/2" x 5-3/4", 7 ga.	1 Carp	1	8	C	2,000	375		2,375	2,775
6494	2-1/2" x 8", 12 ga.		1	8		1,225	375		1,600	1,925
6496	2-7/8" x 6-3/8", 7 ga.		1	8		4,425	375		4,800	5,450
6498	2-7/8" x 12-1/2", 3 ga.		1	8		4,550	375		4,925	5,575
6500	3-3/16" x 9-3/8", 10 ga.		1	8		3,025	375		3,400	3,925
6502	3-1/2" x 11-5/8", 3 ga.		1	8		5,525	375		5,900	6,675
6504	3-1/2" x 14-3/4", 3 ga.		1	8		7,000	375		7,375	8,275
6506	3-1/2" x 16-1/2", 3 ga.		1	8		8,450	375		8,825	9,875
6508	3-1/2" x 20-1/2", 3 ga.		.90	8.889		8,650	415		9,065	10,200
6510	3-1/2" x 24-1/2", 3 ga.		.90	8.889		10,900	415		11,315	12,600
6512	4-1/4" x 20-3/4", 3 ga.		.90	8.889		7,450	415		7,865	8,850
6520	Joist hangers									
6522	Sloped, field adjustable, 18 ga.									
6524	2" x 6"	1 Carp	1.65	4.848	C	545	228		773	950
6526	2" x 8"		1.65	4.848		980	228		1,208	1,425
6528	2" x 10" and up		1.65	4.848		1,625	228		1,853	2,150
6530	3" x 10" and up		1.60	5		1,225	235		1,460	1,700
6532	4" x 10" and up		1.55	5.161		1,475	242		1,717	2,000
6536	Skewed 45°, 16 ga.									
6538	2" x 4"	1 Carp	1.75	4.571	C	870	215		1,085	1,300
6540	2" x 6" or 2" x 8"		1.65	4.848		885	228		1,113	1,325
6542	2" x 10" or 2" x 12"		1.65	4.848		1,025	228		1,253	1,475
6544	2" x 14" or 2" x 16"		1.60	5		1,825	235		2,060	2,350
6546	(2) 2" x 6" or (2) 2" x 8"		1.60	5		1,650	235		1,885	2,150
6548	(2) 2" x 10" or (2) 2" x 12"		1.55	5.161		1,750	242		1,992	2,325
6550	(2) 2" x 14" or (2) 2" x 16"		1.50	5.333		2,775	250		3,025	3,425
6552	4" x 6" or 4" x 8"		1.60	5		1,375	235		1,610	1,850
6554	4" x 10" or 4" x 12"		1.55	5.161		1,600	242		1,842	2,150
6556	4" x 14" or 4" x 16"		1.55	5.161		2,475	242		2,717	3,100
6560	Skewed 45°, 14 ga.									
6562	(2) 2" x 6" or (2) 2" x 8"	1 Carp	1.60	5	C	1,900	235		2,135	2,425
6564	(2) 2" x 10" or (2) 2" x 12"		1.55	5.161		2,625	242		2,867	3,250
6566	(2) 2" x 14" or (2) 2" x 16"		1.50	5.333		3,850	250		4,100	4,600
6568	4" x 6" or 4" x 8"		1.60	5		2,250	235		2,485	2,825
6570	4" x 10" or 4" x 12"		1.55	5.161		2,375	242		2,617	2,975
6572	4" x 14" or 4" x 16"		1.55	5.161		3,175	242		3,417	3,875
6590	Joist hangers, heavy duty 12 ga., galvanized									
6592	2" x 4"	1 Carp	1.75	4.571	C	1,275	215		1,490	1,725
6594	2" x 6"		1.65	4.848		1,400	228		1,628	1,875

06 05 23 – Wood, Plastic, and Composite Fastenings

06 05 23.60 Timber Connectors		Crew	Daily Output	Labor-Hours	Unit	2015 Bare Costs Material	Labor	Equipment	Total	Total Incl O&P
6595	2" x 6", 16 ga.	1 Carp	1.65	4.848	C	1,325	228		1,553	1,800
6596	2" x 8"		1.65	4.848		2,100	228		2,328	2,675
6597	2" x 8", 16 ga.		1.65	4.848		2,000	228		2,228	2,550
6598	2" x 10"		1.65	4.848		2,175	228		2,403	2,725
6600	2" x 12"		1.65	4.848		2,675	228		2,903	3,275
6602	2" x 14"		1.65	4.848		2,775	228		3,003	3,400
6604	2" x 16"		1.65	4.848		2,925	228		3,153	3,575
6606	3" x 4"		1.65	4.848		1,775	228		2,003	2,300
6608	3" x 6"		1.65	4.848		2,375	228		2,603	2,950
6610	3" x 8"		1.65	4.848		2,400	228		2,628	3,000
6612	3" x 10"		1.60	5		2,775	235		3,010	3,425
6614	3" x 12"		1.60	5		3,350	235		3,585	4,025
6616	3" x 14"		1.60	5		3,900	235		4,135	4,650
6618	3" x 16"		1.60	5		4,300	235		4,535	5,100
6620	(2) 2" x 4"		1.75	4.571		2,075	215		2,290	2,600
6622	(2) 2" x 6"		1.60	5		2,500	235		2,735	3,100
6624	(2) 2" x 8"		1.60	5		2,550	235		2,785	3,150
6626	(2) 2" x 10"		1.55	5.161		2,750	242		2,992	3,400
6628	(2) 2" x 12"		1.55	5.161		3,525	242		3,767	4,250
6630	(2) 2" x 14"		1.50	5.333		3,550	250		3,800	4,300
6632	(2) 2" x 16"		1.50	5.333		3,600	250		3,850	4,325
6634	4" x 4"		1.65	4.848		1,500	228		1,728	2,000
6636	4" x 6"		1.60	5		1,650	235		1,885	2,175
6638	4" x 8"		1.60	5		1,900	235		2,135	2,450
6640	4" x 10"		1.55	5.161		2,350	242		2,592	2,950
6642	4" x 12"		1.55	5.161		2,500	242		2,742	3,125
6644	4" x 14"		1.55	5.161		2,975	242		3,217	3,650
6646	4" x 16"		1.55	5.161		3,275	242		3,517	3,975
6648	(3) 2" x 10"		1.50	5.333		3,625	250		3,875	4,350
6650	(3) 2" x 12"		1.50	5.333		4,025	250		4,275	4,825
6652	(3) 2" x 14"		1.45	5.517		4,325	259		4,584	5,150
6654	(3) 2" x 16"		1.45	5.517		4,400	259		4,659	5,250
6656	6" x 6"		1.60	5		1,950	235		2,185	2,500
6658	6" x 8"		1.60	5		2,000	235		2,235	2,550
6660	6" x 10"		1.55	5.161		2,400	242		2,642	3,000
6662	6" x 12"		1.55	5.161		2,725	242		2,967	3,375
6664	6" x 14"		1.50	5.333		3,425	250		3,675	4,125
6666	6" x 16"		1.50	5.333		4,025	250		4,275	4,800
6690	Knee braces, galvanized, 12 ga.									
6692	Beam depth, 10" x 15" x 5' long	1 Carp	1.80	4.444	C	5,225	209		5,434	6,075
6694	15" x 22-1/2" x 7' long		1.70	4.706		6,000	221		6,221	6,950
6696	22-1/2" x 28-1/2" x 8' long		1.60	5		6,450	235		6,685	7,450
6698	28-1/2" x 36" x 10' long		1.55	5.161		6,725	242		6,967	7,775
6700	36" x 42" x 12' long		1.50	5.333		7,425	250		7,675	8,525
6710	Mudsill anchors									
6714	2" x 4" or 3" x 4"	1 Carp	115	.070	C	1,350	3.27		1,353.27	1,475
6716	2" x 6" or 3" x 6"		115	.070		1,350	3.27		1,353.27	1,475
6718	Block wall, 13-1/4" long		115	.070		85	3.27		88.27	98.50
6720	21-1/4" long		115	.070		126	3.27		129.27	144
6730	Post bases, 12 ga. galvanized									
6732	Adjustable, 3-9/16" x 3-9/16"	1 Carp	1.30	6.154	C	1,075	289		1,364	1,625
6734	3-9/16" x 5-1/2"		1.30	6.154		2,025	289		2,314	2,675
6736	4" x 4"		1.30	6.154		930	289		1,219	1,475

06 05 23 – Wood, Plastic, and Composite Fastenings

06 05 23.60 Timber Connectors		Crew	Daily Output	Labor-Hours	Unit	2015 Bare Costs Material	Labor	Equipment	Total	Total Incl O&P
6738	4" x 6"	1 Carp	1.30	6.154	C	2,700	289		2,989	3,425
6740	5-1/2" x 5-1/2"		1.30	6.154		3,325	289		3,614	4,100
6742	6" x 6"		1.30	6.154		3,325	289		3,614	4,100
6744	Elevated, 3-9/16" x 3-1/4"		1.30	6.154		1,150	289		1,439	1,725
6746	5-1/2" x 3-5/16"		1.30	6.154		1,650	289		1,939	2,250
6748	5-1/2" x 5"		1.30	6.154		2,475	289		2,764	3,150
6750	Regular, 3-9/16" x 3-3/8"		1.30	6.154		885	289		1,174	1,425
6752	4" x 3-3/8"		1.30	6.154		1,250	289		1,539	1,825
6754	18 ga., 5-1/4" x 3-1/8"		1.30	6.154		1,300	289		1,589	1,875
6755	5-1/2" x 3-3/8"		1.30	6.154		1,300	289		1,589	1,875
6756	5-1/2" x 5-3/8"		1.30	6.154		1,875	289		2,164	2,500
6758	6" x 3-3/8"		1.30	6.154		2,250	289		2,539	2,925
6760	6" x 5-3/8"		1.30	6.154		2,500	289		2,789	3,225
6762	Post combination cap/bases									
6764	3-9/16" x 3-9/16"	1 Carp	1.20	6.667	C	460	315		775	990
6766	3-9/16" x 5-1/2"		1.20	6.667		1,050	315		1,365	1,625
6768	4" x 4"		1.20	6.667		2,075	315		2,390	2,775
6770	5-1/2" x 5-1/2"		1.20	6.667		1,150	315		1,465	1,750
6772	6" x 6"		1.20	6.667		4,175	315		4,490	5,075
6774	7-1/2" x 7-1/2"		1.20	6.667		4,575	315		4,890	5,500
6776	8" x 8"		1.20	6.667		4,725	315		5,040	5,675
6790	Post-beam connection caps									
6792	Beam size 3-9/16"									
6794	12 ga. post, 4" x 4"	1 Carp	1	8	C	2,900	375		3,275	3,750
6796	4" x 6"		1	8		3,850	375		4,225	4,825
6798	4" x 8"		1	8		5,700	375		6,075	6,850
6800	16 ga. post, 4" x 4"		1	8		1,200	375		1,575	1,875
6802	4" x 6"		1	8		2,000	375		2,375	2,775
6804	4" x 8"		1	8		3,350	375		3,725	4,275
6805	18 ga. post, 2-7/8" x 3"		1	8		3,350	375		3,725	4,275
6806	Beam size 5-1/2"									
6808	12 ga. post, 6" x 4"	1 Carp	1	8	C	3,450	375		3,825	4,375
6810	6" x 6"		1	8		5,450	375		5,825	6,575
6812	6" x 8"		1	8		3,750	375		4,125	4,700
6816	16 ga. post, 6" x 4"		1	8		1,875	375		2,250	2,650
6818	6" x 6"		1	8		2,000	375		2,375	2,775
6820	Beam size 7-1/2"									
6822	12 ga. post, 8" x 4"	1 Carp	1	8	C	4,775	375		5,150	5,850
6824	8" x 6"		1	8		5,025	375		5,400	6,100
6826	8" x 8"		1	8		7,550	375		7,925	8,875
6840	Purlin anchors, embedded									
6842	Heavy duty, 10 ga.									
6844	Straight, 28" long	1 Carp	1.60	5	C	1,450	235		1,685	1,925
6846	35" long		1.50	5.333		1,775	250		2,025	2,325
6848	Twisted, 28" long		1.60	5		1,450	235		1,685	1,925
6850	35" long		1.50	5.333		1,775	250		2,025	2,325
6852	Regular duty, 12 ga.									
6854	Straight, 18-1/2" long	1 Carp	1.80	4.444	C	825	209		1,034	1,225
6856	23-3/4" long		1.70	4.706		1,025	221		1,246	1,500
6858	29" long		1.60	5		1,050	235		1,285	1,525
6860	35" long		1.50	5.333		1,450	250		1,700	1,975
6862	Twisted, 18" long		1.80	4.444		825	209		1,034	1,225
6866	28" long		1.60	5		980	235		1,215	1,425

06 05 23 – Wood, Plastic, and Composite Fastenings

06 05 23.60 Timber Connectors		Crew	Daily Output	Labor-Hours	Unit	Material	Labor	Equipment	Total	Total Incl O&P
						2015 Bare Costs				
6868	35" long	1 Carp	1.50	5.333	C	1,450	250		1,700	1,975
6870	Straight, plastic coated									
6872	23-1/2" long	1 Carp	1.60	5	C	2,025	235		2,260	2,575
6874	26-7/8" long		1.60	5		2,375	235		2,610	2,950
6876	32-1/2" long		1.50	5.333		2,525	250		2,775	3,150
6878	35-7/8" long		1.50	5.333		2,625	250		2,875	3,275
6890	Purlin hangers, painted									
6892	12 ga., 2" x 6"	1 Carp	1.80	4.444	C	1,850	209		2,059	2,375
6894	2" x 8"		1.80	4.444		2,025	209		2,234	2,550
6896	2" x 10"		1.80	4.444		2,175	209		2,384	2,725
6898	2" x 12"		1.75	4.571		2,350	215		2,565	2,900
6900	2" x 14"		1.75	4.571		2,500	215		2,715	3,075
6902	2" x 16"		1.75	4.571		2,650	215		2,865	3,250
6904	3" x 6"		1.70	4.706		1,875	221		2,096	2,400
6906	3" x 8"		1.70	4.706		2,025	221		2,246	2,575
6908	3" x 10"		1.70	4.706		2,175	221		2,396	2,750
6910	3" x 12"		1.65	4.848		2,500	228		2,728	3,100
6912	3" x 14"		1.65	4.848		2,675	228		2,903	3,275
6914	3" x 16"		1.65	4.848		2,825	228		3,053	3,450
6916	4" x 6"		1.65	4.848		1,875	228		2,103	2,425
6918	4" x 8"		1.65	4.848		2,025	228		2,253	2,600
6920	4" x 10"		1.65	4.848		2,200	228		2,428	2,775
6922	4" x 12"		1.60	5		2,600	235		2,835	3,200
6924	4" x 14"		1.60	5		2,750	235		2,985	3,375
6926	4" x 16"		1.60	5		2,925	235		3,160	3,575
6928	6" x 6"		1.60	5		2,475	235		2,710	3,075
6930	6" x 8"		1.60	5		2,625	235		2,860	3,225
6932	6" x 10"		1.55	5.161		2,675	242		2,917	3,325
6934	double 2" x 6"		1.70	4.706		2,025	221		2,246	2,575
6936	double 2" x 8"		1.70	4.706		2,175	221		2,396	2,750
6938	double 2" x 10"		1.70	4.706		2,350	221		2,571	2,925
6940	double 2" x 12"		1.65	4.848		2,500	228		2,728	3,100
6942	double 2" x 14"		1.65	4.848		2,675	228		2,903	3,275
6944	double 2" x 16"		1.65	4.848		2,825	228		3,053	3,450
6960	11 ga., 4" x 6"		1.65	4.848		3,700	228		3,928	4,425
6962	4" x 8"		1.65	4.848		3,975	228		4,203	4,725
6964	4" x 10"		1.65	4.848		4,250	228		4,478	5,025
6966	6" x 6"		1.60	5		3,750	235		3,985	4,475
6968	6" x 8"		1.60	5		4,025	235		4,260	4,775
6970	6" x 10"		1.55	5.161		4,300	242		4,542	5,100
6972	6" x 12"		1.55	5.161		4,575	242		4,817	5,425
6974	6" x 14"		1.55	5.161		4,850	242		5,092	5,725
6976	6" x 16"		1.50	5.333		5,125	250		5,375	6,025
6978	7 ga., 8" x 6"		1.60	5		4,075	235		4,310	4,825
6980	8" x 8"		1.60	5		4,350	235		4,585	5,125
6982	8" x 10"		1.55	5.161		4,625	242		4,867	5,475
6984	8" x 12"		1.55	5.161		4,900	242		5,142	5,775
6986	8" x 14"		1.50	5.333		5,175	250		5,425	6,075
6988	8" x 16"		1.50	5.333		5,450	250		5,700	6,375
7000	Strap connectors, galvanized									
7002	12 ga., 2-1/16" x 36"	1 Carp	1.55	5.161	C	1,175	242		1,417	1,675
7004	2-1/16" x 47"		1.50	5.333		1,650	250		1,900	2,175
7005	10 ga., 2-1/16" x 72"		1.50	5.333		1,725	250		1,975	2,275

06 05 Common Work Results for Wood, Plastics, and Composites

06 05 23 – Wood, Plastic, and Composite Fastenings

06 05 23.60 Timber Connectors		Crew	Daily Output	Labor-Hours	Unit	Material	Labor (2015 Bare Costs)	Equipment	Total	Total Incl O&P
7006	7 ga., 2-1/16" x 34"	1 Carp	1.55	5.161	C	2,925	242		3,167	3,600
7008	2-1/16" x 45"		1.50	5.333		3,825	250		4,075	4,575
7010	3 ga., 3" x 32"		1.55	5.161		4,925	242		5,167	5,800
7012	3" x 41"		1.55	5.161		5,125	242		5,367	6,025
7014	3" x 50"		1.50	5.333		7,825	250		8,075	8,975
7016	3" x 59"		1.50	5.333		9,525	250		9,775	10,900
7018	3-1/2" x 68"		1.45	5.517		9,675	259		9,934	11,000
7030	Tension ties									
7032	19-1/8" long, 16 ga., 3/4" anchor bolt	1 Carp	1.80	4.444	C	1,350	209		1,559	1,825
7034	20" long, 12 ga., 1/2" anchor bolt		1.80	4.444		1,750	209		1,959	2,250
7036	20" long, 12 ga., 3/4" anchor bolt		1.80	4.444		1,750	209		1,959	2,250
7038	27-3/4" long, 12 ga., 3/4" anchor bolt		1.75	4.571		3,100	215		3,315	3,750
7050	Truss connectors, galvanized									
7052	Adjustable hanger									
7054	18 ga., 2" x 6"	1 Carp	1.65	4.848	C	530	228		758	935
7056	4" x 6"		1.65	4.848		700	228		928	1,125
7058	16 ga., 4" x 10"		1.60	5		1,025	235		1,260	1,475
7060	(2) 2" x 10"		1.60	5		1,025	235		1,260	1,475
7062	Connectors to plate									
7064	16 ga., 2" x 4" plate	1 Carp	1.80	4.444	C	525	209		734	895
7066	2" x 6" plate	"	1.80	4.444	"	675	209		884	1,075
7068	Hip jack connector									
7070	14 ga.	1 Carp	1.50	5.333	C	2,750	250		3,000	3,400

06 05 23.80 Metal Bracing		Crew	Daily Output	Labor-Hours	Unit	Material	Labor	Equipment	Total	Total Incl O&P
0010	**METAL BRACING**									
0302	Let-in, "T" shaped, 22 ga. galv. steel, studs at 16" O.C.	1 Carp	580	.014	L.F.	.81	.65		1.46	1.89
0402	Studs at 24" O.C.		600	.013		.81	.63		1.44	1.85
0502	Steel straps, 16 ga. galv. steel, studs at 16" O.C.		600	.013		1.05	.63		1.68	2.12
0602	Studs at 24" O.C.		620	.013		1.05	.61		1.66	2.09

06 11 Wood Framing

06 11 10 – Framing with Dimensional, Engineered or Composite Lumber

06 11 10.01 Forest Stewardship Council Certification		Crew	Daily Output	Labor-Hours	Unit	Material	Labor	Equipment	Total	Total Incl O&P
0010	**FOREST STEWARDSHIP COUNCIL CERTIFICATION**									
0020	For Forest Stewardship Council (FSC) cert dimension lumber, add [G]					65%				

06 11 10.02 Blocking		Crew	Daily Output	Labor-Hours	Unit	Material	Labor	Equipment	Total	Total Incl O&P
0010	**BLOCKING**									
2600	Miscellaneous, to wood construction									
2620	2" x 4"	1 Carp	.17	47.059	M.B.F.	635	2,200		2,835	4,100
2625	Pneumatic nailed		.21	38.095		640	1,800		2,440	3,450
2660	2" x 8"		.27	29.630		690	1,400		2,090	2,900
2665	Pneumatic nailed		.33	24.242		695	1,150		1,845	2,525
2720	To steel construction									
2740	2" x 4"	1 Carp	.14	57.143	M.B.F.	635	2,675		3,310	4,825
2780	2" x 8"	"	.21	38.095	"	690	1,800		2,490	3,500

06 11 10.04 Wood Bracing		Crew	Daily Output	Labor-Hours	Unit	Material	Labor	Equipment	Total	Total Incl O&P
0010	**WOOD BRACING**									
0012	Let-in, with 1" x 6" boards, studs @ 16" O.C.	1 Carp	150	.053	L.F.	.73	2.50		3.23	4.65
0202	Studs @ 24" O.C.	"	230	.035	"	.73	1.63		2.36	3.31

06 11 10 – Framing with Dimensional, Engineered or Composite Lumber

06 11 10.06 Bridging

06 11 10.06 Bridging		Crew	Daily Output	Labor-Hours	Unit	Material	2015 Bare Costs Labor	Equipment	Total	Total Incl O&P
0010	BRIDGING									
0012	Wood, for joists 16" O.C., 1" x 3"	1 Carp	130	.062	Pr.	.63	2.89		3.52	5.15
0017	Pneumatic nailed		170	.047		.68	2.21		2.89	4.15
0102	2" x 3" bridging		130	.062		.69	2.89		3.58	5.20
0107	Pneumatic nailed		170	.047		.69	2.21		2.90	4.16
0302	Steel, galvanized, 18 ga., for 2" x 10" joists at 12" O.C.		130	.062		.92	2.89		3.81	5.45
0352	16" O.C.		135	.059		.93	2.78		3.71	5.30
0402	24" O.C.		140	.057		1.90	2.68		4.58	6.20
0602	For 2" x 14" joists at 16" O.C.		130	.062		1.40	2.89		4.29	6
0902	Compression type, 16" O.C., 2" x 8" joists		200	.040		.93	1.88		2.81	3.91
1002	2" x 12" joists		200	.040		.93	1.88		2.81	3.91

06 11 10.10 Beam and Girder Framing

06 11 10.10 Beam and Girder Framing		Crew	Daily Output	Labor-Hours	Unit	Material	2015 Bare Costs Labor	Equipment	Total	Total Incl O&P
0010	BEAM AND GIRDER FRAMING R061110-30									
3500	Single, 2" x 6"	2 Carp	.70	22.857	M.B.F.	665	1,075		1,740	2,375
3505	Pneumatic nailed		.81	19.704		670	925		1,595	2,175
3520	2" x 8"		.86	18.605		690	875		1,565	2,100
3525	Pneumatic nailed		1	16.048		695	755		1,450	1,925
3540	2" x 10"		1	16		800	750		1,550	2,025
3545	Pneumatic nailed		1.16	13.793		805	650		1,455	1,875
3560	2" x 12"		1.10	14.545		860	685		1,545	2,000
3565	Pneumatic nailed		1.28	12.539		865	590		1,455	1,850
3580	2" x 14"		1.17	13.675		860	640		1,500	1,950
3585	Pneumatic nailed		1.36	11.791		865	555		1,420	1,800
3600	3" x 8"		1.10	14.545		1,275	685		1,960	2,450
3620	3" x 10"		1.25	12.800		1,275	600		1,875	2,350
3640	3" x 12"		1.35	11.852		1,275	555		1,830	2,275
3660	3" x 14"		1.40	11.429		1,275	535		1,810	2,250
3680	4" x 8"	F-3	2.66	15.038		1,075	720	246	2,041	2,575
3700	4" x 10"		3.16	12.658		1,275	605	207	2,087	2,575
3720	4" x 12"		3.60	11.111		1,275	530	182	1,987	2,450
3740	4" x 14"		3.96	10.101		1,275	485	165	1,925	2,350
4000	Double, 2" x 6"	2 Carp	1.25	12.800		665	600		1,265	1,650
4005	Pneumatic nailed		1.45	11.034		670	520		1,190	1,525
4020	2" x 8"		1.60	10		690	470		1,160	1,475
4025	Pneumatic nailed		1.86	8.621		695	405		1,100	1,400
4040	2" x 10"		1.92	8.333		800	390		1,190	1,475
4045	Pneumatic nailed		2.23	7.185		805	335		1,140	1,400
4060	2" x 12"		2.20	7.273		860	340		1,200	1,475
4065	Pneumatic nailed		2.55	6.275		865	295		1,160	1,400
4080	2" x 14"		2.45	6.531		860	305		1,165	1,425
4085	Pneumatic nailed		2.84	5.634		865	265		1,130	1,350
5000	Triple, 2" x 6"		1.65	9.697		665	455		1,120	1,425
5005	Pneumatic nailed		1.91	8.377		670	395		1,065	1,350
5020	2" x 8"		2.10	7.619		690	360		1,050	1,300
5025	Pneumatic nailed		2.44	6.568		695	310		1,005	1,250
5040	2" x 10"		2.50	6.400		800	300		1,100	1,350
5045	Pneumatic nailed		2.90	5.517		805	259		1,064	1,300
5060	2" x 12"		2.85	5.614		860	264		1,124	1,350
5065	Pneumatic nailed		3.31	4.840		865	227		1,092	1,300
5080	2" x 14"		3.15	5.079		860	238		1,098	1,325
5085	Pneumatic nailed		3.35	4.770		865	224		1,089	1,300

06 11 Wood Framing

06 11 10 – Framing with Dimensional, Engineered or Composite Lumber

06 11 10.12 Ceiling Framing

06 11 10.12 Ceiling Framing		Crew	Daily Output	Labor-Hours	Unit	Material	Labor	Equipment	Total	Total Incl O&P
						2015 Bare Costs				
0010	**CEILING FRAMING**									
6400	Suspended, 2" x 3"	2 Carp	.50	32	M.B.F.	785	1,500		2,285	3,175
6450	2" x 4"		.59	27.119		635	1,275		1,910	2,650
6500	2" x 6"		.80	20		665	940		1,605	2,175
6550	2" x 8"		.86	18.605		690	875		1,565	2,100

06 11 10.14 Posts and Columns

06 11 10.14 Posts and Columns		Crew	Daily Output	Labor-Hours	Unit	Material	Labor	Equipment	Total	Total Incl O&P
0010	**POSTS AND COLUMNS**									
0400	4" x 4"	2 Carp	.52	30.769	M.B.F.	1,300	1,450		2,750	3,650
0420	4" x 6"		.55	29.091		1,425	1,375		2,800	3,675
0440	4" x 8"		.59	27.119		1,100	1,275		2,375	3,150
0460	6" x 6"		.65	24.615		1,675	1,150		2,825	3,625
0480	6" x 8"		.70	22.857		1,850	1,075		2,925	3,700
0500	6" x 10"		.75	21.333		1,300	1,000		2,300	2,975

06 11 10.18 Joist Framing

06 11 10.18 Joist Framing		Crew	Daily Output	Labor-Hours	Unit	Material	Labor	Equipment	Total	Total Incl O&P
0010	**JOIST FRAMING** R061110-30									
2650	Joists, 2" x 4"	2 Carp	.83	19.277	M.B.F.	635	905		1,540	2,100
2655	Pneumatic nailed		.96	16.667		640	785		1,425	1,900
2680	2" x 6"		1.25	12.800		665	600		1,265	1,650
2685	Pneumatic nailed		1.44	11.111		670	520		1,190	1,550
2700	2" x 8"		1.46	10.959		690	515		1,205	1,550
2705	Pneumatic nailed		1.68	9.524		695	445		1,140	1,450
2720	2" x 10"		1.49	10.738		800	505		1,305	1,650
2725	Pneumatic nailed		1.71	9.357		805	440		1,245	1,575
2740	2" x 12"		1.75	9.143		860	430		1,290	1,600
2745	Pneumatic nailed		2.01	7.960		865	375		1,240	1,525
2760	2" x 14"		1.79	8.939		860	420		1,280	1,600
2765	Pneumatic nailed		2.06	7.767		865	365		1,230	1,525
2780	3" x 6"		1.39	11.511		1,275	540		1,815	2,225
2790	3" x 8"		1.90	8.421		1,275	395		1,670	2,000
2800	3" x 10"		1.95	8.205		1,275	385		1,660	2,025
2820	3" x 12"		1.80	8.889		1,275	415		1,690	2,075
2840	4" x 6"		1.60	10		1,425	470		1,895	2,275
2860	4" x 10"		2	8		1,275	375		1,650	2,000
2880	4" x 12"		1.80	8.889		1,275	415		1,690	2,075
3000	Composite wood joist 9-1/2" deep		.90	17.778	M.L.F.	1,800	835		2,635	3,250
3010	11-1/2" deep		.88	18.182		1,975	855		2,830	3,500
3020	14" deep		.82	19.512		2,100	915		3,015	3,700
3030	16" deep		.78	20.513		3,475	965		4,440	5,300
4000	Open web joist 12" deep		.88	18.182		3,525	855		4,380	5,200
4002	Per linear foot		880	.018	L.F.	3.52	.85		4.37	5.20
4004	Treated, per linear foot		880	.018	"	4.39	.85		5.24	6.15
4010	14" deep		.82	19.512	M.L.F.	3,725	915		4,640	5,500
4012	Per linear foot		820	.020	L.F.	3.72	.92		4.64	5.50
4014	Treated, per linear foot		820	.020	"	4.74	.92		5.66	6.60
4020	16" deep		.78	20.513	M.L.F.	3,825	965		4,790	5,675
4022	Per linear foot		780	.021	L.F.	3.83	.96		4.79	5.70
4024	Treated, per linear foot		780	.021	"	4.99	.96		5.95	7
4030	18" deep		.74	21.622	M.L.F.	4,000	1,025		5,025	5,950
4032	Per linear foot		740	.022	L.F.	4	1.02		5.02	5.95
4034	Treated, per linear foot		740	.022	"	5.30	1.02		6.32	7.40
6000	Composite rim joist, 1-1/4" x 9-1/2"		.90	17.778	M.L.F.	1,775	835		2,610	3,225
6010	1-1/4" x 11-1/2"		.88	18.182		2,050	855		2,905	3,575

06 11 10 – Framing with Dimensional, Engineered or Composite Lumber

06 11 10.18 Joist Framing		Crew	Daily Output	Labor-Hours	Unit	Material	2015 Bare Costs Labor	Equipment	Total	Total Incl O&P
6020	1-1/4" x 14-1/2"	2 Carp	.82	19.512	M.L.F.	2,750	915		3,665	4,425
6030	1-1/4" x 16-1/2"	↓	.78	20.513	↓	3,375	965		4,340	5,200

06 11 10.24 Miscellaneous Framing		Crew	Daily Output	Labor-Hours	Unit	Material	Labor	Equipment	Total	Total Incl O&P
0010	**MISCELLANEOUS FRAMING**									
8500	Firestops, 2" x 4"	2 Carp	.51	31.373	M.B.F.	635	1,475		2,110	2,975
8505	Pneumatic nailed		.62	25.806		640	1,200		1,840	2,575
8520	2" x 6"		.60	26.667		665	1,250		1,915	2,650
8525	Pneumatic nailed		.73	21.858		670	1,025		1,695	2,325
8540	2" x 8"		.60	26.667		690	1,250		1,940	2,675
8560	2" x 12"		.70	22.857		860	1,075		1,935	2,600
8600	Nailers, treated, wood construction, 2" x 4"		.53	30.189		765	1,425		2,190	3,025
8605	Pneumatic nailed		.64	25.157		770	1,175		1,945	2,675
8620	2" x 6"		.75	21.333		780	1,000		1,780	2,400
8625	Pneumatic nailed		.90	17.778		785	835		1,620	2,150
8640	2" x 8"		.93	17.204		790	810		1,600	2,125
8645	Pneumatic nailed		1.12	14.337		795	675		1,470	1,900
8660	Steel construction, 2" x 4"		.50	32		765	1,500		2,265	3,150
8680	2" x 6"		.70	22.857		780	1,075		1,855	2,500
8700	2" x 8"		.87	18.391		790	865		1,655	2,200
8760	Rough bucks, treated, for doors or windows, 2" x 6"		.40	40		780	1,875		2,655	3,750
8765	Pneumatic nailed		.48	33.333		785	1,575		2,360	3,275
8780	2" x 8"		.51	31.373		790	1,475		2,265	3,150
8785	Pneumatic nailed		.61	26.144		795	1,225		2,020	2,775
8800	Stair stringers, 2" x 10"		.22	72.727		800	3,425		4,225	6,125
8820	2" x 12"		.26	61.538		860	2,900		3,760	5,400
8840	3" x 10"		.31	51.613		1,275	2,425		3,700	5,150
8860	3" x 12"		.38	42.105	↓	1,275	1,975		3,250	4,475
8870	Laminated structural lumber, 1-1/4" x 11-1/2"		130	.123	L.F.	2.05	5.80		7.85	11.15
8880	1-1/4" x 14-1/2"	↓	130	.123	"	2.72	5.80		8.52	11.90

06 11 10.26 Partitions		Crew	Daily Output	Labor-Hours	Unit	Material	Labor	Equipment	Total	Total Incl O&P
0010	**PARTITIONS**									
0020	Single bottom and double top plate, no waste, std. & better lumber									
0180	2" x 4" studs, 8' high, studs 12" O.C.	2 Carp	80	.200	L.F.	4.65	9.40		14.05	19.55
0185	12" O.C., pneumatic nailed		96	.167		4.69	7.85		12.54	17.20
0200	16" O.C.		100	.160		3.81	7.50		11.31	15.75
0205	16" O.C., pneumatic nailed		120	.133		3.84	6.25		10.09	13.85
0300	24" O.C.		125	.128		2.96	6		8.96	12.50
0305	24" O.C., pneumatic nailed		150	.107		2.98	5		7.98	11
0380	10' high, studs 12" O.C.		80	.200		5.50	9.40		14.90	20.50
0385	12" O.C., pneumatic nailed		96	.167		5.55	7.85		13.40	18.15
0400	16" O.C.		100	.160		4.44	7.50		11.94	16.45
0405	16" O.C., pneumatic nailed		120	.133		4.47	6.25		10.72	14.55
0500	24" O.C.		125	.128		3.38	6		9.38	12.95
0505	24" O.C., pneumatic nailed		150	.107		3.41	5		8.41	11.45
0580	12' high, studs 12" O.C.		65	.246		6.35	11.55		17.90	25
0585	12" O.C., pneumatic nailed		78	.205		6.40	9.65		16.05	22
0600	16" O.C.		80	.200		5.10	9.40		14.50	20
0605	16" O.C., pneumatic nailed		96	.167		5.10	7.85		12.95	17.65
0700	24" O.C.		100	.160		3.81	7.50		11.31	15.75
0705	24" O.C., pneumatic nailed		120	.133		3.84	6.25		10.09	13.85
0780	2" x 6" studs, 8' high, studs 12" O.C.		70	.229		7.35	10.75		18.10	24.50
0785	12" O.C., pneumatic nailed	↓	84	.190	↓	7.40	8.95		16.35	22

06 11 10 – Framing with Dimensional, Engineered or Composite Lumber

06 11 10.26 Partitions		Crew	Daily Output	Labor-Hours	Unit	Material	Labor	Equipment	Total	Total Incl O&P
						2015 Bare Costs				
0800	16" O.C.	2 Carp	90	.178	L.F.	6	8.35		14.35	19.45
0805	16" O.C., pneumatic nailed		108	.148		6.05	6.95		13	17.35
0900	24" O.C.		115	.139		4.66	6.55		11.21	15.20
0905	24" O.C., pneumatic nailed		138	.116		4.70	5.45		10.15	13.55
0980	10' high, studs 12" O.C.		70	.229		8.65	10.75		19.40	26
0985	12" O.C., pneumatic nailed		84	.190		8.75	8.95		17.70	23.50
1000	16" O.C.		90	.178		7	8.35		15.35	20.50
1005	16" O.C., pneumatic nailed		108	.148		7.05	6.95		14	18.45
1100	24" O.C.		115	.139		5.35	6.55		11.90	15.90
1105	24" O.C., pneumatic nailed		138	.116		5.35	5.45		10.80	14.30
1180	12' high, studs 12" O.C.		55	.291		10	13.65		23.65	32
1185	12" O.C., pneumatic nailed		66	.242		10.05	11.40		21.45	28.50
1200	16" O.C.		70	.229		8	10.75		18.75	25.50
1205	16" O.C., pneumatic nailed		84	.190		8.05	8.95		17	22.50
1300	24" O.C.		90	.178		6	8.35		14.35	19.45
1305	24" O.C., pneumatic nailed		108	.148		6.05	6.95		13	17.35
1400	For horizontal blocking, 2" x 4", add		600	.027		.42	1.25		1.67	2.40
1500	2" x 6", add		600	.027		.67	1.25		1.92	2.66
1600	For openings, add		250	.064			3		3	4.62
1700	Headers for above openings, material only, add				M.B.F.	695			695	765

06 11 10.28 Porch or Deck Framing		Crew	Daily Output	Labor-Hours	Unit	Material	Labor	Equipment	Total	Total Incl O&P
0010	**PORCH OR DECK FRAMING**									
0100	Treated lumber, posts or columns, 4" x 4"	2 Carp	390	.041	L.F.	1.34	1.93		3.27	4.43
0110	4" x 6"		275	.058		1.91	2.73		4.64	6.30
0120	4" x 8"		220	.073		3.81	3.41		7.22	9.45
0130	Girder, single, 4" x 4"		675	.024		1.34	1.11		2.45	3.18
0140	4" x 6"		600	.027		1.91	1.25		3.16	4.03
0150	4" x 8"		525	.030		3.81	1.43		5.24	6.40
0160	Double, 2" x 4"		625	.026		1.05	1.20		2.25	3
0170	2" x 6"		600	.027		1.60	1.25		2.85	3.69
0180	2" x 8"		575	.028		2.15	1.31		3.46	4.38
0190	2" x 10"		550	.029		2.86	1.37		4.23	5.25
0200	2" x 12"		525	.030		3.95	1.43		5.38	6.55
0210	Triple, 2" x 4"		575	.028		1.57	1.31		2.88	3.73
0220	2" x 6"		550	.029		2.40	1.37		3.77	4.74
0230	2" x 8"		525	.030		3.23	1.43		4.66	5.75
0240	2" x 10"		500	.032		4.29	1.50		5.79	7.05
0250	2" x 12"		475	.034		5.95	1.58		7.53	8.95
0260	Ledger, bolted 4' O.C., 2" x 4"		400	.040		.66	1.88		2.54	3.62
0270	2" x 6"		395	.041		.93	1.90		2.83	3.95
0280	2" x 8"		390	.041		1.20	1.93		3.13	4.28
0290	2" x 10"		385	.042		1.54	1.95		3.49	4.70
0300	2" x 12"		380	.042		2.08	1.98		4.06	5.35
0310	Joists, 2" x 4"		1250	.013		.52	.60		1.12	1.49
0320	2" x 6"		1250	.013		.79	.60		1.39	1.79
0330	2" x 8"		1100	.015		1.07	.68		1.75	2.23
0340	2" x 10"		900	.018		1.42	.83		2.25	2.85
0350	2" x 12"		875	.018		1.71	.86		2.57	3.20
0360	Railings and trim , 1" x 4"	1 Carp	300	.027		.47	1.25		1.72	2.45
0370	2" x 2"		300	.027		.33	1.25		1.58	2.30
0380	2" x 4"		300	.027		.51	1.25		1.76	2.49
0390	2" x 6"		300	.027		.79	1.25		2.04	2.79

06 11 10 – Framing with Dimensional, Engineered or Composite Lumber

06 11 10.28 Porch or Deck Framing		Crew	Daily Output	Labor-Hours	Unit	Material	2015 Bare Costs Labor	Equipment	Total	Total Incl O&P
0400	Decking, 1" x 4"	1 Carp	275	.029	S.F.	2.16	1.37		3.53	4.48
0410	2" x 4"		300	.027		1.73	1.25		2.98	3.83
0420	2" x 6"		320	.025		1.69	1.17		2.86	3.67
0430	5/4" x 6"		320	.025		2.25	1.17		3.42	4.29
0440	Balusters, square, 2" x 2"	2 Carp	660	.024	L.F.	.34	1.14		1.48	2.12
0450	Turned, 2" x 2"		420	.038		.45	1.79		2.24	3.24
0460	Stair stringer, 2" x 10"		130	.123		1.42	5.80		7.22	10.45
0470	2" x 12"		130	.123		1.71	5.80		7.51	10.80
0480	Stair treads, 1" x 4"		140	.114		2.17	5.35		7.52	10.65
0490	2" x 4"		140	.114		.52	5.35		5.87	8.80
0500	2" x 6"		160	.100		.78	4.70		5.48	8.10
0510	5/4" x 6"		160	.100		1.05	4.70		5.75	8.40
0520	Turned handrail post, 4" x 4"		64	.250	Ea.	31	11.75		42.75	52
0530	Lattice panel, 4' x 8', 1/2"		1600	.010	S.F.	.81	.47		1.28	1.61
0535	3/4"		1600	.010	"	1.21	.47		1.68	2.05
0540	Cedar, posts or columns, 4" x 4"		390	.041	L.F.	3.93	1.93		5.86	7.30
0550	4" x 6"		275	.058		7.35	2.73		10.08	12.30
0560	4" x 8"		220	.073		10	3.41		13.41	16.25
0800	Decking, 1" x 4"		550	.029		1.81	1.37		3.18	4.09
0810	2" x 4"		600	.027		3.62	1.25		4.87	5.90
0820	2" x 6"		640	.025		6.60	1.17		7.77	9.05
0830	5/4" x 6"		640	.025		4.45	1.17		5.62	6.70
0840	Railings and trim, 1" x 4"		600	.027		1.81	1.25		3.06	3.92
0860	2" x 4"		600	.027		3.62	1.25		4.87	5.90
0870	2" x 6"		600	.027		6.60	1.25		7.85	9.20
0920	Stair treads, 1" x 4"		140	.114		1.81	5.35		7.16	10.25
0930	2" x 4"		140	.114		3.62	5.35		8.97	12.25
0940	2" x 6"		160	.100		6.60	4.70		11.30	14.50
0950	5/4" x 6"		160	.100		4.45	4.70		9.15	12.15
0980	Redwood, posts or columns, 4" x 4"		390	.041		6.45	1.93		8.38	10
0990	4" x 6"		275	.058		12.60	2.73		15.33	18.05
1000	4" x 8"		220	.073		23.50	3.41		26.91	31.50
1240	Decking, 1" x 4"	1 Carp	275	.029	S.F.	3.97	1.37		5.34	6.45
1260	2" x 6"		340	.024		7.50	1.10		8.60	9.95
1270	5/4" x 6"		320	.025		4.77	1.17		5.94	7.05
1280	Railings and trim, 1" x 4"	2 Carp	600	.027	L.F.	1.17	1.25		2.42	3.22
1310	2" x 6"		600	.027		7.50	1.25		8.75	10.20
1420	Alternative decking, wood/plastic composite, 5/4" x 6" G		640	.025		3.14	1.17		4.31	5.25
1440	1" x 4" square edge fir		550	.029		2.18	1.37		3.55	4.50
1450	1" x 4" tongue and groove fir		450	.036		1.46	1.67		3.13	4.17
1460	1" x 4" mahogany		550	.029		2.01	1.37		3.38	4.31
1462	5/4" x 6" PVC		550	.029		3.73	1.37		5.10	6.20
1465	Framing, porch or deck, alt deck fastening, screws, add	1 Carp	240	.033	S.F.		1.56		1.56	2.41
1470	Accessories, joist hangers, 2" x 4"		160	.050	Ea.	.73	2.35		3.08	4.41
1480	2" x 6" through 2" x 12"		150	.053		1.39	2.50		3.89	5.40
1530	Post footing, incl excav, backfill, tube form & concrete, 4' deep, 8" dia	F-7	12	2.667		12.15	113		125.15	186
1540	10" diameter		11	2.909		17.75	123		140.75	209
1550	12" diameter		10	3.200		23.50	135		158.50	234

06 11 10.30 Roof Framing		Crew	Daily Output	Labor-Hours	Unit	Material	Labor	Equipment	Total	Total Incl O&P
0010	**ROOF FRAMING**									
5250	Composite rafter, 9-1/2" deep	2 Carp	575	.028	L.F.	1.79	1.31		3.10	3.98
5260	11-1/2" deep		575	.028	"	1.97	1.31		3.28	4.18

06 11 10 – Framing with Dimensional, Engineered or Composite Lumber

06 11 10.30 Roof Framing		Crew	Daily Output	Labor-Hours	Unit	Material	2015 Bare Costs Labor	Equipment	Total	Total Incl O&P
6070	Fascia boards, 2″ x 8″	2 Carp	.30	53.333	M.B.F.	690	2,500		3,190	4,600
6080	2″ x 10″		.30	53.333		800	2,500		3,300	4,725
7000	Rafters, to 4 in 12 pitch, 2″ x 6″		1	16		665	750		1,415	1,875
7060	2″ x 8″		1.26	12.698		690	595		1,285	1,675
7300	Hip and valley rafters, 2″ x 6″		.76	21.053		665	990		1,655	2,250
7360	2″ x 8″		.96	16.667		690	785		1,475	1,950
7540	Hip and valley jacks, 2″ x 6″		.60	26.667		665	1,250		1,915	2,650
7600	2″ x 8″		.65	24.615		690	1,150		1,840	2,525
7780	For slopes steeper than 4 in 12, add						30%			
7790	For dormers or complex roofs, add						50%			
7800	Rafter tie, 1″ x 4″, #3	2 Carp	.27	59.259	M.B.F.	1,400	2,775		4,175	5,825
7820	Ridge board, #2 or better, 1″ x 6″		.30	53.333		1,450	2,500		3,950	5,450
7840	1″ x 8″		.37	43.243		1,800	2,025		3,825	5,100
7860	1″ x 10″		.42	38.095		1,875	1,800		3,675	4,800
7880	2″ x 6″		.50	32		665	1,500		2,165	3,025
7900	2″ x 8″		.60	26.667		690	1,250		1,940	2,675
7920	2″ x 10″		.66	24.242		800	1,150		1,950	2,625
7940	Roof cants, split, 4″ x 4″		.86	18.605		1,300	875		2,175	2,775
7960	6″ x 6″		1.80	8.889		1,675	415		2,090	2,500
7980	Roof curbs, untreated, 2″ x 6″		.52	30.769		665	1,450		2,115	2,950
8000	2″ x 12″		.80	20		860	940		1,800	2,400

06 11 10.32 Sill and Ledger Framing		Crew	Daily Output	Labor-Hours	Unit	Material	Labor	Equipment	Total	Total Incl O&P
0010	**SILL AND LEDGER FRAMING**									
4482	Ledgers, nailed, 2″ x 4″	2 Carp	.50	32	M.B.F.	635	1,500		2,135	3,000
4484	2″ x 6″		.60	26.667		665	1,250		1,915	2,650
4486	Bolted, not including bolts, 3″ x 8″		.65	24.615		1,275	1,150		2,425	3,175
4488	3″ x 12″		.70	22.857		1,275	1,075		2,350	3,050
4490	Mud sills, redwood, construction grade, 2″ x 4″		.59	27.119		3,375	1,275		4,650	5,650
4492	2″ x 6″		.78	20.513		3,400	965		4,365	5,225
4500	Sills, 2″ x 4″		.40	40		625	1,875		2,500	3,575
4520	2″ x 6″		.55	29.091		655	1,375		2,030	2,825
4540	2″ x 8″		.67	23.881		680	1,125		1,805	2,475
4600	Treated, 2″ x 4″		.36	44.444		755	2,075		2,830	4,025
4620	2″ x 6″		.50	32		770	1,500		2,270	3,150
4640	2″ x 8″		.60	26.667		780	1,250		2,030	2,775
4700	4″ x 4″		.60	26.667		975	1,250		2,225	3,000
4720	4″ x 6″		.70	22.857		925	1,075		2,000	2,675
4740	4″ x 8″		.80	20		1,400	940		2,340	3,000
4760	4″ x 10″		.87	18.391		1,400	865		2,265	2,875

06 11 10.34 Sleepers		Crew	Daily Output	Labor-Hours	Unit	Material	Labor	Equipment	Total	Total Incl O&P
0010	**SLEEPERS**									
0300	On concrete, treated, 1″ x 2″	2 Carp	.39	41.026	M.B.F.	1,625	1,925		3,550	4,750
0320	1″ x 3″		.50	32		1,800	1,500		3,300	4,300
0340	2″ x 4″		.99	16.162		890	760		1,650	2,150
0360	2″ x 6″		1.30	12.308		905	580		1,485	1,875

06 11 10.36 Soffit and Canopy Framing		Crew	Daily Output	Labor-Hours	Unit	Material	Labor	Equipment	Total	Total Incl O&P
0010	**SOFFIT AND CANOPY FRAMING**									
1300	Canopy or soffit framing, 1″ x 4″	2 Carp	.30	53.333	M.B.F.	1,400	2,500		3,900	5,400
1340	1″ x 8″		.50	32		1,800	1,500		3,300	4,275
1360	2″ x 4″		.41	39.024		635	1,825		2,460	3,525
1400	2″ x 8″		.67	23.881		690	1,125		1,815	2,475
1420	3″ x 4″		.50	32		1,075	1,500		2,575	3,475

06 11 10 – Framing with Dimensional, Engineered or Composite Lumber

06 11 10.36 Soffit and Canopy Framing		Crew	Daily Output	Labor-Hours	Unit	Material	Labor	Equipment	Total	Total Incl O&P
1460	3" x 8"	2 Carp	.60	26.667	M.B.F.	1,275	1,250		2,525	3,325

Column group "2015 Bare Costs" spans Material, Labor, Equipment, Total.

06 11 10.38 Treated Lumber Framing Material

		Crew	Daily Output	Labor-Hours	Unit	Material	Labor	Equipment	Total	Total Incl O&P
0010	**TREATED LUMBER FRAMING MATERIAL**									
0100	2" x 4"				M.B.F.	755			755	830
0110	2" x 6"					770			770	845
0120	2" x 8"					780			780	855
0130	2" x 10"					830			830	910
0140	2" x 12"					960			960	1,050
0200	4" x 4"					975			975	1,075
0210	4" x 6"					925			925	1,025
0220	4" x 8"					1,400			1,400	1,550

06 11 10.40 Wall Framing

		Crew	Daily Output	Labor-Hours	Unit	Material	Labor	Equipment	Total	Total Incl O&P
0010	**WALL FRAMING** R061110-30									
5860	Headers over openings, 2" x 6"	2 Carp	.36	44.444	M.B.F.	665	2,075		2,740	3,925
5865	2" x 6", pneumatic nailed		.43	37.209		670	1,750		2,420	3,450
5880	2" x 8"		.45	35.556		690	1,675		2,365	3,325
5885	2" x 8", pneumatic nailed		.54	29.630		695	1,400		2,095	2,925
5900	2" x 10"		.53	30.189		800	1,425		2,225	3,050
5905	2" x 10", pneumatic nailed		.67	23.881		805	1,125		1,930	2,625
5920	2" x 12"		.60	26.667		860	1,250		2,110	2,875
5925	2" x 12", pneumatic nailed		.72	22.222		865	1,050		1,915	2,550
5940	4" x 12"		.76	21.053		1,275	990		2,265	2,950
5945	4" x 12", pneumatic nailed		.92	17.391		1,300	815		2,115	2,675
5960	6" x 12"		.84	19.048		1,375	895		2,270	2,875
5965	6" x 12", pneumatic nailed		1.01	15.873		1,375	745		2,120	2,650
6000	Plates, untreated, 2" x 3"		.43	37.209		785	1,750		2,535	3,575
6005	2" x 3", pneumatic nailed		.52	30.769		790	1,450		2,240	3,100
6020	2" x 4"		.53	30.189		635	1,425		2,060	2,875
6025	2" x 4", pneumatic nailed		.67	23.881		640	1,125		1,765	2,425
6040	2" x 6"		.75	21.333		665	1,000		1,665	2,275
6045	2" x 6", pneumatic nailed		.90	17.778		670	835		1,505	2,025
6120	Studs, 8' high wall, 2" x 3"		.60	26.667		785	1,250		2,035	2,800
6125	2" x 3", pneumatic nailed		.72	22.222		790	1,050		1,840	2,475
6140	2" x 4"		.92	17.391		635	815		1,450	1,950
6145	2" x 4", pneumatic nailed		1.10	14.493		640	680		1,320	1,750
6160	2" x 6"		1	16		665	750		1,415	1,875
6165	2" x 6", pneumatic nailed		1.20	13.333		670	625		1,295	1,700
6180	3" x 4"		.80	20		1,075	940		2,015	2,625
6185	3" x 4", pneumatic nailed		.96	16.667		1,075	785		1,860	2,375
8200	For 12' high walls, deduct						5%			
8220	For stub wall, 6' high, add						20%			
8240	3' high, add						40%			
8250	For second story & above, add						5%			
8300	For dormer & gable, add						15%			

06 11 10.42 Furring

		Crew	Daily Output	Labor-Hours	Unit	Material	Labor	Equipment	Total	Total Incl O&P
0010	**FURRING**									
0012	Wood strips, 1" x 2", on walls, on wood	1 Carp	550	.015	L.F.	.24	.68		.92	1.31
0015	On wood, pneumatic nailed		710	.011		.24	.53		.77	1.07
0300	On masonry		495	.016		.26	.76		1.02	1.46
0400	On concrete		260	.031		.26	1.44		1.70	2.51
0600	1" x 3", on walls, on wood		550	.015		.39	.68		1.07	1.47
0605	On wood, pneumatic nailed		710	.011		.39	.53		.92	1.23

06 11 Wood Framing

06 11 10 – Framing with Dimensional, Engineered or Composite Lumber

06 11 10.42 Furring		Crew	Daily Output	Labor-Hours	Unit	Material	Labor (2015 Bare Costs)	Equipment	Total	Total Incl O&P
0700	On masonry	1 Carp	495	.016	L.F.	.42	.76		1.18	1.63
0800	On concrete		260	.031		.42	1.44		1.86	2.68
0850	On ceilings, on wood		350	.023		.39	1.07		1.46	2.07
0855	On wood, pneumatic nailed		450	.018		.39	.83		1.22	1.70
0900	On masonry		320	.025		.42	1.17		1.59	2.27
0950	On concrete		210	.038		.42	1.79		2.21	3.21

06 11 10.44 Grounds		Crew	Daily Output	Labor-Hours	Unit	Material	Labor	Equipment	Total	Total Incl O&P
0010	**GROUNDS**									
0020	For casework, 1" x 2" wood strips, on wood	1 Carp	330	.024	L.F.	.24	1.14		1.38	2.01
0100	On masonry		285	.028		.26	1.32		1.58	2.32
0200	On concrete		250	.032		.26	1.50		1.76	2.60
0400	For plaster, 3/4" deep, on wood		450	.018		.24	.83		1.07	1.54
0500	On masonry		225	.036		.26	1.67		1.93	2.86
0600	On concrete		175	.046		.26	2.15		2.41	3.59
0700	On metal lath		200	.040		.26	1.88		2.14	3.18

06 12 Structural Panels

06 12 10 – Structural Insulated Panels

06 12 10.10 OSB Faced Panels			Crew	Daily Output	Labor-Hours	Unit	Material	Labor	Equipment	Total	Total Incl O&P
0010	**OSB FACED PANELS**										
0100	Structural insul. panels, 7/16" OSB both faces, EPS insul, 3-5/8" T	G	F-3	2075	.019	S.F.	3.60	.92	.32	4.84	5.75
0110	5-5/8" thick	G		1725	.023		4.05	1.11	.38	5.54	6.60
0120	7-3/8" thick	G		1425	.028		4.40	1.34	.46	6.20	7.40
0130	9-3/8" thick	G		1125	.036		4.70	1.70	.58	6.98	8.40
0140	7/16" OSB one face, EPS insul, 3-5/8" thick	G		2175	.018		3.70	.88	.30	4.88	5.75
0150	5-5/8" thick	G		1825	.022		4.30	1.05	.36	5.71	6.75
0160	7-3/8" thick	G		1525	.026		4.80	1.26	.43	6.49	7.70
0170	9-3/8" thick	G		1225	.033		5.30	1.56	.53	7.39	8.85
0190	7/16" OSB - 1/2" GWB faces , EPS insul, 3-5/8" T	G		2075	.019		3.29	.92	.32	4.53	5.40
0200	5-5/8" thick	G		1725	.023		3.90	1.11	.38	5.39	6.40
0210	7-3/8" thick	G		1425	.028		4.42	1.34	.46	6.22	7.40
0220	9-3/8" thick	G		1125	.036		5	1.70	.58	7.28	8.75
0240	7/16" OSB - 1/2" MRGWB faces , EPS insul, 3-5/8" T	G		2075	.019		3.39	.92	.32	4.63	5.50
0250	5-5/8" thick	G		1725	.023		4	1.11	.38	5.49	6.50
0260	7-3/8" thick	G		1425	.028		4.52	1.34	.46	6.32	7.55
0270	9-3/8" thick	G		1125	.036		5.10	1.70	.58	7.38	8.90
0300	For 1/2" GWB added to OSB skin, add	G					1.28			1.28	1.41
0310	For 1/2" MRGWB added to OSB skin, add	G					1.28			1.28	1.41
0320	For one T1-11 skin, add to OSB-OSB	G					1.90			1.90	2.09
0330	For one 19/32" CDX skin, add to OSB-OSB	G					1.46			1.46	1.61
0500	Structural insulated panel, 7/16" OSB both sides, straw core										
0510	4-3/8" T, walls (w/sill, splines, plates)	G	F-6	2400	.017	S.F.	7.40	.74	.27	8.41	9.55
0520	Floors (w/splines)	G		2400	.017		7.40	.74	.27	8.41	9.55
0530	Roof (w/splines)	G		2400	.017		7.40	.74	.27	8.41	9.55
0550	7-7/8" T, walls (w/sill, splines, plates)	G		2400	.017		11.20	.74	.27	12.21	13.75
0560	Floors (w/splines)	G		2400	.017		11.20	.74	.27	12.21	13.75
0570	Roof (w/splines)	G		2400	.017		11.20	.74	.27	12.21	13.75

06 12 Structural Panels

06 12 19 – Composite Shearwall Panels

06 12 19.10 Steel and Wood Composite Shearwall Panels		Crew	Daily Output	Labor-Hours	Unit	Material	2015 Bare Costs Labor	Equipment	Total	Total Incl O&P
0010	**STEEL & WOOD COMPOSITE SHEARWALL PANELS**									
0020	Anchor bolts, 36″ long (must be placed in wet concrete)	1 Carp	150	.053	Ea.	32	2.50		34.50	39.50
0030	On concrete, 2 x 4 & 2 x 6 walls, 7′ - 10′ high, 360 lb. shear, 12″ wide	2 Carp	8	2		420	94		514	605
0040	715 lb. shear, 15″ wide		8	2		500	94		594	695
0050	1860 lb. shear, 18″ wide		8	2		520	94		614	715
0060	2780 lb. shear, 21″ wide		8	2		620	94		714	825
0070	3790 lb. shear, 24″ wide		8	2		710	94		804	925
0080	2 x 6 walls, 11′ to 13′ high, 1180 lb. shear, 18″ wide		6	2.667		635	125		760	895
0090	1555 lb. shear, 21″ wide		6	2.667		800	125		925	1,075
0100	2280 lb. shear, 24″ wide		6	2.667		895	125		1,020	1,175
0110	For installing above on wood floor frame, add									
0120	Coupler nuts, threaded rods, bolts, shear transfer plate kit	1 Carp	16	.500	Ea.	63	23.50		86.50	106
0130	Framing anchors, angle (2 required)	″	96	.083	″	2.61	3.91		6.52	8.85
0140	For blocking see Section 06 11 10.02									
0150	For installing above, first floor to second floor, wood floor frame, add									
0160	Add stack option to first floor wall panel				Ea.	69.50			69.50	76
0170	Threaded rods, bolts, shear transfer plate kit	1 Carp	16	.500		76	23.50		99.50	120
0180	Framing anchors, angle (2 required)	″	96	.083		2.61	3.91		6.52	8.85
0190	For blocking see Section 06 11 10.02									
0200	For installing stacked panels, balloon framing									
0210	Add stack option to first floor wall panel				Ea.	69.50			69.50	76
0220	Threaded rods, bolts kit	1 Carp	16	.500	″	44	23.50		67.50	84.50

06 13 Heavy Timber Construction

06 13 23 – Heavy Timber Framing

06 13 23.10 Heavy Framing

		Crew	Daily Output	Labor-Hours	Unit	Material	Labor	Equipment	Total	Total Incl O&P
0010	**HEAVY FRAMING**									
0020	Beams, single 6″ x 10″	2 Carp	1.10	14.545	M.B.F.	1,600	685		2,285	2,800
0100	Single 8″ x 16″		1.20	13.333		2,000	625		2,625	3,175
0200	Built from 2″ lumber, multiple 2″ x 14″		.90	17.778		850	835		1,685	2,200
0210	Built from 3″ lumber, multiple 3″ x 6″		.70	22.857		1,250	1,075		2,325	3,025
0220	Multiple 3″ x 8″		.80	20		1,275	940		2,215	2,850
0230	Multiple 3″ x 10″		.90	17.778		1,275	835		2,110	2,675
0240	Multiple 3″ x 12″		1	16		1,275	750		2,025	2,550
0250	Built from 4″ lumber, multiple 4″ x 6″		.80	20		1,400	940		2,340	3,000
0260	Multiple 4″ x 8″		.90	17.778		1,075	835		1,910	2,450
0270	Multiple 4″ x 10″		1	16		1,275	750		2,025	2,550
0280	Multiple 4″ x 12″		1.10	14.545		1,275	685		1,960	2,450
0290	Columns, structural grade, 1500f, 4″ x 4″		.60	26.667		1,275	1,250		2,525	3,325
0300	6″ x 6″		.65	24.615		1,400	1,150		2,550	3,325
0400	8″ x 8″		.70	22.857		1,350	1,075		2,425	3,125
0500	10″ x 10″		.75	21.333		1,450	1,000		2,450	3,150
0600	12″ x 12″		.80	20		1,525	940		2,465	3,125
0800	Floor planks, 2″ thick, T & G, 2″ x 6″		1.05	15.238		1,600	715		2,315	2,850
0900	2″ x 10″		1.10	14.545		1,600	685		2,285	2,800
1100	3″ thick, 3″ x 6″		1.05	15.238		1,600	715		2,315	2,850
1200	3″ x 10″		1.10	14.545		1,600	685		2,285	2,800
1400	Girders, structural grade, 12″ x 12″		.80	20		1,525	940		2,465	3,125
1500	10″ x 16″		1	16		2,400	750		3,150	3,800
2300	Roof purlins, 4″ thick, structural grade		1.05	15.238		1,075	715		1,790	2,275

06 15 Wood Decking

06 15 16 – Wood Roof Decking

06 15 16.10 Solid Wood Roof Decking		Crew	Daily Output	Labor-Hours	Unit	Material	Labor	Equipment	Total	Total Incl O&P
						2015 Bare Costs				
0010	**SOLID WOOD ROOF DECKING**									
0350	Cedar planks, 2" thick	2 Carp	350	.046	S.F.	6.25	2.15		8.40	10.15
0400	3" thick		320	.050		9.35	2.35		11.70	13.90
0500	4" thick		250	.064		12.45	3		15.45	18.30
0550	6" thick		200	.080		18.70	3.76		22.46	26.50
0650	Douglas fir, 2" thick		350	.046		2.69	2.15		4.84	6.25
0700	3" thick		320	.050		4.04	2.35		6.39	8.05
0800	4" thick		250	.064		5.40	3		8.40	10.50
0850	6" thick		200	.080		8.05	3.76		11.81	14.70
0950	Hemlock, 2" thick		350	.046		2.74	2.15		4.89	6.30
1000	3" thick		320	.050		4.11	2.35		6.46	8.15
1100	4" thick		250	.064		5.50	3		8.50	10.60
1150	6" thick		200	.080		8.20	3.76		11.96	14.85
1250	Western white spruce, 2" thick		350	.046		1.75	2.15		3.90	5.20
1300	3" thick		320	.050		2.62	2.35		4.97	6.50
1400	4" thick		250	.064		3.50	3		6.50	8.45
1450	6" thick		200	.080		5.25	3.76		9.01	11.55

06 15 23 – Laminated Wood Decking

06 15 23.10 Laminated Roof Deck		Crew	Daily Output	Labor-Hours	Unit	Material	Labor	Equipment	Total	Total Incl O&P
0010	**LAMINATED ROOF DECK**									
0020	Pine or hemlock, 3" thick	2 Carp	425	.038	S.F.	6.20	1.77		7.97	9.50
0100	4" thick		325	.049		8.10	2.31		10.41	12.45
0300	Cedar, 3" thick		425	.038		7	1.77		8.77	10.40
0400	4" thick		325	.049		9.40	2.31		11.71	13.90
0600	Fir, 3" thick		425	.038		5.40	1.77		7.17	8.65
0700	4" thick		325	.049		7.40	2.31		9.71	11.70

06 16 Sheathing

06 16 13 – Insulating Sheathing

06 16 13.10 Insulating Sheathing			Crew	Daily Output	Labor-Hours	Unit	Material	Labor	Equipment	Total	Total Incl O&P
0010	**INSULATING SHEATHING**										
0020	Expanded polystyrene, 1#/C.F. density, 3/4" thick R2.89	G	2 Carp	1400	.011	S.F.	.33	.54		.87	1.19
0030	1" thick R3.85	G		1300	.012		.39	.58		.97	1.32
0040	2" thick R7.69	G		1200	.013		.64	.63		1.27	1.67
0050	Extruded polystyrene, 15 PSI compressive strength, 1" thick, R5	G		1300	.012		.69	.58		1.27	1.65
0060	2" thick, R10	G		1200	.013		.86	.63		1.49	1.90
0070	Polyisocyanurate, 2#/C.F. density, 3/4" thick	G		1400	.011		.60	.54		1.14	1.49
0080	1" thick	G		1300	.012		.62	.58		1.20	1.58
0090	1-1/2" thick	G		1250	.013		.78	.60		1.38	1.78
0100	2" thick	G		1200	.013		.97	.63		1.60	2.03

06 16 23 – Subflooring

06 16 23.10 Subfloor			Crew	Daily Output	Labor-Hours	Unit	Material	Labor	Equipment	Total	Total Incl O&P
0010	**SUBFLOOR**	R061636-20									
0011	Plywood, CDX, 1/2" thick		2 Carp	1500	.011	SF Flr.	.66	.50		1.16	1.49
0015	Pneumatic nailed			1860	.009		.66	.40		1.06	1.34
0100	5/8" thick			1350	.012		.78	.56		1.34	1.72
0105	Pneumatic nailed			1674	.010		.78	.45		1.23	1.55
0200	3/4" thick			1250	.013		.93	.60		1.53	1.94
0205	Pneumatic nailed			1550	.010		.93	.48		1.41	1.77
0300	1-1/8" thick, 2-4-1 including underlayment			1050	.015		2.03	.72		2.75	3.33

06 16 Sheathing

06 16 23 – Subflooring

06 16 23.10 Subfloor		Crew	Daily Output	Labor-Hours	Unit	Material	2015 Bare Costs Labor	Equipment	Total	Total Incl O&P
0440	With boards, 1" x 6", S4S, laid regular	2 Carp	900	.018	SF Flr.	1.57	.83		2.40	3.01
0450	1" x 8", laid regular		1000	.016		1.90	.75		2.65	3.25
0460	Laid diagonal		850	.019		1.90	.88		2.78	3.45
0500	1" x 10", laid regular		1100	.015		1.95	.68		2.63	3.19
0600	Laid diagonal		900	.018		1.95	.83		2.78	3.42
8990	Subfloor adhesive, 3/8" bead	1 Carp	2300	.003	L.F.	.12	.16		.28	.38

06 16 26 – Underlayment

06 16 26.10 Wood Product Underlayment

Line	Description	Crew	Daily Output	Labor-Hours	Unit	Material	Labor	Equipment	Total	Total Incl O&P
0010	**WOOD PRODUCT UNDERLAYMENT** R061636-20									
0015	Plywood, underlayment grade, 1/4" thick	2 Carp	1500	.011	S.F.	.83	.50		1.33	1.68
0018	Pneumatic nailed		1860	.009		.83	.40		1.23	1.53
0030	3/8" thick		1500	.011		.92	.50		1.42	1.78
0070	Pneumatic nailed		1860	.009		.92	.40		1.32	1.63
0100	1/2" thick		1450	.011		1.10	.52		1.62	2.01
0105	Pneumatic nailed		1798	.009		1.10	.42		1.52	1.85
0200	5/8" thick		1400	.011		1.20	.54		1.74	2.15
0205	Pneumatic nailed		1736	.009		1.20	.43		1.63	1.99
0300	3/4" thick		1300	.012		1.43	.58		2.01	2.46
0305	Pneumatic nailed		1612	.010		1.43	.47		1.90	2.29
0500	Particle board, 3/8" thick G		1500	.011		.40	.50		.90	1.21
0505	Pneumatic nailed G		1860	.009		.40	.40		.80	1.06
0600	1/2" thick G		1450	.011		.43	.52		.95	1.27
0605	Pneumatic nailed G		1798	.009		.43	.42		.85	1.11
0800	5/8" thick G		1400	.011		.52	.54		1.06	1.40
0805	Pneumatic nailed G		1736	.009		.52	.43		.95	1.24
0900	3/4" thick G		1300	.012		.63	.58		1.21	1.58
0905	Pneumatic nailed G		1612	.010		.63	.47		1.10	1.41
1100	Hardboard, underlayment grade, 4' x 4', .215" thick G		1500	.011		.58	.50		1.08	1.41

06 16 33 – Wood Board Sheathing

06 16 33.10 Board Sheathing

Line	Description	Crew	Daily Output	Labor-Hours	Unit	Material	Labor	Equipment	Total	Total Incl O&P
0009	**BOARD SHEATHING**									
0010	Roof, 1" x 6" boards, laid horizontal	2 Carp	725	.022	S.F.	1.57	1.04		2.61	3.32
0020	On steep roof		520	.031		1.57	1.44		3.01	3.95
0040	On dormers, hips, & valleys		480	.033		1.57	1.56		3.13	4.14
0050	Laid diagonal		650	.025		1.57	1.16		2.73	3.51
0070	1" x 8" boards, laid horizontal		875	.018		1.90	.86		2.76	3.41
0080	On steep roof		635	.025		1.90	1.18		3.08	3.91
0090	On dormers, hips, & valleys		580	.028		1.95	1.30		3.25	4.13
0100	Laid diagonal		725	.022		1.90	1.04		2.94	3.68
0110	Skip sheathing, 1" x 4", 7" OC	1 Carp	1200	.007		.58	.31		.89	1.12
0120	1" x 6", 9" OC		1450	.006		.72	.26		.98	1.19
0180	Tongue and groove sheathing/decking, 1" x 6"		1000	.008		1.80	.38		2.18	2.56
0190	2" x 6"		1000	.008		3.81	.38		4.19	4.77
0200	Walls, 1" x 6" boards, laid regular	2 Carp	650	.025		1.57	1.16		2.73	3.51
0210	Laid diagonal		585	.027		1.57	1.28		2.85	3.71
0220	1" x 8" boards, laid regular		765	.021		1.90	.98		2.88	3.60
0230	Laid diagonal		650	.025		1.90	1.16		3.06	3.87

06 16 36 – Wood Panel Product Sheathing

06 16 36.10 Sheathing		Crew	Daily Output	Labor-Hours	Unit	Material	Labor	Equipment	Total	Total Incl O&P
0010	**SHEATHING** R061636-20									
0012	Plywood on roofs, CDX									
0030	5/16" thick	2 Carp	1600	.010	S.F.	.56	.47		1.03	1.33
0035	Pneumatic nailed R061110-30		1952	.008		.56	.39		.95	1.20
0050	3/8" thick		1525	.010		.60	.49		1.09	1.42
0055	Pneumatic nailed		1860	.009		.60	.40		1	1.28
0100	1/2" thick		1400	.011		.66	.54		1.20	1.55
0105	Pneumatic nailed		1708	.009		.66	.44		1.10	1.40
0200	5/8" thick		1300	.012		.78	.58		1.36	1.75
0205	Pneumatic nailed		1586	.010		.78	.47		1.25	1.59
0300	3/4" thick		1200	.013		.93	.63		1.56	1.98
0305	Pneumatic nailed		1464	.011		.93	.51		1.44	1.81
0500	Plywood on walls, with exterior CDX, 3/8" thick		1200	.013		.60	.63		1.23	1.62
0505	Pneumatic nailed		1488	.011		.60	.50		1.10	1.44
0600	1/2" thick		1125	.014		.66	.67		1.33	1.75
0605	Pneumatic nailed		1395	.011		.66	.54		1.20	1.55
0700	5/8" thick		1050	.015		.78	.72		1.50	1.96
0705	Pneumatic nailed		1302	.012		.78	.58		1.36	1.75
0800	3/4" thick		975	.016		.93	.77		1.70	2.21
0805	Pneumatic nailed		1209	.013		.93	.62		1.55	1.98
1000	For shear wall construction, add						20%			
1200	For structural 1 exterior plywood, add				S.F.	10%				
3000	Wood fiber, regular, no vapor barrier, 1/2" thick	2 Carp	1200	.013		.61	.63		1.24	1.63
3100	5/8" thick		1200	.013		.77	.63		1.40	1.81
3300	No vapor barrier, in colors, 1/2" thick		1200	.013		.74	.63		1.37	1.77
3400	5/8" thick		1200	.013		.78	.63		1.41	1.82
3600	With vapor barrier one side, white, 1/2" thick		1200	.013		.60	.63		1.23	1.62
3700	Vapor barrier 2 sides, 1/2" thick		1200	.013		.83	.63		1.46	1.87
3800	Asphalt impregnated, 25/32" thick		1200	.013		.33	.63		.96	1.32
3850	Intermediate, 1/2" thick		1200	.013		.25	.63		.88	1.24
4500	Oriented strand board, on roof, 7/16" thick G		1460	.011		.50	.51		1.01	1.34
4505	Pneumatic nailed G		1780	.009		.50	.42		.92	1.20
4550	1/2" thick G		1400	.011		.50	.54		1.04	1.38
4555	Pneumatic nailed G		1736	.009		.50	.43		.93	1.22
4600	5/8" thick G		1300	.012		.69	.58		1.27	1.65
4605	Pneumatic nailed G		1586	.010		.69	.47		1.16	1.49
4610	On walls, 7/16" thick		1200	.013		.50	.63		1.13	1.51
4615	Pneumatic nailed		1488	.011		.50	.50		1	1.33
4620	1/2" thick		1195	.013		.50	.63		1.13	1.52
4625	Pneumatic nailed		1325	.012		.50	.57		1.07	1.42
4630	5/8" thick		1050	.015		.69	.72		1.41	1.86
4635	Pneumatic nailed		1302	.012		.69	.58		1.27	1.65
4700	Oriented strand board, factory laminated W.R. barrier, on roof, 1/2" thick G		1400	.011		.78	.54		1.32	1.69
4705	Pneumatic nailed G		1736	.009		.78	.43		1.21	1.53
4720	5/8" thick G		1300	.012		.93	.58		1.51	1.91
4725	Pneumatic nailed G		1586	.010		.93	.47		1.40	1.75
4730	5/8" thick, T&G G		1150	.014		.93	.65		1.58	2.03
4735	Pneumatic nailed, T&G G		1400	.011		.93	.54		1.47	1.85
4740	On walls, 7/16" thick G		1200	.013		.66	.63		1.29	1.69
4745	Pneumatic nailed G		1488	.011		.66	.50		1.16	1.51
4750	1/2" thick G		1195	.013		.78	.63		1.41	1.83
4755	Pneumatic nailed G		1325	.012		.78	.57		1.35	1.73

06 16 Sheathing

06 16 36 – Wood Panel Product Sheathing

06 16 36.10 Sheathing		Crew	Daily Output	Labor-Hours	Unit	Material	2015 Bare Costs Labor	Equipment	Total	Total Incl O&P
4800	Joint sealant tape, 3-1/2"	2 Carp	7600	.002	L.F.	.31	.10		.41	.49
4810	Joint sealant tape, 6"	↓	7600	.002	"	.44	.10		.54	.63

06 16 43 – Gypsum Sheathing

06 16 43.10 Gypsum Sheathing		Crew	Daily Output	Labor-Hours	Unit	Material	Labor	Equipment	Total	Total Incl O&P
0010	**GYPSUM SHEATHING**									
0020	Gypsum, weatherproof, 1/2" thick	2 Carp	1125	.014	S.F.	.45	.67		1.12	1.53
0040	With embedded glass mats	"	1100	.015	"	.72	.68		1.40	1.84

06 17 Shop-Fabricated Structural Wood

06 17 33 – Wood I-Joists

06 17 33.10 Wood and Composite I-Joists		Crew	Daily Output	Labor-Hours	Unit	Material	Labor	Equipment	Total	Total Incl O&P
0010	**WOOD AND COMPOSITE I-JOISTS**									
0100	Plywood webs, incl. bridging & blocking, panels 24" O.C.									
1200	15' to 24' span, 50 psf live load	F-5	2400	.013	SF Flr.	2.02	.63		2.65	3.20
1300	55 psf live load		2250	.014		2.23	.67		2.90	3.49
1400	24' to 30' span, 45 psf live load		2600	.012		2.36	.58		2.94	3.49
1500	55 psf live load	↓	2400	.013	↓	3.93	.63		4.56	5.30

06 17 53 – Shop-Fabricated Wood Trusses

06 17 53.10 Roof Trusses		Crew	Daily Output	Labor-Hours	Unit	Material	Labor	Equipment	Total	Total Incl O&P
0010	**ROOF TRUSSES**									
0100	Fink (W) or King post type, 2'-0" O.C.									
0200	Metal plate connected, 4 in 12 slope									
0210	24' to 29' span	F-3	3000	.013	SF Flr.	1.70	.64	.22	2.56	3.09
0300	30' to 43' span		3000	.013		2.16	.64	.22	3.02	3.59
0400	44' to 60' span	↓	3000	.013		2.29	.64	.22	3.15	3.74
0700	Glued and nailed, add				↓	50%				

06 18 Glued-Laminated Construction

06 18 13 – Glued-Laminated Beams

06 18 13.20 Laminated Framing		Crew	Daily Output	Labor-Hours	Unit	Material	Labor	Equipment	Total	Total Incl O&P
0010	**LAMINATED FRAMING**									
0020	30 lb., short term live load, 15 lb. dead load									
0200	Straight roof beams, 20' clear span, beams 8' O.C.	F-3	2560	.016	SF Flr.	2.04	.75	.26	3.05	3.67
0300	Beams 16' O.C.		3200	.013		1.48	.60	.20	2.28	2.77
0500	40' clear span, beams 8' O.C.		3200	.013		3.90	.60	.20	4.70	5.45
0600	Beams 16' O.C.	↓	3840	.010		3.20	.50	.17	3.87	4.48
0800	60' clear span, beams 8' O.C.	F-4	2880	.017		6.70	.79	.39	7.88	9.05
0900	Beams 16' O.C.	"	3840	.013		5	.59	.29	5.88	6.75
1100	Tudor arches, 30' to 40' clear span, frames 8' O.C.	F-3	1680	.024		8.75	1.14	.39	10.28	11.85
1200	Frames 16' O.C.	"	2240	.018		6.85	.86	.29	8	9.20
1400	50' to 60' clear span, frames 8' O.C.	F-4	2200	.022		9.45	1.04	.51	11	12.55
1500	Frames 16' O.C.		2640	.018		8.05	.86	.43	9.34	10.65
1700	Radial arches, 60' clear span, frames 8' O.C.		1920	.025		8.85	1.19	.59	10.63	12.15
1800	Frames 16' O.C.		2880	.017		6.80	.79	.39	7.98	9.10
2000	100' clear span, frames 8' O.C.		1600	.030		9.15	1.42	.71	11.28	13
2100	Frames 16' O.C.		2400	.020		8.05	.95	.47	9.47	10.80
2300	120' clear span, frames 8' O.C.		1440	.033		12.15	1.58	.78	14.51	16.65
2400	Frames 16' O.C.	↓	1920	.025	↓	11.10	1.19	.59	12.88	14.65

06 18 Glued-Laminated Construction

06 18 13 – Glued-Laminated Beams

06 18 13.20 Laminated Framing		Crew	Daily Output	Labor-Hours	Unit	Material	2015 Bare Costs Labor	Equipment	Total	Total Incl O&P
2600	Bowstring trusses, 20' O.C., 40' clear span	F-3	2400	.017	SF Flr.	5.50	.80	.27	6.57	7.60
2700	60' clear span	F-4	3600	.013		4.92	.63	.31	5.86	6.70
2800	100' clear span		4000	.012		6.95	.57	.28	7.80	8.85
2900	120' clear span		3600	.013		7.45	.63	.31	8.39	9.50
3000	For less than 1000 B.F., add					20%				
3050	For over 5000 B.F., deduct					10%				
3100	For premium appearance, add to S.F. prices					5%				
3300	For industrial type, deduct					15%				
3500	For stain and varnish, add					5%				
3900	For 3/4" laminations, add to straight					25%				
4100	Add to curved					15%				
4300	Alternate pricing method: (use nominal footage of									
4310	components). Straight beams, camber less than 6"	F-3	3.50	11.429	M.B.F.	2,925	545	187	3,657	4,250
4400	Columns, including hardware		2	20		3,125	960	325	4,410	5,275
4600	Curved members, radius over 32'		2.50	16		3,200	765	262	4,227	5,000
4700	Radius 10' to 32'		3	13.333		3,175	640	218	4,033	4,725
4900	For complicated shapes, add maximum					100%				
5100	For pressure treating, add to straight					35%				
5200	Add to curved					45%				
6000	Laminated veneer members, southern pine or western species									
6050	1-3/4" wide x 5-1/2" deep	2 Carp	480	.033	L.F.	3.44	1.56		5	6.20
6100	9-1/2" deep		480	.033		4.74	1.56		6.30	7.60
6150	14" deep		450	.036		7.85	1.67		9.52	11.20
6200	18" deep		450	.036		10.65	1.67		12.32	14.30
6300	Parallel strand members, southern pine or western species									
6350	1-3/4" wide x 9-1/4" deep	2 Carp	480	.033	L.F.	5.10	1.56		6.66	8
6400	11-1/4" deep		450	.036		5.75	1.67		7.42	8.90
6450	14" deep		400	.040		7.85	1.88		9.73	11.55
6500	3-1/2" wide x 9-1/4" deep		480	.033		16.30	1.56		17.86	20.50
6550	11-1/4" deep		450	.036		20.50	1.67		22.17	25
6600	14" deep		400	.040		24	1.88		25.88	29.50
6650	7" wide x 9-1/4" deep		450	.036		34.50	1.67		36.17	40.50
6700	11-1/4" deep		420	.038		43.50	1.79		45.29	50.50
6750	14" deep		400	.040		52	1.88		53.88	60
8000	Straight beams									
8102	20' span									
8104	3-1/8" x 9"	F-3	30	1.333	Ea.	137	64	22	223	273
8106	X 10-1/2"		30	1.333		160	64	22	246	298
8108	X 12"		30	1.333		183	64	22	269	325
8110	X 13-1/2"		30	1.333		205	64	22	291	350
8112	X 15"		29	1.379		228	66	22.50	316.50	375
8114	5-1/8" x 10-1/2"		30	1.333		262	64	22	348	410
8116	X 12"		30	1.333		299	64	22	385	450
8118	X 13-1/2"		30	1.333		335	64	22	421	490
8120	X 15"		29	1.379		375	66	22.50	463.50	535
8122	X 16-1/2"		29	1.379		410	66	22.50	498.50	580
8124	X 18"		29	1.379		450	66	22.50	538.50	620
8126	X 19-1/2"		29	1.379		485	66	22.50	573.50	660
8128	X 21"		28	1.429		525	68.50	23.50	617	705
8130	X 22-1/2"		28	1.429		560	68.50	23.50	652	745
8132	X 24"		28	1.429		600	68.50	23.50	692	790
8134	6-3/4" x 12"		29	1.379		395	66	22.50	483.50	560
8136	X 13-1/2"		29	1.379		445	66	22.50	533.50	615

06 18 13 – Glued-Laminated Beams

06 18 13.20 Laminated Framing		Crew	Daily Output	Labor-Hours	Unit	2015 Bare Costs Material	Labor	Equipment	Total	Total Incl O&P
8138	X 15″	F-3	29	1.379	Ea.	495	66	22.50	583.50	665
8140	X 16-1/2″		28	1.429		540	68.50	23.50	632	725
8142	X 18″		28	1.429		590	68.50	23.50	682	780
8144	X 19-1/2″		28	1.429		640	68.50	23.50	732	835
8146	X 21″		27	1.481		690	71	24	785	895
8148	X 22-1/2″		27	1.481		740	71	24	835	950
8150	X 24″		27	1.481		790	71	24	885	1,000
8152	X 25-1/2″		27	1.481		840	71	24	935	1,050
8154	X 27″		26	1.538		885	73.50	25	983.50	1,125
8156	X 28-1/2″		26	1.538		935	73.50	25	1,033.50	1,175
8158	X 30″		26	1.538		985	73.50	25	1,083.50	1,225
8200	30′ span									
8250	3-1/8″ x 9″	F-3	30	1.333	Ea.	205	64	22	291	350
8252	X 10-1/2″		30	1.333		240	64	22	326	385
8254	X 12″		30	1.333		274	64	22	360	420
8256	X 13-1/2″		30	1.333		310	64	22	396	460
8258	X 15″		29	1.379		340	66	22.50	428.50	500
8260	5-1/8″ x 10-1/2″		30	1.333		395	64	22	481	550
8262	X 12″		30	1.333		450	64	22	536	615
8264	X 13-1/2″		30	1.333		505	64	22	591	675
8266	X 15″		29	1.379		560	66	22.50	648.50	740
8268	X 16-1/2″		29	1.379		615	66	22.50	703.50	805
8270	X 18″		29	1.379		675	66	22.50	763.50	865
8272	X 19-1/2″		29	1.379		730	66	22.50	818.50	930
8274	X 21″		28	1.429		785	68.50	23.50	877	995
8276	X 22-1/2″		28	1.429		840	68.50	23.50	932	1,050
8278	X 24″		28	1.429		900	68.50	23.50	992	1,125
8280	6-3/4″ x 12″		29	1.379		590	66	22.50	678.50	775
8282	X 13-1/2″		29	1.379		665	66	22.50	753.50	855
8284	X 15″		29	1.379		740	66	22.50	828.50	940
8286	X 16-1/2″		28	1.429		815	68.50	23.50	907	1,025
8288	X 18″		28	1.429		885	68.50	23.50	977	1,100
8290	X 19-1/2″		28	1.429		960	68.50	23.50	1,052	1,175
8292	X 21″		27	1.481		1,025	71	24	1,120	1,275
8294	X 22-1/2″		27	1.481		1,100	71	24	1,195	1,350
8296	X 24″		27	1.481		1,175	71	24	1,270	1,425
8298	X 25-1/2″		27	1.481		1,250	71	24	1,345	1,500
8300	X 27″		26	1.538		1,325	73.50	25	1,423.50	1,625
8302	X 28-1/2″		26	1.538		1,400	73.50	25	1,498.50	1,700
8304	X 30″		26	1.538		1,475	73.50	25	1,573.50	1,775
8400	40′ span									
8402	3-1/8″ x 9″	F-3	30	1.333	Ea.	274	64	22	360	420
8404	X 10-1/2″		30	1.333		320	64	22	406	470
8406	X 12″		30	1.333		365	64	22	451	520
8408	X 13-1/2″		30	1.333		410	64	22	496	570
8410	X 15″		29	1.379		455	66	22.50	543.50	625
8412	5-1/8″ x 10-1/2″		30	1.333		525	64	22	611	695
8414	X 12″		30	1.333		600	64	22	686	780
8416	X 13-1/2″		30	1.333		675	64	22	761	860
8418	X 15″		29	1.379		750	66	22.50	838.50	950
8420	X 16-1/2″		29	1.379		825	66	22.50	913.50	1,025
8422	X 18″		29	1.379		900	66	22.50	988.50	1,125
8424	X 19-1/2″		29	1.379		975	66	22.50	1,063.50	1,200

06 18 Glued-Laminated Construction

06 18 13 – Glued-Laminated Beams

06 18 13.20 Laminated Framing		Crew	Daily Output	Labor-Hours	Unit	Material	Labor	Equipment	Total	Total Incl O&P
						2015 Bare Costs				
8426	X 21"	F-3	28	1.429	Ea.	1,050	68.50	23.50	1,142	1,275
8428	X 22-1/2"		28	1.429		1,125	68.50	23.50	1,217	1,350
8430	X 24"		28	1.429		1,200	68.50	23.50	1,292	1,450
8432	6-3/4" x 12"		29	1.379		790	66	22.50	878.50	990
8434	X 13-1/2"		29	1.379		885	66	22.50	973.50	1,100
8436	X 15"		29	1.379		985	66	22.50	1,073.50	1,200
8438	X 16-1/2"		28	1.429		1,075	68.50	23.50	1,167	1,325
8440	X 18"		28	1.429		1,175	68.50	23.50	1,267	1,425
8442	X 19-1/2"		28	1.429		1,275	68.50	23.50	1,367	1,525
8444	X 21"		27	1.481		1,375	71	24	1,470	1,650
8446	X 22-1/2"		27	1.481		1,475	71	24	1,570	1,750
8448	X 24"		27	1.481		1,575	71	24	1,670	1,850
8450	X 25-1/2"		27	1.481		1,675	71	24	1,770	1,975
8452	X 27"		26	1.538		1,775	73.50	25	1,873.50	2,100
8454	X 28-1/2"		26	1.538		1,875	73.50	25	1,973.50	2,200
8456	X 30"		26	1.538		1,975	73.50	25	2,073.50	2,325

06 22 Millwork

06 22 13 – Standard Pattern Wood Trim

06 22 13.15 Moldings, Base

		Crew	Daily Output	Labor-Hours	Unit	Material	Labor	Equipment	Total	Total Incl O&P
0010	**MOLDINGS, BASE**									
5100	Classic profile, 5/8" x 5-1/2", finger jointed and primed	1 Carp	250	.032	L.F.	1.46	1.50		2.96	3.91
5105	Poplar		240	.033		1.54	1.56		3.10	4.10
5110	Red oak		220	.036		2.36	1.71		4.07	5.20
5115	Maple		220	.036		3.57	1.71		5.28	6.55
5120	Cherry		220	.036		4.24	1.71		5.95	7.30
5125	3/4 x 7-1/2", finger jointed and primed		250	.032		1.85	1.50		3.35	4.34
5130	Poplar		240	.033		2.46	1.56		4.02	5.10
5135	Red oak		220	.036		3.43	1.71		5.14	6.40
5140	Maple		220	.036		4.75	1.71		6.46	7.85
5145	Cherry		220	.036		5.65	1.71		7.36	8.85
5150	Modern profile, 5/8" x 3-1/2", finger jointed and primed		250	.032		.89	1.50		2.39	3.29
5155	Poplar		240	.033		.98	1.56		2.54	3.49
5160	Red oak		220	.036		1.62	1.71		3.33	4.41
5165	Maple		220	.036		2.50	1.71		4.21	5.40
5170	Cherry		220	.036		2.70	1.71		4.41	5.60
5175	Ogee profile, 7/16" x 3", finger jointed and primed		250	.032		.59	1.50		2.09	2.96
5180	Poplar		240	.033		.64	1.56		2.20	3.11
5185	Red oak		220	.036		.90	1.71		2.61	3.62
5200	9/16" x 3-1/2", finger jointed and primed		250	.032		.66	1.50		2.16	3.03
5205	Pine		240	.033		1.13	1.56		2.69	3.65
5210	Red oak		220	.036		2.38	1.71		4.09	5.25
5215	9/16" x 4-1/2", red oak		220	.036		3.56	1.71		5.27	6.55
5220	5/8" x 3-1/2", finger jointed and primed		250	.032		.89	1.50		2.39	3.29
5225	Poplar		240	.033		.98	1.56		2.54	3.49
5230	Red oak		220	.036		1.62	1.71		3.33	4.41
5235	Maple		220	.036		2.50	1.71		4.21	5.40
5240	Cherry		220	.036		2.70	1.71		4.41	5.60
5245	5/8" x 4", finger jointed and primed		250	.032		1.15	1.50		2.65	3.57
5250	Poplar		240	.033		1.26	1.56		2.82	3.79
5255	Red oak		220	.036		1.81	1.71		3.52	4.62

06 22 13 – Standard Pattern Wood Trim

06 22 13.15 Moldings, Base		Crew	Daily Output	Labor-Hours	Unit	Material	Labor	Equipment	Total	Total Incl O&P
						2015 Bare Costs				
5260	Maple	1 Carp	220	.036	L.F.	2.79	1.71		4.50	5.70
5265	Cherry		220	.036		2.93	1.71		4.64	5.85
5270	Rectangular profile, oak, 3/8" x 1-1/4"		260	.031		1.25	1.44		2.69	3.60
5275	1/2" x 2-1/2"		255	.031		1.94	1.47		3.41	4.40
5280	1/2" x 3-1/2"		250	.032		2.36	1.50		3.86	4.90
5285	1" x 6"		240	.033		4.09	1.56		5.65	6.90
5290	1" x 8"		240	.033		5.70	1.56		7.26	8.65
5295	Pine, 3/8" x 1-3/4"		260	.031		.46	1.44		1.90	2.72
5300	7/16" x 2-1/2"		255	.031		.74	1.47		2.21	3.08
5305	1" x 6"		240	.033		.83	1.56		2.39	3.32
5310	1" x 8"		240	.033		1.06	1.56		2.62	3.57
5315	Shoe, 1/2" x 3/4", primed		260	.031		.48	1.44		1.92	2.75
5320	Pine		240	.033		.30	1.56		1.86	2.74
5325	Poplar		240	.033		.38	1.56		1.94	2.83
5330	Red oak		220	.036		.52	1.71		2.23	3.20
5335	Maple		220	.036		.69	1.71		2.40	3.39
5340	Cherry		220	.036		.78	1.71		2.49	3.49
5345	11/16" x 1-1/2", pine		240	.033		.70	1.56		2.26	3.18
5350	Caps, 11/16" x 1-3/8", pine		240	.033		.55	1.56		2.11	3.01
5355	3/4" x 1-3/4", finger jointed and primed		260	.031		.70	1.44		2.14	2.99
5360	Poplar		240	.033		.99	1.56		2.55	3.50
5365	Red oak		220	.036		1.11	1.71		2.82	3.85
5370	Maple		220	.036		1.55	1.71		3.26	4.33
5375	Cherry		220	.036		3.21	1.71		4.92	6.15
5380	Combination base & shoe, 9/16" x 3-1/2" & 1/2" x 3/4", pine		125	.064		1.43	3		4.43	6.20
5385	Three piece oak, 6" high		80	.100		5.70	4.70		10.40	13.55
5390	Including 3/4" x 1" base shoe		70	.114		5.90	5.35		11.25	14.75
5395	Flooring cant strip, 3/4" x 3/4", pre-finished pine		260	.031		.51	1.44		1.95	2.78
5400	For pre-finished, stain and clear coat, add					.51			.51	.56
5405	Clear coat only, add					.42			.42	.46

06 22 13.30 Moldings, Casings		Crew	Daily Output	Labor-Hours	Unit	Material	Labor	Equipment	Total	Total Incl O&P
0010	**MOLDINGS, CASINGS**									
0085	Apron, 9/16" x 2-1/2", pine	1 Carp	250	.032	L.F.	1.63	1.50		3.13	4.10
0090	5/8" x 2-1/2", pine		250	.032		1.63	1.50		3.13	4.10
0110	5/8" x 3-1/2", pine		220	.036		1.86	1.71		3.57	4.67
0300	Band, 11/16" x 1-1/8", pine		270	.030		.75	1.39		2.14	2.96
0310	11/16" x 1-1/2", finger jointed and primed		270	.030		.76	1.39		2.15	2.97
0320	Pine		270	.030		.96	1.39		2.35	3.19
0330	11/16" x 1-3/4", finger jointed and primed		270	.030		.95	1.39		2.34	3.18
0350	Pine		270	.030		1.16	1.39		2.55	3.41
0355	Beaded, 3/4" x 3-1/2", finger jointed and primed		220	.036		1.09	1.71		2.80	3.83
0360	Poplar		220	.036		1.09	1.71		2.80	3.83
0365	Red oak		220	.036		1.62	1.71		3.33	4.41
0370	Maple		220	.036		2.50	1.71		4.21	5.40
0375	Cherry		220	.036		3.02	1.71		4.73	5.95
0380	3/4" x 4", finger jointed and primed		220	.036		1.15	1.71		2.86	3.89
0385	Poplar		220	.036		1.41	1.71		3.12	4.18
0390	Red oak		220	.036		1.98	1.71		3.69	4.81
0395	Maple		220	.036		2.60	1.71		4.31	5.50
0400	Cherry		220	.036		3.29	1.71		5	6.25
0405	3/4" x 5-1/2", finger jointed and primed		200	.040		1.15	1.88		3.03	4.15
0410	Poplar		200	.040		1.89	1.88		3.77	4.97

06 22 13 – Standard Pattern Wood Trim

06 22 13.30 Moldings, Casings		Crew	Daily Output	Labor-Hours	Unit	Material	Labor	Equipment	Total	Total Incl O&P
						2015 Bare Costs				
0415	Red oak	1 Carp	200	.040	L.F.	2.78	1.88		4.66	5.95
0420	Maple		200	.040		3.66	1.88		5.54	6.90
0425	Cherry		200	.040		4.22	1.88		6.10	7.55
0430	Classic profile, 3/4" x 2-3/4", finger jointed and primed		250	.032		.77	1.50		2.27	3.16
0435	Poplar		250	.032		.98	1.50		2.48	3.39
0440	Red oak		250	.032		1.43	1.50		2.93	3.88
0445	Maple		250	.032		2.05	1.50		3.55	4.56
0450	Cherry		250	.032		2.32	1.50		3.82	4.86
0455	Fluted, 3/4" x 3-1/2", poplar		220	.036		1.09	1.71		2.80	3.83
0460	Red oak		220	.036		1.62	1.71		3.33	4.41
0465	Maple		220	.036		2.50	1.71		4.21	5.40
0470	Cherry		220	.036		3.02	1.71		4.73	5.95
0475	3/4" x 4", poplar		220	.036		1.41	1.71		3.12	4.18
0480	Red oak		220	.036		1.98	1.71		3.69	4.81
0485	Maple		220	.036		2.60	1.71		4.31	5.50
0490	Cherry		220	.036		3.29	1.71		5	6.25
0495	3/4" x 5-1/2", poplar		200	.040		1.89	1.88		3.77	4.97
0500	Red oak		200	.040		2.78	1.88		4.66	5.95
0505	Maple		200	.040		3.66	1.88		5.54	6.90
0510	Cherry		200	.040		4.22	1.88		6.10	7.55
0515	3/4" x 7-1/2", poplar		190	.042		1.27	1.98		3.25	4.44
0520	Red oak		190	.042		3.85	1.98		5.83	7.25
0525	Maple		190	.042		5.55	1.98		7.53	9.15
0530	Cherry		190	.042		6.65	1.98		8.63	10.35
0535	3/4" x 9-1/2", poplar		180	.044		4.04	2.09		6.13	7.65
0540	Red oak		180	.044		6.45	2.09		8.54	10.25
0545	Maple		180	.044		8.80	2.09		10.89	12.90
0550	Cherry		180	.044		9.60	2.09		11.69	13.80
0555	Modern profile, 9/16" x 2-1/4", poplar		250	.032		.56	1.50		2.06	2.92
0560	Red oak		250	.032		.76	1.50		2.26	3.14
0565	11/16" x 2-1/2", finger jointed & primed		250	.032		.84	1.50		2.34	3.23
0570	Pine		250	.032		1.28	1.50		2.78	3.72
0575	3/4" x 2-1/2", poplar		250	.032		.86	1.50		2.36	3.25
0580	Red oak		250	.032		1.17	1.50		2.67	3.60
0585	Maple		250	.032		1.76	1.50		3.26	4.24
0590	Cherry		250	.032		2.23	1.50		3.73	4.76
0595	Mullion, 5/16" x 2", pine		270	.030		.86	1.39		2.25	3.08
0600	9/16" x 2-1/2", fingerjointed and primed		250	.032		.98	1.50		2.48	3.39
0605	Pine		250	.032		1.28	1.50		2.78	3.72
0610	Red oak		250	.032		2.86	1.50		4.36	5.45
0615	1-1/16" x 3-3/4", red oak		220	.036		6.95	1.71		8.66	10.25
0620	Ogee, 7/16" x 2-1/2", poplar		250	.032		.56	1.50		2.06	2.92
0625	Red oak		250	.032		.77	1.50		2.27	3.16
0630	9/16" x 2-1/4", finger jointed and primed		250	.032		.49	1.50		1.99	2.85
0635	Poplar		250	.032		.56	1.50		2.06	2.92
0640	Red oak		250	.032		.76	1.50		2.26	3.14
0645	11/16" x 2-1/2", finger jointed and primed		250	.032		.72	1.50		2.22	3.10
0700	Pine		250	.032		1.41	1.50		2.91	3.86
0701	Red oak		250	.032		2.70	1.50		4.20	5.30
0730	11/16" x 3-1/2", finger jointed and primed		220	.036		1.11	1.71		2.82	3.85
0750	Pine		220	.036		1.72	1.71		3.43	4.52
0755	3/4" x 2-1/2", finger jointed and primed		250	.032		.89	1.50		2.39	3.29
0760	Poplar		250	.032		.86	1.50		2.36	3.25

06 22 13 – Standard Pattern Wood Trim

06 22 13.30 Moldings, Casings		Crew	Daily Output	Labor-Hours	Unit	Material	Labor	Equipment	Total	Total Incl O&P
						2015 Bare Costs				
0765	Red oak	1 Carp	250	.032	L.F.	1.17	1.50		2.67	3.60
0770	Maple		250	.032		1.76	1.50		3.26	4.24
0775	Cherry		250	.032		2.23	1.50		3.73	4.76
0780	3/4" x 3-1/2", finger jointed and primed		220	.036		.89	1.71		2.60	3.61
0785	Poplar		220	.036		1.09	1.71		2.80	3.83
0790	Red oak		220	.036		1.62	1.71		3.33	4.41
0795	Maple		220	.036		2.50	1.71		4.21	5.40
0800	Cherry		220	.036		3.02	1.71		4.73	5.95
4700	Square profile, 1" x 1", teak		215	.037		2.26	1.75		4.01	5.15
4800	Rectangular profile, 1" x 3", teak		200	.040		6.20	1.88		8.08	9.75

06 22 13.35 Moldings, Ceilings		Crew	Daily Output	Labor-Hours	Unit	Material	Labor	Equipment	Total	Total Incl O&P
0010	**MOLDINGS, CEILINGS**									
0600	Bed, 9/16" x 1-3/4", pine	1 Carp	270	.030	L.F.	1.06	1.39		2.45	3.30
0650	9/16" x 2", pine		270	.030		1.14	1.39		2.53	3.39
0710	9/16" x 1-3/4", oak		270	.030		1.86	1.39		3.25	4.18
1200	Cornice, 9/16" x 1-3/4", pine		270	.030		.96	1.39		2.35	3.19
1300	9/16" x 2-1/4", pine		265	.030		1.27	1.42		2.69	3.57
1350	Cove, 1/2" x 2-1/4", poplar		265	.030		1	1.42		2.42	3.28
1360	Red oak		265	.030		1.33	1.42		2.75	3.64
1370	Hard maple		265	.030		1.83	1.42		3.25	4.19
1380	Cherry		265	.030		2.24	1.42		3.66	4.64
2400	9/16" x 1-3/4", pine		270	.030		.97	1.39		2.36	3.21
2500	11/16" x 2-3/4", pine		265	.030		1.80	1.42		3.22	4.16
2510	Crown, 5/8" x 5/8", poplar		300	.027		.43	1.25		1.68	2.40
2520	Red oak		300	.027		.51	1.25		1.76	2.49
2530	Hard maple		300	.027		.69	1.25		1.94	2.69
2540	Cherry		300	.027		.73	1.25		1.98	2.73
2600	9/16" x 3-5/8", pine		250	.032		2.02	1.50		3.52	4.53
2700	11/16" x 4-1/4", pine		250	.032		2.90	1.50		4.40	5.50
2705	Oak		250	.032		6.40	1.50		7.90	9.35
2710	3/4" x 1-3/4", poplar		270	.030		.70	1.39		2.09	2.91
2720	Red oak		270	.030		1.03	1.39		2.42	3.27
2730	Hard maple		270	.030		1.36	1.39		2.75	3.63
2740	Cherry		270	.030		1.68	1.39		3.07	3.99
2750	3/4" x 2", poplar		270	.030		.92	1.39		2.31	3.15
2760	Red oak		270	.030		1.24	1.39		2.63	3.50
2770	Hard maple		270	.030		1.65	1.39		3.04	3.95
2780	Cherry		270	.030		1.84	1.39		3.23	4.16
2790	3/4" x 2-3/4", poplar		265	.030		1	1.42		2.42	3.28
2800	Red oak		265	.030		1.54	1.42		2.96	3.87
2810	Hard maple		265	.030		2.51	1.42		3.93	4.94
2820	Cherry		265	.030		2.35	1.42		3.77	4.76
2830	3/4" x 3-1/2", poplar		250	.032		1.27	1.50		2.77	3.70
2840	Red oak		250	.032		1.92	1.50		3.42	4.42
2850	Hard maple		250	.032		2.51	1.50		4.01	5.05
2860	Cherry		250	.032		2.94	1.50		4.44	5.55
2870	FJP poplar		250	.032		.90	1.50		2.40	3.30
2880	3/4" x 5", poplar		245	.033		1.89	1.53		3.42	4.44
2890	Red oak		245	.033		2.79	1.53		4.32	5.45
2900	Hard maple		245	.033		3.67	1.53		5.20	6.40
2910	Cherry		245	.033		4.24	1.53		5.77	7
2920	FJP poplar		245	.033		1.29	1.53		2.82	3.78

06 22 Millwork

06 22 13 – Standard Pattern Wood Trim

06 22 13.35 Moldings, Ceilings		Crew	Daily Output	Labor-Hours	Unit	Material	2015 Bare Costs Labor	Equipment	Total	Total Incl O&P
2930	3/4" x 6-1/4", poplar	1 Carp	240	.033	L.F.	2.26	1.56		3.82	4.89
2940	Red oak		240	.033		3.40	1.56		4.96	6.15
2950	Hard maple		240	.033		4.44	1.56		6	7.30
2960	Cherry		240	.033		5.25	1.56		6.81	8.20
2970	7/8" x 8-3/4", poplar		220	.036		3.51	1.71		5.22	6.50
2980	Red oak		220	.036		5.35	1.71		7.06	8.50
2990	Hard maple		220	.036		7.35	1.71		9.06	10.75
3000	Cherry		220	.036		8.40	1.71		10.11	11.90
3010	1" x 7-1/4", poplar		220	.036		3.71	1.71		5.42	6.70
3020	Red oak		220	.036		5.75	1.71		7.46	9
3030	Hard maple		220	.036		8	1.71		9.71	11.40
3040	Cherry		220	.036		9.30	1.71		11.01	12.90
3050	1-1/16" x 4-1/4", poplar		250	.032		2.25	1.50		3.75	4.78
3060	Red oak		250	.032		2.64	1.50		4.14	5.20
3070	Hard maple		250	.032		3.49	1.50		4.99	6.15
3080	Cherry		250	.032		4.98	1.50		6.48	7.75
3090	Dentil crown, 3/4" x 5", poplar		250	.032		1.89	1.50		3.39	4.39
3100	Red oak		250	.032		2.79	1.50		4.29	5.40
3110	Hard maple		250	.032		3.67	1.50		5.17	6.35
3120	Cherry		250	.032		4.24	1.50		5.74	6.95
3130	Dentil piece for above, 1/2" x 1/2", poplar		300	.027		2.68	1.25		3.93	4.87
3140	Red oak		300	.027		3.08	1.25		4.33	5.30
3150	Hard maple		300	.027		3.50	1.25		4.75	5.80
3160	Cherry	▼	300	.027	▼	3.81	1.25		5.06	6.10

06 22 13.40 Moldings, Exterior		Crew	Daily Output	Labor-Hours	Unit	Material	Labor	Equipment	Total	Total Incl O&P
0010	**MOLDINGS, EXTERIOR**									
0100	Band board, cedar, rough sawn, 1" x 2"	1 Carp	300	.027	L.F.	.54	1.25		1.79	2.53
0110	1" x 3"		300	.027		.81	1.25		2.06	2.82
0120	1" x 4"		250	.032		1.07	1.50		2.57	3.49
0130	1" x 6"		250	.032		1.61	1.50		3.11	4.08
0140	1" x 8"		225	.036		2.15	1.67		3.82	4.94
0150	1" x 10"		225	.036		2.67	1.67		4.34	5.50
0160	1" x 12"		200	.040		3.21	1.88		5.09	6.40
0240	STK, 1" x 2"		300	.027		.51	1.25		1.76	2.49
0250	1" x 3"		300	.027		.56	1.25		1.81	2.55
0260	1" x 4"		250	.032		.65	1.50		2.15	3.02
0270	1" x 6"		250	.032		1.03	1.50		2.53	3.44
0280	1" x 8"		225	.036		1.60	1.67		3.27	4.33
0290	1" x 10"		225	.036		2.14	1.67		3.81	4.92
0300	1" x 12"		200	.040		3.54	1.88		5.42	6.80
0310	Pine, #2, 1" x 2"		300	.027		.25	1.25		1.50	2.21
0320	1" x 3"		300	.027		.40	1.25		1.65	2.37
0330	1" x 4"		250	.032		.49	1.50		1.99	2.85
0340	1" x 6"		250	.032		.75	1.50		2.25	3.13
0350	1" x 8"		225	.036		1.22	1.67		2.89	3.91
0360	1" x 10"		225	.036		1.59	1.67		3.26	4.32
0370	1" x 12"		200	.040		2.05	1.88		3.93	5.15
0380	D & better, 1" x 2"		300	.027		.41	1.25		1.66	2.38
0390	1" x 3"		300	.027		.61	1.25		1.86	2.60
0400	1" x 4"		250	.032		.82	1.50		2.32	3.21
0410	1" x 6"		250	.032		1.12	1.50		2.62	3.54
0420	1" x 8"	▼	225	.036	▼	1.62	1.67		3.29	4.35

06 22 13.40 Moldings, Exterior		Crew	Daily Output	Labor-Hours	Unit	2015 Bare Costs Material	Labor	Equipment	Total	Total Incl O&P
0430	1" x 10"	1 Carp	225	.036	L.F.	2.03	1.67		3.70	4.80
0440	1" x 12"		200	.040		2.63	1.88		4.51	5.80
0450	Redwood, clear all heart, 1" x 2"		300	.027		.63	1.25		1.88	2.63
0460	1" x 3"		300	.027		.94	1.25		2.19	2.97
0470	1" x 4"		250	.032		1.19	1.50		2.69	3.62
0480	1" x 6"		252	.032		1.78	1.49		3.27	4.24
0490	1" x 8"		225	.036		2.36	1.67		4.03	5.15
0500	1" x 10"		225	.036		3.96	1.67		5.63	6.95
0510	1" x 12"		200	.040		4.75	1.88		6.63	8.15
0530	Corner board, cedar, rough sawn, 1" x 2"		225	.036		.54	1.67		2.21	3.17
0540	1" x 3"		225	.036		.81	1.67		2.48	3.46
0550	1" x 4"		200	.040		1.07	1.88		2.95	4.07
0560	1" x 6"		200	.040		1.61	1.88		3.49	4.66
0570	1" x 8"		200	.040		2.15	1.88		4.03	5.25
0580	1" x 10"		175	.046		2.67	2.15		4.82	6.25
0590	1" x 12"		175	.046		3.21	2.15		5.36	6.85
0670	STK, 1" x 2"		225	.036		.51	1.67		2.18	3.13
0680	1" x 3"		225	.036		.56	1.67		2.23	3.19
0690	1" x 4"		200	.040		.63	1.88		2.51	3.59
0700	1" x 6"		200	.040		1.03	1.88		2.91	4.02
0710	1" x 8"		200	.040		1.60	1.88		3.48	4.65
0720	1" x 10"		175	.046		2.14	2.15		4.29	5.65
0730	1" x 12"		175	.046		3.54	2.15		5.69	7.20
0740	Pine, #2, 1" x 2"		225	.036		.25	1.67		1.92	2.85
0750	1" x 3"		225	.036		.40	1.67		2.07	3.01
0760	1" x 4"		200	.040		.49	1.88		2.37	3.43
0770	1" x 6"		200	.040		.75	1.88		2.63	3.71
0780	1" x 8"		200	.040		1.22	1.88		3.10	4.23
0790	1" x 10"		175	.046		1.59	2.15		3.74	5.05
0800	1" x 12"		175	.046		2.05	2.15		4.20	5.55
0810	D & better, 1" x 2"		225	.036		.41	1.67		2.08	3.02
0820	1" x 3"		225	.036		.61	1.67		2.28	3.24
0830	1" x 4"		200	.040		.82	1.88		2.70	3.79
0840	1" x 6"		200	.040		1.12	1.88		3	4.12
0850	1" x 8"		200	.040		1.62	1.88		3.50	4.67
0860	1" x 10"		175	.046		2.03	2.15		4.18	5.55
0870	1" x 12"		175	.046		2.63	2.15		4.78	6.20
0880	Redwood, clear all heart, 1" x 2"		225	.036		.63	1.67		2.30	3.27
0890	1" x 3"		225	.036		.94	1.67		2.61	3.61
0900	1" x 4"		200	.040		1.19	1.88		3.07	4.20
0910	1" x 6"		200	.040		1.78	1.88		3.66	4.84
0920	1" x 8"		200	.040		2.36	1.88		4.24	5.50
0930	1" x 10"		175	.046		3.96	2.15		6.11	7.65
0940	1" x 12"		175	.046		4.75	2.15		6.90	8.55
0950	Cornice board, cedar, rough sawn, 1" x 2"		330	.024		.54	1.14		1.68	2.35
0960	1" x 3"		290	.028		.81	1.30		2.11	2.88
0970	1" x 4"		250	.032		1.07	1.50		2.57	3.49
0980	1" x 6"		250	.032		1.61	1.50		3.11	4.08
0990	1" x 8"		200	.040		2.15	1.88		4.03	5.25
1000	1" x 10"		180	.044		2.67	2.09		4.76	6.15
1010	1" x 12"		180	.044		3.21	2.09		5.30	6.75
1020	STK, 1" x 2"		330	.024		.51	1.14		1.65	2.31
1030	1" x 3"		290	.028		.56	1.30		1.86	2.61

06 22 Millwork

06 22 13 – Standard Pattern Wood Trim

06 22 13.40 Moldings, Exterior		Crew	Daily Output	Labor-Hours	Unit	Material	Labor	Equipment	Total	Total Incl O&P
						2015 Bare Costs				
1040	1" x 4"	1 Carp	250	.032	L.F.	.65	1.50		2.15	3.02
1050	1" x 6"		250	.032		1.03	1.50		2.53	3.44
1060	1" x 8"		200	.040		1.60	1.88		3.48	4.65
1070	1" x 10"		180	.044		2.14	2.09		4.23	5.55
1080	1" x 12"		180	.044		3.54	2.09		5.63	7.10
1500	Pine, #2, 1" x 2"		330	.024		.25	1.14		1.39	2.03
1510	1" x 3"		290	.028		.27	1.30		1.57	2.29
1600	1" x 4"		250	.032		.49	1.50		1.99	2.85
1700	1" x 6"		250	.032		.75	1.50		2.25	3.13
1800	1" x 8"		200	.040		1.22	1.88		3.10	4.23
1900	1" x 10"		180	.044		1.59	2.09		3.68	4.96
2000	1" x 12"		180	.044		2.05	2.09		4.14	5.45
2020	D & better, 1" x 2"		330	.024		.41	1.14		1.55	2.20
2030	1" x 3"		290	.028		.61	1.30		1.91	2.66
2040	1" x 4"		250	.032		.82	1.50		2.32	3.21
2050	1" x 6"		250	.032		1.12	1.50		2.62	3.54
2060	1" x 8"		200	.040		1.62	1.88		3.50	4.67
2070	1" x 10"		180	.044		2.03	2.09		4.12	5.45
2080	1" x 12"		180	.044		2.63	2.09		4.72	6.10
2090	Redwood, clear all heart, 1" x 2"		330	.024		.63	1.14		1.77	2.45
2100	1" x 3"		290	.028		.94	1.30		2.24	3.03
2110	1" x 4"		250	.032		1.19	1.50		2.69	3.62
2120	1" x 6"		250	.032		1.78	1.50		3.28	4.26
2130	1" x 8"		200	.040		2.36	1.88		4.24	5.50
2140	1" x 10"		180	.044		3.96	2.09		6.05	7.55
2150	1" x 12"		180	.044		4.75	2.09		6.84	8.45
2160	3 piece, 1" x 2", 1" x 4", 1" x 6", rough sawn cedar		80	.100		3.24	4.70		7.94	10.80
2180	STK cedar		80	.100		2.19	4.70		6.89	9.65
2200	#2 pine		80	.100		1.49	4.70		6.19	8.90
2210	D & better pine		80	.100		2.35	4.70		7.05	9.85
2220	Clear all heart redwood		80	.100		3.60	4.70		8.30	11.20
2230	1" x 8", 1" x 10", 1" x 12", rough sawn cedar		65	.123		8	5.80		13.80	17.70
2240	STK cedar		65	.123		7.25	5.80		13.05	16.90
2300	#2 pine		65	.123		4.82	5.80		10.62	14.20
2320	D & better pine		65	.123		6.25	5.80		12.05	15.80
2330	Clear all heart redwood		65	.123		11.05	5.80		16.85	21
2340	Door/window casing, cedar, rough sawn, 1" x 2"		275	.029		.54	1.37		1.91	2.70
2350	1" x 3"		275	.029		.81	1.37		2.18	2.99
2360	1" x 4"		250	.032		1.07	1.50		2.57	3.49
2370	1" x 6"		250	.032		1.61	1.50		3.11	4.08
2380	1" x 8"		230	.035		2.15	1.63		3.78	4.88
2390	1" x 10"		230	.035		2.67	1.63		4.30	5.45
2395	1" x 12"		210	.038		3.21	1.79		5	6.30
2410	STK, 1" x 2"		275	.029		.51	1.37		1.88	2.66
2420	1" x 3"		275	.029		.56	1.37		1.93	2.72
2430	1" x 4"		250	.032		.65	1.50		2.15	3.02
2440	1" x 6"		250	.032		1.03	1.50		2.53	3.44
2450	1" x 8"		230	.035		1.60	1.63		3.23	4.27
2460	1" x 10"		230	.035		2.14	1.63		3.77	4.86
2470	1" x 12"		210	.038		3.54	1.79		5.33	6.65
2550	Pine, #2, 1" x 2"		275	.029		.25	1.37		1.62	2.38
2560	1" x 3"		275	.029		.40	1.37		1.77	2.54
2570	1" x 4"	↓	250	.032	↓	.49	1.50		1.99	2.85

06 22 13 – Standard Pattern Wood Trim

06 22 13.40 Moldings, Exterior		Crew	Daily Output	Labor-Hours	Unit	Material	Labor	Equipment	Total	Total Incl O&P
						2015 Bare Costs				
2580	1" x 6"	1 Carp	250	.032	L.F.	.75	1.50		2.25	3.13
2590	1" x 8"		230	.035		1.22	1.63		2.85	3.85
2600	1" x 10"		230	.035		1.59	1.63		3.22	4.26
2610	1" x 12"		210	.038		2.05	1.79		3.84	5
2620	Pine, D & better, 1" x 2"		275	.029		.41	1.37		1.78	2.55
2630	1" x 3"		275	.029		.61	1.37		1.98	2.77
2640	1" x 4"		250	.032		.82	1.50		2.32	3.21
2650	1" x 6"		250	.032		1.12	1.50		2.62	3.54
2660	1" x 8"		230	.035		1.62	1.63		3.25	4.29
2670	1" x 10"		230	.035		2.03	1.63		3.66	4.74
2680	1" x 12"		210	.038		2.63	1.79		4.42	5.65
2690	Redwood, clear all heart, 1" x 2"		275	.029		.63	1.37		2	2.80
2695	1" x 3"		275	.029		.94	1.37		2.31	3.14
2710	1" x 4"		250	.032		1.19	1.50		2.69	3.62
2715	1" x 6"		250	.032		1.78	1.50		3.28	4.26
2730	1" x 8"		230	.035		2.36	1.63		3.99	5.10
2740	1" x 10"		230	.035		3.96	1.63		5.59	6.85
2750	1" x 12"		210	.038		4.75	1.79		6.54	8
3500	Bellyband, pine, 11/16" x 4-1/4"		250	.032		3.23	1.50		4.73	5.85
3610	Brickmold, pine, 1-1/4" x 2"		200	.040		2.43	1.88		4.31	5.55
3620	FJP, 1-1/4" x 2"		200	.040		1.09	1.88		2.97	4.09
5100	Fascia, cedar, rough sawn, 1" x 2"		275	.029		.54	1.37		1.91	2.70
5110	1" x 3"		275	.029		.81	1.37		2.18	2.99
5120	1" x 4"		250	.032		1.07	1.50		2.57	3.49
5200	1" x 6"		250	.032		1.61	1.50		3.11	4.08
5300	1" x 8"		230	.035		2.15	1.63		3.78	4.88
5310	1" x 10"		230	.035		2.67	1.63		4.30	5.45
5320	1" x 12"		210	.038		3.21	1.79		5	6.30
5400	2" x 4"		220	.036		1.05	1.71		2.76	3.79
5500	2" x 6"		220	.036		1.58	1.71		3.29	4.37
5600	2" x 8"		200	.040		2.11	1.88		3.99	5.20
5700	2" x 10"		180	.044		2.62	2.09		4.71	6.10
5800	2" x 12"		170	.047		6.25	2.21		8.46	10.30
6120	STK, 1" x 2"		275	.029		.51	1.37		1.88	2.66
6130	1" x 3"		275	.029		.56	1.37		1.93	2.72
6140	1" x 4"		250	.032		.65	1.50		2.15	3.02
6150	1" x 6"		250	.032		1.03	1.50		2.53	3.44
6160	1" x 8"		230	.035		1.60	1.63		3.23	4.27
6170	1" x 10"		230	.035		2.14	1.63		3.77	4.86
6180	1" x 12"		210	.038		3.54	1.79		5.33	6.65
6185	2" x 2"		260	.031		.63	1.44		2.07	2.92
6190	Pine, #2, 1" x 2"		275	.029		.25	1.37		1.62	2.38
6200	1" x 3"		275	.029		.40	1.37		1.77	2.54
6210	1" x 4"		250	.032		.49	1.50		1.99	2.85
6220	1" x 6"		250	.032		.75	1.50		2.25	3.13
6230	1" x 8"		230	.035		1.22	1.63		2.85	3.85
6240	1" x 10"		230	.035		1.59	1.63		3.22	4.26
6250	1" x 12"		210	.038		2.05	1.79		3.84	5
6260	D & better, 1" x 2"		275	.029		.41	1.37		1.78	2.55
6270	1" x 3"		275	.029		.61	1.37		1.98	2.77
6280	1" x 4"		250	.032		.82	1.50		2.32	3.21
6290	1" x 6"		250	.032		1.12	1.50		2.62	3.54
6300	1" x 8"		230	.035		1.62	1.63		3.25	4.29

06 22 13.40 Moldings, Exterior		Crew	Daily Output	Labor-Hours	Unit	Material	Labor	Equipment	Total	Total Incl O&P
						2015 Bare Costs				
6310	1" x 10"	1 Carp	230	.035	L.F.	2.03	1.63		3.66	4.74
6312	1" x 12"		210	.038		2.63	1.79		4.42	5.65
6330	Southern yellow, 1-1/4" x 5"		240	.033		2.27	1.56		3.83	4.90
6340	1-1/4" x 6"		240	.033		2.52	1.56		4.08	5.20
6350	1-1/4" x 8"		215	.037		3.43	1.75		5.18	6.45
6360	1-1/4" x 12"		190	.042		5.05	1.98		7.03	8.60
6370	Redwood, clear all heart, 1" x 2"		275	.029		.63	1.37		2	2.80
6380	1" x 3"		275	.029		1.19	1.37		2.56	3.41
6390	1" x 4"		250	.032		1.19	1.50		2.69	3.62
6400	1" x 6"		250	.032		1.78	1.50		3.28	4.26
6410	1" x 8"		230	.035		2.36	1.63		3.99	5.10
6420	1" x 10"		230	.035		3.96	1.63		5.59	6.85
6430	1" x 12"		210	.038		4.75	1.79		6.54	8
6440	1-1/4" x 5"		240	.033		1.86	1.56		3.42	4.45
6450	1-1/4" x 6"		240	.033		2.22	1.56		3.78	4.85
6460	1-1/4" x 8"		215	.037		3.53	1.75		5.28	6.55
6470	1-1/4" x 12"		190	.042		7.10	1.98		9.08	10.90
6580	Frieze, cedar, rough sawn, 1" x 2"		275	.029		.54	1.37		1.91	2.70
6590	1" x 3"		275	.029		.81	1.37		2.18	2.99
6600	1" x 4"		250	.032		1.07	1.50		2.57	3.49
6610	1" x 6"		250	.032		1.61	1.50		3.11	4.08
6620	1" x 8"		250	.032		2.15	1.50		3.65	4.68
6630	1" x 10"		225	.036		2.67	1.67		4.34	5.50
6640	1" x 12"		200	.040		3.18	1.88		5.06	6.40
6650	STK, 1" x 2"		275	.029		.51	1.37		1.88	2.66
6660	1" x 3"		275	.029		.56	1.37		1.93	2.72
6670	1" x 4"		250	.032		.65	1.50		2.15	3.02
6680	1" x 6"		250	.032		1.03	1.50		2.53	3.44
6690	1" x 8"		250	.032		1.60	1.50		3.10	4.07
6700	1" x 10"		225	.036		2.14	1.67		3.81	4.92
6710	1" x 12"		200	.040		3.54	1.88		5.42	6.80
6790	Pine, #2, 1" x 2"		275	.029		.25	1.37		1.62	2.38
6800	1" x 3"		275	.029		.40	1.37		1.77	2.54
6810	1" x 4"		250	.032		.49	1.50		1.99	2.85
6820	1" x 6"		250	.032		.75	1.50		2.25	3.13
6830	1" x 8"		250	.032		1.22	1.50		2.72	3.65
6840	1" x 10"		225	.036		1.59	1.67		3.26	4.32
6850	1" x 12"		200	.040		2.05	1.88		3.93	5.15
6860	D & better, 1" x 2"		275	.029		.41	1.37		1.78	2.55
6870	1" x 3"		275	.029		.61	1.37		1.98	2.77
6880	1" x 4"		250	.032		.82	1.50		2.32	3.21
6890	1" x 6"		250	.032		1.12	1.50		2.62	3.54
6900	1" x 8"		250	.032		1.62	1.50		3.12	4.09
6910	1" x 10"		225	.036		2.03	1.67		3.70	4.80
6920	1" x 12"		200	.040		2.63	1.88		4.51	5.80
6930	Redwood, clear all heart, 1" x 2"		275	.029		.63	1.37		2	2.80
6940	1" x 3"		275	.029		.94	1.37		2.31	3.14
6950	1" x 4"		250	.032		1.19	1.50		2.69	3.62
6960	1" x 6"		250	.032		1.78	1.50		3.28	4.26
6970	1" x 8"		250	.032		2.36	1.50		3.86	4.90
6980	1" x 10"		225	.036		3.96	1.67		5.63	6.95
6990	1" x 12"		200	.040		4.75	1.88		6.63	8.15
7000	Grounds, 1" x 1", cedar, rough sawn	↓	300	.027	↓	.28	1.25		1.53	2.23

06 22 13.40 Moldings, Exterior		Crew	Daily Output	Labor-Hours	Unit	Material	Labor	Equipment	Total	Total Incl O&P
						2015 Bare Costs				
7010	STK	1 Carp	300	.027	L.F.	.31	1.25		1.56	2.27
7020	Pine, #2		300	.027		.16	1.25		1.41	2.10
7030	D & better		300	.027		.25	1.25		1.50	2.21
7050	Redwood		300	.027		.39	1.25		1.64	2.35
7060	Rake/verge board, cedar, rough sawn, 1" x 2"		225	.036		.54	1.67		2.21	3.17
7070	1" x 3"		225	.036		.81	1.67		2.48	3.46
7080	1" x 4"		200	.040		1.07	1.88		2.95	4.07
7090	1" x 6"		200	.040		1.61	1.88		3.49	4.66
7100	1" x 8"		190	.042		2.15	1.98		4.13	5.40
7110	1" x 10"		190	.042		2.67	1.98		4.65	6
7120	1" x 12"		180	.044		3.21	2.09		5.30	6.75
7130	STK, 1" x 2"		225	.036		.51	1.67		2.18	3.13
7140	1" x 3"		225	.036		.56	1.67		2.23	3.19
7150	1" x 4"		200	.040		.65	1.88		2.53	3.60
7160	1" x 6"		200	.040		1.03	1.88		2.91	4.02
7170	1" x 8"		190	.042		1.60	1.98		3.58	4.80
7180	1" x 10"		190	.042		2.14	1.98		4.12	5.40
7190	1" x 12"		180	.044		3.54	2.09		5.63	7.10
7200	Pine, #2, 1" x 2"		225	.036		.25	1.67		1.92	2.85
7210	1" x 3"		225	.036		.40	1.67		2.07	3.01
7220	1" x 4"		200	.040		.49	1.88		2.37	3.43
7230	1" x 6"		200	.040		.75	1.88		2.63	3.71
7240	1" x 8"		190	.042		1.22	1.98		3.20	4.38
7250	1" x 10"		190	.042		1.59	1.98		3.57	4.79
7260	1" x 12"		180	.044		2.05	2.09		4.14	5.45
7340	D & better, 1" x 2"		225	.036		.41	1.67		2.08	3.02
7350	1" x 3"		225	.036		.61	1.67		2.28	3.24
7360	1" x 4"		200	.040		.82	1.88		2.70	3.79
7370	1" x 6"		200	.040		1.12	1.88		3	4.12
7380	1" x 8"		190	.042		1.62	1.98		3.60	4.82
7390	1" x 10"		190	.042		2.03	1.98		4.01	5.25
7400	1" x 12"		180	.044		2.63	2.09		4.72	6.10
7410	Redwood, clear all heart, 1" x 2"		225	.036		.63	1.67		2.30	3.27
7420	1" x 3"		225	.036		.94	1.67		2.61	3.61
7430	1" x 4"		200	.040		1.19	1.88		3.07	4.20
7440	1" x 6"		200	.040		1.78	1.88		3.66	4.84
7450	1" x 8"		190	.042		2.36	1.98		4.34	5.65
7460	1" x 10"		190	.042		3.96	1.98		5.94	7.40
7470	1" x 12"		180	.044		4.75	2.09		6.84	8.45
7480	2" x 4"		200	.040		2.28	1.88		4.16	5.40
7490	2" x 6"		182	.044		3.43	2.06		5.49	6.95
7500	2" x 8"		165	.048		4.56	2.28		6.84	8.50
7630	Soffit, cedar, rough sawn, 1" x 2"	2 Carp	440	.036		.54	1.71		2.25	3.23
7640	1" x 3"		440	.036		.81	1.71		2.52	3.52
7650	1" x 4"		420	.038		1.07	1.79		2.86	3.93
7660	1" x 6"		420	.038		1.61	1.79		3.40	4.52
7670	1" x 8"		420	.038		2.15	1.79		3.94	5.10
7680	1" x 10"		400	.040		2.67	1.88		4.55	5.85
7690	1" x 12"		400	.040		3.21	1.88		5.09	6.40
7700	STK, 1" x 2"		440	.036		.51	1.71		2.22	3.19
7710	1" x 3"		440	.036		.56	1.71		2.27	3.25
7720	1" x 4"		420	.038		.65	1.79		2.44	3.46
7730	1" x 6"		420	.038		1.03	1.79		2.82	3.88

06 22 Millwork

06 22 13 – Standard Pattern Wood Trim

06 22 13.40 Moldings, Exterior		Crew	Daily Output	Labor-Hours	Unit	Material (2015 Bare Costs)	Labor (2015 Bare Costs)	Equipment (2015 Bare Costs)	Total (2015 Bare Costs)	Total Incl O&P
7740	1" x 8"	2 Carp	420	.038	L.F.	1.60	1.79		3.39	4.51
7750	1" x 10"		400	.040		2.14	1.88		4.02	5.25
7760	1" x 12"		400	.040		3.54	1.88		5.42	6.80
7770	Pine, #2, 1" x 2"		440	.036		.25	1.71		1.96	2.91
7780	1" x 3"		440	.036		.40	1.71		2.11	3.07
7790	1" x 4"		420	.038		.49	1.79		2.28	3.29
7800	1" x 6"		420	.038		.75	1.79		2.54	3.57
7810	1" x 8"		420	.038		1.22	1.79		3.01	4.09
7820	1" x 10"		400	.040		1.59	1.88		3.47	4.64
7830	1" x 12"		400	.040		2.05	1.88		3.93	5.15
7840	D & better, 1" x 2"		440	.036		.41	1.71		2.12	3.08
7850	1" x 3"		440	.036		.61	1.71		2.32	3.30
7860	1" x 4"		420	.038		.82	1.79		2.61	3.65
7870	1" x 6"		420	.038		1.12	1.79		2.91	3.98
7880	1" x 8"		420	.038		1.62	1.79		3.41	4.53
7890	1" x 10"		400	.040		2.03	1.88		3.91	5.10
7900	1" x 12"		400	.040		2.63	1.88		4.51	5.80
7910	Redwood, clear all heart, 1" x 2"		440	.036		.63	1.71		2.34	3.33
7920	1" x 3"		440	.036		.94	1.71		2.65	3.67
7930	1" x 4"		420	.038		1.19	1.79		2.98	4.06
7940	1" x 6"		420	.038		1.78	1.79		3.57	4.70
7950	1" x 8"		420	.038		2.36	1.79		4.15	5.35
7960	1" x 10"		400	.040		3.96	1.88		5.84	7.25
7970	1" x 12"		400	.040		4.75	1.88		6.63	8.15
8050	Trim, crown molding, pine, 11/16" x 4-1/4"	1 Carp	250	.032		4.54	1.50		6.04	7.30
8060	Back band, 11/16" x 1-1/16"		250	.032		.99	1.50		2.49	3.40
8070	Insect screen frame stock, 1-1/16" x 1-3/4"		395	.020		2.39	.95		3.34	4.09
8080	Dentils, 2-1/2" x 2-1/2" x 4", 6" O.C.		30	.267		1.22	12.50		13.72	20.50
8100	Fluted, 5-1/2"		165	.048		5.05	2.28		7.33	9.05
8110	Stucco bead, 1-3/8" x 1-5/8"		250	.032		2.50	1.50		4	5.05

06 22 13.45 Moldings, Trim		Crew	Daily Output	Labor-Hours	Unit	Material	Labor	Equipment	Total	Total Incl O&P
0010	**MOLDINGS, TRIM**									
0200	Astragal, stock pine, 11/16" x 1-3/4"	1 Carp	255	.031	L.F.	1.42	1.47		2.89	3.83
0250	1-5/16" x 2-3/16"		240	.033		2.10	1.56		3.66	4.72
0800	Chair rail, stock pine, 5/8" x 2-1/2"		270	.030		1.58	1.39		2.97	3.88
0900	5/8" x 3-1/2"		240	.033		2.40	1.56		3.96	5.05
1000	Closet pole, stock pine, 1-1/8" diameter		200	.040		1.16	1.88		3.04	4.17
1100	Fir, 1-5/8" diameter		200	.040		2.20	1.88		4.08	5.30
3300	Half round, stock pine, 1/4" x 1/2"		270	.030		.24	1.39		1.63	2.40
3350	1/2" x 1"		255	.031		.73	1.47		2.20	3.07
3400	Handrail, fir, single piece, stock, hardware not included									
3450	1-1/2" x 1-3/4"	1 Carp	80	.100	L.F.	2.56	4.70		7.26	10.05
3470	Pine, 1-1/2" x 1-3/4"		80	.100		2.40	4.70		7.10	9.90
3500	1-1/2" x 2-1/2"		76	.105		2.46	4.94		7.40	10.30
3600	Lattice, stock pine, 1/4" x 1-1/8"		270	.030		.35	1.39		1.74	2.53
3700	1/4" x 1-3/4"		250	.032		.92	1.50		2.42	3.32
3800	Miscellaneous, custom, pine, 1" x 1"		270	.030		.44	1.39		1.83	2.62
3850	1" x 2"		265	.030		.87	1.42		2.29	3.14
3900	1" x 3"		240	.033		1.31	1.56		2.87	3.85
4100	Birch or oak, nominal 1" x 1"		240	.033		.42	1.56		1.98	2.87
4200	Nominal 1" x 3"		215	.037		1.26	1.75		3.01	4.07
4400	Walnut, nominal 1" x 1"		215	.037		.66	1.75		2.41	3.42

06 22 13 – Standard Pattern Wood Trim

06 22 13.45 Moldings, Trim

	06 22 13.45 Moldings, Trim	Crew	Daily Output	Labor-Hours	Unit	Material	Labor	Equipment	Total	Total Incl O&P
						2015 Bare Costs				
4500	Nominal 1" x 3"	1 Carp	200	.040	L.F.	1.99	1.88		3.87	5.10
4700	Teak, nominal 1" x 1"		215	.037		2.80	1.75		4.55	5.75
4800	Nominal 1" x 3"		200	.040		8.40	1.88		10.28	12.15
4900	Quarter round, stock pine, 1/4" x 1/4"		275	.029		.24	1.37		1.61	2.36
4950	3/4" x 3/4"		255	.031		.50	1.47		1.97	2.82
5600	Wainscot moldings, 1-1/8" x 9/16", 2' high, minimum		76	.105	S.F.	11.65	4.94		16.59	20.50
5700	Maximum		65	.123	"	15.75	5.80		21.55	26

06 22 13.50 Moldings, Window and Door

	06 22 13.50 Moldings, Window and Door	Crew	Daily Output	Labor-Hours	Unit	Material	Labor	Equipment	Total	Total Incl O&P
0010	**MOLDINGS, WINDOW AND DOOR**									
2800	Door moldings, stock, decorative, 1-1/8" wide, plain	1 Carp	17	.471	Set	47.50	22		69.50	86.50
2900	Detailed		17	.471	"	91.50	22		113.50	135
2960	Clear pine door jamb, no stops, 11/16" x 4-9/16"		240	.033	L.F.	5.20	1.56		6.76	8.10
3150	Door trim set, 1 head and 2 sides, pine, 2-1/2 wide		12	.667	Opng.	24	31.50		55.50	74
3170	3-1/2" wide		11	.727	"	29	34		63	84.50
3250	Glass beads, stock pine, 3/8" x 1/2"		275	.029	L.F.	.33	1.37		1.70	2.46
3270	3/8" x 7/8"		270	.030		.42	1.39		1.81	2.60
4850	Parting bead, stock pine, 3/8" x 3/4"		275	.029		.44	1.37		1.81	2.58
4870	1/2" x 3/4"		255	.031		.42	1.47		1.89	2.73
5000	Stool caps, stock pine, 11/16" x 3-1/2"		200	.040		2.07	1.88		3.95	5.15
5100	1-1/16" x 3-1/4"		150	.053		3.24	2.50		5.74	7.40
5300	Threshold, oak, 3' long, inside, 5/8" x 3-5/8"		32	.250	Ea.	6.50	11.75		18.25	25
5400	Outside, 1-1/2" x 7-5/8"		16	.500	"	45	23.50		68.50	85.50
5900	Window trim sets, including casings, header, stops,									
5910	stool and apron, 2-1/2" wide, FJP	1 Carp	13	.615	Opng.	31.50	29		60.50	79
5950	Pine		10	.800		37	37.50		74.50	99
6000	Oak		6	1.333		64.50	62.50		127	168

06 22 13.60 Moldings, Soffits

	06 22 13.60 Moldings, Soffits	Crew	Daily Output	Labor-Hours	Unit	Material	Labor	Equipment	Total	Total Incl O&P
0010	**MOLDINGS, SOFFITS**									
0200	Soffits, pine, 1" x 4"	2 Carp	420	.038	L.F.	.47	1.79		2.26	3.26
0210	1" x 6"		420	.038		.72	1.79		2.51	3.54
0220	1" x 8"		420	.038		1.19	1.79		2.98	4.06
0230	1" x 10"		400	.040		1.55	1.88		3.43	4.59
0240	1" x 12"		400	.040		2.01	1.88		3.89	5.10
0250	STK cedar, 1" x 4"		420	.038		.62	1.79		2.41	3.43
0260	1" x 6"		420	.038		1	1.79		2.79	3.85
0270	1" x 8"		420	.038		1.57	1.79		3.36	4.48
0280	1" x 10"		400	.040		2.10	1.88		3.98	5.20
0290	1" x 12"		400	.040		3.50	1.88		5.38	6.75
1000	Exterior AC plywood, 1/4" thick		400	.040	S.F.	.88	1.88		2.76	3.86
1050	3/8" thick		400	.040		.92	1.88		2.80	3.90
1100	1/2" thick		400	.040		1.10	1.88		2.98	4.10
1150	Polyvinyl chloride, white, solid	1 Carp	230	.035		2.10	1.63		3.73	4.82
1160	Perforated	"	230	.035		2.10	1.63		3.73	4.82
1170	Accessories, "J" channel 5/8"	2 Carp	700	.023	L.F.	.45	1.07		1.52	2.15

06 25 Prefinished Paneling

06 25 13 – Prefinished Hardboard Paneling

06 25 13.10 Paneling, Hardboard		Crew	Daily Output	Labor-Hours	Unit	2015 Bare Costs Material	Labor	Equipment	Total	Total Incl O&P
0010	**PANELING, HARDBOARD**									
0050	Not incl. furring or trim, hardboard, tempered, 1/8" thick G	2 Carp	500	.032	S.F.	.41	1.50		1.91	2.76
0100	1/4" thick G		500	.032		.64	1.50		2.14	3.01
0300	Tempered pegboard, 1/8" thick G		500	.032		.40	1.50		1.90	2.75
0400	1/4" thick G		500	.032		.68	1.50		2.18	3.06
0600	Untempered hardboard, natural finish, 1/8" thick G		500	.032		.42	1.50		1.92	2.77
0700	1/4" thick G		500	.032		.51	1.50		2.01	2.87
0900	Untempered pegboard, 1/8" thick G		500	.032		.44	1.50		1.94	2.79
1000	1/4" thick G		500	.032		.48	1.50		1.98	2.84
1200	Plastic faced hardboard, 1/8" thick G		500	.032		.66	1.50		2.16	3.04
1300	1/4" thick G		500	.032		.89	1.50		2.39	3.29
1500	Plastic faced pegboard, 1/8" thick G		500	.032		.67	1.50		2.17	3.05
1600	1/4" thick G		500	.032		.85	1.50		2.35	3.25
1800	Wood grained, plain or grooved, 1/8" thick G		500	.032		.69	1.50		2.19	3.07
1900	1/4" thick G		425	.038	↓	1.40	1.77		3.17	4.26
2100	Moldings, wood grained MDF		500	.032	L.F.	.41	1.50		1.91	2.76
2200	Pine	↓	425	.038	"	1.40	1.77		3.17	4.26

06 25 16 – Prefinished Plywood Paneling

06 25 16.10 Paneling, Plywood		Crew	Daily Output	Labor-Hours	Unit	Material	Labor	Equipment	Total	Total Incl O&P
0010	**PANELING, PLYWOOD** R061636-20									
2400	Plywood, prefinished, 1/4" thick, 4' x 8' sheets									
2410	with vertical grooves. Birch faced, economy	2 Carp	500	.032	S.F.	1.40	1.50		2.90	3.85
2420	Average		420	.038		1.20	1.79		2.99	4.07
2430	Custom		350	.046		1.25	2.15		3.40	4.68
2600	Mahogany, African		400	.040		2.75	1.88		4.63	5.90
2700	Philippine (Lauan)		500	.032		.65	1.50		2.15	3.03
2900	Oak		500	.032		1.40	1.50		2.90	3.85
3000	Cherry		400	.040		2.05	1.88		3.93	5.15
3200	Rosewood		320	.050		2.90	2.35		5.25	6.80
3400	Teak		400	.040		2.90	1.88		4.78	6.10
3600	Chestnut		375	.043		4.80	2		6.80	8.40
3800	Pecan		400	.040		2.50	1.88		4.38	5.65
3900	Walnut, average		500	.032		2.40	1.50		3.90	4.95
3950	Custom		400	.040		5.25	1.88		7.13	8.70
4000	Plywood, prefinished, 3/4" thick, stock grades, economy		320	.050		1.56	2.35		3.91	5.35
4100	Average		224	.071		4.68	3.35		8.03	10.30
4300	Architectural grade, custom		224	.071		5.20	3.35		8.55	10.85
4400	Luxury		160	.100		5.20	4.70		9.90	12.95
4600	Plywood, "A" face, birch, VC, 1/2" thick, natural		450	.036		2.05	1.67		3.72	4.83
4700	Select		450	.036		2.15	1.67		3.82	4.94
4900	Veneer core, 3/4" thick, natural		320	.050		2.24	2.35		4.59	6.05
5000	Select		320	.050		2.44	2.35		4.79	6.30
5200	Lumber core, 3/4" thick, natural		320	.050		3.05	2.35		5.40	6.95
5500	Plywood, knotty pine, 1/4" thick, A2 grade		450	.036		1.70	1.67		3.37	4.44
5600	A3 grade		450	.036		2.10	1.67		3.77	4.88
5800	3/4" thick, veneer core, A2 grade		320	.050		2.15	2.35		4.50	6
5900	A3 grade		320	.050		2.42	2.35		4.77	6.25
6100	Aromatic cedar, 1/4" thick, plywood		400	.040		2.20	1.88		4.08	5.30
6200	1/4" thick, particle board	↓	400	.040	↓	1.05	1.88		2.93	4.05

06 25 Prefinished Paneling

06 25 26 – Panel System

06 25 26.10 Panel Systems		Crew	Daily Output	Labor-Hours	Unit	2015 Bare Costs Material	Labor	Equipment	Total	Total Incl O&P
0010	**PANEL SYSTEMS**									
0100	Raised panel, eng. wood core w/wood veneer, std., paint grade	2 Carp	300	.053	S.F.	11.20	2.50		13.70	16.15
0110	Oak veneer		300	.053		25.50	2.50		28	32
0120	Maple veneer		300	.053		32.50	2.50		35	40
0130	Cherry veneer		300	.053		37	2.50		39.50	45
0300	Class I fire rated, paint grade		300	.053		13	2.50		15.50	18.15
0310	Oak veneer		300	.053		30	2.50		32.50	37
0320	Maple veneer		300	.053		40.50	2.50		43	48.50
0330	Cherry veneer		300	.053		49	2.50		51.50	57.50
0510	Beadboard, 5/8" MDF, standard, primed		300	.053		8.45	2.50		10.95	13.15
0520	Oak veneer, unfinished		300	.053		13.35	2.50		15.85	18.55
0530	Maple veneer, unfinished		300	.053		14.65	2.50		17.15	19.95
0610	Rustic paneling, 5/8" MDF, standard, maple veneer, unfinished		300	.053		18.15	2.50		20.65	24

06 26 Board Paneling

06 26 13 – Profile Board Paneling

06 26 13.10 Paneling, Boards		Crew	Daily Output	Labor-Hours	Unit	Material	Labor	Equipment	Total	Total Incl O&P
0010	**PANELING, BOARDS**									
6400	Wood board paneling, 3/4" thick, knotty pine	2 Carp	300	.053	S.F.	1.91	2.50		4.41	5.95
6500	Rough sawn cedar		300	.053		3.20	2.50		5.70	7.35
6700	Redwood, clear, 1" x 4" boards		300	.053		4.96	2.50		7.46	9.30
6900	Aromatic cedar, closet lining, boards		275	.058		2.32	2.73		5.05	6.75

06 43 Wood Stairs and Railings

06 43 13 – Wood Stairs

06 43 13.20 Prefabricated Wood Stairs		Crew	Daily Output	Labor-Hours	Unit	Material	Labor	Equipment	Total	Total Incl O&P
0010	**PREFABRICATED WOOD STAIRS**									
0100	Box stairs, prefabricated, 3'-0" wide									
0110	Oak treads, up to 14 risers	2 Carp	39	.410	Riser	92.50	19.25		111.75	132
0600	With pine treads for carpet, up to 14 risers	"	39	.410	"	59.50	19.25		78.75	95
1100	For 4' wide stairs, add				Flight	25%				
1550	Stairs, prefabricated stair handrail with balusters	1 Carp	30	.267	L.F.	78.50	12.50		91	105
1700	Basement stairs, prefabricated, pine treads									
1710	Pine risers, 3' wide, up to 14 risers	2 Carp	52	.308	Riser	59.50	14.45		73.95	87.50
4000	Residential, wood, oak treads, prefabricated		1.50	10.667	Flight	1,200	500		1,700	2,100
4200	Built in place		.44	36.364	"	2,175	1,700		3,875	5,000
4400	Spiral, oak, 4'-6" diameter, unfinished, prefabricated,									
4500	incl. railing, 9' high	2 Carp	1.50	10.667	Flight	3,425	500		3,925	4,550

06 43 13.40 Wood Stair Parts		Crew	Daily Output	Labor-Hours	Unit	Material	Labor	Equipment	Total	Total Incl O&P
0010	**WOOD STAIR PARTS**									
0020	Pin top balusters, 1-1/4", oak, 34"	1 Carp	96	.083	Ea.	5.10	3.91		9.01	11.60
0030	38"		96	.083		5.85	3.91		9.76	12.45
0040	42"		96	.083		6.35	3.91		10.26	13
0050	Poplar, 34"		96	.083		3.05	3.91		6.96	9.35
0060	38"		96	.083		3.88	3.91		7.79	10.25
0070	42"		96	.083		8.85	3.91		12.76	15.70
0080	Maple, 34"		96	.083		4.90	3.91		8.81	11.40
0090	38"		96	.083		5.60	3.91		9.51	12.20

06 43 13.40 Wood Stair Parts		Crew	Daily Output	Labor-Hours	Unit	Material	Labor	Equipment	Total	Total Incl O&P
						2015 Bare Costs				
0100	42"	1 Carp	96	.083	Ea.	6.50	3.91		10.41	13.15
0130	Primed, 34"		96	.083		3	3.91		6.91	9.30
0140	38"		96	.083		3.62	3.91		7.53	10
0150	42"		96	.083		4.42	3.91		8.33	10.85
0180	Box top balusters, 1-1/4", oak, 34"		60	.133		8.95	6.25		15.20	19.50
0190	38"		60	.133		9.95	6.25		16.20	20.50
0200	42"		60	.133		10.95	6.25		17.20	21.50
0210	Poplar, 34"		60	.133		6.25	6.25		12.50	16.55
0220	38"		60	.133		6.95	6.25		13.20	17.30
0230	42"		60	.133		7.50	6.25		13.75	17.90
0240	Maple, 34"		60	.133		8.25	6.25		14.50	18.75
0250	38"		60	.133		9	6.25		15.25	19.55
0260	42"		60	.133		10	6.25		16.25	20.50
0290	Primed, 34"		60	.133		7	6.25		13.25	17.35
0300	38"		60	.133		8	6.25		14.25	18.45
0310	42"		60	.133		8.35	6.25		14.60	18.85
0340	Square balusters, cut from lineal stock, pine, 1-1/16" x 1-1/16"		180	.044	L.F.	1.50	2.09		3.59	4.86
0350	1-5/16" x 1-5/16"		180	.044		2.10	2.09		4.19	5.50
0360	1-5/8" x 1-5/8"		180	.044		3.30	2.09		5.39	6.85
0370	Turned newel, oak, 3-1/2" square, 48" high		8	1	Ea.	92	47		139	174
0380	62" high		8	1		92	47		139	174
0390	Poplar, 3-1/2" square, 48" high		8	1		54	47		101	132
0400	62" high		8	1		68	47		115	148
0410	Maple, 3-1/2" square, 48" high		8	1		72	47		119	152
0420	62" high		8	1		92	47		139	174
0430	Square newel, oak, 3-1/2" square, 48" high		8	1		54	47		101	132
0440	58" high		8	1		68	47		115	148
0450	Poplar, 3-1/2" square, 48" high		8	1		35	47		82	111
0460	58" high		8	1		42	47		89	119
0470	Maple, 3" square, 48" high		8	1		52	47		99	130
0480	58" high		8	1		64	47		111	143
0490	Railings, oak, economy		96	.083	L.F.	8.65	3.91		12.56	15.50
0500	Average		96	.083		13	3.91		16.91	20.50
0510	Custom		96	.083		16.50	3.91		20.41	24
0520	Maple, economy		96	.083		11	3.91		14.91	18.10
0530	Average		96	.083		13.50	3.91		17.41	21
0540	Custom		96	.083		16.95	3.91		20.86	24.50
0550	Oak, for bending rail, economy		48	.167		23.50	7.85		31.35	38
0560	Average		48	.167		26	7.85		33.85	40.50
0570	Custom		48	.167		29.50	7.85		37.35	44
0580	Maple, for bending rail, economy		48	.167		32	7.85		39.85	47
0590	Average		48	.167		32	7.85		39.85	47
0600	Custom		48	.167		32	7.85		39.85	47
0610	Risers, oak, 3/4" x 8", 36" long		80	.100	Ea.	13	4.70		17.70	21.50
0620	42" long		70	.114		15.15	5.35		20.50	25
0630	48" long		63	.127		17.30	5.95		23.25	28
0640	54" long		56	.143		19.50	6.70		26.20	32
0650	60" long		50	.160		21.50	7.50		29	35.50
0660	72" long		42	.190		26	8.95		34.95	42.50
0670	Poplar, 3/4" x 8", 36" long		80	.100		12.50	4.70		17.20	21
0680	42" long		71	.113		14.55	5.30		19.85	24
0690	48" long		63	.127		16.65	5.95		22.60	27.50
0700	54" long		56	.143		18.70	6.70		25.40	31

06 43 13 – Wood Stairs

06 43 13.40 Wood Stair Parts		Crew	Daily Output	Labor-Hours	Unit	Material	2015 Bare Costs Labor	Equipment	Total	Total Incl O&P
0710	60" long	1 Carp	50	.160	Ea.	21	7.50		28.50	34.50
0720	72" long		42	.190		25	8.95		33.95	41.50
0730	Pine, 1" x 8", 36" long		80	.100		3.57	4.70		8.27	11.20
0740	42" long		70	.114		4.17	5.35		9.52	12.85
0750	48" long		63	.127		4.76	5.95		10.71	14.40
0760	54" long		56	.143		5.35	6.70		12.05	16.20
0770	60" long		50	.160		5.95	7.50		13.45	18.10
0780	72" long		42	.190		7.15	8.95		16.10	21.50
0790	Treads, oak, no returns, 1-1/32" x 11-1/2" x 36" long		32	.250		27	11.75		38.75	47.50
0800	42" long		32	.250		31.50	11.75		43.25	52.50
0810	48" long		32	.250		36	11.75		47.75	57.50
0820	54" long		32	.250		40.50	11.75		52.25	62.50
0830	60" long		32	.250		45	11.75		56.75	67.50
0840	72" long		32	.250		54	11.75		65.75	77.50
0850	Mitred return one end, 1-1/32" x 11-1/2" x 36" long		24	.333		36	15.65		51.65	63.50
0860	42" long		24	.333		42	15.65		57.65	70
0870	48" long		24	.333		48	15.65		63.65	77
0880	54" long		24	.333		54	15.65		69.65	83.50
0890	60" long		24	.333		60	15.65		75.65	90
0900	72" long		24	.333		72	15.65		87.65	103
0910	Mitred return two ends, 1-1/32" x 11-1/2" x 36" long		12	.667		46	31.50		77.50	98.50
0920	42" long		12	.667		53.50	31.50		85	107
0930	48" long		12	.667		61.50	31.50		93	116
0940	54" long		12	.667		69	31.50		100.50	124
0950	60" long		12	.667		76.50	31.50		108	133
0960	72" long		12	.667		92	31.50		123.50	149
0970	Starting step, oak, 48", bullnose		8	1		172	47		219	262
0980	Double end bullnose		8	1		254	47		301	350
1030	Skirt board, pine, 1" x 10"		55	.145	L.F.	1.55	6.85		8.40	12.20
1040	1" x 12"		52	.154	"	2.01	7.20		9.21	13.30
1050	Oak landing tread, 1-1/16" thick		54	.148	S.F.	9	6.95		15.95	20.50
1060	Oak cove molding		96	.083	L.F.	1	3.91		4.91	7.10
1070	Oak stringer molding		96	.083	"	4	3.91		7.91	10.40
1090	Rail bolt, 5/16" x 3-1/2"		48	.167	Ea.	2.75	7.85		10.60	15.10
1100	5/16" x 4-1/2"		48	.167		2.75	7.85		10.60	15.10
1120	Newel post anchor		16	.500		13	23.50		36.50	50.50
1130	Tapered plug, 1/2"		240	.033		1	1.56		2.56	3.51
1140	1"		240	.033		.99	1.56		2.55	3.50

06 43 16 – Wood Railings

06 43 16.10 Wood Handrails and Railings		Crew	Daily Output	Labor-Hours	Unit	Material	Labor	Equipment	Total	Total Incl O&P
0010	**WOOD HANDRAILS AND RAILINGS**									
0020	Custom design, architectural grade, hardwood, plain	1 Carp	38	.211	L.F.	12.05	9.90		21.95	28.50
0100	Shaped		30	.267		62.50	12.50		75	88.50
0300	Stock interior railing with spindles 4" O.C., 4' long		40	.200		38.50	9.40		47.90	57
0400	8' long		48	.167		38.50	7.85		46.35	54.50

06 44 Ornamental Woodwork

06 44 19 – Wood Grilles

06 44 19.10 Grilles		Crew	Daily Output	Labor-Hours	Unit	Material	2015 Bare Costs Labor	Equipment	Total	Total Incl O&P
0010	GRILLES and panels, hardwood, sanded									
0020	2′ x 4′ to 4′ x 8′, custom designs, unfinished, economy	1 Carp	38	.211	S.F.	62	9.90		71.90	83.50
0050	Average		30	.267		69	12.50		81.50	95.50
0100	Custom		19	.421		72	19.75		91.75	110

06 44 33 – Wood Mantels

06 44 33.10 Fireplace Mantels		Crew	Daily Output	Labor-Hours	Unit	Material	Labor	Equipment	Total	Total Incl O&P
0010	FIREPLACE MANTELS									
0015	6″ molding, 6′ x 3′-6″ opening, plain, paint grade	1 Carp	5	1.600	Opng.	440	75		515	595
0100	Ornate, oak		5	1.600		590	75		665	765
0300	Prefabricated pine, colonial type, stock, deluxe		2	4		1,500	188		1,688	1,950
0400	Economy		3	2.667		690	125		815	955

06 44 33.20 Fireplace Mantel Beam		Crew	Daily Output	Labor-Hours	Unit	Material	Labor	Equipment	Total	Total Incl O&P
0010	FIREPLACE MANTEL BEAM									
0020	Rough texture wood, 4″ x 8″	1 Carp	36	.222	L.F.	8.30	10.45		18.75	25
0100	4″ x 10″		35	.229	″	10.90	10.75		21.65	28.50
0300	Laminated hardwood, 2-1/4″ x 10-1/2″ wide, 6′ long		5	1.600	Ea.	110	75		185	237
0400	8′ long		5	1.600	″	150	75		225	281
0600	Brackets for above, rough sawn		12	.667	Pr.	10	31.50		41.50	59
0700	Laminated		12	.667	″	15	31.50		46.50	64.50

06 44 39 – Wood Posts and Columns

06 44 39.10 Decorative Beams		Crew	Daily Output	Labor-Hours	Unit	Material	Labor	Equipment	Total	Total Incl O&P
0010	DECORATIVE BEAMS									
0020	Rough sawn cedar, non-load bearing, 4″ x 4″	2 Carp	180	.089	L.F.	1.19	4.17		5.36	7.70
0100	4″ x 6″		170	.094		1.79	4.42		6.21	8.75
0200	4″ x 8″		160	.100		2.39	4.70		7.09	9.90
0300	4″ x 10″		150	.107		3.63	5		8.63	11.70
0400	4″ x 12″		140	.114		4.61	5.35		9.96	13.30
0500	8″ x 8″		130	.123		4.78	5.80		10.58	14.15

06 44 39.20 Columns		Crew	Daily Output	Labor-Hours	Unit	Material	Labor	Equipment	Total	Total Incl O&P
0010	COLUMNS									
0050	Aluminum, round colonial, 6″ diameter	2 Carp	80	.200	V.L.F.	19	9.40		28.40	35.50
0100	8″ diameter		62.25	.257		22	12.05		34.05	42.50
0200	10″ diameter		55	.291		22.50	13.65		36.15	46
0250	Fir, stock units, hollow round, 6″ diameter		80	.200		29.50	9.40		38.90	47
0300	8″ diameter		80	.200		35.50	9.40		44.90	53.50
0350	10″ diameter		70	.229		44.50	10.75		55.25	65.50
0360	12″ diameter		65	.246		54.50	11.55		66.05	78
0400	Solid turned, to 8′ high, 3-1/2″ diameter		80	.200		9.60	9.40		19	25
0500	4-1/2″ diameter		75	.213		11.90	10		21.90	28.50
0600	5-1/2″ diameter		70	.229		16	10.75		26.75	34
0800	Square columns, built-up, 5″ x 5″		65	.246		14.20	11.55		25.75	33.50
0900	Solid, 3-1/2″ x 3-1/2″		130	.123		9.60	5.80		15.40	19.45
1600	Hemlock, tapered, T&G, 12″ diam., 10′ high		100	.160		39.50	7.50		47	55
1700	16′ high		65	.246		73	11.55		84.55	98
1900	14″ diameter, 10′ high		100	.160		113	7.50		120.50	137
2000	18′ high		65	.246		103	11.55		114.55	131
2200	18″ diameter, 12′ high		65	.246		165	11.55		176.55	200
2300	20′ high		50	.320		118	15		133	153
2500	20″ diameter, 14′ high		40	.400		180	18.80		198.80	227
2600	20′ high		35	.457		170	21.50		191.50	220
2800	For flat pilasters, deduct					33%				

06 44 Ornamental Woodwork

06 44 39 – Wood Posts and Columns

06 44 39.20 Columns		Crew	Daily Output	Labor-Hours	Unit	Material	2015 Bare Costs Labor	Equipment	Total	Total Incl O&P
3000	For splitting into halves, add				Ea.	106			106	117
4000	Rough sawn cedar posts, 4" x 4"	2 Carp	250	.064	V.L.F.	3.90	3		6.90	8.90
4100	4" x 6"		235	.068		6.80	3.20		10	12.40
4200	6" x 6"		220	.073		9.90	3.41		13.31	16.15
4300	8" x 8"	↓	200	.080	↓	19.20	3.76		22.96	27

06 48 Wood Frames

06 48 13 – Exterior Wood Door Frames

06 48 13.10 Exterior Wood Door Frames and Accessories		Crew	Daily Output	Labor-Hours	Unit	Material	Labor	Equipment	Total	Total Incl O&P
0010	EXTERIOR WOOD DOOR FRAMES AND ACCESSORIES									
0400	Exterior frame, incl. ext. trim, pine, 5/4 x 4-9/16" deep	2 Carp	375	.043	L.F.	6.60	2		8.60	10.40
0420	5-3/16" deep		375	.043		7.90	2		9.90	11.75
0440	6-9/16" deep		375	.043		8.80	2		10.80	12.75
0600	Oak, 5/4 x 4-9/16" deep		350	.046		19.75	2.15		21.90	25
0620	5-3/16" deep		350	.046		21.50	2.15		23.65	27.50
0640	6-9/16" deep		350	.046		19.50	2.15		21.65	25
1000	Sills, 8/4 x 8" deep, oak, no horns		100	.160		6.65	7.50		14.15	18.85
1020	2" horns		100	.160		20.50	7.50		28	34
1040	3" horns		100	.160		20.50	7.50		28	34
1100	8/4 x 10" deep, oak, no horns		90	.178		6.40	8.35		14.75	19.90
1120	2" horns		90	.178		26.50	8.35		34.85	42
1140	3" horns		90	.178	↓	26.50	8.35		34.85	42
2000	Wood frame & trim, ext, colonial, 3' opng, fluted pilasters, flat head		22	.727	Ea.	505	34		539	610
2010	Dentil head		21	.762		580	36		616	690
2020	Ram's head		20	.800		695	37.50		732.50	820
2100	5'-4" opening, in-swing, fluted pilasters, flat head		17	.941		440	44		484	555
2120	Ram's head		15	1.067		1,400	50		1,450	1,600
2140	Out swing, fluted pilasters, flat head		17	.941		520	44		564	640
2160	Ram's head		15	1.067		1,475	50		1,525	1,700
2400	6'-0" opening, in-swing, fluted pilasters, flat head		16	1		520	47		567	645
2420	Ram's head		10	1.600		1,475	75		1,550	1,750
2460	Out-swing, fluted pilasters, flat head		16	1		520	47		567	645
2480	Ram's head		10	1.600	↓	1,475	75		1,550	1,750
2600	For two sidelights, flat head, add		30	.533	Opng.	226	25		251	288
2620	Ram's head, add		20	.800	"	840	37.50		877.50	985
2700	Custom birch frame, 3'-0" opening		16	1	Ea.	240	47		287	335
2750	6'-0" opening		16	1		360	47		407	470
2900	Exterior, modern, plain trim, 3' opng., in-swing, FJP		26	.615		46.50	29		75.50	95.50
2920	Fir		24	.667		55	31.50		86.50	108
2940	Oak	↓	22	.727	↓	63	34		97	122

06 48 16 – Interior Wood Door Frames

06 48 16.10 Interior Wood Door Jamb and Frames		Crew	Daily Output	Labor-Hours	Unit	Material	Labor	Equipment	Total	Total Incl O&P
0010	INTERIOR WOOD DOOR JAMB AND FRAMES									
3000	Interior frame, pine, 11/16" x 3-5/8" deep	2 Carp	375	.043	L.F.	4.44	2		6.44	7.95
3020	4-9/16" deep		375	.043		4.92	2		6.92	8.50
3200	Oak, 11/16" x 3-5/8" deep		350	.046		9.85	2.15		12	14.10
3220	4-9/16" deep		350	.046		9.95	2.15		12.10	14.25
3240	5-3/16" deep		350	.046		13.90	2.15		16.05	18.55
3400	Walnut, 11/16" x 3-5/8" deep		350	.046		9	2.15		11.15	13.20
3420	4-9/16" deep	↓	350	.046	↓	9.45	2.15		11.60	13.70

06 48 Wood Frames

06 48 16 – Interior Wood Door Frames

06 48 16.10 Interior Wood Door Jamb and Frames		Crew	Daily Output	Labor-Hours	Unit	2015 Bare Costs Material	Labor	Equipment	Total	Total Incl O&P
3440	5-3/16" deep	2 Carp	350	.046	L.F.	9.55	2.15		11.70	13.80
3600	Pocket door frame		16	1	Ea.	81	47		128	162
3800	Threshold, oak, 5/8" x 3-5/8" deep		200	.080	L.F.	3.56	3.76		7.32	9.70
3820	4-5/8" deep		190	.084		4.13	3.95		8.08	10.65
3840	5-5/8" deep		180	.089		6.50	4.17		10.67	13.55

06 49 Wood Screens and Exterior Wood Shutters

06 49 19 – Exterior Wood Shutters

06 49 19.10 Shutters, Exterior

		Crew	Daily Output	Labor-Hours	Unit	Material	Labor	Equipment	Total	Total Incl O&P
0010	**SHUTTERS, EXTERIOR**									
0012	Aluminum, louvered, 1'-4" wide, 3'-0" long	1 Carp	10	.800	Pr.	200	37.50		237.50	278
0400	6'-8" long		9	.889		355	41.50		396.50	455
1000	Pine, louvered, primed, each 1'-2" wide, 3'-3" long		10	.800		216	37.50		253.50	296
1100	4'-7" long		10	.800		270	37.50		307.50	355
1500	Each 1'-6" wide, 3'-3" long		10	.800		244	37.50		281.50	325
1600	4'-7" long		10	.800		325	37.50		362.50	415
1620	Cedar, louvered, 1'-2" wide, 5'-7" long		10	.800		315	37.50		352.50	405
1630	Each 1'-4" wide, 2'-2" long		10	.800		175	37.50		212.50	251
1670	4'-3" long		10	.800		275	37.50		312.50	365
1690	5'-11" long		10	.800		350	37.50		387.50	445
1700	Door blinds, 6'-9" long, each 1'-3" wide		9	.889		380	41.50		421.50	480
1710	1'-6" wide		9	.889		435	41.50		476.50	545
1720	Cedar, solid raised panel, each 1'-4" wide, 3'-3" long		10	.800		315	37.50		352.50	405
1740	4'-3" long		10	.800		365	37.50		402.50	460
1770	5'-11" long		10	.800		520	37.50		557.50	630
1800	Door blinds, 6'-9" long, each 1'-3" wide		9	.889		535	41.50		576.50	655
1900	1'-6" wide		9	.889		630	41.50		671.50	760
2500	Polystyrene, solid raised panel, each 1'-4" wide, 3'-3" long		10	.800		71.50	37.50		109	137
2700	4'-7" long		10	.800		105	37.50		142.50	174
4500	Polystyrene, louvered, each 1'-2" wide, 3'-3" long		10	.800		36	37.50		73.50	97.50
4750	5'-3" long		10	.800		59	37.50		96.50	123
6000	Vinyl, louvered, each 1'-2" x 4'-7" long		10	.800		64	37.50		101.50	129
6200	Each 1'-4" x 6'-8" long		9	.889		76	41.50		117.50	148

06 51 Structural Plastic Shapes and Plates

06 51 13 – Plastic Lumber

06 51 13.10 Recycled Plastic Lumber

			Crew	Daily Output	Labor-Hours	Unit	Material	Labor	Equipment	Total	Total Incl O&P
0010	**RECYCLED PLASTIC LUMBER**										
4000	Sheeting, recycled plastic, black or white, 4' x 8' x 1/8"	G	2 Carp	1100	.015	S.F.	1.26	.68		1.94	2.44
4010	4' x 8' x 3/16"	G		1100	.015		1.90	.68		2.58	3.14
4020	4' x 8' x 1/4"	G		950	.017		2.25	.79		3.04	3.70
4030	4' x 8' x 3/8"	G		950	.017		3.80	.79		4.59	5.40
4040	4' x 8' x 1/2"	G		900	.018		5	.83		5.83	6.80
4050	4' x 8' x 5/8"	G		900	.018		7.50	.83		8.33	9.55
4060	4' x 8' x 3/4"	G		850	.019		9.40	.88		10.28	11.70
4070	Add for colors	G				Ea.	5%				
8500	100% recycled plastic, var colors, NLB, 2" x 2"	G				L.F.	1.76			1.76	1.94
8510	2" x 4"	G					3.65			3.65	4.02
8520	2" x 6"	G					5.75			5.75	6.35

06 51 13 – Plastic Lumber

06 51 13.10 Recycled Plastic Lumber		Crew	Daily Output	Labor-Hours	Unit	Material	2015 Bare Costs Labor	Equipment	Total	Total Incl O&P
8530	2" x 8" G				L.F.	7.90			7.90	8.70
8540	2" x 10" G					11.50			11.50	12.65
8550	5/4" x 4" G					4.45			4.45	4.90
8560	5/4" x 6" G					5			5	5.50
8570	1" x 6" G					2.87			2.87	3.16
8580	1/2" x 8" G					3.05			3.05	3.36
8590	2" x 10" T & G G					11.50			11.50	12.65
8600	3" x 10" T & G G					15.30			15.30	16.85
8610	Add for premium colors G				↓	20%				

06 51 13.12 Structural Plastic Lumber		Crew	Daily Output	Labor-Hours	Unit	Material	Labor	Equipment	Total	Total Incl O&P
0010	**STRUCTURAL PLASTIC LUMBER**									
1320	Plastic lumber, posts or columns, 4" x 4"	2 Carp	390	.041	L.F.	9	1.93		10.93	12.85
1325	4" x 6"		275	.058		13.15	2.73		15.88	18.70
1330	4" x 8"		220	.073		19.20	3.41		22.61	26.50
1340	Girder, single, 4" x 4"		675	.024		9	1.11		10.11	11.60
1345	4" x 6"		600	.027		13.15	1.25		14.40	16.45
1350	4" x 8"		525	.030		19.20	1.43		20.63	23
1352	Double, 2" x 4"		625	.026		8.25	1.20		9.45	10.90
1354	2" x 6"		600	.027		12.65	1.25		13.90	15.90
1356	2" x 8"		575	.028		16.10	1.31		17.41	19.70
1358	2" x 10"		550	.029		20	1.37		21.37	24
1360	2" x 12"		525	.030		25	1.43		26.43	29.50
1362	Triple, 2" x 4"		575	.028		12.35	1.31		13.66	15.60
1364	2" x 6"		550	.029		19	1.37		20.37	23
1366	2" x 8"		525	.030		24	1.43		25.43	28.50
1368	2" x 10"		500	.032		30	1.50		31.50	35.50
1370	2" x 12"		475	.034		37	1.58		38.58	43.50
1372	Ledger, bolted 4' O.C., 2" x 4"		400	.040		4.26	1.88		6.14	7.60
1374	2" x 6"		550	.029		6.40	1.37		7.77	9.15
1376	2" x 8"		390	.041		8.15	1.93		10.08	11.95
1378	2" x 10"		385	.042		10.10	1.95		12.05	14.10
1380	2" x 12"		380	.042		12.50	1.98		14.48	16.80
1382	Joists, 2" x 4"		1250	.013		4.12	.60		4.72	5.45
1384	2" x 6"		1250	.013		6.35	.60		6.95	7.85
1386	2" x 8"		1100	.015		8.05	.68		8.73	9.90
1388	2" x 10"		500	.032		10.10	1.50		11.60	13.40
1390	2" x 12"	↓	875	.018		12.40	.86		13.26	14.95
1392	Railings and trim , 5/4" x 4"	1 Carp	300	.027		4.25	1.25		5.50	6.60
1394	2" x 2"		300	.027		2.35	1.25		3.60	4.52
1396	2" x 4"		300	.027		4.10	1.25		5.35	6.45
1398	2" x 6"	↓	300	.027	↓	6.30	1.25		7.55	8.90

06 52 10 – Fiberglass Structural Assemblies

06 52 10.10 Castings, Fiberglass		Crew	Daily Output	Labor-Hours	Unit	2015 Bare Costs Material	Labor	Equipment	Total	Total Incl O&P
0010	CASTINGS, FIBERGLASS									
0100	Angle, 1" x 1" x 1/8" thick	2 Sswk	240	.067	L.F.	2.32	3.51		5.83	8.65
0120	3" x 3" x 1/4" thick		200	.080		6.75	4.21		10.96	14.70
0140	4" x 4" x 1/4" thick		200	.080		9.30	4.21		13.51	17.50
0160	4" x 4" x 3/8" thick		200	.080		13.65	4.21		17.86	22.50
0180	6" x 6" x 1/2" thick		160	.100		34.50	5.25		39.75	47
1000	Flat sheet, 1/8" thick		140	.114	S.F.	6.85	6		12.85	18
1020	1/4" thick		120	.133		11.80	7		18.80	25
1040	3/8" thick		100	.160		18.80	8.40		27.20	35
1060	1/2" thick		80	.200		21.50	10.55		32.05	42.50
2000	Handrail, 42" high, 2" diam. rails pickets 5' O.C.		32	.500	L.F.	50.50	26.50		77	101
3000	Round bar, 1/4" diam.		240	.067		.53	3.51		4.04	6.70
3020	1/2" diam.		200	.080		1.40	4.21		5.61	8.85
3040	3/4" diam.		200	.080		3.75	4.21		7.96	11.45
3060	1" diam.		160	.100		6.40	5.25		11.65	16.20
3080	1-1/4" diam.		160	.100		11.65	5.25		16.90	22
3100	1-1/2" diam.		140	.114		15.60	6		21.60	27.50
3500	Round tube, 1" diam. x 1/8" thick		240	.067		2.77	3.51		6.28	9.15
3520	2" diam. x 1/4" thick		200	.080		8.85	4.21		13.06	17.05
3540	3" diam. x 1/4" thick		160	.100		11.05	5.25		16.30	21.50
4000	Square bar, 1/2" square		240	.067		4.33	3.51		7.84	10.85
4020	1" square		200	.080		6	4.21		10.21	13.90
4040	1-1/2" square		160	.100		11.95	5.25		17.20	22.50
4500	Square tube, 1" x 1" x 1/8" thick		240	.067		3.13	3.51		6.64	9.55
4520	2" x 2" x 1/8" thick		200	.080		6.10	4.21		10.31	14
4540	3" x 3" x 1/4" thick		160	.100		17.05	5.25		22.30	28
5000	Threaded rod, 3/8" diam.		320	.050		4.40	2.63		7.03	9.40
5020	1/2" diam.		320	.050		5.25	2.63		7.88	10.35
5040	5/8" diam.		280	.057		5.60	3.01		8.61	11.40
5060	3/4" diam.		280	.057		6.35	3.01		9.36	12.25
6000	Wide flange beam, 4" x 4" x 1/4" thick		120	.133		13.45	7		20.45	27
6020	6" x 6" x 1/4" thick		100	.160		23.50	8.40		31.90	40.50
6040	8" x 8" x 3/8" thick		80	.200		37	10.55		47.55	59

06 52 10.20 Fiberglass Stair Treads		Crew	Daily Output	Labor-Hours	Unit	Material	Labor	Equipment	Total	Total Incl O&P
0010	FIBERGLASS STAIR TREADS									
0100	24" wide	2 Sswk	52	.308	Ea.	31.50	16.20		47.70	63
0140	30" wide		52	.308		39.50	16.20		55.70	71.50
0180	36" wide		52	.308		47.50	16.20		63.70	80
0220	42" wide		52	.308		55.50	16.20		71.70	89

06 52 10.30 Fiberglass Grating		Crew	Daily Output	Labor-Hours	Unit	Material	Labor	Equipment	Total	Total Incl O&P
0010	FIBERGLASS GRATING									
0100	Molded, green (for mod. corrosive environment)									
0140	1" x 4" mesh, 1" thick	2 Sswk	400	.040	S.F.	8.50	2.11		10.61	13
0180	1-1/2" square mesh, 1" thick		400	.040		18.70	2.11		20.81	24
0220	1-1/4" thick		400	.040		9.35	2.11		11.46	13.90
0260	1-1/2" thick		400	.040		23	2.11		25.11	28.50
0300	2" square mesh, 2" thick		320	.050		26	2.63		28.63	33.50
1000	Orange (for highly corrosive environment)									
1040	1" x 4" mesh, 1" thick	2 Sswk	400	.040	S.F.	16.50	2.11		18.61	22
1080	1-1/2" square mesh, 1" thick		400	.040		19.55	2.11		21.66	25
1120	1-1/4" thick		400	.040		20	2.11		22.11	25.50
1160	1-1/2" thick		400	.040		20.50	2.11		22.61	26

06 52 Plastic Structural Assemblies

06 52 10 – Fiberglass Structural Assemblies

06 52 10.30 Fiberglass Grating		Crew	Daily Output	Labor-Hours	Unit	Material	2015 Bare Costs Labor	Equipment	Total	Total Incl O&P
1200	2" square mesh, 2" thick	2 Sswk	320	.050	S.F.	20.50	2.63		23.13	27
3000	Pultruded, green (for mod. corrosive environment)									
3040	1" O.C. bar spacing, 1" thick	2 Sswk	400	.040	S.F.	14.85	2.11		16.96	20
3080	1-1/2" thick		320	.050		15.30	2.63		17.93	21.50
3120	1-1/2" O.C. bar spacing, 1" thick		400	.040		13.70	2.11		15.81	18.70
3160	1-1/2" thick		400	.040		18.05	2.11		20.16	23.50
4000	Grating support legs, fixed height, no base				Ea.	46.50			46.50	51.50
4040	With base					41			41	45
4080	Adjustable to 60"					61.50			61.50	67.50

06 52 10.40 Fiberglass Floor Grating		Crew	Daily Output	Labor-Hours	Unit	Material	Labor	Equipment	Total	Total Incl O&P
0010	**FIBERGLASS FLOOR GRATING**									
0100	Reinforced polyester, fire retardant, 1" x 4" grid, 1" thick	E-4	510	.063	S.F.	14	3.34	.29	17.63	21.50
0200	1-1/2" x 6" mesh, 1-1/2" thick		500	.064		16.50	3.40	.29	20.19	24.50
0300	With grit surface, 1-1/2" x 6" grid, 1-1/2" thick		500	.064		16.80	3.40	.29	20.49	24.50

06 63 Plastic Railings

06 63 10 – Plastic (PVC) Railings

06 63 10.10 Plastic Railings		Crew	Daily Output	Labor-Hours	Unit	Material	Labor	Equipment	Total	Total Incl O&P
0010	**PLASTIC RAILINGS**									
0100	Horizontal PVC handrail with balusters, 3-1/2" wide, 36" high	1 Carp	96	.083	L.F.	24.50	3.91		28.41	33
0150	42" high		96	.083		28	3.91		31.91	37
0200	Angled PVC handrail with balusters, 3-1/2" wide, 36" high		72	.111		28.50	5.20		33.70	39
0250	42" high		72	.111		31.50	5.20		36.70	42.50
0300	Post sleeve for 4 x 4 post		96	.083		12.80	3.91		16.71	20
0400	Post cap for 4 x 4 post, flat profile		48	.167	Ea.	12.80	7.85		20.65	26
0450	Newel post style profile		48	.167		23.50	7.85		31.35	38
0500	Raised corbeled profile		48	.167		35.50	7.85		43.35	51
0550	Post base trim for 4 x 4 post		96	.083		18.25	3.91		22.16	26

06 65 Plastic Trim

06 65 10 – PVC Trim

06 65 10.10 PVC Trim, Exterior		Crew	Daily Output	Labor-Hours	Unit	Material	Labor	Equipment	Total	Total Incl O&P
0010	**PVC TRIM, EXTERIOR**									
0100	Cornerboards, 5/4" x 6" x 6"	1 Carp	240	.033	L.F.	8.45	1.56		10.01	11.70
0110	Door/window casing, 1" x 4"		200	.040		1.39	1.88		3.27	4.42
0120	1" x 6"		200	.040		2.11	1.88		3.99	5.20
0130	1" x 8"		195	.041		2.78	1.93		4.71	6
0140	1" x 10"		195	.041		3.50	1.93		5.43	6.80
0150	1" x 12"		190	.042		4.33	1.98		6.31	7.80
0160	5/4" x 4"		195	.041		1.69	1.93		3.62	4.82
0170	5/4" x 6"		195	.041		2.72	1.93		4.65	5.95
0180	5/4" x 8"		190	.042		3.56	1.98		5.54	6.95
0190	5/4" x 10"		190	.042		4.56	1.98		6.54	8.05
0200	5/4" x 12"		185	.043		5.25	2.03		7.28	8.90
0210	Fascia, 1" x 4"		250	.032		1.39	1.50		2.89	3.84
0220	1" x 6"		250	.032		2.11	1.50		3.61	4.63
0230	1" x 8"		225	.036		2.78	1.67		4.45	5.65
0240	1" x 10"		225	.036		3.50	1.67		5.17	6.40
0250	1" x 12"		200	.040		4.33	1.88		6.21	7.65

06 65 Plastic Trim

06 65 10 – PVC Trim

06 65 10.10 PVC Trim, Exterior		Crew	Daily Output	Labor-Hours	Unit	2015 Bare Costs Material	Labor	Equipment	Total	Total Incl O&P
0260	5/4" x 4"	1 Carp	240	.033	L.F.	1.69	1.56		3.25	4.27
0270	5/4" x 6"		240	.033		2.72	1.56		4.28	5.40
0280	5/4" x 8"		215	.037		3.56	1.75		5.31	6.60
0290	5/4" x 10"		215	.037		4.56	1.75		6.31	7.70
0300	5/4" x 12"		190	.042		5.25	1.98		7.23	8.85
0310	Frieze, 1" x 4"		250	.032		1.39	1.50		2.89	3.84
0320	1" x 6"		250	.032		2.11	1.50		3.61	4.63
0330	1" x 8"		225	.036		2.78	1.67		4.45	5.65
0340	1" x 10"		225	.036		3.50	1.67		5.17	6.40
0350	1" x 12"		200	.040		4.33	1.88		6.21	7.65
0360	5/4" x 4"		240	.033		1.69	1.56		3.25	4.27
0370	5/4" x 6"		240	.033		2.72	1.56		4.28	5.40
0380	5/4" x 8"		215	.037		3.56	1.75		5.31	6.60
0390	5/4" x 10"		215	.037		4.56	1.75		6.31	7.70
0400	5/4" x 12"		190	.042		5.25	1.98		7.23	8.85
0410	Rake, 1" x 4"		200	.040		1.39	1.88		3.27	4.42
0420	1" x 6"		200	.040		2.11	1.88		3.99	5.20
0430	1" x 8"		190	.042		2.78	1.98		4.76	6.10
0440	1" x 10"		190	.042		3.50	1.98		5.48	6.90
0450	1" x 12"		180	.044		4.33	2.09		6.42	7.95
0460	5/4" x 4"		195	.041		1.69	1.93		3.62	4.82
0470	5/4" x 6"		195	.041		2.72	1.93		4.65	5.95
0480	5/4" x 8"		185	.043		3.56	2.03		5.59	7.05
0490	5/4" x 10"		185	.043		4.56	2.03		6.59	8.10
0500	5/4" x 12"		175	.046		5.25	2.15		7.40	9.10
0510	Rake trim, 1" x 4"		225	.036		1.39	1.67		3.06	4.10
0520	1" x 6"		225	.036		2.11	1.67		3.78	4.89
0560	5/4" x 4"		220	.036		1.69	1.71		3.40	4.49
0570	5/4" x 6"		220	.036		2.72	1.71		4.43	5.60
0610	Soffit, 1" x 4"	2 Carp	420	.038		1.39	1.79		3.18	4.28
0620	1" x 6"		420	.038		2.11	1.79		3.90	5.05
0630	1" x 8"		420	.038		2.78	1.79		4.57	5.80
0640	1" x 10"		400	.040		3.50	1.88		5.38	6.75
0650	1" x 12"		400	.040		4.33	1.88		6.21	7.65
0660	5/4" x 4"		410	.039		1.69	1.83		3.52	4.68
0670	5/4" x 6"		410	.039		2.72	1.83		4.55	5.80
0680	5/4" x 8"		410	.039		3.56	1.83		5.39	6.75
0690	5/4" x 10"		390	.041		4.56	1.93		6.49	7.95
0700	5/4" x 12"		390	.041		5.25	1.93		7.18	8.75

06 80 Composite Fabrications

06 80 10 – Composite Decking

06 80 10.10 Woodgrained Composite Decking		Crew	Daily Output	Labor-Hours	Unit	Material	Labor	Equipment	Total	Total Incl O&P
0010	**WOODGRAINED COMPOSITE DECKING**									
0100	Woodgrained composite decking, 1" x 6"	2 Carp	640	.025	L.F.	3.54	1.17		4.71	5.70
0110	Grooved edge		660	.024		3.68	1.14		4.82	5.80
0120	2" x 6"		640	.025		3.67	1.17		4.84	5.85
0130	Encased, 1" x 6"		640	.025		3.67	1.17		4.84	5.85
0140	Grooved edge		660	.024		3.81	1.14		4.95	5.95
0150	2" x 6"		640	.025		4.98	1.17		6.15	7.30

06 81 Composite Railings

06 81 10 – Encased Railings

06 81 10.10 Encased Composite Railings		Crew	Daily Output	Labor-Hours	Unit	Material	2015 Bare Costs Labor	Equipment	Total	Total Incl O&P
0010	**ENCASED COMPOSITE RAILINGS**									
0100	Encased composite railing, 6′ long, 36″ high, incl. balusters	1 Carp	16	.500	Ea.	156	23.50		179.50	207
0110	42″ high, incl. balusters		16	.500		172	23.50		195.50	226
0120	8′ long, 36″ high, incl. balusters		12	.667		156	31.50		187.50	220
0130	42″ high, incl. balusters		12	.667		172	31.50		203.50	238
0140	Accessories, post sleeve, 4″ x 4″, 39″ long		32	.250		23	11.75		34.75	43.50
0150	96″ long		24	.333		86.50	15.65		102.15	119
0160	6″ x 6″, 39″ long		32	.250		47	11.75		58.75	69.50
0170	96″ long		24	.333		137	15.65		152.65	174
0180	Accessories, post skirt, 4″ x 4″		96	.083		4	3.91		7.91	10.40
0190	6″ x 6″		96	.083		4.85	3.91		8.76	11.35
0200	Post cap, 4″ x 4″, flat		48	.167		6	7.85		13.85	18.65
0210	Pyramid		48	.167		6	7.85		13.85	18.65
0220	Post cap, 6″ x 6″, flat		48	.167		9.60	7.85		17.45	22.50
0230	Pyramid	↓	48	.167	↓	9.60	7.85		17.45	22.50

Division Notes

		CREW	DAILY OUTPUT	LABOR-HOURS	UNIT	BARE COSTS MAT.	LABOR	EQUIP.	TOTAL	TOTAL INCL O&P

Estimating Tips

07 10 00 Dampproofing and Waterproofing

- Be sure of the job specifications before pricing this subdivision. The difference in cost between waterproofing and dampproofing can be great. Waterproofing will hold back standing water. Dampproofing prevents the transmission of water vapor. Also included in this section are vapor retarding membranes.

07 20 00 Thermal Protection

- Insulation and fireproofing products are measured by area, thickness, volume or R-value. Specifications may give only what the specific R-value should be in a certain situation. The estimator may need to choose the type of insulation to meet that R-value.

07 30 00 Steep Slope Roofing
07 40 00 Roofing and Siding Panels

- Many roofing and siding products are bought and sold by the square. One square is equal to an area that measures 100 square feet.

  This simple change in unit of measure could create a large error if the estimator is not observant. Accessories necessary for a complete installation must be figured into any calculations for both material and labor.

07 50 00 Membrane Roofing
07 60 00 Flashing and Sheet Metal
07 70 00 Roofing and Wall Specialties and Accessories

- The items in these subdivisions compose a roofing system. No one component completes the installation, and all must be estimated. Built-up or single-ply membrane roofing systems are made up of many products and installation trades. Wood blocking at roof perimeters or penetrations, parapet coverings, reglets, roof drains, gutters, downspouts, sheet metal flashing, skylights, smoke vents, and roof hatches all need to be considered along with the roofing material. Several different installation trades will need to work together on the roofing system. Inherent difficulties in the scheduling and coordination of various trades must be accounted for when estimating labor costs.

07 90 00 Joint Protection

- To complete the weather-tight shell, the sealants and caulkings must be estimated. Where different materials meet—at expansion joints, at flashing penetrations, and at hundreds of other locations throughout a construction project—they provide another line of defense against water penetration. Often, an entire system is based on the proper location and placement of caulking or sealants. The detailed drawings that are included as part of a set of architectural plans show typical locations for these materials. When caulking or sealants are shown at typical locations, this means the estimator must include them for all the locations where this detail is applicable. Be careful to keep different types of sealants separate, and remember to consider backer rods and primers if necessary.

Reference Numbers

Reference numbers are shown in shaded boxes at the beginning of some major classifications. These numbers refer to related items in the Reference Section. The reference information may be an estimating procedure, an alternate pricing method, or technical information.

Note: Not all subdivisions listed here necessarily appear in this publication. ■

07 01 Operation and Maint. of Thermal and Moisture Protection

07 01 50 – Maintenance of Membrane Roofing

07 01 50.10 Roof Coatings		Crew	Daily Output	Labor-Hours	Unit	Material	2015 Bare Costs Labor	Equipment	Total	Total Incl O&P
0010	**ROOF COATINGS**									
0012	Asphalt, brush grade, material only				Gal.	8.95			8.95	9.80
0200	Asphalt base, fibered aluminum coating G					8.25			8.25	9.10
0300	Asphalt primer, 5 gallon					7.50			7.50	8.25
0600	Coal tar pitch, 200 lb. barrels				Ton	1,525			1,525	1,675
0700	Tar roof cement, 5 gal. lots				Gal.	14.05			14.05	15.45
0800	Glass fibered roof & patching cement, 5 gallon				"	8.25			8.25	9.10
0900	Reinforcing glass membrane, 450 S.F./roll				Ea.	59.50			59.50	65.50
1000	Neoprene roof coating, 5 gal., 2 gal./sq.				Gal.	30.50			30.50	33.50
1100	Roof patch & flashing cement, 5 gallon					8.70			8.70	9.55
1200	Roof resaturant, glass fibered, 3 gal./sq.					8.85			8.85	9.75

07 01 90 – Maintenance of Joint Protection

07 01 90.81 Joint Sealant Replacement		Crew	Daily Output	Labor-Hours	Unit	Material	Labor	Equipment	Total	Total Incl O&P
0010	**JOINT SEALANT REPLACEMENT**									
0050	Control joints in concrete floors/slabs									
0100	Option 1 for joints with hard dry sealant									
0110	Step 1: Sawcut to remove 95% of old sealant									
0112	1/4" wide x 1/2" deep, with single saw blade	C-27	4800	.003	L.F.	.02	.15	.03	.20	.28
0114	3/8" wide x 3/4" deep, with single saw blade		4000	.004		.03	.18	.04	.25	.36
0116	1/2" wide x 1" deep, with double saw blades		3600	.004		.06	.20	.05	.31	.42
0118	3/4" wide x 1-1/2" deep, with double saw blades		3200	.005		.13	.23	.05	.41	.53
0120	Step 2: Water blast joint faces and edges	C-29	2500	.003			.12	.03	.15	.22
0130	Step 3: Air blast joint faces and edges	C-28	2000	.004			.18	.01	.19	.28
0140	Step 4: Sand blast joint faces and edges	E-11	2000	.016			.68	.12	.80	1.24
0150	Step 5: Air blast joint faces and edges	C-28	2000	.004			.18	.01	.19	.28
0200	Option 2 for joints with soft pliable sealant									
0210	Step 1: Plow joint with rectangular blade	B-62	2600	.009	L.F.		.38	.07	.45	.65
0220	Step 2: Sawcut to re-face joint faces									
0222	1/4" wide x 1/2" deep, with single saw blade	C-27	2400	.007	L.F.	.02	.30	.07	.39	.55
0224	3/8" wide x 3/4" deep, with single saw blade		2000	.008		.04	.36	.08	.48	.67
0226	1/2" wide x 1" deep, with double saw blades		1800	.009		.09	.40	.09	.58	.78
0228	3/4" wide x 1-1/2" deep, with double saw blades		1600	.010		.17	.45	.10	.72	.98
0230	Step 3: Water blast joint faces and edges	C-29	2500	.003			.12	.03	.15	.22
0240	Step 4: Air blast joint faces and edges	C-28	2000	.004			.18	.01	.19	.28
0250	Step 5: Sand blast joint faces and edges	E-11	2000	.016			.68	.12	.80	1.24
0260	Step 6: Air blast joint faces and edges	C-28	2000	.004			.18	.01	.19	.28
0290	For saw cutting new control joints, see Section 03 15 16.20									
8910	For backer rod, see Section 07 91 23.10									
8920	For joint sealant, see Sections 03 15 16.30 or 07 92 13.20									

07 05 Common Work Results for Thermal and Moisture Protection

07 05 05 – Selective Demolition for Thermal and Moisture Protection

07 05 05.10 Selective Demo., Thermal and Moist. Protection		Crew	Daily Output	Labor-Hours	Unit	Material	Labor	Equipment	Total	Total Incl O&P
0010	**SELECTIVE DEMO., THERMAL AND MOISTURE PROTECTION**									
0020	Caulking/sealant, to 1" x 1" joint R024119-10	1 Clab	600	.013	L.F.		.50		.50	.77
0120	Downspouts, including hangers		350	.023	"		.86		.86	1.32
0220	Flashing, sheet metal		290	.028	S.F.		1.04		1.04	1.60
0420	Gutters, aluminum or wood, edge hung		240	.033	L.F.		1.25		1.25	1.93
0520	Built-in		100	.080	"		3.01		3.01	4.63
0620	Insulation, air/vapor barrier		3500	.002	S.F.		.09		.09	.13

07 05 05 – Selective Demolition for Thermal and Moisture Protection

07 05 05.10 Selective Demo., Thermal and Moist. Protection		Crew	Daily Output	Labor-Hours	Unit	Material	2015 Bare Costs Labor	Equipment	Total	Total Incl O&P
0670	Batts or blankets	1 Clab	1400	.006	C.F.		.21		.21	.33
0720	Foamed or sprayed in place	2 Clab	1000	.016	B.F.		.60		.60	.93
0770	Loose fitting	1 Clab	3000	.003	C.F.		.10		.10	.15
0870	Rigid board		3450	.002	B.F.		.09		.09	.13
1120	Roll roofing, cold adhesive		12	.667	Sq.		25		25	38.50
1170	Roof accessories, adjustable metal chimney flashing		9	.889	Ea.		33.50		33.50	51.50
1325	Plumbing vent flashing		32	.250	"		9.40		9.40	14.45
1375	Ridge vent strip, aluminum		310	.026	L.F.		.97		.97	1.49
1620	Skylight to 10 S.F.		8	1	Ea.		37.50		37.50	58
2120	Roof edge, aluminum soffit and fascia		570	.014	L.F.		.53		.53	.81
2170	Concrete coping, up to 12" wide	2 Clab	160	.100			3.76		3.76	5.80
2220	Drip edge	1 Clab	1000	.008			.30		.30	.46
2270	Gravel stop		950	.008			.32		.32	.49
2370	Sheet metal coping, up to 12" wide		240	.033			1.25		1.25	1.93
2470	Roof insulation board, over 2" thick	B-2	7800	.005	B.F.		.19		.19	.30
2520	Up to 2" thick	"	3900	.010	S.F.		.39		.39	.60
2620	Roof ventilation, louvered gable vent	1 Clab	16	.500	Ea.		18.80		18.80	29
2670	Remove, roof hatch	G-3	15	2.133			100		100	153
2675	Rafter vents	1 Clab	960	.008			.31		.31	.48
2720	Soffit vent and/or fascia vent		575	.014	L.F.		.52		.52	.80
2775	Soffit vent strip, aluminum, 3" to 4" wide		160	.050			1.88		1.88	2.89
2820	Roofing accessories, shingle moulding, to 1" x 4"		1600	.005			.19		.19	.29
2870	Cant strip	B-2	2000	.020			.76		.76	1.17
2920	Concrete block walkway	1 Clab	230	.035			1.31		1.31	2.01
3070	Roofing, felt paper, 15#		70	.114	Sq.		4.30		4.30	6.60
3125	#30 felt		30	.267	"		10.05		10.05	15.45
3170	Asphalt shingles, 1 layer	B-2	3500	.011	S.F.		.43		.43	.67
3180	2 layers		1750	.023	"		.87		.87	1.34
3370	Modified bitumen		26	1.538	Sq.		58.50		58.50	90
3420	Built-up, no gravel, 3 ply		25	1.600			61		61	93.50
3470	4 ply		21	1.905			72.50		72.50	111
3620	5 ply		1600	.025	S.F.		.95		.95	1.46
3720	5 ply, with gravel		890	.045			1.71		1.71	2.63
3725	Loose gravel removal		5000	.008			.30		.30	.47
3730	Embedded gravel removal		2000	.020			.76		.76	1.17
3870	Fiberglass sheet		1200	.033			1.27		1.27	1.95
4120	Slate shingles		1900	.021			.80		.80	1.23
4170	Ridge shingles, clay or slate		2000	.020	L.F.		.76		.76	1.17
4320	Single ply membrane, attached at seams		52	.769	Sq.		29		29	45
4370	Ballasted		75	.533			20.50		20.50	31
4420	Fully adhered		39	1.026			39		39	60
4550	Roof hatch, 2'-6" x 3'-0"	1 Clab	10	.800	Ea.		30		30	46.50
4670	Wood shingles	B-2	2200	.018	S.F.		.69		.69	1.06
4820	Sheet metal roofing	"	2150	.019			.71		.71	1.09
4970	Siding, horizontal wood clapboards	1 Clab	380	.021			.79		.79	1.22
5025	Exterior insulation finish system	"	120	.067			2.51		2.51	3.86
5070	Tempered hardboard, remove and reset	1 Carp	380	.021			.99		.99	1.52
5120	Tempered hardboard sheet siding	"	375	.021			1		1	1.54
5170	Metal, corner strips	1 Clab	850	.009	L.F.		.35		.35	.54
5225	Horizontal strips		444	.018	S.F.		.68		.68	1.04
5320	Vertical strips		400	.020			.75		.75	1.16
5520	Wood shingles		350	.023			.86		.86	1.32
5620	Stucco siding		360	.022			.84		.84	1.29

07 05 Common Work Results for Thermal and Moisture Protection

07 05 05 – Selective Demolition for Thermal and Moisture Protection

07 05 05.10 Selective Demo., Thermal and Moist. Protection		Crew	Daily Output	Labor-Hours	Unit	Material	Labor	Equipment	Total	Total Incl O&P
						2015 Bare Costs				
5670	Textured plywood	1 Clab	725	.011	S.F.		.41		.41	.64
5720	Vinyl siding		510	.016			.59		.59	.91
5770	Corner strips		900	.009	L.F.		.33		.33	.51
5870	Wood, boards, vertical		400	.020	S.F.		.75		.75	1.16
5920	Waterproofing, protection/drain board	2 Clab	3900	.004	B.F.		.15		.15	.24
5970	Over 1/2" thick		1750	.009	S.F.		.34		.34	.53
6020	To 1/2" thick		2000	.008	"		.30		.30	.46

07 11 Dampproofing

07 11 13 – Bituminous Dampproofing

07 11 13.10 Bituminous Asphalt Coating		Crew	Daily Output	Labor-Hours	Unit	Material	Labor	Equipment	Total	Total Incl O&P
0010	**BITUMINOUS ASPHALT COATING**									
0030	Brushed on, below grade, 1 coat	1 Rofc	665	.012	S.F.	.22	.48		.70	1.07
0100	2 coat		500	.016		.45	.64		1.09	1.59
0300	Sprayed on, below grade, 1 coat		830	.010		.22	.39		.61	.91
0400	2 coat		500	.016		.44	.64		1.08	1.58
0500	Asphalt coating, with fibers				Gal.	8.25			8.25	9.10
0600	Troweled on, asphalt with fibers, 1/16" thick	1 Rofc	500	.016	S.F.	.36	.64		1	1.50
0700	1/8" thick		400	.020		.64	.80		1.44	2.07
1000	1/2" thick		350	.023		2.07	.92		2.99	3.83

07 11 16 – Cementitious Dampproofing

07 11 16.20 Cementitious Parging		Crew	Daily Output	Labor-Hours	Unit	Material	Labor	Equipment	Total	Total Incl O&P
0010	**CEMENTITIOUS PARGING**									
0020	Portland cement, 2 coats, 1/2" thick	D-1	250	.064	S.F.	.33	2.69		3.02	4.47
0100	Waterproofed Portland cement, 1/2" thick, 2 coats	"	250	.064	"	3.93	2.69		6.62	8.45

07 12 Built-up Bituminous Waterproofing

07 12 13 – Built-Up Asphalt Waterproofing

07 12 13.20 Membrane Waterproofing		Crew	Daily Output	Labor-Hours	Unit	Material	Labor	Equipment	Total	Total Incl O&P
0010	**MEMBRANE WATERPROOFING**									
0012	On slabs, 1 ply, felt, mopped	G-1	3000	.019	S.F.	.42	.70	.19	1.31	1.87
0100	On slabs, 1 ply, glass fiber fabric, mopped		2100	.027		.46	1	.27	1.73	2.51
0300	On slabs, 2 ply, felt, mopped		2500	.022		.83	.84	.23	1.90	2.59
0400	On slabs, 2 ply, glass fiber fabric, mopped		1650	.034		1.01	1.27	.34	2.62	3.66
0600	On slabs, 3 ply, felt, mopped		2100	.027		1.25	1	.27	2.52	3.38
0700	On slabs, 3 ply, glass fiber fabric, mopped		1550	.036		1.37	1.36	.37	3.10	4.22
0710	Asphaltic hardboard protection board, 1/8" thick	2 Rofc	500	.032		.61	1.28		1.89	2.86
1000	EPS membrane protection board, 1/4"		3500	.005		.33	.18		.51	.68
1050	3/8" thick		3500	.005		.36	.18		.54	.71
1060	1/2" thick		3500	.005		.40	.18		.58	.74
1070	Fiberglass fabric, black, 20/10 mesh		116	.138	Sq.	15.40	5.55		20.95	26.50

07 13 Sheet Waterproofing

07 13 53 – Elastomeric Sheet Waterproofing

07 13 53.10 Elastomeric Sheet Waterproofing and Access.		Crew	Daily Output	Labor-Hours	Unit	2015 Bare Costs Material	Labor	Equipment	Total	Total Incl O&P
0010	**ELASTOMERIC SHEET WATERPROOFING AND ACCESS.**									
0090	EPDM, plain, 45 mils thick	2 Rofc	580	.028	S.F.	1.41	1.11		2.52	3.44
0100	60 mils thick		570	.028		1.47	1.13		2.60	3.54
0300	Nylon reinforced sheets, 45 mils thick		580	.028		1.55	1.11		2.66	3.59
0400	60 mils thick		570	.028		1.64	1.13		2.77	3.73
0600	Vulcanizing splicing tape for above, 2" wide				C.L.F.	60.50			60.50	66.50
0700	4" wide				"	121			121	133
0900	Adhesive, bonding, 60 S.F. per gal.				Gal.	27			27	29.50
1000	Splicing, 75 S.F. per gal.				"	41			41	45
1200	Neoprene sheets, plain, 45 mils thick	2 Rofc	580	.028	S.F.	1.70	1.11		2.81	3.76
1300	60 mils thick		570	.028		2.17	1.13		3.30	4.31
1500	Nylon reinforced, 45 mils thick		580	.028		1.96	1.11		3.07	4.05
1600	60 mils thick		570	.028		3.05	1.13		4.18	5.30
1800	120 mils thick		500	.032		6.05	1.28		7.33	8.85
1900	Adhesive, splicing, 150 S.F. per gal. per coat				Gal.	41			41	45
2100	Fiberglass reinforced, fluid applied, 1/8" thick	2 Rofc	500	.032	S.F.	1.63	1.28		2.91	3.98
2200	Polyethylene and rubberized asphalt sheets, 60 mils thick		550	.029		.92	1.17		2.09	3
2400	Polyvinyl chloride sheets, plain, 10 mils thick		580	.028		.15	1.11		1.26	2.06
2500	20 mils thick		570	.028		.19	1.13		1.32	2.13
2700	30 mils thick		560	.029		.24	1.15		1.39	2.22
3000	Adhesives, trowel grade, 40-100 S.F. per gal.				Gal.	24			24	26.50
3100	Brush grade, 100-250 S.F. per gal.				"	21.50			21.50	24
3300	Bitumen modified polyurethane, fluid applied, 55 mils thick	2 Rofc	665	.024	S.F.	.93	.96		1.89	2.67

07 16 Cementitious and Reactive Waterproofing

07 16 16 – Crystalline Waterproofing

07 16 16.20 Cementitious Waterproofing		Crew	Daily Output	Labor-Hours	Unit	Material	Labor	Equipment	Total	Total Incl O&P
0010	**CEMENTITIOUS WATERPROOFING**									
0020	1/8" application, sprayed on	G-2A	1000	.024	S.F.	.73	.86	.72	2.31	3.01
0050	4 coat cementitious metallic slurry	1 Cefi	1.20	6.667	C.S.F.	32	300		332	480

07 17 Bentonite Waterproofing

07 17 13 – Bentonite Panel Waterproofing

07 17 13.10 Bentonite		Crew	Daily Output	Labor-Hours	Unit	Material	Labor	Equipment	Total	Total Incl O&P
0010	**BENTONITE**									
0020	Panels, 4' x 4', 3/16" thick	1 Rofc	625	.013	S.F.	1.57	.51		2.08	2.61
0100	Rolls, 3/8" thick, with geotextile fabric both sides	"	550	.015	"	1.47	.58		2.05	2.62
0300	Granular bentonite, 50 lb. bags (.625 C.F.)				Bag	20.50			20.50	22.50
0400	3/8" thick, troweled on	1 Rofc	475	.017	S.F.	1.02	.68		1.70	2.27
0500	Drain board, expanded polystyrene, 1-1/2" thick	1 Rohe	1600	.005		.38	.15		.53	.67
0510	2" thick		1600	.005		.50	.15		.65	.81
0520	3" thick		1600	.005		.75	.15		.90	1.09
0530	4" thick		1600	.005		1	.15		1.15	1.36
0600	With filter fabric, 1-1/2" thick		1600	.005		.44	.15		.59	.74
0625	2" thick		1600	.005		.56	.15		.71	.88
0650	3" thick		1600	.005		.81	.15		.96	1.16
0675	4" thick		1600	.005		1.06	.15		1.21	1.43

07 19 Water Repellents

07 19 19 – Silicone Water Repellents

07 19 19.10 Silicone Based Water Repellents		Crew	Daily Output	Labor-Hours	Unit	Material	Labor	Equipment	Total	Total Incl O&P
						2015 Bare Costs				
0010	**SILICONE BASED WATER REPELLENTS**									
0020	Water base liquid, roller applied	2 Rofc	7000	.002	S.F.	.53	.09		.62	.74
0200	Silicone or stearate, sprayed on CMU, 1 coat	1 Rofc	4000	.002		.34	.08		.42	.51
0300	2 coats	"	3000	.003		.68	.11		.79	.93

07 21 Thermal Insulation

07 21 13 – Board Insulation

07 21 13.10 Rigid Insulation

			Crew	Daily Output	Labor-Hours	Unit	Material	Labor	Equipment	Total	Total Incl O&P
0010	**RIGID INSULATION**, for walls										
0040	Fiberglass, 1.5#/C.F., unfaced, 1" thick, R4.1	G	1 Carp	1000	.008	S.F.	.27	.38		.65	.88
0060	1-1/2" thick, R6.2	G		1000	.008		.40	.38		.78	1.02
0080	2" thick, R8.3	G		1000	.008		.45	.38		.83	1.08
0120	3" thick, R12.4	G		800	.010		.56	.47		1.03	1.34
0370	3#/C.F., unfaced, 1" thick, R4.3	G		1000	.008		.52	.38		.90	1.15
0390	1-1/2" thick, R6.5	G		1000	.008		.78	.38		1.16	1.44
0400	2" thick, R8.7	G		890	.009		1.05	.42		1.47	1.81
0420	2-1/2" thick, R10.9	G		800	.010		1.10	.47		1.57	1.93
0440	3" thick, R13	G		800	.010		1.59	.47		2.06	2.47
0520	Foil faced, 1" thick, R4.3	G		1000	.008		.90	.38		1.28	1.57
0540	1-1/2" thick, R6.5	G		1000	.008		1.35	.38		1.73	2.07
0560	2" thick, R8.7	G		890	.009		1.69	.42		2.11	2.51
0580	2-1/2" thick, R10.9	G		800	.010		1.98	.47		2.45	2.90
0600	3" thick, R13	G		800	.010		2.18	.47		2.65	3.12
1600	Isocyanurate, 4' x 8' sheet, foil faced, both sides										
1610	1/2" thick	G	1 Carp	800	.010	S.F.	.31	.47		.78	1.06
1620	5/8" thick	G		800	.010		.33	.47		.80	1.08
1630	3/4" thick	G		800	.010		.36	.47		.83	1.12
1640	1" thick	G		800	.010		.52	.47		.99	1.29
1650	1-1/2" thick	G		730	.011		.63	.51		1.14	1.48
1660	2" thick	G		730	.011		.80	.51		1.31	1.67
1670	3" thick	G		730	.011		1.80	.51		2.31	2.77
1680	4" thick	G		730	.011		2.05	.51		2.56	3.05
1700	Perlite, 1" thick, R2.77	G		800	.010		.42	.47		.89	1.18
1750	2" thick, R5.55	G		730	.011		.75	.51		1.26	1.62
1900	Extruded polystyrene, 25 PSI compressive strength, 1" thick, R5	G		800	.010		.53	.47		1	1.30
1940	2" thick R10	G		730	.011		1.04	.51		1.55	1.93
1960	3" thick, R15	G		730	.011		1.50	.51		2.01	2.44
2100	Expanded polystyrene, 1" thick, R3.85	G		800	.010		.25	.47		.72	1
2120	2" thick, R7.69	G		730	.011		.50	.51		1.01	1.34
2140	3" thick, R11.49	G		730	.011		.75	.51		1.26	1.62

07 21 13.13 Foam Board Insulation

			Crew	Daily Output	Labor-Hours	Unit	Material	Labor	Equipment	Total	Total Incl O&P
0010	**FOAM BOARD INSULATION**										
0600	Polystyrene, expanded, 1" thick, R4	G	1 Carp	680	.012	S.F.	.25	.55		.80	1.13
0700	2" thick, R8	G	"	675	.012	"	.50	.56		1.06	1.41

07 21 16 – Blanket Insulation

07 21 16.10 Blanket Insulation for Floors/Ceilings

			Crew	Daily Output	Labor-Hours	Unit	Material	Labor	Equipment	Total	Total Incl O&P
0010	**BLANKET INSULATION FOR FLOORS/CEILINGS**										
0020	Including spring type wire fasteners										
2000	Fiberglass, blankets or batts, paper or foil backing										
2100	3-1/2" thick, R13	G	1 Carp	700	.011	S.F.	.37	.54		.91	1.24

07 21 16 – Blanket Insulation

07 21 16.10 Blanket Insulation for Floors/Ceilings			Crew	Daily Output	Labor-Hours	Unit	Material	Labor	Equipment	Total	Total Incl O&P
2150	6-1/4" thick, R19	G	1 Carp	600	.013	S.F.	.49	.63		1.12	1.50
2210	9-1/2" thick, R30	G		500	.016		.71	.75		1.46	1.94
2220	12" thick, R38	G		475	.017		.96	.79		1.75	2.28
3000	Unfaced, 3-1/2" thick, R13	G		600	.013		.31	.63		.94	1.30
3010	6-1/4" thick, R19	G		500	.016		.36	.75		1.11	1.56
3020	9-1/2" thick, R30	G		450	.018		.58	.83		1.41	1.92
3030	12" thick, R38	G		425	.019		.74	.88		1.62	2.17

07 21 16.20 Blanket Insulation for Walls			Crew	Daily Output	Labor-Hours	Unit	Material	Labor	Equipment	Total	Total Incl O&P
0010	BLANKET INSULATION FOR WALLS										
0020	Kraft faced fiberglass, 3-1/2" thick, R11, 15" wide	G	1 Carp	1350	.006	S.F.	.27	.28		.55	.73
0030	23" wide	G		1600	.005		.27	.23		.50	.66
0060	R13, 11" wide	G		1150	.007		.30	.33		.63	.83
0080	15" wide	G		1350	.006		.30	.28		.58	.76
0100	23" wide	G		1600	.005		.30	.23		.53	.69
0110	R15, 11" wide	G		1150	.007		.45	.33		.78	1
0120	15" wide	G		1350	.006		.45	.28		.73	.93
0130	23" wide	G		1600	.005		.45	.23		.68	.86
0140	6" thick, R19, 11" wide	G		1150	.007		.42	.33		.75	.96
0160	15" wide	G		1350	.006		.42	.28		.70	.89
0180	23" wide	G		1600	.005		.42	.23		.65	.82
0182	R21, 11" wide	G		1150	.007		.61	.33		.94	1.17
0184	15" wide	G		1350	.006		.61	.28		.89	1.10
0186	23" wide	G		1600	.005		.61	.23		.84	1.03
0188	9" thick, R30, 11" wide	G		985	.008		.71	.38		1.09	1.37
0200	15" wide	G		1150	.007		.71	.33		1.04	1.28
0220	23" wide	G		1350	.006		.71	.28		.99	1.21
0230	12" thick, R38, 11" wide	G		985	.008		.96	.38		1.34	1.65
0240	15" wide	G		1150	.007		.96	.33		1.29	1.56
0260	23" wide	G		1350	.006		.96	.28		1.24	1.49
0410	Foil faced fiberglass, 3-1/2" thick, R13, 11" wide	G		1150	.007		.45	.33		.78	1
0420	15" wide	G		1350	.006		.45	.28		.73	.93
0440	23" wide	G		1600	.005		.45	.23		.68	.86
0442	R15, 11" wide	G		1150	.007		.47	.33		.80	1.02
0444	15" wide	G		1350	.006		.47	.28		.75	.95
0446	23" wide	G		1600	.005		.47	.23		.70	.88
0448	6" thick, R19, 11" wide	G		1150	.007		.60	.33		.93	1.16
0460	15" wide	G		1350	.006		.60	.28		.88	1.09
0480	23" wide	G		1600	.005		.60	.23		.83	1.02
0482	R21, 11" wide	G		1150	.007		.62	.33		.95	1.18
0484	15" wide	G		1350	.006		.62	.28		.90	1.11
0486	23" wide	G		1600	.005		.62	.23		.85	1.04
0488	9" thick, R30, 11" wide	G		985	.008		.90	.38		1.28	1.58
0500	15" wide	G		1150	.007		.90	.33		1.23	1.49
0550	23" wide	G		1350	.006		.90	.28		1.18	1.42
0560	12" thick, R38, 11" wide	G		985	.008		1.05	.38		1.43	1.75
0570	15" wide	G		1150	.007		1.05	.33		1.38	1.66
0580	23" wide	G		1350	.006		1.05	.28		1.33	1.59
0620	Unfaced fiberglass, 3-1/2" thick, R13, 11" wide	G		1150	.007		.31	.33		.64	.84
0820	15" wide	G		1350	.006		.31	.28		.59	.77
0830	23" wide	G		1600	.005		.31	.23		.54	.70
0832	R15, 11" wide	G		1150	.007		.42	.33		.75	.96
0834	15" wide	G		1350	.006		.42	.28		.70	.89

07 21 Thermal Insulation

07 21 16 – Blanket Insulation

07 21 16.20 Blanket Insulation for Walls			Crew	Daily Output	Labor-Hours	Unit	Material	2015 Bare Costs Labor	Equipment	Total	Total Incl O&P
0836	23″ wide	G	1 Carp	1600	.005	S.F.	.42	.23		.65	.82
0838	6″ thick, R19, 11″ wide	G		1150	.007		.36	.33		.69	.90
0860	15″ wide	G		1150	.007		.36	.33		.69	.90
0880	23″ wide	G		1350	.006		.36	.28		.64	.83
0882	R21, 11″ wide	G		1150	.007		.54	.33		.87	1.09
0886	15″ wide	G		1350	.006		.54	.28		.82	1.02
0888	23″ wide	G		1600	.005		.54	.23		.77	.95
0890	9″ thick, R30, 11″ wide	G		985	.008		.58	.38		.96	1.23
0900	15″ wide	G		1150	.007		.58	.33		.91	1.14
0920	23″ wide	G		1350	.006		.58	.28		.86	1.07
0930	12″ thick, R38, 11″ wide	G		985	.008		.74	.38		1.12	1.40
0940	15″ wide	G		1000	.008		.74	.38		1.12	1.39
0960	23″ wide	G		1150	.007		.74	.33		1.07	1.31
1300	Wall or ceiling insulation, mineral wool batts										
1320	3-1/2″ thick, R15	G	1 Carp	1600	.005	S.F.	.60	.23		.83	1.02
1340	5-1/2″ thick, R23	G		1600	.005		.94	.23		1.17	1.40
1380	7-1/4″ thick, R30	G		1350	.006		1.24	.28		1.52	1.80
1700	Non-rigid insul, recycled blue cotton fiber, unfaced batts, R13, 16″ wide	G		1600	.005		1.01	.23		1.24	1.47
1710	R19, 16″ wide	G		1600	.005		1.38	.23		1.61	1.88
1850	Friction fit wire insulation supports, 16″ O.C.			960	.008	Ea.	.07	.39		.46	.68

07 21 19 – Foamed In Place Insulation

07 21 19.10 Masonry Foamed In Place Insulation			Crew	Daily Output	Labor-Hours	Unit	Material	Labor	Equipment	Total	Total Incl O&P
0010	**MASONRY FOAMED IN PLACE INSULATION**										
0100	Amino-plast foam, injected into block core, 6″ block	G	G-2A	6000	.004	Ea.	.15	.14	.12	.41	.54
0110	8″ block	G		5000	.005		.19	.17	.14	.50	.64
0120	10″ block	G		4000	.006		.23	.22	.18	.63	.82
0130	12″ block	G		3000	.008		.31	.29	.24	.84	1.07
0140	Injected into cavity wall	G		13000	.002	B.F.	.05	.07	.06	.18	.23
0150	Preparation, drill holes into mortar joint every 4 VLF, 5/8″ dia		1 Clab	960	.008	Ea.		.31		.31	.48
0160	7/8″ dia			680	.012			.44		.44	.68
0170	Patch drilled holes, 5/8″ diameter			1800	.004		.03	.17		.20	.30
0180	7/8″ diameter			1200	.007		.05	.25		.30	.44

07 21 23 – Loose-Fill Insulation

07 21 23.10 Poured Loose-Fill Insulation			Crew	Daily Output	Labor-Hours	Unit	Material	Labor	Equipment	Total	Total Incl O&P
0010	**POURED LOOSE-FILL INSULATION**										
0020	Cellulose fiber, R3.8 per inch	G	1 Carp	200	.040	C.F.	.69	1.88		2.57	3.65
0021	4″ thick	G		1000	.008	S.F.	.17	.38		.55	.76
0022	6″ thick	G		800	.010	″	.28	.47		.75	1.03
0080	Fiberglass wool, R4 per inch	G		200	.040	C.F.	.55	1.88		2.43	3.49
0081	4″ thick	G		600	.013	S.F.	.19	.63		.82	1.17
0082	6″ thick	G		400	.020	″	.26	.94		1.20	1.74
0100	Mineral wool, R3 per inch	G		200	.040	C.F.	.41	1.88		2.29	3.34
0101	4″ thick	G		600	.013	S.F.	.14	.63		.77	1.11
0102	6″ thick	G		400	.020	″	.21	.94		1.15	1.68
0300	Polystyrene, R4 per inch	G		200	.040	C.F.	1.60	1.88		3.48	4.65
0301	4″ thick	G		600	.013	S.F.	.53	.63		1.16	1.54
0302	6″ thick	G		400	.020	″	.80	.94		1.74	2.33
0400	Perlite, R2.78 per inch	G		200	.040	C.F.	5.20	1.88		7.08	8.65
0401	4″ thick	G		1000	.008	S.F.	1.73	.38		2.11	2.49
0402	6″ thick	G		800	.010	″	2.61	.47		3.08	3.59

07 21 Thermal Insulation

07 21 23 – Loose-Fill Insulation

07 21 23.20 Masonry Loose-Fill Insulation			Crew	Daily Output	Labor-Hours	Unit	Material	2015 Bare Costs Labor	Equipment	Total	Total Incl O&P
0010	**MASONRY LOOSE-FILL INSULATION**, vermiculite or perlite										
0100	In cores of concrete block, 4" thick wall, .115 C.F./S.F.	G	D-1	4800	.003	S.F.	.60	.14		.74	.87
0200	6" thick wall, .175 C.F./S.F.	G		3000	.005		.91	.22		1.13	1.34
0300	8" thick wall, .258 C.F./S.F.	G		2400	.007		1.34	.28		1.62	1.91
0400	10" thick wall, .340 C.F./S.F.	G		1850	.009		1.77	.36		2.13	2.51
0500	12" thick wall, .422 C.F./S.F.	G		1200	.013		2.20	.56		2.76	3.28
0600	Poured cavity wall, vermiculite or perlite, water repellant	G		250	.064	C.F.	5.20	2.69		7.89	9.85
0700	Foamed in place, urethane in 2-5/8" cavity	G	G-2A	1035	.023	S.F.	1.38	.83	.69	2.90	3.65
0800	For each 1" added thickness, add	G	"	2372	.010	"	.53	.36	.30	1.19	1.51

07 21 26 – Blown Insulation

07 21 26.10 Blown Insulation			Crew	Daily Output	Labor-Hours	Unit	Material	Labor	Equipment	Total	Total Incl O&P
0010	**BLOWN INSULATION** Ceilings, with open access										
0020	Cellulose, 3-1/2" thick, R13	G	G-4	5000	.005	S.F.	.24	.18	.08	.50	.63
0030	5-3/16" thick, R19	G		3800	.006		.35	.24	.11	.70	.88
0050	6-1/2" thick, R22	G		3000	.008		.45	.31	.13	.89	1.12
0100	8-11/16" thick, R30	G		2600	.009		.61	.35	.15	1.11	1.38
0120	10-7/8" thick, R38	G		1800	.013		.78	.51	.22	1.51	1.88
1000	Fiberglass, 5.5" thick, R11	G		3800	.006		.19	.24	.11	.54	.69
1050	6" thick, R12	G		3000	.008		.26	.31	.13	.70	.91
1100	8.8" thick, R19	G		2200	.011		.33	.42	.18	.93	1.20
1200	10" thick, R22	G		1800	.013		.38	.51	.22	1.11	1.45
1300	11.5" thick, R26	G		1500	.016		.46	.61	.27	1.34	1.73
1350	13" thick, R30	G		1400	.017		.53	.66	.29	1.48	1.91
1450	16" thick, R38	G		1145	.021		.68	.80	.35	1.83	2.36
1500	20" thick, R49	G		920	.026		.89	1	.44	2.33	3

07 21 27 – Reflective Insulation

07 21 27.10 Reflective Insulation Options			Crew	Daily Output	Labor-Hours	Unit	Material	Labor	Equipment	Total	Total Incl O&P
0010	**REFLECTIVE INSULATION OPTIONS**										
0020	Aluminum foil on reinforced scrim	G	1 Carp	19	.421	C.S.F.	14.20	19.75		33.95	46
0100	Reinforced with woven polyolefin	G		19	.421		22	19.75		41.75	54.50
0500	With single bubble air space, R8.8	G		15	.533		28	25		53	69.50
0600	With double bubble air space, R9.8	G		15	.533		32	25		57	73.50

07 21 29 – Sprayed Insulation

07 21 29.10 Sprayed-On Insulation			Crew	Daily Output	Labor-Hours	Unit	Material	Labor	Equipment	Total	Total Incl O&P
0010	**SPRAYED-ON INSULATION**										
0020	Fibrous/cementitious, finished wall, 1" thick, R3.7	G	G-2	2050	.012	S.F.	.31	.46	.07	.84	1.11
0100	Attic, 5.2" thick, R19	G		1550	.015	"	.42	.61	.09	1.12	1.48
0200	Fiberglass, R4 per inch, vertical	G		1600	.015	B.F.	.18	.59	.08	.85	1.19
0210	Horizontal	G		1200	.020	"	.18	.79	.11	1:08	1.52
0300	Closed cell, spray polyurethane foam, 2 pounds per cubic foot density										
0310	1" thick	G	G-2A	6000	.004	S.F.	.53	.14	.12	.79	.95
0320	2" thick	G		3000	.008		1.05	.29	.24	1.58	1.89
0330	3" thick	G		2000	.012		1.58	.43	.36	2.37	2.83
0335	3-1/2" thick	G		1715	.014		1.84	.50	.42	2.76	3.31
0340	4" thick	G		1500	.016		2.10	.57	.48	3.15	3.79
0350	5" thick	G		1200	.020		2.63	.72	.60	3.95	4.73
0355	5-1/2" thick	G		1090	.022		2.89	.79	.66	4.34	5.20
0360	6" thick	G		1000	.024		3.15	.86	.72	4.73	5.70

07 22 16 – Roof Board Insulation

07 22 16.10 Roof Deck Insulation			Crew	Daily Output	Labor-Hours	Unit	Material	Labor	Equipment	Total	Total Incl O&P
							2015 Bare Costs				
0010	**ROOF DECK INSULATION**, fastening excluded										
0016	Asphaltic cover board, fiberglass lined, 1/8" thick		1 Rofc	1400	.006	S.F.	.47	.23		.70	.91
0018	1/4" thick			1400	.006		.94	.23		1.17	1.42
0020	Fiberboard low density, 1/2" thick R1.39	G		1300	.006		.30	.25		.55	.75
0030	1" thick R2.78	G		1040	.008		.52	.31		.83	1.10
0080	1-1/2" thick R4.17	G		1040	.008		.80	.31		1.11	1.41
0100	2" thick R5.56	G		1040	.008		1.06	.31		1.37	1.70
0110	Fiberboard high density, 1/2" thick R1.3	G		1300	.006		.30	.25		.55	.75
0120	1" thick R2.5	G		1040	.008		.58	.31		.89	1.17
0130	1-1/2" thick R3.8	G		1040	.008		.88	.31		1.19	1.50
0200	Fiberglass, 3/4" thick R2.78	G		1300	.006		.61	.25		.86	1.09
0400	15/16" thick R3.70	G		1300	.006		.81	.25		1.06	1.31
0460	1-1/16" thick R4.17	G		1300	.006		1.01	.25		1.26	1.53
0600	1-5/16" thick R5.26	G		1300	.006		1.37	.25		1.62	1.93
0650	2-1/16" thick R8.33	G		1040	.008		1.45	.31		1.76	2.13
0700	2-7/16" thick R10	G		1040	.008		1.68	.31		1.99	2.38
0800	Gypsum cover board, fiberglass mat facer, 1/4" thick			1400	.006		.47	.23		.70	.91
0810	1/2" thick			1300	.006		.57	.25		.82	1.05
0820	5/8" thick			1200	.007		.61	.27		.88	1.13
0830	Primed fiberglass mat facer, 1/4" thick			1400	.006		.51	.23		.74	.95
0840	1/2" thick			1300	.006		.62	.25		.87	1.10
0850	5/8" thick			1200	.007		.65	.27		.92	1.18
1650	Perlite, 1/2" thick R1.32	G		1365	.006		.28	.24		.52	.71
1655	3/4" thick R2.08	G		1040	.008		.34	.31		.65	.90
1660	1" thick R2.78	G		1040	.008		.50	.31		.81	1.08
1670	1-1/2" thick R4.17	G		1040	.008		.73	.31		1.04	1.33
1680	2" thick R5.56	G		910	.009		1	.35		1.35	1.70
1685	2-1/2" thick R6.67	G		910	.009		1.30	.35		1.65	2.03
1690	Tapered for drainage	G		1040	.008	B.F.	1.01	.31		1.32	1.64
1700	Polyisocyanurate, 2#/C.F. density, 3/4" thick	G		1950	.004	S.F.	.46	.16		.62	.79
1705	1" thick	G		1820	.004		.48	.18		.66	.83
1715	1-1/2" thick	G		1625	.005		.64	.20		.84	1.04
1725	2" thick	G		1430	.006		.83	.22		1.05	1.29
1735	2-1/2" thick	G		1365	.006		1.06	.24		1.30	1.57
1745	3" thick	G		1300	.006		1.27	.25		1.52	1.82
1755	3-1/2" thick	G		1300	.006		1.95	.25		2.20	2.57
1765	Tapered for drainage	G		1820	.004	B.F.	.64	.18		.82	1
1900	Extruded Polystyrene										
1910	15 PSI compressive strength, 1" thick, R5	G	1 Rofc	1950	.004	S.F.	.55	.16		.71	.89
1920	2" thick, R10	G		1625	.005		.72	.20		.92	1.13
1930	3" thick, R15	G		1300	.006		1.43	.25		1.68	1.99
1932	4" thick, R20	G		1300	.006		1.93	.25		2.18	2.54
1934	Tapered for drainage	G		1950	.004	B.F.	.59	.16		.75	.93
1940	25 PSI compressive strength, 1" thick, R5	G		1950	.004	S.F.	.69	.16		.85	1.04
1942	2" thick, R10	G		1625	.005		1.31	.20		1.51	1.78
1944	3" thick, R15	G		1300	.006		2	.25		2.25	2.62
1946	4" thick, R20	G		1300	.006		2.76	.25		3.01	3.46
1948	Tapered for drainage	G		1950	.004	B.F.	.61	.16		.77	.95
1950	40 psi compressive strength, 1" thick, R5	G		1950	.004	S.F.	.53	.16		.69	.86
1952	2" thick, R10	G		1625	.005		1.01	.20		1.21	1.45
1954	3" thick, R15	G		1300	.006		1.46	.25		1.71	2.02
1956	4" thick, R20	G		1300	.006		1.91	.25		2.16	2.52
1958	Tapered for drainage	G		1820	.004	B.F.	.76	.18		.94	1.14

07 22 Roof and Deck Insulation

07 22 16 – Roof Board Insulation

07 22 16.10 Roof Deck Insulation			Crew	Daily Output	Labor-Hours	Unit	Material	2015 Bare Costs Labor	Equipment	Total	Total Incl O&P
1960	60 PSI compressive strength, 1″ thick, R5	G	1 Rofc	1885	.004	S.F.	.74	.17		.91	1.10
1962	2″ thick, R10	G		1560	.005		1.41	.21		1.62	1.90
1964	3″ thick, R15	G		1270	.006		2.29	.25		2.54	2.95
1966	4″ thick, R20	G		1235	.006		2.85	.26		3.11	3.57
1968	Tapered for drainage	G		1820	.004	B.F.	.96	.18		1.14	1.36
2010	Expanded polystyrene, 1#/C.F. density, 3/4″ thick, R2.89	G		1950	.004	S.F.	.19	.16		.35	.49
2020	1″ thick, R3.85	G		1950	.004		.25	.16		.41	.56
2100	2″ thick, R7.69	G		1625	.005		.50	.20		.70	.89
2110	3″ thick, R11.49	G		1625	.005		.75	.20		.95	1.17
2120	4″ thick, R15.38	G		1625	.005		1	.20		1.20	1.44
2130	5″ thick, R19.23	G		1495	.005		1.25	.21		1.46	1.75
2140	6″ thick, R23.26	G		1495	.005		1.50	.21		1.71	2.02
2150	Tapered for drainage	G		1950	.004	B.F.	.53	.16		.69	.86
2400	Composites with 2″ EPS										
2410	1″ fiberboard	G	1 Rofc	1325	.006	S.F.	1.40	.24		1.64	1.95
2420	7/16″ oriented strand board	G		1040	.008		1.13	.31		1.44	1.77
2430	1/2″ plywood	G		1040	.008		1.39	.31		1.70	2.06
2440	1″ perlite	G		1040	.008		1.14	.31		1.45	1.78
2450	Composites with 1-1/2″ polyisocyanurate										
2460	1″ fiberboard	G	1 Rofc	1040	.008	S.F.	1.19	.31		1.50	1.84
2470	1″ perlite	G		1105	.007		1.06	.29		1.35	1.67
2480	7/16″ oriented strand board	G		1040	.008		.93	.31		1.24	1.55
3000	Fastening alternatives, coated screws, 2″ long			3744	.002	Ea.	.05	.09		.14	.21
3010	4″ long			3120	.003		.10	.10		.20	.29
3020	6″ long			2675	.003		.17	.12		.29	.39
3030	8″ long			2340	.003		.25	.14		.39	.51
3040	10″ long			1872	.004		.43	.17		.60	.76
3050	Pre-drill and drive wedge spike, 2-1/2″			1248	.006		.37	.26		.63	.85
3060	3-1/2″			1101	.007		.48	.29		.77	1.03
3070	4-1/2″			936	.009		.60	.34		.94	1.25
3075	3″ galvanized deck plates			7488	.001		.07	.04		.11	.15
3080	Spot mop asphalt		G-1	295	.190	Sq.	5.55	7.10	1.93	14.58	20.50
3090	Full mop asphalt		″	192	.292		11.10	10.95	2.96	25.01	34
3110	Low-rise polyurethane adhesive, from 5 gallon kit, 12″ OC beads		1 Rofc	45	.178		33	7.15		40.15	48
3120	6″ OC beads			32	.250		65.50	10.05		75.55	89
3130	4″ OC beads			30	.267		98.50	10.70		109.20	126

07 24 Exterior Insulation and Finish Systems

07 24 13 – Polymer-Based Exterior Insulation and Finish System

07 24 13.10 Exterior Insulation and Finish Systems			Crew	Daily Output	Labor-Hours	Unit	Material	Labor	Equipment	Total	Total Incl O&P
0010	**EXTERIOR INSULATION AND FINISH SYSTEMS**										
0095	Field applied, 1″ EPS insulation	G	J-1	390	.103	S.F.	1.91	4.21	.36	6.48	8.80
0100	With 1/2″ cement board sheathing	G		268	.149		2.65	6.10	.52	9.27	12.70
0105	2″ EPS insulation	G		390	.103		2.16	4.21	.36	6.73	9.10
0110	With 1/2″ cement board sheathing	G		268	.149		2.90	6.10	.52	9.52	12.95
0115	3″ EPS insulation	G		390	.103		2.41	4.21	.36	6.98	9.35
0120	With 1/2″ cement board sheathing	G		268	.149		3.15	6.10	.52	9.77	13.25
0125	4″ EPS insulation	G		390	.103		2.66	4.21	.36	7.23	9.65
0130	With 1/2″ cement board sheathing	G		268	.149		4.14	6.10	.52	10.76	14.35
0140	Premium finish add			1265	.032		.33	1.30	.11	1.74	2.42
0150	Heavy duty reinforcement add			914	.044		.83	1.79	.15	2.77	3.77

07 24 Exterior Insulation and Finish Systems

07 24 13 – Polymer-Based Exterior Insulation and Finish System

07 24 13.10 Exterior Insulation and Finish Systems		Crew	Daily Output	Labor-Hours	Unit	2015 Bare Costs Material	Labor	Equipment	Total	Total Incl O&P
0160	2.5#/S.Y. metal lath substrate add	1 Lath	75	.107	S.Y.	2.72	4.59		7.31	9.75
0170	3.4#/S.Y. metal lath substrate add	"	75	.107	"	4.13	4.59		8.72	11.30
0180	Color or texture change,	J-1	1265	.032	S.F.	.77	1.30	.11	2.18	2.91
0190	With substrate leveling base coat	1 Plas	530	.015		.77	.65		1.42	1.82
0210	With substrate sealing base coat	1 Pord	1224	.007	↓	.10	.26		.36	.51
0370	V groove shape in panel face				L.F.	.62			.62	.68
0380	U groove shape in panel face				"	.80			.80	.88
0440	For higher than one story, add						25%			

07 25 Weather Barriers

07 25 10 – Weather Barriers or Wraps

07 25 10.10 Weather Barriers

Line	Description	Crew	Daily Output	Labor-Hours	Unit	Material	Labor	Equipment	Total	Total Incl O&P
0010	**WEATHER BARRIERS**									
0400	Asphalt felt paper, 15#	1 Carp	37	.216	Sq.	5.40	10.15		15.55	21.50
0401	Per square foot	"	3700	.002	S.F.	.05	.10		.15	.22
0450	Housewrap, exterior, spun bonded polypropylene									
0470	Small roll	1 Carp	3800	.002	S.F.	.15	.10		.25	.31
0480	Large roll	"	4000	.002	"	.14	.09		.23	.29
2100	Asphalt felt roof deck vapor barrier, class 1 metal decks	1 Rofc	37	.216	Sq.	22	8.65		30.65	39
2200	For all other decks	"	37	.216		16.50	8.65		25.15	33
2800	Asphalt felt, 50% recycled content, 15 lb., 4 sq. per roll	1 Carp	36	.222		5.40	10.45		15.85	22
2810	30 lb., 2 sq. per roll	"	36	.222	↓	10.80	10.45		21.25	28
3000	Building wrap, spunbonded polyethylene	2 Carp	8000	.002	S.F.	.15	.09		.24	.31

07 26 Vapor Retarders

07 26 10 – Above-Grade Vapor Retarders

07 26 10.10 Vapor Retarders

Line	Description	Crew	Daily Output	Labor-Hours	Unit	Material	Labor	Equipment	Total	Total Incl O&P
0010	**VAPOR RETARDERS**									
0020	Aluminum and kraft laminated, foil 1 side G	1 Carp	37	.216	Sq.	12.50	10.15		22.65	29.50
0100	Foil 2 sides G		37	.216		14	10.15		24.15	31
0600	Polyethylene vapor barrier, standard, 2 mil G		37	.216		1.50	10.15		11.65	17.25
0700	4 mil G		37	.216		2.90	10.15		13.05	18.80
0900	6 mil G		37	.216		4.02	10.15		14.17	20
1200	10 mil G		37	.216		8.85	10.15		19	25.50
1300	Clear reinforced, fire retardant, 8 mil G		37	.216		10.85	10.15		21	27.50
1350	Cross laminated type, 3 mil G		37	.216		7.60	10.15		17.75	24
1400	4 mil G		37	.216		7.95	10.15		18.10	24.50
1800	Reinf. waterproof, 2 mil polyethylene backing, 1 side		37	.216		6.05	10.15		16.20	22.50
1900	2 sides		37	.216		7.95	10.15		18.10	24.50
2400	Waterproofed kraft with sisal or fiberglass fibers	↓	37	.216	↓	12.75	10.15		22.90	29.50

07 27 Air Barriers

07 27 26 – Fluid-Applied Membrane Air Barriers

07 27 26.10 Fluid Applied Membrane Air Barrier		Crew	Daily Output	Labor-Hours	Unit	2015 Bare Costs Material	Labor	Equipment	Total	Total Incl O&P
0010	**FLUID APPLIED MEMBRANE AIR BARRIER**									
0100	Spray applied vapor barrier, 25 S.F./gallon	1 Pord	1375	.006	S.F.	.01	.23		.24	.37

07 31 Shingles and Shakes

07 31 13 – Asphalt Shingles

07 31 13.10 Asphalt Roof Shingles		Crew	Daily Output	Labor-Hours	Unit	Material	Labor	Equipment	Total	Total Incl O&P
0010	**ASPHALT ROOF SHINGLES**									
0100	Standard strip shingles									
0150	Inorganic, class A, 25 year	1 Rofc	5.50	1.455	Sq.	79.50	58.50		138	187
0155	Pneumatic nailed		7	1.143		79.50	46		125.50	166
0200	30 year		5	1.600		94	64		158	213
0205	Pneumatic nailed		6.25	1.280		94	51.50		145.50	191
0250	Standard laminated multi-layered shingles									
0300	Class A, 240-260 lb./square	1 Rofc	4.50	1.778	Sq.	110	71.50		181.50	243
0305	Pneumatic nailed		5.63	1.422		110	57		167	219
0350	Class A, 250-270 lb./square		4	2		110	80		190	258
0355	Pneumatic nailed		5	1.600		110	64		174	231
0400	Premium, laminated multi-layered shingles									
0450	Class A, 260-300 lb./square	1 Rofc	3.50	2.286	Sq.	153	91.50		244.50	325
0455	Pneumatic nailed		4.37	1.831		153	73.50		226.50	293
0500	Class A, 300-385 lb./square		3	2.667		230	107		337	435
0505	Pneumatic nailed		3.75	2.133		230	85.50		315.50	400
0800	#15 felt underlayment		64	.125		5.40	5		10.40	14.50
0825	#30 felt underlayment		58	.138		10.60	5.55		16.15	21
0850	Self adhering polyethylene and rubberized asphalt underlayment		22	.364		75	14.60		89.60	108
0900	Ridge shingles		330	.024	L.F.	2.10	.97		3.07	3.97
0905	Pneumatic nailed		412.50	.019	"	2.10	.78		2.88	3.64
1000	For steep roofs (7 to 12 pitch or greater), add						50%			

07 31 16 – Metal Shingles

07 31 16.10 Aluminum Shingles		Crew	Daily Output	Labor-Hours	Unit	Material	Labor	Equipment	Total	Total Incl O&P
0010	**ALUMINUM SHINGLES**									
0020	Mill finish, .019 thick	1 Carp	5	1.600	Sq.	221	75		296	360
0100	.020" thick	"	5	1.600		224	75		299	360
0300	For colors, add					21			21	23
0600	Ridge cap, .024" thick	1 Carp	170	.047	L.F.	3.63	2.21		5.84	7.40
0700	End wall flashing, .024" thick		170	.047		2.10	2.21		4.31	5.70
0900	Valley section, .024" thick		170	.047		3.57	2.21		5.78	7.35
1000	Starter strip, .024" thick		400	.020		1.67	.94		2.61	3.29
1200	Side wall flashing, .024" thick		170	.047		2.03	2.21		4.24	5.65
1500	Gable flashing, .024" thick		400	.020		1.63	.94		2.57	3.24

07 31 16.20 Steel Shingles		Crew	Daily Output	Labor-Hours	Unit	Material	Labor	Equipment	Total	Total Incl O&P
0010	**STEEL SHINGLES**									
0012	Galvanized, 26 ga.	1 Rots	2.20	3.636	Sq.	335	146		481	620
0200	24 ga.	"	2.20	3.636		335	146		481	620
0300	For colored galvanized shingles, add					56			56	61.50

07 31 26 – Slate Shingles

07 31 26.10 Slate Roof Shingles		Crew	Daily Output	Labor-Hours	Unit	Material	Labor	Equipment	Total	Total Incl O&P
0010	**SLATE ROOF SHINGLES** R073126-20									
0100	Buckingham Virginia black, 3/16" - 1/4" thick G	1 Rots	1.75	4.571	Sq.	470	183		653	830
0200	1/4" thick G		1.75	4.571		470	183		653	830

07 31 Shingles and Shakes

07 31 26 – Slate Shingles

07 31 26.10 Slate Roof Shingles		Crew	Daily Output	Labor-Hours	Unit	Material	2015 Bare Costs Labor	Equipment	Total	Total Incl O&P
0900	Pennsylvania black, Bangor, #1 clear G	1 Rots	1.75	4.571	Sq.	490	183		673	855
1200	Vermont, unfading, green, mottled green G		1.75	4.571		475	183		658	840
1300	Semi-weathering green & gray G		1.75	4.571		345	183		528	695
1400	Purple G		1.75	4.571		425	183		608	780
1500	Black or gray G		1.75	4.571		455	183		638	815
1600	Red G		1.75	4.571		1,175	183		1,358	1,600
1700	Variegated purple		1.75	4.571		415	183		598	770
2700	Ridge shingles, slate		200	.040	L.F.	10	1.60		11.60	13.75

07 31 29 – Wood Shingles and Shakes

07 31 29.13 Wood Shingles		Crew	Daily Output	Labor-Hours	Unit	Material	Labor	Equipment	Total	Total Incl O&P
0010	**WOOD SHINGLES**									
0012	16″ No. 1 red cedar shingles, 5″ exposure, on roof	1 Carp	2.50	3.200	Sq.	287	150		437	545
0015	Pneumatic nailed		3.25	2.462		287	116		403	495
0200	7-1/2″ exposure, on walls		2.05	3.902		191	183		374	490
0205	Pneumatic nailed		2.67	2.996		191	141		332	425
0300	18″ No. 1 red cedar perfections, 5-1/2″ exposure, on roof		2.75	2.909		246	137		383	480
0305	Pneumatic nailed		3.57	2.241		246	105		351	435
0500	7-1/2″ exposure, on walls		2.25	3.556		181	167		348	455
0505	Pneumatic nailed		2.92	2.740		181	129		310	395
0600	Resquared, and rebutted, 5-1/2″ exposure, on roof		3	2.667		279	125		404	500
0605	Pneumatic nailed		3.90	2.051		279	96.50		375.50	455
0900	7-1/2″ exposure, on walls		2.45	3.265		205	153		358	460
0905	Pneumatic nailed		3.18	2.516		205	118		323	410
1000	Add to above for fire retardant shingles					56			56	61.50
1060	Preformed ridge shingles	1 Carp	400	.020	L.F.	5	.94		5.94	6.95
2000	White cedar shingles, 16″ long, extras, 5″ exposure, on roof		2.40	3.333	Sq.	192	157		349	450
2005	Pneumatic nailed		3.12	2.564		192	120		312	395
2050	5″ exposure on walls		2	4		192	188		380	500
2055	Pneumatic nailed		2.60	3.077		192	144		336	435
2100	7-1/2″ exposure, on walls		2	4		137	188		325	440
2105	Pneumatic nailed		2.60	3.077		137	144		281	375
2150	"B" grade, 5″ exposure on walls		2	4		165	188		353	470
2155	Pneumatic nailed		2.60	3.077		165	144		309	405
2300	For 15# organic felt underlayment on roof, 1 layer, add		64	.125		5.40	5.85		11.25	15
2400	2 layers, add		32	.250		10.80	11.75		22.55	30
2600	For steep roofs (7/12 pitch or greater), add to above						50%			
2700	Panelized systems, No.1 cedar shingles on 5/16″ CDX plywood									
2800	On walls, 8′ strips, 7″ or 14″ exposure	2 Carp	700	.023	S.F.	6	1.07		7.07	8.25
3500	On roofs, 8′ strips, 7″ or 14″ exposure	1 Carp	3	2.667	Sq.	600	125		725	855
3505	Pneumatic nailed	"	4	2	"	600	94		694	805

07 31 29.16 Wood Shakes		Crew	Daily Output	Labor-Hours	Unit	Material	Labor	Equipment	Total	Total Incl O&P
0010	**WOOD SHAKES**									
1100	Hand-split red cedar shakes, 1/2″ thick x 24″ long, 10″ exp. on roof	1 Carp	2.50	3.200	Sq.	272	150		422	530
1105	Pneumatic nailed		3.25	2.462		272	116		388	480
1110	3/4″ thick x 24″ long, 10″ exp. on roof		2.25	3.556		272	167		439	555
1115	Pneumatic nailed		2.92	2.740		272	129		401	500
1200	1/2″ thick, 18″ long, 8-1/2″ exp. on roof		2	4		180	188		368	485
1205	Pneumatic nailed		2.60	3.077		180	144		324	420
1210	3/4″ thick x 18″ long, 8 1/2″ exp. on roof		1.80	4.444		180	209		389	520
1215	Pneumatic nailed		2.34	3.419		180	161		341	445
1255	10″ exp. on walls		2	4		174	188		362	480
1260	10″ exposure on walls, pneumatic nailed		2.60	3.077		174	144		318	415

07 31 Shingles and Shakes

07 31 29 – Wood Shingles and Shakes

07 31 29.16 Wood Shakes		Crew	Daily Output	Labor-Hours	Unit	Material (2015 Bare Costs)	Labor (2015 Bare Costs)	Equipment (2015 Bare Costs)	Total (2015 Bare Costs)	Total Incl O&P
1700	Add to above for fire retardant shakes, 24" long				Sq.	56			56	61.50
1800	18" long				↓	56			56	61.50
1810	Ridge shakes	1 Carp	350	.023	L.F.	5	1.07		6.07	7.15

07 32 Roof Tiles

07 32 13 – Clay Roof Tiles

07 32 13.10 Clay Tiles		Crew	Daily Output	Labor-Hours	Unit	Material	Labor	Equipment	Total	Total Incl O&P
0010	**CLAY TILES**, including accessories									
0300	Flat shingle, interlocking, 15", 166 pcs/sq, fireflashed blend	3 Rots	6	4	Sq.	505	160		665	830
0500	Terra cotta red		6	4		510	160		670	835
0600	Roman pan and top, 18", 102 pcs/sq, fireflashed blend	↓	5.50	4.364		555	175		730	910
0640	Terra cotta red	1 Rots	2.40	3.333		555	134		689	840
1100	Barrel mission tile, 18", 166 pcs/sq, fireflashed blend	3 Rots	5.50	4.364		410	175		585	750
1140	Terra cotta red		5.50	4.364		410	175		585	750
1700	Scalloped edge flat shingle, 14", 145 pcs/sq, fireflashed blend		6	4		1,125	160		1,285	1,525
1800	Terra cotta red	↓	6	4		1,100	160		1,260	1,500
3010	#15 felt underlayment	1 Rofc	64	.125		5.40	5		10.40	14.50
3020	#30 felt underlayment		58	.138		10.60	5.55		16.15	21
3040	Polyethylene and rubberized asph. underlayment	↓	22	.364	↓	75	14.60		89.60	108

07 32 16 – Concrete Roof Tiles

07 32 16.10 Concrete Tiles		Crew	Daily Output	Labor-Hours	Unit	Material	Labor	Equipment	Total	Total Incl O&P
0010	**CONCRETE TILES**									
0020	Corrugated, 13" x 16-1/2", 90 per sq., 950 lb. per sq.									
0050	Earthtone colors, nailed to wood deck	1 Rots	1.35	5.926	Sq.	106	238		344	520
0150	Blues		1.35	5.926		106	238		344	520
0200	Greens		1.35	5.926		106	238		344	520
0250	Premium colors	↓	1.35	5.926	↓	106	238		344	520
0500	Shakes, 13" x 16-1/2", 90 per sq., 950 lb. per sq.									
0600	All colors, nailed to wood deck	1 Rots	1.50	5.333	Sq.	108	214		322	485
1500	Accessory pieces, ridge & hip, 10" x 16-1/2", 8 lb. each	"	120	.067	Ea.	3.20	2.67		5.87	8.10
1700	Rake, 6-1/2" x 16-3/4", 9 lb. each					3.20			3.20	3.52
1800	Mansard hip, 10" x 16-1/2", 9.2 lb. each					3.20			3.20	3.52
1900	Hip starter, 10" x 16-1/2", 10.5 lb. each					9.90			9.90	10.90
2000	3 or 4 way apex, 10" each side, 11.5 lb. each				↓	11.40			11.40	12.55

07 32 19 – Metal Roof Tiles

07 32 19.10 Metal Roof Tiles		Crew	Daily Output	Labor-Hours	Unit	Material	Labor	Equipment	Total	Total Incl O&P
0010	**METAL ROOF TILES**									
0020	Accessories included, .032" thick aluminum, mission tile	1 Carp	2.50	3.200	Sq.	825	150		975	1,150
0200	Spanish tiles	"	3	2.667	"	570	125		695	820

07 33 Natural Roof Coverings

07 33 63 – Vegetated Roofing

07 33 63.10 Green Roof Systems			Crew	Daily Output	Labor-Hours	Unit	Material	2015 Bare Costs Labor	Equipment	Total	Total Incl O&P
0010	GREEN ROOF SYSTEMS										
0020	Soil mixture for green roof 30% sand, 55% gravel, 15% soil										
0100	Hoist and spread soil mixture 4 inch depth up to five stories tall roof	G	B-13B	4000	.014	S.F.	.25	.57	.28	1.10	1.47
0150	6 inch depth	G		2667	.021		.38	.86	.42	1.66	2.21
0200	8 inch depth	G		2000	.028		.50	1.15	.56	2.21	2.93
0250	10 inch depth	G		1600	.035		.63	1.43	.70	2.76	3.67
0300	12 inch depth	G		1335	.042		.76	1.72	.84	3.32	4.39
0310	Alt. man-made soil mix, hoist & spread, 4″ deep up to 5 stories tall roof	G		4000	.014		1.86	.57	.28	2.71	3.24
0350	Mobilization 55 ton crane to site	G	1 Eqhv	3.60	2.222	Ea.		115		115	174
0355	Hoisting cost to five stories per day (Avg. 28 picks per day)	G	B-13B	1	56	Day		2,300	1,125	3,425	4,775
0360	Mobilization or demobilization, 100 ton crane to site driver & escort	G	A-3E	2.50	6.400	Ea.		294	62.50	356.50	515
0365	Hoisting cost six to ten stories per day (Avg. 21 picks per day)	G	B-13C	1	56	Day		2,300	1,675	3,975	5,350
0370	Hoist and spread soil mixture 4 inch depth six to ten stories tall roof	G		4000	.014	S.F.	.25	.57	.42	1.24	1.62
0375	6 inch depth	G		2667	.021		.38	.86	.63	1.87	2.43
0380	8 inch depth	G		2000	.028		.50	1.15	.83	2.48	3.23
0385	10 inch depth	G		1600	.035		.63	1.43	1.04	3.10	4.04
0390	12 inch depth	G		1335	.042		.76	1.72	1.25	3.73	4.83
0400	Green roof edging treated lumber 4″ x 4″ no hoisting included	G	2 Carp	400	.040	L.F.	1.34	1.88		3.22	4.36
0410	4″ x 6″	G		400	.040		1.91	1.88		3.79	4.99
0420	4″ x 8″	G		360	.044		3.81	2.09		5.90	7.40
0430	4″ x 6″ double stacked	G		300	.053		3.83	2.50		6.33	8.05
0500	Green roof edging redwood lumber 4″ x 4″ no hoisting included	G		400	.040		6.45	1.88		8.33	9.95
0510	4″ x 6″	G		400	.040		12.60	1.88		14.48	16.75
0520	4″ x 8″	G		360	.044		23.50	2.09		25.59	29
0530	4″ x 6″ double stacked	G		300	.053		25	2.50		27.50	31.50
0550	Components, not including membrane or insulation:										
0560	Fluid applied rubber membrane, reinforced, 215 mil thick	G	G-5	350	.114	S.F.	.28	4.17	.55	5	8
0570	Root barrier	G	2 Rofc	775	.021		.60	.83		1.43	2.07
0580	Moisture retention barrier and reservoir	G	"	900	.018		2.45	.71		3.16	3.92
0600	Planting sedum, light soil, potted, 2-1/4″ diameter, two per S.F.	G	1 Clab	420	.019		4.90	.72		5.62	6.50
0610	one per S.F.	G	"	840	.010		2.45	.36		2.81	3.25
0630	Planting sedum mat per S.F. including shipping (4000 S.F. min)	G	4 Clab	4000	.008		5.90	.30		6.20	6.90
0640	Installation sedum mat system (no soil required) per S.F. (4000 S.F. min)	G	"	4000	.008		8.20	.30		8.50	9.45
0645	Note: pricing of sedum mats shipped in full truck loads (4000-5000 S.F.)										

07 41 Roof Panels

07 41 13 – Metal Roof Panels

07 41 13.10 Aluminum Roof Panels		Crew	Daily Output	Labor-Hours	Unit	Material	Labor	Equipment	Total	Total Incl O&P
0010	ALUMINUM ROOF PANELS									
0020	Corrugated or ribbed, .0155″ thick, natural	G-3	1200	.027	S.F.	.99	1.25		2.24	3
0300	Painted		1200	.027		1.44	1.25		2.69	3.49
0400	Corrugated, .018″ thick, on steel frame, natural finish		1200	.027		1.30	1.25		2.55	3.34
0600	Painted		1200	.027		1.60	1.25		2.85	3.67
0700	Corrugated, on steel frame, natural, .024″ thick		1200	.027		1.85	1.25		3.10	3.95
0800	Painted		1200	.027		2.25	1.25		3.50	4.39
0900	.032″ thick, natural		1200	.027		2.39	1.25		3.64	4.54
1200	Painted		1200	.027		3.06	1.25		4.31	5.30
1300	V-Beam, on steel frame construction, .032″ thick, natural		1200	.027		2.53	1.25		3.78	4.69
1500	Painted		1200	.027		3.10	1.25		4.35	5.30
1600	.040″ thick, natural		1200	.027		3.07	1.25		4.32	5.30
1800	Painted		1200	.027		3.70	1.25		4.95	6

07 41 Roof Panels

07 41 13 – Metal Roof Panels

07 41 13.10 Aluminum Roof Panels		Crew	Daily Output	Labor-Hours	Unit	Material	Labor	Equipment	Total	Total Incl O&P
						2015 Bare Costs				
1900	.050" thick, natural	G-3	1200	.027	S.F.	3.69	1.25		4.94	5.95
2100	Painted		1200	.027		4.45	1.25		5.70	6.80
2200	For roofing on wood frame, deduct		4600	.007		.08	.33		.41	.59
2400	Ridge cap, .032" thick, natural		800	.040	L.F.	2.91	1.87		4.78	6.05

07 41 13.20 Steel Roofing Panels		Crew	Daily Output	Labor-Hours	Unit	Material	Labor	Equipment	Total	Total Incl O&P
0010	**STEEL ROOFING PANELS**									
0012	Corrugated or ribbed, on steel framing, 30 ga. galv	G-3	1100	.029	S.F.	1.60	1.36		2.96	3.85
0100	28 ga.		1050	.030		1.65	1.43		3.08	4
0300	26 ga.		1000	.032		1.68	1.50		3.18	4.14
0400	24 ga.		950	.034		2.90	1.58		4.48	5.60
0600	Colored, 28 ga.		1050	.030		1.67	1.43		3.10	4.02
0700	26 ga.		1000	.032		1.98	1.50		3.48	4.47
0710	Flat profile, 1-3/4" standing seams, 10" wide, standard finish, 26 ga.		1000	.032		3.75	1.50		5.25	6.40
0715	24 ga.		950	.034		4.35	1.58		5.93	7.20
0720	22 ga.		900	.036		5.40	1.66		7.06	8.50
0725	Zinc aluminum alloy finish, 26 ga.		1000	.032		3	1.50		4.50	5.60
0730	24 ga.		950	.034		3.55	1.58		5.13	6.30
0735	22 ga.		900	.036		4.05	1.66		5.71	7
0740	12" wide, standard finish, 26 ga.		1000	.032		3.75	1.50		5.25	6.40
0745	24 ga.		950	.034		4.90	1.58		6.48	7.80
0750	Zinc aluminum alloy finish, 26 ga.		1000	.032		4.26	1.50		5.76	7
0755	24 ga.		950	.034		3.50	1.58		5.08	6.25
0840	Flat profile, 1" x 3/8" batten, 12" wide, standard finish, 26 ga.		1000	.032		3.30	1.50		4.80	5.90
0845	24 ga.		950	.034		3.90	1.58		5.48	6.70
0850	22 ga.		900	.036		4.65	1.66		6.31	7.65
0855	Zinc aluminum alloy finish, 26 ga.		1000	.032		3.20	1.50		4.70	5.80
0860	24 ga.		950	.034		3.60	1.58		5.18	6.35
0865	22 ga.		900	.036		4.15	1.66		5.81	7.10
0870	16-1/2" wide, standard finish, 24 ga.		950	.034		3.85	1.58		5.43	6.65
0875	22 ga.		900	.036		4.30	1.66		5.96	7.30
0880	Zinc aluminum alloy finish, 24 ga.		950	.034		3.35	1.58		4.93	6.10
0885	22 ga.		900	.036		3.75	1.66		5.41	6.70
0890	Flat profile, 2" x 2" batten, 12" wide, standard finish, 26 ga.		1000	.032		3.85	1.50		5.35	6.55
0895	24 ga.		950	.034		4.55	1.58		6.13	7.40
0900	22 ga.		900	.036		5.55	1.66		7.21	8.65
0905	Zinc aluminum alloy finish, 26 ga.		1000	.032		3.55	1.50		5.05	6.20
0910	24 ga.		950	.034		4.05	1.58		5.63	6.85
0915	22 ga.		900	.036		4.70	1.66		6.36	7.70
0920	16-1/2" wide, standard finish, 24 ga.		950	.034		4.20	1.58		5.78	7.05
0925	22 ga.		900	.036		4.90	1.66		6.56	7.95
0930	Zinc aluminum alloy finish, 24 ga.		950	.034		3.80	1.58		5.38	6.60
0935	22 ga.		900	.036		4.30	1.66		5.96	7.30
1200	Ridge, galvanized, 10" wide G		800	.040	L.F.	3.10	1.87		4.97	6.30
1203	14" wide G	2 Shee	316	.051		3.72	2.83		6.55	8.40
1205	18" wide G	"	308	.052		4.34	2.91		7.25	9.20
1210	20" wide G	G-3	750	.043		4.10	2		6.10	7.55

07 41 33 – Plastic Roof Panels

07 41 33.10 Fiberglass Panels		Crew	Daily Output	Labor-Hours	Unit	Material	Labor	Equipment	Total	Total Incl O&P
0010	**FIBERGLASS PANELS**									
0012	Corrugated panels, roofing, 8 oz. per S.F.	G-3	1000	.032	S.F.	2.10	1.50		3.60	4.60
0100	12 oz. per S.F.		1000	.032		4.18	1.50		5.68	6.90
0300	Corrugated siding, 6 oz. per S.F.		880	.036		2.10	1.70		3.80	4.92

07 41 Roof Panels

07 41 33 – Plastic Roof Panels

07 41 33.10 Fiberglass Panels		Crew	Daily Output	Labor-Hours	Unit	Material	Labor	Equipment	Total	Total Incl O&P
						2015 Bare Costs				
0400	8 oz. per S.F.	G-3	880	.036	S.F.	2.10	1.70		3.80	4.92
0500	Fire retardant		880	.036		3.80	1.70		5.50	6.80
0600	12 oz. siding, textured		880	.036		3.75	1.70		5.45	6.75
0700	Fire retardant		880	.036		4.35	1.70		6.05	7.40
0900	Flat panels, 6 oz. per S.F., clear or colors		880	.036		2.40	1.70		4.10	5.25
1100	Fire retardant, class A		880	.036		3.45	1.70		5.15	6.40
1300	8 oz. per S.F., clear or colors		880	.036		2.52	1.70		4.22	5.40
1700	Sandwich panels, fiberglass, 1-9/16" thick, panels to 20 S.F.		180	.178		36	8.30		44.30	52.50
1900	As above, but 2-3/4" thick, panels to 100 S.F.	↓	265	.121	↓	26	5.65		31.65	37

07 42 Wall Panels

07 42 13 – Metal Wall Panels

07 42 13.10 Mansard Panels		Crew	Daily Output	Labor-Hours	Unit	Material	Labor	Equipment	Total	Total Incl O&P
0010	**MANSARD PANELS**									
0600	Aluminum, stock units, straight surfaces	1 Shee	115	.070	S.F.	3.90	3.89		7.79	10.25
0700	Concave or convex surfaces		75	.107	"	2.06	5.95		8.01	11.35
0800	For framing, to 5' high, add		115	.070	L.F.	3.60	3.89		7.49	9.90
0900	Soffits, to 1' wide	↓	125	.064	S.F.	2.20	3.58		5.78	7.85

07 42 13.20 Aluminum Siding Panels		Crew	Daily Output	Labor-Hours	Unit	Material	Labor	Equipment	Total	Total Incl O&P
0010	**ALUMINUM SIDING PANELS**									
0012	Corrugated, on steel framing, .019" thick, natural finish	G-3	775	.041	S.F.	1.52	1.93		3.45	4.63
0100	Painted		775	.041		1.66	1.93		3.59	4.79
0400	Farm type, .021" thick on steel frame, natural		775	.041		1.55	1.93		3.48	4.67
0600	Painted		775	.041		1.65	1.93		3.58	4.78
0700	Industrial type, corrugated, on steel, .024" thick, mill		775	.041		2.15	1.93		4.08	5.35
0900	Painted		775	.041		2.31	1.93		4.24	5.50
1000	.032" thick, mill		775	.041		2.45	1.93		4.38	5.65
1200	Painted		775	.041		3	1.93		4.93	6.25
1300	V-Beam, on steel frame, .032" thick, mill		775	.041		2.80	1.93		4.73	6.05
1500	Painted		775	.041		3.20	1.93		5.13	6.50
1600	.040" thick, mill		775	.041		3.38	1.93		5.31	6.70
1800	Painted		775	.041		3.94	1.93		5.87	7.30
1900	.050" thick, mill		775	.041		3.98	1.93		5.91	7.35
2100	Painted		775	.041		4.73	1.93		6.66	8.15
2200	Ribbed, 3" profile, on steel frame, .032" thick, natural		775	.041		2.40	1.93		4.33	5.60
2400	Painted		775	.041		3.05	1.93		4.98	6.30
2500	.040" thick, natural		775	.041		2.75	1.93		4.68	6
2700	Painted		775	.041		3.20	1.93		5.13	6.50
2750	.050" thick, natural		775	.041		3.15	1.93		5.08	6.45
2760	Painted		775	.041		3.63	1.93		5.56	6.95
3300	For siding on wood frame, deduct from above	↓	2800	.011	↓	.09	.53		.62	.92
3400	Screw fasteners, aluminum, self tapping, neoprene washer, 1"				M	210			210	231
3600	Stitch screws, self tapping, with neoprene washer, 5/8"				"	158			158	174
3630	Flashing, sidewall, .032" thick	G-3	800	.040	L.F.	2.96	1.87		4.83	6.15
3650	End wall, .040" thick		800	.040		3.46	1.87		5.33	6.70
3670	Closure strips, corrugated, .032" thick		800	.040		.88	1.87		2.75	3.84
3680	Ribbed, 4" or 8", .032" thick		800	.040		.88	1.87		2.75	3.84
3690	V-beam, .040" thick	↓	800	.040	↓	1.18	1.87		3.05	4.17
3800	Horizontal, colored clapboard, 8" wide, plain	2 Carp	515	.031	S.F.	2.45	1.46		3.91	4.94
3810	Insulated		515	.031		2.67	1.46		4.13	5.20
4000	Vertical board & batten, colored, non-insulated	↓	515	.031	↓	1.90	1.46		3.36	4.33

07 42 Wall Panels

07 42 13 – Metal Wall Panels

07 42 13.20 Aluminum Siding Panels		Crew	Daily Output	Labor-Hours	Unit	Material	2015 Bare Costs Labor	Equipment	Total	Total Incl O&P
4200	For simulated wood design, add				S.F.	.12			.12	.13
4300	Corners for above, outside	2 Carp	515	.031	V.L.F.	3.36	1.46		4.82	5.95
4500	Inside corners	"	515	.031	"	1.61	1.46		3.07	4.01

07 42 13.30 Steel Siding

		Crew	Daily Output	Labor-Hours	Unit	Material	Labor	Equipment	Total	Total Incl O&P
0010	**STEEL SIDING**									
0020	Beveled, vinyl coated, 8" wide	1 Carp	265	.030	S.F.	1.85	1.42		3.27	4.22
0050	10" wide	"	275	.029		1.88	1.37		3.25	4.17
0080	Galv, corrugated or ribbed, on steel frame, 30 ga.	G-3	800	.040		1.20	1.87		3.07	4.19
0100	28 ga.		795	.040		1.28	1.88		3.16	4.30
0300	26 ga.		790	.041		1.80	1.90		3.70	4.88
0400	24 ga.		785	.041		1.85	1.91		3.76	4.96
0600	22 ga.		770	.042		2.15	1.94		4.09	5.35
0700	Colored, corrugated/ribbed, on steel frame, 10 yr. finish, 28 ga.		800	.040		1.95	1.87		3.82	5
0900	26 ga.		795	.040		2.09	1.88		3.97	5.20
1000	24 ga.		790	.041		2.37	1.90		4.27	5.50
1020	20 ga.		785	.041		3	1.91		4.91	6.20
1200	Factory sandwich panel, 26 ga., 1" insulation, galvanized		380	.084		5.35	3.94		9.29	11.90
1300	Colored 1 side		380	.084		6.60	3.94		10.54	13.30
1500	Galvanized 2 sides		380	.084		7.95	3.94		11.89	14.80
1600	Colored 2 sides		380	.084		8.25	3.94		12.19	15.10
1800	Acrylic paint face, regular paint liner	↓	380	.084		6.20	3.94		10.14	12.85
1900	For 2" thick polystyrene, add					1			1	1.10
2000	22 ga., galv, 2" insulation, baked enamel exterior	G-3	360	.089		12.15	4.16		16.31	19.75
2100	Polyvinylidene exterior finish	"	360	.089	↓	12.80	4.16		16.96	20.50

07 44 Faced Panels

07 44 73 – Metal Faced Panels

07 44 73.10 Metal Faced Panels and Accessories

		Crew	Daily Output	Labor-Hours	Unit	Material	Labor	Equipment	Total	Total Incl O&P
0010	**METAL FACED PANELS AND ACCESSORIES**									
0400	Textured aluminum, 4' x 8' x 5/16" plywood backing, single face	2 Shee	375	.043	S.F.	4.07	2.39		6.46	8.15
0600	Double face		375	.043		5.25	2.39		7.64	9.45
0700	4' x 10' x 5/16" plywood backing, single face		375	.043		4.26	2.39		6.65	8.35
0900	Double face		375	.043		5.80	2.39		8.19	10.05
1000	4' x 12' x 5/16" plywood backing, single face		375	.043		4.26	2.39		6.65	8.35
1300	Smooth aluminum, 1/4" plywood panel, fluoropolymer finish, double face		375	.043		6.05	2.39		8.44	10.30
1350	Clear anodized finish, double face		375	.043		10.05	2.39		12.44	14.70
1400	Double face textured aluminum, structural panel, 1" EPS insulation	↓	375	.043	↓	5.70	2.39		8.09	9.90
1500	Accessories, outside corner	1 Shee	175	.046	L.F.	1.90	2.56		4.46	6
1600	Inside corner		175	.046		1.34	2.56		3.90	5.40
1800	Batten mounting clip		200	.040		.49	2.24		2.73	3.96
1900	Low profile batten		480	.017		.61	.93		1.54	2.10
2100	High profile batten		480	.017		1.41	.93		2.34	2.98
2200	Water table		200	.040		2.12	2.24		4.36	5.75
2400	Horizontal joint connector		200	.040		1.67	2.24		3.91	5.25
2500	Corner cap		200	.040		1.84	2.24		4.08	5.45
2700	H - moulding	↓	480	.017	↓	1.25	.93		2.18	2.81

07 46 Siding

07 46 23 – Wood Siding

07 46 23.10 Wood Board Siding		Crew	Daily Output	Labor-Hours	Unit	Material	Labor	Equipment	Total	Total Incl O&P
						2015 Bare Costs				
0010	**WOOD BOARD SIDING**									
3200	Wood, cedar bevel, A grade, 1/2" x 6"	1 Carp	295	.027	S.F.	4.14	1.27		5.41	6.50
3300	1/2" x 8"		330	.024		5.90	1.14		7.04	8.25
3500	3/4" x 10", clear grade		375	.021		6.30	1		7.30	8.45
3600	"B" grade		375	.021		3.79	1		4.79	5.70
3800	Cedar, rough sawn, 1" x 4", A grade, natural		220	.036		6.55	1.71		8.26	9.90
3900	Stained		220	.036		6.70	1.71		8.41	10
4100	1" x 12", board & batten, #3 & Btr., natural		420	.019		4.48	.89		5.37	6.30
4200	Stained		420	.019		4.81	.89		5.70	6.70
4400	1" x 8" channel siding, #3 & Btr., natural		330	.024		4.57	1.14		5.71	6.75
4500	Stained		330	.024		4.92	1.14		6.06	7.15
4700	Redwood, clear, beveled, vertical grain, 1/2" x 4"		220	.036		4.11	1.71		5.82	7.15
4750	1/2" x 6"		295	.027		4.49	1.27		5.76	6.90
4800	1/2" x 8"		330	.024		4.89	1.14		6.03	7.15
5000	3/4" x 10"		375	.021		4.50	1		5.50	6.50
5200	Channel siding, 1" x 10", B grade		375	.021		4.12	1		5.12	6.05
5250	Redwood, T&G boards, B grade, 1" x 4"		220	.036		2.54	1.71		4.25	5.40
5270	1" x 8"		330	.024		4.10	1.14		5.24	6.25
5400	White pine, rough sawn, 1" x 8", natural		330	.024		2.19	1.14		3.33	4.16
5500	Stained	↓	330	.024	↓	2.19	1.14		3.33	4.16

07 46 29 – Plywood Siding

07 46 29.10 Plywood Siding Options		Crew	Daily Output	Labor-Hours	Unit	Material	Labor	Equipment	Total	Total Incl O&P
0010	**PLYWOOD SIDING OPTIONS**									
0900	Plywood, medium density overlaid, 3/8" thick	2 Carp	750	.021	S.F.	1.25	1		2.25	2.92
1000	1/2" thick		700	.023		1.44	1.07		2.51	3.23
1100	3/4" thick		650	.025		1.85	1.16		3.01	3.82
1600	Texture 1-11, cedar, 5/8" thick, natural		675	.024		2.57	1.11		3.68	4.54
1700	Factory stained		675	.024		2.80	1.11		3.91	4.79
1900	Texture 1-11, fir, 5/8" thick, natural		675	.024		1.57	1.11		2.68	3.44
2000	Factory stained		675	.024		1.80	1.11		2.91	3.69
2050	Texture 1-11, S.Y.P., 5/8" thick, natural		675	.024		1.37	1.11		2.48	3.22
2100	Factory stained		675	.024		1.44	1.11		2.55	3.29
2200	Rough sawn cedar, 3/8" thick, natural		675	.024		1.25	1.11		2.36	3.09
2300	Factory stained		675	.024		1.50	1.11		2.61	3.36
2500	Rough sawn fir, 3/8" thick, natural		675	.024		.84	1.11		1.95	2.63
2600	Factory stained		675	.024		1.04	1.11		2.15	2.85
2800	Redwood, textured siding, 5/8" thick		675	.024		1.92	1.11		3.03	3.82
3000	Polyvinyl chloride coated, 3/8" thick	↓	750	.021	↓	1.11	1		2.11	2.76

07 46 33 – Plastic Siding

07 46 33.10 Vinyl Siding		Crew	Daily Output	Labor-Hours	Unit	Material	Labor	Equipment	Total	Total Incl O&P
0010	**VINYL SIDING**									
3995	Clapboard profile, woodgrain texture, .048 thick, double 4	2 Carp	495	.032	S.F.	1.04	1.52		2.56	3.49
4000	Double 5		550	.029		1.04	1.37		2.41	3.25
4005	Single 8		495	.032		1.26	1.52		2.78	3.73
4010	Single 10		550	.029		1.51	1.37		2.88	3.77
4015	.044 thick, double 4		495	.032		.99	1.52		2.51	3.43
4020	Double 5		550	.029		.99	1.37		2.36	3.19
4025	.042 thick, double 4		495	.032		.92	1.52		2.44	3.36
4030	Double 5		550	.029		.92	1.37		2.29	3.12
4035	Cross sawn texture, .040 thick, double 4		495	.032		.67	1.52		2.19	3.08
4040	Double 5		550	.029		.67	1.37		2.04	2.84
4045	Smooth texture, .042 thick, double 4	↓	495	.032	↓	.75	1.52		2.27	3.17

07 46 Siding

07 46 33 – Plastic Siding

07 46 33.10 Vinyl Siding		Crew	Daily Output	Labor-Hours	Unit	Material	2015 Bare Costs Labor	Equipment	Total	Total Incl O&P
4050	Double 5	2 Carp	550	.029	S.F.	.75	1.37		2.12	2.93
4055	Single 8		495	.032		.75	1.52		2.27	3.17
4060	Cedar texture, .044 thick, double 4		495	.032		1.01	1.52		2.53	3.46
4065	Double 6		600	.027		1.01	1.25		2.26	3.05
4070	Dutch lap profile, woodgrain texture, .048 thick, double 5		550	.029		1.04	1.37		2.41	3.25
4075	.044 thick, double 4.5		525	.030		1.01	1.43		2.44	3.32
4080	.042 thick, double 4.5		525	.030		.84	1.43		2.27	3.13
4085	.040 thick, double 4.5		525	.030		.67	1.43		2.10	2.94
4100	Shake profile, 10″ wide		400	.040		3.45	1.88		5.33	6.70
4105	Vertical pattern, .046 thick, double 5		550	.029		1.42	1.37		2.79	3.67
4110	.044 thick, triple 3		550	.029		1.56	1.37		2.93	3.82
4115	.040 thick, triple 4		550	.029		1.50	1.37		2.87	3.76
4120	.040 thick, triple 2.66		550	.029		1.81	1.37		3.18	4.10
4125	Insulation, fan folded extruded polystyrene, 1/4″		2000	.008		.28	.38		.66	.89
4130	3/8″		2000	.008		.31	.38		.69	.92
4135	Accessories, J channel, 5/8″ pocket		700	.023	L.F.	.46	1.07		1.53	2.15
4140	3/4″ pocket		695	.023		.50	1.08		1.58	2.21
4145	1-1/4″ pocket		680	.024		.77	1.10		1.87	2.54
4150	Flexible, 3/4″ pocket		600	.027		2.33	1.25		3.58	4.49
4155	Under sill finish trim		500	.032		.50	1.50		2	2.86
4160	Vinyl starter strip		700	.023		.58	1.07		1.65	2.29
4165	Aluminum starter strip		700	.023		.28	1.07		1.35	1.96
4170	Window casing, 2-1/2″ wide, 3/4″ pocket		510	.031		1.57	1.47		3.04	3.99
4175	Outside corner, woodgrain finish, 4″ face, 3/4″ pocket		700	.023		1.92	1.07		2.99	3.77
4180	5/8″ pocket		700	.023		1.92	1.07		2.99	3.77
4185	Smooth finish, 4″ face, 3/4″ pocket		700	.023		1.92	1.07		2.99	3.77
4190	7/8″ pocket		690	.023		1.89	1.09		2.98	3.76
4195	1-1/4″ pocket		700	.023		1.33	1.07		2.40	3.12
4200	Soffit and fascia, 1′ overhang, solid		120	.133		4.29	6.25		10.54	14.35
4205	Vented		120	.133		4.29	6.25		10.54	14.35
4207	18″ overhang, solid		110	.145		5	6.85		11.85	16
4208	Vented		110	.145		5	6.85		11.85	16
4210	2′ overhang, solid		100	.160		5.70	7.50		13.20	17.85
4215	Vented		100	.160		5.70	7.50		13.20	17.85
4217	3′ overhang, solid		100	.160		7.10	7.50		14.60	19.40
4218	Vented		100	.160		7.10	7.50		14.60	19.40
4220	Colors for siding and soffits, add				S.F.	.14			.14	.15
4225	Colors for accessories and trim, add				L.F.	.30			.30	.33

07 46 33.20 Polypropylene Siding		Crew	Daily Output	Labor-Hours	Unit	Material	Labor	Equipment	Total	Total Incl O&P
0010	**POLYPROPYLENE SIDING**									
4090	Shingle profile, random grooves, double 7	2 Carp	400	.040	S.F.	2.99	1.88		4.87	6.20
4092	Cornerpost for above	1 Carp	365	.022	L.F.	12	1.03		13.03	14.80
4095	Triple 5	2 Carp	400	.040	S.F.	2.99	1.88		4.87	6.20
4097	Cornerpost for above	1 Carp	365	.022	L.F.	11.05	1.03		12.08	13.75
5000	Staggered butt, double 7″	2 Carp	400	.040	S.F.	3.45	1.88		5.33	6.70
5002	Cornerpost for above	1 Carp	365	.022	L.F.	12	1.03		13.03	14.80
5010	Half round, double 6-1/4″	2 Carp	360	.044	S.F.	3.40	2.09		5.49	6.95
5020	Shake profile, staggered butt, double 9″	″	510	.031	″	3.45	1.47		4.92	6.05
5022	Cornerpost for above	1 Carp	365	.022	L.F.	9	1.03		10.03	11.50
5030	Straight butt, double 7″	2 Carp	400	.040	S.F.	3.40	1.88		5.28	6.65
5032	Cornerpost for above	1 Carp	365	.022	L.F.	11.90	1.03		12.93	14.65
6000	Accessories, J channel, 5/8″ pocket	2 Carp	700	.023		.46	1.07		1.53	2.15

07 46 Siding

07 46 33 – Plastic Siding

07 46 33.20 Polypropylene Siding		Crew	Daily Output	Labor-Hours	Unit	Material	2015 Bare Costs Labor	Equipment	Total	Total Incl O&P
6010	3/4" pocket	2 Carp	695	.023	L.F.	.50	1.08		1.58	2.21
6020	1-1/4" pocket		680	.024		.77	1.10		1.87	2.54
6030	Aluminum starter strip		700	.023		.28	1.07		1.35	1.96

07 46 46 – Fiber Cement Siding

07 46 46.10 Fiber Cement Siding		Crew	Daily Output	Labor-Hours	Unit	Material	Labor	Equipment	Total	Total Incl O&P
0010	FIBER CEMENT SIDING									
0020	Lap siding, 5/16" thick, 6" wide, 4-3/4" exposure, smooth texture	2 Carp	415	.039	S.F.	1.18	1.81		2.99	4.09
0025	Woodgrain texture		415	.039		1.18	1.81		2.99	4.09
0030	7-1/2" wide, 6-1/4" exposure, smooth texture		425	.038		1.46	1.77		3.23	4.33
0035	Woodgrain texture		425	.038		1.46	1.77		3.23	4.33
0040	8" wide, 6-3/4" exposure, smooth texture		425	.038		1.42	1.77		3.19	4.28
0045	Roughsawn texture		425	.038		1.42	1.77		3.19	4.28
0050	9-1/2" wide, 8-1/4" exposure, smooth texture		440	.036		1.22	1.71		2.93	3.97
0055	Woodgrain texture		440	.036		1.22	1.71		2.93	3.97
0060	12" wide, 10-3/8" exposure, smooth texture		455	.035		2.01	1.65		3.66	4.75
0065	Woodgrain texture		455	.035		2.01	1.65		3.66	4.75
0070	Panel siding, 5/16" thick, smooth texture		750	.021		1.25	1		2.25	2.92
0075	Stucco texture		750	.021		1.25	1		2.25	2.92
0080	Grooved woodgrain texture		750	.021		1.25	1		2.25	2.92
0085	V - grooved woodgrain texture		750	.021		1.25	1		2.25	2.92
0088	Shingle siding, 48" x 15-1/4" panels, 7" exposure		700	.023		3.94	1.07		5.01	6
0090	Wood starter strip		400	.040	L.F.	.42	1.88		2.30	3.35

07 46 73 – Soffit

07 46 73.10 Soffit Options		Crew	Daily Output	Labor-Hours	Unit	Material	Labor	Equipment	Total	Total Incl O&P
0010	SOFFIT OPTIONS									
0012	Aluminum, residential, .020" thick	1 Carp	210	.038	S.F.	2	1.79		3.79	4.95
0100	Baked enamel on steel, 16 or 18 ga.		105	.076		5.90	3.58		9.48	12
0300	Polyvinyl chloride, white, solid		230	.035		2.10	1.63		3.73	4.82
0400	Perforated		230	.035		2.10	1.63		3.73	4.82
0500	For colors, add					.15			.15	.17

07 51 Built-Up Bituminous Roofing

07 51 13 – Built-Up Asphalt Roofing

07 51 13.10 Built-Up Roofing Components		Crew	Daily Output	Labor-Hours	Unit	Material	Labor	Equipment	Total	Total Incl O&P
0010	BUILT-UP ROOFING COMPONENTS									
0012	Asphalt saturated felt, #30, 2 square per roll	1 Rofc	58	.138	Sq.	10.60	5.55		16.15	21
0200	#15, 4 sq. per roll, plain or perforated, not mopped		58	.138		5.40	5.55		10.95	15.40
0300	Roll roofing, smooth, #65		15	.533		9.65	21.50		31.15	47
0500	#90		12	.667		37	26.50		63.50	86
0520	Mineralized		12	.667		35	26.50		61.50	84
0540	D.C. (Double coverage), 19" selvage edge		10	.800		52.50	32		84.50	113
0580	Adhesive (lap cement)				Gal.	8.85			8.85	9.75
0800	Steep, flat or dead level asphalt, 10 ton lots, packaged				Ton	925			925	1,025

07 51 13.13 Cold-Applied Built-Up Asphalt Roofing		Crew	Daily Output	Labor-Hours	Unit	Material	Labor	Equipment	Total	Total Incl O&P
0010	COLD-APPLIED BUILT-UP ASPHALT ROOFING									
0020	3 ply system, installation only (components listed below)	G-5	50	.800	Sq.		29	3.86	32.86	54.50
0100	Spunbond poly. fabric, 1.35 oz./S.Y., 36"W, 10.8 sq./roll				Ea.	135			135	149
0500	Base & finish coat, 3 gal./sq., 5 gal./can				Gal.	7.75			7.75	8.55
0600	Coating, ceramic granules, 1/2 sq./bag				Ea.	21.50			21.50	23.50
0700	Aluminum, 2 gal./sq.				Gal.	13.50			13.50	14.85

07 51 13.13 Cold-Applied Built-Up Asphalt Roofing		Crew	Daily Output	Labor-Hours	Unit	2015 Bare Costs Material	Labor	Equipment	Total	Total Incl O&P
0800	Emulsion, fibered or non-fibered, 4 gal./sq.				Gal.	6.50			6.50	7.15

07 51 13.20 Built-Up Roofing Systems

		Crew	Daily Output	Labor-Hours	Unit	Material	Labor	Equipment	Total	Total Incl O&P
0010	**BUILT-UP ROOFING SYSTEMS** R075113-20									
0120	Asphalt flood coat with gravel/slag surfacing, not including									
0140	Insulation, flashing or wood nailers									
0200	Asphalt base sheet, 3 plies #15 asphalt felt, mopped	G-1	22	2.545	Sq.	100	95.50	26	221.50	300
0350	On nailable decks		21	2.667		102	100	27	229	315
0500	4 plies #15 asphalt felt, mopped		20	2.800		136	105	28.50	269.50	360
0550	On nailable decks		19	2.947		120	111	30	261	355
0700	Coated glass base sheet, 2 plies glass (type IV), mopped		22	2.545		107	95.50	26	228.50	310
0850	3 plies glass, mopped		20	2.800		130	105	28.50	263.50	355
0950	On nailable decks		19	2.947		122	111	30	263	355
1100	4 plies glass fiber felt (type IV), mopped		20	2.800		160	105	28.50	293.50	385
1150	On nailable decks		19	2.947		145	111	30	286	380
1200	Coated & saturated base sheet, 3 plies #15 asph. felt, mopped		20	2.800		112	105	28.50	245.50	335
1250	On nailable decks		19	2.947		104	111	30	245	335
1300	4 plies #15 asphalt felt, mopped		22	2.545		130	95.50	26	251.50	335
2000	Asphalt flood coat, smooth surface									
2200	Asphalt base sheet & 3 plies #15 asphalt felt, mopped	G-1	24	2.333	Sq.	105	87.50	23.50	216	290
2400	On nailable decks		23	2.435		96.50	91.50	24.50	212.50	289
2600	4 plies #15 asphalt felt, mopped		24	2.333		123	87.50	23.50	234	310
2700	On nailable decks		23	2.435		115	91.50	24.50	231	310
2900	Coated glass fiber base sheet, mopped, and 2 plies of									
2910	glass fiber felt (type IV)	G-1	25	2.240	Sq.	102	84	23	209	281
3100	On nailable decks		24	2.333		96.50	87.50	23.50	207.50	281
3200	3 plies, mopped		23	2.435		125	91.50	24.50	241	320
3300	On nailable decks		22	2.545		117	95.50	26	238.50	320
3800	4 plies glass fiber felt (type IV), mopped		23	2.435		147	91.50	24.50	263	345
3900	On nailable decks		22	2.545		140	95.50	26	261.50	345
4000	Coated & saturated base sheet, 3 plies #15 asph. felt, mopped		24	2.333		107	87.50	23.50	218	293
4200	On nailable decks		23	2.435		99	91.50	24.50	215	292
4300	4 plies #15 organic felt, mopped		22	2.545		125	95.50	26	246.50	330
4500	Coal tar pitch with gravel/slag surfacing									
4600	4 plies #15 tarred felt, mopped	G-1	21	2.667	Sq.	233	100	27	360	460
4800	3 plies glass fiber felt (type IV), mopped	"	19	2.947	"	193	111	30	334	435
5000	Coated glass fiber base sheet, and 2 plies of									
5010	glass fiber felt, (type IV), mopped	G-1	19	2.947	Sq.	196	111	30	337	440
5300	On nailable decks		18	3.111		170	117	31.50	318.50	420
5600	4 plies glass fiber felt (type IV), mopped		21	2.667		264	100	27	391	490
5800	On nailable decks		20	2.800		237	105	28.50	370.50	470

07 51 13.30 Cants

		Crew	Daily Output	Labor-Hours	Unit	Material	Labor	Equipment	Total	Total Incl O&P
0010	**CANTS**									
0012	Lumber, treated, 4" x 4" cut diagonally	1 Rofc	325	.025	L.F.	1.75	.99		2.74	3.62
0300	Mineral or fiber, trapezoidal, 1" x 4" x 48"		325	.025		.25	.99		1.24	1.97
0400	1-1/2" x 5-5/8" x 48"		325	.025		.44	.99		1.43	2.17

07 51 13.40 Felts

		Crew	Daily Output	Labor-Hours	Unit	Material	Labor	Equipment	Total	Total Incl O&P
0010	**FELTS**									
0012	Glass fibered roofing felt, #15, not mopped	1 Rofc	58	.138	Sq.	9.65	5.55		15.20	20
0300	Base sheet, #80, channel vented		58	.138		42.50	5.55		48.05	56.50
0400	#70, coated		58	.138		17.20	5.55		22.75	28.50
0500	Cap, #87, mineral surfaced		58	.138		80	5.55		85.55	97.50
0600	Flashing membrane, #65		16	.500		9.65	20		29.65	44.50

07 51 Built-Up Bituminous Roofing

07 51 13 – Built-Up Asphalt Roofing

07 51 13.40 Felts		Crew	Daily Output	Labor-Hours	Unit	2015 Bare Costs Material	Labor	Equipment	Total	Total Incl O&P
0800	Coal tar fibered, #15, no mopping	1 Rofc	58	.138	Sq.	18.10	5.55		23.65	29.50
0900	Asphalt felt, #15, 4 sq. per roll, no mopping		58	.138		5.40	5.55		10.95	15.40
1100	#30, 2 sq. per roll		58	.138		10.60	5.55		16.15	21
1200	Double coated, #33		58	.138		11	5.55		16.55	21.50
1400	#40, base sheet		58	.138		9.65	5.55		15.20	20
1450	Coated and saturated		58	.138		12	5.55		17.55	22.50
1500	Tarred felt, organic, #15, 4 sq. rolls		58	.138		13.95	5.55		19.50	25
1550	#30, 2 sq. roll		58	.138		28	5.55		33.55	40
1700	Add for mopping above felts, per ply, asphalt, 24 lb. per sq.	G-1	192	.292		11.10	10.95	2.96	25.01	34
1800	Coal tar mopping, 30 lb. per sq.		186	.301		23	11.30	3.06	37.36	47.50
1900	Flood coat, with asphalt, 60 lb. per sq.		60	.933		27.50	35	9.50	72	101
2000	With coal tar, 75 lb. per sq.		56	1		57	37.50	10.15	104.65	138

07 51 13.50 Walkways for Built-Up Roofs		Crew	Daily Output	Labor-Hours	Unit	Material	Labor	Equipment	Total	Total Incl O&P
0010	**WALKWAYS FOR BUILT-UP ROOFS**									
0020	Asphalt impregnated, 3' x 6' x 1/2" thick	1 Rofc	400	.020	S.F.	1.56	.80		2.36	3.08
0100	3' x 3' x 3/4" thick	"	400	.020		4.48	.80		5.28	6.30
0300	Concrete patio blocks, 2" thick, natural	1 Clab	115	.070		3.37	2.62		5.99	7.75
0400	Colors	"	115	.070		3.73	2.62		6.35	8.10

07 52 Modified Bituminous Membrane Roofing

07 52 13 – Atactic-Polypropylene-Modified Bituminous Membrane Roofing

07 52 13.10 APP Modified Bituminous Membrane		Crew	Daily Output	Labor-Hours	Unit	Material	Labor	Equipment	Total	Total Incl O&P
0010	**APP MODIFIED BITUMINOUS MEMBRANE** R075213-30									
0020	Base sheet, #15 glass fiber felt, nailed to deck	1 Rofc	58	.138	Sq.	10.65	5.55		16.20	21
0030	Spot mopped to deck	G-1	295	.190		15.20	7.10	1.93	24.23	31
0040	Fully mopped to deck	"	192	.292		20.50	10.95	2.96	34.41	45
0050	#15 organic felt, nailed to deck	1 Rofc	58	.138		6.40	5.55		11.95	16.50
0060	Spot mopped to deck	G-1	295	.190		10.95	7.10	1.93	19.98	26.50
0070	Fully mopped to deck	"	192	.292		16.50	10.95	2.96	30.41	40
2100	APP mod., smooth surf. cap sheet, poly. reinf., torched, 160 mils	G-5	2100	.019	S.F.	.68	.69	.09	1.46	2.04
2150	170 mils		2100	.019		.70	.69	.09	1.48	2.06
2200	Granule surface cap sheet, poly. reinf., torched, 180 mils		2000	.020		.89	.73	.10	1.72	2.34
2250	Smooth surface flashing, torched, 160 mils		1260	.032		.68	1.16	.15	1.99	2.90
2300	170 mils		1260	.032		.70	1.16	.15	2.01	2.92
2350	Granule surface flashing, torched, 180 mils		1260	.032		.89	1.16	.15	2.20	3.13
2400	Fibrated aluminum coating	1 Rofc	3800	.002		.08	.08		.16	.23
2450	Seam heat welding	"	205	.039	L.F.	.08	1.56		1.64	2.76

07 52 16 – Styrene-Butadiene-Styrene Modified Bituminous Membrane Roofing

07 52 16.10 SBS Modified Bituminous Membrane		Crew	Daily Output	Labor-Hours	Unit	Material	Labor	Equipment	Total	Total Incl O&P
0010	**SBS MODIFIED BITUMINOUS MEMBRANE**									
0080	Mod bit rfng, SBS mod, gran surf cap sheet, poly reinf									
0650	120 to 149 mils thick	G-1	2000	.028	S.F.	1.25	1.05	.28	2.58	3.48
0750	150 to 160 mils	"	2000	.028		1.72	1.05	.28	3.05	3.99
1150	For reflective granules, add					.71	1.02	.28	2.01	2.82
1600	Smooth surface cap sheet, mopped, 145 mils	G-1	2100	.027		.78	1	.27	2.05	2.87
1620	Lightweight base sheet, fiberglass reinforced, 35 to 47 mil		2100	.027		.27	1	.27	1.54	2.31
1625	Heavyweight base/ply sheet, reinforced, 87 to 120 mil thick		2100	.027		.89	1	.27	2.16	2.99
1650	Granulated walkpad, 180 – 220 mils	1 Rofc	400	.020		1.82	.80		2.62	3.37
1700	Smooth surface flashing, 145 mils	G-1	1260	.044		.78	1.67	.45	2.90	4.21
1800	150 mils		1260	.044		.49	1.67	.45	2.61	3.89

07 52 Modified Bituminous Membrane Roofing

07 52 16 – Styrene-Butadiene-Styrene Modified Bituminous Membrane Roofing

07 52 16.10 SBS Modified Bituminous Membrane		Crew	Daily Output	Labor-Hours	Unit	Material	2015 Bare Costs Labor	Equipment	Total	Total Incl O&P
1900	Granular surface flashing, 150 mils	G-1	1260	.044	S.F.	.65	1.67	.45	2.77	4.07
2000	160 mils		1260	.044		.71	1.67	.45	2.83	4.13
2010	Elastomeric asphalt primer	1 Rofc	2600	.003		.17	.12		.29	.40
2015	Roofing asphalt, 30 lb. per square	G-1	19000	.003		.14	.11	.03	.28	.37
2020	Cold process adhesive, 20 – 30 mils thick	1 Rofc	750	.011		.24	.43		.67	.99
2025	Self adhering vapor retarder, 30 to 45 mils thick	G-5	2150	.019		1.02	.68	.09	1.79	2.38
2050	Seam heat welding	1 Rofc	205	.039	L.F.	.08	1.56		1.64	2.76

07 53 Elastomeric Membrane Roofing

07 53 16 – Chlorosulfonate-Polyethylene Roofing

07 53 16.10 Chlorosulfonated Polyethylene Roofing		Crew	Daily Output	Labor-Hours	Unit	Material	Labor	Equipment	Total	Total Incl O&P
0010	**CHLOROSULFONATED POLYETHYLENE ROOFING**									
0800	Chlorosulfonated polyethylene (CSPE)									
0900	45 mils, heat welded seams, plate attachment	G-5	35	1.143	Sq.	310	41.50	5.50	357	415
1100	Heat welded seams, plate attachment and ballasted		26	1.538		320	56	7.45	383.45	460
1200	60 mils, heat welded seams, plate attachment		35	1.143		410	41.50	5.50	457	525
1300	Heat welded seams, plate attachment and ballasted		26	1.538		420	56	7.45	483.45	565

07 53 23 – Ethylene-Propylene-Diene-Monomer Roofing

07 53 23.20 Ethylene-Propylene-Diene-Monomer Roofing		Crew	Daily Output	Labor-Hours	Unit	Material	Labor	Equipment	Total	Total Incl O&P
0010	**ETHYLENE-PROPYLENE-DIENE-MONOMER ROOFING (EPDM)**									
3500	Ethylene-propylene-diene-monomer (EPDM), 45 mils, 0.28 psf									
3600	Loose-laid & ballasted with stone (10 psf)	G-5	51	.784	Sq.	85	28.50	3.79	117.29	147
3700	Mechanically attached		35	1.143		79	41.50	5.50	126	164
3800	Fully adhered with adhesive		26	1.538		111	56	7.45	174.45	226
4500	60 mils, 0.40 psf									
4600	Loose-laid & ballasted with stone (10 psf)	G-5	51	.784	Sq.	100	28.50	3.79	132.29	163
4700	Mechanically attached		35	1.143		93	41.50	5.50	140	179
4800	Fully adhered with adhesive		26	1.538		125	56	7.45	188.45	242
4810	45 mil, .28 psf, membrane only					50.50			50.50	55.50
4820	60 mil, .40 psf, membrane only					63			63	69.50
4850	Seam tape for membrane, 3" x 100' roll				Ea.	40			40	43.50
4900	Batten strips, 10' sections					3.43			3.43	3.77
4910	Cover tape for batten strips, 6" x 100' roll					183			183	202
4930	Plate anchors				M	78			78	86
4970	Adhesive for fully adhered systems, 60 S.F./gal.				Gal.	19.65			19.65	21.50

07 53 29 – Polyisobutylene Roofing

07 53 29.10 Polyisobutylene Roofing		Crew	Daily Output	Labor-Hours	Unit	Material	Labor	Equipment	Total	Total Incl O&P
0010	**POLYISOBUTYLENE ROOFING**									
7500	Polyisobutylene (PIB), 100 mils, 0.57 psf									
7600	Loose-laid & ballasted with stone/gravel (10 psf)	G-5	51	.784	Sq.	200	28.50	3.79	232.29	273
7700	Partially adhered with adhesive		35	1.143		240	41.50	5.50	287	340
7800	Hot asphalt attachment		35	1.143		230	41.50	5.50	277	330
7900	Fully adhered with contact cement		26	1.538		250	56	7.45	313.45	380

07 54 Thermoplastic Membrane Roofing

07 54 19 – Polyvinyl-Chloride Roofing

07 54 19.10 Polyvinyl-Chloride Roofing (PVC)		Crew	Daily Output	Labor-Hours	Unit	Material	2015 Bare Costs Labor	Equipment	Total	Total Incl O&P
0010	**POLYVINYL-CHLORIDE ROOFING (PVC)**									
8200	Heat welded seams									
8700	Reinforced, 48 mils, 0.33 psf									
8750	Loose-laid & ballasted with stone/gravel (12 psf)	G-5	51	.784	Sq.	118	28.50	3.79	150.29	183
8800	Mechanically attached		35	1.143		111	41.50	5.50	158	199
8850	Fully adhered with adhesive		26	1.538		153	56	7.45	216.45	272
8860	Reinforced, 60 mils, .40 psf									
8870	Loose-laid & ballasted with stone/gravel (12 psf)	G-5	51	.784	Sq.	120	28.50	3.79	152.29	185
8880	Mechanically attached		35	1.143		112	41.50	5.50	159	201
8890	Fully adhered with adhesive		26	1.538		154	56	7.45	217.45	274

07 54 23 – Thermoplastic-Polyolefin Roofing

07 54 23.10 Thermoplastic Polyolefin Roofing (T.P.O.)		Crew	Daily Output	Labor-Hours	Unit	Material	Labor	Equipment	Total	Total Incl O&P
0010	**THERMOPLASTIC POLYOLEFIN ROOFING (T.P.O.)**									
0100	45 mil, loose laid & ballasted with stone (1/2 ton/sq.)	G-5	51	.784	Sq.	87.50	28.50	3.79	119.79	149
0120	Fully adhered		25	1.600		78	58.50	7.75	144.25	194
0140	Mechanically attached		34	1.176		77	43	5.70	125.70	165
0160	Self adhered		35	1.143		87	41.50	5.50	134	173
0180	60 mil membrane, heat welded seams, ballasted		50	.800		100	29	3.86	132.86	164
0200	Fully adhered		25	1.600		90	58.50	7.75	156.25	207
0220	Mechanically attached		34	1.176		93.50	43	5.70	142.20	183
0240	Self adhered		35	1.143		107	41.50	5.50	154	195

07 54 30 – Ketone Ethylene Ester Roofing

07 54 30.10 Ketone Ethylene Ester Roofing		Crew	Daily Output	Labor-Hours	Unit	Material	Labor	Equipment	Total	Total Incl O&P
0010	**KETONE ETHYLENE ESTER ROOFING**									
0100	Ketone ethylene ester roofing, 50 mil, fully adhered	G-5	26	1.538	Sq.	215	56	7.45	278.45	340
0120	Mechanically attached		35	1.143		141	41.50	5.50	188	232
0140	Ballasted with stone		51	.784		149	28.50	3.79	181.29	216
0160	50 mil, fleece backed, adhered w/hot asphalt	G-1	26	2.154		174	81	22	277	355
0180	Accessories, pipe boot	1 Rofc	32	.250	Ea.	24.50	10.05		34.55	43.50
0200	Pre-formed corners		32	.250	"	8.95	10.05		19	27
0220	Ketone clad metal, including up to 4 bends		330	.024	S.F.	3.95	.97		4.92	6
0240	Walkway pad	2 Rofc	800	.020	"	4.27	.80		5.07	6.05
0260	Stripping material	1 Rofc	310	.026	L.F.	.95	1.04		1.99	2.82

07 55 Protected Membrane Roofing

07 55 10 – Protected Membrane Roofing Components

07 55 10.10 Protected Membrane Roofing Components		Crew	Daily Output	Labor-Hours	Unit	Material	Labor	Equipment	Total	Total Incl O&P
0010	**PROTECTED MEMBRANE ROOFING COMPONENTS**									
0100	Choose roofing membrane from 07 50									
0120	Then choose roof deck insulation from 07 22									
0130	Filter fabric	2 Rofc	10000	.002	S.F.	.12	.06		.18	.25
0140	Ballast, 3/8" - 1/2" in place	G-1	36	1.556	Ton	20	58.50	15.80	94.30	139
0150	3/4" - 1-1/2" in place	"	36	1.556	"	20	58.50	15.80	94.30	139
0200	2" concrete blocks, natural	1 Clab	115	.070	S.F.	3.37	2.62		5.99	7.75
0210	Colors	"	115	.070	"	3.73	2.62		6.35	8.10

07 56 Fluid-Applied Roofing

07 56 10 – Fluid-Applied Roofing Elastomers

07 56 10.10 Elastomeric Roofing		Crew	Daily Output	Labor-Hours	Unit	2015 Bare Costs Material	Labor	Equipment	Total	Total Incl O&P
0010	**ELASTOMERIC ROOFING**									
0020	Acrylic, 44% solids, 2 coats, on corrugated metal	2 Rofc	2400	.007	S.F.	.58	.27		.85	1.09
0025	On smooth metal		3000	.005		.46	.21		.67	.87
0030	On foam or modified bitumen		1500	.011		.92	.43		1.35	1.74
0035	On concrete		1500	.011		.92	.43		1.35	1.74
0040	On tar and gravel		1500	.011		.92	.43		1.35	1.74
0045	36% solids, 2 coats, on corrugated metal		2400	.007		.56	.27		.83	1.08
0050	On smooth metal		3000	.005		.45	.21		.66	.85
0055	On foam or modified bitumen		1500	.011		.90	.43		1.33	1.72
0060	On concrete		1500	.011		.90	.43		1.33	1.72
0065	On tar and gravel		1500	.011		.90	.43		1.33	1.72
0070	Primer if required, 2 coats on corrugated metal		2400	.007		.52	.27		.79	1.03
0075	On smooth metal		3000	.005		.52	.21		.73	.93
0080	On foam or modified bitumen		1500	.011		.75	.43		1.18	1.56
0085	On concrete		1500	.011		.56	.43		.99	1.35
0090	On tar & gravel/rolled roof		1500	.011		.75	.43		1.18	1.56
0110	Acrylic rubber, fluid applied, 20 mils thick	G-5	2000	.020		2	.73	.10	2.83	3.56
0120	50 mils, reinforced		1200	.033		3	1.22	.16	4.38	5.55
0130	For walking surface, add		900	.044		1.05	1.62	.21	2.88	4.17
0300	Neoprene, fluid applied, 20 mil thick, not-reinforced	G-1	1135	.049		1.32	1.85	.50	3.67	5.15
0600	Non-woven polyester, reinforced		960	.058		1.45	2.19	.59	4.23	6
0700	5 coat neoprene deck, 60 mil thick, under 10,000 S.F.		325	.172		4.40	6.45	1.75	12.60	17.80
0900	Over 10,000 S.F.		625	.090		4.40	3.36	.91	8.67	11.60

07 57 Coated Foamed Roofing

07 57 13 – Sprayed Polyurethane Foam Roofing

07 57 13.10 Sprayed Polyurethane Foam Roofing (S.P.F.)		Crew	Daily Output	Labor-Hours	Unit	Material	Labor	Equipment	Total	Total Incl O&P
0010	**SPRAYED POLYURETHANE FOAM ROOFING (S.P.F.)**									
0100	Primer for metal substrate (when required)	G-2A	3000	.008	S.F.	.44	.29	.24	.97	1.21
0200	Primer for non-metal substrate (when required)		3000	.008		.17	.29	.24	.70	.92
0300	Closed cell spray, polyurethane foam, 3 lb. per C.F. density, 1", R6.7		15000	.002		.66	.06	.05	.77	.87
0400	2", R13.4		13125	.002		1.32	.07	.05	1.44	1.62
0500	3", R18.6		11485	.002		1.98	.08	.06	2.12	2.37
0550	4", R24.8		10080	.002		2.64	.09	.07	2.80	3.13
0700	Spray-on silicone coating		2500	.010		1.13	.34	.29	1.76	2.13
0800	Warranty 5-20 year manufacturer's								.15	.15
0900	Warranty 20 year, no dollar limit								.20	.20

07 58 Roll Roofing

07 58 10 – Asphalt Roll Roofing

07 58 10.10 Roll Roofing		Crew	Daily Output	Labor-Hours	Unit	Material	Labor	Equipment	Total	Total Incl O&P
0010	**ROLL ROOFING**									
0100	Asphalt, mineral surface									
0200	1 ply #15 organic felt, 1 ply mineral surfaced									
0300	Selvage roofing, lap 19", nailed & mopped	G-1	27	2.074	Sq.	74.50	78	21	173.50	238
0400	3 plies glass fiber felt (type IV), 1 ply mineral surfaced									
0500	Selvage roofing, lapped 19", mopped	G-1	25	2.240	Sq.	126	84	23	233	305
0600	Coated glass fiber base sheet, 2 plies of glass fiber									
0700	Felt (type IV), 1 ply mineral surfaced selvage									

07 58 Roll Roofing

07 58 10 – Asphalt Roll Roofing

07 58 10.10 Roll Roofing		Crew	Daily Output	Labor-Hours	Unit	Material	Labor	Equipment	Total	Total Incl O&P
						2015 Bare Costs				
0800	Roofing, lapped 19", mopped	G-1	25	2.240	Sq.	133	84	23	240	315
0900	On nailable decks	"	24	2.333	"	122	87.50	23.50	233	310
1000	3 plies glass fiber felt (type III), 1 ply mineral surfaced									
1100	Selvage roofing, lapped 19", mopped	G-1	25	2.240	Sq.	126	84	23	233	305

07 61 Sheet Metal Roofing

07 61 13 – Standing Seam Sheet Metal Roofing

07 61 13.10 Standing Seam Sheet Metal Roofing, Field Fab.

Line	Description	Crew	Daily Output	Labor-Hours	Unit	Material	Labor	Equipment	Total	Total Incl O&P
0010	**STANDING SEAM SHEET METAL ROOFING, FIELD FABRICATED**									
0400	Copper standing seam roofing, over 10 squares, 16 oz., 125 lb. per sq.	1 Shee	1.30	6.154	Sq.	960	345		1,305	1,575
0600	18 oz., 140 lb. per sq.		1.20	6.667		1,075	375		1,450	1,750
0700	20 oz., 150 lb. per sq.	↓	1.10	7.273		1,175	405		1,580	1,925
1200	For abnormal conditions or small areas, add					25%	100%			
1300	For lead-coated copper, add				↓	25%				

07 61 16 – Batten Seam Sheet Metal Roofing

07 61 16.10 Batten Seam Sheet Metal Roofing, Field Fabricated

Line	Description	Crew	Daily Output	Labor-Hours	Unit	Material	Labor	Equipment	Total	Total Incl O&P
0010	**BATTEN SEAM SHEET METAL ROOFING, FIELD FABRICATED**									
0012	Copper batten seam roofing, over 10 sq., 16 oz., 130 lb. per sq.	1 Shee	1.10	7.273	Sq.	1,225	405		1,630	1,950
0020	Lead batten seam roofing, 5 lb. per S.F.		1.20	6.667		1,350	375		1,725	2,075
0100	Zinc/copper alloy batten seam roofing, .020 thick		1.20	6.667		1,225	375		1,600	1,925
0200	Copper roofing, batten seam, over 10 sq., 18 oz., 145 lb. per sq.		1	8		1,350	450		1,800	2,175
0300	20 oz., 160 lb. per sq.		1	8		1,500	450		1,950	2,300
0500	Stainless steel batten seam roofing, type 304, 28 ga.		1.20	6.667		585	375		960	1,225
0600	26 ga.		1.15	6.957		545	390		935	1,200
0800	Zinc, copper alloy roofing, batten seam, .027" thick		1.15	6.957		1,550	390		1,940	2,300
0900	.032" thick		1.10	7.273		1,975	405		2,380	2,800
1000	.040" thick	↓	1.05	7.619	↓	2,475	425		2,900	3,375

07 61 19 – Flat Seam Sheet Metal Roofing

07 61 19.10 Flat Seam Sheet Metal Roofing, Field Fabricated

Line	Description	Crew	Daily Output	Labor-Hours	Unit	Material	Labor	Equipment	Total	Total Incl O&P
0010	**FLAT SEAM SHEET METAL ROOFING, FIELD FABRICATED**									
0900	Copper flat seam roofing, over 10 squares, 16 oz., 115 lb./sq.	1 Shee	1.20	6.667	Sq.	890	375		1,265	1,550
0950	18 oz., 130 lb./sq.		1.15	6.957		995	390		1,385	1,700
1000	20 oz., 145 lb./sq.		1.10	7.273		1,100	405		1,505	1,825
1008	Zinc flat seam roofing, .020" thick		1.20	6.667		1,050	375		1,425	1,725
1010	.027" thick		1.15	6.957		1,325	390		1,715	2,050
1020	.032" thick		1.12	7.143		1,675	400		2,075	2,450
1030	.040" thick		1.05	7.619		2,100	425		2,525	2,975
1100	Lead flat seam roofing, 5 lb. per S.F.	↓	1.30	6.154	↓	1,150	345		1,495	1,800

07 65 Flexible Flashing

07 65 10 – Sheet Metal Flashing

07 65 10.10 Sheet Metal Flashing and Counter Flashing		Crew	Daily Output	Labor-Hours	Unit	2015 Bare Costs Material	Labor	Equipment	Total	Total Incl O&P
0010	SHEET METAL FLASHING AND COUNTER FLASHING									
0011	Including up to 4 bends									
0020	Aluminum, mill finish, .013" thick	1 Rofc	145	.055	S.F.	.79	2.21		3	4.65
0030	.016" thick		145	.055		.91	2.21		3.12	4.78
0060	.019" thick		145	.055		1.15	2.21		3.36	5.05
0100	.032" thick		145	.055		1.35	2.21		3.56	5.25
0200	.040" thick		145	.055		2.31	2.21		4.52	6.30
0300	.050" thick		145	.055		2.43	2.21		4.64	6.45
0325	Mill finish 5" x 7" step flashing, .016" thick		1920	.004	Ea.	.15	.17		.32	.46
0350	Mill finish 12" x 12" step flashing, .016" thick		1600	.005	"	.50	.20		.70	.89
0400	Painted finish, add				S.F.	.29			.29	.32
1000	Mastic-coated 2 sides, .005" thick	1 Rofc	330	.024		1.82	.97		2.79	3.66
1100	.016" thick		330	.024		2.01	.97		2.98	3.87
1600	Copper, 16 oz., sheets, under 1000 lb.		115	.070		7.65	2.79		10.44	13.20
1700	Over 4000 lb.		155	.052		7.65	2.07		9.72	12
1900	20 oz. sheets, under 1000 lb.		110	.073		9.40	2.92		12.32	15.35
2000	Over 4000 lb.		145	.055		8.95	2.21		11.16	13.60
2200	24 oz. sheets, under 1000 lb.		105	.076		13.30	3.06		16.36	19.85
2300	Over 4000 lb.		135	.059		12.65	2.38		15.03	17.95
2500	32 oz. sheets, under 1000 lb.		100	.080		17.80	3.21		21.01	25
2600	Over 4000 lb.		130	.062		16.90	2.47		19.37	23
2700	W shape for valleys, 16 oz., 24" wide		100	.080	L.F.	15.40	3.21		18.61	22.50
5800	Lead, 2.5 lb. per S.F., up to 12" wide		135	.059	S.F.	4.99	2.38		7.37	9.55
5900	Over 12" wide		135	.059		3.99	2.38		6.37	8.45
8900	Stainless steel sheets, 32 ga.,		155	.052		3	2.07		5.07	6.85
9000	28 ga.		155	.052		3.99	2.07		6.06	7.90
9100	26 ga.		155	.052		4.25	2.07		6.32	8.20
9200	24 ga.		155	.052		4.75	2.07		6.82	8.80
9290	For mechanically keyed flashing, add					40%				
9320	Steel sheets, galvanized, 20 ga.	1 Rofc	130	.062	S.F.	1.25	2.47		3.72	5.60
9322	22 ga.		135	.059		1.21	2.38		3.59	5.40
9324	24 ga.		140	.057		.92	2.29		3.21	4.92
9326	26 ga.		148	.054		.80	2.17		2.97	4.58
9328	28 ga.		155	.052		.69	2.07		2.76	4.29
9340	30 ga.		160	.050		.58	2.01		2.59	4.06
9400	Terne coated stainless steel, .015" thick, 28 ga		155	.052		7.35	2.07		9.42	11.65
9500	.018" thick, 26 ga		155	.052		8.30	2.07		10.37	12.65
9600	Zinc and copper alloy (brass), .020" thick		155	.052		9.30	2.07		11.37	13.80
9700	.027" thick		155	.052		11	2.07		13.07	15.65
9800	.032" thick		155	.052		14	2.07		16.07	18.95
9900	.040" thick		155	.052		18.50	2.07		20.57	24

07 65 12 – Fabric and Mastic Flashings

07 65 12.10 Fabric and Mastic Flashing and Counter Flashing		Crew	Daily Output	Labor-Hours	Unit	Material	Labor	Equipment	Total	Total Incl O&P
0010	FABRIC AND MASTIC FLASHING AND COUNTER FLASHING									
1300	Asphalt flashing cement, 5 gallon				Gal.	9.20			9.20	10.10
4900	Fabric, asphalt-saturated cotton, specification grade	1 Rofc	35	.229	S.Y.	2.96	9.15		12.11	18.90
5000	Utility grade		35	.229		1.42	9.15		10.57	17.20
5300	Close-mesh fabric, saturated, 17 oz. per S.Y.		35	.229		2.15	9.15		11.30	18
5500	Fiberglass, resin-coated		35	.229		.97	9.15		10.12	16.70
8500	Shower pan, bituminous membrane, 7 oz.		155	.052	S.F.	1.62	2.07		3.69	5.30

07 65 Flexible Flashing

07 65 13 – Laminated Sheet Flashing

07 65 13.10 Laminated Sheet Flashing		Crew	Daily Output	Labor-Hours	Unit	Material	2015 Bare Costs Labor	Equipment	Total	Total Incl O&P
0010	**LAMINATED SHEET FLASHING**, Including up to 4 bends									
0500	Aluminum, fabric-backed 2 sides, mill finish, .004" thick	1 Rofc	330	.024	S.F.	1.54	.97		2.51	3.35
0700	.005" thick		330	.024		1.82	.97		2.79	3.66
0750	Mastic-backed, self adhesive		460	.017		3.51	.70		4.21	5.05
0800	Mastic-coated 2 sides, .004" thick		330	.024		1.54	.97		2.51	3.35
2800	Copper, paperbacked 1 side, 2 oz.		330	.024		1.84	.97		2.81	3.68
2900	3 oz.		330	.024		2.15	.97		3.12	4.03
3100	Paperbacked 2 sides, 2 oz.		330	.024		1.97	.97		2.94	3.83
3150	3 oz.		330	.024		2.25	.97		3.22	4.14
3200	5 oz.		330	.024		3.03	.97		4	4.99
3250	7 oz.		330	.024		5.30	.97		6.27	7.50
3400	Mastic-backed 2 sides, copper, 2 oz.		330	.024		1.82	.97		2.79	3.66
3500	3 oz.		330	.024		2.20	.97		3.17	4.08
3700	5 oz.		330	.024		3.30	.97		4.27	5.30
3800	Fabric-backed 2 sides, copper, 2 oz.		330	.024		2.10	.97		3.07	3.97
4000	3 oz.		330	.024		2.40	.97		3.37	4.30
4100	5 oz.		330	.024		3.40	.97		4.37	5.40
4300	Copper-clad stainless steel, .015" thick, under 500 lb.		115	.070		6.40	2.79		9.19	11.80
4400	Over 2000 lb.		155	.052		6.10	2.07		8.17	10.25
4600	.018" thick, under 500 lb.		100	.080		7.20	3.21		10.41	13.40
4700	Over 2000 lb.		145	.055		6.80	2.21		9.01	11.30
8550	Shower pan, 3 ply copper and fabric, 3 oz.		155	.052		3.50	2.07		5.57	7.40
8600	7 oz.		155	.052		4.20	2.07		6.27	8.15
9300	Stainless steel, paperbacked 2 sides, .005" thick	↓	330	.024	↓	3.62	.97		4.59	5.65

07 65 19 – Plastic Sheet Flashing

07 65 19.10 Plastic Sheet Flashing and Counter Flashing		Crew	Daily Output	Labor-Hours	Unit	Material	Labor	Equipment	Total	Total Incl O&P
0010	**PLASTIC SHEET FLASHING AND COUNTER FLASHING**									
7300	Polyvinyl chloride, black, 10 mil	1 Rofc	285	.028	S.F.	.21	1.13		1.34	2.15
7400	20 mil		285	.028		.28	1.13		1.41	2.23
7600	30 mil		285	.028		.35	1.13		1.48	2.31
7700	60 mil		285	.028		.85	1.13		1.98	2.86
7900	Black or white for exposed roofs, 60 mil	↓	285	.028	↓	1.80	1.13		2.93	3.90
8060	PVC tape, 5" x 45 mils, for joint covers, 100 L.F./roll				Ea.	170			170	187
8850	Polyvinyl chloride, 30 mil	1 Rofc	160	.050	S.F.	1.30	2.01		3.31	4.85

07 65 23 – Rubber Sheet Flashing

07 65 23.10 Rubber Sheet Flashing and Counterflashing		Crew	Daily Output	Labor-Hours	Unit	Material	Labor	Equipment	Total	Total Incl O&P
0010	**RUBBER SHEET FLASHING AND COUNTERFLASHING**									
4810	EPDM 90 mils, 1" diameter pipe flashing	1 Rofc	32	.250	Ea.	19.35	10.05		29.40	38.50
4820	2" diameter		30	.267		19.25	10.70		29.95	39.50
4830	3" diameter		28	.286		20.50	11.45		31.95	42.50
4840	4" diameter		24	.333		23.50	13.35		36.85	49
4850	6" diameter		22	.364	↓	23.50	14.60		38.10	51
8100	Rubber, butyl, 1/32" thick		285	.028	S.F.	1.92	1.13		3.05	4.03
8200	1/16" thick		285	.028		2.60	1.13		3.73	4.78
8300	Neoprene, cured, 1/16" thick		285	.028		2.23	1.13		3.36	4.37
8400	1/8" thick	↓	285	.028	↓	6.05	1.13		7.18	8.55

07 71 Roof Specialties

07 71 19 – Manufactured Gravel Stops and Fasciae

07 71 19.10 Gravel Stop

	07 71 19.10 Gravel Stop	Crew	Daily Output	Labor-Hours	Unit	2015 Bare Costs Material	Labor	Equipment	Total	Total Incl O&P
0010	**GRAVEL STOP**									
0020	Aluminum, .050" thick, 4" face height, mill finish	1 Shee	145	.055	L.F.	6.30	3.09		9.39	11.60
0080	Duranodic finish		145	.055		6.75	3.09		9.84	12.15
0100	Painted		145	.055		6.95	3.09		10.04	12.35
0300	6" face height		135	.059		6.45	3.32		9.77	12.15
0350	Duranodic finish		135	.059		7.25	3.32		10.57	13.05
0400	Painted		135	.059		8.20	3.32		11.52	14.05
0600	8" face height		125	.064		7.35	3.58		10.93	13.55
0650	Duranodic finish		125	.064		8.40	3.58		11.98	14.70
0700	Painted		125	.064		8.75	3.58		12.33	15.10
0900	12" face height, .080" thick, 2 piece		100	.080		9.95	4.48		14.43	17.75
0950	Duranodic finish		100	.080		10.25	4.48		14.73	18.15
1000	Painted		100	.080		12	4.48		16.48	20
1200	Copper, 16 oz., 3" face height		145	.055		24	3.09		27.09	31
1300	6" face height		135	.059		33	3.32		36.32	41.50
1350	Galv steel, 24 ga., 4" leg, plain, with continuous cleat, 4" face		145	.055		6.25	3.09		9.34	11.60
1360	6" face height		145	.055		6.50	3.09		9.59	11.85
1500	Polyvinyl chloride, 6" face height		135	.059		5.20	3.32		8.52	10.75
1600	9" face height		125	.064		6	3.58		9.58	12.05
1800	Stainless steel, 24 ga., 6" face height		135	.059		15.30	3.32		18.62	22
1900	12" face height		100	.080		22.50	4.48		26.98	32
2100	20 ga., 6" face height		135	.059		17.80	3.32		21.12	24.50
2200	12" face height		100	.080		27	4.48		31.48	36.50

07 71 19.30 Fascia

	07 71 19.30 Fascia	Crew	Daily Output	Labor-Hours	Unit	Material	Labor	Equipment	Total	Total Incl O&P
0010	**FASCIA**									
0100	Aluminum, reverse board and batten, .032" thick, colored, no furring incl	1 Shee	145	.055	S.F.	6.75	3.09		9.84	12.15
0300	Steel, galv and enameled, stock, no furring, long panels		145	.055		4.25	3.09		7.34	9.40
0600	Short panels		115	.070		5.80	3.89		9.69	12.35

07 71 23 – Manufactured Gutters and Downspouts

07 71 23.10 Downspouts

	07 71 23.10 Downspouts	Crew	Daily Output	Labor-Hours	Unit	Material	Labor	Equipment	Total	Total Incl O&P
0010	**DOWNSPOUTS**									
0020	Aluminum, embossed, .020" thick, 2" x 3"	1 Shee	190	.042	L.F.	.98	2.36		3.34	4.68
0100	Enameled		190	.042		1.37	2.36		3.73	5.10
0300	.024" thick, 2' x 3"		180	.044		2.05	2.49		4.54	6.05
0400	3" x 4"		140	.057		2.33	3.20		5.53	7.45
0600	Round, corrugated aluminum, 3" diameter, .020" thick		190	.042		1.95	2.36		4.31	5.75
0700	4" diameter, .025" thick		140	.057		2.53	3.20		5.73	7.65
0900	Wire strainer, round, 2" diameter		155	.052	Ea.	2.67	2.89		5.56	7.35
1000	4" diameter		155	.052		3.67	2.89		6.56	8.45
1200	Rectangular, perforated, 2" x 3"		145	.055		2.30	3.09		5.39	7.25
1300	3" x 4"		145	.055		3.32	3.09		6.41	8.35
1500	Copper, round, 16 oz., stock, 2" diameter		190	.042	L.F.	7.60	2.36		9.96	11.95
1600	3" diameter		190	.042		7.50	2.36		9.86	11.85
1800	4" diameter		145	.055		9.70	3.09		12.79	15.35
1900	5" diameter		130	.062		17.05	3.44		20.49	24
2100	Rectangular, corrugated copper, stock, 2" x 3"		190	.042		8.55	2.36		10.91	13
2200	3" x 4"		145	.055		10.40	3.09		13.49	16.15
2400	Rectangular, plain copper, stock, 2" x 3"		190	.042		11.05	2.36		13.41	15.75
2500	3" x 4"		145	.055		13.95	3.09		17.04	20
2700	Wire strainers, rectangular, 2" x 3"		145	.055	Ea.	17.05	3.09		20.14	23.50
2800	3" x 4"		145	.055		17.80	3.09		20.89	24.50
3000	Round, 2" diameter		145	.055		6.40	3.09		9.49	11.75

07 71 23 – Manufactured Gutters and Downspouts

07 71 23.10 Downspouts		Crew	Daily Output	Labor-Hours	Unit	Material	Labor	Equipment	Total	Total Incl O&P
						2015 Bare Costs				
3100	3" diameter	1 Shee	145	.055	Ea.	7.10	3.09		10.19	12.50
3300	4" diameter		145	.055		12.10	3.09		15.19	18
3400	5" diameter		115	.070		22.50	3.89		26.39	30.50
3600	Lead-coated copper, round, stock, 2" diameter		190	.042	L.F.	22	2.36		24.36	28
3700	3" diameter		190	.042		22	2.36		24.36	28
3900	4" diameter		145	.055		23	3.09		26.09	30
4000	5" diameter, corrugated		130	.062		22	3.44		25.44	30
4200	6" diameter, corrugated		105	.076		30	4.26		34.26	39.50
4300	Rectangular, corrugated, stock, 2" x 3"		190	.042		15.40	2.36		17.76	20.50
4500	Plain, stock, 2" x 3"		190	.042		24	2.36		26.36	30
4600	3" x 4"		145	.055		33	3.09		36.09	41
4800	Steel, galvanized, round, corrugated, 2" or 3" diameter, 28 ga.		190	.042		2.08	2.36		4.44	5.90
4900	4" diameter, 28 ga.		145	.055		2.52	3.09		5.61	7.50
5100	5" diameter, 26 ga.		130	.062		4	3.44		7.44	9.65
5400	6" diameter, 28 ga.		105	.076		4.20	4.26		8.46	11.10
5500	26 ga.		105	.076		4	4.26		8.26	10.90
5700	Rectangular, corrugated, 28 ga., 2" x 3"		190	.042		1.94	2.36		4.30	5.75
5800	3" x 4"		145	.055		1.82	3.09		4.91	6.70
6000	Rectangular, plain, 28 ga., galvanized, 2" x 3"		190	.042		3.65	2.36		6.01	7.60
6100	3" x 4"		145	.055		4.06	3.09		7.15	9.20
6300	Epoxy painted, 24 ga., corrugated, 2" x 3"		190	.042		2.25	2.36		4.61	6.10
6400	3" x 4"		145	.055		2.80	3.09		5.89	7.80
6600	Wire strainers, rectangular, 2" x 3"		145	.055	Ea.	17.05	3.09		20.14	23.50
6700	3" x 4"		145	.055		17.80	3.09		20.89	24.50
6900	Round strainers, 2" or 3" diameter		145	.055		3.86	3.09		6.95	8.95
7000	4" diameter		145	.055		5.80	3.09		8.89	11.05
7800	Stainless steel tubing, schedule 5, 2" x 3" or 3" diameter		190	.042	L.F.	38.50	2.36		40.86	46
7900	3" x 4" or 4" diameter		145	.055		49	3.09		52.09	58.50
8100	4" x 5" or 5" diameter		135	.059		101	3.32		104.32	116
8200	Vinyl, rectangular, 2" x 3"		210	.038		2.08	2.13		4.21	5.55
8300	Round, 2-1/2"		220	.036		1.32	2.03		3.35	4.56

07 71 23.20 Downspout Elbows		Crew	Daily Output	Labor-Hours	Unit	Material	Labor	Equipment	Total	Total Incl O&P
0010	**DOWNSPOUT ELBOWS**									
0020	Aluminum, embossed, 2" x 3", .020" thick	1 Shee	100	.080	Ea.	.95	4.48		5.43	7.90
0100	Enameled		100	.080		1.75	4.48		6.23	8.80
0200	Embossed, 3" x 4", .025" thick		100	.080		4.51	4.48		8.99	11.80
0300	Enameled		100	.080		3.75	4.48		8.23	11
0400	Embossed, corrugated, 3" diameter, .020" thick		100	.080		2.74	4.48		7.22	9.85
0500	4" diameter, .025" thick		100	.080		5.80	4.48		10.28	13.20
0600	Copper, 16 oz., 2" diameter		100	.080		9.45	4.48		13.93	17.25
0700	3" diameter		100	.080		9	4.48		13.48	16.75
0800	4" diameter		100	.080		13.90	4.48		18.38	22
1000	2" x 3" corrugated		100	.080		9.35	4.48		13.83	17.15
1100	3" x 4" corrugated		100	.080		13.25	4.48		17.73	21.50
1300	Vinyl, 2-1/2" diameter, 45 or 75 degree bend		100	.080		3.88	4.48		8.36	11.10
1400	Tee Y junction		75	.107		12.90	5.95		18.85	23.50

07 71 23.30 Gutters		Crew	Daily Output	Labor-Hours	Unit	Material	Labor	Equipment	Total	Total Incl O&P
0010	**GUTTERS**									
0012	Aluminum, stock units, 5" K type, .027" thick, plain	1 Shee	125	.064	L.F.	2.72	3.58		6.30	8.45
0100	Enameled		125	.064		2.66	3.58		6.24	8.35
0300	5" K type type, .032" thick, plain		125	.064		3.23	3.58		6.81	9
0400	Enameled		125	.064		3.02	3.58		6.60	8.75

07 71 23 – Manufactured Gutters and Downspouts

07 71 23.30 Gutters		Crew	Daily Output	Labor-Hours	Unit	Material	2015 Bare Costs Labor	Equipment	Total	Total Incl O&P
0700	Copper, half round, 16 oz., stock units, 4" wide	1 Shee	125	.064	L.F.	8.90	3.58		12.48	15.25
0900	5" wide		125	.064		7.25	3.58		10.83	13.40
1000	6" wide		118	.068		11.85	3.79		15.64	18.80
1200	K type, 16 oz., stock, 5" wide		125	.064		8.05	3.58		11.63	14.30
1300	6" wide		125	.064		8.50	3.58		12.08	14.80
1500	Lead coated copper, 16 oz., half round, stock, 4" wide		125	.064		13.80	3.58		17.38	20.50
1600	6" wide		118	.068		21.50	3.79		25.29	30
1800	K type, stock, 5" wide		125	.064		18.35	3.58		21.93	25.50
1900	6" wide		125	.064		17.90	3.58		21.48	25
2100	Copper clad stainless steel, K type, 5" wide		125	.064		7.50	3.58		11.08	13.70
2200	6" wide		125	.064		9.45	3.58		13.03	15.85
2400	Steel, galv, half round or box, 28 ga., 5" wide, plain		125	.064		2.10	3.58		5.68	7.75
2500	Enameled		125	.064		2.10	3.58		5.68	7.75
2700	26 ga., stock, 5" wide		125	.064		2.18	3.58		5.76	7.85
2800	6" wide	↓	125	.064		2.71	3.58		6.29	8.45
3000	Vinyl, O.G., 4" wide	1 Carp	115	.070		1.25	3.27		4.52	6.45
3100	5" wide		115	.070		1.55	3.27		4.82	6.75
3200	4" half round, stock units	↓	115	.070	↓	1.30	3.27		4.57	6.50
3250	Joint connectors				Ea.	2.90			2.90	3.19
3300	Wood, clear treated cedar, fir or hemlock, 3" x 4"	1 Carp	100	.080	L.F.	10	3.76		13.76	16.80
3400	4" x 5"	"	100	.080	"	16.75	3.76		20.51	24.50
5000	Accessories, end cap, K type, aluminum 5"	1 Shee	625	.013	Ea.	.66	.72		1.38	1.82
5010	6"		625	.013		1.44	.72		2.16	2.67
5020	Copper, 5"		625	.013		3.22	.72		3.94	4.63
5030	6"		625	.013		3.54	.72		4.26	4.98
5040	Lead coated copper, 5"		625	.013		12.50	.72		13.22	14.85
5050	6"		625	.013		13.35	.72		14.07	15.80
5060	Copper clad stainless steel, 5"		625	.013		3	.72		3.72	4.39
5070	6"		625	.013		3.60	.72		4.32	5.05
5080	Galvanized steel, 5"		625	.013		1.28	.72		2	2.50
5090	6"	↓	625	.013		2.18	.72		2.90	3.49
5100	Vinyl, 4"	1 Carp	625	.013		13.15	.60		13.75	15.40
5110	5"	"	625	.013		13.15	.60		13.75	15.40
5120	Half round, copper, 4"	1 Shee	625	.013		4.53	.72		5.25	6.05
5130	5"		625	.013		4.53	.72		5.25	6.05
5140	6"		625	.013		7.65	.72		8.37	9.55
5150	Lead coated copper, 5"		625	.013		14.15	.72		14.87	16.65
5160	6"		625	.013		21	.72		21.72	24.50
5170	Copper clad stainless steel, 5"		625	.013		4.50	.72		5.22	6.05
5180	6"		625	.013		4.50	.72		5.22	6.05
5190	Galvanized steel, 5"		625	.013		2.25	.72		2.97	3.57
5200	6"		625	.013		2.80	.72		3.52	4.17
5210	Outlet, aluminum, 2" x 3"		420	.019		.62	1.07		1.69	2.31
5220	3" x 4"		420	.019		1.05	1.07		2.12	2.79
5230	2-3/8" round		420	.019		.56	1.07		1.63	2.25
5240	Copper, 2" x 3"		420	.019		6.50	1.07		7.57	8.80
5250	3" x 4"		420	.019		7.85	1.07		8.92	10.30
5260	2-3/8" round		420	.019		4.55	1.07		5.62	6.65
5270	Lead coated copper, 2" x 3"		420	.019		25.50	1.07		26.57	29.50
5280	3" x 4"		420	.019		28.50	1.07		29.57	33
5290	2-3/8" round		420	.019		25.50	1.07		26.57	29.50
5300	Copper clad stainless steel, 2" x 3"		420	.019		6.50	1.07		7.57	8.80
5310	3" x 4"	↓	420	.019	↓	7.85	1.07		8.92	10.30

07 71 Roof Specialties

07 71 23 – Manufactured Gutters and Downspouts

07 71 23.30 Gutters		Crew	Daily Output	Labor-Hours	Unit	2015 Bare Costs Material	Labor	Equipment	Total	Total Incl O&P
5320	2-3/8" round	1 Shee	420	.019	Ea.	4.55	1.07		5.62	6.65
5330	Galvanized steel, 2" x 3"		420	.019		3.18	1.07		4.25	5.15
5340	3" x 4"		420	.019		4.90	1.07		5.97	7.05
5350	2-3/8" round		420	.019		3.98	1.07		5.05	6
5360	K type mitres, aluminum		65	.123		3	6.90		9.90	13.80
5370	Copper		65	.123		17.40	6.90		24.30	29.50
5380	Lead coated copper		65	.123		54.50	6.90		61.40	70.50
5390	Copper clad stainless steel		65	.123		27.50	6.90		34.40	41
5400	Galvanized steel		65	.123		22.50	6.90		29.40	35.50
5420	Half round mitres, copper		65	.123		63.50	6.90		70.40	80.50
5430	Lead coated copper		65	.123		89.50	6.90		96.40	109
5440	Copper clad stainless steel		65	.123		56	6.90		62.90	72.50
5450	Galvanized steel		65	.123		25.50	6.90		32.40	38.50
5460	Vinyl mitres and outlets		65	.123		9.75	6.90		16.65	21.50
5470	Sealant		940	.009	L.F.	.01	.48		.49	.74
5480	Soldering		96	.083	"	.19	4.66		4.85	7.30

07 71 23.35 Gutter Guard		Crew	Daily Output	Labor-Hours	Unit	Material	Labor	Equipment	Total	Total Incl O&P
0010	**GUTTER GUARD**									
0020	6" wide strip, aluminum mesh	1 Carp	500	.016	L.F.	2.42	.75		3.17	3.82
0100	Vinyl mesh	"	500	.016	"	2.61	.75		3.36	4.03

07 71 26 – Reglets

07 71 26.10 Reglets and Accessories		Crew	Daily Output	Labor-Hours	Unit	Material	Labor	Equipment	Total	Total Incl O&P
0010	**REGLETS AND ACCESSORIES**									
0020	Reglet, aluminum, .025" thick, in parapet	1 Carp	225	.036	L.F.	1.93	1.67		3.60	4.69
0300	16 oz. copper		225	.036		6.40	1.67		8.07	9.60
0400	Galvanized steel, 24 ga.		225	.036		1.34	1.67		3.01	4.04
0600	Stainless steel, .020" thick		225	.036		3.75	1.67		5.42	6.70
0900	Counter flashing for above, 12" wide, .032" aluminum	1 Shee	150	.053		2.40	2.98		5.38	7.20
1200	16 oz. copper		150	.053		6.10	2.98		9.08	11.25
1300	Galvanized steel, 26 ga.		150	.053		1.42	2.98		4.40	6.10
1500	Stainless steel, .020" thick		150	.053		6.15	2.98		9.13	11.30

07 71 29 – Manufactured Roof Expansion Joints

07 71 29.10 Expansion Joints		Crew	Daily Output	Labor-Hours	Unit	Material	Labor	Equipment	Total	Total Incl O&P
0010	**EXPANSION JOINTS**									
0300	Butyl or neoprene center with foam insulation, metal flanges									
0400	Aluminum, .032" thick for openings to 2-1/2"	1 Rofc	165	.048	L.F.	11.30	1.94		13.24	15.75
0600	For joint openings to 3-1/2"		165	.048		11.30	1.94		13.24	15.75
0610	For joint openings to 5"		165	.048		13.10	1.94		15.04	17.70
0620	For joint openings to 8"		165	.048		15.80	1.94		17.74	20.50
0700	Copper, 16 oz. for openings to 2-1/2"		165	.048		16.70	1.94		18.64	21.50
0900	For joint openings to 3-1/2"		165	.048		16.70	1.94		18.64	21.50
0910	For joint openings to 5"		165	.048		18.90	1.94		20.84	24.50
0920	For joint openings to 8"		165	.048		22	1.94		23.94	27.50
1000	Galvanized steel, 26 ga. for openings to 2-1/2"		165	.048		9.80	1.94		11.74	14.10
1200	For joint openings to 3-1/2"		165	.048		9.80	1.94		11.74	14.10
1210	For joint openings to 5"		165	.048		11.30	1.94		13.24	15.75
1220	For joint openings to 8"		165	.048		14.80	1.94		16.74	19.60
1300	Lead-coated copper, 16 oz. for openings to 2-1/2"		165	.048		32	1.94		33.94	38.50
1500	For joint openings to 3-1/2"		165	.048		32	1.94		33.94	38.50
1600	Stainless steel, .018", for openings to 2-1/2"		165	.048		15.40	1.94		17.34	20.50
1800	For joint openings to 3-1/2"		165	.048		15.40	1.94		17.34	20.50

07 71 Roof Specialties

07 71 29 – Manufactured Roof Expansion Joints

07 71 29.10 Expansion Joints		Crew	Daily Output	Labor-Hours	Unit	Material	2015 Bare Costs Labor	Equipment	Total	Total Incl O&P
1810	For joint openings to 5″	1 Rofc	165	.048	L.F.	17.20	1.94		19.14	22
1820	For joint openings to 8″		165	.048		21	1.94		22.94	26.50
1900	Neoprene, double-seal type with thick center, 4-1/2″ wide		125	.064		14.50	2.57		17.07	20.50
1950	Polyethylene bellows, with galv steel flat flanges		100	.080		6.20	3.21		9.41	12.30
1960	With galvanized angle flanges		100	.080		6.40	3.21		9.61	12.55
2000	Roof joint with extruded aluminum cover, 2″	1 Shee	115	.070		28	3.89		31.89	37
2100	Roof joint, plastic curbs, foam center, standard	1 Rofc	100	.080		13	3.21		16.21	19.80
2200	Large	″	100	.080		17	3.21		20.21	24
2500	Roof to wall joint with extruded aluminum cover	1 Shee	115	.070		28	3.89		31.89	37
2700	Wall joint, closed cell foam on PVC cover, 9″ wide	1 Rofc	125	.064		5	2.57		7.57	9.90
2800	12″ wide	″	115	.070		6	2.79		8.79	11.35

07 71 43 – Drip Edge

07 71 43.10 Drip Edge, Rake Edge, Ice Belts

		Crew	Daily Output	Labor-Hours	Unit	Material	Labor	Equipment	Total	Total Incl O&P
0010	**DRIP EDGE, RAKE EDGE, ICE BELTS**									
0020	Aluminum, .016″ thick, 5″ wide, mill finish	1 Carp	400	.020	L.F.	.55	.94		1.49	2.06
0100	White finish		400	.020		.63	.94		1.57	2.14
0200	8″ wide, mill finish		400	.020		1.45	.94		2.39	3.05
0300	Ice belt, 28″ wide, mill finish		100	.080		7.65	3.76		11.41	14.25
0310	Vented, mill finish		400	.020		1.99	.94		2.93	3.64
0320	Painted finish		400	.020		2.24	.94		3.18	3.91
0400	Galvanized, 5″ wide		400	.020		.53	.94		1.47	2.03
0500	8″ wide, mill finish		400	.020		.80	.94		1.74	2.33
0510	Rake edge, aluminum, 1-1/2″ x 1-1/2″		400	.020		.30	.94		1.24	1.78
0520	3-1/2″ x 1-1/2″		400	.020		.42	.94		1.36	1.91

07 72 Roof Accessories

07 72 23 – Relief Vents

07 72 23.10 Roof Vents

		Crew	Daily Output	Labor-Hours	Unit	Material	Labor	Equipment	Total	Total Incl O&P
0010	**ROOF VENTS**									
0020	Mushroom shape, for built-up roofs, aluminum	1 Rofc	30	.267	Ea.	65	10.70		75.70	90
0100	PVC, 6″ high	″	30	.267	″	21	10.70		31.70	41.50

07 72 26 – Ridge Vents

07 72 26.10 Ridge Vents and Accessories

		Crew	Daily Output	Labor-Hours	Unit	Material	Labor	Equipment	Total	Total Incl O&P
0010	**RIDGE VENTS AND ACCESSORIES**									
0100	Aluminum strips, mill finish	1 Rofc	160	.050	L.F.	2.75	2.01		4.76	6.45
0150	Painted finish		160	.050	″	4.05	2.01		6.06	7.90
0200	Connectors		48	.167	Ea.	4.77	6.70		11.47	16.65
0300	End caps		48	.167	″	2.03	6.70		8.73	13.65
0400	Galvanized strips		160	.050	L.F.	3.58	2.01		5.59	7.35
0430	Molded polyethylene, shingles not included		160	.050	″	2.74	2.01		4.75	6.45
0440	End plugs		48	.167	Ea.	2.03	6.70		8.73	13.65
0450	Flexible roll, shingles not included		160	.050	L.F.	2.38	2.01		4.39	6.05
2300	Ridge vent strip, mill finish	1 Shee	155	.052	″	3.85	2.89		6.74	8.65

07 72 33 – Roof Hatches

07 72 33.10 Roof Hatch Options

		Crew	Daily Output	Labor-Hours	Unit	Material	Labor	Equipment	Total	Total Incl O&P
0010	**ROOF HATCH OPTIONS**									
0500	2′-6″ x 3′, aluminum curb and cover	G-3	10	3.200	Ea.	1,025	150		1,175	1,350
0520	Galvanized steel curb and aluminum cover		10	3.200		640	150		790	935
0540	Galvanized steel curb and cover		10	3.200		715	150		865	1,025

07 72 Roof Accessories

07 72 33 – Roof Hatches

07 72 33.10 Roof Hatch Options		Crew	Daily Output	Labor-Hours	Unit	2015 Bare Costs Material	Labor	Equipment	Total	Total Incl O&P
0600	2'-6" x 4'-6", aluminum curb and cover	G-3	9	3.556	Ea.	1,375	166		1,541	1,775
0800	Galvanized steel curb and aluminum cover		9	3.556		885	166		1,051	1,225
0900	Galvanized steel curb and cover		9	3.556		965	166		1,131	1,300
1100	4' x 4' aluminum curb and cover		8	4		1,675	187		1,862	2,125
1120	Galvanized steel curb and aluminum cover		8	4		1,725	187		1,912	2,175
1140	Galvanized steel curb and cover		8	4		1,200	187		1,387	1,575
1200	2'-6" x 8'-0", aluminum curb and cover		6.60	4.848		1,900	227		2,127	2,450
1400	Galvanized steel curb and aluminum cover		6.60	4.848		1,800	227		2,027	2,325
1500	Galvanized steel curb and cover	↓	6.60	4.848		1,525	227		1,752	2,025
1800	For plexiglass panels, 2'-6" x 3'-0", add to above				↓	440			440	485

07 72 36 – Smoke Vents

07 72 36.10 Smoke Hatches

		Crew	Daily Output	Labor-Hours	Unit	Material	Labor	Equipment	Total	Total Incl O&P
0010	**SMOKE HATCHES**									
0200	For 3'-0" long, add to roof hatches from Section 07 72 33.10				Ea.	25%	5%			
0250	For 4'-0" long, add to roof hatches from Section 07 72 33.10					20%	5%			
0300	For 8'-0" long, add to roof hatches from Section 07 72 33.10				↓	10%	5%			

07 72 36.20 Smoke Vent Options

		Crew	Daily Output	Labor-Hours	Unit	Material	Labor	Equipment	Total	Total Incl O&P
0010	**SMOKE VENT OPTIONS**									
0100	4' x 4' aluminum cover and frame	G-3	13	2.462	Ea.	2,050	115		2,165	2,425
0200	Galvanized steel cover and frame		13	2.462		1,800	115		1,915	2,150
0300	4' x 8' aluminum cover and frame		8	4		2,800	187		2,987	3,350
0400	Galvanized steel cover and frame	↓	8	4	↓	2,375	187		2,562	2,900

07 72 53 – Snow Guards

07 72 53.10 Snow Guard Options

		Crew	Daily Output	Labor-Hours	Unit	Material	Labor	Equipment	Total	Total Incl O&P
0010	**SNOW GUARD OPTIONS**									
0100	Slate & asphalt shingle roofs, fastened with nails	1 Rofc	160	.050	Ea.	12.55	2.01		14.56	17.20
0200	Standing seam metal roofs, fastened with set screws		48	.167		16.95	6.70		23.65	30
0300	Surface mount for metal roofs, fastened with solder		48	.167	↓	6.05	6.70		12.75	18.05
0400	Double rail pipe type, including pipe	↓	130	.062	L.F.	32	2.47		34.47	39

07 72 73 – Pitch Pockets

07 72 73.10 Pitch Pockets, Variable Sizes

		Crew	Daily Output	Labor-Hours	Unit	Material	Labor	Equipment	Total	Total Incl O&P
0010	**PITCH POCKETS, VARIABLE SIZES**									
0100	Adjustable, 4" to 7", welded corners, 4" deep	1 Rofc	48	.167	Ea.	16.25	6.70		22.95	29.50
0200	Side extenders, 6"	"	240	.033	"	2.50	1.34		3.84	5.05

07 72 80 – Vents

07 72 80.30 Vent Options

		Crew	Daily Output	Labor-Hours	Unit	Material	Labor	Equipment	Total	Total Incl O&P
0010	**VENT OPTIONS**									
0020	Plastic, for insulated decks, 1 per M.S.F.	1 Rofc	40	.200	Ea.	21	8		29	36.50
0100	Heavy duty		20	.400		38	16.05		54.05	69.50
0300	Aluminum	↓	30	.267		22	10.70		32.70	42.50
0800	Polystyrene baffles, 12" wide for 16" O.C. rafter spacing	1 Carp	90	.089		.42	4.17		4.59	6.85
0900	For 24" O.C. rafter spacing	"	110	.073	↓	.77	3.41		4.18	6.10

07 76 Roof Pavers

07 76 16 – Roof Decking Pavers

07 76 16.10 Roof Pavers and Supports	Crew	Daily Output	Labor-Hours	Unit	Material	Labor (2015 Bare Costs)	Equipment	Total	Total Incl O&P
0010 **ROOF PAVERS AND SUPPORTS**									
1000 Roof decking pavers, concrete blocks, 2" thick, natural	1 Clab	115	.070	S.F.	3.37	2.62		5.99	7.75
1100 Colors		115	.070	"	3.73	2.62		6.35	8.10
1200 Support pedestal, bottom cap		960	.008	Ea.	3	.31		3.31	3.78
1300 Top cap		960	.008		4.80	.31		5.11	5.80
1400 Leveling shims, 1/16"		1920	.004		1.20	.16		1.36	1.56
1500 1/8"		1920	.004		1.20	.16		1.36	1.56
1600 Buffer pad		960	.008		2.50	.31		2.81	3.23
1700 PVC legs (4" SDR 35)		2880	.003	Inch	.12	.10		.22	.29
2000 Alternate pricing method, system in place		101	.079	S.F.	7	2.98		9.98	12.30

07 81 Applied Fireproofing

07 81 16 – Cementitious Fireproofing

07 81 16.10 Sprayed Cementitious Fireproofing	Crew	Daily Output	Labor-Hours	Unit	Material	Labor	Equipment	Total	Total Incl O&P
0010 **SPRAYED CEMENTITIOUS FIREPROOFING**									
0050 Not including canvas protection, normal density									
0100 Per 1" thick, on flat plate steel	G-2	3000	.008	S.F.	.53	.32	.04	.89	1.11
0200 Flat decking		2400	.010		.53	.40	.06	.99	1.24
0400 Beams		1500	.016		.53	.63	.09	1.25	1.64
0500 Corrugated or fluted decks		1250	.019		.79	.76	.11	1.66	2.14
0700 Columns, 1-1/8" thick		1100	.022		.59	.86	.12	1.57	2.08
0800 2-3/16" thick		700	.034		1.25	1.36	.19	2.80	3.64
0900 For canvas protection, add		5000	.005		.08	.19	.03	.30	.41
1000 Not including canvas protection, high density									
1100 Per 1" thick, on flat plate steel	G-2	3000	.008	S.F.	1.85	.32	.04	2.21	2.57
1110 On flat decking		2400	.010		1.85	.40	.06	2.31	2.70
1120 On beams		1500	.016		1.85	.63	.09	2.57	3.10
1130 Corrugated or fluted decks		1250	.019		1.85	.76	.11	2.72	3.31
1140 Columns, 1-1/8" thick		1100	.022		2.08	.86	.12	3.06	3.72
1150 2-3/16" thick		1100	.022		4.16	.86	.12	5.14	6
1170 For canvas protection, add		5000	.005		.08	.19	.03	.30	.41
1200 Not including canvas protection, retrofitting									
1210 Per 1" thick, on flat plate steel	G-2	1500	.016	S.F.	.38	.63	.09	1.10	1.48
1220 On flat decking		1200	.020		.38	.79	.11	1.28	1.74
1230 On beams		750	.032		.38	1.27	.18	1.83	2.53
1240 Corrugated or fluted decks		625	.038		.58	1.52	.21	2.31	3.16
1250 Columns, 1-1/8" thick		550	.044		.43	1.73	.24	2.40	3.36
1260 2-3/16" thick		500	.048		.86	1.90	.27	3.03	4.11
1400 Accessories, preliminary spattered texture coat		4500	.005		.02	.21	.03	.26	.37
1410 Bonding agent	1 Plas	1000	.008		.09	.34		.43	.62
1500 Intumescent epoxy fireproofing on wire mesh, 3/16" thick									
1550 1 hour rating, exterior use	G-2	136	.176	S.F.	7.35	7	.98	15.33	19.75
1600 Magnesium oxychloride, 35# to 40# density, 1/4" thick		3000	.008		1.55	.32	.04	1.91	2.24
1650 1/2" thick		2000	.012		3.10	.47	.07	3.64	4.20
1700 60# to 70# density, 1/4" thick		3000	.008		2.05	.32	.04	2.41	2.79
1750 1/2" thick		2000	.012		4.15	.47	.07	4.69	5.35
2000 Vermiculite cement, troweled or sprayed, 1/4" thick		3000	.008		1.40	.32	.04	1.76	2.07
2050 1/2" thick		2000	.012		2.75	.47	.07	3.29	3.82

07 84 13 – Penetration Firestopping

07 84 13.10 Firestopping		Crew	Daily Output	Labor-Hours	Unit	Material	Labor	Equipment	Total	Total Incl O&P
						2015 Bare Costs				
0010	**FIRESTOPPING** R078413-30									
0100	Metallic piping, non insulated									
0110	Through walls, 2" diameter	1 Carp	16	.500	Ea.	17.20	23.50		40.70	55
0120	4" diameter		14	.571		26	27		53	70.50
0130	6" diameter		12	.667		35.50	31.50		67	87
0140	12" diameter		10	.800		62.50	37.50		100	127
0150	Through floors, 2" diameter		32	.250		9.80	11.75		21.55	29
0160	4" diameter		28	.286		14.20	13.40		27.60	36
0170	6" diameter		24	.333		18.50	15.65		34.15	44.50
0180	12" diameter		20	.400		31.50	18.80		50.30	63.50
0190	Metallic piping, insulated									
0200	Through walls, 2" diameter	1 Carp	16	.500	Ea.	24	23.50		47.50	62.50
0210	4" diameter		14	.571		33	27		60	77.50
0220	6" diameter		12	.667		42	31.50		73.50	94
0230	12" diameter		10	.800		68.50	37.50		106	134
0240	Through floors, 2" diameter		32	.250		16.60	11.75		28.35	36.50
0250	4" diameter		28	.286		21	13.40		34.40	43.50
0260	6" diameter		24	.333		25.50	15.65		41.15	52
0270	12" diameter		20	.400		31.50	18.80		50.30	63.50
0280	Non metallic piping, non insulated									
0290	Through walls, 2" diameter	1 Carp	12	.667	Ea.	68.50	31.50		100	124
0300	4" diameter		10	.800		86	37.50		123.50	153
0310	6" diameter		8	1		120	47		167	205
0330	Through floors, 2" diameter		16	.500		54	23.50		77.50	95.50
0340	4" diameter		6	1.333		66.50	62.50		129	170
0350	6" diameter		6	1.333		80.50	62.50		143	185
0370	Ductwork, insulated & non insulated, round									
0380	Through walls, 6" diameter	1 Carp	12	.667	Ea.	35	31.50		66.50	86.50
0390	12" diameter		10	.800		69.50	37.50		107	135
0400	18" diameter		8	1		113	47		160	198
0410	Through floors, 6" diameter		16	.500		18.60	23.50		42.10	56.50
0420	12" diameter		14	.571		34	27		61	79
0430	18" diameter		12	.667		59	31.50		90.50	113
0440	Ductwork, insulated & non insulated, rectangular									
0450	With stiffener/closure angle, through walls, 6" x 12"	1 Carp	8	1	Ea.	28	47		75	104
0460	12" x 24"		6	1.333		37.50	62.50		100	138
0470	24" x 48"		4	2		107	94		201	262
0480	With stiffener/closure angle, through floors, 6" x 12"		10	.800		15.40	37.50		52.90	75
0490	12" x 24"		8	1		28	47		75	103
0500	24" x 48"		6	1.333		54	62.50		116.50	156
0510	Multi trade openings									
0520	Through walls, 6" x 12"	1 Carp	2	4	Ea.	59	188		247	355
0530	12" x 24"	"	1	8		238	375		613	840
0540	24" x 48"	2 Carp	1	16		950	750		1,700	2,200
0550	48" x 96"	"	.75	21.333		3,700	1,000		4,700	5,625
0560	Through floors, 6" x 12"	1 Carp	2	4		39	188		227	330
0570	12" x 24"	"	1	8		157	375		532	755
0580	24" x 48"	2 Carp	.75	21.333		625	1,000		1,625	2,250
0590	48" x 96"	"	.50	32		2,450	1,500		3,950	4,975
0600	Structural penetrations, through walls									
0610	Steel beams, W8 x 10	1 Carp	8	1	Ea.	37.50	47		84.50	114
0620	W12 x 14		6	1.333		59	62.50		121.50	162
0630	W21 x 44		5	1.600		118	75		193	246

07 84 Firestopping

07 84 13 – Penetration Firestopping

07 84 13.10 Firestopping		Crew	Daily Output	Labor-Hours	Unit	Material	2015 Bare Costs Labor	Equipment	Total	Total Incl O&P
0640	W36 x 135	1 Carp	3	2.667	Ea.	287	125		412	510
0650	Bar joists, 18" deep		6	1.333		54.50	62.50		117	156
0660	24" deep		6	1.333		67.50	62.50		130	171
0670	36" deep		5	1.600		101	75		176	228
0680	48" deep		4	2		118	94		212	275
0690	Construction joints, floor slab at exterior wall									
0700	Precast, brick, block or drywall exterior									
0710	2" wide joint	1 Carp	125	.064	L.F.	8.50	3		11.50	13.95
0720	4" wide joint	"	75	.107	"	17	5		22	26.50
0730	Metal panel, glass or curtain wall exterior									
0740	2" wide joint	1 Carp	40	.200	L.F.	20	9.40		29.40	36.50
0750	4" wide joint	"	25	.320	"	28	15		43	53.50
0760	Floor slab to drywall partition									
0770	Flat joint	1 Carp	100	.080	L.F.	8.40	3.76		12.16	15.05
0780	Fluted joint		50	.160		17.40	7.50		24.90	30.50
0790	Etched fluted joint		75	.107		11.10	5		16.10	19.90
0800	Floor slab to concrete/masonry partition									
0810	Flat joint	1 Carp	75	.107	L.F.	18.75	5		23.75	28
0820	Fluted joint	"	50	.160	"	22.50	7.50		30	36
0830	Concrete/CMU wall joints									
0840	1" wide	1 Carp	100	.080	L.F.	10.25	3.76		14.01	17.10
0850	2" wide		75	.107		18.75	5		23.75	28
0860	4" wide		50	.160		38	7.50		45.50	53
0870	Concrete/CMU floor joints									
0880	1" wide	1 Carp	200	.040	L.F.	5.15	1.88		7.03	8.55
0890	2" wide		150	.053		9.40	2.50		11.90	14.20
0900	4" wide		100	.080		17.90	3.76		21.66	25.50

07 91 Preformed Joint Seals

07 91 13 – Compression Seals

07 91 13.10 Compression Seals

		Crew	Daily Output	Labor-Hours	Unit	Material	Labor	Equipment	Total	Total Incl O&P
0010	**COMPRESSION SEALS**									
4900	O-ring type cord, 1/4"	1 Bric	472	.017	L.F.	.37	.78		1.15	1.60
4910	1/2"		440	.018		.95	.84		1.79	2.33
4920	3/4"		424	.019		1.80	.87		2.67	3.31
4930	1"		408	.020		3.40	.91		4.31	5.10
4940	1-1/4"		384	.021		6.30	.96		7.26	8.40
4950	1-1/2"		368	.022		7.95	1		8.95	10.30
4960	1-3/4"		352	.023		13.25	1.05		14.30	16.20
4970	2"		344	.023		18.50	1.07		19.57	22

07 91 16 – Joint Gaskets

07 91 16.10 Joint Gaskets

		Crew	Daily Output	Labor-Hours	Unit	Material	Labor	Equipment	Total	Total Incl O&P
0010	**JOINT GASKETS**									
4400	Joint gaskets, neoprene, closed cell w/adh, 1/8" x 3/8"	1 Bric	240	.033	L.F.	.28	1.54		1.82	2.66
4500	1/4" x 3/4"		215	.037		.59	1.72		2.31	3.27
4700	1/2" x 1"		200	.040		1.30	1.85		3.15	4.25
4800	3/4" x 1-1/2"		165	.048		1.45	2.24		3.69	5

07 91 Preformed Joint Seals

07 91 23 – Backer Rods

07 91 23.10 Backer Rods		Crew	Daily Output	Labor-Hours	Unit	Material	Labor	Equipment	Total	Total Incl O&P
						2015 Bare Costs				
0010	**BACKER RODS**									
0030	Backer rod, polyethylene, 1/4″ diameter	1 Bric	4.60	1.739	C.L.F.	2.13	80.50		82.63	125
0050	1/2″ diameter		4.60	1.739		3.40	80.50		83.90	127
0070	3/4″ diameter		4.60	1.739		5.55	80.50		86.05	129
0090	1″ diameter		4.60	1.739		8.85	80.50		89.35	133

07 91 26 – Joint Fillers

07 91 26.10 Joint Fillers

Line	Description	Crew	Daily Output	Labor-Hours	Unit	Material	Labor	Equipment	Total	Total Incl O&P
0010	**JOINT FILLERS**									
4360	Butyl rubber filler, 1/4″ x 1/4″	1 Bric	290	.028	L.F.	.22	1.27		1.49	2.19
4365	1/2″ x 1/2″		250	.032		.89	1.48		2.37	3.23
4370	1/2″ x 3/4″		210	.038		1.34	1.76		3.10	4.15
4375	3/4″ x 3/4″		230	.035		2.01	1.61		3.62	4.66
4380	1″ x 1″		180	.044		2.68	2.05		4.73	6.05
4390	For coloring, add					12%				
4980	Polyethylene joint backing, 1/4″ x 2″	1 Bric	2.08	3.846	C.L.F.	12	178		190	284
4990	1/4″ x 6″		1.28	6.250	″	28	288		316	470
5600	Silicone, room temp vulcanizing foam seal, 1/4″ x 1/2″		1312	.006	L.F.	.33	.28		.61	.80
5610	1/2″ x 1/2″		656	.012		.67	.56		1.23	1.59
5620	1/2″ x 3/4″		442	.018		1	.84		1.84	2.38
5630	3/4″ x 3/4″		328	.024		1.50	1.13		2.63	3.37
5640	1/8″ x 1″		1312	.006		.33	.28		.61	.80
5650	1/8″ x 3″		442	.018		1	.84		1.84	2.38
5670	1/4″ x 3″		295	.027		2	1.25		3.25	4.11
5680	1/4″ x 6″		148	.054		3.99	2.49		6.48	8.20
5690	1/2″ x 6″		82	.098		8	4.50		12.50	15.65
5700	1/2″ x 9″		52.50	.152		12	7.05		19.05	24
5710	1/2″ x 12″		33	.242		15.95	11.20		27.15	34.50

07 92 Joint Sealants

07 92 13 – Elastomeric Joint Sealants

07 92 13.20 Caulking and Sealant Options

Line	Description	Crew	Daily Output	Labor-Hours	Unit	Material	Labor	Equipment	Total	Total Incl O&P
0010	**CAULKING AND SEALANT OPTIONS**									
0050	Latex acrylic based, bulk				Gal.	26.50			26.50	29.50
0055	Bulk in place 1/4″ x 1/4″ bead	1 Bric	300	.027	L.F.	.08	1.23		1.31	1.97
0060	1/4″ x 3/8″		294	.027		.14	1.26		1.40	2.07
0065	1/4″ x 1/2″		288	.028		.19	1.28		1.47	2.17
0075	3/8″ x 3/8″		284	.028		.21	1.30		1.51	2.21
0080	3/8″ x 1/2″		280	.029		.28	1.32		1.60	2.32
0085	3/8″ x 5/8″		276	.029		.35	1.34		1.69	2.43
0095	3/8″ x 3/4″		272	.029		.42	1.36		1.78	2.53
0100	1/2″ x 1/2″		275	.029		.37	1.34		1.71	2.46
0105	1/2″ x 5/8″		269	.030		.47	1.37		1.84	2.61
0110	1/2″ x 3/4″		263	.030		.56	1.40		1.96	2.76
0115	1/2″ x 7/8″		256	.031		.65	1.44		2.09	2.92
0120	1/2″ x 1″		250	.032		.75	1.48		2.23	3.07
0125	3/4″ x 3/4″		244	.033		.84	1.51		2.35	3.24
0130	3/4″ x 1″		225	.036		1.12	1.64		2.76	3.74
0135	1″ x 1″		200	.040		1.50	1.85		3.35	4.47
0190	Cartridges				Gal.	32			32	35.50
0200	11 fl. oz. cartridge				Ea.	2.76			2.76	3.04

07 92 Joint Sealants

07 92 13 – Elastomeric Joint Sealants

07 92 13.20 Caulking and Sealant Options		Crew	Daily Output	Labor-Hours	Unit	Material	Labor	Equipment	Total	Total Incl O&P
						2015 Bare Costs				
0500	1/4" x 1/2"	1 Bric	288	.028	L.F.	.23	1.28		1.51	2.21
0600	1/2" x 1/2"		275	.029		.45	1.34		1.79	2.55
0800	3/4" x 3/4"		244	.033		1.01	1.51		2.52	3.43
0900	3/4" x 1"		225	.036		1.35	1.64		2.99	4
1000	1" x 1"	↓	200	.040	↓	1.69	1.85		3.54	4.68
1400	Butyl based, bulk				Gal.	35			35	38.50
1500	Cartridges				"	38.50			38.50	42
1700	1/4" x 1/2", 154 L.F./gal.	1 Bric	288	.028	L.F.	.23	1.28		1.51	2.21
1800	1/2" x 1/2", 77 L.F./gal.	"	275	.029	"	.45	1.34		1.79	2.55
2300	Polysulfide compounds, 1 component, bulk				Gal.	72			72	79
2600	1 or 2 component, in place, 1/4" x 1/4", 308 L.F./gal.	1 Bric	300	.027	L.F.	.23	1.23		1.46	2.14
2700	1/2" x 1/4", 154 L.F./gal.		288	.028		.47	1.28		1.75	2.47
2900	3/4" x 3/8", 68 L.F./gal.		272	.029		1.06	1.36		2.42	3.23
3000	1" x 1/2", 38 L.F./gal.	↓	250	.032	↓	1.89	1.48		3.37	4.33
3200	Polyurethane, 1 or 2 component				Gal.	49			49	54
3500	Bulk, in place, 1/4" x 1/4"	1 Bric	300	.027	L.F.	.16	1.23		1.39	2.06
3655	1/2" x 1/4"		288	.028		.32	1.28		1.60	2.31
3800	3/4" x 3/8"		272	.029		.72	1.36		2.08	2.87
3900	1" x 1/2"	↓	250	.032	↓	1.28	1.48		2.76	3.66
4100	Silicone rubber, bulk				Gal.	49			49	54
4200	Cartridges				"	51			51	56

07 92 16 – Rigid Joint Sealants

07 92 16.10 Rigid Joint Sealants		Crew	Daily Output	Labor-Hours	Unit	Material	Labor	Equipment	Total	Total Incl O&P
0010	**RIGID JOINT SEALANTS**									
5800	Tapes, sealant, PVC foam adhesive, 1/16" x 1/4"				C.L.F.	10			10	11
5900	1/16" x 1/2"					10			10	11
5950	1/16" x 1"					16			16	17.60
6000	1/8" x 1/2"				↓	9.50			9.50	10.45

07 92 19 – Acoustical Joint Sealants

07 92 19.10 Acoustical Sealant		Crew	Daily Output	Labor-Hours	Unit	Material	Labor	Equipment	Total	Total Incl O&P
0010	**ACOUSTICAL SEALANT**									
0020	Acoustical sealant, elastomeric, cartridges				Ea.	8.50			8.50	9.35
0025	In place, 1/4" x 1/4"	1 Bric	300	.027	L.F.	.35	1.23		1.58	2.26
0030	1/4" x 1/2"		288	.028		.69	1.28		1.97	2.72
0035	1/2" x 1/2"		275	.029		1.39	1.34		2.73	3.57
0040	1/2" x 3/4"		263	.030		2.08	1.40		3.48	4.43
0045	3/4" x 3/4"		244	.033		3.12	1.51		4.63	5.75
0050	1" x 1"	↓	200	.040	↓	5.55	1.85		7.40	8.90

07 95 Expansion Control

07 95 13 – Expansion Joint Cover Assemblies

07 95 13.50 Expansion Joint Assemblies		Crew	Daily Output	Labor-Hours	Unit	Material	Labor	Equipment	Total	Total Incl O&P
0010	**EXPANSION JOINT ASSEMBLIES**									
0200	Floor cover assemblies, 1" space, aluminum	1 Sswk	38	.211	L.F.	16.85	11.10		27.95	38
0300	Bronze		38	.211		54.50	11.10		65.60	79.50
0500	2" space, aluminum		38	.211		16.85	11.10		27.95	38
0600	Bronze		38	.211		54.50	11.10		65.60	79.50
0800	Wall and ceiling assemblies, 1" space, aluminum		38	.211		14.85	11.10		25.95	35.50
0900	Bronze		38	.211		48.50	11.10		59.60	73
1100	2" space, aluminum	↓	38	.211	↓	15.15	11.10		26.25	36

07 95 Expansion Control

07 95 13 – Expansion Joint Cover Assemblies

07 95 13.50 Expansion Joint Assemblies		Crew	Daily Output	Labor-Hours	Unit	2015 Bare Costs Material	Labor	Equipment	Total	Total Incl O&P
1200	Bronze	1 Sswk	38	.211	L.F.	48.50	11.10		59.60	73
1400	Floor to wall assemblies, 1" space, aluminum		38	.211		18	11.10		29.10	39
1500	Bronze or stainless		38	.211		59	11.10		70.10	84.50
1700	Gym floor angle covers, aluminum, 3" x 3" angle		46	.174		18	9.15		27.15	35.50
1800	3" x 4" angle		46	.174		21	9.15		30.15	39
2000	Roof closures, aluminum, flat roof, low profile, 1" space		57	.140		22.50	7.40		29.90	38
2100	High profile		57	.140		24	7.40		31.40	39.50
2300	Roof to wall, low profile, 1" space		57	.140		21.50	7.40		28.90	36.50
2400	High profile	↓	57	.140	↓	24	7.40		31.40	39.50

Estimating Tips

08 10 00 Doors and Frames

All exterior doors should be addressed for their energy conservation (insulation and seals).

- Most metal doors and frames look alike, but there may be significant differences among them. When estimating these items, be sure to choose the line item that most closely compares to the specification or door schedule requirements regarding:
  - type of metal
  - metal gauge
  - door core material
  - fire rating
  - finish
- Wood and plastic doors vary considerably in price. The primary determinant is the veneer material. Lauan, birch, and oak are the most common veneers. Other variables include the following:
  - hollow or solid core
  - fire rating
  - flush or raised panel
  - finish
- Door pricing includes bore for cylindrical lockset and mortise for hinges.

08 30 00 Specialty Doors and Frames

- There are many varieties of special doors, and they are usually priced per each. Add frames, hardware, or operators required for a complete installation.

08 40 00 Entrances, Storefronts, and Curtain Walls

- Glazed curtain walls consist of the metal tube framing and the glazing material. The cost data in this subdivision is presented for the metal tube framing alone or the composite wall. If your estimate requires a detailed takeoff of the framing, be sure to add the glazing cost and any tints.

08 50 00 Windows

- Most metal windows are delivered preglazed. However, some metal windows are priced without glass. Refer to 08 80 00 Glazing for glass pricing. The grade C indicates commercial grade windows, usually ASTM C-35.
- All wood windows and vinyl are priced preglazed. The glazing is insulating glass. Add the cost of screens and grills if required, and not already included.

08 70 00 Hardware

- Hardware costs add considerably to the cost of a door. The most efficient method to determine the hardware requirements for a project is to review the door and hardware schedule together. One type of door may have different hardware, depending on the door usage.
- Door hinges are priced by the pair, with most doors requiring 1-1/2 pairs per door. The hinge prices do not include installation labor, because it is included in door installation. Hinges are classified according to the frequency of use, base material, and finish.

08 80 00 Glazing

- Different openings require different types of glass. The most common types are:
  - float
  - tempered
  - insulating
  - impact-resistant
  - ballistic-resistant
- Most exterior windows are glazed with insulating glass. Entrance doors and window walls, where the glass is less than 18" from the floor, are generally glazed with tempered glass. Interior windows and some residential windows are glazed with float glass.
- Coastal communities require the use of impact-resistant glass, dependant on wind speed.
- The insulation or 'u' value is a strong consideration, along with solar heat gain, to determine total energy efficiency.

Reference Numbers

Reference numbers are shown in shaded boxes at the beginning of some major classifications. These numbers refer to related items in the Reference Section. The reference information may be an estimating procedure, an alternate pricing method, or technical information.

Note: Not all subdivisions listed here necessarily appear in this publication. ■

08 05 Common Work Results for Openings

08 05 05 – Selective Demolition for Openings

08 05 05.10 Selective Demolition Doors		Crew	Daily Output	Labor-Hours	Unit	Material	2015 Bare Costs Labor	Equipment	Total	Total Incl O&P
0010	SELECTIVE DEMOLITION DOORS R024119-10									
0200	Doors, exterior, 1-3/4" thick, single, 3' x 7' high	1 Clab	16	.500	Ea.		18.80		18.80	29
0220	Double, 6' x 7' high		12	.667			25		25	38.50
0500	Interior, 1-3/8" thick, single, 3' x 7' high		20	.400			15.05		15.05	23
0520	Double, 6' x 7' high		16	.500			18.80		18.80	29
0700	Bi-folding, 3' x 6'-8" high		20	.400			15.05		15.05	23
0720	6' x 6'-8" high		18	.444			16.70		16.70	25.50
0900	Bi-passing, 3' x 6'-8" high		16	.500			18.80		18.80	29
0940	6' x 6'-8" high		14	.571			21.50		21.50	33
1500	Remove and reset, hollow core	1 Carp	8	1			47		47	72.50
1520	Solid		6	1.333			62.50		62.50	96.50
2000	Frames, including trim, metal		8	1			47		47	72.50
2200	Wood	2 Carp	32	.500			23.50		23.50	36
3000	Special doors, counter doors		6	2.667			125		125	193
3100	Double acting		10	1.600			75		75	116
3200	Floor door (trap type), or access type		8	2			94		94	145
3300	Glass, sliding, including frames		12	1.333			62.50		62.50	96.50
3400	Overhead, commercial, 12' x 12' high		4	4			188		188	289
3440	up to 20' x 16' high		3	5.333			250		250	385
3445	up to 35' x 30' high		1	16			750		750	1,150
3500	Residential, 9' x 7' high		8	2			94		94	145
3540	16' x 7' high		7	2.286			107		107	165
3620	Large		2.50	6.400			300		300	460
3700	Roll-up grille		5	3.200			150		150	231
3800	Revolving door		2	8			375		375	580
3900	Storefront swing door		3	5.333			250		250	385
6600	Demo flexible transparent strip entrance	3 Shee	115	.209	SF Surf		11.70		11.70	17.85
7100	Remove double swing pneumatic doors, openers and sensors	2 Skwk	.50	32	Opng.		1,550		1,550	2,400
7110	Remove automatic operators, industrial, sliding doors, to 12' wide	"	.40	40	"		1,950		1,950	3,000
7550	Hangar door demo	2 Sswk	330	.048	S.F.		2.55		2.55	4.43
7570	Remove shock absorbing door	"	1.90	8.421	Opng.		445		445	770

08 05 05.20 Selective Demolition of Windows		Crew	Daily Output	Labor-Hours	Unit	Material	2015 Bare Costs Labor	Equipment	Total	Total Incl O&P
0010	SELECTIVE DEMOLITION OF WINDOWS R024119-10									
0200	Aluminum, including trim, to 12 S.F.	1 Clab	16	.500	Ea.		18.80		18.80	29
0240	To 25 S.F.		11	.727			27.50		27.50	42
0280	To 50 S.F.		5	1.600			60		60	92.50
0320	Storm windows/screens, to 12 S.F.		27	.296			11.15		11.15	17.15
0360	To 25 S.F.		21	.381			14.30		14.30	22
0400	To 50 S.F.		16	.500			18.80		18.80	29
0600	Glass, up to 10 SF per window		200	.040	S.F.		1.50		1.50	2.31
0620	Over 10 SF per window		150	.053	"		2.01		2.01	3.09
1000	Steel, including trim, to 12 S.F.		13	.615	Ea.		23		23	35.50
1020	To 25 S.F.		9	.889			33.50		33.50	51.50
1040	To 50 S.F.		4	2			75		75	116
2000	Wood, including trim, to 12 S.F.		22	.364			13.65		13.65	21
2020	To 25 S.F.		18	.444			16.70		16.70	25.50
2060	To 50 S.F.		13	.615			23		23	35.50
2065	To 180 S.F.		8	1			37.50		37.50	58
4300	Remove bay/bow window	2 Carp	6	2.667			125		125	193
4410	Remove skylight, plstc domes, flush/curb mtd	G-3	395	.081	S.F.		3.79		3.79	5.80
5020	Remove and reset window, up to a 2'x2' widow	1 Carp	6	1.333	Ea.		62.50		62.50	96.50
5040	Up to a 3'x3' window		4	2			94		94	145

08 05 Common Work Results for Openings

08 05 05 – Selective Demolition for Openings

08 05 05.20 Selective Demolition of Windows		Crew	Daily Output	Labor-Hours	Unit	Material	2015 Bare Costs Labor	Equipment	Total	Total Incl O&P
5080	Up to a 4'x5' window	1 Carp	2	4	Ea.		188		188	289

08 11 Metal Doors and Frames

08 11 16 – Aluminum Doors and Frames

08 11 16.10 Entrance Doors

		Crew	Daily Output	Labor-Hours	Unit	Material	Labor	Equipment	Total	Total Incl O&P
0010	**ENTRANCE DOORS** and frame, Aluminum, narrow stile									
0011	Including standard hardware, clear finish, no glass									
0012	Top and bottom offset pivots, 1/4" beveled glass stops, threshold									
0013	Dead bolt lock with inside thumb screw, standard push pull									
0020	3'-0" x 7'-0" opening	2 Sswk	2	8	Ea.	915	420		1,335	1,725
0025	Anodizing aluminum entr. door & frame, add					104			104	114
0030	3'-6" x 7'-0" opening	2 Sswk	2	8		840	420		1,260	1,650
0100	3'-0" x 10'-0" opening, 3' high transom		1.80	8.889		1,300	470		1,770	2,250
0200	3'-6" x 10'-0" opening, 3' high transom		1.80	8.889		1,350	470		1,820	2,325
0280	5'-0" x 7'-0" opening		2	8		1,425	420		1,845	2,300
0300	6'-0" x 7'-0" opening		1.30	12.308		1,200	650		1,850	2,450
0301	6'-0" x 7'-0" opening		1.30	12.308	Pr.	1,200	650		1,850	2,450
0400	6'-0" x 10'-0" opening, 3' high transom		1.10	14.545	"	1,675	765		2,440	3,150
0520	3'-0" x 7'-0" opening, wide stile		2	8	Ea.	1,000	420		1,420	1,825
0540	3'-6" x 7'-0" opening		2	8		1,200	420		1,620	2,050
0560	5'-0" x 7'-0" opening		2	8		1,525	420		1,945	2,400
0580	6'-0" x 7'-0" opening		1.30	12.308	Pr.	1,575	650		2,225	2,875
0600	7'-0" x 7'-0" opening		1	16	"	1,725	840		2,565	3,350
1200	For non-standard size, add				Leaf	80%				
1250	For installation of non-standard size, add						20%			
1300	Light bronze finish, add				Leaf	36%				
1400	Dark bronze finish, add					25%				
1500	For black finish, add					40%				
1600	Concealed panic device, add					940			940	1,025
1700	Electric striker release, add				Opng.	280			280	310
1800	Floor check, add				Leaf	650			650	710
1900	Concealed closer, add				"	530			530	580

08 11 63 – Metal Screen and Storm Doors and Frames

08 11 63.23 Aluminum Screen and Storm Doors and Frames

		Crew	Daily Output	Labor-Hours	Unit	Material	Labor	Equipment	Total	Total Incl O&P
0010	**ALUMINUM SCREEN AND STORM DOORS AND FRAMES**									
0020	Combination storm and screen									
0420	Clear anodic coating, 2'-8" wide	2 Carp	14	1.143	Ea.	205	53.50		258.50	310
0440	3'-0" wide	"	14	1.143	"	178	53.50		231.50	278
0500	For 7' door height, add					8%				
1020	Mill finish, 2'-8" wide	2 Carp	14	1.143	Ea.	235	53.50		288.50	340
1040	3'-0" wide	"	14	1.143		258	53.50		311.50	365
1100	For 7'-0" door, add					8%				
1520	White painted, 2'-8" wide	2 Carp	14	1.143		286	53.50		339.50	400
1540	3'-0" wide	"	14	1.143		310	53.50		363.50	425
1600	For 7'-0" door, add					8%				
2000	Wood door & screen, see Section 08 14 33.20									

08 11 Metal Doors and Frames

08 11 74 – Sliding Metal Grilles

08 11 74.10 Rolling Grille Supports		Crew	Daily Output	Labor-Hours	Unit	Material	2015 Bare Costs Labor	Equipment	Total	Total Incl O&P
0010	ROLLING GRILLE SUPPORTS									

08 12 Metal Frames

08 12 13 – Hollow Metal Frames

08 12 13.13 Standard Hollow Metal Frames

			Crew	Daily Output	Labor-Hours	Unit	Material	Labor	Equipment	Total	Total Incl O&P
0010	STANDARD HOLLOW METAL FRAMES										
0020	16 ga., up to 5-3/4" jamb depth										
0025	3'-0" x 6'-8" single	G	2 Carp	16	1	Ea.	146	47		193	233
0028	3'-6" wide, single	G		16	1		153	47		200	242
0030	4'-0" wide, single	G		16	1		152	47		199	240
0040	6'-0" wide, double	G		14	1.143		204	53.50		257.50	305
0045	8'-0" wide, double	G		14	1.143		213	53.50		266.50	315
0100	3'-0" x 7'-0" single	G		16	1		151	47		198	239
0110	3'-6" wide, single	G		16	1		159	47		206	247
0112	4'-0" wide, single	G		16	1		159	47		206	247
0140	6'-0" wide, double	G		14	1.143		194	53.50		247.50	297
0145	8'-0" wide, double	G		14	1.143		229	53.50		282.50	335
1000	16 ga., up to 4-7/8" deep, 3'-0" x 7'-0" single	G		16	1		166	47		213	256
1140	6'-0" wide, double	G		14	1.143		188	53.50		241.50	290
1200	16 ga., 8-3/4" deep, 3'-0" x 7'-0" single	G		16	1		199	47		246	292
1240	6'-0" wide, double	G		14	1.143		234	53.50		287.50	340
2800	14 ga., up to 3-7/8" deep, 3'-0" x 7'-0" single	G		16	1		181	47		228	272
2840	6'-0" wide, double	G		14	1.143		217	53.50		270.50	320
3000	14 ga., up to 5-3/4" deep, 3'-0" x 6'-8" single	G		16	1		155	47		202	244
3002	3'-6" wide, single	G		16	1		198	47		245	290
3005	4'-0" wide, single	G		16	1		203	47		250	297
3600	up to 5-3/4" jamb depth, 4'-0" x 7'-0" single	G		15	1.067		185	50		235	280
3620	6'-0" wide, double	G		12	1.333		235	62.50		297.50	355
3640	8'-0" wide, double	G		12	1.333		246	62.50		308.50	370
3700	8'-0" high, 4'-0" wide, single	G		15	1.067		235	50		285	335
3740	8'-0" wide, double	G		12	1.333		290	62.50		352.50	415
4000	6-3/4" deep, 4'-0" x 7'-0" single	G		15	1.067		222	50		272	320
4020	6'-0" wide, double	G		12	1.333		276	62.50		338.50	400
4040	8'-0", wide double	G		12	1.333		284	62.50		346.50	405
4100	8'-0" high, 4'-0" wide, single	G		15	1.067		276	50		326	380
4140	8'-0" wide, double	G		12	1.333		320	62.50		382.50	450
4400	8-3/4" deep, 4'-0" x 7'-0" , single	G		15	1.067		252	50		302	355
4440	8'-0" wide, double	G		12	1.333		315	62.50		377.50	440
4500	4'-0" x 8'-0" , single	G		15	1.067		283	50		333	385
4540	8'-0" wide, double	G		12	1.333		350	62.50		412.50	480
4900	For welded frames, add						63			63	69.50
5400	14 ga., "B" label, up to 5-3/4" deep, 4'-0" x 7'-0" single	G	2 Carp	15	1.067		210	50		260	310
5440	8'-0" wide, double	G		12	1.333		272	62.50		334.50	395
5800	6-3/4" deep, 7'-0" high, 4'-0" wide, single	G		15	1.067		218	50		268	315
5840	8'-0" wide, double	G		12	1.333		385	62.50		447.50	520
6200	8-3/4" deep, 4'-0" x 7'-0" single	G		15	1.067		291	50		341	395
6240	8'-0" wide, double	G		12	1.333		380	62.50		442.50	510
6300	For "A" label use same price as "B" label										
6400	For baked enamel finish, add						30%	15%			
6500	For galvanizing, add						20%				

08 12 Metal Frames

08 12 13 – Hollow Metal Frames

08 12 13.13 Standard Hollow Metal Frames			Crew	Daily Output	Labor-Hours	Unit	Material	2015 Bare Costs Labor	Equipment	Total	Total Incl O&P
6600	For hospital stop, add					Ea.	295			295	325
6620	For hospital stop, stainless steel add					"	380			380	420
7900	Transom lite frames, fixed, add		2 Carp	155	.103	S.F.	54	4.85		58.85	67
8000	Movable, add		"	130	.123	"	68	5.80		73.80	84

08 12 13.25 Channel Metal Frames			Crew	Daily Output	Labor-Hours	Unit	Material	Labor	Equipment	Total	Total Incl O&P
0010	**CHANNEL METAL FRAMES**										
0020	Steel channels with anchors and bar stops										
0100	6" channel @ 8.2#/L.F., 3' x 7' door, weighs 150#	G	E-4	13	2.462	Ea.	239	131	11.20	381.20	500
0200	8" channel @ 11.5#/L.F., 6' x 8' door, weighs 275#	G		9	3.556		435	189	16.20	640.20	830
0300	8' x 12' door, weighs 400#	G		6.50	4.923		635	262	22.50	919.50	1,175
0400	10" channel @ 15.3#/L.F., 10' x 10' door, weighs 500#	G		6	5.333		795	283	24.50	1,102.50	1,400
0500	12' x 12' door, weighs 600#	G		5.50	5.818		955	310	26.50	1,291.50	1,625
0600	12" channel @ 20.7#/L.F., 12' x 12' door, weighs 825#	G		4.50	7.111		1,300	380	32.50	1,712.50	2,150
0700	12' x 16' door, weighs 1000#	G	↓	4	8		1,600	425	36.50	2,061.50	2,525
0800	For frames without bar stops, light sections, deduct						15%				
0900	Heavy sections, deduct					↓	10%				

08 13 Metal Doors

08 13 13 – Hollow Metal Doors

08 13 13.13 Standard Hollow Metal Doors			Crew	Daily Output	Labor-Hours	Unit	Material	Labor	Equipment	Total	Total Incl O&P
0010	**STANDARD HOLLOW METAL DOORS**	R081313-20									
0015	Flush, full panel, hollow core										
0017	When noted doors are prepared but do not include glass or louvers										
0020	1-3/8" thick, 20 ga., 2'-0" x 6'-8"	G	2 Carp	20	.800	Ea.	335	37.50		372.50	430
0040	2'-8" x 6'-8"	G		18	.889		350	41.50		391.50	450
0060	3'-0" x 6'-8"	G		17	.941		350	44		394	455
0100	3'-0" x 7'-0"	G	↓	17	.941		360	44		404	465
0120	For vision lite, add						94.50			94.50	104
0140	For narrow lite, add						102			102	113
0320	Half glass, 20 ga., 2'-0" x 6'-8"	G	2 Carp	20	.800		490	37.50		527.50	600
0340	2'-8" x 6'-8"	G		18	.889		515	41.50		556.50	630
0360	3'-0" x 6'-8"	G		17	.941		510	44		554	630
0400	3'-0" x 7'-0"	G		17	.941		625	44		669	760
0410	1-3/8" thick, 18 ga., 2'-0" x 6'-8"	G		20	.800		400	37.50		437.50	500
0420	3'-0" x 6'-8"	G		17	.941		405	44		449	515
0425	3'-0" x 7'-0"	G	↓	17	.941		415	44		459	530
0450	For vision lite, add						94.50			94.50	104
0452	For narrow lite, add						102			102	113
0460	Half glass, 18 ga., 2'-0" x 6'-8"	G	2 Carp	20	.800		555	37.50		592.50	675
0465	2'-8" x 6'-8"	G		18	.889		575	41.50		616.50	700
0470	3'-0" x 6'-8"	G		17	.941		565	44		609	690
0475	3'-0" x 7'-0"	G		17	.941		565	44		609	695
0500	Hollow core, 1-3/4" thick, full panel, 20 ga., 2'-8" x 6'-8"	G		18	.889		420	41.50		461.50	525
0520	3'-0" x 6'-8"	G		17	.941		420	44		464	530
0640	3'-0" x 7'-0"	G		17	.941		435	44		479	550
0680	4'-0" x 7'-0"	G		15	1.067		630	50		680	770
0700	4'-0" x 8'-0"	G		13	1.231		735	58		793	900
1000	18 ga., 2'-8" x 6'-8"	G		17	.941		490	44		534	610
1020	3'-0" x 6'-8"	G		16	1		475	47		522	600
1120	3'-0" x 7'-0"	G		17	.941		515	44		559	635
1180	4'-0" x 7'-0"	G	↓	14	1.143	↓	630	53.50		683.50	780

08 13 13 – Hollow Metal Doors

08 13 13.13 Standard Hollow Metal Doors			Crew	Daily Output	Labor-Hours	Unit	2015 Bare Costs Material	Labor	Equipment	Total	Total Incl O&P
1200	4'-0" x 8'-0"	G	2 Carp	17	.941	Ea.	735	44		779	880
1212	For vision lite, add						94.50			94.50	104
1214	For narrow lite, add						102			102	113
1230	Half glass, 20 ga., 2'-8" x 6'-8"	G	2 Carp	20	.800		575	37.50		612.50	695
1240	3'-0" x 6'-8"	G		18	.889		575	41.50		616.50	700
1260	3'-0" x 7'-0"	G		18	.889		595	41.50		636.50	715
1280	Embossed panel, 1-3/4" thick, poly core, 20 ga., 3'-0" x 7'-0"			18	.889		415	41.50		456.50	520
1290	Half glass, 1-3/4" thick, poly core, 20 ga., 3'-0" x 7'-0"	G		18	.889		610	41.50		651.50	735
1320	18 ga., 2'-8" x 6'-8"	G		18	.889		640	41.50		681.50	770
1340	3'-0" x 6'-8"	G		17	.941		635	44		679	765
1360	3'-0" x 7'-0"	G		17	.941		645	44		689	780
1380	4'-0" x 7'-0"	G		15	1.067		785	50		835	940
1400	4'-0" x 8'-0"	G		14	1.143		890	53.50		943.50	1,075
1500	Flush full panel, 16 ga., steel hollow core										
1520	2'-0" x 6'-8"	G	2 Carp	20	.800	Ea.	555	37.50		592.50	670
1530	2'-8" x 6'-8"	G		20	.800		555	37.50		592.50	675
1540	3'-0" x 6'-8"	G		20	.800		550	37.50		587.50	665
1560	2'-8" x 7'-0"	G		18	.889		575	41.50		616.50	695
1570	3'-0" x 7'-0"	G		18	.889		560	41.50		601.50	680
1580	3'-6" x 7'-0"	G		18	.889		655	41.50		696.50	785
1590	4'-0" x 7'-0"	G		18	.889		725	41.50		766.50	860
1600	2'-8" x 8'-0"	G		18	.889		700	41.50		741.50	835
1620	3'-0" x 8'-0"	G		18	.889		720	41.50		761.50	855
1630	3'-6" x 8'-0"	G		18	.889		800	41.50		841.50	945
1640	4'-0" x 8'-0"	G		18	.889		850	41.50		891.50	995
1720	Insulated, 1-3/4" thick, full panel, 18 ga., 3'-0" x 6'-8"	G		15	1.067		475	50		525	600
1740	2'-8" x 7'-0"	G		16	1		505	47		552	630
1760	3'-0" x 7'-0"	G		15	1.067		490	50		540	615
1800	4'-0" x 8'-0"	G		13	1.231		740	58		798	900
1820	Half glass, 18 ga., 3'-0" x 6'-8"	G		16	1		635	47		682	770
1840	2'-8" x 7'-0"	G		17	.941		660	44		704	795
1860	3'-0" x 7'-0"	G		16	1		685	47		732	830
1900	4'-0" x 8'-0"	G		14	1.143		670	53.50		723.50	820
2000	For vision lite, add						94.50			94.50	104
2010	For narrow lite, add						102			102	113
8100	For bottom louver, add						282			282	310
8110	For baked enamel finish, add						30%	15%			
8120	For galvanizing, add						20%				

08 13 13.15 Metal Fire Doors			Crew	Daily Output	Labor-Hours	Unit	Material	Labor	Equipment	Total	Total Incl O&P
0010	**METAL FIRE DOORS**	R081313-20									
0015	Steel, flush, "B" label, 90 minute										
0020	Full panel, 20 ga., 2'-0" x 6'-8"		2 Carp	20	.800	Ea.	420	37.50		457.50	520
0040	2'-8" x 6'-8"			18	.889		435	41.50		476.50	545
0060	3'-0" x 6'-8"			17	.941		435	44		479	550
0080	3'-0" x 7'-0"			17	.941		455	44		499	570
0140	18 ga., 3'-0" x 6'-8"			16	1		495	47		542	620
0160	2'-8" x 7'-0"			17	.941		520	44		564	640
0180	3'-0" x 7'-0"			16	1		505	47		552	630
0200	4'-0" x 7'-0"			15	1.067		650	50		700	790
0220	For "A" label, 3 hour, 18 ga., use same price as "B" label										
0240	For vision lite, add					Ea.	156			156	172
0300	Full panel, 16 ga., 2'-0" x 6'-8"		2 Carp	20	.800		545	37.50		582.50	660

08 13 13 – Hollow Metal Doors

08 13 13.15 Metal Fire Doors		Crew	Daily Output	Labor-Hours	Unit	Material	2015 Bare Costs Labor	Equipment	Total	Total Incl O&P
0310	2'-8" x 6'-8"	2 Carp	18	.889	Ea.	545	41.50		586.50	665
0320	3'-0" x 6'-8"		17	.941		540	44		584	665
0350	2'-8" x 7'-0"		17	.941		565	44		609	690
0360	3'-0" x 7'-0"		16	1		545	47		592	675
0370	4'-0" x 7'-0"		15	1.067		705	50		755	855
0520	Flush, "B" label 90 min., egress core, 20 ga., 2'-0" x 6'-8"		18	.889		655	41.50		696.50	785
0540	2'-8" x 6'-8"		17	.941		665	44		709	800
0560	3'-0" x 6'-8"		16	1		665	47		712	805
0580	3'-0" x 7'-0"		16	1		685	47		732	825
0640	Flush, "A" label 3 hour, egress core, 18 ga., 3'-0" x 6'-8"		15	1.067		720	50		770	870
0660	2'-8" x 7'-0"		16	1		750	47		797	895
0680	3'-0" x 7'-0"		15	1.067		740	50		790	890
0700	4'-0" x 7'-0"		14	1.143		885	53.50		938.50	1,050

08 13 13.20 Residential Steel Doors		Crew	Daily Output	Labor-Hours	Unit	Material	Labor	Equipment	Total	Total Incl O&P
0010	**RESIDENTIAL STEEL DOORS**									
0020	Prehung, insulated, exterior									
0030	Embossed, full panel, 2'-8" x 6'-8" G	2 Carp	17	.941	Ea.	315	44		359	415
0040	3'-0" x 6'-8" G		15	1.067		270	50		320	375
0060	3'-0" x 7'-0" G		15	1.067		330	50		380	440
0070	5'-4" x 6'-8", double G		8	2		630	94		724	840
0220	Half glass, 2'-8" x 6'-8" G		17	.941		320	44		364	420
0240	3'-0" x 6'-8" G		16	1		320	47		367	425
0260	3'-0" x 7'-0" G		16	1		370	47		417	480
0270	5'-4" x 6'-8", double G		8	2		650	94		744	860
1320	Flush face, full panel, 2'-8" x 6'-8" G		16	1		264	47		311	365
1340	3'-0" x 6'-8" G		15	1.067		264	50		314	365
1360	3'-0" x 7'-0" G		15	1.067		296	50		346	400
1380	5'-4" x 6'-8", double G		8	2		560	94		654	760
1420	Half glass, 2'-8" x 6'-8" G		17	.941		325	44		369	425
1440	3'-0" x 6'-8" G		16	1		325	47		372	430
1460	3'-0" x 7'-0" G		16	1		360	47		407	470
1480	5'-4" x 6'-8", double G		8	2		640	94		734	850
1500	Sidelight, full lite, 1'-0" x 6'-8" with grille G					252			252	277
1510	1'-0" x 6'-8", low e G					268			268	295
1520	1'-0" x 6'-8", half lite G					280			280	310
1530	1'-0" x 6'-8", half lite, low e G					284			284	310
2300	Interior, residential, closet, bi-fold, 2'-0" x 6'-8" G	2 Carp	16	1		161	47		208	250
2330	3'-0" wide G		16	1		200	47		247	293
2360	4'-0" wide G		15	1.067		260	50		310	365
2400	5'-0" wide G		14	1.143		315	53.50		368.50	435
2420	6'-0" wide G		13	1.231		335	58		393	460

08 13 13.25 Doors Hollow Metal		Crew	Daily Output	Labor-Hours	Unit	Material	Labor	Equipment	Total	Total Incl O&P
0010	**DOORS HOLLOW METAL**									
0500	Exterior, commercial, flush, 20 ga., 1-3/4" x 7'-0" x 2'-6" wide G	2 Carp	15	1.067	Ea.	395	50		445	505
0530	2'-8" wide G		15	1.067		435	50		485	550
0560	3'-0" wide G		14	1.143		435	53.50		488.50	560
1000	18 ga., 1-3/4" x 7'-0" x 2'-6" wide G		15	1.067		495	50		545	620
1030	2'-8" wide G		15	1.067		500	50		550	625
1060	3'-0" wide G		14	1.143		485	53.50		538.50	620
1500	16 ga., 1-3/4 x 7'-0" x 2'-6" wide G		15	1.067		560	50		610	690
1530	2'-8" wide G		15	1.067		565	50		615	700
1560	3'-0" wide G		14	1.143		555	53.50		608.50	695

08 13 Metal Doors

08 13 13 – Hollow Metal Doors

08 13 13.25 Doors Hollow Metal			Crew	Daily Output	Labor-Hours	Unit	Material	2015 Bare Costs Labor	Equipment	Total	Total Incl O&P
1590	3'-6" wide	G	2 Carp	14	1.143	Ea.	645	53.50		698.50	795
2900	Fire door, "A" label, 18 gauge, 1-3/4" x 2'-6" x 7'-0"	G		15	1.067		645	50		695	785
2930	2'-8" wide	G		15	1.067		665	50		715	805
2960	3'-0" wide	G		14	1.143		640	53.50		693.50	790
2990	3'-6" wide	G		14	1.143		735	53.50		788.50	890
3100	"B" label, 2'-6" wide	G		15	1.067		585	50		635	720
3130	2'-8" wide	G		15	1.067		590	50		640	725
3160	3'-0" wide	G	↓	14	1.143	↓	580	53.50		633.50	725

08 13 16 – Aluminum Doors

08 13 16.10 Commercial Aluminum Doors		Crew	Daily Output	Labor-Hours	Unit	Material	Labor	Equipment	Total	Total Incl O&P
0010	**COMMERCIAL ALUMINUM DOORS**, flush, no glazing									
5000	Flush panel doors, pair of 2'-6" x 7'-0"	2 Sswk	2	8	Pr.	1,475	420		1,895	2,325
5050	3'-0" x 7'-0", single		2.50	6.400	Ea.	825	335		1,160	1,500
5100	Pair of 3'-0" x 7'-0"		2	8	Pr.	1,550	420		1,970	2,450
5150	3'-6" x 7'-0", single	↓	2.50	6.400	Ea.	985	335		1,320	1,650

08 14 Wood Doors

08 14 13 – Carved Wood Doors

08 14 13.10 Types of Wood Doors, Carved		Crew	Daily Output	Labor-Hours	Unit	Material	Labor	Equipment	Total	Total Incl O&P
0010	**TYPES OF WOOD DOORS, CARVED**									
3000	Solid wood, 1-3/4" thick stile and rail									
3020	Mahogany, 3'-0" x 7'-0", six panel	2 Carp	14	1.143	Ea.	1,125	53.50		1,178.50	1,325
3030	With two lites		10	1.600		1,850	75		1,925	2,150
3040	3'-6" x 8'-0", six panel		10	1.600		1,500	75		1,575	1,775
3050	With two lites		8	2		2,500	94		2,594	2,900
3100	Pine, 3'-0" x 7'-0", six panel		14	1.143		525	53.50		578.50	665
3110	With two lites		10	1.600		825	75		900	1,025
3120	3'-6" x 8'-0", six panel		10	1.600		920	75		995	1,125
3130	With two lites		8	2		1,825	94		1,919	2,175
3200	Red oak, 3'-0" x 7'-0", six panel		14	1.143		1,750	53.50		1,803.50	1,975
3210	With two lites		10	1.600		2,300	75		2,375	2,650
3220	3'-6" x 8'-0", six panel		10	1.600		2,600	75		2,675	2,975
3230	With two lites	↓	8	2	↓	3,300	94		3,394	3,775
4000	Hand carved door, mahogany									
4020	3'-0" x 7'-0", simple design	2 Carp	14	1.143	Ea.	1,750	53.50		1,803.50	2,000
4030	Intricate design		11	1.455		3,700	68.50		3,768.50	4,175
4040	3'-6" x 8'-0", simple design		10	1.600		3,000	75		3,075	3,425
4050	Intricate design	↓	8	2		3,700	94		3,794	4,225
4400	For custom finish, add					475			475	525
4600	Side light, mahogany, 7'-0" x 1'-6" wide, 4 lites	2 Carp	18	.889		1,100	41.50		1,141.50	1,275
4610	6 lites		14	1.143		2,625	53.50		2,678.50	2,975
4620	8'-0" x 1'-6" wide, 4 lites		14	1.143		1,800	53.50		1,853.50	2,050
4630	6 lites		10	1.600		2,100	75		2,175	2,425
4640	Side light, oak, 7'-0" x 1'-6" wide, 4 lites		18	.889		1,200	41.50		1,241.50	1,400
4650	6 lites		14	1.143		2,100	53.50		2,153.50	2,375
4660	8'-0" x 1'-6" wide, 4 lites		14	1.143		1,100	53.50		1,153.50	1,275
4670	6 lites	↓	10	1.600	↓	2,100	75		2,175	2,425

08 14 16 – Flush Wood Doors

08 14 16.09 Smooth Wood Doors		Crew	Daily Output	Labor-Hours	Unit	Material	2015 Bare Costs Labor	Equipment	Total	Total Incl O&P
0010	**SMOOTH WOOD DOORS**									
0015	Flush, interior, hollow core									
0025	Lauan face, 1-3/8", 3'-0" x 6'-8"	2 Carp	17	.941	Ea.	54.50	44		98.50	128
0030	4'-0" x 6'-8"		16	1		126	47		173	212
0080	1-3/4", 2'-0" x 6'-8"		17	.941		39	44		83	111
0108	3'-0" x 7'-0"		16	1		145	47		192	233
0112	Pair of 3'-0" x 7'-0"		9	1.778	Pr.	104	83.50		187.50	243
0140	Birch face, 1-3/8", 2'-6" x 6'-8"		17	.941	Ea.	85	44		129	161
0180	3'-0" x 6'-8"		17	.941		95.50	44		139.50	173
0200	4'-0" x 6'-8"		16	1		158	47		205	247
0202	1-3/4", 2'-0" x 6'-8"		17	.941		50.50	44		94.50	124
0210	3'-0" x 7'-0"		16	1		110	47		157	194
0214	Pair of 3'-0" x 7'-0"		9	1.778	Pr.	213	83.50		296.50	360
0220	Oak face, 1-3/8", 2'-0" x 6'-8"		17	.941	Ea.	104	44		148	182
0280	3'-0" x 6'-8"		17	.941		115	44		159	195
0300	4'-0" x 6'-8"		16	1		139	47		186	226
0305	1-3/4", 2'-6" x 6'-8"		17	.941		110	44		154	189
0310	3'-0" x 7'-0"		16	1		211	47		258	305
0320	Walnut face, 1-3/8", 2'-0" x 6'-8"		17	.941		182	44		226	269
0340	2'-6" x 6'-8"		17	.941		189	44		233	276
0380	3'-0" x 6'-8"		17	.941		197	44		241	284
0400	4'-0" x 6'-8"		16	1		216	47		263	310
0430	For 7'-0" high, add					26			26	28.50
0440	For 8'-0" high, add					36			36	39.50
0480	For prefinishing, clear, add					46			46	50.50
0500	For prefinishing, stain, add					57			57	62.50
1320	M.D. overlay on hardboard, 1-3/8", 2'-0" x 6'-8"	2 Carp	17	.941		115	44		159	195
1340	2'-6" x 6'-8"		17	.941		115	44		159	195
1380	3'-0" x 6'-8"		17	.941		126	44		170	207
1400	4'-0" x 6'-8"		16	1		178	47		225	269
1420	For 7'-0" high, add					15.75			15.75	17.35
1440	For 8'-0" high, add					31.50			31.50	34.50
1720	H.P. plastic laminate, 1-3/8", 2'-0" x 6'-8"	2 Carp	16	1		260	47		307	360
1740	2'-6" x 6'-8"		16	1		260	47		307	360
1780	3'-0" x 6'-8"		15	1.067		290	50		340	395
1800	4'-0" x 6'-8"		14	1.143		385	53.50		438.50	505
1820	For 7'-0" high, add					15.75			15.75	17.35
1840	For 8'-0" high, add					31.50			31.50	34.50
2020	Particle core, lauan face, 1-3/8", 2'-6" x 6'-8"	2 Carp	15	1.067		92	50		142	178
2040	3'-0" x 6'-8"		14	1.143		94	53.50		147.50	186
2080	3'-0" x 7'-0"		13	1.231		101	58		159	200
2085	4'-0" x 7'-0"		12	1.333		123	62.50		185.50	232
2110	1-3/4", 3'-0" x 7'-0"		13	1.231		145	58		203	249
2120	Birch face, 1-3/8", 2'-6" x 6'-8"		15	1.067		103	50		153	190
2140	3'-0" x 6'-8"		14	1.143		113	53.50		166.50	207
2180	3'-0" x 7'-0"		13	1.231		123	58		181	224
2200	4'-0" x 7'-0"		12	1.333		139	62.50		201.50	250
2205	1-3/4", 3'-0" x 7'-0"		13	1.231		123	58		181	224
2220	Oak face, 1-3/8", 2'-6" x 6'-8"		15	1.067		116	50		166	205
2240	3'-0" x 6'-8"		14	1.143		128	53.50		181.50	224
2280	3'-0" x 7'-0"		13	1.231		133	58		191	235
2300	4'-0" x 7'-0"		12	1.333		156	62.50		218.50	269

08 14 16.09 Smooth Wood Doors		Crew	Daily Output	Labor-Hours	Unit	Material	Labor	Equipment	Total	Total Incl O&P
						2015 Bare Costs				
2305	1-3/4" 3'-0" x 7'-0"	2 Carp	13	1.231	Ea.	190	58		248	298
2320	Walnut face, 1-3/8", 2'-0" x 6'-8"		15	1.067		123	50		173	212
2340	2'-6" x 6'-8"		14	1.143		139	53.50		192.50	236
2380	3'-0" x 6'-8"		13	1.231		156	58		214	261
2400	4'-0" x 6'-8"		12	1.333		206	62.50		268.50	325
2440	For 8'-0" high, add					41			41	45
2460	For 8'-0" high walnut, add					36			36	39.50
2720	For prefinishing, clear, add					36			36	39.50
2740	For prefinishing, stain, add					53			53	58.50
3320	M.D. overlay on hardboard, 1-3/8", 2'-6" x 6'-8"	2 Carp	14	1.143		106	53.50		159.50	200
3340	3'-0" x 6'-8"		13	1.231		115	58		173	216
3380	3'-0" x 7'-0"		12	1.333		117	62.50		179.50	226
3400	4'-0" x 7'-0"		10	1.600		155	75		230	287
3440	For 8'-0" height, add					37			37	40.50
3460	For solid wood core, add					42			42	46
3720	H.P. plastic laminate, 1-3/8", 2'-6" x 6'-8"	2 Carp	13	1.231		160	58		218	265
3740	3'-0" x 6'-8"		12	1.333		185	62.50		247.50	300
3780	3'-0" x 7'-0"		11	1.455		190	68.50		258.50	315
3800	4'-0" x 7'-0"		8	2		225	94		319	395
3840	For 8'-0" height, add					37			37	40.50
3860	For solid wood core, add					42			42	46
4000	Exterior, flush, solid core, birch, 1-3/4" x 2'-6" x 7'-0"	2 Carp	15	1.067		173	50		223	267
4020	2'-8" wide		15	1.067		159	50		209	252
4040	3'-0" wide		14	1.143		209	53.50		262.50	315
4100	Oak faced 1-3/4" x 2'-6" x 7'-0"		15	1.067		215	50		265	315
4120	2'-8" wide		15	1.067		225	50		275	325
4140	3'-0" wide		14	1.143		230	53.50		283.50	335
4200	Walnut faced, 1-3/4" x 2'-6" x 7'-0"		15	1.067		310	50		360	415
4220	2'-8" wide		15	1.067		315	50		365	420
4240	3'-0" wide		14	1.143		320	53.50		373.50	435
4300	For 6'-8" high door, deduct from 7'-0" door					16.80			16.80	18.50
5000	Wood doors, for vision lite, add					94.50			94.50	104
5010	Wood doors, for narrow lite, add					102			102	113
5015	Wood doors, for bottom (or top) louver, add					282			282	310

08 14 16.20 Wood Fire Doors		Crew	Daily Output	Labor-Hours	Unit	Material	Labor	Equipment	Total	Total Incl O&P
0010	**WOOD FIRE DOORS**									
0020	Particle core, 7 face plys, "B" label,									
0040	1 hour, birch face, 1-3/4" x 2'-6" x 6'-8"	2 Carp	14	1.143	Ea.	420	53.50		473.50	545
0080	3'-0" x 6'-8"		13	1.231		400	58		458	530
0090	3'-0" x 7'-0"		12	1.333		435	62.50		497.50	575
0100	4'-0" x 7'-0"		12	1.333		540	62.50		602.50	690
0140	Oak face, 2'-6" x 6'-8"		14	1.143		450	53.50		503.50	580
0180	3'-0" x 6'-8"		13	1.231		455	58		513	590
0190	3'-0" x 7'-0"		12	1.333		460	62.50		522.50	600
0200	4'-0" x 7'-0"		12	1.333		590	62.50		652.50	745
0240	Walnut face, 2'-6" x 6'-8"		14	1.143		470	53.50		523.50	600
0280	3'-0" x 6'-8"		13	1.231		490	58		548	630
0290	3'-0" x 7'-0"		12	1.333		510	62.50		572.50	655
0300	4'-0" x 7'-0"		12	1.333		630	62.50		692.50	790
0440	M.D. overlay on hardboard, 2'-6" x 6'-8"		15	1.067		315	50		365	420
0480	3'-0" x 6'-8"		14	1.143		375	53.50		428.50	500
0490	3'-0" x 7'-0"		13	1.231		395	58		453	525

08 14 Wood Doors

08 14 16 – Flush Wood Doors

08 14 16.20 Wood Fire Doors		Crew	Daily Output	Labor-Hours	Unit	Material	2015 Bare Costs Labor	Equipment	Total	Total Incl O&P
0500	4'-0" x 7'-0"	2 Carp	12	1.333	Ea.	415	62.50		477.50	550
0740	90 minutes, birch face, 1-3/4" x 2'-6" x 6'-8"		14	1.143		325	53.50		378.50	440
0780	3'-0" x 6'-8"		13	1.231		320	58		378	440
0790	3'-0" x 7'-0"		12	1.333		390	62.50		452.50	525
0800	4'-0" x 7'-0"		12	1.333		545	62.50		607.50	690
0840	Oak face, 2'-6" x 6'-8"		14	1.143		430	53.50		483.50	560
0880	3'-0" x 6'-8"		13	1.231		440	58		498	575
0890	3'-0" x 7'-0"		12	1.333		455	62.50		517.50	595
0900	4'-0" x 7'-0"		12	1.333		590	62.50		652.50	745
0940	Walnut face, 2'-6" x 6'-8"		14	1.143		400	53.50		453.50	525
0980	3'-0" x 6'-8"		13	1.231		410	58		468	540
0990	3'-0" x 7'-0"		12	1.333		470	62.50		532.50	610
1000	4'-0" x 7'-0"		12	1.333		620	62.50		682.50	775
1140	M.D. overlay on hardboard, 2'-6" x 6'-8"		15	1.067		355	50		405	465
1180	3'-0" x 6'-8"		14	1.143		375	53.50		428.50	500
1190	3'-0" x 7'-0"		13	1.231		405	58		463	535
1200	4'-0" x 7'-0"		12	1.333		455	62.50		517.50	595
1240	For 8'-0" height, add					75			75	82.50
1260	For 8'-0" height walnut, add					90			90	99
2200	Custom architectural "B" label, flush, 1-3/4" thick, birch,									
2210	Solid core									
2220	2'-6" x 7'-0"	2 Carp	15	1.067	Ea.	305	50		355	410
2260	3'-0" x 7'-0"		14	1.143		315	53.50		368.50	430
2300	4'-0" x 7'-0"		13	1.231		400	58		458	530
2420	4'-0" x 8'-0"		11	1.455		430	68.50		498.50	580
2480	For oak veneer, add					50%				
2500	For walnut veneer, add					75%				

08 14 33 – Stile and Rail Wood Doors

08 14 33.10 Wood Doors Paneled		Crew	Daily Output	Labor-Hours	Unit	Material	2015 Bare Costs Labor	Equipment	Total	Total Incl O&P
0010	**WOOD DOORS PANELED**									
0020	Interior, six panel, hollow core, 1-3/8" thick									
0040	Molded hardboard, 2'-0" x 6'-8"	2 Carp	17	.941	Ea.	62	44		106	136
0060	2'-6" x 6'-8"		17	.941		64	44		108	139
0070	2'-8" x 6'-8"		17	.941		67	44		111	142
0080	3'-0" x 6'-8"		17	.941		72	44		116	147
0140	Embossed print, molded hardboard, 2'-0" x 6'-8"		17	.941		64	44		108	139
0160	2'-6" x 6'-8"		17	.941		64	44		108	139
0180	3'-0" x 6'-8"		17	.941		72	44		116	147
0540	Six panel, solid, 1-3/8" thick, pine, 2'-0" x 6'-8"		15	1.067		155	50		205	248
0560	2'-6" x 6'-8"		14	1.143		170	53.50		223.50	270
0580	3'-0" x 6'-8"		13	1.231		145	58		203	249
1020	Two panel, bored rail, solid, 1-3/8" thick, pine, 1'-6" x 6'-8"		16	1		270	47		317	370
1040	2'-0" x 6'-8"		15	1.067		355	50		405	465
1060	2'-6" x 6'-8"		14	1.143		400	53.50		453.50	525
1340	Two panel, solid, 1-3/8" thick, fir, 2'-0" x 6'-8"		15	1.067		160	50		210	253
1360	2'-6" x 6'-8"		14	1.143		210	53.50		263.50	315
1380	3'-0" x 6'-8"		13	1.231		415	58		473	545
1740	Five panel, solid, 1-3/8" thick, fir, 2'-0" x 6'-8"		15	1.067		280	50		330	385
1760	2'-6" x 6'-8"		14	1.143		420	53.50		473.50	545
1780	3'-0" x 6'-8"		13	1.231		420	58		478	550

08 14 33 – Stile and Rail Wood Doors

08 14 33.20 Wood Doors Residential		Crew	Daily Output	Labor-Hours	Unit	Material	2015 Bare Costs Labor	Equipment	Total	Total Incl O&P
0010	**WOOD DOORS RESIDENTIAL**									
0200	Exterior, combination storm & screen, pine									
0260	2′-8″ wide	2 Carp	10	1.600	Ea.	300	75		375	445
0280	3′-0″ wide		9	1.778		310	83.50		393.50	470
0300	7′-1″ x 3′-0″ wide		9	1.778		335	83.50		418.50	500
0400	Full lite, 6′-9″ x 2′-6″ wide		11	1.455		320	68.50		388.50	455
0420	2′-8″ wide		10	1.600		320	75		395	465
0440	3′-0″ wide		9	1.778		325	83.50		408.50	485
0500	7′-1″ x 3′-0″ wide		9	1.778		355	83.50		438.50	520
0700	Dutch door, pine, 1-3/4″ x 2′-8″ x 6′-8″ , 6 panel		12	1.333		790	62.50		852.50	960
0720	Half glass		10	1.600		900	75		975	1,100
0800	3′-0″ wide, 6 panel		12	1.333		790	62.50		852.50	960
0820	Half glass		10	1.600		900	75		975	1,100
1000	Entrance door, colonial, 1-3/4″ x 6′-8″ x 2′-8″ wide		16	1		500	47		547	625
1020	6 panel pine, 3′-0″ wide		15	1.067		445	50		495	565
1100	8 panel pine, 2′-8″ wide		16	1		580	47		627	710
1120	3′-0″ wide		15	1.067		570	50		620	700
1200	For tempered safety glass lites, (min of 2) add					79			79	87
1300	Flush, birch, solid core, 1-3/4″ x 6′-8″ x 2′-8″ wide	2 Carp	16	1		113	47		160	198
1320	3′-0″ wide		15	1.067		119	50		169	207
1350	7′-0″ x 2′-8″ wide		16	1		122	47		169	207
1360	3′-0″ wide		15	1.067		143	50		193	234
1380	For tempered safety glass lites, add					105			105	116
1920	For fixed wood louvers, add					80			80	88
1930	For dutch door with shelf, add					140%				
2700	Interior, closet, bi-fold, w/hardware, no frame or trim incl.									
2720	Flush, birch, 2′-6″ x 6′-8″	2 Carp	13	1.231	Ea.	67.50	58		125.50	164
2740	3′-0″ wide		13	1.231		71.50	58		129.50	168
2760	4′-0″ wide		12	1.333		109	62.50		171.50	217
2780	5′-0″ wide		11	1.455		105	68.50		173.50	221
2800	6′-0″ wide		10	1.600		128	75		203	257
3000	Raised panel pine, 6′-6″ or 6′-8″ x 2′-6″ wide		13	1.231		198	58		256	305
3020	3′-0″ wide		13	1.231		278	58		336	395
3040	4′-0″ wide		12	1.333		305	62.50		367.50	430
3060	5′-0″ wide		11	1.455		365	68.50		433.50	505
3080	6′-0″ wide		10	1.600		400	75		475	555
3200	Louvered, pine 6′-6″ or 6′-8″ x 2′-6″ wide		13	1.231		144	58		202	247
3220	3′-0″ wide		13	1.231		208	58		266	315
3240	4′-0″ wide		12	1.333		235	62.50		297.50	355
3260	5′-0″ wide		11	1.455		262	68.50		330.50	395
3280	6′-0″ wide		10	1.600		289	75		364	435
4400	Bi-passing closet, incl. hardware and frame, no trim incl.									
4420	Flush, lauan, 6′-8″ x 4′-0″ wide	2 Carp	12	1.333	Opng.	176	62.50		238.50	291
4440	5′-0″ wide		11	1.455		187	68.50		255.50	310
4460	6′-0″ wide		10	1.600		209	75		284	345
4600	Flush, birch, 6′-8″ x 4′-0″ wide		12	1.333		223	62.50		285.50	340
4620	5′-0″ wide		11	1.455		212	68.50		280.50	340
4640	6′-0″ wide		10	1.600		260	75		335	400
4800	Louvered, pine, 6′-8″ x 4′-0″ wide		12	1.333		410	62.50		472.50	550
4820	5′-0″ wide		11	1.455		395	68.50		463.50	540
4840	6′-0″ wide		10	1.600		535	75		610	700
5000	Paneled, pine, 6′-8″ x 4′-0″ wide		12	1.333		515	62.50		577.50	660

08 14 33 – Stile and Rail Wood Doors

08 14 33.20 Wood Doors Residential		Crew	Daily Output	Labor-Hours	Unit	Material	2015 Bare Costs Labor	Equipment	Total	Total Incl O&P
5020	5'-0" wide	2 Carp	11	1.455	Opng.	410	68.50		478.50	555
5040	6'-0" wide		10	1.600		610	75		685	785
5042	8'-0" wide		12	1.333		1,025	62.50		1,087.50	1,225
6100	Folding accordion, closet, including track and frame									
6120	Vinyl, 2 layer, stock	2 Carp	10	1.600	Ea.	66	75		141	189
6140	Woven mahogany and vinyl, stock		10	1.600		57.50	75		132.50	179
6160	Wood slats with vinyl overlay, stock		10	1.600		150	75		225	281
6180	Economy vinyl, stock		10	1.600		41.50	75		116.50	162
6200	Rigid PVC		10	1.600		55.50	75		130.50	177
7310	Passage doors, flush, no frame included									
7320	Hardboard, hollow core, 1-3/8" x 6'-8" x 1'-6" wide	2 Carp	18	.889	Ea.	40.50	41.50		82	109
7330	2'-0" wide		18	.889		43	41.50		84.50	112
7340	2'-6" wide		18	.889		47.50	41.50		89	117
7350	2'-8" wide		18	.889		49	41.50		90.50	118
7360	3'-0" wide		17	.941		52	44		96	126
7420	Lauan, hollow core, 1-3/8" x 6'-8" x 1'-6" wide		18	.889		35	41.50		76.50	103
7440	2'-0" wide		18	.889		35	41.50		76.50	103
7450	2'-4" wide		18	.889		38.50	41.50		80	107
7460	2'-6" wide		18	.889		38.50	41.50		80	107
7480	2'-8" wide		18	.889		40	41.50		81.50	108
7500	3'-0" wide		17	.941		42.50	44		86.50	115
7700	Birch, hollow core, 1-3/8" x 6'-8" x 1'-6" wide		18	.889		41	41.50		82.50	109
7720	2'-0" wide		18	.889		42.50	41.50		84	111
7740	2'-6" wide		18	.889		50	41.50		91.50	119
7760	2'-8" wide		18	.889		50.50	41.50		92	120
7780	3'-0" wide		17	.941		52	44		96	126
8000	Pine louvered, 1-3/8" x 6'-8" x 1'-6" wide		19	.842		120	39.50		159.50	193
8020	2'-0" wide		18	.889		131	41.50		172.50	208
8040	2'-6" wide		18	.889		150	41.50		191.50	229
8060	2'-8" wide		18	.889		161	41.50		202.50	241
8080	3'-0" wide		17	.941		175	44		219	260
8300	Pine paneled, 1-3/8" x 6'-8" x 1'-6" wide		19	.842		128	39.50		167.50	202
8320	2'-0" wide		18	.889		153	41.50		194.50	232
8330	2'-4" wide		18	.889		182	41.50		223.50	264
8340	2'-6" wide		18	.889		182	41.50		223.50	264
8360	2'-8" wide		18	.889		185	41.50		226.50	268
8380	3'-0" wide		17	.941		204	44		248	292

08 14 35 – Torrified Doors

08 14 35.10 Torrified Exterior Doors		Crew	Daily Output	Labor-Hours	Unit	Material	Labor	Equipment	Total	Total Incl O&P
0010	**TORRIFIED EXTERIOR DOORS**									
0020	Wood doors made from torrified wood, exterior									
0030	All doors require a finish be applied, all glass is insulated									
0040	All doors require pilot holes for all fasteners									
0100	6 panel, Paint grade poplar, 1-3/4" x 3'-0"x 6'-8"	2 Carp	12	1.333	Ea.	1,075	62.50		1,137.50	1,275
0120	Half glass 3'-0"x 6'-8"	"	12	1.333		1,175	62.50		1,237.50	1,400
0200	Side lite, full glass, 1-3/4" x 1'-2" x 6'-8"					905			905	995
0220	Side lite, half glass, 1-3/4" x 1'-2" x 6'-8"					905			905	995
0300	Raised Face, 2 Panel, Paint grade poplar, 1-3/4" x 3'-0"x 7'-0"	2 Carp	12	1.333		1,275	62.50		1,337.50	1,500
0320	Side lite, raised face, half glass, 1-3/4" x 1'-2" x 7'-0"					1,050			1,050	1,175
0500	6 panel, Fir, 1-3/4" x 3'-0"x 6'-8"	2 Carp	12	1.333		1,550	62.50		1,612.50	1,800
0520	Half glass 3'-0"x 6'-8"	"	12	1.333		1,650	62.50		1,712.50	1,925
0600	Side lite, full glass, 1-3/4" x 1'-2" x 6'-8"					1,150			1,150	1,275

08 14 Wood Doors

08 14 35 – Torrified Doors

08 14 35.10 Torrified Exterior Doors		Crew	Daily Output	Labor-Hours	Unit	2015 Bare Costs Material	Labor	Equipment	Total	Total Incl O&P
0620	Side lite, half glass, 1-3/4″ x 1′-2″ x 6′-8″				Ea.	1,450			1,450	1,600
0700	6 panel, Mahogany, 1-3/4″ x 3′-0″x 6′-8″	2 Carp	12	1.333		1,625	62.50		1,687.50	1,875
0800	Side lite, full glass, 1-3/4″ x 1′-2″ x 6′-8″					1,225			1,225	1,350
0820	Side lite, half glass, 1-3/4″ x 1′-2″ x 6′-8″					1,200			1,200	1,325

08 14 40 – Interior Cafe Doors

08 14 40.10 Cafe Style Doors		Crew	Daily Output	Labor-Hours	Unit	Material	Labor	Equipment	Total	Total Incl O&P
0010	**CAFE STYLE DOORS**									
6520	Interior cafe doors, 2′-6″ opening, stock, panel pine	2 Carp	16	1	Ea.	218	47		265	310
6540	3′-0″ opening	″	16	1	″	240	47		287	335
6550	Louvered pine									
6560	2′-6″ opening	2 Carp	16	1	Ea.	180	47		227	271
8000	3′-0″ opening		16	1		193	47		240	285
8010	2′-6″ opening, hardwood		16	1		264	47		311	365
8020	3′-0″ opening		16	1		281	47		328	385

08 16 Composite Doors

08 16 13 – Fiberglass Doors

08 16 13.10 Entrance Doors, Fiberous Glass			Crew	Daily Output	Labor-Hours	Unit	Material	Labor	Equipment	Total	Total Incl O&P
0010	**ENTRANCE DOORS, FIBEROUS GLASS**										
0020	Exterior, fiberglass, door, 2′-8″ wide x 6′-8″ high	G	2 Carp	15	1.067	Ea.	270	50		320	375
0040	3′-0″ wide x 6′-8″ high	G		15	1.067		270	50		320	375
0060	3′-0″ wide x 7′-0″ high	G		15	1.067		460	50		510	580
0080	3′-0″ wide x 6′-8″ high, with two lites	G		15	1.067		315	50		365	420
0100	3′-0″ wide x 8′-0″ high, with two lites	G		15	1.067		525	50		575	655
0110	Half glass, 3′-0″ wide x 6′-8″ high	G		15	1.067		435	50		485	555
0120	3′-0″ wide x 6′-8″ high, low e	G		15	1.067		465	50		515	585
0130	3′-0″ wide x 8′-0″ high	G		15	1.067		585	50		635	720
0140	3′-0″ wide x 8′-0″ high, low e	G		15	1.067		655	50		705	795
0150	Side lights, 1′-0″ wide x 6′-8″ high	G					266			266	293
0160	1′-0″ wide x 6′-8″ high, low e	G					281			281	310
0180	1′-0″ wide x 6′-8″ high, full glass	G					310			310	340
0190	1′-0″ wide x 6′-8″ high, low e	G					340			340	370

08 16 14 – French Doors

08 16 14.10 Exterior Doors With Glass Lites		Crew	Daily Output	Labor-Hours	Unit	Material	Labor	Equipment	Total	Total Incl O&P
0010	**EXTERIOR DOORS WITH GLASS LITES**									
0020	French, Fir, 1-3/4″, 3′-0″wide x 6′-8″ high	2 Carp	12	1.333	Ea.	600	62.50		662.50	755
0025	Double		12	1.333		1,200	62.50		1,262.50	1,425
0030	Maple, 1-3/4″, 3′-0″wide x 6′-8″ high		12	1.333		675	62.50		737.50	840
0035	Double		12	1.333		1,350	62.50		1,412.50	1,575
0040	Cherry, 1-3/4″, 3′-0″wide x 6′-8″ high		12	1.333		790	62.50		852.50	960
0045	Double		12	1.333		1,575	62.50		1,637.50	1,825
0100	Mahogany, 1-3/4″, 3′-0″wide x 8′-0″ high		10	1.600		800	75		875	995
0105	Double		10	1.600		1,600	75		1,675	1,875
0110	Fir, 1-3/4″, 3′-0″wide x 8′-0″ high		10	1.600		1,200	75		1,275	1,450
0115	Double		10	1.600		2,400	75		2,475	2,775
0120	Oak, 1-3/4″, 3′-0″wide x 8′-0″ high		10	1.600		1,825	75		1,900	2,125
0125	Double		10	1.600		3,650	75		3,725	4,125

08 17 Integrated Door Opening Assemblies

08 17 13 – Integrated Metal Door Opening Assemblies

08 17 13.20 Stainless Steel Doors and Frames		Crew	Daily Output	Labor-Hours	Unit	2015 Bare Costs Material	Labor	Equipment	Total	Total Incl O&P
0010	**STAINLESS STEEL DOORS AND FRAMES**									
0500	Stainless steel, prehung door, foam core, 14 ga, 3'-0" x 7'-0" G	2 Carp	5	3.200	Ea.	3,400	150		3,550	3,975
0600	Stainless steel, prehung double door, foam core, 14 ga, 3'-0" x 7'-0" G	"	4	4	"	6,700	188		6,888	7,675

08 17 23 – Integrated Wood Door Opening Assemblies

08 17 23.10 Pre-Hung Doors		Crew	Daily Output	Labor-Hours	Unit	Material	Labor	Equipment	Total	Total Incl O&P
0010	**PRE-HUNG DOORS**									
0300	Exterior, wood, comb. storm & screen, 6'-9" x 2'-6" wide	2 Carp	15	1.067	Ea.	296	50		346	400
0320	2'-8" wide		15	1.067		296	50		346	400
0340	3'-0" wide		15	1.067		305	50		355	410
0360	For 7'-0" high door, add					30			30	33
1600	Entrance door, flush, birch, solid core									
1620	4-5/8" solid jamb, 1-3/4" x 6'-8" x 2'-8" wide	2 Carp	16	1	Ea.	289	47		336	395
1640	3'-0" wide		16	1		375	47		422	490
1642	5-5/8" jamb		16	1		325	47		372	435
1680	For 7'-0" high door, add					25			25	27.50
2000	Entrance door, colonial, 6 panel pine									
2020	4-5/8" solid jamb, 1-3/4" x 6'-8" x 2'-8" wide	2 Carp	16	1	Ea.	640	47		687	780
2040	3'-0" wide	"	16	1		675	47		722	815
2060	For 7'-0" high door, add					54			54	59
2200	For 5-5/8" solid jamb, add					41.50			41.50	46
4000	Interior, passage door, 4-5/8" solid jamb									
4400	Lauan, flush, solid core, 1-3/8" x 6'-8" x 2'-6" wide	2 Carp	17	.941	Ea.	186	44		230	273
4420	2'-8" wide		17	.941		186	44		230	273
4440	3'-0" wide		16	1		201	47		248	294
4600	Hollow core, 1-3/8" x 6'-8" x 2'-6" wide		17	.941		125	44		169	206
4620	2'-8" wide		17	.941		125	44		169	205
4640	3'-0" wide		16	1		140	47		187	227
4700	For 7'-0" high door, add					35.50			35.50	39
5000	Birch, flush, solid core, 1-3/8" x 6'-8" x 2'-6" wide	2 Carp	17	.941		275	44		319	375
5020	2'-8" wide		17	.941		201	44		245	289
5040	3'-0" wide		16	1		300	47		347	410
5200	Hollow core, 1-3/8" x 6'-8" x 2'-6" wide		17	.941		222	44		266	310
5220	2'-8" wide		17	.941		266	44		310	360
5240	3'-0" wide		16	1		230	47		277	325
5280	For 7'-0" high door, add					30.50			30.50	33.50
5500	Hardboard paneled, 1-3/8" x 6'-8" x 2'-6" wide	2 Carp	17	.941		145	44		189	227
5520	2'-8" wide		17	.941		153	44		197	236
5540	3'-0" wide		16	1		150	47		197	238
6000	Pine paneled, 1-3/8" x 6'-8" x 2'-6" wide		17	.941		255	44		299	350
6020	2'-8" wide		17	.941		274	44		318	370
6040	3'-0" wide		16	1		282	47		329	385

08 31 13.10 Types of Framed Access Doors		Crew	Daily Output	Labor-Hours	Unit	Material	2015 Bare Costs Labor	Equipment	Total	Total Incl O&P
0010	TYPES OF FRAMED ACCESS DOORS									
1000	Fire rated door with lock									
1100	Metal, 12" x 12"	1 Carp	10	.800	Ea.	165	37.50		202.50	240
1150	18" x 18"		9	.889		220	41.50		261.50	305
1200	24" x 24"		9	.889		325	41.50		366.50	425
1250	24" x 36"		8	1		335	47		382	440
1300	24" x 48"		8	1		430	47		477	550
1350	36" x 36"		7.50	1.067		495	50		545	620
1400	48" x 48"		7.50	1.067		635	50		685	770
1600	Stainless steel, 12" x 12"		10	.800		282	37.50		319.50	370
1650	18" x 18"		9	.889		410	41.50		451.50	515
1700	24" x 24"		9	.889		500	41.50		541.50	615
1750	24" x 36"		8	1		640	47		687	780
2000	Flush door for finishing									
2100	Metal 8" x 8"	1 Carp	10	.800	Ea.	38	37.50		75.50	100
2150	12" x 12"	"	10	.800	"	45	37.50		82.50	108
3000	Recessed door for acoustic tile									
3100	Metal, 12" x 12"	1 Carp	4.50	1.778	Ea.	82	83.50		165.50	218
3150	12" x 24"		4.50	1.778		100	83.50		183.50	238
3200	24" x 24"		4	2		130	94		224	288
3250	24" x 36"		4	2		180	94		274	345
4000	Recessed door for drywall									
4100	Metal 12" x 12"	1 Carp	6	1.333	Ea.	80	62.50		142.50	185
4150	12" x 24"		5.50	1.455		114	68.50		182.50	230
4200	24" x 36"		5	1.600		182	75		257	315
6000	Standard door									
6100	Metal, 8" x 8"	1 Carp	10	.800	Ea.	45	37.50		82.50	108
6150	12" x 12"		10	.800		50	37.50		87.50	113
6200	18" x 18"		9	.889		70	41.50		111.50	141
6250	24" x 24"		9	.889		85	41.50		126.50	158
6300	24" x 36"		8	1		125	47		172	211
6350	36" x 36"		8	1		145	47		192	233
6500	Stainless steel, 8" x 8"		10	.800		85	37.50		122.50	152
6550	12" x 12"		10	.800		110	37.50		147.50	179
6600	18" x 18"		9	.889		205	41.50		246.50	290
6650	24" x 24"		9	.889		265	41.50		306.50	355

08 31 13.20 Bulkhead/Cellar Doors		Crew	Daily Output	Labor-Hours	Unit	Material	Labor	Equipment	Total	Total Incl O&P
0010	BULKHEAD/CELLAR DOORS									
0020	Steel, not incl. sides, 44" x 62"	1 Carp	5.50	1.455	Ea.	560	68.50		628.50	720
0100	52" x 73"		5.10	1.569		790	73.50		863.50	985
0500	With sides and foundation plates, 57" x 45" x 24"		4.70	1.702		845	80		925	1,050
0600	42" x 49" x 51"		4.30	1.860		915	87.50		1,002.50	1,125

08 31 13.30 Commercial Floor Doors		Crew	Daily Output	Labor-Hours	Unit	Material	Labor	Equipment	Total	Total Incl O&P
0010	COMMERCIAL FLOOR DOORS									
0020	Aluminum tile, steel frame, one leaf, 2' x 2' opng.	2 Sswk	3.50	4.571	Opng.	880	241		1,121	1,400
0050	3'-6" x 3'-6" opening		3.50	4.571		1,600	241		1,841	2,175
0500	Double leaf, 4' x 4' opening		3	5.333		1,750	281		2,031	2,425
0550	5' x 5' opening		3	5.333		3,100	281		3,381	3,900

08 31 13.35 Industrial Floor Doors		Crew	Daily Output	Labor-Hours	Unit	Material	Labor	Equipment	Total	Total Incl O&P
0010	INDUSTRIAL FLOOR DOORS									
0020	Steel 300 psf L.L., single leaf, 2' x 2', 175#	2 Sswk	6	2.667	Opng.	760	140		900	1,075
0050	3' x 3' opening, 300#		5.50	2.909		1,075	153		1,228	1,450

08 31 Access Doors and Panels

08 31 13 – Access Doors and Frames

08 31 13.35 Industrial Floor Doors		Crew	Daily Output	Labor-Hours	Unit	Material	Labor (2015 Bare Costs)	Equipment	Total	Total Incl O&P
0300	Double leaf, 4′ x 4′ opening, 455#	2 Sswk	5	3.200	Opng.	2,275	168		2,443	2,800
0350	5′ x 5′ opening, 645#		4.50	3.556		3,300	187		3,487	3,950
1000	Aluminum, 300 psf L.L., single leaf, 2′ x 2′, 60#		6	2.667		800	140		940	1,125
1050	3′ x 3′ opening, 100#		5.50	2.909		1,300	153		1,453	1,700
1500	Double leaf, 4′ x 4′ opening, 160#		5	3.200		2,100	168		2,268	2,600
1550	5′ x 5′ opening, 235#		4.50	3.556		2,800	187		2,987	3,400
2000	Aluminum, 150 psf L.L., single leaf, 2′ x 2′, 60#		6	2.667		720	140		860	1,025
2050	3′ x 3′ opening, 95#		5.50	2.909		1,200	153		1,353	1,600
2500	Double leaf, 4′ x 4′ opening, 150#		5	3.200		1,450	168		1,618	1,900
2550	5′ x 5′ opening, 230#		4.50	3.556		1,950	187		2,137	2,450

08 31 13.40 Kennel Doors		Crew	Daily Output	Labor-Hours	Unit	Material	Labor	Equipment	Total	Total Incl O&P
0010	**KENNEL DOORS**									
0020	2 way, swinging type, 13″ x 19″ opening	2 Carp	11	1.455	Opng.	90	68.50		158.50	204
0100	17″ x 29″ opening		11	1.455		110	68.50		178.50	226
0200	9″ x 9″ opening, electronic with accessories		11	1.455		144	68.50		212.50	263

08 32 Sliding Glass Doors

08 32 13 – Sliding Aluminum-Framed Glass Doors

08 32 13.10 Sliding Aluminum Doors		Crew	Daily Output	Labor-Hours	Unit	Material	Labor	Equipment	Total	Total Incl O&P
0010	**SLIDING ALUMINUM DOORS**									
0350	Aluminum, 5/8″ tempered insulated glass, 6′ wide									
0400	Premium	2 Carp	4	4	Ea.	1,575	188		1,763	2,025
0450	Economy		4	4		815	188		1,003	1,175
0500	8′ wide, premium		3	5.333		1,650	250		1,900	2,200
0550	Economy		3	5.333		1,425	250		1,675	1,925
0600	12′ wide, premium		2.50	6.400		2,975	300		3,275	3,725
0650	Economy		2.50	6.400		1,550	300		1,850	2,150
4000	Aluminum, baked on enamel, temp glass, 6′-8″ x 10′-0″ wide		4	4		1,075	188		1,263	1,475
4020	Insulating glass, 6′-8″ x 6′-0″ wide		4	4		935	188		1,123	1,325
4040	8′-0″ wide		3	5.333		1,100	250		1,350	1,575
4060	10′-0″ wide		2	8		1,350	375		1,725	2,050
4080	Anodized, temp glass, 6′-8″ x 6′-0″ wide		4	4		455	188		643	790
4100	8′-0″ wide		3	5.333		575	250		825	1,025
4120	10′-0″ wide		2	8		645	375		1,020	1,300
5000	Aluminum sliding glass door system									
5010	Sliding door 4′ wide opening single side	2 Carp	2	8	Ea.	5,400	375		5,775	6,525
5015	8′ wide opening single side		2	8		8,100	375		8,475	9,475
5020	Telescoping glass door system, 4′ wide opening biparting		2	8		4,500	375		4,875	5,525
5025	8′ wide opening biparting		2	8		5,400	375		5,775	6,525
5030	Folding glass door, 4′ wide opening biparting		2	8		7,200	375		7,575	8,500
5035	8′ wide opening biparting		2	8		9,000	375		9,375	10,500
5040	ICU-CCU Sliding telescoping glass door, 4′ x 7′, single side opening		2	8		2,650	375		3,025	3,500
5045	8′ x 7′, single side opening		2	8		4,125	375		4,500	5,100

08 32 19 – Sliding Wood-Framed Glass Doors

08 32 19.15 Sliding Glass Vinyl-Clad Wood Doors			Crew	Daily Output	Labor-Hours	Unit	Material	Labor	Equipment	Total	Total Incl O&P
0010	**SLIDING GLASS VINYL-CLAD WOOD DOORS**										
0020	Glass, sliding vinyl clad, insul. glass, 6′-0″ x 6′-8″	G	2 Carp	4	4	Opng.	1,500	188		1,688	1,975
0025	6′-0″ x 6′-10″ high	G		4	4		1,650	188		1,838	2,125
0030	6′-0″ x 8′-0″ high	G		4	4		1,975	188		2,163	2,475
0050	5′-0″ x 6′-8″ high	G		4	4		1,500	188		1,688	1,950

08 32 Sliding Glass Doors

08 32 19 – Sliding Wood-Framed Glass Doors

08 32 19.15 Sliding Glass Vinyl-Clad Wood Doors			Crew	Daily Output	Labor-Hours	Unit	Material	2015 Bare Costs Labor	Equipment	Total	Total Incl O&P
0100	8'-0" x 6'-10" high	G	2 Carp	4	4	Opng.	2,025	188		2,213	2,525
0500	4 leaf, 9'-0" x 6'-10" high	G		3	5.333		3,250	250		3,500	3,950
0600	12'-0" x 6'-10" high	G	↓	3	5.333	↓	3,875	250		4,125	4,650

08 33 Coiling Doors and Grilles

08 33 13 – Coiling Counter Doors

08 33 13.10 Counter Doors, Coiling Type

		Crew	Daily Output	Labor-Hours	Unit	Material	Labor	Equipment	Total	Total Incl O&P
0010	**COUNTER DOORS, COILING TYPE**									
0020	Manual, incl. frame and hardware, galv. stl., 4' roll-up, 6' long	2 Carp	2	8	Opng.	1,225	375		1,600	1,925
0300	Galvanized steel, UL label		1.80	8.889		1,225	415		1,640	2,000
0600	Stainless steel, 4' high roll-up, 6' long		2	8		2,150	375		2,525	2,925
0700	10' long		1.80	8.889		2,500	415		2,915	3,400
2000	Aluminum, 4' high, 4' long		2.20	7.273		1,525	340		1,865	2,200
2020	6' long		2	8		1,800	375		2,175	2,575
2040	8' long		1.90	8.421		2,050	395		2,445	2,850
2060	10' long		1.80	8.889		2,100	415		2,515	2,950
2080	14' long		1.40	11.429		2,675	535		3,210	3,775
2100	6' high, 4' long		2	8		1,850	375		2,225	2,600
2120	6' long		1.60	10		1,675	470		2,145	2,550
2140	10' long	↓	1.40	11.429	↓	2,200	535		2,735	3,250

08 33 16 – Coiling Counter Grilles

08 33 16.10 Coiling Grilles

		Crew	Daily Output	Labor-Hours	Unit	Material	Labor	Equipment	Total	Total Incl O&P
0010	**COILING GRILLES**									
0015	Aluminum, manual, incl. frame, mill finish									
0020	Top coiling, 4' high, 4' long	2 Sswk	3.20	5	Opng.	1,575	263		1,838	2,175
0030	6' long		3.20	5		1,825	263		2,088	2,475
0040	8' long		2.40	6.667		2,050	350		2,400	2,850
0050	12' long		2.40	6.667		2,475	350		2,825	3,325
0060	16' long		1.60	10		2,650	525		3,175	3,850
0070	6' high, 4' long		3.20	5		1,825	263		2,088	2,475
0080	6' long		3.20	5		1,800	263		2,063	2,425
0090	8' long		2.40	6.667		1,800	350		2,150	2,575
0100	12' long		1.60	10		2,375	525		2,900	3,550
0110	16' long		1.20	13.333		2,825	700		3,525	4,350
0200	Side coiling, 8' high, 12' long		.60	26.667		2,350	1,400		3,750	5,025
0220	18' long		.50	32		4,450	1,675		6,125	7,825
0240	24' long		.40	40		3,775	2,100		5,875	7,800
0260	12' high, 12' long		.50	32		3,400	1,675		5,075	6,650
0280	18' long		.40	40		5,300	2,100		7,400	9,475
0300	24' long	↓	.28	57.143	↓	6,725	3,000		9,725	12,600

08 33 23 – Overhead Coiling Doors

08 33 23.10 Coiling Service Doors

		Crew	Daily Output	Labor-Hours	Unit	Material	Labor	Equipment	Total	Total Incl O&P
0010	**COILING SERVICE DOORS** Steel, manual, 20 ga., incl. hardware									
0050	8' x 8' high	2 Sswk	1.60	10	Ea.	1,150	525		1,675	2,200
0100	10' x 10' high		1.40	11.429		1,925	600		2,525	3,150
0200	20' x 10' high		1	16		3,150	840		3,990	4,950
0300	12' x 12' high		1.20	13.333		1,950	700		2,650	3,350
0400	20' x 12' high		.90	17.778		2,175	935		3,110	4,025
0500	14' x 14' high		.80	20		3,100	1,050		4,150	5,250
0600	20' x 16' high	↓	.60	26.667	↓	3,725	1,400		5,125	6,550

08 33 Coiling Doors and Grilles

08 33 23 – Overhead Coiling Doors

08 33 23.10 Coiling Service Doors		Crew	Daily Output	Labor-Hours	Unit	Material	Labor	Equipment	Total	Total Incl O&P
						2015 Bare Costs				
0700	10′ x 20′ high	2 Sswk	.50	32	Ea.	2,625	1,675		4,300	5,825
1000	12′ x 12′, crank operated, crank on door side		.80	20		1,775	1,050		2,825	3,800
1100	Crank thru wall		.70	22.857		2,050	1,200		3,250	4,375
1300	For vision panel, add					360			360	400
1600	3′ x 7′ pass door within rolling steel door, new construction					1,925			1,925	2,125
1700	Existing construction	2 Sswk	2	8		2,050	420		2,470	2,975
2000	Class A fire doors, manual, 20 ga., 8′ x 8′ high		1.40	11.429		1,600	600		2,200	2,800
2100	10′ x 10′ high		1.10	14.545		2,175	765		2,940	3,725
2200	20′ x 10′ high		.80	20		4,450	1,050		5,500	6,725
2300	12′ x 12′ high		1	16		3,425	840		4,265	5,250
2400	20′ x 12′ high		.80	20		4,825	1,050		5,875	7,150
2500	14′ x 14′ high		.60	26.667		3,750	1,400		5,150	6,575
2600	20′ x 16′ high		.50	32		5,900	1,675		7,575	9,425
2700	10′ x 20′ high		.40	40		4,675	2,100		6,775	8,775
3000	For 18 ga. doors, add				S.F.	1.65			1.65	1.82
3300	For enamel finish, add				"	1.90			1.90	2.09
3600	For safety edge bottom bar, pneumatic, add				L.F.	23			23	25
3700	Electric, add					43			43	47.50
4000	For weatherstripping, extruded rubber, jambs, add					14.40			14.40	15.85
4100	Hood, add					8.20			8.20	9
4200	Sill, add					5.15			5.15	5.65
4500	Motor operators, to 14′ x 14′ opening	2 Sswk	5	3.200	Ea.	1,150	168		1,318	1,575
4600	Over 14′ x 14′, jack shaft type	"	5	3.200		1,075	168		1,243	1,475
4700	For fire door, additional fusible link, add					28			28	31

08 34 Special Function Doors

08 34 13 – Cold Storage Doors

08 34 13.10 Doors for Cold Area Storage		Crew	Daily Output	Labor-Hours	Unit	Material	Labor	Equipment	Total	Total Incl O&P
0010	**DOORS FOR COLD AREA STORAGE**									
0020	Single, 20 ga. galvanized steel									
0300	Horizontal sliding, 5′ x 7′, manual operation, 3.5″ thick	2 Carp	2	8	Ea.	3,250	375		3,625	4,150
0400	4″ thick		2	8		3,250	375		3,625	4,150
0500	6″ thick		2	8		3,425	375		3,800	4,350
0800	5′ x 7′, power operation, 2″ thick		1.90	8.421		5,600	395		5,995	6,750
0900	4″ thick		1.90	8.421		5,700	395		6,095	6,875
1000	6″ thick		1.90	8.421		6,500	395		6,895	7,750
1300	9′ x 10′, manual operation, 2″ insulation		1.70	9.412		4,550	440		4,990	5,675
1400	4″ insulation		1.70	9.412		4,675	440		5,115	5,825
1500	6″ insulation		1.70	9.412		5,575	440		6,015	6,825
1800	Power operation, 2″ insulation		1.60	10		7,675	470		8,145	9,175
1900	4″ insulation		1.60	10		7,900	470		8,370	9,400
2000	6″ insulation		1.70	9.412		8,850	440		9,290	10,400
2300	For stainless steel face, add					25%				
3000	Hinged, lightweight, 3′ x 7′-0″, 2″ thick	2 Carp	2	8	Ea.	1,425	375		1,800	2,125
3050	4″ thick		1.90	8.421		1,775	395		2,170	2,550
3300	Polymer doors, 3′ x 7′-0″		1.90	8.421		1,350	395		1,745	2,075
3350	6″ thick		1.40	11.429		2,350	535		2,885	3,400
3600	Stainless steel, 3′ x 7′-0″, 4″ thick		1.90	8.421		1,725	395		2,120	2,500
3650	6″ thick		1.40	11.429		2,950	535		3,485	4,050
3900	Painted, 3′ x 7′-0″, 4″ thick		1.90	8.421		1,275	395		1,670	2,000
3950	6″ thick		1.40	11.429		2,350	535		2,885	3,400

08 34 Special Function Doors

08 34 13 – Cold Storage Doors

08 34 13.10 Doors for Cold Area Storage		Crew	Daily Output	Labor-Hours	Unit	Material	2015 Bare Costs Labor	Equipment	Total	Total Incl O&P
5000	Bi-parting, electric operated									
5010	6′ x 8′ opening, galv. faces, 4″ thick for cooler	2 Carp	.80	20	Opng.	7,225	940		8,165	9,375
5050	For freezer, 4″ thick		.80	20		7,950	940		8,890	10,200
5300	For door buck framing and door protection, add		2.50	6.400		640	300		940	1,175
6000	Galvanized batten door, galvanized hinges, 4′ x 7′		2	8		1,825	375		2,200	2,600
6050	6′ x 8′		1.80	8.889		2,475	415		2,890	3,375
6500	Fire door, 3 hr., 6′ x 8′, single slide		.80	20		8,275	940		9,215	10,600
6550	Double, bi-parting		.70	22.857		12,200	1,075		13,275	15,100

08 34 16 – Hangar Doors

08 34 16.10 Aircraft Hangar Doors		Crew	Daily Output	Labor-Hours	Unit	Material	Labor	Equipment	Total	Total Incl O&P
0010	**AIRCRAFT HANGAR DOORS**									
0020	Bi-fold, ovhd., 20 psf wind load, incl. elec. oper.									
0100	12′ high x 40′	2 Sswk	240	.067	S.F.	16.45	3.51		19.96	24
0200	16′ high x 60′		230	.070		19.80	3.66		23.46	28.50
0300	20′ high x 80′		220	.073		21.50	3.83		25.33	30

08 34 36 – Darkroom Doors

08 34 36.10 Various Types of Darkroom Doors		Crew	Daily Output	Labor-Hours	Unit	Material	Labor	Equipment	Total	Total Incl O&P
0010	**VARIOUS TYPES OF DARKROOM DOORS**									
0015	Revolving, standard, 2 way, 36″ diameter	2 Carp	3.10	5.161	Opng.	2,900	242		3,142	3,550
0020	41″ diameter		3.10	5.161		3,000	242		3,242	3,700
0050	3 way, 51″ diameter		1.40	11.429		3,825	535		4,360	5,025
1000	4 way, 49″ diameter		1.40	11.429		3,950	535		4,485	5,175
2000	Hinged safety, 2 way, 41″ diameter		2.30	6.957		3,750	325		4,075	4,600
2500	3 way, 51″ diameter		1.40	11.429		4,050	535		4,585	5,300
3000	Pop out safety, 2 way, 41″ diameter		3.10	5.161		4,025	242		4,267	4,800
4000	3 way, 51″ diameter		1.40	11.429		4,875	535		5,410	6,200
5000	Wheelchair-type, pop out, 51″ diameter		1.40	11.429		5,500	535		6,035	6,875
5020	72″ diameter		.90	17.778		8,950	835		9,785	11,100
9300	For complete darkrooms, see Section 13 21 53.50									

08 34 53 – Security Doors and Frames

08 34 53.20 Steel Door		Crew	Daily Output	Labor-Hours	Unit	Material	Labor	Equipment	Total	Total Incl O&P
0010	**STEEL DOOR** with ballistic core and welded frame both 14 ga.									
0050	Flush, UL 752 Level 3, 1-3/4″, 3′-0″ x 6′-8″	2 Carp	1.50	10.667	Opng.	2,325	500		2,825	3,350
0055	1-3/4″, 3′-6″ x 6′-8″		1.50	10.667		2,425	500		2,925	3,450
0060	1-3/4″, 4′-0″ x 6′-8″		1.20	13.333		2,650	625		3,275	3,900
0100	UL 752 Level 8, 1-3/4″, 3′-0″ x 6′-8″		1.50	10.667		10,200	500		10,700	12,000
0105	1-3/4″, 3′-6″ x 6′-8″		1.50	10.667		10,300	500		10,800	12,100
0110	1-3/4″, 4′-0″ x 6′-8″		1.20	13.333		11,700	625		12,325	13,900
0120	UL 752 Level 3, 1-3/4″, 3′-0″ x 7′-0″		1.50	10.667		2,350	500		2,850	3,375
0125	1-3/4″, 3′-6″ x 7′-0″		1.50	10.667		2,475	500		2,975	3,500
0130	1-3/4″, 4′-0″ x 7′-0″		1.20	13.333		2,675	625		3,300	3,925
0150	UL 752 Level 8, 1-3/4″, 3′-0″ x 7′-0″		1.50	10.667		10,200	500		10,700	12,000
0155	1-3/4″, 3′-6″ x 7′-0″		1.50	10.667		10,300	500		10,800	12,100
0160	1-3/4″, 4′-0″ x 7′-0″		1.20	13.333		11,700	625		12,325	13,900
1000	Safe Room sliding door and hardware, 1-3/4″, 3′-0″ x 7′-0″ UL 752 Level 3		.50	32		23,500	1,500		25,000	28,200
1050	Safe Room swinging door and hardware, 1-3/4″, 3′-0″ x 7′-0″ UL 752 Level 3		.50	32		28,100	1,500		29,600	33,200

08 34 Special Function Doors

08 34 53 – Security Doors and Frames

08 34 53.30 Wood Ballistic Doors		Crew	Daily Output	Labor-Hours	Unit	Material	2015 Bare Costs Labor	Equipment	Total	Total Incl O&P
0010	**WOOD BALLISTIC DOORS** with frames and hardware									
0050	Wood, 1-3/4", 3'-0" x 7'-0" UL 752 Level 3	2 Carp	1.50	10.667	Opng.	2,250	500		2,750	3,250

08 34 56 – Security Gates

08 34 56.10 Gates		Crew	Daily Output	Labor-Hours	Unit	Material	Labor	Equipment	Total	Total Incl O&P
0010	**GATES**									
0015	Driveway Gates include mounting hardware									
0500	Wood, security gate, driveway, dual, 10' wide	H-4	.80	25	Opng.	3,900	1,100		5,000	5,975
0505	12' wide		.80	25		4,300	1,100		5,400	6,400
0510	15' wide		.80	25		5,100	1,100		6,200	7,275
0600	Steel, security gate, driveway, single, 10' wide		.80	25		2,150	1,100		3,250	4,025
0605	12' wide		.80	25		2,225	1,100		3,325	4,100
0620	Steel, security gate, driveway, dual, 12' wide		.80	25		2,050	1,100		3,150	3,950
0625	14' wide		.80	25		2,200	1,100		3,300	4,100
0630	16' wide		.80	25		2,525	1,100		3,625	4,450
0700	Aluminum, security gate, driveway, dual, 10' wide		.80	25		3,000	1,100		4,100	4,975
0705	12' wide		.80	25		3,200	1,100		4,300	5,200
0710	16' wide		.80	25		3,700	1,100		4,800	5,750
1000	Security gate, driveway, opener 12 VDC				Ea.	800			800	880
1010	Wireless					1,800			1,800	1,975
1020	Security gate, driveway, opener 24 VDC					1,300			1,300	1,425
1030	Wireless					1,450			1,450	1,600
1040	Security gate, driveway, opener 12 VDC, solar panel 10 watt	1 Elec	2	4		300	219		519	660
1050	20 watt	"	2	4		450	219		669	825

08 34 59 – Vault Doors and Day Gates

08 34 59.10 Secure Storage Doors		Crew	Daily Output	Labor-Hours	Unit	Material	Labor	Equipment	Total	Total Incl O&P
0010	**SECURE STORAGE DOORS**									
0020	Door and frame, 32" x 78", clear opening									
0100	1 hour test, 32" door, weighs 750 lb.	2 Sswk	1.50	10.667	Opng.	6,625	560		7,185	8,275
0200	2 hour test, 32" door, weighs 950 lb.		1.30	12.308		8,200	650		8,850	10,100
0250	40" door, weighs 1130 lb.		1	16		9,250	840		10,090	11,700
0300	4 hour test, 32" door, weighs 1025 lb.		1.20	13.333		8,650	700		9,350	10,800
0350	40" door, weighs 1140 lb.		.90	17.778		9,975	935		10,910	12,600
0600	For time lock, two movement, add	1 Elec	2	4	Ea.	1,800	219		2,019	2,300
0800	Day gate, painted, steel, 32" wide	2 Sswk	1.50	10.667		2,000	560		2,560	3,175
0850	40" wide		1.40	11.429		2,100	600		2,700	3,350
0900	Aluminum, 32" wide		1.50	10.667		3,200	560		3,760	4,500
0950	40" wide		1.40	11.429		3,400	600		4,000	4,800
2050	Security vault door, class I, 3' wide, 3 1/2" thick	E-24	.19	166	Opng.	15,700	8,700	3,825	28,225	36,100
2100	Class II, 3' wide, 7" thick		.19	166		18,600	8,700	3,825	31,125	39,300
2150	Class III, 9R, 3' wide, 10" thick		.13	250		23,500	13,000	5,750	42,250	54,000
2160	Class V, type 1, 40" door		2.48	12.903	Ea.	7,125	675	297	8,097	9,275
2170	Class V, type 2, 40" door		2.48	12.903		6,850	675	297	7,822	8,975
2180	Day gate for class V vault	2 Sswk	2	8		2,025	420		2,445	2,950

08 34 63 – Detention Doors and Frames

08 34 63.13 Steel Detention Doors and Frames		Crew	Daily Output	Labor-Hours	Unit	Material	Labor	Equipment	Total	Total Incl O&P
0010	**STEEL DETENTION DOORS AND FRAMES**									
0500	Rolling cell door, bar front, 7/8" bars, 4" O.C., 7' H, 5' W, with hardware G	E-4	2	16	Ea.	5,325	850	73	6,248	7,425
0550	Actuator for rolling cell door, bar front	2 Skwk	2	8		4,250	390		4,640	5,275
1000	Doors & frames, 3' x 7', complete, with hardware, single plate	E-4	4	8		4,600	425	36.50	5,061.50	5,825
1650	Double plate	"	4	8		5,600	425	36.50	6,061.50	6,925

08 34 Special Function Doors

08 34 73 – Sound Control Door Assemblies

08 34 73.10 Acoustical Doors		Crew	Daily Output	Labor-Hours	Unit	Material	2015 Bare Costs Labor	Equipment	Total	Total Incl O&P
0010	ACOUSTICAL DOORS									
0020	Including framed seals, 3' x 7', wood, 40 STC rating	2 Carp	1.50	10.667	Ea.	1,300	500		1,800	2,200
0100	Steel, 41 STC rating		1.50	10.667		3,275	500		3,775	4,375
0200	45 STC rating		1.50	10.667		3,675	500		4,175	4,825
0300	48 STC rating		1.50	10.667		4,275	500		4,775	5,500
0400	52 STC rating		1.50	10.667		4,900	500		5,400	6,175

08 36 Panel Doors

08 36 13 – Sectional Doors

08 36 13.10 Overhead Commercial Doors

		Crew	Daily Output	Labor-Hours	Unit	Material	Labor	Equipment	Total	Total Incl O&P
0010	OVERHEAD COMMERCIAL DOORS									
1000	Stock, sectional, heavy duty, wood, 1-3/4" thick, 8' x 8' high	2 Carp	2	8	Ea.	1,050	375		1,425	1,725
1100	10' x 10' high		1.80	8.889		1,500	415		1,915	2,300
1200	12' x 12' high		1.50	10.667		2,050	500		2,550	3,025
1300	Chain hoist, 14' x 14' high		1.30	12.308		3,300	580		3,880	4,525
1400	12' x 16' high		1	16		3,275	750		4,025	4,750
1500	20' x 8' high		1.30	12.270		2,675	575		3,250	3,825
1600	20' x 16' high		.65	24.615		5,475	1,150		6,625	7,800
1800	Center mullion openings, 8' high		4	4		1,225	188		1,413	1,650
1900	20' high		2	8		2,050	375		2,425	2,825
2100	For medium duty custom door, deduct					5%	5%			
2150	For medium duty stock doors, deduct					10%	5%			
2300	Fiberglass and aluminum, heavy duty, sectional, 12' x 12' high	2 Carp	1.50	10.667	Ea.	2,700	500		3,200	3,750
2450	Chain hoist, 20' x 20' high		.50	32		6,675	1,500		8,175	9,650
2600	Steel, 24 ga. sectional, manual, 8' x 8' high		2	8		820	375		1,195	1,475
2650	10' x 10' high		1.80	8.889		1,100	415		1,515	1,875
2700	12' x 12' high		1.50	10.667		1,350	500		1,850	2,250
2800	Chain hoist, 20' x 14' high		.70	22.857		3,725	1,075		4,800	5,750
2850	For 1-1/4" rigid insulation and 26 ga. galv.									
2860	back panel, add				S.F.	4.75			4.75	5.25
2900	For electric trolley operator, 1/3 H.P., to 12' x 12', add	1 Carp	2	4	Ea.	950	188		1,138	1,350
2950	Over 12' x 12', 1/2 H.P., add		1	8		1,125	375		1,500	1,825
2980	Overhead, for row of clear lites add		1	8		110	375		485	700

08 36 13.20 Residential Garage Doors

		Crew	Daily Output	Labor-Hours	Unit	Material	Labor	Equipment	Total	Total Incl O&P
0010	RESIDENTIAL GARAGE DOORS									
0050	Hinged, wood, custom, double door, 9' x 7'	2 Carp	4	4	Ea.	800	188		988	1,175
0070	16' x 7'		3	5.333		1,200	250		1,450	1,700
0200	Overhead, sectional, incl. hardware, fiberglass, 9' x 7', standard		5.28	3.030		945	142		1,087	1,275
0220	Deluxe		5.28	3.030		1,150	142		1,292	1,475
0300	16' x 7', standard		6	2.667		1,575	125		1,700	1,925
0320	Deluxe		6	2.667		2,150	125		2,275	2,550
0500	Hardboard, 9' x 7', standard		8	2		625	94		719	835
0520	Deluxe		8	2		815	94		909	1,050
0600	16' x 7', standard		6	2.667		1,225	125		1,350	1,550
0620	Deluxe		6	2.667		1,425	125		1,550	1,775
0700	Metal, 9' x 7', standard		5.28	3.030		740	142		882	1,025
0720	Deluxe		8	2		915	94		1,009	1,150
0800	16' x 7', standard		3	5.333		940	250		1,190	1,400
0820	Deluxe		6	2.667		1,400	125		1,525	1,750
0900	Wood, 9' x 7', standard		8	2		970	94		1,064	1,225

08 36 Panel Doors

08 36 13 – Sectional Doors

08 36 13.20 Residential Garage Doors		Crew	Daily Output	Labor-Hours	Unit	Material	2015 Bare Costs Labor	Equipment	Total	Total Incl O&P
0920	Deluxe	2 Carp	8	2	Ea.	2,150	94		2,244	2,500
1000	16' x 7', standard		6	2.667		1,625	125		1,750	2,000
1020	Deluxe		6	2.667		3,025	125		3,150	3,525
1800	Door hardware, sectional	1 Carp	4	2		350	94		444	530
1810	Door tracks only		4	2		163	94		257	325
1820	One side only		7	1.143		120	53.50		173.50	215
3000	Swing-up, including hardware, fiberglass, 9' x 7', standard	2 Carp	8	2		1,000	94		1,094	1,250
3020	Deluxe		8	2		1,100	94		1,194	1,350
3100	16' x 7', standard		6	2.667		1,250	125		1,375	1,575
3120	Deluxe		6	2.667		1,600	125		1,725	1,950
3200	Hardboard, 9' x 7', standard		8	2		550	94		644	750
3220	Deluxe		8	2		650	94		744	860
3300	16' x 7', standard		6	2.667		670	125		795	930
3320	Deluxe		6	2.667		850	125		975	1,125
3400	Metal, 9' x 7', standard		8	2		600	94		694	805
3420	Deluxe		8	2		965	94		1,059	1,200
3500	16' x 7', standard		6	2.667		800	125		925	1,075
3520	Deluxe		6	2.667		1,100	125		1,225	1,400
3600	Wood, 9' x 7', standard		8	2		700	94		794	915
3620	Deluxe		8	2		1,125	94		1,219	1,375
3700	16' x 7', standard		6	2.667		900	125		1,025	1,175
3720	Deluxe		6	2.667		2,100	125		2,225	2,500
3900	Door hardware only, swing up	1 Carp	4	2		168	94		262	330
3920	One side only		7	1.143		90	53.50		143.50	182
4000	For electric operator, economy, add		8	1		425	47		472	545
4100	Deluxe, including remote control		8	1		610	47		657	750
4500	For transmitter/receiver control , add to operator				Total	110			110	121
4600	Transmitters, additional				"	60			60	66

08 36 19 – Multi-Leaf Vertical Lift Doors

08 36 19.10 Sectional Vertical Lift Doors		Crew	Daily Output	Labor-Hours	Unit	Material	Labor	Equipment	Total	Total Incl O&P
0010	**SECTIONAL VERTICAL LIFT DOORS**									
0020	Motorized, 14 ga. steel, incl. frame and control panel									
0050	16' x 16' high	L-10	.50	48	Ea.	21,500	2,550	1,300	25,350	29,400
0100	10' x 20' high		1.30	18.462		34,800	980	505	36,285	40,500
0120	15' x 20' high		1.30	18.462		42,600	980	505	44,085	49,100
0140	20' x 20' high		1	24		49,900	1,275	655	51,830	58,000
0160	25' x 20' high		1	24		56,000	1,275	655	57,930	64,500
0170	32' x 24' high		.75	32		48,700	1,700	870	51,270	57,500
0180	20' x 25' high		1	24		57,500	1,275	655	59,430	66,000
0200	25' x 25' high		.70	34.286		66,000	1,825	935	68,760	77,000
0220	25' x 30' high		.70	34.286		71,000	1,825	935	73,760	82,500
0240	30' x 30' high		.70	34.286		82,500	1,825	935	85,260	94,500
0260	35' x 30' high		.70	34.286		92,500	1,825	935	95,260	106,000

08 38 Traffic Doors

08 38 13 – Flexible Strip Doors

	08 38 13.10 Flexible Transparent Strip Doors	Crew	Daily Output	Labor-Hours	Unit	Material	2015 Bare Costs Labor	Equipment	Total	Total Incl O&P
0010	**FLEXIBLE TRANSPARENT STRIP DOORS**									
0100	12" strip width, 2/3 overlap	3 Shee	135	.178	SF Surf	7.60	9.95		17.55	23.50
0200	Full overlap		115	.209		9.60	11.70		21.30	28.50
0220	8" strip width, 1/2 overlap		140	.171		6.20	9.60		15.80	21.50
0240	Full overlap		120	.200		7.80	11.20		19	25.50
0300	Add for suspension system, header mount				L.F.	9.15			9.15	10.05
0400	Wall mount				"	9.45			9.45	10.40

08 38 19 – Rigid Traffic Doors

08 38 19.20 Double Acting Swing Doors

	08 38 19.20 Double Acting Swing Doors	Crew	Daily Output	Labor-Hours	Unit	Material	Labor	Equipment	Total	Total Incl O&P
0010	**DOUBLE ACTING SWING DOORS**									
0020	Including frame, closer, hardware and vision panel									
1000	Polymer, 7'-0" high, 4'-0" wide	2 Carp	4.20	3.810	Pr.	2,100	179		2,279	2,575
1025	6'-0" wide		4	4		2,200	188		2,388	2,725
1050	6'-8" wide		4	4		2,400	188		2,588	2,950
2000	3/4" thick, stainless steel									
2010	Stainless steel, 7' high opening, 4' wide	2 Carp	4	4	Pr.	2,600	188		2,788	3,150
2050	7' wide	"	3.80	4.211	"	2,800	198		2,998	3,375

08 38 19.30 Shock Absorbing Doors

	08 38 19.30 Shock Absorbing Doors	Crew	Daily Output	Labor-Hours	Unit	Material	Labor	Equipment	Total	Total Incl O&P
0010	**SHOCK ABSORBING DOORS**									
0020	Rigid, no frame, 1-1/2" thick, 5' x 7'	2 Sswk	1.90	8.421	Opng.	1,525	445		1,970	2,450
0100	8' x 8'		1.80	8.889		2,000	470		2,470	3,025
0500	Flexible, no frame, insulated, .16" thick, economy, 5' x 7'		2	8		1,750	420		2,170	2,650
0600	Deluxe		1.90	8.421		2,625	445		3,070	3,650
1000	8' x 8' opening, economy		2	8		2,750	420		3,170	3,750
1100	Deluxe		1.90	8.421		3,500	445		3,945	4,625

08 41 Entrances and Storefronts

08 41 13 – Aluminum-Framed Entrances and Storefronts

08 41 13.20 Tube Framing

	08 41 13.20 Tube Framing	Crew	Daily Output	Labor-Hours	Unit	Material	Labor	Equipment	Total	Total Incl O&P
0010	**TUBE FRAMING**, For window walls and store fronts, aluminum stock									
0050	Plain tube frame, mill finish, 1-3/4" x 1-3/4"	2 Glaz	103	.155	L.F.	9.90	7		16.90	21.50
0150	1-3/4" x 4"		98	.163		13.35	7.35		20.70	26
0200	1-3/4" x 4-1/2"		95	.168		16	7.60		23.60	29
0250	2" x 6"		89	.180		23.50	8.10		31.60	38.50
0350	4" x 4"		87	.184		26.50	8.30		34.80	42
0400	4-1/2" x 4-1/2"		85	.188		28	8.50		36.50	44
0450	Glass bead		240	.067		3.07	3.01		6.08	7.95
1000	Flush tube frame, mill finish, 1/4" glass, 1-3/4" x 4", open header		80	.200		13.25	9		22.25	28.50
1050	Open sill		82	.195		10.85	8.80		19.65	25.50
1100	Closed back header		83	.193		19	8.70		27.70	34.50
1150	Closed back sill		85	.188		18.15	8.50		26.65	33
1160	Tube fmg., spandrel cover both sides, alum 1" wide	1 Sswk	85	.094	S.F.	98	4.96		102.96	117
1170	Tube fmg., spandrel cover both sides, alum 2" wide	"	85	.094	"	38.50	4.96		43.46	50.50
1200	Vertical mullion, one piece	2 Glaz	75	.213	L.F.	19.90	9.60		29.50	36.50
1250	Two piece		73	.219		21	9.90		30.90	38.50
1300	90° or 180° vertical corner post		75	.213		32.50	9.60		42.10	50
1400	1-3/4" x 4-1/2", open header		80	.200		16	9		25	31.50
1450	Open sill		82	.195		13.55	8.80		22.35	28.50
1500	Closed back header		83	.193		19.10	8.70		27.80	34.50
1550	Closed back sill		85	.188		18.95	8.50		27.45	34

08 41 Entrances and Storefronts

08 41 13 – Aluminum-Framed Entrances and Storefronts

08 41 13.20 Tube Framing		Crew	Daily Output	Labor-Hours	Unit	Material	2015 Bare Costs Labor	Equipment	Total	Total Incl O&P
1600	Vertical mullion, one piece	2 Glaz	75	.213	L.F.	21.50	9.60		31.10	38
1650	Two piece		73	.219		22.50	9.90		32.40	39.50
1700	90° or 180° vertical corner post		75	.213		23	9.60		32.60	40
2000	Flush tube frame, mil fin.,ins. glass w/thml brk, 2" x 4-1/2", open header		75	.213		16.30	9.60		25.90	32.50
2050	Open sill		77	.208		13.70	9.35		23.05	29.50
2100	Closed back header		78	.205		15.50	9.25		24.75	31
2150	Closed back sill		80	.200		14.95	9		23.95	30
2200	Vertical mullion, one piece		70	.229		17.20	10.30		27.50	34.50
2250	Two piece		68	.235		18.60	10.60		29.20	36.50
2300	90° or 180° vertical corner post		70	.229		17.70	10.30		28	35
5000	Flush tube frame, mill fin., thermal brk., 2-1/4" x 4-1/2", open header		74	.216		17.15	9.75		26.90	33.50
5050	Open sill		75	.213		15.10	9.60		24.70	31.50
5100	Vertical mullion, one piece		69	.232		17.75	10.45		28.20	35.50
5150	Two piece		67	.239		21.50	10.75		32.25	40
5200	90° or 180° vertical corner post		69	.232		19.10	10.45		29.55	37
6980	Door stop (snap in)		380	.042		3.40	1.90		5.30	6.65
7000	For joints, 90°, clip type, add				Ea.	25.50			25.50	28.50
7050	Screw spline joint, add					24			24	26
7100	For joint other than 90°, add					49.50			49.50	54.50
8000	For bronze anodized aluminum, add					15%				
8020	For black finish, add					30%				
8050	For stainless steel materials, add					350%				
8100	For monumental grade, add					53%				
8150	For steel stiffener, add	2 Glaz	200	.080	L.F.	11.40	3.61		15.01	18.05
8200	For 2 to 5 stories, add per story				Story		8%			

08 41 19 – Stainless-Steel-Framed Entrances and Storefronts

08 41 19.10 Stainless-Steel and Glass Entrance Unit		Crew	Daily Output	Labor-Hours	Unit	Material	Labor	Equipment	Total	Total Incl O&P
0010	**STAINLESS-STEEL AND GLASS ENTRANCE UNIT**, narrow stiles									
0020	3' x 7' opening, including hardware, minimum	2 Sswk	1.60	10	Opng.	6,800	525		7,325	8,400
0050	Average		1.40	11.429		7,250	600		7,850	9,025
0100	Maximum		1.20	13.333		7,825	700		8,525	9,850
1000	For solid bronze entrance units, statuary finish, add					64%				
1100	Without statuary finish, add					45%				
2000	Balanced doors, 3' x 7', economy	2 Sswk	.90	17.778	Ea.	9,200	935		10,135	11,700
2100	Premium	"	.70	22.857	"	15,500	1,200		16,700	19,200

08 41 26 – All-Glass Entrances and Storefronts

08 41 26.10 Window Walls Aluminum, Stock		Crew	Daily Output	Labor-Hours	Unit	Material	Labor	Equipment	Total	Total Incl O&P
0010	**WINDOW WALLS ALUMINUM, STOCK**, including glazing									
0020	Minimum	H-2	160	.150	S.F.	46.50	6.40		52.90	61.50
0050	Average		140	.171		64	7.30		71.30	81.50
0100	Maximum		110	.218		172	9.30		181.30	203
0500	For translucent sandwich wall systems, see Section 07 41 33.10									
0850	Cost of the above walls depends on material,									
0860	finish, repetition, and size of units.									
0870	The larger the opening, the lower the S.F. cost									
1200	Double glazed acoustical window wall for airports,									

08 42 Entrances

08 42 26 – All-Glass Entrances

08 42 26.10 Swinging Glass Doors		Crew	Daily Output	Labor-Hours	Unit	Material	2015 Bare Costs Labor	Equipment	Total	Total Incl O&P
0010	**SWINGING GLASS DOORS**									
0020	Including hardware, 1/2" thick, tempered, 3' x 7' opening	2 Glaz	2	8	Opng.	2,200	360		2,560	2,975
0100	6' x 7' opening	"	1.40	11.429	"	4,400	515		4,915	5,600

08 42 33 – Revolving Door Entrances

08 42 33.10 Circular Rotating Entrance Doors		Crew	Daily Output	Labor-Hours	Unit	Material	Labor	Equipment	Total	Total Incl O&P
0010	**CIRCULAR ROTATING ENTRANCE DOORS**, Aluminum									
0020	6'-10" to 7' high, stock units, minimum	4 Sswk	.75	42.667	Opng.	22,300	2,250		24,550	28,400
0050	Average		.60	53.333		26,300	2,800		29,100	33,800
0100	Maximum		.45	71.111		32,700	3,750		36,450	42,500
1000	Stainless steel		.30	105		41,000	5,575		46,575	55,000
1100	Solid bronze		.15	213		48,000	11,200		59,200	72,500
1500	For automatic controls, add	2 Elec	2	8		15,100	440		15,540	17,300

08 42 36 – Balanced Door Entrances

08 42 36.10 Balanced Entrance Doors		Crew	Daily Output	Labor-Hours	Unit	Material	Labor	Equipment	Total	Total Incl O&P
0010	**BALANCED ENTRANCE DOORS**									
0020	Hardware & frame, alum. & glass, 3' x 7', econ.	2 Sswk	.90	17.778	Ea.	6,600	935		7,535	8,900
0150	Premium	"	.70	22.857	"	7,925	1,200		9,125	10,800

08 43 Storefronts

08 43 13 – Aluminum-Framed Storefronts

08 43 13.10 Aluminum-Framed Entrance Doors and Frames		Crew	Daily Output	Labor-Hours	Unit	Material	Labor	Equipment	Total	Total Incl O&P
0010	**ALUMINUM-FRAMED ENTRANCE DOORS AND FRAMES**									
0015	Standard hardware and glass stops but no glass									
0020	Entrance door, 3' x 7' opening, clear anodized finish	2 Sswk	7	2.286	Opng.	525	120		645	790
0040	Bronze finish		7	2.286		530	120		650	795
0060	Black finish		7	2.286		575	120		695	840
0200	3'-6" x 7'-0", mill finish		7	2.286		635	120		755	905
0220	Bronze finish		7	2.286		655	120		775	930
0240	Black finish		7	2.286		750	120		870	1,025
0500	6' x 7' opening, clear finish		6	2.667		840	140		980	1,175
0520	Bronze finish		6	2.667		910	140		1,050	1,250
0540	Black finish		6	2.667		990	140		1,130	1,350
1000	With 3' high transom above, 3' x 7' opening, clear finish		5.50	2.909		490	153		643	805
1050	Bronze finish		5.50	2.909		510	153		663	825
1100	Black finish		5.50	2.909		535	153		688	850
1300	3'-6" x 7'-0" opening, clear finish		5.50	2.909		340	153		493	640
1320	Bronze finish		5.50	2.909		355	153		508	660
1340	Black finish		5.50	2.909		365	153		518	665
1500	6' x 7' opening, clear finish		5.50	2.909		595	153		748	920
1550	Bronze finish		5.50	2.909		615	153		768	940
1600	Black finish		5.50	2.909		675	153		828	1,000

08 43 13.20 Storefront Systems		Crew	Daily Output	Labor-Hours	Unit	Material	Labor	Equipment	Total	Total Incl O&P
0010	**STOREFRONT SYSTEMS**, aluminum frame clear 3/8" plate glass									
0020	incl. 3' x 7' door with hardware (400 sq. ft. max. wall)									
0500	Wall height to 12' high, commercial grade	2 Glaz	150	.107	S.F.	22.50	4.81		27.31	32.50
0600	Institutional grade		130	.123		28	5.55		33.55	39.50
0700	Monumental grade		115	.139		40.50	6.25		46.75	54
1000	6' x 7' door with hardware, commercial grade		135	.119		30	5.35		35.35	41
1100	Institutional grade		115	.139		28.50	6.25		34.75	41
1200	Monumental grade		100	.160		54.50	7.20		61.70	71

08 43 Storefronts

08 43 13 – Aluminum-Framed Storefronts

08 43 13.20 Storefront Systems		Crew	Daily Output	Labor-Hours	Unit	Material	2015 Bare Costs Labor	Equipment	Total	Total Incl O&P
1500	For bronze anodized finish, add				S.F.	15%				
1600	For black anodized finish, add					36%				
1700	For stainless steel framing, add to monumental				↓	78%				

08 43 29 – Sliding Storefronts

08 43 29.10 Sliding Panels		Crew	Daily Output	Labor-Hours	Unit	Material	Labor	Equipment	Total	Total Incl O&P
0010	**SLIDING PANELS**									
0020	Mall fronts, aluminum & glass, 15′ x 9′ high	2 Glaz	1.30	12.308	Opng.	3,475	555		4,030	4,675
0100	24′ x 9′ high		.70	22.857		4,925	1,025		5,950	7,000
0200	48′ x 9′ high, with fixed panels	↓	.90	17.778		9,100	800		9,900	11,200
0500	For bronze finish, add				↓	17%				

08 44 Curtain Wall and Glazed Assemblies

08 44 13 – Glazed Aluminum Curtain Walls

08 44 13.10 Glazed Curtain Walls		Crew	Daily Output	Labor-Hours	Unit	Material	Labor	Equipment	Total	Total Incl O&P
0010	**GLAZED CURTAIN WALLS**, aluminum, stock, including glazing									
0020	Minimum	H-1	205	.156	S.F.	37.50	7.65		45.15	54
0050	Average, single glazed		195	.164		53.50	8		61.50	71.50
0150	Average, double glazed		180	.178		69	8.70		77.70	90.50
0200	Maximum	↓	160	.200	↓	181	9.80		190.80	215

08 45 Translucent Wall and Roof Assemblies

08 45 10 – Translucent Roof Assemblies

08 45 10.10 Skyroofs		Crew	Daily Output	Labor-Hours	Unit	Material	Labor	Equipment	Total	Total Incl O&P
0010	**SKYROOFS**									
1200	Skylights, circular, clear, double glazed acrylic									
1230	30″ diameter	2 Carp	3	5.333	Ea.	3,000	250		3,250	3,675
1250	60″ diameter		3	5.333		4,000	250		4,250	4,775
1290	96″ diameter	↓	2	8	↓	5,000	375		5,375	6,075
1300	Skylight Barrel Vault, clear, double glazed, acrylic									
1330	3′-0″ X 12′-0″	G-3	3	10.667	Ea.	5,000	500		5,500	6,275
1350	4′-0″ X 12′-0″		3	10.667		5,500	500		6,000	6,825
1390	5′-0″ X 12′-0″	↓	2	16	↓	6,000	750		6,750	7,750
1400	Skylight Pyramid, Aluminum frame, clear low-E laminated glass									
1410	The glass is installed in the frame except where noted									
1430	Square, 3′ X 3′	G-3	3	10.667	Ea.	6,000	500		6,500	7,375
1440	4′ X 4′		3	10.667		7,000	500		7,500	8,475
1450	5′ X 5′, glass must be field installed		3	10.667		8,000	500		8,500	9,575
1460	6′ X 6′, glass must be field installed	↓	2	16	↓	10,000	750		10,750	12,200
1550	Install pre-cut laminated glass in aluminum frame on a flat roof	2 Glaz	55	.291	SF Surf		13.10		13.10	20
1560	Install pre-cut laminated glass in aluminum frame on a sloped roof	″	40	.400	″		18.05		18.05	27.50

08 51 Metal Windows

08 51 13 – Aluminum Windows

08 51 13.10 Aluminum Sash		Crew	Daily Output	Labor-Hours	Unit	Material	2015 Bare Costs Labor	Equipment	Total	Total Incl O&P
0010	**ALUMINUM SASH**									
0020	Stock, grade C, glaze & trim not incl., casement	2 Sswk	200	.080	S.F.	39	4.21		43.21	50.50
0050	Double hung		200	.080		39.50	4.21		43.71	51
0100	Fixed casement		200	.080		17.35	4.21		21.56	26.50
0150	Picture window		200	.080		18.50	4.21		22.71	28
0200	Projected window		200	.080		35.50	4.21		39.71	46.50
0250	Single hung		200	.080		16.55	4.21		20.76	25.50
0300	Sliding		200	.080		21.50	4.21		25.71	31
1000	Mullions for above, tubular		240	.067	L.F.	6.15	3.51		9.66	12.90
2000	Custom aluminum sash, grade HC, glazing not included		140	.114	S.F.	40	6		46	54.50

08 51 13.20 Aluminum Windows		Crew	Daily Output	Labor-Hours	Unit	Material	Labor	Equipment	Total	Total Incl O&P
0010	**ALUMINUM WINDOWS**, incl. frame and glazing, commercial grade									
1000	Stock units, casement, 3′-1″ x 3′-2″ opening	2 Sswk	10	1.600	Ea.	375	84		459	555
1050	Add for storms					120			120	132
1600	Projected, with screen, 3′-1″ x 3′-2″ opening	2 Sswk	10	1.600		355	84		439	535
1700	Add for storms					117			117	129
2000	4′-5″ x 5′-3″ opening	2 Sswk	8	2		400	105		505	625
2100	Add for storms					126			126	139
2500	Enamel finish windows, 3′-1″ x 3′-2″	2 Sswk	10	1.600		360	84		444	540
2600	4′-5″ x 5′-3″		8	2		405	105		510	630
3000	Single hung, 2′ x 3′ opening, enameled, standard glazed		10	1.600		208	84		292	375
3100	Insulating glass		10	1.600		252	84		336	425
3300	2′-8″ x 6′-8″ opening, standard glazed		8	2		365	105		470	585
3400	Insulating glass		8	2		475	105		580	705
3700	3′-4″ x 5′-0″ opening, standard glazed		9	1.778		300	93.50		393.50	495
3800	Insulating glass		9	1.778		335	93.50		428.50	530
3890	Awning type, 3′ x 3′ opening standard glass		14	1.143		425	60		485	575
3900	Insulating glass		14	1.143		450	60		510	600
3910	3′ x 4′ opening, standard glass		10	1.600		490	84		574	685
3920	Insulating glass		10	1.600		565	84		649	765
3930	3′ x 5′-4″ opening, standard glass		10	1.600		590	84		674	795
3940	Insulating glass		10	1.600		695	84		779	910
3950	4′ x 5′-4″ opening, standard glass		9	1.778		650	93.50		743.50	880
3960	Insulating glass		9	1.778		775	93.50		868.50	1,025
4000	Sliding aluminum, 3′ x 2′ opening, standard glazed		10	1.600		217	84		301	385
4100	Insulating glass		10	1.600		232	84		316	400
4300	5′ x 3′ opening, standard glazed		9	1.778		330	93.50		423.50	530
4400	Insulating glass		9	1.778		385	93.50		478.50	590
4600	8′ x 4′ opening, standard glazed		6	2.667		350	140		490	630
4700	Insulating glass		6	2.667		565	140		705	865
5000	9′ x 5′ opening, standard glazed		4	4		530	211		741	950
5100	Insulating glass		4	4		850	211		1,061	1,300
5500	Sliding, with thermal barrier and screen, 6′ x 4′, 2 track		8	2		725	105		830	980
5700	4 track		8	2		910	105		1,015	1,175
6000	For above units with bronze finish, add					15%				
6200	For installation in concrete openings, add					8%				

08 51 23 – Steel Windows

08 51 23.10 Steel Sash		Crew	Daily Output	Labor-Hours	Unit	Material	Labor	Equipment	Total	Total Incl O&P
0010	**STEEL SASH** Custom units, glazing and trim not included R085123-10									
0100	Casement, 100% vented	2 Sswk	200	.080	S.F.	66.50	4.21		70.71	80.50
0200	50% vented		200	.080		54.50	4.21		58.71	67
0300	Fixed		200	.080		29	4.21		33.21	39.50

08 51 Metal Windows

08 51 23 – Steel Windows

08 51 23.10 Steel Sash		Crew	Daily Output	Labor-Hours	Unit	Material	2015 Bare Costs Labor	Equipment	Total	Total Incl O&P
1000	Projected, commercial, 40% vented	2 Sswk	200	.080	S.F.	51.50	4.21		55.71	64
1100	Intermediate, 50% vented		200	.080		58.50	4.21		62.71	71.50
1500	Industrial, horizontally pivoted		200	.080		53	4.21		57.21	65.50
1600	Fixed		200	.080		31	4.21		35.21	41.50
2000	Industrial security sash, 50% vented		200	.080		57.50	4.21		61.71	70.50
2100	Fixed		200	.080		47	4.21		51.21	59
2500	Picture window		200	.080		30	4.21		34.21	40.50
3000	Double hung		200	.080		59.50	4.21		63.71	73
5000	Mullions for above, open interior face		240	.067	L.F.	10.45	3.51		13.96	17.60
5100	With interior cover		240	.067	"	17.30	3.51		20.81	25

08 51 23.20 Steel Windows		Crew	Daily Output	Labor-Hours	Unit	Material	Labor	Equipment	Total	Total Incl O&P
0010	**STEEL WINDOWS** Stock, including frame, trim and insul. glass									
0020	See Section 13 34 19.50									
1000	Custom units, double hung, 2′-8″ x 4′-6″ opening R085123-10	2 Sswk	12	1.333	Ea.	705	70		775	900
1100	2′-4″ x 3′-9″ opening		12	1.333		585	70		655	760
1500	Commercial projected, 3′-9″ x 5′-5″ opening		10	1.600		1,225	84		1,309	1,500
1600	6′-9″ x 4′-1″ opening		7	2.286		1,625	120		1,745	1,975
2000	Intermediate projected, 2′-9″ x 4′-1″ opening		12	1.333		685	70		755	875
2100	4′-1″ x 5′-5″ opening		10	1.600		1,400	84		1,484	1,700

08 51 23.40 Basement Utility Windows		Crew	Daily Output	Labor-Hours	Unit	Material	Labor	Equipment	Total	Total Incl O&P
0010	**BASEMENT UTILITY WINDOWS**									
0015	1′-3″ x 2′-8″	1 Carp	16	.500	Ea.	128	23.50		151.50	176
1100	1′-7″ x 2′-8″	"	16	.500	"	140	23.50		163.50	190

08 51 66 – Metal Window Screens

08 51 66.10 Screens		Crew	Daily Output	Labor-Hours	Unit	Material	Labor	Equipment	Total	Total Incl O&P
0010	**SCREENS**									
0020	For metal sash, aluminum or bronze mesh, flat screen	2 Sswk	1200	.013	S.F.	4.30	.70		5	5.95
0500	Wicket screen, inside window		1000	.016		6.65	.84		7.49	8.75
0800	Security screen, aluminum frame with stainless steel cloth		1200	.013		24	.70		24.70	27
0900	Steel grate, painted, on steel frame		1600	.010		13.10	.53		13.63	15.30
1000	Screens for solar louvers		160	.100		24.50	5.25		29.75	36
4000	See Section 05 58 23.90									

08 52 Wood Windows

08 52 10 – Plain Wood Windows

08 52 10.20 Awning Window		Crew	Daily Output	Labor-Hours	Unit	Material	Labor	Equipment	Total	Total Incl O&P
0010	**AWNING WINDOW**, Including frame, screens and grilles									
0100	34″ x 22″, insulated glass	1 Carp	10	.800	Ea.	270	37.50		307.50	355
0200	Low E glass		10	.800		271	37.50		308.50	355
0300	40″ x 28″, insulated glass		9	.889		315	41.50		356.50	410
0400	Low E Glass		9	.889		340	41.50		381.50	440
0500	48″ x 36″, insulated glass		8	1		465	47		512	590
0600	Low E glass		8	1		490	47		537	615

08 52 10.40 Casement Window		Crew	Daily Output	Labor-Hours	Unit	Material	Labor	Equipment	Total	Total Incl O&P
0010	**CASEMENT WINDOW**, including frame, screen and grilles									
0100	2′-0″ x 3′-0″ H, dbl. insulated glass G	1 Carp	10	.800	Ea.	268	37.50		305.50	355
0150	Low E glass G		10	.800		259	37.50		296.50	345
0200	2′-0″ x 4′-6″ high, double insulated glass G		9	.889		360	41.50		401.50	460
0250	Low E glass G		9	.889		370	41.50		411.50	470
0260	Casement 4′-2″ x 4′-2″ double insulated glass G		11	.727		875	34		909	1,025

08 52 10 – Plain Wood Windows

08 52 10.40 Casement Window		Crew	Daily Output	Labor-Hours	Unit	Material	Labor	Equipment	Total	Total Incl O&P
0270	4'-0" x 4'-0" Low E glass G	1 Carp	11	.727	Ea.	535	34		569	645
0290	6'-4" x 5'-7" Low E glass G		9	.889		1,125	41.50		1,166.50	1,300
0300	2'-4" x 6'-0" high, double insulated glass G		8	1		440	47		487	560
0350	Low E glass G		8	1		475	47		522	600
0522	Vinyl clad, premium, double insulated glass, 2'-0" x 3'-0" G		10	.800		271	37.50		308.50	355
0524	2'-0" x 4'-0" G		9	.889		315	41.50		356.50	415
0525	2'-0" x 5'-0" G		8	1		360	47		407	475
0528	2'-0" x 6'-0" G		8	1		380	47		427	495
0600	3'-0" x 5'-0" G		8	1		665	47		712	805
0700	4'-0" x 3'-0" G		8	1		730	47		777	880
0710	4'-0" x 4'-0" G		8	1		625	47		672	760
0720	4'-8" x 4'-0" G		8	1		690	47		737	835
0730	4'-8" x 5'-0" G		6	1.333		790	62.50		852.50	960
0740	4'-8" x 6'-0" G		6	1.333		880	62.50		942.50	1,075
0750	6'-0" x 4'-0" G		6	1.333		805	62.50		867.50	980
0800	6'-0" x 5'-0" G		6	1.333		895	62.50		957.50	1,075
0900	5'-6" x 5'-6" G	2 Carp	15	1.067		1,425	50		1,475	1,650
2000	Bay, casement units, 8' x 5', w/screens, dbl. insul. glass		2.50	6.400	Opng.	1,600	300		1,900	2,200
2100	Low E glass		2.50	6.400	"	1,675	300		1,975	2,300
8190	For installation, add per leaf				Ea.		15%			
8200	For multiple leaf units, deduct for stationary sash									
8220	2' high				Ea.	23			23	25.50
8240	4'-6" high					26			26	28.50
8260	6' high					34.50			34.50	38

08 52 10.50 Double Hung		Crew	Daily Output	Labor-Hours	Unit	Material	Labor	Equipment	Total	Total Incl O&P
0010	**DOUBLE HUNG**, Including frame, screens and grilles R085216-10									
0100	2'-0" x 3'-0" high, low E insul. glass G	1 Carp	10	.800	Ea.	210	37.50		247.50	289
0200	3'-0" x 4'-0" high, double insulated glass G		9	.889		279	41.50		320.50	370
0300	4'-0" x 4'-6" high, low E insulated glass G		8	1		320	47		367	430

08 52 10.55 Picture Window		Crew	Daily Output	Labor-Hours	Unit	Material	Labor	Equipment	Total	Total Incl O&P
0010	**PICTURE WINDOW**, Including frame and grilles									
0100	3'-6" x 4'-0" high, dbl. insulated glass	2 Carp	12	1.333	Ea.	420	62.50		482.50	560
0150	Low E glass		12	1.333		435	62.50		497.50	570
0200	4'-0" x 4'-6" high, double insulated glass		11	1.455		550	68.50		618.50	710
0250	Low E glass		11	1.455		530	68.50		598.50	690
0300	5'-0" x 4'-0" high, double insulated glass		11	1.455		580	68.50		648.50	745
0350	Low E glass		11	1.455		605	68.50		673.50	770
0400	6'-0" x 4'-6" high, double insulated glass		10	1.600		625	75		700	805
0450	Low E glass		10	1.600		635	75		710	815

08 52 10.65 Wood Sash		Crew	Daily Output	Labor-Hours	Unit	Material	Labor	Equipment	Total	Total Incl O&P
0010	**WOOD SASH**, Including glazing but not trim									
0050	Custom, 5'-0" x 4'-0", 1" dbl. glazed, 3/16" thick lites	2 Carp	3.20	5	Ea.	230	235		465	615
0100	1/4" thick lites		5	3.200		245	150		395	500
0200	1" thick, triple glazed		5	3.200		415	150		565	685
0300	7'-0" x 4'-6" high, 1" double glazed, 3/16" thick lites		4.30	3.721		420	175		595	730
0400	1/4" thick lites		4.30	3.721		475	175		650	790
0500	1" thick, triple glazed		4.30	3.721		540	175		715	865
0600	8'-6" x 5'-0" high, 1" double glazed, 3/16" thick lites		3.50	4.571		565	215		780	955
0700	1/4" thick lites		3.50	4.571		620	215		835	1,000
0800	1" thick, triple glazed		3.50	4.571		625	215		840	1,025
0900	Window frames only, based on perimeter length				L.F.	4.02			4.02	4.42
1200	Window sill, stock, per lineal foot					8.50			8.50	9.35

Material, Labor, Equipment and Total columns: 2015 Bare Costs.

08 52 Wood Windows

08 52 10 – Plain Wood Windows

08 52 10.65 Wood Sash		Crew	Daily Output	Labor-Hours	Unit	Material	2015 Bare Costs Labor	Equipment	Total	Total Incl O&P
1250	Casing, stock				L.F.	3.30			3.30	3.63
08 52 10.70 Sliding Windows										
0010	**SLIDING WINDOWS**									
0100	3'-0" x 3'-0" high, double insulated G	1 Carp	10	.800	Ea.	284	37.50		321.50	370
0120	Low E glass G		10	.800		310	37.50		347.50	400
0200	4'-0" x 3'-6" high, double insulated G		9	.889		355	41.50		396.50	455
0220	Low E glass G		9	.889		360	41.50		401.50	465
0300	6'-0" x 5'-0" high, double insulated G		8	1		490	47		537	610
0320	Low E glass G		8	1		530	47		577	655

08 52 13 – Metal-Clad Wood Windows

08 52 13.10 Awning Windows, Metal-Clad		Crew	Daily Output	Labor-Hours	Unit	Material	Labor	Equipment	Total	Total Incl O&P
0010	**AWNING WINDOWS, METAL-CLAD**									
2000	Metal clad, awning deluxe, double insulated glass, 34" x 22"	1 Carp	9	.889	Ea.	247	41.50		288.50	335
2050	36" x 25"		9	.889		272	41.50		313.50	365
2100	40" x 22"		9	.889		291	41.50		332.50	385
2150	40" x 30"		9	.889		340	41.50		381.50	440
2200	48" x 28"		8	1		345	47		392	455
2250	60" x 36"		8	1		370	47		417	485
08 52 13.20 Casement Windows, Metal-Clad										
0010	**CASEMENT WINDOWS, METAL-CLAD**									
0100	Metal clad, deluxe, dbl. insul. glass, 2'-0" x 3'-0" high G	1 Carp	10	.800	Ea.	279	37.50		316.50	365
0120	2'-0" x 4'-0" high G		9	.889		315	41.50		356.50	410
0130	2'-0" x 5'-0" high G		8	1		325	47		372	435
0140	2'-0" x 6'-0" high G		8	1		365	47		412	475
0300	Metal clad, casement, bldrs mdl, 6'-0" x 4'-0", dbl. insltd gls, 3 panels	2 Carp	10	1.600		1,200	75		1,275	1,450
0310	9'-0" x 4'-0", 4 panels		8	2		1,550	94		1,644	1,850
0320	10'-0" x 5'-0", 5 panels		7	2.286		2,100	107		2,207	2,500
0330	12'-0" x 6'-0", 6 panels		6	2.667		2,700	125		2,825	3,150
08 52 13.30 Double-Hung Windows, Metal-clad										
0010	**DOUBLE-HUNG WINDOWS, METAL-CLAD**									
0100	Metal clad, deluxe, dbl. insul. glass, 2'-6" x 3'-0" high G	1 Carp	10	.800	Ea.	272	37.50		309.50	355
0120	3'-0" x 3'-6" high G		10	.800		315	37.50		352.50	405
0140	3'-0" x 4'-0" high G		9	.889		325	41.50		366.50	425
0160	3'-0" x 4'-6" high G		9	.889		345	41.50		386.50	445
0180	3'-0" x 5'-0" high G		8	1		370	47		417	485
0200	3'-6" x 6'-0" high G		8	1		450	47		497	570
08 52 13.35 Picture and Sliding Windows Metal-Clad										
0010	**PICTURE AND SLIDING WINDOWS METAL-CLAD**									
2000	Metal clad, dlx picture, dbl. insul. glass, 4'-0" x 4'-0" high	2 Carp	12	1.333	Ea.	375	62.50		437.50	505
2100	4'-0" x 6'-0" high		11	1.455		545	68.50		613.50	705
2200	5'-0" x 6'-0" high		10	1.600		610	75		685	785
2300	6'-0" x 6'-0" high		10	1.600		695	75		770	880
2400	Metal clad, dlx sliding, double insulated glass, 3'-0" x 3'-0" high G	1 Carp	10	.800		330	37.50		367.50	420
2420	4'-0" x 3'-6" high G		9	.889		400	41.50		441.50	505
2440	5'-0" x 4'-0" high G		9	.889		480	41.50		521.50	590
2460	6'-0" x 5'-0" high G		8	1		730	47		777	875
08 52 13.40 Bow and Bay Windows, Metal-Clad										
0010	**BOW AND BAY WINDOWS, METAL-CLAD**									
0100	Metal clad, deluxe, dbl. insul. glass, 8'-0" x 5'-0" high, 4 panels	2 Carp	10	1.600	Ea.	1,675	75		1,750	1,950
0120	10'-0" x 5'-0" high, 5 panels		8	2		1,800	94		1,894	2,125
0140	10'-0" x 6'-0" high, 5 panels		7	2.286		2,100	107		2,207	2,500

08 52 Wood Windows

08 52 13 – Metal-Clad Wood Windows

08 52 13.40 Bow and Bay Windows, Metal-Clad		Crew	Daily Output	Labor-Hours	Unit	Material	Labor	Equipment	Total	Total Incl O&P
						2015 Bare Costs				
0160	12'-0" x 6'-0" high, 6 panels	2 Carp	6	2.667	Ea.	2,925	125		3,050	3,425
0400	Double hung, bldrs. model, bay, 8' x 4' high, dbl. insulated glass		10	1.600		1,325	75		1,400	1,575
0440	Low E glass		10	1.600		1,425	75		1,500	1,700
0460	9'-0" x 5'-0" high, double insulated glass		6	2.667		1,425	125		1,550	1,775
0480	Low E glass		6	2.667		1,500	125		1,625	1,850
0500	Metal clad, deluxe, dbl. insul. glass, 7'-0" x 4'-0" high		10	1.600		1,275	75		1,350	1,525
0520	8'-0" x 4'-0" high		8	2		1,300	94		1,394	1,600
0540	8'-0" x 5'-0" high		7	2.286		1,350	107		1,457	1,675
0560	9'-0" x 5'-0" high	↓	6	2.667	↓	1,450	125		1,575	1,775

08 52 16 – Plastic-Clad Wood Windows

08 52 16.10 Bow Window		Crew	Daily Output	Labor-Hours	Unit	Material	Labor	Equipment	Total	Total Incl O&P
0010	**BOW WINDOW** Including frames, screens, and grilles									
0020	End panels operable									
1000	Bow type, casement, wood, bldrs. mdl., 8' x 5' dbl. insltd glass, 4 panel	2 Carp	10	1.600	Ea.	1,500	75		1,575	1,775
1050	Low E glass		10	1.600		1,325	75		1,400	1,575
1100	10'-0" x 5'-0" , double insulated glass, 6 panels		6	2.667		1,350	125		1,475	1,700
1200	Low E glass, 6 panels		6	2.667		1,450	125		1,575	1,800
1300	Vinyl clad, bldrs. model, double insulated glass, 6'-0" x 4'-0", 3 panel		10	1.600		1,025	75		1,100	1,250
1340	9'-0" x 4'-0", 4 panel		8	2		1,350	94		1,444	1,650
1380	10'-0" x 6'-0", 5 panels		7	2.286		2,250	107		2,357	2,650
1420	12'-0" x 6'-0", 6 panels		6	2.667		2,925	125		3,050	3,425
2000	Bay window, 8' x 5', dbl. insul glass		10	1.600		1,875	75		1,950	2,200
2050	Low E glass		10	1.600		2,275	75		2,350	2,625
2100	12'-0" x 6'-0" , double insulated glass, 6 panels		6	2.667		2,350	125		2,475	2,775
2200	Low E glass		6	2.667		2,375	125		2,500	2,825
2280	6'-0" x 4'-0"		11	1.455		1,250	68.50		1,318.50	1,475
2300	Vinyl clad, premium, double insulated glass, 8'-0" x 5'-0"		10	1.600		1,750	75		1,825	2,050
2340	10'-0" x 5'-0"		8	2		2,300	94		2,394	2,675
2380	10'-0" x 6'-0"		7	2.286		2,625	107		2,732	3,050
2420	12'-0" x 6'-0"		6	2.667		3,200	125		3,325	3,725
3300	Vinyl clad, premium, double insulated glass, 7'-0" x 4'-6"		10	1.600		1,375	75		1,450	1,625
3340	8'-0" x 4'-6"		8	2		1,400	94		1,494	1,675
3380	8'-0" x 5'-0"		7	2.286		1,450	107		1,557	1,775
3420	9'-0" x 5'-0"	↓	6	2.667	↓	1,500	125		1,625	1,850

08 52 16.15 Awning Window Vinyl-Clad		Crew	Daily Output	Labor-Hours	Unit	Material	Labor	Equipment	Total	Total Incl O&P
0010	**AWNING WINDOW VINYL-CLAD** Including frames, screens, and grilles									
0240	Vinyl clad, 34" x 22"	1 Carp	10	.800	Ea.	264	37.50		301.50	350
0280	36" x 28"		9	.889		305	41.50		346.50	400
0300	36" x 36"		9	.889		340	41.50		381.50	440
0340	40" x 22"		10	.800		288	37.50		325.50	375
0360	48" x 28"		8	1		370	47		417	480
0380	60" x 36"	↓	8	1	↓	500	47		547	625

08 52 16.30 Palladian Windows		Crew	Daily Output	Labor-Hours	Unit	Material	Labor	Equipment	Total	Total Incl O&P
0010	**PALLADIAN WINDOWS**									
0020	Vinyl clad, double insulated glass, including frame and grilles									
0040	3'-2" x 2'-6" high	2 Carp	11	1.455	Ea.	1,250	68.50		1,318.50	1,475
0060	3'-2" x 4'-10"		11	1.455		1,750	68.50		1,818.50	2,000
0080	3'-2" x 6'-4"		10	1.600		1,700	75		1,775	2,000
0100	4'-0" x 4'-0"	↓	10	1.600		1,525	75		1,600	1,800
0120	4'-0" x 5'-4"	3 Carp	10	2.400		1,875	113		1,988	2,225
0140	4'-0" x 6'-0"		9	2.667		1,950	125		2,075	2,350
0160	4'-0" x 7'-4"	↓	9	2.667	↓	2,125	125		2,250	2,525

08 52 Wood Windows

08 52 16 – Plastic-Clad Wood Windows

08 52 16.30 Palladian Windows

	08 52 16.30 Palladian Windows	Crew	Daily Output	Labor-Hours	Unit	Material (2015 Bare Costs)	Labor (2015 Bare Costs)	Equipment (2015 Bare Costs)	Total (2015 Bare Costs)	Total Incl O&P
0180	5′-5″ x 4′-10″	3 Carp	9	2.667	Ea.	2,275	125		2,400	2,700
0200	5′-5″ x 6′-10″		9	2.667		2,575	125		2,700	3,050
0220	5′-5″ x 7′-9″		9	2.667		2,800	125		2,925	3,275
0240	6′-0″ x 7′-11″		8	3		3,475	141		3,616	4,050
0260	8′-0″ x 6′-0″		8	3		3,075	141		3,216	3,600

08 52 16.35 Double-Hung Window

	08 52 16.35 Double-Hung Window	Crew	Daily Output	Labor-Hours	Unit	Material	Labor	Equipment	Total	Total Incl O&P
0010	**DOUBLE-HUNG WINDOW** Including frames, screens, and grilles									
0300	Vinyl clad, premium, double insulated glass, 2′-6″ x 3′-0″ G	1 Carp	10	.800	Ea.	310	37.50		347.50	400
0305	2′-6″ x 4′-0″ G		10	.800		360	37.50		397.50	455
0400	3′-0″ x 3′-6″ G		10	.800		335	37.50		372.50	430
0500	3′-0″ x 4′-0″ G		9	.889		395	41.50		436.50	500
0600	3′-0″ x 4′-6″ G		9	.889		410	41.50		451.50	515
0700	3′-0″ x 5′-0″ G		8	1		445	47		492	560
0790	3′-4″ x 5′-0″ G		8	1		455	47		502	575
0800	3′-6″ x 6′-0″ G		8	1		490	47		537	615
0820	4′-0″ x 5′-0″ G		7	1.143		560	53.50		613.50	700
0830	4′-0″ x 6′-0″ G		7	1.143		700	53.50		753.50	850

08 52 16.40 Transom Windows

	08 52 16.40 Transom Windows	Crew	Daily Output	Labor-Hours	Unit	Material	Labor	Equipment	Total	Total Incl O&P
0010	**TRANSOM WINDOWS**									
0050	Vinyl clad, premium, double insulated glass, 32″ x 8″	1 Carp	16	.500	Ea.	188	23.50		211.50	242
0100	36″ x 8″		16	.500		199	23.50		222.50	255
0110	36″ x 12″		16	.500		212	23.50		235.50	269
0200	44″ x 48″		12	.667		580	31.50		611.50	690
1000	Vinyl clad, premium, dbl. insul. glass, 4′-0″ x 4′-0″	2 Carp	12	1.333		510	62.50		572.50	655
1100	4′-0″ x 6′-0″		11	1.455		935	68.50		1,003.50	1,125
1200	5′-0″ x 6′-0″		10	1.600		1,050	75		1,125	1,275
1300	6′-0″ x 6′-0″		10	1.600		1,050	75		1,125	1,275

08 52 16.70 Vinyl Clad, Premium, Dbl. Insulated Glass

	08 52 16.70 Vinyl Clad, Premium, Dbl. Insulated Glass	Crew	Daily Output	Labor-Hours	Unit	Material	Labor	Equipment	Total	Total Incl O&P
0010	**VINYL CLAD, PREMIUM, DBL. INSULATED GLASS**									
1000	Sliding, 3′-0″ x 3′-0″ G	1 Carp	10	.800	Ea.	605	37.50		642.50	725
1050	4′-0″ x 3′-6″ G		9	.889		685	41.50		726.50	820
1100	5′-0″ x 4′-0″ G		9	.889		900	41.50		941.50	1,050
1150	6′-0″ x 5′-0″ G		8	1		1,125	47		1,172	1,325

08 52 50 – Window Accessories

08 52 50.10 Window Grille or Muntin

	08 52 50.10 Window Grille or Muntin	Crew	Daily Output	Labor-Hours	Unit	Material	Labor	Equipment	Total	Total Incl O&P
0010	**WINDOW GRILLE OR MUNTIN**, snap in type									
0020	Standard pattern interior grilles									
2000	Wood, awning window, glass size 28″ x 16″ high	1 Carp	30	.267	Ea.	28	12.50		40.50	50.50
2060	44″ x 24″ high		32	.250		40	11.75		51.75	62
2100	Casement, glass size, 20″ x 36″ high		30	.267		32	12.50		44.50	54.50
2180	20″ x 56″ high		32	.250		43	11.75		54.75	65.50
2200	Double hung, glass size, 16″ x 24″ high		24	.333	Set	51	15.65		66.65	80.50
2280	32″ x 32″ high		34	.235	″	131	11.05		142.05	161
2500	Picture, glass size, 48″ x 48″ high		30	.267	Ea.	120	12.50		132.50	151
2580	60″ x 68″ high		28	.286	″	183	13.40		196.40	223
2600	Sliding, glass size, 14″ x 36″ high		24	.333	Set	35.50	15.65		51.15	63
2680	36″ x 36″ high		22	.364	″	43.50	17.05		60.55	74

08 52 Wood Windows

08 52 66 – Wood Window Screens

08 52 66.10 Wood Screens		Crew	Daily Output	Labor-Hours	Unit	Material	Labor	Equipment	Total	Total Incl O&P
						2015 Bare Costs				
0010	**WOOD SCREENS**									
0020	Over 3 S.F., 3/4" frames	2 Carp	375	.043	S.F.	4.89	2		6.89	8.50
0100	1-1/8" frames	"	375	.043	"	8.20	2		10.20	12.10

08 52 69 – Wood Storm Windows

08 52 69.10 Storm Windows

Line	Description	Crew	Daily Output	Labor-Hours	Unit	Material	Labor	Equipment	Total	Total Incl O&P
0010	**STORM WINDOWS**, aluminum residential									
0300	Basement, mill finish, incl. fiberglass screen									
0320	1'-10" x 1'-0" high G	2 Carp	30	.533	Ea.	35	25		60	77
0340	2'-9" x 1'-6" high G		30	.533		38	25		63	80.50
0360	3'-4" x 2'-0" high G	↓	30	.533	↓	45	25		70	88
1600	Double-hung, combination, storm & screen									
2000	Clear anodic coating, 2'-0" x 3'-5" high G	2 Carp	30	.533	Ea.	95	25		120	144
2020	2'-6" x 5'-0" high G		28	.571		117	27		144	170
2040	4'-0" x 6'-0" high G		25	.640		130	30		160	189
2400	White painted, 2'-0" x 3'-5" high G		30	.533		90	25		115	138
2420	2'-6" x 5'-0" high G		28	.571		95	27		122	147
2440	4'-0" x 6'-0" high G		25	.640		110	30		140	167
2600	Mill finish, 2'-0" x 3'-5" high G		30	.533		85	25		110	132
2620	2'-6" x 5'-0" high G		28	.571		90	27		117	141
2640	4'-0" x 6'-8" high G	↓	25	.640	↓	110	30		140	167

08 53 Plastic Windows

08 53 13 – Vinyl Windows

08 53 13.20 Vinyl Single Hung Windows

Line	Description	Crew	Daily Output	Labor-Hours	Unit	Material	Labor	Equipment	Total	Total Incl O&P
0010	**VINYL SINGLE HUNG WINDOWS**, insulated glass									
0100	Grids, low E, J fin, ext. jambs, 21" x 53" G	2 Carp	18	.889	Ea.	198	41.50		239.50	282
0110	21" x 57" G		17	.941		200	44		244	288
0120	21" x 65" G		16	1		205	47		252	299
0130	25" x 41" G		20	.800		190	37.50		227.50	267
0140	25" x 49" G		18	.889		200	41.50		241.50	284
0150	25" x 57" G		17	.941		205	44		249	294
0160	25" x 65" G		16	1		240	47		287	335
0170	29" x 41" G		18	.889		195	41.50		236.50	279
0180	29" x 53" G		18	.889		205	41.50		246.50	290
0190	29" x 57" G		17	.941		210	44		254	299
0200	29" x 65" G		16	1		215	47		262	310
0210	33" x 41" G		20	.800		200	37.50		237.50	278
0220	33" x 53" G		18	.889		215	41.50		256.50	300
0230	33" x 57" G		17	.941		215	44		259	305
0240	33" x 65" G		16	1		220	47		267	315
0250	37" x 41" G		20	.800		225	37.50		262.50	305
0260	37" x 53" G		18	.889		235	41.50		276.50	325
0270	37" x 57" G		17	.941		240	44		284	330
0280	37" x 65" G	↓	16	1	↓	256	47		303	355

08 53 13.30 Vinyl Double Hung Windows

Line	Description	Crew	Daily Output	Labor-Hours	Unit	Material	Labor	Equipment	Total	Total Incl O&P
0010	**VINYL DOUBLE HUNG WINDOWS**, insulated glass									
0100	Grids, low E, J fin, ext. jambs, 21" x 53" G	2 Carp	18	.889	Ea.	215	41.50		256.50	300
0102	21" x 37" G		18	.889		223	41.50		264.50	310
0104	21" x 41" G		18	.889		232	41.50		273.50	320
0106	21" x 49" G	↓	18	.889	↓	247	41.50		288.50	335

08 53 13.30 Vinyl Double Hung Windows		Crew	Daily Output	Labor-Hours	Unit	2015 Bare Costs Material	Labor	Equipment	Total	Total Incl O&P
0110	21" x 57" G	2 Carp	17	.941	Ea.	263	44		307	355
0120	21" x 65" G		16	1		279	47		326	380
0128	25" x 37" G		20	.800		234	37.50		271.50	315
0130	25" x 41" G		20	.800		242	37.50		279.50	325
0140	25" x 49" G		18	.889		258	41.50		299.50	350
0145	25" x 53" G		18	.889		265	41.50		306.50	355
0150	25" x 57" G		17	.941		274	44		318	370
0160	25" x 65" G		16	1		290	47		337	395
0162	25" x 69" G		16	1		297	47		344	400
0164	25" x 77" G		16	1		325	47		372	430
0168	29" x 37" G		18	.889		242	41.50		283.50	330
0170	29" x 41" G		18	.889		250	41.50		291.50	340
0172	29" x 49" G		18	.889		267	41.50		308.50	360
0180	29" x 53" G		18	.889		271	41.50		312.50	360
0190	29" x 57" G		17	.941		280	44		324	380
0200	29" x 65" G		16	1		300	47		347	405
0202	29" x 69" G		16	1		310	47		357	415
0205	29" x 77" G		16	1		335	47		382	440
0208	33" x 37" G		20	.800		255	37.50		292.50	340
0210	33" x 41" G		20	.800		263	37.50		300.50	345
0215	33" x 49" G		20	.800		280	37.50		317.50	370
0220	33" x 53" G		18	.889		285	41.50		326.50	380
0230	33" x 57" G		17	.941		297	44		341	395
0240	33" x 65" G		16	1		315	47		362	420
0242	33" x 69" G		16	1		330	47		377	435
0246	33" x 77" G		16	1		350	47		397	460
0250	37" x 41" G		20	.800		289	37.50		326.50	380
0255	37" x 49" G		20	.800		282	37.50		319.50	370
0260	37" x 53" G		18	.889		315	41.50		356.50	410
0270	37" x 57" G		17	.941		325	44		369	430
0280	37" x 65" G		16	1		350	47		397	460
0282	37" x 69" G		16	1		360	47		407	470
0286	37" x 77" G		16	1		380	47		427	490
0300	Solid vinyl, average quality, double insulated glass, 2'-0" x 3'-0" G	1 Carp	10	.800		291	37.50		328.50	380
0310	3'-0" x 4'-0" G		9	.889		207	41.50		248.50	291
0320	4'-0" x 4'-6" G		8	1		335	47		382	440
0330	Premium, double insulated glass, 2'-6" x 3'-0" G		10	.800		271	37.50		308.50	355
0340	3'-0" x 3'-6" G		9	.889		292	41.50		333.50	385
0350	3'-0" x 4"-0" G		9	.889		315	41.50		356.50	415
0360	3'-0" x 4'-6" G		9	.889		330	41.50		371.50	430
0370	3'-0" x 5'-0" G		8	1		355	47		402	465
0380	3'-6" x 6"-0" G		8	1		380	47		427	495

08 53 13.40 Vinyl Casement Windows		Crew	Daily Output	Labor-Hours	Unit	Material	Labor	Equipment	Total	Total Incl O&P
0010	**VINYL CASEMENT WINDOWS**, insulated glass									
0015	Grids, low E, J fin, extension jambs, screens									
0100	One lite, 21" x 41" G	2 Carp	20	.800	Ea.	295	37.50		332.50	385
0110	21" x 47" G		20	.800		320	37.50		357.50	415
0120	21" x 53" G		20	.800		345	37.50		382.50	440
0128	24" x 35" G		19	.842		283	39.50		322.50	370
0130	24" x 41" G		19	.842		310	39.50		349.50	400
0140	24" x 47" G		19	.842		335	39.50		374.50	425
0150	24" x 53" G		19	.842		360	39.50		399.50	455

08 53 Plastic Windows

08 53 13 – Vinyl Windows

08 53 13.40 Vinyl Casement Windows			Crew	Daily Output	Labor-Hours	Unit	Material	2015 Bare Costs Labor	Equipment	Total	Total Incl O&P
0158	28″ x 35″	G	2 Carp	19	.842	Ea.	300	39.50		339.50	390
0160	28″ x 41″	G		19	.842		330	39.50		369.50	425
0170	28″ x 47″	G		19	.842		350	39.50		389.50	445
0180	28″ x 53″	G		19	.842		385	39.50		424.50	485
0184	28″ x 59″	G		19	.842		395	39.50		434.50	495
0188	Two lites, 33″ x 35″	G		18	.889		470	41.50		511.50	580
0190	33″ x 41″	G		18	.889		500	41.50		541.50	615
0200	33″ x 47″	G		18	.889		540	41.50		581.50	655
0210	33″ x 53″	G		18	.889		575	41.50		616.50	695
0212	33″ x 59″	G		18	.889		610	41.50		651.50	735
0215	33″ x 72″	G		18	.889		635	41.50		676.50	760
0220	41″ x 41″	G		18	.889		545	41.50		586.50	665
0230	41″ x 47″	G		18	.889		585	41.50		626.50	705
0240	41″ x 53″	G		17	.941		620	44		664	750
0242	41″ x 59″	G		17	.941		650	44		694	785
0246	41″ x 72″	G		17	.941		685	44		729	825
0250	47″ x 41″	G		17	.941		555	44		599	680
0260	47″ x 47″	G		17	.941		590	44		634	720
0270	47″ x 53″	G		17	.941		625	44		669	760
0272	47″ x 59″	G		17	.941		680	44		724	820
0280	56″ x 41″	G		15	1.067		595	50		645	730
0290	56″ x 47″	G		15	1.067		625	50		675	765
0300	56″ x 53″	G		15	1.067		680	50		730	825
0302	56″ x 59″	G		15	1.067		710	50		760	855
0310	56″ x 72″	G		15	1.067		770	50		820	925
0340	Solid vinyl, premium, double insulated glass, 2′-0″ x 3′-0″ high	G	1 Carp	10	.800		270	37.50		307.50	355
0360	2′-0″ x 4′-0″ high	G		9	.889		299	41.50		340.50	395
0380	2′-0″ x 5′-0″ high	G		8	1		335	47		382	445

08 53 13.50 Vinyl Picture Windows		Crew	Daily Output	Labor-Hours	Unit	Material	Labor	Equipment	Total	Total Incl O&P
0010	**VINYL PICTURE WINDOWS**, insulated glass									
0100	Grids, low E, J fin, ext. jambs, 33″ x 47″	2 Carp	12	1.333	Ea.	280	62.50		342.50	405
0110	35″ x 71″		12	1.333		375	62.50		437.50	510
0120	47″ x 35″		12	1.333		300	62.50		362.50	425
0130	47″ x 41″		12	1.333		390	62.50		452.50	525
0140	47″ x 47″		12	1.333		345	62.50		407.50	470
0150	47″ x 53″		11	1.455		370	68.50		438.50	510
0160	71″ x 35″		11	1.455		390	68.50		458.50	535
0170	71″ x 41″		11	1.455		410	68.50		478.50	555
0180	71″ x 47″		11	1.455		440	68.50		508.50	590

08 54 Composite Windows

08 54 13 – Fiberglass Windows

08 54 13.10 Fiberglass Single Hung Windows			Crew	Daily Output	Labor-Hours	Unit	Material	Labor	Equipment	Total	Total Incl O&P
0010	**FIBERGLASS SINGLE HUNG WINDOWS**										
0100	Grids, low E, 18″ x 24″	G	2 Carp	18	.889	Ea.	335	41.50		376.50	435
0110	18″ x 40″	G		17	.941		340	44		384	445
0130	24″ x 40″	G		20	.800		360	37.50		397.50	455
0230	36″ x 36″	G		17	.941		370	44		414	475
0250	36″ x 48″	G		20	.800		405	37.50		442.50	505
0260	36″ x 60″	G		18	.889		445	41.50		486.50	555
0280	36″ x 72″	G		16	1		470	47		517	590

08 54 Composite Windows

08 54 13 – Fiberglass Windows

08 54 13.10 Fiberglass Single Hung Windows			Crew	Daily Output	Labor-Hours	Unit	Material	2015 Bare Costs Labor	Equipment	Total	Total Incl O&P
0290	48″ x 40″	G	2 Carp	16	1	Ea.	470	47		517	590

08 56 Special Function Windows

08 56 63 – Detention Windows

08 56 63.13 Visitor Cubicle Windows

			Crew	Daily Output	Labor-Hours	Unit	Material	2015 Bare Costs Labor	Equipment	Total	Total Incl O&P
0010	**VISITOR CUBICLE WINDOWS**										
4000	Visitor cubicle, vision panel, no intercom		E-4	2	16	Ea.	3,200	850	73	4,123	5,075

08 62 Unit Skylights

08 62 13 – Domed Unit Skylights

08 62 13.20 Skylights

			Crew	Daily Output	Labor-Hours	Unit	Material	2015 Bare Costs Labor	Equipment	Total	Total Incl O&P
0010	**SKYLIGHTS**, flush or curb mounted										
2120	Ventilating insulated plexiglass dome with										
2130	curb mounting, 36″ x 36″	G	G-3	12	2.667	Ea.	480	125		605	720
2150	52″ x 52″	G		12	2.667		670	125		795	925
2160	28″ x 52″	G		10	3.200		490	150		640	770
2170	36″ x 52″	G		10	3.200		545	150		695	825
2180	For electric opening system, add	G					315			315	345
2300	Insulated safety glass with aluminum frame	G	G-3	160	.200	S.F.	94	9.35		103.35	117

08 63 Metal-Framed Skylights

08 63 13 – Domed Metal-Framed Skylights

08 63 13.20 Skylight Rigid Metal-Framed

		Crew	Daily Output	Labor-Hours	Unit	Material	2015 Bare Costs Labor	Equipment	Total	Total Incl O&P
0010	**SKYLIGHT RIGID METAL-FRAMED** Skylght framing is aluminum									
0050	Fixed acrylic double domes, curb mount, 25-1/2″ x 25-1/2″	G-3	10	3.200	Ea.	200	150		350	450
0060	25-1/2″ x 33-1/2″		10	3.200		230	150		380	480
0070	25-1/2″ x 49-1/2″		6	5.333		265	249		514	670
0080	33-1/2″ x 33-1/2″		8	4		290	187		477	605
0090	37-1/2″ x 25-1/2″		8	4		240	187		427	550
0100	37-1/2″ x 37-1/2″		8	4		235	187		422	545
0110	37-1/2″ x 49-1/2″		6	5.333		400	249		649	820
0120	49-1/2″ x 33-1/2″		6	5.333		355	249		604	770
0130	49-1/2″ x 49-1/2″		6	5.333		445	249		694	870
1000	Fixed tempered glass, curb mount, 17-1/2″ x 33-1/2″		10	3.200		170	150		320	415
1020	17-1/2″ x 49-1/2″		6	5.333		190	249		439	590
1030	25-1/2″ x 25-1/2″		10	3.200		170	150		320	415
1040	25-1/2″ x 33-1/2″		10	3.200		200	150		350	450
1050	25-1/2″ x 37-1/2″		8	4		210	187		397	520
1060	25-1/2″ x 49-1/2″		6	5.333		222	249		471	625
1070	25-1/2″ x 73-1/2″		6	5.333		375	249		624	790
1080	33-1/2″ x 33-1/2″		8	4		235	187		422	545
2000	Manual vent tempered glass & screen, curb, 25-1/2″ x 25-1/2″		10	3.200		410	150		560	680
2020	25-1/2″ x 37-1/2″		10	3.200		465	150		615	740
2030	25-1/2″ x 49-1/2″		8	4		505	187		692	840
2040	33-1/2″ x 33-1/2″		8	4		535	187		722	875
2050	33-1/2″ x 49-1/2″		6	5.333		675	249		924	1,125
2060	37-1/2″ x 37-1/2″		6	5.333		615	249		864	1,050

08 63 Metal-Framed Skylights

08 63 13 – Domed Metal-Framed Skylights

08 63 13.20 Skylight Rigid Metal-Framed		Crew	Daily Output	Labor-Hours	Unit	Material	2015 Bare Costs Labor	Equipment	Total	Total Incl O&P
2070	49-1/2" x 49-1/2"	G-3	6	5.333	Ea.	790	249		1,039	1,250
3000	Electric vent tempered glass , curb mount, 25-1/2" x 25-1/2"		10	3.200		960	150		1,110	1,275
3020	25-1/2" x 37-1/2"		10	3.200		1,050	150		1,200	1,375
3030	25-1/2" x 49-1/2"		8	4		1,125	187		1,312	1,500
3040	33-1/2" x 33-1/2"		8	4		1,125	187		1,312	1,525
3050	33-1/2" x 49-1/2"		6	5.333		1,225	249		1,474	1,700
3060	37-1/2" x 37-1/2"		6	5.333		1,200	249		1,449	1,675
3070	49-1/2" x 49-1/2"		6	5.333		1,325	249		1,574	1,825

08 71 Door Hardware

08 71 13 – Automatic Door Operators

08 71 13.10 Automatic Openers Commercial

Line	Description	Crew	Daily Output	Labor-Hours	Unit	Material	Labor	Equipment	Total	Total Incl O&P
0010	**AUTOMATIC OPENERS COMMERCIAL**									
0020	Pneumatic, incl opener, motion sens, control box, tubing, compressor									
0050	For single swing door, per opening	2 Skwk	.80	20	Ea.	4,550	975		5,525	6,500
0100	Pair, per opening		.50	32	Opng.	7,375	1,550		8,925	10,500
1000	For single sliding door, per opening		.60	26.667		4,975	1,300		6,275	7,475
1300	Bi-parting pair		.50	32		7,475	1,550		9,025	10,600
1420	Electronic door opener incl motion sens, 12 V control box, motor									
1450	For single swing door, per opening	2 Skwk	.80	20	Opng.	3,600	975		4,575	5,475
1500	Pair, per opening		.50	32		6,650	1,550		8,200	9,725
1600	For single sliding door, per opening		.60	26.667		4,400	1,300		5,700	6,850
1700	Bi-parting pair		.50	32		5,350	1,550		6,900	8,300
1750	Handicap actuator buttons, 2, including 12 V DC wiring, add	1 Carp	1.50	5.333	Pr.	470	250		720	900

08 71 13.20 Automatic Openers Industrial

Line	Description	Crew	Daily Output	Labor-Hours	Unit	Material	Labor	Equipment	Total	Total Incl O&P
0010	**AUTOMATIC OPENERS INDUSTRIAL**									
0015	Sliding doors up to 6' wide	2 Skwk	.60	26.667	Opng.	5,900	1,300		7,200	8,500
0200	To 12' wide	"	.40	40	"	7,075	1,950		9,025	10,800
0400	Over 12' wide, add per L.F. of excess				L.F.	800			800	880
1000	Swing doors, to 5' wide	2 Skwk	.80	20	Ea.	3,500	975		4,475	5,350
1860	Add for controls, wall pushbutton, 3 button		4	4		240	195		435	565
1870	Control pull cord		4.30	3.721		195	181		376	495

08 71 20 – Hardware

08 71 20.10 Bolts, Flush

Line	Description	Crew	Daily Output	Labor-Hours	Unit	Material	Labor	Equipment	Total	Total Incl O&P
0010	**BOLTS, FLUSH**									
0020	Standard, concealed	1 Carp	7	1.143	Ea.	23.50	53.50		77	108
0800	Automatic fire exit	"	5	1.600		278	75		353	420
1600	Electrified dead bolt	1 Elec	3	2.667		140	146		286	375
3000	Barrel, brass, 2" long	1 Carp	40	.200		7.80	9.40		17.20	23
3020	4" long		40	.200		13	9.40		22.40	29
3060	6" long		40	.200		26	9.40		35.40	43

08 71 20.15 Hardware

Line	Description	Crew	Daily Output	Labor-Hours	Unit	Material	Labor	Equipment	Total	Total Incl O&P
0010	**HARDWARE**									
0020	Average percentage for hardware, total job cost									
0500	Total hardware for building, average distribution				Job	85%	15%			
1000	Door hardware, apartment, interior	1 Carp	4	2	Door	455	94		549	645
1300	Average, door hardware, motel/hotel interior, with access card		4	2		580	94		674	785
1500	Hospital bedroom, average quality		4	2		640	94		734	850
2000	High quality		3	2.667		740	125		865	1,000
2100	Pocket door		6	1.333	Ea.	100	62.50		162.50	207

08 71 Door Hardware

08 71 20 – Hardware

08 71 20.15 Hardware		Crew	Daily Output	Labor-Hours	Unit	2015 Bare Costs Material	Labor	Equipment	Total	Total Incl O&P
2250	School, single exterior, incl. lever, incl. panic device	1 Carp	3	2.667	Door	1,300	125		1,425	1,650
2500	Single interior, regular use, lever included		3	2.667		635	125		760	895
2550	Avg., door hdwe., school, classroom, ANSI F84, lever handle		3	2.667		850	125		975	1,125
2600	Avg., door hdwe.set, school, classroom, ANSI F88, incl. lever		3	2.667		905	125		1,030	1,200
2850	Stairway, single interior		3	2.667		590	125		715	840
3100	Double exterior, with panic device		2	4	Pr.	2,425	188		2,613	2,975
6020	Add for fire alarm door holder, electro-magnetic	1 Elec	4	2	Ea.	103	109		212	277

08 71 20.20 Door Protectors		Crew	Daily Output	Labor-Hours	Unit	Material	Labor	Equipment	Total	Total Incl O&P
0010	**DOOR PROTECTORS**									
0020	1-3/4" x 3/4" U channel	2 Carp	80	.200	L.F.	28	9.40		37.40	45.50
0021	1-3/4" x 1-1/4" U channel		80	.200	"	30	9.40		39.40	47.50
1000	Tear drop, spring-stl, 8" high x 19" long		15	1.067	Ea.	120	50		170	209
1010	8" high x 32" long		15	1.067		180	50		230	275
1100	Tear drop, stainless stl., 8" high x 19" long		15	1.067		315	50		365	420
1200	8" high x 32" long		15	1.067		400	50		450	515

08 71 20.30 Door Closers		Crew	Daily Output	Labor-Hours	Unit	Material	Labor	Equipment	Total	Total Incl O&P
0010	**DOOR CLOSERS** Adjustible backcheck, multiple mounting									
0015	and rack and pinion									
0020	Standard Regular Arm	1 Carp	6	1.333	Ea.	192	62.50		254.50	310
0040	Hold open arm		6	1.333		200	62.50		262.50	315
0100	Fusible link		6.50	1.231		165	58		223	271
0210	Light duty, regular arm		6	1.333		109	62.50		171.50	217
0220	Parallel arm		6	1.333		133	62.50		195.50	243
0230	Hold open arm		6	1.333		117	62.50		179.50	226
0240	Fusible link arm		6	1.333		148	62.50		210.50	260
0250	Medium duty, regular arm		6	1.333		117	62.50		179.50	226
0500	Surface mount regular arm		6.50	1.231		153	58		211	257
0550	Fusible link		6.50	1.231		145	58		203	249
1520	Overhead concealed, all sizes, regular arm		5.50	1.455		200	68.50		268.50	325
1525	Concealed arm		5	1.600		315	75		390	460
1530	Concealed in door, all sizes, regular arm		5.50	1.455		325	68.50		393.50	465
1535	Concealed arm		5	1.600		255	75		330	395
1560	Floor concealed, all sizes, single acting		2.20	3.636		500	171		671	815
1565	Double acting		2.20	3.636		475	171		646	790
1610	Hold open arm		6	1.333		420	62.50		482.50	555
1620	Double acting, standard arm		6	1.333		760	62.50		822.50	930
1630	Hold open arm		6	1.333		765	62.50		827.50	935
1640	Floor, center hung, single acting, bottom arm		6	1.333		410	62.50		472.50	545
1650	Double acting		6	1.333		465	62.50		527.50	605
1660	Offset hung, single acting, bottom arm		6	1.333		550	62.50		612.50	700
2000	Backcheck and adjustable power, hinge face mount									
5000	For cast aluminum cylinder, deduct				Ea.	35			35	38.50
5010	For delayed action add	1 Carp	6	1.333		34	62.50		96.50	134
5040	For delayed action, add					46			46	50.50
5080	For fusible link arm, add					35			35	38.50
5120	For shock absorbing arm, add					50			50	55
5160	For spring power adjustment, add					40			40	44
6000	Closer-holder, hinge face mount, all sizes, exposed arm	1 Carp	6.50	1.231		205	58		263	315
6500	Electro magnetic closer/holder									
6510	Single point, no detector	1 Carp	4	2	Ea.	505	94		599	705
6515	Including detector		4	2		665	94		759	880
6520	Multi-point, no detector		4	2		885	94		979	1,125

08 71 20.30 Door Closers		Crew	Daily Output	Labor-Hours	Unit	Material	2015 Bare Costs Labor	Equipment	Total	Total Incl O&P
6524	Including detector	1 Carp	4	2	Ea.	1,325	94		1,419	1,600
6550	Electric automatic operators									
6555	Operator	1 Carp	4	2	Ea.	1,975	94		2,069	2,325
6570	Wall plate actuator		4	2		217	94		311	385
7000	Electronic closer-holder, hinge facemount, concealed arm		5	1.600		420	75		495	575
7400	With built-in detector		5	1.600		605	75		680	780
8000	Surface mounted, stand. duty, parallel arm, primed, traditional		6	1.333		199	62.50		261.50	315
8030	Light duty		6	1.333		143	62.50		205.50	255
8042	Extra duty parallel arm		6	1.333		117	62.50		179.50	226
8044	Hold open arm		6	1.333		156	62.50		218.50	269
8046	Positive stop arm		6	1.333		212	62.50		274.50	330
8050	Heavy duty		6	1.333		234	62.50		296.50	355
8052	Heavy duty, regular arm		6	1.333		215	62.50		277.50	335
8054	Top jamb mount		6	1.333		215	62.50		277.50	335
8056	Extra duty parallel arm		6	1.333		215	62.50		277.50	335
8058	Hold open arm		6	1.333		261	62.50		323.50	385
8060	Positive stop arm		6	1.333		238	62.50		300.50	360
8062	Fusible link arm		6	1.333		257	62.50		319.50	380
8080	Universal heavy duty, regular arm		6	1.333		225	62.50		287.50	345
8084	Parallel arm		6	1.333		225	62.50		287.50	345
8088	Extra duty, parallel arm		6	1.333		257	62.50		319.50	380
8090	Hold open arm		6	1.333		265	62.50		327.50	390
8094	Positive stop arm		6	1.333		275	62.50		337.50	395
8100	Standard duty, parallel arm, modern		6	1.333		243	62.50		305.50	365
8150	Heavy duty		6	1.333		274	62.50		336.50	395

08 71 20.35 Panic Devices		Crew	Daily Output	Labor-Hours	Unit	Material	Labor	Equipment	Total	Total Incl O&P
0010	**PANIC DEVICES**									
0015	For rim locks, single door exit only	1 Carp	6	1.333	Ea.	460	62.50		522.50	600
0020	Outside key and pull		5	1.600		570	75		645	745
0200	Bar and vertical rod, exit only		5	1.600		865	75		940	1,075
0210	Outside key and pull		4	2		975	94		1,069	1,225
0400	Bar and concealed rod		4	2		745	94		839	965
0600	Touch bar, exit only		6	1.333		555	62.50		617.50	705
0610	Outside key and pull		5	1.600		685	75		760	865
0700	Touch bar and vertical rod, exit only		5	1.600		845	75		920	1,050
0710	Outside key and pull		4	2		970	94		1,064	1,225
1000	Mortise, bar, exit only		4	2		725	94		819	945
1600	Touch bar, exit only		4	2		760	94		854	980
2000	Narrow stile, rim mounted, bar, exit only		6	1.333		610	62.50		672.50	765
2010	Outside key and pull		5	1.600		890	75		965	1,100
2200	Bar and vertical rod, exit only		5	1.600		920	75		995	1,125
2210	Outside key and pull		4	2		920	94		1,014	1,150
2400	Bar and concealed rod, exit only		3	2.667		1,075	125		1,200	1,375
3000	Mortise, bar, exit only		4	2		600	94		694	805
3600	Touch bar, exit only		4	2		770	94		864	990

08 71 20.40 Lockset		Crew	Daily Output	Labor-Hours	Unit	Material	Labor	Equipment	Total	Total Incl O&P
0010	**LOCKSET**, Standard duty									
0020	Non-keyed, passage, w/sect.trim	1 Carp	12	.667	Ea.	70	31.50		101.50	125
0100	Privacy		12	.667		75	31.50		106.50	131
0400	Keyed, single cylinder function		10	.800		108	37.50		145.50	176
0420	Hotel (see also Section 08 71 20.15)		8	1		200	47		247	293
0500	Lever handled, keyed, single cylinder function		10	.800		128	37.50		165.50	199

08 71 Door Hardware

08 71 20 – Hardware

08 71 20.40 Lockset

	08 71 20.40 Lockset	Crew	Daily Output	Labor-Hours	Unit	2015 Bare Costs Material	Labor	Equipment	Total	Total Incl O&P
1000	Heavy duty with sectional trim, non-keyed, passages	1 Carp	12	.667	Ea.	125	31.50		156.50	186
1100	Privacy		12	.667		155	31.50		186.50	219
1400	Keyed, single cylinder function		10	.800		185	37.50		222.50	262
1420	Hotel		8	1		500	47		547	625
1600	Communicating		10	.800		300	37.50		337.50	390
1690	For re-core cylinder, add					50			50	55
3800	Cipher lockset w/key pad (security item)	1 Carp	13	.615		920	29		949	1,075
3900	Cipher lockset with dial for swinging doors (security item)		13	.615		1,800	29		1,829	2,050
3920	with dial for swinging doors & drill resistant plate (security item)		12	.667		2,200	31.50		2,231.50	2,475
3950	Cipher lockset with dial for safe/vault door (security item)		12	.667		1,500	31.50		1,531.50	1,700
3980	Keyless, pushbutton type									
4000	Residential/light commercial, deadbolt, standard	1 Carp	9	.889	Ea.	140	41.50		181.50	218
4010	Heavy duty		9	.889		230	41.50		271.50	315
4020	Industrial, heavy duty, with deadbolt		9	.889		380	41.50		421.50	480
4030	Key override		9	.889		390	41.50		431.50	495
4040	Lever activated handle		9	.889		415	41.50		456.50	520
4050	Key override		9	.889		440	41.50		481.50	550
4060	Double sided pushbutton type		8	1		755	47		802	905
4070	Key override		8	1		790	47		837	945

08 71 20.41 Dead Locks

	08 71 20.41 Dead Locks	Crew	Daily Output	Labor-Hours	Unit	Material	Labor	Equipment	Total	Total Incl O&P
0010	**DEAD LOCKS**									
0011	Mortise heavy duty outside key (security item)	1 Carp	9	.889	Ea.	175	41.50		216.50	257
0020	Double cylinder		9	.889		175	41.50		216.50	257
0100	Medium duty, outside key		10	.800		110	37.50		147.50	179
0110	Double cylinder		10	.800		120	37.50		157.50	190
1000	Tubular, standard duty, outside key		10	.800		45	37.50		82.50	108
1010	Double cylinder		10	.800		60	37.50		97.50	124
1200	Night latch, outside key		10	.800		45	37.50		82.50	108

08 71 20.42 Mortise Locksets

	08 71 20.42 Mortise Locksets	Crew	Daily Output	Labor-Hours	Unit	Material	Labor	Equipment	Total	Total Incl O&P
0010	**MORTISE LOCKSETS**, Comm., wrought knobs & full escutcheon trim									
0015	Assumes mortise is cut									
0020	Non-keyed, passage, grade 3	1 Carp	9	.889	Ea.	149	41.50		190.50	228
0030	Grade 1		8	1		405	47		452	520
0040	Privacy set Grade 3		9	.889		169	41.50		210.50	250
0050	Grade 1		8	1		455	47		502	575
0100	Keyed, office/entrance/apartment, Grade 2		8	1		197	47		244	290
0110	Grade 1		7	1.143		515	53.50		568.50	655
0120	Single cylinder, typical, Grade 3		8	1		189	47		236	281
0130	Grade 1		7	1.143		500	53.50		553.50	635
0200	Hotel, room, Grade 3		7	1.143		190	53.50		243.50	292
0210	Grade 1 (see also Section 08 71 20.15)		6	1.333		510	62.50		572.50	655
0300	Double cylinder, Grade 3		8	1		225	47		272	320
0310	Grade 1		7	1.143		515	53.50		568.50	655
1000	Wrought knobs and sectional trim, non-keyed, passage, Grade 3		10	.800		130	37.50		167.50	201
1010	Grade 1		9	.889		405	41.50		446.50	510
1040	Privacy, Grade 3		10	.800		145	37.50		182.50	218
1050	Grade 1		9	.889		455	41.50		496.50	565
1100	Keyed, entrance, office/apartment, Grade 3		9	.889		220	41.50		261.50	305
1103	Install lockset		6.92	1.156		220	54.50		274.50	325
1110	Grade 1		8	1		520	47		567	645
1120	Single cylinder, Grade 3		9	.889		225	41.50		266.50	310
1130	Grade 1		8	1		500	47		547	625

08 71 Door Hardware

08 71 20 – Hardware

08 71 20.42 Mortise Locksets		Crew	Daily Output	Labor-Hours	Unit	2015 Bare Costs Material	Labor	Equipment	Total	Total Incl O&P
2000	Cast knobs and full escutcheon trim									
2010	Non-keyed, passage, Grade 3	1 Carp	9	.889	Ea.	275	41.50		316.50	370
2020	Grade 1		8	1		380	47		427	495
2040	Privacy, Grade 3		9	.889		320	41.50		361.50	415
2050	Grade 1		8	1		440	47		487	560
2120	Keyed, single cylinder, Grade 3		8	1		330	47		377	440
2123	Mortise lock		6.15	1.301		330	61		391	460
2130	Grade 1		7	1.143		525	53.50		578.50	665
3000	Cast knob and sectional trim, non-keyed, passage, Grade 3		10	.800		210	37.50		247.50	289
3010	Grade 1		10	.800		380	37.50		417.50	480
3040	Privacy, Grade 3		10	.800		225	37.50		262.50	305
3050	Grade 1		10	.800		440	37.50		477.50	545
3100	Keyed, office/entrance/apartment, Grade 3		9	.889		255	41.50		296.50	345
3110	Grade 1		9	.889		580	41.50		621.50	705
3120	Single cylinder, Grade 3		9	.889		260	41.50		301.50	350
3130	Grade 1		9	.889		525	41.50		566.50	645
3190	For re-core cylinder, add					75			75	82.50

08 71 20.45 Peepholes		Crew	Daily Output	Labor-Hours	Unit	Material	Labor	Equipment	Total	Total Incl O&P
0010	**PEEPHOLES**									
2010	Peephole	1 Carp	32	.250	Ea.	15.80	11.75		27.55	35.50
2020	Peephole, wide view	"	32	.250	"	16.50	11.75		28.25	36

08 71 20.50 Door Stops		Crew	Daily Output	Labor-Hours	Unit	Material	Labor	Equipment	Total	Total Incl O&P
0010	**DOOR STOPS**									
0020	Holder & bumper, floor or wall	1 Carp	32	.250	Ea.	34.50	11.75		46.25	55.50
1300	Wall bumper, 4" diameter, with rubber pad, aluminum		32	.250		11.70	11.75		23.45	31
1600	Door bumper, floor type, aluminum		32	.250		8.10	11.75		19.85	27
1900	Plunger type, door mounted		32	.250		28	11.75		39.75	48.50

08 71 20.55 Push-Pull Plates		Crew	Daily Output	Labor-Hours	Unit	Material	Labor	Equipment	Total	Total Incl O&P
0010	**PUSH-PULL PLATES**									
0090	Push plate, 0.050 thick, 3" x 12", aluminum	1 Carp	12	.667	Ea.	6.25	31.50		37.75	55
0100	4" x 16"		12	.667		12.80	31.50		44.30	62
0110	6" x 16"		12	.667		8.50	31.50		40	57.50
0120	8" x 16"		12	.667		9.90	31.50		41.40	59
0200	Push plate, 0.050 thick, 3" x 12", brass		12	.667		14	31.50		45.50	63.50
0210	4" x 16"		12	.667		17.50	31.50		49	67.50
0220	6" x 16"		12	.667		27	31.50		58.50	77.50
0230	8" x 16"		12	.667		35	31.50		66.50	86.50
0250	Push plate, 0.050 thick, 3" x 12", satin brass		12	.667		14.25	31.50		45.75	63.50
0260	4" x 16"		12	.667		17.70	31.50		49.20	67.50
0270	6" x 16"		12	.667		27	31.50		58.50	77.50
0280	8" x 16"		12	.667		35	31.50		66.50	86.50
0490	Push plate, 0.050 thick, 3" x 12", bronze		12	.667		17	31.50		48.50	66.50
0500	4" x 16"		12	.667		25	31.50		56.50	75.50
0510	6" x 16"		12	.667		32	31.50		63.50	83
0520	8" x 16"		12	.667		38.50	31.50		70	90.50
0740	Push plate, 0.050 thick, 3" x 12", stainless steel		12	.667		12	31.50		43.50	61
0760	6" x 16"		12	.667		18.65	31.50		50.15	68.50
0780	8" x 16"		12	.667		23.50	31.50		55	74
0790	Push plate, 0.050 thick, 3" x 12", satin stainless steel		12	.667		7	31.50		38.50	55.50
0820	6" x 16"		12	.667		11.70	31.50		43.20	61
0830	8" x 16"		12	.667		16	31.50		47.50	65.50
0980	Pull plate, 0.050 thick, 3" x 12", aluminum		12	.667		25.50	31.50		57	76

08 71 Door Hardware

08 71 20 – Hardware

08 71 20.55 Push-Pull Plates		Crew	Daily Output	Labor-Hours	Unit	Material	Labor	Equipment	Total	Total Incl O&P
						2015 Bare Costs				
1050	Pull plate, 0.050 thick, 3" x 12", brass	1 Carp	12	.667	Ea.	39.50	31.50		71	91.50
1080	Pull plate, 0.050 thick, 3" x 12", bronze		12	.667		50	31.50		81.50	103
1180	Pull plate, 0.050 thick, 3" x 12", stainless steel		12	.667		45.50	31.50		77	98
1250	Pull plate, 0.050 thick, 3" x 12", chrome		12	.667		44.50	31.50		76	97
1500	Pull handle and push bar, aluminum		11	.727		118	34		152	183
2000	Bronze		10	.800		159	37.50		196.50	233

08 71 20.60 Entrance Locks

Line	Description	Crew	Daily Output	Labor-Hours	Unit	Material	Labor	Equipment	Total	Total Incl O&P
0010	**ENTRANCE LOCKS**									
0015	Cylinder, grip handle deadlocking latch	1 Carp	9	.889	Ea.	175	41.50		216.50	257
0020	Deadbolt		8	1		175	47		222	266
0100	Push and pull plate, dead bolt		8	1		225	47		272	320
0900	For handicapped lever, add					150			150	165

08 71 20.65 Thresholds

Line	Description	Crew	Daily Output	Labor-Hours	Unit	Material	Labor	Equipment	Total	Total Incl O&P
0010	**THRESHOLDS**									
0011	Threshold 3' long saddles aluminum	1 Carp	48	.167	L.F.	9.40	7.85		17.25	22.50
0100	Aluminum, 8" wide, 1/2" thick		12	.667	Ea.	49	31.50		80.50	102
0500	Bronze		60	.133	L.F.	42	6.25		48.25	56
0600	Bronze, panic threshold, 5" wide, 1/2" thick		12	.667	Ea.	158	31.50		189.50	221
0700	Rubber, 1/2" thick, 5-1/2" wide		20	.400		40	18.80		58.80	73
0800	2-3/4" wide		20	.400		45	18.80		63.80	78.50
1950	ADA Compliant Thresholds									
2300	Threshold, aluminum 4" wide x 36" long	1 Carp	12	.667	Ea.	33	31.50		64.50	84.50
2310	4" wide x 48" long		12	.667		41	31.50		72.50	93
2360	6" wide x 36" long		12	.667		53	31.50		84.50	107
2370	6" wide x 48" long		12	.667		68	31.50		99.50	123

08 71 20.70 Floor Checks

Line	Description	Crew	Daily Output	Labor-Hours	Unit	Material	Labor	Equipment	Total	Total Incl O&P
0010	**FLOOR CHECKS**									
0020	For over 3' wide doors single acting	1 Carp	2.50	3.200	Ea.	745	150		895	1,050
0500	Double acting	"	2.50	3.200	"	860	150		1,010	1,175

08 71 20.75 Door Hardware Accessories

Line	Description	Crew	Daily Output	Labor-Hours	Unit	Material	Labor	Equipment	Total	Total Incl O&P
0010	**DOOR HARDWARE ACCESSORIES**									
0050	Door closing coordinator, 36" (for paired openings up to 56")	1 Carp	8	1	Ea.	98	47		145	181
0060	48" (for paired openings up to 84")		8	1		105	47		152	189
0070	56" (for paired openings up to 96")		8	1		116	47		163	200

08 71 20.80 Hasps

Line	Description	Crew	Daily Output	Labor-Hours	Unit	Material	Labor	Equipment	Total	Total Incl O&P
0010	**HASPS**, steel assembly									
0015	3"	1 Carp	26	.308	Ea.	4.90	14.45		19.35	27.50
0020	4-1/2"		13	.615		6.60	29		35.60	52
0040	6"		12.50	.640		9.35	30		39.35	56.50

08 71 20.90 Hinges

Line	Description	Crew	Daily Output	Labor-Hours	Unit	Material	Labor	Equipment	Total	Total Incl O&P
0010	**HINGES**									
0012	Full mortise, avg. freq., steel base, USP, 4-1/2" x 4-1/2"				Pr.	36			36	40
0100	5" x 5", USP					57.50			57.50	63
0200	6" x 6", USP					119			119	131
0400	Brass base, 4-1/2" x 4-1/2", US10					58.50			58.50	64.50
0500	5" x 5", US10					86			86	94.50
0600	6" x 6", US10					162			162	178
0800	Stainless steel base, 4-1/2" x 4-1/2", US32					74			74	81.50
0900	For non removable pin, add (security item)				Ea.	4.88			4.88	5.35
0910	For floating pin, driven tips, add					3.30			3.30	3.63
0930	For hospital type tip on pin, add					13.95			13.95	15.35

08 71 20 – Hardware

08 71 20.90 Hinges		Crew	Daily Output	Labor-Hours	Unit	Material	Labor	Equipment	Total	Total Incl O&P
						2015 Bare Costs				
0940	For steeple type tip on pin, add				Ea.	18.95			18.95	21
0950	Full mortise, high frequency, steel base, 3-1/2" x 3-1/2", US26D				Pr.	30			30	33
1000	4-1/2" x 4-1/2", USP					63			63	69
1100	5" x 5", USP					50			50	55
1200	6" x 6", USP					134			134	148
1400	Brass base, 3-1/2" x 3-1/2", US4					51			51	56
1430	4-1/2" x 4-1/2", US10					73.50			73.50	81
1500	5" x 5", US10					122			122	134
1600	6" x 6", US10					165			165	181
1800	Stainless steel base, 4-1/2" x 4-1/2", US32					103			103	114
1810	5" x 4-1/2", US32					137			137	151
1930	For hospital type tip on pin, add				Ea.	13.15			13.15	14.45
1950	Full mortise, low frequency, steel base, 3-1/2" x 3-1/2", US26D				Pr.	23.50			23.50	26
2000	4-1/2" x 4-1/2", USP					22			22	24.50
2100	5" x 5", USP					47.50			47.50	52
2200	6" x 6", USP					89			89	98
2300	4-1/2" x 4-1/2", US3					16.85			16.85	18.50
2310	5" x 5", US3					41			41	45
2400	Brass bass, 4-1/2" x 4-1/2", US10					53			53	58.50
2500	5" x 5", US10					76.50			76.50	84
2800	Stainless steel base, 4-1/2" x 4-1/2", US32					74.50			74.50	82

08 71 20.91 Special Hinges		Crew	Daily Output	Labor-Hours	Unit	Material	Labor	Equipment	Total	Total Incl O&P
0010	**SPECIAL HINGES**									
0015	Paumelle, high frequency									
0020	Steel base, 6" x 4-1/2", US10				Pr.	145			145	160
0100	Brass base, 5" x 4-1/2", US10				Ea.	248			248	273
0200	Paumelle, average frequency, steel base, 4-1/2" x 3-1/2", US10				Pr.	98.50			98.50	108
0400	Olive knuckle, low frequency, brass base, 6" x 4-1/2", US10				Ea.	144			144	159
1000	Electric hinge with concealed conductor, average frequency									
1010	Steel base, 4-1/2" x 4-1/2", US26D				Pr.	310			310	340
1100	Bronze base, 4-1/2" x 4-1/2", US26D				"	320			320	350
1200	Electric hinge with concealed conductor, high frequency									
1210	Steel base, 4-1/2" x 4-1/2", US26D				Pr.	262			262	288
1600	Double weight, 800 lb., steel base, removable pin, 5" x 6", USP					410			410	450
1700	Steel base-welded pin, 5" x 6", USP					166			166	183
1800	Triple weight, 2000 lb., steel base, welded pin, 5" x 6", USP					535			535	590
2000	Pivot reinf., high frequency, steel base, 7-3/4" door plate, USP					149			149	164
2200	Bronze base, 7-3/4" door plate, US10					232			232	255
3000	Swing clear, full mortise, full or half surface, high frequency,									
3010	Steel base, 5" high, USP				Pr.	143			143	157
3200	Swing clear, full mortise, average frequency									
3210	Steel base, 4-1/2" high, USP				Pr.	128			128	141
4000	Wide throw, average frequency, steel base, 4-1/2" x 6", USP					94.50			94.50	104
4200	High frequency, steel base, 4-1/2" x 6", USP					112			112	124
4600	Spring hinge, single acting, 6" flange, steel				Ea.	50			50	55
4700	Brass					94.50			94.50	104
4900	Double acting, 6" flange, steel					80			80	88
4950	Brass					133			133	146
8000	Continuous hinges									
8010	Steel, piano, 2" x 72"	1 Carp	20	.400	Ea.	22	18.80		40.80	53
8020	Brass, piano, 1-1/16" x 30"		30	.267		8	12.50		20.50	28
8030	Acrylic, piano, 1-3/4" x 12"		40	.200		15	9.40		24.40	31

08 71 20.91 Special Hinges		Crew	Daily Output	Labor-Hours	Unit	Material	Labor	Equipment	Total	Total Incl O&P
						2015 Bare Costs				
9000	Continuous hinge, steel, full mortise, heavy duty, 96 inch	1 Carp	2	4	Ea.	460	188		648	795

08 71 20.95 Kick Plates		Crew	Daily Output	Labor-Hours	Unit	Material	Labor	Equipment	Total	Total Incl O&P
0010	**KICK PLATES**									
0020	Stainless steel, .050, 16 ga., 8″ x 28″, US32	1 Carp	15	.533	Ea.	38	25		63	80.50
0030	8″ x 30″		15	.533		41	25		66	83.50
0040	8″ x 34″		15	.533		46	25		71	89
0050	10″ x 28″		15	.533		76	25		101	122
0060	10″ x 30″		15	.533		82	25		107	129
0070	10″ x 34″		15	.533		92	25		117	140
0080	Mop/Kick, 4″ x 28″		15	.533		34	25		59	76
0090	4″ x 30″		15	.533		36	25		61	78
0100	4″ x 34″		15	.533		41	25		66	83.50
0110	6″ x 28″		15	.533		43	25		68	86
0120	6″ x 30″		15	.533		47	25		72	90
0130	6″ x 34″		15	.533		53	25		78	97
0500	Bronze, .050″, 8″ x 28″		15	.533		66.50	25		91.50	112
0510	8″ x 30″		15	.533		65	25		90	110
0520	8″ x 34″		15	.533		73	25		98	119
0530	10″ x 28″		15	.533		75	25		100	121
0540	10″ x 30″		15	.533		80	25		105	127
0550	10″ x 34″		15	.533		91	25		116	139
0560	Mop/Kick, 4″ x 28″		15	.533		33	25		58	75
0570	4″ x 30″		15	.533		36	25		61	78
0580	4″ x 34″		15	.533		37	25		62	79
0590	6″ x 28″		15	.533		46	25		71	89
0600	6″ x 30″		15	.533		52	25		77	95.50
0610	6″ x 34″		15	.533		56	25		81	100
1000	Acrylic, .125″, 8″ x 26″		15	.533		28	25		53	69.50
1010	8″ x 36″		15	.533		38	25		63	80.50
1020	8″ x 42″		15	.533		45	25		70	88
1030	10″ x 26″		15	.533		35	25		60	77
1040	10″ x 36″		15	.533		48	25		73	91.50
1050	10″ x 42″		15	.533		68.50	25		93.50	114
1060	Mop/Kick, 4″ x 26″		15	.533		17	25		42	57
1070	4″ x 36″		15	.533		24	25		49	65
1080	4″ x 42″		15	.533		27	25		52	68
1090	6″ x 26″		15	.533		23	25		48	64
1100	6″ x 36″		15	.533		34	25		59	76
1110	6″ x 42″		15	.533		39	25		64	81.50
1220	Brass, .050″, 8″ x 26″		15	.533		56	25		81	100
1230	8″ x 36″		15	.533		75	25		100	121
1240	8″ x 42″		15	.533		86	25		111	133
1250	10″ x 26″		15	.533		71	25		96	117
1260	10″ x 36″		15	.533		91	25		116	139
1270	10″ x 42″		15	.533		105	25		130	155
1320	Mop/Kick, 4″ x 26″		15	.533		28	25		53	69.50
1330	4″ x 36″		15	.533		39	25		64	81.50
1340	4″ x 42″		15	.533		44	25		69	87
1350	6″ x 26″		15	.533		38	25		63	80.50
1360	6″ x 36″		15	.533		48	25		73	91.50
1370	6″ x 42″		15	.533		55	25		80	99
1800	Aluminum, .050″, 8″ x 26″		15	.533		35	25		60	77

08 71 Door Hardware

08 71 20 – Hardware

08 71 20.95 Kick Plates		Crew	Daily Output	Labor-Hours	Unit	Material	2015 Bare Costs Labor	Equipment	Total	Total Incl O&P
1810	8" x 36"	1 Carp	15	.533	Ea.	40	25		65	82.50
1820	8" x 42"		15	.533		47	25		72	90
1830	10" x 26"		15	.533		36	25		61	78
1840	10" x 36"		15	.533		50	25		75	93.50
1850	10" x 42"		15	.533		59	25		84	104
1860	Mop/Kick, 4" x 26"		15	.533		15	25		40	55
1870	4" x 36"		15	.533		20	25		45	60.50
1880	4" x 42"		15	.533		24	25		49	65
1890	6" x 26"		15	.533		22	25		47	62.50
1900	6" x 36"		15	.533		30	25		55	71.50
1910	6" x 42"		15	.533		35	25		60	77

08 71 21 – Astragals

08 71 21.10 Exterior Mouldings, Astragals		Crew	Daily Output	Labor-Hours	Unit	Material	Labor	Equipment	Total	Total Incl O&P
0010	**EXTERIOR MOULDINGS, ASTRAGALS**									
0400	One piece, overlapping cadmium plated steel, flat, 3/16" x 2"	1 Carp	90	.089	L.F.	4	4.17		8.17	10.80
0600	Prime coated steel, flat, 1/8" x 3"		90	.089		5.75	4.17		9.92	12.75
0800	Stainless steel, flat, 3/32" x 1-5/8"		90	.089		16	4.17		20.17	24
1000	Aluminum, flat, 1/8" x 2"		90	.089		4.10	4.17		8.27	10.90
1200	Nail on, "T" extrusion		120	.067		1.90	3.13		5.03	6.90
1300	Vinyl bulb insert		105	.076		2.50	3.58		6.08	8.25
1600	Screw on, "T" extrusion		90	.089		3.75	4.17		7.92	10.55
1700	Vinyl insert		75	.107		4.50	5		9.50	12.65
2000	"L" extrusion, neoprene bulbs		75	.107		4.10	5		9.10	12.20
2100	Neoprene sponge insert		75	.107		6.90	5		11.90	15.30
2200	Magnetic		75	.107		10.60	5		15.60	19.35
2400	Spring hinged security seal, with cam		75	.107		6.80	5		11.80	15.20
2600	Spring loaded locking bolt, vinyl insert		45	.178		9.20	8.35		17.55	23
2800	Neoprene sponge strip, "Z" shaped, aluminum		60	.133		8.30	6.25		14.55	18.80
2900	Solid neoprene strip, nail on aluminum strip		90	.089		4.05	4.17		8.22	10.85
3000	One piece stile protection									
3020	Neoprene fabric loop, nail on aluminum strips	1 Carp	60	.133	L.F.	1.10	6.25		7.35	10.85
3110	Flush mounted aluminum extrusion, 1/2" x 1-1/4"		60	.133		6.90	6.25		13.15	17.25
3140	3/4" x 1-3/8"		60	.133		4.10	6.25		10.35	14.15
3160	1-1/8" x 1-3/4"		60	.133		4.70	6.25		10.95	14.80
3300	Mortise, 9/16" x 3/4"		60	.133		4.10	6.25		10.35	14.15
3320	13/16" x 1-3/8"		60	.133		4.30	6.25		10.55	14.40
3600	Spring bronze strip, nail on type		105	.076		1.85	3.58		5.43	7.55
3620	Screw on, with retainer		75	.107		2.70	5		7.70	10.65
3800	Flexible stainless steel housing, pile insert, 1/2" door		105	.076		7.25	3.58		10.83	13.50
3820	3/4" door		105	.076		8.10	3.58		11.68	14.40
4000	Extruded aluminum retainer, flush mount, pile insert		105	.076		2.25	3.58		5.83	8
4080	Mortise, felt insert		90	.089		4.55	4.17		8.72	11.40
4160	Mortise with spring, pile insert		90	.089		3.40	4.17		7.57	10.15
4400	Rigid vinyl retainer, mortise, pile insert		105	.076		2.70	3.58		6.28	8.45
4600	Wool pile filler strip, aluminum backing		105	.076		2.70	3.58		6.28	8.45
5000	Two piece overlapping astragal, extruded aluminum retainer									
5010	Pile insert	1 Carp	60	.133	L.F.	3.35	6.25		9.60	13.35
5020	Vinyl bulb insert		60	.133		1.85	6.25		8.10	11.70
5040	Vinyl flap insert		60	.133		3.65	6.25		9.90	13.65
5060	Solid neoprene flap insert		60	.133		6.55	6.25		12.80	16.85
5080	Hypalon rubber flap insert		60	.133		6.65	6.25		12.90	16.95
5090	Snap on cover, pile insert		60	.133		9.45	6.25		15.70	20

08 71 Door Hardware

08 71 21 – Astragals

08 71 21.10 Exterior Mouldings, Astragals		Crew	Daily Output	Labor-Hours	Unit	Material	2015 Bare Costs Labor	Equipment	Total	Total Incl O&P
5400	Magnetic aluminum, surface mounted	1 Carp	60	.133	L.F.	23	6.25		29.25	34.50
5500	Interlocking aluminum, 5/8" x 1" neoprene bulb insert		45	.178		5.70	8.35		14.05	19.10
5600	Adjustable aluminum, 9/16" x 21/32", pile insert		45	.178		17.55	8.35		25.90	32
5800	Magnetic, adjustable, 9/16" x 21/32"	↓	45	.178	↓	23	8.35		31.35	38
6000	Two piece stile protection									
6010	Cloth backed rubber loop, 1" gap, nail on aluminum strips	1 Carp	45	.178	L.F.	4.35	8.35		12.70	17.65
6040	Screw on aluminum strips		45	.178		6.55	8.35		14.90	20
6100	1-1/2" gap, screw on aluminum extrusion		45	.178		5.85	8.35		14.20	19.30
6240	Vinyl fabric loop, slotted aluminum extrusion, 1" gap		45	.178		2.20	8.35		10.55	15.25
6300	1-1/4" gap	↓	45	.178	↓	6.20	8.35		14.55	19.65

08 71 25 – Weatherstripping

08 71 25.10 Mechanical Seals, Weatherstripping

		Crew	Daily Output	Labor-Hours	Unit	Material	Labor	Equipment	Total	Total Incl O&P
0010	**MECHANICAL SEALS, WEATHERSTRIPPING**									
1000	Doors, wood frame, interlocking, for 3' x 7' door, zinc	1 Carp	3	2.667	Opng.	44	125		169	242
1100	Bronze		3	2.667		56	125		181	255
1300	6' x 7' opening, zinc		2	4		54	188		242	350
1400	Bronze	↓	2	4	↓	65	188		253	360
1700	Wood frame, spring type, bronze									
1800	3' x 7' door	1 Carp	7.60	1.053	Opng.	23	49.50		72.50	102
1900	6' x 7' door	"	7	1.143	"	29.50	53.50		83	115
2200	Metal frame, spring type, bronze									
2300	3' x 7' door	1 Carp	3	2.667	Opng.	46.50	125		171.50	244
2400	6' x 7' door	"	2.50	3.200	"	52	150		202	288
2500	For stainless steel, spring type, add					133%				
2700	Metal frame, extruded sections, 3' x 7' door, aluminum	1 Carp	3	2.667	Opng.	28	125		153	224
2800	Bronze		3	2.667		82	125		207	283
3100	6' x 7' door, aluminum		1.50	5.333		35	250		285	425
3200	Bronze	↓	1.50	5.333	↓	137	250		387	535
3500	Threshold weatherstripping									
3650	Door sweep, flush mounted, aluminum	1 Carp	25	.320	Ea.	19	15		34	44
3700	Vinyl		25	.320		18	15		33	43
5000	Garage door bottom weatherstrip, 12' aluminum, clear		14	.571		25	27		52	69
5010	Bronze		14	.571		90	27		117	141
5050	Bottom protection, Rubber		14	.571		37	27		64	82
5100	Threshold	↓	14	.571	↓	72	27		99	121

08 74 Access Control Hardware

08 74 13 – Card Key Access Control Hardware

08 74 13.50 Card Key Access

		Crew	Daily Output	Labor-Hours	Unit	Material	Labor	Equipment	Total	Total Incl O&P
0010	**CARD KEY ACCESS**									
0020	Computerized system , processor, proximity reader and cards									
0030	Does not inculde door hardware, lockset or wiring									
0040	Card key system for 1 door				Ea.	1,225			1,225	1,350
0060	Card key system for 2 doors					2,125			2,125	2,350
0080	Card key system for 4 doors					2,650			2,650	2,900
0100	Processor for card key access system					850			850	935
0160	Magnetic lock for electric access, 600 Pound holding force					190			190	209
0170	Magnetic lock for electric access, 1200 Pound holding force					190			190	209
0200	Proximity card reader				↓	130			130	143

08 74 Access Control Hardware

08 74 19 – Biometric Identity Access Control Hardware

08 74 19.50 Biometric Identity Access		Crew	Daily Output	Labor-Hours	Unit	2015 Bare Costs Material	Labor	Equipment	Total	Total Incl O&P
0010	**BIOMETRIC IDENTITY ACCESS**									
0220	Hand geometry scanner, mem of 512 users, excl striker/power	1 Elec	3	2.667	Ea.	2,100	146		2,246	2,525
0230	Memory upgrade for, adds 9,700 user profiles		8	1		300	54.50		354.50	410
0240	Adds 32,500 user profiles		8	1		600	54.50		654.50	740
0250	Prison type, memory of 256 users, excl striker, power		3	2.667		2,600	146		2,746	3,075
0260	Memory upgrade for, adds 3,300 user profiles		8	1		250	54.50		304.50	355
0270	Adds 9,700 user profiles		8	1		460	54.50		514.50	585
0280	Adds 27,900 user profiles		8	1		610	54.50		664.50	750
0290	All weather, mem of 512 users, excl striker/power		3	2.667		3,900	146		4,046	4,525
0300	Facial & fingerprint scanner, combination unit, excl striker/power		3	2.667		4,300	146		4,446	4,950
0310	Access for, for initial setup, excl striker/power	↓	3	2.667	↓	1,100	146		1,246	1,425

08 75 Window Hardware

08 75 30 – Weatherstripping

08 75 30.10 Mechanical Weather Seals		Crew	Daily Output	Labor-Hours	Unit	Material	Labor	Equipment	Total	Total Incl O&P
0010	**MECHANICAL WEATHER SEALS**, Window, double hung, 3′ X 5′									
0020	Zinc	1 Carp	7.20	1.111	Opng.	20	52		72	103
0100	Bronze		7.20	1.111		40	52		92	125
0500	As above but heavy duty, zinc		4.60	1.739		20	81.50		101.50	148
0600	Bronze	↓	4.60	1.739	↓	70	81.50		151.50	203

08 79 Hardware Accessories

08 79 13 – Key Storage Equipment

08 79 13.10 Key Cabinets		Crew	Daily Output	Labor-Hours	Unit	Material	Labor	Equipment	Total	Total Incl O&P
0010	**KEY CABINETS**									
0020	Wall mounted, 60 key capacity	1 Carp	20	.400	Ea.	93.50	18.80		112.30	132
0200	Drawer type, 600 key capacity	1 Clab	15	.533		800	20		820	910
0300	2,400 key capacity		20	.400		4,200	15.05		4,215.05	4,650
0400	Tray type, 20 key capacity		50	.160		67.50	6		73.50	84
0500	50 key capacity	↓	40	.200	↓	105	7.50		112.50	128

08 79 20 – Door Accessories

08 79 20.10 Door Hardware Accessories		Crew	Daily Output	Labor-Hours	Unit	Material	Labor	Equipment	Total	Total Incl O&P
0010	**DOOR HARDWARE ACCESSORIES**									
0140	Door bolt, surface, 4″	1 Carp	32	.250	Ea.	11.70	11.75		23.45	31
0160	Door latch	″	12	.667	″	8.55	31.50		40.05	57.50
0200	Sliding closet door									
0220	Track and hanger, single	1 Carp	10	.800	Ea.	58.50	37.50		96	123
0240	Double		8	1		80	47		127	161
0260	Door guide, single		48	.167		30	7.85		37.85	45
0280	Double		48	.167		40	7.85		47.85	56
0600	Deadbolt and lock cover plate, brass or stainless steel		30	.267		28	12.50		40.50	50.50
0620	Hole cover plate, brass or chrome		35	.229		8	10.75		18.75	25.50
2240	Mortise lockset, passage, lever handle		9	.889		160	41.50		201.50	240
4000	Security chain, standard	↓	18	.444	↓	10	21		31	43

08 81 Glass Glazing

08 81 10 – Float Glass

08 81 10.10 Various Types and Thickness of Float Glass		Crew	Daily Output	Labor-Hours	Unit	Material	2015 Bare Costs Labor	Equipment	Total	Total Incl O&P
0010	VARIOUS TYPES AND THICKNESS OF FLOAT GLASS R088110-10									
0020	3/16" Plain	2 Glaz	130	.123	S.F.	5.05	5.55		10.60	14
0200	Tempered, clear		130	.123		6.95	5.55		12.50	16.10
0300	Tinted		130	.123		8	5.55		13.55	17.25
0600	1/4" thick, clear, plain		120	.133		5.95	6		11.95	15.70
0700	Tinted		120	.133		9.10	6		15.10	19.15
0800	Tempered, clear		120	.133		8.85	6		14.85	18.90
0900	Tinted		120	.133		10.90	6		16.90	21
1600	3/8" thick, clear, plain		75	.213		10.20	9.60		19.80	26
1700	Tinted		75	.213		15.70	9.60		25.30	32
1800	Tempered, clear		75	.213		16.75	9.60		26.35	33
1900	Tinted		75	.213		18.85	9.60		28.45	35
2200	1/2" thick, clear, plain		55	.291		17.30	13.10		30.40	39
2300	Tinted		55	.291		27.50	13.10		40.60	50
2400	Tempered, clear		55	.291		25	13.10		38.10	47.50
2500	Tinted		55	.291		26	13.10		39.10	48.50
2800	5/8" thick, clear, plain		45	.356		27.50	16.05		43.55	54.50
2900	Tempered, clear		45	.356		31.50	16.05		47.55	59
3200	3/4" thick, clear, plain		35	.457		35.50	20.50		56	70.50
3300	Tempered, clear		35	.457		41	20.50		61.50	77
3600	1" thick, clear, plain	↓	30	.533		59	24		83	101
8900	For low emissivity coating for 3/16" & 1/4" only, add to above				↓	18%				

08 81 13 – Decorative Glass Glazing

08 81 13.10 Beveled Glass		Crew	Daily Output	Labor-Hours	Unit	Material	Labor	Equipment	Total	Total Incl O&P
0010	**BEVELED GLASS**, with design patterns									
0020	Simple pattern	2 Glaz	150	.107	S.F.	60.50	4.81		65.31	74
0050	Intricate pattern	"	125	.128	"	133	5.75		138.75	155

08 81 13.30 Sandblasted Glass		Crew	Daily Output	Labor-Hours	Unit	Material	Labor	Equipment	Total	Total Incl O&P
0010	**SANDBLASTED GLASS**, float glass									
0020	1/8" thick	2 Glaz	160	.100	S.F.	11	4.51		15.51	18.95
0100	3/16" thick		130	.123		12.15	5.55		17.70	22
0500	1/4" thick		120	.133		12.65	6		18.65	23
0600	3/8" thick	↓	75	.213	↓	13.60	9.60		23.20	29.50

08 81 17 – Fire Glass

08 81 17.10 Fire Resistant Glass		Crew	Daily Output	Labor-Hours	Unit	Material	Labor	Equipment	Total	Total Incl O&P
0010	**FIRE RESISTANT GLASS**									
0020	Fire Glass Minimum	2 Glaz	40	.400	S.F.	35	18.05		53.05	66
0030	Mid Range		40	.400		76	18.05		94.05	111
0050	High End	↓	40	.400	↓	340	18.05		358.05	405

08 81 20 – Vision Panels

08 81 20.10 Full Vision		Crew	Daily Output	Labor-Hours	Unit	Material	Labor	Equipment	Total	Total Incl O&P
0010	**FULL VISION**, window system with 3/4" glass mullions									
0020	Up to 10' high	H-2	130	.185	S.F.	63.50	7.85		71.35	81.50
0100	10' to 20' high, minimum		110	.218		67.50	9.30		76.80	88.50
0150	Average		100	.240		73	10.20		83.20	95.50
0200	Maximum	↓	80	.300	↓	81.50	12.80		94.30	110

08 81 Glass Glazing

08 81 25 – Glazing Variables

08 81 25.10 Applications of Glazing		Crew	Daily Output	Labor-Hours	Unit	2015 Bare Costs Material	Labor	Equipment	Total	Total Incl O&P
0010	**APPLICATIONS OF GLAZING** R088110-10									
0600	For glass replacement, add				S.F.		100%			
0700	For gasket settings, add				L.F.	5.75			5.75	6.35
0900	For sloped glazing, add				S.F.		26%			
2000	Fabrication, polished edges, 1/4" thick				Inch	.55			.55	.61
2100	1/2" thick					1.30			1.30	1.43
2500	Mitered edges, 1/4" thick					1.30			1.30	1.43
2600	1/2" thick				↓	2.15			2.15	2.37

08 81 30 – Insulating Glass

08 81 30.10 Reduce Heat Transfer Glass		Crew	Daily Output	Labor-Hours	Unit	Material	Labor	Equipment	Total	Total Incl O&P
0010	**REDUCE HEAT TRANSFER GLASS** R088110-10									
0015	2 lites 1/8" float, 1/2" thk under 15 S.F.									
0020	Clear G	2 Glaz	95	.168	S.F.	10.05	7.60		17.65	22.50
0100	Tinted G		95	.168		14.05	7.60		21.65	27
0200	2 lites 3/16" float, for 5/8" thk unit, 15 to 30 S.F., clear G		90	.178		13.70	8		21.70	27.50
0300	Tinted G		90	.178		13.75	8		21.75	27.50
0400	1" thk, dbl. glazed, 1/4" float, 30-70 S.F., clear G		75	.213		16.80	9.60		26.40	33
0500	Tinted G		75	.213		23.50	9.60		33.10	40.50
0600	1" thick double glazed, 1/4" float, 1/4" wire		75	.213		23.50	9.60		33.10	40.50
0700	1/4" float, 1/4" tempered		75	.213		31	9.60		40.60	48.50
0800	1/4" wire, 1/4" tempered		75	.213		29.50	9.60		39.10	47
2000	Both lites, light & heat reflective G		85	.188		31.50	8.50		40	47.50
2500	Heat reflective, film inside, 1" thick unit, clear G		85	.188		27.50	8.50		36	43.50
2600	Tinted G		85	.188		28.50	8.50		37	44.50
3000	Film on weatherside, clear, 1/2" thick unit G		95	.168		19.60	7.60		27.20	33
3100	5/8" thick unit G		90	.178		20	8		28	34
3200	1" thick unit G	↓	85	.188	↓	27	8.50		35.50	43

08 81 35 – Translucent Glass

08 81 35.10 Obscure Glass		Crew	Daily Output	Labor-Hours	Unit	Material	Labor	Equipment	Total	Total Incl O&P
0010	**OBSCURE GLASS**									
0020	1/8" thick, textured	2 Glaz	140	.114	S.F.	11.60	5.15		16.75	20.50
0100	Color		125	.128		13.70	5.75		19.45	24
0300	7/32" thick, textured		120	.133		12.70	6		18.70	23
0400	Color	↓	105	.152	↓	15.95	6.85		22.80	28

08 81 35.20 Patterned Glass		Crew	Daily Output	Labor-Hours	Unit	Material	Labor	Equipment	Total	Total Incl O&P
0010	**PATTERNED GLASS**, colored									
0020	1/8" thick	2 Glaz	140	.114	S.F.	9.35	5.15		14.50	18.15
0300	7/32" thick	"	120	.133	"	11.70	6		17.70	22

08 81 45 – Sheet Glass

08 81 45.10 Window Glass, Sheet		Crew	Daily Output	Labor-Hours	Unit	Material	Labor	Equipment	Total	Total Incl O&P
0010	**WINDOW GLASS, SHEET** gray									
0020	1/8" thick	2 Glaz	160	.100	S.F.	5.80	4.51		10.31	13.25
0200	1/4" thick	"	130	.123	"	7.10	5.55		12.65	16.25

08 81 50 – Spandrel Glass

08 81 50.10 Glass for Non Vision Areas		Crew	Daily Output	Labor-Hours	Unit	Material	Labor	Equipment	Total	Total Incl O&P
0010	**GLASS FOR NON VISION AREAS**, 1/4" thick standard colors									
0020	Up to 1000 S.F.	2 Glaz	110	.145	S.F.	16.95	6.55		23.50	28.50
0200	1,000 to 2,000 S.F.	"	120	.133	"	15.70	6		21.70	26.50
0300	For custom colors, add				Total	10%				
0500	For 3/8" thick, add				S.F.	11.90			11.90	13.10

08 81 Glass Glazing

08 81 50 – Spandrel Glass

08 81 50.10 Glass for Non Vision Areas		Crew	Daily Output	Labor-Hours	Unit	Material	2015 Bare Costs Labor	Equipment	Total	Total Incl O&P
1000	For double coated, 1/4" thick, add				S.F.	4.25			4.25	4.68
1200	For insulation on panels, add					6.95			6.95	7.65
2000	Panels, insulated, with aluminum backed fiberglass, 1" thick	2 Glaz	120	.133		16.85	6		22.85	27.50
2100	2" thick	"	120	.133	↓	20	6		26	31

08 81 55 – Window Glass

08 81 55.10 Sheet Glass		Crew	Daily Output	Labor-Hours	Unit	Material	Labor	Equipment	Total	Total Incl O&P
0010	**SHEET GLASS** (window), clear float, stops, putty bed									
0015	1/8" thick, clear float	2 Glaz	480	.033	S.F.	3.65	1.50		5.15	6.30
0500	3/16" thick, clear		480	.033		5.90	1.50		7.40	8.75
0600	Tinted		480	.033		7.45	1.50		8.95	10.50
0700	Tempered	↓	480	.033	↓	9.20	1.50		10.70	12.40

08 81 65 – Wire Glass

08 81 65.10 Glass Reinforced With Wire		Crew	Daily Output	Labor-Hours	Unit	Material	Labor	Equipment	Total	Total Incl O&P
0010	**GLASS REINFORCED WITH WIRE**									
0012	1/4" thick rough obscure	2 Glaz	135	.119	S.F.	23.50	5.35		28.85	34
1000	Polished wire, 1/4" thick, diamond, clear		135	.119		28	5.35		33.35	38.50
1500	Pinstripe, obscure	↓	135	.119	↓	41	5.35		46.35	53

08 83 Mirrors

08 83 13 – Mirrored Glass Glazing

08 83 13.10 Mirrors		Crew	Daily Output	Labor-Hours	Unit	Material	Labor	Equipment	Total	Total Incl O&P
0010	**MIRRORS**, No frames, wall type, 1/4" plate glass, polished edge									
0100	Up to 5 S.F.	2 Glaz	125	.128	S.F.	9.50	5.75		15.25	19.25
0200	Over 5 S.F.		160	.100		9.25	4.51		13.76	17.05
0500	Door type, 1/4" plate glass, up to 12 S.F.		160	.100		8.70	4.51		13.21	16.45
1000	Float glass, up to 10 S.F., 1/8" thick		160	.100		5.90	4.51		10.41	13.35
1100	3/16" thick		150	.107		7.30	4.81		12.11	15.40
1500	12" x 12" wall tiles, square edge, clear		195	.082		2.22	3.70		5.92	8.10
1600	Veined		195	.082		5.85	3.70		9.55	12.05
2000	1/4" thick, stock sizes, one way transparent		125	.128		19.60	5.75		25.35	30.50
2010	Bathroom, unframed, laminated	↓	160	.100	↓	13.90	4.51		18.41	22

08 83 13.15 Reflective Glass		Crew	Daily Output	Labor-Hours	Unit	Material	Labor	Equipment	Total	Total Incl O&P
0010	**REFLECTIVE GLASS**									
0100	1/4" float with fused metallic oxide fixed G	2 Glaz	115	.139	S.F.	16.95	6.25		23.20	28
0500	1/4" float glass with reflective applied coating G	"	115	.139	"	13.70	6.25		19.95	24.50

08 84 Plastic Glazing

08 84 10 – Plexiglass Glazing

08 84 10.10 Plexiglass Acrylic		Crew	Daily Output	Labor-Hours	Unit	Material	Labor	Equipment	Total	Total Incl O&P
0010	**PLEXIGLASS ACRYLIC**, clear, masked,									
0020	1/8" thick, cut sheets	2 Glaz	170	.094	S.F.	12	4.24		16.24	19.65
0200	Full sheets		195	.082		5	3.70		8.70	11.15
0500	1/4" thick, cut sheets		165	.097		14	4.37		18.37	22
0600	Full sheets		185	.086		9	3.90		12.90	15.85
0900	3/8" thick, cut sheets		155	.103		20	4.66		24.66	29
1000	Full sheets		180	.089		15	4.01		19.01	22.50
1300	1/2" thick, cut sheets		135	.119		28	5.35		33.35	39
1400	Full sheets	↓	150	.107	↓	20	4.81		24.81	29.50

08 84 Plastic Glazing

08 84 10 – Plexiglass Glazing

08 84 10.10 Plexiglass Acrylic		Crew	Daily Output	Labor-Hours	Unit	Material	Labor	Equipment	Total	Total Incl O&P
						2015 Bare Costs				
1700	3/4" thick, cut sheets	2 Glaz	115	.139	S.F.	69	6.25		75.25	85.50
1800	Full sheets		130	.123		40	5.55		45.55	52.50
2100	1" thick, cut sheets		105	.152		77.50	6.85		84.35	96
2200	Full sheets		125	.128		48	5.75		53.75	62
3000	Colored, 1/8" thick, cut sheets		170	.094		18	4.24		22.24	26.50
3200	Full sheets		195	.082		11	3.70		14.70	17.75
3500	1/4" thick, cut sheets		165	.097		20	4.37		24.37	28.50
3600	Full sheets		185	.086		14	3.90		17.90	21.50
4000	Mirrors, untinted, cut sheets, 1/8" thick		185	.086		12	3.90		15.90	19.15
4200	1/4" thick	↓	180	.089	↓	16	4.01		20.01	23.50

08 84 20 – Polycarbonate

08 84 20.10 Thermoplastic		Crew	Daily Output	Labor-Hours	Unit	Material	Labor	Equipment	Total	Total Incl O&P
0010	**THERMOPLASTIC**, clear, masked, cut sheets									
0020	1/8" thick	2 Glaz	170	.094	S.F.	14	4.24		18.24	22
0500	3/16" thick		165	.097		16	4.37		20.37	24.50
1000	1/4" thick		155	.103		17	4.66		21.66	26
1500	3/8" thick	↓	150	.107	↓	26	4.81		30.81	36

08 87 Glazing Surface Films

08 87 13 – Solar Control Films

08 87 13.10 Solar Films On Glass		Crew	Daily Output	Labor-Hours	Unit	Material	Labor	Equipment	Total	Total Incl O&P
0010	**SOLAR FILMS ON GLASS** (glass not included)									
2000	Minimum G	2 Glaz	180	.089	S.F.	6.80	4.01		10.81	13.60
2050	Maximum G	"	225	.071	"	15.30	3.21		18.51	21.50

08 87 23 – Safety and Security Films

08 87 23.16 Security Films		Crew	Daily Output	Labor-Hours	Unit	Material	Labor	Equipment	Total	Total Incl O&P
0010	**SECURITY FILMS**, clear, 32000 psi tensile strength, adhered to glass									
0100	.002" thick, daylight installation	H-2	950	.025	S.F.	2.70	1.08		3.78	4.61
0150	.004" thick, daylight installation R088110-10		800	.030		3.35	1.28		4.63	5.65
0200	.006" thick, daylight installation		700	.034		3.60	1.46		5.06	6.20
0210	Install for anchorage		600	.040		4	1.70		5.70	7
0400	.007" thick, daylight installation		600	.040		4.45	1.70		6.15	7.50
0410	Install for anchorage		500	.048		4.94	2.04		6.98	8.55
0500	.008" thick, daylight installation		500	.048		5	2.04		7.04	8.60
0510	Install for anchorage		500	.048		5.55	2.04		7.59	9.20
0600	.015" thick, daylight installation		400	.060		8.80	2.56		11.36	13.60
0610	Install for anchorage	↓	400	.060	↓	5.55	2.56		8.11	10
0900	Security film anchorage, mechanical attachment and cover plate	H-3	370	.043	L.F.	9.90	1.74		11.64	13.55
0950	Security film anchorage, wet glaze structural caulking	1 Glaz	225	.036	"	.99	1.60		2.59	3.53
1000	Adhered security film removal	1 Clab	275	.029	S.F.		1.09		1.09	1.68

08 88 Special Function Glazing

08 88 40 – Acoustical Glass Units

08 88 40.10 Sound Reduction Units		Crew	Daily Output	Labor-Hours	Unit	Material	2015 Bare Costs Labor	Equipment	Total	Total Incl O&P
0010	**SOUND REDUCTION UNITS**, 1lite at 3/8", 1 lite at 3/16"									
0020	For 1" thick	2 Glaz	100	.160	S.F.	34	7.20		41.20	48.50
0100	For 4" thick	"	80	.200	"	58.50	9		67.50	78

08 88 56 – Ballistics-Resistant Glazing

08 88 56.10 Laminated Glass		Crew	Daily Output	Labor-Hours	Unit	Material	Labor	Equipment	Total	Total Incl O&P
0010	**LAMINATED GLASS**									
0020	Clear float .03" vinyl 1/4"	2 Glaz	90	.178	S.F.	12.40	8		20.40	26
0100	3/8" thick		78	.205		22.50	9.25		31.75	38.50
0200	.06" vinyl, 1/2" thick		65	.246		25.50	11.10		36.60	45
1000	5/8" thick		90	.178		29.50	8		37.50	44.50
2000	Bullet-resisting, 1-3/16" thick, to 15 S.F.		16	1		105	45		150	185
2100	Over 15 S.F.		16	1		119	45		164	200
2500	2-1/4" thick, to 15 S.F.		12	1.333		179	60		239	289
2600	Over 15 S.F.		12	1.333		168	60		228	277
2700	Level 2 (.357 magnum), NIJ and UL		12	1.333		77	60		137	176
2750	Level 3A (.44 magnum) NIJ, UL 3		12	1.333		82	60		142	182
2800	Level 4 (AK-47) NIJ, UL 7 & 8		12	1.333		113	60		173	216
2850	Level 5 (M-16) UL		12	1.333		116	60		176	220
2900	Level 3 (7.62 Armor Piercing) NIJ, UL 4 & 5	↓	12	1.333	↓	137	60		197	243

08 91 Louvers

08 91 19 – Fixed Louvers

08 91 19.10 Aluminum Louvers		Crew	Daily Output	Labor-Hours	Unit	Material	Labor	Equipment	Total	Total Incl O&P
0010	**ALUMINUM LOUVERS**									
0020	Aluminum with screen, residential, 8" x 8"	1 Carp	38	.211	Ea.	20	9.90		29.90	37
0100	12" x 12"		38	.211		16	9.90		25.90	33
0200	12" x 18"		35	.229		20	10.75		30.75	38.50
0250	14" x 24"		30	.267		29	12.50		41.50	51.50
0300	18" x 24"		27	.296		32	13.90		45.90	56.50
0500	24" x 30"		24	.333		60	15.65		75.65	90
0700	Triangle, adjustable, small		20	.400		55	18.80		73.80	89.50
0800	Large	↓	15	.533	↓	76	25		101	122
1200	Extruded aluminum, see Section 23 37 15.40									
2100	Midget, aluminum, 3/4" deep, 1" diameter	1 Carp	85	.094	Ea.	.75	4.42		5.17	7.65
2150	3" diameter		60	.133		2.47	6.25		8.72	12.35
2200	4" diameter		50	.160		4.95	7.50		12.45	17
2250	6" diameter	↓	30	.267	↓	4.10	12.50		16.60	24

08 91 26 – Door Louvers

08 91 26.10 Steel Louvers, 18 Gauge, Fixed Blade		Crew	Daily Output	Labor-Hours	Unit	Material	Labor	Equipment	Total	Total Incl O&P
0010	**STEEL LOUVERS, 18 GAUGE, FIXED BLADE**									
0050	12" x 12", with enamel or powder coat	1 Carp	20	.400	Ea.	82	18.80		100.80	119
0055	18" x 12"		20	.400		87.50	18.80		106.30	125
0060	18" x 18"		20	.400		103	18.80		121.80	143
0065	24" x 12"		20	.400		108	18.80		126.80	148
0070	24" x 18"		20	.400		117	18.80		135.80	158
0075	24" x 24"		20	.400		144	18.80		162.80	187
0100	12" x 12", galvanized		20	.400		72.50	18.80		91.30	109
0105	18" x 12"		20	.400		85.50	18.80		104.30	123
0115	24" x 12"		20	.400		102	18.80		120.80	141
0125	24" x 24"	↓	20	.400	↓	143	18.80		161.80	186

08 95 Vents

08 95 13 – Soffit Vents

08 95 13.10 Wall Louvers		Crew	Daily Output	Labor-Hours	Unit	Material	2015 Bare Costs Labor	Equipment	Total	Total Incl O&P
0010	**WALL LOUVERS**									
2330	Soffit vent, continuous, 3" wide, aluminum, mill finish	1 Carp	200	.040	L.F.	.67	1.88		2.55	3.63
2340	Baked enamel finish		200	.040	"	5.50	1.88		7.38	8.95
2400	Under eaves vent, aluminum, mill finish, 16" x 4"		48	.167	Ea.	1.90	7.85		9.75	14.15
2500	16" x 8"	↓	48	.167	"	2.18	7.85		10.03	14.45

08 95 16 – Wall Vents

08 95 16.10 Louvers

08 95 16.10 Louvers		Crew	Daily Output	Labor-Hours	Unit	Material	Labor	Equipment	Total	Total Incl O&P
0010	**LOUVERS**									
0020	Redwood, 2'-0" diameter, full circle	1 Carp	16	.500	Ea.	190	23.50		213.50	245
0100	Half circle		16	.500		180	23.50		203.50	234
0200	Octagonal		16	.500		142	23.50		165.50	192
0300	Triangular, 5/12 pitch, 5'-0" at base		16	.500		200	23.50		223.50	256
7000	Vinyl gable vent, 8" x 8"		38	.211		14	9.90		23.90	30.50
7020	12" x 12"		38	.211		27	9.90		36.90	44.50
7080	12" x 18"		35	.229		35	10.75		45.75	55
7200	18" x 24"	↓	30	.267	↓	45	12.50		57.50	69

Estimating Tips

General

- Room Finish Schedule: A complete set of plans should contain a room finish schedule. If one is not available, it would be well worth the time and effort to obtain one.

09 20 00 Plaster and Gypsum Board

- Lath is estimated by the square yard plus a 5% allowance for waste. Furring, channels, and accessories are measured by the linear foot. An extra foot should be allowed for each accessory miter or stop.
- Plaster is also estimated by the square yard. Deductions for openings vary by preference, from zero deduction to 50% of all openings over 2 feet in width. The estimator should allow one extra square foot for each linear foot of horizontal interior or exterior angle located below the ceiling level. Also, double the areas of small radius work.
- Drywall accessories, studs, track, and acoustical caulking are all measured by the linear foot. Drywall taping is figured by the square foot. Gypsum wallboard is estimated by the square foot. No material deductions should be made for door or window openings under 32 S.F.

09 60 00 Flooring

- Tile and terrazzo areas are taken off on a square foot basis. Trim and base materials are measured by the linear foot. Accent tiles are listed per each. Two basic methods of installation are used. Mud set is approximately 30% more expensive than thin set. In terrazzo work, be sure to include the linear footage of embedded decorative strips, grounds, machine rubbing, and power cleanup.
- Wood flooring is available in strip, parquet, or block configuration. The latter two types are set in adhesives with quantities estimated by the square foot. The laying pattern will influence labor costs and material waste. In addition to the material and labor for laying wood floors, the estimator must make allowances for sanding and finishing these areas, unless the flooring is prefinished.
- Sheet flooring is measured by the square yard. Roll widths vary, so consideration should be given to use the most economical width, as waste must be figured into the total quantity. Consider also the installation methods available, direct glue down or stretched.

09 70 00 Wall Finishes

- Wall coverings are estimated by the square foot. The area to be covered is measured, length by height of wall above baseboards, to calculate the square footage of each wall. This figure is divided by the number of square feet in the single roll which is being used. Deduct, in full, the areas of openings such as doors and windows. Where a pattern match is required allow 25%–30% waste.

09 80 00 Acoustic Treatment

- Acoustical systems fall into several categories. The takeoff of these materials should be by the square foot of area with a 5% allowance for waste. Do not forget about scaffolding, if applicable, when estimating these systems.

09 90 00 Painting and Coating

- A major portion of the work in painting involves surface preparation. Be sure to include cleaning, sanding, filling, and masking costs in the estimate.
- Protection of adjacent surfaces is not included in painting costs. When considering the method of paint application, an important factor is the amount of protection and masking required. These must be estimated separately and may be the determining factor in choosing the method of application.

Reference Numbers

Reference numbers are shown in shaded boxes at the beginning of some major classifications. These numbers refer to related items in the Reference Section. The reference information may be an estimating procedure, an alternate pricing method, or technical information.

Note: Not all subdivisions listed here necessarily appear in this publication. ■

09 01 Maintenance of Finishes

09 01 60 – Maintenance of Flooring

09 01 60.10 Carpet Maintenance		Crew	Daily Output	Labor-Hours	Unit	Material	2015 Bare Costs Labor	Equipment	Total	Total Incl O&P
0010	**CARPET MAINTENANCE**									
0020	Steam clean, per cleaning, routine maintenance	1 Clab	3000	.003	S.F.	.05	.10		.15	.21
0500	Stain removal	″	2000	.004	″	.07	.15		.22	.31

09 01 70 – Maintenance of Wall Finishes

09 01 70.10 Gypsum Wallboard Repairs

		Crew	Daily Output	Labor-Hours	Unit	Material	Labor	Equipment	Total	Total Incl O&P
0010	**GYPSUM WALLBOARD REPAIRS**									
0100	Fill and sand, pin/nail holes	1 Carp	960	.008	Ea.		.39		.39	.60
0110	Screw head pops		480	.017			.78		.78	1.20
0120	Dents, up to 2″ square		48	.167		.01	7.85		7.86	12.05
0130	2″ to 4″ square		24	.333		.03	15.65		15.68	24
0140	Cut square, patch, sand and finish, holes, up to 2″ square		12	.667		.03	31.50		31.53	48
0150	2″ to 4″ square		11	.727		.09	34		34.09	52.50
0160	4″ to 8″ square		10	.800		.23	37.50		37.73	58.50
0170	8″ to 12″ square		8	1		.46	47		47.46	73
0180	12″ to 32″ square		6	1.333		1.55	62.50		64.05	98
0210	16″ by 48″		5	1.600		2.65	75		77.65	119
0220	32″ by 48″		4	2		4.09	94		98.09	150
0230	48″ square		3.50	2.286		5.75	107		112.75	171
0240	60″ square		3.20	2.500		9.50	117		126.50	191
0500	Skim coat surface with joint compound		1600	.005	S.F.	.03	.23		.26	.40
0510	Prepare, retape and refinish joints		60	.133	L.F.	.64	6.25		6.89	10.35

09 05 Common Work Results for Finishes

09 05 05 – Selective Demolition for Finishes

09 05 05.10 Selective Demolition, Ceilings

		Crew	Daily Output	Labor-Hours	Unit	Material	Labor	Equipment	Total	Total Incl O&P
0010	**SELECTIVE DEMOLITION, CEILINGS** R024119-10									
0200	Ceiling, drywall, furred and nailed or screwed	2 Clab	800	.020	S.F.		.75		.75	1.16
0220	On metal frame		760	.021			.79		.79	1.22
0240	On suspension system, including system		720	.022			.84		.84	1.29
1000	Plaster, lime and horse hair, on wood lath, incl. lath		700	.023			.86		.86	1.32
1020	On metal lath		570	.028			1.06		1.06	1.62
1100	Gypsum, on gypsum lath		720	.022			.84		.84	1.29
1120	On metal lath		500	.032			1.20		1.20	1.85
1200	Suspended ceiling, mineral fiber, 2′ x 2′ or 2′ x 4′		1500	.011			.40		.40	.62
1250	On suspension system, incl. system		1200	.013			.50		.50	.77
1500	Tile, wood fiber, 12″ x 12″, glued		900	.018			.67		.67	1.03
1540	Stapled		1500	.011			.40		.40	.62
1580	On suspension system, incl. system		760	.021			.79		.79	1.22
2000	Wood, tongue and groove, 1″ x 4″		1000	.016			.60		.60	.93
2040	1″ x 8″		1100	.015			.55		.55	.84
2400	Plywood or wood fiberboard, 4′ x 8′ sheets		1200	.013			.50		.50	.77

09 05 05.20 Selective Demolition, Flooring

		Crew	Daily Output	Labor-Hours	Unit	Material	Labor	Equipment	Total	Total Incl O&P
0010	**SELECTIVE DEMOLITION, FLOORING** R024119-10									
0200	Brick with mortar	2 Clab	475	.034	S.F.		1.27		1.27	1.95
0400	Carpet, bonded, including surface scraping		2000	.008			.30		.30	.46
0440	Scrim applied		8000	.002			.08		.08	.12
0480	Tackless		9000	.002			.07		.07	.10
0550	Carpet tile, releasable adhesive		5000	.003			.12		.12	.19
0560	Permanent adhesive		1850	.009			.33		.33	.50
0600	Composition, acrylic or epoxy		400	.040			1.50		1.50	2.31

09 05 Common Work Results for Finishes

09 05 05 – Selective Demolition for Finishes

09 05 05.20 Selective Demolition, Flooring		Crew	Daily Output	Labor-Hours	Unit	Material	2015 Bare Costs Labor	Equipment	Total	Total Incl O&P
0700	Concrete, scarify skin	A-1A	225	.036	S.F.		1.73	.95	2.68	3.72
0800	Resilient, sheet goods	2 Clab	1400	.011			.43		.43	.66
0820	For gym floors	"	900	.018			.67		.67	1.03
0850	Vinyl or rubber cove base	1 Clab	1000	.008	L.F.		.30		.30	.46
0860	Vinyl or rubber cove base, molded corner	"	1000	.008	Ea.		.30		.30	.46
0870	For glued and caulked installation, add to labor						50%			
0900	Vinyl composition tile, 12" x 12"	2 Clab	1000	.016	S.F.		.60		.60	.93
2000	Tile, ceramic, thin set		675	.024			.89		.89	1.37
2020	Mud set		625	.026			.96		.96	1.48
2200	Marble, slate, thin set		675	.024			.89		.89	1.37
2220	Mud set		625	.026			.96		.96	1.48
2600	Terrazzo, thin set		450	.036			1.34		1.34	2.06
2620	Mud set		425	.038			1.42		1.42	2.18
2640	Terrazzo, cast in place		300	.053			2.01		2.01	3.09
3000	Wood, block, on end	1 Carp	400	.020			.94		.94	1.45
3200	Parquet		450	.018			.83		.83	1.28
3400	Strip flooring, interior, 2-1/4" x 25/32" thick		325	.025			1.16		1.16	1.78
3500	Exterior, porch flooring, 1" x 4"		220	.036			1.71		1.71	2.63
3800	Subfloor, tongue and groove, 1" x 6"		325	.025			1.16		1.16	1.78
3820	1" x 8"		430	.019			.87		.87	1.34
3840	1" x 10"		520	.015			.72		.72	1.11
4000	Plywood, nailed		600	.013			.63		.63	.96
4100	Glued and nailed		400	.020			.94		.94	1.45
4200	Hardboard, 1/4" thick		760	.011			.49		.49	.76
8000	Remove flooring, bead blast, simple floor plan	A-1A	1000	.008			.39	.21	.60	.84
8100	complex floor plan		400	.020			.97	.54	1.51	2.09
8150	Mastic only		1500	.005			.26	.14	.40	.56

09 05 05.30 Selective Demolition, Walls and Partitions		Crew	Daily Output	Labor-Hours	Unit	Material	Labor	Equipment	Total	Total Incl O&P
0010	**SELECTIVE DEMOLITION, WALLS AND PARTITIONS** R024119-10									
0020	Walls, concrete, reinforced	B-39	120	.400	C.F.		15.90	1.94	17.84	26.50
0025	Plain	"	160	.300			11.95	1.46	13.41	19.90
0100	Brick, 4" to 12" thick	B-9	220	.182			6.90	1.06	7.96	11.80
0200	Concrete block, 4" thick		1150	.035	S.F.		1.32	.20	1.52	2.25
0280	8" thick		1050	.038			1.45	.22	1.67	2.47
0300	Exterior stucco 1" thick over mesh		3200	.013			.48	.07	.55	.81
1000	Drywall, nailed or screwed	1 Clab	1000	.008			.30		.30	.46
1010	2 layers		400	.020			.75		.75	1.16
1020	Glued and nailed		900	.009			.33		.33	.51
1500	Fiberboard, nailed		900	.009			.33		.33	.51
1520	Glued and nailed		800	.010			.38		.38	.58
1568	Plenum barrier, sheet lead		300	.027			1		1	1.54
2000	Movable walls, metal, 5' high		300	.027			1		1	1.54
2020	8' high		400	.020			.75		.75	1.16
2200	Metal or wood studs, finish 2 sides, fiberboard	B-1	520	.046			1.77		1.77	2.72
2250	Lath and plaster		260	.092			3.53		3.53	5.45
2300	Plasterboard (drywall)		520	.046			1.77		1.77	2.72
2350	Plywood		450	.053			2.04		2.04	3.14
2800	Paneling, 4' x 8' sheets	1 Clab	475	.017			.63		.63	.97
3000	Plaster, lime and horsehair, on wood lath		400	.020			.75		.75	1.16
3020	On metal lath		335	.024			.90		.90	1.38
3400	Gypsum or perlite, on gypsum lath		410	.020			.73		.73	1.13
3420	On metal lath		300	.027			1		1	1.54

09 05 Common Work Results for Finishes

09 05 05 – Selective Demolition for Finishes

09 05 05.30 Selective Demolition, Walls and Partitions		Crew	Daily Output	Labor-Hours	Unit	Material	2015 Bare Costs Labor	Equipment	Total	Total Incl O&P
3450	Plaster, interior gypsum, acoustic, or cement	1 Clab	60	.133	S.Y.		5		5	7.70
3500	Stucco, on masonry		145	.055			2.07		2.07	3.19
3510	Commercial 3-coat		80	.100			3.76		3.76	5.80
3520	Interior stucco	↓	25	.320	↓		12.05		12.05	18.50
3600	Plywood, one side	B-1	1500	.016	S.F.		.61		.61	.94
3750	Terra cotta block and plaster, to 6" thick	"	175	.137			5.25		5.25	8.05
3760	Tile, ceramic, on walls, thin set	1 Clab	300	.027			1		1	1.54
3765	Mud set		250	.032	↓		1.20		1.20	1.85
3800	Toilet partitions, slate or marble		5	1.600	Ea.		60		60	92.50
3820	Metal or plastic	↓	8	1	"		37.50		37.50	58
5000	Wallcovering, vinyl	1 Pape	700	.011	S.F.		.47		.47	.70
5010	With release agent		1500	.005			.22		.22	.33
5025	Wallpaper, 2 layers or less, by hand		250	.032			1.31		1.31	1.97
5035	3 layers or more		165	.048			1.98		1.98	2.99
5040	Designer	↓	480	.017	↓		.68		.68	1.03

09 05 71 – Acoustic Underlayment

09 05 71.10 Acoustical Underlayment		Crew	Daily Output	Labor-Hours	Unit	Material	Labor	Equipment	Total	Total Incl O&P
0010	**ACOUSTICAL UNDERLAYMENT**									
4000	Nylon matting 0.4" thick, with carbon black spinerette									
4010	plus polyester fabric, on floor	D-7	1600	.010	S.F.	1.36	.38		1.74	2.06
4200	Fiberglass reinf. backer board underlayment, 7/16" thick, on floor	"	1500	.011	"	2.77	.41		3.18	3.65

09 21 Plaster and Gypsum Board Assemblies

09 21 13 – Plaster Assemblies

09 21 13.10 Plaster Partition Wall		Crew	Daily Output	Labor-Hours	Unit	Material	Labor	Equipment	Total	Total Incl O&P
0010	**PLASTER PARTITION WALL**									
0400	Stud walls, 3.4 lb. metal lath, 3 coat gypsum plaster, 2 sides									
0600	2" x 4" wood studs, 16" O.C.	J-2	315	.152	S.F.	3.39	6.30	.44	10.13	13.60
0700	2-1/2" metal studs, 25 ga., 12" O.C.		325	.148		3.14	6.10	.43	9.67	13.05
0800	3-5/8" metal studs, 25 ga., 16" O.C.	↓	320	.150	↓	3.16	6.20	.44	9.80	13.20
0900	Gypsum lath, 2 coat vermiculite plaster, 2 sides									
1000	2" x 4" wood studs, 16" O.C.	J-2	355	.135	S.F.	3.77	5.60	.39	9.76	12.95
1200	2-1/2" metal studs, 25 ga., 12" O.C.		365	.132		3.35	5.45	.38	9.18	12.25
1300	3-5/8" metal studs, 25 ga., 16" O.C.	↓	360	.133	↓	3.44	5.50	.39	9.33	12.45

09 21 16 – Gypsum Board Assemblies

09 21 16.23 Gypsum Board Shaft Wall Assemblies		Crew	Daily Output	Labor-Hours	Unit	Material	Labor	Equipment	Total	Total Incl O&P
0010	**GYPSUM BOARD SHAFT WALL ASSEMBLIES**									
0020	Cavity type on 25 ga. J-track & C-H studs, 24" O.C.									
0030	1" thick coreboard wall liner on shaft side									
0040	2-hour assembly with double layer									
0060	5/8" fire rated gypsum board on room side	2 Carp	220	.073	S.F.	2.10	3.41		5.51	7.55
0100	3-hour assembly with triple layer									
0300	5/8" fire rated gypsum board on room side	2 Carp	180	.089	S.F.	1.77	4.17		5.94	8.35
0400	4-hour assembly, 1" coreboard, 5/8" fire rated gypsum board									
0600	and 3/4" galv. metal furring channels, 24" O.C., with									
0700	Double layer 5/8" fire rated gypsum board on room side	2 Carp	110	.145	S.F.	1.67	6.85		8.52	12.35
0900	For taping & finishing, add per side	1 Carp	1050	.008	"	.05	.36		.41	.60
1000	For insulation, see Section 07 21									
5200	For work over 8' high, add	2 Carp	3060	.005	S.F.		.25		.25	.38
5300	For distribution cost over 3 stories high, add per story	"	6100	.003	"		.12		.12	.19

09 21 16 – Gypsum Board Assemblies

09 21 16.33 Partition Wall		Crew	Daily Output	Labor-Hours	Unit	Material	2015 Bare Costs Labor	Equipment	Total	Total Incl O&P
0010	**PARTITION WALL** Stud wall, 8′ to 12′ high									
0050	1/2″, interior, gypsum board, std, tape & finish 2 sides									
0500	Installed on and incl., 2″ x 4″ wood studs, 16″ O.C.	2 Carp	310	.052	S.F.	1.16	2.42		3.58	5
1000	Metal studs, NLB, 25 ga., 16″ O.C., 3-5/8″ wide		350	.046		1.06	2.15		3.21	4.47
1200	6″ wide		330	.048		1.19	2.28		3.47	4.81
1400	Water resistant, on 2″ x 4″ wood studs, 16″ O.C.		310	.052		1.34	2.42		3.76	5.20
1600	Metal studs, NLB, 25 ga., 16″ O.C., 3-5/8″ wide		350	.046		1.24	2.15		3.39	4.67
1800	6″ wide		330	.048		1.37	2.28		3.65	5
2000	Fire res., 2 layers, 1-1/2 hr., on 2″ x 4″ wood studs, 16″ O.C.		210	.076		1.96	3.58		5.54	7.65
2200	Metal studs, NLB, 25 ga., 16″ O.C., 3-5/8″ wide		250	.064		1.86	3		4.86	6.65
2400	6″ wide		230	.070		1.99	3.27		5.26	7.25
2600	Fire & water res., 2 layers, 1-1/2 hr., 2″ x 4″ studs, 16″ O.C.		210	.076		1.96	3.58		5.54	7.65
2800	Metal studs, NLB, 25 ga., 16″ O.C., 3-5/8″ wide		250	.064		1.86	3		4.86	6.65
3000	6″ wide		230	.070		1.99	3.27		5.26	7.25
3200	5/8″, interior, gypsum board, standard, tape & finish 2 sides									
3400	Installed on and including 2″ x 4″ wood studs, 16″ O.C.	2 Carp	300	.053	S.F.	1.22	2.50		3.72	5.20
3600	24″ O.C.		330	.048		1.12	2.28		3.40	4.73
3800	Metal studs, NLB, 25 ga., 16″ O.C., 3-5/8″ wide		340	.047		1.12	2.21		3.33	4.63
4000	6″ wide		320	.050		1.25	2.35		3.60	4.98
4200	24″ O.C., 3-5/8″ wide		360	.044		1.02	2.09		3.11	4.34
4400	6″ wide		340	.047		1.12	2.21		3.33	4.63
4800	Water resistant, on 2″ x 4″ wood studs, 16″ O.C.		300	.053		1.40	2.50		3.90	5.40
5000	24″ O.C.		330	.048		1.30	2.28		3.58	4.93
5200	Metal studs, NLB, 25 ga. 16″ O.C., 3-5/8″ wide		340	.047		1.30	2.21		3.51	4.83
5400	6″ wide		320	.050		1.43	2.35		3.78	5.20
5600	24″ O.C., 3-5/8″ wide		360	.044		1.20	2.09		3.29	4.54
5800	6″ wide		340	.047		1.30	2.21		3.51	4.83
6000	Fire resistant, 2 layers, 2 hr., on 2″ x 4″ wood studs, 16″ O.C.		205	.078		1.82	3.66		5.48	7.65
6200	24″ O.C.		235	.068		1.82	3.20		5.02	6.90
6400	Metal studs, NLB, 25 ga., 16″ O.C., 3-5/8″ wide		245	.065		1.85	3.07		4.92	6.75
6600	6″ wide		225	.071		1.95	3.34		5.29	7.30
6800	24″ O.C., 3-5/8″ wide		265	.060		1.72	2.83		4.55	6.25
7000	6″ wide		245	.065		1.82	3.07		4.89	6.70
7200	Fire & water resistant, 2 layers, 2 hr., 2″ x 4″ studs, 16″ O.C.		205	.078		1.92	3.66		5.58	7.75
7400	24″ O.C.		235	.068		1.82	3.20		5.02	6.90
7600	Metal studs, NLB, 25 ga., 16″ O.C., 3-5/8″ wide		245	.065		1.82	3.07		4.89	6.70
7800	6″ wide		225	.071		1.95	3.34		5.29	7.30
8000	24″ O.C., 3-5/8″ wide		265	.060		1.72	2.83		4.55	6.25
8200	6″ wide		245	.065		1.82	3.07		4.89	6.70
8600	1/2″ blueboard, mesh tape both sides									
8620	Installed on and including 2″ x 4″ wood studs, 16″ O.C.	2 Carp	300	.053	S.F.	1.22	2.50		3.72	5.20
8640	Metal studs, NLB, 25 ga., 16″ O.C., 3-5/8″ wide		340	.047		1.12	2.21		3.33	4.63
8660	6″ wide		320	.050		1.25	2.35		3.60	4.98
8800	Hospital security partition, 5/8″ fiber reinf. high abuse gyp. bd.									
8810	Mtl. studs, NLB, 20 ga., 16″ O.C., 3-5/8″ wide, w/sec. mesh, gyp. bd.	2 Carp	208	.077	S.F.	4.18	3.61		7.79	10.15
9000	Exterior, 1/2″ gypsum sheathing, 1/2″ gypsum finished, interior,									
9100	including foil faced insulation, metal studs, 20 ga.									
9200	16″ O.C., 3-5/8″ wide	2 Carp	290	.055	S.F.	1.70	2.59		4.29	5.85
9400	6″ wide		270	.059		1.88	2.78		4.66	6.35
9600	Partitions, for work over 8′ high, add		1530	.010			.49		.49	.76

09 22 03 – Fastening Methods for Finishes

09 22 03.20 Drilling Plaster/Drywall		Crew	Daily Output	Labor-Hours	Unit	Material	2015 Bare Costs Labor	Equipment	Total	Total Incl O&P
0010	**DRILLING PLASTER/DRYWALL**									
1100	Drilling & layout for drywall/plaster walls, up to 1" deep, no anchor									
1200	Holes, 1/4" diameter	1 Carp	150	.053	Ea.	.01	2.50		2.51	3.86
1300	3/8" diameter		140	.057		.01	2.68		2.69	4.14
1400	1/2" diameter		130	.062		.01	2.89		2.90	4.46
1500	3/4" diameter		120	.067		.01	3.13		3.14	4.84
1600	1" diameter		110	.073		.02	3.41		3.43	5.25
1700	1-1/4" diameter		100	.080		.03	3.76		3.79	5.85
1800	1-1/2" diameter		90	.089		.05	4.17		4.22	6.45
1900	For ceiling installations, add						40%			

09 22 13 – Metal Furring

09 22 13.13 Metal Channel Furring		Crew	Daily Output	Labor-Hours	Unit	Material	Labor	Equipment	Total	Total Incl O&P
0010	**METAL CHANNEL FURRING**									
0030	Beams and columns, 7/8" channels, galvanized, 12" O.C.	1 Lath	155	.052	S.F.	.40	2.22		2.62	3.72
0050	16" O.C.		170	.047		.33	2.02		2.35	3.34
0070	24" O.C.		185	.043		.22	1.86		2.08	2.98
0100	Ceilings, on steel, 7/8" channels, galvanized, 12" O.C.		210	.038		.37	1.64		2.01	2.82
0300	16" O.C.		290	.028		.33	1.19		1.52	2.11
0400	24" O.C.		420	.019		.22	.82		1.04	1.45
0600	1-5/8" channels, galvanized, 12" O.C.		190	.042		.49	1.81		2.30	3.21
0700	16" O.C.		260	.031		.44	1.32		1.76	2.43
0900	24" O.C.		390	.021		.29	.88		1.17	1.62
0930	7/8" channels with sound isolation clips, 12" O.C.		120	.067		1.74	2.87		4.61	6.15
0940	16" O.C.		100	.080		1.31	3.44		4.75	6.50
0950	24" O.C.		165	.048		.87	2.08		2.95	4.03
0960	1-5/8" channels, galvanized, 12" O.C.		110	.073		1.86	3.13		4.99	6.65
0970	16" O.C.		100	.080		1.40	3.44		4.84	6.60
0980	24" O.C.		155	.052		.93	2.22		3.15	4.30
1000	Walls, 7/8" channels, galvanized, 12" O.C.		235	.034		.37	1.46		1.83	2.56
1200	16" O.C.		265	.030		.33	1.30		1.63	2.27
1300	24" O.C.		350	.023		.22	.98		1.20	1.69
1500	1-5/8" channels, galvanized, 12" O.C.		210	.038		.49	1.64		2.13	2.96
1600	16" O.C.		240	.033		.44	1.43		1.87	2.59
1800	24" O.C.		305	.026		.29	1.13		1.42	1.98
1920	7/8" channels with sound isolation clips, 12" O.C.		125	.064		1.74	2.75		4.49	6
1940	16" O.C.		100	.080		1.31	3.44		4.75	6.50
1950	24" O.C.		150	.053		.87	2.29		3.16	4.34
1960	1-5/8" channels, galvanized, 12" O.C.		115	.070		1.86	2.99		4.85	6.45
1970	16" O.C.		95	.084		1.40	3.62		5.02	6.90
1980	24" O.C.		140	.057		.93	2.46		3.39	4.65

09 22 16 – Non-Structural Metal Framing

09 22 16.13 Non-Structural Metal Stud Framing		Crew	Daily Output	Labor-Hours	Unit	Material	Labor	Equipment	Total	Total Incl O&P
0010	**NON-STRUCTURAL METAL STUD FRAMING**									
1600	Non-load bearing, galv., 8' high, 25 ga. 1-5/8" wide, 16" O.C.	1 Carp	619	.013	S.F.	.27	.61		.88	1.22
1610	24" O.C.		950	.008		.20	.40		.60	.83
1620	2-1/2" wide, 16" O.C.		613	.013		.33	.61		.94	1.30
1630	24" O.C.		938	.009		.24	.40		.64	.89
1640	3-5/8" wide, 16" O.C.		600	.013		.39	.63		1.02	1.39
1650	24" O.C.		925	.009		.29	.41		.70	.95
1660	4" wide, 16" O.C.		594	.013		.43	.63		1.06	1.44
1670	24" O.C.		925	.009		.32	.41		.73	.99
1680	6" wide, 16" O.C.		588	.014		.52	.64		1.16	1.55

09 22 16.13 Non-Structural Metal Stud Framing		Crew	Daily Output	Labor-Hours	Unit	Material	2015 Bare Costs Labor	Equipment	Total	Total Incl O&P
1690	24" O.C.	1 Carp	906	.009	S.F.	.39	.41		.80	1.07
1700	20 ga. studs, 1-5/8" wide, 16" O.C.		494	.016		.34	.76		1.10	1.54
1710	24" O.C.		763	.010		.25	.49		.74	1.04
1720	2-1/2" wide, 16" O.C.		488	.016		.42	.77		1.19	1.64
1730	24" O.C.		750	.011		.31	.50		.81	1.12
1740	3-5/8" wide, 16" O.C.		481	.017		.48	.78		1.26	1.73
1750	24" O.C.		738	.011		.36	.51		.87	1.18
1760	4" wide, 16" O.C.		475	.017		.57	.79		1.36	1.85
1770	24" O.C.		738	.011		.43	.51		.94	1.25
1780	6" wide, 16" O.C.		469	.017		.66	.80		1.46	1.96
1790	24" O.C.		725	.011		.50	.52		1.02	1.35
2000	Non-load bearing, galv., 10' high, 25 ga. 1-5/8" wide, 16" O.C.		495	.016		.25	.76		1.01	1.45
2100	24" O.C.		760	.011		.19	.49		.68	.96
2200	2-1/2" wide, 16" O.C.		490	.016		.31	.77		1.08	1.52
2250	24" O.C.		750	.011		.23	.50		.73	1.02
2300	3-5/8" wide, 16" O.C.		480	.017		.37	.78		1.15	1.60
2350	24" O.C.		740	.011		.27	.51		.78	1.08
2400	4" wide, 16" O.C.		475	.017		.41	.79		1.20	1.67
2450	24" O.C.		740	.011		.30	.51		.81	1.11
2500	6" wide, 16" O.C.		470	.017		.49	.80		1.29	1.77
2550	24" O.C.		725	.011		.36	.52		.88	1.20
2600	20 ga. studs, 1-5/8" wide, 16" O.C.		395	.020		.32	.95		1.27	1.81
2650	24" O.C.		610	.013		.23	.62		.85	1.21
2700	2-1/2" wide, 16" O.C.		390	.021		.39	.96		1.35	1.91
2750	24" O.C.		600	.013		.29	.63		.92	1.28
2800	3-5/8" wide, 16" OC		385	.021		.46	.98		1.44	2
2850	24" O.C.		590	.014		.34	.64		.98	1.35
2900	4" wide, 16" O.C.		380	.021		.54	.99		1.53	2.11
2950	24" O.C.		590	.014		.40	.64		1.04	1.42
3000	6" wide, 16" O.C.		375	.021		.63	1		1.63	2.23
3050	24" O.C.		580	.014		.46	.65		1.11	1.51
3060	Non-load bearing, galv., 12' high, 25 ga. 1-5/8" wide, 16" O.C.		413	.019		.24	.91		1.15	1.67
3070	24" O.C.		633	.013		.18	.59		.77	1.10
3080	2-1/2" wide, 16" O.C.		408	.020		.30	.92		1.22	1.75
3090	24" O.C.		625	.013		.22	.60		.82	1.16
3100	3-5/8" wide, 16" O.C.		400	.020		.35	.94		1.29	1.84
3110	24" O.C.		617	.013		.26	.61		.87	1.22
3120	4" wide, 16" O.C.		396	.020		.39	.95		1.34	1.89
3130	24" O.C.		617	.013		.28	.61		.89	1.25
3140	6" wide, 16" O.C.		392	.020		.47	.96		1.43	1.99
3150	24" O.C.		604	.013		.34	.62		.96	1.34
3160	20 ga. studs, 1-5/8" wide, 16" O.C.		329	.024		.30	1.14		1.44	2.09
3170	24" O.C.		508	.016		.22	.74		.96	1.38
3180	2-1/2" wide, 16" O.C.		325	.025		.38	1.16		1.54	2.19
3190	24" O.C.		500	.016		.27	.75		1.02	1.46
3200	3-5/8" wide, 16" O.C.		321	.025		.44	1.17		1.61	2.28
3210	24" O.C.		492	.016		.32	.76		1.08	1.52
3220	4" wide, 16" O.C.		317	.025		.52	1.19		1.71	2.39
3230	24" O.C.		492	.016		.38	.76		1.14	1.58
3240	6" wide, 16" O.C.		313	.026		.60	1.20		1.80	2.51
3250	24" O.C.	▼	483	.017	▼	.44	.78		1.22	1.68
5000	Load bearing studs, see Section 05 41 13.30									

09 22 Supports for Plaster and Gypsum Board

09 22 26 – Suspension Systems

09 22 26.13 Ceiling Suspension Systems		Crew	Daily Output	Labor-Hours	Unit	2015 Bare Costs Material	Labor	Equipment	Total	Total Incl O&P
0010	**CEILING SUSPENSION SYSTEMS** for gypsum board or plaster									
8000	Suspended ceilings, including carriers									
8200	1-1/2" carriers, 24" O.C. with:									
8300	7/8" channels, 16" O.C.	1 Lath	275	.029	S.F.	.54	1.25		1.79	2.43
8320	24" O.C.		310	.026		.42	1.11		1.53	2.11
8400	1-5/8" channels, 16" O.C.		205	.039		.64	1.68		2.32	3.18
8420	24" O.C.		250	.032		.50	1.38		1.88	2.58
8600	2" carriers, 24" O.C. with:									
8700	7/8" channels, 16" O.C.	1 Lath	250	.032	S.F.	.59	1.38		1.97	2.68
8720	24" O.C.		285	.028		.48	1.21		1.69	2.31
8800	1-5/8" channels, 16" O.C.		190	.042		.70	1.81		2.51	3.44
8820	24" O.C.		225	.036		.55	1.53		2.08	2.86

09 22 36 – Lath

09 22 36.13 Gypsum Lath

Line	Description	Crew	Daily Output	Labor-Hours	Unit	Material	Labor	Equipment	Total	Total Incl O&P
0010	**GYPSUM LATH** R092000-50									
0020	Plain or perforated, nailed, 3/8" thick	1 Lath	85	.094	S.Y.	3.06	4.05		7.11	9.30
0100	1/2" thick		80	.100		2.43	4.30		6.73	9
0300	Clipped to steel studs, 3/8" thick		75	.107		3.06	4.59		7.65	10.10
0400	1/2" thick		70	.114		2.43	4.91		7.34	9.90
1500	For ceiling installations, add		216	.037			1.59		1.59	2.35
1600	For columns and beams, add		170	.047			2.02		2.02	2.98

09 22 36.23 Metal Lath

Line	Description	Crew	Daily Output	Labor-Hours	Unit	Material	Labor	Equipment	Total	Total Incl O&P
0010	**METAL LATH** R092000-50									
0020	Diamond, expanded, 2.5 lb. per S.Y., painted				S.Y.	3.61			3.61	3.97
0100	Galvanized					2.72			2.72	2.99
0300	3.4 lb. per S.Y., painted					4.08			4.08	4.49
0400	Galvanized					4.13			4.13	4.54
0600	For 15# asphalt sheathing paper, add					.49			.49	.53
0900	Flat rib, 1/8" high, 2.75 lb., painted					3.40			3.40	3.74
1000	Foil backed					3.58			3.58	3.94
1200	3.4 lb. per S.Y., painted					4.24			4.24	4.66
1300	Galvanized					4.35			4.35	4.79
1500	For 15# asphalt sheathing paper, add					.49			.49	.53
1800	High rib, 3/8" high, 3.4 lb. per S.Y., painted					4.12			4.12	4.53
1900	Galvanized					3.70			3.70	4.07
2400	3/4" high, painted, .60 lb. per S.F.				S.F.	.62			.62	.68
2500	.75 lb. per S.F.				"	1.33			1.33	1.46
2800	Stucco mesh, painted, 3.6 lb.				S.Y.	3.69			3.69	4.06
3000	K-lath, perforated, absorbent paper, regular					4.42			4.42	4.86
3100	Heavy duty					5.20			5.20	5.75
3300	Waterproof, heavy duty, grade B backing					5.10			5.10	5.60
3400	Fire resistant backing					5.65			5.65	6.20
3600	2.5 lb. diamond painted, on wood framing, on walls	1 Lath	85	.094		3.61	4.05		7.66	9.90
3700	On ceilings		75	.107		3.61	4.59		8.20	10.70
3900	3.4 lb. diamond painted, on wood framing, on walls		80	.100		4.24	4.30		8.54	11
4000	On ceilings		70	.114		4.24	4.91		9.15	11.90
4200	3.4 lb. diamond painted, wired to steel framing		75	.107		4.24	4.59		8.83	11.40
4300	On ceilings		60	.133		4.24	5.75		9.99	13.10
4500	Columns and beams, wired to steel		40	.200		4.24	8.60		12.84	17.35
4600	Cornices, wired to steel		35	.229		4.24	9.85		14.09	19.15
4800	Screwed to steel studs, 2.5 lb.		80	.100		3.61	4.30		7.91	10.30

09 22 Supports for Plaster and Gypsum Board

09 22 36 – Lath

09 22 36.23 Metal Lath		Crew	Daily Output	Labor-Hours	Unit	Material	2015 Bare Costs Labor	Equipment	Total	Total Incl O&P
4900	3.4 lb.	1 Lath	75	.107	S.Y.	4.08	4.59		8.67	11.25
5100	Rib lath, painted, wired to steel, on walls, 2.5 lb.		75	.107		3.40	4.59		7.99	10.50
5200	3.4 lb.		70	.114		4.12	4.91		9.03	11.80
5400	4.0 lb.		65	.123		5.65	5.30		10.95	14
5500	For self-furring lath, add					.11			.11	.12
5700	Suspended ceiling system, incl. 3.4 lb. diamond lath, painted	1 Lath	15	.533		4.31	23		27.31	38.50
5800	Galvanized	"	15	.533		4.24	23		27.24	38.50
6000	Hollow metal stud partitions, 3.4 lb. painted lath both sides									
6010	Non-load bearing, 25 ga., w/rib lath 2-1/2" studs, 12" O.C.	1 Lath	20.30	.394	S.Y.	11.75	16.95		28.70	38
6300	16" O.C.		21.10	.379		11	16.30		27.30	36
6350	24" O.C.		22.70	.352		10.30	15.15		25.45	34
6400	3-5/8" studs, 16" O.C.		19.50	.410		11.55	17.65		29.20	38.50
6600	24" O.C.		20.40	.392		10.65	16.85		27.50	37
6700	4" studs, 16" O.C.		20.40	.392		11.90	16.85		28.75	38
6900	24" O.C.		21.60	.370		10.95	15.95		26.90	35.50
7000	6" studs, 16" O.C.		19.50	.410		12.65	17.65		30.30	40
7100	24" O.C.		21.10	.379		11.50	16.30		27.80	36.50
7200	L.B. partitions, 16 ga., w/rib lath, 2-1/2" studs, 16" O.C.		20	.400		12.45	17.20		29.65	39
7300	3-5/8" studs, 16 ga.		19.70	.406		14.15	17.45		31.60	41.50
7500	4" studs, 16 ga.		19.50	.410		14.65	17.65		32.30	42
7600	6" studs, 16 ga.		18.70	.428		17.30	18.40		35.70	46

09 22 36.43 Security Mesh		Crew	Daily Output	Labor-Hours	Unit	Material	Labor	Equipment	Total	Total Incl O&P
0010	**SECURITY MESH**, expanded metal, flat, screwed to framing									
0100	On walls, 3/4", 1.76 lb./S.F.	2 Carp	1500	.011	S.F.	1.84	.50		2.34	2.79
0110	1-1/2", 1.14 lb./S.F.		1600	.010		1.41	.47		1.88	2.27
0200	On ceilings, 3/4", 1.76 lb./S.F.		1350	.012		1.84	.56		2.40	2.88
0210	1-1/2", 1.14 lb./S.F.		1450	.011		1.41	.52		1.93	2.35

09 22 36.83 Accessories, Plaster		Crew	Daily Output	Labor-Hours	Unit	Material	Labor	Equipment	Total	Total Incl O&P
0010	**ACCESSORIES, PLASTER**									
0020	Casing bead, expanded flange, galvanized	1 Lath	2.70	2.963	C.L.F.	55	127		182	249
0200	Foundation weep screed, galvanized	"	2.70	2.963		52	127		179	245
0900	Channels, cold rolled, 16 ga., 3/4" deep, galvanized					37			37	40.50
1200	1-1/2" deep, 16 ga., galvanized					49			49	54
1620	Corner bead, expanded bullnose, 3/4" radius, #10, galvanized	1 Lath	2.60	3.077		24.50	132		156.50	222
1650	#1, galvanized		2.55	3.137		48.50	135		183.50	252
1670	Expanded wing, 2-3/4" wide, #1, galvanized		2.65	3.019		37	130		167	232
1700	Inside corner (corner rite), 3" x 3", painted		2.60	3.077		20.50	132		152.50	218
1750	Strip-ex, 4" wide, painted		2.55	3.137		24	135		159	226
1800	Expansion joint, 3/4" grounds, limited expansion, galv., 1 piece		2.70	2.963		75	127		202	271
2100	Extreme expansion, galvanized, 2 piece		2.60	3.077		140	132		272	350

09 23 Gypsum Plastering

09 23 13 – Acoustical Gypsum Plastering

09 23 13.10 Perlite or Vermiculite Plaster		Crew	Daily Output	Labor-Hours	Unit	Material	2015 Bare Costs Labor	Equipment	Total	Total Incl O&P
0010	**PERLITE OR VERMICULITE PLASTER** R092000-50									
0020	In 100 lb. bags, under 200 bags				Bag	17.55			17.55	19.35
0100	Over 200 bags				"	16.80			16.80	18.45
0300	2 coats, no lath included, on walls	J-1	92	.435	S.Y.	5.85	17.85	1.53	25.23	34.50
0400	On ceilings	"	79	.506		5.85	21	1.78	28.63	39.50
0600	On and incl. 3/8" gypsum lath, on metal studs	J-2	84	.571		9.55	23.50	1.67	34.72	48
0700	On ceilings	"	70	.686		9.55	28.50	2	40.05	55
0900	3 coats, no lath included, on walls	J-1	74	.541		6.40	22	1.90	30.30	42
1000	On ceilings	"	63	.635		6.40	26	2.23	34.63	48.50
1200	On and incl. painted metal lath, on metal studs	J-2	72	.667		10.50	27.50	1.95	39.95	54.50
1300	On ceilings		61	.787		10.50	32.50	2.30	45.30	62.50
1500	On and incl. suspended metal lath ceiling		37	1.297		10.70	53.50	3.79	67.99	96
1700	For irregular or curved surfaces, add to above						30%			
1800	For columns and beams, add to above						50%			
1900	For soffits, add to ceiling prices						40%			

09 23 20 – Gypsum Plaster

09 23 20.10 Gypsum Plaster On Walls and Ceilings

Line	Description	Crew	Daily Output	Labor-Hours	Unit	Material	Labor	Equipment	Total	Total Incl O&P
0010	**GYPSUM PLASTER ON WALLS AND CEILINGS** R092000-50									
0020	80# bag, less than 1 ton				Bag	15.95			15.95	17.55
0100	Over 1 ton				"	13.95			13.95	15.35
0300	2 coats, no lath included, on walls	J-1	105	.381	S.Y.	3.61	15.60	1.34	20.55	29
0400	On ceilings	"	92	.435		3.61	17.85	1.53	22.99	32
0600	On and incl. 3/8" gypsum lath on steel, on walls	J-2	97	.495		6.65	20.50	1.45	28.60	39.50
0700	On ceilings	"	83	.578		6.65	24	1.69	32.34	44.50
0900	3 coats, no lath included, on walls	J-1	87	.460		5.20	18.85	1.61	25.66	36
1000	On ceilings	"	78	.513		5.20	21	1.80	28	39
1200	On and including painted metal lath, on wood studs	J-2	86	.558		10.20	23	1.63	34.83	47.50
1300	On ceilings	"	76.50	.627		10.20	26	1.83	38.03	52.50
1600	For irregular or curved surfaces, add						30%			
1800	For columns & beams, add						50%			

09 23 20.20 Gauging Plaster

Line	Description	Crew	Daily Output	Labor-Hours	Unit	Material	Labor	Equipment	Total	Total Incl O&P
0010	**GAUGING PLASTER** R092000-50									
0020	100 lb. bags, less than 1 ton				Bag	19.35			19.35	21.50
0100	Over 1 ton				"	18.35			18.35	20

09 23 20.30 Keenes Cement

Line	Description	Crew	Daily Output	Labor-Hours	Unit	Material	Labor	Equipment	Total	Total Incl O&P
0010	**KEENES CEMENT** R092000-50									
0020	In 100 lb. bags, less than 1 ton				Bag	22			22	24
0100	Over 1 ton				"	20			20	22.50
0300	Finish only, add to plaster prices, standard	J-1	215	.186	S.Y.	1.95	7.65	.65	10.25	14.30
0400	High quality	"	144	.278	"	1.97	11.40	.98	14.35	20.50

09 24 Cement Plastering

09 24 23 – Cement Stucco

09 24 23.40 Stucco		Crew	Daily Output	Labor-Hours	Unit	Material	2015 Bare Costs Labor	Equipment	Total	Total Incl O&P
0010	**STUCCO** R092000-50									
0015	3 coats 1" thick, float finish, with mesh, on wood frame	J-2	63	.762	S.Y.	6.25	31.50	2.22	39.97	56.50
0100	On masonry construction, no mesh incl.	J-1	67	.597		2.55	24.50	2.10	29.15	41.50
0300	For trowel finish, add	1 Plas	170	.047			2.02		2.02	3.03
0400	For 3/4" thick, on masonry, deduct	J-1	880	.045		.63	1.86	.16	2.65	3.66
0600	For coloring add		685	.058		.40	2.39	.20	2.99	4.26
0700	For special texture add		200	.200		1.41	8.20	.70	10.31	14.60
0900	For soffits, add	J-2	155	.310		2.18	12.80	.90	15.88	22.50
1000	Exterior stucco, with bonding agent, 3 coats, on walls, no mesh incl.	J-1	200	.200		3.65	8.20	.70	12.55	17.10
1200	Ceilings		180	.222		3.65	9.10	.78	13.53	18.55
1300	Beams		80	.500		3.65	20.50	1.76	25.91	36.50
1500	Columns		100	.400		3.65	16.40	1.40	21.45	30
1600	Mesh, painted, nailed to wood, 1.8 lb.	1 Lath	60	.133		6.20	5.75		11.95	15.25
1800	3.6 lb.		55	.145		3.69	6.25		9.94	13.25
1900	Wired to steel, painted, 1.8 lb.		53	.151		6.20	6.50		12.70	16.35
2100	3.6 lb.		50	.160		3.69	6.90		10.59	14.20

09 25 Other Plastering

09 25 23 – Lime Based Plastering

09 25 23.10 Venetian Plaster

Line	Description	Crew	Daily Output	Labor-Hours	Unit	Material	Labor	Equipment	Total	Total Incl O&P
0010	**VENETIAN PLASTER**									
0100	Walls, 1 coat primer, roller applied	1 Plas	950	.008	S.F.	.16	.36		.52	.72
0200	Plaster, 3 coats, incl. sanding	2 Plas	700	.023	"	.46	.98		1.44	1.97
0210	For pigment, light colors add per ea.				Ea.	3			3	3.30
0220	For pigment, dark colors add per ea.				"	9			9	9.90
0300	For sealer/wax coat incl. burnishing, add	1 Plas	300	.027	S.F.	.41	1.15		1.56	2.17

09 26 Veneer Plastering

09 26 13 – Gypsum Veneer Plastering

09 26 13.20 Blueboard

Line	Description	Crew	Daily Output	Labor-Hours	Unit	Material	Labor	Equipment	Total	Total Incl O&P
0010	**BLUEBOARD** For use with thin coat									
0100	plaster application see Section 09 26 13.80									
1000	3/8" thick, on walls or ceilings, standard, no finish included	2 Carp	1900	.008	S.F.	.34	.40		.74	.98
1100	With thin coat plaster finish		875	.018		.45	.86		1.31	1.82
1400	On beams, columns, or soffits, standard, no finish included		675	.024		.39	1.11		1.50	2.14
1450	With thin coat plaster finish		475	.034		.50	1.58		2.08	2.98
3000	1/2" thick, on walls or ceilings, standard, no finish included		1900	.008		.33	.40		.73	.97
3100	With thin coat plaster finish		875	.018		.44	.86		1.30	1.81
3300	Fire resistant, no finish included		1900	.008		.33	.40		.73	.97
3400	With thin coat plaster finish		875	.018		.44	.86		1.30	1.81
3450	On beams, columns, or soffits, standard, no finish included		675	.024		.38	1.11		1.49	2.13
3500	With thin coat plaster finish		475	.034		.49	1.58		2.07	2.97
3700	Fire resistant, no finish included		675	.024		.38	1.11		1.49	2.13
3800	With thin coat plaster finish		475	.034		.49	1.58		2.07	2.97
5000	5/8" thick, on walls or ceilings, fire resistant, no finish included		1900	.008		.34	.40		.74	.98
5100	With thin coat plaster finish		875	.018		.45	.86		1.31	1.82
5500	On beams, columns, or soffits, no finish included		675	.024		.39	1.11		1.50	2.14
5600	With thin coat plaster finish		475	.034		.50	1.58		2.08	2.98
6000	For high ceilings, over 8' high, add		3060	.005			.25		.25	.38

09 26 Veneer Plastering

09 26 13 – Gypsum Veneer Plastering

09 26 13.20 Blueboard		Crew	Daily Output	Labor-Hours	Unit	Material	2015 Bare Costs Labor	Equipment	Total	Total Incl O&P
6500	For over 3 stories high, add per story	2 Carp	6100	.003	S.F.		.12		.12	.19

09 26 13.80 Thin Coat Plaster

		Crew	Daily Output	Labor-Hours	Unit	Material	Labor	Equipment	Total	Total Incl O&P
0010	**THIN COAT PLASTER**									
0012	1 coat veneer, not incl. lath	J-1	3600	.011	S.F.	.11	.46	.04	.61	.84
1000	In 50 lb. bags				Bag	15.25			15.25	16.80

09 28 Backing Boards and Underlayments

09 28 13 – Cementitious Backing Boards

09 28 13.10 Cementitious Backerboard

		Crew	Daily Output	Labor-Hours	Unit	Material	Labor	Equipment	Total	Total Incl O&P
0010	**CEMENTITIOUS BACKERBOARD**									
0070	Cementitious backerboard, on floor, 3′ x 4′ x 1/2″ sheets	2 Carp	525	.030	S.F.	.78	1.43		2.21	3.06
0080	3′ x 5′ x 1/2″ sheets		525	.030		.76	1.43		2.19	3.04
0090	3′ x 6′ x 1/2″ sheets		525	.030		.74	1.43		2.17	3.01
0100	3′ x 4′ x 5/8″ sheets		525	.030		.99	1.43		2.42	3.29
0110	3′ x 5′ x 5/8″ sheets		525	.030		.99	1.43		2.42	3.29
0120	3′ x 6′ x 5/8″ sheets		525	.030		.96	1.43		2.39	3.26
0150	On wall, 3′ x 4′ x 1/2″ sheets		350	.046		.78	2.15		2.93	4.16
0160	3′ x 5′ x 1/2″ sheets		350	.046		.76	2.15		2.91	4.14
0170	3′ x 6′ x 1/2″ sheets		350	.046		.74	2.15		2.89	4.11
0180	3′ x 4′ x 5/8″ sheets		350	.046		.99	2.15		3.14	4.39
0190	3′ x 5′ x 5/8″ sheets		350	.046		.99	2.15		3.14	4.39
0200	3′ x 6′ x 5/8″ sheets		350	.046		.96	2.15		3.11	4.36
0250	On counter, 3′ x 4′ x 1/2″ sheets		180	.089		.78	4.17		4.95	7.25
0260	3′ x 5′ x 1/2″ sheets		180	.089		.76	4.17		4.93	7.25
0270	3′ x 6′ x 1/2″ sheets		180	.089		.74	4.17		4.91	7.20
0300	3′ x 4′ x 5/8″ sheets		180	.089		.99	4.17		5.16	7.50
0310	3′ x 5′ x 5/8″ sheets		180	.089		.99	4.17		5.16	7.50
0320	3′ x 6′ x 5/8″ sheets		180	.089		.96	4.17		5.13	7.45

09 29 Gypsum Board

09 29 10 – Gypsum Board Panels

09 29 10.30 Gypsum Board

		Crew	Daily Output	Labor-Hours	Unit	Material	Labor	Equipment	Total	Total Incl O&P
0010	**GYPSUM BOARD** on walls & ceilings R092910-10									
0100	Nailed or screwed to studs unless otherwise noted									
0110	1/4″ thick, on walls or ceilings, standard, no finish included	2 Carp	1330	.012	S.F.	.35	.56		.91	1.26
0115	1/4″ thick, on walls or ceilings, flexible, no finish included		1050	.015		.50	.72		1.22	1.65
0117	1/4″ thick, on columns or soffits, flexible, no finish included		1050	.015		.50	.72		1.22	1.65
0130	1/4″ thick, standard, no finish included, less than 800 S.F.		510	.031		.35	1.47		1.82	2.66
0150	3/8″ thick, on walls, standard, no finish included		2000	.008		.34	.38		.72	.95
0200	On ceilings, standard, no finish included		1800	.009		.34	.42		.76	1.01
0250	On beams, columns, or soffits, no finish included		675	.024		.34	1.11		1.45	2.08
0300	1/2″ thick, on walls, standard, no finish included		2000	.008		.30	.38		.68	.91
0350	Taped and finished (level 4 finish)		965	.017		.35	.78		1.13	1.58
0390	With compound skim coat (level 5 finish)		775	.021		.40	.97		1.37	1.93
0400	Fire resistant, no finish included		2000	.008		.35	.38		.73	.97
0450	Taped and finished (level 4 finish)		965	.017		.40	.78		1.18	1.64
0490	With compound skim coat (level 5 finish)		775	.021		.45	.97		1.42	1.99
0500	Water resistant, no finish included		2000	.008		.39	.38		.77	1.01
0550	Taped and finished (level 4 finish)		965	.017		.44	.78		1.22	1.68

09 29 10.30 Gypsum Board		Crew	Daily Output	Labor-Hours	Unit	Material	2015 Bare Costs Labor	Equipment	Total	Total Incl O&P
0590	With compound skim coat (level 5 finish)	2 Carp	775	.021	S.F.	.49	.97		1.46	2.03
0600	Prefinished, vinyl, clipped to studs		900	.018		.48	.83		1.31	1.81
0700	Mold resistant, no finish included		2000	.008		.44	.38		.82	1.06
0710	Taped and finished (level 4 finish)		965	.017		.49	.78		1.27	1.74
0720	With compound skim coat (level 5 finish)		775	.021		.54	.97		1.51	2.08
1000	On ceilings, standard, no finish included		1800	.009		.30	.42		.72	.97
1050	Taped and finished (level 4 finish)		765	.021		.35	.98		1.33	1.89
1090	With compound skim coat (level 5 finish)		610	.026		.40	1.23		1.63	2.34
1100	Fire resistant, no finish included		1800	.009		.35	.42		.77	1.03
1150	Taped and finished (level 4 finish)		765	.021		.40	.98		1.38	1.95
1195	With compound skim coat (level 5 finish)		610	.026		.45	1.23		1.68	2.40
1200	Water resistant, no finish included		1800	.009		.39	.42		.81	1.07
1250	Taped and finished (level 4 finish)		765	.021		.44	.98		1.42	1.99
1290	With compound skim coat (level 5 finish)		610	.026		.49	1.23		1.72	2.44
1310	Mold resistant, no finish included		1800	.009		.44	.42		.86	1.12
1320	Taped and finished (level 4 finish)		765	.021		.49	.98		1.47	2.05
1330	With compound skim coat (level 5 finish)		610	.026		.54	1.23		1.77	2.49
1350	Sag resistant, no finish included		1600	.010		.34	.47		.81	1.09
1360	Taped and finished (level 4 finish)		765	.021		.39	.98		1.37	1.94
1370	With compound skim coat (level 5 finish)		610	.026		.44	1.23		1.67	2.38
1500	On beams, columns, or soffits, standard, no finish included		675	.024		.35	1.11		1.46	2.09
1550	Taped and finished (level 4 finish)		540	.030		.35	1.39		1.74	2.52
1590	With compound skim coat (level 5 finish)		475	.034		.40	1.58		1.98	2.87
1600	Fire resistant, no finish included		675	.024		.35	1.11		1.46	2.10
1650	Taped and finished (level 4 finish)		540	.030		.40	1.39		1.79	2.58
1690	With compound skim coat (level 5 finish)		475	.034		.45	1.58		2.03	2.93
1700	Water resistant, no finish included		675	.024		.45	1.11		1.56	2.20
1750	Taped and finished (level 4 finish)		540	.030		.44	1.39		1.83	2.62
1790	With compound skim coat (level 5 finish)		475	.034		.49	1.58		2.07	2.97
1800	Mold resistant, no finish included		675	.024		.51	1.11		1.62	2.27
1810	Taped and finished (level 4 finish)		540	.030		.49	1.39		1.88	2.68
1820	With compound skim coat (level 5 finish)		475	.034		.54	1.58		2.12	3.02
1850	Sag resistant, no finish included		675	.024		.39	1.11		1.50	2.14
1860	Taped and finished (level 4 finish)		540	.030		.39	1.39		1.78	2.57
1870	With compound skim coat (level 5 finish)		475	.034		.44	1.58		2.02	2.91
2000	5/8" thick, on walls, standard, no finish included		2000	.008		.33	.38		.71	.94
2050	Taped and finished (level 4 finish)		965	.017		.38	.78		1.16	1.61
2090	With compound skim coat (level 5 finish)		775	.021		.43	.97		1.40	1.96
2100	Fire resistant, no finish included		2000	.008		.34	.38		.72	.95
2150	Taped and finished (level 4 finish)		965	.017		.39	.78		1.17	1.63
2195	With compound skim coat (level 5 finish)		775	.021		.44	.97		1.41	1.97
2200	Water resistant, no finish included		2000	.008		.42	.38		.80	1.04
2250	Taped and finished (level 4 finish)		965	.017		.47	.78		1.25	1.71
2290	With compound skim coat (level 5 finish)		775	.021		.52	.97		1.49	2.06
2300	Prefinished, vinyl, clipped to studs		900	.018		.76	.83		1.59	2.12
2510	Mold resistant, no finish included		2000	.008		.46	.38		.84	1.09
2520	Taped and finished (level 4 finish)		965	.017		.51	.78		1.29	1.76
2530	With compound skim coat (level 5 finish)		775	.021		.56	.97		1.53	2.11
3000	On ceilings, standard, no finish included		1800	.009		.33	.42		.75	1
3050	Taped and finished (level 4 finish)		765	.021		.38	.98		1.36	1.92
3090	With compound skim coat (level 5 finish)		615	.026		.43	1.22		1.65	2.35
3100	Fire resistant, no finish included		1800	.009		.34	.42		.76	1.01
3150	Taped and finished (level 4 finish)		765	.021		.39	.98		1.37	1.94

09 29 10.30 Gypsum Board		Crew	Daily Output	Labor-Hours	Unit	2015 Bare Costs Material	Labor	Equipment	Total	Total Incl O&P
3190	With compound skim coat (level 5 finish)	2 Carp	615	.026	S.F.	.44	1.22		1.66	2.36
3200	Water resistant, no finish included		1800	.009		.42	.42		.84	1.10
3250	Taped and finished (level 4 finish)		765	.021		.47	.98		1.45	2.02
3290	With compound skim coat (level 5 finish)		615	.026		.52	1.22		1.74	2.45
3300	Mold resistant, no finish included		1800	.009		.46	.42		.88	1.15
3310	Taped and finished (level 4 finish)		765	.021		.51	.98		1.49	2.07
3320	With compound skim coat (level 5 finish)		615	.026		.56	1.22		1.78	2.50
3500	On beams, columns, or soffits, no finish included		675	.024		.38	1.11		1.49	2.13
3550	Taped and finished (level 4 finish)		475	.034		.43	1.58		2.01	2.91
3590	With compound skim coat (level 5 finish)		380	.042		.49	1.98		2.47	3.58
3600	Fire resistant, no finish included		675	.024		.39	1.11		1.50	2.14
3650	Taped and finished (level 4 finish)		475	.034		.45	1.58		2.03	2.92
3690	With compound skim coat (level 5 finish)		380	.042		.44	1.98		2.42	3.52
3700	Water resistant, no finish included		675	.024		.48	1.11		1.59	2.24
3750	Taped and finished (level 4 finish)		475	.034		.52	1.58		2.10	3
3790	With compound skim coat (level 5 finish)		380	.042		.54	1.98		2.52	3.63
3800	Mold resistant, no finish included		675	.024		.53	1.11		1.64	2.29
3810	Taped and finished (level 4 finish)		475	.034		.56	1.58		2.14	3.05
3820	With compound skim coat (level 5 finish)		380	.042		.58	1.98		2.56	3.68
4000	Fireproofing, beams or columns, 2 layers, 1/2" thick, incl finish		330	.048		.79	2.28		3.07	4.37
4010	Mold resistant		330	.048		.97	2.28		3.25	4.57
4050	5/8" thick		300	.053		.77	2.50		3.27	4.70
4060	Mold resistant		300	.053		1.01	2.50		3.51	4.97
4100	3 layers, 1/2" thick		225	.071		1.19	3.34		4.53	6.45
4110	Mold resistant		225	.071		1.46	3.34		4.80	6.75
4150	5/8" thick		210	.076		1.16	3.58		4.74	6.80
4160	Mold resistant		210	.076		1.52	3.58		5.10	7.15
5050	For 1" thick coreboard on columns		480	.033		.78	1.56		2.34	3.27
5100	For foil-backed board, add					.15			.15	.17
5200	For work over 8' high, add	2 Carp	3060	.005			.25		.25	.38
5270	For textured spray, add	2 Lath	1600	.010		.04	.43		.47	.67
5300	For distribution cost over 3 stories high, add per story	2 Carp	6100	.003			.12		.12	.19
5350	For finishing inner corners, add		950	.017	L.F.	.10	.79		.89	1.33
5355	For finishing outer corners, add		1250	.013		.22	.60		.82	1.17
5500	For acoustical sealant, add per bead	1 Carp	500	.016		.04	.75		.79	1.21
5550	Sealant, 1 quart tube				Ea.	7.05			7.05	7.80
6000	Gypsum sound dampening panels									
6010	1/2" thick on walls, multi-layer, light weight, no finish included	2 Carp	1500	.011	S.F.	1.87	.50		2.37	2.83
6015	Taped and finished (level 4 finish)		725	.022		1.92	1.04		2.96	3.70
6020	With compound skim coat (level 5 finish)		580	.028		1.97	1.30		3.27	4.16
6025	5/8" thick on walls, for wood studs, no finish included		1500	.011		2.17	.50		2.67	3.16
6030	Taped and finished (level 4 finish)		725	.022		2.22	1.04		3.26	4.03
6035	With compound skim coat (level 5 finish)		580	.028		2.27	1.30		3.57	4.49
6040	For metal stud, no finish included		1500	.011		2.06	.50		2.56	3.04
6045	Taped and finished (level 4 finish)		725	.022		2.11	1.04		3.15	3.91
6050	With compound skim coat (level 5 finish)		580	.028		2.16	1.30		3.46	4.37
6055	Abuse resist, no finish included		1500	.011		3.75	.50		4.25	4.90
6060	Taped and finished (level 4 finish)		725	.022		3.80	1.04		4.84	5.75
6065	With compound skim coat (level 5 finish)		580	.028		3.85	1.30		5.15	6.25
6070	Shear rated, no finish included		1500	.011		4.30	.50		4.80	5.50
6075	Taped and finished (level 4 finish)		725	.022		4.35	1.04		5.39	6.35
6080	With compound skim coat (level 5 finish)		580	.028		4.40	1.30		5.70	6.85
6085	For SCIF applications, no finish included		1500	.011		4.72	.50		5.22	5.95

09 29 10 – Gypsum Board Panels

09 29 10.30 Gypsum Board		Crew	Daily Output	Labor-Hours	Unit	Material	Labor	Equipment	Total	Total Incl O&P
						2015 Bare Costs				
6090	Taped and finished (level 4 finish)	2 Carp	725	.022	S.F.	4.77	1.04		5.81	6.85
6095	With compound skim coat (level 5 finish)		580	.028		4.82	1.30		6.12	7.30
6100	1-3/8" thick on walls, THX Certified, no finish included		1500	.011		8.40	.50		8.90	10
6105	Taped and finished (level 4 finish)		725	.022		8.45	1.04		9.49	10.90
6110	With compound skim coat (level 5 finish)		580	.028		8.50	1.30		9.80	11.35
6115	5/8" thick on walls, score & snap installation, no finish included		2000	.008		1.69	.38		2.07	2.44
6120	Taped and finished (level 4 finish)		965	.017		1.74	.78		2.52	3.11
6125	With compound skim coat (level 5 finish)		775	.021		1.79	.97		2.76	3.46
7020	5/8" thick on ceilings, for wood joists, no finish included		1200	.013		2.17	.63		2.80	3.35
7025	Taped and finished (level 4 finish)		510	.031		2.22	1.47		3.69	4.71
7030	With compound skim coat (level 5 finish)		410	.039		2.27	1.83		4.10	5.30
7035	For metal joists, no finish included		1200	.013		2.06	.63		2.69	3.23
7040	Taped and finished (level 4 finish)		510	.031		2.11	1.47		3.58	4.59
7045	With compound skim coat (level 5 finish)		410	.039		2.16	1.83		3.99	5.20
7050	Abuse resist, no finish included		1200	.013		3.75	.63		4.38	5.10
7055	Taped and finished (level 4 finish)		510	.031		3.80	1.47		5.27	6.45
7060	With compound skim coat (level 5 finish)		410	.039		3.85	1.83		5.68	7.05
7065	Shear rated, no finish included		1200	.013		4.30	.63		4.93	5.70
7070	Taped and finished (level 4 finish)		510	.031		4.35	1.47		5.82	7.05
7075	With compound skim coat (level 5 finish)		410	.039		4.40	1.83		6.23	7.65
7080	For SCIF applications, no finish included		1200	.013		4.72	.63		5.35	6.15
7085	Taped and finished (level 4 finish)		510	.031		4.77	1.47		6.24	7.50
7090	With compound skim coat (level 5 finish)		410	.039		4.82	1.83		6.65	8.10
8010	5/8" thick on ceilings, score & snap installation, no finish included		1600	.010		1.69	.47		2.16	2.58
8015	Taped and finished (level 4 finish)		680	.024		1.74	1.10		2.84	3.61
8020	With compound skim coat (level 5 finish)		545	.029		1.79	1.38		3.17	4.09

09 29 10.50 High Abuse Gypsum Board		Crew	Daily Output	Labor-Hours	Unit	Material	Labor	Equipment	Total	Total Incl O&P
0010	**HIGH ABUSE GYPSUM BOARD**, fiber reinforced, nailed or									
0100	screwed to studs unless otherwise noted									
0110	1/2" thick, on walls, no finish included	2 Carp	1800	.009	S.F.	.70	.42		1.12	1.41
0120	Taped and finished (level 4 finish)		870	.018		.75	.86		1.61	2.15
0130	With compound skim coat (level 5 finish)		700	.023		.80	1.07		1.87	2.53
0150	On ceilings, no finish included		1620	.010		.70	.46		1.16	1.48
0160	Taped and finished (level 4 finish)		690	.023		.75	1.09		1.84	2.50
0170	With compound skim coat (level 5 finish)		550	.029		.80	1.37		2.17	2.98
0210	5/8" thick, on walls, no finish included		1800	.009		.85	.42		1.27	1.58
0220	Taped and finished (level 4 finish)		870	.018		.90	.86		1.76	2.32
0230	With compound skim coat (level 5 finish)		700	.023		.95	1.07		2.02	2.70
0250	On ceilings, no finish included		1620	.010		.85	.46		1.31	1.65
0260	Taped and finished (level 4 finish)		690	.023		.90	1.09		1.99	2.67
0270	With compound skim coat (level 5 finish)		550	.029		.95	1.37		2.32	3.15
0310	5/8" thick, on walls, very high impact, no finish included		1800	.009		.98	.42		1.40	1.72
0320	Taped and finished (level 4 finish)		870	.018		1.03	.86		1.89	2.46
0330	With compound skim coat (level 5 finish)		700	.023		1.08	1.07		2.15	2.84
0350	On ceilings, no finish included		1620	.010		.98	.46		1.44	1.79
0360	Taped and finished (level 4 finish)		690	.023		1.03	1.09		2.12	2.81
0370	With compound skim coat (level 5 finish)		550	.029		1.08	1.37		2.45	3.29
0400	High abuse, gypsum core, paper face									
0410	1/2" thick, on walls, no finish included	2 Carp	1800	.009	S.F.	.62	.42		1.04	1.32
0420	Taped and finished (level 4 finish)		870	.018		.67	.86		1.53	2.06
0430	With compound skim coat (level 5 finish)		700	.023		.72	1.07		1.79	2.44
0450	On ceilings, no finish included		1620	.010		.62	.46		1.08	1.39

09 29 Gypsum Board

09 29 10 – Gypsum Board Panels

09 29 10.50 High Abuse Gypsum Board		Crew	Daily Output	Labor-Hours	Unit	Material	2015 Bare Costs Labor	Equipment	Total	Total Incl O&P
0460	Taped and finished (level 4 finish)	2 Carp	690	.023	S.F.	.67	1.09		1.76	2.41
0470	With compound skim coat (level 5 finish)		550	.029		.72	1.37		2.09	2.89
0510	5/8" thick, on walls, no finish included		1800	.009		.66	.42		1.08	1.37
0520	Taped and finished (level 4 finish)		870	.018		.71	.86		1.57	2.11
0530	With compound skim coat (level 5 finish)		700	.023		.76	1.07		1.83	2.49
0550	On ceilings, no finish included		1620	.010		.66	.46		1.12	1.44
0560	Taped and finished (level 4 finish)		690	.023		.71	1.09		1.80	2.46
0570	With compound skim coat (level 5 finish)		550	.029		.76	1.37		2.13	2.94
1000	For high ceilings, over 8' high, add		2750	.006			.27		.27	.42
1010	For distribution cost over 3 stories high, add per story		5500	.003			.14		.14	.21

09 29 15 – Gypsum Board Accessories

09 29 15.10 Accessories, Gypsum Board		Crew	Daily Output	Labor-Hours	Unit	Material	Labor	Equipment	Total	Total Incl O&P
0010	**ACCESSORIES, GYPSUM BOARD**									
0020	Casing bead, galvanized steel	1 Carp	2.90	2.759	C.L.F.	24	130		154	226
0100	Vinyl		3	2.667		22	125		147	218
0300	Corner bead, galvanized steel, 1" x 1"		4	2		14.70	94		108.70	161
0400	1-1/4" x 1-1/4"		3.50	2.286		16.15	107		123.15	183
0600	Vinyl		4	2		20	94		114	167
0900	Furring channel, galv. steel, 7/8" deep, standard		2.60	3.077		33.50	144		177.50	259
1000	Resilient		2.55	3.137		25.50	147		172.50	255
1100	J trim, galvanized steel, 1/2" wide		3	2.667		22	125		147	217
1120	5/8" wide		2.95	2.712		31	127		158	230
1140	L trim, galvanized		3	2.667		19.30	125		144.30	214
1150	U trim, galvanized		2.95	2.712		22.50	127		149.50	221
1160	Screws #6 x 1" A				M	10.05			10.05	11.05
1170	#6 x 1-5/8" A				"	15.10			15.10	16.60
1200	For stud partitions, see Section 05 41 13.30 and 09 22 16.13									
1500	Z stud, galvanized steel, 1-1/2" wide	1 Carp	2.60	3.077	C.L.F.	38.50	144		182.50	265
1600	2" wide	"	2.55	3.137	"	62	147		209	295

09 30 Tiling

09 30 13 – Ceramic Tiling

09 30 13.10 Ceramic Tile		Crew	Daily Output	Labor-Hours	Unit	Material	Labor	Equipment	Total	Total Incl O&P
0010	**CERAMIC TILE**									
0020	Backsplash, thinset, average grade tiles	1 Tilf	50	.160	S.F.	2.46	6.85		9.31	12.85
0022	Custom grade tiles		50	.160		4.93	6.85		11.78	15.55
0024	Luxury grade tiles		50	.160		9.85	6.85		16.70	21
0026	Economy grade tiles		50	.160		2.25	6.85		9.10	12.65
0050	Base, using 1' x 4" high pc. with 1" x 1" tiles, mud set	D-7	82	.195	L.F.	5.20	7.45		12.65	16.70
0100	Thin set	"	128	.125		4.84	4.77		9.61	12.35
0300	For 6" high base, 1" x 1" tile face, add					.77			.77	.85
0400	For 2" x 2" tile face, add to above					.42			.42	.46
0600	Cove base, 4-1/4" x 4-1/4" high, mud set	D-7	91	.176		3.99	6.70		10.69	14.35
0700	Thin set		128	.125		3.87	4.77		8.64	11.30
0900	6" x 4-1/4" high, mud set		100	.160		4.54	6.10		10.64	14.05
1000	Thin set		137	.117		4.42	4.46		8.88	11.45
1200	Sanitary cove base, 6" x 4-1/4" high, mud set		93	.172		4.36	6.55		10.91	14.50
1300	Thin set		124	.129		4.24	4.93		9.17	11.95
1500	6" x 6" high, mud set		84	.190		5.35	7.25		12.60	16.65
1600	Thin set		117	.137		5.20	5.20		10.40	13.45

09 30 13 – Ceramic Tiling

09 30 13.10 Ceramic Tile		Crew	Daily Output	Labor-Hours	Unit	Material	Labor	Equipment	Total	Total Incl O&P
						2015 Bare Costs				
1800	Bathroom accessories, average (soap dish, tooth brush holder)	D-7	82	.195	Ea.	12	7.45		19.45	24
1900	Bathtub, 5′, rec. 4-1/4″ x 4-1/4″ tile wainscot, adhesive set 6′ high		2.90	5.517		156	211		367	480
2100	7′ high wainscot		2.50	6.400		179	244		423	555
2200	8′ high wainscot		2.20	7.273		190	278		468	620
2400	Bullnose trim, 4-1/4″ x 4-1/4″, mud set		82	.195	L.F.	3.92	7.45		11.37	15.30
2500	Thin set		128	.125		3.84	4.77		8.61	11.25
2700	2″ x 6″ bullnose trim, mud set		84	.190		4.05	7.25		11.30	15.20
2800	Thin set		124	.129		3.99	4.93		8.92	11.70
3000	Floors, natural clay, random or uniform, thin set, color group 1		183	.087	S.F.	4.15	3.34		7.49	9.50
3100	Color group 2		183	.087		5.85	3.34		9.19	11.40
3255	Floors, glazed, thin set, 6″ x 6″, color group 1		300	.053		4.45	2.04		6.49	7.90
3260	8″ x 8″ tile		300	.053		4.45	2.04		6.49	7.90
3270	12″ x 12″ tile		290	.055		6.25	2.11		8.36	10
3280	16″ x 16″ tile		280	.057		6.70	2.18		8.88	10.65
3281	18″ x 18″ tile		270	.059		8.65	2.26		10.91	12.85
3282	20″ x 20″ tile		260	.062		9.90	2.35		12.25	14.35
3283	24″ x 24″ tile		250	.064		11.25	2.44		13.69	15.95
3285	Border, 6″ x 12″ tile		200	.080		12.55	3.05		15.60	18.30
3290	3″ x 12″ tile		200	.080		40	3.05		43.05	48.50
3300	Porcelain type, 1 color, color group 2, 1″ x 1″		183	.087		5.20	3.34		8.54	10.65
3310	2″ x 2″ or 2″ x 1″, thin set		190	.084		6.10	3.22		9.32	11.45
3350	For random blend, 2 colors, add					1			1	1.10
3360	4 colors, add					1.50			1.50	1.65
3370	For color group 3, add					.65			.65	.72
3380	For abrasive non-slip tile, add					.44			.44	.48
4300	Specialty tile, 4-1/4″ x 4-1/4″ x 1/2″, decorator finish	D-7	183	.087		10.40	3.34		13.74	16.40
4500	Add for epoxy grout, 1/16″ joint, 1″ x 1″ tile		800	.020		.67	.76		1.43	1.87
4600	2″ x 2″ tile		820	.020		.62	.74		1.36	1.78
4610	Add for epoxy grout, 1/8″ joint, 8″ x 8″ x 3/8″ tile, add		900	.018		1.44	.68		2.12	2.59
4800	Pregrouted sheets, walls, 4-1/4″ x 4-1/4″, 6″ x 4-1/4″									
4810	and 8-1/2″ x 4-1/4″, 4 S.F. sheets, silicone grout	D-7	240	.067	S.F.	5.10	2.55		7.65	9.40
5100	Floors, unglazed, 2 S.F. sheets,									
5110	Urethane adhesive	D-7	180	.089	S.F.	5.10	3.39		8.49	10.60
5400	Walls, interior, thin set, 4-1/4″ x 4-1/4″ tile		190	.084		2.26	3.22		5.48	7.25
5500	6″ x 4-1/4″ tile		190	.084		2.92	3.22		6.14	7.95
5700	8-1/2″ x 4-1/4″ tile		190	.084		4.86	3.22		8.08	10.10
5800	6″ x 6″ tile		175	.091		3.28	3.49		6.77	8.75
5810	8″ x 8″ tile		170	.094		4.44	3.59		8.03	10.20
5820	12″ x 12″ tile		160	.100		4.35	3.82		8.17	10.45
5830	16″ x 16″ tile		150	.107		4.77	4.07		8.84	11.25
6000	Decorated wall tile, 4-1/4″ x 4-1/4″, color group 1		270	.059		3.18	2.26		5.44	6.85
6100	Color group 4		180	.089		49.50	3.39		52.89	59.50
6300	Exterior walls, frostproof, mud set, 4-1/4″ x 4-1/4″		102	.157		7.15	6		13.15	16.70
6400	1-3/8″ x 1-3/8″		93	.172		6.10	6.55		12.65	16.40
6600	Crystalline glazed, 4-1/4″ x 4-1/4″, mud set, plain		100	.160		4.36	6.10		10.46	13.85
6700	4-1/4″ x 4-1/4″, scored tile		100	.160		5.80	6.10		11.90	15.45
6900	6″ x 6″ plain		93	.172		6.70	6.55		13.25	17.05
7000	For epoxy grout, 1/16″ joints, 4-1/4″ tile, add		800	.020		.41	.76		1.17	1.58
7200	For tile set in dry mortar, add		1735	.009			.35		.35	.52
7300	For tile set in Portland cement mortar, add		290	.055		.16	2.11		2.27	3.29
9300	Ceramic tiles, recycled glass, standard colors, 2″ x 2″ thru 6″ x 6″ G		190	.084		21	3.22		24.22	28.50
9310	6″ x 6″ G		175	.091		21.50	3.49		24.99	28.50
9320	8″ x 8″ G		170	.094		22.50	3.59		26.09	30.50

09 30 Tiling

09 30 13 – Ceramic Tiling

09 30 13.10 Ceramic Tile			Crew	Daily Output	Labor-Hours	Unit	Material	2015 Bare Costs Labor	Equipment	Total	Total Incl O&P
9330	12″ x 12″	G	D-7	160	.100	S.F.	22.50	3.82		26.32	30.50
9340	Earthtones, 2″ x 2″ to 4″ x 8″	G		190	.084		25	3.22		28.22	32.50
9350	6″ x 6″	G		175	.091		25	3.49		28.49	32.50
9360	8″ x 8″	G		170	.094		26	3.59		29.59	34
9370	12″ x 12″	G		160	.100		26	3.82		29.82	34
9380	Deep colors, 2″ x 2″ to 4″ x 8″	G		190	.084		29.50	3.22		32.72	37.50
9390	6″ x 6″	G		175	.091		29.50	3.49		32.99	37.50
9400	8″ x 8″	G		170	.094		31	3.59		34.59	39.50
9410	12″ x 12″	G		160	.100		31	3.82		34.82	39.50
09 30 13.20 Ceramic Tile Repairs											
0010	**CERAMIC TILE REPAIRS**										
1000	Grout removal, carbide tipped, rotary grinder		1 Clab	240	.033	L.F.		1.25		1.25	1.93
1100	Regrout tile 4-1/2 x 4-1/2, or larger, wall		1 Tilf	100	.080	S.F.	.14	3.42		3.56	5.20
1150	Floor			125	.064		.15	2.74		2.89	4.22
1200	Seal tile and grout			360	.022			.95		.95	1.41
09 30 13.45 Ceramic Tile Accessories											
0010	**CERAMIC TILE ACCESSORIES**										
0100	Spacers, 1/8″					C	1.98			1.98	2.18
1310	Sealer for natural stone tile, installed		1 Tilf	650	.012	S.F.	.05	.53		.58	.84

09 30 16 – Quarry Tiling

09 30 16.10 Quarry Tile		Crew	Daily Output	Labor-Hours	Unit	Material	Labor	Equipment	Total	Total Incl O&P
0010	**QUARRY TILE**									
0100	Base, cove or sanitary, mud set, to 5″ high, 1/2″ thick	D-7	110	.145	L.F.	5.35	5.55		10.90	14.10
0300	Bullnose trim, red, mud set, 6″ x 6″ x 1/2″ thick		120	.133		4.39	5.10		9.49	12.40
0400	4″ x 4″ x 1/2″ thick		110	.145		4.50	5.55		10.05	13.15
0600	4″ x 8″ x 1/2″ thick, using 8″ as edge		130	.123		4.50	4.70		9.20	11.90
0700	Floors, mud set, 1,000 S.F. lots, red, 4″ x 4″ x 1/2″ thick		120	.133	S.F.	8.05	5.10		13.15	16.40
0900	6″ x 6″ x 1/2″ thick		140	.114		7.55	4.36		11.91	14.75
1000	4″ x 8″ x 1/2″ thick		130	.123		5.55	4.70		10.25	13.05
1300	For waxed coating, add					.75			.75	.83
1500	For non-standard colors, add					.46			.46	.51
1600	For abrasive surface, add					.52			.52	.57
1800	Brown tile, imported, 6″ x 6″ x 3/4″	D-7	120	.133		7.95	5.10		13.05	16.30
1900	8″ x 8″ x 1″		110	.145		8.70	5.55		14.25	17.75
2100	For thin set mortar application, deduct		700	.023			.87		.87	1.29
2200	For epoxy grout & mortar, 6″ x 6″ x 1/2″, add		350	.046		2.04	1.75		3.79	4.82
2700	Stair tread, 6″ x 6″ x 3/4″, plain		50	.320		7.20	12.20		19.40	26
2800	Abrasive		47	.340		6.05	13		19.05	26
3000	Wainscot, 6″ x 6″ x 1/2″, thin set, red		105	.152		4.68	5.80		10.48	13.75
3100	Non-standard colors		105	.152		5.20	5.80		11	14.30
3300	Window sill, 6″ wide, 3/4″ thick		90	.178	L.F.	9.20	6.80		16	20
3400	Corners		80	.200	Ea.	6.65	7.65		14.30	18.60

09 30 23 – Glass Mosaic Tiling

09 30 23.10 Glass Mosaics		Crew	Daily Output	Labor-Hours	Unit	Material	Labor	Equipment	Total	Total Incl O&P
0010	**GLASS MOSAICS** 3/4″ tile on 12″ sheets, standard grout									
0300	Color group 1 & 2	D-7	73	.219	S.F.	17.20	8.35		25.55	31.50
0350	Color group 3		73	.219		20.50	8.35		28.85	35
0400	Color group 4		73	.219		26.50	8.35		34.85	42
0450	Color group 5		73	.219		29	8.35		37.35	44.50
0500	Color group 6		73	.219		40	8.35		48.35	56.50
0600	Color group 7		73	.219		40.50	8.35		48.85	57

09 30 Tiling

09 30 23 – Glass Mosaic Tiling

	09 30 23.10 Glass Mosaics	Crew	Daily Output	Labor-Hours	Unit	Material	2015 Bare Costs Labor	Equipment	Total	Total Incl O&P
0700	Color group 8, golds, silvers & specialties	D-7	64	.250	S.F.	41	9.55		50.55	59.50
1020	1" tile on 12" sheets, opalescent finish		73	.219		16.85	8.35		25.20	31
1040	1" x 2" tile on 12" sheet, blend		73	.219		18.15	8.35		26.50	32.50
1060	2" tile on 12" sheet, blend		73	.219		16.50	8.35		24.85	30.50
1080	5/8" x random tile, linear, on 12" sheet, blend		73	.219		26	8.35		34.35	41
1600	Dots on 12" sheet		73	.219		26	8.35		34.35	41
1700	For glass mosaic tiles set in dry mortar, add		290	.055		.45	2.11		2.56	3.61
1720	For glass mosaic tile set in Portland cement mortar, add		290	.055		.01	2.11		2.12	3.12
1730	For polyblend sanded tile grout		96.15	.166	Lb.	2.19	6.35		8.54	11.80

09 30 29 – Metal Tiling

	09 30 29.10 Metal Tile	Crew	Daily Output	Labor-Hours	Unit	Material	Labor	Equipment	Total	Total Incl O&P
0010	**METAL TILE** 4' x 4' sheet, 24 ga., tile pattern, nailed									
0200	Stainless steel	2 Carp	512	.031	S.F.	28	1.47		29.47	33.50
0400	Aluminized steel	"	512	.031	"	15.10	1.47		16.57	18.85

09 34 Waterproofing-Membrane Tiling

09 34 13 – Waterproofing-Membrane Ceramic Tiling

	09 34 13.10 Ceramic Tile Waterproofing Membrane	Crew	Daily Output	Labor-Hours	Unit	Material	Labor	Equipment	Total	Total Incl O&P
0010	**CERAMIC TILE WATERPROOFING MEMBRANE**									
0020	On floors, including thinset									
0030	Fleece laminated polyethylene grid, 1/8" thick	D-7	250	.064	S.F.	2.26	2.44		4.70	6.10
0040	5/16" thick	"	250	.064	"	2.58	2.44		5.02	6.45
0050	On walls, including thinset									
0060	Fleece laminated polyethylene sheet, 8 mil thick	D-7	480	.033	S.F.	2.26	1.27		3.53	4.37
0070	Accessories, including thinset									
0080	Joint and corner sheet, 4 mils thick, 5" wide	1 Tilf	240	.033	L.F.	1.33	1.43		2.76	3.58
0090	7-1/4" wide		180	.044		1.69	1.90		3.59	4.67
0100	10" wide		120	.067		2.06	2.85		4.91	6.50
0110	Pre-formed corners, inside		32	.250	Ea.	6.90	10.70		17.60	23.50
0120	Outside		32	.250		7.65	10.70		18.35	24.50
0130	2" flanged floor drain with 6" stainless steel grate		16	.500		370	21.50		391.50	440
0140	EPS, sloped shower floor		480	.017	S.F.	4.95	.71		5.66	6.50
0150	Curb		32	.250	L.F.	14	10.70		24.70	31.50

09 51 Acoustical Ceilings

09 51 23 – Acoustical Tile Ceilings

	09 51 23.10 Suspended Acoustic Ceiling Tiles	Crew	Daily Output	Labor-Hours	Unit	Material	Labor	Equipment	Total	Total Incl O&P
0010	**SUSPENDED ACOUSTIC CEILING TILES**, not including									
0100	suspension system									
0300	Fiberglass boards, film faced, 2' x 2' or 2' x 4', 5/8" thick	1 Carp	625	.013	S.F.	1.24	.60		1.84	2.28
0400	3/4" thick		600	.013		2.63	.63		3.26	3.85
0500	3" thick, thermal, R11		450	.018		2.42	.83		3.25	3.94
0600	Glass cloth faced fiberglass, 3/4" thick		500	.016		2.65	.75		3.40	4.08
0700	1" thick		485	.016		3.13	.77		3.90	4.63
0820	1-1/2" thick, nubby face		475	.017		2.53	.79		3.32	4
1110	Mineral fiber tile, lay-in, 2' x 2' or 2' x 4', 5/8" thick, fine texture		625	.013		.91	.60		1.51	1.92
1115	Rough textured		625	.013		.85	.60		1.45	1.86
1125	3/4" thick, fine textured		600	.013		1.94	.63		2.57	3.09
1130	Rough textured		600	.013		1.56	.63		2.19	2.68

09 51 Acoustical Ceilings

09 51 23 – Acoustical Tile Ceilings

09 51 23.10 Suspended Acoustic Ceiling Tiles		Crew	Daily Output	Labor-Hours	Unit	2015 Bare Costs Material	Labor	Equipment	Total	Total Incl O&P
1135	Fissured	1 Carp	600	.013	S.F.	1.96	.63		2.59	3.12
1150	Tegular, 5/8" thick, fine textured		470	.017		1.03	.80		1.83	2.36
1155	Rough textured		470	.017		1.14	.80		1.94	2.48
1165	3/4" thick, fine textured		450	.018		2.14	.83		2.97	3.63
1170	Rough textured		450	.018		1.43	.83		2.26	2.85
1175	Fissured		450	.018		2.16	.83		2.99	3.66
1185	For plastic film face, add					.75			.75	.83
1190	For fire rating, add					.44			.44	.48
1300	Metal panel, lay-in, 2' x 2', sq. edge	1 Carp	500	.016		9.55	.75		10.30	11.65
1350	Tegular edge		500	.016		13.20	.75		13.95	15.65
1400	2' x 4', sq. edge		500	.016		12.90	.75		13.65	15.35
1450	Tegular edge		500	.016		13.20	.75		13.95	15.65
1500	Perforated alum. clip-in, 2' x 2'		500	.016		13.45	.75		14.20	15.90
1550	2' x 4'		500	.016		10.85	.75		11.60	13.10
1600	Solid alum. planks, 3-1/4"x12', open reveal		500	.016		2.35	.75		3.10	3.75
1650	Closed reveal		500	.016		3	.75		3.75	4.46
1700	7-1/4"x12', open reveal		500	.016		4	.75		4.75	5.55
1750	Closed reveal		500	.016		5.10	.75		5.85	6.75
1775	Metal, open cell, 2'x2', 6" cell		500	.016		8	.75		8.75	9.95
1800	8" cell		500	.016		8.85	.75		9.60	10.90
1825	2'x4', 6" cell		500	.016		5.10	.75		5.85	6.75
1850	8" cell		500	.016		5.10	.75		5.85	6.75
1870	Translucent lay-in panels, 2'x2'		500	.016		23	.75		23.75	26
1890	2'x6'		500	.016		17.20	.75		17.95	20
3720	Mineral fiber, 24" x 24" or 48", reveal edge, painted, 5/8" thick		600	.013		1.15	.63		1.78	2.23
3740	3/4" thick		575	.014		1.52	.65		2.17	2.68
5020	66 – 78% recycled content, 3/4" thick G		600	.013		1.93	.63		2.56	3.08
5040	Mylar, 42% recycled content, 3/4" thick G		600	.013		4.54	.63		5.17	5.95
6000	Remove and replace ceiling tiles, min fiber, 2x2 or 2x4, 5/8"thk.		335	.024		.91	1.12		2.03	2.73

09 51 23.30 Suspended Ceilings, Complete		Crew	Daily Output	Labor-Hours	Unit	Material	Labor	Equipment	Total	Total Incl O&P
0010	**SUSPENDED CEILINGS, COMPLETE**, including standard									
0100	suspension system but not incl. 1-1/2" carrier channels									
0600	Fiberglass ceiling board, 2' x 4' x 5/8", plain faced	1 Carp	500	.016	S.F.	1.97	.75		2.72	3.33
0700	Offices, 2' x 4' x 3/4"		380	.021		3.36	.99		4.35	5.20
0800	Mineral fiber, on 15/16" T bar susp. 2' x 2' x 3/4" lay-in board		345	.023		2.89	1.09		3.98	4.86
0810	2' x 4' x 5/8" tile		380	.021		2.31	.99		3.30	4.07
0820	Tegular, 2' x 2' x 5/8" tile on 9/16" grid		250	.032		2.40	1.50		3.90	4.95
0830	2' x 4' x 3/4" tile		275	.029		2.60	1.37		3.97	4.97
0900	Luminous panels, prismatic, acrylic		255	.031		3.37	1.47		4.84	6
1200	Metal pan with acoustic pad, steel		75	.107		4.51	5		9.51	12.65
1300	Painted aluminum		75	.107		3.07	5		8.07	11.10
1500	Aluminum, degreased finish		75	.107		5.20	5		10.20	13.40
1600	Stainless steel		75	.107		9.75	5		14.75	18.40
1800	Tile, Z bar suspension, 5/8" mineral fiber tile		150	.053		2.32	2.50		4.82	6.40
1900	3/4" mineral fiber tile		150	.053		2.48	2.50		4.98	6.60
2400	For strip lighting, see Section 26 51 13.50									
2500	For rooms under 500 S.F., add				S.F.		25%			

09 51 53 – Direct-Applied Acoustical Ceilings

09 51 53.10 Ceiling Tile		Crew	Daily Output	Labor-Hours	Unit	Material	Labor	Equipment	Total	Total Incl O&P
0010	**CEILING TILE**, stapled or cemented									
0100	12" x 12" or 12" x 24", not including furring									
0600	Mineral fiber, vinyl coated, 5/8" thick	1 Carp	300	.027	S.F.	2.18	1.25		3.43	4.33

09 51 Acoustical Ceilings

09 51 53 – Direct-Applied Acoustical Ceilings

09 51 53.10 Ceiling Tile		Crew	Daily Output	Labor-Hours	Unit	Material	2015 Bare Costs Labor	Equipment	Total	Total Incl O&P
0700	3/4" thick	1 Carp	300	.027	S.F.	2.40	1.25		3.65	4.57
0900	Fire rated, 3/4" thick, plain faced		300	.027		1.27	1.25		2.52	3.33
1000	Plastic coated face		300	.027		1.84	1.25		3.09	3.95
1200	Aluminum faced, 5/8" thick, plain		300	.027		1.66	1.25		2.91	3.76
3700	Wall application of above, add	↓	1000	.008			.38		.38	.58
3900	For ceiling primer, add					.13			.13	.14
4000	For ceiling cement, add				↓	.39			.39	.43

09 53 Acoustical Ceiling Suspension Assemblies

09 53 23 – Metal Acoustical Ceiling Suspension Assemblies

09 53 23.30 Ceiling Suspension Systems		Crew	Daily Output	Labor-Hours	Unit	Material	Labor	Equipment	Total	Total Incl O&P
0010	**CEILING SUSPENSION SYSTEMS** for boards and tile									
0050	Class A suspension system, 15/16" T bar, 2' x 4' grid	1 Carp	800	.010	S.F.	.73	.47		1.20	1.53
0300	2' x 2' grid		650	.012		.95	.58		1.53	1.93
0310	25% recycled steel, 2' x 4' grid G		800	.010		.77	.47		1.24	1.57
0320	2' x 2' grid G	↓	650	.012		.96	.58		1.54	1.95
0350	For 9/16" grid, add					.16			.16	.18
0360	For fire rated grid, add					.09			.09	.10
0370	For colored grid, add					.21			.21	.23
0400	Concealed Z bar suspension system, 12" module	1 Carp	520	.015		.84	.72		1.56	2.03
0600	1-1/2" carrier channels, 4' O.C., add	"	470	.017	↓	.11	.80		.91	1.35
0700	Carrier channels for ceilings with									
0900	recessed lighting fixtures, add	1 Carp	460	.017	S.F.	.20	.82		1.02	1.48
1040	Hanging wire, 12 ga., 4' long		65	.123	C.S.F.	.37	5.80		6.17	9.30
1080	8' long	↓	65	.123	"	.74	5.80		6.54	9.70
3000	Seismic ceiling bracing, IBC Site Class D, Occupancy Category II									
3050	For ceilings less than 2500 S.F.									
3060	Seismic clips at attached walls	1 Carp	180	.044	Ea.	1.12	2.09		3.21	4.44
3100	For ceilings greater than 2500 S.F., add									
3120	Seismic clips, joints at cross tees	1 Carp	120	.067	Ea.	3.29	3.13		6.42	8.45
3140	At cross tees and mains, mains field cut	"	60	.133	"	3.29	6.25		9.54	13.25
3200	Compression posts, telescopic, attached to structure above									
3210	To 30" high	1 Carp	26	.308	Ea.	39	14.45		53.45	65
3220	30" to 48" high		25.50	.314		43.50	14.75		58.25	70.50
3230	48" to 84" high		25	.320		52.50	15		67.50	80.50
3240	84" to 102" high		24.50	.327		60	15.35		75.35	89.50
3250	102" to 120" high		24	.333		85.50	15.65		101.15	119
3260	120" to 144" high	↓	24	.333	↓	95	15.65		110.65	129
3300	Stabilizer bars									
3310	12" long	1 Carp	240	.033	Ea.	.97	1.56		2.53	3.48
3320	24" long		235	.034		.92	1.60		2.52	3.47
3330	36" long		230	.035		.89	1.63		2.52	3.49
3340	48" long	↓	220	.036	↓	.73	1.71		2.44	3.44
3400	Wire support for light fixtures, per L.F. height to structure above									
3410	Less than 10 lb.	1 Carp	400	.020	L.F.	.28	.94		1.22	1.76
3420	10 lb. to 56 lb.	"	240	.033	"	.56	1.56		2.12	3.02

09 54 Specialty Ceilings

09 54 26 – Suspended Wood Ceilings

09 54 26.10 Wood Ceilings		Crew	Daily Output	Labor-Hours	Unit	Material	2015 Bare Costs Labor	Equipment	Total	Total Incl O&P
0010	**WOOD CEILINGS**									
1000	4" - 6" wood slats on heavy duty 15/16" T-bar grid	2 Carp	250	.064	S.F.	24	3		27	31

09 54 33 – Decorative Panel Ceilings

09 54 33.20 Metal Panel Ceilings		Crew	Daily Output	Labor-Hours	Unit	Material	Labor	Equipment	Total	Total Incl O&P
0010	**METAL PANEL CEILINGS**									
0020	Lay-in or screwed to furring, not including grid									
0100	Tin ceilings, 2' x 2' or 2' x 4', bare steel finish	2 Carp	300	.053	S.F.	2.46	2.50		4.96	6.55
0120	Painted white finish		300	.053	"	3.71	2.50		6.21	7.95
0140	Copper, chrome or brass finish		300	.053	L.F.	6.55	2.50		9.05	11.05
0200	Cornice molding, 2-1/2" to 3-1/2" wide, 4' long, bare steel finish		200	.080	S.F.	2.21	3.76		5.97	8.25
0220	Painted white finish		200	.080		2.75	3.76		6.51	8.80
0240	Copper, chrome or brass finish		200	.080		3.91	3.76		7.67	10.10
0320	5" to 6-1/2" wide, 4' long, bare steel finish		150	.107		3.19	5		8.19	11.20
0340	Painted white finish		150	.107		4.05	5		9.05	12.15
0360	Copper, chrome or brass finish		150	.107		6.45	5		11.45	14.80
0420	Flat molding, 3-1/2" to 5" wide, 4' long, bare steel finish		250	.064		3.71	3		6.71	8.70
0440	Painted white finish		250	.064		3.96	3		6.96	9
0460	Copper, chrome or brass finish	↓	250	.064	↓	7.50	3		10.50	12.85

09 61 Flooring Treatment

09 61 19 – Concrete Floor Staining

09 61 19.40 Floors, Interior		Crew	Daily Output	Labor-Hours	Unit	Material	Labor	Equipment	Total	Total Incl O&P
0010	**FLOORS, INTERIOR**									
0300	Acid stain and sealer									
0310	Stain, one coat	1 Pord	650	.012	S.F.	.12	.50		.62	.88
0320	Two coats		570	.014		.23	.57		.80	1.11
0330	Acrylic sealer, one coat		2600	.003		.23	.12		.35	.44
0340	Two coats	↓	1400	.006	↓	.46	.23		.69	.86

09 62 Specialty Flooring

09 62 19 – Laminate Flooring

09 62 19.10 Floating Floor		Crew	Daily Output	Labor-Hours	Unit	Material	Labor	Equipment	Total	Total Incl O&P
0010	**FLOATING FLOOR**									
8300	Floating floor, laminate, wood pattern strip, complete	1 Clab	133	.060	S.F.	4.35	2.26		6.61	8.25
8310	Components, T&G wood composite strips					3.91			3.91	4.30
8320	Film					.14			.14	.15
8330	Foam					.25			.25	.28
8340	Adhesive					.65			.65	.72
8350	Installation kit				↓	.17			.17	.19
8360	Trim, 2" wide x 3' long				L.F.	4.30			4.30	4.73
8370	Reducer moulding				"	5.70			5.70	6.25

09 62 23 – Bamboo Flooring

09 62 23.10 Flooring, Bamboo		Crew	Daily Output	Labor-Hours	Unit	Material	Labor	Equipment	Total	Total Incl O&P
0010	**FLOORING, BAMBOO**									
8600	Flooring, wood, bamboo strips, unfinished, 5/8" x 4" x 3' G	1 Carp	255	.031	S.F.	4.60	1.47		6.07	7.30
8610	5/8" x 4" x 4' G		275	.029		4.77	1.37		6.14	7.35
8620	5/8" x 4" x 6' G		295	.027		5.25	1.27		6.52	7.70
8630	Finished, 5/8" x 4" x 3' G	↓	255	.031	↓	5.05	1.47		6.52	7.80

09 62 Specialty Flooring

09 62 23 – Bamboo Flooring

09 62 23.10 Flooring, Bamboo			Crew	Daily Output	Labor-Hours	Unit	Material	2015 Bare Costs Labor	Equipment	Total	Total Incl O&P
8640	5/8″ x 4″ x 4′	G	1 Carp	275	.029	S.F.	5.30	1.37		6.67	7.95
8650	5/8″ x 4″ x 6′	G		295	.027		4.61	1.27		5.88	7
8660	Stair treads, unfinished, 1-1/16″ x 11-1/2″ x 4′	G		18	.444	Ea.	44	21		65	80.50
8670	Finished, 1-1/16″ x 11-1/2″ x 4′	G		18	.444		78	21		99	118
8680	Stair risers, unfinished, 5/8″ x 7-1/2″ x 4′	G		18	.444		16.30	21		37.30	50
8690	Finished, 5/8″ x 7-1/2″ x 4′	G		18	.444		31	21		52	66
8700	Stair nosing, unfinished, 6′ long	G		16	.500		36	23.50		59.50	75.50
8710	Finished, 6′ long	G		16	.500		42.50	23.50		66	83

09 63 Masonry Flooring

09 63 13 – Brick Flooring

09 63 13.10 Miscellaneous Brick Flooring			Crew	Daily Output	Labor-Hours	Unit	Material	Labor	Equipment	Total	Total Incl O&P
0010	**MISCELLANEOUS BRICK FLOORING**										
0020	Acid-proof shales, red, 8″ x 3-3/4″ x 1-1/4″ thick		D-7	.43	37.209	M	695	1,425		2,120	2,875
0050	2-1/4″ thick		D-1	.40	40		965	1,675		2,640	3,625
0200	Acid-proof clay brick, 8″ x 3-3/4″ x 2-1/4″ thick	G		.40	40		935	1,675		2,610	3,600
0250	9″ x 4-1/2″ x 3″	G		95	.168	S.F.	4.14	7.10		11.24	15.40
0260	Cast ceramic, pressed, 4″ x 8″ x 1/2″, unglazed		D-7	100	.160		6.35	6.10		12.45	16
0270	Glazed			100	.160		8.45	6.10		14.55	18.35
0280	Hand molded flooring, 4″ x 8″ x 3/4″, unglazed			95	.168		8.35	6.45		14.80	18.70
0290	Glazed			95	.168		10.50	6.45		16.95	21
0300	8″ hexagonal, 3/4″ thick, unglazed			85	.188		9.20	7.20		16.40	21
0310	Glazed			85	.188		16.60	7.20		23.80	29
0400	Heavy duty industrial, cement mortar bed, 2″ thick, not incl. brick		D-1	80	.200		1.06	8.40		9.46	14
0450	Acid-proof joints, 1/4″ wide		″	65	.246		1.46	10.35		11.81	17.40
0500	Pavers, 8″ x 4″, 1″ to 1-1/4″ thick, red		D-7	95	.168		3.69	6.45		10.14	13.55
0510	Ironspot		″	95	.168		5.20	6.45		11.65	15.25
0540	1-3/8″ to 1-3/4″ thick, red		D-1	95	.168		3.56	7.10		10.66	14.75
0560	Ironspot			95	.168		5.15	7.10		12.25	16.55
0580	2-1/4″ thick, red			90	.178		3.62	7.50		11.12	15.45
0590	Ironspot			90	.178		5.60	7.50		13.10	17.65
0700	Paver, adobe brick, 6″ x 12″, 1/2″ joint	G		42	.381		1.30	16.05		17.35	26
0710	Mexican red, 12″ x 12″	G	1 Tilf	48	.167		1.66	7.15		8.81	12.40
0720	Saltillo, 12″ x 12″	G	″	48	.167		1.40	7.15		8.55	12.10
0800	For sidewalks and patios with pavers, see Section 32 14 16.10										
0870	For epoxy joints, add		D-1	600	.027	S.F.	2.81	1.12		3.93	4.80
0880	For Furan underlayment, add		″	600	.027		2.33	1.12		3.45	4.27
0890	For waxed surface, steam cleaned, add		A-1H	1000	.008		.20	.30	.08	.58	.76

09 63 40 – Stone Flooring

09 63 40.10 Marble		Crew	Daily Output	Labor-Hours	Unit	Material	Labor	Equipment	Total	Total Incl O&P
0010	**MARBLE**									
0020	Thin gauge tile, 12″ x 6″, 3/8″, white Carara	D-7	60	.267	S.F.	14.40	10.20		24.60	31
0100	Travertine		60	.267		11.60	10.20		21.80	28
0200	12″ x 12″ x 3/8″, thin set, floors		60	.267		10.15	10.20		20.35	26
0300	On walls		52	.308		9.85	11.75		21.60	28
1000	Marble threshold, 4″ wide x 36″ long x 5/8″ thick, white		60	.267	Ea.	10.15	10.20		20.35	26

09 63 40.20 Slate Tile		Crew	Daily Output	Labor-Hours	Unit	Material	Labor	Equipment	Total	Total Incl O&P
0010	**SLATE TILE**									
0020	Vermont, 6″ x 6″ x 1/4″ thick, thin set	D-7	180	.089	S.F.	7.50	3.39		10.89	13.25
0200	See also Section 32 14 40.10									

09 64 Wood Flooring

09 64 16 – Wood Block Flooring

09 64 16.10 End Grain Block Flooring		Crew	Daily Output	Labor-Hours	Unit	2015 Bare Costs Material	Labor	Equipment	Total	Total Incl O&P
0010	**END GRAIN BLOCK FLOORING**									
0020	End grain flooring, coated, 2" thick	1 Carp	295	.027	S.F.	3.57	1.27		4.84	5.90
0400	Natural finish, 1" thick, fir		125	.064		3.69	3		6.69	8.70
0600	1-1/2" thick, pine		125	.064		3.62	3		6.62	8.60
0700	2" thick, pine		125	.064		4.44	3		7.44	9.50

09 64 19 – Wood Composition Flooring

09 64 19.10 Wood Composition		Crew	Daily Output	Labor-Hours	Unit	Material	Labor	Equipment	Total	Total Incl O&P
0010	**WOOD COMPOSITION** Gym floors									
0100	2-1/4" x 6-7/8" x 3/8", on 2" grout setting bed	D-7	150	.107	S.F.	6.15	4.07		10.22	12.75
0200	Thin set, on concrete	"	250	.064		5.60	2.44		8.04	9.75
0300	Sanding and finishing, add	1 Carp	200	.040		.84	1.88		2.72	3.81

09 64 23 – Wood Parquet Flooring

09 64 23.10 Wood Parquet		Crew	Daily Output	Labor-Hours	Unit	Material	Labor	Equipment	Total	Total Incl O&P
0010	**WOOD PARQUET** flooring									
5200	Parquetry, 5/16" thk, no finish, oak, plain pattern	1 Carp	160	.050	S.F.	5.25	2.35		7.60	9.35
5300	Intricate pattern		100	.080		9.60	3.76		13.36	16.35
5500	Teak, plain pattern		160	.050		5.85	2.35		8.20	10.05
5600	Intricate pattern		100	.080		10.05	3.76		13.81	16.85
5650	13/16" thick, select grade oak, plain pattern		160	.050		9.75	2.35		12.10	14.30
5700	Intricate pattern		100	.080		16.40	3.76		20.16	24
5800	Custom parquetry, including finish, plain pattern		100	.080		16.80	3.76		20.56	24.50
5900	Intricate pattern		50	.160		24	7.50		31.50	37.50
6700	Parquetry, prefinished white oak, 5/16" thick, plain pattern		160	.050		7.90	2.35		10.25	12.30
6800	Intricate pattern		100	.080		8.45	3.76		12.21	15.05
7000	Walnut or teak, parquetry, plain pattern		160	.050		8	2.35		10.35	12.40
7100	Intricate pattern		100	.080		11.50	3.76		15.26	18.45
7200	Acrylic wood parquet blocks, 12" x 12" x 5/16",									
7210	Irradiated, set in epoxy	1 Carp	160	.050	S.F.	10	2.35		12.35	14.60

09 64 29 – Wood Strip and Plank Flooring

09 64 29.10 Wood		Crew	Daily Output	Labor-Hours	Unit	Material	Labor	Equipment	Total	Total Incl O&P
0010	**WOOD**									
0020	Fir, vertical grain, 1" x 4", not incl. finish, grade B & better	1 Carp	255	.031	S.F.	2.79	1.47		4.26	5.35
0100	C grade & better		255	.031		2.63	1.47		4.10	5.15
4000	Maple, strip, 25/32" x 2-1/4", not incl. finish, select		170	.047		4.95	2.21		7.16	8.85
4100	#2 & better		170	.047		4.29	2.21		6.50	8.10
4300	33/32" x 3-1/4", not incl. finish, #1 grade		170	.047		4.56	2.21		6.77	8.40
4400	#2 & better		170	.047		4.06	2.21		6.27	7.85
4600	Oak, white or red, 25/32" x 2-1/4", not incl. finish									
4700	#1 common	1 Carp	170	.047	S.F.	3.19	2.21		5.40	6.90
4900	Select quartered, 2-1/4" wide		170	.047		3.89	2.21		6.10	7.70
5000	Clear		170	.047		4.01	2.21		6.22	7.80
6100	Prefinished, white oak, prime grade, 2-1/4" wide		170	.047		4.69	2.21		6.90	8.55
6200	3-1/4" wide		185	.043		5.10	2.03		7.13	8.70
6400	Ranch plank		145	.055		7.15	2.59		9.74	11.85
6500	Hardwood blocks, 9" x 9", 25/32" thick		160	.050		6	2.35		8.35	10.20
7400	Yellow pine, 3/4" x 3-1/8", T & G, C & better, not incl. finish		200	.040		1.49	1.88		3.37	4.53
7500	Refinish wood floor, sand, 2 coats poly, wax, soft wood	1 Clab	400	.020		.90	.75		1.65	2.15
7600	Hard wood		130	.062		1.34	2.31		3.65	5.05
7800	Sanding and finishing, 2 coats polyurethane		295	.027		.90	1.02		1.92	2.56
7900	Subfloor and underlayment, see Section 06 16									
8015	Transition molding, 2 1/4" wide, 5' long	1 Carp	19.20	.417	Ea.	10.85	19.55		30.40	42

09 64 Wood Flooring

09 64 66 – Wood Athletic Flooring

	09 64 66.10 Gymnasium Flooring	Crew	Daily Output	Labor-Hours	Unit	Material	2015 Bare Costs Labor	Equipment	Total	Total Incl O&P
0010	**GYMNASIUM FLOORING**									
0600	Gym floor, in mastic, over 2 ply felt, #2 & better									
0700	25/32" thick maple	1 Carp	100	.080	S.F.	3.99	3.76		7.75	10.20
0900	33/32" thick maple		98	.082		4.99	3.83		8.82	11.40
1000	For 1/2" corkboard underlayment, add		750	.011		.97	.50		1.47	1.84
1300	For #1 grade maple, add					.51			.51	.56
1600	Maple flooring, over sleepers, #2 & better									
1700	25/32" thick	1 Carp	85	.094	S.F.	4.70	4.42		9.12	11.95
1900	33/32" thick	"	83	.096		5.45	4.53		9.98	12.95
2000	For #1 grade, add					.55			.55	.61
2200	For 3/4" subfloor, add	1 Carp	350	.023		1.22	1.07		2.29	2.99
2300	With two 1/2" subfloors, 25/32" thick	"	69	.116		5.90	5.45		11.35	14.85
2500	Maple, incl. finish, #2 & btr., 25/32" thick, on rubber									
2600	Sleepers, with two 1/2" subfloors	1 Carp	76	.105	S.F.	6.30	4.94		11.24	14.55
2800	With steel spline, double connection to channels	"	73	.110		6.75	5.15		11.90	15.30
2900	For 33/32" maple, add					.72			.72	.79
3100	For #1 grade maple, add					.55			.55	.61
3500	For termite proofing all of the above, add					.29			.29	.32
3700	Portable hardwood, prefinished panels	1 Carp	83	.096		8.40	4.53		12.93	16.20
3720	Insulated with polystyrene, 1" thick, add		165	.048		.72	2.28		3	4.29
3750	Running tracks, Sitka spruce surface, 25/32" x 2-1/4"		62	.129		15.40	6.05		21.45	26.50
3770	3/4" plywood surface, finished		100	.080		3.65	3.76		7.41	9.80

09 65 Resilient Flooring

09 65 10 – Resilient Tile Underlayment

	09 65 10.10 Latex Underlayment	Crew	Daily Output	Labor-Hours	Unit	Material	Labor	Equipment	Total	Total Incl O&P
0010	**LATEX UNDERLAYMENT**									
3600	Latex underlayment, 1/8" thk., cementitious for resilient flooring	1 Tilf	160	.050	S.F.	1.19	2.14		3.33	4.48
4000	Liquid, fortified				Gal.	33			33	36

09 65 13 – Resilient Base and Accessories

	09 65 13.13 Resilient Base	Crew	Daily Output	Labor-Hours	Unit	Material	Labor	Equipment	Total	Total Incl O&P
0010	**RESILIENT BASE**									
0690	1/8" vinyl base, 2 1/2" H, straight or cove, standard colors	1 Tilf	315	.025	L.F.	.67	1.09		1.76	2.35
0700	4" high		315	.025		1.32	1.09		2.41	3.06
0710	6" high		315	.025		1.40	1.09		2.49	3.15
0720	Corners, 2 1/2" high		315	.025	Ea.	2.02	1.09		3.11	3.83
0730	4" high		315	.025		2.14	1.09		3.23	3.96
0740	6" high		315	.025		2.45	1.09		3.54	4.31
0800	1/8" rubber base, 2 1/2" H, straight or cove, standard colors		315	.025	L.F.	1.13	1.09		2.22	2.85
1100	4" high		315	.025		1.02	1.09		2.11	2.73
1110	6" high		315	.025		1.77	1.09		2.86	3.56
1150	Corners, 2 1/2" high		315	.025	Ea.	2	1.09		3.09	3.81
1153	4" high		315	.025		2.05	1.09		3.14	3.87
1155	6" high		315	.025		2.51	1.09		3.60	4.37
1450	For premium color/finish add					50%				
1500	Millwork profile	1 Tilf	315	.025	L.F.	5.85	1.09		6.94	8.05

	09 65 13.23 Resilient Stair Treads and Risers	Crew	Daily Output	Labor-Hours	Unit	Material	Labor	Equipment	Total	Total Incl O&P
0010	**RESILIENT STAIR TREADS AND RISERS**									
0300	Rubber, molded tread, 12" wide, 5/16" thick, black	1 Tilf	115	.070	L.F.	14.75	2.98		17.73	20.50
0400	Colors		115	.070		15.35	2.98		18.33	21.50

09 65 Resilient Flooring

09 65 13 – Resilient Base and Accessories

09 65 13.23 Resilient Stair Treads and Risers		Crew	Daily Output	Labor-Hours	Unit	Material	Labor	Equipment	Total	Total Incl O&P
						2015 Bare Costs				
0600	1/4" thick, black	1 Tilf	115	.070	L.F.	13.35	2.98		16.33	19.10
0700	Colors		115	.070		14.95	2.98		17.93	21
0900	Grip strip safety tread, colors, 5/16" thick		115	.070		20.50	2.98		23.48	27
1000	3/16" thick		120	.067		15.25	2.85		18.10	21
1200	Landings, smooth sheet rubber, 1/8" thick		120	.067	S.F.	7.80	2.85		10.65	12.80
1300	3/16" thick		120	.067	"	8.30	2.85		11.15	13.35
1500	Nosings, 3" wide, 3/16" thick, black		140	.057	L.F.	4.23	2.45		6.68	8.25
1600	Colors		140	.057		4.88	2.45		7.33	8.95
1800	Risers, 7" high, 1/8" thick, flat		250	.032		7.70	1.37		9.07	10.50
1900	Coved		250	.032		8.55	1.37		9.92	11.45
2100	Vinyl, molded tread, 12" wide, colors, 1/8" thick		115	.070		5.35	2.98		8.33	10.25
2200	1/4" thick		115	.070		6.70	2.98		9.68	11.80
2300	Landing material, 1/8" thick		200	.040	S.F.	6.05	1.71		7.76	9.20
2400	Riser, 7" high, 1/8" thick, coved		175	.046	L.F.	2.70	1.96		4.66	5.85
2500	Tread and riser combined, 1/8" thick		80	.100	"	9.95	4.28		14.23	17.25

09 65 16 – Resilient Sheet Flooring

09 65 16.10 Rubber and Vinyl Sheet Flooring			Crew	Daily Output	Labor-Hours	Unit	Material	Labor	Equipment	Total	Total Incl O&P
0010	**RUBBER AND VINYL SHEET FLOORING**										
5500	Linoleum, sheet goods	G	1 Tilf	360	.022	S.F.	3.59	.95		4.54	5.35
5900	Rubber, sheet goods, 36" wide, 1/8" thick			120	.067		7.75	2.85		10.60	12.75
5950	3/16" thick			100	.080		10.50	3.42		13.92	16.60
6000	1/4" thick			90	.089		12.45	3.80		16.25	19.30
8000	Vinyl sheet goods, backed, .065" thick, plain pattern/colors			250	.032		4.09	1.37		5.46	6.55
8050	Intricate pattern/ colors			200	.040		4.57	1.71		6.28	7.60
8100	.080" thick, plain pattern/colors			230	.035		4.10	1.49		5.59	6.70
8150	Intricate pattern/colors			200	.040		5.90	1.71		7.61	9.05
8200	.125" thick, plain pattern/colors			230	.035		4.25	1.49		5.74	6.90
8250	intricate pattern/colors			200	.040		7.50	1.71		9.21	10.80
8400	For welding seams, add			100	.080	L.F.	.25	3.42		3.67	5.35
8450	For integral cove base, add			175	.046	"	.75	1.96		2.71	3.72
8700	Adhesive cement, 1 gallon per 200 to 300 S.F.					Gal.	27			27	29.50
8800	Asphalt primer, 1 gallon per 300 S.F.						14			14	15.40
8900	Emulsion, 1 gallon per 140 S.F.						18			18	19.80

09 65 19 – Resilient Tile Flooring

09 65 19.10 Miscellaneous Resilient Tile Flooring			Crew	Daily Output	Labor-Hours	Unit	Material	Labor	Equipment	Total	Total Incl O&P
0010	**MISCELLANEOUS RESILIENT TILE FLOORING**										
2200	Cork tile, standard finish, 1/8" thick	G	1 Tilf	315	.025	S.F.	6.95	1.09		8.04	9.25
2250	3/16" thick	G		315	.025		6.65	1.09		7.74	8.90
2300	5/16" thick	G		315	.025		8.30	1.09		9.39	10.70
2350	1/2" thick	G		315	.025		10.95	1.09		12.04	13.60
2500	Urethane finish, 1/8" thick	G		315	.025		8.10	1.09		9.19	10.50
2550	3/16" thick	G		315	.025		8.25	1.09		9.34	10.65
2600	5/16" thick	G		315	.025		8.85	1.09		9.94	11.35
2650	1/2" thick	G		315	.025		12.15	1.09		13.24	14.95
6700	Synthetic turf, 3/8" thick			90	.089		4.54	3.80		8.34	10.65
6750	Interlocking 2' x 2' squares, 1/2" thick, not										
6810	cemented, for playgrounds, 3/8" thick		1 Tilf	210	.038	S.F.	4.69	1.63		6.32	7.55
6850	1/2" thick		"	190	.042	"	5.10	1.80		6.90	8.25

09 65 19.19 Vinyl Composition Tile Flooring		Crew	Daily Output	Labor-Hours	Unit	Material	Labor	Equipment	Total	Total Incl O&P
0010	**VINYL COMPOSITION TILE FLOORING**									
7000	Vinyl composition tile, 12" x 12", 1/16" thick	1 Tilf	500	.016	S.F.	1.18	.68		1.86	2.31

09 65 Resilient Flooring

09 65 19 – Resilient Tile Flooring

09 65 19.19 Vinyl Composition Tile Flooring		Crew	Daily Output	Labor-Hours	Unit	2015 Bare Costs Material	Labor	Equipment	Total	Total Incl O&P
7050	Embossed	1 Tilf	500	.016	S.F.	2.21	.68		2.89	3.44
7100	Marbleized		500	.016		2.21	.68		2.89	3.44
7150	Solid		500	.016		2.85	.68		3.53	4.15
7200	3/32" thick, embossed		500	.016		1.51	.68		2.19	2.67
7250	Marbleized		500	.016		2.54	.68		3.22	3.80
7300	Solid		500	.016		2.36	.68		3.04	3.61
7350	1/8" thick, marbleized		500	.016		2.40	.68		3.08	3.65
7400	Solid		500	.016		1.53	.68		2.21	2.69
7450	Conductive		500	.016		6.05	.68		6.73	7.65

09 65 19.23 Vinyl Tile Flooring		Crew	Daily Output	Labor-Hours	Unit	Material	Labor	Equipment	Total	Total Incl O&P
0010	**VINYL TILE FLOORING**									
7500	Vinyl tile, 12" x 12", 3/32" thick,	1 Tilf	500	.016	S.F.	3.59	.68		4.27	4.96
7550	3/32" thick, premium colors/patterns		500	.016		7.25	.68		7.93	9
7600	1/8" thick, standard colors/patterns		500	.016		5.65	.68		6.33	7.20
7650	Solid colors		500	.016		3.25	.68		3.93	4.59
7700	Marbleized or Travertine pattern		500	.016		5.90	.68		6.58	7.50
7750	Florentine pattern		500	.016		6.30	.68		6.98	7.95
7800	Premium colors/patterns		500	.016		6.20	.68		6.88	7.85

09 65 19.33 Rubber Tile Flooring		Crew	Daily Output	Labor-Hours	Unit	Material	Labor	Equipment	Total	Total Incl O&P
0010	**RUBBER TILE FLOORING**									
6050	Rubber tile, marbleized colors, 12" x 12", 1/8" thick	1 Tilf	400	.020	S.F.	5.70	.86		6.56	7.50
6100	3/16" thick		400	.020		9.25	.86		10.11	11.45
6300	Special tile, plain colors, 1/8" thick		400	.020		7.70	.86		8.56	9.75
6350	3/16" thick		400	.020		9.05	.86		9.91	11.20
6410	Raised, radial or square, .5 mm black		400	.020		8.25	.86		9.11	10.30
6430	.5 mm colored		400	.020		9.65	.86		10.51	11.85
6450	For golf course, skating rink, etc., 1/4" thick		275	.029		9.95	1.25		11.20	12.80

09 65 33 – Conductive Resilient Flooring

09 65 33.10 Conductive Rubber and Vinyl Flooring		Crew	Daily Output	Labor-Hours	Unit	Material	Labor	Equipment	Total	Total Incl O&P
0010	**CONDUCTIVE RUBBER AND VINYL FLOORING**									
1700	Conductive flooring, rubber tile, 1/8" thick	1 Tilf	315	.025	S.F.	6.95	1.09		8.04	9.25
1800	Homogeneous vinyl tile, 1/8" thick	"	315	.025	"	6.40	1.09		7.49	8.65

09 66 Terrazzo Flooring

09 66 13 – Portland Cement Terrazzo Flooring

09 66 13.10 Portland Cement Terrazzo		Crew	Daily Output	Labor-Hours	Unit	Material	Labor	Equipment	Total	Total Incl O&P
0010	**PORTLAND CEMENT TERRAZZO**, cast-in-place R096613-10									
0020	Cove base, 6" high, 16 ga. zinc	1 Mstz	20	.400	L.F.	3.28	17.20		20.48	29
0100	Curb, 6" high and 6" wide		6	1.333		5.65	57.50		63.15	90.50
0300	Divider strip for floors, 14 ga., 1-1/4" deep, zinc		375	.021		1.42	.92		2.34	2.91
0400	Brass		375	.021		2.46	.92		3.38	4.06
0600	Heavy top strip 1/4" thick, 1-1/4" deep, zinc		300	.027		2.17	1.15		3.32	4.08
0900	Galv. bottoms, brass		300	.027		3.25	1.15		4.40	5.25
1200	For thin set floors, 16 ga., 1/2" x 1/2", zinc		350	.023		1.12	.98		2.10	2.68
1300	Brass		350	.023		2.32	.98		3.30	4
1500	Floor, bonded to concrete, 1-3/4" thick, gray cement	J-3	75	.213	S.F.	3.35	8.35	4.05	15.75	20.50
1600	White cement, mud set		75	.213		3.73	8.35	4.05	16.13	21
1800	Not bonded, 3" total thickness, gray cement		70	.229		4.17	8.95	4.34	17.46	22.50
1900	White cement, mud set		70	.229		4.86	8.95	4.34	18.15	23.50
2100	For Venetian terrazzo, 1" topping, add					50%	50%			

09 66 Terrazzo Flooring

09 66 13 – Portland Cement Terrazzo Flooring

09 66 13.10 Portland Cement Terrazzo		Crew	Daily Output	Labor-Hours	Unit	2015 Bare Costs Material	Labor	Equipment	Total	Total Incl O&P
2200	For heavy duty abrasive terrazzo, add					50%	50%			
2700	Monolithic terrazzo, 1/2″ thick									
2710	10′ panels	J-3	125	.128	S.F.	3.14	5	2.43	10.57	13.50
3000	Stairs, cast in place, pan filled treads		30	.533	L.F.	3.57	21	10.10	34.67	46
3100	Treads and risers		14	1.143	″	6.15	45	21.50	72.65	97.50
3300	For stair landings, add to floor prices						50%			
3400	Stair stringers and fascia	J-3	30	.533	S.F.	5.25	21	10.10	36.35	48
3600	For abrasive metal nosings on stairs, add		150	.107	L.F.	9.15	4.18	2.02	15.35	18.55
3700	For abrasive surface finish, add		600	.027	S.F.	1.54	1.05	.51	3.10	3.80
3900	For raised abrasive strips, add		150	.107	L.F.	1.34	4.18	2.02	7.54	9.90
4000	Wainscot, bonded, 1-1/2″ thick		30	.533	S.F.	3.93	21	10.10	35.03	46.50
4200	1/4″ thick		40	.400	″	5.55	15.70	7.60	28.85	37.50
4300	Stone chips, onyx gemstone, per 50 lb. bag				Bag	16.80			16.80	18.50

09 66 16 – Terrazzo Floor Tile

09 66 16.10 Tile or Terrazzo Base

Code	Description	Crew	Daily Output	Labor-Hours	Unit	Material	Labor	Equipment	Total	Total Incl O&P
0010	**TILE OR TERRAZZO BASE**									
0020	Scratch coat only	1 Mstz	150	.053	S.F.	.43	2.29		2.72	3.86
0500	Scratch and brown coat only	″	75	.107	″	.82	4.58		5.40	7.65

09 66 16.13 Portland Cement Terrazzo Floor Tile

Code	Description	Crew	Daily Output	Labor-Hours	Unit	Material	Labor	Equipment	Total	Total Incl O&P
0010	**PORTLAND CEMENT TERRAZZO FLOOR TILE**									
1200	Floor tiles, non-slip, 1″ thick, 12″ x 12″	D-1	60	.267	S.F.	20	11.25		31.25	39
1300	1-1/4″ thick, 12″ x 12″		60	.267		20.50	11.25		31.75	40
1500	16″ x 16″		50	.320		22.50	13.45		35.95	45
1600	1-1/2″ thick, 16″ x 16″		45	.356		20.50	14.95		35.45	45.50
1800	For Venetian terrazzo, add					6.15			6.15	6.80
1900	For white cement, add					.58			.58	.64

09 66 16.16 Plastic Matrix Terrazzo Floor Tile

Code	Description	Crew	Daily Output	Labor-Hours	Unit	Material	Labor	Equipment	Total	Total Incl O&P
0010	**PLASTIC MATRIX TERRAZZO FLOOR TILE**									
0100	12″ x 12″, 3/16″ thick, Floor tiles w/marble chips	1 Tilf	500	.016	S.F.	7.30	.68		7.98	9.05
0200	12″ x 12″, 3/16″ thick, Floor tiles w/glass chips		500	.016		7.90	.68		8.58	9.65
0300	12″ x 12″, 3/16″ thick, Floor tiles w/recycled content		500	.016		6.20	.68		6.88	7.85

09 66 16.30 Terrazzo, Precast

Code	Description	Crew	Daily Output	Labor-Hours	Unit	Material	Labor	Equipment	Total	Total Incl O&P
0010	**TERRAZZO, PRECAST**									
0020	Base, 6″ high, straight	1 Mstz	70	.114	L.F.	11.95	4.91		16.86	20.50
0100	Cove		60	.133		12.80	5.75		18.55	22.50
0300	8″ high, straight		60	.133		11.45	5.75		17.20	21
0400	Cove		50	.160		16.85	6.85		23.70	28.50
0600	For white cement, add					.45			.45	.50
0700	For 16 ga. zinc toe strip, add					1.72			1.72	1.89
0900	Curbs, 4″ x 4″ high	1 Mstz	40	.200		32.50	8.60		41.10	48
1000	8″ x 8″ high	″	30	.267		38	11.45		49.45	59
2400	Stair treads, 1-1/2″ thick, non-slip, three line pattern	2 Mstz	70	.229		41	9.80		50.80	59.50
2500	Nosing and two lines		70	.229		41	9.80		50.80	59.50
2700	2″ thick treads, straight		60	.267		45.50	11.45		56.95	67
2800	Curved		50	.320		59	13.75		72.75	85.50
3000	Stair risers, 1″ thick, to 6″ high, straight sections		60	.267		10.80	11.45		22.25	29
3100	Cove		50	.320		15.65	13.75		29.40	37.50
3300	Curved, 1″ thick, to 6″ high, vertical		48	.333		21.50	14.30		35.80	44.50
3400	Cove		38	.421		40.50	18.10		58.60	71
3600	Stair tread and riser, single piece, straight, smooth surface		60	.267		52.50	11.45		63.95	75
3700	Non skid surface		40	.400		68	17.20		85.20	101

09 66 Terrazzo Flooring

09 66 16 – Terrazzo Floor Tile

09 66 16.30 Terrazzo, Precast		Crew	Daily Output	Labor-Hours	Unit	Material	Labor	Equipment	Total	Total Incl O&P
						2015 Bare Costs				
3900	Curved tread and riser, smooth surface	2 Mstz	40	.400	L.F.	74	17.20		91.20	107
4000	Non skid surface		32	.500		93	21.50		114.50	134
4200	Stair stringers, notched, 1" thick		25	.640		30	27.50		57.50	73.50
4300	2" thick		22	.727		35.50	31		66.50	85
4500	Stair landings, structural, non-slip, 1-1/2" thick		85	.188	S.F.	33	8.10		41.10	48
4600	3" thick		75	.213		46.50	9.15		55.65	65
4800	Wainscot, 12" x 12" x 1" tiles	1 Mstz	12	.667		6.85	28.50		35.35	50
4900	16" x 16" x 1-1/2" tiles	"	8	1		14.25	43		57.25	79

09 66 23 – Resinous Matrix Terrazzo Flooring

09 66 23.13 Polyacrylate Modified Cementitious Terrazzo Flooring		Crew	Daily Output	Labor-Hours	Unit	Material	Labor	Equipment	Total	Total Incl O&P
0010	**POLYACRYLATE MODIFIED CEMENTITIOUS TERRAZZO FLOORING**									
3150	Polyacrylate, 1/4" thick, granite chips	C-6	735	.065	S.F.	3.50	2.56	.08	6.14	7.85
3170	Recycled porcelain		480	.100		4.51	3.92	.13	8.56	11.10
3200	3/8" thick, granite chips		620	.077		4.58	3.03	.10	7.71	9.80
3220	Recycled porcelain		480	.100		6.40	3.92	.13	10.45	13.20

09 66 23.16 Epoxy-Resin Terrazzo Flooring		Crew	Daily Output	Labor-Hours	Unit	Material	Labor	Equipment	Total	Total Incl O&P
0010	**EPOXY-RESIN TERRAZZO FLOORING**									
1800	Epoxy terrazzo, 1/4" thick, chemical resistant, granite chips	J-3	200	.080	S.F.	5.95	3.14	1.52	10.61	12.85
1900	Recycled porcelain		150	.107		9.05	4.18	2.02	15.25	18.40
2500	Epoxy terrazzo, 1/4" thick, granite chips		200	.080		5.15	3.14	1.52	9.81	12
2550	Average		175	.091		5.45	3.59	1.74	10.78	13.20
2600	Recycled porcelain		150	.107		6.35	4.18	2.02	12.55	15.45

09 66 33 – Conductive Terrazzo Flooring

09 66 33.10 Conductive Terrazzo		Crew	Daily Output	Labor-Hours	Unit	Material	Labor	Equipment	Total	Total Incl O&P
0010	**CONDUCTIVE TERRAZZO**									
2400	Bonded conductive floor for hospitals	J-3	90	.178	S.F.	4.84	6.95	3.37	15.16	19.30

09 66 33.13 Conductive Epoxy-Resin Terrazzo		Crew	Daily Output	Labor-Hours	Unit	Material	Labor	Equipment	Total	Total Incl O&P
0010	**CONDUCTIVE EPOXY-RESIN TERRAZZO**									
2100	Epoxy terrazzo, 1/4" thick, conductive, granite chips	J-3	100	.160	S.F.	8	6.30	3.04	17.34	21.50
2200	Recycled porcelain	"	90	.178	"	10.40	6.95	3.37	20.72	25.50

09 66 33.19 Conductive Plastic-Matrix Terrazzo Flooring		Crew	Daily Output	Labor-Hours	Unit	Material	Labor	Equipment	Total	Total Incl O&P
0010	**CONDUCTIVE PLASTIC-MATRIX TERRAZZO FLOORING**									
3300	Conductive, 1/4" thick, granite chips	C-6	450	.107	S.F.	7.45	4.18	.14	11.77	14.75
3330	Recycled porcelain		305	.157		10	6.15	.20	16.35	20.50
3350	3/8" thick, granite chips		365	.132		9.80	5.15	.17	15.12	18.85
3370	Recycled porcelain		255	.188		12.75	7.35	.24	20.34	25.50
3450	Granite, conductive, 1/4" thick, 20% chip		695	.069		9.25	2.71	.09	12.05	14.40
3470	50% chip		420	.114		11.90	4.48	.15	16.53	20
3500	3/8" thick, 20% chip		695	.069		13.50	2.71	.09	16.30	19.10
3520	50% chip		380	.126		16.30	4.95	.16	21.41	25.50

09 67 Fluid-Applied Flooring

09 67 13 – Elastomeric Liquid Flooring

09 67 13.13 Elastomeric Liquid Flooring		Crew	Daily Output	Labor-Hours	Unit	2015 Bare Costs Material	Labor	Equipment	Total	Total Incl O&P
0010	**ELASTOMERIC LIQUID FLOORING**									
0020	Cementitious acrylic, 1/4" thick	C-6	520	.092	S.F.	1.66	3.62	.12	5.40	7.45
0100	3/8" thick	"	450	.107		2.10	4.18	.14	6.42	8.85
0200	Methyl methachrylate, 1/4" thick	C-8A	3000	.016		6.50	.65		7.15	8.15
0210	1/8" thick	"	3000	.016		5.50	.65		6.15	7.05
0300	Cupric oxychloride, on bond coat, simple configs and patterns	C-6	480	.100		3.58	3.92	.13	7.63	10.10
0400	Complex configurations and patterns		420	.114		6	4.48	.15	10.63	13.60
2400	Mastic, hot laid, 2 coat, 1-1/2" thick, standard, simple configurations and		690	.070		4.15	2.73	.09	6.97	8.85
2500	Maximum		520	.092		5.35	3.62	.12	9.09	11.50
2700	Acid-proof, minimum		605	.079		5.35	3.11	.10	8.56	10.70
2800	Maximum		350	.137		7.40	5.35	.18	12.93	16.50
3000	Neoprene, troweled on, 1/4" thick, minimum		545	.088		4.09	3.45	.11	7.65	9.90
3100	Maximum		430	.112		5.55	4.37	.15	10.07	13
4300	Polyurethane, with suspended vinyl chips, clear		1065	.045		7.60	1.77	.06	9.43	11.10
4500	Pigmented		860	.056		11.05	2.19	.07	13.31	15.55

09 67 26 – Quartz Flooring

09 67 26.26 Quartz Flooring		Crew	Daily Output	Labor-Hours	Unit	Material	Labor	Equipment	Total	Total Incl O&P
0010	**QUARTZ FLOORING**									
0600	Epoxy, with colored quartz chips, broadcast, 3/8" thick	C-6	675	.071	S.F.	2.81	2.79	.09	5.69	7.45
0700	1/2" thick		490	.098		4.05	3.84	.13	8.02	10.45
0900	Troweled, minimum		560	.086		3.57	3.36	.11	7.04	9.20
1000	Maximum		480	.100		5.40	3.92	.13	9.45	12.05
1200	Heavy duty epoxy topping, 1/4" thick,									
1300	500 to 1,000 S.F.	C-6	420	.114	S.F.	5.55	4.48	.15	10.18	13.15
1500	1,000 to 2,000 S.F.		450	.107		5.05	4.18	.14	9.37	12.10
1600	Over 10,000 S.F.		480	.100		4.66	3.92	.13	8.71	11.30
3600	Polyester, with colored quartz chips, 1/16" thick, minimum		1065	.045		3.16	1.77	.06	4.99	6.25
3700	Maximum		560	.086		4.19	3.36	.11	7.66	9.90
3900	1/8" thick, minimum		810	.059		3.67	2.32	.08	6.07	7.65
4000	Maximum		675	.071		4.73	2.79	.09	7.61	9.55
4200	Polyester, heavy duty, compared to epoxy, add		2590	.019		1.51	.73	.02	2.26	2.80

09 67 66 – Fluid-Applied Athletic Flooring

09 67 66.10 Polyurethane		Crew	Daily Output	Labor-Hours	Unit	Material	Labor	Equipment	Total	Total Incl O&P
0010	**POLYURETHANE**									
4400	Thermoset, prefabricated in place, indoor									
4500	3/8" thick for basketball, gyms, etc.	1 Tilf	100	.080	S.F.	5.40	3.42		8.82	11
4600	1/2" thick for professional sports		95	.084		7.25	3.60		10.85	13.30
4700	Outdoor, 1/4" thick, smooth, for tennis		100	.080		5.60	3.42		9.02	11.25
5000	Poured in place, indoor, with finish, 1/4" thick		80	.100		3.99	4.28		8.27	10.75
5050	3/8" thick		65	.123		4.84	5.25		10.09	13.10
5100	1/2" thick		50	.160		6.75	6.85		13.60	17.60

09 68 Carpeting

09 68 05 – Carpet Accessories

09 68 05.11 Flooring Transition Strip		Crew	Daily Output	Labor-Hours	Unit	Material	2015 Bare Costs Labor	Equipment	Total	Total Incl O&P
0010	**FLOORING TRANSITION STRIP**									
0107	Clamp down brass divider, 12′ strip, vinyl to carpet	1 Tilf	31.25	.256	Ea.	13.30	10.95		24.25	31
0117	Vinyl to hard surface	″	31.25	.256	″	13.30	10.95		24.25	31

09 68 10 – Carpet Pad

09 68 10.10 Commercial Grade Carpet Pad

Line	Description	Crew	Daily Output	Labor-Hours	Unit	Material	Labor	Equipment	Total	Total Incl O&P
0010	**COMMERCIAL GRADE CARPET PAD**									
9000	Sponge rubber pad, 20 oz./sq. yd.	1 Tilf	150	.053	S.Y.	4.51	2.28		6.79	8.35
9100	40-62 oz./sq. yd.		150	.053		8.75	2.28		11.03	13.05
9200	Felt pad, 20 oz./sq. yd.		150	.053		5.30	2.28		7.58	9.25
9300	32 to 56 oz./sq. yd.		150	.053		8.95	2.28		11.23	13.25
9400	Bonded urethane pad, 2.7 density		150	.053		5.95	2.28		8.23	9.95
9500	13.0 Density		150	.053		8	2.28		10.28	12.20
9600	Prime urethane pad, 2.7 density		150	.053		3.15	2.28		5.43	6.85
9700	13.0 density		150	.053		4.95	2.28		7.23	8.85

09 68 13 – Tile Carpeting

09 68 13.10 Carpet Tile

Line	Description	Crew	Daily Output	Labor-Hours	Unit	Material	Labor	Equipment	Total	Total Incl O&P
0010	**CARPET TILE**									
0100	Tufted nylon, 18″ x 18″, hard back, 20 oz.	1 Tilf	80	.100	S.Y.	24	4.28		28.28	33
0110	26 oz.		80	.100		32	4.28		36.28	41.50
0200	Cushion back, 20 oz.		80	.100		28	4.28		32.28	37
0210	26 oz.		80	.100		43	4.28		47.28	53.50
1100	Tufted, 24″ x 24″, hard back, 24 oz. nylon		80	.100		30	4.28		34.28	39.50
1180	35 oz.		80	.100		35	4.28		39.28	45
5060	42 oz.		80	.100		46	4.28		50.28	57.50

09 68 16 – Sheet Carpeting

09 68 16.10 Sheet Carpet

Line	Description	Crew	Daily Output	Labor-Hours	Unit	Material	Labor	Equipment	Total	Total Incl O&P
0010	**SHEET CARPET**									
0700	Nylon, level loop, 26 oz., light to medium traffic	1 Tilf	75	.107	S.Y.	25	4.57		29.57	34.50
0720	28 oz., light to medium traffic		75	.107		32	4.57		36.57	42
0900	32 oz., medium traffic		75	.107		42.50	4.57		47.07	54
1100	40 oz., medium to heavy traffic		75	.107		59.50	4.57		64.07	72.50
2920	Nylon plush, 30 oz., medium traffic		57	.140		29.50	6		35.50	41.50
3000	36 oz., medium traffic		75	.107		34.50	4.57		39.07	45
3100	42 oz., medium to heavy traffic		70	.114		45	4.89		49.89	57
3200	46 oz., medium to heavy traffic		70	.114		49	4.89		53.89	61.50
3300	54 oz., heavy traffic		70	.114		55	4.89		59.89	68
3340	60 oz., heavy traffic		70	.114		62.50	4.89		67.39	76.50
3665	Olefin, 24 oz., light to medium traffic		75	.107		18.65	4.57		23.22	27.50
3670	26 oz., medium traffic		75	.107		13.10	4.57		17.67	21
3680	28 oz., medium to heavy traffic		75	.107		23.50	4.57		28.07	33
3700	32 oz., medium to heavy traffic		75	.107		27	4.57		31.57	36.50
3730	42 oz., heavy traffic		70	.114		28.50	4.89		33.39	38.50
4110	Wool, level loop, 40 oz., medium traffic		75	.107		106	4.57		110.57	124
4500	50 oz., medium to heavy traffic		75	.107		100	4.57		104.57	117
4700	Patterned, 32 oz., medium to heavy traffic		70	.114		53.50	4.89		58.39	66.50
4900	48 oz., heavy traffic		70	.114		100	4.89		104.89	117
5000	For less than full roll (approx. 1500 S.F.), add					25%				
5100	For small rooms, less than 12′ wide, add						25%			
5200	For large open areas (no cuts), deduct						25%			
5600	For bound carpet baseboard, add	1 Tilf	300	.027	L.F.	3	1.14		4.14	4.99
5610	For stairs, not incl. price of carpet, add	″	30	.267	Riser		11.40		11.40	16.90

09 68 Carpeting

09 68 16 – Sheet Carpeting

09 68 16.10 Sheet Carpet		Crew	Daily Output	Labor-Hours	Unit	Material	2015 Bare Costs Labor	Equipment	Total	Total Incl O&P
5620	For borders and patterns, add to labor						18%			
8950	For tackless, stretched installation, add padding from 09 68 10.10 to above									
9850	For brand-named specific fiber, add				S.Y.	25%				

09 68 20 – Athletic Carpet

09 68 20.10 Indoor Athletic Carpet		Crew	Daily Output	Labor-Hours	Unit	Material	Labor	Equipment	Total	Total Incl O&P
0010	**INDOOR ATHLETIC CARPET**									
3700	Polyethylene, in rolls, no base incl., landscape surfaces	1 Tilf	275	.029	S.F.	4.09	1.25		5.34	6.35
3800	Nylon action surface, 1/8" thick		275	.029		3.76	1.25		5.01	6
3900	1/4" thick		275	.029		5.45	1.25		6.70	7.80
4000	3/8" thick		275	.029		6.80	1.25		8.05	9.35
4100	Golf tee surface with foam back		235	.034		6.75	1.46		8.21	9.60
4200	Practice putting, knitted nylon surface		235	.034		5.70	1.46		7.16	8.45
5500	Polyvinyl chloride, sheet goods for gyms, 1/4" thick		80	.100		7.90	4.28		12.18	15.05
5600	3/8" thick		60	.133		11.80	5.70		17.50	21.50

09 69 Access Flooring

09 69 13 – Rigid-Grid Access Flooring

09 69 13.10 Access Floors		Crew	Daily Output	Labor-Hours	Unit	Material	Labor	Equipment	Total	Total Incl O&P
0010	**ACCESS FLOORS**									
0015	Access floor package including panel, pedestal, stringers & laminate cover									
0100	Computer room, greater than 6,000 S.F.	4 Carp	750	.043	S.F.	7.50	2		9.50	11.35
0110	Less than 6,000 S.F.	2 Carp	375	.043		8.15	2		10.15	12.10
0120	Office, greater than 6,000 S.F.	4 Carp	1050	.030		5.35	1.43		6.78	8.10
0250	Panels, particle board or steel, 1250# load, no covering, under 6,000 S.F.	2 Carp	600	.027		3.75	1.25		5	6.05
0300	Over 6,000 S.F.		640	.025		3.21	1.17		4.38	5.35
0400	Aluminum, 24" panels		500	.032		33	1.50		34.50	38.50
0600	For carpet covering, add					8.75			8.75	9.65
0700	For vinyl floor covering, add					9.05			9.05	9.95
0900	For high pressure laminate covering, add					7.50			7.50	8.25
0910	For snap on stringer system, add	2 Carp	1000	.016		1.56	.75		2.31	2.88
0950	Office applications, steel or concrete panels,									
0960	no covering, over 6,000 S.F.	2 Carp	960	.017	S.F.	10.55	.78		11.33	12.80
1000	Machine cutouts after initial installation	1 Carp	50	.160	Ea.	20	7.50		27.50	33.50
1050	Pedestals, 6" to 12"	2 Carp	85	.188		8.40	8.85		17.25	23
1100	Air conditioning grilles, 4" x 12"	1 Carp	17	.471		68.50	22		90.50	109
1150	4" x 18"	"	14	.571		93.50	27		120.50	145
1200	Approach ramps, steel	2 Carp	60	.267	S.F.	24.50	12.50		37	46.50
1300	Aluminum	"	40	.400	"	34	18.80		52.80	66.50
1500	Handrail, 2 rail, aluminum	1 Carp	15	.533	L.F.	109	25		134	158

09 72 Wall Coverings

09 72 19 – Textile Wall Covering

09 72 19.10 Textile Wall Covering		Crew	Daily Output	Labor-Hours	Unit	Material	2015 Bare Costs Labor	Equipment	Total	Total Incl O&P
0010	**TEXTILE WALL COVERING**, including sizing; add 10-30% waste @ takeoff									
0020	Silk	1 Pape	640	.013	S.F.	4.15	.51		4.66	5.35
0030	Cotton		640	.013		6.65	.51		7.16	8.05
0040	Linen		640	.013		1.85	.51		2.36	2.81
0050	Blend	↓	640	.013	↓	2.97	.51		3.48	4.04

09 72 20 – Natural Fiber Wall Covering

09 72 20.10 Natural Fiber Wall Covering		Crew	Daily Output	Labor-Hours	Unit	Material	Labor	Equipment	Total	Total Incl O&P
0010	**NATURAL FIBER WALL COVERING**, including sizing; add 10-30% waste @ takeoff									
0015	Bamboo	1 Pape	640	.013	S.F.	2.26	.51		2.77	3.26
0030	Burlap		640	.013		2.03	.51		2.54	3
0045	Jute		640	.013		1.29	.51		1.80	2.19
0060	Sisal	↓	640	.013	↓	1.46	.51		1.97	2.38

09 72 23 – Wallpapering

09 72 23.10 Wallpaper		Crew	Daily Output	Labor-Hours	Unit	Material	Labor	Equipment	Total	Total Incl O&P
0010	**WALLPAPER** including sizing; add 10-30 percent waste @ takeoff R097223-10									
0050	Aluminum foil	1 Pape	275	.029	S.F.	1.04	1.19		2.23	2.93
0100	Copper sheets, .025" thick, vinyl backing		240	.033		5.55	1.36		6.91	8.15
0300	Phenolic backing		240	.033		7.20	1.36		8.56	10
0600	Cork tiles, light or dark, 12" x 12" x 3/16"		240	.033		4.51	1.36		5.87	7
0700	5/16" thick		235	.034		3.13	1.39		4.52	5.55
0900	1/4" basketweave		240	.033		3.50	1.36		4.86	5.90
1000	1/2" natural, non-directional pattern		240	.033		6.90	1.36		8.26	9.65
1100	3/4" natural, non-directional pattern		240	.033		11.35	1.36		12.71	14.55
1200	Granular surface, 12" x 36", 1/2" thick		385	.021		1.31	.85		2.16	2.72
1300	1" thick		370	.022		1.68	.88		2.56	3.18
1500	Polyurethane coated, 12" x 12" x 3/16" thick		240	.033		4.08	1.36		5.44	6.55
1600	5/16" thick		235	.034		6.60	1.39		7.99	9.35
1800	Cork wallpaper, paperbacked, natural		480	.017		2.06	.68		2.74	3.30
1900	Colors		480	.017		2.87	.68		3.55	4.19
2100	Flexible wood veneer, 1/32" thick, plain woods		100	.080		2.44	3.27		5.71	7.60
2200	Exotic woods	↓	95	.084	↓	3.69	3.44		7.13	9.25
2400	Gypsum-based, fabric-backed, fire resistant									
2500	for masonry walls, 21 oz./S.Y.	1 Pape	800	.010	S.F.	.85	.41		1.26	1.56
2700	Small quantities	"	640	.013		.76	.51		1.27	1.61
2750	Acrylic, modified, semi-rigid PVC, .028" thick	2 Carp	330	.048		1.37	2.28		3.65	5
2800	.040" thick	"	320	.050		1.81	2.35		4.16	5.60
3000	Vinyl wall covering, fabric-backed, lightweight, type 1 (12-15 oz./S.Y.)	1 Pape	640	.013		.97	.51		1.48	1.84
3300	Medium weight, type 2 (20-24 oz./S.Y.)		480	.017		.90	.68		1.58	2.02
3400	Heavy weight, type 3 (28 oz./S.Y.)	↓	435	.018	↓	1.37	.75		2.12	2.64
3600	Adhesive, 5 gal. lots (18 S.Y./gal.)				Gal.	14.20			14.20	15.60
3700	Wallpaper, average workmanship, solid pattern, low cost paper	1 Pape	640	.013	S.F.	.61	.51		1.12	1.44
3900	basic patterns (matching required), avg. cost paper		535	.015		1.11	.61		1.72	2.14
4000	Paper at $85 per double roll, quality workmanship	↓	435	.018	↓	2.20	.75		2.95	3.55
4100	Linen wall covering, paper backed									
4150	Flame treatment				S.F.	1.01			1.01	1.11
4180	Stain resistance treatment					1.84			1.84	2.02
4190	Grass cloth, natural fabric G	1 Pape	400	.020		2.01	.82		2.83	3.44
4200	Grass cloths with lining paper G		400	.020		.91	.82		1.73	2.23
4300	Premium texture/color G	↓	350	.023	↓	2.91	.94		3.85	4.61

09 77 Special Wall Surfacing

09 77 30 – Fiberglass Reinforced Panels

09 77 30.10 Fiberglass Reinforced Plastic Panels		Crew	Daily Output	Labor-Hours	Unit	2015 Bare Costs Material	Labor	Equipment	Total	Total Incl O&P
0010	**FIBERGLASS REINFORCED PLASTIC PANELS**, .090" thick									
0020	On walls, adhesive mounted, embossed surface	2 Carp	640	.025	S.F.	1.11	1.17		2.28	3.03
0030	Smooth surface		640	.025		1.37	1.17		2.54	3.32
0040	Fire rated, embossed surface		640	.025		1.99	1.17		3.16	4
0050	Nylon rivet mounted, on drywall, embossed surface		480	.033		1.11	1.56		2.67	3.63
0060	Smooth surface		480	.033		1.37	1.56		2.93	3.92
0070	Fire rated, embossed surface		480	.033		1.99	1.56		3.55	4.60
0080	On masonry, embossed surface		320	.050		1.11	2.35		3.46	4.83
0090	Smooth surface		320	.050		1.37	2.35		3.72	5.10
0100	Fire rated, embossed surface		320	.050		1.99	2.35		4.34	5.80
0110	Nylon rivet and adhesive mounted, on drywall, embossed surface		240	.067		1.26	3.13		4.39	6.20
0120	Smooth surface		240	.067		1.26	3.13		4.39	6.20
0130	Fire rated, embossed surface		240	.067		2.19	3.13		5.32	7.25
0140	On masonry, embossed surface		190	.084		1.26	3.95		5.21	7.50
0150	Smooth surface		190	.084		1.26	3.95		5.21	7.50
0160	Fire rated, embossed surface		190	.084		2.19	3.95		6.14	8.50
0170	For moldings add	1 Carp	250	.032	L.F.	.26	1.50		1.76	2.60
0180	On ceilings, for lay in grid system, embossed surface		400	.020	S.F.	1.11	.94		2.05	2.67
0190	Smooth surface		400	.020		1.37	.94		2.31	2.96
0200	Fire rated, embossed surface		400	.020		1.99	.94		2.93	3.64

09 77 43 – Panel Systems

09 77 43.20 Slatwall Panels and Accessories		Crew	Daily Output	Labor-Hours	Unit	Material	Labor	Equipment	Total	Total Incl O&P
0010	**SLATWALL PANELS AND ACCESSORIES**									
0100	Slatwall panel, 4' x 8' x 3/4" T, MDF, paint grade	1 Carp	500	.016	S.F.	1.41	.75		2.16	2.71
0110	Melamine finish		500	.016		2.12	.75		2.87	3.49
0120	High pressure plastic laminate finish		500	.016		3.72	.75		4.47	5.25
0125	Wood veneer		500	.016		3.41	.75		4.16	4.91
0130	Aluminum channel inserts, add					2.43			2.43	2.67
0200	Accessories, corner forms, 8' L				L.F.	5.80			5.80	6.40
0210	T-connector, 8' L					8.40			8.40	9.25
0220	J-mold, 8' L					1.26			1.26	1.39
0230	Edge cap, 8' L					1.33			1.33	1.46
0240	Finish end cap, 8' L					3.39			3.39	3.73
0300	Display hook, metal, 4" L				Ea.	.43			.43	.47
0310	6" L					.48			.48	.53
0320	8" L					.52			.52	.57
0330	10" L					.59			.59	.65
0340	12" L					.64			.64	.70
0350	Acrylic, 4" L					.92			.92	1.01
0360	6" L					1.06			1.06	1.17
0370	8" L					1.11			1.11	1.22
0380	10" L					1.25			1.25	1.38
0400	Waterfall hanger, metal, 12" - 16"					3.84			3.84	4.22
0410	Acrylic					11.30			11.30	12.40
0500	Shelf bracket, metal, 8"					1.79			1.79	1.97
0510	10"					1.97			1.97	2.17
0520	12"					2.17			2.17	2.39
0530	14"					2.39			2.39	2.63
0540	16"					2.81			2.81	3.09
0550	Acrylic, 8"					3.79			3.79	4.17
0560	10"					4.03			4.03	4.43
0570	12"					4.26			4.26	4.69

09 77 Special Wall Surfacing

09 77 43 – Panel Systems

09 77 43.20 Slatwall Panels and Accessories		Crew	Daily Output	Labor-Hours	Unit	Material	Labor	Equipment	Total	Total Incl O&P
						2015 Bare Costs				
0580	14"				Ea.	4.48			4.48	4.93
0600	Shelf, acrylic, 12" x 16" x 1/4"					17.30			17.30	19
0610	12" x 24" x 1/4"				↓	24			24	26

09 81 Acoustic Insulation

09 81 16 – Acoustic Blanket Insulation

09 81 16.10 Sound Attenuation Blanket

		Crew	Daily Output	Labor-Hours	Unit	Material	Labor	Equipment	Total	Total Incl O&P
0010	**SOUND ATTENUATION BLANKET**									
0020	Blanket, 1" thick	1 Carp	925	.009	S.F.	.25	.41		.66	.91
0500	1-1/2" thick		920	.009		.25	.41		.66	.91
1000	2" thick		915	.009		.36	.41		.77	1.03
1500	3" thick	↓	910	.009	↓	.52	.41		.93	1.21
2000	Wall hung, STC 18 – 21, 1" thick, 4' x 20'	2 Carp	22	.727	Ea.	385	34		419	475
2010	10' x 20'	"	19	.842		960	39.50		999.50	1,100
2020	Wall hung, STC 27 – 28, 3" thick, 4' x 20'	3 Carp	12	2		545	94		639	745
2030	10' x 20'	"	9	2.667	↓	1,375	125		1,500	1,700
3000	Thermal or acoustical batt above ceiling, 2" thick	1 Carp	900	.009	S.F.	.50	.42		.92	1.19
3100	3" thick		900	.009		.75	.42		1.17	1.47
3200	4" thick	↓	900	.009		.94	.42		1.36	1.67
3400	Urethane plastic foam, open cell, on wall, 2" thick	2 Carp	2050	.008		3.11	.37		3.48	3.98
3500	3" thick		1550	.010		4.13	.48		4.61	5.30
3600	4" thick		1050	.015		5.80	.72		6.52	7.45
3700	On ceiling, 2" thick		1700	.009		3.10	.44		3.54	4.09
3800	3" thick		1300	.012		4.13	.58		4.71	5.45
3900	4" thick	↓	900	.018	↓	5.80	.83		6.63	7.65

09 84 Acoustic Room Components

09 84 13 – Fixed Sound-Absorptive Panels

09 84 13.10 Fixed Panels

		Crew	Daily Output	Labor-Hours	Unit	Material	Labor	Equipment	Total	Total Incl O&P
0010	**FIXED PANELS** Perforated steel facing, painted with									
0100	Fiberglass or mineral filler, no backs, 2-1/4" thick, modular									
0200	space units, ceiling or wall hung, white or colored	1 Carp	100	.080	S.F.	8.80	3.76		12.56	15.45
0300	Fiberboard sound deadening panels, 1/2" thick	"	600	.013	"	.33	.63		.96	1.32
0500	Fiberglass panels, 4' x 8' x 1" thick, with									
0600	glass cloth face for walls, cemented	1 Carp	155	.052	S.F.	8.35	2.42		10.77	12.95
0700	1-1/2" thick, dacron covered, inner aluminum frame,									
0710	wall mounted	1 Carp	300	.027	S.F.	8.90	1.25		10.15	11.75
0900	Mineral fiberboard panels, fabric covered, 30" x 108",									
1000	3/4" thick, concealed spline, wall mounted	1 Carp	150	.053	S.F.	6.40	2.50		8.90	10.90

09 84 36 – Sound-Absorbing Ceiling Units

09 84 36.10 Barriers

		Crew	Daily Output	Labor-Hours	Unit	Material	Labor	Equipment	Total	Total Incl O&P
0010	**BARRIERS** Plenum									
0600	Aluminum foil, fiberglass reinf., parallel with joists	1 Carp	275	.029	S.F.	1.08	1.37		2.45	3.29
0700	Perpendicular to joists		180	.044		1.08	2.09		3.17	4.40
0900	Aluminum mesh, kraft paperbacked		275	.029		.79	1.37		2.16	2.97
0970	Fiberglass batts, kraft faced, 3-1/2" thick		1400	.006		.37	.27		.64	.82
0980	6" thick		1300	.006		.66	.29		.95	1.17
1000	Sheet lead, 1 lb., 1/64" thick, perpendicular to joists		150	.053		6.40	2.50		8.90	10.90
1100	Vinyl foam reinforced, 1/8" thick, 1.0 lb. per S.F.	↓	150	.053	↓	4.49	2.50		6.99	8.80

09 91 Painting

09 91 03 – Paint Restoration

09 91 03.20 Sanding		Crew	Daily Output	Labor-Hours	Unit	Material	Labor (2015 Bare Costs)	Equipment (2015 Bare Costs)	Total (2015 Bare Costs)	Total Incl O&P
0010	**SANDING** and puttying interior trim, compared to R099100-10									
0100	Painting 1 coat, on quality work				L.F.		100%			
0300	Medium work						50%			
0400	Industrial grade						25%			
0500	Surface protection, placement and removal									
0510	Basic drop cloths	1 Pord	6400	.001	S.F.		.05		.05	.08
0520	Masking with paper		800	.010		.07	.40		.47	.69
0530	Volume cover up (using plastic sheathing, or building paper)		16000	.001			.02		.02	.03

09 91 03.30 Exterior Surface Preparation

		Crew	Daily Output	Labor-Hours	Unit	Material	Labor	Equipment	Total	Total Incl O&P
0010	**EXTERIOR SURFACE PREPARATION** R099100-10									
0015	Doors, per side, not incl. frames or trim									
0020	Scrape & sand									
0030	Wood, flush	1 Pord	616	.013	S.F.		.52		.52	.79
0040	Wood, detail		496	.016			.65		.65	.98
0050	Wood, louvered		280	.029			1.15		1.15	1.74
0060	Wood, overhead		616	.013			.52		.52	.79
0070	Wire brush									
0080	Metal, flush	1 Pord	640	.013	S.F.		.50		.50	.76
0090	Metal, detail		520	.015			.62		.62	.94
0100	Metal, louvered		360	.022			.90		.90	1.35
0110	Metal or fibr., overhead		640	.013			.50		.50	.76
0120	Metal, roll up		560	.014			.58		.58	.87
0130	Metal, bulkhead		640	.013			.50		.50	.76
0140	Power wash, based on 2500 lb. operating pressure									
0150	Metal, flush	A-1H	2240	.004	S.F.		.13	.03	.16	.25
0160	Metal, detail		2120	.004			.14	.04	.18	.26
0170	Metal, louvered		2000	.004			.15	.04	.19	.27
0180	Metal or fibr., overhead		2400	.003			.13	.03	.16	.22
0190	Metal, roll up		2400	.003			.13	.03	.16	.22
0200	Metal, bulkhead		2200	.004			.14	.03	.17	.25
0400	Windows, per side, not incl. trim									
0410	Scrape & sand									
0420	Wood, 1-2 lite	1 Pord	320	.025	S.F.		1.01		1.01	1.52
0430	Wood, 3-6 lite		280	.029			1.15		1.15	1.74
0440	Wood, 7-10 lite		240	.033			1.34		1.34	2.03
0450	Wood, 12 lite		200	.040			1.61		1.61	2.43
0460	Wood, Bay/Bow		320	.025			1.01		1.01	1.52
0470	Wire brush									
0480	Metal, 1-2 lite	1 Pord	480	.017	S.F.		.67		.67	1.01
0490	Metal, 3-6 lite		400	.020			.81		.81	1.22
0500	Metal, Bay/Bow		480	.017			.67		.67	1.01
0510	Power wash, based on 2500 lb. operating pressure									
0520	1-2 lite	A-1H	4400	.002	S.F.		.07	.02	.09	.13
0530	3-6 lite		4320	.002			.07	.02	.09	.13
0540	7-10 lite		4240	.002			.07	.02	.09	.13
0550	12 lite		4160	.002			.07	.02	.09	.13
0560	Bay/Bow		4400	.002			.07	.02	.09	.13
0600	Siding, scrape and sand, light=10-30%, med.=30-70%									
0610	Heavy=70-100% of surface to sand									
0650	Texture 1-11, light	1 Pord	480	.017	S.F.		.67		.67	1.01
0660	Med.		440	.018			.73		.73	1.11
0670	Heavy		360	.022			.90		.90	1.35

09 91 03 – Paint Restoration

09 91 03.30 Exterior Surface Preparation		Crew	Daily Output	Labor-Hours	Unit	Material	2015 Bare Costs Labor	Equipment	Total	Total Incl O&P
0680	Wood shingles, shakes, light	1 Pord	440	.018	S.F.		.73		.73	1.11
0690	Med.		360	.022			.90		.90	1.35
0700	Heavy		280	.029			1.15		1.15	1.74
0710	Clapboard, light		520	.015			.62		.62	.94
0720	Med.		480	.017			.67		.67	1.01
0730	Heavy		400	.020			.81		.81	1.22
0740	Wire brush									
0750	Aluminum, light	1 Pord	600	.013	S.F.		.54		.54	.81
0760	Med.		520	.015			.62		.62	.94
0770	Heavy		440	.018			.73		.73	1.11
0780	Pressure wash, based on 2500 lb. operating pressure									
0790	Stucco	A-1H	3080	.003	S.F.		.10	.02	.12	.18
0800	Aluminum or vinyl		3200	.003			.09	.02	.11	.17
0810	Siding, masonry, brick & block		2400	.003			.13	.03	.16	.22
1300	Miscellaneous, wire brush									
1310	Metal, pedestrian gate	1 Pord	100	.080	S.F.		3.23		3.23	4.86
1320	Aluminum chain link, both sides		250	.032			1.29		1.29	1.95
1400	Existing galvanized surface, clean and prime, prep for painting		380	.021		.11	.85		.96	1.41
8000	For chemical washing, see Section 04 01 30									
8010	For steam cleaning, see Section 04 01 30.20									
8020	For sand blasting, see Section 03 35 29.60 and 05 01 10.51									

09 91 03.40 Interior Surface Preparation		Crew	Daily Output	Labor-Hours	Unit	Material	Labor	Equipment	Total	Total Incl O&P
0010	**INTERIOR SURFACE PREPARATION** R099100-10									
0020	Doors, per side, not incl. frames or trim									
0030	Scrape & sand									
0040	Wood, flush	1 Pord	616	.013	S.F.		.52		.52	.79
0050	Wood, detail		496	.016			.65		.65	.98
0060	Wood, louvered		280	.029			1.15		1.15	1.74
0070	Wire brush									
0080	Metal, flush	1 Pord	640	.013	S.F.		.50		.50	.76
0090	Metal, detail		520	.015			.62		.62	.94
0100	Metal, louvered		360	.022			.90		.90	1.35
0110	Hand wash									
0120	Wood, flush	1 Pord	2160	.004	S.F.		.15		.15	.23
0130	Wood, detailed		2000	.004			.16		.16	.24
0140	Wood, louvered		1360	.006			.24		.24	.36
0150	Metal, flush		2160	.004			.15		.15	.23
0160	Metal, detail		2000	.004			.16		.16	.24
0170	Metal, louvered		1360	.006			.24		.24	.36
0400	Windows, per side, not incl. trim									
0410	Scrape & sand									
0420	Wood, 1-2 lite	1 Pord	360	.022	S.F.		.90		.90	1.35
0430	Wood, 3-6 lite		320	.025			1.01		1.01	1.52
0440	Wood, 7-10 lite		280	.029			1.15		1.15	1.74
0450	Wood, 12 lite		240	.033			1.34		1.34	2.03
0460	Wood, Bay/Bow		360	.022			.90		.90	1.35
0470	Wire brush									
0480	Metal, 1-2 lite	1 Pord	520	.015	S.F.		.62		.62	.94
0490	Metal, 3-6 lite		440	.018			.73		.73	1.11
0500	Metal, Bay/Bow		520	.015			.62		.62	.94
0600	Walls, sanding, light=10-30%, medium - 30-70%,									
0610	heavy=70-100% of surface to sand									

09 91 Painting

09 91 03 – Paint Restoration

09 91 03.40 Interior Surface Preparation		Crew	Daily Output	Labor-Hours	Unit	Material	2015 Bare Costs Labor	Equipment	Total	Total Incl O&P
0650	Walls, sand									
0660	Gypsum board or plaster, light	1 Pord	3077	.003	S.F.		.10		.10	.16
0670	Gypsum board or plaster, medium		2160	.004			.15		.15	.23
0680	Gypsum board or plaster, heavy		923	.009			.35		.35	.53
0690	Wood, T&G, light		2400	.003			.13		.13	.20
0700	Wood, T&G, med.		1600	.005			.20		.20	.30
0710	Wood, T&G, heavy		800	.010			.40		.40	.61
0720	Walls, wash									
0730	Gypsum board or plaster	1 Pord	3200	.003	S.F.		.10		.10	.15
0740	Wood, T&G		3200	.003			.10		.10	.15
0750	Masonry, brick & block, smooth		2800	.003			.12		.12	.17
0760	Masonry, brick & block, coarse		2000	.004			.16		.16	.24
8000	For chemical washing, see Section 04 01 30									
8010	For steam cleaning, see Section 04 01 30.20									
8020	For sand blasting, see Section 03 35 29.60 and 05 01 10.51									

09 91 13 – Exterior Painting

09 91 13.30 Fences

Line	Description	Crew	Daily Output	Labor-Hours	Unit	Material	Labor	Equipment	Total	Total Incl O&P
0010	**FENCES** R099100-20									
0100	Chain link or wire metal, one side, water base									
0110	Roll & brush, first coat	1 Pord	960	.008	S.F.	.08	.34		.42	.60
0120	Second coat		1280	.006		.07	.25		.32	.46
0130	Spray, first coat		2275	.004		.08	.14		.22	.30
0140	Second coat		2600	.003		.08	.12		.20	.28
0150	Picket, water base									
0160	Roll & brush, first coat	1 Pord	865	.009	S.F.	.08	.37		.45	.65
0170	Second coat		1050	.008		.08	.31		.39	.55
0180	Spray, first coat		2275	.004		.08	.14		.22	.30
0190	Second coat		2600	.003		.08	.12		.20	.28
0200	Stockade, water base									
0210	Roll & brush, first coat	1 Pord	1040	.008	S.F.	.08	.31		.39	.56
0220	Second coat		1200	.007		.08	.27		.35	.50
0230	Spray, first coat		2275	.004		.08	.14		.22	.30
0240	Second coat		2600	.003		.08	.12		.20	.28

09 91 13.42 Miscellaneous, Exterior

Line	Description	Crew	Daily Output	Labor-Hours	Unit	Material	Labor	Equipment	Total	Total Incl O&P
0010	**MISCELLANEOUS, EXTERIOR** R099100-20									
0015	For painting metals, see Section 09 97 13.23									
0100	Railing, ext., decorative wood, incl. cap & baluster									
0110	Newels & spindles @ 12" O.C.									
0120	Brushwork, stain, sand, seal & varnish									
0130	First coat	1 Pord	90	.089	L.F.	.81	3.59		4.40	6.30
0140	Second coat	"	120	.067	"	.81	2.69		3.50	4.94
0150	Rough sawn wood, 42" high, 2" x 2" verticals, 6" O.C.									
0160	Brushwork, stain, each coat	1 Pord	90	.089	L.F.	.26	3.59		3.85	5.70
0170	Wrought iron, 1" rail, 1/2" sq. verticals									
0180	Brushwork, zinc chromate, 60" high, bars 6" O.C.									
0190	Primer	1 Pord	130	.062	L.F.	.86	2.48		3.34	4.69
0200	Finish coat		130	.062		1.13	2.48		3.61	4.98
0210	Additional coat		190	.042		1.32	1.70		3.02	4.01
0220	Shutters or blinds, single panel, 2' x 4', paint all sides									
0230	Brushwork, primer	1 Pord	20	.400	Ea.	.66	16.15		16.81	25
0240	Finish coat, exterior latex		20	.400		.62	16.15		16.77	25
0250	Primer & 1 coat, exterior latex		13	.615		1.14	25		26.14	39

09 91 13 – Exterior Painting

09 91 13.42 Miscellaneous, Exterior			Crew	Daily Output	Labor-Hours	Unit	Material	2015 Bare Costs Labor	Equipment	Total	Total Incl O&P
0260	Spray, primer		1 Pord	35	.229	Ea.	.96	9.20		10.16	14.95
0270	Finish coat, exterior latex			35	.229		1.33	9.20		10.53	15.35
0280	Primer & 1 coat, exterior latex		↓	20	.400	↓	1.04	16.15		17.19	25.50
0290	For louvered shutters, add					S.F.	10%				
0300	Stair stringers, exterior, metal										
0310	Roll & brush, zinc chromate, to 14", each coat		1 Pord	320	.025	L.F.	.38	1.01		1.39	1.93
0320	Rough sawn wood, 4" x 12"										
0330	Roll & brush, exterior latex, each coat		1 Pord	215	.037	L.F.	.09	1.50		1.59	2.36
0340	Trellis/lattice, 2" x 2" @ 3" O.C. with 2" x 8" supports										
0350	Spray, latex, per side, each coat		1 Pord	475	.017	S.F.	.09	.68		.77	1.12
0450	Decking, ext., sealer, alkyd, brushwork, sealer coat			1140	.007		.10	.28		.38	.54
0460	1st coat			1140	.007		.10	.28		.38	.54
0470	2nd coat			1300	.006		.07	.25		.32	.45
0500	Paint, alkyd, brushwork, primer coat			1140	.007		.11	.28		.39	.56
0510	1st coat			1140	.007		.13	.28		.41	.57
0520	2nd coat			1300	.006		.09	.25		.34	.47
0600	Sand paint, alkyd, brushwork, 1 coat		↓	150	.053	↓	.14	2.15		2.29	3.39

09 91 13.60 Siding Exterior			Crew	Daily Output	Labor-Hours	Unit	Material	Labor	Equipment	Total	Total Incl O&P
0010	**SIDING EXTERIOR**, Alkyd (oil base)	R099100-10									
0450	Steel siding, oil base, paint 1 coat, brushwork		2 Pord	2015	.008	S.F.	.10	.32		.42	.59
0500	Spray	R099100-20		4550	.004		.15	.14		.29	.38
0800	Paint 2 coats, brushwork			1300	.012		.20	.50		.70	.97
1000	Spray			2750	.006		.17	.23		.40	.54
1200	Stucco, rough, oil base, paint 2 coats, brushwork			1300	.012		.20	.50		.70	.97
1400	Roller			1625	.010		.21	.40		.61	.83
1600	Spray			2925	.005		.22	.22		.44	.58
1800	Texture 1-11 or clapboard, oil base, primer coat, brushwork			1300	.012		.15	.50		.65	.91
2000	Spray			4550	.004		.15	.14		.29	.37
2400	Paint 2 coats, brushwork			810	.020		.30	.80		1.10	1.52
2600	Spray			2600	.006		.33	.25		.58	.73
3400	Stain 2 coats, brushwork			950	.017		.17	.68		.85	1.21
4000	Spray			3050	.005		.19	.21		.40	.53
4200	Wood shingles, oil base primer coat, brushwork			1300	.012		.14	.50		.64	.90
4400	Spray			3900	.004		.13	.17		.30	.39
5000	Paint 2 coats, brushwork			810	.020		.25	.80		1.05	1.47
5200	Spray			2275	.007		.23	.28		.51	.69
6500	Stain 2 coats, brushwork			950	.017		.17	.68		.85	1.21
7000	Spray		↓	2660	.006		.24	.24		.48	.63
8000	For latex paint, deduct						10%				
8100	For work over 12' H, from pipe scaffolding, add							15%			
8200	For work over 12' H, from extension ladder, add							25%			
8300	For work over 12' H, from swing staging, add					↓		35%			

09 91 13.62 Siding, Misc.			Crew	Daily Output	Labor-Hours	Unit	Material	Labor	Equipment	Total	Total Incl O&P
0010	**SIDING, MISC.**, latex paint	R099100-10									
0100	Aluminum siding										
0110	Brushwork, primer	R099100-20	2 Pord	2275	.007	S.F.	.06	.28		.34	.50
0120	Finish coat, exterior latex			2275	.007		.06	.28		.34	.49
0130	Primer & 1 coat exterior latex			1300	.012		.13	.50		.63	.89
0140	Primer & 2 coats exterior latex		↓	975	.016	↓	.19	.66		.85	1.20
0150	Mineral fiber shingles										
0160	Brushwork, primer		2 Pord	1495	.011	S.F.	.15	.43		.58	.81
0170	Finish coat, industrial enamel		↓	1495	.011	↓	.18	.43		.61	.85

09 91 13 – Exterior Painting

09 91 13.62 Siding, Misc.		Crew	Daily Output	Labor-Hours	Unit	2015 Bare Costs Material	Labor	Equipment	Total	Total Incl O&P
0180	Primer & 1 coat enamel	2 Pord	810	.020	S.F.	.33	.80		1.13	1.56
0190	Primer & 2 coats enamel		540	.030		.51	1.20		1.71	2.37
0200	Roll, primer		1625	.010		.17	.40		.57	.78
0210	Finish coat, industrial enamel		1625	.010		.20	.40		.60	.82
0220	Primer & 1 coat enamel		975	.016		.36	.66		1.02	1.40
0230	Primer & 2 coats enamel		650	.025		.56	.99		1.55	2.12
0240	Spray, primer		3900	.004		.13	.17		.30	.39
0250	Finish coat, industrial enamel		3900	.004		.16	.17		.33	.43
0260	Primer & 1 coat enamel		2275	.007		.29	.28		.57	.75
0270	Primer & 2 coats enamel		1625	.010		.46	.40		.86	1.10
0280	Waterproof sealer, first coat		4485	.004		.09	.14		.23	.31
0290	Second coat	↓	5235	.003	↓	.08	.12		.20	.28
0300	Rough wood incl. shingles, shakes or rough sawn siding									
0310	Brushwork, primer	2 Pord	1280	.013	S.F.	.13	.50		.63	.91
0320	Finish coat, exterior latex		1280	.013		.10	.50		.60	.87
0330	Primer & 1 coat exterior latex		960	.017		.24	.67		.91	1.27
0340	Primer & 2 coats exterior latex		700	.023		.34	.92		1.26	1.76
0350	Roll, primer		2925	.005		.18	.22		.40	.53
0360	Finish coat, exterior latex		2925	.005		.12	.22		.34	.46
0370	Primer & 1 coat exterior latex		1790	.009		.30	.36		.66	.87
0380	Primer & 2 coats exterior latex		1300	.012		.42	.50		.92	1.21
0390	Spray, primer		3900	.004		.15	.17		.32	.41
0400	Finish coat, exterior latex		3900	.004		.09	.17		.26	.35
0410	Primer & 1 coat exterior latex		2600	.006		.24	.25		.49	.64
0420	Primer & 2 coats exterior latex		2080	.008		.34	.31		.65	.84
0430	Waterproof sealer, first coat		4485	.004		.16	.14		.30	.39
0440	Second coat	↓	4485	.004	↓	.09	.14		.23	.31
0450	Smooth wood incl. butt, T&G, beveled, drop or B&B siding									
0460	Brushwork, primer	2 Pord	2325	.007	S.F.	.10	.28		.38	.53
0470	Finish coat, exterior latex		1280	.013		.10	.50		.60	.87
0480	Primer & 1 coat exterior latex		800	.020		.20	.81		1.01	1.44
0490	Primer & 2 coats exterior latex		630	.025		.30	1.02		1.32	1.87
0500	Roll, primer		2275	.007		.11	.28		.39	.55
0510	Finish coat, exterior latex		2275	.007		.11	.28		.39	.55
0520	Primer & 1 coat exterior latex		1300	.012		.22	.50		.72	.99
0530	Primer & 2 coats exterior latex		975	.016		.33	.66		.99	1.36
0540	Spray, primer		4550	.004		.08	.14		.22	.30
0550	Finish coat, exterior latex		4550	.004		.09	.14		.23	.31
0560	Primer & 1 coat exterior latex		2600	.006		.18	.25		.43	.57
0570	Primer & 2 coats exterior latex		1950	.008		.27	.33		.60	.80
0580	Waterproof sealer, first coat		5230	.003		.09	.12		.21	.28
0590	Second coat	↓	5980	.003		.09	.11		.20	.25
0600	For oil base paint, add				↓	10%				

09 91 13.70 Doors and Windows, Exterior		Crew	Daily Output	Labor-Hours	Unit	Material	Labor	Equipment	Total	Total Incl O&P
0010	**DOORS AND WINDOWS, EXTERIOR** R099100-10									
0100	Door frames & trim, only									
0110	Brushwork, primer R099100-20	1 Pord	512	.016	L.F.	.06	.63		.69	1.02
0120	Finish coat, exterior latex		512	.016		.08	.63		.71	1.04
0130	Primer & 1 coat, exterior latex		300	.027		.14	1.08		1.22	1.77
0140	Primer & 2 coats, exterior latex	↓	265	.030	↓	.22	1.22		1.44	2.08
0150	Doors, flush, both sides, incl. frame & trim									
0160	Roll & brush, primer	1 Pord	10	.800	Ea.	4.55	32.50		37.05	53.50

09 91 13.70 Doors and Windows, Exterior		Crew	Daily Output	Labor-Hours	Unit	Material	2015 Bare Costs Labor	Equipment	Total	Total Incl O&P
0170	Finish coat, exterior latex	1 Pord	10	.800	Ea.	5.95	32.50		38.45	55
0180	Primer & 1 coat, exterior latex		7	1.143		10.50	46		56.50	81
0190	Primer & 2 coats, exterior latex		5	1.600		16.40	64.50		80.90	116
0200	Brushwork, stain, sealer & 2 coats polyurethane		4	2		28.50	80.50		109	154
0210	Doors, French, both sides, 10-15 lite, incl. frame & trim									
0220	Brushwork, primer	1 Pord	6	1.333	Ea.	2.27	54		56.27	83.50
0230	Finish coat, exterior latex		6	1.333		2.97	54		56.97	84.50
0240	Primer & 1 coat, exterior latex		3	2.667		5.25	108		113.25	168
0250	Primer & 2 coats, exterior latex		2	4		8.05	161		169.05	252
0260	Brushwork, stain, sealer & 2 coats polyurethane		2.50	3.200		10.30	129		139.30	206
0270	Doors, louvered, both sides, incl. frame & trim									
0280	Brushwork, primer	1 Pord	7	1.143	Ea.	4.55	46		50.55	74.50
0290	Finish coat, exterior latex		7	1.143		5.95	46		51.95	76
0300	Primer & 1 coat, exterior latex		4	2		10.50	80.50		91	134
0310	Primer & 2 coats, exterior latex		3	2.667		16.10	108		124.10	180
0320	Brushwork, stain, sealer & 2 coats polyurethane		4.50	1.778		28.50	71.50		100	140
0330	Doors, panel, both sides, incl. frame & trim									
0340	Roll & brush, primer	1 Pord	6	1.333	Ea.	4.55	54		58.55	86
0350	Finish coat, exterior latex		6	1.333		5.95	54		59.95	87.50
0360	Primer & 1 coat, exterior latex		3	2.667		10.50	108		118.50	174
0370	Primer & 2 coats, exterior latex		2.50	3.200		16.10	129		145.10	213
0380	Brushwork, stain, sealer & 2 coats polyurethane		3	2.667		28.50	108		136.50	194
0400	Windows, per ext. side, based on 15 S.F.									
0410	1 to 6 lite									
0420	Brushwork, primer	1 Pord	13	.615	Ea.	.90	25		25.90	38.50
0430	Finish coat, exterior latex		13	.615		1.17	25		26.17	39
0440	Primer & 1 coat, exterior latex		8	1		2.07	40.50		42.57	63.50
0450	Primer & 2 coats, exterior latex		6	1.333		3.17	54		57.17	84.50
0460	Stain, sealer & 1 coat varnish		7	1.143		4.06	46		50.06	74
0470	7 to 10 lite									
0480	Brushwork, primer	1 Pord	11	.727	Ea.	.90	29.50		30.40	45
0490	Finish coat, exterior latex		11	.727		1.17	29.50		30.67	45.50
0500	Primer & 1 coat, exterior latex		7	1.143		2.07	46		48.07	72
0510	Primer & 2 coats, exterior latex		5	1.600		3.17	64.50		67.67	101
0520	Stain, sealer & 1 coat varnish		6	1.333		4.06	54		58.06	85.50
0530	12 lite									
0540	Brushwork, primer	1 Pord	10	.800	Ea.	.90	32.50		33.40	49.50
0550	Finish coat, exterior latex		10	.800		1.17	32.50		33.67	50
0560	Primer & 1 coat, exterior latex		6	1.333		2.07	54		56.07	83.50
0570	Primer & 2 coats, exterior latex		5	1.600		3.17	64.50		67.67	101
0580	Stain, sealer & 1 coat varnish		6	1.333		4.15	54		58.15	85.50
0590	For oil base paint, add					10%				

09 91 13.80 Trim, Exterior		Crew	Daily Output	Labor-Hours	Unit	Material	Labor	Equipment	Total	Total Incl O&P
0010	**TRIM, EXTERIOR** R099100-10									
0100	Door frames & trim (see Doors, interior or exterior)									
0110	Fascia, latex paint, one coat coverage R099100-20									
0120	1" x 4", brushwork	1 Pord	640	.013	L.F.	.02	.50		.52	.79
0130	Roll		1280	.006		.03	.25		.28	.41
0140	Spray		2080	.004		.02	.16		.18	.25
0150	1" x 6" to 1" x 10", brushwork		640	.013		.08	.50		.58	.85
0160	Roll		1230	.007		.09	.26		.35	.50
0170	Spray		2100	.004		.07	.15		.22	.30

09 91 Painting

09 91 13 – Exterior Painting

09 91 13.80 Trim, Exterior		Crew	Daily Output	Labor-Hours	Unit	2015 Bare Costs Material	Labor	Equipment	Total	Total Incl O&P
0180	1" x 12", brushwork	1 Pord	640	.013	L.F.	.08	.50		.58	.85
0190	Roll		1050	.008		.09	.31		.40	.56
0200	Spray		2200	.004		.07	.15		.22	.29
0210	Gutters & downspouts, metal, zinc chromate paint									
0220	Brushwork, gutters, 5", first coat	1 Pord	640	.013	L.F.	.40	.50		.90	1.20
0230	Second coat		960	.008		.38	.34		.72	.92
0240	Third coat		1280	.006		.30	.25		.55	.72
0250	Downspouts, 4", first coat		640	.013		.40	.50		.90	1.20
0260	Second coat		960	.008		.38	.34		.72	.92
0270	Third coat		1280	.006		.30	.25		.55	.72
0280	Gutters & downspouts, wood									
0290	Brushwork, gutters, 5", primer	1 Pord	640	.013	L.F.	.06	.50		.56	.83
0300	Finish coat, exterior latex		640	.013		.07	.50		.57	.84
0310	Primer & 1 coat exterior latex		400	.020		.14	.81		.95	1.37
0320	Primer & 2 coats exterior latex		325	.025		.22	.99		1.21	1.74
0330	Downspouts, 4", primer		640	.013		.06	.50		.56	.83
0340	Finish coat, exterior latex		640	.013		.07	.50		.57	.84
0350	Primer & 1 coat exterior latex		400	.020		.14	.81		.95	1.37
0360	Primer & 2 coats exterior latex		325	.025		.11	.99		1.10	1.62
0370	Molding, exterior, up to 14" wide									
0380	Brushwork, primer	1 Pord	640	.013	L.F.	.07	.50		.57	.84
0390	Finish coat, exterior latex		640	.013		.08	.50		.58	.85
0400	Primer & 1 coat exterior latex		400	.020		.17	.81		.98	1.40
0410	Primer & 2 coats exterior latex		315	.025		.17	1.02		1.19	1.72
0420	Stain & fill		1050	.008		.10	.31		.41	.57
0430	Shellac		1850	.004		.13	.17		.30	.40
0440	Varnish		1275	.006		.11	.25		.36	.50

09 91 13.90 Walls, Masonry (CMU), Exterior		Crew	Daily Output	Labor-Hours	Unit	Material	Labor	Equipment	Total	Total Incl O&P
0010	**WALLS, MASONRY (CMU), EXTERIOR** R099100-10									
0360	Concrete masonry units (CMU), smooth surface									
0370	Brushwork, latex, first coat	1 Pord	640	.013	S.F.	.07	.50		.57	.84
0380	Second coat		960	.008		.06	.34		.40	.57
0390	Waterproof sealer, first coat		736	.011		.25	.44		.69	.93
0400	Second coat		1104	.007		.25	.29		.54	.71
0410	Roll, latex, paint, first coat		1465	.005		.09	.22		.31	.42
0420	Second coat		1790	.004		.07	.18		.25	.34
0430	Waterproof sealer, first coat		1680	.005		.25	.19		.44	.56
0440	Second coat		2060	.004		.25	.16		.41	.51
0450	Spray, latex, paint, first coat		1950	.004		.07	.17		.24	.32
0460	Second coat		2600	.003		.05	.12		.17	.25
0470	Waterproof sealer, first coat		2245	.004		.25	.14		.39	.49
0480	Second coat		2990	.003		.25	.11		.36	.43
0490	Concrete masonry unit (CMU), porous									
0500	Brushwork, latex, first coat	1 Pord	640	.013	S.F.	.14	.50		.64	.92
0510	Second coat		960	.008		.07	.34		.41	.59
0520	Waterproof sealer, first coat		736	.011		.25	.44		.69	.93
0530	Second coat		1104	.007		.25	.29		.54	.71
0540	Roll latex, first coat		1465	.005		.11	.22		.33	.45
0550	Second coat		1790	.004		.07	.18		.25	.35
0560	Waterproof sealer, first coat		1680	.005		.25	.19		.44	.56
0570	Second coat		2060	.004		.25	.16		.41	.51
0580	Spray latex, first coat		1950	.004		.08	.17		.25	.34

09 91 Painting

09 91 13 – Exterior Painting

09 91 13.90 Walls, Masonry (CMU), Exterior		Crew	Daily Output	Labor-Hours	Unit	Material	Labor (2015 Bare Costs)	Equipment	Total	Total Incl O&P
0590	Second coat	1 Pord	2600	.003	S.F.	.05	.12		.17	.25
0600	Waterproof sealer, first coat		2245	.004		.25	.14		.39	.49
0610	Second coat		2990	.003		.25	.11		.36	.43

09 91 23 – Interior Painting

09 91 23.20 Cabinets and Casework

		Crew	Daily Output	Labor-Hours	Unit	Material	Labor	Equipment	Total	Total Incl O&P
0010	CABINETS AND CASEWORK R099100-10									
1000	Primer coat, oil base, brushwork	1 Pord	650	.012	S.F.	.07	.50		.57	.83
2000	Paint, oil base, brushwork, 1 coat R099100-20		650	.012		.11	.50		.61	.87
3000	Stain, brushwork, wipe off		650	.012		.09	.50		.59	.84
4000	Shellac, 1 coat, brushwork		650	.012		.11	.50		.61	.87
4500	Varnish, 3 coats, brushwork, sand after 1st coat		325	.025		.28	.99		1.27	1.81
5000	For latex paint, deduct					10%				
6300	Strip, prep and refinish wood furniture									
6310	Remove paint using chemicals, wood furniture	1 Pord	28	.286	S.F.	1.58	11.55		13.13	19.10
6320	Prep for painting, sanding		75	.107		.23	4.30		4.53	6.75
6350	Stain and wipe, brushwork		600	.013		.09	.54		.63	.90
6355	Spray applied		900	.009		.08	.36		.44	.62
6360	Sealer or varnish, brushwork		1080	.007		.09	.30		.39	.55
6365	Spray applied		2100	.004		.09	.15		.24	.33
6370	Paint, primer, brushwork		720	.011		.07	.45		.52	.76
6375	Spray applied		2100	.004		.07	.15		.22	.30
6380	Finish coat, brushwork		810	.010		.11	.40		.51	.72
6385	Spray applied		2100	.004		.10	.15		.25	.34

09 91 23.33 Doors and Windows, Interior Alkyd (Oil Base)

		Crew	Daily Output	Labor-Hours	Unit	Material	Labor	Equipment	Total	Total Incl O&P
0010	DOORS AND WINDOWS, INTERIOR ALKYD (OIL BASE) R099100-10									
0500	Flush door & frame, 3′ x 7′, oil, primer, brushwork	1 Pord	10	.800	Ea.	3.45	32.50		35.95	52.50
1000	Paint, 1 coat R099100-20		10	.800		4.14	32.50		36.64	53
1400	Stain, brushwork, wipe off		18	.444		1.81	17.95		19.76	29
1600	Shellac, 1 coat, brushwork		25	.320		2.25	12.90		15.15	22
1800	Varnish, 3 coats, brushwork, sand after 1st coat		9	.889		5.85	36		41.85	60.50
2000	Panel door & frame, 3′ x 7′, oil, primer, brushwork		6	1.333		2.66	54		56.66	84
2200	Paint, 1 coat		6	1.333		4.14	54		58.14	85.50
2600	Stain, brushwork, panel door, 3′ x 7′, not incl. frame		16	.500		1.81	20		21.81	32.50
2800	Shellac, 1 coat, brushwork		22	.364		2.25	14.65		16.90	24.50
3000	Varnish, 3 coats, brushwork, sand after 1st coat		7.50	1.067		5.85	43		48.85	71.50
4400	Windows, including frame and trim, per side									
4600	Colonial type, 6/6 lites, 2′ x 3′, oil, primer, brushwork	1 Pord	14	.571	Ea.	.42	23		23.42	35
5800	Paint, 1 coat		14	.571		.65	23		23.65	35
6200	3′ x 5′ opening, 6/6 lites, primer coat, brushwork		12	.667		1.05	27		28.05	41.50
6400	Paint, 1 coat		12	.667		1.64	27		28.64	42.50
6800	4′ x 8′ opening, 6/6 lites, primer coat, brushwork		8	1		2.24	40.50		42.74	63.50
7000	Paint, 1 coat		8	1		3.49	40.50		43.99	65
8000	Single lite type, 2′ x 3′, oil base, primer coat, brushwork		33	.242		.42	9.80		10.22	15.20
8200	Paint, 1 coat		33	.242		.65	9.80		10.45	15.45
8600	3′ x 5′ opening, primer coat, brushwork		20	.400		1.05	16.15		17.20	25.50
8800	Paint, 1 coat		20	.400		1.64	16.15		17.79	26.50
9200	4′ x 8′ opening, primer coat, brushwork		14	.571		2.24	23		25.24	37
9400	Paint, 1 coat		14	.571		3.49	23		26.49	38.50

09 91 23.35 Doors and Windows, Interior Latex

		Crew	Daily Output	Labor-Hours	Unit	Material	Labor	Equipment	Total	Total Incl O&P
0010	DOORS & WINDOWS, INTERIOR LATEX R099100-10									
0100	Doors, flush, both sides, incl. frame & trim									
0110	Roll & brush, primer R099100-20	1 Pord	10	.800	Ea.	4.13	32.50		36.63	53

09 91 23 – Interior Painting

09 91 23.35 Doors and Windows, Interior Latex		Crew	Daily Output	Labor-Hours	Unit	Material	Labor	Equipment	Total	Total Incl O&P
0120	Finish coat, latex	1 Pord	10	.800	Ea.	5.30	32.50		37.80	54.50
0130	Primer & 1 coat latex		7	1.143		9.40	46		55.40	80
0140	Primer & 2 coats latex		5	1.600		14.40	64.50		78.90	113
0160	Spray, both sides, primer		20	.400		4.35	16.15		20.50	29.50
0170	Finish coat, latex		20	.400		5.55	16.15		21.70	30.50
0180	Primer & 1 coat latex		11	.727		9.95	29.50		39.45	55
0190	Primer & 2 coats latex		8	1		15.25	40.50		55.75	78
0200	Doors, French, both sides, 10-15 lite, incl. frame & trim									
0210	Roll & brush, primer	1 Pord	6	1.333	Ea.	2.06	54		56.06	83.50
0220	Finish coat, latex		6	1.333		2.65	54		56.65	84
0230	Primer & 1 coat latex		3	2.667		4.71	108		112.71	167
0240	Primer & 2 coats latex		2	4		7.20	161		168.20	251
0260	Doors, louvered, both sides, incl. frame & trim									
0270	Roll & brush, primer	1 Pord	7	1.143	Ea.	4.13	46		50.13	74
0280	Finish coat, latex		7	1.143		5.30	46		51.30	75.50
0290	Primer & 1 coat, latex		4	2		9.20	80.50		89.70	132
0300	Primer & 2 coats, latex		3	2.667		14.70	108		122.70	178
0320	Spray, both sides, primer		20	.400		4.35	16.15		20.50	29.50
0330	Finish coat, latex		20	.400		5.55	16.15		21.70	30.50
0340	Primer & 1 coat, latex		11	.727		9.95	29.50		39.45	55
0350	Primer & 2 coats, latex		8	1		15.60	40.50		56.10	78
0360	Doors, panel, both sides, incl. frame & trim									
0370	Roll & brush, primer	1 Pord	6	1.333	Ea.	4.35	54		58.35	86
0380	Finish coat, latex		6	1.333		5.30	54		59.30	87
0390	Primer & 1 coat, latex		3	2.667		9.40	108		117.40	172
0400	Primer & 2 coats, latex		2.50	3.200		14.70	129		143.70	211
0420	Spray, both sides, primer		10	.800		4.35	32.50		36.85	53.50
0430	Finish coat, latex		10	.800		5.55	32.50		38.05	54.50
0440	Primer & 1 coat, latex		5	1.600		9.95	64.50		74.45	108
0450	Primer & 2 coats, latex		4	2		15.60	80.50		96.10	139
0460	Windows, per interior side, based on 15 S.F.									
0470	1 to 6 lite									
0480	Brushwork, primer	1 Pord	13	.615	Ea.	.81	25		25.81	38.50
0490	Finish coat, enamel		13	.615		1.04	25		26.04	38.50
0500	Primer & 1 coat enamel		8	1		1.86	40.50		42.36	63
0510	Primer & 2 coats enamel		6	1.333		2.90	54		56.90	84
0530	7 to 10 lite									
0540	Brushwork, primer	1 Pord	11	.727	Ea.	.81	29.50		30.31	45
0550	Finish coat, enamel		11	.727		1.04	29.50		30.54	45
0560	Primer & 1 coat enamel		7	1.143		1.86	46		47.86	71.50
0570	Primer & 2 coats enamel		5	1.600		2.90	64.50		67.40	101
0590	12 lite									
0600	Brushwork, primer	1 Pord	10	.800	Ea.	.81	32.50		33.31	49.50
0610	Finish coat, enamel		10	.800		1.04	32.50		33.54	49.50
0620	Primer & 1 coat enamel		6	1.333		1.86	54		55.86	83
0630	Primer & 2 coats enamel		5	1.600		2.90	64.50		67.40	101
0650	For oil base paint, add					10%				

09 91 23.39 Doors and Windows, Interior Latex, Zero Voc		Crew	Daily Output	Labor-Hours	Unit	Material	Labor	Equipment	Total	Total Incl O&P
0010	**DOORS & WINDOWS, INTERIOR LATEX, ZERO VOC**									
0100	Doors flush, both sides, incl. frame & trim									
0110	Roll & brush, primer G	1 Pord	10	.800	Ea.	4.90	32.50		37.40	54
0120	Finish coat, latex G		10	.800		5.70	32.50		38.20	55

Costs columns (Material, Labor, Equipment, Total) are 2015 Bare Costs.

09 91 23 – Interior Painting

09 91 23.39 Doors and Windows, Interior Latex, Zero Voc

	09 91 23.39 Doors and Windows, Interior Latex, Zero Voc		Crew	Daily Output	Labor-Hours	Unit	2015 Bare Costs Material	Labor	Equipment	Total	Total Incl O&P
0130	Primer & 1 coat latex	G	1 Pord	7	1.143	Ea.	10.60	46		56.60	81
0140	Primer & 2 coats latex	G		5	1.600		16	64.50		80.50	115
0160	Spray, both sides, primer	G		20	.400		5.15	16.15		21.30	30
0170	Finish coat, latex	G		20	.400		6	16.15		22.15	31
0180	Primer & 1 coat latex	G		11	.727		11.25	29.50		40.75	56.50
0190	Primer & 2 coats latex	G		8	1		16.95	40.50		57.45	79.50
0200	Doors, French, both sides, 10-15 lite, incl. frame & trim										
0210	Roll & brush, primer	G	1 Pord	6	1.333	Ea.	2.45	54		56.45	83.50
0220	Finish coat, latex	G		6	1.333		2.86	54		56.86	84
0230	Primer & 1 coat latex	G		3	2.667		5.30	108		113.30	168
0240	Primer & 2 coats latex	G		2	4		8	161		169	252
0360	Doors, panel, both sides, incl. frame & trim										
0370	Roll & brush, primer	G	1 Pord	6	1.333	Ea.	5.15	54		59.15	86.50
0380	Finish coat, latex	G		6	1.333		5.70	54		59.70	87.50
0390	Primer & 1 coat, latex	G		3	2.667		10.60	108		118.60	174
0400	Primer & 2 coats, latex	G		2.50	3.200		16.35	129		145.35	213
0420	Spray, both sides, primer	G		10	.800		5.15	32.50		37.65	54
0430	Finish coat, latex	G		10	.800		6	32.50		38.50	55
0440	Primer & 1 coat, latex	G		5	1.600		11.25	64.50		75.75	110
0450	Primer & 2 coats, latex	G		4	2		17.30	80.50		97.80	141
0460	Windows, per interior side, based on 15 S.F.										
0470	1 to 6 lite										
0480	Brushwork, primer	G	1 Pord	13	.615	Ea.	.97	25		25.97	38.50
0490	Finish coat, enamel	G		13	.615		1.13	25		26.13	38.50
0500	Primer & 1 coat enamel	G		8	1		2.09	40.50		42.59	63.50
0510	Primer & 2 coats enamel	G		6	1.333		3.22	54		57.22	84.50

09 91 23.40 Floors, Interior

	09 91 23.40 Floors, Interior	Crew	Daily Output	Labor-Hours	Unit	Material	Labor	Equipment	Total	Total Incl O&P
0010	**FLOORS, INTERIOR** R099100-10									
0100	Concrete paint, latex									
0110	Brushwork									
0120	1st coat	1 Pord	975	.008	S.F.	.15	.33		.48	.66
0130	2nd coat		1150	.007		.10	.28		.38	.53
0140	3rd coat		1300	.006		.08	.25		.33	.46
0150	Roll									
0160	1st coat	1 Pord	2600	.003	S.F.	.20	.12		.32	.41
0170	2nd coat		3250	.002		.12	.10		.22	.28
0180	3rd coat		3900	.002		.09	.08		.17	.22
0190	Spray									
0200	1st coat	1 Pord	2600	.003	S.F.	.17	.12		.29	.38
0210	2nd coat		3250	.002		.09	.10		.19	.25
0220	3rd coat		3900	.002		.07	.08		.15	.20

09 91 23.44 Anti-Slip Floor Treatments

	09 91 23.44 Anti-Slip Floor Treatments	Crew	Daily Output	Labor-Hours	Unit	Material	Labor	Equipment	Total	Total Incl O&P
0010	**ANTI-SLIP FLOOR TREATMENTS**									
1000	Walking surface treatment, ADA compliant, mop on and rinse									
1100	For tile, terrazzo, stone or smooth concrete	1 Pord	4000	.002	S.F.	.32	.08		.40	.48
1110	For marble		4000	.002		.21	.08		.29	.35
1120	For wood		4000	.002		.20	.08		.28	.34
1130	For baths and showers		500	.016		.37	.65		1.02	1.38
2000	Granular additive for paint or sealer, add to paint cost					.02			.02	.02

09 91 Painting

09 91 23 – Interior Painting

09 91 23.52 Miscellaneous, Interior		Crew	Daily Output	Labor-Hours	Unit	2015 Bare Costs Material	Labor	Equipment	Total	Total Incl O&P
0010	**MISCELLANEOUS, INTERIOR** R099100-10									
2400	Floors, conc./wood, oil base, primer/sealer coat, brushwork	2 Pord	1950	.008	S.F.	.09	.33		.42	.60
2450	Roller		5200	.003		.10	.12		.22	.30
2600	Spray		6000	.003		.10	.11		.21	.27
2650	Paint 1 coat, brushwork		1950	.008		.10	.33		.43	.61
2800	Roller		5200	.003		.10	.12		.22	.30
2850	Spray		6000	.003		.11	.11		.22	.28
3000	Stain, wood floor, brushwork, 1 coat		4550	.004		.09	.14		.23	.30
3200	Roller		5200	.003		.09	.12		.21	.29
3250	Spray		6000	.003		.09	.11		.20	.26
3400	Varnish, wood floor, brushwork		4550	.004		.09	.14		.23	.31
3450	Roller		5200	.003		.10	.12		.22	.30
3600	Spray		6000	.003		.10	.11		.21	.27
3650	For dust proofing or anti skid, see Section 03 35 29.30									
3800	Grilles, per side, oil base, primer coat, brushwork	1 Pord	520	.015	S.F.	.14	.62		.76	1.09
3850	Spray		1140	.007		.15	.28		.43	.59
3920	Paint 2 coats, brushwork		325	.025		.42	.99		1.41	1.97
3940	Spray		650	.012		.48	.50		.98	1.28
4600	Miscellaneous surfaces, metallic paint, spray applied									
4610	Water based, non-tintable, warm silver	1 Pord	1140	.007	S.F.	.75	.28		1.03	1.26
4620	Rusted iron		1140	.007		.97	.28		1.25	1.50
4630	Low VOC, tintable		1140	.007		.33	.28		.61	.79
5000	Pipe, 1" - 4" diameter, primer or sealer coat, oil base, brushwork	2 Pord	1250	.013	L.F.	.10	.52		.62	.89
5100	Spray		2165	.007		.09	.30		.39	.55
5350	Paint 2 coats, brushwork		775	.021		.19	.83		1.02	1.47
5400	Spray		1240	.013		.22	.52		.74	1.02
6300	13" - 16" diameter, primer or sealer coat, brushwork		310	.052		.39	2.08		2.47	3.57
6350	Spray		540	.030		.44	1.20		1.64	2.28
6500	Paint 2 coats, brushwork		195	.082		.78	3.31		4.09	5.85
6550	Spray		310	.052		.86	2.08		2.94	4.09
7000	Trim, wood, incl. puttying, under 6" wide									
7200	Primer coat, oil base, brushwork	1 Pord	650	.012	L.F.	.03	.50		.53	.79
7250	Paint, 1 coat, brushwork		650	.012		.05	.50		.55	.81
7450	3 coats		325	.025		.16	.99		1.15	1.67
7500	Over 6" wide, primer coat, brushwork		650	.012		.07	.50		.57	.83
7550	Paint, 1 coat, brushwork		650	.012		.11	.50		.61	.87
7650	3 coats		325	.025		.31	.99		1.30	1.85
8000	Cornice, simple design, primer coat, oil base, brushwork		650	.012	S.F.	.07	.50		.57	.83
8250	Paint, 1 coat		650	.012		.11	.50		.61	.87
8350	Ornate design, primer coat		350	.023		.07	.92		.99	1.47
8400	Paint, 1 coat		350	.023		.11	.92		1.03	1.51
8600	Balustrades, primer coat, oil base, brushwork		520	.015		.07	.62		.69	1.02
8650	Paint, 1 coat		520	.015		.11	.62		.73	1.06
8900	Trusses and wood frames, primer coat, oil base, brushwork		800	.010		.07	.40		.47	.69
8950	Spray		1200	.007		.07	.27		.34	.49
9220	Paint 2 coats, brushwork		500	.016		.21	.65		.86	1.20
9240	Spray		600	.013		.24	.54		.78	1.07
9260	Stain, brushwork, wipe off		600	.013		.09	.54		.63	.90
9280	Varnish, 3 coats, brushwork		275	.029		.28	1.17		1.45	2.08
9350	For latex paint, deduct					10%				

09 91 Painting

09 91 23 – Interior Painting

09 91 23.62 Electrostatic Painting		Crew	Daily Output	Labor-Hours	Unit	Material	Labor	Equipment	Total	Total Incl O&P
						2015 Bare Costs				
0010	**ELECTROSTATIC PAINTING**									
0100	In shop									
0200	Flat surfaces (lockers, casework, elevator doors. etc.)									
0300	One coat	1 Pord	200	.040	S.F.	.53	1.61		2.14	3.01
0400	Two coats	"	120	.067	"	.77	2.69		3.46	4.89
0500	Irregular surfaces (furniture, door frames, etc.)									
0600	One coat	1 Pord	150	.053	S.F.	.53	2.15		2.68	3.82
0700	Two coats	"	100	.080	"	.77	3.23		4	5.70
0800	On site									
0900	Flat surfaces (lockers, casework, elevator doors, etc.)									
1000	One coat	1 Pord	150	.053	S.F.	.53	2.15		2.68	3.82
1100	Two coats	"	100	.080	"	.77	3.23		4	5.70
1200	Irregular surfaces (furniture, door frames, etc)									
1300	One coat	1 Pord	115	.070	S.F.	.53	2.81		3.34	4.81
1400	Two coats		70	.114		.77	4.61		5.38	7.80
2000	Anti-microbial coating, hospital application		150	.053		.06	2.15		2.21	3.31

09 91 23.72 Walls and Ceilings, Interior		Crew	Daily Output	Labor-Hours	Unit	Material	Labor	Equipment	Total	Total Incl O&P
0010	**WALLS AND CEILINGS, INTERIOR** R099100-10									
0100	Concrete, drywall or plaster, latex , primer or sealer coat R099100-20									
0200	Smooth finish, brushwork	1 Pord	1150	.007	S.F.	.06	.28		.34	.49
0240	Roller		1350	.006		.06	.24		.30	.43
0280	Spray		2750	.003		.05	.12		.17	.24
0300	Sand finish, brushwork		975	.008		.06	.33		.39	.57
0340	Roller		1150	.007		.06	.28		.34	.49
0380	Spray		2275	.004		.05	.14		.19	.27
0800	Paint 2 coats, smooth finish, brushwork		680	.012		.13	.47		.60	.87
0840	Roller		800	.010		.13	.40		.53	.76
0880	Spray		1625	.005		.12	.20		.32	.44
0900	Sand finish, brushwork		605	.013		.13	.53		.66	.95
0940	Roller		1020	.008		.13	.32		.45	.63
0980	Spray		1700	.005		.12	.19		.31	.43
1200	Paint 3 coats, smooth finish, brushwork		510	.016		.20	.63		.83	1.17
1240	Roller		650	.012		.20	.50		.70	.97
1280	Spray		1625	.005		.19	.20		.39	.50
1600	Glaze coating, 2 coats, spray, clear		1200	.007		.49	.27		.76	.95
1640	Multicolor		1200	.007		.99	.27		1.26	1.50
1660	Painting walls, complete, including surface prep, primer &									
1670	2 coats finish, on drywall or plaster, with roller	1 Pord	325	.025	S.F.	.20	.99		1.19	1.72
1700	For oil base paint, add					10%				
1800	For ceiling installations, add						25%			
2000	Masonry or concrete block, primer/sealer, latex paint									
2100	Primer, smooth finish, brushwork	1 Pord	1000	.008	S.F.	.11	.32		.43	.61
2110	Roller		1150	.007		.11	.28		.39	.54
2180	Spray		2400	.003		.10	.13		.23	.31
2200	Sand finish, brushwork		850	.009		.11	.38		.49	.69
2210	Roller		975	.008		.11	.33		.44	.62
2280	Spray		2050	.004		.10	.16		.26	.35
2400	Finish coat, smooth finish, brush		1100	.007		.08	.29		.37	.53
2410	Roller		1300	.006		.08	.25		.33	.46
2480	Spray		2400	.003		.07	.13		.20	.28
2500	Sand finish, brushwork		950	.008		.08	.34		.42	.60
2510	Roller		1090	.007		.08	.30		.38	.54

09 91 Painting

09 91 23 – Interior Painting

09 91 23.72 Walls and Ceilings, Interior		Crew	Daily Output	Labor-Hours	Unit	Material	Labor	Equipment	Total	Total Incl O&P
2580	Spray	1 Pord	2040	.004	S.F.	.07	.16		.23	.32
2800	Primer plus one finish coat, smooth brush		525	.015		.30	.61		.91	1.26
2810	Roller		615	.013		.19	.53		.72	1
2880	Spray		1200	.007		.17	.27		.44	.59
2900	Sand finish, brushwork		450	.018		.19	.72		.91	1.29
2910	Roller		515	.016		.19	.63		.82	1.15
2980	Spray		1025	.008		.17	.31		.48	.65
3200	Primer plus 2 finish coats, smooth, brush		355	.023		.27	.91		1.18	1.66
3210	Roller		415	.019		.27	.78		1.05	1.46
3280	Spray		800	.010		.23	.40		.63	.87
3300	Sand finish, brushwork		305	.026		.27	1.06		1.33	1.88
3310	Roller		350	.023		.27	.92		1.19	1.68
3380	Spray		675	.012		.23	.48		.71	.98
3600	Glaze coating, 3 coats, spray, clear		900	.009		.70	.36		1.06	1.31
3620	Multicolor		900	.009		1.15	.36		1.51	1.81
4000	Block filler, 1 coat, brushwork		425	.019		.12	.76		.88	1.27
4100	Silicone, water repellent, 2 coats, spray		2000	.004		.33	.16		.49	.61
4120	For oil base paint, add					10%				
8200	For work 8' - 15' H, add						10%			
8300	For work over 15' H, add						20%			
8400	For light textured surfaces, add						10%			
8410	Heavy textured, add						25%			

09 91 23.74 Walls and Ceilings, Interior, Zero VOC Latex			Crew	Daily Output	Labor-Hours	Unit	Material	Labor	Equipment	Total	Total Incl O&P
0010	**WALLS AND CEILINGS, INTERIOR, ZERO VOC LATEX**										
0100	Concrete, dry wall or plaster, latex, primer or sealer coat										
0200	Smooth finish, brushwork	G	1 Pord	1150	.007	S.F.	.06	.28		.34	.49
0240	Roller	G		1350	.006		.06	.24		.30	.43
0280	Spray	G		2750	.003		.05	.12		.17	.23
0300	Sand finish, brushwork	G		975	.008		.06	.33		.39	.57
0340	Roller	G		1150	.007		.07	.28		.35	.49
0380	Spray	G		2275	.004		.05	.14		.19	.27
0800	Paint 2 coats, smooth finish, brushwork	G		680	.012		.15	.47		.62	.89
0840	Roller	G		800	.010		.16	.40		.56	.78
0880	Spray	G		1625	.005		.13	.20		.33	.45
0900	Sand finish, brushwork	G		605	.013		.15	.53		.68	.96
0940	Roller	G		1020	.008		.16	.32		.48	.65
0980	Spray	G		1700	.005		.13	.19		.32	.44
1200	Paint 3 coats, smooth finish, brushwork	G		510	.016		.22	.63		.85	1.19
1240	Roller	G		650	.012		.23	.50		.73	1.01
1280	Spray	G		1625	.005		.20	.20		.40	.52
1800	For ceiling installations, add	G						25%			
8200	For work 8' - 15' H, add							10%			
8300	For work over 15' H, add							20%			

09 91 23.75 Dry Fall Painting			Crew	Daily Output	Labor-Hours	Unit	Material	Labor	Equipment	Total	Total Incl O&P
0010	**DRY FALL PAINTING**	R099100-10									
0100	Sprayed on walls, gypsum board or plaster										
0220	One coat	R099100-20	1 Pord	2600	.003	S.F.	.06	.12		.18	.25
0250	Two coats			1560	.005		.11	.21		.32	.43
0280	Concrete or textured plaster, one coat			1560	.005		.06	.21		.27	.37
0310	Two coats			1300	.006		.11	.25		.36	.49
0340	Concrete block, one coat			1560	.005		.06	.21		.27	.37
0370	Two coats			1300	.006		.11	.25		.36	.49

09 91 Painting

09 91 23 – Interior Painting

09 91 23.75 Dry Fall Painting		Crew	Daily Output	Labor-Hours	Unit	Material	2015 Bare Costs Labor	Equipment	Total	Total Incl O&P
0400	Wood, one coat	1 Pord	877	.009	S.F.	.06	.37		.43	.61
0430	Two coats	↓	650	.012	↓	.11	.50		.61	.87
0440	On ceilings, gypsum board or plaster									
0470	One coat	1 Pord	1560	.005	S.F.	.06	.21		.27	.37
0500	Two coats		1300	.006		.11	.25		.36	.49
0530	Concrete or textured plaster, one coat		1560	.005		.06	.21		.27	.37
0560	Two coats		1300	.006		.11	.25		.36	.49
0570	Structural steel, bar joists or metal deck, one coat		1560	.005		.06	.21		.27	.37
0580	Two coats	↓	1040	.008	↓	.11	.31		.42	.59

09 93 Staining and Transparent Finishing

09 93 23 – Interior Staining and Finishing

09 93 23.10 Varnish		Crew	Daily Output	Labor-Hours	Unit	Material	Labor	Equipment	Total	Total Incl O&P
0010	**VARNISH**									
0012	1 coat + sealer, on wood trim, brush, no sanding included	1 Pord	400	.020	S.F.	.07	.81		.88	1.30
0020	1 coat + sealer, on wood trim, brush, no sanding included, no VOC		400	.020		.21	.81		1.02	1.45
0100	Hardwood floors, 2 coats, no sanding included, roller	↓	1890	.004	↓	.15	.17		.32	.43

09 96 High-Performance Coatings

09 96 23 – Graffiti-Resistant Coatings

09 96 23.10 Graffiti Resistant Treatments		Crew	Daily Output	Labor-Hours	Unit	Material	Labor	Equipment	Total	Total Incl O&P
0010	**GRAFFITI RESISTANT TREATMENTS**, sprayed on walls									
0100	Non-sacrificial, permanent non-stick coating, clear, on metals	1 Pord	2000	.004	S.F.	2.06	.16		2.22	2.51
0200	Concrete		2000	.004		2.35	.16		2.51	2.82
0300	Concrete block		2000	.004		3.03	.16		3.19	3.58
0400	Brick		2000	.004		3.44	.16		3.60	4.02
0500	Stone		2000	.004		3.44	.16		3.60	4.02
0600	Unpainted wood		2000	.004		3.97	.16		4.13	4.61
2000	Semi-permanent cross linking polymer primer, on metals		2000	.004		.70	.16		.86	1.01
2100	Concrete		2000	.004		.84	.16		1	1.16
2200	Concrete block		2000	.004		1.05	.16		1.21	1.39
2300	Brick		2000	.004		.84	.16		1	1.16
2400	Stone		2000	.004		.84	.16		1	1.16
2500	Unpainted wood		2000	.004		1.16	.16		1.32	1.52
3000	Top coat, on metals		2000	.004		.55	.16		.71	.84
3100	Concrete		2000	.004		.62	.16		.78	.93
3200	Concrete block		2000	.004		.87	.16		1.03	1.20
3300	Brick		2000	.004		.73	.16		.89	1.04
3400	Stone		2000	.004		.73	.16		.89	1.04
3500	Unpainted wood		2000	.004		.87	.16		1.03	1.20
5000	Sacrificial, water based, on metal		2000	.004		.32	.16		.48	.60
5100	Concrete		2000	.004		.32	.16		.48	.60
5200	Concrete block		2000	.004		.32	.16		.48	.60
5300	Brick		2000	.004		.32	.16		.48	.60
5400	Stone		2000	.004		.32	.16		.48	.60
5500	Unpainted wood	↓	2000	.004	↓	.32	.16		.48	.60
8000	Cleaner for use after treatment									
8100	Towels or wipes, per package of 30				Ea.	.63			.63	.70
8200	Aerosol spray, 24 oz. can				"	18			18	19.80

09 96 High-Performance Coatings

09 96 46 – Intumescent Coatings

09 96 46.10 Coatings, Intumescent		Crew	Daily Output	Labor-Hours	Unit	Material	2015 Bare Costs Labor	Equipment	Total	Total Incl O&P
0010	**COATINGS, INTUMESCENT**, spray applied									
0100	On exterior structural steel, 0.25" d.f.t.	1 Pord	475	.017	S.F.	.41	.68		1.09	1.47
0150	0.51" d.f.t.		350	.023		.41	.92		1.33	1.84
0200	0.98" d.f.t.		280	.029		.41	1.15		1.56	2.19
0300	On interior structural steel, 0.108" d.f.t.		300	.027		.41	1.08		1.49	2.07
0350	0.310" d.f.t.		150	.053		.41	2.15		2.56	3.69
0400	0.670" d.f.t.		100	.080		.41	3.23		3.64	5.30

09 96 53 – Elastomeric Coatings

09 96 53.10 Coatings, Elastomeric		Crew	Daily Output	Labor-Hours	Unit	Material	Labor	Equipment	Total	Total Incl O&P
0010	**COATINGS, ELASTOMERIC**									
0020	High build, water proof, one coat system									
0100	Concrete, brush	1 Pord	650	.012	S.F.	.27	.50		.77	1.05
0110	Roll		1650	.005		.27	.20		.47	.59
0120	Spray		2600	.003		.27	.12		.39	.49
0200	Concrete block, brush		600	.013		.32	.54		.86	1.17
0210	Roll		1400	.006		.32	.23		.55	.71
0220	Spray		1900	.004		.32	.17		.49	.62
0300	Stucco, brush		400	.020		.44	.81		1.25	1.71
0310	Roll		1000	.008		.44	.32		.76	.98
0320	Spray		1500	.005		.44	.22		.66	.81

09 96 56 – Epoxy Coatings

09 96 56.20 Wall Coatings		Crew	Daily Output	Labor-Hours	Unit	Material	Labor	Equipment	Total	Total Incl O&P
0010	**WALL COATINGS**									
0100	Acrylic glazed coatings, matte	1 Pord	525	.015	S.F.	.31	.61		.92	1.27
0200	Gloss		305	.026		.65	1.06		1.71	2.31
0300	Epoxy coatings, solvent based		525	.015		.40	.61		1.01	1.37
0400	Water based		170	.047		1.20	1.90		3.10	4.18
0600	Exposed aggregate, troweled on, 1/16" to 1/4", solvent based		235	.034		.62	1.37		1.99	2.75
0700	Water based (epoxy or polyacrylate)		130	.062		1.33	2.48		3.81	5.20
0900	1/2" to 5/8" aggregate, solvent based		130	.062		1.20	2.48		3.68	5.05
1000	Water based		80	.100		2.08	4.04		6.12	8.40
1500	Exposed aggregate, sprayed on, 1/8" aggregate, solvent based		295	.027		.57	1.09		1.66	2.28
1600	Water based		145	.055		1.05	2.23		3.28	4.51
1800	High build epoxy, 50 mil, solvent based		390	.021		.68	.83		1.51	2
1900	Water based		95	.084		1.15	3.40		4.55	6.35
2100	Laminated epoxy with fiberglass, solvent based		295	.027		.73	1.09		1.82	2.45
2200	Water based		145	.055		1.32	2.23		3.55	4.80
2400	Sprayed perlite or vermiculite, 1/16" thick, solvent based		2935	.003		.27	.11		.38	.47
2500	Water based		640	.013		.74	.50		1.24	1.57
2700	Vinyl plastic wall coating, solvent based		735	.011		.33	.44		.77	1.02
2800	Water based		240	.033		.82	1.34		2.16	2.93
3000	Urethane on smooth surface, 2 coats, solvent based		1135	.007		.27	.28		.55	.73
3100	Water based		665	.012		.59	.49		1.08	1.38
3600	Ceramic-like glazed coating, cementitious, solvent based		440	.018		.48	.73		1.21	1.64
3700	Water based		345	.023		.81	.94		1.75	2.30
3900	Resin base, solvent based		640	.013		.33	.50		.83	1.12
4000	Water based		330	.024		.54	.98		1.52	2.06

09 97 13 – Steel Coatings

09 97 13.23 Exterior Steel Coatings		Crew	Daily Output	Labor-Hours	Unit	Material	Labor	Equipment	Total	Total Incl O&P
						2015 Bare Costs				
0010	**EXTERIOR STEEL COATINGS** R050516-30									
6100	Cold galvanizing, brush in field	1 Psst	1100	.007	S.F.	.23	.30		.53	.79
6510	Paints & protective coatings, sprayed in field									
6520	Alkyds, primer	2 Psst	3600	.004	S.F.	.09	.18		.27	.42
6540	Gloss topcoats		3200	.005		.08	.21		.29	.45
6560	Silicone alkyd		3200	.005		.15	.21		.36	.52
6610	Epoxy, primer		3000	.005		.29	.22		.51	.71
6630	Intermediate or topcoat		2800	.006		.26	.24		.50	.71
6650	Enamel coat		2800	.006		.33	.24		.57	.78
6700	Epoxy ester, primer		2800	.006		.42	.24		.66	.88
6720	Topcoats		2800	.006		.22	.24		.46	.66
6810	Latex primer		3600	.004		.06	.18		.24	.39
6830	Topcoats		3200	.005		.07	.21		.28	.44
6910	Universal primers, one part, phenolic, modified alkyd		2000	.008		.37	.33		.70	.99
6940	Two part, epoxy spray		2000	.008		.33	.33		.66	.95
7000	Zinc rich primers, self cure, spray, inorganic		1800	.009		.86	.37		1.23	1.60
7010	Epoxy, spray, organic	▼	1800	.009	▼	.26	.37		.63	.94
7020	Above one story, spray painting simple structures, add						25%			
7030	Intricate structures, add						50%			

Division Notes

I	CREW	DAILY OUTPUT	LABOR-HOURS	UNIT	BARE COSTS MAT.	LABOR	EQUIP.	TOTAL	TOTAL INCL O&P

Estimating Tips

General

- The items in this division are usually priced per square foot or each.
- Many items in Division 10 require some type of support system or special anchors that are not usually furnished with the item. The required anchors must be added to the estimate in the appropriate division.
- Some items in Division 10, such as lockers, may require assembly before installation. Verify the amount of assembly required. Assembly can often exceed installation time.

10 20 00 Interior Specialties

- Support angles and blocking are not included in the installation of toilet compartments, shower/ dressing compartments, or cubicles. Appropriate line items from Divisions 5 or 6 may need to be added to support the installations.
- Toilet partitions are priced by the stall. A stall consists of a side wall, pilaster, and door with hardware. Toilet tissue holders and grab bars are extra.
- The required acoustical rating of a folding partition can have a significant impact on costs. Verify the sound transmission coefficient rating of the panel priced to the specification requirements.
- Grab bar installation does not include supplemental blocking or backing to support the required load. When grab bars are installed at an existing facility, provisions must be made to attach the grab bars to solid structure.

Reference Numbers

Reference numbers are shown in shaded boxes at the beginning of some major classifications. These numbers refer to related items in the Reference Section. The reference information may be an estimating procedure, an alternate pricing method, or technical information.

Note: Not all subdivisions listed here necessarily appear in this publication. ■

10 05 Common Work Results for Specialties

10 05 05 – Selective Demolition for Specialties

10 05 05.10 Selective Demolition, Specialties		Crew	Daily Output	Labor-Hours	Unit	Material	2015 Bare Costs Labor	Equipment	Total	Total Incl O&P
0010	**SELECTIVE DEMOLITION, SPECIALTIES**									
1100	Boards and panels, wall mounted	2 Clab	15	1.067	Ea.		40		40	61.50
1200	Cases, for directory and/or bulletin boards, including doors		24	.667			25		25	38.50
1850	Shower partitions, cabinet or stall, including base and door		8	2			75		75	116
1855	Shower receptor, terrazzo or concrete	1 Clab	14	.571			21.50		21.50	33
1900	Curtain track or rod, hospital type, ceiling mounted or suspended	"	220	.036	L.F.		1.37		1.37	2.10
1910	Toilet cubicles, remove	2 Clab	8	2	Ea.		75		75	116
1930	Urinal screen, remove	1 Clab	12	.667	"		25		25	38.50
2650	Wall guard, misc. wall or corner protection	"	320	.025	L.F.		.94		.94	1.45
2750	Access floor, metal panel system, including pedestals, covering	2 Clab	850	.019	S.F.		.71		.71	1.09
3050	Fireplace, prefab, freestanding or wall hung, including hood and screen	1 Clab	2	4	Ea.		150		150	231
3054	Chimney top, simulated brick, 4′ high	"	15	.533			20		20	31
3200	Stove, woodburning, cast iron	2 Clab	2	8			300		300	465
3440	Weathervane, residential	1 Clab	12	.667			25		25	38.50
3500	Flagpole, groundset, to 70′ high, excluding base/foundation	K-1	1	16			690	305	995	1,375
3555	To 30′ high	"	2.50	6.400			275	121	396	555
4300	Letter, signs or plaques, exterior on wall	1 Clab	20	.400			15.05		15.05	23
4310	Signs, street, reflective aluminum, including post and bracket		60	.133			5		5	7.70
4320	Door signs interior on door 6″ x 6″, selective demolition		20	.400			15.05		15.05	23
4550	Turnstiles, manual or electric	2 Clab	2	8			300		300	465
5050	Lockers	1 Clab	15	.533	Opng.		20		20	31
5250	Cabinets, recessed	Q-12	12	1.333	Ea.		67.50		67.50	102
5260	Mail boxes, Horiz., Key Lock, front loading, Remove	1 Carp	34	.235	"		11.05		11.05	17
5350	Awning, fabric, including frame	2 Clab	100	.160	S.F.		6		6	9.25
6050	Partition, woven wire		1400	.011			.43		.43	.66
6100	Folding gate, security, door or window		500	.032			1.20		1.20	1.85
6580	Acoustic air wall		650	.025			.93		.93	1.42
7550	Telephone enclosure, exterior, post mounted		3	5.333	Ea.		201		201	310
8850	Scale, platform, excludes foundation or pit		.25	64	"		2,400		2,400	3,700

10 11 Visual Display Units

10 11 13 – Chalkboards

10 11 13.13 Fixed Chalkboards		Crew	Daily Output	Labor-Hours	Unit	Material	Labor	Equipment	Total	Total Incl O&P
0010	**FIXED CHALKBOARDS** Porcelain enamel steel									
3900	Wall hung									
4000	Aluminum frame and chalktrough									
4200	3′ x 4′	2 Carp	16	1	Ea.	248	47		295	345
4300	3′ x 5′		15	1.067		320	50		370	425
4500	4′ x 8′		14	1.143		430	53.50		483.50	560
4600	4′ x 12′		13	1.231		600	58		658	750
4700	Wood frame and chalktrough									
4800	3′ x 4′	2 Carp	16	1	Ea.	194	47		241	287
5000	3′ x 5′		15	1.067		243	50		293	345
5100	4′ x 5′		14	1.143		255	53.50		308.50	365
5300	4′ x 8′		13	1.231		345	58		403	470
5400	Liquid chalk, white porcelain enamel, wall hung									
5420	Deluxe units, aluminum trim and chalktrough									
5450	4′ x 4′	2 Carp	16	1	Ea.	253	47		300	350
5500	4′ x 8′		14	1.143		390	53.50		443.50	515
5550	4′ x 12′		12	1.333		535	62.50		597.50	685
5700	Wood trim and chalktrough									

10 11 13 – Chalkboards

10 11 13.13 Fixed Chalkboards		Crew	Daily Output	Labor-Hours	Unit	Material	2015 Bare Costs Labor	Equipment	Total	Total Incl O&P
5900	4′ x 4′	2 Carp	16	1	Ea.	715	47		762	860
6000	4′ x 6′		15	1.067		810	50		860	965
6200	4′ x 8′		14	1.143		970	53.50		1,023.50	1,150
6300	Liquid chalk, felt tip markers					2.12			2.12	2.33
6500	Erasers					1.93			1.93	2.12
6600	Board cleaner, 8 oz. bottle					6.20			6.20	6.80

10 11 13.23 Modular-Support-Mounted Chalkboards		Crew	Daily Output	Labor-Hours	Unit	Material	Labor	Equipment	Total	Total Incl O&P
0010	**MODULAR-SUPPORT-MOUNTED CHALKBOARDS**									
0400	Sliding chalkboards									
0450	Vertical, one sliding board with back panel, wall mounted									
0500	8′ x 4′	2 Carp	8	2	Ea.	2,225	94		2,319	2,600
0520	8′ x 8′		7.50	2.133		3,225	100		3,325	3,700
0540	8′ x 12′		7	2.286		4,175	107		4,282	4,775
0600	Two sliding boards, with back panel									
0620	8′ x 4′	2 Carp	8	2	Ea.	3,425	94		3,519	3,925
0640	8′ x 8′		7.50	2.133		5,025	100		5,125	5,700
0660	8′ x 12′		7	2.286		8,275	107		8,382	9,275
0700	Horizontal, two track									
0800	4′ x 8′, 2 sliding panels	2 Carp	8	2	Ea.	1,950	94		2,044	2,300
0820	4′ x 12′, 2 sliding panels		7.50	2.133		2,550	100		2,650	2,950
0840	4′ x 16′, 4 sliding panels		7	2.286		3,425	107		3,532	3,950
0900	Four track, four sliding panels									
0920	4′ x 8′	2 Carp	8	2	Ea.	3,150	94		3,244	3,600
0940	4′ x 12′		7.50	2.133		4,100	100		4,200	4,675
0960	4′ x 16′		7	2.286		5,350	107		5,457	6,050
1200	Vertical, motor operated									
1400	One sliding panel with back panel									
1450	10′ x 4′	2 Carp	4	4	Ea.	5,325	188		5,513	6,175
1500	10′ x 10′		3.75	4.267		6,425	200		6,625	7,375
1550	10′ x 16′		3.50	4.571		7,575	215		7,790	8,675
1700	Two sliding panels with back panel									
1750	10′ x 4′	2 Carp	4	4	Ea.	9,475	188		9,663	10,700
1800	10′ x 10′		3.75	4.267		10,600	200		10,800	12,000
1850	10′ x 16′		3.50	4.571		12,600	215		12,815	14,200
2000	Three sliding panels with back panel									
2100	10′ x 4′	2 Carp	4	4	Ea.	13,200	188		13,388	14,800
2150	10′ x 10′		3.75	4.267		14,700	200		14,900	16,400
2200	10′ x 16′		3.50	4.571		17,500	215		17,715	19,500
2400	For projection screen, glass beaded, add				S.F.	4.57			4.57	5.05
2500	For remote control, 1 panel control, add				Ea.	360			360	395
2600	2 panel control, add				"	620			620	680
2800	For units without back panels, deduct				S.F.	4.84			4.84	5.30
2850	For liquid chalk porcelain panels, add				"	5.15			5.15	5.70
3000	Swing leaf, any comb. of chalkboard & cork, aluminum frame									
3100	Floor style, 6 panels									
3150	30″ x 40″ panels				Ea.	1,550			1,550	1,700
3200	48″ x 40″ panels				"	2,600			2,600	2,875
3300	Wall mounted, 6 panels									
3400	30″ x 40″ panels	2 Carp	16	1	Ea.	1,475	47		1,522	1,700
3450	48″ x 40″ panels	"	16	1	"	1,800	47		1,847	2,075
3600	Extra panels for swing leaf units									
3700	30″ x 40″ panels				Ea.	294			294	325

10 11 Visual Display Units

10 11 13 – Chalkboards

10 11 13.23 Modular-Support-Mounted Chalkboards		Crew	Daily Output	Labor-Hours	Unit	2015 Bare Costs Material	Labor	Equipment	Total	Total Incl O&P
3750	48″ x 40″ panels				Ea.	365			365	400

10 11 13.43 Portable Chalkboards

Code	Description	Crew	Daily Output	Labor-Hours	Unit	Material	Labor	Equipment	Total	Total Incl O&P
0010	**PORTABLE CHALKBOARDS**									
0100	Freestanding, reversible									
0120	Economy, wood frame, 4′ x 6′									
0140	Chalkboard both sides				Ea.	610			610	670
0160	Chalkboard one side, cork other side				″	575			575	630
0200	Standard, lightweight satin finished aluminum, 4′ x 6′									
0220	Chalkboard both sides				Ea.	635			635	695
0240	Chalkboard one side, cork other side				″	640			640	705
0300	Deluxe, heavy duty extruded aluminum, 4′ x 6′									
0320	Chalkboard both sides				Ea.	1,050			1,050	1,150
0340	Chalkboard one side, cork other side				″	970			970	1,075

10 11 16 – Markerboards

10 11 16.53 Electronic Markerboards

Code	Description	Crew	Daily Output	Labor-Hours	Unit	Material	Labor	Equipment	Total	Total Incl O&P
0010	**ELECTRONIC MARKERBOARDS**									
0100	Wall hung or free standing, 3′ x 4′ to 4′ x 6′	2 Carp	8	2	S.F.	87.50	94		181.50	241
0150	5′ x 6′ to 4′ x 8′		8	2	″	61	94		155	212
0500	Interactive projection module for existing whiteboards		8	2	Ea.	1,300	94		1,394	1,575

10 11 23 – Tackboards

10 11 23.10 Fixed Tackboards

Code	Description	Crew	Daily Output	Labor-Hours	Unit	Material	Labor	Equipment	Total	Total Incl O&P
0010	**FIXED TACKBOARDS**									
0020	Cork sheets, unbacked, no frame, 1/4″ thick	2 Carp	290	.055	S.F.	1.54	2.59		4.13	5.70
0100	1/2″ thick		290	.055		4.14	2.59		6.73	8.55
0300	Fabric-face, no frame, on 7/32″ cork underlay		290	.055		6.85	2.59		9.44	11.55
0400	On 1/4″ cork on 1/4″ hardboard		290	.055		8.25	2.59		10.84	13.10
0600	With edges wrapped		290	.055		9.90	2.59		12.49	14.85
0700	On 7/16″ fire retardant core		290	.055		6.50	2.59		9.09	11.15
0900	With edges wrapped		290	.055		8.20	2.59		10.79	13.05
1000	Designer fabric only, cut to size					2.70			2.70	2.97
1200	1/4″ vinyl cork, on 1/4″ hardboard, no frame	2 Carp	290	.055		8.45	2.59		11.04	13.25
1300	On 1/4″ coreboard		290	.055		5.45	2.59		8.04	9.95
2000	For map and display rail, economy, add		385	.042	L.F.	3.06	1.95		5.01	6.35
2100	Deluxe, add		350	.046	″	4.70	2.15		6.85	8.45
2120	Prefabricated, 1/4″ cork, 3′ x 5′ with aluminum frame		16	1	Ea.	132	47		179	218
2140	Wood frame		16	1		156	47		203	244
2160	4′ x 4′ with aluminum frame		16	1		135	47		182	221
2180	Wood frame		16	1		177	47		224	268
2200	4′ x 8′ with aluminum frame		14	1.143		270	53.50		323.50	380
2210	With wood frame		14	1.143		241	53.50		294.50	350
2220	4′ x 12′ with aluminum frame		12	1.333		400	62.50		462.50	535
2230	Bulletin board case, single glass door, with lock									
2240	36″ x 24″, economy	2 Carp	12	1.333	Ea.	315	62.50		377.50	440
2250	Deluxe		12	1.333		370	62.50		432.50	500
2260	42″ x 30″, economy		12	1.333		380	62.50		442.50	515
2270	Deluxe		12	1.333		510	62.50		572.50	655
2300	Glass enclosed cabinets, alum., cork panel, hinged doors									
2400	3′ x 3′, 1 door	2 Carp	12	1.333	Ea.	605	62.50		667.50	760
2500	4′ x 4′, 2 door		11	1.455		1,000	68.50		1,068.50	1,200
2600	4′ x 7′, 3 door		10	1.600		1,775	75		1,850	2,075
2800	4′ x 10′, 4 door		8	2		2,350	94		2,444	2,750

10 11 Visual Display Units

10 11 23 – Tackboards

10 11 23.10 Fixed Tackboards		Crew	Daily Output	Labor-Hours	Unit	Material	2015 Bare Costs Labor	Equipment	Total	Total Incl O&P
2900	For lights, add per door opening	1 Elec	13	.615	Ea.	165	33.50		198.50	233
3100	Horizontal sliding units, 4 doors, 4′ x 8′, 8′ x 4′	2 Carp	9	1.778		1,950	83.50		2,033.50	2,275
3200	4′ x 12′		7	2.286		2,550	107		2,657	2,975
3400	8 doors, 4′ x 16′		5	3.200		3,450	150		3,600	4,000
3500	4′ x 24′	↓	4	4	↓	4,650	188		4,838	5,400

10 11 23.20 Control Boards		Crew	Daily Output	Labor-Hours	Unit	Material	Labor	Equipment	Total	Total Incl O&P
0010	**CONTROL BOARDS**									
0020	Magnetic, porcelain finish, 18″ x 24″, framed	2 Carp	8	2	Ea.	194	94		288	360
0100	24″ x 36″		7.50	2.133		268	100		368	450
0200	36″ x 48″		7	2.286		370	107		477	570
0300	48″ x 72″		6	2.667		650	125		775	910
0400	48″ x 96″	↓	5	3.200	↓	1,075	150		1,225	1,400
1000	Hospital patient display board, 4-color custom design									
1010	Porcelain steel dry erase board, 36″ x 24″	2 Carp	7.50	2.133	Ea.	246	100		346	425

10 13 Directories

10 13 10 – Building Directories

10 13 10.10 Directory Boards		Crew	Daily Output	Labor-Hours	Unit	Material	Labor	Equipment	Total	Total Incl O&P
0010	**DIRECTORY BOARDS**									
0050	Plastic, glass covered, 30″ x 20″	2 Carp	3	5.333	Ea.	198	250		448	600
0100	36″ x 48″		2	8		845	375		1,220	1,500
0300	Grooved cork, 30″ x 20″		3	5.333		405	250		655	830
0400	36″ x 48″		2	8		555	375		930	1,200
0600	Black felt, 30″ x 20″		3	5.333		239	250		489	650
0700	36″ x 48″		2	8		465	375		840	1,100
0900	Outdoor, weatherproof, black plastic, 36″ x 24″		2	8		760	375		1,135	1,425
1000	36″ x 36″		1.50	10.667		880	500		1,380	1,750
1800	Indoor, economy, open face, 18″ x 24″		7	2.286		163	107		270	345
1900	24″ x 36″		7	2.286		154	107		261	335
2000	36″ x 24″		6	2.667		154	125		279	360
2100	36″ x 48″		6	2.667		251	125		376	470
2400	Building directory, alum., black felt panels, 1 door, 24″ x 18″		4	4		315	188		503	635
2500	36″ x 24″		3.50	4.571		385	215		600	755
2600	48″ x 32″		3	5.333		610	250		860	1,050
2700	2 door, 36″ x 48″		2.50	6.400		680	300		980	1,200
2800	36″ x 60″		2	8		860	375		1,235	1,525
2900	48″ x 60″	↓	1	16		970	750		1,720	2,225
3100	For bronze enamel finish, add					15%				
3200	For bronze anodized finish, add					25%				
3400	For illuminated directory, single door unit, add				↓	138			138	151
3500	For 6″ header panel, 6 letters per foot, add				L.F.	21.50			21.50	23.50
6050	Building directory, electronic display, alum. frame, wall mounted	2 Carp	32	.500	S.F.	2,625	23.50		2,648.50	2,900
6100	Free standing	″	60	.267	″	3,700	12.50		3,712.50	4,100

10 14 Signage

10 14 19 – Dimensional Letter Signage

10 14 19.10 Exterior Signs		Crew	Daily Output	Labor-Hours	Unit	Material	Labor	Equipment	Total	Total Incl O&P
						2015 Bare Costs				
0010	**EXTERIOR SIGNS**									
0020	Letters, 2″ high, 3/8″ deep, cast bronze	1 Carp	24	.333	Ea.	25	15.65		40.65	51.50
0140	1/2″ deep, cast aluminum		18	.444		25	21		46	59.50
0160	Cast bronze		32	.250		30	11.75		41.75	51
0300	6″ high, 5/8″ deep, cast aluminum		24	.333		29	15.65		44.65	56
0400	Cast bronze		24	.333		62.50	15.65		78.15	93
0600	8″ high, 3/4″ deep, cast aluminum		14	.571		36	27		63	81
0700	Cast bronze		20	.400		88	18.80		106.80	126
0900	10″ high, 1″ deep, cast aluminum		18	.444		53	21		74	90
1000	Bronze		18	.444		104	21		125	146
1200	12″ high, 1-1/4″ deep, cast aluminum		12	.667		53.50	31.50		85	107
1500	Cast bronze		18	.444		127	21		148	171
1600	14″ high, 2-5/16″ deep, cast aluminum		12	.667		101	31.50		132.50	159
1800	Fabricated stainless steel, 6″ high, 2″ deep		20	.400		41.50	18.80		60.30	74.50
1900	12″ high, 3″ deep		18	.444		67	21		88	106
2100	18″ high, 3″ deep		12	.667		109	31.50		140.50	167
2200	24″ high, 4″ deep		10	.800		212	37.50		249.50	291
2700	Acrylic, on high density foam, 12″ high, 2″ deep		20	.400		19.80	18.80		38.60	51
2800	18″ high, 2″ deep		18	.444		37.50	21		58.50	73
3900	Plaques, custom, 20″ x 30″, for up to 450 letters, cast aluminum	2 Carp	4	4		1,850	188		2,038	2,325
4000	Cast bronze		4	4		1,750	188		1,938	2,200
4200	30″ x 36″, up to 900 letters cast aluminum		3	5.333		2,625	250		2,875	3,275
4300	Cast bronze		3	5.333		4,025	250		4,275	4,800
4500	36″ x 48″, for up to 1300 letters, cast bronze		2	8		4,650	375		5,025	5,700
4800	Signs, reflective alum. directional signs, dbl. face, 2-way, w/bracket		30	.533		144	25		169	197
4900	4-way		30	.533		231	25		256	293
5100	Exit signs, 24 ga. alum., 14″ x 12″ surface mounted	1 Carp	30	.267		47.50	12.50		60	72
5200	10″ x 7″		20	.400		25.50	18.80		44.30	57
5400	Bracket mounted, double face, 12″ x 10″		30	.267		56	12.50		68.50	81.50
5500	Sticky back, stock decals, 14″ x 10″	1 Clab	50	.160		26.50	6		32.50	38.50
6000	Interior elec., wall mount, fiberglass panels, 2 lamps, 6″	1 Elec	8	1		92	54.50		146.50	183
6100	8″	″	8	1		114	54.50		168.50	208
6400	Replacement sign faces, 6″ or 8″	1 Clab	50	.160		62.50	6		68.50	78

10 14 23 – Panel Signage

10 14 23.13 Engraved Panel Signage

Line	Description	Crew	Daily Output	Labor-Hours	Unit	Material	Labor	Equipment	Total	Total Incl O&P
0010	**ENGRAVED PANEL SIGNAGE**, interior									
1010	Flexible door sign, adhesive back, w/Braille, 5/8″ letters, 4″ x 4″	1 Clab	32	.250	Ea.	33	9.40		42.40	51
1050	6″ x 6″		32	.250		48.50	9.40		57.90	68
1100	8″ x 2″		32	.250		33	9.40		42.40	51
1150	8″ x 4″		32	.250		43.50	9.40		52.90	62
1200	8″ x 8″		32	.250		53	9.40		62.40	73
1250	12″ x 2″		32	.250		36	9.40		45.40	54
1300	12″ x 6″		32	.250		39	9.40		48.40	57.50
1350	12″ x 12″		32	.250		150	9.40		159.40	179
1500	Graphic symbols, 2″ x 2″		32	.250		12	9.40		21.40	27.50
1550	6″ x 6″		32	.250		31	9.40		40.40	48.50
1600	8″ x 8″		32	.250		39	9.40		48.40	57
2010	Corridor, stock acrylic, 2-sided, with mounting bracket, 2″ x 8″	1 Carp	24	.333		24.50	15.65		40.15	51
2020	2″ x 10″		24	.333		35.50	15.65		51.15	63
2050	3″ x 8″		24	.333		28.50	15.65		44.15	55.50
2060	3″ x 10″		24	.333		40	15.65		55.65	68.50
2070	3″ x 12″		24	.333		37	15.65		52.65	64.50

10 14 Signage

10 14 23 – Panel Signage

10 14 23.13 Engraved Panel Signage		Crew	Daily Output	Labor-Hours	Unit	Material	2015 Bare Costs Labor	Equipment	Total	Total Incl O&P
2100	4" x 8"	1 Carp	24	.333	Ea.	21	15.65		36.65	47
2110	4" x 10"		24	.333		38	15.65		53.65	66
2120	4" x 12"	↓	24	.333	↓	51	15.65		66.65	80.50

10 14 53 – Traffic Signage

10 14 53.20 Traffic Signs		Crew	Daily Output	Labor-Hours	Unit	Material	Labor	Equipment	Total	Total Incl O&P
0010	**TRAFFIC SIGNS**									
0012	Stock, 24" x 24", no posts, .080" alum. reflectorized	B-80	70	.457	Ea.	85	18.85	10.30	114.15	134
0100	High intensity		70	.457		97.50	18.85	10.30	126.65	148
0300	30" x 30", reflectorized		70	.457		123	18.85	10.30	152.15	175
0400	High intensity		70	.457		135	18.85	10.30	164.15	188
0600	Guide and directional signs, 12" x 18", reflectorized		70	.457		34.50	18.85	10.30	63.65	78
0700	High intensity		70	.457		52	18.85	10.30	81.15	97.50
0900	18" x 24", stock signs, reflectorized		70	.457		47	18.85	10.30	76.15	92
1000	High intensity		70	.457		52	18.85	10.30	81.15	97.50
1200	24" x 24", stock signs, reflectorized		70	.457		57	18.85	10.30	86.15	103
1300	High intensity		70	.457		62	18.85	10.30	91.15	108
1500	Add to above for steel posts, galvanized, 10'-0" upright, bolted		200	.160		32.50	6.60	3.60	42.70	50
1600	12'-0" upright, bolted		140	.229	↓	39	9.40	5.15	53.55	63
1800	Highway road signs, aluminum, over 20 S.F., reflectorized		350	.091	S.F.	33.50	3.77	2.06	39.33	45
2000	High intensity		350	.091		33.50	3.77	2.06	39.33	45
2200	Highway, suspended over road, 80 S.F. min., reflectorized		165	.194		32	8	4.37	44.37	52.50
2300	High intensity	↓	165	.194	↓	30.50	8	4.37	42.87	51

10 17 Telephone Specialties

10 17 16 – Telephone Enclosures

10 17 16.10 Commercial Telephone Enclosures		Crew	Daily Output	Labor-Hours	Unit	Material	Labor	Equipment	Total	Total Incl O&P
0010	**COMMERCIAL TELEPHONE ENCLOSURES**									
0300	Shelf type, wall hung, recessed	2 Carp	5	3.200	Ea.	745	150		895	1,050
0400	Surface mount	"	5	3.200	"	1,625	150		1,775	2,000

10 21 Compartments and Cubicles

10 21 13 – Toilet Compartments

10 21 13.13 Metal Toilet Compartments		Crew	Daily Output	Labor-Hours	Unit	Material	Labor	Equipment	Total	Total Incl O&P
0010	**METAL TOILET COMPARTMENTS**									
0110	Cubicles, ceiling hung									
0200	Powder coated steel	2 Carp	4	4	Ea.	525	188		713	865
0500	Stainless steel	"	4	4		1,075	188		1,263	1,475
0600	For handicap units, incl. 52" grab bars, add				↓	450			450	495
0900	Floor and ceiling anchored									
1000	Powder coated steel	2 Carp	5	3.200	Ea.	590	150		740	875
1300	Stainless steel	"	5	3.200		1,275	150		1,425	1,625
1400	For handicap units, incl. 52" grab bars, add				↓	315			315	345
1610	Floor anchored									
1700	Powder coated steel	2 Carp	7	2.286	Ea.	590	107		697	815
2000	Stainless steel	"	7	2.286		1,400	107		1,507	1,725
2100	For handicap units, incl. 52" grab bars, add					310			310	345
2200	For juvenile units, deduct				↓	41.50			41.50	45.50
2450	Floor anchored, headrail braced									
2500	Powder coated steel	2 Carp	6	2.667	Ea.	390	125		515	625

10 21 Compartments and Cubicles

10 21 13 – Toilet Compartments

10 21 13.13 Metal Toilet Compartments		Crew	Daily Output	Labor-Hours	Unit	2015 Bare Costs Material	Labor	Equipment	Total	Total Incl O&P
2804	Stainless steel	2 Carp	4.60	3.478	Ea.	1,025	163		1,188	1,375
2900	For handicap units, incl. 52" grab bars, add					370			370	410
3000	Wall hung partitions, powder coated steel	2 Carp	7	2.286		635	107		742	865
3300	Stainless steel	"	7	2.286		1,650	107		1,757	2,000
3400	For handicap units, incl. 52" grab bars, add					370			370	410
4000	Screens, entrance, floor mounted, 58" high, 48" wide									
4200	Powder coated steel	2 Carp	15	1.067	Ea.	242	50		292	345
4500	Stainless steel	"	15	1.067	"	910	50		960	1,075
4650	Urinal screen, 18" wide									
4704	Powder coated steel	2 Carp	6.15	2.602	Ea.	211	122		333	420
5004	Stainless steel	"	6.15	2.602	"	580	122		702	830
5100	Floor mounted, head rail braced									
5300	Powder coated steel	2 Carp	8	2	Ea.	230	94		324	400
5600	Stainless steel	"	8	2	"	570	94		664	770
5750	Pilaster, flush									
5800	Powder coated steel	2 Carp	10	1.600	Ea.	278	75		353	420
6100	Stainless steel		10	1.600		625	75		700	805
6300	Post braced, powder coated steel		10	1.600		163	75		238	295
6600	Stainless steel		10	1.600		450	75		525	610
6700	Wall hung, bracket supported									
6800	Powder coated steel	2 Carp	10	1.600	Ea.	163	75		238	295
7100	Stainless steel		10	1.600		278	75		353	420
7400	Flange supported, powder coated steel		10	1.600		106	75		181	232
7700	Stainless steel		10	1.600		310	75		385	455
7800	Wedge type, powder coated steel		10	1.600		134	75		209	264
8100	Stainless steel		10	1.600		575	75		650	750

10 21 13.14 Metal Toilet Compartment Components		Crew	Daily Output	Labor-Hours	Unit	Material	Labor	Equipment	Total	Total Incl O&P
0010	**METAL TOILET COMPARTMENT COMPONENTS**									
0100	Pilasters									
0110	Overhead braced, powder coated steel, 7" wide x 82" high	2 Carp	22.20	.721	Ea.	73.50	34		107.50	133
0120	Stainless steel		22.20	.721		125	34		159	189
0130	Floor braced, powder coated steel, 7" wide x 70" high		23.30	.687		128	32		160	191
0140	Stainless steel		23.30	.687		245	32		277	320
0150	Ceiling hung, powder coated steel, 7" wide x 83" high		13.30	1.203		136	56.50		192.50	236
0160	Stainless steel		13.30	1.203		274	56.50		330.50	385
0170	Wall hung, powder coated steel, 3" wide x 58" high		18.90	.847		133	40		173	207
0180	Stainless steel		18.90	.847		193	40		233	274
0200	Panels									
0210	Powder coated steel, 31" wide x 58" high	2 Carp	18.90	.847	Ea.	137	40		177	211
0220	Stainless steel		18.90	.847		365	40		405	460
0230	Powder coated steel, 53" wide x 58" high		18.90	.847		169	40		209	247
0240	Stainless steel		18.90	.847		475	40		515	580
0250	Powder coated steel, 63" wide x 58" high		18.90	.847		205	40		245	287
0260	Stainless steel		18.90	.847		520	40		560	635
0300	Doors									
0310	Powder coated steel, 24" wide x 58" high	2 Carp	14.10	1.135	Ea.	139	53.50		192.50	235
0320	Stainless steel		14.10	1.135		290	53.50		343.50	400
0330	Powder coated steel, 26" wide x 58" high		14.10	1.135		141	53.50		194.50	238
0340	Stainless steel		14.10	1.135		300	53.50		353.50	410
0350	Powder coated steel, 28" wide x 58" high		14.10	1.135		162	53.50		215.50	260
0360	Stainless steel		14.10	1.135		335	53.50		388.50	445
0370	Powder coated steel, 36" wide x 58" high		14.10	1.135		174	53.50		227.50	274

10 21 13.14 Metal Toilet Compartment Components		Crew	Daily Output	Labor-Hours	Unit	Material	Labor	Equipment	Total	Total Incl O&P
						2015 Bare Costs				
0380	Stainless steel	2 Carp	14.10	1.135	Ea.	375	53.50		428.50	490
0400	Headrails									
0410	For powder coated steel, 62" long	2 Carp	65	.246	Ea.	22	11.55		33.55	42.50
0420	Stainless steel		65	.246		22	11.55		33.55	42.50
0430	For powder coated steel, 84" long		50	.320		31.50	15		46.50	58
0440	Stainless steel		50	.320		31.50	15		46.50	58
0450	For powder coated steel, 120" long		30	.533		43	25		68	86
0460	Stainless steel		30	.533		42.50	25		67.50	85.50

10 21 13.16 Plastic-Laminate-Clad Toilet Compartments		Crew	Daily Output	Labor-Hours	Unit	Material	Labor	Equipment	Total	Total Incl O&P
0010	**PLASTIC-LAMINATE-CLAD TOILET COMPARTMENTS**									
0110	Cubicles, ceiling hung									
0300	Plastic laminate on particle board	2 Carp	4	4	Ea.	520	188		708	860
0600	For handicap units, incl. 52" grab bars, add				"	450			450	495
0900	Floor and ceiling anchored									
1100	Plastic laminate on particle board	2 Carp	5	3.200	Ea.	790	150		940	1,100
1400	For handicap units, incl. 52" grab bars, add				"	315			315	345
1610	Floor mounted									
1800	Plastic laminate on particle board	2 Carp	7	2.286	Ea.	535	107		642	750
2450	Floor mounted, headrail braced									
2600	Plastic laminate on particle board	2 Carp	6	2.667	Ea.	750	125		875	1,025
3400	For handicap units, incl. 52" grab bars, add					370			370	410
4300	Entrance screen, floor mtd., plas. lam., 58" high, 48" wide	2 Carp	15	1.067		610	50		660	745
4800	Urinal screen, 18" wide, ceiling braced, plastic laminate		8	2		194	94		288	360
5400	Floor mounted, headrail braced		8	2		200	94		294	365
5900	Pilaster, flush, plastic laminate		10	1.600		505	75		580	675
6400	Post braced, plastic laminate		10	1.600		305	75		380	450
6700	Wall hung, bracket supported									
6900	Plastic laminate on particle board	2 Carp	10	1.600	Ea.	94.50	75		169.50	220
7450	Flange supported									
7500	Plastic laminate on particle board	2 Carp	10	1.600	Ea.	230	75		305	370

10 21 13.17 Plastic-Lam. Clad Toilet Compart. Components		Crew	Daily Output	Labor-Hours	Unit	Material	Labor	Equipment	Total	Total Incl O&P
0010	**PLASTIC-LAMINATE CLAD TOILET COMPARTMENT COMPONENTS**									
0100	Pilasters									
0110	Overhead braced, 7" wide x 82" high	2 Carp	22.20	.721	Ea.	99.50	34		133.50	161
0130	Floor anchored, 7" wide x 70" high		23.30	.687		99.50	32		131.50	159
0150	Ceiling hung, 7" wide x 83" high		13.30	1.203		104	56.50		160.50	201
0180	Wall hung, 3" wide x 58" high		18.90	.847		93	40		133	163
0200	Panels									
0210	31" wide x 58" high	2 Carp	18.90	.847	Ea.	148	40		188	223
0230	51" wide x 58" high		18.90	.847		202	40		242	283
0250	63" wide x 58" high		18.90	.847		235	40		275	320
0300	Doors									
0310	24" wide x 58" high	2 Carp	14.10	1.135	Ea.	142	53.50		195.50	238
0330	26" wide x 58" high		14.10	1.135		147	53.50		200.50	244
0350	28" wide x 58" high		14.10	1.135		152	53.50		205.50	249
0370	36" wide x 58" high		14.10	1.135		182	53.50		235.50	282
0400	Headrails									
0410	62" long	2 Carp	65	.246	Ea.	22.50	11.55		34.05	43
0430	84" long		60	.267		31	12.50		43.50	53.50
0450	120" long		30	.533		42	25		67	84.50

10 21 13.19 Plastic Toilet Compartments

	10 21 13.19 Plastic Toilet Compartments	Crew	Daily Output	Labor-Hours	Unit	Material	Labor (2015 Bare Costs)	Equipment (2015 Bare Costs)	Total (2015 Bare Costs)	Total Incl O&P
0010	**PLASTIC TOILET COMPARTMENTS**									
0110	Cubicles, ceiling hung									
0250	Phenolic	2 Carp	4	4	Ea.	865	188		1,053	1,250
0600	For handicap units, incl. 52″ grab bars, add				″	450			450	495
0900	Floor and ceiling anchored									
1050	Phenolic	2 Carp	5	3.200	Ea.	810	150		960	1,125
1400	For handicap units, incl. 52″ grab bars, add				″	315			315	345
1610	Floor mounted									
1750	Phenolic	2 Carp	7	2.286	Ea.	750	107		857	990
2100	For handicap units, incl. 52″ grab bars, add					310			310	345
2200	For juvenile units, deduct				↓	41.50			41.50	45.50
2450	Floor mounted, headrail braced									
2550	Phenolic	2 Carp	6	2.667	Ea.	750	125		875	1,025

10 21 13.20 Plastic Toilet Compartment Components

	10 21 13.20 Plastic Toilet Compartment Components	Crew	Daily Output	Labor-Hours	Unit	Material	Labor (2015 Bare Costs)	Equipment (2015 Bare Costs)	Total (2015 Bare Costs)	Total Incl O&P
0010	**PLASTIC TOILET COMPARTMENT COMPONENTS**									
0100	Pilasters									
0110	Overhead braced, polymer plastic, 7″ wide x 82″ high	2 Carp	22.20	.721	Ea.	121	34		155	185
0120	Phenolic		22.20	.721		151	34		185	218
0130	Floor braced, polymer plastic, 7″ wide x 70″ high		23.30	.687		171	32		203	239
0140	Phenolic		23.30	.687		142	32		174	206
0150	Ceiling hung, polymer plastic, 7″ wide x 83″ high		13.30	1.203		171	56.50		227.50	275
0160	Phenolic		13.30	1.203		161	56.50		217.50	264
0180	Wall hung, phenolic, 3″ wide x 58″ high	↓	18.90	.847	↓	96	40		136	166
0200	Panels									
0203	Polymer plastic, 18″ high x 55″ high	2 Carp	18.90	.847	Ea.	252	40		292	340
0206	Phenolic, 18″ wide x 58″ high		18.90	.847		222	40		262	305
0210	Polymer plastic, 31″ high x 55″ high		18.90	.847		305	40		345	395
0220	Phenolic, 31″ wide x 58″ high		18.90	.847		263	40		303	350
0223	Polymer plastic, 48″ high x 55″ high		18.90	.847		470	40		510	575
0226	Phenolic, 48″ wide x 58″ high		18.90	.847		430	40		470	530
0230	Polymer plastic, 51″ wide x 55″ high		18.90	.847		410	40		450	510
0240	Phenolic, 51″ wide x 58″ high		18.90	.847		400	40		440	500
0250	Polymer plastic, 63″ wide x 55″ high		18.90	.847		560	40		600	675
0260	Phenolic, 63″ wide x 58″ high	↓	18.90	.847	↓	420	40		460	520
0300	Doors									
0310	Polymer plastic, 24″ wide x 55″ high	2 Carp	14.10	1.135	Ea.	211	53.50		264.50	315
0320	Phenolic, 24″ wide x 58″ high		14.10	1.135		296	53.50		349.50	405
0330	Polymer plastic, 26″ high x 55″ high		14.10	1.135		225	53.50		278.50	330
0340	Phenolic, 26″ wide x 58″ high		14.10	1.135		315	53.50		368.50	425
0350	Polymer plastic, 28″ wide x 55″ high		14.10	1.135		250	53.50		303.50	355
0360	Phenolic, 28″ wide x 58″ high		14.10	1.135		330	53.50		383.50	445
0370	Polymer plastic, 36″ wide x 55″ high		14.10	1.135		291	53.50		344.50	400
0380	Phenolic, 36″ wide x 58″ high	↓	14.10	1.135	↓	415	53.50		468.50	535
0400	Headrails									
0410	For polymer plastic, 62″ long	2 Carp	65	.246	Ea.	22.50	11.55		34.05	43
0420	Phenolic		65	.246		22.50	11.55		34.05	43
0430	For polymer plastic, 84″ long		50	.320		32	15		47	58.50
0440	Phenolic		50	.320		32	15		47	58.50
0450	For polymer plastic, 120″ long		30	.533		43	25		68	86
0460	Phenolic	↓	30	.533	↓	43	25		68	86

10 21 Compartments and Cubicles

10 21 13 – Toilet Compartments

10 21 13.40 Stone Toilet Compartments		Crew	Daily Output	Labor-Hours	Unit	Material	2015 Bare Costs Labor	Equipment	Total	Total Incl O&P
0010	**STONE TOILET COMPARTMENTS**									
0100	Cubicles, ceiling hung, marble	2 Marb	2	8	Ea.	1,800	350		2,150	2,500
0600	For handicap units, incl. 52″ grab bars, add					450			450	495
0800	Floor & ceiling anchored, marble	2 Marb	2.50	6.400		1,975	278		2,253	2,600
1400	For handicap units, incl. 52″ grab bars, add					315			315	345
1600	Floor mounted, marble	2 Marb	3	5.333		1,225	232		1,457	1,700
2400	Floor mounted, headrail braced, marble	″	3	5.333		1,150	232		1,382	1,625
2900	For handicap units, incl. 52″ grab bars, add					370			370	410
4100	Entrance screen, floor mounted marble, 58″ high, 48″ wide	2 Marb	9	1.778		795	77.50		872.50	990
4600	Urinal screen, 18″ wide, ceiling braced, marble	D-1	6	2.667		755	112		867	1,000
5100	Floor mounted, head rail braced									
5200	Marble	D-1	6	2.667	Ea.	645	112		757	880
5700	Pilaster, flush, marble		9	1.778		840	75		915	1,050
6200	Post braced, marble		9	1.778		825	75		900	1,025

10 21 23 – Cubicle Curtains and Track

10 21 23.16 Cubicle Track and Hardware		Crew	Daily Output	Labor-Hours	Unit	Material	Labor	Equipment	Total	Total Incl O&P
0010	**CUBICLE TRACK AND HARDWARE**									
0020	Curtain track, box channel, ceiling mounted	1 Carp	135	.059	L.F.	6.25	2.78		9.03	11.15
0100	Suspended	″	100	.080	″	8.30	3.76		12.06	14.90
0300	Curtains, nylon mesh tops, fire resistant, 11 oz. per lineal yard									
0310	Polyester oxford cloth, 9′ ceiling height	1 Carp	425	.019	L.F.	16.15	.88		17.03	19.10
0500	8′ ceiling height		425	.019		7.35	.88		8.23	9.45
0550	Polyester, antimicrobial, 9′ ceiling height		425	.019		19.65	.88		20.53	23
0560	8′ ceiling height		425	.019		17.50	.88		18.38	20.50
0700	Designer oxford cloth		425	.019		7.35	.88		8.23	9.45
0800	I.V. track systems									
0820	I.V. track, oval	1 Carp	135	.059	L.F.	7.80	2.78		10.58	12.85
0830	I.V. trolley		32	.250	Ea.	41	11.75		52.75	63
0840	I.V. pendent, (tree, 5 hook)		32	.250	″	171	11.75		182.75	206

10 22 Partitions

10 22 13 – Wire Mesh Partitions

10 22 13.10 Partitions, Woven Wire		Crew	Daily Output	Labor-Hours	Unit	Material	Labor	Equipment	Total	Total Incl O&P
0010	**PARTITIONS, WOVEN WIRE** for tool or stockroom enclosures									
0100	Channel frame, 1-1/2″ diamond mesh, 10 ga. wire, painted									
0300	Wall panels, 4′-0″ wide, 7′ high	2 Carp	25	.640	Ea.	144	30		174	204
0400	8′ high		23	.696		162	32.50		194.50	230
0600	10′ high		18	.889		190	41.50		231.50	273
0700	For 5′ wide panels, add					5%				
0900	Ceiling panels, 10′ long, 2′ wide	2 Carp	25	.640		130	30		160	189
1000	4′ wide		15	1.067		197	50		247	294
1200	Panel with service window & shelf, 5′ wide, 7′ high		20	.800		375	37.50		412.50	470
1300	8′ high		15	1.067		450	50		500	570
1500	Sliding doors, full height, 3′ wide, 7′ high		6	2.667		490	125		615	730
1600	10′ high		5	3.200		525	150		675	805
1800	6′ wide sliding door, 7′ full height		5	3.200		670	150		820	970
1900	10′ high		4	4		830	188		1,018	1,200
2100	Swinging doors, 3′ wide, 7′ high, no transom		6	2.667		310	125		435	540
2200	7′ high, 3′ transom		5	3.200		375	150		525	640

10 22 16 – Folding Gates

	10 22 16.10 Security Gates	Crew	Daily Output	Labor-Hours	Unit	Material	Labor	Equipment	Total	Total Incl O&P
0010	**SECURITY GATES** for roll up type, see Section 08 33 13.10									
0300	Scissors type folding gate, ptd. steel, single, 6-1/2′ high, 5-1/2′ wide	2 Sswk	4	4	Opng.	226	211		437	615
0350	6-1/2′ wide		4	4		246	211		457	635
0400	7-1/2′ wide		4	4		237	211		448	625
0600	Double gate, 8′ high, 8′ wide		2.50	6.400		390	335		725	1,025
0650	10′ wide		2.50	6.400		425	335		760	1,050
0700	12′ wide		2	8		620	420		1,040	1,400
0750	14′ wide		2	8		630	420		1,050	1,425
0900	Door gate, folding steel, 4′ wide, 61″ high		4	4		139	211		350	520
1000	71″ high		4	4		169	211		380	550
1200	81″ high		4	4		195	211		406	580
1300	Window gates, 2′ to 4′ wide, 31″ high		4	4		80	211		291	455
1500	55″ high		3.75	4.267		122	225		347	525
1600	79″ high		3.50	4.571		144	241		385	580

10 22 19 – Demountable Partitions

	10 22 19.43 Demountable Composite Partitions	Crew	Daily Output	Labor-Hours	Unit	Material	Labor	Equipment	Total	Total Incl O&P
0010	**DEMOUNTABLE COMPOSITE PARTITIONS**, add for doors									
0100	Do not deduct door openings from total L.F.									
0900	Demountable gypsum system on 2″ to 2-1/2″									
1000	steel studs, 9′ high, 3″ to 3-3/4″ thick									
1200	Vinyl clad gypsum	2 Carp	48	.333	L.F.	60	15.65		75.65	90
1300	Fabric clad gypsum		44	.364		150	17.05		167.05	192
1500	Steel clad gypsum		40	.400		167	18.80		185.80	213
1600	1.75 system, aluminum framing, vinyl clad hardboard,									
1800	paper honeycomb core panel, 1-3/4″ to 2-1/2″ thick									
1900	9′ high	2 Carp	48	.333	L.F.	101	15.65		116.65	135
2100	7′ high		60	.267		90.50	12.50		103	119
2200	5′ high		80	.200		76.50	9.40		85.90	98.50
2250	Unitized gypsum system									
2300	Unitized panel, 9′ high, 2″ to 2-1/2″ thick									
2350	Vinyl clad gypsum	2 Carp	48	.333	L.F.	130	15.65		145.65	167
2400	Fabric clad gypsum	″	44	.364	″	214	17.05		231.05	262
2500	Unitized mineral fiber system									
2510	Unitized panel, 9′ high, 2-1/4″ thick, aluminum frame									
2550	Vinyl clad mineral fiber	2 Carp	48	.333	L.F.	129	15.65		144.65	166
2600	Fabric clad mineral fiber	″	44	.364	″	193	17.05		210.05	239
2800	Movable steel walls, modular system									
2900	Unitized panels, 9′ high, 48″ wide									
3100	Baked enamel, pre-finished	2 Carp	60	.267	L.F.	146	12.50		158.50	180
3200	Fabric clad steel		56	.286	″	212	13.40		225.40	254
5310	Trackless wall, cork finish, semi-acoustic, 1-5/8″ thick, unsealed		325	.049	S.F.	38.50	2.31		40.81	46
5320	Sealed		190	.084		42.50	3.95		46.45	53
5330	Acoustic, 2″ thick, unsealed		305	.052		36.50	2.46		38.96	44
5340	Sealed		225	.071		56	3.34		59.34	66.50
5500	For acoustical partitions, add, unsealed					2.36			2.36	2.60
5550	Sealed					11			11	12.10
5700	For doors, see Sections 08 11 & 08 16									
5800	For door hardware, see Section 08 71									
6100	In-plant modular office system, w/prehung hollow core door									
6200	3″ thick polystyrene core panels									
6250	12′ x 12′, 2 wall	2 Clab	3.80	4.211	Ea.	4,025	158		4,183	4,700
6300	4 wall		1.90	8.421		6,150	315		6,465	7,225

10 22 Partitions

10 22 19 – Demountable Partitions

10 22 19.43 Demountable Composite Partitions		Crew	Daily Output	Labor-Hours	Unit	Material	2015 Bare Costs Labor	Equipment	Total	Total Incl O&P
6350	16′ x 16′, 2 wall	2 Clab	3.60	4.444	Ea.	6,125	167		6,292	7,000
6400	4 wall	↓	1.80	8.889	↓	8,325	335		8,660	9,675

10 22 23 – Portable Partitions, Screens, and Panels

10 22 23.13 Wall Screens		Crew	Daily Output	Labor-Hours	Unit	Material	Labor	Equipment	Total	Total Incl O&P
0010	**WALL SCREENS**, divider panels, free standing, fiber core									
0020	Fabric face straight									
0100	3′-0″ long, 4′-0″ high	2 Carp	100	.160	L.F.	123	7.50		130.50	147
0200	5′-0″ high		90	.178		103	8.35		111.35	126
0500	6′-0″ high		75	.213		104	10		114	130
0900	5′-0″ long, 4′-0″ high		175	.091		67.50	4.29		71.79	81
1000	5′-0″ high		150	.107		75.50	5		80.50	90.50
1500	6″-0″ high		125	.128		90.50	6		96.50	109
1600	6′-0″ long, 5′-0″ high		162	.099		75.50	4.64		80.14	90
3200	Economical panels, fabric face, 4′-0″ long, 5′-0″ high		132	.121		50	5.70		55.70	64
3250	6′-0″ high		112	.143		55.50	6.70		62.20	71.50
3300	5′-0″ long, 5′-0″ high		150	.107		54.50	5		59.50	67
3350	6′-0″ high		125	.128		50	6		56	64.50
3450	Acoustical panels, 60 to 90 NRC, 3′-0″ long, 5′-0″ high		90	.178		76.50	8.35		84.85	97
3550	6′-0″ high		75	.213		89.50	10		99.50	114
3600	5′-0″ long, 5′-0″ high		150	.107		61	5		66	74.50
3650	6′-0″ high		125	.128		69	6		75	85
3700	6′-0″ long, 5′-0″ high		162	.099		53	4.64		57.64	65.50
3750	6′-0″ high		138	.116		82	5.45		87.45	98.50
3800	Economy acoustical panels, 40 N.R.C., 4′-0″ long, 5′-0″ high		132	.121		50	5.70		55.70	64
3850	6′-0″ high		112	.143		55.50	6.70		62.20	71.50
3900	5′-0″ long, 6′-0″ high		125	.128		50	6		56	64.50
3950	6′-0″ long, 5′-0″ high		162	.099		47	4.64		51.64	59
4000	Metal chalkboard, 6′-6″ high, chalkboard, 1 side		125	.128		120	6		126	141
4100	Metal chalkboard, 2 sides		120	.133		137	6.25		143.25	161
4300	Tackboard, both sides	↓	123	.130	↓	109	6.10		115.10	128

10 22 33 – Accordion Folding Partitions

10 22 33.10 Partitions, Accordion Folding		Crew	Daily Output	Labor-Hours	Unit	Material	Labor	Equipment	Total	Total Incl O&P
0010	**PARTITIONS, ACCORDION FOLDING**									
0100	Vinyl covered, over 150 S.F., frame not included									
0300	Residential, 1.25 lb. per S.F., 8′ maximum height	2 Carp	300	.053	S.F.	25	2.50		27.50	31.50
0400	Commercial, 1.75 lb. per S.F., 8′ maximum height		225	.071		28.50	3.34		31.84	36.50
0600	2 lb. per S.F., 17′ maximum height		150	.107		29.50	5		34.50	40
0700	Industrial, 4 lb. per S.F., 20′ maximum height		75	.213		44.50	10		54.50	64.50
0900	Acoustical, 3 lb. per S.F., 17′ maximum height		100	.160		31.50	7.50		39	46.50
1200	5 lb. per S.F., 20′ maximum height		95	.168		44	7.90		51.90	60.50
1300	5.5 lb. per S.F., 17′ maximum height		90	.178		51.50	8.35		59.85	69.50
1400	Fire rated, 4.5 psf, 20′ maximum height		160	.100		51.50	4.70		56.20	64
1500	Vinyl clad wood or steel, electric operation, 5.0 psf		160	.100		63	4.70		67.70	76.50
1900	Wood, non-acoustic, birch or mahogany, to 10′ high	↓	300	.053	↓	34	2.50		36.50	41

10 22 39 – Folding Panel Partitions

10 22 39.10 Partitions, Folding Panel		Crew	Daily Output	Labor-Hours	Unit	Material	Labor	Equipment	Total	Total Incl O&P
0010	**PARTITIONS, FOLDING PANEL**, acoustic, wood									
0100	Vinyl faced, to 18′ high, 6 psf, economy trim	2 Carp	60	.267	S.F.	57	12.50		69.50	82
0150	Standard trim		45	.356		68	16.70		84.70	100
0200	Premium trim		30	.533		87.50	25		112.50	135
0400	Plastic laminate or hardwood finish, standard trim	↓	60	.267	↓	58.50	12.50		71	84

10 22 Partitions

10 22 39 – Folding Panel Partitions

10 22 39.10 Partitions, Folding Panel		Crew	Daily Output	Labor-Hours	Unit	Material	Labor	Equipment	Total	Total Incl O&P
						2015 Bare Costs				
0500	Premium trim	2 Carp	30	.533	S.F.	62.50	25		87.50	107
0600	Wood, low acoustical type, 4.5 psf, to 14′ high		50	.320		42.50	15		57.50	70
1100	Steel, acoustical, 9 to 12 lb. per S.F., vinyl faced, standard trim		60	.267		60.50	12.50		73	86
1200	Premium trim		30	.533		74	25		99	120
1700	Aluminum framed, acoustical, to 12′ high, 5.5 psf, standard trim		60	.267		41	12.50		53.50	64.50
1800	Premium trim		30	.533		49.50	25		74.50	92.50
2000	6.5 lb. per S.F., standard trim		60	.267		43	12.50		55.50	66.50
2100	Premium trim		30	.533		53	25		78	97

10 22 43 – Sliding Partitions

10 22 43.10 Partitions, Sliding		Crew	Daily Output	Labor-Hours	Unit	Material	Labor	Equipment	Total	Total Incl O&P
0010	**PARTITIONS, SLIDING**									
0020	Acoustic air wall, 1-5/8″ thick, standard trim	2 Carp	375	.043	S.F.	33	2		35	39.50
0100	Premium trim		365	.044		56.50	2.06		58.56	65
0300	2-1/4″ thick, standard trim		360	.044		37	2.09		39.09	44
0400	Premium trim		330	.048		65	2.28		67.28	75
0600	For track type, add to above				L.F.	121			121	133
0700	Overhead track type, acoustical, 3″ thick, 11 psf, standard trim	2 Carp	350	.046	S.F.	83.50	2.15		85.65	95
0800	Premium trim	″	300	.053	″	100	2.50		102.50	114

10 26 Wall and Door Protection

10 26 13 – Corner Guards

10 26 13.10 Metal Corner Guards		Crew	Daily Output	Labor-Hours	Unit	Material	Labor	Equipment	Total	Total Incl O&P
0010	**METAL CORNER GUARDS**									
0020	Steel angle w/anchors, 1″ x 1″ x 1/4″, 1.5#/L.F.	2 Carp	160	.100	L.F.	7.10	4.70		11.80	15.05
0100	2″ x 2″ x 1/4″ angles, 3.2#/L.F.		150	.107		10.20	5		15.20	18.90
0200	3″ x 3″ x 5/16″ angles, 6.1#/L.F.		140	.114		14.75	5.35		20.10	24.50
0300	4″ x 4″ x 5/16″ angles, 8.2#/L.F.		120	.133		19.95	6.25		26.20	31.50
0350	For angles drilled and anchored to masonry, add					15%	120%			
0370	Drilled and anchored to concrete, add					20%	170%			
0400	For galvanized angles, add					35%				
0450	For stainless steel angles, add					100%				
0500	Steel door track/wheel guards, 4′ - 0″ high	E-4	22	1.455	Ea.	109	77.50	6.65	193.15	261
0800	Pipe bumper for truck doors, 8′ long, 6″ diameter, filled		20	1.600		625	85	7.30	717.30	845
0900	8″ diameter		20	1.600		725	85	7.30	817.30	955
1000	Wall protection, stainless steel, 16 ga, 48″ x 36″ tall, screwed to studs	2 Skwk	500	.032	S.F.	8.25	1.56		9.81	11.50
1050	Wall end guard, stainless steel, 16 ga, 36″ tall, screwed to studs	1 Skwk	30	.267	Ea.	25	12.95		37.95	47.50

10 26 13.20 Corner Protection		Crew	Daily Output	Labor-Hours	Unit	Material	Labor	Equipment	Total	Total Incl O&P
0010	**CORNER PROTECTION**									
0100	Stainless steel, 16 ga., adhesive mount, 3-1/2″ leg	1 Carp	80	.100	L.F.	24	4.70		28.70	34
0200	12 ga. stainless, adhesive mount	″	80	.100		26.50	4.70		31.20	36.50
0300	For screw mount, add						10%			
0500	Vinyl acrylic, adhesive mount, 3″ leg	1 Carp	128	.063		9.25	2.93		12.18	14.70
0550	1-1/2″ leg		160	.050		4.83	2.35		7.18	8.90
0600	Screw mounted, 3″ leg		80	.100		10.10	4.70		14.80	18.35
0650	1-1/2″ leg		100	.080		4.57	3.76		8.33	10.85
0700	Clear plastic, screw mounted, 2-1/2″		60	.133		4.37	6.25		10.62	14.45
1000	Vinyl cover, alum. retainer, surface mount, 3″ x 3″		48	.167		10.45	7.85		18.30	23.50
1050	2″ x 2″		48	.167		9.45	7.85		17.30	22.50
1100	Flush mounted, 3″ x 3″		32	.250		20.50	11.75		32.25	40.50
1150	2″ x 2″		32	.250		16.70	11.75		28.45	36.50

10 26 Wall and Door Protection

10 26 16 – Bumper Guards

10 26 16.10 Wallguard		Crew	Daily Output	Labor-Hours	Unit	Material	2015 Bare Costs Labor	Equipment	Total	Total Incl O&P
0010	**WALLGUARD**									
0400	Rub rail, vinyl, adhesive mounted	1 Carp	185	.043	L.F.	8.65	2.03		10.68	12.65
0500	Neoprene, aluminum backing, 1-1/2" x 2"		110	.073		8.80	3.41		12.21	14.95
1000	Trolley rail, PVC, clipped to wall, 5" high		185	.043		8.80	2.03		10.83	12.80
1050	8" high		180	.044		14.70	2.09		16.79	19.35
1200	Bed bumper, vinyl acrylic, alum. retainer, 21" long		10	.800	Ea.	40.50	37.50		78	103
1300	53" long with aligner		9	.889	"	102	41.50		143.50	176
1400	Bumper, vinyl cover, alum. retain., cush. mnt., 1-1/2" x 2-3/4"		80	.100	L.F.	14.20	4.70		18.90	23
1500	2" x 4-1/4"		80	.100		20.50	4.70		25.20	30.50
1600	Surface mounted, 1-3/4" x 3-5/8"		80	.100		11.95	4.70		16.65	20.50
1700	Bumper rail, stainless steel, flat bar on brackets, 4" x 1/4"	2 Skwk	120	.133		42	6.50		48.50	56.50
1750	Wallguard stainless steel baseboard, 12" tall, adhesive applied	"	260	.062		33	2.99		35.99	41
2000	Crash rail, vinyl cover, alum. retainer, 1" x 4"	1 Carp	110	.073		11.05	3.41		14.46	17.40
2100	1" x 8"		90	.089		18.15	4.17		22.32	26.50
2150	Vinyl inserts, aluminum plate, 1" x 2-1/2"		110	.073		14.90	3.41		18.31	21.50
2200	1" x 5"		90	.089		23	4.17		27.17	32
3000	Handrail/bumper, vinyl cover, alum. retainer									
3010	Bracket mounted, flat rail, 5-1/2"	1 Carp	80	.100	L.F.	18.30	4.70		23	27.50
3100	6-1/2"		80	.100		23	4.70		27.70	32.50
3200	Bronze bracket, 1-3/4" diam. rail		80	.100		16.50	4.70		21.20	25.50
4000	Handrail, with antimicrobial copper alloy, #6 finish, 1-1/2" OD		80	.100		7.50	4.70		12.20	15.50

10 28 Toilet, Bath, and Laundry Accessories

10 28 13 – Toilet Accessories

10 28 13.13 Commercial Toilet Accessories		Crew	Daily Output	Labor-Hours	Unit	Material	Labor	Equipment	Total	Total Incl O&P
0010	**COMMERCIAL TOILET ACCESSORIES**									
0200	Curtain rod, stainless steel, 5' long, 1" diameter	1 Carp	13	.615	Ea.	28.50	29		57.50	75.50
0300	1-1/4" diameter		13	.615		29	29		58	76
0350	Chrome, 1" diameter		13	.615		32.50	29		61.50	80
0360	For vinyl curtain, add		1950	.004	S.F.	.91	.19		1.10	1.30
0400	Diaper changing station, horizontal, wall mounted, plastic		10	.800	Ea.	229	37.50		266.50	310
0420	Vertical		10	.800		229	37.50		266.50	310
0430	Oval shaped		10	.800		225	37.50		262.50	305
0440	Recessed, with stainless steel flange		6	1.333		565	62.50		627.50	720
0500	Dispenser units, combined soap & towel dispensers,									
0510	mirror and shelf, flush mounted	1 Carp	10	.800	Ea.	310	37.50		347.50	400
0600	Towel dispenser and waste receptacle,									
0610	18 gallon capacity	1 Carp	10	.800	Ea.	300	37.50		337.50	390
0800	Grab bar, straight, 1-1/4" diameter, stainless steel, 18" long		24	.333		29	15.65		44.65	56
0900	24" long		23	.348		29	16.35		45.35	56.50
1000	30" long		22	.364		31.50	17.05		48.55	61
1100	36" long		20	.400		38.50	18.80		57.30	71.50
1105	42" long		20	.400		46	18.80		64.80	79.50
1120	Corner, 36" long		20	.400		85.50	18.80		104.30	123
1200	1-1/2" diameter, 24" long		23	.348		31	16.35		47.35	59
1300	36" long		20	.400		33.50	18.80		52.30	65.50
1310	42" long		18	.444		38	21		59	73.50
1500	Tub bar, 1-1/4" diameter, 24" x 36"		14	.571		92.50	27		119.50	144
1600	Plus vertical arm		12	.667		97.50	31.50		129	155
1900	End tub bar, 1" diameter, 90° angle, 16" x 32"		12	.667		109	31.50		140.50	168

10 28 13 – Toilet Accessories

10 28 13.13 Commercial Toilet Accessories		Crew	Daily Output	Labor-Hours	Unit	Material	2015 Bare Costs Labor	Equipment	Total	Total Incl O&P
2010	Tub/shower/toilet, 2-wall, 36" x 24"	1 Carp	12	.667	Ea.	91	31.50		122.50	148
2110	Antimicrobial copper alloy finish, straight, 18" long		24	.333		68.50	15.65		84.15	99.50
2120	24" long		23	.348		75.50	16.35		91.85	108
2130	36" long		20	.400		90	18.80		108.80	128
2140	48" long		19	.421		103	19.75		122.75	144
2300	Hand dryer, surface mounted, electric, 115 volt, 20 amp		4	2		445	94		539	630
2400	230 volt, 10 amp		4	2		745	94		839	965
2450	Hand dryer, touch free, 1400 watt, 81,000 rpm		4	2		1,025	94		1,119	1,275
2600	Hat and coat strip, stainless steel, 4 hook, 36" long		24	.333		68	15.65		83.65	99
2700	6 hook, 60" long		20	.400		124	18.80		142.80	166
3000	Mirror, with stainless steel 3/4" square frame, 18" x 24"		20	.400		49	18.80		67.80	83
3100	36" x 24"		15	.533		109	25		134	159
3200	48" x 24"		10	.800		150	37.50		187.50	223
3300	72" x 24"		6	1.333		289	62.50		351.50	415
3500	With 5" stainless steel shelf, 18" x 24"		20	.400		194	18.80		212.80	242
3600	36" x 24"		15	.533		236	25		261	298
3700	48" x 24"		10	.800		245	37.50		282.50	325
3800	72" x 24"		6	1.333		300	62.50		362.50	425
4100	Mop holder strip, stainless steel, 5 holders, 48" long		20	.400		89	18.80		107.80	127
4200	Napkin/tampon dispenser, recessed		15	.533		590	25		615	690
4220	Semi-recessed		6.50	1.231		310	58		368	430
4250	Napkin receptacle, recessed		6.50	1.231		165	58		223	271
4300	Robe hook, single, regular		36	.222		18.95	10.45		29.40	37
4400	Heavy duty, concealed mounting		36	.222		19.60	10.45		30.05	37.50
4600	Soap dispenser, chrome, surface mounted, liquid		20	.400		46.50	18.80		65.30	80
4700	Powder		20	.400		56.50	18.80		75.30	91.50
5000	Recessed stainless steel, liquid		10	.800		154	37.50		191.50	227
5600	Shelf, stainless steel, 5" wide, 18 ga., 24" long		24	.333		80.50	15.65		96.15	113
5700	48" long		16	.500		157	23.50		180.50	208
5800	8" wide shelf, 18 ga., 24" long		22	.364		69	17.05		86.05	103
5900	48" long		14	.571		121	27		148	175
6000	Toilet seat cover dispenser, stainless steel, recessed		20	.400		167	18.80		185.80	212
6050	Surface mounted		15	.533		34.50	25		59.50	76.50
6100	Toilet tissue dispenser, surface mounted, SS, single roll		30	.267		17.80	12.50		30.30	39
6200	Double roll		24	.333		23.50	15.65		39.15	50
6240	Plastic, twin/jumbo dbl. roll		24	.333		28.50	15.65		44.15	55
6400	Towel bar, stainless steel, 18" long		23	.348		41.50	16.35		57.85	71
6500	30" long		21	.381		111	17.90		128.90	151
6610	Antimicrobial copper alloy finish, 3/4" round, straight, w/o mounting				L.F.	10.95			10.95	12.05
6620	Antimicrobial copper alloy finish, 1" round, straight, w/o mounting				"	13.15			13.15	14.50
6630	24" long, including mounting	1 Carp	23	.348	Ea.	93.50	16.35		109.85	128
6700	Towel dispenser, stainless steel, surface mounted		16	.500		44.50	23.50		68	85
6800	Flush mounted, recessed		10	.800		257	37.50		294.50	340
6900	Plastic, touchless, battery operated		16	.500		88	23.50		111.50	133
7000	Towel holder, hotel type, 2 guest size		20	.400		52	18.80		70.80	86.50
7200	Towel shelf, stainless steel, 24" long, 8" wide		20	.400		60	18.80		78.80	95
7400	Tumbler holder, for tumbler only		30	.267		17.80	12.50		30.30	39
7410	Tumbler holder, recessed		20	.400		9.80	18.80		28.60	40
7500	Soap, tumbler & toothbrush		30	.267		19.60	12.50		32.10	41
7510	Tumbler & toothbrush holder		20	.400		13.60	18.80		32.40	44
7700	Wall urn ash receiver, surface mount, 11" long		12	.667		95	31.50		126.50	152
7800	7-1/2", long		18	.444		102	21		123	144
8000	Waste receptacles, stainless steel, with top, 13 gallon		10	.800		295	37.50		332.50	385

10 28 13 – Toilet Accessories

10 28 13.13 Commercial Toilet Accessories		Crew	Daily Output	Labor-Hours	Unit	Material	Labor	Equipment	Total	Total Incl O&P
						2015 Bare Costs				
8100	36 gallon	1 Carp	8	1	Ea.	405	47		452	520

10 28 16 – Bath Accessories

10 28 16.20 Medicine Cabinets		Crew	Daily Output	Labor-Hours	Unit	Material	Labor	Equipment	Total	Total Incl O&P
0010	**MEDICINE CABINETS**									
0020	With mirror, sst frame, 16" x 22", unlighted	1 Carp	14	.571	Ea.	98	27		125	150
0100	Wood frame		14	.571		128	27		155	183
0300	Sliding mirror doors, 20" x 16" x 4-3/4", unlighted		7	1.143		124	53.50		177.50	219
0400	24" x 19" x 8-1/2", lighted		5	1.600		179	75		254	315
0600	Triple door, 30" x 32", unlighted, plywood body		7	1.143		325	53.50		378.50	445
0700	Steel body		7	1.143		375	53.50		428.50	495
0900	Oak door, wood body, beveled mirror, single door		7	1.143		199	53.50		252.50	300
1000	Double door		6	1.333		380	62.50		442.50	515
1200	Hotel cabinets, stainless, with lower shelf, unlighted		10	.800		200	37.50		237.50	277
1300	Lighted		5	1.600		305	75		380	450

10 28 19 – Tub and Shower Enclosures

10 28 19.10 Partitions, Shower		Crew	Daily Output	Labor-Hours	Unit	Material	Labor	Equipment	Total	Total Incl O&P
0010	**PARTITIONS, SHOWER** floor mounted, no plumbing									
0400	Cabinet, one piece, fiberglass, 32" x 32"	2 Carp	5	3.200	Ea.	530	150		680	815
0420	36" x 36"		5	3.200		570	150		720	855
0440	36" x 48"		5	3.200		1,375	150		1,525	1,725
0460	Acrylic, 32" x 32"		5	3.200		310	150		460	570
0480	36" x 36"		5	3.200		1,025	150		1,175	1,350
0500	36" x 48"		5	3.200		1,500	150		1,650	1,900
0520	Shower door for above, clear plastic, 24" wide	1 Carp	8	1		186	47		233	277
0540	28" wide		8	1		206	47		253	300
0560	Tempered glass, 24" wide		8	1		200	47		247	293
0580	28" wide		8	1		226	47		273	320
2400	Glass stalls, with doors, no receptors, chrome on brass	2 Shee	3	5.333		1,650	298		1,948	2,275
2700	Anodized aluminum	"	4	4		1,150	224		1,374	1,625
2900	Marble shower stall, stock design, with shower door	2 Marb	1.20	13.333		2,475	580		3,055	3,600
3000	With curtain		1.30	12.308		2,200	535		2,735	3,225
3200	Receptors, precast terrazzo, 32" x 32"		14	1.143		360	49.50		409.50	470
3300	48" x 34"		9.50	1.684		465	73.50		538.50	620
3500	Plastic, simulated terrazzo receptor, 32" x 32"		14	1.143		148	49.50		197.50	239
3600	32" x 48"		12	1.333		215	58		273	325
3800	Precast concrete, colors, 32" x 32"		14	1.143		186	49.50		235.50	281
3900	48" x 48"		8	2		251	87		338	410
4100	Shower doors, economy plastic, 24" wide	1 Shee	9	.889		144	49.50		193.50	234
4200	Tempered glass door, economy		8	1		258	56		314	370
4400	Folding, tempered glass, aluminum frame		6	1.333		390	74.50		464.50	545
4500	Sliding, tempered glass, 48" opening		6	1.333		540	74.50		614.50	705
4700	Deluxe, tempered glass, chrome on brass frame, 42" to 44"		8	1		385	56		441	505
4800	39" to 48" wide		1	8		625	450		1,075	1,375
4850	On anodized aluminum frame, obscure glass		2	4		540	224		764	930
4900	Clear glass		1	8		625	450		1,075	1,375
5100	Shower enclosure, tempered glass, anodized alum. frame									
5120	2 panel & door, corner unit, 32" x 32"	1 Shee	2	4	Ea.	1,025	224		1,249	1,500
5140	Neo-angle corner unit, 16" x 24" x 16"	"	2	4		1,075	224		1,299	1,550
5200	Shower surround, 3 wall, polypropylene, 32" x 32"	1 Carp	4	2		455	94		549	645
5220	PVC, 32" x 32"		4	2		380	94		474	565
5240	Fiberglass		4	2		420	94		514	605
5250	2 wall, polypropylene, 32" x 32"		4	2		315	94		409	495

10 28 Toilet, Bath, and Laundry Accessories

10 28 19 – Tub and Shower Enclosures

10 28 19.10 Partitions, Shower		Crew	Daily Output	Labor-Hours	Unit	Material	Labor	Equipment	Total	Total Incl O&P
5270	PVC	1 Carp	4	2	Ea.	395	94		489	580
5290	Fiberglass	↓	4	2		400	94		494	585
5300	Tub doors, tempered glass & frame, obscure glass	1 Shee	8	1		229	56		285	340
5400	Clear glass		6	1.333		535	74.50		609.50	705
5600	Chrome plated, brass frame, obscure glass		8	1		305	56		361	420
5700	Clear glass		6	1.333		745	74.50		819.50	930
5900	Tub/shower enclosure, temp. glass, alum. frame, obscure glass		2	4		410	224		634	790
6200	Clear glass		1.50	5.333		855	298		1,153	1,400
6500	On chrome-plated brass frame, obscure glass		2	4		565	224		789	965
6600	Clear glass	↓	1.50	5.333		1,200	298		1,498	1,775
6800	Tub surround, 3 wall, polypropylene	1 Carp	4	2		256	94		350	425
6900	PVC		4	2		390	94		484	575
7000	Fiberglass, obscure glass		4	2		400	94		494	585
7100	Clear glass	↓	3	2.667	↓	680	125		805	945

10 28 23 – Laundry Accessories

10 28 23.13 Built-In Ironing Boards		Crew	Daily Output	Labor-Hours	Unit	Material	Labor	Equipment	Total	Total Incl O&P
0010	**BUILT-IN IRONING BOARDS**									
0020	Including cabinet, board & light, 42″	1 Carp	2	4	Ea.	395	188		583	725
0100	46″, see also Section 11 23 23.13	″	1.50	5.333	″	515	250		765	950

10 31 Manufactured Fireplaces

10 31 13 – Manufactured Fireplace Chimneys

10 31 13.10 Fireplace Chimneys		Crew	Daily Output	Labor-Hours	Unit	Material	Labor	Equipment	Total	Total Incl O&P
0010	**FIREPLACE CHIMNEYS**									
0500	Chimney dbl. wall, all stainless, over 8′-6″, 7″ diam., add to fireplace	1 Carp	33	.242	V.L.F.	83.50	11.40		94.90	109
0600	10″ diameter, add to fireplace		32	.250		132	11.75		143.75	163
0700	12″ diameter, add to fireplace		31	.258		158	12.10		170.10	193
0800	14″ diameter, add to fireplace		30	.267	↓	227	12.50		239.50	269
1000	Simulated brick chimney top, 4′ high, 16″ x 16″		10	.800	Ea.	425	37.50		462.50	530
1100	24″ x 24″	↓	7	1.143	″	540	53.50		593.50	680

10 31 13.20 Chimney Accessories		Crew	Daily Output	Labor-Hours	Unit	Material	Labor	Equipment	Total	Total Incl O&P
0010	**CHIMNEY ACCESSORIES**									
0020	Chimney screens, galv., 13″ x 13″ flue	1 Bric	8	1	Ea.	58.50	46		104.50	135
0050	24″ x 24″ flue		5	1.600		124	74		198	250
0200	Stainless steel, 13″ x 13″ flue		8	1		97.50	46		143.50	178
0250	20″ x 20″ flue		5	1.600		152	74		226	280
2400	Squirrel and bird screens, galvanized, 8″ x 8″ flue		16	.500		53	23		76	93.50
2450	13″ x 13″ flue	↓	12	.667	↓	55	31		86	108

10 31 16 – Manufactured Fireplace Forms

10 31 16.10 Fireplace Forms		Crew	Daily Output	Labor-Hours	Unit	Material	Labor	Equipment	Total	Total Incl O&P
0010	**FIREPLACE FORMS**									
1800	Fireplace forms, no accessories, 32″ opening	1 Bric	3	2.667	Ea.	670	123		793	930
1900	36″ opening		2.50	3.200		855	148		1,003	1,175
2000	40″ opening		2	4		1,125	185		1,310	1,525
2100	78″ opening	↓	1.50	5.333	↓	1,650	246		1,896	2,200

10 31 Manufactured Fireplaces

10 31 23 – Prefabricated Fireplaces

10 31 23.10 Fireplace, Prefabricated		Crew	Daily Output	Labor-Hours	Unit	Material	2015 Bare Costs Labor	Equipment	Total	Total Incl O&P
0010	**FIREPLACE, PREFABRICATED**, free standing or wall hung									
0100	With hood & screen, painted	1 Carp	1.30	6.154	Ea.	1,375	289		1,664	1,975
0150	Average		1	8		1,625	375		2,000	2,350
0200	Stainless steel		.90	8.889		3,050	415		3,465	4,025
1500	Simulated logs, gas fired, 40,000 BTU, 2′ long, manual safety pilot		7	1.143	Set	425	53.50		478.50	550
1600	Adjustable flame remote pilot		6	1.333		1,100	62.50		1,162.50	1,325
1700	Electric, 1,500 BTU, 1′-6″ long, incandescent flame		7	1.143		197	53.50		250.50	299
1800	1,500 BTU, LED flame		6	1.333		300	62.50		362.50	430
2000	Fireplace, built-in, 36″ hearth, radiant		1.30	6.154	Ea.	660	289		949	1,175
2100	Recirculating, small fan		1	8		880	375		1,255	1,550
2150	Large fan		.90	8.889		1,900	415		2,315	2,750
2200	42″ hearth, radiant		1.20	6.667		890	315		1,205	1,450
2300	Recirculating, small fan		.90	8.889		1,175	415		1,590	1,950
2350	Large fan		.80	10		1,300	470		1,770	2,150
2400	48″ hearth, radiant		1.10	7.273		2,075	340		2,415	2,800
2500	Recirculating, small fan		.80	10		2,350	470		2,820	3,325
2550	Large fan		.70	11.429		2,375	535		2,910	3,425
3000	See through, including doors		.80	10		2,225	470		2,695	3,175
3200	Corner (2 wall)		1	8		3,250	375		3,625	4,150

10 32 Fireplace Specialties

10 32 13 – Fireplace Dampers

10 32 13.10 Dampers

		Crew	Daily Output	Labor-Hours	Unit	Material	Labor	Equipment	Total	Total Incl O&P
0010	**DAMPERS**									
0800	Damper, rotary control, steel, 30″ opening	1 Bric	6	1.333	Ea.	119	61.50		180.50	225
0850	Cast iron, 30″ opening		6	1.333		125	61.50		186.50	231
0880	36″ opening		6	1.333		127	61.50		188.50	234
0900	48″ opening		6	1.333		167	61.50		228.50	278
0920	60″ opening		6	1.333		355	61.50		416.50	485
0950	72″ opening		5	1.600		425	74		499	580
1000	84″ opening, special order		5	1.600		910	74		984	1,125
1050	96″ opening, special order		4	2		925	92.50		1,017.50	1,175
1200	Steel plate, poker control, 60″ opening		8	1		320	46		366	425
1250	84″ opening, special order		5	1.600		585	74		659	760
1400	"Universal" type, chain operated, 32″ x 20″ opening		8	1		250	46		296	345
1450	48″ x 24″ opening		5	1.600		375	74		449	525

10 32 23 – Fireplace Doors

10 32 23.10 Doors

		Crew	Daily Output	Labor-Hours	Unit	Material	Labor	Equipment	Total	Total Incl O&P
0010	**DOORS**									
0400	Cleanout doors and frames, cast iron, 8″ x 8″	1 Bric	12	.667	Ea.	41	31		72	92
0450	12″ x 12″		10	.800		108	37		145	176
0500	18″ x 24″		8	1		150	46		196	236
0550	Cast iron frame, steel door, 24″ x 30″		5	1.600		315	74		389	460
1600	Dutch Oven door and frame, cast iron, 12″ x 15″ opening		13	.615		131	28.50		159.50	189
1650	Copper plated, 12″ x 15″ opening		13	.615		257	28.50		285.50	325

10 35 Stoves

10 35 13 – Heating Stoves

10 35 13.10 Woodburning Stoves		Crew	Daily Output	Labor-Hours	Unit	Material	2015 Bare Costs Labor	Equipment	Total	Total Incl O&P
0010	**WOODBURNING STOVES**									
0015	Cast iron, °1500sf	2 Carp	1.30	12.308	Ea.	1,175	580		1,755	2,200
0020	1500-2000sf		1	16		2,025	750		2,775	3,375
0030	2,000 sf		.80	20		2,775	940		3,715	4,500
0050	For gas log lighter, add					45			45	49.50

10 43 Emergency Aid Specialties

10 43 13 – Defibrillator Cabinets

10 43 13.05 Defibrillator Cabinets		Crew	Daily Output	Labor-Hours	Unit	Material	Labor	Equipment	Total	Total Incl O&P
0010	**DEFIBRILLATOR CABINETS**, not equipped, stainless steel									
0050	Defibrillator cabinet, stainless steel with strobe & alarm 12″ x 27″	1 Carp	10	.800	Ea.	430	37.50		467.50	530
0100	Automatic External Defibrillator	″	30	.267	″	1,350	12.50		1,362.50	1,500

10 44 Fire Protection Specialties

10 44 13 – Fire Protection Cabinets

10 44 13.53 Fire Equipment Cabinets		Crew	Daily Output	Labor-Hours	Unit	Material	Labor	Equipment	Total	Total Incl O&P
0010	**FIRE EQUIPMENT CABINETS**, not equipped, 20 ga. steel box									
0040	recessed, D.S. glass in door, box size given									
1000	Portable extinguisher, single, 8″ x 12″ x 27″, alum. door & frame	Q-12	8	2	Ea.	155	101		256	325
1100	Steel door and frame		8	2		116	101		217	281
2700	Fire blanket & extinguisher cab, inc blanket, rec stl., 14″ x 40″ x 8″		7	2.286		188	116		304	380
2800	Fire blanket cab, inc blanket, surf mtd, stl, 15″x10″x5″, w/pwdr coat fin		8	2		94	101		195	257
3000	Hose rack assy., 1-1/2″ valve & 100′ hose, 24″ x 40″ x 5-1/2″									
3100	Aluminum door and frame	Q-12	6	2.667	Ea.	370	135		505	610
3200	Steel door and frame		6	2.667		246	135		381	475
3300	Stainless steel door and frame		6	2.667		455	135		590	710
4000	Hose rack assy., 2-1/2″ x 1-1/2″ valve, 100′ hose, 24″ x 40″ x 8″									
4100	Aluminum door and frame	Q-12	6	2.667	Ea.	375	135		510	615
4200	Steel door and frame		6	2.667		253	135		388	480
4300	Stainless steel door and frame		6	2.667		495	135		630	750
5000	Hose rack assy., 2-1/2″ x 1-1/2″ valve, 100′ hose									
5010	and extinguisher, 30″ x 40″ x 8″									
5100	Aluminum door and frame	Q-12	5	3.200	Ea.	475	162		637	770
5200	Steel door and frame		5	3.200		264	162		426	535
5300	Stainless steel door and frame		5	3.200		535	162		697	835
8000	Valve cabinet for 2-1/2″ FD angle valve, 18″ x 18″ x 8″									
8100	Aluminum door and frame	Q-12	12	1.333	Ea.	164	67.50		231.50	282
8200	Steel door and frame		12	1.333		136	67.50		203.50	251
8300	Stainless steel door and frame		12	1.333		221	67.50		288.50	345

10 44 16 – Fire Extinguishers

10 44 16.13 Portable Fire Extinguishers		Crew	Daily Output	Labor-Hours	Unit	Material	Labor	Equipment	Total	Total Incl O&P
0010	**PORTABLE FIRE EXTINGUISHERS**									
0140	CO_2, with hose and "H" horn, 10 lb.				Ea.	275			275	305
0160	15 lb.					355			355	390
0180	20 lb.					395			395	435
1000	Dry chemical, pressurized									
1040	Standard type, portable, painted, 2-1/2 lb.				Ea.	37.50			37.50	41
1060	5 lb.					51			51	56.50
1080	10 lb.					81			81	89.50

10 44 Fire Protection Specialties

10 44 16 – Fire Extinguishers

10 44 16.13 Portable Fire Extinguishers		Crew	Daily Output	Labor-Hours	Unit	Material	2015 Bare Costs Labor	Equipment	Total	Total Incl O&P
1100	20 lb.				Ea.	136			136	150
1120	30 lb.					425			425	470
1300	Standard type, wheeled, 150 lb.					2,425			2,425	2,650
2000	ABC all purpose type, portable, 2-1/2 lb.					21			21	23.50
2060	5 lb.					26.50			26.50	29
2080	9-1/2 lb.					44			44	48
2100	20 lb.					79.50			79.50	87.50
3500	Halotron 1, 2-1/2 lb.					126			126	139
3600	5 lb.					198			198	218
3700	11 lb.					390			390	430
5000	Pressurized water, 2-1/2 gallon, stainless steel					102			102	112
5060	With anti-freeze					106			106	116
9400	Installation of extinguishers, 12 or more, on nailable surface	1 Carp	30	.267			12.50		12.50	19.25
9420	On masonry or concrete	"	15	.533			25		25	38.50

10 44 16.16 Wheeled Fire Extinguisher Units		Crew	Daily Output	Labor-Hours	Unit	Material	Labor	Equipment	Total	Total Incl O&P
0010	**WHEELED FIRE EXTINGUISHER UNITS**									
0350	CO_2, portable, with swivel horn									
0360	Wheeled type, cart mounted, 50 lb.				Ea.	1,125			1,125	1,250
0400	100 lb.				"	3,850			3,850	4,250
2200	ABC all purpose type									
2300	Wheeled, 45 lb.				Ea.	735			735	810
2360	150 lb.				"	1,850			1,850	2,025

10 51 Lockers

10 51 13 – Metal Lockers

10 51 13.10 Lockers		Crew	Daily Output	Labor-Hours	Unit	Material	Labor	Equipment	Total	Total Incl O&P
0011	**LOCKERS** steel, baked enamel, pre-assembled									
0110	Single tier box locker, 12" x 15" x 72"	1 Shee	20	.400	Ea.	223	22.50		245.50	280
0120	18" x 15" x 72"		20	.400		240	22.50		262.50	298
0130	12" x 18" x 72"		20	.400		228	22.50		250.50	285
0140	18" x 18" x 72"		20	.400		283	22.50		305.50	345
0410	Double tier, 12" x 15" x 36"		30	.267		232	14.90		246.90	278
0420	18" x 15" x 36"		30	.267		235	14.90		249.90	282
0430	12" x 18" x 36"		30	.267		264	14.90		278.90	315
0440	18" x 18" x 36"		30	.267		241	14.90		255.90	288
0500	Two person, 18" x 15" x 72"		20	.400		291	22.50		313.50	355
0510	18" x 18" x 72"		20	.400		325	22.50		347.50	390
0520	Duplex, 15" x 15" x 72"		20	.400		325	22.50		347.50	390
0530	15" x 21" x 72"		20	.400		365	22.50		387.50	435
0600	5 tier box lockers, unassembled		30	.267	Opng.	49	14.90		63.90	77
0700	Set up		24	.333		52.50	18.65		71.15	86.50
0900	6 tier box lockers, unassembled		36	.222		37.50	12.45		49.95	60
1000	Set up		30	.267		46	14.90		60.90	73.50
1100	Wire meshed wardrobe, floor. mtd., open front varsity type		7.50	1.067	Ea.	272	59.50		331.50	390
2400	16-person locker unit with clothing rack									
2500	72 wide x 15" deep x 72" high	1 Shee	15	.533	Ea.	455	30		485	545
2550	18" deep	"	15	.533	"	605	30		635	710
3000	Wall mounted lockers, 4 person, with coat bar									
3100	48" wide x 18" deep x 12" high	1 Shee	20	.400	Ea.	325	22.50		347.50	390
3250	Rack w/24 wire mesh baskets		1.50	5.333	Set	400	298		698	895
3260	30 baskets		1.25	6.400		350	360		710	930

10 51 Lockers

10 51 13 – Metal Lockers

10 51 13.10 Lockers		Crew	Daily Output	Labor-Hours	Unit	Material	2015 Bare Costs Labor	Equipment	Total	Total Incl O&P
3270	36 baskets	1 Shee	.95	8.421	Set	510	470		980	1,275
3280	42 baskets		.80	10		555	560		1,115	1,475
3300	For built-in lock with 2 keys, add				Ea.	13.20			13.20	14.50
3600	For hanger rods, add					1.90			1.90	2.09
3650	For number plate kit, 100 plates #1 - #100, add	1 Shee	4	2		82	112		194	261
3700	For locker base, closed front panel		90	.089		7.25	4.97		12.22	15.60
3710	End panel, bolted		36	.222		9	12.45		21.45	29
3800	For sloping top, 12″ wide		24	.333		31.50	18.65		50.15	63
3810	15″ wide		24	.333		34	18.65		52.65	66
3820	18″ wide		24	.333		35.50	18.65		54.15	68
3850	Sloping top end panel, 12″ deep		72	.111		12.20	6.20		18.40	23
3860	15″ deep		72	.111		12.20	6.20		18.40	23
3870	18″ deep		72	.111		12.20	6.20		18.40	23
3900	For finish end panels, steel, 60″ high, 15″ deep		12	.667		36	37.50		73.50	96.50
3910	72″ high, 12″ deep		12	.667		29	37.50		66.50	89
3920	18″ deep		12	.667		43	37.50		80.50	104
5000	For 'ready to assemble' lockers,									
5010	Add to labor						75%			
5020	Deduct from material					20%				
6000	Heavy duty for detention facility, tamper proof, 14 ga. welded steel, solid									
6100	24″ W x 24″ D x 74″ H, single tier	1 Shee	18	.444	Ea.	515	25		540	610
6110	Double tier		18	.444		505	25		530	595
6120	Triple tier		18	.444		520	25		545	615

10 51 26 – Plastic Lockers

10 51 26.13 Recycled Plastic Lockers			Crew	Daily Output	Labor-Hours	Unit	Material	Labor	Equipment	Total	Total Incl O&P
0011	**RECYCLED PLASTIC LOCKERS**, 30% recycled										
0110	Single tier box locker, 12″ x 12″ x 72″	G	1 Shee	8	1	Ea.	450	56		506	580
0120	12″ x 15″ x 72″	G		8	1		470	56		526	605
0130	12″ x 18″ x 72″	G		8	1		460	56		516	590
0410	Double tier, 12″ x 12″ x 72″	G		21	.381		470	21.50		491.50	550
0420	12″ x 15″ x 72″	G		21	.381		495	21.50		516.50	580
0430	12″ x 18″ x 72″	G		21	.381		485	21.50		506.50	565

10 51 53 – Locker Room Benches

10 51 53.10 Benches		Crew	Daily Output	Labor-Hours	Unit	Material	Labor	Equipment	Total	Total Incl O&P
0010	**BENCHES**									
2100	Locker bench, laminated maple, top only	1 Shee	100	.080	L.F.	24.50	4.48		28.98	34
2200	Pedestals, steel pipe		25	.320	Ea.	44	17.90		61.90	76
2250	Plastic, 9.5″ top with PVC pedestals		80	.100	L.F.	59.50	5.60		65.10	73.50

10 55 Postal Specialties

10 55 23 – Mail Boxes

10 55 23.10 Commercial Mail Boxes		Crew	Daily Output	Labor-Hours	Unit	Material	Labor	Equipment	Total	Total Incl O&P
0010	**COMMERCIAL MAIL BOXES**									
0020	Horiz., key lock, 5″H x 6″W x 15″D, alum., rear load	1 Carp	34	.235	Ea.	40	11.05		51.05	61
0100	Front loading		34	.235		40	11.05		51.05	61
0200	Double, 5″H x 12″W x 15″D, rear loading		26	.308		66.50	14.45		80.95	95.50
0300	Front loading		26	.308		70	14.45		84.45	99
0500	Quadruple, 10″H x 12″W x 15″D, rear loading		20	.400		106	18.80		124.80	146
0600	Front loading		20	.400		88.50	18.80		107.30	126
0800	Vertical, front load, 15″H x 5″W x 6″D, alum., per compartment		34	.235		40	11.05		51.05	61

10 55 Postal Specialties

10 55 23 – Mail Boxes

10 55 23.10 Commercial Mail Boxes		Crew	Daily Output	Labor-Hours	Unit	Material	2015 Bare Costs Labor	Equipment	Total	Total Incl O&P
0900	Bronze, duranodic finish	1 Carp	34	.235	Ea.	46.50	11.05		57.55	68
1000	Steel, enameled		34	.235		40	11.05		51.05	61
1700	Alphabetical directories, 120 names		10	.800		125	37.50		162.50	196
1800	Letter collection box	↓	6	1.333		715	62.50		777.50	880
1830	Lobby collection boxes, aluminum	2 Shee	5	3.200		1,750	179		1,929	2,200
1840	Bronze or stainless	"	4.50	3.556		1,800	199		1,999	2,300
1900	Letter slot, residential	1 Carp	20	.400		80	18.80		98.80	117
2000	Post office type		8	1		104	47		151	187
2250	Key keeper, single key, aluminum		26	.308		39.50	14.45		53.95	65.50
2300	Steel, enameled	↓	26	.308	↓	75	14.45		89.45	105

10 56 Storage Assemblies

10 56 13 – Metal Storage Shelving

10 56 13.10 Shelving		Crew	Daily Output	Labor-Hours	Unit	Material	Labor	Equipment	Total	Total Incl O&P
0010	**SHELVING**									
0020	Metal, industrial, cross-braced, 3′ wide, 12″ deep	1 Sswk	175	.046	SF Shlf	7.65	2.41		10.06	12.60
0100	24″ deep		330	.024		5.05	1.28		6.33	7.75
0300	4′ wide, 12″ deep		185	.043		6.50	2.28		8.78	11.10
0400	24″ deep		380	.021		4.62	1.11		5.73	7.05
1200	Enclosed sides, cross-braced back, 3′ wide, 12″ deep		175	.046		11.85	2.41		14.26	17.25
1300	24″ deep		290	.028		8.20	1.45		9.65	11.55
1500	Fully enclosed, sides and back, 3′ wide, 12″ deep		150	.053		15.60	2.81		18.41	22
1600	24″ deep		255	.031		10.40	1.65		12.05	14.30
1800	4′ wide, 12″ deep		150	.053		9.95	2.81		12.76	15.85
1900	24″ deep		290	.028		8.30	1.45		9.75	11.65
2200	Wide span, 1600 lb. capacity per shelf, 6′ wide, 24″ deep		380	.021		7	1.11		8.11	9.65
2400	36″ deep		440	.018		6.85	.96		7.81	9.15
2600	8′ wide, 24″ deep		440	.018		6.45	.96		7.41	8.75
2800	36″ deep	↓	520	.015		6.20	.81		7.01	8.20
4000	Pallet racks, steel frame 5,000 lb. capacity, 8′ long, 36″ deep	2 Sswk	450	.036		9.25	1.87		11.12	13.45
4200	42″ deep		500	.032		8.10	1.68		9.78	11.90
4400	48″ deep	↓	520	.031	↓	7.55	1.62		9.17	11.10

10 56 13.20 Parts Bins		Crew	Daily Output	Labor-Hours	Unit	Material	Labor	Equipment	Total	Total Incl O&P
0010	**PARTS BINS** metal, gray baked enamel finish									
0100	6′-3″ high, 3′ wide									
0300	12 bins, 18″ wide x 12″ high, 12″ deep	2 Clab	10	1.600	Ea.	320	60		380	445
0400	24″ deep		10	1.600		405	60		465	540
0600	72 bins, 6″ wide x 6″ high, 12″ deep		8	2		540	75		615	710
0700	18″ deep	↓	8	2	↓	850	75		925	1,050
1000	7′-3″ high, 3′ wide									
1200	14 bins, 18″ wide x 12″ high, 12″ deep	2 Clab	10	1.600	Ea.	350	60		410	480
1300	24″ deep		10	1.600		440	60		500	575
1500	84 bins, 6″ wide x 6″ high, 12″ deep		8	2		925	75		1,000	1,150
1600	24″ deep	↓	8	2	↓	1,125	75		1,200	1,350

10 57 Wardrobe and Closet Specialties

10 57 13 – Hat and Coat Racks

10 57 13.10 Coat Racks and Wardrobes		Crew	Daily Output	Labor-Hours	Unit	2015 Bare Costs Material	Labor	Equipment	Total	Total Incl O&P
0010	**COAT RACKS AND WARDROBES**									
0020	Hat & coat rack, floor model, 6 hangers									
0050	Standing, beech wood, 21″ x 21″ x 72″, chrome				Ea.	237			237	260
0100	18 ga. tubular steel, 21″ x 21″ x 69″, wood walnut				″	299			299	330
0500	16 ga. steel frame, 22 ga. steel shelves									
0650	Single pedestal, 30″ x 18″ x 63″				Ea.	262			262	289
0800	Single face rack, 29″ x 18-1/2″ x 62″					315			315	345
0900	51″ x 18-1/2″ x 70″					410			410	450
0910	Double face rack, 39″ x 26″ x 70″					390			390	430
0920	63″ x 26″ x 70″					465			465	515
0940	For 2″ ball casters, add				Set	88			88	97
1400	Utility hook strips, 3/8″ x 2-1/2″ x 18″, 6 hooks	1 Carp	48	.167	Ea.	62	7.85		69.85	80
1500	34″ long, 12 hooks	″	48	.167	″	66	7.85		73.85	84.50
1650	Wall mounted racks, 16 ga. steel frame, 22 ga. steel shelves									
1850	12″ x 15″ x 26″, 6 hangers	1 Carp	32	.250	Ea.	155	11.75		166.75	188
2000	12″ x 15″ x 50″, 12 hangers	″	32	.250	″	181	11.75		192.75	217
2150	Wardrobe cabinet, steel, baked enamel finish									
2300	36″ x 21″ x 78″, incl. top shelf & hanger rod				Ea.	315			315	350
2400	Wardrobe, 24″ x 24″ x 76″, KD, w/door, hospital, baked enamel steel	1 Carp	2	4		635	188		823	985
2500	Hardwood	″	2	4		1,150	188		1,338	1,575

10 57 23 – Closet and Utility Shelving

10 57 23.19 Wood Closet and Utility Shelving		Crew	Daily Output	Labor-Hours	Unit	Material	Labor	Equipment	Total	Total Incl O&P
0010	**WOOD CLOSET AND UTILITY SHELVING**									
0020	Pine, clear grade, no edge band, 1″ x 8″	1 Carp	115	.070	L.F.	3.49	3.27		6.76	8.90
0100	1″ x 10″		110	.073		4.34	3.41		7.75	10
0200	1″ x 12″		105	.076		5.25	3.58		8.83	11.25
0600	Plywood, 3/4″ thick with lumber edge, 12″ wide		75	.107		1.87	5		6.87	9.75
0700	24″ wide		70	.114		3.30	5.35		8.65	11.90
0900	Bookcase, clear grade pine, shelves 12″ O.C., 8″ deep, per S.F. shelf		70	.114	S.F.	11.35	5.35		16.70	20.50
1000	12″ deep shelves		65	.123	″	17	5.80		22.80	27.50
1200	Adjustable closet rod and shelf, 12″ wide, 3′ long		20	.400	Ea.	11.15	18.80		29.95	41.50
1300	8′ long		15	.533	″	21.50	25		46.50	62
1500	Prefinished shelves with supports, stock, 8″ wide		75	.107	L.F.	4.58	5		9.58	12.75
1600	10″ wide		70	.114	″	6.50	5.35		11.85	15.40

10 71 Exterior Protection

10 71 13 – Exterior Sun Control Devices

10 71 13.19 Rolling Exterior Shutters		Crew	Daily Output	Labor-Hours	Unit	Material	Labor	Equipment	Total	Total Incl O&P
0010	**ROLLING EXTERIOR SHUTTERS**									
0020	Roll-up, manual operation, aluminum, 3′ x 4′, incl. frame	2 Carp	8	2	Ea.	635	94		729	845
0030	6′ x 7′	″	8	2	″	1,375	94		1,469	1,675

10 73 13 – Awnings

10 73 13.10 Awnings, Fabric		Crew	Daily Output	Labor-Hours	Unit	Material	2015 Bare Costs Labor	Equipment	Total	Total Incl O&P
0010	**AWNINGS, FABRIC**									
0020	Including acrylic canvas and frame, standard design									
0100	Door and window, slope, 3′ high, 4′ wide	1 Carp	4.50	1.778	Ea.	720	83.50		803.50	920
0110	6′ wide		3.50	2.286		925	107		1,032	1,200
0120	8′ wide		3	2.667		1,125	125		1,250	1,450
0200	Quarter round convex, 4′ wide		3	2.667		1,125	125		1,250	1,425
0210	6′ wide		2.25	3.556		1,450	167		1,617	1,850
0220	8′ wide		1.80	4.444		1,775	209		1,984	2,275
0300	Dome, 4′ wide		7.50	1.067		430	50		480	550
0310	6′ wide		3.50	2.286		970	107		1,077	1,250
0320	8′ wide		2	4		1,725	188		1,913	2,200
0350	Elongated dome, 4′ wide		1.33	6.015		1,625	282		1,907	2,200
0360	6′ wide		1.11	7.207		1,925	340		2,265	2,650
0370	8′ wide		1	8		2,275	375		2,650	3,075
1000	Entry or walkway, peak, 12′ long, 4′ wide	2 Carp	.90	17.778		5,225	835		6,060	7,025
1010	6′ wide		.60	26.667		8,050	1,250		9,300	10,800
1020	8′ wide		.40	40		11,100	1,875		12,975	15,100
1100	Radius with dome end, 4′ wide		1.10	14.545		3,950	685		4,635	5,400
1110	6′ wide		.70	22.857		6,350	1,075		7,425	8,650
1120	8′ wide		.50	32		9,050	1,500		10,550	12,300
2000	Retractable lateral arm awning, manual									
2010	To 12′ wide, 8′ - 6″ projection	2 Carp	1.70	9.412	Ea.	1,175	440		1,615	1,975
2020	To 14′ wide, 8′ - 6″ projection		1.10	14.545		1,375	685		2,060	2,575
2030	To 19′ wide, 8′ - 6″ projection		.85	18.824		1,875	885		2,760	3,400
2040	To 24′ wide, 8′ - 6″ projection		.67	23.881		2,350	1,125		3,475	4,325
2050	Motor for above, add	1 Carp	2.67	3		1,000	141		1,141	1,350
3000	Patio/deck canopy with frame									
3010	12′ wide, 12′ projection	2 Carp	2	8	Ea.	1,675	375		2,050	2,400
3020	16′ wide, 14′ projection	"	1.20	13.333		2,600	625		3,225	3,825
9000	For fire retardant canvas, add					7%				
9010	For lettering or graphics, add					35%				
9020	For painted or coated acrylic canvas, deduct					8%				
9030	For translucent or opaque vinyl canvas, add					10%				
9040	For 6 or more units, deduct					20%	15%			

10 73 16 – Canopies

10 73 16.20 Metal Canopies		Crew	Daily Output	Labor-Hours	Unit	Material	Labor	Equipment	Total	Total Incl O&P
0010	**METAL CANOPIES**									
0020	Wall hung, .032″, aluminum, prefinished, 8′ x 10′	K-2	1.30	18.462	Ea.	2,150	900	233	3,283	4,125
0300	8′ x 20′		1.10	21.818		3,650	1,075	276	5,001	6,075
0500	10′ x 10′		1.30	18.462		2,825	900	233	3,958	4,850
0700	10′ x 20′		1.10	21.818		4,600	1,075	276	5,951	7,125
1000	12′ x 20′		1	24		5,250	1,175	305	6,730	8,075
1360	12′ x 30′		.80	30		7,875	1,475	380	9,730	11,500
1700	12′ x 40′		.60	40		10,500	1,950	505	12,955	15,400
1900	For free standing units, add					20%	10%			
2300	Aluminum entrance canopies, flat soffit, .032″									
2500	3′-6″ x 4′-0″, clear anodized	2 Carp	4	4	Ea.	915	188		1,103	1,300
2700	Bronze anodized		4	4		1,625	188		1,813	2,075
3000	Polyurethane painted		4	4		1,300	188		1,488	1,750
3300	4′-6″ x 10′-0″, clear anodized		2	8		2,525	375		2,900	3,350
3500	Bronze anodized		2	8		3,225	375		3,600	4,125
3700	Polyurethane painted		2	8		2,700	375		3,075	3,550

10 73 Protective Covers

10 73 16 – Canopies

10 73 16.20 Metal Canopies		Crew	Daily Output	Labor-Hours	Unit	2015 Bare Costs Material	Labor	Equipment	Total	Total Incl O&P
4000	Wall downspout, 10 L.F., clear anodized	1 Carp	7	1.143	Ea.	157	53.50		210.50	256
4300	Bronze anodized		7	1.143		274	53.50		327.50	385
4500	Polyurethane painted		7	1.143		236	53.50		289.50	340
7000	Carport, baked vinyl finish, .032", 20' x 10', no foundations, flat panel	K-2	4	6	Car	4,000	293	76	4,369	4,975
7250	Insulated flat panel		2	12	"	6,300	585	152	7,037	8,075
7500	Walkway cover, to 12' wide, stl., vinyl finish, .032",no fndtns., flat		250	.096	S.F.	22.50	4.68	1.21	28.39	33.50
7750	Arched		200	.120	"	55	5.85	1.52	62.37	72

10 74 Manufactured Exterior Specialties

10 74 23 – Cupolas

10 74 23.10 Wood Cupolas		Crew	Daily Output	Labor-Hours	Unit	Material	Labor	Equipment	Total	Total Incl O&P
0010	**WOOD CUPOLAS**									
0020	Stock units, pine, painted, 18" sq., 28" high, alum. roof	1 Carp	4.10	1.951	Ea.	184	91.50		275.50	345
0100	Copper roof		3.80	2.105		255	99		354	435
0300	23" square, 33" high, aluminum roof		3.70	2.162		360	102		462	550
0400	Copper roof		3.30	2.424		470	114		584	690
0600	30" square, 37" high, aluminum roof		3.70	2.162		560	102		662	770
0700	Copper roof		3.30	2.424		670	114		784	910
0900	Hexagonal, 31" wide, 46" high, copper roof		4	2		930	94		1,024	1,175
1000	36" wide, 50" high, copper roof		3.50	2.286		1,400	107		1,507	1,725
1200	For deluxe stock units, add to above					25%				
1400	For custom built units, add to above					50%	50%			

10 74 29 – Steeples

10 74 29.10 Prefabricated Steeples		Crew	Daily Output	Labor-Hours	Unit	Material	Labor	Equipment	Total	Total Incl O&P
0010	**PREFABRICATED STEEPLES**									
4000	Steeples, translucent fiberglass, 30" square, 15' high	F-3	2	20	Ea.	8,375	960	325	9,660	11,000
4150	25' high		1.80	22.222		9,750	1,075	365	11,190	12,700
4350	Opaque fiberglass, 24" square, 14' high		2	20		6,975	960	325	8,260	9,475
4500	28' high		1.80	22.222		6,375	1,075	365	7,815	9,025
4600	Aluminum, baked finish, 16" square, 14' high					5,700			5,700	6,275
4620	20' high, 3'-6" base					9,550			9,550	10,500
4640	35' high, 8' base					36,700			36,700	40,400
4660	60' high, 14' base					77,500			77,500	85,000
4680	152' high, custom					632,000			632,000	695,000
4700	Porcelain enamel steeples, custom, 40' high	F-3	.50	80		14,300	3,825	1,300	19,425	23,000
4800	60' high	"	.30	133		24,800	6,375	2,175	33,350	39,500

10 74 46 – Window Wells

10 74 46.10 Area Window Wells		Crew	Daily Output	Labor-Hours	Unit	Material	Labor	Equipment	Total	Total Incl O&P
0010	**AREA WINDOW WELLS**, Galvanized steel									
0020	20 ga., 3'-2" wide, 1' deep	1 Sswk	29	.276	Ea.	17.45	14.50		31.95	44
0100	2' deep		23	.348		31	18.30		49.30	66.50
0300	16 ga., 3'-2" wide, 1' deep		29	.276		23	14.50		37.50	50.50
0400	3' deep		23	.348		47	18.30		65.30	83.50
0600	Welded grating for above, 15 lb., painted		45	.178		87.50	9.35		96.85	112
0700	Galvanized		45	.178		118	9.35		127.35	146
0900	Translucent plastic cap for above		60	.133		19	7		26	33

10 75 Flagpoles

10 75 16 – Ground-Set Flagpoles

10 75 16.10 Flagpoles		Crew	Daily Output	Labor-Hours	Unit	Material	2015 Bare Costs Labor	Equipment	Total	Total Incl O&P
0010	**FLAGPOLES**, ground set									
0050	Not including base or foundation									
0100	Aluminum, tapered, ground set 20′ high	K-1	2	8	Ea.	1,050	345	152	1,547	1,850
0200	25′ high		1.70	9.412		1,100	405	178	1,683	2,050
0300	30′ high		1.50	10.667		1,300	460	202	1,962	2,350
0400	35′ high		1.40	11.429		1,800	490	217	2,507	3,000
0500	40′ high		1.20	13.333		2,775	575	253	3,603	4,200
0600	50′ high		1	16		3,175	690	305	4,170	4,875
0700	60′ high		.90	17.778		4,950	765	335	6,050	6,975
0800	70′ high		.80	20		8,375	860	380	9,615	11,000
1100	Counterbalanced, internal halyard, 20′ high		1.80	8.889		2,450	380	168	2,998	3,475
1200	30′ high		1.50	10.667		2,675	460	202	3,337	3,875
1300	40′ high		1.30	12.308		6,475	530	233	7,238	8,200
1400	50′ high		1	16		9,025	690	305	10,020	11,300
2820	Aluminum, electronically operated, 30′ high		1.40	11.429		4,150	490	217	4,857	5,575
2840	35′ high		1.30	12.308		4,950	530	233	5,713	6,500
2860	39′ high		1.10	14.545		6,025	625	276	6,926	7,875
2880	45′ high		1	16		6,350	690	305	7,345	8,375
2900	50′ high		.90	17.778		8,175	765	335	9,275	10,500
3000	Fiberglass, tapered, ground set, 23′ high		2	8		580	345	152	1,077	1,325
3100	29′-7″ high		1.50	10.667		1,525	460	202	2,187	2,600
3200	36′-1″ high		1.40	11.429		2,000	490	217	2,707	3,200
3300	39′-5″ high		1.20	13.333		2,100	575	253	2,928	3,450
3400	49′-2″ high		1	16		3,900	690	305	4,895	5,675
3500	59′ high		.90	17.778		4,825	765	335	5,925	6,850
4300	Steel, direct imbedded installation									
4400	Internal halyard, 20′ high	K-1	2.50	6.400	Ea.	1,375	275	121	1,771	2,050
4500	25′ high		2.50	6.400		2,050	275	121	2,446	2,800
4600	30′ high		2.30	6.957		2,450	299	132	2,881	3,275
4700	40′ high		2.10	7.619		3,725	330	144	4,199	4,750
4800	50′ high		1.90	8.421		4,325	360	160	4,845	5,475
5000	60′ high		1.80	8.889		7,200	380	168	7,748	8,700
5100	70′ high		1.60	10		7,850	430	190	8,470	9,500
5200	80′ high		1.40	11.429		10,100	490	217	10,807	12,100
5300	90′ high		1.20	13.333		15,500	575	253	16,328	18,200
5500	100′ high		1	16		17,600	690	305	18,595	20,800
6400	Wood poles, tapered, clear vertical grain fir with tilting									
6410	base, not incl. foundation, 4″ butt, 25′ high	K-1	1.90	8.421	Ea.	1,400	360	160	1,920	2,250
6800	6″ butt, 30′ high	″	1.30	12.308	″	2,600	530	233	3,363	3,925
7300	Foundations for flagpoles, including									
7400	excavation and concrete, to 35′ high poles	C-1	10	3.200	Ea.	685	143		828	975
7600	40′ to 50′ high		3.50	9.143		1,275	410		1,685	2,025
7700	Over 60′ high		2	16		1,575	715		2,290	2,825

10 75 23 – Wall-Mounted Flagpoles

10 75 23.10 Flagpoles		Crew	Daily Output	Labor-Hours	Unit	Material	Labor	Equipment	Total	Total Incl O&P
0010	**FLAGPOLES**, structure mounted									
0100	Fiberglass, vertical wall set, 19′-8″ long	K-1	1.50	10.667	Ea.	1,150	460	202	1,812	2,200
0200	23′ long		1.40	11.429		1,425	490	217	2,132	2,550
0300	26′-3″ long		1.30	12.308		2,050	530	233	2,813	3,325
0800	19′-8″ long outrigger		1.30	12.308		1,325	530	233	2,088	2,525
1300	Aluminum, vertical wall set, tapered, with base, 20′ high		1.20	13.333		1,075	575	253	1,903	2,325
1400	29′-6″ high		1	16		2,650	690	305	3,645	4,300

10 75 Flagpoles

10 75 23 – Wall-Mounted Flagpoles

10 75 23.10 Flagpoles		Crew	Daily Output	Labor-Hours	Unit	Material	2015 Bare Costs Labor	Equipment	Total	Total Incl O&P
2400	Outrigger poles with base, 12′ long	K-1	1.30	12.308	Ea.	1,100	530	233	1,863	2,275
2500	14′ long	↓	1	16	↓	1,425	690	305	2,420	2,950

10 81 Pest Control Devices

10 81 13 – Bird Control Devices

10 81 13.10 Bird Control Netting

		Crew	Daily Output	Labor-Hours	Unit	Material	Labor	Equipment	Total	Total Incl O&P
0010	**BIRD CONTROL NETTING**									
0020	1/8″ square mesh	4 Clab	4000	.008	S.F.	.67	.30		.97	1.20
0100	1/4″ square mesh		4000	.008		.78	.30		1.08	1.32
0120	1/2″ square mesh		4000	.008		.14	.30		.44	.61
0140	5/8″ x 3/4″ mesh		4000	.008		.07	.30		.37	.54
0160	1-1/4″ x 1-1/2″ mesh		4000	.008		.14	.30		.44	.61
0200	4″ square mesh	↓	4000	.008	↓	.10	.30		.40	.57
1000	Poly clips				Ea.	.09			.09	.10

10 86 Security Mirrors and Domes

10 86 10 – Security Mirrors

10 86 10.10 Exterior Traffic Control Mirrors

		Crew	Daily Output	Labor-Hours	Unit	Material	Labor	Equipment	Total	Total Incl O&P
0010	**EXTERIOR TRAFFIC CONTROL MIRRORS**									
0100	Convex, stainless steel, 20 ga., 26″ diameter	1 Carp	12	.667	Ea.	213	31.50		244.50	282

10 86 20 – Security Domes

10 86 20.10 Domes

		Crew	Daily Output	Labor-Hours	Unit	Material	Labor	Equipment	Total	Total Incl O&P
0010	**DOMES** for security cameras (CCTV)									
0100	Ceiling mounted, 10″ diameter	1 Carp	30	.267	Ea.	11.60	12.50		24.10	32
0110	12″ diameter	″	30	.267	″	7.55	12.50		20.05	27.50

10 88 Scales

10 88 05 – Commercial Scales

10 88 05.10 Scales

		Crew	Daily Output	Labor-Hours	Unit	Material	Labor	Equipment	Total	Total Incl O&P
0010	**SCALES**									
0700	Truck scales, incl. steel weigh bridge,									
0800	not including foundation, pits									
1550	Digital, electronic, 100 ton capacity, steel deck 12′ x 10′ platform	3 Carp	.20	120	Ea.	14,100	5,625		19,725	24,200
1600	40′ x 10′ platform		.14	171		28,100	8,050		36,150	43,300
1640	60′ x 10′ platform		.13	184		37,100	8,675		45,775	54,000
1680	70′ x 10′ platform	↓	.12	200		39,400	9,400		48,800	58,000
2000	For standard automatic printing device, add					1,250			1,250	1,375
2100	For remote reading electronic system, add					2,725			2,725	3,000
2300	Concrete foundation pits for above, 8′ x 6′, 5 C.Y. required	C-1	.50	64		1,075	2,850		3,925	5,575
2400	14′ x 6′ platform, 10 C.Y. required		.35	91.429		1,575	4,075		5,650	8,000
2600	50′ x 10′ platform, 30 C.Y. required		.25	128		2,125	5,700		7,825	11,100
2700	70′ x 10′ platform, 40 C.Y. required	↓	.15	213		4,625	9,525		14,150	19,700
2750	Crane scales, dial, 1 ton capacity					1,150			1,150	1,275
2780	5 ton capacity					1,550			1,550	1,700
2800	Digital, 1 ton capacity					1,900			1,900	2,075
2850	10 ton capacity				↓	4,800			4,800	5,300
2900	Low profile electronic warehouse scale,									

10 88 05 – Commercial Scales

10 88 05.10 Scales		Crew	Daily Output	Labor-Hours	Unit	Material	Labor	Equipment	Total	Total Incl O&P
						2015 Bare Costs				
3000	not incl. printer, 4' x 4' platform, 10,000 lb. capacity	2 Carp	.30	53.333	Ea.	1,425	2,500		3,925	5,400
3300	5' x 7' platform, 10,000 lb. capacity		.25	64		4,825	3,000		7,825	9,925
3400	20,000 lb. capacity		.20	80		5,925	3,750		9,675	12,300
3500	For printers, incl. time, date & numbering, add					890			890	980
3800	Portable, beam type, capacity 1000#, platform 18" x 24"					785			785	860
3900	Dial type, capacity 2000#, platform 24" x 24"					1,425			1,425	1,550
4000	Digital type, capacity 1000#, platform 24" x 30"					2,275			2,275	2,500
4100	Portable contractor truck scales, 50 ton cap., 40' x 10' platform					33,600			33,600	37,000
4200	60' x 10' platform					31,600			31,600	34,800

Division Notes

	CREW	DAILY OUTPUT	LABOR-HOURS	UNIT	BARE COSTS MAT.	BARE COSTS LABOR	BARE COSTS EQUIP.	BARE COSTS TOTAL	TOTAL INCL O&P

Estimating Tips

General

- The items in this division are usually priced per square foot or each. Many of these items are purchased by the owner for installation by the contractor. Check the specifications for responsibilities and include time for receiving, storage, installation, and mechanical and electrical hookups in the appropriate divisions.
- Many items in Division 11 require some type of support system that is not usually furnished with the item. Examples of these systems include blocking for the attachment of casework and support angles for ceiling-hung projection screens. The required blocking or supports must be added to the estimate in the appropriate division.
- Some items in Division 11 may require assembly or electrical hookups. Verify the amount of assembly required or the need for a hard electrical connection and add the appropriate costs.

Reference Numbers

Reference numbers are shown in shaded boxes at the beginning of some major classifications. These numbers refer to related items in the Reference Section. The reference information may be an estimating procedure, an alternate pricing method, or technical information.

Note: Not all subdivisions listed here necessarily appear in this publication. ■

11 05 Common Work Results for Equipment

11 05 05 – Selective Demolition for Equipment

11 05 05.10 Selective Demolition		Crew	Daily Output	Labor-Hours	Unit	2015 Bare Costs Material	Labor	Equipment	Total	Total Incl O&P
0010	**SELECTIVE DEMOLITION**									
0130	Central vacuum, motor unit, residential or commercial	1 Clab	2	4	Ea.		150		150	231
0210	Vault door and frame	2 Skwk	2	8			390		390	600
0215	Day gate, for vault	″	3	5.333			259		259	400
0380	Bank equipment, teller window, bullet resistant	1 Clab	1.20	6.667			251		251	385
0381	Counter	2 Clab	1.50	10.667	Station		400		400	615
0382	Drive-up window, including drawer and glass		1.50	10.667	″		400		400	615
0383	Thru-wall boxes and chests, selective demolition		2.50	6.400	Ea.		241		241	370
0384	Bullet resistant partitions		20	.800	L.F.		30		30	46.50
0385	Pneumatic tube system, 2 lane drive-up	L-3	.45	35.556	Ea.		1,825		1,825	2,775
0386	Safety deposit box	1 Clab	50	.160	Opng.		6		6	9.25
0387	Surveillance system, video, complete	2 Elec	2	8	Ea.		440		440	655
0410	Church equipment, misc moveable fixtures	2 Clab	1	16			600		600	925
0412	Steeple, to 28′ high	F-3	3	13.333			640	218	858	1,225
0414	40′ to 60′ high	″	.80	50			2,400	820	3,220	4,575
0510	Library equipment, bookshelves, wood, to 90″ high	1 Clab	20	.400	L.F.		15.05		15.05	23
0515	Carrels, hardwood, 36″ x 24″	″	9	.889	Ea.		33.50		33.50	51.50
0630	Stage equipment, light control panel	1 Elec	1	8	″		440		440	655
0632	Border lights		40	.200	L.F.		10.95		10.95	16.40
0634	Spotlights		8	1	Ea.		54.50		54.50	82
0636	Telescoping platforms and risers	2 Clab	175	.091	SF Stg.		3.44		3.44	5.30
1020	Barber equipment, hydraulic chair	1 Clab	40	.200	Ea.		7.50		7.50	11.55
1030	Checkout counter, supermarket or warehouse conveyor	2 Clab	18	.889			33.50		33.50	51.50
1040	Food cases, refrigerated or frozen	Q-5	6	2.667			143		143	217
1190	Laundry equipment, commercial	L-6	3	4			229		229	345
1360	Movie equipment, lamphouse, to 4000 watt, incl rectifier	1 Elec	4	2			109		109	164
1365	Sound system, incl amplifier	″	1.25	6.400			350		350	525
1410	Air compressor, to 5 H.P.	2 Clab	2.50	6.400			241		241	370
1412	Lubrication equipment, automotive, 3 reel type, incl pump, excl piping	L-4	1	24	Set		1,050		1,050	1,650
1414	Booth, spray paint, complete, to 26′ long	″	.80	30	″		1,325		1,325	2,050
1560	Parking equipment, cashier booth	B-22	2	15	Ea.		660	103	763	1,125
1600	Loading dock equipment, dock bumpers, rubber	1 Clab	50	.160	″		6		6	9.25
1610	Door seal for door perimeter	″	50	.160	L.F.		6		6	9.25
1620	Platform lifter, fixed, 6′ x 8′, 5000 lb. capacity	E-16	1.50	10.667	Ea.		570	97.50	667.50	1,100
1630	Dock leveller	″	2	8			430	73	503	825
1640	Lights, single or double arm	1 Elec	8	1			54.50		54.50	82
1650	Shelter, fabric, truck or train	1 Clab	1.50	5.333			201		201	310
1790	Waste handling equipment, commercial compactor	L-4	2	12			530		530	820
1792	Commercial or municipal incinerator, gas	″	2	12			530		530	820
1795	Crematory, excluding building	Q-3	.25	128			7,150		7,150	10,800
1910	Detection equipment, cell bar front	E-4	4	8			425	36.50	461.50	780
1912	Cell door and frame		8	4			213	18.25	231.25	390
1914	Prefab cell, 4′ to 5′ wide, 7′ to 8′ high, 7′ deep		8	4			213	18.25	231.25	390
1916	Cot, bolted, single		40	.800			42.50	3.65	46.15	78
1918	Visitor cubicle		4	8			425	36.50	461.50	780
2850	Hydraulic gates, canal, flap, knife, slide or sluice, to 18″ diameter	L-5A	8	4			212	75	287	440
2852	19″ to 36″ diameter		6	5.333			282	100	382	585
2854	37″ to 48″ diameter		2	16			845	300	1,145	1,750
2856	49″ to 60″ diameter		1	32			1,700	600	2,300	3,500
2858	Over 60″ diameter		.30	106			5,650	2,000	7,650	11,700
3100	Sewage pumping system, prefabricated, to 1000 GPM	C-17D	.20	420			20,700	4,050	24,750	36,400
3110	Sewage treatment, holding tank for recirc chemical water closet	1 Plum	8	1			58.50		58.50	88.50
3900	Wastewater treatment system, to 1500 gallons	B-21	2	14			605	69	674	1,000

11 05 Common Work Results for Equipment

11 05 05 – Selective Demolition for Equipment

11 05 05.10 Selective Demolition		Crew	Daily Output	Labor-Hours	Unit	Material	2015 Bare Costs Labor	Equipment	Total	Total Incl O&P
4050	Food storage equipment, walk-in refrigerator/freezer	2 Clab	64	.250	S.F.		9.40		9.40	14.45
4052	Shelving, stainless steel, 4 tier or dunnage rack	1 Clab	12	.667	Ea.		25		25	38.50
4100	Food preparation equipment, small countertop		18	.444			16.70		16.70	25.50
4150	Food delivery carts, heated cabinets		18	.444			16.70		16.70	25.50
4200	Cooking equipment, commercial range	Q-1	12	1.333			70.50		70.50	106
4250	Hood and ventilation equipment, kitchen exhaust hood, excl fire prot	1 Clab	3	2.667			100		100	154
4255	Fire protection system	Q-1	3	5.333			282		282	425
4300	Food dispensing equipment, countertop items	1 Clab	15	.533			20		20	31
4310	Serving counter	"	65	.123	L.F.		4.63		4.63	7.10
4350	Ice machine, ice cube maker, flakers and storage bins, to 2000 lb./day	Q-1	1.60	10	Ea.		530		530	800
4400	Cleaning and disposal, commercial dishwasher, to 50 racks per hour	L-6	1	12			690		690	1,025
4405	To 275 racks per hour	L-4	1	24			1,050		1,050	1,650
4410	Dishwasher hood	2 Clab	5	3.200			120		120	185
4420	Garbage disposal, commercial, to 5 H.P.	L-1	8	2			113		113	171
4540	Water heater, residential, to 80 gal/day	"	5	3.200			181		181	273
4542	Water softener, automatic	2 Plum	10	1.600			94		94	142
4544	Disappearing stairway, to 15′ floor height	2 Clab	6	2.667			100		100	154
4710	Darkroom equipment, light	L-7	10	2.800			127		127	195
4712	Heavy	"	1.50	18.667			845		845	1,300
4720	Doors	2 Clab	3.50	4.571	Opng.		172		172	264
4830	Bowling alley, complete, incl pinsetter, scorer, counters, misc supplies	4 Clab	.40	80	Lane		3,000		3,000	4,625
4840	Health club equipment, circuit training apparatus	2 Clab	2	8	Set		300		300	465
4842	Squat racks	"	10	1.600	Ea.		60		60	92.50
4860	School equipment, basketball backstop	L-2	2	8			330		330	510
4862	Table and benches, folding, in wall, 14′ long	L-4	4	6			266		266	410
4864	Bleachers, telescoping, to 30 tier	F-5	120	.267	Seat		12.65		12.65	19.45
4866	Boxing ring, elevated	L-4	.20	120	Ea.		5,300		5,300	8,200
4867	Boxing ring, floor level	"	2	12			530		530	820
4868	Exercise equipment	1 Clab	6	1.333			50		50	77
4870	Gym divider	L-4	1000	.024	S.F.		1.06		1.06	1.64
4875	Scoreboard	R-3	2	10	Ea.		545	69	614	890
4880	Shooting range, incl bullet traps, targets, excl structure	L-9	1	36	Point		1,550		1,550	2,450
5200	Vocational shop equipment	2 Clab	8	2	Ea.		75		75	116
6200	Fume hood, incl countertop, excl HVAC	"	6	2.667	L.F.		100		100	154
7100	Medical sterilizing, distiller, water, steam heated, 50 gal. capacity	1 Plum	2.80	2.857	Ea.		168		168	253
7200	Medical equipment, surgery table, minor	1 Clab	1	8			300		300	465
7210	Surgical lights, doctors office, single or double arm	2 Elec	3	5.333			292		292	435
7300	Physical therapy, table	2 Clab	4	4			150		150	231
7310	Whirlpool bath, fixed, incl mixing valves	1 Plum	4	2			117		117	177
7400	Dental equipment, chair, electric or hydraulic	1 Clab	.75	10.667			400		400	615
7410	Central suction system	1 Plum	2	4			235		235	355
7420	Drill console with accessories	1 Clab	3.20	2.500			94		94	145
7430	X-ray unit	"	4	2			75		75	116
7440	X-ray developer	1 Plum	10	.800			47		47	71

11 05 10 – Equipment Installation

11 05 10.10 Industrial Equipment Installation		Crew	Daily Output	Labor-Hours	Unit	Material	Labor	Equipment	Total	Total Incl O&P
0010	**INDUSTRIAL EQUIPMENT INSTALLATION**									
0020	Industrial equipment, minimum	E-2	12	4.667	Ton		241	126	367	545
0200	Maximum	"	2	28	"		1,450	755	2,205	3,250

11 11 Vehicle Service Equipment

11 11 13 – Compressed-Air Vehicle Service Equipment

11 11 13.10 Compressed Air Equipment		Crew	Daily Output	Labor-Hours	Unit	2015 Bare Costs Material	Labor	Equipment	Total	Total Incl O&P
0010	**COMPRESSED AIR EQUIPMENT**									
0030	Compressors, electric, 1-1/2 H.P., standard controls	L-4	1.50	16	Ea.	455	710		1,165	1,600
0550	Dual controls		1.50	16		810	710		1,520	2,000
0600	5 H.P., 115/230 volt, standard controls		1	24		2,575	1,050		3,625	4,475
0650	Dual controls		1	24		3,400	1,050		4,450	5,375

11 11 19 – Vehicle Lubrication Equipment

11 11 19.10 Lubrication Equipment		Crew	Daily Output	Labor-Hours	Unit	Material	Labor	Equipment	Total	Total Incl O&P
0010	**LUBRICATION EQUIPMENT**									
3000	Lube equipment, 3 reel type, with pumps, not including piping	L-4	.50	48	Set	8,800	2,125		10,925	13,000
3100	Hose reel, including hose, oil/lube, 1000 PSI	2 Sswk	2	8	Ea.	740	420		1,160	1,550
3200	Grease, 5000 PSI		2	8		800	420		1,220	1,600
3300	Air, 50 feet, 160 PSI		2	8		875	420		1,295	1,700
3350	25 feet, 160 PSI		2	8		525	420		945	1,300

11 11 33 – Vehicle Spray Painting Equipment

11 11 33.10 Spray Painting Equipment		Crew	Daily Output	Labor-Hours	Unit	Material	Labor	Equipment	Total	Total Incl O&P
0010	**SPRAY PAINTING EQUIPMENT**									
4000	Spray painting booth, 26' long, complete	L-4	.40	60	Ea.	16,400	2,650		19,050	22,100

11 12 Parking Control Equipment

11 12 13 – Parking Key and Card Control Units

11 12 13.10 Parking Control Units		Crew	Daily Output	Labor-Hours	Unit	Material	Labor	Equipment	Total	Total Incl O&P
0010	**PARKING CONTROL UNITS**									
5100	Card reader	1 Elec	2	4	Ea.	1,950	219		2,169	2,475
5120	Proximity with customer display	2 Elec	1	16		5,575	875		6,450	7,425
6000	Parking control software, basic functionality	1 Elec	.50	16		23,900	875		24,775	27,500
6020	multi-function	"	.20	40		104,500	2,200		106,700	118,500

11 12 16 – Parking Ticket Dispensers

11 12 16.10 Ticket Dispensers		Crew	Daily Output	Labor-Hours	Unit	Material	Labor	Equipment	Total	Total Incl O&P
0010	**TICKET DISPENSERS**									
5900	Ticket spitter with time/date stamp, standard	2 Elec	2	8	Ea.	6,200	440		6,640	7,475
5920	Mag stripe encoding	"	2	8	"	18,800	440		19,240	21,400

11 12 26 – Parking Fee Collection Equipment

11 12 26.13 Parking Fee Coin Collection Equipment		Crew	Daily Output	Labor-Hours	Unit	Material	Labor	Equipment	Total	Total Incl O&P
0010	**PARKING FEE COIN COLLECTION EQUIPMENT**									
5200	Cashier booth, average	B-22	1	30	Ea.	10,400	1,325	207	11,932	13,800
5300	Collector station, pay on foot	2 Elec	.20	80		111,000	4,375		115,375	128,500
5320	Credit card only	"	.50	32		20,500	1,750		22,250	25,100

11 12 26.23 Fee Equipment		Crew	Daily Output	Labor-Hours	Unit	Material	Labor	Equipment	Total	Total Incl O&P
0010	**FEE EQUIPMENT**									
5600	Fee computer	1 Elec	1.50	5.333	Ea.	14,200	292		14,492	16,000

11 12 33 – Parking Gates

11 12 33.13 Lift Arm Parking Gates		Crew	Daily Output	Labor-Hours	Unit	Material	Labor	Equipment	Total	Total Incl O&P
0010	**LIFT ARM PARKING GATES**									
5000	Barrier gate with programmable controller	2 Elec	3	5.333	Ea.	3,400	292		3,692	4,150
5020	Industrial		3	5.333		5,050	292		5,342	5,975
5050	Non-programmable, with reader and 12' arm		3	5.333		1,875	292		2,167	2,475
5500	Exit verifier		1	16		17,600	875		18,475	20,700
5700	Full sign, 4" letters	1 Elec	2	4		1,225	219		1,444	1,675

11 12 Parking Control Equipment

11 12 33 – Parking Gates

11 12 33.13 Lift Arm Parking Gates		Crew	Daily Output	Labor-Hours	Unit	Material	2015 Bare Costs Labor	Equipment	Total	Total Incl O&P
5800	Inductive loop	2 Elec	4	4	Ea.	171	219		390	520
5950	Vehicle detector, microprocessor based	1 Elec	3	2.667		425	146		571	690
7100	Traffic spike unit, flush mount, spring loaded, 72″ L	B-89	4	4		1,250	175	122	1,547	1,775
7200	Surface mount, 72″ L	2 Skwk	10	1.600		2,275	78		2,353	2,625

11 13 Loading Dock Equipment

11 13 13 – Loading Dock Bumpers

11 13 13.10 Dock Bumpers

		Crew	Daily Output	Labor-Hours	Unit	Material	Labor	Equipment	Total	Total Incl O&P
0010	**DOCK BUMPERS** Bolts not included									
0020	2″ x 6″ to 4″ x 8″, average	1 Carp	.30	26.667	M.B.F.	1,350	1,250		2,600	3,400
0050	Bumpers, rubber blocks 4-1/2″ thick, 10″ high, 14″ long		26	.308	Ea.	64	14.45		78.45	92
0200	24″ long		22	.364		96	17.05		113.05	133
0300	36″ long		17	.471		78	22		100	120
0500	12″ high, 14″ long		25	.320		101	15		116	135
0550	24″ long		20	.400		112	18.80		130.80	152
0600	36″ long		15	.533		125	25		150	176
0800	Rubber blocks 6″ thick, 10″ high, 14″ long		22	.364		104	17.05		121.05	141
0850	24″ long		18	.444		123	21		144	167
0900	36″ long		13	.615		188	29		217	252
0910	20″ high, 11″ long		13	.615		148	29		177	208
0920	Extruded rubber bumpers, T section, 22″ x 22″ x 3″ thick		41	.195		65.50	9.15		74.65	86
0940	Molded rubber bumpers, 24″ x 12″ x 3″ thick		20	.400		57.50	18.80		76.30	92.50
1000	Welded installation of above bumpers	E-14	8	1		3.78	54.50	18.25	76.53	119
1100	For drilled anchors, add per anchor	1 Carp	36	.222		6.80	10.45		17.25	23.50
1300	Steel bumpers, see Section 10 26 13.10									

11 13 16 – Loading Dock Seals and Shelters

11 13 16.10 Dock Seals and Shelters

		Crew	Daily Output	Labor-Hours	Unit	Material	Labor	Equipment	Total	Total Incl O&P
0010	**DOCK SEALS AND SHELTERS**									
3600	Door seal for door perimeter, 12″ x 12″, vinyl covered	1 Carp	26	.308	L.F.	32.50	14.45		46.95	57.50
3900	Folding gates, see Section 10 22 16.10									
6200	Shelters, fabric, for truck or train, scissor arms, minimum	1 Carp	1	8	Ea.	1,775	375		2,150	2,525
6300	Maximum	″	.50	16	″	2,425	750		3,175	3,825

11 13 19 – Stationary Loading Dock Equipment

11 13 19.10 Dock Equipment

		Crew	Daily Output	Labor-Hours	Unit	Material	Labor	Equipment	Total	Total Incl O&P
0010	**DOCK EQUIPMENT**									
2200	Dock boards, heavy duty, 60″ x 60″, aluminum, 5,000 lb. capacity				Ea.	1,450			1,450	1,600
2700	9,000 lb. capacity					1,625			1,625	1,775
3200	15,000 lb. capacity					1,750			1,750	1,925
4200	Platform lifter, 6′ x 6′, portable, 3,000 lb. capacity					9,250			9,250	10,200
4250	4,000 lb. capacity					11,400			11,400	12,500
4400	Fixed, 6′ x 8′, 5,000 lb. capacity	E-16	.70	22.857		9,975	1,225	208	11,408	13,400
4500	Levelers, hinged for trucks, 10 ton capacity, 6′ x 8′		1.08	14.815		5,025	795	135	5,955	7,075
4650	7′ x 8′		1.08	14.815		5,975	795	135	6,905	8,100
4670	Air bag power operated, 10 ton cap., 6′ x 8′		1.08	14.815		5,850	795	135	6,780	7,950
4680	7′ x 8′		1.08	14.815		5,875	795	135	6,805	7,975
4700	Hydraulic, 10 ton capacity, 6′ x 8′		1.08	14.815		8,750	795	135	9,680	11,200
4800	7′ x 8′		1.08	14.815		9,425	795	135	10,355	11,900
5800	Loading dock safety restraints, manual style		1.08	14.815		3,200	795	135	4,130	5,050
5900	Automatic style		1.08	14.815		5,400	795	135	6,330	7,450
6000	Dock leveler, 15 ton capacity									

11 13 Loading Dock Equipment

11 13 19 – Stationary Loading Dock Equipment

11 13 19.10 Dock Equipment		Crew	Daily Output	Labor-Hours	Unit	Material	2015 Bare Costs Labor	Equipment	Total	Total Incl O&P
6100	I beam construction, mechanical, 6′ x 8′	E-16	.50	32	Ea.	4,400	1,725	292	6,417	8,125
6150	Hydraulic		.50	32		5,225	1,725	292	7,242	9,050
6200	Formed beam deck construction, mechanical, 6′ x 8′		.50	32		3,250	1,725	292	5,267	6,875
6250	Hydraulic		.50	32		4,450	1,725	292	6,467	8,200
6300	Edge of dock leveler, mechanical, 15 ton capacity	↓	2	8	↓	970	430	73	1,473	1,900
7000	22.5 ton capacity									
7100	Vertical storing dock leveler, hydraulic, 6′ x 6′	E-16	.40	40	Ea.	7,200	2,150	365	9,715	12,100

11 13 26 – Loading Dock Lights

11 13 26.10 Dock Lights		Crew	Daily Output	Labor-Hours	Unit	Material	Labor	Equipment	Total	Total Incl O&P
0010	**DOCK LIGHTS**									
5000	Lights for loading docks, single arm, 24″ long	1 Elec	3.80	2.105	Ea.	138	115		253	325
5700	Double arm, 60″ long	″	3.80	2.105	″	211	115		326	405

11 14 Pedestrian Control Equipment

11 14 13 – Pedestrian Gates

11 14 13.13 Portable Posts and Railings		Crew	Daily Output	Labor-Hours	Unit	Material	Labor	Equipment	Total	Total Incl O&P
0010	**PORTABLE POSTS AND RAILINGS**									
0020	Portable for pedestrian traffic control, standard				Ea.	134			134	147
0300	Deluxe posts				″	210			210	231
0600	Ropes for above posts, plastic covered, 1-1/2″ diameter				L.F.	17.25			17.25	19
0700	Chain core				″	11.65			11.65	12.80
1500	Portable security or safety barrier, black with 7′ yellow strap				Ea.	213			213	234
1510	12′ yellow strap					238			238	262
1550	Sign holder, standard design				↓	76			76	83.50

11 14 13.19 Turnstiles		Crew	Daily Output	Labor-Hours	Unit	Material	Labor	Equipment	Total	Total Incl O&P
0010	**TURNSTILES**									
0020	One way, 4 arm, 46″ diameter, economy, manual	2 Carp	5	3.200	Ea.	1,700	150		1,850	2,100
0100	Electric		1.20	13.333		2,050	625		2,675	3,250
0300	High security, galv., 5′-5″ diameter, 7′ high, manual		1	16		5,875	750		6,625	7,625
0350	Electric		.60	26.667		8,225	1,250		9,475	11,000
0420	Three arm, 24″ opening, light duty, manual		2	8		3,500	375		3,875	4,425
0450	Heavy duty		1.50	10.667		5,000	500		5,500	6,275
0460	Manual, with registering & controls, light duty		2	8		3,750	375		4,125	4,725
0470	Heavy duty		1.50	10.667		4,100	500		4,600	5,275
0480	Electric, heavy duty	↓	1.10	14.545		4,775	685		5,460	6,300
0500	For coin or token operating, add				↓	710			710	780
1200	One way gate with horizontal bars, 5′-5″ diameter									
1300	7′ high, recreation or transit type	2 Carp	.80	20	Ea.	5,350	940		6,290	7,350
1500	For electronic counter, add				″	211			211	232

11 21 Retail and Service Equipment

11 21 13 – Cash Registers and Checking Equipment

11 21 13.10 Checkout Counter		Crew	Daily Output	Labor-Hours	Unit	Material	2015 Bare Costs Labor	Equipment	Total	Total Incl O&P
0010	CHECKOUT COUNTER									
0020	Supermarket conveyor, single belt	2 Clab	10	1.600	Ea.	3,175	60		3,235	3,600
0100	Double belt, power take-away		9	1.778		4,575	67		4,642	5,125
0400	Double belt, power take-away, incl. side scanning		7	2.286		5,375	86		5,461	6,025
0800	Warehouse or bulk type	↓	6	2.667	↓	6,350	100		6,450	7,125
1000	Scanning system, 2 lanes, w/registers, scan gun & memory				System	16,600			16,600	18,300
1100	10 lanes, single processor, full scan, with scales				"	158,000			158,000	173,500
2000	Register, restaurant, minimum				Ea.	740			740	815
2100	Maximum					3,125			3,125	3,425
2150	Store, minimum					740			740	815
2200	Maximum				↓	3,125			3,125	3,425

11 21 33 – Checkroom Equipment

11 21 33.10 Clothes Check Equipment		Crew	Daily Output	Labor-Hours	Unit	Material	Labor	Equipment	Total	Total Incl O&P
0010	CLOTHES CHECK EQUIPMENT									
0030	Clothes check rack, free standing, st. stl., 2-tier, 90 bag capacity				Ea.	1,525			1,525	1,700
0050	Wall mounted, 45 bag capacity	L-2	8	2		750	82.50		832.50	950
0100	Garment checking bag, green mesh fabric, 21" H x 17" W with 4.5" hook				↓	19.10			19.10	21

11 21 53 – Barber and Beauty Shop Equipment

11 21 53.10 Barber Equipment		Crew	Daily Output	Labor-Hours	Unit	Material	Labor	Equipment	Total	Total Incl O&P
0010	BARBER EQUIPMENT									
0020	Chair, hydraulic, movable, minimum	1 Carp	24	.333	Ea.	560	15.65		575.65	640
0050	Maximum	"	16	.500		3,475	23.50		3,498.50	3,850
0200	Wall hung styling station with mirrors, minimum	L-2	8	2		460	82.50		542.50	630
0300	Maximum	"	4	4		2,325	165		2,490	2,800
0500	Sink, hair washing basin, rough plumbing not incl.	1 Plum	8	1		495	58.50		553.50	635
1000	Sterilizer, liquid solution for tools					161			161	177
1100	Total equipment, rule of thumb, per chair, minimum	L-8	1	20		1,925	985		2,910	3,625
1150	Maximum	"	1	20	↓	5,150	985		6,135	7,175

11 21 73 – Commercial Laundry and Dry Cleaning Equipment

11 21 73.13 Dry Cleaning Equipment		Crew	Daily Output	Labor-Hours	Unit	Material	Labor	Equipment	Total	Total Incl O&P
0010	DRY CLEANING EQUIPMENT									
2000	Dry cleaners, electric, 20 lb. capacity, not incl. rough-in	L-1	.20	80	Ea.	33,700	4,525		38,225	43,800
2050	25 lb. capacity		.17	94.118		48,600	5,325		53,925	61,500
2100	30 lb. capacity		.15	106		51,000	6,050		57,050	65,500
2150	60 lb. capacity	↓	.09	177	↓	79,000	10,100		89,100	102,000

11 21 73.16 Drying and Conditioning Equipment		Crew	Daily Output	Labor-Hours	Unit	Material	Labor	Equipment	Total	Total Incl O&P
0010	DRYING AND CONDITIONING EQUIPMENT									
0100	Dryers, Not including rough-in									
1500	Industrial, 30 lb. capacity	1 Plum	2	4	Ea.	3,175	235		3,410	3,825
1600	50 lb. capacity	"	1.70	4.706		3,400	276		3,676	4,175
4700	Lint collector, ductwork not included, 8,000 to 10,000 CFM	Q-10	.30	80	↓	9,200	4,175		13,375	16,500

11 21 73.19 Finishing Equipment		Crew	Daily Output	Labor-Hours	Unit	Material	Labor	Equipment	Total	Total Incl O&P
0010	FINISHING EQUIPMENT									
3500	Folders, blankets & sheets, minimum	1 Elec	.17	47.059	Ea.	33,100	2,575		35,675	40,300
3700	King size with automatic stacker		.10	80		60,000	4,375		64,375	72,500
3800	For conveyor delivery, add	↓	.45	17.778		14,500	970		15,470	17,400
4900	Spreader feeders, 240V, 2 station	L-6	.70	17.143		56,000	985		56,985	63,500
4920	4 station	"	.35	34.286	↓	68,000	1,975		69,975	78,000

11 21 Retail and Service Equipment

11 21 73 – Commercial Laundry and Dry Cleaning Equipment

	11 21 73.23 Commercial Ironing Equipment	Crew	Daily Output	Labor-Hours	Unit	2015 Bare Costs Material	Labor	Equipment	Total	Total Incl O&P
0010	**COMMERCIAL IRONING EQUIPMENT**									
4500	Ironers, institutional, 110″, single roll	1 Elec	.20	40	Ea.	32,100	2,200		34,300	38,600
4800	Pressers, low capacity air operated	L-6	1.75	6.857		9,650	395		10,045	11,200
4820	Hand operated		1.75	6.857		8,900	395		9,295	10,400
4840	Ironer 48″, 240V		3.50	3.429		115,000	197		115,197	127,000
6600	Hand operated presser		.70	17.143		6,175	985		7,160	8,275
6620	Mushroom press 115 V		.70	17.143		7,700	985		8,685	9,950

11 21 73.26 Commercial Washers and Extractors

		Crew	Daily Output	Labor-Hours	Unit	Material	Labor	Equipment	Total	Total Incl O&P
0010	**COMMERCIAL WASHERS AND EXTRACTORS**, not including rough-in									
6000	Combination washer/extractor, 20 lb. capacity	L-6	1.50	8	Ea.	5,950	460		6,410	7,225
6100	30 lb. capacity		.80	15		9,275	860		10,135	11,500
6200	50 lb. capacity		.68	17.647		10,900	1,000		11,900	13,500
6300	75 lb. capacity		.30	40		20,400	2,300		22,700	26,000
6350	125 lb. capacity		.16	75		27,700	4,300		32,000	36,900
6380	Washer extractor/dryer, 110 lb., 240V		1	12		8,900	690		9,590	10,800
6400	Washer extractor, 135 lb., 240V		1	12		25,900	690		26,590	29,500
6450	Pass through		1	12		69,500	690		70,190	77,500
6500	200 lb. washer extractor		1	12		72,000	690		72,690	80,000
6550	Pass through		1	12		75,500	690		76,190	84,000
6600	Extractor, low capacity		1.75	6.857		7,375	395		7,770	8,700

11 21 73.33 Coin-Operated Laundry Equipment

		Crew	Daily Output	Labor-Hours	Unit	Material	Labor	Equipment	Total	Total Incl O&P
0010	**COIN-OPERATED LAUNDRY EQUIPMENT**									
0990	Dryer, gas fired									
1000	Commercial, 30 lb. capacity, coin operated, single	1 Plum	3	2.667	Ea.	3,300	157		3,457	3,850
1100	Double stacked	″	2	4		7,275	235		7,510	8,350
4860	Coin dry cleaner 20 lb.	L-6	1.75	6.857		28,800	395		29,195	32,300
5290	Clothes washer									
5300	Commercial, coin operated, average	1 Plum	3	2.667	Ea.	1,250	157		1,407	1,600

11 21 83 – Photo Processing Equipment

11 21 83.13 Darkroom Equipment

		Crew	Daily Output	Labor-Hours	Unit	Material	Labor	Equipment	Total	Total Incl O&P
0010	**DARKROOM EQUIPMENT**									
0020	Developing sink, 5″ deep, 24″ x 48″	Q-1	2	8	Ea.	4,150	425		4,575	5,225
0050	48″ x 52″		1.70	9.412		4,425	495		4,920	5,625
0200	10″ deep, 24″ x 48″		1.70	9.412		1,550	495		2,045	2,450
0250	24″ x 108″		1.50	10.667		3,650	565		4,215	4,875
0500	Dryers, dehumidified filtered air, 36″ x 25″ x 68″ high	L-7	6	4.667		4,300	212		4,512	5,050
0550	48″ x 25″ x 68″ high		5	5.600		10,100	254		10,354	11,500
2000	Processors, automatic, color print, minimum		4	7		15,900	320		16,220	18,000
2050	Maximum		.60	46.667		25,000	2,125		27,125	30,800
2300	Black and white print, minimum		2	14		12,100	635		12,735	14,300
2350	Maximum		.80	35		62,000	1,600		63,600	70,500
2600	Manual processor, 16″ x 20″ maximum print size		2	14		9,575	635		10,210	11,500
2650	20″ x 24″ maximum print size		1	28		9,100	1,275		10,375	12,000
3000	Viewing lights, 20″ x 24″		6	4.667		288	212		500	640
3100	20″ x 24″ with color correction		6	4.667		345	212		557	705
3500	Washers, round, minimum sheet 11″ x 14″	Q-1	2	8		3,275	425		3,700	4,250
3550	Maximum sheet 20″ x 24″		1	16		3,600	845		4,445	5,250
3800	Square, minimum sheet 20″ x 24″		1	16		3,150	845		3,995	4,750
3900	Maximum sheet 50″ x 56″		.80	20		4,900	1,050		5,950	7,000
4500	Combination tank sink, tray sink, washers, with									
4510	Dry side tables, average	Q-1	.45	35.556	Ea.	10,800	1,875		12,675	14,600

11 22 Banking Equipment

11 22 13 – Vault Equipment

11 22 13.16 Safes		Crew	Daily Output	Labor-Hours	Unit	Material	2015 Bare Costs Labor	Equipment	Total	Total Incl O&P
0010	**SAFES**									
0200	Office, 1 hr. rating, 30" x 18" x 18"				Ea.	2,275			2,275	2,500
0250	40" x 18" x 18"					4,875			4,875	5,350
0300	60" x 36" x 18", double door					7,375			7,375	8,125
0600	Data, 1 hr. rating, 27" x 19" x 16"					4,550			4,550	5,025
0700	63" x 34" x 16"					14,500			14,500	16,000
0750	Diskette, 1 hr., 14" x 12" x 11", inside					4,050			4,050	4,450
0800	Money, "B" label, 9" x 14" x 14"					540			540	595
0900	Tool resistive, 24" x 24" x 20"					4,150			4,150	4,550
1050	Tool and torch resistive, 24" x 24" x 20"					7,875			7,875	8,675
1150	Jewelers, 23" x 20" x 18"					8,775			8,775	9,650
1200	63" x 25" x 18"					14,300			14,300	15,800
1300	For handling into building, add, minimum	A-2	8.50	2.824			108	29	137	197
1400	Maximum	"	.78	30.769			1,175	315	1,490	2,150

11 22 16 – Teller and Service Equipment

11 22 16.13 Teller Equipment Systems		Crew	Daily Output	Labor-Hours	Unit	Material	2015 Bare Costs Labor	Equipment	Total	Total Incl O&P
0010	**TELLER EQUIPMENT SYSTEMS**									
0020	Alarm system, police	2 Elec	1.60	10	Ea.	4,975	545		5,520	6,300
0100	With vault alarm	"	.40	40		19,700	2,200		21,900	25,000
0400	Bullet resistant teller window, 44" x 60"	1 Glaz	.60	13.333		3,850	600		4,450	5,175
0500	48" x 60"	"	.60	13.333		5,600	600		6,200	7,075
3000	Counters for banks, frontal only	2 Carp	1	16	Station	1,850	750		2,600	3,175
3100	Complete with steel undercounter	"	.50	32	"	3,625	1,500		5,125	6,275
4600	Door and frame, bullet-resistant, with vision panel, minimum	2 Sswk	1.10	14.545	Ea.	5,575	765		6,340	7,450
4700	Maximum		1.10	14.545		7,550	765		8,315	9,625
4800	Drive-up window, drawer & mike, not incl. glass, minimum		1	16		6,900	840		7,740	9,075
4900	Maximum		.50	32		9,225	1,675		10,900	13,000
5000	Night depository, with chest, minimum		1	16		7,550	840		8,390	9,775
5100	Maximum		.50	32		10,700	1,675		12,375	14,700
5200	Package receiver, painted		3.20	5		1,375	263		1,638	1,975
5300	Stainless steel		3.20	5		2,350	263		2,613	3,025
5400	Partitions, bullet-resistant, 1-3/16" glass, 8' high	2 Carp	10	1.600	L.F.	201	75		276	335
5450	Acrylic	"	10	1.600	"	380	75		455	535
5500	Pneumatic tube systems, 2 lane drive-up, complete	L-3	.25	64	Total	26,100	3,275		29,375	33,800
5550	With T.V. viewer	"	.20	80	"	50,500	4,100		54,600	62,000
5570	Safety deposit boxes, minimum	1 Sswk	44	.182	Opng.	58	9.55		67.55	80.50
5580	Maximum, 10" x 15" opening		19	.421		123	22		145	174
5590	Teller locker, average		15	.533		1,600	28		1,628	1,800
5600	Pass thru, bullet-res. window, painted steel, 24" x 36"	2 Sswk	1.60	10	Ea.	2,650	525		3,175	3,850
5700	48" x 48"		1.20	13.333		2,700	700		3,400	4,200
5800	72" x 40"		.80	20		4,250	1,050		5,300	6,500
5900	For stainless steel frames, add					20%				
6100	Surveillance system, video camera, complete	2 Elec	1	16	Ea.	9,675	875		10,550	12,000
6110	For each additional camera, add				"	1,000			1,000	1,100
6120	CCTV system, see Section 27 41 33.10									
6200	Twenty-four hour teller, single unit,									
6300	automated deposit, cash and memo	L-3	.25	64	Ea.	45,200	3,275		48,475	54,500
7000	Vault front, see Section 08 34 59.10									

11 30 13 – Residential Appliances

11 30 13.15 Cooking Equipment		Crew	Daily Output	Labor-Hours	Unit	Material	Labor	Equipment	Total	Total Incl O&P
						2015 Bare Costs				
0010	**COOKING EQUIPMENT**									
0020	Cooking range, 30" free standing, 1 oven, minimum	2 Clab	10	1.600	Ea.	440	60		500	580
0050	Maximum		4	4		2,025	150		2,175	2,450
0150	2 oven, minimum		10	1.600		1,225	60		1,285	1,425
0200	Maximum		10	1.600		2,550	60		2,610	2,900
0350	Built-in, 30" wide, 1 oven, minimum	1 Elec	6	1.333		680	73		753	860
0400	Maximum	2 Carp	2	8		1,350	375		1,725	2,075
0500	2 oven, conventional, minimum		4	4		1,325	188		1,513	1,750
0550	1 conventional, 1 microwave, maximum		2	8		1,800	375		2,175	2,550
0700	Free-standing, 1 oven, 21" wide range, minimum	2 Clab	10	1.600		455	60		515	595
0750	21" wide, maximum	"	4	4		450	150		600	725
0900	Countertop cooktops, 4 burner, standard, minimum	1 Elec	6	1.333		296	73		369	435
0950	Maximum		3	2.667		1,300	146		1,446	1,675
1050	As above, but with grill and griddle attachment, minimum		6	1.333		1,250	73		1,323	1,475
1100	Maximum		3	2.667		3,700	146		3,846	4,300
1200	Induction cooktop, 30" wide		3	2.667		1,200	146		1,346	1,550
1250	Microwave oven, minimum		4	2		120	109		229	296
1300	Maximum		2	4		465	219		684	840

11 30 13.16 Refrigeration Equipment		Crew	Daily Output	Labor-Hours	Unit	Material	Labor	Equipment	Total	Total Incl O&P
0010	**REFRIGERATION EQUIPMENT**									
2000	Deep freeze, 15 to 23 C.F., minimum	2 Clab	10	1.600	Ea.	610	60		670	765
2050	Maximum		5	3.200		820	120		940	1,075
2200	30 C.F., minimum		8	2		705	75		780	890
2250	Maximum		3	5.333		910	201		1,111	1,300
5200	Icemaker, automatic, 20 lb. per day	1 Plum	7	1.143		895	67		962	1,075
5350	51 lb. per day	"	2	4		1,575	235		1,810	2,100
5500	Refrigerator, no frost, 10 C.F. to 12 C.F., minimum	2 Clab	10	1.600		425	60		485	560
5600	Maximum		6	2.667		450	100		550	650
5750	14 C.F. to 16 C.F., minimum		9	1.778		505	67		572	660
5800	Maximum		5	3.200		620	120		740	870
5950	18 C.F. to 20 C.F., minimum		8	2		630	75		705	805
6000	Maximum		4	4		1,425	150		1,575	1,800
6150	21 C.F. to 29 C.F., minimum		7	2.286		965	86		1,051	1,200
6200	Maximum		3	5.333		3,375	201		3,576	4,000
6790	Energy-star qualified, 18 C.F., minimum G	2 Carp	4	4		510	188		698	850
6795	Maximum G		2	8		1,175	375		1,550	1,850
6797	21.7 C.F., minimum G		4	4		1,025	188		1,213	1,425
6799	Maximum G		4	4		2,100	188		2,288	2,625

11 30 13.17 Kitchen Cleaning Equipment		Crew	Daily Output	Labor-Hours	Unit	Material	Labor	Equipment	Total	Total Incl O&P
0010	**KITCHEN CLEANING EQUIPMENT**									
2750	Dishwasher, built-in, 2 cycles, minimum	L-1	4	4	Ea.	238	227		465	600
2800	Maximum		2	8		435	455		890	1,150
2950	4 or more cycles, minimum		4	4		375	227		602	755
2960	Average		4	4		500	227		727	890
3000	Maximum		2	8		1,100	455		1,555	1,875
3100	Energy-star qualified, minimum G		4	4		370	227		597	745
3110	Maximum G		2	8		1,500	455		1,955	2,325

11 30 13.18 Waste Disposal Equipment		Crew	Daily Output	Labor-Hours	Unit	Material	Labor	Equipment	Total	Total Incl O&P
0010	**WASTE DISPOSAL EQUIPMENT**									
1750	Compactor, residential size, 4 to 1 compaction, minimum	1 Carp	5	1.600	Ea.	645	75		720	825
1800	Maximum	"	3	2.667		1,025	125		1,150	1,325
3300	Garbage disposal, sink type, minimum	L-1	10	1.600		87.50	90.50		178	232

11 30 Residential Equipment

11 30 13 – Residential Appliances

11 30 13.18 Waste Disposal Equipment			Crew	Daily Output	Labor-Hours	Unit	Material	2015 Bare Costs Labor	Equipment	Total	Total Incl O&P
3350	Maximum		L-1	10	1.600	Ea.	220	90.50		310.50	380

11 30 13.19 Kitchen Ventilation Equipment

			Crew	Daily Output	Labor-Hours	Unit	Material	Labor	Equipment	Total	Total Incl O&P
0010	**KITCHEN VENTILATION EQUIPMENT**										
4150	Hood for range, 2 speed, vented, 30" wide, minimum		L-3	5	3.200	Ea.	70.50	164		234.50	330
4200	Maximum			3	5.333		780	273		1,053	1,275
4300	42" wide, minimum			5	3.200		168	164		332	435
4330	Custom			5	3.200		1,650	164		1,814	2,075
4350	Maximum			3	5.333		2,025	273		2,298	2,650
4500	For ventless hood, 2 speed, add						18.65			18.65	20.50
4650	For vented 1 speed, deduct from maximum						50			50	55

11 30 13.24 Washers

			Crew	Daily Output	Labor-Hours	Unit	Material	Labor	Equipment	Total	Total Incl O&P
0010	**WASHERS**										
5000	Residential, 4 cycle, average		1 Plum	3	2.667	Ea.	875	157		1,032	1,200
6650	Washing machine, automatic, minimum			3	2.667		485	157		642	765
6700	Maximum			1	8		1,525	470		1,995	2,375
6750	Energy star qualified, front loading, minimum	G		3	2.667		655	157		812	955
6760	Maximum	G		1	8		1,600	470		2,070	2,450
6764	Top loading, minimum	G		3	2.667		450	157		607	730
6766	Maximum	G		3	2.667		1,250	157		1,407	1,600

11 30 13.25 Dryers

			Crew	Daily Output	Labor-Hours	Unit	Material	Labor	Equipment	Total	Total Incl O&P
0010	**DRYERS**										
0500	Gas fired residential, 16 lb. capacity, average		1 Plum	3	2.667	Ea.	675	157		832	975
6770	Electric, front loading, energy-star qualified, minimum	G	L-2	3	5.333		385	220		605	765
6780	Maximum	G	"	2	8		1,925	330		2,255	2,600
7450	Vent kits for dryers		1 Carp	10	.800		39	37.50		76.50	101

11 30 15 – Miscellaneous Residential Appliances

11 30 15.13 Sump Pumps

		Crew	Daily Output	Labor-Hours	Unit	Material	Labor	Equipment	Total	Total Incl O&P
0010	**SUMP PUMPS**									
6400	Cellar drainer, pedestal, 1/3 H.P., molded PVC base	1 Plum	3	2.667	Ea.	135	157		292	385
6450	Solid brass	"	2	4	"	289	235		524	675
6460	Sump pump, see also Section 22 14 29.16									

11 30 15.23 Water Heaters

		Crew	Daily Output	Labor-Hours	Unit	Material	Labor	Equipment	Total	Total Incl O&P
0010	**WATER HEATERS**									
6900	Electric, glass lined, 30 gallon, minimum	L-1	5	3.200	Ea.	430	181		611	745
6950	Maximum		3	5.333		595	300		895	1,100
7100	80 gallon, minimum		2	8		1,225	455		1,680	2,025
7150	Maximum		1	16		1,700	905		2,605	3,225
7180	Gas, glass lined, 30 gallon, minimum	2 Plum	5	3.200		805	188		993	1,175
7220	Maximum		3	5.333		1,125	315		1,440	1,700
7260	50 gallon, minimum		2.50	6.400		845	375		1,220	1,500
7300	Maximum		1.50	10.667		1,175	625		1,800	2,225
7310	Water heater, see also Section 22 33 30.13									

11 30 15.43 Air Quality

		Crew	Daily Output	Labor-Hours	Unit	Material	Labor	Equipment	Total	Total Incl O&P
0010	**AIR QUALITY**									
2450	Dehumidifier, portable, automatic, 15 pint	1 Elec	4	2	Ea.	152	109		261	330
2550	40 pint		3.75	2.133		209	117		326	405
3550	Heater, electric, built-in, 1250 watt, ceiling type, minimum		4	2		107	109		216	282
3600	Maximum		3	2.667		175	146		321	410
3700	Wall type, minimum		4	2		172	109		281	355
3750	Maximum		3	2.667		185	146		331	420
3900	1500 watt wall type, with blower		4	2		172	109		281	355

11 30 Residential Equipment

11 30 15 – Miscellaneous Residential Appliances

11 30 15.43 Air Quality		Crew	Daily Output	Labor-Hours	Unit	Material	2015 Bare Costs Labor	Equipment	Total	Total Incl O&P
3950	3000 watt	1 Elec	3	2.667	Ea.	350	146		496	605
4850	Humidifier, portable, 8 gallons per day					133			133	146
5000	15 gallons per day					211			211	232

11 30 33 – Retractable Stairs

11 30 33.10 Disappearing Stairway		Crew	Daily Output	Labor-Hours	Unit	Material	Labor	Equipment	Total	Total Incl O&P
0010	**DISAPPEARING STAIRWAY** No trim included									
0100	Custom grade, pine, 8'-6" ceiling, minimum	1 Carp	4	2	Ea.	177	94		271	340
0150	Average		3.50	2.286		253	107		360	445
0200	Maximum		3	2.667		325	125		450	555
0500	Heavy duty, pivoted, from 7'-7" to 12'-10" floor to floor		3	2.667		740	125		865	1,000
0600	16'-0" ceiling		2	4		1,525	188		1,713	1,975
0800	Economy folding, pine, 8'-6" ceiling		4	2		176	94		270	340
0900	9'-6" ceiling		4	2		196	94		290	360
1100	Automatic electric, aluminum, floor to floor height, 8' to 9'	2 Carp	1	16		8,550	750		9,300	10,600
1400	11' to 12'		.90	17.778		9,075	835		9,910	11,300
1700	14' to 15'		.70	22.857		9,775	1,075		10,850	12,500

11 32 Unit Kitchens

11 32 13 – Metal Unit Kitchens

11 32 13.10 Commercial Unit Kitchens		Crew	Daily Output	Labor-Hours	Unit	Material	Labor	Equipment	Total	Total Incl O&P
0010	**COMMERCIAL UNIT KITCHENS**									
1500	Combination range, refrigerator and sink, 30" wide, minimum	L-1	2	8	Ea.	1,100	455		1,555	1,900
1550	Maximum		1	16		1,475	905		2,380	3,000
1570	60" wide, average		1.40	11.429		1,525	650		2,175	2,675
1590	72" wide, average		1.20	13.333		1,625	755		2,380	2,900
1600	Office model, 48" wide		2	8		2,025	455		2,480	2,925
1620	Refrigerator and sink only		2.40	6.667		2,525	380		2,905	3,350
1640	Combination range, refrigerator, sink, microwave									
1660	Oven and ice maker	L-1	.80	20	Ea.	4,550	1,125		5,675	6,700

11 41 Foodservice Storage Equipment

11 41 13 – Refrigerated Food Storage Cases

11 41 13.10 Refrigerated Food Cases		Crew	Daily Output	Labor-Hours	Unit	Material	Labor	Equipment	Total	Total Incl O&P
0010	**REFRIGERATED FOOD CASES**									
0030	Dairy, multi-deck, 12' long	Q-5	3	5.333	Ea.	11,300	287		11,587	12,800
0100	For rear sliding doors, add					1,800			1,800	2,000
0200	Delicatessen case, service deli, 12' long, single deck	Q-5	3.90	4.103		7,750	221		7,971	8,850
0300	Multi-deck, 18 S.F. shelf display		3	5.333		7,150	287		7,437	8,300
0400	Freezer, self-contained, chest-type, 30 C.F.		3.90	4.103		8,525	221		8,746	9,725
0500	Glass door, upright, 78 C.F.		3.30	4.848		10,200	261		10,461	11,600
0600	Frozen food, chest type, 12' long		3.30	4.848		8,225	261		8,486	9,450
0700	Glass door, reach-in, 5 door		3	5.333		14,100	287		14,387	15,900
0800	Island case, 12' long, single deck		3.30	4.848		7,275	261		7,536	8,425
0900	Multi-deck		3	5.333		8,500	287		8,787	9,775
1000	Meat case, 12' long, single deck		3.30	4.848		7,275	261		7,536	8,425
1050	Multi-deck		3.10	5.161		10,400	278		10,678	11,800
1100	Produce, 12' long, single deck		3.30	4.848		6,875	261		7,136	7,950
1200	Multi-deck		3.10	5.161		8,725	278		9,003	10,000

11 41 13.20 Refrigerated Food Storage Equipment

	11 41 13.20 Refrigerated Food Storage Equipment	Crew	Daily Output	Labor-Hours	Unit	2015 Bare Costs Material	Labor	Equipment	Total	Total Incl O&P
0010	**REFRIGERATED FOOD STORAGE EQUIPMENT**									
2350	Cooler, reach-in, beverage, 6' long	Q-1	6	2.667	Ea.	3,600	141		3,741	4,175
4300	Freezers, reach-in, 44 C.F.		4	4		4,425	211		4,636	5,175
4500	68 C.F.		3	5.333		4,825	282		5,107	5,725
4600	Freezer, pre-fab, 8' x 8' w/refrigeration	2 Carp	.45	35.556		11,100	1,675		12,775	14,800
4620	8' x 12'		.35	45.714		11,200	2,150		13,350	15,600
4640	8' x 16'		.25	64		14,300	3,000		17,300	20,300
4660	8' x 20'		.17	94.118		19,400	4,425		23,825	28,100
4680	Reach-in, 1 compartment	Q-1	4	4		2,500	211		2,711	3,100
4685	Energy star rated G	R-18	7.80	3.333		2,575	143		2,718	3,050
4700	2 compartment	Q-1	3	5.333		4,100	282		4,382	4,950
4705	Energy star rated G	R-18	6.20	4.194		3,100	180		3,280	3,675
4710	3 compartment	Q-1	3	5.333		5,300	282		5,582	6,250
4715	Energy star rated G	R-18	5.60	4.643		4,175	199		4,374	4,900
8320	Refrigerator, reach-in, 1 compartment		7.80	3.333		2,425	143		2,568	2,900
8325	Energy star rated G		7.80	3.333		2,575	143		2,718	3,050
8330	2 compartment		6.20	4.194		3,800	180		3,980	4,450
8335	Energy star rated G		6.20	4.194		3,100	180		3,280	3,675
8340	3 compartment		5.60	4.643		4,775	199		4,974	5,550
8345	Energy star rated G		5.60	4.643		4,175	199		4,374	4,900
8350	Pre-fab, with refrigeration, 8' x 8'	2 Carp	.45	35.556		7,050	1,675		8,725	10,300
8360	8' x 12'		.35	45.714		8,000	2,150		10,150	12,100
8370	8' x 16'		.25	64		12,300	3,000		15,300	18,200
8380	8' x 20'		.17	94.118		15,700	4,425		20,125	24,000
8390	Pass-thru/roll-in, 1 compartment	R-18	7.80	3.333		4,500	143		4,643	5,175
8400	2 compartment		6.24	4.167		6,325	179		6,504	7,225
8410	3 compartment		5.60	4.643		8,625	199		8,824	9,800
8420	Walk-in, alum, door & floor only, no refrig, 6' x 6' x 7'-6"	2 Carp	1.40	11.429		8,775	535		9,310	10,500
8430	10' x 6' x 7'-6"		.55	29.091		12,700	1,375		14,075	16,100
8440	12' x 14' x 7'-6"		.25	64		17,500	3,000		20,500	23,800
8450	12' x 20' x 7'-6"		.17	94.118		19,300	4,425		23,725	28,000
8460	Refrigerated cabinets, mobile					3,925			3,925	4,325
8470	Refrigerator/freezer, reach-in, 1 compartment	R-18	5.60	4.643		5,800	199		5,999	6,675
8480	2 compartment	"	4.80	5.417		7,575	232		7,807	8,675

11 41 13.30 Wine Cellar

	11 41 13.30 Wine Cellar	Crew	Daily Output	Labor-Hours	Unit	Material	Labor	Equipment	Total	Total Incl O&P
0010	**WINE CELLAR**, refrigerated, Redwood interior, carpeted, walk-in type									
0020	6'-8" high, including racks									
0200	80" W x 48" D for 900 bottles	2 Carp	1.50	10.667	Ea.	4,300	500		4,800	5,500
0250	80" W x 72" D for 1300 bottles		1.33	12.030		5,225	565		5,790	6,625
0300	80" W x 94" D for 1900 bottles		1.17	13.675		6,350	640		6,990	7,975
0400	80" W x 124" D for 2500 bottles		1	16		7,450	750		8,200	9,350
0600	Portable cabinets, red oak, reach-in temp.& humidity controlled									
0650	26-5/8"W x 26-1/2"D x 68"H for 235 bottles				Ea.	3,575			3,575	3,925
0660	32"W x 21-1/2"D x 73-1/2"H for 144 bottles					2,900			2,900	3,200
0670	32"W x 29-1/2"D x 73-1/2"H for 288 bottles					3,875			3,875	4,275
0680	39-1/2"W x 29-1/2"D x 86-1/2"H for 440 bottles					4,100			4,100	4,500
0690	52-1/2"W x 29-1/2"D x 73-1/2"H for 468 bottles					4,375			4,375	4,825
0700	52-1/2"W x 29-1/2"D x 86-1/2"H for 572 bottles					4,475			4,475	4,925
0730	Portable, red oak, can be built-in with glass door									
0750	23-7/8"W x 24"D x 34-1/2"H for 50 bottles				Ea.	940			940	1,025

11 41 Foodservice Storage Equipment

11 41 33 – Foodservice Shelving

11 41 33.20 Metal Food Storage Shelving		Crew	Daily Output	Labor-Hours	Unit	2015 Bare Costs Material	Labor	Equipment	Total	Total Incl O&P
0010	**METAL FOOD STORAGE SHELVING**									
8600	Stainless steel shelving, louvered 4-tier, 20" x 3'	1 Clab	6	1.333	Ea.	1,400	50		1,450	1,625
8605	20" x 4'		6	1.333		1,550	50		1,600	1,775
8610	20" x 6'		6	1.333		2,200	50		2,250	2,500
8615	24" x 3'		6	1.333		1,975	50		2,025	2,250
8620	24" x 4'		6	1.333		2,350	50		2,400	2,675
8625	24" x 6'		6	1.333		3,275	50		3,325	3,675
8630	Flat 4-tier, 20" x 3'		6	1.333		1,150	50		1,200	1,350
8635	20" x 4'		6	1.333		1,400	50		1,450	1,625
8640	20" x 5'		6	1.333		1,625	50		1,675	1,850
8645	24" x 3'		6	1.333		1,275	50		1,325	1,475
8650	24" x 4'		6	1.333		2,225	50		2,275	2,525
8655	24" x 6'		6	1.333		2,675	50		2,725	3,000
8700	Galvanized shelving, louvered 4-tier, 20" x 3'		6	1.333		760	50		810	910
8705	20" x 4'		6	1.333		860	50		910	1,025
8710	20" x 6'		6	1.333		1,000	50		1,050	1,175
8715	24" x 3'		6	1.333		705	50		755	855
8720	24" x 4'		6	1.333		995	50		1,045	1,175
8725	24" x 6'		6	1.333		1,325	50		1,375	1,525
8730	Flat 4-tier, 20" x 3'		6	1.333		700	50		750	845
8735	20" x 4'		6	1.333		695	50		745	840
8740	20" x 6'		6	1.333		905	50		955	1,075
8745	24" x 3'		6	1.333		670	50		720	810
8750	24" x 4'		6	1.333		755	50		805	905
8755	24" x 6'		6	1.333		950	50		1,000	1,125
8760	Stainless steel dunnage rack, 24" x 3'		8	1		330	37.50		367.50	425
8765	24" x 4'		8	1		425	37.50		462.50	525
8770	Galvanized dunnage rack, 24" x 3'		8	1		168	37.50		205.50	243
8775	24" x 4'		8	1		190	37.50		227.50	267

11 42 Food Preparation Equipment

11 42 10 – Commercial Food Preparation Equipment

11 42 10.10 Choppers, Mixers and Misc. Equipment

		Crew	Daily Output	Labor-Hours	Unit	Material	Labor	Equipment	Total	Total Incl O&P
0010	**CHOPPERS, MIXERS AND MISC. EQUIPMENT**									
1700	Choppers, 5 pounds	R-18	7	3.714	Ea.	2,150	159		2,309	2,625
1720	16 pounds		5	5.200		2,200	223		2,423	2,775
1740	35 to 40 pounds		4	6.500		3,550	279		3,829	4,325
1840	Coffee brewer, 5 burners	1 Plum	3	2.667		1,325	157		1,482	1,675
1850	Coffee urn, twin 6 gallon urns		2	4		2,400	235		2,635	3,000
1860	Single, 3 gallon		3	2.667		1,825	157		1,982	2,225
3000	Fast food equipment, total package, minimum	6 Skwk	.08	600		203,500	29,200		232,700	269,000
3100	Maximum	"	.07	685		277,500	33,400		310,900	356,500
3800	Food mixers, bench type, 20 quarts	L-7	7	4		2,775	182		2,957	3,325
3850	40 quarts		5.40	5.185		6,725	235		6,960	7,750
3900	60 quarts		5	5.600		11,300	254		11,554	12,900
4040	80 quarts		3.90	7.179		12,700	325		13,025	14,500
4100	Floor type, 20 quarts		15	1.867		3,225	84.50		3,309.50	3,675
4120	60 quarts		14	2		9,950	91		10,041	11,000
4140	80 quarts		12	2.333		15,400	106		15,506	17,200
4160	140 quarts		8.60	3.256		25,600	148		25,748	28,400

11 42 Food Preparation Equipment

11 42 10 – Commercial Food Preparation Equipment

11 42 10.10 Choppers, Mixers and Misc. Equipment		Crew	Daily Output	Labor-Hours	Unit	Material	Labor	Equipment	Total	Total Incl O&P
						2015 Bare Costs				
6700	Peelers, small	R-18	8	3.250	Ea.	1,900	139		2,039	2,300
6720	Large	"	6	4.333		4,825	186		5,011	5,575
6800	Pulper/extractor, close coupled, 5 HP	1 Plum	1.90	4.211		3,525	247		3,772	4,250
8580	Slicer with table	R-18	9	2.889	↓	4,675	124		4,799	5,350

11 43 Food Delivery Carts and Conveyors

11 43 13 – Food Delivery Carts

11 43 13.10 Mobile Carts, Racks and Trays

		Crew	Daily Output	Labor-Hours	Unit	Material	Labor	Equipment	Total	Total Incl O&P
0010	**MOBILE CARTS, RACKS AND TRAYS**									
1650	Cabinet, heated, 1 compartment, reach-in	R-18	5.60	4.643	Ea.	3,300	199		3,499	3,925
1655	Pass-thru roll-in		5.60	4.643		3,825	199		4,024	4,500
1660	2 compartment, reach-in	↓	4.80	5.417		9,050	232		9,282	10,300
1670	Mobile					3,525			3,525	3,875
2000	Hospital food cart, hot and cold service, 20 tray capacity					15,300			15,300	16,800
6850	Mobile rack w/pan slide					1,400			1,400	1,525
9180	Tray and silver dispenser, mobile	1 Clab	16	.500	↓	915	18.80		933.80	1,025

11 44 Food Cooking Equipment

11 44 13 – Commercial Ranges

11 44 13.10 Cooking Equipment

		Crew	Daily Output	Labor-Hours	Unit	Material	Labor	Equipment	Total	Total Incl O&P
0010	**COOKING EQUIPMENT**									
0020	Bake oven, gas, one section	Q-1	8	2	Ea.	5,450	106		5,556	6,150
0300	Two sections		7	2.286		9,075	121		9,196	10,200
0600	Three sections	↓	6	2.667		11,200	141		11,341	12,500
0900	Electric convection, single deck	L-7	4	7		6,225	320		6,545	7,325
1300	Broiler, without oven, standard	Q-1	8	2		3,550	106		3,656	4,050
1550	Infrared	L-7	4	7		7,425	320		7,745	8,625
4750	Fryer, with twin baskets, modular model	Q-1	7	2.286		1,250	121		1,371	1,550
5000	Floor model, on 6" legs	"	5	3.200		2,425	169		2,594	2,925
5100	Extra single basket, large					100			100	110
5170	Energy star rated, 50 lb. capacity G	R-18	4	6.500		4,650	279		4,929	5,525
5175	85 lb. capacity G	"	4	6.500		8,525	279		8,804	9,800
5300	Griddle, SS, 24" plate, w/4" legs, elec, 208 V, 3 phase, 3' long	Q-1	7	2.286		1,400	121		1,521	1,725
5550	4' long	"	6	2.667		2,250	141		2,391	2,700
6200	Iced tea brewer	1 Plum	3.44	2.326		750	137		887	1,025
6350	Kettle, w/steam jacket, tilting, w/positive lock, SS, 20 gallons	L-7	7	4		8,225	182		8,407	9,325
6600	60 gallons	"	6	4.667		11,000	212		11,212	12,400
6900	Range, restaurant type, 6 burners and 1 standard oven, 36" wide	Q-1	7	2.286		2,500	121		2,621	2,925
6950	Convection		7	2.286		4,450	121		4,571	5,075
7150	2 standard ovens, 24" griddle, 60" wide		6	2.667		4,575	141		4,716	5,250
7200	1 standard, 1 convection oven		6	2.667		9,475	141		9,616	10,600
7450	Heavy duty, single 34" standard oven, open top		5	3.200		5,050	169		5,219	5,800
7500	Convection oven		5	3.200		5,450	169		5,619	6,250
7700	Griddle top		6	2.667		2,850	141		2,991	3,375
7750	Convection oven	↓	6	2.667		7,775	141		7,916	8,775
7760	Induction cooker, electric	L-7	7	4		1,850	182		2,032	2,325
8850	Steamer, electric 27 KW		7	4		10,700	182		10,882	12,100
9100	Electric, 10 KW or gas 100,000 BTU	↓	5	5.600		6,325	254		6,579	7,350
9150	Toaster, conveyor type, 16-22 slices per minute				↓	1,075			1,075	1,200

11 44 Food Cooking Equipment

11 44 13 – Commercial Ranges

11 44 13.10 Cooking Equipment		Crew	Daily Output	Labor-Hours	Unit	Material	2015 Bare Costs Labor	Equipment	Total	Total Incl O&P
9160	Pop-up, 2 slot				Ea.	615			615	680
9200	For deluxe models of above equipment, add					75%				
9400	Rule of thumb: Equipment cost based									
9410	on kitchen work area									
9420	Office buildings, minimum	L-7	77	.364	S.F.	90.50	16.50		107	125
9450	Maximum		58	.483		153	22		175	202
9550	Public eating facilities, minimum		77	.364		119	16.50		135.50	157
9600	Maximum		46	.609		193	27.50		220.50	255
9750	Hospitals, minimum		58	.483		122	22		144	168
9800	Maximum		39	.718		225	32.50		257.50	297

11 46 Food Dispensing Equipment

11 46 16 – Service Line Equipment

11 46 16.10 Commercial Food Dispensing Equipment

		Crew	Daily Output	Labor-Hours	Unit	Material	Labor	Equipment	Total	Total Incl O&P
0010	COMMERCIAL FOOD DISPENSING EQUIPMENT									
1050	Butter pat dispenser	1 Clab	13	.615	Ea.	1,000	23		1,023	1,125
1100	Bread dispenser, counter top		13	.615		890	23		913	1,025
1900	Cup and glass dispenser, drop in		4	2		1,075	75		1,150	1,300
1920	Disposable cup, drop in		16	.500		495	18.80		513.80	575
2650	Dish dispenser, drop in, 12"		11	.727		2,250	27.50		2,277.50	2,525
2660	Mobile		10	.800		2,625	30		2,655	2,950
3300	Food warmer, counter, 1.2 KW					665			665	735
3550	1.6 KW					2,150			2,150	2,350
3600	Well, hot food, built-in, rectangular, 12" x 20"	R-30	10	2.600		815	115		930	1,075
3610	Circular, 7 qt.		10	2.600		420	115		535	635
3620	Refrigerated, 2 compartments		10	2.600		3,075	115		3,190	3,550
3630	3 compartments		9	2.889		3,700	128		3,828	4,275
3640	4 compartments		8	3.250		4,350	144		4,494	5,025
4720	Frost cold plate		9	2.889		18,400	128		18,528	20,500
5700	Hot chocolate dispenser	1 Plum	4	2		1,175	117		1,292	1,450
5750	Ice dispenser 567 pound	Q-1	6	2.667		5,200	141		5,341	5,950
6250	Jet spray dispenser	R-18	4.50	5.778		3,175	248		3,423	3,875
6300	Juice dispenser, concentrate	"	4.50	5.778		1,900	248		2,148	2,450
6690	Milk dispenser, bulk, 2 flavor	R-30	8	3.250		1,850	144		1,994	2,250
6695	3 flavor	"	8	3.250		2,475	144		2,619	2,950
8800	Serving counter, straight	1 Carp	40	.200	L.F.	925	9.40		934.40	1,050
8820	Curved section	"	30	.267	"	1,125	12.50		1,137.50	1,275
8825	Solid surface, see Section 12 36 61.16									
8860	Sneeze guard with lights, 60" L	1 Clab	16	.500	Ea.	395	18.80		413.80	465
8900	Sneeze guard, stainless steel and glass, single sided									
8910	Portable, 48" W				Ea.	305			305	335
8920	Portable, 72" W					325			325	360
8930	Adjustable, 36" W	1 Carp	24	.333		253	15.65		268.65	300
8940	Adjustable, 48" W	"	20	.400		320	18.80		338.80	385
9100	Soft serve ice cream machine, medium	R-18	11	2.364		12,100	101		12,201	13,500
9110	Large	"	9	2.889		21,800	124		21,924	24,100

11 46 Food Dispensing Equipment

11 46 83 – Ice Machines

11 46 83.10 Commercial Ice Equipment		Crew	Daily Output	Labor-Hours	Unit	Material	Labor (2015 Bare Costs)	Equipment (2015 Bare Costs)	Total (2015 Bare Costs)	Total Incl O&P
0010	**COMMERCIAL ICE EQUIPMENT**									
5800	Ice cube maker, 50 pounds per day	Q-1	6	2.667	Ea.	1,600	141		1,741	2,000
5810	65 pounds per day, energy star rated		6	2.667		1,475	141		1,616	1,850
5900	250 pounds per day		1.20	13.333		2,675	705		3,380	4,025
5950	300 pounds per day, remote condensing		1.20	13.333		2,350	705		3,055	3,675
6050	500 pounds per day		4	4		2,775	211		2,986	3,375
6060	With bin		1.20	13.333		3,400	705		4,105	4,800
6070	Modular, with bin and condenser		1.20	13.333		3,875	705		4,580	5,350
6090	1000 pounds per day, with bin		1	16		4,950	845		5,795	6,700
6100	Ice flakers, 300 pounds per day		1.60	10		2,800	530		3,330	3,875
6120	600 pounds per day		.95	16.842		3,950	890		4,840	5,675
6130	1000 pounds per day		.75	21.333		4,750	1,125		5,875	6,925
6140	2000 pounds per day		.65	24.615		21,600	1,300		22,900	25,700
6160	Ice storage bin, 500 pound capacity	Q-5	1	16		1,050	860		1,910	2,450
6180	1000 pound	"	.56	28.571		2,525	1,525		4,050	5,100

11 48 Foodservice Cleaning and Disposal Equipment

11 48 13 – Commercial Dishwashers

11 48 13.10 Dishwashers		Crew	Daily Output	Labor-Hours	Unit	Material	Labor (2015 Bare Costs)	Equipment (2015 Bare Costs)	Total (2015 Bare Costs)	Total Incl O&P
0010	**DISHWASHERS**									
2700	Dishwasher, commercial, rack type									
2720	10 to 12 racks per hour	Q-1	3.20	5	Ea.	3,525	264		3,789	4,275
2730	Energy star rated, 35 to 40 racks/hour G		1.30	12.308		4,700	650		5,350	6,150
2740	50 to 60 racks/hour G		1.30	12.308		10,200	650		10,850	12,200
2800	Automatic, 190 to 230 racks per hour	L-6	.35	34.286		13,900	1,975		15,875	18,300
2820	235 to 275 racks per hour		.25	48		31,500	2,750		34,250	38,800
2840	8,750 to 12,500 dishes per hour		.10	120		55,500	6,875		62,375	71,500
2950	Dishwasher hood, canopy type	L-3A	10	1.200	L.F.	865	61.50		926.50	1,050
2960	Pant leg type	"	2.50	4.800	Ea.	8,700	246		8,946	9,950
5200	Garbage disposal 1.5 HP, 100 GPH	L-1	4.80	3.333		2,175	189		2,364	2,675
5210	3 HP, 120 GPH		4.60	3.478		2,625	197		2,822	3,200
5220	5 HP, 250 GPH		4.50	3.556		3,600	202		3,802	4,275
6750	Pot sink, 3 compartment	1 Plum	7.25	1.103	L.F.	980	65		1,045	1,175
6760	Pot washer, low temp wash/rinse		1.60	5	Ea.	4,050	294		4,344	4,900
6770	High pressure wash, high temperature rinse		1.20	6.667		36,900	390		37,290	41,200
9170	Trash compactor, small, up to 125 lb. compacted weight	L-4	4	6		24,200	266		24,466	27,000
9175	Large, up to 175 lb. compacted weight	"	3	8		29,500	355		29,855	33,000

11 52 Audio-Visual Equipment

11 52 13 – Projection Screens

11 52 13.10 Projection Screens, Wall or Ceiling Hung		Crew	Daily Output	Labor-Hours	Unit	Material	Labor (2015 Bare Costs)	Equipment (2015 Bare Costs)	Total (2015 Bare Costs)	Total Incl O&P
0010	**PROJECTION SCREENS, WALL OR CEILING HUNG**, matte white									
0100	Manually operated, economy	2 Carp	500	.032	S.F.	5.90	1.50		7.40	8.80
0300	Intermediate		450	.036		6.90	1.67		8.57	10.15
0400	Deluxe		400	.040		9.55	1.88		11.43	13.40
0600	Electric operated, matte white, 25 S.F., economy		5	3.200	Ea.	865	150		1,015	1,175
0700	Deluxe		4	4		1,800	188		1,988	2,275
0900	50 S.F., economy		3	5.333		705	250		955	1,150
1000	Deluxe		2	8		2,000	375		2,375	2,775

11 52 Audio-Visual Equipment

11 52 13 – Projection Screens

11 52 13.10 Projection Screens, Wall or Ceiling Hung		Crew	Daily Output	Labor-Hours	Unit	2015 Bare Costs Material	Labor	Equipment	Total	Total Incl O&P
1200	Heavy duty, electric operated, 200 S.F.	2 Carp	1.50	10.667	Ea.	3,950	500		4,450	5,125
1300	400 S.F.		1	16		4,875	750		5,625	6,500
1500	Rigid acrylic in wall, for rear projection, 1/4" thick	2 Glaz	30	.533	S.F.	47.50	24		71.50	89
1600	1/2" thick (maximum size 10' x 20')	"	25	.640	"	84.50	29		113.50	137

11 52 16 – Projectors

11 52 16.10 Movie Equipment

		Crew	Daily Output	Labor-Hours	Unit	Material	Labor	Equipment	Total	Total Incl O&P
0010	**MOVIE EQUIPMENT**									
0020	Changeover, minimum				Ea.	470			470	520
0100	Maximum					915			915	1,000
0400	Film transport, incl. platters and autowind, minimum					5,075			5,075	5,600
0500	Maximum					14,400			14,400	15,900
0800	Lamphouses, incl. rectifiers, xenon, 1,000 watt	1 Elec	2	4		6,725	219		6,944	7,725
0900	1,600 watt		2	4		7,175	219		7,394	8,225
1000	2,000 watt		1.50	5.333		7,700	292		7,992	8,900
1100	4,000 watt		1.50	5.333		9,550	292		9,842	10,900
1400	Lenses, anamorphic, minimum					1,300			1,300	1,425
1500	Maximum					2,900			2,900	3,200
1800	Flat 35 mm, minimum					1,125			1,125	1,225
1900	Maximum					1,750			1,750	1,925
2200	Pedestals, for projectors					1,500			1,500	1,650
2300	Console type					10,800			10,800	11,900
2600	Projector mechanisms, incl. soundhead, 35 mm, minimum					11,200			11,200	12,300
2700	Maximum					15,400			15,400	16,900
3000	Projection screens, rigid, in wall, acrylic, 1/4" thick	2 Glaz	195	.082	S.F.	42.50	3.70		46.20	52
3100	1/2" thick	"	130	.123	"	49	5.55		54.55	62.50
3300	Electric operated, heavy duty, 400 S.F.	2 Carp	1	16	Ea.	3,000	750		3,750	4,425
3320	Theater projection screens, matte white, including frames	"	200	.080	S.F.	6.70	3.76		10.46	13.20
3400	Also see Section 11 52 13.10									
3700	Sound systems, incl. amplifier, mono, minimum	1 Elec	.90	8.889	Ea.	3,350	485		3,835	4,400
3800	Dolby/Super Sound, maximum		.40	20		18,300	1,100		19,400	21,800
4100	Dual system, 2 channel, front surround, minimum		.70	11.429		4,675	625		5,300	6,075
4200	Dolby/Super Sound, 4 channel, maximum		.40	20		16,700	1,100		17,800	20,100
4500	Sound heads, 35 mm					5,350			5,350	5,900
4900	Splicer, tape system, minimum					750			750	825
5000	Tape type, maximum					1,350			1,350	1,475
5300	Speakers, recessed behind screen, minimum	1 Elec	2	4		1,075	219		1,294	1,500
5400	Maximum	"	1	8		3,125	440		3,565	4,100
5700	Seating, painted steel, upholstered, minimum	2 Carp	35	.457		133	21.50		154.50	180
5800	Maximum	"	28	.571		425	27		452	510
6100	Rewind tables, minimum					2,675			2,675	2,950
6200	Maximum					4,775			4,775	5,250
7000	For automation, varying sophistication, minimum	1 Elec	1	8	System	2,425	440		2,865	3,300
7100	Maximum	2 Elec	.30	53.333	"	5,625	2,925		8,550	10,600

11 52 16.20 Movie Equipment- Digital

		Crew	Daily Output	Labor-Hours	Unit	Material	Labor	Equipment	Total	Total Incl O&P
0010	**MOVIE EQUIPMENT- DIGITAL**									
1000	Digital 2K projection system, 98" DMD	1 Elec	2	4	Ea.	44,500	219		44,719	49,300
1100	OEM lens		2	4		5,425	219		5,644	6,300
2000	Pedestal with power distribution		2	4		2,075	219		2,294	2,600
3000	Software		2	4		1,750	219		1,969	2,250

11 53 Laboratory Equipment

11 53 03 – Laboratory Test Equipment

11 53 03.13 Test Equipment		Crew	Daily Output	Labor-Hours	Unit	Material	2015 Bare Costs Labor	Equipment	Total	Total Incl O&P
0010	TEST EQUIPMENT									
1700	Thermometer, electric, portable				Ea.	500			500	550
1800	Titration unit, four 2000 ml reservoirs				"	5,550			5,550	6,125

11 53 13 – Laboratory Fume Hoods

11 53 13.13 Recirculating Laboratory Fume Hoods		Crew	Daily Output	Labor-Hours	Unit	Material	Labor	Equipment	Total	Total Incl O&P
0010	RECIRCULATING LABORATORY FUME HOODS									
0600	Fume hood, with countertop & base, not including HVAC									
0610	Simple, minimum	2 Carp	5.40	2.963	L.F.	500	139		639	765
0620	Complex, including fixtures		2.40	6.667		805	315		1,120	1,375
0630	Special, maximum	↓	1.70	9.412	↓	840	440		1,280	1,600
0670	Service fixtures, average				Ea.	240			240	264
0680	For sink assembly with hot and cold water, add	1 Plum	1.40	5.714		735	335		1,070	1,325
0750	Glove box, fiberglass, bacteriological					17,100			17,100	18,800
0760	Controlled atmosphere					19,600			19,600	21,600
0770	Radioisotope					17,100			17,100	18,800
0780	Carcinogenic				↓	17,100			17,100	18,800

11 53 13.23 Exhaust Hoods		Crew	Daily Output	Labor-Hours	Unit	Material	Labor	Equipment	Total	Total Incl O&P
0010	EXHAUST HOODS									
0650	Ductwork, minimum	2 Shee	1	16	Hood	4,075	895		4,970	5,850
0660	Maximum	"	.50	32	"	6,275	1,800		8,075	9,625

11 53 16 – Laboratory Incubators

11 53 16.13 Incubators		Crew	Daily Output	Labor-Hours	Unit	Material	Labor	Equipment	Total	Total Incl O&P
0010	INCUBATORS									
1000	Incubators, minimum				Ea.	3,050			3,050	3,350
1010	Maximum				"	11,800			11,800	13,000

11 53 19 – Laboratory Sterilizers

11 53 19.13 Sterilizers		Crew	Daily Output	Labor-Hours	Unit	Material	Labor	Equipment	Total	Total Incl O&P
0010	STERILIZERS									
0700	Glassware washer, undercounter, minimum	L-1	1.80	8.889	Ea.	6,325	505		6,830	7,700
0710	Maximum	"	1	16		13,300	905		14,205	16,000
1850	Utensil washer-sanitizer	1 Plum	2	4	↓	11,300	235		11,535	12,800

11 53 23 – Laboratory Refrigerators

11 53 23.13 Refrigerators		Crew	Daily Output	Labor-Hours	Unit	Material	Labor	Equipment	Total	Total Incl O&P
0010	REFRIGERATORS									
1200	Blood bank, 28.6 C.F. emergency signal				Ea.	9,650			9,650	10,600
1210	Reach-in, 16.9 C.F.				"	8,525			8,525	9,375

11 53 33 – Emergency Safety Appliances

11 53 33.13 Emergency Equipment		Crew	Daily Output	Labor-Hours	Unit	Material	Labor	Equipment	Total	Total Incl O&P
0010	EMERGENCY EQUIPMENT									
1400	Safety equipment, eye wash, hand held				Ea.	410			410	450
1450	Deluge shower				"	770			770	850

11 53 43 – Service Fittings and Accessories

11 53 43.13 Fittings		Crew	Daily Output	Labor-Hours	Unit	Material	Labor	Equipment	Total	Total Incl O&P
0010	FITTINGS									
1600	Sink, one piece plastic, flask wash, hose, free standing	1 Plum	1.60	5	Ea.	1,950	294		2,244	2,600
1610	Epoxy resin sink, 25" x 16" x 10"	"	2	4	"	221	235		456	600
1950	Utility table, acid resistant top with drawers	2 Carp	30	.533	L.F.	164	25		189	219
8000	Alternate pricing method: as percent of lab furniture									
8050	Installation, not incl. plumbing & duct work				% Furn.				22%	22%

11 53 Laboratory Equipment

11 53 43 – Service Fittings and Accessories

	11 53 43.13 Fittings	Crew	Daily Output	Labor-Hours	Unit	Material	2015 Bare Costs Labor	Equipment	Total	Total Incl O&P
8100	Plumbing, final connections, simple system				% Furn.				10%	10%
8110	Moderately complex system								15%	15%
8120	Complex system								20%	20%
8150	Electrical, simple system								10%	10%
8160	Moderately complex system								20%	20%
8170	Complex system								35%	35%

11 53 53 – Biological Safety Cabinets

	11 53 53.10 Pharmacy Cabinets	Crew	Daily Output	Labor-Hours	Unit	Material	Labor	Equipment	Total	Total Incl O&P
0010	**PHARMACY CABINETS**, vertical flow									
0100	Class II, type B2, 6′ L	2 Carp	1.50	10.667	Ea.	13,200	500		13,700	15,300

11 57 Vocational Shop Equipment

11 57 10 – Shop Equipment

	11 57 10.10 Vocational School Shop Equipment	Crew	Daily Output	Labor-Hours	Unit	Material	Labor	Equipment	Total	Total Incl O&P
0010	**VOCATIONAL SCHOOL SHOP EQUIPMENT**									
0020	Benches, work, wood, average	2 Carp	5	3.200	Ea.	635	150		785	930
0100	Metal, average		5	3.200		550	150		700	835
0400	Combination belt & disc sander, 6″		4	4		1,650	188		1,838	2,125
0700	Drill press, floor mounted, 12″, 1/2 H.P.		4	4		415	188		603	745
0800	Dust collector, not incl. ductwork, 6″ diameter	1 Shee	1.10	7.273		4,650	405		5,055	5,750
0810	Dust collector bag, 20″ diameter	″	5	1.600		440	89.50		529.50	620
1000	Grinders, double wheel, 1/2 H.P.	2 Carp	5	3.200		217	150		367	470
1300	Jointer, 4″, 3/4 H.P.		4	4		1,375	188		1,563	1,800
1600	Kilns, 16 C.F., to 2000°		4	4		1,475	188		1,663	1,925
1900	Lathe, woodworking, 10″, 1/2 H.P.		4	4		545	188		733	890
2200	Planer, 13″ x 6″		4	4		1,100	188		1,288	1,500
2500	Potter's wheel, motorized		4	4		1,125	188		1,313	1,550
2800	Saws, band, 14″, 3/4 H.P.		4	4		930	188		1,118	1,325
3100	Metal cutting band saw, 14″		4	4		2,525	188		2,713	3,075
3400	Radial arm saw, 10″, 2 H.P.		4	4		1,375	188		1,563	1,800
3700	Scroll saw, 24″		4	4		585	188		773	935
4000	Table saw, 10″, 3 H.P.		4	4		2,725	188		2,913	3,300
4300	Welder AC arc, 30 amp capacity		4	4		3,175	188		3,363	3,800

11 61 Broadcast, Theater, and Stage Equipment

11 61 23 – Folding and Portable Stages

	11 61 23.10 Portable Stages	Crew	Daily Output	Labor-Hours	Unit	Material	Labor	Equipment	Total	Total Incl O&P
0010	**PORTABLE STAGES**									
1500	Flooring, portable oak parquet, 3′ x 3′ sections				S.F.	13.55			13.55	14.90
1600	Cart to carry 225 S.F. of flooring				Ea.	405			405	445
5000	Stages, portable with steps, folding legs, stock, 8″ high				SF Stg.	32.50			32.50	36
5100	16″ high					49.50			49.50	54.50
5200	32″ high					53.50			53.50	59
5300	40″ high					60			60	66
6000	Telescoping platforms, extruded alum., straight, minimum	4 Carp	157	.204		34.50	9.55		44.05	52.50
6100	Maximum		77	.416		48	19.50		67.50	82.50
6500	Pie-shaped, minimum		150	.213		74	10		84	97
6600	Maximum		70	.457		83	21.50		104.50	124
6800	For 3/4″ plywood covered deck, deduct					4.21			4.21	4.63

11 61 Broadcast, Theater, and Stage Equipment

11 61 23 – Folding and Portable Stages

11 61 23.10 Portable Stages		Crew	Daily Output	Labor-Hours	Unit	Material	2015 Bare Costs Labor	Equipment	Total	Total Incl O&P
7000	Band risers, steel frame, plywood deck, minimum	4 Carp	275	.116	SF Stg.	31	5.45		36.45	42.50
7100	Maximum	"	138	.232	↓	69.50	10.90		80.40	93.50
7500	Chairs for above, self-storing, minimum	2 Carp	43	.372	Ea.	110	17.45		127.45	148
7600	Maximum	"	40	.400	"	194	18.80		212.80	243

11 61 33 – Rigging Systems and Controls

11 61 33.10 Controls		Crew	Daily Output	Labor-Hours	Unit	Material	Labor	Equipment	Total	Total Incl O&P
0010	**CONTROLS**									
0050	Control boards with dimmers and breakers, minimum	1 Elec	1	8	Ea.	12,600	440		13,040	14,600
0100	Average		.50	16		39,700	875		40,575	45,000
0150	Maximum	↓	.20	40	↓	129,000	2,200		131,200	145,500
8000	Rule of thumb: total stage equipment, minimum	4 Carp	100	.320	SF Stg.	98.50	15		113.50	131
8100	Maximum	"	25	1.280	"	555	60		615	705

11 61 43 – Stage Curtains

11 61 43.10 Curtains		Crew	Daily Output	Labor-Hours	Unit	Material	Labor	Equipment	Total	Total Incl O&P
0010	**CURTAINS**									
0500	Curtain track, straight, light duty	2 Carp	20	.800	L.F.	27.50	37.50		65	88.50
0600	Heavy duty		18	.889		62	41.50		103.50	133
0700	Curved sections		12	1.333	↓	177	62.50		239.50	292
1000	Curtains, velour, medium weight		600	.027	S.F.	8	1.25		9.25	10.75
1150	Silica based yarn, inherently fire retardant	↓	50	.320	"	15.40	15		30.40	40

11 62 Musical Equipment

11 62 16 – Carillons

11 62 16.10 Bell Tower Equipment		Crew	Daily Output	Labor-Hours	Unit	Material	Labor	Equipment	Total	Total Incl O&P
0010	**BELL TOWER EQUIPMENT**									
0300	Carillon, 4 octave (48 bells), with keyboard				System	996,000			996,000	1,095,500
0320	2 octave (24 bells)					468,500			468,500	515,500
0340	3 to 4 bell peal, minimum					117,000			117,000	129,000
0360	Maximum				↓	703,000			703,000	773,000
0380	Cast bronze bell, average				Ea.	105,500			105,500	116,000
0400	Electronic, digital, minimum					17,600			17,600	19,300
0410	With keyboard, maximum				↓	88,000			88,000	96,500

11 66 Athletic Equipment

11 66 13 – Exercise Equipment

11 66 13.10 Physical Training Equipment		Crew	Daily Output	Labor-Hours	Unit	Material	Labor	Equipment	Total	Total Incl O&P
0010	**PHYSICAL TRAINING EQUIPMENT**									
0020	Abdominal rack, 2 board capacity				Ea.	490			490	535
0050	Abdominal board, upholstered					665			665	730
0200	Bicycle trainer, minimum					485			485	535
0300	Deluxe, electric					4,350			4,350	4,775
0400	Barbell set, chrome plated steel, 25 lb.					252			252	277
0420	100 lb.					465			465	510
0450	200 lb.				↓	705			705	780
0500	Weight plates, cast iron, per lb.				Lb.	5.25			5.25	5.80
0520	Storage rack, 10 station				Ea.	910			910	1,000
0600	Circuit training apparatus, 12 machines minimum	2 Clab	1.25	12.800	Set	29,000	480		29,480	32,600
0700	Average	↓	1	16	↓	35,600	600		36,200	40,100

11 66 13 – Exercise Equipment

11 66 13.10 Physical Training Equipment		Crew	Daily Output	Labor-Hours	Unit	2015 Bare Costs Material	Labor	Equipment	Total	Total Incl O&P
0800	Maximum	2 Clab	.75	21.333	Set	42,200	800		43,000	47,600
0820	Dumbbell set, cast iron, with rack and 5 pair					625			625	690
0900	Squat racks	2 Clab	5	3.200	Ea.	900	120		1,020	1,175
1200	Multi-station gym machine, 5 station					5,050			5,050	5,575
1250	9 station					11,700			11,700	12,900
1280	Rowing machine, hydraulic					1,775			1,775	1,950
1300	Treadmill, manual					1,100			1,100	1,225
1320	Motorized					3,550			3,550	3,900
1340	Electronic					3,750			3,750	4,125
1360	Cardio-testing					4,600			4,600	5,050
1400	Treatment/massage tables, minimum					595			595	655
1420	Deluxe, with accessories					730			730	800
4150	Exercise equipment, bicycle trainer					760			760	835
4180	Chinning bar, adjustable, wall mounted	1 Carp	5	1.600		212	75		287	350
4200	Exercise ladder, 16′ x 1′-7″, suspended	L-2	3	5.333		1,375	220		1,595	1,850
4210	High bar, floor plate attached	1 Carp	4	2		2,250	94		2,344	2,625
4240	Parallel bars, adjustable		4	2		1,700	94		1,794	2,025
4270	Uneven parallel bars, adjustable		4	2		3,225	94		3,319	3,700
4280	Wall mounted, adjustable	L-2	1.50	10.667	Set	865	440		1,305	1,625
4300	Rope, ceiling mounted, 18′ long	1 Carp	3.66	2.186	Ea.	190	103		293	365
4330	Side horse, vaulting		5	1.600		1,375	75		1,450	1,625
4360	Treadmill, motorized, deluxe, training type		5	1.600		3,850	75		3,925	4,350
4390	Weight lifting multi-station, minimum	2 Clab	1	16		335	600		935	1,300
4450	Maximum	"	.50	32		14,900	1,200		16,100	18,300

11 66 23 – Gymnasium Equipment

11 66 23.13 Basketball Equipment		Crew	Daily Output	Labor-Hours	Unit	Material	Labor	Equipment	Total	Total Incl O&P
0010	**BASKETBALL EQUIPMENT**									
1000	Backstops, wall mtd., 6′ extended, fixed, minimum	L-2	1	16	Ea.	1,375	660		2,035	2,550
1100	Maximum		1	16		1,900	660		2,560	3,125
1200	Swing up, minimum		1	16		1,450	660		2,110	2,600
1250	Maximum		1	16		2,850	660		3,510	4,150
1300	Portable, manual, heavy duty, spring operated		1.90	8.421		12,700	345		13,045	14,500
1400	Ceiling suspended, stationary, minimum		.78	20.513		4,050	845		4,895	5,750
1450	Fold up, with accessories, maximum		.40	40		6,075	1,650		7,725	9,225
1600	For electrically operated, add	1 Elec	1	8		2,350	440		2,790	3,225
5800	Wall pads, 1-1/2″ thick, standard (not fire rated)	2 Carp	640	.025	S.F.	6.25	1.17		7.42	8.70

11 66 23.19 Boxing Ring		Crew	Daily Output	Labor-Hours	Unit	Material	Labor	Equipment	Total	Total Incl O&P
0010	**BOXING RING**									
4100	Elevated, 22′ x 22′	L-4	.10	240	Ea.	7,075	10,600		17,675	24,200
4110	For cellular plastic foam padding, add		.10	240		1,025	10,600		11,625	17,600
4120	Floor level, including posts and ropes only, 20′ x 20′		.80	30		4,550	1,325		5,875	7,050
4130	Canvas, 30′ x 30′		5	4.800		1,275	212		1,487	1,725

11 66 23.47 Gym Mats		Crew	Daily Output	Labor-Hours	Unit	Material	Labor	Equipment	Total	Total Incl O&P
0010	**GYM MATS**									
5500	2″ thick, naugahyde covered				S.F.	3.71			3.71	4.08
5600	Vinyl/nylon covered					8.05			8.05	8.85
6000	Wrestling mats, 1″ thick, heavy duty					5.60			5.60	6.15

11 66 Athletic Equipment

11 66 43 – Interior Scoreboards

11 66 43.10 Scoreboards		Crew	Daily Output	Labor-Hours	Unit	Material	2015 Bare Costs Labor	Equipment	Total	Total Incl O&P
0010	**SCOREBOARDS**									
7000	Baseball, minimum	R-3	1.30	15.385	Ea.	4,125	835	106	5,066	5,925
7200	Maximum		.05	400		18,100	21,700	2,750	42,550	55,500
7300	Football, minimum		.86	23.256		5,325	1,275	160	6,760	7,950
7400	Maximum		.20	100		15,900	5,425	690	22,015	26,400
7500	Basketball (one side), minimum		2.07	9.662		2,400	525	66.50	2,991.50	3,525
7600	Maximum		.30	66.667		3,525	3,625	460	7,610	9,800
7700	Hockey-basketball (four sides), minimum		.25	80		5,675	4,350	550	10,575	13,400
7800	Maximum		.15	133		5,750	7,250	920	13,920	18,200

11 66 53 – Gymnasium Dividers

11 66 53.10 Divider Curtains

		Crew	Daily Output	Labor-Hours	Unit	Material	Labor	Equipment	Total	Total Incl O&P
0010	**DIVIDER CURTAINS**									
4500	Gym divider curtain, mesh top, vinyl bottom, manual	L-4	500	.048	S.F.	9	2.12		11.12	13.20
4700	Electric roll up	L-7	400	.070	"	12.05	3.18		15.23	18.10

11 67 Recreational Equipment

11 67 13 – Bowling Alley Equipment

11 67 13.10 Bowling Alleys

		Crew	Daily Output	Labor-Hours	Unit	Material	Labor	Equipment	Total	Total Incl O&P
0010	**BOWLING ALLEYS** Including alley, pinsetter, scorer,									
0020	Counters and misc. supplies, minimum	4 Carp	.20	160	Lane	42,200	7,500		49,700	58,000
0150	Average		.19	168		46,400	7,900		54,300	63,000
0300	Maximum		.18	177		53,500	8,350		61,850	71,500
0400	Combo table ball rack, add					1,150			1,150	1,275
0600	For automatic scorer, add, minimum					8,325			8,325	9,150
0700	Maximum					10,000			10,000	11,000

11 67 23 – Shooting Range Equipment

11 67 23.10 Shooting Range

		Crew	Daily Output	Labor-Hours	Unit	Material	Labor	Equipment	Total	Total Incl O&P
0010	**SHOOTING RANGE** Incl. bullet traps, target provisions, controls,									
0100	Separators, ceiling system, etc. Not incl. structural shell									
0200	Commercial	L-9	.64	56.250	Point	28,300	2,425		30,725	35,000
0300	Law enforcement		.28	128		38,500	5,525		44,025	51,000
0400	National Guard armories		.71	50.704		20,700	2,175		22,875	26,300
0500	Reserve training centers		.71	50.704		15,900	2,175		18,075	21,000
0600	Schools and colleges		.32	112		35,400	4,825		40,225	46,700
0700	Major academies		.19	189		52,500	8,150		60,650	71,000
0800	For acoustical treatment, add					10%	10%			
0900	For lighting, add					28%	25%			
1000	For plumbing, add					5%	5%			
1100	For ventilating system, add, minimum					40%	40%			
1200	Add, average					25%	25%			
1300	Add, maximum					35%	35%			

11 68 13 – Playground Equipment

11 68 13.10 Free-Standing Playground Equipment

	11 68 13.10 Free-Standing Playground Equipment	Crew	Daily Output	Labor-Hours	Unit	Material	2015 Bare Costs Labor	Equipment	Total	Total Incl O&P
0010	**FREE-STANDING PLAYGROUND EQUIPMENT** See also individual items									
0200	Bike rack, 10′ long, permanent G	B-1	12	2	Ea.	425	76.50		501.50	590
0392	Upper body warm-up station		2.60	9.231		2,025	355		2,380	2,775
0394	Bench stepper station		2.60	9.231		2,875	355		3,230	3,700
0396	Standing push up station		2.60	9.231		1,075	355		1,430	1,750
0398	Upper body stretch station		2.60	9.231		1,700	355		2,055	2,425
0400	Horizontal monkey ladder, 14′ long, 6′ high		4	6		905	230		1,135	1,350
0590	Parallel bars, 10′ long		4	6		320	230		550	705
0600	Posts, tether ball set, 2-3/8″ O.D.		12	2		455	76.50		531.50	620
0800	Poles, multiple purpose, 10′-6″ long		12	2	Pr.	179	76.50		255.50	315
1000	Ground socket for movable posts, 2-3/8″ post		10	2.400		102	92		194	253
1100	3-1/2″ post		10	2.400		167	92		259	325
1300	See-saw, spring, steel, 2 units		6	4	Ea.	740	153		893	1,050
1400	4 units		4	6		1,275	230		1,505	1,775
1500	6 units		3	8		1,675	305		1,980	2,325
1700	Shelter, fiberglass golf tee, 3 person		4.60	5.217		4,375	200		4,575	5,100
1900	Slides, stainless steel bed, 12′ long, 6′ high		3	8		3,325	305		3,630	4,125
2000	20′ long, 10′ high		2	12		3,675	460		4,135	4,750
2200	Swings, plain seats, 8′ high, 4 seats		2	12		1,150	460		1,610	1,975
2300	8 seats		1.30	18.462		2,175	705		2,880	3,450
2500	12′ high, 4 seats		2	12		2,150	460		2,610	3,075
2600	8 seats		1.30	18.462		3,525	705		4,230	4,950
2800	Whirlers, 8′ diameter		3	8		2,775	305		3,080	3,525
2900	10′ diameter		3	8		6,275	305		6,580	7,375

11 68 13.20 Modular Playground

	11 68 13.20 Modular Playground	Crew	Daily Output	Labor-Hours	Unit	Material	Labor	Equipment	Total	Total Incl O&P
0010	**MODULAR PLAYGROUND** Basic components									
0100	Deck, square, steel, 48″ x 48″	B-1	1	24	Ea.	520	920		1,440	2,000
0110	Recycled polyurethane		1	24		515	920		1,435	2,000
0120	Triangular, steel, 48″ side		1	24		680	920		1,600	2,175
0130	Post, steel, 5″ square		18	1.333	L.F.	40.50	51		91.50	123
0140	Aluminum, 2-3/8″ square		20	1.200		39.50	46		85.50	114
0150	5″ square		18	1.333		40.50	51		91.50	123
0160	Roof, square poly, 54″ side		18	1.333	Ea.	1,400	51		1,451	1,625
0170	Wheelchair transfer module, for 3′ high deck		3	8	″	2,950	305		3,255	3,725
0180	Guardrail, pipe, 36″ high		60	.400	L.F.	198	15.30		213.30	242
0190	Steps, deck-to-deck, 3 – 8″ steps		8	3	Ea.	1,125	115		1,240	1,400
0200	Activity panel, crawl through panel		2	12		500	460		960	1,250
0210	Alphabet/spelling panel		2	12		555	460		1,015	1,325
0360	With guardrails		3	8		1,850	305		2,155	2,500
0370	Crawl tunnel, straight, 56″ long		4	6		1,200	230		1,430	1,675
0380	90°, 4′ long		4	6		1,400	230		1,630	1,900
1200	Slide, tunnel, for 56″ high deck		8	3		1,750	115		1,865	2,100
1210	Straight, poly		8	3		390	115		505	605
1220	Stainless steel, 54″ high deck		6	4		690	153		843	995
1230	Curved, poly, 40″ high deck		6	4		865	153		1,018	1,200
1240	Spiral slide, 56″ - 72″ high		5	4.800		4,475	184		4,659	5,200
1300	Ladder, vertical, for 24″ - 72″ high deck		5	4.800		555	184		739	895
1310	Horizontal, 8′ long		5	4.800		695	184		879	1,050
1320	Corkscrew climber, 6′ high		3	8		1,125	305		1,430	1,725
1330	Fire pole for 72″ high deck		6	4		655	153		808	955
1340	Bridge, ring climber, 8′ long		4	6		1,900	230		2,130	2,425
1350	Suspension		4	6	L.F.	360	230		590	750

11 68 Play Field Equipment and Structures

11 68 16 – Play Structures

11 68 16.10 Handball/Squash Court		Crew	Daily Output	Labor-Hours	Unit	2015 Bare Costs Material	Labor	Equipment	Total	Total Incl O&P
0010	**HANDBALL/SQUASH COURT**, outdoor									
0900	Handball or squash court, outdoor, wood	2 Carp	.50	32	Ea.	5,050	1,500		6,550	7,875
1000	Masonry handball/squash court	D-1	.30	53.333	"	25,100	2,250		27,350	31,000

11 68 16.30 Platform/Paddle Tennis Court

		Crew	Daily Output	Labor-Hours	Unit	Material	Labor	Equipment	Total	Total Incl O&P
0010	**PLATFORM/PADDLE TENNIS COURT** Complete with lighting, etc.									
0100	Aluminum slat deck with aluminum frame	B-1	.08	300	Court	60,500	11,500		72,000	84,500
0500	Aluminum slat deck with wood frame	C-1	.12	266		63,500	11,900		75,400	88,000
0800	Aluminum deck heater, add	B-1	1.18	20.339		2,575	780		3,355	4,050
0900	Douglas fir planking with wood frame 2" x 6" x 30'	C-1	.12	266		59,500	11,900		71,400	84,000
1000	Plywood deck with steel frame		.12	266		59,500	11,900		71,400	84,000
1100	Steel slat deck with wood frame		.12	266		39,300	11,900		51,200	61,500

11 68 33 – Athletic Field Equipment

11 68 33.13 Football Field Equipment

		Crew	Daily Output	Labor-Hours	Unit	Material	Labor	Equipment	Total	Total Incl O&P
0010	**FOOTBALL FIELD EQUIPMENT**									
0020	Goal posts, steel, football, double post	B-1	1.50	16	Pr.	3,800	610		4,410	5,125
0100	Deluxe, single post		1.50	16		2,625	610		3,235	3,850
0300	Football, convertible to soccer		1.50	16		2,825	610		3,435	4,075
0500	Soccer, regulation		2	12		1,425	460		1,885	2,250

11 71 Medical Sterilizing Equipment

11 71 10 – Medical Sterilizers & Distillers

11 71 10.10 Sterilizers and Distillers

		Crew	Daily Output	Labor-Hours	Unit	Material	Labor	Equipment	Total	Total Incl O&P
0010	**STERILIZERS AND DISTILLERS**									
0700	Distiller, water, steam heated, 50 gal. capacity	1 Plum	1.40	5.714	Ea.	19,200	335		19,535	21,700
3010	Portable, top loading, 105 – 135 degree C, 3 to 30 psi, 50 L chamber					10,000			10,000	11,000
3020	Stainless steel basket, 10.7" diam. x 11.8" H					223			223	245
3025	Stainless steel pail, 10.7" diam. x 10.7" H					243			243	267
3050	85 L chamber					14,300			14,300	15,700
3060	Stainless steel basket, 15.3" diam. x 11.5" H					305			305	335
3065	Stainless steel pail, 15.3" diam. x 11" H					425			425	470
5600	Sterilizers, floor loading, 26" x 62" x 42", single door, steam					122,000			122,000	134,500
5650	Double door, steam					206,000			206,000	227,000
5800	General purpose, 20" x 20" x 38", single door					13,100			13,100	14,400
6000	Portable, counter top, steam, minimum					3,550			3,550	3,900
6020	Maximum					4,325			4,325	4,750
6050	Portable, counter top, gas, 17" x 15" x 32-1/2"					39,800			39,800	43,700
6150	Manual washer/sterilizer, 16" x 16" x 26"	1 Plum	2	4		54,500	235		54,735	60,500
6200	Steam generators, electric 10 kW to 180 kW, freestanding									
6250	Minimum	1 Elec	3	2.667	Ea.	8,775	146		8,921	9,875
6300	Maximum	"	.70	11.429		29,200	625		29,825	33,000
8200	Bed pan washer-sanitizer	1 Plum	2	4		7,500	235		7,735	8,600

11 72 Examination and Treatment Equipment

11 72 13 – Examination Equipment

11 72 13.13 Examination Equipment		Crew	Daily Output	Labor-Hours	Unit	Material	2015 Bare Costs Labor	Equipment	Total	Total Incl O&P
0010	**EXAMINATION EQUIPMENT**									
0300	Blood pressure unit, mercurial, wall				Ea.	152			152	167
0400	Diagnostic set, wall					780			780	860
4400	Scale, physician's, with height rod				↓	320			320	350

11 72 53 – Treatment Equipment

11 72 53.13 Medical Treatment Equipment		Crew	Daily Output	Labor-Hours	Unit	Material	Labor	Equipment	Total	Total Incl O&P
0010	**MEDICAL TREATMENT EQUIPMENT**									
6300	Exam light, portable, 14" flexible arm				Ea.	197			197	217
6500	Surgery table, minor minimum	1 Sswk	.70	11.429		12,100	600		12,700	14,500
6520	Maximum	"	.50	16		20,200	840		21,040	23,800
6700	Surgical lights, doctor's office, single arm	2 Elec	2	8		2,475	440		2,915	3,375
6750	Dual arm	"	1	16	↓	4,575	875		5,450	6,325

11 73 Patient Care Equipment

11 73 10 – Patient Treatment Equipment

11 73 10.10 Treatment Equipment		Crew	Daily Output	Labor-Hours	Unit	Material	Labor	Equipment	Total	Total Incl O&P
0010	**TREATMENT EQUIPMENT**									
0750	Exam room furnishings, average per room				Ea.	7,050			7,050	7,750
1800	Heat therapy unit, humidified, 26" x 78" x 28"				"	3,600			3,600	3,975
2100	Hubbard tank with accessories, stainless steel,									
2110	125 GPM at 45 psi water pressure				Ea.	27,200			27,200	29,900
2150	For electric overhead hoist, add					2,975			2,975	3,250
2900	K-Module for heat therapy, 20 oz. capacity, 75°F to 110°F					425			425	465
3600	Paraffin bath, 126°F, auto controlled					1,100			1,100	1,200
3900	Parallel bars for walking training, 12'-0"					1,450			1,450	1,600
4600	Station, dietary, medium, with ice					16,700			16,700	18,400
4700	Medicine					7,550			7,550	8,300
7000	Tables, physical therapy, walk off, electric	2 Carp	3	5.333		3,375	250		3,625	4,075
7150	Standard, vinyl top with base cabinets, minimum		3	5.333		1,025	250		1,275	1,500
7200	Maximum	↓	2	8		5,625	375		6,000	6,750
7250	Table, hospital, adjustable height					1,200			1,200	1,325
8400	Whirlpool bath, mobile, sst, 18" x 24" x 60"					4,900			4,900	5,400
8450	Fixed, incl. mixing valves	1 Plum	2	4	↓	9,700	235		9,935	11,100

11 73 10.20 Bariatric Equipment		Crew	Daily Output	Labor-Hours	Unit	Material	Labor	Equipment	Total	Total Incl O&P
0010	**BARIATRIC EQUIPMENT**									
5000	Patient lift, electric operated, arm style									
5110	400 lb. capacity				Ea.	2,000			2,000	2,200
5120	450 lb. capacity					3,050			3,050	3,350
5130	600 lb. capacity					3,200			3,200	3,525
5140	700 lb. capacity					5,050			5,050	5,550
5150	1,000 lb. capacity					10,800			10,800	11,900
5200	Overhead, 4-post, 1,000 lb. capacity					10,400			10,400	11,500
5300	Overhead, track type, 450 lb. capacity, not including track					3,000			3,000	3,275
5500	For fabric sling, add					315			315	345
5550	For digital scale, add				↓	780			780	855

11 74 Dental Equipment

11 74 10 – Dental Office Equipment

11 74 10.10 Diagnostic and Treatment Equipment		Crew	Daily Output	Labor-Hours	Unit	Material	2015 Bare Costs Labor	Equipment	Total	Total Incl O&P
0010	DIAGNOSTIC AND TREATMENT EQUIPMENT									
0020	Central suction system, minimum	1 Plum	1.20	6.667	Ea.	1,575	390		1,965	2,325
0100	Maximum	"	.90	8.889		4,450	520		4,970	5,700
0300	Air compressor, minimum	1 Skwk	.80	10		3,025	485		3,510	4,075
0400	Maximum		.50	16		8,950	780		9,730	11,100
0600	Chair, electric or hydraulic, minimum		.50	16		2,250	780		3,030	3,675
0700	Maximum		.25	32		7,825	1,550		9,375	11,000
0800	Doctor's/assistant's stool, minimum					254			254	279
0850	Maximum					715			715	785
1000	Drill console with accessories, minimum	1 Skwk	1.60	5		2,100	243		2,343	2,675
1100	Maximum		1.60	5		4,900	243		5,143	5,775
2000	Light, ceiling mounted, minimum		8	1		1,175	48.50		1,223.50	1,350
2100	Maximum		8	1		2,025	48.50		2,073.50	2,300
2200	Unit light, minimum	2 Skwk	5.33	3.002		735	146		881	1,025
2210	Maximum		5.33	3.002		1,575	146		1,721	1,950
2220	Track light, minimum		3.20	5		1,575	243		1,818	2,100
2230	Maximum		3.20	5		2,650	243		2,893	3,300
2300	Sterilizers, steam portable, minimum					1,300			1,300	1,425
2350	Maximum					10,500			10,500	11,600
2600	Steam, institutional					3,275			3,275	3,600
2650	Dry heat, electric, portable, 3 trays					1,225			1,225	1,350
2700	Ultra-sonic cleaner, portable, minimum					445			445	490
2750	Maximum (institutional)					1,325			1,325	1,475
3000	X-ray unit, wall, minimum	1 Skwk	4	2		2,350	97.50		2,447.50	2,725
3010	Maximum		4	2		4,050	97.50		4,147.50	4,600
3100	Panoramic unit		.60	13.333		15,700	650		16,350	18,300
3105	Deluxe, minimum	2 Skwk	1.60	10		16,700	485		17,185	19,100
3110	Maximum	"	1.60	10		41,000	485		41,485	45,900
3500	Developers, X-ray, average	1 Plum	5.33	1.501		5,000	88		5,088	5,650
3600	Maximum	"	5.33	1.501		8,200	88		8,288	9,125

11 76 Operating Room Equipment

11 76 10 – Operating Room Equipment

11 76 10.10 Surgical Equipment		Crew	Daily Output	Labor-Hours	Unit	Material	Labor	Equipment	Total	Total Incl O&P
0010	SURGICAL EQUIPMENT									
5000	Scrub, surgical, stainless steel, single station, minimum	1 Plum	3	2.667	Ea.	4,300	157		4,457	4,950
5100	Maximum					6,975			6,975	7,675
6550	Major surgery table, minimum	1 Sswk	.50	16		26,300	840		27,140	30,400
6570	Maximum		.50	16		28,300	840		29,140	32,700
6600	Hydraulic, hand-held control, general surgery		.60	13.333		30,400	700		31,100	34,600
6650	Stationary, universal		.50	16		39,200	840		40,040	44,600
6800	Surgical lights, major operating room, dual head, minimum	2 Elec	1	16		4,375	875		5,250	6,100
6850	Maximum		1	16		30,000	875		30,875	34,400
6900	Ceiling mount articulation, single arm		1	16		3,800	875		4,675	5,475

11 77 Radiology Equipment

11 77 10 – Radiology Equipment

11 77 10.10 X-Ray Equipment		Crew	Daily Output	Labor-Hours	Unit	Material	2015 Bare Costs Labor	Equipment	Total	Total Incl O&P
0010	**X-RAY EQUIPMENT**									
8700	X-ray, mobile, minimum				Ea.	16,800			16,800	18,500
8750	Maximum					77,500			77,500	85,500
8900	Stationary, minimum					43,200			43,200	47,500
8950	Maximum					224,500			224,500	247,000
9150	Developing processors, minimum					4,750			4,750	5,225
9200	Maximum					12,900			12,900	14,100

11 78 Mortuary Equipment

11 78 13 – Mortuary Refrigerators

11 78 13.10 Mortuary and Autopsy Equipment		Crew	Daily Output	Labor-Hours	Unit	Material	Labor	Equipment	Total	Total Incl O&P
0010	**MORTUARY AND AUTOPSY EQUIPMENT**									
0015	Autopsy table, standard	1 Plum	1	8	Ea.	9,750	470		10,220	11,400
0020	Deluxe	"	.60	13.333		15,100	785		15,885	17,800
3200	Mortuary refrigerator, end operated, 2 capacity					12,300			12,300	13,500
3300	6 capacity					22,200			22,200	24,400

11 78 16 – Crematorium Equipment

11 78 16.10 Crematory		Crew	Daily Output	Labor-Hours	Unit	Material	Labor	Equipment	Total	Total Incl O&P
0010	**CREMATORY**									
1500	Crematory, not including building, 1 place	Q-3	.20	160	Ea.	72,500	8,950		81,450	93,000
1750	2 place	"	.10	320	"	103,500	17,900		121,400	141,000

11 81 Facility Maintenance Equipment

11 81 19 – Vacuum Cleaning Systems

11 81 19.10 Vacuum Cleaning		Crew	Daily Output	Labor-Hours	Unit	Material	Labor	Equipment	Total	Total Incl O&P
0010	**VACUUM CLEANING**									
0020	Central, 3 inlet, residential	1 Skwk	.90	8.889	Total	1,075	430		1,505	1,850
0200	Commercial		.70	11.429		1,225	555		1,780	2,200
0400	5 inlet system, residential		.50	16		1,500	780		2,280	2,850
0600	7 inlet system, commercial		.40	20		1,700	975		2,675	3,375
0800	9 inlet system, residential		.30	26.667		3,750	1,300		5,050	6,125
4010	Rule of thumb: First 1200 S.F., installed								1,425	1,575
4020	For each additional S.F., add				S.F.				.26	.26

11 82 Facility Solid Waste Handling Equipment

11 82 19 – Packaged Incinerators

11 82 19.10 Packaged Gas Fired Incinerators		Crew	Daily Output	Labor-Hours	Unit	Material	Labor	Equipment	Total	Total Incl O&P
0010	**PACKAGED GAS FIRED INCINERATORS**									
4400	Incinerator, gas, not incl. chimney, elec. or pipe, 50#/hr., minimum	Q-3	.80	40	Ea.	38,500	2,225		40,725	45,700
4420	Maximum		.70	45.714		40,500	2,550		43,050	48,400
4440	200 lb. per hr., minimum (batch type)		.60	53.333		68,000	2,975		70,975	79,000
4460	Maximum (with feeder)		.50	64		76,000	3,575		79,575	89,000
4480	400 lb. per hr., minimum (batch type)		.30	106		79,000	5,950		84,950	96,000
4500	Maximum (with feeder)		.25	128		101,000	7,150		108,150	122,500
4520	800 lb. per hr., with feeder, minimum		.20	160		121,500	8,950		130,450	147,000
4540	Maximum		.17	188		182,000	10,500		192,500	216,500
4560	1,200 lb. per hr., with feeder, minimum		.15	213		154,000	11,900		165,900	187,000

11 82 Facility Solid Waste Handling Equipment

11 82 19 – Packaged Incinerators

11 82 19.10 Packaged Gas Fired Incinerators		Crew	Daily Output	Labor-Hours	Unit	2015 Bare Costs Material	Labor	Equipment	Total	Total Incl O&P
4580	Maximum	Q-3	.11	290	Ea.	200,000	16,300		216,300	244,500
4600	2,000 lb. per hr., with feeder, minimum		.10	320		405,000	17,900		422,900	472,500
4620	Maximum		.05	640		607,000	35,800		642,800	722,000
4700	For heat recovery system, add, minimum		.25	128		81,000	7,150		88,150	100,000
4710	Add, maximum		.11	290		253,000	16,300		269,300	303,000
4720	For automatic ash conveyer, add		.50	64		33,700	3,575		37,275	42,500
4750	Large municipal incinerators, incl. stack, minimum		.25	128	Ton/day	20,500	7,150		27,650	33,300
4850	Maximum		.10	320	"	54,500	17,900		72,400	87,000

11 82 26 – Facility Waste Compactors

11 82 26.10 Compactors		Crew	Daily Output	Labor-Hours	Unit	Material	Labor	Equipment	Total	Total Incl O&P
0010	**COMPACTORS**									
0020	Compactors, 115 volt, 250#/hr., chute fed	L-4	1	24	Ea.	12,100	1,050		13,150	15,000
0100	Hand fed		2.40	10		15,100	445		15,545	17,300
0300	Multi-bag, 230 volt, 600#/hr., chute fed		1	24		15,100	1,050		16,150	18,300
0400	Hand fed		1	24		15,300	1,050		16,350	18,500
0500	Containerized, hand fed, 2 to 6 C.Y. containers, 250#/hr.		1	24		15,100	1,050		16,150	18,300
0550	For chute fed, add per floor		1	24		1,400	1,050		2,450	3,175
1000	Heavy duty industrial compactor, 0.5 C.Y. capacity		1	24		10,100	1,050		11,150	12,800
1050	1.0 C.Y. capacity		1	24		15,200	1,050		16,250	18,400
1100	3.0 C.Y. capacity		.50	48		25,900	2,125		28,025	31,800
1150	5.0 C.Y. capacity		.50	48		32,600	2,125		34,725	39,100
1200	Combination shredder/compactor (5,000 lb./hr.)		.50	48		63,500	2,125		65,625	73,000
1400	For handling hazardous waste materials, 55 gallon drum packer, std.					19,700			19,700	21,700
1410	55 gallon drum packer w/HEPA filter					24,600			24,600	27,100
1420	55 gallon drum packer w/charcoal & HEPA filter					32,800			32,800	36,100
1430	All of the above made explosion proof, add					1,450			1,450	1,575
5500	Shredder, municipal use, 35 ton per hour					304,500			304,500	335,000
5600	60 ton per hour					648,500			648,500	713,500
5750	Shredder & baler, 50 ton per day					608,000			608,000	669,000
5800	Shredder, industrial, minimum					24,000			24,000	26,400
5850	Maximum					128,500			128,500	141,500
5900	Baler, industrial, minimum					9,625			9,625	10,600
5950	Maximum					560,500			560,500	616,500
6000	Transfer station compactor, with power unit									
6050	and pedestal, not including pit, 50 ton per hour				Ea.	192,500			192,500	211,500

11 82 39 – Medical Waste Disposal Systems

11 82 39.10 Off-Site Disposal		Crew	Daily Output	Labor-Hours	Unit	Material	Labor	Equipment	Total	Total Incl O&P
0010	**OFF-SITE DISPOSAL**									
0100	Medical waste disposal, Red Bag system, pick up & treat, 200 lb. per week				Week	177			177	195
0110	Per month				Month	700			700	770
0150	Red bags, 7-10 gal., 1.2 mil, pkg of 500				Ea.	65.50			65.50	72.50
0200	15 gal., package of 250					61.50			61.50	68
0250	33 gal., package of 250					64			64	70
0300	45 gal., package of 100					54.50			54.50	60

11 82 39.20 Disposal Carts		Crew	Daily Output	Labor-Hours	Unit	Material	Labor	Equipment	Total	Total Incl O&P
0010	**DISPOSAL CARTS**									
2010	Medical waste disposal cart, HDPE, w/lid, 28 gal. capacity				Ea.	261			261	287
2020	96 gal. capacity					305			305	335
2030	150 gal. capacity, low profile					470			470	515
2040	200 gal. capacity					910			910	1,000

11 82 Facility Solid Waste Handling Equipment

11 82 39 – Medical Waste Disposal Systems

11 82 39.30 Medical Waste Sanitizers		Crew	Daily Output	Labor-Hours	Unit	Material	2015 Bare Costs Labor	Equipment	Total	Total Incl O&P
0010	MEDICAL WASTE SANITIZERS									
2010	Small, hand loaded, 1.5 C.Y., 225 lb. capacity				Ea.	78,000			78,000	86,000
2020	Medium, cart loaded, 6.25 C.Y., 938 lb. capacity					104,000			104,000	114,500
2030	Large, cart loaded, 15 C.Y., 2250 lb. capacity					130,500			130,500	143,500
3010	Cart, aluminum, 75 lb. capacity					2,150			2,150	2,375
3020	95 lb. capacity					2,350			2,350	2,575
4010	Stainless steel, 173 lb. capacity					3,075			3,075	3,375
4020	232 lb. capacity					3,450			3,450	3,775
4030	Cart lift, hydraulic scissor type					6,100			6,100	6,700
4040	Portable aluminum ramp					1,725			1,725	1,900
4050	Fold-down steel tracks					1,375			1,375	1,525
4060	Pull-out drawer, small					6,575			6,575	7,225
4070	Medium					9,475			9,475	10,400
4080	Large				▼	13,400			13,400	14,800
5000	Medical waste treatment, sanitize, on-site									
5010	Less than 15,000 lb. per month				Lb.	.20			.20	.22
5020	Over 15,000 lb. per month				"	.16			.16	.18

11 91 Religious Equipment

11 91 13 – Baptisteries

11 91 13.10 Baptistry		Crew	Daily Output	Labor-Hours	Unit	Material	Labor	Equipment	Total	Total Incl O&P
0010	BAPTISTRY									
0150	Fiberglass, 3'-6" deep, x 13'-7" long,									
0160	steps at both ends, incl. plumbing, minimum	L-8	1	20	Ea.	5,725	985		6,710	7,800
0200	Maximum	"	.70	28.571		9,325	1,400		10,725	12,500
0250	Add for filter, heater and lights				▼	1,850			1,850	2,050

11 91 23 – Sanctuary Equipment

11 91 23.10 Sanctuary Furnishings		Crew	Daily Output	Labor-Hours	Unit	Material	Labor	Equipment	Total	Total Incl O&P
0010	SANCTUARY FURNISHINGS									
0020	Altar, wood, custom design, plain	1 Carp	1.40	5.714	Ea.	2,550	268		2,818	3,250
0050	Deluxe	"	.20	40		12,400	1,875		14,275	16,500
0070	Granite or marble, average	2 Marb	.50	32		13,300	1,400		14,700	16,800
0090	Deluxe	"	.20	80		38,000	3,475		41,475	47,100
0100	Arks, prefabricated, plain	2 Carp	.80	20		9,750	940		10,690	12,200
0130	Deluxe, maximum	"	.20	80		138,500	3,750		142,250	158,500
0500	Reconciliation room, wood, prefabricated, single, plain	1 Carp	.60	13.333		3,175	625		3,800	4,475
0550	Deluxe		.40	20		8,750	940		9,690	11,100
0650	Double, plain		.40	20		6,375	940		7,315	8,450
0700	Deluxe		.20	40		19,000	1,875		20,875	23,800
1000	Lecterns, wood, plain		5	1.600		845	75		920	1,050
1100	Deluxe		2	4		6,125	188		6,313	7,050
2000	Pulpits, hardwood, prefabricated, plain		2	4		1,475	188		1,663	1,925
2100	Deluxe		1.60	5	▼	10,100	235		10,335	11,500
2500	Railing, hardwood, average		25	.320	L.F.	208	15		223	252
3000	Seating, individual, oak, contour, laminated		21	.381	Person	178	17.90		195.90	224
3100	Cushion seat		21	.381		162	17.90		179.90	206
3200	Fully upholstered		21	.381		158	17.90		175.90	202
3300	Combination, self-rising	▼	21	.381		305	17.90		322.90	365
3500	For cherry, add				▼	30%				
5000	Wall cross, aluminum, extruded, 2" x 2" section	1 Carp	34	.235	L.F.	218	11.05		229.05	257

11 91 Religious Equipment

11 91 23 – Sanctuary Equipment

11 91 23.10 Sanctuary Furnishings		Crew	Daily Output	Labor-Hours	Unit	Material	2015 Bare Costs Labor	Equipment	Total	Total Incl O&P
5150	4" x 4" section	1 Carp	29	.276	L.F.	315	12.95		327.95	365
5300	Bronze, extruded, 1" x 2" section		31	.258		430	12.10		442.10	490
5350	2-1/2" x 2-1/2" section		34	.235		650	11.05		661.05	730
5450	Solid bar stock, 1/2" x 3" section		29	.276		855	12.95		867.95	960
5600	Fiberglass, stock		34	.235		147	11.05		158.05	178
5700	Stainless steel, 4" deep, channel section		29	.276		690	12.95		702.95	780
5800	4" deep box section	↓	29	.276	↓	940	12.95		952.95	1,050

11 97 Security Equipment

11 97 30 – Security Drawers

11 97 30.10 Pass Through Drawer		Crew	Daily Output	Labor-Hours	Unit	Material	Labor	Equipment	Total	Total Incl O&P
0010	**PASS THROUGH DRAWER**									
0100	Pass-thru drawer for personal items, 18" x 15" x 24"	1 Skwk	2	4	Ea.	2,800	195		2,995	3,375
0110	Including speakers	"	1.50	5.333	"	3,150	259		3,409	3,875

11 98 Detention Equipment

11 98 30 – Detention Cell Equipment

11 98 30.10 Cell Equipment		Crew	Daily Output	Labor-Hours	Unit	Material	Labor	Equipment	Total	Total Incl O&P
0010	**CELL EQUIPMENT**									
3000	Toilet apparatus including wash basin, average	L-8	1.50	13.333	Ea.	3,400	655		4,055	4,750

Division Notes

I	CREW	DAILY OUTPUT	LABOR-HOURS	UNIT	BARE COSTS MAT.	BARE COSTS LABOR	BARE COSTS EQUIP.	BARE COSTS TOTAL	TOTAL INCL O&P

Estimating Tips

General

- The items in this division are usually priced per square foot or each. Most of these items are purchased by the owner and installed by the contractor. Do not assume the items in Division 12 will be purchased and installed by the contractor. Check the specifications for responsibilities and include receiving, storage, installation, and mechanical and electrical hookups in the appropriate divisions.
- Some items in this division require some type of support system that is not usually furnished with the item. Examples of these systems include blocking for the attachment of casework and heavy drapery rods. The required blocking must be added to the estimate in the appropriate division.

Reference Numbers

Reference numbers are shown in shaded boxes at the beginning of some major classifications. These numbers refer to related items in the Reference Section. The reference information may be an estimating procedure, an alternate pricing method, or technical information.

Note: Not all subdivisions listed here necessarily appear in this publication. ■

12 21 Window Blinds

12 21 13 – Horizontal Louver Blinds

12 21 13.13 Metal Horizontal Louver Blinds		Crew	Daily Output	Labor-Hours	Unit	2015 Bare Costs Material	Labor	Equipment	Total	Total Incl O&P
0010	**METAL HORIZONTAL LOUVER BLINDS**									
0020	Horizontal, 1" aluminum slats, solid color, stock	1 Carp	590	.014	S.F.	4.90	.64		5.54	6.40
0070	Horizontal, 1" aluminum slats, custom color		590	.014		5.40	.64		6.04	6.95
0250	2" aluminum slats, solid color, stock		590	.014		5.40	.64		6.04	6.95
0275	2" aluminum slats, custom color		590	.014		5.70	.64		6.34	7.30

12 21 13.33 Vinyl Horizontal Louver Blinds		Crew	Daily Output	Labor-Hours	Unit	Material	Labor	Equipment	Total	Total Incl O&P
0010	**VINYL HORIZONTAL LOUVER BLINDS**									
0100	2" composite, 48" wide, 48" high	1 Carp	30	.267	Ea.	92	12.50		104.50	120
0120	72" high		29	.276		131	12.95		143.95	164
0140	96" high		28	.286		180	13.40		193.40	219
0200	60" wide, 60" high		27	.296		99.50	13.90		113.40	131
0220	72" high		25	.320		114	15		129	149
0240	96" high		24	.333		182	15.65		197.65	224
0300	72" wide, 72" high		25	.320		194	15		209	237
0320	96" high		23	.348		271	16.35		287.35	325
0400	96" wide, 96" high		20	.400		315	18.80		333.80	375
1000	2" faux wood, 48" wide, 48" high		30	.267		59	12.50		71.50	84.50
1020	72" high		29	.276		81	12.95		93.95	109
1040	96" high		28	.286		100	13.40		113.40	131
1300	72" wide, 72" high		25	.320		125	15		140	160
1320	96" high		23	.348		196	16.35		212.35	240
1400	96" wide, 96" high		20	.400		217	18.80		235.80	267

12 21 16 – Vertical Louver Blinds

12 21 16.13 Metal Vertical Louver Blinds		Crew	Daily Output	Labor-Hours	Unit	Material	Labor	Equipment	Total	Total Incl O&P
0010	**METAL VERTICAL LOUVER BLINDS**									
1500	Vertical, 3" PVC strips, minimum	1 Carp	460	.017	S.F.	7.95	.82		8.77	10
1600	Maximum		400	.020		23.50	.94		24.44	27.50
1800	4" aluminum slats, minimum		460	.017		8.10	.82		8.92	10.15
1900	Maximum		400	.020		14.95	.94		15.89	17.90

12 22 Curtains and Drapes

12 22 16 – Drapery Track and Accessories

12 22 16.10 Drapery Hardware		Crew	Daily Output	Labor-Hours	Unit	Material	Labor	Equipment	Total	Total Incl O&P
0010	**DRAPERY HARDWARE**									
0030	Standard traverse, per foot, minimum	1 Carp	59	.136	L.F.	7	6.35		13.35	17.50
0100	Maximum		51	.157	"	10.20	7.35		17.55	22.50
4000	Traverse rods, adjustable, 28" to 48"		22	.364	Ea.	22.50	17.05		39.55	51.50
4020	48" to 84"		20	.400		29	18.80		47.80	60.50
4040	66" to 120"		18	.444		35	21		56	70.50
4060	84" to 156"		16	.500		40	23.50		63.50	79.50
4080	100" to 180"		14	.571		46.50	27		73.50	92.50
4090	156" to 228"		13	.615		55.50	29		84.50	106
4100	228" to 312"		13	.615		64.50	29		93.50	115
4200	Double rods, adjustable, 30" to 48"		9	.889		42.50	41.50		84	111
4220	48" to 86"		9	.889		59	41.50		100.50	129
4240	86" to 150"		8	1		65	47		112	144
4260	100" to 180"		7	1.143		68.50	53.50		122	158
4300	Curtain rod & brackets, adjustable, 30" to 48"		9	.889		29.50	41.50		71	96.50
4320	48" to 86"		9	.889		42	41.50		83.50	111
4340	86" to 150"		8	1		52	47		99	130

12 22 Curtains and Drapes

12 22 16 – Drapery Track and Accessories

12 22 16.10 Drapery Hardware		Crew	Daily Output	Labor-Hours	Unit	2015 Bare Costs Material	Labor	Equipment	Total	Total Incl O&P
4360	100" to 180"	1 Carp	7	1.143	Ea.	64.50	53.50		118	153
4600	Valance, pinch pleated fabric, 12" deep, up to 54" long, minimum					39			39	43
4610	Maximum					98			98	108
4620	Up to 77" long, minimum					60.50			60.50	66.50
4630	Maximum					158			158	174
5000	Stationary rods, first 2'					8.25			8.25	9.10
5020	Each additional foot, add				L.F.	3.90			3.90	4.29

12 22 16.20 Blast Curtains		Crew	Daily Output	Labor-Hours	Unit	Material	Labor	Equipment	Total	Total Incl O&P
0010	**BLAST CURTAINS** per L.F. horizontal opening width, off-white or gray fabric									
0100	Blast curtains, drapery system, complete, including hardware, minimum	1 Carp	10.25	.780	L.F.	189	36.50		225.50	265
0120	Average		10.25	.780		204	36.50		240.50	281
0140	Maximum		10.25	.780		235	36.50		271.50	315

12 23 Interior Shutters

12 23 10 – Wood Interior Shutters

12 23 10.10 Wood Interior Shutters		Crew	Daily Output	Labor-Hours	Unit	Material	Labor	Equipment	Total	Total Incl O&P
0010	**WOOD INTERIOR SHUTTERS**, louvered									
0200	Two panel, 27" wide, 36" high	1 Carp	5	1.600	Set	150	75		225	281
0300	33" wide, 36" high		5	1.600		194	75		269	330
0500	47" wide, 36" high		5	1.600		260	75		335	400
1000	Four panel, 27" wide, 36" high		5	1.600		220	75		295	360
1100	33" wide, 36" high		5	1.600		282	75		357	425
1300	47" wide, 36" high		5	1.600		375	75		450	530

12 23 10.13 Wood Panels		Crew	Daily Output	Labor-Hours	Unit	Material	Labor	Equipment	Total	Total Incl O&P
0010	**WOOD PANELS**									
3000	Wood folding panels with movable louvers, 7" x 20" each	1 Carp	17	.471	Pr.	79.50	22		101.50	122
3300	8" x 28" each		17	.471		79.50	22		101.50	122
3450	9" x 36" each		17	.471		91.50	22		113.50	135
3600	10" x 40" each		17	.471		100	22		122	144
4000	Fixed louver type, stock units, 8" x 20" each		17	.471		94	22		116	137
4150	10" x 28" each		17	.471		79.50	22		101.50	122
4300	12" x 36" each		17	.471		94	22		116	137
4450	18" x 40" each		17	.471		134	22		156	182
5000	Insert panel type, stock, 7" x 20" each		17	.471		21	22		43	57
5150	8" x 28" each		17	.471		38	22		60	76
5300	9" x 36" each		17	.471		48.50	22		70.50	87
5450	10" x 40" each		17	.471		52	22		74	91
5600	Raised panel type, stock, 10" x 24" each		17	.471		247	22		269	305
5650	12" x 26" each		17	.471		247	22		269	305
5700	14" x 30" each		17	.471		273	22		295	335
5750	16" x 36" each		17	.471		300	22		322	370
6000	For custom built pine, add					22%				
6500	For custom built hardwood blinds, add					42%				

12 24 Window Shades

12 24 13 – Roller Window Shades

12 24 13.10 Shades		Crew	Daily Output	Labor-Hours	Unit	Material	2015 Bare Costs Labor	Equipment	Total	Total Incl O&P
0010	**SHADES**									
0020	Basswood, roll-up, stain finish, 3/8″ slats	1 Carp	300	.027	S.F.	14.95	1.25		16.20	18.35
0200	7/8″ slats		300	.027		14.10	1.25		15.35	17.45
0300	Vertical side slide, stain finish, 3/8″ slats		300	.027		19.30	1.25		20.55	23.50
0400	7/8″ slats		300	.027		19.30	1.25		20.55	23.50
0500	For fire retardant finishes, add					16%				
0600	For ″B″ rated finishes, add					20%				
0900	Mylar, single layer, non-heat reflective	1 Carp	685	.012		5.05	.55		5.60	6.40
0910	Mylar, single layer, heat reflective		685	.012		5.45	.55		6	6.85
1000	Double layered, heat reflective		685	.012		5.85	.55		6.40	7.30
1100	Triple layered, heat reflective		685	.012		6.40	.55		6.95	7.85
1200	For metal roller instead of wood, add per				Shade	4.51			4.51	4.96
1300	Vinyl coated cotton, standard	1 Carp	685	.012	S.F.	2.90	.55		3.45	4.03
1400	Lightproof decorator shades		685	.012		3	.55		3.55	4.14
1500	Vinyl, lightweight, 4 ga.		685	.012		.61	.55		1.16	1.51
1600	Heavyweight, 6 ga.		685	.012		1.87	.55		2.42	2.90
1700	Vinyl laminated fiberglass, 6 ga., translucent		685	.012		2.60	.55		3.15	3.70
1800	Lightproof		685	.012		4.27	.55		4.82	5.55
2000	Polyester, room darkening, with continuous cord, GEI									
2010	36″ x 72″	1 Carp	38	.211	Ea.	224	9.90		233.90	262
2020	48″ x 72″		28	.286		293	13.40		306.40	340
2030	60″ x 72″		23	.348		345	16.35		361.35	405
2040	72″ x 72″		19	.421		395	19.75		414.75	465
6011	Solar screening, fiberglass G		85	.094	S.F.	6.75	4.42		11.17	14.20

12 32 Manufactured Wood Casework

12 32 16 – Manufactured Plastic-Laminate-Clad Casework

12 32 16.20 Plastic Laminate Casework Doors

		Crew	Daily Output	Labor-Hours	Unit	Material	Labor	Equipment	Total	Total Incl O&P
0010	**PLASTIC LAMINATE CASEWORK DOORS**									
1000	For casework frames, see Section 12 32 23.15									
1100	For casework hardware, see Section 12 32 23.35									
6000	Plastic laminate on particle board									
6100	12″ wide, 18″ high	1 Carp	25	.320	Ea.	24	15		39	49.50
6140	30″ high		23	.348		40	16.35		56.35	69
6500	18″ wide, 18″ high		24	.333		36	15.65		51.65	63.50
6600	30″ high		22	.364		60	17.05		77.05	92.50

12 32 16.25 Plastic Laminate Drawer Fronts

		Crew	Daily Output	Labor-Hours	Unit	Material	Labor	Equipment	Total	Total Incl O&P
0010	**PLASTIC LAMINATE DRAWER FRONTS**									
2800	Plastic laminate on particle board front									
3000	4″ high, 12″ wide	1 Carp	17	.471	Ea.	4.51	22		26.51	39
3200	18″ wide	″	16	.500	″	6.75	23.50		30.25	43.50

12 32 23 – Hardwood Casework

12 32 23.10 Manufactured Wood Casework, Stock Units

		Crew	Daily Output	Labor-Hours	Unit	Material	Labor	Equipment	Total	Total Incl O&P
0010	**MANUFACTURED WOOD CASEWORK, STOCK UNITS**									
0300	Built-in drawer units, pine, 18″ deep, 32″ high, unfinished									
0400	Minimum	2 Carp	53	.302	L.F.	120	14.15		134.15	154
0500	Maximum	″	40	.400	″	141	18.80		159.80	184
0700	Kitchen base cabinets, hardwood, not incl. counter tops,									
0710	24″ deep, 35″ high, prefinished									
0800	One top drawer, one door below, 12″ wide	2 Carp	24.80	.645	Ea.	265	30.50		295.50	340

12 32 23 – Hardwood Casework

12 32 23.10 Manufactured Wood Casework, Stock Units		Crew	Daily Output	Labor-Hours	Unit	Material	2015 Bare Costs Labor	Equipment	Total	Total Incl O&P
0840	18" wide	2 Carp	23.30	.687	Ea.	300	32		332	380
0880	24" wide		22.30	.717		365	33.50		398.50	450
1000	Four drawers, 12" wide		24.80	.645		279	30.50		309.50	350
1040	18" wide		23.30	.687		310	32		342	395
1060	24" wide		22.30	.717		345	33.50		378.50	430
1200	Two top drawers, two doors below, 27" wide		22	.727		390	34		424	485
1260	36" wide		20.30	.788		455	37		492	555
1300	48" wide		18.90	.847		515	40		555	630
1500	Range or sink base, two doors below, 30" wide		21.40	.748		350	35		385	440
1540	36" wide		20.30	.788		395	37		432	490
1580	48" wide		18.90	.847		435	40		475	540
1800	For sink front units, deduct					161			161	177
2000	Corner base cabinets, 36" wide, standard	2 Carp	18	.889		625	41.50		666.50	755
2100	Lazy Susan with revolving door	"	16.50	.970		840	45.50		885.50	995
4000	Kitchen wall cabinets, hardwood, 12" deep with two doors									
4050	12" high, 30" wide	2 Carp	24.80	.645	Ea.	237	30.50		267.50	310
4100	36" wide		24	.667		282	31.50		313.50	360
4400	15" high, 30" wide		24	.667		241	31.50		272.50	315
4440	36" wide		22.70	.705		290	33		323	370
4700	24" high, 30" wide		23.30	.687		325	32		357	405
4720	36" wide		22.70	.705		355	33		388	440
5000	30" high, one door, 12" wide		22	.727		216	34		250	291
5040	18" wide		20.90	.766		265	36		301	350
5060	24" wide		20.30	.788		310	37		347	395
5300	Two doors, 27" wide		19.80	.808		340	38		378	435
5340	36" wide		18.80	.851		405	40		445	510
5380	48" wide		18.40	.870		500	41		541	615
6000	Corner wall, 30" high, 24" wide		18	.889		355	41.50		396.50	455
6050	30" wide		17.20	.930		380	43.50		423.50	480
6100	36" wide		16.50	.970		430	45.50		475.50	540
6500	Revolving Lazy Susan		15.20	1.053		480	49.50		529.50	600
7000	Broom cabinet, 84" high, 24" deep, 18" wide		10	1.600		650	75		725	830
7500	Oven cabinets, 84" high, 24" deep, 27" wide		8	2		1,000	94		1,094	1,250
7750	Valance board trim		396	.040	L.F.	13	1.90		14.90	17.20
7780	Toe kick trim	1 Carp	256	.031	"	2.79	1.47		4.26	5.35
7790	Base cabinet corner filler		16	.500	Ea.	41.50	23.50		65	81.50
7800	Cabinet filler, 3" x 24"		20	.400		18.05	18.80		36.85	49
7810	3" x 30"		20	.400		22.50	18.80		41.30	54
7820	3" x 42"		18	.444		31.50	21		52.50	66.50
7830	3" x 80"		16	.500		60	23.50		83.50	102
7850	Cabinet panel		50	.160	S.F.	8.65	7.50		16.15	21
9000	For deluxe models of all cabinets, add					40%				
9500	For custom built in place, add					25%	10%			
9558	Rule of thumb, kitchen cabinets not including									
9560	appliances & counter top, minimum	2 Carp	30	.533	L.F.	176	25		201	232
9600	Maximum	"	25	.640	"	395	30		425	480
9610	For metal cabinets, see Section 12 35 70.13									

12 32 23.15 Manufactured Wood Casework Frames		Crew	Daily Output	Labor-Hours	Unit	Material	Labor	Equipment	Total	Total Incl O&P
0010	**MANUFACTURED WOOD CASEWORK FRAMES**									
0050	Base cabinets, counter storage, 36" high									
0100	One bay, 18" wide	1 Carp	2.70	2.963	Ea.	171	139		310	400
0400	Two bay, 36" wide		2.20	3.636		261	171		432	550

12 32 23 – Hardwood Casework

12 32 23.15 Manufactured Wood Casework Frames		Crew	Daily Output	Labor-Hours	Unit	2015 Bare Costs Material	Labor	Equipment	Total	Total Incl O&P
1100	Three bay, 54" wide	1 Carp	1.50	5.333	Ea.	310	250		560	725
2800	Bookcases, one bay, 7' high, 18" wide		2.40	3.333		201	157		358	460
3500	Two bay, 36" wide		1.60	5		292	235		527	680
4100	Three bay, 54" wide		1.20	6.667		485	315		800	1,000
5100	Coat racks, one bay, 7' high, 24" wide		4.50	1.778		201	83.50		284.50	350
5300	Two bay, 48" wide		2.75	2.909		279	137		416	515
5800	Three bay, 72" wide		2.10	3.810		410	179		589	725
6100	Wall mounted cabinet, one bay, 24" high, 18" wide		3.60	2.222		110	104		214	282
6800	Two bay, 36" wide		2.20	3.636		161	171		332	440
7400	Three bay, 54" wide		1.70	4.706		201	221		422	560
8400	30" high, one bay, 18" wide		3.60	2.222		120	104		224	293
9000	Two bay, 36" wide		2.15	3.721		160	175		335	445
9400	Three bay, 54" wide		1.60	5		199	235		434	580
9800	Wardrobe, 7' high, single, 24" wide		2.70	2.963		222	139		361	460
9880	Partition & adjustable shelves, 48" wide		1.70	4.706		282	221		503	650
9950	Partition, adjustable shelves & drawers, 48" wide		1.40	5.714		425	268		693	880

12 32 23.20 Manufactured Hardwood Casework Doors

		Crew	Daily Output	Labor-Hours	Unit	Material	Labor	Equipment	Total	Total Incl O&P
0010	**MANUFACTURED HARDWOOD CASEWORK DOORS**									
2000	Glass panel, hardwood frame									
2200	12" wide, 18" high	1 Carp	34	.235	Ea.	27	11.05		38.05	46.50
2600	30" high		32	.250		45	11.75		56.75	67.50
4450	18" wide, 18" high		32	.250		40.50	11.75		52.25	62.50
4550	30" high		29	.276		67.50	12.95		80.45	94.50
5000	Hardwood, raised panel									
5100	12" wide, 18" high	1 Carp	16	.500	Ea.	28.50	23.50		52	67.50
5200	30" high		15	.533		47.50	25		72.50	91
5500	18" wide, 18" high		15	.533		43	25		68	85.50
5600	30" high		14	.571		71.50	27		98.50	120

12 32 23.25 Manufactured Wood Casework Drawer Fronts

		Crew	Daily Output	Labor-Hours	Unit	Material	Labor	Equipment	Total	Total Incl O&P
0010	**MANUFACTURED WOOD CASEWORK DRAWER FRONTS**									
0100	Solid hardwood front									
1000	4" high, 12" wide	1 Carp	17	.471	Ea.	4.33	22		26.33	39
1200	18" wide	"	16	.500	"	6.50	23.50		30	43

12 32 23.30 Manufactured Wood Casework Vanities

		Crew	Daily Output	Labor-Hours	Unit	Material	Labor	Equipment	Total	Total Incl O&P
0010	**MANUFACTURED WOOD CASEWORK VANITIES**									
8000	Vanity bases, 2 doors, 30" high, 21" deep, 24" wide	2 Carp	20	.800	Ea.	310	37.50		347.50	400
8050	30" wide		16	1		370	47		417	480
8100	36" wide		13.33	1.200		360	56.50		416.50	480
8150	48" wide		11.43	1.400		470	65.50		535.50	620
9000	For deluxe models of all vanities, add to above					40%				
9500	For custom built in place, add to above					25%	10%			

12 32 23.35 Manufactured Wood Casework Hardware

		Crew	Daily Output	Labor-Hours	Unit	Material	Labor	Equipment	Total	Total Incl O&P
0010	**MANUFACTURED WOOD CASEWORK HARDWARE**									
1000	Catches, minimum	1 Carp	235	.034	Ea.	1.22	1.60		2.82	3.80
1040	Maximum	"	80	.100	"	7.55	4.70		12.25	15.55
2000	Door/drawer pulls, handles									
2200	Handles and pulls, projecting, metal, minimum	1 Carp	48	.167	Ea.	5	7.85		12.85	17.55
2240	Maximum		36	.222		10.60	10.45		21.05	27.50
2300	Wood, minimum		48	.167		5.25	7.85		13.10	17.85
2340	Maximum		36	.222		9.65	10.45		20.10	26.50
2400	Drawer pulls, antimicrobial copper alloy finish		50	.160		18.75	7.50		26.25	32
2600	Flush, metal, minimum		48	.167		5.25	7.85		13.10	17.85

12 32 Manufactured Wood Casework

12 32 23 – Hardwood Casework

12 32 23.35 Manufactured Wood Casework Hardware		Crew	Daily Output	Labor-Hours	Unit	2015 Bare Costs Material	Labor	Equipment	Total	Total Incl O&P
2640	Maximum	1 Carp	36	.222	Ea.	9.65	10.45		20.10	26.50
2900	Drawer knobs, antimicrobial copper alloy finish		50	.160	↓	12.50	7.50		20	25.50
3000	Drawer tracks/glides, minimum		48	.167	Pr.	8.95	7.85		16.80	22
3040	Maximum		24	.333		26	15.65		41.65	52.50
4000	Cabinet hinges, minimum		160	.050		3.02	2.35		5.37	6.95
4040	Maximum		68	.118	↓	11.45	5.50		16.95	21
7000	Appliance pulls, antimicrobial copper alloy finish	↓	50	.160	L.F.	62.50	7.50		70	80.50

12 35 Specialty Casework

12 35 50 – Educational/Library Casework

12 35 50.13 Educational Casework

		Crew	Daily Output	Labor-Hours	Unit	Material	Labor	Equipment	Total	Total Incl O&P
0010	**EDUCATIONAL CASEWORK**									
5000	School, 24" deep, metal, 84" high units	2 Carp	15	1.067	L.F.	430	50		480	545
5150	Counter height units		20	.800		288	37.50		325.50	375
5450	Wood, custom fabricated, 32" high counter		20	.800		240	37.50		277.50	320
5600	Add for counter top		56	.286		25.50	13.40		38.90	48.50
5800	84" high wall units	↓	15	1.067	↓	465	50		515	585
6000	Laminated plastic finish is same price as wood									

12 35 53 – Laboratory Casework

12 35 53.13 Metal Laboratory Casework

		Crew	Daily Output	Labor-Hours	Unit	Material	Labor	Equipment	Total	Total Incl O&P
0010	**METAL LABORATORY CASEWORK**									
0020	Cabinets, base, door units, metal	2 Carp	18	.889	L.F.	231	41.50		272.50	320
0300	Drawer units		18	.889		515	41.50		556.50	630
0700	Tall storage cabinets, open, 7' high		20	.800		495	37.50		532.50	605
0900	With glazed doors		20	.800		740	37.50		777.50	875
1300	Wall cabinets, metal, 12-1/2" deep, open		20	.800		166	37.50		203.50	241
1500	With doors	↓	20	.800	↓	345	37.50		382.50	440
6300	Rule of thumb: lab furniture including installation & connection									
6320	High school				S.F.				35	39
6340	College								52	57
6360	Clinical, health care								45	49.50
6380	Industrial				↓				72.50	79.50

12 35 59 – Display Casework

12 35 59.10 Display Cases

		Crew	Daily Output	Labor-Hours	Unit	Material	Labor	Equipment	Total	Total Incl O&P
0010	**DISPLAY CASES** Free standing, all glass									
0020	Aluminum frame, 42" high x 36" x 12" deep	2 Carp	8	2	Ea.	1,225	94		1,319	1,500
0100	70" high x 48" x 18" deep	"	6	2.667		3,775	125		3,900	4,350
0500	For wood bases, add					9%				
0600	For hardwood frames, deduct					8%				
0700	For bronze, baked enamel finish, add				↓	10%				
2000	Wall mounted, glass front, aluminum frame									
2010	Non-illuminated, one section 3' x 4' x 1'-4"	2 Carp	5	3.200	Ea.	2,175	150		2,325	2,625
2100	5' x 4' x 1'-4"		5	3.200		2,525	150		2,675	3,000
2200	6' x 4' x 1'-4"		4	4		3,050	188		3,238	3,650
2500	Two sections, 8' x 4' x 1'-4"		2	8		2,225	375		2,600	3,025
2600	10' x 4' x 1'-4"		2	8		2,725	375		3,100	3,575
3000	Three sections, 16' x 4' x 1'-4"	↓	1.50	10.667	↓	4,125	500		4,625	5,325
3500	For fluorescent lights, add				Section	330			330	365
4000	Table exhibit cases, 2' wide, 3' high, 4' long, flat top	2 Carp	5	3.200	Ea.	1,375	150		1,525	1,725
4100	3' wide, 3' high, 4' long, sloping top	"	3	5.333	"	825	250		1,075	1,300

12 35 Specialty Casework

12 35 70 – Healthcare Casework

12 35 70.13 Hospital Casework

	12 35 70.13 Hospital Casework	Crew	Daily Output	Labor-Hours	Unit	Material	2015 Bare Costs Labor	Equipment	Total	Total Incl O&P
0010	**HOSPITAL CASEWORK**									
0500	Base cabinets, laminated plastic	2 Carp	10	1.600	L.F.	275	75		350	415
1000	Stainless steel	"	10	1.600		505	75		580	670
1200	For all drawers, add					28.50			28.50	31.50
1300	Cabinet base trim, 4" high, enameled steel	2 Carp	200	.080		46	3.76		49.76	56.50
1400	Stainless steel		200	.080		92	3.76		95.76	107
1450	Countertop, laminated plastic, no backsplash		40	.400		47.50	18.80		66.30	81
1650	With backsplash		40	.400		59	18.80		77.80	94
1800	For sink cutout, add		12.20	1.311	Ea.		61.50		61.50	95
1900	Stainless steel counter top		40	.400	L.F.	153	18.80		171.80	198
2000	For drop-in stainless 43" x 21" sink, add				Ea.	1,000			1,000	1,100
2050	Laminate with antimicrobial finish #4	2 Carp	40	.400	L.F.	32	18.80		50.80	64
2500	Wall cabinets, laminated plastic		15	1.067		206	50		256	305
2600	Enameled steel		15	1.067		253	50		303	355
2700	Stainless steel		15	1.067		505	50		555	630
3000	Hospital cabinets, stainless steel with glass door(s), lockable									
3010	One door, 24" W x 18" D x 60" H	2 Clab	18	.889	Ea.	2,500	33.50		2,533.50	2,775
3020	Two doors, 36" W x 18" D x 60" H		15	1.067		2,825	40		2,865	3,150
3030	36" W x 24" D x 67" H		15	1.067		4,250	40		4,290	4,725
3040	48" W x 24" D x 66" H		12	1.333		4,000	50		4,050	4,475
3050	48" W x 24" D x 72" H		12	1.333		5,050	50		5,100	5,625
3060	60" W x 24" D x 72" H		9	1.778		5,500	67		5,567	6,150

12 35 70.16 Nurse Station Casework

	12 35 70.16 Nurse Station Casework	Crew	Daily Output	Labor-Hours	Unit	Material	Labor	Equipment	Total	Total Incl O&P
0010	**NURSE STATION CASEWORK**									
2100	Door type, laminated plastic	2 Carp	10	1.600	L.F.	320	75		395	465
2200	Enameled steel		10	1.600		305	75		380	450
2300	Stainless steel		10	1.600		610	75		685	785
2400	For drawer type, add					258			258	284

12 35 80 – Commercial Kitchen Casework

12 35 80.13 Metal Kitchen Casework

	12 35 80.13 Metal Kitchen Casework	Crew	Daily Output	Labor-Hours	Unit	Material	Labor	Equipment	Total	Total Incl O&P
0010	**METAL KITCHEN CASEWORK**									
3500	Base cabinets, metal, minimum	2 Carp	30	.533	L.F.	74	25		99	120
3600	Maximum		25	.640		188	30		218	253
3700	Wall cabinets, metal, minimum		30	.533		74	25		99	120
3800	Maximum		25	.640		170	30		200	233

12 36 Countertops

12 36 16 – Metal Countertops

12 36 16.10 Stainless Steel Countertops

	12 36 16.10 Stainless Steel Countertops	Crew	Daily Output	Labor-Hours	Unit	Material	Labor	Equipment	Total	Total Incl O&P
0010	**STAINLESS STEEL COUNTERTOPS**									
3200	Stainless steel, custom	1 Carp	24	.333	S.F.	153	15.65		168.65	192

12 36 19 – Wood Countertops

12 36 19.10 Maple Countertops

	12 36 19.10 Maple Countertops	Crew	Daily Output	Labor-Hours	Unit	Material	Labor	Equipment	Total	Total Incl O&P
0010	**MAPLE COUNTERTOPS**									
2900	Solid, laminated, 1-1/2" thick, no splash	1 Carp	28	.286	L.F.	75.50	13.40		88.90	104
3000	With square splash		28	.286	"	90	13.40		103.40	119
3400	Recessed cutting block with trim, 16" x 20" x 1"		8	1	Ea.	92	47		139	174

12 36 Countertops

12 36 23 – Plastic Countertops

12 36 23.13 Plastic-Laminate-Clad Countertops		Crew	Daily Output	Labor-Hours	Unit	Material	2015 Bare Costs Labor	Equipment	Total	Total Incl O&P
0010	**PLASTIC-LAMINATE-CLAD COUNTERTOPS**									
0020	Stock, 24" wide w/backsplash, minimum	1 Carp	30	.267	L.F.	17	12.50		29.50	38
0100	Maximum		25	.320		34.50	15		49.50	61
0300	Custom plastic, 7/8" thick, aluminum molding, no splash		30	.267		30	12.50		42.50	52.50
0400	Cove splash		30	.267		29	12.50		41.50	51.50
0600	1-1/4" thick, no splash		28	.286		35.50	13.40		48.90	59.50
0700	Square splash		28	.286		42.50	13.40		55.90	67
0900	Square edge, plastic face, 7/8" thick, no splash		30	.267		33	12.50		45.50	56
1000	With splash		30	.267		39.50	12.50		52	62.50
1200	For stainless channel edge, 7/8" thick, add					3.12			3.12	3.43
1300	1-1/4" thick, add					3.72			3.72	4.09
1500	For solid color suede finish, add					4.08			4.08	4.49
1700	For end splash, add				Ea.	18.35			18.35	20
1900	For cut outs, standard, add, minimum	1 Carp	32	.250		12.25	11.75		24	31.50
2000	Maximum		8	1		6.10	47		53.10	79.50
2100	Postformed, including backsplash and front edge		30	.267	L.F.	10.20	12.50		22.70	30.50
2110	Mitred, add		12	.667	Ea.		31.50		31.50	48
2200	Built-in place, 25" wide, plastic laminate		25	.320	L.F.	41.50	15		56.50	68.50

12 36 33 – Tile Countertops

12 36 33.10 Ceramic Tile Countertops

		Crew	Daily Output	Labor-Hours	Unit	Material	Labor	Equipment	Total	Total Incl O&P
0010	**CERAMIC TILE COUNTERTOPS**									
2300	Ceramic tile mosaic	1 Carp	25	.320	L.F.	33.50	15		48.50	60

12 36 40 – Stone Countertops

12 36 40.10 Natural Stone Countertops

		Crew	Daily Output	Labor-Hours	Unit	Material	Labor	Equipment	Total	Total Incl O&P
0010	**NATURAL STONE COUNTERTOPS**									
2500	Marble, stock, with splash, 1/2" thick, minimum	1 Bric	17	.471	L.F.	42	21.50		63.50	79
2700	3/4" thick, maximum		13	.615		105	28.50		133.50	160
2800	Granite, average, 1-1/4" thick, 24" wide, no splash		13.01	.615		138	28.50		166.50	195

12 36 53 – Laboratory Countertops

12 36 53.10 Laboratory Countertops and Sinks

		Crew	Daily Output	Labor-Hours	Unit	Material	Labor	Equipment	Total	Total Incl O&P
0010	**LABORATORY COUNTERTOPS AND SINKS**									
0020	Countertops, epoxy resin, not incl. base cabinets, acid-proof, minimum	2 Carp	82	.195	S.F.	40.50	9.15		49.65	58.50
0030	Maximum		70	.229		50	10.75		60.75	71.50
0040	Stainless steel		82	.195		131	9.15		140.15	158

12 36 61 – Simulated Stone Countertops

12 36 61.16 Solid Surface Countertops

		Crew	Daily Output	Labor-Hours	Unit	Material	Labor	Equipment	Total	Total Incl O&P
0010	**SOLID SURFACE COUNTERTOPS**, Acrylic polymer									
0020	Pricing for orders of 100 L.F. or greater									
0100	25" wide, solid colors	2 Carp	28	.571	L.F.	54.50	27		81.50	102
0200	Patterned colors		28	.571		69	27		96	118
0300	Premium patterned colors		28	.571		86.50	27		113.50	137
0400	With silicone attached 4" backsplash, solid colors		27	.593		60	28		88	109
0500	Patterned colors		27	.593		76	28		104	127
0600	Premium patterned colors		27	.593		94.50	28		122.50	147
0700	With hard seam attached 4" backsplash, solid colors		23	.696		60	32.50		92.50	117
0800	Patterned colors		23	.696		76	32.50		108.50	134
0900	Premium patterned colors		23	.696		94.50	32.50		127	155
1000	Pricing for order of 51 – 99 L.F.									
1100	25" wide, solid colors	2 Carp	24	.667	L.F.	63	31.50		94.50	117
1200	Patterned colors		24	.667		79.50	31.50		111	136

12 36 61 – Simulated Stone Countertops

12 36 61.16 Solid Surface Countertops		Crew	Daily Output	Labor-Hours	Unit	Material	Labor	Equipment	Total	Total Incl O&P
						2015 Bare Costs				
1300	Premium patterned colors	2 Carp	24	.667	L.F.	99.50	31.50		131	157
1400	With silicone attached 4″ backsplash, solid colors		23	.696		69	32.50		101.50	127
1500	Patterned colors		23	.696		87.50	32.50		120	147
1600	Premium patterned colors		23	.696		109	32.50		141.50	171
1700	With hard seam attached 4″ backsplash, solid colors		20	.800		69	37.50		106.50	134
1800	Patterned colors		20	.800		87.50	37.50		125	154
1900	Premium patterned colors		20	.800		109	37.50		146.50	178
2000	Pricing for order of 1 – 50 L.F.									
2100	25″ wide, solid colors	2 Carp	20	.800	L.F.	73.50	37.50		111	139
2200	Patterned colors		20	.800		93.50	37.50		131	161
2300	Premium patterned colors		20	.800		117	37.50		154.50	187
2400	With silicone attached 4″ backsplash, solid colors		19	.842		81	39.50		120.50	150
2500	Patterned colors		19	.842		102	39.50		141.50	174
2600	Premium patterned colors		19	.842		128	39.50		167.50	201
2700	With hard seam attached 4″ backsplash, solid colors		15	1.067		81	50		131	166
2800	Patterned colors		15	1.067		102	50		152	190
2900	Premium patterned colors		15	1.067		128	50		178	217
3000	Sinks, pricing for order of 100 or greater units									
3100	Single bowl, hard seamed, solid colors, 13″ x 17″	1 Carp	3	2.667	Ea.	370	125		495	600
3200	10″ x 15″		7	1.143		170	53.50		223.50	270
3300	Cutouts for sinks		8	1			47		47	72.50
3400	Sinks, pricing for order of 51 – 99 units									
3500	Single bowl, hard seamed, solid colors, 13″ x 17″	1 Carp	2.55	3.137	Ea.	425	147		572	690
3600	10″ x 15″		6	1.333		196	62.50		258.50	315
3700	Cutouts for sinks		7	1.143			53.50		53.50	82.50
3800	Sinks, pricing for order of 1 – 50 units									
3900	Single bowl, hard seamed, solid colors, 13″ x 17″	1 Carp	2	4	Ea.	500	188		688	840
4000	10″ x 15″		4.55	1.758		230	82.50		312.50	380
4100	Cutouts for sinks		5.25	1.524			71.50		71.50	110
4200	Cooktop cutouts, pricing for 100 or greater units		4	2		27	94		121	175
4300	51 – 99 units		3.40	2.353		31.50	110		141.50	205
4400	1 – 50 units		3	2.667		36.50	125		161.50	234

12 36 61.17 Solid Surface Vanity Tops		Crew	Daily Output	Labor-Hours	Unit	Material	Labor	Equipment	Total	Total Incl O&P
0010	**SOLID SURFACE VANITY TOPS**									
0015	Solid surface, center bowl, 17″ x 19″	1 Carp	12	.667	Ea.	190	31.50		221.50	257
0020	19″ x 25″		12	.667		194	31.50		225.50	262
0030	19″ x 31″		12	.667		227	31.50		258.50	298
0040	19″ x 37″		12	.667		264	31.50		295.50	340
0050	22″ x 25″		10	.800		345	37.50		382.50	440
0060	22″ x 31″		10	.800		405	37.50		442.50	505
0070	22″ x 37″		10	.800		470	37.50		507.50	580
0080	22″ x 43″		10	.800		535	37.50		572.50	650
0090	22″ x 49″		10	.800		595	37.50		632.50	715
0110	22″ x 55″		8	1		675	47		722	820
0120	22″ x 61″		8	1		770	47		817	925
0220	Double bowl, 22″ x 61″		8	1		870	47		917	1,025
0230	Double bowl, 22″ x 73″		8	1		950	47		997	1,125
0240	For aggregate colors, add					35%				
0250	For faucets and fittings, see Section 22 41 39.10									

12 36 Countertops

12 36 61 – Simulated Stone Countertops

12 36 61.19 Quartz Agglomerate Countertops		Crew	Daily Output	Labor-Hours	Unit	2015 Bare Costs Material	Labor	Equipment	Total	Total Incl O&P
0010	**QUARTZ AGGLOMERATE COUNTERTOPS**									
0100	25" wide, 4" backsplash, color group A, minimum	2 Carp	15	1.067	L.F.	64.50	50		114.50	148
0110	Maximum		15	1.067		90	50		140	176
0120	Color group B, minimum		15	1.067		66.50	50		116.50	151
0130	Maximum		15	1.067		94.50	50		144.50	181
0140	Color group C, minimum		15	1.067		78	50		128	163
0150	Maximum		15	1.067		107	50		157	194
0160	Color group D, minimum		15	1.067		84.50	50		134.50	170
0170	Maximum	↓	15	1.067	↓	115	50		165	203

12 46 Furnishing Accessories

12 46 13 – Ash Receptacles

12 46 13.10 Ash/Trash Receivers

		Crew	Daily Output	Labor-Hours	Unit	Material	Labor	Equipment	Total	Total Incl O&P
0010	**ASH/TRASH RECEIVERS**									
1000	Ash urn, cylindrical metal									
1020	8" diameter, 20" high	1 Clab	60	.133	Ea.	158	5		163	182
1060	10" diameter, 26" high	"	60	.133	"	126	5		131	147
2000	Combination ash/trash urn, metal									
2020	8" diameter, 20" high	1 Clab	60	.133	Ea.	158	5		163	182
2050	10" diameter, 26" high	"	60	.133	"	126	5		131	147

12 46 19 – Clocks

12 46 19.50 Wall Clocks

		Crew	Daily Output	Labor-Hours	Unit	Material	Labor	Equipment	Total	Total Incl O&P
0010	**WALL CLOCKS**									
0080	12" diameter, single face	1 Elec	8	1	Ea.	130	54.50		184.50	225
0100	Double face	"	6.20	1.290	"	300	70.50		370.50	435

12 46 33 – Waste Receptacles

12 46 33.13 Trash Receptacles

		Crew	Daily Output	Labor-Hours	Unit	Material	Labor	Equipment	Total	Total Incl O&P
0010	**TRASH RECEPTACLES**									
4000	Trash receptacle, metal									
4020	8" diameter, 15" high	1 Clab	60	.133	Ea.	73.50	5		78.50	88.50
4040	10" diameter, 18" high		60	.133		137	5		142	159
5040	16" x 8" x 14" high	↓	60	.133	↓	31	5		36	41.50
5500	Plastic, with lid									
5520	35 gallon	1 Clab	60	.133	Ea.	145	5		150	167
5540	45 gallon		60	.133		233	5		238	265
5550	Plastic recycling barrel, w/lid & wheels, 32 gal. G		60	.133		84.50	5		89.50	101
5560	65 gal. G		60	.133		545	5		550	610
5570	95 gal. G	↓	60	.133	↓	1,025	5		1,030	1,125

12 48 Rugs and Mats

12 48 13 – Entrance Floor Mats and Frames

12 48 13.13 Entrance Floor Mats		Crew	Daily Output	Labor-Hours	Unit	Material	Labor	Equipment	Total	Total Incl O&P
						2015 Bare Costs				
0010	**ENTRANCE FLOOR MATS**									
0020	Recessed, black rubber, 3/8" thick, solid	1 Clab	155	.052	S.F.	26	1.94		27.94	31.50
0050	Perforated		155	.052		16.15	1.94		18.09	20.50
0100	1/2" thick, solid		155	.052		19.40	1.94		21.34	24.50
0150	Perforated		155	.052		23.50	1.94		25.44	28.50
0200	In colors, 3/8" thick, solid		155	.052		21	1.94		22.94	26
0250	Perforated		155	.052		21.50	1.94		23.44	26.50
0300	1/2" thick, solid		155	.052		27	1.94		28.94	32.50
0350	Perforated		155	.052		27.50	1.94		29.44	33.50
1225	Recessed, alum. rail, hinged mat, 7/16" thk									
1250	Carpet insert	1 Clab	360	.022	S.F.	49.50	.84		50.34	56
1275	Vinyl insert		360	.022		49.50	.84		50.34	56
1300	Abrasive insert		360	.022		49.50	.84		50.34	56
1325	Recessed, vinyl rail, hinged mat, 7/16" thk									
1350	Carpet insert	1 Clab	360	.022	S.F.	55	.84		55.84	62
1375	Vinyl insert		360	.022		55	.84		55.84	62
1400	Abrasive insert		360	.022		55	.84		55.84	62
2000	Recycled rubber tire tile, 12" x 12" x 3/8" thick G		125	.064		10.10	2.41		12.51	14.80
2510	Natural cocoa fiber, 1/2" thick G		125	.064		8.35	2.41		10.76	12.90
2520	3/4" thick G		125	.064		6.90	2.41		9.31	11.30
2530	1" thick G		125	.064		9.60	2.41		12.01	14.25
3000	Hospital tacky mats, package of 30 with frame				Ea.	56			56	61.50
3010	4 packages of 30				"	86.50			86.50	95.50

12 51 Office Furniture

12 51 16 – Case Goods

12 51 16.13 Metal Case Goods		Crew	Daily Output	Labor-Hours	Unit	Material	Labor	Equipment	Total	Total Incl O&P
0010	**METAL CASE GOODS**									
0020	Desks, 29" high, double pedestal, 30" x 60", metal, minimum				Ea.	595			595	655
0030	Maximum					1,550			1,550	1,700
0600	Desks, single pedestal, 30" x 60", metal, minimum					540			540	595
0620	Maximum					1,250			1,250	1,400
0720	Desks, secretarial, 30" x 60", metal, minimum					485			485	535
0730	Maximum					860			860	945
0740	Return, 20" x 42", minimum					360			360	395
0750	Maximum					555			555	610
0940	59" x 12" x 23" high, steel, minimum					305			305	340
0960	Maximum					390			390	430

12 51 16.16 Wood Case Goods		Crew	Daily Output	Labor-Hours	Unit	Material	Labor	Equipment	Total	Total Incl O&P
0010	**WOOD CASE GOODS**									
0150	Desk, 29" high, double pedestal, 30" x 60"									
0160	Wood, minimum				Ea.	740			740	815
0180	Maximum				"	3,025			3,025	3,325
0630	Single pedestal, 30" x 60"									
0640	Wood, minimum				Ea.	600			600	660
0650	Maximum					945			945	1,050
0670	Executive return, 24" x 42", with box, file, wood, minimum					390			390	430
0680	Maximum					945			945	1,050
0790	Desk, 29" high, secretarial, 30" x 60"									
0800	Wood, minimum				Ea.	510			510	560
0810	Maximum					3,075			3,075	3,375

12 51 Office Furniture

12 51 16 – Case Goods

12 51 16.16 Wood Case Goods		Crew	Daily Output	Labor-Hours	Unit	2015 Bare Costs Material	Labor	Equipment	Total	Total Incl O&P
0820	Return, 20" x 42", minimum				Ea.	320			320	350
0830	Maximum					1,150			1,150	1,250
0900	Desktop organizer, 72" x 14" x 36" high, wood, minimum					180			180	198
0920	Maximum				↓	465			465	510
1110	Furniture, credenza, 29" high, 18" to 22" x 60" to 72"									
1120	Wood, minimum				Ea.	725			725	795
1140	Maximum				"	2,675			2,675	2,950

12 51 23 – Office Tables

12 51 23.33 Conference Tables		Crew	Daily Output	Labor-Hours	Unit	Material	Labor	Equipment	Total	Total Incl O&P
0010	**CONFERENCE TABLES**									
6050	Boat, 96" x 42", minimum				Ea.	800			800	875
6150	Maximum					3,675			3,675	4,050
6720	Rectangle, 96" x 42", minimum					1,325			1,325	1,450
6740	Maximum				↓	3,675			3,675	4,050

12 52 Seating

12 52 23 – Office Seating

12 52 23.13 Office Chairs		Crew	Daily Output	Labor-Hours	Unit	Material	Labor	Equipment	Total	Total Incl O&P
0010	**OFFICE CHAIRS**									
2000	Standard office chair, executive, minimum				Ea.	305			305	335
2150	Maximum					2,075			2,075	2,300
2200	Management, minimum					231			231	254
2250	Maximum					2,225			2,225	2,450
2280	Task, minimum					169			169	186
2290	Maximum					560			560	615
2300	Arm kit, minimum					77			77	85
2320	Maximum				↓	117			117	129

12 54 Hospitality Furniture

12 54 13 – Hotel and Motel Furniture

12 54 13.10 Hotel Furniture		Crew	Daily Output	Labor-Hours	Unit	Material	Labor	Equipment	Total	Total Incl O&P
0010	**HOTEL FURNITURE**									
0020	Standard quality set, minimum				Room	2,400			2,400	2,650
0200	Maximum				"	8,600			8,600	9,450

12 54 16 – Restaurant Furniture

12 54 16.10 Tables, Folding		Crew	Daily Output	Labor-Hours	Unit	Material	Labor	Equipment	Total	Total Incl O&P
0010	**TABLES, FOLDING** Laminated plastic tops									
1000	Tubular steel legs with glides									
1020	18" x 60", minimum				Ea.	272			272	299
1040	Maximum					1,550			1,550	1,700
1840	36" x 96", minimum					335			335	370
1860	Maximum					3,125			3,125	3,425
2000	Round, wood stained, plywood top, 60" diameter, minimum					213			213	234
2020	Maximum				↓	295			295	325

12 54 16.20 Furniture, Restaurant		Crew	Daily Output	Labor-Hours	Unit	Material	Labor	Equipment	Total	Total Incl O&P
0010	**FURNITURE, RESTAURANT**									
0020	Bars, built-in, front bar	1 Carp	5	1.600	L.F.	280	75		355	425
0200	Back bar	"	5	1.600	"	203	75		278	340

12 54 Hospitality Furniture

12 54 16 – Restaurant Furniture

12 54 16.20 Furniture, Restaurant		Crew	Daily Output	Labor-Hours	Unit	Material	2015 Bare Costs Labor	Equipment	Total	Total Incl O&P
0300	Booth seating, see Section 12 54 16.70									
2000	Chair, bentwood side chair, metal, minimum				Ea.	100			100	110
2020	Maximum					117			117	129
2600	Upholstered seat & back, arms, minimum					163			163	179
2620	Maximum					460			460	510

12 54 16.70 Booths		Crew	Daily Output	Labor-Hours	Unit	Material	Labor	Equipment	Total	Total Incl O&P
0010	**BOOTHS**									
1000	Banquet, upholstered seat and back, custom									
1500	Straight, minimum	2 Carp	40	.400	L.F.	201	18.80		219.80	250
1520	Maximum		36	.444		390	21		411	460
1600	"L" or "U" shape, minimum		35	.457		205	21.50		226.50	259
1620	Maximum		30	.533		365	25		390	440
1800	Upholstered outside finished backs for									
1810	single booths and custom banquets									
1820	Minimum	2 Carp	44	.364	L.F.	23	17.05		40.05	52
1840	Maximum	"	40	.400	"	69.50	18.80		88.30	106
3000	Fixed seating, one piece plastic chair and									
3010	plastic laminate table top									
3100	Two seat, 24" x 24" table, minimum	F-7	30	1.067	Ea.	810	45		855	960
3120	Maximum		26	1.231		1,150	52		1,202	1,350
3200	Four seat, 24" x 48" table, minimum		28	1.143		805	48.50		853.50	960
3220	Maximum		24	1.333		1,375	56.50		1,431.50	1,575
5000	Mount in floor, wood fiber core with									
5010	plastic laminate face, single booth									
5050	24" wide	F-7	30	1.067	Ea.	310	45		355	410
5100	48" wide	"	28	1.143	"	395	48.50		443.50	510

12 55 Detention Furniture

12 55 13 – Detention Bunks

12 55 13.13 Cots		Crew	Daily Output	Labor-Hours	Unit	Material	Labor	Equipment	Total	Total Incl O&P
0010	**COTS**									
2500	Bolted, single, painted steel	E-4	20	1.600	Ea.	335	85	7.30	427.30	525
2700	Stainless steel	"	20	1.600	"	975	85	7.30	1,067.30	1,225

12 56 Institutional Furniture

12 56 33 – Classroom Furniture

12 56 33.10 Furniture, School		Crew	Daily Output	Labor-Hours	Unit	Material	Labor	Equipment	Total	Total Incl O&P
0010	**FURNITURE, SCHOOL**									
0500	Classroom, movable chair & desk type, minimum				Set				73.50	81
0600	Maximum				"				155	171
1000	Chair, molded plastic									
1100	Integral tablet arm, minimum				Ea.	100			100	110
1150	Maximum					189			189	207
2000	Desk, single pedestal, top book compartment, minimum					91.50			91.50	101
2020	Maximum					188			188	206
2200	Flip top, minimum					228			228	251
2220	Maximum					277			277	305

12 56 Institutional Furniture

12 56 43 – Dormitory Furniture

12 56 43.10 Dormitory Furnishings		Crew	Daily Output	Labor-Hours	Unit	Material	2015 Bare Costs Labor	Equipment	Total	Total Incl O&P
0010	**DORMITORY FURNISHINGS**									
0300	Bunkable bed, twin, minimum				Ea.	375			375	415
0320	Maximum					550			550	605
1000	Chest, four drawer, minimum					360			360	395
1020	Maximum					690			690	760
1050	Built-in, minimum	2 Carp	13	1.231	L.F.	120	58		178	221
1150	Maximum		10	1.600		221	75		296	360
1200	Desk top, built-in, laminated plastic, 24″ deep, minimum		50	.320		44	15		59	71.50
1300	Maximum		40	.400		132	18.80		150.80	175
1450	30″ deep, minimum		50	.320		56.50	15		71.50	85
1550	Maximum		40	.400		247	18.80		265.80	300
1750	Dressing unit, built-in, minimum		12	1.333		179	62.50		241.50	294
1850	Maximum		8	2		540	94		634	735
8000	Rule of thumb: total cost for furniture, minimum				Student				2,525	2,800
8050	Maximum				″				4,850	5,350

12 56 51 – Library Furniture

12 56 51.10 Library Furnishings		Crew	Daily Output	Labor-Hours	Unit	Material	Labor	Equipment	Total	Total Incl O&P
0010	**LIBRARY FURNISHINGS**									
0100	Attendant desk, 36″ x 62″ x 29″ high	1 Carp	16	.500	Ea.	1,900	23.50		1,923.50	2,125
0200	Book display, "A" frame display, both sides, 42″ x 42″ x 60″ high		16	.500		1,250	23.50		1,273.50	1,400
0220	Table with bulletin board, 42″ x 24″ x 49″ high		16	.500		740	23.50		763.50	850
0800	Card catalogue, 30 tray unit		16	.500		3,350	23.50		3,373.50	3,700
0840	60 tray unit		16	.500		6,675	23.50		6,698.50	7,375
0880	72 tray unit	2 Carp	16	1		8,700	47		8,747	9,625
1000	Carrels, single face, initial unit	1 Carp	16	.500		820	23.50		843.50	935
1500	Double face, initial unit	2 Carp	16	1		1,300	47		1,347	1,500
1710	Carrels, hardwood, 36″ x 24″, minimum	1 Carp	5	1.600		785	75		860	975
1720	Maximum	″	4	2		2,000	94		2,094	2,350
2700	Card catalog file, 60 trays, complete					7,925			7,925	8,700
2720	Alternate method: each tray					132			132	145
3800	Charging desk, built-in, with counter, plastic laminated top	1 Carp	7	1.143	L.F.	305	53.50		358.50	420
4000	Dictionary stand, stationary		16	.500	Ea.	690	23.50		713.50	790
4020	Revolving		16	.500		214	23.50		237.50	272
4200	Exhibit case, table style, 60″ x 28″ x 36″		11	.727		2,625	34		2,659	2,925
6010	Bookshelf, metal, 90″ high, 10″ shelf, double face		11.50	.696	L.F.	150	32.50		182.50	216
6020	Single face		12	.667	″	124	31.50		155.50	185
6050	For 8″ shelving, subtract from above					10%				
6060	For 12″ shelving, add to above					10%				
6070	For 42″ high with countertop, subtract from above					20%				
6100	Mobile compacted shelving, hand crank, 9′-0″ high									
6110	Double face, including track, 3′ section				Ea.	1,175			1,175	1,275
6150	For electrical operation, add					25%				
6200	Magazine shelving, 82″ high, 12″ deep, single face	1 Carp	11.50	.696	L.F.	151	32.50		183.50	217
6210	Double face	″	11.50	.696	″	248	32.50		280.50	325
7200	Reading table, laminated top, 60″ x 36″				Ea.	705			705	775

12 56 70 – Healthcare Furniture

12 56 70.10 Furniture, Hospital		Crew	Daily Output	Labor-Hours	Unit	Material	Labor	Equipment	Total	Total Incl O&P
0010	**FURNITURE, HOSPITAL**									
0020	Beds, manual, minimum				Ea.	800			800	880
0100	Maximum					2,550			2,550	2,800
0600	All electric hospital beds, minimum					1,625			1,625	1,800

12 56 Institutional Furniture

12 56 70 – Healthcare Furniture

12 56 70.10 Furniture, Hospital		Crew	Daily Output	Labor-Hours	Unit	2015 Bare Costs Material	Labor	Equipment	Total	Total Incl O&P
0700	Maximum				Ea.	4,500			4,500	4,925
0900	Manual, nursing home beds, minimum					770			770	850
1000	Maximum					2,200			2,200	2,400
1020	Overbed table, laminated top, minimum					460			460	505
1040	Maximum					885			885	975
1100	Patient wall systems, not incl. plumbing, minimum				Room	1,425			1,425	1,575
1200	Maximum				"	1,925			1,925	2,125
2000	Geriatric chairs, minimum				Ea.	440			440	480
2020	Maximum				"	765			765	840

12 61 Fixed Audience Seating

12 61 13 – Upholstered Audience Seating

12 61 13.13 Auditorium Chairs

		Crew	Daily Output	Labor-Hours	Unit	Material	Labor	Equipment	Total	Total Incl O&P
0010	**AUDITORIUM CHAIRS**									
2000	All veneer construction	2 Carp	22	.727	Ea.	232	34		266	310
2200	Veneer back, padded seat		22	.727		242	34		276	320
2350	Fully upholstered, spring seat		22	.727		242	34		276	320
2450	For tablet arms, add					68			68	74.50
2500	For fire retardancy, CATB-133, add					30			30	33

12 61 13.23 Lecture Hall Seating

		Crew	Daily Output	Labor-Hours	Unit	Material	Labor	Equipment	Total	Total Incl O&P
0010	**LECTURE HALL SEATING**									
1000	Pedestal type, minimum	2 Carp	22	.727	Ea.	191	34		225	263
1200	Maximum	"	14.50	1.103	"	490	52		542	620

12 63 Stadium and Arena Seating

12 63 13 – Stadium and Arena Bench Seating

12 63 13.13 Bleachers

		Crew	Daily Output	Labor-Hours	Unit	Material	Labor	Equipment	Total	Total Incl O&P
0010	**BLEACHERS**									
3000	Telescoping, manual to 15 tier, minimum	F-5	65	.492	Seat	91.50	23.50		115	137
3100	Maximum		60	.533		137	25.50		162.50	190
3300	16 to 20 tier, minimum		60	.533		220	25.50		245.50	281
3400	Maximum		55	.582		275	27.50		302.50	345
3600	21 to 30 tier, minimum		50	.640		229	30.50		259.50	299
3700	Maximum		40	.800		300	38		338	390
3900	For integral power operation, add, minimum	2 Elec	300	.053		46	2.92		48.92	55
4000	Maximum	"	250	.064		73.50	3.50		77	86
5000	Benches, folding, in wall, 14' table, 2 benches	L-4	2	12	Set	775	530		1,305	1,675

12 67 Pews and Benches

12 67 13 – Pews

12 67 13.13 Sanctuary Pews		Crew	Daily Output	Labor-Hours	Unit	Material	2015 Bare Costs Labor	Equipment	Total	Total Incl O&P
0010	**SANCTUARY PEWS**									
1500	Bench type, hardwood, minimum	1 Carp	20	.400	L.F.	94.50	18.80		113.30	133
1550	Maximum	″	15	.533		187	25		212	245
1570	For kneeler, add					22.50			22.50	24.50

12 92 Interior Planters and Artificial Plants

12 92 33 – Interior Planters

12 92 33.10 Planters		Crew	Daily Output	Labor-Hours	Unit	Material	2015 Bare Costs Labor	Equipment	Total	Total Incl O&P
0010	**PLANTERS**									
1000	Fiberglass, hanging, 12″ diameter, 7″ high				Ea.	118			118	129
1500	Rectangular, 48″ long, 16″ high x 15″ wide					665			665	735
1650	60″ long, 30″ high, 28″ wide					1,025			1,025	1,125
2000	Round, 12″ diameter, 13″ high					154			154	170
2050	25″ high					203			203	223
5000	Square, 10″ side, 20″ high					193			193	212
5100	14″ side, 15″ high					234			234	257
6000	Metal bowl, 32″ diameter, 8″ high, minimum					555			555	610
6050	Maximum					755			755	830
8750	Wood, fiberglass liner, square									
8780	14″ square, 15″ high, minimum				Ea.	455			455	500
8800	Maximum					560			560	615
9400	Plastic cylinder, molded, 10″ diameter, 10″ high					18.45			18.45	20.50
9500	11″ diameter, 11″ high					35			35	38.50

12 93 Interior Public Space Furnishings

12 93 23 – Trash and Litter Receptacles

12 93 23.10 Trash Receptacles		Crew	Daily Output	Labor-Hours	Unit	Material	2015 Bare Costs Labor	Equipment	Total	Total Incl O&P
0010	**TRASH RECEPTACLES**									
0020	Fiberglass, 2′ square, 18″ high	2 Clab	30	.533	Ea.	560	20		580	645
0100	2′ square, 2′-6″ high		30	.533		795	20		815	905
0300	Circular , 2′ diameter, 18″ high		30	.533		475	20		495	550
0400	2′ diameter, 2′-6″ high		30	.533		530	20		550	615
0500	Recycled plastic, var colors, round, 32 gal., 28″ x 38″ H G		5	3.200		510	120		630	750
0510	32 gal., 31″ x 32″ H G		5	3.200		585	120		705	830
9110	Plastic, with dome lid, 32 gal. capacity		35	.457		58	17.20		75.20	90.50
9120	Recycled plastic slats, plastic dome lid, 32 gal. capacity		35	.457		284	17.20		301.20	335

12 93 23.20 Trash Closure		Crew	Daily Output	Labor-Hours	Unit	Material	2015 Bare Costs Labor	Equipment	Total	Total Incl O&P
0010	**TRASH CLOSURE**									
0020	Steel with pullover cover, 2′-3″ wide, 4′-7″ high, 6′-2″ long	2 Clab	5	3.200	Ea.	1,950	120		2,070	2,325
0100	10′-1″ long		4	4		2,425	150		2,575	2,900
0300	Wood, 10′ wide, 6′ high, 10′ long		1.20	13.333		1,750	500		2,250	2,700

Division Notes

		CREW	DAILY OUTPUT	LABOR-HOURS	UNIT	BARE COSTS MAT.	BARE COSTS LABOR	BARE COSTS EQUIP.	BARE COSTS TOTAL	TOTAL INCL O&P

Estimating Tips

General

- The items and systems in this division are usually estimated, purchased, supplied, and installed as a unit by one or more subcontractors. The estimator must ensure that all parties are operating from the same set of specifications and assumptions, and that all necessary items are estimated and will be provided. Many times the complex items and systems are covered, but the more common ones, such as excavation or a crane, are overlooked for the very reason that everyone assumes nobody could miss them. The estimator should be the central focus and be able to ensure that all systems are complete.
- Another area where problems can develop in this division is at the interface between systems. The estimator must ensure, for instance, that anchor bolts, nuts, and washers are estimated and included for the air-supported structures and pre-engineered buildings to be bolted to their foundations. Utility supply is a common area where essential items or pieces of equipment can be missed or overlooked, because each subcontractor may feel it is another's responsibility. The estimator should also be aware of certain items which may be supplied as part of a package but installed by others, and ensure that the installing contractor's estimate includes the cost of installation. Conversely, the estimator must also ensure that items are not costed by two different subcontractors, resulting in an inflated overall estimate.

13 30 00 Special Structures

- The foundations and floor slab, as well as rough mechanical and electrical, should be estimated, as this work is required for the assembly and erection of the structure. Generally, as noted in the book, the pre-engineered building comes as a shell. Pricing is based on the size and structural design parameters stated in the reference section. Additional features, such as windows and doors with their related structural framing, must also be included by the estimator. Here again, the estimator must have a clear understanding of the scope of each portion of the work and all the necessary interfaces.

Reference Numbers

Reference numbers are shown in shaded boxes at the beginning of some major classifications. These numbers refer to related items in the Reference Section. The reference information may be an estimating procedure, an alternate pricing method, or technical information.

Note: Not all subdivisions listed here necessarily appear in this publication. ■

13 05 Common Work Results for Special Construction

13 05 05 – Selective Demolition for Special Construction

13 05 05.10 Selective Demolition, Air Supported Structures		Crew	Daily Output	Labor-Hours	Unit	Material	2015 Bare Costs Labor	Equipment	Total	Total Incl O&P
0010	SELECTIVE DEMOLITION, AIR SUPPORTED STRUCTURES									
0020	Tank covers, scrim, dbl. layer, vinyl poly w/hdwe., blower & controls									
0050	Round and rectangular R024119-10	B-2	9000	.004	S.F.		.17		.17	.26
0100	Warehouse structures									
0120	Poly/vinyl fabric, 28 oz., incl. tension cables & inflation system	4 Clab	9000	.004	SF Flr.		.13		.13	.21
0150	Reinforced vinyl, 12 oz., 3000 S.F.	″	5000	.006			.24		.24	.37
0200	12,000 to 24,000 S.F.	8 Clab	20000	.003			.12		.12	.19
0250	Tedlar vinyl fabric, 28 oz. w/liner, to 3000 S.F.	4 Clab	5000	.006			.24		.24	.37
0300	12,000 to 24,000 S.F.	8 Clab	20000	.003			.12		.12	.19
0350	Greenhouse/shelter, woven polyethylene with liner									
0400	3000 S.F.	4 Clab	5000	.006	SF Flr.		.24		.24	.37
0450	12,000 to 24,000 S.F.	8 Clab	20000	.003			.12		.12	.19
0500	Tennis/gymnasium, poly/vinyl fabric, 28 oz., incl. thermal liner	4 Clab	9000	.004			.13		.13	.21
0600	Stadium/convention center, teflon coated fiberglass, incl. thermal liner	9 Clab	40000	.002			.07		.07	.10
0700	Doors, air lock, 15′ long, 10′ x 10′	2 Carp	1.50	10.667	Ea.		500		500	770
0720	15′ x 15′		.80	20			940		940	1,450
0750	Revolving personnel door, 6′ diam. x 6′-6″ high		1.50	10.667			500		500	770

13 05 05.20 Selective Demolition, Garden Houses		Crew	Daily Output	Labor-Hours	Unit	Material	Labor	Equipment	Total	Total Incl O&P
0010	SELECTIVE DEMOLITION, GARDEN HOUSES R024119-10									
0020	Prefab, wood, excl foundation, average	2 Clab	400	.040	SF Flr.		1.50		1.50	2.31

13 05 05.25 Selective Demolition, Geodesic Domes		Crew	Daily Output	Labor-Hours	Unit	Material	Labor	Equipment	Total	Total Incl O&P
0010	SELECTIVE DEMOLITION, GEODESIC DOMES									
0050	Shell only, interlocking plywood panels, 30′ diameter	F-5	3.20	10	Ea.		475		475	730
0060	34′ diameter		2.30	13.913			660		660	1,025
0070	39′ diameter		2	16			760		760	1,175
0080	45′ diameter	F-3	2.20	18.182			870	297	1,167	1,650
0090	55′ diameter		2	20			960	325	1,285	1,825
0100	60′ diameter		2	20			960	325	1,285	1,825
0110	65′ diameter		1.60	25			1,200	410	1,610	2,275

13 05 05.30 Selective Demolition, Greenhouses		Crew	Daily Output	Labor-Hours	Unit	Material	Labor	Equipment	Total	Total Incl O&P
0010	SELECTIVE DEMOLITION, GREENHOUSES R024119-10									
0020	Resi-type, free standing, excl. foundations, 9′ long x 8′ wide	2 Clab	160	.100	SF Flr.		3.76		3.76	5.80
0030	9′ long x 11′ wide		170	.094			3.54		3.54	5.45
0040	9′ long x 14′ wide		220	.073			2.73		2.73	4.21
0050	9′ long x 17′ wide		320	.050			1.88		1.88	2.89
0060	Lean-to type, 4′ wide		64	.250			9.40		9.40	14.45
0070	7′ wide		120	.133			5		5	7.70
0080	Geodesic hemisphere, 1/8″ plexiglass glazing, 8′ diam.		4	4	Ea.		150		150	231
0090	24′ diam.		.80	20			750		750	1,150
0100	48′ diam.		.40	40			1,500		1,500	2,325

13 05 05.35 Selective Demolition, Hangars		Crew	Daily Output	Labor-Hours	Unit	Material	Labor	Equipment	Total	Total Incl O&P
0010	SELECTIVE DEMOLITION, HANGARS									
0020	T type hangars, prefab, steel , galv roof & walls, incl doors, excl fndtn	E-2	2550	.022	SF Flr.		1.14	.59	1.73	2.56
0030	Circular type, prefab, steel frame, plastic skin, incl foundation, 80′ diam	″	.50	112	Total		5,800	3,025	8,825	13,100

13 05 05.45 Selective Demolition, Lightning Protection		Crew	Daily Output	Labor-Hours	Unit	Material	Labor	Equipment	Total	Total Incl O&P
0010	SELECTIVE DEMOLITION, LIGHTNING PROTECTION									
0020	Air terminal & base, copper, 3/8″ diam. x 10″, to 75′ h	1 Clab	16	.500	Ea.		18.80		18.80	29
0030	1/2″ diam. x 12″, over 75′ h		16	.500			18.80		18.80	29
0050	Aluminum, 1/2″ diam. x 12″, to 75′ h		16	.500			18.80		18.80	29
0060	5/8″ diam. x 12″, over 75′ h		16	.500			18.80		18.80	29
0070	Cable, copper, 220 lb. per thousand feet, to 75′ high		640	.013	L.F.		.47		.47	.72

13 05 05 – Selective Demolition for Special Construction

13 05 05.45 Selective Demolition, Lightning Protection

	13 05 05.45 Selective Demolition, Lightning Protection	Crew	Daily Output	Labor-Hours	Unit	Material	2015 Bare Costs Labor	Equipment	Total	Total Incl O&P
0080	375 lb. per thousand feet, over 75′ high	1 Clab	460	.017	L.F.		.65		.65	1.01
0090	Aluminum, 101 lb. per thousand feet, to 75′ high		560	.014			.54		.54	.83
0100	199 lb. per thousand feet, over 75′ high		480	.017			.63		.63	.96
0110	Arrester, 175 V AC, to ground		16	.500	Ea.		18.80		18.80	29
0120	650 V AC, to ground		13	.615	″		23		23	35.50

13 05 05.50 Selective Demolition, Pre-Engineered Steel Buildings

		Crew	Daily Output	Labor-Hours	Unit	Material	Labor	Equipment	Total	Total Incl O&P
0010	**SELECTIVE DEMOLITION, PRE-ENGINEERED STEEL BUILDINGS**									
0500	Pre-engd. steel bldgs., rigid frame, clear span & multi post, excl. salvage									
0550	3,500 to 7,500 S.F.	L-10	1000	.024	SF Flr.		1.27	.65	1.92	2.84
0600	7,501 to 12,500 S.F.		1500	.016			.85	.44	1.29	1.89
0650	12,500 S.F. or greater		1650	.015			.77	.40	1.17	1.72
0700	Pre-engd. steel building components									
0710	Entrance canopy, including frame 4′ x 4′	E-24	8	4	Ea.		209	92	301	450
0720	4′ x 8′	″	7	4.571			238	105	343	515
0730	HM doors, self framing, single leaf	2 Skwk	8	2			97.50		97.50	150
0740	Double leaf		5	3.200			156		156	240
0760	Gutter, eave type		600	.027	L.F.		1.30		1.30	2
0770	Sash, single slide, double slide or fixed		24	.667	Ea.		32.50		32.50	50
0780	Skylight, fiberglass, to 30 S.F.		16	1			48.50		48.50	75
0785	Roof vents, circular, 12″ to 24″ diameter		12	1.333			65		65	100
0790	Continuous, 10′ long		8	2			97.50		97.50	150
0900	Shelters, aluminum frame									
0910	Acrylic glazing, 3′ x 9′ x 8′ high	2 Skwk	2	8	Ea.		390		390	600
0920	9′ x 12′ x 8′ high	″	1.50	10.667	″		520		520	800

13 05 05.60 Selective Demolition, Silos

		Crew	Daily Output	Labor-Hours	Unit	Material	Labor	Equipment	Total	Total Incl O&P
0010	**SELECTIVE DEMOLITION, SILOS**									
0020	Conc stave, indstrl, conical/sloping bott, excl fndtn, 12′ diam., 35′ h	E-24	.18	177	Ea.		9,275	4,100	13,375	20,100
0030	16′ diam., 45′ h		.12	266			13,900	6,150	20,050	30,200
0040	25′ diam., 75′ h		.08	400			20,900	9,200	30,100	45,200
0050	Steel, factory fabricated, 30,000 gal. cap, painted or epoxy lined	L-5	2	28			1,475	370	1,845	2,925

13 05 05.65 Selective Demolition, Sound Control

		Crew	Daily Output	Labor-Hours	Unit	Material	Labor	Equipment	Total	Total Incl O&P
0010	**SELECTIVE DEMOLITION, SOUND CONTROL** R024119-10									
0120	Acoustical enclosure, 4″ thick walls & ceiling panels, 8 lb./S.F.	3 Carp	144	.167	SF Surf		7.85		7.85	12.05
0130	10.5 lb./S.F.		128	.188			8.80		8.80	13.55
0140	Reverb chamber, parallel walls, 4″ thick		120	.200			9.40		9.40	14.45
0150	Skewed walls, parallel roof, 4″ thick		110	.218			10.25		10.25	15.75
0160	Skewed walls/roof, 4″ layer/air space		96	.250			11.75		11.75	18.05
0170	Sound-absorbing panels, painted metal, 2′-6″ x 8′, under 1,000 S.F.		430	.056			2.62		2.62	4.03
0180	Over 1,000 S.F.		480	.050			2.35		2.35	3.61
0190	Flexible transparent curtain, clear	3 Shee	430	.056			3.12		3.12	4.77
0192	50% clear, 50% foam		430	.056			3.12		3.12	4.77
0194	25% clear, 75% foam		430	.056			3.12		3.12	4.77
0196	100% foam		430	.056			3.12		3.12	4.77
0200	Audio-masking sys., incl. speakers, amplfr., signal gnrtr.									
0205	Ceiling mounted, 5,000 S.F.	2 Elec	4800	.003	S.F.		.18		.18	.27
0210	10,000 S.F.		5600	.003			.16		.16	.23
0220	Plenum mounted, 5,000 S.F.		7600	.002			.12		.12	.17
0230	10,000 S.F.		8800	.002			.10		.10	.15

13 05 05.70 Selective Demolition, Special Purpose Rooms

		Crew	Daily Output	Labor-Hours	Unit	Material	Labor	Equipment	Total	Total Incl O&P
0010	**SELECTIVE DEMOLITION, SPECIAL PURPOSE ROOMS** R024119-10									
0100	Audiometric rooms, under 500 S.F. surface	4 Carp	200	.160	SF Surf		7.50		7.50	11.55
0110	Over 500 S.F. surface	″	240	.133	″		6.25		6.25	9.65

13 05 Common Work Results for Special Construction

13 05 05 – Selective Demolition for Special Construction

13 05 05.70 Selective Demolition, Special Purpose Rooms		Crew	Daily Output	Labor-Hours	Unit	Material	2015 Bare Costs Labor	Equipment	Total	Total Incl O&P
0200	Clean rooms, 12′ x 12′ soft wall, class 100	1 Carp	.30	26.667	Ea.		1,250		1,250	1,925
0210	Class 1000		.30	26.667			1,250		1,250	1,925
0220	Class 10,000		.35	22.857			1,075		1,075	1,650
0230	Class 100,000		.35	22.857			1,075		1,075	1,650
0300	Darkrooms, shell complete, 8′ high	2 Carp	220	.073	SF Flr.		3.41		3.41	5.25
0310	12′ high		110	.145	"		6.85		6.85	10.50
0350	Darkrooms doors, mini-cylindrical, revolving		4	4	Ea.		188		188	289
0400	Music room, practice modular		140	.114	SF Surf		5.35		5.35	8.25
0500	Refrigeration structures and finishes									
0510	Wall finish, 2 coat portland cement plaster, 1/2″ thick	1 Clab	200	.040	S.F.		1.50		1.50	2.31
0520	Fiberglass panels, 1/8″ thick		400	.020			.75		.75	1.16
0530	Ceiling finish, polystyrene plastic, 1″ to 2″ thick		500	.016			.60		.60	.93
0540	4″ thick		450	.018			.67		.67	1.03
0550	Refrigerator, prefab aluminum walk-in, 7′-6″ high, 6′ x 6′ OD	2 Carp	100	.160	SF Flr.		7.50		7.50	11.55
0560	10′ x 10′ OD		160	.100			4.70		4.70	7.25
0570	Over 150 S.F.		200	.080			3.76		3.76	5.80
0600	Sauna, prefabricated, including heater & controls, 7′ high, to 30 S.F.		120	.133			6.25		6.25	9.65
0610	To 40 S.F.		140	.114			5.35		5.35	8.25
0620	To 60 S.F.		175	.091			4.29		4.29	6.60
0630	To 100 S.F.		220	.073			3.41		3.41	5.25
0640	To 130 S.F.		250	.064			3		3	4.62
0650	Steam bath, heater, timer, head, single, to 140 C.F.	1 Plum	2.20	3.636	Ea.		213		213	320
0660	To 300 C.F.		2.20	3.636			213		213	320
0670	Steam bath, comm. size, w/blow-down assembly, to 800 C.F.		1.80	4.444			261		261	395
0680	To 2500 C.F.		1.60	5			294		294	445
0690	Steam bath, comm. size, multiple, for motels, apts, 500 C.F., 2 baths		2	4			235		235	355
0700	1,000 C.F., 4 baths		1.40	5.714			335		335	505

13 05 05.75 Selective Demolition, Storage Tanks		Crew	Daily Output	Labor-Hours	Unit	Material	Labor	Equipment	Total	Total Incl O&P
0010	**SELECTIVE DEMOLITION, STORAGE TANKS**									
0500	Steel tank, single wall, above ground, not incl. fdn., pumps or piping									
0510	Single wall, 275 gallon R024119-10	Q-1	3	5.333	Ea.		282		282	425
0520	550 thru 2,000 gallon	B-34P	2	12			600	335	935	1,275
0530	5,000 thru 10,000 gallon	B-34Q	2	12			600	630	1,230	1,600
0540	15,000 thru 30,000 gallon	B-34S	2	16			845	1,775	2,620	3,225
0600	Steel tank, double wall, above ground not incl. fdn., pumps & piping									
0620	500 thru 2,000 gallon	B-34P	2	12	Ea.		600	335	935	1,275

13 05 05.85 Selective Demolition, Swimming Pool Equip		Crew	Daily Output	Labor-Hours	Unit	Material	Labor	Equipment	Total	Total Incl O&P
0010	**SELECTIVE DEMOLITION, SWIMMING POOL EQUIP**									
0020	Diving stand, stainless steel, 3 meter	2 Clab	3	5.333	Ea.		201		201	310
0030	1 meter		5	3.200			120		120	185
0040	Diving board, 16′ long, aluminum		5.40	2.963			111		111	171
0050	Fiberglass		5.40	2.963			111		111	171
0070	Ladders, heavy duty, stainless steel, 2 tread		14	1.143			43		43	66
0080	4 tread		12	1.333			50		50	77
0090	Lifeguard chair, stainless steel, fixed		5	3.200			120		120	185
0100	Slide, tubular, fiberglass, aluminum handrails & ladder, 5′, straight		4	4			150		150	231
0110	8′, curved		6	2.667			100		100	154
0120	10′, curved		3	5.333			201		201	310
0130	12′ straight, with platform		2.50	6.400			241		241	370
0140	Removable access ramp, stainless steel		4	4			150		150	231
0150	Removable stairs, stainless steel, collapsible		4	4			150		150	231

13 05 Common Work Results for Special Construction

13 05 05 – Selective Demolition for Special Construction

13 05 05.90 Selective Demolition, Tension Structures		Crew	Daily Output	Labor-Hours	Unit	Material	2015 Bare Costs Labor	Equipment	Total	Total Incl O&P
0010	**SELECTIVE DEMOLITION, TENSION STRUCTURES**									
0020	Steel/alum. frame, fabric shell, 60′ clear span, 6,000 S.F.	B-41	2000	.022	SF Flr.		.86	.14	1	1.48
0030	12,000 S.F.		2200	.020			.78	.13	.91	1.34
0040	80′ clear span, 20,800 S.F.		2440	.018			.70	.12	.82	1.21
0050	100′ clear span, 10,000 S.F.	L-5	4350	.013			.68	.17	.85	1.35
0060	26,000 S.F.		4600	.012			.64	.16	.80	1.28
0070	36,000 S.F.		5000	.011			.59	.15	.74	1.17

13 05 05.95 Selective Demo, X-Ray/Radio Freq Protection

Line	Description	Crew	Daily Output	Labor-Hours	Unit	Material	Labor	Equipment	Total	Total Incl O&P
0010	**SELECTIVE DEMO, X-RAY/RADIO FREQ PROTECTION**									
0020	Shielding lead, lined door frame, excl. hdwe., 1/16″ thick	1 Clab	4.80	1.667	Ea.		62.50		62.50	96.50
0030	Lead sheets, 1/16″ thick R024119-10	2 Clab	270	.059	S.F.		2.23		2.23	3.43
0040	1/8″ thick		240	.067			2.51		2.51	3.86
0050	Lead shielding, 1/4″ thick		270	.059			2.23		2.23	3.43
0060	1/2″ thick		240	.067			2.51		2.51	3.86
0070	Lead glass, 1/4″ thick, 2.0 mm LE, 12″ x 16″	2 Glaz	16	1	Ea.		45		45	68.50
0080	24″ x 36″		8	2			90		90	137
0090	36″ x 60″		4	4			180		180	275
0100	Lead glass window frame, with 1/16″ lead & voice passage, 36″ x 60″		4	4			180		180	275
0110	Lead glass window frame, 24″ x 36″		8	2			90		90	137
0120	Lead gypsum board, 5/8″ thick with 1/16″ lead	2 Clab	320	.050	S.F.		1.88		1.88	2.89
0130	1/8″ lead		280	.057			2.15		2.15	3.31
0140	1/32″ lead		400	.040			1.50		1.50	2.31
0150	Butt joints, 1/8″ lead or thicker, 2″ x 7′ long batten strip		480	.033	Ea.		1.25		1.25	1.93
0160	X-ray protection, average radiography room, up to 300 S.F., 1/16″ lead, min		.50	32	Total		1,200		1,200	1,850
0170	Maximum		.30	53.333			2,000		2,000	3,075
0180	Deep therapy X-ray room, 250 kV cap, up to 300 S.F., 1/4″ lead, min		.20	80			3,000		3,000	4,625
0190	Maximum		.12	133			5,025		5,025	7,725
0880	Radio frequency shielding, prefab or screen-type copper or steel, minimum		360	.044	SF Surf		1.67		1.67	2.57
0890	Average		310	.052			1.94		1.94	2.99
0895	Maximum		290	.055			2.07		2.07	3.19

13 11 Swimming Pools

13 11 13 – Below-Grade Swimming Pools

13 11 13.50 Swimming Pools

Line	Description	Crew	Daily Output	Labor-Hours	Unit	Material	Labor	Equipment	Total	Total Incl O&P
0010	**SWIMMING POOLS** Residential in-ground, vinyl lined, concrete									
0020	Swimming pools,resi in-ground,vyl lined,conc sides,W/ equip ,sand bot	B-52	300	.187	SF Surf	23.50	8.15	1.98	33.63	40
0100	Metal or polystyrene sides R131113-20	B-14	410	.117		19.55	4.66	.89	25.10	29.50
0200	Add for vermiculite bottom					1.49			1.49	1.64
0500	Gunite bottom and sides, white plaster finish									
0600	12′ x 30′ pool	B-52	145	.386	SF Surf	43.50	16.85	4.10	64.45	78.50
0720	16′ x 32′ pool		155	.361		39	15.75	3.83	58.58	71
0750	20′ x 40′ pool		250	.224		35	9.75	2.38	47.13	56
0810	Concrete bottom and sides, tile finish									
0820	12′ x 30′ pool	B-52	80	.700	SF Surf	44	30.50	7.45	81.95	103
0830	16′ x 32′ pool		95	.589		36.50	25.50	6.25	68.25	86.50
0840	20′ x 40′ pool		130	.431		29	18.80	4.57	52.37	66
1100	Motel, gunite with plaster finish, incl. medium									
1150	capacity filtration & chlorination	B-52	115	.487	SF Surf	53.50	21	5.15	79.65	97
1200	Municipal, gunite with plaster finish, incl. high									
1250	capacity filtration & chlorination	B-52	100	.560	SF Surf	69.50	24.50	5.95	99.95	120

13 11 Swimming Pools

13 11 13 – Below-Grade Swimming Pools

13 11 13.50 Swimming Pools		Crew	Daily Output	Labor-Hours	Unit	Material (2015 Bare Costs)	Labor (2015 Bare Costs)	Equipment (2015 Bare Costs)	Total (2015 Bare Costs)	Total Incl O&P
1350	Add for formed gutters				L.F.	102			102	112
1360	Add for stainless steel gutters				"	300			300	330
1600	For water heating system, see Section 23 52 28.10									
1700	Filtration and deck equipment only, as % of total				Total				20%	20%
1800	Deck equipment, rule of thumb, 20′ x 40′ pool				SF Pool				1.18	1.30
1900	5000 S.F. pool				"				1.73	1.90
3000	Painting pools, preparation + 3 coats, 20′ x 40′ pool, epoxy	2 Pord	.33	48.485	Total	1,775	1,950		3,725	4,900
3100	Rubber base paint, 18 gallons	"	.33	48.485		1,225	1,950		3,175	4,300
3500	42′ x 82′ pool, 75 gallons, epoxy paint	3 Pord	.14	171		7,500	6,925		14,425	18,700
3600	Rubber base paint	"	.14	171		5,075	6,925		12,000	16,000

13 11 46 – Swimming Pool Accessories

13 11 46.50 Swimming Pool Equipment		Crew	Daily Output	Labor-Hours	Unit	Material	Labor	Equipment	Total	Total Incl O&P
0010	**SWIMMING POOL EQUIPMENT**									
0020	Diving stand, stainless steel, 3 meter	2 Carp	.40	40	Ea.	15,100	1,875		16,975	19,500
0300	1 meter		2.70	5.926		9,175	278		9,453	10,500
0600	Diving boards, 16′ long, aluminum		2.70	5.926		3,950	278		4,228	4,775
0700	Fiberglass		2.70	5.926		3,250	278		3,528	4,000
0800	14′ long, aluminum		2.70	5.926		3,575	278		3,853	4,350
0850	Fiberglass		2.70	5.926		3,225	278		3,503	3,950
1200	Ladders, heavy duty, stainless steel, 2 tread		7	2.286		810	107		917	1,050
1500	4 tread		6	2.667		1,075	125		1,200	1,375
1800	Lifeguard chair, stainless steel, fixed		2.70	5.926		3,225	278		3,503	3,975
1900	Portable					2,775			2,775	3,050
2100	Lights, underwater, 12 volt, with transformer, 300 watt	1 Elec	1	8		330	440		770	1,025
2200	110 volt, 500 watt, standard		1	8		294	440		734	980
2400	Low water cutoff type		1	8		300	440		740	985
2800	Heaters, see Section 23 52 28.10									
3000	Pool covers, reinforced vinyl	3 Clab	1800	.013	S.F.	1.13	.50		1.63	2.01
3050	Automatic, electric								8.75	9.65
3100	Vinyl, for winter, 400 SF max pool surface	3 Clab	3200	.008		.26	.28		.54	.72
3200	With water tubes, 400SF max pool surface	"	3000	.008		.29	.30		.59	.78
3250	Sealed air bubble polyethylene solar blanket, 16 mils					.30			.30	.33
3300	Slides, tubular, fiberglass, aluminum handrails & ladder, 5′-0″, straight	2 Carp	1.60	10	Ea.	3,575	470		4,045	4,675
3320	8′-0″, curved		3	5.333		7,225	250		7,475	8,325
3400	10′-0″, curved		1	16		22,000	750		22,750	25,400
3420	12′-0″, straight with platform		1.20	13.333		13,800	625		14,425	16,200
4500	Hydraulic lift, movable pool bottom, single ram									
4520	Under 1,000 S.F. area	L-9	72	.500	S.F.	160	21.50		181.50	210
4600	Four ram lift, over 1,000 S.F.	"	109	.330	"	130	14.20		144.20	166
5000	Removable access ramp, stainless steel	2 Clab	2	8	Ea.	5,800	300		6,100	6,850

13 17 Tubs and Pools

13 17 33 – Whirlpool Tubs

13 17 33.10 Whirlpool Bath		Crew	Daily Output	Labor-Hours	Unit	Material	2015 Bare Costs Labor	Equipment	Total	Total Incl O&P
0010	**WHIRLPOOL BATH**									
6000	Whirlpool, bath with vented overflow, molded fiberglass									
6100	66" x 36" x 24"	Q-1	1	16	Ea.	3,475	845		4,320	5,100

13 18 Ice Rinks

13 18 13 – Ice Rink Floor Systems

13 18 13.50 Ice Skating

		Crew	Daily Output	Labor-Hours	Unit	Material	Labor	Equipment	Total	Total Incl O&P
0010	**ICE SKATING** Equipment incl. refrigeration, plumbing & cooling									
0020	coils & concrete slab, 85′ x 200′ rink									
0300	55° system, 5 mos., 100 ton				Total	575,000			575,000	632,500
0700	90° system, 12 mos., 135 ton				"	650,000			650,000	715,000
1200	Subsoil heating system (recycled from compressor), 85′ x 200′	Q-7	.27	118	Ea.	40,000	6,750		46,750	54,000
1300	Subsoil insulation, 2 lb. polystyrene with vapor barrier, 85′ x 200′	2 Carp	.14	114	"	30,000	5,375		35,375	41,300

13 18 16 – Ice Rink Dasher Boards

13 18 16.50 Ice Rink Dasher Boards

		Crew	Daily Output	Labor-Hours	Unit	Material	Labor	Equipment	Total	Total Incl O&P
0010	**ICE RINK DASHER BOARDS**									
1000	Dasher boards, 1/2" H.D. polyethylene faced steel frame, 3′ acrylic									
1020	screen at sides, 5′ acrylic ends, 85′ x 200′	F-5	.06	533	Ea.	135,000	25,300		160,300	187,500
1100	Fiberglass & aluminum construction, same sides and ends	"	.06	533	"	155,000	25,300		180,300	209,500

13 21 Controlled Environment Rooms

13 21 13 – Clean Rooms

13 21 13.50 Clean Room Components

		Crew	Daily Output	Labor-Hours	Unit	Material	Labor	Equipment	Total	Total Incl O&P
0010	**CLEAN ROOM COMPONENTS**									
1100	Clean room, soft wall, 12′ x 12′, Class 100	1 Carp	.18	44.444	Ea.	18,600	2,075		20,675	23,700
1110	Class 1,000		.18	44.444		15,400	2,075		17,475	20,200
1120	Class 10,000		.21	38.095		13,000	1,800		14,800	17,100
1130	Class 100,000		.21	38.095		12,000	1,800		13,800	16,000
2800	Ceiling grid support, slotted channel struts 4′-0" O.C., ea. way				S.F.				5.90	6.50
3000	Ceiling panel, vinyl coated foil on mineral substrate									
3020	Sealed, non-perforated				S.F.				1.27	1.40
4000	Ceiling panel seal, silicone sealant, 150 L.F./gal.	1 Carp	150	.053	L.F.	.34	2.50		2.84	4.22
4100	Two sided adhesive tape	"	240	.033	"	.12	1.56		1.68	2.54
4200	Clips, one per panel				Ea.	.99			.99	1.09
6000	HEPA filter, 2′ x 4′, 99.97% eff., 3" dp beveled frame (silicone seal)					470			470	515
6040	6" deep skirted frame (channel seal)					440			440	485
6100	99.99% efficient, 3" deep beveled frame (silicone seal)					525			525	580
6140	6" deep skirted frame (channel seal)					455			455	500
6200	99.999% efficient, 3" deep beveled frame (silicone seal)					605			605	665
6240	6" deep skirted frame (channel seal)					485			485	530
7000	Wall panel systems, including channel strut framing									
7020	Polyester coated aluminum, particle board				S.F.				18.20	20
7100	Porcelain coated aluminum, particle board								32	35
7400	Wall panel support, slotted channel struts, to 12′ high								16.35	18

13 21 Controlled Environment Rooms

13 21 26 – Cold Storage Rooms

13 21 26.50 Refrigeration		Crew	Daily Output	Labor-Hours	Unit	Material	2015 Bare Costs Labor	Equipment	Total	Total Incl O&P
0010	**REFRIGERATION**									
0020	Curbs, 12" high, 4" thick, concrete	2 Carp	58	.276	L.F.	5.15	12.95		18.10	25.50
1000	Doors, see Section 08 34 13.10									
2400	Finishes, 2 coat portland cement plaster, 1/2" thick	1 Plas	48	.167	S.F.	1.41	7.15		8.56	12.30
2500	For galvanized reinforcing mesh, add	1 Lath	335	.024		.94	1.03		1.97	2.54
2700	3/16" thick latex cement	1 Plas	88	.091		2.40	3.90		6.30	8.50
2900	For glass cloth reinforced ceilings, add	"	450	.018		.57	.76		1.33	1.78
3100	Fiberglass panels, 1/8" thick	1 Carp	149.45	.054		3.18	2.51		5.69	7.35
3200	Polystyrene, plastic finish ceiling, 1" thick		274	.029		2.94	1.37		4.31	5.35
3400	2" thick		274	.029		3.36	1.37		4.73	5.80
3500	4" thick		219	.037		3.71	1.72		5.43	6.70
3800	Floors, concrete, 4" thick	1 Cefi	93	.086		1.32	3.87		5.19	7.20
3900	6" thick	"	85	.094		2.07	4.24		6.31	8.50
4000	Insulation, 1" to 6" thick, cork				B.F.	1.35			1.35	1.49
4100	Urethane					.52			.52	.57
4300	Polystyrene, regular					.53			.53	.58
4400	Bead board					.25			.25	.28
4600	Installation of above, add per layer	2 Carp	657.60	.024	S.F.	.45	1.14		1.59	2.26
4700	Wall and ceiling juncture		298.90	.054	L.F.	2.19	2.51		4.70	6.30
4900	Partitions, galvanized sandwich panels, 4" thick, stock		219.20	.073	S.F.	9.20	3.43		12.63	15.35
5000	Aluminum or fiberglass		219.20	.073	"	10.05	3.43		13.48	16.35
5200	Prefab walk-in, 7'-6" high, aluminum, incl. refrigeration, door & floor									
5210	not incl. partitions, 6' x 6'	2 Carp	54.80	.292	SF Flr.	164	13.70		177.70	202
5500	10' x 10'		82.20	.195		132	9.15		141.15	159
5700	12' x 14'		109.60	.146		119	6.85		125.85	142
5800	12' x 20'		109.60	.146		103	6.85		109.85	125
6100	For 8'-6" high, add					5%				
6300	Rule of thumb for complete units, w/o doors & refrigeration, cooler	2 Carp	146	.110		149	5.15		154.15	172
6400	Freezer		109.60	.146		176	6.85		182.85	205
6600	Shelving, plated or galvanized, steel wire type		360	.044	SF Hor.	13.05	2.09		15.14	17.55
6700	Slat shelf type		375	.043		16.10	2		18.10	21
6900	For stainless steel shelving, add					300%				
7000	Vapor barrier, on wood walls	2 Carp	1644	.010	S.F.	.20	.46		.66	.92
7200	On masonry walls	"	1315	.012	"	.49	.57		1.06	1.42
7500	For air curtain doors, see Section 23 34 33.10									

13 21 48 – Sound-Conditioned Rooms

13 21 48.10 Anechoic Chambers		Crew	Daily Output	Labor-Hours	Unit	Material	Labor	Equipment	Total	Total Incl O&P
0010	**ANECHOIC CHAMBERS** Standard units, 7' ceiling heights									
0100	Area for pricing is net inside dimensions									
0300	200 cycles per second cutoff, 25 S.F. floor area				SF Flr.	1,625			1,625	1,800
0400	50 S.F.								1,050	1,150
0600	75 S.F.								1,000	1,100
0700	100 S.F.					1,225			1,225	1,350
0900	For 150 cycles per second cutoff, add to 100 S.F. room								30%	30%
1000	For 100 cycles per second cutoff, add to 100 S.F. room								45%	45%

13 21 48.15 Audiometric Rooms		Crew	Daily Output	Labor-Hours	Unit	Material	Labor	Equipment	Total	Total Incl O&P
0010	**AUDIOMETRIC ROOMS**									
0020	Under 500 S.F. surface	4 Carp	98	.327	SF Surf	52.50	15.35		67.85	81
0100	Over 500 S.F. surface	"	120	.267	"	50	12.50		62.50	74.50

13 21 Controlled Environment Rooms

13 21 53 – Darkrooms

13 21 53.50 Darkrooms		Crew	Daily Output	Labor-Hours	Unit	Material	2015 Bare Costs Labor	Equipment	Total	Total Incl O&P
0010	**DARKROOMS**									
0020	Shell, complete except for door, 64 S.F., 8′ high	2 Carp	128	.125	SF Flr.	51	5.85		56.85	65
0100	12′ high		64	.250		66.50	11.75		78.25	91
0500	120 S.F. floor, 8′ high		120	.133		37	6.25		43.25	50.50
0600	12′ high		60	.267		50.50	12.50		63	75
0800	240 S.F. floor, 8′ high		120	.133		27	6.25		33.25	39
0900	12′ high		60	.267		37	12.50		49.50	60
1200	Mini-cylindrical, revolving, unlined, 4′ diameter		3.50	4.571	Ea.	2,725	215		2,940	3,325
1400	5′-6″ diameter		2.50	6.400		5,675	300		5,975	6,675
1600	Add for lead lining, inner cylinder, 1/32″ thick					1,650			1,650	1,825
1700	1/16″ thick					4,450			4,450	4,875
1800	Add for lead lining, inner and outer cylinder, 1/32″ thick					3,050			3,050	3,375
1900	1/16″ thick					6,825			6,825	7,500
2000	For darkroom door, see Section 08 34 36.10									

13 21 56 – Music Rooms

13 21 56.50 Music Rooms		Crew	Daily Output	Labor-Hours	Unit	Material	Labor	Equipment	Total	Total Incl O&P
0010	**MUSIC ROOMS**									
0020	Practice room, modular, perforated steel, under 500 S.F.	2 Carp	70	.229	SF Surf	32	10.75		42.75	52
0100	Over 500 S.F.	"	80	.200	"	27	9.40		36.40	44.50

13 24 Special Activity Rooms

13 24 16 – Saunas

13 24 16.50 Saunas and Heaters		Crew	Daily Output	Labor-Hours	Unit	Material	Labor	Equipment	Total	Total Incl O&P
0010	**SAUNAS AND HEATERS**									
0020	Prefabricated, incl. heater & controls, 7′ high, 6′ x 4′, C/C	L-7	2.20	12.727	Ea.	5,175	580		5,755	6,575
0050	6′ x 4′, C/P		2	14		4,700	635		5,335	6,125
0400	6′ x 5′, C/C		2	14		5,875	635		6,510	7,450
0450	6′ x 5′, C/P		2	14		5,325	635		5,960	6,825
0600	6′ x 6′, C/C		1.80	15.556		6,225	705		6,930	7,925
0650	6′ x 6′, C/P		1.80	15.556		5,675	705		6,380	7,300
0800	6′ x 9′, C/C		1.60	17.500		7,975	795		8,770	10,000
0850	6′ x 9′, C/P		1.60	17.500		7,225	795		8,020	9,150
1000	8′ x 12′, C/C		1.10	25.455		11,700	1,150		12,850	14,700
1050	8′ x 12′, C/P		1.10	25.455		10,500	1,150		11,650	13,400
1200	8′ x 8′, C/C		1.40	20		9,175	910		10,085	11,500
1250	8′ x 8′, C/P		1.40	20		8,450	910		9,360	10,700
1400	8′ x 10′, C/C		1.20	23.333		10,200	1,050		11,250	12,800
1450	8′ x 10′, C/P		1.20	23.333		9,250	1,050		10,300	11,800
1600	10′ x 12′, C/C		1	28		12,200	1,275		13,475	15,500
1650	10′ x 12′, C/P		1	28		11,000	1,275		12,275	14,100
1700	Door only, cedar, 2′x6′, with 1′x4′ tempered insulated glass window	2 Carp	3.40	4.706		750	221		971	1,175
1800	Prehung, incl. jambs, pulls & hardware	"	12	1.333		745	62.50		807.50	915
2500	Heaters only (incl. above), wall mounted, to 200 C.F.					685			685	755
2750	To 300 C.F.					930			930	1,025
3000	Floor standing, to 720 C.F., 10,000 watts, w/controls	1 Elec	3	2.667		2,950	146		3,096	3,475
3250	To 1,000 C.F., 16,000 watts	"	3	2.667		3,825	146		3,971	4,425

13 24 Special Activity Rooms

13 24 26 – Steam Baths

13 24 26.50 Steam Baths and Components		Crew	Daily Output	Labor-Hours	Unit	2015 Bare Costs Material	Labor	Equipment	Total	Total Incl O&P
0010	**STEAM BATHS AND COMPONENTS**									
0020	Heater, timer & head, single, to 140 C.F.	1 Plum	1.20	6.667	Ea.	2,200	390		2,590	3,000
0500	To 300 C.F.		1.10	7.273		2,425	425		2,850	3,325
1000	Commercial size, with blow-down assembly, to 800 C.F.		.90	8.889		5,725	520		6,245	7,100
1500	To 2500 C.F.	↓	.80	10		7,650	585		8,235	9,300
2000	Multiple, motels, apts., 2 baths, w/blow-down assm., 500 C.F.	Q-1	1.30	12.308		6,325	650		6,975	7,950
2500	4 baths	"	.70	22.857		10,200	1,200		11,400	13,000
2700	Conversion unit for residential tub, including door				↓	3,550			3,550	3,925

13 28 Athletic and Recreational Special Construction

13 28 33 – Athletic and Recreational Court Walls

13 28 33.50 Sport Court

		Crew	Daily Output	Labor-Hours	Unit	Material	Labor	Equipment	Total	Total Incl O&P
0010	**SPORT COURT**									
0020	Floors, No. 2 & better maple, 25/32" thick				SF Flr.				6.05	6.65
0100	Walls, laminated plastic bonded to galv. steel studs				SF Wall				7.70	8.45
0300	Squash, regulation court in existing building, minimum				Court	36,800			36,800	40,400
0400	Maximum				"	41,000			41,000	45,000
0450	Rule of thumb for components:									
0470	Walls	3 Carp	.15	160	Court	11,000	7,500		18,500	23,700
0500	Floor	"	.25	96		8,725	4,500		13,225	16,500
0550	Lighting	2 Elec	.60	26.667		2,100	1,450		3,550	4,475
0600	Handball, racquetball court in existing building, minimum	C-1	.20	160		39,800	7,150		46,950	55,000
0800	Maximum	"	.10	320		43,100	14,300		57,400	69,500
0900	Rule of thumb for components: walls	3 Carp	.12	200		12,600	9,400		22,000	28,400
1000	Floor		.25	96		8,725	4,500		13,225	16,500
1100	Ceiling	↓	.33	72.727		4,200	3,425		7,625	9,875
1200	Lighting	2 Elec	.60	26.667	↓	2,200	1,450		3,650	4,600

13 31 Fabric Structures

13 31 13 – Air-Supported Fabric Structures

13 31 13.09 Air Supported Tank Covers

		Crew	Daily Output	Labor-Hours	Unit	Material	Labor	Equipment	Total	Total Incl O&P
0010	**AIR SUPPORTED TANK COVERS**, vinyl polyester									
0100	Scrim, double layer, with hardware, blower, standby & controls									
0200	Round, 75' diameter	B-2	4500	.009	S.F.	11.75	.34		12.09	13.45
0300	100' diameter		5000	.008		10.70	.30		11	12.20
0400	150' diameter		5000	.008		8.45	.30		8.75	9.75
0500	Rectangular, 20' x 20'		4500	.009		23	.34		23.34	26
0600	30' x 40'		4500	.009		23	.34		23.34	26
0700	50' x 60'	↓	4500	.009		23	.34		23.34	26
0800	For single wall construction, deduct, minimum					.79			.79	.87
0900	Maximum					2.33			2.33	2.56
1000	For maximum resistance to atmosphere or cold, add				↓	1.14			1.14	1.25
1100	For average shipping charges, add				Total	1,975			1,975	2,175

13 31 13.13 Single-Walled Air-Supported Structures

		Crew	Daily Output	Labor-Hours	Unit	Material	Labor	Equipment	Total	Total Incl O&P
0010	**SINGLE-WALLED AIR-SUPPORTED STRUCTURES** R133113-10									
0020	Site preparation, incl. anchor placement and utilities	B-11B	1000	.016	SF Flr.	1.16	.69	.30	2.15	2.66
0030	For concrete, see Section 03 30 53.40									
0050	Warehouse, polyester/vinyl fabric, 28 oz., over 10 yr. life, welded									

13 31 Fabric Structures

13 31 13 – Air-Supported Fabric Structures

13 31 13.13 Single-Walled Air-Supported Structures		Crew	Daily Output	Labor-Hours	Unit	Material	2015 Bare Costs Labor	Equipment	Total	Total Incl O&P
0060	Seams, tension cables, primary & auxiliary inflation system,									
0070	airlock, personnel doors and liner									
0100	5,000 S.F.	4 Clab	5000	.006	SF Flr.	26.50	.24		26.74	29.50
0250	12,000 S.F.	"	6000	.005		18.85	.20		19.05	21.50
0400	24,000 S.F.	8 Clab	12000	.005		13.30	.20		13.50	14.90
0500	50,000 S.F.	"	12500	.005		12.30	.19		12.49	13.85
0700	12 oz. reinforced vinyl fabric, 5 yr. life, sewn seams,									
0710	accordion door, including liner									
0750	3000 S.F.	4 Clab	3000	.011	SF Flr.	13.20	.40		13.60	15.15
0800	12,000 S.F.	"	6000	.005		11.25	.20		11.45	12.65
0850	24,000 S.F.	8 Clab	12000	.005		9.50	.20		9.70	10.75
0950	Deduct for single layer					1.03			1.03	1.13
1000	Add for welded seams					1.50			1.50	1.65
1050	Add for double layer, welded seams included					3			3	3.30
1250	Tedlar/vinyl fabric, 28 oz., with liner, over 10 yr. life,									
1260	incl. overhead and personnel doors									
1300	3000 S.F.	4 Clab	3000	.011	SF Flr.	24.50	.40		24.90	27.50
1450	12,000 S.F.	"	6000	.005		17.20	.20		17.40	19.25
1550	24,000 S.F.	8 Clab	12000	.005		13.30	.20		13.50	14.95
1700	Deduct for single layer					2			2	2.20
2250	Greenhouse/shelter, woven polyethylene with liner, 2 yr. life,									
2260	sewn seams, including doors									
2300	3000 S.F.	4 Clab	3000	.011	SF Flr.	16	.40		16.40	18.20
2350	12,000 S.F.	"	6000	.005		14	.20		14.20	15.70
2450	24,000 S.F.	8 Clab	12000	.005		12	.20		12.20	13.50
2550	Deduct for single layer					.98			.98	1.08
2600	Tennis/gymnasium, polyester/vinyl fabric, 28 oz., over 10 yr. life,									
2610	including thermal liner, heat and lights									
2650	7,200 S.F.	4 Clab	6000	.005	SF Flr.	23.50	.20		23.70	26.50
2750	13,000 S.F.	"	6500	.005		18	.19		18.19	20
2850	Over 24,000 S.F.	8 Clab	12000	.005		16.45	.20		16.65	18.35
2860	For low temperature conditions, add					1.14			1.14	1.25
2870	For average shipping charges, add				Total	5,600			5,600	6,150
2900	Thermal liner, translucent reinforced vinyl				SF Flr.	1.14			1.14	1.25
2950	Metalized mylar fabric and mesh, double liner				"	2.33			2.33	2.56
3050	Stadium/convention center, teflon coated fiberglass, heavy weight,									
3060	over 20 yr. life, incl. thermal liner and heating system									
3100	Minimum	9 Clab	26000	.003	SF Flr.	57.50	.10		57.60	63
3110	Maximum	"	19000	.004	"	68	.14		68.14	75
3400	Doors, air lock, 15′ long, 10′ x 10′	2 Carp	.80	20	Ea.	20,400	940		21,340	24,000
3600	15′ x 15′	"	.50	32		30,300	1,500		31,800	35,700
3700	For each added 5′ length, add					5,425			5,425	5,950
3900	Revolving personnel door, 6′ diameter, 6′-6″ high	2 Carp	.80	20		15,200	940		16,140	18,200

13 31 23 – Tensioned Fabric Structures

13 31 23.50 Tension Structures		Crew	Daily Output	Labor-Hours	Unit	Material	Labor	Equipment	Total	Total Incl O&P
0010	**TENSION STRUCTURES** Rigid steel/alum. frame, vinyl coated poly									
0100	Fabric shell, 60′ clear span, not incl. foundations or floors									
0200	6,000 S.F.	B-41	1000	.044	SF Flr.	13.80	1.71	.29	15.80	18.15
0300	12,000 S.F.		1100	.040		13.20	1.56	.26	15.02	17.25
0400	80′ to 99′ clear span, 20,800 S.F.		1220	.036		13	1.40	.24	14.64	16.70
0410	100′ to 119′ clear span, 10,000 S.F.	L-5	2175	.026		13.70	1.36	.34	15.40	17.75
0430	26,000 S.F.		2300	.024		12.75	1.29	.32	14.36	16.60

13 31 Fabric Structures

13 31 23 – Tensioned Fabric Structures

	13 31 23.50 Tension Structures	Crew	Daily Output	Labor-Hours	Unit	Material	2015 Bare Costs Labor	Equipment	Total	Total Incl O&P
0450	36,000 S.F.	L-5	2500	.022	SF Flr.	12.60	1.18	.29	14.07	16.20
0460	120′ to 149′ clear span, 24,000 S.F.		3000	.019		13.90	.99	.25	15.14	17.20
0470	150′ to 199′ clear span, 30,000 S.F.		6000	.009		14.45	.49	.12	15.06	16.85
0480	200′ clear span, 40,000 S.F.	E-6	8000	.016		17.75	.84	.23	18.82	21
0500	For roll-up door, 12′ x 14′, add	L-2	1	16	Ea.	5,500	660		6,160	7,075

13 34 Fabricated Engineered Structures

13 34 13 – Glazed Structures

13 34 13.13 Greenhouses

	13 34 13.13 Greenhouses	Crew	Daily Output	Labor-Hours	Unit	Material	Labor	Equipment	Total	Total Incl O&P
0010	**GREENHOUSES**, Shell only, stock units, not incl. 2′ stub walls,									
0020	foundation, floors, heat or compartments									
0300	Residential type, free standing, 8′-6″ long x 7′-6″ wide	2 Carp	59	.271	SF Flr.	20	12.75		32.75	41.50
0400	10′-6″ wide		85	.188		37	8.85		45.85	54
0600	13′-6″ wide		108	.148		39	6.95		45.95	53
0700	17′-0″ wide		160	.100		43.50	4.70		48.20	55.50
0900	Lean-to type, 3′-10″ wide		34	.471		41.50	22		63.50	80
1000	6′-10″ wide		58	.276		50	12.95		62.95	75
1500	Commercial, custom, truss frame, incl. equip., plumbing, elec.,									
1550	benches and controls, under 2,000 S.F.				SF Flr.	13			13	14.30
1700	Over 5,000 S.F.				"	11.95			11.95	13.15
2000	Institutional, custom, rigid frame, including compartments and									
2050	multi-controls, under 500 S.F.				SF Flr.	25.50			25.50	28
2150	Over 2,000 S.F.				"	10.70			10.70	11.75
3700	For 1/4″ tempered glass, add				SF Surf	1.34			1.34	1.47
3900	Cooling, 1200 CFM exhaust fan, add				Ea.	310			310	340
4000	7850 CFM					1,050			1,050	1,150
4200	For heaters, 10 MBH, add					215			215	237
4300	60 MBH, add					780			780	855
4500	For benches, 2′ x 8′, add					160			160	176
4600	4′ x 10′, add					195			195	214
4800	For ventilation & humidity control w/ 4 integrated outlets, add				Total	240			240	264
4900	For environmental controls and automation, 8 outputs, 9 stages, add				"	765			765	840
5100	For humidification equipment, add				Ea.	299			299	330
5200	For vinyl shading, add				S.F.	.24			.24	.26
6000	Geodesic hemisphere, 1/8″ plexiglass glazing									
6050	8′ diameter	2 Carp	2	8	Ea.	6,250	375		6,625	7,450
6150	24′ diameter		.35	45.714		14,000	2,150		16,150	18,600
6250	48′ diameter		.20	80		33,000	3,750		36,750	42,000

13 34 13.19 Swimming Pool Enclosures

	13 34 13.19 Swimming Pool Enclosures	Crew	Daily Output	Labor-Hours	Unit	Material	Labor	Equipment	Total	Total Incl O&P
0010	**SWIMMING POOL ENCLOSURES** Translucent, free standing									
0020	not including foundations, heat or light									
0200	Economy	2 Carp	200	.080	SF Hor.	37	3.76		40.76	47
0600	Deluxe	"	70	.229		93	10.75		103.75	119
0700	For motorized roof, 40% opening, solid roof, add					21			21	23
0800	Skylight type roof, add					13.50			13.50	14.85

13 34 16 – Grandstands and Bleachers

13 34 16.13 Grandstands

	13 34 16.13 Grandstands	Crew	Daily Output	Labor-Hours	Unit	Material	Labor	Equipment	Total	Total Incl O&P
0010	**GRANDSTANDS** Permanent, municipal, including foundation									
0300	Steel, economy				Seat	19.90			19.90	22
0400	Steel, deluxe					22.50			22.50	24.50

13 34 Fabricated Engineered Structures

13 34 16 – Grandstands and Bleachers

13 34 16.13 Grandstands		Crew	Daily Output	Labor-Hours	Unit	2015 Bare Costs Material	Labor	Equipment	Total	Total Incl O&P
0900	Composite, steel, wood and plastic, stock design, economy				Seat	38.50			38.50	42
1000	Deluxe					68.50			68.50	75
13 34 16.53 Bleachers										
0010	**BLEACHERS**									
0020	Bleachers, outdoor, portable, 5 tiers, 42 seats	2 Sswk	120	.133	Seat	90	7		97	111
0100	5 tiers, 54 seats		80	.200		81.50	10.55		92.05	108
0200	10 tiers, 104 seats		120	.133		92.50	7		99.50	114
0300	10 tiers, 144 seats		80	.200		83	10.55		93.55	109
0500	Permanent bleachers, aluminum seat, steel frame, 24" row									
0600	8 tiers, 80 seats	2 Sswk	60	.267	Seat	67	14.05		81.05	98.50
0700	8 tiers, 160 seats		48	.333		58	17.55		75.55	94.50
0925	15 tiers, 154 to 165 seats		60	.267		94	14.05		108.05	128
0975	15 tiers, 214 to 225 seats		60	.267		84.50	14.05		98.55	118
1050	15 tiers, 274 to 285 seats		60	.267		77	14.05		91.05	109
1200	Seat backs only, 30" row, fiberglass		160	.100		23	5.25		28.25	34.50
1300	Steel and wood		160	.100		22	5.25		27.25	33.50
1400	NOTE: average seating is 1.5' in width									

13 34 19 – Metal Building Systems

13 34 19.50 Pre-Engineered Steel Buildings		Crew	Daily Output	Labor-Hours	Unit	Material	Labor	Equipment	Total	Total Incl O&P
0010	**PRE-ENGINEERED STEEL BUILDINGS** R133419-10									
0100	Clear span rigid frame, 26 ga. colored roofing and siding									
0150	20' to 29' wide, 10' eave height	E-2	425	.132	SF Flr.	8.80	6.80	3.55	19.15	25
0160	14' eave height		350	.160		9.50	8.30	4.32	22.12	29
0170	16' eave height		320	.175		10.20	9.05	4.72	23.97	31.50
0180	20' eave height		275	.204		11.15	10.55	5.50	27.20	36
0190	24' eave height		240	.233		12.30	12.05	6.30	30.65	41
0200	30' to 49' wide, 10' eave height		535	.105		6.75	5.40	2.82	14.97	19.65
0300	14' eave height		450	.124		7.30	6.45	3.36	17.11	22.50
0400	16' eave height		415	.135		7.80	7	3.64	18.44	24.50
0500	20' eave height		360	.156		8.45	8.05	4.20	20.70	27.50
0600	24' eave height		320	.175		9.30	9.05	4.72	23.07	30.50
0700	50' to 100' wide, 10' eave height		770	.073		5.75	3.76	1.96	11.47	14.75
0900	16' eave height		600	.093		6.60	4.83	2.52	13.95	18.10
1000	20' eave height		490	.114		7.15	5.90	3.08	16.13	21
1100	24' eave height		435	.129		7.90	6.65	3.47	18.02	23.50
1200	Clear span tapered beam frame, 26 ga. colored roofing/siding									
1300	30' to 39' wide, 10' eave height	E-2	535	.105	SF Flr.	7.65	5.40	2.82	15.87	20.50
1400	14' eave height		450	.124		8.45	6.45	3.36	18.26	24
1500	16' eave height		415	.135		8.90	7	3.64	19.54	25.50
1600	20' eave height		360	.156		9.80	8.05	4.20	22.05	29
1700	40' wide, 10' eave height		600	.093		6.80	4.83	2.52	14.15	18.35
1800	14' eave height		510	.110		7.55	5.70	2.96	16.21	21
1900	16' eave height		475	.118		7.90	6.10	3.18	17.18	22.50
2000	20' eave height		415	.135		8.70	7	3.64	19.34	25.50
2100	50' to 79' wide, 10' eave height		770	.073		6.40	3.76	1.96	12.12	15.45
2200	14' eave height		675	.083		6.95	4.29	2.24	13.48	17.25
2300	16' eave height		635	.088		7.20	4.56	2.38	14.14	18.20
2400	20' eave height		490	.114		8	5.90	3.08	16.98	22
2410	80' to 100' wide, 10' eave height		935	.060		5.65	3.10	1.62	10.37	13.25
2420	14' eave height		750	.075		6.25	3.86	2.01	12.12	15.55
2430	16' eave height		685	.082		6.50	4.23	2.21	12.94	16.70
2440	20' eave height		560	.100		6.95	5.15	2.70	14.80	19.30

13 34 19.50 Pre-Engineered Steel Buildings		Crew	Daily Output	Labor-Hours	Unit	Material	Labor	Equipment	Total	Total Incl O&P
						2015 Bare Costs				
2460	101′ to 120′ wide, 10′ eave height	E-2	950	.059	SF Flr.	5.20	3.05	1.59	9.84	12.55
2470	14′ eave height		770	.073		5.75	3.76	1.96	11.47	14.80
2480	16′ eave height		675	.083		6.15	4.29	2.24	12.68	16.40
2490	20′ eave height		560	.100		6.55	5.15	2.70	14.40	18.90
2500	Single post 2-span frame, 26 ga. colored roofing and siding									
2600	80′ wide, 14′ eave height	E-2	740	.076	SF Flr.	5.75	3.92	2.04	11.71	15.10
2700	16′ eave height		695	.081		6.10	4.17	2.17	12.44	16.10
2800	20′ eave height		625	.090		6.60	4.64	2.42	13.66	17.70
2900	24′ eave height		570	.098		7.25	5.10	2.65	15	19.40
3000	100′ wide, 14′ eave height		835	.067		5.60	3.47	1.81	10.88	13.95
3100	16′ eave height		795	.070		5.20	3.64	1.90	10.74	13.95
3200	20′ eave height		730	.077		6.30	3.97	2.07	12.34	15.90
3300	24′ eave height		670	.084		7	4.32	2.26	13.58	17.45
3400	120′ wide, 14′ eave height		870	.064		6.45	3.33	1.74	11.52	14.60
3500	16′ eave height		830	.067		5.80	3.49	1.82	11.11	14.20
3600	20′ eave height		765	.073		6.30	3.79	1.97	12.06	15.45
3700	24′ eave height		705	.079		6.90	4.11	2.14	13.15	16.85
3800	Double post 3-span frame, 26 ga. colored roofing and siding									
3900	150′ wide, 14′ eave height	E-2	925	.061	SF Flr.	4.57	3.13	1.63	9.33	12.10
4000	16′ eave height		890	.063		4.77	3.26	1.70	9.73	12.55
4100	20′ eave height		820	.068		5.20	3.53	1.84	10.57	13.70
4200	24′ eave height		765	.073		5.75	3.79	1.97	11.51	14.85
4300	Triple post 4-span frame, 26 ga. colored roofing and siding									
4400	160′ wide, 14′ eave height	E-2	970	.058	SF Flr.	4.50	2.99	1.56	9.05	11.65
4500	16′ eave height		930	.060		4.69	3.12	1.62	9.43	12.20
4600	20′ eave height		870	.064		4.62	3.33	1.74	9.69	12.60
4700	24′ eave height		815	.069		5.25	3.56	1.85	10.66	13.80
4800	200′ wide, 14′ eave height		1030	.054		4.13	2.81	1.47	8.41	10.85
4900	16′ eave height		995	.056		4.28	2.91	1.52	8.71	11.25
5000	20′ eave height		935	.060		4.72	3.10	1.62	9.44	12.20
5100	24′ eave height		885	.063		5.30	3.27	1.71	10.28	13.25
5200	Accessory items: add to the basic building cost above									
5250	Eave overhang, 2′ wide, 26 ga., with soffit	E-2	360	.156	L.F.	31.50	8.05	4.20	43.75	52.50
5300	4′ wide, without soffit		300	.187		27.50	9.65	5.05	42.20	52
5350	With soffit		250	.224		40	11.60	6.05	57.65	70
5400	6′ wide, without soffit		250	.224		35.50	11.60	6.05	53.15	65
5450	With soffit		200	.280		48	14.50	7.55	70.05	85.50
5500	Entrance canopy, incl. frame, 4′ x 4′		25	2.240	Ea.	475	116	60.50	651.50	780
5550	4′ x 8′		19	2.947	"	550	153	79.50	782.50	950
5600	End wall roof overhang, 4′ wide, without soffit		850	.066	L.F.	17.60	3.41	1.78	22.79	27
5650	With soffit		500	.112	"	28	5.80	3.02	36.82	44
5700	Doors, HM self-framing, incl. butts, lockset and trim									
5750	Single leaf, 3070 (3′ x 7′), economy	2 Sswk	5	3.200	Opng.	585	168		753	940
5800	Deluxe		4	4		640	211		851	1,075
5825	Glazed		4	4		745	211		956	1,175
5850	3670 (3′-6″ x 7′)		4	4		815	211		1,026	1,275
5900	4070 (4′ x 7′)		3	5.333		885	281		1,166	1,450
5950	Double leaf, 6070 (6′ x 7′)		2	8		1,100	420		1,520	1,925
6000	Glazed		2	8		1,400	420		1,820	2,275
6050	Framing only, for openings, 3′ x 7′		4	4		185	211		396	570
6100	10′ x 10′		3	5.333		610	281		891	1,150
6150	For windows below, 2020 (2′ x 2′)		6	2.667		196	140		336	460
6200	4030 (4′ x 3′)		5	3.200		239	168		407	555

13 34 19.50 Pre-Engineered Steel Buildings		Crew	Daily Output	Labor-Hours	Unit	2015 Bare Costs Material	Labor	Equipment	Total	Total Incl O&P
6250	Flashings, 26 ga., corner or eave, painted	2 Sswk	240	.067	L.F.	4.47	3.51		7.98	11
6300	Galvanized		240	.067		4.10	3.51		7.61	10.60
6350	Rake flashing, painted		240	.067		4.83	3.51		8.34	11.40
6400	Galvanized		240	.067		4.40	3.51		7.91	10.95
6450	Ridge flashing, 18" wide, painted		240	.067		6.45	3.51		9.96	13.20
6500	Galvanized		240	.067		7.10	3.51		10.61	13.90
6550	Gutter, eave type, 26 ga., painted		320	.050		6.95	2.63		9.58	12.20
6650	Valley type, between buildings, painted		120	.133		12.85	7		19.85	26.50
6710	Insulation, rated .6 lb. density, unfaced 4" thick, R13	2 Carp	2300	.007	S.F.	.42	.33		.75	.96
6720	6" thick, R19		2300	.007		.58	.33		.91	1.14
6730	10" thick, R30		2300	.007		1.10	.33		1.43	1.71
6750	Insulation, rated .6 lb. density, poly/scrim/foil (PSF) faced									
6760	4" thick R13	2 Carp	2300	.007	S.F.	.63	.33		.96	1.19
6770	6" thick, R19		2300	.007		.89	.33		1.22	1.48
6780	9 1/2" thick, R30		2300	.007		.98	.33		1.31	1.58
6800	Insulation, rated .6 lb. density, vinyl faced 1-1/2" thick, R5		2300	.007		.31	.33		.64	.84
6850	3" thick, R10		2300	.007		.32	.33		.65	.85
6900	4" thick, R13		2300	.007		.42	.33		.75	.96
6920	6" thick, R19		2300	.007		.55	.33		.88	1.11
6930	10" thick, R30		2300	.007		1.44	.33		1.77	2.08
6950	Foil/scrim/kraft (FSK) faced, 1-1/2" thick, R5		2300	.007		.33	.33		.66	.86
7000	2" thick, R6		2300	.007		.43	.33		.76	.97
7050	3" thick, R10		2300	.007		.43	.33		.76	.97
7100	4" thick, R13		2300	.007		.45	.33		.78	1
7110	6" thick, R19		2300	.007		.66	.33		.99	1.23
7120	10" thick, R30		2300	.007		.92	.33		1.25	1.51
7150	Metalized polyester/scrim/kraft (PSK) facing,1-1/2" thk, R5		2300	.007		.53	.33		.86	1.08
7200	2" thick, R6		2300	.007		.63	.33		.96	1.19
7250	3" thick, R11		2300	.007		.72	.33		1.05	1.29
7300	4" thick, R13		2300	.007		.81	.33		1.14	1.39
7310	6" thick, R19		2300	.007		1.06	.33		1.39	1.67
7320	10" thick, R30		2300	.007		1.18	.33		1.51	1.80
7350	Vinyl/scrim/foil (VSF), 1-1/2" thick, R5		2300	.007		.47	.33		.80	1.02
7400	2" thick, R6		2300	.007		.60	.33		.93	1.16
7450	3" thick, R10		2300	.007		.65	.33		.98	1.22
7500	4" thick, R13		2300	.007		.79	.33		1.12	1.37
7510	Vinyl/scrim/vinyl (VSV) 4" thick, R13		2300	.007		.53	.33		.86	1.08
7520	6" thick, R19		2300	.007		.68	.33		1.01	1.25
7530	9 1/2" thick, R30		2300	.007		.92	.33		1.25	1.51
7540	Polyprop/scrim/polyester (PSP), 4", R13		2300	.007		.58	.33		.91	1.14
7550	6" thick, R19		2300	.007		.73	.33		1.06	1.30
7555	10" thick, R30		2300	.007		1.28	.33		1.61	1.91
7560	Polyprop/scrim/kraft/polyester (PSKP), 4" thick, R13		2300	.007		.54	.33		.87	1.09
7570	6" thick, R19		2300	.007		.74	.33		1.07	1.31
7580	10" thick, R19		2300	.007		1.41	.33		1.74	2.05
7585	Vinyl/scrim/polyester (VSP), 4" thick, R13		2300	.007		.57	.33		.90	1.13
7590	6" thick, R19		2300	.007		.70	.33		1.03	1.27
7600	10" thick, R30		2300	.007		1.09	.33		1.42	1.70
7635	Insulation installation, over the purlin, second layer, up to 4" thick, add						90%			
7640	Insulation installation, between the purlins, up to 4" thick, add						100%			
7650	Sash, single slide, glazed, with screens, 2020 (2' x 2')	E-1	22	1.091	Opng.	127	56.50	6.65	190.15	242
7700	3030 (3' x 3')		14	1.714		285	89	10.40	384.40	475
7750	4030 (4' x 3')		13	1.846		380	96	11.20	487.20	590

13 34 Fabricated Engineered Structures

13 34 19 – Metal Building Systems

13 34 19.50 Pre-Engineered Steel Buildings		Crew	Daily Output	Labor-Hours	Unit	2015 Bare Costs Material	Labor	Equipment	Total	Total Incl O&P
7800	6040 (6′ x 4′)	E-1	12	2	Opng.	760	104	12.15	876.15	1,025
7850	Double slide sash, 3030 (3′ x 3′)		14	1.714		224	89	10.40	323.40	405
7900	6040 (6′ x 4′)		12	2		595	104	12.15	711.15	840
7950	Fixed glass, no screens, 3030 (3′ x 3′)		14	1.714		220	89	10.40	319.40	400
8000	6040 (6′ x 4′)		12	2		585	104	12.15	701.15	830
8050	Prefinished storm sash, 3030 (3′ x 3′)		70	.343		80	17.80	2.08	99.88	120
8100	Siding and roofing, see Sections 07 41 13.00 & 07 42 13.00									
8200	Skylight, fiberglass panels, to 30 S.F.	E-1	10	2.400	Ea.	125	125	14.60	264.60	360
8250	Larger sizes, add for excess over 30 S.F.	″	300	.080	S.F.	4.18	4.16	.49	8.83	12.10
8300	Roof vents, turbine ventilator, wind driven									
8350	No damper, includes base, galvanized									
8400	12″ diameter	Q-9	10	1.600	Ea.	91	80.50		171.50	223
8450	20″ diameter		8	2		256	101		357	435
8500	24″ diameter		8	2		390	101		491	580
8600	Continuous, 26 ga., 10′ long, 9″ wide	2 Sswk	4	4		36	211		247	405
8650	12″ wide	″	4	4		36	211		247	405

13 34 23 – Fabricated Structures

13 34 23.10 Comfort Stations		Crew	Daily Output	Labor-Hours	Unit	Material	Labor	Equipment	Total	Total Incl O&P
0010	**COMFORT STATIONS** Prefab., stock, w/doors, windows & fixt.									
0100	Not incl. interior finish or electrical									
0300	Mobile, on steel frame, 2 unit				S.F.	163			163	180
0350	7 unit					271			271	298
0400	Permanent, including concrete slab, 2 unit	B-12J	50	.320		249	14.30	17.65	280.95	315
0500	6 unit	″	43	.372		191	16.60	20.50	228.10	259
0600	Alternate pricing method, mobile, 2 fixture				Fixture	5,650			5,650	6,225
0650	7 fixture					9,750			9,750	10,700
0700	Permanent, 2 unit	B-12J	.70	22.857		20,500	1,025	1,250	22,775	25,500
0750	6 unit	″	.50	32		17,400	1,425	1,775	20,600	23,300

13 34 23.15 Domes		Crew	Daily Output	Labor-Hours	Unit	Material	Labor	Equipment	Total	Total Incl O&P
0010	**DOMES**									
0020	Domes, rev. alum., elec. drive, for astronomy obsv. shell only, stock units									
0600	10′-6″ diameter	2 Carp	.25	64	Ea.	38,000	3,000		41,000	46,400
0900	18′-6″ diameter		.17	94.118		78,000	4,425		82,425	93,000
1200	24′-6″ diameter		.08	200		103,000	9,400		112,400	128,000
1500	Domes, bulk storage, shell only, dual radius hemisphere, arch, steel									
1600	framing, corrugated steel covering, 150′ diameter	E-2	550	.102	SF Flr.	35	5.25	2.75	43	50
1700	400′ diameter	″	720	.078		28.50	4.02	2.10	34.62	40
1800	Wood framing, wood decking, to 400′ diameter	F-4	400	.120		37.50	5.70	2.82	46.02	53.50
1900	Radial framed wood (2″ x 6″), 1/2″ thick									
2000	plywood, asphalt shingles, 50′ diameter	F-3	2000	.020	SF Flr.	72.50	.96	.33	73.79	81.50
2100	60′ diameter		1900	.021		62	1.01	.34	63.35	70
2200	72′ diameter		1800	.022		51.50	1.06	.36	52.92	58.50
2300	116′ diameter		1730	.023		35	1.11	.38	36.49	40.50
2400	150′ diameter		1500	.027		37.50	1.28	.44	39.22	44

13 34 23.16 Fabricated Control Booths		Crew	Daily Output	Labor-Hours	Unit	Material	Labor	Equipment	Total	Total Incl O&P
0010	**FABRICATED CONTROL BOOTHS**									
0100	Guard House, prefab conc. w/bullet resistant doors & windows, roof & wiring									
0110	8′ x 8′, Level III	L-10	1	24	Ea.	44,200	1,275	655	46,130	51,500
0120	8′ x 8′, Level IV	″	1	24	″	50,500	1,275	655	52,430	58,500

13 34 Fabricated Engineered Structures

13 34 23 – Fabricated Structures

13 34 23.25 Garage Costs

	13 34 23.25 Garage Costs	Crew	Daily Output	Labor-Hours	Unit	2015 Bare Costs Material	Labor	Equipment	Total	Total Incl O&P
0010	**GARAGE COSTS**									
0020	Public parking, average				Car				18,400	20,200
0100	See also Square Foot Costs in Reference Section									
0300	Residential, wood, 12′ x 20′, one car prefab shell, stock, economy	2 Carp	1	16	Total	5,375	750		6,125	7,075
0350	Custom		.67	23.881		5,975	1,125		7,100	8,300
0400	Two car, 24′ x 20′, economy		.67	23.881		10,100	1,125		11,225	12,800
0450	Custom		.50	32		12,100	1,500		13,600	15,600

13 34 23.30 Garden House

	13 34 23.30 Garden House	Crew	Daily Output	Labor-Hours	Unit	Material	Labor	Equipment	Total	Total Incl O&P
0010	**GARDEN HOUSE** Prefab wood, no floors or foundations									
0100	6′ x 6′	2 Carp	200	.080	SF Flr.	51.50	3.76		55.26	62.50
0300	8′ x 12′	″	48	.333	″	38	15.65		53.65	66

13 34 23.35 Geodesic Domes

	13 34 23.35 Geodesic Domes	Crew	Daily Output	Labor-Hours	Unit	Material	Labor	Equipment	Total	Total Incl O&P
0010	**GEODESIC DOMES** Shell only, interlocking plywood panels R133423-30									
0400	30′ diameter	F-5	1.60	20	Ea.	23,000	950		23,950	26,800
0500	33′ diameter		1.14	28.070		24,300	1,325		25,625	28,800
0600	40′ diameter		1	32		28,200	1,525		29,725	33,300
0700	45′ diameter	F-3	1.13	35.556		29,500	1,700	580	31,780	35,800
0750	56′ diameter		1	40		53,000	1,925	655	55,580	62,000
0800	60′ diameter		1	40		60,000	1,925	655	62,580	69,500
0850	67′ diameter		.80	50		84,500	2,400	820	87,720	97,500
1100	Aluminum panel, with 6″ insulation									
1200	100′ diameter				SF Flr.	30.50			30.50	33.50
1300	500′ diameter				″	29.50			29.50	32.50
1600	Aluminum framed, plexiglass closure panels									
1700	40′ diameter				SF Flr.	75.50			75.50	83
1800	200′ diameter				″	69.50			69.50	76.50
2100	Aluminum framed, aluminum closure panels									
2200	40′ diameter				SF Flr.	24.50			24.50	27
2300	100′ diameter					23.50			23.50	26
2400	200′ diameter					23.50			23.50	26
2500	For VRP faced bonded fiberglass insulation, add								10	10
2700	Aluminum framed, fiberglass sandwich panel closure									
2800	6′ diameter	2 Carp	150	.107	SF Flr.	33	5		38	44
2900	28′ diameter	″	350	.046	″	30	2.15		32.15	36.50

13 34 23.45 Kiosks

	13 34 23.45 Kiosks	Crew	Daily Output	Labor-Hours	Unit	Material	Labor	Equipment	Total	Total Incl O&P
0010	**KIOSKS**									
0020	Round, advertising type, 5′ diameter, 7′ high, aluminum wall, illuminated				Ea.	23,500			23,500	25,800
0100	Aluminum wall, non-illuminated					22,500			22,500	24,700
0500	Rectangular, 5′ x 9′, 7′-6″ high, aluminum wall, illuminated					25,500			25,500	28,000
0600	Aluminum wall, non-illuminated					24,000			24,000	26,400

13 34 23.60 Portable Booths

	13 34 23.60 Portable Booths	Crew	Daily Output	Labor-Hours	Unit	Material	Labor	Equipment	Total	Total Incl O&P
0010	**PORTABLE BOOTHS** Prefab. aluminum with doors, windows, ext. roof									
0100	lights wiring & insulation, 15 S.F. building, O.D., painted				S.F.	266			266	293
0300	30 S.F. building					214			214	235
0400	50 S.F. building					169			169	186
0600	80 S.F. building					147			147	161
0700	100 S.F. building					123			123	135
0900	Acoustical booth, 27 Db @ 1,000 Hz, 15 S.F. floor				Ea.	3,550			3,550	3,925
1000	7′ x 7′-6″, including light & ventilation					7,325			7,325	8,050
1200	Ticket booth, galv. steel, not incl. foundations., 4′ x 4′					4,625			4,625	5,100
1300	4′ x 6′					6,825			6,825	7,500

13 34 Fabricated Engineered Structures

13 34 23 – Fabricated Structures

13 34 23.70 Shelters		Crew	Daily Output	Labor-Hours	Unit	Material	2015 Bare Costs Labor	Equipment	Total	Total Incl O&P
0010	**SHELTERS**									
0020	Aluminum frame, acrylic glazing, 3' x 9' x 8' high	2 Sswk	1.14	14.035	Ea.	3,075	740		3,815	4,650
0100	9' x 12' x 8' high	"	.73	21.918	"	7,300	1,150		8,450	10,000

13 34 43 – Aircraft Hangars

13 34 43.50 Hangars		Crew	Daily Output	Labor-Hours	Unit	Material	Labor	Equipment	Total	Total Incl O&P
0010	**HANGARS** Prefabricated steel T hangars, Galv. steel roof &									
0100	walls, incl. electric bi-folding doors									
0110	not including floors or foundations, 4 unit	E-2	1275	.044	SF Flr.	12.75	2.27	1.19	16.21	19.15
0130	8 unit		1063	.053		11.60	2.73	1.42	15.75	18.90
0900	With bottom rolling doors, 4 unit		1386	.040		11.75	2.09	1.09	14.93	17.60
1000	8 unit	↓	966	.058	↓	10.55	3	1.56	15.11	18.35
1200	Alternate pricing method:									
1300	Galv. roof and walls, electric bi-folding doors, 4 plane	E-2	1.06	52.830	Plane	16,900	2,725	1,425	21,050	24,800
1500	8 plane		.91	61.538		13,800	3,175	1,650	18,625	22,400
1600	With bottom rolling doors, 4 plane		1.25	44.800		15,600	2,325	1,200	19,125	22,300
1800	8 plane	↓	.97	57.732	↓	12,600	2,975	1,550	17,125	20,600
2000	Circular type, prefab., steel frame, plastic skin, electric									
2010	door, including foundations, 80' diameter,									

13 34 53 – Agricultural Structures

13 34 53.50 Silos		Crew	Daily Output	Labor-Hours	Unit	Material	Labor	Equipment	Total	Total Incl O&P
0010	**SILOS**									
0500	Steel, factory fab., 30,000 gallon cap., painted, economy	L-5	1	56	Ea.	22,100	2,950	735	25,785	30,200
0700	Deluxe		.50	112		35,100	5,925	1,475	42,500	50,500
0800	Epoxy lined, economy		1	56		36,100	2,950	735	39,785	45,600
1000	Deluxe	↓	.50	112	↓	45,700	5,925	1,475	53,100	62,000

13 34 63 – Natural Fiber Construction

13 34 63.50 Straw Bale Construction		Crew	Daily Output	Labor-Hours	Unit	Material	Labor	Equipment	Total	Total Incl O&P
0010	**STRAW BALE CONSTRUCTION**									
2020	Straw bales in walls w/modified post and beam frame G	2 Carp	320	.050	S.F.	6.15	2.35		8.50	10.35

13 36 Towers

13 36 13 – Metal Towers

13 36 13.50 Control Towers		Crew	Daily Output	Labor-Hours	Unit	Material	Labor	Equipment	Total	Total Incl O&P
0010	**CONTROL TOWERS**									
0020	Modular 12' x 10', incl. instruments				Ea.	757,000			757,000	832,500
0500	With standard 40' tower				"	1,191,000			1,191,000	1,310,000
1000	Temporary portable control towers, 8' x 12',									
1010	complete with one position communications				Ea.				266,000	293,000

13 42 Building Modules

13 42 63 – Detention Cell Modules

13 42 63.16 Steel Detention Cell Modules		Crew	Daily Output	Labor-Hours	Unit	2015 Bare Costs Material	Labor	Equipment	Total	Total Incl O&P
0010	**STEEL DETENTION CELL MODULES**									
2000	Cells, prefab., 5′ to 6′ wide, 7′ to 8′ high, 7′ to 8′ deep,									
2010	bar front, cot, not incl. plumbing	E-4	1.50	21.333	Ea.	9,775	1,125	97.50	10,997.50	12,800

13 47 Facility Protection

13 47 13 – Cathodic Protection

13 47 13.16 Cathodic Prot. for Underground Storage Tanks

		Crew	Daily Output	Labor-Hours	Unit	Material	Labor	Equipment	Total	Total Incl O&P
0010	**CATHODIC PROTECTION FOR UNDERGROUND STORAGE TANKS**									
1000	Anodes, magnesium type, 9 #	R-15	18.50	2.595	Ea.	41.50	140	16.30	197.80	273
1010	17 #		13	3.692		71.50	199	23	293.50	400
1020	32 #		10	4.800		124	258	30	412	555
1030	48 #		7.20	6.667		164	360	42	566	765
1100	Graphite type w/epoxy cap, 3″ x 60″ (32 #)	R-22	8.40	4.438		128	206		334	455
1110	4″ x 80″ (68 #)		6	6.213		241	289		530	705
1120	6″ x 72″ (80 #)		5.20	7.169		1,500	335		1,835	2,150
1130	6″ x 36″ (45 #)		9.60	3.883		760	181		941	1,100
2000	Rectifiers, silicon type, air cooled, 28 V/10 A	R-19	3.50	5.714		2,300	315		2,615	3,025
2010	20 V/20 A		3.50	5.714		2,375	315		2,690	3,075
2100	Oil immersed, 28 V/10 A		3	6.667		3,175	365		3,540	4,025
2110	20 V/20 A		3	6.667		3,175	365		3,540	4,050
3000	Anode backfill, coke breeze	R-22	3850	.010	Lb.	.22	.45		.67	.92
4000	Cable, HMWPE, No. 8		2.40	15.533	M.L.F.	500	720		1,220	1,650
4010	No. 6		2.40	15.533		745	720		1,465	1,925
4020	No. 4		2.40	15.533		1,125	720		1,845	2,350
4030	No. 2		2.40	15.533		1,750	720		2,470	3,025
4040	No. 1		2.20	16.945		2,400	790		3,190	3,850
4050	No. 1/0		2.20	16.945		2,975	790		3,765	4,475
4060	No. 2/0		2.20	16.945		4,625	790		5,415	6,300
4070	No. 4/0		2	18.640		6,000	865		6,865	7,925
5000	Test station, 7 terminal box, flush curb type w/lockable cover	R-19	12	1.667	Ea.	71.50	91.50		163	216
5010	Reference cell, 2″ dia PVC conduit, cplg., plug, set flush	″	4.80	4.167	″	151	228		379	505

13 48 Sound, Vibration, and Seismic Control

13 48 13 – Manufactured Sound and Vibration Control Components

13 48 13.50 Audio Masking

		Crew	Daily Output	Labor-Hours	Unit	Material	Labor	Equipment	Total	Total Incl O&P
0010	**AUDIO MASKING**, acoustical enclosure, 4″ thick wall and ceiling									
0020	8# per S.F., up to 12′ span	3 Carp	72	.333	SF Surf	33	15.65		48.65	60
0300	Better quality panels, 10.5# per S.F.		64	.375		37	17.60		54.60	68
0400	Reverb-chamber, 4″ thick, parallel walls		60	.400		46.50	18.80		65.30	80
0600	Skewed wall, parallel roof, 4″ thick panels		55	.436		53	20.50		73.50	90
0700	Skewed walls, skewed roof, 4″ layers, 4″ air space		48	.500		59.50	23.50		83	102
0900	Sound-absorbing panels, pntd. mtl., 2′-6″ x 8′, under 1,000 S.F.		215	.112		12.30	5.25		17.55	21.50
1100	Over 1000 S.F.		240	.100		11.85	4.70		16.55	20.50
1200	Fabric faced		240	.100		9.60	4.70		14.30	17.85
1500	Flexible transparent curtain, clear	3 Shee	215	.112		7.45	6.25		13.70	17.75
1600	50% foam		215	.112		10.40	6.25		16.65	21
1700	75% foam		215	.112		10.40	6.25		16.65	21
1800	100% foam		215	.112		10.40	6.25		16.65	21
3100	Audio masking system, including speakers, amplification									

13 48 Sound, Vibration, and Seismic Control

13 48 13 – Manufactured Sound and Vibration Control Components

13 48 13.50 Audio Masking		Crew	Daily Output	Labor-Hours	Unit	Material	Labor	Equipment	Total	Total Incl O&P
						2015 Bare Costs				
3110	and signal generator									
3200	Ceiling mounted, 5,000 S.F.	2 Elec	2400	.007	S.F.	1.26	.36		1.62	1.94
3300	10,000 S.F.		2800	.006		1.02	.31		1.33	1.59
3400	Plenum mounted, 5,000 S.F.		3800	.004		1.08	.23		1.31	1.54
3500	10,000 S.F.		4400	.004		.73	.20		.93	1.10

13 49 Radiation Protection

13 49 13 – Integrated X-Ray Shielding Assemblies

13 49 13.50 Lead Sheets

		Crew	Daily Output	Labor-Hours	Unit	Material	Labor	Equipment	Total	Total Incl O&P
0010	**LEAD SHEETS**									
0300	Lead sheets, 1/16″ thick	2 Lath	135	.119	S.F.	10.50	5.10		15.60	19.05
0400	1/8″ thick		120	.133		31	5.75		36.75	42.50
0500	Lead shielding, 1/4″ thick		135	.119		41.50	5.10		46.60	53.50
0550	1/2″ thick		120	.133		73.50	5.75		79.25	89.50
0950	Lead headed nails (average 1 lb. per sheet)				Lb.	8			8	8.80
1000	Butt joints in 1/8″ lead or thicker, 2″ batten strip x 7′ long	2 Lath	240	.067	Ea.	28	2.87		30.87	35
1200	X-ray protection, average radiography or fluoroscopy									
1210	room, up to 300 S.F. floor, 1/16″ lead, economy	2 Lath	.25	64	Total	10,100	2,750		12,850	15,200
1500	7′-0″ walls, deluxe	″	.15	106	″	12,200	4,575		16,775	20,200
1600	Deep therapy X-ray room, 250 kV capacity,									
1800	up to 300 S.F. floor, 1/4″ lead, economy	2 Lath	.08	200	Total	28,300	8,600		36,900	43,800
1900	7′-0″ walls, deluxe	″	.06	266	″	34,900	11,500		46,400	55,500

13 49 19 – Lead-Lined Materials

13 49 19.50 Shielding Lead

		Crew	Daily Output	Labor-Hours	Unit	Material	Labor	Equipment	Total	Total Incl O&P
0010	**SHIELDING LEAD**									
0100	Laminated lead in wood doors, 1/16″ thick, no hardware				S.F.	52.50			52.50	57.50
0200	Lead lined door frame, not incl. hardware,									
0210	1/16″ thick lead, butt prepared for hardware	1 Lath	2.40	3.333	Ea.	810	143		953	1,100
0850	Window frame with 1/16″ lead and voice passage, 36″ x 60″	2 Glaz	2	8		4,200	360		4,560	5,150
0870	24″ x 36″ frame		4	4		2,175	180		2,355	2,675
0900	Lead gypsum board, 5/8″ thick with 1/16″ lead		160	.100	S.F.	10.95	4.51		15.46	18.90
0910	1/8″ lead		140	.114		23	5.15		28.15	33.50
0930	1/32″ lead	2 Lath	200	.080		7.95	3.44		11.39	13.80

13 49 21 – Lead Glazing

13 49 21.50 Lead Glazing

		Crew	Daily Output	Labor-Hours	Unit	Material	Labor	Equipment	Total	Total Incl O&P
0010	**LEAD GLAZING**									
0600	Lead glass, 1/4″ thick, 2.0 mm LE, 12″ x 16″	2 Glaz	13	1.231	Ea.	380	55.50		435.50	500
0700	24″ x 36″		8	2		1,325	90		1,415	1,575
0800	36″ x 60″		2	8		3,675	360		4,035	4,575
2000	X-ray viewing panels, clear lead plastic									
2010	7 mm thick, 0.3 mm LE, 2.3 lb./S.F.	H-3	139	.115	S.F.	241	4.64		245.64	272
2020	12 mm thick, 0.5 mm LE, 3.9 lb./S.F.		82	.195		355	7.85		362.85	400
2030	18 mm thick, 0.8 mm LE, 5.9 lb./S.F.		54	.296		405	11.95		416.95	465
2040	22 mm thick, 1.0 mm LE, 7.2 lb./S.F.		44	.364		530	14.65		544.65	610
2050	35 mm thick, 1.5 mm LE, 11.5 lb./S.F.		28	.571		815	23		838	930
2060	46 mm thick, 2.0 mm LE, 15.0 lb./S.F.		21	.762		1,050	30.50		1,080.50	1,225
2090	For panels 12 S.F. to 48 S.F., add crating charge				Ea.				50	50

13 49 23 – Integrated RFI/EMI Shielding Assemblies

13 49 23.50 Modular Shielding Partitions		Crew	Daily Output	Labor-Hours	Unit	Material	2015 Bare Costs Labor	Equipment	Total	Total Incl O&P
0010	**MODULAR SHIELDING PARTITIONS**									
4000	X-ray barriers, modular, panels mounted within framework for									
4002	attaching to floor, wall or ceiling, upper portion is clear lead									
4005	plastic window panels 48"H, lower portion is opaque leaded									
4008	steel panels 36"H, structural supports not incl.									
4010	1-section barrier, 36"W x 84"H overall									
4020	0.5 mm LE panels	H-3	6.40	2.500	Ea.	8,100	101		8,201	9,050
4030	0.8 mm LE panels		6.40	2.500		8,725	101		8,826	9,750
4040	1.0 mm LE panels		5.33	3.002		10,200	121		10,321	11,400
4050	1.5 mm LE panels		5.33	3.002		13,600	121		13,721	15,200
4060	2-section barrier, 72"W x 84"H overall									
4070	0.5 mm LE panels	H-3	4	4	Ea.	11,800	161		11,961	13,200
4080	0.8 mm LE panels		4	4		13,100	161		13,261	14,600
4090	1.0 mm LE panels		3.56	4.494		16,000	181		16,181	17,900
5000	1.5 mm LE panels		3.20	5		22,800	201		23,001	25,400
5010	3-section barrier, 108"W x 84"H overall									
5020	0.5 mm LE panels	H-3	3.20	5	Ea.	17,700	201		17,901	19,800
5030	0.8 mm LE panels		3.20	5		19,600	201		19,801	21,800
5040	1.0 mm LE panels		2.67	5.993		24,000	241		24,241	26,800
5050	1.5 mm LE panels		2.46	6.504		34,200	262		34,462	38,000
7000	X-ray barriers, mobile, mounted within framework w/casters on									
7005	bottom, clear lead plastic window panels on upper portion,									
7010	opaque on lower, 30"W x 75"H overall, incl. framework									
7020	24"H upper w/0.5 mm LE, 48"H lower w/0.8 mm LE	1 Carp	16	.500	Ea.	3,800	23.50		3,823.50	4,200
7030	48"W x 75"H overall, incl. framework									
7040	36"H upper w/0.5 mm LE, 36"H lower w/0.8 mm LE	1 Carp	16	.500	Ea.	6,175	23.50		6,198.50	6,825
7050	36"H upper w/1.0 mm LE, 36"H lower w/1.5 mm LE	"	16	.500	"	7,300	23.50		7,323.50	8,075
7060	72"W x 75"H overall, incl. framework									
7070	36"H upper w/0.5 mm LE, 36"H lower w/0.8 mm LE	1 Carp	16	.500	Ea.	7,300	23.50		7,323.50	8,075
7080	36"H upper w/1.0 mm LE, 36"H lower w/1.5 mm LE	"	16	.500	"	9,150	23.50		9,173.50	10,100

13 49 33 – Radio Frequency Shielding

13 49 33.50 Shielding, Radio Frequency		Crew	Daily Output	Labor-Hours	Unit	Material	Labor	Equipment	Total	Total Incl O&P
0010	**SHIELDING, RADIO FREQUENCY**									
0020	Prefabricated, galvanized steel	2 Carp	375	.043	SF Surf	4.46	2		6.46	8
0040	5 oz., copper floor panel		480	.033		3.68	1.56		5.24	6.45
0050	5 oz., copper wall/ceiling panel		155	.103		3.68	4.85		8.53	11.50
0100	12 oz., copper floor panel		470	.034		7.85	1.60		9.45	11.10
0110	12 oz., copper wall/ceiling panel		140	.114		7.85	5.35		13.20	16.90
0150	Door, copper/wood laminate, 4' x 7'		1.50	10.667	Ea.	7,400	500		7,900	8,900

13 53 Meteorological Instrumentation

13 53 09 – Weather Instrumentation

13 53 09.50 Weather Station		Crew	Daily Output	Labor-Hours	Unit	2015 Bare Costs Material	Labor	Equipment	Total	Total Incl O&P
0010	**WEATHER STATION**									
0020	Remote recording, solar powered, with rain gauge & display, 400 ft range				Ea.	775			775	850
0100	1 mile range				"	1,900			1,900	2,075

Estimating Tips

General

- Many products in Division 14 will require some type of support or blocking for installation not included with the item itself. Examples are supports for conveyors or tube systems, attachment points for lifts, and footings for hoists or cranes. Add these supports in the appropriate division.

14 10 00 Dumbwaiters
14 20 00 Elevators

- Dumbwaiters and elevators are estimated and purchased in a method similar to buying a car. The manufacturer has a base unit with standard features. Added to this base unit price will be whatever options the owner or specifications require. Increased load capacity, additional vertical travel, additional stops, higher speed, and cab finish options are items to be considered. When developing an estimate for dumbwaiters and elevators, remember that some items needed by the installers may have to be included as part of the general contract.

  Examples are:
  - shaftway
  - rail support brackets
  - machine room
  - electrical supply
  - sill angles
  - electrical connections
  - pits
  - roof penthouses
  - pit ladders

  Check the job specifications and drawings before pricing.
- Installation of elevators and handicapped lifts in historic structures can require significant additional costs. The associated structural requirements may involve cutting into and repairing finishes, moldings, flooring, etc. The estimator must account for these special conditions.

14 30 00 Escalators and Moving Walks

- Escalators and moving walks are specialty items installed by specialty contractors. There are numerous options associated with these items. For specific options, contact a manufacturer or contractor. In a method similar to estimating dumbwaiters and elevators, you should verify the extent of general contract work and add items as necessary.

14 40 00 Lifts
14 90 00 Other Conveying Equipment

- Products such as correspondence lifts, chutes, and pneumatic tube systems, as well as other items specified in this subdivision, may require trained installers. The general contractor might not have any choice as to who will perform the installation, or when it will be performed. Long lead times are often required for these products, making early decisions in scheduling necessary.

Reference Numbers

Reference numbers are shown in shaded boxes at the beginning of some major classifications. These numbers refer to related items in the Reference Section. The reference information may be an estimating procedure, an alternate pricing method, or technical information.

Note: Not all subdivisions listed here necessarily appear in this publication. ■

14 11 Manual Dumbwaiters

14 11 10 – Hand Operated Dumbwaiters

14 11 10.20 Manual Dumbwaiters		Crew	Daily Output	Labor-Hours	Unit	Material	2015 Bare Costs Labor	Equipment	Total	Total Incl O&P
0010	**MANUAL DUMBWAITERS**									
0020	2 stop, hand powered, up to 75 lb. capacity	2 Elev	.75	21.333	Ea.	3,000	1,625		4,625	5,725
0100	76 lb capacity and up		.50	32	"	6,700	2,450		9,150	11,000
0300	For each additional stop, add	↓	.75	21.333	Stop	1,100	1,625		2,725	3,625

14 12 Electric Dumbwaiters

14 12 10 – Dumbwaiters

14 12 10.10 Electric Dumbwaiters		Crew	Daily Output	Labor-Hours	Unit	Material	Labor	Equipment	Total	Total Incl O&P
0010	**ELECTRIC DUMBWAITERS**									
0020	2 stop, up to 75 lb capacity	2 Elev	.13	123	Ea.	7,400	9,425		16,825	22,300
0100	76 lb capacity and up		.11	145	"	22,300	11,100		33,400	41,100
0600	For each additional stop, add	↓	.54	29.630	Stop	3,300	2,275		5,575	7,000

14 21 Electric Traction Elevators

14 21 13 – Electric Traction Freight Elevators

14 21 13.10 Electric Traction Freight Elevators and Options		Crew	Daily Output	Labor-Hours	Unit	Material	Labor	Equipment	Total	Total Incl O&P
0010	**ELECTRIC TRACTION FREIGHT ELEVATORS AND OPTIONS** R142000-10									
0425	Electric freight, base unit, 4000 lb., 200 fpm, 4 stop, std. fin.	2 Elev	.05	320	Ea.	110,000	24,500		134,500	158,000
0450	For 5000 lb. capacity, add					5,850			5,850	6,425
0500	For 6000 lb. capacity, add					14,500			14,500	15,900
0525	For 7000 lb. capacity, add					18,100			18,100	19,900
0550	For 8000 lb. capacity, add					22,300			22,300	24,500
0575	For 10000 lb. capacity, add					30,500			30,500	33,600
0600	For 12000 lb. capacity, add					37,500			37,500	41,300
0625	For 16000 lb. capacity, add					45,000			45,000	49,500
0650	For 20000 lb. capacity, add					50,000			50,000	55,000
0675	For increased speed, 250 fpm, add					17,000			17,000	18,700
0700	300 fpm, geared electric, add					20,600			20,600	22,700
0725	350 fpm, geared electric, add					25,100			25,100	27,600
0750	400 fpm, geared electric, add					29,400			29,400	32,300
0775	500 fpm, gearless electric, add					36,400			36,400	40,100
0800	600 fpm, gearless electric, add					43,600			43,600	47,900
0825	700 fpm, gearless electric, add					52,000			52,000	57,000
0850	800 fpm, gearless electric, add					59,500			59,500	65,500
0875	For class "B" loading, add					4,800			4,800	5,275
0900	For class "C-1" loading, add					6,950			6,950	7,650
0925	For class "C-2" loading, add					8,000			8,000	8,800
0950	For class "C-3" loading, add				↓	10,700			10,700	11,700
0975	For travel over 40 V.L.F., add	2 Elev	7.25	2.207	V.L.F.	645	169		814	960
1000	For number of stops over 4, add	"	.27	59.259	Stop	4,225	4,525		8,750	11,400

14 21 23 – Electric Traction Passenger Elevators

14 21 23.10 Electric Traction Passenger Elevators and Options		Crew	Daily Output	Labor-Hours	Unit	Material	Labor	Equipment	Total	Total Incl O&P
0010	**ELECTRIC TRACTION PASSENGER ELEVATORS AND OPTIONS**									
1625	Electric pass., base unit, 2000 lb., 200 fpm, 4 stop, std. fin.	2 Elev	.05	320	Ea.	92,000	24,500		116,500	137,500
1650	For 2500 lb. capacity, add					3,800			3,800	4,175
1675	For 3000 lb. capacity, add					4,400			4,400	4,825
1700	For 3500 lb. capacity, add					5,675			5,675	6,250
1725	For 4000 lb. capacity, add					6,825			6,825	7,500
1750	For 4500 lb. capacity, add				↓	9,150			9,150	10,100

14 21 Electric Traction Elevators

14 21 23 – Electric Traction Passenger Elevators

14 21 23.10 Electric Traction Passenger Elevators and Options		Crew	Daily Output	Labor-Hours	Unit	2015 Bare Costs Material	Labor	Equipment	Total	Total Incl O&P
1775	For 5000 lb. capacity, add				Ea.	11,400			11,400	12,600
1800	For increased speed, 250 fpm, geared electric, add					4,800			4,800	5,275
1825	300 fpm, geared electric, add					6,575			6,575	7,225
1850	350 fpm, geared electric, add					8,775			8,775	9,650
1875	400 fpm, geared electric, add					11,500			11,500	12,600
1900	500 fpm, gearless electric, add					28,000			28,000	30,800
1925	600 fpm, gearless electric, add					43,700			43,700	48,000
1950	700 fpm, gearless electric, add					50,000			50,000	55,500
1975	800 fpm, gearless electric, add					55,500			55,500	61,000
2000	For travel over 40 V.L.F., add	2 Elev	7.25	2.207	V.L.F.	695	169		864	1,025
2025	For number of stops over 4, add		.27	59.259	Stop	3,075	4,525		7,600	10,200
2400	Electric hospital, base unit, 4000 lb., 200 fpm, 4 stop, std fin.		.05	320	Ea.	79,500	24,500		104,000	124,000
2425	For 4500 lb. capacity, add					5,475			5,475	6,025
2450	For 5000 lb. capacity, add					7,175			7,175	7,900
2475	For increased speed, 250 fpm, geared electric, add					4,925			4,925	5,425
2500	300 fpm, geared electric, add					7,600			7,600	8,375
2525	350 fpm, geared electric, add					8,775			8,775	9,650
2550	400 fpm, geared electric, add					11,500			11,500	12,700
2575	500 fpm, gearless electric, add					32,600			32,600	35,900
2600	600 fpm, gearless electric, add					47,600			47,600	52,500
2625	700 fpm, gearless electric, add					52,500			52,500	58,000
2650	800 fpm, gearless electric, add					59,000			59,000	65,000
2675	For travel over 40 V.L.F., add	2 Elev	7.25	2.207	V.L.F.	435	169		604	730
2700	For number of stops over 4, add	″	.27	59.259	Stop	4,325	4,525		8,850	11,600

14 21 33 – Electric Traction Residential Elevators

14 21 33.20 Residential Elevators

Line	Description	Crew	Daily Output	Labor-Hours	Unit	Material	Labor	Equipment	Total	Total Incl O&P
0010	**RESIDENTIAL ELEVATORS**									
7000	Residential, cab type, 1 floor, 2 stop, economy model	2 Elev	.20	80	Ea.	11,400	6,125		17,525	21,700
7100	Custom model		.10	160		19,300	12,200		31,500	39,500
7200	2 floor, 3 stop, economy model		.12	133		17,000	10,200		27,200	33,900
7300	Custom model		.06	266		27,700	20,400		48,100	61,000

14 24 Hydraulic Elevators

14 24 13 – Hydraulic Freight Elevators

14 24 13.10 Hydraulic Freight Elevators and Options

Line	Description	Crew	Daily Output	Labor-Hours	Unit	Material	Labor	Equipment	Total	Total Incl O&P
0010	**HYDRAULIC FREIGHT ELEVATORS AND OPTIONS**									
1025	Hydraulic freight, base unit, 2000 lb., 50 fpm, 2 stop, std. fin.	2 Elev	.10	160	Ea.	79,500	12,200		91,700	106,000
1050	For 2500 lb. capacity, add					3,875			3,875	4,250
1075	For 3000 lb. capacity, add					5,250			5,250	5,775
1100	For 3500 lb. capacity, add					8,350			8,350	9,200
1125	For 4000 lb. capacity, add					9,750			9,750	10,700
1150	For 4500 lb. capacity, add					14,300			14,300	15,700
1175	For 5000 lb. capacity, add					15,000			15,000	16,500
1200	For 6000 lb. capacity, add					15,800			15,800	17,400
1225	For 7000 lb. capacity, add					23,500			23,500	25,800
1250	For 8000 lb. capacity, add					30,900			30,900	33,900
1275	For 10000 lb. capacity, add					37,800			37,800	41,600
1300	For 12000 lb. capacity, add					49,100			49,100	54,000
1325	For 16000 lb. capacity, add					70,000			70,000	77,000
1350	For 20000 lb. capacity, add					80,500			80,500	89,000

14 24 Hydraulic Elevators

14 24 13 – Hydraulic Freight Elevators

14 24 13.10 Hydraulic Freight Elevators and Options		Crew	Daily Output	Labor-Hours	Unit	2015 Bare Costs Material	Labor	Equipment	Total	Total Incl O&P
1375	For increased speed, 100 fpm, add				Ea.	1,550			1,550	1,700
1400	125 fpm, add					3,425			3,425	3,775
1425	150 fpm, add					4,675			4,675	5,150
1450	175 fpm, add					7,150			7,150	7,875
1475	For class "B" loading, add					4,525			4,525	4,975
1500	For class "C-1" loading, add					6,900			6,900	7,600
1525	For class "C-2" loading, add					7,950			7,950	8,750
1550	For class "C-3" loading, add					10,700			10,700	11,800
1575	For travel over 20 V.L.F., add	2 Elev	7.25	2.207	V.L.F.	845	169		1,014	1,175
1600	For number of stops over 2, add	"	.27	59.259	Stop	2,150	4,525		6,675	9,125

14 24 23 – Hydraulic Passenger Elevators

14 24 23.10 Hydraulic Passenger Elevators and Options		Crew	Daily Output	Labor-Hours	Unit	Material	Labor	Equipment	Total	Total Incl O&P
0010	**HYDRAULIC PASSENGER ELEVATORS AND OPTIONS**									
2050	Hyd. pass., base unit, 1500 lb., 100 fpm, 2 stop, std. fin.	2 Elev	.10	160	Ea.	37,700	12,200		49,900	59,500
2075	For 2000 lb. capacity, add					810			810	890
2100	For 2500 lb. capacity, add					2,800			2,800	3,100
2125	For 3000 lb. capacity, add					3,975			3,975	4,375
2150	For 3500 lb. capacity, add					6,875			6,875	7,550
2175	For 4000 lb. capacity, add					8,225			8,225	9,050
2200	For 4500 lb. capacity, add					10,800			10,800	11,900
2225	For 5000 lb. capacity, add					15,200			15,200	16,700
2250	For increased speed, 125 fpm, add					1,975			1,975	2,175
2275	150 fpm, add					2,550			2,550	2,825
2300	175 fpm, add					4,850			4,850	5,325
2325	200 fpm, add					9,275			9,275	10,200
2350	For travel over 12 V.L.F., add	2 Elev	7.25	2.207	V.L.F.	690	169		859	1,000
2375	For number of stops over 2, add		.27	59.259	Stop	960	4,525		5,485	7,825
2725	Hydraulic hospital, base unit, 4000 lb., 100 fpm, 2 stop, std. fin.		.10	160	Ea.	59,500	12,200		71,700	84,000
2775	For 4500 lb. capacity, add					6,700			6,700	7,350
2800	For 5000 lb. capacity, add					9,800			9,800	10,800
2825	For increased speed, 125 fpm, add					2,450			2,450	2,700
2850	150 fpm, add					3,450			3,450	3,800
2875	175 fpm, add					5,775			5,775	6,350
2900	200 fpm, add					8,450			8,450	9,300
2925	For travel over 12 V.L.F., add	2 Elev	7.25	2.207	V.L.F.	530	169		699	830
2950	For number of stops over 2, add	"	.27	59.259	Stop	4,175	4,525		8,700	11,400

14 27 Custom Elevator Cabs and Doors

14 27 13 – Custom Elevator Cab Finishes

14 27 13.10 Cab Finishes		Crew	Daily Output	Labor-Hours	Unit	Material	Labor	Equipment	Total	Total Incl O&P
0010	**CAB FINISHES**									
3325	Passenger elevator cab finishes (based on 3500 lb. cab size)									
3350	Acrylic panel ceiling				Ea.	755			755	830
3375	Aluminum eggcrate ceiling					865			865	950
3400	Stainless steel doors					3,950			3,950	4,350
3425	Carpet flooring					610			610	670
3450	Epoxy flooring					465			465	510
3475	Quarry tile flooring					875			875	960
3500	Slate flooring					1,600			1,600	1,750
3525	Textured rubber flooring					650			650	715

14 27 Custom Elevator Cabs and Doors

14 27 13 – Custom Elevator Cab Finishes

14 27 13.10 Cab Finishes		Crew	Daily Output	Labor-Hours	Unit	2015 Bare Costs Material	Labor	Equipment	Total	Total Incl O&P
3550	Stainless steel walls				Ea.	4,100			4,100	4,500
3575	Stainless steel returns at door				↓	1,175			1,175	1,275
4450	Hospital elevator cab finishes (based on 3500 lb. cab size)									
4475	Aluminum eggcrate ceiling				Ea.	900			900	990
4500	Stainless steel doors					3,950			3,950	4,350
4525	Epoxy flooring					465			465	510
4550	Quarry tile flooring					875			875	960
4575	Textured rubber flooring					650			650	715
4600	Stainless steel walls					4,550			4,550	5,000
4625	Stainless steel returns at door				↓	960			960	1,050

14 28 Elevator Equipment and Controls

14 28 10 – Elevator Equipment and Control Options

14 28 10.10 Elevator Controls and Doors		Crew	Daily Output	Labor-Hours	Unit	Material	Labor	Equipment	Total	Total Incl O&P
0010	**ELEVATOR CONTROLS AND DOORS**									
2975	Passenger elevator options									
3000	2 car group automatic controls	2 Elev	.66	24.242	Ea.	4,625	1,850		6,475	7,850
3025	3 car group automatic controls		.44	36.364		8,825	2,775		11,600	13,900
3050	4 car group automatic controls		.33	48.485		17,600	3,700		21,300	24,900
3075	5 car group automatic controls		.26	61.538		31,800	4,700		36,500	41,900
3100	6 car group automatic controls		.22	72.727		64,500	5,575		70,075	79,000
3125	Intercom service		3	5.333		955	410		1,365	1,650
3150	Duplex car selective collective		.66	24.242		8,100	1,850		9,950	11,700
3175	Center opening 1 speed doors		2	8		1,950	610		2,560	3,075
3200	Center opening 2 speed doors		2	8		2,750	610		3,360	3,950
3225	Rear opening doors (opposite front)		2	8		4,225	610		4,835	5,575
3250	Side opening 2 speed doors		2	8		4,350	610		4,960	5,700
3275	Automatic emergency power switching		.66	24.242		1,200	1,850		3,050	4,075
3300	Manual emergency power switching	↓	8	2		515	153		668	795
3625	Hall finishes, stainless steel doors					1,425			1,425	1,575
3650	Stainless steel frames					1,450			1,450	1,600
3675	12 month maintenance contract								3,600	3,950
3700	Signal devices, hall lanterns	2 Elev	8	2		520	153		673	800
3725	Position indicators, up to 3		9.40	1.702		475	130		605	720
3750	Position indicators, per each over 3	↓	32	.500		420	38.50		458.50	515
3775	High speed heavy duty door opener					3,250			3,250	3,575
3800	Variable voltage, O.H. gearless machine, min.	2 Elev	.16	100		34,000	7,650		41,650	48,800
3815	Maximum		.07	228		81,000	17,500		98,500	115,000
3825	Basement installed geared machine	↓	.33	48.485	↓	49,200	3,700		52,900	59,500
3850	Freight elevator options									
3875	Doors, bi-parting	2 Elev	.66	24.242	Ea.	7,550	1,850		9,400	11,100
3900	Power operated door and gate	"	.66	24.242		24,300	1,850		26,150	29,500
3925	Finishes, steel plate floor					1,825			1,825	2,000
3950	14 ga. 1/4" x 4' steel plate walls					2,000			2,000	2,200
3975	12 month maintenance contract								3,600	3,950
4000	Signal devices, hall lanterns	2 Elev	8	2		505	153		658	785
4025	Position indicators, up to 3		9.40	1.702		490	130		620	730
4050	Position indicators, per each over 3		32	.500		425	38.50		463.50	525
4075	Variable voltage basement installed geared machine	↓	.66	24.242	↓	20,000	1,850		21,850	24,800
4100	Hospital elevator options									
4125	2 car group automatic controls	2 Elev	.66	24.242	Ea.	4,850	1,850		6,700	8,125

14 28 Elevator Equipment and Controls

14 28 10 – Elevator Equipment and Control Options

14 28 10.10 Elevator Controls and Doors		Crew	Daily Output	Labor-Hours	Unit	Material	2015 Bare Costs Labor	Equipment	Total	Total Incl O&P
4150	3 car group automatic controls	2 Elev	.44	36.364	Ea.	9,275	2,775		12,050	14,400
4175	4 car group automatic controls		.33	48.485		12,400	3,700		16,100	19,200
4200	5 car group automatic controls		.26	61.538		33,500	4,700		38,200	43,900
4225	6 car group automatic controls		.22	72.727		67,500	5,575		73,075	83,000
4250	Intercom service		3	5.333		1,000	410		1,410	1,700
4275	Duplex car selective collective		.66	24.242		8,100	1,850		9,950	11,700
4300	Center opening 1 speed doors		2	8		1,950	610		2,560	3,075
4325	Center opening 2 speed doors		2	8		2,575	610		3,185	3,775
4350	Rear opening doors (opposite front)		2	8		4,250	610		4,860	5,600
4375	Side opening 2 speed doors		2	8		6,350	610		6,960	7,900
4400	Automatic emergency power switching		.66	24.242		1,175	1,850		3,025	4,050
4425	Manual emergency power switching		8	2		505	153		658	785
4675	Hall finishes, stainless steel doors					1,600			1,600	1,750
4700	Stainless steel frames					1,475			1,475	1,625
4725	12 month maintenance contract								3,600	3,950
4750	Signal devices, hall lanterns	2 Elev	8	2		490	153		643	765
4775	Position indicators, up to 3		9.40	1.702		475	130		605	715
4800	Position indicators, per each over 3		32	.500		415	38.50		453.50	515
4825	High speed heavy duty door opener					3,250			3,250	3,575
4850	Variable voltage, O.H. gearless machine, min.	2 Elev	.16	100		51,500	7,650		59,150	68,000
4865	Maximum		.07	228		79,500	17,500		97,000	113,500
4875	Basement installed geared machine		.33	48.485		20,300	3,700		24,000	27,900
5000	Drilling for piston, casing included, 18" diameter	B-48	80	.700	V.L.F.	56.50	30	36.50	123	148

14 31 Escalators

14 31 10 – Glass and Steel Escalators

14 31 10.10 Escalators		Crew	Daily Output	Labor-Hours	Unit	Material	Labor	Equipment	Total	Total Incl O&P
0010	**ESCALATORS**									
1000	Glass, 32" wide x 10' floor to floor height	M-1	.07	457	Ea.	86,500	33,200	660	120,360	145,500
1010	48" wide x 10' floor to floor height		.07	457		93,500	33,200	660	127,360	153,000
1020	32" wide x 15' floor to floor height		.06	533		91,000	38,800	770	130,570	159,000
1030	48" wide x 15' floor to floor height		.06	533		96,500	38,800	770	136,070	165,000
1040	32" wide x 20' floor to floor height		.05	653		96,500	47,500	940	144,940	178,500
1050	48" wide x 20' floor to floor height		.05	653		105,000	47,500	940	153,440	187,500
1060	32" wide x 25' floor to floor height		.04	800		105,500	58,000	1,150	164,650	204,500
1070	48" wide x 25' floor to floor height		.04	800		122,000	58,000	1,150	181,150	222,500
1080	Enameled steel, 32" wide x 10' floor to floor height		.07	457		93,500	33,200	660	127,360	153,500
1090	48" wide x 10' floor to floor height		.07	457		101,500	33,200	660	135,360	162,000
1110	32" wide x 15' floor to floor height		.06	533		98,500	38,800	770	138,070	167,500
1120	48" wide x 15' floor to floor height		.06	533		104,500	38,800	770	144,070	174,000
1130	32" wide x 20' floor to floor height		.05	653		105,000	47,500	940	153,440	187,500
1140	48" wide x 20' floor to floor height		.05	653		113,500	47,500	940	161,940	197,000
1150	32" wide x 25' floor to floor height		.04	800		114,000	58,000	1,150	173,150	214,000
1160	48" wide x 25' floor to floor height		.04	800		131,500	58,000	1,150	190,650	233,000
1170	Stainless steel, 32" wide x 10' floor to floor height		.07	457		99,000	33,200	660	132,860	159,000
1180	48" wide x 10' floor to floor height		.07	457		106,500	33,200	660	140,360	167,500
1500	32" wide x 15' floor to floor height		.06	533		104,000	38,800	770	143,570	173,500
1700	48" wide x 15' floor to floor height		.06	533		110,000	38,800	770	149,570	180,000
1750	32" wide x 18' floor to floor height		.05	615		102,500	44,700	885	148,085	180,500
1775	48" wide x 18' floor to floor height		.05	615		111,500	44,700	885	157,085	190,500
2300	32" wide x 25' floor to floor height		.04	800		119,500	58,000	1,150	178,650	220,000

14 31 Escalators

14 31 10 – Glass and Steel Escalators

14 31 10.10 Escalators		Crew	Daily Output	Labor-Hours	Unit	Material	2015 Bare Costs Labor	Equipment	Total	Total Incl O&P
2500	48″ wide x 25′ floor to floor height	M-1	.04	800	Ea.	138,000	58,000	1,150	197,150	240,000

14 32 Moving Walks

14 32 10 – Moving Walkways

14 32 10.10 Moving Walks

		Crew	Daily Output	Labor-Hours	Unit	Material	Labor	Equipment	Total	Total Incl O&P
0010	MOVING WALKS R143210-20									
0020	Walk, 27″ tread width, minimum	M-1	6.50	4.923	L.F.	870	360	7.10	1,237.10	1,500
0100	300′ to 500′, maximum		4.43	7.223		1,200	525	10.40	1,735.40	2,125
0300	48″ tread width walk, minimum		4.43	7.223		1,950	525	10.40	2,485.40	2,950
0400	100′ to 350′, maximum		3.82	8.377		2,300	610	12.05	2,922.05	3,450
0600	Ramp, 12° incline, 36″ tread width, minimum		5.27	6.072		1,600	440	8.75	2,048.75	2,450
0700	70′ to 90′ maximum		3.82	8.377		2,300	610	12.05	2,922.05	3,450
0900	48″ tread width, minimum		3.57	8.964		2,350	650	12.90	3,012.90	3,550
1000	40′ to 70′, maximum	↓	2.91	10.997	↓	2,950	800	15.85	3,765.85	4,475

14 42 Wheelchair Lifts

14 42 13 – Inclined Wheelchair Lifts

14 42 13.10 Inclined Wheelchair Lifts and Stairclimbers

		Crew	Daily Output	Labor-Hours	Unit	Material	Labor	Equipment	Total	Total Incl O&P
0010	INCLINED WHEELCHAIR LIFTS AND STAIRCLIMBERS									
7700	Stair climber (chair lift), single seat, minimum	2 Elev	1	16	Ea.	5,325	1,225		6,550	7,700
7800	Maximum		.20	80		7,350	6,125		13,475	17,200
8700	Stair lift, minimum		1	16		14,500	1,225		15,725	17,700
8900	Maximum	↓	.20	80	↓	22,900	6,125		29,025	34,300

14 42 16 – Vertical Wheelchair Lifts

14 42 16.10 Wheelchair Lifts

		Crew	Daily Output	Labor-Hours	Unit	Material	Labor	Equipment	Total	Total Incl O&P
0010	WHEELCHAIR LIFTS									
8000	Wheelchair lift, minimum	2 Elev	1	16	Ea.	7,325	1,225		8,550	9,875
8500	Maximum	″	.50	32	″	17,300	2,450		19,750	22,700

14 45 Vehicle Lifts

14 45 10 – Hydraulic Vehicle Lifts

14 45 10.10 Hydraulic Lifts

		Crew	Daily Output	Labor-Hours	Unit	Material	Labor	Equipment	Total	Total Incl O&P
0010	HYDRAULIC LIFTS									
2200	Single post, 8000 lb. capacity	L-4	.40	60	Ea.	5,550	2,650		8,200	10,200
2810	Double post, 6000 lb. capacity		2.67	8.989		7,825	400		8,225	9,250
2815	9000 lb. capacity		2.29	10.480		18,600	465		19,065	21,200
2820	15,000 lb. capacity		2	12		21,000	530		21,530	23,900
2822	Four post, 26,000 lb. capacity		1.80	13.333		14,200	590		14,790	16,500
2825	30,000 lb. capacity		1.60	15		46,400	665		47,065	52,000
2830	Ramp style, 4 post, 25,000 lb. capacity		2	12		18,200	530		18,730	20,800
2835	35,000 lb. capacity		1	24		84,500	1,050		85,550	94,500
2840	50,000 lb. capacity		1	24		94,500	1,050		95,550	105,500
2845	75,000 lb. capacity	↓	1	24		110,000	1,050		111,050	122,500
2850	For drive thru tracks, add, minimum					1,175			1,175	1,275
2855	Maximum					2,000			2,000	2,200
2860	Ramp extensions, 3′ (set of 2)					960			960	1,050
2865	Rolling jack platform				↓	3,325			3,325	3,675

14 45 Vehicle Lifts

14 45 10 – Hydraulic Vehicle Lifts

14 45 10.10 Hydraulic Lifts		Crew	Daily Output	Labor-Hours	Unit	Material	2015 Bare Costs Labor	Equipment	Total	Total Incl O&P
2870	Electric/hydraulic jacking beam				Ea.	8,925			8,925	9,825
2880	Scissor lift, portable, 6000 lb. capacity				↓	8,750			8,750	9,625

14 91 Facility Chutes

14 91 33 – Laundry and Linen Chutes

14 91 33.10 Chutes

		Crew	Daily Output	Labor-Hours	Unit	Material	Labor	Equipment	Total	Total Incl O&P
0011	**CHUTES**, linen, trash or refuse									
0050	Aluminized steel, 16 ga., 18" diameter	2 Shee	3.50	4.571	Floor	1,700	256		1,956	2,275
0100	24" diameter		3.20	5		1,775	280		2,055	2,375
0200	30" diameter		3	5.333		2,125	298		2,423	2,775
0300	36" diameter		2.80	5.714		2,625	320		2,945	3,400
0400	Galvanized steel, 16 ga., 18" diameter		3.50	4.571		1,000	256		1,256	1,500
0500	24" diameter		3.20	5		1,125	280		1,405	1,675
0600	30" diameter		3	5.333		1,275	298		1,573	1,850
0700	36" diameter		2.80	5.714		1,500	320		1,820	2,150
0800	Stainless steel, 18" diameter		3.50	4.571		3,000	256		3,256	3,700
0900	24" diameter		3.20	5		3,150	280		3,430	3,875
1000	30" diameter		3	5.333		3,750	298		4,048	4,575
1005	36" diameter		2.80	5.714	↓	3,950	320		4,270	4,825
1200	Linen chute bottom collector, aluminized steel		4	4	Ea.	1,400	224		1,624	1,900
1300	Stainless steel		4	4		1,800	224		2,024	2,325
1500	Refuse, bottom hopper, aluminized steel, 18" diameter		3	5.333		1,025	298		1,323	1,575
1600	24" diameter		3	5.333		1,250	298		1,548	1,825
1800	36" diameter	↓	3	5.333	↓	2,500	298		2,798	3,200

14 91 82 – Trash Chutes

14 91 82.10 Trash Chutes and Accessories

		Crew	Daily Output	Labor-Hours	Unit	Material	Labor	Equipment	Total	Total Incl O&P
0010	**TRASH CHUTES AND ACCESSORIES**									
2900	Package chutes, spiral type, minimum	2 Shee	4.50	3.556	Floor	2,425	199		2,624	2,975
3000	Maximum	"	1.50	10.667	"	6,325	595		6,920	7,875

14 92 Pneumatic Tube Systems

14 92 10 – Conventional, Automatic and Computer Controlled Pneumatic Tube Systems

14 92 10.10 Pneumatic Tube Systems

		Crew	Daily Output	Labor-Hours	Unit	Material	Labor	Equipment	Total	Total Incl O&P
0010	**PNEUMATIC TUBE SYSTEMS**									
0020	100' long, single tube, 2 stations, stock									
0100	3" diameter	2 Stpi	.12	133	Total	3,375	7,975		11,350	15,700
0300	4" diameter	"	.09	177	"	4,275	10,600		14,875	20,700
0400	Twin tube, two stations or more, conventional system									
0600	2-1/2" round	2 Stpi	62.50	.256	L.F.	38	15.30		53.30	65
0700	3" round		46	.348		38	21		59	73.50
0900	4" round		49.60	.323		48	19.25		67.25	82
1000	4" x 7" oval		37.60	.426	↓	89	25.50		114.50	137
1050	Add for blower		2	8	System	5,200	480		5,680	6,450
1110	Plus for each round station, add		7.50	2.133	Ea.	1,350	127		1,477	1,700
1150	Plus for each oval station, add		7.50	2.133	"	1,350	127		1,477	1,700
1200	Alternate pricing method: base cost, economy model		.75	21.333	Total	5,750	1,275		7,025	8,250
1300	Custom model		.25	64	"	11,500	3,825		15,325	18,500
1500	Plus total system length, add, for economy model		93.40	.171	L.F.	8.25	10.25		18.50	24.50
1600	For custom model	↓	37.60	.426	"	25	25.50		50.50	66

14 92 Pneumatic Tube Systems

14 92 10 – Conventional, Automatic and Computer Controlled Pneumatic Tube Systems

14 92 10.10 Pneumatic Tube Systems		Crew	Daily Output	Labor-Hours	Unit	2015 Bare Costs Material	Labor	Equipment	Total	Total Incl O&P
1800	Completely automatic system, 4" round, 15 to 50 stations	2 Stpi	.29	55.172	Station	20,100	3,300		23,400	27,100
2200	51 to 144 stations		.32	50		15,600	3,000		18,600	21,700
2400	6" round or 4" x 7" oval, 15 to 50 stations		.24	66.667		25,200	3,975		29,175	33,700
2800	51 to 144 stations		.23	69.565		21,100	4,150		25,250	29,500

Division Notes

	CREW	DAILY OUTPUT	LABOR-HOURS	UNIT	BARE COSTS MAT.	LABOR	EQUIP.	TOTAL	TOTAL INCL O&P

Division 21 — Fire Suppression

Estimating Tips

Pipe for fire protection and all uses is located in Subdivisions 21 11 13 and 22 11 13.

The labor adjustment factors listed in Subdivision 22 01 02.20 also apply to Division 21.

Many, but not all, areas in the U.S. require backflow protection in the fire system. It is advisable to check local building codes for specific requirements.

For your reference, the following is a list of the most applicable Fire Codes and Standards which may be purchased from the NFPA, 1 Batterymarch Park, Quincy, MA 02169-7471.

- NFPA 1: Uniform Fire Code
- NFPA 10: Portable Fire Extinguishers
- NFPA 11: Low-, Medium-, and High-Expansion Foam
- NFPA 12: Carbon Dioxide Extinguishing Systems (Also companion 12A)
- NFPA 13: Installation of Sprinkler Systems (Also companion 13D, 13E, and 13R)
- NFPA 14: Installation of Standpipe and Hose Systems
- NFPA 15: Water Spray Fixed Systems for Fire Protection
- NFPA 16: Installation of Foam-Water Sprinkler and Foam-Water Spray Systems
- NFPA 17: Dry Chemical Extinguishing Systems (Also companion 17A)
- NFPA 18: Wetting Agents
- NFPA 20: Installation of Stationary Pumps for Fire Protection
- NFPA 22: Water Tanks for Private Fire Protection
- NFPA 24: Installation of Private Fire Service Mains and their Appurtenances
- NFPA 25: Inspection, Testing and Maintenance of Water-Based Fire Protection

Reference Numbers

Reference numbers are shown in shaded boxes at the beginning of some major classifications. These numbers refer to related items in the Reference Section. The reference information may be an estimating procedure, an alternate pricing method, or technical information.

Note: Not all subdivisions listed here necessarily appear in this publication. ■

21 05 Common Work Results for Fire Suppression

21 05 23 – General-Duty Valves for Water-Based Fire-Suppression Piping

21 05 23.50 General-Duty Valves		Crew	Daily Output	Labor-Hours	Unit	2015 Bare Costs Material	Labor	Equipment	Total	Total Incl O&P
0010	**GENERAL-DUTY VALVES**, for water-based fire suppression									
6200	Valves and components									
6500	Check, swing, C.I. body, brass fittings, auto. ball drip									
6520	4" size	Q-12	3	5.333	Ea.	350	269		619	795
6800	Check, wafer, butterfly type, C.I. body, bronze fittings									
6820	4" size	Q-12	4	4	Ea.	1,000	202		1,202	1,400

21 11 Facility Fire-Suppression Water-Service Piping

21 11 16 – Facility Fire Hydrants

21 11 16.50 Fire Hydrants for Buildings

		Crew	Daily Output	Labor-Hours	Unit	Material	Labor	Equipment	Total	Total Incl O&P
0010	**FIRE HYDRANTS FOR BUILDINGS**									
3750	Hydrants, wall, w/caps, single, flush, polished brass									
3800	2-1/2" x 2-1/2"	Q-12	5	3.200	Ea.	213	162		375	480
3840	2-1/2" x 3"	"	5	3.200		430	162		592	715
3900	For polished chrome, add				↓	20%				
3950	Double, flush, polished brass									
4000	2-1/2" x 2-1/2" x 4"	Q-12	5	3.200	Ea.	570	162		732	875
4040	2-1/2" x 2-1/2" x 6"	"	4.60	3.478		825	176		1,001	1,175
4200	For polished chrome, add				↓	10%				
4350	Double, projecting, polished brass									
4400	2-1/2" x 2-1/2" x 4"	Q-12	5	3.200	Ea.	254	162		416	525
4450	2-1/2" x 2-1/2" x 6"	"	4.60	3.478	"	520	176		696	835
4460	Valve control, dbl. flush/projecting hydrant, cap &									
4470	chain, extension rod & cplg., escutcheon, polished brass	Q-12	8	2	Ea.	300	101		401	485

21 11 19 – Fire-Department Connections

21 11 19.50 Connections for the Fire-Department

		Crew	Daily Output	Labor-Hours	Unit	Material	Labor	Equipment	Total	Total Incl O&P
0010	**CONNECTIONS FOR THE FIRE-DEPARTMENT**									
7140	Standpipe connections, wall, w/plugs & chains									
7160	Single, flush, brass, 2-1/2" x 2-1/2"	Q-12	5	3.200	Ea.	159	162		321	420
7180	2-1/2" x 3"	"	5	3.200	"	164	162		326	425
7240	For polished chrome, add					15%				
7280	Double, flush, polished brass									
7300	2-1/2" x 2-1/2" x 4"	Q-12	5	3.200	Ea.	520	162		682	815
7330	2-1/2" x 2-1/2" x 6"	"	4.60	3.478	"	725	176		901	1,050
7400	For polished chrome, add					15%				
7440	For sill cock combination, add				Ea.	90.50			90.50	99.50
7900	Three way, flush, polished brass									
7920	2-1/2" (3) x 4"	Q-12	4.80	3.333	Ea.	1,650	168		1,818	2,075
7930	2-1/2" (3) x 6"	"	4.80	3.333		1,650	168		1,818	2,075
8000	For polished chrome, add				↓	9%				
8020	Three way, projecting, polished brass									
8040	2-1/2"(3) x 4"	Q-12	4.80	3.333	Ea.	1,575	168		1,743	1,975

21 12 Fire-Suppression Standpipes

21 12 13 – Fire-Suppression Hoses and Nozzles

21 12 13.50 Fire Hoses and Nozzles		Crew	Daily Output	Labor-Hours	Unit	Material	2015 Bare Costs Labor	Equipment	Total	Total Incl O&P
0010	FIRE HOSES AND NOZZLES									
0200	Adapters, rough brass, straight hose threads									
0220	One piece, female to male, rocker lugs									
0240	1" x 1"				Ea.	48			48	53
0320	2" x 2"					42			42	46
0380	2-1/2" x 2-1/2"					19			19	21
2200	Hose, less couplings									
2260	Synthetic jacket, lined, 300 lb. test, 1-1/2" diameter	Q-12	2600	.006	L.F.	3.24	.31		3.55	4.03
2280	2-1/2" diameter		2200	.007		5.60	.37		5.97	6.70
2360	High strength, 500 lb. test, 1-1/2" diameter		2600	.006		3.35	.31		3.66	4.16
2380	2-1/2" diameter		2200	.007		5.90	.37		6.27	7.05
5600	Nozzles, brass									
5620	Adjustable fog, 3/4" booster line				Ea.	111			111	122
5630	1" booster line					137			137	150
5640	1-1/2" leader line					101			101	112
5660	2-1/2" direct connection					151			151	166
5680	2-1/2" playpipe nozzle					223			223	246
5780	For chrome plated, add					8%				
5850	Electrical fire, adjustable fog, no shock									
5900	1-1/2"				Ea.	415			415	455
5920	2-1/2"					560			560	615
5980	For polished chrome, add					6%				
6200	Heavy duty, comb. adj. fog and str. stream, with handle									
6210	1" booster line				Ea.	375			375	410

21 12 19 – Fire-Suppression Hose Racks

21 12 19.50 Fire Hose Racks		Crew	Daily Output	Labor-Hours	Unit	Material	Labor	Equipment	Total	Total Incl O&P
0010	FIRE HOSE RACKS									
2600	Hose rack, swinging, for 1-1/2" diameter hose,									
2620	Enameled steel, 50' & 75' lengths of hose	Q-12	20	.800	Ea.	61	40.50		101.50	128
2640	100' and 125' lengths of hose	"	20	.800	"	61	40.50		101.50	128

21 12 23 – Fire-Suppression Hose Valves

21 12 23.70 Fire Hose Valves		Crew	Daily Output	Labor-Hours	Unit	Material	Labor	Equipment	Total	Total Incl O&P
0010	FIRE HOSE VALVES									
0080	Wheel handle, 300 lb., 1-1/2"	1 Spri	12	.667	Ea.	95	37.50		132.50	161
0090	2-1/2"	"	7	1.143	"	175	64		239	289
0100	For polished brass, add					35%				
0110	For polished chrome, add					50%				

21 13 Fire-Suppression Sprinkler Systems

21 13 13 – Wet-Pipe Sprinkler Systems

21 13 13.50 Wet-Pipe Sprinkler System Components		Crew	Daily Output	Labor-Hours	Unit	Material	Labor	Equipment	Total	Total Incl O&P
0010	WET-PIPE SPRINKLER SYSTEM COMPONENTS									
2600	Sprinkler heads, not including supply piping									
3700	Standard spray, pendent or upright, brass, 135°F to 286°F									
3730	1/2" NPT, 7/16" orifice	1 Spri	16	.500	Ea.	15.20	28		43.20	59.50
3740	1/2" NPT, 1/2" orifice	"	16	.500		9.90	28		37.90	53.50
3860	For wax and lead coating, add					35			35	38
3880	For wax coating, add					21			21	23
3900	For lead coating, add					22.50			22.50	25
3920	For 360°F, same cost									

21 13 Fire-Suppression Sprinkler Systems

21 13 13 – Wet-Pipe Sprinkler Systems

21 13 13.50 Wet-Pipe Sprinkler System Components		Crew	Daily Output	Labor-Hours	Unit	Material	2015 Bare Costs Labor	Equipment	Total	Total Incl O&P
3930	For 400°F				Ea.	93			93	102
3940	For 500°F				"	93			93	102
4500	Sidewall, horizontal, brass, 135°F to 286°F									
4520	1/2" NPT, 1/2" orifice	1 Spri	16	.500	Ea.	25	28		53	70
4540	For 360°F, same cost									
4800	Recessed pendent, brass, 135°F to 286°F									
4820	1/2" NPT, 3/8" orifice	1 Spri	10	.800	Ea.	42.50	45		87.50	115
4830	1/2" NPT, 7/16" orifice		10	.800		18.40	45		63.40	88
4840	1/2" NPT, 1/2" orifice	↓	10	.800	↓	14.25	45		59.25	83.50

21 13 16 – Dry-Pipe Sprinkler Systems

21 13 16.50 Dry-Pipe Sprinkler System Components		Crew	Daily Output	Labor-Hours	Unit	Material	Labor	Equipment	Total	Total Incl O&P
0010	**DRY-PIPE SPRINKLER SYSTEM COMPONENTS**									
0600	Accelerator	1 Spri	8	1	Ea.	755	56		811	915
2600	Sprinkler heads, not including supply piping									
2640	Dry, pendent, 1/2" orifice, 3/4" or 1" NPT									
2700	15-1/4" to 18" length	1 Spri	14	.571	Ea.	145	32		177	209
2710	18-1/4" to 21" length		13	.615		151	34.50		185.50	219
2720	21-1/4" to 24" length		13	.615		156	34.50		190.50	225
2730	24-1/4" to 27" length	↓	13	.615	↓	162	34.50		196.50	231

21 13 26 – Deluge Fire-Suppression Sprinkler Systems

21 13 26.50 Deluge Fire-Suppression Sprinkler Sys. Comp.		Crew	Daily Output	Labor-Hours	Unit	Material	Labor	Equipment	Total	Total Incl O&P
0010	**DELUGE FIRE-SUPPRESSION SPRINKLER SYSTEM COMPONENTS**									
1400	Deluge system, monitoring panel w/deluge valve & trim	1 Spri	18	.444	Ea.	10,800	25		10,825	11,900
6200	Valves and components									
7000	Deluge, assembly, incl. trim, pressure									
7020	operated relief, emergency release, gauges									
7040	2" size	Q-12	2	8	Ea.	3,675	405		4,080	4,625
7060	3" size	"	1.50	10.667	"	4,100	540		4,640	5,350

21 13 39 – Foam-Water Systems

21 13 39.50 Foam-Water System Components		Crew	Daily Output	Labor-Hours	Unit	Material	Labor	Equipment	Total	Total Incl O&P
0010	**FOAM-WATER SYSTEM COMPONENTS**									
2600	Sprinkler heads, not including supply piping									
3600	Foam-water, pendent or upright, 1/2" NPT	1 Spri	12	.667	Ea.	210	37.50		247.50	287

21 21 Carbon-Dioxide Fire-Extinguishing Systems

21 21 16 – Carbon-Dioxide Fire-Extinguishing Equipment

21 21 16.50 CO2 Fire Extinguishing System		Crew	Daily Output	Labor-Hours	Unit	Material	Labor	Equipment	Total	Total Incl O&P
0010	**CO_2 FIRE EXTINGUISHING SYSTEM**									
0042	For detectors and control stations, see Section 28 31 23.50									
0100	Control panel, single zone with batteries (2 zones det., 1 suppr.)	1 Elec	1	8	Ea.	1,725	440		2,165	2,550
0150	Multizone (4) with batteries (8 zones det., 4 suppr.)	"	.50	16		3,275	875		4,150	4,900
1000	Dispersion nozzle, CO_2, 3" x 5"	1 Plum	18	.444		67	26		93	113
2000	Extinguisher, CO_2 system, high pressure, 75 lb. cylinder	Q-1	6	2.667		1,275	141		1,416	1,625
2100	100 lb. cylinder	"	5	3.200		1,300	169		1,469	1,700
3000	Electro/mechanical release	L-1	4	4		167	227		394	525
3400	Manual pull station	1 Plum	6	1.333		60.50	78.50		139	185
4000	Pneumatic damper release	"	8	1	↓	223	58.50		281.50	335

21 22 Clean-Agent Fire-Extinguishing Systems

21 22 16 – Clean-Agent Fire-Extinguishing Equipment

21 22 16.50 FM200 Fire Extinguishing System		Crew	Daily Output	Labor-Hours	Unit	Material	2015 Bare Costs Labor	Equipment	Total	Total Incl O&P
0010	**FM200 FIRE EXTINGUISHING SYSTEM**									
1100	Dispersion nozzle FM200, 1-1/2"	1 Plum	14	.571	Ea.	67	33.50		100.50	124
2400	Extinguisher, FM200 system, filled, with mounting bracket									
2460	26 lb. container	Q-1	8	2	Ea.	2,300	106		2,406	2,675
2480	44 lb. container		7	2.286		3,050	121		3,171	3,550
2500	63 lb. container		6	2.667		3,575	141		3,716	4,150
2520	101 lb. container		5	3.200		4,775	169		4,944	5,500
2540	196 lb. container	↓	4	4	↓	7,775	211		7,986	8,875
6000	FM200 system, simple nozzle layout, with broad dispersion				C.F.	1.76			1.76	1.94
6020	Complex nozzle layout and/or including underfloor dispersion				"	3.50			3.50	3.85

21 31 Centrifugal Fire Pumps

21 31 13 – Electric-Drive, Centrifugal Fire Pumps

21 31 13.50 Electric-Drive Fire Pumps		Crew	Daily Output	Labor-Hours	Unit	Material	Labor	Equipment	Total	Total Incl O&P
0010	**ELECTRIC-DRIVE FIRE PUMPS** Including controller, fittings and relief valve									
3100	250 GPM, 55 psi, 15 HP, 3550 RPM, 2" pump	Q-13	.70	45.714	Ea.	14,500	2,450		16,950	19,600
3200	500 GPM, 50 psi, 27 HP, 1770 RPM, 4" pump		.68	47.059		14,900	2,525		17,425	20,200
3350	750 GPM, 50 psi, 44 HP, 1770 RPM, 5" pump		.64	50		15,500	2,675		18,175	21,200
3400	750 GPM, 100 psi, 66 HP, 3550 RPM, 4" pump	↓	.58	55.172		17,900	2,950		20,850	24,200
5000	For jockey pump 1", 3 HP, with control, add	Q-12	2	8	↓	2,600	405		3,005	3,450

21 31 16 – Diesel-Drive, Centrifugal Fire Pumps

21 31 16.50 Diesel-Drive Fire Pumps		Crew	Daily Output	Labor-Hours	Unit	Material	Labor	Equipment	Total	Total Incl O&P
0010	**DIESEL-DRIVE FIRE PUMPS** Including controller, fittings and relief valve									
0050	500 GPM, 50 psi, 27 HP, 4" pump	Q-13	.64	50	Ea.	33,800	2,675		36,475	41,200
0200	750 GPM, 50 psi, 44 HP, 5" pump		.60	53.333		34,800	2,850		37,650	42,500
0400	1000 GPM, 100 psi, 89 HP, 4" pump		.56	57.143		39,400	3,050		42,450	47,900
0700	2000 GPM, 100 psi, 167 HP, 6" pump		.34	94.118		49,900	5,025		54,925	62,500
0950	3500 GPM, 100 psi, 300 HP, 10" pump	↓	.24	133	↓	71,000	7,125		78,125	89,000

Division Notes

		CREW	DAILY OUTPUT	LABOR-HOURS	UNIT	BARE COSTS MAT.	LABOR	EQUIP.	TOTAL	TOTAL INCL O&P

Estimating Tips

22 10 00 Plumbing Piping and Pumps

This subdivision is primarily basic pipe and related materials. The pipe may be used by any of the mechanical disciplines, i.e., plumbing, fire protection, heating, and air conditioning.

Note: CPVC plastic piping approved for fire protection is located in 21 11 13.

- The labor adjustment factors listed in Subdivision 22 01 02.20 apply throughout Divisions 21, 22, and 23. CAUTION: the correct percentage may vary for the same items. For example, the percentage add for the basic pipe installation should be based on the maximum height that the craftsman must install for that particular section. If the pipe is to be located 14' above the floor but it is suspended on threaded rod from beams, the bottom flange of which is 18' high (4' rods), then the height is actually 18' and the add is 20%. The pipe coverer, however, does not have to go above the 14', and so the add should be 10%.
- Most pipe is priced first as straight pipe with a joint (coupling, weld, etc.) every 10' and a hanger usually every 10'. There are exceptions with hanger spacing such as for cast iron pipe (5') and plastic pipe (3 per 10'). Following each type of pipe there are several lines listing sizes and the amount to be subtracted to delete couplings and hangers. This is for pipe that is to be buried or supported together on trapeze hangers. The reason that the couplings are deleted is that these runs are usually long, and frequently longer lengths of pipe are used. By deleting the couplings, the estimator is expected to look up and add back the correct reduced number of couplings.
- When preparing an estimate, it may be necessary to approximate the fittings. Fittings usually run between 25% and 50% of the cost of the pipe. The lower percentage is for simpler runs, and the higher number is for complex areas, such as mechanical rooms.
- For historic restoration projects, the systems must be as invisible as possible, and pathways must be sought for pipes, conduit, and ductwork. While installations in accessible spaces (such as basements and attics) are relatively straightforward to estimate, labor costs may be more difficult to determine when delivery systems must be concealed.

22 40 00 Plumbing Fixtures

- Plumbing fixture costs usually require two lines: the fixture itself and its "rough-in, supply, and waste."
- In the Assemblies Section (Plumbing D2010) for the desired fixture, the System Components Group at the center of the page shows the fixture on the first line. The rest of the list (fittings, pipe, tubing, etc.) will total up to what we refer to in the Unit Price section as "Rough-in, supply, waste, and vent." Note that for most fixtures we allow a nominal 5' of tubing to reach from the fixture to a main or riser.
- Remember that gas- and oil-fired units need venting.

Reference Numbers

Reference numbers are shown in shaded boxes at the beginning of some major classifications. These numbers refer to related items in the Reference Section. The reference information may be an estimating procedure, an alternate pricing method, or technical information.

Note: Not all subdivisions listed here necessarily appear in this publication. ■

22 01 Operation and Maintenance of Plumbing

22 01 02 – Labor Adjustments

22 01 02.10 Boilers, General		Crew	Daily Output	Labor-Hours	Unit	Material	2015 Bare Costs Labor	Equipment	Total	Total Incl O&P
0010	**BOILERS, GENERAL**, Prices do not include flue piping, elec. wiring,									
0020	gas or oil piping, boiler base, pad, or tankless unless noted									
0100	Boiler H.P.: 10 KW = 34 lb./steam/hr. = 33,475 BTU/hr.									
0150	To convert SFR to BTU rating: Hot water, 150 x SFR;									
0160	Forced hot water, 180 x SFR; steam, 240 x SFR									

22 01 02.20 Labor Adjustment Factors

		Crew	Daily Output	Labor-Hours	Unit	Material	Labor	Equipment	Total	Total Incl O&P
0010	**LABOR ADJUSTMENT FACTORS**, (For Div. 21, 22 and 23) R220102-20									
0100	Labor factors, The below are reasonable suggestions, however									
0110	each project must be evaluated for its own peculiarities, and									
0120	the adjustments be increased or decreased depending on the									
0130	severity of the special conditions.									
1000	Add to labor for elevated installation (Above floor level)									
1080	10′ to 14.5′ high						10%			
1100	15′ to 19.5′ high						20%			
1120	20′ to 24.5′ high						25%			
1140	25′ to 29.5′ high						35%			
1160	30′ to 34.5′ high						40%			
1180	35′ to 39.5′ high						50%			
1200	40′ and higher						55%			
2000	Add to labor for crawl space									
2100	3′ high						40%			
2140	4′ high						30%			
3000	Add to labor for multi-story building									
3100	Add per floor for floors 3 thru 19						2%			
3140	Add per floor for floors 20 and up						4%			
4000	Add to labor for working in existing occupied buildings									
4100	Hospital						35%			
4140	Office building						25%			
4180	School						20%			
4220	Factory or warehouse						15%			
4260	Multi dwelling						15%			
5000	Add to labor, miscellaneous									
5100	Cramped shaft						35%			
5140	Congested area						15%			
5180	Excessive heat or cold						30%			
9000	Labor factors, The above are reasonable suggestions, however									
9010	each project should be evaluated for its own peculiarities.									
9100	Other factors to be considered are:									
9140	Movement of material and equipment through finished areas									
9180	Equipment room									
9220	Attic space									
9260	No service road									
9300	Poor unloading/storage area									
9340	Congested site area/heavy traffic									

22 05 05 – Selective Demolition for Plumbing

22 05 05.10 Plumbing Demolition		Crew	Daily Output	Labor-Hours	Unit	Material	2015 Bare Costs Labor	Equipment	Total	Total Incl O&P
0010	**PLUMBING DEMOLITION**									
1020	Fixtures, including 10' piping									
1100	Bathtubs, cast iron	1 Plum	4	2	Ea.		117		117	177
1120	Fiberglass		6	1.333			78.50		78.50	118
1140	Steel		5	1.600			94		94	142
1200	Lavatory, wall hung		10	.800			47		47	71
1220	Counter top		8	1			58.50		58.50	88.50
1300	Sink, single compartment		8	1			58.50		58.50	88.50
1320	Double compartment		7	1.143			67		67	101
1400	Water closet, floor mounted		8	1			58.50		58.50	88.50
1420	Wall mounted		7	1.143			67		67	101
1500	Urinal, floor mounted		4	2			117		117	177
1520	Wall mounted		7	1.143			67		67	101
1600	Water fountains, free standing		8	1			58.50		58.50	88.50
1620	Wall or deck mounted		6	1.333			78.50		78.50	118
2000	Piping, metal, up thru 1-1/2" diameter		200	.040	L.F.		2.35		2.35	3.55
2050	2" thru 3-1/2" diameter		150	.053			3.13		3.13	4.73
2100	4" thru 6" diameter	2 Plum	100	.160			9.40		9.40	14.20
2150	8" thru 14" diameter	"	60	.267			15.65		15.65	23.50
2153	16" thru 20" diameter	Q-18	70	.343			19.10	.82	19.92	30
2155	24" thru 26" diameter		55	.436			24.50	1.05	25.55	37.50
2156	30" thru 36" diameter		40	.600			33.50	1.44	34.94	52
2160	Plastic pipe with fittings, up thru 1-1/2" diameter	1 Plum	250	.032			1.88		1.88	2.84
2162	2" thru 3" diameter	"	200	.040			2.35		2.35	3.55
2164	4" thru 6" diameter	Q-1	200	.080			4.23		4.23	6.40
2166	8" thru 14" diameter		150	.107			5.65		5.65	8.50
2168	16" diameter		100	.160			8.45		8.45	12.75
2212	Deduct for salvage, aluminum scrap				Ton				700	770
2214	Brass scrap								2,450	2,675
2216	Copper scrap								3,200	3,525
2218	Lead scrap								520	570
2220	Steel scrap								180	200
2250	Water heater, 40 gal.	1 Plum	6	1.333	Ea.		78.50		78.50	118
9470	Water softener	Q-1	2	8	"		425		425	640

22 05 23 – General-Duty Valves for Plumbing Piping

22 05 23.10 Valves, Brass		Crew	Daily Output	Labor-Hours	Unit	Material	Labor	Equipment	Total	Total Incl O&P
0010	**VALVES, BRASS**									
0500	Gas cocks, threaded									
0530	1/2"	1 Plum	24	.333	Ea.	13.15	19.55		32.70	44
0540	3/4"		22	.364		15.50	21.50		37	49
0550	1"		19	.421		31	24.50		55.50	71.50
0560	1-1/4"		15	.533		44.50	31.50		76	96.50

22 05 23.20 Valves, Bronze		Crew	Daily Output	Labor-Hours	Unit	Material	Labor	Equipment	Total	Total Incl O&P
0010	**VALVES, BRONZE**									
1020	Angle, 150 lb., rising stem, threaded									
1030	1/8"	1 Plum	24	.333	Ea.	128	19.55		147.55	170
1040	1/4"		24	.333		128	19.55		147.55	170
1050	3/8"		24	.333		130	19.55		149.55	173
1060	1/2"		22	.364		142	21.50		163.50	188
1070	3/4"		20	.400		193	23.50		216.50	249
1080	1"		19	.421		279	24.50		303.50	345
1100	1-1/2"		13	.615		470	36		506	570

22 05 23 – General-Duty Valves for Plumbing Piping

22 05 23.20 Valves, Bronze		Crew	Daily Output	Labor-Hours	Unit	Material	2015 Bare Costs Labor	Equipment	Total	Total Incl O&P
1110	2"	1 Plum	11	.727	Ea.	755	42.50		797.50	895
1300	Ball									
1398	Threaded, 150 psi									
1400	1/4"	1 Plum	24	.333	Ea.	14.15	19.55		33.70	45
1430	3/8"		24	.333		14.15	19.55		33.70	45
1450	1/2"		22	.364		14.15	21.50		35.65	47.50
1460	3/4"		20	.400		23.50	23.50		47	61
1470	1"		19	.421		33.50	24.50		58	74.50
1480	1-1/4"		15	.533		58.50	31.50		90	112
1490	1-1/2"		13	.615		76.50	36		112.50	139
1500	2"		11	.727		93	42.50		135.50	167
1750	Check, swing, class 150, regrinding disc, threaded									
1800	1/8"	1 Plum	24	.333	Ea.	66.50	19.55		86.05	103
1830	1/4"		24	.333		66.50	19.55		86.05	103
1840	3/8"		24	.333		70.50	19.55		90.05	107
1850	1/2"		24	.333		75.50	19.55		95.05	113
1860	3/4"		20	.400		100	23.50		123.50	146
1870	1"		19	.421		144	24.50		168.50	196
1880	1-1/4"		15	.533		208	31.50		239.50	277
1890	1-1/2"		13	.615		242	36		278	320
1900	2"		11	.727		355	42.50		397.50	455
1910	2-1/2"	Q-1	15	1.067		800	56.50		856.50	965
2000	For 200 lb., add					5%	10%			
2040	For 300 lb., add					15%	15%			
2850	Gate, N.R.S., soldered, 125 psi									
2900	3/8"	1 Plum	24	.333	Ea.	61	19.55		80.55	96.50
2920	1/2"		24	.333		54	19.55		73.55	89
2940	3/4"		20	.400		63.50	23.50		87	106
2950	1"		19	.421		76.50	24.50		101	122
2960	1-1/4"		15	.533		126	31.50		157.50	187
2970	1-1/2"		13	.615		142	36		178	212
2980	2"		11	.727		185	42.50		227.50	269
2990	2-1/2"	Q-1	15	1.067		450	56.50		506.50	580
3000	3"	"	13	1.231		560	65		625	715
3850	Rising stem, soldered, 300 psi									
3950	1"	1 Plum	19	.421	Ea.	179	24.50		203.50	235
3980	2"	"	11	.727		480	42.50		522.50	595
4000	3"	Q-1	13	1.231		1,575	65		1,640	1,850
4250	Threaded, class 150									
4310	1/4"	1 Plum	24	.333	Ea.	68.50	19.55		88.05	105
4320	3/8"		24	.333		68.50	19.55		88.05	105
4330	1/2"		24	.333		63	19.55		82.55	98.50
4340	3/4"		20	.400		73.50	23.50		97	116
4350	1"		19	.421		98.50	24.50		123	146
4360	1-1/4"		15	.533		134	31.50		165.50	195
4370	1-1/2"		13	.615		169	36		205	240
4380	2"		11	.727		227	42.50		269.50	315
4390	2-1/2"	Q-1	15	1.067		530	56.50		586.50	670
4400	3"	"	13	1.231		740	65		805	910
4500	For 300 psi, threaded, add					100%	15%			
4540	For chain operated type, add					15%				
4850	Globe, class 150, rising stem, threaded									
4920	1/4"	1 Plum	24	.333	Ea.	97	19.55		116.55	136

22 05 23.20 Valves, Bronze		Crew	Daily Output	Labor-Hours	Unit	Material	2015 Bare Costs Labor	Equipment	Total	Total Incl O&P
4940	3/8"	1 Plum	24	.333	Ea.	95.50	19.55		115.05	135
4950	1/2"		24	.333		95.50	19.55		115.05	135
4960	3/4"		20	.400		99	23.50		122.50	145
4970	1"		19	.421		154	24.50		178.50	207
4980	1-1/4"		15	.533		245	31.50		276.50	315
4990	1-1/2"		13	.615		320	36		356	410
5000	2"		11	.727		465	42.50		507.50	580
5010	2-1/2"	Q-1	15	1.067		1,175	56.50		1,231.50	1,350
5020	3"	"	13	1.231		1,675	65		1,740	1,925
5120	For 300 lb. threaded, add					50%	15%			
5600	Relief, pressure & temperature, self-closing, ASME, threaded									
5640	3/4"	1 Plum	28	.286	Ea.	253	16.75		269.75	305
5650	1"		24	.333		405	19.55		424.55	475
5660	1-1/4"		20	.400		895	23.50		918.50	1,025
5670	1-1/2"		18	.444		1,250	26		1,276	1,425
5680	2"		16	.500		1,350	29.50		1,379.50	1,550
5950	Pressure, poppet type, threaded									
6000	1/2"	1 Plum	30	.267	Ea.	78.50	15.65		94.15	110
6040	3/4"	"	28	.286	"	73.50	16.75		90.25	106
6400	Pressure, water, ASME, threaded									
6440	3/4"	1 Plum	28	.286	Ea.	116	16.75		132.75	153
6450	1"		24	.333		260	19.55		279.55	315
6460	1-1/4"		20	.400		390	23.50		413.50	465
6470	1-1/2"		18	.444		570	26		596	665
6480	2"		16	.500		820	29.50		849.50	950
6490	2-1/2"		15	.533		3,150	31.50		3,181.50	3,525
6900	Reducing, water pressure									
6920	300 psi to 25-75 psi, threaded or sweat									
6940	1/2"	1 Plum	24	.333	Ea.	395	19.55		414.55	465
6950	3/4"		20	.400		405	23.50		428.50	485
6960	1"		19	.421		630	24.50		654.50	730
6970	1-1/4"		15	.533		1,100	31.50		1,131.50	1,250
6980	1-1/2"		13	.615		1,650	36		1,686	1,875
8350	Tempering, water, sweat connections									
8400	1/2"	1 Plum	24	.333	Ea.	98	19.55		117.55	138
8440	3/4"	"	20	.400	"	126	23.50		149.50	175
8650	Threaded connections									
8700	1/2"	1 Plum	24	.333	Ea.	126	19.55		145.55	169
8740	3/4"		20	.400		770	23.50		793.50	880
8750	1"		19	.421		865	24.50		889.50	995
8760	1-1/4"		15	.533		1,350	31.50		1,381.50	1,525
8770	1-1/2"		13	.615		1,475	36		1,511	1,650
8780	2"		11	.727		2,200	42.50		2,242.50	2,500

22 05 23.60 Valves, Plastic		Crew	Daily Output	Labor-Hours	Unit	Material	Labor	Equipment	Total	Total Incl O&P
0010	**VALVES, PLASTIC**									
1100	Angle, PVC, threaded									
1110	1/4"	1 Plum	26	.308	Ea.	39.50	18.05		57.55	71
1120	1/2"		26	.308		56.50	18.05		74.55	89.50
1130	3/4"		25	.320		67	18.80		85.80	103
1140	1"		23	.348		81.50	20.50		102	121
1150	Ball, PVC, socket or threaded, true union									
1230	1/2"	1 Plum	26	.308	Ea.	40.50	18.05		58.55	72

22 05 23 – General-Duty Valves for Plumbing Piping

22 05 23.60 Valves, Plastic		Crew	Daily Output	Labor-Hours	Unit	Material	Labor	Equipment	Total	Total Incl O&P
						2015 Bare Costs				
1240	3/4"	1 Plum	25	.320	Ea.	40.50	18.80		59.30	73
1250	1"		23	.348		48	20.50		68.50	84
1260	1-1/4"		21	.381		83.50	22.50		106	126
1270	1-1/2"		20	.400		83.50	23.50		107	128
1280	2"		17	.471		110	27.50		137.50	163
1360	For PVC, flanged, add					100%	15%			
1650	CPVC, socket or threaded, single union									
1700	1/2"	1 Plum	26	.308	Ea.	58	18.05		76.05	91
1720	3/4"		25	.320		77	18.80		95.80	113
1730	1"		23	.348		87.50	20.50		108	127
1750	1-1/4"		21	.381		140	22.50		162.50	188
1760	1-1/2"		20	.400		140	23.50		163.50	190
1840	For CPVC, flanged, add					65%	15%			
1880	For true union, socket or threaded, add					50%	5%			
2050	Polypropylene, threaded									
2100	1/4"	1 Plum	26	.308	Ea.	45	18.05		63.05	77
2120	3/8"		26	.308		45	18.05		63.05	77
2130	1/2"		26	.308		45	18.05		63.05	77
2140	3/4"		25	.320		54.50	18.80		73.30	88.50
2150	1"		23	.348		62.50	20.50		83	99.50
2160	1-1/4"		21	.381		84	22.50		106.50	127
2170	1-1/2"		20	.400		103	23.50		126.50	150
2180	2"		17	.471		138	27.50		165.50	194
4850	Foot valve, PVC, socket or threaded									
4900	1/2"	1 Plum	34	.235	Ea.	73	13.80		86.80	101
4930	3/4"		32	.250		83	14.70		97.70	114
4940	1"		28	.286		107	16.75		123.75	144
4950	1-1/4"		27	.296		206	17.40		223.40	254
4960	1-1/2"		26	.308		206	18.05		224.05	255
6350	Y sediment strainer, PVC, socket or threaded									
6400	1/2"	1 Plum	26	.308	Ea.	55	18.05		73.05	88
6440	3/4"		24	.333		58.50	19.55		78.05	93.50
6450	1"		23	.348		69.50	20.50		90	108
6460	1-1/4"		21	.381		117	22.50		139.50	162
6470	1-1/2"		20	.400		117	23.50		140.50	164

22 05 48 – Vibration and Seismic Controls for Plumbing Piping and Equipment

22 05 48.10 Seismic Bracing Supports		Crew	Daily Output	Labor-Hours	Unit	Material	Labor	Equipment	Total	Total Incl O&P
0010	**SEISMIC BRACING SUPPORTS**									
0020	Clamps									
0030	C-clamp, for mounting on steel beam									
0040	3/8" threaded rod	1 Skwk	160	.050	Ea.	2.05	2.43		4.48	6
0050	1/2" threaded rod		160	.050		2.20	2.43		4.63	6.20
0060	5/8" threaded rod		160	.050		3.70	2.43		6.13	7.85
0070	3/4" threaded rod		160	.050		4.55	2.43		6.98	8.75
0100	Brackets									
0110	Beam side or wall mallable iron									
0120	3/8" threaded rod	1 Skwk	48	.167	Ea.	3.11	8.10		11.21	15.95
0130	1/2" threaded rod		48	.167		4.39	8.10		12.49	17.40
0140	5/8" threaded rod		48	.167		8.20	8.10		16.30	21.50
0150	3/4" threaded rod		48	.167		11.30	8.10		19.40	25
0160	7/8" threaded rod		48	.167		11.70	8.10		19.80	25.50
0170	For concrete installation, add						30%			

22 05 48 – Vibration and Seismic Controls for Plumbing Piping and Equipment

22 05 48.10 Seismic Bracing Supports		Crew	Daily Output	Labor-Hours	Unit	2015 Bare Costs Material	Labor	Equipment	Total	Total Incl O&P
0180	Wall, welded steel									
0190	0 size 12" wide 18" deep	1 Skwk	34	.235	Ea.	172	11.45		183.45	207
0200	1 size 18" wide 24" deep		34	.235		211	11.45		222.45	250
0210	2 size 24" wide 30" deep		34	.235		290	11.45		301.45	340
0300	Rod, carbon steel									
0310	Continuous thread									
0320	1/4" thread	1 Skwk	144	.056	L.F.	1.70	2.70		4.40	6.05
0330	3/8" thread		144	.056		1.81	2.70		4.51	6.15
0340	1/2" thread		144	.056		2.86	2.70		5.56	7.35
0350	5/8" thread		144	.056		4.05	2.70		6.75	8.65
0360	3/4" thread		144	.056		7.15	2.70		9.85	12.05
0370	7/8" thread		144	.056		8.95	2.70		11.65	14.05
0380	For galvanized, add					30%				
0400	Channel, steel									
0410	3/4" x 1-1/2"	1 Skwk	80	.100	L.F.	4.10	4.87		8.97	12
0420	1-1/2" x 1-1/2"		70	.114		5.40	5.55		10.95	14.55
0430	1-7/8" x 1-1/2"		60	.133		23	6.50		29.50	35.50
0440	3" x 1-1/2"		50	.160		39.50	7.80		47.30	55.50
0450	Spring nuts									
0460	3/8"	1 Skwk	100	.080	Ea.	1.20	3.89		5.09	7.30
0470	1/2"	"	80	.100	"	1.67	4.87		6.54	9.35
0500	Welding, field									
0510	Cleaning and welding plates, bars, or rods									
0520	To existing beams, columns, or trusses									
0530	1" weld	1 Skwk	144	.056	Ea.	.23	2.70		2.93	4.43
0540	2" weld		72	.111		.39	5.40		5.79	8.80
0550	3" weld		54	.148		.61	7.20		7.81	11.80
0560	4" weld		36	.222		.83	10.80		11.63	17.60
0570	5" weld		30	.267		1.05	12.95		14	21
0580	6" weld		24	.333		1.18	16.20		17.38	26.50
0600	Vibration absorbers									
0610	Hangers, neoprene flex									
0620	10-120 lb. capacity	1 Skwk	8	1	Ea.	26	48.50		74.50	104
0630	75-550 lb. capacity		8	1		37	48.50		85.50	116
0640	250-1100 lb. capacity		6	1.333		72	65		137	179
0650	1000-4000 lb. capacity		6	1.333		134	65		199	248
0660	Spring flex									
0670	60-450 lb. capacity	1 Skwk	8	1	Ea.	70.50	48.50		119	153
0680	85-450 lb. capacity		8	1		96.50	48.50		145	181
0690	600-900 lb. capacity		8	1		108	48.50		156.50	194
0700	1100-1300 lb. capacity		6	1.333		117	65		182	229
0710	Mounts, neoprene									
0720	135-380 lb. capacity	1 Skwk	7	1.143	Ea.	17.25	55.50		72.75	105
0730	250-1100 lb. capacity		7	1.143		49.50	55.50		105	140
0740	1000-4000 lb. capacity		5	1.600		111	78		189	242

22 05 76 – Facility Drainage Piping Cleanouts

22 05 76.10 Cleanouts		Crew	Daily Output	Labor-Hours	Unit	Material	Labor	Equipment	Total	Total Incl O&P
0010	**CLEANOUTS**									
0060	Floor type									
0080	Round or square, scoriated nickel bronze top									
0100	2" pipe size	1 Plum	10	.800	Ea.	197	47		244	288
0120	3" pipe size		8	1		295	58.50		353.50	415

22 05 Common Work Results for Plumbing

22 05 76 – Facility Drainage Piping Cleanouts

22 05 76.10 Cleanouts		Crew	Daily Output	Labor-Hours	Unit	Material	2015 Bare Costs Labor	Equipment	Total	Total Incl O&P
0140	4" pipe size	1 Plum	6	1.333	Ea.	295	78.50		373.50	445
0980	Round top, recessed for terrazzo									
1000	2" pipe size	1 Plum	9	.889	Ea.	197	52		249	296
1080	3" pipe size		6	1.333		295	78.50		373.50	445
1100	4" pipe size		4	2		295	117		412	500
1120	5" pipe size	Q-1	6	2.667		375	141		516	630

22 05 76.20 Cleanout Tees		Crew	Daily Output	Labor-Hours	Unit	Material	Labor	Equipment	Total	Total Incl O&P
0010	**CLEANOUT TEES**									
0100	Cast iron, B&S, with countersunk plug									
0200	2" pipe size	1 Plum	4	2	Ea.	267	117		384	470
0220	3" pipe size		3.60	2.222		292	130		422	515
0240	4" pipe size		3.30	2.424		365	142		507	615
0280	6" pipe size	Q-1	5	3.200		980	169		1,149	1,325
0500	For round smooth access cover, same price									
4000	Plastic, tees and adapters. Add plugs									
4010	ABS, DWV									
4020	Cleanout tee, 1-1/2" pipe size	1 Plum	15	.533	Ea.	23	31.50		54.50	72.50
4030	2" pipe size	Q-1	27	.593		25	31.50		56.50	75
4040	3" pipe size		21	.762		47	40.50		87.50	113
4050	4" pipe size		16	1		101	53		154	191
4100	Cleanout plug, 1-1/2" pipe size	1 Plum	32	.250		3.94	14.70		18.64	26.50
4110	2" pipe size	Q-1	56	.286		5.20	15.10		20.30	28.50
4120	3" pipe size		36	.444		8.35	23.50		31.85	44.50
4130	4" pipe size		30	.533		14.65	28		42.65	58.50
4180	Cleanout adapter fitting, 1-1/2" pipe size	1 Plum	32	.250		6.25	14.70		20.95	29
4190	2" pipe size	Q-1	56	.286		9.60	15.10		24.70	33.50
4200	3" pipe size		36	.444		24.50	23.50		48	62.50
4210	4" pipe size		30	.533		45	28		73	92
5000	PVC, DWV									
5010	Cleanout tee, 1-1/2" pipe size	1 Plum	15	.533	Ea.	17.10	31.50		48.60	66.50
5020	2" pipe size	Q-1	27	.593		19.95	31.50		51.45	69.50
5030	3" pipe size		21	.762		35.50	40.50		76	100
5040	4" pipe size		16	1		69.50	53		122.50	157
5090	Cleanout plug, 1-1/2" pipe size	1 Plum	32	.250		4.08	14.70		18.78	26.50
5100	2" pipe size	Q-1	56	.286		4.53	15.10		19.63	28
5110	3" pipe size		36	.444		8.10	23.50		31.60	44.50
5120	4" pipe size		30	.533		11.95	28		39.95	55.50
5130	6" pipe size		24	.667		39	35		74	96
5170	Cleanout adapter fitting, 1-1/2" pipe size	1 Plum	32	.250		5.45	14.70		20.15	28
5180	2" pipe size	Q-1	56	.286		7	15.10		22.10	30.50
5190	3" pipe size		36	.444		19.75	23.50		43.25	57
5200	4" pipe size		30	.533		32.50	28		60.50	78
5210	6" pipe size		24	.667		94.50	35		129.50	157

22 07 19 – Plumbing Piping Insulation

22 07 19.10 Piping Insulation		Crew	Daily Output	Labor-Hours	Unit	Material	Labor	Equipment	Total	Total Incl O&P
						2015 Bare Costs				
0010	**PIPING INSULATION**									
0100	Rule of thumb, as a percentage of total mechanical costs				Job				10%	10%
0110	Insulation req'd. is based on the surface size/area to be covered									
0600	Pipe covering (price copper tube one size less than IPS)									
6600	Fiberglass, with all service jacket									
6840	1" wall, 1/2" iron pipe size G	Q-14	240	.067	L.F.	.83	3.14		3.97	5.80
6870	1" iron pipe size G		220	.073		.97	3.43		4.40	6.40
6900	2" iron pipe size G		200	.080		1.22	3.77		4.99	7.20
6920	3" iron pipe size G		180	.089		1.49	4.19		5.68	8.15
6940	4" iron pipe size G		150	.107		1.97	5.05		7.02	10
7320	2" wall, 1/2" iron pipe size G		220	.073		2.43	3.43		5.86	8
7440	6" iron pipe size G		100	.160		4.88	7.55		12.43	17.10
7460	8" iron pipe size G		80	.200		5.95	9.45		15.40	21
7480	10" iron pipe size G		70	.229		7.10	10.75		17.85	24.50
7490	12" iron pipe size G		65	.246		7.95	11.60		19.55	27
7800	For fiberglass with standard canvas jacket, deduct					5%				
7802	For fittings, add 3 L.F. for each fitting									
7804	plus 4 L.F. for each flange of the fitting									
7810	Finishes									
7812	For .016" aluminum jacket, add G	Q-14	200	.080	S.F.	.92	3.77		4.69	6.85
7813	For .010" stainless steel, add G	"	160	.100	"	2.88	4.71		7.59	10.50
7814	For single layer of felt, add					10%	10%			
7816	For roofing paper, 45 lb. to 55 lb., add					25%	10%			
7879	Rubber tubing, flexible closed cell foam									
7880	3/8" wall, 1/4" iron pipe size G	1 Asbe	120	.067	L.F.	.33	3.49		3.82	5.80
7910	1/2" iron pipe size G		115	.070		.41	3.64		4.05	6.10
7920	3/4" iron pipe size G		115	.070		.46	3.64		4.10	6.15
7930	1" iron pipe size G		110	.073		.52	3.81		4.33	6.50
7950	1-1/2" iron pipe size G		110	.073		.73	3.81		4.54	6.75
8100	1/2" wall, 1/4" iron pipe size G		90	.089		.54	4.65		5.19	7.85
8130	1/2" iron pipe size G		89	.090		.67	4.71		5.38	8.10
8140	3/4" iron pipe size G		89	.090		.75	4.71		5.46	8.20
8150	1" iron pipe size G		88	.091		.82	4.76		5.58	8.30
8170	1-1/2" iron pipe size G		87	.092		1.15	4.81		5.96	8.75
8180	2" iron pipe size G		86	.093		1.47	4.87		6.34	9.20
8200	3" iron pipe size G		85	.094		2.06	4.93		6.99	9.90
8300	3/4" wall, 1/4" iron pipe size G		90	.089		.85	4.65		5.50	8.20
8330	1/2" iron pipe size G		89	.090		1.10	4.71		5.81	8.55
8340	3/4" iron pipe size G		89	.090		1.35	4.71		6.06	8.85
8350	1" iron pipe size G		88	.091		1.54	4.76		6.30	9.10
8370	1-1/2" iron pipe size G		87	.092		2.32	4.81		7.13	10.05
8380	2" iron pipe size G		86	.093		2.68	4.87		7.55	10.55
8400	3" iron pipe size G		85	.094		4.08	4.93		9.01	12.15
8444	1" wall, 1/2" iron pipe size G		86	.093		2.05	4.87		6.92	9.85
8445	3/4" iron pipe size G		84	.095		2.48	4.99		7.47	10.50
8446	1" iron pipe size G		84	.095		2.89	4.99		7.88	10.95
8447	1-1/4" iron pipe size G		82	.098		3.23	5.10		8.33	11.50
8448	1-1/2" iron pipe size G		82	.098		3.76	5.10		8.86	12.10
8449	2" iron pipe size G		80	.100		4.95	5.25		10.20	13.60
8450	2-1/2" iron pipe size G		80	.100		6.45	5.25		11.70	15.25
8456	Rubber insulation tape, 1/8" x 2" x 30' G				Ea.	12.05			12.05	13.25

22 11 13 – Facility Water Distribution Piping

22 11 13.14 Pipe, Brass		Crew	Daily Output	Labor-Hours	Unit	2015 Bare Costs Material	Labor	Equipment	Total	Total Incl O&P
0010	**PIPE, BRASS**, Plain end									
0900	Field threaded, coupling & clevis hanger assembly 10′ O.C.									
0920	Regular weight									
1120	1/2″ diameter	1 Plum	48	.167	L.F.	7.15	9.80		16.95	22.50
1140	3/4″ diameter		46	.174		9.30	10.20		19.50	25.50
1160	1″ diameter		43	.186		13.30	10.90		24.20	31
1180	1-1/4″ diameter	Q-1	72	.222		19.90	11.75		31.65	40
1200	1-1/2″ diameter		65	.246		23.50	13		36.50	45.50
1220	2″ diameter		53	.302		33	15.95		48.95	60.50

22 11 13.23 Pipe/Tube, Copper		Crew	Daily Output	Labor-Hours	Unit	Material	Labor	Equipment	Total	Total Incl O&P
0010	**PIPE/TUBE, COPPER**, Solder joints									
1000	Type K tubing, couplings & clevis hanger assemblies 10′ O.C.									
1100	1/4″ diameter	1 Plum	84	.095	L.F.	3.71	5.60		9.31	12.55
1200	1″ diameter		66	.121		10.35	7.10		17.45	22
1260	2″ diameter		40	.200		23.50	11.75		35.25	43.50
2000	Type L tubing, couplings & clevis hanger assemblies 10′ O.C.									
2100	1/4″ diameter	1 Plum	88	.091	L.F.	2.77	5.35		8.12	11.10
2120	3/8″ diameter		84	.095		3.50	5.60		9.10	12.30
2140	1/2″ diameter		81	.099		3.70	5.80		9.50	12.80
2160	5/8″ diameter		79	.101		5.35	5.95		11.30	14.85
2180	3/4″ diameter		76	.105		5.20	6.20		11.40	15.05
2200	1″ diameter		68	.118		7.85	6.90		14.75	19.05
2220	1-1/4″ diameter		58	.138		10.35	8.10		18.45	23.50
2240	1-1/2″ diameter		52	.154		12.95	9.05		22	28
2260	2″ diameter		42	.190		18.80	11.20		30	37.50
2280	2-1/2″ diameter	Q-1	62	.258		29.50	13.65		43.15	53
2300	3″ diameter		56	.286		37	15.10		52.10	63.50
2320	3-1/2″ diameter		43	.372		53	19.65		72.65	87.50
2340	4″ diameter		39	.410		65.50	21.50		87	105
2360	5″ diameter		34	.471		120	25		145	170
2380	6″ diameter	Q-2	40	.600		166	33		199	232
2400	8″ diameter	″	36	.667		285	36.50		321.50	370
2410	For other than full hard temper, add					21%				
2590	For silver solder, add						15%			
4000	Type DWV tubing, couplings & clevis hanger assemblies 10′ O.C.									
4100	1-1/4″ diameter	1 Plum	60	.133	L.F.	9.15	7.85		17	22
4120	1-1/2″ diameter		54	.148		11.15	8.70		19.85	25.50
4140	2″ diameter		44	.182		14.60	10.65		25.25	32
4160	3″ diameter	Q-1	58	.276		26.50	14.55		41.05	51
4180	4″ diameter		40	.400		44	21		65	80
4200	5″ diameter		36	.444		107	23.50		130.50	154
4220	6″ diameter	Q-2	42	.571		153	31.50		184.50	216

22 11 13.44 Pipe, Steel		Crew	Daily Output	Labor-Hours	Unit	Material	Labor	Equipment	Total	Total Incl O&P
0010	**PIPE, STEEL**									
0012	The steel pipe in this section does not include fittings such as ells, tees									
0014	For fittings either add a % (usually 25 to 35%) or see									
0015	the Mechanical or Plumbing Cost Data									
0020	All pipe sizes are to Spec. A-53 unless noted otherwise R221113-50									
0050	Schedule 40, threaded, with couplings, and clevis hanger									
0060	assemblies sized for covering, 10′ O.C.									
0540	Black, 1/4″ diameter	1 Plum	66	.121	L.F.	6.55	7.10		13.65	17.95
0550	3/8″ diameter		65	.123		7.30	7.20		14.50	18.95

22 11 13 – Facility Water Distribution Piping

22 11 13.44 Pipe, Steel		Crew	Daily Output	Labor-Hours	Unit	Material	2015 Bare Costs Labor	Equipment	Total	Total Incl O&P
0560	1/2" diameter	1 Plum	63	.127	L.F.	3.32	7.45		10.77	14.90
0570	3/4" diameter		61	.131		3.91	7.70		11.61	15.95
0580	1" diameter		53	.151		5.05	8.85		13.90	18.95
0590	1-1/4" diameter	Q-1	89	.180		6.20	9.50		15.70	21
0600	1-1/2" diameter		80	.200		7.10	10.55		17.65	24
0610	2" diameter		64	.250		9	13.20		22.20	30
0620	2-1/2" diameter		50	.320		14	16.90		30.90	41
0630	3" diameter		43	.372		17.80	19.65		37.45	49
0640	3-1/2" diameter		40	.400		24.50	21		45.50	59
0650	4" diameter		36	.444		27	23.50		50.50	65
1280	All pipe sizes are to Spec. A-53 unless noted otherwise									
1281	Schedule 40, threaded, with couplings and clevis hanger									
1282	assemblies sized for covering, 10' O. C.									
1290	Galvanized, 1/4" diameter	1 Plum	66	.121	L.F.	9.10	7.10		16.20	21
1300	3/8" diameter		65	.123		9.95	7.20		17.15	22
1310	1/2" diameter		63	.127		3.69	7.45		11.14	15.30
1320	3/4" diameter		61	.131		4.24	7.70		11.94	16.30
1330	1" diameter		53	.151		5.75	8.85		14.60	19.70
1340	1-1/4" diameter	Q-1	89	.180		7	9.50		16.50	22
1350	1-1/2" diameter		80	.200		8.05	10.55		18.60	25
1360	2" diameter		64	.250		10.35	13.20		23.55	31.50
1370	2-1/2" diameter		50	.320		16.30	16.90		33.20	43.50
1380	3" diameter		43	.372		20.50	19.65		40.15	52.50
1390	3-1/2" diameter		40	.400		26	21		47	61
1400	4" diameter		36	.444		30	23.50		53.50	68.50
2000	Welded, sch. 40, on yoke & roll hanger assy's, sized for covering, 10' O.C.									
2040	Black, 1" diameter	Q-15	93	.172	L.F.	4.72	9.10	.62	14.44	19.65
2070	2" diameter		61	.262		7.85	13.85	.95	22.65	30.50
2090	3" diameter		43	.372		15.45	19.65	1.34	36.44	48
2110	4" diameter		37	.432		21.50	23	1.56	46.06	60
2120	5" diameter		32	.500		34.50	26.50	1.81	62.81	80
2130	6" diameter	Q-16	36	.667		43	36.50	1.60	81.10	104
2140	8" diameter		29.	.828		68.50	45.50	1.99	115.99	146
2150	10" diameter		24	1		89.50	55	2.40	146.90	183
2160	12" diameter		19	1.263		105	69	3.03	177.03	224

22 11 13.48 Pipe, Fittings and Valves, Steel, Grooved-Joint		Crew	Daily Output	Labor-Hours	Unit	Material	Labor	Equipment	Total	Total Incl O&P
0010	**PIPE, FITTINGS AND VALVES, STEEL, GROOVED-JOINT**									
0012	Fittings are ductile iron. Steel fittings noted.									
0020	Pipe includes coupling & clevis type hanger assemblies, 10' O.C.									
1000	Schedule 40, black									
1040	3/4" diameter	1 Plum	71	.113	L.F.	5.50	6.60		12.10	16
1050	1" diameter		63	.127		5.30	7.45		12.75	17.05
1060	1-1/4" diameter		58	.138		6.60	8.10		14.70	19.50
1070	1-1/2" diameter		51	.157		7.35	9.20		16.55	22
1080	2" diameter		40	.200		8.70	11.75		20.45	27.50
1090	2-1/2" diameter	Q-1	57	.281		13.65	14.85		28.50	37.50
1100	3" diameter		50	.320		16.75	16.90		33.65	44
1110	4" diameter		45	.356		23.50	18.80		42.30	54.50
1120	5" diameter		37	.432		38	23		61	76.50
1130	6" diameter	Q-2	42	.571		48.50	31.50		80	101
1800	Galvanized									
1840	3/4" diameter	1 Plum	71	.113	L.F.	5.80	6.60		12.40	16.35

22 11 13 – Facility Water Distribution Piping

22 11 13.48 Pipe, Fittings and Valves, Steel, Grooved-Joint		Crew	Daily Output	Labor-Hours	Unit	Material	2015 Bare Costs Labor	Equipment	Total	Total Incl O&P
1850	1" diameter	1 Plum	63	.127	L.F.	7	7.45		14.45	18.95
1860	1-1/4" diameter		58	.138		8.95	8.10		17.05	22
1870	1-1/2" diameter		51	.157		10.10	9.20		19.30	25
1880	2" diameter		40	.200		12.50	11.75		24.25	31.50
1890	2-1/2" diameter	Q-1	57	.281		17.60	14.85		32.45	42
1900	3" diameter		50	.320		22.50	16.90		39.40	50
1910	4" diameter		45	.356		32	18.80		50.80	63.50
1920	5" diameter		37	.432		60	23		83	100
1930	6" diameter	Q-2	42	.571		64.50	31.50		96	119
3990	Fittings: coupling material required at joints not incl. in fitting price.									
3994	Add 1 selected coupling, material only, per joint for installed price.									
4000	Elbow, 90° or 45°, painted									
4030	3/4" diameter	1 Plum	50	.160	Ea.	60	9.40		69.40	80
4040	1" diameter		50	.160		32	9.40		41.40	49
4050	1-1/4" diameter		40	.200		32	11.75		43.75	53
4060	1-1/2" diameter		33	.242		32	14.25		46.25	56.50
4070	2" diameter		25	.320		32	18.80		50.80	63.50
4080	2-1/2" diameter	Q-1	40	.400		32	21		53	67
4090	3" diameter		33	.485		56.50	25.50		82	101
4100	4" diameter		25	.640		61.50	34		95.50	119
4110	5" diameter		20	.800		146	42.50		188.50	225
4120	6" diameter	Q-2	25	.960		172	52.50		224.50	269
4250	For galvanized elbows, add					26%				
4690	Tee, painted									
4700	3/4" diameter	1 Plum	38	.211	Ea.	64.50	12.35		76.85	89.50
4740	1" diameter		33	.242		50	14.25		64.25	76.50
4750	1-1/4" diameter		27	.296		50	17.40		67.40	81.50
4760	1-1/2" diameter		22	.364		50	21.50		71.50	87
4770	2" diameter		17	.471		50	27.50		77.50	96.50
4780	2-1/2" diameter	Q-1	27	.593		50	31.50		81.50	103
4790	3" diameter		22	.727		68	38.50		106.50	133
4800	4" diameter		17	.941		103	49.50		152.50	189
4810	5" diameter		13	1.231		241	65		306	365
4820	6" diameter	Q-2	17	1.412		278	77.50		355.50	420
4900	For galvanized tees, add					24%				
4906	Couplings, rigid style, painted									
4908	1" diameter	1 Plum	100	.080	Ea.	25	4.70		29.70	34.50
4909	1-1/4" diameter		100	.080		25	4.70		29.70	34.50
4910	1-1/2" diameter		67	.119		25	7		32	38
4912	2" diameter		50	.160		31.50	9.40		40.90	48.50
4914	2-1/2" diameter	Q-1	80	.200		36	10.55		46.55	55.50
4916	3" diameter		67	.239		41.50	12.60		54.10	65
4918	4" diameter		50	.320		58	16.90		74.90	89.50
4920	5" diameter		40	.400		75	21		96	115
4922	6" diameter	Q-2	50	.480		99	26.50		125.50	149
4940	Flexible, standard, painted									
4950	3/4" diameter	1 Plum	100	.080	Ea.	17.80	4.70		22.50	26.50
4960	1" diameter		100	.080		17.80	4.70		22.50	26.50
4970	1-1/4" diameter		80	.100		23.50	5.85		29.35	34.50
4980	1-1/2" diameter		67	.119		25.50	7		32.50	38.50
4990	2" diameter		50	.160		27	9.40		36.40	44
5000	2-1/2" diameter	Q-1	80	.200		31.50	10.55		42.05	51
5010	3" diameter		67	.239		35	12.60		47.60	57.50

22 11 13.48 Pipe, Fittings and Valves, Steel, Grooved-Joint		Crew	Daily Output	Labor-Hours	Unit	Material	2015 Bare Costs Labor	Equipment	Total	Total Incl O&P
5020	3-1/2" diameter	Q-1	57	.281	Ea.	50	14.85		64.85	77.50
5030	4" diameter		50	.320		50.50	16.90		67.40	81
5040	5" diameter		40	.400		76.50	21		97.50	117
5050	6" diameter	Q-2	50	.480		90.50	26.50		117	139
5200	For galvanized couplings, add					33%				

22 11 13.64 Pipe, Stainless Steel

		Crew	Daily Output	Labor-Hours	Unit	Material	Labor	Equipment	Total	Total Incl O&P
0010	**PIPE, STAINLESS STEEL**									
3500	Threaded, couplings and clevis hanger assemblies, 10' O.C.									
3520	Schedule 40, type 304									
3540	1/4" diameter	1 Plum	54	.148	L.F.	10.85	8.70		19.55	25
3550	3/8" diameter		53	.151		11.05	8.85		19.90	25.50
3560	1/2" diameter		52	.154		12.85	9.05		21.90	28
3580	1" diameter		45	.178		19.30	10.45		29.75	37
3610	2" diameter	Q-1	57	.281		45	14.85		59.85	72
3640	4" diameter	Q-2	51	.471		138	26		164	190
3740	For small quantities, add					10%				
4250	Schedule 40, type 316									
4290	1/4" diameter	1 Plum	54	.148	L.F.	11.65	8.70		20.35	26
4300	3/8" diameter		53	.151		12.70	8.85		21.55	27.50
4310	1/2" diameter		52	.154		15.50	9.05		24.55	30.50
4320	3/4" diameter		51	.157		17.95	9.20		27.15	33.50
4330	1" diameter		45	.178		25	10.45		35.45	43.50
4360	2" diameter	Q-1	57	.281		53	14.85		67.85	81
4390	4" diameter	Q-2	51	.471		153	26		179	207
4490	For small quantities, add					10%				

22 11 13.74 Pipe, Plastic

		Crew	Daily Output	Labor-Hours	Unit	Material	Labor	Equipment	Total	Total Incl O&P
0010	**PIPE, PLASTIC**									
1800	PVC, couplings 10' O.C., clevis hanger assemblies, 3 per 10'									
1820	Schedule 40									
1860	1/2" diameter	1 Plum	54	.148	L.F.	4.88	8.70		13.58	18.50
1870	3/4" diameter		51	.157		5.20	9.20		14.40	19.65
1880	1" diameter		46	.174		5.85	10.20		16.05	22
1890	1-1/4" diameter		42	.190		6.55	11.20		17.75	24
1900	1-1/2" diameter		36	.222		6.85	13.05		19.90	27.50
1910	2" diameter	Q-1	59	.271		8.05	14.35		22.40	30.50
1920	2-1/2" diameter		56	.286		10.35	15.10		25.45	34.50
1930	3" diameter		53	.302		12.60	15.95		28.55	38
1940	4" diameter		48	.333		16.10	17.60		33.70	44
1950	5" diameter		43	.372		26.50	19.65		46.15	58.50
1960	6" diameter		39	.410		27.50	21.50		49	62.50
4100	DWV type, schedule 40, couplings 10' O.C., clevis hanger assy's, 3 per 10'									
4210	ABS, schedule 40, foam core type									
4212	Plain end black									
4214	1-1/2" diameter	1 Plum	39	.205	L.F.	5.20	12.05		17.25	24
4216	2" diameter	Q-1	62	.258		5.60	13.65		19.25	26.50
4218	3" diameter		56	.286		8	15.10		23.10	32
4220	4" diameter		51	.314		10.20	16.55		26.75	36
4222	6" diameter		42	.381		18	20		38	50.50
4240	To delete coupling & hangers, subtract									
4244	1-1/2" diam. to 6" diam.					43%	48%			
4400	PVC									
4410	1-1/4" diameter	1 Plum	42	.190	L.F.	5.35	11.20		16.55	23

22 11 13.74 Pipe, Plastic		Crew	Daily Output	Labor-Hours	Unit	2015 Bare Costs Material	Labor	Equipment	Total	Total Incl O&P
4420	1-1/2" diameter	1 Plum	36	.222	L.F.	5.30	13.05		18.35	25.50
4460	2" diameter	Q-1	59	.271		5.65	14.35		20	27.50
4470	3" diameter		53	.302		8.05	15.95		24	33
4480	4" diameter		48	.333		10	17.60		27.60	37.50
4490	6" diameter		39	.410		16.40	21.50		37.90	50.50
5300	CPVC, socket joint, couplings 10' O.C., clevis hanger assemblies, 3 per 10'									
5302	Schedule 40									
5304	1/2" diameter	1 Plum	54	.148	L.F.	5.90	8.70		14.60	19.65
5305	3/4" diameter		51	.157		6.75	9.20		15.95	21.50
5306	1" diameter		46	.174		8.10	10.20		18.30	24.50
5307	1-1/4" diameter		42	.190		9.65	11.20		20.85	27.50
5308	1-1/2" diameter		36	.222		10.85	13.05		23.90	31.50
5309	2" diameter	Q-1	59	.271		12.85	14.35		27.20	35.50
5310	2-1/2" diameter		56	.286		19.90	15.10		35	45
5311	3" diameter		53	.302		23.50	15.95		39.45	50
5360	CPVC, threaded, couplings 10' O.C., clevis hanger assemblies, 3 per 10'									
5380	Schedule 40									
5460	1/2" diameter	1 Plum	54	.148	L.F.	6.70	8.70		15.40	20.50
5470	3/4" diameter		51	.157		8.15	9.20		17.35	23
5480	1" diameter		46	.174		9.55	10.20		19.75	26
5490	1-1/4" diameter		42	.190		10.80	11.20		22	29
5500	1-1/2" diameter		36	.222		11.80	13.05		24.85	32.50
5510	2" diameter	Q-1	59	.271		14	14.35		28.35	37
5520	2-1/2" diameter		56	.286		21	15.10		36.10	46
5530	3" diameter		53	.302		25.50	15.95		41.45	52
7280	PEX, flexible, no couplings or hangers									
7282	Note: For labor costs add 25% to the couplings and fittings labor total.									
7285	For fittings see section 23 83 16.10 7000									
7300	Non-barrier type, hot/cold tubing rolls									
7310	1/4" diameter x 100'				L.F.	.49			.49	.54
7350	3/8" diameter x 100'					.55			.55	.61
7360	1/2" diameter x 100'					.61			.61	.67
7370	1/2" diameter x 500'					.61			.61	.67
7380	1/2" diameter x 1000'					.61			.61	.67
7400	3/4" diameter x 100'					1.11			1.11	1.22
7410	3/4" diameter x 500'					1.11			1.11	1.22
7420	3/4" diameter x 1000'					1.11			1.11	1.22
7460	1" diameter x 100'					1.90			1.90	2.09
7470	1" diameter x 300'					1.90			1.90	2.09
7480	1" diameter x 500'					1.90			1.90	2.09
7500	1-1/4" diameter x 100'					3.23			3.23	3.55
7510	1-1/4" diameter x 300'					3.23			3.23	3.55
7540	1-1/2" diameter x 100'					4.39			4.39	4.83
7550	1-1/2" diameter x 300'					4.39			4.39	4.83
7596	Most sizes available in red or blue									
7700	Non-barrier type, hot/cold tubing straight lengths									
7710	1/2" diameter x 20'				L.F.	.60			.60	.66
7750	3/4" diameter x 20'					1.10			1.10	1.21
7760	1" diameter x 20'					1.90			1.90	2.09
7770	1-1/4" diameter x 20'					3.23			3.23	3.55
7780	1-1/2" diameter x 20'					4.39			4.39	4.83
7790	2" diameter					8.60			8.60	9.45
7796	Most sizes available in red or blue									

22 11 19 – Domestic Water Piping Specialties

22 11 19.10 Flexible Connectors

Line	22 11 19.10 Flexible Connectors	Crew	Daily Output	Labor-Hours	Unit	2015 Bare Costs Material	2015 Bare Costs Labor	2015 Bare Costs Equipment	2015 Bare Costs Total	Total Incl O&P
0010	**FLEXIBLE CONNECTORS**, Corrugated, 7/8″ O.D., 1/2″ I.D.									
0050	Gas, seamless brass, steel fittings									
0200	12″ long	1 Plum	36	.222	Ea.	17.40	13.05		30.45	39
0220	18″ long		36	.222		21.50	13.05		34.55	43.50
0240	24″ long		34	.235		25.50	13.80		39.30	49
0280	36″ long		32	.250		30.50	14.70		45.20	55.50
0340	60″ long		30	.267		46	15.65		61.65	74
2000	Water, copper tubing, dielectric separators									
2100	12″ long	1 Plum	36	.222	Ea.	17.35	13.05		30.40	39
2260	24″ long	″	34	.235	″	26	13.80		39.80	49.50

22 11 19.14 Flexible Metal Hose

Line	22 11 19.14 Flexible Metal Hose	Crew	Daily Output	Labor-Hours	Unit	Material	Labor	Equipment	Total	Total Incl O&P
0010	**FLEXIBLE METAL HOSE**, Connectors, standard lengths									
0100	Bronze braided, bronze ends									
0120	3/8″ diameter x 12″	1 Stpi	26	.308	Ea.	21	18.40		39.40	51
0160	3/4″ diameter x 12″		20	.400		32	24		56	71
0180	1″ diameter x 18″		19	.421		38	25		63	80
0200	1-1/2″ diameter x 18″		13	.615		52	37		89	113
0220	2″ diameter x 18″		11	.727		68	43.50		111.50	141

22 11 19.26 Pressure Regulators

Line	22 11 19.26 Pressure Regulators	Crew	Daily Output	Labor-Hours	Unit	Material	Labor	Equipment	Total	Total Incl O&P
0010	**PRESSURE REGULATORS**									
3000	Steam, high capacity, bronze body, stainless steel trim									
3020	Threaded, 1/2″ diameter	1 Stpi	24	.333	Ea.	1,975	19.90		1,994.90	2,200
3030	3/4″ diameter		24	.333		2,150	19.90		2,169.90	2,375
3040	1″ diameter		19	.421		2,400	25		2,425	2,700
3060	1-1/4″ diameter		15	.533		2,500	32		2,532	2,800
3080	1-1/2″ diameter		13	.615		3,025	37		3,062	3,375
3100	2″ diameter		11	.727		3,725	43.50		3,768.50	4,150
3120	2-1/2″ diameter	Q-5	12	1.333		4,650	71.50		4,721.50	5,225
3140	3″ diameter	″	11	1.455		5,075	78		5,153	5,700
3500	Flanged connection, iron body, 125 lb. W.S.P.									
3520	3″ diameter	Q-5	11	1.455	Ea.	5,800	78		5,878	6,525
3540	4″ diameter	″	5	3.200	″	7,325	172		7,497	8,300

22 11 19.38 Water Supply Meters

Line	22 11 19.38 Water Supply Meters	Crew	Daily Output	Labor-Hours	Unit	Material	Labor	Equipment	Total	Total Incl O&P
0010	**WATER SUPPLY METERS**									
2000	Domestic/commercial, bronze									
2020	Threaded									
2060	5/8″ diameter, to 20 GPM	1 Plum	16	.500	Ea.	50	29.50		79.50	99.50
2080	3/4″ diameter, to 30 GPM		14	.571		91	33.50		124.50	151
2100	1″ diameter, to 50 GPM		12	.667		138	39		177	211
2300	Threaded/flanged									
2340	1-1/2″ diameter, to 100 GPM	1 Plum	8	1	Ea.	340	58.50		398.50	460
2360	2″ diameter, to 160 GPM	″	6	1.333	″	460	78.50		538.50	625
2600	Flanged, compound									
2640	3″ diameter, 320 GPM	Q-1	3	5.333	Ea.	3,125	282		3,407	3,850
2660	4″ diameter, to 500 GPM		1.50	10.667		5,000	565		5,565	6,350
2680	6″ diameter, to 1,000 GPM		1	16		7,975	845		8,820	10,000
2700	8″ diameter, to 1,800 GPM		.80	20		12,500	1,050		13,550	15,300

22 11 19.42 Backflow Preventers

Line	22 11 19.42 Backflow Preventers	Crew	Daily Output	Labor-Hours	Unit	Material	Labor	Equipment	Total	Total Incl O&P
0010	**BACKFLOW PREVENTERS**, Includes valves									
0020	and four test cocks, corrosion resistant, automatic operation									
4000	Reduced pressure principle									

22 11 Facility Water Distribution

22 11 19 – Domestic Water Piping Specialties

22 11 19.42 Backflow Preventers		Crew	Daily Output	Labor-Hours	Unit	2015 Bare Costs Material	Labor	Equipment	Total	Total Incl O&P
4100	Threaded, bronze, valves are ball									
4120	3/4" pipe size	1 Plum	16	.500	Ea.	445	29.50		474.50	535
4140	1" pipe size		14	.571		480	33.50		513.50	575
4150	1-1/4" pipe size		12	.667		845	39		884	990
4160	1-1/2" pipe size		10	.800		960	47		1,007	1,125
4180	2" pipe size		7	1.143		1,075	67		1,142	1,300
5000	Flanged, bronze, valves are OS&Y									
5060	2-1/2" pipe size	Q-1	5	3.200	Ea.	4,025	169		4,194	4,675
5080	3" pipe size		4.50	3.556		4,650	188		4,838	5,375
5100	4" pipe size		3	5.333		5,500	282		5,782	6,475
5120	6" pipe size	Q-2	3	8		8,750	440		9,190	10,300
5600	Flanged, iron, valves are OS&Y									
5660	2-1/2" pipe size	Q-1	5	3.200	Ea.	3,025	169		3,194	3,575
5680	3" pipe size		4.50	3.556		3,175	188		3,363	3,775
5700	4" pipe size		3	5.333		3,975	282		4,257	4,800
5720	6" pipe size	Q-2	3	8		5,775	440		6,215	7,000
5740	8" pipe size		2	12		10,100	655		10,755	12,100
5760	10" pipe size		1	24		13,600	1,325		14,925	16,900

22 11 19.50 Vacuum Breakers		Crew	Daily Output	Labor-Hours	Unit	Material	Labor	Equipment	Total	Total Incl O&P
0010	**VACUUM BREAKERS**									
0013	See also backflow preventers Section 22 11 19.42									
1000	Anti-siphon continuous pressure type									
1010	Max. 150 PSI - 210°F									
1020	Bronze body									
1030	1/2" size	1 Stpi	24	.333	Ea.	187	19.90		206.90	236
1040	3/4" size		20	.400		187	24		211	242
1050	1" size		19	.421		194	25		219	251
1060	1-1/4" size		15	.533		380	32		412	470
1070	1-1/2" size		13	.615		470	37		507	570
1080	2" size		11	.727		485	43.50		528.50	595
1200	Max. 125 PSI with atmospheric vent									
1210	Brass, in-line construction									
1220	1/4" size	1 Stpi	24	.333	Ea.	117	19.90		136.90	159
1230	3/8" size	"	24	.333		117	19.90		136.90	159
1260	For polished chrome finish, add					13%				
2000	Anti-siphon, non-continuous pressure type									
2010	Hot or cold water 125 PSI - 210°F									
2020	Bronze body									
2030	1/4" size	1 Stpi	24	.333	Ea.	64	19.90		83.90	101
2040	3/8" size		24	.333		64	19.90		83.90	101
2050	1/2" size		24	.333		72.50	19.90		92.40	110
2060	3/4" size		20	.400		86.50	24		110.50	131
2070	1" size		19	.421		134	25		159	185
2080	1-1/4" size		15	.533		235	32		267	305
2090	1-1/2" size		13	.615		276	37		313	360
2100	2" size		11	.727		430	43.50		473.50	535
2110	2-1/2" size		8	1		1,225	60		1,285	1,450
2120	3" size		6	1.333		1,625	79.50		1,704.50	1,925
2150	For polished chrome finish, add					50%				

22 11 19 – Domestic Water Piping Specialties

22 11 19.54 Water Hammer Arresters/Shock Absorbers		Crew	Daily Output	Labor-Hours	Unit	Material	2015 Bare Costs Labor	Equipment	Total	Total Incl O&P
0010	**WATER HAMMER ARRESTERS/SHOCK ABSORBERS**									
0490	Copper									
0500	3/4" male I.P.S. For 1 to 11 fixtures	1 Plum	12	.667	Ea.	28	39		67	90
0600	1" male I.P.S. For 12 to 32 fixtures		8	1		45.50	58.50		104	139
0700	1-1/4" male I.P.S. For 33 to 60 fixtures		8	1		47	58.50		105.50	140
0800	1-1/2" male I.P.S. For 61 to 113 fixtures		8	1		67.50	58.50		126	163
0900	2" male I.P.S. For 114 to 154 fixtures		8	1		98.50	58.50		157	197
1000	2-1/2" male I.P.S. For 155 to 330 fixtures		4	2		305	117		422	510

22 11 19.64 Hydrants

		Crew	Daily Output	Labor-Hours	Unit	Material	Labor	Equipment	Total	Total Incl O&P
0010	**HYDRANTS**									
0050	Wall type, moderate climate, bronze, encased									
0200	3/4" IPS connection	1 Plum	16	.500	Ea.	745	29.50		774.50	860
0300	1" IPS connection		14	.571		850	33.50		883.50	985
0500	Anti-siphon type, 3/4" connection		16	.500		640	29.50		669.50	750
1000	Non-freeze, bronze, exposed									
1100	3/4" IPS connection, 4" to 9" thick wall	1 Plum	14	.571	Ea.	500	33.50		533.50	600
1120	10" to 14" thick wall		12	.667		545	39		584	660
1140	15" to 19" thick wall		12	.667		605	39		644	725
1160	20" to 24" thick wall		10	.800		640	47		687	775
1200	For 1" IPS connection, add					15%	10%			
1240	For 3/4" adapter type vacuum breaker, add				Ea.	63			63	69.50
1280	For anti-siphon type, add				"	132			132	145
2000	Non-freeze bronze, encased, anti-siphon type									
2100	3/4" IPS connection, 5" to 9" thick wall	1 Plum	14	.571	Ea.	1,275	33.50		1,308.50	1,450
2120	10" to 14" thick wall		12	.667		1,300	39		1,339	1,500
2140	15" to 19" thick wall		12	.667		1,375	39		1,414	1,550
3000	Ground box type, bronze frame, 3/4" IPS connection									
3080	Non-freeze, all bronze, polished face, set flush									
3100	2 feet depth of bury	1 Plum	8	1	Ea.	950	58.50		1,008.50	1,150
3140	4 feet depth of bury		8	1		1,075	58.50		1,133.50	1,300
3180	6 feet depth of bury		7	1.143		1,250	67		1,317	1,475
3220	8 feet depth of bury		5	1.600		1,375	94		1,469	1,675
3400	For 1" IPS connection, add					15%	10%			
3550	For 2" connection, add					445%	24%			
3600	For tapped drain port in box, add					86			86	94.50
5000	Moderate climate, all bronze, polished face									
5020	and scoriated cover, set flush									
5100	3/4" IPS connection	1 Plum	16	.500	Ea.	655	29.50		684.50	770
5120	1" IPS connection	"	14	.571		810	33.50		843.50	940
5200	For tapped drain port in box, add					86			86	94.50

22 11 23 – Domestic Water Pumps

22 11 23.10 General Utility Pumps

		Crew	Daily Output	Labor-Hours	Unit	Material	Labor	Equipment	Total	Total Incl O&P
0010	**GENERAL UTILITY PUMPS**									
2000	Single stage									
3000	Double suction,									
3190	75 HP, to 2500 GPM	Q-3	.28	114	Ea.	20,100	6,375		26,475	31,800
3220	100 HP, to 3000 GPM		.26	123		25,500	6,875		32,375	38,500
3240	150 HP, to 4000 GPM		.24	133		36,000	7,450		43,450	51,000

22 13 16.20 Pipe, Cast Iron

	22 13 16.20 Pipe, Cast Iron	Crew	Daily Output	Labor-Hours	Unit	Material	2015 Bare Costs Labor	Equipment	Total	Total Incl O&P
0010	**PIPE, CAST IRON**, Soil, on clevis hanger assemblies, 5′ O.C.									
0020	Single hub, service wt., lead & oakum joints 10′ O.C.									
2120	2″ diameter	Q-1	63	.254	L.F.	11.35	13.40		24.75	33
2140	3″ diameter		60	.267		15	14.10		29.10	38
2160	4″ diameter		55	.291		18.55	15.35		33.90	43.50
2180	5″ diameter	Q-2	76	.316		25.50	17.30		42.80	54
2200	6″ diameter	″	73	.329		31	18		49	61.50
2220	8″ diameter	Q-3	59	.542		47	30.50		77.50	98
2240	10″ diameter		54	.593		74.50	33		107.50	132
2260	12″ diameter		48	.667		105	37.50		142.50	173
2320	For service weight, double hub, add					10%				
2340	For extra heavy, single hub, add					48%	4%			
2360	For extra heavy, double hub, add					71%	4%			
2400	Lead for caulking, (1#/diam. in.)	Q-1	160	.100	Lb.	1.04	5.30		6.34	9.15
2420	Oakum for caulking, (1/8#/diam. in.)	″	40	.400	″	3.60	21		24.60	36
4000	No hub, couplings 10′ O.C.									
4100	1-1/2″ diameter	Q-1	71	.225	L.F.	11.05	11.90		22.95	30
4120	2″ diameter		67	.239		11.25	12.60		23.85	31.50
4140	3″ diameter		64	.250		15.20	13.20		28.40	36.50
4160	4″ diameter		58	.276		19.10	14.55		33.65	43

22 13 16.50 Shower Drains

	22 13 16.50 Shower Drains	Crew	Daily Output	Labor-Hours	Unit	Material	Labor	Equipment	Total	Total Incl O&P
0010	**SHOWER DRAINS**									
2780	Shower, with strainer, uniform diam. trap, bronze top									
2800	2″ and 3″ pipe size	Q-1	8	2	Ea.	480	106		586	690
2820	4″ pipe size	″	7	2.286		485	121		606	710
2840	For galvanized body, add					189			189	208

22 13 16.60 Traps

	22 13 16.60 Traps	Crew	Daily Output	Labor-Hours	Unit	Material	Labor	Equipment	Total	Total Incl O&P
0010	**TRAPS**									
0030	Cast iron, service weight									
0050	Running P trap, without vent									
1100	2″	Q-1	16	1	Ea.	143	53		196	237
1140	3″		14	1.143		143	60.50		203.50	248
1150	4″		13	1.231		143	65		208	255
1160	6″	Q-2	17	1.412		635	77.50		712.50	810
1180	Running trap, single hub, with vent									
2080	3″ pipe size, 3″ vent	Q-1	14	1.143	Ea.	118	60.50		178.50	220
2120	4″ pipe size, 4″ vent	″	13	1.231		154	65		219	267
2300	For double hub, vent, add					10%	20%			
3000	P trap, B&S, 2″ pipe size	Q-1	16	1		34	53		87	118
3040	3″ pipe size	″	14	1.143		50.50	60.50		111	147
3350	Deep seal trap, B&S									
3400	1-1/4″ pipe size	Q-1	14	1.143	Ea.	53	60.50		113.50	149
3410	1-1/2″ pipe size		14	1.143		53	60.50		113.50	149
3420	2″ pipe size		14	1.143		49	60.50		109.50	145
3440	3″ pipe size		12	1.333		62.50	70.50		133	175
4700	Copper, drainage, drum trap									
4800	3″ x 5″ solid, 1-1/2″ pipe size	1 Plum	16	.500	Ea.	106	29.50		135.50	162
4840	3″ x 6″ swivel, 1-1/2″ pipe size	″	16	.500	″	160	29.50		189.50	221
5100	P trap, standard pattern									
5200	1-1/4″ pipe size	1 Plum	18	.444	Ea.	78.50	26		104.50	126
5240	1-1/2″ pipe size		17	.471		72	27.50		99.50	121
5260	2″ pipe size		15	.533		111	31.50		142.50	170

22 13 Facility Sanitary Sewerage

22 13 16 – Sanitary Waste and Vent Piping

22 13 16.60 Traps		Crew	Daily Output	Labor-Hours	Unit	Material	2015 Bare Costs Labor	Equipment	Total	Total Incl O&P
5280	3″ pipe size	1 Plum	11	.727	Ea.	281	42.50		323.50	375
5340	With cleanout, swivel joint and slip joint									
5360	1-1/4″ pipe size	1 Plum	18	.444	Ea.	99.50	26		125.50	149
5400	1-1/2″ pipe size	″	17	.471	″	106	27.50		133.50	159

22 13 16.80 Vent Flashing and Caps

		Crew	Daily Output	Labor-Hours	Unit	Material	Labor	Equipment	Total	Total Incl O&P
0010	**VENT FLASHING AND CAPS**									
0120	Vent caps									
0140	Cast iron									
0180	2-1/2″ - 3-5/8″ pipe	1 Plum	21	.381	Ea.	45	22.50		67.50	83.50
0190	4″ - 4-1/8″ pipe	″	19	.421	″	55	24.50		79.50	98
0900	Vent flashing									
1000	Aluminum with lead ring									
1020	1-1/4″ pipe	1 Plum	20	.400	Ea.	8	23.50		31.50	44.50
1030	1-1/2″ pipe		20	.400		8.45	23.50		31.95	45
1040	2″ pipe		18	.444		8.60	26		34.60	49
1050	3″ pipe		17	.471		9.55	27.50		37.05	52
1060	4″ pipe		16	.500		11.50	29.50		41	57
1350	Copper with neoprene ring									
1400	1-1/4″ pipe	1 Plum	20	.400	Ea.	61.50	23.50		85	103
1430	1-1/2″ pipe		20	.400		61.50	23.50		85	103
1440	2″ pipe		18	.444		61.50	26		87.50	107
1450	3″ pipe		17	.471		74.50	27.50		102	124
1460	4″ pipe		16	.500		74.50	29.50		104	127

22 13 19 – Sanitary Waste Piping Specialties

22 13 19.13 Sanitary Drains

		Crew	Daily Output	Labor-Hours	Unit	Material	Labor	Equipment	Total	Total Incl O&P
0010	**SANITARY DRAINS**									
0400	Deck, auto park, C.I., 13″ top									
0440	3″, 4″, 5″, and 6″ pipe size	Q-1	8	2	Ea.	1,450	106		1,556	1,725
0480	For galvanized body, add				″	780			780	855
2000	Floor, medium duty, C.I., deep flange, 7″ diam. top									
2040	2″ and 3″ pipe size	Q-1	12	1.333	Ea.	208	70.50		278.50	335
2080	For galvanized body, add					96.50			96.50	106
2120	With polished bronze top					315			315	345
2400	Heavy duty, with sediment bucket, C.I., 12″ diam. loose grate									
2420	2″, 3″, 4″, 5″, and 6″ pipe size	Q-1	9	1.778	Ea.	690	94		784	900
2460	With polished bronze top				″	975			975	1,075
2500	Heavy duty, cleanout & trap w/bucket, C.I., 15″ top									
2540	2″, 3″, and 4″ pipe size	Q-1	6	2.667	Ea.	6,575	141		6,716	7,450
2560	For galvanized body, add					1,675			1,675	1,850
2580	With polished bronze top					7,300			7,300	8,025

22 13 23 – Sanitary Waste Interceptors

22 13 23.10 Interceptors

		Crew	Daily Output	Labor-Hours	Unit	Material	Labor	Equipment	Total	Total Incl O&P
0010	**INTERCEPTORS**									
0150	Grease, fabricated steel, 4 GPM, 8 lb. fat capacity	1 Plum	4	2	Ea.	1,150	117		1,267	1,450
0200	7 GPM, 14 lb. fat capacity		4	2		1,600	117		1,717	1,950
1000	10 GPM, 20 lb. fat capacity		4	2		1,875	117		1,992	2,250
1040	15 GPM, 30 lb. fat capacity		4	2		2,800	117		2,917	3,250
1060	20 GPM, 40 lb. fat capacity		3	2.667		3,425	157		3,582	3,975
1120	50 GPM, 100 lb. fat capacity	Q-1	2	8		6,300	425		6,725	7,575
1160	100 GPM, 200 lb. fat capacity	″	2	8		15,100	425		15,525	17,200
1580	For seepage pan, add					7%				

22 13 Facility Sanitary Sewerage

22 13 23 – Sanitary Waste Interceptors

22 13 23.10 Interceptors		Crew	Daily Output	Labor-Hours	Unit	Material	2015 Bare Costs Labor	Equipment	Total	Total Incl O&P
3000	Hair, cast iron, 1-1/4" and 1-1/2" pipe connection	1 Plum	8	1	Ea.	435	58.50		493.50	565
3100	For chrome-plated cast iron, add					266			266	293
4000	Oil, fabricated steel, 10 GPM, 2" pipe size	1 Plum	4	2		2,575	117		2,692	3,000
4100	15 GPM, 2" or 3" pipe size		4	2		3,525	117		3,642	4,075
4120	20 GPM, 2" or 3" pipe size		3	2.667		4,625	157		4,782	5,300
4220	100 GPM, 3" pipe size	Q-1	2	8		14,300	425		14,725	16,300
6000	Solids, precious metals recovery, C.I., 1-1/4" to 2" pipe	1 Plum	4	2		655	117		772	895
6100	Dental Lab., large, C.I., 1-1/2" to 2" pipe	"	3	2.667		2,275	157		2,432	2,750

22 13 29 – Sanitary Sewerage Pumps

22 13 29.13 Wet-Pit-Mounted, Vertical Sewerage Pumps

		Crew	Daily Output	Labor-Hours	Unit	Material	Labor	Equipment	Total	Total Incl O&P
0010	**WET-PIT-MOUNTED, VERTICAL SEWERAGE PUMPS**									
0020	Controls incl. alarm/disconnect panel w/wire. Excavation not included									
0260	Simplex, 9 GPM at 60 PSIG, 91 gal. tank				Ea.	3,325			3,325	3,675
0300	Unit with manway, 26" I.D., 18" high					3,700			3,700	4,075
0340	26" I.D., 36" high					3,750			3,750	4,125
0380	43" I.D., 4' high					3,975			3,975	4,375
3000	Indoor residential type installation									
3020	Simplex, 9 GPM at 60 PSIG, 91 gal. HDPE tank				Ea.	3,350			3,350	3,675

22 13 29.14 Sewage Ejector Pumps

		Crew	Daily Output	Labor-Hours	Unit	Material	Labor	Equipment	Total	Total Incl O&P
0010	**SEWAGE EJECTOR PUMPS**, With operating and level controls									
0100	Simplex system incl. tank, cover, pump 15' head									
0500	37 gal. PE tank, 12 GPM, 1/2 HP, 2" discharge	Q-1	3.20	5	Ea.	480	264		744	925
0510	3" discharge		3.10	5.161		520	273		793	980
0530	87 GPM, .7 HP, 2" discharge		3.20	5		735	264		999	1,200
0540	3" discharge		3.10	5.161		795	273		1,068	1,275
0600	45 gal. coated stl. tank, 12 GPM, 1/2 HP, 2" discharge		3	5.333		855	282		1,137	1,375
0610	3" discharge		2.90	5.517		890	291		1,181	1,425
0630	87 GPM, .7 HP, 2" discharge		3	5.333		1,100	282		1,382	1,625
0640	3" discharge		2.90	5.517		1,150	291		1,441	1,725
0660	134 GPM, 1 HP, 2" discharge		2.80	5.714		1,175	300		1,475	1,750
0680	3" discharge		2.70	5.926		1,250	315		1,565	1,850
0700	70 gal. PE tank, 12 GPM, 1/2 HP, 2" discharge		2.60	6.154		920	325		1,245	1,525
0710	3" discharge		2.40	6.667		980	350		1,330	1,600
0730	87 GPM, 0.7 HP, 2" discharge		2.50	6.400		1,200	340		1,540	1,800
0740	3" discharge		2.30	6.957		1,275	370		1,645	1,950
0760	134 GPM, 1 HP, 2" discharge		2.20	7.273		1,300	385		1,685	2,000
0770	3" discharge		2	8		1,375	425		1,800	2,175

22 14 Facility Storm Drainage

22 14 23 – Storm Drainage Piping Specialties

22 14 23.33 Backwater Valves

		Crew	Daily Output	Labor-Hours	Unit	Material	Labor	Equipment	Total	Total Incl O&P
0010	**BACKWATER VALVES**, C.I. Body									
6980	Bronze gate and automatic flapper valves									
7000	3" and 4" pipe size	Q-1	13	1.231	Ea.	2,050	65		2,115	2,350
7100	5" and 6" pipe size	"	13	1.231	"	3,125	65		3,190	3,550
7240	Bronze flapper valve, bolted cover									
7260	2" pipe size	Q-1	16	1	Ea.	595	53		648	735
7300	4" pipe size	"	13	1.231		1,150	65		1,215	1,375
7340	6" pipe size	Q-2	17	1.412		1,650	77.50		1,727.50	1,950

22 14 26 – Facility Storm Drains

22 14 26.13 Roof Drains		Crew	Daily Output	Labor-Hours	Unit	Material	2015 Bare Costs Labor	Equipment	Total	Total Incl O&P
0010	ROOF DRAINS									
0140	Cornice, C.I., 45° or 90° outlet									
0200	3" and 4" pipe size	Q-1	12	1.333	Ea.	330	70.50		400.50	470
0260	For galvanized body, add					75.50			75.50	83
0280	For polished bronze dome, add				↓	85.50			85.50	94
3860	Roof, flat metal deck, C.I. body, 12" C.I. dome									
3890	3" pipe size	Q-1	14	1.143	Ea.	395	60.50		455.50	525
3920	6" pipe size	"	10	1.600	"	680	84.50		764.50	880
4620	Main, all aluminum, 12" low profile dome									
4640	2", 3" and 4" pipe size	Q-1	14	1.143	Ea.	435	60.50		495.50	570

22 14 26.16 Facility Area Drains		Crew	Daily Output	Labor-Hours	Unit	Material	Labor	Equipment	Total	Total Incl O&P
0010	FACILITY AREA DRAINS									
4980	Scupper floor, oblique strainer, C.I.									
5000	6" x 7" top, 2", 3" and 4" pipe size	Q-1	16	1	Ea.	283	53		336	390
5100	8" x 12" top, 5" and 6" pipe size	"	14	1.143		550	60.50		610.50	695
5160	For galvanized body, add					40%				
5200	For polished bronze strainer, add				↓	85%				

22 14 26.19 Facility Trench Drains		Crew	Daily Output	Labor-Hours	Unit	Material	Labor	Equipment	Total	Total Incl O&P
0010	FACILITY TRENCH DRAINS									
5980	Trench, floor, heavy duty, modular, C.I., 12" x 12" top									
6000	2", 3", 4", 5", & 6" pipe size	Q-1	8	2	Ea.	895	106		1,001	1,150
6100	For unit with polished bronze top	"	8	2	"	1,325	106		1,431	1,625
6600	Trench, floor, for cement concrete encasement									
6610	Not including trenching or concrete									
6640	Polyester polymer concrete									
6650	4" internal width, with grate									
6660	Light duty steel grate	Q-1	120	.133	L.F.	35	7.05		42.05	49
6670	Medium duty steel grate		115	.139		40.50	7.35		47.85	55.50
6680	Heavy duty iron grate	↓	110	.145	↓	62	7.70		69.70	79.50
6700	12" internal width, with grate									
6770	Heavy duty galvanized grate	Q-1	80	.200	L.F.	169	10.55		179.55	201
6800	Fiberglass									
6810	8" internal width, with grate									
6820	Medium duty galvanized grate	Q-1	115	.139	L.F.	111	7.35		118.35	133
6830	Heavy duty iron grate	"	110	.145	"	106	7.70		113.70	128

22 14 29 – Sump Pumps

22 14 29.13 Wet-Pit-Mounted, Vertical Sump Pumps		Crew	Daily Output	Labor-Hours	Unit	Material	Labor	Equipment	Total	Total Incl O&P
0010	WET-PIT-MOUNTED, VERTICAL SUMP PUMPS									
0400	Molded PVC base, 21 GPM at 15' head, 1/3 HP	1 Plum	5	1.600	Ea.	135	94		229	291
0800	Iron base, 21 GPM at 15' head, 1/3 HP		5	1.600		164	94		258	320
1200	Solid brass, 21 GPM at 15' head, 1/3 HP	↓	5	1.600	↓	289	94		383	460
2000	Sump pump, single stage									
2010	25 GPM, 1 HP, 1-1/2" discharge	Q-1	1.80	8.889	Ea.	3,825	470		4,295	4,925
2020	75 GPM, 1-1/2 HP, 2" discharge		1.50	10.667		4,050	565		4,615	5,300
2030	100 GPM, 2 HP, 2-1/2" discharge		1.30	12.308		4,125	650		4,775	5,525
2040	150 GPM, 3 HP, 3" discharge		1.10	14.545		4,125	770		4,895	5,700
2050	200 GPM, 3 HP, 3" discharge	↓	1	16		4,375	845		5,220	6,100
2060	300 GPM, 10 HP, 4" discharge	Q-2	1.20	20		4,725	1,100		5,825	6,850
2070	500 GPM, 15 HP, 5" discharge		1.10	21.818		5,375	1,200		6,575	7,700
2080	800 GPM, 20 HP, 6" discharge		1	24		6,350	1,325		7,675	8,950
2090	1000 GPM, 30 HP, 6" discharge	↓	.85	28.235	↓	6,975	1,550		8,525	10,000

22 14 Facility Storm Drainage

22 14 29 – Sump Pumps

22 14 29.13 Wet-Pit-Mounted, Vertical Sump Pumps		Crew	Daily Output	Labor-Hours	Unit	Material	2015 Bare Costs Labor	Equipment	Total	Total Incl O&P
2100	1600 GPM, 50 HP, 8" discharge	Q-2	.72	33.333	Ea.	10,900	1,825		12,725	14,800
2110	2000 GPM, 60 HP, 8" discharge	Q-3	.85	37.647		11,100	2,100		13,200	15,400
2202	For general purpose float switch, copper coated float, add	Q-1	5	3.200		108	169		277	375

22 14 29.16 Submersible Sump Pumps

		Crew	Daily Output	Labor-Hours	Unit	Material	Labor	Equipment	Total	Total Incl O&P
0010	**SUBMERSIBLE SUMP PUMPS**									
7000	Sump pump, automatic									
7100	Plastic, 1-1/4" discharge, 1/4 HP	1 Plum	6.40	1.250	Ea.	138	73.50		211.50	263
7140	1/3 HP		6	1.333		200	78.50		278.50	340
7160	1/2 HP		5.40	1.481		246	87		333	400
7180	1-1/2" discharge, 1/2 HP		5.20	1.538		281	90.50		371.50	445
7500	Cast iron, 1-1/4" discharge, 1/4 HP		6	1.333		194	78.50		272.50	330
7540	1/3 HP		6	1.333		229	78.50		307.50	370
7560	1/2 HP		5	1.600		277	94		371	445

22 31 Domestic Water Softeners

22 31 13 – Residential Domestic Water Softeners

22 31 13.10 Residential Water Softeners

		Crew	Daily Output	Labor-Hours	Unit	Material	Labor	Equipment	Total	Total Incl O&P
0010	**RESIDENTIAL WATER SOFTENERS**									
7350	Water softener, automatic, to 30 grains per gallon	2 Plum	5	3.200	Ea.	405	188		593	730
7400	To 100 grains per gallon	"	4	4	"	660	235		895	1,075

22 31 16 – Commercial Domestic Water Softeners

22 31 16.10 Water Softeners

		Crew	Daily Output	Labor-Hours	Unit	Material	Labor	Equipment	Total	Total Incl O&P
0010	**WATER SOFTENERS**									
5800	Softener systems, automatic, intermediate sizes									
5820	available, may be used in multiples.									
6000	Hardness capacity between regenerations and flow									
6100	150,000 grains, 37 GPM cont., 51 GPM peak	Q-1	1.20	13.333	Ea.	6,075	705		6,780	7,750
6200	300,000 grains, 81 GPM cont., 113 GPM peak		1	16		9,850	845		10,695	12,100
6300	750,000 grains, 160 GPM cont., 230 GPM peak		.80	20		12,800	1,050		13,850	15,700
6400	900,000 grains, 185 GPM cont., 270 GPM peak		.70	22.857		20,700	1,200		21,900	24,500

22 33 Electric Domestic Water Heaters

22 33 13 – Instantaneous Electric Domestic Water Heaters

22 33 13.10 Hot Water Dispensers

		Crew	Daily Output	Labor-Hours	Unit	Material	Labor	Equipment	Total	Total Incl O&P
0010	**HOT WATER DISPENSERS**									
0160	Commercial, 100 cup, 11.3 amp	1 Plum	14	.571	Ea.	510	33.50		543.50	615
3180	Household, 60 cup	"	14	.571	"	269	33.50		302.50	345

22 33 30 – Residential, Electric Domestic Water Heaters

22 33 30.13 Residential, Small-Capacity Elec. Water Heaters

		Crew	Daily Output	Labor-Hours	Unit	Material	Labor	Equipment	Total	Total Incl O&P
0010	**RESIDENTIAL, SMALL-CAPACITY ELECTRIC DOMESTIC WATER HEATERS**									
1000	Residential, electric, glass lined tank, 5 yr., 10 gal., single element	1 Plum	2.30	3.478	Ea.	330	204		534	670
1040	20 gallon, single element		2.20	3.636		410	213		623	770
1060	30 gallon, double element		2.20	3.636		475	213		688	845
1080	40 gallon, double element		2	4		800	235		1,035	1,225
1100	52 gallon, double element		2	4		895	235		1,130	1,350
1180	120 gallon, double element		1.40	5.714		1,900	335		2,235	2,575

22 33 Electric Domestic Water Heaters

22 33 33 – Light-Commercial Electric Domestic Water Heaters

22 33 33.10 Commercial Electric Water Heaters		Crew	Daily Output	Labor-Hours	Unit	Material	2015 Bare Costs Labor	Equipment	Total	Total Incl O&P
0010	COMMERCIAL ELECTRIC WATER HEATERS									
4000	Commercial, 100° rise. NOTE: for each size tank, a range of									
4010	heaters between the ones shown are available									
4020	Electric									
4100	5 gal., 3 kW, 12 GPH, 208 volt	1 Plum	2	4	Ea.	2,850	235		3,085	3,500
4120	10 gal., 6 kW, 25 GPH, 208 volt		2	4		3,175	235		3,410	3,825
4130	30 gal., 24 kW, 98 GPH, 208 volt		1.92	4.167		5,200	245		5,445	6,100
4136	40 gal., 36 kW, 148 GPH, 208 volt		1.88	4.255		6,225	250		6,475	7,225
4140	50 gal., 9 kW, 37 GPH, 208 volt		1.80	4.444		4,350	261		4,611	5,175
4160	50 gal., 36 kW, 148 GPH, 208 volt		1.80	4.444		6,650	261		6,911	7,700
4300	200 gal., 15 kW, 61 GPH, 480 volt	Q-1	1.70	9.412		20,500	495		20,995	23,300
4320	200 gal., 120 kW , 490 GPH, 480 volt		1.70	9.412		28,000	495		28,495	31,600
4460	400 gal., 30 kW, 123 GPH, 480 volt		1	16		27,800	845		28,645	31,900
5400	Modulating step control for under 90 kW, 2-5 steps	1 Elec	5.30	1.509		810	82.50		892.50	1,025
5440	1 through 5 steps beyond standard		3.20	2.500		221	137		358	450
5460	6 through 10 steps beyond standard		2.70	2.963		455	162		617	745
5480	11 through 18 steps beyond standard		1.60	5		680	274		954	1,150

22 34 Fuel-Fired Domestic Water Heaters

22 34 13 – Instantaneous, Tankless, Gas Domestic Water Heaters

22 34 13.10 Instantaneous, Tankless, Gas Water Heaters		Crew	Daily Output	Labor-Hours	Unit	Material	Labor	Equipment	Total	Total Incl O&P
0010	INSTANTANEOUS, TANKLESS, GAS WATER HEATERS									
9410	Natural gas/propane, 3.2 GPM G	1 Plum	2	4	Ea.	370	235		605	760
9420	6.4 GPM G		1.90	4.211		625	247		872	1,050
9430	8.4 GPM G		1.80	4.444		730	261		991	1,200
9440	9.5 GPM G		1.60	5		930	294		1,224	1,475

22 34 30 – Residential Gas Domestic Water Heaters

22 34 30.13 Residential, Atmos, Gas Domestic Wtr Heaters		Crew	Daily Output	Labor-Hours	Unit	Material	Labor	Equipment	Total	Total Incl O&P
0010	RESIDENTIAL, ATMOSPHERIC, GAS DOMESTIC WATER HEATERS									
2000	Gas fired, foam lined tank, 10 yr., vent not incl.									
2040	30 gallon	1 Plum	2	4	Ea.	895	235		1,130	1,350
2100	75 gallon		1.50	5.333		1,350	315		1,665	1,975
2120	100 gallon		1.30	6.154		1,600	360		1,960	2,325
2900	Water heater, safety-drain pan, 26" round		20	.400		37	23.50		60.50	76.50

22 34 36 – Commercial Gas Domestic Water Heaters

22 34 36.13 Commercial, Atmos., Gas Domestic Water Htrs.		Crew	Daily Output	Labor-Hours	Unit	Material	Labor	Equipment	Total	Total Incl O&P
0010	COMMERCIAL, ATMOSPHERIC, GAS DOMESTIC WATER HEATERS									
6000	Gas fired, flush jacket, std. controls, vent not incl.									
6040	75 MBH input, 73 GPH	1 Plum	1.40	5.714	Ea.	3,500	335		3,835	4,350
6060	98 MBH input, 95 GPH		1.40	5.714		5,300	335		5,635	6,325
6080	120 MBH input, 110 GPH		1.20	6.667		5,500	390		5,890	6,625
6180	200 MBH input, 192 GPH		.60	13.333		8,925	785		9,710	11,000
6200	250 MBH input, 245 GPH		.50	16		9,325	940		10,265	11,700
6900	For low water cutoff, add		8	1		350	58.50		408.50	475
6960	For bronze body hot water circulator, add		4	2		1,925	117		2,042	2,300

22 34 46 – Oil-Fired Domestic Water Heaters

22 34 46.10 Residential Oil-Fired Water Heaters		Crew	Daily Output	Labor-Hours	Unit	Material	Labor	Equipment	Total	Total Incl O&P
0010	RESIDENTIAL OIL-FIRED WATER HEATERS									
3000	Oil fired, glass lined tank, 5 yr., vent not included, 30 gallon	1 Plum	2	4	Ea.	1,175	235		1,410	1,625

22 34 Fuel-Fired Domestic Water Heaters

22 34 46 – Oil-Fired Domestic Water Heaters

22 34 46.10 Residential Oil-Fired Water Heaters		Crew	Daily Output	Labor-Hours	Unit	Material	2015 Bare Costs Labor	Equipment	Total	Total Incl O&P
3040	50 gallon	1 Plum	1.80	4.444	Ea.	1,375	261		1,636	1,925
3060	70 gallon	↓	1.50	5.333	↓	1,975	315		2,290	2,650

22 34 46.20 Commercial Oil-Fired Water Heaters		Crew	Daily Output	Labor-Hours	Unit	Material	Labor	Equipment	Total	Total Incl O&P
0010	**COMMERCIAL OIL-FIRED WATER HEATERS**									
8000	Oil fired, glass lined, UL listed, std. controls, vent not incl.									
8060	140 gal., 140 MBH input, 134 GPH	Q-1	2.13	7.512	Ea.	19,900	395		20,295	22,500
8080	140 gal., 199 MBH input, 191 GPH		2	8		20,600	425		21,025	23,300
8100	140 gal., 255 MBH input, 247 GPH		1.60	10		21,200	530		21,730	24,100
8160	140 gal., 540 MBH input, 519 GPH		.96	16.667		28,100	880		28,980	32,200
8180	140 gal., 720 MBH input, 691 GPH	↓	.92	17.391		28,600	920		29,520	32,900
8280	201 gal., 1250 MBH input, 1200 GPH	Q-2	1.22	19.672		43,900	1,075		44,975	49,900
8300	201 gal., 1500 MBH input, 1441 GPH	"	1.16	20.690		47,800	1,125		48,925	54,000
8900	For low water cutoff, add	1 Plum	8	1		350	58.50		408.50	475
8960	For bronze body hot water circulator, add	"	4	2	↓	710	117		827	955

22 35 Domestic Water Heat Exchangers

22 35 30 – Water Heating by Steam

22 35 30.10 Water Heating Transfer Package		Crew	Daily Output	Labor-Hours	Unit	Material	Labor	Equipment	Total	Total Incl O&P
0010	**WATER HEATING TRANSFER PACKAGE**, Complete controls,									
0020	expansion tank, converter, air separator									
1000	Hot water, 180°F enter, 200°F leaving, 15# steam									
1010	One pump system, 28 GPM	Q-6	.75	32	Ea.	20,000	1,775		21,775	24,800
1020	35 GPM		.70	34.286		21,800	1,900		23,700	26,900
1040	55 GPM		.65	36.923		25,800	2,050		27,850	31,400
1060	130 GPM		.55	43.636		32,500	2,425		34,925	39,500
1080	255 GPM		.40	60		42,900	3,350		46,250	52,000
1100	550 GPM	↓	.30	80	↓	58,000	4,450		62,450	70,500

22 41 Residential Plumbing Fixtures

22 41 06 – Plumbing Fixtures General

22 41 06.10 Plumbing Fixture Notes		Crew	Daily Output	Labor-Hours	Unit	Material	Labor	Equipment	Total	Total Incl O&P
0010	**PLUMBING FIXTURE NOTES**, Incl. trim fittings unless otherwise noted									
0080	For rough-in, supply, waste, and vent, see add for each type									
0122	For electric water coolers, see Section 22 47 16.10									
0160	For color, unless otherwise noted, add				Ea.	20%				

22 41 13 – Residential Water Closets, Urinals, and Bidets

22 41 13.13 Water Closets		Crew	Daily Output	Labor-Hours	Unit	Material	Labor	Equipment	Total	Total Incl O&P
0010	**WATER CLOSETS**									
0032	For automatic flush, see Line 22 42 39.10 0972									
0150	Tank type, vitreous china, incl. seat, supply pipe w/stop, 1.6 gpf or noted									
0200	Wall hung									
0400	Two piece, close coupled	Q-1	5.30	3.019	Ea.	630	159		789	935
0960	For rough-in, supply, waste, vent and carrier	"	2.73	5.861	"	1,025	310		1,335	1,600
0999	Floor mounted									
1100	Two piece, close coupled	Q-1	5.30	3.019	Ea.	237	159		396	500
1102	Economy		5.30	3.019		132	159		291	385
1110	Two piece, close coupled, dual flush		5.30	3.019		310	159		469	585
1140	Two piece, close coupled, 1.28 gpf, ADA G	↓	5.30	3.019	↓	310	159		469	585

22 41 Residential Plumbing Fixtures

22 41 13 – Residential Water Closets, Urinals, and Bidets

22 41 13.13 Water Closets		Crew	Daily Output	Labor-Hours	Unit	Material	Labor	Equipment	Total	Total Incl O&P
						2015 Bare Costs				
1960	For color, add					30%				
1980	For rough-in, supply, waste and vent	Q-1	3.05	5.246	Ea.	330	277		607	785

22 41 16 – Residential Lavatories and Sinks

22 41 16.13 Lavatories		Crew	Daily Output	Labor-Hours	Unit	Material	Labor	Equipment	Total	Total Incl O&P
0010	**LAVATORIES**, With trim, white unless noted otherwise									
0500	Vanity top, porcelain enamel on cast iron									
0600	20" x 18"	Q-1	6.40	2.500	Ea.	335	132		467	570
0640	33" x 19" oval		6.40	2.500		475	132		607	725
0720	19" round		6.40	2.500		435	132		567	680
0860	For color, add					25%				
1000	Cultured marble, 19" x 17", single bowl	Q-1	6.40	2.500	Ea.	175	132		307	390
1040	25" x 19", single bowl	"	6.40	2.500	"	206	132		338	425
1580	For color, same price									
1900	Stainless steel, self-rimming, 25" x 22", single bowl, ledge	Q-1	6.40	2.500	Ea.	365	132		497	605
1960	17" x 22", single bowl		6.40	2.500		355	132		487	590
2600	Steel, enameled, 20" x 17", single bowl		5.80	2.759		161	146		307	395
2660	19" round		5.80	2.759		171	146		317	410
2900	Vitreous china, 20" x 16", single bowl		5.40	2.963		260	157		417	520
2960	20" x 17", single bowl		5.40	2.963		176	157		333	430
3020	19" round, single bowl		5.40	2.963		174	157		331	430
3200	22" x 13", single bowl		5.40	2.963		267	157		424	530
3560	For color, add					50%				
3580	Rough-in, supply, waste and vent for all above lavatories	Q-1	2.30	6.957	Ea.	231	370		601	810
4000	Wall hung									
4040	Porcelain enamel on cast iron, 16" x 14", single bowl	Q-1	8	2	Ea.	520	106		626	730
4180	20" x 18", single bowl		8	2		277	106		383	465
4240	22" x 19", single bowl		8	2		700	106		806	930
4580	For color, add					30%				
6000	Vitreous china, 18" x 15", single bowl with backsplash	Q-1	7	2.286	Ea.	231	121		352	435
6500	For color, add					30%				
6960	Rough-in, supply, waste and vent for above lavatories	Q-1	1.66	9.639	Ea.	455	510		965	1,275
7000	Pedestal type									
7600	Vitreous china, 27" x 21", white	Q-1	6.60	2.424	Ea.	700	128		828	965
7610	27" x 21", colored		6.60	2.424		880	128		1,008	1,150
7620	27" x 21", premium color		6.60	2.424		995	128		1,123	1,300
7660	26" x 20", white		6.60	2.424		795	128		923	1,075
7670	26" x 20", colored		6.60	2.424		1,000	128		1,128	1,300
7680	26" x 20", premium color		6.60	2.424		1,125	128		1,253	1,450
7700	24" x 20", white		6.60	2.424		405	128		533	640
7710	24" x 20", colored		6.60	2.424		475	128		603	720
7720	24" x 20", premium color		6.60	2.424		495	128		623	740
7760	21" x 18", white		6.60	2.424		291	128		419	515
7770	21" x 18", colored		6.60	2.424		291	128		419	515
7990	Rough-in, supply, waste and vent for pedestal lavatories		1.66	9.639		455	510		965	1,275

22 41 16.16 Sinks		Crew	Daily Output	Labor-Hours	Unit	Material	Labor	Equipment	Total	Total Incl O&P
0010	**SINKS**, With faucets and drain									
2000	Kitchen, counter top style, P.E. on C.I., 24" x 21" single bowl	Q-1	5.60	2.857	Ea.	285	151		436	545
2100	31" x 22" single bowl		5.60	2.857		615	151		766	910
2200	32" x 21" double bowl		4.80	3.333		355	176		531	655
3000	Stainless steel, self rimming, 19" x 18" single bowl		5.60	2.857		590	151		741	880
3100	25" x 22" single bowl		5.60	2.857		660	151		811	955
4000	Steel, enameled, with ledge, 24" x 21" single bowl		5.60	2.857		510	151		661	790

22 41 Residential Plumbing Fixtures

22 41 16 – Residential Lavatories and Sinks

22 41 16.16 Sinks		Crew	Daily Output	Labor-Hours	Unit	Material	Labor	Equipment	Total	Total Incl O&P
						2015 Bare Costs				
4100	32″ x 21″ double bowl	Q-1	4.80	3.333	Ea.	495	176		671	810
4960	For color sinks except stainless steel, add					10%				
4980	For rough-in, supply, waste and vent, counter top sinks	Q-1	2.14	7.477		260	395		655	880
5000	Kitchen, raised deck, P.E. on C.I.									
5100	32″ x 21″, dual level, double bowl	Q-1	2.60	6.154	Ea.	420	325		745	950
5700	For color, add					20%				
5790	For rough-in, supply, waste & vent, sinks	Q-1	1.85	8.649		260	455		715	975

22 41 19 – Residential Bathtubs

22 41 19.10 Baths		Crew	Daily Output	Labor-Hours	Unit	Material	Labor	Equipment	Total	Total Incl O&P
0010	**BATHS**									
0100	Tubs, recessed porcelain enamel on cast iron, with trim									
0180	48″ x 42″	Q-1	4	4	Ea.	2,625	211		2,836	3,225
0220	72″ x 36″		3	5.333		2,725	282		3,007	3,425
2000	Enameled formed steel, 4′-6″ long		5.80	2.759		495	146		641	765
4000	Soaking, acrylic, w/pop-up drain 66″ x 36″ x 20″ deep		5.50	2.909		2,300	154		2,454	2,750
4100	60″ x 42″ x 20″ deep		5	3.200		1,175	169		1,344	1,550
9600	Rough-in, supply, waste and vent, for all above tubs, add		2.07	7.729		345	410		755	995

22 41 23 – Residential Showers

22 41 23.20 Showers		Crew	Daily Output	Labor-Hours	Unit	Material	Labor	Equipment	Total	Total Incl O&P
0010	**SHOWERS**									
1500	Stall, with drain only. Add for valve and door/curtain									
3000	Fiberglass, one piece, with 3 walls, 32″ x 32″ square	Q-1	5.50	2.909	Ea.	490	154		644	770
3100	36″ x 36″ square		5.50	2.909		505	154		659	785
3250	64″ x 65-3/4″ x 81-1/2″ fold. seat, whlchr.		3.80	4.211		2,325	222		2,547	2,875
4000	Polypropylene, stall only, w/molded-stone floor, 30″ x 30″		2	8		635	425		1,060	1,350
4200	Rough-in, supply, waste and vent for above showers		2.05	7.805		345	410		755	1,000

22 41 23.40 Shower System Components		Crew	Daily Output	Labor-Hours	Unit	Material	Labor	Equipment	Total	Total Incl O&P
0010	**SHOWER SYSTEM COMPONENTS**									
4500	Receptor only									
4510	For tile, 36″ x 36″	1 Plum	4	2	Ea.	365	117		482	580
4520	Fiberglass receptor only, 32″ x 32″		8	1		114	58.50		172.50	214
4530	34″ x 34″		7.80	1.026		128	60		188	231
4540	36″ x 36″		7.60	1.053		133	62		195	240
4600	Rectangular									
4620	32″ x 48″	1 Plum	7.40	1.081	Ea.	158	63.50		221.50	270
4630	34″ x 54″		7.20	1.111		190	65		255	310
4640	34″ x 60″		7	1.143		201	67		268	320
5000	Built-in, head, arm, 2.5 GPM valve		4	2		80	117		197	265
5200	Head, arm, by-pass, integral stops, handles		3.60	2.222		255	130		385	480

22 41 36 – Residential Laundry Trays

22 41 36.10 Laundry Sinks		Crew	Daily Output	Labor-Hours	Unit	Material	Labor	Equipment	Total	Total Incl O&P
0010	**LAUNDRY SINKS**, With trim									
0020	Porcelain enamel on cast iron, black iron frame									
0050	24″ x 21″, single compartment	Q-1	6	2.667	Ea.	580	141		721	850
0100	26″ x 21″, single compartment	″	6	2.667	″	610	141		751	885
2000	Molded stone, on wall hanger or legs									
2020	22″ x 23″, single compartment	Q-1	6	2.667	Ea.	167	141		308	395
2100	45″ x 21″, double compartment	″	5	3.200	″	330	169		499	620
3000	Plastic, on wall hanger or legs									
3020	18″ x 23″, single compartment	Q-1	6.50	2.462	Ea.	135	130		265	345
3300	40″ x 24″, double compartment		5.50	2.909		278	154		432	535

22 41 Residential Plumbing Fixtures

22 41 36 – Residential Laundry Trays

22 41 36.10 Laundry Sinks		Crew	Daily Output	Labor-Hours	Unit	2015 Bare Costs Material	Labor	Equipment	Total	Total Incl O&P
5000	Stainless steel, counter top, 22" x 17" single compartment	Q-1	6	2.667	Ea.	64	141		205	284
5200	33" x 22", double compartment		5	3.200		79	169		248	340
9600	Rough-in, supply, waste and vent, for all laundry sinks		2.14	7.477		260	395		655	880

22 41 39 – Residential Faucets, Supplies and Trim

22 41 39.10 Faucets and Fittings		Crew	Daily Output	Labor-Hours	Unit	Material	Labor	Equipment	Total	Total Incl O&P
0010	FAUCETS AND FITTINGS									
0150	Bath, faucets, diverter spout combination, sweat	1 Plum	8	1	Ea.	86.50	58.50		145	184
0200	For integral stops, IPS unions, add					109			109	120
0420	Bath, press-bal mix valve w/diverter, spout, shower head, arm/flange	1 Plum	8	1		168	58.50		226.50	274
0810	Bidet									
0812	Fitting, over the rim, swivel spray/pop-up drain	1 Plum	8	1	Ea.	206	58.50		264.50	315
1000	Kitchen sink faucets, top mount, cast spout		10	.800		61.50	47		108.50	139
1100	For spray, add		24	.333		16.15	19.55		35.70	47.50
1300	Single control lever handle									
1310	With pull out spray									
1320	Polished chrome	1 Plum	10	.800	Ea.	196	47		243	286
2000	Laundry faucets, shelf type, IPS or copper unions		12	.667		49.50	39		88.50	114
2100	Lavatory faucet, centerset, without drain		10	.800		44.50	47		91.50	120
2210	Porcelain cross handles and pop-up drain									
2220	Polished chrome	1 Plum	6.66	1.201	Ea.	189	70.50		259.50	315
2230	Polished brass	"	6.66	1.201	"	298	70.50		368.50	435
2260	Single lever handle and pop-up drain									
2280	Satin nickel	1 Plum	6.66	1.201	Ea.	277	70.50		347.50	410
2290	Polished chrome		6.66	1.201		198	70.50		268.50	325
2810	Automatic sensor and operator, with faucet head G		6.15	1.301		450	76.50		526.50	610
4000	Shower by-pass valve with union		18	.444		68.50	26		94.50	115
4200	Shower thermostatic mixing valve, concealed, with shower head trim kit		8	1		345	58.50		403.50	470
4220	Shower pressure balancing mixing valve,									
4230	With shower head, arm, flange and diverter tub spout									
4240	Chrome	1 Plum	6.14	1.303	Ea.	360	76.50		436.50	510
4250	Satin nickel		6.14	1.303		555	76.50		631.50	725
4260	Polished graphite		6.14	1.303		555	76.50		631.50	725
5000	Sillcock, compact, brass, IPS or copper to hose		24	.333		9.70	19.55		29.25	40

22 42 Commercial Plumbing Fixtures

22 42 13 – Commercial Water Closets, Urinals, and Bidets

22 42 13.13 Water Closets		Crew	Daily Output	Labor-Hours	Unit	Material	Labor	Equipment	Total	Total Incl O&P
0010	WATER CLOSETS									
3000	Bowl only, with flush valve, seat, 1.6 gpf unless noted									
3100	Wall hung	Q-1	5.80	2.759	Ea.	945	146		1,091	1,275
3200	For rough-in, supply, waste and vent, single WC		2.56	6.250		1,075	330		1,405	1,675
3300	Floor mounted		5.80	2.759		315	146		461	565
3350	With wall outlet		5.80	2.759		545	146		691	820
3360	With floor outlet, 1.28 gpf G		5.80	2.759		555	146		701	835
3362	With floor outlet, 1.28 gpf, ADA G		5.80	2.759		580	146		726	855
3370	For rough-in, supply, waste and vent, single WC		2.84	5.634		370	298		668	855
3390	Floor mounted children's size, 10-3/4" high									
3392	With automatic flush sensor, 1.6 gpf	Q-1	6.20	2.581	Ea.	615	136		751	885
3396	With automatic flush sensor, 1.28 gpf		6.20	2.581		620	136		756	885
3400	For rough-in, supply, waste and vent, single WC		2.84	5.634		370	298		668	855

22 42 Commercial Plumbing Fixtures

22 42 13 – Commercial Water Closets, Urinals, and Bidets

22 42 13.16 Urinals		Crew	Daily Output	Labor-Hours	Unit	2015 Bare Costs Material	Labor	Equipment	Total	Total Incl O&P
0010	**URINALS**									
3000	Wall hung, vitreous china, with self-closing valve									
3100	Siphon jet type	Q-1	3	5.333	Ea.	282	282		564	735
3120	Blowout type		3	5.333		465	282		747	935
3140	Water saving .5 gpf G		3	5.333		550	282		832	1,025
3300	Rough-in, supply, waste & vent		2.83	5.654		595	299		894	1,100
5000	Stall type, vitreous china, includes valve		2.50	6.400		740	340		1,080	1,325
6980	Rough-in, supply, waste and vent		1.99	8.040		365	425		790	1,050
8000	Waterless (no flush) urinal									
8010	Wall hung									
8014	Fiberglass reinforced polyester									
8020	Standard unit G	Q-1	21.30	.751	Ea.	385	39.50		424.50	480
8030	ADA compliant unit G	"	21.30	.751		400	39.50		439.50	500
8070	For solid color, add G					48			48	53
8080	For 2" brass flange, (new const.), add G	Q-1	96	.167		19.20	8.80		28	34.50
8200	Vitreous china									
8220	ADA compliant unit, 14" G	Q-1	21.30	.751	Ea.	198	39.50		237.50	278
8240	ADA compliant unit, 18" G		21.30	.751		320	39.50		359.50	410
8250	ADA compliant unit, 15.5" G		21.30	.751		272	39.50		311.50	360
8270	For solid color, add G					48			48	53
8290	Rough-in, supply, waste & vent G	Q-1	2.92	5.479		560	289		849	1,050
8400	Trap liquid									
8410	1 quart G				Ea.	15.95			15.95	17.55
8420	1 gallon G				"	58			58	64

22 42 16 – Commercial Lavatories and Sinks

22 42 16.13 Lavatories		Crew	Daily Output	Labor-Hours	Unit	Material	Labor	Equipment	Total	Total Incl O&P
0010	**LAVATORIES**, With trim, white unless noted otherwise									
0020	Commercial lavatories same as residential. See Section 22 41 16									
22 42 16.40 Service Sinks										
0010	**SERVICE SINKS**									
6650	Service, floor, corner, P.E. on C.I., 28" x 28"	Q-1	4.40	3.636	Ea.	1,000	192		1,192	1,400
6790	For rough-in, supply, waste & vent, floor service sinks		1.64	9.756		705	515		1,220	1,550
7000	Service, wall, P.E. on C.I., roll rim, 22" x 18"		4	4		770	211		981	1,175
7100	24" x 20"		4	4		850	211		1,061	1,250
8600	Vitreous china, 22" x 20"		4	4		610	211		821	995
8960	For stainless steel rim guard, front or one side, add					56			56	62
8980	For rough-in, supply, waste & vent, wall service sinks	Q-1	1.30	12.308		1,100	650		1,750	2,175

22 42 23 – Commercial Showers

22 42 23.30 Group Showers		Crew	Daily Output	Labor-Hours	Unit	Material	Labor	Equipment	Total	Total Incl O&P
0010	**GROUP SHOWERS**									
6000	Group, w/pressure balancing valve, rough-in and rigging not included									
6800	Column, 6 heads, no receptors, less partitions	Q-1	3	5.333	Ea.	9,200	282		9,482	10,500
6900	With stainless steel partitions		1	16		11,900	845		12,745	14,400
7600	5 heads, no receptors, less partitions		3	5.333		6,350	282		6,632	7,400
7620	4 heads (1 handicap) no receptors, less partitions		3	5.333		5,650	282		5,932	6,625
7700	With stainless steel partitions		1	16		5,650	845		6,495	7,475
8000	Wall, 2 heads, no receptors, less partitions		4	4		2,725	211		2,936	3,300
8100	With stainless steel partitions		2	8		5,975	425		6,400	7,200

22 42 Commercial Plumbing Fixtures

22 42 33 – Wash Fountains

22 42 33.20 Commercial Wash Fountains		Crew	Daily Output	Labor-Hours	Unit	Material	2015 Bare Costs Labor	Equipment	Total	Total Incl O&P
0010	COMMERCIAL WASH FOUNTAINS									
1900	Group, foot control									
2000	Precast terrazzo, circular, 36" diam., 5 or 6 persons	Q-2	3	8	Ea.	7,275	440		7,715	8,650
2100	54" diameter for 8 or 10 persons		2.50	9.600		9,050	525		9,575	10,800
2400	Semi-circular, 36" diam. for 3 persons		3	8		6,375	440		6,815	7,675
2500	54" diam. for 4 or 5 persons		2.50	9.600		8,575	525		9,100	10,200
2700	Quarter circle (corner), 54" for 3 persons		3.50	6.857		7,850	375		8,225	9,225
3000	Stainless steel, circular, 36" diameter		3.50	6.857		6,550	375		6,925	7,775
3100	54" diameter		2.80	8.571		8,050	470		8,520	9,550
3400	Semi-circular, 36" diameter		3.50	6.857		5,025	375		5,400	6,125
3500	54" diameter		2.80	8.571		7,000	470		7,470	8,425
5610	Group, infrared control, barrier free									
5614	Precast terrazzo									
5620	Semi-circular 36" diam. for 3 persons	Q-2	3	8	Ea.	7,250	440		7,690	8,625
5630	46" diam. for 4 persons		2.80	8.571		7,825	470		8,295	9,300
5640	Circular, 54" diam. for 8 persons, button control		2.50	9.600		9,475	525		10,000	11,200
5700	Rough-in, supply, waste and vent for above wash fountains	Q-1	1.82	8.791		370	465		835	1,100
6200	Duo for small washrooms, stainless steel		2	8		2,725	425		3,150	3,625
6500	Rough-in, supply, waste & vent for duo fountains		2.02	7.921		209	420		629	860

22 42 39 – Commercial Faucets, Supplies, and Trim

22 42 39.10 Faucets and Fittings

		Crew	Daily Output	Labor-Hours	Unit	Material	Labor	Equipment	Total	Total Incl O&P
0010	FAUCETS AND FITTINGS									
0840	Flush valves, with vacuum breaker									
0850	Water closet									
0860	Exposed, rear spud	1 Plum	8	1	Ea.	148	58.50		206.50	252
0870	Top spud		8	1		149	58.50		207.50	252
0880	Concealed, rear spud		8	1		197	58.50		255.50	305
0890	Top spud		8	1		159	58.50		217.50	264
0900	Wall hung		8	1		177	58.50		235.50	284
0920	Urinal									
0930	Exposed, stall	1 Plum	8	1	Ea.	149	58.50		207.50	253
0940	Wall, (washout)		8	1		149	58.50		207.50	252
0950	Pedestal, top spud		8	1		143	58.50		201.50	246
0960	Concealed, stall		8	1		156	58.50		214.50	261
0970	Wall (washout)		8	1		168	58.50		226.50	274
0971	Automatic flush sensor and operator for									
0972	urinals or water closets, standard G	1 Plum	8	1	Ea.	450	58.50		508.50	585
0980	High efficiency water saving									
0984	Water closets, 1.28 gpf G	1 Plum	8	1	Ea.	415	58.50		473.50	550
0988	Urinals, .5 gpf G	"	8	1	"	415	58.50		473.50	550
2790	Faucets for lavatories									
2800	Self-closing, center set	1 Plum	10	.800	Ea.	131	47		178	215
2810	Automatic sensor and operator, with faucet head		6.15	1.301		450	76.50		526.50	610
3000	Service sink faucet, cast spout, pail hook, hose end		14	.571		80	33.50		113.50	139

22 42 39.30 Carriers and Supports

		Crew	Daily Output	Labor-Hours	Unit	Material	Labor	Equipment	Total	Total Incl O&P
0010	CARRIERS AND SUPPORTS, For plumbing fixtures									
0500	Drinking fountain, wall mounted									
0600	Plate type with studs, top back plate	1 Plum	7	1.143	Ea.	95	67		162	206
0700	Top front and back plate		7	1.143		116	67		183	229
0800	Top & bottom, front & back plates, w/bearing jacks		7	1.143		193	67		260	315
3000	Lavatory, concealed arm									

22 42 39.30 Carriers and Supports		Crew	Daily Output	Labor-Hours	Unit	Material	2015 Bare Costs Labor	Equipment	Total	Total Incl O&P
3050	Floor mounted, single									
3100	High back fixture	1 Plum	6	1.333	Ea.	525	78.50		603.50	700
3200	Flat slab fixture		6	1.333		455	78.50		533.50	620
3220	Paraplegic		6	1.333		590	78.50		668.50	770
3250	Floor mounted, back to back									
3300	High back fixtures	1 Plum	5	1.600	Ea.	750	94		844	965
3400	Flat slab fixtures		5	1.600		925	94		1,019	1,175
3430	Paraplegic		5	1.600		860	94		954	1,075
3500	Wall mounted, in stud or masonry									
3600	High back fixture	1 Plum	6	1.333	Ea.	320	78.50		398.50	470
3700	Flat slab fixture	"	6	1.333	"	270	78.50		348.50	415
4600	Sink, floor mounted									
4650	Exposed arm system									
4700	Single heavy fixture	1 Plum	5	1.600	Ea.	880	94		974	1,100
4750	Single heavy sink with slab		5	1.600		1,075	94		1,169	1,350
4800	Back to back, standard fixtures		5	1.600		650	94		744	850
4850	Back to back, heavy fixtures		5	1.600		920	94		1,014	1,150
4900	Back to back, heavy sink with slab		5	1.600		920	94		1,014	1,150
4950	Exposed offset arm system									
5000	Single heavy deep fixture	1 Plum	5	1.600	Ea.	840	94		934	1,075
5100	Plate type system									
5200	With bearing jacks, single fixture	1 Plum	5	1.600	Ea.	885	94		979	1,125
5300	With exposed arms, single heavy fixture		5	1.600		1,225	94		1,319	1,500
5400	Wall mounted, exposed arms, single heavy fixture		5	1.600		450	94		544	635
6000	Urinal, floor mounted, 2" or 3" coupling, blowout type		6	1.333		560	78.50		638.50	735
6100	With fixture or hanger bolts, blowout or washout		6	1.333		390	78.50		468.50	550
6200	With bearing plate		6	1.333		445	78.50		523.50	610
6300	Wall mounted, plate type system		6	1.333		345	78.50		423.50	500
6980	Water closet, siphon jet									
7000	Horizontal, adjustable, caulk									
7040	Single, 4" pipe size	1 Plum	5.33	1.501	Ea.	890	88		978	1,100
7050	4" pipe size, paraplegic		5.33	1.501		890	88		978	1,100
7060	5" pipe size		5.33	1.501		965	88		1,053	1,175
7100	Double, 4" pipe size		5	1.600		1,575	94		1,669	1,875
7110	4" pipe size, paraplegic		5	1.600		1,575	94		1,669	1,875
7120	5" pipe size		5	1.600		1,700	94		1,794	2,025
7160	Horizontal, adjustable, extended, caulk									
7180	Single, 4" pipe size	1 Plum	5.33	1.501	Ea.	1,000	88		1,088	1,225
7200	5" pipe size		5.33	1.501		1,275	88		1,363	1,525
7240	Double, 4" pipe size		5	1.600		1,750	94		1,844	2,075
7260	5" pipe size		5	1.600		2,125	94		2,219	2,500
7400	Vertical, adjustable, caulk or thread									
7440	Single, 4" pipe size	1 Plum	5.33	1.501	Ea.	890	88		978	1,100
7460	5" pipe size		5.33	1.501		1,125	88		1,213	1,350
7480	6" pipe size		5	1.600		1,325	94		1,419	1,600
7520	Double, 4" pipe size		5	1.600		1,525	94		1,619	1,850
7540	5" pipe size		5	1.600		1,775	94		1,869	2,100
7560	6" pipe size		4	2		1,950	117		2,067	2,325
7600	Vertical, adjustable, extended, caulk									
7620	Single, 4" pipe size	1 Plum	5.33	1.501	Ea.	1,025	88		1,113	1,250
7640	5" pipe size		5.33	1.501		1,275	88		1,363	1,525
7680	6" pipe size		5	1.600		1,450	94		1,544	1,750
7720	Double, 4" pipe size		5	1.600		1,675	94		1,769	2,000

22 42 Commercial Plumbing Fixtures

22 42 39 – Commercial Faucets, Supplies, and Trim

22 42 39.30 Carriers and Supports		Crew	Daily Output	Labor-Hours	Unit	Material	Labor (2015 Bare Costs)	Equipment (2015 Bare Costs)	Total (2015 Bare Costs)	Total Incl O&P
7740	5″ pipe size	1 Plum	5	1.600	Ea.	1,925	94		2,019	2,275
7760	6″ pipe size		4	2		2,100	117		2,217	2,500
7780	Water closet, blow out									
7800	Vertical offset, caulk or thread									
7820	Single, 4″ pipe size	1 Plum	5.33	1.501	Ea.	760	88		848	970
7840	Double, 4″ pipe size	"	5	1.600	"	1,300	94		1,394	1,575
7880	Vertical offset, extended, caulk									
7900	Single, 4″ pipe size	1 Plum	5.33	1.501	Ea.	950	88		1,038	1,175
7920	Double, 4″ pipe size	"	5	1.600	"	1,500	94		1,594	1,800
7960	Vertical, for floor mounted back-outlet									
7980	Single, 4″ thread, 2″ vent	1 Plum	5.33	1.501	Ea.	670	88		758	870
8000	Double, 4″ thread, 2″ vent	"	6	1.333	"	1,975	78.50		2,053.50	2,275
8040	Vertical, for floor mounted back-outlet, extended									
8060	Single, 4″ caulk, 2″ vent	1 Plum	6	1.333	Ea.	670	78.50		748.50	855
8080	Double, 4″ caulk, 2″ vent	"	6	1.333	"	1,975	78.50		2,053.50	2,275
8200	Water closet, residential									
8220	Vertical centerline, floor mount									
8240	Single, 3″ caulk, 2″ or 3″ vent	1 Plum	6	1.333	Ea.	610	78.50		688.50	790
8260	4″ caulk, 2″ or 4″ vent		6	1.333		785	78.50		863.50	980
8280	3″ copper sweat, 3″ vent		6	1.333		545	78.50		623.50	720
8300	4″ copper sweat, 4″ vent		6	1.333		660	78.50		738.50	845
8400	Vertical offset, floor mount									
8420	Single, 3″ or 4″ caulk, vent	1 Plum	4	2	Ea.	760	117		877	1,000
8440	3″ or 4″ copper sweat, vent		5	1.600		760	94		854	975
8460	Double, 3″ or 4″ caulk, vent		4	2		1,300	117		1,417	1,600
8480	3″ or 4″ copper sweat, vent		5	1.600		1,300	94		1,394	1,575
9000	Water cooler (electric), floor mounted									
9100	Plate type with bearing plate, single	1 Plum	6	1.333	Ea.	385	78.50		463.50	545

22 45 Emergency Plumbing Fixtures

22 45 13 – Emergency Showers

22 45 13.10 Emergency Showers		Crew	Daily Output	Labor-Hours	Unit	Material	Labor	Equipment	Total	Total Incl O&P
0010	**EMERGENCY SHOWERS**, Rough-in not included									
5000	Shower, single head, drench, ball valve, pull, freestanding	Q-1	4	4	Ea.	380	211		591	735
5200	Horizontal or vertical supply		4	4		555	211		766	930
6000	Multi-nozzle, eye/face wash combination		4	4		660	211		871	1,050
6400	Multi-nozzle, 12 spray, shower only		4	4		2,000	211		2,211	2,525
6600	For freeze-proof, add		6	2.667		465	141		606	730

22 45 16 – Eyewash Equipment

22 45 16.10 Eyewash Safety Equipment		Crew	Daily Output	Labor-Hours	Unit	Material	Labor	Equipment	Total	Total Incl O&P
0010	**EYEWASH SAFETY EQUIPMENT**, Rough-in not included									
1000	Eye wash fountain									
1400	Plastic bowl, pedestal mounted	Q-1	4	4	Ea.	282	211		493	630
1600	Unmounted		4	4		247	211		458	590
1800	Wall mounted		4	4		455	211		666	820
2000	Stainless steel, pedestal mounted		4	4		350	211		561	705
2200	Unmounted		4	4		272	211		483	620
2400	Wall mounted		4	4		288	211		499	635

22 45 Emergency Plumbing Fixtures

22 45 19 – Self-Contained Eyewash Equipment

22 45 19.10 Self-Contained Eyewash Safety Equipment		Crew	Daily Output	Labor-Hours	Unit	2015 Bare Costs Material	Labor	Equipment	Total	Total Incl O&P
0010	**SELF-CONTAINED EYEWASH SAFETY EQUIPMENT**									
3000	Eye wash, portable, self-contained				Ea.	1,025			1,025	1,125

22 45 26 – Eye/Face Wash Equipment

22 45 26.10 Eye/Face Wash Safety Equipment		Crew	Daily Output	Labor-Hours	Unit	Material	Labor	Equipment	Total	Total Incl O&P
0010	**EYE/FACE WASH SAFETY EQUIPMENT**, Rough-in not included									
4000	Eye and face wash, combination fountain									
4200	Stainless steel, pedestal mounted	Q-1	4	4	Ea.	1,075	211		1,286	1,500
4400	Unmounted		4	4		272	211		483	620
4600	Wall mounted		4	4		246	211		457	590

22 47 Drinking Fountains and Water Coolers

22 47 13 – Drinking Fountains

22 47 13.10 Drinking Water Fountains		Crew	Daily Output	Labor-Hours	Unit	Material	Labor	Equipment	Total	Total Incl O&P
0010	**DRINKING WATER FOUNTAINS**, For connection to cold water supply									
1000	Wall mounted, non-recessed									
1400	Bronze, with no back	1 Plum	4	2	Ea.	1,000	117		1,117	1,275
1800	Cast aluminum, enameled, for correctional institutions		4	2		1,675	117		1,792	2,000
2000	Fiberglass, 12" back, single bubbler unit		4	2		1,925	117		2,042	2,300
2040	Dual bubbler		3.20	2.500		2,225	147		2,372	2,650
2400	Precast stone, no back		4	2		915	117		1,032	1,175
2700	Stainless steel, single bubbler, no back		4	2		1,050	117		1,167	1,325
2740	With back		4	2		565	117		682	795
2780	Dual handle & wheelchair projection type		4	2		745	117		862	995
2820	Dual level for handicapped type		3.20	2.500		1,575	147		1,722	1,950
3300	Vitreous china									
3340	7" back	1 Plum	4	2	Ea.	620	117		737	855
3940	For vandal-resistant bottom plate, add					77			77	84.50
3960	For freeze-proof valve system, add	1 Plum	2	4		705	235		940	1,125
3980	For rough-in, supply and waste, add	"	2.21	3.620		174	212		386	510
4000	Wall mounted, semi-recessed									
4200	Poly-marble, single bubbler	1 Plum	4	2	Ea.	955	117		1,072	1,225
4600	Stainless steel, satin finish, single bubbler		4	2		1,150	117		1,267	1,450
4900	Vitreous china, single bubbler		4	2		895	117		1,012	1,150
5980	For rough-in, supply and waste, add		1.83	4.372		174	257		431	580
6000	Wall mounted, fully recessed									
6400	Poly-marble, single bubbler	1 Plum	4	2	Ea.	1,600	117		1,717	1,925
6800	Stainless steel, single bubbler		4	2		1,550	117		1,667	1,875
7560	For freeze-proof valve system, add		2	4		795	235		1,030	1,225
7580	For rough-in, supply and waste, add		1.83	4.372		174	257		431	580
7600	Floor mounted, pedestal type									
7700	Aluminum, architectural style, C.I. base	1 Plum	2	4	Ea.	2,250	235		2,485	2,825
7780	Wheelchair handicap unit		2	4		1,625	235		1,860	2,125
8400	Stainless steel, architectural style		2	4		1,750	235		1,985	2,275
8600	Enameled iron, heavy duty service, 2 bubblers		2	4		2,575	235		2,810	3,200
8660	4 bubblers		2	4		3,950	235		4,185	4,675
8880	For freeze-proof valve system, add		2	4		705	235		940	1,125
8900	For rough-in, supply and waste, add		1.83	4.372		174	257		431	580
9100	Deck mounted									
9500	Stainless steel, circular receptor	1 Plum	4	2	Ea.	435	117		552	655
9760	White enameled steel, 14" x 9" receptor		4	2		375	117		492	590

22 47 Drinking Fountains and Water Coolers

22 47 13 – Drinking Fountains

22 47 13.10 Drinking Water Fountains		Crew	Daily Output	Labor-Hours	Unit	Material	2015 Bare Costs Labor	Equipment	Total	Total Incl O&P
9860	White enameled cast iron, 24″ x 16″ receptor	1 Plum	3	2.667	Ea.	460	157		617	740
9980	For rough-in, supply and waste, add		1.83	4.372		174	257		431	580

22 47 16 – Pressure Water Coolers

22 47 16.10 Electric Water Coolers		Crew	Daily Output	Labor-Hours	Unit	Material	Labor	Equipment	Total	Total Incl O&P
0010	**ELECTRIC WATER COOLERS**									
0100	Wall mounted, non-recessed									
0140	4 GPH	Q-1	4	4	Ea.	680	211		891	1,075
0160	8 GPH, barrier free, sensor operated		4	4		1,025	211		1,236	1,475
0180	8.2 GPH		4	4		750	211		961	1,150
0600	8 GPH hot and cold water		4	4		1,050	211		1,261	1,500
0640	For stainless steel cabinet, add					90			90	99.50
1000	Dual height, 8.2 GPH	Q-1	3.80	4.211		2,075	222		2,297	2,600
1040	14.3 GPH	″	3.80	4.211		975	222		1,197	1,400
1240	For stainless steel cabinet, add					171			171	188
2600	Wheelchair type, 8 GPH	Q-1	4	4		915	211		1,126	1,325
3300	Semi-recessed, 8.1 GPH		4	4		750	211		961	1,150
3320	12 GPH		4	4		865	211		1,076	1,275
4600	Floor mounted, flush-to-wall									
4640	4 GPH	1 Plum	3	2.667	Ea.	740	157		897	1,050
4680	8.2 GPH		3	2.667		780	157		937	1,100
4720	14.3 GPH		3	2.667		890	157		1,047	1,225
4960	14 GPH hot and cold water		3	2.667		1,100	157		1,257	1,450
4980	For stainless steel cabinet, add					134			134	148
5000	Dual height, 8.2 GPH	1 Plum	2	4		1,175	235		1,410	1,625
5040	14.3 GPH	″	2	4		1,200	235		1,435	1,675
5120	For stainless steel cabinet, add					196			196	215
9800	For supply, waste & vent, all coolers	1 Plum	2.21	3.620		174	212		386	510

22 51 Swimming Pool Plumbing Systems

22 51 19 – Swimming Pool Water Treatment Equipment

22 51 19.50 Swimming Pool Filtration Equipment		Crew	Daily Output	Labor-Hours	Unit	Material	Labor	Equipment	Total	Total Incl O&P
0010	**SWIMMING POOL FILTRATION EQUIPMENT**									
0900	Filter system, sand or diatomite type, incl. pump, 6,000 gal./hr.	2 Plum	1.80	8.889	Total	1,975	520		2,495	2,975
1020	Add for chlorination system, 800 S.F. pool		3	5.333	Ea.	181	315		496	675
1040	5,000 S.F. pool		3	5.333	″	1,850	315		2,165	2,500

22 52 Fountain Plumbing Systems

22 52 16 – Fountain Pumps

22 52 16.10 Fountain Water Pumps		Crew	Daily Output	Labor-Hours	Unit	Material	Labor	Equipment	Total	Total Incl O&P
0010	**FOUNTAIN WATER PUMPS**									
0100	Pump w/controls									
0200	Single phase, 100′ cord, 1/2 H.P. pump	2 Skwk	4.40	3.636	Ea.	1,275	177		1,452	1,675
0300	3/4 H.P. pump		4.30	3.721		2,175	181		2,356	2,675
0400	1 H.P. pump		4.20	3.810		2,350	185		2,535	2,875
0500	1-1/2 H.P. pump		4.10	3.902		2,800	190		2,990	3,375
0600	2 H.P. pump		4	4		3,775	195		3,970	4,475
0700	Three phase, 200′ cord, 5 H.P. pump		3.90	4.103		5,175	200		5,375	6,000
0800	7-1/2 H.P. pump		3.80	4.211		9,000	205		9,205	10,200
0900	10 H.P. pump		3.70	4.324		13,300	210		13,510	15,000

22 52 Fountain Plumbing Systems

22 52 16 – Fountain Pumps

22 52 16.10 Fountain Water Pumps		Crew	Daily Output	Labor-Hours	Unit	Material	2015 Bare Costs Labor	Equipment	Total	Total Incl O&P
1000	15 H.P. pump	2 Skwk	3.60	4.444	Ea.	16,800	216		17,016	18,800
2000	DESIGN NOTE: Use two horsepower per surface acre.									

22 52 33 – Fountain Ancillary

22 52 33.10 Fountain Miscellaneous		Crew	Daily Output	Labor-Hours	Unit	Material	Labor	Equipment	Total	Total Incl O&P
0010	**FOUNTAIN MISCELLANEOUS**									
1300	Lights w/mounting kits, 200 watt	2 Skwk	18	.889	Ea.	1,050	43		1,093	1,225
1400	300 watt		18	.889		1,275	43		1,318	1,475
1500	500 watt		18	.889		1,425	43		1,468	1,650
1600	Color blender	↓	12	1.333	↓	555	65		620	710

22 66 Chemical-Waste Systems for Lab. and Healthcare Facilities

22 66 53 – Laboratory Chemical-Waste and Vent Piping

22 66 53.30 Glass Pipe		Crew	Daily Output	Labor-Hours	Unit	Material	Labor	Equipment	Total	Total Incl O&P
0010	**GLASS PIPE**, Borosilicate, couplings & clevis hanger assemblies, 10′ O.C.									
0020	Drainage									
1100	1-1/2″ diameter	Q-1	52	.308	L.F.	11.40	16.25		27.65	37
1120	2″ diameter		44	.364		14.65	19.20		33.85	45
1140	3″ diameter		39	.410		19.50	21.50		41	54
1160	4″ diameter		30	.533		34	28		62	80
1180	6″ diameter	↓	26	.615	↓	58.50	32.50		91	114

22 66 53.60 Corrosion Resistant Pipe		Crew	Daily Output	Labor-Hours	Unit	Material	Labor	Equipment	Total	Total Incl O&P
0010	**CORROSION RESISTANT PIPE**, No couplings or hangers									
0020	Iron alloy, drain, mechanical joint									
1000	1-1/2″ diameter	Q-1	70	.229	L.F.	46	12.10		58.10	69
1100	2″ diameter		66	.242		47	12.80		59.80	71
1120	3″ diameter		60	.267		60.50	14.10		74.60	88
1140	4″ diameter	↓	52	.308	↓	77.50	16.25		93.75	110
2980	Plastic, epoxy, fiberglass filament wound, B&S joint									
3000	2″ diameter	Q-1	62	.258	L.F.	12	13.65		25.65	33.50
3100	3″ diameter		51	.314		14	16.55		30.55	40.50
3120	4″ diameter		45	.356		20	18.80		38.80	50.50
3140	6″ diameter	↓	32	.500	↓	28	26.50		54.50	71
3980	Polyester, fiberglass filament wound, B&S joint									
4000	2″ diameter	Q-1	62	.258	L.F.	13.05	13.65		26.70	35
4100	3″ diameter		51	.314		17	16.55		33.55	43.50
4120	4″ diameter		45	.356		25	18.80		43.80	56
4140	6″ diameter	↓	32	.500	↓	36	26.50		62.50	79.50
4980	Polypropylene, acid resistant, fire retardant, schedule 40									
5000	1-1/2″ diameter	Q-1	68	.235	L.F.	7.90	12.45		20.35	27.50
5100	2″ diameter		62	.258		12.45	13.65		26.10	34
5120	3″ diameter		51	.314		22	16.55		38.55	49.50
5140	4″ diameter	↓	45	.356	↓	28	18.80		46.80	59.50
5980	Proxylene, fire retardant, Schedule 40									
6000	1-1/2″ diameter	Q-1	68	.235	L.F.	12.40	12.45		24.85	32.50
6100	2″ diameter		62	.258		17	13.65		30.65	39
6120	3″ diameter		51	.314		30.50	16.55		47.05	59
6140	4″ diameter	↓	45	.356	↓	43.50	18.80		62.30	76

Estimating Tips

The labor adjustment factors listed in Subdivision 22 01 02.20 also apply to Division 23.

23 10 00 Facility Fuel Systems

- The prices in this subdivision for above- and below-ground storage tanks do not include foundations or hold-down slabs, unless noted. The estimator should refer to Divisions 3 and 31 for foundation system pricing. In addition to the foundations, required tank accessories, such as tank gauges, leak detection devices, and additional manholes and piping, must be added to the tank prices.

23 50 00 Central Heating Equipment

- When estimating the cost of an HVAC system, check to see who is responsible for providing and installing the temperature control system. It is possible to overlook controls, assuming that they would be included in the electrical estimate.
- When looking up a boiler, be careful on specified capacity. Some manufacturers rate their products on output while others use input.
- Include HVAC insulation for pipe, boiler, and duct (wrap and liner).
- Be careful when looking up mechanical items to get the correct pressure rating and connection type (thread, weld, flange).

23 70 00 Central HVAC Equipment

- Combination heating and cooling units are sized by the air conditioning requirements. (See Reference No. R236000-20 for preliminary sizing guide.)
- A ton of air conditioning is nominally 400 CFM.
- Rectangular duct is taken off by the linear foot for each size, but its cost is usually estimated by the pound. Remember that SMACNA standards now base duct on internal pressure.
- Prefabricated duct is estimated and purchased like pipe: straight sections and fittings.
- Note that cranes or other lifting equipment are not included on any lines in Division 23. For example, if a crane is required to lift a heavy piece of pipe into place high above a gym floor, or to put a rooftop unit on the roof of a four-story building, etc., it must be added. Due to the potential for extreme variation—from nothing additional required to a major crane or helicopter—we feel that including a nominal amount for "lifting contingency" would be useless and detract from the accuracy of the estimate. When using equipment rental cost data from RSMeans, do not forget to include the cost of the operator(s).

Reference Numbers

Reference numbers are shown in shaded boxes at the beginning of some major classifications. These numbers refer to related items in the Reference Section. The reference information may be an estimating procedure, an alternate pricing method, or technical information.

Note: Not all subdivisions listed here necessarily appear in this publication. ■

*Note: **Trade Service**, in part, has been used as a reference source for some of the material prices used in Division 23.*

23 05 Common Work Results for HVAC

23 05 02 – HVAC General

23 05 02.10 Air Conditioning, General		Crew	Daily Output	Labor-Hours	Unit	Material	2015 Bare Costs Labor	Equipment	Total	Total Incl O&P
0010	**AIR CONDITIONING, GENERAL** Prices are for standard efficiencies (SEER 13)									
0020	for upgrade to SEER 14 add					10%				

23 05 05 – Selective Demolition for HVAC

23 05 05.10 HVAC Demolition

		Crew	Daily Output	Labor-Hours	Unit	Material	Labor	Equipment	Total	Total Incl O&P
0010	**HVAC DEMOLITION**									
0100	Air conditioner, split unit, 3 ton	Q-5	2	8	Ea.		430		430	650
0150	Package unit, 3 ton	Q-6	3	8	"		445		445	675
0298	Boilers									
0300	Electric, up thru 148 kW	Q-19	2	12	Ea.		650		650	975
0310	150 thru 518 kW	"	1	24			1,300		1,300	1,950
0320	550 thru 2000 kW	Q-21	.40	80			4,450		4,450	6,700
0330	2070 kW and up	"	.30	106			5,925		5,925	8,925
0340	Gas and/or oil, up thru 150 MBH	Q-7	2.20	14.545			825		825	1,250
0350	160 thru 2000 MBH		.80	40			2,275		2,275	3,425
0360	2100 thru 4500 MBH		.50	64			3,650		3,650	5,500
0370	4600 thru 7000 MBH		.30	106			6,075		6,075	9,175
0380	7100 thru 12,000 MBH		.16	200			11,400		11,400	17,200
0390	12,200 thru 25,000 MBH		.12	266			15,200		15,200	22,900
1000	Ductwork, 4" high, 8" wide	1 Clab	200	.040	L.F.		1.50		1.50	2.31
1100	6" high, 8" wide		165	.048			1.82		1.82	2.80
1200	10" high, 12" wide		125	.064			2.41		2.41	3.70
1300	12"-14" high, 16"-18" wide		85	.094			3.54		3.54	5.45
1400	18" high, 24" wide		67	.119			4.49		4.49	6.90
1500	30" high, 36" wide		56	.143			5.35		5.35	8.25
1540	72" wide		50	.160			6		6	9.25
3000	Mechanical equipment, light items. Unit is weight, not cooling.	Q-5	.90	17.778	Ton		955		955	1,450
3600	Heavy items	"	1.10	14.545	"		780		780	1,175
5090	Remove refrigerant from system	1 Stpi	40	.200	Lb.		11.95		11.95	18.05

23 05 23 – General-Duty Valves for HVAC Piping

23 05 23.30 Valves, Iron Body

		Crew	Daily Output	Labor-Hours	Unit	Material	Labor	Equipment	Total	Total Incl O&P
0010	**VALVES, IRON BODY**									
1020	Butterfly, wafer type, gear actuator, 200 lb.									
1030	2"	1 Plum	14	.571	Ea.	96	33.50		129.50	156
1040	2-1/2"	Q-1	9	1.778		97.50	94		191.50	249
1050	3"		8	2		101	106		207	271
1060	4"		5	3.200		113	169		282	380
1070	5"	Q-2	5	4.800		126	263		389	535
1080	6"	"	5	4.800		143	263		406	555
1650	Gate, 125 lb., N.R.S.									
2150	Flanged									
2200	2"	1 Plum	5	1.600	Ea.	715	94		809	930
2240	2-1/2"	Q-1	5	3.200		735	169		904	1,075
2260	3"		4.50	3.556		825	188		1,013	1,200
2280	4"		3	5.333		1,175	282		1,457	1,725
2300	6"	Q-2	3	8		2,025	440		2,465	2,875
3550	OS&Y, 125 lb., flanged									
3600	2"	1 Plum	5	1.600	Ea.	475	94		569	665
3660	3"	Q-1	4.50	3.556		530	188		718	865
3680	4"	"	3	5.333		770	282		1,052	1,275
3700	6"	Q-2	3	8		1,250	440		1,690	2,025
3900	For 175 lb., flanged, add					200%	10%			
5450	Swing check, 125 lb., threaded									

23 05 Common Work Results for HVAC

23 05 23 – General-Duty Valves for HVAC Piping

23 05 23.30 Valves, Iron Body		Crew	Daily Output	Labor-Hours	Unit	2015 Bare Costs Material	Labor	Equipment	Total	Total Incl O&P
5500	2″	1 Plum	11	.727	Ea.	430	42.50		472.50	540
5540	2-1/2″	Q-1	15	1.067		555	56.50		611.50	695
5550	3″		13	1.231		590	65		655	750
5560	4″		10	1.600		955	84.50		1,039.50	1,175
5950	Flanged									
6000	2″	1 Plum	5	1.600	Ea.	395	94		489	575
6040	2-1/2″	Q-1	5	3.200		380	169		549	675
6050	3″		4.50	3.556		410	188		598	735
6060	4″		3	5.333		605	282		887	1,100
6070	6″	Q-2	3	8		1,025	440		1,465	1,800

23 05 23.80 Valves, Steel

		Crew	Daily Output	Labor-Hours	Unit	Material	Labor	Equipment	Total	Total Incl O&P
0010	**VALVES, STEEL**									
0800	Cast									
1350	Check valve, swing type, 150 lb., flanged									
1370	1″	1 Plum	10	.800	Ea.	375	47		422	485
1400	2″	″	8	1		685	58.50		743.50	840
1440	2-1/2″	Q-1	5	3.200		890	169		1,059	1,225
1450	3″		4.50	3.556		805	188		993	1,175
1460	4″		3	5.333		1,225	282		1,507	1,750
1540	For 300 lb., flanged, add					50%	15%			
1548	For 600 lb., flanged, add					110%	20%			
1950	Gate valve, 150 lb., flanged									
2000	2″	1 Plum	8	1	Ea.	780	58.50		838.50	950
2040	2-1/2″	Q-1	5	3.200		1,100	169		1,269	1,475
2050	3″		4.50	3.556		1,100	188		1,288	1,500
2060	4″		3	5.333		1,350	282		1,632	1,925
2070	6″	Q-2	3	8		2,250	440		2,690	3,125
3650	Globe valve, 150 lb., flanged									
3700	2″	1 Plum	8	1	Ea.	980	58.50		1,038.50	1,175
3740	2-1/2″	Q-1	5	3.200		1,250	169		1,419	1,625
3750	3″		4.50	3.556		1,250	188		1,438	1,650
3760	4″		3	5.333		1,825	282		2,107	2,425
3770	6″	Q-2	3	8		2,875	440		3,315	3,800
5150	Forged									
5650	Check valve, class 800, horizontal, socket									
5698	Threaded									
5700	1/4″	1 Plum	24	.333	Ea.	94.50	19.55		114.05	134
5720	3/8″		24	.333		94.50	19.55		114.05	134
5730	1/2″		24	.333		94.50	19.55		114.05	134
5740	3/4″		20	.400		101	23.50		124.50	147
5750	1″		19	.421		119	24.50		143.50	169
5760	1-1/4″		15	.533		233	31.50		264.50	305

23 05 93 – Testing, Adjusting, and Balancing for HVAC

23 05 93.10 Balancing, Air

		Crew	Daily Output	Labor-Hours	Unit	Material	Labor	Equipment	Total	Total Incl O&P
0010	**BALANCING, AIR** (Subcontractor's quote incl. material and labor)									
0900	Heating and ventilating equipment									
1000	Centrifugal fans, utility sets				Ea.				410	410
1100	Heating and ventilating unit								615	615
1200	In-line fan								615	615
1300	Propeller and wall fan								116	116
1400	Roof exhaust fan								274	274
2000	Air conditioning equipment, central station								890	890

23 05 Common Work Results for HVAC

23 05 93 – Testing, Adjusting, and Balancing for HVAC

23 05 93.10 Balancing, Air		Crew	Daily Output	Labor-Hours	Unit	Material	2015 Bare Costs Labor	Equipment	Total	Total Incl O&P
2100	Built-up low pressure unit				Ea.				820	820
2200	Built-up high pressure unit								960	960
2500	Multi-zone A.C. and heating unit								615	615
2600	For each zone over one, add								137	137
2700	Package A.C. unit								340	340
2800	Rooftop heating and cooling unit								480	480
3000	Supply, return, exhaust, registers & diffusers, avg. height ceiling								82	82
3100	High ceiling								123	123
3200	Floor height				↓				68.50	68.50

23 05 93.20 Balancing, Water		Crew	Daily Output	Labor-Hours	Unit	Material	Labor	Equipment	Total	Total Incl O&P
0010	**BALANCING, WATER** (Subcontractor's quote incl. material and labor)									
0050	Air cooled condenser				Ea.				253	253
0080	Boiler								510	510
0100	Cabinet unit heater								86.50	86.50
0200	Chiller								615	615
0300	Convector								72	72
0500	Cooling tower								470	470
0600	Fan coil unit, unit ventilator								130	130
0700	Fin tube and radiant panels								144	144
0800	Main and duct re-heat coils								134	134
0810	Heat exchanger								134	134
1000	Pumps								320	320
1100	Unit heater				↓				101	101

23 07 HVAC Insulation

23 07 13 – Duct Insulation

23 07 13.10 Duct Thermal Insulation		Crew	Daily Output	Labor-Hours	Unit	Material	Labor	Equipment	Total	Total Incl O&P
0010	**DUCT THERMAL INSULATION**									
0100	Rule of thumb, as a percentage of total mechanical costs				Job				10%	10%
0110	Insulation req'd. is based on the surface size/area to be covered									
3000	Ductwork									
3020	Blanket type, fiberglass, flexible									
3030	Fire rated for grease and hazardous exhaust ducts									
3060	1-1/2" thick	Q-14	300	.053	S.F.	4.54	2.51		7.05	8.90
3090	Fire rated for plenums									
3100	1/2" x24" x 25'	Q-14	7.20	2.222	Roll	167	105		272	345
3110	1/2" x24" x 25'		360	.044	S.F.	3.35	2.09		5.44	6.95
3120	1/2" x 48" x 25'		3.80	4.211	Roll	335	198		533	680
3126	1/2" x 48" x 25'	↓	380	.042	S.F.	3.35	1.98		5.33	6.75
3140	FSK vapor barrier wrap, .75 lb. density									
3160	1" thick G	Q-14	350	.046	S.F.	.18	2.15		2.33	3.55
3170	1-1/2" thick G		320	.050		.22	2.36		2.58	3.91
3180	2" thick G		300	.053		.26	2.51		2.77	4.20
3190	3" thick G		260	.062		.36	2.90		3.26	4.92
3200	4" thick G	↓	242	.066	↓	.51	3.12		3.63	5.40
3210	Vinyl jacket, same as FSK									
3280	Unfaced, 1 lb. density									
3310	1" thick G	Q-14	360	.044	S.F.	.19	2.09		2.28	3.47
3320	1-1/2" thick G		330	.048		.29	2.28		2.57	3.88
3330	2" thick G	↓	310	.052	↓	.35	2.43		2.78	4.18
3400	FSK facing, 1 lb. density									

23 07 HVAC Insulation

23 07 13 – Duct Insulation

23 07 13.10 Duct Thermal Insulation			Crew	Daily Output	Labor-Hours	Unit	2015 Bare Costs Material	Labor	Equipment	Total	Total Incl O&P
3420	1-1/2" thick	G	Q-14	310	.052	S.F.	.23	2.43		2.66	4.04
3430	2" thick	G	"	300	.053	"	.35	2.51		2.86	4.30
3450	FSK facing, 1.5 lb. density										
3470	1-1/2" thick	G	Q-14	300	.053	S.F.	.36	2.51		2.87	4.31
3480	2" thick	G	"	290	.055	"	.43	2.60		3.03	4.52
3795	Finishes										
3800	Stainless steel woven mesh		Q-14	100	.160	S.F.	.74	7.55		8.29	12.55
3810	For .010" stainless steel, add			160	.100		2.88	4.71		7.59	10.50
3820	18 oz. fiberglass cloth, pasted on			170	.094		.67	4.44		5.11	7.65
3900	8 oz. canvas, pasted on			180	.089		.21	4.19		4.40	6.75
3940	For .016" aluminum jacket, add		↓	200	.080	↓	.92	3.77		4.69	6.85
7878	Contact cement, quart can					Ea.	10.50			10.50	11.55

23 07 16 – HVAC Equipment Insulation

23 07 16.10 HVAC Equipment Thermal Insulation			Crew	Daily Output	Labor-Hours	Unit	Material	Labor	Equipment	Total	Total Incl O&P
0010	**HVAC EQUIPMENT THERMAL INSULATION**										
0100	Rule of thumb, as a percentage of total mechanical costs					Job				10%	10%
0110	Insulation req'd. is based on the surface size/area to be covered										
1000	Boiler, 1-1/2" calcium silicate only	G	Q-14	110	.145	S.F.	4.64	6.85		11.49	15.75
1020	Plus 2" fiberglass	G	"	80	.200	"	5.65	9.45		15.10	21
2000	Breeching, 2" calcium silicate										
2020	Rectangular	G	Q-14	42	.381	S.F.	9.10	17.95		27.05	38
2040	Round	G	"	38.70	.413	"	9.50	19.50		29	41

23 09 Instrumentation and Control for HVAC

23 09 33 – Electric and Electronic Control System for HVAC

23 09 33.10 Electronic Control Systems			Crew	Daily Output	Labor-Hours	Unit	Material	Labor	Equipment	Total	Total Incl O&P
0010	**ELECTRONIC CONTROL SYSTEMS**										
0020	For electronic costs, add to Section 23 09 43.10					Ea.				15%	15%

23 09 43 – Pneumatic Control System for HVAC

23 09 43.10 Pneumatic Control Systems			Crew	Daily Output	Labor-Hours	Unit	Material	Labor	Equipment	Total	Total Incl O&P
0010	**PNEUMATIC CONTROL SYSTEMS**										
0011	Including a nominal 50 ft. of tubing. Add control panelboard if req'd.										
0100	Heating and ventilating, split system										
0200	Mixed air control, economizer cycle, panel readout, tubing										
0220	Up to 10 tons	G	Q-19	.68	35.294	Ea.	4,200	1,900		6,100	7,500
0240	For 10 to 20 tons	G		.63	37.915		4,500	2,050		6,550	8,050
0260	For over 20 tons	G	↓	.58	41.096	↓	4,875	2,225		7,100	8,700
0300	Heating coil, hot water, 3 way valve,										
0320	Freezestat, limit control on discharge, readout		Q-5	.69	23.088	Ea.	3,125	1,250		4,375	5,300
0500	Cooling coil, chilled water, room										
0520	Thermostat, 3 way valve		Q-5	2	8	Ea.	1,400	430		1,830	2,175
0600	Cooling tower, fan cycle, damper control,										
0620	Control system including water readout in/out at panel		Q-19	.67	35.821	Ea.	5,525	1,925		7,450	9,000
1000	Unit ventilator, day/night operation,										
1100	freezestat, ASHRAE, cycle 2		Q-19	.91	26.374	Ea.	3,050	1,425		4,475	5,525
2000	Compensated hot water from boiler, valve control,										
2100	readout and reset at panel, up to 60 GPM		Q-19	.55	43.956	Ea.	5,725	2,375		8,100	9,875
2120	For 120 GPM			.51	47.059		6,125	2,550		8,675	10,600
2140	For 240 GPM			.49	49.180		6,400	2,650		9,050	11,100
3000	Boiler room combustion air, damper to 5 S.F., controls		↓	1.37	17.582	↓	2,750	950		3,700	4,450

23 09 Instrumentation and Control for HVAC

23 09 43 – Pneumatic Control System for HVAC

23 09 43.10 Pneumatic Control Systems		Crew	Daily Output	Labor-Hours	Unit	2015 Bare Costs Material	Labor	Equipment	Total	Total Incl O&P
3500	Fan coil, heating and cooling valves, 4 pipe control system	Q-19	3	8	Ea.	1,250	435		1,685	2,025
3600	Heat exchanger system controls		.86	27.907		2,675	1,500		4,175	5,225
4000	Pneumatic thermostat, including controlling room radiator valve	Q-5	2.43	6.593		830	355		1,185	1,450
4060	Pump control system	Q-19	3	8		1,275	435		1,710	2,050
4500	Air supply for pneumatic control system									
4600	Tank mounted duplex compressor, starter, alternator,									
4620	piping, dryer, PRV station and filter									
4630	1/2 HP	Q-19	.68	35.139	Ea.	10,300	1,900		12,200	14,200
4660	1-1/2 HP		.58	41.739		12,500	2,250		14,750	17,200
4690	5 HP		.42	57.143		29,800	3,100		32,900	37,500

23 13 Facility Fuel-Storage Tanks

23 13 13 – Facility Underground Fuel-Oil, Storage Tanks

23 13 13.09 Single-Wall Steel Fuel-Oil Tanks		Crew	Daily Output	Labor-Hours	Unit	Material	Labor	Equipment	Total	Total Incl O&P
0010	**SINGLE-WALL STEEL FUEL-OIL TANKS**									
5000	Tanks, steel ugnd., sti-p3, not incl. hold-down bars									
5500	Excavation, pad, pumps and piping not included									
5510	Single wall, 500 gallon capacity, 7 ga. shell	Q-5	2.70	5.926	Ea.	2,050	320		2,370	2,725
5520	1,000 gallon capacity, 7 ga. shell	"	2.50	6.400		3,825	345		4,170	4,725
5530	2,000 gallon capacity, 1/4" thick shell	Q-7	4.60	6.957		6,200	395		6,595	7,425
5535	2,500 gallon capacity, 7 ga. shell	Q-5	3	5.333		6,850	287		7,137	7,950
5540	5,000 gallon capacity, 1/4" thick shell	Q-7	3.20	10		11,500	570		12,070	13,600
5580	15,000 gallon capacity, 5/16" thick shell		1.70	18.824		19,000	1,075		20,075	22,500
5600	20,000 gallon capacity, 5/16" thick shell		1.50	21.333		26,700	1,225		27,925	31,100
5610	25,000 gallon capacity, 3/8" thick shell		1.30	24.615		37,900	1,400		39,300	43,800
5620	30,000 gallon capacity, 3/8" thick shell		1.10	29.091		38,400	1,650		40,050	44,700
5630	40,000 gallon capacity, 3/8" thick shell		.90	35.556		42,000	2,025		44,025	49,300
5640	50,000 gallon capacity, 3/8" thick shell		.80	40		46,600	2,275		48,875	55,000

23 13 13.23 Glass-Fiber-Reinfcd-Plastic, Fuel-Oil, Storage		Crew	Daily Output	Labor-Hours	Unit	Material	Labor	Equipment	Total	Total Incl O&P
0010	**GLASS-FIBER-REINFCD-PLASTIC, UNDERGRND. FUEL-OIL, STORAGE**									
0210	Fiberglass, underground, single wall, U.L. listed, not including									
0220	manway or hold-down strap									
0240	2,000 gallon capacity	Q-7	4.57	7.002	Ea.	5,950	400		6,350	7,150
0245	3,000 gallon capacity		3.90	8.205		7,125	465		7,590	8,550
0250	4,000 gallon capacity		3.55	9.014		8,275	515		8,790	9,875
0255	5,000 gallon capacity		3.20	10		9,200	570		9,770	11,000
0260	6,000 gallon capacity		2.67	11.985		9,400	680		10,080	11,300
0280	10,000 gallon capacity		2	16		12,800	910		13,710	15,500
0284	15,000 gallon capacity		1.68	19.048		20,500	1,075		21,575	24,100
0290	20,000 gallon capacity		1.45	22.069		26,300	1,250		27,550	30,900
0500	For manway, fittings and hold-downs, add					20%	15%			
1020	Fiberglass, underground, double wall, U.L. listed									
1030	includes manways, not incl. hold-down straps									
1040	600 gallon capacity	Q-5	2.42	6.612	Ea.	7,200	355		7,555	8,450
1050	1,000 gallon capacity	"	2.25	7.111		9,825	380		10,205	11,400
1060	2,500 gallon capacity	Q-7	4.16	7.692		14,800	440		15,240	17,000
1070	3,000 gallon capacity		3.90	8.205		16,000	465		16,465	18,300
1080	4,000 gallon capacity		3.64	8.791		16,200	500		16,700	18,700
1090	6,000 gallon capacity		2.42	13.223		21,100	750		21,850	24,400
1100	8,000 gallon capacity		2.08	15.385		23,500	875		24,375	27,200
1110	10,000 gallon capacity		1.82	17.582		27,300	1,000		28,300	31,500

23 13 Facility Fuel-Storage Tanks

23 13 13 – Facility Underground Fuel-Oil, Storage Tanks

23 13 13.23 Glass-Fiber-Reinfcd-Plastic, Fuel-Oil, Storage		Crew	Daily Output	Labor-Hours	Unit	Material	2015 Bare Costs Labor	Equipment	Total	Total Incl O&P
1120	12,000 gallon capacity	Q-7	1.70	18.824	Ea.	34,100	1,075		35,175	39,100
2210	Fiberglass, underground, single wall, U.L. listed, including									
2220	hold-down straps, no manways									
2240	2,000 gallon capacity	Q-7	3.55	9.014	Ea.	6,400	515		6,915	7,825
2250	4,000 gallon capacity		2.90	11.034		8,725	630		9,355	10,600
2260	6,000 gallon capacity		2	16		10,300	910		11,210	12,700
2280	10,000 gallon capacity		1.60	20		13,700	1,150		14,850	16,800
2284	15,000 gallon capacity		1.39	23.022		21,400	1,300		22,700	25,500
2290	20,000 gallon capacity		1.14	28.070		27,700	1,600		29,300	32,800
3020	Fiberglass, underground, double wall, U.L. listed									
3030	includes manways and hold-down straps									
3040	600 gallon capacity	Q-5	1.86	8.602	Ea.	7,650	465		8,115	9,100
3050	1,000 gallon capacity	"	1.70	9.412		10,300	505		10,805	12,100
3060	2,500 gallon capacity	Q-7	3.29	9.726		15,200	555		15,755	17,500
3070	3,000 gallon capacity		3.13	10.224		16,500	580		17,080	19,000
3080	4,000 gallon capacity		2.93	10.922		16,700	620		17,320	19,300
3090	6,000 gallon capacity		1.86	17.204		22,000	980		22,980	25,700
3100	8,000 gallon capacity		1.65	19.394		24,400	1,100		25,500	28,600
3110	10,000 gallon capacity		1.48	21.622		28,200	1,225		29,425	32,900
3120	12,000 gallon capacity		1.40	22.857		34,900	1,300		36,200	40,400

23 13 23 – Facility Aboveground Fuel-Oil, Storage Tanks

23 13 23.13 Vertical, Steel, Abvground Fuel-Oil, Stor. Tanks		Crew	Daily Output	Labor-Hours	Unit	Material	Labor	Equipment	Total	Total Incl O&P
0010	**VERTICAL, STEEL, ABOVEGROUND FUEL-OIL, STORAGE TANKS**									
4000	Fixed roof oil storage tanks, steel, (1 BBL=42 gal. w/foundation 3′D x 1′W)									
4200	5,000 barrels				Ea.				194,000	213,500
4300	24,000 barrels								333,500	367,000
4500	56,000 barrels								729,000	802,000
4600	110,000 barrels								1,060,000	1,166,000
4800	143,000 barrels								1,250,000	1,375,000
4900	225,000 barrels								1,360,000	1,496,000
5100	Floating roof gasoline tanks, steel, 5,000 barrels (w/foundation 3′D x 1′W)								204,000	225,000
5200	25,000 barrels								381,000	419,000
5400	55,000 barrels								839,000	923,000
5500	100,000 barrels								1,253,000	1,379,000
5700	150,000 barrels								1,532,000	1,685,000
5800	225,000 barrels								2,300,000	2,783,000

23 13 23.16 Horizontal, Stl, Abvgrd Fuel-Oil, Storage Tanks		Crew	Daily Output	Labor-Hours	Unit	Material	Labor	Equipment	Total	Total Incl O&P
0010	**HORIZONTAL, STEEL, ABOVEGROUND FUEL-OIL, STORAGE TANKS**									
3000	Steel, storage, above ground, including cradles, coating,									
3020	fittings, not including foundation, pumps or piping									
3040	Single wall, 275 gallon	Q-5	5	3.200	Ea.	490	172		662	800
3060	550 gallon	"	2.70	5.926		3,750	320		4,070	4,600
3080	1,000 gallon	Q-7	5	6.400		4,025	365		4,390	4,975
3100	1,500 gallon		4.75	6.737		8,775	385		9,160	10,300
3120	2,000 gallon		4.60	6.957		10,600	395		10,995	12,300
3140	5,000 gallon		3.20	10		19,100	570		19,670	21,900
3150	10,000 gallon		2	16		35,100	910		36,010	40,000
3160	15,000 gallon		1.70	18.824		45,100	1,075		46,175	51,000
3170	20,000 gallon		1.45	22.069		58,500	1,250		59,750	66,000
3180	25,000 gallon		1.30	24.615		68,000	1,400		69,400	77,000
3190	30,000 gallon		1.10	29.091		81,500	1,650		83,150	92,500
3320	Double wall, 500 gallon capacity	Q-5	2.40	6.667		2,625	360		2,985	3,425

23 13 Facility Fuel-Storage Tanks

23 13 23 – Facility Aboveground Fuel-Oil, Storage Tanks

23 13 23.16 Horizontal, Stl, Abvgrd Fuel-Oil, Storage Tanks		Crew	Daily Output	Labor-Hours	Unit	Material	2015 Bare Costs Labor	Equipment	Total	Total Incl O&P
3330	2000 gallon capacity	Q-7	4.15	7.711	Ea.	9,975	440		10,415	11,700
3340	4000 gallon capacity		3.60	8.889		17,800	505		18,305	20,400
3350	6000 gallon capacity		2.40	13.333		21,000	760		21,760	24,300
3360	8000 gallon capacity		2	16		27,000	910		27,910	31,100
3370	10000 gallon capacity		1.80	17.778		30,200	1,000		31,200	34,700
3380	15000 gallon capacity		1.50	21.333		45,900	1,225		47,125	52,500
3390	20000 gallon capacity		1.30	24.615		52,500	1,400		53,900	59,500
3400	25000 gallon capacity		1.15	27.826		63,500	1,575		65,075	72,500
3410	30000 gallon capacity	↓	1	32	↓	69,500	1,825		71,325	79,500
23 13 23.26 Horizontal, Conc., Abvgrd Fuel-Oil, Stor. Tanks										
0010	**HORIZONTAL, CONCRETE, ABOVEGROUND FUEL-OIL, STORAGE TANKS**									
0050	Concrete, storage, above ground, including pad & pump									
0100	500 gallon	F-3	2	20	Ea.	10,000	960	325	11,285	12,800
0200	1,000 gallon	"	2	20		14,000	960	325	15,285	17,200
0300	2,000 gallon	F-4	2	24		18,000	1,150	565	19,715	22,200
0400	4,000 gallon		2	24		23,000	1,150	565	24,715	27,700
0500	8,000 gallon		2	24		36,000	1,150	565	37,715	42,000
0600	12,000 gallon	↓	2	24	↓	48,000	1,150	565	49,715	55,500

23 21 Hydronic Piping and Pumps

23 21 20 – Hydronic HVAC Piping Specialties

23 21 20.10 Air Control		Crew	Daily Output	Labor-Hours	Unit	Material	Labor	Equipment	Total	Total Incl O&P
0010	**AIR CONTROL**									
0030	Air separator, with strainer									
0040	2" diameter	Q-5	6	2.667	Ea.	1,175	143		1,318	1,525
0080	2-1/2" diameter		5	3.200		1,325	172		1,497	1,700
0100	3" diameter		4	4		2,050	215		2,265	2,575
0120	4" diameter	↓	3	5.333		2,950	287		3,237	3,650
0130	5" diameter	Q-6	3.60	6.667		3,750	370		4,120	4,675
0140	6" diameter	"	3.40	7.059	↓	4,500	395		4,895	5,550
23 21 20.18 Automatic Air Vent										
0010	**AUTOMATIC AIR VENT**									
0020	Cast iron body, stainless steel internals, float type									
0060	1/2" NPT inlet, 300 psi	1 Stpi	12	.667	Ea.	109	40		149	180
0220	3/4" NPT inlet, 250 psi	"	10	.800		350	48		398	455
0340	1-1/2" NPT inlet, 250 psi	Q-5	12	1.333	↓	1,075	71.50		1,146.50	1,300
23 21 20.42 Expansion Joints										
0010	**EXPANSION JOINTS**									
0100	Bellows type, neoprene cover, flanged spool									
0140	6" face to face, 1-1/4" diameter	1 Stpi	11	.727	Ea.	255	43.50		298.50	345
0160	1-1/2" diameter	"	10.60	.755		255	45		300	350
0180	2" diameter	Q-5	13.30	1.203		258	64.50		322.50	380
0190	2-1/2" diameter		12.40	1.290		267	69.50		336.50	400
0200	3" diameter		11.40	1.404		299	75.50		374.50	445
0480	10" face to face, 2" diameter		13	1.231		370	66		436	510
0500	2-1/2" diameter		12	1.333		390	71.50		461.50	540
0520	3" diameter		11	1.455		400	78		478	560
0540	4" diameter		8	2		455	108		563	660
0560	5" diameter		7	2.286		540	123		663	780
0580	6" diameter	↓	6	2.667	↓	560	143		703	830

23 21 20 – Hydronic HVAC Piping Specialties

23 21 20.46 Expansion Tanks		Crew	Daily Output	Labor-Hours	Unit	Material	2015 Bare Costs Labor	Equipment	Total	Total Incl O&P
0010	EXPANSION TANKS									
1507	Underground fuel-oil storage tanks, see Section 23 13 13									
1512	Tank leak detection systems, see Section 28 33 33.50									
2000	Steel, liquid expansion, ASME, painted, 15 gallon capacity	Q-5	17	.941	Ea.	640	50.50		690.50	775
2020	24 gallon capacity		14	1.143		715	61.50		776.50	880
2040	30 gallon capacity		12	1.333		715	71.50		786.50	895
2060	40 gallon capacity		10	1.600		835	86		921	1,050
2080	60 gallon capacity		8	2		1,000	108		1,108	1,250
2100	80 gallon capacity		7	2.286		1,075	123		1,198	1,350
2120	100 gallon capacity		6	2.667		1,450	143		1,593	1,825
3000	Steel ASME expansion, rubber diaphragm, 19 gal. cap. accept.		12	1.333		2,425	71.50		2,496.50	2,775
3020	31 gallon capacity		8	2		2,700	108		2,808	3,125
3040	61 gallon capacity		6	2.667		3,800	143		3,943	4,400
3080	119 gallon capacity		4	4		4,100	215		4,315	4,825
3100	158 gallon capacity		3.80	4.211		5,675	226		5,901	6,600
3140	317 gallon capacity		2.80	5.714		8,575	305		8,880	9,900
3180	528 gallon capacity	↓	2.40	6.667	↓	13,900	360		14,260	15,800

23 21 20.58 Hydronic Heating Control Valves		Crew	Daily Output	Labor-Hours	Unit	Material	Labor	Equipment	Total	Total Incl O&P
0010	HYDRONIC HEATING CONTROL VALVES									
0050	Hot water, nonelectric, thermostatic									
0100	Radiator supply, 1/2" diameter	1 Stpi	24	.333	Ea.	68.50	19.90		88.40	105
0120	3/4" diameter		20	.400		71.50	24		95.50	115
0140	1" diameter		19	.421		88	25		113	135
0160	1-1/4" diameter	↓	15	.533	↓	120	32		152	180
0500	For low pressure steam, add					25%				

23 21 20.70 Steam Traps		Crew	Daily Output	Labor-Hours	Unit	Material	Labor	Equipment	Total	Total Incl O&P
0010	STEAM TRAPS									
0030	Cast iron body, threaded									
0040	Inverted bucket									
0050	1/2" pipe size	1 Stpi	12	.667	Ea.	157	40		197	232
0070	3/4" pipe size		10	.800		278	48		326	375
0100	1" pipe size		9	.889		420	53		473	545
0120	1-1/4" pipe size	↓	8	1	↓	635	60		695	790
1000	Float & thermostatic, 15 psi									
1010	3/4" pipe size	1 Stpi	16	.500	Ea.	141	30		171	200
1020	1" pipe size		15	.533		169	32		201	234
1040	1-1/2" pipe size		9	.889		298	53		351	410
1060	2" pipe size	↓	6	1.333	↓	590	79.50		669.50	770

23 21 20.76 Strainers, Y Type, Bronze Body		Crew	Daily Output	Labor-Hours	Unit	Material	Labor	Equipment	Total	Total Incl O&P
0010	STRAINERS, Y TYPE, BRONZE BODY									
0050	Screwed, 125 lb., 1/4" pipe size	1 Stpi	24	.333	Ea.	23	19.90		42.90	55.50
0070	3/8" pipe size		24	.333		27.50	19.90		47.40	60.50
0100	1/2" pipe size		20	.400		27.50	24		51.50	66.50
0140	1" pipe size		17	.471		50	28		78	97.50
0160	1-1/2" pipe size		14	.571		108	34		142	171
0180	2" pipe size		13	.615		144	37		181	214
0182	3" pipe size	↓	12	.667		845	40		885	990
0200	300 lb., 2-1/2" pipe size	Q-5	17	.941		510	50.50		560.50	635
0220	3" pipe size		16	1		1,000	54		1,054	1,175
0240	4" pipe size	↓	15	1.067	↓	2,300	57.50		2,357.50	2,600
0500	For 300 lb. rating 1/4" thru 2", add					15%				

23 21 Hydronic Piping and Pumps

23 21 20 – Hydronic HVAC Piping Specialties

23 21 20.76 Strainers, Y Type, Bronze Body		Crew	Daily Output	Labor-Hours	Unit	Material	2015 Bare Costs Labor	Equipment	Total	Total Incl O&P
1000	Flanged, 150 lb., 1-1/2" pipe size	1 Stpi	11	.727	Ea.	520	43.50		563.50	640
1020	2" pipe size	"	8	1		705	60		765	865
1030	2-1/2" pipe size	Q-5	5	3.200		950	172		1,122	1,300
1040	3" pipe size		4.50	3.556		1,175	191		1,366	1,575
1060	4" pipe size		3	5.333		1,775	287		2,062	2,375
1100	6" pipe size	Q-6	3	8		3,400	445		3,845	4,400
1106	8" pipe size	"	2.60	9.231		3,725	515		4,240	4,875
1500	For 300 lb. rating, add					40%				

23 21 20.78 Strainers, Y Type, Iron Body		Crew	Daily Output	Labor-Hours	Unit	Material	Labor	Equipment	Total	Total Incl O&P
0010	**STRAINERS, Y TYPE, IRON BODY**									
0050	Screwed, 250 lb., 1/4" pipe size	1 Stpi	20	.400	Ea.	11.05	24		35.05	48
0070	3/8" pipe size		20	.400		11.05	24		35.05	48
0100	1/2" pipe size		20	.400		11.05	24		35.05	48
0140	1" pipe size		16	.500		18.25	30		48.25	65
0160	1-1/2" pipe size		12	.667		30	40		70	92.50
0180	2" pipe size		8	1		44.50	60		104.50	139
0220	3" pipe size	Q-5	11	1.455		289	78		367	440
0240	4" pipe size	"	5	3.200		490	172		662	800
0500	For galvanized body, add					50%				
1000	Flanged, 125 lb., 1-1/2" pipe size	1 Stpi	11	.727	Ea.	107	43.50		150.50	183
1020	2" pipe size	"	8	1		113	60		173	214
1040	3" pipe size	Q-5	4.50	3.556		186	191		377	495
1060	4" pipe size	"	3	5.333		305	287		592	770
1080	5" pipe size	Q-6	3.40	7.059		385	395		780	1,025
1100	6" pipe size	"	3	8		615	445		1,060	1,350
1500	For 250 lb. rating, add					20%				
2000	For galvanized body, add					50%				
2500	For steel body, add					40%				

23 21 20.88 Venturi Flow		Crew	Daily Output	Labor-Hours	Unit	Material	Labor	Equipment	Total	Total Incl O&P
0010	**VENTURI FLOW**, Measuring device									
0050	1/2" diameter	1 Stpi	24	.333	Ea.	281	19.90		300.90	340
0120	1" diameter		19	.421		276	25		301	345
0140	1-1/4" diameter		15	.533		340	32		372	425
0160	1-1/2" diameter		13	.615		355	37		392	445
0180	2" diameter		11	.727		365	43.50		408.50	470
0220	3" diameter	Q-5	14	1.143		515	61.50		576.50	665
0240	4" diameter	"	11	1.455		775	78		853	970
0280	6" diameter	Q-6	3.50	6.857		1,125	380		1,505	1,825
0500	For meter, add					2,125			2,125	2,350

23 21 23 – Hydronic Pumps

23 21 23.13 In-Line Centrifugal Hydronic Pumps		Crew	Daily Output	Labor-Hours	Unit	Material	Labor	Equipment	Total	Total Incl O&P
0010	**IN-LINE CENTRIFUGAL HYDRONIC PUMPS**									
0600	Bronze, sweat connections, 1/40 HP, in line									
0640	3/4" size	Q-1	16	1	Ea.	218	53		271	320
1000	Flange connection, 3/4" to 1-1/2" size									
1040	1/12 HP	Q-1	6	2.667	Ea.	565	141		706	835
1060	1/8 HP		6	2.667		970	141		1,111	1,300
1100	1/3 HP		6	2.667		1,075	141		1,216	1,425
1140	2" size, 1/6 HP		5	3.200		1,400	169		1,569	1,775
1180	2-1/2" size, 1/4 HP		5	3.200		1,775	169		1,944	2,200
2000	Cast iron, flange connection									
2040	3/4" to 1-1/2" size, in line, 1/12 HP	Q-1	6	2.667	Ea.	365	141		506	615

23 21 Hydronic Piping and Pumps

23 21 23 – Hydronic Pumps

23 21 23.13 In-Line Centrifugal Hydronic Pumps		Crew	Daily Output	Labor-Hours	Unit	Material	2015 Bare Costs Labor	Equipment	Total	Total Incl O&P
2100	1/3 HP	Q-1	6	2.667	Ea.	680	141		821	960
2101	Pumps, circulating, 3/4" to 1-1/2" size, 1/3 HP		6	2.667		790	141		931	1,075
2140	2" size, 1/6 HP		5	3.200		745	169		914	1,075
2180	2-1/2" size, 1/4 HP		5	3.200		960	169		1,129	1,300
2220	3" size, 1/4 HP	↓	4	4	↓	975	211		1,186	1,400
2600	For nonferrous impeller, add					3%				

23 21 29 – Automatic Condensate Pump Units

23 21 29.10 Condensate Removal Pump System		Crew	Daily Output	Labor-Hours	Unit	Material	Labor	Equipment	Total	Total Incl O&P
0010	**CONDENSATE REMOVAL PUMP SYSTEM**									
0020	Pump with 1 gal. ABS tank									
0100	115 V									
0120	1/50 HP, 200 GPH G	1 Stpi	12	.667	Ea.	197	40		237	277
0140	1/18 HP, 270 GPH G		10	.800		210	48		258	305
0160	1/5 HP, 450 GPH G	↓	8	1	↓	470	60		530	605
0200	230 V									
0260	1/5 HP, 450 GPH G	1 Stpi	8	1	Ea.	520	60		580	660

23 22 Steam and Condensate Piping and Pumps

23 22 13 – Steam and Condensate Heating Piping

23 22 13.23 Aboveground Steam and Condensate Piping		Crew	Daily Output	Labor-Hours	Unit	Material	Labor	Equipment	Total	Total Incl O&P
0010	**ABOVEGROUND STEAM AND CONDENSATE HEATING PIPING**									
0020	Condensate meter									
0100	500 lb. per hour	1 Stpi	14	.571	Ea.	3,275	34		3,309	3,650
0140	1500 lb. per hour	"	7	1.143	"	4,100	68.50		4,168.50	4,600

23 22 23 – Steam Condensate Pumps

23 22 23.10 Condensate Return System		Crew	Daily Output	Labor-Hours	Unit	Material	Labor	Equipment	Total	Total Incl O&P
0010	**CONDENSATE RETURN SYSTEM**									
2000	Simplex									
2010	With pump, motor, CI receiver, float switch									
2020	3/4 HP, 15 GPM	Q-1	1.80	8.889	Ea.	6,425	470		6,895	7,750
2100	Duplex									
2110	With 2 pumps and motors, CI receiver, float switch, alternator									
2120	3/4 HP, 15 GPM, 15 gal. CI rcvr.	Q-1	1.40	11.429	Ea.	7,000	605		7,605	8,600
2130	1 HP, 25 GPM		1.20	13.333		8,400	705		9,105	10,300
2140	1-1/2 HP, 45 GPM		1	16		9,750	845		10,595	12,000
2150	1-1/2 HP, 60 GPM	↓	1	16	↓	10,900	845		11,745	13,300

23 31 HVAC Ducts and Casings

23 31 13 – Metal Ducts

23 31 13.13 Rectangular Metal Ducts		Crew	Daily Output	Labor-Hours	Unit	Material	Labor	Equipment	Total	Total Incl O&P
0010	**RECTANGULAR METAL DUCTS**									
0020	Fabricated rectangular, includes fittings, joints, supports,									
0021	allowance for flexible connections and field sketches.									
0030	Does not include "as-built dwgs." or insulation.									
0031	NOTE: Fabrication and installation are combined									
0040	as LABOR cost. Approx. 25% fittings assumed.									
0042	Fabrication/Inst. is to commercial quality standards									
0043	(SMACNA or equiv.) for structure, sealing, leak testing, etc.									

23 31 13 – Metal Ducts

23 31 13.13 Rectangular Metal Ducts		Crew	Daily Output	Labor-Hours	Unit	Material	2015 Bare Costs Labor	Equipment	Total	Total Incl O&P
0050	Add to labor for elevated installation									
0051	of fabricated ductwork									
0052	10' to 15' high						6%			
0053	15' to 20' high						12%			
0054	20' to 25' high						15%			
0055	25' to 30' high						21%			
0056	30' to 35' high						24%			
0057	35' to 40' high						30%			
0058	Over 40' high						33%			
0072	For duct insulation see Line 23 07 13.10 3000									
0100	Aluminum, alloy 3003-H14, under 100 lb.	Q-10	75	.320	Lb.	3.20	16.70		19.90	29
0110	100 to 500 lb.		80	.300		1.88	15.65		17.53	26
0120	500 to 1,000 lb.		95	.253		1.82	13.20		15.02	22
0140	1,000 to 2,000 lb.		120	.200		1.77	10.45		12.22	17.90
0150	2,000 to 5,000 lb.		130	.185		1.77	9.65		11.42	16.70
0160	Over 5,000 lb.		145	.166		1.77	8.65		10.42	15.15
0500	Galvanized steel, under 200 lb.		235	.102		.65	5.35		6	8.85
0520	200 to 500 lb.		245	.098		.64	5.10		5.74	8.50
0540	500 to 1,000 lb.		255	.094		.62	4.91		5.53	8.20
0560	1,000 to 2,000 lb.		265	.091		.61	4.73		5.34	7.90
0570	2,000 to 5,000 lb.		275	.087		.61	4.56		5.17	7.60
0580	Over 5,000 lb.		285	.084		.61	4.40		5.01	7.35
1000	Stainless steel, type 304, under 100 lb.		165	.145		6.35	7.60		13.95	18.55
1020	100 to 500 lb.		175	.137		4.05	7.15		11.20	15.40
1030	500 to 1,000 lb.		190	.126		2.95	6.60		9.55	13.35
1040	1,000 to 2,000 lb.		200	.120		2.89	6.25		9.14	12.80
1050	2,000 to 5,000 lb.		225	.107		2.39	5.55		7.94	11.15
1060	Over 5,000 lb.		235	.102		1.96	5.35		7.31	10.30
1100	For medium pressure ductwork, add						15%			
1200	For high pressure ductwork, add						40%			
1210	For welded ductwork, add						85%			
1220	For 30% fittings, add						11%			
1224	For 40% fittings, add						34%			
1228	For 50% fittings, add						56%			
1232	For 60% fittings, add						79%			
1236	For 70% fittings, add						101%			
1240	For 80% fittings, add						124%			
1244	For 90% fittings, add						147%			
1248	For 100% fittings, add						169%			
1252	Note: Fittings add includes time for detailing and installation.									

23 31 13.19 Metal Duct Fittings

		Crew	Daily Output	Labor-Hours	Unit	Material	Labor	Equipment	Total	Total Incl O&P
0010	**METAL DUCT FITTINGS**									
2000	Fabrics for flexible connections, with metal edge	1 Shee	100	.080	L.F.	3.44	4.48		7.92	10.65
2100	Without metal edge	"	160	.050	"	2.46	2.80		5.26	7

23 31 16 – Nonmetal Ducts

23 31 16.13 Fibrous-Glass Ducts

		Crew	Daily Output	Labor-Hours	Unit	Material	Labor	Equipment	Total	Total Incl O&P
0010	**FIBROUS-GLASS DUCTS**									
3490	Rigid fiberglass duct board, foil reinf. kraft facing									
3500	Rectangular, 1" thick, alum. faced, (FRK), std. weight	Q-10	350	.069	SF Surf	.79	3.58		4.37	6.30

23 33 Air Duct Accessories

23 33 13 – Dampers

23 33 13.13 Volume-Control Dampers		Crew	Daily Output	Labor-Hours	Unit	Material	2015 Bare Costs Labor	Equipment	Total	Total Incl O&P
0010	**VOLUME-CONTROL DAMPERS**									
5990	Multi-blade dampers, opposed blade, 8″ x 6″	1 Shee	24	.333	Ea.	21.50	18.65		40.15	52.50
5994	8″ x 8″		22	.364		22.50	20.50		43	55.50
5996	10″ x 10″		21	.381		26.50	21.50		48	62
6000	12″ x 12″		21	.381		29.50	21.50		51	65
6020	12″ x 18″		18	.444		39.50	25		64.50	81.50
6030	14″ x 10″		20	.400		29	22.50		51.50	65.50
6031	14″ x 14″		17	.471		35.50	26.50		62	79
6033	16″ x 12″		17	.471		35.50	26.50		62	79
6035	16″ x 16″		16	.500		44	28		72	91.50
6037	18″ x 16″		15	.533		48	30		78	98.50
6038	18″ x 18″		15	.533		52	30		82	103
6070	20″ x 16″		14	.571		52	32		84	106
6072	20″ x 20″		13	.615		62.50	34.50		97	122
6074	22″ x 18″		14	.571		62.50	32		94.50	118
6076	24″ x 16″		11	.727		61	40.50		101.50	130
6078	24″ x 20″		8	1		72.50	56		128.50	166
6080	24″ x 24″		8	1		85	56		141	179
6110	26″ x 26″	↓	6	1.333		95	74.50		169.50	219
6133	30″ x 30″	Q-9	6.60	2.424		137	122		259	335
6135	32″ x 32″		6.40	2.500		154	126		280	360
6180	48″ x 36″	↓	5.60	2.857	↓	253	144		397	500
8000	Multi-blade dampers, parallel blade									
8100	8″ x 8″	1 Shee	24	.333	Ea.	82	18.65		100.65	119
8140	16″ x 10″		20	.400		103	22.50		125.50	148
8200	24″ x 16″		11	.727		134	40.50		174.50	210
8260	30″ x 18″	↓	7	1.143	↓	186	64		250	300

23 33 13.16 Fire Dampers		Crew	Daily Output	Labor-Hours	Unit	Material	Labor	Equipment	Total	Total Incl O&P
0010	**FIRE DAMPERS**									
3000	Fire damper, curtain type, 1-1/2 hr. rated, vertical, 6″ x 6″	1 Shee	24	.333	Ea.	24	18.65		42.65	55
3020	8″ x 6″		22	.364		24	20.50		44.50	57.50
3240	16″ x 14″		18	.444		44	25		69	86.50
3400	24″ x 20″	↓	8	1	↓	54	56		110	145

23 33 13.28 Splitter Damper Assembly		Crew	Daily Output	Labor-Hours	Unit	Material	Labor	Equipment	Total	Total Incl O&P
0010	**SPLITTER DAMPER ASSEMBLY**									
7000	Self locking, 1′ rod	1 Shee	24	.333	Ea.	23.50	18.65		42.15	54
7020	3′ rod		22	.364		30	20.50		50.50	64
7040	4′ rod		20	.400		33.50	22.50		56	71
7060	6′ rod	↓	18	.444	↓	41	25		66	83

23 33 19 – Duct Silencers

23 33 19.10 Duct Silencers		Crew	Daily Output	Labor-Hours	Unit	Material	Labor	Equipment	Total	Total Incl O&P
0010	**DUCT SILENCERS**									
9000	Silencers, noise control for air flow, duct				MCFM	57			57	63

23 33 33 – Duct-Mounting Access Doors

23 33 33.13 Duct Access Doors		Crew	Daily Output	Labor-Hours	Unit	Material	Labor	Equipment	Total	Total Incl O&P
0010	**DUCT ACCESS DOORS**									
1000	Duct access door, insulated, 6″ x 6″	1 Shee	14	.571	Ea.	16.05	32		48.05	66.50
1020	10″ x 10″		11	.727		18.45	40.50		58.95	82.50
1040	12″ x 12″		10	.800		20	45		65	90.50
1050	12″ x 18″		9	.889		37	49.50		86.50	117
1070	18″ x 18″	↓	8	1	↓	32.50	56		88.50	121

23 33 33 – Duct-Mounting Access Doors

23 33 33.13 Duct Access Doors		Crew	Daily Output	Labor-Hours	Unit	Material	2015 Bare Costs Labor	Equipment	Total	Total Incl O&P
1074	24" x 18"	1 Shee	8	1	Ea.	44	56		100	134

23 33 46 – Flexible Ducts

23 33 46.10 Flexible Air Ducts			Crew	Daily Output	Labor-Hours	Unit	Material	Labor	Equipment	Total	Total Incl O&P
0010	**FLEXIBLE AIR DUCTS**										
1280	Add to labor for elevated installation										
1282	of prefabricated (purchased) ductwork										
1283	10' to 15' high							10%			
1284	15' to 20' high							20%			
1285	20' to 25' high							25%			
1286	25' to 30' high							35%			
1287	30' to 35' high							40%			
1288	35' to 40' high							50%			
1289	Over 40' high							55%			
1300	Flexible, coated fiberglass fabric on corr. resist. metal helix										
1400	pressure to 12" (WG) UL-181										
1500	Noninsulated, 3" diameter		Q-9	400	.040	L.F.	1.12	2.01		3.13	4.31
1540	5" diameter			320	.050		1.30	2.52		3.82	5.30
1560	6" diameter			280	.057		1.50	2.88		4.38	6.05
1580	7" diameter			240	.067		1.53	3.36		4.89	6.85
1600	8" diameter			200	.080		1.91	4.03		5.94	8.25
1640	10" diameter			160	.100		2.46	5.05		7.51	10.40
1660	12" diameter			120	.133		2.94	6.70		9.64	13.50
1900	Insulated, 1" thick, PE jacket, 3" diameter	G		380	.042		2.60	2.12		4.72	6.10
1910	4" diameter	G		340	.047		2.60	2.37		4.97	6.50
1920	5" diameter	G		300	.053		2.60	2.69		5.29	6.95
1940	6" diameter	G		260	.062		2.94	3.10		6.04	7.95
1960	7" diameter	G		220	.073		3.20	3.66		6.86	9.10
1980	8" diameter	G		180	.089		3.49	4.48		7.97	10.70
2020	10" diameter	G		140	.114		4.25	5.75		10	13.50
2040	12" diameter	G		100	.160		4.90	8.05		12.95	17.70

23 33 53 – Duct Liners

23 33 53.10 Duct Liner Board			Crew	Daily Output	Labor-Hours	Unit	Material	Labor	Equipment	Total	Total Incl O&P
0010	**DUCT LINER BOARD**										
3340	Board type fiberglass liner, FSK, 1-1/2 lb. density										
3344	1" thick	G	Q-14	150	.107	S.F.	.62	5.05		5.67	8.55
3345	1-1/2" thick	G		130	.123		.68	5.80		6.48	9.80
3346	2" thick	G		120	.133		.79	6.30		7.09	10.65
3348	3" thick	G		110	.145		1.02	6.85		7.87	11.75
3350	4" thick	G		100	.160		1.25	7.55		8.80	13.15
3356	3 lb. density, 1" thick	G		150	.107		.79	5.05		5.84	8.70
3358	1-1/2" thick	G		130	.123		1	5.80		6.80	10.15
3360	2" thick	G		120	.133		1.22	6.30		7.52	11.15
3362	2-1/2" thick	G		110	.145		1.43	6.85		8.28	12.20
3364	3" thick	G		100	.160		1.64	7.55		9.19	13.55
3366	4" thick	G		90	.178		2.06	8.40		10.46	15.30
3370	6 lb. density, 1" thick	G		140	.114		1.12	5.40		6.52	9.65
3374	1-1/2" thick	G		120	.133		1.50	6.30		7.80	11.45
3378	2" thick	G		100	.160		1.88	7.55		9.43	13.80
3490	Board type, fiberglass liner, 3 lb. density										
3680	No finish										
3700	1" thick	G	Q-14	170	.094	S.F.	.44	4.44		4.88	7.40
3710	1-1/2" thick	G		140	.114		.66	5.40		6.06	9.15

23 33 Air Duct Accessories

23 33 53 – Duct Liners

23 33 53.10 Duct Liner Board		Crew	Daily Output	Labor-Hours	Unit	Material	Labor	Equipment	Total	Total Incl O&P
3720	2" thick G	Q-14	130	.123	S.F.	.88	5.80		6.68	10
3940	Board type, non-fibrous foam									
3950	Temperature, bacteria and fungi resistant									
3960	1" thick G	Q-14	150	.107	S.F.	2.38	5.05		7.43	10.45
3970	1-1/2" thick G		130	.123		3.28	5.80		9.08	12.65
3980	2" thick G	↓	120	.133	↓	3.98	6.30		10.28	14.20

(Material, Labor, Equipment and Total are 2015 Bare Costs.)

23 34 HVAC Fans

23 34 13 – Axial HVAC Fans

23 34 13.10 Axial Flow HVAC Fans

Line	Description	Crew	Daily Output	Labor-Hours	Unit	Material	Labor	Equipment	Total	Total Incl O&P
0010	**AXIAL FLOW HVAC FANS**									
0020	Air conditioning and process air handling									
1500	Vaneaxial, low pressure, 2000 CFM, 1/2 HP	Q-20	3.60	5.556	Ea.	2,225	285		2,510	2,875
1520	4,000 CFM, 1 HP		3.20	6.250		2,600	320		2,920	3,325
1540	8,000 CFM, 2 HP	↓	2.80	7.143	↓	3,275	365		3,640	4,175

23 34 14 – Blower HVAC Fans

23 34 14.10 Blower Type HVAC Fans

Line	Description	Crew	Daily Output	Labor-Hours	Unit	Material	Labor	Equipment	Total	Total Incl O&P
0010	**BLOWER TYPE HVAC FANS**									
2500	Ceiling fan, right angle, extra quiet, 0.10" S.P.									
2520	95 CFM	Q-20	20	1	Ea.	300	51		351	410
2540	210 CFM		19	1.053		355	54		409	470
2560	385 CFM		18	1.111		450	57		507	580
2580	885 CFM		16	1.250		890	64		954	1,075
2600	1,650 CFM		13	1.538		1,225	79		1,304	1,475
2620	2,960 CFM	↓	11	1.818		1,650	93		1,743	1,950
2640	For wall or roof cap, add	1 Shee	16	.500		300	28		328	375
2660	For straight thru fan, add					10%				
2680	For speed control switch, add	1 Elec	16	.500	↓	164	27.50		191.50	222
7500	Utility set, steel construction, pedestal, 1/4" S.P.									
7520	Direct drive, 150 CFM, 1/8 HP	Q-20	6.40	3.125	Ea.	870	160		1,030	1,200
7540	485 CFM, 1/6 HP		5.80	3.448		1,100	177		1,277	1,475
7560	1950 CFM, 1/2 HP		4.80	4.167		1,275	213		1,488	1,725
7580	2410 CFM, 3/4 HP		4.40	4.545		2,375	233		2,608	2,950
7600	3328 CFM, 1-1/2 HP	↓	3	6.667	↓	2,625	340		2,965	3,425
7680	V-belt drive, drive cover, 3 phase									
7700	800 CFM, 1/4 HP	Q-20	6	3.333	Ea.	980	171		1,151	1,325
7720	1,300 CFM, 1/3 HP		5	4		1,025	205		1,230	1,425
7740	2,000 CFM, 1 HP		4.60	4.348		1,225	223		1,448	1,700
7760	2,900 CFM, 3/4 HP	↓	4.20	4.762	↓	1,650	244		1,894	2,175

23 34 16 – Centrifugal HVAC Fans

23 34 16.10 Centrifugal Type HVAC Fans

Line	Description	Crew	Daily Output	Labor-Hours	Unit	Material	Labor	Equipment	Total	Total Incl O&P
0010	**CENTRIFUGAL TYPE HVAC FANS**									
0200	In-line centrifugal, supply/exhaust booster									
0220	aluminum wheel/hub, disconnect switch, 1/4" S.P.									
0240	500 CFM, 10" diameter connection	Q-20	3	6.667	Ea.	1,300	340		1,640	1,950
0260	1,380 CFM, 12" diameter connection		2	10		1,375	510		1,885	2,300
0280	1,520 CFM, 16" diameter connection		2	10		1,500	510		2,010	2,425
0300	2,560 CFM, 18" diameter connection		1	20		1,625	1,025		2,650	3,350
0320	3,480 CFM, 20" diameter connection		.80	25		1,925	1,275		3,200	4,075
0326	5,080 CFM, 20" diameter connection	↓	.75	26.667	↓	2,100	1,375		3,475	4,400

23 34 16 – Centrifugal HVAC Fans

23 34 16.10 Centrifugal Type HVAC Fans		Crew	Daily Output	Labor-Hours	Unit	2015 Bare Costs Material	Labor	Equipment	Total	Total Incl O&P
3500	Centrifugal, airfoil, motor and drive, complete									
3520	1000 CFM, 1/2 HP	Q-20	2.50	8	Ea.	1,900	410		2,310	2,700
3540	2,000 CFM, 1 HP		2	10		2,150	510		2,660	3,125
3560	4,000 CFM, 3 HP		1.80	11.111		2,725	570		3,295	3,875
3580	8,000 CFM, 7-1/2 HP		1.40	14.286		4,100	730		4,830	5,625
3600	12,000 CFM, 10 HP		1	20		5,450	1,025		6,475	7,550
5000	Utility set, centrifugal, V belt drive, motor									
5020	1/4" S.P., 1200 CFM, 1/4 HP	Q-20	6	3.333	Ea.	1,750	171		1,921	2,175
5040	1520 CFM, 1/3 HP		5	4		2,225	205		2,430	2,750
5060	1850 CFM, 1/2 HP		4	5		2,200	256		2,456	2,825
5080	2180 CFM, 3/4 HP		3	6.667		2,600	340		2,940	3,400
5100	1/2" S.P., 3600 CFM, 1 HP		2	10		2,700	510		3,210	3,750
5120	4250 CFM, 1-1/2 HP		1.60	12.500		3,300	640		3,940	4,600
5140	4800 CFM, 2 HP		1.40	14.286		4,000	730		4,730	5,500
7000	Roof exhauster, centrifugal, aluminum housing, 12" galvanized									
7020	curb, bird screen, back draft damper, 1/4" S.P.									
7100	Direct drive, 320 CFM, 11" sq. damper	Q-20	7	2.857	Ea.	705	146		851	1,000
7120	600 CFM, 11" sq. damper		6	3.333		900	171		1,071	1,250
7140	815 CFM, 13" sq. damper		5	4		900	205		1,105	1,300
7160	1450 CFM, 13" sq. damper		4.20	4.762		1,450	244		1,694	1,975
7180	2050 CFM, 16" sq. damper		4	5		1,725	256		1,981	2,300
7200	V-belt drive, 1650 CFM, 12" sq. damper		6	3.333		1,300	171		1,471	1,675
7220	2750 CFM, 21" sq. damper		5	4		1,550	205		1,755	2,000
7230	3500 CFM, 21" sq. damper		4.50	4.444		1,725	228		1,953	2,250
7240	4910 CFM, 23" sq. damper		4	5		2,125	256		2,381	2,725
7260	8525 CFM, 28" sq. damper		3	6.667		2,800	340		3,140	3,600
7280	13,760 CFM, 35" sq. damper		2	10		3,925	510		4,435	5,100
7300	20,558 CFM, 43" sq. damper		1	20		7,875	1,025		8,900	10,200
7320	For 2 speed winding, add					15%				
7340	For explosionproof motor, add					600			600	660
7360	For belt driven, top discharge, add					15%				
8500	Wall exhausters, centrifugal, auto damper, 1/8" S.P.									
8520	Direct drive, 610 CFM, 1/20 HP	Q-20	14	1.429	Ea.	425	73		498	575
8540	796 CFM, 1/12 HP		13	1.538		880	79		959	1,075
8560	822 CFM, 1/6 HP		12	1.667		1,075	85.50		1,160.50	1,300
8580	1,320 CFM, 1/4 HP		12	1.667		1,250	85.50		1,335.50	1,500
9500	V-belt drive, 3 phase									
9520	2,800 CFM, 1/4 HP	Q-20	9	2.222	Ea.	1,925	114		2,039	2,300
9540	3,740 CFM, 1/2 HP	"	8	2.500	"	2,000	128		2,128	2,400

23 34 23 – HVAC Power Ventilators

23 34 23.10 HVAC Power Circulators and Ventilators		Crew	Daily Output	Labor-Hours	Unit	Material	Labor	Equipment	Total	Total Incl O&P
0010	**HVAC POWER CIRCULATORS AND VENTILATORS**									
3000	Paddle blade air circulator, 3 speed switch									
3020	42", 5,000 CFM high, 3000 CFM low G	1 Elec	2.40	3.333	Ea.	163	182		345	450
3040	52", 6,500 CFM high, 4000 CFM low G	"	2.20	3.636	"	170	199		369	485
3100	For antique white motor, same cost									
3200	For brass plated motor, same cost									
3300	For light adaptor kit, add G				Ea.	41			41	45
6000	Propeller exhaust, wall shutter									
6020	Direct drive, one speed, .075" S.P.									
6100	653 CFM, 1/30 HP	Q-20	10	2	Ea.	192	102		294	365
6120	1033 CFM, 1/20 HP		9	2.222		287	114		401	490

23 34 23 – HVAC Power Ventilators

23 34 23.10 HVAC Power Circulators and Ventilators		Crew	Daily Output	Labor-Hours	Unit	Material	2015 Bare Costs Labor	Equipment	Total	Total Incl O&P
6140	1323 CFM, 1/15 HP	Q-20	8	2.500	Ea.	320	128		448	545
6300	V-belt drive, 3 phase									
6320	6175 CFM, 3/4 HP	Q-20	5	4	Ea.	2,950	205		3,155	3,550
6340	7500 CFM, 3/4 HP		5	4		3,025	205		3,230	3,625
6360	10,100 CFM, 1 HP		4.50	4.444		3,175	228		3,403	3,850
6380	14,300 CFM, 1-1/2 HP	↓	4	5	↓	3,450	256		3,706	4,175
6650	Residential, bath exhaust, grille, back draft damper									
6660	50 CFM	Q-20	24	.833	Ea.	63.50	42.50		106	135
6670	110 CFM		22	.909		98	46.50		144.50	179
6680	Light combination, squirrel cage, 100 watt, 70 CFM	↓	24	.833	↓	112	42.50		154.50	188
6700	Light/heater combination, ceiling mounted									
6710	70 CFM, 1450 watt	Q-20	24	.833	Ea.	162	42.50		204.50	243
6800	Heater combination, recessed, 70 CFM		24	.833		67.50	42.50		110	139
6820	With 2 infrared bulbs		23	.870		105	44.50		149.50	184
6900	Kitchen exhaust, grille, complete, 160 CFM		22	.909		108	46.50		154.50	190
6910	180 CFM		20	1		90.50	51		141.50	178
6920	270 CFM		18	1.111		171	57		228	275
6930	350 CFM	↓	16	1.250	↓	129	64		193	240
6940	Residential roof jacks and wall caps									
6944	Wall cap with back draft damper									
6946	3" & 4" diam. round duct	1 Shee	11	.727	Ea.	26	40.50		66.50	90.50
6948	6" diam. round duct	"	11	.727	"	64	40.50		104.50	133
6958	Roof jack with bird screen and back draft damper									
6960	3" & 4" diam. round duct	1 Shee	11	.727	Ea.	26	40.50		66.50	90.50
6962	3-1/4" x 10" rectangular duct	"	10	.800	"	48.50	45		93.50	122
6980	Transition									
6982	3-1/4" x 10" to 6" diam. round	1 Shee	20	.400	Ea.	32	22.50		54.50	69.50

23 34 33 – Air Curtains

23 34 33.10 Air Barrier Curtains		Crew	Daily Output	Labor-Hours	Unit	Material	Labor	Equipment	Total	Total Incl O&P
0010	**AIR BARRIER CURTAINS**, Incl. motor starters, transformers,									
0050	and door switches									
2450	Conveyor openings or service windows									
3000	Service window, 5' high x 25" wide	2 Shee	5	3.200	Ea.	305	179		484	610
3100	Environmental separation									
3110	Door heights up to 8', low profile, super quiet									
3120	Unheated, variable speed									
3130	36" wide	2 Shee	4	4	Ea.	610	224		834	1,000
3134	42" wide		3.80	4.211		635	236		871	1,050
3138	48" wide		3.60	4.444		655	249		904	1,100
3142	60" wide	↓	3.40	4.706		685	263		948	1,150
3146	72" wide	Q-3	4.60	6.957		850	390		1,240	1,525
3150	96" wide		4.40	7.273		1,300	405		1,705	2,050
3154	120" wide		4.20	7.619		1,400	425		1,825	2,175
3158	144" wide	↓	4	8	↓	1,700	445		2,145	2,550
3200	Door heights up to 10'									
3210	Unheated									
3230	36" wide	2 Shee	3.80	4.211	Ea.	645	236		881	1,075
3234	42" wide		3.60	4.444		665	249		914	1,100
3238	48" wide		3.40	4.706		685	263		948	1,150
3242	60" wide	↓	3.20	5		1,025	280		1,305	1,550
3246	72" wide	Q-3	4.40	7.273		1,075	405		1,480	1,825
3250	96" wide	↓	4.20	7.619	↓	1,275	425		1,700	2,050

23 34 HVAC Fans

23 34 33 – Air Curtains

23 34 33.10 Air Barrier Curtains		Crew	Daily Output	Labor-Hours	Unit	Material	2015 Bare Costs Labor	Equipment	Total	Total Incl O&P
3254	120″ wide	Q-3	4	8	Ea.	1,775	445		2,220	2,625
3258	144″ wide	↓	3.80	8.421	↓	1,925	470		2,395	2,800
3300	Door heights up to 12′									
3310	Unheated									
3334	42″ wide	2 Shee	3.40	4.706	Ea.	920	263		1,183	1,400
3338	48″ wide		3.20	5		925	280		1,205	1,450
3342	60″ wide	↓	3	5.333		945	298		1,243	1,500
3346	72″ wide	Q-3	4.20	7.619		1,675	425		2,100	2,475
3350	96″ wide		4	8		1,825	445		2,270	2,675
3354	120″ wide		3.80	8.421		2,225	470		2,695	3,150
3358	144″ wide	↓	3.60	8.889	↓	2,375	495		2,870	3,375
3400	Door heights up to 16′									
3410	Unheated									
3438	48″ wide	2 Shee	3	5.333	Ea.	1,100	298		1,398	1,675
3442	60″ wide	″	2.80	5.714		1,175	320		1,495	1,775
3446	72″ wide	Q-3	3.80	8.421		2,000	470		2,470	2,900
3450	96″ wide		3.60	8.889		2,075	495		2,570	3,050
3454	120″ wide		3.40	9.412		2,825	525		3,350	3,900
3458	144″ wide	↓	3.20	10	↓	2,925	560		3,485	4,050
3470	Heated, electric									
3474	48″ wide	2 Shee	2.90	5.517	Ea.	1,850	310		2,160	2,500
3478	60″ wide	″	2.70	5.926		1,875	330		2,205	2,575
3482	72″ wide	Q-3	3.70	8.649		3,175	485		3,660	4,225
3486	96″ wide		3.50	9.143		3,275	510		3,785	4,375
3490	120″ wide		3.30	9.697		3,325	540		3,865	4,500
3494	144″ wide	↓	3.10	10.323	↓	4,475	575		5,050	5,800

23 37 Air Outlets and Inlets

23 37 13 – Diffusers, Registers, and Grilles

23 37 13.10 Diffusers		Crew	Daily Output	Labor-Hours	Unit	Material	Labor	Equipment	Total	Total Incl O&P
0010	**DIFFUSERS**, Aluminum, opposed blade damper unless noted									
0100	Ceiling, linear, also for sidewall									
0500	Perforated, 24″ x 24″ lay-in panel size, 6″ x 6″	1 Shee	16	.500	Ea.	151	28		179	209
0520	8″ x 8″		15	.533		159	30		189	221
0530	9″ x 9″		14	.571		161	32		193	226
0540	10″ x 10″		14	.571		162	32		194	227
0560	12″ x 12″		12	.667		168	37.50		205.50	242
0590	16″ x 16″		11	.727		189	40.50		229.50	269
0600	18″ x 18″		10	.800		202	45		247	291
0610	20″ x 20″		10	.800		218	45		263	310
0620	24″ x 24″		9	.889		239	49.50		288.50	340
1000	Rectangular, 1 to 4 way blow, 6″ x 6″		16	.500		50	28		78	98
1010	8″ x 8″		15	.533		58	30		88	109
1014	9″ x 9″		15	.533		66	30		96	118
1016	10″ x 10″		15	.533		80	30		110	133
1020	12″ x 6″		15	.533		71.50	30		101.50	124
1040	12″ x 9″		14	.571		75.50	32		107.50	133
1060	12″ x 12″		12	.667		84.50	37.50		122	150
1070	14″ x 6″		13	.615		77.50	34.50		112	138
1074	14″ x 14″		12	.667		128	37.50		165.50	197
1150	18″ x 18″	↓	9	.889	↓	138	49.50		187.50	227

23 37 Air Outlets and Inlets

23 37 13 – Diffusers, Registers, and Grilles

23 37 13.10 Diffusers		Crew	Daily Output	Labor-Hours	Unit	2015 Bare Costs Material	Labor	Equipment	Total	Total Incl O&P
1160	21" x 21"	1 Shee	8	1	Ea.	215	56		271	320
1170	24" x 12"		10	.800		163	45		208	249
1500	Round, butterfly damper, steel, diffuser size, 6" diameter		18	.444		9	25		34	48
1520	8" diameter		16	.500		9.55	28		37.55	53.50
1540	10" diameter		14	.571		11.85	32		43.85	62
1560	12" diameter		12	.667		15.65	37.50		53.15	74
1580	14" diameter		10	.800		19.55	45		64.55	90
2000	T bar mounting, 24" x 24" lay-in frame, 6" x 6"		16	.500		61	28		89	110
2020	8" x 8"		14	.571		62	32		94	117
2040	12" x 12"		12	.667		74	37.50		111.50	138
2060	16" x 16"		11	.727		94.50	40.50		135	166
2080	18" x 18"		10	.800		106	45		151	186
6000	For steel diffusers instead of aluminum, deduct					10%				

23 37 13.30 Grilles		Crew	Daily Output	Labor-Hours	Unit	Material	Labor	Equipment	Total	Total Incl O&P
0010	**GRILLES**									
0020	Aluminum, unless noted otherwise									
1000	Air return, steel, 6" x 6"	1 Shee	26	.308	Ea.	18.70	17.20		35.90	47
1020	10" x 6"		24	.333		18.70	18.65		37.35	49
1080	16" x 8"		22	.364		26.50	20.50		47	60
1100	12" x 12"		22	.364		26.50	20.50		47	60
1120	24" x 12"		18	.444		36	25		61	77.50
1220	24" x 18"		16	.500		43.50	28		71.50	91
1280	36" x 24"		14	.571		75	32		107	132
3000	Filter grille with filter, 12" x 12"		24	.333		53	18.65		71.65	86.50
3020	18" x 12"		20	.400		71	22.50		93.50	112
3040	24" x 18"		18	.444		83.50	25		108.50	130
3060	24" x 24"		16	.500		98	28		126	151
6000	For steel grilles instead of aluminum in above, deduct					10%				

23 37 13.60 Registers		Crew	Daily Output	Labor-Hours	Unit	Material	Labor	Equipment	Total	Total Incl O&P
0010	**REGISTERS**									
0980	Air supply									
1000	Ceiling/wall, O.B. damper, anodized aluminum									
1010	One or two way deflection, adj. curved face bars									
1020	8" x 4"	1 Shee	26	.308	Ea.	11.55	17.20		28.75	39.50
1120	12" x 12"		18	.444		21	25		46	61.50
1240	20" x 6"		18	.444		18.90	25		43.90	59
1340	24" x 8"		13	.615		25.50	34.50		60	80.50
1350	24" x 18"		12	.667		47.50	37.50		85	109
2700	Above registers in steel instead of aluminum, deduct					10%				
4000	Floor, toe operated damper, enameled steel									
4020	4" x 8"	1 Shee	32	.250	Ea.	9	14		23	31.50
4100	8" x 10"		22	.364		11	20.50		31.50	43
4140	10" x 10"		20	.400		13.15	22.50		35.65	48.50
4220	14" x 14"		16	.500		42	28		70	89.50
4240	14" x 20"		15	.533		49.50	30		79.50	99.50
4980	Air return									
5000	Ceiling or wall, fixed 45° face blades									
5010	Adjustable O.B. damper, anodized aluminum									
5020	4" x 8"	1 Shee	26	.308	Ea.	11.85	17.20		29.05	39.50
5060	6" x 10"		19	.421		14.20	23.50		37.70	51.50
5280	24" x 24"		11	.727		65	40.50		105.50	134
5300	24" x 36"		8	1		110	56		166	207

23 37 Air Outlets and Inlets

23 37 13 – Diffusers, Registers, and Grilles

23 37 13.60 Registers		Crew	Daily Output	Labor-Hours	Unit	Material	Labor	Equipment	Total	Total Incl O&P
						2015 Bare Costs				
6000	For steel construction instead of aluminum, deduct					10%				

23 37 15 – Louvers

23 37 15.40 HVAC Louvers

		Crew	Daily Output	Labor-Hours	Unit	Material	Labor	Equipment	Total	Total Incl O&P
0010	**HVAC LOUVERS**									
0100	Aluminum, extruded, with screen, mill finish									
1002	Brick vent, see also Section 04 05 23.19									
1100	Standard, 4" deep, 8" wide, 5" high	1 Shee	24	.333	Ea.	34.50	18.65		53.15	66
1200	Modular, 4" deep, 7-3/4" wide, 5" high		24	.333		36	18.65		54.65	68
1300	Speed brick, 4" deep, 11-5/8" wide, 3-7/8" high		24	.333		36	18.65		54.65	68
1400	Fuel oil brick, 4" deep, 8" wide, 5" high		24	.333		61.50	18.65		80.15	96.50
2000	Cooling tower and mechanical equip., screens, light weight		40	.200	S.F.	15.40	11.20		26.60	34
2020	Standard weight		35	.229		41	12.80		53.80	64.50
2500	Dual combination, automatic, intake or exhaust		20	.400		56	22.50		78.50	95.50
2520	Manual operation		20	.400		41.50	22.50		64	80
2540	Electric or pneumatic operation		20	.400		41.50	22.50		64	80
2560	Motor, for electric or pneumatic		14	.571	Ea.	480	32		512	580
3000	Fixed blade, continuous line									
3100	Mullion type, stormproof	1 Shee	28	.286	S.F.	41.50	16		57.50	70.50
3200	Stormproof		28	.286		41.50	16		57.50	70.50
3300	Vertical line		28	.286		49.50	16		65.50	78.50
3500	For damper to use with above, add					50%	30%			
3520	Motor, for damper, electric or pneumatic	1 Shee	14	.571	Ea.	480	32		512	580
4000	Operating, 45°, manual, electric or pneumatic		24	.333	S.F.	50	18.65		68.65	83.50
4100	Motor, for electric or pneumatic		14	.571	Ea.	480	32		512	580
4200	Penthouse, roof		56	.143	S.F.	24.50	8		32.50	39
4300	Walls		40	.200		58	11.20		69.20	81
5000	Thinline, under 4" thick, fixed blade		40	.200		24	11.20		35.20	43.50
5010	Finishes, applied by mfr. at additional cost, available in colors									
5020	Prime coat only, add				S.F.	3.30			3.30	3.63
5040	Baked enamel finish coating, add					6.10			6.10	6.70
5060	Anodized finish, add					6.60			6.60	7.25
5080	Duranodic finish, add					12			12	13.20
5100	Fluoropolymer finish coating, add					18.90			18.90	21
9980	For small orders (under 10 pieces), add					25%				

23 37 23 – HVAC Gravity Ventilators

23 37 23.10 HVAC Gravity Air Ventilators

		Crew	Daily Output	Labor-Hours	Unit	Material	Labor	Equipment	Total	Total Incl O&P
0010	**HVAC GRAVITY AIR VENTILATORS**, Includes base									
1280	Rotary ventilators, wind driven, galvanized									
1300	4" neck diameter	Q-9	20	.800	Ea.	64.50	40.50		105	133
1340	6" neck diameter		16	1		64.50	50.50		115	148
1400	12" neck diameter		10	1.600		91	80.50		171.50	223
1500	24" neck diameter		8	2		390	101		491	580
1540	36" neck diameter		6	2.667		655	134		789	925
2000	Stationary, gravity, syphon, galvanized									
2160	6" neck diameter, 66 CFM	Q-9	16	1	Ea.	45	50.50		95.50	127
2240	12" neck diameter, 160 CFM		10	1.600		106	80.50		186.50	240
2340	24" neck diameter, 900 CFM		8	2		360	101		461	550
2380	36" neck diameter, 2,000 CFM		6	2.667		440	134		574	690
4200	Stationary mushroom, aluminum, 16" orifice diameter		10	1.600		620	80.50		700.50	810
4220	26" orifice diameter		6.15	2.602		915	131		1,046	1,200
4230	30" orifice diameter		5.71	2.802		1,350	141		1,491	1,700
4240	38" orifice diameter		5	3.200		1,925	161		2,086	2,350

23 37 Air Outlets and Inlets

23 37 23 – HVAC Gravity Ventilators

23 37 23.10 HVAC Gravity Air Ventilators		Crew	Daily Output	Labor-Hours	Unit	2015 Bare Costs Material	Labor	Equipment	Total	Total Incl O&P
4250	42" orifice diameter	Q-9	4.70	3.404	Ea.	2,550	171		2,721	3,050
4260	50" orifice diameter		4.44	3.604		3,025	181		3,206	3,600
5000	Relief vent									
5500	Rectangular, aluminum, galvanized curb									
5510	intake/exhaust, 0.033" SP									
5580	500 CFM, 12" x 12"	Q-9	8.60	1.860	Ea.	700	93.50		793.50	915
5600	600 CFM, 12" x 16"		8	2		785	101		886	1,025
5640	1000 CFM, 12" x 24"		6.60	2.424		880	122		1,002	1,150
5680	3000 CFM, 20" x 42"		4	4		1,550	201		1,751	2,000
5880	Size is throat area, volume is at 500 fpm									
7000	Note: sizes based on exhaust. Intake, with 0.125" SP									
7100	loss, approximately twice listed capacity.									

23 38 Ventilation Hoods

23 38 13 – Commercial-Kitchen Hoods

23 38 13.10 Hood and Ventilation Equipment

Line	Description	Crew	Daily Output	Labor-Hours	Unit	Material	Labor	Equipment	Total	Total Incl O&P
0010	**HOOD AND VENTILATION EQUIPMENT**									
2970	Exhaust hood, sst, gutter on all sides, 4' x 4' x 2'	1 Carp	1.80	4.444	Ea.	4,725	209		4,934	5,525
2980	4' x 4' x 7'	"	1.60	5	"	7,525	235		7,760	8,625

23 41 Particulate Air Filtration

23 41 13 – Panel Air Filters

23 41 13.10 Panel Type Air Filters

Line	Description	Crew	Daily Output	Labor-Hours	Unit	Material	Labor	Equipment	Total	Total Incl O&P
0010	**PANEL TYPE AIR FILTERS**									
2950	Mechanical media filtration units									
3000	High efficiency type, with frame, non-supported [G]				MCFM	35			35	38.50
3100	Supported type [G]				"	45			45	49.50
5500	Throwaway glass or paper media type				Ea.	3.42			3.42	3.76

23 41 16 – Renewable-Media Air Filters

23 41 16.10 Disposable Media Air Filters

Line	Description	Crew	Daily Output	Labor-Hours	Unit	Material	Labor	Equipment	Total	Total Incl O&P
0010	**DISPOSABLE MEDIA AIR FILTERS**									
5000	Renewable disposable roll				C.S.F.	1.54			1.54	1.70

23 41 19 – Washable Air Filters

23 41 19.10 Permanent Air Filters

Line	Description	Crew	Daily Output	Labor-Hours	Unit	Material	Labor	Equipment	Total	Total Incl O&P
0010	**PERMANENT AIR FILTERS**									
4500	Permanent washable [G]				MCFM	20			20	22

23 41 23 – Extended Surface Filters

23 41 23.10 Expanded Surface Filters

Line	Description	Crew	Daily Output	Labor-Hours	Unit	Material	Labor	Equipment	Total	Total Incl O&P
0010	**EXPANDED SURFACE FILTERS**									
4000	Medium efficiency, extended surface [G]				MCFM	5.50			5.50	6.05

23 42 Gas-Phase Air Filtration

23 42 13 – Activated-Carbon Air Filtration

23 42 13.10 Charcoal Type Air Filtration		Crew	Daily Output	Labor-Hours	Unit	Material	2015 Bare Costs Labor	Equipment	Total	Total Incl O&P
0010	**CHARCOAL TYPE AIR FILTRATION**									
0050	Activated charcoal type, full flow				MCFM	600			600	660
0060	Full flow, impregnated media 12" deep					225			225	248
0070	HEPA filter & frame for field erection					350			350	385
0080	HEPA filter-diffuser, ceiling install.				↓	300			300	330

23 43 Electronic Air Cleaners

23 43 13 – Washable Electronic Air Cleaners

23 43 13.10 Electronic Air Cleaners		Crew	Daily Output	Labor-Hours	Unit	Material	Labor	Equipment	Total	Total Incl O&P
0010	**ELECTRONIC AIR CLEANERS**									
2000	Electronic air cleaner, duct mounted									
2150	1000 CFM	1 Shee	4	2	Ea.	420	112		532	635
2200	1200 CFM		3.80	2.105		505	118		623	735
2250	1400 CFM	↓	3.60	2.222	↓	520	124		644	765

23 51 Breechings, Chimneys, and Stacks

23 51 13 – Draft Control Devices

23 51 13.13 Draft-Induction Fans		Crew	Daily Output	Labor-Hours	Unit	Material	Labor	Equipment	Total	Total Incl O&P
0010	**DRAFT-INDUCTION FANS**									
1000	Breeching installation									
1800	Hot gas, 600°F, variable pitch pulley and motor									
1860	8" diam. inlet, 1/4 H.P., 1phase, 1120 CFM	Q-9	4	4	Ea.	2,150	201		2,351	2,650
1900	12" diam. inlet, 3/4 H.P., 3 phase, 2960 CFM		3	5.333		2,925	269		3,194	3,600
1980	24" diam. inlet, 7-1/2 H.P., 3 phase, 17,760 CFM	↓	.80	20		8,375	1,000		9,375	10,800
2300	For multi-blade damper at fan inlet, add				↓	20%				

23 51 23 – Gas Vents

23 51 23.10 Gas Chimney Vents		Crew	Daily Output	Labor-Hours	Unit	Material	Labor	Equipment	Total	Total Incl O&P
0010	**GAS CHIMNEY VENTS**, Prefab metal, U.L. listed									
0020	Gas, double wall, galvanized steel									
0080	3" diameter	Q-9	72	.222	V.L.F.	5.55	11.20		16.75	23
0100	4" diameter		68	.235		7.20	11.85		19.05	26
0120	5" diameter		64	.250		7.80	12.60		20.40	28
0140	6" diameter		60	.267		9.45	13.45		22.90	31
0160	7" diameter		56	.286		15.10	14.40		29.50	38.50
0180	8" diameter		52	.308		17.40	15.50		32.90	42.50
0200	10" diameter		48	.333		36	16.80		52.80	65.50
0220	12" diameter		44	.364		43	18.30		61.30	75.50
0260	16" diameter	↓	40	.400		103	20		123	145
0300	20" diameter	Q-10	36	.667	↓	150	35		185	218

23 51 26 – All-Fuel Vent Chimneys

23 51 26.30 All-Fuel Vent Chimneys, Double Wall, St. Stl.		Crew	Daily Output	Labor-Hours	Unit	Material	Labor	Equipment	Total	Total Incl O&P
0010	**ALL-FUEL VENT CHIMNEYS, DOUBLE WALL, STAINLESS STEEL**									
7780	All fuel, pressure tight, double wall, 4" insulation, U.L. listed, 1400°F.									
7790	304 stainless steel liner, aluminized steel outer jacket									
7800	6" diameter	Q-9	60	.267	V.L.F.	64	13.45		77.45	90.50
7804	8" diameter		52	.308		73	15.50		88.50	104
7806	10" diameter		48	.333		81.50	16.80		98.30	116
7808	12" diameter	↓	44	.364	↓	93.50	18.30		111.80	131

23 51 Breechings, Chimneys, and Stacks

23 51 26 – All-Fuel Vent Chimneys

23 51 26.30 All-Fuel Vent Chimneys, Double Wall, St. Stl.		Crew	Daily Output	Labor-Hours	Unit	Material	2015 Bare Costs Labor	Equipment	Total	Total Incl O&P
7810	14" diameter	Q-9	42	.381	V.L.F.	105	19.20		124.20	145
7880	For 316 stainless steel liner add				L.F.	30%				

23 52 Heating Boilers

23 52 13 – Electric Boilers

23 52 13.10 Electric Boilers, ASME

Line	Description	Crew	Daily Output	Labor-Hours	Unit	Material	Labor	Equipment	Total	Total Incl O&P
0010	**ELECTRIC BOILERS, ASME**, Standard controls and trim									
1000	Steam, 6 KW, 20.5 MBH	Q-19	1.20	20	Ea.	3,950	1,075		5,025	5,975
1160	60 KW, 205 MBH		1	24		6,650	1,300		7,950	9,275
1220	112 KW, 382 MBH		.75	32		9,375	1,725		11,100	12,900
1280	222 KW, 758 MBH		.55	43.636		23,800	2,350		26,150	29,800
1380	518 KW, 1768 MBH	Q-21	.36	88.889		32,600	4,925		37,525	43,300
1480	814 KW, 2778 MBH		.25	128		40,600	7,100		47,700	55,500
1600	2,340 KW, 7984 MBH		.16	200		86,500	11,100		97,600	111,500
2000	Hot water, 7.5 KW, 25.6 MBH	Q-19	1.30	18.462		4,975	1,000		5,975	6,975
2100	90 KW, 307 MBH		1.10	21.818		6,000	1,175		7,175	8,375
2220	296 KW, 1010 MBH		.55	43.636		16,100	2,350		18,450	21,300
2500	1036 KW, 3536 MBH	Q-21	.34	94.118		35,500	5,225		40,725	47,000
2680	2400 KW, 8191 MBH		.25	128		68,000	7,100		75,100	85,000
2820	3600 KW, 12,283 MBH		.16	200		94,000	11,100		105,100	120,000

23 52 23 – Cast-Iron Boilers

23 52 23.20 Gas-Fired Boilers

Line	Description	Crew	Daily Output	Labor-Hours	Unit	Material	Labor	Equipment	Total	Total Incl O&P
0010	**GAS-FIRED BOILERS**, Natural or propane, standard controls, packaged									
1000	Cast iron, with insulated jacket									
2000	Steam, gross output, 81 MBH	Q-7	1.40	22.857	Ea.	2,450	1,300		3,750	4,675
2080	203 MBH		.90	35.556		3,825	2,025		5,850	7,250
2180	400 MBH		.56	56.838		5,900	3,225		9,125	11,400
2240	765 MBH		.43	74.419		12,200	4,225		16,425	19,800
2320	1,875 MBH		.30	106		25,200	6,075		31,275	36,900
2440	4,720 MBH		.15	207		68,000	11,800		79,800	93,000
2480	6,100 MBH		.13	246		88,000	14,000		102,000	118,000
2540	6,970 MBH		.10	320		98,500	18,200		116,700	136,000
3000	Hot water, gross output, 80 MBH		1.46	21.918		1,975	1,250		3,225	4,050
3140	320 MBH		.80	40		4,675	2,275		6,950	8,575
3260	1,088 MBH		.40	80		13,500	4,550		18,050	21,800
3360	2,856 MBH		.20	160		30,800	9,100		39,900	47,600
3380	3,264 MBH		.18	179		32,700	10,200		42,900	51,500
3480	6,100 MBH		.13	250		115,500	14,200		129,700	148,500
3540	6,970 MBH		.09	359		118,500	20,500		139,000	161,500
7000	For tankless water heater, add					10%				

23 52 23.30 Gas/Oil Fired Boilers

Line	Description	Crew	Daily Output	Labor-Hours	Unit	Material	Labor	Equipment	Total	Total Incl O&P
0010	**GAS/OIL FIRED BOILERS**, Combination with burners and controls, packaged									
1000	Cast iron with insulated jacket									
2000	Steam, gross output, 720 MBH	Q-7	.43	74.074	Ea.	14,700	4,225		18,925	22,600
2080	1,600 MBH		.30	107		20,900	6,100		27,000	32,200
2140	2,700 MBH		.19	165		28,100	9,425		37,525	45,100
2280	5,520 MBH		.14	235		90,500	13,400		103,900	119,500
2340	6,390 MBH		.11	296		97,000	16,900		113,900	132,000
2380	6,970 MBH		.09	372		102,500	21,200		123,700	145,000
2900	Hot water, gross output									

23 52 Heating Boilers

23 52 23 – Cast-Iron Boilers

23 52 23.30 Gas/Oil Fired Boilers		Crew	Daily Output	Labor-Hours	Unit	Material	Labor	Equipment	Total	Total Incl O&P
						2015 Bare Costs				
2910	200 MBH	Q-6	.62	39.024	Ea.	10,500	2,175		12,675	14,900
2920	300 MBH		.49	49.080		10,500	2,725		13,225	15,700
2930	400 MBH		.41	57.971		12,300	3,225		15,525	18,400
2940	500 MBH		.36	67.039		13,300	3,750		17,050	20,300
3000	584 MBH	Q-7	.44	72.072		14,700	4,100		18,800	22,300
3060	1,460 MBH		.28	113		36,400	6,425		42,825	49,700
3160	4,088 MBH		.16	195		61,500	11,100		72,600	84,500
3300	13,500 MBH, 403.3 BHP		.04	727		190,500	41,400		231,900	272,000

23 52 23.40 Oil-Fired Boilers

		Crew	Daily Output	Labor-Hours	Unit	Material	Labor	Equipment	Total	Total Incl O&P
0010	**OIL-FIRED BOILERS**, Standard controls, flame retention burner, packaged									
1000	Cast iron, with insulated flush jacket									
2000	Steam, gross output, 109 MBH	Q-7	1.20	26.667	Ea.	2,325	1,525		3,850	4,850
2060	207 MBH		.90	35.556		3,175	2,025		5,200	6,550
2180	1,084 MBH		.38	85.106		10,900	4,850		15,750	19,300
2280	3,000 MBH		.19	170		23,700	9,675		33,375	40,700
2380	5,520 MBH		.14	235		78,500	13,400		91,900	106,500
2460	6,970 MBH		.09	363		101,000	20,700		121,700	142,000
3000	Hot water, same price as steam									
4000	For tankless coil in smaller sizes, add				Ea.	15%				

23 52 26 – Steel Boilers

23 52 26.40 Oil-Fired Boilers

		Crew	Daily Output	Labor-Hours	Unit	Material	Labor	Equipment	Total	Total Incl O&P
0010	**OIL-FIRED BOILERS**, Standard controls, flame retention burner									
5000	Steel, with insulated flush jacket									
7000	Hot water, gross output, 103 MBH	Q-6	1.60	15	Ea.	1,775	835		2,610	3,225
7120	420 MBH		.70	34.483		6,900	1,925		8,825	10,500
7320	3,150 MBH		.13	184		30,000	10,300		40,300	48,500
7340	For tankless coil in steam or hot water, add					7%				

23 52 28 – Swimming Pool Boilers

23 52 28.10 Swimming Pool Heaters

		Crew	Daily Output	Labor-Hours	Unit	Material	Labor	Equipment	Total	Total Incl O&P
0010	**SWIMMING POOL HEATERS**, Not including wiring, external									
0020	piping, base or pad,									
0160	Gas fired, input, 155 MBH	Q-6	1.50	16	Ea.	2,000	890		2,890	3,550
0200	199 MBH		1	24		2,125	1,350		3,475	4,375
0280	500 MBH		.40	60		8,900	3,350		12,250	14,800
0400	1,800 MBH		.14	171		19,300	9,550		28,850	35,700
2000	Electric, 12 KW, 4,800 gallon pool	Q-19	3	8		2,075	435		2,510	2,925
2020	15 KW, 7,200 gallon pool		2.80	8.571		2,100	465		2,565	3,025
2040	24 KW, 9,600 gallon pool		2.40	10		2,425	540		2,965	3,500
2100	57 KW, 24,000 gallon pool		1.20	20		3,575	1,075		4,650	5,550

23 54 Furnaces

23 54 13 – Electric-Resistance Furnaces

23 54 13.10 Electric Furnaces		Crew	Daily Output	Labor-Hours	Unit	2015 Bare Costs Material	Labor	Equipment	Total	Total Incl O&P
0010	**ELECTRIC FURNACES**, Hot air, blowers, std. controls									
0011	not including gas, oil or flue piping									
1000	Electric, UL listed									
1100	34.1 MBH	Q-20	4.40	4.545	Ea.	455	233		688	855

23 54 16 – Fuel-Fired Furnaces

23 54 16.13 Gas-Fired Furnaces		Crew	Daily Output	Labor-Hours	Unit	Material	Labor	Equipment	Total	Total Incl O&P
0010	**GAS-FIRED FURNACES**									
3000	Gas, AGA certified, upflow, direct drive models									
3020	45 MBH input	Q-9	4	4	Ea.	535	201		736	900
3040	60 MBH input		3.80	4.211		535	212		747	915
3060	75 MBH input		3.60	4.444		575	224		799	975
3100	100 MBH input	↓	3.20	5	↓	625	252		877	1,075

23 54 16.16 Oil-Fired Furnaces		Crew	Daily Output	Labor-Hours	Unit	Material	Labor	Equipment	Total	Total Incl O&P
0010	**OIL-FIRED FURNACES**									
6000	Oil, UL listed, atomizing gun type burner									
6020	56 MBH output	Q-9	3.60	4.444	Ea.	1,725	224		1,949	2,250
6030	84 MBH output		3.50	4.571		1,850	230		2,080	2,400
6040	95 MBH output		3.40	4.706		1,875	237		2,112	2,425
6060	134 MBH output		3.20	5		2,175	252		2,427	2,775
6080	151 MBH output	↓	3	5.333	↓	2,250	269		2,519	2,875

23 55 Fuel-Fired Heaters

23 55 13 – Fuel-Fired Duct Heaters

23 55 13.16 Gas-Fired Duct Heaters		Crew	Daily Output	Labor-Hours	Unit	Material	Labor	Equipment	Total	Total Incl O&P
0010	**GAS-FIRED DUCT HEATERS**, Includes burner, controls, stainless steel									
0020	heat exchanger. Gas fired, electric ignition									
0030	Indoor installation									
0100	120 MBH output	Q-5	4	4	Ea.	3,175	215		3,390	3,825
0130	200 MBH output		2.70	5.926		4,025	320		4,345	4,900
0140	240 MBH output		2.30	6.957		4,225	375		4,600	5,225
0180	320 MBH output	↓	1.60	10		5,150	540		5,690	6,475
0300	For powered venter and adapter, add				↓	525			525	575
0502	For required flue pipe, see Section 23 51 23.10									
1000	Outdoor installation, with power venter									
1020	75 MBH output	Q-5	4	4	Ea.	3,525	215		3,740	4,200
1060	120 MBH output		4	4		3,875	215		4,090	4,575
1100	187 MBH output		3	5.333		4,800	287		5,087	5,700
1140	300 MBH output		1.80	8.889		7,750	480		8,230	9,250
1180	450 MBH output	↓	1.40	11.429	↓	9,350	615		9,965	11,200

23 55 33 – Fuel-Fired Unit Heaters

23 55 33.13 Oil-Fired Unit Heaters		Crew	Daily Output	Labor-Hours	Unit	Material	Labor	Equipment	Total	Total Incl O&P
0010	**OIL-FIRED UNIT HEATERS**, Cabinet, grilles, fan, ctrl., burner, no piping									
6000	Oil fired, suspension mounted, 94 MBH output	Q-5	4	4	Ea.	4,825	215		5,040	5,625
6040	140 MBH output		3	5.333		5,075	287		5,362	6,000
6060	184 MBH output	↓	3	5.333	↓	5,375	287		5,662	6,350

23 55 33.16 Gas-Fired Unit Heaters		Crew	Daily Output	Labor-Hours	Unit	Material	Labor	Equipment	Total	Total Incl O&P
0010	**GAS-FIRED UNIT HEATERS**, Cabinet, grilles, fan, ctrls., burner, no piping									
0022	thermostat, no piping. For flue see Section 23 51 23.10									
1000	Gas fired, floor mounted									

23 55 Fuel-Fired Heaters

23 55 33 – Fuel-Fired Unit Heaters

23 55 33.16 Gas-Fired Unit Heaters		Crew	Daily Output	Labor-Hours	Unit	Material (2015 Bare Costs)	Labor (2015 Bare Costs)	Equipment (2015 Bare Costs)	Total (2015 Bare Costs)	Total Incl O&P
1100	60 MBH output	Q-5	10	1.600	Ea.	870	86		956	1,075
1140	100 MBH output		8	2		960	108		1,068	1,200
1180	180 MBH output		6	2.667		1,375	143		1,518	1,725
2000	Suspension mounted, propeller fan, 20 MBH output		8.50	1.882		1,125	101		1,226	1,400
2040	60 MBH output		7	2.286		1,700	123		1,823	2,050
2060	80 MBH output		6	2.667		1,875	143		2,018	2,275
2100	130 MBH output		5	3.200		2,225	172		2,397	2,700
2240	320 MBH output		2	8		4,100	430		4,530	5,175
2500	For powered venter and adapter, add					490			490	540
5000	Wall furnace, 17.5 MBH output	Q-5	6	2.667		745	143		888	1,025
5020	24 MBH output		5	3.200		745	172		917	1,075
5040	35 MBH output		4	4		790	215		1,005	1,200

23 56 Solar Energy Heating Equipment

23 56 16 – Packaged Solar Heating Equipment

23 56 16.40 Solar Heating Systems			Crew	Daily Output	Labor-Hours	Unit	Material	Labor	Equipment	Total	Total Incl O&P
0010	**SOLAR HEATING SYSTEMS**	R235616-60									
0020	System/Package prices, not including connecting										
0030	pipe, insulation, or special heating/plumbing fixtures										
0500	Hot water, standard package, low temperature										
0540	1 collector, circulator, fittings, 65 gal. tank	G	Q-1	.50	32	Ea.	3,675	1,700		5,375	6,600
0580	2 collectors, circulator, fittings, 120 gal. tank	G		.40	40		5,050	2,125		7,175	8,750
0620	3 collectors, circulator, fittings, 120 gal. tank	G		.34	47.059		6,875	2,475		9,350	11,300
0700	Medium temperature package										
0720	1 collector, circulator, fittings, 80 gal. tank	G	Q-1	.50	32	Ea.	5,025	1,700		6,725	8,075
0740	2 collectors, circulator, fittings, 120 gal. tank	G		.40	40		6,450	2,125		8,575	10,300
0780	3 collectors, circulator, fittings, 120 gal. tank	G		.30	53.333		7,325	2,825		10,150	12,300
0980	For each additional 120 gal. tank, add	G					1,750			1,750	1,925

23 56 19 – Solar Heating Components

23 56 19.50 Solar Heating Ancillary			Crew	Daily Output	Labor-Hours	Unit	Material	Labor	Equipment	Total	Total Incl O&P
0010	**SOLAR HEATING ANCILLARY**										
2300	Circulators, air										
2310	Blowers										
2400	Reversible fan, 20" diameter, 2 speed	G	Q-9	18	.889	Ea.	113	45		158	193
2870	1/12 HP, 30 GPM	G	Q-1	10	1.600	"	345	84.50		429.50	510
3000	Collector panels, air with aluminum absorber plate										
3010	Wall or roof mount										
3040	Flat black, plastic glazing										
3080	4' x 8'	G	Q-9	6	2.667	Ea.	660	134		794	935
3200	Flush roof mount, 10' to 16' x 22" wide	G	"	96	.167	L.F.	132	8.40		140.40	158
3300	Collector panels, liquid with copper absorber plate										
3330	Alum. frame, 4' x 8', 5/32" single glazing	G	Q-1	9.50	1.684	Ea.	995	89		1,084	1,225
3390	Alum. frame, 4' x 10', 5/32" single glazing	G		6	2.667		1,150	141		1,291	1,475
3450	Flat black, alum. frame, 3.5' x 7.5'	G		9	1.778		880	94		974	1,100
3500	4' x 8'	G		5.50	2.909		1,050	154		1,204	1,375
3520	4' x 10'	G		10	1.600		1,250	84.50		1,334.50	1,500
3540	4' x 12.5'	G		5	3.200		1,250	169		1,419	1,625
3600	Liquid, full wetted, plastic, alum. frame, 4' x 10'	G		5	3.200		320	169		489	610
3650	Collector panel mounting, flat roof or ground rack	G		7	2.286		244	121		365	450
3670	Roof clamps	G		70	.229	Set	2.80	12.10		14.90	21.50

23 56 Solar Energy Heating Equipment

23 56 19 – Solar Heating Components

23 56 19.50 Solar Heating Ancillary			Crew	Daily Output	Labor-Hours	Unit	Material	2015 Bare Costs Labor	Equipment	Total	Total Incl O&P
3700	Roof strap, teflon	G	1 Plum	205	.039	L.F.	23.50	2.29		25.79	29.50
3900	Differential controller with two sensors										
3930	Thermostat, hard wired	G	1 Plum	8	1	Ea.	101	58.50		159.50	200
4100	Five station with digital read-out	G	"	3	2.667	"	263	157		420	525
4300	Heat exchanger										
4580	Fluid to fluid package includes two circulating pumps										
4590	expansion tank, check valve, relief valve										
4600	controller, high temperature cutoff and sensors	G	Q-1	2.50	6.400	Ea.	800	340		1,140	1,400
4650	Heat transfer fluid										
4700	Propylene glycol, inhibited anti-freeze	G	1 Plum	28	.286	Gal.	15.55	16.75		32.30	42.50
8250	Water storage tank with heat exchanger and electric element										
8300	80 gal. with 2" x 2 lb. density insulation	G	1 Plum	1.60	5	Ea.	1,600	294		1,894	2,225
8380	120 gal. with 2" x 2 lb. density insulation	G		1.40	5.714		1,825	335		2,160	2,525
8400	120 gal. with 2" x 2 lb. density insul., 40 S.F. heat coil	G	↓	1.40	5.714	↓	2,325	335		2,660	3,050

23 57 Heat Exchangers for HVAC

23 57 16 – Steam-to-Water Heat Exchangers

23 57 16.10 Shell/Tube Type Steam-to-Water Heat Exch.

		Crew	Daily Output	Labor-Hours	Unit	Material	Labor	Equipment	Total	Total Incl O&P
0010	**SHELL AND TUBE TYPE STEAM-TO-WATER HEAT EXCHANGERS**									
0016	Shell & tube type, 2 or 4 pass, 3/4" O.D. copper tubes,									
0020	C.I. heads, C.I. tube sheet, steel shell									
0100	Hot water 40°F to 180°F, by steam at 10 PSI									
0120	8 GPM	Q-5	6	2.667	Ea.	2,150	143		2,293	2,600
0140	10 GPM		5	3.200		3,250	172		3,422	3,825
0160	40 GPM		4	4		5,025	215		5,240	5,850
0180	64 GPM		2	8		7,700	430		8,130	9,100
0200	96 GPM	↓	1	16		10,300	860		11,160	12,600
0220	120 GPM	Q-6	1.50	16	↓	13,500	890		14,390	16,300

23 57 19 – Liquid-to-Liquid Heat Exchangers

23 57 19.13 Plate-Type, Liquid-to-Liquid Heat Exchangers

		Crew	Daily Output	Labor-Hours	Unit	Material	Labor	Equipment	Total	Total Incl O&P
0010	**PLATE-TYPE, LIQUID-TO-LIQUID HEAT EXCHANGERS**									
3000	Plate type,									
3100	400 GPM	Q-6	.80	30	Ea.	36,700	1,675		38,375	42,900
3120	800 GPM	"	.50	48		63,500	2,675		66,175	73,500
3140	1200 GPM	Q-7	.34	94.118		94,000	5,350		99,350	111,500
3160	1800 GPM	"	.24	133	↓	125,000	7,575		132,575	149,000

23 57 19.16 Shell-Type, Liquid-to-Liquid Heat Exchangers

		Crew	Daily Output	Labor-Hours	Unit	Material	Labor	Equipment	Total	Total Incl O&P
0010	**SHELL-TYPE, LIQUID-TO-LIQUID HEAT EXCHANGERS**									
1000	Hot water 40°F to 140°F, by water at 200°F									
1020	7 GPM	Q-5	6	2.667	Ea.	2,650	143		2,793	3,125
1040	16 GPM		5	3.200		3,750	172		3,922	4,375
1060	34 GPM		4	4		5,675	215		5,890	6,550
1100	74 GPM	↓	1.50	10.667	↓	10,300	575		10,875	12,200

23 62 Packaged Compressor and Condenser Units

23 62 13 – Packaged Air-Cooled Refrigerant Compressor and Condenser Units

	23 62 13.10 Packaged Air-Cooled Refrig. Condensing Units	Crew	Daily Output	Labor-Hours	Unit	Material	Labor	Equipment	Total	Total Incl O&P
						2015 Bare Costs				
0010	**PACKAGED AIR-COOLED REFRIGERANT CONDENSING UNITS**									
0020	Condensing unit									
0030	Air cooled, compressor, standard controls									
0050	1.5 ton	Q-5	2.50	6.400	Ea.	1,250	345		1,595	1,900
0500	5 ton		.60	26.667		2,375	1,425		3,800	4,775
0600	10 ton		.50	32		5,000	1,725		6,725	8,100
0700	20 ton	Q-6	.40	60		11,600	3,350		14,950	17,800

23 63 Refrigerant Condensers

23 63 13 – Air-Cooled Refrigerant Condensers

	23 63 13.10 Air-Cooled Refrig. Condensers	Crew	Daily Output	Labor-Hours	Unit	Material	Labor	Equipment	Total	Total Incl O&P
0010	**AIR-COOLED REFRIG. CONDENSERS**									
0080	Air cooled, belt drive, propeller fan									
0240	50 ton	Q-6	.69	34.985	Ea.	10,600	1,950		12,550	14,600
0280	59 ton		.58	41.308		12,700	2,300		15,000	17,500
0320	73 ton		.47	51.173		16,600	2,850		19,450	22,600
0360	86 ton		.40	60.302		19,200	3,375		22,575	26,200
0380	88 ton		.39	61.697		20,600	3,450		24,050	27,900
1550	Air cooled, direct drive, propeller fan									
1590	1 ton	Q-5	3.80	4.211	Ea.	1,650	226		1,876	2,175
1600	1-1/2 ton		3.60	4.444		1,975	239		2,214	2,525
1620	2 ton		3.20	5		2,175	269		2,444	2,800
1640	5 ton		2	8		5,400	430		5,830	6,575
1660	10 ton		1.40	11.429		6,325	615		6,940	7,875
1690	16 ton		1.10	14.545		10,200	780		10,980	12,400
1720	26 ton		.84	19.002		12,500	1,025		13,525	15,400
1760	41 ton	Q-6	.77	31.008		17,600	1,725		19,325	22,000
1800	63 ton	″	.55	44.037		27,700	2,450		30,150	34,200

23 64 Packaged Water Chillers

23 64 13 – Absorption Water Chillers

	23 64 13.16 Indirect-Fired Absorption Water Chillers	Crew	Daily Output	Labor-Hours	Unit	Material	Labor	Equipment	Total	Total Incl O&P
0010	**INDIRECT-FIRED ABSORPTION WATER CHILLERS**									
0020	Steam or hot water, water cooled									
0050	100 ton	Q-7	.13	240	Ea.	125,000	13,700		138,700	158,000
0400	420 ton	″	.10	323	″	366,500	18,400		384,900	431,000

23 64 16 – Centrifugal Water Chillers

	23 64 16.10 Centrifugal Type Water Chillers	Crew	Daily Output	Labor-Hours	Unit	Material	Labor	Equipment	Total	Total Incl O&P
0010	**CENTRIFUGAL TYPE WATER CHILLERS**, With standard controls									
0020	Centrifugal liquid chiller, water cooled									
0030	not including water tower									
0100	2000 ton (twin 1000 ton units)	Q-7	.07	477	Ea.	699,000	27,200		726,200	810,000

23 64 19 – Reciprocating Water Chillers

	23 64 19.10 Reciprocating Type Water Chillers	Crew	Daily Output	Labor-Hours	Unit	Material	Labor	Equipment	Total	Total Incl O&P
0010	**RECIPROCATING TYPE WATER CHILLERS**, With standard controls									
0494	Water chillers, integral air cooled condenser									
0600	100 ton cooling	Q-7	.25	129	Ea.	73,500	7,350		80,850	92,000
0980	Water cooled, multiple compressor, semi-hermetic, tower not incl.									

23 64 19 – Reciprocating Water Chillers

23 64 19.10 Reciprocating Type Water Chillers		Crew	Daily Output	Labor-Hours	Unit	Material	2015 Bare Costs Labor	Equipment	Total	Total Incl O&P
1000	15 ton cooling	Q-6	.36	65.934	Ea.	18,900	3,675		22,575	26,400
1020	25 ton cooling	Q-7	.41	78.049		19,500	4,450		23,950	28,200
1060	35 ton cooling		.31	101		24,300	5,800		30,100	35,600
1090	45 ton cooling		.29	111		28,200	6,325		34,525	40,600
1100	50 ton cooling		.28	113		35,800	6,475		42,275	49,200
1160	100 ton cooling		.18	179		60,500	10,200		70,700	82,000
1180	125 ton cooling		.16	196		64,000	11,200		75,200	87,000
1200	145 ton cooling		.16	202		71,500	11,500		83,000	96,000
1451	Water cooled, dual compressors, semi-hermetic, tower not incl.									
1500	80 ton cooling	Q-7	.14	222	Ea.	29,600	12,600		42,200	51,500
1520	100 ton cooling		.14	228		40,000	13,000		53,000	63,500
1540	120 ton cooling		.14	231		48,500	13,200		61,700	73,500

23 64 23 – Scroll Water Chillers

23 64 23.10 Scroll Water Chillers		Crew	Daily Output	Labor-Hours	Unit	Material	Labor	Equipment	Total	Total Incl O&P
0010	**SCROLL WATER CHILLERS**, With standard controls									
0480	Packaged w/integral air cooled condenser									
0482	10 ton cooling	Q-7	.34	94.118	Ea.	20,300	5,350		25,650	30,500
0490	15 ton cooling		.37	86.486		20,800	4,925		25,725	30,300
0500	20 ton cooling		.34	94.118		21,500	5,350		26,850	31,700
0520	40 ton cooling		.30	108		32,900	6,150		39,050	45,500
0680	Scroll water cooled, single compressor, hermetic, tower not incl.									
0700	2 ton cooling	Q-5	.57	28.070	Ea.	3,500	1,500		5,000	6,125
0710	5 ton cooling		.57	28.070		4,175	1,500		5,675	6,875
0740	8 ton cooling		.31	52.117		5,750	2,800		8,550	10,600
0760	10 ton cooling	Q-6	.36	67.039		6,625	3,750		10,375	12,900
0800	20 ton cooling	Q-7	.38	83.990		11,800	4,775		16,575	20,200
0820	30 ton cooling	"	.33	96.096		13,200	5,475		18,675	22,900

23 64 26 – Rotary-Screw Water Chillers

23 64 26.10 Rotary-Screw Type Water Chillers		Crew	Daily Output	Labor-Hours	Unit	Material	Labor	Equipment	Total	Total Incl O&P
0010	**ROTARY-SCREW TYPE WATER CHILLERS**, With standard controls									
0110	Screw, liquid chiller, air cooled, insulated evaporator									
0120	130 ton	Q-7	.14	228	Ea.	91,500	13,000		104,500	120,000
0124	160 ton		.13	246		112,500	14,000		126,500	144,500
0128	180 ton		.13	250		126,500	14,200		140,700	160,500
0132	210 ton		.12	258		139,000	14,700		153,700	175,000
0136	270 ton		.12	266		159,000	15,200		174,200	198,000
0140	320 ton		.12	275		199,500	15,700		215,200	243,000
0200	Packaged unit, water cooled, not incl. tower									
0210	80 ton	Q-7	.14	223	Ea.	46,000	12,700		58,700	69,500
0240	200 ton		.13	251		82,000	14,300		96,300	111,500
0270	350 ton		.12	275		133,500	15,700		149,200	170,000
1450	Water cooled, tower not included									
1580	150 ton cooling, screw compressors	Q-7	.13	240	Ea.	69,000	13,700		82,700	96,000
1620	200 ton cooling, screw compressors		.13	250		95,000	14,200		109,200	126,000
1660	291 ton cooling, screw compressors		.12	260		99,000	14,800		113,800	131,500

23 65 Cooling Towers

23 65 13 – Forced-Draft Cooling Towers

23 65 13.10 Forced-Draft Type Cooling Towers		Crew	Daily Output	Labor-Hours	Unit	2015 Bare Costs Material	Labor	Equipment	Total	Total Incl O&P
0010	**FORCED-DRAFT TYPE COOLING TOWERS**, Packaged units									
0070	Galvanized steel									
0080	Induced draft, crossflow									
0100	Vertical, belt drive, 61 tons	Q-6	90	.267	TonAC	215	14.85		229.85	260
0150	100 ton		100	.240		208	13.40		221.40	248
0200	115 ton		109	.220		180	12.30		192.30	218
0250	131 ton		120	.200		217	11.15		228.15	255
0260	162 ton		132	.182		175	10.15		185.15	208
1000	For higher capacities, use multiples									
1500	Induced air, double flow									
1900	Vertical, gear drive, 167 ton	Q-6	126	.190	TonAC	170	10.60		180.60	203
2000	297 ton		129	.186		105	10.40		115.40	131
2100	582 ton		132	.182		58.50	10.15		68.65	79.50
2150	849 ton		142	.169		78.50	9.45		87.95	101
2200	1016 ton		150	.160		79	8.90		87.90	100
3000	For higher capacities, use multiples									
3500	For pumps and piping, add	Q-6	38	.632	TonAC	108	35		143	171
4000	For absorption systems, add				"	75%	75%			
4100	Cooling water chemical feeder	Q-5	3	5.333	Ea.	365	287		652	835
5000	Fiberglass tower on galvanized steel support structure									
5010	Draw thru									
5100	100 ton	Q-6	1.40	17.143	Ea.	13,900	955		14,855	16,800
5120	120 ton		1.20	20		16,200	1,125		17,325	19,500
5140	140 ton		1	24		17,500	1,350		18,850	21,200
5160	160 ton		.80	30		19,400	1,675		21,075	23,900
5180	180 ton		.65	36.923		22,100	2,050		24,150	27,400
5200	200 ton		.48	50		25,000	2,800		27,800	31,700
5300	For stainless steel support structure, add					30%				
5360	For higher capacities, use multiples of each size									
6000	Stainless steel									
6010	Induced draft, crossflow, horizontal, belt drive									
6100	57 ton	Q-6	1.50	16	Ea.	27,100	890		27,990	31,200
6120	91 ton		.99	24.242		32,700	1,350		34,050	38,100
6140	111 ton		.43	55.814		41,600	3,125		44,725	50,500
6160	126 ton		.22	109		41,600	6,075		47,675	55,000

23 73 Indoor Central-Station Air-Handling Units

23 73 13 – Modular Indoor Central-Station Air-Handling Units

23 73 13.10 Air-Handling Units		Crew	Daily Output	Labor-Hours	Unit	Material	Labor	Equipment	Total	Total Incl O&P
0010	**AIR-HANDLING UNITS**, Built-Up									
0100	With cooling/heating coil section, filters, mixing box									
0880	Single zone, horizontal/vertical									
0890	Constant volume									
0900	1600 CFM	Q-5	1.20	13.333	Ea.	5,000	715		5,715	6,575
0920	5000 CFM	Q-6	1.40	17.143		14,000	955		14,955	16,900
0940	11,500 CFM		1	24		25,000	1,350		26,350	29,500
0970	22,000 CFM		.60	40		49,000	2,225		51,225	57,500
1000	40,000 CFM		.30	80		90,000	4,450		94,450	105,500

23 73 Indoor Central-Station Air-Handling Units

23 73 39 – Indoor, Direct Gas-Fired Heating and Ventilating Units

23 73 39.10 Make-Up Air Unit		Crew	Daily Output	Labor-Hours	Unit	Material	2015 Bare Costs Labor	Equipment	Total	Total Incl O&P
0010	**MAKE-UP AIR UNIT**									
0020	Indoor suspension, natural/LP gas, direct fired,									
0032	standard control. For flue see Section 23 51 23.10									
0040	70°F temperature rise, MBH is input									
0100	75 MBH input	Q-6	3.60	6.667	Ea.	4,925	370		5,295	5,975
0160	150 MBH input		3	8		5,975	445		6,420	7,250
0220	225 MBH input		2.40	10		6,750	560		7,310	8,250
0300	400 MBH input		1.60	15		15,400	835		16,235	18,200
0600	For discharge louver assembly, add					5%				
0700	For filters, add					10%				
0800	For air shut-off damper section, add					30%				

23 74 Packaged Outdoor HVAC Equipment

23 74 33 – Dedicated Outdoor-Air Units

23 74 33.10 Rooftop Air Conditioners		Crew	Daily Output	Labor-Hours	Unit	Material	Labor	Equipment	Total	Total Incl O&P
0010	**ROOFTOP AIR CONDITIONERS**, Standard controls, curb, economizer									
1000	Single zone, electric cool, gas heat									
1100	3 ton cooling, 60 MBH heating R236000-20	Q-5	.70	22.857	Ea.	3,925	1,225		5,150	6,175
1120	4 ton cooling, 95 MBH heating		.61	26.403		4,575	1,425		6,000	7,200
1140	5 ton cooling, 112 MBH heating		.56	28.521		5,125	1,525		6,650	7,950
1145	6 ton cooling, 140 MBH heating		.52	30.769		5,875	1,650		7,525	8,975
1150	7.5 ton cooling, 170 MBH heating		.50	32.258		6,825	1,725		8,550	10,200
1156	8.5 ton cooling, 170 MBH heating		.46	34.783		8,125	1,875		10,000	11,800
1160	10 ton cooling, 200 MBH heating	Q-6	.67	35.982		9,425	2,000		11,425	13,400
1170	12.5 ton cooling, 230 MBH heating		.63	37.975		11,900	2,125		14,025	16,300
1190	17.5 ton cooling, 330 MBH heating		.52	45.889		16,100	2,550		18,650	21,700
1200	20 ton cooling, 360 MBH heating	Q-7	.67	47.976		23,800	2,725		26,525	30,300
1210	25 ton cooling, 450 MBH heating		.56	57.554		27,100	3,275		30,375	34,800
1220	30 ton cooling, 540 MBH heating		.47	68.376		30,800	3,900		34,700	39,800
1240	40 ton cooling, 675 MBH heating		.35	91.168		40,200	5,175		45,375	52,000
2000	Multizone, electric cool, gas heat, economizer									
2100	15 ton cooling, 360 MBH heating	Q-7	.61	52.545	Ea.	65,000	3,000		68,000	76,000
2120	20 ton cooling, 360 MBH heating		.53	60.038		70,000	3,425		73,425	82,000
2200	40 ton cooling, 540 MBH heating		.28	113		122,500	6,475		128,975	144,500
2210	50 ton cooling, 540 MBH heating		.23	142		152,000	8,100		160,100	179,500
2220	70 ton cooling, 1500 MBH heating		.16	198		164,500	11,300		175,800	198,000
2240	80 ton cooling, 1500 MBH heating		.14	228		188,000	13,000		201,000	226,500
2260	90 ton cooling, 1500 MBH heating		.13	256		197,500	14,600		212,100	239,000
2280	105 ton cooling, 1500 MBH heating		.11	290		217,500	16,500		234,000	264,000
2400	For hot water heat coil, deduct					5%				
2500	For steam heat coil, deduct					2%				
2600	For electric heat, deduct					3%	5%			

23 81 Decentralized Unitary HVAC Equipment

23 81 13 – Packaged Terminal Air-Conditioners

23 81 13.10 Packaged Cabinet Type Air-Conditioners		Crew	Daily Output	Labor-Hours	Unit	2015 Bare Costs Material	Labor	Equipment	Total	Total Incl O&P
0010	**PACKAGED CABINET TYPE AIR-CONDITIONERS**, Cabinet, wall sleeve,									
0100	louver, electric heat, thermostat, manual changeover, 208 V									
0200	6,000 BTUH cooling, 8800 BTU heat	Q-5	6	2.667	Ea.	775	143		918	1,075
0220	9,000 BTUH cooling, 13,900 BTU heat		5	3.200		1,200	172		1,372	1,575
0240	12,000 BTUH cooling, 13,900 BTU heat		4	4		1,350	215		1,565	1,825
0260	15,000 BTUH cooling, 13,900 BTU heat		3	5.333		1,450	287		1,737	2,025
0500	For hot water coil, increase heat by 10%, add					5%	10%			
1000	For steam, increase heat output by 30%, add					8%	10%			

23 81 19 – Self-Contained Air-Conditioners

23 81 19.20 Self-Contained Single Package		Crew	Daily Output	Labor-Hours	Unit	Material	Labor	Equipment	Total	Total Incl O&P
0010	**SELF-CONTAINED SINGLE PACKAGE**									
0100	Air cooled, for free blow or duct, not incl. remote condenser									
0110	Constant volume									
0200	3 ton cooling	Q-5	1	16	Ea.	3,750	860		4,610	5,400
0220	5 ton cooling	Q-6	1.20	20		4,450	1,125		5,575	6,575
0240	10 ton cooling	Q-7	1	32		7,125	1,825		8,950	10,600
0260	20 ton cooling		.90	35.556		12,800	2,025		14,825	17,200
0280	30 ton cooling		.80	40		25,300	2,275		27,575	31,200
0340	60 ton cooling	Q-8	.40	80		56,000	4,550	144	60,694	68,500
0490	For duct mounting no price change									
0500	For steam heating coils, add				Ea.	10%	10%			
1000	Water cooled for free blow or duct, not including tower									
1010	Constant volume									
1100	3 ton cooling	Q-6	1	24	Ea.	3,725	1,350		5,075	6,125
1120	5 ton cooling	"	1	24		4,850	1,350		6,200	7,350
1140	10 ton cooling	Q-7	.90	35.556		9,450	2,025		11,475	13,500
1160	20 ton cooling		.80	40		28,600	2,275		30,875	34,800
1180	30 ton cooling		.70	45.714		37,900	2,600		40,500	45,600

23 81 23 – Computer-Room Air-Conditioners

23 81 23.10 Computer Room Units		Crew	Daily Output	Labor-Hours	Unit	Material	Labor	Equipment	Total	Total Incl O&P
0010	**COMPUTER ROOM UNITS**									
1000	Air cooled, includes remote condenser but not									
1020	interconnecting tubing or refrigerant									
1080	3 ton	Q-5	.50	32	Ea.	18,800	1,725		20,525	23,300
1120	5 ton		.45	35.556		20,100	1,900		22,000	25,000
1160	6 ton		.30	53.333		37,000	2,875		39,875	45,000
1200	8 ton		.27	59.259		37,400	3,175		40,575	46,000
1240	10 ton		.25	64		39,100	3,450		42,550	48,300
1260	12 ton		.24	66.667		40,500	3,575		44,075	50,000
1280	15 ton		.22	72.727		43,100	3,900		47,000	53,500
1290	18 ton		.20	80		49,300	4,300		53,600	61,000
1300	20 ton	Q-6	.26	92.308		51,500	5,150		56,650	64,500
1320	22 ton		.24	100		52,000	5,575		57,575	66,000
1360	30 ton		.21	114		64,500	6,375		70,875	80,000
2200	Chilled water, for connection to									
2220	existing chiller system of adequate capacity									
2260	5 ton	Q-5	.74	21.622	Ea.	14,300	1,175		15,475	17,600

23 81 Decentralized Unitary HVAC Equipment

23 81 43 – Air-Source Unitary Heat Pumps

23 81 43.10 Air-Source Heat Pumps		Crew	Daily Output	Labor-Hours	Unit	Material	2015 Bare Costs Labor	Equipment	Total	Total Incl O&P
0010	**AIR-SOURCE HEAT PUMPS**, Not including interconnecting tubing									
1000	Air to air, split system, not including curbs, pads, fan coil and ductwork									
1012	Outside condensing unit only, for fan coil see Section 23 82 19.10									
1020	2 ton cooling, 8.5 MBH heat @ 0°F	Q-5	2	8	Ea.	2,400	430		2,830	3,300
1060	5 ton cooling, 27 MBH heat @ 0°F		.50	32		3,600	1,725		5,325	6,550
1080	7.5 ton cooling, 33 MBH heat @ 0°F		.45	35.556		6,425	1,900		8,325	9,950
1100	10 ton cooling, 50 MBH heat @ 0°F	Q-6	.64	37.500		8,500	2,100		10,600	12,500
1120	15 ton cooling, 64 MBH heat @ 0°F		.50	48		11,800	2,675		14,475	17,100
1130	20 ton cooling, 85 MBH heat @ 0°F		.35	68.571		17,000	3,825		20,825	24,500
1140	25 ton cooling, 119 MBH heat @ 0°F		.25	96		20,400	5,350		25,750	30,600
1500	Single package, not including curbs, pads, or plenums									
1520	2 ton cooling, 6.5 MBH heat @ 0°F	Q-5	1.50	10.667	Ea.	3,100	575		3,675	4,300
1580	4 ton cooling, 13 MBH heat @ 0°F		.96	16.667		4,200	895		5,095	5,975
1640	7.5 ton cooling, 35 MBH heat @ 0°F		.40	40		7,325	2,150		9,475	11,300

23 81 46 – Water-Source Unitary Heat Pumps

23 81 46.10 Water Source Heat Pumps		Crew	Daily Output	Labor-Hours	Unit	Material	Labor	Equipment	Total	Total Incl O&P
0010	**WATER SOURCE HEAT PUMPS**, Not incl. connecting tubing or water source									
2000	Water source to air, single package									
2100	1 ton cooling, 13 MBH heat @ 75°F	Q-5	2	8	Ea.	2,075	430		2,505	2,950
2140	2 ton cooling, 19 MBH heat @ 75°F		1.70	9.412		2,575	505		3,080	3,600
2220	5 ton cooling, 29 MBH heat @ 75°F		.90	17.778		3,825	955		4,780	5,650
3960	For supplementary heat coil, add					10%				
4000	For increase in capacity thru use									
4020	of solar collector, size boiler at 60%									

23 82 Convection Heating and Cooling Units

23 82 16 – Air Coils

23 82 16.10 Flanged Coils		Crew	Daily Output	Labor-Hours	Unit	Material	Labor	Equipment	Total	Total Incl O&P
0010	**FLANGED COILS**									
0500	Chilled water cooling, 6 rows, 24" x 48"	Q-5	3.20	5	Ea.	4,100	269		4,369	4,900
1000	Direct expansion cooling, 6 rows, 24" x 48"		2.80	5.714		4,450	305		4,755	5,375
1500	Hot water heating, 1 row, 24" x 48"		4	4		1,625	215		1,840	2,125
2000	Steam heating, 1 row, 24" x 48"		3.06	5.229		2,325	281		2,606	2,975

23 82 16.20 Duct Heaters		Crew	Daily Output	Labor-Hours	Unit	Material	Labor	Equipment	Total	Total Incl O&P
0010	**DUCT HEATERS**, Electric, 480 V, 3 Ph.									
0020	Finned tubular insert, 500°F									
0100	8" wide x 6" high, 4.0 kW	Q-20	16	1.250	Ea.	765	64		829	945
0120	12" high, 8.0 kW		15	1.333		1,275	68.50		1,343.50	1,500
0140	18" high, 12.0 kW		14	1.429		1,775	73		1,848	2,050
0160	24" high, 16.0 kW		13	1.538		2,300	79		2,379	2,650
0180	30" high, 20.0 kW		12	1.667		2,800	85.50		2,885.50	3,200
0300	12" wide x 6" high, 6.7 kW		15	1.333		815	68.50		883.50	1,000
0360	24" high, 26.7 kW		12	1.667		2,375	85.50		2,460.50	2,725
0700	24" wide x 6" high, 17.8 kW		13	1.538		965	79		1,044	1,200
0760	24" high, 71.1 kW		10	2		2,950	102		3,052	3,400
8000	To obtain BTU multiply kW by 3413									

23 82 Convection Heating and Cooling Units

23 82 19 – Fan Coil Units

	23 82 19.10 Fan Coil Air Conditioning	Crew	Daily Output	Labor-Hours	Unit	Material	Labor (2015 Bare Costs)	Equipment	Total	Total Incl O&P
0010	**FAN COIL AIR CONDITIONING**									
0030	Fan coil AC, cabinet mounted, filters and controls									
0100	Chilled water, 1/2 ton cooling	Q-5	8	2	Ea.	815	108		923	1,050
0120	1 ton cooling		6	2.667		950	143		1,093	1,275
0140	1.5 ton cooling		5.50	2.909		1,050	156		1,206	1,375
0150	2 ton cooling		5.25	3.048		1,350	164		1,514	1,725
0180	3 ton cooling		4	4		2,175	215		2,390	2,725
0262	For hot water coil, add					40%	10%			
0940	Direct expansion, for use w/air cooled condensing unit, 1.5 ton cooling	Q-5	5	3.200	Ea.	680	172		852	1,000
1000	5 ton cooling	"	3	5.333		1,300	287		1,587	1,875
1040	10 ton cooling	Q-6	2.60	9.231		2,800	515		3,315	3,850
1060	20 ton cooling	"	.70	34.286		5,400	1,900		7,300	8,825
1500	For hot water coil, add					40%	10%			

	23 82 19.20 Heating and Ventilating Units	Crew	Daily Output	Labor-Hours	Unit	Material	Labor	Equipment	Total	Total Incl O&P
0010	**HEATING AND VENTILATING UNITS**, Classroom units									
0020	Includes filter, heating/cooling coils, standard controls									
0080	750 CFM, 2 tons cooling	Q-6	2	12	Ea.	4,050	670		4,720	5,475
0120	1250 CFM, 3 tons cooling		1.40	17.143		4,950	955		5,905	6,900
0140	1500 CFM, 4 tons cooling		.80	30		5,300	1,675		6,975	8,350
0500	For electric heat, add					35%				
1000	For no cooling, deduct					25%	10%			

23 82 27 – Infrared Units

	23 82 27.10 Infrared Type Heating Units	Crew	Daily Output	Labor-Hours	Unit	Material	Labor	Equipment	Total	Total Incl O&P
0010	**INFRARED TYPE HEATING UNITS**									
0020	Gas fired, unvented, electric ignition, 100% shutoff.									
0030	Piping and wiring not included									
0120	45 MBH	Q-5	5	3.200	Ea.	950	172		1,122	1,300
0160	60 MBH		4	4		950	215		1,165	1,375
0240	120 MBH		2	8		1,575	430		2,005	2,375
1000	Gas fired, vented, electric ignition, tubular									
1020	Piping and wiring not included, 20' to 80' lengths									
1030	Single stage, input, 60 MBH	Q-6	4.50	5.333	Ea.	1,425	297		1,722	2,025
1040	80 MBH		3.90	6.154		1,425	345		1,770	2,100
1050	100 MBH		3.40	7.059		1,425	395		1,820	2,175
1060	125 MBH		2.90	8.276		1,425	460		1,885	2,275
1070	150 MBH		2.70	8.889		1,425	495		1,920	2,325
1080	170 MBH		2.50	9.600		1,425	535		1,960	2,375
1090	200 MBH		2.20	10.909		1,625	610		2,235	2,725
1100	Note: Final pricing may vary due to									
1110	tube length and configuration package selected									
1130	Two stage, input, 60 MBH high, 45 MBH low	Q-6	4.50	5.333	Ea.	1,750	297		2,047	2,375
1140	80 MBH high, 60 MBH low		3.90	6.154		1,750	345		2,095	2,450
1150	100 MBH high, 65 MBH low		3.40	7.059		1,750	395		2,145	2,525
1160	125 MBH high, 95 MBH low		2.90	8.276		1,750	460		2,210	2,625
1170	150 MBH high, 100 MBH low		2.70	8.889		1,750	495		2,245	2,675
1180	170 MBH high, 125 MBH low		2.50	9.600		1,950	535		2,485	2,950
1190	200 MBH high, 150 MBH low		2.20	10.909		1,950	610		2,560	3,075
1220	Note: Final pricing may vary due to									
1230	tube length and configuration package selected									

23 82 29 – Radiators

23 82 29.10 Hydronic Heating		Crew	Daily Output	Labor-Hours	Unit	Material	2015 Bare Costs Labor	Equipment	Total	Total Incl O&P
0010	**HYDRONIC HEATING**, Terminal units, not incl. main supply pipe									
1000	Radiation									
1100	Panel, baseboard, C.I., including supports, no covers	Q-5	46	.348	L.F.	38	18.70		56.70	69.50
3000	Radiators, cast iron									
3100	Free standing or wall hung, 6 tube, 25" high	Q-5	96	.167	Section	44	8.95		52.95	62
3200	4 tube, 19" high	"	96	.167	"	31.50	8.95		40.45	48
3250	Adj. brackets, 2 per wall radiator up to 30 sections	1 Stpi	32	.250	Ea.	53	14.95		67.95	81
9500	To convert SFR to BTU rating: Hot water, 150 x SFR									
9510	Forced hot water, 180 x SFR; steam, 240 x SFR									

23 82 33 – Convectors

23 82 33.10 Convector Units		Crew	Daily Output	Labor-Hours	Unit	Material	Labor	Equipment	Total	Total Incl O&P
0010	**CONVECTOR UNITS**, Terminal units, not incl. main supply pipe									
2204	Convector, multifin, 2 pipe w/cabinet									
2210	17" H x 24" L	Q-5	10	1.600	Ea.	99	86		185	239
2214	17" H x 36" L		8.60	1.860		148	100		248	315
2218	17" H x 48" L		7.40	2.162		198	116		314	395
2222	21" H x 24" L		9	1.778		101	95.50		196.50	255
2226	21" H x 36" L		8.20	1.951		151	105		256	325
2228	21" H x 48" L	↓	6.80	2.353	↓	201	127		328	410
2240	For knob operated damper, add					140%				
2241	For metal trim strips, add	Q-5	64	.250	Ea.	12.90	13.45		26.35	34.50
2243	For snap-on inlet grille, add					10%	10%			
2245	For hinged access door, add	Q-5	64	.250	Ea.	35.50	13.45		48.95	59.50
2246	For air chamber, auto-venting, add	"	58	.276	"	7.75	14.85		22.60	31

23 82 36 – Finned-Tube Radiation Heaters

23 82 36.10 Finned Tube Radiation		Crew	Daily Output	Labor-Hours	Unit	Material	Labor	Equipment	Total	Total Incl O&P
0010	**FINNED TUBE RADIATION**, Terminal units, not incl. main supply pipe									
1150	Fin tube, wall hung, 14" slope top cover, with damper									
1200	1-1/4" copper tube, 4-1/4" alum. fin	Q-5	38	.421	L.F.	43	22.50		65.50	81
1250	1-1/4" steel tube, 4-1/4" steel fin	"	36	.444	"	39	24		63	79
1500	Note: fin tube may also require corners, caps, etc.									

23 82 39 – Unit Heaters

23 82 39.16 Propeller Unit Heaters		Crew	Daily Output	Labor-Hours	Unit	Material	Labor	Equipment	Total	Total Incl O&P
0010	**PROPELLER UNIT HEATERS**									
3950	Unit heaters, propeller, 115 V 2 psi steam, 60°F entering air									
4000	Horizontal, 12 MBH	Q-5	12	1.333	Ea.	360	71.50		431.50	505
4060	43.9 MBH		8	2		545	108		653	760
4140	96.8 MBH		6	2.667		795	143		938	1,100
4180	157.6 MBH		4	4		1,050	215		1,265	1,500
4240	286.9 MBH		2	8		1,650	430		2,080	2,475
4260	364 MBH		1.80	8.889		2,075	480		2,555	3,000
4270	404 MBH	↓	1.60	10	↓	2,150	540		2,690	3,175
4300	Vertical diffuser same price									
4310	Vertical flow, 40 MBH	Q-5	11	1.455	Ea.	540	78		618	715
4314	58.5 MBH		8	2		560	108		668	775
4326	131.0 MBH		4	4		845	215		1,060	1,250
4346	297.0 MBH	↓	1.80	8.889		1,600	480		2,080	2,500
4354	420 MBH,(460 V)	Q-6	1.80	13.333		2,150	745		2,895	3,500
4358	500 MBH, (460 V)		1.71	14.035		2,875	785		3,660	4,325
4362	570 MBH, (460 V)	↓	1.40	17.143	↓	3,925	955		4,880	5,775

23 82 Convection Heating and Cooling Units

23 82 39 – Unit Heaters

23 82 39.16 Propeller Unit Heaters		Crew	Daily Output	Labor-Hours	Unit	Material	2015 Bare Costs Labor	Equipment	Total	Total Incl O&P
4366	620 MBH, (460 V)	Q-6	1.30	18.462	Ea.	3,925	1,025		4,950	5,875
4370	960 MBH, (460 V)	↓	1.10	21.818	↓	7,450	1,225		8,675	10,000

23 83 Radiant Heating Units

23 83 16 – Radiant-Heating Hydronic Piping

23 83 16.10 Radiant Floor Heating		Crew	Daily Output	Labor-Hours	Unit	Material	Labor	Equipment	Total	Total Incl O&P
0010	**RADIANT FLOOR HEATING**									
0100	Tubing, PEX (cross-linked polyethylene)									
0110	Oxygen barrier type for systems with ferrous materials									
0120	1/2"	Q-5	800	.020	L.F.	.96	1.08		2.04	2.68
0130	3/4"		535	.030		1.35	1.61		2.96	3.92
0140	1"	↓	400	.040	↓	2.11	2.15		4.26	5.55
0200	Non barrier type for ferrous free systems									
0210	1/2"	Q-5	800	.020	L.F.	.61	1.08		1.69	2.29
0220	3/4"		535	.030		1.11	1.61		2.72	3.65
0230	1"	↓	400	.040	↓	1.90	2.15		4.05	5.35
1000	Manifolds									
1110	Brass									
1120	With supply and return valves, flow meter, thermometer,									
1122	auto air vent and drain/fill valve.									
1130	1", 2 circuit	Q-5	14	1.143	Ea.	257	61.50		318.50	375
1140	1", 3 circuit		13.50	1.185		294	63.50		357.50	420
1150	1", 4 circuit		13	1.231		320	66		386	450
1154	1", 5 circuit		12.50	1.280		365	69		434	510
1158	1", 6 circuit		12	1.333		415	71.50		486.50	565
1162	1", 7 circuit		11.50	1.391		455	75		530	615
1166	1", 8 circuit		11	1.455		500	78		578	670
1172	1", 9 circuit		10.50	1.524		540	82		622	720
1174	1", 10 circuit		10	1.600		580	86		666	770
1178	1", 11 circuit		9.50	1.684		605	90.50		695.50	800
1182	1", 12 circuit	↓	9	1.778	↓	670	95.50		765.50	880
1610	Copper manifold header, (cut to size)									
1620	1" header, 12 – 1/2" sweat outlets	Q-5	3.33	4.805	Ea.	91	258		349	490
1630	1-1/4" header, 12 – 1/2" sweat outlets		3.20	5		106	269		375	520
1640	1-1/4" header, 12 – 3/4" sweat outlets		3	5.333		114	287		401	560
1650	1-1/2" header, 12 – 3/4" sweat outlets		3.10	5.161		137	278		415	570
1660	2" header, 12 – 3/4" sweat outlets	↓	2.90	5.517	↓	201	297		498	670
3000	Valves									
3110	Thermostatic zone valve actuator with end switch	Q-5	40	.400	Ea.	38.50	21.50		60	75
3114	Thermostatic zone valve actuator	"	36	.444	"	81	24		105	125
3120	Motorized straight zone valve with operator complete									
3130	3/4"	Q-5	35	.457	Ea.	130	24.50		154.50	180
3140	1"		32	.500		141	27		168	196
3150	1-1/4"	↓	29.60	.541	↓	179	29		208	241
3500	4 Way mixing valve, manual, brass									
3530	1"	Q-5	13.30	1.203	Ea.	180	64.50		244.50	296
3540	1-1/4"		11.40	1.404		195	75.50		270.50	330
3550	1-1/2"		11	1.455		249	78		327	390
3560	2"		10.60	1.509		350	81		431	515
3800	Mixing valve motor, 4 way for valves, 1" and 1-1/4"		34	.471		310	25.50		335.50	380
3810	Mixing valve motor, 4 way for valves, 1-1/2" and 2"	↓	30	.533	↓	355	28.50		383.50	435

23 83 Radiant Heating Units

23 83 16 – Radiant-Heating Hydronic Piping

23 83 16.10 Radiant Floor Heating		Crew	Daily Output	Labor-Hours	Unit	Material	2015 Bare Costs Labor	Equipment	Total	Total Incl O&P
5000	Radiant floor heating, zone control panel									
5120	4 Zone actuator valve control, expandable	Q-5	20	.800	Ea.	163	43		206	244
5130	6 Zone actuator valve control, expandable	↓	18	.889		226	48		274	320
6070	Thermal track, straight panel for long continuous runs, 5.333 S.F.		40	.400		26.50	21.50		48	62
6080	Thermal track, utility panel, for direction reverse at run end, 5.333 S.F.		40	.400		26.50	21.50		48	62
6090	Combination panel, for direction reverse plus straight run, 5.333 S.F.	↓	40	.400	↓	26.50	21.50		48	62
7000	PEX tubing fittings									
7100	Compression type									
7116	Coupling									
7120	1/2" x 1/2"	1 Stpi	27	.296	Ea.	6.70	17.70		24.40	34
7124	3/4" x 3/4"	"	23	.348	"	10.60	21		31.60	43
7130	Adapter									
7132	1/2" x female sweat 1/2"	1 Stpi	27	.296	Ea.	4.36	17.70		22.06	31.50
7134	1/2" x female sweat 3/4"		26	.308		4.88	18.40		23.28	33.50
7136	5/8" x female sweat 3/4"	↓	24	.333	↓	7	19.90		26.90	37.50
7140	Elbow									
7142	1/2" x female sweat 1/2"	1 Stpi	27	.296	Ea.	6.65	17.70		24.35	34
7144	1/2" x female sweat 3/4"		26	.308		7.80	18.40		26.20	36.50
7146	5/8" x female sweat 3/4"	↓	24	.333	↓	8.75	19.90		28.65	39.50
7200	Insert type									
7206	PEX x male NPT									
7210	1/2" x 1/2"	1 Stpi	29	.276	Ea.	2.59	16.50		19.09	28
7220	3/4" x 3/4"		27	.296		3.81	17.70		21.51	30.50
7230	1" x 1"	↓	26	.308	↓	6.45	18.40		24.85	35
7300	PEX coupling									
7310	1/2" x 1/2"	1 Stpi	30	.267	Ea.	1.76	15.95		17.71	26
7320	3/4" x 3/4"		29	.276		2.15	16.50		18.65	27.50
7330	1" x 1"	↓	28	.286	↓	5.85	17.05		22.90	32.50
7400	PEX stainless crimp ring									
7410	1/2" x 1/2"	1 Stpi	86	.093	Ea.	.37	5.55		5.92	8.80
7420	3/4" x 3/4"		84	.095		.51	5.70		6.21	9.15
7430	1" x 1"	↓	82	.098	↓	.73	5.85		6.58	9.60

23 83 33 – Electric Radiant Heaters

23 83 33.10 Electric Heating		Crew	Daily Output	Labor-Hours	Unit	Material	Labor	Equipment	Total	Total Incl O&P
0010	**ELECTRIC HEATING**, not incl. conduit or feed wiring									
1100	Rule of thumb: Baseboard units, including control	1 Elec	4.40	1.818	kW	104	99.50		203.50	263
1300	Baseboard heaters, 2' long, 350 watt		8	1	Ea.	28.50	54.50		83	114
1400	3' long, 750 watt		8	1		33	54.50		87.50	119
1600	4' long, 1000 watt		6.70	1.194		39	65.50		104.50	141
1800	5' long, 935 watt		5.70	1.404		48	77		125	168
2000	6' long, 1500 watt		5	1.600		54.50	87.50		142	191
2400	8' long, 2000 watt	↓	4	2	↓	66	109		175	237
2950	Wall heaters with fan, 120 to 277 volt									
3170	1000 watt	1 Elec	6	1.333	Ea.	134	73		207	256
3180	1250 watt		5	1.600		134	87.50		221.50	278
3190	1500 watt		4	2		134	109		243	310
3600	Thermostats, integral		16	.500		30	27.50		57.50	74
3800	Line voltage, 1 pole	↓	8	1	↓	19.20	54.50		73.70	103

23 84 Humidity Control Equipment

23 84 13 – Humidifiers

23 84 13.10 Humidifier Units		Crew	Daily Output	Labor-Hours	Unit	2015 Bare Costs Material	Labor	Equipment	Total	Total Incl O&P
0010	**HUMIDIFIER UNITS**									
0520	Steam, room or duct, filter, regulators, auto. controls, 220 V									
0540	11 lb. per hour	Q-5	6	2.667	Ea.	2,700	143		2,843	3,200
0560	22 lb. per hour		5	3.200		2,975	172		3,147	3,525
0580	33 lb. per hour		4	4		3,050	215		3,265	3,700
0600	50 lb. per hour		4	4		3,550	215		3,765	4,225
0620	100 lb. per hour		3	5.333		4,475	287		4,762	5,350

Estimating Tips

26 05 00 Common Work Results for Electrical

- Conduit should be taken off in three main categories—power distribution, branch power, and branch lighting—so the estimator can concentrate on systems and components, therefore making it easier to ensure all items have been accounted for.
- For cost modifications for elevated conduit installation, add the percentages to labor according to the height of installation, and only to the quantities exceeding the different height levels, not to the total conduit quantities.
- Remember that aluminum wiring of equal ampacity is larger in diameter than copper and may require larger conduit.
- If more than three wires at a time are being pulled, deduct percentages from the labor hours of that grouping of wires.
- When taking off grounding systems, identify separately the type and size of wire, and list each unique type of ground connection.
- The estimator should take the weights of materials into consideration when completing a takeoff. Topics to consider include: How will the materials be supported? What methods of support are available? How high will the support structure have to reach? Will the final support structure be able to withstand the total burden? Is the support material included or separate from the fixture, equipment, and material specified?
- Do not overlook the costs for equipment used in the installation. If scaffolding or highlifts are available in the field, contractors may use them in lieu of the proposed ladders and rolling staging.

26 20 00 Low-Voltage Electrical Transmission

- Supports and concrete pads may be shown on drawings for the larger equipment, or the support system may be only a piece of plywood for the back of a panelboard. In either case, it must be included in the costs.

26 40 00 Electrical and Cathodic Protection

- When taking off cathodic protections systems, identify the type and size of cable, and list each unique type of anode connection.

26 50 00 Lighting

- Fixtures should be taken off room by room, using the fixture schedule, specifications, and the ceiling plan. For large concentrations of lighting fixtures in the same area, deduct the percentages from labor hours.

Reference Numbers

Reference numbers are shown in shaded boxes at the beginning of some major classifications. These numbers refer to related items in the Reference Section. The reference information may be an estimating procedure, an alternate pricing method, or technical information.

Note: Not all subdivisions listed here necessarily appear in this publication. ■

*Note: **Trade Service**, in part, has been used as a reference source for some of the material prices used in Division 26.*

26 05 05 – Selective Demolition for Electrical

26 05 05.10 Electrical Demolition	Crew	Daily Output	Labor-Hours	Unit	Material	2015 Bare Costs Labor	Equipment	Total	Total Incl O&P
0010 **ELECTRICAL DEMOLITION**									
0020 Conduit to 15′ high, including fittings & hangers									
0100 Rigid galvanized steel, 1/2″ to 1″ diameter	1 Elec	242	.033	L.F.		1.81		1.81	2.71
0120 1-1/4″ to 2″	″	200	.040			2.19		2.19	3.28
0140 2-1/2″ to 3-1/2″	2 Elec	302	.053			2.90		2.90	4.34
0160 4″ to 6″	″	160	.100			5.45		5.45	8.20
0200 Electric metallic tubing (EMT), 1/2″ to 1″	1 Elec	394	.020			1.11		1.11	1.66
0220 1-1/4″ to 1-1/2″		326	.025			1.34		1.34	2.01
0240 2″ to 3″		236	.034			1.85		1.85	2.78
0260 3-1/2″ to 4″	2 Elec	310	.052			2.82		2.82	4.23
0270 Armored cable, (BX) avg. 50′ runs									
0280 #14, 2 wire	1 Elec	690	.012	L.F.		.63		.63	.95
0290 #14, 3 wire		571	.014			.77		.77	1.15
0300 #12, 2 wire		605	.013			.72		.72	1.08
0310 #12, 3 wire		514	.016			.85		.85	1.28
0320 #10, 2 wire		514	.016			.85		.85	1.28
0330 #10, 3 wire		425	.019			1.03		1.03	1.54
0340 #8, 3 wire		342	.023			1.28		1.28	1.92
0350 Non metallic sheathed cable (Romex)									
0360 #14, 2 wire	1 Elec	720	.011	L.F.		.61		.61	.91
0370 #14, 3 wire		657	.012			.67		.67	1
0380 #12, 2 wire		629	.013			.70		.70	1.04
0390 #10, 3 wire		450	.018			.97		.97	1.46
0400 Wiremold raceway, including fittings & hangers									
0420 No. 3000	1 Elec	250	.032	L.F.		1.75		1.75	2.62
0440 No. 4000		217	.037			2.02		2.02	3.02
0460 No. 6000		166	.048			2.64		2.64	3.95
0462 Plugmold with receptacle		114	.070			3.84		3.84	5.75
0465 Telephone/power pole		12	.667	Ea.		36.50		36.50	54.50
0470 Non-metallic, straight section		480	.017	L.F.		.91		.91	1.37
0500 Channels, steel, including fittings & hangers									
0520 3/4″ x 1-1/2″	1 Elec	308	.026	L.F.		1.42		1.42	2.13
0540 1-1/2″ x 1-1/2″		269	.030			1.63		1.63	2.44
0560 1-1/2″ x 1-7/8″		229	.035			1.91		1.91	2.86
0600 Copper bus duct, indoor, 3 phase									
0610 Including hangers & supports									
0620 225 amp	2 Elec	135	.119	L.F.		6.50		6.50	9.70
0640 400 amp		106	.151			8.25		8.25	12.35
0660 600 amp		86	.186			10.20		10.20	15.25
0680 1000 amp		60	.267			14.60		14.60	22
0700 1600 amp		40	.400			22		22	33
0720 3000 amp		10	1.600			87.50		87.50	131
1300 Transformer, dry type, 1 phase, incl. removal of									
1320 supports, wire & conduit terminations									
1340 1 kVA	1 Elec	7.70	1.039	Ea.		57		57	85
1420 75 kVA	2 Elec	2.50	6.400	″		350		350	525
1440 3 phase to 600V, primary									
1460 3 kVA	1 Elec	3.87	2.067	Ea.		113		113	169
1520 75 kVA	2 Elec	2.69	5.948			325		325	485
1550 300 kVA	R-3	1.80	11.111			605	76.50	681.50	990
1570 750 kVA	″	1.10	18.182			985	125	1,110	1,625
1800 Wire, THW-THWN-THHN, removed from									
1810 in place conduit, to 15′ high									

26 05 Common Work Results for Electrical

26 05 05 – Selective Demolition for Electrical

26 05 05.10 Electrical Demolition		Crew	Daily Output	Labor-Hours	Unit	Material	2015 Bare Costs Labor	Equipment	Total	Total Incl O&P
1830	#14	1 Elec	65	.123	C.L.F.		6.75		6.75	10.10
1840	#12		55	.145			7.95		7.95	11.90
1850	#10		45.50	.176			9.60		9.60	14.40
1860	#8		40.40	.198			10.85		10.85	16.25
1870	#6		32.60	.245			13.40		13.40	20
1880	#4	2 Elec	53	.302			16.50		16.50	24.50
1890	#3		50	.320			17.50		17.50	26
1900	#2		44.60	.359			19.60		19.60	29.50
1910	1/0		33.20	.482			26.50		26.50	39.50
1920	2/0		29.20	.548			30		30	45
1930	3/0		25	.640			35		35	52.50
1940	4/0		22	.727			40		40	59.50
1950	250 kcmil		20	.800			44		44	65.50
1960	300 kcmil		19	.842			46		46	69
1970	350 kcmil		18	.889			48.50		48.50	73
1980	400 kcmil		17	.941			51.50		51.50	77
1990	500 kcmil		16.20	.988			54		54	81
2000	Interior fluorescent fixtures, incl. supports									
2010	& whips, to 15′ high									
2100	Recessed drop-in 2′ x 2′, 2 lamp	2 Elec	35	.457	Ea.		25		25	37.50
2120	2′ x 4′, 2 lamp		33	.485			26.50		26.50	39.50
2140	2′ x 4′, 4 lamp		30	.533			29		29	43.50
2160	4′ x 4′, 4 lamp		20	.800			44		44	65.50
2180	Surface mount, acrylic lens & hinged frame									
2200	1′ x 4′, 2 lamp	2 Elec	44	.364	Ea.		19.90		19.90	30
2220	2′ x 2′, 2 lamp		44	.364			19.90		19.90	30
2260	2′ x 4′, 4 lamp		33	.485			26.50		26.50	39.50
2280	4′ x 4′, 4 lamp		23	.696			38		38	57
2300	Strip fixtures, surface mount									
2320	4′ long, 1 lamp	2 Elec	53	.302	Ea.		16.50		16.50	24.50
2340	4′ long, 2 lamp		50	.320			17.50		17.50	26
2360	8′ long, 1 lamp		42	.381			21		21	31
2380	8′ long, 2 lamp		40	.400			22		22	33
2400	Pendant mount, industrial, incl. removal									
2410	of chain or rod hangers, to 15′ high									
2420	4′ long, 2 lamp	2 Elec	35	.457	Ea.		25		25	37.50
2440	8′ long, 2 lamp	"	27	.593	"		32.50		32.50	48.50

26 05 13 – Medium-Voltage Cables

26 05 13.16 Medium-Voltage, Single Cable		Crew	Daily Output	Labor-Hours	Unit	Material	2015 Bare Costs Labor	Equipment	Total	Total Incl O&P
0010	**MEDIUM-VOLTAGE, SINGLE CABLE** Splicing & terminations not included									
0040	Copper, XLP shielding, 5 kV, #6	2 Elec	4.40	3.636	C.L.F.	160	199		359	475
0050	#4		4.40	3.636		207	199		406	525
0100	#2		4	4		228	219		447	580
0200	#1		4	4		281	219		500	640
0400	1/0		3.80	4.211		315	230		545	690
0600	2/0		3.60	4.444		385	243		628	785
0800	4/0		3.20	5		520	274		794	980
1000	250 kcmil	3 Elec	4.50	5.333		595	292		887	1,100
1200	350 kcmil		3.90	6.154		780	335		1,115	1,375
1400	500 kcmil		3.60	6.667		955	365		1,320	1,600
1600	15 kV, ungrounded neutral, #1	2 Elec	4	4		340	219		559	705
1800	1/0		3.80	4.211		410	230		640	795

26 05 13 – Medium-Voltage Cables

26 05 13.16 Medium-Voltage, Single Cable		Crew	Daily Output	Labor-Hours	Unit	Material	2015 Bare Costs Labor	Equipment	Total	Total Incl O&P
2000	2/0	2 Elec	3.60	4.444	C.L.F.	465	243		708	875
2200	4/0		3.20	5		620	274		894	1,100
2400	250 kcmil	3 Elec	4.50	5.333		685	292		977	1,200
2600	350 kcmil		3.90	6.154		865	335		1,200	1,450
2800	500 kcmil		3.60	6.667		1,075	365		1,440	1,725

26 05 19 – Low-Voltage Electrical Power Conductors and Cables

26 05 19.20 Armored Cable

		Crew	Daily Output	Labor-Hours	Unit	Material	Labor	Equipment	Total	Total Incl O&P
0010	**ARMORED CABLE**									
0050	600 volt, copper (BX), #14, 2 conductor, solid	1 Elec	2.40	3.333	C.L.F.	46	182		228	325
0100	3 conductor, solid		2.20	3.636		71	199		270	375
0150	#12, 2 conductor, solid		2.30	3.478		46	190		236	335
0200	3 conductor, solid		2	4		75.50	219		294.50	415
0250	#10, 2 conductor, solid		2	4		85	219		304	425
0300	3 conductor, solid		1.60	5		118	274		392	540
0340	#8, 2 conductor, stranded		1.50	5.333		222	292		514	680
0350	3 conductor, stranded		1.30	6.154		222	335		557	750
0400	3 conductor with PVC jacket, in cable tray, #6		3.10	2.581		540	141		681	800
0450	#4	2 Elec	5.40	2.963		655	162		817	965
0500	#2		4.60	3.478		795	190		985	1,150
0550	#1		4	4		1,025	219		1,244	1,450
0600	1/0		3.60	4.444		1,050	243		1,293	1,525
0650	2/0		3.40	4.706		1,250	257		1,507	1,750
0700	3/0		3.20	5		1,700	274		1,974	2,250
0750	4/0		3	5.333		2,000	292		2,292	2,625
0800	250 kcmil	3 Elec	3.60	6.667		2,325	365		2,690	3,125
0850	350 kcmil		3.30	7.273		3,150	400		3,550	4,050
0900	500 kcmil		3	8		4,350	440		4,790	5,425
1050	5 kV, copper, 3 conductor with PVC jacket,									
1060	non-shielded, in cable tray, #4	2 Elec	380	.042	L.F.	7.40	2.30		9.70	11.60
1100	#2		360	.044		9.65	2.43		12.08	14.25
1200	#1		300	.053		12.30	2.92		15.22	17.85
1400	1/0		290	.055		14.20	3.02		17.22	20
1600	2/0		260	.062		16.40	3.37		19.77	23
2000	4/0		240	.067		22	3.65		25.65	29.50
2100	250 kcmil	3 Elec	330	.073		30	3.98		33.98	39
2150	350 kcmil		315	.076		37	4.17		41.17	47
2200	500 kcmil		270	.089		51.50	4.86		56.36	64.50
2400	15 kV, copper, 3 conductor with PVC jacket galv., steel armored									
2500	grounded neutral, in cable tray, #2	2 Elec	300	.053	L.F.	15.60	2.92		18.52	21.50
2600	#1		280	.057		16.60	3.13		19.73	23
2800	1/0		260	.062		19	3.37		22.37	26
2900	2/0		220	.073		25	3.98		28.98	34
3000	4/0		190	.084		28.50	4.61		33.11	38.50
3100	250 kcmil	3 Elec	270	.089		32	4.86		36.86	42.50
3150	350 kcmil		240	.100		37.50	5.45		42.95	49.50
3200	500 kcmil		210	.114		50.50	6.25		56.75	65
3400	15 kV, copper, 3 conductor with PVC jacket,									
3450	ungrounded neutral, in cable tray, #2	2 Elec	260	.062	L.F.	16.80	3.37		20.17	23.50
3500	#1		230	.070		18.60	3.81		22.41	26
3600	1/0		200	.080		21.50	4.38		25.88	30
3700	2/0		190	.084		26	4.61		30.61	36
3800	4/0		160	.100		31.50	5.45		36.95	42.50

26 05 Common Work Results for Electrical

26 05 19 – Low-Voltage Electrical Power Conductors and Cables

26 05 19.20 Armored Cable		Crew	Daily Output	Labor-Hours	Unit	Material	Labor	Equipment	Total	Total Incl O&P
						2015 Bare Costs				
4000	250 kcmil	3 Elec	210	.114	L.F.	37	6.25		43.25	50
4050	350 kcmil		195	.123		48.50	6.75		55.25	63.50
4100	500 kcmil		180	.133		59	7.30		66.30	76
9010	600 volt, copper (MC) steel clad, #14, 2 wire	1 Elec	2.40	3.333	C.L.F.	46.50	182		228.50	325
9020	3 wire		2.20	3.636		72	199		271	380
9030	4 wire		2	4		100	219		319	440
9040	#12, 2 wire		2.30	3.478		47	190		237	335
9050	3 wire		2	4		79.50	219		298.50	420
9070	#10, 2 wire		2	4		98.50	219		317.50	440
9080	3 wire		1.60	5		138	274		412	560
9100	#8, 2 wire, stranded		1.80	4.444		190	243		433	575
9110	3 wire, stranded		1.30	6.154		267	335		602	800
9200	600 volt, copper (MC) aluminum clad, #14, 2 wire		2.65	3.019		45.50	165		210.50	297
9210	3 wire		2.45	3.265		71	179		250	345
9220	4 wire		2.20	3.636		99	199		298	405
9230	#12, 2 wire		2.55	3.137		47	172		219	310
9240	3 wire		2.20	3.636		78.50	199		277.50	385
9250	4 wire		2	4		106	219		325	445
9260	#10, 2 wire		2.20	3.636		97.50	199		296.50	405
9270	3 wire		1.80	4.444		137	243		380	515
9280	4 wire		1.55	5.161		215	282		497	660

26 05 19.35 Cable Terminations

Line	Description	Crew	Daily Output	Labor-Hours	Unit	Material	Labor	Equipment	Total	Total Incl O&P
0010	**CABLE TERMINATIONS**									
0015	Wire connectors, screw type, #22 to #14	1 Elec	260	.031	Ea.	.06	1.68		1.74	2.59
0020	#18 to #12		240	.033		.07	1.82		1.89	2.81
0025	#18 to #10		240	.033		.12	1.82		1.94	2.86
0030	Screw-on connectors, insulated, #18 to #12		240	.033		.25	1.82		2.07	3.01
0035	#16 to #10		230	.035		.27	1.90		2.17	3.15
0040	#14 to #8		210	.038		.30	2.08		2.38	3.45
0045	#12 to #6		180	.044		.62	2.43		3.05	4.32
0050	Terminal lugs, solderless, #16 to #10		50	.160		.37	8.75		9.12	13.50
0100	#8 to #4		30	.267		.73	14.60		15.33	23
0150	#2 to #1		22	.364		1.03	19.90		20.93	31
0200	1/0 to 2/0		16	.500		1.64	27.50		29.14	43
0250	3/0		12	.667		3.30	36.50		39.80	58
0300	4/0		11	.727		4.05	40		44.05	64
0350	250 kcmil		9	.889		3.41	48.50		51.91	77
0400	350 kcmil		7	1.143		4.44	62.50		66.94	98.50
0450	500 kcmil		6	1.333		8.20	73		81.20	118
1600	Crimp 1 hole lugs, copper or aluminum, 600 volt									
1620	#14	1 Elec	60	.133	Ea.	.56	7.30		7.86	11.55
1630	#12		50	.160		.90	8.75		9.65	14.10
1640	#10		45	.178		.90	9.70		10.60	15.55
1780	#8		36	.222		1.86	12.15		14.01	20.50
1800	#6		30	.267		2.11	14.60		16.71	24.50
2000	#4		27	.296		2.88	16.20		19.08	27.50
2200	#2		24	.333		4.64	18.25		22.89	32.50
2400	#1		20	.400		4.84	22		26.84	38.50
2500	1/0		17.50	.457		5.20	25		30.20	43
2600	2/0		15	.533		6.30	29		35.30	50.50
2800	3/0		12	.667		7	36.50		43.50	62
3000	4/0		11	.727		7.80	40		47.80	68

26 05 19.35 Cable Terminations

	26 05 19.35 Cable Terminations	Crew	Daily Output	Labor-Hours	Unit	2015 Bare Costs Material	Labor	Equipment	Total	Total Incl O&P
3200	250 kcmil	1 Elec	9	.889	Ea.	9.10	48.50		57.60	83
3400	300 kcmil		8	1		11.05	54.50		65.55	94
3500	350 kcmil		7	1.143		11.50	62.50		74	106
3600	400 kcmil		6.50	1.231		13.75	67.50		81.25	116
3800	500 kcmil		6	1.333		15.70	73		88.70	126

26 05 19.50 Mineral Insulated Cable

	26 05 19.50 Mineral Insulated Cable	Crew	Daily Output	Labor-Hours	Unit	Material	Labor	Equipment	Total	Total Incl O&P
0010	**MINERAL INSULATED CABLE** 600 volt									
0100	1 conductor, #12	1 Elec	1.60	5	C.L.F.	365	274		639	810
0200	#10		1.60	5		470	274		744	925
0400	#8		1.50	5.333		520	292		812	1,000
0500	#6		1.40	5.714		620	315		935	1,150
0600	#4	2 Elec	2.40	6.667		835	365		1,200	1,475
0800	#2		2.20	7.273		1,175	400		1,575	1,900
0900	#1		2.10	7.619		1,350	415		1,765	2,125
1000	1/0		2	8		1,600	440		2,040	2,400
1100	2/0		1.90	8.421		1,900	460		2,360	2,800
1200	3/0		1.80	8.889		2,275	485		2,760	3,225
1400	4/0		1.60	10		2,625	545		3,170	3,725
1410	250 kcmil	3 Elec	2.40	10		2,975	545		3,520	4,100
1420	350 kcmil		1.95	12.308		3,400	675		4,075	4,725
1430	500 kcmil		1.95	12.308		4,375	675		5,050	5,800

26 05 19.55 Non-Metallic Sheathed Cable

	26 05 19.55 Non-Metallic Sheathed Cable	Crew	Daily Output	Labor-Hours	Unit	Material	Labor	Equipment	Total	Total Incl O&P
0010	**NON-METALLIC SHEATHED CABLE** 600 volt									
0100	Copper with ground wire, (Romex)									
0150	#14, 2 conductor	1 Elec	2.70	2.963	C.L.F.	24	162		186	270
0200	3 conductor		2.40	3.333		34	182		216	310
0220	4 conductor		2.20	3.636		49	199		248	350
0250	#12, 2 conductor		2.50	3.200		36.50	175		211.50	300
0300	3 conductor		2.20	3.636		52	199		251	355
0320	4 conductor		2	4		76.50	219		295.50	415
0350	#10, 2 conductor		2.20	3.636		56.50	199		255.50	360
0400	3 conductor		1.80	4.444		82.50	243		325.50	455
0430	#8, 2 conductor		1.60	5		88.50	274		362.50	510
0450	3 conductor		1.50	5.333		133	292		425	580
0500	#6, 3 conductor		1.40	5.714		215	315		530	705
0550	SE type SER aluminum cable, 3 RHW and									
0600	1 bare neutral, 3 #8 & 1 #8	1 Elec	1.60	5	C.L.F.	158	274		432	585
0650	3 #6 & 1 #6	"	1.40	5.714		179	315		494	665
0700	3 #4 & 1 #6	2 Elec	2.40	6.667		165	365		530	725
0750	3 #2 & 1 #4		2.20	7.273		296	400		696	920
0800	3 #1/0 & 1 #2		2	8		450	440		890	1,150
0850	3 #2/0 & 1 #1		1.80	8.889		530	485		1,015	1,300
0900	3 #4/0 & 1 #2/0		1.60	10		755	545		1,300	1,650

26 05 19.90 Wire

	26 05 19.90 Wire	Crew	Daily Output	Labor-Hours	Unit	Material	Labor	Equipment	Total	Total Incl O&P
0010	**WIRE** R260519-92									
0020	600 volt, copper type THW, solid, #14	1 Elec	13	.615	C.L.F.	7.80	33.50		41.30	59
0030	#12		11	.727		11.95	40		51.95	72.50
0040	#10		10	.800		18.65	44		62.65	86
0050	Stranded, #14 R260533-22		13	.615		9.35	33.50		42.85	61
0100	#12		11	.727		14.35	40		54.35	75.50
0120	#10		10	.800		22.50	44		66.50	90.50
0140	#8		8	1		37.50	54.50		92	123

26 05 19 – Low-Voltage Electrical Power Conductors and Cables

26 05 19.90 Wire		Crew	Daily Output	Labor-Hours	Unit	Material	2015 Bare Costs Labor	Equipment	Total	Total Incl O&P
0160	#6	1 Elec	6.50	1.231	C.L.F.	63.50	67.50		131	171
0180	#4	2 Elec	10.60	1.509		100	82.50		182.50	234
0200	#3		10	1.600		126	87.50		213.50	269
0220	#2		9	1.778		158	97		255	320
0240	#1		8	2		200	109		309	385
0260	1/0		6.60	2.424		250	133		383	475
0280	2/0		5.80	2.759		315	151		466	570
0300	3/0		5	3.200		395	175		570	695
0350	4/0		4.40	3.636		500	199		699	850
0400	250 kcmil	3 Elec	6	4		585	219		804	975
0420	300 kcmil		5.70	4.211		700	230		930	1,125
0450	350 kcmil		5.40	4.444		855	243		1,098	1,300
0480	400 kcmil		5.10	4.706		985	257		1,242	1,450
0490	500 kcmil		4.80	5		1,150	274		1,424	1,675
0540	600 volt, aluminum type THHN, stranded, #6	1 Elec	8	1		43	54.50		97.50	130
0560	#4	2 Elec	13	1.231		53.50	67.50		121	160
0580	#2		10.60	1.509		72.50	82.50		155	204
0600	#1		9	1.778		106	97		203	262
0620	1/0		8	2		127	109		236	305
0640	2/0		7.20	2.222		150	122		272	345
0680	3/0		6.60	2.424		186	133		319	405
0700	4/0		6.20	2.581		207	141		348	440
0720	250 kcmil	3 Elec	8.70	2.759		253	151		404	505
0740	300 kcmil		8.10	2.963		350	162		512	630
0760	350 kcmil		7.50	3.200		355	175		530	650
0780	400 kcmil		6.90	3.478		415	190		605	740
0800	500 kcmil		6	4		460	219		679	835
0850	600 kcmil		5.70	4.211		580	230		810	985
0880	700 kcmil		5.10	4.706		670	257		927	1,125
0900	750 kcmil		4.80	5		695	274		969	1,175
0910	1000 kcmil		3.78	6.349		1,025	345		1,370	1,650
0920	600 volt, copper type THWN-THHN, solid, #14	1 Elec	13	.615		7.80	33.50		41.30	59
0940	#12		11	.727		11.95	40		51.95	72.50
0960	#10		10	.800		18.65	44		62.65	86
1000	Stranded, #14		13	.615		8.90	33.50		42.40	60.50
1200	#12		11	.727		13.30	40		53.30	74
1250	#10		10	.800		20.50	44		64.50	88
1300	#8		8	1		33.50	54.50		88	119
1350	#6		6.50	1.231		57.50	67.50		125	165
1400	#4	2 Elec	10.60	1.509		87.50	82.50		170	220

26 05 23 – Control-Voltage Electrical Power Cables

26 05 23.10 Control Cable		Crew	Daily Output	Labor-Hours	Unit	Material	Labor	Equipment	Total	Total Incl O&P
0010	**CONTROL CABLE**									
0020	600 volt, copper, #14 THWN wire with PVC jacket, 2 wires	1 Elec	9	.889	C.L.F.	30.50	48.50		79	107
0030	3 wires		8	1		42.50	54.50		97	129
0100	4 wires		7	1.143		52	62.50		114.50	151
0150	5 wires		6.50	1.231		65	67.50		132.50	173
0200	6 wires		6	1.333		85.50	73		158.50	203
0300	8 wires		5.30	1.509		106	82.50		188.50	241
0400	10 wires		4.80	1.667		126	91		217	276
0500	12 wires		4.30	1.860		148	102		250	315
0600	14 wires		3.80	2.105		176	115		291	365

26 05 Common Work Results for Electrical

26 05 23 – Control-Voltage Electrical Power Cables

26 05 23.10 Control Cable		Crew	Daily Output	Labor-Hours	Unit	2015 Bare Costs Material	Labor	Equipment	Total	Total Incl O&P
0700	16 wires	1 Elec	3.50	2.286	C.L.F.	182	125		307	385
0800	18 wires		3.30	2.424		199	133		332	420
0810	19 wires		3.10	2.581		224	141		365	460
0900	20 wires		3	2.667		242	146		388	485
1000	22 wires		2.80	2.857		249	156		405	510

26 05 26 – Grounding and Bonding for Electrical Systems

26 05 26.80 Grounding		Crew	Daily Output	Labor-Hours	Unit	Material	Labor	Equipment	Total	Total Incl O&P
0010	**GROUNDING**									
0030	Rod, copper clad, 8′ long, 1/2″ diameter	1 Elec	5.50	1.455	Ea.	20	79.50		99.50	141
0050	3/4″ diameter		5.30	1.509		33.50	82.50		116	161
0080	10′ long, 1/2″ diameter		4.80	1.667		22	91		113	162
0100	3/4″ diameter		4.40	1.818		38.50	99.50		138	192
0130	15′ long, 3/4″ diameter		4	2		54	109		163	224
0390	Bare copper wire, stranded, #8		11	.727	C.L.F.	29.50	40		69.50	92
0400	#6		10	.800		52	44		96	123
0600	#2	2 Elec	10	1.600		138	87.50		225.50	283
0800	3/0		6.60	2.424		310	133		443	540
1000	4/0		5.70	2.807		395	154		549	665
1200	250 kcmil	3 Elec	7.20	3.333		465	182		647	785
1800	Water pipe ground clamps, heavy duty									
2000	Bronze, 1/2″ to 1″ diameter	1 Elec	8	1	Ea.	24	54.50		78.50	108
2100	1-1/4″ to 2″ diameter		8	1		33	54.50		87.50	119
2200	2-1/2″ to 3″ diameter		6	1.333		44.50	73		117.50	158
2800	Brazed connections, #6 wire		12	.667		16.15	36.50		52.65	72.50
3000	#2 wire		10	.800		21.50	44		65.50	89.50
3100	3/0 wire		8	1		32.50	54.50		87	118
3200	4/0 wire		7	1.143		37	62.50		99.50	135
3400	250 kcmil wire		5	1.600		43.50	87.50		131	179
3600	500 kcmil wire		4	2		53.50	109		162.50	223

26 05 33 – Raceway and Boxes for Electrical Systems

26 05 33.13 Conduit		Crew	Daily Output	Labor-Hours	Unit	Material	Labor	Equipment	Total	Total Incl O&P
0010	**CONDUIT** To 15′ high, includes 2 terminations, 2 elbows, R260533-22									
0020	11 beam clamps, and 11 couplings per 100 L.F.									
0300	Aluminum, 1/2″ diameter	1 Elec	100	.080	L.F.	1.75	4.38		6.13	8.45
0500	3/4″ diameter		90	.089		2.41	4.86		7.27	9.95
0700	1″ diameter		80	.100		3.46	5.45		8.91	12
1000	1-1/4″ diameter		70	.114		4.27	6.25		10.52	14.05
1030	1-1/2″ diameter		65	.123		5.45	6.75		12.20	16.10
1050	2″ diameter		60	.133		7.70	7.30		15	19.40
1070	2-1/2″ diameter		50	.160		11.50	8.75		20.25	26
1100	3″ diameter	2 Elec	90	.178		15.70	9.70		25.40	32
1130	3-1/2″ diameter		80	.200		21.50	10.95		32.45	40
1140	4″ diameter		70	.229		25	12.50		37.50	46.50
1750	Rigid galvanized steel, 1/2″ diameter	1 Elec	90	.089		2.49	4.86		7.35	10.05
1770	3/4″ diameter		80	.100		2.74	5.45		8.19	11.20
1800	1″ diameter		65	.123		3.95	6.75		10.70	14.45
1830	1-1/4″ diameter		60	.133		5.15	7.30		12.45	16.60
1850	1-1/2″ diameter		55	.145		6.10	7.95		14.05	18.65
1870	2″ diameter		45	.178		7.90	9.70		17.60	23.50
1900	2-1/2″ diameter		35	.229		13.75	12.50		26.25	34
1930	3″ diameter	2 Elec	50	.320		15.95	17.50		33.45	43.50
1950	3-1/2″ diameter		44	.364		20	19.90		39.90	52

26 05 33 – Raceway and Boxes for Electrical Systems

26 05 33.13 Conduit		Crew	Daily Output	Labor-Hours	Unit	Material	2015 Bare Costs Labor	Equipment	Total	Total Incl O&P
1970	4" diameter	2 Elec	40	.400	L.F.	23	22		45	58.50
2500	Steel, intermediate conduit (IMC), 1/2" diameter	1 Elec	100	.080		1.79	4.38		6.17	8.50
2530	3/4" diameter		90	.089		2.22	4.86		7.08	9.75
2550	1" diameter		70	.114		3.28	6.25		9.53	12.95
2570	1-1/4" diameter		65	.123		4.01	6.75		10.76	14.50
2600	1-1/2" diameter		60	.133		5.35	7.30		12.65	16.85
2630	2" diameter		50	.160		6.40	8.75		15.15	20
2650	2-1/2" diameter		40	.200		11.35	10.95		22.30	29
2670	3" diameter	2 Elec	60	.267		15.40	14.60		30	39
2700	3-1/2" diameter		54	.296		20.50	16.20		36.70	47
2730	4" diameter		50	.320		22	17.50		39.50	50.50
5000	Electric metallic tubing (EMT), 1/2" diameter	1 Elec	170	.047		.67	2.57		3.24	4.60
5020	3/4" diameter		130	.062		.94	3.37		4.31	6.10
5040	1" diameter		115	.070		1.60	3.81		5.41	7.45
5060	1-1/4" diameter		100	.080		2.61	4.38		6.99	9.40
5080	1-1/2" diameter		90	.089		3.35	4.86		8.21	11
5100	2" diameter		80	.100		4.20	5.45		9.65	12.80
5120	2-1/2" diameter		60	.133		9.05	7.30		16.35	21
5140	3" diameter	2 Elec	100	.160		10.60	8.75		19.35	25
5160	3-1/2" diameter		90	.178		13.25	9.70		22.95	29
5180	4" diameter		80	.200		14.50	10.95		25.45	32.50
9900	Add to labor for higher elevated installation									
9910	15' to 20' high, add						10%			
9920	20' to 25' high, add						20%			
9930	25' to 30' high, add						25%			
9940	30' to 35' high, add						30%			
9950	35' to 40' high, add						35%			
9960	Over 40' high, add						40%			

26 05 33.16 Outlet Boxes		Crew	Daily Output	Labor-Hours	Unit	Material	Labor	Equipment	Total	Total Incl O&P
0010	**OUTLET BOXES**									
0020	Pressed steel, octagon, 4"	1 Elec	20	.400	Ea.	2.61	22		24.61	36
0060	Covers, blank		64	.125		1.10	6.85		7.95	11.45
0100	Extension rings		40	.200		4.33	10.95		15.28	21
0150	Square, 4"		20	.400		2.42	22		24.42	35.50
0200	Extension rings		40	.200		4.37	10.95		15.32	21
0250	Covers, blank		64	.125		1.18	6.85		8.03	11.55
0300	Plaster rings		64	.125		2.40	6.85		9.25	12.90
0650	Switchbox		27	.296		4.18	16.20		20.38	29
1100	Concrete, floor, 1 gang		5.30	1.509		89.50	82.50		172	223
2000	Poke-thru fitting, fire rated, for 3-3/4" floor		6.80	1.176		131	64.50		195.50	241
2040	For 7" floor		6.80	1.176		165	64.50		229.50	278
2100	Pedestal, 15 amp, duplex receptacle & blank plate		5.25	1.524		138	83.50		221.50	276
2120	Duplex receptacle and telephone plate		5.25	1.524		138	83.50		221.50	276
2140	Pedestal, 20 amp, duplex recept. & phone plate		5	1.600		139	87.50		226.50	283
2200	Abandonment plate		32	.250		38.50	13.70		52.20	63

26 05 33.18 Pull Boxes		Crew	Daily Output	Labor-Hours	Unit	Material	Labor	Equipment	Total	Total Incl O&P
0010	**PULL BOXES**									
0100	Steel, pull box, NEMA 1, type SC, 6" W x 6" H x 4" D	1 Elec	8	1	Ea.	9.65	54.50		64.15	92.50
0200	8" W x 8" H x 4" D		8	1		14.50	54.50		69	98
0300	10" W x 12" H x 6" D		5.30	1.509		24.50	82.50		107	151
0400	16" W x 20" H x 8" D		4	2		84.50	109		193.50	257
0500	20" W x 24" H x 8" D		3.20	2.500		98	137		235	315

26 05 33 – Raceway and Boxes for Electrical Systems

26 05 33.18 Pull Boxes		Crew	Daily Output	Labor-Hours	Unit	Material	2015 Bare Costs Labor	Equipment	Total	Total Incl O&P
0600	24" W x 36" H x 8" D	1 Elec	2.70	2.963	Ea.	151	162		313	410
0650	Pull box, hinged, NEMA 1, 6" W x 6" H x 4" D		8	1		11.60	54.50		66.10	95
0800	12" W x 16" H x 6" D		4.70	1.702		49.50	93		142.50	194
1000	20" W x 20" H x 6" D		3.60	2.222		86	122		208	277
1200	20" W x 20" H x 8" D		3.20	2.500		149	137		286	370
1400	24" W x 36" H x 8" D		2.70	2.963		237	162		399	505
1600	24" W x 42" H x 8" D	↓	2	4	↓	350	219		569	715
2100	Pull box, NEMA 3R, type SC, raintight & weatherproof									
2150	6" L x 6" W x 6" D	1 Elec	10	.800	Ea.	14.75	44		58.75	82
2200	8" L x 6" W x 6" D		8	1		20.50	54.50		75	105
2250	10" L x 6" W x 6" D		7	1.143		32	62.50		94.50	129
2300	12" L x 12" W x 6" D		5	1.600		54.50	87.50		142	191
2350	16" L x 16" W x 6" D		4.50	1.778		77.50	97		174.50	232
2400	20" L x 20" W x 6" D		4	2		101	109		210	275
2450	24" L x 18" W x 8" D		3	2.667		123	146		269	355
2500	24" L x 24" W x 10" D		2.50	3.200		291	175		466	580
2550	30" L x 24" W x 12" D		2	4		385	219		604	750
2600	36" L x 36" W x 12" D	↓	1.50	5.333	↓	430	292		722	910
2800	Cast iron, pull boxes for surface mounting									
3000	NEMA 4, watertight & dust tight									
3050	6" L x 6" W x 6" D	1 Elec	4	2	Ea.	257	109		366	445
3100	8" L x 6" W x 6" D		3.20	2.500		360	137		497	600
3150	10" L x 6" W x 6" D		2.50	3.200		400	175		575	700
3200	12" L x 12" W x 6" D		2.30	3.478		730	190		920	1,075
3250	16" L x 16" W x 6" D		1.30	6.154		960	335		1,295	1,550
3300	20" L x 20" W x 6" D		.80	10		1,800	545		2,345	2,800
3350	24" L x 18" W x 8" D		.70	11.429		2,625	625		3,250	3,825
3400	24" L x 24" W x 10" D		.50	16		4,975	875		5,850	6,775
3450	30" L x 24" W x 12" D		.40	20		5,450	1,100		6,550	7,650
3500	36" L x 36" W x 12" D	↓	.20	40	↓	5,975	2,200		8,175	9,850
6000	J.I.C. wiring boxes, NEMA 12, dust tight & drip tight									
6050	6" L x 8" W x 4" D	1 Elec	10	.800	Ea.	51.50	44		95.50	122
6100	8" L x 10" W x 4" D		8	1		64	54.50		118.50	153
6150	12" L x 14" W x 6" D		5.30	1.509		124	82.50		206.50	261
6200	14" L x 16" W x 6" D		4.70	1.702		147	93		240	300
6250	16" L x 20" W x 6" D		4.40	1.818		226	99.50		325.50	395
6300	24" L x 30" W x 6" D		3.20	2.500		330	137		467	570
6350	24" L x 30" W x 8" D		2.90	2.759		340	151		491	600
6400	24" L x 36" W x 8" D		2.70	2.963		380	162		542	660
6450	24" L x 42" W x 8" D		2.30	3.478		425	190		615	755
6500	24" L x 48" W x 8" D	↓	2	4	↓	465	219		684	840

26 05 33.23 Wireway		Crew	Daily Output	Labor-Hours	Unit	Material	Labor	Equipment	Total	Total Incl O&P
0010	**WIREWAY** to 15' high									
0100	NEMA 1, Screw cover w/fittings and supports, 2-1/2" x 2-1/2"	1 Elec	45	.178	L.F.	10.85	9.70		20.55	26.50
0200	4" x 4"	"	40	.200		11.35	10.95		22.30	29
0400	6" x 6"	2 Elec	60	.267		19.60	14.60		34.20	43.50
0600	8" x 8"	"	40	.400		33.50	22		55.50	70
4475	NEMA 3R, Screw cover w/fittings and supports, 4" x 4"	1 Elec	36	.222		16.50	12.15		28.65	36.50
4480	6" x 6"	2 Elec	55	.291		22	15.90		37.90	48.50
4485	8" x 8"		36	.444		34	24.50		58.50	74
4490	12" x 12"	↓	18	.889	↓	54.50	48.50		103	133

26 05 Common Work Results for Electrical

26 05 36 – Cable Trays for Electrical Systems

26 05 36.10 Cable Tray Ladder Type

	26 05 36.10 Cable Tray Ladder Type	Crew	Daily Output	Labor-Hours	Unit	Material (2015 Bare Costs)	Labor (2015 Bare Costs)	Equipment (2015 Bare Costs)	Total (2015 Bare Costs)	Total Incl O&P
0010	**CABLE TRAY LADDER TYPE** w/ftngs. & supports, 4″ dp., to 15′ elev.									
0160	Galvanized steel tray									
0170	4″ rung spacing, 6″ wide	2 Elec	98	.163	L.F.	18.15	8.95		27.10	33.50
0200	12″ wide		86	.186		22	10.20		32.20	39.50
0400	18″ wide		82	.195		25.50	10.65		36.15	44
0600	24″ wide		78	.205		29	11.20		40.20	49
3200	Aluminum tray, 4″ deep, 6″ rung spacing, 6″ wide		134	.119		18	6.55		24.55	29.50
3220	12″ wide		124	.129		20	7.05		27.05	32.50
3230	18″ wide		114	.140		22.50	7.70		30.20	36
3240	24″ wide		106	.151		26	8.25		34.25	41

26 05 39 – Underfloor Raceways for Electrical Systems

26 05 39.30 Conduit In Concrete Slab

	26 05 39.30 Conduit In Concrete Slab	Crew	Daily Output	Labor-Hours	Unit	Material	Labor	Equipment	Total	Total Incl O&P
0010	**CONDUIT IN CONCRETE SLAB** Including terminations,									
0020	fittings and supports									
3230	PVC, schedule 40, 1/2″ diameter	1 Elec	270	.030	L.F.	.57	1.62		2.19	3.06
3250	3/4″ diameter		230	.035		.66	1.90		2.56	3.57
3270	1″ diameter		200	.040		.87	2.19		3.06	4.23
3300	1-1/4″ diameter		170	.047		1.17	2.57		3.74	5.15
3330	1-1/2″ diameter		140	.057		1.42	3.13		4.55	6.25
3350	2″ diameter		120	.067		1.78	3.65		5.43	7.40
4350	Rigid galvanized steel, 1/2″ diameter		200	.040		2.48	2.19		4.67	6
4400	3/4″ diameter		170	.047		2.62	2.57		5.19	6.75
4450	1″ diameter		130	.062		3.60	3.37		6.97	9
4500	1-1/4″ diameter		110	.073		4.94	3.98		8.92	11.40
4600	1-1/2″ diameter		100	.080		5.50	4.38		9.88	12.60
4800	2″ diameter		90	.089		6.80	4.86		11.66	14.80

26 05 39.40 Conduit In Trench

	26 05 39.40 Conduit In Trench	Crew	Daily Output	Labor-Hours	Unit	Material	Labor	Equipment	Total	Total Incl O&P
0010	**CONDUIT IN TRENCH** Includes terminations and fittings									
0020	Does not include excavation or backfill, see Section 31 23 16.00									
0200	Rigid galvanized steel, 2″ diameter	1 Elec	150	.053	L.F.	6.40	2.92		9.32	11.40
0400	2-1/2″ diameter	″	100	.080		12.45	4.38		16.83	20.50
0600	3″ diameter	2 Elec	160	.100		14.55	5.45		20	24
0800	3-1/2″ diameter		140	.114		19.20	6.25		25.45	30.50
1000	4″ diameter		100	.160		21	8.75		29.75	36.50
1200	5″ diameter		80	.200		44	10.95		54.95	64.50
1400	6″ diameter		60	.267		61	14.60		75.60	89.50

26 05 43 – Underground Ducts and Raceways for Electrical Systems

26 05 43.10 Trench Duct

	26 05 43.10 Trench Duct	Crew	Daily Output	Labor-Hours	Unit	Material	Labor	Equipment	Total	Total Incl O&P
0010	**TRENCH DUCT** Steel with cover									
0020	Standard adjustable, depths to 4″									
0100	Straight, single compartment, 9″ wide	2 Elec	40	.400	L.F.	112	22		134	156
0200	12″ wide		32	.500		136	27.50		163.50	190
0400	18″ wide		26	.615		167	33.50		200.50	234
0600	24″ wide		22	.727		199	40		239	279
0800	30″ wide		20	.800		234	44		278	325
1000	36″ wide		16	1		265	54.50		319.50	375
1200	Horizontal elbow, 9″ wide		5.40	2.963	Ea.	385	162		547	670
1400	12″ wide		4.60	3.478		445	190		635	775
1600	18″ wide		4	4		570	219		789	960
1800	24″ wide		3.20	5		800	274		1,074	1,300
2000	30″ wide		2.60	6.154		1,075	335		1,410	1,675

26 05 43 – Underground Ducts and Raceways for Electrical Systems

26 05 43.10 Trench Duct		Crew	Daily Output	Labor-Hours	Unit	2015 Bare Costs Material	Labor	Equipment	Total	Total Incl O&P
2200	36″ wide	2 Elec	2.40	6.667	Ea.	1,400	365		1,765	2,100
2400	Vertical elbow, 9″ wide		5.40	2.963		135	162		297	390
2600	12″ wide		4.60	3.478		146	190		336	445
2800	18″ wide		4	4		168	219		387	515
3000	24″ wide		3.20	5		209	274		483	640
3200	30″ wide		2.60	6.154		230	335		565	760
3400	36″ wide		2.40	6.667		253	365		618	825
3600	Cross, 9″ wide		4	4		635	219		854	1,025
3800	12″ wide		3.20	5		670	274		944	1,150
4000	18″ wide		2.60	6.154		800	335		1,135	1,375
4200	24″ wide		2.20	7.273		1,025	400		1,425	1,725
4400	30″ wide		2	8		1,325	440		1,765	2,100
4600	36″ wide		1.80	8.889		1,650	485		2,135	2,550
4800	End closure, 9″ wide		14.40	1.111		39.50	61		100.50	135
5000	12″ wide		12	1.333		45.50	73		118.50	159
5200	18″ wide		10	1.600		69.50	87.50		157	208
5400	24″ wide		8	2		91.50	109		200.50	265
5600	30″ wide		6.60	2.424		115	133		248	325
5800	36″ wide		5.80	2.759		136	151		287	375
6000	Tees, 9″ wide		4	4		385	219		604	755
6200	12″ wide		3.60	4.444		445	243		688	855
6400	18″ wide		3.20	5		570	274		844	1,050
6600	24″ wide		3	5.333		820	292		1,112	1,350
6800	30″ wide		2.60	6.154		1,075	335		1,410	1,675
7000	36″ wide		2	8		1,400	440		1,840	2,200
7200	Riser, and cabinet connector, 9″ wide		5.40	2.963		168	162		330	430
7400	12″ wide		4.60	3.478		196	190		386	500
7600	18″ wide		4	4		241	219		460	595
7800	24″ wide		3.20	5		291	274		565	730
8000	30″ wide		2.60	6.154		335	335		670	875
8200	36″ wide		2	8		390	440		830	1,075
8400	Insert assembly, cell to conduit adapter, 1-1/4″	1 Elec	16	.500		66.50	27.50		94	114

26 05 43.20 Underfloor Duct		Crew	Daily Output	Labor-Hours	Unit	Material	Labor	Equipment	Total	Total Incl O&P
0010	**UNDERFLOOR DUCT**									
0100	Duct, 1-3/8″ x 3-1/8″ blank, standard	2 Elec	160	.100	L.F.	14.25	5.45		19.70	24
0200	1-3/8″ x 7-1/4″ blank, super duct		120	.133		28.50	7.30		35.80	42.50
0400	7/8″ or 1-3/8″ insert type, 24″ O.C., 1-3/8″ x 3-1/8″, std.		140	.114		19.05	6.25		25.30	30.50
0600	1-3/8″ x 7-1/4″, super duct		100	.160		33.50	8.75		42.25	49.50
0800	Junction box, single duct, 1 level, 3-1/8″	1 Elec	4	2	Ea.	430	109		539	635
1000	Junction box, single duct, 1 level, 7-1/4″		2.70	2.963		500	162		662	795
1200	1 level, 2 duct, 3-1/8″		3.20	2.500		570	137		707	835
1400	Junction box, 1 level, 2 duct, 7-1/4″		2.30	3.478		1,475	190		1,665	1,900
1580	Junction box, 1 level, one 3-1/8″ + one 7-1/4″ x same		2.30	3.478		970	190		1,160	1,350
1600	Triple duct, 3-1/8″		2.30	3.478		970	190		1,160	1,350
1800	Insert to conduit adapter, 3/4″ & 1″		32	.250		35	13.70		48.70	59
2000	Support, single cell		27	.296		52.50	16.20		68.70	82.50
2200	Super duct		16	.500		52.50	27.50		80	99
2400	Double cell		16	.500		52.50	27.50		80	99
2600	Triple cell		11	.727		52.50	40		92.50	118
2800	Vertical elbow, standard duct		10	.800		94.50	44		138.50	170
3000	Super duct		8	1		94.50	54.50		149	186
3200	Cabinet connector, standard duct		32	.250		71	13.70		84.70	98.50

26 05 43 – Underground Ducts and Raceways for Electrical Systems

26 05 43.20 Underfloor Duct		Crew	Daily Output	Labor-Hours	Unit	Material	2015 Bare Costs Labor	Equipment	Total	Total Incl O&P
3400	Super duct	1 Elec	27	.296	Ea.	71	16.20		87.20	103
3600	Conduit adapter, 1" to 1-1/4"		32	.250		71	13.70		84.70	98.50
3800	2" to 1-1/4"		27	.296		85	16.20		101.20	119
4000	Outlet, low tension (tele, computer, etc.)		8	1		99.50	54.50		154	192
4200	High tension, receptacle (120 volt)		8	1		99.50	54.50		154	192

26 05 80 – Wiring Connections

26 05 80.10 Motor Connections		Crew	Daily Output	Labor-Hours	Unit	Material	Labor	Equipment	Total	Total Incl O&P
0010	**MOTOR CONNECTIONS**									
0020	Flexible conduit and fittings, 115 volt, 1 phase, up to 1 HP motor	1 Elec	8	1	Ea.	6.35	54.50		60.85	89
0110	230 volt, 3 phase, 3 HP motor		6.78	1.180		7.65	64.50		72.15	105
0112	5 HP motor		5.47	1.463		6.50	80		86.50	127
0114	7-1/2 HP motor		4.61	1.735		9.55	95		104.55	153
0120	10 HP motor		4.20	1.905		21	104		125	180
0200	25 HP motor		2.70	2.963		29	162		191	275
0400	50 HP motor		2.20	3.636		59	199		258	365
0600	100 HP motor		1.50	5.333		140	292		432	590

26 05 90 – Residential Applications

26 05 90.10 Residential Wiring		Crew	Daily Output	Labor-Hours	Unit	Material	Labor	Equipment	Total	Total Incl O&P
0010	**RESIDENTIAL WIRING**									
0020	20' avg. runs and #14/2 wiring incl. unless otherwise noted									
1000	Service & panel, includes 24' SE-AL cable, service eye, meter,									
1010	Socket, panel board, main bkr., ground rod, 15 or 20 amp									
1020	1-pole circuit breakers, and misc. hardware									
1100	100 amp, with 10 branch breakers	1 Elec	1.19	6.723	Ea.	560	370		930	1,175
1110	With PVC conduit and wire		.92	8.696		605	475		1,080	1,375
1120	With RGS conduit and wire		.73	10.959		765	600		1,365	1,750
1150	150 amp, with 14 branch breakers		1.03	7.767		860	425		1,285	1,575
1170	With PVC conduit and wire		.82	9.756		955	535		1,490	1,850
1180	With RGS conduit and wire		.67	11.940		1,275	655		1,930	2,375
1200	200 amp, with 18 branch breakers	2 Elec	1.80	8.889		1,150	485		1,635	2,000
1220	With PVC conduit and wire		1.46	10.959		1,250	600		1,850	2,275
1230	With RGS conduit and wire		1.24	12.903		1,650	705		2,355	2,850
1800	Lightning surge suppressor	1 Elec	32	.250		50.50	13.70		64.20	76.50
2000	Switch devices									
2100	Single pole, 15 amp, Ivory, with a 1-gang box, cover plate,									
2110	Type NM (Romex) cable	1 Elec	17.10	.468	Ea.	12.35	25.50		37.85	52
2120	Type MC (BX) cable		14.30	.559		21.50	30.50		52	69.50
2130	EMT & wire		5.71	1.401		30	76.50		106.50	148
2150	3-way, #14/3, type NM cable		14.55	.550		14.85	30		44.85	61.50
2170	Type MC cable		12.31	.650		27	35.50		62.50	83
2180	EMT & wire		5	1.600		32	87.50		119.50	166
2200	4-way, #14/3, type NM cable		14.55	.550		21	30		51	68
2220	Type MC cable		12.31	.650		33	35.50		68.50	90
2230	EMT & wire		5	1.600		38.50	87.50		126	173
2250	S.P., 20 amp, #12/2, type NM cable		13.33	.600		22	33		55	73
2270	Type MC cable		11.43	.700		28.50	38.50		67	89
2280	EMT & wire		4.85	1.649		41	90		131	180
2290	S.P. rotary dimmer, 600W, no wiring		17	.471		29.50	25.50		55	71
2300	S.P. rotary dimmer, 600W, type NM cable		14.55	.550		34.50	30		64.50	83
2320	Type MC cable		12.31	.650		43.50	35.50		79	102
2330	EMT & wire		5	1.600		53.50	87.50		141	190
2350	3-way rotary dimmer, type NM cable		13.33	.600		28.50	33		61.50	80.50

26 05 90 – Residential Applications

26 05 90.10 Residential Wiring		Crew	Daily Output	Labor-Hours	Unit	2015 Bare Costs Material	Labor	Equipment	Total	Total Incl O&P
2370	Type MC cable	1 Elec	11.43	.700	Ea.	37.50	38.50		76	98.50
2380	EMT & wire		4.85	1.649		47.50	90		137.50	188
2400	Interval timer wall switch, 20 amp, 1-30 min., #12/2									
2410	Type NM cable	1 Elec	14.55	.550	Ea.	56	30		86	107
2420	Type MC cable		12.31	.650		60	35.50		95.50	120
2430	EMT & wire		5	1.600		75	87.50		162.50	214
2500	Decorator style									
2510	S.P., 15 amp, type NM cable	1 Elec	17.10	.468	Ea.	16.25	25.50		41.75	56.50
2520	Type MC cable		14.30	.559		25	30.50		55.50	73.50
2530	EMT & wire		5.71	1.401		34	76.50		110.50	152
2550	3-way, #14/3, type NM cable		14.55	.550		18.75	30		48.75	65.50
2570	Type MC cable		12.31	.650		30.50	35.50		66	87
2580	EMT & wire		5	1.600		36	87.50		123.50	171
2600	4-way, #14/3, type NM cable		14.55	.550		25	30		55	72.50
2620	Type MC cable		12.31	.650		37	35.50		72.50	94
2630	EMT & wire		5	1.600		42	87.50		129.50	178
2650	S.P., 20 amp, #12/2, type NM cable		13.33	.600		26	33		59	77.50
2670	Type MC cable		11.43	.700		32.50	38.50		71	93
2680	EMT & wire		4.85	1.649		45	90		135	185
2700	S.P., slide dimmer, type NM cable		17.10	.468		28.50	25.50		54	70
2720	Type MC cable		14.30	.559		37.50	30.50		68	87.50
2730	EMT & wire		5.71	1.401		47.50	76.50		124	168
2750	S.P., touch dimmer, type NM cable		17.10	.468		32	25.50		57.50	74
2770	Type MC cable		14.30	.559		41	30.50		71.50	91.50
2780	EMT & wire		5.71	1.401		51.50	76.50		128	172
2800	3-way touch dimmer, type NM cable		13.33	.600		52	33		85	106
2820	Type MC cable		11.43	.700		60.50	38.50		99	125
2830	EMT & wire		4.85	1.649		71	90		161	213
3000	Combination devices									
3100	S.P. switch/15 amp recpt., Ivory, 1-gang box, plate									
3110	Type NM cable	1 Elec	11.43	.700	Ea.	24.50	38.50		63	84.50
3120	Type MC cable		10	.800		33.50	44		77.50	102
3130	EMT & wire		4.40	1.818		43.50	99.50		143	197
3150	S.P. switch/pilot light, type NM cable		11.43	.700		24.50	38.50		63	84.50
3170	Type MC cable		10	.800		33.50	44		77.50	103
3180	EMT & wire		4.43	1.806		43.50	99		142.50	196
3190	2-S.P. switches, 2-#14/2, no wiring		14	.571		6.90	31.50		38.40	54.50
3200	2-S.P. switches, 2-#14/2, type NM cables		10	.800		27.50	44		71.50	95.50
3220	Type MC cable		8.89	.900		41	49		90	119
3230	EMT & wire		4.10	1.951		46.50	107		153.50	211
3250	3-way switch/15 amp recpt., #14/3, type NM cable		10	.800		32	44		76	101
3270	Type MC cable		8.89	.900		44	49		93	122
3280	EMT & wire		4.10	1.951		49	107		156	214
3300	2-3 way switches, 2-#14/3, type NM cables		8.89	.900		41.50	49		90.50	120
3320	Type MC cable		8	1		60.50	54.50		115	149
3330	EMT & wire		4	2		56.50	109		165.50	226
3350	S.P. switch/20 amp recpt., #12/2, type NM cable		10	.800		32	44		76	101
3370	Type MC cable		8.89	.900		36	49		85	114
3380	EMT & wire		4.10	1.951		51	107		158	216
3400	Decorator style									
3410	S.P. switch/15 amp recpt., type NM cable	1 Elec	11.43	.700	Ea.	28.50	38.50		67	88.50
3420	Type MC cable		10	.800		37	44		81	107
3430	EMT & wire		4.40	1.818		47.50	99.50		147	201

26 05 90.10 Residential Wiring		Crew	Daily Output	Labor-Hours	Unit	2015 Bare Costs Material	Labor	Equipment	Total	Total Incl O&P
3450	S.P. switch/pilot light, type NM cable	1 Elec	11.43	.700	Ea.	28.50	38.50		67	88.50
3470	Type MC cable		10	.800		37.50	44		81.50	107
3480	EMT & wire		4.40	1.818		47.50	99.50		147	202
3500	2-S.P. switches, 2-#14/2, type NM cables		10	.800		31.50	44		75.50	100
3520	Type MC cable		8.89	.900		44.50	49		93.50	123
3530	EMT & wire		4.10	1.951		50.50	107		157.50	216
3550	3-way/15 amp recpt., #14/3, type NM cable		10	.800		36	44		80	105
3570	Type MC cable		8.89	.900		47.50	49		96.50	127
3580	EMT & wire		4.10	1.951		53	107		160	218
3650	2-3 way switches, 2-#14/3, type NM cables		8.89	.900		45.50	49		94.50	124
3670	Type MC cable		8	1		64.50	54.50		119	153
3680	EMT & wire		4	2		60.50	109		169.50	231
3700	S.P. switch/20 amp recpt., #12/2, type NM cable		10	.800		35.50	44		79.50	105
3720	Type MC cable		8.89	.900		39.50	49		88.50	118
3730	EMT & wire		4.10	1.951		55	107		162	221
4000	Receptacle devices									
4010	Duplex outlet, 15 amp recpt., Ivory, 1-gang box, plate									
4015	Type NM cable	1 Elec	14.55	.550	Ea.	10.80	30		40.80	57
4020	Type MC cable		12.31	.650		19.75	35.50		55.25	75
4030	EMT & wire		5.33	1.501		28.50	82		110.50	155
4050	With #12/2, type NM cable		12.31	.650		13.30	35.50		48.80	68
4070	Type MC cable		10.67	.750		19.85	41		60.85	83.50
4080	EMT & wire		4.71	1.699		32.50	93		125.50	175
4100	20 amp recpt., #12/2, type NM cable		12.31	.650		19.95	35.50		55.45	75.50
4120	Type MC cable		10.67	.750		26.50	41		67.50	90.50
4130	EMT & wire		4.71	1.699		39	93		132	182
4140	For GFI see Section 26 05 90.10 line 4300 below									
4150	Decorator style, 15 amp recpt., type NM cable	1 Elec	14.55	.550	Ea.	14.70	30		44.70	61
4170	Type MC cable		12.31	.650		23.50	35.50		59	79.50
4180	EMT & wire		5.33	1.501		32.50	82		114.50	159
4200	With #12/2, type NM cable		12.31	.650		17.20	35.50		52.70	72.50
4220	Type MC cable		10.67	.750		23.50	41		64.50	87.50
4230	EMT & wire		4.71	1.699		36.50	93		129.50	179
4250	20 amp recpt. #12/2, type NM cable		12.31	.650		24	35.50		59.50	79.50
4270	Type MC cable		10.67	.750		30.50	41		71.50	95
4280	EMT & wire		4.71	1.699		43	93		136	187
4300	GFI, 15 amp recpt., type NM cable		12.31	.650		41.50	35.50		77	99
4320	Type MC cable		10.67	.750		50.50	41		91.50	117
4330	EMT & wire		4.71	1.699		59	93		152	204
4350	GFI with #12/2, type NM cable		10.67	.750		44	41		85	110
4370	Type MC cable		9.20	.870		50.50	47.50		98	127
4380	EMT & wire		4.21	1.900		63	104		167	226
4400	20 amp recpt., #12/2 type NM cable		10.67	.750		52	41		93	119
4420	Type MC cable		9.20	.870		58.50	47.50		106	136
4430	EMT & wire		4.21	1.900		71.50	104		175.50	235
4500	Weather-proof cover for above receptacles, add		32	.250		3.46	13.70		17.16	24.50
4550	Air conditioner outlet, 20 amp-240 volt recpt.									
4560	30' of #12/2, 2 pole circuit breaker									
4570	Type NM cable	1 Elec	10	.800	Ea.	63	44		107	135
4580	Type MC cable		9	.889		70.50	48.50		119	151
4590	EMT & wire		4	2		82.50	109		191.50	255
4600	Decorator style, type NM cable		10	.800		68	44		112	141
4620	Type MC cable		9	.889		75.50	48.50		124	156

26 05 90.10 Residential Wiring		Crew	Daily Output	Labor-Hours	Unit	2015 Bare Costs Material	Labor	Equipment	Total	Total Incl O&P
4630	EMT & wire	1 Elec	4	2	Ea.	87	109		196	260
4650	Dryer outlet, 30 amp-240 volt recpt., 20′ of #10/3									
4660	2 pole circuit breaker									
4670	Type NM cable	1 Elec	6.41	1.248	Ea.	61.50	68.50		130	170
4680	Type MC cable		5.71	1.401		64.50	76.50		141	186
4690	EMT & wire		3.48	2.299		76	126		202	272
4700	Range outlet, 50 amp-240 volt recpt., 30′ of #8/3									
4710	Type NM cable	1 Elec	4.21	1.900	Ea.	90.50	104		194.50	256
4720	Type MC cable		4	2		124	109		233	300
4730	EMT & wire		2.96	2.703		108	148		256	340
4750	Central vacuum outlet, Type NM cable		6.40	1.250		58.50	68.50		127	167
4770	Type MC cable		5.71	1.401		69.50	76.50		146	192
4780	EMT & wire		3.48	2.299		83.50	126		209.50	280
4800	30 amp-110 volt locking recpt., #10/2 circ. bkr.									
4810	Type NM cable	1 Elec	6.20	1.290	Ea.	67	70.50		137.50	180
4820	Type MC cable		5.40	1.481		80	81		161	209
4830	EMT & wire		3.20	2.500		94	137		231	310
4900	Low voltage outlets									
4910	Telephone recpt., 20′ of 4/C phone wire	1 Elec	26	.308	Ea.	9.30	16.85		26.15	35.50
4920	TV recpt., 20′ of RG59U coax wire, F type connector	"	16	.500	"	18.75	27.50		46.25	61.50
4950	Door bell chime, transformer, 2 buttons, 60′ of bellwire									
4970	Economy model	1 Elec	11.50	.696	Ea.	51.50	38		89.50	114
4980	Custom model		11.50	.696		99	38		137	166
4990	Luxury model, 3 buttons		9.50	.842		281	46		327	380
6000	Lighting outlets									
6050	Wire only (for fixture), type NM cable	1 Elec	32	.250	Ea.	6.90	13.70		20.60	28
6070	Type MC cable		24	.333		12.30	18.25		30.55	41
6080	EMT & wire		10	.800		20	44		64	87.50
6100	Box (4″), and wire (for fixture), type NM cable		25	.320		14.45	17.50		31.95	42
6120	Type MC cable		20	.400		19.90	22		41.90	55
6130	EMT & wire		11	.727		27.50	40		67.50	90
6200	Fixtures (use with lines 6050 or 6100 above)									
6210	Canopy style, economy grade	1 Elec	40	.200	Ea.	32	10.95		42.95	51.50
6220	Custom grade		40	.200		53.50	10.95		64.45	75.50
6250	Dining room chandelier, economy grade		19	.421		79.50	23		102.50	122
6260	Custom grade		19	.421		315	23		338	380
6270	Luxury grade		15	.533		715	29		744	830
6310	Kitchen fixture (fluorescent), economy grade		30	.267		71.50	14.60		86.10	101
6320	Custom grade		25	.320		219	17.50		236.50	267
6350	Outdoor, wall mounted, economy grade		30	.267		30	14.60		44.60	55
6360	Custom grade		30	.267		120	14.60		134.60	154
6370	Luxury grade		25	.320		248	17.50		265.50	299
6410	Outdoor PAR floodlights, 1 lamp, 150 watt		20	.400		32.50	22		54.50	68.50
6420	2 lamp, 150 watt each		20	.400		53.50	22		75.50	92
6425	Motion sensing, 2 lamp, 150 watt each		20	.400		87	22		109	129
6430	For infrared security sensor, add		32	.250		132	13.70		145.70	166
6450	Outdoor, quartz-halogen, 300 watt flood		20	.400		39	22		61	75.50
6600	Recessed downlight, round, pre-wired, 50 or 75 watt trim		30	.267		80	14.60		94.60	110
6610	With shower light trim		30	.267		89	14.60		103.60	120
6620	With wall washer trim		28	.286		99.50	15.65		115.15	133
6630	With eye-ball trim		28	.286		99.50	15.65		115.15	133
6700	Porcelain lamp holder		40	.200		2.96	10.95		13.91	19.65
6710	With pull switch		40	.200		6.55	10.95		17.50	23.50

26 05 90 – Residential Applications

26 05 90.10 Residential Wiring		Crew	Daily Output	Labor-Hours	Unit	Material	2015 Bare Costs Labor	Equipment	Total	Total Incl O&P
6750	Fluorescent strip, 2-20 watt tube, wrap around diffuser, 24"	1 Elec	24	.333	Ea.	49.50	18.25		67.75	82
6760	1-34 watt tube, 48"		24	.333		87	18.25		105.25	124
6770	2-34 watt tubes, 48"		20	.400		103	22		125	146
6800	Bathroom heat lamp, 1-250 watt		28	.286		44	15.65		59.65	71.50
6810	2-250 watt lamps	↓	28	.286	↓	70	15.65		85.65	101
6820	For timer switch, see Section 26 05 90.10 line 2400									
6900	Outdoor post lamp, incl. post, fixture, 35' of #14/2									
6910	Type NMC cable	1 Elec	3.50	2.286	Ea.	258	125		383	470
6920	Photo-eye, add		27	.296		32	16.20		48.20	60
6950	Clock dial time switch, 24 hr., w/enclosure, type NM cable		11.43	.700		73.50	38.50		112	139
6970	Type MC cable		11	.727		82.50	40		122.50	150
6980	EMT & wire	↓	4.85	1.649	↓	91	90		181	235
7000	Alarm systems									
7050	Smoke detectors, box, #14/3, type NM cable	1 Elec	14.55	.550	Ea.	34.50	30		64.50	83
7070	Type MC cable		12.31	.650		44	35.50		79.50	102
7080	EMT & wire	↓	5	1.600		49.50	87.50		137	186
7090	For relay output to security system, add				↓	12			12	13.20
8000	Residential equipment									
8050	Disposal hook-up, incl. switch, outlet box, 3' of flex									
8060	20 amp-1 pole circ. bkr., and 25' of #12/2									
8070	Type NM cable	1 Elec	10	.800	Ea.	32.50	44		76.50	101
8080	Type MC cable		8	1		39.50	54.50		94	126
8090	EMT & wire	↓	5	1.600	↓	54.50	87.50		142	191
8100	Trash compactor or dishwasher hook-up, incl. outlet box,									
8110	3' of flex, 15 amp-1 pole circ. bkr., and 25' of #14/2									
8120	Type NM cable	1 Elec	10	.800	Ea.	24	44		68	91.50
8130	Type MC cable		8	1		34	54.50		88.50	120
8140	EMT & wire	↓	5	1.600	↓	46	87.50		133.50	182
8150	Hot water sink dispensor hook-up, use line 8100									
8200	Vent/exhaust fan hook-up, type NM cable	1 Elec	32	.250	Ea.	6.90	13.70		20.60	28
8220	Type MC cable		24	.333		12.30	18.25		30.55	41
8230	EMT & wire	↓	10	.800	↓	20	44		64	87.50
8250	Bathroom vent fan, 50 CFM (use with above hook-up)									
8260	Economy model	1 Elec	15	.533	Ea.	23	29		52	69
8270	Low noise model		15	.533		41	29		70	88.50
8280	Custom model	↓	12	.667	↓	127	36.50		163.50	195
8300	Bathroom or kitchen vent fan, 110 CFM									
8310	Economy model	1 Elec	15	.533	Ea.	67.50	29		96.50	118
8320	Low noise model	"	15	.533	"	92	29		121	145
8350	Paddle fan, variable speed (w/o lights)									
8360	Economy model (AC motor)	1 Elec	10	.800	Ea.	109	44		153	186
8362	With light kit		10	.800		150	44		194	231
8370	Custom model (AC motor)		10	.800		227	44		271	315
8372	With light kit		10	.800		268	44		312	360
8380	Luxury model (DC motor)		8	1		330	54.50		384.50	445
8382	With light kit		8	1		375	54.50		429.50	490
8390	Remote speed switch for above, add	↓	12	.667	↓	37.50	36.50		74	95.50
8500	Whole house exhaust fan, ceiling mount, 36", variable speed									
8510	Remote switch, incl. shutters, 20 amp-1 pole circ. bkr.									
8520	30' of #12/2, type NM cable	1 Elec	4	2	Ea.	1,300	109		1,409	1,600
8530	Type MC cable		3.50	2.286		1,300	125		1,425	1,600
8540	EMT & wire	↓	3	2.667	↓	1,325	146		1,471	1,675
8600	Whirlpool tub hook-up, incl. timer switch, outlet box									

26 05 Common Work Results for Electrical

26 05 90 – Residential Applications

26 05 90.10 Residential Wiring		Crew	Daily Output	Labor-Hours	Unit	Material	Labor	Equipment	Total	Total Incl O&P
						2015 Bare Costs				
8610	3′ of flex, 20 amp-1 pole GFI circ. bkr.									
8620	30′ of #12/2, type NM cable	1 Elec	5	1.600	Ea.	142	87.50		229.50	287
8630	Type MC cable		4.20	1.905		145	104		249	315
8640	EMT & wire		3.40	2.353		158	129		287	365
8650	Hot water heater hook-up, incl. 1-2 pole circ. bkr., box;									
8660	3′ of flex, 20′ of #10/2, type NM cable	1 Elec	5	1.600	Ea.	33	87.50		120.50	168
8670	Type MC cable		4.20	1.905		43	104		147	204
8680	EMT & wire		3.40	2.353		48.50	129		177.50	246
9000	Heating/air conditioning									
9050	Furnace/boiler hook-up, incl. firestat, local on-off switch									
9060	Emergency switch, and 40′ of type NM cable	1 Elec	4	2	Ea.	52.50	109		161.50	222
9070	Type MC cable		3.50	2.286		65.50	125		190.50	260
9080	EMT & wire		1.50	5.333		83	292		375	525
9100	Air conditioner hook-up, incl. local 60 amp disc. switch									
9110	3′ sealtite, 40 amp, 2 pole circuit breaker									
9130	40′ of #8/2, type NM cable	1 Elec	3.50	2.286	Ea.	163	125		288	365
9140	Type MC cable		3	2.667		216	146		362	455
9150	EMT & wire		1.30	6.154		204	335		539	730
9200	Heat pump hook-up, 1-40 & 1-100 amp 2 pole circ. bkr.									
9210	Local disconnect switch, 3′ sealtite									
9220	40′ of #8/2 & 30′ of #3/2									
9230	Type NM cable	1 Elec	1.30	6.154	Ea.	520	335		855	1,075
9240	Type MC cable		1.08	7.407		535	405		940	1,200
9250	EMT & wire		.94	8.511		585	465		1,050	1,350
9500	Thermostat hook-up, using low voltage wire									
9520	Heating only, 25′ of #18-3	1 Elec	24	.333	Ea.	8.80	18.25		27.05	37
9530	Heating/cooling, 25′ of #18-4	"	20	.400	"	11.45	22		33.45	45.50

26 09 Instrumentation and Control for Electrical Systems

26 09 13 – Electrical Power Monitoring

26 09 13.10 Switchboard Instruments

Line	Description	Crew	Daily Output	Labor-Hours	Unit	Material	Labor	Equipment	Total	Total Incl O&P
0010	**SWITCHBOARD INSTRUMENTS** 3 phase, 4 wire									
0100	AC indicating, ammeter & switch	1 Elec	8	1	Ea.	2,500	54.50		2,554.50	2,825
0200	Voltmeter & switch		8	1		2,500	54.50		2,554.50	2,825
0300	Wattmeter		8	1		4,250	54.50		4,304.50	4,750
0400	AC recording, ammeter		4	2		7,550	109		7,659	8,500
0500	Voltmeter		4	2		7,550	109		7,659	8,500
0600	Ground fault protection, zero sequence		2.70	2.963		6,675	162		6,837	7,600
0700	Ground return path		2.70	2.963		6,675	162		6,837	7,600
0800	3 current transformers, 5 to 800 amp		2	4		3,100	219		3,319	3,750
0900	1000 to 1500 amp		1.30	6.154		4,475	335		4,810	5,425
1200	2000 to 4000 amp		1	8		5,275	440		5,715	6,450
1300	Fused potential transformer, maximum 600 volt		8	1		1,175	54.50		1,229.50	1,350

26 09 13.30 Smart Metering

Line	Description	Crew	Daily Output	Labor-Hours	Unit	Material	Labor	Equipment	Total	Total Incl O&P
0010	**SMART METERING**, In panel									
0100	Single phase, 120/208 volt, 100 amp G	1 Elec	8.78	.911	Ea.	375	50		425	485
0120	200 amp G		8.78	.911		375	50		425	485
0200	277 volt, 100 amp G		8.78	.911		400	50		450	515
0220	200 amp G		8.78	.911		400	50		450	515
1100	Three phase, 120/208 volt, 100 amp G		4.69	1.706		690	93.50		783.50	900
1120	200 amp G		4.69	1.706		690	93.50		783.50	900

26 09 Instrumentation and Control for Electrical Systems

26 09 13 – Electrical Power Monitoring

26 09 13.30 Smart Metering			Crew	Daily Output	Labor-Hours	Unit	Material	2015 Bare Costs Labor	Equipment	Total	Total Incl O&P
1130	400 amp	G	1 Elec	4.69	1.706	Ea.	690	93.50		783.50	900
1140	800 amp	G		4.69	1.706		690	93.50		783.50	900
1150	1600 amp	G		4.69	1.706		690	93.50		783.50	900
1200	277/480 volt, 100 amp	G		4.69	1.706		775	93.50		868.50	990
1220	200 amp	G		4.69	1.706		775	93.50		868.50	990
1230	400 amp	G		4.69	1.706		775	93.50		868.50	990
1240	800 amp	G		4.69	1.706		775	93.50		868.50	990
1250	1600 amp	G		4.69	1.706		785	93.50		878.50	1,000
2000	Data recorder, 8 meters	G		10.97	.729		1,400	40		1,440	1,575
2100	16 meters	G		8.53	.938		1,950	51.50		2,001.50	2,225
3000	Software package, per meter, basic	G					236			236	260
3100	Premium	G					610			610	675

26 09 23 – Lighting Control Devices

26 09 23.10 Energy Saving Lighting Devices			Crew	Daily Output	Labor-Hours	Unit	Material	Labor	Equipment	Total	Total Incl O&P
0010	**ENERGY SAVING LIGHTING DEVICES**										
0100	Occupancy sensors, passive infrared ceiling mounted	G	1 Elec	7	1.143	Ea.	81	62.50		143.50	183
0110	Ultrasonic ceiling mounted	G		7	1.143		89.50	62.50		152	192
0120	Dual technology ceiling mounted	G		6.50	1.231		131	67.50		198.50	245
0150	Automatic wall switches	G		24	.333		64.50	18.25		82.75	98.50
0160	Daylighting sensor, manual control, ceiling mounted	G		7	1.143		111	62.50		173.50	216
0170	Remote and dimming control with remote controller	G		6.50	1.231		152	67.50		219.50	268
0200	Remote power pack	G		10	.800		31	44		75	99.50
0250	Photoelectric control, S.P.S.T. 120 V	G		8	1		19.35	54.50		73.85	104
0300	S.P.S.T. 208 V/277 V	G		8	1		28.50	54.50		83	114
0350	D.P.S.T. 120 V	G		6	1.333		217	73		290	350
0400	D.P.S.T. 208 V/277 V	G		6	1.333		176	73		249	305
0450	S.P.D.T. 208 V/277 V	G		6	1.333		211	73		284	340
0460	Daylight level sensor, wall mounted, on/off or dimming	G		8	1		135	54.50		189.50	231

26 12 Medium-Voltage Transformers

26 12 19 – Pad-Mounted, Liquid-Filled, Medium-Voltage Transformers

26 12 19.10 Transformer, Oil-Filled		Crew	Daily Output	Labor-Hours	Unit	Material	Labor	Equipment	Total	Total Incl O&P
0010	**TRANSFORMER, OIL-FILLED** primary delta or Y,									
0050	Pad mounted 5 kV or 15 kV, with taps, 277/480 V secondary, 3 phase									
0100	150 kVA	R-3	.65	30.769	Ea.	9,250	1,675	212	11,137	12,900
0200	300 kVA		.45	44.444		13,200	2,425	305	15,930	18,500
0300	500 kVA		.40	50		18,700	2,725	345	21,770	25,100
0400	750 kVA		.38	52.632		23,700	2,850	365	26,915	30,800
0500	1000 kVA		.26	76.923		28,100	4,175	530	32,805	37,800
0600	1500 kVA		.23	86.957		33,400	4,725	600	38,725	44,500
0700	2000 kVA		.20	100		42,100	5,425	690	48,215	55,500
0800	3750 kVA		.16	125		79,000	6,800	860	86,660	98,000

26 22 Low-Voltage Transformers

26 22 13 – Low-Voltage Distribution Transformers

26 22 13.10 Transformer, Dry-Type		Crew	Daily Output	Labor-Hours	Unit	Material	2015 Bare Costs Labor	Equipment	Total	Total Incl O&P
0010	**TRANSFORMER, DRY-TYPE**									
0050	Single phase, 240/480 volt primary, 120/240 volt secondary									
0100	1 kVA	1 Elec	2	4	Ea.	330	219		549	690
0300	2 kVA		1.60	5		490	274		764	950
0500	3 kVA		1.40	5.714		610	315		925	1,150
0700	5 kVA		1.20	6.667		835	365		1,200	1,475
0900	7.5 kVA	2 Elec	2.20	7.273		1,175	400		1,575	1,875
1100	10 kVA		1.60	10		1,450	545		1,995	2,425
1300	15 kVA		1.20	13.333		1,700	730		2,430	2,975
1500	25 kVA		1	16		2,125	875		3,000	3,625
1700	37.5 kVA		.80	20		2,750	1,100		3,850	4,675
1900	50 kVA		.70	22.857		3,250	1,250		4,500	5,475
2100	75 kVA		.65	24.615		4,325	1,350		5,675	6,775
2190	480 V primary 120/240 V secondary, nonvent., 15 kVA		1.20	13.333		1,575	730		2,305	2,825
2200	25 kVA		.90	17.778		2,300	970		3,270	4,000
2210	37 kVA		.75	21.333		2,750	1,175		3,925	4,775
2220	50 kVA		.65	24.615		3,250	1,350		4,600	5,625
2300	3 phase, 480 volt primary 120/208 volt secondary									
2310	Ventilated, 3 kVA	1 Elec	1	8	Ea.	910	440		1,350	1,650
2700	6 kVA		.80	10		1,025	545		1,570	1,950
2900	9 kVA		.70	11.429		1,075	625		1,700	2,125
3100	15 kVA	2 Elec	1.10	14.545		1,300	795		2,095	2,625
3300	30 kVA		.90	17.778		1,425	970		2,395	3,025
3500	45 kVA		.80	20		1,700	1,100		2,800	3,525
3700	75 kVA		.70	22.857		2,375	1,250		3,625	4,475
3900	112.5 kVA	R-3	.90	22.222		3,400	1,200	153	4,753	5,725
4100	150 kVA		.85	23.529		4,450	1,275	162	5,887	7,000
4300	225 kVA		.65	30.769		6,075	1,675	212	7,962	9,400
4500	300 kVA		.55	36.364		7,600	1,975	251	9,826	11,600
4700	500 kVA		.45	44.444		12,600	2,425	305	15,330	17,900
4800	750 kVA		.35	57.143		21,100	3,100	395	24,595	28,400

26 24 Switchboards and Panelboards

26 24 13 – Switchboards

26 24 13.10 Incoming Switchboards										
0010	**INCOMING SWITCHBOARDS** main service section									
0100	Aluminum bus bars, not including CT's or PT's									
0200	No main disconnect, includes CT compartment									
0300	120/208 volt, 4 wire, 600 amp	2 Elec	1	16	Ea.	4,225	875		5,100	5,950
0400	800 amp		.88	18.182		4,225	995		5,220	6,150
0500	1000 amp		.80	20		5,075	1,100		6,175	7,225
0600	1200 amp		.72	22.222		5,075	1,225		6,300	7,400
0700	1600 amp		.66	24.242		5,075	1,325		6,400	7,550
0800	2000 amp		.62	25.806		5,450	1,400		6,850	8,125
1000	3000 amp		.56	28.571		7,200	1,575		8,775	10,300
2000	Fused switch & CT compartment									
2100	120/208 volt, 4 wire, 400 amp	2 Elec	1.12	14.286	Ea.	2,825	780		3,605	4,275
2200	600 amp		.94	17.021		3,350	930		4,280	5,075
2300	800 amp		.84	19.048		11,400	1,050		12,450	14,200
2400	1200 amp		.68	23.529		14,800	1,275		16,075	18,200
2900	Pressure switch & CT compartment									

26 24 Switchboards and Panelboards

26 24 13 – Switchboards

26 24 13.10 Incoming Switchboards

	26 24 13.10 Incoming Switchboards	Crew	Daily Output	Labor-Hours	Unit	Material	2015 Bare Costs Labor	Equipment	Total	Total Incl O&P
3000	120/208 volt, 4 wire, 800 amp	2 Elec	.80	20	Ea.	10,200	1,100		11,300	13,000
3100	1200 amp		.66	24.242		19,800	1,325		21,125	23,800
3200	1600 amp		.62	25.806		21,100	1,400		22,500	25,300
3300	2000 amp		.56	28.571		22,400	1,575		23,975	27,100
4400	Circuit breaker, molded case & CT compartment									
4600	3 pole, 4 wire, 600 amp	2 Elec	.94	17.021	Ea.	8,825	930		9,755	11,100
4800	800 amp		.84	19.048		10,600	1,050		11,650	13,200
5000	1200 amp		.68	23.529		14,400	1,275		15,675	17,700
5100	Copper bus bars, not incl. CT's or PT's, add, minimum					15%				

26 24 13.30 Distribution Switchboards Section

	26 24 13.30 Distribution Switchboards Section	Crew	Daily Output	Labor-Hours	Unit	Material	Labor	Equipment	Total	Total Incl O&P
0010	**DISTRIBUTION SWITCHBOARDS SECTION**									
0100	Aluminum bus bars, not including breakers									
0195	120/208 or 277/480 volt, 4 wire, 400 amp	2 Elec	1.10	14.545	Ea.	1,300	795		2,095	2,650
0200	600 amp		1	16		1,625	875		2,500	3,075
0300	800 amp		.88	18.182		2,100	995		3,095	3,825
0400	1000 amp		.80	20		2,625	1,100		3,725	4,550
0500	1200 amp		.72	22.222		3,125	1,225		4,350	5,275
0600	1600 amp		.66	24.242		3,575	1,325		4,900	5,925
0700	2000 amp		.62	25.806		4,200	1,400		5,600	6,725
0800	2500 amp		.60	26.667		4,725	1,450		6,175	7,375
0900	3000 amp		.56	28.571		5,725	1,575		7,300	8,650
0950	4000 amp		.52	30.769		8,375	1,675		10,050	11,800

26 24 13.40 Switchboards Feeder Section

	26 24 13.40 Switchboards Feeder Section	Crew	Daily Output	Labor-Hours	Unit	Material	Labor	Equipment	Total	Total Incl O&P
0010	**SWITCHBOARDS FEEDER SECTION** group mounted devices									
0030	Circuit breakers									
0160	FA frame, 15 to 60 amp, 240 volt, 1 pole	1 Elec	8	1	Ea.	120	54.50		174.50	214
0280	FA frame, 70 to 100 amp, 240 volt, 1 pole		7	1.143		191	62.50		253.50	305
0420	KA frame, 70 to 225 amp		3.20	2.500		1,250	137		1,387	1,575
0430	LA frame, 125 to 400 amp		2.30	3.478		2,475	190		2,665	3,000
0460	MA frame, 450 to 600 amp		1.60	5		5,000	274		5,274	5,900
0470	700 to 800 amp		1.30	6.154		6,500	335		6,835	7,650
0480	MAL frame, 1000 amp		1	8		6,725	440		7,165	8,050
0490	PA frame, 1200 amp		.80	10		13,700	545		14,245	15,900
0500	Branch circuit, fusible switch, 600 volt, double 30/30 amp		4	2		895	109		1,004	1,150
0550	60/60 amp		3.20	2.500		920	137		1,057	1,200
0600	100/100 amp		2.70	2.963		1,150	162		1,312	1,525
0650	Single, 30 amp		5.30	1.509		735	82.50		817.50	935
0700	60 amp		4.70	1.702		815	93		908	1,025
0750	100 amp		4	2		1,075	109		1,184	1,350
0800	200 amp		2.70	2.963		1,425	162		1,587	1,825
0850	400 amp		2.30	3.478		2,625	190		2,815	3,150
0900	600 amp		1.80	4.444		3,200	243		3,443	3,900
0950	800 amp		1.30	6.154		5,375	335		5,710	6,400
1000	1200 amp		.80	10		6,150	545		6,695	7,600

26 24 16 – Panelboards

26 24 16.20 Panelboard and Load Center Circuit Breakers

	26 24 16.20 Panelboard and Load Center Circuit Breakers	Crew	Daily Output	Labor-Hours	Unit	Material	Labor	Equipment	Total	Total Incl O&P
0010	**PANELBOARD AND LOAD CENTER CIRCUIT BREAKERS**									
0050	Bolt-on, 10,000 amp I.C., 120 volt, 1 pole									
0100	15 to 50 amp	1 Elec	10	.800	Ea.	16.65	44		60.65	84
0200	60 amp		8	1		19.20	54.50		73.70	103
0300	70 amp		8	1		28	54.50		82.50	113
0350	240 volt, 2 pole									

26 24 16.20 Panelboard and Load Center Circuit Breakers		Crew	Daily Output	Labor-Hours	Unit	Material	2015 Bare Costs Labor	Equipment	Total	Total Incl O&P
0400	15 to 50 amp	1 Elec	8	1	Ea.	36	54.50		90.50	122
0500	60 amp		7.50	1.067		49	58.50		107.50	141
0600	80 to 100 amp		5	1.600		93.50	87.50		181	234
0700	3 pole, 15 to 60 amp		6.20	1.290		115	70.50		185.50	233
0800	70 amp		5	1.600		146	87.50		233.50	292
0900	80 to 100 amp		3.60	2.222		166	122		288	365
1000	22,000 amp I.C., 240 volt, 2 pole, 70 – 225 amp		2.70	2.963		630	162		792	940
1100	3 pole, 70 – 225 amp		2.30	3.478		700	190		890	1,050
1200	14,000 amp I.C., 277 volts, 1 pole, 15 – 30 amp		8	1		44	54.50		98.50	131
1300	22,000 amp I.C., 480 volts, 2 pole, 70 – 225 amp		2.70	2.963		630	162		792	940
1400	3 pole, 70 – 225 amp	↓	2.30	3.478	↓	780	190		970	1,150

26 24 16.30 Panelboards Commercial Applications

		Crew	Daily Output	Labor-Hours	Unit	Material	Labor	Equipment	Total	Total Incl O&P
0010	**PANELBOARDS COMMERCIAL APPLICATIONS**									
0050	NQOD, w/20 amp 1 pole bolt-on circuit breakers									
0100	3 wire, 120/240 volts, 100 amp main lugs									
0150	10 circuits	1 Elec	1	8	Ea.	545	440		985	1,250
0200	14 circuits		.88	9.091		660	495		1,155	1,475
0250	18 circuits		.75	10.667		720	585		1,305	1,675
0300	20 circuits	↓	.65	12.308		805	675		1,480	1,875
0350	225 amp main lugs, 24 circuits	2 Elec	1.20	13.333		910	730		1,640	2,100
0400	30 circuits		.90	17.778		1,050	970		2,020	2,600
0450	36 circuits		.80	20		1,200	1,100		2,300	2,975
0500	38 circuits		.72	22.222		1,275	1,225		2,500	3,225
0550	42 circuits	↓	.66	24.242		1,350	1,325		2,675	3,450
0600	4 wire, 120/208 volts, 100 amp main lugs, 12 circuits	1 Elec	1	8		640	440		1,080	1,350
0650	16 circuits		.75	10.667		730	585		1,315	1,675
0700	20 circuits		.65	12.308		845	675		1,520	1,925
0750	24 circuits		.60	13.333		875	730		1,605	2,075
0800	30 circuits	↓	.53	15.094		1,050	825		1,875	2,375
0850	225 amp main lugs, 32 circuits	2 Elec	.90	17.778		1,175	970		2,145	2,750
0900	34 circuits		.84	19.048		1,200	1,050		2,250	2,875
0950	36 circuits		.80	20		1,225	1,100		2,325	3,000
1000	42 circuits	↓	.68	23.529	↓	1,375	1,275		2,650	3,425
1200	NEHB,w/20 amp, 1 pole bolt-on circuit breakers									
1250	4 wire, 277/480 volts, 100 amp main lugs, 12 circuits	1 Elec	.88	9.091	Ea.	1,225	495		1,720	2,100
1300	20 circuits	"	.60	13.333		1,825	730		2,555	3,100
1350	225 amp main lugs, 24 circuits	2 Elec	.90	17.778		2,075	970		3,045	3,725
1400	30 circuits		.80	20		2,475	1,100		3,575	4,375
1450	36 circuits	↓	.72	22.222	↓	2,875	1,225		4,100	5,000
1600	NQOD panel, w/20 amp, 1 pole, circuit breakers									
1650	3 wire, 120/240 volt with main circuit breaker									
1700	100 amp main, 12 circuits	1 Elec	.80	10	Ea.	800	545		1,345	1,700
1750	20 circuits	"	.60	13.333		1,025	730		1,755	2,225
1800	225 amp main, 30 circuits	2 Elec	.68	23.529		1,900	1,275		3,175	4,025
1850	42 circuits		.52	30.769		2,200	1,675		3,875	4,950
1900	400 amp main, 30 circuits		.54	29.630		2,625	1,625		4,250	5,325
1950	42 circuits	↓	.50	32	↓	2,925	1,750		4,675	5,850
2000	4 wire, 120/208 volts with main circuit breaker									
2050	100 amp main, 24 circuits	1 Elec	.47	17.021	Ea.	1,175	930		2,105	2,700
2100	30 circuits	"	.40	20		1,325	1,100		2,425	3,100
2200	225 amp main, 32 circuits	2 Elec	.72	22.222		2,225	1,225		3,450	4,250
2250	42 circuits	↓	.56	28.571	↓	2,425	1,575		4,000	5,025

26 24 Switchboards and Panelboards

26 24 16 – Panelboards

26 24 16.30 Panelboards Commercial Applications		Crew	Daily Output	Labor-Hours	Unit	Material	2015 Bare Costs Labor	Equipment	Total	Total Incl O&P
2300	400 amp main, 42 circuits	2 Elec	.48	33.333	Ea.	3,250	1,825		5,075	6,300
2350	600 amp main, 42 circuits		.40	40		4,825	2,200		7,025	8,575
2400	NEHB, with 20 amp, 1 pole circuit breaker									
2450	4 wire, 277/480 volts with main circuit breaker									
2500	100 amp main, 24 circuits	1 Elec	.42	19.048	Ea.	2,375	1,050		3,425	4,175
2550	30 circuits	″	.38	21.053		2,775	1,150		3,925	4,800
2600	225 amp main, 30 circuits	2 Elec	.72	22.222		3,500	1,225		4,725	5,675
2650	42 circuits	″	.56	28.571		4,300	1,575		5,875	7,100

26 24 19 – Motor-Control Centers

26 24 19.40 Motor Starters and Controls		Crew	Daily Output	Labor-Hours	Unit	Material	Labor	Equipment	Total	Total Incl O&P
0010	**MOTOR STARTERS AND CONTROLS**									
0050	Magnetic, FVNR, with enclosure and heaters, 480 volt									
0080	2 HP, size 00	1 Elec	3.50	2.286	Ea.	203	125		328	410
0100	5 HP, size 0		2.30	3.478		272	190		462	585
0200	10 HP, size 1		1.60	5		275	274		549	715
0300	25 HP, size 2	2 Elec	2.20	7.273		520	400		920	1,175
0400	50 HP, size 3		1.80	8.889		845	485		1,330	1,650
0500	100 HP, size 4		1.20	13.333		1,875	730		2,605	3,150
0600	200 HP, size 5		.90	17.778		4,375	970		5,345	6,275
0700	Combination, with motor circuit protectors, 5 HP, size 0	1 Elec	1.80	4.444		880	243		1,123	1,325
0800	10 HP, size 1	″	1.30	6.154		915	335		1,250	1,500
0900	25 HP, size 2	2 Elec	2	8		1,275	440		1,715	2,050
1000	50 HP, size 3		1.32	12.121		1,850	665		2,515	3,025
1200	100 HP, size 4		.80	20		4,000	1,100		5,100	6,050
1400	Combination, with fused switch, 5 HP, size 0	1 Elec	1.80	4.444		610	243		853	1,025
1600	10 HP, size 1	″	1.30	6.154		650	335		985	1,225
1800	25 HP, size 2	2 Elec	2	8		1,050	440		1,490	1,800
2000	50 HP, size 3		1.32	12.121		1,775	665		2,440	2,975
2200	100 HP, size 4		.80	20		3,125	1,100		4,225	5,075

26 25 Enclosed Bus Assemblies

26 25 13 – Bus Duct/Busway and Fittings

26 25 13.40 Copper Bus Duct		Crew	Daily Output	Labor-Hours	Unit	Material	Labor	Equipment	Total	Total Incl O&P
0010	**COPPER BUS DUCT** 10 ft. long									
0050	Indoor 3 pole 4 wire, plug-in, straight section, 225 amp	2 Elec	40	.400	L.F.	230	22		252	286
1000	400 amp		32	.500		230	27.50		257.50	294
1500	600 amp		26	.615		230	33.50		263.50	305
2400	800 amp		20	.800		273	44		317	365
2450	1000 amp		18	.889		300	48.50		348.50	405
2500	1350 amp		16	1		415	54.50		469.50	535
2510	1600 amp		12	1.333		470	73		543	625
2520	2000 amp		10	1.600		595	87.50		682.50	785
2550	Feeder, 600 amp		28	.571		204	31.50		235.50	271
2600	800 amp		22	.727		247	40		287	330
2700	1000 amp		20	.800		276	44		320	370
2800	1350 amp		18	.889		385	48.50		433.50	500
2900	1600 amp		14	1.143		440	62.50		502.50	580
3000	2000 amp		12	1.333		565	73		638	735
3100	Elbows, 225 amp		4	4	Ea.	1,375	219		1,594	1,825
3200	400 amp		3.60	4.444		1,375	243		1,618	1,875

26 25 13 – Bus Duct/Busway and Fittings

26 25 13.40 Copper Bus Duct		Crew	Daily Output	Labor-Hours	Unit	Material	Labor	Equipment	Total	Total Incl O&P
						2015 Bare Costs				
3300	600 amp	2 Elec	3.20	5	Ea.	1,375	274		1,649	1,900
3400	800 amp		2.80	5.714		1,475	315		1,790	2,100
3500	1000 amp		2.60	6.154		1,650	335		1,985	2,325
3600	1350 amp		2.40	6.667		1,850	365		2,215	2,600
3700	1600 amp		2.20	7.273		2,025	400		2,425	2,825
3800	2000 amp		1.80	8.889		2,500	485		2,985	3,475
4000	End box, 225 amp		34	.471		165	25.50		190.50	221
4100	400 amp		32	.500		186	27.50		213.50	246
4200	600 amp		28	.571		186	31.50		217.50	252
4300	800 amp		26	.615		186	33.50		219.50	256
4400	1000 amp		24	.667		186	36.50		222.50	260
4500	1350 amp		22	.727		177	40		217	254
4600	1600 amp		20	.800		177	44		221	260
4700	2000 amp		18	.889		217	48.50		265.50	310
4800	Cable tap box end, 225 amp		3.20	5		1,100	274		1,374	1,600
5000	400 amp		2.60	6.154		1,200	335		1,535	1,825
5100	600 amp		2.20	7.273		1,400	400		1,800	2,150
5200	800 amp		2	8		1,475	440		1,915	2,275
5300	1000 amp		1.60	10		1,500	545		2,045	2,475
5400	1350 amp		1.40	11.429		2,200	625		2,825	3,350
5500	1600 amp		1.20	13.333		2,475	730		3,205	3,825
5600	2000 amp		1	16		2,750	875		3,625	4,325
5700	Switchboard stub, 225 amp		5.40	2.963		1,250	162		1,412	1,650
5800	400 amp		4.60	3.478		1,325	190		1,515	1,725
5900	600 amp		4	4		1,375	219		1,594	1,850
6000	800 amp		3.20	5		1,675	274		1,949	2,225
6100	1000 amp		3	5.333		1,925	292		2,217	2,550
6200	1350 amp		2.60	6.154		2,400	335		2,735	3,125
6300	1600 amp		2.40	6.667		2,700	365		3,065	3,525
6400	2000 amp		2	8		3,275	440		3,715	4,250
6490	Tee fittings, 225 amp		2.40	6.667		1,900	365		2,265	2,625
6500	400 amp		2	8		1,900	440		2,340	2,725
6600	600 amp		1.80	8.889		1,900	485		2,385	2,800
6700	800 amp		1.60	10		2,175	545		2,720	3,225
6800	1350 amp		1.20	13.333		3,000	730		3,730	4,400
7000	1600 amp		1	16		3,400	875		4,275	5,050
7100	2000 amp		.80	20		4,025	1,100		5,125	6,100
7200	Plug-in fusible switches w/3 fuses, 600 volt, 3 pole, 30 amp	1 Elec	4	2		850	109		959	1,100
7300	60 amp		3.60	2.222		955	122		1,077	1,225
7400	100 amp		2.70	2.963		1,450	162		1,612	1,850
7500	200 amp	2 Elec	3.20	5		2,600	274		2,874	3,275
7600	400 amp		1.40	11.429		7,625	625		8,250	9,300
7700	600 amp		.90	17.778		8,650	970		9,620	11,000
7800	800 amp		.66	24.242		12,100	1,325		13,425	15,300
7900	1200 amp		.50	32		22,800	1,750		24,550	27,600
8000	Plug-in circuit breakers, molded case, 15 to 50 amp	1 Elec	4.40	1.818		805	99.50		904.50	1,025
8100	70 to 100 amp	"	3.10	2.581		895	141		1,036	1,200
8200	150 to 225 amp	2 Elec	3.40	4.706		2,425	257		2,682	3,050
8300	250 to 400 amp		1.40	11.429		4,250	625		4,875	5,600
8400	500 to 600 amp		1	16		5,750	875		6,625	7,600
8500	700 to 800 amp		.64	25		7,075	1,375		8,450	9,825
8600	900 to 1000 amp		.56	28.571		10,100	1,575		11,675	13,500
8700	1200 amp		.44	36.364		12,200	2,000		14,200	16,400

26 27 16 – Electrical Cabinets and Enclosures

26 27 16.10 Cabinets		Crew	Daily Output	Labor-Hours	Unit	Material	2015 Bare Costs Labor	Equipment	Total	Total Incl O&P
0010	**CABINETS**									
7000	Cabinets, current transformer									
7050	Single door, 24" H x 24" W x 10" D	1 Elec	1.60	5	Ea.	152	274		426	575
7100	30" H x 24" W x 10" D		1.30	6.154		165	335		500	685
7150	36" H x 24" W x 10" D		1.10	7.273		177	400		577	790
7200	30" H x 30" W x 10" D		1	8		223	440		663	900
7250	36" H x 30" W x 10" D		.90	8.889		262	485		747	1,025
7300	36" H x 36" W x 10" D		.80	10		270	545		815	1,125
7500	Double door, 48" H x 36" W x 10" D		.60	13.333		590	730		1,320	1,750
7550	24" H x 24" W x 12" D		1	8		173	440		613	845

26 27 23 – Indoor Service Poles

26 27 23.40 Surface Raceway		Crew	Daily Output	Labor-Hours	Unit	Material	Labor	Equipment	Total	Total Incl O&P
0010	**SURFACE RACEWAY**									
0090	Metal, straight section									
0100	No. 500	1 Elec	100	.080	L.F.	1.04	4.38		5.42	7.70
0110	No. 700		100	.080		1.17	4.38		5.55	7.85
0400	No. 1500, small pancake		90	.089		2.15	4.86		7.01	9.65
0600	No. 2000, base & cover, blank		90	.089		2.19	4.86		7.05	9.70
0800	No. 3000, base & cover, blank		75	.107		4.18	5.85		10.03	13.35
1000	No. 4000, base & cover, blank		65	.123		6.80	6.75		13.55	17.60
1200	No. 6000, base & cover, blank		50	.160		11.40	8.75		20.15	25.50
2400	Fittings, elbows, No. 500		40	.200	Ea.	1.90	10.95		12.85	18.50
2800	Elbow cover, No. 2000		40	.200		3.57	10.95		14.52	20.50
2880	Tee, No. 500		42	.190		3.66	10.40		14.06	19.65
2900	No. 2000		27	.296		11.85	16.20		28.05	37.50
3000	Switch box, No. 500		16	.500		11	27.50		38.50	53
3400	Telephone outlet, No. 1500		16	.500		14.10	27.50		41.60	56.50
3600	Junction box, No. 1500		16	.500		9.65	27.50		37.15	51.50
3800	Plugmold wired sections, No. 2000									
4000	1 circuit, 6 outlets, 3 ft. long	1 Elec	8	1	Ea.	36	54.50		90.50	122
4100	2 circuits, 8 outlets, 6 ft. long	"	5.30	1.509	"	53	82.50		135.50	183

26 27 26 – Wiring Devices

26 27 26.10 Low Voltage Switching		Crew	Daily Output	Labor-Hours	Unit	Material	Labor	Equipment	Total	Total Incl O&P
0010	**LOW VOLTAGE SWITCHING**									
3600	Relays, 120 V or 277 V standard	1 Elec	12	.667	Ea.	41.50	36.50		78	101
3800	Flush switch, standard		40	.200		11.50	10.95		22.45	29
4000	Interchangeable		40	.200		15.05	10.95		26	33
4100	Surface switch, standard		40	.200		8.15	10.95		19.10	25.50
4200	Transformer 115 V to 25 V		12	.667		130	36.50		166.50	198
4400	Master control, 12 circuit, manual		4	2		126	109		235	305
4500	25 circuit, motorized		4	2		140	109		249	320
4600	Rectifier, silicon		12	.667		45.50	36.50		82	105
4800	Switchplates, 1 gang, 1, 2 or 3 switch, plastic		80	.100		5	5.45		10.45	13.70
5000	Stainless steel		80	.100		11.35	5.45		16.80	20.50
5400	2 gang, 3 switch, stainless steel		53	.151		23	8.25		31.25	37.50
5500	4 switch, plastic		53	.151		10.35	8.25		18.60	23.50
5800	3 gang, 9 switch, stainless steel		32	.250		64.50	13.70		78.20	91.50

26 27 26.20 Wiring Devices Elements		Crew	Daily Output	Labor-Hours	Unit	Material	Labor	Equipment	Total	Total Incl O&P
0010	**WIRING DEVICES ELEMENTS**									
0200	Toggle switch, quiet type, single pole, 15 amp	1 Elec	40	.200	Ea.	6.55	10.95		17.50	23.50
0600	3 way, 15 amp		23	.348		5.05	19.05		24.10	34

26 27 Low-Voltage Distribution Equipment

26 27 26 – Wiring Devices

26 27 26.20 Wiring Devices Elements		Crew	Daily Output	Labor-Hours	Unit	Material	Labor	Equipment	Total	Total Incl O&P
						2015 Bare Costs				
0900	4 way, 15 amp	1 Elec	15	.533	Ea.	9.10	29		38.10	53.50
1650	Dimmer switch, 120 volt, incandescent, 600 watt, 1 pole G		16	.500		21	27.50		48.50	64
2460	Receptacle, duplex, 120 volt, grounded, 15 amp		40	.200		1.26	10.95		12.21	17.80
2470	20 amp		27	.296		7.90	16.20		24.10	33
2490	Dryer, 30 amp		15	.533		4.39	29		33.39	48.50
2500	Range, 50 amp		11	.727		12.15	40		52.15	73
2600	Wall plates, stainless steel, 1 gang		80	.100		2.56	5.45		8.01	11
2800	2 gang		53	.151		4.33	8.25		12.58	17.10
3200	Lampholder, keyless		26	.308		12.65	16.85		29.50	39
3400	Pullchain with receptacle	↓	22	.364	↓	20.50	19.90		40.40	52.50

26 27 73 – Door Chimes

26 27 73.10 Doorbell System		Crew	Daily Output	Labor-Hours	Unit	Material	Labor	Equipment	Total	Total Incl O&P
0010	**DOORBELL SYSTEM**, incl. transformer, button & signal									
0100	6" bell	1 Elec	4	2	Ea.	129	109		238	305
0200	Buzzer	"	4	2	"	106	109		215	281

26 28 Low-Voltage Circuit Protective Devices

26 28 16 – Enclosed Switches and Circuit Breakers

26 28 16.10 Circuit Breakers		Crew	Daily Output	Labor-Hours	Unit	Material	Labor	Equipment	Total	Total Incl O&P
0010	**CIRCUIT BREAKERS** (in enclosure)									
0100	Enclosed (NEMA 1), 600 volt, 3 pole, 30 amp	1 Elec	3.20	2.500	Ea.	500	137		637	755
0200	60 amp		2.80	2.857		615	156		771	915
0400	100 amp		2.30	3.478		705	190		895	1,050
0500	200 amp		1.50	5.333		1,475	292		1,767	2,050
0600	225 amp	↓	1.50	5.333		1,625	292		1,917	2,225
0700	400 amp	2 Elec	1.60	10		2,775	545		3,320	3,900
0800	600 amp		1.20	13.333		4,025	730		4,755	5,550
1000	800 amp	↓	.94	17.021	↓	5,250	930		6,180	7,175

26 28 16.20 Safety Switches		Crew	Daily Output	Labor-Hours	Unit	Material	Labor	Equipment	Total	Total Incl O&P
0010	**SAFETY SWITCHES**									
0100	General duty 240 volt, 3 pole NEMA 1, fusible, 30 amp	1 Elec	3.20	2.500	Ea.	73.50	137		210.50	286
0200	60 amp		2.30	3.478		124	190		314	420
0300	100 amp		1.90	4.211		213	230		443	580
0400	200 amp	↓	1.30	6.154		455	335		790	1,000
0500	400 amp	2 Elec	1.80	8.889		1,150	485		1,635	2,000
0600	600 amp	"	1.20	13.333	↓	2,150	730		2,880	3,475
2900	Heavy duty, 240 volt, 3 pole NEMA 1 fusible									
2910	30 amp	1 Elec	3.20	2.500	Ea.	118	137		255	335
3000	60 amp		2.30	3.478		199	190		389	505
3300	100 amp		1.90	4.211		315	230		545	690
3500	200 amp	↓	1.30	6.154		540	335		875	1,100
3700	400 amp	2 Elec	1.80	8.889		1,400	485		1,885	2,250
3900	600 amp	"	1.20	13.333	↓	2,800	730		3,530	4,175

26 29 Low-Voltage Controllers

26 29 13 – Enclosed Controllers

26 29 13.20 Control Stations		Crew	Daily Output	Labor-Hours	Unit	Material	2015 Bare Costs Labor	Equipment	Total	Total Incl O&P
0010	**CONTROL STATIONS**									
0050	NEMA 1, heavy duty, stop/start	1 Elec	8	1	Ea.	135	54.50		189.50	231
0100	Stop/start, pilot light		6.20	1.290		184	70.50		254.50	310
0200	Hand/off/automatic		6.20	1.290		100	70.50		170.50	216
0400	Stop/start/reverse		5.30	1.509		182	82.50		264.50	325

26 32 Packaged Generator Assemblies

26 32 13 – Engine Generators

26 32 13.13 Diesel-Engine-Driven Generator Sets

		Crew	Daily Output	Labor-Hours	Unit	Material	Labor	Equipment	Total	Total Incl O&P
0010	**DIESEL-ENGINE-DRIVEN GENERATOR SETS**									
2000	Diesel engine, including battery, charger,									
2010	muffler, & day tank, 30 kW	R-3	.55	36.364	Ea.	14,000	1,975	251	16,226	18,700
2100	50 kW		.42	47.619		19,800	2,575	330	22,705	26,000
2200	75 kW		.35	57.143		21,300	3,100	395	24,795	28,500
2300	100 kW		.31	64.516		29,900	3,500	445	33,845	38,600
2400	125 kW		.29	68.966		31,700	3,750	475	35,925	40,900
2500	150 kW		.26	76.923		33,800	4,175	530	38,505	44,100
2600	175 kW		.25	80		39,100	4,350	550	44,000	50,000
2700	200 kW		.24	83.333		42,200	4,525	575	47,300	54,000
2800	250 kW		.23	86.957		45,200	4,725	600	50,525	57,500
2900	300 kW		.22	90.909		49,100	4,925	625	54,650	62,000
3000	350 kW		.20	100		55,500	5,425	690	61,615	70,000
3100	400 kW		.19	105		68,500	5,725	725	74,950	85,000
3200	500 kW		.18	111		86,500	6,025	765	93,290	105,000

26 32 13.16 Gas-Engine-Driven Generator Sets

		Crew	Daily Output	Labor-Hours	Unit	Material	Labor	Equipment	Total	Total Incl O&P
0010	**GAS-ENGINE-DRIVEN GENERATOR SETS**									
0020	Gas or gasoline operated, includes battery,									
0050	charger, & muffler									
0200	3 phase 4 wire, 277/480 volt, 7.5 kW	R-3	.83	24.096	Ea.	7,350	1,300	166	8,816	10,200
0300	11.5 kW		.71	28.169		10,400	1,525	194	12,119	14,000
0400	20 kW		.63	31.746		12,300	1,725	219	14,244	16,300
0500	35 kW		.55	36.364		14,600	1,975	251	16,826	19,400
0600	80 kW		.40	50		24,000	2,725	345	27,070	30,900
0700	100 kW		.33	60.606		26,300	3,300	420	30,020	34,300
0800	125 kW		.28	71.429		54,000	3,875	490	58,365	65,500
0900	185 kW		.25	80		71,000	4,350	550	75,900	85,000

26 33 Battery Equipment

26 33 43 – Battery Chargers

26 33 43.55 Electric Vehicle Charging

		Crew	Daily Output	Labor-Hours	Unit	Material	Labor	Equipment	Total	Total Incl O&P
0010	**ELECTRIC VEHICLE CHARGING**									
0020	Level 2, wall mounted									
2200	Heavy duty G	R-1A	15.36	1.042	Ea.	2,525	47		2,572	2,850
2210	with RFID G		12.29	1.302		2,700	58.50		2,758.50	3,075
2300	Free standing, single connector G		10.24	1.563		2,900	70.50		2,970.50	3,275
2310	with RFID G		8.78	1.822		3,625	82		3,707	4,100
2320	Double connector G		7.68	2.083		4,850	94		4,944	5,475
2330	with RFID G		6.83	2.343		6,300	106		6,406	7,075

26 35 Power Filters and Conditioners

26 35 13 – Capacitors

26 35 13.10 Capacitors Indoor		Crew	Daily Output	Labor-Hours	Unit	Material	2015 Bare Costs Labor	Equipment	Total	Total Incl O&P
0010	**CAPACITORS INDOOR**									
0020	240 volts, single & 3 phase, 0.5 kVAR	1 Elec	2.70	2.963	Ea.	430	162		592	715
0100	1.0 kVAR		2.70	2.963		515	162		677	815
0150	2.5 kVAR		2	4		580	219		799	970
0200	5.0 kVAR		1.80	4.444		790	243		1,033	1,225
0250	7.5 kVAR		1.60	5		885	274		1,159	1,375
0300	10 kVAR		1.50	5.333		1,025	292		1,317	1,550
0350	15 kVAR		1.30	6.154		1,325	335		1,660	1,950
0400	20 kVAR		1.10	7.273		1,600	400		2,000	2,350
0450	25 kVAR		1	8		1,825	440		2,265	2,675
1000	480 volts, single & 3 phase, 1 kVAR		2.70	2.963		390	162		552	675
1050	2 kVAR		2.70	2.963		450	162		612	740
1100	5 kVAR		2	4		565	219		784	950
1150	7.5 kVAR		2	4		610	219		829	1,000
1200	10 kVAR		2	4		710	219		929	1,100
1250	15 kVAR		2	4		835	219		1,054	1,250
1300	20 kVAR		1.60	5		915	274		1,189	1,400
1350	30 kVAR		1.50	5.333		1,100	292		1,392	1,650
1400	40 kVAR		1.20	6.667		1,375	365		1,740	2,050
1450	50 kVAR	↓	1.10	7.273	↓	1,600	400		2,000	2,375

26 51 Interior Lighting

26 51 13 – Interior Lighting Fixtures, Lamps, and Ballasts

26 51 13.50 Interior Lighting Fixtures		Crew	Daily Output	Labor-Hours	Unit	Material	Labor	Equipment	Total	Total Incl O&P
0010	**INTERIOR LIGHTING FIXTURES** Including lamps, mounting									
0030	hardware and connections									
0100	Fluorescent, C.W. lamps, troffer, recess mounted in grid, RS									
0130	Grid ceiling mount									
0200	Acrylic lens, 1'W x 4'L, two 40 watt	1 Elec	5.70	1.404	Ea.	47.50	77		124.50	167
0210	1'W x 4'L, three 40 watt		5.40	1.481		54	81		135	180
0300	2'W x 2'L, two U40 watt		5.70	1.404		51	77		128	172
0400	2'W x 4'L, two 40 watt		5.30	1.509		50	82.50		132.50	179
0500	2'W x 4'L, three 40 watt		5	1.600		55	87.50		142.50	192
0600	2'W x 4'L, four 40 watt	↓	4.70	1.702		57.50	93		150.50	203
0700	4'W x 4'L, four 40 watt	2 Elec	6.40	2.500		293	137		430	525
0800	4'W x 4'L, six 40 watt		6.20	2.581		305	141		446	545
0900	4'W x 4'L, eight 40 watt	↓	5.80	2.759		315	151		466	570
0910	Acrylic lens, 1'W x 4'L, two 32 watt T8 G	1 Elec	5.70	1.404		59.50	77		136.50	181
0930	2'W x 2'L, two U32 watt T8 G		5.70	1.404		81	77		158	204
0940	2'W x 4'L, two 32 watt T8 G		5.30	1.509		67	82.50		149.50	198
0950	2'W x 4'L, three 32 watt T8 G		5	1.600		68	87.50		155.50	206
0960	2'W x 4'L, four 32 watt T8 G	↓	4.70	1.702	↓	71	93		164	217
1000	Surface mounted, RS									
1030	Acrylic lens with hinged & latched door frame									
1100	1'W x 4'L, two 40 watt	1 Elec	7	1.143	Ea.	64	62.50		126.50	164
1110	1'W x 4'L, three 40 watt		6.70	1.194		66	65.50		131.50	171
1200	2'W x 2'L, two U40 watt		7	1.143		68.50	62.50		131	169
1300	2'W x 4'L, two 40 watt		6.20	1.290		78	70.50		148.50	192
1400	2'W x 4'L, three 40 watt		5.70	1.404		79	77		156	202
1500	2'W x 4'L, four 40 watt	↓	5.30	1.509		81	82.50		163.50	213
1600	4'W x 4'L, four 40 watt	2 Elec	7.20	2.222	↓	400	122		522	620

26 51 13 – Interior Lighting Fixtures, Lamps, and Ballasts

26 51 13.50 Interior Lighting Fixtures		Crew	Daily Output	Labor-Hours	Unit	Material	Labor	Equipment	Total	Total Incl O&P
						2015 Bare Costs				
1700	4'W x 4'L, six 40 watt	2 Elec	6.60	2.424	Ea.	435	133		568	675
1800	4'W x 4'L, eight 40 watt		6.20	2.581		450	141		591	705
1900	2'W x 8'L, four 40 watt		6.40	2.500		159	137		296	380
2000	2'W x 8'L, eight 40 watt		6.20	2.581		171	141		312	400
2100	Strip fixture									
2130	Surface mounted									
2200	4' long, one 40 watt, RS	1 Elec	8.50	.941	Ea.	27.50	51.50		79	107
2300	4' long, two 40 watt, RS		8	1		38.50	54.50		93	124
2400	4' long, one 40 watt, SL		8	1		46.50	54.50		101	133
2500	4' long, two 40 watt, SL		7	1.143		63	62.50		125.50	163
2600	8' long, one 75 watt, SL	2 Elec	13.40	1.194		48	65.50		113.50	151
2700	8' long, two 75 watt, SL	"	12.40	1.290		58	70.50		128.50	170
2800	4' long, two 60 watt, HO	1 Elec	6.70	1.194		93.50	65.50		159	201
2900	8' long, two 110 watt, HO	2 Elec	10.60	1.509		98.50	82.50		181	232
2950	High bay pendent mounted, 16" W x 4' L, four 54 watt, T5HO G		8.90	1.798		229	98.50		327.50	400
2952	2' W x 4' L, six 54 watt, T5HO G		8.50	1.882		305	103		408	490
2954	2' W x 4' L, six 32 watt, T8 G		8.50	1.882		179	103		282	350
3000	Strip, pendent mounted, industrial, white porcelain enamel									
3100	4' long, two 40 watt, RS	1 Elec	5.70	1.404	Ea.	52.50	77		129.50	173
3200	4' long, two 60 watt, HO	"	5	1.600		82.50	87.50		170	222
3300	8' long, two 75 watt, SL	2 Elec	8.80	1.818		98	99.50		197.50	257
3400	8' long, two 110 watt, HO	"	8	2		125	109		234	300
3470	Troffer, air handling, 2'W x 4'L with four 32 watt T8 G	1 Elec	4	2		112	109		221	287
3480	2'W x 2'L with two U32 watt T8 G		5.50	1.455		108	79.50		187.50	238
3490	Air connector insulated, 5" diameter		20	.400		67.50	22		89.50	107
3500	6" diameter		20	.400		69	22		91	109
3510	Troffer parabolic lay-in, 1'W x 4'L with one 32 W T8 G		5.70	1.404		114	77		191	241
3520	1'W x 4'L with two 32 W T8 G		5.30	1.509		137	82.50		219.50	274
3525	2'W x 2'L with two U32 W T8 G		5.70	1.404		115	77		192	242
3530	2'W x 4'L with three 32 W T8 G		5	1.600		128	87.50		215.50	271
4450	Incandescent, high hat can, round alzak reflector, prewired									
4470	100 watt	1 Elec	8	1	Ea.	71.50	54.50		126	161
4480	150 watt		8	1		104	54.50		158.50	196
4500	300 watt		6.70	1.194		241	65.50		306.50	365
4600	Square glass lens with metal trim, prewired									
4630	100 watt	1 Elec	6.70	1.194	Ea.	55	65.50		120.50	159
4700	200 watt		6.70	1.194		97	65.50		162.50	205
4800	300 watt		5.70	1.404		145	77		222	274
4900	Ceiling/wall, surface mounted, metal cylinder, 75 watt		10	.800		56	44		100	127
4920	150 watt		10	.800		80.50	44		124.50	154
5200	Ceiling, surface mounted, opal glass drum									
5300	8", one 60 watt lamp	1 Elec	10	.800	Ea.	44.50	44		88.50	115
5400	10", two 60 watt lamps		8	1		50	54.50		104.50	137
5500	12", four 60 watt lamps		6.70	1.194		70.50	65.50		136	176
6010	Vapor tight, incandescent, ceiling mounted, 200 watt		6.20	1.290		82.50	70.50		153	197
6100	Fluorescent, surface mounted, 2 lamps, 4'L, RS, 40 watt		3.20	2.500		115	137		252	330
6850	Vandalproof, surface mounted, fluorescent, two 32 watt T8 G		3.20	2.500		252	137		389	485
6860	Incandescent, one 150 watt		8	1		98.50	54.50		153	190
7500	Ballast replacement, by weight of ballast, to 15' high									
7520	Indoor fluorescent, less than 2 lb.	1 Elec	10	.800	Ea.	26	44		70	94
7540	Two 40W, watt reducer, 2 to 5 lb.		9.40	.851		41	46.50		87.50	115
7560	Two F96 slimline, over 5 lb.		8	1		77	54.50		131.50	167
7580	Vaportite ballast, less than 2 lb.		9.40	.851		26	46.50		72.50	98

26 51 Interior Lighting

26 51 13 – Interior Lighting Fixtures, Lamps, and Ballasts

26 51 13.50 Interior Lighting Fixtures			Crew	Daily Output	Labor-Hours	Unit	Material	2015 Bare Costs Labor	Equipment	Total	Total Incl O&P
7600	2 lb. to 5 lb.		1 Elec	8.90	.899	Ea.	41	49		90	119
7620	Over 5 lb.			7.60	1.053		77	57.50		134.50	171
7630	Electronic ballast for two tubes			8	1		37	54.50		91.50	123
7640	Dimmable ballast one lamp	G		8	1		106	54.50		160.50	199
7650	Dimmable ballast two-lamp	G	↓	7.60	1.053	↓	104	57.50		161.50	201

26 51 13.55 Interior LED Fixtures

Line	Description		Crew	Daily Output	Labor-Hours	Unit	Material	Labor	Equipment	Total	Total Incl O&P
0010	**INTERIOR LED FIXTURES** Incl. lamps, and mounting hardware										
0100	Downlight, recess mounted, 7.5" diameter, 25 watt	G	1 Elec	8	1	Ea.	335	54.50		389.50	445
0120	10" diameter, 36 watt	G		8	1		360	54.50		414.50	475
0160	cylinder, 10 watts	G		8	1		102	54.50		156.50	194
0180	20 watts	G		8	1		585	54.50		639.50	725
1000	Troffer, recess mounted, 2' x 4', 3200 Lumens	G		5.30	1.509		138	82.50		220.50	275
1010	4800 Lumens	G		5	1.600		179	87.50		266.50	325
1020	6400 Lumens	G		4.70	1.702		198	93		291	355
1100	Troffer retrofit lamp, 38 watt	G		21	.381		238	21		259	292
1110	60 watt	G		20	.400		340	22		362	410
1120	100 watt	G		18	.444		510	24.50		534.50	595
1200	Troffer, volumetric recess mounted, 2' x 2'	G		5.70	1.404		251	77		328	390
2000	Strip, surface mounted, one light bar 4' long, 3500K	G		8.50	.941		299	51.50		350.50	405
2010	5000K	G		8	1		299	54.50		353.50	410
2020	Two light bar 4' long, 5000K	G		7	1.143		470	62.50		532.50	610
3000	Linear, suspended mounted, one light bar 4' long, 37 watt	G	↓	6.70	1.194		195	65.50		260.50	315
3010	One light bar 8' long, 74 watt	G	2 Elec	12.20	1.311		360	71.50		431.50	505
3020	Two light bar 4' long, 74 watt	G	1 Elec	5.70	1.404		390	77		467	540
3030	Two light bar 8' long, 148 watt	G	2 Elec	8.80	1.818		450	99.50		549.50	645
4000	High bay, surface mounted, round, 150 watts	G		5.41	2.959		605	162		767	905
4010	2 bars,164 watts	G		5.41	2.959		570	162		732	865
4020	3 bars, 246 watts	G		5.01	3.197		730	175		905	1,075
4030	4 bars, 328 watts	G	↓	4.60	3.478		895	190		1,085	1,275
4040	5 bars, 410 watts	G	3 Elec	4.20	5.716		1,050	315		1,365	1,625
4050	6 bars, 492 watts	G		3.80	6.324		1,200	345		1,545	1,825
4060	7 bars, 574 watts	G		3.39	7.075		1,350	385		1,735	2,075
4070	8 bars, 656 watts	G	↓	2.99	8.029		1,500	440		1,940	2,300
5000	track, lighthead, 6 watt	G	1 Elec	32	.250		54.50	13.70		68.20	80.50
5010	9 watt	G	"	32	.250		61.50	13.70		75.20	88
6000	Garage, surface mount, 103 watts	G	2 Elec	6.50	2.462		970	135		1,105	1,275
6100	pendent mount, 80 watts	G		6.50	2.462		565	135		700	820
6200	95 watts	G		6.50	2.462		635	135		770	900
6300	125 watts	G	↓	6.50	2.462	↓	690	135		825	960

26 52 Emergency Lighting

26 52 13 – Emergency Lighting Equipments

26 52 13.10 Emergency Lighting and Battery Units

Line	Description	Crew	Daily Output	Labor-Hours	Unit	Material	Labor	Equipment	Total	Total Incl O&P
0010	**EMERGENCY LIGHTING AND BATTERY UNITS**									
0300	Emergency light units, battery operated									
0350	Twin sealed beam light, 25 watt, 6 volt each									
0500	Lead battery operated	1 Elec	4	2	Ea.	153	109		262	330
0700	Nickel cadmium battery operated		4	2		560	109		669	780
0900	Self-contained fluorescent lamp pack	↓	10	.800	↓	156	44		200	237

26 53 Exit Signs

26 53 13 – Exit Lighting

26 53 13.10 Exit Lighting Fixtures		Crew	Daily Output	Labor-Hours	Unit	Material	2015 Bare Costs Labor	Equipment	Total	Total Incl O&P
0010	**EXIT LIGHTING FIXTURES**									
0080	Exit light ceiling or wall mount, incandescent, single face	1 Elec	8	1	Ea.	39	54.50		93.50	125
0100	Double face		6.70	1.194		39	65.50		104.50	141
0200	LED standard, single face G		8	1		77	54.50		131.50	167
0220	Double face G		6.70	1.194		77	65.50		142.50	183
0230	LED vandal-resistant, single face G		7.27	1.100		212	60		272	325
0240	LED w/battery unit, single face G		4.40	1.818		160	99.50		259.50	325
0260	Double face G		4	2		163	109		272	345
0262	LED w/battery unit, vandal-resistant, single face G		4.40	1.818		245	99.50		344.50	420
0270	Combination emergency light units and exit sign		4	2		179	109		288	360

26 54 Classified Location Lighting

26 54 13 – Classified Lighting

26 54 13.20 Explosionproof		Crew	Daily Output	Labor-Hours	Unit	Material	Labor	Equipment	Total	Total Incl O&P
0010	**EXPLOSIONPROOF**, incl lamps, mounting hardware and connections									
6510	Incandescent, ceiling mounted, 200 watt	1 Elec	4	2	Ea.	1,075	109		1,184	1,350
6600	Fluorescent, RS, 4′ long, ceiling mounted, two 40 watt	"	2.70	2.963	"	2,850	162		3,012	3,375

26 55 Special Purpose Lighting

26 55 61 – Theatrical Lighting

26 55 61.10 Lights		Crew	Daily Output	Labor-Hours	Unit	Material	Labor	Equipment	Total	Total Incl O&P
0010	**LIGHTS**									
2000	Lights, border, quartz, reflector, vented,									
2100	colored or white	1 Elec	20	.400	L.F.	175	22		197	225
2500	Spotlight, follow spot, with transformer, 2,100 watt	"	4	2	Ea.	3,100	109		3,209	3,600
2600	For no transformer, deduct					920			920	1,025
3000	Stationary spot, fresnel quartz, 6″ lens	1 Elec	4	2		133	109		242	310
3100	8″ lens		4	2		234	109		343	420
3500	Ellipsoidal quartz, 1,000W, 6″ lens		4	2		340	109		449	540
3600	12″ lens		4	2		605	109		714	830
4000	Strobe light, 1 to 15 flashes per second, quartz		3	2.667		760	146		906	1,050
4500	Color wheel, portable, five hole, motorized		4	2		197	109		306	380

26 56 Exterior Lighting

26 56 13 – Lighting Poles and Standards

26 56 13.10 Lighting Poles		Crew	Daily Output	Labor-Hours	Unit	Material	Labor	Equipment	Total	Total Incl O&P
0010	**LIGHTING POLES**									
2800	Light poles, anchor base									
2820	not including concrete bases									
2840	Aluminum pole, 8′ high	1 Elec	4	2	Ea.	705	109		814	940
3000	20′ high	R-3	2.90	6.897		935	375	47.50	1,357.50	1,625
3200	30′ high		2.60	7.692		1,775	420	53	2,248	2,625
3400	35′ high		2.30	8.696		1,925	470	60	2,455	2,900
3600	40′ high		2	10		2,200	545	69	2,814	3,325
3800	Bracket arms, 1 arm	1 Elec	8	1		121	54.50		175.50	215
4000	2 arms		8	1		243	54.50		297.50	350
4200	3 arms		5.30	1.509		365	82.50		447.50	525

26 56 Exterior Lighting

26 56 13 – Lighting Poles and Standards

26 56 13.10 Lighting Poles

26 56 13.10 Lighting Poles		Crew	Daily Output	Labor-Hours	Unit	2015 Bare Costs Material	Labor	Equipment	Total	Total Incl O&P
4400	4 arms	1 Elec	5.30	1.509	Ea.	485	82.50		567.50	660
4500	Steel pole, galvanized, 8′ high		3.80	2.105		610	115		725	845
4600	20′ high	R-3	2.60	7.692		1,100	420	53	1,573	1,875
4800	30′ high		2.30	8.696		1,300	470	60	1,830	2,200
5000	35′ high		2.20	9.091		1,425	495	62.50	1,982.50	2,375
5200	40′ high		1.70	11.765		1,775	640	81	2,496	3,000
5400	Bracket arms, 1 arm	1 Elec	8	1		180	54.50		234.50	280
5600	2 arms		8	1		216	54.50		270.50	320
5800	3 arms		5.30	1.509		254	82.50		336.50	405
6000	4 arms		5.30	1.509		310	82.50		392.50	465

26 56 16 – Parking Lighting

26 56 16.55 Parking LED Lighting

Line	Description		Crew	Daily Output	Labor-Hours	Unit	Material	Labor	Equipment	Total	Total Incl O&P
0010	**PARKING LED LIGHTING**										
0100	Round pole mounting, 88 lamp watts	G	1 Elec	2	4	Ea.	995	219		1,214	1,425
0110	Square pole mounting, 223 lamp watts	G	"	2	4	"	1,750	219		1,969	2,250

26 56 19 – Roadway Lighting

26 56 19.20 Roadway Luminaire

Line	Description		Crew	Daily Output	Labor-Hours	Unit	Material	Labor	Equipment	Total	Total Incl O&P
0010	**ROADWAY LUMINAIRE**										
2650	Roadway area luminaire, low pressure sodium, 135 watt		1 Elec	2	4	Ea.	650	219		869	1,050
2700	180 watt		"	2	4		700	219		919	1,100
2750	Metal halide, 400 watt		2 Elec	4.40	3.636		555	199		754	910
2760	1000 watt			4	4		625	219		844	1,025
2780	High pressure sodium, 400 watt			4.40	3.636		580	199		779	940
2790	1000 watt			4	4		660	219		879	1,050

26 56 19.55 Roadway LED Luminaire

Line	Description		Crew	Daily Output	Labor-Hours	Unit	Material	Labor	Equipment	Total	Total Incl O&P
0010	**ROADWAY LED LUMINAIRE**										
0100	LED fixture, 72 LEDs, 120 V AC or 12 V DC, equal to 60 watt	G	1 Elec	2.70	2.963	Ea.	595	162		757	895
0110	108 LEDs, 120 V AC or 12 V DC, equal to 90 watt	G		2.70	2.963		695	162		857	1,000
0120	144 LEDs, 120 V AC or 12 V DC, equal to 120 watt	G		2.70	2.963		855	162		1,017	1,175
0130	252 LEDs, 120 V AC or 12 V DC, equal to 210 watt	G	2 Elec	4.40	3.636		1,175	199		1,374	1,600
0140	Replaces high pressure sodium fixture, 75 watt	G	1 Elec	2.70	2.963		720	162		882	1,050
0150	125 watt	G		2.70	2.963		825	162		987	1,150
0160	150 watt	G		2.70	2.963		1,025	162		1,187	1,375
0170	175 watt	G		2.70	2.963		1,225	162		1,387	1,600
0180	200 watt	G		2.70	2.963		1,450	162		1,612	1,825
0190	250 watt	G	2 Elec	4.40	3.636		1,650	199		1,849	2,125
0200	320 watt	G	"	4.40	3.636		1,800	199		1,999	2,275

26 56 23 – Area Lighting

26 56 23.10 Exterior Fixtures

Line	Description		Crew	Daily Output	Labor-Hours	Unit	Material	Labor	Equipment	Total	Total Incl O&P
0010	**EXTERIOR FIXTURES** With lamps										
0200	Wall mounted, incandescent, 100 watt		1 Elec	8	1	Ea.	35	54.50		89.50	121
0400	Quartz, 500 watt			5.30	1.509		50	82.50		132.50	179
1100	Wall pack, low pressure sodium, 35 watt			4	2		227	109		336	415
1150	55 watt			4	2		270	109		379	460

26 56 23.55 Exterior LED Fixtures

Line	Description		Crew	Daily Output	Labor-Hours	Unit	Material	Labor	Equipment	Total	Total Incl O&P
0010	**EXTERIOR LED FIXTURES**										
0100	Wall mounted, indoor/outdoor, 12 watt	G	1 Elec	10	.800	Ea.	258	44		302	350
0110	32 watt	G		10	.800		350	44		394	450
0120	66 watt	G		10	.800		500	44		544	615
0200	outdoor, 110 watt	G		10	.800		765	44		809	905

26 56 Exterior Lighting

26 56 23 – Area Lighting

26 56 23.55 Exterior LED Fixtures			Crew	Daily Output	Labor-Hours	Unit	2015 Bare Costs Material	Labor	Equipment	Total	Total Incl O&P
0210	220 watt	G	1 Elec	10	.800	Ea.	1,350	44		1,394	1,550
0300	modular, type IV, 120 V, 50 lamp watts	G		9	.889		930	48.50		978.50	1,100
0310	101 lamp watts	G		9	.889		1,050	48.50		1,098.50	1,225
0320	126 lamp watts	G		9	.889		1,325	48.50		1,373.50	1,525
0330	202 lamp watts	G		9	.889		1,500	48.50		1,548.50	1,725
0340	240 V, 50 lamp watts	G		8	1		970	54.50		1,024.50	1,150
0350	101 lamp watts	G		8	1		1,100	54.50		1,154.50	1,275
0360	126 lamp watts	G		8	1		1,350	54.50		1,404.50	1,575
0370	202 lamp watts	G		8	1		1,550	54.50		1,604.50	1,775
0400	wall pack, glass, 13 lamp watts	G		4	2		360	109		469	560
0410	poly w/photocell, 26 lamp watts	G		4	2		226	109		335	410
0420	50 lamp watts	G		4	2		530	109		639	745
0430	replacement, 40 watts	G		4	2		460	109		569	670
0440	60 watts	G		4	2		580	109		689	805

26 56 36 – Flood Lighting

26 56 36.20 Floodlights		Crew	Daily Output	Labor-Hours	Unit	Material	Labor	Equipment	Total	Total Incl O&P
0010	**FLOODLIGHTS** with ballast and lamp,									
1400	Pole mounted, pole not included									
1950	Metal halide, 175 watt	1 Elec	2.70	2.963	Ea.	340	162		502	615
2000	400 watt	2 Elec	4.40	3.636		420	199		619	765
2200	1000 watt	"	4	4		580	219		799	965
2340	High pressure sodium, 70 watt	1 Elec	2.70	2.963		246	162		408	515
2400	400 watt	2 Elec	4.40	3.636		380	199		579	720
2600	1000 watt	"	4	4		650	219		869	1,050

26 56 36.55 LED Floodlights			Crew	Daily Output	Labor-Hours	Unit	Material	Labor	Equipment	Total	Total Incl O&P
0010	**LED FLOODLIGHTS** with ballast and lamp,										
0020	Pole mounted, pole not included										
0100	11 watt	G	1 Elec	4	2	Ea.	345	109		454	545
0110	46 watt	G		4	2		1,050	109		1,159	1,325
0120	90 watt	G		4	2		1,725	109		1,834	2,075
0130	288 watt	G		4	2		2,125	109		2,234	2,500

26 61 Lighting Systems and Accessories

26 61 23 – Lamps Applications

26 61 23.10 Lamps			Crew	Daily Output	Labor-Hours	Unit	Material	Labor	Equipment	Total	Total Incl O&P
0010	**LAMPS**										
0080	Fluorescent, rapid start, cool white, 2' long, 20 watt		1 Elec	1	8	C	310	440		750	995
0100	4' long, 40 watt			.90	8.889		236	485		721	990
0200	Slimline, 4' long, 40 watt			.90	8.889		1,250	485		1,735	2,100
0210	4' long, 30 watt energy saver	G		.90	8.889		1,250	485		1,735	2,100
0400	High output, 4' long, 60 watt			.90	8.889		650	485		1,135	1,450
0410	8' long, 95 watt energy saver	G		.80	10		640	545		1,185	1,525
0500	8' long, 110 watt			.80	10		640	545		1,185	1,525
0512	2' long, T5, 14 watt energy saver	G		1	8		860	440		1,300	1,600
0514	3' long, T5, 21 watt energy saver	G		.90	8.889		1,000	485		1,485	1,850
0516	4' long, T5, 28 watt energy saver	G		.90	8.889		1,300	485		1,785	2,150
0517	4' long, T5, 54 watt energy saver	G		.90	8.889		1,575	485		2,060	2,450
0560	Twin tube compact lamp	G		.90	8.889		360	485		845	1,125
0570	Double twin tube compact lamp	G		.80	10		805	545		1,350	1,700
0600	Mercury vapor, mogul base, deluxe white, 100 watt			.30	26.667		3,650	1,450		5,100	6,200

26 61 Lighting Systems and Accessories

26 61 23 – Lamps Applications

26 61 23.10 Lamps		Crew	Daily Output	Labor-Hours	Unit	Material	Labor (2015 Bare Costs)	Equipment (2015 Bare Costs)	Total	Total Incl O&P
0700	250 watt	1 Elec	.30	26.667	C	3,000	1,450		4,450	5,475
0800	400 watt		.30	26.667		4,350	1,450		5,800	6,950
0900	1000 watt		.20	40		7,350	2,200		9,550	11,400
1000	Metal halide, mogul base, 175 watt		.30	26.667		2,550	1,450		4,000	5,000
1200	400 watt		.30	26.667		4,750	1,450		6,200	7,400
1300	1000 watt		.20	40		6,625	2,200		8,825	10,600
1350	High pressure sodium, 70 watt		.30	26.667		3,125	1,450		4,575	5,600
1380	250 watt		.30	26.667		4,225	1,450		5,675	6,825
1400	400 watt		.30	26.667		5,525	1,450		6,975	8,250
1450	1000 watt		.20	40		10,000	2,200		12,200	14,300
3000	Guards, fluorescent lamp, 4′ long		1	8		1,400	440		1,840	2,200
3200	8′ long		.90	8.889		2,800	485		3,285	3,825

26 61 23.55 LED Lamps		Crew	Daily Output	Labor-Hours	Unit	Material	Labor	Equipment	Total	Total Incl O&P
0010	**LED LAMPS**									
0100	LED lamp, interior, shape A60, equal to 60 watt G	1 Elec	160	.050	Ea.	22.50	2.74		25.24	28.50
0200	Globe frosted A60, equal to 60 watt G		160	.050		22.50	2.74		25.24	28.50
0300	Globe earth, equal to 100 watt G		160	.050		74	2.74		76.74	85.50
1100	MR16, 3 watt, replacement of halogen lamp 25 watt G		130	.062		21.50	3.37		24.87	28.50
1200	6 watt replacement of halogen lamp 45 watt G		130	.062		47	3.37		50.37	56.50
2100	10 watt, PAR20, equal to 60 watt G		130	.062		47.50	3.37		50.87	57.50
2200	15 watt, PAR30, equal to 100 watt G		130	.062		79	3.37		82.37	92

26 71 Electrical Machines

26 71 13 – Motors Applications

26 71 13.20 Motors		Crew	Daily Output	Labor-Hours	Unit	Material	Labor (2015 Bare Costs)	Equipment (2015 Bare Costs)	Total	Total Incl O&P
0010	**MOTORS** 230/460 volts, 60 HZ									
0050	Dripproof, premium efficiency, 1.15 service factor									
0060	1800 RPM, 1/4 HP	1 Elec	5.33	1.501	Ea.	190	82		272	330
0070	1/3 HP		5.33	1.501		195	82		277	340
0080	1/2 HP		5.33	1.501		218	82		300	365
0090	3/4 HP		5.33	1.501		248	82		330	395
0100	1 HP		4.50	1.778		279	97		376	450
0250	5 HP		4.50	1.778		610	97		707	815
0350	10 HP		4	2		990	109		1,099	1,275
0450	20 HP	2 Elec	5.20	3.077		1,750	168		1,918	2,175

Estimating Tips

27 20 00 Data Communications
27 30 00 Voice Communications
27 40 00 Audio-Video Communications

When estimating material costs for special systems, it is always prudent to obtain manufacturers' quotations for equipment prices and special installation requirements which will affect the total costs.

Reference Numbers

Reference numbers are shown in shaded boxes at the beginning of some major classifications. These numbers refer to related items in the Reference Section. The reference information may be an estimating procedure, an alternate pricing method, or technical information.

Note: Not all subdivisions listed here necessarily appear in this publication. ■

*Note: **Trade Service**, in part, has been used as a reference source for some of the material prices used in Division 27.*

27 13 Communications Backbone Cabling

27 13 23 – Communications Optical Fiber Backbone Cabling

27 13 23.13 Communications Optical Fiber		Crew	Daily Output	Labor-Hours	Unit	2015 Bare Costs Material	Labor	Equipment	Total	Total Incl O&P
0010	**COMMUNICATIONS OPTICAL FIBER**									
0040	Specialized tools & techniques cause installation costs to vary.									
0070	Fiber optic, cable, bulk simplex, single mode	1 Elec	8	1	C.L.F.	22.50	54.50		77	107
0080	Multi mode		8	1		42	54.50		96.50	129
0090	4 strand, single mode		7.34	1.090		38	59.50		97.50	131
0095	Multi mode		7.34	1.090		50.50	59.50		110	145
0100	12 strand, single mode		6.67	1.199		90	65.50		155.50	198
0105	Multi mode		6.67	1.199		96.50	65.50		162	205
0150	Jumper				Ea.	33			33	36.50
0200	Pigtail					33.50			33.50	37
0300	Connector	1 Elec	24	.333		23.50	18.25		41.75	53
0350	Finger splice		32	.250		32.50	13.70		46.20	56.50
0400	Transceiver (low cost bi-directional)		8	1		420	54.50		474.50	540
0450	Rack housing, 4 rack spaces, 12 panels (144 fibers)		2	4		500	219		719	880
0500	Patch panel, 12 ports		6	1.333		247	73		320	380

27 41 Audio-Video Systems

27 41 33 – Master Antenna Television Systems

27 41 33.10 T.V. Systems		Crew	Daily Output	Labor-Hours	Unit	Material	Labor	Equipment	Total	Total Incl O&P
0010	**T.V. SYSTEMS**, not including rough-in wires, cables & conduits									
0100	Master TV antenna system									
0200	VHF reception & distribution, 12 outlets	1 Elec	6	1.333	Outlet	133	73		206	255
0400	30 outlets		10	.800		141	44		185	221
0600	100 outlets		13	.615		144	33.50		177.50	209
0800	VHF & UHF reception & distribution, 12 outlets		6	1.333		213	73		286	345
1000	30 outlets		10	.800		141	44		185	221
1200	100 outlets		13	.615		144	33.50		177.50	209
1400	School and deluxe systems, 12 outlets		2.40	3.333		281	182		463	585
1600	30 outlets		4	2		246	109		355	435
1800	80 outlets		5.30	1.509		237	82.50		319.50	385

27 51 Distributed Audio-Video Communications Systems

27 51 16 – Public Address and Mass Notification Systems

27 51 16.10 Public Address System		Crew	Daily Output	Labor-Hours	Unit	Material	Labor	Equipment	Total	Total Incl O&P
0010	**PUBLIC ADDRESS SYSTEM**									
0100	Conventional, office	1 Elec	5.33	1.501	Speaker	135	82		217	272
0200	Industrial	"	2.70	2.963	"	261	162		423	530

27 51 19 – Sound Masking Systems

27 51 19.10 Sound System		Crew	Daily Output	Labor-Hours	Unit	Material	Labor	Equipment	Total	Total Incl O&P
0010	**SOUND SYSTEM**, not including rough-in wires, cables & conduits									
0100	Components, projector outlet	1 Elec	8	1	Ea.	46	54.50		100.50	133
0200	Microphone		4	2		81.50	109		190.50	254
0400	Speakers, ceiling or wall		8	1		119	54.50		173.50	213
0600	Trumpets		4	2		222	109		331	410
0800	Privacy switch		8	1		88.50	54.50		143	180
1000	Monitor panel		4	2		395	109		504	600
1200	Antenna, AM/FM		4	2		138	109		247	315
1400	Volume control		8	1		90	54.50		144.50	181
1600	Amplifier, 250 watts		1	8		1,275	440		1,715	2,050

27 51 Distributed Audio-Video Communications Systems

27 51 19 – Sound Masking Systems

27 51 19.10 Sound System		Crew	Daily Output	Labor-Hours	Unit	2015 Bare Costs Material	Labor	Equipment	Total	Total Incl O&P
1800	Cabinets	1 Elec	1	8	Ea.	860	440		1,300	1,600
2000	Intercom, 30 station capacity, master station	2 Elec	2	8		2,350	440		2,790	3,250
2200	Remote station	1 Elec	8	1		166	54.50		220.50	264
2400	Intercom outlets		8	1		97.50	54.50		152	189
2600	Handset		4	2		320	109		429	520
2800	Emergency call system, 12 zones, annunciator		1.30	6.154		970	335		1,305	1,575
3000	Bell		5.30	1.509		100	82.50		182.50	234
3200	Light or relay		8	1		50	54.50		104.50	137
3400	Transformer		4	2		220	109		329	405
3600	House telephone, talking station		1.60	5		475	274		749	930
3800	Press to talk, release to listen		5.30	1.509		110	82.50		192.50	245
4000	System-on button					66			66	72.50
4200	Door release	1 Elec	4	2		118	109		227	294
4400	Combination speaker and microphone		8	1		201	54.50		255.50	305
4600	Termination box		3.20	2.500		63	137		200	275
4800	Amplifier or power supply		5.30	1.509		725	82.50		807.50	925
5000	Vestibule door unit		16	.500	Name	133	27.50		160.50	188
5200	Strip cabinet		27	.296	Ea.	252	16.20		268.20	300
5400	Directory		16	.500	″	119	27.50		146.50	172

27 52 Healthcare Communications and Monitoring Systems

27 52 23 – Nurse Call/Code Blue Systems

27 52 23.10 Nurse Call Systems		Crew	Daily Output	Labor-Hours	Unit	Material	Labor	Equipment	Total	Total Incl O&P
0010	**NURSE CALL SYSTEMS**									
0100	Single bedside call station	1 Elec	8	1	Ea.	231	54.50		285.50	335
0200	Ceiling speaker station		8	1		67.50	54.50		122	156
0400	Emergency call station		8	1		72	54.50		126.50	162
0600	Pillow speaker		8	1		177	54.50		231.50	277
0800	Double bedside call station		4	2		142	109		251	320
1000	Duty station		4	2		148	109		257	325
1200	Standard call button		8	1		87	54.50		141.50	178
1400	Lights, corridor, dome or zone indicator		8	1		49	54.50		103.50	136
1600	Master control station for 20 stations	2 Elec	.65	24.615	Total	3,975	1,350		5,325	6,400

27 53 Distributed Systems

27 53 13 – Clock Systems

27 53 13.50 Clock Equipments		Crew	Daily Output	Labor-Hours	Unit	Material	Labor	Equipment	Total	Total Incl O&P
0010	**CLOCK EQUIPMENTS**, not including wires & conduits									
0100	Time system components, master controller	1 Elec	.33	24.242	Ea.	1,850	1,325		3,175	4,025
0200	Program bell		8	1		86.50	54.50		141	178
0400	Combination clock & speaker		3.20	2.500		214	137		351	440
0600	Frequency generator		2	4		2,400	219		2,619	2,975
0800	Job time automatic stamp recorder		4	2		540	109		649	760
1600	Master time clock system, clocks & bells, 20 room	4 Elec	.20	160		6,200	8,750		14,950	19,900
1800	50 room	″	.08	400		12,700	21,900		34,600	46,800
1900	Time clock	1 Elec	3.20	2.500		445	137		582	695
2000	100 cards in & out, 1 color					9.15			9.15	10.10
2200	2 colors					9.15			9.15	10.10
2800	Metal rack for 25 cards	1 Elec	7	1.143		44	62.50		106.50	142

Division Notes

		CREW	DAILY OUTPUT	LABOR-HOURS	UNIT	BARE COSTS MAT.	LABOR	EQUIP.	TOTAL	TOTAL INCL O&P

Estimating Tips

- When estimating material costs for electronic safety and security systems, it is always prudent to obtain manufacturers' quotations for equipment prices and special installation requirements that affect the total cost.
- Fire alarm systems consist of control panels, annunciator panels, battery with rack, charger, and fire alarm actuating and indicating devices. Some fire alarm systems include speakers, telephone lines, door closer controls, and other components. Be careful not to overlook the costs related to installation for these items. Also be aware of costs for integrated automation instrumentation and terminal devices, control equipment, control wiring, and programming.
- Security equipment includes items such as CCTV, access control, and other detection and identification systems to perform alert and alarm functions. Be sure to consider the costs related to installation for this security equipment, such as for integrated automation instrumentation and terminal devices, control equipment, control wiring, and programming.

Reference Numbers

Reference numbers are shown in shaded boxes at the beginning of some major classifications. These numbers refer to related items in the Reference Section. The reference information may be an estimating procedure, an alternate pricing method, or technical information.

Note: Not all subdivisions listed here necessarily appear in this publication. ■

28 13 Access Control

28 13 53 – Security Access Detection

28 13 53.13 Security Access Metal Detectors		Crew	Daily Output	Labor-Hours	Unit	2015 Bare Costs Material	Labor	Equipment	Total	Total Incl O&P
0010	**SECURITY ACCESS METAL DETECTORS**									
0240	Metal detector, hand-held, wand type, unit only				Ea.	89			89	98
0250	Metal detector, walk through portal type, single zone	1 Elec	2	4		3,700	219		3,919	4,400
0260	Multi-zone	″	2	4		4,700	219		4,919	5,500

28 13 53.16 Security Access X-Ray Equipment		Crew	Daily Output	Labor-Hours	Unit	Material	Labor	Equipment	Total	Total Incl O&P
0010	**SECURITY ACCESS X-RAY EQUIPMENT**									
0290	X-ray machine, desk top, for mail/small packages/letters	1 Elec	4	2	Ea.	3,425	109		3,534	3,950
0300	Conveyor type, incl monitor		2	4		16,000	219		16,219	17,900
0310	Includes additional features		2	4		28,600	219		28,819	31,700
0320	X-ray machine, large unit, for airports, incl monitor	2 Elec	1	16		40,000	875		40,875	45,300
0330	Full console	″	.50	32		68,500	1,750		70,250	78,000

28 13 53.23 Security Access Explosive Detection Equipment		Crew	Daily Output	Labor-Hours	Unit	Material	Labor	Equipment	Total	Total Incl O&P
0010	**SECURITY ACCESS EXPLOSIVE DETECTION EQUIPMENT**									
0270	Explosives detector, walk through portal type	1 Elec	2	4	Ea.	44,200	219		44,419	48,900
0280	Hand-held, battery operated				″				25,500	28,100

28 16 Intrusion Detection

28 16 16 – Intrusion Detection Systems Infrastructure

28 16 16.50 Intrusion Detection		Crew	Daily Output	Labor-Hours	Unit	Material	Labor	Equipment	Total	Total Incl O&P
0010	**INTRUSION DETECTION**, not including wires & conduits									
0100	Burglar alarm, battery operated, mechanical trigger	1 Elec	4	2	Ea.	278	109		387	470
0200	Electrical trigger		4	2		330	109		439	530
0400	For outside key control, add		8	1		84	54.50		138.50	175
0600	For remote signaling circuitry, add		8	1		125	54.50		179.50	219
0800	Card reader, flush type, standard		2.70	2.963		930	162		1,092	1,275
1000	Multi-code		2.70	2.963		1,200	162		1,362	1,575
1200	Door switches, hinge switch		5.30	1.509		58.50	82.50		141	189
1400	Magnetic switch		5.30	1.509		69	82.50		151.50	200
1600	Exit control locks, horn alarm		4	2		277	109		386	470
1800	Flashing light alarm		4	2		305	109		414	500
2000	Indicating panels, 1 channel		2.70	2.963		370	162		532	650
2200	10 channel	2 Elec	3.20	5		1,050	274		1,324	1,550
2400	20 channel		2	8		2,450	440		2,890	3,350
2600	40 channel		1.14	14.035		4,450	770		5,220	6,050
2800	Ultrasonic motion detector, 12 volt	1 Elec	2.30	3.478		230	190		420	540
3000	Infrared photoelectric detector	″	4	2		189	109		298	370

28 23 Video Surveillance

28 23 13 – Video Surveillance Control and Management Systems

28 23 13.10 Closed Circuit Television System		Crew	Daily Output	Labor-Hours	Unit	Material	Labor	Equipment	Total	Total Incl O&P
0010	**CLOSED CIRCUIT TELEVISION SYSTEM**									
2000	Surveillance, one station (camera & monitor)	2 Elec	2.60	6.154	Total	1,350	335		1,685	1,975
2200	For additional camera stations, add	1 Elec	2.70	2.963	Ea.	755	162		917	1,075
2400	Industrial quality, one station (camera & monitor)	2 Elec	2.60	6.154	Total	2,800	335		3,135	3,575
2600	For additional camera stations, add	1 Elec	2.70	2.963	Ea.	1,725	162		1,887	2,125
2610	For low light, add		2.70	2.963		1,375	162		1,537	1,775
2620	For very low light, add		2.70	2.963		10,200	162		10,362	11,400
2800	For weatherproof camera station, add		1.30	6.154		1,050	335		1,385	1,675
3000	For pan and tilt, add		1.30	6.154		2,725	335		3,060	3,500

28 23 Video Surveillance

28 23 13 – Video Surveillance Control and Management Systems

28 23 13.10 Closed Circuit Television System		Crew	Daily Output	Labor-Hours	Unit	Material	2015 Bare Costs Labor	Equipment	Total	Total Incl O&P
3200	For zoom lens - remote control, add	1 Elec	2	4	Ea.	2,525	219		2,744	3,100
3400	Extended zoom lens		2	4		9,200	219		9,419	10,400
3410	For automatic iris for low light, add		2	4		2,200	219		2,419	2,750
3600	Educational T.V. studio, basic 3 camera system, black & white,									
3800	electrical & electronic equip. only	4 Elec	.80	40	Total	13,100	2,200		15,300	17,700
4000	Full console		.28	114		56,000	6,250		62,250	71,000
4100	As above, but color system		.28	114		74,000	6,250		80,250	90,500
4120	Full console		.12	266		321,000	14,600		335,600	375,000
4200	For film chain, black & white, add	1 Elec	1	8	Ea.	15,000	440		15,440	17,200
4250	Color, add		.25	32		18,200	1,750		19,950	22,600
4400	For video recorders, add		1	8		3,150	440		3,590	4,125
4600	Premium	4 Elec	.40	80		26,200	4,375		30,575	35,400

28 23 23 – Video Surveillance Systems Infrastructure

28 23 23.50 Video Surveillance Equipments		Crew	Daily Output	Labor-Hours	Unit	Material	Labor	Equipment	Total	Total Incl O&P
0010	**VIDEO SURVEILLANCE EQUIPMENTS**									
0200	Video cameras, wireless, hidden in exit signs, clocks, etc., incl. receiver	1 Elec	3	2.667	Ea.	108	146		254	340
0210	Accessories for video recorder, single camera		3	2.667		183	146		329	420
0220	For multiple cameras		3	2.667		1,750	146		1,896	2,150
0230	Video cameras, wireless, for under vehicle searching, complete		2	4		10,200	219		10,419	11,500

28 31 Fire Detection and Alarm

28 31 23 – Fire Detection and Alarm Annunciation Panels and Fire Stations

28 31 23.50 Alarm Panels and Devices		Crew	Daily Output	Labor-Hours	Unit	Material	Labor	Equipment	Total	Total Incl O&P
0010	**ALARM PANELS AND DEVICES**, not including wires & conduits									
3594	Fire, alarm control panel									
3600	4 zone	2 Elec	2	8	Ea.	400	440		840	1,100
3800	8 zone		1	16		780	875		1,655	2,150
4000	12 zone		.67	23.988		2,400	1,300		3,700	4,600
4020	Alarm device	1 Elec	8	1		238	54.50		292.50	345
4050	Actuating device		8	1		335	54.50		389.50	450
4200	Battery and rack		4	2		410	109		519	620
4400	Automatic charger		8	1		585	54.50		639.50	720
4600	Signal bell		8	1		78.50	54.50		133	169
4800	Trouble buzzer or manual station		8	1		83.50	54.50		138	174
5600	Strobe and horn		5.30	1.509		152	82.50		234.50	291
5800	Fire alarm horn		6.70	1.194		61	65.50		126.50	165
6000	Door holder, electro-magnetic		4	2		103	109		212	277
6200	Combination holder and closer		3.20	2.500		123	137		260	340
6600	Drill switch		8	1		370	54.50		424.50	490
6800	Master box		2.70	2.963		6,400	162		6,562	7,300
7000	Break glass station		8	1		55.50	54.50		110	143
7800	Remote annunciator, 8 zone lamp		1.80	4.444		209	243		452	595
8000	12 zone lamp	2 Elec	2.60	6.154		335	335		670	875
8200	16 zone lamp	"	2.20	7.273		420	400		820	1,050

28 31 43 – Fire Detection Sensors

28 31 43.50 Fire and Heat Detectors		Crew	Daily Output	Labor-Hours	Unit	Material	Labor	Equipment	Total	Total Incl O&P
0010	**FIRE & HEAT DETECTORS**									
5000	Detector, rate of rise	1 Elec	8	1	Ea.	51	54.50		105.50	138

28 31 Fire Detection and Alarm

28 31 46 – Smoke Detection Sensors

28 31 46.50 Smoke Detectors		Crew	Daily Output	Labor-Hours	Unit	2015 Bare Costs Material	Labor	Equipment	Total	Total Incl O&P
0010	**SMOKE DETECTORS**									
5200	Smoke detector, ceiling type	1 Elec	6.20	1.290	Ea.	110	70.50		180.50	227
5400	Duct type	"	3.20	2.500	"	325	137		462	565

28 33 Gas Detection and Alarm

28 33 33 – Gas Detection Sensors

28 33 33.50 Tank Leak Detection Systems		Crew	Daily Output	Labor-Hours	Unit	Material	Labor	Equipment	Total	Total Incl O&P
0010	**TANK LEAK DETECTION SYSTEMS** Liquid and vapor									
0100	For hydrocarbons and hazardous liquids/vapors									
0120	Controller, data acquisition, incl. printer, modem, RS232 port									
0140	24 channel, for use with all probes				Ea.	4,175			4,175	4,575
0160	9 channel, for external monitoring				"	915			915	1,000
0200	Probes									
0210	Well monitoring									
0220	Liquid phase detection				Ea.	475			475	525
0230	Hydrocarbon vapor, fixed position					480			480	530
0240	Hydrocarbon vapor, float mounted					480			480	530
0250	Both liquid and vapor hydrocarbon				↓	480			480	530
0300	Secondary containment, liquid phase									
0310	Pipe trench/manway sump				Ea.	820			820	905
0320	Double wall pipe and manual sump					825			825	905
0330	Double wall fiberglass annular space					325			325	355
0340	Double wall steel tank annular space				↓	335			335	370
0500	Accessories									
0510	Modem, non-dedicated phone line				Ea.	305			305	335
0600	Monitoring, internal									
0610	Automatic tank gauge, incl. overfill				Ea.	1,150			1,150	1,275
0620	Product line				"	1,150			1,150	1,275
0700	Monitoring, special									
0710	Cathodic protection				Ea.	725			725	795
0720	Annular space chemical monitor				"	985			985	1,075

28 39 Mass Notification Systems

28 39 10 – Notification Systems

28 39 10.10 Mass Notification System		Crew	Daily Output	Labor-Hours	Unit	Material	Labor	Equipment	Total	Total Incl O&P
0010	**MASS NOTIFICATION SYSTEM**									
0100	Wireless command center, 10,000 devices	2 Elec	1.33	12.030	Ea.	3,600	660		4,260	4,925
0200	Option, email notification					1,750			1,750	1,925
0210	Remote device supervision & monitor					2,450			2,450	2,675
0300	Antenna VHF or UHF, for medium range	1 Elec	4	2		129	109		238	305
0310	For high-power transmitter		2	4		670	219		889	1,075
0400	Transmitter, 25 watt		4	2		2,025	109		2,134	2,400
0410	40 watt		2.66	3.008		2,700	165		2,865	3,225
0420	100 watt		1.33	6.015		6,775	330		7,105	7,950
0500	Wireless receiver/control module for speaker		8	1		265	54.50		319.50	375
0600	Desktop paging controller, stand alone	↓	4	2	↓	370	109		479	575

Estimating Tips

31 05 00 Common Work Results for Earthwork

- Estimating the actual cost of performing earthwork requires careful consideration of the variables involved. This includes items such as type of soil, whether water will be encountered, dewatering, whether banks need bracing, disposal of excavated earth, and length of haul to fill or spoil sites, etc. If the project has large quantities of cut or fill, consider raising or lowering the site to reduce costs, while paying close attention to the effect on site drainage and utilities.
- If the project has large quantities of fill, creating a borrow pit on the site can significantly lower the costs.
- It is very important to consider what time of year the project is scheduled for completion. Bad weather can create large cost overruns from dewatering, site repair, and lost productivity from cold weather.

Reference Numbers

Reference numbers are shown in shaded boxes at the beginning of some major classifications. These numbers refer to related items in the Reference Section. The reference information may be an estimating procedure, an alternate pricing method, or technical information.

Note: Not all subdivisions listed here necessarily appear in this publication. ■

31 05 Common Work Results for Earthwork

31 05 13 – Soils for Earthwork

31 05 13.10 Borrow		Crew	Daily Output	Labor-Hours	Unit	Material	Labor	Equipment	Total	Total Incl O&P
						2015 Bare Costs				
0010	**BORROW** R312316-40									
0020	Spread, 200 H.P. dozer, no compaction, 2 mi. RT haul									
0200	Common borrow	B-15	600	.047	C.Y.	12.40	1.99	4.62	19.01	22
0700	Screened loam		600	.047		27.50	1.99	4.62	34.11	38.50
0800	Topsoil, weed free		600	.047		24.50	1.99	4.62	31.11	35
0900	For 5 mile haul, add	B-34B	200	.040			1.60	3.46	5.06	6.20

31 05 16 – Aggregates for Earthwork

31 05 16.10 Borrow		Crew	Daily Output	Labor-Hours	Unit	Material	Labor	Equipment	Total	Total Incl O&P
0010	**BORROW** R312316-40									
0020	Spread, with 200 H.P. dozer, no compaction, 2 mi. RT haul									
0100	Bank run gravel	B-15	600	.047	L.C.Y.	22	1.99	4.62	28.61	32.50
0300	Crushed stone (1.40 tons per CY) , 1-1/2"		600	.047		23.50	1.99	4.62	30.11	34
0320	3/4"		600	.047		23.50	1.99	4.62	30.11	34
0340	1/2"		600	.047		26.50	1.99	4.62	33.11	37
0360	3/8"		600	.047		27.50	1.99	4.62	34.11	38.50
0400	Sand, washed, concrete		600	.047		41	1.99	4.62	47.61	53
0500	Dead or bank sand		600	.047		17.85	1.99	4.62	24.46	27.50
0600	Select structural fill		600	.047		21	1.99	4.62	27.61	31
0900	For 5 mile haul, add	B-34B	200	.040			1.60	3.46	5.06	6.20

31 05 23 – Cement and Concrete for Earthwork

31 05 23.30 Plant Mixed Bituminous Concrete		Crew	Daily Output	Labor-Hours	Unit	Material	Labor	Equipment	Total	Total Incl O&P
0010	**PLANT MIXED BITUMINOUS CONCRETE**									
0020	Asphaltic concrete plant mix (145 lb. per C.F.)				Ton	70			70	77
0040	Asphaltic concrete less than 300 tons add trucking costs									
0050	See Section 31 23 23.20 for hauling costs									
0200	All weather patching mix, hot				Ton	69			69	76
0250	Cold patch					79.50			79.50	87.50
0300	Berm mix					69			69	76

31 06 Schedules for Earthwork

31 06 60 – Schedules for Special Foundations and Load Bearing Elements

31 06 60.14 Piling Special Costs		Crew	Daily Output	Labor-Hours	Unit	Material	Labor	Equipment	Total	Total Incl O&P
0010	**PILING SPECIAL COSTS**									
0011	Piling special costs, pile caps, see Section 03 30 53.40									
0500	Cutoffs, concrete piles, plain	1 Pile	5.50	1.455	Ea.		67		67	106
0600	With steel thin shell, add		38	.211			9.70		9.70	15.35
0700	Steel pile or "H" piles		19	.421			19.40		19.40	30.50
0800	Wood piles		38	.211			9.70		9.70	15.35
0900	Pre-augering up to 30' deep, average soil, 24" diameter	B-43	180	.267	L.F.		11.10	14.10	25.20	32.50
0920	36" diameter		115	.417			17.35	22	39.35	51
0960	48" diameter		70	.686			28.50	36.50	65	83.50
0980	60" diameter		50	.960			40	51	91	117
1000	Testing, any type piles, test load is twice the design load									
1050	50 ton design load, 100 ton test				Ea.				14,000	15,500
1100	100 ton design load, 200 ton test								20,000	22,000
1150	150 ton design load, 300 ton test								26,000	28,500
1200	200 ton design load, 400 ton test								28,000	31,000
1250	400 ton design load, 800 ton test								32,000	35,000
1500	Wet conditions, soft damp ground									
1600	Requiring mats for crane, add								40%	40%

31 06 Schedules for Earthwork

31 06 60 – Schedules for Special Foundations and Load Bearing Elements

31 06 60.14 Piling Special Costs		Crew	Daily Output	Labor-Hours	Unit	Material	2015 Bare Costs Labor	Equipment	Total	Total Incl O&P
1700	Barge mounted driving rig, add								30%	30%

31 06 60.15 Mobilization		Crew	Daily Output	Labor-Hours	Unit	Material	2015 Bare Costs Labor	Equipment	Total	Total Incl O&P
0010	**MOBILIZATION**									
0020	Set up & remove, air compressor, 600 CFM	A-5	3.30	5.455	Ea.		206	18.60	224.60	335
0100	1200 CFM	"	2.20	8.182			310	28	338	505
0200	Crane, with pile leads and pile hammer, 75 ton	B-19	.60	106			5,075	2,900	7,975	11,100
0300	150 ton	"	.36	177			8,475	4,825	13,300	18,500
0500	Drill rig, for caissons, to 36", minimum	B-43	2	24			995	1,275	2,270	2,925
0600	Up to 84"	"	1	48			2,000	2,550	4,550	5,850
0800	Auxiliary boiler, for steam small	A-5	1.66	10.843			410	37	447	670
0900	Large	"	.83	21.687			820	74	894	1,325
1100	Rule of thumb: complete pile driving set up, small	B-19	.45	142			6,775	3,850	10,625	14,900
1200	Large	"	.27	237	↓		11,300	6,425	17,725	24,700
1500	Mobilization, barge, by tug boat	B-83	25	.640	Mile		28	35	63	81.50

31 11 Clearing and Grubbing

31 11 10 – Clearing and Grubbing Land

31 11 10.10 Clear and Grub Site		Crew	Daily Output	Labor-Hours	Unit	Material	2015 Bare Costs Labor	Equipment	Total	Total Incl O&P
0010	**CLEAR AND GRUB SITE**									
0020	Cut & chip light trees to 6" diam.	B-7	1	48	Acre		1,925	1,675	3,600	4,775
0150	Grub stumps and remove	B-30	2	12			525	1,200	1,725	2,125
0200	Cut & chip medium, trees to 12" diam.	B-7	.70	68.571			2,750	2,375	5,125	6,850
0250	Grub stumps and remove	B-30	1	24			1,050	2,400	3,450	4,225
0300	Cut & chip heavy, trees to 24" diam.	B-7	.30	160			6,425	5,575	12,000	16,000
0350	Grub stumps and remove	B-30	.50	48			2,100	4,825	6,925	8,475
0400	If burning is allowed, deduct cut & chip				↓				40%	40%
3000	Chipping stumps, to 18" deep, 12" diam.	B-86	20	.400	Ea.		20	9.30	29.30	40.50
3040	18" diameter		16	.500			25.50	11.60	37.10	51.50
3080	24" diameter		14	.571			29	13.25	42.25	58.50
3100	30" diameter		12	.667			33.50	15.45	48.95	68
3120	36" diameter		10	.800			40.50	18.55	59.05	82
3160	48" diameter	↓	8	1	↓		50.50	23	73.50	103
5000	Tree thinning, feller buncher, conifer									
5080	Up to 8" diameter	B-93	240	.033	Ea.		1.69	3.49	5.18	6.40
5120	12" diameter		160	.050			2.53	5.25	7.78	9.60
5240	Hardwood, up to 4" diameter		240	.033			1.69	3.49	5.18	6.40
5280	8" diameter		180	.044			2.25	4.66	6.91	8.50
5320	12" diameter	↓	120	.067	↓		3.37	7	10.37	12.80
7000	Tree removal, congested area, aerial lift truck									
7040	8" diameter	B-85	7	5.714	Ea.		233	147	380	515
7080	12" diameter		6	6.667			271	172	443	605
7120	18" diameter		5	8			325	206	531	720
7160	24" diameter		4	10			405	258	663	905
7240	36" diameter		3	13.333			545	345	890	1,200
7280	48" diameter	↓	2	20	↓		815	515	1,330	1,825

31 13 13 – Selective Tree and Shrub Removal

31 13 13.10 Selective Clearing		Crew	Daily Output	Labor-Hours	Unit	Material	2015 Bare Costs Labor	Equipment	Total	Total Incl O&P
0010	**SELECTIVE CLEARING**									
0020	Clearing brush with brush saw	A-1C	.25	32	Acre		1,200	125	1,325	1,975
0100	By hand	1 Clab	.12	66.667			2,500		2,500	3,850
0300	With dozer, ball and chain, light clearing	B-11A	2	8			355	695	1,050	1,300
0400	Medium clearing		1.50	10.667			470	925	1,395	1,750
0500	With dozer and brush rake, light		10	1.600			70.50	139	209.50	261
0550	Medium brush to 4" diameter		8	2			88	173	261	325
0600	Heavy brush to 4" diameter		6.40	2.500			110	217	327	405
1000	Brush mowing, tractor w/rotary mower, no removal									
1020	Light density	B-84	2	4	Acre		202	185	387	510
1040	Medium density		1.50	5.333			270	247	517	680
1080	Heavy density		1	8			405	370	775	1,025

31 13 13.20 Selective Tree Removal		Crew	Daily Output	Labor-Hours	Unit	Material	2015 Bare Costs Labor	Equipment	Total	Total Incl O&P
0010	**SELECTIVE TREE REMOVAL**									
0011	With tractor, large tract, firm									
0020	level terrain, no boulders, less than 12" diam. trees									
0300	300 HP dozer, up to 400 trees/acre, 0 to 25% hardwoods	B-10M	.75	16	Acre		740	2,525	3,265	3,900
0340	25% to 50% hardwoods		.60	20			925	3,150	4,075	4,875
0370	75% to 100% hardwoods		.45	26.667			1,225	4,225	5,450	6,500
0400	500 trees/acre, 0% to 25% hardwoods		.60	20			925	3,150	4,075	4,875
0440	25% to 50% hardwoods		.48	25			1,150	3,950	5,100	6,100
0470	75% to 100% hardwoods		.36	33.333			1,550	5,275	6,825	8,150
0500	More than 600 trees/acre, 0 to 25% hardwoods		.52	23.077			1,075	3,650	4,725	5,650
0540	25% to 50% hardwoods		.42	28.571			1,325	4,525	5,850	7,000
0570	75% to 100% hardwoods		.31	38.710			1,800	6,125	7,925	9,450
0900	Large tract clearing per tree									
1500	300 HP dozer, to 12" diameter, softwood	B-10M	320	.038	Ea.		1.74	5.95	7.69	9.15
1550	Hardwood		100	.120			5.55	18.95	24.50	29.50
1600	12" to 24" diameter, softwood		200	.060			2.78	9.50	12.28	14.70
1650	Hardwood		80	.150			6.95	23.50	30.45	36.50
1700	24" to 36" diameter, softwood		100	.120			5.55	18.95	24.50	29.50
1750	Hardwood		50	.240			11.10	38	49.10	58.50
1800	36" to 48" diameter, softwood		70	.171			7.95	27	34.95	42
1850	Hardwood		35	.343			15.85	54	69.85	83.50
2000	Stump removal on site by hydraulic backhoe, 1-1/2 C.Y.									
2040	4" to 6" diameter	B-17	60	.533	Ea.		22	12.90	34.90	47.50
2050	8" to 12" diameter	B-30	33	.727			31.50	73	104.50	129
2100	14" to 24" diameter		25	.960			42	96.50	138.50	170
2150	26" to 36" diameter		16	1.500			65.50	151	216.50	265
3000	Remove selective trees, on site using chain saws and chipper,									
3050	not incl. stumps, up to 6" diameter	B-7	18	2.667	Ea.		107	93	200	266
3100	8" to 12" diameter		12	4			160	139	299	400
3150	14" to 24" diameter		10	4.800			192	167	359	480
3200	26" to 36" diameter		8	6			241	209	450	600
3300	Machine load, 2 mile haul to dump, 12" diam. tree	A-3B	8	2			90.50	151	241.50	305

31 14 Earth Stripping and Stockpiling

31 14 13 – Soil Stripping and Stockpiling

	31 14 13.23 Topsoil Stripping and Stockpiling	Crew	Daily Output	Labor-Hours	Unit	Material	2015 Bare Costs Labor	Equipment	Total	Total Incl O&P
0010	**TOPSOIL STRIPPING AND STOCKPILING**									
0020	200 H.P. dozer, ideal conditions	B-10B	2300	.005	C.Y.		.24	.60	.84	1.03
0100	Adverse conditions	"	1150	.010			.48	1.21	1.69	2.06
0200	300 H.P. dozer, ideal conditions	B-10M	3000	.004			.19	.63	.82	.98
0300	Adverse conditions	"	1650	.007			.34	1.15	1.49	1.77
0400	400 H.P. dozer, ideal conditions	B-10X	3900	.003			.14	.62	.76	.90
0500	Adverse conditions	"	2000	.006			.28	1.20	1.48	1.74
0600	Clay, dry and soft, 200 H.P. dozer, ideal conditions	B-10B	1600	.008			.35	.87	1.22	1.48
0700	Adverse conditions	"	800	.015			.69	1.73	2.42	2.97
1000	Medium hard, 300 H.P. dozer, ideal conditions	B-10M	2000	.006			.28	.95	1.23	1.46
1100	Adverse conditions	"	1100	.011			.50	1.72	2.22	2.67
1200	Very hard, 400 H.P. dozer, ideal conditions	B-10X	2600	.005			.21	.93	1.14	1.35
1300	Adverse conditions	"	1340	.009			.41	1.80	2.21	2.61
1400	Loam or topsoil, remove and stockpile on site									
1420	6" deep, 200' haul	B-10B	865	.014	C.Y.		.64	1.60	2.24	2.74
1430	300' haul		520	.023			1.07	2.67	3.74	4.56
1440	500' haul		225	.053			2.47	6.15	8.62	10.55
1450	Alternate method: 6" deep, 200' haul		5090	.002	S.Y.		.11	.27	.38	.47
1460	500' haul		1325	.009	"		.42	1.05	1.47	1.79
1500	Loam or topsoil, remove/stockpile on site									
1510	By hand, 6" deep, 50' haul, less than 100 S.Y.	B-1	100	.240	S.Y.		9.20		9.20	14.15
1520	By skid steer, 6" deep, 100' haul, 101-500 S.Y.	B-62	500	.048			1.98	.35	2.33	3.41
1530	100' haul, 501-900 S.Y.	"	900	.027			1.10	.19	1.29	1.89
1540	200' haul, 901-1100 S.Y.	B-63	1000	.040			1.59	.17	1.76	2.63
1550	By dozer, 200' haul, 1101-4000 S.Y.	B-10B	4000	.003			.14	.35	.49	.59

31 22 Grading

31 22 13 – Rough Grading

31 22 13.20 Rough Grading Sites

	31 22 13.20 Rough Grading Sites	Crew	Daily Output	Labor-Hours	Unit	Material	Labor	Equipment	Total	Total Incl O&P
0010	**ROUGH GRADING SITES**									
0100	Rough grade sites 400 S.F. or less	B-1	2	12	Ea.		460		460	705
0120	410-1000 S.F.	"	1	24			920		920	1,425
0130	1100-3000 S.F.	B-62	1.50	16			660	116	776	1,125
0140	3100-5000 S.F.	"	1	24			990	174	1,164	1,725
0150	5100-8000 S.F.	B-63	1	40			1,600	174	1,774	2,650
0160	8100-10000 S.F.	"	.75	53.333			2,125	231	2,356	3,500
0170	8100-10000 S.F.	B-10L	1	12			555	470	1,025	1,375
0200	Rough grade open sites 10000-20000 S.F.	B-11L	1.80	8.889			390	410	800	1,050
0210	20100-25000 S.F.		1.40	11.429			505	525	1,030	1,350
0220	25100-30000 S.F.		1.20	13.333			590	615	1,205	1,575
0230	30100-35000 S.F.		1	16			705	735	1,440	1,875
0240	35100-40000 S.F.		.90	17.778			785	820	1,605	2,100
0250	40100-45000 S.F.		.80	20			880	920	1,800	2,350
0260	45100-50000 S.F.		.72	22.222			980	1,025	2,005	2,625
0270	50100-75000 S.F.		.50	32			1,400	1,475	2,875	3,775
0280	75100-100000 S.F.		.36	44.444			1,950	2,050	4,000	5,250

31 22 Grading

31 22 16 – Fine Grading

31 22 16.10 Finish Grading		Crew	Daily Output	Labor-Hours	Unit	Material	2015 Bare Costs Labor	Equipment	Total	Total Incl O&P
0010	**FINISH GRADING**									
0012	Finish grading area to be paved with grader, small area	B-11L	400	.040	S.Y.		1.76	1.84	3.60	4.71
0100	Large area		2000	.008			.35	.37	.72	.94
1100	Fine grade for slab on grade, machine		1040	.015			.68	.71	1.39	1.82
1150	Hand grading	B-18	700	.034			1.31	.07	1.38	2.09
3500	Finish grading lagoon bottoms	B-11L	4	4	M.S.F.		176	184	360	470

31 23 Excavation and Fill

31 23 16 – Excavation

31 23 16.13 Excavating, Trench

		Crew	Daily Output	Labor-Hours	Unit	Material	Labor	Equipment	Total	Total Incl O&P
0010	**EXCAVATING, TRENCH**									
0011	Or continuous footing									
0020	Common earth with no sheeting or dewatering included									
0050	1′ to 4′ deep, 3/8 C.Y. excavator	B-11C	150	.107	B.C.Y.		4.70	2.43	7.13	9.85
0060	1/2 C.Y. excavator	B-11M	200	.080			3.53	1.96	5.49	7.55
0090	4′ to 6′ deep, 1/2 C.Y. excavator	″	200	.080			3.53	1.96	5.49	7.55
0100	5/8 C.Y. excavator	B-12Q	250	.064			2.86	2.36	5.22	6.95
0110	3/4 C.Y. excavator	B-12F	300	.053			2.38	2.18	4.56	6.05
0300	1/2 C.Y. excavator, truck mounted	B-12J	200	.080			3.57	4.41	7.98	10.30
0500	6′ to 10′ deep, 3/4 C.Y. excavator	B-12F	225	.071			3.18	2.91	6.09	8.05
0510	1 C.Y. excavator	B-12A	400	.040			1.79	2.03	3.82	4.96
0600	1 C.Y. excavator, truck mounted	B-12K	400	.040			1.79	2.51	4.30	5.50
0610	1-1/2 C.Y. excavator	B-12B	600	.027			1.19	1.72	2.91	3.71
0900	10′ to 14′ deep, 3/4 C.Y. excavator	B-12F	200	.080			3.57	3.27	6.84	9.05
0910	1 C.Y. excavator	B-12A	360	.044			1.98	2.26	4.24	5.50
1000	1-1/2 C.Y. excavator	B-12B	540	.030			1.32	1.91	3.23	4.12
1300	14′ to 20′ deep, 1 C.Y. excavator	B-12A	320	.050			2.23	2.54	4.77	6.20
1310	1-1/2 C.Y. excavator	B-12B	480	.033			1.49	2.15	3.64	4.63
1320	2-1/2 C.Y. excavator	B-12S	765	.021			.93	2.10	3.03	3.74
1340	20′ to 24′ deep, 1 C.Y. excavator	B-12A	288	.056			2.48	2.82	5.30	6.90
1342	1-1/2 C.Y. excavator	B-12B	432	.037			1.65	2.38	4.03	5.15
1344	2-1/2 C.Y. excavator	B-12S	685	.023			1.04	2.34	3.38	4.17
1352	4′ to 6′ deep, 1/2 C.Y. excavator w/trench box	B-13H	188	.085			3.80	5.10	8.90	11.45
1354	5/8 C.Y. excavator	″	235	.068			3.04	4.10	7.14	9.15
1356	3/4 C.Y. excavator	B-13G	282	.057			2.53	2.61	5.14	6.75
1362	6′ to 10′ deep, 3/4 C.Y. excavator w/trench box	″	212	.075			3.37	3.47	6.84	8.95
1370	1 C.Y. excavator	B-13D	376	.043			1.90	2.38	4.28	5.50
1371	1-1/2 C.Y. excavator	B-13E	564	.028			1.27	1.97	3.24	4.10
1374	10′ to 14′ deep, 3/4 C.Y. excavator w/trench box	B-13G	188	.085			3.80	3.91	7.71	10.10
1375	1 C.Y. excavator	B-13D	338	.047			2.11	2.64	4.75	6.15
1376	1-1/2 C.Y. excavator	B-13E	508	.032			1.41	2.19	3.60	4.56
1381	14′ to 20′ deep, 1 C.Y. excavator w/trench box	B-13D	301	.053			2.37	2.97	5.34	6.90
1382	1-1/2 C.Y. excavator	B-13E	451	.035			1.58	2.46	4.04	5.15
1383	2-1/2 C.Y. excavator	B-13J	720	.022			.99	2.34	3.33	4.09
1386	20′ to 24′ deep, 1 C.Y. excavator w/trench box	B-13D	271	.059			2.64	3.30	5.94	7.65
1387	1-1/2 C.Y. excavator	B-13E	406	.039			1.76	2.74	4.50	5.70
1388	2-1/2 C.Y. excavator	B-13J	645	.025			1.11	2.61	3.72	4.57
1391	Shoring by S.F./day trench wall protected loose mat., 4′ W	B-6	3200	.008	SF Wall	.53	.31	.11	.95	1.18
1392	Rent shoring per week per S.F. wall protected, loose mat., 4′ W					1.46			1.46	1.60
1395	Hydraulic shoring, S.F. trench wall protected stable mat., 4′ W	2 Clab	2700	.006		.19	.22		.41	.55

31 23 Excavation and Fill

31 23 16 – Excavation

31 23 16.13 Excavating, Trench		Crew	Daily Output	Labor-Hours	Unit	Material	2015 Bare Costs Labor	Equipment	Total	Total Incl O&P
1397	semi-stable material, 4′ W	2 Clab	2400	.007	SF Wall	.27	.25		.52	.69
1398	Rent hydraulic shoring per day/S.F. wall, stable mat., 4′ W					.32			.32	.35
1399	semi-stable material					.40			.40	.44
1400	By hand with pick and shovel 2′ to 6′ deep, light soil	1 Clab	8	1	B.C.Y.		37.50		37.50	58
1500	Heavy soil	″	4	2	″		75		75	116
1700	For tamping backfilled trenches, air tamp, add	A-1G	100	.080	E.C.Y.		3.01	.56	3.57	5.25
1900	Vibrating plate, add	B-18	180	.133	″		5.10	.26	5.36	8.15
2100	Trim sides and bottom for concrete pours, common earth		1500	.016	S.F.		.61	.03	.64	.97
2300	Hardpan		600	.040	″		1.53	.08	1.61	2.45
2400	Pier and spread footing excavation, add to above				B.C.Y.				30%	30%
3000	Backfill trench, F.E. loader, wheel mtd., 1 C.Y. bucket									
3020	Minimal haul	B-10R	400	.030	L.C.Y.		1.39	.75	2.14	2.93
3040	100′ haul	″	200	.060			2.78	1.50	4.28	5.85
3080	2-1/4 C.Y. bucket, minimum haul	B-10T	600	.020			.93	.87	1.80	2.36
3090	100′ haul	″	300	.040			1.85	1.73	3.58	4.73
5020	Loam & Sandy clay with no sheeting or dewatering included									
5050	1′ to 4′ deep, 3/8 C.Y. tractor loader/backhoe	B-11C	162	.099	B.C.Y.		4.36	2.25	6.61	9.10
5060	1/2 C.Y. excavator	B-11M	216	.074			3.27	1.81	5.08	7
5080	4′ to 6′ deep, 1/2 C.Y. excavator	″	216	.074			3.27	1.81	5.08	7
5090	5/8 C.Y. excavator	B-12Q	276	.058			2.59	2.14	4.73	6.30
5100	3/4 C.Y. excavator	B-12F	324	.049			2.20	2.02	4.22	5.60
5130	1/2 C.Y. excavator, truck mounted	B-12J	216	.074			3.31	4.08	7.39	9.55
5140	6′ to 10′ deep, 3/4 C.Y. excavator	B-12F	243	.066			2.94	2.69	5.63	7.45
5150	1 C.Y. excavator	B-12A	432	.037			1.65	1.88	3.53	4.59
5160	1 C.Y. excavator, truck mounted	B-12K	432	.037			1.65	2.32	3.97	5.05
5170	1-1/2 C.Y. excavator	B-12B	648	.025			1.10	1.59	2.69	3.43
5190	10′ to 14′ deep, 3/4 C.Y. excavator	B-12F	216	.074			3.31	3.03	6.34	8.40
5200	1 C.Y. excavator	B-12A	389	.041			1.84	2.09	3.93	5.10
5210	1-1/2 C.Y. excavator	B-12B	583	.027			1.23	1.77	3	3.81
5250	14′ to 20′ deep, 1 C.Y. excavator	B-12A	346	.046			2.06	2.35	4.41	5.75
5260	1-1/2 C.Y. excavator	B-12B	518	.031			1.38	1.99	3.37	4.30
5270	2-1/2 C.Y. excavator	B-12S	826	.019			.86	1.94	2.80	3.46
5300	20′ to 24′ deep, 1 C.Y. excavator	B-12A	311	.051			2.30	2.61	4.91	6.40
5310	1-1/2 C.Y. excavator	B-12B	467	.034			1.53	2.21	3.74	4.76
5320	2-1/2 C.Y. excavator	B-12S	740	.022			.97	2.17	3.14	3.86
5352	4′ to 6′ deep, 1/2 C.Y. excavator w/trench box	B-13H	205	.078			3.48	4.70	8.18	10.45
5354	5/8 C.Y. excavator	″	257	.062			2.78	3.75	6.53	8.35
5356	3/4 C.Y. excavator	B-13G	308	.052			2.32	2.39	4.71	6.15
5362	6′ to 10′ deep, 3/4 C.Y. excavator w/trench box	″	231	.069			3.09	3.19	6.28	8.20
5364	1 C.Y. excavator	B-13D	410	.039			1.74	2.18	3.92	5.05
5366	1-1/2 C.Y. excavator	B-13E	616	.026			1.16	1.80	2.96	3.75
5370	10′ to 14′ deep, 3/4 C.Y. excavator w/trench box	B-13G	205	.078			3.48	3.59	7.07	9.25
5372	1 C.Y. excavator	B-13D	370	.043			1.93	2.42	4.35	5.60
5374	1-1/2 C.Y. excavator	B-13E	554	.029			1.29	2.01	3.30	4.18
5382	14′ to 20′ deep, 1 C.Y. excavator w/trench box	B-13D	329	.049			2.17	2.72	4.89	6.30
5384	1-1/2 C.Y. excavator	B-13E	492	.033			1.45	2.26	3.71	4.70
5386	2-1/2 C.Y. excavator	B-13J	780	.021			.92	2.16	3.08	3.78
5392	20′ to 24′ deep, 1 C.Y. excavator w/trench box	B-13D	295	.054			2.42	3.03	5.45	7.05
5394	1-1/2 C.Y. excavator	B-13E	444	.036			1.61	2.50	4.11	5.20
5396	2-1/2 C.Y. excavator	B-13J	695	.023			1.03	2.43	3.46	4.24
6020	Sand & gravel with no sheeting or dewatering included									
6050	1′ to 4′ deep, 3/8 C.Y. excavator	B-11C	165	.097	B.C.Y.		4.28	2.21	6.49	9
6060	1/2 C.Y. excavator	B-11M	220	.073			3.21	1.78	4.99	6.85

31 23 16.13 Excavating, Trench		Crew	Daily Output	Labor-Hours	Unit	Material	2015 Bare Costs Labor	Equipment	Total	Total Incl O&P
6080	4' to 6' deep, 1/2 C.Y. excavator	B-11M	220	.073	B.C.Y.		3.21	1.78	4.99	6.85
6090	5/8 C.Y. excavator	B-12Q	275	.058			2.60	2.14	4.74	6.35
6100	3/4 C.Y. excavator	B-12F	330	.048			2.16	1.98	4.14	5.50
6130	1/2 C.Y. excavator, truck mounted	B-12J	220	.073			3.25	4.01	7.26	9.35
6140	6' to 10' deep, 3/4 C.Y. excavator	B-12F	248	.065			2.88	2.64	5.52	7.30
6150	1 C.Y. excavator	B-12A	440	.036			1.62	1.85	3.47	4.51
6160	1 C.Y. excavator, truck mounted	B-12K	440	.036			1.62	2.28	3.90	4.98
6170	1-1/2 C.Y. excavator	B-12B	660	.024			1.08	1.56	2.64	3.37
6190	10' to 14' deep, 3/4 C.Y. excavator	B-12F	220	.073			3.25	2.98	6.23	8.25
6200	1 C.Y. excavator	B-12A	396	.040			1.80	2.05	3.85	5
6210	1-1/2 C.Y. excavator	B-12B	594	.027			1.20	1.73	2.93	3.75
6250	14' to 20' deep, 1 C.Y. excavator	B-12A	352	.045			2.03	2.31	4.34	5.65
6260	1-1/2 C.Y. excavator	B-12B	528	.030			1.35	1.95	3.30	4.21
6270	2-1/2 C.Y. excavator	B-12S	840	.019			.85	1.91	2.76	3.40
6300	20' to 24' deep, 1 C.Y. excavator	B-12A	317	.050			2.25	2.56	4.81	6.25
6310	1-1/2 C.Y. excavator	B-12B	475	.034			1.50	2.17	3.67	4.68
6320	2-1/2 C.Y. excavator	B-12S	755	.021			.95	2.13	3.08	3.78
6352	4' to 6' deep, 1/2 C.Y. excavator w/trench box	B-13H	209	.077			3.42	4.61	8.03	10.25
6354	5/8 C.Y. excavator	"	261	.061			2.74	3.69	6.43	8.25
6356	3/4 C.Y. excavator	B-13G	314	.051			2.28	2.34	4.62	6.05
6362	6' to 10' deep, 3/4 C.Y. excavator w/trench box	"	236	.068			3.03	3.12	6.15	8.05
6364	1 C.Y. excavator	B-13D	418	.038			1.71	2.14	3.85	4.96
6366	1-1/2 C.Y. excavator	B-13E	627	.026			1.14	1.77	2.91	3.69
6370	10' to 14' deep, 3/4 C.Y. excavator w/trench box	B-13G	209	.077			3.42	3.52	6.94	9.05
6372	1 C.Y. excavator	B-13D	376	.043			1.90	2.38	4.28	5.50
6374	1-1/2 C.Y. excavator	B-13E	564	.028			1.27	1.97	3.24	4.10
6382	14' to 20' deep, 1 C.Y. excavator w/trench box	B-13D	334	.048			2.14	2.68	4.82	6.20
6384	1-1/2 C.Y. excavator	B-13E	502	.032			1.42	2.21	3.63	4.60
6386	2-1/2 C.Y. excavator	B-13J	790	.020			.90	2.13	3.03	3.73
6392	20' to 24' deep, 1 C.Y. excavator w/trench box	B-13D	301	.053			2.37	2.97	5.34	6.90
6394	1-1/2 C.Y. excavator	B-13E	452	.035			1.58	2.46	4.04	5.10
6396	2-1/2 C.Y. excavator	B-13J	710	.023	▼		1.01	2.38	3.39	4.15
7020	Dense hard clay with no sheeting or dewatering included									
7050	1' to 4' deep, 3/8 C.Y. excavator	B-11C	132	.121	B.C.Y.		5.35	2.76	8.11	11.20
7060	1/2 C.Y. excavator	B-11M	176	.091			4.01	2.23	6.24	8.55
7080	4' to 6' deep, 1/2 C.Y. excavator	"	176	.091			4.01	2.23	6.24	8.55
7090	5/8 C.Y. excavator	B-12Q	220	.073			3.25	2.68	5.93	7.90
7100	3/4 C.Y. excavator	B-12F	264	.061			2.71	2.48	5.19	6.85
7130	1/2 C.Y. excavator, truck mounted	B-12J	176	.091			4.06	5	9.06	11.70
7140	6' to 10' deep, 3/4 C.Y. excavator	B-12F	198	.081			3.61	3.31	6.92	9.15
7150	1 C.Y. excavator	B-12A	352	.045			2.03	2.31	4.34	5.65
7160	1 C.Y. excavator, truck mounted	B-12K	352	.045			2.03	2.85	4.88	6.25
7170	1-1/2 C.Y. excavator	B-12B	528	.030			1.35	1.95	3.30	4.21
7190	10' to 14' deep, 3/4 C.Y. excavator	B-12F	176	.091			4.06	3.72	7.78	10.30
7200	1 C.Y. excavator	B-12A	317	.050			2.25	2.56	4.81	6.25
7210	1-1/2 C.Y. excavator	B-12B	475	.034			1.50	2.17	3.67	4.68
7250	14' to 20' deep, 1 C.Y. excavator	B-12A	282	.057			2.53	2.88	5.41	7.05
7260	1-1/2 C.Y. excavator	B-12B	422	.038			1.69	2.44	4.13	5.25
7270	2-1/2 C.Y. excavator	B-12S	675	.024			1.06	2.38	3.44	4.24
7300	20' to 24' deep, 1 C.Y. excavator	B-12A	254	.063			2.81	3.20	6.01	7.80
7310	1-1/2 C.Y. excavator	B-12B	380	.042			1.88	2.71	4.59	5.85
7320	2-1/2 C.Y. excavator	B-12S	605	.026	▼		1.18	2.65	3.83	4.72

31 23 16.14 Excavating, Utility Trench		Crew	Daily Output	Labor-Hours	Unit	Material	2015 Bare Costs Labor	Equipment	Total	Total Incl O&P
0010	**EXCAVATING, UTILITY TRENCH**									
0011	Common earth									
0050	Trenching with chain trencher, 12 H.P., operator walking									
0100	4″ wide trench, 12″ deep	B-53	800	.010	L.F.		.49	.09	.58	.83
0150	18″ deep		750	.011			.52	.09	.61	.89
0200	24″ deep		700	.011			.56	.10	.66	.95
0300	6″ wide trench, 12″ deep		650	.012			.60	.10	.70	1.03
0350	18″ deep		600	.013			.65	.11	.76	1.10
0400	24″ deep		550	.015			.71	.12	.83	1.21
0450	36″ deep		450	.018			.86	.15	1.01	1.48
0600	8″ wide trench, 12″ deep		475	.017			.82	.14	.96	1.40
0650	18″ deep		400	.020			.97	.17	1.14	1.67
0700	24″ deep		350	.023			1.11	.19	1.30	1.90
0750	36″ deep		300	.027			1.30	.23	1.53	2.22
1000	Backfill by hand including compaction, add									
1050	4″ wide trench, 12″ deep	A-1G	800	.010	L.F.		.38	.07	.45	.66
1100	18″ deep		530	.015			.57	.11	.68	.99
1150	24″ deep		400	.020			.75	.14	.89	1.32
1300	6″ wide trench, 12″ deep		540	.015			.56	.10	.66	.97
1350	18″ deep		405	.020			.74	.14	.88	1.29
1400	24″ deep		270	.030			1.11	.21	1.32	1.94
1450	36″ deep		180	.044			1.67	.31	1.98	2.91
1600	8″ wide trench, 12″ deep		400	.020			.75	.14	.89	1.32
1650	18″ deep		265	.030			1.14	.21	1.35	1.98
1700	24″ deep		200	.040			1.50	.28	1.78	2.62
1750	36″ deep		135	.059			2.23	.42	2.65	3.89
2000	Chain trencher, 40 H.P. operator riding									
2050	6″ wide trench and backfill, 12″ deep	B-54	1200	.007	L.F.		.32	.28	.60	.80
2100	18″ deep		1000	.008			.39	.34	.73	.96
2150	24″ deep		975	.008			.40	.34	.74	.99
2200	36″ deep		900	.009			.43	.37	.80	1.07
2250	48″ deep		750	.011			.52	.45	.97	1.28
2300	60″ deep		650	.012			.60	.52	1.12	1.48
2400	8″ wide trench and backfill, 12″ deep		1000	.008			.39	.34	.73	.96
2450	18″ deep		950	.008			.41	.35	.76	1.01
2500	24″ deep		900	.009			.43	.37	.80	1.07
2550	36″ deep		800	.010			.49	.42	.91	1.20
2600	48″ deep		650	.012			.60	.52	1.12	1.48
2700	12″ wide trench and backfill, 12″ deep		975	.008			.40	.34	.74	.99
2750	18″ deep		860	.009			.45	.39	.84	1.12
2800	24″ deep		800	.010			.49	.42	.91	1.20
2850	36″ deep		725	.011			.54	.46	1	1.32
3000	16″ wide trench and backfill, 12″ deep		835	.010			.47	.40	.87	1.15
3050	18″ deep		750	.011			.52	.45	.97	1.28
3100	24″ deep		700	.011			.56	.48	1.04	1.37
3200	Compaction with vibratory plate, add								35%	35%
5100	Hand excavate and trim for pipe bells after trench excavation									
5200	8″ pipe	1 Clab	155	.052	L.F.		1.94		1.94	2.99
5300	18″ pipe	″	130	.062	″		2.31		2.31	3.56

31 23 16.16 Structural Excavation for Minor Structures		Crew	Daily Output	Labor-Hours	Unit	Material	2015 Bare Costs Labor	Equipment	Total	Total Incl O&P
0010	STRUCTURAL EXCAVATION FOR MINOR STRUCTURES R312316-40									
0015	Hand, pits to 6' deep, sandy soil	1 Clab	8	1	B.C.Y.		37.50		37.50	58
0100	Heavy soil or clay		4	2			75		75	116
0300	Pits 6' to 12' deep, sandy soil		5	1.600			60		60	92.50
0500	Heavy soil or clay		3	2.667			100		100	154
0700	Pits 12' to 18' deep, sandy soil		4	2			75		75	116
0900	Heavy soil or clay		2	4			150		150	231
1100	Hand loading trucks from stock pile, sandy soil		12	.667			25		25	38.50
1300	Heavy soil or clay		8	1			37.50		37.50	58
1500	For wet or muck hand excavation, add to above								50%	50%
6000	Machine excavation, for spread and mat footings, elevator pits,									
6001	and small building foundations									
6030	Common earth, hydraulic backhoe, 1/2 C.Y. bucket	B-12E	55	.291	B.C.Y.		13	8.15	21.15	29
6035	3/4 C.Y. bucket	B-12F	90	.178			7.95	7.30	15.25	20
6040	1 C.Y. bucket	B-12A	108	.148			6.60	7.50	14.10	18.40
6050	1-1/2 C.Y. bucket	B-12B	144	.111			4.96	7.15	12.11	15.40
6060	2 C.Y. bucket	B-12C	200	.080			3.57	5.90	9.47	11.90
6070	Sand and gravel, 3/4 C.Y. bucket	B-12F	100	.160			7.15	6.55	13.70	18.10
6080	1 C.Y. bucket	B-12A	120	.133			5.95	6.75	12.70	16.55
6090	1-1/2 C.Y. bucket	B-12B	160	.100			4.47	6.45	10.92	13.90
6100	2 C.Y. bucket	B-12C	220	.073			3.25	5.35	8.60	10.85
6110	Clay, till, or blasted rock, 3/4 C.Y. bucket	B-12F	80	.200			8.95	8.20	17.15	22.50
6120	1 C.Y. bucket	B-12A	95	.168			7.50	8.55	16.05	21
6130	1-1/2 C.Y. bucket	B-12B	130	.123			5.50	7.90	13.40	17.10
6140	2 C.Y. bucket	B-12C	175	.091			4.08	6.70	10.78	13.65
6230	Sandy clay & loam, hydraulic backhoe, 1/2 C.Y. bucket	B-12E	60	.267			11.90	7.45	19.35	26.50
6235	3/4 C.Y. bucket	B-12F	98	.163			7.30	6.70	14	18.50
6240	1 C.Y. bucket	B-12A	116	.138			6.15	7	13.15	17.10
6250	1-1/2 C.Y. bucket	B-12B	156	.103			4.58	6.60	11.18	14.25
9010	For mobilization or demobilization, see Section 01 54 36.50									
9020	For dewatering, see Section 31 23 19.20									
9022	For larger structures, see Bulk Excavation, Section 31 23 16.42									
9024	For loading onto trucks, add								15%	15%
9026	For hauling, see Section 31 23 23.20									
9030	For sheeting or soldier bms/lagging, see Section 31 52 16.10									
9040	For trench excavation of strip ftgs, see Section 31 23 16.13									

31 23 16.26 Rock Removal		Crew	Daily Output	Labor-Hours	Unit	Material	Labor	Equipment	Total	Total Incl O&P
0010	ROCK REMOVAL R312316-40									
0015	Drilling only rock, 2" hole for rock bolts	B-47	316	.076	L.F.		3.18	5.10	8.28	10.45
0800	2-1/2" hole for pre-splitting		600	.040			1.68	2.68	4.36	5.50
4600	Quarry operations, 2-1/2" to 3-1/2" diameter		715	.034			1.41	2.25	3.66	4.62

31 23 16.30 Drilling and Blasting Rock		Crew	Daily Output	Labor-Hours	Unit	Material	Labor	Equipment	Total	Total Incl O&P
0010	DRILLING AND BLASTING ROCK									
0020	Rock, open face, under 1500 C.Y.	B-47	225	.107	B.C.Y.	3.20	4.47	7.15	14.82	18.20
0100	Over 1500 C.Y.		300	.080		3.20	3.35	5.35	11.90	14.55
0200	Areas where blasting mats are required, under 1500 C.Y.		175	.137		3.20	5.75	9.20	18.15	22.50
0250	Over 1500 C.Y.		250	.096		3.20	4.03	6.45	13.68	16.75
0300	Bulk drilling and blasting, can vary greatly, average								9.65	12.20
0500	Pits, average								25.50	31.50
1300	Deep hole method, up to 1500 C.Y.	B-47	50	.480		3.20	20	32	55.20	70
1400	Over 1500 C.Y.		66	.364		3.20	15.25	24.50	42.95	54
1900	Restricted areas, up to 1500 C.Y.		13	1.846		3.20	77.50	124	204.70	258

31 23 16.30 Drilling and Blasting Rock		Crew	Daily Output	Labor-Hours	Unit	Material	2015 Bare Costs Labor	Equipment	Total	Total Incl O&P
2000	Over 1500 C.Y.	B-47	20	1.200	B.C.Y.	3.20	50.50	80.50	134.20	169
2200	Trenches, up to 1500 C.Y.		22	1.091		9.30	45.50	73	127.80	161
2300	Over 1500 C.Y.		26	.923		9.30	38.50	62	109.80	137
2500	Pier holes, up to 1500 C.Y.		22	1.091		3.20	45.50	73	121.70	154
2600	Over 1500 C.Y.		31	.774		3.20	32.50	52	87.70	110
2800	Boulders under 1/2 C.Y., loaded on truck, no hauling	B-10O	80	.150			6.95	11.95	18.90	23.50
2900	Boulders, drilled, blasted	B-47	100	.240		3.20	10.05	16.10	29.35	36.50
3100	Jackhammer operators with foreman compressor, air tools	B-9	1	40	Day		1,525	233	1,758	2,600
3300	Track drill, compressor, operator and foreman	B-47	1	24	"		1,000	1,600	2,600	3,325
3500	Blasting caps				Ea.	6.25			6.25	6.90
3700	Explosives					.48			.48	.52
3900	Blasting mats, rent, for first day					137			137	151
4000	Per added day					47			47	51.50
4200	Preblast survey for 6 room house, individual lot, minimum	A-6	2.40	6.667			315	23	338	505
4300	Maximum	"	1.35	11.852			555	40.50	595.50	900
4500	City block within zone of influence, minimum	A-8	25200	.001	S.F.		.06		.06	.10
4600	Maximum	"	15100	.002	"		.11		.11	.16

31 23 16.42 Excavating, Bulk Bank Measure		Crew	Daily Output	Labor-Hours	Unit	Material	Labor	Equipment	Total	Total Incl O&P
0010	**EXCAVATING, BULK BANK MEASURE** R312316-40									
0011	Common earth piled									
0020	For loading onto trucks, add								15%	15%
0050	For mobilization and demobilization, see Section 01 54 36.50 R312316-45									
0100	For hauling, see Section 31 23 23.20									
0200	Excavator, hydraulic, crawler mtd., 1 C.Y. cap. = 100 C.Y./hr.	B-12A	800	.020	B.C.Y.		.89	1.02	1.91	2.48
0250	1-1/2 C.Y. cap. = 125 C.Y./hr.	B-12B	1000	.016			.71	1.03	1.74	2.22
0260	2 C.Y. cap. = 165 C.Y./hr.	B-12C	1320	.012			.54	.89	1.43	1.81
0300	3 C.Y. cap. = 260 C.Y./hr.	B-12D	2080	.008			.34	1.17	1.51	1.81
0305	3-1/2 C.Y. cap. = 300 C.Y./hr.	"	2400	.007			.30	1.02	1.32	1.57
0310	Wheel mounted, 1/2 C.Y. cap. = 40 C.Y./hr.	B-12E	320	.050			2.23	1.40	3.63	4.95
0360	3/4 C.Y. cap. = 60 C.Y./hr.	B-12F	480	.033			1.49	1.36	2.85	3.77
0500	Clamshell, 1/2 C.Y. cap. = 20 C.Y./hr.	B-12G	160	.100			4.47	4.43	8.90	11.65
0550	1 C.Y. cap. = 35 C.Y./hr.	B-12H	280	.057			2.55	4.28	6.83	8.60
0950	Dragline, 1/2 C.Y. cap. = 30 C.Y./hr.	B-12I	240	.067			2.98	3.68	6.66	8.60
1000	3/4 C.Y. cap. = 35 C.Y./hr.	"	280	.057			2.55	3.15	5.70	7.35
1050	1-1/2 C.Y. cap. = 65 C.Y./hr.	B-12P	520	.031			1.37	2.29	3.66	4.62
1200	Front end loader, track mtd., 1-1/2 C.Y. cap. = 70 C.Y./hr.	B-10N	560	.021			.99	.93	1.92	2.54
1250	2-1/2 C.Y. cap. = 95 C.Y./hr.	B-10O	760	.016			.73	1.26	1.99	2.49
1300	3 C.Y. cap. = 130 C.Y./hr.	B-10P	1040	.012			.53	1.14	1.67	2.07
1350	5 C.Y. cap. = 160 C.Y./hr.	B-10Q	1280	.009			.43	1.22	1.65	2
1500	Wheel mounted, 3/4 C.Y. cap. = 45 C.Y./hr.	B-10R	360	.033			1.54	.83	2.37	3.26
1550	1-1/2 C.Y. cap. = 80 C.Y./hr.	B-10S	640	.019			.87	.59	1.46	1.97
1600	2-1/4 C.Y. cap. = 100 C.Y./hr.	B-10T	800	.015			.69	.65	1.34	1.77
1650	5 C.Y. cap. = 185 C.Y./hr.	B-10U	1480	.008			.38	.73	1.11	1.37
1800	Hydraulic excavator, truck mtd. 1/2 C.Y. = 30 C.Y./hr.	B-12J	240	.067			2.98	3.68	6.66	8.60
1850	48 inch bucket, 1 C.Y. = 45 C.Y./hr.	B-12K	360	.044			1.98	2.78	4.76	6.10
3700	Shovel, 1/2 C.Y. capacity = 55 C.Y./hr.	B-12L	440	.036			1.62	1.66	3.28	4.30
3750	3/4 C.Y. capacity = 85 C.Y./hr.	B-12M	680	.024			1.05	1.36	2.41	3.10
3800	1 C.Y. capacity = 120 C.Y./hr.	B-12N	960	.017			.74	1.27	2.01	2.54
3850	1-1/2 C.Y. capacity = 160 C.Y./hr.	B-12O	1280	.013			.56	.97	1.53	1.91
3900	3 C.Y. cap. = 250 C.Y./hr.	B-12T	2000	.008			.36	.78	1.14	1.41
4000	For soft soil or sand, deduct								15%	15%
4100	For heavy soil or stiff clay, add								60%	60%

31 23 16.42 Excavating, Bulk Bank Measure

	31 23 16.42 Excavating, Bulk Bank Measure	Crew	Daily Output	Labor-Hours	Unit	Material	2015 Bare Costs Labor	Equipment	Total	Total Incl O&P
4200	For wet excavation with clamshell or dragline, add				B.C.Y.				100%	100%
4250	All other equipment, add								50%	50%
4400	Clamshell in sheeting or cofferdam, minimum	B-12H	160	.100			4.47	7.50	11.97	15.05
4450	Maximum	"	60	.267			11.90	19.95	31.85	40
5000	Excavating, bulk bank measure, sandy clay & loam piled									
5020	For loading onto trucks, add								15%	15%
5100	Excavator, hydraulic, crawler mtd., 1 C.Y. cap. = 120 C.Y./hr.	B-12A	960	.017	B.C.Y.		.74	.85	1.59	2.07
5150	1-1/2 C.Y. cap. = 150 C.Y./hr.	B-12B	1200	.013			.60	.86	1.46	1.85
5300	2 C.Y. cap. = 195 C.Y./hr.	B-12C	1560	.010			.46	.75	1.21	1.53
5400	3 C.Y. cap. = 300 C.Y./hr.	B-12D	2400	.007			.30	1.02	1.32	1.57
5500	3.5 C.Y. cap. = 350 C.Y./hr.	"	2800	.006			.26	.87	1.13	1.35
5610	Wheel mounted, 1/2 C.Y. cap. = 44 C.Y./hr.	B-12E	352	.045			2.03	1.27	3.30	4.50
5660	3/4 C.Y. cap. = 66 C.Y./hr.	B-12F	528	.030			1.35	1.24	2.59	3.42
8000	For hauling excavated material, see Section 31 23 23.20									

31 23 16.46 Excavating, Bulk, Dozer

	31 23 16.46 Excavating, Bulk, Dozer	Crew	Daily Output	Labor-Hours	Unit	Material	Labor	Equipment	Total	Total Incl O&P
0010	**EXCAVATING, BULK, DOZER**									
0011	Open site									
2000	80 H.P., 50' haul, sand & gravel	B-10L	460	.026	B.C.Y.		1.21	1.03	2.24	2.97
2010	Sandy clay & loam		440	.027			1.26	1.07	2.33	3.10
2020	Common earth		400	.030			1.39	1.18	2.57	3.41
2040	Clay		250	.048			2.22	1.89	4.11	5.45
2200	150' haul, sand & gravel		230	.052			2.41	2.05	4.46	5.95
2210	Sandy clay & loam		220	.055			2.52	2.15	4.67	6.20
2220	Common earth		200	.060			2.78	2.36	5.14	6.85
2240	Clay		125	.096			4.44	3.78	8.22	10.90
2400	300' haul, sand & gravel		120	.100			4.63	3.94	8.57	11.40
2410	Sandy clay & loam		115	.104			4.83	4.11	8.94	11.85
2420	Common earth		100	.120			5.55	4.72	10.27	13.65
2440	Clay		65	.185			8.55	7.25	15.80	21
3000	105 H.P., 50' haul, sand & gravel	B-10W	700	.017			.79	.86	1.65	2.16
3010	Sandy clay & loam		680	.018			.82	.89	1.71	2.22
3020	Common earth		610	.020			.91	.99	1.90	2.48
3040	Clay		385	.031			1.44	1.57	3.01	3.92
3200	150' haul, sand & gravel		310	.039			1.79	1.94	3.73	4.87
3210	Sandy clay & loam		300	.040			1.85	2.01	3.86	5.05
3220	Common earth		270	.044			2.06	2.23	4.29	5.60
3240	Clay		170	.071			3.27	3.55	6.82	8.85
3300	300' haul, sand & gravel		140	.086			3.97	4.31	8.28	10.80
3310	Sandy clay & loam		135	.089			4.11	4.46	8.57	11.15
3320	Common earth		120	.100			4.63	5	9.63	12.60
3340	Clay		100	.120			5.55	6.05	11.60	15.10
4000	200 H.P., 50' haul, sand & gravel	B-10B	1400	.009			.40	.99	1.39	1.69
4010	Sandy clay & loam		1360	.009			.41	1.02	1.43	1.74
4020	Common earth		1230	.010			.45	1.13	1.58	1.93
4040	Clay		770	.016			.72	1.80	2.52	3.08
4200	150' haul, sand & gravel		595	.020			.93	2.33	3.26	3.98
4210	Sandy clay & loam		580	.021			.96	2.39	3.35	4.09
4220	Common earth		516	.023			1.08	2.69	3.77	4.60
4240	Clay		325	.037			1.71	4.27	5.98	7.30
4400	300' haul, sand & gravel		310	.039			1.79	4.47	6.26	7.65
4410	Sandy clay & loam		300	.040			1.85	4.62	6.47	7.90
4420	Common earth		270	.044			2.06	5.15	7.21	8.80

31 23 16.46 Excavating, Bulk, Dozer

	31 23 16.46 Excavating, Bulk, Dozer	Crew	Daily Output	Labor-Hours	Unit	Material	2015 Bare Costs Labor	Equipment	Total	Total Incl O&P
4440	Clay	B-10B	170	.071	B.C.Y.		3.27	8.15	11.42	13.90
5000	300 H.P., 50′ haul, sand & gravel	B-10M	1900	.006			.29	1	1.29	1.55
5010	Sandy clay & loam		1850	.006			.30	1.03	1.33	1.59
5020	Common earth		1650	.007			.34	1.15	1.49	1.77
5040	Clay		1025	.012			.54	1.85	2.39	2.87
5200	150′ haul, sand & gravel		920	.013			.60	2.06	2.66	3.19
5210	Sandy clay & loam		895	.013			.62	2.12	2.74	3.27
5220	Common earth		800	.015			.69	2.37	3.06	3.67
5240	Clay		500	.024			1.11	3.79	4.90	5.85
5400	300′ haul, sand & gravel		470	.026			1.18	4.04	5.22	6.25
5410	Sandy clay & loam		455	.026			1.22	4.17	5.39	6.45
5420	Common earth		410	.029			1.35	4.63	5.98	7.15
5440	Clay		250	.048			2.22	7.60	9.82	11.75
5500	460 H.P., 50′ haul, sand & gravel	B-10X	1930	.006			.29	1.25	1.54	1.81
5506	Sandy clay & loam		1880	.006			.30	1.28	1.58	1.86
5510	Common earth		1680	.007			.33	1.43	1.76	2.07
5520	Clay		1050	.011			.53	2.29	2.82	3.33
5530	150′ haul, sand & gravel		1290	.009			.43	1.86	2.29	2.71
5535	Sandy clay & loam		1250	.010			.44	1.92	2.36	2.80
5540	Common earth		1120	.011			.50	2.15	2.65	3.11
5550	Clay		700	.017			.79	3.44	4.23	4.99
5560	300′ haul, sand & gravel		660	.018			.84	3.64	4.48	5.30
5565	Sandy clay & loam		640	.019			.87	3.76	4.63	5.45
5570	Common earth		575	.021			.97	4.18	5.15	6.05
5580	Clay		350	.034			1.59	6.85	8.44	9.95
6000	700 H.P., 50′ haul, sand & gravel	B-10V	3500	.003			.16	1.42	1.58	1.81
6006	Sandy clay & loam		3400	.004			.16	1.46	1.62	1.86
6010	Common earth		3035	.004			.18	1.64	1.82	2.08
6020	Clay		1925	.006			.29	2.59	2.88	3.28
6030	150′ haul, sand & gravel		2025	.006			.27	2.46	2.73	3.13
6035	Sandy clay & loam		1960	.006			.28	2.54	2.82	3.22
6040	Common earth		1750	.007			.32	2.85	3.17	3.61
6050	Clay		1100	.011			.50	4.53	5.03	5.75
6060	300′ haul, sand & gravel		1030	.012			.54	4.83	5.37	6.10
6065	Sandy clay & loam		1005	.012			.55	4.95	5.50	6.30
6070	Common earth		900	.013			.62	5.55	6.17	7.05
6080	Clay		550	.022			1.01	9.05	10.06	11.50

31 23 16.50 Excavation, Bulk, Scrapers

	31 23 16.50 Excavation, Bulk, Scrapers	Crew	Daily Output	Labor-Hours	Unit	Material	Labor	Equipment	Total	Total Incl O&P
0010	**EXCAVATION, BULK, SCRAPERS** R312316-40									
0100	Elev. scraper 11 C.Y., sand & gravel 1500′ haul, 1/4 dozer	B-33F	690	.020	B.C.Y.		.95	2.37	3.32	4.06
0150	3000′ haul		610	.023			1.08	2.68	3.76	4.59
0200	5000′ haul		505	.028			1.30	3.24	4.54	5.55
0300	Common earth, 1500′ haul		600	.023			1.09	2.73	3.82	4.66
0350	3000′ haul		530	.026			1.24	3.09	4.33	5.30
0400	5000′ haul		440	.032			1.49	3.72	5.21	6.35
0410	Sandy clay & loam, 1500′ haul		648	.022			1.01	2.53	3.54	4.32
0420	3000′ haul		572	.024			1.15	2.86	4.01	4.90
0430	5000′ haul		475	.029			1.38	3.45	4.83	5.90
0500	Clay, 1500′ haul		375	.037			1.75	4.37	6.12	7.45
0550	3000′ haul		330	.042			1.99	4.96	6.95	8.50
0600	5000′ haul		275	.051			2.39	5.95	8.34	10.20
1000	Self propelled scraper, 14 C.Y. 1/4 push dozer, sand									

31 23 16 – Excavation

31 23 16.50 Excavation, Bulk, Scrapers		Crew	Daily Output	Labor-Hours	Unit	Material	2015 Bare Costs Labor	Equipment	Total	Total Incl O&P
1050	Sand and gravel, 1500' haul	B-33D	920	.015	B.C.Y.		.71	2.56	3.27	3.91
1100	3000' haul		805	.017			.82	2.93	3.75	4.46
1200	5000' haul		645	.022			1.02	3.66	4.68	5.55
1300	Common earth, 1500' haul		800	.018			.82	2.95	3.77	4.49
1350	3000' haul		700	.020			.94	3.37	4.31	5.15
1400	5000' haul		560	.025			1.17	4.21	5.38	6.40
1420	Sandy clay & loam, 1500' haul		864	.016			.76	2.73	3.49	4.16
1430	3000' haul		786	.018			.84	3	3.84	4.57
1440	5000' haul		605	.023			1.09	3.90	4.99	5.95
1500	Clay, 1500' haul		500	.028			1.31	4.72	6.03	7.20
1550	3000' haul		440	.032			1.49	5.35	6.84	8.15
1600	5000' haul		350	.040			1.88	6.75	8.63	10.25
2000	21 C.Y., 1/4 push dozer, sand & gravel, 1500' haul	B-33E	1180	.012			.56	2.68	3.24	3.80
2100	3000' haul		910	.015			.72	3.47	4.19	4.92
2200	5000' haul		750	.019			.88	4.22	5.10	5.95
2300	Common earth, 1500' haul		1030	.014			.64	3.07	3.71	4.35
2350	3000' haul		790	.018			.83	4	4.83	5.65
2400	5000' haul		650	.022			1.01	4.87	5.88	6.90
2420	Sandy clay & loam, 1500' haul		1112	.013			.59	2.84	3.43	4.03
2430	3000' haul		854	.016			.77	3.70	4.47	5.25
2440	5000' haul		702	.020			.94	4.50	5.44	6.35
2500	Clay, 1500' haul		645	.022			1.02	4.90	5.92	6.95
2550	3000' haul		495	.028			1.33	6.40	7.73	9.05
2600	5000' haul		405	.035			1.62	7.80	9.42	11.05
2700	Towed, 10 C.Y., 1/4 push dozer, sand & gravel, 1500' haul	B-33B	560	.025			1.17	4.50	5.67	6.75
2720	3000' haul		450	.031			1.46	5.60	7.06	8.35
2730	5000' haul		365	.038			1.80	6.90	8.70	10.35
2750	Common earth, 1500' haul		420	.033			1.56	6	7.56	9
2770	3000' haul		400	.035			1.64	6.30	7.94	9.40
2780	5000' haul		310	.045			2.12	8.10	10.22	12.15
2785	Sandy clay & Loam, 1500' haul		454	.031			1.45	5.55	7	8.30
2790	3000' haul		432	.032			1.52	5.85	7.37	8.70
2795	5000' haul		340	.041			1.93	7.40	9.33	11.10
2800	Clay, 1500' haul		315	.044			2.08	8	10.08	11.95
2820	3000' haul		300	.047			2.19	8.40	10.59	12.60
2840	5000' haul		225	.062			2.92	11.20	14.12	16.75
2900	15 C.Y., 1/4 push dozer, sand & gravel, 1500' haul	B-33C	800	.018			.82	3.17	3.99	4.74
2920	3000' haul		640	.022			1.03	3.96	4.99	5.90
2940	5000' haul		520	.027			1.26	4.88	6.14	7.25
2960	Common earth, 1500' haul		600	.023			1.09	4.23	5.32	6.30
2980	3000' haul		560	.025			1.17	4.53	5.70	6.75
3000	5000' haul		440	.032			1.49	5.75	7.24	8.60
3005	Sandy clay & Loam, 1500' haul		648	.022			1.01	3.91	4.92	5.85
3010	3000' haul		605	.023			1.09	4.19	5.28	6.25
3015	5000' haul		475	.029			1.38	5.35	6.73	7.95
3020	Clay, 1500' haul		450	.031			1.46	5.65	7.11	8.40
3040	3000' haul		420	.033			1.56	6.05	7.61	9.05
3060	5000' haul		320	.044			2.05	7.95	10	11.80

31 23 Excavation and Fill

31 23 19 – Dewatering

31 23 19.20 Dewatering Systems		Crew	Daily Output	Labor-Hours	Unit	Material	2015 Bare Costs Labor	Equipment	Total	Total Incl O&P
0010	**DEWATERING SYSTEMS**									
0020	Excavate drainage trench, 2′ wide, 2′ deep	B-11C	90	.178	C.Y.		7.85	4.05	11.90	16.40
0100	2′ wide, 3′ deep, with backhoe loader	″	135	.119			5.25	2.70	7.95	10.95
0200	Excavate sump pits by hand, light soil	1 Clab	7.10	1.127			42.50		42.50	65
0300	Heavy soil	″	3.50	2.286			86		86	132
0500	Pumping 8 hr., attended 2 hrs. per day, including 20 L.F.									
0550	of suction hose & 100 L.F. discharge hose									
0600	2″ diaphragm pump used for 8 hours	B-10H	4	3	Day		139	18.70	157.70	232
0650	4″ diaphragm pump used for 8 hours	B-10I	4	3			139	30.50	169.50	245
0800	8 hrs. attended, 2″ diaphragm pump	B-10H	1	12			555	75	630	925
0900	3″ centrifugal pump	B-10J	1	12			555	84	639	940
1000	4″ diaphragm pump	B-10I	1	12			555	122	677	980
1100	6″ centrifugal pump	B-10K	1	12			555	370	925	1,250
1300	CMP, incl. excavation 3′ deep, 12″ diameter	B-6	115	.209	L.F.	11.05	8.60	3.17	22.82	29
1400	18″ diameter		100	.240	″	16.70	9.90	3.64	30.24	37.50
1600	Sump hole construction, incl. excavation and gravel, pit		1250	.019	C.F.	1.09	.79	.29	2.17	2.73
1700	With 12″ gravel collar, 12″ pipe, corrugated, 16 ga.		70	.343	L.F.	21.50	14.15	5.20	40.85	50.50
1800	15″ pipe, corrugated, 16 ga.		55	.436		28	18	6.60	52.60	65.50
1900	18″ pipe, corrugated, 16 ga.		50	.480		32	19.80	7.30	59.10	74
2000	24″ pipe, corrugated, 14 ga.		40	.600		38.50	25	9.10	72.60	90.50
2200	Wood lining, up to 4′ x 4′, add		300	.080	SFCA	16.25	3.30	1.21	20.76	24.50
9950	See Section 31 23 19.40 for wellpoints									
9960	See Section 31 23 19.30 for deep well systems									

31 23 19.30 Wells

		Crew	Daily Output	Labor-Hours	Unit	Material	Labor	Equipment	Total	Total Incl O&P
0010	**WELLS**									
0011	For dewatering 10′ to 20′ deep, 2′ diameter									
0020	with steel casing, minimum	B-6	165	.145	V.L.F.	38	6	2.21	46.21	53
0050	Average		98	.245		43	10.10	3.72	56.82	66.50
0100	Maximum		49	.490		47.50	20	7.45	74.95	91.50
0300	For dewatering pumps see 01 54 33 in Reference Section									
0500	For domestic water wells, see Section 33 21 13.10									

31 23 19.40 Wellpoints

		Crew	Daily Output	Labor-Hours	Unit	Material	Labor	Equipment	Total	Total Incl O&P
0010	**WELLPOINTS** R312319-90									
0011	For equipment rental, see 01 54 33 in Reference Section									
0100	Installation and removal of single stage system									
0110	Labor only, .75 labor-hours per L.F.	1 Clab	10.70	.748	LF Hdr		28		28	43.50
0200	2.0 labor-hours per L.F.	″	4	2	″		75		75	116
0400	Pump operation, 4 @ 6 hr. shifts									
0410	Per 24 hour day	4 Eqlt	1.27	25.197	Day		1,225		1,225	1,850
0500	Per 168 hour week, 160 hr. straight, 8 hr. double time		.18	177	Week		8,650		8,650	13,100
0550	Per 4.3 week month		.04	800	Month		38,900		38,900	59,000
0600	Complete installation, operation, equipment rental, fuel &									
0610	removal of system with 2″ wellpoints 5′ O.C.									
0700	100′ long header, 6″ diameter, first month	4 Eqlt	3.23	9.907	LF Hdr	159	480		639	905
0800	Thereafter, per month		4.13	7.748		127	375		502	710
1000	200′ long header, 8″ diameter, first month		6	5.333		145	259		404	555
1100	Thereafter, per month		8.39	3.814		71.50	185		256.50	360
1300	500′ long header, 8″ diameter, first month		10.63	3.010		55.50	146		201.50	283
1400	Thereafter, per month		20.91	1.530		39.50	74.50		114	157
1600	1,000′ long header, 10″ diameter, first month		11.62	2.754		47.50	134		181.50	256
1700	Thereafter, per month		41.81	.765		24	37		61	82.50
1900	Note: above figures include pumping 168 hrs. per week									

31 23 19.40 Wellpoints		Crew	Daily Output	Labor-Hours	Unit	Material	2015 Bare Costs Labor	Equipment	Total	Total Incl O&P
1910	and include the pump operator and one stand-by pump.									

31 23 23 – Fill

31 23 23.13 Backfill

		Crew	Daily Output	Labor-Hours	Unit	Material	Labor	Equipment	Total	Total Incl O&P
0010	**BACKFILL** R312323-30									
0015	By hand, no compaction, light soil	1 Clab	14	.571	L.C.Y.		21.50		21.50	33
0100	Heavy soil		11	.727	"		27.50		27.50	42
0300	Compaction in 6" layers, hand tamp, add to above		20.60	.388	E.C.Y.		14.60		14.60	22.50
0400	Roller compaction operator walking, add	B-10A	100	.120			5.55	1.80	7.35	10.45
0500	Air tamp, add	B-9D	190	.211			8	1.40	9.40	13.85
0600	Vibrating plate, add	A-1D	60	.133			5	.60	5.60	8.35
0800	Compaction in 12" layers, hand tamp, add to above	1 Clab	34	.235			8.85		8.85	13.60
0900	Roller compaction operator walking, add	B-10A	150	.080			3.70	1.20	4.90	6.95
1000	Air tamp, add	B-9	285	.140			5.35	.82	6.17	9.10
1100	Vibrating plate, add	A-1E	90	.089			3.34	.52	3.86	5.70
1300	Dozer backfilling, bulk, up to 300' haul, no compaction	B-10B	1200	.010	L.C.Y.		.46	1.16	1.62	1.97
1400	Air tamped, add	B-11B	80	.200	E.C.Y.		8.60	3.77	12.37	17.30
1600	Compacting backfill, 6" to 12" lifts, vibrating roller	B-10C	800	.015			.69	2.23	2.92	3.51
1700	Sheepsfoot roller	B-10D	750	.016			.74	2.42	3.16	3.79
1900	Dozer backfilling, trench, up to 300' haul, no compaction	B-10B	900	.013	L.C.Y.		.62	1.54	2.16	2.63
2000	Air tamped, add	B-11B	80	.200	E.C.Y.		8.60	3.77	12.37	17.30
2200	Compacting backfill, 6" to 12" lifts, vibrating roller	B-10C	700	.017			.79	2.55	3.34	4.01
2300	Sheepsfoot roller	B-10D	650	.018			.85	2.79	3.64	4.37
2350	Spreading in 8" layers, small dozer	B-10B	1060	.011	L.C.Y.		.52	1.31	1.83	2.24

31 23 23.14 Backfill, Structural

		Crew	Daily Output	Labor-Hours	Unit	Material	Labor	Equipment	Total	Total Incl O&P
0010	**BACKFILL, STRUCTURAL**									
0011	Dozer or F.E. loader									
0020	From existing stockpile, no compaction									
2000	80 H.P., 50' haul, sand & gravel	B-10L	1100	.011	L.C.Y.		.50	.43	.93	1.24
2010	Sandy clay & loam		1070	.011			.52	.44	.96	1.28
2020	Common earth		975	.012			.57	.48	1.05	1.40
2040	Clay		850	.014			.65	.56	1.21	1.60
2400	300' haul, sand & gravel		370	.032			1.50	1.28	2.78	3.68
2410	Sandy clay & loam		360	.033			1.54	1.31	2.85	3.79
2420	Common earth		330	.036			1.68	1.43	3.11	4.13
2440	Clay		290	.041			1.91	1.63	3.54	4.71
3000	105 H.P., 50' haul, sand & gravel	B-10W	1350	.009			.41	.45	.86	1.12
3010	Sandy clay & loam		1325	.009			.42	.46	.88	1.14
3020	Common earth		1225	.010			.45	.49	.94	1.23
3040	Clay		1100	.011			.50	.55	1.05	1.37
3300	300' haul, sand & gravel		465	.026			1.19	1.30	2.49	3.25
3310	Sandy clay & loam		455	.026			1.22	1.32	2.54	3.32
3320	Common earth		415	.029			1.34	1.45	2.79	3.64
3340	Clay		370	.032			1.50	1.63	3.13	4.07
4000	200 H.P., 50' haul, sand & gravel	B-10B	2500	.005			.22	.55	.77	.95
4010	Sandy clay & loam		2435	.005			.23	.57	.80	.98
4020	Common earth		2200	.005			.25	.63	.88	1.07
4040	Clay		1950	.006			.28	.71	.99	1.21
4400	300' haul, sand & gravel		805	.015			.69	1.72	2.41	2.95
4410	Sandy clay & loam		790	.015			.70	1.76	2.46	3
4420	Common earth		735	.016			.76	1.89	2.65	3.23
4440	Clay		660	.018			.84	2.10	2.94	3.59
5000	300 H.P., 50' haul, sand & gravel	B-10M	3170	.004			.18	.60	.78	.93

31 23 23.14 Backfill, Structural

	31 23 23.14 Backfill, Structural	Crew	Daily Output	Labor-Hours	Unit	Material	2015 Bare Costs Labor	Equipment	Total	Total Incl O&P
5010	Sandy clay & loam	B-10M	3110	.004	L.C.Y.		.18	.61	.79	.94
5020	Common earth		2900	.004			.19	.65	.84	1.01
5040	Clay		2700	.004			.21	.70	.91	1.08
5400	300′ haul, sand & gravel		1500	.008			.37	1.26	1.63	1.95
5410	Sandy clay & loam		1470	.008			.38	1.29	1.67	1.99
5420	Common earth		1350	.009			.41	1.41	1.82	2.18
5440	Clay		1225	.010			.45	1.55	2	2.39
6010	For trench backfill, see Section 31 23 16.13 and 31 23 16.14									
6100	For compaction, see Section 31 23 23.24									

31 23 23.16 Fill By Borrow and Utility Bedding

	31 23 23.16 Fill By Borrow and Utility Bedding	Crew	Daily Output	Labor-Hours	Unit	Material	Labor	Equipment	Total	Total Incl O&P
0010	**FILL BY BORROW AND UTILITY BEDDING**									
0015	Fill by borrow, load, 1 mile haul, spread with dozer									
0020	for embankments	B-15	1200	.023	L.C.Y.	12.40	1	2.31	15.71	17.70
0035	Select fill for shoulders & embankments	″	1200	.023	″	21	1	2.31	24.31	27
0040	Fill,for hauling over 1 mile,add to above per C.Y., see Section 31 23 23.20				Mile				1.41	1.73
0049	Utility bedding, for pipe & conduit, not incl. compaction									
0050	Crushed or screened bank run gravel	B-6	150	.160	L.C.Y.	25.50	6.60	2.43	34.53	41
0100	Crushed stone 3/4″ to 1/2″		150	.160		23.50	6.60	2.43	32.53	39
0200	Sand, dead or bank		150	.160		17.85	6.60	2.43	26.88	32.50
0500	Compacting bedding in trench	A-1D	90	.089	E.C.Y.		3.34	.40	3.74	5.60
0600	If material source exceeds 2 miles, add for extra mileage.									
0610	See Section 31 23 23.20 for hauling mileage add.									

31 23 23.17 General Fill

	31 23 23.17 General Fill	Crew	Daily Output	Labor-Hours	Unit	Material	Labor	Equipment	Total	Total Incl O&P
0010	**GENERAL FILL**									
0011	Spread dumped material, no compaction									
0020	By dozer, no compaction	B-10B	1000	.012	L.C.Y.		.56	1.39	1.95	2.38
0100	By hand	1 Clab	12	.667	″		25		25	38.50
0500	Gravel fill, compacted, under floor slabs, 4″ deep	B-37	10000	.005	S.F.	.42	.19	.02	.63	.77
0600	6″ deep		8600	.006		.63	.22	.02	.87	1.05
0700	9″ deep		7200	.007		1.05	.27	.02	1.34	1.58
0800	12″ deep		6000	.008		1.47	.32	.03	1.82	2.14
1000	Alternate pricing method, 4″ deep		120	.400	E.C.Y.	31.50	15.90	1.31	48.71	60.50
1100	6″ deep		160	.300		31.50	11.95	.98	44.43	54
1200	9″ deep		200	.240		31.50	9.55	.79	41.84	50
1300	12″ deep		220	.218		31.50	8.70	.72	40.92	48.50
1500	For fill under exterior paving, see Section 32 11 23.23									

31 23 23.20 Hauling

	31 23 23.20 Hauling	Crew	Daily Output	Labor-Hours	Unit	Material	Labor	Equipment	Total	Total Incl O&P
0010	**HAULING**									
0011	Excavated or borrow, loose cubic yards									
0012	no loading equipment, including hauling,waiting, loading/dumping									
0013	time per cycle (wait, load, travel, unload or dump & return)									
0014	8 C.Y. truck, 15 MPH ave, cycle 0.5 miles, 10 min. wait/Ld./Uld.	B-34A	320	.025	L.C.Y.		1	1.29	2.29	2.92
0016	cycle 1 mile		272	.029			1.18	1.51	2.69	3.44
0018	cycle 2 miles		208	.038			1.54	1.98	3.52	4.50
0020	cycle 4 miles		144	.056			2.23	2.86	5.09	6.50
0022	cycle 6 miles		112	.071			2.86	3.67	6.53	8.35
0024	cycle 8 miles		88	.091			3.64	4.67	8.31	10.65
0026	20 MPH ave,cycle 0.5 mile		336	.024			.95	1.22	2.17	2.79
0028	cycle 1 mile		296	.027			1.08	1.39	2.47	3.17
0030	cycle 2 miles		240	.033			1.33	1.71	3.04	3.90
0032	cycle 4 miles		176	.045			1.82	2.34	4.16	5.30
0034	cycle 6 miles		136	.059			2.36	3.02	5.38	6.90

31 23 23.20 Hauling		Crew	Daily Output	Labor-Hours	Unit	Material	2015 Bare Costs Labor	Equipment	Total	Total Incl O&P
0036	cycle 8 miles	B-34A	112	.071	L.C.Y.		2.86	3.67	6.53	8.35
0044	25 MPH ave, cycle 4 miles		192	.042			1.67	2.14	3.81	4.88
0046	cycle 6 miles		160	.050			2	2.57	4.57	5.85
0048	cycle 8 miles		128	.063			2.50	3.21	5.71	7.30
0050	30 MPH ave, cycle 4 miles		216	.037			1.48	1.90	3.38	4.33
0052	cycle 6 miles		176	.045			1.82	2.34	4.16	5.30
0054	cycle 8 miles		144	.056			2.23	2.86	5.09	6.50
0114	15 MPH ave, cycle 0.5 mile, 15 min. wait/Ld./Uld.		224	.036			1.43	1.84	3.27	4.18
0116	cycle 1 mile		200	.040			1.60	2.06	3.66	4.68
0118	cycle 2 miles		168	.048			1.91	2.45	4.36	5.55
0120	cycle 4 miles		120	.067			2.67	3.43	6.10	7.80
0122	cycle 6 miles		96	.083			3.34	4.28	7.62	9.75
0124	cycle 8 miles		80	.100			4.01	5.15	9.16	11.70
0126	20 MPH ave, cycle 0.5 mile		232	.034			1.38	1.77	3.15	4.04
0128	cycle 1 mile		208	.038			1.54	1.98	3.52	4.50
0130	cycle 2 miles		184	.043			1.74	2.23	3.97	5.10
0132	cycle 4 miles		144	.056			2.23	2.86	5.09	6.50
0134	cycle 6 miles		112	.071			2.86	3.67	6.53	8.35
0136	cycle 8 miles		96	.083			3.34	4.28	7.62	9.75
0144	25 MPH ave, cycle 4 miles		152	.053			2.11	2.71	4.82	6.15
0146	cycle 6 miles		128	.063			2.50	3.21	5.71	7.30
0148	cycle 8 miles		112	.071			2.86	3.67	6.53	8.35
0150	30 MPH ave, cycle 4 miles		168	.048			1.91	2.45	4.36	5.55
0152	cycle 6 miles		144	.056			2.23	2.86	5.09	6.50
0154	cycle 8 miles		120	.067			2.67	3.43	6.10	7.80
0214	15 MPH ave, cycle 0.5 mile, 20 min wait/Ld./Uld.		176	.045			1.82	2.34	4.16	5.30
0216	cycle 1 mile		160	.050			2	2.57	4.57	5.85
0218	cycle 2 miles		136	.059			2.36	3.02	5.38	6.90
0220	cycle 4 miles		104	.077			3.08	3.95	7.03	9
0222	cycle 6 miles		88	.091			3.64	4.67	8.31	10.65
0224	cycle 8 miles		72	.111			4.45	5.70	10.15	13.05
0226	20 MPH ave, cycle 0.5 mile		176	.045			1.82	2.34	4.16	5.30
0228	cycle 1 mile		168	.048			1.91	2.45	4.36	5.55
0230	cycle 2 miles		144	.056			2.23	2.86	5.09	6.50
0232	cycle 4 miles		120	.067			2.67	3.43	6.10	7.80
0234	cycle 6 miles		96	.083			3.34	4.28	7.62	9.75
0236	cycle 8 miles		88	.091			3.64	4.67	8.31	10.65
0244	25 MPH ave, cycle 4 miles		128	.063			2.50	3.21	5.71	7.30
0246	cycle 6 miles		112	.071			2.86	3.67	6.53	8.35
0248	cycle 8 miles		96	.083			3.34	4.28	7.62	9.75
0250	30 MPH ave, cycle 4 miles		136	.059			2.36	3.02	5.38	6.90
0252	cycle 6 miles		120	.067			2.67	3.43	6.10	7.80
0254	cycle 8 miles		104	.077			3.08	3.95	7.03	9
0314	15 MPH ave, cycle 0.5 mile, 25 min wait/Ld./Uld.		144	.056			2.23	2.86	5.09	6.50
0316	cycle 1 mile		128	.063			2.50	3.21	5.71	7.30
0318	cycle 2 miles		112	.071			2.86	3.67	6.53	8.35
0320	cycle 4 miles		96	.083			3.34	4.28	7.62	9.75
0322	cycle 6 miles		80	.100			4.01	5.15	9.16	11.70
0324	cycle 8 miles		64	.125			5	6.45	11.45	14.60
0326	20 MPH ave, cycle 0.5 mile		144	.056			2.23	2.86	5.09	6.50
0328	cycle 1 mile		136	.059			2.36	3.02	5.38	6.90
0330	cycle 2 miles		120	.067			2.67	3.43	6.10	7.80
0332	cycle 4 miles		104	.077			3.08	3.95	7.03	9

31 23 23 – Fill

31 23 23.20 Hauling		Crew	Daily Output	Labor-Hours	Unit	Material	2015 Bare Costs Labor	Equipment	Total	Total Incl O&P
0334	cycle 6 miles	B-34A	88	.091	L.C.Y.		3.64	4.67	8.31	10.65
0336	cycle 8 miles		80	.100			4.01	5.15	9.16	11.70
0344	25 MPH ave, cycle 4 miles		112	.071			2.86	3.67	6.53	8.35
0346	cycle 6 miles		96	.083			3.34	4.28	7.62	9.75
0348	cycle 8 miles		88	.091			3.64	4.67	8.31	10.65
0350	30 MPH ave, cycle 4 miles		112	.071			2.86	3.67	6.53	8.35
0352	cycle 6 miles		104	.077			3.08	3.95	7.03	9
0354	cycle 8 miles		96	.083			3.34	4.28	7.62	9.75
0414	15 MPH ave, cycle 0.5 mile, 30 min wait/Ld./Uld.		120	.067			2.67	3.43	6.10	7.80
0416	cycle 1 mile		112	.071			2.86	3.67	6.53	8.35
0418	cycle 2 miles		96	.083			3.34	4.28	7.62	9.75
0420	cycle 4 miles		80	.100			4.01	5.15	9.16	11.70
0422	cycle 6 miles		72	.111			4.45	5.70	10.15	13.05
0424	cycle 8 miles		64	.125			5	6.45	11.45	14.60
0426	20 MPH ave, cycle 0.5 mile		120	.067			2.67	3.43	6.10	7.80
0428	cycle 1 mile		112	.071			2.86	3.67	6.53	8.35
0430	cycle 2 miles		104	.077			3.08	3.95	7.03	9
0432	cycle 4 miles		88	.091			3.64	4.67	8.31	10.65
0434	cycle 6 miles		80	.100			4.01	5.15	9.16	11.70
0436	cycle 8 miles		72	.111			4.45	5.70	10.15	13.05
0444	25 MPH ave, cycle 4 miles		96	.083			3.34	4.28	7.62	9.75
0446	cycle 6 miles		88	.091			3.64	4.67	8.31	10.65
0448	cycle 8 miles		80	.100			4.01	5.15	9.16	11.70
0450	30 MPH ave, cycle 4 miles		96	.083			3.34	4.28	7.62	9.75
0452	cycle 6 miles		88	.091			3.64	4.67	8.31	10.65
0454	cycle 8 miles		80	.100			4.01	5.15	9.16	11.70
0514	15 MPH ave, cycle 0.5 mile, 35 min wait/Ld./Uld.		104	.077			3.08	3.95	7.03	9
0516	cycle 1 mile		96	.083			3.34	4.28	7.62	9.75
0518	cycle 2 miles		88	.091			3.64	4.67	8.31	10.65
0520	cycle 4 miles		72	.111			4.45	5.70	10.15	13.05
0522	cycle 6 miles		64	.125			5	6.45	11.45	14.60
0524	cycle 8 miles		56	.143			5.70	7.35	13.05	16.75
0526	20 MPH ave, cycle 0.5 mile		104	.077			3.08	3.95	7.03	9
0528	cycle 1 mile		96	.083			3.34	4.28	7.62	9.75
0530	cycle 2 miles		96	.083			3.34	4.28	7.62	9.75
0532	cycle 4 miles		80	.100			4.01	5.15	9.16	11.70
0534	cycle 6 miles		72	.111			4.45	5.70	10.15	13.05
0536	cycle 8 miles		64	.125			5	6.45	11.45	14.60
0544	25 MPH ave, cycle 4 miles		88	.091			3.64	4.67	8.31	10.65
0546	cycle 6 miles		80	.100			4.01	5.15	9.16	11.70
0548	cycle 8 miles		72	.111			4.45	5.70	10.15	13.05
0550	30 MPH ave, cycle 4 miles		88	.091			3.64	4.67	8.31	10.65
0552	cycle 6 miles		80	.100			4.01	5.15	9.16	11.70
0554	cycle 8 miles		72	.111			4.45	5.70	10.15	13.05
1014	12 C.Y. truck, cycle 0.5 mile, 15 MPH ave, 15 min. wait/Ld./Uld.	B-34B	336	.024			.95	2.06	3.01	3.70
1016	cycle 1 mile		300	.027			1.07	2.30	3.37	4.14
1018	cycle 2 miles		252	.032			1.27	2.74	4.01	4.94
1020	cycle 4 miles		180	.044			1.78	3.84	5.62	6.90
1022	cycle 6 miles		144	.056			2.23	4.80	7.03	8.65
1024	cycle 8 miles		120	.067			2.67	5.75	8.42	10.40
1025	cycle 10 miles		96	.083			3.34	7.20	10.54	12.95
1026	20 MPH ave, cycle 0.5 mile		348	.023			.92	1.99	2.91	3.57
1028	cycle 1 mile		312	.026			1.03	2.21	3.24	3.99

31 23 23.20 Hauling		Crew	Daily Output	Labor-Hours	Unit	Material	2015 Bare Costs Labor	Equipment	Total	Total Incl O&P
1030	cycle 2 miles	B-34B	276	.029	L.C.Y.		1.16	2.50	3.66	4.51
1032	cycle 4 miles		216	.037			1.48	3.20	4.68	5.75
1034	cycle 6 miles		168	.048			1.91	4.11	6.02	7.40
1036	cycle 8 miles		144	.056			2.23	4.80	7.03	8.65
1038	cycle 10 miles		120	.067			2.67	5.75	8.42	10.40
1040	25 MPH ave, cycle 4 miles		228	.035			1.41	3.03	4.44	5.45
1042	cycle 6 miles		192	.042			1.67	3.60	5.27	6.50
1044	cycle 8 miles		168	.048			1.91	4.11	6.02	7.40
1046	cycle 10 miles		144	.056			2.23	4.80	7.03	8.65
1050	30 MPH ave, cycle 4 miles		252	.032			1.27	2.74	4.01	4.94
1052	cycle 6 miles		216	.037			1.48	3.20	4.68	5.75
1054	cycle 8 miles		180	.044			1.78	3.84	5.62	6.90
1056	cycle 10 miles		156	.051			2.05	4.43	6.48	8
1060	35 MPH ave, cycle 4 miles		264	.030			1.21	2.62	3.83	4.71
1062	cycle 6 miles		228	.035			1.41	3.03	4.44	5.45
1064	cycle 8 miles		204	.039			1.57	3.39	4.96	6.10
1066	cycle 10 miles		180	.044			1.78	3.84	5.62	6.90
1068	cycle 20 miles		120	.067			2.67	5.75	8.42	10.40
1069	cycle 30 miles		84	.095			3.81	8.25	12.06	14.80
1070	cycle 40 miles		72	.111			4.45	9.60	14.05	17.30
1072	40 MPH ave, cycle 6 miles		240	.033			1.33	2.88	4.21	5.20
1074	cycle 8 miles		216	.037			1.48	3.20	4.68	5.75
1076	cycle 10 miles		192	.042			1.67	3.60	5.27	6.50
1078	cycle 20 miles		120	.067			2.67	5.75	8.42	10.40
1080	cycle 30 miles		96	.083			3.34	7.20	10.54	12.95
1082	cycle 40 miles		72	.111			4.45	9.60	14.05	17.30
1084	cycle 50 miles		60	.133			5.35	11.50	16.85	20.50
1094	45 MPH ave, cycle 8 miles		216	.037			1.48	3.20	4.68	5.75
1096	cycle 10 miles		204	.039			1.57	3.39	4.96	6.10
1098	cycle 20 miles		132	.061			2.43	5.25	7.68	9.40
1100	cycle 30 miles		108	.074			2.97	6.40	9.37	11.55
1102	cycle 40 miles		84	.095			3.81	8.25	12.06	14.80
1104	cycle 50 miles		72	.111			4.45	9.60	14.05	17.30
1106	50 MPH ave, cycle 10 miles		216	.037			1.48	3.20	4.68	5.75
1108	cycle 20 miles		144	.056			2.23	4.80	7.03	8.65
1110	cycle 30 miles		108	.074			2.97	6.40	9.37	11.55
1112	cycle 40 miles		84	.095			3.81	8.25	12.06	14.80
1114	cycle 50 miles		72	.111			4.45	9.60	14.05	17.30
1214	15 MPH ave, cycle 0.5 mile, 20 min. wait/Ld./Uld.		264	.030			1.21	2.62	3.83	4.71
1216	cycle 1 mile		240	.033			1.33	2.88	4.21	5.20
1218	cycle 2 miles		204	.039			1.57	3.39	4.96	6.10
1220	cycle 4 miles		156	.051			2.05	4.43	6.48	8
1222	cycle 6 miles		132	.061			2.43	5.25	7.68	9.40
1224	cycle 8 miles		108	.074			2.97	6.40	9.37	11.55
1225	cycle 10 miles		96	.083			3.34	7.20	10.54	12.95
1226	20 MPH ave, cycle 0.5 mile		264	.030			1.21	2.62	3.83	4.71
1228	cycle 1 mile		252	.032			1.27	2.74	4.01	4.94
1230	cycle 2 miles		216	.037			1.48	3.20	4.68	5.75
1232	cycle 4 miles		180	.044			1.78	3.84	5.62	6.90
1234	cycle 6 miles		144	.056			2.23	4.80	7.03	8.65
1236	cycle 8 miles		132	.061			2.43	5.25	7.68	9.40
1238	cycle 10 miles		108	.074			2.97	6.40	9.37	11.55
1240	25 MPH ave, cycle 4 miles	↓	192	.042	↓		1.67	3.60	5.27	6.50

31 23 23.20 Hauling		Crew	Daily Output	Labor-Hours	Unit	Material	2015 Bare Costs Labor	Equipment	Total	Total Incl O&P
1242	cycle 6 miles	B-34B	168	.048	L.C.Y.		1.91	4.11	6.02	7.40
1244	cycle 8 miles		144	.056			2.23	4.80	7.03	8.65
1246	cycle 10 miles		132	.061			2.43	5.25	7.68	9.40
1250	30 MPH ave, cycle 4 miles		204	.039			1.57	3.39	4.96	6.10
1252	cycle 6 miles		180	.044			1.78	3.84	5.62	6.90
1254	cycle 8 miles		156	.051			2.05	4.43	6.48	8
1256	cycle 10 miles		144	.056			2.23	4.80	7.03	8.65
1260	35 MPH ave, cycle 4 miles		216	.037			1.48	3.20	4.68	5.75
1262	cycle 6 miles		192	.042			1.67	3.60	5.27	6.50
1264	cycle 8 miles		168	.048			1.91	4.11	6.02	7.40
1266	cycle 10 miles		156	.051			2.05	4.43	6.48	8
1268	cycle 20 miles		108	.074			2.97	6.40	9.37	11.55
1269	cycle 30 miles		72	.111			4.45	9.60	14.05	17.30
1270	cycle 40 miles		60	.133			5.35	11.50	16.85	20.50
1272	40 MPH ave, cycle 6 miles		192	.042			1.67	3.60	5.27	6.50
1274	cycle 8 miles		180	.044			1.78	3.84	5.62	6.90
1276	cycle 10 miles		156	.051			2.05	4.43	6.48	8
1278	cycle 20 miles		108	.074			2.97	6.40	9.37	11.55
1280	cycle 30 miles		84	.095			3.81	8.25	12.06	14.80
1282	cycle 40 miles		72	.111			4.45	9.60	14.05	17.30
1284	cycle 50 miles		60	.133			5.35	11.50	16.85	20.50
1294	45 MPH ave, cycle 8 miles		180	.044			1.78	3.84	5.62	6.90
1296	cycle 10 miles		168	.048			1.91	4.11	6.02	7.40
1298	cycle 20 miles		120	.067			2.67	5.75	8.42	10.40
1300	cycle 30 miles		96	.083			3.34	7.20	10.54	12.95
1302	cycle 40 miles		72	.111			4.45	9.60	14.05	17.30
1304	cycle 50 miles		60	.133			5.35	11.50	16.85	20.50
1306	50 MPH ave, cycle 10 miles		180	.044			1.78	3.84	5.62	6.90
1308	cycle 20 miles		132	.061			2.43	5.25	7.68	9.40
1310	cycle 30 miles		96	.083			3.34	7.20	10.54	12.95
1312	cycle 40 miles		84	.095			3.81	8.25	12.06	14.80
1314	cycle 50 miles		72	.111			4.45	9.60	14.05	17.30
1414	15 MPH ave, cycle 0.5 mile, 25 min. wait/Ld./Uld.		204	.039			1.57	3.39	4.96	6.10
1416	cycle 1 mile		192	.042			1.67	3.60	5.27	6.50
1418	cycle 2 miles		168	.048			1.91	4.11	6.02	7.40
1420	cycle 4 miles		132	.061			2.43	5.25	7.68	9.40
1422	cycle 6 miles		120	.067			2.67	5.75	8.42	10.40
1424	cycle 8 miles		96	.083			3.34	7.20	10.54	12.95
1425	cycle 10 miles		84	.095			3.81	8.25	12.06	14.80
1426	20 MPH ave, cycle 0.5 mile		216	.037			1.48	3.20	4.68	5.75
1428	cycle 1 mile		204	.039			1.57	3.39	4.96	6.10
1430	cycle 2 miles		180	.044			1.78	3.84	5.62	6.90
1432	cycle 4 miles		156	.051			2.05	4.43	6.48	8
1434	cycle 6 miles		132	.061			2.43	5.25	7.68	9.40
1436	cycle 8 miles		120	.067			2.67	5.75	8.42	10.40
1438	cycle 10 miles		96	.083			3.34	7.20	10.54	12.95
1440	25 MPH ave, cycle 4 miles		168	.048			1.91	4.11	6.02	7.40
1442	cycle 6 miles		144	.056			2.23	4.80	7.03	8.65
1444	cycle 8 miles		132	.061			2.43	5.25	7.68	9.40
1446	cycle 10 miles		108	.074			2.97	6.40	9.37	11.55
1450	30 MPH ave, cycle 4 miles		168	.048			1.91	4.11	6.02	7.40
1452	cycle 6 miles		156	.051			2.05	4.43	6.48	8
1454	cycle 8 miles		132	.061			2.43	5.25	7.68	9.40

31 23 23.20 Hauling		Crew	Daily Output	Labor-Hours	Unit	Material	2015 Bare Costs Labor	Equipment	Total	Total Incl O&P
1456	cycle 10 miles	B-34B	120	.067	L.C.Y.		2.67	5.75	8.42	10.40
1460	35 MPH ave, cycle 4 miles		180	.044			1.78	3.84	5.62	6.90
1462	cycle 6 miles		156	.051			2.05	4.43	6.48	8
1464	cycle 8 miles		144	.056			2.23	4.80	7.03	8.65
1466	cycle 10 miles		132	.061			2.43	5.25	7.68	9.40
1468	cycle 20 miles		96	.083			3.34	7.20	10.54	12.95
1469	cycle 30 miles		72	.111			4.45	9.60	14.05	17.30
1470	cycle 40 miles		60	.133			5.35	11.50	16.85	20.50
1472	40 MPH ave, cycle 6 miles		168	.048			1.91	4.11	6.02	7.40
1474	cycle 8 miles		156	.051			2.05	4.43	6.48	8
1476	cycle 10 miles		144	.056			2.23	4.80	7.03	8.65
1478	cycle 20 miles		96	.083			3.34	7.20	10.54	12.95
1480	cycle 30 miles		84	.095			3.81	8.25	12.06	14.80
1482	cycle 40 miles		60	.133			5.35	11.50	16.85	20.50
1484	cycle 50 miles		60	.133			5.35	11.50	16.85	20.50
1494	45 MPH ave, cycle 8 miles		156	.051			2.05	4.43	6.48	8
1496	cycle 10 miles		144	.056			2.23	4.80	7.03	8.65
1498	cycle 20 miles		108	.074			2.97	6.40	9.37	11.55
1500	cycle 30 miles		84	.095			3.81	8.25	12.06	14.80
1502	cycle 40 miles		72	.111			4.45	9.60	14.05	17.30
1504	cycle 50 miles		60	.133			5.35	11.50	16.85	20.50
1506	50 MPH ave, cycle 10 miles		156	.051			2.05	4.43	6.48	8
1508	cycle 20 miles		120	.067			2.67	5.75	8.42	10.40
1510	cycle 30 miles		96	.083			3.34	7.20	10.54	12.95
1512	cycle 40 miles		72	.111			4.45	9.60	14.05	17.30
1514	cycle 50 miles		60	.133			5.35	11.50	16.85	20.50
1614	15 MPH, cycle 0.5 mile, 30 min. wait/Ld./Uld.		180	.044			1.78	3.84	5.62	6.90
1616	cycle 1 mile		168	.048			1.91	4.11	6.02	7.40
1618	cycle 2 miles		144	.056			2.23	4.80	7.03	8.65
1620	cycle 4 miles		120	.067			2.67	5.75	8.42	10.40
1622	cycle 6 miles		108	.074			2.97	6.40	9.37	11.55
1624	cycle 8 miles		84	.095			3.81	8.25	12.06	14.80
1625	cycle 10 miles		84	.095			3.81	8.25	12.06	14.80
1626	20 MPH ave, cycle 0.5 mile		180	.044			1.78	3.84	5.62	6.90
1628	cycle 1 mile		168	.048			1.91	4.11	6.02	7.40
1630	cycle 2 miles		156	.051			2.05	4.43	6.48	8
1632	cycle 4 miles		132	.061			2.43	5.25	7.68	9.40
1634	cycle 6 miles		120	.067			2.67	5.75	8.42	10.40
1636	cycle 8 miles		108	.074			2.97	6.40	9.37	11.55
1638	cycle 10 miles		96	.083			3.34	7.20	10.54	12.95
1640	25 MPH ave, cycle 4 miles		144	.056			2.23	4.80	7.03	8.65
1642	cycle 6 miles		132	.061			2.43	5.25	7.68	9.40
1644	cycle 8 miles		108	.074			2.97	6.40	9.37	11.55
1646	cycle 10 miles		108	.074			2.97	6.40	9.37	11.55
1650	30 MPH ave, cycle 4 miles		144	.056			2.23	4.80	7.03	8.65
1652	cycle 6 miles		132	.061			2.43	5.25	7.68	9.40
1654	cycle 8 miles		120	.067			2.67	5.75	8.42	10.40
1656	cycle 10 miles		108	.074			2.97	6.40	9.37	11.55
1660	35 MPH ave, cycle 4 miles		156	.051			2.05	4.43	6.48	8
1662	cycle 6 miles		144	.056			2.23	4.80	7.03	8.65
1664	cycle 8 miles		132	.061			2.43	5.25	7.68	9.40
1666	cycle 10 miles		120	.067			2.67	5.75	8.42	10.40
1668	cycle 20 miles		84	.095			3.81	8.25	12.06	14.80

31 23 23.20 Hauling		Crew	Daily Output	Labor-Hours	Unit	Material	2015 Bare Costs Labor	Equipment	Total	Total Incl O&P
1669	cycle 30 miles	B-34B	72	.111	L.C.Y.		4.45	9.60	14.05	17.30
1670	cycle 40 miles		60	.133			5.35	11.50	16.85	20.50
1672	40 MPH, cycle 6 miles		144	.056			2.23	4.80	7.03	8.65
1674	cycle 8 miles		132	.061			2.43	5.25	7.68	9.40
1676	cycle 10 miles		120	.067			2.67	5.75	8.42	10.40
1678	cycle 20 miles		96	.083			3.34	7.20	10.54	12.95
1680	cycle 30 miles		72	.111			4.45	9.60	14.05	17.30
1682	cycle 40 miles		60	.133			5.35	11.50	16.85	20.50
1684	cycle 50 miles		48	.167			6.70	14.40	21.10	26
1694	45 MPH ave, cycle 8 miles		144	.056			2.23	4.80	7.03	8.65
1696	cycle 10 miles		132	.061			2.43	5.25	7.68	9.40
1698	cycle 20 miles		96	.083			3.34	7.20	10.54	12.95
1700	cycle 30 miles		84	.095			3.81	8.25	12.06	14.80
1702	cycle 40 miles		60	.133			5.35	11.50	16.85	20.50
1704	cycle 50 miles		60	.133			5.35	11.50	16.85	20.50
1706	50 MPH ave, cycle 10 miles		132	.061			2.43	5.25	7.68	9.40
1708	cycle 20 miles		108	.074			2.97	6.40	9.37	11.55
1710	cycle 30 miles		84	.095			3.81	8.25	12.06	14.80
1712	cycle 40 miles		72	.111			4.45	9.60	14.05	17.30
1714	cycle 50 miles		60	.133			5.35	11.50	16.85	20.50
2000	Hauling, 8 C.Y. truck, small project cost per hour	B-34A	8	1	Hr.		40	51.50	91.50	117
2100	12 C.Y. Truck	B-34B	8	1			40	86.50	126.50	156
2150	16.5 C.Y. Truck	B-34C	8	1			40	91	131	161
2175	18 C.Y. 8 wheel Truck	B-34I	8	1			40	109	149	180
2200	20 C.Y. Truck	B-34D	8	1			40	92.50	132.50	163
2300	Grading at dump, or embankment if required, by dozer	B-10B	1000	.012	L.C.Y.		.56	1.39	1.95	2.38
2310	Spotter at fill or cut, if required	1 Clab	8	1	Hr.		37.50		37.50	58
9014	18 C.Y. truck, 8 wheels,15 min. wait/Ld./Uld.,15 MPH, cycle 0.5 mi.	B-34I	504	.016	L.C.Y.		.64	1.72	2.36	2.86
9016	cycle 1 mile		450	.018			.71	1.93	2.64	3.20
9018	cycle 2 miles		378	.021			.85	2.30	3.15	3.81
9020	cycle 4 miles		270	.030			1.19	3.22	4.41	5.35
9022	cycle 6 miles		216	.037			1.48	4.02	5.50	6.65
9024	cycle 8 miles		180	.044			1.78	4.83	6.61	8
9025	cycle 10 miles		144	.056			2.23	6.05	8.28	10
9026	20 MPH ave, cycle 0.5 mile		522	.015			.61	1.67	2.28	2.76
9028	cycle 1 mile		468	.017			.68	1.86	2.54	3.07
9030	cycle 2 miles		414	.019			.77	2.10	2.87	3.48
9032	cycle 4 miles		324	.025			.99	2.68	3.67	4.45
9034	cycle 6 miles		252	.032			1.27	3.45	4.72	5.70
9036	cycle 8 miles		216	.037			1.48	4.02	5.50	6.65
9038	cycle 10 miles		180	.044			1.78	4.83	6.61	8
9040	25 MPH ave, cycle 4 miles		342	.023			.94	2.54	3.48	4.21
9042	cycle 6 miles		288	.028			1.11	3.02	4.13	5
9044	cycle 8 miles		252	.032			1.27	3.45	4.72	5.70
9046	cycle 10 miles		216	.037			1.48	4.02	5.50	6.65
9050	30 MPH ave, cycle 4 miles		378	.021			.85	2.30	3.15	3.81
9052	cycle 6 miles		324	.025			.99	2.68	3.67	4.45
9054	cycle 8 miles		270	.030			1.19	3.22	4.41	5.35
9056	cycle 10 miles		234	.034			1.37	3.71	5.08	6.15
9060	35 MPH ave, cycle 4 miles		396	.020			.81	2.19	3	3.63
9062	cycle 6 miles		342	.023			.94	2.54	3.48	4.21
9064	cycle 8 miles		288	.028			1.11	3.02	4.13	5
9066	cycle 10 miles		270	.030			1.19	3.22	4.41	5.35

31 23 23 – Fill

31 23 23.20 Hauling		Crew	Daily Output	Labor-Hours	Unit	Material	2015 Bare Costs Labor	Equipment	Total	Total Incl O&P
9068	cycle 20 miles	B-34I	162	.049	L.C.Y.		1.98	5.35	7.33	8.90
9070	cycle 30 miles		126	.063			2.54	6.90	9.44	11.45
9072	cycle 40 miles		90	.089			3.56	9.65	13.21	16
9074	40 MPH ave, cycle 6 miles		360	.022			.89	2.41	3.30	4.01
9076	cycle 8 miles		324	.025			.99	2.68	3.67	4.45
9078	cycle 10 miles		288	.028			1.11	3.02	4.13	5
9080	cycle 20 miles		180	.044			1.78	4.83	6.61	8
9082	cycle 30 miles		144	.056			2.23	6.05	8.28	10
9084	cycle 40 miles		108	.074			2.97	8.05	11.02	13.35
9086	cycle 50 miles		90	.089			3.56	9.65	13.21	16
9094	45 MPH ave, cycle 8 miles		324	.025			.99	2.68	3.67	4.45
9096	cycle 10 miles		306	.026			1.05	2.84	3.89	4.70
9098	cycle 20 miles		198	.040			1.62	4.39	6.01	7.30
9100	cycle 30 miles		144	.056			2.23	6.05	8.28	10
9102	cycle 40 miles		126	.063			2.54	6.90	9.44	11.45
9104	cycle 50 miles		108	.074			2.97	8.05	11.02	13.35
9106	50 MPH ave, cycle 10 miles		324	.025			.99	2.68	3.67	4.45
9108	cycle 20 miles		216	.037			1.48	4.02	5.50	6.65
9110	cycle 30 miles		162	.049			1.98	5.35	7.33	8.90
9112	cycle 40 miles		126	.063			2.54	6.90	9.44	11.45
9114	cycle 50 miles		108	.074			2.97	8.05	11.02	13.35
9214	20 min. wait/Ld./Uld.,15 MPH, cycle 0.5 mi.		396	.020			.81	2.19	3	3.63
9216	cycle 1 mile		360	.022			.89	2.41	3.30	4.01
9218	cycle 2 miles		306	.026			1.05	2.84	3.89	4.70
9220	cycle 4 miles		234	.034			1.37	3.71	5.08	6.15
9222	cycle 6 miles		198	.040			1.62	4.39	6.01	7.30
9224	cycle 8 miles		162	.049			1.98	5.35	7.33	8.90
9225	cycle 10 miles		144	.056			2.23	6.05	8.28	10
9226	20 MPH ave, cycle 0.5 mile		396	.020			.81	2.19	3	3.63
9228	cycle 1 mile		378	.021			.85	2.30	3.15	3.81
9230	cycle 2 miles		324	.025			.99	2.68	3.67	4.45
9232	cycle 4 miles		270	.030			1.19	3.22	4.41	5.35
9234	cycle 6 miles		216	.037			1.48	4.02	5.50	6.65
9236	cycle 8 miles		198	.040			1.62	4.39	6.01	7.30
9238	cycle 10 miles		162	.049			1.98	5.35	7.33	8.90
9240	25 MPH ave, cycle 4 miles		288	.028			1.11	3.02	4.13	5
9242	cycle 6 miles		252	.032			1.27	3.45	4.72	5.70
9244	cycle 8 miles		216	.037			1.48	4.02	5.50	6.65
9246	cycle 10 miles		198	.040			1.62	4.39	6.01	7.30
9250	30 MPH ave, cycle 4 miles		306	.026			1.05	2.84	3.89	4.70
9252	cycle 6 miles		270	.030			1.19	3.22	4.41	5.35
9254	cycle 8 miles		234	.034			1.37	3.71	5.08	6.15
9256	cycle 10 miles		216	.037			1.48	4.02	5.50	6.65
9260	35 MPH ave, cycle 4 miles		324	.025			.99	2.68	3.67	4.45
9262	cycle 6 miles		288	.028			1.11	3.02	4.13	5
9264	cycle 8 miles		252	.032			1.27	3.45	4.72	5.70
9266	cycle 10 miles		234	.034			1.37	3.71	5.08	6.15
9268	cycle 20 miles		162	.049			1.98	5.35	7.33	8.90
9270	cycle 30 miles		108	.074			2.97	8.05	11.02	13.35
9272	cycle 40 miles		90	.089			3.56	9.65	13.21	16
9274	40 MPH ave, cycle 6 miles		288	.028			1.11	3.02	4.13	5
9276	cycle 8 miles		270	.030			1.19	3.22	4.41	5.35
9278	cycle 10 miles		234	.034			1.37	3.71	5.08	6.15

31 23 23.20 Hauling		Crew	Daily Output	Labor-Hours	Unit	Material	2015 Bare Costs Labor	Equipment	Total	Total Incl O&P
9280	cycle 20 miles	B-34I	162	.049	L.C.Y.		1.98	5.35	7.33	8.90
9282	cycle 30 miles		126	.063			2.54	6.90	9.44	11.45
9284	cycle 40 miles		108	.074			2.97	8.05	11.02	13.35
9286	cycle 50 miles		90	.089			3.56	9.65	13.21	16
9294	45 MPH ave, cycle 8 miles		270	.030			1.19	3.22	4.41	5.35
9296	cycle 10 miles		252	.032			1.27	3.45	4.72	5.70
9298	cycle 20 miles		180	.044			1.78	4.83	6.61	8
9300	cycle 30 miles		144	.056			2.23	6.05	8.28	10
9302	cycle 40 miles		108	.074			2.97	8.05	11.02	13.35
9304	cycle 50 miles		90	.089			3.56	9.65	13.21	16
9306	50 MPH ave, cycle 10 miles		270	.030			1.19	3.22	4.41	5.35
9308	cycle 20 miles		198	.040			1.62	4.39	6.01	7.30
9310	cycle 30 miles		144	.056			2.23	6.05	8.28	10
9312	cycle 40 miles		126	.063			2.54	6.90	9.44	11.45
9314	cycle 50 miles		108	.074			2.97	8.05	11.02	13.35
9414	25 min. wait/Ld./Uld.,15 MPH, cycle 0.5 mi.		306	.026			1.05	2.84	3.89	4.70
9416	cycle 1 mile		288	.028			1.11	3.02	4.13	5
9418	cycle 2 miles		252	.032			1.27	3.45	4.72	5.70
9420	cycle 4 miles		198	.040			1.62	4.39	6.01	7.30
9422	cycle 6 miles		180	.044			1.78	4.83	6.61	8
9424	cycle 8 miles		144	.056			2.23	6.05	8.28	10
9425	cycle 10 miles		126	.063			2.54	6.90	9.44	11.45
9426	20 MPH ave, cycle 0.5 mile		324	.025			.99	2.68	3.67	4.45
9428	cycle 1 mile		306	.026			1.05	2.84	3.89	4.70
9430	cycle 2 miles		270	.030			1.19	3.22	4.41	5.35
9432	cycle 4 miles		234	.034			1.37	3.71	5.08	6.15
9434	cycle 6 miles		198	.040			1.62	4.39	6.01	7.30
9436	cycle 8 miles		180	.044			1.78	4.83	6.61	8
9438	cycle 10 miles		144	.056			2.23	6.05	8.28	10
9440	25 MPH ave, cycle 4 miles		252	.032			1.27	3.45	4.72	5.70
9442	cycle 6 miles		216	.037			1.48	4.02	5.50	6.65
9444	cycle 8 miles		198	.040			1.62	4.39	6.01	7.30
9446	cycle 10 miles		180	.044			1.78	4.83	6.61	8
9450	30 MPH ave, cycle 4 miles		252	.032			1.27	3.45	4.72	5.70
9452	cycle 6 miles		234	.034			1.37	3.71	5.08	6.15
9454	cycle 8 miles		198	.040			1.62	4.39	6.01	7.30
9456	cycle 10 miles		180	.044			1.78	4.83	6.61	8
9460	35 MPH ave, cycle 4 miles		270	.030			1.19	3.22	4.41	5.35
9462	cycle 6 miles		234	.034			1.37	3.71	5.08	6.15
9464	cycle 8 miles		216	.037			1.48	4.02	5.50	6.65
9466	cycle 10 miles		198	.040			1.62	4.39	6.01	7.30
9468	cycle 20 miles		144	.056			2.23	6.05	8.28	10
9470	cycle 30 miles		108	.074			2.97	8.05	11.02	13.35
9472	cycle 40 miles		90	.089			3.56	9.65	13.21	16
9474	40 MPH ave, cycle 6 miles		252	.032			1.27	3.45	4.72	5.70
9476	cycle 8 miles		234	.034			1.37	3.71	5.08	6.15
9478	cycle 10 miles		216	.037			1.48	4.02	5.50	6.65
9480	cycle 20 miles		144	.056			2.23	6.05	8.28	10
9482	cycle 30 miles		126	.063			2.54	6.90	9.44	11.45
9484	cycle 40 miles		90	.089			3.56	9.65	13.21	16
9486	cycle 50 miles		90	.089			3.56	9.65	13.21	16
9494	45 MPH ave, cycle 8 miles		234	.034			1.37	3.71	5.08	6.15
9496	cycle 10 miles		216	.037			1.48	4.02	5.50	6.65

31 23 23.20 Hauling		Crew	Daily Output	Labor-Hours	Unit	Material	2015 Bare Costs Labor	Equipment	Total	Total Incl O&P
9498	cycle 20 miles	B-34I	162	.049	L.C.Y.		1.98	5.35	7.33	8.90
9500	cycle 30 miles		126	.063			2.54	6.90	9.44	11.45
9502	cycle 40 miles		108	.074			2.97	8.05	11.02	13.35
9504	cycle 50 miles		90	.089			3.56	9.65	13.21	16
9506	50 MPH ave, cycle 10 miles		234	.034			1.37	3.71	5.08	6.15
9508	cycle 20 miles		180	.044			1.78	4.83	6.61	8
9510	cycle 30 miles		144	.056			2.23	6.05	8.28	10
9512	cycle 40 miles		108	.074			2.97	8.05	11.02	13.35
9514	cycle 50 miles		90	.089			3.56	9.65	13.21	16
9614	30 min. wait/Ld./Uld.,15 MPH, cycle 0.5 mi.		270	.030			1.19	3.22	4.41	5.35
9616	cycle 1 mile		252	.032			1.27	3.45	4.72	5.70
9618	cycle 2 miles		216	.037			1.48	4.02	5.50	6.65
9620	cycle 4 miles		180	.044			1.78	4.83	6.61	8
9622	cycle 6 miles		162	.049			1.98	5.35	7.33	8.90
9624	cycle 8 miles		126	.063			2.54	6.90	9.44	11.45
9625	cycle 10 miles		126	.063			2.54	6.90	9.44	11.45
9626	20 MPH ave, cycle 0.5 mile		270	.030			1.19	3.22	4.41	5.35
9628	cycle 1 mile		252	.032			1.27	3.45	4.72	5.70
9630	cycle 2 miles		234	.034			1.37	3.71	5.08	6.15
9632	cycle 4 miles		198	.040			1.62	4.39	6.01	7.30
9634	cycle 6 miles		180	.044			1.78	4.83	6.61	8
9636	cycle 8 miles		162	.049			1.98	5.35	7.33	8.90
9638	cycle 10 miles		144	.056			2.23	6.05	8.28	10
9640	25 MPH ave, cycle 4 miles		216	.037			1.48	4.02	5.50	6.65
9642	cycle 6 miles		198	.040			1.62	4.39	6.01	7.30
9644	cycle 8 miles		180	.044			1.78	4.83	6.61	8
9646	cycle 10 miles		162	.049			1.98	5.35	7.33	8.90
9650	30 MPH ave, cycle 4 miles		216	.037			1.48	4.02	5.50	6.65
9652	cycle 6 miles		198	.040			1.62	4.39	6.01	7.30
9654	cycle 8 miles		180	.044			1.78	4.83	6.61	8
9656	cycle 10 miles		162	.049			1.98	5.35	7.33	8.90
9660	35 MPH ave, cycle 4 miles		234	.034			1.37	3.71	5.08	6.15
9662	cycle 6 miles		216	.037			1.48	4.02	5.50	6.65
9664	cycle 8 miles		198	.040			1.62	4.39	6.01	7.30
9666	cycle 10 miles		180	.044			1.78	4.83	6.61	8
9668	cycle 20 miles		126	.063			2.54	6.90	9.44	11.45
9670	cycle 30 miles		108	.074			2.97	8.05	11.02	13.35
9672	cycle 40 miles		90	.089			3.56	9.65	13.21	16
9674	40 MPH ave, cycle 6 miles		216	.037			1.48	4.02	5.50	6.65
9676	cycle 8 miles		198	.040			1.62	4.39	6.01	7.30
9678	cycle 10 miles		180	.044			1.78	4.83	6.61	8
9680	cycle 20 miles		144	.056			2.23	6.05	8.28	10
9682	cycle 30 miles		108	.074			2.97	8.05	11.02	13.35
9684	cycle 40 miles		90	.089			3.56	9.65	13.21	16
9686	cycle 50 miles		72	.111			4.45	12.05	16.50	20
9694	45 MPH ave, cycle 8 miles		216	.037			1.48	4.02	5.50	6.65
9696	cycle 10 miles		198	.040			1.62	4.39	6.01	7.30
9698	cycle 20 miles		144	.056			2.23	6.05	8.28	10
9700	cycle 30 miles		126	.063			2.54	6.90	9.44	11.45
9702	cycle 40 miles		108	.074			2.97	8.05	11.02	13.35
9704	cycle 50 miles		90	.089			3.56	9.65	13.21	16
9706	50 MPH ave, cycle 10 miles		198	.040			1.62	4.39	6.01	7.30
9708	cycle 20 miles		162	.049			1.98	5.35	7.33	8.90

31 23 Excavation and Fill

31 23 23 – Fill

31 23 23.20 Hauling			Crew	Daily Output	Labor-Hours	Unit	Material	2015 Bare Costs Labor	Equipment	Total	Total Incl O&P
9710	cycle 30 miles		B-34I	126	.063	L.C.Y.		2.54	6.90	9.44	11.45
9712	cycle 40 miles			108	.074			2.97	8.05	11.02	13.35
9714	cycle 50 miles			90	.089			3.56	9.65	13.21	16
31 23 23.24 Compaction, Structural											
0010	**COMPACTION, STRUCTURAL**	R312323-30									
0020	Steel wheel tandem roller, 5 tons		B-10E	8	1.500	Hr.		69.50	19.70	89.20	128
0100	10 tons		B-10F	8	1.500	″		69.50	29.50	99	139
0300	Sheepsfoot or wobbly wheel roller, 8″ lifts, common fill		B-10G	1300	.009	E.C.Y.		.43	.93	1.36	1.67
0400	Select fill		″	1500	.008			.37	.80	1.17	1.44
0600	Vibratory plate, 8″ lifts, common fill		A-1D	200	.040			1.50	.18	1.68	2.51
0700	Select fill		″	216	.037			1.39	.17	1.56	2.32

31 25 Erosion and Sedimentation Controls

31 25 14 – Stabilization Measures for Erosion and Sedimentation Control

31 25 14.16 Rolled Erosion Control Mats and Blankets			Crew	Daily Output	Labor-Hours	Unit	Material	Labor	Equipment	Total	Total Incl O&P
0010	**ROLLED EROSION CONTROL MATS AND BLANKETS**										
0020	Jute mesh, 100 S.Y. per roll, 4′ wide, stapled	G	B-80A	2400	.010	S.Y.	1.03	.38	.13	1.54	1.85
0070	Paper biodegradable mesh	G	B-1	2500	.010		.11	.37		.48	.69
0080	Paper mulch	G	B-64	20000	.001		.11	.03	.02	.16	.19
0100	Plastic netting, stapled, 2″ x 1″ mesh, 20 mil	G	B-1	2500	.010		.24	.37		.61	.83
0200	Polypropylene mesh, stapled, 6.5 oz./S.Y.	G		2500	.010		2.30	.37		2.67	3.10
0300	Tobacco netting, or jute mesh #2, stapled	G		2500	.010		.19	.37		.56	.78
1000	Silt fence, Install and Maintain, Remove	G	B-62	1300	.018	L.F.	.72	.76	.13	1.61	2.11
1100	Allow 25% per month for maintenance; 6-month max life										
1200	Place and remove hay bales	G	A-2	3	8	Ton	246	305	82	633	825
1250	Hay bales, staked	G	″	2500	.010	L.F.	9.85	.37	.10	10.32	11.45

31 31 Soil Treatment

31 31 16 – Termite Control

31 31 16.13 Chemical Termite Control		Crew	Daily Output	Labor-Hours	Unit	Material	Labor	Equipment	Total	Total Incl O&P
0010	**CHEMICAL TERMITE CONTROL**									
0020	Slab and walls, residential	1 Skwk	1200	.007	SF Flr.	.32	.32		.64	.85
0100	Commercial, minimum		2496	.003		.33	.16		.49	.60
0200	Maximum		1645	.005		.50	.24		.74	.91
0400	Insecticides for termite control, minimum		14.20	.563	Gal.	68.50	27.50		96	118
0500	Maximum		11	.727	″	117	35.50		152.50	184

31 32 Soil Stabilization

31 32 13 – Soil Mixing Stabilization

31 32 13.30 Calcium Chloride		Crew	Daily Output	Labor-Hours	Unit	Material	Labor	Equipment	Total	Total Incl O&P
0010	**CALCIUM CHLORIDE**									
0020	Calcium chloride delivered, 100 lb. bags, truckload lots				Ton	390			390	430
0030	Solution, 4 lb. flake per gallon, tank truck delivery				Gal.	1.48			1.48	1.63

31 33 Rock Stabilization

31 33 13 – Rock Bolting and Grouting

31 33 13.10 Rock Bolting		Crew	Daily Output	Labor-Hours	Unit	Material	2015 Bare Costs Labor	Equipment	Total	Total Incl O&P
0010	**ROCK BOLTING**									
2020	Hollow core, prestressable anchor, 1" diameter, 5' long	2 Skwk	32	.500	Ea.	182	24.50		206.50	238
2025	10' long		24	.667		305	32.50		337.50	390
2060	2" diameter, 5' long		32	.500		590	24.50		614.50	690
2065	10' long		24	.667		1,225	32.50		1,257.50	1,400
2100	Super high-tensile, 3/4" diameter, 5' long		32	.500		49.50	24.50		74	92
2105	10' long		24	.667		136	32.50		168.50	199
2160	2" diameter, 5' long		32	.500		410	24.50		434.50	490
2165	10' long	↓	24	.667		690	32.50		722.50	805
4400	Drill hole for rock bolt, 1-3/4" diam., 5' long (for 3/4" bolt)	B-56	17	.941			40.50	93.50	134	165
4405	10' long		9	1.778			76.50	177	253.50	310
4420	2" diameter, 5' long (for 1" bolt)		13	1.231			53	123	176	216
4425	10' long		7	2.286			98.50	228	326.50	400
4460	3-1/2" diameter, 5' long (for 2" bolt)		10	1.600			69	159	228	280
4465	10' long	↓	5	3.200	↓		138	320	458	560

31 36 Gabions

31 36 13 – Gabion Boxes

31 36 13.10 Gabion Box Systems		Crew	Daily Output	Labor-Hours	Unit	Material	Labor	Equipment	Total	Total Incl O&P
0010	**GABION BOX SYSTEMS**									
0400	Gabions, galvanized steel mesh mats or boxes, stone filled, 6" deep	B-13	200	.280	S.Y.	26.50	11.50	3.68	41.68	50.50
0500	9" deep		163	.344		39	14.10	4.52	57.62	69.50
0600	12" deep		153	.366		52.50	15	4.81	72.31	86
0700	18" deep		102	.549		71.50	22.50	7.20	101.20	121
0800	36" deep	↓	60	.933	↓	98	38.50	12.25	148.75	180

31 37 Riprap

31 37 13 – Machined Riprap

31 37 13.10 Riprap and Rock Lining		Crew	Daily Output	Labor-Hours	Unit	Material	Labor	Equipment	Total	Total Incl O&P
0010	**RIPRAP AND ROCK LINING**									
0011	Random, broken stone									
0100	Machine placed for slope protection	B-12G	62	.258	L.C.Y.	30	11.50	11.40	52.90	63
0110	3/8 to 1/4 C.Y. pieces, grouted	B-13	80	.700	S.Y.	59.50	28.50	9.20	97.20	120
0200	18" minimum thickness, not grouted	"	53	1.057	"	18.90	43.50	13.90	76.30	103
0300	Dumped, 50 lb. average	B-11A	800	.020	Ton	26.50	.88	1.73	29.11	32.50
0350	100 lb. average		700	.023		26.50	1.01	1.98	29.49	32.50
0370	300 lb. average	↓	600	.027	↓	26.50	1.18	2.31	29.99	33.50

31 41 Shoring

31 41 13 – Timber Shoring

31 41 13.10 Building Shoring		Crew	Daily Output	Labor-Hours	Unit	Material	2015 Bare Costs Labor	Equipment	Total	Total Incl O&P
0010	**BUILDING SHORING**									
0020	Shoring, existing building, with timber, no salvage allowance	B-51	2.20	21.818	M.B.F.	850	835	111	1,796	2,325
1000	On cribbing with 35 ton screw jacks, per box and jack	"	3.60	13.333	Jack	68.50	510	68	646.50	930

31 41 16 – Sheet Piling

31 41 16.10 Sheet Piling Systems		Crew	Daily Output	Labor-Hours	Unit	Material	2015 Bare Costs Labor	Equipment	Total	Total Incl O&P
0010	**SHEET PILING SYSTEMS**									
0020	Sheet piling steel, not incl. wales, 22 psf, 15′ excav., left in place	B-40	10.81	5.920	Ton	1,600	282	345	2,227	2,575
0100	Drive, extract & salvage R314116-45		6	10.667		510	510	625	1,645	2,025
0300	20′ deep excavation, 27 psf, left in place		12.95	4.942		1,600	235	289	2,124	2,425
0400	Drive, extract & salvage		6.55	9.771		510	465	570	1,545	1,925
0600	25′ deep excavation, 38 psf, left in place		19	3.368		1,600	160	197	1,957	2,225
0700	Drive, extract & salvage		10.50	6.095		510	290	355	1,155	1,400
0900	40′ deep excavation, 38 psf, left in place		21.20	3.019		1,600	144	177	1,921	2,175
1000	Drive, extract & salvage		12.25	5.224		510	249	305	1,064	1,275
1200	15′ deep excavation, 22 psf, left in place		983	.065	S.F.	18.60	3.10	3.81	25.51	29.50
1300	Drive, extract & salvage		545	.117		5.70	5.60	6.85	18.15	22.50
1500	20′ deep excavation, 27 psf, left in place		960	.067		23.50	3.18	3.90	30.58	34.50
1600	Drive, extract & salvage		485	.132		7.45	6.30	7.70	21.45	26.50
1800	25′ deep excavation, 38 psf, left in place		1000	.064		34.50	3.05	3.74	41.29	47
1900	Drive, extract & salvage		553	.116		10.15	5.50	6.75	22.40	27.50
2100	Rent steel sheet piling and wales, first month				Ton	310			310	340
2200	Per added month					31			31	34
2300	Rental piling left in place, add to rental					1,150			1,150	1,275
2500	Wales, connections & struts, 2/3 salvage					480			480	530
2700	High strength piling, 50,000 psi, add					63.50			63.50	70
2800	55,000 psi, add					82			82	90
3000	Tie rod, not upset, 1-1/2″ to 4″ diameter with turnbuckle					2,125			2,125	2,325
3100	No turnbuckle					1,650			1,650	1,825
3300	Upset, 1-3/4″ to 4″ diameter with turnbuckle					2,400			2,400	2,625
3400	No turnbuckle					2,075			2,075	2,275
3600	Lightweight, 18″ to 28″ wide, 7 ga., 9.22 psf, and									
3610	9 ga., 8.6 psf, minimum				Lb.	.77			.77	.85
3700	Average					.87			.87	.96
3750	Maximum					1.04			1.04	1.14
3900	Wood, solid sheeting, incl. wales, braces and spacers, R314116-40									
3910	drive, extract & salvage, 8′ deep excavation	B-31	330	.121	S.F.	1.89	4.83	.66	7.38	10.25
4000	10′ deep, 50 S.F./hr. in & 150 S.F./hr. out		300	.133		1.95	5.30	.73	7.98	11.15
4100	12′ deep, 45 S.F./hr. in & 135 S.F./hr. out		270	.148		2	5.90	.81	8.71	12.20
4200	14′ deep, 42 S.F./hr. in & 126 S.F./hr. out		250	.160		2.06	6.40	.88	9.34	13.05
4300	16′ deep, 40 S.F./hr. in & 120 S.F./hr. out		240	.167		2.13	6.65	.91	9.69	13.60
4400	18′ deep, 38 S.F./hr. in & 114 S.F./hr. out		230	.174		2.20	6.95	.95	10.10	14.10
4500	20′ deep, 35 S.F./hr. in & 105 S.F./hr. out		210	.190		2.27	7.60	1.04	10.91	15.35
4520	Left in place, 8′ deep, 55 S.F./hr.		440	.091		3.41	3.62	.50	7.53	9.90
4540	10′ deep, 50 S.F./hr.		400	.100		3.58	3.99	.55	8.12	10.70
4560	12′ deep, 45 S.F./hr.		360	.111		3.78	4.43	.61	8.82	11.65
4565	14′ deep, 42 S.F./hr.		335	.119		4.01	4.76	.65	9.42	12.50
4570	16′ deep, 40 S.F./hr.		320	.125		4.26	4.98	.68	9.92	13.10
4580	18′ deep, 38 S.F./hr.		305	.131		4.54	5.25	.72	10.51	13.85
4590	20′ deep, 35 S.F./hr.		280	.143		4.86	5.70	.78	11.34	14.95
4700	Alternate pricing, left in place, 8′ deep		1.76	22.727	M.B.F.	765	905	124	1,794	2,375
4800	Drive, extract and salvage, 8′ deep		1.32	30.303	"	680	1,200	166	2,046	2,775
5000	For treated lumber add cost of treatment to lumber									

31 43 Concrete Raising

31 43 13 – Pressure Grouting

31 43 13.13 Concrete Pressure Grouting		Crew	Daily Output	Labor-Hours	Unit	Material	2015 Bare Costs Labor	Equipment	Total	Total Incl O&P
0010	**CONCRETE PRESSURE GROUTING**									
0020	Grouting, pressure, cement & sand, 1:1 mix, minimum	B-61	124	.323	Bag	12.10	12.95	2.83	27.88	36.50
0100	Maximum		51	.784	"	12.10	31.50	6.90	50.50	69.50
0200	Cement and sand, 1:1 mix, minimum		250	.160	C.F.	24	6.45	1.40	31.85	38
0300	Maximum		100	.400		36.50	16.10	3.51	56.11	68.50
0400	Epoxy cement grout, minimum		137	.292		700	11.75	2.56	714.31	790
0500	Maximum		57	.702		700	28	6.15	734.15	820
0700	Alternate pricing method: (Add for materials)									
0710	5 person crew and equipment	B-61	1	40	Day		1,600	350	1,950	2,850

31 45 Vibroflotation and Densification

31 45 13 – Vibroflotation

31 45 13.10 Vibroflotation Densification

		Crew	Daily Output	Labor-Hours	Unit	Material	Labor	Equipment	Total	Total Incl O&P
0010	**VIBROFLOTATION DENSIFICATION** R314513-90									
0900	Vibroflotation compacted sand cylinder, minimum	B-60	750	.075	V.L.F.		3.30	2.91	6.21	8.25
0950	Maximum		325	.172			7.60	6.70	14.30	19
1100	Vibro replacement compacted stone cylinder, minimum		500	.112			4.94	4.36	9.30	12.35
1150	Maximum		250	.224			9.90	8.70	18.60	24.50
1300	Mobilization and demobilization, minimum		.47	119	Total		5,250	4,650	9,900	13,100
1400	Maximum		.14	400	"		17,700	15,600	33,300	44,000

31 46 Needle Beams

31 46 13 – Cantilever Needle Beams

31 46 13.10 Needle Beams

		Crew	Daily Output	Labor-Hours	Unit	Material	Labor	Equipment	Total	Total Incl O&P
0010	**NEEDLE BEAMS**									
0011	Incl. wood shoring 10′ x 10′ opening									
0400	Block, concrete, 8″ thick	B-9	7.10	5.634	Ea.	50	214	33	297	420
0420	12″ thick		6.70	5.970		60	227	35	322	455
0800	Brick, 4″ thick with 8″ backup block		5.70	7.018		60	267	41	368	520
1000	Brick, solid, 8″ thick		6.20	6.452		50	245	37.50	332.50	470
1040	12″ thick		4.90	8.163		60	310	47.50	417.50	595
1080	16″ thick		4.50	8.889		80.50	340	51.50	472	665
2000	Add for additional floors of shoring	B-1	6	4		50	153		203	291

31 48 Underpinning

31 48 13 – Underpinning Piers

31 48 13.10 Underpinning Foundations

		Crew	Daily Output	Labor-Hours	Unit	Material	Labor	Equipment	Total	Total Incl O&P
0010	**UNDERPINNING FOUNDATIONS**									
0011	Including excavation,									
0020	forming, reinforcing, concrete and equipment									
0100	5′ to 16′ below grade, 100 to 500 C.Y.	B-52	2.30	24.348	C.Y.	283	1,050	258	1,591	2,225
0200	Over 500 C.Y.		2.50	22.400		255	975	238	1,468	2,050
0400	16′ to 25′ below grade, 100 to 500 C.Y.		2	28		310	1,225	297	1,832	2,550
0500	Over 500 C.Y.		2.10	26.667		295	1,175	283	1,753	2,400
0700	26′ to 40′ below grade, 100 to 500 C.Y.		1.60	35		340	1,525	370	2,235	3,100
0800	Over 500 C.Y.		1.80	31.111		310	1,350	330	1,990	2,775
0900	For under 50 C.Y., add					10%	40%			

31 48 Underpinning

31 48 13 – Underpinning Piers

31 48 13.10 Underpinning Foundations		Crew	Daily Output	Labor-Hours	Unit	Material	2015 Bare Costs Labor	Equipment	Total	Total Incl O&P
1000	For 50 C.Y. to 100 C.Y., add				C.Y.	5%	20%			

31 52 Cofferdams

31 52 16 – Timber Cofferdams

31 52 16.10 Cofferdams		Crew	Daily Output	Labor-Hours	Unit	Material	2015 Bare Costs Labor	Equipment	Total	Total Incl O&P
0010	**COFFERDAMS**									
0011	Incl. mobilization and temporary sheeting									
0080	Soldier beams & lagging H piles with 3" wood sheeting									
0090	horizontal between piles, including removal of wales & braces									
0100	No hydrostatic head, 15' deep, 1 line of braces, minimum	B-50	545	.206	S.F.	8.50	9.30	4.48	22.28	29
0200	Maximum		495	.226		9.45	10.25	4.93	24.63	32
0400	15' to 22' deep with 2 lines of braces, 10" H, minimum		360	.311		10	14.10	6.80	30.90	40.50
0500	Maximum		330	.339		11.35	15.35	7.40	34.10	44.50
0700	23' to 35' deep with 3 lines of braces, 12" H, minimum		325	.345		13.10	15.60	7.50	36.20	47
0800	Maximum		295	.380		14.20	17.20	8.30	39.70	51.50
1000	36' to 45' deep with 4 lines of braces, 14" H, minimum		290	.386		14.70	17.50	8.40	40.60	53
1100	Maximum		265	.423		15.50	19.15	9.20	43.85	57
1300	No hydrostatic head, left in place, 15' dp., 1 line of braces, min.		635	.176		11.35	8	3.85	23.20	29
1400	Maximum		575	.195		12.15	8.80	4.25	25.20	32
1600	15' to 22' deep with 2 lines of braces, minimum		455	.246		17.05	11.15	5.35	33.55	42
1700	Maximum		415	.270		18.95	12.25	5.90	37.10	46.50
1900	23' to 35' deep with 3 lines of braces, minimum		420	.267		20.50	12.10	5.80	38.40	48
2000	Maximum		380	.295		22.50	13.35	6.45	42.30	52.50
2200	36' to 45' deep with 4 lines of braces, minimum		385	.291		24.50	13.20	6.35	44.05	54.50
2300	Maximum		350	.320		28.50	14.50	7	50	61
2350	Lagging only, 3" thick wood between piles 8' O.C., minimum	B-46	400	.120		1.89	5.05	.11	7.05	10.10
2370	Maximum		250	.192		2.84	8.10	.18	11.12	15.95
2400	Open sheeting no bracing, for trenches to 10' deep, min.		1736	.028		.85	1.17	.03	2.05	2.79
2450	Maximum		1510	.032		.95	1.34	.03	2.32	3.17
2500	Tie-back method, add to open sheeting, add, minimum								20%	20%
2550	Maximum								60%	60%
2700	Tie-backs only, based on tie-backs total length, minimum	B-46	86.80	.553	L.F.	14.60	23.50	.52	38.62	53
2750	Maximum		38.50	1.247	"	25.50	52.50	1.17	79.17	112
3500	Tie-backs only, typical average, 25' long		2	24	Ea.	645	1,000	22.50	1,667.50	2,300
3600	35' long		1.58	30.380	"	855	1,275	28.50	2,158.50	2,975

31 56 Slurry Walls

31 56 23 – Lean Concrete Slurry Walls

31 56 23.20 Slurry Trench		Crew	Daily Output	Labor-Hours	Unit	Material	2015 Bare Costs Labor	Equipment	Total	Total Incl O&P
0010	**SLURRY TRENCH**									
0011	Excavated slurry trench in wet soils									
0020	backfilled with 3000 PSI concrete, no reinforcing steel									
0050	Minimum	C-7	333	.216	C.F.	7.55	8.85	3.65	20.05	26
0100	Maximum		200	.360	"	12.60	14.75	6.05	33.40	43
0200	Alternate pricing method, minimum		150	.480	S.F.	15.05	19.65	8.10	42.80	55.50
0300	Maximum		120	.600		22.50	24.50	10.10	57.10	73.50
0500	Reinforced slurry trench, minimum	B-48	177	.316		11.30	13.45	16.40	41.15	51
0600	Maximum	"	69	.812		37.50	34.50	42	114	141
0800	Haul for disposal, 2 mile haul, excavated material, add	B-34B	99	.081	C.Y.		3.24	7	10.24	12.60

31 56 Slurry Walls

31 56 23 – Lean Concrete Slurry Walls

31 56 23.20 Slurry Trench		Crew	Daily Output	Labor-Hours	Unit	Material	2015 Bare Costs Labor	Equipment	Total	Total Incl O&P
0900	Haul bentonite castings for disposal, add	B-34B	40	.200	C.Y.		8	17.30	25.30	31

31 62 Driven Piles

31 62 13 – Concrete Piles

31 62 13.23 Prestressed Concrete Piles

		Crew	Daily Output	Labor-Hours	Unit	Material	Labor	Equipment	Total	Total Incl O&P
0010	**PRESTRESSED CONCRETE PILES**, 200 piles									
0020	Unless specified otherwise, not incl. pile caps or mobilization									
2200	Precast, prestressed, 50′ long, 12″ diam., 2-3/8″ wall	B-19	720	.089	V.L.F.	23.50	4.23	2.41	30.14	35.50
2300	14″ diameter, 2-1/2″ wall		680	.094		32	4.48	2.55	39.03	45
2500	16″ diameter, 3″ wall		640	.100		45.50	4.76	2.71	52.97	60.50
2600	18″ diameter, 3″ wall	B-19A	600	.107		57	5.10	3.61	65.71	74.50
2800	20″ diameter, 3-1/2″ wall		560	.114		67.50	5.45	3.87	76.82	87
2900	24″ diameter, 3-1/2″ wall		520	.123		76	5.85	4.16	86.01	97
3100	Precast, prestressed, 40′ long, 10″ thick, square	B-19	700	.091		18.35	4.36	2.48	25.19	29.50
3200	12″ thick, square		680	.094		22.50	4.48	2.55	29.53	34.50
3400	14″ thick, square		600	.107		25.50	5.10	2.90	33.50	39
3500	Octagonal		640	.100		34.50	4.76	2.71	41.97	48.50
3700	16″ thick, square		560	.114		41	5.45	3.10	49.55	57
3800	Octagonal		600	.107		41.50	5.10	2.90	49.50	57
4000	18″ thick, square	B-19A	520	.123		47.50	5.85	4.16	57.51	65.50
4100	Octagonal	B-19	560	.114		49.50	5.45	3.10	58.05	66
4300	20″ thick, square	B-19A	480	.133		60.50	6.35	4.51	71.36	81.50
4400	Octagonal	B-19	520	.123		54.50	5.85	3.34	63.69	73
4600	24″ thick, square	B-19A	440	.145		72	6.95	4.92	83.87	95
4700	Octagonal	B-19	480	.133		78.50	6.35	3.62	88.47	100
4730	Precast, prestressed, 60′ long, 10″ thick, square		700	.091		19.25	4.36	2.48	26.09	30.50
4740	12″ thick, square (60′ long)		680	.094		23	4.48	2.55	30.03	35.50
4750	Mobilization for 10,000 L.F. pile job, add		3300	.019			.92	.53	1.45	2.02
4800	25,000 L.F. pile job, add		8500	.008			.36	.20	.56	.78

31 62 16 – Steel Piles

31 62 16.13 Steel Piles

		Crew	Daily Output	Labor-Hours	Unit	Material	Labor	Equipment	Total	Total Incl O&P
0010	**STEEL PILES**									
0100	Step tapered, round, concrete filled									
0110	8″ tip, 60 ton capacity, 30′ depth	B-19	760	.084	V.L.F.	11.45	4.01	2.29	17.75	21.50
0120	60′ depth		740	.086		12.95	4.12	2.35	19.42	23
0130	80′ depth		700	.091		13.40	4.36	2.48	20.24	24.50
0150	10″ tip, 90 ton capacity, 30′ depth		700	.091		13.75	4.36	2.48	20.59	24.50
0160	60′ depth		690	.093		14.20	4.42	2.52	21.14	25.50
0170	80′ depth		670	.096		15.30	4.55	2.59	22.44	27
0190	12″ tip, 120 ton capacity, 30′ depth		660	.097		18.35	4.62	2.63	25.60	30
0200	60′ depth, 12″ diameter		630	.102		17.65	4.84	2.76	25.25	30
0210	80′ depth		590	.108		17.15	5.15	2.94	25.24	30
0250	"H" Sections, 50′ long, HP8 x 36		640	.100		16.05	4.76	2.71	23.52	28
0400	HP10 X 42		610	.105		18.35	5	2.85	26.20	31
0500	HP10 X 57		610	.105		25.50	5	2.85	33.35	39
0700	HP12 X 53		590	.108		25	5.15	2.94	33.09	39
0800	HP12 X 74	B-19A	590	.108		34.50	5.15	3.67	43.32	50
1000	HP14 X 73		540	.119		35	5.65	4.01	44.66	51.50
1100	HP14 X 89		540	.119		41	5.65	4.01	50.66	58
1300	HP14 X 102		510	.125		47	6	4.24	57.24	65.50

31 62 Driven Piles

31 62 16 – Steel Piles

31 62 16.13 Steel Piles		Crew	Daily Output	Labor-Hours	Unit	Material	2015 Bare Costs Labor	Equipment	Total	Total Incl O&P
1400	HP14 X 117	B-19A	510	.125	V.L.F.	54	6	4.24	64.24	73
1600	Splice on standard points, not in leads, 8″ or 10″	1 Sswl	5	1.600	Ea.	120	84		204	278
1700	12″ or 14″		4	2		177	105		282	375
1900	Heavy duty points, not in leads, 10″ wide		4	2		190	105		295	390
2100	14″ wide	↓	3.50	2.286	↓	238	120		358	470

31 62 19 – Timber Piles

31 62 19.10 Wood Piles		Crew	Daily Output	Labor-Hours	Unit	Material	Labor	Equipment	Total	Total Incl O&P
0010	**WOOD PILES**									
0011	Friction or end bearing, not including									
0050	mobilization or demobilization									
0100	Untreated piles, up to 30′ long, 12″ butts, 8″ points	B-19	625	.102	V.L.F.	16.85	4.88	2.78	24.51	29
0200	30′ to 39′ long, 12″ butts, 8″ points		700	.091		16.85	4.36	2.48	23.69	28
0300	40′ to 49′ long, 12″ butts, 7″ points		720	.089		16.85	4.23	2.41	23.49	28
0400	50′ to 59′ long, 13″butts, 7″ points		800	.080		16.85	3.81	2.17	22.83	27
0500	60′ to 69′ long, 13″ butts, 7″ points		840	.076		18.95	3.63	2.07	24.65	29
0600	70′ to 80′ long, 13″ butts, 6″ points	↓	840	.076	↓	21	3.63	2.07	26.70	31
0800	Treated piles, 12 lb. per C.F.,									
0810	friction or end bearing, ASTM class B									
1000	Up to 30′ long, 12″ butts, 8″ points	B-19	625	.102	V.L.F.	15.85	4.88	2.78	23.51	28
1100	30′ to 39′ long, 12″ butts, 8″ points		700	.091		17.80	4.36	2.48	24.64	29
1200	40′ to 49′ long, 12″ butts, 7″ points		720	.089		19.35	4.23	2.41	25.99	31
1300	50′ to 59′ long, 13″ butts, 7″ points	↓	800	.080		22	3.81	2.17	27.98	32.50
1400	60′ to 69′ long, 13″ butts, 6″ points	B-19A	840	.076		26	3.63	2.58	32.21	37
1500	70′ to 80′ long, 13″ butts, 6″ points	″	840	.076	↓	28	3.63	2.58	34.21	39.50
1600	Treated piles, C.C.A., 2.5# per C.F.									
1610	8″ butts, 10′ long	B-19	400	.160	V.L.F.	18.40	7.60	4.34	30.34	36.50
1620	11′ to 16′ long		500	.128		18.40	6.10	3.47	27.97	33.50
1630	17′ to 20′ long		575	.111		18.40	5.30	3.02	26.72	31.50
1640	10″ butts, 10′ to 16′ long		500	.128		18.40	6.10	3.47	27.97	33.50
1650	17′ to 20′ long		575	.111		18.40	5.30	3.02	26.72	31.50
1660	21′ to 40′ long		700	.091		18.40	4.36	2.48	25.24	29.50
1670	12″ butts, 10′ to 20′ long		575	.111		18.40	5.30	3.02	26.72	31.50
1680	21′ to 35′ long		650	.098		18.40	4.69	2.67	25.76	30
1690	36′ to 40′ long		700	.091		18.40	4.36	2.48	25.24	29.50
1695	14″ butts, to 40′ long	↓	700	.091	↓	18.40	4.36	2.48	25.24	29.50
1700	Boot for pile tip, minimum	1 Pile	27	.296	Ea.	23	13.65		36.65	47
1800	Maximum		21	.381		69.50	17.55		87.05	105
2000	Point for pile tip, minimum		20	.400		23	18.45		41.45	54.50
2100	Maximum	↓	15	.533		83.50	24.50		108	131
2300	Splice for piles over 50′ long, minimum	B-46	35	1.371		57	58	1.29	116.29	155
2400	Maximum		20	2.400	↓	68.50	101	2.26	171.76	236
2600	Concrete encasement with wire mesh and tube	↓	331	.145	V.L.F.	10.60	6.10	.14	16.84	21.50
2700	Mobilization for 10,000 L.F. pile job, add	B-19	3300	.019			.92	.53	1.45	2.02
2800	25,000 L.F. pile job, add	″	8500	.008	↓		.36	.20	.56	.78

31 62 23 – Composite Piles

31 62 23.13 Concrete-Filled Steel Piles		Crew	Daily Output	Labor-Hours	Unit	Material	Labor	Equipment	Total	Total Incl O&P
0010	**CONCRETE-FILLED STEEL PILES** no mobilization or demobilization									
2600	Pipe piles, 50′ lg. 8″ diam., 29 lb. per L.F., no concrete	B-19	500	.128	V.L.F.	20.50	6.10	3.47	30.07	36
2700	Concrete filled		460	.139		21.50	6.65	3.78	31.93	38
2900	10″ diameter, 34 lb. per L.F., no concrete		500	.128		27	6.10	3.47	36.57	43.50
3000	Concrete filled		450	.142		29	6.80	3.86	39.66	47
3200	12″ diameter, 44 lb. per L.F., no concrete	↓	475	.135	↓	33	6.40	3.66	43.06	50.50

31 62 Driven Piles

31 62 23 – Composite Piles

	31 62 23.13 Concrete-Filled Steel Piles	Crew	Daily Output	Labor-Hours	Unit	Material	2015 Bare Costs Labor	Equipment	Total	Total Incl O&P
3300	Concrete filled	B-19	415	.154	V.L.F.	37	7.35	4.19	48.54	57
3500	14" diameter, 46 lb. per L.F., no concrete		430	.149		35.50	7.10	4.04	46.64	55
3600	Concrete filled		355	.180		41.50	8.60	4.89	54.99	64.50
3800	16" diameter, 52 lb. per L.F., no concrete		385	.166		39	7.90	4.51	51.41	60.50
3900	Concrete filled		335	.191		45.50	9.10	5.20	59.80	70
4100	18" diameter, 59 lb. per L.F., no concrete		355	.180		48.50	8.60	4.89	61.99	72.50
4200	Concrete filled		310	.206		54.50	9.85	5.60	69.95	81.50
4400	Splices for pipe piles, not in leads, 8" diameter	1 Sswl	4.67	1.713	Ea.	90.50	90		180.50	257
4500	14" diameter		3.79	2.111		119	111		230	325
4600	16" diameter		3.03	2.640		147	139		286	400
4800	Points, standard, 8" diameter		4.61	1.735		148	91.50		239.50	320
4900	14" diameter		4.05	1.975		206	104		310	405
5000	16" diameter		3.37	2.374		251	125		376	495
5200	Points, heavy duty, 10" diameter		2.89	2.768		296	146		442	580
5300	14" or 16" diameter		2.02	3.960		475	209		684	880
5500	For reinforcing steel, add		1150	.007	Lb.	.93	.37		1.30	1.66
5700	For thick wall sections, add				"	.96			.96	1.06

31 63 Bored Piles

31 63 26 – Drilled Caissons

31 63 26.13 Fixed End Caisson Piles

	31 63 26.13 Fixed End Caisson Piles	Crew	Daily Output	Labor-Hours	Unit	Material	Labor	Equipment	Total	Total Incl O&P
0010	**FIXED END CAISSON PILES** R316326-60									
0015	Including excavation, concrete, 50 lb. reinforcing									
0020	per C.Y., not incl. mobilization, boulder removal, disposal									
0100	Open style, machine drilled, to 50' deep, in stable ground, no									
0110	casings or ground water, 18" diam., 0.065 C.Y./L.F.	B-43	200	.240	V.L.F.	8.15	9.95	12.70	30.80	38
0200	24" diameter, 0.116 C.Y./L.F.		190	.253		14.60	10.50	13.40	38.50	47
0300	30" diameter, 0.182 C.Y./L.F.		150	.320		23	13.30	16.95	53.25	64
0400	36" diameter, 0.262 C.Y./L.F.		125	.384		33	15.95	20.50	69.45	83
0500	48" diameter, 0.465 C.Y./L.F.		100	.480		58.50	19.95	25.50	103.95	123
0600	60" diameter, 0.727 C.Y./L.F.		90	.533		91.50	22	28.50	142	166
0700	72" diameter, 1.05 C.Y./L.F.		80	.600		132	25	32	189	218
0800	84" diameter, 1.43 C.Y./L.F.		75	.640		180	26.50	34	240.50	276
1000	For bell excavation and concrete, add									
1020	4' bell diameter, 24" shaft, 0.444 C.Y.	B-43	20	2.400	Ea.	45	99.50	127	271.50	345
1040	6' bell diameter, 30" shaft, 1.57 C.Y.		5.70	8.421		160	350	445	955	1,200
1060	8' bell diameter, 36" shaft, 3.72 C.Y.		2.40	20		380	830	1,050	2,260	2,875
1080	9' bell diameter, 48" shaft, 4.48 C.Y.		2	24		455	995	1,275	2,725	3,425
1100	10' bell diameter, 60" shaft, 5.24 C.Y.		1.70	28.235		535	1,175	1,500	3,210	4,025
1120	12' bell diameter, 72" shaft, 8.74 C.Y.		1	48		890	2,000	2,550	5,440	6,825
1140	14' bell diameter, 84" shaft, 13.6 C.Y.		.70	68.571		1,375	2,850	3,625	7,850	9,875
1200	Open style, machine drilled, to 50' deep, in wet ground, pulled									
1300	casing and pumping, 18" diameter, 0.065 C.Y./L.F.	B-48	160	.350	V.L.F.	8.15	14.90	18.15	41.20	52
1400	24" diameter, 0.116 C.Y./L.F.		125	.448		14.60	19.05	23.50	57.15	70.50
1500	30" diameter, 0.182 C.Y./L.F.		85	.659		23	28	34	85	106
1600	36" diameter, 0.262 C.Y./L.F.		60	.933		33	39.50	48.50	121	150
1700	48" diameter, 0.465 C.Y./L.F.	B-49	55	1.600		58.50	71	66	195.50	247
1800	60" diameter, 0.727 C.Y./L.F.		35	2.514		91.50	111	104	306.50	385
1900	72" diameter, 1.05 C.Y./L.F.		30	2.933		132	130	121	383	480
2000	84" diameter, 1.43 C.Y./L.F.		25	3.520		180	156	146	482	595
2100	For bell excavation and concrete, add									

31 63 26 – Drilled Caissons

31 63 26.13 Fixed End Caisson Piles		Crew	Daily Output	Labor-Hours	Unit	Material	2015 Bare Costs Labor	Equipment	Total	Total Incl O&P
2120	4' bell diameter, 24" shaft, 0.444 C.Y.	B-48	19.80	2.828	Ea.	45	120	147	312	395
2140	6' bell diameter, 30" shaft, 1.57 C.Y.		5.70	9.825		160	420	510	1,090	1,375
2160	8' bell diameter, 36" shaft, 3.72 C.Y.	↓	2.40	23.333		380	995	1,200	2,575	3,275
2180	9' bell diameter, 48" shaft, 4.48 C.Y.	B-49	3.30	26.667		455	1,175	1,100	2,730	3,550
2200	10' bell diameter, 60" shaft, 5.24 C.Y.		2.80	31.429		535	1,400	1,300	3,235	4,150
2220	12' bell diameter, 72" shaft, 8.74 C.Y.		1.60	55		890	2,425	2,275	5,590	7,225
2240	14' bell diameter, 84" shaft, 13.6 C.Y.	↓	1	88	↓	1,375	3,900	3,650	8,925	11,500
2300	Open style, machine drilled, to 50' deep, in soft rocks and									
2400	medium hard shales, 18" diameter, 0.065 C.Y./L.F.	B-49	50	1.760	V.L.F.	8.15	78	73	159.15	209
2500	24" diameter, 0.116 C.Y./L.F.		30	2.933		14.60	130	121	265.60	350
2600	30" diameter, 0.182 C.Y./L.F.		20	4.400		23	195	182	400	525
2700	36" diameter, 0.262 C.Y./L.F.		15	5.867		33	260	243	536	705
2800	48" diameter, 0.465 C.Y./L.F.		10	8.800		58.50	390	365	813.50	1,075
2900	60" diameter, 0.727 C.Y./L.F.		7	12.571		91.50	555	520	1,166.50	1,525
3000	72" diameter, 1.05 C.Y./L.F.		6	14.667		132	650	605	1,387	1,825
3100	84" diameter, 1.43 C.Y./L.F.	↓	5	17.600	↓	180	780	730	1,690	2,200
3200	For bell excavation and concrete, add									
3220	4' bell diameter, 24" shaft, 0.444 C.Y.	B-49	10.90	8.073	Ea.	45	355	335	735	970
3240	6' bell diameter, 30" shaft, 1.57 C.Y.		3.10	28.387		160	1,250	1,175	2,585	3,400
3260	8' bell diameter, 36" shaft, 3.72 C.Y.		1.30	67.692		380	3,000	2,800	6,180	8,100
3280	9' bell diameter, 48" shaft, 4.48 C.Y.		1.10	80		455	3,550	3,300	7,305	9,600
3300	10' bell diameter, 60" shaft, 5.24 C.Y.		.90	97.778		535	4,325	4,050	8,910	11,700
3320	12' bell diameter, 72" shaft, 8.74 C.Y.		.60	146		890	6,500	6,075	13,465	17,600
3340	14' bell diameter, 84" shaft, 13.6 C.Y.		.40	220	↓	1,375	9,750	9,100	20,225	26,500
3600	For rock excavation, sockets, add, minimum		120	.733	C.F.		32.50	30.50	63	83.50
3650	Average		95	.926			41	38.50	79.50	105
3700	Maximum	↓	48	1.833	↓		81	76	157	209
3900	For 50' to 100' deep, add				V.L.F.				7%	7%
4000	For 100' to 150' deep, add								25%	25%
4100	For 150' to 200' deep, add				↓				30%	30%
4200	For casings left in place, add				Lb.	1.19			1.19	1.31
4300	For other than 50 lb. reinf. per C.Y., add or deduct				"	1.16			1.16	1.28
4400	For steel "I" beam cores, add	B-49	8.30	10.602	Ton	2,125	470	440	3,035	3,525
4500	Load and haul excess excavation, 2 miles	B-34B	178	.045	L.C.Y.		1.80	3.88	5.68	7
4600	For mobilization, 50 mile radius, rig to 36"	B-43	2	24	Ea.		995	1,275	2,270	2,925
4650	Rig to 84"	B-48	1.75	32			1,350	1,650	3,000	3,900
4700	For low headroom, add								50%	50%
5000	Bottom inspection	1 Skwk	1.20	6.667	↓		325		325	500

31 63 26.16 Concrete Caissons for Marine Construction		Crew	Daily Output	Labor-Hours	Unit	Material	Labor	Equipment	Total	Total Incl O&P
0010	**CONCRETE CAISSONS FOR MARINE CONSTRUCTION**									
0100	Caissons, incl. mobilization and demobilization, up to 50 miles									
0200	Uncased shafts, 30 to 80 tons cap., 17" diam., 10' depth	B-44	88	.727	V.L.F.	20.50	34	20.50	75	98
0300	25' depth		165	.388		14.55	18.10	11	43.65	56.50
0400	80-150 ton capacity, 22" diameter, 10' depth		80	.800		25.50	37.50	22.50	85.50	112
0500	20' depth		130	.492		20.50	23	14	57.50	74
0700	Cased shafts, 10 to 30 ton capacity, 10-5/8" diam., 20' depth		175	.366		14.55	17.10	10.40	42.05	54
0800	30' depth		240	.267		13.55	12.45	7.55	33.55	42.50
0850	30 to 60 ton capacity, 12" diameter, 20' depth		160	.400		20.50	18.70	11.35	50.55	64
0900	40' depth		230	.278		15.65	13	7.90	36.55	46.50
1000	80 to 100 ton capacity, 16" diameter, 20' depth		160	.400		29	18.70	11.35	59.05	73.50
1100	40' depth		230	.278		27	13	7.90	47.90	59
1200	110 to 140 ton capacity, 17-5/8" diameter, 20' depth	↓	160	.400	↓	31.50	18.70	11.35	61.55	76

31 63 Bored Piles

31 63 26 – Drilled Caissons

31 63 26.16 Concrete Caissons for Marine Construction		Crew	Daily Output	Labor-Hours	Unit	Material	Labor	Equipment	Total	Total Incl O&P
						2015 Bare Costs				
1300	40′ depth	B-44	230	.278	V.L.F.	29	13	7.90	49.90	61
1400	140 to 175 ton capacity, 19″ diameter, 20′ depth		130	.492		34	23	14	71	89
1500	40′ depth	↓	210	.305	↓	31.50	14.25	8.65	54.40	66
1700	Over 30′ long, L.F. cost tends to be lower									
1900	Maximum depth is about 90′									

31 63 29 – Drilled Concrete Piers and Shafts

31 63 29.13 Uncased Drilled Concrete Piers

		Crew	Daily Output	Labor-Hours	Unit	Material	Labor	Equipment	Total	Total Incl O&P
0010	**UNCASED DRILLED CONCRETE PIERS**									
0020	Unless specified otherwise, not incl. pile caps or mobilization									
0050	Cast in place augered piles, no casing or reinforcing									
0060	8″ diameter	B-43	540	.089	V.L.F.	4.06	3.69	4.71	12.46	15.30
0065	10″ diameter		480	.100		6.45	4.16	5.30	15.91	19.30
0070	12″ diameter		420	.114		9.10	4.75	6.05	19.90	24
0075	14″ diameter		360	.133		12.25	5.55	7.05	24.85	30
0080	16″ diameter		300	.160		16.50	6.65	8.45	31.60	37.50
0085	18″ diameter	↓	240	.200	↓	20.50	8.30	10.60	39.40	47
0100	Cast in place, thin wall shell pile, straight sided,									
0110	not incl. reinforcing, 8″ diam., 16 ga., 5.8 lb./L.F.	B-19	700	.091	V.L.F.	9.10	4.36	2.48	15.94	19.55
0200	10″ diameter, 16 ga. corrugated, 7.3 lb./L.F.		650	.098		11.90	4.69	2.67	19.26	23.50
0300	12″ diameter, 16 ga. corrugated, 8.7 lb./L.F.		600	.107		15.45	5.10	2.90	23.45	28
0400	14″ diameter, 16 ga. corrugated, 10.0 lb./L.F.		550	.116		18.20	5.55	3.16	26.91	32
0500	16″ diameter, 16 ga. corrugated, 11.6 lb./L.F.	↓	500	.128	↓	22.50	6.10	3.47	32.07	38
0800	Cast in place friction pile, 50′ long, fluted,									
0810	tapered steel, 4000 psi concrete, no reinforcing									
0900	12″ diameter, 7 ga.	B-19	600	.107	V.L.F.	29	5.10	2.90	37	43
1000	14″ diameter, 7 ga.		560	.114		31.50	5.45	3.10	40.05	47
1100	16″ diameter, 7 ga.		520	.123		37.50	5.85	3.34	46.69	54
1200	18″ diameter, 7 ga.	↓	480	.133	↓	43.50	6.35	3.62	53.47	62
1300	End bearing, fluted, constant diameter,									
1320	4000 psi concrete, no reinforcing									
1340	12″ diameter, 7 ga.	B-19	600	.107	V.L.F.	30.50	5.10	2.90	38.50	44.50
1360	14″ diameter, 7 ga.		560	.114		38	5.45	3.10	46.55	54
1380	16″ diameter, 7 ga.		520	.123		44	5.85	3.34	53.19	61.50
1400	18″ diameter, 7 ga.	↓	480	.133	↓	48.50	6.35	3.62	58.47	67.50

31 63 29.20 Cast In Place Piles, Adds

		Crew	Daily Output	Labor-Hours	Unit	Material	Labor	Equipment	Total	Total Incl O&P
0010	**CAST IN PLACE PILES, ADDS**									
1500	For reinforcing steel, add				Lb.	.97			.97	1.06
1700	For ball or pedestal end, add	B-19	11	5.818	C.Y.	146	277	158	581	765
1900	For lengths above 60′, concrete, add	″	11	5.818	″	152	277	158	587	770
2000	For steel thin shell, pipe only				Lb.	1.26			1.26	1.39

Estimating Tips

32 01 00 Operations and Maintenance of Exterior Improvements

- Recycling of asphalt pavement is becoming very popular and is an alternative to removal and replacement. It can be a good value engineering proposal if removed pavement can be recycled, either at the project site or at another site that is reasonably close to the project site. Sections on repair of flexible and rigid pavement are included.

32 10 00 Bases, Ballasts, and Paving

- When estimating paving, keep in mind the project schedule. Also note that prices for asphalt and concrete are generally higher in the cold seasons. Lines for pavement markings, including tactile warning systems and fence lines, are included.

32 90 00 Planting

- The timing of planting and guarantee specifications often dictate the costs for establishing tree and shrub growth and a stand of grass or ground cover. Establish the work performance schedule to coincide with the local planting season. Maintenance and growth guarantees can add from 20%–100% to the total landscaping cost and can be contractually cumbersome. The cost to replace trees and shrubs can be as high as 5% of the total cost, depending on the planting zone, soil conditions, and time of year.

Reference Numbers

Reference numbers are shown in shaded boxes at the beginning of some major classifications. These numbers refer to related items in the Reference Section. The reference information may be an estimating procedure, an alternate pricing method, or technical information.

Note: Not all subdivisions listed here necessarily appear in this publication. ■

32 01 Operation and Maintenance of Exterior Improvements

32 01 13 – Flexible Paving Surface Treatment

32 01 13.61 Slurry Seal (Latex Modified)		Crew	Daily Output	Labor-Hours	Unit	2015 Bare Costs Material	Labor	Equipment	Total	Total Incl O&P
0010	**SLURRY SEAL (LATEX MODIFIED)**									
3780	Rubberized asphalt (latex) seal	B-45	5000	.003	S.Y.	2.81	.15	.18	3.14	3.51

32 01 13.64 Sand Seal

		Crew	Daily Output	Labor-Hours	Unit	Material	Labor	Equipment	Total	Total Incl O&P
0010	**SAND SEAL**									
2080	Sand sealing, sharp sand, asphalt emulsion, small area	B-91	10000	.006	S.Y.	1.48	.29	.23	2	2.33
2120	Roadway or large area	"	18000	.004	"	1.27	.16	.13	1.56	1.78

32 01 13.66 Fog Seal

		Crew	Daily Output	Labor-Hours	Unit	Material	Labor	Equipment	Total	Total Incl O&P
0010	**FOG SEAL**									
0012	Sealcoating, 2 coat coal tar pitch emulsion over 10,000 S.Y.	B-45	5000	.003	S.Y.	.90	.15	.18	1.23	1.41
0030	1000 to 10,000 S.Y.	"	3000	.005		.90	.24	.30	1.44	1.69
0100	Under 1000 S.Y.	B-1	1050	.023		.90	.87		1.77	2.34
0300	Petroleum resistant, over 10,000 S.Y.	B-45	5000	.003		1.30	.15	.18	1.63	1.85
0320	1000 to 10,000 S.Y.	"	3000	.005		1.30	.24	.30	1.84	2.13
0400	Under 1000 S.Y.	B-1	1050	.023		1.30	.87		2.17	2.78
0600	Non-skid pavement renewal, over 10,000 S.Y.	B-45	5000	.003		1.37	.15	.18	1.70	1.93
0620	1000 to 10,000 S.Y.	"	3000	.005		1.37	.24	.30	1.91	2.21
0700	Under 1000 S.Y.	B-1	1050	.023		1.37	.87		2.24	2.86
0800	Prepare and clean surface for above	A-2	8545	.003			.11	.03	.14	.19
1000	Hand seal asphalt curbing	B-1	4420	.005	L.F.	.62	.21		.83	1
1900	Asphalt surface treatment, single course, small area									
1901	0.30 gal/S.Y. asphalt material, 20#/S.Y. aggregate	B-91	5000	.013	S.Y.	1.30	.57	.46	2.33	2.81
1910	Roadway or large area		10000	.006		1.20	.29	.23	1.72	2.02
1950	Asphalt surface treatment, dbl. course for small area		3000	.021		2.94	.95	.77	4.66	5.55
1960	Roadway or large area		6000	.011		2.65	.48	.39	3.52	4.07
1980	Asphalt surface treatment, single course, for shoulders		7500	.009		1.46	.38	.31	2.15	2.53

32 06 Schedules for Exterior Improvements

32 06 10 – Schedules for Bases, Ballasts, and Paving

32 06 10.10 Sidewalks, Driveways and Patios

		Crew	Daily Output	Labor-Hours	Unit	Material	Labor	Equipment	Total	Total Incl O&P
0010	**SIDEWALKS, DRIVEWAYS AND PATIOS** No base									
0020	Asphaltic concrete, 2" thick	B-37	720	.067	S.Y.	7.40	2.65	.22	10.27	12.45
0100	2-1/2" thick	"	660	.073	"	9.40	2.89	.24	12.53	15.05
0300	Concrete, 3000 psi, CIP, 6 x 6 - W1.4 x W1.4 mesh,									
0310	broomed finish, no base, 4" thick	B-24	600	.040	S.F.	1.69	1.73		3.42	4.48
0350	5" thick		545	.044		2.26	1.90		4.16	5.40
0400	6" thick		510	.047		2.64	2.03		4.67	6
0450	For bank run gravel base, 4" thick, add	B-18	2500	.010		.52	.37	.02	.91	1.16
0520	8" thick, add	"	1600	.015		1.04	.57	.03	1.64	2.06
0550	Exposed aggregate finish, add to above, minimum	B-24	1875	.013		.11	.55		.66	.96
0600	Maximum	"	455	.053		.36	2.28		2.64	3.86
1000	Crushed stone, 1" thick, white marble	2 Clab	1700	.009		.44	.35		.79	1.03
1050	Bluestone	"	1700	.009		.13	.35		.48	.68
1700	Redwood, prefabricated, 4' x 4' sections	2 Carp	316	.051		4.77	2.38		7.15	8.90
1750	Redwood planks, 1" thick, on sleepers	"	240	.067		4.77	3.13		7.90	10.05
2250	Stone dust, 4" thick	B-62	900	.027	S.Y.	3.33	1.10	.19	4.62	5.55

32 06 10.20 Steps

		Crew	Daily Output	Labor-Hours	Unit	Material	Labor	Equipment	Total	Total Incl O&P
0010	**STEPS**									
0011	Incl. excav., borrow & concrete base as required									
0100	Brick steps	B-24	35	.686	LF Riser	15.35	29.50		44.85	62
0200	Railroad ties	2 Clab	25	.640		3.59	24		27.59	41

32 06 Schedules for Exterior Improvements

32 06 10 – Schedules for Bases, Ballasts, and Paving

32 06 10.20 Steps		Crew	Daily Output	Labor-Hours	Unit	2015 Bare Costs Material	Labor	Equipment	Total	Total Incl O&P
0300	Bluestone treads, 12" x 2" or 12" x 1-1/2"	B-24	30	.800	LF Riser	33.50	34.50		68	89
0500	Concrete, cast in place, see Section 03 30 53.40									
0600	Precast concrete, see Section 03 41 23.50									
4025	Steel edge strips, incl. stakes, 1/4" x 5"	B-1	390	.062	L.F.	3.92	2.36		6.28	7.95
4050	Edging, landscape timber or railroad ties, 6" x 8"	2 Carp	170	.094	"	2.79	4.42		7.21	9.85

32 11 Base Courses

32 11 23 – Aggregate Base Courses

32 11 23.23 Base Course Drainage Layers

Code	Description	Crew	Daily Output	Labor-Hours	Unit	Material	Labor	Equipment	Total	Total Incl O&P
0010	**BASE COURSE DRAINAGE LAYERS**									
0011	For roadways and large areas									
0050	Crushed 3/4" stone base, compacted, 3" deep	B-36C	5200	.008	S.Y.	2.30	.36	.81	3.47	3.96
0100	6" deep		5000	.008		4.61	.37	.84	5.82	6.55
0200	9" deep		4600	.009		6.90	.40	.91	8.21	9.20
0300	12" deep		4200	.010		9.20	.44	1	10.64	11.90
0301	Crushed 1-1/2" stone base, compacted to 4" deep	B-36B	6000	.011		4.46	.48	.80	5.74	6.50
0302	6" deep		5400	.012		6.70	.53	.89	8.12	9.15
0303	8" deep		4500	.014		8.90	.64	1.07	10.61	11.95
0304	12" deep		3800	.017		13.35	.75	1.27	15.37	17.25
0350	Bank run gravel, spread and compacted									
0370	6" deep	B-32	6000	.005	S.Y.	4.33	.25	.39	4.97	5.55
0390	9" deep		4900	.007		6.50	.31	.48	7.29	8.15
0400	12" deep		4200	.008		8.65	.36	.56	9.57	10.65
6000	Stabilization fabric, polypropylene, 6 oz./S.Y.	B-6	10000	.002		1.28	.10	.04	1.42	1.60
6900	For small and irregular areas, add						50%	50%		
7000	Prepare and roll sub-base, small areas to 2500 S.Y.	B-32A	1500	.016	S.Y.		.74	.95	1.69	2.17
8000	Large areas over 2500 S.Y.	"	3500	.007			.32	.41	.73	.93
8050	For roadways	B-32	4000	.008			.38	.59	.97	1.23

32 11 26 – Asphaltic Base Courses

32 11 26.19 Bituminous-Stabilized Base Courses

Code	Description	Crew	Daily Output	Labor-Hours	Unit	Material	Labor	Equipment	Total	Total Incl O&P
0010	**BITUMINOUS-STABILIZED BASE COURSES**									
0020	And large paved areas									
0700	Liquid application to gravel base, asphalt emulsion	B-45	6000	.003	Gal.	4.29	.12	.15	4.56	5.05
0800	Prime and seal, cut back asphalt		6000	.003	"	5.05	.12	.15	5.32	5.90
1000	Macadam penetration crushed stone, 2 gal. per S.Y., 4" thick		6000	.003	S.Y.	8.60	.12	.15	8.87	9.80
1100	6" thick, 3 gal. per S.Y.		4000	.004		12.85	.18	.23	13.26	14.65
1200	8" thick, 4 gal. per S.Y.		3000	.005		17.15	.24	.30	17.69	19.60
8900	For small and irregular areas, add						50%	50%		

32 12 16 – Asphalt Paving

32 12 16.13 Plant-Mix Asphalt Paving

	32 12 16.13 Plant-Mix Asphalt Paving	Crew	Daily Output	Labor-Hours	Unit	2015 Bare Costs Material	Labor	Equipment	Total	Total Incl O&P
0010	**PLANT-MIX ASPHALT PAVING**									
0020	And large paved areas with no hauling included									
0025	See Section 31 23 23.20 for hauling costs									
0080	Binder course, 1-1/2" thick	B-25	7725	.011	S.Y.	5.70	.47	.35	6.52	7.40
0120	2" thick		6345	.014		7.60	.57	.43	8.60	9.70
0130	2-1/2" thick		5620	.016		9.50	.65	.48	10.63	11.95
0160	3" thick		4905	.018		11.40	.74	.55	12.69	14.30
0170	3-1/2" thick		4520	.019		13.30	.80	.60	14.70	16.55
0200	4" thick		4140	.021		15.25	.88	.66	16.79	18.80
0300	Wearing course, 1" thick	B-25B	10575	.009		3.78	.38	.28	4.44	5.05
0340	1-1/2" thick		7725	.012		6.35	.52	.38	7.25	8.20
0380	2" thick		6345	.015		8.50	.64	.46	9.60	10.90
0420	2-1/2" thick		5480	.018		10.50	.74	.54	11.78	13.25
0460	3" thick		4900	.020		12.55	.82	.60	13.97	15.70
0470	3-1/2" thick		4520	.021		14.70	.89	.65	16.24	18.30
0480	4" thick		4140	.023		16.80	.98	.71	18.49	21
0500	Open graded friction course	B-25C	5000	.010		2.13	.41	.47	3.01	3.48
0800	Alternate method of figuring paving costs									
0810	Binder course, 1-1/2" thick	B-25	630	.140	Ton	70	5.75	4.31	80.06	90.50
0811	2" thick		690	.128		70	5.25	3.93	79.18	89.50
0812	3" thick		800	.110		70	4.55	3.39	77.94	87.50
0813	4" thick		900	.098		70	4.04	3.01	77.05	86.50
0850	Wearing course, 1" thick	B-25B	575	.167		69	7.05	5.15	81.20	92.50
0851	1-1/2" thick		630	.152		69	6.40	4.68	80.08	91
0852	2" thick		690	.139		69	5.85	4.28	79.13	89.50
0853	2-1/2" thick		765	.125		69	5.30	3.86	78.16	88.50
0854	3" thick		800	.120		69	5.05	3.69	77.74	88
1000	Pavement replacement over trench, 2" thick	B-37	90	.533	S.Y.	7.85	21	1.75	30.60	43
1050	4" thick		70	.686		15.55	27.50	2.25	45.30	61.50
1080	6" thick		55	.873		25	34.50	2.86	62.36	83.50

32 12 16.14 Asphaltic Concrete Paving

	32 12 16.14 Asphaltic Concrete Paving	Crew	Daily Output	Labor-Hours	Unit	Material	Labor	Equipment	Total	Total Incl O&P
0011	**ASPHALTIC CONCRETE PAVING**, parking lots & driveways									
0015	No asphalt hauling included									
0018	Use 6.05 C.Y. per inch per M.S.F. for hauling									
0020	6" stone base, 2" binder course, 1" topping	B-25C	9000	.005	S.F.	1.86	.23	.26	2.35	2.68
0025	2" binder course, 2" topping		9000	.005		2.28	.23	.26	2.77	3.14
0030	3" binder course, 2" topping		9000	.005		2.71	.23	.26	3.20	3.61
0035	4" binder course, 2" topping		9000	.005		3.13	.23	.26	3.62	4.07
0040	1.5" binder course, 1" topping		9000	.005		1.65	.23	.26	2.14	2.45
0042	3" binder course, 1" topping		9000	.005		2.29	.23	.26	2.78	3.15
0045	3" binder course, 3" topping		9000	.005		3.13	.23	.26	3.62	4.07
0050	4" binder course, 3" topping		9000	.005		3.55	.23	.26	4.04	4.53
0055	4" binder course, 4" topping		9000	.005		3.96	.23	.26	4.45	4.99
0300	Binder course, 1-1/2" thick		35000	.001		.64	.06	.07	.77	.86
0400	2" thick		25000	.002		.82	.08	.09	.99	1.13
0500	3" thick		15000	.003		1.27	.14	.16	1.57	1.78
0600	4" thick		10800	.004		1.67	.19	.22	2.08	2.36
0800	Sand finish course, 3/4" thick		41000	.001		.32	.05	.06	.43	.49
0900	1" thick		34000	.001		.40	.06	.07	.53	.61
1000	Fill pot holes, hot mix, 2" thick	B-16	4200	.008		.85	.30	.16	1.31	1.57
1100	4" thick		3500	.009		1.25	.35	.20	1.80	2.13
1120	6" thick		3100	.010		1.67	.40	.22	2.29	2.70

32 12 Flexible Paving

32 12 16 – Asphalt Paving

32 12 16.14 Asphaltic Concrete Paving		Crew	Daily Output	Labor-Hours	Unit	Material	Labor	Equipment	Total	Total Incl O&P
						2015 Bare Costs				
1140	Cold patch, 2″ thick	B-51	3000	.016	S.F.	.99	.61	.08	1.68	2.12
1160	4″ thick		2700	.018		1.89	.68	.09	2.66	3.22
1180	6″ thick		1900	.025		2.94	.96	.13	4.03	4.86

32 13 Rigid Paving

32 13 13 – Concrete Paving

32 13 13.23 Concrete Paving Surface Treatment		Crew	Daily Output	Labor-Hours	Unit	Material	Labor	Equipment	Total	Total Incl O&P
0010	**CONCRETE PAVING SURFACE TREATMENT**									
0015	Including joints, finishing and curing									
0020	Fixed form, 12′ pass, unreinforced, 6″ thick	B-26	3000	.029	S.Y.	22	1.24	1.18	24.42	27
0100	8″ thick		2750	.032		30	1.35	1.28	32.63	36.50
0110	8″ thick, small area		1375	.064		30	2.70	2.57	35.27	40
0200	9″ thick		2500	.035		34.50	1.48	1.41	37.39	41.50
0300	10″ thick		2100	.042		37.50	1.77	1.68	40.95	46
0310	10″ thick, small area		1050	.084		37.50	3.54	3.36	44.40	50.50
0400	12″ thick		1800	.049		43	2.06	1.96	47.02	53
0410	Conc. pavement, w/jt.,fnsh.&curing,fix form,24′ pass,unreinforced,6″T		6000	.015		21	.62	.59	22.21	24.50
0430	8″ thick		5500	.016		28.50	.67	.64	29.81	33
0440	9″ thick		5000	.018		32.50	.74	.71	33.95	38
0450	10″ thick		4200	.021		36	.88	.84	37.72	42
0460	12″ thick		3600	.024		41.50	1.03	.98	43.51	48
0470	15″ thick		3000	.029		54.50	1.24	1.18	56.92	62.50
0500	Fixed form 12′ pass 15″ thick		1500	.059		55	2.47	2.35	59.82	67
0510	For small irregular areas, add				%	10%	100%	100%		
0520	Welded wire fabric, sheets for rigid paving 2.33 lb./S.Y.	2 Rodm	389	.041	S.Y.	1.31	2.16		3.47	4.82
0530	Reinforcing steel for rigid paving 12 lb./S.Y.		666	.024		6.10	1.26		7.36	8.65
0540	Reinforcing steel for rigid paving 18 lb./S.Y.		444	.036		9.15	1.89		11.04	13
0620	Slip form, 12′ pass, unreinforced, 6″ thick	B-26A	5600	.016		21	.66	.66	22.32	24.50
0624	8″ thick		5300	.017		29	.70	.70	30.40	34
0626	9″ thick		4820	.018		33	.77	.77	34.54	38.50
0628	10″ thick		4050	.022		36.50	.92	.91	38.33	42.50
0630	12″ thick		3470	.025		42	1.07	1.07	44.14	49
0632	15″ thick		2890	.030		53	1.28	1.28	55.56	62
0640	Slip form, 24′ pass, unreinforced, 6″ thick		11200	.008		20.50	.33	.33	21.16	24
0644	8″ thick		10600	.008		28	.35	.35	28.70	31.50
0646	9″ thick		9640	.009		32	.39	.38	32.77	36
0648	10″ thick		8100	.011		35	.46	.46	35.92	40
0650	12″ thick		6940	.013		41	.53	.53	42.06	46.50
0652	15″ thick		5780	.015		51.50	.64	.64	52.78	58
0700	Finishing, broom finish small areas	2 Cefi	120	.133			6		6	8.90
1000	Curing, with sprayed membrane by hand	2 Clab	1500	.011		1	.40		1.40	1.71
1650	For integral coloring, see Section 03 05 13.20									

32 14 Unit Paving

32 14 13 – Precast Concrete Unit Paving

32 14 13.13 Interlocking Precast Concrete Unit Paving		Crew	Daily Output	Labor-Hours	Unit	Material	2015 Bare Costs Labor	Equipment	Total	Total Incl O&P
0010	**INTERLOCKING PRECAST CONCRETE UNIT PAVING**									
0020	"V" blocks for retaining soil	D-1	205	.078	S.F.	10	3.29		13.29	16

32 14 13.16 Precast Concrete Unit Paving Slabs

Line	Description	Crew	Daily Output	Labor-Hours	Unit	Material	Labor	Equipment	Total	Total Incl O&P
0010	**PRECAST CONCRETE UNIT PAVING SLABS**									
0710	Precast concrete patio blocks, 2-3/8" thick, colors, 8" x 16"	D-1	265	.060	S.F.	1.85	2.54		4.39	5.90
0750	Exposed local aggregate, natural	2 Bric	250	.064		7.30	2.95		10.25	12.55
0800	Colors		250	.064		7.30	2.95		10.25	12.55
0850	Exposed granite or limestone aggregate		250	.064		7.30	2.95		10.25	12.55
0900	Exposed white tumblestone aggregate		250	.064		5.70	2.95		8.65	10.75

32 14 13.18 Precast Concrete Plantable Pavers

Line	Description	Crew	Daily Output	Labor-Hours	Unit	Material	Labor	Equipment	Total	Total Incl O&P
0010	**PRECAST CONCRETE PLANTABLE PAVERS** (50% grass)									
0015	Subgrade preparation and grass planting not included									
0100	Precast concrete plantable pavers with topsoil, 24" x 16"	B-63	800	.050	S.F.	4.15	1.99	.22	6.36	7.85
0200	Less than 600 Square Feet or irregular area	"	500	.080	"	4.15	3.18	.35	7.68	9.85
0300	3/4" crushed stone base for plantable pavers, 6 inch depth	B-62	1000	.024	S.Y.	4.61	.99	.17	5.77	6.75
0400	8 inch depth		900	.027		6.15	1.10	.19	7.44	8.70
0500	10 inch depth		800	.030		7.70	1.24	.22	9.16	10.60
0600	12 inch depth		700	.034		9.20	1.42	.25	10.87	12.60
0700	Hydro seeding plantable pavers	B-81A	20	.800	M.S.F.	12.45	30.50	21.50	64.45	84
0800	Apply fertilizer and seed to plantable pavers	1 Clab	8	1	"	50.50	37.50		88	114

32 14 16 – Brick Unit Paving

32 14 16.10 Brick Paving

Line	Description	Crew	Daily Output	Labor-Hours	Unit	Material	Labor	Equipment	Total	Total Incl O&P
0010	**BRICK PAVING**									
0012	4" x 8" x 1-1/2", without joints (4.5 brick/S.F.)	D-1	110	.145	S.F.	2.21	6.10		8.31	11.80
0100	Grouted, 3/8" joint (3.9 brick/S.F.)		90	.178		1.95	7.50		9.45	13.60
0200	4" x 8" x 2-1/4", without joints (4.5 bricks/S.F.)		110	.145		2.25	6.10		8.35	11.85
0300	Grouted, 3/8" joint (3.9 brick/S.F.)		90	.178		1.95	7.50		9.45	13.60
0455	Pervious brick paving, 4" x 8" x 3-1/4", without joints (4.5 bricks/S.F.)		110	.145		3.38	6.10		9.48	13.05
0500	Bedding, asphalt, 3/4" thick	B-25	5130	.017		.70	.71	.53	1.94	2.44
0540	Course washed sand bed, 1" thick	B-18	5000	.005		.32	.18	.01	.51	.64
0580	Mortar, 1" thick	D-1	300	.053		.74	2.25		2.99	4.24
0620	2" thick		200	.080		1.48	3.37		4.85	6.75
1500	Brick on 1" thick sand bed laid flat, 4.5 per S.F.		100	.160		3.18	6.75		9.93	13.80
2000	Brick pavers, laid on edge, 7.2 per S.F.		70	.229		3.86	9.60		13.46	18.95
2500	For 4" thick concrete bed and joints, add		595	.027		1.19	1.13		2.32	3.04
2800	For steam cleaning, add	A-1H	950	.008		.09	.32	.08	.49	.68

32 14 23 – Asphalt Unit Paving

32 14 23.10 Asphalt Blocks

Line	Description	Crew	Daily Output	Labor-Hours	Unit	Material	Labor	Equipment	Total	Total Incl O&P
0010	**ASPHALT BLOCKS**									
0020	Rectangular, 6" x 12" x 1-1/4", w/bed & neopr. adhesive	D-1	135	.119	S.F.	9.30	4.99		14.29	17.80
0100	3" thick		130	.123		13	5.20		18.20	22
0300	Hexagonal tile, 8" wide, 1-1/4" thick		135	.119		9.30	4.99		14.29	17.80
0400	2" thick		130	.123		13	5.20		18.20	22
0500	Square, 8" x 8", 1-1/4" thick		135	.119		9.30	4.99		14.29	17.80
0600	2" thick		130	.123		13	5.20		18.20	22
0900	For exposed aggregate (ground finish) add					.62			.62	.68
0910	For colors, add					.47			.47	.52

32 14 Unit Paving

32 14 40 – Stone Paving

32 14 40.10 Stone Pavers		Crew	Daily Output	Labor-Hours	Unit	Material	Labor	Equipment	Total	Total Incl O&P
						2015 Bare Costs				
0010	**STONE PAVERS**									
1100	Flagging, bluestone, irregular, 1" thick,	D-1	81	.198	S.F.	6.40	8.30		14.70	19.75
1150	Snapped random rectangular, 1" thick		92	.174		9.70	7.30		17	22
1200	1-1/2" thick		85	.188		11.65	7.90		19.55	25
1250	2" thick		83	.193		13.60	8.10		21.70	27.50
1300	Slate, natural cleft, irregular, 3/4" thick		92	.174		9.20	7.30		16.50	21.50
1350	Random rectangular, gauged, 1/2" thick		105	.152		19.85	6.40		26.25	32
1400	Random rectangular, butt joint, gauged, 1/4" thick		150	.107		21.50	4.49		25.99	30.50
1500	For interior setting, add								25%	25%
1550	Granite blocks, 3-1/2" x 3-1/2" x 3-1/2"	D-1	92	.174	S.F.	12	7.30		19.30	24.50
1600	4" to 12" long, 3" to 5" wide, 3" to 5" thick		98	.163		10	6.85		16.85	21.50
1650	6" to 15" long, 3" to 6" wide, 3" to 5" thick		105	.152		5.35	6.40		11.75	15.70

32 16 Curbs, Gutters, Sidewalks, and Driveways

32 16 13 – Curbs and Gutters

32 16 13.13 Cast-in-Place Concrete Curbs and Gutters

		Crew	Daily Output	Labor-Hours	Unit	Material	Labor	Equipment	Total	Total Incl O&P
0010	**CAST-IN-PLACE CONCRETE CURBS AND GUTTERS**									
0290	Forms only, no concrete									
0300	Concrete, wood forms, 6" x 18", straight	C-2	500	.096	L.F.	2.99	4.39		7.38	10.05
0400	6" x 18", radius	"	200	.240	"	3.11	11		14.11	20.50
0402	Forms and concrete complete									
0404	Concrete, wood forms, 6" x 18", straight & concrete	C-2A	500	.096	L.F.	5.85	4.36		10.21	13.05
0406	6" x 18", radius		200	.240		5.95	10.90		16.85	23
0410	Steel forms, 6" x 18", straight		700	.069		4.40	3.11		7.51	9.60
0411	6" x 18", radius		400	.120		3.61	5.45		9.06	12.35
0415	Machine formed, 6" x 18", straight	B-69A	2000	.024		3.52	.99	.50	5.01	5.95
0416	6" x 18", radius	"	900	.053		3.55	2.20	1.11	6.86	8.50
0421	Curb and gutter, straight									
0422	with 6" high curb and 6" thick gutter, wood forms									
0430	24" wide, .055 C.Y. per L.F.	C-2A	375	.128	L.F.	15.35	5.80		21.15	26
0435	30" wide, .066 C.Y. per L.F.		340	.141		16.85	6.40		23.25	28.50
0440	Steel forms, 24" wide, straight		700	.069		6.80	3.11		9.91	12.20
0441	Radius		500	.096		6.75	4.36		11.11	14.10
0442	30" wide, straight		700	.069		7.90	3.11		11.01	13.45
0443	Radius		500	.096		7.70	4.36		12.06	15.10
0445	Machine formed, 24" wide, straight	B-69A	2000	.024		5.70	.99	.50	7.19	8.30
0446	Radius		900	.053		5.70	2.20	1.11	9.01	10.85
0447	30" wide, straight		2000	.024		6.60	.99	.50	8.09	9.30
0448	Radius		900	.053		6.60	2.20	1.11	9.91	11.85

32 16 13.23 Precast Concrete Curbs and Gutters

		Crew	Daily Output	Labor-Hours	Unit	Material	Labor	Equipment	Total	Total Incl O&P
0010	**PRECAST CONCRETE CURBS AND GUTTERS**									
0550	Precast, 6" x 18", straight	B-29	700	.080	L.F.	10.30	3.28	1.26	14.84	17.75
0600	6" x 18", radius	"	325	.172	"	10.80	7.05	2.71	20.56	25.50

32 16 13.33 Asphalt Curbs

		Crew	Daily Output	Labor-Hours	Unit	Material	Labor	Equipment	Total	Total Incl O&P
0010	**ASPHALT CURBS**									
0012	Curbs, asphaltic, machine formed, 8" wide, 6" high, 40 L.F./ton	B-27	1000	.032	L.F.	1.82	1.22	.30	3.34	4.21
0100	8" wide, 8" high, 30 L.F. per ton		900	.036		2.43	1.35	.33	4.11	5.10
0150	Asphaltic berm, 12" W, 3"-6" H, 35 L.F./ton, before pavement		700	.046		.04	1.74	.42	2.20	3.19
0200	12" W, 1-1/2" to 4" H, 60 L.F. per ton, laid with pavement	B-2	1050	.038		.02	1.45		1.47	2.26

32 16 Curbs, Gutters, Sidewalks, and Driveways

32 16 13 – Curbs and Gutters

32 16 13.43 Stone Curbs		Crew	Daily Output	Labor-Hours	Unit	2015 Bare Costs Material	Labor	Equipment	Total	Total Incl O&P
0010	**STONE CURBS**									
1000	Granite, split face, straight, 5″ x 16″	D-13	275	.175	L.F.	11.65	7.85	1.72	21.22	26.50
1100	6″ x 18″	″	250	.192		15.35	8.60	1.89	25.84	32
1300	Radius curbing, 6″ x 18″, over 10′ radius	B-29	260	.215		18.75	8.85	3.39	30.99	37.50
1400	Corners, 2′ radius	″	80	.700	Ea.	63	28.50	11.05	102.55	126
1600	Edging, 4-1/2″ x 12″, straight	d-13	300	.160	L.F.	5.85	7.15	1.58	14.58	19.10
1800	Curb inlets, (guttermouth) straight	B-29	41	1.366	Ea.	140	56	21.50	217.50	263
2000	Indian granite (belgian block)									
2100	Jumbo, 10-1/2″ x 7-1/2″ x 4″, grey	D-1	150	.107	L.F.	6.15	4.49		10.64	13.60
2150	Pink		150	.107		7.75	4.49		12.24	15.35
2200	Regular, 9″ x 4-1/2″ x 4-1/2″, grey		160	.100		4.67	4.21		8.88	11.60
2250	Pink		160	.100		6.15	4.21		10.36	13.20
2300	Cubes, 4″ x 4″ x 4″, grey		175	.091		3.96	3.85		7.81	10.25
2350	Pink		175	.091		3.81	3.85		7.66	10.10
2400	6″ x 6″ x 6″, pink		155	.103		12.60	4.35		16.95	20.50
2500	Alternate pricing method for indian granite									
2550	Jumbo, 10-1/2″ x 7-1/2″ x 4″ (30 lb.), grey				Ton	350			350	385
2600	Pink					450			450	495
2650	Regular, 9″ x 4-1/2″ x 4-1/2″ (20 lb.), grey					330			330	365
2700	Pink					430			430	475
2750	Cubes, 4″ x 4″ x 4″ (5 lb.), grey					480			480	530
2800	Pink					490			490	540
2850	6″ x 6″ x 6″ (25 lb.), pink					490			490	540
2900	For pallets, add					22			22	24

32 17 Paving Specialties

32 17 13 – Parking Bumpers

32 17 13.13 Metal Parking Bumpers		Crew	Daily Output	Labor-Hours	Unit	Material	Labor	Equipment	Total	Total Incl O&P
0010	**METAL PARKING BUMPERS**									
0015	Bumper rails for garages, 12 Ga. rail, 6″ wide, with steel									
0020	posts 12′-6″ O.C., minimum	E-4	190	.168	L.F.	18.25	8.95	.77	27.97	36.50
0030	Average		165	.194		23	10.30	.88	34.18	44
0100	Maximum		140	.229		27.50	12.15	1.04	40.69	52
0300	12″ channel rail, minimum		160	.200		23	10.65	.91	34.56	44.50
0400	Maximum		120	.267		34	14.15	1.22	49.37	63.50
1300	Pipe bollards, conc. filled/paint, 8′ L x 4′ D hole, 6″ diam.	B-6	20	1.200	Ea.	600	49.50	18.20	667.70	755
1400	8″ diam.		15	1.600		685	66	24.50	775.50	880
1500	12″ diam.		12	2		965	82.50	30.50	1,078	1,200
2030	Folding with individual padlocks	B-2	50	.800		605	30.50		635.50	710
8000	Parking lot control, see Section 11 12 13.10									
8900	Security bollards, SS, lighted, hyd., incl. controls, group of 3	L-7	.06	509	Ea.	48,400	23,100		71,500	89,000
8910	Group of 5	″	.04	682	″	65,000	31,000		96,000	119,000

32 17 13.16 Plastic Parking Bumpers		Crew	Daily Output	Labor-Hours	Unit	Material	Labor	Equipment	Total	Total Incl O&P
0010	**PLASTIC PARKING BUMPERS**									
1200	Thermoplastic, 6″ x 10″ x 6′-0″	B-2	120	.333	Ea.	52.50	12.65		65.15	77.50

32 17 13.19 Precast Concrete Parking Bumpers		Crew	Daily Output	Labor-Hours	Unit	Material	Labor	Equipment	Total	Total Incl O&P
0010	**PRECAST CONCRETE PARKING BUMPERS**									
1000	Wheel stops, precast concrete incl. dowels, 6″ x 10″ x 6′-0″	B-2	120	.333	Ea.	39.50	12.65		52.15	63
1100	8″ x 13″ x 6′-0″	″	120	.333	″	46	12.65		58.65	70

32 17 13 – Parking Bumpers

32 17 13.26 Wood Parking Bumpers		Crew	Daily Output	Labor-Hours	Unit	Material	2015 Bare Costs Labor	Equipment	Total	Total Incl O&P
0010	**WOOD PARKING BUMPERS**									
0020	Parking barriers, timber w/saddles, treated type									
0100	4" x 4" for cars	B-2	520	.077	L.F.	2.92	2.92		5.84	7.70
0200	6" x 6" for trucks		520	.077	"	6.10	2.92		9.02	11.20
0600	Flexible fixed stanchion, 2' high, 3" diameter		100	.400	Ea.	40.50	15.20		55.70	68

32 17 23 – Pavement Markings

32 17 23.13 Painted Pavement Markings

		Crew	Daily Output	Labor-Hours	Unit	Material	Labor	Equipment	Total	Total Incl O&P
0010	**PAINTED PAVEMENT MARKINGS**									
0020	Acrylic waterborne, white or yellow, 4" wide, less than 3000 L.F.	B-78	20000	.002	L.F.	.15	.09	.03	.27	.34
0200	6" wide, less than 3000 L.F.		11000	.004		.23	.17	.05	.45	.57
0500	8" wide, less than 3000 L.F.		10000	.005		.31	.18	.06	.55	.69
0600	12" wide, less than 3000 L.F.		4000	.012		.46	.46	.15	1.07	1.37
0620	Arrows or gore lines		2300	.021	S.F.	.22	.80	.26	1.28	1.75
0640	Temporary paint, white or yellow, less than 3000 L.F.		15000	.003	L.F.	.08	.12	.04	.24	.32
0660	Removal	1 Clab	300	.027			1		1	1.54
0680	Temporary tape	2 Clab	1500	.011		.48	.40		.88	1.15
0710	Thermoplastic, white or yellow, 4" wide, less than 6000 L.F.	B-79	15000	.003		.31	.10	.10	.51	.61
0730	6" wide, less than 6000 L.F.		14000	.003		.47	.11	.11	.69	.80
0740	8" wide, less than 6000 L.F.		12000	.003		.62	.13	.12	.87	1.02
0750	12" wide, less than 6000 L.F.		6000	.007		.91	.26	.25	1.42	1.67
0760	Arrows		660	.061	S.F.	.61	2.32	2.24	5.17	6.70
0770	Gore lines		2500	.016		.61	.61	.59	1.81	2.26
0780	Letters		660	.061		.61	2.32	2.24	5.17	6.70
1000	Airport painted markings									
1050	Traffic safety flashing truck for airport painting	A-2B	1	8	Day		315	245	560	745
1100	Painting, white or yellow, taxiway markings	B-78	4000	.012	S.F.	.27	.46	.15	.88	1.16
1110	with 12 lb. beads per 100 S.F.		4000	.012		.53	.46	.15	1.14	1.44
1200	Runway markings		3500	.014		.27	.52	.17	.96	1.29
1210	with 12 lb. beads per 100 S.F.		3500	.014		.53	.52	.17	1.22	1.57
1300	Pavement location or direction signs		2500	.019		.27	.73	.24	1.24	1.68
1310	with 12 lb. beads per 100 S.F.		2500	.019		.53	.73	.24	1.50	1.96
1350	Mobilization airport pavement painting		4	12	Ea.		460	150	610	870
1400	Paint markings or pavement signs removal daytime	B-78B	400	.045	S.F.		1.75	.94	2.69	3.71
1500	Removal nighttime		335	.054	"		2.09	1.12	3.21	4.43
1600	Mobilization pavement paint removal		4	4.500	Ea.		175	94	269	370

32 17 23.14 Pavement Parking Markings

		Crew	Daily Output	Labor-Hours	Unit	Material	Labor	Equipment	Total	Total Incl O&P
0010	**PAVEMENT PARKING MARKINGS**									
0790	Layout of pavement marking	A-2	25000	.001	L.F.		.04	.01	.05	.07
0800	Lines on pvmt., parking stall, paint, white, 4" wide	B-78B	400	.045	Stall	4.44	1.75	.94	7.13	8.60
0825	Parking stall, small quantities	2 Pord	80	.200		8.90	8.05		16.95	22
0830	Lines on pvmt., parking stall, thermoplastic, white, 4" wide	B-79	300	.133		13.50	5.10	4.92	23.52	28
1000	Street letters and numbers	B-78B	1600	.011	S.F.	.67	.44	.23	1.34	1.67

32 18 Athletic and Recreational Surfacing

32 18 13 – Synthetic Grass Surfacing

32 18 13.10 Artificial Grass Surfacing		Crew	Daily Output	Labor-Hours	Unit	2015 Bare Costs Material	Labor	Equipment	Total	Total Incl O&P
0010	**ARTIFICIAL GRASS SURFACING**									
0015	Not including asphalt base or drainage,									
0020	but including cushion pad, over 50,000 S.F.									
0200	1/2" pile and 5/16" cushion pad, standard	C-17	3200	.025	S.F.	10.15	1.23		11.38	13.10
0300	Deluxe		2560	.031		15	1.53		16.53	18.85
0500	1/2" pile and 5/8" cushion pad, standard		2844	.028		14.60	1.38		15.98	18.20
0600	Deluxe	↓	2327	.034	↓	16.10	1.69		17.79	20.50
0800	For asphaltic concrete base, 2-1/2" thick,									
0900	with 6" crushed stone sub-base, add	B-25	12000	.007	S.F.	1.69	.30	.23	2.22	2.57

32 18 16 – Synthetic Resilient Surfacing

32 18 16.13 Playground Protective Surfacing		Crew	Daily Output	Labor-Hours	Unit	Material	Labor	Equipment	Total	Total Incl O&P
0010	**PLAYGROUND PROTECTIVE SURFACING**									
0100	Resilient rubber surface, poured in place, 4" thick, black	2 Skwk	300	.053	S.F.	13.35	2.59		15.94	18.70
0150	2" thick topping, colors	"	2800	.006		7.25	.28		7.53	8.40
0200	Wood chip mulch, 6" deep	1 Clab	300	.027	↓	.76	1		1.76	2.38

32 18 23 – Athletic Surfacing

32 18 23.33 Running Track Surfacing		Crew	Daily Output	Labor-Hours	Unit	Material	Labor	Equipment	Total	Total Incl O&P
0010	**RUNNING TRACK SURFACING**									
0020	Running track, asphalt, incl base, 3" thick	B-37	300	.160	S.Y.	13	6.35	.52	19.87	24.50
0102	Surface, latex rubber system, 1/2" thick, black	B-20	115	.209		41	8.75		49.75	59
0152	Colors		115	.209		50.50	8.75		59.25	69
0302	Urethane rubber system, 1/2" thick, black		110	.218		30.50	9.15		39.65	47.50
0402	Color coating	↓	110	.218	↓	37.50	9.15		46.65	55

32 18 23.53 Tennis Court Surfacing		Crew	Daily Output	Labor-Hours	Unit	Material	Labor	Equipment	Total	Total Incl O&P
0010	**TENNIS COURT SURFACING**									
0020	Tennis court, asphalt, incl. base, 2-1/2" thick, one court	B-37	450	.107	S.Y.	37.50	4.24	.35	42.09	48.50
0200	Two courts		675	.071		15.85	2.83	.23	18.91	22
0300	Clay courts		360	.133		42	5.30	.44	47.74	55
0400	Pulverized natural greenstone with 4" base, fast dry		250	.192		39.50	7.65	.63	47.78	56
0800	Rubber-acrylic base resilient pavement	↓	600	.080		56	3.18	.26	59.44	66.50
1000	Colored sealer, acrylic emulsion, 3 coats	2 Clab	800	.020		6.05	.75		6.80	7.80
1100	3 coat, 2 colors	"	900	.018		8.45	.67		9.12	10.30
1200	For preparing old courts, add	1 Clab	825	.010	↓		.36		.36	.56
1400	Posts for nets, 3-1/2" diameter with eye bolts	B-1	3.40	7.059	Pr.	305	270		575	750
1500	With pulley & reel		3.40	7.059	"	780	270		1,050	1,275
1700	Net, 42' long, nylon thread with binder		50	.480	Ea.	254	18.35		272.35	310
1800	All metal	↓	6.50	3.692	"	490	141		631	755
2000	Paint markings on asphalt, 2 coats	1 Pord	1.78	4.494	Court	187	181		368	480
2200	Complete court with fence, etc., asphaltic conc., minimum	B-37	.20	240		28,700	9,550	785	39,035	47,000
2300	Maximum		.16	300		56,500	11,900	985	69,385	82,000
2800	Clay courts, minimum		.20	240		31,500	9,550	785	41,835	50,000
2900	Maximum	↓	.16	300	↓	58,000	11,900	985	70,885	83,500

32 31 13.20 Fence, Chain Link Industrial		Crew	Daily Output	Labor-Hours	Unit	Material	2015 Bare Costs Labor	Equipment	Total	Total Incl O&P
0010	**FENCE, CHAIN LINK INDUSTRIAL**									
0011	Schedule 40, including concrete									
0020	3 strands barb wire, 2" post @ 10′ O.C., set in concrete, 6′ H									
0200	9 ga. wire, galv. steel, in concrete	B-80C	240	.100	L.F.	19.10	3.81	1.06	23.97	28
0248	Fence, add for vinyl coated fabric				S.F.	.66			.66	.73
0300	Aluminized steel	B-80C	240	.100	L.F.	20.50	3.81	1.06	25.37	29.50
0500	6 ga. wire, galv. steel		240	.100		21	3.81	1.06	25.87	30
0600	Aluminized steel		240	.100		30	3.81	1.06	34.87	40
0800	6 ga. wire, 6′ high but omit barbed wire, galv. steel		250	.096		19.55	3.66	1.02	24.23	28
0900	Aluminized steel, in concrete		250	.096		23.50	3.66	1.02	28.18	32.50
0920	8′ H, 6 ga. wire, 2-1/2" line post, galv. steel, in concrete		180	.133		31	5.10	1.41	37.51	43.50
0940	Aluminized steel, in concrete		180	.133		38	5.10	1.41	44.51	51
1100	Add for corner posts, 3" diam., galv. steel, in concrete		40	.600	Ea.	86.50	23	6.35	115.85	138
1200	Aluminized steel, in concrete		40	.600		86.50	23	6.35	115.85	138
1300	Add for braces, galv. steel		80	.300		35.50	11.45	3.17	50.12	60
1350	Aluminized steel		80	.300		46	11.45	3.17	60.62	71.50
1400	Gate for 6′ high fence, 1-5/8" frame, 3′ wide, galv. steel		10	2.400		203	91.50	25.50	320	390
1500	Aluminized steel, in concrete		10	2.400		205	91.50	25.50	322	395
2000	5′-0" high fence, 9 ga., no barbed wire, 2" line post, in concrete									
2010	10′ O.C., 1-5/8" top rail, in concrete									
2100	Galvanized steel, in concrete	B-80C	300	.080	L.F.	18.45	3.05	.85	22.35	26
2200	Aluminized steel, in concrete		300	.080	"	18.75	3.05	.85	22.65	26
2400	Gate, 4′ wide, 5′ high, 2" frame, galv. steel, in concrete		10	2.400	Ea.	188	91.50	25.50	305	375
2500	Aluminized steel, in concrete		10	2.400	"	200	91.50	25.50	317	390
3100	Overhead slide gate, chain link, 6′ high, to 18′ wide, in concrete		38	.632	L.F.	97	24	6.70	127.70	151
3110	Cantilever type, in concrete	B-80	48	.667		129	27.50	15	171.50	201
3120	8′ high, in concrete		24	1.333		155	55	30	240	288
3130	10′ high, in concrete		18	1.778		190	73.50	40	303.50	365
5000	Double swing gates, incl. posts & hardware, in concrete									
5010	5′ high, 12′ opening, in concrete	B-80C	3.40	7.059	Opng.	395	269	74.50	738.50	925
5020	20′ opening, in concrete		2.80	8.571		520	325	90.50	935.50	1,175
5060	6′ high, 12′ opening, in concrete		3.20	7.500		455	286	79.50	820.50	1,025
5070	20′ opening, in concrete		2.60	9.231		655	350	97.50	1,102.50	1,375
5080	8′ high, 12′ opening, in concrete	B-80	2.13	15.002		460	620	340	1,420	1,825
5090	20′ opening, in concrete		1.45	22.069		685	910	495	2,090	2,700
5100	10′ high, 12′ opening, in concrete		1.31	24.427		825	1,000	550	2,375	3,025
5110	20′ opening, in concrete		1.03	31.068		865	1,275	700	2,840	3,675
5120	12′ high, 12′ opening, in concrete		1.05	30.476		1,175	1,250	685	3,110	3,975
5130	20′ opening, in concrete		.85	37.647		1,225	1,550	845	3,620	4,650
5190	For aluminized steel add					20%				

32 31 13.25 Fence, Chain Link Residential		Crew	Daily Output	Labor-Hours	Unit	Material	Labor	Equipment	Total	Total Incl O&P
0010	**FENCE, CHAIN LINK RESIDENTIAL**									
0011	Schedule 20, 11 ga. wire, 1-5/8" post									
0020	10′ O.C., 1-3/8" top rail, 2" corner post, galv. stl. 3′ high	B-80C	500	.048	L.F.	2.12	1.83	.51	4.46	5.70
0050	4′ high		400	.060		7.10	2.29	.63	10.02	12
0100	6′ high		200	.120		9.45	4.57	1.27	15.29	18.80
0150	Add for gate 3′ wide, 1-3/8" frame, 3′ high		12	2	Ea.	81.50	76	21	178.50	230
0170	4′ high		10	2.400		87.50	91.50	25.50	204.50	264
0190	6′ high		10	2.400		109	91.50	25.50	226	288
0200	Add for gate 4′ wide, 1-3/8" frame, 3′ high		9	2.667		91.50	102	28	221.50	287
0220	4′ high		9	2.667		97.50	102	28	227.50	293
0240	6′ high		8	3		123	114	31.50	268.50	345

32 31 13 – Chain Link Fences and Gates

32 31 13.25 Fence, Chain Link Residential		Crew	Daily Output	Labor-Hours	Unit	Material	Labor	Equipment	Total	Total Incl O&P
						2015 Bare Costs				
0350	Aluminized steel, 11 ga. wire, 3′ high	B-80C	500	.048	L.F.	8.80	1.83	.51	11.14	13.05
0380	4′ high		400	.060		9.25	2.29	.63	12.17	14.35
0400	6′ high		200	.120		11.15	4.57	1.27	16.99	20.50
0450	Add for gate 3′ wide, 1-3/8″ frame, 3′ high		12	2	Ea.	95	76	21	192	245
0470	4′ high		10	2.400		101	91.50	25.50	218	279
0490	6′ high		10	2.400		125	91.50	25.50	242	305
0500	Add for gate 4′ wide, 1-3/8″ frame, 3′ high		10	2.400		105	91.50	25.50	222	284
0520	4′ high		9	2.667		121	102	28	251	320
0540	6′ high		8	3		130	114	31.50	275.50	355
0620	Vinyl covered, 9 ga. wire, 3′ high		500	.048	L.F.	7.80	1.83	.51	10.14	11.95
0640	4′ high		400	.060		8.15	2.29	.63	11.07	13.20
0660	6′ high		200	.120		10.15	4.57	1.27	15.99	19.55
0720	Add for gate 3′ wide, 1-3/8″ frame, 3′ high		12	2	Ea.	94.50	76	21	191.50	245
0740	4′ high		10	2.400		101	91.50	25.50	218	279
0760	6′ high		10	2.400		120	91.50	25.50	237	300
0780	Add for gate 4′ wide, 1-3/8″ frame, 3′ high		10	2.400		99.50	91.50	25.50	216.50	277
0800	4′ high		9	2.667		103	102	28	233	299
0820	6′ high		8	3		128	114	31.50	273.50	350
7076	Fence, for small jobs 100 L.F. fence or less w/or wo gate, add				S.F.	20%				

32 31 13.26 Tennis Court Fences and Gates		Crew	Daily Output	Labor-Hours	Unit	Material	Labor	Equipment	Total	Total Incl O&P
0010	**TENNIS COURT FENCES AND GATES**									
0860	Tennis courts, 11 ga. wire, 2-1/2″ post set									
0870	in concrete, 10′ O.C., 1-5/8″ top rail									
0900	10′ high	B-80	190	.168	L.F.	22.50	6.95	3.79	33.24	39.50
0920	12′ high		170	.188	″	23.50	7.75	4.24	35.49	42.50
1000	Add for gate 4′ wide, 1-5/8″ frame 7′ high		10	3.200	Ea.	242	132	72	446	545
1040	Aluminized steel, 11 ga. wire 10′ high		190	.168	L.F.	21	6.95	3.79	31.74	38
1100	12′ high		170	.188	″	23	7.75	4.24	34.99	41.50
1140	Add for gate 4′ wide, 1-5/8″ frame, 7′ high		10	3.200	Ea.	262	132	72	466	570
1250	Vinyl covered, 9 ga. wire, 10′ high		190	.168	L.F.	21.50	6.95	3.79	32.24	38.50
1300	12′ high		170	.188	″	25.50	7.75	4.24	37.49	44.50
1310	Fence, CL, tennis court, transom gate, single, galv., 4′ x 7′	B-80A	8.72	2.752	Ea.	310	103	35	448	540
1400	Add for gate 4′ wide, 1-5/8″ frame, 7′ high	B-80	10	3.200	″	315	132	72	519	625

32 31 13.33 Chain Link Backstops		Crew	Daily Output	Labor-Hours	Unit	Material	Labor	Equipment	Total	Total Incl O&P
0010	**CHAIN LINK BACKSTOPS**									
0015	Backstops, baseball, prefabricated, 30′ wide, 12′ high & 1 overhang	B-1	1	24	Ea.	2,575	920		3,495	4,275
0100	40′ wide, 12′ high & 2 overhangs	″	.75	32		6,775	1,225		8,000	9,325
0300	Basketball, steel, single goal	B-13	3.04	18.421		1,425	755	242	2,422	3,000
0400	Double goal	″	1.92	29.167		1,925	1,200	385	3,510	4,375
0600	Tennis, wire mesh with pair of ends	B-1	2.48	9.677	Set	2,650	370		3,020	3,475
0700	Enclosed court	″	1.30	18.462	Ea.	8,975	705		9,680	10,900

32 31 13.53 High-Security Chain Link Fences, Gates and Sys.		Crew	Daily Output	Labor-Hours	Unit	Material	Labor	Equipment	Total	Total Incl O&P
0010	**HIGH-SECURITY CHAIN LINK FENCES, GATES AND SYSTEMS**									
0100	Fence, chain link, security, 7′ H, standard FE-7, incl excavation & posts	B-80C	480	.050	L.F.	44	1.91	.53	46.44	52
0200	Fence, barbed wire, security, 7′ high, with 3 wire barbed wire arm	″	400	.060	″	7.75	2.29	.63	10.67	12.70
0300	Complete systems, including material and installation									
0310	Taunt wire fence detection system				M.L.F.				25,100	27,600
0410	Microwave fence detection system								41,300	45,400
0510	Passive magnetic fence detection system								19,500	21,400
0610	Infrared fence detection system								12,900	14,400
0710	Strain relief fence detection system								25,100	27,600
0810	Electro-shock fence detection system								35,900	39,500

32 31 Fences and Gates

32 31 13 – Chain Link Fences and Gates

32 31 13.53 High-Security Chain Link Fences, Gates and Sys.		Crew	Daily Output	Labor-Hours	Unit	Material	2015 Bare Costs Labor	Equipment	Total	Total Incl O&P
0910	Photo-electric fence detection system				M.L.F.				16,300	18,000

32 31 19 – Decorative Metal Fences and Gates

32 31 19.10 Decorative Fence		Crew	Daily Output	Labor-Hours	Unit	Material	Labor	Equipment	Total	Total Incl O&P
0010	**DECORATIVE FENCE**									
5300	Tubular picket, steel, 6′ sections, 1-9/16″ posts, 4′ high	B-80C	300	.080	L.F.	31	3.05	.85	34.90	39.50
5400	2″ posts, 5′ high		240	.100		35	3.81	1.06	39.87	45.50
5600	2″ posts, 6′ high		200	.120		42	4.57	1.27	47.84	55
5700	Staggered picket 1-9/16″ posts, 4′ high		300	.080		31	3.05	.85	34.90	39.50
5800	2″ posts, 5′ high		240	.100		35	3.81	1.06	39.87	45.50
5900	2″ posts, 6′ high		200	.120		42	4.57	1.27	47.84	55
6200	Gates, 4′ high, 3′ wide	B-1	10	2.400	Ea.	282	92		374	450
6300	5′ high, 3′ wide		10	2.400		350	92		442	525
6400	6′ high, 3′ wide		10	2.400		415	92		507	595
6500	4′ wide		10	2.400		420	92		512	605

32 31 26 – Wire Fences and Gates

32 31 26.10 Fences, Misc. Metal		Crew	Daily Output	Labor-Hours	Unit	Material	Labor	Equipment	Total	Total Incl O&P
0010	**FENCES, MISC. METAL**									
0012	Chicken wire, posts @ 4′, 1″ mesh, 4′ high	B-80C	410	.059	L.F.	3.30	2.23	.62	6.15	7.70
0100	2″ mesh, 6′ high		350	.069		3.82	2.61	.73	7.16	9
0200	Galv. steel, 12 ga., 2″ x 4″ mesh, posts 5′ O.C., 3′ high		300	.080		2.78	3.05	.85	6.68	8.65
0300	5′ high		300	.080		3.38	3.05	.85	7.28	9.30
0400	14 ga., 1″ x 2″ mesh, 3′ high		300	.080		3.42	3.05	.85	7.32	9.35
0500	5′ high		300	.080		4.57	3.05	.85	8.47	10.65
1000	Kennel fencing, 1-1/2″ mesh, 6′ long, 3′-6″ wide, 6′-2″ high	2 Clab	4	4	Ea.	510	150		660	790
1050	12′ long		4	4		720	150		870	1,025
1200	Top covers, 1-1/2″ mesh, 6′ long		15	1.067		136	40		176	212
1250	12′ long		12	1.333		191	50		241	287
1300	For kennel doors, see Section 08 31 13.40									
4500	Security fence, prison grade, set in concrete, 12′ high	B-80	25	1.280	L.F.	61.50	53	29	143.50	180
4600	16′ high	″	20	1.600	″	79	66	36	181	228

32 31 26.20 Wire Fencing, General		Crew	Daily Output	Labor-Hours	Unit	Material	Labor	Equipment	Total	Total Incl O&P
0010	**WIRE FENCING, GENERAL**									
0015	Barbed wire, galvanized, domestic steel, hi-tensile 15-1/2 ga.				M.L.F.	98.50			98.50	108
0020	Standard, 12-3/4 ga.					111			111	122
0210	Barbless wire, 2-strand galvanized, 12-1/2 ga.					111			111	122
0500	Helical razor ribbon, stainless steel, 18″ dia x 18″ spacing				C.L.F.	164			164	180
0600	Hardware cloth galv., 1/4″ mesh, 23 ga., 2′ wide				C.S.F.	60			60	66
0700	3′ wide					43.50			43.50	48
0900	1/2″ mesh, 19 ga., 2′ wide					35.50			35.50	39
1000	4′ wide					23.50			23.50	26
1200	Chain link fabric, steel, 2″ mesh, 6 ga., galvanized					152			152	167
1300	9 ga., galvanized					86			86	94.50
1350	Vinyl coated					82			82	90
1360	Aluminized					79.50			79.50	87.50
1400	2-1/4″ mesh, 11.5 ga., galvanized					54.50			54.50	60
1600	1-3/4″ mesh (tennis courts), 11.5 ga. (core), vinyl coated					61			61	67
1700	9 ga., galvanized					82.50			82.50	90.50
2100	Welded wire fabric, galvanized, 1″ x 2″, 14 ga.					59.50			59.50	65.50
2200	2″ x 4″, 12-1/2 ga.					57			57	62.50

32 31 Fences and Gates

32 31 29 – Wood Fences and Gates

32 31 29.20 Fence, Wood Rail		Crew	Daily Output	Labor-Hours	Unit	Material	2015 Bare Costs Labor	Equipment	Total	Total Incl O&P
0010	**FENCE, WOOD RAIL**									
0012	Picket, No. 2 cedar, Gothic, 2 rail, 3′ high	B-1	160	.150	L.F.	7.60	5.75		13.35	17.20
0050	Gate, 3′-6″ wide	B-80C	9	2.667	Ea.	77	102	28	207	271
0400	3 rail, 4′ high		150	.160	L.F.	8.50	6.10	1.69	16.29	20.50
0500	Gate, 3′-6″ wide		9	2.667	Ea.	95	102	28	225	290
0600	Open rail, rustic, No. 1 cedar, 2 rail, 3′ high		160	.150	L.F.	5.90	5.70	1.59	13.19	17
0650	Gate, 3′ wide		9	2.667	Ea.	82	102	28	212	276
0700	3 rail, 4′ high		150	.160	L.F.	7.10	6.10	1.69	14.89	18.95
0900	Gate, 3′ wide		9	2.667	Ea.	102	102	28	232	298
1200	Stockade, No. 2 cedar, treated wood rails, 6′ high		160	.150	L.F.	8.90	5.70	1.59	16.19	20.50
1250	Gate, 3′ wide		9	2.667	Ea.	96	102	28	226	292
1300	No. 1 cedar, 3-1/4″ cedar rails, 6′ high		160	.150	L.F.	20.50	5.70	1.59	27.79	33
1500	Gate, 3′ wide		9	2.667	Ea.	217	102	28	347	425
1520	Open rail, split, No. 1 cedar, 2 rail, 3′ high		160	.150	L.F.	5.90	5.70	1.59	13.19	17
1540	3 rail, 4′-0″ high		150	.160		7.85	6.10	1.69	15.64	19.75
3300	Board, shadow box, 1″ x 6″, treated pine, 6′ high		160	.150		12.45	5.70	1.59	19.74	24
3400	No. 1 cedar, 6′ high		150	.160		24.50	6.10	1.69	32.29	38
3900	Basket weave, No. 1 cedar, 6′ high		160	.150		34	5.70	1.59	41.29	47.50
3950	Gate, 3′-6″ wide	B-1	8	3	Ea.	232	115		347	430
4000	Treated pine, 6′ high		150	.160	L.F.	16	6.10		22.10	27
4200	Gate, 3′-6″ wide		9	2.667	Ea.	176	102		278	350
5000	Fence rail, redwood, 2″ x 4″, merch. grade 8′		2400	.010	L.F.	2.42	.38		2.80	3.25
5050	Select grade, 8′		2400	.010	″	5.35	.38		5.73	6.45
6000	Fence post, select redwood, earthpacked & treated, 4″ x 4″ x 6′		96	.250	Ea.	13.60	9.55		23.15	29.50
6010	4″ x 4″ x 8′		96	.250		18.70	9.55		28.25	35
6020	Set in concrete, 4″ x 4″ x 6′		50	.480		21	18.35		39.35	52
6030	4″ x 4″ x 8′		50	.480		22.50	18.35		40.85	53
6040	Wood post, 4′ high, set in concrete, incl. concrete		50	.480		13.80	18.35		32.15	43.50
6050	Earth packed		96	.250		16.70	9.55		26.25	33
6060	6′ high, set in concrete, incl. concrete		50	.480		17.20	18.35		35.55	47.50
6070	Earth packed		96	.250		13.10	9.55		22.65	29

32 32 Retaining Walls

32 32 13 – Cast-in-Place Concrete Retaining Walls

32 32 13.10 Retaining Walls, Cast Concrete		Crew	Daily Output	Labor-Hours	Unit	Material	Labor	Equipment	Total	Total Incl O&P
0010	**RETAINING WALLS, CAST CONCRETE**									
1800	Concrete gravity wall with vertical face including excavation & backfill									
1850	No reinforcing									
1900	6′ high, level embankment	C-17C	36	2.306	L.F.	74.50	113	16.90	204.40	276
2000	33° slope embankment		32	2.594		86.50	127	19.05	232.55	315
2200	8′ high, no surcharge		27	3.074		92.50	151	22.50	266	360
2300	33° slope embankment		24	3.458		112	170	25.50	307.50	415
2500	10′ high, level embankment		19	4.368		132	215	32	379	510
2600	33° slope embankment		18	4.611		183	227	34	444	590
2800	Reinforced concrete cantilever, incl. excavation, backfill & reinf.									
2900	6′ high, 33° slope embankment	C-17C	35	2.371	L.F.	68	117	17.40	202.40	274
3000	8′ high, 33° slope embankment		29	2.862		78	141	21	240	325
3100	10′ high, 33° slope embankment		20	4.150		102	204	30.50	336.50	460
3200	20′ high, 500 lb. per L.F. surcharge		7.50	11.067		305	545	81	931	1,275
3500	Concrete cribbing, incl. excavation and backfill									

32 32 Retaining Walls

32 32 13 – Cast-in-Place Concrete Retaining Walls

32 32 13.10 Retaining Walls, Cast Concrete		Crew	Daily Output	Labor-Hours	Unit	Material	2015 Bare Costs Labor	Equipment	Total	Total Incl O&P
3700	12′ high, open face	B-13	210	.267	S.F.	34	10.95	3.51	48.46	58
3900	Closed face	″	210	.267	″	32	10.95	3.51	46.46	55.50
4100	Concrete filled slurry trench, see Section 31 56 23.20									

32 32 23 – Segmental Retaining Walls

32 32 23.13 Segmental Conc. Unit Masonry Retaining Walls		Crew	Daily Output	Labor-Hours	Unit	Material	Labor	Equipment	Total	Total Incl O&P
0010	**SEGMENTAL CONC. UNIT MASONRY RETAINING WALLS**									
7100	Segmental Retaining Wall system, incl. pins, and void fill									
7120	base and backfill not included									
7140	Large unit, 8″ high x 18″ wide x 20″ deep, 3 plane split	B-62	300	.080	S.F.	13.50	3.30	.58	17.38	20.50
7150	Straight split		300	.080		13.60	3.30	.58	17.48	20.50
7160	Medium, lt. wt., 8″ high x 18″ wide x 12″ deep, 3 plane split		400	.060		10.50	2.48	.43	13.41	15.80
7170	Straight split		400	.060		10.40	2.48	.43	13.31	15.70
7180	Small unit, 4″ x 18″ x 10″ deep, 3 plane split		400	.060		13.40	2.48	.43	16.31	19
7190	Straight split		400	.060		13.20	2.48	.43	16.11	18.75
7200	Cap unit, 3 plane split		300	.080		13.80	3.30	.58	17.68	21
7210	Cap unit, straight split		300	.080		13.80	3.30	.58	17.68	21
7250	Geo-grid soil reinforcement 4′ x 50′	2 Clab	22500	.001		.76	.03		.79	.88
7255	Geo-grid soil reinforcement 6′ x 150′	″	22500	.001		.60	.03		.63	.70
8000	For higher walls, add components as necessary									

32 32 26 – Metal Crib Retaining Walls

32 32 26.10 Metal Bin Retaining Walls		Crew	Daily Output	Labor-Hours	Unit	Material	Labor	Equipment	Total	Total Incl O&P
0010	**METAL BIN RETAINING WALLS**									
0011	Aluminized steel bin, excavation									
0020	and backfill not included, 10′ wide									
0100	4′ high, 5.5′ deep	B-13	650	.086	S.F.	27	3.53	1.13	31.66	36
0200	8′ high, 5.5′ deep		615	.091		31	3.73	1.20	35.93	41
0300	10′ high, 7.7′ deep		580	.097		34.50	3.96	1.27	39.73	45.50
0400	12′ high, 7.7′ deep		530	.106		37	4.33	1.39	42.72	49
0500	16′ high, 7.7′ deep		515	.109		39.50	4.46	1.43	45.39	51.50
0600	16′ high, 9.9′ deep		500	.112		41.50	4.59	1.47	47.56	54
0700	20′ high, 9.9′ deep		470	.119		46.50	4.88	1.57	52.95	60.50
0800	20′ high, 12.1′ deep		460	.122		42	4.99	1.60	48.59	55.50
0900	24′ high, 12.1′ deep		455	.123		44.50	5.05	1.62	51.17	58.50
1000	24′ high, 14.3′ deep		450	.124		52.50	5.10	1.64	59.24	67
1100	28′ high, 14.3′ deep		440	.127		54.50	5.20	1.67	61.37	70
1300	For plain galvanized bin type walls, deduct					10%				

32 32 29 – Timber Retaining Walls

32 32 29.10 Landscape Timber Retaining Walls		Crew	Daily Output	Labor-Hours	Unit	Material	Labor	Equipment	Total	Total Incl O&P
0010	**LANDSCAPE TIMBER RETAINING WALLS**									
0100	Treated timbers, 6″ x 6″	1 Clab	265	.030	L.F.	2.01	1.14		3.15	3.96
0110	6″ x 8″	″	200	.040	″	2.65	1.50		4.15	5.25
0120	Drilling holes in timbers for fastening, 1/2″	1 Carp	450	.018	Inch		.83		.83	1.28
0130	5/8″	″	450	.018	″		.83		.83	1.28
0140	Reinforcing rods for fastening, 1/2″	1 Clab	312	.026	L.F.	.36	.96		1.32	1.87
0150	5/8″	″	312	.026	″	.56	.96		1.52	2.09
0160	Reinforcing fabric	2 Clab	2500	.006	S.Y.	1.90	.24		2.14	2.46
0170	Gravel backfill		28	.571	C.Y.	20	21.50		41.50	55
0180	Perforated pipe, 4″ diameter with silt sock		1200	.013	L.F.	1.25	.50		1.75	2.15
0190	Galvanized 60d common nails	1 Clab	625	.013	Ea.	.16	.48		.64	.92
0200	20d common nails	″	3800	.002	″	.04	.08		.12	.16

32 32 Retaining Walls

32 32 36 – Gabion Retaining Walls

32 32 36.10 Stone Gabion Retaining Walls		Crew	Daily Output	Labor-Hours	Unit	Material	Labor	Equipment	Total	Total Incl O&P
						2015 Bare Costs				
0010	**STONE GABION RETAINING WALLS**									
4300	Stone filled gabions, not incl. excavation,									
4310	Stone, delivered, 3′ wide									
4350	Galvanized, 6′ high, 33° slope embankment	B-13	49	1.143	L.F.	45	47	15.05	107.05	138
4500	Highway surcharge		27	2.074		88	85	27.50	200.50	257
4600	9′ high, up to 33° slope embankment		24	2.333		101	95.50	30.50	227	291
4700	Highway surcharge		16	3.500		154	143	46	343	440
4900	12′ high, up to 33° slope embankment		14	4		157	164	52.50	373.50	480
5000	Highway surcharge	↓	11	5.091		221	209	67	497	635
5950	For PVC coating, add				↓	12%				

32 32 53 – Stone Retaining Walls

32 32 53.10 Retaining Walls, Stone		Crew	Daily Output	Labor-Hours	Unit	Material	Labor	Equipment	Total	Total Incl O&P
0010	**RETAINING WALLS, STONE**									
0015	Including excavation, concrete footing and									
0020	stone 3′ below grade. Price is exposed face area.									
0200	Decorative random stone, to 6′ high, 1′-6″ thick, dry set	D-1	35	.457	S.F.	58.50	19.25		77.75	94
0300	Mortar set		40	.400		60.50	16.85		77.35	92
0500	Cut stone, to 6′ high, 1′-6″ thick, dry set		35	.457		60.50	19.25		79.75	96.50
0600	Mortar set		40	.400		61.50	16.85		78.35	93
0800	Random stone, 6′ to 10′ high, 2′ thick, dry set		45	.356		66.50	14.95		81.45	96
0900	Mortar set		50	.320		69	13.45		82.45	96
1100	Cut stone, 6′ to 10′ high, 2′ thick, dry set		45	.356		67	14.95		81.95	96.50
1200	Mortar set	↓	50	.320	↓	69	13.45		82.45	96.50

32 33 Site Furnishings

32 33 33 – Site Manufactured Planters

32 33 33.10 Planters		Crew	Daily Output	Labor-Hours	Unit	Material	Labor	Equipment	Total	Total Incl O&P
0010	**PLANTERS**									
0012	Concrete, sandblasted, precast, 48″ diameter, 24″ high	2 Clab	15	1.067	Ea.	630	40		670	755
0100	Fluted, precast, 7′ diameter, 36″ high		10	1.600		1,575	60		1,635	1,825
0300	Fiberglass, circular, 36″ diameter, 24″ high		15	1.067		725	40		765	855
0320	36″ diameter, 27″ high		12	1.333		725	50		775	870
0330	33″ high		15	1.067		770	40		810	905
0335	24″ diameter, 36″ high		15	1.067		425	40		465	530
0340	60″ diameter, 39″ high		8	2		1,275	75		1,350	1,525
0400	60″ diameter, 24″ high		10	1.600		1,125	60		1,185	1,325
0600	Square, 24″ side, 36″ high		15	1.067		620	40		660	740
0610	24″ side, 27″ high		12	1.333		675	50		725	815
0620	24″ side, 16″ high		20	.800		320	30		350	400
0700	48″ side, 36″ high		15	1.067		1,025	40		1,065	1,175
0900	Planter/bench, 72″ square, 36″ high		5	3.200		1,800	120		1,920	2,150
1000	96″ square, 27″ high		5	3.200		2,225	120		2,345	2,625
1200	Wood, square, 48″ side, 24″ high		15	1.067		1,400	40		1,440	1,600
1300	Circular, 48″ diameter, 30″ high		10	1.600		985	60		1,045	1,175
1500	72″ diameter, 30″ high		10	1.600		1,725	60		1,785	2,000
1600	Planter/bench, 72″	↓	5	3.200	↓	3,275	120		3,395	3,775

32 33 Site Furnishings

32 33 43 – Site Seating and Tables

	32 33 43.13 Site Seating	Crew	Daily Output	Labor-Hours	Unit	Material	2015 Bare Costs Labor	Equipment	Total	Total Incl O&P
0010	**SITE SEATING**									
0012	Seating, benches, park, precast conc., w/backs, wood rails, 4' long	2 Clab	5	3.200	Ea.	595	120		715	840
0100	8' long		4	4		955	150		1,105	1,275
0300	Fiberglass, without back, one piece, 4' long		10	1.600		620	60		680	775
0400	8' long		7	2.286		815	86		901	1,025
0500	Steel barstock pedestals w/backs, 2" x 3" wood rails, 4' long		10	1.600		1,125	60		1,185	1,325
0510	8' long		7	2.286		1,425	86		1,511	1,700
0515	Powder coated steel, 4" x 4" plastic slats, 6' L		8	2		490	75		565	650
0520	3" x 8" wood plank, 4' long		10	1.600		1,200	60		1,260	1,425
0530	8' long		7	2.286		1,500	86		1,586	1,775
0540	Backless, 4" x 4" wood plank, 4' square		10	1.600		940	60		1,000	1,125
0550	8' long		7	2.286		1,000	86		1,086	1,225
0560	Powder coated steel, with back and 2 anti-vagrant dividers, 6' long		8	2		1,075	75		1,150	1,325
0600	Aluminum pedestals, with backs, aluminum slats, 8' long		8	2		480	75		555	640
0610	15' long		5	3.200		975	120		1,095	1,250
0620	Portable, aluminum slats, 8' long		8	2		465	75		540	625
0630	15' long		5	3.200		570	120		690	815
0800	Cast iron pedestals, back & arms, wood slats, 4' long		8	2		385	75		460	535
0820	8' long		5	3.200		1,075	120		1,195	1,350
0840	Backless, wood slats, 4' long		8	2		590	75		665	760
0860	8' long		5	3.200		1,150	120		1,270	1,450
1700	Steel frame, fir seat, 10' long		10	1.600		355	60		415	490
2000	Benches, park, with back, galv. stl. frame, 4" x 4" plastic slats, 6' L	▼	7	2.286	▼	490	86		576	665

32 34 Fabricated Bridges

32 34 20 – Fabricated Pedestrian Bridges

	32 34 20.10 Bridges, Pedestrian	Crew	Daily Output	Labor-Hours	Unit	Material	Labor	Equipment	Total	Total Incl O&P
0010	**BRIDGES, PEDESTRIAN**									
0011	Spans over streams, roadways, etc.									
0020	including erection, not including foundations									
0050	Precast concrete, complete in place, 8' wide, 60' span	E-2	215	.260	S.F.	116	13.50	7.05	136.55	158
0100	100' span		185	.303		127	15.65	8.15	150.80	176
0150	120' span		160	.350		138	18.10	9.45	165.55	193
0200	150' span		145	.386		144	20	10.40	174.40	203
0300	Steel, trussed or arch spans, compl. in place, 8' wide, 40' span		320	.175		117	9.05	4.72	130.77	149
0400	50' span		395	.142		105	7.35	3.83	116.18	133
0500	60' span		465	.120		105	6.25	3.25	114.50	130
0600	80' span		570	.098		125	5.10	2.65	132.75	149
0700	100' span		465	.120		176	6.25	3.25	185.50	208
0800	120' span		365	.153		223	7.95	4.14	235.09	263
0900	150' span		310	.181		237	9.35	4.87	251.22	281
1000	160' span		255	.220		237	11.35	5.95	254.30	286
1100	10' wide, 80' span		640	.088		125	4.53	2.36	131.89	147
1200	120' span		415	.135		162	7	3.64	172.64	194
1300	150' span		445	.126		182	6.50	3.40	191.90	215
1400	200' span	▼	205	.273		194	14.15	7.35	215.50	245
1600	Wood, laminated type, complete in place, 80' span	C-12	203	.236		86.50	11	3.22	100.72	116
1700	130' span	"	153	.314	▼	90.50	14.60	4.27	109.37	127

32 35 Screening Devices

32 35 16 – Sound Barriers

32 35 16.10 Traffic Barriers, Highway Sound Barriers		Crew	Daily Output	Labor-Hours	Unit	Material	2015 Bare Costs Labor	Equipment	Total	Total Incl O&P
0010	**TRAFFIC BARRIERS, HIGHWAY SOUND BARRIERS**									
0020	Highway sound barriers, not including footing									
0100	Precast concrete, concrete columns @ 30' OC, 8" T, 8' H	C-12	400	.120	L.F.	131	5.60	1.63	138.23	154
0110	12' H		265	.181		197	8.45	2.47	207.92	232
0120	16' H		200	.240		262	11.15	3.27	276.42	310
0130	20' H		160	.300		330	13.95	4.09	348.04	385
0400	Lt. Wt. composite panel, cementitious face, St. posts @ 12' OC, 8' H	B-80B	190	.168		156	6.80	1.28	164.08	184
0410	12' H		125	.256		235	10.35	1.95	247.30	276
0420	16' H		95	.337		315	13.60	2.57	331.17	370
0430	20' H		75	.427		390	17.20	3.25	410.45	460

32 84 Planting Irrigation

32 84 23 – Underground Sprinklers

32 84 23.10 Sprinkler Irrigation System

		Crew	Daily Output	Labor-Hours	Unit	Material	Labor	Equipment	Total	Total Incl O&P
0010	**SPRINKLER IRRIGATION SYSTEM**									
0011	For lawns									
0100	Golf course with fully automatic system	C-17	.05	1600	9 holes	100,000	78,500		178,500	231,000
0200	24' diam. head at 15' O.C. incl. piping, auto oper., minimum	B-20	70	.343	Head	27	14.40		41.40	51.50
0300	Maximum		40	.600		45	25		70	88.50
0600	Sprinkler irrigation sys, golf course, auto sys, 60' dia HD		23	1.043		150	44		194	233
0800	Residential system, custom, 1" supply		2000	.012	S.F.	.26	.50		.76	1.07
0900	1-1/2" supply		1800	.013	"	.49	.56		1.05	1.40
1020	Pop up spray head w/risers, hi-pop, full circle pattern, 4"	2 Skwk	76	.211	Ea.	5.15	10.25		15.40	21.50
1030	1/2 circle pattern, 4"		76	.211		5.15	10.25		15.40	21.50
1040	6", full circle pattern		76	.211		9.50	10.25		19.75	26.50
1050	1/2 circle pattern, 6"		76	.211		9.50	10.25		19.75	26.50
1060	12", full circle pattern		76	.211		11.15	10.25		21.40	28
1070	1/2 circle pattern, 12"		76	.211		13.75	10.25		24	31
1080	Pop up bubbler head w/risers, hi-pop bubbler head, 4"		76	.211		4.40	10.25		14.65	20.50
1090	6"		76	.211		9.20	10.25		19.45	26
1100	12"		76	.211		11	10.25		21.25	28
1110	Impact full/part circle sprinklers, 28'-54' 25-60 PSI		37	.432		17.65	21		38.65	52
1120	Spaced 37'-49' @ 25-50 PSI		37	.432		22	21		43	57
1130	Spaced 43'-61' @ 30-60 PSI		37	.432		61.50	21		82.50	100
1140	Spaced 54'-78' @ 40-80 PSI		37	.432		106	21		127	150
1145	Impact rotor pop-up full/part commercial circle sprinklers									
1150	Spaced 42'-65' 35-80 PSI	2 Skwk	25	.640	Ea.	15.30	31		46.30	65
1160	Spaced 48'-76' 45-85 PSI	"	25	.640	"	16.85	31		47.85	66.50
1165	Impact rotor pop-up part. circle comm., 53'-75', 55-100 PSI, w/accessories									
1170	Plastic case, metal cover	2 Skwk	25	.640	Ea.	74	31		105	129
1180	Rubber cover		25	.640		56.50	31		87.50	111
1190	Iron case, metal cover		22	.727		122	35.50		157.50	189
1200	Rubber cover		22	.727		129	35.50		164.50	197
1250	Plastic case, 2 nozzle, metal cover		25	.640		89.50	31		120.50	146
1260	Rubber cover		25	.640		91.50	31		122.50	148
1270	Iron case, 2 nozzle, metal cover		22	.727		130	35.50		165.50	197
1280	Rubber cover		22	.727		130	35.50		165.50	197
1282	Impact rotor pop-up full circle commercial, 39'-99', 30-100 PSI									
1284	Plastic case, metal cover	2 Skwk	25	.640	Ea.	83	31		114	139
1286	Rubber cover		25	.640		94.50	31		125.50	152
1288	Iron case, metal cover		22	.727		122	35.50		157.50	189

32 84 Planting Irrigation

32 84 23 – Underground Sprinklers

32 84 23.10 Sprinkler Irrigation System		Crew	Daily Output	Labor-Hours	Unit	Material	2015 Bare Costs Labor	Equipment	Total	Total Incl O&P
1290	Rubber cover	2 Skwk	22	.727	Ea.	126	35.50		161.50	194
1292	Plastic case, 2 nozzle, metal cover		22	.727		90	35.50		125.50	154
1294	Rubber cover		22	.727		90	35.50		125.50	154
1296	Iron case, 2 nozzle, metal cover		20	.800		119	39		158	190
1298	Rubber cover		20	.800		123	39		162	195
1305	Electric remote control valve, plastic, 3/4"		18	.889		23.50	43		66.50	93
1310	1"		18	.889		23.50	43		66.50	93
1320	1-1/2"		18	.889		91	43		134	167
1330	2"		18	.889		109	43		152	187
1335	Quick coupling valves, brass, locking cover									
1340	Inlet coupling valve, 3/4"	2 Skwk	18.75	.853	Ea.	21.50	41.50		63	87.50
1350	1"		18.75	.853		30	41.50		71.50	97
1360	Controller valve boxes, 6" round boxes		18.75	.853		7.10	41.50		48.60	72
1370	10" round boxes		14.25	1.123		11.40	54.50		65.90	97
1380	12" square box		9.75	1.641		16.05	80		96.05	141
1388	Electromech. control, 14 day 3-60 min., auto start to 23/day									
1390	4 station	2 Skwk	1.04	15.385	Ea.	75	750		825	1,225
1400	7 station		.64	25		140	1,225		1,365	2,025
1410	12 station		.40	40		170	1,950		2,120	3,175
1420	Dual programs, 18 station		.24	66.667		200	3,250		3,450	5,225
1430	23 station		.16	100		220	4,875		5,095	7,775
1435	Backflow preventer, bronze, 0-175 PSI, w/valves, test cocks									
1440	3/4"	2 Skwk	6	2.667	Ea.	82	130		212	290
1450	1"		6	2.667		94	130		224	305
1460	1-1/2"		6	2.667		224	130		354	445
1470	2"		6	2.667		276	130		406	505
1475	Pressure vacuum breaker, brass, 15-150 PSI									
1480	3/4"	2 Skwk	6	2.667	Ea.	25	130		155	228
1490	1"		6	2.667		30	130		160	233
1500	1-1/2"		6	2.667		70	130		200	277
1510	2"		6	2.667		120	130		250	330

32 91 Planting Preparation

32 91 13 – Soil Preparation

32 91 13.16 Mulching		Crew	Daily Output	Labor-Hours	Unit	Material	Labor	Equipment	Total	Total Incl O&P
0010	**MULCHING**									
0100	Aged barks, 3" deep, hand spread	1 Clab	100	.080	S.Y.	3.55	3.01		6.56	8.55
0150	Skid steer loader	B-63	13.50	2.963	M.S.F.	395	118	12.85	525.85	630
0200	Hay, 1" deep, hand spread	1 Clab	475	.017	S.Y.	.52	.63		1.15	1.54
0250	Power mulcher, small	B-64	180	.089	M.S.F.	58	3.41	2.21	63.62	71
0350	Large	B-65	530	.030	"	58	1.16	1.07	60.23	66.50
0400	Humus peat, 1" deep, hand spread	1 Clab	700	.011	S.Y.	2.42	.43		2.85	3.32
0450	Push spreader	"	2500	.003	"	2.42	.12		2.54	2.85
0550	Tractor spreader	B-66	700	.011	M.S.F.	269	.56	.38	269.94	297
0600	Oat straw, 1" deep, hand spread	1 Clab	475	.017	S.Y.	.60	.63		1.23	1.63
0650	Power mulcher, small	B-64	180	.089	M.S.F.	66.50	3.41	2.21	72.12	81
0700	Large	B-65	530	.030	"	66.50	1.16	1.07	68.73	76.50
0750	Add for asphaltic emulsion	B-45	1770	.009	Gal.	5.80	.41	.51	6.72	7.60
0800	Peat moss, 1" deep, hand spread	1 Clab	900	.009	S.Y.	2.80	.33		3.13	3.59
0850	Push spreader	"	2500	.003	"	2.80	.12		2.92	3.27
0950	Tractor spreader	B-66	700	.011	M.S.F.	310	.56	.38	310.94	340

32 91 Planting Preparation

32 91 13 – Soil Preparation

32 91 13.16 Mulching		Crew	Daily Output	Labor-Hours	Unit	Material	2015 Bare Costs Labor	Equipment	Total	Total Incl O&P
1000	Polyethylene film, 6 mil	2 Clab	2000	.008	S.Y.	.46	.30		.76	.97
1100	Redwood nuggets, 3" deep, hand spread	1 Clab	150	.053	"	2.77	2.01		4.78	6.15
1150	Skid steer loader	B-63	13.50	2.963	M.S.F.	310	118	12.85	440.85	535
1200	Stone mulch, hand spread, ceramic chips, economy	1 Clab	125	.064	S.Y.	6.85	2.41		9.26	11.25
1250	Deluxe	"	95	.084	"	10.60	3.17		13.77	16.50
1300	Granite chips	B-1	10	2.400	C.Y.	64	92		156	212
1400	Marble chips		10	2.400		152	92		244	310
1600	Pea gravel		28	.857		108	33		141	169
1700	Quartz		10	2.400		188	92		280	350
1800	Tar paper, 15 lb. felt	1 Clab	800	.010	S.Y.	.49	.38		.87	1.11
1900	Wood chips, 2" deep, hand spread	"	220	.036	"	1.60	1.37		2.97	3.86
1950	Skid steer loader	B-63	20.30	1.970	M.S.F.	178	78.50	8.55	265.05	325

32 91 13.26 Planting Beds		Crew	Daily Output	Labor-Hours	Unit	Material	Labor	Equipment	Total	Total Incl O&P
0010	**PLANTING BEDS**									
0100	Backfill planting pit, by hand, on site topsoil	2 Clab	18	.889	C.Y.		33.50		33.50	51.50
0200	Prepared planting mix, by hand	"	24	.667			25		25	38.50
0300	Skid steer loader, on site topsoil	B-62	340	.071			2.91	.51	3.42	5
0400	Prepared planting mix	"	410	.059			2.42	.42	2.84	4.17
1000	Excavate planting pit, by hand, sandy soil	2 Clab	16	1			37.50		37.50	58
1100	Heavy soil or clay	"	8	2			75		75	116
1200	1/2 C.Y. backhoe, sandy soil	B-11C	150	.107			4.70	2.43	7.13	9.85
1300	Heavy soil or clay	"	115	.139			6.15	3.17	9.32	12.85
2000	Mix planting soil, incl. loam, manure, peat, by hand	2 Clab	60	.267		43	10.05		53.05	62.50
2100	Skid steer loader	B-62	150	.160		43	6.60	1.16	50.76	58.50
3000	Pile sod, skid steer loader	"	2800	.009	S.Y.		.35	.06	.41	.61
3100	By hand	2 Clab	400	.040			1.50		1.50	2.31
4000	Remove sod, F.E. loader	B-10S	2000	.006			.28	.19	.47	.63
4100	Sod cutter	B-12K	3200	.005			.22	.31	.53	.68
4200	By hand	2 Clab	240	.067			2.51		2.51	3.86

32 91 19 – Landscape Grading

32 91 19.13 Topsoil Placement and Grading		Crew	Daily Output	Labor-Hours	Unit	Material	Labor	Equipment	Total	Total Incl O&P
0010	**TOPSOIL PLACEMENT AND GRADING**									
0400	Spread from pile to rough finish grade, F.E. loader, 1.5 C.Y.	B-10S	200	.060	C.Y.		2.78	1.89	4.67	6.30
0500	Up to 200' radius, by hand	1 Clab	14	.571			21.50		21.50	33
0600	Top dress by hand, 1 C.Y. for 600 S.F.	"	11.50	.696		27.50	26		53.50	70.50
0700	Furnish and place, truck dumped, screened, 4" deep	B-10S	1300	.009	S.Y.	3.44	.43	.29	4.16	4.75
0800	6" deep	"	820	.015	"	4.40	.68	.46	5.54	6.40

32 92 Turf and Grasses

32 92 19 – Seeding

32 92 19.13 Mechanical Seeding		Crew	Daily Output	Labor-Hours	Unit	Material	Labor	Equipment	Total	Total Incl O&P
0010	**MECHANICAL SEEDING**									
0020	Mechanical seeding, 215 lb./acre	B-66	1.50	5.333	Acre	560	259	175	994	1,200
0100	44 lb./M.S.Y.	"	2500	.003	S.Y.	.18	.16	.11	.45	.56
0101	$2.00/lb., 44 lb./M.S.Y.	1 Clab	13950	.001	S.F.	.02	.02		.04	.05
0300	Fine grading and seeding incl. lime, fertilizer & seed,									
0310	with equipment	B-14	1000	.048	S.Y.	.44	1.91	.36	2.71	3.81
0400	Fertilizer hand push spreader, 35 lb. per M.S.F.	1 Clab	200	.040	M.S.F.	9.80	1.50		11.30	13.10
0600	Limestone hand push spreader, 50 lb. per M.S.F.		180	.044		5.40	1.67		7.07	8.45
0800	Grass seed hand push spreader, 4.5 lb. per M.S.F.		180	.044		20	1.67		21.67	24.50

32 92 Turf and Grasses

32 92 19 – Seeding

32 92 19.13 Mechanical Seeding		Crew	Daily Output	Labor-Hours	Unit	2015 Bare Costs Material	Labor	Equipment	Total	Total Incl O&P
1000	Hydro or air seeding for large areas, incl. seed and fertilizer	B-81	8900	.003	S.Y.	.43	.12	.08	.63	.74
1100	With wood fiber mulch added	"	8900	.003	"	1.76	.12	.08	1.96	2.21
1300	Seed only, over 100 lb., field seed, minimum				Lb.	1.75			1.75	1.93
1400	Maximum					1.75			1.75	1.93
1500	Lawn seed, minimum					1.41			1.41	1.55
1600	Maximum					2.46			2.46	2.71
1800	Aerial operations, seeding only, field seed	B-58	50	.480	Acre	570	19.80	63	652.80	725
1900	Lawn seed		50	.480		460	19.80	63	542.80	605
2100	Seed and liquid fertilizer, field seed		50	.480		690	19.80	63	772.80	860
2200	Lawn seed		50	.480		580	19.80	63	662.80	735

32 92 23 – Sodding

32 92 23.10 Sodding Systems

		Crew	Daily Output	Labor-Hours	Unit	Material	Labor	Equipment	Total	Total Incl O&P
0010	**SODDING SYSTEMS**									
0020	Sodding, 1" deep, bluegrass sod, on level ground, over 8 M.S.F.	B-63	22	1.818	M.S.F.	243	72.50	7.90	323.40	390
0200	4 M.S.F.		17	2.353		254	93.50	10.20	357.70	435
0300	1000 S.F.		13.50	2.963		295	118	12.85	425.85	520
0500	Sloped ground, over 8 M.S.F.		6	6.667		243	265	29	537	705
0600	4 M.S.F.		5	8		254	320	34.50	608.50	805
0700	1000 S.F.		4	10		295	400	43.50	738.50	985
1000	Bent grass sod, on level ground, over 6 M.S.F.		20	2		252	79.50	8.70	340.20	410
1100	3 M.S.F.		18	2.222		261	88.50	9.65	359.15	435
1200	Sodding 1000 S.F. or less		14	2.857		286	114	12.40	412.40	505
1500	Sloped ground, over 6 M.S.F.		15	2.667		252	106	11.55	369.55	455
1600	3 M.S.F.		13.50	2.963		261	118	12.85	391.85	480
1700	1000 S.F.		12	3.333		286	133	14.45	433.45	535

32 93 Plants

32 93 10 – General Planting Costs

32 93 10.12 Travel

		Crew	Daily Output	Labor-Hours	Unit	Material	Labor	Equipment	Total	Total Incl O&P
0010	**TRAVEL** add to all nursery items									
0015	10 to 20 miles one way, add				All				5%	5%
0100	30 to 50 miles one way, add				"				10%	10%

32 93 13 – Ground Covers

32 93 13.10 Ground Cover Plants

		Crew	Daily Output	Labor-Hours	Unit	Material	Labor	Equipment	Total	Total Incl O&P
0010	**GROUND COVER PLANTS**									
0012	Plants, pachysandra, in prepared beds	B-1	15	1.600	C	62.50	61		123.50	163
0200	Vinca minor, 1 yr., bare root, in prepared beds		12	2	"	111	76.50		187.50	240
0600	Stone chips, in 50 lb. bags, Georgia marble		520	.046	Bag	3.76	1.77		5.53	6.85
0700	Onyx gemstone		260	.092		16.90	3.53		20.43	24
0800	Quartz		260	.092		16.80	3.53		20.33	24
0900	Pea gravel, truckload lots		28	.857	Ton	25.50	33		58.50	78.50

32 93 33 – Shrubs

32 93 33.10 Shrubs and Trees

		Crew	Daily Output	Labor-Hours	Unit	Material	Labor	Equipment	Total	Total Incl O&P
0010	**SHRUBS AND TREES**									
0011	Evergreen, in prepared beds, B & B									
0100	Arborvitae pyramidal, 4'-5'	B-17	30	1.067	Ea.	100	43.50	26	169.50	206
0150	Globe, 12"-15"	B-1	96	.250		21.50	9.55		31.05	38
0300	Cedar, blue, 8'-10'	B-17	18	1.778		231	73	43	347	415
0500	Hemlock, Canadian, 2-1/2'-3'	B-1	36	.667		31	25.50		56.50	74

32 93 33 – Shrubs

32 93 33.10 Shrubs and Trees		Crew	Daily Output	Labor-Hours	Unit	2015 Bare Costs Material	Labor	Equipment	Total	Total Incl O&P
0550	Holly, Savannah, 8′ - 10′ H	B-1	9.68	2.479	Ea.	257	95		352	430
0600	Juniper, andorra, 18″-24″		80	.300		36.50	11.50		48	57.50
0620	Wiltoni, 15″-18″		80	.300		26	11.50		37.50	46
0640	Skyrocket, 4-1/2′-5′	B-17	55	.582		107	24	14.10	145.10	169
0660	Blue pfitzer, 2′-2-1/2′	B-1	44	.545		38.50	21		59.50	74.50
0680	Ketleerie, 2-1/2′-3′		50	.480		52	18.35		70.35	85.50
0700	Pine, black, 2-1/2′-3′		50	.480		59.50	18.35		77.85	93.50
0720	Mugo, 18″-24″		60	.400		59	15.30		74.30	88.50
0740	White, 4′-5′	B-17	75	.427		51	17.50	10.35	78.85	94
0800	Spruce, blue, 18″-24″	B-1	60	.400		66.50	15.30		81.80	97
0840	Norway, 4′-5′	B-17	75	.427		83	17.50	10.35	110.85	129
0900	Yew, denisforma, 12″-15″	B-1	60	.400		35	15.30		50.30	62
1000	Capitata, 18″-24″		30	.800		32.50	30.50		63	83
1100	Hicksi, 2′-2-1/2′		30	.800		78.50	30.50		109	134

32 93 33.20 Shrubs		Crew	Daily Output	Labor-Hours	Unit	Material	Labor	Equipment	Total	Total Incl O&P
0010	**SHRUBS**									
0011	Broadleaf Evergreen, planted in prepared beds									
0100	Andromeda, 15″-18″, container	B-1	96	.250	Ea.	32.50	9.55		42.05	50.50
0200	Azalea, 15″ - 18″, container		96	.250		29.50	9.55		39.05	47
0300	Barberry, 9″-12″, container		130	.185		17.60	7.05		24.65	30
0400	Boxwood, 15″-18″, B&B		96	.250		43	9.55		52.55	61.50
0500	Euonymus, emerald gaiety, 12″ to 15″, container		115	.209		24	8		32	38.50
0600	Holly, 15″-18″, B & B		96	.250		36	9.55		45.55	54.50
0900	Mount laurel, 18″ - 24″, B & B		80	.300		70.50	11.50		82	95
1000	Paxistema, 9 – 12″ high		130	.185		21	7.05		28.05	34.50
1100	Rhododendron, 18″-24″, container		48	.500		37.50	19.15		56.65	71
1200	Rosemary, 1 gal. container		600	.040		17.70	1.53		19.23	22
2000	Deciduous, planted in prepared beds, amelanchier, 2′-3′, B & B		57	.421		119	16.10		135.10	156
2100	Azalea, 15″-18″, B & B		96	.250		29	9.55		38.55	46.50
2300	Bayberry, 2′-3′, B & B		57	.421		24	16.10		40.10	51.50
2600	Cotoneaster, 15″-18″, B & B		80	.300		26.50	11.50		38	46.50
2800	Dogwood, 3′-4′, B & B	B-17	40	.800		32	33	19.40	84.40	107
2900	Euonymus, alatus compacta, 15″ to 18″, container	B-1	80	.300		26	11.50		37.50	46.50
3200	Forsythia, 2′-3′, container	″	60	.400		18.45	15.30		33.75	44
3300	Hibiscus, 3′-4′, B & B	B-17	75	.427		48	17.50	10.35	75.85	91
3400	Honeysuckle, 3′-4′, B & B	B-1	60	.400		26.50	15.30		41.80	52.50
3500	Hydrangea, 2′-3′, B & B	″	57	.421		30	16.10		46.10	58
3600	Lilac, 3′-4′, B & B	B-17	40	.800		28	33	19.40	80.40	102
3900	Privet, bare root, 18″-24″	B-1	80	.300		14.50	11.50		26	33.50
4100	Quince, 2′-3′, B & B	″	57	.421		28	16.10		44.10	56
4200	Russian olive, 3′-4′, B & B	B-17	75	.427		29	17.50	10.35	56.85	69.50
4400	Spirea, 3′-4′, B & B	B-1	70	.343		20.50	13.10		33.60	42.50
4500	Viburnum, 3′-4′, B & B	B-17	40	.800		25	33	19.40	77.40	99

32 93 43 – Trees

32 93 43.20 Trees			Crew	Daily Output	Labor-Hours	Unit	Material	Labor	Equipment	Total	Total Incl O&P
0010	**TREES**										
0011	Deciduous, in prep. beds, balled & burlapped (B&B)										
0100	Ash, 2″ caliper	G	B-17	8	4	Ea.	186	164	97	447	560
0200	Beech, 5′-6′	G		50	.640		187	26	15.50	228.50	263
0300	Birch, 6′-8′, 3 stems	G		20	1.600		168	65.50	39	272.50	330
0500	Crabapple, 6′-8′	G		20	1.600		139	65.50	39	243.50	295
0600	Dogwood, 4′-5′	G		40	.800		131	33	19.40	183.40	216

32 93 Plants

32 93 43 – Trees

32 93 43.20 Trees			Crew	Daily Output	Labor-Hours	Unit	Material	Labor	Equipment	Total	Total Incl O&P
							2015 Bare Costs				
0700	Eastern redbud 4'-5'	G	B-17	40	.800	Ea.	149	33	19.40	201.40	236
0800	Elm, 8'-10'	G		20	1.600		258	65.50	39	362.50	425
0900	Ginkgo, 6'-7'	G		24	1.333		148	54.50	32.50	235	282
1000	Hawthorn, 8'-10', 1" caliper	G		20	1.600		162	65.50	39	266.50	320
1100	Honeylocust, 10'-12', 1-1/2" caliper	G		10	3.200		206	131	77.50	414.50	515
1300	Larch, 8'	G		32	1		127	41	24	192	229
1400	Linden, 8'-10', 1" caliper	G		20	1.600		144	65.50	39	248.50	300
1500	Magnolia, 4'-5'	G		20	1.600		101	65.50	39	205.50	254
1600	Maple, red, 8'-10', 1-1/2" caliper	G		10	3.200		202	131	77.50	410.50	510
1700	Mountain ash, 8'-10', 1" caliper	G		16	2		176	82	48.50	306.50	375
1800	Oak, 2-1/2"-3" caliper	G		6	5.333		325	218	129	672	835
2100	Planetree, 9'-11', 1-1/4" caliper	G		10	3.200		236	131	77.50	444.50	545
2200	Plum, 6'-8', 1" caliper	G		20	1.600		82	65.50	39	186.50	233
2300	Poplar, 9'-11', 1-1/4" caliper	G		10	3.200		144	131	77.50	352.50	445
2500	Sumac, 2'-3'	G		75	.427		44	17.50	10.35	71.85	86.50
2700	Tulip, 5'-6'	G		40	.800		47	33	19.40	99.40	123
2800	Willow, 6'-8', 1" caliper	G	↓	20	1.600	↓	96	65.50	39	200.50	249

32 94 Planting Accessories

32 94 13 – Landscape Edging

32 94 13.20 Edging		Crew	Daily Output	Labor-Hours	Unit	Material	Labor	Equipment	Total	Total Incl O&P
0010	**EDGING**									
0050	Aluminum alloy, including stakes, 1/8" x 4", mill finish	B-1	390	.062	L.F.	2.30	2.36		4.66	6.15
0051	Black paint		390	.062		2.67	2.36		5.03	6.55
0052	Black anodized	↓	390	.062		3.08	2.36		5.44	7
0100	Brick, set horizontally, 1-1/2 bricks per L.F.	D-1	370	.043		1.22	1.82		3.04	4.13
0150	Set vertically, 3 bricks per L.F.	"	135	.119		3.68	4.99		8.67	11.65
0200	Corrugated aluminum, roll, 4" wide	1 Carp	650	.012		1.93	.58		2.51	3.01
0250	6" wide	"	550	.015		2.41	.68		3.09	3.70
0600	Railroad ties, 6" x 8"	2 Carp	170	.094		2.79	4.42		7.21	9.85
0650	7" x 9"		136	.118		3.10	5.50		8.60	11.90
0750	Redwood 2" x 4"	↓	330	.048		2.25	2.28		4.53	5.95
0800	Steel edge strips, incl. stakes, 1/4" x 5"	B-1	390	.062		3.92	2.36		6.28	7.95
0850	3/16" x 4"	"	390	.062	↓	3.10	2.36		5.46	7.05

32 94 50 – Tree Guying

32 94 50.10 Tree Guying Systems		Crew	Daily Output	Labor-Hours	Unit	Material	Labor	Equipment	Total	Total Incl O&P
0010	**TREE GUYING SYSTEMS**									
0015	Tree guying Including stakes, guy wire and wrap									
0100	Less than 3" caliper, 2 stakes	2 Clab	35	.457	Ea.	14.25	17.20		31.45	42
0200	3" to 4" caliper, 3 stakes	"	21	.762	"	19.25	28.50		47.75	65
1000	Including arrowhead anchor, cable, turnbuckles and wrap									
1100	Less than 3" caliper, 3" anchors	2 Clab	20	.800	Ea.	44.50	30		74.50	95.50
1200	3" to 6" caliper, 4" anchors		15	1.067		38.50	40		78.50	104
1300	6" caliper, 6" anchors		12	1.333		44.50	50		94.50	126
1400	8" caliper, 8" anchors	↓	9	1.778	↓	129	67		196	245

32 96 Transplanting

32 96 23 – Plant and Bulb Transplanting

32 96 23.23 Planting		Crew	Daily Output	Labor-Hours	Unit	Material	2015 Bare Costs Labor	Equipment	Total	Total Incl O&P
0010	**PLANTING**									
0012	Moving shrubs on site, 12" ball	B-62	28	.857	Ea.		35.50	6.20	41.70	61
0100	24" ball	"	22	1.091	"		45	7.90	52.90	77.50

32 96 23.43 Moving Trees		Crew	Daily Output	Labor-Hours	Unit	Material	Labor	Equipment	Total	Total Incl O&P
0010	**MOVING TREES**, On site									
0300	Moving trees on site, 36" ball	B-6	3.75	6.400	Ea.		264	97	361	510
0400	60" ball	"	1	24	"		990	365	1,355	1,925

Estimating Tips

33 10 00 Water Utilities
33 30 00 Sanitary Sewerage Utilities
33 40 00 Storm Drainage Utilities

- Never assume that the water, sewer, and drainage lines will go in at the early stages of the project. Consider the site access needs before dividing the site in half with open trenches, loose pipe, and machinery obstructions. Always inspect the site to establish that the site drawings are complete. Check off all existing utilities on your drawings as you locate them. Be especially careful with underground utilities because appurtenances are sometimes buried during regrading or repaving operations. If you find any discrepancies, mark up the site plan for further research. Differing site conditions can be very costly if discovered later in the project.
- See also Section 33 01 00 for restoration of pipe where removal/replacement may be undesirable. Use of new types of piping materials can reduce the overall project cost. Owners/design engineers should consider the installing contractor as a valuable source of current information on utility products and local conditions that could lead to significant cost savings.

Reference Numbers

Reference numbers are shown in shaded boxes at the beginning of some major classifications. These numbers refer to related items in the Reference Section. The reference information may be an estimating procedure, an alternate pricing method, or technical information.

Note: Not all subdivisions listed here necessarily appear in this publication. ■

*Note: **Trade Service**, in part, has been used as a reference source for some of the material prices used in Division 33.*

33 01 Operation and Maintenance of Utilities

33 01 10 – Operation and Maintenance of Water Utilities

33 01 10.10 Corrosion Resistance		Crew	Daily Output	Labor-Hours	Unit	Material	Labor	Equipment	Total	Total Incl O&P
						2015 Bare Costs				
0010	**CORROSION RESISTANCE**									
0012	Wrap & coat, add to pipe, 4" diameter				L.F.	2.21			2.21	2.43
0040	6" diameter					3.28			3.28	3.61
0060	8" diameter					4.04			4.04	4.44
0100	12" diameter					6.20			6.20	6.80
0200	24" diameter					12.95			12.95	14.25
0500	Coating, bituminous, per diameter inch, 1 coat, add					.63			.63	.69
0540	3 coat					1.92			1.92	2.11
0560	Coal tar epoxy, per diameter inch, 1 coat, add					.22			.22	.24
0600	3 coat				↓	.68			.68	.75

33 01 30 – Operation and Maintenance of Sewer Utilities

33 01 30.72 Relining Sewers		Crew	Daily Output	Labor-Hours	Unit	Material	Labor	Equipment	Total	Total Incl O&P
0010	**RELINING SEWERS**									
0011	With cement incl. bypass & cleaning									
0020	Less than 10,000 L.F., urban, 6" to 10"	C-17E	130	.615	L.F.	9.55	30	.74	40.29	58
0200	24" to 36"		90	.889		15.25	43.50	1.08	59.83	85.50
0300	48" to 72"	↓	80	1	↓	24.50	49	1.21	74.71	104

33 05 Common Work Results for Utilities

33 05 16 – Utility Structures

33 05 16.13 Precast Concrete Utility Boxes		Crew	Daily Output	Labor-Hours	Unit	Material	Labor	Equipment	Total	Total Incl O&P
0010	**PRECAST CONCRETE UTILITY BOXES**, 6" thick									
0050	5' x 10' x 6' high, I.D.	B-13	2	28	Ea.	3,750	1,150	370	5,270	6,275
0100	6' x 10' x 6' high, I.D.		2	28		3,900	1,150	370	5,420	6,425
0150	5' x 12' x 6' high, I.D.		2	28		4,125	1,150	370	5,645	6,675
0200	6' x 12' x 6' high, I.D.		1.80	31.111		4,600	1,275	410	6,285	7,475
0250	6' x 13' x 6' high, I.D.		1.50	37.333		6,050	1,525	490	8,065	9,550
0300	8' x 14' x 7' high, I.D.	↓	1	56	↓	6,525	2,300	735	9,560	11,500
0350	Hand hole, precast concrete, 1-1/2" thick									
0400	1'-0" x 2'-0" x 1'-9", I.D., light duty	B-1	4	6	Ea.	400	230		630	795
0450	4'-6" x 3'-2" x 2'-0", O.D., heavy duty	B-6	3	8	"	1,475	330	121	1,926	2,250

33 05 23 – Trenchless Utility Installation

33 05 23.19 Microtunneling		Crew	Daily Output	Labor-Hours	Unit	Material	Labor	Equipment	Total	Total Incl O&P
0010	**MICROTUNNELING**									
0011	Not including excavation, backfill, shoring,									
0020	or dewatering, average 50'/day, slurry method									
0100	24" to 48" outside diameter, minimum				L.F.				875	965
0110	Adverse conditions, add				%				50%	50%
1000	Rent microtunneling machine, average monthly lease				Month				97,500	107,000
1010	Operating technician				Day				630	705
1100	Mobilization and demobilization, minimum				Job				41,200	45,900
1110	Maximum				"				445,500	490,500

33 05 23.20 Horizontal Boring		Crew	Daily Output	Labor-Hours	Unit	Material	Labor	Equipment	Total	Total Incl O&P
0010	**HORIZONTAL BORING**									
0011	Casing only, 100' minimum,									
0020	not incl. jacking pits or dewatering									
0100	Roadwork, 1/2" thick wall, 24" diameter casing	B-42	20	3.200	L.F.	121	136	68	325	420
0200	36" diameter		16	4		223	170	85	478	605
0300	48" diameter		15	4.267		310	181	90.50	581.50	725
0500	Railroad work, 24" diameter	↓	15	4.267	↓	121	181	90.50	392.50	515

33 05 Common Work Results for Utilities

33 05 23 – Trenchless Utility Installation

33 05 23.20 Horizontal Boring		Crew	Daily Output	Labor-Hours	Unit	Material (2015 Bare Costs)	Labor (2015 Bare Costs)	Equipment (2015 Bare Costs)	Total (2015 Bare Costs)	Total Incl O&P
0600	36″ diameter	B-42	14	4.571	L.F.	223	194	97	514	655
0700	48″ diameter		12	5.333		310	226	113	649	820
0900	For ledge, add								20%	20%

33 05 26 – Utility Identification

33 05 26.05 Utility Connection		Crew	Daily Output	Labor-Hours	Unit	Material	Labor	Equipment	Total	Total Incl O&P
0010	**UTILITY CONNECTION**									
0020	Water, sanitary, stormwater, gas, single connection	B-14	1	48	Ea.	3,225	1,900	365	5,490	6,875
0030	Telecommunication	″	1	48	″	395	1,900	365	2,660	3,750

33 05 26.10 Utility Accessories		Crew	Daily Output	Labor-Hours	Unit	Material	Labor	Equipment	Total	Total Incl O&P
0010	**UTILITY ACCESSORIES**									
0400	Underground tape, detectable, reinforced, alum. foil core, 2″	1 Clab	150	.053	C.L.F.	6.50	2.01		8.51	10.25
0500	6″	″	140	.057	″	27.50	2.15		29.65	33.50

33 11 Water Utility Distribution Piping

33 11 13 – Public Water Utility Distribution Piping

33 11 13.15 Water Supply, Ductile Iron Pipe		Crew	Daily Output	Labor-Hours	Unit	Material	Labor	Equipment	Total	Total Incl O&P
0010	**WATER SUPPLY, DUCTILE IRON PIPE** R331113-80									
0020	Not including excavation or backfill									
2000	Pipe, class 50 water piping, 18′ lengths									
2020	Mechanical joint, 4″ diameter	B-21A	200	.200	L.F.	30.50	9.40	2.37	42.27	50.50
2040	6″ diameter		160	.250		32	11.75	2.96	46.71	56
2060	8″ diameter		133.33	.300		44.50	14.05	3.55	62.10	74.50
2080	10″ diameter		114.29	.350		58.50	16.40	4.14	79.04	93.50
2100	12″ diameter		105.26	.380		79	17.85	4.50	101.35	119
2120	14″ diameter		100	.400		93	18.75	4.74	116.49	136
2140	16″ diameter		72.73	.550		94.50	26	6.50	127	151
2160	18″ diameter		68.97	.580		126	27	6.85	159.85	188
2170	20″ diameter		57.14	.700		127	33	8.30	168.30	199
2180	24″ diameter		47.06	.850		141	40	10.05	191.05	227
3000	Push-on joint, 4″ diameter		400	.100		21	4.69	1.18	26.87	31.50
3020	6″ diameter		333.33	.120		17	5.65	1.42	24.07	29
3040	8″ diameter		200	.200		23	9.40	2.37	34.77	42.50
3060	10″ diameter		181.82	.220		36.50	10.30	2.60	49.40	59
3080	12″ diameter		160	.250		38.50	11.75	2.96	53.21	63.50
3100	14″ diameter		133.33	.300		42.50	14.05	3.55	60.10	72.50
3120	16″ diameter		114.29	.350		51.50	16.40	4.14	72.04	86
3140	18″ diameter		100	.400		57	18.75	4.74	80.49	96
3160	20″ diameter		88.89	.450		59.50	21	5.35	85.85	103
3180	24″ diameter		76.92	.520		59	24.50	6.15	89.65	109
8000	Piping, fittings, mechanical joint, AWWA C110									
8006	90° bend, 4″ diameter	B-20A	16	2	Ea.	155	91.50		246.50	310
8020	6″ diameter		12.80	2.500		229	114		343	425
8040	8″ diameter		10.67	2.999		450	137		587	705
8060	10″ diameter	B-21A	11.43	3.500		620	164	41.50	825.50	975
8080	12″ diameter		10.53	3.799		880	178	45	1,103	1,300
8100	14″ diameter		10	4		1,200	188	47.50	1,435.50	1,650
8120	16″ diameter		7.27	5.502		1,525	258	65	1,848	2,150
8140	18″ diameter		6.90	5.797		2,125	272	68.50	2,465.50	2,850
8160	20″ diameter		5.71	7.005		2,650	330	83	3,063	3,525
8180	24″ diameter		4.70	8.511		4,200	400	101	4,701	5,325

33 11 13.15 Water Supply, Ductile Iron Pipe

	33 11 13.15 Water Supply, Ductile Iron Pipe	Crew	Daily Output	Labor-Hours	Unit	2015 Bare Costs Material	Labor	Equipment	Total	Total Incl O&P
8200	Wye or tee, 4" diameter	B-20A	10.67	2.999	Ea.	355	137		492	600
8220	6" diameter		8.53	3.751		535	171		706	850
8240	8" diameter		7.11	4.501		850	206		1,056	1,250
8260	10" diameter	B-21A	7.62	5.249		1,225	246	62	1,533	1,800
8280	12" diameter		7.02	5.698		1,625	267	67.50	1,959.50	2,250
8300	14" diameter		6.67	5.997		2,600	281	71	2,952	3,375
8320	16" diameter		4.85	8.247		2,900	385	97.50	3,382.50	3,875
8340	18" diameter		4.60	8.696		3,900	410	103	4,413	5,025
8360	20" diameter		3.81	10.499		5,475	490	124	6,089	6,900
8380	24" diameter		3.14	12.739		9,275	600	151	10,026	11,300
8450	Decreaser, 6" x 4" diameter	B-20A	14.22	2.250		207	103		310	385
8460	8" x 6" diameter	"	11.64	2.749		310	126		436	530
8470	10" x 6" diameter	B-21A	13.33	3.001		390	141	35.50	566.50	685
8480	12" x 6" diameter		12.70	3.150		550	148	37.50	735.50	870
8490	16" x 6" diameter		10	4		890	188	47.50	1,125.50	1,325
8500	20" x 6" diameter		8.42	4.751		1,625	223	56.50	1,904.50	2,175
8550	Piping, butterfly valves, cast iron									
8560	4" diameter	B-20	6	4	Ea.	470	168		638	775
8570	6" diameter	"	5	4.800		645	201		846	1,025
8580	8" diameter	B-21	4	7		835	305	34.50	1,174.50	1,425
8590	10" diameter		3.50	8		1,175	345	39.50	1,559.50	1,850
8600	12" diameter		3	9.333		1,600	405	46	2,051	2,450
8610	14" diameter		2	14		3,075	605	69	3,749	4,375
8620	16" diameter		2	14		4,700	605	69	5,374	6,150

33 11 13.25 Water Supply, Polyvinyl Chloride Pipe

	33 11 13.25 Water Supply, Polyvinyl Chloride Pipe	Crew	Daily Output	Labor-Hours	Unit	Material	Labor	Equipment	Total	Total Incl O&P
0010	**WATER SUPPLY, POLYVINYL CHLORIDE PIPE**									
0020	Not including excavation or backfill, unless specified									
2100	PVC pipe, Class 150, 1-1/2" diameter	Q-1A	750	.013	L.F.	.45	.79		1.24	1.69
2120	2" diameter		686	.015		.68	.86		1.54	2.05
2140	2-1/2" diameter		500	.020		1.30	1.18		2.48	3.22
2160	3" diameter	B-20	430	.056		1.40	2.34		3.74	5.15
3010	AWWA C905, PR 100, DR 25									
3030	14" diameter	B-20A	213	.150	L.F.	13.65	6.85		20.50	25.50
3040	16" diameter		200	.160		19.25	7.30		26.55	32
3050	18" diameter		160	.200		24.50	9.15		33.65	40.50
3060	20" diameter		133	.241		30	11		41	50
3070	24" diameter		107	.299		43	13.65		56.65	68.50
3080	30" diameter		80	.400		81	18.30		99.30	117
3090	36" diameter		80	.400		126	18.30		144.30	166
3100	42" diameter		60	.533		168	24.50		192.50	222
3200	48" diameter		60	.533		220	24.50		244.50	279
4520	Pressure pipe Class 150, SDR 18, AWWA C900, 4" diameter		380	.084		2.73	3.85		6.58	8.85
4530	6" diameter		316	.101		5.35	4.63		9.98	12.95
4540	8" diameter		264	.121		8.80	5.55		14.35	18.10
4550	10" diameter		220	.145		14.05	6.65		20.70	25.50
4560	12" diameter		186	.172		19.90	7.85		27.75	34
8000	Fittings with rubber gasket									
8003	Class 150, DR 18									
8006	90° Bend , 4" diameter	B-20	100	.240	Ea.	42.50	10.05		52.55	62
8020	6" diameter		90	.267		75	11.20		86.20	100
8040	8" diameter		80	.300		145	12.60		157.60	178
8060	10" diameter		50	.480		330	20		350	395

33 11 Water Utility Distribution Piping

33 11 13 – Public Water Utility Distribution Piping

33 11 13.25 Water Supply, Polyvinyl Chloride Pipe		Crew	Daily Output	Labor-Hours	Unit	Material	Labor	Equipment	Total	Total Incl O&P
						2015 Bare Costs				
8080	12″ diameter	B-20	30	.800	Ea.	425	33.50		458.50	515
8100	Tee, 4″ diameter		90	.267		58.50	11.20		69.70	82
8120	6″ diameter		80	.300		144	12.60		156.60	177
8140	8″ diameter		70	.343		185	14.40		199.40	225
8160	10″ diameter		40	.600		590	25		615	690
8180	12″ diameter		20	1.200		765	50.50		815.50	920
8200	45° Bend, 4″ diameter		100	.240		42	10.05		52.05	62
8220	6″ diameter		90	.267		73	11.20		84.20	97.50
8240	8″ diameter		50	.480		139	20		159	183
8260	10″ diameter		50	.480		278	20		298	335
8280	12″ diameter		30	.800		360	33.50		393.50	445
8300	Reducing tee 6″ x 4″		100	.240		104	10.05		114.05	131
8320	8″ x 6″		90	.267		166	11.20		177.20	199
8330	10″ x 6″		90	.267		196	11.20		207.20	232
8340	10″ x 8″		90	.267		216	11.20		227.20	255
8350	12″ x 6″		90	.267		245	11.20		256.20	286
8360	12″ x 8″		90	.267		265	11.20		276.20	310
8400	Tapped service tee (threaded type) 6″ x 6″ x 3/4″		100	.240		95.50	10.05		105.55	121
8430	6″ x 6″ x 1″		90	.267		95.50	11.20		106.70	122
8440	6″ x 6″ x 1-1/2″		90	.267		95.50	11.20		106.70	122
8450	6″ x 6″ x 2″		90	.267		95.50	11.20		106.70	122
8460	8″ x 8″ x 3/4″		90	.267		140	11.20		151.20	171
8470	8″ x 8″ x 1″		90	.267		140	11.20		151.20	171
8480	8″ x 8″ x 1-1/2″		90	.267		140	11.20		151.20	171
8490	8″ x 8″ x 2″		90	.267		140	11.20		151.20	171
8500	Repair coupling 4″		100	.240		26	10.05		36.05	44
8520	6″ diameter		90	.267		40.50	11.20		51.70	62
8540	8″ diameter		50	.480		96.50	20		116.50	137
8560	10″ diameter		50	.480		203	20		223	254
8580	12″ diameter		50	.480		296	20		316	355
8600	Plug end 4″		100	.240		23	10.05		33.05	41
8620	6″ diameter		90	.267		41	11.20		52.20	62.50
8640	8″ diameter		50	.480		69	20		89	107
8660	10″ diameter		50	.480		97	20		117	137
8680	12″ diameter	▼	50	.480	▼	119	20		139	162

33 12 Water Utility Distribution Equipment

33 12 19 – Water Utility Distribution Fire Hydrants

33 12 19.10 Fire Hydrants		Crew	Daily Output	Labor-Hours	Unit	Material	Labor	Equipment	Total	Total Incl O&P
0010	**FIRE HYDRANTS**									
0020	Mechanical joints unless otherwise noted									
1000	Fire hydrants, two way; excavation and backfill not incl.									
1100	4-1/2″ valve size, depth 2′-0″	B-21	10	2.800	Ea.	1,650	121	13.80	1,784.80	2,025
1120	2′-6″		10	2.800		1,750	121	13.80	1,884.80	2,125
1140	3′-0″		10	2.800		1,875	121	13.80	2,009.80	2,275
1300	7′-0″	▼	6	4.667		2,150	202	23	2,375	2,700
2400	Lower barrel extensions with stems, 1′-0″	B-20	14	1.714		320	72		392	460
2480	3′-0″	″	12	2	▼	770	84		854	975

33 16 Water Utility Storage Tanks

33 16 13 – Aboveground Water Utility Storage Tanks

33 16 13.13 Steel Water Storage Tanks		Crew	Daily Output	Labor-Hours	Unit	2015 Bare Costs Material	Labor	Equipment	Total	Total Incl O&P
0010	**STEEL WATER STORAGE TANKS**									
0910	Steel, ground level, ht./diam. less than 1, not incl. fdn., 100,000 gallons				Ea.				202,000	244,500
1000	250,000 gallons								295,500	324,000
1200	500,000 gallons								417,000	458,500
1250	750,000 gallons								538,000	591,500
1300	1,000,000 gallons								558,000	725,500
1500	2,000,000 gallons								1,043,000	1,148,000
1600	4,000,000 gallons								2,121,000	2,333,000
1800	6,000,000 gallons								3,095,000	3,405,000
1850	8,000,000 gallons								4,068,000	4,475,000
1910	10,000,000 gallons				↓				5,050,000	5,554,500
2100	Steel standpipes, ht./diam. more than 1, 100′ to overflow, no fdn.									
2200	500,000 gallons				Ea.				546,500	600,500
2400	750,000 gallons								722,500	794,500
2500	1,000,000 gallons								1,060,500	1,167,000
2700	1,500,000 gallons								1,749,000	1,923,000
2800	2,000,000 gallons				↓				2,327,000	2,559,000

33 16 13.16 Prestressed Conc. Water Storage Tanks

Line	Description	Crew	Daily Output	Labor-Hours	Unit	Material	Labor	Equipment	Total	Total Incl O&P
0010	**PRESTRESSED CONC. WATER STORAGE TANKS**									
0020	Not including fdn., pipe or pumps, 250,000 gallons				Ea.				299,000	329,500
0100	500,000 gallons								487,000	536,000
0300	1,000,000 gallons								707,000	807,500
0400	2,000,000 gallons								1,072,000	1,179,000
0600	4,000,000 gallons								1,706,000	1,877,000
0700	6,000,000 gallons								2,266,000	2,493,000
0750	8,000,000 gallons								2,924,000	3,216,000
0800	10,000,000 gallons				↓				3,533,000	3,886,000

33 16 13.23 Plastic-Coated Fabric Pillow Water Tanks

Line	Description	Crew	Daily Output	Labor-Hours	Unit	Material	Labor	Equipment	Total	Total Incl O&P
0010	**PLASTIC-COATED FABRIC PILLOW WATER TANKS**									
7000	Water tanks, vinyl coated fabric pillow tanks, freestanding, 5,000 gallons	4 Clab	4	8	Ea.	3,675	300		3,975	4,525
7100	Supporting embankment not included, 25,000 gallons	6 Clab	2	24		13,200	900		14,100	15,900
7200	50,000 gallons	8 Clab	1.50	42.667		18,500	1,600		20,100	22,800
7300	100,000 gallons	9 Clab	.90	80		42,300	3,000		45,300	51,000
7400	150,000 gallons		.50	144		60,500	5,425		65,925	75,500
7500	200,000 gallons		.40	180		75,000	6,775		81,775	93,000
7600	250,000 gallons	↓	.30	240	↓	105,500	9,025		114,525	130,000

33 16 19 – Elevated Water Utility Storage Tanks

33 16 19.50 Elevated Water Storage Tanks

Line	Description	Crew	Daily Output	Labor-Hours	Unit	Material	Labor	Equipment	Total	Total Incl O&P
0010	**ELEVATED WATER STORAGE TANKS**									
0011	Not incl. pipe, pumps or foundation									
3000	Elevated water tanks, 100′ to bottom capacity line, incl. painting									
3010	50,000 gallons				Ea.				185,000	204,000
3300	100,000 gallons								280,000	307,500
3400	250,000 gallons								751,500	826,500
3600	500,000 gallons								1,336,000	1,470,000
3700	750,000 gallons								1,622,000	1,783,500
3900	1,000,000 gallons				↓				2,322,000	2,556,000

33 21 Water Supply Wells

33 21 13 – Public Water Supply Wells

33 21 13.10 Wells and Accessories		Crew	Daily Output	Labor-Hours	Unit	2015 Bare Costs Material	Labor	Equipment	Total	Total Incl O&P
0010	**WELLS & ACCESSORIES**									
0011	Domestic									
0100	Drilled, 4″ to 6″ diameter	B-23	120	.333	L.F.		12.65	23.50	36.15	45.50
0200	8″ diameter	″	95.20	.420	″		15.95	30	45.95	57.50
0400	Gravel pack well, 40′ deep, incl. gravel & casing, complete									
0500	24″ diameter casing x 18″ diameter screen	B-23	.13	307	Total	38,800	11,700	21,900	72,400	85,000
0600	36″ diameter casing x 18″ diameter screen		.12	333	″	40,000	12,700	23,700	76,400	89,500
0800	Observation wells, 1-1/4″ riser pipe		163	.245	V.L.F.	20.50	9.35	17.45	47.30	56
0900	For flush Buffalo roadway box, add	1 Skwk	16.60	.482	Ea.	51	23.50		74.50	92
1200	Test well, 2-1/2″ diameter, up to 50′ deep (15 to 50 GPM)	B-23	1.51	26.490	″	805	1,000	1,875	3,680	4,525
1300	Over 50′ deep, add	″	121.80	.328	L.F.	21.50	12.50	23.50	57.50	68
1500	Pumps, installed in wells to 100′ deep, 4″ submersible									
1510	1/2 H.P.	Q-1	3.22	4.969	Ea.	350	263		613	780
1520	3/4 H.P.		2.66	6.015		400	320		720	920
1600	1 H.P.		2.29	6.987		425	370		795	1,025
1700	1-1/2 H.P.	Q-22	1.60	10		800	530	410	1,740	2,125
1800	2 H.P.		1.33	12.030		830	635	490	1,955	2,425
1900	3 H.P.		1.14	14.035		1,150	740	575	2,465	3,025
2000	5 H.P.		1.14	14.035		1,600	740	575	2,915	3,500
3000	Pump, 6″ submersible, 25′ to 150′ deep, 25 H.P., 249 to 297 GPM		.89	17.978		3,350	950	735	5,035	5,900
3100	25′ to 500′ deep, 30 H.P., 100 to 300 GPM		.73	21.918		4,100	1,150	895	6,145	7,225
8000	Steel well casing	B-23A	3020	.008	Lb.	1.21	.34	.89	2.44	2.83
9950	See Section 31 23 19.40 for wellpoints									
9960	See Section 31 23 19.30 for drainage wells									

33 21 13.20 Water Supply Wells, Pumps		Crew	Daily Output	Labor-Hours	Unit	Material	Labor	Equipment	Total	Total Incl O&P
0010	**WATER SUPPLY WELLS, PUMPS**									
0011	With pressure control									
1000	Deep well, jet, 42 gal. galvanized tank									
1040	3/4 HP	1 Plum	.80	10	Ea.	1,100	585		1,685	2,100
3000	Shallow well, jet, 30 gal. galvanized tank									
3040	1/2 HP	1 Plum	2	4	Ea.	895	235		1,130	1,350

33 31 Sanitary Utility Sewerage Piping

33 31 13 – Public Sanitary Utility Sewerage Piping

33 31 13.15 Sewage Collection, Concrete Pipe		Crew	Daily Output	Labor-Hours	Unit	Material	Labor	Equipment	Total	Total Incl O&P
0010	**SEWAGE COLLECTION, CONCRETE PIPE**									
0020	See Section 33 41 13.60 for sewage/drainage collection, concrete pipe									

33 31 13.25 Sewage Collection, Polyvinyl Chloride Pipe		Crew	Daily Output	Labor-Hours	Unit	Material	Labor	Equipment	Total	Total Incl O&P
0010	**SEWAGE COLLECTION, POLYVINYL CHLORIDE PIPE**									
0020	Not including excavation or backfill									
2000	20′ lengths, SDR 35, B&S, 4″ diameter	B-20	375	.064	L.F.	1.45	2.68		4.13	5.75
2040	6″ diameter		350	.069		3.28	2.88		6.16	8.05
2080	13′ lengths , SDR 35, B&S, 8″ diameter		335	.072		6.90	3.01		9.91	12.25
2120	10″ diameter	B-21	330	.085		11.40	3.68	.42	15.50	18.60
2160	12″ diameter		320	.088		12.75	3.79	.43	16.97	20.50
2200	15″ diameter		240	.117		12.95	5.05	.57	18.57	22.50
4000	Piping, DWV PVC, no exc./bkfill., 10′ L, Sch 40, 4″ diameter	B-20	375	.064		3.84	2.68		6.52	8.35
4010	6″ diameter		350	.069		7.65	2.88		10.53	12.85
4020	8″ diameter		335	.072		12	3.01		15.01	17.85

33 36 Utility Septic Tanks

33 36 13 – Utility Septic Tank and Effluent Wet Wells

33 36 13.13 Concrete Utility Septic Tank		Crew	Daily Output	Labor-Hours	Unit	Material	2015 Bare Costs Labor	Equipment	Total	Total Incl O&P
0010	**CONCRETE UTILITY SEPTIC TANK**									
0011	Not including excavation or piping									
0015	Septic tanks, precast, 1,000 gallon	B-21	8	3.500	Ea.	1,000	152	17.20	1,169.20	1,350
0060	1,500 gallon		7	4		1,325	173	19.70	1,517.70	1,775
0100	2,000 gallon		5	5.600		1,925	243	27.50	2,195.50	2,500
0200	5,000 gallon	B-13	3.50	16		9,500	655	210	10,365	11,700
0300	15,000 gallon, 4 piece	B-13B	1.70	32.941		20,000	1,350	665	22,015	24,800
0400	25,000 gallon, 4 piece		1.10	50.909		38,800	2,075	1,025	41,900	47,000
0500	40,000 gallon, 4 piece		.80	70		49,700	2,875	1,400	53,975	60,500
0520	50,000 gallon, 5 piece	B-13C	.60	93.333		57,000	3,825	2,775	63,600	72,000
0640	75,000 gallon, cast in place	C-14C	.25	448		69,500	20,200	125	89,825	107,500
0660	100,000 gallon	″	.15	746		86,000	33,700	209	119,909	146,500
1150	Leaching field chambers, 13′ x 3′-7″ x 1′-4″, standard	B-13	16	3.500		485	143	46	674	800
1200	Heavy duty, 8′ x 4′ x 1′-6″		14	4		284	164	52.50	500.50	620
1300	13′ x 3′-9″ x 1′-6″		12	4.667		1,125	191	61.50	1,377.50	1,575
1350	20′ x 4′ x 1′-6″		5	11.200		1,175	460	147	1,782	2,150
1400	Leaching pit, precast concrete, 3′ diameter, 3′ deep	B-21	8	3.500		710	152	17.20	879.20	1,025
1500	6′ diameter, 3′ section		4.70	5.957		885	258	29.50	1,172.50	1,400
2000	Velocity reducing pit, precast conc., 6′ diameter, 3′ deep		4.70	5.957		1,600	258	29.50	1,887.50	2,175

33 36 13.19 Polyethylene Utility Septic Tank		Crew	Daily Output	Labor-Hours	Unit	Material	Labor	Equipment	Total	Total Incl O&P
0010	**POLYETHYLENE UTILITY SEPTIC TANK**									
0015	High density polyethylene, 1,000 gallon	B-21	8	3.500	Ea.	1,550	152	17.20	1,719.20	1,950
0020	1,250 gallon		8	3.500		1,200	152	17.20	1,369.20	1,575
0025	1,500 gallon		7	4		1,475	173	19.70	1,667.70	1,925

33 36 19 – Utility Septic Tank Effluent Filter

33 36 19.13 Utility Septic Tank Effluent Tube Filter		Crew	Daily Output	Labor-Hours	Unit	Material	Labor	Equipment	Total	Total Incl O&P
0010	**UTILITY SEPTIC TANK EFFLUENT TUBE FILTER**									
3000	Effluent filter, 4″ diameter	1 Skwk	8	1	Ea.	42	48.50		90.50	121
3020	6″ diameter		7	1.143		249	55.50		304.50	360
3030	8″ diameter		7	1.143		225	55.50		280.50	335
3040	8″ diameter, very fine		7	1.143		460	55.50		515.50	590
3050	10″ diameter, very fine		6	1.333		235	65		300	360
3060	10″ diameter		6	1.333		280	65		345	410
3080	12″ diameter		6	1.333		670	65		735	840
3090	15″ diameter		5	1.600		1,150	78		1,228	1,375

33 36 33 – Utility Septic Tank Drainage Field

33 36 33.13 Utility Septic Tank Tile Drainage Field		Crew	Daily Output	Labor-Hours	Unit	Material	Labor	Equipment	Total	Total Incl O&P
0010	**UTILITY SEPTIC TANK TILE DRAINAGE FIELD**									
0015	Distribution box, concrete, 5 outlets	2 Clab	20	.800	Ea.	80.50	30		110.50	135
0020	7 outlets		16	1		80.50	37.50		118	147
0025	9 outlets		8	2		490	75		565	655
0115	Distribution boxes, HDPE, 5 outlets		20	.800		64.50	30		94.50	118
0117	6 outlets		15	1.067		65	40		105	133
0118	7 outlets		15	1.067		65	40		105	133
0120	8 outlets		10	1.600		68	60		128	168
0240	Distribution boxes, Outlet Flow Leveler	1 Clab	50	.160		2.25	6		8.25	11.75
0300	Precast concrete, galley, 4′ x 4′ x 4′	B-21	16	1.750		250	76	8.60	334.60	400
0350	HDPE infiltration chamber 12″ H X 15″ W	2 Clab	300	.053	L.F.	7.30	2.01		9.31	11.15
0351	12″ H X 15″ W End Cap	1 Clab	32	.250	Ea.	19.75	9.40		29.15	36
0355	chamber 12″ H X 22″ W	2 Clab	300	.053	L.F.	6.70	2.01		8.71	10.45
0356	12″ H X 22″ W End Cap	1 Clab	32	.250	Ea.	16.85	9.40		26.25	33

33 36 Utility Septic Tanks

33 36 33 – Utility Septic Tank Drainage Field

33 36 33.13 Utility Septic Tank Tile Drainage Field		Crew	Daily Output	Labor-Hours	Unit	Material	2015 Bare Costs Labor	Equipment	Total	Total Incl O&P
0360	chamber 13" H X 34" W	2 Clab	300	.053	L.F.	15	2.01		17.01	19.60
0361	13" H X 34" W End Cap	1 Clab	32	.250	Ea.	53.50	9.40		62.90	73
0365	chamber 16" H X 34" W	2 Clab	300	.053	L.F.	10.70	2.01		12.71	14.90
0366	16" H X 34" W End Cap	1 Clab	32	.250	Ea.	8	9.40		17.40	23.50
0370	chamber 8" H X 16" W	2 Clab	300	.053	L.F.	9.65	2.01		11.66	13.70
0371	8" H X 16" W End Cap	1 Clab	32	.250	Ea.	10.80	9.40		20.20	26.50

33 36 50 – Drainage Field Systems

33 36 50.10 Drainage Field Excavation and Fill		Crew	Daily Output	Labor-Hours	Unit	Material	Labor	Equipment	Total	Total Incl O&P
0010	**DRAINAGE FIELD EXCAVATION AND FILL**									
2200	Septic tank & drainage field excavation with 3/4 c.y backhoe	B-12F	145	.110	C.Y.		4.93	4.52	9.45	12.45
2400	4' trench for disposal field, 3/4 C.Y. backhoe	"	335	.048	L.F.		2.13	1.95	4.08	5.40
2600	Gravel fill, run of bank	B-6	150	.160	C.Y.	20	6.60	2.43	29.03	35
2800	Crushed stone, 3/4"	"	150	.160	"	36.50	6.60	2.43	45.53	53

33 41 Storm Utility Drainage Piping

33 41 13 – Public Storm Utility Drainage Piping

33 41 13.40 Piping, Storm Drainage, Corrugated Metal		Crew	Daily Output	Labor-Hours	Unit	Material	Labor	Equipment	Total	Total Incl O&P
0010	**PIPING, STORM DRAINAGE, CORRUGATED METAL**									
0020	Not including excavation or backfill									
2000	Corrugated metal pipe, galvanized									
2020	Bituminous coated with paved invert, 20' lengths									
2040	8" diameter, 16 ga.	B-14	330	.145	L.F.	8.65	5.80	1.10	15.55	19.60
2060	10" diameter, 16 ga.		260	.185		9	7.35	1.40	17.75	22.50
2080	12" diameter, 16 ga.		210	.229		11.05	9.10	1.73	21.88	28
2100	15" diameter, 16 ga.		200	.240		15.15	9.55	1.82	26.52	33.50
2120	18" diameter, 16 ga.		190	.253		16.70	10.05	1.92	28.67	36
2140	24" diameter, 14 ga.		160	.300		21.50	11.95	2.28	35.73	44.50
2160	30" diameter, 14 ga.	B-13	120	.467		27.50	19.15	6.15	52.80	67
2180	36" diameter, 12 ga.		120	.467		35.50	19.15	6.15	60.80	75.50
2200	48" diameter, 12 ga.		100	.560		52.50	23	7.35	82.85	101
2220	60" diameter, 10 ga.	B-13B	75	.747		80	30.50	15.05	125.55	151
2240	72" diameter, 8 ga.	"	45	1.244		95.50	51	25	171.50	211
2500	Galvanized, uncoated, 20' lengths									
2520	8" diameter, 16 ga.	B-14	355	.135	L.F.	7.80	5.40	1.03	14.23	18
2540	10" diameter, 16 ga.		280	.171		8.95	6.80	1.30	17.05	21.50
2560	12" diameter, 16 ga.		220	.218		9.95	8.70	1.66	20.31	26
2580	15" diameter, 16 ga.		220	.218		12.45	8.70	1.66	22.81	29
2600	18" diameter, 16 ga.		205	.234		15.05	9.30	1.78	26.13	33
2620	24" diameter, 14 ga.		175	.274		18.95	10.90	2.08	31.93	40
2640	30" diameter, 14 ga.	B-13	130	.431		25	17.65	5.65	48.30	61
2660	36" diameter, 12 ga.		130	.431		32	17.65	5.65	55.30	68.50
2680	48" diameter, 12 ga.		110	.509		47.50	21	6.70	75.20	91.50
2690	60" diameter, 10 ga.	B-13B	78	.718		72	29.50	14.45	115.95	140
2780	End sections, 8" diameter	B-14	35	1.371	Ea.	73	54.50	10.40	137.90	175
2785	10" diameter		35	1.371		77	54.50	10.40	141.90	179
2790	12" diameter		35	1.371		114	54.50	10.40	178.90	220
2800	18" diameter		30	1.600		115	63.50	12.15	190.65	237
2810	24" diameter	B-13	25	2.240		215	92	29.50	336.50	410
2820	30" diameter		25	2.240		330	92	29.50	451.50	535
2825	36" diameter		20	2.800		480	115	37	632	745

33 41 13.40 Piping, Storm Drainage, Corrugated Metal		Crew	Daily Output	Labor-Hours	Unit	Material	2015 Bare Costs Labor	Equipment	Total	Total Incl O&P
2830	48" diameter	B-13	10	5.600	Ea.	950	230	73.50	1,253.50	1,475
2835	60" diameter	B-13B	5	11.200		1,650	460	226	2,336	2,750
2840	72" diameter	"	4	14		1,975	575	282	2,832	3,350

33 41 13.60 Sewage/Drainage Collection, Concrete Pipe		Crew	Daily Output	Labor-Hours	Unit	Material	2015 Bare Costs Labor	Equipment	Total	Total Incl O&P
0010	**SEWAGE/DRAINAGE COLLECTION, CONCRETE PIPE**									
0020	Not including excavation or backfill									
1000	Non-reinforced pipe, extra strength, B&S or T&G joints									
1010	6" diameter	B-14	265.04	.181	L.F.	6	7.20	1.37	14.57	19.15
1020	8" diameter		224	.214		6.60	8.50	1.63	16.73	22
1030	10" diameter		216	.222		7.30	8.85	1.69	17.84	23.50
1040	12" diameter		200	.240		8.20	9.55	1.82	19.57	25.50
1050	15" diameter		180	.267		12	10.60	2.02	24.62	31.50
1060	18" diameter		144	.333		15	13.25	2.53	30.78	40
1070	21" diameter		112	.429		17	17.05	3.25	37.30	48.50
1080	24" diameter		100	.480		22	19.10	3.64	44.74	57.50
2000	Reinforced culvert, class 3, no gaskets									
2010	12" diameter	B-14	150	.320	L.F.	11	12.75	2.43	26.18	34.50
2020	15" diameter		150	.320		14	12.75	2.43	29.18	37.50
2030	18" diameter		132	.364		18	14.45	2.76	35.21	45
2035	21" diameter		120	.400		22	15.90	3.04	40.94	52
2040	24" diameter		100	.480		26	19.10	3.64	48.74	62
2045	27" diameter	B-13	92	.609		37	25	8	70	87.50
2050	30" diameter		88	.636		42	26	8.35	76.35	95
2060	36" diameter		72	.778		56	32	10.25	98.25	122
2070	42" diameter	B-13B	72	.778		75	32	15.65	122.65	149
2080	48" diameter		64	.875		89	36	17.60	142.60	172
2090	60" diameter		48	1.167		136	48	23.50	207.50	249
2100	72" diameter		40	1.400		206	57.50	28	291.50	345
2120	84" diameter		32	1.750		275	71.50	35.50	382	455
2140	96" diameter		24	2.333		330	95.50	47	472.50	565
2200	With gaskets, class 3, 12" diameter	B-21	168	.167		12.10	7.20	.82	20.12	25.50
2220	15" diameter		160	.175		15.40	7.60	.86	23.86	29.50
2230	18" diameter		152	.184		19.80	8	.91	28.71	35.50
2240	24" diameter		136	.206		33	8.90	1.01	42.91	51.50
2260	30" diameter	B-13	88	.636		50	26	8.35	84.35	104
2270	36" diameter	"	72	.778		65.50	32	10.25	107.75	132
2290	48" diameter	B-13B	64	.875		102	36	17.60	155.60	186
2310	72" diameter	"	40	1.400		225	57.50	28	310.50	365
2330	Flared ends, 12" diameter	B-21	31	.903	Ea.	226	39	4.44	269.44	315
2340	15" diameter		25	1.120		267	48.50	5.50	321	375
2400	18" diameter		20	1.400		305	60.50	6.90	372.40	440
2420	24" diameter		14	2		370	86.50	9.85	466.35	555
2440	36" diameter	B-13	10	5.600		790	230	73.50	1,093.50	1,300
3080	Radius pipe, add to pipe prices, 12" to 60" diameter				L.F.	50%				
3090	Over 60" diameter, add				"	20%				
3500	Reinforced elliptical, 8' lengths, C507 class 3									
3520	14" x 23" inside, round equivalent 18" diameter	B-21	82	.341	L.F.	41	14.80	1.68	57.48	70
3530	24" x 38" inside, round equivalent 30" diameter	B-13	58	.966		62	39.50	12.70	114.20	142
3540	29" x 45" inside, round equivalent 36" diameter		52	1.077		78	44	14.15	136.15	169
3550	38" x 60" inside, round equivalent 48" diameter		38	1.474		134	60.50	19.40	213.90	261
3560	48" x 76" inside, round equivalent 60" diameter		26	2.154		186	88.50	28.50	303	370
3570	58" x 91" inside, round equivalent 72" diameter		22	2.545		272	104	33.50	409.50	495

33 41 Storm Utility Drainage Piping

33 41 13 – Public Storm Utility Drainage Piping

33 41 13.60 Sewage/Drainage Collection, Concrete Pipe		Crew	Daily Output	Labor-Hours	Unit	Material	2015 Bare Costs Labor	Equipment	Total	Total Incl O&P
3780	Concrete slotted pipe, class 4 mortar joint									
3800	12" diameter	B-21	168	.167	L.F.	28	7.20	.82	36.02	43
3840	18" diameter	"	152	.184	"	32	8	.91	40.91	48.50
3900	Concrete slotted pipe, Class 4 O-ring joint									
3940	12" diameter	B-21	168	.167	L.F.	28	7.20	.82	36.02	43
3960	18" diameter	"	152	.184	"	32	8	.91	40.91	48.50

33 42 Culverts

33 42 16 – Concrete Culverts

33 42 16.15 Oval Arch Culverts

		Crew	Daily Output	Labor-Hours	Unit	Material	Labor	Equipment	Total	Total Incl O&P
0010	**OVAL ARCH CULVERTS**									
3000	Corrugated galvanized or aluminum, coated & paved									
3020	17" x 13", 16 ga., 15" equivalent	B-14	200	.240	L.F.	13.20	9.55	1.82	24.57	31
3040	21" x 15", 16 ga., 18" equivalent		150	.320		16.05	12.75	2.43	31.23	40
3060	28" x 20", 14 ga., 24" equivalent		125	.384		24	15.25	2.91	42.16	53
3080	35" x 24", 14 ga., 30" equivalent		100	.480		29.50	19.10	3.64	52.24	66
3100	42" x 29", 12 ga., 36" equivalent	B-13	100	.560		35.50	23	7.35	65.85	82
3120	49" x 33", 12 ga., 42" equivalent		90	.622		41	25.50	8.20	74.70	93.50
3140	57" x 38", 12 ga., 48" equivalent		75	.747		57	30.50	9.80	97.30	120
3160	Steel, plain oval arch culverts, plain									
3180	17" x 13", 16 ga., 15" equivalent	B-14	225	.213	L.F.	11.95	8.50	1.62	22.07	28
3200	21" x 15", 16 ga., 18" equivalent		175	.274		14.45	10.90	2.08	27.43	35
3220	28" x 20", 14 ga., 24" equivalent		150	.320		21.50	12.75	2.43	36.68	45.50
3240	35" x 24", 14 ga., 30" equivalent	B-13	108	.519		26.50	21.50	6.80	54.80	69
3260	42" x 29", 12 ga., 36" equivalent		108	.519		32	21.50	6.80	60.30	75.50
3280	49" x 33", 12 ga., 42" equivalent		92	.609		37	25	8	70	88
3300	57" x 38", 12 ga., 48" equivalent		75	.747		51.50	30.50	9.80	91.80	114
3320	End sections, 17" x 13"		22	2.545	Ea.	144	104	33.50	281.50	355
3340	42" x 29"		17	3.294	"	395	135	43.50	573.50	690
3360	Multi-plate arch, steel	B-20	1690	.014	Lb.	1.25	.60		1.85	2.30

33 44 Storm Utility Water Drains

33 44 13 – Utility Area Drains

33 44 13.13 Catchbasins

		Crew	Daily Output	Labor-Hours	Unit	Material	Labor	Equipment	Total	Total Incl O&P
0010	**CATCHBASINS**									
0011	Not including footing & excavation									
1600	Frames & grates, C.I., 24" square, 500 lb.	B-6	7.80	3.077	Ea.	340	127	46.50	513.50	620
1700	26" D shape, 600 lb.		7	3.429		505	142	52	699	830
1800	Light traffic, 18" diameter, 100 lb.		10	2.400		123	99	36.50	258.50	325
1900	24" diameter, 300 lb.		8.70	2.759		196	114	42	352	435
2000	36" diameter, 900 lb.		5.80	4.138		570	171	63	804	955
2100	Heavy traffic, 24" diameter, 400 lb.		7.80	3.077		244	127	46.50	417.50	515
2200	36" diameter, 1150 lb.		3	8		795	330	121	1,246	1,525
2300	Mass. State standard, 26" diameter, 475 lb.		7	3.429		266	142	52	460	565
2400	30" diameter, 620 lb.		7	3.429		345	142	52	539	655
2500	Watertight, 24" diameter, 350 lb.		7.80	3.077		320	127	46.50	493.50	595
2600	26" diameter, 500 lb.		7	3.429		425	142	52	619	745
2700	32" diameter, 575 lb.		6	4		850	165	60.50	1,075.50	1,250
2800	3 piece cover & frame, 10" deep,									

33 44 Storm Utility Water Drains

33 44 13 – Utility Area Drains

33 44 13.13 Catchbasins		Crew	Daily Output	Labor-Hours	Unit	2015 Bare Costs Material	Labor	Equipment	Total	Total Incl O&P
2900	1200 lb., for heavy equipment	B-6	3	8	Ea.	1,050	330	121	1,501	1,825
3000	Raised for paving 1-1/4" to 2" high									
3100	4 piece expansion ring									
3200	20" to 26" diameter	1 Clab	3	2.667	Ea.	158	100		258	330
3300	30" to 36" diameter	"	3	2.667	"	217	100		317	395
3320	Frames and covers, existing, raised for paving, 2", including									
3340	row of brick, concrete collar, up to 12" wide frame	B-6	18	1.333	Ea.	45	55	20	120	156
3360	20" to 26" wide frame		11	2.182		67.50	90	33	190.50	249
3380	30" to 36" wide frame		9	2.667		83.50	110	40.50	234	305
3400	Inverts, single channel brick	D-1	3	5.333		97	225		322	450
3500	Concrete		5	3.200		104	135		239	320
3600	Triple channel, brick		2	8		148	335		483	680
3700	Concrete		3	5.333		139	225		364	500

33 44 13.50 Stormwater Management		Crew	Daily Output	Labor-Hours	Unit	Material	Labor	Equipment	Total	Total Incl O&P
0010	**STORMWATER MANAGEMENT**									
0020	Allowance, add per SF of impervious surface				S.F.				3	3

33 46 Subdrainage

33 46 16 – Subdrainage Piping

33 46 16.25 Piping, Subdrainage, Corrugated Metal		Crew	Daily Output	Labor-Hours	Unit	Material	Labor	Equipment	Total	Total Incl O&P
0010	**PIPING, SUBDRAINAGE, CORRUGATED METAL**									
0021	Not including excavation and backfill									
2010	Aluminum, perforated									
2020	6" diameter, 18 ga.	B-20	380	.063	L.F.	6.50	2.65		9.15	11.25
2200	8" diameter, 16 ga.	"	370	.065		8.55	2.72		11.27	13.60
2220	10" diameter, 16 ga.	B-21	360	.078		10.70	3.37	.38	14.45	17.35
2240	12" diameter, 16 ga.		285	.098		11.95	4.26	.48	16.69	20
2260	18" diameter, 16 ga.		205	.137		17.95	5.90	.67	24.52	29.50
3000	Uncoated galvanized, perforated									
3020	6" diameter, 18 ga.	B-20	380	.063	L.F.	6.05	2.65		8.70	10.75
3200	8" diameter, 16 ga.	"	370	.065		8.30	2.72		11.02	13.35
3220	10" diameter, 16 ga.	B-21	360	.078		8.80	3.37	.38	12.55	15.30
3240	12" diameter, 16 ga.		285	.098		9.80	4.26	.48	14.54	17.85
3260	18" diameter, 16 ga.		205	.137		15	5.90	.67	21.57	26.50
4000	Steel, perforated, asphalt coated									
4020	6" diameter 18 ga.	B-20	380	.063	L.F.	6.50	2.65		9.15	11.25
4030	8" diameter 18 ga.	"	370	.065		8.55	2.72		11.27	13.60
4040	10" diameter 16 ga.	B-21	360	.078		10.05	3.37	.38	13.80	16.65
4050	12" diameter 16 ga.		285	.098		11.05	4.26	.48	15.79	19.25
4060	18" diameter 16 ga.		205	.137		17.10	5.90	.67	23.67	28.50

33 46 16.40 Piping, Subdrainage, Polyvinyl Chloride		Crew	Daily Output	Labor-Hours	Unit	Material	Labor	Equipment	Total	Total Incl O&P
0010	**PIPING, SUBDRAINAGE, POLYVINYL CHLORIDE**									
0020	Perforated, price as solid pipe, Section 33 31 13.25									

33 49 Storm Drainage Structures

33 49 13 – Storm Drainage Manholes, Frames, and Covers

33 49 13.10 Storm Drainage Manholes, Frames and Covers		Crew	Daily Output	Labor-Hours	Unit	Material	2015 Bare Costs Labor	Equipment	Total	Total Incl O&P
0010	**STORM DRAINAGE MANHOLES, FRAMES & COVERS**									
0020	Excludes footing, excavation, backfill (See line items for frame & cover)									
0050	Brick, 4′ inside diameter, 4′ deep	D-1	1	16	Ea.	520	675		1,195	1,600
0100	6′ deep		.70	22.857		740	960		1,700	2,300
0150	8′ deep		.50	32		955	1,350		2,305	3,100
0200	For depths over 8′, add		4	4	V.L.F.	83.50	168		251.50	350
0400	Concrete blocks (radial), 4′ I.D., 4′ deep		1.50	10.667	Ea.	415	450		865	1,150
0500	6′ deep		1	16		565	675		1,240	1,650
0600	8′ deep		.70	22.857		710	960		1,670	2,250
0700	For depths over 8′, add		5.50	2.909	V.L.F.	77	122		199	272
0800	Concrete, cast in place, 4′ x 4′, 8″ thick, 4′ deep	C-14H	2	24	Ea.	505	1,100	15.60	1,620.60	2,275
0900	6′ deep		1.50	32		730	1,475	21	2,226	3,100
1000	8′ deep		1	48		1,050	2,225	31	3,306	4,575
1100	For depths over 8′, add		8	6	V.L.F.	119	278	3.90	400.90	560
1110	Precast, 4′ I.D., 4′ deep	B-22	4.10	7.317	Ea.	725	320	50.50	1,095.50	1,350
1120	6′ deep		3	10		925	440	69	1,434	1,775
1130	8′ deep		2	15		1,075	660	103	1,838	2,300
1140	For depths over 8′, add		16	1.875	V.L.F.	127	82.50	12.90	222.40	280
1150	5′ I.D., 4′ deep	B-6	3	8	Ea.	1,650	330	121	2,101	2,450
1160	6′ deep		2	12		1,925	495	182	2,602	3,075
1170	8′ deep		1.50	16		2,400	660	243	3,303	3,925
1180	For depths over 8′, add		12	2	V.L.F.	280	82.50	30.50	393	470
1190	6′ I.D., 4′ deep		2	12	Ea.	2,150	495	182	2,827	3,325
1200	6′ deep		1.50	16		2,600	660	243	3,503	4,125
1210	8′ deep		1	24		3,200	990	365	4,555	5,425
1220	For depths over 8′, add		8	3	V.L.F.	380	124	45.50	549.50	655
1250	Slab tops, precast, 8″ thick									
1300	4′ diameter manhole	B-6	8	3	Ea.	252	124	45.50	421.50	515
1400	5′ diameter manhole		7.50	3.200		410	132	48.50	590.50	710
1500	6′ diameter manhole		7	3.429		635	142	52	829	975
3800	Steps, heavyweight cast iron, 7″ x 9″	1 Bric	40	.200		19.25	9.25		28.50	35
3900	8″ x 9″		40	.200		23	9.25		32.25	39.50
3928	12″ x 10-1/2″		40	.200		27	9.25		36.25	43.50
4000	Standard sizes, galvanized steel		40	.200		22	9.25		31.25	38
4100	Aluminum		40	.200		24	9.25		33.25	40.50
4150	Polyethylene		40	.200		26	9.25		35.25	42.50

33 51 Natural-Gas Distribution

33 51 13 – Natural-Gas Piping

33 51 13.10 Piping, Gas Service and Distribution, P.E.		Crew	Daily Output	Labor-Hours	Unit	Material	Labor	Equipment	Total	Total Incl O&P
0010	**PIPING, GAS SERVICE AND DISTRIBUTION, POLYETHYLENE**									
0020	Not including excavation or backfill									
1000	60 psi coils, compression coupling @ 100′, 1/2″ diameter, SDR 11	B-20A	608	.053	L.F.	.45	2.41		2.86	4.16
1010	1″ diameter, SDR 11		544	.059		1.05	2.69		3.74	5.25
1040	1-1/4″ diameter, SDR 11		544	.059		1.62	2.69		4.31	5.85
1100	2″ diameter, SDR 11		488	.066		2.82	3		5.82	7.65
1160	3″ diameter, SDR 11		408	.078		6.45	3.59		10.04	12.55
1500	60 PSI 40′ joints with coupling, 3″ diameter, SDR 11	B-21A	408	.098		10.05	4.60	1.16	15.81	19.35
1540	4″ diameter, SDR 11		352	.114		14.25	5.35	1.35	20.95	25
1600	6″ diameter, SDR 11		328	.122		33.50	5.70	1.44	40.64	47.50
1640	8″ diameter, SDR 11		272	.147		51.50	6.90	1.74	60.14	69

33 52 16 – Gasoline Distribution

33 52 16.13 Gasoline Piping		Crew	Daily Output	Labor-Hours	Unit	Material	Labor (2015 Bare Costs)	Equipment (2015 Bare Costs)	Total (2015 Bare Costs)	Total Incl O&P
0010	**GASOLINE PIPING**									
0020	Primary containment pipe, fiberglass-reinforced									
0030	Plastic pipe 15′ & 30′ lengths									
0040	2″ diameter	Q-6	425	.056	L.F.	5.95	3.15		9.10	11.25
0050	3″ diameter		400	.060		10.35	3.35		13.70	16.45
0060	4″ diameter	↓	375	.064	↓	13.65	3.57		17.22	20.50
0100	Fittings									
0110	Elbows, 90° & 45°, bell-ends, 2″	Q-6	24	1	Ea.	43.50	56		99.50	132
0120	3″ diameter		22	1.091		55	61		116	153
0130	4″ diameter		20	1.200		70	67		137	178
0200	Tees, bell ends, 2″		21	1.143		60.50	63.50		124	163
0210	3″ diameter		18	1.333		64	74.50		138.50	183
0220	4″ diameter		15	1.600		84	89		173	228
0230	Flanges bell ends, 2″		24	1		33.50	56		89.50	121
0240	3″ diameter		22	1.091		39	61		100	135
0250	4″ diameter		20	1.200		45	67		112	151
0260	Sleeve couplings, 2″		21	1.143		12.40	63.50		75.90	110
0270	3″ diameter		18	1.333		17.80	74.50		92.30	132
0280	4″ diameter		15	1.600		23	89		112	160
0290	Threaded adapters 2″		21	1.143		17.95	63.50		81.45	116
0300	3″ diameter		18	1.333		34	74.50		108.50	150
0310	4″ diameter		15	1.600		38	89		127	177
0320	Reducers, 2″		27	.889		27.50	49.50		77	105
0330	3″ diameter		22	1.091		27.50	61		88.50	122
0340	4″ diameter	↓	20	1.200	↓	37	67		104	142
1010	Gas station product line for secondary containment (double wall)									
1100	Fiberglass reinforced plastic pipe 25′ lengths									
1120	Pipe, plain end, 3″ diameter	Q-6	375	.064	L.F.	26.50	3.57		30.07	34.50
1130	4″ diameter		350	.069		32	3.82		35.82	41.50
1140	5″ diameter		325	.074		35.50	4.12		39.62	45
1150	6″ diameter	↓	300	.080	↓	39	4.46		43.46	50
1200	Fittings									
1230	Elbows, 90° & 45°, 3″ diameter	Q-6	18	1.333	Ea.	136	74.50		210.50	262
1240	4″ diameter		16	1.500		167	83.50		250.50	310
1250	5″ diameter		14	1.714		184	95.50		279.50	345
1260	6″ diameter		12	2		204	112		316	390
1270	Tees, 3″ diameter		15	1.600		166	89		255	315
1280	4″ diameter		12	2		202	112		314	390
1290	5″ diameter		9	2.667		315	149		464	575
1300	6″ diameter		6	4		375	223		598	750
1310	Couplings, 3″ diameter		18	1.333		55	74.50		129.50	173
1320	4″ diameter		16	1.500		119	83.50		202.50	257
1330	5″ diameter		14	1.714		214	95.50		309.50	380
1340	6″ diameter		12	2		315	112		427	520
1350	Cross-over nipples, 3″ diameter		18	1.333		10.60	74.50		85.10	124
1360	4″ diameter		16	1.500		12.70	83.50		96.20	140
1370	5″ diameter		14	1.714		15.90	95.50		111.40	162
1380	6″ diameter		12	2		19.05	112		131.05	189
1400	Telescoping, reducers, concentric 4″ x 3″		18	1.333		48	74.50		122.50	165
1410	5″ x 4″		17	1.412		95.50	78.50		174	224
1420	6″ x 5″	↓	16	1.500	↓	234	83.50		317.50	385

33 71 16 – Electrical Utility Poles

33 71 16.33 Wood Electrical Utility Poles		Crew	Daily Output	Labor-Hours	Unit	Material	2015 Bare Costs Labor	Equipment	Total	Total Incl O&P
0010	WOOD ELECTRICAL UTILITY POLES									
0011	Excludes excavation, backfill and cast-in-place concrete									
6200	Electric & tel sitework, 20′ high, treated wd., see Section 26 56 13.10	R-3	3.10	6.452	Ea.	224	350	44.50	618.50	820
6400	25′ high		2.90	6.897		265	375	47.50	687.50	905
6600	30′ high		2.60	7.692		435	420	53	908	1,150
6800	35′ high		2.40	8.333		495	455	57.50	1,007.50	1,300
7000	40′ high		2.30	8.696		710	470	60	1,240	1,550
7200	45′ high		1.70	11.765		865	640	81	1,586	2,000
7400	Cross arms with hardware & insulators									
7600	4′ long	1 Elec	2.50	3.200	Ea.	150	175		325	425
7800	5′ long		2.40	3.333		172	182		354	460
8000	6′ long		2.20	3.636		166	199		365	480

33 71 19 – Electrical Underground Ducts and Manholes

33 71 19.17 Electric and Telephone Underground		Crew	Daily Output	Labor-Hours	Unit	Material	Labor	Equipment	Total	Total Incl O&P
0010	ELECTRIC AND TELEPHONE UNDERGROUND									
0011	Not including excavation									
0200	backfill and cast in place concrete									
0400	Hand holes, precast concrete, with concrete cover									
0600	2′ x 2′ x 3′ deep	R-3	2.40	8.333	Ea.	405	455	57.50	917.50	1,200
0800	3′ x 3′ x 3′ deep		1.90	10.526		525	570	72.50	1,167.50	1,525
1000	4′ x 4′ x 4′ deep		1.40	14.286		1,400	775	98.50	2,273.50	2,825
1200	Manholes, precast with iron racks & pulling irons, C.I. frame									
1400	and cover, 4′ x 6′ x 7′ deep	B-13	2	28	Ea.	6,050	1,150	370	7,570	8,800
1600	6′ x 8′ x 7′ deep		1.90	29.474		6,800	1,200	390	8,390	9,750
1800	6′ x 10′ x 7′ deep		1.80	31.111		7,625	1,275	410	9,310	10,800
4200	Underground duct, banks ready for concrete fill, min. of 7.5″									
4400	between conduits, center to center									
4580	PVC, type EB, 1 @ 2″ diameter	2 Elec	480	.033	L.F.	.71	1.82		2.53	3.51
4600	2 @ 2″ diameter		240	.067		1.41	3.65		5.06	7
4800	4 @ 2″ diameter		120	.133		2.82	7.30		10.12	14.05
5000	2 @ 3″ diameter		200	.080		1.96	4.38		6.34	8.70
5200	4 @ 3″ diameter		100	.160		3.91	8.75		12.66	17.40
5400	2 @ 4″ diameter		160	.100		3.05	5.45		8.50	11.55
5600	4 @ 4″ diameter		80	.200		6.10	10.95		17.05	23
5800	6 @ 4″ diameter		54	.296		9.15	16.20		25.35	34.50
6200	Rigid galvanized steel, 2 @ 2″ diameter		180	.089		12.90	4.86		17.76	21.50
6400	4 @ 2″ diameter		90	.178		26	9.70		35.70	43
6800	2 @ 3″ diameter		100	.160		29	8.75		37.75	44.50
7000	4 @ 3″ diameter		50	.320		57.50	17.50		75	89.50
7200	2 @ 4″ diameter		70	.229		41.50	12.50		54	64.50
7400	4 @ 4″ diameter		34	.471		83	25.50		108.50	130
7600	6 @ 4″ diameter		22	.727		124	40		164	197

33 81 13 – Communications Transmission Towers

33 81 13.10 Radio Towers		Crew	Daily Output	Labor-Hours	Unit	2015 Bare Costs Material	Labor	Equipment	Total	Total Incl O&P
0010	**RADIO TOWERS**									
0020	Guyed, 50′ H, 40 lb. sect., 70 MPH basic wind spd.	2 Sswk	1	16	Ea.	2,750	840		3,590	4,500
0100	Wind load 90 MPH basic wind speed	"	1	16		3,700	840		4,540	5,550
0300	190′ high, 40 lb. section, wind load 70 MPH basic wind speed	K-2	.33	72.727		9,550	3,550	920	14,020	17,500
0400	200′ high, 70 lb. section, wind load 90 MPH basic wind speed		.33	72.727		15,900	3,550	920	20,370	24,500
0600	300′ high, 70 lb. section, wind load 70 MPH basic wind speed		.20	120		25,500	5,850	1,525	32,875	39,500
0700	270′ high, 90 lb. section, wind load 90 MPH basic wind speed		.20	120		27,600	5,850	1,525	34,975	41,800
0800	400′ high, 100 lb. section, wind load 70 MPH basic wind speed		.14	171		38,100	8,375	2,175	48,650	58,500
0900	Self-supporting, 60′ high, wind load 70 MPH basic wind speed		.80	30		4,500	1,475	380	6,355	7,825
0910	60′ high, wind load 90 MPH basic wind speed		.45	53.333		5,275	2,600	675	8,550	10,900
1000	120′ high, wind load 70 MPH basic wind speed		.40	60		10,000	2,925	760	13,685	16,700
1200	190′ high, wind load 90 MPH basic wind speed		.20	120		28,300	5,850	1,525	35,675	42,600
2000	For states west of Rocky Mountains, add for shipping					10%				

Estimating Tips

34 11 00 Rail Tracks

This subdivision includes items that may involve either repair of existing, or construction of new, railroad tracks. Additional preparation work, such as the roadbed earthwork, would be found in Division 31. Additional new construction siding and turnouts are found in Subdivision 34 72. Maintenance of railroads is found under 34 01 23 Operation and Maintenance of Railways.

34 40 00 Traffic Signals

This subdivision includes traffic signal systems. Other traffic control devices such as traffic signs are found in Subdivision 10 14 53 Traffic Signage.

34 70 00 Vehicle Barriers

This subdivision includes security vehicle barriers, guide and guard rails, crash barriers, and delineators. The actual maintenance and construction of concrete and asphalt pavement is found in Division 32.

Reference Numbers

Reference numbers are shown in shaded boxes at the beginning of some major classifications. These numbers refer to related items in the Reference Section. The reference information may be an estimating procedure, an alternate pricing method, or technical information.

Note: Not all subdivisions listed here necessarily appear in this publication. ■

34 01 Operation and Maintenance of Transportation

34 01 23 – Operation and Maintenance of Railways

34 01 23.51 Maintenance of Railroads		Crew	Daily Output	Labor-Hours	Unit	Material	2015 Bare Costs Labor	Equipment	Total	Total Incl O&P
0010	MAINTENANCE OF RAILROADS									
0400	Resurface and realign existing track	B-14	200	.240	L.F.		9.55	1.82	11.37	16.65
0600	For crushed stone ballast, add	"	500	.096	"	13.40	3.82	.73	17.95	21.50

34 11 Rail Tracks

34 11 13 – Track Rails

34 11 13.23 Heavy Rail Track

		Crew	Daily Output	Labor-Hours	Unit	Material	Labor	Equipment	Total	Total Incl O&P
0010	HEAVY RAIL TRACK R347216-10									
1000	Rail, 100 lb. prime grade				L.F.	35			35	38.50
1500	Relay rail				"	17.55			17.55	19.30

34 11 33 – Track Cross Ties

34 11 33.13 Concrete Track Cross Ties

		Crew	Daily Output	Labor-Hours	Unit	Material	Labor	Equipment	Total	Total Incl O&P
0010	CONCRETE TRACK CROSS TIES									
1400	Ties, concrete, 8'-6" long, 30" O.C.	B-14	80	.600	Ea.	173	24	4.55	201.55	232

34 11 33.16 Timber Track Cross Ties

		Crew	Daily Output	Labor-Hours	Unit	Material	Labor	Equipment	Total	Total Incl O&P
0010	TIMBER TRACK CROSS TIES									
1600	Wood, pressure treated, 6" x 8" x 8'-6", C.L. lots	B-14	90	.533	Ea.	52	21	4.05	77.05	94
1700	L.C.L. lots		90	.533		54.50	21	4.05	79.55	97
1900	Heavy duty, 7" x 9" x 8'-6", C.L. lots		70	.686		57	27.50	5.20	89.70	110
2000	L.C.L. lots		70	.686		57	27.50	5.20	89.70	110

34 11 33.17 Timber Switch Ties

		Crew	Daily Output	Labor-Hours	Unit	Material	Labor	Equipment	Total	Total Incl O&P
0010	TIMBER SWITCH TIES									
1200	Switch timber, for a #8 switch, pressure treated	B-14	3.70	12.973	M.B.F.	3,100	515	98.50	3,713.50	4,325
1300	Complete set of timbers, 3.7 MBF for #8 switch	"	1	48	Total	12,000	1,900	365	14,265	16,500

34 11 93 – Track Appurtenances and Accessories

34 11 93.50 Track Accessories

		Crew	Daily Output	Labor-Hours	Unit	Material	Labor	Equipment	Total	Total Incl O&P
0010	TRACK ACCESSORIES									
0020	Car bumpers, test	B-14	2	24	Ea.	3,750	955	182	4,887	5,800
0100	Heavy duty R347216-20		2	24		7,100	955	182	8,237	9,500
0200	Derails hand throw (sliding)		10	4.800		1,250	191	36.50	1,477.50	1,700
0300	Hand throw with standard timbers, open stand & target		8	6		1,350	239	45.50	1,634.50	1,900
2400	Wheel stops, fixed		18	2.667	Pr.	920	106	20	1,046	1,200
2450	Hinged		14	3.429	"	1,250	136	26	1,412	1,625

34 11 93.60 Track Material

		Crew	Daily Output	Labor-Hours	Unit	Material	Labor	Equipment	Total	Total Incl O&P
0010	TRACK MATERIAL									
0020	Track bolts				Ea.	4.17			4.17	4.59
0100	Joint bars				Pr.	93			93	103
0200	Spikes				Ea.	2			2	2.20
0300	Tie plates				"	15.60			15.60	17.15

34 41 Roadway Signaling and Control Equipment

34 41 13 – Traffic Signals

34 41 13.10 Traffic Signals Systems		Crew	Daily Output	Labor-Hours	Unit	Material	2015 Bare Costs Labor	Equipment	Total	Total Incl O&P
0010	**TRAFFIC SIGNALS SYSTEMS**									
0020	Component costs									
0600	Crew employs crane/directional driller as required									
1000	Vertical mast with foundation									
1010	Mast sized for single arm to 40'; no lighting or power function	R-11	.50	112	Signal	10,000	5,825	1,725	17,550	21,700
1100	Horizontal Arm									
1110	Per linear foot of arm	R-11	50	1.120	Signal	200	58	17.20	275.20	325
1200	Traffic Signal									
1210	Includes signal, bracket, sensor, and wiring	R-11	2.50	22.400	Signal	1,000	1,175	345	2,520	3,225
1300	Pedestrian Signals and Callers									
1310	Includes four signals with brackets and two call buttons	R-11	2.50	22.400	Signal	3,000	1,175	345	4,520	5,425
1400	Controller, Design, and underground conduit									
1410	Includes miscellaneous signage and adjacent surface work	R-11	.25	224	Signal	20,000	11,600	3,450	35,050	43,300

34 71 Roadway Construction

34 71 13 – Vehicle Barriers

34 71 13.17 Security Vehicle Barriers		Crew	Daily Output	Labor-Hours	Unit	Material	2015 Bare Costs Labor	Equipment	Total	Total Incl O&P
0010	**SECURITY VEHICLE BARRIERS**									
0020	Security planters excludes filling material									
0100	Concrete security planter, exposed aggregate 36" diam. x 30" high	B-11M	8	2	Ea.	565	88	49	702	810
0200	48" diam. x 36" high		8	2		845	88	49	982	1,125
0300	53" diam. x 18" high		8	2		945	88	49	1,082	1,250
0400	72" diam. x 18" high with seats		8	2		2,275	88	49	2,412	2,700
0450	84" diam. x 36" high		8	2		2,250	88	49	2,387	2,675
0500	36" x 36" x 24" high square		8	2		510	88	49	647	755
0600	36" x 36" x 30" high square		8	2		820	88	49	957	1,100
0700	48" L x 24" W x 30" H rectangle		8	2		525	88	49	662	765
0800	72" L x 24" W x 30" H rectangle		8	2		610	88	49	747	865
0900	96" L x 24" W x 30" H rectangle		8	2		780	88	49	917	1,050
0950	Decorative geometric concrete barrier, 96" L x 24" W x 36" H		8	2		830	88	49	967	1,100
1000	Concrete security planter, filling with washed sand or gravel <1 C.Y.		8	2		39	88	49	176	232
1050	2 C.Y. or less	↓	6	2.667		78	118	65.50	261.50	335
1200	Jersey concrete barrier, 10' L x 2' by 0.5' W x 30" H	B-21B	16	2.500		385	102	41	528	620
1300	10 or more same site		24	1.667		385	68	27.50	480.50	555
1400	10' L x 2' by 0.5' W x 32" H		16	2.500		405	102	41	548	645
1500	10 or more same site		24	1.667		405	68	27.50	500.50	580
1600	20' L x 2' by 0.5' W x 30" H		12	3.333		565	136	54.50	755.50	890
1700	10 or more same site		18	2.222		565	90.50	36.50	692	800
1800	20' L x 2' by 0.5' W x 32" H		12	3.333		680	136	54.50	870.50	1,025
1900	10 or more same site		18	2.222		680	90.50	36.50	807	930
2000	GFRC decorative security barrier per 10 feet section including concrete		4	10		3,200	410	164	3,774	4,300
2100	Per 12 feet section including concrete	↓	4	10	↓	3,825	410	164	4,399	5,025
2210	GFRC decorative security barrier will stop 30 MPH, 4000 lb. vehicle									
2300	High security barrier base prep per 12 feet section on bare ground	B-11C	4	4	Ea.	21.50	176	91	288.50	395
2310	GFRC barrier base prep does not include haul away of excavated matl.									
2400	GFRC decorative high security barrier per 12 feet section w/concrete	B-21B	3	13.333	Ea.	6,025	545	218	6,788	7,700
2410	GFRC decorative high security barrier will stop 50 MPH, 15000 lb. vehicle									
2500	GFRC decorative impaler security barrier per 10 feet section w/prep	B-6	4	6	Ea.	2,300	248	91	2,639	3,000
2600	Per 12 feet section w/prep	"	4	6	"	2,750	248	91	3,089	3,500
2610	Impaler barrier should stop 50 MPH, 15000 lb. vehicle w/some penetr.									
2700	Pipe bollards, steel, concrete filled/painted, 8' L x 4' D hole, 8" diam.	B-6	10	2.400	Ea.	875	99	36.50	1,010.50	1,150

34 71 Roadway Construction

34 71 13 – Vehicle Barriers

34 71 13.17 Security Vehicle Barriers

	34 71 13.17 Security Vehicle Barriers	Crew	Daily Output	Labor-Hours	Unit	2015 Bare Costs Material	Labor	Equipment	Total	Total Incl O&P
2710	Schedule 80 concrete bollards will stop 4000 lb. vehicle @ 30 MPH									
2800	GFRC decorative jersey barrier cover per 10 feet section excludes soil	B-6	8	3	Ea.	990	124	45.50	1,159.50	1,350
2900	Per 12 feet section excludes soil		8	3		1,175	124	45.50	1,344.50	1,550
3000	GFRC decorative 8″ diameter bollard cover		12	2		515	82.50	30.50	628	725
3100	GFRC decorative barrier face 10 foot section excludes earth backing		10	2.400		1,075	99	36.50	1,210.50	1,400
3200	Drop arm crash barrier, 15000 lb. vehicle @ 50 MPH									
3205	Includes all material, labor for complete installation									
3210	12′ width				Ea.				64,500	71,000
3310	24′ width								86,000	95,000
3410	Wedge crash barrier, 10′ width								98,000	108,000
3510	12.5′ width								108,000	119,000
3520	15′ width								118,500	130,000
3610	Sliding crash barrier, 12′ width								178,500	196,000
3710	Sliding roller crash barrier, 20′ width								198,000	217,500
3810	Sliding cantilever crash barrier, 20′ width								198,000	217,500
3890	Note: Raised bollard crash barriers should be used w/tire shredders									
3910	Raised bollard crash barrier, 10′ width				Ea.				38,800	42,900
4010	12′ width								44,900	49,000
4110	Raised bollard crash barrier, 10′ width, solar powered								46,900	52,000
4210	12′ width								53,000	58,000
4310	In ground tire shredder, 16′ width								43,900	47,900

34 71 13.26 Vehicle Guide Rails

	34 71 13.26 Vehicle Guide Rails	Crew	Daily Output	Labor-Hours	Unit	Material	Labor	Equipment	Total	Total Incl O&P
0010	**VEHICLE GUIDE RAILS**									
0012	Corrugated stl., galv. stl. posts, 6′-3″ O.C.	B-80	850	.038	L.F.	23	1.55	.85	25.40	28.50
0200	End sections, galvanized, flared		50	.640	Ea.	98.50	26.50	14.40	139.40	164
0300	Wrap around end		50	.640	″	141	26.50	14.40	181.90	211
0400	Timber guide rail, 4″ x 8″ with 6″ x 8″ wood posts, treated		960	.033	L.F.	12.05	1.37	.75	14.17	16.20
0600	Cable guide rail, 3 at 3/4″ cables, steel posts, single face		900	.036		10.75	1.47	.80	13.02	14.95
0700	Wood posts		950	.034		12.95	1.39	.76	15.10	17.20
0900	Guide rail, steel box beam, 6″ x 6″		120	.267		34	11	6	51	61
1100	Median barrier, steel box beam, 6″ x 8″		215	.149		45	6.15	3.35	54.50	62.50
1400	Resilient guide fence and light shield, 6′ high	B-2	130	.308		38	11.70		49.70	60
1500	Concrete posts, individual, 6′-5″, triangular	B-80	110	.291	Ea.	72.50	12	6.55	91.05	106
1550	Square	″	110	.291	″	78	12	6.55	96.55	112
2000	Median, precast concrete, 3′-6″ high, 2′ wide, single face	B-29	380	.147	L.F.	54.50	6.05	2.32	62.87	71.50
2200	Double face	″	340	.165	″	62.50	6.75	2.59	71.84	81.50
2400	Speed bumps, thermoplastic, 10-1/2″ x 2-1/4″ x 48″ long	B-2	120	.333	Ea.	138	12.65		150.65	171
3030	Impact barrier, UTMCD, barrel type	B-16	30	1.067	″	510	41.50	23	574.50	655

34 71 19 – Vehicle Delineators

34 71 19.13 Fixed Vehicle Delineators

	34 71 19.13 Fixed Vehicle Delineators	Crew	Daily Output	Labor-Hours	Unit	Material	Labor	Equipment	Total	Total Incl O&P
0010	**FIXED VEHICLE DELINEATORS**									
0020	Crash barriers									
0100	Traffic channelizing pavement markers, layout only	A-7	2000	.012	Ea.		.62	.03	.65	.98
0110	13″ x 7-1/2″ x 2-1/2″ high, non-plowable install	2 Clab	96	.167		24.50	6.25		30.75	36
0200	8″ x 8″ x 3-1/4″ high, non-plowable, install		96	.167		22.50	6.25		28.75	34.50
0230	4″ x 4″ x 3/4″ high, non-plowable, install		120	.133		1.79	5		6.79	9.65
0240	9-1/4″ x 5-7/8″ x 1/4″ high, plowable, concrete pavmt.	A-2A	70	.343		18.20	13.05	5.90	37.15	46.50
0250	9-1/4″ x 5-7/8″ x 1/4″ high, plowable, asphalt pav′t	″	120	.200		3.83	7.60	3.44	14.87	19.65
0300	Barrier and curb delineators, reflectorized, 2″ x 4″	2 Clab	150	.107		2.04	4.01		6.05	8.40
0310	3″ x 5″	″	150	.107		3.98	4.01		7.99	10.55
0500	Rumble strip, polycarbonate									
0510	24″ x 3-1/2″ x 1/2″ high	2 Clab	50	.320	Ea.	8.40	12.05		20.45	28

34 72 16 – Railway Siding

34 72 16.50 Railroad Sidings		Crew	Daily Output	Labor-Hours	Unit	Material	2015 Bare Costs Labor	Equipment	Total	Total Incl O&P
0010	**RAILROAD SIDINGS** R347216-10									
0800	Siding, yard spur, level grade									
0820	100 lb. new rail	B-14	57	.842	L.F.	133	33.50	6.40	172.90	205
1002	Steel ties in concrete, incl. fasteners & plates									
1020	100 lb. new rail	B-14	22	2.182	L.F.	181	87	16.55	284.55	350

34 72 16.60 Railroad Turnouts		Crew	Daily Output	Labor-Hours	Unit	Material	Labor	Equipment	Total	Total Incl O&P
0010	**RAILROAD TURNOUTS**									
2200	Turnout, #8 complete, w/rails, plates, bars, frog, switch point,									
2250	timbers, and ballast to 6" below bottom of ties									
2280	90 lb. rails	B-13	.25	224	Ea.	40,300	9,175	2,950	52,425	61,500
2290	90 lb. relay rails		.25	224		27,100	9,175	2,950	39,225	47,200
2300	100 lb. rails		.25	224		44,900	9,175	2,950	57,025	67,000
2310	100 lb. relay rails		.25	224		30,200	9,175	2,950	42,325	50,500
2320	110 lb. rails		.25	224		49,500	9,175	2,950	61,625	72,000
2330	110 lb. relay rails		.25	224		33,300	9,175	2,950	45,425	54,000
2340	115 lb. rails		.25	224		54,500	9,175	2,950	66,625	77,000
2350	115 lb. relay rails		.25	224		36,100	9,175	2,950	48,225	57,000
2360	132 lb. rails		.25	224		62,000	9,175	2,950	74,125	85,500
2370	132 lb. relay rails	▼	.25	224	▼	41,100	9,175	2,950	53,225	62,500

Division Notes

		CREW	DAILY OUTPUT	LABOR-HOURS	UNIT	BARE COSTS MAT.	LABOR	EQUIP.	TOTAL	TOTAL INCL O&P

Estimating Tips

35 01 50 Operation and Maintenance of Marine Construction

Includes unit price lines for pile cleaning and pile wrapping for protection.

35 20 16 Hydraulic Gates

This subdivision includes various types of gates that are commonly used in waterway and canal construction. Various earthwork items and structural support is found in Division 31, and concrete work in Division 3.

35 20 23 Dredging

This subdivision includes barge and shore dredging systems for rivers, canals, and channels.

35 31 00 Shoreline Protection

This subdivision includes breakwaters, bulkheads, and revetments for ocean and river inlets. Additional earthwork may be required from Division 31, and concrete work from Division 3.

35 41 00 Levees

Information on levee construction, including estimated cost of clay cone material.

35 49 00 Waterway Structures

This subdivision includes breakwaters and bulkheads for canals.

35 51 00 Floating Construction

This section includes floating piers, docks, and dock accessories. Fixed Pier Timber Construction is found in 06 13 33. Driven piles are found in Division 31, as well as sheet piling, cofferdams, and riprap.

Reference Numbers

Reference numbers are shown in shaded boxes at the beginning of some major classifications. These numbers refer to related items in the Reference Section. The reference information may be an estimating procedure, an alternate pricing method, or technical information.

Note: Not all subdivisions listed here necessarily appear in this publication. ■

35 20 Waterway and Marine Construction and Equipment

35 20 23 – Dredging

35 20 23.13 Mechanical Dredging		Crew	Daily Output	Labor-Hours	Unit	Material	2015 Bare Costs Labor	Equipment	Total	Total Incl O&P
0010	MECHANICAL DREDGING									
0020	Dredging mobilization and demobilization, add to below, minimum	B-8	.53	120	Total		5,150	6,250	11,400	14,700
0100	Maximum	"	.10	640	"		27,300	33,100	60,400	78,000
0300	Barge mounted clamshell excavation into scows									
0310	Dumped 20 miles at sea, minimum	B-57	310	.155	B.C.Y.		6.70	5.40	12.10	16.20
0400	Maximum	"	213	.225	"		9.80	7.90	17.70	23.50
0500	Barge mounted dragline or clamshell, hopper dumped,									
0510	pumped 1000′ to shore dump, minimum	B-57	340	.141	B.C.Y.		6.10	4.94	11.04	14.80
0525	All pumping uses 2000 gallons of water per cubic yard									
0600	Maximum	B-57	243	.198	B.C.Y.		8.55	6.90	15.45	20.50

35 20 23.23 Hydraulic Dredging		Crew	Daily Output	Labor-Hours	Unit	Material	Labor	Equipment	Total	Total Incl O&P
0010	HYDRAULIC DREDGING									
1000	Hydraulic method, pumped 1000′ to shore dump, minimum	B-57	460	.104	B.C.Y.		4.53	3.65	8.18	10.90
1100	Maximum		310	.155			6.70	5.40	12.10	16.20
1400	Into scows dumped 20 miles, minimum		425	.113			4.90	3.95	8.85	11.85
1500	Maximum		243	.198			8.55	6.90	15.45	20.50
1600	For inland rivers and canals in South, deduct								30%	30%

35 51 Floating Construction

35 51 13 – Floating Piers

35 51 13.23 Floating Wood Piers		Crew	Daily Output	Labor-Hours	Unit	Material	Labor	Equipment	Total	Total Incl O&P
0010	FLOATING WOOD PIERS									
0020	Polyethylene encased polystyrene, no pilings included	F-3	330	.121	S.F.	29	5.80	1.98	36.78	43
0200	Pile supported, shore constructed, bare, 3″ decking		130	.308		28	14.75	5.05	47.80	58.50
0250	4″ decking		120	.333		29	15.95	5.45	50.40	62.50
0400	Floating, small boat, prefab, no shore facilities, minimum		250	.160		24.50	7.65	2.62	34.77	41.50
0500	Maximum		150	.267		52	12.75	4.36	69.11	81.50
0700	Per slip, minimum (180 S.F. each)		1.59	25.157	Ea.	5,000	1,200	410	6,610	7,800
0800	Maximum		1.40	28.571	"	8,400	1,375	465	10,240	11,900

Estimating Tips

Products such as conveyors, material handling cranes and hoists, as well as other items specified in this division, require trained installers. The general contractor may not have any choice as to who will perform the installation or when it will be performed. Long lead times are often required for these products, making early decisions in purchasing and scheduling necessary. The installation of this type of equipment may require the embedment of mounting hardware during construction of floors, structural walls, or interior walls/partitions. Electrical connections will require coordination with the electrical contractor.

Reference Numbers

Reference numbers are shown in shaded boxes at the beginning of some major classifications. These numbers refer to related items in the Reference Section. The reference information may be an estimating procedure, an alternate pricing method, or technical information.

Note: Not all subdivisions listed here necessarily appear in this publication. ■

41 21 Conveyors

41 21 23 – Piece Material Conveyors

41 21 23.16 Container Piece Material Conveyors		Crew	Daily Output	Labor-Hours	Unit	Material	2015 Bare Costs Labor	Equipment	Total	Total Incl O&P
0010	**CONTAINER PIECE MATERIAL CONVEYORS**									
0020	Gravity fed, 2" rollers, 3" O.C.									
0050	10' sections with 2 supports, 600 lb. capacity, 18" wide				Ea.	480			480	530
0100	24" wide					545			545	595
0150	1400 lb. capacity, 18" wide					640			640	705
0200	24" wide					545			545	595
0350	Horizontal belt, center drive and takeup, 60 fpm									
0400	16" belt, 26.5' length	2 Mill	.50	32	Ea.	3,350	1,575		4,925	6,025
0450	24" belt, 41.5' length		.40	40		5,100	1,975		7,075	8,525
0500	61.5' length		.30	53.333		7,200	2,625		9,825	11,800
0600	Inclined belt, 10' rise with horizontal loader and									
0620	End idler assembly, 27.5' length, 18" belt	2 Mill	.30	53.333	Ea.	7,300	2,625		9,925	11,900
0700	24" belt	"	.15	106	"	8,925	5,250		14,175	17,600
3600	Monorail, overhead, manual, channel type									
3700	125 lb. per L.F.	1 Mill	26	.308	L.F.	19.40	15.10		34.50	44
3900	500 lb. per L.F.	"	21	.381	"	20.50	18.70		39.20	50
4000	Trolleys for above, 2 wheel, 125 lb. capacity				Ea.	81.50			81.50	89.50
4200	4 wheel, 250 lb. capacity					330			330	365
4300	8 wheel, 500 lb. capacity					770			770	845

41 22 Cranes and Hoists

41 22 13 – Cranes

41 22 13.10 Crane Rail

41 22 13.10 Crane Rail		Crew	Daily Output	Labor-Hours	Unit	Material	Labor	Equipment	Total	Total Incl O&P
0010	**CRANE RAIL**									
0020	Box beam bridge, no equipment included	E-4	3400	.009	Lb.	1.33	.50	.04	1.87	2.38
0200	Running track only, 104 lb. per yard	"	5600	.006	"	.66	.30	.03	.99	1.29

41 22 13.13 Bridge Cranes

41 22 13.13 Bridge Cranes		Crew	Daily Output	Labor-Hours	Unit	Material	Labor	Equipment	Total	Total Incl O&P
0010	**BRIDGE CRANES**									
0100	1 girder, 20' span, 3 ton	M-3	1	34	Ea.	25,400	1,950	139	27,489	31,100
0125	5 ton		1	34		27,900	1,950	139	29,989	33,800
0150	7.5 ton		1	34		33,100	1,950	139	35,189	39,500
0175	10 ton		.80	42.500		43,900	2,450	173	46,523	52,000
0200	15 ton		.80	42.500		56,500	2,450	173	59,123	66,000
0225	30' span, 3 ton		1	34		26,500	1,950	139	28,589	32,200
0250	5 ton		1	34		29,100	1,950	139	31,189	35,100
0275	7.5 ton		1	34		34,800	1,950	139	36,889	41,400
0300	10 ton		.80	42.500		45,400	2,450	173	48,023	54,000
0325	15 ton		.80	42.500		59,000	2,450	173	61,623	69,000
0350	2 girder, 40' span, 3 ton	M-4	.50	72		43,600	4,100	350	48,050	54,500
0375	5 ton		.50	72		45,600	4,100	350	50,050	56,500
0400	7.5 ton		.50	72		50,000	4,100	350	54,450	61,500
0425	10 ton		.40	90		58,500	5,125	440	64,065	72,500
0450	15 ton		.40	90		80,000	5,125	440	85,565	95,500
0475	25 ton		.30	120		94,000	6,825	585	101,410	114,500
0500	50' span, 3 ton		.50	72		49,700	4,100	350	54,150	61,000
0525	5 ton		.50	72		51,500	4,100	350	55,950	63,500
0550	7.5 ton		.50	72		55,500	4,100	350	59,950	67,500
0575	10 ton		.40	90		64,000	5,125	440	69,565	78,000
0600	15 ton		.40	90		83,500	5,125	440	89,065	100,000
0625	25 ton		.30	120		98,500	6,825	585	105,910	119,500

41 22 Cranes and Hoists

41 22 13 – Cranes

41 22 13.19 Jib Cranes		Crew	Daily Output	Labor-Hours	Unit	Material	2015 Bare Costs Labor	Equipment	Total	Total Incl O&P
0010	**JIB CRANES**									
0020	Jib crane, wall cantilever, 500 lb. capacity, 8′ span	2 Mill	1	16	Ea.	1,400	785		2,185	2,675
0040	12′ span		1	16		1,525	785		2,310	2,825
0060	16′ span		1	16		1,700	785		2,485	3,025
0080	20′ span		1	16		2,150	785		2,935	3,525
0100	1000 lb. capacity, 8′ span		1	16		1,500	785		2,285	2,800
0120	12′ span		1	16		1,625	785		2,410	2,950
0130	16′ span		1	16		2,150	785		2,935	3,525
0150	20′ span		1	16		2,525	785		3,310	3,925

41 22 23 – Hoists

41 22 23.10 Material Handling		Crew	Daily Output	Labor-Hours	Unit	Material	Labor	Equipment	Total	Total Incl O&P
0010	**MATERIAL HANDLING**, cranes, hoists and lifts									
1500	Cranes, portable hydraulic, floor type, 2,000 lb. capacity				Ea.	3,500			3,500	3,875
1600	4,000 lb. capacity					4,025			4,025	4,425
1800	Movable gantry type, 12′ to 15′ range, 2,000 lb. capacity					3,950			3,950	4,325
1900	6,000 lb. capacity					5,975			5,975	6,575
2100	Hoists, electric overhead, chain, hook hung, 15′ lift, 1 ton cap.					2,500			2,500	2,750
2200	3 ton capacity					3,425			3,425	3,775
2500	5 ton capacity					6,925			6,925	7,625
2600	For hand-pushed trolley, add					15%				
2700	For geared trolley, add					30%				
2800	For motor trolley, add					75%				
3000	For lifts over 15′, 1 ton, add				L.F.	24.50			24.50	27
3100	5 ton, add				″	56.50			56.50	62
3300	Lifts, scissor type, portable, electric, 36″ high, 2,000 lb.				Ea.	3,475			3,475	3,825
3400	48″ high, 4,000 lb.				″	4,175			4,175	4,600

Division Notes

		CREW	DAILY OUTPUT	LABOR-HOURS	UNIT	BARE COSTS MAT.	LABOR	EQUIP.	TOTAL	TOTAL INCL O&P

Estimating Tips

This section involves equipment and construction costs for air noise and odor pollution control systems. These systems may be interrelated and care must be taken that the complete systems are estimated. For example, air pollution equipment may include dust and air-entrained particles that have to be collected. The vacuum systems could be noisy, requiring silencers to reduce noise pollution, and the collected solids have to be disposed of to prevent solid pollution.

Reference Numbers

Reference numbers are shown in shaded boxes at the beginning of some major classifications. These numbers refer to related items in the Reference Section. The reference information may be an estimating procedure, an alternate pricing method, or technical information.

Note: Not all subdivisions listed here necessarily appear in this publication. ■

44 11 16 – Fugitive Dust Barrier Systems

44 11 16.10 Dust Collection Systems		Crew	Daily Output	Labor-Hours	Unit	2015 Bare Costs Material	Labor	Equipment	Total	Total Incl O&P
0010	**DUST COLLECTION SYSTEMS** Commercial/Industrial									
0120	Central vacuum units									
0130	Includes stand, filters and motorized shaker									
0200	500 CFM, 10" inlet, 2 HP	Q-20	2.40	8.333	Ea.	4,175	425		4,600	5,250
0220	1000 CFM, 10" inlet, 3 HP		2.20	9.091		4,400	465		4,865	5,525
0240	1500 CFM, 10" inlet, 5 HP		2	10		4,625	510		5,135	5,875
0260	3000 CFM, 13" inlet, 10 HP		1.50	13.333		12,900	685		13,585	15,300
0280	5000 CFM, 16" inlet, 2 @ 10 HP	↓	1	20	↓	13,700	1,025		14,725	16,700
1000	Vacuum tubing, galvanized									
1100	2-1/8" OD, 16 ga.	Q-9	440	.036	L.F.	2.95	1.83		4.78	6.05
1110	2-1/2" OD, 16 ga.		420	.038		3.24	1.92		5.16	6.50
1120	3" OD, 16 ga.		400	.040		4.13	2.01		6.14	7.60
1130	3-1/2" OD, 16 ga.		380	.042		5.85	2.12		7.97	9.70
1140	4" OD, 16 ga.		360	.044		6	2.24		8.24	10
1150	5" OD, 14 ga.		320	.050		10.80	2.52		13.32	15.70
1160	6" OD, 14 ga.		280	.057		12.10	2.88		14.98	17.70
1170	8" OD, 14 ga.		200	.080		18.30	4.03		22.33	26
1180	10" OD, 12 ga.		160	.100		35.50	5.05		40.55	46.50
1190	12" OD, 12 ga.		120	.133		47	6.70		53.70	62
1200	14" OD, 12 ga.	↓	80	.200	↓	53.50	10.05		63.55	74
1940	Hose, flexible wire reinforced rubber									
1956	3" diam.	Q-9	400	.040	L.F.	9.15	2.01		11.16	13.15
1960	4" diam.		360	.044		10.60	2.24		12.84	15.05
1970	5" diam.		320	.050		13.75	2.52		16.27	19
1980	6" diam.	↓	280	.057	↓	14.40	2.88		17.28	20
2000	90° Elbow, slip fit									
2110	2-1/8" diam.	Q-9	70	.229	Ea.	10.85	11.50		22.35	29.50
2120	2-1/2" diam.		65	.246		15.20	12.40		27.60	35.50
2130	3" diam.		60	.267		20.50	13.45		33.95	43
2140	3-1/2" diam.		55	.291		25.50	14.65		40.15	50.50
2150	4" diam.		50	.320		31.50	16.10		47.60	59
2160	5" diam.		45	.356		60	17.90		77.90	93.50
2170	6" diam.		40	.400		82.50	20		102.50	122
2180	8" diam.	↓	30	.533	↓	157	27		184	213
2400	45° Elbow, slip fit									
2410	2-1/8" diam.	Q-9	70	.229	Ea.	9.50	11.50		21	28
2420	2-1/2" diam.		65	.246		14	12.40		26.40	34.50
2430	3" diam.		60	.267		16.90	13.45		30.35	39
2440	3-1/2" diam.		55	.291		21.50	14.65		36.15	46
2450	4" diam.		50	.320		27.50	16.10		43.60	55
2460	5" diam.		45	.356		46.50	17.90		64.40	78.50
2470	6" diam.		40	.400		63	20		83	100
2480	8" diam.	↓	35	.457	↓	123	23		146	171
2800	90° TY, slip fit thru 6" diam.									
2810	2-1/8" diam.	Q-9	42	.381	Ea.	17.75	19.20		36.95	49
2820	2-1/2" diam.		39	.410		26.50	20.50		47	60.50
2830	3" diam.		36	.444		37	22.50		59.50	74.50
2840	3-1/2" diam.		33	.485		39	24.50		63.50	80
2850	4" diam.		30	.533		41.50	27		68.50	87
2860	5" diam.		27	.593		70.50	30		100.50	123
2870	6" diam.	↓	24	.667		101	33.50		134.50	163
2880	8" diam., butt end					155			155	170
2890	10" diam., butt end				↓	475			475	525

44 11 16 – Fugitive Dust Barrier Systems

44 11 16.10 Dust Collection Systems		Crew	Daily Output	Labor-Hours	Unit	Material	Labor	Equipment	Total	Total Incl O&P
						2015 Bare Costs				
2900	12" diam., butt end				Ea.	475			475	525
2910	14" diam., butt end					855			855	940
2920	6" x 4" diam., butt end					106			106	116
2930	8" x 4" diam., butt end					200			200	220
2940	10" x 4" diam., butt end					263			263	289
2950	12" x 4" diam., butt end					300			300	330
3100	90° Elbow, butt end segmented									
3110	8" diam., butt end, segmented				Ea.	157			157	172
3120	10" diam., butt end, segmented					390			390	430
3130	12" diam., butt end, segmented					495			495	545
3140	14" diam., butt end, segmented					620			620	680
3200	45° Elbow, butt end segmented									
3210	8" diam., butt end, segmented				Ea.	123			123	136
3220	10" diam., butt end, segmented					282			282	310
3230	12" diam., butt end, segmented					288			288	315
3240	14" diam., butt end, segmented					585			585	645
3400	All butt end fittings require one coupling per joint.									
3410	Labor for fitting included with couplings.									
3460	Compression coupling, galvanized, neoprene gasket									
3470	2-1/8" diam.	Q-9	44	.364	Ea.	10.85	18.30		29.15	40
3480	2-1/2" diam.		44	.364		10.85	18.30		29.15	40
3490	3" diam.		38	.421		15.60	21		36.60	49.50
3500	3-1/2" diam.		35	.457		17.45	23		40.45	54
3510	4" diam.		33	.485		18.95	24.50		43.45	58.50
3520	5" diam.		29	.552		21.50	28		49.50	66.50
3530	6" diam.		26	.615		25.50	31		56.50	75.50
3540	8" diam.		22	.727		46	36.50		82.50	107
3550	10" diam.		20	.800		67.50	40.50		108	136
3560	12" diam.		18	.889		84	45		129	161
3570	14" diam.		16	1		119	50.50		169.50	208
3800	Air gate valves, galvanized									
3810	2-1/8" diam.	Q-9	30	.533	Ea.	111	27		138	163
3820	2-1/2" diam.		28	.571		119	29		148	175
3830	3" diam.		26	.615		127	31		158	188
3840	4" diam.		23	.696		147	35		182	216
3850	6" diam.		18	.889		205	45		250	294

Division Notes

I		CREW	DAILY OUTPUT	LABOR-HOURS	UNIT	BARE COSTS MAT.	LABOR	EQUIP.	TOTAL	TOTAL INCL O&P

Estimating Tips

This division contains information about water and wastewater equipment and systems, which was formerly located in Division 44. The main areas of focus are total wastewater treatment plants and components of wastewater treatment plants. In addition, there are assemblies such as sewage treatment lagoons that can be found in some publications under G30 Site Mechanical Utilities. Also included in this section are oil/water separators for wastewater treatment.

Reference Numbers

Reference numbers are shown in shaded boxes at the beginning of some major classifications. These numbers refer to related items in the Reference Section. The reference information may be an estimating procedure, an alternate pricing method, or technical information.

Note: Not all subdivisions listed here necessarily appear in this publication. ■

46 07 53 – Packaged Wastewater Treatment Equipment

46 07 53.10 Biological Pkg. Wastewater Treatment Plants		Crew	Daily Output	Labor-Hours	Unit	2015 Bare Costs Material	Labor	Equipment	Total	Total Incl O&P
0010	BIOLOGICAL PACKAGED WASTEWATER TREATMENT PLANTS									
0011	Not including fencing or external piping									
0020	Steel packaged, blown air aeration plants									
0100	1,000 GPD				Gal.				55	60.50
0200	5,000 GPD								22	24
0300	15,000 GPD								22	24
0400	30,000 GPD								15.40	16.95
0500	50,000 GPD								11	12.10
0600	100,000 GPD								9.90	10.90
0700	200,000 GPD								8.80	9.70
0800	500,000 GPD				↓				7.70	8.45
1000	Concrete, extended aeration, primary and secondary treatment									
1010	10,000 GPD				Gal.				22	24
1100	30,000 GPD								15.40	16.95
1200	50,000 GPD								11	12.10
1400	100,000 GPD								9.90	10.90
1500	500,000 GPD				↓				7.70	8.45
1700	Municipal wastewater treatment facility									
1720	1.0 MGD				Gal.				11	12.10
1740	1.5 MGD								10.60	11.65
1760	2.0 MGD								10	11
1780	3.0 MGD								7.80	8.60
1800	5.0 MGD				↓				5.80	6.70
2000	Holding tank system, not incl. excavation or backfill									
2010	Recirculating chemical water closet	2 Plum	4	4	Ea.	530	235		765	940
2100	For voltage converter, add	"	16	1		300	58.50		358.50	420
2200	For high level alarm, add	1 Plum	7.80	1.026	↓	117	60		177	219

46 07 53.20 Wastewater Treatment System		Crew	Daily Output	Labor-Hours	Unit	Material	Labor	Equipment	Total	Total Incl O&P
0010	WASTEWATER TREATMENT SYSTEM									
0020	Fiberglass, 1,000 gallon	B-21	1.29	21.705	Ea.	4,250	940	107	5,297	6,250
0100	1,500 gallon	"	1.03	27.184	"	8,475	1,175	134	9,784	11,300

Estimating Tips

- When estimating costs for the installation of electrical power generation equipment, factors to review include access to the job site, access and setting at the installation site, required connections, uncrating pads, anchors, leveling, final assembly of the components, and temporary protection from physical damage, including from exposure to the environment.
- Be aware of the cost of equipment supports, concrete pads, and vibration isolators, and cross-reference to other trades' specifications. Also, review site and structural drawings for items that must be included in the estimates.
- It is important to include items that are not documented in the plans and specifications but must be priced. These items include, but are not limited to, testing, dust protection, roof penetration, core drilling concrete floors and walls, patching, cleanup, and final adjustments. Add a contingency or allowance for utility company fees for power hookups, if needed.
- The project size and scope of electrical power generation equipment will have a significant impact on cost. The intent of RSMeans cost data is to provide a benchmark cost so that owners, engineers, and electrical contractors will have a comfortable number with which to start a project. Additionally, there are many websites available to use for research and to obtain a vendor's quote to finalize costs.

Reference Numbers

Reference numbers are shown in shaded boxes at the beginning of some major classifications. These numbers refer to related items in the Reference Section. The reference information may be an estimating procedure, an alternate pricing method, or technical information.

Note: Not all subdivisions listed here necessarily appear in this publication. ■

48 15 Wind Energy Electrical Power Generation Equipment

48 15 13 – Wind Turbines

48 15 13.50 Wind Turbines and Components			Crew	Daily Output	Labor-Hours	Unit	2015 Bare Costs Material	Labor	Equipment	Total	Total Incl O&P
0010	**WIND TURBINES & COMPONENTS**										
0500	Complete system, grid connected										
1000	20 kW, 31′ dia, incl. labor & material	G				System				49,900	49,900
1010	Enhanced	G								92,000	92,000
1500	10 kW, 23′ dia, incl. labor & material	G								74,000	74,000
2000	2.4 kW, 12′ dia, incl. labor & material	G				↓				18,000	18,000
2900	Component system										
3000	Turbine, 400 watt, 3′ dia	G	1 Elec	3.41	2.346	Ea.	650	128		778	905
3100	600 watt, 3′ dia	G		2.56	3.125		880	171		1,051	1,225
3200	1000 watt, 9′ dia	G	↓	2.05	3.902	↓	3,650	213		3,863	4,350
3400	Mounting hardware										
3500	30′ guyed tower kit	G	2 Clab	5.12	3.125	Ea.	415	118		533	640
3505	3′ galvanized helical earth screw	G	1 Clab	8	1		49.50	37.50		87	113
3510	Attic mount kit	G	1 Rofc	2.56	3.125		154	125		279	385
3520	Roof mount kit	G	1 Clab	3.41	2.346	↓	136	88		224	286
8900	Equipment										
9100	DC to AC inverter for, 48 V, 4,000 watt	G	1 Elec	2	4	Ea.	2,325	219		2,544	2,900

Reference Section

All the reference information is in one section, making it easy to find what you need to know . . . and easy to use the book on a daily basis. This section is visually identified by a vertical black bar on the page edges.

In this Reference Section, we've included Equipment Rental Costs, a listing of rental and operating costs; Crew Listings, a full listing of all crews and equipment, and their costs; Historical Cost Indexes for cost comparisons over time; City Cost Indexes and Location Factors for adjusting costs to the region you are in; Reference Tables, where you will find explanations, estimating information and procedures, or technical data; Change Orders, information on pricing changes to contract documents; Square Foot Costs that allow you to make a rough estimate for the overall cost of a project; and an explanation of all the Abbreviations in the book.

Table of Contents

Estimating Tips

- This section contains the average costs to rent and operate hundreds of pieces of construction equipment. This is useful information when estimating the time and material requirements of any particular operation in order to establish a unit or total cost. Equipment costs include not only rental, but also operating costs for equipment under normal use.

Rental Costs

- Equipment rental rates are obtained from the following industry sources throughout North America: contractors, suppliers, dealers, manufacturers, and distributors.
- Rental rates vary throughout the country, with larger cities generally having lower rates. Lease plans for new equipment are available for periods in excess of six months, with a percentage of payments applying toward purchase.
- Monthly rental rates vary from 2% to 5% of the purchase price of the equipment depending on the anticipated life of the equipment and its wearing parts.
- Weekly rental rates are about 1/3 the monthly rates, and daily rental rates are about 1/3 the weekly rate.
- Rental rates can also be treated as reimbursement costs for contractor-owned equipment. Owned equipment costs include depreciation, loan payments, interest, taxes, insurance, storage, and major repairs.

Operating Costs

- The operating costs include parts and labor for routine servicing, such as repair and replacement of pumps, filters and worn lines. Normal operating expendables, such as fuel, lubricants, tires and electricity (where applicable), are also included.
- Extraordinary operating expendables with highly variable wear patterns, such as diamond bits and blades, are excluded. These costs can be found as material costs in the Unit Price section.
- The hourly operating costs listed do not include the operator's wages.

Equipment Cost/Day

- Any power equipment required by a crew is shown in the Crew Listings with a daily cost.
- The daily cost of equipment needed by a crew is based on dividing the weekly rental rate by 5 (number of working days in the week), and then adding the hourly operating cost times 8 (the number of hours in a day). This "Equipment Cost/ Day" is shown in the far right column of the Equipment Rental pages.
- If equipment is needed for only one or two days, it is best to develop your own cost by including components for daily rent and hourly operating cost. This is important when the listed Crew for a task does not contain the equipment needed, such as a crane for lifting mechanical heating/ cooling equipment up onto a roof.
- If the quantity of work is less than the crew's Daily Output shown for a Unit Price line item that includes a bare unit equipment cost, it is recommended to estimate one day's rental cost and operating cost for equipment shown in the Crew Listing for that line item.

Mobilization/ Demobilization

- The cost to move construction equipment from an equipment yard or rental company to the job site and back again is not included in equipment rental costs listed in the Reference Section, nor in the bare equipment cost of any unit price line item, nor in any equipment costs shown in the Crew listings.
- Mobilization (to the site) and demobilization (from the site) costs can be found in the Unit Price Section.
- If a piece of equipment is already at the job site, it is not appropriate to utilize mobilization/demobilization. costs again in an estimate. ■

01 54 33 | Equipment Rental

		01 54 33 Equipment Rental	UNIT	HOURLY OPER. COST	RENT PER DAY	RENT PER WEEK	RENT PER MONTH	EQUIPMENT COST/DAY	
10	0010	**CONCRETE EQUIPMENT RENTAL** without operators R015433-10							10
	0200	Bucket, concrete lightweight, 1/2 C.Y.	Ea.	.80	23.50	70	210	20.40	
	0300	1 C.Y.		.90	27.50	83	249	23.80	
	0400	1-1/2 C.Y.		1.10	36.50	110	330	30.80	
	0500	2 C.Y.		1.25	45	135	405	37	
	0580	8 C.Y.		6.10	257	770	2,300	202.80	
	0600	Cart, concrete, self-propelled, operator walking, 10 C.F.		3.25	56.50	170	510	60	
	0700	Operator riding, 18 C.F.		5.35	95	285	855	99.80	
	0800	Conveyer for concrete, portable, gas, 16" wide, 26' long		12.60	123	370	1,100	174.80	
	0900	46' long		13.00	148	445	1,325	193	
	1000	56' long		13.10	157	470	1,400	198.80	
	1100	Core drill, electric, 2-1/2 H.P., 1" to 8" bit diameter		1.77	68.50	205	615	55.15	
	1150	11 H.P., 8" to 18" cores		5.95	113	340	1,025	115.60	
	1200	Finisher, concrete floor, gas, riding trowel, 96" wide		12.75	145	435	1,300	189	
	1300	Gas, walk-behind, 3 blade, 36" trowel		2.15	20	60	180	29.20	
	1400	4 blade, 48" trowel		4.20	27.50	83	249	50.20	
	1500	Float, hand-operated (Bull float) 48" wide		.08	13.35	40	120	8.65	
	1570	Curb builder, 14 H.P., gas, single screw		14.65	248	745	2,225	266.20	
	1590	Double screw		15.35	290	870	2,600	296.80	
	1600	Floor grinder, concrete and terrazzo, electric, 22" path		2.57	158	475	1,425	115.55	
	1700	Edger, concrete, electric, 7" path		1.04	51.50	155	465	39.30	
	1750	Vacuum pick-up system for floor grinders, wet/dry		1.49	81.50	245	735	60.90	
	1800	Mixer, powered, mortar and concrete, gas, 6 C.F., 18 H.P.		8.65	118	355	1,075	140.20	
	1900	10 C.F., 25 H.P.		10.80	143	430	1,300	172.40	
	2000	16 C.F.		11.15	165	495	1,475	188.20	
	2100	Concrete, stationary, tilt drum, 2 C.Y.		6.90	232	695	2,075	194.20	
	2120	Pump, concrete, truck mounted 4" line 80' boom		25.00	865	2,600	7,800	720	
	2140	5" line, 110' boom		32.40	1,150	3,435	10,300	946.20	
	2160	Mud jack, 50 C.F. per hr.		7.30	125	375	1,125	133.40	
	2180	225 C.F. per hr.		9.50	145	435	1,300	163	
	2190	Shotcrete pump rig, 12 C.Y./hr.		14.85	220	660	1,975	250.80	
	2200	35 C.Y./hr.		17.50	235	705	2,125	281	
	2600	Saw, concrete, manual, gas, 18 H.P.		6.75	45	135	405	81	
	2650	Self-propelled, gas, 30 H.P.		13.30	102	305	915	167.40	
	2675	V-groove crack chaser, manual, gas, 6 H.P.		2.35	17.35	52	156	29.20	
	2700	Vibrators, concrete, electric, 60 cycle, 2 H.P.		.46	8.65	26	78	8.90	
	2800	3 H.P.		.60	11.65	35	105	11.80	
	2900	Gas engine, 5 H.P.		2.00	16	48	144	25.60	
	3000	8 H.P.		2.75	15.35	46	138	31.20	
	3050	Vibrating screed, gas engine, 8 H.P.		2.91	71.50	215	645	66.30	
	3120	Concrete transit mixer, 6 x 4, 250 H.P., 8 C.Y., rear discharge		61.55	570	1,715	5,150	835.40	
	3200	Front discharge		72.45	700	2,095	6,275	998.60	
	3300	6 x 6, 285 H.P., 12 C.Y., rear discharge		71.50	660	1,980	5,950	968	
	3400	Front discharge	↓	74.00	705	2,120	6,350	1,016	
20	0010	**EARTHWORK EQUIPMENT RENTAL** without operators R015433-10							20
	0040	Aggregate spreader, push type 8' to 12' wide	Ea.	3.05	25.50	76	228	39.60	
	0045	Tailgate type, 8' wide		2.90	32.50	98	294	42.80	
	0055	Earth auger, truck-mounted, for fence & sign posts, utility poles		19.15	440	1,320	3,950	417.20	
	0060	For borings and monitoring wells		46.65	675	2,030	6,100	779.20	
	0070	Portable, trailer mounted		3.30	32.50	98	294	46	
	0075	Truck-mounted, for caissons, water wells		100.15	2,900	8,705	26,100	2,542	
	0080	Horizontal boring machine, 12" to 36" diameter, 45 H.P.		24.20	192	575	1,725	308.60	
	0090	12" to 48" diameter, 65 H.P.		34.35	335	1,000	3,000	474.80	
	0095	Auger, for fence posts, gas engine, hand held		.60	6	18	54	8.40	
	0100	Excavator, diesel hydraulic, crawler mounted, 1/2 C.Y. cap.		24.65	420	1,255	3,775	448.20	
	0120	5/8 C.Y. capacity		32.55	550	1,645	4,925	589.40	
	0140	3/4 C.Y. capacity		35.60	615	1,850	5,550	654.80	
	0150	1 C.Y. capacity	↓	49.70	690	2,075	6,225	812.60	

01 54 33 | Equipment Rental

			UNIT	HOURLY OPER. COST	RENT PER DAY	RENT PER WEEK	RENT PER MONTH	EQUIPMENT COST/DAY	
20	0200	1-1/2 C.Y. capacity	Ea.	59.80	920	2,760	8,275	1,030	20
	0300	2 C.Y. capacity		67.35	1,050	3,180	9,550	1,175	
	0320	2-1/2 C.Y. capacity		102.55	1,300	3,925	11,800	1,605	
	0325	3-1/2 C.Y. capacity		144.45	2,150	6,420	19,300	2,440	
	0330	4-1/2 C.Y. capacity		173.40	2,625	7,880	23,600	2,963	
	0335	6 C.Y. capacity		219.80	2,900	8,680	26,000	3,494	
	0340	7 C.Y. capacity		222.05	3,025	9,070	27,200	3,590	
	0342	Excavator attachments, bucket thumbs		3.20	245	735	2,200	172.60	
	0345	Grapples		2.75	193	580	1,750	138	
	0346	Hydraulic hammer for boom mounting, 4000 ft lb.		12.40	350	1,045	3,125	308.20	
	0347	5000 ft lb.		14.40	425	1,280	3,850	371.20	
	0348	8000 ft lb.		21.30	625	1,875	5,625	545.40	
	0349	12,000 ft lb.		23.35	750	2,245	6,725	635.80	
	0350	Gradall type, truck mounted, 3 ton @ 15′ radius, 5/8 C.Y.		43.65	890	2,665	8,000	882.20	
	0370	1 C.Y. capacity		47.85	1,025	3,095	9,275	1,002	
	0400	Backhoe-loader, 40 to 45 H.P., 5/8 C.Y. capacity		14.90	240	720	2,150	263.20	
	0450	45 H.P. to 60 H.P., 3/4 C.Y. capacity		23.40	295	885	2,650	364.20	
	0460	80 H.P., 1-1/4 C.Y. capacity		25.60	310	935	2,800	391.80	
	0470	112 H.P., 1-1/2 C.Y. capacity		40.95	645	1,930	5,800	713.60	
	0482	Backhoe-loader attachment, compactor, 20,000 lb.		5.80	142	425	1,275	131.40	
	0485	Hydraulic hammer, 750 ft lb.		3.30	96.50	290	870	84.40	
	0486	Hydraulic hammer, 1200 ft lb.		6.30	217	650	1,950	180.40	
	0500	Brush chipper, gas engine, 6″ cutter head, 35 H.P.		11.10	105	315	945	151.80	
	0550	Diesel engine, 12″ cutter head, 130 H.P.		27.55	285	855	2,575	391.40	
	0600	15″ cutter head, 165 H.P.		33.10	335	1,005	3,025	465.80	
	0750	Bucket, clamshell, general purpose, 3/8 C.Y.		1.30	38.50	115	345	33.40	
	0800	1/2 C.Y.		1.40	45	135	405	38.20	
	0850	3/4 C.Y.		1.55	55	165	495	45.40	
	0900	1 C.Y.		1.60	58.50	175	525	47.80	
	0950	1-1/2 C.Y.		2.55	81.50	245	735	69.40	
	1000	2 C.Y.		2.70	90	270	810	75.60	
	1010	Bucket, dragline, medium duty, 1/2 C.Y.		.75	23.50	70	210	20	
	1020	3/4 C.Y.		.75	24.50	73	219	20.60	
	1030	1 C.Y.		.80	26	78	234	22	
	1040	1-1/2 C.Y.		1.20	40	120	360	33.60	
	1050	2 C.Y.		1.30	45	135	405	37.40	
	1070	3 C.Y.		1.95	61.50	185	555	52.60	
	1200	Compactor, manually guided 2-drum vibratory smooth roller, 7.5 H.P.		7.25	203	610	1,825	180	
	1250	Rammer/tamper, gas, 8″		2.75	46.50	140	420	50	
	1260	15″		3.05	53.50	160	480	56.40	
	1300	Vibratory plate, gas, 18″ plate, 3000 lb. blow		2.70	24.50	73	219	36.20	
	1350	21″ plate, 5000 lb. blow		3.40	32.50	98	294	46.80	
	1370	Curb builder/extruder, 14 H.P., gas, single screw		14.65	248	745	2,225	266.20	
	1390	Double screw		15.35	290	870	2,600	296.80	
	1500	Disc harrow attachment, for tractor		.44	73.50	221	665	47.70	
	1810	Feller buncher, shearing & accumulating trees, 100 H.P.		48.30	755	2,260	6,775	838.40	
	1860	Grader, self-propelled, 25,000 lb.		39.50	640	1,925	5,775	701	
	1910	30,000 lb.		43.15	650	1,955	5,875	736.20	
	1920	40,000 lb.		64.90	1,100	3,275	9,825	1,174	
	1930	55,000 lb.		84.15	1,700	5,130	15,400	1,699	
	1950	Hammer, pavement breaker, self-propelled, diesel, 1000 to 1250 lb.		29.90	350	1,055	3,175	450.20	
	2000	1300 to 1500 lb.		44.85	705	2,110	6,325	780.80	
	2050	Pile driving hammer, steam or air, 4150 ft lb. @ 225 bpm		10.40	475	1,420	4,250	367.20	
	2100	8750 ft lb. @ 145 bpm		12.40	660	1,975	5,925	494.20	
	2150	15,000 ft lb. @ 60 bpm		14.00	790	2,370	7,100	586	
	2200	24,450 ft lb. @ 111 bpm		15.00	875	2,630	7,900	646	
	2250	Leads, 60′ high for pile driving hammers up to 20,000 ft lb.		3.30	80.50	242	725	74.80	
	2300	90′ high for hammers over 20,000 ft lb.		4.95	141	424	1,275	124.40	

01 54 33 | Equipment Rental

			UNIT	HOURLY OPER. COST	RENT PER DAY	RENT PER WEEK	RENT PER MONTH	EQUIPMENT COST/DAY	
20	2350	Diesel type hammer, 22,400 ft lb.	Ea.	22.55	460	1,380	4,150	456.40	20
	2400	41,300 ft lb.		32.00	540	1,625	4,875	581	
	2450	141,000 ft lb.		51.65	960	2,875	8,625	988.20	
	2500	Vib. elec. hammer/extractor, 200 kW diesel generator, 34 H.P.		54.90	635	1,900	5,700	819.20	
	2550	80 H.P.		100.65	925	2,775	8,325	1,360	
	2600	150 H.P.		190.90	1,775	5,300	15,900	2,587	
	2800	Log chipper, up to 22" diameter, 600 H.P.		63.30	640	1,915	5,750	889.40	
	2850	Logger, for skidding & stacking logs, 150 H.P.		52.75	815	2,445	7,325	911	
	2860	Mulcher, diesel powered, trailer mounted		24.05	212	635	1,900	319.40	
	2900	Rake, spring tooth, with tractor		18.23	345	1,042	3,125	354.25	
	3000	Roller, vibratory, tandem, smooth drum, 20 H.P.		8.80	145	435	1,300	157.40	
	3050	35 H.P.		11.10	247	740	2,225	236.80	
	3100	Towed type vibratory compactor, smooth drum, 50 H.P.		26.00	315	940	2,825	396	
	3150	Sheepsfoot, 50 H.P.		27.35	345	1,035	3,100	425.80	
	3170	Landfill compactor, 220 H.P.		91.00	1,475	4,440	13,300	1,616	
	3200	Pneumatic tire roller, 80 H.P.		16.80	350	1,050	3,150	344.40	
	3250	120 H.P.		25.20	600	1,800	5,400	561.60	
	3300	Sheepsfoot vibratory roller, 240 H.P.		68.95	1,100	3,270	9,800	1,206	
	3320	340 H.P.		100.85	1,575	4,730	14,200	1,753	
	3350	Smooth drum vibratory roller, 75 H.P.		24.85	585	1,755	5,275	549.80	
	3400	125 H.P.		32.55	710	2,135	6,400	687.40	
	3410	Rotary mower, brush, 60", with tractor		22.60	315	950	2,850	370.80	
	3420	Rototiller, walk-behind, gas, 5 H.P.		1.96	51.50	155	465	46.70	
	3422	8 H.P.		3.08	80	240	720	72.65	
	3440	Scrapers, towed type, 7 C.Y. capacity		5.80	113	340	1,025	114.40	
	3450	10 C.Y. capacity		6.60	157	470	1,400	146.80	
	3500	15 C.Y. capacity		7.10	180	540	1,625	164.80	
	3525	Self-propelled, single engine, 14 C.Y. capacity		114.76	1,600	4,830	14,500	1,884	
	3550	Dual engine, 21 C.Y. capacity		173.80	2,175	6,490	19,500	2,688	
	3600	31 C.Y. capacity		231.70	3,075	9,195	27,600	3,693	
	3640	44 C.Y. capacity		287.80	3,950	11,885	35,700	4,679	
	3650	Elevating type, single engine, 11 C.Y. capacity		71.45	985	2,955	8,875	1,163	
	3700	22 C.Y. capacity		139.55	2,350	7,025	21,100	2,521	
	3710	Screening plant 110 H.P. w/5' x 10' screen		25.35	400	1,200	3,600	442.80	
	3720	5' x 16' screen		29.65	505	1,515	4,550	540.20	
	3850	Shovel, crawler-mounted, front-loading, 7 C.Y. capacity		258.85	3,200	9,610	28,800	3,993	
	3855	12 C.Y. capacity		386.45	4,400	13,220	39,700	5,736	
	3860	Shovel/backhoe bucket, 1/2 C.Y.		2.50	66.50	200	600	60	
	3870	3/4 C.Y.		2.55	75	225	675	65.40	
	3880	1 C.Y.		2.65	83.50	250	750	71.20	
	3890	1-1/2 C.Y.		2.75	96.50	290	870	80	
	3910	3 C.Y.		3.15	130	390	1,175	103.20	
	3950	Stump chipper, 18" deep, 30 H.P.		7.82	205	615	1,850	185.55	
	4110	Dozer, crawler, torque converter, diesel 80 H.P.		29.05	400	1,200	3,600	472.40	
	4150	105 H.P.		35.60	530	1,590	4,775	602.80	
	4200	140 H.P.		51.25	795	2,380	7,150	886	
	4260	200 H.P.		77.25	1,275	3,845	11,500	1,387	
	4310	300 H.P.		100.65	1,825	5,460	16,400	1,897	
	4360	410 H.P.		132.60	2,250	6,720	20,200	2,405	
	4370	500 H.P.		171.15	3,225	9,650	29,000	3,299	
	4380	700 H.P.		280.90	4,550	13,660	41,000	4,979	
	4400	Loader, crawler, torque conv., diesel, 1-1/2 C.Y., 80 H.P.		31.25	455	1,360	4,075	522	
	4450	1-1/2 to 1-3/4 C.Y., 95 H.P.		34.60	590	1,765	5,300	629.80	
	4510	1-3/4 to 2-1/4 C.Y., 130 H.P.		53.80	875	2,630	7,900	956.40	
	4530	2-1/2 to 3-1/4 C.Y., 190 H.P.		65.95	1,100	3,300	9,900	1,188	
	4560	3-1/2 to 5 C.Y., 275 H.P.		87.80	1,425	4,295	12,900	1,561	
	4610	Front end loader, 4WD, articulated frame, diesel, 1 to 1-1/4 C.Y., 70 H.P.		19.75	235	705	2,125	299	
	4620	1-1/2 to 1-3/4 C.Y., 95 H.P.	↓	24.80	300	900	2,700	378.40	

01 54 33 | Equipment Rental

		01 54 33 Equipment Rental	UNIT	HOURLY OPER. COST	RENT PER DAY	RENT PER WEEK	RENT PER MONTH	EQUIPMENT COST/DAY	
20	4650	1-3/4 to 2 C.Y., 130 H.P.	Ea.	29.15	380	1,140	3,425	461.20	20
	4710	2-1/2 to 3-1/2 C.Y., 145 H.P.		33.10	425	1,275	3,825	519.80	
	4730	3 to 4-1/2 C.Y., 185 H.P.		42.40	550	1,650	4,950	669.20	
	4760	5-1/4 to 5-3/4 C.Y., 270 H.P.		67.50	905	2,710	8,125	1,082	
	4810	7 to 9 C.Y., 475 H.P.		114.05	1,700	5,125	15,400	1,937	
	4870	9 - 11 C.Y., 620 H.P.		160.55	2,775	8,290	24,900	2,942	
	4880	Skid steer loader, wheeled, 10 C.F., 30 H.P. gas		10.45	150	450	1,350	173.60	
	4890	1 C.Y., 78 H.P., diesel		20.20	262	785	2,350	318.60	
	4892	Skid-steer attachment, auger		.48	80.50	242	725	52.25	
	4893	Backhoe		.66	111	332	995	71.70	
	4894	Broom		.74	124	372	1,125	80.30	
	4895	Forks		.22	36	108	325	23.35	
	4896	Grapple		.56	92.50	278	835	60.10	
	4897	Concrete hammer		1.06	177	531	1,600	114.70	
	4898	Tree spade		1.03	172	515	1,550	111.25	
	4899	Trencher		.54	90	270	810	58.30	
	4900	Trencher, chain, boom type, gas, operator walking, 12 H.P.		5.00	46.50	140	420	68	
	4910	Operator riding, 40 H.P.		20.15	290	870	2,600	335.20	
	5000	Wheel type, diesel, 4′ deep, 12″ wide		85.95	845	2,535	7,600	1,195	
	5100	6′ deep, 20″ wide		92.75	1,925	5,740	17,200	1,890	
	5150	Chain type, diesel, 5′ deep, 8″ wide		38.35	560	1,675	5,025	641.80	
	5200	Diesel, 8′ deep, 16″ wide		155.95	3,600	10,770	32,300	3,402	
	5202	Rock trencher, wheel type, 6″ wide x 18″ deep		21.50	335	1,000	3,000	372	
	5206	Chain type, 18″ wide x 7′ deep		112.00	2,800	8,395	25,200	2,575	
	5210	Tree spade, self-propelled		14.38	267	800	2,400	275.05	
	5250	Truck, dump, 2-axle, 12 ton, 8 C.Y. payload, 220 H.P.		34.40	227	680	2,050	411.20	
	5300	Three axle dump, 16 ton, 12 C.Y. payload, 400 H.P.		61.00	340	1,015	3,050	691	
	5310	Four axle dump, 25 ton, 18 C.Y. payload, 450 H.P.		71.75	490	1,475	4,425	869	
	5350	Dump trailer only, rear dump, 16-1/2 C.Y.		5.45	138	415	1,250	126.60	
	5400	20 C.Y.		5.90	157	470	1,400	141.20	
	5450	Flatbed, single axle, 1-1/2 ton rating		25.55	68.50	205	615	245.40	
	5500	3 ton rating		30.65	96.50	290	870	303.20	
	5550	Off highway rear dump, 25 ton capacity		73.90	1,275	3,800	11,400	1,351	
	5600	35 ton capacity		82.50	1,425	4,260	12,800	1,512	
	5610	50 ton capacity		102.65	1,700	5,080	15,200	1,837	
	5620	65 ton capacity		105.65	1,700	5,090	15,300	1,863	
	5630	100 ton capacity		151.40	2,850	8,570	25,700	2,925	
	6000	Vibratory plow, 25 H.P., walking	↓	8.45	60	180	540	103.60	
40	0010	**GENERAL EQUIPMENT RENTAL** without operators R015433-10							40
	0150	Aerial lift, scissor type, to 15′ high, 1000 lb. cap., electric	Ea.	3.05	51.50	155	465	55.40	
	0160	To 25′ high, 2000 lb. capacity		3.45	66.50	200	600	67.60	
	0170	Telescoping boom to 40′ high, 500 lb. capacity, diesel		13.55	320	965	2,900	301.40	
	0180	To 45′ high, 500 lb. capacity		14.75	350	1,055	3,175	329	
	0190	To 60′ high, 600 lb. capacity		17.20	495	1,490	4,475	435.60	
	0195	Air compressor, portable, 6.5 CFM, electric		.67	12.65	38	114	12.95	
	0196	Gasoline		.81	19	57	171	17.90	
	0200	Towed type, gas engine, 60 CFM		13.55	48.50	145	435	137.40	
	0300	160 CFM		15.80	50	150	450	156.40	
	0400	Diesel engine, rotary screw, 250 CFM		17.05	108	325	975	201.40	
	0500	365 CFM		23.10	132	395	1,175	263.80	
	0550	450 CFM		29.35	165	495	1,475	333.80	
	0600	600 CFM		51.70	228	685	2,050	550.60	
	0700	750 CFM	↓	51.90	237	710	2,125	557.20	
	0800	For silenced models, small sizes, add to rent		3%	5%	5%	5%		
	0900	Large sizes, add to rent		5%	7%	7%	7%		
	0930	Air tools, breaker, pavement, 60 lb.	Ea.	.50	9.65	29	87	9.80	
	0940	80 lb.		.50	10.35	31	93	10.20	
	0950	Drills, hand (jackhammer) 65 lb.	↓	.60	17.35	52	156	15.20	

01 54 33 | Equipment Rental

			UNIT	HOURLY OPER. COST	RENT PER DAY	RENT PER WEEK	RENT PER MONTH	EQUIPMENT COST/DAY	
40	0960	Track or wagon, swing boom, 4" drifter	Ea.	62.50	880	2,640	7,925	1,028	40
	0970	5" drifter		74.45	1,075	3,235	9,700	1,243	
	0975	Track mounted quarry drill, 6" diameter drill		128.10	1,600	4,765	14,300	1,978	
	0980	Dust control per drill		1.02	23.50	71	213	22.35	
	0990	Hammer, chipping, 12 lb.		.55	26	78	234	20	
	1000	Hose, air with couplings, 50' long, 3/4" diameter		.03	5	15	45	3.25	
	1100	1" diameter		.04	6.35	19	57	4.10	
	1200	1-1/2" diameter		.05	9	27	81	5.80	
	1300	2" diameter		.07	12	36	108	7.75	
	1400	2-1/2" diameter		.11	19	57	171	12.30	
	1410	3" diameter		.14	23	69	207	14.90	
	1450	Drill, steel, 7/8" x 2'		.05	8.65	26	78	5.60	
	1460	7/8" x 6'		.05	9	27	81	5.80	
	1520	Moil points		.02	3.33	10	30	2.15	
	1525	Pneumatic nailer w/accessories		.58	38.50	115	345	27.65	
	1530	Sheeting driver for 60 lb. breaker		.04	6	18	54	3.90	
	1540	For 90 lb. breaker		.12	8	24	72	5.75	
	1550	Spade, 25 lb.		.45	6.65	20	60	7.60	
	1560	Tamper, single, 35 lb.		.55	36.50	109	325	26.20	
	1570	Triple, 140 lb.		.82	54.50	164	490	39.35	
	1580	Wrenches, impact, air powered, up to 3/4" bolt		.40	12.65	38	114	10.80	
	1590	Up to 1-1/4" bolt		.50	23.50	70	210	18	
	1600	Barricades, barrels, reflectorized, 1 to 99 barrels		.03	4.60	13.80	41.50	3	
	1610	100 to 200 barrels		.02	3.53	10.60	32	2.30	
	1620	Barrels with flashers, 1 to 99 barrels		.03	5.25	15.80	47.50	3.40	
	1630	100 to 200 barrels		.03	4.20	12.60	38	2.75	
	1640	Barrels with steady burn type C lights		.04	7	21	63	4.50	
	1650	Illuminated board, trailer mounted, with generator		3.50	130	390	1,175	106	
	1670	Portable barricade, stock, with flashers, 1 to 6 units		.03	5.25	15.80	47.50	3.40	
	1680	25 to 50 units		.03	4.90	14.70	44	3.20	
	1685	Butt fusion machine, wheeled, 1.5 HP electric, 2" - 8" diameter pipe		2.63	167	500	1,500	121.05	
	1690	Tracked, 20 HP diesel, 4"-12" diameter pipe		11.14	490	1,465	4,400	382.10	
	1695	83 HP diesel, 8" - 24" diameter pipe		30.46	975	2,930	8,800	829.70	
	1700	Carts, brick, hand powered, 1000 lb. capacity		.50	83.50	251	755	54.20	
	1800	Gas engine, 1500 lb., 7-1/2' lift		4.17	115	345	1,025	102.35	
	1822	Dehumidifier, medium, 6 lb./hr., 150 CFM		1.00	62	186	560	45.20	
	1824	Large, 18 lb./hr., 600 CFM		2.02	126	378	1,125	91.75	
	1830	Distributor, asphalt, trailer mounted, 2000 gal., 38 H.P. diesel		9.95	325	980	2,950	275.60	
	1840	3000 gal., 38 H.P. diesel		11.45	355	1,070	3,200	305.60	
	1850	Drill, rotary hammer, electric		1.02	26	78	234	23.75	
	1860	Carbide bit, 1-1/2" diameter, add to electric rotary hammer		.02	3.61	10.84	32.50	2.35	
	1865	Rotary, crawler, 250 H.P.		148.50	2,050	6,185	18,600	2,425	
	1870	Emulsion sprayer, 65 gal., 5 H.P. gas engine		2.91	97	291	875	81.50	
	1880	200 gal., 5 H.P. engine		7.85	162	485	1,450	159.80	
	1900	Floor auto-scrubbing machine, walk-behind, 28" path		4.98	325	970	2,900	233.85	
	1930	Floodlight, mercury vapor, or quartz, on tripod, 1000 watt		.44	20.50	62	186	15.90	
	1940	2000 watt		.83	41	123	370	31.25	
	1950	Floodlights, trailer mounted with generator, 1 - 300 watt light		3.65	71.50	215	645	72.20	
	1960	2 - 1000 watt lights		4.85	96.50	290	870	96.80	
	2000	4 - 300 watt lights		4.55	93.50	280	840	92.40	
	2005	Foam spray rig, incl. box trailer, compressor, generator, proportioner		35.03	490	1,465	4,400	573.25	
	2020	Forklift, straight mast, 12' lift, 5000 lb., 2 wheel drive, gas		26.35	202	605	1,825	331.80	
	2040	21' lift, 5000 lb., 4 wheel drive, diesel		20.90	240	720	2,150	311.20	
	2050	For rough terrain, 42' lift, 35' reach, 9000 lb., 110 H.P.		29.55	485	1,450	4,350	526.40	
	2060	For plant, 4 ton capacity, 80 H.P., 2 wheel drive, gas		16.15	93.50	280	840	185.20	
	2080	10 ton capacity, 120 H.P., 2 wheel drive, diesel		24.10	162	485	1,450	289.80	
	2100	Generator, electric, gas engine, 1.5 kW to 3 kW		3.75	11.35	34	102	36.80	
	2200	5 kW	↓	4.80	15.35	46	138	47.60	

01 54 33 | Equipment Rental

			UNIT	HOURLY OPER. COST	RENT PER DAY	RENT PER WEEK	RENT PER MONTH	EQUIPMENT COST/DAY	
40	2300	10 kW	Ea.	9.10	36.50	110	330	94.80	40
	2400	25 kW		10.60	85	255	765	135.80	
	2500	Diesel engine, 20 kW		12.10	66.50	200	600	136.80	
	2600	50 kW		23.35	103	310	930	248.80	
	2700	100 kW		43.30	128	385	1,150	423.40	
	2800	250 kW		85.15	235	705	2,125	822.20	
	2850	Hammer, hydraulic, for mounting on boom, to 500 ft lb.		2.55	75	225	675	65.40	
	2860	1000 ft lb.		4.35	127	380	1,150	110.80	
	2900	Heaters, space, oil or electric, 50 MBH		2.01	7.65	23	69	20.70	
	3000	100 MBH		3.76	10.65	32	96	36.50	
	3100	300 MBH		11.03	38.50	115	345	111.25	
	3150	500 MBH		18.15	45	135	405	172.20	
	3200	Hose, water, suction with coupling, 20′ long, 2″ diameter		.02	3	9	27	1.95	
	3210	3″ diameter		.03	4.33	13	39	2.85	
	3220	4″ diameter		.03	5	15	45	3.25	
	3230	6″ diameter		.11	17.65	53	159	11.50	
	3240	8″ diameter		.27	44.50	133	400	28.75	
	3250	Discharge hose with coupling, 50′ long, 2″ diameter		.01	1.33	4	12	.90	
	3260	3″ diameter		.01	2.33	7	21	1.50	
	3270	4″ diameter		.02	3.67	11	33	2.35	
	3280	6″ diameter		.06	9.35	28	84	6.10	
	3290	8″ diameter		.18	30	90	270	19.45	
	3295	Insulation blower		.78	6	18	54	9.85	
	3300	Ladders, extension type, 16′ to 36′ long		.14	22.50	68	204	14.70	
	3400	40′ to 60′ long		.19	31	93	279	20.10	
	3405	Lance for cutting concrete		2.35	71.50	215	645	61.80	
	3407	Lawn mower, rotary, 22″, 5 H.P.		1.80	49	147	440	43.80	
	3408	48″ self propelled		2.98	75	225	675	68.85	
	3410	Level, electronic, automatic, with tripod and leveling rod		1.20	75.50	226	680	54.80	
	3430	Laser type, for pipe and sewer line and grade		.73	48.50	145	435	34.85	
	3440	Rotating beam for interior control		.78	52	156	470	37.45	
	3460	Builder's optical transit, with tripod and rod		.09	14.65	44	132	9.50	
	3500	Light towers, towable, with diesel generator, 2000 watt		4.55	93.50	280	840	92.40	
	3600	4000 watt		4.85	96.50	290	870	96.80	
	3700	Mixer, powered, plaster and mortar, 6 C.F., 7 H.P.		2.70	20	60	180	33.60	
	3800	10 C.F., 9 H.P.		2.80	31.50	94	282	41.20	
	3850	Nailer, pneumatic		.58	38.50	115	345	27.65	
	3900	Paint sprayers complete, 8 CFM		1.04	69.50	208	625	49.90	
	4000	17 CFM		1.90	127	380	1,150	91.20	
	4020	Pavers, bituminous, rubber tires, 8′ wide, 50 H.P., diesel		31.15	490	1,465	4,400	542.20	
	4030	10′ wide, 150 H.P.		106.10	1,800	5,370	16,100	1,923	
	4050	Crawler, 8′ wide, 100 H.P., diesel		88.65	1,800	5,385	16,200	1,786	
	4060	10′ wide, 150 H.P.		113.15	2,200	6,610	19,800	2,227	
	4070	Concrete paver, 12′ to 24′ wide, 250 H.P.		101.40	1,550	4,675	14,000	1,746	
	4080	Placer-spreader-trimmer, 24′ wide, 300 H.P.		150.25	2,650	7,960	23,900	2,794	
	4100	Pump, centrifugal gas pump, 1-1/2″ diam., 65 GPM		3.90	48.50	145	435	60.20	
	4200	2″ diameter, 130 GPM		5.35	53.50	160	480	74.80	
	4300	3″ diameter, 250 GPM		5.65	55	165	495	78.20	
	4400	6″ diameter, 1500 GPM		30.45	172	515	1,550	346.60	
	4500	Submersible electric pump, 1-1/4″ diameter, 55 GPM		.39	16.35	49	147	12.90	
	4600	1-1/2″ diameter, 83 GPM		.43	18.65	56	168	14.65	
	4700	2″ diameter, 120 GPM		1.50	23.50	70	210	26	
	4800	3″ diameter, 300 GPM		2.69	41.50	125	375	46.50	
	4900	4″ diameter, 560 GPM		12.34	158	475	1,425	193.70	
	5000	6″ diameter, 1590 GPM		18.40	212	635	1,900	274.20	
	5100	Diaphragm pump, gas, single, 1-1/2″ diameter		1.18	49.50	148	445	39.05	
	5200	2″ diameter		4.25	61.50	185	555	71	
	5300	3″ diameter		4.25	61.50	185	555	71	

01 54 33 | Equipment Rental

			UNIT	HOURLY OPER. COST	RENT PER DAY	RENT PER WEEK	RENT PER MONTH	EQUIPMENT COST/DAY	
40	5400	Double, 4″ diameter	Ea.	6.40	105	315	945	114.20	40
	5450	Pressure washer 5 GPM, 3000 psi		4.90	51.50	155	465	70.20	
	5460	7 GPM, 3000 psi		6.45	60	180	540	87.60	
	5500	Trash pump, self-priming, gas, 2″ diameter		4.60	21	63	189	49.40	
	5600	Diesel, 4″ diameter		9.15	88.50	265	795	126.20	
	5650	Diesel, 6″ diameter		25.25	147	440	1,325	290	
	5655	Grout Pump		26.35	268	805	2,425	371.80	
	5700	Salamanders, L.P. gas fired, 100,000 Btu		3.96	13.65	41	123	39.90	
	5705	50,000 Btu		2.21	10.35	31	93	23.90	
	5720	Sandblaster, portable, open top, 3 C.F. capacity		.55	26.50	80	240	20.40	
	5730	6 C.F. capacity		.95	40	120	360	31.60	
	5740	Accessories for above		.13	21.50	65	195	14.05	
	5750	Sander, floor		.70	14.35	43	129	14.20	
	5760	Edger		.50	14.35	43	129	12.60	
	5800	Saw, chain, gas engine, 18″ long		2.30	21.50	64	192	31.20	
	5900	Hydraulic powered, 36″ long		.75	65	195	585	45	
	5950	60″ long		.75	66.50	200	600	46	
	6000	Masonry, table mounted, 14″ diameter, 5 H.P.		1.32	56.50	170	510	44.55	
	6050	Portable cut-off, 8 H.P.		2.50	33.50	100	300	40	
	6100	Circular, hand held, electric, 7-1/4″ diameter		.23	4.67	14	42	4.65	
	6200	12″ diameter		.23	8	24	72	6.65	
	6250	Wall saw, w/hydraulic power, 10 H.P.		9.70	61.50	185	555	114.60	
	6275	Shot blaster, walk-behind, 20″ wide		4.75	293	880	2,650	214	
	6280	Sidewalk broom, walk-behind		2.52	78.50	235	705	67.15	
	6300	Steam cleaner, 100 gallons per hour		3.70	76.50	230	690	75.60	
	6310	200 gallons per hour		5.35	95	285	855	99.80	
	6340	Tar Kettle/Pot, 400 gallons		15.60	75	225	675	169.80	
	6350	Torch, cutting, acetylene-oxygen, 150′ hose, excludes gases		.30	15	45	135	11.40	
	6360	Hourly operating cost includes tips and gas		19.00				152	
	6410	Toilet, portable chemical		.13	21	63	189	13.65	
	6420	Recycle flush type		.15	25	75	225	16.20	
	6430	Toilet, fresh water flush, garden hose,		.18	30.50	91	273	19.65	
	6440	Hoisted, non-flush, for high rise		.15	24.50	74	222	16	
	6465	Tractor, farm with attachment		21.50	297	890	2,675	350	
	6480	Trailers, platform, flush deck, 2 axle, 3 ton capacity		1.45	20	60	180	23.60	
	6500	25 ton capacity		5.45	117	350	1,050	113.60	
	6600	40 ton capacity		7.00	163	490	1,475	154	
	6700	3 axle, 50 ton capacity		7.55	180	540	1,625	168.40	
	6800	75 ton capacity		9.40	235	705	2,125	216.20	
	6810	Trailer mounted cable reel for high voltage line work		5.45	260	779	2,325	199.40	
	6820	Trailer mounted cable tensioning rig		10.85	515	1,550	4,650	396.80	
	6830	Cable pulling rig		73.05	2,900	8,680	26,000	2,320	
	6900	Water tank trailer, engine driven discharge, 5000 gallons		7.00	142	425	1,275	141	
	6925	10,000 gallons		9.50	197	590	1,775	194	
	6950	Water truck, off highway, 6000 gallons		88.15	770	2,315	6,950	1,168	
	7010	Tram car for high voltage line work, powered, 2 conductor		6.60	141	423	1,275	137.40	
	7020	Transit (builder's level) with tripod		.09	14.65	44	132	9.50	
	7030	Trench box, 3000 lb., 6′ x 8′		.56	93.50	280	840	60.50	
	7040	7200 lb., 6′ x 20′		.75	125	375	1,125	81	
	7050	8000 lb., 8′ x 16′		1.08	180	540	1,625	116.65	
	7060	9500 lb., 8′ x 20′		1.21	201	603	1,800	130.30	
	7065	11,000 lb., 8′ x 24′		1.27	211	633	1,900	136.75	
	7070	12,000 lb., 10′ x 20′		1.50	249	748	2,250	161.60	
	7100	Truck, pickup, 3/4 ton, 2 wheel drive		13.65	58.50	175	525	144.20	
	7200	4 wheel drive		13.95	73.50	220	660	155.60	
	7250	Crew carrier, 9 passenger		19.30	86.50	260	780	206.40	
	7290	Flat bed truck, 20,000 lb. GVW		21.70	125	375	1,125	248.60	
	7300	Tractor, 4 x 2, 220 H.P.	↓	30.20	197	590	1,775	359.60	

01 54 33 | Equipment Rental

		01 54 33 | Equipment Rental	UNIT	HOURLY OPER. COST	RENT PER DAY	RENT PER WEEK	RENT PER MONTH	EQUIPMENT COST/DAY	
40	7410	330 H.P.	Ea.	44.75	270	810	2,425	520	40
	7500	6 x 4, 380 H.P.		51.35	315	945	2,825	599.80	
	7600	450 H.P.		62.30	380	1,145	3,425	727.40	
	7610	Tractor, with A frame, boom and winch, 225 H.P.		33.00	272	815	2,450	427	
	7620	Vacuum truck, hazardous material, 2500 gallons		12.05	290	870	2,600	270.40	
	7625	5,000 gallons		14.26	405	1,220	3,650	358.10	
	7650	Vacuum, HEPA, 16 gallon, wet/dry		.85	18	54	162	17.60	
	7655	55 gallon, wet/dry		.80	27	81	243	22.60	
	7660	Water tank, portable		.16	26.50	80	240	17.30	
	7690	Sewer/catch basin vacuum, 14 C.Y., 1500 gallons		18.53	610	1,830	5,500	514.25	
	7700	Welder, electric, 200 amp		3.88	16.35	49	147	40.85	
	7800	300 amp		5.71	20	60	180	57.70	
	7900	Gas engine, 200 amp		14.65	23.50	70	210	131.20	
	8000	300 amp		16.38	24.50	74	222	145.85	
	8100	Wheelbarrow, any size		.08	13	39	117	8.45	
	8200	Wrecking ball, 4000 lb.	↓	2.35	68.50	205	615	59.80	
50	0010	**HIGHWAY EQUIPMENT RENTAL** without operators R015433 -10							50
	0050	Asphalt batch plant, portable drum mixer, 100 ton/hr.	Ea.	78.06	1,425	4,270	12,800	1,478	
	0060	200 ton/hr.		89.18	1,500	4,530	13,600	1,619	
	0070	300 ton/hr.		106.34	1,775	5,325	16,000	1,916	
	0100	Backhoe attachment, long stick, up to 185 H.P., 10.5′ long		.35	23.50	70	210	16.80	
	0140	Up to 250 H.P., 12′ long		.38	25	75	225	18.05	
	0180	Over 250 H.P., 15′ long		.53	35	105	315	25.25	
	0200	Special dipper arm, up to 100 H.P., 32′ long		1.08	71.50	215	645	51.65	
	0240	Over 100 H.P., 33′ long		1.34	89.50	268	805	64.30	
	0280	Catch basin/sewer cleaning truck, 3 ton, 9 C.Y., 1000 gal.		42.70	385	1,160	3,475	573.60	
	0300	Concrete batch plant, portable, electric, 200 C.Y./hr.		22.87	515	1,550	4,650	492.95	
	0520	Grader/dozer attachment, ripper/scarifier, rear mounted, up to 135 H.P.		2.90	56.50	170	510	57.20	
	0540	Up to 180 H.P.		3.90	88.50	265	795	84.20	
	0580	Up to 250 H.P.		5.00	117	350	1,050	110	
	0700	Pvmt. removal bucket, for hyd. excavator, up to 90 H.P.		1.90	53.50	160	480	47.20	
	0740	Up to 200 H.P.		2.15	73.50	220	660	61.20	
	0780	Over 200 H.P.		2.25	85	255	765	69	
	0900	Aggregate spreader, self-propelled, 187 H.P.		53.00	685	2,050	6,150	834	
	1000	Chemical spreader, 3 C.Y.		3.30	43.50	130	390	52.40	
	1900	Hammermill, traveling, 250 H.P.		78.83	2,075	6,240	18,700	1,879	
	2000	Horizontal borer, 3″ diameter, 13 H.P. gas driven		6.35	55	165	495	83.80	
	2150	Horizontal directional drill, 20,000 lb. thrust, 78 H.P. diesel		31.15	675	2,025	6,075	654.20	
	2160	30,000 lb. thrust, 115 H.P.		38.90	1,025	3,100	9,300	931.20	
	2170	50,000 lb. thrust, 170 H.P.		55.65	1,325	3,960	11,900	1,237	
	2190	Mud trailer for HDD, 1500 gallons, 175 H.P., gas		31.90	152	455	1,375	346.20	
	2200	Hydromulcher, diesel, 3000 gallon, for truck mounting		23.40	245	735	2,200	334.20	
	2300	Gas, 600 gallon		8.60	98.50	295	885	127.80	
	2400	Joint & crack cleaner, walk behind, 25 H.P.		3.95	50	150	450	61.60	
	2500	Filler, trailer mounted, 400 gallons, 20 H.P.		9.50	210	630	1,900	202	
	3000	Paint striper, self-propelled, 40 gallon, 22 H.P.		7.30	155	465	1,400	151.40	
	3100	120 gallon, 120 H.P.		23.00	395	1,185	3,550	421	
	3200	Post drivers, 6″ I-Beam frame, for truck mounting		19.05	380	1,135	3,400	379.40	
	3400	Road sweeper, self-propelled, 8′ wide, 90 H.P.		39.80	575	1,720	5,150	662.40	
	3450	Road sweeper, vacuum assisted, 4 C.Y., 220 gallons		75.75	635	1,900	5,700	986	
	4000	Road mixer, self-propelled, 130 H.P.		46.85	765	2,295	6,875	833.80	
	4100	310 H.P.		83.25	2,100	6,315	18,900	1,929	
	4220	Cold mix paver, incl. pug mill and bitumen tank, 165 H.P.		97.00	2,275	6,855	20,600	2,147	
	4240	Pavement brush, towed		3.20	93.50	280	840	81.60	
	4250	Paver, asphalt, wheel or crawler, 130 H.P., diesel		96.65	2,275	6,795	20,400	2,132	
	4300	Paver, road widener, gas 1′ to 6′, 67 H.P.		48.50	895	2,680	8,050	924	
	4400	Diesel, 2′ to 14′, 88 H.P.		62.70	1,050	3,135	9,400	1,129	
	4600	Slipform pavers, curb and gutter, 2 track, 75 H.P.	↓	57.20	900	2,700	8,100	997.60	

01 54 33 | Equipment Rental

			UNIT	HOURLY OPER. COST	RENT PER DAY	RENT PER WEEK	RENT PER MONTH	EQUIPMENT COST/DAY	
50	4700	4 track, 165 H.P.	Ea.	41.50	740	2,220	6,650	776	50
	4800	Median barrier, 215 H.P.		60.50	1,050	3,150	9,450	1,114	
	4901	Trailer, low bed, 75 ton capacity		10.10	235	705	2,125	221.80	
	5000	Road planer, walk behind, 10″ cutting width, 10 H.P.		3.70	33.50	100	300	49.60	
	5100	Self-propelled, 12″ cutting width, 64 H.P.		10.40	113	340	1,025	151.20	
	5120	Traffic line remover, metal ball blaster, truck mounted, 115 H.P.		47.95	730	2,190	6,575	821.60	
	5140	Grinder, truck mounted, 115 H.P.		54.60	790	2,375	7,125	911.80	
	5160	Walk-behind, 11 H.P.		4.20	53.50	160	480	65.60	
	5200	Pavement profiler, 4′ to 6′ wide, 450 H.P.		254.50	3,325	9,960	29,900	4,028	
	5300	8′ to 10′ wide, 750 H.P.		399.20	4,350	13,055	39,200	5,805	
	5400	Roadway plate, steel, 1″ x 8′ x 20′		.08	13.35	40	120	8.65	
	5600	Stabilizer, self-propelled, 150 H.P.		49.30	600	1,795	5,375	753.40	
	5700	310 H.P.		99.05	1,675	5,030	15,100	1,798	
	5800	Striper, truck mounted, 120 gallon paint, 460 H.P.		64.20	485	1,455	4,375	804.60	
	5900	Thermal paint heating kettle, 115 gallons		8.20	25.50	77	231	81	
	6000	Tar kettle, 330 gallon, trailer mounted		12.25	56.50	170	510	132	
	7000	Tunnel locomotive, diesel, 8 to 12 ton		31.85	585	1,750	5,250	604.80	
	7005	Electric, 10 ton		26.00	665	1,995	5,975	607	
	7010	Muck cars, 1/2 C.Y. capacity		2.00	24.50	73	219	30.60	
	7020	1 C.Y. capacity		2.25	32.50	97	291	37.40	
	7030	2 C.Y. capacity		2.35	36.50	110	330	40.80	
	7040	Side dump, 2 C.Y. capacity		2.60	45	135	405	47.80	
	7050	3 C.Y. capacity		3.45	50	150	450	57.60	
	7060	5 C.Y. capacity		4.95	63.50	190	570	77.60	
	7100	Ventilating blower for tunnel, 7-1/2 H.P.		2.05	51.50	155	465	47.40	
	7110	10 H.P.		2.28	51.50	155	465	49.25	
	7120	20 H.P.		3.48	67.50	202	605	68.25	
	7140	40 H.P.		5.76	96.50	290	870	104.10	
	7160	60 H.P.		8.77	152	455	1,375	161.15	
	7175	75 H.P.		11.66	207	620	1,850	217.30	
	7180	200 H.P.		23.65	305	910	2,725	371.20	
	7800	Windrow loader, elevating	↓	55.00	1,300	3,935	11,800	1,227	
60	0010	**LIFTING AND HOISTING EQUIPMENT RENTAL** without operators R015433-10							60
	0120	Aerial lift truck, 2 person, to 80′	Ea.	26.30	705	2,120	6,350	634.40	
	0140	Boom work platform, 40′ snorkel R015433-15		12.00	275	825	2,475	261	
	0150	Crane, flatbed mounted, 3 ton capacity		16.35	188	565	1,700	243.80	
	0200	Crane, climbing, 106′ jib, 6000 lb. capacity, 410 fpm R312316-45		39.35	1,650	4,960	14,900	1,307	
	0300	101′ jib, 10,250 lb. capacity, 270 fpm		45.95	2,100	6,280	18,800	1,624	
	0500	Tower, static, 130′ high, 106′ jib, 6200 lb. capacity at 400 fpm		43.20	1,900	5,730	17,200	1,492	
	0600	Crawler mounted, lattice boom, 1/2 C.Y., 15 tons at 12′ radius		36.98	625	1,870	5,600	669.85	
	0700	3/4 C.Y., 20 tons at 12′ radius		49.31	780	2,340	7,025	862.50	
	0800	1 C.Y., 25 tons at 12′ radius		65.75	1,050	3,120	9,350	1,150	
	0900	1-1/2 C.Y., 40 tons at 12′ radius		65.75	1,050	3,150	9,450	1,156	
	1000	2 C.Y., 50 tons at 12′ radius		69.70	1,225	3,680	11,000	1,294	
	1100	3 C.Y., 75 tons at 12′ radius		74.70	1,425	4,305	12,900	1,459	
	1200	100 ton capacity, 60′ boom		84.60	1,650	4,950	14,900	1,667	
	1300	165 ton capacity, 60′ boom		107.75	1,925	5,785	17,400	2,019	
	1400	200 ton capacity, 70′ boom		130.55	2,400	7,210	21,600	2,486	
	1500	350 ton capacity, 80′ boom		182.90	3,625	10,845	32,500	3,632	
	1600	Truck mounted, lattice boom, 6 x 4, 20 tons at 10′ radius		37.11	1,075	3,260	9,775	948.90	
	1700	25 tons at 10′ radius		40.09	1,175	3,550	10,700	1,031	
	1800	8 x 4, 30 tons at 10′ radius		43.61	1,250	3,780	11,300	1,105	
	1900	40 tons at 12′ radius		46.70	1,325	3,950	11,900	1,164	
	2000	60 tons at 15′ radius		53.07	1,400	4,180	12,500	1,261	
	2050	82 tons at 15′ radius		59.80	1,475	4,460	13,400	1,370	
	2100	90 tons at 15′ radius		67.36	1,625	4,860	14,600	1,511	
	2200	115 tons at 15′ radius		76.11	1,800	5,430	16,300	1,695	
	2300	150 tons at 18′ radius	↓	83.65	1,900	5,720	17,200	1,813	

01 54 33 | Equipment Rental

			UNIT	HOURLY OPER. COST	RENT PER DAY	RENT PER WEEK	RENT PER MONTH	EQUIPMENT COST/DAY
60	2350	165 tons at 18′ radius	Ea.	90.05	2,025	6,060	18,200	1,932
	2400	Truck mounted, hydraulic, 12 ton capacity		42.60	520	1,565	4,700	653.80
	2500	25 ton capacity		44.95	630	1,885	5,650	736.60
	2550	33 ton capacity		45.50	645	1,930	5,800	750
	2560	40 ton capacity		58.85	750	2,250	6,750	920.80
	2600	55 ton capacity		76.80	855	2,570	7,700	1,128
	2700	80 ton capacity		100.45	1,375	4,105	12,300	1,625
	2720	100 ton capacity		94.20	1,425	4,250	12,800	1,604
	2740	120 ton capacity		100.05	1,525	4,575	13,700	1,715
	2760	150 ton capacity		127.60	2,000	5,995	18,000	2,220
	2800	Self-propelled, 4 x 4, with telescoping boom, 5 ton		17.55	225	675	2,025	275.40
	2900	12-1/2 ton capacity		32.30	360	1,075	3,225	473.40
	3000	15 ton capacity		33.00	375	1,130	3,400	490
	3050	20 ton capacity		35.95	445	1,335	4,000	554.60
	3100	25 ton capacity		37.55	500	1,495	4,475	599.40
	3150	40 ton capacity		46.00	560	1,675	5,025	703
	3200	Derricks, guy, 20 ton capacity, 60′ boom, 75′ mast		27.55	405	1,214	3,650	463.20
	3300	100′ boom, 115′ mast		43.21	695	2,090	6,275	763.70
	3400	Stiffleg, 20 ton capacity, 70′ boom, 37′ mast		30.11	525	1,580	4,750	556.90
	3500	100′ boom, 47′ mast		46.29	845	2,530	7,600	876.30
	3550	Helicopter, small, lift to 1250 lb. maximum, w/pilot		103.70	3,275	9,800	29,400	2,790
	3600	Hoists, chain type, overhead, manual, 3/4 ton		.10	.33	1	3	1
	3900	10 ton		.70	6	18	54	9.20
	4000	Hoist and tower, 5000 lb. cap., portable electric, 40′ high		4.95	233	699	2,100	179.40
	4100	For each added 10′ section, add		.11	18.35	55	165	11.90
	4200	Hoist and single tubular tower, 5000 lb. electric, 100′ high		6.70	325	976	2,925	248.80
	4300	For each added 6′-6″ section, add		.19	31.50	95	285	20.50
	4400	Hoist and double tubular tower, 5000 lb., 100′ high		7.19	360	1,075	3,225	272.50
	4500	For each added 6′-6″ section, add		.21	35	105	315	22.70
	4550	Hoist and tower, mast type, 6000 lb., 100′ high		7.76	370	1,115	3,350	285.10
	4570	For each added 10′ section, add		.13	21.50	65	195	14.05
	4600	Hoist and tower, personnel, electric, 2000 lb., 100′ @ 125 fpm		16.31	990	2,970	8,900	724.50
	4700	3000 lb., 100′ @ 200 fpm		18.62	1,125	3,360	10,100	820.95
	4800	3000 lb., 150′ @ 300 fpm		20.62	1,250	3,760	11,300	916.95
	4900	4000 lb., 100′ @ 300 fpm		21.38	1,275	3,840	11,500	939.05
	5000	6000 lb., 100′ @ 275 fpm	↓	23.06	1,350	4,030	12,100	990.50
	5100	For added heights up to 500′, add	L.F.	.01	1.67	5	15	1.10
	5200	Jacks, hydraulic, 20 ton	Ea.	.05	2	6	18	1.60
	5500	100 ton		.40	11.65	35	105	10.20
	6100	Jacks, hydraulic, climbing w/50′ jackrods, control console, 30 ton cap.		2.01	134	402	1,200	96.50
	6150	For each added 10′ jackrod section, add		.05	3.33	10	30	2.40
	6300	50 ton capacity		3.23	215	646	1,950	155.05
	6350	For each added 10′ jackrod section, add		.06	4	12	36	2.90
	6500	125 ton capacity		8.45	565	1,690	5,075	405.60
	6550	For each added 10′ jackrod section, add		.58	38.50	115	345	27.65
	6600	Cable jack, 10 ton capacity with 200′ cable		1.68	112	336	1,000	80.65
	6650	For each added 50′ of cable, add	↓	.20	13.35	40	120	9.60
70	0010	**WELLPOINT EQUIPMENT RENTAL** without operators R015433-10						
	0020	Based on 2 months rental						
	0100	Combination jetting & wellpoint pump, 60 H.P. diesel	Ea.	18.40	330	996	3,000	346.40
	0200	High pressure gas jet pump, 200 H.P., 300 psi	"	44.06	284	851	2,550	522.70
	0300	Discharge pipe, 8″ diameter	L.F.	.01	.54	1.62	4.86	.40
	0350	12″ diameter		.01	.79	2.38	7.15	.55
	0400	Header pipe, flows up to 150 GPM, 4″ diameter		.01	.49	1.47	4.41	.35
	0500	400 GPM, 6″ diameter		.01	.57	1.72	5.15	.40
	0600	800 GPM, 8″ diameter		.01	.79	2.38	7.15	.55
	0700	1500 GPM, 10″ diameter	↓	.01	.84	2.51	7.55	.60

01 54 33 | Equipment Rental

			UNIT	HOURLY OPER. COST	RENT PER DAY	RENT PER WEEK	RENT PER MONTH	EQUIPMENT COST/DAY	
70	0800	2500 GPM, 12″ diameter	L.F.	.02	1.58	4.74	14.20	1.10	70
	0900	4500 GPM, 16″ diameter		.03	2.02	6.07	18.20	1.45	
	0950	For quick coupling aluminum and plastic pipe, add	↓	.03	2.09	6.28	18.85	1.50	
	1100	Wellpoint, 25′ long, with fittings & riser pipe, 1-1/2″ or 2″ diameter	Ea.	.06	4.18	12.54	37.50	3	
	1200	Wellpoint pump, diesel powered, 4″ suction, 20 H.P.		7.83	191	574	1,725	177.45	
	1300	6″ suction, 30 H.P.		10.70	237	712	2,125	228	
	1400	8″ suction, 40 H.P.		14.45	325	976	2,925	310.80	
	1500	10″ suction, 75 H.P.		22.27	380	1,141	3,425	406.35	
	1600	12″ suction, 100 H.P.		31.72	605	1,810	5,425	615.75	
	1700	12″ suction, 175 H.P.	↓	47.41	670	2,010	6,025	781.30	
80	0010	**MARINE EQUIPMENT RENTAL** without operators R015433-10							80
	0200	Barge, 400 Ton, 30′ wide x 90′ long	Ea.	17.50	1,050	3,180	9,550	776	
	0240	800 Ton, 45′ wide x 90′ long		21.25	1,300	3,870	11,600	944	
	2000	Tugboat, diesel, 100 H.P.		38.20	217	650	1,950	435.60	
	2040	250 H.P.		79.25	395	1,180	3,550	870	
	2080	380 H.P.		156.80	1,175	3,535	10,600	1,961	
	3000	Small work boat, gas, 16-foot, 50 H.P.		17.15	60	180	540	173.20	
	4000	Large, diesel, 48-foot, 200 H.P.	↓	89.40	1,250	3,740	11,200	1,463	

Crew A-1	Bare Costs Hr.	Bare Costs Daily	Incl. Subs O&P Hr.	Incl. Subs O&P Daily	Cost Per Labor-Hour Bare Costs	Cost Per Labor-Hour Incl. O&P
1 Building Laborer	$37.60	$300.80	$57.85	$462.80	$37.60	$57.85
1 Concrete Saw, Gas Manual		81.00		89.10	10.13	11.14
8 L.H., Daily Totals		$381.80		$551.90	$47.73	$68.99

Crew A-1A	Hr.	Daily	Hr.	Daily	Bare Costs	Incl. O&P
1 Skilled Worker	$48.65	$389.20	$75.15	$601.20	$48.65	$75.15
1 Shot Blaster, 20"		214.00		235.40	26.75	29.43
8 L.H., Daily Totals		$603.20		$836.60	$75.40	$104.58

Crew A-1B	Hr.	Daily	Hr.	Daily	Bare Costs	Incl. O&P
1 Building Laborer	$37.60	$300.80	$57.85	$462.80	$37.60	$57.85
1 Concrete Saw		167.40		184.14	20.93	23.02
8 L.H., Daily Totals		$468.20		$646.94	$58.52	$80.87

Crew A-1C	Hr.	Daily	Hr.	Daily	Bare Costs	Incl. O&P
1 Building Laborer	$37.60	$300.80	$57.85	$462.80	$37.60	$57.85
1 Chain Saw, Gas, 18"		31.20		34.32	3.90	4.29
8 L.H., Daily Totals		$332.00		$497.12	$41.50	$62.14

Crew A-1D	Hr.	Daily	Hr.	Daily	Bare Costs	Incl. O&P
1 Building Laborer	$37.60	$300.80	$57.85	$462.80	$37.60	$57.85
1 Vibrating Plate, Gas, 18"		36.20		39.82	4.53	4.98
8 L.H., Daily Totals		$337.00		$502.62	$42.13	$62.83

Crew A-1E	Hr.	Daily	Hr.	Daily	Bare Costs	Incl. O&P
1 Building Laborer	$37.60	$300.80	$57.85	$462.80	$37.60	$57.85
1 Vibrating Plate, Gas, 21"		46.80		51.48	5.85	6.43
8 L.H., Daily Totals		$347.60		$514.28	$43.45	$64.28

Crew A-1F	Hr.	Daily	Hr.	Daily	Bare Costs	Incl. O&P
1 Building Laborer	$37.60	$300.80	$57.85	$462.80	$37.60	$57.85
1 Rammer/Tamper, Gas, 8"		50.00		55.00	6.25	6.88
8 L.H., Daily Totals		$350.80		$517.80	$43.85	$64.72

Crew A-1G	Hr.	Daily	Hr.	Daily	Bare Costs	Incl. O&P
1 Building Laborer	$37.60	$300.80	$57.85	$462.80	$37.60	$57.85
1 Rammer/Tamper, Gas, 15"		56.40		62.04	7.05	7.75
8 L.H., Daily Totals		$357.20		$524.84	$44.65	$65.61

Crew A-1H	Hr.	Daily	Hr.	Daily	Bare Costs	Incl. O&P
1 Building Laborer	$37.60	$300.80	$57.85	$462.80	$37.60	$57.85
1 Exterior Steam Cleaner		75.60		83.16	9.45	10.40
8 L.H., Daily Totals		$376.40		$545.96	$47.05	$68.25

Crew A-1J	Hr.	Daily	Hr.	Daily	Bare Costs	Incl. O&P
1 Building Laborer	$37.60	$300.80	$57.85	$462.80	$37.60	$57.85
1 Cultivator, Walk-Behind, 5 H.P.		46.70		51.37	5.84	6.42
8 L.H., Daily Totals		$347.50		$514.17	$43.44	$64.27

Crew A-1K	Hr.	Daily	Hr.	Daily	Bare Costs	Incl. O&P
1 Building Laborer	$37.60	$300.80	$57.85	$462.80	$37.60	$57.85
1 Cultivator, Walk-Behind, 8 H.P.		72.65		79.92	9.08	9.99
8 L.H., Daily Totals		$373.45		$542.72	$46.68	$67.84

Crew A-1M	Hr.	Daily	Hr.	Daily	Bare Costs	Incl. O&P
1 Building Laborer	$37.60	$300.80	$57.85	$462.80	$37.60	$57.85
1 Snow Blower, Walk-Behind		67.15		73.86	8.39	9.23
8 L.H., Daily Totals		$367.95		$536.66	$45.99	$67.08

Crew A-2	Bare Costs Hr.	Bare Costs Daily	Incl. Subs O&P Hr.	Incl. Subs O&P Daily	Cost Per Labor-Hour Bare Costs	Cost Per Labor-Hour Incl. O&P
2 Laborers	$37.60	$601.60	$57.85	$925.60	$38.10	$58.27
1 Truck Driver (light)	39.10	312.80	59.10	472.80		
1 Flatbed Truck, Gas, 1.5 Ton		245.40		269.94	10.23	11.25
24 L.H., Daily Totals		$1159.80		$1668.34	$48.33	$69.51

Crew A-2A	Hr.	Daily	Hr.	Daily	Bare Costs	Incl. O&P
2 Laborers	$37.60	$601.60	$57.85	$925.60	$38.10	$58.27
1 Truck Driver (light)	39.10	312.80	59.10	472.80		
1 Flatbed Truck, Gas, 1.5 Ton		245.40		269.94		
1 Concrete Saw		167.40		184.14	17.20	18.92
24 L.H., Daily Totals		$1327.20		$1852.48	$55.30	$77.19

Crew A-2B	Hr.	Daily	Hr.	Daily	Bare Costs	Incl. O&P
1 Truck Driver (light)	$39.10	$312.80	$59.10	$472.80	$39.10	$59.10
1 Flatbed Truck, Gas, 1.5 Ton		245.40		269.94	30.68	33.74
8 L.H., Daily Totals		$558.20		$742.74	$69.78	$92.84

Crew A-3A	Hr.	Daily	Hr.	Daily	Bare Costs	Incl. O&P
1 Equip. Oper. (light)	$48.60	$388.80	$73.75	$590.00	$48.60	$73.75
1 Pickup Truck, 4 x 4, 3/4 Ton		155.60		171.16	19.45	21.40
8 L.H., Daily Totals		$544.40		$761.16	$68.05	$95.14

Crew A-3B	Hr.	Daily	Hr.	Daily	Bare Costs	Incl. O&P
1 Equip. Oper. (medium)	$50.60	$404.80	$76.75	$614.00	$45.33	$68.65
1 Truck Driver (heavy)	40.05	320.40	60.55	484.40		
1 Dump Truck, 12 C.Y., 400 H.P.		691.00		760.10		
1 F.E. Loader, W.M.,2.5 C.Y.		519.80		571.78	75.67	83.24
16 L.H., Daily Totals		$1936.00		$2430.28	$121.00	$151.89

Crew A-3C	Hr.	Daily	Hr.	Daily	Bare Costs	Incl. O&P
1 Equip. Oper. (light)	$48.60	$388.80	$73.75	$590.00	$48.60	$73.75
1 Loader, Skid Steer, 78 H.P.		318.60		350.46	39.83	43.81
8 L.H., Daily Totals		$707.40		$940.46	$88.42	$117.56

Crew A-3D	Hr.	Daily	Hr.	Daily	Bare Costs	Incl. O&P
1 Truck Driver (light)	$39.10	$312.80	$59.10	$472.80	$39.10	$59.10
1 Pickup Truck, 4 x 4, 3/4 Ton		155.60		171.16		
1 Flatbed Trailer, 25 Ton		113.60		124.96	33.65	37.02
8 L.H., Daily Totals		$582.00		$768.92	$72.75	$96.11

Crew A-3E	Hr.	Daily	Hr.	Daily	Bare Costs	Incl. O&P
1 Equip. Oper. (crane)	$51.70	$413.60	$78.45	$627.60	$45.88	$69.50
1 Truck Driver (heavy)	40.05	320.40	60.55	484.40		
1 Pickup Truck, 4 x 4, 3/4 Ton		155.60		171.16	9.72	10.70
16 L.H., Daily Totals		$889.60		$1283.16	$55.60	$80.20

Crew A-3F	Hr.	Daily	Hr.	Daily	Bare Costs	Incl. O&P
1 Equip. Oper. (crane)	$51.70	$413.60	$78.45	$627.60	$45.88	$69.50
1 Truck Driver (heavy)	40.05	320.40	60.55	484.40		
1 Pickup Truck, 4 x 4, 3/4 Ton		155.60		171.16		
1 Truck Tractor, 6x4, 380 H.P.		599.80		659.78		
1 Lowbed Trailer, 75 Ton		221.80		243.98	61.08	67.18
16 L.H., Daily Totals		$1711.20		$2186.92	$106.95	$136.68

Crew No.	Bare Costs		Incl. Subs O&P		Cost Per Labor-Hour	
Crew A-3G	**Hr.**	**Daily**	**Hr.**	**Daily**	**Bare Costs**	**Incl. O&P**
1 Equip. Oper. (crane)	$51.70	$413.60	$78.45	$627.60	$45.88	$69.50
1 Truck Driver (heavy)	40.05	320.40	60.55	484.40		
1 Pickup Truck, 4 x 4, 3/4 Ton		155.60		171.16		
1 Truck Tractor, 6x4, 450 H.P.		727.40		800.14		
1 Lowbed Trailer, 75 Ton		221.80		243.98	69.05	75.95
16 L.H., Daily Totals		$1838.80		$2327.28	$114.93	$145.46

Crew A-3H	Hr.	Daily	Hr.	Daily	Bare Costs	Incl. O&P
1 Equip. Oper. (crane)	$51.70	$413.60	$78.45	$627.60	$51.70	$78.45
1 Hyd. Crane, 12 Ton (Daily)		860.80		946.88	107.60	118.36
8 L.H., Daily Totals		$1274.40		$1574.48	$159.30	$196.81

Crew A-3I	Hr.	Daily	Hr.	Daily	Bare Costs	Incl. O&P
1 Equip. Oper. (crane)	$51.70	$413.60	$78.45	$627.60	$51.70	$78.45
1 Hyd. Crane, 25 Ton (Daily)		989.60		1088.56	123.70	136.07
8 L.H., Daily Totals		$1403.20		$1716.16	$175.40	$214.52

Crew A-3J	Hr.	Daily	Hr.	Daily	Bare Costs	Incl. O&P
1 Equip. Oper. (crane)	$51.70	$413.60	$78.45	$627.60	$51.70	$78.45
1 Hyd. Crane, 40 Ton (Daily)		1221.00		1343.10	152.63	167.89
8 L.H., Daily Totals		$1634.60		$1970.70	$204.32	$246.34

Crew A-3K	Hr.	Daily	Hr.	Daily	Bare Costs	Incl. O&P
1 Equip. Oper. (crane)	$51.70	$413.60	$78.45	$627.60	$48.45	$73.50
1 Equip. Oper. (oiler)	45.20	361.60	68.55	548.40		
1 Hyd. Crane, 55 Ton (Daily)		1469.00		1615.90		
1 P/U Truck, 3/4 Ton (Daily)		167.20		183.92	102.26	112.49
16 L.H., Daily Totals		$2411.40		$2975.82	$150.71	$185.99

Crew A-3L	Hr.	Daily	Hr.	Daily	Bare Costs	Incl. O&P
1 Equip. Oper. (crane)	$51.70	$413.60	$78.45	$627.60	$48.45	$73.50
1 Equip. Oper. (oiler)	45.20	361.60	68.55	548.40		
1 Hyd. Crane, 80 Ton (Daily)		2174.00		2391.40		
1 P/U Truck, 3/4 Ton (Daily)		167.20		183.92	146.32	160.96
16 L.H., Daily Totals		$3116.40		$3751.32	$194.78	$234.46

Crew A-3M	Hr.	Daily	Hr.	Daily	Bare Costs	Incl. O&P
1 Equip. Oper. (crane)	$51.70	$413.60	$78.45	$627.60	$48.45	$73.50
1 Equip. Oper. (oiler)	45.20	361.60	68.55	548.40		
1 Hyd. Crane, 100 Ton (Daily)		2169.00		2385.90		
1 P/U Truck, 3/4 Ton (Daily)		167.20		183.92	146.01	160.61
16 L.H., Daily Totals		$3111.40		$3745.82	$194.46	$234.11

Crew A-3N	Hr.	Daily	Hr.	Daily	Bare Costs	Incl. O&P
1 Equip. Oper. (crane)	$51.70	$413.60	$78.45	$627.60	$51.70	$78.45
1 Tower Cane (monthly)		1128.00		1240.80	141.00	155.10
8 L.H., Daily Totals		$1541.60		$1868.40	$192.70	$233.55

Crew A-3P	Hr.	Daily	Hr.	Daily	Bare Costs	Incl. O&P
1 Equip. Oper. (light)	$48.60	$388.80	$73.75	$590.00	$48.60	$73.75
1 A.T. Forklift, 42' lift		526.40		579.04	65.80	72.38
8 L.H., Daily Totals		$915.20		$1169.04	$114.40	$146.13

Crew A-3Q	Hr.	Daily	Hr.	Daily	Bare Costs	Incl. O&P
1 Equip. Oper. (light)	$48.60	$388.80	$73.75	$590.00	$48.60	$73.75
1 Pickup Truck, 4 x 4, 3/4 Ton		155.60		171.16		
1 Flatbed trailer, 3 Ton		23.60		25.96	22.40	24.64
8 L.H., Daily Totals		$568.00		$787.12	$71.00	$98.39

Crew No.	Bare Costs		Incl. Subs O&P		Cost Per Labor-Hour	
Crew A-4	**Hr.**	**Daily**	**Hr.**	**Daily**	**Bare Costs**	**Incl. O&P**
2 Carpenters	$46.95	$751.20	$72.25	$1156.00	$44.75	$68.43
1 Painter, Ordinary	40.35	322.80	60.80	486.40		
24 L.H., Daily Totals		$1074.00		$1642.40	$44.75	$68.43

Crew A-5	Hr.	Daily	Hr.	Daily	Bare Costs	Incl. O&P
2 Laborers	$37.60	$601.60	$57.85	$925.60	$37.77	$57.99
.25 Truck Driver (light)	39.10	78.20	59.10	118.20		
.25 Flatbed Truck, Gas, 1.5 Ton		61.35		67.48	3.41	3.75
18 L.H., Daily Totals		$741.15		$1111.29	$41.17	$61.74

Crew A-6	Hr.	Daily	Hr.	Daily	Bare Costs	Incl. O&P
1 Instrument Man	$48.65	$389.20	$75.15	$601.20	$46.88	$71.92
1 Rodman/Chainman	45.10	360.80	68.70	549.60		
1 Level, Electronic		54.80		60.28	3.42	3.77
16 L.H., Daily Totals		$804.80		$1211.08	$50.30	$75.69

Crew A-7	Hr.	Daily	Hr.	Daily	Bare Costs	Incl. O&P
1 Chief of Party	$60.90	$487.20	$93.20	$745.60	$51.55	$79.02
1 Instrument Man	48.65	389.20	75.15	601.20		
1 Rodman/Chainman	45.10	360.80	68.70	549.60		
1 Level, Electronic		54.80		60.28	2.28	2.51
24 L.H., Daily Totals		$1292.00		$1956.68	$53.83	$81.53

Crew A-8	Hr.	Daily	Hr.	Daily	Bare Costs	Incl. O&P
1 Chief of Party	$60.90	$487.20	$93.20	$745.60	$49.94	$76.44
1 Instrument Man	48.65	389.20	75.15	601.20		
2 Rodmen/Chainmen	45.10	721.60	68.70	1099.20		
1 Level, Electronic		54.80		60.28	1.71	1.88
32 L.H., Daily Totals		$1652.80		$2506.28	$51.65	$78.32

Crew A-9	Hr.	Daily	Hr.	Daily	Bare Costs	Incl. O&P
1 Asbestos Foreman	$52.85	$422.80	$82.30	$658.40	$52.41	$81.60
7 Asbestos Workers	52.35	2931.60	81.50	4564.00		
64 L.H., Daily Totals		$3354.40		$5222.40	$52.41	$81.60

Crew A-10A	Hr.	Daily	Hr.	Daily	Bare Costs	Incl. O&P
1 Asbestos Foreman	$52.85	$422.80	$82.30	$658.40	$52.52	$81.77
2 Asbestos Workers	52.35	837.60	81.50	1304.00		
24 L.H., Daily Totals		$1260.40		$1962.40	$52.52	$81.77

Crew A-10B	Hr.	Daily	Hr.	Daily	Bare Costs	Incl. O&P
1 Asbestos Foreman	$52.85	$422.80	$82.30	$658.40	$52.48	$81.70
3 Asbestos Workers	52.35	1256.40	81.50	1956.00		
32 L.H., Daily Totals		$1679.20		$2614.40	$52.48	$81.70

Crew A-10C	Hr.	Daily	Hr.	Daily	Bare Costs	Incl. O&P
3 Asbestos Workers	$52.35	$1256.40	$81.50	$1956.00	$52.35	$81.50
1 Flatbed Truck, Gas, 1.5 Ton		245.40		269.94	10.23	11.25
24 L.H., Daily Totals		$1501.80		$2225.94	$62.58	$92.75

Crew A-10D	Hr.	Daily	Hr.	Daily	Bare Costs	Incl. O&P
2 Asbestos Workers	$52.35	$837.60	$81.50	$1304.00	$50.40	$77.50
1 Equip. Oper. (crane)	51.70	413.60	78.45	627.60		
1 Equip. Oper. (oiler)	45.20	361.60	68.55	548.40		
1 Hydraulic Crane, 33 Ton		750.00		825.00	23.44	25.78
32 L.H., Daily Totals		$2362.80		$3305.00	$73.84	$103.28

Crew No.	Bare Costs		Incl. Subs O&P		Cost Per Labor-Hour	
Crew A-11	**Hr.**	**Daily**	**Hr.**	**Daily**	**Bare Costs**	**Incl. O&P**
1 Asbestos Foreman	$52.85	$422.80	$82.30	$658.40	$52.41	$81.60
7 Asbestos Workers	52.35	2931.60	81.50	4564.00		
2 Chip. Hammers, 12 Lb., Elec.		40.00		44.00	.63	.69
64 L.H., Daily Totals		$3394.40		$5266.40	$53.04	$82.29
Crew A-12	**Hr.**	**Daily**	**Hr.**	**Daily**	**Bare Costs**	**Incl. O&P**
1 Asbestos Foreman	$52.85	$422.80	$82.30	$658.40	$52.41	$81.60
7 Asbestos Workers	52.35	2931.60	81.50	4564.00		
1 Trk-Mtd Vac, 14 CY, 1500 Gal.		514.25		565.67		
1 Flatbed Truck, 20,000 GVW		248.60		273.46	11.92	13.11
64 L.H., Daily Totals		$4117.25		$6061.53	$64.33	$94.71
Crew A-13	**Hr.**	**Daily**	**Hr.**	**Daily**	**Bare Costs**	**Incl. O&P**
1 Equip. Oper. (light)	$48.60	$388.80	$73.75	$590.00	$48.60	$73.75
1 Trk-Mtd Vac, 14 CY, 1500 Gal.		514.25		565.67		
1 Flatbed Truck, 20,000 GVW		248.60		273.46	95.36	104.89
8 L.H., Daily Totals		$1151.65		$1429.14	$143.96	$178.64
Crew B-1	**Hr.**	**Daily**	**Hr.**	**Daily**	**Bare Costs**	**Incl. O&P**
1 Labor Foreman (outside)	$39.60	$316.80	$60.95	$487.60	$38.27	$58.88
2 Laborers	37.60	601.60	57.85	925.60		
24 L.H., Daily Totals		$918.40		$1413.20	$38.27	$58.88
Crew B-1A	**Hr.**	**Daily**	**Hr.**	**Daily**	**Bare Costs**	**Incl. O&P**
1 Labor Foreman (outside)	$39.60	$316.80	$60.95	$487.60	$38.27	$58.88
2 Laborers	37.60	601.60	57.85	925.60		
2 Cutting Torches		22.80		25.08		
2 Sets of Gases		304.00		334.40	13.62	14.98
24 L.H., Daily Totals		$1245.20		$1772.68	$51.88	$73.86
Crew B-1B	**Hr.**	**Daily**	**Hr.**	**Daily**	**Bare Costs**	**Incl. O&P**
1 Labor Foreman (outside)	$39.60	$316.80	$60.95	$487.60	$41.63	$63.77
2 Laborers	37.60	601.60	57.85	925.60		
1 Equip. Oper. (crane)	51.70	413.60	78.45	627.60		
2 Cutting Torches		22.80		25.08		
2 Sets of Gases		304.00		334.40		
1 Hyd. Crane, 12 Ton		653.80		719.18	30.64	33.71
32 L.H., Daily Totals		$2312.60		$3119.46	$72.27	$97.48
Crew B-1C	**Hr.**	**Daily**	**Hr.**	**Daily**	**Bare Costs**	**Incl. O&P**
1 Labor Foreman (outside)	$39.60	$316.80	$60.95	$487.60	$38.27	$58.88
2 Laborers	37.60	601.60	57.85	925.60		
1 Aerial Lift Truck, 60' Boom		435.60		479.16	18.15	19.97
24 L.H., Daily Totals		$1354.00		$1892.36	$56.42	$78.85
Crew B-1D	**Hr.**	**Daily**	**Hr.**	**Daily**	**Bare Costs**	**Incl. O&P**
2 Laborers	$37.60	$601.60	$57.85	$925.60	$37.60	$57.85
1 Small Work Boat, Gas, 50 H.P.		173.20		190.52		
1 Pressure Washer, 7 GPM		87.60		96.36	16.30	17.93
16 L.H., Daily Totals		$862.40		$1212.48	$53.90	$75.78
Crew B-1E	**Hr.**	**Daily**	**Hr.**	**Daily**	**Bare Costs**	**Incl. O&P**
1 Labor Foreman (outside)	$39.60	$316.80	$60.95	$487.60	$38.10	$58.63
3 Laborers	37.60	902.40	57.85	1388.40		
1 Work Boat, Diesel, 200 H.P.		1463.00		1609.30		
2 Pressure Washer, 7 GPM		175.20		192.72	51.19	56.31
32 L.H., Daily Totals		$2857.40		$3678.02	$89.29	$114.94

Crew No.	Bare Costs		Incl. Subs O&P		Cost Per Labor-Hour	
Crew B-1F	**Hr.**	**Daily**	**Hr.**	**Daily**	**Bare Costs**	**Incl. O&P**
2 Skilled Workers	$48.65	$778.40	$75.15	$1202.40	$44.97	$69.38
1 Laborer	37.60	300.80	57.85	462.80		
1 Small Work Boat, Gas, 50 H.P.		173.20		190.52		
1 Pressure Washer, 7 GPM		87.60		96.36	10.87	11.95
24 L.H., Daily Totals		$1340.00		$1952.08	$55.83	$81.34
Crew B-1G	**Hr.**	**Daily**	**Hr.**	**Daily**	**Bare Costs**	**Incl. O&P**
2 Laborers	$37.60	$601.60	$57.85	$925.60	$37.60	$57.85
1 Small Work Boat, Gas, 50 H.P.		173.20		190.52	10.82	11.91
16 L.H., Daily Totals		$774.80		$1116.12	$48.42	$69.76
Crew B-1H	**Hr.**	**Daily**	**Hr.**	**Daily**	**Bare Costs**	**Incl. O&P**
2 Skilled Workers	$48.65	$778.40	$75.15	$1202.40	$44.97	$69.38
1 Laborer	37.60	300.80	57.85	462.80		
1 Small Work Boat, Gas, 50 H.P.		173.20		190.52	7.22	7.94
24 L.H., Daily Totals		$1252.40		$1855.72	$52.18	$77.32
Crew B-1J	**Hr.**	**Daily**	**Hr.**	**Daily**	**Bare Costs**	**Incl. O&P**
1 Labor Foreman (inside)	$38.10	$304.80	$58.65	$469.20	$37.85	$58.25
1 Laborer	37.60	300.80	57.85	462.80		
16 L.H., Daily Totals		$605.60		$932.00	$37.85	$58.25
Crew B-1K	**Hr.**	**Daily**	**Hr.**	**Daily**	**Bare Costs**	**Incl. O&P**
1 Carpenter Foreman (inside)	$47.45	$379.60	$73.05	$584.40	$47.20	$72.65
1 Carpenter	46.95	375.60	72.25	578.00		
16 L.H., Daily Totals		$755.20		$1162.40	$47.20	$72.65
Crew B-2	**Hr.**	**Daily**	**Hr.**	**Daily**	**Bare Costs**	**Incl. O&P**
1 Labor Foreman (outside)	$39.60	$316.80	$60.95	$487.60	$38.00	$58.47
4 Laborers	37.60	1203.20	57.85	1851.20		
40 L.H., Daily Totals		$1520.00		$2338.80	$38.00	$58.47
Crew B-2A	**Hr.**	**Daily**	**Hr.**	**Daily**	**Bare Costs**	**Incl. O&P**
1 Labor Foreman (outside)	$39.60	$316.80	$60.95	$487.60	$38.27	$58.88
2 Laborers	37.60	601.60	57.85	925.60		
1 Aerial Lift Truck, 60' Boom		435.60		479.16	18.15	19.97
24 L.H., Daily Totals		$1354.00		$1892.36	$56.42	$78.85
Crew B-3	**Hr.**	**Daily**	**Hr.**	**Daily**	**Bare Costs**	**Incl. O&P**
1 Labor Foreman (outside)	$39.60	$316.80	$60.95	$487.60	$40.92	$62.42
2 Laborers	37.60	601.60	57.85	925.60		
1 Equip. Oper. (medium)	50.60	404.80	76.75	614.00		
2 Truck Drivers (heavy)	40.05	640.80	60.55	968.80		
1 Crawler Loader, 3 C.Y.		1188.00		1306.80		
2 Dump Trucks, 12 C.Y., 400 H.P.		1382.00		1520.20	53.54	58.90
48 L.H., Daily Totals		$4534.00		$5823.00	$94.46	$121.31
Crew B-3A	**Hr.**	**Daily**	**Hr.**	**Daily**	**Bare Costs**	**Incl. O&P**
4 Laborers	$37.60	$1203.20	$57.85	$1851.20	$40.20	$61.63
1 Equip. Oper. (medium)	50.60	404.80	76.75	614.00		
1 Hyd. Excavator, 1.5 C.Y.		1030.00		1133.00	25.75	28.32
40 L.H., Daily Totals		$2638.00		$3598.20	$65.95	$89.95
Crew B-3B	**Hr.**	**Daily**	**Hr.**	**Daily**	**Bare Costs**	**Incl. O&P**
2 Laborers	$37.60	$601.60	$57.85	$925.60	$41.46	$63.25
1 Equip. Oper. (medium)	50.60	404.80	76.75	614.00		
1 Truck Driver (heavy)	40.05	320.40	60.55	484.40		
1 Backhoe Loader, 80 H.P.		391.80		430.98		
1 Dump Truck, 12 C.Y., 400 H.P.		691.00		760.10	33.84	37.22
32 L.H., Daily Totals		$2409.60		$3215.08	$75.30	$100.47

Crew No.	Bare Costs		Incl. Subs O&P		Cost Per Labor-Hour	
Crew B-3C	**Hr.**	**Daily**	**Hr.**	**Daily**	**Bare Costs**	**Incl. O&P**
3 Laborers	$37.60	$902.40	$57.85	$1388.40	$40.85	$62.58
1 Equip. Oper. (medium)	50.60	404.80	76.75	614.00		
1 Crawler Loader, 4 C.Y.		1561.00		1717.10	48.78	53.66
32 L.H., Daily Totals		$2868.20		$3719.50	$89.63	$116.23

Crew No.	Bare Costs		Incl. Subs O&P		Cost Per Labor-Hour	
Crew B-4	**Hr.**	**Daily**	**Hr.**	**Daily**	**Bare Costs**	**Incl. O&P**
1 Labor Foreman (outside)	$39.60	$316.80	$60.95	$487.60	$38.34	$58.82
4 Laborers	37.60	1203.20	57.85	1851.20		
1 Truck Driver (heavy)	40.05	320.40	60.55	484.40		
1 Truck Tractor, 220 H.P.		359.60		395.56		
1 Flatbed Trailer, 40 Ton		154.00		169.40	10.70	11.77
48 L.H., Daily Totals		$2354.00		$3388.16	$49.04	$70.59

Crew No.	Bare Costs		Incl. Subs O&P		Cost Per Labor-Hour	
Crew B-5	**Hr.**	**Daily**	**Hr.**	**Daily**	**Bare Costs**	**Incl. O&P**
1 Labor Foreman (outside)	$39.60	$316.80	$60.95	$487.60	$41.60	$63.69
4 Laborers	37.60	1203.20	57.85	1851.20		
2 Equip. Oper. (medium)	50.60	809.60	76.75	1228.00		
1 Air Compressor, 250 cfm		201.40		221.54		
2 Breakers, Pavement, 60 lb.		19.60		21.56		
2 -50' Air Hoses, 1.5"		11.60		12.76		
1 Crawler Loader, 3 C.Y.		1188.00		1306.80	25.37	27.90
56 L.H., Daily Totals		$3750.20		$5129.46	$66.97	$91.60

Crew No.	Bare Costs		Incl. Subs O&P		Cost Per Labor-Hour	
Crew B-5A	**Hr.**	**Daily**	**Hr.**	**Daily**	**Bare Costs**	**Incl. O&P**
1 Labor Foreman (outside)	$39.60	$316.80	$60.95	$487.60	$41.26	$63.03
6 Laborers	37.60	1804.80	57.85	2776.80		
2 Equip. Oper. (medium)	50.60	809.60	76.75	1228.00		
1 Equip. Oper. (light)	48.60	388.80	73.75	590.00		
2 Truck Drivers (heavy)	40.05	640.80	60.55	968.80		
1 Air Compressor, 365 cfm		263.80		290.18		
2 Breakers, Pavement, 60 lb.		19.60		21.56		
8 -50' Air Hoses, 1"		32.80		36.08		
2 Dump Trucks, 8 C.Y., 220 H.P.		822.40		904.64	11.86	13.05
96 L.H., Daily Totals		$5099.40		$7303.66	$53.12	$76.08

Crew No.	Bare Costs		Incl. Subs O&P		Cost Per Labor-Hour	
Crew B-5B	**Hr.**	**Daily**	**Hr.**	**Daily**	**Bare Costs**	**Incl. O&P**
1 Powderman	$48.65	$389.20	$75.15	$601.20	$45.00	$68.38
2 Equip. Oper. (medium)	50.60	809.60	76.75	1228.00		
3 Truck Drivers (heavy)	40.05	961.20	60.55	1453.20		
1 F.E. Loader, W.M.,2.5 C.Y.		519.80		571.78		
3 Dump Trucks, 12 C.Y., 400 H.P.		2073.00		2280.30		
1 Air Compressor, 365 CFM		263.80		290.18	59.51	65.46
48 L.H., Daily Totals		$5016.60		$6424.66	$104.51	$133.85

Crew No.	Bare Costs		Incl. Subs O&P		Cost Per Labor-Hour	
Crew B-5C	**Hr.**	**Daily**	**Hr.**	**Daily**	**Bare Costs**	**Incl. O&P**
3 Laborers	$37.60	$902.40	$57.85	$1388.40	$42.55	$64.80
1 Equip. Oper. (medium)	50.60	404.80	76.75	614.00		
2 Truck Drivers (heavy)	40.05	640.80	60.55	968.80		
1 Equip. Oper. (crane)	51.70	413.60	78.45	627.60		
1 Equip. Oper. (oiler)	45.20	361.60	68.55	548.40		
2 Dump Trucks, 12 C.Y., 400 H.P.		1382.00		1520.20		
1 Crawler Loader, 4 C.Y.		1561.00		1717.10		
1 S.P. Crane, 4x4, 25 Ton		599.40		659.34	55.35	60.88
64 L.H., Daily Totals		$6265.60		$8043.84	$97.90	$125.69

Crew No.	Bare Costs		Incl. Subs O&P		Cost Per Labor-Hour	
Crew B-5D	**Hr.**	**Daily**	**Hr.**	**Daily**	**Bare Costs**	**Incl. O&P**
1 Labor Foreman (outside)	$39.60	$316.80	$60.95	$487.60	$41.41	$63.30
4 Laborers	37.60	1203.20	57.85	1851.20		
2 Equip. Oper. (medium)	50.60	809.60	76.75	1228.00		
1 Truck Driver (heavy)	40.05	320.40	60.55	484.40		
1 Air Compressor, 250 cfm		201.40		221.54		
2 Breakers, Pavement, 60 lb.		19.60		21.56		
2 -50' Air Hoses, 1.5"		11.60		12.76		
1 Crawler Loader, 3 C.Y.		1188.00		1306.80		
1 Dump Truck, 12 C.Y., 400 H.P.		691.00		760.10	32.99	36.29
64 L.H., Daily Totals		$4761.60		$6373.96	$74.40	$99.59

Crew No.	Bare Costs		Incl. Subs O&P		Cost Per Labor-Hour	
Crew B-6	**Hr.**	**Daily**	**Hr.**	**Daily**	**Bare Costs**	**Incl. O&P**
2 Laborers	$37.60	$601.60	$57.85	$925.60	$41.27	$63.15
1 Equip. Oper. (light)	48.60	388.80	73.75	590.00		
1 Backhoe Loader, 48 H.P.		364.20		400.62	15.18	16.69
24 L.H., Daily Totals		$1354.60		$1916.22	$56.44	$79.84

Crew No.	Bare Costs		Incl. Subs O&P		Cost Per Labor-Hour	
Crew B-6A	**Hr.**	**Daily**	**Hr.**	**Daily**	**Bare Costs**	**Incl. O&P**
.5 Labor Foreman (outside)	$39.60	$158.40	$60.95	$243.80	$43.20	$66.03
1 Laborer	37.60	300.80	57.85	462.80		
1 Equip. Oper. (medium)	50.60	404.80	76.75	614.00		
1 Vacuum Truck, 5000 Gal.		358.10		393.91	17.91	19.70
20 L.H., Daily Totals		$1222.10		$1714.51	$61.10	$85.73

Crew No.	Bare Costs		Incl. Subs O&P		Cost Per Labor-Hour	
Crew B-6B	**Hr.**	**Daily**	**Hr.**	**Daily**	**Bare Costs**	**Incl. O&P**
2 Labor Foremen (outside)	$39.60	$633.60	$60.95	$975.20	$38.27	$58.88
4 Laborers	37.60	1203.20	57.85	1851.20		
1 S.P. Crane, 4x4, 5 Ton		275.40		302.94		
1 Flatbed Truck, Gas, 1.5 Ton		245.40		269.94		
1 Butt Fusion Mach., 4"-12" diam.		382.10		420.31	18.81	20.69
48 L.H., Daily Totals		$2739.70		$3819.59	$57.08	$79.57

Crew No.	Bare Costs		Incl. Subs O&P		Cost Per Labor-Hour	
Crew B-6C	**Hr.**	**Daily**	**Hr.**	**Daily**	**Bare Costs**	**Incl. O&P**
2 Labor Foremen (outside)	$39.60	$633.60	$60.95	$975.20	$38.27	$58.88
4 Laborers	37.60	1203.20	57.85	1851.20		
1 S.P. Crane, 4x4, 12 Ton		473.40		520.74		
1 Flatbed Truck, Gas, 3 Ton		303.20		333.52		
1 Butt Fusion Mach., 8"-24" diam.		829.70		912.67	33.46	36.81
48 L.H., Daily Totals		$3443.10		$4593.33	$71.73	$95.69

Crew No.	Bare Costs		Incl. Subs O&P		Cost Per Labor-Hour	
Crew B-7	**Hr.**	**Daily**	**Hr.**	**Daily**	**Bare Costs**	**Incl. O&P**
1 Labor Foreman (outside)	$39.60	$316.80	$60.95	$487.60	$40.10	$61.52
4 Laborers	37.60	1203.20	57.85	1851.20		
1 Equip. Oper. (medium)	50.60	404.80	76.75	614.00		
1 Brush Chipper, 12", 130 H.P.		391.40		430.54		
1 Crawler Loader, 3 C.Y.		1188.00		1306.80		
2 Chain Saws, Gas, 36" Long		90.00		99.00	34.78	38.26
48 L.H., Daily Totals		$3594.20		$4789.14	$74.88	$99.77

Crew No.	Bare Costs		Incl. Subs O&P		Cost Per Labor-Hour	
Crew B-7A	**Hr.**	**Daily**	**Hr.**	**Daily**	**Bare Costs**	**Incl. O&P**
2 Laborers	$37.60	$601.60	$57.85	$925.60	$41.27	$63.15
1 Equip. Oper. (light)	48.60	388.80	73.75	590.00		
1 Rake w/Tractor		354.25		389.68		
2 Chain Saw, Gas, 18"		62.40		68.64	17.36	19.10
24 L.H., Daily Totals		$1407.05		$1973.92	$58.63	$82.25

Crew No.	Bare Costs		Incl. Subs O&P		Cost Per Labor-Hour	
Crew B-7B	**Hr.**	**Daily**	**Hr.**	**Daily**	**Bare Costs**	**Incl. O&P**
1 Labor Foreman (outside)	$39.60	$316.80	$60.95	$487.60	$40.09	$61.38
4 Laborers	37.60	1203.20	57.85	1851.20		
1 Equip. Oper. (medium)	50.60	404.80	76.75	614.00		
1 Truck Driver (heavy)	40.05	320.40	60.55	484.40		
1 Brush Chipper, 12", 130 H.P.		391.40		430.54		
1 Crawler Loader, 3 C.Y.		1188.00		1306.80		
2 Chain Saws, Gas, 36" Long		90.00		99.00		
1 Dump Truck, 8 C.Y., 220 H.P.		411.20		452.32	37.15	40.87
56 L.H., Daily Totals		$4325.80		$5725.86	$77.25	$102.25
Crew B-7C	**Hr.**	**Daily**	**Hr.**	**Daily**	**Bare Costs**	**Incl. O&P**
1 Labor Foreman (outside)	$39.60	$316.80	$60.95	$487.60	$40.09	$61.38
4 Laborers	37.60	1203.20	57.85	1851.20		
1 Equip. Oper. (medium)	50.60	404.80	76.75	614.00		
1 Truck Driver (heavy)	40.05	320.40	60.55	484.40		
1 Brush Chipper, 12", 130 H.P.		391.40		430.54		
1 Crawler Loader, 3 C.Y.		1188.00		1306.80		
2 Chain Saws, Gas, 36" Long		90.00		99.00		
1 Dump Truck, 12 C.Y., 400 H.P.		691.00		760.10	42.15	46.37
56 L.H., Daily Totals		$4605.60		$6033.64	$82.24	$107.74
Crew B-8	**Hr.**	**Daily**	**Hr.**	**Daily**	**Bare Costs**	**Incl. O&P**
1 Labor Foreman (outside)	$39.60	$316.80	$60.95	$487.60	$42.66	$64.97
2 Laborers	37.60	601.60	57.85	925.60		
2 Equip. Oper. (medium)	50.60	809.60	76.75	1228.00		
1 Equip. Oper. (oiler)	45.20	361.60	68.55	548.40		
2 Truck Drivers (heavy)	40.05	640.80	60.55	968.80		
1 Hyd. Crane, 25 Ton		736.60		810.26		
1 Crawler Loader, 3 C.Y.		1188.00		1306.80		
2 Dump Trucks, 12 C.Y., 400 H.P.		1382.00		1520.20	51.67	56.83
64 L.H., Daily Totals		$6037.00		$7795.66	$94.33	$121.81
Crew B-9	**Hr.**	**Daily**	**Hr.**	**Daily**	**Bare Costs**	**Incl. O&P**
1 Labor Foreman (outside)	$39.60	$316.80	$60.95	$487.60	$38.00	$58.47
4 Laborers	37.60	1203.20	57.85	1851.20		
1 Air Compressor, 250 cfm		201.40		221.54		
2 Breakers, Pavement, 60 lb.		19.60		21.56		
2 -50' Air Hoses, 1.5"		11.60		12.76	5.82	6.40
40 L.H., Daily Totals		$1752.60		$2594.66	$43.81	$64.87
Crew B-9A	**Hr.**	**Daily**	**Hr.**	**Daily**	**Bare Costs**	**Incl. O&P**
2 Laborers	$37.60	$601.60	$57.85	$925.60	$38.42	$58.75
1 Truck Driver (heavy)	40.05	320.40	60.55	484.40		
1 Water Tank Trailer, 5000 Gal.		141.00		155.10		
1 Truck Tractor, 220 H.P.		359.60		395.56		
2 -50' Discharge Hoses, 3"		3.00		3.30	20.98	23.08
24 L.H., Daily Totals		$1425.60		$1963.96	$59.40	$81.83
Crew B-9B	**Hr.**	**Daily**	**Hr.**	**Daily**	**Bare Costs**	**Incl. O&P**
2 Laborers	$37.60	$601.60	$57.85	$925.60	$38.42	$58.75
1 Truck Driver (heavy)	40.05	320.40	60.55	484.40		
2 -50' Discharge Hoses, 3"		3.00		3.30		
1 Water Tank Trailer, 5000 Gal.		141.00		155.10		
1 Truck Tractor, 220 H.P.		359.60		395.56		
1 Pressure Washer		70.20		77.22	23.91	26.30
24 L.H., Daily Totals		$1495.80		$2041.18	$62.33	$85.05

Crew No.	Bare Costs		Incl. Subs O&P		Cost Per Labor-Hour	
Crew B-9D	**Hr.**	**Daily**	**Hr.**	**Daily**	**Bare Costs**	**Incl. O&P**
1 Labor Foreman (outside)	$39.60	$316.80	$60.95	$487.60	$38.00	$58.47
4 Common Laborers	37.60	1203.20	57.85	1851.20		
1 Air Compressor, 250 cfm		201.40		221.54		
2 -50' Air Hoses, 1.5"		11.60		12.76		
2 Air Powered Tampers		52.40		57.64	6.63	7.30
40 L.H., Daily Totals		$1785.40		$2630.74	$44.63	$65.77
Crew B-10	**Hr.**	**Daily**	**Hr.**	**Daily**	**Bare Costs**	**Incl. O&P**
1 Equip. Oper. (medium)	$50.60	$404.80	$76.75	$614.00	$46.27	$70.45
.5 Laborer	37.60	150.40	57.85	231.40		
12 L.H., Daily Totals		$555.20		$845.40	$46.27	$70.45
Crew B-10A	**Hr.**	**Daily**	**Hr.**	**Daily**	**Bare Costs**	**Incl. O&P**
1 Equip. Oper. (medium)	$50.60	$404.80	$76.75	$614.00	$46.27	$70.45
.5 Laborer	37.60	150.40	57.85	231.40		
1 Roller, 2-Drum, W.B., 7.5 H.P.		180.00		198.00	15.00	16.50
12 L.H., Daily Totals		$735.20		$1043.40	$61.27	$86.95
Crew B-10B	**Hr.**	**Daily**	**Hr.**	**Daily**	**Bare Costs**	**Incl. O&P**
1 Equip. Oper. (medium)	$50.60	$404.80	$76.75	$614.00	$46.27	$70.45
.5 Laborer	37.60	150.40	57.85	231.40		
1 Dozer, 200 H.P.		1387.00		1525.70	115.58	127.14
12 L.H., Daily Totals		$1942.20		$2371.10	$161.85	$197.59
Crew B-10C	**Hr.**	**Daily**	**Hr.**	**Daily**	**Bare Costs**	**Incl. O&P**
1 Equip. Oper. (medium)	$50.60	$404.80	$76.75	$614.00	$46.27	$70.45
.5 Laborer	37.60	150.40	57.85	231.40		
1 Dozer, 200 H.P.		1387.00		1525.70		
1 Vibratory Roller, Towed, 23 Ton		396.00		435.60	148.58	163.44
12 L.H., Daily Totals		$2338.20		$2806.70	$194.85	$233.89
Crew B-10D	**Hr.**	**Daily**	**Hr.**	**Daily**	**Bare Costs**	**Incl. O&P**
1 Equip. Oper. (medium)	$50.60	$404.80	$76.75	$614.00	$46.27	$70.45
.5 Laborer	37.60	150.40	57.85	231.40		
1 Dozer, 200 H.P.		1387.00		1525.70		
1 Sheepsft. Roller, Towed		425.80		468.38	151.07	166.17
12 L.H., Daily Totals		$2368.00		$2839.48	$197.33	$236.62
Crew B-10E	**Hr.**	**Daily**	**Hr.**	**Daily**	**Bare Costs**	**Incl. O&P**
1 Equip. Oper. (medium)	$50.60	$404.80	$76.75	$614.00	$46.27	$70.45
.5 Laborer	37.60	150.40	57.85	231.40		
1 Tandem Roller, 5 Ton		157.40		173.14	13.12	14.43
12 L.H., Daily Totals		$712.60		$1018.54	$59.38	$84.88
Crew B-10F	**Hr.**	**Daily**	**Hr.**	**Daily**	**Bare Costs**	**Incl. O&P**
1 Equip. Oper. (medium)	$50.60	$404.80	$76.75	$614.00	$46.27	$70.45
.5 Laborer	37.60	150.40	57.85	231.40		
1 Tandem Roller, 10 Ton		236.80		260.48	19.73	21.71
12 L.H., Daily Totals		$792.00		$1105.88	$66.00	$92.16
Crew B-10G	**Hr.**	**Daily**	**Hr.**	**Daily**	**Bare Costs**	**Incl. O&P**
1 Equip. Oper. (medium)	$50.60	$404.80	$76.75	$614.00	$46.27	$70.45
.5 Laborer	37.60	150.40	57.85	231.40		
1 Sheepsfoot Roller, 240 H.P.		1206.00		1326.60	100.50	110.55
12 L.H., Daily Totals		$1761.20		$2172.00	$146.77	$181.00

Crew No.	Bare Costs		Incl. Subs O&P		Cost Per Labor-Hour	
Crew B-10H	**Hr.**	**Daily**	**Hr.**	**Daily**	**Bare Costs**	**Incl. O&P**
1 Equip. Oper. (medium)	$50.60	$404.80	$76.75	$614.00	$46.27	$70.45
.5 Laborer	37.60	150.40	57.85	231.40		
1 Diaphragm Water Pump, 2"		71.00		78.10		
1 -20' Suction Hose, 2"		1.95		2.15		
2 -50' Discharge Hoses, 2"		1.80		1.98	6.23	6.85
12 L.H., Daily Totals		$629.95		$927.63	$52.50	$77.30

Crew B-10I	Hr.	Daily	Hr.	Daily	Bare Costs	Incl. O&P
1 Equip. Oper. (medium)	$50.60	$404.80	$76.75	$614.00	$46.27	$70.45
.5 Laborer	37.60	150.40	57.85	231.40		
1 Diaphragm Water Pump, 4"		114.20		125.62		
1 -20' Suction Hose, 4"		3.25		3.58		
2 -50' Discharge Hoses, 4"		4.70		5.17	10.18	11.20
12 L.H., Daily Totals		$677.35		$979.76	$56.45	$81.65

Crew B-10J	Hr.	Daily	Hr.	Daily	Bare Costs	Incl. O&P
1 Equip. Oper. (medium)	$50.60	$404.80	$76.75	$614.00	$46.27	$70.45
.5 Laborer	37.60	150.40	57.85	231.40		
1 Centrifugal Water Pump, 3"		78.20		86.02		
1 -20' Suction Hose, 3"		2.85		3.13		
2 -50' Discharge Hoses, 3"		3.00		3.30	7.00	7.70
12 L.H., Daily Totals		$639.25		$937.86	$53.27	$78.15

Crew B-10K	Hr.	Daily	Hr.	Daily	Bare Costs	Incl. O&P
1 Equip. Oper. (medium)	$50.60	$404.80	$76.75	$614.00	$46.27	$70.45
.5 Laborer	37.60	150.40	57.85	231.40		
1 Centr. Water Pump, 6"		346.60		381.26		
1 -20' Suction Hose, 6"		11.50		12.65		
2 -50' Discharge Hoses, 6"		12.20		13.42	30.86	33.94
12 L.H., Daily Totals		$925.50		$1252.73	$77.13	$104.39

Crew B-10L	Hr.	Daily	Hr.	Daily	Bare Costs	Incl. O&P
1 Equip. Oper. (medium)	$50.60	$404.80	$76.75	$614.00	$46.27	$70.45
.5 Laborer	37.60	150.40	57.85	231.40		
1 Dozer, 80 H.P.		472.40		519.64	39.37	43.30
12 L.H., Daily Totals		$1027.60		$1365.04	$85.63	$113.75

Crew B-10M	Hr.	Daily	Hr.	Daily	Bare Costs	Incl. O&P
1 Equip. Oper. (medium)	$50.60	$404.80	$76.75	$614.00	$46.27	$70.45
.5 Laborer	37.60	150.40	57.85	231.40		
1 Dozer, 300 H.P.		1897.00		2086.70	158.08	173.89
12 L.H., Daily Totals		$2452.20		$2932.10	$204.35	$244.34

Crew B-10N	Hr.	Daily	Hr.	Daily	Bare Costs	Incl. O&P
1 Equip. Oper. (medium)	$50.60	$404.80	$76.75	$614.00	$46.27	$70.45
.5 Laborer	37.60	150.40	57.85	231.40		
1 F.E. Loader, T.M., 1.5 C.Y		522.00		574.20	43.50	47.85
12 L.H., Daily Totals		$1077.20		$1419.60	$89.77	$118.30

Crew B-10O	Hr.	Daily	Hr.	Daily	Bare Costs	Incl. O&P
1 Equip. Oper. (medium)	$50.60	$404.80	$76.75	$614.00	$46.27	$70.45
.5 Laborer	37.60	150.40	57.85	231.40		
1 F.E. Loader, T.M., 2.25 C.Y.		956.40		1052.04	79.70	87.67
12 L.H., Daily Totals		$1511.60		$1897.44	$125.97	$158.12

Crew B-10P	Hr.	Daily	Hr.	Daily	Bare Costs	Incl. O&P
1 Equip. Oper. (medium)	$50.60	$404.80	$76.75	$614.00	$46.27	$70.45
.5 Laborer	37.60	150.40	57.85	231.40		
1 Crawler Loader, 3 C.Y.		1188.00		1306.80	99.00	108.90
12 L.H., Daily Totals		$1743.20		$2152.20	$145.27	$179.35

Crew B-10Q	Hr.	Daily	Hr.	Daily	Bare Costs	Incl. O&P
1 Equip. Oper. (medium)	$50.60	$404.80	$76.75	$614.00	$46.27	$70.45
.5 Laborer	37.60	150.40	57.85	231.40		
1 Crawler Loader, 4 C.Y.		1561.00		1717.10	130.08	143.09
12 L.H., Daily Totals		$2116.20		$2562.50	$176.35	$213.54

Crew B-10R	Hr.	Daily	Hr.	Daily	Bare Costs	Incl. O&P
1 Equip. Oper. (medium)	$50.60	$404.80	$76.75	$614.00	$46.27	$70.45
.5 Laborer	37.60	150.40	57.85	231.40		
1 F.E. Loader, W.M., 1 C.Y.		299.00		328.90	24.92	27.41
12 L.H., Daily Totals		$854.20		$1174.30	$71.18	$97.86

Crew B-10S	Hr.	Daily	Hr.	Daily	Bare Costs	Incl. O&P
1 Equip. Oper. (medium)	$50.60	$404.80	$76.75	$614.00	$46.27	$70.45
.5 Laborer	37.60	150.40	57.85	231.40		
1 F.E. Loader, W.M., 1.5 C.Y.		378.40		416.24	31.53	34.69
12 L.H., Daily Totals		$933.60		$1261.64	$77.80	$105.14

Crew B-10T	Hr.	Daily	Hr.	Daily	Bare Costs	Incl. O&P
1 Equip. Oper. (medium)	$50.60	$404.80	$76.75	$614.00	$46.27	$70.45
.5 Laborer	37.60	150.40	57.85	231.40		
1 F.E. Loader, W.M.,2.5 C.Y.		519.80		571.78	43.32	47.65
12 L.H., Daily Totals		$1075.00		$1417.18	$89.58	$118.10

Crew B-10U	Hr.	Daily	Hr.	Daily	Bare Costs	Incl. O&P
1 Equip. Oper. (medium)	$50.60	$404.80	$76.75	$614.00	$46.27	$70.45
.5 Laborer	37.60	150.40	57.85	231.40		
1 F.E. Loader, W.M., 5.5 C.Y.		1082.00		1190.20	90.17	99.18
12 L.H., Daily Totals		$1637.20		$2035.60	$136.43	$169.63

Crew B-10V	Hr.	Daily	Hr.	Daily	Bare Costs	Incl. O&P
1 Equip. Oper. (medium)	$50.60	$404.80	$76.75	$614.00	$46.27	$70.45
.5 Laborer	37.60	150.40	57.85	231.40		
1 Dozer, 700 H.P.		4979.00		5476.90	414.92	456.41
12 L.H., Daily Totals		$5534.20		$6322.30	$461.18	$526.86

Crew B-10W	Hr.	Daily	Hr.	Daily	Bare Costs	Incl. O&P
1 Equip. Oper. (medium)	$50.60	$404.80	$76.75	$614.00	$46.27	$70.45
.5 Laborer	37.60	150.40	57.85	231.40		
1 Dozer, 105 H.P.		602.80		663.08	50.23	55.26
12 L.H., Daily Totals		$1158.00		$1508.48	$96.50	$125.71

Crew B-10X	Hr.	Daily	Hr.	Daily	Bare Costs	Incl. O&P
1 Equip. Oper. (medium)	$50.60	$404.80	$76.75	$614.00	$46.27	$70.45
.5 Laborer	37.60	150.40	57.85	231.40		
1 Dozer, 410 H.P.		2405.00		2645.50	200.42	220.46
12 L.H., Daily Totals		$2960.20		$3490.90	$246.68	$290.91

Crew B-10Y	Hr.	Daily	Hr.	Daily	Bare Costs	Incl. O&P
1 Equip. Oper. (medium)	$50.60	$404.80	$76.75	$614.00	$46.27	$70.45
.5 Laborer	37.60	150.40	57.85	231.40		
1 Vibr. Roller, Towed, 12 Ton		549.80		604.78	45.82	50.40
12 L.H., Daily Totals		$1105.00		$1450.18	$92.08	$120.85

Crew B-11A	Hr.	Daily	Hr.	Daily	Bare Costs	Incl. O&P
1 Equipment Oper. (med.)	$50.60	$404.80	$76.75	$614.00	$44.10	$67.30
1 Laborer	37.60	300.80	57.85	462.80		
1 Dozer, 200 H.P.		1387.00		1525.70	86.69	95.36
16 L.H., Daily Totals		$2092.60		$2602.50	$130.79	$162.66

Crew No.	Bare Costs		Incl. Subs O&P		Cost Per Labor-Hour	

Crew B-11B	Hr.	Daily	Hr.	Daily	Bare Costs	Incl. O&P
1 Equipment Oper. (light)	$48.60	$388.80	$73.75	$590.00	$43.10	$65.80
1 Laborer	37.60	300.80	57.85	462.80		
1 Air Powered Tamper		26.20		28.82		
1 Air Compressor, 365 cfm		263.80		290.18		
2 -50' Air Hoses, 1.5"		11.60		12.76	18.85	20.73
16 L.H., Daily Totals		$991.20		$1384.56	$61.95	$86.53

Crew B-11C	Hr.	Daily	Hr.	Daily	Bare Costs	Incl. O&P
1 Equipment Oper. (med.)	$50.60	$404.80	$76.75	$614.00	$44.10	$67.30
1 Laborer	37.60	300.80	57.85	462.80		
1 Backhoe Loader, 48 H.P.		364.20		400.62	22.76	25.04
16 L.H., Daily Totals		$1069.80		$1477.42	$66.86	$92.34

Crew B-11J	Hr.	Daily	Hr.	Daily	Bare Costs	Incl. O&P
1 Equipment Oper. (med.)	$50.60	$404.80	$76.75	$614.00	$44.10	$67.30
1 Laborer	37.60	300.80	57.85	462.80		
1 Grader, 30,000 Lbs.		736.20		809.82		
1 Ripper, Beam & 1 Shank		84.20		92.62	51.27	56.40
16 L.H., Daily Totals		$1526.00		$1979.24	$95.38	$123.70

Crew B-11K	Hr.	Daily	Hr.	Daily	Bare Costs	Incl. O&P
1 Equipment Oper. (med.)	$50.60	$404.80	$76.75	$614.00	$44.10	$67.30
1 Laborer	37.60	300.80	57.85	462.80		
1 Trencher, Chain Type, 8' D		3402.00		3742.20	212.63	233.89
16 L.H., Daily Totals		$4107.60		$4819.00	$256.73	$301.19

Crew B-11L	Hr.	Daily	Hr.	Daily	Bare Costs	Incl. O&P
1 Equipment Oper. (med.)	$50.60	$404.80	$76.75	$614.00	$44.10	$67.30
1 Laborer	37.60	300.80	57.85	462.80		
1 Grader, 30,000 Lbs.		736.20		809.82	46.01	50.61
16 L.H., Daily Totals		$1441.80		$1886.62	$90.11	$117.91

Crew B-11M	Hr.	Daily	Hr.	Daily	Bare Costs	Incl. O&P
1 Equipment Oper. (med.)	$50.60	$404.80	$76.75	$614.00	$44.10	$67.30
1 Laborer	37.60	300.80	57.85	462.80		
1 Backhoe Loader, 80 H.P.		391.80		430.98	24.49	26.94
16 L.H., Daily Totals		$1097.40		$1507.78	$68.59	$94.24

Crew B-11N	Hr.	Daily	Hr.	Daily	Bare Costs	Incl. O&P
1 Labor Foreman (outside)	$39.60	$316.80	$60.95	$487.60	$42.34	$64.19
2 Equipment Operators (med.)	50.60	809.60	76.75	1228.00		
6 Truck Drivers (heavy)	40.05	1922.40	60.55	2906.40		
1 F.E. Loader, W.M., 5.5 C.Y.		1082.00		1190.20		
1 Dozer, 410 H.P.		2405.00		2645.50		
6 Dump Trucks, Off Hwy., 50 Ton		11022.00		12124.20	201.51	221.67
72 L.H., Daily Totals		$17557.80		$20581.90	$243.86	$285.86

Crew B-11Q	Hr.	Daily	Hr.	Daily	Bare Costs	Incl. O&P
1 Equipment Operator (med.)	$50.60	$404.80	$76.75	$614.00	$46.27	$70.45
.5 Laborer	37.60	150.40	57.85	231.40		
1 Dozer, 140 H.P.		886.00		974.60	73.83	81.22
12 L.H., Daily Totals		$1441.20		$1820.00	$120.10	$151.67

Crew B-11R	Hr.	Daily	Hr.	Daily	Bare Costs	Incl. O&P
1 Equipment Operator (med.)	$50.60	$404.80	$76.75	$614.00	$46.27	$70.45
.5 Laborer	37.60	150.40	57.85	231.40		
1 Dozer, 200 H.P.		1387.00		1525.70	115.58	127.14
12 L.H., Daily Totals		$1942.20		$2371.10	$161.85	$197.59

Crew B-11S	Hr.	Daily	Hr.	Daily	Bare Costs	Incl. O&P
1 Equipment Operator (med.)	$50.60	$404.80	$76.75	$614.00	$46.27	$70.45
.5 Laborer	37.60	150.40	57.85	231.40		
1 Dozer, 300 H.P.		1897.00		2086.70		
1 Ripper, Beam & 1 Shank		84.20		92.62	165.10	181.61
12 L.H., Daily Totals		$2536.40		$3024.72	$211.37	$252.06

Crew B-11T	Hr.	Daily	Hr.	Daily	Bare Costs	Incl. O&P
1 Equipment Operator (med.)	$50.60	$404.80	$76.75	$614.00	$46.27	$70.45
.5 Laborer	37.60	150.40	57.85	231.40		
1 Dozer, 410 H.P.		2405.00		2645.50		
1 Ripper, Beam & 2 Shanks		110.00		121.00	209.58	230.54
12 L.H., Daily Totals		$3070.20		$3611.90	$255.85	$300.99

Crew B-11U	Hr.	Daily	Hr.	Daily	Bare Costs	Incl. O&P
1 Equipment Operator (med.)	$50.60	$404.80	$76.75	$614.00	$46.27	$70.45
.5 Laborer	37.60	150.40	57.85	231.40		
1 Dozer, 520 H.P.		3299.00		3628.90	274.92	302.41
12 L.H., Daily Totals		$3854.20		$4474.30	$321.18	$372.86

Crew B-11V	Hr.	Daily	Hr.	Daily	Bare Costs	Incl. O&P
3 Laborers	$37.60	$902.40	$57.85	$1388.40	$37.60	$57.85
1 Roller, 2-Drum, W.B., 7.5 H.P.		180.00		198.00	7.50	8.25
24 L.H., Daily Totals		$1082.40		$1586.40	$45.10	$66.10

Crew B-11W	Hr.	Daily	Hr.	Daily	Bare Costs	Incl. O&P
1 Equipment Operator (med.)	$50.60	$404.80	$76.75	$614.00	$40.73	$61.67
1 Common Laborer	37.60	300.80	57.85	462.80		
10 Truck Drivers (heavy)	40.05	3204.00	60.55	4844.00		
1 Dozer, 200 H.P.		1387.00		1525.70		
1 Vibratory Roller, Towed, 23 Ton		396.00		435.60		
10 Dump Trucks, 8 C.Y., 220 H.P.		4112.00		4523.20	61.41	67.55
96 L.H., Daily Totals		$9804.60		$12405.30	$102.13	$129.22

Crew B-11Y	Hr.	Daily	Hr.	Daily	Bare Costs	Incl. O&P
1 Labor Foreman (outside)	$39.60	$316.80	$60.95	$487.60	$42.16	$64.49
5 Common Laborers	37.60	1504.00	57.85	2314.00		
3 Equipment Operators (med.)	50.60	1214.40	76.75	1842.00		
1 Dozer, 80 H.P.		472.40		519.64		
2 Roller, 2-Drum, W.B., 7.5 H.P.		360.00		396.00		
4 Vibrating Plate, Gas, 21"		187.20		205.92	14.16	15.58
72 L.H., Daily Totals		$4054.80		$5765.16	$56.32	$80.07

Crew B-12A	Hr.	Daily	Hr.	Daily	Bare Costs	Incl. O&P
1 Equip. Oper. (crane)	$51.70	$413.60	$78.45	$627.60	$44.65	$68.15
1 Laborer	37.60	300.80	57.85	462.80		
1 Hyd. Excavator, 1 C.Y.		812.60		893.86	50.79	55.87
16 L.H., Daily Totals		$1527.00		$1984.26	$95.44	$124.02

Crew B-12B	Hr.	Daily	Hr.	Daily	Bare Costs	Incl. O&P
1 Equip. Oper. (crane)	$51.70	$413.60	$78.45	$627.60	$44.65	$68.15
1 Laborer	37.60	300.80	57.85	462.80		
1 Hyd. Excavator, 1.5 C.Y.		1030.00		1133.00	64.38	70.81
16 L.H., Daily Totals		$1744.40		$2223.40	$109.03	$138.96

Crew B-12C	Hr.	Daily	Hr.	Daily	Bare Costs	Incl. O&P
1 Equip. Oper. (crane)	$51.70	$413.60	$78.45	$627.60	$44.65	$68.15
1 Laborer	37.60	300.80	57.85	462.80		
1 Hyd. Excavator, 2 C.Y.		1175.00		1292.50	73.44	80.78
16 L.H., Daily Totals		$1889.40		$2382.90	$118.09	$148.93

Crew B-12D	Hr.	Daily	Hr.	Daily	Bare Costs	Incl. O&P
1 Equip. Oper. (crane)	$51.70	$413.60	$78.45	$627.60	$44.65	$68.15
1 Laborer	37.60	300.80	57.85	462.80		
1 Hyd. Excavator, 3.5 C.Y.		2440.00		2684.00	152.50	167.75
16 L.H., Daily Totals		$3154.40		$3774.40	$197.15	$235.90

Crew B-12E	Hr.	Daily	Hr.	Daily	Bare Costs	Incl. O&P
1 Equip. Oper. (crane)	$51.70	$413.60	$78.45	$627.60	$44.65	$68.15
1 Laborer	37.60	300.80	57.85	462.80		
1 Hyd. Excavator, .5 C.Y.		448.20		493.02	28.01	30.81
16 L.H., Daily Totals		$1162.60		$1583.42	$72.66	$98.96

Crew B-12F	Hr.	Daily	Hr.	Daily	Bare Costs	Incl. O&P
1 Equip. Oper. (crane)	$51.70	$413.60	$78.45	$627.60	$44.65	$68.15
1 Laborer	37.60	300.80	57.85	462.80		
1 Hyd. Excavator, .75 C.Y.		654.80		720.28	40.92	45.02
16 L.H., Daily Totals		$1369.20		$1810.68	$85.58	$113.17

Crew B-12G	Hr.	Daily	Hr.	Daily	Bare Costs	Incl. O&P
1 Equip. Oper. (crane)	$51.70	$413.60	$78.45	$627.60	$44.65	$68.15
1 Laborer	37.60	300.80	57.85	462.80		
1 Crawler Crane, 15 Ton		669.85		736.84		
1 Clamshell Bucket, .5 C.Y.		38.20		42.02	44.25	48.68
16 L.H., Daily Totals		$1422.45		$1869.26	$88.90	$116.83

Crew B-12H	Hr.	Daily	Hr.	Daily	Bare Costs	Incl. O&P
1 Equip. Oper. (crane)	$51.70	$413.60	$78.45	$627.60	$44.65	$68.15
1 Laborer	37.60	300.80	57.85	462.80		
1 Crawler Crane, 25 Ton		1150.00		1265.00		
1 Clamshell Bucket, 1 C.Y.		47.80		52.58	74.86	82.35
16 L.H., Daily Totals		$1912.20		$2407.98	$119.51	$150.50

Crew B-12I	Hr.	Daily	Hr.	Daily	Bare Costs	Incl. O&P
1 Equip. Oper. (crane)	$51.70	$413.60	$78.45	$627.60	$44.65	$68.15
1 Laborer	37.60	300.80	57.85	462.80		
1 Crawler Crane, 20 Ton		862.50		948.75		
1 Dragline Bucket, .75 C.Y.		20.60		22.66	55.19	60.71
16 L.H., Daily Totals		$1597.50		$2061.81	$99.84	$128.86

Crew B-12J	Hr.	Daily	Hr.	Daily	Bare Costs	Incl. O&P
1 Equip. Oper. (crane)	$51.70	$413.60	$78.45	$627.60	$44.65	$68.15
1 Laborer	37.60	300.80	57.85	462.80		
1 Gradall, 5/8 C.Y.		882.20		970.42	55.14	60.65
16 L.H., Daily Totals		$1596.60		$2060.82	$99.79	$128.80

Crew B-12K	Hr.	Daily	Hr.	Daily	Bare Costs	Incl. O&P
1 Equip. Oper. (crane)	$51.70	$413.60	$78.45	$627.60	$44.65	$68.15
1 Laborer	37.60	300.80	57.85	462.80		
1 Gradall, 3 Ton, 1 C.Y.		1002.00		1102.20	62.63	68.89
16 L.H., Daily Totals		$1716.40		$2192.60	$107.28	$137.04

Crew B-12L	Hr.	Daily	Hr.	Daily	Bare Costs	Incl. O&P
1 Equip. Oper. (crane)	$51.70	$413.60	$78.45	$627.60	$44.65	$68.15
1 Laborer	37.60	300.80	57.85	462.80		
1 Crawler Crane, 15 Ton		669.85		736.84		
1 F.E. Attachment, .5 C.Y.		60.00		66.00	45.62	50.18
16 L.H., Daily Totals		$1444.25		$1893.23	$90.27	$118.33

Crew B-12M	Hr.	Daily	Hr.	Daily	Bare Costs	Incl. O&P
1 Equip. Oper. (crane)	$51.70	$413.60	$78.45	$627.60	$44.65	$68.15
1 Laborer	37.60	300.80	57.85	462.80		
1 Crawler Crane, 20 Ton		862.50		948.75		
1 F.E. Attachment, .75 C.Y.		65.40		71.94	57.99	63.79
16 L.H., Daily Totals		$1642.30		$2111.09	$102.64	$131.94

Crew B-12N	Hr.	Daily	Hr.	Daily	Bare Costs	Incl. O&P
1 Equip. Oper. (crane)	$51.70	$413.60	$78.45	$627.60	$44.65	$68.15
1 Laborer	37.60	300.80	57.85	462.80		
1 Crawler Crane, 25 Ton		1150.00		1265.00		
1 F.E. Attachment, 1 C.Y.		71.20		78.32	76.33	83.96
16 L.H., Daily Totals		$1935.60		$2433.72	$120.97	$152.11

Crew B-12O	Hr.	Daily	Hr.	Daily	Bare Costs	Incl. O&P
1 Equip. Oper. (crane)	$51.70	$413.60	$78.45	$627.60	$44.65	$68.15
1 Laborer	37.60	300.80	57.85	462.80		
1 Crawler Crane, 40 Ton		1156.00		1271.60		
1 F.E. Attachment, 1.5 C.Y.		80.00		88.00	77.25	84.97
16 L.H., Daily Totals		$1950.40		$2450.00	$121.90	$153.13

Crew B-12P	Hr.	Daily	Hr.	Daily	Bare Costs	Incl. O&P
1 Equip. Oper. (crane)	$51.70	$413.60	$78.45	$627.60	$44.65	$68.15
1 Laborer	37.60	300.80	57.85	462.80		
1 Crawler Crane, 40 Ton		1156.00		1271.60		
1 Dragline Bucket, 1.5 C.Y.		33.60		36.96	74.35	81.78
16 L.H., Daily Totals		$1904.00		$2398.96	$119.00	$149.94

Crew B-12Q	Hr.	Daily	Hr.	Daily	Bare Costs	Incl. O&P
1 Equip. Oper. (crane)	$51.70	$413.60	$78.45	$627.60	$44.65	$68.15
1 Laborer	37.60	300.80	57.85	462.80		
1 Hyd. Excavator, 5/8 C.Y.		589.40		648.34	36.84	40.52
16 L.H., Daily Totals		$1303.80		$1738.74	$81.49	$108.67

Crew B-12S	Hr.	Daily	Hr.	Daily	Bare Costs	Incl. O&P
1 Equip. Oper. (crane)	$51.70	$413.60	$78.45	$627.60	$44.65	$68.15
1 Laborer	37.60	300.80	57.85	462.80		
1 Hyd. Excavator, 2.5 C.Y.		1605.00		1765.50	100.31	110.34
16 L.H., Daily Totals		$2319.40		$2855.90	$144.96	$178.49

Crew B-12T	Hr.	Daily	Hr.	Daily	Bare Costs	Incl. O&P
1 Equip. Oper. (crane)	$51.70	$413.60	$78.45	$627.60	$44.65	$68.15
1 Laborer	37.60	300.80	57.85	462.80		
1 Crawler Crane, 75 Ton		1459.00		1604.90		
1 F.E. Attachment, 3 C.Y.		103.20		113.52	97.64	107.40
16 L.H., Daily Totals		$2276.60		$2808.82	$142.29	$175.55

Crew B-12V	Hr.	Daily	Hr.	Daily	Bare Costs	Incl. O&P
1 Equip. Oper. (crane)	$51.70	$413.60	$78.45	$627.60	$44.65	$68.15
1 Laborer	37.60	300.80	57.85	462.80		
1 Crawler Crane, 75 Ton		1459.00		1604.90		
1 Dragline Bucket, 3 C.Y.		52.60		57.86	94.47	103.92
16 L.H., Daily Totals		$2226.00		$2753.16	$139.13	$172.07

Crew B-12Y	Hr.	Daily	Hr.	Daily	Bare Costs	Incl. O&P
1 Equip. Oper. (crane)	$51.70	$413.60	$78.45	$627.60	$42.30	$64.72
2 Laborers	37.60	601.60	57.85	925.60		
1 Hyd. Excavator, 3.5 C.Y.		2440.00		2684.00	101.67	111.83
24 L.H., Daily Totals		$3455.20		$4237.20	$143.97	$176.55

Crew No.	Bare Costs		Incl. Subs O&P		Cost Per Labor-Hour	
Crew B-12Z	**Hr.**	**Daily**	**Hr.**	**Daily**	**Bare Costs**	**Incl. O&P**
1 Equip. Oper. (crane)	$51.70	$413.60	$78.45	$627.60	$42.30	$64.72
2 Laborers	37.60	601.60	57.85	925.60		
1 Hyd. Excavator, 2.5 C.Y.		1605.00		1765.50	66.88	73.56
24 L.H., Daily Totals		$2620.20		$3318.70	$109.18	$138.28
Crew B-13	**Hr.**	**Daily**	**Hr.**	**Daily**	**Bare Costs**	**Incl. O&P**
1 Labor Foreman (outside)	$39.60	$316.80	$60.95	$487.60	$40.99	$62.76
4 Laborers	37.60	1203.20	57.85	1851.20		
1 Equip. Oper. (crane)	51.70	413.60	78.45	627.60		
1 Equip. Oper. (oiler)	45.20	361.60	68.55	548.40		
1 Hyd. Crane, 25 Ton		736.60		810.26	13.15	14.47
56 L.H., Daily Totals		$3031.80		$4325.06	$54.14	$77.23
Crew B-13A	**Hr.**	**Daily**	**Hr.**	**Daily**	**Bare Costs**	**Incl. O&P**
1 Labor Foreman (outside)	$39.60	$316.80	$60.95	$487.60	$42.30	$64.46
2 Laborers	37.60	601.60	57.85	925.60		
2 Equipment Operators (med.)	50.60	809.60	76.75	1228.00		
2 Truck Drivers (heavy)	40.05	640.80	60.55	968.80		
1 Crawler Crane, 75 Ton		1459.00		1604.90		
1 Crawler Loader, 4 C.Y.		1561.00		1717.10		
2 Dump Trucks, 8 C.Y., 220 H.P.		822.40		904.64	68.61	75.48
56 L.H., Daily Totals		$6211.20		$7836.64	$110.91	$139.94
Crew B-13B	**Hr.**	**Daily**	**Hr.**	**Daily**	**Bare Costs**	**Incl. O&P**
1 Labor Foreman (outside)	$39.60	$316.80	$60.95	$487.60	$40.99	$62.76
4 Laborers	37.60	1203.20	57.85	1851.20		
1 Equip. Oper. (crane)	51.70	413.60	78.45	627.60		
1 Equip. Oper. (oiler)	45.20	361.60	68.55	548.40		
1 Hyd. Crane, 55 Ton		1128.00		1240.80	20.14	22.16
56 L.H., Daily Totals		$3423.20		$4755.60	$61.13	$84.92
Crew B-13C	**Hr.**	**Daily**	**Hr.**	**Daily**	**Bare Costs**	**Incl. O&P**
1 Labor Foreman (outside)	$39.60	$316.80	$60.95	$487.60	$40.99	$62.76
4 Laborers	37.60	1203.20	57.85	1851.20		
1 Equip. Oper. (crane)	51.70	413.60	78.45	627.60		
1 Equip. Oper. (oiler)	45.20	361.60	68.55	548.40		
1 Crawler Crane, 100 Ton		1667.00		1833.70	29.77	32.74
56 L.H., Daily Totals		$3962.20		$5348.50	$70.75	$95.51
Crew B-13D	**Hr.**	**Daily**	**Hr.**	**Daily**	**Bare Costs**	**Incl. O&P**
1 Laborer	$37.60	$300.80	$57.85	$462.80	$44.65	$68.15
1 Equip. Oper. (crane)	51.70	413.60	78.45	627.60		
1 Hyd. Excavator, 1 C.Y.		812.60		893.86		
1 Trench Box		81.00		89.10	55.85	61.44
16 L.H., Daily Totals		$1608.00		$2073.36	$100.50	$129.59
Crew B-13E	**Hr.**	**Daily**	**Hr.**	**Daily**	**Bare Costs**	**Incl. O&P**
1 Laborer	$37.60	$300.80	$57.85	$462.80	$44.65	$68.15
1 Equip. Oper. (crane)	51.70	413.60	78.45	627.60		
1 Hyd. Excavator, 1.5 C.Y.		1030.00		1133.00		
1 Trench Box		81.00		89.10	69.44	76.38
16 L.H., Daily Totals		$1825.40		$2312.50	$114.09	$144.53
Crew B-13F	**Hr.**	**Daily**	**Hr.**	**Daily**	**Bare Costs**	**Incl. O&P**
1 Laborer	$37.60	$300.80	$57.85	$462.80	$44.65	$68.15
1 Equip. Oper. (crane)	51.70	413.60	78.45	627.60		
1 Hyd. Excavator, 3.5 C.Y.		2440.00		2684.00		
1 Trench Box		81.00		89.10	157.56	173.32
16 L.H., Daily Totals		$3235.40		$3863.50	$202.21	$241.47

Crew No.	Bare Costs		Incl. Subs O&P		Cost Per Labor-Hour	
Crew B-13G	**Hr.**	**Daily**	**Hr.**	**Daily**	**Bare Costs**	**Incl. O&P**
1 Laborer	$37.60	$300.80	$57.85	$462.80	$44.65	$68.15
1 Equip. Oper. (crane)	51.70	413.60	78.45	627.60		
1 Hyd. Excavator, .75 C.Y.		654.80		720.28		
1 Trench Box		81.00		89.10	45.99	50.59
16 L.H., Daily Totals		$1450.20		$1899.78	$90.64	$118.74
Crew B-13H	**Hr.**	**Daily**	**Hr.**	**Daily**	**Bare Costs**	**Incl. O&P**
1 Laborer	$37.60	$300.80	$57.85	$462.80	$44.65	$68.15
1 Equip. Oper. (crane)	51.70	413.60	78.45	627.60		
1 Gradall, 5/8 C.Y.		882.20		970.42		
1 Trench Box		81.00		89.10	60.20	66.22
16 L.H., Daily Totals		$1677.60		$2149.92	$104.85	$134.37
Crew B-13I	**Hr.**	**Daily**	**Hr.**	**Daily**	**Bare Costs**	**Incl. O&P**
1 Laborer	$37.60	$300.80	$57.85	$462.80	$44.65	$68.15
1 Equip. Oper. (crane)	51.70	413.60	78.45	627.60		
1 Gradall, 3 Ton, 1 C.Y.		1002.00		1102.20		
1 Trench Box		81.00		89.10	67.69	74.46
16 L.H., Daily Totals		$1797.40		$2281.70	$112.34	$142.61
Crew B-13J	**Hr.**	**Daily**	**Hr.**	**Daily**	**Bare Costs**	**Incl. O&P**
1 Laborer	$37.60	$300.80	$57.85	$462.80	$44.65	$68.15
1 Equip. Oper. (crane)	51.70	413.60	78.45	627.60		
1 Hyd. Excavator, 2.5 C.Y.		1605.00		1765.50		
1 Trench Box		81.00		89.10	105.38	115.91
16 L.H., Daily Totals		$2400.40		$2945.00	$150.03	$184.06
Crew B-13K	**Hr.**	**Daily**	**Hr.**	**Daily**	**Bare Costs**	**Incl. O&P**
2 Equip. Opers. (crane)	$51.70	$827.20	$78.45	$1255.20	$51.70	$78.45
1 Hyd. Excavator, .75 C.Y.		654.80		720.28		
1 Hyd. Hammer, 4000 ft-lb		308.20		339.02		
1 Hyd. Excavator, .75 C.Y.		654.80		720.28	101.11	111.22
16 L.H., Daily Totals		$2445.00		$3034.78	$152.81	$189.67
Crew B-13L	**Hr.**	**Daily**	**Hr.**	**Daily**	**Bare Costs**	**Incl. O&P**
2 Equip. Opers. (crane)	$51.70	$827.20	$78.45	$1255.20	$51.70	$78.45
1 Hyd. Excavator, 1.5 C.Y.		1030.00		1133.00		
1 Hyd. Hammer, 5000 ft-lb		371.20		408.32		
1 Hyd. Excavator, .75 C.Y.		654.80		720.28	128.50	141.35
16 L.H., Daily Totals		$2883.20		$3516.80	$180.20	$219.80
Crew B-13M	**Hr.**	**Daily**	**Hr.**	**Daily**	**Bare Costs**	**Incl. O&P**
2 Equip. Opers. (crane)	$51.70	$827.20	$78.45	$1255.20	$51.70	$78.45
1 Hyd. Excavator, 2.5 C.Y.		1605.00		1765.50		
1 Hyd. Hammer, 8000 ft-lb		545.40		599.94		
1 Hyd. Excavator, 1.5 C.Y.		1030.00		1133.00	198.78	218.65
16 L.H., Daily Totals		$4007.60		$4753.64	$250.47	$297.10
Crew B-13N	**Hr.**	**Daily**	**Hr.**	**Daily**	**Bare Costs**	**Incl. O&P**
2 Equip. Opers. (crane)	$51.70	$827.20	$78.45	$1255.20	$51.70	$78.45
1 Hyd. Excavator, 3.5 C.Y.		2440.00		2684.00		
1 Hyd. Hammer, 12,000 ft-lb		635.80		699.38		
1 Hyd. Excavator, 1.5 C.Y.		1030.00		1133.00	256.61	282.27
16 L.H., Daily Totals		$4933.00		$5771.58	$308.31	$360.72

Crew No.	Bare Costs		Incl. Subs O&P		Cost Per Labor-Hour	
Crew B-14	**Hr.**	**Daily**	**Hr.**	**Daily**	**Bare Costs**	**Incl. O&P**
1 Labor Foreman (outside)	$39.60	$316.80	$60.95	$487.60	$39.77	$61.02
4 Laborers	37.60	1203.20	57.85	1851.20		
1 Equip. Oper. (light)	48.60	388.80	73.75	590.00		
1 Backhoe Loader, 48 H.P.		364.20		400.62	7.59	8.35
48 L.H., Daily Totals		$2273.00		$3329.42	$47.35	$69.36
Crew B-14A	**Hr.**	**Daily**	**Hr.**	**Daily**	**Bare Costs**	**Incl. O&P**
1 Equip. Oper. (crane)	$51.70	$413.60	$78.45	$627.60	$47.00	$71.58
.5 Laborer	37.60	150.40	57.85	231.40		
1 Hyd. Excavator, 4.5 C.Y.		2963.00		3259.30	246.92	271.61
12 L.H., Daily Totals		$3527.00		$4118.30	$293.92	$343.19
Crew B-14B	**Hr.**	**Daily**	**Hr.**	**Daily**	**Bare Costs**	**Incl. O&P**
1 Equip. Oper. (crane)	$51.70	$413.60	$78.45	$627.60	$47.00	$71.58
.5 Laborer	37.60	150.40	57.85	231.40		
1 Hyd. Excavator, 6 C.Y.		3494.00		3843.40	291.17	320.28
12 L.H., Daily Totals		$4058.00		$4702.40	$338.17	$391.87
Crew B-14C	**Hr.**	**Daily**	**Hr.**	**Daily**	**Bare Costs**	**Incl. O&P**
1 Equip. Oper. (crane)	$51.70	$413.60	$78.45	$627.60	$47.00	$71.58
.5 Laborer	37.60	150.40	57.85	231.40		
1 Hyd. Excavator, 7 C.Y.		3590.00		3949.00	299.17	329.08
12 L.H., Daily Totals		$4154.00		$4808.00	$346.17	$400.67
Crew B-14F	**Hr.**	**Daily**	**Hr.**	**Daily**	**Bare Costs**	**Incl. O&P**
1 Equip. Oper. (crane)	$51.70	$413.60	$78.45	$627.60	$47.00	$71.58
.5 Laborer	37.60	150.40	57.85	231.40		
1 Hyd. Shovel, 7 C.Y.		3993.00		4392.30	332.75	366.02
12 L.H., Daily Totals		$4557.00		$5251.30	$379.75	$437.61
Crew B-14G	**Hr.**	**Daily**	**Hr.**	**Daily**	**Bare Costs**	**Incl. O&P**
1 Equip. Oper. (crane)	$51.70	$413.60	$78.45	$627.60	$47.00	$71.58
.5 Laborer	37.60	150.40	57.85	231.40		
1 Hyd. Shovel, 12 C.Y.		5736.00		6309.60	478.00	525.80
12 L.H., Daily Totals		$6300.00		$7168.60	$525.00	$597.38
Crew B-14J	**Hr.**	**Daily**	**Hr.**	**Daily**	**Bare Costs**	**Incl. O&P**
1 Equip. Oper. (medium)	$50.60	$404.80	$76.75	$614.00	$46.27	$70.45
.5 Laborer	37.60	150.40	57.85	231.40		
1 F.E. Loader, 8 C.Y.		1937.00		2130.70	161.42	177.56
12 L.H., Daily Totals		$2492.20		$2976.10	$207.68	$248.01
Crew B-14K	**Hr.**	**Daily**	**Hr.**	**Daily**	**Bare Costs**	**Incl. O&P**
1 Equip. Oper. (medium)	$50.60	$404.80	$76.75	$614.00	$46.27	$70.45
.5 Laborer	37.60	150.40	57.85	231.40		
1 F.E. Loader, 10 C.Y.		2942.00		3236.20	245.17	269.68
12 L.H., Daily Totals		$3497.20		$4081.60	$291.43	$340.13
Crew B-15	**Hr.**	**Daily**	**Hr.**	**Daily**	**Bare Costs**	**Incl. O&P**
1 Equipment Oper. (med.)	$50.60	$404.80	$76.75	$614.00	$42.71	$64.79
.5 Laborer	37.60	150.40	57.85	231.40		
2 Truck Drivers (heavy)	40.05	640.80	60.55	968.80		
2 Dump Trucks, 12 C.Y., 400 H.P.		1382.00		1520.20		
1 Dozer, 200 H.P.		1387.00		1525.70	98.89	108.78
28 L.H., Daily Totals		$3965.00		$4860.10	$141.61	$173.57

Crew No.	Bare Costs		Incl. Subs O&P		Cost Per Labor-Hour	
Crew B-16	**Hr.**	**Daily**	**Hr.**	**Daily**	**Bare Costs**	**Incl. O&P**
1 Labor Foreman (outside)	$39.60	$316.80	$60.95	$487.60	$38.71	$59.30
2 Laborers	37.60	601.60	57.85	925.60		
1 Truck Driver (heavy)	40.05	320.40	60.55	484.40		
1 Dump Truck, 12 C.Y., 400 H.P.		691.00		760.10	21.59	23.75
32 L.H., Daily Totals		$1929.80		$2657.70	$60.31	$83.05
Crew B-17	**Hr.**	**Daily**	**Hr.**	**Daily**	**Bare Costs**	**Incl. O&P**
2 Laborers	$37.60	$601.60	$57.85	$925.60	$40.96	$62.50
1 Equip. Oper. (light)	48.60	388.80	73.75	590.00		
1 Truck Driver (heavy)	40.05	320.40	60.55	484.40		
1 Backhoe Loader, 48 H.P.		364.20		400.62		
1 Dump Truck, 8 C.Y., 220 H.P.		411.20		452.32	24.23	26.65
32 L.H., Daily Totals		$2086.20		$2852.94	$65.19	$89.15
Crew B-17A	**Hr.**	**Daily**	**Hr.**	**Daily**	**Bare Costs**	**Incl. O&P**
2 Labor Foremen (outside)	$39.60	$633.60	$60.95	$975.20	$40.41	$62.24
6 Laborers	37.60	1804.80	57.85	2776.80		
1 Skilled Worker Foreman (out)	50.65	405.20	78.25	626.00		
1 Skilled Worker	48.65	389.20	75.15	601.20		
80 L.H., Daily Totals		$3232.80		$4979.20	$40.41	$62.24
Crew B-17B	**Hr.**	**Daily**	**Hr.**	**Daily**	**Bare Costs**	**Incl. O&P**
2 Laborers	$37.60	$601.60	$57.85	$925.60	$40.96	$62.50
1 Equip. Oper. (light)	48.60	388.80	73.75	590.00		
1 Truck Driver (heavy)	40.05	320.40	60.55	484.40		
1 Backhoe Loader, 48 H.P.		364.20		400.62		
1 Dump Truck, 12 C.Y., 400 H.P.		691.00		760.10	32.98	36.27
32 L.H., Daily Totals		$2366.00		$3160.72	$73.94	$98.77
Crew B-18	**Hr.**	**Daily**	**Hr.**	**Daily**	**Bare Costs**	**Incl. O&P**
1 Labor Foreman (outside)	$39.60	$316.80	$60.95	$487.60	$38.27	$58.88
2 Laborers	37.60	601.60	57.85	925.60		
1 Vibrating Plate, Gas, 21"		46.80		51.48	1.95	2.15
24 L.H., Daily Totals		$965.20		$1464.68	$40.22	$61.03
Crew B-19	**Hr.**	**Daily**	**Hr.**	**Daily**	**Bare Costs**	**Incl. O&P**
1 Pile Driver Foreman (outside)	$48.10	$384.80	$76.15	$609.20	$47.64	$74.20
4 Pile Drivers	46.10	1475.20	73.00	2336.00		
2 Equip. Oper. (crane)	51.70	827.20	78.45	1255.20		
1 Equip. Oper. (oiler)	45.20	361.60	68.55	548.40		
1 Crawler Crane, 40 Ton		1156.00		1271.60		
1 Lead, 90' High		124.40		136.84		
1 Hammer, Diesel, 22k ft-lb		456.40		502.04	27.14	29.85
64 L.H., Daily Totals		$4785.60		$6659.28	$74.78	$104.05
Crew B-19A	**Hr.**	**Daily**	**Hr.**	**Daily**	**Bare Costs**	**Incl. O&P**
1 Pile Driver Foreman (outside)	$48.10	$384.80	$76.15	$609.20	$47.64	$74.20
4 Pile Drivers	46.10	1475.20	73.00	2336.00		
2 Equip. Oper. (crane)	51.70	827.20	78.45	1255.20		
1 Equip. Oper. (oiler)	45.20	361.60	68.55	548.40		
1 Crawler Crane, 75 Ton		1459.00		1604.90		
1 Lead, 90' high		124.40		136.84		
1 Hammer, Diesel, 41k ft-lb		581.00		639.10	33.82	37.20
64 L.H., Daily Totals		$5213.20		$7129.64	$81.46	$111.40

Crew No.	Bare Costs		Incl. Subs O&P		Cost Per Labor-Hour	
Crew B-19B	**Hr.**	**Daily**	**Hr.**	**Daily**	**Bare Costs**	**Incl. O&P**
1 Pile Driver Foreman (outside)	$48.10	$384.80	$76.15	$609.20	$47.64	$74.20
4 Pile Drivers	46.10	1475.20	73.00	2336.00		
2 Equip. Oper. (crane)	51.70	827.20	78.45	1255.20		
1 Equip. Oper. (oiler)	45.20	361.60	68.55	548.40		
1 Crawler Crane, 40 Ton		1156.00		1271.60		
1 Lead, 90' High		124.40		136.84		
1 Hammer, Diesel, 22k ft-lb		456.40		502.04		
1 Barge, 400 Ton		776.00		853.60	39.26	43.19
64 L.H., Daily Totals		$5561.60		$7512.88	$86.90	$117.39

Crew B-19C	Hr.	Daily	Hr.	Daily	Bare Costs	Incl. O&P
1 Pile Driver Foreman (outside)	$48.10	$384.80	$76.15	$609.20	$47.64	$74.20
4 Pile Drivers	46.10	1475.20	73.00	2336.00		
2 Equip. Oper. (crane)	51.70	827.20	78.45	1255.20		
1 Equip. Oper. (oiler)	45.20	361.60	68.55	548.40		
1 Crawler Crane, 75 Ton		1459.00		1604.90		
1 Lead, 90' High		124.40		136.84		
1 Hammer, Diesel, 41k ft-lb		581.00		639.10		
1 Barge, 400 Ton		776.00		853.60	45.94	50.54
64 L.H., Daily Totals		$5989.20		$7983.24	$93.58	$124.74

Crew B-20	Hr.	Daily	Hr.	Daily	Bare Costs	Incl. O&P
1 Labor Foreman (outside)	$39.60	$316.80	$60.95	$487.60	$41.95	$64.65
1 Skilled Worker	48.65	389.20	75.15	601.20		
1 Laborer	37.60	300.80	57.85	462.80		
24 L.H., Daily Totals		$1006.80		$1551.60	$41.95	$64.65

Crew B-20A	Hr.	Daily	Hr.	Daily	Bare Costs	Incl. O&P
1 Labor Foreman (outside)	$39.60	$316.80	$60.95	$487.60	$45.71	$69.59
1 Laborer	37.60	300.80	57.85	462.80		
1 Plumber	58.70	469.60	88.65	709.20		
1 Plumber Apprentice	46.95	375.60	70.90	567.20		
32 L.H., Daily Totals		$1462.80		$2226.80	$45.71	$69.59

Crew B-21	Hr.	Daily	Hr.	Daily	Bare Costs	Incl. O&P
1 Labor Foreman (outside)	$39.60	$316.80	$60.95	$487.60	$43.34	$66.62
1 Skilled Worker	48.65	389.20	75.15	601.20		
1 Laborer	37.60	300.80	57.85	462.80		
.5 Equip. Oper. (crane)	51.70	206.80	78.45	313.80		
.5 S.P. Crane, 4x4, 5 Ton		137.70		151.47	4.92	5.41
28 L.H., Daily Totals		$1351.30		$2016.87	$48.26	$72.03

Crew B-21A	Hr.	Daily	Hr.	Daily	Bare Costs	Incl. O&P
1 Labor Foreman (outside)	$39.60	$316.80	$60.95	$487.60	$46.91	$71.36
1 Laborer	37.60	300.80	57.85	462.80		
1 Plumber	58.70	469.60	88.65	709.20		
1 Plumber Apprentice	46.95	375.60	70.90	567.20		
1 Equip. Oper. (crane)	51.70	413.60	78.45	627.60		
1 S.P. Crane, 4x4, 12 Ton		473.40		520.74	11.84	13.02
40 L.H., Daily Totals		$2349.80		$3375.14	$58.74	$84.38

Crew B-21B	Hr.	Daily	Hr.	Daily	Bare Costs	Incl. O&P
1 Labor Foreman (outside)	$39.60	$316.80	$60.95	$487.60	$40.82	$62.59
3 Laborers	37.60	902.40	57.85	1388.40		
1 Equip. Oper. (crane)	51.70	413.60	78.45	627.60		
1 Hyd. Crane, 12 Ton		653.80		719.18	16.34	17.98
40 L.H., Daily Totals		$2286.60		$3222.78	$57.16	$80.57

Crew B-21C	Hr.	Daily	Hr.	Daily	Bare Costs	Incl. O&P
1 Labor Foreman (outside)	$39.60	$316.80	$60.95	$487.60	$40.99	$62.76
4 Laborers	37.60	1203.20	57.85	1851.20		
1 Equip. Oper. (crane)	51.70	413.60	78.45	627.60		
1 Equip. Oper. (oiler)	45.20	361.60	68.55	548.40		
2 Cutting Torches		22.80		25.08		
2 Sets of Gases		304.00		334.40		
1 Lattice Boom Crane, 90 Ton		1511.00		1662.10	32.82	36.10
56 L.H., Daily Totals		$4133.00		$5536.38	$73.80	$98.86

Crew B-22	Hr.	Daily	Hr.	Daily	Bare Costs	Incl. O&P
1 Labor Foreman (outside)	$39.60	$316.80	$60.95	$487.60	$43.90	$67.41
1 Skilled Worker	48.65	389.20	75.15	601.20		
1 Laborer	37.60	300.80	57.85	462.80		
.75 Equip. Oper. (crane)	51.70	310.20	78.45	470.70		
.75 S.P. Crane, 4x4, 5 Ton		206.55		227.21	6.88	7.57
30 L.H., Daily Totals		$1523.55		$2249.51	$50.78	$74.98

Crew B-22A	Hr.	Daily	Hr.	Daily	Bare Costs	Incl. O&P
1 Labor Foreman (outside)	$39.60	$316.80	$60.95	$487.60	$43.03	$66.05
1 Skilled Worker	48.65	389.20	75.15	601.20		
2 Laborers	37.60	601.60	57.85	925.60		
1 Equipment Operator, Crane	51.70	413.60	78.45	627.60		
1 S.P. Crane, 4x4, 5 Ton		275.40		302.94		
1 Butt Fusion Mach., 4"-12" diam.		382.10		420.31	16.44	18.08
40 L.H., Daily Totals		$2378.70		$3365.25	$59.47	$84.13

Crew B-22B	Hr.	Daily	Hr.	Daily	Bare Costs	Incl. O&P
1 Labor Foreman (outside)	$39.60	$316.80	$60.95	$487.60	$43.03	$66.05
1 Skilled Worker	48.65	389.20	75.15	601.20		
2 Laborers	37.60	601.60	57.85	925.60		
1 Equip. Oper. (crane)	51.70	413.60	78.45	627.60		
1 S.P. Crane, 4x4, 5 Ton		275.40		302.94		
1 Butt Fusion Mach., 8"-24" diam.		829.70		912.67	27.63	30.39
40 L.H., Daily Totals		$2826.30		$3857.61	$70.66	$96.44

Crew B-22C	Hr.	Daily	Hr.	Daily	Bare Costs	Incl. O&P
1 Skilled Worker	$48.65	$389.20	$75.15	$601.20	$43.13	$66.50
1 Laborer	37.60	300.80	57.85	462.80		
1 Butt Fusion Mach., 2"-8" diam.		121.05		133.16	7.57	8.32
16 L.H., Daily Totals		$811.05		$1197.16	$50.69	$74.82

Crew B-23	Hr.	Daily	Hr.	Daily	Bare Costs	Incl. O&P
1 Labor Foreman (outside)	$39.60	$316.80	$60.95	$487.60	$38.00	$58.47
4 Laborers	37.60	1203.20	57.85	1851.20		
1 Drill Rig, Truck-Mounted		2542.00		2796.20		
1 Flatbed Truck, Gas, 3 Ton		303.20		333.52	71.13	78.24
40 L.H., Daily Totals		$4365.20		$5468.52	$109.13	$136.71

Crew B-23A	Hr.	Daily	Hr.	Daily	Bare Costs	Incl. O&P
1 Labor Foreman (outside)	$39.60	$316.80	$60.95	$487.60	$42.60	$65.18
1 Laborer	37.60	300.80	57.85	462.80		
1 Equip. Oper. (medium)	50.60	404.80	76.75	614.00		
1 Drill Rig, Truck-Mounted		2542.00		2796.20		
1 Pickup Truck, 3/4 Ton		144.20		158.62	111.93	123.12
24 L.H., Daily Totals		$3708.60		$4519.22	$154.53	$188.30

Crew B-23B	Bare Costs Hr.	Bare Costs Daily	Incl. Subs O&P Hr.	Incl. Subs O&P Daily	Cost Per Labor-Hour Bare Costs	Cost Per Labor-Hour Incl. O&P
1 Labor Foreman (outside)	$39.60	$316.80	$60.95	$487.60	$42.60	$65.18
1 Laborer	37.60	300.80	57.85	462.80		
1 Equip. Oper. (medium)	50.60	404.80	76.75	614.00		
1 Drill Rig, Truck-Mounted		2542.00		2796.20		
1 Pickup Truck, 3/4 Ton		144.20		158.62		
1 Centr. Water Pump, 6"		346.60		381.26	126.37	139.00
24 L.H., Daily Totals		$4055.20		$4900.48	$168.97	$204.19

Crew B-24	Hr.	Daily	Hr.	Daily	Bare Costs	Incl. O&P
1 Cement Finisher	$45.00	$360.00	$66.65	$533.20	$43.18	$65.58
1 Laborer	37.60	300.80	57.85	462.80		
1 Carpenter	46.95	375.60	72.25	578.00		
24 L.H., Daily Totals		$1036.40		$1574.00	$43.18	$65.58

Crew B-25	Hr.	Daily	Hr.	Daily	Bare Costs	Incl. O&P
1 Labor Foreman (outside)	$39.60	$316.80	$60.95	$487.60	$41.33	$63.29
7 Laborers	37.60	2105.60	57.85	3239.60		
3 Equip. Oper. (medium)	50.60	1214.40	76.75	1842.00		
1 Asphalt Paver, 130 H.P.		2132.00		2345.20		
1 Tandem Roller, 10 Ton		236.80		260.48		
1 Roller, Pneum. Whl., 12 Ton		344.40		378.84	30.83	33.91
88 L.H., Daily Totals		$6350.00		$8553.72	$72.16	$97.20

Crew B-25B	Hr.	Daily	Hr.	Daily	Bare Costs	Incl. O&P
1 Labor Foreman (outside)	$39.60	$316.80	$60.95	$487.60	$42.10	$64.41
7 Laborers	37.60	2105.60	57.85	3239.60		
4 Equip. Oper. (medium)	50.60	1619.20	76.75	2456.00		
1 Asphalt Paver, 130 H.P.		2132.00		2345.20		
2 Tandem Rollers, 10 Ton		473.60		520.96		
1 Roller, Pneum. Whl., 12 Ton		344.40		378.84	30.73	33.80
96 L.H., Daily Totals		$6991.60		$9428.20	$72.83	$98.21

Crew B-25C	Hr.	Daily	Hr.	Daily	Bare Costs	Incl. O&P
1 Labor Foreman (outside)	$39.60	$316.80	$60.95	$487.60	$42.27	$64.67
3 Laborers	37.60	902.40	57.85	1388.40		
2 Equip. Oper. (medium)	50.60	809.60	76.75	1228.00		
1 Asphalt Paver, 130 H.P.		2132.00		2345.20		
1 Tandem Roller, 10 Ton		236.80		260.48	49.35	54.28
48 L.H., Daily Totals		$4397.60		$5709.68	$91.62	$118.95

Crew B-25D	Hr.	Daily	Hr.	Daily	Bare Costs	Incl. O&P
1 Labor Foreman (outside)	$39.60	$316.80	$60.95	$487.60	$42.39	$64.83
3 Laborers	37.60	902.40	57.85	1388.40		
2.125 Equip. Oper. (medium)	50.60	860.20	76.75	1304.75		
.125 Truck Driver (heavy)	40.05	40.05	60.55	60.55		
.125 Truck Tractor, 6x4, 380 H.P.		74.97		82.47		
.125 Dist. Tanker, 3000 Gallon		38.20		42.02		
1 Asphalt Paver, 130 H.P.		2132.00		2345.20		
1 Tandem Roller, 10 Ton		236.80		260.48	49.64	54.60
50 L.H., Daily Totals		$4601.43		$5971.47	$92.03	$119.43

Crew B-25E	Hr.	Daily	Hr.	Daily	Bare Costs	Incl. O&P
1 Labor Foreman (outside)	$39.60	$316.80	$60.95	$487.60	$42.50	$64.97
3 Laborers	37.60	902.40	57.85	1388.40		
2.250 Equip. Oper. (medium)	50.60	910.80	76.75	1381.50		
.25 Truck Driver (heavy)	40.05	80.10	60.55	121.10		
.25 Truck Tractor, 6x4, 380 H.P.		149.95		164.94		
.25 Dist. Tanker, 3000 Gallon		76.40		84.04		
1 Asphalt Paver, 130 H.P.		2132.00		2345.20		
1 Tandem Roller, 10 Ton		236.80		260.48	49.91	54.90
52 L.H., Daily Totals		$4805.25		$6233.27	$92.41	$119.87

Crew B-26	Hr.	Daily	Hr.	Daily	Bare Costs	Incl. O&P
1 Labor Foreman (outside)	$39.60	$316.80	$60.95	$487.60	$42.18	$64.57
6 Laborers	37.60	1804.80	57.85	2776.80		
2 Equip. Oper. (medium)	50.60	809.60	76.75	1228.00		
1 Rodman (reinf.)	52.55	420.40	82.10	656.80		
1 Cement Finisher	45.00	360.00	66.65	533.20		
1 Grader, 30,000 Lbs.		736.20		809.82		
1 Paving Mach. & Equip.		2794.00		3073.40	40.12	44.13
88 L.H., Daily Totals		$7241.80		$9565.62	$82.29	$108.70

Crew B-26A	Hr.	Daily	Hr.	Daily	Bare Costs	Incl. O&P
1 Labor Foreman (outside)	$39.60	$316.80	$60.95	$487.60	$42.18	$64.57
6 Laborers	37.60	1804.80	57.85	2776.80		
2 Equip. Oper. (medium)	50.60	809.60	76.75	1228.00		
1 Rodman (reinf.)	52.55	420.40	82.10	656.80		
1 Cement Finisher	45.00	360.00	66.65	533.20		
1 Grader, 30,000 Lbs.		736.20		809.82		
1 Paving Mach. & Equip.		2794.00		3073.40		
1 Concrete Saw		167.40		184.14	42.02	46.22
88 L.H., Daily Totals		$7409.20		$9749.76	$84.20	$110.79

Crew B-26B	Hr.	Daily	Hr.	Daily	Bare Costs	Incl. O&P
1 Labor Foreman (outside)	$39.60	$316.80	$60.95	$487.60	$42.88	$65.59
6 Laborers	37.60	1804.80	57.85	2776.80		
3 Equip. Oper. (medium)	50.60	1214.40	76.75	1842.00		
1 Rodman (reinf.)	52.55	420.40	82.10	656.80		
1 Cement Finisher	45.00	360.00	66.65	533.20		
1 Grader, 30,000 Lbs.		736.20		809.82		
1 Paving Mach. & Equip.		2794.00		3073.40		
1 Concrete Pump, 110' Boom		946.20		1040.82	46.63	51.29
96 L.H., Daily Totals		$8592.80		$11220.44	$89.51	$116.88

Crew B-26C	Hr.	Daily	Hr.	Daily	Bare Costs	Incl. O&P
1 Labor Foreman (outside)	$39.60	$316.80	$60.95	$487.60	$41.34	$63.35
6 Laborers	37.60	1804.80	57.85	2776.80		
1 Equip. Oper. (medium)	50.60	404.80	76.75	614.00		
1 Rodman (reinf.)	52.55	420.40	82.10	656.80		
1 Cement Finisher	45.00	360.00	66.65	533.20		
1 Paving Mach. & Equip.		2794.00		3073.40		
1 Concrete Saw		167.40		184.14	37.02	40.72
80 L.H., Daily Totals		$6268.20		$8325.94	$78.35	$104.07

Crew B-27	Hr.	Daily	Hr.	Daily	Bare Costs	Incl. O&P
1 Labor Foreman (outside)	$39.60	$316.80	$60.95	$487.60	$38.10	$58.63
3 Laborers	37.60	902.40	57.85	1388.40		
1 Berm Machine		296.80		326.48	9.28	10.20
32 L.H., Daily Totals		$1516.00		$2202.48	$47.38	$68.83

Crew B-28	Hr.	Daily	Hr.	Daily	Bare Costs	Incl. O&P
2 Carpenters	$46.95	$751.20	$72.25	$1156.00	$43.83	$67.45
1 Laborer	37.60	300.80	57.85	462.80		
24 L.H., Daily Totals		$1052.00		$1618.80	$43.83	$67.45

Crew B-29	Hr.	Daily	Hr.	Daily	Bare Costs	Incl. O&P
1 Labor Foreman (outside)	$39.60	$316.80	$60.95	$487.60	$40.99	$62.76
4 Laborers	37.60	1203.20	57.85	1851.20		
1 Equip. Oper. (crane)	51.70	413.60	78.45	627.60		
1 Equip. Oper. (oiler)	45.20	361.60	68.55	548.40		
1 Gradall, 5/8 C.Y.		882.20		970.42	15.75	17.33
56 L.H., Daily Totals		$3177.40		$4485.22	$56.74	$80.09

Crew No.	Bare Costs		Incl. Subs O&P		Cost Per Labor-Hour	
Crew B-30	**Hr.**	**Daily**	**Hr.**	**Daily**	**Bare Costs**	**Incl. O&P**
1 Equip. Oper. (medium)	$50.60	$404.80	$76.75	$614.00	$43.57	$65.95
2 Truck Drivers (heavy)	40.05	640.80	60.55	968.80		
1 Hyd. Excavator, 1.5 C.Y.		1030.00		1133.00		
2 Dump Trucks, 12 C.Y., 400 H.P.		1382.00		1520.20	100.50	110.55
24 L.H., Daily Totals		$3457.60		$4236.00	$144.07	$176.50
Crew B-31	**Hr.**	**Daily**	**Hr.**	**Daily**	**Bare Costs**	**Incl. O&P**
1 Labor Foreman (outside)	$39.60	$316.80	$60.95	$487.60	$39.87	$61.35
3 Laborers	37.60	902.40	57.85	1388.40		
1 Carpenter	46.95	375.60	72.25	578.00		
1 Air Compressor, 250 cfm		201.40		221.54		
1 Sheeting Driver		5.75		6.33		
2 -50' Air Hoses, 1.5"		11.60		12.76	5.47	6.02
40 L.H., Daily Totals		$1813.55		$2694.63	$45.34	$67.37
Crew B-32	**Hr.**	**Daily**	**Hr.**	**Daily**	**Bare Costs**	**Incl. O&P**
1 Laborer	$37.60	$300.80	$57.85	$462.80	$47.35	$72.03
3 Equip. Oper. (medium)	50.60	1214.40	76.75	1842.00		
1 Grader, 30,000 Lbs.		736.20		809.82		
1 Tandem Roller, 10 Ton		236.80		260.48		
1 Dozer, 200 H.P.		1387.00		1525.70	73.75	81.13
32 L.H., Daily Totals		$3875.20		$4900.80	$121.10	$153.15
Crew B-32A	**Hr.**	**Daily**	**Hr.**	**Daily**	**Bare Costs**	**Incl. O&P**
1 Laborer	$37.60	$300.80	$57.85	$462.80	$46.27	$70.45
2 Equip. Oper. (medium)	50.60	809.60	76.75	1228.00		
1 Grader, 30,000 Lbs.		736.20		809.82		
1 Roller, Vibratory, 25 Ton		687.40		756.14	59.32	65.25
24 L.H., Daily Totals		$2534.00		$3256.76	$105.58	$135.70
Crew B-32B	**Hr.**	**Daily**	**Hr.**	**Daily**	**Bare Costs**	**Incl. O&P**
1 Laborer	$37.60	$300.80	$57.85	$462.80	$46.27	$70.45
2 Equip. Oper. (medium)	50.60	809.60	76.75	1228.00		
1 Dozer, 200 H.P.		1387.00		1525.70		
1 Roller, Vibratory, 25 Ton		687.40		756.14	86.43	95.08
24 L.H., Daily Totals		$3184.80		$3972.64	$132.70	$165.53
Crew B-32C	**Hr.**	**Daily**	**Hr.**	**Daily**	**Bare Costs**	**Incl. O&P**
1 Labor Foreman (outside)	$39.60	$316.80	$60.95	$487.60	$44.43	$67.82
2 Laborers	37.60	601.60	57.85	925.60		
3 Equip. Oper. (medium)	50.60	1214.40	76.75	1842.00		
1 Grader, 30,000 Lbs.		736.20		809.82		
1 Tandem Roller, 10 Ton		236.80		260.48		
1 Dozer, 200 H.P.		1387.00		1525.70	49.17	54.08
48 L.H., Daily Totals		$4492.80		$5851.20	$93.60	$121.90
Crew B-33A	**Hr.**	**Daily**	**Hr.**	**Daily**	**Bare Costs**	**Incl. O&P**
1 Equip. Oper. (medium)	$50.60	$404.80	$76.75	$614.00	$46.89	$71.35
.5 Laborer	37.60	150.40	57.85	231.40		
.25 Equip. Oper. (medium)	50.60	101.20	76.75	153.50		
1 Scraper, Towed, 7 C.Y.		114.40		125.84		
1.25 Dozers, 300 H.P.		2371.25		2608.38	177.55	195.30
14 L.H., Daily Totals		$3142.05		$3733.11	$224.43	$266.65
Crew B-33B	**Hr.**	**Daily**	**Hr.**	**Daily**	**Bare Costs**	**Incl. O&P**
1 Equip. Oper. (medium)	$50.60	$404.80	$76.75	$614.00	$46.89	$71.35
.5 Laborer	37.60	150.40	57.85	231.40		
.25 Equip. Oper. (medium)	50.60	101.20	76.75	153.50		
1 Scraper, Towed, 10 C.Y.		146.80		161.48		
1.25 Dozers, 300 H.P.		2371.25		2608.38	179.86	197.85
14 L.H., Daily Totals		$3174.45		$3768.76	$226.75	$269.20

Crew No.	Bare Costs		Incl. Subs O&P		Cost Per Labor-Hour	
Crew B-33C	**Hr.**	**Daily**	**Hr.**	**Daily**	**Bare Costs**	**Incl. O&P**
1 Equip. Oper. (medium)	$50.60	$404.80	$76.75	$614.00	$46.89	$71.35
.5 Laborer	37.60	150.40	57.85	231.40		
.25 Equip. Oper. (medium)	50.60	101.20	76.75	153.50		
1 Scraper, Towed, 15 C.Y.		164.80		181.28		
1.25 Dozers, 300 H.P.		2371.25		2608.38	181.15	199.26
14 L.H., Daily Totals		$3192.45		$3788.55	$228.03	$270.61
Crew B-33D	**Hr.**	**Daily**	**Hr.**	**Daily**	**Bare Costs**	**Incl. O&P**
1 Equip. Oper. (medium)	$50.60	$404.80	$76.75	$614.00	$46.89	$71.35
.5 Laborer	37.60	150.40	57.85	231.40		
.25 Equip. Oper. (medium)	50.60	101.20	76.75	153.50		
1 S.P. Scraper, 14 C.Y.		1884.00		2072.40		
.25 Dozer, 300 H.P.		474.25		521.67	168.45	185.29
14 L.H., Daily Totals		$3014.65		$3592.97	$215.33	$256.64
Crew B-33E	**Hr.**	**Daily**	**Hr.**	**Daily**	**Bare Costs**	**Incl. O&P**
1 Equip. Oper. (medium)	$50.60	$404.80	$76.75	$614.00	$46.89	$71.35
.5 Laborer	37.60	150.40	57.85	231.40		
.25 Equip. Oper. (medium)	50.60	101.20	76.75	153.50		
1 S.P. Scraper, 21 C.Y.		2688.00		2956.80		
.25 Dozer, 300 H.P.		474.25		521.67	225.88	248.46
14 L.H., Daily Totals		$3818.65		$4477.38	$272.76	$319.81
Crew B-33F	**Hr.**	**Daily**	**Hr.**	**Daily**	**Bare Costs**	**Incl. O&P**
1 Equip. Oper. (medium)	$50.60	$404.80	$76.75	$614.00	$46.89	$71.35
.5 Laborer	37.60	150.40	57.85	231.40		
.25 Equip. Oper. (medium)	50.60	101.20	76.75	153.50		
1 Elev. Scraper, 11 C.Y.		1163.00		1279.30		
.25 Dozer, 300 H.P.		474.25		521.67	116.95	128.64
14 L.H., Daily Totals		$2293.65		$2799.88	$163.83	$199.99
Crew B-33G	**Hr.**	**Daily**	**Hr.**	**Daily**	**Bare Costs**	**Incl. O&P**
1 Equip. Oper. (medium)	$50.60	$404.80	$76.75	$614.00	$46.89	$71.35
.5 Laborer	37.60	150.40	57.85	231.40		
.25 Equip. Oper. (medium)	50.60	101.20	76.75	153.50		
1 Elev. Scraper, 22 C.Y.		2521.00		2773.10		
.25 Dozer, 300 H.P.		474.25		521.67	213.95	235.34
14 L.H., Daily Totals		$3651.65		$4293.68	$260.83	$306.69
Crew B-33H	**Hr.**	**Daily**	**Hr.**	**Daily**	**Bare Costs**	**Incl. O&P**
.5 Laborer	$37.60	$150.40	$57.85	$231.40	$46.89	$71.35
1 Equipment Operator (med.)	50.60	404.80	76.75	614.00		
.25 Equipment Operator (med.)	50.60	101.20	76.75	153.50		
1 S.P. Scraper, 44 C.Y.		4679.00		5146.90		
.25 Dozer, 410 H.P.		601.25		661.38	377.16	414.88
14 L.H., Daily Totals		$5936.65		$6807.18	$424.05	$486.23
Crew B-33J	**Hr.**	**Daily**	**Hr.**	**Daily**	**Bare Costs**	**Incl. O&P**
1 Equipment Operator (med.)	$50.60	$404.80	$76.75	$614.00	$50.60	$76.75
1 S.P. Scraper, 14 C.Y.		1884.00		2072.40	235.50	259.05
8 L.H., Daily Totals		$2288.80		$2686.40	$286.10	$335.80
Crew B-33K	**Hr.**	**Daily**	**Hr.**	**Daily**	**Bare Costs**	**Incl. O&P**
1 Equipment Operator (med.)	$50.60	$404.80	$76.75	$614.00	$46.89	$71.35
.25 Equipment Operator (med.)	50.60	101.20	76.75	153.50		
.5 Laborer	37.60	150.40	57.85	231.40		
1 S.P. Scraper, 31 C.Y.		3693.00		4062.30		
.25 Dozer, 410 H.P.		601.25		661.38	306.73	337.41
14 L.H., Daily Totals		$4950.65		$5722.57	$353.62	$408.76

Crew No.	Bare Costs		Incl. Subs O&P		Cost Per Labor-Hour	
Crew B-34A	**Hr.**	**Daily**	**Hr.**	**Daily**	**Bare Costs**	**Incl. O&P**
1 Truck Driver (heavy)	$40.05	$320.40	$60.55	$484.40	$40.05	$60.55
1 Dump Truck, 8 C.Y., 220 H.P.		411.20		452.32	51.40	56.54
8 L.H., Daily Totals		$731.60		$936.72	$91.45	$117.09
Crew B-34B	**Hr.**	**Daily**	**Hr.**	**Daily**	**Bare Costs**	**Incl. O&P**
1 Truck Driver (heavy)	$40.05	$320.40	$60.55	$484.40	$40.05	$60.55
1 Dump Truck, 12 C.Y., 400 H.P.		691.00		760.10	86.38	95.01
8 L.H., Daily Totals		$1011.40		$1244.50	$126.43	$155.56
Crew B-34C	**Hr.**	**Daily**	**Hr.**	**Daily**	**Bare Costs**	**Incl. O&P**
1 Truck Driver (heavy)	$40.05	$320.40	$60.55	$484.40	$40.05	$60.55
1 Truck Tractor, 6x4, 380 H.P.		599.80		659.78		
1 Dump Trailer, 16.5 C.Y.		126.60		139.26	90.80	99.88
8 L.H., Daily Totals		$1046.80		$1283.44	$130.85	$160.43
Crew B-34D	**Hr.**	**Daily**	**Hr.**	**Daily**	**Bare Costs**	**Incl. O&P**
1 Truck Driver (heavy)	$40.05	$320.40	$60.55	$484.40	$40.05	$60.55
1 Truck Tractor, 6x4, 380 H.P.		599.80		659.78		
1 Dump Trailer, 20 C.Y.		141.20		155.32	92.63	101.89
8 L.H., Daily Totals		$1061.40		$1299.50	$132.68	$162.44
Crew B-34E	**Hr.**	**Daily**	**Hr.**	**Daily**	**Bare Costs**	**Incl. O&P**
1 Truck Driver (heavy)	$40.05	$320.40	$60.55	$484.40	$40.05	$60.55
1 Dump Truck, Off Hwy., 25 Ton		1351.00		1486.10	168.88	185.76
8 L.H., Daily Totals		$1671.40		$1970.50	$208.93	$246.31
Crew B-34F	**Hr.**	**Daily**	**Hr.**	**Daily**	**Bare Costs**	**Incl. O&P**
1 Truck Driver (heavy)	$40.05	$320.40	$60.55	$484.40	$40.05	$60.55
1 Dump Truck, Off Hwy., 35 Ton		1512.00		1663.20	189.00	207.90
8 L.H., Daily Totals		$1832.40		$2147.60	$229.05	$268.45
Crew B-34G	**Hr.**	**Daily**	**Hr.**	**Daily**	**Bare Costs**	**Incl. O&P**
1 Truck Driver (heavy)	$40.05	$320.40	$60.55	$484.40	$40.05	$60.55
1 Dump Truck, Off Hwy., 50 Ton		1837.00		2020.70	229.63	252.59
8 L.H., Daily Totals		$2157.40		$2505.10	$269.68	$313.14
Crew B-34H	**Hr.**	**Daily**	**Hr.**	**Daily**	**Bare Costs**	**Incl. O&P**
1 Truck Driver (heavy)	$40.05	$320.40	$60.55	$484.40	$40.05	$60.55
1 Dump Truck, Off Hwy., 65 Ton		1863.00		2049.30	232.88	256.16
8 L.H., Daily Totals		$2183.40		$2533.70	$272.93	$316.71
Crew B-34I	**Hr.**	**Daily**	**Hr.**	**Daily**	**Bare Costs**	**Incl. O&P**
1 Truck Driver (heavy)	$40.05	$320.40	$60.55	$484.40	$40.05	$60.55
1 Dump Truck, 18 C.Y., 450 H.P.		869.00		955.90	108.63	119.49
8 L.H., Daily Totals		$1189.40		$1440.30	$148.68	$180.04
Crew B-34J	**Hr.**	**Daily**	**Hr.**	**Daily**	**Bare Costs**	**Incl. O&P**
1 Truck Driver (heavy)	$40.05	$320.40	$60.55	$484.40	$40.05	$60.55
1 Dump Truck, Off Hwy., 100 Ton		2925.00		3217.50	365.63	402.19
8 L.H., Daily Totals		$3245.40		$3701.90	$405.68	$462.74
Crew B-34K	**Hr.**	**Daily**	**Hr.**	**Daily**	**Bare Costs**	**Incl. O&P**
1 Truck Driver (heavy)	$40.05	$320.40	$60.55	$484.40	$40.05	$60.55
1 Truck Tractor, 6x4, 450 H.P.		727.40		800.14		
1 Lowbed Trailer, 75 Ton		221.80		243.98	118.65	130.51
8 L.H., Daily Totals		$1269.60		$1528.52	$158.70	$191.07

Crew No.	Bare Costs		Incl. Subs O&P		Cost Per Labor-Hour	
Crew B-34L	**Hr.**	**Daily**	**Hr.**	**Daily**	**Bare Costs**	**Incl. O&P**
1 Equip. Oper. (light)	$48.60	$388.80	$73.75	$590.00	$48.60	$73.75
1 Flatbed Truck, Gas, 1.5 Ton		245.40		269.94	30.68	33.74
8 L.H., Daily Totals		$634.20		$859.94	$79.28	$107.49
Crew B-34M	**Hr.**	**Daily**	**Hr.**	**Daily**	**Bare Costs**	**Incl. O&P**
1 Equip. Oper. (light)	$48.60	$388.80	$73.75	$590.00	$48.60	$73.75
1 Flatbed Truck, Gas, 3 Ton		303.20		333.52	37.90	41.69
8 L.H., Daily Totals		$692.00		$923.52	$86.50	$115.44
Crew B-34N	**Hr.**	**Daily**	**Hr.**	**Daily**	**Bare Costs**	**Incl. O&P**
1 Truck Driver (heavy)	$40.05	$320.40	$60.55	$484.40	$45.33	$68.65
1 Equip. Oper. (medium)	50.60	404.80	76.75	614.00		
1 Truck Tractor, 6x4, 380 H.P.		599.80		659.78		
1 Flatbed Trailer, 40 Ton		154.00		169.40	47.11	51.82
16 L.H., Daily Totals		$1479.00		$1927.58	$92.44	$120.47
Crew B-34P	**Hr.**	**Daily**	**Hr.**	**Daily**	**Bare Costs**	**Incl. O&P**
1 Pipe Fitter	$59.75	$478.00	$90.20	$721.60	$49.82	$75.35
1 Truck Driver (light)	39.10	312.80	59.10	472.80		
1 Equip. Oper. (medium)	50.60	404.80	76.75	614.00		
1 Flatbed Truck, Gas, 3 Ton		303.20		333.52		
1 Backhoe Loader, 48 H.P.		364.20		400.62	27.81	30.59
24 L.H., Daily Totals		$1863.00		$2542.54	$77.63	$105.94
Crew B-34Q	**Hr.**	**Daily**	**Hr.**	**Daily**	**Bare Costs**	**Incl. O&P**
1 Pipe Fitter	$59.75	$478.00	$90.20	$721.60	$50.18	$75.92
1 Truck Driver (light)	39.10	312.80	59.10	472.80		
1 Equip. Oper. (crane)	51.70	413.60	78.45	627.60		
1 Flatbed Trailer, 25 Ton		113.60		124.96		
1 Dump Truck, 8 C.Y., 220 H.P.		411.20		452.32		
1 Hyd. Crane, 25 Ton		736.60		810.26	52.56	57.81
24 L.H., Daily Totals		$2465.80		$3209.54	$102.74	$133.73
Crew B-34R	**Hr.**	**Daily**	**Hr.**	**Daily**	**Bare Costs**	**Incl. O&P**
1 Pipe Fitter	$59.75	$478.00	$90.20	$721.60	$50.18	$75.92
1 Truck Driver (light)	39.10	312.80	59.10	472.80		
1 Equip. Oper. (crane)	51.70	413.60	78.45	627.60		
1 Flatbed Trailer, 25 Ton		113.60		124.96		
1 Dump Truck, 8 C.Y., 220 H.P.		411.20		452.32		
1 Hyd. Crane, 25 Ton		736.60		810.26		
1 Hyd. Excavator, 1 C.Y.		812.60		893.86	86.42	95.06
24 L.H., Daily Totals		$3278.40		$4103.40	$136.60	$170.97
Crew B-34S	**Hr.**	**Daily**	**Hr.**	**Daily**	**Bare Costs**	**Incl. O&P**
2 Pipe Fitters	$59.75	$956.00	$90.20	$1443.20	$52.81	$79.85
1 Truck Driver (heavy)	40.05	320.40	60.55	484.40		
1 Equip. Oper. (crane)	51.70	413.60	78.45	627.60		
1 Flatbed Trailer, 40 Ton		154.00		169.40		
1 Truck Tractor, 6x4, 380 H.P.		599.80		659.78		
1 Hyd. Crane, 80 Ton		1625.00		1787.50		
1 Hyd. Excavator, 2 C.Y.		1175.00		1292.50	111.06	122.16
32 L.H., Daily Totals		$5243.80		$6464.38	$163.87	$202.01
Crew B-34T	**Hr.**	**Daily**	**Hr.**	**Daily**	**Bare Costs**	**Incl. O&P**
2 Pipe Fitters	$59.75	$956.00	$90.20	$1443.20	$52.81	$79.85
1 Truck Driver (heavy)	40.05	320.40	60.55	484.40		
1 Equip. Oper. (crane)	51.70	413.60	78.45	627.60		
1 Flatbed Trailer, 40 Ton		154.00		169.40		
1 Truck Tractor, 6x4, 380 H.P.		599.80		659.78		
1 Hyd. Crane, 80 Ton		1625.00		1787.50	74.34	81.77
32 L.H., Daily Totals		$4068.80		$5171.88	$127.15	$161.62

Crew No.	Bare Costs		Incl. Subs O&P		Cost Per Labor-Hour	
Crew B-34U	**Hr.**	**Daily**	**Hr.**	**Daily**	**Bare Costs**	**Incl. O&P**
1 Truck Driver (heavy)	$40.05	$320.40	$60.55	$484.40	$44.33	$67.15
1 Equip. Oper. (light)	48.60	388.80	73.75	590.00		
1 Truck Tractor, 220 H.P.		359.60		395.56		
1 Flatbed Trailer, 25 Ton		113.60		124.96	29.57	32.53
16 L.H., Daily Totals		$1182.40		$1594.92	$73.90	$99.68
Crew B-34V	**Hr.**	**Daily**	**Hr.**	**Daily**	**Bare Costs**	**Incl. O&P**
1 Truck Driver (heavy)	$40.05	$320.40	$60.55	$484.40	$46.78	$70.92
1 Equip. Oper. (crane)	51.70	413.60	78.45	627.60		
1 Equip. Oper. (light)	48.60	388.80	73.75	590.00		
1 Truck Tractor, 6x4, 450 H.P.		727.40		800.14		
1 Equipment Trailer, 50 Ton		168.40		185.24		
1 Pickup Truck, 4 x 4, 3/4 Ton		155.60		171.16	43.81	48.19
24 L.H., Daily Totals		$2174.20		$2858.54	$90.59	$119.11
Crew B-34W	**Hr.**	**Daily**	**Hr.**	**Daily**	**Bare Costs**	**Incl. O&P**
5 Truck Drivers (heavy)	$40.05	$1602.00	$60.55	$2422.00	$43.66	$66.21
2 Equip. Opers. (crane)	51.70	827.20	78.45	1255.20		
1 Equip. Oper. (mechanic)	51.65	413.20	78.35	626.80		
1 Laborer	37.60	300.80	57.85	462.80		
4 Truck Tractor, 6x4, 380 H.P.		2399.20		2639.12		
2 Equipment Trailer, 50 Ton		336.80		370.48		
2 Flatbed Trailer, 40 Ton		308.00		338.80		
1 Pickup Truck, 4 x 4, 3/4 Ton		155.60		171.16		
1 S.P. Crane, 4x4, 20 Ton		554.60		610.06	52.14	57.36
72 L.H., Daily Totals		$6897.40		$8896.42	$95.80	$123.56
Crew B-35	**Hr.**	**Daily**	**Hr.**	**Daily**	**Bare Costs**	**Incl. O&P**
1 Labor Foreman (outside)	$39.60	$316.80	$60.95	$487.60	$46.91	$71.60
1 Skilled Worker	48.65	389.20	75.15	601.20		
1 Welder (plumber)	58.70	469.60	88.65	709.20		
1 Laborer	37.60	300.80	57.85	462.80		
1 Equip. Oper. (crane)	51.70	413.60	78.45	627.60		
1 Equip. Oper. (oiler)	45.20	361.60	68.55	548.40		
1 Welder, Electric, 300 amp		57.70		63.47		
1 Hyd. Excavator, .75 C.Y.		654.80		720.28	14.84	16.33
48 L.H., Daily Totals		$2964.10		$4220.55	$61.75	$87.93
Crew B-35A	**Hr.**	**Daily**	**Hr.**	**Daily**	**Bare Costs**	**Incl. O&P**
1 Labor Foreman (outside)	$39.60	$316.80	$60.95	$487.60	$45.58	$69.64
2 Laborers	37.60	601.60	57.85	925.60		
1 Skilled Worker	48.65	389.20	75.15	601.20		
1 Welder (plumber)	58.70	469.60	88.65	709.20		
1 Equip. Oper. (crane)	51.70	413.60	78.45	627.60		
1 Equip. Oper. (oiler)	45.20	361.60	68.55	548.40		
1 Welder, Gas Engine, 300 amp		145.85		160.44		
1 Crawler Crane, 75 Ton		1459.00		1604.90	28.66	31.52
56 L.H., Daily Totals		$4157.25		$5664.94	$74.24	$101.16
Crew B-36	**Hr.**	**Daily**	**Hr.**	**Daily**	**Bare Costs**	**Incl. O&P**
1 Labor Foreman (outside)	$39.60	$316.80	$60.95	$487.60	$43.20	$66.03
2 Laborers	37.60	601.60	57.85	925.60		
2 Equip. Oper. (medium)	50.60	809.60	76.75	1228.00		
1 Dozer, 200 H.P.		1387.00		1525.70		
1 Aggregate Spreader		39.60		43.56		
1 Tandem Roller, 10 Ton		236.80		260.48	41.59	45.74
40 L.H., Daily Totals		$3391.40		$4470.94	$84.78	$111.77

Crew No.	Bare Costs		Incl. Subs O&P		Cost Per Labor-Hour	
Crew B-36A	**Hr.**	**Daily**	**Hr.**	**Daily**	**Bare Costs**	**Incl. O&P**
1 Labor Foreman (outside)	$39.60	$316.80	$60.95	$487.60	$45.31	$69.09
2 Laborers	37.60	601.60	57.85	925.60		
4 Equip. Oper. (medium)	50.60	1619.20	76.75	2456.00		
1 Dozer, 200 H.P.		1387.00		1525.70		
1 Aggregate Spreader		39.60		43.56		
1 Tandem Roller, 10 Ton		236.80		260.48		
1 Roller, Pneum. Whl., 12 Ton		344.40		378.84	35.85	39.44
56 L.H., Daily Totals		$4545.40		$6077.78	$81.17	$108.53
Crew B-36B	**Hr.**	**Daily**	**Hr.**	**Daily**	**Bare Costs**	**Incl. O&P**
1 Labor Foreman (outside)	$39.60	$316.80	$60.95	$487.60	$44.66	$68.03
2 Laborers	37.60	601.60	57.85	925.60		
4 Equip. Oper. (medium)	50.60	1619.20	76.75	2456.00		
1 Truck Driver (heavy)	40.05	320.40	60.55	484.40		
1 Grader, 30,000 Lbs.		736.20		809.82		
1 F.E. Loader, Crl, 1.5 C.Y.		629.80		692.78		
1 Dozer, 300 H.P.		1897.00		2086.70		
1 Roller, Vibratory, 25 Ton		687.40		756.14		
1 Truck Tractor, 6x4, 450 H.P.		727.40		800.14		
1 Water Tank Trailer, 5000 Gal.		141.00		155.10	75.29	82.82
64 L.H., Daily Totals		$7676.80		$9654.28	$119.95	$150.85
Crew B-36C	**Hr.**	**Daily**	**Hr.**	**Daily**	**Bare Costs**	**Incl. O&P**
1 Labor Foreman (outside)	$39.60	$316.80	$60.95	$487.60	$46.29	$70.35
3 Equip. Oper. (medium)	50.60	1214.40	76.75	1842.00		
1 Truck Driver (heavy)	40.05	320.40	60.55	484.40		
1 Grader, 30,000 Lbs.		736.20		809.82		
1 Dozer, 300 H.P.		1897.00		2086.70		
1 Roller, Vibratory, 25 Ton		687.40		756.14		
1 Truck Tractor, 6x4, 450 H.P.		727.40		800.14		
1 Water Tank Trailer, 5000 Gal.		141.00		155.10	104.72	115.20
40 L.H., Daily Totals		$6040.60		$7421.90	$151.01	$185.55
Crew B-36D	**Hr.**	**Daily**	**Hr.**	**Daily**	**Bare Costs**	**Incl. O&P**
1 Labor Foreman (outside)	$39.60	$316.80	$60.95	$487.60	$47.85	$72.80
3 Equip. Oper. (medium)	50.60	1214.40	76.75	1842.00		
1 Grader, 30,000 Lbs.		736.20		809.82		
1 Dozer, 300 H.P.		1897.00		2086.70		
1 Roller, Vibratory, 25 Ton		687.40		756.14	103.77	114.15
32 L.H., Daily Totals		$4851.80		$5982.26	$151.62	$186.95
Crew B-37	**Hr.**	**Daily**	**Hr.**	**Daily**	**Bare Costs**	**Incl. O&P**
1 Labor Foreman (outside)	$39.60	$316.80	$60.95	$487.60	$39.77	$61.02
4 Laborers	37.60	1203.20	57.85	1851.20		
1 Equip. Oper. (light)	48.60	388.80	73.75	590.00		
1 Tandem Roller, 5 Ton		157.40		173.14	3.28	3.61
48 L.H., Daily Totals		$2066.20		$3101.94	$43.05	$64.62
Crew B-37A	**Hr.**	**Daily**	**Hr.**	**Daily**	**Bare Costs**	**Incl. O&P**
2 Laborers	$37.60	$601.60	$57.85	$925.60	$38.10	$58.27
1 Truck Driver (light)	39.10	312.80	59.10	472.80		
1 Flatbed Truck, Gas, 1.5 Ton		245.40		269.94		
1 Tar Kettle, T.M.		132.00		145.20	15.73	17.30
24 L.H., Daily Totals		$1291.80		$1813.54	$53.83	$75.56
Crew B-37B	**Hr.**	**Daily**	**Hr.**	**Daily**	**Bare Costs**	**Incl. O&P**
3 Laborers	$37.60	$902.40	$57.85	$1388.40	$37.98	$58.16
1 Truck Driver (light)	39.10	312.80	59.10	472.80		
1 Flatbed Truck, Gas, 1.5 Ton		245.40		269.94		
1 Tar Kettle, T.M.		132.00		145.20	11.79	12.97
32 L.H., Daily Totals		$1592.60		$2276.34	$49.77	$71.14

Crew No.	Bare Costs		Incl. Subs O&P		Cost Per Labor-Hour	
Crew B-37C	**Hr.**	**Daily**	**Hr.**	**Daily**	**Bare Costs**	**Incl. O&P**
2 Laborers	$37.60	$601.60	$57.85	$925.60	$38.35	$58.48
2 Truck Drivers (light)	39.10	625.60	59.10	945.60		
2 Flatbed Trucks, Gas, 1.5 Ton		490.80		539.88		
1 Tar Kettle, T.M.		132.00		145.20	19.46	21.41
32 L.H., Daily Totals		$1850.00		$2556.28	$57.81	$79.88
Crew B-37D	**Hr.**	**Daily**	**Hr.**	**Daily**	**Bare Costs**	**Incl. O&P**
1 Laborer	$37.60	$300.80	$57.85	$462.80	$38.35	$58.48
1 Truck Driver (light)	39.10	312.80	59.10	472.80		
1 Pickup Truck, 3/4 Ton		144.20		158.62	9.01	9.91
16 L.H., Daily Totals		$757.80		$1094.22	$47.36	$68.39
Crew B-37E	**Hr.**	**Daily**	**Hr.**	**Daily**	**Bare Costs**	**Incl. O&P**
3 Laborers	$37.60	$902.40	$57.85	$1388.40	$41.46	$63.18
1 Equip. Oper. (light)	48.60	388.80	73.75	590.00		
1 Equip. Oper. (medium)	50.60	404.80	76.75	614.00		
2 Truck Drivers (light)	39.10	625.60	59.10	945.60		
4 Barrels w/ Flasher		13.60		14.96		
1 Concrete Saw		167.40		184.14		
1 Rotary Hammer Drill		23.75		26.13		
1 Hammer Drill Bit		2.35		2.59		
1 Loader, Skid Steer, 30 H.P.		173.60		190.96		
1 Conc. Hammer Attach.		114.70		126.17		
1 Vibrating Plate, Gas, 18"		36.20		39.82		
2 Flatbed Trucks, Gas, 1.5 Ton		490.80		539.88	18.26	20.08
56 L.H., Daily Totals		$3344.00		$4662.64	$59.71	$83.26
Crew B-37F	**Hr.**	**Daily**	**Hr.**	**Daily**	**Bare Costs**	**Incl. O&P**
3 Laborers	$37.60	$902.40	$57.85	$1388.40	$37.98	$58.16
1 Truck Driver (light)	39.10	312.80	59.10	472.80		
4 Barrels w/ Flasher		13.60		14.96		
1 Concrete Mixer, 10 C.F.		172.40		189.64		
1 Air Compressor, 60 cfm		137.40		151.14		
1 -50' Air Hose, 3/4"		3.25		3.58		
1 Spade (Chipper)		7.60		8.36		
1 Flatbed Truck, Gas, 1.5 Ton		245.40		269.94	18.11	19.93
32 L.H., Daily Totals		$1794.85		$2498.82	$56.09	$78.09
Crew B-37G	**Hr.**	**Daily**	**Hr.**	**Daily**	**Bare Costs**	**Incl. O&P**
1 Labor Foreman (outside)	$39.60	$316.80	$60.95	$487.60	$39.77	$61.02
4 Laborers	37.60	1203.20	57.85	1851.20		
1 Equip. Oper. (light)	48.60	388.80	73.75	590.00		
1 Berm Machine		296.80		326.48		
1 Tandem Roller, 5 Ton		157.40		173.14	9.46	10.41
48 L.H., Daily Totals		$2363.00		$3428.42	$49.23	$71.43
Crew B-37H	**Hr.**	**Daily**	**Hr.**	**Daily**	**Bare Costs**	**Incl. O&P**
1 Labor Foreman (outside)	$39.60	$316.80	$60.95	$487.60	$39.77	$61.02
4 Laborers	37.60	1203.20	57.85	1851.20		
1 Equip. Oper. (light)	48.60	388.80	73.75	590.00		
1 Tandem Roller, 5 Ton		157.40		173.14		
1 Flatbed Trucks, Gas, 1.5 Ton		245.40		269.94		
1 Tar Kettle, T.M.		132.00		145.20	11.14	12.26
48 L.H., Daily Totals		$2443.60		$3517.08	$50.91	$73.27

Crew No.	Bare Costs		Incl. Subs O&P		Cost Per Labor-Hour	
Crew B-37I	**Hr.**	**Daily**	**Hr.**	**Daily**	**Bare Costs**	**Incl. O&P**
3 Laborers	$37.60	$902.40	$57.85	$1388.40	$41.46	$63.18
1 Equip. Oper. (light)	48.60	388.80	73.75	590.00		
1 Equip. Oper. (medium)	50.60	404.80	76.75	614.00		
2 Truck Drivers (light)	39.10	625.60	59.10	945.60		
4 Barrels w/ Flasher		13.60		14.96		
1 Concrete Saw		167.40		184.14		
1 Rotary Hammer Drill		23.75		26.13		
1 Hammer Drill Bit		2.35		2.59		
1 Air Compressor, 60 cfm		137.40		151.14		
1 -50' Air Hose, 3/4"		3.25		3.58		
1 Spade (Chipper)		7.60		8.36		
1 Loader, Skid Steer, 30 H.P.		173.60		190.96		
1 Conc. Hammer Attach.		114.70		126.17		
1 Concrete Mixer, 10 C.F.		172.40		189.64		
1 Vibrating Plate, Gas, 18"		36.20		39.82		
2 Flatbed Trucks, Gas, 1.5 Ton		490.80		539.88	23.98	26.38
56 L.H., Daily Totals		$3664.65		$5015.35	$65.44	$89.56
Crew B-37J	**Hr.**	**Daily**	**Hr.**	**Daily**	**Bare Costs**	**Incl. O&P**
1 Labor Foreman (outside)	$39.60	$316.80	$60.95	$487.60	$39.77	$61.02
4 Laborers	37.60	1203.20	57.85	1851.20		
1 Equip. Oper. (light)	48.60	388.80	73.75	590.00		
1 Air Compressor, 60 cfm		137.40		151.14		
1 -50' Air Hose, 3/4"		3.25		3.58		
2 Concrete Mixer, 10 C.F.		344.80		379.28		
2 Flatbed Trucks, Gas, 1.5 Ton		490.80		539.88		
1 Shot Blaster, 20"		214.00		235.40	24.80	27.28
48 L.H., Daily Totals		$3099.05		$4238.07	$64.56	$88.29
Crew B-37K	**Hr.**	**Daily**	**Hr.**	**Daily**	**Bare Costs**	**Incl. O&P**
1 Labor Foreman (outside)	$39.60	$316.80	$60.95	$487.60	$39.77	$61.02
4 Laborers	37.60	1203.20	57.85	1851.20		
1 Equip. Oper. (light)	48.60	388.80	73.75	590.00		
1 Air Compressor, 60 cfm		137.40		151.14		
1 -50' Air Hose, 3/4"		3.25		3.58		
2 Flatbed Trucks, Gas, 1.5 Ton		490.80		539.88		
1 Shot Blaster, 20"		214.00		235.40	17.61	19.37
48 L.H., Daily Totals		$2754.25		$3858.80	$57.38	$80.39
Crew B-38	**Hr.**	**Daily**	**Hr.**	**Daily**	**Bare Costs**	**Incl. O&P**
1 Labor Foreman (outside)	$39.60	$316.80	$60.95	$487.60	$42.80	$65.43
2 Laborers	37.60	601.60	57.85	925.60		
1 Equip. Oper. (light)	48.60	388.80	73.75	590.00		
1 Equip. Oper. (medium)	50.60	404.80	76.75	614.00		
1 Backhoe Loader, 48 H.P.		364.20		400.62		
1 Hyd. Hammer, (1200 lb.)		180.40		198.44		
1 F.E. Loader, W.M., 4 C.Y.		669.20		736.12		
1 Pvmt. Rem. Bucket		61.20		67.32	31.88	35.06
40 L.H., Daily Totals		$2987.00		$4019.70	$74.67	$100.49
Crew B-39	**Hr.**	**Daily**	**Hr.**	**Daily**	**Bare Costs**	**Incl. O&P**
1 Labor Foreman (outside)	$39.60	$316.80	$60.95	$487.60	$39.77	$61.02
4 Laborers	37.60	1203.20	57.85	1851.20		
1 Equip. Oper. (light)	48.60	388.80	73.75	590.00		
1 Air Compressor, 250 cfm		201.40		221.54		
2 Breakers, Pavement, 60 lb.		19.60		21.56		
2 -50' Air Hoses, 1.5"		11.60		12.76	4.85	5.33
48 L.H., Daily Totals		$2141.40		$3184.66	$44.61	$66.35

Crew B-40	Bare Costs Hr.	Bare Costs Daily	Incl. Subs O&P Hr.	Incl. Subs O&P Daily	Cost Per Labor-Hour Bare Costs	Cost Per Labor-Hour Incl. O&P
1 Pile Driver Foreman (outside)	$48.10	$384.80	$76.15	$609.20	$47.64	$74.20
4 Pile Drivers	46.10	1475.20	73.00	2336.00		
2 Equip. Oper. (crane)	51.70	827.20	78.45	1255.20		
1 Equip. Oper. (oiler)	45.20	361.60	68.55	548.40		
1 Crawler Crane, 40 Ton		1156.00		1271.60		
1 Vibratory Hammer & Gen.		2587.00		2845.70	58.48	64.33
64 L.H., Daily Totals		$6791.80		$8866.10	$106.12	$138.53

Crew B-40B	Hr.	Daily	Hr.	Daily	Bare Costs	Incl. O&P
1 Labor Foreman (outside)	$39.60	$316.80	$60.95	$487.60	$41.55	$63.58
3 Laborers	37.60	902.40	57.85	1388.40		
1 Equip. Oper. (crane)	51.70	413.60	78.45	627.60		
1 Equip. Oper. (oiler)	45.20	361.60	68.55	548.40		
1 Lattice Boom Crane, 40 Ton		1164.00		1280.40	24.25	26.68
48 L.H., Daily Totals		$3158.40		$4332.40	$65.80	$90.26

Crew B-41	Hr.	Daily	Hr.	Daily	Bare Costs	Incl. O&P
1 Labor Foreman (outside)	$39.60	$316.80	$60.95	$487.60	$38.95	$59.84
4 Laborers	37.60	1203.20	57.85	1851.20		
.25 Equip. Oper. (crane)	51.70	103.40	78.45	156.90		
.25 Equip. Oper. (oiler)	45.20	90.40	68.55	137.10		
.25 Crawler Crane, 40 Ton		289.00		317.90	6.57	7.22
44 L.H., Daily Totals		$2002.80		$2950.70	$45.52	$67.06

Crew B-42	Hr.	Daily	Hr.	Daily	Bare Costs	Incl. O&P
1 Labor Foreman (outside)	$39.60	$316.80	$60.95	$487.60	$42.44	$66.35
4 Laborers	37.60	1203.20	57.85	1851.20		
1 Equip. Oper. (crane)	51.70	413.60	78.45	627.60		
1 Equip. Oper. (oiler)	45.20	361.60	68.55	548.40		
1 Welder	52.65	421.20	91.45	731.60		
1 Hyd. Crane, 25 Ton		736.60		810.26		
1 Welder, Gas Engine, 300 amp		145.85		160.44		
1 Horz. Boring Csg. Mch.		474.80		522.28	21.21	23.33
64 L.H., Daily Totals		$4073.65		$5739.38	$63.65	$89.68

Crew B-43	Hr.	Daily	Hr.	Daily	Bare Costs	Incl. O&P
1 Labor Foreman (outside)	$39.60	$316.80	$60.95	$487.60	$41.55	$63.58
3 Laborers	37.60	902.40	57.85	1388.40		
1 Equip. Oper. (crane)	51.70	413.60	78.45	627.60		
1 Equip. Oper. (oiler)	45.20	361.60	68.55	548.40		
1 Drill Rig, Truck-Mounted		2542.00		2796.20	52.96	58.25
48 L.H., Daily Totals		$4536.40		$5848.20	$94.51	$121.84

Crew B-44	Hr.	Daily	Hr.	Daily	Bare Costs	Incl. O&P
1 Pile Driver Foreman (outside)	$48.10	$384.80	$76.15	$609.20	$46.69	$72.86
4 Pile Drivers	46.10	1475.20	73.00	2336.00		
2 Equip. Oper. (crane)	51.70	827.20	78.45	1255.20		
1 Laborer	37.60	300.80	57.85	462.80		
1 Crawler Crane, 40 Ton		1156.00		1271.60		
1 Lead, 60' High		74.80		82.28		
1 Hammer, Diesel, 15K ft.-lbs.		586.00		644.60	28.39	31.23
64 L.H., Daily Totals		$4804.80		$6661.68	$75.08	$104.09

Crew B-45	Hr.	Daily	Hr.	Daily	Bare Costs	Incl. O&P
1 Equip. Oper. (medium)	$50.60	$404.80	$76.75	$614.00	$45.33	$68.65
1 Truck Driver (heavy)	40.05	320.40	60.55	484.40		
1 Dist. Tanker, 3000 Gallon		305.60		336.16		
1 Truck Tractor, 6x4, 380 H.P.		599.80		659.78	56.59	62.25
16 L.H., Daily Totals		$1630.60		$2094.34	$101.91	$130.90

Crew B-46	Hr.	Daily	Hr.	Daily	Bare Costs	Incl. O&P
1 Pile Driver Foreman (outside)	$48.10	$384.80	$76.15	$609.20	$42.18	$65.95
2 Pile Drivers	46.10	737.60	73.00	1168.00		
3 Laborers	37.60	902.40	57.85	1388.40		
1 Chain Saw, Gas, 36" Long		45.00		49.50	.94	1.03
48 L.H., Daily Totals		$2069.80		$3215.10	$43.12	$66.98

Crew B-47	Hr.	Daily	Hr.	Daily	Bare Costs	Incl. O&P
1 Blast Foreman (outside)	$39.60	$316.80	$60.95	$487.60	$41.93	$64.18
1 Driller	37.60	300.80	57.85	462.80		
1 Equip. Oper. (light)	48.60	388.80	73.75	590.00		
1 Air Track Drill, 4"		1028.00		1130.80		
1 Air Compressor, 600 cfm		550.60		605.66		
2 -50' Air Hoses, 3"		29.80		32.78	67.02	73.72
24 L.H., Daily Totals		$2614.80		$3309.64	$108.95	$137.90

Crew B-47A	Hr.	Daily	Hr.	Daily	Bare Costs	Incl. O&P
1 Drilling Foreman (outside)	$39.60	$316.80	$60.95	$487.60	$45.50	$69.32
1 Equip. Oper. (heavy)	51.70	413.60	78.45	627.60		
1 Equip. Oper. (oiler)	45.20	361.60	68.55	548.40		
1 Air Track Drill, 5"		1243.00		1367.30	51.79	56.97
24 L.H., Daily Totals		$2335.00		$3030.90	$97.29	$126.29

Crew B-47C	Hr.	Daily	Hr.	Daily	Bare Costs	Incl. O&P
1 Laborer	$37.60	$300.80	$57.85	$462.80	$43.10	$65.80
1 Equip. Oper. (light)	48.60	388.80	73.75	590.00		
1 Air Compressor, 750 cfm		557.20		612.92		
2 -50' Air Hoses, 3"		29.80		32.78		
1 Air Track Drill, 4"		1028.00		1130.80	100.94	111.03
16 L.H., Daily Totals		$2304.60		$2829.30	$144.04	$176.83

Crew B-47E	Hr.	Daily	Hr.	Daily	Bare Costs	Incl. O&P
1 Labor Foreman (outside)	$39.60	$316.80	$60.95	$487.60	$38.10	$58.63
3 Laborers	37.60	902.40	57.85	1388.40		
1 Flatbed Truck, Gas, 3 Ton		303.20		333.52	9.47	10.42
32 L.H., Daily Totals		$1522.40		$2209.52	$47.58	$69.05

Crew B-47G	Hr.	Daily	Hr.	Daily	Bare Costs	Incl. O&P
1 Labor Foreman (outside)	$39.60	$316.80	$60.95	$487.60	$40.85	$62.60
2 Laborers	37.60	601.60	57.85	925.60		
1 Equip. Oper. (light)	48.60	388.80	73.75	590.00		
1 Air Track Drill, 4"		1028.00		1130.80		
1 Air Compressor, 600 cfm		550.60		605.66		
2 -50' Air Hoses, 3"		29.80		32.78		
1 Gunite Pump Rig		371.80		408.98	61.88	68.07
32 L.H., Daily Totals		$3287.40		$4181.42	$102.73	$130.67

Crew B-47H	Hr.	Daily	Hr.	Daily	Bare Costs	Incl. O&P
1 Skilled Worker Foreman (out)	$50.65	$405.20	$78.25	$626.00	$49.15	$75.92
3 Skilled Workers	48.65	1167.60	75.15	1803.60		
1 Flatbed Truck, Gas, 3 Ton		303.20		333.52	9.47	10.42
32 L.H., Daily Totals		$1876.00		$2763.12	$58.63	$86.35

Crew B-48	Bare Costs Hr.	Bare Costs Daily	Incl. Subs O&P Hr.	Incl. Subs O&P Daily	Cost Per Labor-Hour Bare Costs	Cost Per Labor-Hour Incl. O&P
1 Labor Foreman (outside)	$39.60	$316.80	$60.95	$487.60	$42.56	$65.04
3 Laborers	37.60	902.40	57.85	1388.40		
1 Equip. Oper. (crane)	51.70	413.60	78.45	627.60		
1 Equip. Oper. (oiler)	45.20	361.60	68.55	548.40		
1 Equip. Oper. (light)	48.60	388.80	73.75	590.00		
1 Centr. Water Pump, 6"		346.60		381.26		
1 -20' Suction Hose, 6"		11.50		12.65		
1 -50' Discharge Hose, 6"		6.10		6.71		
1 Drill Rig, Truck-Mounted		2542.00		2796.20	51.90	57.09
56 L.H., Daily Totals		$5289.40		$6838.82	$94.45	$122.12

Crew B-49	Hr.	Daily	Hr.	Daily	Bare Costs	Incl. O&P
1 Labor Foreman (outside)	$39.60	$316.80	$60.95	$487.60	$44.27	$68.02
3 Laborers	37.60	902.40	57.85	1388.40		
2 Equip. Oper. (crane)	51.70	827.20	78.45	1255.20		
2 Equip. Oper. (oilers)	45.20	723.20	68.55	1096.80		
1 Equip. Oper. (light)	48.60	388.80	73.75	590.00		
2 Pile Drivers	46.10	737.60	73.00	1168.00		
1 Hyd. Crane, 25 Ton		736.60		810.26		
1 Centr. Water Pump, 6"		346.60		381.26		
1 -20' Suction Hose, 6"		11.50		12.65		
1 -50' Discharge Hose, 6"		6.10		6.71		
1 Drill Rig, Truck-Mounted		2542.00		2796.20	41.40	45.53
88 L.H., Daily Totals		$7538.80		$9993.08	$85.67	$113.56

Crew B-50	Hr.	Daily	Hr.	Daily	Bare Costs	Incl. O&P
2 Pile Driver Foremen (outside)	$48.10	$769.60	$76.15	$1218.40	$45.30	$70.66
6 Pile Drivers	46.10	2212.80	73.00	3504.00		
2 Equip. Oper. (crane)	51.70	827.20	78.45	1255.20		
1 Equip. Oper. (oiler)	45.20	361.60	68.55	548.40		
3 Laborers	37.60	902.40	57.85	1388.40		
1 Crawler Crane, 40 Ton		1156.00		1271.60		
1 Lead, 60' High		74.80		82.28		
1 Hammer, Diesel, 15K ft.-lbs.		586.00		644.60		
1 Air Compressor, 600 cfm		550.60		605.66		
2 -50' Air Hoses, 3"		29.80		32.78		
1 Chain Saw, Gas, 36" Long		45.00		49.50	21.81	23.99
112 L.H., Daily Totals		$7515.80		$10600.82	$67.11	$94.65

Crew B-51	Hr.	Daily	Hr.	Daily	Bare Costs	Incl. O&P
1 Labor Foreman (outside)	$39.60	$316.80	$60.95	$487.60	$38.18	$58.58
4 Laborers	37.60	1203.20	57.85	1851.20		
1 Truck Driver (light)	39.10	312.80	59.10	472.80		
1 Flatbed Truck, Gas, 1.5 Ton		245.40		269.94	5.11	5.62
48 L.H., Daily Totals		$2078.20		$3081.54	$43.30	$64.20

Crew B-52	Hr.	Daily	Hr.	Daily	Bare Costs	Incl. O&P
1 Carpenter Foreman (outside)	$48.95	$391.60	$75.35	$602.80	$43.61	$66.75
1 Carpenter	46.95	375.60	72.25	578.00		
3 Laborers	37.60	902.40	57.85	1388.40		
1 Cement Finisher	45.00	360.00	66.65	533.20		
.5 Rodman (reinf.)	52.55	210.20	82.10	328.40		
.5 Equip. Oper. (medium)	50.60	202.40	76.75	307.00		
.5 Crawler Loader, 3 C.Y.		594.00		653.40	10.61	11.67
56 L.H., Daily Totals		$3036.20		$4391.20	$54.22	$78.41

Crew B-53	Hr.	Daily	Hr.	Daily	Bare Costs	Incl. O&P
1 Equip. Oper. (light)	$48.60	$388.80	$73.75	$590.00	$48.60	$73.75
1 Trencher, Chain, 12 H.P.		68.00		74.80	8.50	9.35
8 L.H., Daily Totals		$456.80		$664.80	$57.10	$83.10

Crew B-54	Hr.	Daily	Hr.	Daily	Bare Costs	Incl. O&P
1 Equip. Oper. (light)	$48.60	$388.80	$73.75	$590.00	$48.60	$73.75
1 Trencher, Chain, 40 H.P.		335.20		368.72	41.90	46.09
8 L.H., Daily Totals		$724.00		$958.72	$90.50	$119.84

Crew B-54A	Hr.	Daily	Hr.	Daily	Bare Costs	Incl. O&P
.17 Labor Foreman (outside)	$39.60	$53.86	$60.95	$82.89	$49.00	$74.45
1 Equipment Operator (med.)	50.60	404.80	76.75	614.00		
1 Wheel Trencher, 67 H.P.		1195.00		1314.50	127.67	140.44
9.36 L.H., Daily Totals		$1653.66		$2011.39	$176.67	$214.89

Crew B-54B	Hr.	Daily	Hr.	Daily	Bare Costs	Incl. O&P
.25 Labor Foreman (outside)	$39.60	$79.20	$60.95	$121.90	$48.40	$73.59
1 Equipment Operator (med.)	50.60	404.80	76.75	614.00		
1 Wheel Trencher, 150 H.P.		1890.00		2079.00	189.00	207.90
10 L.H., Daily Totals		$2374.00		$2814.90	$237.40	$281.49

Crew B-54C	Hr.	Daily	Hr.	Daily	Bare Costs	Incl. O&P
1 Laborer	$37.60	$300.80	$57.85	$462.80	$44.10	$67.30
1 Equipment Operator (med.)	50.60	404.80	76.75	614.00		
1 Wheel Trencher, 67 H.P.		1195.00		1314.50	74.69	82.16
16 L.H., Daily Totals		$1900.60		$2391.30	$118.79	$149.46

Crew B-54D	Hr.	Daily	Hr.	Daily	Bare Costs	Incl. O&P
1 Laborer	$37.60	$300.80	$57.85	$462.80	$44.10	$67.30
1 Equipment Operator (med.)	50.60	404.80	76.75	614.00		
1 Rock Trencher, 6" Width		372.00		409.20	23.25	25.57
16 L.H., Daily Totals		$1077.60		$1486.00	$67.35	$92.88

Crew B-54E	Hr.	Daily	Hr.	Daily	Bare Costs	Incl. O&P
1 Laborer	$37.60	$300.80	$57.85	$462.80	$44.10	$67.30
1 Equipment Operator (med.)	50.60	404.80	76.75	614.00		
1 Rock Trencher, 18" Width		2575.00		2832.50	160.94	177.03
16 L.H., Daily Totals		$3280.60		$3909.30	$205.04	$244.33

Crew B-55	Hr.	Daily	Hr.	Daily	Bare Costs	Incl. O&P
2 Laborers	$37.60	$601.60	$57.85	$925.60	$38.10	$58.27
1 Truck Driver (light)	39.10	312.80	59.10	472.80		
1 Truck-Mounted Earth Auger		779.20		857.12		
1 Flatbed Truck, Gas, 3 Ton		303.20		333.52	45.10	49.61
24 L.H., Daily Totals		$1996.80		$2589.04	$83.20	$107.88

Crew B-56	Hr.	Daily	Hr.	Daily	Bare Costs	Incl. O&P
1 Laborer	$37.60	$300.80	$57.85	$462.80	$43.10	$65.80
1 Equip. Oper. (light)	48.60	388.80	73.75	590.00		
1 Air Track Drill, 4"		1028.00		1130.80		
1 Air Compressor, 600 cfm		550.60		605.66		
1 -50' Air Hose, 3"		14.90		16.39	99.59	109.55
16 L.H., Daily Totals		$2283.10		$2805.65	$142.69	$175.35

Crew No.	Bare Costs		Incl. Subs O&P		Cost Per Labor-Hour	
Crew B-57	**Hr.**	**Daily**	**Hr.**	**Daily**	**Bare Costs**	**Incl. O&P**
1 Labor Foreman (outside)	$39.60	$316.80	$60.95	$487.60	$43.38	$66.23
2 Laborers	37.60	601.60	57.85	925.60		
1 Equip. Oper. (crane)	51.70	413.60	78.45	627.60		
1 Equip. Oper. (light)	48.60	388.80	73.75	590.00		
1 Equip. Oper. (oiler)	45.20	361.60	68.55	548.40		
1 Crawler Crane, 25 Ton		1150.00		1265.00		
1 Clamshell Bucket, 1 C.Y.		47.80		52.58		
1 Centr. Water Pump, 6"		346.60		381.26		
1 -20' Suction Hose, 6"		11.50		12.65		
20 -50' Discharge Hoses, 6"		122.00		134.20	34.96	38.45
48 L.H., Daily Totals		$3760.30		$5024.89	$78.34	$104.69

Crew B-58	Hr.	Daily	Hr.	Daily	Bare Costs	Incl. O&P
2 Laborers	$37.60	$601.60	$57.85	$925.60	$41.27	$63.15
1 Equip. Oper. (light)	48.60	388.80	73.75	590.00		
1 Backhoe Loader, 48 H.P.		364.20		400.62		
1 Small Helicopter, w/ Pilot		2790.00		3069.00	131.43	144.57
24 L.H., Daily Totals		$4144.60		$4985.22	$172.69	$207.72

Crew B-59	Hr.	Daily	Hr.	Daily	Bare Costs	Incl. O&P
1 Truck Driver (heavy)	$40.05	$320.40	$60.55	$484.40	$40.05	$60.55
1 Truck Tractor, 220 H.P.		359.60		395.56		
1 Water Tank Trailer, 5000 Gal.		141.00		155.10	62.58	68.83
8 L.H., Daily Totals		$821.00		$1035.06	$102.63	$129.38

Crew B-59A	Hr.	Daily	Hr.	Daily	Bare Costs	Incl. O&P
2 Laborers	$37.60	$601.60	$57.85	$925.60	$38.42	$58.75
1 Truck Driver (heavy)	40.05	320.40	60.55	484.40		
1 Water Tank Trailer, 5000 Gal.		141.00		155.10		
1 Truck Tractor, 220 H.P.		359.60		395.56	20.86	22.94
24 L.H., Daily Totals		$1422.60		$1960.66	$59.27	$81.69

Crew B-60	Hr.	Daily	Hr.	Daily	Bare Costs	Incl. O&P
1 Labor Foreman (outside)	$39.60	$316.80	$60.95	$487.60	$44.13	$67.31
2 Laborers	37.60	601.60	57.85	925.60		
1 Equip. Oper. (crane)	51.70	413.60	78.45	627.60		
2 Equip. Oper. (light)	48.60	777.60	73.75	1180.00		
1 Equip. Oper. (oiler)	45.20	361.60	68.55	548.40		
1 Crawler Crane, 40 Ton		1156.00		1271.60		
1 Lead, 60' High		74.80		82.28		
1 Hammer, Diesel, 15K ft.-lbs.		586.00		644.60		
1 Backhoe Loader, 48 H.P.		364.20		400.62	38.95	42.84
56 L.H., Daily Totals		$4652.20		$6168.30	$83.08	$110.15

Crew B-61	Hr.	Daily	Hr.	Daily	Bare Costs	Incl. O&P
1 Labor Foreman (outside)	$39.60	$316.80	$60.95	$487.60	$40.20	$61.65
3 Laborers	37.60	902.40	57.85	1388.40		
1 Equip. Oper. (light)	48.60	388.80	73.75	590.00		
1 Cement Mixer, 2 C.Y.		194.20		213.62		
1 Air Compressor, 160 cfm		156.40		172.04	8.77	9.64
40 L.H., Daily Totals		$1958.60		$2851.66	$48.97	$71.29

Crew B-62	Hr.	Daily	Hr.	Daily	Bare Costs	Incl. O&P
2 Laborers	$37.60	$601.60	$57.85	$925.60	$41.27	$63.15
1 Equip. Oper. (light)	48.60	388.80	73.75	590.00		
1 Loader, Skid Steer, 30 H.P.		173.60		190.96	7.23	7.96
24 L.H., Daily Totals		$1164.00		$1706.56	$48.50	$71.11

Crew No.	Bare Costs		Incl. Subs O&P		Cost Per Labor-Hour	
Crew B-62A	**Hr.**	**Daily**	**Hr.**	**Daily**	**Bare Costs**	**Incl. O&P**
2 Laborers	$37.60	$601.60	$57.85	$925.60	$41.27	$63.15
1 Equip. Oper. (light)	48.60	388.80	73.75	590.00		
1 Loader, Skid Steer, 30 H.P.		173.60		190.96		
1 Trencher Attachment		58.30		64.13	9.66	10.63
24 L.H., Daily Totals		$1222.30		$1770.69	$50.93	$73.78

Crew B-63	Hr.	Daily	Hr.	Daily	Bare Costs	Incl. O&P
4 Laborers	$37.60	$1203.20	$57.85	$1851.20	$39.80	$61.03
1 Equip. Oper. (light)	48.60	388.80	73.75	590.00		
1 Loader, Skid Steer, 30 H.P.		173.60		190.96	4.34	4.77
40 L.H., Daily Totals		$1765.60		$2632.16	$44.14	$65.80

Crew B-63B	Hr.	Daily	Hr.	Daily	Bare Costs	Incl. O&P
1 Labor Foreman (inside)	$38.10	$304.80	$58.65	$469.20	$40.48	$62.02
2 Laborers	37.60	601.60	57.85	925.60		
1 Equip. Oper. (light)	48.60	388.80	73.75	590.00		
1 Loader, Skid Steer, 78 H.P.		318.60		350.46	9.96	10.95
32 L.H., Daily Totals		$1613.80		$2335.26	$50.43	$72.98

Crew B-64	Hr.	Daily	Hr.	Daily	Bare Costs	Incl. O&P
1 Laborer	$37.60	$300.80	$57.85	$462.80	$38.35	$58.48
1 Truck Driver (light)	39.10	312.80	59.10	472.80		
1 Power Mulcher (small)		151.80		166.98		
1 Flatbed Truck, Gas, 1.5 Ton		245.40		269.94	24.82	27.31
16 L.H., Daily Totals		$1010.80		$1372.52	$63.17	$85.78

Crew B-65	Hr.	Daily	Hr.	Daily	Bare Costs	Incl. O&P
1 Laborer	$37.60	$300.80	$57.85	$462.80	$38.35	$58.48
1 Truck Driver (light)	39.10	312.80	59.10	472.80		
1 Power Mulcher (Large)		319.40		351.34		
1 Flatbed Truck, Gas, 1.5 Ton		245.40		269.94	35.30	38.83
16 L.H., Daily Totals		$1178.40		$1556.88	$73.65	$97.31

Crew B-66	Hr.	Daily	Hr.	Daily	Bare Costs	Incl. O&P
1 Equip. Oper. (light)	$48.60	$388.80	$73.75	$590.00	$48.60	$73.75
1 Loader-Backhoe, 40 H.P.		263.20		289.52	32.90	36.19
8 L.H., Daily Totals		$652.00		$879.52	$81.50	$109.94

Crew B-67	Hr.	Daily	Hr.	Daily	Bare Costs	Incl. O&P
1 Millwright	$49.15	$393.20	$72.35	$578.80	$48.88	$73.05
1 Equip. Oper. (light)	48.60	388.80	73.75	590.00		
1 Forklift, R/T, 4,000 Lb.		311.20		342.32	19.45	21.40
16 L.H., Daily Totals		$1093.20		$1511.12	$68.33	$94.44

Crew B-67B	Hr.	Daily	Hr.	Daily	Bare Costs	Incl. O&P
1 Millwright Foreman (inside)	$49.65	$397.20	$73.10	$584.80	$49.40	$72.72
1 Millwright	49.15	393.20	72.35	578.80		
16 L.H., Daily Totals		$790.40		$1163.60	$49.40	$72.72

Crew B-68	Hr.	Daily	Hr.	Daily	Bare Costs	Incl. O&P
2 Millwrights	$49.15	$786.40	$72.35	$1157.60	$48.97	$72.82
1 Equip. Oper. (light)	48.60	388.80	73.75	590.00		
1 Forklift, R/T, 4,000 Lb.		311.20		342.32	12.97	14.26
24 L.H., Daily Totals		$1486.40		$2089.92	$61.93	$87.08

Crew B-68A	Hr.	Daily	Hr.	Daily	Bare Costs	Incl. O&P
1 Millwright Foreman (inside)	$49.65	$397.20	$73.10	$584.80	$49.32	$72.60
2 Millwrights	49.15	786.40	72.35	1157.60		
1 Forklift, 8,000 Lb.		185.20		203.72	7.72	8.49
24 L.H., Daily Totals		$1368.80		$1946.12	$57.03	$81.09

Crew B-68B	Bare Costs Hr.	Bare Costs Daily	Incl. Subs O&P Hr.	Incl. Subs O&P Daily	Cost Per Labor-Hour Bare Costs	Cost Per Labor-Hour Incl. O&P
1 Millwright Foreman (inside)	$49.65	$397.20	$73.10	$584.80	$53.54	$79.86
2 Millwrights	49.15	786.40	72.35	1157.60		
2 Electricians	54.70	875.20	81.95	1311.20		
2 Plumbers	58.70	939.20	88.65	1418.40		
1 Forklift, 5,000 Lb.		331.80		364.98	5.92	6.52
56 L.H., Daily Totals		$3329.80		$4836.98	$59.46	$86.37

Crew B-68C	Hr.	Daily	Hr.	Daily	Bare Costs	Incl. O&P
1 Millwright Foreman (inside)	$49.65	$397.20	$73.10	$584.80	$53.05	$79.01
1 Millwright	49.15	393.20	72.35	578.80		
1 Electrician	54.70	437.60	81.95	655.60		
1 Plumber	58.70	469.60	88.65	709.20		
1 Forklift, 5,000 Lb.		331.80		364.98	10.37	11.41
32 L.H., Daily Totals		$2029.40		$2893.38	$63.42	$90.42

Crew B-68D	Hr.	Daily	Hr.	Daily	Bare Costs	Incl. O&P
1 Labor Foreman (inside)	$38.10	$304.80	$58.65	$469.20	$41.43	$63.42
1 Laborer	37.60	300.80	57.85	462.80		
1 Equip. Oper. (light)	48.60	388.80	73.75	590.00		
1 Forklift, 5,000 Lb.		331.80		364.98	13.82	15.21
24 L.H., Daily Totals		$1326.20		$1886.98	$55.26	$78.62

Crew B-68E	Hr.	Daily	Hr.	Daily	Bare Costs	Incl. O&P
1 Struc. Steel Foreman (inside)	$53.15	$425.20	$92.30	$738.40	$52.75	$91.62
3 Struc. Steel Workers	52.65	1263.60	91.45	2194.80		
1 Welder	52.65	421.20	91.45	731.60		
1 Forklift, 8,000 Lb.		185.20		203.72	4.63	5.09
40 L.H., Daily Totals		$2295.20		$3868.52	$57.38	$96.71

Crew B-68F	Hr.	Daily	Hr.	Daily	Bare Costs	Incl. O&P
1 Skilled Worker Foreman (out)	$50.65	$405.20	$78.25	$626.00	$49.32	$76.18
2 Skilled Workers	48.65	778.40	75.15	1202.40		
1 Forklift, 5,000 Lb.		331.80		364.98	13.82	15.21
24 L.H., Daily Totals		$1515.40		$2193.38	$63.14	$91.39

Crew B-68G	Hr.	Daily	Hr.	Daily	Bare Costs	Incl. O&P
2 Structural Steel Workers	$52.65	$842.40	$91.45	$1463.20	$52.65	$91.45
1 Forklift, 5,000 Lb.		331.80		364.98	20.74	22.81
16 L.H., Daily Totals		$1174.20		$1828.18	$73.39	$114.26

Crew B-69	Hr.	Daily	Hr.	Daily	Bare Costs	Incl. O&P
1 Labor Foreman (outside)	$39.60	$316.80	$60.95	$487.60	$41.55	$63.58
3 Laborers	37.60	902.40	57.85	1388.40		
1 Equip. Oper. (crane)	51.70	413.60	78.45	627.60		
1 Equip. Oper. (oiler)	45.20	361.60	68.55	548.40		
1 Hyd. Crane, 80 Ton		1625.00		1787.50	33.85	37.24
48 L.H., Daily Totals		$3619.40		$4839.50	$75.40	$100.82

Crew B-69A	Hr.	Daily	Hr.	Daily	Bare Costs	Incl. O&P
1 Labor Foreman (outside)	$39.60	$316.80	$60.95	$487.60	$41.33	$62.98
3 Laborers	37.60	902.40	57.85	1388.40		
1 Equip. Oper. (medium)	50.60	404.80	76.75	614.00		
1 Concrete Finisher	45.00	360.00	66.65	533.20		
1 Curb/Gutter Paver, 2-Track		997.60		1097.36	20.78	22.86
48 L.H., Daily Totals		$2981.60		$4120.56	$62.12	$85.84

Crew B-69B	Hr.	Daily	Hr.	Daily	Bare Costs	Incl. O&P
1 Labor Foreman (outside)	$39.60	$316.80	$60.95	$487.60	$41.33	$62.98
3 Laborers	37.60	902.40	57.85	1388.40		
1 Equip. Oper. (medium)	50.60	404.80	76.75	614.00		
1 Cement Finisher	45.00	360.00	66.65	533.20		
1 Curb/Gutter Paver, 4-Track		776.00		853.60	16.17	17.78
48 L.H., Daily Totals		$2760.00		$3876.80	$57.50	$80.77

Crew B-70	Hr.	Daily	Hr.	Daily	Bare Costs	Incl. O&P
1 Labor Foreman (outside)	$39.60	$316.80	$60.95	$487.60	$43.46	$66.39
3 Laborers	37.60	902.40	57.85	1388.40		
3 Equip. Oper. (medium)	50.60	1214.40	76.75	1842.00		
1 Grader, 30,000 Lbs.		736.20		809.82		
1 Ripper, Beam & 1 Shank		84.20		92.62		
1 Road Sweeper, S.P., 8' wide		662.40		728.64		
1 F.E. Loader, W.M., 1.5 C.Y.		378.40		416.24	33.24	36.56
56 L.H., Daily Totals		$4294.80		$5765.32	$76.69	$102.95

Crew B-70A	Hr.	Daily	Hr.	Daily	Bare Costs	Incl. O&P
1 Laborer	$37.60	$300.80	$57.85	$462.80	$48.00	$72.97
4 Equip. Oper. (medium)	50.60	1619.20	76.75	2456.00		
1 Grader, 40,000 Lbs.		1174.00		1291.40		
1 F.E. Loader, W.M.,2.5 C.Y.		519.80		571.78		
1 Dozer, 80 H.P.		472.40		519.64		
1 Roller, Pneum. Whl., 12 Ton		344.40		378.84	62.77	69.04
40 L.H., Daily Totals		$4430.60		$5680.46	$110.77	$142.01

Crew B-71	Hr.	Daily	Hr.	Daily	Bare Costs	Incl. O&P
1 Labor Foreman (outside)	$39.60	$316.80	$60.95	$487.60	$43.46	$66.39
3 Laborers	37.60	902.40	57.85	1388.40		
3 Equip. Oper. (medium)	50.60	1214.40	76.75	1842.00		
1 Pvmt. Profiler, 750 H.P.		5805.00		6385.50		
1 Road Sweeper, S.P., 8' wide		662.40		728.64		
1 F.E. Loader, W.M., 1.5 C.Y.		378.40		416.24	122.25	134.47
56 L.H., Daily Totals		$9279.40		$11248.38	$165.70	$200.86

Crew B-72	Hr.	Daily	Hr.	Daily	Bare Costs	Incl. O&P
1 Labor Foreman (outside)	$39.60	$316.80	$60.95	$487.60	$44.35	$67.69
3 Laborers	37.60	902.40	57.85	1388.40		
4 Equip. Oper. (medium)	50.60	1619.20	76.75	2456.00		
1 Pvmt. Profiler, 750 H.P.		5805.00		6385.50		
1 Hammermill, 250 H.P.		1879.00		2066.90		
1 Windrow Loader		1227.00		1349.70		
1 Mix Paver 165 H.P.		2147.00		2361.70		
1 Roller, Pneum. Whl., 12 Ton		344.40		378.84	178.16	195.98
64 L.H., Daily Totals		$14240.80		$16874.64	$222.51	$263.67

Crew B-73	Hr.	Daily	Hr.	Daily	Bare Costs	Incl. O&P
1 Labor Foreman (outside)	$39.60	$316.80	$60.95	$487.60	$45.98	$70.05
2 Laborers	37.60	601.60	57.85	925.60		
5 Equip. Oper. (medium)	50.60	2024.00	76.75	3070.00		
1 Road Mixer, 310 H.P.		1929.00		2121.90		
1 Tandem Roller, 10 Ton		236.80		260.48		
1 Hammermill, 250 H.P.		1879.00		2066.90		
1 Grader, 30,000 Lbs.		736.20		809.82		
.5 F.E. Loader, W.M., 1.5 C.Y.		189.20		208.12		
.5 Truck Tractor, 220 H.P.		179.80		197.78		
.5 Water Tank Trailer, 5000 Gal.		70.50		77.55	81.57	89.73
64 L.H., Daily Totals		$8162.90		$10225.75	$127.55	$159.78

Crew No.	Bare Costs		Incl. Subs O&P		Cost Per Labor-Hour	

Crew B-74	Hr.	Daily	Hr.	Daily	Bare Costs	Incl. O&P
1 Labor Foreman (outside)	$39.60	$316.80	$60.95	$487.60	$44.96	$68.36
1 Laborer	37.60	300.80	57.85	462.80		
4 Equip. Oper. (medium)	50.60	1619.20	76.75	2456.00		
2 Truck Drivers (heavy)	40.05	640.80	60.55	968.80		
1 Grader, 30,000 Lbs.		736.20		809.82		
1 Ripper, Beam & 1 Shank		84.20		92.62		
2 Stabilizers, 310 H.P.		3596.00		3955.60		
1 Flatbed Truck, Gas, 3 Ton		303.20		333.52		
1 Chem. Spreader, Towed		52.40		57.64		
1 Roller, Vibratory, 25 Ton		687.40		756.14		
1 Water Tank Trailer, 5000 Gal.		141.00		155.10		
1 Truck Tractor, 220 H.P.		359.60		395.56	93.13	102.44
64 L.H., Daily Totals		$8837.60		$10931.20	$138.09	$170.80

Crew B-75	Hr.	Daily	Hr.	Daily	Bare Costs	Incl. O&P
1 Labor Foreman (outside)	$39.60	$316.80	$60.95	$487.60	$45.66	$69.48
1 Laborer	37.60	300.80	57.85	462.80		
4 Equip. Oper. (medium)	50.60	1619.20	76.75	2456.00		
1 Truck Driver (heavy)	40.05	320.40	60.55	484.40		
1 Grader, 30,000 Lbs.		736.20		809.82		
1 Ripper, Beam & 1 Shank		84.20		92.62		
2 Stabilizers, 310 H.P.		3596.00		3955.60		
1 Dist. Tanker, 3000 Gallon		305.60		336.16		
1 Truck Tractor, 6x4, 380 H.P.		599.80		659.78		
1 Roller, Vibratory, 25 Ton		687.40		756.14	107.31	118.04
56 L.H., Daily Totals		$8566.40		$10500.92	$152.97	$187.52

Crew B-76	Hr.	Daily	Hr.	Daily	Bare Costs	Incl. O&P
1 Dock Builder Foreman (outside)	$48.10	$384.80	$76.15	$609.20	$47.47	$74.07
5 Dock Builders	46.10	1844.00	73.00	2920.00		
2 Equip. Oper. (crane)	51.70	827.20	78.45	1255.20		
1 Equip. Oper. (oiler)	45.20	361.60	68.55	548.40		
1 Crawler Crane, 50 Ton		1294.00		1423.40		
1 Barge, 400 Ton		776.00		853.60		
1 Hammer, Diesel, 15K ft.-lbs.		586.00		644.60		
1 Lead, 60' High		74.80		82.28		
1 Air Compressor, 600 cfm		550.60		605.66		
2 -50' Air Hoses, 3"		29.80		32.78	45.99	50.59
72 L.H., Daily Totals		$6728.80		$8975.12	$93.46	$124.65

Crew B-76A	Hr.	Daily	Hr.	Daily	Bare Costs	Incl. O&P
1 Labor Foreman (outside)	$39.60	$316.80	$60.95	$487.60	$40.56	$62.15
5 Laborers	37.60	1504.00	57.85	2314.00		
1 Equip. Oper. (crane)	51.70	413.60	78.45	627.60		
1 Equip. Oper. (oiler)	45.20	361.60	68.55	548.40		
1 Crawler Crane, 50 Ton		1294.00		1423.40		
1 Barge, 400 Ton		776.00		853.60	32.34	35.58
64 L.H., Daily Totals		$4666.00		$6254.60	$72.91	$97.73

Crew B-77	Hr.	Daily	Hr.	Daily	Bare Costs	Incl. O&P
1 Labor Foreman (outside)	$39.60	$316.80	$60.95	$487.60	$38.30	$58.72
3 Laborers	37.60	902.40	57.85	1388.40		
1 Truck Driver (light)	39.10	312.80	59.10	472.80		
1 Crack Cleaner, 25 H.P.		61.60		67.76		
1 Crack Filler, Trailer Mtd.		202.00		222.20		
1 Flatbed Truck, Gas, 3 Ton		303.20		333.52	14.17	15.59
40 L.H., Daily Totals		$2098.80		$2972.28	$52.47	$74.31

Crew B-78	Hr.	Daily	Hr.	Daily	Bare Costs	Incl. O&P
1 Labor Foreman (outside)	$39.60	$316.80	$60.95	$487.60	$38.18	$58.58
4 Laborers	37.60	1203.20	57.85	1851.20		
1 Truck Driver (light)	39.10	312.80	59.10	472.80		
1 Paint Striper, S.P., 40 Gallon		151.40		166.54		
1 Flatbed Truck, Gas, 3 Ton		303.20		333.52		
1 Pickup Truck, 3/4 Ton		144.20		158.62	12.48	13.72
48 L.H., Daily Totals		$2431.60		$3470.28	$50.66	$72.30

Crew B-78A	Hr.	Daily	Hr.	Daily	Bare Costs	Incl. O&P
1 Equip. Oper. (light)	$48.60	$388.80	$73.75	$590.00	$48.60	$73.75
1 Line Rem. (Metal Balls) 115 H.P.		821.60		903.76	102.70	112.97
8 L.H., Daily Totals		$1210.40		$1493.76	$151.30	$186.72

Crew B-78B	Hr.	Daily	Hr.	Daily	Bare Costs	Incl. O&P
2 Laborers	$37.60	$601.60	$57.85	$925.60	$38.82	$59.62
.25 Equip. Oper. (light)	48.60	97.20	73.75	147.50		
1 Pickup Truck, 3/4 Ton		144.20		158.62		
1 Line Rem.,11 H.P.,Walk Behind		65.60		72.16		
.25 Road Sweeper, S.P., 8' wide		165.60		182.16	20.86	22.94
18 L.H., Daily Totals		$1074.20		$1486.04	$59.68	$82.56

Crew B-78C	Hr.	Daily	Hr.	Daily	Bare Costs	Incl. O&P
1 Labor Foreman (outside)	$39.60	$316.80	$60.95	$487.60	$38.18	$58.58
4 Laborers	37.60	1203.20	57.85	1851.20		
1 Truck Driver (light)	39.10	312.80	59.10	472.80		
1 Paint Striper, T.M., 120 Gal.		804.60		885.06		
1 Flatbed Truck, Gas, 3 Ton		303.20		333.52		
1 Pickup Truck, 3/4 Ton		144.20		158.62	26.08	28.69
48 L.H., Daily Totals		$3084.80		$4188.80	$64.27	$87.27

Crew B-78D	Hr.	Daily	Hr.	Daily	Bare Costs	Incl. O&P
2 Labor Foremen (outside)	$39.60	$633.60	$60.95	$975.20	$38.15	$58.59
7 Laborers	37.60	2105.60	57.85	3239.60		
1 Truck Driver (light)	39.10	312.80	59.10	472.80		
1 Paint Striper, T.M., 120 Gal.		804.60		885.06		
1 Flatbed Truck, Gas, 3 Ton		303.20		333.52		
3 Pickup Trucks, 3/4 Ton		432.60		475.86		
1 Air Compressor, 60 cfm		137.40		151.14		
1 -50' Air Hose, 3/4"		3.25		3.58		
1 Breakers, Pavement, 60 lb.		9.80		10.78	21.14	23.25
80 L.H., Daily Totals		$4742.85		$6547.53	$59.29	$81.84

Crew B-78E	Hr.	Daily	Hr.	Daily	Bare Costs	Incl. O&P
2 Labor Foremen (outside)	$39.60	$633.60	$60.95	$975.20	$38.06	$58.47
9 Laborers	37.60	2707.20	57.85	4165.20		
1 Truck Driver (light)	39.10	312.80	59.10	472.80		
1 Paint Striper, T.M., 120 Gal.		804.60		885.06		
1 Flatbed Truck, Gas, 3 Ton		303.20		333.52		
4 Pickup Trucks, 3/4 Ton		576.80		634.48		
2 Air Compressor, 60 cfm		274.80		302.28		
2 -50' Air Hose, 3/4"		6.50		7.15		
2 Breakers, Pavement, 60 lb.		19.60		21.56	20.68	22.75
96 L.H., Daily Totals		$5639.10		$7797.25	$58.74	$81.22

Crew No.	Bare Costs		Incl. Subs O&P		Cost Per Labor-Hour	
Crew B-78F	**Hr.**	**Daily**	**Hr.**	**Daily**	**Bare Costs**	**Incl. O&P**
2 Labor Foremen (outside)	$39.60	$633.60	$60.95	$975.20	$37.99	$58.38
11 Laborers	37.60	3308.80	57.85	5090.80		
1 Truck Driver (light)	39.10	312.80	59.10	472.80		
1 Paint Striper, T.M., 120 Gal.		804.60		885.06		
1 Flatbed Truck, Gas, 3 Ton		303.20		333.52		
7 Pickup Trucks, 3/4 Ton		1009.40		1110.34		
3 Air Compressor, 60 cfm		412.20		453.42		
3 -50' Air Hose, 3/4"		9.75		10.73		
3 Breakers, Pavement, 60 lb.		29.40		32.34	22.93	25.23
112 L.H., Daily Totals		$6823.75		$9364.20	$60.93	$83.61

Crew B-79	Hr.	Daily	Hr.	Daily	Bare Costs	Incl. O&P
1 Labor Foreman (outside)	$39.60	$316.80	$60.95	$487.60	$38.30	$58.72
3 Laborers	37.60	902.40	57.85	1388.40		
1 Truck Driver (light)	39.10	312.80	59.10	472.80		
1 Paint Striper, T.M., 120 Gal.		804.60		885.06		
1 Heating Kettle, 115 Gallon		81.00		89.10		
1 Flatbed Truck, Gas, 3 Ton		303.20		333.52		
2 Pickup Trucks, 3/4 Ton		288.40		317.24	36.93	40.62
40 L.H., Daily Totals		$3009.20		$3973.72	$75.23	$99.34

Crew B-79A	Hr.	Daily	Hr.	Daily	Bare Costs	Incl. O&P
1.5 Equip. Oper. (light)	$48.60	$583.20	$73.75	$885.00	$48.60	$73.75
.5 Line Remov. (Grinder) 115 H.P.		455.90		501.49		
1 Line Rem. (Metal Balls) 115 H.P.		821.60		903.76	106.46	117.10
12 L.H., Daily Totals		$1860.70		$2290.25	$155.06	$190.85

Crew B-79B	Hr.	Daily	Hr.	Daily	Bare Costs	Incl. O&P
1 Laborer	$37.60	$300.80	$57.85	$462.80	$37.60	$57.85
1 Set of Gases		152.00		167.20	19.00	20.90
8 L.H., Daily Totals		$452.80		$630.00	$56.60	$78.75

Crew B-79C	Hr.	Daily	Hr.	Daily	Bare Costs	Incl. O&P
1 Labor Foreman (outside)	$39.60	$316.80	$60.95	$487.60	$38.10	$58.47
5 Laborers	37.60	1504.00	57.85	2314.00		
1 Truck Driver (light)	39.10	312.80	59.10	472.80		
1 Paint Striper, T.M., 120 Gal.		804.60		885.06		
1 Heating Kettle, 115 Gallon		81.00		89.10		
1 Flatbed Truck, Gas, 3 Ton		303.20		333.52		
3 Pickup Trucks, 3/4 Ton		432.60		475.86		
1 Air Compressor, 60 cfm		137.40		151.14		
1 -50' Air Hose, 3/4"		3.25		3.58		
1 Breakers, Pavement, 60 lb.		9.80		10.78	31.64	34.80
56 L.H., Daily Totals		$3905.45		$5223.44	$69.74	$93.28

Crew B-79D	Hr.	Daily	Hr.	Daily	Bare Costs	Incl. O&P
2 Labor Foremen (outside)	$39.60	$633.60	$60.95	$975.20	$38.29	$58.78
5 Laborers	37.60	1504.00	57.85	2314.00		
1 Truck Driver (light)	39.10	312.80	59.10	472.80		
1 Paint Striper, T.M., 120 Gal.		804.60		885.06		
1 Heating Kettle, 115 Gallon		81.00		89.10		
1 Flatbed Truck, Gas, 3 Ton		303.20		333.52		
4 Pickup Trucks, 3/4 Ton		576.80		634.48		
1 Air Compressor, 60 cfm		137.40		151.14		
1 -50' Air Hose, 3/4"		3.25		3.58		
1 Breakers, Pavement, 60 lb.		9.80		10.78	29.94	32.93
64 L.H., Daily Totals		$4366.45		$5869.65	$68.23	$91.71

Crew No.	Bare Costs		Incl. Subs O&P		Cost Per Labor-Hour	
Crew B-79E	**Hr.**	**Daily**	**Hr.**	**Daily**	**Bare Costs**	**Incl. O&P**
2 Labor Foremen (outside)	$39.60	$633.60	$60.95	$975.20	$38.15	$58.59
7 Laborers	37.60	2105.60	57.85	3239.60		
1 Truck Driver (light)	39.10	312.80	59.10	472.80		
1 Paint Striper, T.M., 120 Gal.		804.60		885.06		
1 Heating Kettle, 115 Gallon		81.00		89.10		
1 Flatbed Truck, Gas, 3 Ton		303.20		333.52		
5 Pickup Trucks, 3/4 Ton		721.00		793.10		
2 Air Compressors, 60 cfm		274.80		302.28		
2 -50' Air Hoses, 3/4"		6.50		7.15		
2 Breakers, Pavement, 60 lb.		19.60		21.56	27.63	30.40
80 L.H., Daily Totals		$5262.70		$7119.37	$65.78	$88.99

Crew B-80	Hr.	Daily	Hr.	Daily	Bare Costs	Incl. O&P
1 Labor Foreman (outside)	$39.60	$316.80	$60.95	$487.60	$41.23	$62.91
1 Laborer	37.60	300.80	57.85	462.80		
1 Truck Driver (light)	39.10	312.80	59.10	472.80		
1 Equip. Oper. (light)	48.60	388.80	73.75	590.00		
1 Flatbed Truck, Gas, 3 Ton		303.20		333.52		
1 Earth Auger, Truck-Mtd.		417.20		458.92	22.51	24.76
32 L.H., Daily Totals		$2039.60		$2805.64	$63.74	$87.68

Crew B-80A	Hr.	Daily	Hr.	Daily	Bare Costs	Incl. O&P
3 Laborers	$37.60	$902.40	$57.85	$1388.40	$37.60	$57.85
1 Flatbed Truck, Gas, 3 Ton		303.20		333.52	12.63	13.90
24 L.H., Daily Totals		$1205.60		$1721.92	$50.23	$71.75

Crew B-80B	Hr.	Daily	Hr.	Daily	Bare Costs	Incl. O&P
3 Laborers	$37.60	$902.40	$57.85	$1388.40	$40.35	$61.83
1 Equip. Oper. (light)	48.60	388.80	73.75	590.00		
1 Crane, Flatbed Mounted, 3 Ton		243.80		268.18	7.62	8.38
32 L.H., Daily Totals		$1535.00		$2246.58	$47.97	$70.21

Crew B-80C	Hr.	Daily	Hr.	Daily	Bare Costs	Incl. O&P
2 Laborers	$37.60	$601.60	$57.85	$925.60	$38.10	$58.27
1 Truck Driver (light)	39.10	312.80	59.10	472.80		
1 Flatbed Truck, Gas, 1.5 Ton		245.40		269.94		
1 Manual Fence Post Auger, Gas		8.40		9.24	10.57	11.63
24 L.H., Daily Totals		$1168.20		$1677.58	$48.67	$69.90

Crew B-81	Hr.	Daily	Hr.	Daily	Bare Costs	Incl. O&P
1 Laborer	$37.60	$300.80	$57.85	$462.80	$42.75	$65.05
1 Equip. Oper. (medium)	50.60	404.80	76.75	614.00		
1 Truck Driver (heavy)	40.05	320.40	60.55	484.40		
1 Hydromulcher, T.M., 3000 Gal.		334.20		367.62		
1 Truck Tractor, 220 H.P.		359.60		395.56	28.91	31.80
24 L.H., Daily Totals		$1719.80		$2324.38	$71.66	$96.85

Crew B-81A	Hr.	Daily	Hr.	Daily	Bare Costs	Incl. O&P
1 Laborer	$37.60	$300.80	$57.85	$462.80	$38.35	$58.48
1 Truck Driver (light)	39.10	312.80	59.10	472.80		
1 Hydromulcher, T.M., 600 Gal.		127.80		140.58		
1 Flatbed Truck, Gas, 3 Ton		303.20		333.52	26.94	29.63
16 L.H., Daily Totals		$1044.60		$1409.70	$65.29	$88.11

Crew B-82	Hr.	Daily	Hr.	Daily	Bare Costs	Incl. O&P
1 Laborer	$37.60	$300.80	$57.85	$462.80	$43.10	$65.80
1 Equip. Oper. (light)	48.60	388.80	73.75	590.00		
1 Horiz. Borer, 6 H.P.		83.80		92.18	5.24	5.76
16 L.H., Daily Totals		$773.40		$1144.98	$48.34	$71.56

Crew No.	Bare Costs		Incl. Subs O&P		Cost Per Labor-Hour	
Crew B-82A	**Hr.**	**Daily**	**Hr.**	**Daily**	**Bare Costs**	**Incl. O&P**
2 Laborers	$37.60	$601.60	$57.85	$925.60	$43.10	$65.80
2 Equip. Opers. (light)	48.60	777.60	73.75	1180.00		
2 Dump Truck, 8 C.Y., 220 H.P.		822.40		904.64		
1 Flatbed Trailer, 25 Ton		113.60		124.96		
1 Horiz. Dir. Drill, 20k lb. Thrust		654.20		719.62		
1 Mud Trailer for HDD, 1500 Gal.		346.20		380.82		
1 Pickup Truck, 4 x 4, 3/4 Ton		155.60		171.16		
1 Flatbed trailer, 3 Ton		23.60		25.96		
1 Loader, Skid Steer, 78 H.P.		318.60		350.46	76.07	83.68
32 L.H., Daily Totals		$3813.40		$4783.22	$119.17	$149.48
Crew B-82B	**Hr.**	**Daily**	**Hr.**	**Daily**	**Bare Costs**	**Incl. O&P**
2 Laborers	$37.60	$601.60	$57.85	$925.60	$43.10	$65.80
2 Equip. Opers. (light)	48.60	777.60	73.75	1180.00		
2 Dump Truck, 8 C.Y., 220 H.P.		822.40		904.64		
1 Flatbed Trailer, 25 Ton		113.60		124.96		
1 Horiz. Dir. Drill, 30k lb. Thrust		931.20		1024.32		
1 Mud Trailer for HDD, 1500 Gal.		346.20		380.82		
1 Pickup Truck, 4 x 4, 3/4 Ton		155.60		171.16		
1 Flatbed trailer, 3 Ton		23.60		25.96		
1 Loader, Skid Steer, 78 H.P.		318.60		350.46	84.72	93.20
32 L.H., Daily Totals		$4090.40		$5087.92	$127.83	$159.00
Crew B-82C	**Hr.**	**Daily**	**Hr.**	**Daily**	**Bare Costs**	**Incl. O&P**
2 Laborers	$37.60	$601.60	$57.85	$925.60	$43.10	$65.80
2 Equip. Opers. (light)	48.60	777.60	73.75	1180.00		
2 Dump Truck, 8 C.Y., 220 H.P.		822.40		904.64		
1 Flatbed Trailer, 25 Ton		113.60		124.96		
1 Horiz. Dir. Drill, 50k lb. Thrust		1237.00		1360.70		
1 Mud Trailer for HDD, 1500 Gal.		346.20		380.82		
1 Pickup Truck, 4 x 4, 3/4 Ton		155.60		171.16		
1 Flatbed trailer, 3 Ton		23.60		25.96		
1 Loader, Skid Steer, 78 H.P.		318.60		350.46	94.28	103.71
32 L.H., Daily Totals		$4396.20		$5424.30	$137.38	$169.51
Crew B-82D	**Hr.**	**Daily**	**Hr.**	**Daily**	**Bare Costs**	**Incl. O&P**
1 Equip. Oper. (light)	$48.60	$388.80	$73.75	$590.00	$48.60	$73.75
1 Mud Trailer for HDD, 1500 Gal.		346.20		380.82	43.27	47.60
8 L.H., Daily Totals		$735.00		$970.82	$91.88	$121.35
Crew B-83	**Hr.**	**Daily**	**Hr.**	**Daily**	**Bare Costs**	**Incl. O&P**
1 Tugboat Captain	$50.60	$404.80	$76.75	$614.00	$44.10	$67.30
1 Tugboat Hand	37.60	300.80	57.85	462.80		
1 Tugboat, 250 H.P.		870.00		957.00	54.38	59.81
16 L.H., Daily Totals		$1575.60		$2033.80	$98.47	$127.11
Crew B-84	**Hr.**	**Daily**	**Hr.**	**Daily**	**Bare Costs**	**Incl. O&P**
1 Equip. Oper. (medium)	$50.60	$404.80	$76.75	$614.00	$50.60	$76.75
1 Rotary Mower/Tractor		370.80		407.88	46.35	50.98
8 L.H., Daily Totals		$775.60		$1021.88	$96.95	$127.74
Crew B-85	**Hr.**	**Daily**	**Hr.**	**Daily**	**Bare Costs**	**Incl. O&P**
3 Laborers	$37.60	$902.40	$57.85	$1388.40	$40.69	$62.17
1 Equip. Oper. (medium)	50.60	404.80	76.75	614.00		
1 Truck Driver (heavy)	40.05	320.40	60.55	484.40		
1 Aerial Lift Truck, 80'		634.40		697.84		
1 Brush Chipper, 12", 130 H.P.		391.40		430.54		
1 Pruning Saw, Rotary		6.65		7.32	25.81	28.39
40 L.H., Daily Totals		$2660.05		$3622.49	$66.50	$90.56

Crew No.	Bare Costs		Incl. Subs O&P		Cost Per Labor-Hour	
Crew B-86	**Hr.**	**Daily**	**Hr.**	**Daily**	**Bare Costs**	**Incl. O&P**
1 Equip. Oper. (medium)	$50.60	$404.80	$76.75	$614.00	$50.60	$76.75
1 Stump Chipper, S.P.		185.55		204.10	23.19	25.51
8 L.H., Daily Totals		$590.35		$818.11	$73.79	$102.26
Crew B-86A	**Hr.**	**Daily**	**Hr.**	**Daily**	**Bare Costs**	**Incl. O&P**
1 Equip. Oper. (medium)	$50.60	$404.80	$76.75	$614.00	$50.60	$76.75
1 Grader, 30,000 Lbs.		736.20		809.82	92.03	101.23
8 L.H., Daily Totals		$1141.00		$1423.82	$142.63	$177.98
Crew B-86B	**Hr.**	**Daily**	**Hr.**	**Daily**	**Bare Costs**	**Incl. O&P**
1 Equip. Oper. (medium)	$50.60	$404.80	$76.75	$614.00	$50.60	$76.75
1 Dozer, 200 H.P.		1387.00		1525.70	173.38	190.71
8 L.H., Daily Totals		$1791.80		$2139.70	$223.97	$267.46
Crew B-87	**Hr.**	**Daily**	**Hr.**	**Daily**	**Bare Costs**	**Incl. O&P**
1 Laborer	$37.60	$300.80	$57.85	$462.80	$48.00	$72.97
4 Equip. Oper. (medium)	50.60	1619.20	76.75	2456.00		
2 Feller Bunchers, 100 H.P.		1676.80		1844.48		
1 Log Chipper, 22" Tree		889.40		978.34		
1 Dozer, 105 H.P.		602.80		663.08		
1 Chain Saw, Gas, 36" Long		45.00		49.50	80.35	88.39
40 L.H., Daily Totals		$5134.00		$6454.20	$128.35	$161.35
Crew B-88	**Hr.**	**Daily**	**Hr.**	**Daily**	**Bare Costs**	**Incl. O&P**
1 Laborer	$37.60	$300.80	$57.85	$462.80	$48.74	$74.05
6 Equip. Oper. (medium)	50.60	2428.80	76.75	3684.00		
2 Feller Bunchers, 100 H.P.		1676.80		1844.48		
1 Log Chipper, 22" Tree		889.40		978.34		
2 Log Skidders, 50 H.P.		1822.00		2004.20		
1 Dozer, 105 H.P.		602.80		663.08		
1 Chain Saw, Gas, 36" Long		45.00		49.50	89.93	98.92
56 L.H., Daily Totals		$7765.60		$9686.40	$138.67	$172.97
Crew B-89	**Hr.**	**Daily**	**Hr.**	**Daily**	**Bare Costs**	**Incl. O&P**
1 Equip. Oper. (light)	$48.60	$388.80	$73.75	$590.00	$43.85	$66.42
1 Truck Driver (light)	39.10	312.80	59.10	472.80		
1 Flatbed Truck, Gas, 3 Ton		303.20		333.52		
1 Concrete Saw		167.40		184.14		
1 Water Tank, 65 Gal.		17.30		19.03	30.49	33.54
16 L.H., Daily Totals		$1189.50		$1599.49	$74.34	$99.97
Crew B-89A	**Hr.**	**Daily**	**Hr.**	**Daily**	**Bare Costs**	**Incl. O&P**
1 Skilled Worker	$48.65	$389.20	$75.15	$601.20	$43.13	$66.50
1 Laborer	37.60	300.80	57.85	462.80		
1 Core Drill (Large)		115.60		127.16	7.22	7.95
16 L.H., Daily Totals		$805.60		$1191.16	$50.35	$74.45
Crew B-89B	**Hr.**	**Daily**	**Hr.**	**Daily**	**Bare Costs**	**Incl. O&P**
1 Equip. Oper. (light)	$48.60	$388.80	$73.75	$590.00	$43.85	$66.42
1 Truck Driver (light)	39.10	312.80	59.10	472.80		
1 Wall Saw, Hydraulic, 10 H.P.		114.60		126.06		
1 Generator, Diesel, 100 kW		423.40		465.74		
1 Water Tank, 65 Gal.		17.30		19.03		
1 Flatbed Truck, Gas, 3 Ton		303.20		333.52	53.66	59.02
16 L.H., Daily Totals		$1560.10		$2007.15	$97.51	$125.45

Crew No.	Bare Costs		Incl. Subs O&P		Cost Per Labor-Hour	
Crew B-90	**Hr.**	**Daily**	**Hr.**	**Daily**	**Bare Costs**	**Incl. O&P**
1 Labor Foreman (outside)	$39.60	$316.80	$60.95	$487.60	$41.21	$62.89
3 Laborers	37.60	902.40	57.85	1388.40		
2 Equip. Oper. (light)	48.60	777.60	73.75	1180.00		
2 Truck Drivers (heavy)	40.05	640.80	60.55	968.80		
1 Road Mixer, 310 H.P.		1929.00		2121.90		
1 Dist. Truck, 2000 Gal.		275.60		303.16	34.45	37.89
64 L.H., Daily Totals		$4842.20		$6449.86	$75.66	$100.78
Crew B-90A	**Hr.**	**Daily**	**Hr.**	**Daily**	**Bare Costs**	**Incl. O&P**
1 Labor Foreman (outside)	$39.60	$316.80	$60.95	$487.60	$45.31	$69.09
2 Laborers	37.60	601.60	57.85	925.60		
4 Equip. Oper. (medium)	50.60	1619.20	76.75	2456.00		
2 Graders, 30,000 Lbs.		1472.40		1619.64		
1 Tandem Roller, 10 Ton		236.80		260.48		
1 Roller, Pneum. Whl., 12 Ton		344.40		378.84	36.67	40.34
56 L.H., Daily Totals		$4591.20		$6128.16	$81.99	$109.43
Crew B-90B	**Hr.**	**Daily**	**Hr.**	**Daily**	**Bare Costs**	**Incl. O&P**
1 Labor Foreman (outside)	$39.60	$316.80	$60.95	$487.60	$44.43	$67.82
2 Laborers	37.60	601.60	57.85	925.60		
3 Equip. Oper. (medium)	50.60	1214.40	76.75	1842.00		
1 Roller, Pneum. Whl., 12 Ton		344.40		378.84		
1 Road Mixer, 310 H.P.		1929.00		2121.90	47.36	52.10
48 L.H., Daily Totals		$4406.20		$5755.94	$91.80	$119.92
Crew B-90C	**Hr.**	**Daily**	**Hr.**	**Daily**	**Bare Costs**	**Incl. O&P**
1 Labor Foreman (outside)	$39.60	$316.80	$60.95	$487.60	$42.00	$64.02
4 Laborers	37.60	1203.20	57.85	1851.20		
3 Equip. Oper. (medium)	50.60	1214.40	76.75	1842.00		
3 Truck Drivers (heavy)	40.05	961.20	60.55	1453.20		
3 Road Mixers, 310 H.P.		5787.00		6365.70	65.76	72.34
88 L.H., Daily Totals		$9482.60		$11999.70	$107.76	$136.36
Crew B-90D	**Hr.**	**Daily**	**Hr.**	**Daily**	**Bare Costs**	**Incl. O&P**
1 Labor Foreman (outside)	$39.60	$316.80	$60.95	$487.60	$41.32	$63.07
6 Laborers	37.60	1804.80	57.85	2776.80		
3 Equip. Oper. (medium)	50.60	1214.40	76.75	1842.00		
3 Truck Drivers (heavy)	40.05	961.20	60.55	1453.20		
3 Road Mixers, 310 H.P.		5787.00		6365.70	55.64	61.21
104 L.H., Daily Totals		$10084.20		$12925.30	$96.96	$124.28
Crew B-90E	**Hr.**	**Daily**	**Hr.**	**Daily**	**Bare Costs**	**Incl. O&P**
1 Labor Foreman (outside)	$39.60	$316.80	$60.95	$487.60	$42.43	$64.79
4 Laborers	37.60	1203.20	57.85	1851.20		
3 Equip. Oper. (medium)	50.60	1214.40	76.75	1842.00		
1 Truck Driver (heavy)	40.05	320.40	60.55	484.40		
1 Road Mixers, 310 H.P.		1929.00		2121.90	26.79	29.47
72 L.H., Daily Totals		$4983.80		$6787.10	$69.22	$94.27
Crew B-91	**Hr.**	**Daily**	**Hr.**	**Daily**	**Bare Costs**	**Incl. O&P**
1 Labor Foreman (outside)	$39.60	$316.80	$60.95	$487.60	$44.66	$68.03
2 Laborers	37.60	601.60	57.85	925.60		
4 Equip. Oper. (medium)	50.60	1619.20	76.75	2456.00		
1 Truck Driver (heavy)	40.05	320.40	60.55	484.40		
1 Dist. Tanker, 3000 Gallon		305.60		336.16		
1 Truck Tractor, 6x4, 380 H.P.		599.80		659.78		
1 Aggreg. Spreader, S.P.		834.00		917.40		
1 Roller, Pneum. Whl., 12 Ton		344.40		378.84		
1 Tandem Roller, 10 Ton		236.80		260.48	36.26	39.89
64 L.H., Daily Totals		$5178.60		$6906.26	$80.92	$107.91

Crew No.	Bare Costs		Incl. Subs O&P		Cost Per Labor-Hour	
Crew B-91B	**Hr.**	**Daily**	**Hr.**	**Daily**	**Bare Costs**	**Incl. O&P**
1 Laborer	$37.60	$300.80	$57.85	$462.80	$44.10	$67.30
1 Equipment Oper. (med.)	50.60	404.80	76.75	614.00		
1 Road Sweeper, Vac. Assist.		986.00		1084.60	61.63	67.79
16 L.H., Daily Totals		$1691.60		$2161.40	$105.72	$135.09
Crew B-91C	**Hr.**	**Daily**	**Hr.**	**Daily**	**Bare Costs**	**Incl. O&P**
1 Laborer	$37.60	$300.80	$57.85	$462.80	$38.35	$58.48
1 Truck Driver (light)	39.10	312.80	59.10	472.80		
1 Catch Basin Cleaning Truck		573.60		630.96	35.85	39.44
16 L.H., Daily Totals		$1187.20		$1566.56	$74.20	$97.91
Crew B-91D	**Hr.**	**Daily**	**Hr.**	**Daily**	**Bare Costs**	**Incl. O&P**
1 Labor Foreman (outside)	$39.60	$316.80	$60.95	$487.60	$43.13	$65.77
5 Laborers	37.60	1504.00	57.85	2314.00		
5 Equip. Oper. (medium)	50.60	2024.00	76.75	3070.00		
2 Truck Drivers (heavy)	40.05	640.80	60.55	968.80		
1 Aggreg. Spreader, S.P.		834.00		917.40		
2 Truck Tractor, 6x4, 380 H.P.		1199.60		1319.56		
2 Dist. Tanker, 3000 Gallon		611.20		672.32		
2 Pavement Brush, Towed		163.20		179.52		
2 Roller, Pneum. Whl., 12 Ton		688.80		757.68	33.62	36.99
104 L.H., Daily Totals		$7982.40		$10686.88	$76.75	$102.76
Crew B-92	**Hr.**	**Daily**	**Hr.**	**Daily**	**Bare Costs**	**Incl. O&P**
1 Labor Foreman (outside)	$39.60	$316.80	$60.95	$487.60	$38.10	$58.63
3 Laborers	37.60	902.40	57.85	1388.40		
1 Crack Cleaner, 25 H.P.		61.60		67.76		
1 Air Compressor, 60 cfm		137.40		151.14		
1 Tar Kettle, T.M.		132.00		145.20		
1 Flatbed Truck, Gas, 3 Ton		303.20		333.52	19.82	21.80
32 L.H., Daily Totals		$1853.40		$2573.62	$57.92	$80.43
Crew B-93	**Hr.**	**Daily**	**Hr.**	**Daily**	**Bare Costs**	**Incl. O&P**
1 Equip. Oper. (medium)	$50.60	$404.80	$76.75	$614.00	$50.60	$76.75
1 Feller Buncher, 100 H.P.		838.40		922.24	104.80	115.28
8 L.H., Daily Totals		$1243.20		$1536.24	$155.40	$192.03
Crew B-94A	**Hr.**	**Daily**	**Hr.**	**Daily**	**Bare Costs**	**Incl. O&P**
1 Laborer	$37.60	$300.80	$57.85	$462.80	$37.60	$57.85
1 Diaphragm Water Pump, 2"		71.00		78.10		
1 -20' Suction Hose, 2"		1.95		2.15		
2 -50' Discharge Hoses, 2"		1.80		1.98	9.34	10.28
8 L.H., Daily Totals		$375.55		$545.02	$46.94	$68.13
Crew B-94B	**Hr.**	**Daily**	**Hr.**	**Daily**	**Bare Costs**	**Incl. O&P**
1 Laborer	$37.60	$300.80	$57.85	$462.80	$37.60	$57.85
1 Diaphragm Water Pump, 4"		114.20		125.62		
1 -20' Suction Hose, 4"		3.25		3.58		
2 -50' Discharge Hoses, 4"		4.70		5.17	15.27	16.80
8 L.H., Daily Totals		$422.95		$597.16	$52.87	$74.65
Crew B-94C	**Hr.**	**Daily**	**Hr.**	**Daily**	**Bare Costs**	**Incl. O&P**
1 Laborer	$37.60	$300.80	$57.85	$462.80	$37.60	$57.85
1 Centrifugal Water Pump, 3"		78.20		86.02		
1 -20' Suction Hose, 3"		2.85		3.13		
2 -50' Discharge Hoses, 3"		3.00		3.30	10.51	11.56
8 L.H., Daily Totals		$384.85		$555.26	$48.11	$69.41

Crew No.	Bare Costs		Incl. Subs O&P		Cost Per Labor-Hour	
Crew B-94D	**Hr.**	**Daily**	**Hr.**	**Daily**	**Bare Costs**	**Incl. O&P**
1 Laborer	$37.60	$300.80	$57.85	$462.80	$37.60	$57.85
1 Centr. Water Pump, 6"		346.60		381.26		
1 -20' Suction Hose, 6"		11.50		12.65		
2 -50' Discharge Hoses, 6"		12.20		13.42	46.29	50.92
8 L.H., Daily Totals		$671.10		$870.13	$83.89	$108.77

Crew C-1	Hr.	Daily	Hr.	Daily	Bare Costs	Incl. O&P
3 Carpenters	$46.95	$1126.80	$72.25	$1734.00	$44.61	$68.65
1 Laborer	37.60	300.80	57.85	462.80		
32 L.H., Daily Totals		$1427.60		$2196.80	$44.61	$68.65

Crew C-2	Hr.	Daily	Hr.	Daily	Bare Costs	Incl. O&P
1 Carpenter Foreman (outside)	$48.95	$391.60	$75.35	$602.80	$45.73	$70.37
4 Carpenters	46.95	1502.40	72.25	2312.00		
1 Laborer	37.60	300.80	57.85	462.80		
48 L.H., Daily Totals		$2194.80		$3377.60	$45.73	$70.37

Crew C-2A	Hr.	Daily	Hr.	Daily	Bare Costs	Incl. O&P
1 Carpenter Foreman (outside)	$48.95	$391.60	$75.35	$602.80	$45.40	$69.43
3 Carpenters	46.95	1126.80	72.25	1734.00		
1 Cement Finisher	45.00	360.00	66.65	533.20		
1 Laborer	37.60	300.80	57.85	462.80		
48 L.H., Daily Totals		$2179.20		$3332.80	$45.40	$69.43

Crew C-3	Hr.	Daily	Hr.	Daily	Bare Costs	Incl. O&P
1 Rodman Foreman (outside)	$54.55	$436.40	$85.20	$681.60	$48.57	$75.38
4 Rodmen (reinf.)	52.55	1681.60	82.10	2627.20		
1 Equip. Oper. (light)	48.60	388.80	73.75	590.00		
2 Laborers	37.60	601.60	57.85	925.60		
3 Stressing Equipment		30.60		33.66		
.5 Grouting Equipment		81.50		89.65	1.75	1.93
64 L.H., Daily Totals		$3220.50		$4947.71	$50.32	$77.31

Crew C-4	Hr.	Daily	Hr.	Daily	Bare Costs	Incl. O&P
1 Rodman Foreman (outside)	$54.55	$436.40	$85.20	$681.60	$53.05	$82.88
3 Rodmen (reinf.)	52.55	1261.20	82.10	1970.40		
3 Stressing Equipment		30.60		33.66	.96	1.05
32 L.H., Daily Totals		$1728.20		$2685.66	$54.01	$83.93

Crew C-4A	Hr.	Daily	Hr.	Daily	Bare Costs	Incl. O&P
2 Rodmen (reinf.)	$52.55	$840.80	$82.10	$1313.60	$52.55	$82.10
4 Stressing Equipment		40.80		44.88	2.55	2.81
16 L.H., Daily Totals		$881.60		$1358.48	$55.10	$84.91

Crew C-5	Hr.	Daily	Hr.	Daily	Bare Costs	Incl. O&P
1 Rodman Foreman (outside)	$54.55	$436.40	$85.20	$681.60	$51.66	$80.09
4 Rodmen (reinf.)	52.55	1681.60	82.10	2627.20		
1 Equip. Oper. (crane)	51.70	413.60	78.45	627.60		
1 Equip. Oper. (oiler)	45.20	361.60	68.55	548.40		
1 Hyd. Crane, 25 Ton		736.60		810.26	13.15	14.47
56 L.H., Daily Totals		$3629.80		$5295.06	$64.82	$94.55

Crew C-6	Hr.	Daily	Hr.	Daily	Bare Costs	Incl. O&P
1 Labor Foreman (outside)	$39.60	$316.80	$60.95	$487.60	$39.17	$59.83
4 Laborers	37.60	1203.20	57.85	1851.20		
1 Cement Finisher	45.00	360.00	66.65	533.20		
2 Gas Engine Vibrators		62.40		68.64	1.30	1.43
48 L.H., Daily Totals		$1942.40		$2940.64	$40.47	$61.26

Crew C-7	Hr.	Daily	Hr.	Daily	Bare Costs	Incl. O&P
1 Labor Foreman (outside)	$39.60	$316.80	$60.95	$487.60	$40.93	$62.46
5 Laborers	37.60	1504.00	57.85	2314.00		
1 Cement Finisher	45.00	360.00	66.65	533.20		
1 Equip. Oper. (medium)	50.60	404.80	76.75	614.00		
1 Equip. Oper. (oiler)	45.20	361.60	68.55	548.40		
2 Gas Engine Vibrators		62.40		68.64		
1 Concrete Bucket, 1 C.Y.		23.80		26.18		
1 Hyd. Crane, 55 Ton		1128.00		1240.80	16.86	18.55
72 L.H., Daily Totals		$4161.40		$5832.82	$57.80	$81.01

Crew C-7A	Hr.	Daily	Hr.	Daily	Bare Costs	Incl. O&P
1 Labor Foreman (outside)	$39.60	$316.80	$60.95	$487.60	$38.46	$58.91
5 Laborers	37.60	1504.00	57.85	2314.00		
2 Truck Drivers (heavy)	40.05	640.80	60.55	968.80		
2 Conc. Transit Mixers		2032.00		2235.20	31.75	34.92
64 L.H., Daily Totals		$4493.60		$6005.60	$70.21	$93.84

Crew C-7B	Hr.	Daily	Hr.	Daily	Bare Costs	Incl. O&P
1 Labor Foreman (outside)	$39.60	$316.80	$60.95	$487.60	$40.56	$62.15
5 Laborers	37.60	1504.00	57.85	2314.00		
1 Equipment Operator, Crane	51.70	413.60	78.45	627.60		
1 Equipment Oiler	45.20	361.60	68.55	548.40		
1 Conc. Bucket, 2 C.Y.		37.00		40.70		
1 Lattice Boom Crane, 165 Ton		1932.00		2125.20	30.77	33.84
64 L.H., Daily Totals		$4565.00		$6143.50	$71.33	$95.99

Crew C-7C	Hr.	Daily	Hr.	Daily	Bare Costs	Incl. O&P
1 Labor Foreman (outside)	$39.60	$316.80	$60.95	$487.60	$41.10	$62.96
5 Laborers	37.60	1504.00	57.85	2314.00		
2 Equipment Operators (med.)	50.60	809.60	76.75	1228.00		
2 F.E. Loaders, W.M., 4 C.Y.		1338.40		1472.24	20.91	23.00
64 L.H., Daily Totals		$3968.80		$5501.84	$62.01	$85.97

Crew C-7D	Hr.	Daily	Hr.	Daily	Bare Costs	Incl. O&P
1 Labor Foreman (outside)	$39.60	$316.80	$60.95	$487.60	$39.74	$60.99
5 Laborers	37.60	1504.00	57.85	2314.00		
1 Equip. Oper. (medium)	50.60	404.80	76.75	614.00		
1 Concrete Conveyer		198.80		218.68	3.55	3.90
56 L.H., Daily Totals		$2424.40		$3634.28	$43.29	$64.90

Crew C-8	Hr.	Daily	Hr.	Daily	Bare Costs	Incl. O&P
1 Labor Foreman (outside)	$39.60	$316.80	$60.95	$487.60	$41.86	$63.51
3 Laborers	37.60	902.40	57.85	1388.40		
2 Cement Finishers	45.00	720.00	66.65	1066.40		
1 Equip. Oper. (medium)	50.60	404.80	76.75	614.00		
1 Concrete Pump (Small)		720.00		792.00	12.86	14.14
56 L.H., Daily Totals		$3064.00		$4348.40	$54.71	$77.65

Crew C-8A	Hr.	Daily	Hr.	Daily	Bare Costs	Incl. O&P
1 Labor Foreman (outside)	$39.60	$316.80	$60.95	$487.60	$40.40	$61.30
3 Laborers	37.60	902.40	57.85	1388.40		
2 Cement Finishers	45.00	720.00	66.65	1066.40		
48 L.H., Daily Totals		$1939.20		$2942.40	$40.40	$61.30

Crew No.	Bare Costs		Incl. Subs O&P		Cost Per Labor-Hour	
Crew C-8B	**Hr.**	**Daily**	**Hr.**	**Daily**	**Bare Costs**	**Incl. O&P**
1 Labor Foreman (outside)	$39.60	$316.80	$60.95	$487.60	$40.60	$62.25
3 Laborers	37.60	902.40	57.85	1388.40		
1 Equip. Oper. (medium)	50.60	404.80	76.75	614.00		
1 Vibrating Power Screed		66.30		72.93		
1 Roller, Vibratory, 25 Ton		687.40		756.14		
1 Dozer, 200 H.P.		1387.00		1525.70	53.52	58.87
40 L.H., Daily Totals		$3764.70		$4844.77	$94.12	$121.12
Crew C-8C	**Hr.**	**Daily**	**Hr.**	**Daily**	**Bare Costs**	**Incl. O&P**
1 Labor Foreman (outside)	$39.60	$316.80	$60.95	$487.60	$41.33	$62.98
3 Laborers	37.60	902.40	57.85	1388.40		
1 Cement Finisher	45.00	360.00	66.65	533.20		
1 Equip. Oper. (medium)	50.60	404.80	76.75	614.00		
1 Shotcrete Rig, 12 C.Y./hr		250.80		275.88		
1 Air Compressor, 160 cfm		156.40		172.04		
4 -50' Air Hoses, 1"		16.40		18.04		
4 -50' Air Hoses, 2"		31.00		34.10	9.47	10.42
48 L.H., Daily Totals		$2438.60		$3523.26	$50.80	$73.40
Crew C-8D	**Hr.**	**Daily**	**Hr.**	**Daily**	**Bare Costs**	**Incl. O&P**
1 Labor Foreman (outside)	$39.60	$316.80	$60.95	$487.60	$42.70	$64.80
1 Laborer	37.60	300.80	57.85	462.80		
1 Cement Finisher	45.00	360.00	66.65	533.20		
1 Equipment Oper. (light)	48.60	388.80	73.75	590.00		
1 Air Compressor, 250 cfm		201.40		221.54		
2 -50' Air Hoses, 1"		8.20		9.02	6.55	7.21
32 L.H., Daily Totals		$1576.00		$2304.16	$49.25	$72.00
Crew C-8E	**Hr.**	**Daily**	**Hr.**	**Daily**	**Bare Costs**	**Incl. O&P**
1 Labor Foreman (outside)	$39.60	$316.80	$60.95	$487.60	$41.00	$62.48
3 Laborers	37.60	902.40	57.85	1388.40		
1 Cement Finisher	45.00	360.00	66.65	533.20		
1 Equipment Oper. (light)	48.60	388.80	73.75	590.00		
1 Shotcrete Rig, 35 C.Y./hr		281.00		309.10		
1 Air Compressor, 250 cfm		201.40		221.54		
4 -50' Air Hoses, 1"		16.40		18.04		
4 -50' Air Hoses, 2"		31.00		34.10	11.04	12.14
48 L.H., Daily Totals		$2497.80		$3581.98	$52.04	$74.62
Crew C-10	**Hr.**	**Daily**	**Hr.**	**Daily**	**Bare Costs**	**Incl. O&P**
1 Laborer	$37.60	$300.80	$57.85	$462.80	$42.53	$63.72
2 Cement Finishers	45.00	720.00	66.65	1066.40		
24 L.H., Daily Totals		$1020.80		$1529.20	$42.53	$63.72
Crew C-10B	**Hr.**	**Daily**	**Hr.**	**Daily**	**Bare Costs**	**Incl. O&P**
3 Laborers	$37.60	$902.40	$57.85	$1388.40	$40.56	$61.37
2 Cement Finishers	45.00	720.00	66.65	1066.40		
1 Concrete Mixer, 10 C.F.		172.40		189.64		
2 Trowels, 48" Walk-Behind		100.40		110.44	6.82	7.50
40 L.H., Daily Totals		$1895.20		$2754.88	$47.38	$68.87
Crew C-10C	**Hr.**	**Daily**	**Hr.**	**Daily**	**Bare Costs**	**Incl. O&P**
1 Laborer	$37.60	$300.80	$57.85	$462.80	$42.53	$63.72
2 Cement Finishers	45.00	720.00	66.65	1066.40		
1 Trowel, 48" Walk-Behind		50.20		55.22	2.09	2.30
24 L.H., Daily Totals		$1071.00		$1584.42	$44.63	$66.02

Crew No.	Bare Costs		Incl. Subs O&P		Cost Per Labor-Hour	
Crew C-10D	**Hr.**	**Daily**	**Hr.**	**Daily**	**Bare Costs**	**Incl. O&P**
1 Laborer	$37.60	$300.80	$57.85	$462.80	$42.53	$63.72
2 Cement Finishers	45.00	720.00	66.65	1066.40		
1 Vibrating Power Screed		66.30		72.93		
1 Trowel, 48" Walk-Behind		50.20		55.22	4.85	5.34
24 L.H., Daily Totals		$1137.30		$1657.35	$47.39	$69.06
Crew C-10E	**Hr.**	**Daily**	**Hr.**	**Daily**	**Bare Costs**	**Incl. O&P**
1 Laborer	$37.60	$300.80	$57.85	$462.80	$42.53	$63.72
2 Cement Finishers	45.00	720.00	66.65	1066.40		
1 Vibrating Power Screed		66.30		72.93		
1 Cement Trowel, 96" Ride-On		189.00		207.90	10.64	11.70
24 L.H., Daily Totals		$1276.10		$1810.03	$53.17	$75.42
Crew C-10F	**Hr.**	**Daily**	**Hr.**	**Daily**	**Bare Costs**	**Incl. O&P**
1 Laborer	$37.60	$300.80	$57.85	$462.80	$42.53	$63.72
2 Cement Finishers	45.00	720.00	66.65	1066.40		
1 Aerial Lift Truck, 60' Boom		435.60		479.16	18.15	19.97
24 L.H., Daily Totals		$1456.40		$2008.36	$60.68	$83.68
Crew C-11	**Hr.**	**Daily**	**Hr.**	**Daily**	**Bare Costs**	**Incl. O&P**
1 Struc. Steel Foreman (outside)	$54.65	$437.20	$94.95	$759.60	$51.94	$87.85
6 Struc. Steel Workers	52.65	2527.20	91.45	4389.60		
1 Equip. Oper. (crane)	51.70	413.60	78.45	627.60		
1 Equip. Oper. (oiler)	45.20	361.60	68.55	548.40		
1 Lattice Boom Crane, 150 Ton		1813.00		1994.30	25.18	27.70
72 L.H., Daily Totals		$5552.60		$8319.50	$77.12	$115.55
Crew C-12	**Hr.**	**Daily**	**Hr.**	**Daily**	**Bare Costs**	**Incl. O&P**
1 Carpenter Foreman (outside)	$48.95	$391.60	$75.35	$602.80	$46.52	$71.40
3 Carpenters	46.95	1126.80	72.25	1734.00		
1 Laborer	37.60	300.80	57.85	462.80		
1 Equip. Oper. (crane)	51.70	413.60	78.45	627.60		
1 Hyd. Crane, 12 Ton		653.80		719.18	13.62	14.98
48 L.H., Daily Totals		$2886.60		$4146.38	$60.14	$86.38
Crew C-13	**Hr.**	**Daily**	**Hr.**	**Daily**	**Bare Costs**	**Incl. O&P**
1 Struc. Steel Worker	$52.65	$421.20	$91.45	$731.60	$50.75	$85.05
1 Welder	52.65	421.20	91.45	731.60		
1 Carpenter	46.95	375.60	72.25	578.00		
1 Welder, Gas Engine, 300 amp		145.85		160.44	6.08	6.68
24 L.H., Daily Totals		$1363.85		$2201.64	$56.83	$91.73
Crew C-14	**Hr.**	**Daily**	**Hr.**	**Daily**	**Bare Costs**	**Incl. O&P**
1 Carpenter Foreman (outside)	$48.95	$391.60	$75.35	$602.80	$46.18	$70.93
5 Carpenters	46.95	1878.00	72.25	2890.00		
4 Laborers	37.60	1203.20	57.85	1851.20		
4 Rodmen (reinf.)	52.55	1681.60	82.10	2627.20		
2 Cement Finishers	45.00	720.00	66.65	1066.40		
1 Equip. Oper. (crane)	51.70	413.60	78.45	627.60		
1 Equip. Oper. (oiler)	45.20	361.60	68.55	548.40		
1 Hyd. Crane, 80 Ton		1625.00		1787.50	11.28	12.41
144 L.H., Daily Totals		$8274.60		$12001.10	$57.46	$83.34

Crew No.	Bare Costs		Incl. Subs O&P		Cost Per Labor-Hour	
Crew C-14A	**Hr.**	**Daily**	**Hr.**	**Daily**	**Bare Costs**	**Incl. O&P**
1 Carpenter Foreman (outside)	$48.95	$391.60	$75.35	$602.80	$47.25	$72.75
16 Carpenters	46.95	6009.60	72.25	9248.00		
4 Rodmen (reinf.)	52.55	1681.60	82.10	2627.20		
2 Laborers	37.60	601.60	57.85	925.60		
1 Cement Finisher	45.00	360.00	66.65	533.20		
1 Equip. Oper. (medium)	50.60	404.80	76.75	614.00		
1 Gas Engine Vibrator		31.20		34.32		
1 Concrete Pump (Small)		720.00		792.00	3.76	4.13
200 L.H., Daily Totals		$10200.40		$15377.12	$51.00	$76.89

Crew No.	Bare Costs		Incl. Subs O&P		Cost Per Labor-Hour	
Crew C-14B	**Hr.**	**Daily**	**Hr.**	**Daily**	**Bare Costs**	**Incl. O&P**
1 Carpenter Foreman (outside)	$48.95	$391.60	$75.35	$602.80	$47.16	$72.52
16 Carpenters	46.95	6009.60	72.25	9248.00		
4 Rodmen (reinf.)	52.55	1681.60	82.10	2627.20		
2 Laborers	37.60	601.60	57.85	925.60		
2 Cement Finishers	45.00	720.00	66.65	1066.40		
1 Equip. Oper. (medium)	50.60	404.80	76.75	614.00		
1 Gas Engine Vibrator		31.20		34.32		
1 Concrete Pump (Small)		720.00		792.00	3.61	3.97
208 L.H., Daily Totals		$10560.40		$15910.32	$50.77	$76.49

Crew No.	Bare Costs		Incl. Subs O&P		Cost Per Labor-Hour	
Crew C-14C	**Hr.**	**Daily**	**Hr.**	**Daily**	**Bare Costs**	**Incl. O&P**
1 Carpenter Foreman (outside)	$48.95	$391.60	$75.35	$602.80	$45.08	$69.36
6 Carpenters	46.95	2253.60	72.25	3468.00		
2 Rodmen (reinf.)	52.55	840.80	82.10	1313.60		
4 Laborers	37.60	1203.20	57.85	1851.20		
1 Cement Finisher	45.00	360.00	66.65	533.20		
1 Gas Engine Vibrator		31.20		34.32	.28	.31
112 L.H., Daily Totals		$5080.40		$7803.12	$45.36	$69.67

Crew No.	Bare Costs		Incl. Subs O&P		Cost Per Labor-Hour	
Crew C-14D	**Hr.**	**Daily**	**Hr.**	**Daily**	**Bare Costs**	**Incl. O&P**
1 Carpenter Foreman (outside)	$48.95	$391.60	$75.35	$602.80	$46.80	$71.97
18 Carpenters	46.95	6760.80	72.25	10404.00		
2 Rodmen (reinf.)	52.55	840.80	82.10	1313.60		
2 Laborers	37.60	601.60	57.85	925.60		
1 Cement Finisher	45.00	360.00	66.65	533.20		
1 Equip. Oper. (medium)	50.60	404.80	76.75	614.00		
1 Gas Engine Vibrator		31.20		34.32		
1 Concrete Pump (Small)		720.00		792.00	3.76	4.13
200 L.H., Daily Totals		$10110.80		$15219.52	$50.55	$76.10

Crew No.	Bare Costs		Incl. Subs O&P		Cost Per Labor-Hour	
Crew C-14E	**Hr.**	**Daily**	**Hr.**	**Daily**	**Bare Costs**	**Incl. O&P**
1 Carpenter Foreman (outside)	$48.95	$391.60	$75.35	$602.80	$46.44	$71.68
2 Carpenters	46.95	751.20	72.25	1156.00		
4 Rodmen (reinf.)	52.55	1681.60	82.10	2627.20		
3 Laborers	37.60	902.40	57.85	1388.40		
1 Cement Finisher	45.00	360.00	66.65	533.20		
1 Gas Engine Vibrator		31.20		34.32	.35	.39
88 L.H., Daily Totals		$4118.00		$6341.92	$46.80	$72.07

Crew No.	Bare Costs		Incl. Subs O&P		Cost Per Labor-Hour	
Crew C-14F	**Hr.**	**Daily**	**Hr.**	**Daily**	**Bare Costs**	**Incl. O&P**
1 Labor Foreman (outside)	$39.60	$316.80	$60.95	$487.60	$42.76	$64.06
2 Laborers	37.60	601.60	57.85	925.60		
6 Cement Finishers	45.00	2160.00	66.65	3199.20		
1 Gas Engine Vibrator		31.20		34.32	.43	.48
72 L.H., Daily Totals		$3109.60		$4646.72	$43.19	$64.54

Crew No.	Bare Costs		Incl. Subs O&P		Cost Per Labor-Hour	
Crew C-14G	**Hr.**	**Daily**	**Hr.**	**Daily**	**Bare Costs**	**Incl. O&P**
1 Labor Foreman (outside)	$39.60	$316.80	$60.95	$487.60	$42.11	$63.32
2 Laborers	37.60	601.60	57.85	925.60		
4 Cement Finishers	45.00	1440.00	66.65	2132.80		
1 Gas Engine Vibrator		31.20		34.32	.56	.61
56 L.H., Daily Totals		$2389.60		$3580.32	$42.67	$63.93

Crew No.	Bare Costs		Incl. Subs O&P		Cost Per Labor-Hour	
Crew C-14H	**Hr.**	**Daily**	**Hr.**	**Daily**	**Bare Costs**	**Incl. O&P**
1 Carpenter Foreman (outside)	$48.95	$391.60	$75.35	$602.80	$46.33	$71.08
2 Carpenters	46.95	751.20	72.25	1156.00		
1 Rodman (reinf.)	52.55	420.40	82.10	656.80		
1 Laborer	37.60	300.80	57.85	462.80		
1 Cement Finisher	45.00	360.00	66.65	533.20		
1 Gas Engine Vibrator		31.20		34.32	.65	.71
48 L.H., Daily Totals		$2255.20		$3445.92	$46.98	$71.79

Crew No.	Bare Costs		Incl. Subs O&P		Cost Per Labor-Hour	
Crew C-14L	**Hr.**	**Daily**	**Hr.**	**Daily**	**Bare Costs**	**Incl. O&P**
1 Carpenter Foreman (outside)	$48.95	$391.60	$75.35	$602.80	$43.84	$67.24
6 Carpenters	46.95	2253.60	72.25	3468.00		
4 Laborers	37.60	1203.20	57.85	1851.20		
1 Cement Finisher	45.00	360.00	66.65	533.20		
1 Gas Engine Vibrator		31.20		34.32	.33	.36
96 L.H., Daily Totals		$4239.60		$6489.52	$44.16	$67.60

Crew No.	Bare Costs		Incl. Subs O&P		Cost Per Labor-Hour	
Crew C-14M	**Hr.**	**Daily**	**Hr.**	**Daily**	**Bare Costs**	**Incl. O&P**
1 Carpenter Foreman (outside)	$48.95	$391.60	$75.35	$602.80	$45.77	$70.13
2 Carpenters	46.95	751.20	72.25	1156.00		
1 Rodman (reinf.)	52.55	420.40	82.10	656.80		
2 Laborers	37.60	601.60	57.85	925.60		
1 Cement Finisher	45.00	360.00	66.65	533.20		
1 Equip. Oper. (medium)	50.60	404.80	76.75	614.00		
1 Gas Engine Vibrator		31.20		34.32		
1 Concrete Pump (Small)		720.00		792.00	11.74	12.91
64 L.H., Daily Totals		$3680.80		$5314.72	$57.51	$83.04

Crew No.	Bare Costs		Incl. Subs O&P		Cost Per Labor-Hour	
Crew C-15	**Hr.**	**Daily**	**Hr.**	**Daily**	**Bare Costs**	**Incl. O&P**
1 Carpenter Foreman (outside)	$48.95	$391.60	$75.35	$602.80	$44.24	$67.64
2 Carpenters	46.95	751.20	72.25	1156.00		
3 Laborers	37.60	902.40	57.85	1388.40		
2 Cement Finishers	45.00	720.00	66.65	1066.40		
1 Rodman (reinf.)	52.55	420.40	82.10	656.80		
72 L.H., Daily Totals		$3185.60		$4870.40	$44.24	$67.64

Crew No.	Bare Costs		Incl. Subs O&P		Cost Per Labor-Hour	
Crew C-16	**Hr.**	**Daily**	**Hr.**	**Daily**	**Bare Costs**	**Incl. O&P**
1 Labor Foreman (outside)	$39.60	$316.80	$60.95	$487.60	$41.86	$63.51
3 Laborers	37.60	902.40	57.85	1388.40		
2 Cement Finishers	45.00	720.00	66.65	1066.40		
1 Equip. Oper. (medium)	50.60	404.80	76.75	614.00		
1 Gunite Pump Rig		371.80		408.98		
2 -50' Air Hoses, 3/4"		6.50		7.15		
2 -50' Air Hoses, 2"		15.50		17.05	7.03	7.74
56 L.H., Daily Totals		$2737.80		$3989.58	$48.89	$71.24

Crew No.	Bare Costs		Incl. Subs O&P		Cost Per Labor-Hour	
Crew C-16A	**Hr.**	**Daily**	**Hr.**	**Daily**	**Bare Costs**	**Incl. O&P**
1 Laborer	$37.60	$300.80	$57.85	$462.80	$44.55	$66.97
2 Cement Finishers	45.00	720.00	66.65	1066.40		
1 Equip. Oper. (medium)	50.60	404.80	76.75	614.00		
1 Gunite Pump Rig		371.80		408.98		
2 -50' Air Hoses, 3/4"		6.50		7.15		
2 -50' Air Hoses, 2"		15.50		17.05		
1 Aerial Lift Truck, 60' Boom		435.60		479.16	25.92	28.51
32 L.H., Daily Totals		$2255.00		$3055.54	$70.47	$95.49

Crew C-17	Bare Costs Hr.	Bare Costs Daily	Incl. Subs O&P Hr.	Incl. Subs O&P Daily	Cost Per Labor-Hour Bare Costs	Cost Per Labor-Hour Incl. O&P
2 Skilled Worker Foremen (out)	$50.65	$810.40	$78.25	$1252.00	$49.05	$75.77
8 Skilled Workers	48.65	3113.60	75.15	4809.60		
80 L.H., Daily Totals		$3924.00		$6061.60	$49.05	$75.77

Crew C-17A	Hr.	Daily	Hr.	Daily	Bare Costs	Incl. O&P
2 Skilled Worker Foremen (out)	$50.65	$810.40	$78.25	$1252.00	$49.08	$75.80
8 Skilled Workers	48.65	3113.60	75.15	4809.60		
.125 Equip. Oper. (crane)	51.70	51.70	78.45	78.45		
.125 Hyd. Crane, 80 Ton		203.13		223.44	2.51	2.76
81 L.H., Daily Totals		$4178.82		$6363.49	$51.59	$78.56

Crew C-17B	Hr.	Daily	Hr.	Daily	Bare Costs	Incl. O&P
2 Skilled Worker Foremen (out)	$50.65	$810.40	$78.25	$1252.00	$49.11	$75.84
8 Skilled Workers	48.65	3113.60	75.15	4809.60		
.25 Equip. Oper. (crane)	51.70	103.40	78.45	156.90		
.25 Hyd. Crane, 80 Ton		406.25		446.88		
.25 Trowel, 48" Walk-Behind		12.55		13.81	5.11	5.62
82 L.H., Daily Totals		$4446.20		$6679.18	$54.22	$81.45

Crew C-17C	Hr.	Daily	Hr.	Daily	Bare Costs	Incl. O&P
2 Skilled Worker Foremen (out)	$50.65	$810.40	$78.25	$1252.00	$49.15	$75.87
8 Skilled Workers	48.65	3113.60	75.15	4809.60		
.375 Equip. Oper. (crane)	51.70	155.10	78.45	235.35		
.375 Hyd. Crane, 80 Ton		609.38		670.31	7.34	8.08
83 L.H., Daily Totals		$4688.48		$6967.26	$56.49	$83.94

Crew C-17D	Hr.	Daily	Hr.	Daily	Bare Costs	Incl. O&P
2 Skilled Worker Foremen (out)	$50.65	$810.40	$78.25	$1252.00	$49.18	$75.90
8 Skilled Workers	48.65	3113.60	75.15	4809.60		
.5 Equip. Oper. (crane)	51.70	206.80	78.45	313.80		
.5 Hyd. Crane, 80 Ton		812.50		893.75	9.67	10.64
84 L.H., Daily Totals		$4943.30		$7269.15	$58.85	$86.54

Crew C-17E	Hr.	Daily	Hr.	Daily	Bare Costs	Incl. O&P
2 Skilled Worker Foremen (out)	$50.65	$810.40	$78.25	$1252.00	$49.05	$75.77
8 Skilled Workers	48.65	3113.60	75.15	4809.60		
1 Hyd. Jack with Rods		96.50		106.15	1.21	1.33
80 L.H., Daily Totals		$4020.50		$6167.75	$50.26	$77.10

Crew C-18	Hr.	Daily	Hr.	Daily	Bare Costs	Incl. O&P
.125 Labor Foreman (outside)	$39.60	$39.60	$60.95	$60.95	$37.82	$58.19
1 Laborer	37.60	300.80	57.85	462.80		
1 Concrete Cart, 10 C.F.		60.00		66.00	6.67	7.33
9 L.H., Daily Totals		$400.40		$589.75	$44.49	$65.53

Crew C-19	Hr.	Daily	Hr.	Daily	Bare Costs	Incl. O&P
.125 Labor Foreman (outside)	$39.60	$39.60	$60.95	$60.95	$37.82	$58.19
1 Laborer	37.60	300.80	57.85	462.80		
1 Concrete Cart, 18 C.F.		99.80		109.78	11.09	12.20
9 L.H., Daily Totals		$440.20		$633.53	$48.91	$70.39

Crew C-20	Hr.	Daily	Hr.	Daily	Bare Costs	Incl. O&P
1 Labor Foreman (outside)	$39.60	$316.80	$60.95	$487.60	$40.40	$61.70
5 Laborers	37.60	1504.00	57.85	2314.00		
1 Cement Finisher	45.00	360.00	66.65	533.20		
1 Equip. Oper. (medium)	50.60	404.80	76.75	614.00		
2 Gas Engine Vibrators		62.40		68.64		
1 Concrete Pump (Small)		720.00		792.00	12.23	13.45
64 L.H., Daily Totals		$3368.00		$4809.44	$52.63	$75.15

Crew C-21	Hr.	Daily	Hr.	Daily	Bare Costs	Incl. O&P
1 Labor Foreman (outside)	$39.60	$316.80	$60.95	$487.60	$40.40	$61.70
5 Laborers	37.60	1504.00	57.85	2314.00		
1 Cement Finisher	45.00	360.00	66.65	533.20		
1 Equip. Oper. (medium)	50.60	404.80	76.75	614.00		
2 Gas Engine Vibrators		62.40		68.64		
1 Concrete Conveyer		198.80		218.68	4.08	4.49
64 L.H., Daily Totals		$2846.80		$4236.12	$44.48	$66.19

Crew C-22	Hr.	Daily	Hr.	Daily	Bare Costs	Incl. O&P
1 Rodman Foreman (outside)	$54.55	$436.40	$85.20	$681.60	$52.74	$82.28
4 Rodmen (reinf.)	52.55	1681.60	82.10	2627.20		
.125 Equip. Oper. (crane)	51.70	51.70	78.45	78.45		
.125 Equip. Oper. (oiler)	45.20	45.20	68.55	68.55		
.125 Hyd. Crane, 25 Ton		92.08		101.28	2.19	2.41
42 L.H., Daily Totals		$2306.97		$3557.08	$54.93	$84.69

Crew C-23	Hr.	Daily	Hr.	Daily	Bare Costs	Incl. O&P
2 Skilled Worker Foremen (out)	$50.65	$810.40	$78.25	$1252.00	$49.01	$75.44
6 Skilled Workers	48.65	2335.20	75.15	3607.20		
1 Equip. Oper. (crane)	51.70	413.60	78.45	627.60		
1 Equip. Oper. (oiler)	45.20	361.60	68.55	548.40		
1 Lattice Boom Crane, 90 Ton		1511.00		1662.10	18.89	20.78
80 L.H., Daily Totals		$5431.80		$7697.30	$67.90	$96.22

Crew C-23A	Hr.	Daily	Hr.	Daily	Bare Costs	Incl. O&P
1 Labor Foreman (outside)	$39.60	$316.80	$60.95	$487.60	$42.34	$64.73
2 Laborers	37.60	601.60	57.85	925.60		
1 Equip. Oper. (crane)	51.70	413.60	78.45	627.60		
1 Equip. Oper. (oiler)	45.20	361.60	68.55	548.40		
1 Crawler Crane, 100 Ton		1667.00		1833.70		
3 Conc. Buckets, 8 C.Y.		608.40		669.24	56.88	62.57
40 L.H., Daily Totals		$3969.00		$5092.14	$99.22	$127.30

Crew C-24	Hr.	Daily	Hr.	Daily	Bare Costs	Incl. O&P
2 Skilled Worker Foremen (out)	$50.65	$810.40	$78.25	$1252.00	$49.01	$75.44
6 Skilled Workers	48.65	2335.20	75.15	3607.20		
1 Equip. Oper. (crane)	51.70	413.60	78.45	627.60		
1 Equip. Oper. (oiler)	45.20	361.60	68.55	548.40		
1 Lattice Boom Crane, 150 Ton		1813.00		1994.30	22.66	24.93
80 L.H., Daily Totals		$5733.80		$8029.50	$71.67	$100.37

Crew C-25	Hr.	Daily	Hr.	Daily	Bare Costs	Incl. O&P
2 Rodmen (reinf.)	$52.55	$840.80	$82.10	$1313.60	$41.30	$66.70
2 Rodmen Helpers	30.05	480.80	51.30	820.80		
32 L.H., Daily Totals		$1321.60		$2134.40	$41.30	$66.70

Crew C-27	Hr.	Daily	Hr.	Daily	Bare Costs	Incl. O&P
2 Cement Finishers	$45.00	$720.00	$66.65	$1066.40	$45.00	$66.65
1 Concrete Saw		167.40		184.14	10.46	11.51
16 L.H., Daily Totals		$887.40		$1250.54	$55.46	$78.16

Crew C-28	Hr.	Daily	Hr.	Daily	Bare Costs	Incl. O&P
1 Cement Finisher	$45.00	$360.00	$66.65	$533.20	$45.00	$66.65
1 Portable Air Compressor, Gas		17.90		19.69	2.24	2.46
8 L.H., Daily Totals		$377.90		$552.89	$47.24	$69.11

Crew C-29	Hr.	Daily	Hr.	Daily	Bare Costs	Incl. O&P
1 Laborer	$37.60	$300.80	$57.85	$462.80	$37.60	$57.85
1 Pressure Washer		70.20		77.22	8.78	9.65
8 L.H., Daily Totals		$371.00		$540.02	$46.38	$67.50

Crew No.	Bare Costs		Incl. Subs O&P		Cost Per Labor-Hour	
Crew C-30	**Hr.**	**Daily**	**Hr.**	**Daily**	**Bare Costs**	**Incl. O&P**
1 Laborer	$37.60	$300.80	$57.85	$462.80	$37.60	$57.85
1 Concrete Mixer, 10 C.F.		172.40		189.64	21.55	23.70
8 L.H., Daily Totals		$473.20		$652.44	$59.15	$81.56
Crew C-31	**Hr.**	**Daily**	**Hr.**	**Daily**	**Bare Costs**	**Incl. O&P**
1 Cement Finisher	$45.00	$360.00	$66.65	$533.20	$45.00	$66.65
1 Grout Pump		371.80		408.98	46.48	51.12
8 L.H., Daily Totals		$731.80		$942.18	$91.47	$117.77
Crew C-32	**Hr.**	**Daily**	**Hr.**	**Daily**	**Bare Costs**	**Incl. O&P**
1 Cement Finisher	$45.00	$360.00	$66.65	$533.20	$41.30	$62.25
1 Laborer	37.60	300.80	57.85	462.80		
1 Crack Chaser Saw, Gas, 6 H.P.		29.20		32.12		
1 Vacuum Pick-Up System		60.90		66.99	5.63	6.19
16 L.H., Daily Totals		$750.90		$1095.11	$46.93	$68.44
Crew D-1	**Hr.**	**Daily**	**Hr.**	**Daily**	**Bare Costs**	**Incl. O&P**
1 Bricklayer	$46.15	$369.20	$70.45	$563.60	$42.10	$64.28
1 Bricklayer Helper	38.05	304.40	58.10	464.80		
16 L.H., Daily Totals		$673.60		$1028.40	$42.10	$64.28
Crew D-2	**Hr.**	**Daily**	**Hr.**	**Daily**	**Bare Costs**	**Incl. O&P**
3 Bricklayers	$46.15	$1107.60	$70.45	$1690.80	$43.28	$66.12
2 Bricklayer Helpers	38.05	608.80	58.10	929.60		
.5 Carpenter	46.95	187.80	72.25	289.00		
44 L.H., Daily Totals		$1904.20		$2909.40	$43.28	$66.12
Crew D-3	**Hr.**	**Daily**	**Hr.**	**Daily**	**Bare Costs**	**Incl. O&P**
3 Bricklayers	$46.15	$1107.60	$70.45	$1690.80	$43.10	$65.83
2 Bricklayer Helpers	38.05	608.80	58.10	929.60		
.25 Carpenter	46.95	93.90	72.25	144.50		
42 L.H., Daily Totals		$1810.30		$2764.90	$43.10	$65.83
Crew D-4	**Hr.**	**Daily**	**Hr.**	**Daily**	**Bare Costs**	**Incl. O&P**
1 Bricklayer	$46.15	$369.20	$70.45	$563.60	$42.71	$65.10
2 Bricklayer Helpers	38.05	608.80	58.10	929.60		
1 Equip. Oper. (light)	48.60	388.80	73.75	590.00		
1 Grout Pump, 50 C.F./hr.		133.40		146.74	4.17	4.59
32 L.H., Daily Totals		$1500.20		$2229.94	$46.88	$69.69
Crew D-5	**Hr.**	**Daily**	**Hr.**	**Daily**	**Bare Costs**	**Incl. O&P**
1 Bricklayer	46.15	369.20	70.45	563.60	46.15	70.45
8 L.H., Daily Totals		$369.20		$563.60	$46.15	$70.45
Crew D-6	**Hr.**	**Daily**	**Hr.**	**Daily**	**Bare Costs**	**Incl. O&P**
3 Bricklayers	$46.15	$1107.60	$70.45	$1690.80	$42.29	$64.59
3 Bricklayer Helpers	38.05	913.20	58.10	1394.40		
.25 Carpenter	46.95	93.90	72.25	144.50		
50 L.H., Daily Totals		$2114.70		$3229.70	$42.29	$64.59
Crew D-7	**Hr.**	**Daily**	**Hr.**	**Daily**	**Bare Costs**	**Incl. O&P**
1 Tile Layer	$42.80	$342.40	$63.30	$506.40	$38.17	$56.45
1 Tile Layer Helper	33.55	268.40	49.60	396.80		
16 L.H., Daily Totals		$610.80		$903.20	$38.17	$56.45
Crew D-8	**Hr.**	**Daily**	**Hr.**	**Daily**	**Bare Costs**	**Incl. O&P**
3 Bricklayers	$46.15	$1107.60	$70.45	$1690.80	$42.91	$65.51
2 Bricklayer Helpers	38.05	608.80	58.10	929.60		
40 L.H., Daily Totals		$1716.40		$2620.40	$42.91	$65.51

Crew No.	Bare Costs		Incl. Subs O&P		Cost Per Labor-Hour	
Crew D-9	**Hr.**	**Daily**	**Hr.**	**Daily**	**Bare Costs**	**Incl. O&P**
3 Bricklayers	$46.15	$1107.60	$70.45	$1690.80	$42.10	$64.28
3 Bricklayer Helpers	38.05	913.20	58.10	1394.40		
48 L.H., Daily Totals		$2020.80		$3085.20	$42.10	$64.28
Crew D-10	**Hr.**	**Daily**	**Hr.**	**Daily**	**Bare Costs**	**Incl. O&P**
1 Bricklayer Foreman (outside)	$48.15	$385.20	$73.55	$588.40	$46.01	$70.14
1 Bricklayer	46.15	369.20	70.45	563.60		
1 Bricklayer Helper	38.05	304.40	58.10	464.80		
1 Equip. Oper. (crane)	51.70	413.60	78.45	627.60		
1 S.P. Crane, 4x4, 12 Ton		473.40		520.74	14.79	16.27
32 L.H., Daily Totals		$1945.80		$2765.14	$60.81	$86.41
Crew D-11	**Hr.**	**Daily**	**Hr.**	**Daily**	**Bare Costs**	**Incl. O&P**
1 Bricklayer Foreman (outside)	$48.15	$385.20	$73.55	$588.40	$44.12	$67.37
1 Bricklayer	46.15	369.20	70.45	563.60		
1 Bricklayer Helper	38.05	304.40	58.10	464.80		
24 L.H., Daily Totals		$1058.80		$1616.80	$44.12	$67.37
Crew D-12	**Hr.**	**Daily**	**Hr.**	**Daily**	**Bare Costs**	**Incl. O&P**
1 Bricklayer Foreman (outside)	$48.15	$385.20	$73.55	$588.40	$42.60	$65.05
1 Bricklayer	46.15	369.20	70.45	563.60		
2 Bricklayer Helpers	38.05	608.80	58.10	929.60		
32 L.H., Daily Totals		$1363.20		$2081.60	$42.60	$65.05
Crew D-13	**Hr.**	**Daily**	**Hr.**	**Daily**	**Bare Costs**	**Incl. O&P**
1 Bricklayer Foreman (outside)	$48.15	$385.20	$73.55	$588.40	$44.84	$68.48
1 Bricklayer	46.15	369.20	70.45	563.60		
2 Bricklayer Helpers	38.05	608.80	58.10	929.60		
1 Carpenter	46.95	375.60	72.25	578.00		
1 Equip. Oper. (crane)	51.70	413.60	78.45	627.60		
1 S.P. Crane, 4x4, 12 Ton		473.40		520.74	9.86	10.85
48 L.H., Daily Totals		$2625.80		$3807.94	$54.70	$79.33
Crew E-1	**Hr.**	**Daily**	**Hr.**	**Daily**	**Bare Costs**	**Incl. O&P**
1 Welder Foreman (outside)	$54.65	$437.20	$94.95	$759.60	$51.97	$86.72
1 Welder	52.65	421.20	91.45	731.60		
1 Equip. Oper. (light)	48.60	388.80	73.75	590.00		
1 Welder, Gas Engine, 300 amp		145.85		160.44	6.08	6.68
24 L.H., Daily Totals		$1393.05		$2241.64	$58.04	$93.40
Crew E-2	**Hr.**	**Daily**	**Hr.**	**Daily**	**Bare Costs**	**Incl. O&P**
1 Struc. Steel Foreman (outside)	$54.65	$437.20	$94.95	$759.60	$51.74	$86.82
4 Struc. Steel Workers	52.65	1684.80	91.45	2926.40		
1 Equip. Oper. (crane)	51.70	413.60	78.45	627.60		
1 Equip. Oper. (oiler)	45.20	361.60	68.55	548.40		
1 Lattice Boom Crane, 90 Ton		1511.00		1662.10	26.98	29.68
56 L.H., Daily Totals		$4408.20		$6524.10	$78.72	$116.50
Crew E-3	**Hr.**	**Daily**	**Hr.**	**Daily**	**Bare Costs**	**Incl. O&P**
1 Struc. Steel Foreman (outside)	$54.65	$437.20	$94.95	$759.60	$53.32	$92.62
1 Struc. Steel Worker	52.65	421.20	91.45	731.60		
1 Welder	52.65	421.20	91.45	731.60		
1 Welder, Gas Engine, 300 amp		145.85		160.44	6.08	6.68
24 L.H., Daily Totals		$1425.45		$2383.24	$59.39	$99.30

Crew No.	Bare Costs		Incl. Subs O&P		Cost Per Labor-Hour	
Crew E-3A	**Hr.**	**Daily**	**Hr.**	**Daily**	**Bare Costs**	**Incl. O&P**
1 Struc. Steel Foreman (outside)	$54.65	$437.20	$94.95	$759.60	$53.32	$92.62
1 Struc. Steel Worker	52.65	421.20	91.45	731.60		
1 Welder	52.65	421.20	91.45	731.60		
1 Welder, Gas Engine, 300 amp		145.85		160.44		
1 Aerial Lift Truck, 40' Boom		301.40		331.54	18.64	20.50
24 L.H., Daily Totals		$1726.85		$2714.78	$71.95	$113.12
Crew E-4	**Hr.**	**Daily**	**Hr.**	**Daily**	**Bare Costs**	**Incl. O&P**
1 Struc. Steel Foreman (outside)	$54.65	$437.20	$94.95	$759.60	$53.15	$92.33
3 Struc. Steel Workers	52.65	1263.60	91.45	2194.80		
1 Welder, Gas Engine, 300 amp		145.85		160.44	4.56	5.01
32 L.H., Daily Totals		$1846.65		$3114.84	$57.71	$97.34
Crew E-5	**Hr.**	**Daily**	**Hr.**	**Daily**	**Bare Costs**	**Incl. O&P**
2 Struc. Steel Foremen (outside)	$54.65	$874.40	$94.95	$1519.20	$52.21	$88.56
5 Struc. Steel Workers	52.65	2106.00	91.45	3658.00		
1 Equip. Oper. (crane)	51.70	413.60	78.45	627.60		
1 Welder	52.65	421.20	91.45	731.60		
1 Equip. Oper. (oiler)	45.20	361.60	68.55	548.40		
1 Lattice Boom Crane, 90 Ton		1511.00		1662.10		
1 Welder, Gas Engine, 300 amp		145.85		160.44	20.71	22.78
80 L.H., Daily Totals		$5833.65		$8907.33	$72.92	$111.34
Crew E-6	**Hr.**	**Daily**	**Hr.**	**Daily**	**Bare Costs**	**Incl. O&P**
3 Struc. Steel Foremen (outside)	$54.65	$1311.60	$94.95	$2278.80	$52.25	$88.76
9 Struc. Steel Workers	52.65	3790.80	91.45	6584.40		
1 Equip. Oper. (crane)	51.70	413.60	78.45	627.60		
1 Welder	52.65	421.20	91.45	731.60		
1 Equip. Oper. (oiler)	45.20	361.60	68.55	548.40		
1 Equip. Oper. (light)	48.60	388.80	73.75	590.00		
1 Lattice Boom Crane, 90 Ton		1511.00		1662.10		
1 Welder, Gas Engine, 300 amp		145.85		160.44		
1 Air Compressor, 160 cfm		156.40		172.04		
2 Impact Wrenches		36.00		39.60	14.45	15.89
128 L.H., Daily Totals		$8536.85		$13394.98	$66.69	$104.65
Crew E-7	**Hr.**	**Daily**	**Hr.**	**Daily**	**Bare Costs**	**Incl. O&P**
1 Struc. Steel Foreman (outside)	$54.65	$437.20	$94.95	$759.60	$52.21	$88.56
4 Struc. Steel Workers	52.65	1684.80	91.45	2926.40		
1 Equip. Oper. (crane)	51.70	413.60	78.45	627.60		
1 Equip. Oper. (oiler)	45.20	361.60	68.55	548.40		
1 Welder Foreman (outside)	54.65	437.20	94.95	759.60		
2 Welders	52.65	842.40	91.45	1463.20		
1 Lattice Boom Crane, 90 Ton		1511.00		1662.10		
2 Welder, Gas Engine, 300 amp		291.70		320.87	22.53	24.79
80 L.H., Daily Totals		$5979.50		$9067.77	$74.74	$113.35
Crew E-8	**Hr.**	**Daily**	**Hr.**	**Daily**	**Bare Costs**	**Incl. O&P**
1 Struc. Steel Foreman (outside)	$54.65	$437.20	$94.95	$759.60	$52.00	$87.87
4 Struc. Steel Workers	52.65	1684.80	91.45	2926.40		
1 Welder Foreman (outside)	54.65	437.20	94.95	759.60		
4 Welders	52.65	1684.80	91.45	2926.40		
1 Equip. Oper. (crane)	51.70	413.60	78.45	627.60		
1 Equip. Oper. (oiler)	45.20	361.60	68.55	548.40		
1 Equip. Oper. (light)	48.60	388.80	73.75	590.00		
1 Lattice Boom Crane, 90 Ton		1511.00		1662.10		
4 Welder, Gas Engine, 300 amp		583.40		641.74	20.14	22.15
104 L.H., Daily Totals		$7502.40		$11441.84	$72.14	$110.02

Crew No.	Bare Costs		Incl. Subs O&P		Cost Per Labor-Hour	
Crew E-9	**Hr.**	**Daily**	**Hr.**	**Daily**	**Bare Costs**	**Incl. O&P**
2 Struc. Steel Foremen (outside)	$54.65	$874.40	$94.95	$1519.20	$52.25	$88.76
5 Struc. Steel Workers	52.65	2106.00	91.45	3658.00		
1 Welder Foreman (outside)	54.65	437.20	94.95	759.60		
5 Welders	52.65	2106.00	91.45	3658.00		
1 Equip. Oper. (crane)	51.70	413.60	78.45	627.60		
1 Equip. Oper. (oiler)	45.20	361.60	68.55	548.40		
1 Equip. Oper. (light)	48.60	388.80	73.75	590.00		
1 Lattice Boom Crane, 90 Ton		1511.00		1662.10		
5 Welder, Gas Engine, 300 amp		729.25		802.17	17.50	19.25
128 L.H., Daily Totals		$8927.85		$13825.08	$69.75	$108.01
Crew E-10	**Hr.**	**Daily**	**Hr.**	**Daily**	**Bare Costs**	**Incl. O&P**
1 Welder Foreman (outside)	$54.65	$437.20	$94.95	$759.60	$53.65	$93.20
1 Welder	52.65	421.20	91.45	731.60		
1 Welder, Gas Engine, 300 amp		145.85		160.44		
1 Flatbed Truck, Gas, 3 Ton		303.20		333.52	28.07	30.87
16 L.H., Daily Totals		$1307.45		$1985.16	$81.72	$124.07
Crew E-11	**Hr.**	**Daily**	**Hr.**	**Daily**	**Bare Costs**	**Incl. O&P**
2 Painters, Struc. Steel	$41.45	$663.20	$72.95	$1167.20	$42.27	$69.38
1 Building Laborer	37.60	300.80	57.85	462.80		
1 Equip. Oper. (light)	48.60	388.80	73.75	590.00		
1 Air Compressor, 250 cfm		201.40		221.54		
1 Sandblaster, Portable, 3 C.F.		20.40		22.44		
1 Set Sand Blasting Accessories		14.05		15.46	7.37	8.11
32 L.H., Daily Totals		$1588.65		$2479.43	$49.65	$77.48
Crew E-11A	**Hr.**	**Daily**	**Hr.**	**Daily**	**Bare Costs**	**Incl. O&P**
2 Painters, Struc. Steel	$41.45	$663.20	$72.95	$1167.20	$42.27	$69.38
1 Building Laborer	37.60	300.80	57.85	462.80		
1 Equip. Oper. (light)	48.60	388.80	73.75	590.00		
1 Air Compressor, 250 cfm		201.40		221.54		
1 Sandblaster, Portable, 3 C.F.		20.40		22.44		
1 Set Sand Blasting Accessories		14.05		15.46		
1 Aerial Lift Truck, 60' Boom		435.60		479.16	20.98	23.08
32 L.H., Daily Totals		$2024.25		$2958.59	$63.26	$92.46
Crew E-11B	**Hr.**	**Daily**	**Hr.**	**Daily**	**Bare Costs**	**Incl. O&P**
2 Painters, Struc. Steel	$41.45	$663.20	$72.95	$1167.20	$40.17	$67.92
1 Building Laborer	37.60	300.80	57.85	462.80		
2 Paint Sprayer, 8 C.F.M.		99.80		109.78		
1 Aerial Lift Truck, 60' Boom		435.60		479.16	22.31	24.54
24 L.H., Daily Totals		$1499.40		$2218.94	$62.48	$92.46
Crew E-12	**Hr.**	**Daily**	**Hr.**	**Daily**	**Bare Costs**	**Incl. O&P**
1 Welder Foreman (outside)	$54.65	$437.20	$94.95	$759.60	$51.63	$84.35
1 Equip. Oper. (light)	48.60	388.80	73.75	590.00		
1 Welder, Gas Engine, 300 amp		145.85		160.44	9.12	10.03
16 L.H., Daily Totals		$971.85		$1510.04	$60.74	$94.38
Crew E-13	**Hr.**	**Daily**	**Hr.**	**Daily**	**Bare Costs**	**Incl. O&P**
1 Welder Foreman (outside)	$54.65	$437.20	$94.95	$759.60	$52.63	$87.88
.5 Equip. Oper. (light)	48.60	194.40	73.75	295.00		
1 Welder, Gas Engine, 300 amp		145.85		160.44	12.15	13.37
12 L.H., Daily Totals		$777.45		$1215.04	$64.79	$101.25
Crew E-14	**Hr.**	**Daily**	**Hr.**	**Daily**	**Bare Costs**	**Incl. O&P**
1 Welder Foreman (outside)	$54.65	$437.20	$94.95	$759.60	$54.65	$94.95
1 Welder, Gas Engine, 300 amp		145.85		160.44	18.23	20.05
8 L.H., Daily Totals		$583.05		$920.03	$72.88	$115.00

Crew No.	Bare Costs		Incl. Subs O&P		Cost Per Labor-Hour	
Crew E-16	**Hr.**	**Daily**	**Hr.**	**Daily**	**Bare Costs**	**Incl. O&P**
1 Welder Foreman (outside)	$54.65	$437.20	$94.95	$759.60	$53.65	$93.20
1 Welder	52.65	421.20	91.45	731.60		
1 Welder, Gas Engine, 300 amp		145.85		160.44	9.12	10.03
16 L.H., Daily Totals		$1004.25		$1651.64	$62.77	$103.23
Crew E-17	**Hr.**	**Daily**	**Hr.**	**Daily**	**Bare Costs**	**Incl. O&P**
1 Struc. Steel Foreman (outside)	$54.65	$437.20	$94.95	$759.60	$53.65	$93.20
1 Structural Steel Worker	52.65	421.20	91.45	731.60		
16 L.H., Daily Totals		$858.40		$1491.20	$53.65	$93.20
Crew E-18	**Hr.**	**Daily**	**Hr.**	**Daily**	**Bare Costs**	**Incl. O&P**
1 Struc. Steel Foreman (outside)	$54.65	$437.20	$94.95	$759.60	$52.64	$89.21
3 Structural Steel Workers	52.65	1263.60	91.45	2194.80		
1 Equipment Operator (med.)	50.60	404.80	76.75	614.00		
1 Lattice Boom Crane, 20 Ton		948.90		1043.79	23.72	26.09
40 L.H., Daily Totals		$3054.50		$4612.19	$76.36	$115.30
Crew E-19	**Hr.**	**Daily**	**Hr.**	**Daily**	**Bare Costs**	**Incl. O&P**
1 Struc. Steel Foreman (outside)	$54.65	$437.20	$94.95	$759.60	$51.97	$86.72
1 Structural Steel Worker	52.65	421.20	91.45	731.60		
1 Equip. Oper. (light)	48.60	388.80	73.75	590.00		
1 Lattice Boom Crane, 20 Ton		948.90		1043.79	39.54	43.49
24 L.H., Daily Totals		$2196.10		$3124.99	$91.50	$130.21
Crew E-20	**Hr.**	**Daily**	**Hr.**	**Daily**	**Bare Costs**	**Incl. O&P**
1 Struc. Steel Foreman (outside)	$54.65	$437.20	$94.95	$759.60	$51.85	$87.40
5 Structural Steel Workers	52.65	2106.00	91.45	3658.00		
1 Equip. Oper. (crane)	51.70	413.60	78.45	627.60		
1 Equip. Oper. (oiler)	45.20	361.60	68.55	548.40		
1 Lattice Boom Crane, 40 Ton		1164.00		1280.40	18.19	20.01
64 L.H., Daily Totals		$4482.40		$6874.00	$70.04	$107.41
Crew E-22	**Hr.**	**Daily**	**Hr.**	**Daily**	**Bare Costs**	**Incl. O&P**
1 Skilled Worker Foreman (out)	$50.65	$405.20	$78.25	$626.00	$49.32	$76.18
2 Skilled Workers	48.65	778.40	75.15	1202.40		
24 L.H., Daily Totals		$1183.60		$1828.40	$49.32	$76.18
Crew E-24	**Hr.**	**Daily**	**Hr.**	**Daily**	**Bare Costs**	**Incl. O&P**
3 Structural Steel Workers	$52.65	$1263.60	$91.45	$2194.80	$52.14	$87.78
1 Equipment Operator (med.)	50.60	404.80	76.75	614.00		
1 Hyd. Crane, 25 Ton		736.60		810.26	23.02	25.32
32 L.H., Daily Totals		$2405.00		$3619.06	$75.16	$113.10
Crew E-25	**Hr.**	**Daily**	**Hr.**	**Daily**	**Bare Costs**	**Incl. O&P**
1 Welder Foreman (outside)	$54.65	$437.20	$94.95	$759.60	$54.65	$94.95
1 Cutting Torch		11.40		12.54	1.43	1.57
8 L.H., Daily Totals		$448.60		$772.14	$56.08	$96.52
Crew F-3	**Hr.**	**Daily**	**Hr.**	**Daily**	**Bare Costs**	**Incl. O&P**
4 Carpenters	$46.95	$1502.40	$72.25	$2312.00	$47.90	$73.49
1 Equip. Oper. (crane)	51.70	413.60	78.45	627.60		
1 Hyd. Crane, 12 Ton		653.80		719.18	16.34	17.98
40 L.H., Daily Totals		$2569.80		$3658.78	$64.25	$91.47
Crew F-4	**Hr.**	**Daily**	**Hr.**	**Daily**	**Bare Costs**	**Incl. O&P**
4 Carpenters	$46.95	$1502.40	$72.25	$2312.00	$47.45	$72.67
1 Equip. Oper. (crane)	51.70	413.60	78.45	627.60		
1 Equip. Oper. (oiler)	45.20	361.60	68.55	548.40		
1 Hyd. Crane, 55 Ton		1128.00		1240.80	23.50	25.85
48 L.H., Daily Totals		$3405.60		$4728.80	$70.95	$98.52

Crew No.	Bare Costs		Incl. Subs O&P		Cost Per Labor-Hour	
Crew F-5	**Hr.**	**Daily**	**Hr.**	**Daily**	**Bare Costs**	**Incl. O&P**
1 Carpenter Foreman (outside)	$48.95	$391.60	$75.35	$602.80	$47.45	$73.03
3 Carpenters	46.95	1126.80	72.25	1734.00		
32 L.H., Daily Totals		$1518.40		$2336.80	$47.45	$73.03
Crew F-6	**Hr.**	**Daily**	**Hr.**	**Daily**	**Bare Costs**	**Incl. O&P**
2 Carpenters	$46.95	$751.20	$72.25	$1156.00	$44.16	$67.73
2 Building Laborers	37.60	601.60	57.85	925.60		
1 Equip. Oper. (crane)	51.70	413.60	78.45	627.60		
1 Hyd. Crane, 12 Ton		653.80		719.18	16.34	17.98
40 L.H., Daily Totals		$2420.20		$3428.38	$60.51	$85.71
Crew F-7	**Hr.**	**Daily**	**Hr.**	**Daily**	**Bare Costs**	**Incl. O&P**
2 Carpenters	$46.95	$751.20	$72.25	$1156.00	$42.27	$65.05
2 Building Laborers	37.60	601.60	57.85	925.60		
32 L.H., Daily Totals		$1352.80		$2081.60	$42.27	$65.05
Crew G-1	**Hr.**	**Daily**	**Hr.**	**Daily**	**Bare Costs**	**Incl. O&P**
1 Roofer Foreman (outside)	$42.10	$336.80	$71.85	$574.80	$37.51	$64.04
4 Roofers Composition	40.10	1283.20	68.45	2190.40		
2 Roofer Helpers	30.05	480.80	51.30	820.80		
1 Application Equipment		193.00		212.30		
1 Tar Kettle/Pot		169.80		186.78		
1 Crew Truck		206.40		227.04	10.16	11.18
56 L.H., Daily Totals		$2670.00		$4212.12	$47.68	$75.22
Crew G-2	**Hr.**	**Daily**	**Hr.**	**Daily**	**Bare Costs**	**Incl. O&P**
1 Plasterer	$42.95	$343.60	$64.40	$515.20	$39.55	$59.78
1 Plasterer Helper	38.10	304.80	57.10	456.80		
1 Building Laborer	37.60	300.80	57.85	462.80		
1 Grout Pump, 50 C.F./hr.		133.40		146.74	5.56	6.11
24 L.H., Daily Totals		$1082.60		$1581.54	$45.11	$65.90
Crew G-2A	**Hr.**	**Daily**	**Hr.**	**Daily**	**Bare Costs**	**Incl. O&P**
1 Roofer Composition	$40.10	$320.80	$68.45	$547.60	$35.92	$59.20
1 Roofer Helper	30.05	240.40	51.30	410.40		
1 Building Laborer	37.60	300.80	57.85	462.80		
1 Foam Spray Rig, Trailer-Mtd.		573.25		630.58		
1 Pickup Truck, 3/4 Ton		144.20		158.62	29.89	32.88
24 L.H., Daily Totals		$1579.45		$2209.99	$65.81	$92.08
Crew G-3	**Hr.**	**Daily**	**Hr.**	**Daily**	**Bare Costs**	**Incl. O&P**
2 Sheet Metal Workers	$55.95	$895.20	$85.50	$1368.00	$46.77	$71.67
2 Building Laborers	37.60	601.60	57.85	925.60		
32 L.H., Daily Totals		$1496.80		$2293.60	$46.77	$71.67
Crew G-4	**Hr.**	**Daily**	**Hr.**	**Daily**	**Bare Costs**	**Incl. O&P**
1 Labor Foreman (outside)	$39.60	$316.80	$60.95	$487.60	$38.27	$58.88
2 Building Laborers	37.60	601.60	57.85	925.60		
1 Flatbed Truck, Gas, 1.5 Ton		245.40		269.94		
1 Air Compressor, 160 cfm		156.40		172.04	16.74	18.42
24 L.H., Daily Totals		$1320.20		$1855.18	$55.01	$77.30
Crew G-5	**Hr.**	**Daily**	**Hr.**	**Daily**	**Bare Costs**	**Incl. O&P**
1 Roofer Foreman (outside)	$42.10	$336.80	$71.85	$574.80	$36.48	$62.27
2 Roofers Composition	40.10	641.60	68.45	1095.20		
2 Roofer Helpers	30.05	480.80	51.30	820.80		
1 Application Equipment		193.00		212.30	4.83	5.31
40 L.H., Daily Totals		$1652.20		$2703.10	$41.31	$67.58

Crew No.	Bare Costs		Incl. Subs O&P		Cost Per Labor-Hour	
Crew G-6A	**Hr.**	**Daily**	**Hr.**	**Daily**	**Bare Costs**	**Incl. O&P**
2 Roofers Composition	$40.10	$641.60	$68.45	$1095.20	$40.10	$68.45
1 Small Compressor, Electric		12.95		14.24		
2 Pneumatic Nailers		55.30		60.83	4.27	4.69
16 L.H., Daily Totals		$709.85		$1170.28	$44.37	$73.14
Crew G-7	**Hr.**	**Daily**	**Hr.**	**Daily**	**Bare Costs**	**Incl. O&P**
1 Carpenter	$46.95	$375.60	$72.25	$578.00	$46.95	$72.25
1 Small Compressor, Electric		12.95		14.24		
1 Pneumatic Nailer		27.65		30.41	5.08	5.58
8 L.H., Daily Totals		$416.20		$622.66	$52.02	$77.83
Crew H-1	**Hr.**	**Daily**	**Hr.**	**Daily**	**Bare Costs**	**Incl. O&P**
2 Glaziers	$45.10	$721.60	$68.70	$1099.20	$48.88	$80.08
2 Struc. Steel Workers	52.65	842.40	91.45	1463.20		
32 L.H., Daily Totals		$1564.00		$2562.40	$48.88	$80.08
Crew H-2	**Hr.**	**Daily**	**Hr.**	**Daily**	**Bare Costs**	**Incl. O&P**
2 Glaziers	$45.10	$721.60	$68.70	$1099.20	$42.60	$65.08
1 Building Laborer	37.60	300.80	57.85	462.80		
24 L.H., Daily Totals		$1022.40		$1562.00	$42.60	$65.08
Crew H-3	**Hr.**	**Daily**	**Hr.**	**Daily**	**Bare Costs**	**Incl. O&P**
1 Glazier	$45.10	$360.80	$68.70	$549.60	$40.27	$61.83
1 Helper	35.45	283.60	54.95	439.60		
16 L.H., Daily Totals		$644.40		$989.20	$40.27	$61.83
Crew H-4	**Hr.**	**Daily**	**Hr.**	**Daily**	**Bare Costs**	**Incl. O&P**
1 Carpenter	$46.95	$375.60	$72.25	$578.00	$43.90	$67.27
1 Carpenter Helper	35.45	283.60	54.95	439.60		
.5 Electrician	54.70	218.80	81.95	327.80		
20 L.H., Daily Totals		$878.00		$1345.40	$43.90	$67.27
Crew J-1	**Hr.**	**Daily**	**Hr.**	**Daily**	**Bare Costs**	**Incl. O&P**
3 Plasterers	$42.95	$1030.80	$64.40	$1545.60	$41.01	$61.48
2 Plasterer Helpers	38.10	609.60	57.10	913.60		
1 Mixing Machine, 6 C.F.		140.20		154.22	3.50	3.86
40 L.H., Daily Totals		$1780.60		$2613.42	$44.52	$65.34
Crew J-2	**Hr.**	**Daily**	**Hr.**	**Daily**	**Bare Costs**	**Incl. O&P**
3 Plasterers	$42.95	$1030.80	$64.40	$1545.60	$41.34	$61.80
2 Plasterer Helpers	38.10	609.60	57.10	913.60		
1 Lather	43.00	344.00	63.40	507.20		
1 Mixing Machine, 6 C.F.		140.20		154.22	2.92	3.21
48 L.H., Daily Totals		$2124.60		$3120.62	$44.26	$65.01
Crew J-3	**Hr.**	**Daily**	**Hr.**	**Daily**	**Bare Costs**	**Incl. O&P**
1 Terrazzo Worker	$42.95	$343.60	$63.50	$508.00	$39.23	$58.00
1 Terrazzo Helper	35.50	284.00	52.50	420.00		
1 Floor Grinder, 22" Path		115.55		127.11		
1 Terrazzo Mixer		188.20		207.02	18.98	20.88
16 L.H., Daily Totals		$931.35		$1262.13	$58.21	$78.88
Crew J-4	**Hr.**	**Daily**	**Hr.**	**Daily**	**Bare Costs**	**Incl. O&P**
2 Cement Finishers	$45.00	$720.00	$66.65	$1066.40	$42.53	$63.72
1 Laborer	37.60	300.80	57.85	462.80		
1 Floor Grinder, 22" Path		115.55		127.11		
1 Floor Edger, 7" Path		39.30		43.23		
1 Vacuum Pick-Up System		60.90		66.99	8.99	9.89
24 L.H., Daily Totals		$1236.55		$1766.53	$51.52	$73.61

Crew No.	Bare Costs		Incl. Subs O&P		Cost Per Labor-Hour	
Crew J-4A	**Hr.**	**Daily**	**Hr.**	**Daily**	**Bare Costs**	**Incl. O&P**
2 Cement Finishers	$45.00	$720.00	$66.65	$1066.40	$41.30	$62.25
2 Laborers	37.60	601.60	57.85	925.60		
1 Floor Grinder, 22" Path		115.55		127.11		
1 Floor Edger, 7" Path		39.30		43.23		
1 Vacuum Pick-Up System		60.90		66.99		
1 Floor Auto Scrubber		233.85		257.24	14.05	15.46
32 L.H., Daily Totals		$1771.20		$2486.56	$55.35	$77.70
Crew J-4B	**Hr.**	**Daily**	**Hr.**	**Daily**	**Bare Costs**	**Incl. O&P**
1 Laborer	$37.60	$300.80	$57.85	$462.80	$37.60	$57.85
1 Floor Auto Scrubber		233.85		257.24	29.23	32.15
8 L.H., Daily Totals		$534.65		$720.03	$66.83	$90.00
Crew J-6	**Hr.**	**Daily**	**Hr.**	**Daily**	**Bare Costs**	**Incl. O&P**
2 Painters	$40.35	$645.60	$60.80	$972.80	$41.73	$63.30
1 Building Laborer	37.60	300.80	57.85	462.80		
1 Equip. Oper. (light)	48.60	388.80	73.75	590.00		
1 Air Compressor, 250 cfm		201.40		221.54		
1 Sandblaster, Portable, 3 C.F.		20.40		22.44		
1 Set Sand Blasting Accessories		14.05		15.46	7.37	8.11
32 L.H., Daily Totals		$1571.05		$2285.03	$49.10	$71.41
Crew J-7	**Hr.**	**Daily**	**Hr.**	**Daily**	**Bare Costs**	**Incl. O&P**
2 Painters	$40.35	$645.60	$60.80	$972.80	$40.35	$60.80
1 Floor Belt Sander		14.20		15.62		
1 Floor Sanding Edger		12.60		13.86	1.68	1.84
16 L.H., Daily Totals		$672.40		$1002.28	$42.02	$62.64
Crew K-1	**Hr.**	**Daily**	**Hr.**	**Daily**	**Bare Costs**	**Incl. O&P**
1 Carpenter	$46.95	$375.60	$72.25	$578.00	$43.02	$65.67
1 Truck Driver (light)	39.10	312.80	59.10	472.80		
1 Flatbed Truck, Gas, 3 Ton		303.20		333.52	18.95	20.84
16 L.H., Daily Totals		$991.60		$1384.32	$61.98	$86.52
Crew K-2	**Hr.**	**Daily**	**Hr.**	**Daily**	**Bare Costs**	**Incl. O&P**
1 Struc. Steel Foreman (outside)	$54.65	$437.20	$94.95	$759.60	$48.80	$81.83
1 Struc. Steel Worker	52.65	421.20	91.45	731.60		
1 Truck Driver (light)	39.10	312.80	59.10	472.80		
1 Flatbed Truck, Gas, 3 Ton		303.20		333.52	12.63	13.90
24 L.H., Daily Totals		$1474.40		$2297.52	$61.43	$95.73
Crew L-1	**Hr.**	**Daily**	**Hr.**	**Daily**	**Bare Costs**	**Incl. O&P**
1 Electrician	$54.70	$437.60	$81.95	$655.60	$56.70	$85.30
1 Plumber	58.70	469.60	88.65	709.20		
16 L.H., Daily Totals		$907.20		$1364.80	$56.70	$85.30
Crew L-2	**Hr.**	**Daily**	**Hr.**	**Daily**	**Bare Costs**	**Incl. O&P**
1 Carpenter	$46.95	$375.60	$72.25	$578.00	$41.20	$63.60
1 Carpenter Helper	35.45	283.60	54.95	439.60		
16 L.H., Daily Totals		$659.20		$1017.60	$41.20	$63.60
Crew L-3	**Hr.**	**Daily**	**Hr.**	**Daily**	**Bare Costs**	**Incl. O&P**
1 Carpenter	$46.95	$375.60	$72.25	$578.00	$51.14	$77.99
.5 Electrician	54.70	218.80	81.95	327.80		
.5 Sheet Metal Worker	55.95	223.80	85.50	342.00		
16 L.H., Daily Totals		$818.20		$1247.80	$51.14	$77.99

Crew L-3A	Bare Costs Hr.	Bare Costs Daily	Incl. Subs O&P Hr.	Incl. Subs O&P Daily	Cost Per Labor-Hour Bare Costs	Cost Per Labor-Hour Incl. O&P
1 Carpenter Foreman (outside)	$48.95	$391.60	$75.35	$602.80	$51.28	$78.73
.5 Sheet Metal Worker	55.95	223.80	85.50	342.00		
12 L.H., Daily Totals		$615.40		$944.80	$51.28	$78.73

Crew L-4	Hr.	Daily	Hr.	Daily	Bare Costs	Incl. O&P
2 Skilled Workers	$48.65	$778.40	$75.15	$1202.40	$44.25	$68.42
1 Helper	35.45	283.60	54.95	439.60		
24 L.H., Daily Totals		$1062.00		$1642.00	$44.25	$68.42

Crew L-5	Hr.	Daily	Hr.	Daily	Bare Costs	Incl. O&P
1 Struc. Steel Foreman (outside)	$54.65	$437.20	$94.95	$759.60	$52.80	$90.09
5 Struc. Steel Workers	52.65	2106.00	91.45	3658.00		
1 Equip. Oper. (crane)	51.70	413.60	78.45	627.60		
1 Hyd. Crane, 25 Ton		736.60		810.26	13.15	14.47
56 L.H., Daily Totals		$3693.40		$5855.46	$65.95	$104.56

Crew L-5A	Hr.	Daily	Hr.	Daily	Bare Costs	Incl. O&P
1 Struc. Steel Foreman (outside)	$54.65	$437.20	$94.95	$759.60	$52.91	$89.08
2 Structural Steel Workers	52.65	842.40	91.45	1463.20		
1 Equip. Oper. (crane)	51.70	413.60	78.45	627.60		
1 S.P. Crane, 4x4, 25 Ton		599.40		659.34	18.73	20.60
32 L.H., Daily Totals		$2292.60		$3509.74	$71.64	$109.68

Crew L-5B	Hr.	Daily	Hr.	Daily	Bare Costs	Incl. O&P
1 Struc. Steel Foreman (outside)	$54.65	$437.20	$94.95	$759.60	$53.97	$85.46
2 Structural Steel Workers	52.65	842.40	91.45	1463.20		
2 Electricians	54.70	875.20	81.95	1311.20		
2 Steamfitters/Pipefitters	59.75	956.00	90.20	1443.20		
1 Equip. Oper. (crane)	51.70	413.60	78.45	627.60		
1 Equip. Oper. (oiler)	45.20	361.60	68.55	548.40		
1 Hyd. Crane, 80 Ton		1625.00		1787.50	22.57	24.83
72 L.H., Daily Totals		$5511.00		$7940.70	$76.54	$110.29

Crew L-6	Hr.	Daily	Hr.	Daily	Bare Costs	Incl. O&P
1 Plumber	$58.70	$469.60	$88.65	$709.20	$57.37	$86.42
.5 Electrician	54.70	218.80	81.95	327.80		
12 L.H., Daily Totals		$688.40		$1037.00	$57.37	$86.42

Crew L-7	Hr.	Daily	Hr.	Daily	Bare Costs	Incl. O&P
2 Carpenters	$46.95	$751.20	$72.25	$1156.00	$45.39	$69.52
1 Building Laborer	37.60	300.80	57.85	462.80		
.5 Electrician	54.70	218.80	81.95	327.80		
28 L.H., Daily Totals		$1270.80		$1946.60	$45.39	$69.52

Crew L-8	Hr.	Daily	Hr.	Daily	Bare Costs	Incl. O&P
2 Carpenters	$46.95	$751.20	$72.25	$1156.00	$49.30	$75.53
.5 Plumber	58.70	234.80	88.65	354.60		
20 L.H., Daily Totals		$986.00		$1510.60	$49.30	$75.53

Crew L-9	Hr.	Daily	Hr.	Daily	Bare Costs	Incl. O&P
1 Labor Foreman (inside)	$38.10	$304.80	$58.65	$469.20	$42.96	$68.17
2 Building Laborers	37.60	601.60	57.85	925.60		
1 Struc. Steel Worker	52.65	421.20	91.45	731.60		
.5 Electrician	54.70	218.80	81.95	327.80		
36 L.H., Daily Totals		$1546.40		$2454.20	$42.96	$68.17

Crew L-10	Hr.	Daily	Hr.	Daily	Bare Costs	Incl. O&P
1 Struc. Steel Foreman (outside)	$54.65	$437.20	$94.95	$759.60	$53.00	$88.28
1 Structural Steel Worker	52.65	421.20	91.45	731.60		
1 Equip. Oper. (crane)	51.70	413.60	78.45	627.60		
1 Hyd. Crane, 12 Ton		653.80		719.18	27.24	29.97
24 L.H., Daily Totals		$1925.80		$2837.98	$80.24	$118.25

Crew L-11	Hr.	Daily	Hr.	Daily	Bare Costs	Incl. O&P
2 Wreckers	$37.60	$601.60	$62.10	$993.60	$43.88	$69.10
1 Equip. Oper. (crane)	51.70	413.60	78.45	627.60		
1 Equip. Oper. (light)	48.60	388.80	73.75	590.00		
1 Hyd. Excavator, 2.5 C.Y.		1605.00		1765.50		
1 Loader, Skid Steer, 78 H.P.		318.60		350.46	60.11	66.12
32 L.H., Daily Totals		$3327.60		$4327.16	$103.99	$135.22

Crew M-1	Hr.	Daily	Hr.	Daily	Bare Costs	Incl. O&P
3 Elevator Constructors	$76.50	$1836.00	$114.20	$2740.80	$72.67	$108.49
1 Elevator Apprentice	61.20	489.60	91.35	730.80		
5 Hand Tools		46.00		50.60	1.44	1.58
32 L.H., Daily Totals		$2371.60		$3522.20	$74.11	$110.07

Crew M-3	Hr.	Daily	Hr.	Daily	Bare Costs	Incl. O&P
1 Electrician Foreman (outside)	$56.70	$453.60	$84.95	$679.60	$57.56	$86.48
1 Common Laborer	37.60	300.80	57.85	462.80		
.25 Equipment Operator (med.)	50.60	101.20	76.75	153.50		
1 Elevator Constructor	76.50	612.00	114.20	913.60		
1 Elevator Apprentice	61.20	489.60	91.35	730.80		
.25 S.P. Crane, 4x4, 20 Ton		138.65		152.51	4.08	4.49
34 L.H., Daily Totals		$2095.85		$3092.82	$61.64	$90.97

Crew M-4	Hr.	Daily	Hr.	Daily	Bare Costs	Incl. O&P
1 Electrician Foreman (outside)	$56.70	$453.60	$84.95	$679.60	$56.94	$85.58
1 Common Laborer	37.60	300.80	57.85	462.80		
.25 Equipment Operator, Crane	51.70	103.40	78.45	156.90		
.25 Equip. Oper. (oiler)	45.20	90.40	68.55	137.10		
1 Elevator Constructor	76.50	612.00	114.20	913.60		
1 Elevator Apprentice	61.20	489.60	91.35	730.80		
.25 S.P. Crane, 4x4, 40 Ton		175.75		193.32	4.88	5.37
36 L.H., Daily Totals		$2225.55		$3274.13	$61.82	$90.95

Crew Q-1	Hr.	Daily	Hr.	Daily	Bare Costs	Incl. O&P
1 Plumber	$58.70	$469.60	$88.65	$709.20	$52.83	$79.78
1 Plumber Apprentice	46.95	375.60	70.90	567.20		
16 L.H., Daily Totals		$845.20		$1276.40	$52.83	$79.78

Crew Q-1A	Hr.	Daily	Hr.	Daily	Bare Costs	Incl. O&P
.25 Plumber Foreman (outside)	$60.70	$121.40	$91.65	$183.30	$59.10	$89.25
1 Plumber	58.70	469.60	88.65	709.20		
10 L.H., Daily Totals		$591.00		$892.50	$59.10	$89.25

Crew Q-1C	Hr.	Daily	Hr.	Daily	Bare Costs	Incl. O&P
1 Plumber	$58.70	$469.60	$88.65	$709.20	$52.08	$78.77
1 Plumber Apprentice	46.95	375.60	70.90	567.20		
1 Equip. Oper. (medium)	50.60	404.80	76.75	614.00		
1 Trencher, Chain Type, 8' D		3402.00		3742.20	141.75	155.93
24 L.H., Daily Totals		$4652.00		$5632.60	$193.83	$234.69

Crew Q-2	Hr.	Daily	Hr.	Daily	Bare Costs	Incl. O&P
2 Plumbers	$58.70	$939.20	$88.65	$1418.40	$54.78	$82.73
1 Plumber Apprentice	46.95	375.60	70.90	567.20		
24 L.H., Daily Totals		$1314.80		$1985.60	$54.78	$82.73

Crew Q-3	Hr.	Daily	Hr.	Daily	Bare Costs	Incl. O&P
1 Plumber Foreman (inside)	$59.20	$473.60	$89.40	$715.20	$55.89	$84.40
2 Plumbers	58.70	939.20	88.65	1418.40		
1 Plumber Apprentice	46.95	375.60	70.90	567.20		
32 L.H., Daily Totals		$1788.40		$2700.80	$55.89	$84.40

Crew Q-4	Hr.	Daily	Hr.	Daily	Bare Costs	Incl. O&P
1 Plumber Foreman (inside)	$59.20	$473.60	$89.40	$715.20	$55.89	$84.40
1 Plumber	58.70	469.60	88.65	709.20		
1 Welder (plumber)	58.70	469.60	88.65	709.20		
1 Plumber Apprentice	46.95	375.60	70.90	567.20		
1 Welder, Electric, 300 amp		57.70		63.47	1.80	1.98
32 L.H., Daily Totals		$1846.10		$2764.27	$57.69	$86.38

Crew Q-5	Hr.	Daily	Hr.	Daily	Bare Costs	Incl. O&P
1 Steamfitter	$59.75	$478.00	$90.20	$721.60	$53.77	$81.20
1 Steamfitter Apprentice	47.80	382.40	72.20	577.60		
16 L.H., Daily Totals		$860.40		$1299.20	$53.77	$81.20

Crew Q-6	Hr.	Daily	Hr.	Daily	Bare Costs	Incl. O&P
2 Steamfitters	$59.75	$956.00	$90.20	$1443.20	$55.77	$84.20
1 Steamfitter Apprentice	47.80	382.40	72.20	577.60		
24 L.H., Daily Totals		$1338.40		$2020.80	$55.77	$84.20

Crew Q-7	Hr.	Daily	Hr.	Daily	Bare Costs	Incl. O&P
1 Steamfitter Foreman (inside)	$60.25	$482.00	$91.00	$728.00	$56.89	$85.90
2 Steamfitters	59.75	956.00	90.20	1443.20		
1 Steamfitter Apprentice	47.80	382.40	72.20	577.60		
32 L.H., Daily Totals		$1820.40		$2748.80	$56.89	$85.90

Crew Q-8	Hr.	Daily	Hr.	Daily	Bare Costs	Incl. O&P
1 Steamfitter Foreman (inside)	$60.25	$482.00	$91.00	$728.00	$56.89	$85.90
1 Steamfitter	59.75	478.00	90.20	721.60		
1 Welder (steamfitter)	59.75	478.00	90.20	721.60		
1 Steamfitter Apprentice	47.80	382.40	72.20	577.60		
1 Welder, Electric, 300 amp		57.70		63.47	1.80	1.98
32 L.H., Daily Totals		$1878.10		$2812.27	$58.69	$87.88

Crew Q-9	Hr.	Daily	Hr.	Daily	Bare Costs	Incl. O&P
1 Sheet Metal Worker	$55.95	$447.60	$85.50	$684.00	$50.35	$76.95
1 Sheet Metal Apprentice	44.75	358.00	68.40	547.20		
16 L.H., Daily Totals		$805.60		$1231.20	$50.35	$76.95

Crew Q-10	Hr.	Daily	Hr.	Daily	Bare Costs	Incl. O&P
2 Sheet Metal Workers	$55.95	$895.20	$85.50	$1368.00	$52.22	$79.80
1 Sheet Metal Apprentice	44.75	358.00	68.40	547.20		
24 L.H., Daily Totals		$1253.20		$1915.20	$52.22	$79.80

Crew Q-11	Hr.	Daily	Hr.	Daily	Bare Costs	Incl. O&P
1 Sheet Metal Foreman (inside)	$56.45	$451.60	$86.25	$690.00	$53.27	$81.41
2 Sheet Metal Workers	55.95	895.20	85.50	1368.00		
1 Sheet Metal Apprentice	44.75	358.00	68.40	547.20		
32 L.H., Daily Totals		$1704.80		$2605.20	$53.27	$81.41

Crew Q-12	Hr.	Daily	Hr.	Daily	Bare Costs	Incl. O&P
1 Sprinkler Installer	$56.15	$449.20	$84.95	$679.60	$50.52	$76.45
1 Sprinkler Apprentice	44.90	359.20	67.95	543.60		
16 L.H., Daily Totals		$808.40		$1223.20	$50.52	$76.45

Crew Q-13	Hr.	Daily	Hr.	Daily	Bare Costs	Incl. O&P
1 Sprinkler Foreman (inside)	$56.65	$453.20	$85.70	$685.60	$53.46	$80.89
2 Sprinkler Installers	56.15	898.40	84.95	1359.20		
1 Sprinkler Apprentice	44.90	359.20	67.95	543.60		
32 L.H., Daily Totals		$1710.80		$2588.40	$53.46	$80.89

Crew Q-14	Hr.	Daily	Hr.	Daily	Bare Costs	Incl. O&P
1 Asbestos Worker	$52.35	$418.80	$81.50	$652.00	$47.13	$73.38
1 Asbestos Apprentice	41.90	335.20	65.25	522.00		
16 L.H., Daily Totals		$754.00		$1174.00	$47.13	$73.38

Crew Q-15	Hr.	Daily	Hr.	Daily	Bare Costs	Incl. O&P
1 Plumber	$58.70	$469.60	$88.65	$709.20	$52.83	$79.78
1 Plumber Apprentice	46.95	375.60	70.90	567.20		
1 Welder, Electric, 300 amp		57.70		63.47	3.61	3.97
16 L.H., Daily Totals		$902.90		$1339.87	$56.43	$83.74

Crew Q-16	Hr.	Daily	Hr.	Daily	Bare Costs	Incl. O&P
2 Plumbers	$58.70	$939.20	$88.65	$1418.40	$54.78	$82.73
1 Plumber Apprentice	46.95	375.60	70.90	567.20		
1 Welder, Electric, 300 amp		57.70		63.47	2.40	2.64
24 L.H., Daily Totals		$1372.50		$2049.07	$57.19	$85.38

Crew Q-17	Hr.	Daily	Hr.	Daily	Bare Costs	Incl. O&P
1 Steamfitter	$59.75	$478.00	$90.20	$721.60	$53.77	$81.20
1 Steamfitter Apprentice	47.80	382.40	72.20	577.60		
1 Welder, Electric, 300 amp		57.70		63.47	3.61	3.97
16 L.H., Daily Totals		$918.10		$1362.67	$57.38	$85.17

Crew Q-17A	Hr.	Daily	Hr.	Daily	Bare Costs	Incl. O&P
1 Steamfitter	$59.75	$478.00	$90.20	$721.60	$53.08	$80.28
1 Steamfitter Apprentice	47.80	382.40	72.20	577.60		
1 Equip. Oper. (crane)	51.70	413.60	78.45	627.60		
1 Hyd. Crane, 12 Ton		653.80		719.18		
1 Welder, Electric, 300 amp		57.70		63.47	29.65	32.61
24 L.H., Daily Totals		$1985.50		$2709.45	$82.73	$112.89

Crew Q-18	Hr.	Daily	Hr.	Daily	Bare Costs	Incl. O&P
2 Steamfitters	$59.75	$956.00	$90.20	$1443.20	$55.77	$84.20
1 Steamfitter Apprentice	47.80	382.40	72.20	577.60		
1 Welder, Electric, 300 amp		57.70		63.47	2.40	2.64
24 L.H., Daily Totals		$1396.10		$2084.27	$58.17	$86.84

Crew Q-19	Hr.	Daily	Hr.	Daily	Bare Costs	Incl. O&P
1 Steamfitter	$59.75	$478.00	$90.20	$721.60	$54.08	$81.45
1 Steamfitter Apprentice	47.80	382.40	72.20	577.60		
1 Electrician	54.70	437.60	81.95	655.60		
24 L.H., Daily Totals		$1298.00		$1954.80	$54.08	$81.45

Crew Q-20	Hr.	Daily	Hr.	Daily	Bare Costs	Incl. O&P
1 Sheet Metal Worker	$55.95	$447.60	$85.50	$684.00	$51.22	$77.95
1 Sheet Metal Apprentice	44.75	358.00	68.40	547.20		
.5 Electrician	54.70	218.80	81.95	327.80		
20 L.H., Daily Totals		$1024.40		$1559.00	$51.22	$77.95

Crew Q-21	Hr.	Daily	Hr.	Daily	Bare Costs	Incl. O&P
2 Steamfitters	$59.75	$956.00	$90.20	$1443.20	$55.50	$83.64
1 Steamfitter Apprentice	47.80	382.40	72.20	577.60		
1 Electrician	54.70	437.60	81.95	655.60		
32 L.H., Daily Totals		$1776.00		$2676.40	$55.50	$83.64

Crew No.	Bare Costs		Incl. Subs O&P		Cost Per Labor-Hour	
Crew Q-22	**Hr.**	**Daily**	**Hr.**	**Daily**	**Bare Costs**	**Incl. O&P**
1 Plumber	$58.70	$469.60	$88.65	$709.20	$52.83	$79.78
1 Plumber Apprentice	46.95	375.60	70.90	567.20		
1 Hyd. Crane, 12 Ton		653.80		719.18	40.86	44.95
16 L.H., Daily Totals		$1499.00		$1995.58	$93.69	$124.72
Crew Q-22A	**Hr.**	**Daily**	**Hr.**	**Daily**	**Bare Costs**	**Incl. O&P**
1 Plumber	$58.70	$469.60	$88.65	$709.20	$48.74	$73.96
1 Plumber Apprentice	46.95	375.60	70.90	567.20		
1 Laborer	37.60	300.80	57.85	462.80		
1 Equip. Oper. (crane)	51.70	413.60	78.45	627.60		
1 Hyd. Crane, 12 Ton		653.80		719.18	20.43	22.47
32 L.H., Daily Totals		$2213.40		$3085.98	$69.17	$96.44
Crew Q-23	**Hr.**	**Daily**	**Hr.**	**Daily**	**Bare Costs**	**Incl. O&P**
1 Plumber Foreman (outside)	$60.70	$485.60	$91.65	$733.20	$56.67	$85.68
1 Plumber	58.70	469.60	88.65	709.20		
1 Equip. Oper. (medium)	50.60	404.80	76.75	614.00		
1 Lattice Boom Crane, 20 Ton		948.90		1043.79	39.54	43.49
24 L.H., Daily Totals		$2308.90		$3100.19	$96.20	$129.17
Crew R-1	**Hr.**	**Daily**	**Hr.**	**Daily**	**Bare Costs**	**Incl. O&P**
1 Electrician Foreman	$55.20	$441.60	$82.70	$661.60	$48.37	$73.08
3 Electricians	54.70	1312.80	81.95	1966.80		
2 Helpers	35.45	567.20	54.95	879.20		
48 L.H., Daily Totals		$2321.60		$3507.60	$48.37	$73.08
Crew R-1A	**Hr.**	**Daily**	**Hr.**	**Daily**	**Bare Costs**	**Incl. O&P**
1 Electrician	$54.70	$437.60	$81.95	$655.60	$45.08	$68.45
1 Helper	35.45	283.60	54.95	439.60		
16 L.H., Daily Totals		$721.20		$1095.20	$45.08	$68.45
Crew R-2	**Hr.**	**Daily**	**Hr.**	**Daily**	**Bare Costs**	**Incl. O&P**
1 Electrician Foreman	$55.20	$441.60	$82.70	$661.60	$48.84	$73.84
3 Electricians	54.70	1312.80	81.95	1966.80		
2 Helpers	35.45	567.20	54.95	879.20		
1 Equip. Oper. (crane)	51.70	413.60	78.45	627.60		
1 S.P. Crane, 4x4, 5 Ton		275.40		302.94	4.92	5.41
56 L.H., Daily Totals		$3010.60		$4438.14	$53.76	$79.25
Crew R-3	**Hr.**	**Daily**	**Hr.**	**Daily**	**Bare Costs**	**Incl. O&P**
1 Electrician Foreman	$55.20	$441.60	$82.70	$661.60	$54.30	$81.55
1 Electrician	54.70	437.60	81.95	655.60		
.5 Equip. Oper. (crane)	51.70	206.80	78.45	313.80		
.5 S.P. Crane, 4x4, 5 Ton		137.70		151.47	6.88	7.57
20 L.H., Daily Totals		$1223.70		$1782.47	$61.19	$89.12
Crew R-4	**Hr.**	**Daily**	**Hr.**	**Daily**	**Bare Costs**	**Incl. O&P**
1 Struc. Steel Foreman (outside)	$54.65	$437.20	$94.95	$759.60	$53.46	$90.25
3 Struc. Steel Workers	52.65	1263.60	91.45	2194.80		
1 Electrician	54.70	437.60	81.95	655.60		
1 Welder, Gas Engine, 300 amp		145.85		160.44	3.65	4.01
40 L.H., Daily Totals		$2284.25		$3770.43	$57.11	$94.26

Crew No.	Bare Costs		Incl. Subs O&P		Cost Per Labor-Hour	
Crew R-5	**Hr.**	**Daily**	**Hr.**	**Daily**	**Bare Costs**	**Incl. O&P**
1 Electrician Foreman	$55.20	$441.60	$82.70	$661.60	$47.75	$72.20
4 Electrician Linemen	54.70	1750.40	81.95	2622.40		
2 Electrician Operators	54.70	875.20	81.95	1311.20		
4 Electrician Groundmen	35.45	1134.40	54.95	1758.40		
1 Crew Truck		206.40		227.04		
1 Flatbed Truck, 20,000 GVW		248.60		273.46		
1 Pickup Truck, 3/4 Ton		144.20		158.62		
.2 Hyd. Crane, 55 Ton		225.60		248.16		
.2 Hyd. Crane, 12 Ton		130.76		143.84		
.2 Earth Auger, Truck-Mtd.		83.44		91.78		
1 Tractor w/Winch		427.00		469.70	16.66	18.32
88 L.H., Daily Totals		$5667.60		$7966.20	$64.40	$90.53
Crew R-6	**Hr.**	**Daily**	**Hr.**	**Daily**	**Bare Costs**	**Incl. O&P**
1 Electrician Foreman	$55.20	$441.60	$82.70	$661.60	$47.75	$72.20
4 Electrician Linemen	54.70	1750.40	81.95	2622.40		
2 Electrician Operators	54.70	875.20	81.95	1311.20		
4 Electrician Groundmen	35.45	1134.40	54.95	1758.40		
1 Crew Truck		206.40		227.04		
1 Flatbed Truck, 20,000 GVW		248.60		273.46		
1 Pickup Truck, 3/4 Ton		144.20		158.62		
.2 Hyd. Crane, 55 Ton		225.60		248.16		
.2 Hyd. Crane, 12 Ton		130.76		143.84		
.2 Earth Auger, Truck-Mtd.		83.44		91.78		
1 Tractor w/Winch		427.00		469.70		
3 Cable Trailers		598.20		658.02		
.5 Tensioning Rig		198.40		218.24		
.5 Cable Pulling Rig		1160.00		1276.00	38.89	42.78
88 L.H., Daily Totals		$7624.20		$10118.46	$86.64	$114.98
Crew R-7	**Hr.**	**Daily**	**Hr.**	**Daily**	**Bare Costs**	**Incl. O&P**
1 Electrician Foreman	$55.20	$441.60	$82.70	$661.60	$38.74	$59.58
5 Electrician Groundmen	35.45	1418.00	54.95	2198.00		
1 Crew Truck		206.40		227.04	4.30	4.73
48 L.H., Daily Totals		$2066.00		$3086.64	$43.04	$64.31
Crew R-8	**Hr.**	**Daily**	**Hr.**	**Daily**	**Bare Costs**	**Incl. O&P**
1 Electrician Foreman	$55.20	$441.60	$82.70	$661.60	$48.37	$73.08
3 Electrician Linemen	54.70	1312.80	81.95	1966.80		
2 Electrician Groundmen	35.45	567.20	54.95	879.20		
1 Pickup Truck, 3/4 Ton		144.20		158.62		
1 Crew Truck		206.40		227.04	7.30	8.03
48 L.H., Daily Totals		$2672.20		$3893.26	$55.67	$81.11
Crew R-9	**Hr.**	**Daily**	**Hr.**	**Daily**	**Bare Costs**	**Incl. O&P**
1 Electrician Foreman	$55.20	$441.60	$82.70	$661.60	$45.14	$68.54
1 Electrician Lineman	54.70	437.60	81.95	655.60		
2 Electrician Operators	54.70	875.20	81.95	1311.20		
4 Electrician Groundmen	35.45	1134.40	54.95	1758.40		
1 Pickup Truck, 3/4 Ton		144.20		158.62		
1 Crew Truck		206.40		227.04	5.48	6.03
64 L.H., Daily Totals		$3239.40		$4772.46	$50.62	$74.57
Crew R-10	**Hr.**	**Daily**	**Hr.**	**Daily**	**Bare Costs**	**Incl. O&P**
1 Electrician Foreman	$55.20	$441.60	$82.70	$661.60	$51.58	$77.58
4 Electrician Linemen	54.70	1750.40	81.95	2622.40		
1 Electrician Groundman	35.45	283.60	54.95	439.60		
1 Crew Truck		206.40		227.04		
3 Tram Cars		412.20		453.42	12.89	14.18
48 L.H., Daily Totals		$3094.20		$4404.06	$64.46	$91.75

Crew R-11	Bare Costs Hr.	Bare Costs Daily	Incl. Subs O&P Hr.	Incl. Subs O&P Daily	Cost Per Labor-Hour Bare Costs	Cost Per Labor-Hour Incl. O&P
1 Electrician Foreman	$55.20	$441.60	$82.70	$661.60	$51.90	$78.11
4 Electricians	54.70	1750.40	81.95	2622.40		
1 Equip. Oper. (crane)	51.70	413.60	78.45	627.60		
1 Common Laborer	37.60	300.80	57.85	462.80		
1 Crew Truck		206.40		227.04		
1 Hyd. Crane, 12 Ton		653.80		719.18	15.36	16.90
56 L.H., Daily Totals		$3766.60		$5320.62	$67.26	$95.01

Crew R-12	Hr.	Daily	Hr.	Daily	Bare Costs	Incl. O&P
1 Carpenter Foreman (inside)	$47.45	$379.60	$73.05	$584.40	$44.45	$69.24
4 Carpenters	46.95	1502.40	72.25	2312.00		
4 Common Laborers	37.60	1203.20	57.85	1851.20		
1 Equip. Oper. (medium)	50.60	404.80	76.75	614.00		
1 Steel Worker	52.65	421.20	91.45	731.60		
1 Dozer, 200 H.P.		1387.00		1525.70		
1 Pickup Truck, 3/4 Ton		144.20		158.62	17.40	19.14
88 L.H., Daily Totals		$5442.40		$7777.52	$61.85	$88.38

Crew R-13	Hr.	Daily	Hr.	Daily	Bare Costs	Incl. O&P
1 Electrician Foreman	$55.20	$441.60	$82.70	$661.60	$52.84	$79.37
3 Electricians	54.70	1312.80	81.95	1966.80		
.25 Equip. Oper. (crane)	51.70	103.40	78.45	156.90		
1 Equipment Oiler	45.20	361.60	68.55	548.40		
.25 Hydraulic Crane, 33 Ton		187.50		206.25	4.46	4.91
42 L.H., Daily Totals		$2406.90		$3539.95	$57.31	$84.28

Crew R-15	Hr.	Daily	Hr.	Daily	Bare Costs	Incl. O&P
1 Electrician Foreman	$55.20	$441.60	$82.70	$661.60	$53.77	$80.71
4 Electricians	54.70	1750.40	81.95	2622.40		
1 Equipment Oper. (light)	48.60	388.80	73.75	590.00		
1 Aerial Lift Truck, 40' Boom		301.40		331.54	6.28	6.91
48 L.H., Daily Totals		$2882.20		$4205.54	$60.05	$87.62

Crew R-18	Hr.	Daily	Hr.	Daily	Bare Costs	Incl. O&P
.25 Electrician Foreman	$55.20	$110.40	$82.70	$165.40	$42.89	$65.39
1 Electrician	54.70	437.60	81.95	655.60		
2 Helpers	35.45	567.20	54.95	879.20		
26 L.H., Daily Totals		$1115.20		$1700.20	$42.89	$65.39

Crew R-19	Hr.	Daily	Hr.	Daily	Bare Costs	Incl. O&P
.5 Electrician Foreman	$55.20	$220.80	$82.70	$330.80	$54.80	$82.10
2 Electricians	54.70	875.20	81.95	1311.20		
20 L.H., Daily Totals		$1096.00		$1642.00	$54.80	$82.10

Crew R-21	Hr.	Daily	Hr.	Daily	Bare Costs	Incl. O&P
1 Electrician Foreman	$55.20	$441.60	$82.70	$661.60	$54.72	$82.01
3 Electricians	54.70	1312.80	81.95	1966.80		
.1 Equip. Oper. (medium)	50.60	40.48	76.75	61.40		
.1 S.P. Crane, 4x4, 25 Ton		59.94		65.93	1.83	2.01
32.8 L.H., Daily Totals		$1854.82		$2755.73	$56.55	$84.02

Crew R-22	Hr.	Daily	Hr.	Daily	Bare Costs	Incl. O&P
.66 Electrician Foreman	$55.20	$291.46	$82.70	$436.66	$46.51	$70.47
2 Electricians	54.70	875.20	81.95	1311.20		
2 Helpers	35.45	567.20	54.95	879.20		
37.28 L.H., Daily Totals		$1733.86		$2627.06	$46.51	$70.47

Crew R-30	Hr.	Daily	Hr.	Daily	Bare Costs	Incl. O&P
.25 Electrician Foreman (outside)	$56.70	$113.40	$84.95	$169.90	$44.33	$67.35
1 Electrician	54.70	437.60	81.95	655.60		
2 Laborers, (Semi-Skilled)	37.60	601.60	57.85	925.60		
26 L.H., Daily Totals		$1152.60		$1751.10	$44.33	$67.35

Crew R-31	Hr.	Daily	Hr.	Daily	Bare Costs	Incl. O&P
1 Electrician	$54.70	$437.60	$81.95	$655.60	$54.70	$81.95
1 Core Drill, Electric, 2.5 H.P.		55.15		60.66	6.89	7.58
8 L.H., Daily Totals		$492.75		$716.26	$61.59	$89.53

Crew W-41E	Hr.	Daily	Hr.	Daily	Bare Costs	Incl. O&P
.5 Plumber Foreman (outside)	$60.70	$242.80	$91.65	$366.60	$50.66	$76.93
1 Plumber	58.70	469.60	88.65	709.20		
1 Laborer	37.60	300.80	57.85	462.80		
20 L.H., Daily Totals		$1013.20		$1538.60	$50.66	$76.93

Historical Cost Indexes

The table below lists both the RSMeans® historical cost index based on Jan. 1, 1993 = 100 as well as the computed value of an index based on Jan. 1, 2015 costs. Since the Jan. 1, 2015 figure is estimated, space is left to write in the actual index figures as they become available through either the quarterly *RSMeans Construction Cost Indexes* or as printed in the *Engineering News-Record*. To compute the actual index based on Jan. 1, 2015 = 100, divide the historical cost index for a particular year by the actual Jan. 1, 2015 construction cost index. Space has been left to advance the index figures as the year progresses.

Year	Historical Cost Index Jan. 1, 1993 = 100		Current Index Based on Jan. 1, 2015 = 100		Year	Historical Cost Index Jan. 1, 1993 = 100	Current Index Based on Jan. 1, 2015 = 100		Year	Historical Cost Index Jan. 1, 1993 = 100	Current Index Based on Jan. 1, 2015 = 100	
	Est.	Actual	Est.	Actual		Actual	Est.	Actual		Actual	Est.	Actual
Oct 2015*					July 2000	120.9	58.5		July 1982	76.1	36.8	
July 2015*					1999	117.6	56.9		1981	70.0	33.9	
April 2015*					1998	115.1	55.7		1980	62.9	30.4	
Jan 2015*	206.7		100.0	100.0	1997	112.8	54.6		1979	57.8	28.0	
July 2014		204.9	99.1		1996	110.2	53.3		1978	53.5	25.9	
2013		201.2	97.3		1995	107.6	52.1		1977	49.5	23.9	
2012		194.6	94.1		1994	104.4	50.5		1976	46.9	22.7	
2011		191.2	92.5		1993	101.7	49.2		1975	44.8	21.7	
2010		183.5	88.8		1992	99.4	48.1		1974	41.4	20.0	
2009		180.1	87.1		1991	96.8	46.8		1973	37.7	18.2	
2008		180.4	87.3		1990	94.3	45.6		1972	34.8	16.8	
2007		169.4	82.0		1989	92.1	44.6		1971	32.1	15.5	
2006		162.0	78.4		1988	89.9	43.5		1970	28.7	13.9	
2005		151.6	73.3		1987	87.7	42.4		1969	26.9	13.0	
2004		143.7	69.5		1986	84.2	40.7		1968	24.9	12.0	
2003		132.0	63.9		1985	82.6	40.0		1967	23.5	11.4	
2002		128.7	62.3		1984	82.0	39.7		1966	22.7	11.0	
▼ 2001		125.1	60.5		▼ 1983	80.2	38.8		▼ 1965	21.7	10.5	

Adjustments to Costs

The "Historical Cost Index" can be used to convert national average building costs at a particular time to the approximate building costs for some other time.

Time Adjustment Using the Historical Cost Indexes:

$$\frac{\text{Index for Year A}}{\text{Index for Year B}} \times \text{Cost in Year B} = \text{Cost in Year A}$$

Example:

Estimate and compare construction costs for different years in the same city.

To estimate the national average construction cost of a building in 1970, knowing that it cost $900,000 in 2015:

INDEX in 1970 = 28.7

INDEX in 2015 = 206.7

$$\frac{\text{INDEX 1970}}{\text{INDEX 2015}} \times \text{Cost 2015} = \text{Cost 1970}$$

$$\frac{28.7}{206.7} \times \$900{,}000 = .139 \times \$900{,}000 = \$125{,}100$$

The construction cost of the building in 1970 is $125,100.

Note: The city cost indexes for Canada can be used to convert U.S. national averages to local costs in Canadian dollars.

Example:

To estimate and compare the cost of a building in Toronto, ON in 2015 with the known cost of $600,000 (US$) in New York, NY in 2015:

INDEX Toronto = 110.9

INDEX New York = 131.8

$$\frac{\text{INDEX Toronto}}{\text{INDEX New York}} \times \text{Cost New York} = \text{Cost Toronto}$$

$$\frac{110.9}{131.8} \times \$600{,}000 = .841 \times \$600{,}000 = \$504{,}600$$

The construction cost of the building in Toronto is $504,600 (CN$).

*Historical Cost Index updates and other resources are provided on the following website.
http://info.thegordiangroup.com/RSMeans.html

How to Use the City Cost Indexes

What you should know before you begin

RSMeans City Cost Indexes (CCI) are an extremely useful tool to use when you want to compare costs from city to city and region to region.

This publication contains average construction cost indexes for 731 U.S. and Canadian cities covering over 930 three-digit zip code locations, as listed directly under each city.

Keep in mind that a City Cost Index number is a percentage ratio of a specific city's cost to the national average cost of the same item at a stated time period.

In other words, these index figures represent relative construction factors (or, if you prefer, multipliers) for Material and Installation costs, as well as the weighted average for Total In Place costs for each CSI MasterFormat division. Installation costs include both labor and equipment rental costs. When estimating equipment rental rates only, for a specific location, use 01 54 33 EQUIPMENT RENTAL COSTS in the Reference Section at the back of the book.

The 30 City Average Index is the average of 30 major U.S. cities and serves as a National Average.

Index figures for both material and installation are based on the 30 major city average of 100 and represent the cost relationship as of July 1, 2014. The index for each division is computed from representative material and labor quantities for that division. The weighted average for each city is a weighted total of the components listed above it, but does not include relative productivity between trades or cities.

As changes occur in local material prices, labor rates, and equipment rental rates (including fuel costs), the impact of these changes should be accurately measured by the change in the City Cost Index for each particular city (as compared to the 30 City Average).

Therefore, if you know (or have estimated) building costs in one city today, you can easily convert those costs to expected building costs in another city.

In addition, by using the Historical Cost Index, you can easily convert National Average building costs at a particular time to the approximate building costs for some other time. The City Cost Indexes can then be applied to calculate the costs for a particular city.

Quick Calculations

Location Adjustment Using the City Cost Indexes:

$$\frac{\text{Index for City A}}{\text{Index for City B}} \times \text{Cost in City B} = \text{Cost in City A}$$

Time Adjustment for the National Average Using the Historical Cost Index:

$$\frac{\text{Index for Year A}}{\text{Index for Year B}} \times \text{Cost in Year B} = \text{Cost in Year A}$$

Adjustment from the National Average:

$$\frac{\text{Index for City A}}{100} \times \text{National Average Cost} = \text{Cost in City A}$$

Since each of the other RSMeans publications contains many different items, any *one* item multiplied by the particular city index may give incorrect results. However, the larger the number of items compiled, the closer the results should be to actual costs for that particular city.

The City Cost Indexes for Canadian cities are calculated using Canadian material and equipment prices and labor rates, in Canadian dollars. Therefore, indexes for Canadian cities can be used to convert U.S. National Average prices to local costs in Canadian dollars.

How to use this section

1. Compare costs from city to city.

In using the RSMeans Indexes, remember that an index number is not a fixed number but a ratio: It's a percentage ratio of a building component's cost at any stated time to the National Average cost of that same component at the same time period. Put in the form of an equation:

$$\frac{\text{Specific City Cost}}{\text{National Average Cost}} \times 100 = \text{City Index Number}$$

Therefore, when making cost comparisons between cities, do not subtract one city's index number from the index number of another city and read the result as a percentage difference. Instead, divide one city's index number by that of the other city. The resulting number may then be used as a multiplier to calculate cost differences from city to city.

The formula used to find cost differences between cities for the purpose of comparison is as follows:

$$\frac{\text{City A Index}}{\text{City B Index}} \times \text{City B Cost (Known)} = \text{City A Cost (Unknown)}$$

In addition, you can use RSMeans CCI to calculate and compare costs division by division between cities using the same basic formula. (Just be sure that you're comparing similar divisions.)

2. Compare a specific city's construction costs with the National Average.

When you're studying construction location feasibility, it's advisable to compare a prospective project's cost index with an index of the National Average cost.

For example, divide the weighted average index of construction costs of a specific city by that of the 30 City Average, which = 100.

$$\frac{\text{City Index}}{100} = \text{\% of National Average}$$

As a result, you get a ratio that indicates the relative cost of construction in that city in comparison with the National Average.

3. Convert U.S. National Average to actual costs in Canadian City.

$$\frac{\text{Index for Canadian City}}{100} \times \text{National Average Cost} = \text{Cost in Canadian City in \$ CAN}$$

4. **Adjust construction cost data based on a National Average.**

When you use a source of construction cost data which is based on a National Average (such as RSMeans cost data publications), it is necessary to adjust those costs to a specific location.

$$\frac{\text{City Index}}{100} \times \frac{\text{"Book" Cost Based on}}{\text{National Average Costs}} = \frac{\text{City Cost}}{\text{(Unknown)}}$$

5. **When applying the City Cost Indexes to demolition projects, use the appropriate division installation index.** For example, for removal of existing doors and windows, use Division 8 (Openings) index.

What you might like to know about how we developed the Indexes

The information presented in the CCI is organized according to the Construction Specifications Institute (CSI) MasterFormat 2014 classification system.

To create a reliable index, RSMeans researched the building type most often constructed in the United States and Canada. Because it was concluded that no one type of building completely represented the building construction industry, nine different types of buildings were combined to create a composite model.

The exact material, labor, and equipment quantities are based on detailed analyses of these nine building types, and then each quantity is weighted in proportion to expected usage. These various material items, labor hours, and equipment rental rates are thus combined to form a composite building representing as closely as possible the actual usage of materials, labor, and equipment used in the North American building construction industry.

The following structures were chosen to make up that composite model:

1. Factory, 1 story
2. Office, 2-4 story
3. Store, Retail
4. Town Hall, 2-3 story
5. High School, 2-3 story
6. Hospital, 4-8 story
7. Garage, Parking
8. Apartment, 1-3 story
9. Hotel/Motel, 2-3 story

For the purposes of ensuring the timeliness of the data, the components of the index for the composite model have been streamlined. They currently consist of:

- specific quantities of 66 commonly used construction materials;
- specific labor-hours for 21 building construction trades; and
- specific days of equipment rental for 6 types of construction equipment (normally used to install the 66 material items by the 21 trades.) Fuel costs and routine maintenance costs are included in the equipment cost.

A sophisticated computer program handles the updating of all costs for each city on a quarterly basis. Material and equipment price quotations are gathered quarterly from cities in the United States and Canada. These prices and the latest negotiated labor wage rates for 21 different building trades are used to compile the quarterly update of the City Cost Index.

The 30 major U.S. cities used to calculate the National Average are:

Atlanta, GA	Memphis, TN
Baltimore, MD	Milwaukee, WI
Boston, MA	Minneapolis, MN
Buffalo, NY	Nashville, TN
Chicago, IL	New Orleans, LA
Cincinnati, OH	New York, NY
Cleveland, OH	Philadelphia, PA
Columbus, OH	Phoenix, AZ
Dallas, TX	Pittsburgh, PA
Denver, CO	St. Louis, MO
Detroit, MI	San Antonio, TX
Houston, TX	San Diego, CA
Indianapolis, IN	San Francisco, CA
Kansas City, MO	Seattle, WA
Los Angeles, CA	Washington, DC

What the CCI does not indicate

The weighted average for each city is a total of the divisional components weighted to reflect typical usage, but it does not include the productivity variations between trades or cities.

In addition, the CCI does not take into consideration factors such as the following:

- managerial efficiency
- competitive conditions
- automation
- restrictive union practices
- unique local requirements
- regional variations due to specific building codes

DIVISION		UNITED STATES 30 CITY AVERAGE			ALABAMA ANNISTON 362			ALABAMA BIRMINGHAM 350 - 352			ALABAMA BUTLER 369			ALABAMA DECATUR 356			ALABAMA DOTHAN 363		
		MAT.	INST.	TOTAL	MAT.	INST.	TOTAL	MAT.	INST.	TOTAL	MAT.	INST.	TOTAL	MAT.	INST.	TOTAL	MAT.	INST.	TOTAL
015433	CONTRACTOR EQUIPMENT		100.0	100.0		101.0	101.0		101.1	101.1		98.2	98.2		101.0	101.0		98.2	98.2
0241, 31 - 34	SITE & INFRASTRUCTURE, DEMOLITION	100.0	100.0	100.0	90.2	93.1	92.3	97.4	93.8	94.8	103.5	87.1	91.8	88.8	91.5	90.7	101.2	87.1	91.1
0310	Concrete Forming & Accessories	100.0	100.0	100.0	90.1	47.1	53.0	93.1	75.9	78.2	86.4	42.6	48.6	94.6	45.4	52.1	95.5	42.0	49.4
0320	Concrete Reinforcing	100.0	100.0	100.0	87.8	86.3	87.1	94.7	86.9	90.7	92.8	45.9	68.9	88.6	77.0	82.7	92.8	45.4	68.7
0330	Cast-in-Place Concrete	100.0	100.0	100.0	101.8	49.7	80.4	110.0	74.6	95.4	99.3	54.8	81.0	103.4	64.1	87.2	99.3	46.3	77.5
03	CONCRETE	100.0	100.0	100.0	102.1	57.0	79.9	102.2	78.4	90.5	103.0	49.3	76.6	98.3	59.3	79.1	102.3	46.0	74.7
04	MASONRY	100.0	100.0	100.0	100.1	66.8	79.3	98.8	76.0	84.6	105.9	48.6	70.2	97.3	51.3	68.6	107.1	39.7	65.1
05	METALS	100.0	100.0	100.0	104.0	91.5	100.1	104.0	94.1	101.0	102.8	76.6	94.8	106.2	86.1	100.0	102.9	75.3	94.4
06	WOOD, PLASTICS & COMPOSITES	100.0	100.0	100.0	90.5	42.7	63.7	97.4	76.1	85.4	85.0	42.4	61.1	98.9	42.9	67.6	97.4	42.4	66.6
07	THERMAL & MOISTURE PROTECTION	100.0	100.0	100.0	97.9	52.9	79.5	99.7	81.5	92.2	98.0	56.7	81.0	96.7	58.3	81.0	98.0	49.1	77.9
08	OPENINGS	100.0	100.0	100.0	98.6	50.1	87.3	103.2	76.0	96.8	98.6	44.2	85.9	106.5	50.4	93.5	98.6	44.2	86.0
0920	Plaster & Gypsum Board	100.0	100.0	100.0	90.8	41.4	57.4	94.8	75.8	81.9	87.9	41.0	56.2	95.8	41.6	59.2	98.6	41.0	59.7
0950, 0980	Ceilings & Acoustic Treatment	100.0	100.0	100.0	80.4	41.4	54.7	89.8	75.8	80.6	80.4	41.0	54.5	87.2	41.6	57.2	80.4	41.0	54.5
0960	Flooring	100.0	100.0	100.0	91.5	40.2	76.8	102.5	76.7	95.1	95.8	56.0	84.4	99.7	50.6	85.6	100.6	27.5	79.7
0970, 0990	Wall Finishes & Painting/Coating	100.0	100.0	100.0	103.7	31.5	60.1	103.9	66.5	81.3	103.7	53.3	73.3	99.6	66.8	79.8	103.7	53.3	73.3
09	FINISHES	100.0	100.0	100.0	87.0	43.0	62.8	95.4	74.8	84.1	89.9	45.7	65.5	91.9	47.7	67.6	92.5	40.1	63.6
COVERS	DIVS. 10 - 14, 25, 28, 41, 43, 44, 46	100.0	100.0	100.0	100.0	68.3	93.6	100.0	88.4	97.7	100.0	41.4	88.2	100.0	43.2	88.5	100.0	41.5	88.2
21, 22, 23	FIRE SUPPRESSION, PLUMBING & HVAC	100.0	100.0	100.0	100.0	55.4	82.0	100.0	70.4	88.1	97.3	33.3	71.5	100.0	40.8	76.1	97.3	33.2	71.5
26, 27, 3370	ELECTRICAL, COMMUNICATIONS & UTIL.	100.0	100.0	100.0	93.1	57.3	74.2	99.0	61.6	79.3	95.0	40.2	66.0	94.2	64.9	78.7	93.8	57.4	74.6
MF2014	WEIGHTED AVERAGE	100.0	100.0	100.0	98.6	61.8	82.6	100.6	76.7	90.2	98.9	49.4	77.3	99.8	57.9	81.5	99.0	49.4	77.4

DIVISION		ALABAMA EVERGREEN 364			ALABAMA GADSDEN 359			ALABAMA HUNTSVILLE 357 - 358			ALABAMA JASPER 355			ALABAMA MOBILE 365 - 366			ALABAMA MONTGOMERY 360 - 361		
		MAT.	INST.	TOTAL	MAT.	INST.	TOTAL	MAT.	INST.	TOTAL	MAT.	INST.	TOTAL	MAT.	INST.	TOTAL	MAT.	INST.	TOTAL
015433	CONTRACTOR EQUIPMENT		98.2	98.2		101.0	101.0		101.0	101.0		101.0	101.0		98.2	98.2		98.2	98.2
0241, 31 - 34	SITE & INFRASTRUCTURE, DEMOLITION	104.0	87.4	92.2	94.9	92.7	93.3	88.6	92.7	91.5	94.6	92.4	93.1	96.4	88.1	90.5	96.5	88.1	90.5
0310	Concrete Forming & Accessories	83.0	44.2	49.5	86.9	42.5	48.6	94.6	69.9	73.3	92.1	34.6	42.5	94.5	54.0	59.6	94.1	44.1	51.0
0320	Concrete Reinforcing	92.9	45.9	69.0	94.0	86.8	90.3	88.6	81.6	85.1	88.6	85.6	87.1	90.6	83.6	87.0	95.8	86.1	90.9
0330	Cast-in-Place Concrete	99.3	57.3	82.1	103.4	66.7	88.3	100.7	69.5	87.9	114.5	44.1	85.6	104.1	66.4	88.6	105.2	56.5	85.2
03	CONCRETE	103.3	50.9	77.5	102.9	60.8	82.2	97.0	72.9	85.2	106.6	49.4	78.5	98.9	65.3	82.4	100.2	57.9	79.4
04	MASONRY	105.9	53.2	73.0	95.6	62.4	74.9	98.8	67.7	79.4	92.9	41.2	60.7	103.7	54.2	72.9	100.0	49.7	68.6
05	METALS	102.9	76.5	94.7	104.0	93.2	100.7	106.2	90.6	101.4	104.0	90.2	99.7	105.0	91.6	100.9	104.0	91.9	100.3
06	WOOD, PLASTICS & COMPOSITES	81.3	42.4	59.5	89.3	38.7	61.0	98.9	72.6	84.2	95.9	31.6	59.9	95.9	52.8	71.8	95.3	42.4	65.6
07	THERMAL & MOISTURE PROTECTION	97.9	56.9	81.1	96.8	71.7	86.5	96.6	77.6	88.8	96.8	45.4	75.7	97.6	70.1	86.3	97.9	66.6	85.0
08	OPENINGS	98.6	44.2	85.9	102.8	49.0	90.3	106.5	67.4	97.4	102.8	48.0	90.0	102.2	58.8	92.1	103.3	53.1	91.7
0920	Plaster & Gypsum Board	87.2	41.0	56.0	88.2	37.3	53.8	95.8	72.2	79.9	92.1	30.0	50.1	95.2	51.8	65.9	94.6	41.0	58.4
0950, 0980	Ceilings & Acoustic Treatment	80.4	41.0	54.5	83.8	37.3	53.2	89.0	72.2	78.0	83.8	30.0	48.4	85.6	51.8	63.4	88.0	41.0	57.1
0960	Flooring	93.9	56.0	83.1	95.5	76.7	90.1	99.7	76.7	93.1	97.8	40.2	81.3	100.5	58.5	88.5	97.3	58.5	86.2
0970, 0990	Wall Finishes & Painting/Coating	103.7	53.3	73.3	99.6	60.7	76.1	99.6	65.9	79.3	99.6	38.6	62.8	107.3	54.3	75.3	103.3	53.3	73.1
09	FINISHES	89.2	46.8	65.9	89.4	48.7	67.0	92.3	70.7	80.4	90.5	34.4	59.6	92.7	53.5	71.1	92.8	46.1	67.0
COVERS	DIVS. 10 - 14, 25, 28, 41, 43, 44, 46	100.0	42.9	88.5	100.0	79.8	95.9	100.0	85.3	97.0	100.0	39.8	87.8	100.0	83.1	96.6	100.0	79.6	95.9
21, 22, 23	FIRE SUPPRESSION, PLUMBING & HVAC	97.3	35.6	72.4	102.0	36.3	75.5	100.0	61.6	84.5	102.0	61.4	85.6	99.9	60.7	84.1	100.0	34.2	73.4
26, 27, 3370	ELECTRICAL, COMMUNICATIONS & UTIL.	92.5	40.2	64.8	94.3	61.6	77.0	95.3	64.9	79.3	93.8	61.3	76.6	95.7	59.1	76.3	96.3	60.9	77.6
MF2014	WEIGHTED AVERAGE	98.6	50.7	77.7	99.9	60.0	82.5	99.8	72.1	87.8	100.2	57.5	81.6	99.9	65.4	84.9	99.9	57.1	81.2

DIVISION		ALABAMA PHENIX CITY 368			ALABAMA SELMA 367			ALABAMA TUSCALOOSA 354			ALASKA ANCHORAGE 995 - 996			ALASKA FAIRBANKS 997			ALASKA JUNEAU 998		
		MAT.	INST.	TOTAL	MAT.	INST.	TOTAL	MAT.	INST.	TOTAL	MAT.	INST.	TOTAL	MAT.	INST.	TOTAL	MAT.	INST.	TOTAL
015433	CONTRACTOR EQUIPMENT		98.2	98.2		98.2	98.2		101.0	101.0		114.7	114.7		114.7	114.7		114.7	114.7
0241, 31 - 34	SITE & INFRASTRUCTURE, DEMOLITION	107.8	88.2	93.9	101.0	88.1	91.8	89.1	92.9	91.8	127.2	129.6	128.9	119.2	129.6	126.6	135.6	129.6	131.3
0310	Concrete Forming & Accessories	90.1	39.1	46.1	87.6	42.6	48.8	94.5	47.9	54.3	123.7	119.8	120.3	131.6	119.9	121.5	130.1	119.8	121.2
0320	Concrete Reinforcing	92.8	61.4	76.8	92.8	85.2	89.0	88.6	86.8	87.7	149.7	110.7	129.9	150.9	110.7	130.5	135.7	110.7	123.0
0330	Cast-in-Place Concrete	99.3	57.0	81.9	99.3	46.5	77.6	104.9	68.4	89.9	132.0	117.8	126.2	128.3	118.2	124.1	133.2	117.8	126.9
03	CONCRETE	106.2	51.3	79.2	101.7	53.7	78.1	99.0	63.8	81.7	138.6	116.7	127.9	123.8	116.9	120.4	136.0	116.7	126.5
04	MASONRY	105.9	41.1	65.5	109.6	38.7	65.4	97.6	65.5	77.6	184.6	125.3	147.6	189.5	125.3	149.5	176.6	125.3	144.6
05	METALS	102.8	81.9	96.4	102.8	89.3	98.7	105.3	93.2	101.6	113.6	103.4	110.5	117.6	103.5	113.3	116.3	103.4	112.3
06	WOOD, PLASTICS & COMPOSITES	90.2	35.4	59.5	87.0	42.4	62.0	98.9	44.4	68.4	129.2	118.7	123.4	136.1	118.7	126.4	129.4	118.7	123.4
07	THERMAL & MOISTURE PROTECTION	98.3	63.8	84.1	97.8	52.1	79.1	96.7	73.7	87.3	159.5	118.0	142.5	166.5	119.1	147.1	168.0	118.0	147.5
08	OPENINGS	98.6	43.5	85.8	98.6	53.1	88.0	106.5	58.8	95.4	132.7	115.6	128.7	128.9	115.7	125.8	129.8	115.6	126.5
0920	Plaster & Gypsum Board	92.2	33.9	52.8	90.1	41.0	56.9	95.8	43.1	60.2	140.9	119.1	126.2	164.8	119.1	133.9	148.1	119.1	128.5
0950, 0980	Ceilings & Acoustic Treatment	80.4	33.9	49.8	80.4	41.0	54.5	89.0	43.1	58.9	119.6	119.1	119.3	119.2	119.1	119.2	125.9	119.1	121.4
0960	Flooring	97.6	58.5	86.4	96.2	28.5	76.8	99.7	76.7	93.1	134.6	133.0	134.1	127.2	133.0	128.9	131.0	133.0	131.6
0970, 0990	Wall Finishes & Painting/Coating	103.7	53.3	73.3	103.7	53.3	73.3	99.6	45.9	67.2	136.8	116.3	124.4	133.5	121.2	126.1	132.1	116.3	122.6
09	FINISHES	91.4	42.2	64.3	90.0	40.2	62.6	92.3	51.2	69.7	134.3	122.2	127.7	132.4	122.8	127.1	134.4	122.2	127.7
COVERS	DIVS. 10 - 14, 25, 28, 41, 43, 44, 46	100.0	79.3	95.8	100.0	41.4	88.2	100.0	81.4	96.2	100.0	112.8	102.6	100.0	112.8	102.6	100.0	112.8	102.6
21, 22, 23	FIRE SUPPRESSION, PLUMBING & HVAC	97.3	34.5	72.0	97.3	33.6	71.6	100.0	34.0	73.4	100.3	105.0	102.2	100.2	108.0	103.4	100.3	105.0	102.2
26, 27, 3370	ELECTRICAL, COMMUNICATIONS & UTIL.	94.4	69.5	81.3	93.6	40.2	65.4	94.8	61.6	77.3	117.7	117.8	117.7	130.0	117.8	123.5	119.9	117.8	118.8
MF2014	WEIGHTED AVERAGE	99.5	54.6	79.9	98.7	49.9	77.4	99.8	61.2	83.0	121.1	115.6	118.7	121.0	116.4	119.0	121.3	115.6	118.8

DIVISION		ALASKA			ARIZONA															
		KETCHIKAN			CHAMBERS			FLAGSTAFF			GLOBE			KINGMAN			MESA/TEMPE			
		999			865			860			855			864			852			
		MAT.	INST.	TOTAL	MAT.	INST.	TOTAL	MAT.	INST.	TOTAL	MAT.	INST.	TOTAL	MAT.	INST.	TOTAL	MAT.	INST.	TOTAL	
015433	CONTRACTOR EQUIPMENT		114.7	114.7		93.0	93.0		93.0	93.0		92.2	92.2		93.0	93.0		92.2	92.2	
0241, 31 - 34	SITE & INFRASTRUCTURE, DEMOLITION	173.8	129.6	142.4	69.6	96.1	88.4	87.3	96.3	93.7	97.3	95.6	96.1	69.6	96.3	88.6	89.0	95.8	93.8	
0310	Concrete Forming & Accessories	122.9	119.8	120.2	98.8	58.1	63.7	104.4	65.1	70.5	99.0	58.1	63.7	97.0	65.1	69.5	102.2	69.3	73.9	
0320	Concrete Reinforcing	117.4	110.7	114.0	97.1	85.2	91.1	97.0	85.3	91.0	108.0	85.2	96.4	97.2	85.3	91.1	108.7	85.3	96.8	
0330	Cast-in-Place Concrete	259.2	117.8	201.2	90.7	73.2	83.5	90.8	73.4	83.7	94.3	72.0	85.1	90.4	73.4	83.4	95.0	72.3	85.7	
03	CONCRETE	208.7	116.7	163.5	95.0	68.6	82.0	114.1	71.8	93.3	110.0	68.2	89.5	94.6	71.8	83.4	101.8	73.3	87.8	
04	MASONRY	198.4	125.3	152.8	92.5	62.3	73.7	92.6	62.4	73.8	109.9	62.2	80.1	92.5	62.4	73.7	110.1	62.2	80.3	
05	METALS	117.8	103.4	113.4	96.2	75.6	89.8	96.7	76.3	90.4	93.3	76.0	88.0	96.8	76.3	90.5	93.6	76.8	88.4	
06	WOOD, PLASTICS & COMPOSITES	126.3	118.7	122.1	101.0	55.0	75.2	107.3	64.2	83.1	97.6	55.0	73.8	96.1	64.2	78.2	101.1	69.8	83.6	
07	THERMAL & MOISTURE PROTECTION	169.3	118.0	148.3	94.6	65.0	82.5	96.2	68.1	84.7	101.9	63.0	85.9	94.6	65.6	82.7	101.2	65.1	86.4	
08	OPENINGS	130.8	115.6	127.3	108.1	65.3	98.2	108.3	70.3	99.4	100.0	65.3	91.9	108.3	70.3	99.5	100.1	73.3	93.8	
0920	Plaster & Gypsum Board	152.7	119.1	130.0	90.2	53.7	65.5	93.6	63.2	73.0	96.6	53.7	67.6	82.9	63.2	69.6	98.7	68.9	78.6	
0950, 0980	Ceilings & Acoustic Treatment	112.7	119.1	116.9	99.7	53.7	69.5	100.5	63.2	76.0	88.9	53.7	65.8	100.5	63.2	76.0	88.9	68.9	75.8	
0960	Flooring	127.2	133.0	128.9	93.3	39.5	77.9	95.4	39.7	79.5	103.5	39.5	85.2	92.1	53.7	81.1	104.7	49.5	88.9	
0970, 0990	Wall Finishes & Painting/Coating	133.5	116.3	123.1	98.3	55.1	72.2	98.3	55.1	72.2	103.4	55.1	74.3	98.3	55.1	72.2	103.4	55.1	74.3	
09	FINISHES	133.4	122.2	127.2	93.7	52.9	71.2	96.6	58.4	75.6	97.1	53.0	72.8	92.5	60.7	75.0	96.8	63.6	78.5	
COVERS	DIVS. 10 - 14, 25, 28, 41, 43, 44, 46	100.0	112.8	102.6	100.0	82.3	96.4	100.0	83.3	96.6	100.0	82.4	96.4	100.0	83.3	96.6	100.0	84.0	96.8	
21, 22, 23	FIRE SUPPRESSION, PLUMBING & HVAC	98.4	105.0	101.1	97.0	78.9	89.7	100.2	79.0	91.6	95.2	78.9	88.7	97.0	79.0	89.7	100.0	79.0	91.5	
26, 27, 3370	ELECTRICAL, COMMUNICATIONS & UTIL.	129.9	117.8	123.5	104.5	70.9	86.7	103.4	61.3	81.1	97.5	61.2	78.3	104.5	61.3	81.7	94.2	61.3	76.8	
MF2014	WEIGHTED AVERAGE	132.0	115.6	124.8	97.6	71.5	86.2	101.3	71.8	88.4	98.8	70.0	86.3	97.6	72.0	86.4	98.6	72.8	87.4	

DIVISION		ARIZONA												ARKANSAS					
		PHOENIX			PRESCOTT			SHOW LOW			TUCSON			BATESVILLE			CAMDEN		
		850,853			863			859			856 - 857			725			717		
		MAT.	INST.	TOTAL	MAT.	INST.	TOTAL	MAT.	INST.	TOTAL	MAT.	INST.	TOTAL	MAT.	INST.	TOTAL	MAT.	INST.	TOTAL
015433	CONTRACTOR EQUIPMENT		92.7	92.7		93.0	93.0		92.2	92.2		92.2	92.2		89.1	89.1		89.1	89.1
0241, 31 - 34	SITE & INFRASTRUCTURE, DEMOLITION	89.4	96.0	94.1	76.0	96.0	90.2	99.2	95.6	96.6	85.2	95.8	92.7	77.1	84.9	82.6	78.3	84.4	82.7
0310	Concrete Forming & Accessories	103.1	68.3	73.1	100.4	52.5	59.1	106.3	69.1	74.2	102.5	68.1	72.9	85.9	44.9	50.5	83.7	30.7	38.0
0320	Concrete Reinforcing	107.0	85.4	96.0	97.0	85.2	91.0	108.7	85.2	96.8	89.9	85.3	87.5	87.4	67.5	77.3	93.6	67.4	80.3
0330	Cast-in-Place Concrete	95.1	72.4	85.8	90.7	73.1	83.5	94.3	72.1	85.2	97.8	72.3	87.3	77.4	45.3	64.2	82.6	38.3	64.4
03	CONCRETE	101.4	73.0	87.4	100.1	66.1	83.4	112.4	73.1	93.1	100.0	72.8	86.6	83.5	50.1	67.1	87.5	41.3	64.8
04	MASONRY	97.5	64.1	76.7	92.6	62.3	73.7	109.9	62.2	80.1	95.5	62.2	74.8	99.8	40.7	63.0	116.6	31.4	63.5
05	METALS	95.1	77.6	89.7	96.7	75.3	90.1	93.1	76.1	87.8	94.3	76.7	88.9	99.5	66.9	89.4	103.9	66.5	92.4
06	WOOD, PLASTICS & COMPOSITES	102.1	68.3	83.2	102.4	47.4	71.6	105.8	69.8	85.6	101.4	68.3	82.8	89.9	45.5	65.0	90.3	29.7	56.4
07	THERMAL & MOISTURE PROTECTION	100.9	67.0	87.0	95.1	64.2	82.4	102.1	65.0	86.9	102.4	64.2	86.7	98.2	43.9	76.0	94.2	35.7	70.2
08	OPENINGS	102.0	72.5	95.2	108.3	61.1	97.3	99.2	73.3	93.2	96.3	72.5	90.7	97.7	45.8	85.6	102.1	40.6	87.8
0920	Plaster & Gypsum Board	101.0	67.4	78.2	90.3	45.9	60.3	100.9	68.9	79.3	104.0	67.4	79.2	83.3	44.3	56.9	85.0	28.0	46.5
0950, 0980	Ceilings & Acoustic Treatment	96.5	67.4	77.3	98.8	45.9	64.0	88.9	68.9	75.8	89.8	67.4	75.0	83.2	44.3	57.6	83.3	28.0	47.0
0960	Flooring	104.9	51.7	89.7	94.1	39.5	78.5	106.2	45.0	88.7	95.2	44.2	80.6	96.7	59.1	85.9	99.6	39.6	82.4
0970, 0990	Wall Finishes & Painting/Coating	103.4	61.7	78.3	98.3	55.1	72.2	103.4	55.1	74.3	104.5	55.1	74.7	107.4	41.2	67.4	104.4	50.4	71.8
09	FINISHES	98.9	63.8	79.6	94.2	48.5	69.0	98.7	62.7	78.8	94.9	61.6	76.5	86.6	46.7	64.6	88.5	33.5	58.2
COVERS	DIVS. 10 - 14, 25, 28, 41, 43, 44, 46	100.0	83.9	96.7	100.0	81.4	96.3	100.0	84.0	96.8	100.0	83.9	96.7	100.0	39.5	87.8	100.0	35.6	87.0
21, 22, 23	FIRE SUPPRESSION, PLUMBING & HVAC	99.9	79.0	91.5	100.2	78.9	91.6	95.2	79.0	88.7	100.0	79.0	91.5	95.3	51.6	77.7	95.3	53.1	78.3
26, 27, 3370	ELECTRICAL, COMMUNICATIONS & UTIL.	101.0	66.7	82.9	103.0	61.2	80.9	94.6	61.2	77.0	96.5	61.3	77.9	96.5	63.5	79.0	96.8	62.1	78.5
MF2014	WEIGHTED AVERAGE	99.2	73.8	88.1	99.2	68.9	86.0	98.9	72.5	87.4	97.4	72.4	86.5	94.6	54.6	77.1	97.1	50.1	76.6

DIVISION		ARKANSAS																	
		FAYETTEVILLE			FORT SMITH			HARRISON			HOT SPRINGS			JONESBORO			LITTLE ROCK		
		727			729			726			719			724			720 - 722		
		MAT.	INST.	TOTAL	MAT.	INST.	TOTAL	MAT.	INST.	TOTAL	MAT.	INST.	TOTAL	MAT.	INST.	TOTAL	MAT.	INST.	TOTAL
015433	CONTRACTOR EQUIPMENT		89.1	89.1		89.1	89.1		89.1	89.1		89.1	89.1		108.3	108.3		89.1	89.1
0241, 31 - 34	SITE & INFRASTRUCTURE, DEMOLITION	76.4	86.3	83.5	81.8	86.7	85.3	82.1	84.9	84.1	81.4	85.8	84.5	103.2	101.1	101.7	93.1	86.8	88.6
0310	Concrete Forming & Accessories	81.2	38.5	44.4	101.3	57.3	63.4	90.6	44.8	51.1	81.1	35.7	41.9	89.6	48.2	53.9	96.4	60.0	65.0
0320	Concrete Reinforcing	87.4	71.3	79.2	88.4	72.0	80.1	87.0	63.7	75.1	91.8	67.5	79.4	84.5	70.1	77.2	93.4	68.3	80.6
0330	Cast-in-Place Concrete	77.4	49.4	65.9	88.5	70.3	81.0	85.8	43.4	68.4	84.5	38.8	65.7	84.3	55.0	72.2	87.3	70.3	80.3
03	CONCRETE	83.2	49.3	66.5	91.3	65.0	78.4	90.8	48.7	70.1	91.2	43.7	67.9	88.1	56.0	72.3	92.8	65.5	79.4
04	MASONRY	90.5	41.4	59.9	96.9	50.9	68.3	100.1	39.0	62.0	87.6	33.5	53.9	91.9	41.9	60.7	96.8	50.9	68.2
05	METALS	99.5	67.8	89.7	101.8	70.6	92.2	100.6	66.5	90.1	103.8	66.9	92.5	95.9	80.0	91.0	102.0	69.4	92.0
06	WOOD, PLASTICS & COMPOSITES	85.8	35.5	57.6	107.6	59.0	80.4	96.0	45.5	67.7	87.5	35.7	58.5	93.9	48.3	68.4	99.9	62.5	78.9
07	THERMAL & MOISTURE PROTECTION	99.0	45.3	77.0	99.4	57.5	82.2	98.5	42.6	75.6	94.4	37.9	71.2	103.3	49.1	81.1	95.7	57.8	80.2
08	OPENINGS	97.7	44.7	85.4	98.5	56.2	88.7	98.5	46.5	86.4	102.1	42.3	88.2	100.6	54.0	89.7	99.5	58.4	89.9
0920	Plaster & Gypsum Board	81.8	34.0	49.5	89.4	58.3	68.4	88.3	44.3	58.6	84.0	34.2	50.4	97.7	46.9	63.4	94.6	61.8	72.5
0950, 0980	Ceilings & Acoustic Treatment	83.2	34.0	50.9	84.8	58.3	67.4	84.8	44.3	58.2	83.3	34.2	51.0	86.2	46.9	60.3	88.0	61.8	70.8
0960	Flooring	94.0	59.1	84.0	102.9	60.8	90.9	98.8	59.1	87.5	98.5	59.1	87.2	68.9	53.4	64.5	101.0	60.8	89.5
0970, 0990	Wall Finishes & Painting/Coating	107.4	29.4	60.3	107.4	53.8	75.1	107.4	41.2	67.4	104.4	55.2	74.7	94.9	48.4	66.8	108.8	55.2	76.5
09	FINISHES	85.7	40.0	60.5	89.9	57.6	72.1	88.6	46.8	65.6	88.4	41.0	62.3	83.7	48.5	64.3	92.7	59.8	74.6
COVERS	DIVS. 10 - 14, 25, 28, 41, 43, 44, 46	100.0	48.7	89.6	100.0	78.8	95.7	100.0	49.1	89.7	100.0	36.5	87.2	100.0	45.2	88.9	100.0	79.2	95.8
21, 22, 23	FIRE SUPPRESSION, PLUMBING & HVAC	95.3	50.2	77.1	100.1	50.6	80.1	95.3	49.4	76.8	95.3	48.7	76.5	100.3	52.5	81.0	99.9	56.2	82.3
26, 27, 3370	ELECTRICAL, COMMUNICATIONS & UTIL.	90.1	51.2	69.5	93.7	64.5	78.3	95.0	37.1	64.4	99.0	67.8	82.5	100.2	63.5	80.8	100.7	70.2	84.6
MF2014	WEIGHTED AVERAGE	93.4	52.0	75.4	97.2	61.6	81.7	95.9	50.4	76.0	96.4	51.8	76.9	96.7	59.1	80.3	98.5	64.0	83.4

City Cost Indexes

DIVISION		ARKANSAS												CALIFORNIA					
		PINE BLUFF			RUSSELLVILLE			TEXARKANA			WEST MEMPHIS			ALHAMBRA			ANAHEIM		
		716			728			718			723			917 - 918			928		
		MAT.	INST.	TOTAL	MAT.	INST.	TOTAL	MAT.	INST.	TOTAL	MAT.	INST.	TOTAL	MAT.	INST.	TOTAL	MAT.	INST.	TOTAL
015433	CONTRACTOR EQUIPMENT		89.1	89.1		89.1	89.1		89.9	89.9		108.3	108.3		100.1	100.1		100.8	100.8
0241, 31 - 34	SITE & INFRASTRUCTURE, DEMOLITION	83.7	86.8	85.9	78.3	84.9	83.0	94.9	87.1	89.3	110.5	101.1	103.8	97.9	110.9	107.1	99.4	108.3	105.7
0310	Concrete Forming & Accessories	80.7	60.0	62.8	86.6	54.7	59.1	87.3	40.1	46.6	95.5	48.5	54.9	117.1	116.4	116.5	105.4	124.2	121.6
0320	Concrete Reinforcing	93.5	68.2	80.7	88.0	67.3	77.5	93.1	67.5	80.1	84.5	70.1	77.2	106.6	113.3	110.0	93.8	113.2	103.7
0330	Cast-in-Place Concrete	84.5	70.3	78.7	81.1	45.3	66.4	92.1	43.3	72.0	88.4	55.1	74.7	95.1	120.8	105.6	92.9	123.3	105.4
03	CONCRETE	92.1	65.5	79.0	86.7	54.4	70.8	89.8	47.2	68.9	95.5	56.2	76.2	101.1	116.5	108.7	101.8	120.8	111.2
04	MASONRY	125.0	50.9	78.8	96.1	37.3	59.5	102.4	33.1	59.2	79.4	41.9	56.0	122.7	121.1	121.7	80.0	118.8	104.2
05	METALS	104.6	69.4	93.8	99.5	66.5	89.3	96.8	66.9	87.6	95.0	80.3	90.5	85.3	100.9	90.1	102.7	101.1	102.2
06	WOOD, PLASTICS & COMPOSITES	87.0	62.5	73.3	91.3	59.0	73.2	95.5	42.0	65.5	100.1	48.3	71.1	100.0	112.1	106.7	101.9	122.6	113.5
07	THERMAL & MOISTURE PROTECTION	94.5	57.8	79.5	99.2	43.9	76.5	95.2	44.5	74.4	103.7	49.1	81.3	95.1	118.1	104.5	99.4	122.0	108.7
08	OPENINGS	102.1	58.4	91.9	97.7	55.7	87.9	107.3	46.5	93.2	100.6	54.0	89.7	91.8	114.8	97.2	104.2	120.6	108.0
0920	Plaster & Gypsum Board	83.6	61.8	68.9	83.3	58.3	66.4	86.6	40.7	55.6	100.2	46.9	64.2	97.4	112.5	107.6	107.1	123.2	118.0
0950, 0980	Ceilings & Acoustic Treatment	83.3	61.8	69.2	83.2	58.3	66.8	86.6	40.7	56.4	84.3	46.9	59.7	100.2	112.5	108.3	105.0	123.2	117.0
0960	Flooring	98.2	60.8	87.5	96.2	59.1	85.6	100.4	51.0	86.3	71.0	53.4	66.0	95.4	110.2	99.6	101.9	112.0	104.8
0970, 0990	Wall Finishes & Painting/Coating	104.4	55.2	74.7	107.4	35.0	63.7	104.4	30.3	59.7	94.9	50.4	68.0	100.5	111.5	107.1	98.9	106.8	103.7
09	FINISHES	88.3	59.8	72.6	86.7	54.1	68.7	90.7	40.4	62.9	85.0	48.8	65.0	99.8	114.0	107.6	100.5	120.1	111.3
COVERS	DIVS. 10 - 14, 25, 28, 41, 43, 44, 46	100.0	79.2	95.8	100.0	41.0	88.1	100.0	32.6	86.4	100.0	45.2	88.9	100.0	110.5	102.1	100.0	112.0	102.4
21, 22, 23	FIRE SUPPRESSION, PLUMBING & HVAC	100.1	52.9	81.1	95.3	51.5	77.6	100.1	53.5	81.3	95.6	65.8	83.6	95.2	115.6	103.4	100.0	118.6	107.5
26, 27, 3370	ELECTRICAL, COMMUNICATIONS & UTIL.	97.0	70.2	82.8	93.7	44.7	67.8	98.9	36.0	65.7	101.9	65.7	82.8	118.4	120.5	119.5	92.2	105.7	99.3
MF2014	WEIGHTED AVERAGE	99.4	63.3	83.6	94.6	53.7	76.8	98.2	49.2	76.8	96.1	62.4	81.4	98.3	114.8	105.5	99.4	114.9	106.2

DIVISION		CALIFORNIA																	
		BAKERSFIELD			BERKELEY			EUREKA			FRESNO			INGLEWOOD			LONG BEACH		
		932 - 933			947			955			936 - 938			903 - 905			906 - 908		
		MAT.	INST.	TOTAL	MAT.	INST.	TOTAL	MAT.	INST.	TOTAL	MAT.	INST.	TOTAL	MAT.	INST.	TOTAL	MAT.	INST.	TOTAL
015433	CONTRACTOR EQUIPMENT		98.8	98.8		100.3	100.3		98.5	98.5		98.8	98.8		96.5	96.5		96.5	96.5
0241, 31 - 34	SITE & INFRASTRUCTURE, DEMOLITION	100.4	106.0	104.4	118.4	107.6	110.7	110.8	103.9	105.9	103.1	105.4	104.7	89.3	103.5	99.4	95.9	103.5	101.3
0310	Concrete Forming & Accessories	104.0	124.3	121.5	115.9	149.9	145.2	114.8	133.6	131.0	104.3	134.2	130.1	111.8	115.4	114.9	106.5	115.4	114.1
0320	Concrete Reinforcing	103.4	113.2	108.4	94.5	114.4	104.6	102.5	114.1	108.4	80.6	113.7	97.5	102.7	113.3	108.1	101.9	113.3	107.7
0330	Cast-in-Place Concrete	94.0	122.6	105.7	123.4	124.8	124.0	100.8	118.0	107.9	99.4	118.9	107.4	84.5	121.4	99.7	96.2	121.4	106.5
03	CONCRETE	97.9	120.6	109.1	107.3	132.7	119.8	112.8	123.2	117.9	98.6	123.8	111.0	90.9	116.3	103.4	99.8	116.3	107.9
04	MASONRY	100.2	118.3	111.5	113.4	134.3	126.5	104.0	132.6	121.8	103.6	122.0	115.1	76.3	121.2	104.3	85.3	121.2	107.7
05	METALS	105.3	100.9	104.0	106.9	100.5	104.9	102.6	98.9	101.4	105.6	100.5	104.1	94.4	101.5	96.6	94.3	101.5	96.5
06	WOOD, PLASTICS & COMPOSITES	96.0	122.7	111.0	111.4	154.9	135.8	116.6	138.5	128.9	108.1	138.5	125.1	103.8	110.7	107.6	96.9	110.7	104.6
07	THERMAL & MOISTURE PROTECTION	98.1	116.1	105.5	105.2	137.2	118.3	103.4	119.8	110.1	88.7	116.0	99.9	98.3	118.0	106.4	98.5	118.0	106.5
08	OPENINGS	97.8	116.1	102.0	93.0	140.6	104.1	103.5	113.5	105.8	98.1	124.7	104.3	88.8	114.1	94.7	88.8	114.1	94.7
0920	Plaster & Gypsum Board	101.9	123.2	116.3	111.6	156.0	141.6	113.4	139.5	131.0	100.6	139.5	126.9	103.8	111.0	108.7	99.9	111.0	107.4
0950, 0980	Ceilings & Acoustic Treatment	96.3	123.2	114.0	106.2	156.0	139.0	110.1	139.5	129.4	92.8	139.5	123.5	100.3	111.0	107.4	100.3	111.0	107.4
0960	Flooring	106.8	110.2	107.7	108.2	127.6	113.8	106.0	116.5	109.0	113.3	133.5	119.1	104.4	110.2	106.0	101.8	110.2	104.2
0970, 0990	Wall Finishes & Painting/Coating	111.4	102.0	105.7	102.5	143.9	127.5	100.5	46.2	67.7	129.0	107.2	115.8	100.9	111.5	107.3	100.9	111.5	107.3
09	FINISHES	100.0	120.5	111.3	105.4	147.2	128.4	106.1	123.6	115.8	102.1	133.2	119.2	102.7	113.2	108.5	101.8	113.2	108.1
COVERS	DIVS. 10 - 14, 25, 28, 41, 43, 44, 46	100.0	111.9	102.4	100.0	126.4	105.3	100.0	122.2	104.5	100.0	122.2	104.5	100.0	110.4	102.1	100.0	110.4	102.1
21, 22, 23	FIRE SUPPRESSION, PLUMBING & HVAC	100.1	117.1	107.0	95.3	147.8	116.5	95.2	110.6	101.4	100.2	111.3	104.7	94.8	115.6	103.2	94.8	115.6	103.2
26, 27, 3370	ELECTRICAL, COMMUNICATIONS & UTIL.	103.6	102.5	103.1	109.4	145.0	128.2	99.1	116.7	108.4	93.8	100.8	97.5	103.8	120.5	112.6	103.5	120.5	112.5
MF2014	WEIGHTED AVERAGE	100.7	113.6	106.3	102.7	135.1	116.8	102.1	116.4	108.4	100.1	115.3	106.7	94.8	114.1	103.2	96.2	114.1	104.0

DIVISION		CALIFORNIA																	
		LOS ANGELES			MARYSVILLE			MODESTO			MOJAVE			OAKLAND			OXNARD		
		900 - 902			959			953			935			946			930		
		MAT.	INST.	TOTAL	MAT.	INST.	TOTAL	MAT.	INST.	TOTAL	MAT.	INST.	TOTAL	MAT.	INST.	TOTAL	MAT.	INST.	TOTAL
015433	CONTRACTOR EQUIPMENT		100.1	100.1		98.5	98.5		98.5	98.5		98.8	98.8		100.3	100.3		97.7	97.7
0241, 31 - 34	SITE & INFRASTRUCTURE, DEMOLITION	96.6	107.1	104.1	107.2	105.0	105.7	102.2	105.1	104.3	96.4	106.0	103.2	124.7	107.6	112.5	103.3	104.0	103.8
0310	Concrete Forming & Accessories	108.6	124.3	122.2	104.6	134.4	130.3	100.7	134.3	129.7	116.1	112.1	112.6	104.9	149.9	143.7	107.5	124.3	122.0
0320	Concrete Reinforcing	103.6	113.4	108.6	102.5	113.8	108.2	106.2	113.8	110.0	97.3	113.1	105.3	96.7	114.4	105.7	95.5	113.1	104.4
0330	Cast-in-Place Concrete	91.2	122.0	103.9	112.6	119.2	115.3	100.8	119.2	108.4	86.0	122.4	101.0	117.0	124.8	120.2	100.1	122.7	109.4
03	CONCRETE	95.9	120.6	108.0	113.9	124.0	118.9	104.7	124.0	114.2	91.0	115.1	102.8	106.8	132.7	119.5	99.2	120.6	109.7
04	MASONRY	90.5	122.6	110.5	104.9	123.1	116.2	102.9	123.1	115.5	101.9	118.3	112.1	120.6	134.3	129.2	104.8	116.5	112.1
05	METALS	101.1	102.9	101.7	102.0	101.8	102.0	99.1	101.7	99.9	102.8	100.3	102.0	101.4	100.4	101.1	100.7	100.7	100.7
06	WOOD, PLASTICS & COMPOSITES	103.0	122.6	113.9	103.0	138.5	122.9	98.5	138.5	120.9	107.4	106.8	107.1	98.8	154.9	130.2	102.1	122.7	113.6
07	THERMAL & MOISTURE PROTECTION	97.6	121.2	107.3	102.9	121.5	110.5	102.5	118.7	109.1	94.4	110.9	101.2	103.1	137.2	117.1	97.4	119.8	106.6
08	OPENINGS	95.4	120.7	101.3	102.8	125.3	108.0	101.6	125.4	107.1	93.2	107.5	96.5	93.1	140.6	104.1	95.7	120.6	101.5
0920	Plaster & Gypsum Board	103.6	123.2	116.9	105.0	139.5	128.3	107.5	139.5	129.1	110.8	106.8	108.1	105.7	156.0	139.7	104.7	123.2	117.2
0950, 0980	Ceilings & Acoustic Treatment	110.5	123.2	118.8	109.2	139.5	129.1	105.0	139.5	127.7	94.2	106.8	102.5	108.9	156.0	139.9	96.6	123.2	114.1
0960	Flooring	102.0	112.0	104.9	102.0	117.6	106.4	102.4	112.8	105.3	112.6	110.2	111.9	103.4	127.6	110.3	104.8	112.0	106.9
0970, 0990	Wall Finishes & Painting/Coating	100.0	111.5	106.9	100.5	115.8	109.7	100.5	117.0	110.5	111.1	102.0	105.6	102.5	143.9	127.5	111.1	101.1	105.1
09	FINISHES	104.2	120.6	113.3	103.1	131.5	118.8	102.2	130.9	118.0	101.8	109.9	106.2	104.3	147.2	127.9	99.4	119.6	110.5
COVERS	DIVS. 10 - 14, 25, 28, 41, 43, 44, 46	100.0	111.9	102.4	100.0	122.2	104.5	100.0	122.2	104.5	100.0	107.8	101.6	100.0	126.3	105.3	100.0	112.2	102.5
21, 22, 23	FIRE SUPPRESSION, PLUMBING & HVAC	99.9	118.6	107.5	95.2	109.0	100.8	100.0	111.3	104.6	95.3	115.1	103.3	100.1	147.8	119.4	100.1	118.6	107.6
26, 27, 3370	ELECTRICAL, COMMUNICATIONS & UTIL.	102.0	121.3	112.2	95.7	107.4	101.9	98.2	104.0	101.3	92.0	99.9	96.2	108.5	139.3	124.8	97.7	107.3	102.8
MF2014	WEIGHTED AVERAGE	99.2	117.5	107.2	101.3	115.9	107.7	100.8	115.7	107.3	96.8	109.8	102.5	103.1	134.3	116.7	99.6	114.4	106.0

DIVISION		CALIFORNIA																	
		PALM SPRINGS			PALO ALTO			PASADENA			REDDING			RICHMOND			RIVERSIDE		
		922			943			910 - 912			960			948			925		
		MAT.	INST.	TOTAL	MAT.	INST.	TOTAL	MAT.	INST.	TOTAL	MAT.	INST.	TOTAL	MAT.	INST.	TOTAL	MAT.	INST.	TOTAL
015433	CONTRACTOR EQUIPMENT		99.6	99.6		100.3	100.3		100.1	100.1		98.5	98.5		100.3	100.3		99.6	99.6
0241, 31 - 34	SITE & INFRASTRUCTURE, DEMOLITION	90.9	106.3	101.8	114.4	107.6	109.6	94.9	110.9	106.3	125.1	105.0	110.8	123.8	107.6	112.3	97.9	106.3	103.9
0310	Concrete Forming & Accessories	101.9	115.3	113.4	103.0	146.3	140.3	106.2	115.3	114.1	107.8	142.7	137.9	118.7	149.7	145.5	105.8	124.2	121.7
0320	Concrete Reinforcing	107.8	113.1	110.5	94.5	114.4	104.6	107.5	113.3	110.5	118.7	113.8	116.2	94.5	114.3	104.6	104.7	113.1	109.0
0330	Cast-in-Place Concrete	88.6	123.2	102.8	104.4	124.8	112.8	90.2	120.8	102.8	118.0	119.2	118.5	120.0	124.8	122.0	96.3	123.3	107.4
03	CONCRETE	96.6	116.8	106.5	96.5	131.1	113.5	96.6	116.0	106.2	122.7	127.7	125.2	108.8	132.6	120.5	102.6	120.8	111.6
04	MASONRY	77.6	118.5	103.1	97.4	134.3	120.4	106.5	121.1	115.6	130.0	123.1	125.7	113.2	132.8	125.4	78.6	118.1	103.2
05	METALS	103.3	100.8	102.5	98.9	100.3	99.3	85.3	100.8	90.1	101.6	101.8	101.7	98.9	100.2	99.3	102.8	101.0	102.3
06	WOOD, PLASTICS & COMPOSITES	96.6	110.8	104.5	96.1	150.2	126.4	86.5	110.6	100.0	110.6	149.8	132.6	115.3	154.9	137.5	101.9	122.6	113.5
07	THERMAL & MOISTURE PROTECTION	99.1	118.7	107.2	102.7	137.5	117.0	94.9	117.4	104.1	121.6	122.1	121.8	103.3	135.9	116.7	99.6	121.1	108.4
08	OPENINGS	100.1	114.1	103.4	93.1	136.9	103.3	91.8	114.0	97.0	114.1	131.5	118.1	93.1	139.4	103.9	102.8	120.6	107.0
0920	Plaster & Gypsum Board	102.0	111.0	108.1	103.9	151.1	135.9	90.2	111.0	104.3	106.2	151.1	136.6	113.6	156.0	142.3	106.3	123.2	117.7
0950, 0980	Ceilings & Acoustic Treatment	101.7	111.0	107.8	107.1	151.1	136.0	100.2	111.0	107.3	131.5	151.1	144.4	107.1	156.0	139.3	109.2	123.2	118.4
0960	Flooring	104.6	107.2	105.3	102.4	127.6	109.6	90.8	110.2	96.3	98.0	117.6	103.6	109.9	127.6	115.0	106.0	112.0	107.7
0970, 0990	Wall Finishes & Painting/Coating	97.2	111.3	105.7	102.5	143.9	127.5	100.5	111.5	107.1	115.9	115.8	115.8	102.5	143.9	127.5	97.2	106.8	103.0
09	FINISHES	99.1	112.8	106.6	102.7	144.4	125.6	97.3	113.2	106.0	109.3	138.2	125.3	106.9	147.2	129.1	102.1	120.1	112.0
COVERS	DIVS. 10 - 14, 25, 28, 41, 43, 44, 46	100.0	110.7	102.2	100.0	125.9	105.2	100.0	110.3	102.1	100.0	123.4	104.7	100.0	126.3	105.3	100.0	112.0	102.4
21, 22, 23	FIRE SUPPRESSION, PLUMBING & HVAC	95.2	115.5	103.4	95.3	145.5	115.6	95.2	115.6	103.4	100.2	109.0	103.7	95.3	147.8	116.5	100.0	118.6	107.5
26, 27, 3370	ELECTRICAL, COMMUNICATIONS & UTIL.	95.3	105.4	100.7	108.4	150.2	130.5	115.1	120.5	118.0	99.2	107.4	103.5	109.0	132.7	121.5	91.9	105.4	99.1
MF2014	WEIGHTED AVERAGE	97.2	112.0	103.6	98.9	134.5	114.4	96.4	114.6	104.3	107.8	117.7	112.1	101.8	133.1	115.4	99.4	114.6	106.0

DIVISION		CALIFORNIA																	
		SACRAMENTO			SALINAS			SAN BERNARDINO			SAN DIEGO			SAN FRANCISCO			SAN JOSE		
		942,956 - 958			939			923 - 924			919 - 921			940 - 941			951		
		MAT.	INST.	TOTAL	MAT.	INST.	TOTAL	MAT.	INST.	TOTAL	MAT.	INST.	TOTAL	MAT.	INST.	TOTAL	MAT.	INST.	TOTAL
015433	CONTRACTOR EQUIPMENT		99.9	99.9		98.8	98.8		99.6	99.6		100.1	100.1		110.7	110.7		99.3	99.3
0241, 31 - 34	SITE & INFRASTRUCTURE, DEMOLITION	100.6	113.3	109.6	116.3	105.7	108.8	77.7	106.3	98.0	101.7	104.0	103.4	127.0	113.6	117.5	133.5	100.4	109.9
0310	Concrete Forming & Accessories	103.3	137.0	132.4	111.0	137.4	133.7	109.6	115.2	114.4	105.6	113.3	112.2	104.6	150.9	144.6	107.0	149.8	143.9
0320	Concrete Reinforcing	89.4	113.8	101.8	96.0	114.2	105.2	104.7	113.0	108.9	104.2	113.1	108.7	110.1	115.0	112.6	93.5	114.5	104.1
0330	Cast-in-Place Concrete	99.0	120.3	107.7	98.9	119.6	107.4	66.6	123.2	89.8	98.9	107.8	102.6	120.0	126.4	122.6	117.1	124.2	120.0
03	CONCRETE	97.0	125.2	110.9	108.5	125.6	116.9	76.8	116.8	96.4	101.8	110.7	106.2	110.3	134.4	122.1	110.8	132.9	121.7
04	MASONRY	98.4	123.1	113.8	101.6	128.1	118.1	84.9	115.9	104.2	97.4	115.6	108.7	121.1	141.5	133.8	131.4	134.4	133.3
05	METALS	97.0	96.1	96.7	105.5	102.9	104.7	102.8	100.5	102.1	101.1	101.4	101.2	107.6	110.8	108.6	97.5	107.5	100.6
06	WOOD, PLASTICS & COMPOSITES	92.9	141.7	120.2	107.3	141.5	126.4	105.8	110.8	108.6	100.6	110.2	106.0	98.8	155.1	130.3	111.3	154.7	135.6
07	THERMAL & MOISTURE PROTECTION	110.8	121.3	115.1	95.0	126.1	107.7	98.3	117.7	106.3	99.3	107.1	102.5	104.9	141.9	120.1	98.9	139.3	115.4
08	OPENINGS	106.3	127.2	111.1	96.9	133.3	105.4	100.2	114.1	103.4	99.1	111.8	102.1	97.3	140.7	107.4	92.8	140.5	103.9
0920	Plaster & Gypsum Board	101.0	142.6	129.1	105.8	142.6	130.6	108.2	111.0	110.1	97.8	110.3	106.3	108.3	156.0	140.6	102.7	156.0	138.7
0950, 0980	Ceilings & Acoustic Treatment	107.1	142.6	130.4	94.2	142.6	126.0	105.0	111.0	109.0	107.7	110.3	109.4	117.1	156.0	142.7	103.3	156.0	138.0
0960	Flooring	102.6	117.6	106.9	107.1	121.1	111.1	107.8	110.2	108.5	98.5	112.0	102.4	103.4	127.6	110.3	95.4	127.6	104.6
0970, 0990	Wall Finishes & Painting/Coating	100.2	116.7	110.1	112.3	143.9	131.4	97.2	104.0	101.3	97.1	111.5	105.8	102.5	153.2	133.1	100.8	143.9	126.8
09	FINISHES	101.4	133.6	119.1	101.4	137.0	121.0	100.5	112.5	107.1	102.9	112.9	108.4	106.5	148.3	129.6	100.7	147.0	126.2
COVERS	DIVS. 10 - 14, 25, 28, 41, 43, 44, 46	100.0	123.0	104.6	100.0	122.6	104.6	100.0	108.1	101.6	100.0	109.7	102.0	100.0	126.8	105.4	100.0	125.9	105.2
21, 22, 23	FIRE SUPPRESSION, PLUMBING & HVAC	100.0	120.8	108.4	95.3	117.7	104.4	95.2	115.6	103.4	99.9	116.9	106.8	100.1	175.0	130.3	100.0	147.3	119.1
26, 27, 3370	ELECTRICAL, COMMUNICATIONS & UTIL.	103.5	108.7	106.2	93.2	120.9	107.8	95.3	103.4	99.6	101.2	98.5	99.8	108.6	157.8	134.6	101.1	155.9	130.1
MF2014	WEIGHTED AVERAGE	100.5	119.3	108.7	100.2	121.6	109.5	95.1	111.3	102.1	100.5	109.6	104.5	105.2	145.4	122.7	102.5	136.6	117.4

DIVISION		CALIFORNIA																	
		SAN LUIS OBISPO			SAN MATEO			SAN RAFAEL			SANTA ANA			SANTA BARBARA			SANTA CRUZ		
		934			944			949			926 - 927			931			950		
		MAT.	INST.	TOTAL	MAT.	INST.	TOTAL	MAT.	INST.	TOTAL	MAT.	INST.	TOTAL	MAT.	INST.	TOTAL	MAT.	INST.	TOTAL
015433	CONTRACTOR EQUIPMENT		98.8	98.8		100.3	100.3		100.3	100.3		99.6	99.6		98.8	98.8		99.3	99.3
0241, 31 - 34	SITE & INFRASTRUCTURE, DEMOLITION	108.4	106.0	106.7	121.3	107.6	111.6	113.2	113.4	113.4	89.4	106.3	101.4	103.3	106.0	105.2	133.1	100.0	109.6
0310	Concrete Forming & Accessories	117.9	115.4	115.8	109.0	149.8	144.2	112.9	149.9	144.9	109.9	115.3	114.5	108.3	124.2	122.1	107.0	137.6	133.4
0320	Concrete Reinforcing	97.3	113.1	105.4	94.5	114.6	104.7	95.1	114.6	105.1	108.3	113.1	110.8	95.5	113.2	104.5	115.7	114.2	114.9
0330	Cast-in-Place Concrete	106.3	122.5	112.9	116.2	124.8	119.7	135.2	124.1	130.7	85.0	123.2	100.7	99.8	122.6	109.1	116.3	121.4	118.4
03	CONCRETE	106.9	116.6	111.7	105.4	132.7	118.8	126.0	132.4	129.1	94.2	116.8	105.3	99.0	120.6	109.6	113.9	126.5	120.1
04	MASONRY	103.5	118.3	112.7	112.9	137.3	128.1	93.2	137.4	120.7	74.6	118.8	102.1	102.2	119.1	112.7	135.3	128.2	130.9
05	METALS	103.5	100.6	102.6	98.7	100.6	99.3	100.1	98.8	99.7	102.9	100.8	102.2	101.2	100.9	101.1	104.5	106.3	105.0
06	WOOD, PLASTICS & COMPOSITES	109.9	110.9	110.5	104.0	154.9	132.5	101.8	154.7	131.4	107.8	110.8	109.5	102.1	122.7	113.6	111.3	141.6	128.3
07	THERMAL & MOISTURE PROTECTION	95.1	116.9	104.0	103.1	138.8	117.7	107.2	138.9	120.2	99.4	118.3	107.2	94.6	118.9	104.5	98.8	129.2	111.3
08	OPENINGS	95.1	109.7	98.5	93.1	139.4	103.8	103.8	139.3	112.1	99.5	114.1	102.9	96.6	120.6	102.2	94.0	133.4	103.2
0920	Plaster & Gypsum Board	112.2	111.0	111.4	108.9	156.0	140.8	111.5	156.0	141.6	109.6	111.0	110.6	104.7	123.2	117.2	110.1	142.6	132.0
0950, 0980	Ceilings & Acoustic Treatment	94.2	111.0	105.3	107.1	156.0	139.3	115.4	156.0	142.1	105.0	111.0	109.0	96.6	123.2	114.1	106.7	142.6	130.3
0960	Flooring	113.5	110.2	112.6	105.1	127.6	111.5	115.0	121.1	116.8	108.3	110.2	108.8	106.2	108.6	106.9	99.6	121.1	105.7
0970, 0990	Wall Finishes & Painting/Coating	111.1	101.7	105.5	102.5	143.9	127.5	98.9	142.6	125.3	97.2	106.8	103.0	111.1	101.1	105.1	101.0	143.9	126.9
09	FINISHES	103.2	112.3	108.2	104.7	147.2	128.1	107.6	145.8	128.6	102.0	112.8	107.9	100.0	119.0	110.5	103.7	137.1	122.1
COVERS	DIVS. 10 - 14, 25, 28, 41, 43, 44, 46	100.0	120.6	104.2	100.0	126.4	105.3	100.0	125.7	105.2	100.0	110.7	102.2	100.0	112.2	102.5	100.0	122.9	104.6
21, 22, 23	FIRE SUPPRESSION, PLUMBING & HVAC	95.3	115.6	103.5	95.3	142.2	114.2	95.3	169.5	125.2	95.2	114.3	102.9	100.1	118.6	107.6	100.0	117.8	107.2
26, 27, 3370	ELECTRICAL, COMMUNICATIONS & UTIL.	92.0	106.4	99.6	108.4	148.0	129.3	105.4	120.0	113.1	95.4	105.7	100.8	91.0	110.5	101.3	100.3	120.9	111.2
MF2014	WEIGHTED AVERAGE	99.4	112.0	104.9	100.9	134.6	115.6	103.5	136.7	118.0	96.9	111.8	103.4	98.9	115.1	106.0	104.5	121.8	112.0

DIVISION		CALIFORNIA															COLORADO		
		SANTA ROSA			STOCKTON			SUSANVILLE			VALLEJO			VAN NUYS			ALAMOSA		
		954			952			961			945			913 - 916			811		
		MAT.	INST.	TOTAL	MAT.	INST.	TOTAL	MAT.	INST.	TOTAL	MAT.	INST.	TOTAL	MAT.	INST.	TOTAL	MAT.	INST.	TOTAL
015433	CONTRACTOR EQUIPMENT		99.0	99.0		98.5	98.5		98.5	98.5		100.3	100.3		100.1	100.1		93.8	93.8
0241, 31 - 34	SITE & INFRASTRUCTURE, DEMOLITION	103.0	105.1	104.5	102.0	105.1	104.2	131.8	105.0	112.8	100.5	113.2	109.5	111.4	110.9	111.0	134.7	88.0	101.5
0310	Concrete Forming & Accessories	102.9	148.1	141.9	104.8	136.6	132.2	109.1	142.8	138.1	103.4	147.7	141.6	112.7	115.3	115.0	105.0	67.5	72.7
0320	Concrete Reinforcing	103.4	114.8	109.2	106.2	113.8	110.1	118.7	113.2	115.9	96.3	114.5	105.6	107.5	113.3	110.5	105.1	76.4	90.5
0330	Cast-in-Place Concrete	110.6	120.8	114.8	98.2	119.2	106.9	107.4	119.3	112.3	107.7	121.5	113.4	95.1	120.8	105.7	101.1	78.4	91.8
03	CONCRETE	113.7	130.9	122.1	103.7	125.0	114.2	125.2	127.7	126.4	102.2	130.5	116.1	110.2	116.0	113.0	113.5	73.3	93.8
04	MASONRY	103.1	137.2	124.3	102.8	123.1	115.5	127.9	112.7	118.4	71.6	133.1	109.9	122.7	121.1	121.7	126.0	71.9	92.3
05	METALS	103.2	105.0	103.8	99.3	101.8	100.0	100.6	101.1	100.7	100.1	98.1	99.5	84.4	100.8	89.5	98.2	80.2	92.7
06	WOOD, PLASTICS & COMPOSITES	98.0	154.5	129.6	104.3	141.5	125.1	112.5	149.8	133.4	90.6	154.7	126.5	95.0	110.6	103.7	98.0	67.5	80.9
07	THERMAL & MOISTURE PROTECTION	99.8	136.2	114.7	102.6	119.9	109.7	122.2	120.4	121.4	105.0	135.3	117.4	95.7	117.4	104.6	104.9	78.1	93.9
08	OPENINGS	101.0	140.4	110.2	101.6	127.1	107.5	113.9	131.5	118.0	105.6	140.5	113.7	91.6	114.0	96.8	98.3	74.1	92.7
0920	Plaster & Gypsum Board	103.9	156.0	139.2	107.5	142.6	131.2	107.1	151.1	136.9	105.8	156.0	139.8	95.3	111.0	105.9	79.0	66.3	70.4
0950, 0980	Ceilings & Acoustic Treatment	105.0	156.0	138.6	112.5	142.6	132.3	124.1	151.1	141.9	117.3	156.0	142.8	97.7	111.0	106.5	96.3	66.3	76.6
0960	Flooring	105.0	115.2	107.9	102.4	112.8	105.3	98.4	122.1	105.2	110.6	127.6	115.4	93.2	110.2	98.0	113.5	54.8	96.7
0970, 0990	Wall Finishes & Painting/Coating	97.2	142.6	124.6	100.5	117.0	110.5	115.9	115.8	115.8	99.8	142.6	125.6	100.5	111.5	107.1	114.1	39.8	69.3
09	FINISHES	101.1	143.6	124.5	103.8	132.7	119.7	108.9	139.0	125.5	104.6	145.8	127.3	99.3	113.2	107.0	102.5	61.7	80.0
COVERS	DIVS. 10 - 14, 25, 28, 41, 43, 44, 46	100.0	123.8	104.8	100.0	122.5	104.6	100.0	123.5	104.7	100.0	124.3	104.9	100.0	110.3	102.1	100.0	88.5	97.7
21, 22, 23	FIRE SUPPRESSION, PLUMBING & HVAC	95.2	167.3	124.3	100.0	111.3	104.6	95.4	109.0	100.9	100.1	127.5	111.1	95.2	115.6	103.4	95.2	72.0	85.9
26, 27, 3370	ELECTRICAL, COMMUNICATIONS & UTIL.	95.7	114.3	105.5	98.2	110.6	104.8	99.6	118.4	109.5	100.9	122.4	112.3	115.1	120.5	118.0	99.1	72.5	85.0
MF2014	WEIGHTED AVERAGE	100.8	134.7	115.6	100.9	117.2	108.0	106.8	118.1	111.7	100.1	127.2	111.9	99.1	114.6	105.9	102.2	73.7	89.8

DIVISION		COLORADO																	
		BOULDER			COLORADO SPRINGS			DENVER			DURANGO			FORT COLLINS			FORT MORGAN		
		803			808 - 809			800 - 802			813			805			807		
		MAT.	INST.	TOTAL	MAT.	INST.	TOTAL	MAT.	INST.	TOTAL	MAT.	INST.	TOTAL	MAT.	INST.	TOTAL	MAT.	INST.	TOTAL
015433	CONTRACTOR EQUIPMENT		97.5	97.5		95.8	95.8		100.4	100.4		93.8	93.8		97.5	97.5		97.5	97.5
0241, 31 - 34	SITE & INFRASTRUCTURE, DEMOLITION	94.8	96.4	95.9	96.9	94.7	95.3	96.1	102.6	100.7	128.5	88.0	99.7	107.1	96.0	99.2	97.7	95.9	96.4
0310	Concrete Forming & Accessories	103.1	79.9	83.1	93.6	79.4	81.3	100.0	76.0	79.3	111.4	67.7	73.7	100.7	74.7	78.3	103.6	74.9	78.9
0320	Concrete Reinforcing	98.9	76.6	87.5	98.1	80.2	89.0	98.1	80.2	89.0	105.1	76.5	90.5	99.0	76.6	87.6	99.1	76.5	87.6
0330	Cast-in-Place Concrete	101.8	80.0	92.9	104.6	88.3	97.9	99.0	82.8	92.4	116.3	78.5	100.8	114.9	78.9	100.1	99.9	78.9	91.3
03	CONCRETE	103.1	79.6	91.5	106.3	82.9	94.8	100.7	79.5	90.3	116.1	73.4	95.1	113.5	76.7	95.4	101.5	76.8	89.4
04	MASONRY	91.0	72.1	79.2	91.3	81.9	85.4	93.3	72.2	80.1	113.6	71.9	87.6	108.5	75.7	88.1	104.9	72.2	84.5
05	METALS	96.0	83.5	92.2	99.1	85.7	95.0	101.6	85.6	96.7	98.2	80.4	92.7	97.2	80.4	92.0	95.7	80.3	91.0
06	WOOD, PLASTICS & COMPOSITES	102.2	83.0	91.4	91.7	77.3	83.6	99.8	77.2	87.2	107.6	67.5	85.1	99.6	77.2	87.0	102.2	77.2	88.2
07	THERMAL & MOISTURE PROTECTION	103.1	81.7	94.3	103.9	85.6	96.4	102.4	75.8	91.5	104.9	78.1	93.9	103.5	73.5	91.2	103.0	81.2	94.1
08	OPENINGS	98.2	82.5	94.5	102.4	80.5	97.3	103.0	80.4	97.8	105.3	74.1	98.0	98.2	79.4	93.8	98.1	79.4	93.8
0920	Plaster & Gypsum Board	105.1	82.7	89.9	87.9	76.7	80.3	100.7	76.8	84.5	92.5	66.3	74.8	99.1	76.7	84.0	105.1	76.7	85.9
0950, 0980	Ceilings & Acoustic Treatment	96.9	82.7	87.5	104.4	76.7	86.2	107.3	76.8	87.2	96.3	66.3	76.6	96.9	76.7	83.6	96.9	76.7	83.6
0960	Flooring	99.7	83.8	95.2	92.0	68.2	85.2	96.7	84.8	93.3	118.2	54.8	100.0	96.6	54.8	84.6	100.1	54.8	87.1
0970, 0990	Wall Finishes & Painting/Coating	101.7	66.6	80.5	101.4	40.5	64.7	101.7	75.7	86.0	114.1	39.8	69.3	101.7	40.2	64.6	101.7	53.6	72.6
09	FINISHES	101.7	79.4	89.4	98.8	72.8	84.5	102.4	77.3	88.6	105.0	61.7	81.1	100.5	67.4	82.3	101.8	68.9	83.7
COVERS	DIVS. 10 - 14, 25, 28, 41, 43, 44, 46	100.0	89.4	97.9	100.0	92.2	98.4	100.0	88.8	97.7	100.0	88.4	97.7	100.0	88.8	97.7	100.0	88.8	97.7
21, 22, 23	FIRE SUPPRESSION, PLUMBING & HVAC	95.3	77.3	88.1	100.2	88.2	95.4	100.0	79.9	91.9	95.2	83.1	90.3	100.1	77.2	90.9	95.3	77.2	88.0
26, 27, 3370	ELECTRICAL, COMMUNICATIONS & UTIL.	97.4	84.5	90.6	100.8	82.1	90.9	102.6	83.2	92.4	98.5	69.4	83.1	97.4	84.5	90.6	97.8	84.5	90.7
MF2014	WEIGHTED AVERAGE	97.8	81.3	90.6	100.4	84.0	93.2	100.8	81.7	92.5	102.7	75.7	90.9	101.3	79.0	91.6	98.3	79.0	89.9

DIVISION		COLORADO																	
		GLENWOOD SPRINGS			GOLDEN			GRAND JUNCTION			GREELEY			MONTROSE			PUEBLO		
		816			804			815			806			814			810		
		MAT.	INST.	TOTAL	MAT.	INST.	TOTAL	MAT.	INST.	TOTAL	MAT.	INST.	TOTAL	MAT.	INST.	TOTAL	MAT.	INST.	TOTAL
015433	CONTRACTOR EQUIPMENT		96.7	96.7		97.5	97.5		96.7	96.7		97.5	97.5		95.2	95.2		93.8	93.8
0241, 31 - 34	SITE & INFRASTRUCTURE, DEMOLITION	143.2	95.1	109.0	107.3	96.2	99.4	128.4	94.7	104.5	94.0	95.4	95.0	137.1	91.2	104.5	120.6	91.0	99.6
0310	Concrete Forming & Accessories	102.0	75.0	78.7	96.1	74.8	77.7	110.2	74.4	79.3	98.6	78.4	81.2	101.4	74.8	78.4	107.3	79.4	83.3
0320	Concrete Reinforcing	104.0	76.5	90.0	99.1	76.4	87.6	104.3	76.3	90.1	98.9	75.1	86.8	103.9	76.4	89.9	100.6	80.2	90.2
0330	Cast-in-Place Concrete	101.1	77.9	91.6	100.0	78.9	91.3	111.9	77.1	97.6	96.1	59.1	80.9	101.1	77.8	91.5	100.4	87.8	95.2
03	CONCRETE	118.5	76.5	97.9	111.9	76.7	94.6	112.5	75.9	94.5	98.3	71.3	85.0	109.7	76.3	93.3	102.7	82.7	92.9
04	MASONRY	98.9	72.1	82.2	107.7	71.8	85.3	132.6	71.5	94.5	102.6	49.0	69.2	106.2	71.9	84.9	95.4	81.7	86.9
05	METALS	97.9	80.9	92.6	95.9	80.1	91.0	99.5	79.2	93.2	97.2	77.5	91.1	97.1	80.0	91.8	101.1	86.2	96.5
06	WOOD, PLASTICS & COMPOSITES	93.2	77.3	84.3	94.3	77.2	84.7	105.3	77.3	89.6	96.8	83.0	89.1	94.3	77.4	84.8	100.9	77.6	87.8
07	THERMAL & MOISTURE PROTECTION	104.8	79.2	94.3	103.9	75.4	92.2	104.0	68.5	89.5	102.9	66.4	87.9	105.0	79.2	94.4	103.5	83.6	95.3
08	OPENINGS	104.2	79.4	98.5	98.2	78.8	93.7	105.0	78.8	98.9	98.1	82.4	94.5	105.5	79.5	99.4	100.2	80.6	95.6
0920	Plaster & Gypsum Board	117.8	76.7	90.0	96.9	76.7	83.3	131.0	76.7	94.3	97.6	82.7	87.5	78.3	76.7	77.2	83.9	76.7	79.1
0950, 0980	Ceilings & Acoustic Treatment	95.5	76.7	83.2	96.9	76.7	83.6	95.5	76.7	83.2	96.9	82.7	87.5	96.3	76.7	83.4	104.7	76.7	86.3
0960	Flooring	112.8	50.4	95.0	94.7	54.8	83.3	117.5	54.8	99.6	95.7	54.8	84.0	115.8	45.4	95.7	114.6	84.8	106.1
0970, 0990	Wall Finishes & Painting/Coating	114.1	66.5	85.4	101.7	66.6	80.5	114.1	66.6	85.4	101.7	25.0	55.4	114.1	39.8	69.3	114.1	37.5	67.9
09	FINISHES	107.9	69.6	86.8	100.4	71.0	84.2	109.3	70.4	87.9	99.3	69.2	82.7	103.2	65.7	82.5	103.7	75.6	88.2
COVERS	DIVS. 10 - 14, 25, 28, 41, 43, 44, 46	100.0	88.8	97.7	100.0	88.8	97.7	100.0	88.8	97.7	100.0	89.4	97.9	100.0	89.1	97.8	100.0	92.7	98.5
21, 22, 23	FIRE SUPPRESSION, PLUMBING & HVAC	95.2	83.0	90.3	95.3	76.7	87.8	99.9	82.5	92.9	100.1	77.2	90.8	95.2	83.0	90.3	99.9	76.8	90.6
26, 27, 3370	ELECTRICAL, COMMUNICATIONS & UTIL.	95.7	69.4	81.8	97.8	84.5	90.7	98.2	53.8	74.7	97.4	84.5	90.6	98.2	56.4	76.1	99.1	73.0	85.3
MF2014	WEIGHTED AVERAGE	102.3	78.1	91.8	99.8	78.9	90.7	104.8	75.3	91.9	98.9	75.5	88.7	101.4	75.4	90.1	101.1	80.3	92.0

DIVISION		COLORADO			CONNECTICUT															
		SALIDA			BRIDGEPORT			BRISTOL			HARTFORD			MERIDEN			NEW BRITAIN		
		812			066			060			061			064			060		
		MAT.	INST.	TOTAL	MAT.	INST.	TOTAL	MAT.	INST.	TOTAL	MAT.	INST.	TOTAL	MAT.	INST.	TOTAL	MAT.	INST.	TOTAL
015433	CONTRACTOR EQUIPMENT		95.2	95.2		100.9	100.9		100.9	100.9		100.9	100.9		101.3	101.3		100.9	100.9
0241, 31 - 34	SITE & INFRASTRUCTURE, DEMOLITION	128.2	91.5	102.1	109.8	105.1	106.4	108.9	105.0	106.2	104.4	105.0	104.9	106.7	105.8	106.0	109.1	105.0	106.2
0310	Concrete Forming & Accessories	110.1	74.7	79.6	98.9	124.5	121.0	98.9	124.3	120.8	95.7	124.3	120.4	98.6	124.3	120.8	99.2	124.3	120.9
0320	Concrete Reinforcing	103.7	76.4	89.8	105.1	127.3	116.4	105.1	127.3	116.4	104.6	127.3	116.2	105.1	127.3	116.4	105.1	127.3	116.4
0330	Cast-in-Place Concrete	115.8	77.8	100.2	107.9	127.1	115.8	101.1	127.1	111.8	103.0	127.1	112.9	97.3	127.1	109.5	102.8	127.1	112.7
03	CONCRETE	111.2	76.3	94.1	106.6	125.7	116.0	103.4	125.6	114.3	104.0	125.6	114.6	101.6	125.6	113.4	104.2	125.6	114.7
04	MASONRY	133.9	71.9	95.3	111.8	134.0	125.6	102.9	134.0	122.3	107.7	134.0	124.1	102.5	134.0	122.1	105.2	134.0	123.1
05	METALS	96.7	79.9	91.6	100.5	125.5	108.2	100.5	125.4	108.1	105.6	125.4	111.7	97.4	125.4	106.0	96.5	125.4	105.4
06	WOOD, PLASTICS & COMPOSITES	101.9	77.4	88.2	98.8	123.4	112.6	98.8	123.4	112.6	92.3	123.4	109.7	98.8	123.4	112.6	98.8	123.4	112.6
07	THERMAL & MOISTURE PROTECTION	104.0	79.2	93.8	98.6	129.1	111.1	98.7	126.0	109.9	102.8	126.0	112.3	98.7	126.0	109.9	98.7	126.0	109.9
08	OPENINGS	98.4	79.5	94.0	101.2	132.1	108.4	101.2	132.1	108.4	100.3	132.1	107.7	103.6	132.1	110.2	101.2	132.1	108.4
0920	Plaster & Gypsum Board	78.6	76.7	77.3	101.7	123.5	116.4	101.7	123.5	116.4	97.4	123.5	115.0	103.2	123.5	116.9	101.7	123.5	116.4
0950, 0980	Ceilings & Acoustic Treatment	96.3	76.7	83.4	86.4	123.5	110.8	86.4	123.5	110.8	89.7	123.5	111.9	90.4	123.5	112.2	86.4	123.5	110.8
0960	Flooring	120.6	45.4	99.1	101.9	131.2	110.3	101.9	131.2	110.3	100.5	131.2	109.2	101.9	131.2	110.3	101.9	131.2	110.3
0970, 0990	Wall Finishes & Painting/Coating	114.1	41.7	70.4	104.9	122.9	115.8	104.9	122.9	115.8	105.9	122.9	116.2	104.9	122.9	115.8	104.9	122.9	115.8
09	FINISHES	103.6	65.9	82.8	97.2	125.0	112.6	97.3	125.0	112.6	96.3	125.0	112.1	98.4	125.0	113.1	97.3	125.0	112.6
COVERS	DIVS. 10 - 14, 25, 28, 41, 43, 44, 46	100.0	89.2	97.8	100.0	112.4	102.5	100.0	112.4	102.5	100.0	112.4	102.5	100.0	112.4	102.5	100.0	112.4	102.5
21, 22, 23	FIRE SUPPRESSION, PLUMBING & HVAC	95.2	72.0	85.8	100.0	116.4	106.6	100.0	116.4	106.6	100.0	116.4	106.6	95.2	116.4	103.8	100.0	116.4	106.6
26, 27, 3370	ELECTRICAL, COMMUNICATIONS & UTIL.	98.4	72.5	84.7	98.5	112.2	105.7	98.5	111.8	105.6	97.9	112.8	105.8	98.4	111.8	105.5	98.6	111.8	105.6
MF2014	WEIGHTED AVERAGE	101.9	75.3	90.3	101.3	120.8	109.8	100.5	120.6	109.3	101.4	120.8	109.8	98.9	120.7	108.4	100.1	120.6	109.1

DIVISION		CONNECTICUT																	
		NEW HAVEN			NEW LONDON			NORWALK			STAMFORD			WATERBURY			WILLIMANTIC		
		065			063			068			069			067			062		
		MAT.	INST.	TOTAL	MAT.	INST.	TOTAL	MAT.	INST.	TOTAL	MAT.	INST.	TOTAL	MAT.	INST.	TOTAL	MAT.	INST.	TOTAL
015433	CONTRACTOR EQUIPMENT		101.3	101.3		101.3	101.3		100.9	100.9		100.9	100.9		100.9	100.9		100.9	100.9
0241, 31 - 34	SITE & INFRASTRUCTURE, DEMOLITION	108.9	105.8	106.7	101.1	105.8	104.4	109.5	105.1	106.4	110.2	105.1	106.6	109.5	105.0	106.3	109.5	105.0	106.3
0310	Concrete Forming & Accessories	98.6	124.3	120.8	98.6	124.3	120.8	98.9	124.9	121.3	98.9	124.9	121.3	98.9	124.3	120.8	98.9	124.1	120.6
0320	Concrete Reinforcing	105.1	127.3	116.4	82.4	127.3	105.2	105.1	127.4	116.5	105.1	127.4	116.5	105.1	127.3	116.4	105.1	127.2	116.4
0330	Cast-in-Place Concrete	104.5	127.1	113.8	89.1	127.1	104.7	106.2	128.6	115.4	107.9	128.6	116.4	107.9	127.1	115.8	100.8	125.8	111.0
03	CONCRETE	118.7	125.6	122.1	91.4	125.6	108.2	105.8	126.4	115.9	106.6	126.4	116.3	106.6	125.6	115.9	103.3	125.0	114.0
04	MASONRY	103.2	134.0	122.4	101.5	134.0	121.7	102.7	135.4	123.1	103.4	135.4	123.3	103.4	134.0	122.5	102.7	134.0	122.2
05	METALS	96.8	125.4	105.6	96.5	125.4	105.4	100.5	126.1	108.3	100.5	126.1	108.3	100.5	125.4	108.1	100.2	125.2	107.9
06	WOOD, PLASTICS & COMPOSITES	98.8	123.4	112.6	98.8	123.4	112.6	98.8	123.4	112.6	98.8	123.4	112.6	98.8	123.4	112.6	98.8	123.4	112.6
07	THERMAL & MOISTURE PROTECTION	98.8	126.1	110.0	98.7	126.0	109.9	98.8	129.6	111.4	98.7	129.6	111.4	98.7	126.1	110.0	98.9	124.6	109.5
08	OPENINGS	101.2	132.1	108.4	104.0	132.1	110.6	101.2	132.1	108.4	101.2	132.1	108.4	101.2	132.1	108.4	104.0	132.1	110.6
0920	Plaster & Gypsum Board	101.7	123.5	116.4	101.7	123.5	116.4	101.7	123.5	116.4	101.7	123.5	116.4	101.7	123.5	116.4	101.7	123.5	116.4
0950, 0980	Ceilings & Acoustic Treatment	86.4	123.5	110.8	84.5	123.5	110.2	86.4	123.5	110.8	86.4	123.5	110.8	86.4	123.5	110.8	84.5	123.5	110.2
0960	Flooring	101.9	131.2	110.3	101.9	131.2	110.3	101.9	131.2	110.3	101.9	131.2	110.3	101.9	131.2	110.3	101.9	133.2	110.9
0970, 0990	Wall Finishes & Painting/Coating	104.9	122.9	115.8	104.9	122.9	115.8	104.9	122.9	115.8	104.9	122.9	115.8	104.9	122.9	115.8	104.9	122.9	115.8
09	FINISHES	97.3	125.0	112.6	96.4	125.0	112.2	97.3	125.0	112.6	97.4	125.0	112.6	97.2	125.0	112.5	97.0	125.4	112.6
COVERS	DIVS. 10 - 14, 25, 28, 41, 43, 44, 46	100.0	112.4	102.5	100.0	112.4	102.5	100.0	112.6	102.6	100.0	112.6	102.6	100.0	112.4	102.5	100.0	112.4	102.5
21, 22, 23	FIRE SUPPRESSION, PLUMBING & HVAC	100.0	116.4	106.6	95.2	116.4	103.8	100.0	116.4	106.6	100.0	116.4	106.6	100.0	116.4	106.6	100.0	116.1	106.5
26, 27, 3370	ELECTRICAL, COMMUNICATIONS & UTIL.	98.4	111.8	105.5	95.3	111.8	104.0	98.5	167.2	134.8	98.5	167.2	134.8	98.0	112.2	105.5	98.5	109.6	104.4
MF2014	WEIGHTED AVERAGE	101.6	120.7	109.9	97.1	120.7	107.4	100.8	128.7	113.0	100.9	128.7	113.0	100.9	120.7	109.5	100.8	120.2	109.2

DIVISION		D.C.			DELAWARE									FLORIDA					
		WASHINGTON			DOVER			NEWARK			WILMINGTON			DAYTONA BEACH			FORT LAUDERDALE		
		200 - 205			199			197			198			321			333		
		MAT.	INST.	TOTAL	MAT.	INST.	TOTAL	MAT.	INST.	TOTAL	MAT.	INST.	TOTAL	MAT.	INST.	TOTAL	MAT.	INST.	TOTAL
015433	CONTRACTOR EQUIPMENT		105.5	105.5		118.0	118.0		118.0	118.0		118.2	118.2		98.2	98.2		91.0	91.0
0241, 31 - 34	SITE & INFRASTRUCTURE, DEMOLITION	103.8	94.5	97.2	102.3	113.3	110.1	102.6	113.3	110.2	99.5	113.6	109.5	105.9	89.2	94.0	96.4	77.1	82.7
0310	Concrete Forming & Accessories	97.4	78.3	80.9	96.3	102.2	101.4	97.0	102.2	101.5	98.1	102.2	101.6	97.6	68.1	72.1	95.7	68.4	72.1
0320	Concrete Reinforcing	105.4	93.8	99.5	94.0	102.9	98.5	91.5	102.9	97.3	93.4	102.9	98.2	91.2	76.6	83.8	88.2	72.1	80.0
0330	Cast-in-Place Concrete	115.4	88.0	104.2	104.6	104.6	104.6	90.0	104.6	96.0	99.2	104.6	101.4	90.2	70.8	82.2	94.6	77.1	87.4
03	CONCRETE	107.7	85.9	97.0	102.2	104.0	103.1	94.6	104.0	99.2	99.6	104.0	101.8	91.6	71.8	81.9	94.5	73.3	84.0
04	MASONRY	98.2	80.0	86.9	107.5	98.1	101.6	103.2	98.1	100.0	108.5	98.1	102.0	99.1	65.7	78.3	102.8	68.3	81.3
05	METALS	99.9	108.6	102.6	102.0	116.7	106.5	103.5	116.7	107.5	102.0	116.7	106.5	104.7	91.1	100.5	102.0	89.6	98.2
06	WOOD, PLASTICS & COMPOSITES	95.3	76.2	84.6	94.0	101.9	98.4	95.8	101.9	99.2	92.0	101.9	97.6	96.3	69.2	81.1	84.3	67.2	74.7
07	THERMAL & MOISTURE PROTECTION	101.7	85.8	95.2	98.8	110.0	103.4	101.7	110.0	105.1	97.9	110.0	102.9	96.2	73.6	86.9	100.3	77.2	90.8
08	OPENINGS	99.9	87.9	97.1	91.3	110.9	95.9	91.6	110.9	96.1	89.9	110.9	94.8	98.6	67.4	91.3	97.5	65.5	90.1
0920	Plaster & Gypsum Board	108.0	75.5	86.0	97.1	101.8	100.3	99.5	101.8	101.1	101.7	101.8	101.8	95.6	68.7	77.4	105.4	66.6	79.2
0950, 0980	Ceilings & Acoustic Treatment	112.2	75.5	88.1	91.3	101.8	98.2	88.7	101.8	97.3	92.4	101.8	98.6	84.9	68.7	74.2	87.9	66.6	73.9
0960	Flooring	104.7	93.6	101.5	99.8	109.7	102.7	96.8	109.7	100.5	104.3	109.7	105.9	110.5	73.6	100.0	105.1	69.3	94.9
0970, 0990	Wall Finishes & Painting/Coating	108.0	82.2	92.4	97.2	106.4	102.8	97.3	106.4	102.8	95.0	106.4	101.9	104.5	74.1	86.2	98.9	70.2	81.6
09	FINISHES	100.8	80.5	89.6	96.2	103.6	100.3	95.0	103.6	99.7	98.7	103.6	101.4	96.9	69.3	81.7	94.9	68.1	80.1
COVERS	DIVS. 10 - 14, 25, 28, 41, 43, 44, 46	100.0	98.6	99.7	100.0	88.0	97.6	100.0	88.0	97.6	100.0	88.0	97.6	100.0	84.4	96.9	100.0	87.4	97.5
21, 22, 23	FIRE SUPPRESSION, PLUMBING & HVAC	100.1	92.1	96.9	100.0	117.4	107.0	100.1	117.4	107.1	100.1	117.4	107.1	99.9	76.1	90.3	100.0	66.5	86.5
26, 27, 3370	ELECTRICAL, COMMUNICATIONS & UTIL.	101.5	105.8	103.8	96.6	111.9	104.7	98.7	111.9	105.7	99.8	111.9	106.2	95.8	54.9	74.2	96.0	72.7	83.7
MF2014	WEIGHTED AVERAGE	101.1	91.9	97.1	99.3	109.2	103.6	98.8	109.2	103.3	99.3	109.2	103.6	99.0	72.9	87.6	98.6	72.6	87.2

DIVISION		FLORIDA																	
		FORT MYERS			GAINESVILLE			JACKSONVILLE			LAKELAND			MELBOURNE			MIAMI		
		339,341			326,344			320,322			338			329			330 - 332,340		
		MAT.	INST.	TOTAL	MAT.	INST.	TOTAL	MAT.	INST.	TOTAL	MAT.	INST.	TOTAL	MAT.	INST.	TOTAL	MAT.	INST.	TOTAL
015433	CONTRACTOR EQUIPMENT		98.2	98.2		98.2	98.2		98.2	98.2		98.2	98.2		98.2	98.2		91.0	91.0
0241, 31 - 34	SITE & INFRASTRUCTURE, DEMOLITION	107.3	88.3	93.8	113.9	88.3	95.7	106.0	88.6	93.6	109.2	88.7	94.6	112.7	88.5	95.5	99.5	76.9	83.4
0310	Concrete Forming & Accessories	91.5	74.8	77.1	92.7	54.3	59.6	97.4	54.7	60.5	88.0	75.3	77.1	93.8	69.8	73.1	100.7	68.7	73.1
0320	Concrete Reinforcing	89.3	92.7	91.0	96.8	64.8	80.5	91.2	64.9	77.8	91.5	93.6	92.5	92.3	76.6	84.3	94.3	72.2	83.0
0330	Cast-in-Place Concrete	98.7	68.2	86.2	103.6	62.6	86.7	91.1	68.3	81.8	100.9	69.6	88.1	108.7	73.1	94.1	95.5	78.1	88.3
03	CONCRETE	95.1	76.9	86.2	102.6	60.7	82.0	92.1	62.8	77.7	96.8	77.8	87.5	103.0	73.3	88.4	96.1	73.7	85.1
04	MASONRY	95.8	61.9	74.7	114.0	61.2	81.1	99.1	61.2	75.5	114.1	75.8	90.2	97.5	69.8	80.2	103.1	71.9	83.7
05	METALS	104.2	96.8	102.0	103.6	85.0	97.9	103.2	85.4	97.7	104.1	97.8	102.2	113.4	91.3	106.6	102.4	88.5	98.1
06	WOOD, PLASTICS & COMPOSITES	81.1	77.4	79.0	89.9	51.8	68.5	96.3	51.8	71.4	76.5	77.4	77.0	91.5	69.2	79.0	90.6	67.2	77.5
07	THERMAL & MOISTURE PROTECTION	100.1	80.0	91.9	96.5	61.7	82.2	96.4	62.3	82.4	100.1	84.6	93.7	96.6	76.0	88.2	101.6	71.6	89.3
08	OPENINGS	98.4	74.4	92.8	96.9	52.7	86.6	98.6	55.8	88.6	98.4	75.1	93.0	97.8	70.7	91.5	99.7	65.5	91.7
0920	Plaster & Gypsum Board	101.0	77.1	84.9	91.3	50.7	63.9	95.6	50.7	65.3	97.4	77.1	83.7	91.3	68.7	76.0	103.2	66.6	78.4
0950, 0980	Ceilings & Acoustic Treatment	82.7	77.1	79.0	79.5	50.7	60.6	84.9	50.7	62.5	82.7	77.1	79.0	83.3	68.7	73.7	92.2	66.6	75.4
0960	Flooring	102.3	55.1	88.8	108.1	43.1	89.5	110.5	64.5	97.4	100.3	56.5	87.8	108.3	73.6	98.4	107.3	73.0	97.5
0970, 0990	Wall Finishes & Painting/Coating	103.7	67.6	81.9	104.5	67.6	82.3	104.5	67.6	82.3	103.7	67.6	81.9	104.5	91.9	96.9	96.0	70.2	80.5
09	FINISHES	94.4	70.7	81.3	95.3	52.5	71.7	97.0	56.7	74.8	93.5	71.0	81.1	95.9	72.3	82.9	97.1	69.2	81.7
COVERS	DIVS. 10 - 14, 25, 28, 41, 43, 44, 46	100.0	72.6	94.5	100.0	82.8	96.5	100.0	81.0	96.2	100.0	72.6	94.5	100.0	85.8	97.1	100.0	88.0	97.6
21, 22, 23	FIRE SUPPRESSION, PLUMBING & HVAC	97.4	64.3	84.1	98.8	64.1	84.8	99.9	64.2	85.5	97.4	80.8	90.7	99.9	78.2	91.2	100.0	66.4	86.4
26, 27, 3370	ELECTRICAL, COMMUNICATIONS & UTIL.	98.1	62.8	79.5	96.1	71.8	83.3	95.5	62.2	77.9	96.3	61.6	77.9	97.0	67.8	81.6	100.1	74.9	86.8
MF2014	WEIGHTED AVERAGE	98.6	72.6	87.3	100.3	66.7	85.7	98.8	66.4	84.7	99.4	77.8	90.0	101.7	76.3	90.6	99.8	73.1	88.2

DIVISION		FLORIDA																	
		ORLANDO			PANAMA CITY			PENSACOLA			SARASOTA			ST. PETERSBURG			TALLAHASSEE		
		327 - 328,347			324			325			342			337			323		
		MAT.	INST.	TOTAL	MAT.	INST.	TOTAL	MAT.	INST.	TOTAL	MAT.	INST.	TOTAL	MAT.	INST.	TOTAL	MAT.	INST.	TOTAL
015433	CONTRACTOR EQUIPMENT		98.2	98.2		98.2	98.2		98.2	98.2		98.2	98.2		98.2	98.2		98.2	98.2
0241, 31 - 34	SITE & INFRASTRUCTURE, DEMOLITION	107.6	88.4	93.9	117.6	87.4	96.1	117.6	87.9	96.5	114.0	88.4	95.8	110.9	88.1	94.7	105.1	87.7	92.7
0310	Concrete Forming & Accessories	101.4	72.0	76.1	96.7	43.9	51.1	94.6	51.8	57.7	96.0	75.0	77.9	94.9	51.0	57.1	99.9	44.1	51.8
0320	Concrete Reinforcing	96.6	73.9	85.0	95.3	72.5	83.7	97.7	73.0	85.1	92.5	93.5	93.0	91.5	86.1	88.7	98.3	64.7	81.2
0330	Cast-in-Place Concrete	112.1	70.6	95.1	95.7	56.7	79.7	118.2	65.9	96.8	106.6	69.5	91.4	102.0	64.5	86.6	97.2	56.4	80.5
03	CONCRETE	103.4	73.0	88.4	100.8	55.4	78.5	110.7	62.2	86.9	100.2	77.5	89.1	98.5	63.8	81.4	97.3	54.0	76.0
04	MASONRY	100.6	65.7	78.8	103.9	47.3	68.6	124.7	54.5	80.9	99.8	75.8	84.8	157.9	49.0	90.0	103.6	53.8	72.6
05	METALS	102.4	89.6	98.5	104.4	86.5	98.9	105.6	87.9	100.2	105.2	97.5	102.8	105.0	92.9	101.3	102.2	84.6	96.8
06	WOOD, PLASTICS & COMPOSITES	95.8	75.2	84.2	95.1	41.8	65.2	92.7	51.5	69.7	95.7	77.4	85.4	85.5	50.1	65.7	95.0	41.3	64.9
07	THERMAL & MOISTURE PROTECTION	94.9	74.3	86.4	96.7	56.6	80.2	96.6	62.3	82.5	98.1	84.6	92.5	100.3	57.8	82.8	102.6	71.5	89.9
08	OPENINGS	101.4	69.2	93.9	96.5	46.5	84.9	96.5	56.9	87.3	99.7	74.2	93.8	98.4	60.9	89.6	100.2	47.3	87.9
0920	Plaster & Gypsum Board	99.7	74.9	82.9	94.5	40.4	57.9	97.4	50.5	65.7	97.8	77.1	83.8	103.5	49.0	66.7	108.1	40.0	62.0
0950, 0980	Ceilings & Acoustic Treatment	91.4	74.9	80.5	83.3	40.4	55.1	83.3	50.5	61.7	86.2	77.1	80.3	84.5	49.0	61.2	94.1	40.0	58.5
0960	Flooring	104.1	73.6	95.4	110.1	43.0	90.9	106.1	63.0	93.7	111.1	57.9	95.9	104.1	55.2	90.1	111.1	62.0	97.0
0970, 0990	Wall Finishes & Painting/Coating	103.4	70.2	83.4	104.5	64.9	80.6	104.5	67.6	82.3	109.5	67.6	84.2	103.7	60.8	77.8	99.5	67.6	80.3
09	FINISHES	98.2	72.4	84.0	97.5	44.8	68.5	96.4	54.8	73.5	99.9	71.2	84.1	95.9	51.7	71.6	100.7	48.5	72.0
COVERS	DIVS. 10 - 14, 25, 28, 41, 43, 44, 46	100.0	85.1	97.0	100.0	46.2	89.1	100.0	46.3	89.1	100.0	72.6	94.5	100.0	55.8	91.1	100.0	65.8	93.1
21, 22, 23	FIRE SUPPRESSION, PLUMBING & HVAC	99.9	56.6	82.4	99.9	52.3	80.7	99.9	52.5	80.8	99.9	64.8	85.7	100.0	58.4	83.2	100.0	38.9	75.4
26, 27, 3370	ELECTRICAL, COMMUNICATIONS & UTIL.	97.6	60.0	77.7	94.5	59.3	75.9	98.5	55.8	75.9	97.2	61.6	78.4	96.3	61.6	77.9	103.3	60.0	80.4
MF2014	WEIGHTED AVERAGE	100.5	70.0	87.2	100.2	57.8	81.7	102.7	61.2	84.6	100.8	74.3	89.2	102.6	63.3	85.5	100.8	56.9	81.7

DIVISION		FLORIDA						GEORGIA											
		TAMPA			WEST PALM BEACH			ALBANY			ATHENS			ATLANTA			AUGUSTA		
		335 - 336,346			334,349			317,398			306			300 - 303,399			308 - 309		
		MAT.	INST.	TOTAL	MAT.	INST.	TOTAL	MAT.	INST.	TOTAL	MAT.	INST.	TOTAL	MAT.	INST.	TOTAL	MAT.	INST.	TOTAL
015433	CONTRACTOR EQUIPMENT		98.2	98.2		91.0	91.0		91.9	91.9		94.2	94.2		94.7	94.7		94.2	94.2
0241, 31 - 34	SITE & INFRASTRUCTURE, DEMOLITION	111.4	88.6	95.2	93.2	77.1	81.8	98.8	78.7	84.5	100.9	93.4	95.6	97.6	94.8	95.7	94.5	93.2	93.6
0310	Concrete Forming & Accessories	97.7	75.6	78.6	99.2	68.1	72.4	90.4	43.6	50.0	92.9	45.9	52.4	96.7	73.5	76.7	94.2	65.5	69.4
0320	Concrete Reinforcing	88.2	93.6	91.0	90.7	71.8	81.1	90.6	80.2	85.3	95.3	77.9	86.5	94.6	81.2	87.8	95.7	71.8	83.6
0330	Cast-in-Place Concrete	99.7	69.7	87.4	90.0	74.3	83.5	92.4	54.5	76.8	107.8	55.6	86.4	107.8	70.8	92.6	101.9	49.9	80.5
03	CONCRETE	97.0	77.9	87.6	91.3	72.1	81.9	93.1	55.8	74.8	104.8	56.0	80.8	102.2	74.1	88.4	97.5	61.7	79.9
04	MASONRY	103.9	75.8	86.4	102.3	66.6	80.0	102.5	49.1	69.2	81.5	53.8	64.2	95.2	66.2	77.1	95.5	43.3	63.0
05	METALS	104.0	98.1	102.2	101.1	89.3	97.4	105.0	85.5	99.0	91.9	73.3	86.2	92.8	77.0	88.0	91.6	71.6	85.5
06	WOOD, PLASTICS & COMPOSITES	89.3	77.4	82.6	89.2	67.2	76.9	85.0	38.0	58.7	91.8	40.9	63.3	96.1	75.8	84.7	93.3	70.9	80.8
07	THERMAL & MOISTURE PROTECTION	100.5	84.6	94.0	100.1	70.2	87.8	95.9	60.3	81.3	94.1	52.8	77.2	94.0	72.0	85.0	93.7	57.0	78.7
08	OPENINGS	99.7	79.2	94.9	96.8	65.5	89.5	91.1	44.2	80.2	89.5	45.7	79.3	94.7	71.9	89.4	89.5	62.2	83.1
0920	Plaster & Gypsum Board	106.1	77.1	86.5	110.2	66.6	80.7	97.5	36.5	56.3	98.7	39.4	58.7	100.9	75.4	83.6	99.8	70.4	79.9
0950, 0980	Ceilings & Acoustic Treatment	87.9	77.1	80.8	82.7	66.6	72.1	84.0	36.5	52.8	97.2	39.4	59.2	97.2	75.4	82.9	98.1	70.4	79.9
0960	Flooring	105.1	56.5	91.2	106.8	66.4	95.3	110.9	47.8	92.9	96.4	53.9	84.2	97.7	65.2	88.4	96.6	46.6	82.3
0970, 0990	Wall Finishes & Painting/Coating	103.7	67.6	81.9	98.9	70.2	81.6	105.3	52.8	73.6	106.0	46.0	69.8	106.0	85.1	93.4	106.0	46.0	69.8
09	FINISHES	97.4	71.0	82.8	94.7	67.5	79.7	98.1	43.2	67.9	95.4	45.7	68.0	95.7	73.4	83.4	95.2	60.5	76.1
COVERS	DIVS. 10 - 14, 25, 28, 41, 43, 44, 46	100.0	85.1	97.0	100.0	87.4	97.5	100.0	80.2	96.0	100.0	77.3	95.4	100.0	85.6	97.1	100.0	78.5	95.7
21, 22, 23	FIRE SUPPRESSION, PLUMBING & HVAC	100.0	80.8	92.2	97.4	62.7	83.4	99.9	68.0	87.0	95.2	69.6	84.9	99.9	70.6	88.1	100.0	61.2	84.4
26, 27, 3370	ELECTRICAL, COMMUNICATIONS & UTIL.	96.0	61.6	77.8	97.2	72.7	84.3	96.9	58.7	76.7	99.7	69.5	83.8	99.0	71.9	84.7	100.4	61.8	80.0
MF2014	WEIGHTED AVERAGE	100.1	78.4	90.6	97.4	71.1	85.9	98.5	61.3	82.3	95.5	63.9	81.7	97.6	74.4	87.5	96.3	63.6	82.1

	DIVISION	GEORGIA																	
		COLUMBUS			DALTON			GAINESVILLE			MACON			SAVANNAH			STATESBORO		
		318 - 319			307			305			310 - 312			313 - 314			304		
		MAT.	INST.	TOTAL	MAT.	INST.	TOTAL	MAT.	INST.	TOTAL	MAT.	INST.	TOTAL	MAT.	INST.	TOTAL	MAT.	INST.	TOTAL
015433	CONTRACTOR EQUIPMENT		91.9	91.9		106.1	106.1		94.2	94.2		102.3	102.3		92.8	92.8		93.4	93.4
0241, 31 - 34	SITE & INFRASTRUCTURE, DEMOLITION	98.7	78.7	84.5	101.1	97.8	98.8	100.8	93.3	95.5	99.8	93.5	95.3	100.6	80.1	86.1	102.0	77.5	84.6
0310	Concrete Forming & Accessories	90.3	54.3	59.3	85.7	46.7	52.1	96.4	42.9	50.2	89.9	51.9	57.1	92.1	50.0	55.8	80.3	51.4	55.4
0320	Concrete Reinforcing	90.9	80.6	85.7	94.8	73.9	84.2	95.1	77.8	86.3	92.1	80.3	86.1	98.1	71.9	84.7	94.4	41.8	67.6
0330	Cast-in-Place Concrete	92.1	53.9	76.4	104.7	50.2	82.3	113.3	52.9	88.5	90.9	65.6	80.5	100.2	55.2	81.7	107.6	59.7	87.9
03	CONCRETE	93.0	60.4	77.0	103.9	54.6	79.7	106.7	53.7	80.6	92.6	63.3	78.2	98.0	57.3	78.0	104.2	54.0	79.6
04	MASONRY	102.6	55.2	73.1	83.8	36.7	54.4	89.5	54.2	67.5	115.9	44.5	71.4	98.8	50.9	69.0	84.7	40.6	57.2
05	METALS	104.5	86.3	98.9	92.8	82.8	89.7	91.2	72.5	85.4	100.1	85.9	95.7	101.3	82.4	95.5	96.3	72.6	89.0
06	WOOD, PLASTICS & COMPOSITES	85.0	52.1	66.6	75.2	47.5	59.7	95.9	37.9	63.4	91.8	51.2	69.0	96.0	45.9	68.0	68.5	53.9	60.3
07	THERMAL & MOISTURE PROTECTION	95.9	63.5	82.6	96.1	51.5	77.8	94.1	54.8	78.0	94.4	62.2	81.2	95.0	56.6	79.2	94.7	51.1	76.8
08	OPENINGS	91.1	59.0	83.7	90.1	49.5	80.6	89.5	40.0	77.9	90.0	52.8	81.3	94.9	48.4	84.1	91.1	40.6	79.4
0920	Plaster & Gypsum Board	97.5	51.1	66.1	86.2	46.2	59.2	100.9	36.3	57.3	103.0	50.1	67.2	95.6	44.7	61.2	86.9	52.8	63.9
0950, 0980	Ceilings & Acoustic Treatment	84.0	51.1	62.4	109.3	46.2	67.8	97.2	36.3	57.2	79.2	50.1	60.1	91.5	44.7	60.8	105.7	52.8	71.0
0960	Flooring	110.9	50.4	93.6	96.8	46.9	82.5	97.6	46.6	83.0	87.9	47.8	76.5	108.6	46.4	90.8	115.6	44.9	95.4
0970, 0990	Wall Finishes & Painting/Coating	105.3	66.5	81.9	96.3	59.8	74.2	106.0	46.0	69.8	107.7	52.8	74.6	103.8	58.3	76.3	103.5	39.2	64.7
09	FINISHES	98.0	53.5	73.5	103.0	47.1	72.2	95.9	41.6	66.0	86.7	49.9	66.4	98.2	48.9	71.0	106.9	49.0	75.0
COVERS	DIVS. 10 - 14, 25, 28, 41, 43, 44, 46	100.0	81.8	96.3	100.0	22.8	84.4	100.0	39.4	87.8	100.0	80.2	96.0	100.0	78.0	95.6	100.0	43.5	88.6
21, 22, 23	FIRE SUPPRESSION, PLUMBING & HVAC	99.9	63.4	85.2	95.2	57.0	79.8	95.2	69.5	84.8	99.9	66.7	86.6	100.0	61.8	84.6	95.7	56.7	80.0
26, 27, 3370	ELECTRICAL, COMMUNICATIONS & UTIL.	97.1	69.6	82.5	109.7	67.7	87.5	99.7	69.5	83.8	96.1	61.2	77.7	101.7	57.4	78.3	99.9	57.4	77.4
MF2014	WEIGHTED AVERAGE	98.4	65.4	84.0	97.2	59.0	80.5	96.0	61.6	81.0	97.2	64.7	83.0	99.2	60.9	82.5	97.3	55.8	79.2

	DIVISION	GEORGIA						HAWAII									IDAHO		
		VALDOSTA			WAYCROSS			HILO			HONOLULU			STATES & POSS., GUAM			BOISE		
		316			315			967			968			969			836 - 837		
		MAT.	INST.	TOTAL	MAT.	INST.	TOTAL	MAT.	INST.	TOTAL	MAT.	INST.	TOTAL	MAT.	INST.	TOTAL	MAT.	INST.	TOTAL
015433	CONTRACTOR EQUIPMENT		91.9	91.9		91.9	91.9		99.5	99.5		99.5	99.5		164.5	164.5		98.2	98.2
0241, 31 - 34	SITE & INFRASTRUCTURE, DEMOLITION	107.9	78.8	87.2	104.7	77.4	85.3	144.8	106.5	117.6	155.4	106.5	120.6	184.9	103.7	127.2	85.5	96.3	93.2
0310	Concrete Forming & Accessories	81.0	44.1	49.2	82.8	64.7	67.2	111.2	135.6	132.2	123.6	135.6	134.0	113.7	63.6	70.5	100.8	77.4	80.7
0320	Concrete Reinforcing	92.8	76.4	84.4	92.8	72.7	82.5	123.5	117.9	120.7	132.5	117.9	125.1	214.4	30.2	120.7	99.2	80.4	89.6
0330	Cast-in-Place Concrete	90.5	55.9	76.3	102.2	48.2	80.0	198.3	125.7	168.5	163.7	125.7	148.1	171.9	105.5	144.7	91.6	88.4	90.3
03	CONCRETE	98.1	55.8	77.3	101.2	61.4	81.6	158.1	127.7	143.2	150.8	127.7	139.5	159.9	72.3	116.9	99.1	81.9	90.7
04	MASONRY	108.9	50.5	72.5	109.7	39.3	65.8	150.5	128.3	136.7	150.6	128.3	136.7	213.9	43.5	107.7	121.4	84.2	98.2
05	METALS	104.1	84.0	97.9	103.1	77.7	95.3	108.1	107.3	107.8	120.7	107.3	116.5	139.2	76.2	119.8	102.2	80.9	95.6
06	WOOD, PLASTICS & COMPOSITES	73.4	37.6	53.4	75.0	72.4	73.5	114.5	139.2	128.3	134.4	139.2	137.1	127.1	67.4	93.7	93.1	75.9	83.4
07	THERMAL & MOISTURE PROTECTION	96.1	62.4	82.3	95.9	50.3	77.2	120.8	124.6	122.4	138.4	124.6	132.7	140.4	65.2	109.6	94.6	81.4	89.2
08	OPENINGS	87.3	42.5	76.9	87.4	57.8	80.5	111.3	132.8	116.3	121.8	132.8	124.4	116.6	54.1	102.1	99.1	71.2	92.6
0920	Plaster & Gypsum Board	90.4	36.2	53.8	90.4	72.0	78.0	109.7	140.1	130.3	151.3	140.1	143.7	219.9	55.5	108.8	92.4	75.1	80.7
0950, 0980	Ceilings & Acoustic Treatment	81.5	36.2	51.7	79.6	72.0	74.6	121.5	140.1	133.8	132.0	140.1	137.3	237.3	55.5	117.8	99.1	75.1	83.3
0960	Flooring	104.7	47.8	88.4	105.9	29.9	84.1	116.0	139.9	122.8	132.2	139.9	134.4	135.5	46.4	110.0	96.3	83.7	92.7
0970, 0990	Wall Finishes & Painting/Coating	105.3	52.2	73.2	105.3	46.0	69.5	110.4	143.8	130.6	119.4	143.8	134.2	115.9	35.9	67.6	103.2	39.5	64.7
09	FINISHES	95.6	43.6	67.0	95.2	57.4	74.4	113.7	138.6	127.4	128.4	138.6	134.0	188.8	58.9	117.2	96.0	74.8	84.3
COVERS	DIVS. 10 - 14, 25, 28, 41, 43, 44, 46	100.0	77.1	95.4	100.0	51.5	90.2	100.0	116.0	103.2	100.0	116.0	103.2	100.0	75.2	95.0	100.0	87.4	97.5
21, 22, 23	FIRE SUPPRESSION, PLUMBING & HVAC	99.9	70.0	87.9	96.9	57.1	80.8	100.2	107.7	103.2	100.3	107.7	103.3	102.6	37.4	76.3	100.0	71.8	88.6
26, 27, 3370	ELECTRICAL, COMMUNICATIONS & UTIL.	95.0	56.2	74.5	99.6	57.4	77.3	106.5	121.1	114.2	107.9	121.1	114.9	153.5	41.2	94.1	98.5	72.8	84.9
MF2014	WEIGHTED AVERAGE	98.5	61.4	82.3	98.3	59.4	81.4	114.8	120.2	117.2	119.5	120.2	119.8	136.2	58.0	102.1	100.1	78.6	90.7

	DIVISION	IDAHO															ILLINOIS		
		COEUR D'ALENE			IDAHO FALLS			LEWISTON			POCATELLO			TWIN FALLS			BLOOMINGTON		
		838			834			835			832			833			617		
		MAT.	INST.	TOTAL	MAT.	INST.	TOTAL	MAT.	INST.	TOTAL	MAT.	INST.	TOTAL	MAT.	INST.	TOTAL	MAT.	INST.	TOTAL
015433	CONTRACTOR EQUIPMENT		92.8	92.8		98.2	98.2		92.8	92.8		98.2	98.2		98.2	98.2		101.6	101.6
0241, 31 - 34	SITE & INFRASTRUCTURE, DEMOLITION	83.1	91.7	89.2	83.1	96.1	92.4	89.9	92.5	91.7	85.8	96.3	93.3	92.1	97.2	95.8	97.4	98.4	98.1
0310	Concrete Forming & Accessories	112.6	81.1	85.5	94.7	78.1	80.4	117.9	82.2	87.1	101.0	77.1	80.4	102.0	54.7	61.2	84.3	117.2	112.7
0320	Concrete Reinforcing	106.2	96.5	101.3	101.0	80.0	90.3	106.2	96.8	101.4	99.6	80.2	89.7	101.3	79.9	90.4	94.6	110.7	102.8
0330	Cast-in-Place Concrete	99.0	87.2	94.1	87.2	74.3	81.9	102.9	86.0	95.9	94.1	88.3	91.7	96.6	63.9	83.1	102.8	114.8	107.7
03	CONCRETE	105.7	86.1	96.1	91.3	77.2	84.4	109.3	86.2	98.0	98.3	81.7	90.1	105.8	63.3	84.9	100.1	115.2	107.5
04	MASONRY	122.9	83.9	98.6	116.5	81.0	94.4	123.4	85.9	100.0	118.8	81.1	95.3	121.6	81.1	96.4	120.3	118.3	119.1
05	METALS	96.3	86.9	93.4	110.0	79.1	100.5	95.7	88.0	93.4	110.1	80.3	100.9	110.1	78.9	100.5	97.6	113.6	102.5
06	WOOD, PLASTICS & COMPOSITES	96.8	81.0	88.0	86.8	78.8	82.3	102.4	81.0	90.4	93.1	75.9	83.4	94.2	47.2	67.9	86.4	115.4	102.6
07	THERMAL & MOISTURE PROTECTION	148.3	80.8	120.6	94.3	71.7	85.0	148.5	81.4	121.0	94.7	73.7	86.1	95.4	74.7	86.9	98.4	112.8	104.3
08	OPENINGS	117.8	73.0	107.4	102.8	68.0	94.7	117.8	75.6	108.0	99.8	66.3	92.0	102.8	48.8	90.2	94.9	104.9	97.2
0920	Plaster & Gypsum Board	161.4	80.5	106.7	78.5	78.1	78.2	162.5	80.5	107.0	80.7	75.1	76.9	82.3	45.6	57.5	91.3	115.8	107.9
0950, 0980	Ceilings & Acoustic Treatment	129.0	80.5	97.1	97.2	78.1	84.6	129.0	80.5	97.1	104.7	75.1	85.2	99.7	45.6	64.1	88.0	115.8	106.3
0960	Flooring	138.3	45.9	111.9	96.4	43.1	81.2	141.2	93.8	127.7	99.6	83.7	95.1	100.7	43.1	84.3	93.8	118.7	100.9
0970, 0990	Wall Finishes & Painting/Coating	123.8	67.8	90.0	103.3	40.3	65.3	123.8	67.8	90.0	103.1	41.1	65.7	103.3	37.3	63.4	93.8	126.4	113.5
09	FINISHES	161.8	73.3	113.0	93.3	68.7	79.7	162.9	82.8	118.8	96.6	75.0	84.7	96.7	49.7	70.8	93.3	118.9	107.4
COVERS	DIVS. 10 - 14, 25, 28, 41, 43, 44, 46	100.0	87.4	97.5	100.0	47.7	89.4	100.0	87.6	97.5	100.0	87.4	97.5	100.0	44.2	88.7	100.0	105.2	101.0
21, 22, 23	FIRE SUPPRESSION, PLUMBING & HVAC	99.5	82.6	92.7	100.9	71.7	89.1	100.7	85.6	94.6	99.9	71.8	88.6	99.9	69.3	87.6	95.1	108.8	100.7
26, 27, 3370	ELECTRICAL, COMMUNICATIONS & UTIL.	91.0	77.6	83.9	90.8	70.4	80.0	88.9	80.7	84.5	96.2	70.4	82.6	92.3	60.1	75.3	94.2	94.6	94.4
MF2014	WEIGHTED AVERAGE	107.9	82.2	96.7	99.8	74.7	88.9	108.7	84.9	98.3	101.1	77.4	90.8	102.2	67.3	87.0	97.5	109.4	102.7

DIVISION		ILLINOIS																	
		CARBONDALE			CENTRALIA			CHAMPAIGN			CHICAGO			DECATUR			EAST ST. LOUIS		
		629			628			618 - 619			606 - 608			625			620 - 622		
		MAT.	INST.	TOTAL	MAT.	INST.	TOTAL	MAT.	INST.	TOTAL	MAT.	INST.	TOTAL	MAT.	INST.	TOTAL	MAT.	INST.	TOTAL
015433	CONTRACTOR EQUIPMENT		108.0	108.0		108.0	108.0		102.4	102.4		94.3	94.3		102.4	102.4		108.0	108.0
0241, 31 - 34	SITE & INFRASTRUCTURE, DEMOLITION	98.7	99.5	99.2	99.0	100.1	99.8	106.4	99.0	101.2	107.0	95.7	99.0	93.2	99.1	97.4	101.2	99.8	100.2
0310	Concrete Forming & Accessories	90.0	109.0	106.4	91.6	112.8	109.9	90.7	115.2	111.8	96.6	156.7	148.4	91.8	114.9	111.8	87.6	114.8	111.1
0320	Concrete Reinforcing	91.2	111.6	101.6	91.2	111.8	101.7	94.6	105.1	99.9	99.0	158.1	129.1	89.3	102.7	96.1	91.1	109.9	100.6
0330	Cast-in-Place Concrete	97.2	102.5	99.4	97.7	117.3	105.7	119.2	109.5	115.2	107.5	149.4	124.7	106.1	110.5	107.9	99.3	117.0	106.6
03	CONCRETE	90.1	107.9	98.9	90.6	114.8	102.5	113.1	111.3	112.2	102.6	153.2	127.5	101.3	111.1	106.1	91.7	115.2	103.2
04	MASONRY	83.6	108.6	99.2	83.6	116.7	104.2	145.3	117.7	128.1	101.1	156.6	135.7	79.6	114.7	101.5	83.9	116.6	104.3
05	METALS	96.2	119.6	103.4	96.2	120.8	103.8	97.6	108.7	101.0	94.3	134.3	106.6	99.9	108.1	102.4	97.3	119.4	104.1
06	WOOD, PLASTICS & COMPOSITES	91.7	106.1	99.8	94.1	109.7	102.8	93.5	113.9	104.9	104.0	155.9	133.1	91.9	113.9	104.2	89.2	112.1	102.0
07	THERMAL & MOISTURE PROTECTION	97.1	101.5	98.9	97.2	111.7	103.1	99.0	113.2	104.9	98.0	145.1	117.3	103.2	109.3	105.7	97.2	110.2	102.5
08	OPENINGS	89.4	114.2	95.2	89.4	116.2	95.7	95.5	111.2	99.2	105.3	158.7	117.8	100.9	110.6	103.2	89.5	116.9	95.9
0920	Plaster & Gypsum Board	97.2	106.2	103.3	98.3	110.0	106.2	93.7	114.2	107.6	92.7	157.6	136.6	100.0	114.2	109.6	96.1	112.4	107.2
0950, 0980	Ceilings & Acoustic Treatment	91.4	106.2	101.2	91.4	110.0	103.6	88.0	114.2	105.2	99.2	157.6	137.6	98.1	114.2	108.7	91.4	112.4	105.2
0960	Flooring	122.2	120.2	121.6	123.2	115.7	121.0	96.7	120.2	103.4	96.9	148.0	111.5	109.1	116.7	111.3	121.2	115.7	119.6
0970, 0990	Wall Finishes & Painting/Coating	112.6	97.4	103.5	112.6	106.6	109.0	93.8	110.0	103.6	91.9	153.0	128.8	102.5	106.5	104.9	112.6	103.5	107.1
09	FINISHES	101.8	108.6	105.5	102.2	112.6	107.9	95.2	115.9	106.6	98.0	155.6	129.7	101.8	115.0	109.1	101.4	113.8	108.2
COVERS	DIVS. 10 - 14, 25, 28, 41, 43, 44, 46	100.0	102.6	100.5	100.0	104.1	100.8	100.0	104.4	100.9	100.0	124.5	105.0	100.0	104.3	100.9	100.0	104.4	100.9
21, 22, 23	FIRE SUPPRESSION, PLUMBING & HVAC	95.1	106.1	99.5	95.1	95.3	95.2	95.1	105.5	99.3	99.8	133.9	113.6	99.9	98.5	99.4	99.9	98.8	99.4
26, 27, 3370	ELECTRICAL, COMMUNICATIONS & UTIL.	95.4	107.7	101.9	96.8	107.7	102.6	97.4	94.9	96.1	98.3	133.9	117.1	99.5	90.4	94.7	96.4	103.5	100.1
MF2014	WEIGHTED AVERAGE	94.7	107.9	100.5	95.0	108.5	100.9	101.0	107.6	103.9	99.8	139.7	117.2	99.2	104.9	101.7	96.4	108.7	101.8

DIVISION		ILLINOIS																	
		EFFINGHAM			GALESBURG			JOLIET			KANKAKEE			LA SALLE			NORTH SUBURBAN		
		624			614			604			609			613			600 - 603		
		MAT.	INST.	TOTAL	MAT.	INST.	TOTAL	MAT.	INST.	TOTAL	MAT.	INST.	TOTAL	MAT.	INST.	TOTAL	MAT.	INST.	TOTAL
015433	CONTRACTOR EQUIPMENT		102.4	102.4		101.6	101.6		92.5	92.5		92.5	92.5		101.6	101.6		92.5	92.5
0241, 31 - 34	SITE & INFRASTRUCTURE, DEMOLITION	97.6	98.9	98.5	99.9	98.3	98.8	106.9	94.9	98.4	100.5	94.6	96.3	99.2	99.2	99.2	106.1	94.9	98.1
0310	Concrete Forming & Accessories	96.3	114.0	111.6	90.6	116.6	113.1	98.4	159.6	151.2	91.8	143.2	136.2	104.6	125.3	122.4	97.8	154.0	146.3
0320	Concrete Reinforcing	92.2	99.9	96.1	94.1	110.6	102.5	99.0	150.8	125.4	99.8	147.4	124.0	94.3	144.7	119.9	99.0	156.5	128.3
0330	Cast-in-Place Concrete	105.7	108.4	106.8	105.9	106.2	106.0	107.4	146.5	123.5	100.1	132.5	113.4	105.7	122.5	112.6	107.5	141.2	121.3
03	CONCRETE	102.1	109.4	105.7	103.2	112.0	107.5	102.7	152.1	126.9	96.7	139.3	117.7	104.1	127.9	115.8	102.7	148.8	125.4
04	MASONRY	88.2	108.5	100.8	120.5	118.0	118.9	104.3	148.5	131.8	100.6	141.2	125.9	120.5	125.1	123.4	101.1	144.1	127.9
05	METALS	97.0	105.6	99.7	97.6	113.0	102.4	92.2	129.1	103.6	92.2	126.6	102.8	97.7	131.6	108.1	93.4	131.4	105.1
06	WOOD, PLASTICS & COMPOSITES	94.2	113.9	105.2	93.3	115.5	105.7	105.7	161.0	136.7	97.8	142.2	122.7	108.8	123.7	117.1	104.0	155.7	132.9
07	THERMAL & MOISTURE PROTECTION	102.7	106.6	104.3	98.5	107.6	102.2	97.8	141.6	115.8	97.0	135.7	112.9	98.7	117.8	106.5	98.2	139.3	115.1
08	OPENINGS	94.9	109.9	98.4	94.9	111.0	98.6	102.8	159.6	116.0	95.4	148.4	107.7	94.9	129.5	102.9	102.9	158.2	115.8
0920	Plaster & Gypsum Board	99.7	114.2	109.5	93.7	115.9	108.7	89.9	162.9	139.2	86.7	143.4	125.0	100.1	124.3	116.5	92.7	157.3	136.4
0950, 0980	Ceilings & Acoustic Treatment	91.4	114.2	106.4	88.0	115.9	106.3	99.2	162.9	141.0	99.2	143.4	128.3	88.0	124.3	111.9	99.2	157.3	137.4
0960	Flooring	110.1	120.2	113.0	96.6	118.7	102.9	96.5	140.3	109.1	93.7	131.9	104.6	102.8	122.6	108.5	96.9	140.3	109.3
0970, 0990	Wall Finishes & Painting/Coating	102.5	104.2	103.5	93.8	95.1	94.5	90.1	152.5	127.8	90.1	126.4	112.0	93.8	126.4	113.5	91.9	152.5	128.4
09	FINISHES	100.6	115.2	108.7	94.6	115.4	106.1	97.4	156.9	130.2	95.7	137.9	119.0	97.3	124.3	112.2	97.9	152.7	128.1
COVERS	DIVS. 10 - 14, 25, 28, 41, 43, 44, 46	100.0	70.8	94.1	100.0	105.1	101.0	100.0	124.5	105.0	100.0	120.8	104.2	100.0	102.5	100.5	100.0	122.3	104.5
21, 22, 23	FIRE SUPPRESSION, PLUMBING & HVAC	95.2	102.8	98.2	95.1	105.3	99.2	99.9	131.2	112.5	95.1	129.0	108.7	95.1	123.4	106.6	99.8	128.6	111.4
26, 27, 3370	ELECTRICAL, COMMUNICATIONS & UTIL.	97.3	107.6	102.8	95.1	86.6	90.6	97.5	137.9	118.9	92.3	137.1	116.0	92.2	137.1	116.0	97.3	128.5	113.8
MF2014	WEIGHTED AVERAGE	97.3	106.0	101.1	98.2	106.7	101.9	99.2	138.4	116.3	95.6	131.7	111.3	98.4	124.4	109.7	99.3	135.1	114.9

DIVISION		ILLINOIS																	
		PEORIA			QUINCY			ROCK ISLAND			ROCKFORD			SOUTH SUBURBAN			SPRINGFIELD		
		615 - 616			623			612			610 - 611			605			626 - 627		
		MAT.	INST.	TOTAL	MAT.	INST.	TOTAL	MAT.	INST.	TOTAL	MAT.	INST.	TOTAL	MAT.	INST.	TOTAL	MAT.	INST.	TOTAL
015433	CONTRACTOR EQUIPMENT		101.6	101.6		102.4	102.4		101.6	101.6		101.6	101.6		92.5	92.5		102.4	102.4
0241, 31 - 34	SITE & INFRASTRUCTURE, DEMOLITION	100.4	98.3	98.9	96.5	98.7	98.0	98.0	97.4	97.5	99.8	99.7	99.7	106.1	94.9	98.1	99.0	99.1	99.1
0310	Concrete Forming & Accessories	93.7	117.4	114.2	94.2	111.8	109.4	92.3	104.2	102.5	98.1	131.3	126.7	97.8	154.0	146.3	92.6	115.3	112.1
0320	Concrete Reinforcing	91.7	110.8	101.4	91.8	105.3	98.7	94.1	103.7	99.0	86.7	136.9	112.2	99.0	156.5	128.3	94.2	105.1	99.7
0330	Cast-in-Place Concrete	102.7	114.1	107.3	105.9	102.9	104.7	103.6	97.3	101.0	105.1	126.4	113.8	107.5	141.2	121.3	100.8	109.2	104.2
03	CONCRETE	100.2	115.0	107.5	101.7	107.6	104.6	101.0	102.0	101.5	100.9	130.4	115.4	102.7	148.8	125.4	99.1	111.3	105.1
04	MASONRY	120.1	118.1	118.8	111.7	104.1	106.9	120.3	97.0	105.8	94.1	135.8	120.0	101.1	144.1	127.9	91.0	117.8	107.7
05	METALS	100.4	113.8	104.5	97.1	108.0	100.5	97.7	108.4	101.0	100.4	128.2	108.9	93.4	131.4	105.1	97.5	109.3	101.1
06	WOOD, PLASTICS & COMPOSITES	101.2	115.5	109.2	91.8	113.9	104.2	95.0	104.4	100.3	101.1	128.3	116.4	104.0	155.7	132.9	89.2	113.9	103.0
07	THERMAL & MOISTURE PROTECTION	99.3	112.4	104.7	102.7	103.5	103.0	98.5	98.3	98.4	101.8	129.0	113.0	98.2	139.3	115.1	104.4	112.2	107.6
08	OPENINGS	101.3	116.0	104.7	95.8	111.4	99.4	94.9	103.2	96.8	101.3	132.1	108.5	102.9	158.2	115.8	103.0	111.2	104.9
0920	Plaster & Gypsum Board	97.1	115.9	109.8	98.3	114.2	109.0	93.7	104.5	101.0	97.1	129.1	118.8	92.7	157.3	136.4	98.7	114.2	109.2
0950, 0980	Ceilings & Acoustic Treatment	93.0	115.9	108.1	91.4	114.2	106.4	88.0	104.5	98.8	93.0	129.1	116.7	99.2	157.3	137.4	102.1	114.2	110.1
0960	Flooring	99.9	118.7	105.3	109.1	108.1	108.8	97.7	105.6	100.0	99.9	122.6	106.4	96.9	140.3	109.3	113.5	107.8	111.9
0970, 0990	Wall Finishes & Painting/Coating	93.8	126.4	113.5	102.5	106.5	104.9	93.8	95.1	94.5	93.8	133.1	117.5	91.9	152.5	128.4	101.1	106.5	104.3
09	FINISHES	97.2	118.9	109.2	100.1	111.8	106.6	94.9	103.7	99.7	97.2	130.0	115.3	97.9	152.7	128.1	104.9	113.5	109.6
COVERS	DIVS. 10 - 14, 25, 28, 41, 43, 44, 46	100.0	105.2	101.0	100.0	71.2	94.2	100.0	99.0	99.8	100.0	114.3	102.9	100.0	122.3	104.5	100.0	104.5	100.9
21, 22, 23	FIRE SUPPRESSION, PLUMBING & HVAC	99.9	104.9	101.9	95.2	101.0	97.5	95.1	99.9	97.1	100.0	116.5	106.7	99.8	128.6	111.4	99.9	103.3	101.3
26, 27, 3370	ELECTRICAL, COMMUNICATIONS & UTIL.	96.1	96.7	96.4	94.7	80.8	87.3	87.3	95.5	91.6	96.4	131.1	114.7	97.3	128.5	113.8	102.4	92.0	96.9
MF2014	WEIGHTED AVERAGE	100.5	109.3	104.4	98.1	101.0	99.3	97.2	100.5	98.6	99.5	124.8	110.5	99.3	135.1	114.9	100.0	106.5	102.8

	DIVISION	INDIANA																	
		ANDERSON			BLOOMINGTON			COLUMBUS			EVANSVILLE			FORT WAYNE			GARY		
		460			474			472			476 - 477			467 - 468			463 - 464		
		MAT.	INST.	TOTAL	MAT.	INST.	TOTAL	MAT.	INST.	TOTAL	MAT.	INST.	TOTAL	MAT.	INST.	TOTAL	MAT.	INST.	TOTAL
015433	CONTRACTOR EQUIPMENT		97.0	97.0		86.5	86.5		86.5	86.5		116.0	116.0		97.0	97.0		97.0	97.0
0241, 31 - 34	SITE & INFRASTRUCTURE, DEMOLITION	93.9	96.1	95.5	85.8	94.4	91.9	82.2	94.3	90.8	91.0	123.9	114.4	94.8	96.0	95.7	94.5	99.6	98.1
0310	Concrete Forming & Accessories	97.8	81.5	83.7	101.0	80.7	83.5	95.0	78.6	80.8	94.4	82.7	84.3	96.1	75.2	78.1	97.9	116.3	113.8
0320	Concrete Reinforcing	95.8	82.8	89.2	86.8	80.9	83.8	87.2	80.9	84.0	95.1	78.1	86.4	95.8	75.5	85.5	95.8	110.8	103.5
0330	Cast-in-Place Concrete	109.4	80.2	97.4	103.6	78.2	93.1	103.1	71.2	90.0	99.0	87.9	94.4	116.3	82.8	102.5	114.3	113.9	114.1
03	CONCRETE	100.8	81.7	91.5	105.0	79.7	92.6	104.2	76.3	90.5	105.3	83.8	94.7	104.0	78.4	91.4	103.2	114.2	108.6
04	MASONRY	94.1	80.1	85.4	95.6	76.5	83.7	95.4	76.5	83.6	91.2	82.5	85.8	98.5	77.9	85.6	95.6	114.6	107.5
05	METALS	92.8	89.3	91.7	95.8	77.6	90.2	95.9	77.0	90.0	89.3	85.0	87.9	92.8	85.9	90.7	92.8	108.7	97.7
06	WOOD, PLASTICS & COMPOSITES	100.2	81.9	89.9	112.9	80.9	95.0	107.7	78.2	91.2	93.5	82.0	87.0	99.9	74.7	85.8	97.7	115.7	107.8
07	THERMAL & MOISTURE PROTECTION	107.7	75.9	94.7	95.1	78.3	88.2	94.7	78.1	87.9	99.4	83.7	93.0	107.5	77.7	95.3	106.3	108.6	107.2
08	OPENINGS	98.5	82.1	94.7	105.8	81.1	100.1	101.5	79.6	96.4	99.1	80.8	94.9	98.5	74.2	92.9	98.5	120.3	103.6
0920	Plaster & Gypsum Board	102.5	81.6	88.4	97.5	81.0	86.4	95.0	78.2	83.6	93.4	80.8	84.9	101.8	74.2	83.1	96.1	116.4	109.8
0950, 0980	Ceilings & Acoustic Treatment	87.9	81.6	83.7	80.9	81.0	81.0	80.9	78.2	79.1	85.5	80.8	82.4	87.9	74.2	78.9	87.9	116.4	106.6
0960	Flooring	100.6	85.0	96.2	105.5	75.0	96.8	100.8	75.0	93.4	100.3	80.7	94.7	100.6	78.8	94.4	100.6	123.4	107.1
0970, 0990	Wall Finishes & Painting/Coating	105.6	69.4	83.7	95.3	82.4	87.5	95.3	82.4	87.5	101.2	84.9	91.4	105.6	73.9	86.5	105.6	123.9	116.7
09	FINISHES	95.6	81.0	87.6	94.8	79.8	86.5	93.0	78.2	84.9	93.8	82.6	87.6	95.4	75.7	84.5	94.7	118.7	107.9
COVERS	DIVS. 10 - 14, 25, 28, 41, 43, 44, 46	100.0	91.8	98.3	100.0	90.8	98.1	100.0	90.5	98.1	100.0	95.9	99.2	100.0	91.7	98.3	100.0	106.1	101.2
21, 22, 23	FIRE SUPPRESSION, PLUMBING & HVAC	100.0	78.8	91.5	99.7	78.9	91.3	94.9	78.8	88.4	99.9	79.6	91.7	100.0	72.1	88.7	100.0	106.6	102.7
26, 27, 3370	ELECTRICAL, COMMUNICATIONS & UTIL.	87.0	88.4	87.7	98.9	88.0	93.2	98.1	87.9	92.7	95.1	88.3	91.5	87.7	79.2	83.2	98.7	107.6	103.4
MF2014	WEIGHTED AVERAGE	97.0	83.8	91.3	99.3	81.8	91.6	97.2	80.9	90.1	97.1	86.9	92.6	97.6	79.1	89.5	98.3	110.4	103.6

	DIVISION	INDIANA																	
		INDIANAPOLIS			KOKOMO			LAFAYETTE			LAWRENCEBURG			MUNCIE			NEW ALBANY		
		461 - 462			469			479			470			473			471		
		MAT.	INST.	TOTAL	MAT.	INST.	TOTAL	MAT.	INST.	TOTAL	MAT.	INST.	TOTAL	MAT.	INST.	TOTAL	MAT.	INST.	TOTAL
015433	CONTRACTOR EQUIPMENT		93.7	93.7		97.0	97.0		86.5	86.5		104.4	104.4		95.0	95.0		94.5	94.5
0241, 31 - 34	SITE & INFRASTRUCTURE, DEMOLITION	93.5	99.3	97.7	90.3	96.0	94.4	83.1	94.3	91.0	81.2	110.4	102.0	85.6	94.6	92.0	78.2	96.8	91.4
0310	Concrete Forming & Accessories	98.6	85.6	87.4	101.2	76.6	80.0	92.6	82.6	84.0	91.5	75.4	77.6	92.6	81.0	82.6	89.5	73.3	75.5
0320	Concrete Reinforcing	99.4	83.0	91.1	86.7	81.1	83.9	86.8	82.7	84.7	86.1	74.2	80.1	96.0	82.7	89.3	87.4	77.1	82.2
0330	Cast-in-Place Concrete	101.9	86.4	95.5	108.3	82.7	97.8	103.7	81.8	94.7	97.0	74.8	87.9	108.8	78.6	96.4	100.1	74.0	89.4
03	CONCRETE	100.9	85.1	93.1	97.5	80.1	89.0	104.5	82.1	93.5	97.4	75.5	86.6	103.5	81.0	92.4	102.8	74.6	89.0
04	MASONRY	95.8	80.3	86.2	93.7	79.1	84.6	101.2	80.3	88.2	79.9	74.7	76.7	97.7	80.1	86.7	86.9	68.3	75.3
05	METALS	93.5	80.5	89.5	89.3	88.3	89.0	94.3	78.4	89.4	91.0	83.5	88.7	97.6	89.2	95.0	92.9	81.0	89.3
06	WOOD, PLASTICS & COMPOSITES	99.7	86.2	92.2	103.4	75.0	87.5	104.7	83.1	92.6	91.9	75.2	82.6	106.3	81.5	92.4	94.1	74.3	83.0
07	THERMAL & MOISTURE PROTECTION	100.7	80.8	92.5	107.3	75.7	94.3	94.7	80.6	88.9	100.1	76.7	90.5	97.7	76.8	89.2	86.8	69.6	79.7
08	OPENINGS	105.7	84.5	100.8	93.3	77.9	89.8	99.9	82.8	95.9	101.4	75.3	95.3	99.0	81.9	95.0	98.7	76.3	93.5
0920	Plaster & Gypsum Board	97.0	85.8	89.5	107.5	74.5	85.2	92.2	83.3	86.1	71.5	75.0	73.9	93.4	81.6	85.4	91.6	73.8	79.6
0950, 0980	Ceilings & Acoustic Treatment	93.8	85.8	88.6	87.9	74.5	79.1	76.7	83.3	81.0	89.7	75.0	80.0	81.7	81.6	81.6	85.5	73.8	77.8
0960	Flooring	101.4	85.0	96.7	104.6	93.3	101.3	99.7	88.8	96.6	74.1	85.0	77.2	99.9	85.0	95.6	97.9	62.8	87.9
0970, 0990	Wall Finishes & Painting/Coating	103.6	82.4	90.8	105.6	71.7	85.2	95.3	93.1	94.0	96.2	71.6	81.4	95.3	69.4	79.7	101.2	82.7	90.0
09	FINISHES	96.1	85.4	90.2	97.3	78.8	87.1	91.4	85.0	87.9	83.7	77.2	80.1	92.4	80.6	85.9	93.1	72.7	81.8
COVERS	DIVS. 10 - 14, 25, 28, 41, 43, 44, 46	100.0	93.3	98.7	100.0	91.1	98.2	100.0	91.1	98.2	100.0	43.6	88.6	100.0	90.8	98.1	100.0	43.1	88.5
21, 22, 23	FIRE SUPPRESSION, PLUMBING & HVAC	99.9	79.4	91.6	95.2	79.0	88.7	94.9	79.4	88.6	95.8	74.1	87.1	99.7	78.7	91.2	95.2	75.2	87.1
26, 27, 3370	ELECTRICAL, COMMUNICATIONS & UTIL.	101.0	88.4	94.4	91.4	78.7	84.7	97.6	82.5	89.6	93.0	73.5	82.7	90.9	79.6	84.9	93.6	75.8	84.2
MF2014	WEIGHTED AVERAGE	99.1	84.8	92.9	94.8	81.5	89.0	96.9	82.7	90.7	94.0	77.7	86.9	97.9	82.3	91.1	95.1	75.4	86.5

	DIVISION	INDIANA									IOWA								
		SOUTH BEND			TERRE HAUTE			WASHINGTON			BURLINGTON			CARROLL			CEDAR RAPIDS		
		465 - 466			478			475			526			514			522 - 524		
		MAT.	INST.	TOTAL	MAT.	INST.	TOTAL	MAT.	INST.	TOTAL	MAT.	INST.	TOTAL	MAT.	INST.	TOTAL	MAT.	INST.	TOTAL
015433	CONTRACTOR EQUIPMENT		105.3	105.3		116.0	116.0		116.0	116.0		99.4	99.4		99.4	99.4		96.1	96.1
0241, 31 - 34	SITE & INFRASTRUCTURE, DEMOLITION	96.7	96.2	96.3	92.9	124.1	115.1	92.4	121.3	113.0	98.3	96.5	97.0	87.5	96.5	93.9	100.0	95.1	96.5
0310	Concrete Forming & Accessories	99.5	81.4	83.9	95.4	82.2	84.0	96.2	78.8	81.2	94.8	75.5	78.2	82.6	50.4	54.8	100.5	82.3	84.8
0320	Concrete Reinforcing	97.3	79.9	88.4	95.1	82.9	88.9	87.9	48.8	68.0	91.0	82.7	86.8	91.8	78.7	85.1	91.7	84.2	87.9
0330	Cast-in-Place Concrete	106.6	79.8	95.6	95.9	85.6	91.6	104.3	84.8	96.3	111.7	55.1	88.4	111.7	59.7	90.3	111.9	79.7	98.7
03	CONCRETE	98.3	81.8	90.2	108.4	83.7	96.2	114.3	75.3	95.1	102.6	70.8	86.9	101.3	60.2	81.1	102.7	82.3	92.7
04	MASONRY	105.7	77.8	88.3	98.9	79.6	86.9	91.3	80.8	84.8	102.3	64.9	79.0	104.2	74.1	85.4	108.2	78.0	89.4
05	METALS	92.8	99.9	95.0	90.0	87.4	89.2	84.6	68.2	79.5	88.8	93.0	90.1	88.9	88.7	88.8	91.3	92.2	91.5
06	WOOD, PLASTICS & COMPOSITES	94.8	81.6	87.4	95.7	82.1	88.1	96.1	78.9	86.5	95.1	76.6	84.8	81.6	44.9	61.1	101.9	82.6	91.1
07	THERMAL & MOISTURE PROTECTION	101.9	80.6	93.2	99.5	80.8	91.8	99.6	82.2	92.5	102.9	73.9	91.0	103.2	69.4	89.3	103.9	80.9	94.5
08	OPENINGS	96.0	80.5	92.4	99.7	82.2	95.6	96.4	67.6	89.7	94.8	70.4	89.1	99.3	53.4	88.6	100.3	81.9	96.0
0920	Plaster & Gypsum Board	90.8	81.3	84.4	93.4	81.0	85.0	93.4	77.7	82.8	102.3	75.9	84.5	97.7	43.2	60.9	106.9	82.3	90.3
0950, 0980	Ceilings & Acoustic Treatment	88.9	81.3	83.9	85.5	81.0	82.6	79.6	77.7	78.3	98.7	75.9	83.7	98.7	43.2	62.3	101.2	82.3	88.8
0960	Flooring	99.3	91.1	96.9	100.3	85.0	95.9	101.2	76.4	94.1	107.9	37.8	87.8	102.2	33.2	82.5	124.0	78.3	110.9
0970, 0990	Wall Finishes & Painting/Coating	99.2	87.1	91.9	101.2	83.1	90.3	101.2	83.4	90.5	106.3	78.5	89.5	106.3	78.7	89.6	107.6	72.2	86.3
09	FINISHES	95.4	83.9	89.1	93.8	82.8	87.7	93.1	79.6	85.6	102.5	68.9	84.0	98.7	48.3	71.0	108.4	80.6	93.1
COVERS	DIVS. 10 - 14, 25, 28, 41, 43, 44, 46	100.0	92.8	98.5	100.0	93.7	98.7	100.0	95.3	99.0	100.0	86.8	97.3	100.0	65.1	92.9	100.0	91.8	98.3
21, 22, 23	FIRE SUPPRESSION, PLUMBING & HVAC	99.9	77.1	90.7	99.9	79.6	91.7	95.2	77.6	88.1	95.4	75.2	87.3	95.4	72.6	86.2	100.2	80.5	92.2
26, 27, 3370	ELECTRICAL, COMMUNICATIONS & UTIL.	99.6	86.6	92.7	93.3	86.7	89.8	93.8	84.2	88.8	100.6	67.7	83.2	101.3	81.3	90.7	98.1	80.6	88.8
MF2014	WEIGHTED AVERAGE	98.0	84.4	92.1	97.9	86.5	92.9	95.8	81.8	89.7	97.2	75.3	87.6	97.0	71.2	85.8	100.0	83.2	92.7

	DIVISION	IOWA																	
		COUNCIL BLUFFS			CRESTON			DAVENPORT			DECORAH			DES MOINES			DUBUQUE		
		515			508			527 - 528			521			500 - 503,509			520		
		MAT.	INST.	TOTAL	MAT.	INST.	TOTAL	MAT.	INST.	TOTAL	MAT.	INST.	TOTAL	MAT.	INST.	TOTAL	MAT.	INST.	TOTAL
015433	CONTRACTOR EQUIPMENT		95.5	95.5		99.4	99.4		99.4	99.4		99.4	99.4		101.1	101.1		94.9	94.9
0241, 31 - 34	SITE & INFRASTRUCTURE, DEMOLITION	103.8	91.5	95.0	93.9	95.5	95.1	98.7	98.6	98.6	96.9	95.5	95.9	102.9	99.8	100.7	97.9	92.4	94.0
0310	Concrete Forming & Accessories	82.0	71.2	72.7	78.4	65.8	67.5	100.0	91.3	92.5	92.3	45.4	51.8	95.4	81.8	83.6	83.3	74.9	76.1
0320	Concrete Reinforcing	93.6	79.0	86.1	89.3	82.4	85.8	91.7	97.0	94.4	91.0	76.5	83.6	98.2	86.7	92.3	90.4	84.1	87.2
0330	Cast-in-Place Concrete	116.4	73.6	98.8	115.2	64.1	94.2	107.8	90.1	100.5	108.6	58.0	87.8	101.4	87.5	95.7	109.6	98.6	105.1
03	CONCRETE	105.0	74.3	89.9	102.9	69.3	86.4	100.7	92.5	96.7	100.5	57.0	79.1	100.0	85.4	92.8	99.4	85.5	92.6
04	MASONRY	109.8	75.0	88.1	108.1	81.9	91.7	105.3	86.8	93.7	124.0	71.4	91.2	101.0	81.2	88.7	109.2	73.9	87.2
05	METALS	96.3	89.3	94.1	93.8	90.5	92.8	91.3	102.7	94.8	89.0	85.8	88.0	99.9	96.5	98.8	89.8	92.0	90.5
06	WOOD, PLASTICS & COMPOSITES	80.4	70.9	75.1	74.1	61.6	67.1	101.9	91.0	95.8	92.2	37.1	61.3	91.8	81.1	85.8	82.1	73.5	77.3
07	THERMAL & MOISTURE PROTECTION	103.3	67.1	88.5	104.2	77.6	93.3	103.4	87.5	96.8	103.1	53.4	82.7	98.5	79.4	90.6	103.6	73.9	91.4
08	OPENINGS	99.3	75.9	93.9	108.4	63.9	98.1	100.3	90.8	98.1	97.8	48.6	86.4	102.7	86.5	98.9	99.3	79.3	94.7
0920	Plaster & Gypsum Board	97.7	70.3	79.2	93.6	60.5	71.2	106.9	90.8	96.0	101.2	35.2	56.6	90.4	80.6	83.8	97.7	72.9	81.0
0950, 0980	Ceilings & Acoustic Treatment	98.7	70.3	80.0	90.8	60.5	70.9	101.2	90.8	94.4	98.7	35.2	57.0	94.2	80.6	85.2	98.7	72.9	81.8
0960	Flooring	100.8	80.7	95.0	97.4	33.2	79.1	110.3	97.0	106.5	107.6	47.2	90.3	102.5	89.6	98.8	114.9	77.0	104.1
0970, 0990	Wall Finishes & Painting/Coating	102.2	65.8	80.2	101.0	78.7	87.5	106.3	95.5	99.8	106.3	32.9	62.0	96.2	88.6	91.6	106.7	63.8	80.8
09	FINISHES	99.4	72.1	84.4	94.2	60.2	75.4	104.3	92.7	97.9	102.2	42.6	69.4	97.1	83.7	89.7	103.7	73.5	87.1
COVERS	DIVS. 10 - 14, 25, 28, 41, 43, 44, 46	100.0	88.6	97.7	100.0	69.5	93.8	100.0	94.5	98.9	100.0	82.9	96.5	100.0	92.1	98.4	100.0	90.2	98.0
21, 22, 23	FIRE SUPPRESSION, PLUMBING & HVAC	100.2	74.2	89.7	95.2	79.7	89.0	100.2	92.8	97.2	95.4	72.9	86.3	99.9	79.7	91.7	100.2	75.3	90.1
26, 27, 3370	ELECTRICAL, COMMUNICATIONS & UTIL.	103.6	81.5	91.9	93.8	81.3	87.2	95.6	91.0	93.2	98.1	44.3	69.7	105.4	81.3	92.7	102.1	78.2	89.4
MF2014	WEIGHTED AVERAGE	100.7	78.1	90.9	98.2	77.3	89.1	99.0	93.1	96.4	98.0	64.2	83.3	100.5	85.2	93.8	99.1	80.2	90.9

	DIVISION	IOWA																	
		FORT DODGE			MASON CITY			OTTUMWA			SHENANDOAH			SIBLEY			SIOUX CITY		
		505			504			525			516			512			510 - 511		
		MAT.	INST.	TOTAL	MAT.	INST.	TOTAL	MAT.	INST.	TOTAL	MAT.	INST.	TOTAL	MAT.	INST.	TOTAL	MAT.	INST.	TOTAL
015433	CONTRACTOR EQUIPMENT		99.4	99.4		99.4	99.4		94.9	94.9		95.5	95.5		99.4	99.4		99.4	99.4
0241, 31 - 34	SITE & INFRASTRUCTURE, DEMOLITION	102.2	94.4	96.6	102.3	95.4	97.4	98.1	90.5	92.7	102.2	90.5	93.9	108.1	94.2	98.2	109.9	95.0	99.3
0310	Concrete Forming & Accessories	78.9	45.1	49.7	83.1	45.5	50.6	90.3	73.7	75.9	83.6	56.5	60.2	84.0	37.9	44.2	100.5	66.4	71.1
0320	Concrete Reinforcing	89.3	67.2	78.1	89.2	81.8	85.4	91.0	86.1	88.5	93.6	68.4	80.7	93.6	65.3	79.2	91.7	77.9	84.7
0330	Cast-in-Place Concrete	108.2	44.2	81.9	108.2	57.5	87.4	112.4	53.7	88.3	112.5	59.8	90.9	110.3	44.7	83.4	111.0	55.5	88.2
03	CONCRETE	98.2	50.4	74.7	98.5	57.9	78.5	102.1	69.9	86.3	102.4	61.0	82.1	101.4	47.0	74.6	102.0	65.8	84.2
04	MASONRY	106.9	37.9	63.9	120.5	69.8	88.9	105.8	57.8	75.9	109.4	75.0	88.0	128.6	38.1	72.2	102.2	54.6	72.6
05	METALS	93.9	81.3	90.1	94.0	88.7	92.4	88.8	91.7	89.7	95.3	83.3	91.6	89.1	79.7	86.2	91.3	87.4	90.1
06	WOOD, PLASTICS & COMPOSITES	74.5	45.2	58.1	78.4	37.1	55.3	89.3	80.9	84.6	82.1	52.8	65.7	82.9	35.9	56.6	101.9	66.6	82.1
07	THERMAL & MOISTURE PROTECTION	103.5	57.1	84.5	103.0	63.5	86.8	103.7	64.6	87.7	102.6	65.2	87.2	102.9	48.3	80.5	103.4	64.3	87.4
08	OPENINGS	101.9	49.5	89.7	93.5	50.1	83.4	99.3	77.7	94.3	90.3	54.7	82.0	95.9	43.8	83.8	100.3	65.9	92.3
0920	Plaster & Gypsum Board	93.6	43.5	59.8	93.6	35.2	54.1	98.8	80.6	86.5	97.7	51.6	66.5	97.7	33.9	54.6	106.9	65.7	79.0
0950, 0980	Ceilings & Acoustic Treatment	90.8	43.5	59.7	90.8	35.2	54.2	98.7	80.6	86.8	98.7	51.6	67.7	98.7	33.9	56.1	101.2	65.7	77.9
0960	Flooring	98.9	47.2	84.1	100.9	47.2	85.6	117.9	50.2	98.5	101.5	34.3	82.3	103.1	34.0	83.3	110.3	53.7	94.1
0970, 0990	Wall Finishes & Painting/Coating	101.0	61.7	77.3	101.0	30.1	58.2	106.7	78.7	89.8	102.2	65.8	80.2	106.3	61.7	79.4	106.3	63.5	80.5
09	FINISHES	96.1	46.1	68.5	96.7	42.0	66.5	104.9	70.4	85.9	99.5	52.3	73.4	101.9	38.0	66.7	105.8	63.6	82.5
COVERS	DIVS. 10 - 14, 25, 28, 41, 43, 44, 46	100.0	81.7	96.3	100.0	86.5	97.3	100.0	82.4	96.4	100.0	66.1	93.2	100.0	80.8	96.1	100.0	88.2	97.6
21, 22, 23	FIRE SUPPRESSION, PLUMBING & HVAC	95.2	65.5	83.3	95.2	70.5	85.2	95.4	70.9	85.5	95.4	73.5	86.6	95.4	68.0	84.3	100.2	75.8	90.4
26, 27, 3370	ELECTRICAL, COMMUNICATIONS & UTIL.	100.3	41.9	69.4	99.4	55.6	76.3	100.4	72.0	85.4	98.1	81.5	89.3	98.1	41.4	68.1	98.1	73.0	84.8
MF2014	WEIGHTED AVERAGE	97.9	58.2	80.6	97.6	65.8	83.8	97.9	73.7	87.3	97.6	71.2	86.1	98.3	56.4	80.0	99.7	72.6	87.9

	DIVISION	IOWA						KANSAS											
		SPENCER			WATERLOO			BELLEVILLE			COLBY			DODGE CITY			EMPORIA		
		513			506 - 507			669			677			678			668		
		MAT.	INST.	TOTAL	MAT.	INST.	TOTAL	MAT.	INST.	TOTAL	MAT.	INST.	TOTAL	MAT.	INST.	TOTAL	MAT.	INST.	TOTAL
015433	CONTRACTOR EQUIPMENT		99.4	99.4		99.4	99.4		103.4	103.4		103.4	103.4		103.4	103.4		101.6	101.6
0241, 31 - 34	SITE & INFRASTRUCTURE, DEMOLITION	108.1	94.2	98.2	107.6	95.3	98.8	111.0	95.0	99.7	112.3	95.2	100.1	114.9	94.9	100.7	103.0	92.5	95.5
0310	Concrete Forming & Accessories	90.2	37.9	45.1	94.1	55.1	60.5	96.2	54.1	59.8	99.5	58.3	64.0	93.1	58.2	63.0	87.2	65.3	68.3
0320	Concrete Reinforcing	93.6	67.1	80.1	89.9	83.9	86.8	96.6	55.1	75.5	98.9	55.2	76.7	96.5	55.1	75.4	95.3	55.7	75.2
0330	Cast-in-Place Concrete	110.3	44.7	83.4	115.8	59.2	92.6	124.2	55.7	96.1	127.1	55.8	97.8	129.3	55.5	99.0	120.1	53.6	92.8
03	CONCRETE	101.8	47.3	75.0	104.4	63.1	84.1	119.3	56.2	88.3	119.9	58.1	89.5	121.3	57.9	90.2	111.5	60.5	86.5
04	MASONRY	128.6	38.1	72.2	107.7	73.1	86.1	99.2	58.7	73.9	108.1	59.9	78.1	118.9	58.9	81.5	105.5	68.5	82.4
05	METALS	89.0	80.5	86.4	96.4	91.1	94.7	95.6	77.3	90.0	96.0	77.7	90.4	97.4	76.7	91.1	95.3	78.9	90.3
06	WOOD, PLASTICS & COMPOSITES	89.2	35.9	59.3	91.6	48.0	67.2	97.9	52.1	72.3	103.6	57.9	78.0	95.8	57.9	74.5	88.7	66.6	76.3
07	THERMAL & MOISTURE PROTECTION	103.8	48.2	81.0	103.3	71.3	90.2	94.8	60.7	80.8	99.1	61.6	83.7	99.1	60.7	83.4	93.1	75.9	86.0
08	OPENINGS	107.5	44.4	92.9	94.4	58.2	86.0	98.6	48.4	86.9	103.2	51.5	91.2	103.1	51.5	91.1	96.3	56.3	87.0
0920	Plaster & Gypsum Board	98.8	33.9	54.9	101.9	46.5	64.4	97.2	50.6	65.7	99.7	56.5	70.5	94.0	56.5	68.7	94.4	65.5	74.9
0950, 0980	Ceilings & Acoustic Treatment	98.7	33.9	56.1	94.2	46.5	62.8	86.3	50.6	62.8	84.0	56.5	65.9	84.0	56.5	65.9	86.3	65.5	72.7
0960	Flooring	105.7	34.0	85.2	106.2	64.9	94.4	103.0	39.2	84.8	103.1	39.2	84.9	99.6	39.2	82.3	98.0	36.4	80.4
0970, 0990	Wall Finishes & Painting/Coating	106.3	61.7	79.4	101.0	77.8	87.0	100.0	38.9	63.1	104.7	38.9	65.0	104.7	38.9	65.0	100.0	38.9	63.1
09	FINISHES	102.8	38.0	67.1	100.3	57.6	76.8	96.9	49.3	70.6	96.6	52.7	72.4	94.9	52.7	71.6	94.0	57.3	73.8
COVERS	DIVS. 10 - 14, 25, 28, 41, 43, 44, 46	100.0	80.8	96.1	100.0	87.8	97.5	100.0	40.8	88.0	100.0	41.4	88.2	100.0	41.4	88.2	100.0	40.5	88.0
21, 22, 23	FIRE SUPPRESSION, PLUMBING & HVAC	95.4	68.0	84.3	100.0	79.8	91.9	95.2	71.6	85.7	95.2	69.0	84.6	100.0	69.0	87.5	95.2	73.4	86.4
26, 27, 3370	ELECTRICAL, COMMUNICATIONS & UTIL.	99.9	41.4	69.0	95.8	55.7	74.6	96.5	67.7	81.3	100.0	64.5	81.2	96.8	73.8	84.6	93.9	71.4	82.0
MF2014	WEIGHTED AVERAGE	99.9	56.6	81.0	99.5	71.8	87.4	99.6	64.9	84.4	101.1	65.0	85.4	102.7	66.0	86.7	97.9	69.2	85.4

DIVISION		KANSAS																	
		FORT SCOTT			HAYS			HUTCHINSON			INDEPENDENCE			KANSAS CITY			LIBERAL		
		667			676			675			673			660 - 662			679		
		MAT.	INST.	TOTAL	MAT.	INST.	TOTAL	MAT.	INST.	TOTAL	MAT.	INST.	TOTAL	MAT.	INST.	TOTAL	MAT.	INST.	TOTAL
015433	CONTRACTOR EQUIPMENT		102.5	102.5		103.4	103.4		103.4	103.4		103.4	103.4		99.9	99.9		103.4	103.4
0241, 31 - 34	SITE & INFRASTRUCTURE, DEMOLITION	99.8	93.4	95.2	117.7	95.2	101.7	96.2	95.0	95.4	116.0	95.1	101.1	94.4	93.1	93.5	117.3	95.0	101.5
0310	Concrete Forming & Accessories	104.2	76.8	80.6	97.2	58.3	63.7	87.9	58.2	62.2	107.9	66.3	72.0	100.9	94.2	95.1	93.5	58.2	63.0
0320	Concrete Reinforcing	94.6	93.3	94.0	96.5	55.2	75.5	96.5	55.1	75.4	95.9	62.7	79.0	91.8	97.2	94.6	97.8	55.1	76.1
0330	Cast-in-Place Concrete	111.4	54.5	88.0	100.3	55.8	82.0	92.9	53.1	76.6	129.9	53.4	98.5	95.4	97.4	96.3	100.3	53.1	80.9
03	CONCRETE	106.6	72.9	90.1	110.4	58.1	84.7	92.4	57.1	75.0	122.6	62.2	92.9	98.4	96.2	97.3	112.5	57.1	85.2
04	MASONRY	106.7	60.3	77.8	118.0	59.9	81.8	107.9	58.6	77.2	105.3	64.5	79.9	107.8	99.4	102.6	116.6	58.6	80.4
05	METALS	95.3	92.8	94.5	95.6	77.7	90.1	95.4	76.7	89.6	95.3	80.5	90.8	102.9	100.1	102.1	95.9	76.7	90.0
06	WOOD, PLASTICS & COMPOSITES	108.7	83.5	94.6	100.6	57.9	76.7	90.9	57.9	72.4	113.7	68.0	88.1	104.4	94.5	98.8	96.4	57.9	74.8
07	THERMAL & MOISTURE PROTECTION	94.0	80.8	88.6	99.4	61.6	83.9	98.0	60.6	82.7	99.1	80.5	91.5	93.7	98.0	95.5	99.5	60.6	83.6
08	OPENINGS	96.3	78.2	92.1	103.1	51.5	91.1	103.0	51.5	91.0	100.7	58.8	90.9	97.7	89.0	95.7	103.1	51.5	91.1
0920	Plaster & Gypsum Board	99.7	82.9	88.4	97.3	56.5	69.7	93.0	56.5	68.3	106.9	67.0	79.9	92.9	94.2	93.8	94.8	56.5	68.9
0950, 0980	Ceilings & Acoustic Treatment	86.3	82.9	84.1	84.0	56.5	65.9	84.0	56.5	65.9	84.0	67.0	72.8	86.3	94.2	91.5	84.0	56.5	65.9
0960	Flooring	113.7	39.4	92.5	102.0	39.2	84.0	96.8	39.2	80.3	107.3	39.2	87.8	91.4	98.7	93.5	99.8	39.2	82.5
0970, 0990	Wall Finishes & Painting/Coating	101.8	38.9	63.8	104.7	38.9	65.0	104.7	38.9	65.0	104.7	38.9	65.0	108.5	67.7	83.9	104.7	38.9	65.0
09	FINISHES	99.6	66.9	81.6	96.4	52.7	72.3	92.4	52.7	70.5	98.9	58.7	76.8	94.3	91.0	92.5	95.7	52.7	72.0
COVERS	DIVS. 10 - 14, 25, 28, 41, 43, 44, 46	100.0	46.2	89.1	100.0	41.4	88.2	100.0	41.4	88.2	100.0	42.5	88.4	100.0	61.9	92.3	100.0	41.4	88.2
21, 22, 23	FIRE SUPPRESSION, PLUMBING & HVAC	95.2	68.3	84.3	95.2	69.0	84.6	95.2	69.0	84.6	95.2	70.1	85.1	99.9	96.4	98.5	95.2	67.2	83.9
26, 27, 3370	ELECTRICAL, COMMUNICATIONS & UTIL.	93.2	71.4	81.6	98.9	67.7	82.4	93.9	67.7	80.0	96.1	71.4	83.0	98.7	97.2	97.9	96.8	73.8	84.6
MF2014	WEIGHTED AVERAGE	97.9	73.0	87.0	100.5	65.4	85.2	96.5	65.0	82.8	100.9	69.2	87.1	99.5	94.8	97.5	100.4	65.5	85.2

DIVISION		KANSAS									KENTUCKY								
		SALINA			TOPEKA			WICHITA			ASHLAND			BOWLING GREEN			CAMPTON		
		674			664 - 666			670 - 672			411 - 412			421 - 422			413 - 414		
		MAT.	INST.	TOTAL	MAT.	INST.	TOTAL	MAT.	INST.	TOTAL	MAT.	INST.	TOTAL	MAT.	INST.	TOTAL	MAT.	INST.	TOTAL
015433	CONTRACTOR EQUIPMENT		103.4	103.4		101.6	101.6		103.4	103.4		97.3	97.3		94.5	94.5		101.1	101.1
0241, 31 - 34	SITE & INFRASTRUCTURE, DEMOLITION	105.2	95.2	98.1	98.2	91.5	93.4	101.8	93.8	96.1	112.3	86.3	93.8	78.6	97.1	91.8	87.0	98.3	95.0
0310	Concrete Forming & Accessories	89.7	54.9	59.7	98.4	40.9	48.8	95.3	51.0	57.1	87.9	105.2	102.8	85.7	83.0	83.4	89.4	82.6	83.6
0320	Concrete Reinforcing	95.9	59.4	77.3	94.9	99.5	97.2	95.3	77.3	86.1	88.8	108.7	98.9	86.2	91.3	88.8	87.0	101.8	94.5
0330	Cast-in-Place Concrete	112.3	53.7	88.2	100.4	45.5	77.9	105.3	51.7	83.3	91.1	101.1	95.2	90.4	95.5	92.5	100.8	73.6	89.6
03	CONCRETE	107.3	56.7	82.5	99.4	55.1	77.7	101.6	57.4	79.9	98.6	104.8	101.6	96.8	88.9	92.9	100.7	83.2	92.1
04	MASONRY	134.8	59.9	88.2	102.5	55.9	73.4	106.3	50.2	71.4	97.8	103.7	101.5	99.7	80.2	87.5	96.9	69.0	79.5
05	METALS	97.3	80.6	92.1	99.4	96.9	98.6	99.4	84.2	94.7	92.1	110.3	97.7	93.6	87.3	91.7	92.9	92.0	92.6
06	WOOD, PLASTICS & COMPOSITES	92.4	52.1	69.8	97.1	37.3	63.6	95.1	51.3	70.6	76.0	105.2	92.3	88.6	83.1	85.5	87.1	89.0	88.2
07	THERMAL & MOISTURE PROTECTION	98.6	61.4	83.3	96.7	67.4	84.7	97.2	56.1	80.4	91.0	97.5	93.7	86.6	80.2	84.0	99.7	67.5	86.5
08	OPENINGS	103.1	48.5	90.4	100.8	55.4	90.2	105.3	56.2	93.9	97.9	99.6	98.3	98.7	79.3	94.2	100.2	87.6	97.3
0920	Plaster & Gypsum Board	93.0	50.6	64.3	96.8	35.3	55.2	91.8	49.8	63.4	60.9	105.4	91.0	87.4	82.9	84.3	87.4	88.1	87.9
0950, 0980	Ceilings & Acoustic Treatment	84.0	50.6	62.0	92.0	35.3	54.7	88.8	49.8	63.2	80.8	105.4	97.0	85.5	82.9	83.8	85.5	88.1	87.2
0960	Flooring	98.2	61.9	87.8	103.4	73.7	94.9	105.4	63.0	93.3	79.4	102.4	86.0	95.8	86.0	93.0	98.0	41.2	81.7
0970, 0990	Wall Finishes & Painting/Coating	104.7	38.9	65.0	101.8	52.5	72.1	103.6	48.0	70.0	103.1	101.2	102.0	101.2	72.7	84.0	101.2	65.4	79.6
09	FINISHES	93.6	53.6	71.6	98.5	45.9	69.5	97.7	52.3	72.7	80.9	104.8	94.1	91.8	82.8	86.8	92.6	74.0	82.4
COVERS	DIVS. 10 - 14, 25, 28, 41, 43, 44, 46	100.0	86.0	97.2	100.0	47.2	89.3	100.0	83.0	96.6	100.0	94.1	98.8	100.0	62.3	92.4	100.0	55.8	91.1
21, 22, 23	FIRE SUPPRESSION, PLUMBING & HVAC	100.0	69.1	87.5	100.0	69.1	87.5	99.8	64.8	85.7	95.0	92.5	94.0	99.9	82.7	93.0	95.2	79.8	89.0
26, 27, 3370	ELECTRICAL, COMMUNICATIONS & UTIL.	96.6	73.8	84.5	97.5	71.8	83.9	100.9	73.8	86.5	91.4	97.3	94.5	94.0	81.5	87.4	91.4	66.1	78.0
MF2014	WEIGHTED AVERAGE	101.5	67.6	86.7	99.5	66.0	84.9	100.7	66.0	85.6	94.6	99.3	96.6	96.2	83.9	90.8	95.8	78.6	88.3

DIVISION		KENTUCKY																	
		CORBIN			COVINGTON			ELIZABETHTOWN			FRANKFORT			HAZARD			HENDERSON		
		407 - 409			410			427			406			417 - 418			424		
		MAT.	INST.	TOTAL	MAT.	INST.	TOTAL	MAT.	INST.	TOTAL	MAT.	INST.	TOTAL	MAT.	INST.	TOTAL	MAT.	INST.	TOTAL
015433	CONTRACTOR EQUIPMENT		101.1	101.1		104.4	104.4		94.5	94.5		101.1	101.1		101.1	101.1		116.0	116.0
0241, 31 - 34	SITE & INFRASTRUCTURE, DEMOLITION	87.9	98.1	95.1	82.6	112.3	103.7	73.1	97.2	90.2	88.9	99.2	96.2	84.8	99.6	95.3	81.1	124.0	111.6
0310	Concrete Forming & Accessories	84.5	70.6	72.5	85.0	82.0	82.4	80.7	80.5	80.5	94.2	71.8	74.8	86.1	83.5	83.8	92.5	81.5	83.0
0320	Concrete Reinforcing	86.0	69.8	77.7	85.7	91.1	88.5	86.6	90.8	88.8	97.7	90.8	94.2	87.4	107.6	97.7	86.3	84.8	85.6
0330	Cast-in-Place Concrete	93.6	58.6	79.2	96.5	97.6	96.9	81.8	73.0	78.2	89.4	69.1	81.1	97.0	82.5	91.0	79.9	90.6	84.3
03	CONCRETE	93.3	66.8	80.3	98.9	89.7	94.4	88.6	80.0	84.4	93.8	74.8	84.5	97.3	87.7	92.6	93.9	85.4	89.7
04	MASONRY	94.4	61.9	74.1	111.8	101.5	105.4	83.5	75.4	78.5	94.6	79.4	85.1	94.8	71.5	80.3	103.6	92.9	96.9
05	METALS	90.4	78.2	86.6	91.0	96.9	92.8	92.8	87.2	91.1	93.0	87.8	91.4	92.9	94.2	93.3	84.4	86.9	85.1
06	WOOD, PLASTICS & COMPOSITES	76.9	75.9	76.4	85.0	71.7	77.6	83.7	82.0	82.7	91.4	68.7	78.7	84.1	89.0	86.8	91.3	79.0	84.4
07	THERMAL & MOISTURE PROTECTION	99.3	63.8	84.7	100.3	92.3	97.0	86.3	76.9	82.5	101.0	75.1	90.4	99.7	69.5	87.3	99.0	91.3	95.9
08	OPENINGS	97.9	65.6	90.4	102.4	81.7	97.6	98.7	83.2	95.1	104.5	76.7	98.0	100.6	79.9	95.8	96.8	79.9	92.9
0920	Plaster & Gypsum Board	92.3	74.6	80.3	68.8	71.4	70.6	86.7	81.7	83.3	96.5	67.2	76.7	86.7	88.1	87.6	90.2	77.8	81.8
0950, 0980	Ceilings & Acoustic Treatment	80.4	74.6	76.6	88.8	71.4	77.4	85.5	81.7	83.0	89.6	67.2	74.8	85.5	88.1	87.2	79.6	77.8	78.4
0960	Flooring	97.2	41.2	81.2	71.8	88.6	76.6	93.4	86.0	91.3	106.0	54.4	91.2	96.4	42.0	80.9	99.5	86.0	95.6
0970, 0990	Wall Finishes & Painting/Coating	107.2	49.4	72.3	96.2	83.2	88.3	101.2	76.2	86.1	109.1	79.8	91.4	101.2	65.4	79.6	101.2	96.0	98.1
09	FINISHES	90.7	63.6	75.8	82.7	82.0	82.4	90.6	80.7	85.1	97.6	68.8	81.7	91.9	74.8	82.5	91.4	83.8	87.2
COVERS	DIVS. 10 - 14, 25, 28, 41, 43, 44, 46	100.0	48.4	89.6	100.0	98.7	99.7	100.0	84.2	96.8	100.0	65.5	93.0	100.0	56.6	91.2	100.0	65.4	93.0
21, 22, 23	FIRE SUPPRESSION, PLUMBING & HVAC	95.2	71.4	85.6	95.9	92.4	94.5	95.5	80.1	89.3	100.1	82.3	92.9	95.2	80.7	89.4	95.5	77.6	88.3
26, 27, 3370	ELECTRICAL, COMMUNICATIONS & UTIL.	91.4	80.1	85.4	95.2	79.3	86.8	91.2	80.2	85.4	101.3	87.1	93.8	91.4	58.4	73.9	93.4	80.3	86.5
MF2014	WEIGHTED AVERAGE	94.0	71.8	84.3	95.9	91.4	93.9	92.8	82.0	88.1	98.1	80.7	90.5	95.2	78.7	88.0	93.7	86.1	90.4

	DIVISION	KENTUCKY																	
		LEXINGTON			LOUISVILLE			OWENSBORO			PADUCAH			PIKEVILLE			SOMERSET		
		403 - 405			400 - 402			423			420			415 - 416			425 - 426		
		MAT.	INST.	TOTAL	MAT.	INST.	TOTAL	MAT.	INST.	TOTAL	MAT.	INST.	TOTAL	MAT.	INST.	TOTAL	MAT.	INST.	TOTAL
015433	**CONTRACTOR EQUIPMENT**		101.1	101.1		94.5	94.5		116.0	116.0		116.0	116.0		97.3	97.3		101.1	101.1
0241, 31 - 34	**SITE & INFRASTRUCTURE, DEMOLITION**	90.1	100.6	97.5	84.8	97.3	93.7	91.0	124.5	114.8	83.7	123.4	111.9	123.2	85.3	96.3	77.9	98.8	92.8
0310	Concrete Forming & Accessories	96.4	73.0	76.2	95.8	82.0	83.9	90.9	85.1	85.9	88.9	79.9	81.1	96.8	88.9	90.0	87.0	75.9	77.4
0320	Concrete Reinforcing	94.4	92.2	93.3	96.8	92.6	94.7	86.3	91.8	89.1	86.8	86.2	86.5	89.3	108.4	99.0	86.6	90.8	88.7
0330	Cast-in-Place Concrete	95.8	92.7	94.5	96.4	73.9	87.2	93.2	91.1	92.4	85.2	86.5	85.7	100.1	96.7	98.7	79.9	98.3	87.5
03	**CONCRETE**	96.5	83.7	90.2	97.1	81.3	89.4	106.3	88.5	97.5	98.6	83.5	91.2	112.7	96.0	104.5	83.6	86.6	85.1
04	**MASONRY**	93.7	73.3	81.0	91.6	78.8	83.6	95.9	91.5	93.2	98.8	88.4	92.3	95.3	96.8	96.2	89.8	76.7	81.6
05	**METALS**	92.8	89.0	91.6	93.9	88.2	92.1	85.8	90.6	87.3	82.9	86.7	84.1	92.0	108.8	97.2	92.8	87.6	91.2
06	**WOOD, PLASTICS & COMPOSITES**	92.0	68.7	79.0	90.9	83.1	86.5	89.2	83.3	85.9	86.8	79.1	82.5	85.5	88.3	87.1	84.6	75.9	79.7
07	**THERMAL & MOISTURE PROTECTION**	99.5	84.1	93.2	96.3	78.2	88.9	99.4	92.8	96.7	99.1	78.8	90.8	91.7	79.5	86.7	99.0	71.2	87.6
08	**OPENINGS**	98.1	73.9	92.5	93.5	84.3	91.4	96.8	85.7	94.2	96.0	77.5	91.7	98.6	87.8	96.1	99.5	77.7	94.4
0920	Plaster & Gypsum Board	101.9	67.2	78.4	98.0	82.9	87.8	88.7	82.2	84.3	87.7	77.9	81.1	63.8	88.1	80.2	86.7	74.6	78.5
0950, 0980	Ceilings & Acoustic Treatment	84.5	67.2	73.1	88.9	82.9	84.9	79.6	82.2	81.3	79.6	77.9	78.5	80.8	88.1	85.6	85.5	74.6	78.3
0960	Flooring	101.9	70.3	92.9	98.6	86.0	95.0	98.9	86.0	95.2	97.8	59.4	86.8	83.4	102.4	88.8	96.7	41.2	80.8
0970, 0990	Wall Finishes & Painting/Coating	107.2	81.0	91.4	104.7	76.2	87.5	101.2	96.0	98.1	101.2	79.7	88.2	103.1	81.2	89.9	101.2	76.2	86.1
09	**FINISHES**	94.4	72.3	82.2	94.9	82.3	88.0	91.5	86.0	88.5	90.7	76.1	82.7	83.3	90.9	87.5	91.2	69.4	79.2
COVERS	**DIVS. 10 - 14, 25, 28, 41, 43, 44, 46**	100.0	86.5	97.3	100.0	84.9	96.9	100.0	102.9	100.6	100.0	61.8	92.3	100.0	56.3	91.2	100.0	58.5	91.6
21, 22, 23	**FIRE SUPPRESSION, PLUMBING & HVAC**	100.0	79.5	91.7	100.0	80.9	92.3	99.9	78.6	91.4	95.5	81.1	89.7	95.0	87.2	91.8	95.5	77.4	88.2
26, 27, 3370	**ELECTRICAL, COMMUNICATIONS & UTIL.**	94.2	81.5	87.5	99.4	81.5	89.9	93.5	81.4	87.1	95.8	79.9	87.4	94.4	73.4	83.3	91.8	80.2	85.6
MF2014	**WEIGHTED AVERAGE**	96.6	81.4	90.0	96.6	83.3	90.8	96.3	88.9	93.1	93.9	84.5	89.8	96.9	88.6	93.3	93.2	79.9	87.4

	DIVISION	LOUISIANA																	
		ALEXANDRIA			BATON ROUGE			HAMMOND			LAFAYETTE			LAKE CHARLES			MONROE		
		713 - 714			707 - 708			704			705			706			712		
		MAT.	INST.	TOTAL	MAT.	INST.	TOTAL	MAT.	INST.	TOTAL	MAT.	INST.	TOTAL	MAT.	INST.	TOTAL	MAT.	INST.	TOTAL
015433	**CONTRACTOR EQUIPMENT**		89.9	89.9		89.5	89.5		90.1	90.1		90.1	90.1		89.5	89.5		89.9	89.9
0241, 31 - 34	**SITE & INFRASTRUCTURE, DEMOLITION**	101.2	87.1	91.1	104.0	87.2	92.0	102.1	87.8	91.9	103.2	88.1	92.5	103.9	87.0	91.9	101.2	87.0	91.1
0310	Concrete Forming & Accessories	82.9	43.4	48.9	97.6	60.8	65.8	78.4	46.2	50.7	96.0	54.1	59.9	96.8	56.7	62.2	82.4	43.3	48.7
0320	Concrete Reinforcing	94.9	54.3	74.2	101.1	58.9	79.6	97.9	59.3	78.3	99.3	58.9	78.7	99.3	59.1	78.9	93.8	54.0	73.6
0330	Cast-in-Place Concrete	96.2	49.9	77.2	97.2	58.8	81.4	95.8	44.0	74.5	95.3	47.9	75.8	100.3	67.3	86.8	96.2	56.8	80.0
03	**CONCRETE**	95.5	48.8	72.6	98.9	60.4	80.0	95.2	48.9	72.5	96.4	53.7	75.4	98.8	61.5	80.5	95.3	51.1	73.6
04	**MASONRY**	121.3	53.4	79.0	94.9	54.1	69.5	95.7	51.2	68.0	95.7	52.4	68.7	95.0	58.8	72.5	115.6	47.9	73.4
05	**METALS**	95.4	70.4	87.7	100.6	72.1	91.8	95.6	71.2	88.1	94.8	72.1	87.8	94.8	72.6	87.9	95.4	69.9	87.6
06	**WOOD, PLASTICS & COMPOSITES**	90.5	41.7	63.2	99.0	63.8	79.3	84.2	47.0	63.3	105.7	55.7	77.7	103.8	57.7	78.0	89.8	42.2	63.1
07	**THERMAL & MOISTURE PROTECTION**	95.6	62.9	82.2	96.2	65.4	83.5	96.5	61.4	82.1	97.1	63.8	83.4	96.1	66.8	84.1	95.6	60.8	81.3
08	**OPENINGS**	109.1	44.7	94.1	97.9	59.1	88.9	93.0	51.7	83.4	96.7	53.2	86.6	96.7	54.9	87.0	109.1	47.6	94.8
0920	Plaster & Gypsum Board	84.2	40.4	54.6	96.7	63.0	74.0	97.5	45.7	62.5	105.7	54.7	71.2	105.7	56.8	72.6	83.8	40.9	54.8
0950, 0980	Ceilings & Acoustic Treatment	84.1	40.4	55.4	91.3	63.0	72.7	94.1	45.7	62.3	92.4	54.7	67.6	93.3	56.8	69.3	84.1	40.9	55.7
0960	Flooring	98.5	65.7	89.1	101.8	65.7	91.5	96.3	65.7	87.5	104.9	65.7	93.7	104.9	72.9	95.7	98.1	57.1	86.4
0970, 0990	Wall Finishes & Painting/Coating	104.4	64.4	80.3	94.7	46.3	65.5	99.4	48.3	68.6	99.4	57.9	74.4	99.4	50.0	69.6	104.4	48.1	70.4
09	**FINISHES**	89.8	48.9	67.2	94.7	60.2	75.7	94.8	49.7	70.0	98.1	56.3	75.0	98.3	58.6	76.4	89.6	45.5	65.3
COVERS	**DIVS. 10 - 14, 25, 28, 41, 43, 44, 46**	100.0	50.0	89.9	100.0	80.8	96.1	100.0	45.4	89.0	100.0	79.4	95.8	100.0	80.3	96.0	100.0	45.7	89.0
21, 22, 23	**FIRE SUPPRESSION, PLUMBING & HVAC**	100.1	55.3	82.1	100.0	59.5	83.7	95.3	42.8	74.1	100.1	61.1	84.4	100.1	62.0	84.7	100.1	54.3	81.6
26, 27, 3370	**ELECTRICAL, COMMUNICATIONS & UTIL.**	95.7	57.9	75.7	100.8	61.6	80.1	96.7	56.0	75.2	97.9	67.4	81.8	97.4	65.0	80.3	97.7	59.6	77.6
MF2014	**WEIGHTED AVERAGE**	99.4	57.2	81.0	99.1	63.7	83.7	95.8	54.5	77.8	97.9	63.0	82.7	98.1	65.0	83.7	99.3	56.5	80.6

	DIVISION	LOUISIANA									MAINE								
		NEW ORLEANS			SHREVEPORT			THIBODAUX			AUGUSTA			BANGOR			BATH		
		700 - 701			710 - 711			703			043			044			045		
		MAT.	INST.	TOTAL	MAT.	INST.	TOTAL	MAT.	INST.	TOTAL	MAT.	INST.	TOTAL	MAT.	INST.	TOTAL	MAT.	INST.	TOTAL
015433	**CONTRACTOR EQUIPMENT**		90.4	90.4		89.9	89.9		90.1	90.1		100.9	100.9		100.9	100.9		100.9	100.9
0241, 31 - 34	**SITE & INFRASTRUCTURE, DEMOLITION**	103.4	91.1	94.7	105.5	87.1	92.4	104.4	88.1	92.8	91.6	101.1	98.4	93.3	101.3	99.0	91.0	101.1	98.2
0310	Concrete Forming & Accessories	95.1	65.5	69.5	98.5	44.2	51.7	90.4	64.6	68.1	95.9	94.8	94.9	93.1	96.6	96.1	88.9	94.9	94.1
0320	Concrete Reinforcing	100.2	60.1	79.8	98.3	54.4	76.0	97.9	59.1	78.2	96.1	106.7	101.5	87.7	108.1	98.1	86.8	107.9	97.5
0330	Cast-in-Place Concrete	98.0	69.4	86.3	101.1	53.6	81.6	102.9	51.6	81.8	101.5	58.8	84.0	82.2	110.6	93.9	82.2	58.9	72.7
03	**CONCRETE**	99.0	66.3	82.9	100.4	50.5	75.9	100.5	59.7	80.4	103.1	84.1	93.8	94.1	102.9	98.4	94.1	84.4	89.4
04	**MASONRY**	97.7	59.8	74.1	107.1	50.5	71.8	120.5	48.8	75.8	107.4	63.6	80.1	118.0	103.9	109.2	125.6	86.1	101.0
05	**METALS**	110.3	73.3	98.9	99.5	70.6	90.6	95.6	71.6	88.2	106.0	86.6	100.1	96.1	89.3	94.0	94.6	88.4	92.7
06	**WOOD, PLASTICS & COMPOSITES**	97.9	66.3	80.2	101.7	42.7	68.7	92.3	69.7	79.6	91.1	105.3	99.0	90.1	96.4	93.6	84.6	105.3	96.2
07	**THERMAL & MOISTURE PROTECTION**	94.9	68.8	84.2	95.3	59.8	80.8	96.4	61.8	82.2	100.1	65.0	85.7	97.8	84.4	92.3	97.8	71.8	87.1
08	**OPENINGS**	97.8	65.4	90.2	108.0	45.3	93.4	97.6	64.1	89.8	106.3	87.9	102.0	102.8	83.0	98.2	102.8	87.8	99.3
0920	Plaster & Gypsum Board	101.4	65.7	77.2	94.0	41.5	58.5	99.7	69.2	79.0	100.6	104.8	103.5	99.4	95.7	96.9	95.5	104.8	101.8
0950, 0980	Ceilings & Acoustic Treatment	98.2	65.7	76.8	88.1	41.5	57.4	94.1	69.2	77.7	97.4	104.8	102.3	86.2	95.7	92.5	84.3	104.8	97.8
0960	Flooring	105.7	65.7	94.2	101.9	61.3	90.3	102.3	44.1	85.7	103.9	55.3	90.0	97.1	114.3	102.0	95.3	55.3	83.9
0970, 0990	Wall Finishes & Painting/Coating	101.9	64.1	79.1	100.4	44.6	66.7	100.7	49.6	69.9	112.7	63.3	82.9	105.0	44.1	68.3	105.0	44.1	68.3
09	**FINISHES**	101.2	65.1	81.3	94.5	46.3	67.9	97.2	59.8	76.6	100.2	85.8	92.3	95.7	94.8	95.2	94.2	83.7	88.4
COVERS	**DIVS. 10 - 14, 25, 28, 41, 43, 44, 46**	100.0	83.1	96.6	100.0	78.1	95.6	100.0	81.4	96.2	100.0	100.5	100.1	100.0	109.8	102.0	100.0	100.5	100.1
21, 22, 23	**FIRE SUPPRESSION, PLUMBING & HVAC**	100.0	65.1	85.9	99.9	58.6	83.3	95.3	61.7	81.8	99.9	64.2	85.5	100.1	74.3	89.7	95.4	64.3	82.8
26, 27, 3370	**ELECTRICAL, COMMUNICATIONS & UTIL.**	102.1	73.3	86.9	103.5	66.6	84.0	95.3	73.3	83.7	100.9	80.1	89.9	97.2	75.8	85.9	95.2	80.1	87.2
MF2014	**WEIGHTED AVERAGE**	101.4	69.5	87.5	101.0	59.6	83.0	98.2	65.4	83.9	102.1	79.5	92.3	99.0	89.5	94.9	97.5	81.9	90.7

City Cost Indexes

	DIVISION	MAINE																	
		HOULTON			KITTERY			LEWISTON			MACHIAS			PORTLAND			ROCKLAND		
		047			039			042			046			040 - 041			048		
		MAT.	INST.	TOTAL	MAT.	INST.	TOTAL	MAT.	INST.	TOTAL	MAT.	INST.	TOTAL	MAT.	INST.	TOTAL	MAT.	INST.	TOTAL
015433	CONTRACTOR EQUIPMENT		100.9	100.9		100.9	100.9		100.9	100.9		100.9	100.9		100.9	100.9		100.9	100.9
0241, 31 - 34	SITE & INFRASTRUCTURE, DEMOLITION	93.0	102.2	99.5	84.8	101.2	96.5	90.9	101.3	98.3	92.3	102.2	99.4	90.1	101.3	98.1	88.7	102.2	98.3
0310	Concrete Forming & Accessories	96.7	101.6	101.0	89.2	95.5	94.6	98.5	96.7	96.9	93.9	101.6	100.5	98.5	96.7	96.9	94.9	101.6	100.7
0320	Concrete Reinforcing	87.7	106.5	97.2	84.1	107.9	96.2	107.8	108.1	107.9	87.7	106.5	97.2	102.2	108.1	105.2	87.7	106.5	97.2
0330	Cast-in-Place Concrete	82.3	71.7	77.9	81.2	59.8	72.4	83.9	110.7	94.9	82.2	71.1	77.7	96.7	110.7	102.4	83.9	71.1	78.7
03	CONCRETE	95.1	91.5	93.4	89.7	85.0	87.4	94.7	102.9	98.7	94.6	91.3	93.0	101.9	102.9	102.4	92.0	91.3	91.6
04	MASONRY	101.3	76.3	85.7	113.3	87.1	97.0	101.9	103.9	103.2	101.3	76.3	85.7	112.8	103.9	107.2	95.2	76.3	83.4
05	METALS	94.8	86.6	92.3	85.5	88.6	86.4	99.5	89.3	96.4	94.8	86.5	92.3	107.9	89.3	102.2	94.7	86.5	92.2
06	WOOD, PLASTICS & COMPOSITES	94.0	105.3	100.3	89.2	105.3	98.2	96.0	96.4	96.2	91.0	105.3	99.0	93.7	96.4	95.2	91.9	105.3	99.4
07	THERMAL & MOISTURE PROTECTION	97.9	72.6	87.5	99.2	71.1	87.7	97.6	84.4	92.2	97.8	71.6	87.1	100.1	84.4	93.7	97.5	71.7	86.9
08	OPENINGS	102.9	86.1	99.0	102.0	87.8	98.7	106.2	83.0	100.8	102.9	86.1	99.0	104.5	83.0	99.5	102.8	86.1	98.9
0920	Plaster & Gypsum Board	101.6	104.8	103.8	100.0	104.8	103.3	104.9	95.7	98.7	100.1	104.8	103.3	102.5	95.7	97.9	100.1	104.8	103.3
0950, 0980	Ceilings & Acoustic Treatment	84.3	104.8	97.8	94.2	104.8	101.2	95.4	95.7	95.6	84.3	104.8	97.8	99.1	95.7	96.9	84.3	104.8	97.8
0960	Flooring	98.3	51.7	85.0	99.8	57.7	87.7	100.0	114.3	104.1	97.5	51.7	84.4	98.6	114.3	103.1	97.9	51.7	84.7
0970, 0990	Wall Finishes & Painting/Coating	105.0	137.1	124.4	96.4	38.4	61.4	105.0	44.1	68.3	105.0	137.1	124.4	102.0	44.1	67.1	105.0	137.1	124.4
09	FINISHES	95.9	97.8	97.0	96.5	83.7	89.5	98.9	94.8	96.6	95.4	97.8	96.7	98.8	94.8	96.6	95.2	97.8	96.6
COVERS	DIVS. 10 - 14, 25, 28, 41, 43, 44, 46	100.0	106.4	101.3	100.0	100.8	100.2	100.0	109.8	102.0	100.0	106.4	101.3	100.0	109.8	102.0	100.0	106.4	101.3
21, 22, 23	FIRE SUPPRESSION, PLUMBING & HVAC	95.4	73.2	86.4	95.3	76.6	87.8	100.1	74.3	89.7	95.4	73.2	86.4	100.0	74.3	89.6	95.4	73.2	86.4
26, 27, 3370	ELECTRICAL, COMMUNICATIONS & UTIL.	99.2	80.1	89.1	97.0	80.1	88.1	99.3	80.1	89.1	99.2	80.1	89.1	98.9	80.1	89.0	99.1	80.1	89.1
MF2014	WEIGHTED AVERAGE	97.2	85.6	92.1	95.2	84.7	90.6	99.6	90.1	95.5	97.1	85.5	92.0	102.0	90.1	96.8	96.3	85.5	91.6

	DIVISION	MAINE			MARYLAND														
		WATERVILLE			ANNAPOLIS			BALTIMORE			COLLEGE PARK			CUMBERLAND			EASTON		
		049			214			210 - 212			207 - 208			215			216		
		MAT.	INST.	TOTAL	MAT.	INST.	TOTAL	MAT.	INST.	TOTAL	MAT.	INST.	TOTAL	MAT.	INST.	TOTAL	MAT.	INST.	TOTAL
015433	CONTRACTOR EQUIPMENT		100.9	100.9		99.8	99.8		103.4	103.4		105.5	105.5		99.8	99.8		99.8	99.8
0241, 31 - 34	SITE & INFRASTRUCTURE, DEMOLITION	92.8	101.1	98.7	103.0	91.5	94.9	101.5	95.6	97.3	100.5	94.5	96.2	94.3	91.8	92.5	101.4	88.6	92.3
0310	Concrete Forming & Accessories	88.4	94.8	93.9	97.2	73.6	76.8	101.4	71.1	75.3	84.3	70.2	72.2	91.4	81.0	82.4	89.2	70.2	72.8
0320	Concrete Reinforcing	87.7	106.7	97.3	100.3	80.7	90.3	105.6	80.7	92.9	105.4	78.1	91.5	86.6	72.3	79.3	86.0	79.0	82.4
0330	Cast-in-Place Concrete	82.3	58.8	72.6	110.0	75.7	95.9	107.5	76.4	94.7	116.5	77.5	100.4	91.7	85.6	89.2	101.8	48.1	79.8
03	CONCRETE	95.6	84.1	90.0	104.8	76.6	90.9	104.8	75.8	90.6	107.5	75.5	91.8	90.2	81.8	86.1	98.0	65.2	81.9
04	MASONRY	112.2	63.6	81.9	103.5	71.6	83.6	100.1	71.7	82.4	111.8	69.0	85.1	99.1	83.8	89.5	113.4	42.9	69.4
05	METALS	94.8	86.6	92.3	104.0	93.4	100.8	100.6	93.9	98.5	86.8	97.4	90.0	98.5	90.5	96.1	98.8	87.2	95.2
06	WOOD, PLASTICS & COMPOSITES	84.0	105.3	95.9	92.9	75.4	83.1	99.6	72.1	84.2	80.0	70.0	74.4	84.9	80.1	82.2	82.7	77.4	79.7
07	THERMAL & MOISTURE PROTECTION	97.9	65.0	84.4	100.2	78.8	91.4	101.9	78.6	92.3	101.9	78.7	92.4	100.2	78.9	91.5	100.3	59.0	83.4
08	OPENINGS	102.9	87.9	99.4	105.8	80.3	99.9	99.0	78.5	94.2	94.3	74.8	89.7	99.7	78.0	94.7	98.0	71.4	91.8
0920	Plaster & Gypsum Board	95.5	104.8	101.8	100.5	74.9	83.2	102.3	71.3	81.4	99.0	69.0	78.8	102.3	79.8	87.1	102.3	77.0	85.2
0950, 0980	Ceilings & Acoustic Treatment	84.3	104.8	97.8	91.3	74.9	80.6	94.0	71.3	79.1	102.2	69.0	80.4	93.3	79.8	84.4	93.3	77.0	82.6
0960	Flooring	95.0	55.3	83.6	99.8	77.2	93.3	100.7	77.2	94.0	97.8	79.7	92.6	95.5	90.8	94.2	94.7	51.7	82.4
0970, 0990	Wall Finishes & Painting/Coating	105.0	63.3	79.8	92.1	79.2	84.3	97.9	79.2	86.6	108.0	77.6	89.7	96.5	73.4	82.5	96.5	77.6	85.1
09	FINISHES	94.3	85.8	89.6	94.8	74.3	83.5	100.1	72.3	84.8	95.3	71.9	82.4	96.5	81.9	88.5	96.7	69.1	81.5
COVERS	DIVS. 10 - 14, 25, 28, 41, 43, 44, 46	100.0	100.5	100.1	100.0	86.2	97.2	100.0	86.2	97.2	100.0	85.5	97.1	100.0	90.5	98.1	100.0	75.3	95.0
21, 22, 23	FIRE SUPPRESSION, PLUMBING & HVAC	95.4	64.2	82.8	100.1	80.5	92.2	100.0	80.6	92.2	95.4	83.6	90.7	95.2	73.0	86.3	95.2	68.3	84.4
26, 27, 3370	ELECTRICAL, COMMUNICATIONS & UTIL.	99.2	80.1	89.1	99.5	91.9	95.5	102.0	91.9	96.6	101.1	99.7	100.4	98.3	81.8	89.6	97.8	65.2	80.5
MF2014	WEIGHTED AVERAGE	97.5	79.5	89.7	101.5	82.0	93.0	100.8	81.9	92.6	97.2	83.4	91.2	96.8	81.8	90.2	98.3	68.6	85.3

	DIVISION	MARYLAND															MASSACHUSETTS		
		ELKTON			HAGERSTOWN			SALISBURY			SILVER SPRING			WALDORF			BOSTON		
		219			217			218			209			206			020 - 022, 024		
		MAT.	INST.	TOTAL	MAT.	INST.	TOTAL	MAT.	INST.	TOTAL	MAT.	INST.	TOTAL	MAT.	INST.	TOTAL	MAT.	INST.	TOTAL
015433	CONTRACTOR EQUIPMENT		99.8	99.8		99.8	99.8		99.8	99.8		98.1	98.1		98.1	98.1		106.5	106.5
0241, 31 - 34	SITE & INFRASTRUCTURE, DEMOLITION	88.4	89.5	89.2	92.6	92.3	92.4	101.3	88.6	92.3	89.0	87.6	88.0	95.4	87.3	89.7	98.9	109.1	106.2
0310	Concrete Forming & Accessories	95.4	81.0	82.9	90.3	75.1	77.2	103.9	51.7	58.9	92.8	70.4	73.5	100.2	68.7	73.0	102.7	143.8	138.1
0320	Concrete Reinforcing	86.0	105.3	95.8	86.6	72.3	79.3	86.0	63.5	74.6	104.1	77.9	90.8	104.8	78.0	91.1	108.9	155.8	132.8
0330	Cast-in-Place Concrete	82.5	75.2	79.5	87.4	85.7	86.7	101.8	46.7	79.2	119.3	79.3	102.8	133.6	76.5	110.2	104.0	151.2	123.4
03	CONCRETE	83.3	84.3	83.8	86.8	79.2	83.0	99.0	53.7	76.8	105.9	76.1	91.2	116.5	74.4	95.8	105.7	147.5	126.2
04	MASONRY	98.3	58.9	73.8	105.3	83.8	91.9	113.1	47.3	72.1	110.9	72.3	86.8	95.4	67.2	77.9	106.6	164.2	142.5
05	METALS	98.8	101.3	99.6	98.7	90.7	96.2	98.8	81.2	93.4	91.0	93.8	91.9	91.0	93.9	91.9	100.2	132.4	110.1
06	WOOD, PLASTICS & COMPOSITES	90.0	86.8	88.3	84.0	71.9	77.2	100.6	55.1	75.1	86.5	69.2	76.8	94.0	69.2	80.1	98.6	143.3	123.6
07	THERMAL & MOISTURE PROTECTION	99.9	73.5	89.1	100.0	76.8	90.5	100.6	63.7	85.4	105.4	84.9	97.0	105.8	82.7	96.4	103.4	151.9	123.3
08	OPENINGS	98.0	81.3	94.1	98.0	73.0	92.2	98.2	61.7	89.7	85.7	74.3	83.0	86.2	74.3	83.5	101.4	146.2	111.8
0920	Plaster & Gypsum Board	105.2	86.7	92.7	102.3	71.3	81.4	111.6	54.0	72.6	105.1	69.0	80.7	108.3	69.0	81.8	107.5	144.1	132.2
0950, 0980	Ceilings & Acoustic Treatment	93.3	86.7	89.0	94.3	71.3	79.2	93.3	54.0	67.5	112.2	69.0	83.8	112.2	69.0	83.8	106.1	144.1	131.1
0960	Flooring	97.0	58.5	86.0	95.2	90.8	93.9	100.7	64.7	90.4	104.1	79.7	97.2	107.7	79.0	99.5	97.9	182.7	122.2
0970, 0990	Wall Finishes & Painting/Coating	96.5	77.6	85.1	96.5	77.6	85.1	96.5	77.6	85.1	115.8	77.6	92.8	115.8	77.6	92.8	104.4	155.9	135.4
09	FINISHES	96.9	76.9	85.9	96.4	77.5	86.0	99.8	56.3	75.8	96.9	71.4	82.8	98.6	70.7	83.2	104.2	152.7	130.9
COVERS	DIVS. 10 - 14, 25, 28, 41, 43, 44, 46	100.0	59.9	91.9	100.0	89.6	97.9	100.0	38.7	87.6	100.0	84.7	96.9	100.0	83.0	96.6	100.0	119.4	103.9
21, 22, 23	FIRE SUPPRESSION, PLUMBING & HVAC	95.2	80.5	89.3	100.0	86.2	94.4	95.2	59.9	81.0	95.4	84.8	91.2	95.4	82.3	90.1	100.1	132.4	113.1
26, 27, 3370	ELECTRICAL, COMMUNICATIONS & UTIL.	99.6	91.9	95.5	98.1	81.8	89.5	96.5	67.1	81.0	98.4	99.7	99.1	95.7	99.7	97.8	100.8	135.4	119.1
MF2014	WEIGHTED AVERAGE	95.9	81.9	89.8	97.6	83.3	91.4	98.7	62.1	82.7	96.5	83.2	90.7	97.1	81.7	90.4	101.6	139.5	118.1

	DIVISION	MASSACHUSETTS																	
		BROCKTON			BUZZARDS BAY			FALL RIVER			FITCHBURG			FRAMINGHAM			GREENFIELD		
		023			025			027			014			017			013		
		MAT.	INST.	TOTAL	MAT.	INST.	TOTAL	MAT.	INST.	TOTAL	MAT.	INST.	TOTAL	MAT.	INST.	TOTAL	MAT.	INST.	TOTAL
015433	CONTRACTOR EQUIPMENT		102.7	102.7		102.7	102.7		103.7	103.7		100.9	100.9		101.9	101.9		100.9	100.9
0241, 31 - 34	SITE & INFRASTRUCTURE, DEMOLITION	94.5	105.7	102.4	84.8	105.6	99.6	93.5	105.8	102.2	86.3	105.6	100.0	83.0	105.3	98.9	90.0	104.3	100.1
0310	Concrete Forming & Accessories	101.6	138.7	133.6	99.2	138.5	133.1	101.6	138.8	133.6	93.8	130.8	125.7	101.2	138.8	133.6	92.1	113.2	110.3
0320	Concrete Reinforcing	106.4	155.6	131.4	85.3	160.1	123.4	106.4	160.2	133.7	83.9	151.4	118.3	83.9	155.6	120.4	87.3	124.4	106.2
0330	Cast-in-Place Concrete	97.7	151.5	119.8	81.2	151.5	110.1	94.5	152.0	118.1	85.9	150.5	112.4	85.9	148.3	111.5	88.3	131.2	105.9
03	CONCRETE	101.8	145.2	123.1	86.2	145.9	115.5	100.3	146.2	122.9	85.9	140.4	112.7	88.7	144.1	116.0	89.6	120.9	104.9
04	MASONRY	101.7	160.0	138.0	94.4	160.0	135.3	102.6	159.9	138.3	101.0	159.0	137.2	107.2	160.1	140.1	105.4	136.0	124.4
05	METALS	97.4	130.0	107.5	92.3	131.7	104.4	97.4	132.1	108.1	95.5	125.5	104.7	95.5	130.0	106.2	97.8	110.4	101.7
06	WOOD, PLASTICS & COMPOSITES	99.1	138.4	121.1	96.1	138.4	119.8	99.1	138.7	121.3	93.7	128.3	113.1	100.1	138.2	121.4	91.5	110.9	102.4
07	THERMAL & MOISTURE PROTECTION	100.9	148.7	120.5	100.1	147.0	119.4	100.8	146.3	119.5	99.2	141.5	116.5	99.3	149.1	119.7	99.2	121.9	108.5
08	OPENINGS	102.2	143.5	111.9	97.9	139.2	107.5	102.2	139.3	110.9	104.6	137.0	112.2	94.9	143.4	106.2	104.7	115.0	107.1
0920	Plaster & Gypsum Board	93.7	139.0	124.3	89.5	139.0	123.0	93.7	139.0	124.3	100.3	128.6	119.4	102.8	139.0	127.3	101.0	110.7	107.5
0950, 0980	Ceilings & Acoustic Treatment	102.2	139.0	126.4	88.2	139.0	121.6	102.2	139.0	126.4	89.0	128.6	115.0	89.0	139.0	121.9	97.4	110.7	106.1
0960	Flooring	98.6	182.7	122.7	96.7	182.7	121.3	97.7	182.7	122.0	98.1	182.7	122.3	99.7	182.7	123.5	97.3	154.0	113.5
0970, 0990	Wall Finishes & Painting/Coating	99.9	155.1	133.2	99.9	155.1	133.2	99.9	155.1	133.2	99.5	155.1	133.1	100.5	155.1	133.5	99.5	115.8	109.4
09	FINISHES	98.9	149.1	126.6	94.2	149.1	124.4	98.7	149.3	126.6	94.6	143.1	121.3	95.3	148.9	124.8	96.7	121.0	110.0
COVERS	DIVS. 10 - 14, 25, 28, 41, 43, 44, 46	100.0	117.9	103.6	100.0	117.9	103.6	100.0	118.5	103.7	100.0	109.0	101.8	100.0	117.5	103.5	100.0	104.6	100.9
21, 22, 23	FIRE SUPPRESSION, PLUMBING & HVAC	100.1	111.1	104.6	95.4	111.1	101.7	100.1	111.2	104.6	96.0	113.6	103.1	96.0	126.1	108.1	96.0	102.8	98.7
26, 27, 3370	ELECTRICAL, COMMUNICATIONS & UTIL.	99.3	100.1	99.7	96.3	100.1	98.3	99.2	100.1	99.7	100.1	105.1	102.8	96.6	128.9	113.7	100.1	98.9	99.5
MF2014	WEIGHTED AVERAGE	99.9	128.1	112.2	94.4	128.1	109.1	99.7	128.3	112.2	96.5	126.5	109.6	95.7	135.1	112.9	97.7	112.3	104.1

	DIVISION	MASSACHUSETTS																	
		HYANNIS			LAWRENCE			LOWELL			NEW BEDFORD			PITTSFIELD			SPRINGFIELD		
		026			019			018			027			012			010 - 011		
		MAT.	INST.	TOTAL	MAT.	INST.	TOTAL	MAT.	INST.	TOTAL	MAT.	INST.	TOTAL	MAT.	INST.	TOTAL	MAT.	INST.	TOTAL
015433	CONTRACTOR EQUIPMENT		102.7	102.7		102.7	102.7		100.9	100.9		103.7	103.7		100.9	100.9		100.9	100.9
0241, 31 - 34	SITE & INFRASTRUCTURE, DEMOLITION	90.9	105.6	101.3	95.2	105.7	102.7	94.2	105.6	102.3	92.1	105.8	101.8	95.2	104.2	101.6	94.6	104.4	101.6
0310	Concrete Forming & Accessories	93.5	138.5	132.3	102.4	139.0	134.0	99.0	139.0	133.5	101.6	138.8	133.6	99.0	112.2	110.4	99.3	113.6	111.6
0320	Concrete Reinforcing	85.3	160.1	123.4	103.8	150.4	127.5	104.6	150.4	127.9	106.4	160.2	133.7	86.6	122.3	104.8	104.6	124.4	114.7
0330	Cast-in-Place Concrete	89.1	151.5	114.7	99.4	148.9	119.7	90.3	148.9	114.4	83.3	152.0	111.5	98.5	129.7	111.3	94.0	131.7	109.5
03	CONCRETE	92.4	145.9	118.7	103.7	143.5	123.2	95.1	143.3	118.8	95.0	146.2	120.2	96.2	119.5	107.6	96.9	121.2	108.8
04	MASONRY	100.6	160.0	137.6	112.3	160.7	142.5	100.2	160.0	137.5	100.6	159.9	137.6	100.9	133.2	121.0	100.5	136.9	123.2
05	METALS	93.7	131.7	105.4	98.2	128.3	107.5	98.2	126.0	106.7	97.4	132.1	108.1	98.0	109.5	101.5	100.9	110.4	103.8
06	WOOD, PLASTICS & COMPOSITES	89.3	138.4	116.8	100.6	138.4	121.8	99.8	138.4	121.4	99.1	138.7	121.3	99.8	110.9	106.0	99.8	110.9	106.0
07	THERMAL & MOISTURE PROTECTION	100.4	147.0	119.5	99.7	148.5	119.7	99.5	148.6	119.7	100.7	146.3	119.4	99.6	120.7	108.2	99.5	122.2	108.8
08	OPENINGS	98.5	139.2	108.0	99.0	142.1	109.0	105.9	142.1	114.4	102.2	139.3	110.9	105.9	114.4	107.9	105.9	115.0	108.0
0920	Plaster & Gypsum Board	85.2	139.0	121.6	105.6	139.0	128.2	105.6	139.0	128.2	93.7	139.0	124.3	105.6	110.7	109.0	105.6	110.7	109.0
0950, 0980	Ceilings & Acoustic Treatment	93.8	139.0	123.5	99.3	139.0	125.4	99.3	139.0	125.4	102.2	139.0	126.4	99.3	110.7	106.8	99.3	110.7	106.8
0960	Flooring	94.4	182.7	119.6	100.2	182.7	123.8	100.2	182.7	123.8	97.7	182.7	122.0	100.6	154.0	115.9	99.7	154.0	115.3
0970, 0990	Wall Finishes & Painting/Coating	99.9	155.1	133.2	99.6	155.1	133.1	99.5	155.1	133.1	99.9	155.1	133.2	99.5	115.8	109.4	100.9	115.8	109.9
09	FINISHES	94.4	149.1	124.6	98.8	149.1	126.5	98.7	149.1	126.5	98.6	149.3	126.5	98.8	120.3	110.6	98.7	121.2	111.1
COVERS	DIVS. 10 - 14, 25, 28, 41, 43, 44, 46	100.0	117.9	103.6	100.0	118.1	103.6	100.0	118.1	103.6	100.0	118.5	103.7	100.0	103.7	100.7	100.0	104.9	101.0
21, 22, 23	FIRE SUPPRESSION, PLUMBING & HVAC	100.1	111.1	104.6	100.1	124.8	110.0	100.1	125.7	110.4	100.1	111.2	104.6	100.1	101.4	100.6	100.1	103.3	101.4
26, 27, 3370	ELECTRICAL, COMMUNICATIONS & UTIL.	96.8	100.1	98.5	99.1	128.9	114.9	99.6	128.9	115.1	100.0	100.1	100.1	99.6	98.9	99.2	99.7	98.9	99.2
MF2014	WEIGHTED AVERAGE	97.0	128.1	110.6	100.3	134.6	115.3	99.5	134.5	114.8	99.1	128.3	111.8	99.7	111.3	104.8	100.2	112.6	105.6

	DIVISION	MASSACHUSETTS			MICHIGAN														
		WORCESTER			ANN ARBOR			BATTLE CREEK			BAY CITY			DEARBORN			DETROIT		
		015 - 016			481			490			487			481			482		
		MAT.	INST.	TOTAL	MAT.	INST.	TOTAL	MAT.	INST.	TOTAL	MAT.	INST.	TOTAL	MAT.	INST.	TOTAL	MAT.	INST.	TOTAL
015433	CONTRACTOR EQUIPMENT		100.9	100.9		110.7	110.7		102.8	102.8		110.7	110.7		110.7	110.7		98.8	98.8
0241, 31 - 34	SITE & INFRASTRUCTURE, DEMOLITION	94.6	105.6	102.4	82.3	98.8	94.0	93.3	87.5	89.2	73.9	97.8	90.9	82.1	98.9	94.1	94.7	100.6	98.9
0310	Concrete Forming & Accessories	99.6	130.7	126.5	98.3	111.6	109.8	97.6	87.6	89.0	98.4	87.1	88.6	98.2	115.8	113.3	100.8	115.8	113.7
0320	Concrete Reinforcing	104.6	150.6	128.0	92.3	120.1	106.5	91.4	93.3	92.4	92.3	119.2	106.0	92.3	120.2	106.5	94.9	120.1	107.8
0330	Cast-in-Place Concrete	93.5	150.5	116.9	88.8	107.4	96.4	99.0	100.4	99.6	85.0	90.6	87.3	86.8	109.9	96.3	94.1	109.9	100.6
03	CONCRETE	96.6	140.2	118.0	93.5	112.0	102.6	98.1	92.5	95.4	91.7	95.2	93.4	92.6	114.7	103.5	96.6	113.6	105.0
04	MASONRY	100.0	159.0	136.8	104.4	108.4	106.9	105.2	87.2	94.0	104.0	87.0	93.4	104.3	111.6	108.9	97.9	111.6	106.5
05	METALS	101.0	125.2	108.4	94.2	118.8	101.8	98.3	87.6	95.0	94.8	115.4	101.2	94.3	119.1	101.9	94.9	102.1	97.1
06	WOOD, PLASTICS & COMPOSITES	100.3	128.3	116.0	97.5	112.8	106.0	96.5	87.2	91.3	97.5	86.5	91.3	97.5	116.8	108.3	103.0	116.8	110.7
07	THERMAL & MOISTURE PROTECTION	99.5	141.5	116.7	100.4	108.2	103.6	94.6	84.8	90.6	98.2	92.4	95.8	98.9	115.7	105.8	97.7	115.7	105.1
08	OPENINGS	105.9	136.7	113.1	99.4	110.4	101.9	96.6	82.3	93.3	99.4	93.9	98.1	99.4	112.6	102.5	99.2	112.9	102.4
0920	Plaster & Gypsum Board	105.6	128.6	121.2	103.8	112.4	109.6	93.7	82.9	86.4	103.8	85.3	91.3	103.8	116.6	112.4	100.5	116.6	111.3
0950, 0980	Ceilings & Acoustic Treatment	99.3	128.6	118.5	88.7	112.4	104.3	96.3	82.9	87.5	89.7	85.3	86.8	88.7	116.6	107.0	91.2	116.6	107.9
0960	Flooring	100.2	179.8	123.0	97.5	117.6	103.2	105.7	91.9	101.7	97.5	81.5	92.9	96.9	113.9	101.8	95.9	113.9	101.1
0970, 0990	Wall Finishes & Painting/Coating	99.5	155.1	133.1	96.2	98.7	97.7	107.9	81.9	92.2	96.2	82.0	87.6	96.2	100.8	99.0	97.3	100.8	99.4
09	FINISHES	98.7	142.5	122.9	92.8	111.6	103.1	98.0	87.6	92.3	92.6	84.8	88.3	92.6	114.3	104.5	93.9	114.3	105.1
COVERS	DIVS. 10 - 14, 25, 28, 41, 43, 44, 46	100.0	109.0	101.8	100.0	105.8	101.2	100.0	96.4	99.3	100.0	94.1	98.8	100.0	107.1	101.4	100.0	107.1	101.4
21, 22, 23	FIRE SUPPRESSION, PLUMBING & HVAC	100.1	113.6	105.5	100.0	98.8	99.5	100.0	87.4	94.9	100.0	83.6	93.4	100.0	109.2	103.7	100.0	110.2	104.1
26, 27, 3370	ELECTRICAL, COMMUNICATIONS & UTIL.	99.7	105.1	102.5	96.3	108.8	102.9	94.4	83.7	88.8	95.2	90.3	92.6	96.3	107.0	101.9	98.2	107.0	102.9
MF2014	WEIGHTED AVERAGE	100.2	126.4	111.6	97.1	107.5	101.7	98.3	87.6	93.7	96.6	91.7	94.5	97.0	111.0	103.1	97.8	109.6	102.9

DIVISION		MICHIGAN																	
		FLINT			GAYLORD			GRAND RAPIDS			IRON MOUNTAIN			JACKSON			KALAMAZOO		
		484 - 485			497			493,495			498 - 499			492			491		
		MAT.	INST.	TOTAL	MAT.	INST.	TOTAL	MAT.	INST.	TOTAL	MAT.	INST.	TOTAL	MAT.	INST.	TOTAL	MAT.	INST.	TOTAL
015433	CONTRACTOR EQUIPMENT		110.7	110.7		104.7	104.7		102.8	102.8		93.7	93.7		104.7	104.7		102.8	102.8
0241, 31 - 34	SITE & INFRASTRUCTURE, DEMOLITION	71.7	98.0	90.4	88.5	85.0	86.0	93.2	87.6	89.2	96.1	93.7	94.4	109.4	87.3	93.7	93.6	87.5	89.3
0310	Concrete Forming & Accessories	101.3	89.9	91.5	95.5	77.3	79.8	97.1	84.6	86.3	87.8	85.0	85.4	92.5	86.7	87.5	97.6	87.3	88.7
0320	Concrete Reinforcing	92.3	119.6	106.2	85.2	118.6	102.2	97.8	93.2	95.5	85.0	99.8	92.5	82.8	119.4	101.4	91.4	93.4	92.4
0330	Cast-in-Place Concrete	89.4	92.4	90.6	98.7	86.1	93.5	104.1	97.6	101.4	116.8	85.4	103.9	98.6	92.2	96.0	101.0	98.8	100.1
03	CONCRETE	94.0	97.1	95.5	94.6	88.9	91.8	101.6	90.2	96.0	104.3	87.9	96.2	89.3	95.4	92.3	101.6	91.8	96.8
04	MASONRY	104.5	96.5	99.5	116.2	79.1	93.1	102.0	87.4	92.9	101.2	86.1	91.8	95.1	91.6	92.9	103.8	89.1	94.6
05	METALS	94.3	116.3	101.0	99.7	109.9	102.8	95.3	87.1	92.8	99.0	89.6	96.1	99.9	114.1	104.2	98.3	86.5	94.7
06	WOOD, PLASTICS & COMPOSITES	100.9	88.2	93.8	89.4	76.8	82.4	97.7	83.2	89.6	85.0	85.1	85.0	88.1	84.7	86.2	96.5	87.2	91.3
07	THERMAL & MOISTURE PROTECTION	98.2	96.3	97.4	93.2	74.5	85.5	97.4	76.6	88.9	96.7	81.4	90.4	92.5	92.6	92.5	94.6	85.1	90.7
08	OPENINGS	99.4	95.6	98.5	97.5	75.1	92.3	103.4	84.7	99.0	104.6	76.0	98.0	96.5	93.1	95.7	96.6	81.7	93.1
0920	Plaster & Gypsum Board	105.2	87.1	93.0	93.4	74.9	80.9	95.4	78.8	84.2	51.7	85.2	74.3	91.7	82.9	85.8	93.7	82.9	86.4
0950, 0980	Ceilings & Acoustic Treatment	88.7	87.1	87.6	95.4	74.9	81.9	104.3	78.8	87.5	94.4	85.2	88.3	95.4	82.9	87.2	96.3	82.9	87.5
0960	Flooring	97.5	95.8	97.0	97.8	92.4	96.2	105.5	85.7	99.8	122.4	93.8	114.2	96.5	82.7	92.6	105.7	82.7	99.1
0970, 0990	Wall Finishes & Painting/Coating	96.2	85.4	89.7	103.8	82.0	90.6	109.7	80.8	92.3	126.1	72.1	93.5	103.8	96.8	99.6	107.9	81.9	92.2
09	FINISHES	92.1	89.8	90.9	97.4	79.1	87.3	101.6	84.2	92.0	100.0	85.3	91.9	98.3	86.2	91.6	98.0	85.6	91.2
COVERS	DIVS. 10 - 14, 25, 28, 41, 43, 44, 46	100.0	95.5	99.1	100.0	88.6	97.7	100.0	96.4	99.3	100.0	87.2	97.4	100.0	100.9	100.2	100.0	96.4	99.3
21, 22, 23	FIRE SUPPRESSION, PLUMBING & HVAC	100.0	89.6	95.8	95.6	78.4	88.6	100.0	83.0	93.1	95.5	85.3	91.4	95.6	88.4	92.7	100.0	80.8	92.3
26, 27, 3370	ELECTRICAL, COMMUNICATIONS & UTIL.	96.3	96.0	96.1	92.4	79.1	85.4	99.7	79.1	88.8	98.7	84.3	91.1	96.3	108.8	102.9	94.3	81.8	87.7
MF2014	WEIGHTED AVERAGE	96.8	96.0	96.4	97.2	83.6	91.3	99.7	85.1	93.3	99.3	86.2	93.6	96.5	95.1	95.9	98.7	85.7	93.0

DIVISION		MICHIGAN															MINNESOTA		
		LANSING			MUSKEGON			ROYAL OAK			SAGINAW			TRAVERSE CITY			BEMIDJI		
		488 - 489			494			480,483			486			496			566		
		MAT.	INST.	TOTAL	MAT.	INST.	TOTAL	MAT.	INST.	TOTAL	MAT.	INST.	TOTAL	MAT.	INST.	TOTAL	MAT.	INST.	TOTAL
015433	CONTRACTOR EQUIPMENT		110.7	110.7		102.8	102.8		96.2	96.2		110.7	110.7		93.7	93.7		98.2	98.2
0241, 31 - 34	SITE & INFRASTRUCTURE, DEMOLITION	95.6	98.0	97.3	91.2	87.5	88.5	86.5	98.4	95.0	74.9	97.8	91.2	82.4	93.2	90.0	95.4	97.5	96.9
0310	Concrete Forming & Accessories	98.2	89.7	90.9	98.0	84.1	86.0	94.1	112.6	110.1	98.3	87.6	89.1	87.8	75.2	76.9	86.1	89.6	89.1
0320	Concrete Reinforcing	94.3	119.4	107.1	92.1	93.3	92.7	83.7	119.9	102.1	92.3	119.2	106.0	86.3	92.5	89.5	93.7	106.6	100.2
0330	Cast-in-Place Concrete	103.5	92.1	98.8	98.6	94.8	97.1	77.8	106.3	89.5	87.7	90.5	88.9	91.2	78.2	85.9	106.0	103.0	104.8
03	CONCRETE	100.8	96.9	98.9	96.4	89.1	92.8	80.5	110.8	95.4	93.0	95.4	94.2	85.8	79.8	82.9	97.6	98.5	98.1
04	MASONRY	106.6	94.2	98.9	102.2	88.5	93.7	98.2	112.0	106.8	106.0	87.0	94.1	99.2	82.4	88.7	102.5	103.7	103.3
05	METALS	93.9	115.6	100.6	96.0	87.2	93.3	97.3	98.0	97.5	94.3	115.2	100.7	99.0	88.2	95.7	92.1	120.0	100.7
06	WOOD, PLASTICS & COMPOSITES	96.3	88.8	92.1	93.2	83.4	87.7	93.0	113.1	104.3	93.8	87.6	90.3	85.0	75.1	79.4	71.3	85.5	79.3
07	THERMAL & MOISTURE PROTECTION	98.6	89.6	94.9	93.6	78.9	87.6	96.8	111.2	102.7	99.0	92.6	96.4	95.8	71.7	85.9	104.7	95.4	100.9
08	OPENINGS	103.2	95.1	101.3	95.8	84.5	93.2	99.3	109.4	101.7	97.4	94.5	96.7	104.6	70.6	96.7	100.2	107.2	101.8
0920	Plaster & Gypsum Board	96.9	87.7	90.7	75.7	79.0	77.9	101.3	112.7	109.0	103.8	86.4	92.1	51.7	74.9	67.3	102.2	85.3	90.8
0950, 0980	Ceilings & Acoustic Treatment	94.0	87.7	89.8	97.1	79.0	85.2	88.0	112.7	104.3	88.7	86.4	87.2	94.4	74.9	81.6	122.8	85.3	98.2
0960	Flooring	106.6	85.7	100.6	104.5	85.7	99.1	94.8	113.9	100.3	97.5	81.5	92.9	122.4	94.9	114.5	104.4	120.8	109.1
0970, 0990	Wall Finishes & Painting/Coating	108.0	80.8	91.6	106.3	80.8	90.9	97.6	96.8	97.1	96.2	82.0	87.6	126.1	44.7	76.9	102.2	95.9	98.4
09	FINISHES	98.3	87.6	92.4	94.7	83.2	88.4	91.7	111.5	102.6	92.4	85.5	88.6	99.0	74.1	85.3	106.2	95.2	100.1
COVERS	DIVS. 10 - 14, 25, 28, 41, 43, 44, 46	100.0	100.6	100.1	100.0	96.0	99.2	100.0	103.0	100.6	100.0	94.3	98.8	100.0	84.6	96.9	100.0	96.9	99.4
21, 22, 23	FIRE SUPPRESSION, PLUMBING & HVAC	99.9	88.6	95.3	99.9	82.3	92.8	95.6	107.2	100.3	100.0	83.1	93.2	95.5	77.7	88.3	95.6	83.4	90.6
26, 27, 3370	ELECTRICAL, COMMUNICATIONS & UTIL.	98.6	93.3	95.8	94.8	79.1	86.5	98.4	104.4	101.6	93.9	90.5	92.1	94.0	79.1	86.1	104.3	104.6	104.5
MF2014	WEIGHTED AVERAGE	99.3	94.7	97.3	97.2	84.8	91.8	94.9	106.9	100.1	96.4	91.8	94.4	96.3	80.1	89.3	98.2	98.2	98.2

DIVISION		MINNESOTA																	
		BRAINERD			DETROIT LAKES			DULUTH			MANKATO			MINNEAPOLIS			ROCHESTER		
		564			565			556 - 558			560			553 - 555			559		
		MAT.	INST.	TOTAL	MAT.	INST.	TOTAL	MAT.	INST.	TOTAL	MAT.	INST.	TOTAL	MAT.	INST.	TOTAL	MAT.	INST.	TOTAL
015433	CONTRACTOR EQUIPMENT		100.8	100.8		98.2	98.2		99.1	99.1		100.8	100.8		102.6	102.6		99.1	99.1
0241, 31 - 34	SITE & INFRASTRUCTURE, DEMOLITION	96.9	102.2	100.7	93.6	97.9	96.6	100.6	99.4	99.7	93.8	102.9	100.3	97.9	104.7	102.8	97.4	99.0	98.5
0310	Concrete Forming & Accessories	87.7	91.6	91.1	83.1	91.6	90.4	98.7	109.9	108.4	95.6	99.1	98.6	100.7	128.6	124.8	99.3	106.5	105.5
0320	Concrete Reinforcing	92.5	106.7	99.7	93.7	106.6	100.2	97.4	107.1	102.3	92.4	116.2	104.5	92.2	116.6	104.6	91.7	116.4	104.3
0330	Cast-in-Place Concrete	115.2	106.7	111.7	102.9	106.1	104.2	114.8	106.0	111.2	106.2	107.3	106.6	109.6	117.4	112.8	112.3	99.4	107.0
03	CONCRETE	101.7	100.7	101.2	95.2	100.4	97.8	106.5	108.7	107.6	97.2	106.1	101.5	103.4	122.8	112.9	102.5	106.8	104.6
04	MASONRY	126.4	112.0	117.4	126.3	110.1	116.2	107.6	117.6	113.9	114.5	110.4	112.0	109.0	126.7	120.1	104.5	114.1	110.5
05	METALS	93.2	120.0	101.5	92.1	120.0	100.6	101.7	122.4	108.0	93.1	125.5	103.0	99.1	128.9	108.3	100.4	127.3	108.7
06	WOOD, PLASTICS & COMPOSITES	89.4	85.3	87.1	68.4	85.5	78.0	98.2	109.3	104.4	98.8	96.4	97.5	106.7	127.4	118.3	103.2	105.3	104.4
07	THERMAL & MOISTURE PROTECTION	103.4	107.7	105.2	104.6	108.3	106.1	101.6	114.9	107.1	103.8	99.0	101.8	102.8	127.4	112.9	104.9	100.9	103.2
08	OPENINGS	86.9	107.1	91.6	100.2	107.2	101.8	102.5	116.5	105.8	91.5	115.5	97.1	97.1	133.6	105.6	96.4	121.5	102.2
0920	Plaster & Gypsum Board	86.9	85.3	85.8	101.1	85.3	90.5	90.2	110.0	103.6	91.5	96.8	95.1	95.8	128.6	117.9	97.6	105.8	103.2
0950, 0980	Ceilings & Acoustic Treatment	59.3	85.3	76.4	122.8	85.3	98.2	96.1	110.0	105.3	59.3	96.8	84.0	98.2	128.6	118.2	93.2	105.8	101.5
0960	Flooring	103.3	120.8	108.3	103.2	120.8	108.2	101.9	122.1	107.6	105.0	120.8	109.5	97.9	120.8	104.4	102.9	120.8	108.0
0970, 0990	Wall Finishes & Painting/Coating	96.0	95.9	95.9	102.2	95.9	98.4	101.9	110.3	107.0	108.3	104.9	106.3	98.2	126.1	115.1	97.2	104.9	101.9
09	FINISHES	89.4	96.4	93.2	105.5	96.5	100.6	98.3	112.5	106.1	91.1	103.4	97.9	98.6	127.6	114.6	97.4	109.1	103.8
COVERS	DIVS. 10 - 14, 25, 28, 41, 43, 44, 46	100.0	98.3	99.7	100.0	98.6	99.7	100.0	97.6	99.5	100.0	98.6	99.7	100.0	109.7	102.0	100.0	102.8	100.6
21, 22, 23	FIRE SUPPRESSION, PLUMBING & HVAC	94.8	86.1	91.3	95.6	86.0	91.7	99.8	96.5	98.5	94.8	87.7	91.9	99.9	115.6	106.2	99.9	96.2	98.4
26, 27, 3370	ELECTRICAL, COMMUNICATIONS & UTIL.	101.8	102.4	102.1	104.0	67.5	84.7	98.4	102.4	100.6	108.6	89.2	98.3	101.2	111.4	106.6	98.6	89.2	93.6
MF2014	WEIGHTED AVERAGE	96.9	100.5	98.5	98.9	95.1	97.3	101.3	107.3	103.9	97.2	101.3	99.0	100.4	120.0	108.9	99.9	104.7	102.0

		MINNESOTA															MISSISSIPPI		
DIVISION		SAINT PAUL			ST. CLOUD			THIEF RIVER FALLS			WILLMAR			WINDOM			BILOXI		
		550 - 551			563			567			562			561			395		
		MAT.	INST.	TOTAL	MAT.	INST.	TOTAL	MAT.	INST.	TOTAL	MAT.	INST.	TOTAL	MAT.	INST.	TOTAL	MAT.	INST.	TOTAL
015433	CONTRACTOR EQUIPMENT		99.1	99.1		100.8	100.8		98.2	98.2		100.8	100.8		100.8	100.8		98.8	98.8
0241, 31 - 34	SITE & INFRASTRUCTURE, DEMOLITION	95.2	100.1	98.7	92.5	104.3	100.9	94.4	97.5	96.6	91.8	102.0	99.1	86.0	101.0	96.6	100.6	88.4	91.9
0310	Concrete Forming & Accessories	93.9	123.0	119.0	84.9	117.7	113.2	86.8	89.0	88.7	84.7	91.1	90.2	89.1	84.8	85.4	94.7	52.9	58.7
0320	Concrete Reinforcing	98.9	116.6	107.9	92.6	116.4	104.7	94.0	106.5	100.4	92.2	116.0	104.3	92.2	115.3	104.0	95.0	58.3	76.4
0330	Cast-in-Place Concrete	124.0	116.7	121.0	101.7	114.8	107.1	105.1	88.1	98.1	103.3	85.3	95.9	89.4	66.3	79.9	108.4	54.0	86.0
03	CONCRETE	111.2	120.1	115.5	92.9	116.9	104.7	96.3	93.1	94.7	92.9	94.9	93.9	83.6	85.5	84.5	101.6	55.9	79.2
04	MASONRY	111.7	126.7	121.1	111.1	119.9	116.6	102.4	104.5	103.7	115.4	112.0	113.3	125.8	92.1	104.8	102.7	44.3	66.3
05	METALS	101.1	128.6	109.6	93.9	126.8	104.0	92.2	119.4	100.6	93.0	124.5	102.7	92.9	122.4	102.0	94.9	81.7	90.8
06	WOOD, PLASTICS & COMPOSITES	93.4	119.9	108.2	86.7	114.9	102.5	72.3	85.5	79.7	86.4	85.5	85.9	90.7	82.8	86.3	99.1	56.1	75.0
07	THERMAL & MOISTURE PROTECTION	104.1	126.4	113.3	103.6	117.5	109.3	105.6	98.4	102.7	103.4	108.6	105.5	103.4	89.6	97.7	97.4	56.1	80.4
08	OPENINGS	99.8	129.5	106.7	91.5	126.7	99.7	100.2	107.2	101.8	89.0	91.9	89.7	92.7	90.4	92.2	99.6	57.7	89.8
0920	Plaster & Gypsum Board	88.9	120.9	110.6	86.9	115.8	106.4	101.8	85.3	90.7	86.9	85.6	86.0	86.9	82.8	84.1	101.7	55.2	70.3
0950, 0980	Ceilings & Acoustic Treatment	95.6	120.9	112.3	59.3	115.8	96.4	122.8	85.3	98.2	59.3	85.6	76.6	59.3	82.8	74.7	87.3	55.2	66.2
0960	Flooring	103.7	120.8	108.6	100.0	120.8	105.9	104.1	120.8	108.8	101.6	120.8	107.1	103.9	120.8	108.8	101.4	55.1	88.1
0970, 0990	Wall Finishes & Painting/Coating	105.3	126.1	117.8	108.3	126.1	119.1	102.2	95.9	98.4	102.2	95.9	98.4	102.2	104.9	103.8	99.5	43.4	65.7
09	FINISHES	97.9	123.1	111.8	88.9	119.2	105.6	106.0	95.2	100.0	88.9	94.2	91.8	89.1	91.3	90.3	93.3	52.4	70.8
COVERS	DIVS. 10 - 14, 25, 28, 41, 43, 44, 46	100.0	108.6	101.7	100.0	104.3	100.9	100.0	96.8	99.3	100.0	98.2	99.6	100.0	93.7	98.7	100.0	55.4	91.0
21, 22, 23	FIRE SUPPRESSION, PLUMBING & HVAC	99.9	113.8	105.5	99.6	113.8	105.3	95.6	83.1	90.5	94.8	106.8	99.6	94.8	84.1	90.5	100.0	49.1	79.5
26, 27, 3370	ELECTRICAL, COMMUNICATIONS & UTIL.	97.1	111.4	104.7	101.8	111.4	106.9	101.3	67.5	83.5	101.8	85.2	93.0	108.6	89.2	98.3	103.8	56.9	79.0
MF2014	WEIGHTED AVERAGE	101.4	117.9	108.6	96.9	116.0	105.2	97.8	92.4	95.4	95.5	101.3	98.0	95.9	92.3	94.3	99.2	58.1	81.3

		MISSISSIPPI																	
DIVISION		CLARKSDALE			COLUMBUS			GREENVILLE			GREENWOOD			JACKSON			LAUREL		
		386			397			387			389			390 - 392			394		
		MAT.	INST.	TOTAL	MAT.	INST.	TOTAL	MAT.	INST.	TOTAL	MAT.	INST.	TOTAL	MAT.	INST.	TOTAL	MAT.	INST.	TOTAL
015433	CONTRACTOR EQUIPMENT		98.8	98.8		98.8	98.8		98.8	98.8		98.8	98.8		98.8	98.8		98.8	98.8
0241, 31 - 34	SITE & INFRASTRUCTURE, DEMOLITION	98.3	88.2	91.1	98.9	88.4	91.5	103.9	88.4	92.9	101.0	87.9	91.7	97.6	88.4	91.1	104.4	88.3	92.9
0310	Concrete Forming & Accessories	85.0	49.2	54.2	83.3	51.2	55.6	81.7	63.7	66.1	94.2	49.2	55.4	93.6	57.5	62.5	83.3	52.9	57.1
0320	Concrete Reinforcing	99.8	41.6	70.1	101.8	42.5	71.6	100.3	60.3	79.9	99.8	52.5	75.7	100.2	56.1	77.7	102.4	39.1	70.2
0330	Cast-in-Place Concrete	105.6	53.5	84.2	110.5	58.6	89.2	108.7	55.6	86.9	113.5	53.2	88.7	98.7	56.8	81.5	108.1	55.2	86.4
03	CONCRETE	98.7	50.9	75.2	103.2	53.7	78.9	104.2	61.6	83.3	105.2	52.7	79.4	97.7	58.5	78.5	106.0	52.7	79.8
04	MASONRY	100.6	45.0	66.0	131.6	51.6	81.7	149.8	49.5	87.3	101.3	44.9	66.1	109.2	49.5	72.0	127.1	48.1	77.9
05	METALS	95.4	73.4	88.6	91.9	74.2	86.5	96.4	82.5	92.1	95.4	74.8	89.0	101.8	80.8	95.4	92.0	73.3	86.3
06	WOOD, PLASTICS & COMPOSITES	82.1	51.2	64.8	83.9	52.2	66.1	78.7	68.1	72.8	95.0	51.2	70.5	96.3	59.6	75.8	85.1	54.6	68.0
07	THERMAL & MOISTURE PROTECTION	95.7	47.2	75.8	97.3	51.5	78.5	96.0	60.0	81.2	96.1	47.1	76.0	97.5	59.2	81.8	97.4	53.1	79.3
08	OPENINGS	97.0	49.4	85.9	98.9	49.5	87.4	97.0	61.4	88.7	97.0	50.7	86.2	103.7	55.6	92.5	95.6	50.3	85.0
0920	Plaster & Gypsum Board	90.2	50.2	63.1	91.3	51.2	64.2	89.8	67.6	74.8	100.9	50.2	66.6	95.0	58.8	70.5	91.3	53.7	65.8
0950, 0980	Ceilings & Acoustic Treatment	82.9	50.2	61.4	82.0	51.2	61.8	85.7	67.6	73.8	82.9	50.2	61.4	89.7	58.8	69.4	82.0	53.7	63.4
0960	Flooring	105.7	52.8	90.6	95.5	59.3	85.1	104.1	52.8	89.4	111.1	52.8	94.4	99.3	55.1	86.7	94.2	52.8	82.4
0970, 0990	Wall Finishes & Painting/Coating	104.4	46.1	69.2	99.5	53.0	71.4	104.4	65.9	81.2	104.4	46.1	69.2	96.5	65.9	78.0	99.5	60.9	76.2
09	FINISHES	94.0	49.9	69.7	89.1	53.2	69.3	94.6	62.9	77.1	97.4	49.9	71.2	93.2	58.3	74.0	89.2	54.3	70.0
COVERS	DIVS. 10 - 14, 25, 28, 41, 43, 44, 46	100.0	55.7	91.1	100.0	56.8	91.3	100.0	58.5	91.6	100.0	55.7	91.1	100.0	57.5	91.4	100.0	37.5	87.4
21, 22, 23	FIRE SUPPRESSION, PLUMBING & HVAC	97.9	41.3	75.1	97.4	37.5	73.3	100.0	47.6	78.8	97.9	39.3	74.3	100.1	57.0	82.7	97.5	45.8	76.6
26, 27, 3370	ELECTRICAL, COMMUNICATIONS & UTIL.	97.1	46.4	70.3	100.8	49.1	73.5	97.1	62.7	78.9	97.1	43.4	68.8	105.4	62.7	82.8	102.6	56.9	78.5
MF2014	WEIGHTED AVERAGE	97.3	52.6	77.8	98.8	53.9	79.2	101.0	61.7	83.9	98.4	52.2	78.3	100.7	62.2	83.9	98.9	55.8	80.1

		MISSISSIPPI									MISSOURI								
DIVISION		MCCOMB			MERIDIAN			TUPELO			BOWLING GREEN			CAPE GIRARDEAU			CHILLICOTHE		
		396			393			388			633			637			646		
		MAT.	INST.	TOTAL	MAT.	INST.	TOTAL	MAT.	INST.	TOTAL	MAT.	INST.	TOTAL	MAT.	INST.	TOTAL	MAT.	INST.	TOTAL
015433	CONTRACTOR EQUIPMENT		98.8	98.8		98.8	98.8		98.8	98.8		106.9	106.9		106.9	106.9		101.9	101.9
0241, 31 - 34	SITE & INFRASTRUCTURE, DEMOLITION	92.2	88.0	89.2	96.2	88.7	90.8	95.6	88.1	90.3	89.6	95.6	93.9	91.3	95.4	94.2	105.2	94.6	97.7
0310	Concrete Forming & Accessories	83.3	50.9	55.4	80.4	63.3	65.7	82.2	51.2	55.5	92.3	87.3	88.0	85.2	85.1	85.1	89.0	93.0	92.5
0320	Concrete Reinforcing	103.0	41.0	71.4	101.8	56.1	78.5	97.6	52.5	74.7	101.4	102.8	102.1	102.7	88.7	95.5	96.8	106.5	101.8
0330	Cast-in-Place Concrete	96.0	53.0	78.3	102.7	59.2	84.8	105.6	54.9	84.8	91.2	84.7	88.5	90.2	86.0	88.5	103.0	85.8	95.9
03	CONCRETE	92.9	51.3	72.5	97.2	61.9	79.8	98.2	54.2	76.6	95.8	90.5	93.2	94.9	87.3	91.2	105.6	93.6	99.7
04	MASONRY	133.0	44.4	77.7	102.4	53.5	71.9	138.4	47.9	82.0	117.6	98.7	105.8	113.9	80.0	92.8	102.9	98.8	100.3
05	METALS	92.2	71.6	85.8	93.0	81.0	89.3	95.3	75.1	89.1	94.6	114.4	100.7	95.6	107.0	99.1	95.1	108.5	99.2
06	WOOD, PLASTICS & COMPOSITES	83.9	54.0	67.2	81.2	65.1	72.2	79.3	52.2	64.1	90.8	86.0	88.1	83.5	85.5	84.6	93.6	93.3	93.4
07	THERMAL & MOISTURE PROTECTION	96.9	46.4	76.2	97.0	61.6	82.5	95.7	52.9	78.2	100.3	98.9	99.7	100.2	84.2	93.6	98.3	94.5	96.8
08	OPENINGS	99.0	51.0	87.8	98.9	64.1	90.8	97.0	50.6	86.2	97.7	97.5	97.6	97.7	82.9	94.2	95.0	95.7	95.2
0920	Plaster & Gypsum Board	91.3	53.0	65.4	91.3	64.5	73.2	89.8	51.2	63.7	95.2	85.6	88.7	94.2	85.1	88.0	102.0	92.9	95.8
0950, 0980	Ceilings & Acoustic Treatment	82.0	53.0	63.0	83.9	64.5	71.2	82.9	51.2	62.0	92.6	85.6	88.0	92.6	85.1	87.6	89.1	92.9	91.6
0960	Flooring	95.5	52.8	83.3	94.2	55.1	83.0	104.4	52.8	89.6	96.4	96.8	96.5	93.3	84.0	90.6	97.3	102.5	98.8
0970, 0990	Wall Finishes & Painting/Coating	99.5	42.3	65.0	99.5	76.1	85.4	104.4	58.7	76.8	97.1	102.8	100.5	97.1	73.2	82.7	87.9	106.3	99.0
09	FINISHES	88.5	51.0	67.8	88.8	63.7	75.0	93.5	52.6	71.0	97.8	89.7	93.4	96.7	83.0	89.2	100.2	96.1	97.9
COVERS	DIVS. 10 - 14, 25, 28, 41, 43, 44, 46	100.0	59.1	91.7	100.0	59.5	91.8	100.0	56.8	91.3	100.0	88.2	97.6	100.0	93.5	98.7	100.0	88.5	97.7
21, 22, 23	FIRE SUPPRESSION, PLUMBING & HVAC	97.4	34.6	72.1	100.0	59.1	83.5	98.1	43.2	76.0	95.2	96.2	95.6	99.9	96.8	98.7	95.3	97.0	96.0
26, 27, 3370	ELECTRICAL, COMMUNICATIONS & UTIL.	99.2	49.4	72.9	102.6	57.8	78.9	96.9	49.3	71.7	96.9	82.6	89.3	96.9	101.5	99.3	92.4	78.6	85.1
MF2014	WEIGHTED AVERAGE	97.4	51.7	77.5	97.6	64.1	83.0	98.8	54.9	79.7	97.3	94.4	96.0	98.2	92.4	95.7	97.6	94.5	96.3

	DIVISION	MISSOURI																	
		COLUMBIA 652			FLAT RIVER 636			HANNIBAL 634			HARRISONVILLE 647			JEFFERSON CITY 650 - 651			JOPLIN 648		
		MAT.	INST.	TOTAL	MAT.	INST.	TOTAL	MAT.	INST.	TOTAL	MAT.	INST.	TOTAL	MAT.	INST.	TOTAL	MAT.	INST.	TOTAL
015433	CONTRACTOR EQUIPMENT		108.0	108.0		106.9	106.9		106.9	106.9		101.9	101.9		108.0	108.0		105.7	105.7
0241, 31 - 34	SITE & INFRASTRUCTURE, DEMOLITION	102.8	97.2	98.8	92.2	95.6	94.6	87.6	95.7	93.4	96.6	95.7	95.9	102.3	97.2	98.6	105.6	99.7	101.4
0310	Concrete Forming & Accessories	86.9	81.8	82.5	98.3	84.3	86.2	90.5	79.0	80.6	86.2	98.7	97.0	98.3	81.8	84.1	101.3	75.3	78.9
0320	Concrete Reinforcing	96.3	119.0	107.9	102.7	107.7	105.3	100.9	97.7	99.3	96.4	115.3	106.0	98.5	111.9	105.3	99.9	98.5	99.2
0330	Cast-in-Place Concrete	96.1	87.8	92.7	94.2	93.8	94.0	86.3	86.0	86.2	105.5	103.9	104.8	102.0	87.0	95.9	111.5	77.9	97.7
03	CONCRETE	93.2	92.2	92.7	98.8	93.2	96.0	92.1	86.3	89.2	101.8	104.0	102.9	98.9	90.6	94.8	105.2	81.4	93.5
04	MASONRY	154.1	90.0	114.2	114.5	81.4	93.9	109.0	101.3	104.2	97.4	103.4	101.2	112.0	90.0	98.3	96.5	85.9	89.9
05	METALS	98.4	121.2	105.4	94.5	115.6	101.0	94.6	111.9	99.9	95.6	113.2	101.0	98.0	118.3	104.2	98.2	100.5	98.9
06	WOOD, PLASTICS & COMPOSITES	91.7	78.7	84.4	99.1	82.6	89.9	88.9	73.8	80.4	89.9	97.5	94.2	100.0	78.7	88.1	106.0	74.2	88.2
07	THERMAL & MOISTURE PROTECTION	96.2	87.5	92.6	100.5	93.8	97.7	100.1	97.0	98.8	97.6	103.2	99.9	101.8	87.6	96.0	97.7	78.9	90.0
08	OPENINGS	99.5	95.0	98.4	97.7	97.2	97.6	97.7	82.3	94.1	95.1	102.8	96.9	99.1	93.1	97.7	96.2	78.5	92.1
0920	Plaster & Gypsum Board	87.2	78.0	81.0	101.3	82.1	88.3	94.5	73.0	79.9	97.3	97.3	97.3	91.3	78.0	82.3	108.2	73.2	84.5
0950, 0980	Ceilings & Acoustic Treatment	93.2	78.0	83.2	92.6	82.1	85.7	92.6	73.0	79.7	89.1	97.3	94.5	97.2	78.0	84.6	90.0	73.2	78.9
0960	Flooring	97.6	98.6	97.9	99.4	84.0	95.0	95.8	96.8	96.1	92.6	102.5	95.4	104.4	98.6	102.7	123.8	74.9	109.8
0970, 0990	Wall Finishes & Painting/Coating	105.2	77.1	88.2	97.1	78.4	85.8	97.1	102.8	100.5	92.5	103.8	99.3	102.3	77.1	87.1	87.5	74.9	79.9
09	FINISHES	92.8	83.0	87.4	99.7	82.2	90.1	97.4	83.1	89.5	97.8	99.5	98.7	97.2	83.0	89.4	107.0	75.3	89.6
COVERS	DIVS. 10 - 14, 25, 28, 41, 43, 44, 46	100.0	96.2	99.2	100.0	91.8	98.3	100.0	87.8	97.5	100.0	90.6	98.1	100.0	96.2	99.2	100.0	87.9	97.6
21, 22, 23	FIRE SUPPRESSION, PLUMBING & HVAC	99.8	96.4	98.4	95.2	97.8	96.2	95.2	97.5	96.1	95.2	100.0	97.2	99.8	98.1	99.1	100.1	72.1	88.8
26, 27, 3370	ELECTRICAL, COMMUNICATIONS & UTIL.	94.8	86.5	90.4	101.1	101.5	101.3	95.7	82.6	88.8	98.8	99.6	99.2	100.4	86.5	93.0	90.4	77.5	83.6
MF2014	WEIGHTED AVERAGE	100.2	93.9	97.5	98.1	95.0	96.8	96.2	92.4	94.6	97.1	101.5	99.1	99.9	93.7	97.2	99.5	81.7	91.8

	DIVISION	MISSOURI																	
		KANSAS CITY 640 - 641			KIRKSVILLE 635			POPLAR BLUFF 639			ROLLA 654 - 655			SEDALIA 653			SIKESTON 638		
		MAT.	INST.	TOTAL	MAT.	INST.	TOTAL	MAT.	INST.	TOTAL	MAT.	INST.	TOTAL	MAT.	INST.	TOTAL	MAT.	INST.	TOTAL
015433	CONTRACTOR EQUIPMENT		103.3	103.3		99.1	99.1		101.5	101.5		108.0	108.0		99.9	99.9		101.5	101.5
0241, 31 - 34	SITE & INFRASTRUCTURE, DEMOLITION	98.9	98.2	98.4	91.8	91.4	91.5	79.1	95.4	90.7	101.5	97.1	98.3	99.0	92.8	94.6	82.4	95.4	91.7
0310	Concrete Forming & Accessories	100.1	105.4	104.6	83.2	80.6	81.0	83.3	85.0	84.7	95.0	91.9	92.3	92.4	81.4	82.9	84.3	84.9	84.9
0320	Concrete Reinforcing	94.9	119.8	107.6	101.6	98.8	100.2	104.7	94.7	99.6	96.7	94.7	95.7	95.3	106.2	100.8	104.0	98.4	101.2
0330	Cast-in-Place Concrete	103.7	106.6	104.9	94.1	83.8	89.9	72.4	86.5	78.2	98.3	96.2	97.5	102.7	83.8	94.9	77.4	86.5	81.1
03	CONCRETE	101.1	108.7	104.8	110.8	86.0	98.6	85.9	88.1	86.9	95.2	95.0	95.1	109.8	87.7	99.0	89.5	88.7	89.1
04	MASONRY	99.5	107.5	104.5	121.0	88.9	101.0	112.5	80.1	92.3	126.3	87.9	102.4	132.9	88.3	105.1	112.2	80.1	92.2
05	METALS	105.9	115.8	108.9	94.2	103.8	97.2	94.8	101.7	96.9	97.8	110.5	101.7	96.5	108.0	100.1	95.1	103.2	97.6
06	WOOD, PLASTICS & COMPOSITES	105.3	105.0	105.1	77.0	78.7	77.9	76.1	85.5	81.4	100.3	92.8	96.1	93.1	78.7	85.1	77.7	85.5	82.1
07	THERMAL & MOISTURE PROTECTION	97.8	106.8	101.5	106.8	88.2	99.2	105.2	85.9	97.3	96.4	93.4	95.1	102.5	93.1	98.7	105.3	84.5	96.8
08	OPENINGS	102.5	109.2	104.0	102.7	85.8	98.8	103.7	84.5	99.2	99.5	88.3	96.9	104.5	87.4	100.5	103.7	85.5	99.5
0920	Plaster & Gypsum Board	105.2	105.0	105.0	89.8	78.0	81.8	90.2	85.1	86.7	89.3	92.5	91.5	82.5	78.0	79.5	92.0	85.1	87.3
0950, 0980	Ceilings & Acoustic Treatment	96.7	105.0	102.1	90.9	78.0	82.4	92.6	85.1	87.6	93.2	92.5	92.7	93.2	78.0	83.2	92.6	85.1	87.6
0960	Flooring	98.0	109.0	101.2	74.4	95.8	80.5	88.3	82.0	86.5	101.2	95.8	99.6	78.8	95.8	83.7	88.7	82.0	86.8
0970, 0990	Wall Finishes & Painting/Coating	92.5	112.2	104.4	92.5	82.1	86.2	92.0	73.2	80.7	105.2	100.6	102.5	105.2	106.3	105.9	92.0	73.2	80.7
09	FINISHES	102.0	106.6	104.5	96.5	82.7	88.9	96.5	82.7	88.9	94.3	93.4	93.8	91.9	85.4	88.3	97.1	82.7	89.2
COVERS	DIVS. 10 - 14, 25, 28, 41, 43, 44, 46	100.0	100.6	100.1	100.0	87.1	97.4	100.0	93.6	98.7	100.0	89.0	97.8	100.0	90.2	98.0	100.0	94.2	98.8
21, 22, 23	FIRE SUPPRESSION, PLUMBING & HVAC	100.0	104.3	101.7	95.2	97.3	96.1	95.2	96.3	95.6	95.0	98.2	96.3	95.0	95.9	95.4	95.2	96.3	95.6
26, 27, 3370	ELECTRICAL, COMMUNICATIONS & UTIL.	100.3	100.1	100.2	95.8	82.5	88.8	96.1	101.5	99.0	93.2	86.5	89.6	94.7	128.8	112.7	95.3	101.5	98.6
MF2014	WEIGHTED AVERAGE	101.4	105.7	103.3	99.5	89.9	95.3	96.2	92.0	94.4	97.9	94.6	96.5	100.2	97.3	98.9	96.7	92.2	94.8

	DIVISION	MISSOURI									MONTANA								
		SPRINGFIELD 656 - 658			ST. JOSEPH 644 - 645			ST. LOUIS 630 - 631			BILLINGS 590 - 591			BUTTE 597			GREAT FALLS 594		
		MAT.	INST.	TOTAL	MAT.	INST.	TOTAL	MAT.	INST.	TOTAL	MAT.	INST.	TOTAL	MAT.	INST.	TOTAL	MAT.	INST.	TOTAL
015433	CONTRACTOR EQUIPMENT		102.5	102.5		101.9	101.9		107.9	107.9		98.5	98.5		98.2	98.2		98.2	98.2
0241, 31 - 34	SITE & INFRASTRUCTURE, DEMOLITION	101.6	95.1	97.0	100.5	93.9	95.8	91.7	98.9	96.8	95.0	96.3	95.9	102.0	96.1	97.8	105.7	96.2	98.9
0310	Concrete Forming & Accessories	101.4	79.1	82.2	100.0	89.6	91.0	98.4	104.8	103.9	98.4	71.5	75.2	85.5	71.8	73.7	98.4	71.5	75.2
0320	Concrete Reinforcing	92.7	118.7	105.9	93.8	115.0	104.6	93.9	112.7	103.4	90.2	81.7	85.9	97.8	81.9	89.7	90.2	81.8	85.9
0330	Cast-in-Place Concrete	104.3	79.6	94.2	103.6	103.3	103.5	90.2	105.0	96.3	129.3	67.6	103.9	142.3	69.0	112.2	150.4	63.7	114.8
03	CONCRETE	105.8	87.4	96.8	100.9	99.7	100.3	94.4	107.1	100.7	109.8	72.9	91.7	113.8	73.5	94.0	119.5	71.5	96.0
04	MASONRY	102.9	89.8	94.8	99.3	96.1	97.3	95.4	111.9	105.7	128.0	73.7	94.2	123.5	76.6	94.3	128.0	75.4	95.2
05	METALS	102.8	108.8	104.7	101.9	112.4	105.2	100.3	120.2	106.4	105.8	90.0	100.9	100.1	90.1	97.1	103.2	90.0	99.2
06	WOOD, PLASTICS & COMPOSITES	101.8	77.1	88.0	106.0	87.6	95.7	99.0	103.0	101.2	98.6	70.5	82.8	85.7	70.8	77.4	99.9	70.5	83.4
07	THERMAL & MOISTURE PROTECTION	100.7	82.2	93.2	98.2	95.6	97.1	100.2	106.9	103.0	106.5	69.9	91.5	106.3	71.0	91.8	106.9	68.5	91.2
08	OPENINGS	107.2	84.3	101.9	100.6	98.6	100.1	97.6	110.9	100.7	97.3	68.7	90.6	95.6	70.0	89.6	98.5	68.0	91.4
0920	Plaster & Gypsum Board	90.4	76.3	80.9	109.7	87.0	94.4	102.2	103.1	102.8	98.0	69.8	78.9	98.7	70.2	79.4	107.3	69.8	82.0
0950, 0980	Ceilings & Acoustic Treatment	93.2	76.3	82.1	95.8	87.0	90.0	97.6	103.1	101.2	81.5	69.8	73.8	88.3	70.2	76.4	90.0	69.8	76.7
0960	Flooring	99.9	74.9	92.8	101.5	107.3	103.2	99.2	97.6	98.7	106.6	70.2	96.2	104.4	74.2	95.7	111.5	74.2	100.8
0970, 0990	Wall Finishes & Painting/Coating	99.0	74.9	84.4	87.9	86.7	87.2	97.1	106.0	102.5	105.2	86.6	94.0	103.4	51.2	71.9	103.4	86.6	93.3
09	FINISHES	96.7	78.0	86.4	103.2	91.7	96.9	100.8	103.1	102.1	96.6	72.6	83.4	97.2	69.7	82.0	101.1	73.3	85.8
COVERS	DIVS. 10 - 14, 25, 28, 41, 43, 44, 46	100.0	94.8	99.0	100.0	96.9	99.4	100.0	102.4	100.5	100.0	94.7	98.9	100.0	94.8	98.9	100.0	94.7	98.9
21, 22, 23	FIRE SUPPRESSION, PLUMBING & HVAC	99.8	74.6	89.6	100.1	89.0	95.6	100.0	107.4	103.0	100.1	74.7	89.9	100.2	75.6	90.3	100.2	70.7	88.3
26, 27, 3370	ELECTRICAL, COMMUNICATIONS & UTIL.	99.2	69.7	83.6	98.8	78.6	88.2	99.2	108.3	104.0	98.5	72.2	84.6	105.5	71.7	87.7	97.7	71.9	84.1
MF2014	WEIGHTED AVERAGE	101.7	83.7	93.8	100.6	93.5	97.5	98.7	107.8	102.7	102.7	77.1	91.5	102.6	77.3	91.6	104.1	76.2	91.9

DIVISION		MONTANA																	
		HAVRE			HELENA			KALISPELL			MILES CITY			MISSOULA			WOLF POINT		
		595			596			599			593			598			592		
		MAT.	INST.	TOTAL	MAT.	INST.	TOTAL	MAT.	INST.	TOTAL	MAT.	INST.	TOTAL	MAT.	INST.	TOTAL	MAT.	INST.	TOTAL
015433	CONTRACTOR EQUIPMENT		98.2	98.2		98.2	98.2		98.2	98.2		98.2	98.2		98.2	98.2		98.2	98.2
0241, 31 - 34	SITE & INFRASTRUCTURE, DEMOLITION	109.2	95.8	99.7	98.5	95.8	96.6	92.2	96.0	94.9	98.5	95.4	96.3	85.1	95.8	92.7	115.5	95.6	101.3
0310	Concrete Forming & Accessories	78.4	69.0	70.3	101.5	65.2	70.2	88.7	71.4	73.8	96.8	67.6	71.6	88.7	71.4	73.8	89.2	67.9	70.8
0320	Concrete Reinforcing	98.6	81.7	90.0	102.5	81.7	91.9	100.4	83.4	91.8	98.2	81.8	89.8	99.5	83.4	91.3	99.6	81.8	90.5
0330	Cast-in-Place Concrete	153.3	60.4	115.1	115.0	62.9	93.6	123.6	62.7	98.6	135.3	60.3	104.5	104.9	64.9	88.4	151.5	60.6	114.2
03	CONCRETE	122.7	69.3	96.5	106.7	68.4	87.9	102.6	71.5	87.3	110.4	68.6	89.9	90.0	72.2	81.3	126.5	68.9	98.2
04	MASONRY	124.5	69.8	90.4	120.1	75.7	92.4	122.2	77.0	94.0	129.7	65.3	89.6	147.9	76.6	103.4	131.0	66.0	90.5
05	METALS	96.3	89.6	94.2	102.1	88.8	98.0	96.1	90.6	94.4	95.4	89.9	93.7	96.6	90.7	94.8	95.5	89.2	93.6
06	WOOD, PLASTICS & COMPOSITES	77.4	70.5	73.5	102.9	62.3	80.2	89.1	70.5	78.7	96.9	70.5	82.1	89.1	70.5	78.7	88.5	70.5	78.4
07	THERMAL & MOISTURE PROTECTION	106.6	68.0	90.8	103.9	70.7	90.3	105.9	72.8	92.3	106.2	65.7	89.6	105.5	79.1	94.7	107.1	66.0	90.3
08	OPENINGS	95.6	68.0	89.2	99.8	65.2	91.7	95.6	68.5	89.3	95.1	68.0	88.8	95.6	68.3	89.2	95.1	67.4	88.7
0920	Plaster & Gypsum Board	94.8	69.8	77.9	96.0	61.4	72.6	98.7	69.8	79.2	106.7	69.8	81.8	98.7	69.8	79.2	102.1	69.8	80.3
0950, 0980	Ceilings & Acoustic Treatment	88.3	69.8	76.2	90.8	61.4	71.5	88.3	69.8	76.2	87.5	69.8	75.9	88.3	69.8	76.2	87.5	69.8	75.9
0960	Flooring	101.8	74.2	93.9	106.7	59.1	93.1	106.2	74.2	97.1	111.3	70.2	99.5	106.2	74.2	97.1	107.6	70.2	96.9
0970, 0990	Wall Finishes & Painting/Coating	103.4	48.6	70.4	99.7	47.2	68.0	103.4	47.2	69.5	103.4	47.2	69.5	103.4	62.0	78.4	103.4	48.6	70.4
09	FINISHES	96.5	67.7	80.6	98.6	61.6	78.2	97.2	69.1	81.7	100.0	65.7	81.1	96.7	70.7	82.3	99.6	66.0	81.1
COVERS	DIVS. 10 - 14, 25, 28, 41, 43, 44, 46	100.0	70.8	94.1	100.0	94.0	98.8	100.0	93.9	98.8	100.0	90.4	98.1	100.0	93.7	98.7	100.0	90.6	98.1
21, 22, 23	FIRE SUPPRESSION, PLUMBING & HVAC	95.4	67.8	84.3	100.1	70.1	88.0	95.4	69.2	84.8	95.4	69.6	85.0	100.2	69.2	87.7	95.4	70.0	85.1
26, 27, 3370	ELECTRICAL, COMMUNICATIONS & UTIL.	97.7	71.2	83.7	104.6	71.3	87.0	102.2	69.3	84.8	97.7	76.9	86.7	103.2	67.7	84.5	97.7	76.9	86.7
MF2014	WEIGHTED AVERAGE	101.2	73.1	89.0	102.4	73.8	89.9	99.0	75.3	88.7	100.1	74.0	88.7	99.9	75.5	89.3	102.3	74.2	90.0

DIVISION		NEBRASKA																	
		ALLIANCE			COLUMBUS			GRAND ISLAND			HASTINGS			LINCOLN			MCCOOK		
		693			686			688			689			683 - 685			690		
		MAT.	INST.	TOTAL	MAT.	INST.	TOTAL	MAT.	INST.	TOTAL	MAT.	INST.	TOTAL	MAT.	INST.	TOTAL	MAT.	INST.	TOTAL
015433	CONTRACTOR EQUIPMENT		97.8	97.8		101.6	101.6		101.6	101.6		101.6	101.6		101.6	101.6		101.6	101.6
0241, 31 - 34	SITE & INFRASTRUCTURE, DEMOLITION	100.5	98.8	99.3	100.2	92.3	94.6	105.0	93.1	96.5	103.7	92.3	95.6	92.7	93.0	92.9	103.8	92.2	95.6
0310	Concrete Forming & Accessories	87.3	56.4	60.7	95.5	67.4	71.2	95.1	72.4	75.5	98.2	71.9	75.5	95.7	68.2	71.9	93.0	56.1	61.2
0320	Concrete Reinforcing	108.8	85.1	96.7	98.3	76.5	87.2	97.7	76.4	86.9	97.7	77.1	87.2	96.7	76.0	86.1	101.8	76.2	88.8
0330	Cast-in-Place Concrete	113.7	60.0	91.6	115.7	60.0	92.8	122.5	67.2	99.8	122.5	63.2	98.2	92.4	69.6	83.1	122.6	59.3	96.6
03	CONCRETE	123.9	63.5	94.2	108.8	67.3	88.4	113.7	72.1	93.3	113.9	70.6	92.6	95.6	71.0	83.5	112.9	62.0	87.9
04	MASONRY	115.6	74.1	89.7	122.7	77.8	94.7	115.4	76.1	90.9	124.9	88.0	101.9	104.8	67.2	81.4	110.7	73.9	87.8
05	METALS	101.5	78.7	94.5	96.1	84.6	92.6	97.8	86.3	94.3	98.7	85.8	94.7	99.9	85.5	95.5	96.3	83.6	92.4
06	WOOD, PLASTICS & COMPOSITES	87.0	52.2	67.5	101.2	66.7	81.9	100.4	72.2	84.6	104.1	72.2	86.3	98.4	66.7	80.7	95.7	52.2	71.4
07	THERMAL & MOISTURE PROTECTION	104.3	67.8	89.4	100.8	69.8	88.1	100.9	72.8	89.4	100.9	83.4	93.8	98.1	69.9	86.5	98.9	66.6	85.6
08	OPENINGS	94.5	58.6	86.1	95.4	66.1	88.6	95.5	69.7	89.5	95.5	69.1	89.3	107.8	63.0	97.4	95.0	56.6	86.1
0920	Plaster & Gypsum Board	82.7	50.7	61.1	90.9	65.6	73.8	90.1	71.3	77.4	91.9	71.3	78.0	96.2	65.6	75.5	93.7	50.7	64.7
0950, 0980	Ceilings & Acoustic Treatment	91.9	50.7	64.8	80.2	65.6	70.6	80.2	71.3	74.3	80.2	71.3	74.3	90.4	65.6	74.1	88.0	50.7	63.5
0960	Flooring	99.9	82.1	94.8	92.9	100.4	95.0	92.7	107.2	96.8	93.9	100.4	95.7	101.0	83.6	96.0	97.6	82.1	93.1
0970, 0990	Wall Finishes & Painting/Coating	168.5	54.8	99.9	85.5	63.7	72.4	85.5	67.6	74.7	85.5	63.7	72.4	103.8	77.9	88.1	93.8	47.7	66.0
09	FINISHES	98.4	59.5	77.0	89.9	71.9	80.0	90.0	77.1	82.9	90.6	75.2	82.1	96.9	71.7	83.0	95.6	58.6	75.2
COVERS	DIVS. 10 - 14, 25, 28, 41, 43, 44, 46	100.0	65.3	93.0	100.0	86.1	97.2	100.0	88.7	97.7	100.0	86.7	97.3	100.0	88.1	97.6	100.0	64.5	92.8
21, 22, 23	FIRE SUPPRESSION, PLUMBING & HVAC	95.3	70.5	85.3	95.2	76.0	87.4	100.0	78.0	91.1	95.2	76.0	87.5	99.9	78.0	91.0	95.1	75.9	87.4
26, 27, 3370	ELECTRICAL, COMMUNICATIONS & UTIL.	92.3	69.4	80.2	93.6	82.3	87.6	92.2	67.0	78.9	91.6	83.6	87.4	105.4	67.0	85.1	94.4	69.4	81.2
MF2014	WEIGHTED AVERAGE	101.0	70.5	87.7	98.4	77.1	89.1	100.0	77.1	90.0	99.4	79.8	90.9	100.5	74.9	89.4	98.7	71.1	86.7

DIVISION		NEBRASKA												NEVADA					
		NORFOLK			NORTH PLATTE			OMAHA			VALENTINE			CARSON CITY			ELKO		
		687			691			680 - 681			692			897			898		
		MAT.	INST.	TOTAL	MAT.	INST.	TOTAL	MAT.	INST.	TOTAL	MAT.	INST.	TOTAL	MAT.	INST.	TOTAL	MAT.	INST.	TOTAL
015433	CONTRACTOR EQUIPMENT		92.8	92.8		101.6	101.6		92.8	92.8		95.9	95.9		98.2	98.2		98.2	98.2
0241, 31 - 34	SITE & INFRASTRUCTURE, DEMOLITION	83.6	91.5	89.2	105.2	92.3	96.0	90.0	92.1	91.5	88.6	96.5	94.2	84.5	99.0	94.8	66.9	97.6	88.8
0310	Concrete Forming & Accessories	81.7	71.1	72.5	95.4	71.5	74.8	94.3	72.4	75.4	83.7	55.8	59.6	103.3	92.8	94.2	109.2	81.3	85.2
0320	Concrete Reinforcing	98.4	66.5	82.2	101.2	77.1	88.9	99.7	76.6	87.9	101.8	66.1	83.7	103.2	118.4	111.0	102.7	113.2	108.1
0330	Cast-in-Place Concrete	116.7	62.4	94.4	122.6	64.5	98.7	101.1	75.5	90.6	108.4	57.2	87.4	100.4	85.7	94.4	95.1	80.2	89.0
03	CONCRETE	107.3	67.5	87.8	113.0	70.8	92.3	100.1	74.5	87.5	110.2	58.9	85.0	100.9	95.2	98.1	95.4	87.2	91.4
04	MASONRY	129.4	79.1	98.0	98.6	87.9	92.0	106.8	79.0	89.4	110.9	73.9	87.8	117.3	94.6	103.2	115.0	71.9	88.1
05	METALS	99.6	73.5	91.6	95.5	85.4	92.4	99.9	78.9	93.4	107.6	73.4	97.1	94.0	104.7	97.3	96.9	99.4	97.7
06	WOOD, PLASTICS & COMPOSITES	84.0	71.7	77.1	97.8	72.2	83.5	94.1	72.0	81.7	81.7	51.7	64.9	92.1	94.4	93.4	100.6	81.8	90.1
07	THERMAL & MOISTURE PROTECTION	100.8	73.6	89.6	98.9	82.8	92.3	95.5	78.7	88.6	99.6	67.6	86.5	103.5	92.3	99.0	99.9	76.1	90.1
08	OPENINGS	97.2	66.5	90.0	94.3	69.1	88.4	100.0	71.5	93.3	96.9	55.6	87.3	99.5	107.9	101.4	100.9	85.9	97.4
0920	Plaster & Gypsum Board	90.8	71.3	77.6	93.7	71.3	78.6	100.5	71.6	81.0	95.6	50.7	65.3	100.9	94.1	96.3	105.3	81.2	89.0
0950, 0980	Ceilings & Acoustic Treatment	92.0	71.3	78.4	88.0	71.3	77.0	93.0	71.6	78.9	104.9	50.7	69.3	94.0	94.1	94.1	94.4	81.2	85.7
0960	Flooring	117.4	100.4	112.5	98.5	100.4	99.1	113.4	81.9	104.4	127.4	79.8	113.8	101.9	107.9	103.6	105.4	51.1	89.9
0970, 0990	Wall Finishes & Painting/Coating	151.7	63.7	98.6	93.8	63.7	75.6	126.2	67.2	90.6	167.2	66.1	106.1	101.2	80.8	88.9	102.6	95.9	98.5
09	FINISHES	108.8	75.2	90.3	95.9	75.2	84.5	106.0	73.3	88.0	118.2	59.7	86.0	97.9	94.3	95.9	97.6	77.2	86.3
COVERS	DIVS. 10 - 14, 25, 28, 41, 43, 44, 46	100.0	85.5	97.1	100.0	66.8	93.3	100.0	87.6	97.5	100.0	62.5	92.4	100.0	89.8	97.9	100.0	62.8	92.5
21, 22, 23	FIRE SUPPRESSION, PLUMBING & HVAC	95.0	75.8	87.2	99.9	76.0	90.2	99.9	76.8	90.6	94.8	75.7	87.1	100.0	80.1	92.0	97.7	79.6	90.4
26, 27, 3370	ELECTRICAL, COMMUNICATIONS & UTIL.	92.4	82.3	87.1	92.6	77.1	84.4	99.1	82.3	90.2	89.7	86.6	88.0	103.2	96.7	99.7	100.7	96.6	98.6
MF2014	WEIGHTED AVERAGE	100.0	76.7	89.8	99.0	78.3	90.0	100.2	78.6	90.8	101.2	72.5	88.7	99.7	93.5	97.0	98.3	84.9	92.4

DIVISION		NEVADA									NEW HAMPSHIRE								
		ELY			LAS VEGAS			RENO			CHARLESTON			CLAREMONT			CONCORD		
		893			889 - 891			894 - 895			036			037			032 - 033		
		MAT.	INST.	TOTAL	MAT.	INST.	TOTAL	MAT.	INST.	TOTAL	MAT.	INST.	TOTAL	MAT.	INST.	TOTAL	MAT.	INST.	TOTAL
015433	CONTRACTOR EQUIPMENT		98.2	98.2		98.2	98.2		98.2	98.2		100.9	100.9		100.9	100.9		100.9	100.9
0241, 31 - 34	SITE & INFRASTRUCTURE, DEMOLITION	72.2	99.1	91.3	75.4	101.0	93.6	72.2	99.0	91.3	86.9	99.6	96.0	80.9	99.6	94.2	93.7	103.9	100.9
0310	Concrete Forming & Accessories	102.3	105.7	105.2	103.5	109.6	108.8	98.7	92.7	93.5	86.8	81.1	81.9	92.6	81.1	82.7	95.5	93.6	93.8
0320	Concrete Reinforcing	101.5	111.6	106.7	93.7	119.9	107.1	96.0	119.9	108.1	84.1	93.5	88.9	84.1	93.5	88.9	95.3	94.1	94.7
0330	Cast-in-Place Concrete	102.1	105.4	103.4	98.9	107.8	102.5	108.0	85.7	98.8	97.1	70.5	86.2	89.2	70.5	81.6	112.2	90.5	103.3
03	CONCRETE	103.4	106.4	104.9	98.7	110.6	104.6	103.1	95.4	99.3	99.1	80.0	89.7	91.1	80.0	85.6	106.6	92.5	99.6
04	MASONRY	119.7	102.5	109.0	108.3	104.0	105.6	114.3	94.6	102.0	97.5	84.7	89.5	96.9	84.7	89.3	110.9	104.2	106.7
05	METALS	96.9	103.3	98.9	104.5	108.6	105.8	98.4	105.1	100.5	92.8	90.1	92.0	92.8	90.1	92.0	98.8	93.8	97.3
06	WOOD, PLASTICS & COMPOSITES	91.3	106.8	100.0	89.8	107.9	99.9	85.9	94.4	90.6	87.4	90.2	89.0	93.6	90.2	91.7	95.4	92.4	93.7
07	THERMAL & MOISTURE PROTECTION	100.3	95.7	98.4	113.8	102.6	109.2	99.9	92.3	96.8	98.8	72.3	88.0	98.7	72.3	87.9	101.9	91.5	97.6
08	OPENINGS	100.8	98.8	100.3	99.9	115.6	103.5	98.7	102.3	99.6	102.0	76.4	96.0	103.3	76.4	97.0	104.9	87.2	100.8
0920	Plaster & Gypsum Board	101.0	107.0	105.1	97.5	108.1	104.7	92.0	94.1	93.4	100.0	89.3	92.8	101.0	89.3	93.1	103.9	91.6	95.6
0950, 0980	Ceilings & Acoustic Treatment	94.4	107.0	102.7	102.7	108.1	106.3	99.4	94.1	95.9	94.2	89.3	91.0	94.2	89.3	91.0	99.9	91.6	94.4
0960	Flooring	103.0	57.2	89.9	94.6	107.9	98.4	99.9	107.9	102.2	98.4	32.1	79.4	100.6	32.1	81.0	102.7	116.9	106.8
0970, 0990	Wall Finishes & Painting/Coating	102.6	121.0	113.7	105.4	121.0	114.8	102.6	80.8	89.4	96.4	45.7	65.8	96.4	46.0	65.9	95.8	95.4	95.6
09	FINISHES	96.8	98.9	98.0	95.8	110.4	103.9	95.6	94.3	94.9	95.3	69.7	81.2	95.6	69.8	81.4	97.4	97.9	97.7
COVERS	DIVS. 10 - 14, 25, 28, 41, 43, 44, 46	100.0	63.6	92.6	100.0	104.4	100.9	100.0	89.8	97.9	100.0	92.2	98.4	100.0	92.2	98.4	100.0	104.3	100.9
21, 22, 23	FIRE SUPPRESSION, PLUMBING & HVAC	97.7	104.5	100.5	100.1	104.4	101.8	100.0	80.1	92.0	95.3	39.6	72.8	95.3	39.7	72.9	99.9	84.4	93.7
26, 27, 3370	ELECTRICAL, COMMUNICATIONS & UTIL.	101.0	109.1	105.3	105.4	119.9	113.1	101.4	96.7	98.9	98.5	52.9	74.4	98.5	52.9	74.4	99.5	83.6	91.1
MF2014	WEIGHTED AVERAGE	99.4	102.2	100.6	100.9	108.7	104.3	99.6	93.3	96.8	96.7	69.5	84.8	95.9	69.5	84.4	101.1	92.5	97.4

DIVISION		NEW HAMPSHIRE															NEW JERSEY		
		KEENE			LITTLETON			MANCHESTER			NASHUA			PORTSMOUTH			ATLANTIC CITY		
		034			035			031			030			038			082,084		
		MAT.	INST.	TOTAL	MAT.	INST.	TOTAL	MAT.	INST.	TOTAL	MAT.	INST.	TOTAL	MAT.	INST.	TOTAL	MAT.	INST.	TOTAL
015433	CONTRACTOR EQUIPMENT		100.9	100.9		100.9	100.9		100.9	100.9		100.9	100.9		100.9	100.9		99.1	99.1
0241, 31 - 34	SITE & INFRASTRUCTURE, DEMOLITION	94.0	99.9	98.2	81.0	99.9	94.4	93.6	103.9	100.9	95.7	103.9	101.5	89.2	104.0	99.7	96.4	105.2	102.7
0310	Concrete Forming & Accessories	91.3	83.0	84.2	102.9	83.0	85.8	96.9	94.0	94.4	99.3	94.0	94.8	88.2	94.3	93.4	109.0	127.5	125.0
0320	Concrete Reinforcing	84.1	93.6	88.9	84.8	93.6	89.3	103.7	94.1	98.8	105.1	94.1	99.5	84.1	94.2	89.2	76.3	118.1	97.6
0330	Cast-in-Place Concrete	97.6	72.8	87.4	87.6	72.7	81.5	110.6	113.1	111.7	92.4	113.1	100.9	87.6	113.2	98.1	86.1	133.0	105.3
03	CONCRETE	98.7	81.6	90.3	90.5	81.6	86.1	107.2	100.4	103.9	98.9	100.4	99.7	90.6	100.6	95.5	93.1	126.5	109.5
04	MASONRY	101.3	88.3	93.2	107.9	88.3	95.7	102.9	104.2	103.7	101.4	104.2	103.1	97.2	104.2	101.5	111.8	132.0	124.4
05	METALS	93.5	90.7	92.6	93.5	90.7	92.6	101.3	94.3	99.2	98.8	94.3	97.5	95.0	94.9	94.9	95.6	106.1	98.8
06	WOOD, PLASTICS & COMPOSITES	92.0	90.2	91.0	103.5	90.2	96.0	95.3	92.4	93.7	101.9	92.4	96.6	88.7	92.4	90.8	109.6	127.2	119.4
07	THERMAL & MOISTURE PROTECTION	99.2	75.3	89.4	98.8	73.8	88.5	101.5	95.0	98.8	99.6	95.0	97.7	99.3	115.0	105.7	105.9	124.2	113.4
08	OPENINGS	100.4	81.2	96.0	104.4	77.1	98.0	105.2	87.2	101.0	105.4	87.2	101.2	106.2	84.0	101.0	101.7	124.4	107.0
0920	Plaster & Gypsum Board	100.3	89.3	92.9	114.6	89.3	97.5	104.7	91.6	95.8	110.0	91.6	97.5	100.0	91.6	94.3	105.4	127.5	120.3
0950, 0980	Ceilings & Acoustic Treatment	94.2	89.3	91.0	94.2	89.3	91.0	101.7	91.6	95.0	105.1	91.6	96.2	95.1	91.6	92.8	81.3	127.5	111.6
0960	Flooring	100.2	52.6	86.6	109.9	32.1	87.6	98.6	116.9	103.8	103.7	116.9	107.4	98.5	116.9	103.8	105.4	159.9	121.0
0970, 0990	Wall Finishes & Painting/Coating	96.4	45.7	65.8	96.4	59.9	74.4	100.7	95.4	97.5	96.4	95.4	95.8	96.4	95.4	95.8	93.9	130.2	115.8
09	FINISHES	97.0	74.5	84.6	100.2	72.2	84.8	98.6	97.9	98.2	101.7	97.9	99.6	96.2	97.9	97.1	96.0	134.5	117.2
COVERS	DIVS. 10 - 14, 25, 28, 41, 43, 44, 46	100.0	93.4	98.7	100.0	98.6	99.7	100.0	104.3	100.9	100.0	104.3	100.9	100.0	104.3	100.9	100.0	111.5	102.3
21, 22, 23	FIRE SUPPRESSION, PLUMBING & HVAC	95.3	43.3	74.3	95.3	63.8	82.6	99.9	84.4	93.7	100.1	84.4	93.8	100.1	84.4	93.8	99.7	123.6	109.3
26, 27, 3370	ELECTRICAL, COMMUNICATIONS & UTIL.	98.5	63.2	79.9	99.8	55.9	76.6	99.8	83.6	91.2	101.2	83.6	91.9	99.1	83.6	90.9	91.1	138.1	116.0
MF2014	WEIGHTED AVERAGE	97.2	73.3	86.7	97.1	76.3	88.0	101.4	93.8	98.1	100.5	93.8	97.6	97.9	94.3	96.4	98.2	124.8	109.8

DIVISION		NEW JERSEY																	
		CAMDEN			DOVER			ELIZABETH			HACKENSACK			JERSEY CITY			LONG BRANCH		
		081			078			072			076			073			077		
		MAT.	INST.	TOTAL	MAT.	INST.	TOTAL	MAT.	INST.	TOTAL	MAT.	INST.	TOTAL	MAT.	INST.	TOTAL	MAT.	INST.	TOTAL
015433	CONTRACTOR EQUIPMENT		99.1	99.1		100.9	100.9		100.9	100.9		100.9	100.9		99.1	99.1		98.7	98.7
0241, 31 - 34	SITE & INFRASTRUCTURE, DEMOLITION	97.8	105.5	103.3	102.0	106.3	105.1	106.1	106.3	106.3	103.0	106.3	105.3	93.4	106.3	102.5	97.5	106.0	103.6
0310	Concrete Forming & Accessories	100.2	127.5	123.8	97.8	128.3	124.1	110.2	128.4	125.9	97.8	128.3	124.1	101.8	128.3	124.7	102.4	128.0	124.5
0320	Concrete Reinforcing	100.1	118.3	109.4	77.2	137.2	107.7	77.2	137.2	107.7	77.2	137.2	107.7	100.1	137.2	119.0	77.2	137.1	107.7
0330	Cast-in-Place Concrete	83.5	132.9	103.8	101.5	127.7	112.2	87.1	131.4	105.3	99.2	131.4	112.4	79.2	127.7	99.1	88.0	132.9	106.4
03	CONCRETE	94.0	126.5	110.0	98.8	128.7	113.5	94.5	130.1	112.0	97.0	130.0	113.2	92.1	128.6	110.0	96.2	130.2	112.9
04	MASONRY	101.3	132.0	120.4	93.0	132.5	117.6	108.7	132.5	123.5	96.9	132.5	119.1	87.0	132.5	115.4	101.3	132.0	120.5
05	METALS	101.1	106.1	102.6	93.3	116.2	100.4	94.8	116.2	101.4	93.4	116.1	100.4	98.9	113.9	103.5	93.4	113.7	99.7
06	WOOD, PLASTICS & COMPOSITES	98.4	127.2	114.5	96.2	127.2	113.5	111.7	127.2	120.4	96.2	127.2	113.5	97.1	127.2	114.0	98.2	127.1	114.4
07	THERMAL & MOISTURE PROTECTION	105.8	123.3	113.0	100.6	132.8	113.8	100.8	133.3	114.1	100.4	125.5	110.7	100.2	132.8	113.5	100.3	124.4	110.2
08	OPENINGS	104.1	124.5	108.8	105.5	127.7	110.6	103.6	127.7	109.2	102.9	127.7	108.7	101.6	127.7	107.7	97.8	127.6	104.7
0920	Plaster & Gypsum Board	101.5	127.5	119.0	100.9	127.5	118.9	107.7	127.5	121.0	100.9	127.5	118.9	104.3	127.5	120.0	102.7	127.5	119.4
0950, 0980	Ceilings & Acoustic Treatment	91.5	127.5	115.1	83.8	127.5	112.5	85.7	127.5	113.1	83.8	127.5	112.5	94.9	127.5	116.3	83.8	127.5	112.5
0960	Flooring	101.5	159.9	118.2	93.6	178.1	117.8	98.5	178.1	121.3	93.6	178.1	117.8	94.6	178.1	118.5	94.8	178.1	118.6
0970, 0990	Wall Finishes & Painting/Coating	93.9	130.2	115.8	98.4	132.5	119.0	98.4	132.5	119.0	98.4	132.5	119.0	98.5	132.5	119.0	98.5	130.2	117.6
09	FINISHES	96.4	134.5	117.4	93.7	136.9	117.5	96.9	136.9	118.9	93.5	136.9	117.4	96.6	136.9	118.8	94.5	137.5	118.2
COVERS	DIVS. 10 - 14, 25, 28, 41, 43, 44, 46	100.0	111.5	102.3	100.0	118.1	103.7	100.0	118.1	103.7	100.0	118.1	103.7	100.0	118.1	103.7	100.0	111.3	102.3
21, 22, 23	FIRE SUPPRESSION, PLUMBING & HVAC	100.0	123.6	109.5	99.7	127.4	110.9	100.1	125.5	110.3	99.7	127.4	110.9	100.1	127.4	111.1	99.7	127.1	110.8
26, 27, 3370	ELECTRICAL, COMMUNICATIONS & UTIL.	96.0	138.1	118.2	92.8	137.7	116.5	93.4	137.7	116.8	92.8	140.1	117.8	97.6	140.1	120.1	92.5	130.8	112.7
MF2014	WEIGHTED AVERAGE	99.5	124.8	110.5	97.9	127.8	110.9	98.8	127.6	111.3	97.6	128.1	110.9	97.9	127.9	111.0	97.1	126.3	109.8

DIVISION		NEW JERSEY																	
		NEW BRUNSWICK			NEWARK			PATERSON			POINT PLEASANT			SUMMIT			TRENTON		
		088 - 089			070 - 071			074 - 075			087			079			085 - 086		
		MAT.	INST.	TOTAL	MAT.	INST.	TOTAL	MAT.	INST.	TOTAL	MAT.	INST.	TOTAL	MAT.	INST.	TOTAL	MAT.	INST.	TOTAL
015433	CONTRACTOR EQUIPMENT		98.7	98.7		100.9	100.9		100.9	100.9		98.7	98.7		100.9	100.9		98.7	98.7
0241, 31 - 34	SITE & INFRASTRUCTURE, DEMOLITION	109.2	106.0	107.0	107.8	106.3	106.8	105.0	106.3	105.9	110.9	106.0	107.4	103.7	106.3	105.6	95.8	106.0	103.0
0310	Concrete Forming & Accessories	103.3	128.3	124.8	97.6	128.4	124.2	99.9	128.2	124.3	97.7	127.9	123.7	100.6	128.4	124.5	98.6	127.7	123.7
0320	Concrete Reinforcing	77.2	137.2	107.7	99.8	137.2	118.8	100.1	137.2	119.0	77.2	137.1	107.7	77.2	137.2	107.7	99.8	112.4	106.2
0330	Cast-in-Place Concrete	106.3	133.2	117.4	108.8	131.5	118.1	100.8	131.4	113.4	106.3	132.8	117.2	84.3	131.4	103.6	101.8	132.8	114.5
03	CONCRETE	110.4	130.5	120.3	105.8	130.1	117.7	102.2	130.0	115.9	110.1	130.1	119.9	91.5	130.1	110.5	102.5	125.5	113.8
04	MASONRY	109.3	132.5	123.7	99.6	132.5	120.1	93.6	132.5	117.9	97.0	132.0	118.8	95.5	132.5	118.6	103.9	132.0	121.4
05	METALS	95.6	113.8	101.2	100.8	116.3	105.5	94.2	116.1	100.9	95.6	113.5	101.1	93.3	116.2	100.4	100.8	105.2	102.1
06	WOOD, PLASTICS & COMPOSITES	103.0	127.1	116.5	93.9	127.2	112.5	98.8	127.2	114.7	95.9	127.1	113.4	100.1	127.2	115.3	95.6	127.1	113.3
07	THERMAL & MOISTURE PROTECTION	106.1	132.0	116.7	101.1	133.3	114.3	100.7	125.5	110.9	106.2	124.4	113.7	101.0	133.3	114.2	105.5	124.3	113.2
08	OPENINGS	96.1	127.6	103.5	104.4	127.7	109.8	108.3	127.7	112.8	98.2	129.2	105.4	110.2	127.7	114.2	106.1	122.2	109.8
0920	Plaster & Gypsum Board	102.6	127.5	119.4	101.1	127.5	118.9	104.3	127.5	120.0	98.3	127.5	118.0	102.7	127.5	119.4	98.7	127.5	118.1
0950, 0980	Ceilings & Acoustic Treatment	81.3	127.5	111.6	97.6	127.5	117.2	94.9	127.5	116.3	81.3	127.5	111.6	83.8	127.5	112.5	93.1	127.5	115.7
0960	Flooring	103.0	178.1	124.5	95.7	178.1	119.2	94.6	178.1	118.5	100.5	159.9	117.5	94.9	178.1	118.7	102.1	172.2	122.1
0970, 0990	Wall Finishes & Painting/Coating	93.9	132.5	117.2	99.8	132.5	119.5	98.4	132.5	119.0	93.9	130.2	115.8	98.4	132.5	119.0	99.1	130.2	117.8
09	FINISHES	96.0	136.8	118.5	95.6	136.9	118.3	96.7	136.9	118.9	94.7	134.5	116.6	94.6	136.9	117.9	96.6	136.5	118.6
COVERS	DIVS. 10 - 14, 25, 28, 41, 43, 44, 46	100.0	118.0	103.6	100.0	118.1	103.7	100.0	118.1	103.7	100.0	108.8	101.8	100.0	118.1	103.7	100.0	111.3	102.3
21, 22, 23	FIRE SUPPRESSION, PLUMBING & HVAC	99.7	127.4	110.8	100.1	127.4	111.1	100.1	127.4	111.1	99.7	127.0	110.7	99.7	125.5	110.1	100.1	126.9	110.9
26, 27, 3370	ELECTRICAL, COMMUNICATIONS & UTIL.	91.8	136.8	115.6	101.2	140.1	121.8	97.6	137.7	118.8	91.1	130.8	112.1	93.4	137.7	116.8	99.3	135.4	118.4
MF2014	WEIGHTED AVERAGE	99.7	127.6	111.9	101.2	128.4	113.0	99.6	127.8	111.9	99.2	125.9	110.8	97.9	127.6	110.8	100.9	125.1	111.5

DIVISION		NEW JERSEY			NEW MEXICO														
		VINELAND			ALBUQUERQUE			CARRIZOZO			CLOVIS			FARMINGTON			GALLUP		
		080,083			870 - 872			883			881			874			873		
		MAT.	INST.	TOTAL	MAT.	INST.	TOTAL	MAT.	INST.	TOTAL	MAT.	INST.	TOTAL	MAT.	INST.	TOTAL	MAT.	INST.	TOTAL
015433	CONTRACTOR EQUIPMENT		99.1	99.1		110.1	110.1		110.1	110.1		110.1	110.1		110.1	110.1		110.1	110.1
0241, 31 - 34	SITE & INFRASTRUCTURE, DEMOLITION	100.7	105.5	104.1	87.6	103.3	98.7	105.8	103.3	104.0	94.1	103.3	100.6	94.1	103.3	100.6	102.1	103.3	102.9
0310	Concrete Forming & Accessories	95.2	127.6	123.1	101.7	64.6	69.7	99.2	64.6	69.3	99.2	64.5	69.2	101.7	64.6	69.7	101.7	64.6	69.7
0320	Concrete Reinforcing	76.3	116.1	96.6	98.4	68.9	83.4	107.0	68.9	87.6	108.2	68.9	88.2	107.6	68.9	87.9	103.0	68.9	85.6
0330	Cast-in-Place Concrete	92.8	133.0	109.3	98.9	70.4	87.2	95.8	70.4	85.4	95.7	70.3	85.3	99.8	70.4	87.7	93.9	70.4	84.3
03	CONCRETE	97.9	126.2	111.8	100.2	68.5	84.6	117.0	68.5	93.2	105.8	68.4	87.4	103.9	68.5	86.5	110.0	68.5	89.6
04	MASONRY	99.4	132.0	119.7	105.0	58.7	76.1	103.4	58.7	75.5	103.4	58.7	75.5	113.7	58.7	79.4	99.4	58.7	74.0
05	METALS	95.5	105.5	98.6	105.2	86.8	99.5	99.5	86.8	95.6	99.2	86.7	95.3	102.9	86.8	97.9	102.0	86.8	97.3
06	WOOD, PLASTICS & COMPOSITES	92.8	127.2	112.1	97.5	65.5	79.6	93.1	65.5	77.6	93.1	65.5	77.6	97.7	65.5	79.7	97.7	65.5	79.7
07	THERMAL & MOISTURE PROTECTION	105.7	124.2	113.3	94.8	73.5	86.0	100.3	73.5	89.3	99.2	73.5	88.7	94.9	73.5	86.2	95.9	73.5	86.7
08	OPENINGS	97.7	124.1	103.8	101.2	67.7	93.4	98.6	67.7	91.4	98.7	67.7	91.5	103.6	67.7	95.3	103.6	67.7	95.3
0920	Plaster & Gypsum Board	96.9	127.5	117.5	95.8	64.2	74.4	78.6	64.2	68.8	78.6	64.2	68.8	89.4	64.2	72.3	89.4	64.2	72.3
0950, 0980	Ceilings & Acoustic Treatment	81.3	127.5	111.6	95.5	64.2	74.9	96.3	64.2	75.2	96.3	64.2	75.2	92.6	64.2	73.9	92.6	64.2	73.9
0960	Flooring	99.6	159.9	116.8	101.9	66.0	91.6	101.2	66.0	91.2	101.2	66.0	91.2	103.7	66.0	92.9	103.7	66.0	92.9
0970, 0990	Wall Finishes & Painting/Coating	93.9	130.2	115.8	113.2	66.6	85.1	103.3	66.6	81.1	103.3	66.6	81.1	107.1	66.6	82.7	107.1	66.6	82.7
09	FINISHES	93.5	134.5	116.1	95.9	64.8	78.8	96.6	64.8	79.1	95.4	64.8	78.5	94.8	64.8	78.3	96.0	64.8	78.8
COVERS	DIVS. 10 - 14, 25, 28, 41, 43, 44, 46	100.0	111.5	102.3	100.0	82.9	96.5	100.0	82.9	96.5	100.0	82.9	96.5	100.0	82.9	96.5	100.0	82.9	96.5
21, 22, 23	FIRE SUPPRESSION, PLUMBING & HVAC	99.7	123.6	109.3	100.1	69.0	87.6	97.2	69.0	85.8	97.2	68.7	85.7	100.0	69.0	87.5	97.1	69.0	85.8
26, 27, 3370	ELECTRICAL, COMMUNICATIONS & UTIL.	91.1	138.1	116.0	91.2	71.2	80.6	92.8	71.2	81.4	90.2	71.2	80.2	88.8	71.2	79.5	88.0	71.2	79.1
MF2014	WEIGHTED AVERAGE	97.5	124.7	109.4	99.6	72.6	87.8	100.3	72.6	88.2	98.3	72.5	87.1	100.1	72.6	88.1	99.5	72.6	87.8

DIVISION		NEW MEXICO																	
		LAS CRUCES			LAS VEGAS			ROSWELL			SANTA FE			SOCORRO			TRUTH/CONSEQUENCES		
		880			877			882			875			878			879		
		MAT.	INST.	TOTAL	MAT.	INST.	TOTAL	MAT.	INST.	TOTAL	MAT.	INST.	TOTAL	MAT.	INST.	TOTAL	MAT.	INST.	TOTAL
015433	CONTRACTOR EQUIPMENT		86.0	86.0		110.1	110.1		110.1	110.1		110.1	110.1		110.1	110.1		86.0	86.0
0241, 31 - 34	SITE & INFRASTRUCTURE, DEMOLITION	94.0	82.9	86.1	93.6	103.3	100.5	96.3	103.3	101.3	98.8	103.3	102.0	89.8	103.3	99.4	108.3	82.9	90.3
0310	Concrete Forming & Accessories	95.7	63.4	67.8	101.7	64.6	69.7	99.2	64.6	69.3	100.4	64.6	69.5	101.7	64.6	69.7	99.3	63.4	68.3
0320	Concrete Reinforcing	104.6	68.7	86.4	104.7	68.9	86.5	108.2	68.9	88.2	103.8	68.9	86.1	106.8	68.9	87.5	100.8	68.7	84.5
0330	Cast-in-Place Concrete	90.4	62.7	79.0	97.1	70.4	86.1	95.8	70.4	85.3	105.3	70.4	91.0	95.1	70.4	85.0	104.2	62.7	87.1
03	CONCRETE	84.9	65.0	75.1	101.2	68.5	85.1	106.5	68.5	87.8	104.0	68.5	86.5	100.2	68.5	84.6	94.1	65.0	79.8
04	MASONRY	99.2	58.3	73.7	99.7	58.7	74.1	114.3	58.7	79.6	104.1	58.7	75.8	99.6	58.7	74.1	97.2	58.3	73.0
05	METALS	98.1	80.3	92.6	101.7	86.8	97.1	100.4	86.8	96.2	99.0	86.8	95.2	102.0	86.8	97.3	101.6	80.3	95.0
06	WOOD, PLASTICS & COMPOSITES	82.5	64.4	72.4	97.7	65.5	79.7	93.1	65.5	77.6	99.3	65.5	80.4	97.7	65.5	79.7	89.1	64.4	75.3
07	THERMAL & MOISTURE PROTECTION	86.0	68.6	78.8	94.5	73.5	85.9	99.3	73.5	88.7	96.9	73.5	87.3	94.5	73.5	85.9	83.3	68.6	77.2
08	OPENINGS	91.5	67.1	85.8	99.9	67.7	92.4	98.5	67.7	91.4	102.0	67.7	94.0	99.8	67.7	92.3	93.1	67.1	87.0
0920	Plaster & Gypsum Board	77.5	64.2	68.5	89.4	64.2	72.3	78.6	64.2	68.8	98.9	64.2	75.4	89.4	64.2	72.3	91.1	64.2	72.9
0950, 0980	Ceilings & Acoustic Treatment	84.2	64.2	71.0	92.6	64.2	73.9	96.3	64.2	75.2	93.0	64.2	74.0	92.6	64.2	73.9	82.4	64.2	70.4
0960	Flooring	131.2	66.0	112.5	103.7	66.0	92.9	101.2	66.0	91.2	110.8	66.0	98.0	103.7	66.0	92.9	135.3	66.0	115.4
0970, 0990	Wall Finishes & Painting/Coating	90.7	66.6	76.2	107.1	66.6	82.7	103.3	66.6	81.1	111.8	66.6	84.5	107.1	66.6	82.7	97.7	66.6	79.0
09	FINISHES	105.2	63.9	82.5	94.7	64.8	78.2	95.5	64.8	78.6	99.7	64.8	80.5	94.6	64.8	78.2	107.5	63.9	83.5
COVERS	DIVS. 10 - 14, 25, 28, 41, 43, 44, 46	100.0	80.2	96.0	100.0	82.9	96.5	100.0	82.9	96.5	100.0	82.9	96.5	100.0	82.9	96.5	100.0	80.2	96.0
21, 22, 23	FIRE SUPPRESSION, PLUMBING & HVAC	100.3	68.7	87.6	97.1	69.0	85.8	99.9	69.0	87.5	100.0	69.0	87.5	97.1	69.0	85.8	97.1	68.7	85.6
26, 27, 3370	ELECTRICAL, COMMUNICATIONS & UTIL.	92.1	71.2	81.0	90.6	71.2	80.4	91.7	71.2	80.9	102.9	71.2	86.2	88.5	71.2	79.4	92.3	71.2	81.2
MF2014	WEIGHTED AVERAGE	96.1	69.4	84.4	98.0	72.6	86.9	100.0	72.6	88.0	100.8	72.6	88.5	97.6	72.6	86.7	97.5	69.4	85.2

	DIVISION	NEW MEXICO			NEW YORK															
		TUCUMCARI			ALBANY			BINGHAMTON			BRONX			BROOKLYN			BUFFALO			
		884			120 - 122			137 - 139			104			112			140 - 142			
		MAT.	INST.	TOTAL	MAT.	INST.	TOTAL	MAT.	INST.	TOTAL	MAT.	INST.	TOTAL	MAT.	INST.	TOTAL	MAT.	INST.	TOTAL	
015433	CONTRACTOR EQUIPMENT		110.1	110.1		112.9	112.9		114.1	114.1		110.6	110.6		113.3	113.3		97.1	97.1	
0241, 31 - 34	SITE & INFRASTRUCTURE, DEMOLITION	93.8	103.3	100.5	83.8	106.3	99.8	95.9	94.1	94.6	108.8	120.8	117.3	120.3	127.6	125.5	98.3	98.2	98.2	
0310	Concrete Forming & Accessories	99.2	64.5	69.2	100.1	100.6	100.5	100.8	88.5	90.2	98.4	175.3	164.8	107.2	182.9	172.5	97.3	116.5	113.9	
0320	Concrete Reinforcing	106.0	68.9	87.1	104.0	103.8	103.9	93.7	97.8	95.8	103.9	184.4	144.9	95.1	205.9	151.5	97.8	102.1	100.0	
0330	Cast-in-Place Concrete	95.7	70.3	85.3	91.9	111.3	99.9	102.7	127.9	113.1	95.9	173.4	127.7	104.8	172.1	132.4	106.7	119.7	112.0	
03	CONCRETE	105.0	68.4	87.0	99.0	105.4	102.1	95.8	105.2	100.4	96.1	174.8	134.8	107.5	181.3	143.7	103.0	114.1	108.4	
04	MASONRY	114.6	58.7	79.8	101.0	112.6	108.2	109.7	127.8	121.0	92.7	177.5	145.5	120.2	177.4	155.9	105.1	121.0	115.0	
05	METALS	99.2	86.7	95.3	104.1	110.3	106.0	96.3	118.9	103.3	99.9	151.6	115.8	104.4	150.1	118.4	99.6	95.4	98.3	
06	WOOD, PLASTICS & COMPOSITES	93.1	65.5	77.6	98.7	97.9	98.3	104.9	84.7	93.6	94.4	174.9	139.5	106.9	185.1	150.6	99.4	116.4	108.9	
07	THERMAL & MOISTURE PROTECTION	99.1	73.5	88.6	105.9	105.8	105.9	107.3	103.6	105.8	108.5	163.2	131.0	108.4	164.1	131.2	101.8	110.6	105.4	
08	OPENINGS	98.5	67.7	91.3	102.6	94.0	100.6	93.0	85.0	91.1	87.9	175.5	108.3	90.4	180.0	111.2	97.7	103.9	99.1	
0920	Plaster & Gypsum Board	78.6	64.2	68.8	97.2	97.6	97.5	107.4	83.8	91.4	99.2	176.9	151.7	102.8	187.6	160.2	98.3	116.7	110.7	
0950, 0980	Ceilings & Acoustic Treatment	96.3	64.2	75.2	92.4	97.6	95.8	91.1	83.8	86.3	83.8	176.9	145.0	87.1	187.6	153.2	102.0	116.7	111.6	
0960	Flooring	101.2	66.0	91.2	97.1	113.7	101.9	106.8	103.6	105.9	98.3	186.8	123.6	113.9	186.8	134.8	101.3	121.2	107.0	
0970, 0990	Wall Finishes & Painting/Coating	103.3	66.6	81.1	103.8	94.3	98.0	93.3	98.8	96.6	102.9	157.4	135.8	123.4	157.4	143.9	98.8	112.4	107.0	
09	FINISHES	95.4	64.8	78.5	94.7	101.9	98.7	94.9	91.1	92.8	94.0	175.9	139.1	108.4	181.9	148.9	100.6	117.9	110.2	
COVERS	DIVS. 10 - 14, 25, 28, 41, 43, 44, 46	100.0	82.9	96.5	100.0	99.0	99.8	100.0	96.2	99.2	100.0	135.0	107.1	100.0	135.4	107.1	100.0	105.6	101.1	
21, 22, 23	FIRE SUPPRESSION, PLUMBING & HVAC	97.2	68.7	85.7	100.0	102.8	101.1	100.5	90.2	96.3	100.2	165.5	126.5	99.7	165.4	126.2	100.0	96.7	98.7	
26, 27, 3370	ELECTRICAL, COMMUNICATIONS & UTIL.	92.8	71.2	81.4	98.7	104.1	101.6	99.9	105.4	102.8	97.0	181.9	141.9	99.7	181.9	143.1	100.2	102.7	101.5	
MF2014	WEIGHTED AVERAGE	99.0	72.5	87.4	100.1	104.7	102.1	98.5	101.4	99.8	97.7	166.1	127.5	102.8	168.5	131.4	100.3	106.2	102.9	

	DIVISION	NEW YORK																	
		ELMIRA			FAR ROCKAWAY			FLUSHING			GLENS FALLS			HICKSVILLE			JAMAICA		
		148 - 149			116			113			128			115,117,118			114		
		MAT.	INST.	TOTAL	MAT.	INST.	TOTAL	MAT.	INST.	TOTAL	MAT.	INST.	TOTAL	MAT.	INST.	TOTAL	MAT.	INST.	TOTAL
015433	CONTRACTOR EQUIPMENT		116.0	116.0		113.3	113.3		113.3	113.3		112.9	112.9		113.3	113.3		113.3	113.3
0241, 31 - 34	SITE & INFRASTRUCTURE, DEMOLITION	97.1	94.1	95.0	123.4	127.6	126.4	123.4	127.6	126.4	73.6	105.8	96.5	113.3	126.3	122.6	117.7	127.6	124.7
0310	Concrete Forming & Accessories	81.3	93.2	91.6	93.5	175.1	163.9	97.3	175.1	164.4	85.7	90.3	89.7	90.0	154.5	145.6	97.3	175.1	164.4
0320	Concrete Reinforcing	97.4	95.7	96.5	95.1	205.9	151.5	96.7	205.9	152.3	95.4	94.4	94.9	95.1	210.2	153.7	95.1	205.9	151.5
0330	Cast-in-Place Concrete	93.7	103.2	97.6	113.3	172.1	137.5	113.3	172.1	137.5	85.3	106.7	94.1	96.3	165.5	124.7	104.8	172.1	132.4
03	CONCRETE	90.8	98.6	94.6	113.6	177.8	145.1	114.1	177.8	145.4	88.3	97.5	92.9	99.4	167.1	132.6	106.8	177.8	141.7
04	MASONRY	103.5	102.1	102.6	124.1	177.4	157.4	118.1	177.4	155.1	100.7	105.6	103.7	114.6	166.8	147.1	122.2	177.4	156.6
05	METALS	96.7	117.3	103.0	104.4	150.1	118.4	104.4	150.1	118.4	97.7	106.1	100.3	105.8	149.1	119.2	104.4	150.1	118.4
06	WOOD, PLASTICS & COMPOSITES	85.0	92.2	89.0	90.1	174.6	137.4	94.8	174.6	139.5	87.1	87.6	87.3	107.9	156.3	127.8	94.8	174.6	139.5
07	THERMAL & MOISTURE PROTECTION	103.9	94.1	99.9	108.3	163.0	130.7	108.3	163.0	130.7	98.8	97.2	98.2	107.9	156.3	127.8	108.1	163.0	130.6
08	OPENINGS	99.4	88.9	97.0	89.0	174.4	108.8	89.0	174.4	108.8	93.3	86.2	91.7	89.0	163.4	106.3	89.0	174.4	108.8
0920	Plaster & Gypsum Board	97.8	91.7	93.7	91.6	176.9	149.2	94.1	176.9	150.0	89.1	87.0	87.7	91.2	154.2	133.8	94.1	176.9	150.0
0950, 0980	Ceilings & Acoustic Treatment	95.4	91.7	93.0	76.2	176.9	142.4	76.2	176.9	142.4	82.1	87.0	85.3	75.2	154.2	127.2	76.2	176.9	142.4
0960	Flooring	94.6	103.6	97.1	109.2	186.8	131.4	110.6	186.8	132.4	86.2	111.3	93.3	108.3	185.1	130.2	110.6	186.8	132.4
0970, 0990	Wall Finishes & Painting/Coating	98.9	90.0	93.5	123.4	157.4	143.9	123.4	157.4	143.9	101.1	87.4	92.8	123.4	157.4	143.9	123.4	157.4	143.9
09	FINISHES	95.5	94.7	95.0	103.5	175.7	143.3	104.3	175.7	143.6	86.1	93.1	90.0	102.1	159.7	133.8	103.8	175.7	143.4
COVERS	DIVS. 10 - 14, 25, 28, 41, 43, 44, 46	100.0	99.4	99.9	100.0	134.2	106.9	100.0	134.2	106.9	100.0	96.2	99.2	100.0	128.2	105.7	100.0	134.2	106.9
21, 22, 23	FIRE SUPPRESSION, PLUMBING & HVAC	95.5	91.5	93.9	95.0	165.3	123.4	95.0	165.3	123.4	95.5	95.7	95.6	99.7	154.4	121.8	95.0	165.3	123.4
26, 27, 3370	ELECTRICAL, COMMUNICATIONS & UTIL.	96.5	96.2	96.3	107.0	181.9	146.6	107.0	181.9	146.6	93.6	100.5	97.3	99.0	143.4	122.5	97.9	181.9	142.3
MF2014	WEIGHTED AVERAGE	96.7	97.4	97.0	102.6	166.8	130.6	102.5	166.8	130.5	94.0	98.6	96.0	100.9	153.3	123.7	100.8	166.8	129.6

	DIVISION	NEW YORK																	
		JAMESTOWN			KINGSTON			LONG ISLAND CITY			MONTICELLO			MOUNT VERNON			NEW ROCHELLE		
		147			124			111			127			105			108		
		MAT.	INST.	TOTAL	MAT.	INST.	TOTAL	MAT.	INST.	TOTAL	MAT.	INST.	TOTAL	MAT.	INST.	TOTAL	MAT.	INST.	TOTAL
015433	CONTRACTOR EQUIPMENT		93.9	93.9		113.3	113.3		113.3	113.3		113.3	113.3		110.6	110.6		110.6	110.6
0241, 31 - 34	SITE & INFRASTRUCTURE, DEMOLITION	98.4	94.2	95.4	140.6	122.9	128.0	121.3	127.6	125.8	135.5	122.7	126.4	115.2	118.5	117.6	114.6	118.5	117.4
0310	Concrete Forming & Accessories	81.3	86.9	86.1	87.1	104.4	102.0	101.6	175.1	165.0	94.7	102.9	101.8	88.8	139.1	132.1	103.6	139.0	134.2
0320	Concrete Reinforcing	97.6	98.9	98.2	95.8	140.6	118.6	95.1	205.9	151.5	95.1	140.1	118.0	102.9	183.3	143.8	103.0	183.3	143.9
0330	Cast-in-Place Concrete	97.2	102.1	99.2	115.2	135.1	123.4	108.1	172.1	134.4	107.8	124.8	114.8	107.1	141.3	121.1	107.0	141.3	121.1
03	CONCRETE	93.8	94.3	94.0	110.8	121.5	116.0	109.9	177.8	143.3	105.5	117.2	111.2	105.4	146.9	125.8	105.0	146.9	125.6
04	MASONRY	111.8	99.9	104.4	116.0	141.3	131.8	116.9	177.4	154.6	108.6	128.7	121.1	98.1	147.3	128.8	98.1	147.3	128.8
05	METALS	94.0	92.1	93.5	106.1	117.2	109.5	104.4	150.1	118.4	106.0	116.2	109.2	99.6	135.6	110.7	99.9	135.6	110.9
06	WOOD, PLASTICS & COMPOSITES	83.5	84.0	83.8	88.4	96.4	92.9	100.9	174.6	142.2	96.1	96.4	96.3	84.6	136.2	113.5	101.4	136.2	120.9
07	THERMAL & MOISTURE PROTECTION	103.4	93.5	99.3	122.3	137.2	128.4	108.3	163.0	130.7	122.0	132.1	126.1	109.4	144.1	123.6	109.5	144.1	123.7
08	OPENINGS	99.2	85.6	96.1	95.9	117.3	100.9	89.0	174.4	108.8	91.1	117.3	97.2	87.9	148.6	102.0	88.0	148.6	102.1
0920	Plaster & Gypsum Board	88.5	83.2	84.9	91.4	96.3	94.7	98.7	176.9	151.5	92.1	96.3	94.9	94.9	137.0	123.3	106.3	137.0	127.0
0950, 0980	Ceilings & Acoustic Treatment	92.0	83.2	86.2	72.9	96.3	88.3	76.2	176.9	142.4	72.9	96.3	88.3	82.1	137.0	118.2	82.1	137.0	118.2
0960	Flooring	97.7	103.6	99.4	103.8	72.5	94.9	112.1	186.8	133.5	106.1	72.5	96.5	90.2	186.8	117.8	97.1	186.8	122.8
0970, 0990	Wall Finishes & Painting/Coating	100.2	94.2	96.6	134.4	116.8	123.8	123.4	157.4	143.9	134.4	116.8	123.8	101.1	157.4	135.1	101.1	157.4	135.1
09	FINISHES	94.6	89.9	92.0	101.0	97.1	98.8	105.1	175.7	144.0	101.4	96.3	98.6	91.2	149.2	123.2	94.6	149.2	124.7
COVERS	DIVS. 10 - 14, 25, 28, 41, 43, 44, 46	100.0	98.6	99.7	100.0	111.9	102.4	100.0	134.2	106.9	100.0	110.8	102.2	100.0	125.2	105.1	100.0	121.9	104.4
21, 22, 23	FIRE SUPPRESSION, PLUMBING & HVAC	95.4	87.2	92.1	95.4	122.3	106.3	99.7	165.3	126.2	95.4	116.7	104.0	95.5	134.1	111.1	95.5	134.1	111.1
26, 27, 3370	ELECTRICAL, COMMUNICATIONS & UTIL.	95.4	96.3	95.9	95.1	111.0	103.5	98.5	181.9	142.6	95.1	111.0	103.5	95.2	155.2	126.9	95.2	155.2	126.9
MF2014	WEIGHTED AVERAGE	96.8	92.5	94.9	102.6	118.5	109.6	102.4	166.8	130.5	101.1	115.1	107.2	97.5	141.4	116.6	97.9	141.3	116.8

	DIVISION	NEW YORK																	
		NEW YORK			NIAGARA FALLS			PLATTSBURGH			POUGHKEEPSIE			QUEENS			RIVERHEAD		
		100 - 102			143			129			125 - 126			110			119		
		MAT.	INST.	TOTAL	MAT.	INST.	TOTAL	MAT.	INST.	TOTAL	MAT.	INST.	TOTAL	MAT.	INST.	TOTAL	MAT.	INST.	TOTAL
015433	CONTRACTOR EQUIPMENT		111.1	111.1		93.9	93.9		98.9	98.9		113.3	113.3		113.3	113.3		113.3	113.3
0241, 31 - 34	SITE & INFRASTRUCTURE, DEMOLITION	117.3	121.5	120.2	100.5	95.5	97.0	106.1	103.0	103.9	136.5	123.5	127.2	116.6	127.6	124.4	114.5	125.9	122.6
0310	Concrete Forming & Accessories	102.6	183.3	172.2	81.3	114.4	109.8	91.1	95.4	94.9	87.1	165.0	154.3	90.1	175.1	163.4	94.6	153.4	145.3
0320	Concrete Reinforcing	109.9	210.4	161.0	96.3	102.5	99.5	99.9	102.7	101.3	95.8	141.0	118.8	96.7	205.9	152.3	96.9	184.2	141.3
0330	Cast-in-Place Concrete	107.8	177.7	136.5	100.6	120.3	108.7	104.4	104.4	104.4	111.5	137.9	122.4	99.6	172.1	129.4	97.9	164.5	125.3
03	CONCRETE	106.2	184.2	144.5	95.9	113.3	104.4	102.5	99.3	101.0	107.9	149.5	128.3	102.6	177.8	139.6	100.1	161.4	130.2
04	MASONRY	102.5	177.5	149.2	119.9	127.5	124.6	95.0	101.6	99.1	108.6	144.5	131.0	111.1	177.4	152.4	119.9	166.3	148.8
05	METALS	113.1	151.9	125.0	96.7	93.8	95.8	102.0	91.2	98.7	106.1	119.8	110.3	104.4	150.1	118.4	106.3	138.4	116.2
06	WOOD, PLASTICS & COMPOSITES	98.5	185.4	147.2	83.4	110.3	98.5	93.8	93.5	93.6	88.4	174.6	136.7	86.7	174.6	135.9	91.8	152.6	125.9
07	THERMAL & MOISTURE PROTECTION	108.6	164.5	131.5	103.5	112.9	107.4	116.3	100.5	109.8	122.3	146.8	132.3	107.9	163.0	130.5	108.8	156.1	128.2
08	OPENINGS	93.7	180.9	114.0	99.2	100.7	99.6	101.6	90.8	99.1	95.9	159.5	110.7	89.0	174.4	108.8	89.0	157.5	104.9
0920	Plaster & Gypsum Board	106.0	187.6	161.2	88.5	110.3	103.3	109.4	92.6	98.0	91.4	176.9	149.2	91.2	176.9	149.1	92.5	154.2	134.2
0950, 0980	Ceilings & Acoustic Treatment	102.2	187.6	158.4	92.0	110.3	104.1	98.1	92.6	94.5	72.9	176.9	141.2	76.2	176.9	142.4	76.1	154.2	127.5
0960	Flooring	99.5	186.8	124.5	97.7	121.2	104.4	108.9	113.7	110.3	103.8	167.2	122.0	108.3	186.8	130.7	109.2	142.7	118.8
0970, 0990	Wall Finishes & Painting/Coating	102.9	157.4	135.8	100.2	112.4	107.6	130.0	91.4	106.7	134.4	117.2	124.0	123.4	157.4	143.9	123.4	157.4	143.9
09	FINISHES	99.9	182.1	145.2	94.7	115.9	106.4	97.9	98.0	98.0	100.8	163.0	135.1	102.5	175.7	142.8	102.7	151.3	129.5
COVERS	DIVS. 10 - 14, 25, 28, 41, 43, 44, 46	100.0	136.1	107.3	100.0	107.1	101.4	100.0	98.0	99.6	100.0	121.6	104.4	100.0	134.2	106.9	100.0	127.6	105.6
21, 22, 23	FIRE SUPPRESSION, PLUMBING & HVAC	100.1	165.5	126.5	95.4	100.0	97.2	95.4	96.6	95.9	95.4	127.0	108.1	99.7	165.3	126.2	99.9	151.4	120.7
26, 27, 3370	ELECTRICAL, COMMUNICATIONS & UTIL.	104.7	181.9	145.5	94.0	100.3	97.3	91.6	90.3	90.9	95.1	119.9	108.2	99.0	181.9	142.8	100.6	133.6	118.1
MF2014	WEIGHTED AVERAGE	103.4	168.7	131.8	97.8	106.4	101.5	99.1	96.7	98.0	101.8	136.7	117.0	100.9	166.8	129.6	101.6	148.2	121.9

	DIVISION	NEW YORK																	
		ROCHESTER			SCHENECTADY			STATEN ISLAND			SUFFERN			SYRACUSE			UTICA		
		144 - 146			123			103			109			130 - 132			133 - 135		
		MAT.	INST.	TOTAL	MAT.	INST.	TOTAL	MAT.	INST.	TOTAL	MAT.	INST.	TOTAL	MAT.	INST.	TOTAL	MAT.	INST.	TOTAL
015433	CONTRACTOR EQUIPMENT		116.7	116.7		112.9	112.9		110.6	110.6		110.6	110.6		112.9	112.9		112.9	112.9
0241, 31 - 34	SITE & INFRASTRUCTURE, DEMOLITION	85.9	109.9	102.9	84.2	106.3	99.9	119.9	120.8	120.6	111.4	116.3	114.9	94.8	105.6	102.4	73.5	104.3	95.4
0310	Concrete Forming & Accessories	98.7	98.2	98.3	103.0	100.6	100.9	88.3	183.3	170.2	96.9	135.2	129.9	99.8	91.0	92.2	101.0	87.9	89.7
0320	Concrete Reinforcing	100.1	95.7	97.9	94.4	103.8	99.2	103.9	210.4	158.1	103.0	140.9	122.3	94.7	96.4	95.6	94.7	95.7	95.2
0330	Cast-in-Place Concrete	94.2	104.4	98.4	102.9	111.3	106.4	107.1	173.5	134.4	103.8	138.2	117.9	95.4	105.9	99.7	87.3	104.7	94.4
03	CONCRETE	99.3	100.6	99.9	102.9	105.4	104.1	107.1	182.7	144.2	102.0	136.3	118.9	98.9	97.8	98.4	96.9	95.9	96.4
04	MASONRY	107.0	104.9	105.7	97.7	112.6	107.0	104.7	177.5	150.1	97.6	142.6	125.7	101.5	105.2	103.8	93.1	103.9	99.8
05	METALS	104.1	107.3	105.1	101.9	110.3	104.5	97.8	151.8	114.4	97.8	119.4	104.5	99.8	105.6	101.6	97.7	105.1	100.0
06	WOOD, PLASTICS & COMPOSITES	97.6	97.6	97.6	107.4	97.9	102.1	83.3	185.4	140.4	94.0	136.2	117.6	101.3	88.5	94.2	101.3	84.7	92.0
07	THERMAL & MOISTURE PROTECTION	103.5	101.7	102.7	100.3	105.8	102.6	108.9	164.3	131.6	109.3	142.1	122.8	102.3	97.6	100.4	90.6	97.6	93.5
08	OPENINGS	105.3	92.2	102.2	99.4	94.0	98.1	87.9	181.2	109.6	88.0	139.0	99.8	95.0	85.6	92.8	98.0	83.4	94.6
0920	Plaster & Gypsum Board	107.2	97.4	100.6	98.8	97.6	98.0	95.0	187.6	157.6	98.5	137.0	124.5	98.0	88.0	91.2	98.0	84.0	88.5
0950, 0980	Ceilings & Acoustic Treatment	100.4	97.4	98.5	88.7	97.6	94.6	83.8	187.6	152.0	82.1	137.0	118.2	91.1	88.0	89.0	91.1	84.0	86.4
0960	Flooring	93.8	113.3	99.4	92.9	113.7	98.9	94.2	186.8	120.7	93.5	185.1	119.7	94.6	102.1	96.8	92.1	102.2	95.0
0970, 0990	Wall Finishes & Painting/Coating	100.3	99.1	99.6	101.1	94.3	97.0	102.9	157.4	135.8	101.1	124.9	115.5	98.5	99.8	99.3	91.3	99.8	96.4
09	FINISHES	100.4	101.3	100.9	91.6	101.9	97.3	93.2	182.1	142.2	92.3	140.0	118.6	94.0	93.5	93.7	92.1	91.1	91.6
COVERS	DIVS. 10 - 14, 25, 28, 41, 43, 44, 46	100.0	99.8	100.0	100.0	99.0	99.8	100.0	136.1	107.3	100.0	123.6	104.8	100.0	96.5	99.3	100.0	96.2	99.2
21, 22, 23	FIRE SUPPRESSION, PLUMBING & HVAC	100.1	90.3	96.1	100.2	102.8	101.2	100.2	165.5	126.5	95.5	123.3	106.7	100.2	91.9	96.9	100.2	92.2	97.0
26, 27, 3370	ELECTRICAL, COMMUNICATIONS & UTIL.	98.7	94.6	96.6	98.1	104.1	101.3	97.0	181.9	141.9	102.8	119.9	111.8	100.0	102.2	101.1	98.1	102.2	100.3
MF2014	WEIGHTED AVERAGE	101.1	99.1	100.2	99.3	104.7	101.7	99.2	168.4	129.4	97.6	129.0	111.3	98.9	98.0	98.5	97.1	97.1	97.1

	DIVISION	NEW YORK									NORTH CAROLINA								
		WATERTOWN			WHITE PLAINS			YONKERS			ASHEVILLE			CHARLOTTE			DURHAM		
		136			106			107			287 - 288			281 - 282			277		
		MAT.	INST.	TOTAL	MAT.	INST.	TOTAL	MAT.	INST.	TOTAL	MAT.	INST.	TOTAL	MAT.	INST.	TOTAL	MAT.	INST.	TOTAL
015433	CONTRACTOR EQUIPMENT		112.9	112.9		110.6	110.6		110.6	110.6		96.3	96.3		96.3	96.3		101.7	101.7
0241, 31 - 34	SITE & INFRASTRUCTURE, DEMOLITION	80.9	105.7	98.5	108.5	118.5	115.6	116.7	118.5	118.0	102.4	76.9	84.3	105.6	76.9	85.2	100.4	85.7	89.9
0310	Concrete Forming & Accessories	85.9	94.3	93.1	101.9	139.1	134.0	102.2	143.9	138.2	96.0	41.3	48.8	99.7	42.8	50.6	100.2	44.3	52.0
0320	Concrete Reinforcing	95.3	96.5	95.9	103.0	183.3	143.9	107.1	183.3	145.9	93.0	63.0	77.7	98.8	58.5	78.3	91.2	57.9	74.2
0330	Cast-in-Place Concrete	101.5	107.6	104.1	95.1	141.3	114.1	106.3	141.4	120.7	115.4	51.3	89.1	120.2	49.8	91.3	101.6	47.1	79.2
03	CONCRETE	109.3	99.9	104.7	95.7	146.9	120.9	105.2	149.1	126.7	106.0	50.6	78.8	108.6	49.9	79.8	98.7	49.5	74.5
04	MASONRY	94.2	107.9	102.8	97.1	147.3	128.4	102.1	147.3	130.3	93.6	43.7	62.5	101.9	51.2	70.3	86.4	37.9	56.2
05	METALS	97.8	105.6	100.2	99.3	135.6	110.5	108.9	135.7	117.1	103.1	82.3	96.7	104.0	80.9	96.9	121.4	79.9	108.6
06	WOOD, PLASTICS & COMPOSITES	83.0	92.2	88.1	99.5	136.2	120.0	99.3	142.7	123.6	97.9	40.3	65.6	103.0	41.9	68.8	96.7	44.7	67.6
07	THERMAL & MOISTURE PROTECTION	90.9	100.1	94.7	109.2	144.1	123.5	109.5	144.8	124.0	106.7	43.7	80.8	100.8	46.6	78.6	106.9	45.7	81.8
08	OPENINGS	98.0	89.4	96.0	88.0	148.6	102.1	91.1	151.7	105.2	97.1	44.1	84.8	102.5	45.0	89.1	105.8	47.1	92.2
0920	Plaster & Gypsum Board	88.7	91.7	90.8	101.3	137.0	125.4	105.6	143.6	131.3	100.1	38.3	58.3	100.0	39.9	59.4	105.9	42.8	63.3
0950, 0980	Ceilings & Acoustic Treatment	91.1	91.7	91.5	82.1	137.0	118.2	100.5	143.6	128.9	85.7	38.3	54.5	88.8	39.9	56.7	88.1	42.8	58.3
0960	Flooring	86.0	102.2	90.6	95.6	186.8	121.7	95.2	186.8	121.4	102.2	42.7	85.2	101.6	43.4	84.9	103.9	42.7	86.4
0970, 0990	Wall Finishes & Painting/Coating	91.3	96.2	94.3	101.1	157.4	135.1	101.1	157.4	135.1	113.6	40.3	69.3	113.8	49.4	74.9	105.4	37.4	64.3
09	FINISHES	89.7	95.7	93.0	92.9	149.2	123.9	98.0	153.0	128.3	95.6	40.7	65.3	95.5	42.9	66.5	96.0	42.9	66.7
COVERS	DIVS. 10 - 14, 25, 28, 41, 43, 44, 46	100.0	97.5	99.5	100.0	125.2	105.1	100.0	126.3	105.3	100.0	77.7	95.5	100.0	78.1	95.6	100.0	72.1	94.4
21, 22, 23	FIRE SUPPRESSION, PLUMBING & HVAC	100.2	85.9	94.4	100.3	134.1	114.0	100.3	134.1	114.0	100.4	53.4	81.4	100.0	54.1	81.5	100.5	52.9	81.3
26, 27, 3370	ELECTRICAL, COMMUNICATIONS & UTIL.	100.0	90.4	94.9	95.2	155.2	126.9	102.9	163.8	135.1	102.2	55.7	77.6	101.2	58.5	78.6	96.1	55.6	74.7
MF2014	WEIGHTED AVERAGE	98.5	96.3	97.5	97.6	141.4	116.7	102.1	143.7	120.2	100.8	55.3	80.9	101.9	56.8	82.2	102.9	55.3	82.1

	DIVISION	NORTH CAROLINA																	
		ELIZABETH CITY			FAYETTEVILLE			GASTONIA			GREENSBORO			HICKORY			KINSTON		
		279			283			280			270,272 - 274			286			285		
		MAT.	INST.	TOTAL	MAT.	INST.	TOTAL	MAT.	INST.	TOTAL	MAT.	INST.	TOTAL	MAT.	INST.	TOTAL	MAT.	INST.	TOTAL
015433	CONTRACTOR EQUIPMENT		106.0	106.0		101.7	101.7		96.3	96.3		101.7	101.7		101.7	101.7		101.7	101.7
0241, 31 - 34	SITE & INFRASTRUCTURE, DEMOLITION	104.6	87.4	92.4	101.7	85.6	90.3	102.3	76.9	84.2	100.3	85.7	89.9	101.2	85.5	90.1	100.3	85.6	89.8
0310	Concrete Forming & Accessories	85.3	42.5	48.4	95.6	60.3	65.2	103.1	38.7	47.5	99.9	44.4	52.1	92.0	37.1	44.6	88.2	42.1	48.4
0320	Concrete Reinforcing	89.2	45.9	67.2	96.8	58.0	77.0	93.5	56.5	74.7	90.1	58.0	73.8	93.0	56.1	74.2	92.5	45.8	68.7
0330	Cast-in-Place Concrete	101.8	47.0	79.3	121.0	48.3	91.2	112.8	52.7	88.1	100.8	47.6	79.0	115.4	48.1	87.7	111.4	44.8	84.0
03	CONCRETE	98.7	46.4	73.0	108.1	57.1	83.1	104.5	48.7	77.1	98.1	49.8	74.4	105.7	46.3	76.5	102.4	45.5	74.4
04	MASONRY	98.6	48.0	67.1	97.3	38.9	60.9	98.3	50.9	68.7	82.5	41.4	56.9	82.3	43.7	58.2	89.1	48.1	63.6
05	METALS	107.1	75.8	97.5	124.2	80.0	110.6	103.9	80.0	96.5	113.7	80.0	103.4	103.2	78.7	95.7	102.0	74.8	93.6
06	WOOD, PLASTICS & COMPOSITES	80.1	43.6	59.7	97.1	66.8	80.1	107.0	36.6	67.6	96.4	44.9	67.6	92.1	35.2	60.2	88.6	42.9	63.0
07	THERMAL & MOISTURE PROTECTION	106.2	43.3	80.4	106.3	45.9	81.5	106.9	45.2	81.6	106.7	42.8	80.5	107.0	41.7	80.2	106.8	42.8	80.6
08	OPENINGS	102.6	38.4	87.7	97.2	59.1	88.4	100.9	41.0	87.0	105.8	47.2	92.2	97.2	36.8	83.1	97.3	43.1	84.7
0920	Plaster & Gypsum Board	98.6	41.0	59.7	104.7	65.6	78.3	107.0	34.5	58.0	107.5	43.0	63.9	100.1	33.1	54.8	99.8	41.0	60.1
0950, 0980	Ceilings & Acoustic Treatment	88.1	41.0	57.1	86.5	65.6	72.7	89.0	34.5	53.2	88.1	43.0	58.4	85.7	33.1	51.1	89.0	41.0	57.4
0960	Flooring	95.7	23.2	75.0	102.4	42.7	85.3	105.4	42.7	87.4	103.9	39.7	85.6	102.1	32.8	82.3	99.6	22.6	77.6
0970, 0990	Wall Finishes & Painting/Coating	105.4	42.0	67.1	113.6	33.2	65.0	113.6	40.3	69.3	105.4	31.6	60.9	113.6	40.3	69.3	113.6	39.2	68.7
09	FINISHES	93.0	38.7	63.0	96.5	55.3	73.8	98.1	38.6	65.3	96.3	41.9	66.3	95.8	35.4	62.5	95.5	38.0	63.8
COVERS	DIVS. 10 - 14, 25, 28, 41, 43, 44, 46	100.0	79.4	95.8	100.0	74.3	94.8	100.0	77.3	95.4	100.0	75.9	95.1	100.0	77.1	95.4	100.0	71.6	94.3
21, 22, 23	FIRE SUPPRESSION, PLUMBING & HVAC	95.6	51.1	77.7	100.2	52.7	81.0	100.4	52.4	81.0	100.4	53.1	81.3	95.6	52.2	78.1	95.6	51.2	77.7
26, 27, 3370	ELECTRICAL, COMMUNICATIONS & UTIL.	95.9	34.5	63.4	101.4	50.7	74.6	101.6	57.4	78.2	95.2	55.7	74.3	99.6	57.4	77.3	99.4	46.1	71.2
MF2014	WEIGHTED AVERAGE	99.3	51.6	78.5	104.5	58.1	84.2	101.6	55.1	81.3	101.3	55.7	81.4	98.8	53.9	79.2	98.4	52.7	78.5

	DIVISION	NORTH CAROLINA															NORTH DAKOTA		
		MURPHY			RALEIGH			ROCKY MOUNT			WILMINGTON			WINSTON-SALEM			BISMARCK		
		289			275 - 276			278			284			271			585		
		MAT.	INST.	TOTAL	MAT.	INST.	TOTAL	MAT.	INST.	TOTAL	MAT.	INST.	TOTAL	MAT.	INST.	TOTAL	MAT.	INST.	TOTAL
015433	CONTRACTOR EQUIPMENT		96.3	96.3		101.7	101.7		101.7	101.7		96.3	96.3		101.7	101.7		98.2	98.2
0241, 31 - 34	SITE & INFRASTRUCTURE, DEMOLITION	103.5	76.7	84.5	101.3	85.7	90.2	102.6	85.7	90.6	103.7	77.1	84.8	100.6	85.7	90.0	101.4	96.9	98.2
0310	Concrete Forming & Accessories	103.7	39.3	48.2	98.9	46.6	53.8	91.9	44.2	50.8	97.5	48.8	55.5	101.9	48.0	55.4	104.9	40.8	49.6
0320	Concrete Reinforcing	92.6	45.4	68.6	96.2	56.0	75.7	89.2	55.8	72.2	93.7	57.9	75.5	90.1	57.9	73.7	99.9	92.7	96.2
0330	Cast-in-Place Concrete	119.4	44.4	88.6	106.7	51.4	84.0	99.6	47.7	78.3	115.0	49.5	88.1	103.4	50.0	81.4	102.6	47.5	79.9
03	CONCRETE	109.2	44.1	77.2	101.7	51.7	77.1	99.5	49.3	74.9	105.9	52.4	79.6	99.4	52.2	76.2	101.5	54.2	78.3
04	MASONRY	85.3	41.2	57.8	86.6	42.2	58.9	76.7	39.5	53.5	82.8	41.7	57.2	82.7	38.6	55.2	114.6	56.4	78.3
05	METALS	100.9	74.2	92.7	105.7	79.2	97.5	106.3	78.4	97.7	102.7	79.9	95.7	110.8	79.9	101.3	103.4	88.4	98.8
06	WOOD, PLASTICS & COMPOSITES	107.8	39.7	69.7	95.1	47.5	68.5	87.4	44.7	63.5	100.1	49.0	71.5	96.4	49.4	70.1	95.3	34.7	61.4
07	THERMAL & MOISTURE PROTECTION	106.9	40.8	79.8	100.5	47.0	78.5	106.6	40.8	79.7	106.7	46.9	82.2	106.7	43.5	80.8	112.4	51.3	87.4
08	OPENINGS	97.1	38.9	83.5	104.8	47.4	91.4	101.8	41.9	87.9	97.3	51.0	86.5	105.8	49.7	92.8	108.0	50.1	94.5
0920	Plaster & Gypsum Board	106.1	37.6	59.8	99.8	45.7	63.3	100.5	42.8	61.5	102.4	47.2	65.1	107.5	47.6	67.0	98.8	33.0	54.3
0950, 0980	Ceilings & Acoustic Treatment	85.7	37.6	54.1	88.8	45.7	60.5	85.6	42.8	57.5	86.5	47.2	60.7	88.1	47.6	61.5	112.9	33.0	60.4
0960	Flooring	105.7	24.1	82.4	101.2	42.7	84.5	99.4	21.8	77.2	102.9	44.4	86.2	103.9	42.7	86.4	99.4	72.1	91.6
0970, 0990	Wall Finishes & Painting/Coating	113.6	39.4	68.8	104.2	36.9	63.5	105.4	38.5	65.0	113.6	38.7	68.4	105.4	36.7	64.0	101.9	29.4	58.1
09	FINISHES	97.7	36.1	63.7	95.5	44.7	67.5	93.9	39.3	63.8	96.3	46.8	69.0	96.3	45.7	68.4	102.7	42.8	69.7
COVERS	DIVS. 10 - 14, 25, 28, 41, 43, 44, 46	100.0	76.9	95.3	100.0	72.6	94.5	100.0	72.5	94.5	100.0	73.8	94.7	100.0	78.7	95.7	100.0	82.5	96.5
21, 22, 23	FIRE SUPPRESSION, PLUMBING & HVAC	95.6	50.9	77.6	100.0	52.5	80.8	95.6	52.7	78.3	100.4	54.9	82.0	100.4	53.3	81.4	100.1	71.9	88.7
26, 27, 3370	ELECTRICAL, COMMUNICATIONS & UTIL.	103.2	29.3	64.1	98.3	40.4	67.7	97.9	39.7	67.1	102.4	50.7	75.1	95.2	55.7	74.3	102.5	71.6	86.2
MF2014	WEIGHTED AVERAGE	99.6	48.3	77.2	100.5	54.1	80.3	98.5	52.3	78.3	100.3	55.9	81.0	101.0	56.5	81.6	103.1	66.2	87.0

	DIVISION	NORTH DAKOTA																	
		DEVILS LAKE			DICKINSON			FARGO			GRAND FORKS			JAMESTOWN			MINOT		
		583			586			580 - 581			582			584			587		
		MAT.	INST.	TOTAL	MAT.	INST.	TOTAL	MAT.	INST.	TOTAL	MAT.	INST.	TOTAL	MAT.	INST.	TOTAL	MAT.	INST.	TOTAL
015433	CONTRACTOR EQUIPMENT		98.2	98.2		98.2	98.2		98.2	98.2		98.2	98.2		98.2	98.2		98.2	98.2
0241, 31 - 34	SITE & INFRASTRUCTURE, DEMOLITION	105.7	94.4	97.7	113.6	92.9	98.9	102.6	96.9	98.5	109.6	92.9	97.7	104.7	92.9	96.3	107.1	96.9	99.8
0310	Concrete Forming & Accessories	101.2	35.4	44.4	90.8	34.9	42.5	99.3	41.5	49.4	94.5	34.2	42.5	92.3	34.1	42.1	90.5	67.0	70.2
0320	Concrete Reinforcing	100.2	93.1	96.6	101.1	92.8	96.9	96.9	92.8	94.8	98.7	92.9	95.7	100.8	91.3	95.9	102.1	93.2	97.6
0330	Cast-in-Place Concrete	126.3	46.0	93.3	114.3	44.4	85.6	113.6	49.2	87.2	114.3	44.1	85.5	124.7	44.1	91.6	114.3	46.5	86.4
03	CONCRETE	110.6	51.3	81.5	109.6	50.2	80.4	105.9	55.1	80.9	106.8	49.8	78.8	109.1	48.7	79.5	105.5	65.6	85.9
04	MASONRY	121.7	64.9	86.3	124.3	60.4	84.5	114.2	56.4	78.2	115.9	64.4	83.8	135.1	32.8	71.3	114.2	65.4	83.8
05	METALS	99.9	87.4	96.0	99.8	82.5	94.5	103.0	88.7	98.6	99.8	82.1	94.4	99.8	63.7	88.7	100.1	89.3	96.8
06	WOOD, PLASTICS & COMPOSITES	95.2	32.0	59.8	83.1	32.0	54.5	92.9	35.2	60.6	87.4	32.0	56.4	85.0	32.0	55.3	82.8	70.1	75.7
07	THERMAL & MOISTURE PROTECTION	107.6	49.8	83.9	108.1	48.5	83.7	107.4	51.6	84.5	107.8	49.8	84.0	107.4	41.5	80.4	107.5	57.4	87.0
08	OPENINGS	100.7	42.7	87.2	100.7	42.7	87.2	100.5	50.3	88.9	100.7	42.7	87.2	100.7	32.2	84.8	100.9	69.3	93.5
0920	Plaster & Gypsum Board	118.4	30.2	58.8	109.1	30.2	55.8	96.6	33.5	54.0	110.5	30.2	56.2	110.2	30.2	56.1	109.1	69.4	82.3
0950, 0980	Ceilings & Acoustic Treatment	114.3	30.2	59.0	114.3	30.2	59.0	112.1	33.5	60.5	114.3	30.2	59.0	114.3	30.2	59.0	114.3	69.4	84.8
0960	Flooring	107.9	35.9	87.3	101.4	35.9	82.7	102.8	72.1	94.0	103.3	35.9	84.0	102.1	35.9	83.2	101.1	89.3	97.8
0970, 0990	Wall Finishes & Painting/Coating	104.4	22.5	54.9	104.4	31.5	60.4	101.2	70.7	82.8	104.4	28.2	58.4	104.4	22.5	54.9	104.4	27.9	58.2
09	FINISHES	108.2	32.2	66.3	105.9	33.2	65.8	103.8	47.7	72.9	106.1	32.8	65.7	105.3	32.2	65.0	105.0	67.0	84.1
COVERS	DIVS. 10 - 14, 25, 28, 41, 43, 44, 46	100.0	32.4	86.3	100.0	32.6	86.4	100.0	82.6	96.5	100.0	32.5	86.4	100.0	80.6	96.1	100.0	86.4	97.3
21, 22, 23	FIRE SUPPRESSION, PLUMBING & HVAC	95.6	74.0	86.9	95.6	66.2	83.7	100.1	77.0	90.8	100.4	33.3	73.3	95.6	35.0	71.2	100.4	60.2	84.2
26, 27, 3370	ELECTRICAL, COMMUNICATIONS & UTIL.	98.3	35.8	65.3	107.7	72.5	89.1	102.4	68.4	84.4	102.1	53.5	76.4	98.3	35.7	65.2	105.5	75.7	89.8
MF2014	WEIGHTED AVERAGE	102.0	58.6	83.1	102.8	60.9	84.6	102.7	67.6	87.4	102.7	51.6	80.4	102.1	45.2	77.3	102.7	71.3	89.0

		NORTH DAKOTA			OHIO														
DIVISION		WILLISTON			AKRON			ATHENS			CANTON			CHILLICOTHE			CINCINNATI		
		588			442 - 443			457			446 - 447			456			451 - 452		
		MAT.	INST.	TOTAL	MAT.	INST.	TOTAL	MAT.	INST.	TOTAL	MAT.	INST.	TOTAL	MAT.	INST.	TOTAL	MAT.	INST.	TOTAL
015433	CONTRACTOR EQUIPMENT		98.2	98.2		94.4	94.4		90.4	90.4		94.4	94.4		99.7	99.7		99.5	99.5
0241, 31 - 34	SITE & INFRASTRUCTURE, DEMOLITION	107.5	92.9	97.1	98.1	101.6	100.6	111.0	91.6	97.2	98.2	101.0	100.2	97.0	102.6	101.0	94.1	102.4	100.0
0310	Concrete Forming & Accessories	96.3	34.9	43.3	99.0	94.3	95.0	95.0	84.3	85.8	99.0	83.7	85.8	97.4	92.3	93.0	99.3	79.2	82.0
0320	Concrete Reinforcing	103.1	92.8	97.9	99.5	92.1	95.7	93.1	87.0	90.0	99.5	75.8	87.5	90.1	80.4	85.1	95.5	79.2	87.2
0330	Cast-in-Place Concrete	114.3	44.4	85.6	94.0	98.0	95.7	111.5	96.5	105.4	94.9	95.0	95.0	101.2	99.3	100.4	93.2	92.4	92.9
03	CONCRETE	106.9	50.2	79.1	97.3	94.4	95.9	107.6	88.6	98.3	97.7	85.7	91.8	101.9	92.4	97.2	96.3	84.0	90.3
04	MASONRY	108.8	60.4	78.7	93.3	96.2	95.1	86.4	90.4	88.9	94.0	84.3	87.9	94.0	100.5	98.1	93.7	85.0	88.2
05	METALS	100.0	82.5	94.6	95.4	81.6	91.1	103.0	80.0	95.9	95.4	74.5	88.9	94.9	85.7	92.1	97.2	84.6	93.3
06	WOOD, PLASTICS & COMPOSITES	88.8	32.0	57.0	99.1	93.6	96.0	85.7	84.9	85.3	99.5	82.8	90.2	98.4	89.6	93.5	100.9	76.8	87.4
07	THERMAL & MOISTURE PROTECTION	107.7	48.5	83.5	110.8	96.5	104.9	99.4	94.8	97.5	112.0	91.4	103.5	101.2	97.4	99.7	99.2	88.7	94.9
08	OPENINGS	100.8	42.7	87.3	110.7	93.3	106.7	101.7	82.3	97.2	104.3	77.9	98.2	93.8	83.5	91.4	102.0	77.4	96.3
0920	Plaster & Gypsum Board	110.5	30.2	56.2	96.9	93.2	94.4	91.4	84.1	86.5	98.0	82.1	87.2	93.3	89.5	90.7	94.9	76.3	82.3
0950, 0980	Ceilings & Acoustic Treatment	114.3	30.2	59.0	93.0	93.2	93.1	103.3	84.1	90.7	93.0	82.1	85.8	97.9	89.5	92.4	98.7	76.3	84.0
0960	Flooring	104.1	35.9	84.6	96.9	93.4	95.9	121.3	99.0	114.9	97.1	81.7	92.7	98.3	97.7	98.1	99.3	90.7	96.8
0970, 0990	Wall Finishes & Painting/Coating	104.4	31.5	60.4	96.3	106.3	102.3	100.5	99.1	99.6	96.3	83.2	88.4	97.8	92.9	94.9	97.8	83.9	89.4
09	FINISHES	106.2	33.2	66.0	98.0	95.3	96.5	101.8	89.0	94.8	98.2	82.9	89.8	98.8	93.3	95.8	99.2	81.1	89.3
COVERS	DIVS. 10 - 14, 25, 28, 41, 43, 44, 46	100.0	32.6	86.4	100.0	98.2	99.6	100.0	52.0	90.3	100.0	95.4	99.1	100.0	92.6	98.5	100.0	89.3	97.8
21, 22, 23	FIRE SUPPRESSION, PLUMBING & HVAC	95.6	66.2	83.7	100.1	94.6	97.9	95.3	51.3	77.6	100.1	81.4	92.5	95.8	95.2	95.6	100.0	83.3	93.3
26, 27, 3370	ELECTRICAL, COMMUNICATIONS & UTIL.	102.5	72.5	86.6	99.3	93.9	96.5	95.5	98.1	96.9	98.5	92.7	95.5	95.0	85.6	90.0	93.9	79.4	86.3
MF2014	WEIGHTED AVERAGE	101.2	60.9	83.7	99.9	94.2	97.4	99.6	80.4	91.2	99.3	85.6	93.3	96.8	93.0	95.1	98.3	84.5	92.2

		OHIO																	
DIVISION		CLEVELAND			COLUMBUS			DAYTON			HAMILTON			LIMA			LORAIN		
		441			430 - 432			453 - 454			450			458			440		
		MAT.	INST.	TOTAL	MAT.	INST.	TOTAL	MAT.	INST.	TOTAL	MAT.	INST.	TOTAL	MAT.	INST.	TOTAL	MAT.	INST.	TOTAL
015433	CONTRACTOR EQUIPMENT		94.7	94.7		93.6	93.6		94.7	94.7		99.7	99.7		92.9	92.9		94.4	94.4
0241, 31 - 34	SITE & INFRASTRUCTURE, DEMOLITION	98.0	102.2	101.0	95.5	98.3	97.5	92.9	101.8	99.2	92.8	102.1	99.4	104.6	91.6	95.4	97.5	102.9	101.4
0310	Concrete Forming & Accessories	99.1	99.6	99.5	100.3	84.5	86.7	99.3	79.0	81.8	99.4	79.4	82.1	95.0	86.9	88.1	99.1	86.8	88.5
0320	Concrete Reinforcing	100.0	92.5	96.2	102.7	82.6	92.5	95.5	81.3	88.3	95.5	79.2	87.2	93.1	81.5	87.2	99.5	92.4	95.9
0330	Cast-in-Place Concrete	92.2	106.5	98.1	94.1	93.7	93.9	86.7	86.4	86.6	92.9	92.7	92.8	102.4	97.5	100.4	89.5	104.1	95.5
03	CONCRETE	96.5	99.8	98.2	97.9	87.1	92.6	93.3	81.8	87.6	96.2	84.2	90.3	100.4	89.3	94.9	95.2	93.2	94.2
04	MASONRY	97.8	105.6	102.7	96.6	94.0	95.0	93.2	83.5	87.1	93.5	85.5	88.5	117.7	86.4	98.2	89.9	104.6	99.1
05	METALS	96.9	84.4	93.1	96.9	80.7	91.9	96.4	78.6	91.0	96.5	84.5	92.8	103.0	81.5	96.4	96.0	82.8	91.9
06	WOOD, PLASTICS & COMPOSITES	98.2	97.1	97.6	97.1	82.3	88.8	102.2	77.1	88.1	100.9	76.8	87.4	85.6	86.3	86.0	99.1	80.9	88.9
07	THERMAL & MOISTURE PROTECTION	109.5	108.6	109.1	100.9	94.9	98.4	104.7	87.6	97.7	101.4	88.8	96.2	99.0	95.0	97.3	111.9	103.0	108.2
08	OPENINGS	100.6	95.2	99.3	102.4	80.0	97.2	101.3	78.1	95.9	98.9	77.4	93.9	101.7	79.8	96.6	104.3	86.4	100.2
0920	Plaster & Gypsum Board	96.2	96.8	96.6	93.5	81.8	85.6	94.9	76.6	82.6	94.9	76.3	82.3	91.4	85.5	87.4	96.9	80.1	85.5
0950, 0980	Ceilings & Acoustic Treatment	91.3	96.8	94.9	97.7	81.8	87.3	99.7	76.6	84.5	98.7	76.3	84.0	102.4	85.5	91.3	93.0	80.1	84.5
0960	Flooring	96.7	105.0	99.1	92.0	90.3	91.5	102.0	80.7	95.9	99.3	90.7	96.8	120.3	92.2	112.3	97.1	105.0	99.3
0970, 0990	Wall Finishes & Painting/Coating	96.3	105.1	101.6	95.5	92.9	94.0	97.8	85.0	90.1	97.8	84.7	89.9	100.5	82.7	89.8	96.3	105.1	101.6
09	FINISHES	97.5	100.9	99.4	94.1	85.9	89.6	100.2	79.3	88.7	99.1	81.4	89.3	100.8	87.2	93.3	98.0	91.1	94.2
COVERS	DIVS. 10 - 14, 25, 28, 41, 43, 44, 46	100.0	102.5	100.5	100.0	93.3	98.6	100.0	89.1	97.8	100.0	89.5	97.9	100.0	95.2	99.0	100.0	100.3	100.1
21, 22, 23	FIRE SUPPRESSION, PLUMBING & HVAC	100.0	100.5	100.2	100.1	92.0	96.8	100.9	86.6	95.2	100.6	83.6	93.7	95.3	89.0	92.8	100.1	92.1	96.9
26, 27, 3370	ELECTRICAL, COMMUNICATIONS & UTIL.	98.8	105.0	102.1	97.6	87.8	92.4	92.6	83.0	87.5	93.0	83.3	87.9	95.8	80.3	87.6	98.7	88.0	93.0
MF2014	WEIGHTED AVERAGE	99.0	100.3	99.6	98.6	89.2	94.5	98.0	84.4	92.1	97.9	85.1	92.3	100.0	86.8	94.3	98.9	93.0	96.3

		OHIO																	
DIVISION		MANSFIELD			MARION			SPRINGFIELD			STEUBENVILLE			TOLEDO			YOUNGSTOWN		
		448 - 449			433			455			439			434 - 436			444 - 445		
		MAT.	INST.	TOTAL	MAT.	INST.	TOTAL	MAT.	INST.	TOTAL	MAT.	INST.	TOTAL	MAT.	INST.	TOTAL	MAT.	INST.	TOTAL
015433	CONTRACTOR EQUIPMENT		94.4	94.4		93.2	93.2		94.7	94.7		98.1	98.1		96.0	96.0		94.4	94.4
0241, 31 - 34	SITE & INFRASTRUCTURE, DEMOLITION	93.9	101.5	99.3	91.8	97.1	95.6	93.2	100.5	98.4	133.1	106.9	114.5	94.8	99.0	97.8	98.0	102.0	100.8
0310	Concrete Forming & Accessories	89.1	83.7	84.5	96.7	81.2	83.3	99.3	83.5	85.7	98.3	89.4	90.7	100.3	97.1	97.6	99.0	87.5	89.1
0320	Concrete Reinforcing	90.6	76.4	83.4	94.7	82.5	88.5	95.5	81.3	88.3	92.3	85.5	88.8	102.7	85.4	93.9	99.5	85.6	92.4
0330	Cast-in-Place Concrete	87.1	94.7	90.2	85.9	92.7	88.7	89.1	86.3	87.9	93.3	94.1	93.6	94.1	100.0	96.5	93.1	96.9	94.7
03	CONCRETE	89.7	85.7	87.7	89.9	85.1	87.6	94.4	83.8	89.2	93.8	89.7	91.8	97.9	95.5	96.7	96.9	89.8	93.4
04	MASONRY	92.4	97.0	95.3	98.8	95.5	96.7	93.4	83.5	87.2	86.2	94.2	91.2	104.8	99.9	101.7	93.6	93.6	93.6
05	METALS	96.2	75.5	89.8	96.0	78.7	90.6	96.4	78.4	90.9	92.5	79.5	88.5	96.7	85.5	93.3	95.4	78.7	90.3
06	WOOD, PLASTICS & COMPOSITES	86.7	80.9	83.4	92.8	79.8	85.5	103.6	83.6	92.4	88.4	88.3	88.3	97.1	96.9	97.0	99.1	86.1	91.8
07	THERMAL & MOISTURE PROTECTION	111.3	94.4	104.4	100.5	85.2	94.2	104.6	88.2	97.9	112.8	96.4	106.1	102.5	102.7	102.6	112.1	94.3	104.8
08	OPENINGS	104.9	76.0	98.2	96.6	76.5	91.9	99.3	79.0	94.6	97.0	83.7	94.0	99.7	91.3	97.7	104.3	85.1	99.8
0920	Plaster & Gypsum Board	90.7	80.1	83.5	91.4	79.3	83.2	94.9	83.3	87.1	90.7	87.5	88.6	93.5	96.9	95.8	96.9	85.5	89.2
0950, 0980	Ceilings & Acoustic Treatment	93.9	80.1	84.8	97.7	79.3	85.6	99.7	83.3	88.9	94.8	87.5	90.0	97.7	96.9	97.2	93.0	85.5	88.1
0960	Flooring	92.5	107.5	96.8	90.9	107.5	95.6	102.0	80.7	95.9	118.6	100.4	113.4	91.2	99.7	93.6	97.1	93.9	96.2
0970, 0990	Wall Finishes & Painting/Coating	96.3	88.5	91.6	95.5	47.5	66.5	97.8	85.0	90.1	107.8	103.1	104.9	95.6	102.0	99.4	96.3	92.6	94.1
09	FINISHES	95.8	88.2	91.6	93.1	82.6	87.3	100.2	83.2	90.8	110.0	92.6	100.4	93.8	98.1	96.2	98.1	88.8	93.0
COVERS	DIVS. 10 - 14, 25, 28, 41, 43, 44, 46	100.0	96.2	99.2	100.0	54.4	90.8	100.0	89.8	97.9	100.0	95.0	99.0	100.0	98.7	99.7	100.0	96.4	99.3
21, 22, 23	FIRE SUPPRESSION, PLUMBING & HVAC	95.3	88.9	92.7	95.3	91.0	93.6	100.9	81.3	93.0	95.7	92.0	94.2	100.1	100.4	100.2	100.1	87.1	94.8
26, 27, 3370	ELECTRICAL, COMMUNICATIONS & UTIL.	96.2	79.0	87.1	91.9	79.0	85.1	92.6	87.8	90.1	86.9	114.1	101.3	97.6	104.7	101.4	98.7	86.0	92.0
[illegible]	WEIGHTED AVERAGE	96.7	87.3	92.6	95.1	85.3	90.8	98.0	84.7	92.2	96.8	94.9	96.0	98.7	98.1	98.4	99.2	89.0	94.7

DIVISION		OHIO			OKLAHOMA															
		ZANESVILLE			ARDMORE			CLINTON			DURANT			ENID			GUYMON		
		437 - 438			734			736			747			737			739		
		MAT.	INST.	TOTAL	MAT.	INST.	TOTAL	MAT.	INST.	TOTAL	MAT.	INST.	TOTAL	MAT.	INST.	TOTAL	MAT.	INST.	TOTAL
015433	CONTRACTOR EQUIPMENT		93.2	93.2		82.7	82.7		81.8	81.8		81.8	81.8		81.8	81.8		81.8	81.8
0241, 31 - 34	SITE & INFRASTRUCTURE, DEMOLITION	94.5	98.7	97.5	96.2	91.6	92.9	97.6	90.2	92.3	95.5	89.9	91.5	99.3	90.2	92.8	101.6	90.0	93.3
0310	Concrete Forming & Accessories	93.7	81.3	83.0	94.0	40.4	47.8	92.5	46.6	53.0	85.5	44.1	49.8	96.2	34.1	42.7	99.8	44.9	52.4
0320	Concrete Reinforcing	94.1	85.1	89.5	90.8	79.8	85.2	91.3	79.8	85.5	95.8	80.2	87.9	90.7	79.8	85.2	91.3	79.8	85.5
0330	Cast-in-Place Concrete	90.5	91.7	91.0	97.5	43.4	75.3	94.3	45.1	74.1	91.7	42.8	71.6	94.3	46.4	74.6	94.3	43.1	73.3
03	CONCRETE	93.5	85.3	89.5	95.2	49.3	72.7	94.7	52.7	74.1	92.3	50.8	71.9	95.3	47.6	71.8	98.2	51.2	75.1
04	MASONRY	95.5	85.9	89.5	103.4	57.1	74.5	129.7	57.1	84.4	95.6	62.3	74.9	110.1	57.1	77.0	105.9	53.7	73.3
05	METALS	97.4	80.4	92.1	102.5	70.1	92.5	102.6	70.1	92.6	93.5	70.7	86.5	104.0	70.0	93.6	103.2	69.8	92.9
06	WOOD, PLASTICS & COMPOSITES	88.5	79.8	83.6	97.4	37.9	64.0	96.4	46.2	68.3	88.1	43.0	62.9	100.1	29.3	60.5	104.0	46.0	71.5
07	THERMAL & MOISTURE PROTECTION	100.6	92.6	97.3	108.9	60.6	89.1	109.1	61.5	89.6	99.7	61.8	84.2	109.2	59.9	89.0	109.5	57.3	88.1
08	OPENINGS	96.6	79.3	92.5	102.9	49.0	90.3	102.9	53.3	91.4	96.3	52.0	86.0	102.9	43.9	89.1	103.0	49.2	90.5
0920	Plaster & Gypsum Board	88.2	79.3	82.2	89.8	36.6	53.8	89.4	45.2	59.6	79.4	41.9	54.1	90.5	27.8	48.1	90.7	44.9	59.8
0950, 0980	Ceilings & Acoustic Treatment	97.7	79.3	85.6	85.7	36.6	53.4	85.7	45.2	59.1	82.4	41.9	55.8	85.7	27.8	47.6	86.5	44.9	59.2
0960	Flooring	89.3	90.3	89.6	107.0	43.2	88.7	105.8	41.1	87.3	102.8	61.9	91.1	107.7	41.1	88.6	109.3	24.4	85.0
0970, 0990	Wall Finishes & Painting/Coating	95.5	92.9	94.0	104.7	51.3	72.5	104.7	51.3	72.5	103.9	51.3	72.1	104.7	51.3	72.5	104.7	33.0	61.5
09	FINISHES	92.4	83.6	87.5	92.9	40.1	63.8	92.7	44.7	66.3	90.2	46.6	66.2	93.5	34.6	61.1	94.5	38.9	63.9
COVERS	DIVS. 10 - 14, 25, 28, 41, 43, 44, 46	100.0	88.5	97.7	100.0	77.6	95.5	100.0	78.6	95.7	100.0	77.9	95.5	100.0	76.7	95.3	100.0	77.4	95.4
21, 22, 23	FIRE SUPPRESSION, PLUMBING & HVAC	95.3	89.9	93.1	95.6	65.8	83.6	95.6	65.9	83.6	95.5	65.4	83.4	100.3	65.8	86.4	95.6	63.8	82.7
26, 27, 3370	ELECTRICAL, COMMUNICATIONS & UTIL.	92.3	83.4	87.5	93.6	70.9	81.6	94.6	70.9	82.1	96.5	70.9	83.0	94.6	70.9	82.1	96.2	61.5	77.9
MF2014	WEIGHTED AVERAGE	95.6	86.5	91.6	98.3	61.9	82.4	99.6	63.1	83.7	95.1	63.5	81.3	100.2	60.5	82.9	99.4	59.7	82.1

DIVISION		OKLAHOMA																	
		LAWTON			MCALESTER			MIAMI			MUSKOGEE			OKLAHOMA CITY			PONCA CITY		
		735			745			743			744			730 - 731			746		
		MAT.	INST.	TOTAL	MAT.	INST.	TOTAL	MAT.	INST.	TOTAL	MAT.	INST.	TOTAL	MAT.	INST.	TOTAL	MAT.	INST.	TOTAL
015433	CONTRACTOR EQUIPMENT		82.7	82.7		81.8	81.8		89.9	89.9		89.9	89.9		83.0	83.0		81.8	81.8
0241, 31 - 34	SITE & INFRASTRUCTURE, DEMOLITION	95.7	91.6	92.8	89.1	90.1	89.9	90.5	87.3	88.2	90.8	87.1	88.2	95.1	92.0	92.9	95.9	90.1	91.8
0310	Concrete Forming & Accessories	99.8	44.3	52.0	83.5	43.6	49.1	96.9	66.2	70.4	101.4	33.4	42.7	97.7	57.9	63.4	92.2	46.3	52.6
0320	Concrete Reinforcing	91.0	79.8	85.3	95.5	79.8	87.5	94.0	79.9	86.8	94.9	79.3	87.0	96.2	79.9	87.9	94.9	79.9	87.2
0330	Cast-in-Place Concrete	91.2	46.4	72.8	80.4	45.2	65.9	84.3	47.5	69.2	85.3	45.8	69.1	92.9	46.9	74.0	94.2	44.7	73.8
03	CONCRETE	91.5	52.1	72.1	82.9	51.3	67.4	87.6	63.0	75.5	89.4	47.7	68.9	94.9	58.3	76.9	94.4	52.4	73.8
04	MASONRY	105.8	57.1	75.4	114.3	56.1	78.1	98.4	56.3	72.1	116.9	45.2	72.2	109.8	56.4	76.5	90.9	56.1	69.2
05	METALS	108.0	70.1	96.4	93.4	70.0	86.2	93.4	82.1	89.9	94.8	80.4	90.4	100.7	70.2	91.3	93.4	70.3	86.3
06	WOOD, PLASTICS & COMPOSITES	103.0	43.1	69.5	85.6	43.0	61.7	101.0	72.9	85.3	105.6	30.7	63.6	97.0	61.8	77.3	96.3	46.0	68.1
07	THERMAL & MOISTURE PROTECTION	108.9	61.3	89.4	99.4	60.6	83.5	99.8	64.3	85.2	99.9	48.6	78.8	101.1	62.9	85.4	99.9	66.2	86.1
08	OPENINGS	104.6	51.4	92.2	96.2	51.8	85.9	96.2	68.3	89.7	96.2	42.4	83.7	104.0	61.6	94.2	96.2	53.1	86.2
0920	Plaster & Gypsum Board	92.4	42.0	58.3	78.4	41.9	53.7	84.8	72.6	76.5	86.9	29.0	47.8	98.0	61.2	73.1	83.7	44.9	57.5
0950, 0980	Ceilings & Acoustic Treatment	93.2	42.0	59.6	82.4	41.9	55.8	82.4	72.6	75.9	90.8	29.0	50.2	96.4	61.2	73.3	82.4	44.9	57.8
0960	Flooring	109.7	41.1	90.1	101.8	41.1	84.4	108.8	61.9	95.4	111.3	39.8	90.8	106.2	41.1	87.6	105.9	41.1	87.4
0970, 0990	Wall Finishes & Painting/Coating	104.7	51.3	72.5	103.9	37.8	64.0	103.9	77.5	87.9	103.9	34.2	61.8	106.4	51.3	73.2	103.9	51.3	72.1
09	FINISHES	95.5	42.8	66.5	89.2	41.0	62.6	92.0	67.3	78.4	94.9	33.0	60.8	97.6	53.7	73.4	91.8	44.3	65.6
COVERS	DIVS. 10 - 14, 25, 28, 41, 43, 44, 46	100.0	78.2	95.6	100.0	77.9	95.5	100.0	81.6	96.3	100.0	76.1	95.2	100.0	80.0	96.0	100.0	78.2	95.6
21, 22, 23	FIRE SUPPRESSION, PLUMBING & HVAC	100.3	65.9	86.4	95.5	61.6	81.8	95.5	61.9	81.9	100.3	60.6	84.3	100.1	65.5	86.2	95.5	61.8	81.9
26, 27, 3370	ELECTRICAL, COMMUNICATIONS & UTIL.	96.2	70.9	82.8	94.9	64.5	78.8	96.3	64.6	79.5	94.5	52.4	72.2	101.7	70.9	85.4	94.5	70.4	81.7
MF2014	WEIGHTED AVERAGE	100.7	62.8	84.2	94.5	60.4	79.7	94.8	67.5	82.9	97.4	55.7	79.2	100.3	65.7	85.2	95.1	62.1	80.7

DIVISION		OKLAHOMA												OREGON					
		POTEAU			SHAWNEE			TULSA			WOODWARD			BEND			EUGENE		
		749			748			740 - 741			738			977			974		
		MAT.	INST.	TOTAL	MAT.	INST.	TOTAL	MAT.	INST.	TOTAL	MAT.	INST.	TOTAL	MAT.	INST.	TOTAL	MAT.	INST.	TOTAL
015433	CONTRACTOR EQUIPMENT		89.1	89.1		81.8	81.8		89.9	89.9		81.8	81.8		98.8	98.8		98.8	98.8
0241, 31 - 34	SITE & INFRASTRUCTURE, DEMOLITION	77.2	85.8	83.3	99.0	90.1	92.7	97.1	87.5	90.3	97.9	90.2	92.4	105.4	102.9	103.6	96.1	102.9	100.9
0310	Concrete Forming & Accessories	90.1	41.0	47.8	85.4	43.7	49.5	101.5	40.3	48.7	92.6	46.8	53.1	110.1	98.8	100.4	106.5	98.7	99.8
0320	Concrete Reinforcing	95.9	79.9	87.8	94.9	79.8	87.2	95.1	79.8	87.3	90.7	79.9	85.2	93.3	99.6	96.5	97.4	99.6	98.5
0330	Cast-in-Place Concrete	84.3	44.9	68.1	97.2	42.9	74.9	92.9	46.2	73.7	94.3	45.3	74.2	104.4	102.7	103.7	101.1	102.6	101.7
03	CONCRETE	89.9	50.9	70.7	96.0	50.6	73.7	94.7	51.0	73.2	95.0	52.8	74.3	107.1	100.0	103.6	98.8	100.0	99.4
04	MASONRY	98.7	56.2	72.2	115.6	56.1	78.5	99.3	60.0	74.8	98.6	57.1	72.7	104.1	103.3	103.6	101.1	103.3	102.5
05	METALS	93.4	81.7	89.8	93.3	69.9	86.1	98.1	81.4	93.0	102.7	70.5	92.8	91.9	96.4	93.3	92.6	96.2	93.7
06	WOOD, PLASTICS & COMPOSITES	93.0	39.1	62.8	88.0	43.0	62.8	104.8	36.6	66.6	96.5	46.0	68.2	101.7	98.4	99.9	97.6	98.4	98.1
07	THERMAL & MOISTURE PROTECTION	99.9	60.5	83.7	99.9	59.3	83.2	99.8	62.2	84.4	109.1	66.6	91.7	107.9	94.9	102.6	107.2	91.6	100.8
08	OPENINGS	96.2	49.9	85.5	96.2	51.8	85.9	97.8	47.5	86.1	102.9	53.1	91.3	96.9	102.3	98.1	97.1	102.3	98.4
0920	Plaster & Gypsum Board	82.3	37.8	52.2	79.4	41.9	54.1	86.9	35.2	51.9	89.6	44.9	59.4	107.4	98.2	101.2	106.0	98.2	100.7
0950, 0980	Ceilings & Acoustic Treatment	82.4	37.8	53.1	82.4	41.9	55.8	90.8	35.2	54.3	86.5	44.9	59.2	91.3	98.2	95.8	92.3	98.2	96.1
0960	Flooring	105.2	61.9	92.8	102.8	32.7	82.8	110.0	42.8	90.8	105.8	43.2	87.9	110.5	103.4	108.5	108.9	103.4	107.3
0970, 0990	Wall Finishes & Painting/Coating	103.9	51.3	72.1	103.9	34.1	61.8	103.9	40.5	65.6	104.7	51.3	72.5	106.4	74.8	87.3	106.4	74.8	87.3
09	FINISHES	89.9	44.3	64.8	90.4	39.2	62.2	94.7	38.9	63.9	93.0	44.9	66.5	102.7	96.8	99.5	101.2	96.8	98.8
COVERS	DIVS. 10 - 14, 25, 28, 41, 43, 44, 46	100.0	77.7	95.5	100.0	77.9	95.5	100.0	78.8	95.7	100.0	78.5	95.7	100.0	99.6	99.9	100.0	99.6	99.9
21, 22, 23	FIRE SUPPRESSION, PLUMBING & HVAC	95.5	61.8	81.9	95.5	65.4	83.4	100.3	63.5	85.4	95.6	65.9	83.6	95.1	100.8	97.4	99.9	100.8	100.2
26, 27, 3370	ELECTRICAL, COMMUNICATIONS & UTIL.	94.6	64.6	78.7	96.6	70.9	83.0	96.5	64.6	79.6	96.1	70.9	82.8	101.6	96.1	98.7	100.1	96.1	98.0
MF2014	WEIGHTED AVERAGE	94.3	61.4	80.0	96.5	61.8	81.4	98.1	61.6	82.2	98.3	63.4	83.1	98.9	99.4	99.1	98.6	99.3	98.9

	DIVISION	OREGON																	
		KLAMATH FALLS			MEDFORD			PENDLETON			PORTLAND			SALEM			VALE		
		976			975			978			970 - 972			973			979		
		MAT.	INST.	TOTAL	MAT.	INST.	TOTAL	MAT.	INST.	TOTAL	MAT.	INST.	TOTAL	MAT.	INST.	TOTAL	MAT.	INST.	TOTAL
015433	CONTRACTOR EQUIPMENT		98.8	98.8		98.8	98.8		96.2	96.2		98.8	98.8		98.8	98.8		96.2	96.2
0241, 31 - 34	SITE & INFRASTRUCTURE, DEMOLITION	109.2	102.9	104.7	103.4	102.9	103.0	102.5	96.4	98.2	98.5	102.9	101.6	91.9	102.9	99.7	90.5	96.4	94.7
0310	Concrete Forming & Accessories	102.9	98.6	99.2	102.0	98.6	99.1	103.3	99.0	99.6	107.7	98.9	100.1	106.0	98.8	99.8	109.9	98.7	100.2
0320	Concrete Reinforcing	93.3	99.6	96.5	94.9	99.6	97.3	92.6	99.7	96.2	98.1	99.7	98.9	103.6	99.6	101.6	90.4	99.6	95.1
0330	Cast-in-Place Concrete	104.4	102.6	103.7	104.4	102.6	103.7	105.2	103.9	104.7	103.9	102.7	103.4	97.9	102.7	99.9	82.9	103.8	91.5
03	CONCRETE	109.7	99.9	104.9	104.6	99.9	102.3	91.4	100.6	95.9	100.3	100.1	100.2	98.2	100.0	99.1	76.9	100.4	88.5
04	MASONRY	117.5	103.3	108.6	98.2	103.3	101.4	107.8	103.4	105.0	102.5	103.3	103.0	109.3	103.3	105.6	106.1	103.4	104.4
05	METALS	91.9	96.2	93.2	92.2	96.2	93.4	98.1	96.9	97.7	93.5	96.5	94.4	98.7	96.4	98.0	98.0	96.6	97.5
06	WOOD, PLASTICS & COMPOSITES	92.6	98.4	95.9	91.5	98.4	95.3	94.7	98.5	96.8	98.6	98.4	98.5	95.5	98.4	97.1	103.2	98.5	100.6
07	THERMAL & MOISTURE PROTECTION	108.2	92.9	101.9	107.8	92.9	101.7	101.1	93.5	98.0	107.1	96.7	102.9	104.9	94.9	100.8	100.5	94.4	98.0
08	OPENINGS	96.9	102.3	98.1	99.7	102.3	100.3	93.2	102.4	95.4	95.0	102.3	96.7	98.7	102.3	99.5	93.2	94.1	93.4
0920	Plaster & Gypsum Board	102.9	98.2	99.7	102.2	98.2	99.5	90.2	98.2	95.6	105.5	98.2	100.5	102.1	98.2	99.4	95.9	98.2	97.4
0950, 0980	Ceilings & Acoustic Treatment	98.9	98.2	98.4	105.4	98.2	100.7	64.4	98.2	86.6	94.2	98.2	96.8	99.3	98.2	98.5	64.4	98.2	86.6
0960	Flooring	107.6	103.4	106.4	107.1	103.4	106.1	74.7	103.4	82.9	106.4	103.4	105.5	107.2	103.4	106.1	76.7	103.4	84.4
0970, 0990	Wall Finishes & Painting/Coating	106.4	70.4	84.6	106.4	70.4	84.6	96.6	72.8	82.2	106.2	72.8	86.0	104.6	74.8	86.6	96.6	74.8	83.4
09	FINISHES	103.4	96.3	99.5	103.8	96.3	99.6	72.3	96.7	85.7	100.8	96.6	98.5	100.0	96.8	98.2	72.7	96.9	86.0
COVERS	DIVS. 10 - 14, 25, 28, 41, 43, 44, 46	100.0	99.5	99.9	100.0	99.5	99.9	100.0	90.1	98.0	100.0	99.6	99.9	100.0	99.6	99.9	100.0	99.9	100.0
21, 22, 23	FIRE SUPPRESSION, PLUMBING & HVAC	95.1	100.7	97.4	99.9	100.7	100.2	97.0	113.2	103.5	99.9	100.8	100.2	99.9	100.8	100.2	97.0	98.0	97.4
26, 27, 3370	ELECTRICAL, COMMUNICATIONS & UTIL.	100.2	80.8	89.9	103.9	80.8	91.6	92.3	97.1	94.8	100.4	103.2	101.9	108.0	96.1	101.7	92.3	97.0	94.8
MF2014	WEIGHTED AVERAGE	99.8	97.1	98.6	100.0	97.1	98.8	94.8	101.4	97.7	98.8	100.4	99.5	100.5	99.4	100.0	92.9	98.1	95.2

	DIVISION	PENNSYLVANIA																	
		ALLENTOWN			ALTOONA			BEDFORD			BRADFORD			BUTLER			CHAMBERSBURG		
		181			166			155			167			160			172		
		MAT.	INST.	TOTAL	MAT.	INST.	TOTAL	MAT.	INST.	TOTAL	MAT.	INST.	TOTAL	MAT.	INST.	TOTAL	MAT.	INST.	TOTAL
015433	CONTRACTOR EQUIPMENT		112.9	112.9		112.9	112.9		109.6	109.6		112.9	112.9		112.9	112.9		112.1	112.1
0241, 31 - 34	SITE & INFRASTRUCTURE, DEMOLITION	93.6	104.5	101.3	96.8	104.4	102.2	101.4	100.1	100.5	92.4	103.5	100.3	88.1	105.8	100.7	89.2	101.2	97.8
0310	Concrete Forming & Accessories	99.2	113.0	111.1	84.2	80.4	80.9	84.1	80.8	81.3	86.4	81.9	82.5	85.7	96.0	94.6	90.2	80.0	81.4
0320	Concrete Reinforcing	94.7	108.1	101.5	91.8	102.7	97.3	93.1	80.9	86.9	93.7	103.2	98.5	92.4	109.1	100.9	91.7	103.0	97.4
0330	Cast-in-Place Concrete	86.5	104.9	94.0	96.3	86.1	92.1	106.7	70.1	91.6	92.0	92.6	92.3	85.1	96.7	89.9	91.1	69.8	82.3
03	CONCRETE	93.6	110.1	101.7	89.1	88.1	88.6	102.1	78.5	90.5	95.4	91.0	93.2	81.3	99.9	90.4	101.9	82.2	92.3
04	MASONRY	97.4	101.0	99.7	100.5	64.5	78.0	114.1	87.0	97.2	97.6	88.4	91.9	102.6	98.9	100.3	102.6	87.4	93.1
05	METALS	100.0	123.0	107.1	93.9	116.8	101.0	97.6	104.0	99.6	97.9	116.5	103.6	93.6	121.9	102.3	97.5	114.6	102.7
06	WOOD, PLASTICS & COMPOSITES	100.7	115.8	109.1	79.2	83.2	81.4	84.0	80.7	82.2	85.5	80.1	82.5	80.6	95.5	89.0	88.0	80.8	84.0
07	THERMAL & MOISTURE PROTECTION	102.3	119.1	109.2	101.4	91.0	97.1	103.0	87.7	96.7	102.3	91.7	97.9	101.1	100.7	100.9	99.3	72.5	88.3
08	OPENINGS	95.0	115.0	99.6	88.7	89.3	88.8	98.0	81.6	94.2	95.2	91.0	94.2	88.7	104.8	92.4	94.1	82.8	91.4
0920	Plaster & Gypsum Board	95.9	116.0	109.5	87.7	82.5	84.2	92.3	80.0	84.0	88.5	79.2	82.2	87.7	95.2	92.7	96.7	80.0	85.4
0950, 0980	Ceilings & Acoustic Treatment	82.8	116.0	104.6	87.0	82.5	84.0	95.0	80.0	85.1	85.4	79.2	81.3	87.9	95.2	92.7	84.0	80.0	81.4
0960	Flooring	94.6	95.9	95.0	88.2	97.5	90.9	93.1	88.2	91.7	89.4	105.3	94.0	89.0	81.0	86.7	97.9	46.2	83.1
0970, 0990	Wall Finishes & Painting/Coating	98.5	70.5	81.6	94.0	109.2	103.2	100.2	81.6	89.0	98.5	94.6	96.1	94.0	109.2	103.2	105.9	81.6	91.2
09	FINISHES	91.8	105.5	99.4	90.0	85.8	87.7	95.0	80.8	87.2	89.9	85.5	87.5	89.9	93.6	91.9	92.1	73.7	82.0
COVERS	DIVS. 10 - 14, 25, 28, 41, 43, 44, 46	100.0	105.6	101.1	100.0	95.4	99.1	100.0	98.3	99.7	100.0	99.2	99.8	100.0	102.4	100.5	100.0	96.1	99.2
21, 22, 23	FIRE SUPPRESSION, PLUMBING & HVAC	100.2	112.1	105.0	99.7	82.5	92.8	95.2	87.3	92.0	95.5	90.7	93.6	95.0	93.9	94.6	95.5	87.1	92.1
26, 27, 3370	ELECTRICAL, COMMUNICATIONS & UTIL.	99.1	98.4	98.8	89.3	112.2	101.4	94.3	112.2	103.8	92.7	112.2	103.0	89.9	111.2	101.1	89.6	86.3	87.8
MF2014	WEIGHTED AVERAGE	97.9	108.6	102.5	94.6	91.9	93.4	98.2	91.3	95.2	95.7	96.4	96.0	92.5	102.0	96.6	96.2	88.0	92.6

	DIVISION	PENNSYLVANIA																	
		DOYLESTOWN			DUBOIS			ERIE			GREENSBURG			HARRISBURG			HAZLETON		
		189			158			164 - 165			156			170 - 171			182		
		MAT.	INST.	TOTAL	MAT.	INST.	TOTAL	MAT.	INST.	TOTAL	MAT.	INST.	TOTAL	MAT.	INST.	TOTAL	MAT.	INST.	TOTAL
015433	CONTRACTOR EQUIPMENT		93.9	93.9		109.6	109.6		112.9	112.9		109.6	109.6		112.1	112.1		112.9	112.9
0241, 31 - 34	SITE & INFRASTRUCTURE, DEMOLITION	106.9	90.2	95.0	106.1	100.5	102.1	93.7	105.2	101.9	97.8	102.9	101.4	89.6	103.5	99.5	86.9	104.7	99.6
0310	Concrete Forming & Accessories	83.2	128.2	122.0	83.5	84.6	84.4	98.5	88.1	89.5	90.7	95.9	95.2	100.4	87.9	89.6	80.8	88.7	87.6
0320	Concrete Reinforcing	91.6	130.9	111.6	92.6	103.5	98.1	93.7	103.3	98.6	92.6	109.1	101.0	101.4	106.2	103.9	91.9	107.1	99.6
0330	Cast-in-Place Concrete	81.7	86.7	83.8	102.9	92.9	98.8	94.7	81.8	89.4	99.0	96.3	97.9	93.2	95.8	94.3	81.7	93.1	86.4
03	CONCRETE	89.2	114.2	101.5	103.6	92.2	98.0	88.2	90.0	89.1	97.2	99.6	98.4	97.7	95.4	96.6	86.3	95.0	90.6
04	MASONRY	100.7	128.1	117.7	114.4	91.1	99.9	89.6	92.5	91.4	124.8	98.9	108.7	102.5	89.6	94.5	110.2	95.7	101.2
05	METALS	97.5	123.8	105.6	97.6	116.1	103.3	94.1	117.0	101.2	97.5	120.6	104.6	104.0	121.4	109.4	99.7	120.6	106.1
06	WOOD, PLASTICS & COMPOSITES	81.0	130.7	108.8	82.9	83.1	83.0	96.9	86.6	91.1	90.9	95.4	93.4	95.8	86.7	90.7	79.6	86.6	83.5
07	THERMAL & MOISTURE PROTECTION	100.1	131.4	112.9	103.2	96.5	100.5	101.9	92.6	98.1	102.8	100.7	102.0	102.5	110.2	105.7	101.8	106.6	103.8
08	OPENINGS	97.2	137.9	106.7	98.0	92.7	96.7	88.8	92.2	89.6	97.9	104.7	99.5	100.6	94.0	99.0	95.6	92.3	94.8
0920	Plaster & Gypsum Board	86.2	131.4	116.7	91.2	82.5	85.3	95.9	86.0	89.2	93.5	95.2	94.6	100.6	86.0	90.7	86.6	86.0	86.2
0950, 0980	Ceilings & Acoustic Treatment	82.0	131.4	114.5	95.0	82.5	86.7	82.8	86.0	84.9	94.2	95.2	94.8	92.6	86.0	88.3	83.7	86.0	85.2
0960	Flooring	79.4	134.3	95.1	92.9	105.3	96.4	93.2	91.8	92.8	96.2	68.0	88.2	103.2	91.7	99.9	87.0	94.6	89.2
0970, 0990	Wall Finishes & Painting/Coating	98.0	69.0	80.5	100.2	106.6	104.1	103.5	94.6	98.1	100.2	106.6	104.1	106.4	89.6	96.3	98.5	106.4	103.2
09	FINISHES	83.2	122.4	104.8	95.2	89.5	92.1	92.2	88.8	90.4	95.6	91.7	93.4	96.9	88.0	92.0	88.1	89.5	88.9
COVERS	DIVS. 10 - 14, 25, 28, 41, 43, 44, 46	100.0	70.6	94.1	100.0	99.1	99.8	100.0	100.8	100.2	100.0	102.2	100.4	100.0	97.4	99.5	100.0	101.0	100.2
21, 22, 23	FIRE SUPPRESSION, PLUMBING & HVAC	95.0	127.4	108.1	95.2	88.4	92.4	99.7	92.9	97.0	95.2	91.1	93.5	100.1	92.1	96.9	95.5	99.2	97.0
26, 27, 3370	ELECTRICAL, COMMUNICATIONS & UTIL.	92.1	128.1	111.1	94.9	112.2	104.0	91.1	96.7	94.0	94.9	112.2	104.1	96.8	88.3	92.3	93.6	91.7	92.6
MF2014	WEIGHTED AVERAGE	94.9	120.6	106.1	98.5	96.8	97.8	94.5	95.8	95.1	98.2	100.9	99.4	99.8	95.6	98.0	95.4	98.3	96.6

DIVISION		PENNSYLVANIA																	
		INDIANA			JOHNSTOWN			KITTANNING			LANCASTER			LEHIGH VALLEY			MONTROSE		
		157			159			162			175 - 176			180			188		
		MAT.	INST.	TOTAL	MAT.	INST.	TOTAL	MAT.	INST.	TOTAL	MAT.	INST.	TOTAL	MAT.	INST.	TOTAL	MAT.	INST.	TOTAL
015433	CONTRACTOR EQUIPMENT		109.6	109.6		109.6	109.6		112.9	112.9		112.1	112.1		112.9	112.9		112.9	112.9
0241, 31 - 34	SITE & INFRASTRUCTURE, DEMOLITION	95.9	101.2	99.7	101.9	102.2	102.1	90.7	105.7	101.4	81.6	103.5	97.2	90.7	104.3	100.4	89.4	102.1	98.4
0310	Concrete Forming & Accessories	84.7	86.3	86.1	83.5	84.8	84.6	85.7	95.9	94.5	92.3	87.5	88.2	92.8	112.7	110.0	81.8	88.5	87.6
0320	Concrete Reinforcing	91.8	109.2	100.6	93.1	108.9	101.2	92.4	109.2	101.0	91.4	106.1	98.9	91.9	108.1	100.1	96.3	106.3	101.4
0330	Cast-in-Place Concrete	97.1	95.9	96.6	107.6	92.4	101.4	88.4	96.5	91.7	77.5	97.9	85.9	88.4	104.7	95.1	86.7	90.9	88.4
03	CONCRETE	94.7	95.2	94.9	102.9	93.2	98.2	83.8	99.7	91.6	89.8	96.0	92.8	92.6	109.8	101.1	91.3	93.7	92.5
04	MASONRY	110.1	100.1	103.9	110.9	90.6	98.3	105.4	100.1	102.1	108.2	90.1	97.0	97.3	101.0	99.6	97.3	95.1	95.9
05	METALS	97.7	120.1	104.6	97.6	118.9	104.2	93.7	121.9	102.4	97.5	120.8	104.7	99.7	122.0	106.5	98.0	114.1	102.9
06	WOOD, PLASTICS & COMPOSITES	84.8	83.1	83.8	82.9	83.1	83.0	80.6	95.5	89.0	90.8	86.7	88.5	92.2	115.8	105.4	80.4	88.4	84.9
07	THERMAL & MOISTURE PROTECTION	102.7	98.2	100.9	103.0	95.1	99.7	101.1	101.0	101.1	98.8	97.1	98.1	102.2	102.8	102.4	101.8	91.7	97.7
08	OPENINGS	98.0	94.4	97.1	98.0	90.9	96.3	88.7	101.2	91.6	94.1	98.6	95.1	95.6	115.0	100.1	92.2	93.1	92.4
0920	Plaster & Gypsum Board	92.7	82.5	85.8	91.0	82.5	85.2	87.7	95.2	92.7	98.5	86.0	90.1	89.5	116.0	107.4	87.1	87.8	87.6
0950, 0980	Ceilings & Acoustic Treatment	95.0	82.5	86.7	94.2	82.5	86.5	87.9	95.2	92.7	84.0	86.0	85.3	83.7	116.0	105.0	85.4	87.8	87.0
0960	Flooring	93.7	105.3	97.0	92.9	98.2	94.4	89.0	105.3	93.6	98.8	91.7	96.8	91.9	95.9	93.0	87.6	57.9	79.1
0970, 0990	Wall Finishes & Painting/Coating	100.2	106.6	104.1	100.2	109.2	105.6	94.0	106.6	101.6	105.9	57.1	76.5	98.5	66.4	79.1	98.5	106.4	103.2
09	FINISHES	94.8	90.0	92.1	94.6	88.8	91.4	90.1	97.4	94.1	92.0	84.5	87.9	90.2	104.1	97.9	88.9	84.7	86.6
COVERS	DIVS. 10 - 14, 25, 28, 41, 43, 44, 46	100.0	100.6	100.1	100.0	99.3	99.9	100.0	102.2	100.5	100.0	97.4	99.5	100.0	106.0	101.2	100.0	101.2	100.2
21, 22, 23	FIRE SUPPRESSION, PLUMBING & HVAC	95.2	90.8	93.4	95.2	88.7	92.5	95.0	96.9	95.8	95.5	92.3	94.2	95.5	112.0	102.2	95.5	98.4	96.7
26, 27, 3370	ELECTRICAL, COMMUNICATIONS & UTIL.	94.9	112.2	104.1	94.9	112.2	104.0	89.3	112.2	101.4	91.0	44.0	66.2	93.6	143.4	119.9	92.7	97.6	95.3
MF2014	WEIGHTED AVERAGE	97.1	99.3	98.0	98.1	97.2	97.7	92.9	103.3	97.4	95.1	89.0	92.4	95.8	114.0	103.7	94.7	96.9	95.7

DIVISION		PENNSYLVANIA																	
		NEW CASTLE			NORRISTOWN			OIL CITY			PHILADELPHIA			PITTSBURGH			POTTSVILLE		
		161			194			163			190 - 191			150 - 152			179		
		MAT.	INST.	TOTAL	MAT.	INST.	TOTAL	MAT.	INST.	TOTAL	MAT.	INST.	TOTAL	MAT.	INST.	TOTAL	MAT.	INST.	TOTAL
015433	CONTRACTOR EQUIPMENT		112.9	112.9		98.8	98.8		112.9	112.9		98.2	98.2		110.8	110.8		112.1	112.1
0241, 31 - 34	SITE & INFRASTRUCTURE, DEMOLITION	88.5	105.8	100.8	95.8	100.6	99.2	87.1	103.7	98.9	101.6	100.0	100.5	101.3	104.9	103.8	84.4	103.4	97.9
0310	Concrete Forming & Accessories	85.7	95.6	94.3	83.2	129.7	123.3	85.7	83.4	83.7	98.9	140.1	134.4	98.5	96.7	97.0	83.3	89.7	88.8
0320	Concrete Reinforcing	91.3	92.5	91.9	90.2	137.9	114.5	92.4	92.3	92.4	100.6	137.9	119.6	93.6	109.4	101.6	90.7	102.7	96.8
0330	Cast-in-Place Concrete	85.9	96.5	90.2	85.0	127.1	102.3	83.4	95.5	88.4	98.0	132.2	112.1	102.9	96.5	100.3	82.6	100.9	90.1
03	CONCRETE	81.6	96.4	88.9	89.2	130.1	109.3	80.1	90.5	85.2	99.4	136.4	117.6	100.5	100.1	100.3	93.3	97.3	95.3
04	MASONRY	101.9	97.7	99.3	110.4	124.9	119.4	101.8	95.8	98.0	96.7	130.7	117.9	103.0	102.1	102.4	102.1	92.2	95.9
05	METALS	93.7	113.6	99.8	99.2	130.0	108.7	93.7	111.5	99.2	102.0	130.2	110.7	99.0	121.3	105.9	97.7	118.8	104.2
06	WOOD, PLASTICS & COMPOSITES	80.6	95.9	89.2	79.9	130.6	108.3	80.6	80.1	80.3	98.6	141.9	122.9	100.6	95.8	97.9	80.2	88.1	84.6
07	THERMAL & MOISTURE PROTECTION	101.1	98.0	99.8	101.4	131.1	113.5	101.0	94.4	98.3	102.8	135.2	116.1	103.1	101.6	102.4	98.9	106.7	102.1
08	OPENINGS	88.7	96.6	90.5	86.9	140.1	99.2	88.7	80.3	86.7	98.6	146.2	109.7	101.8	105.0	102.5	94.1	91.8	93.6
0920	Plaster & Gypsum Board	87.7	95.6	93.0	85.6	131.4	116.5	87.7	79.2	82.0	96.6	143.1	128.0	99.8	95.6	96.9	93.5	87.5	89.5
0950, 0980	Ceilings & Acoustic Treatment	87.9	95.6	92.9	84.5	131.4	115.3	87.9	79.2	82.2	94.0	143.1	126.2	95.0	95.6	95.4	84.0	87.5	86.3
0960	Flooring	89.0	57.5	80.0	92.9	139.2	106.1	89.0	105.3	93.6	98.4	139.2	110.1	99.5	106.6	101.5	95.2	91.7	94.2
0970, 0990	Wall Finishes & Painting/Coating	94.0	109.2	103.2	98.6	147.8	128.3	94.0	106.6	101.6	99.8	154.7	133.0	100.2	119.2	111.7	105.9	106.4	106.2
09	FINISHES	90.0	89.6	89.8	90.7	133.0	114.0	89.8	87.8	88.7	98.9	141.8	122.5	97.7	100.2	99.1	90.5	91.3	90.9
COVERS	DIVS. 10 - 14, 25, 28, 41, 43, 44, 46	100.0	102.4	100.5	100.0	117.5	103.5	100.0	100.6	100.1	100.0	120.8	104.2	100.0	102.2	100.4	100.0	98.7	99.7
21, 22, 23	FIRE SUPPRESSION, PLUMBING & HVAC	95.0	93.6	94.4	95.2	127.3	108.2	95.0	92.9	94.2	100.0	131.8	112.9	99.9	100.4	100.1	95.5	98.6	96.8
26, 27, 3370	ELECTRICAL, COMMUNICATIONS & UTIL.	89.9	98.1	94.2	93.0	148.4	122.3	91.8	111.1	102.0	97.1	148.4	124.2	97.2	112.2	105.1	89.2	95.5	92.5
MF2014	WEIGHTED AVERAGE	92.5	97.9	94.8	95.0	129.5	110.0	92.4	96.9	94.4	99.7	133.6	114.5	99.9	104.6	102.0	94.9	98.5	96.5

DIVISION		PENNSYLVANIA																	
		READING			SCRANTON			STATE COLLEGE			STROUDSBURG			SUNBURY			UNIONTOWN		
		195 - 196			184 - 185			168			183			178			154		
		MAT.	INST.	TOTAL	MAT.	INST.	TOTAL	MAT.	INST.	TOTAL	MAT.	INST.	TOTAL	MAT.	INST.	TOTAL	MAT.	INST.	TOTAL
015433	CONTRACTOR EQUIPMENT		118.0	118.0		112.9	112.9		112.1	112.1		112.9	112.9		112.9	112.9		109.6	109.6
0241, 31 - 34	SITE & INFRASTRUCTURE, DEMOLITION	100.3	112.7	109.1	94.1	104.8	101.7	84.4	103.1	97.7	88.5	102.2	98.2	96.6	104.5	102.2	96.5	102.9	101.0
0310	Concrete Forming & Accessories	98.8	90.2	91.4	99.3	88.6	90.0	84.4	80.3	80.8	87.2	89.4	89.1	96.1	88.5	89.6	77.6	96.0	93.5
0320	Concrete Reinforcing	91.5	104.7	98.2	94.7	106.6	100.8	93.0	103.4	98.3	95.0	111.6	103.4	93.2	106.2	99.8	92.6	109.1	101.0
0330	Cast-in-Place Concrete	76.2	97.3	84.9	90.3	93.1	91.5	87.1	65.6	78.2	85.1	72.6	80.0	90.2	95.5	92.4	97.1	96.4	96.8
03	CONCRETE	88.2	96.7	92.4	95.4	94.9	95.1	95.5	81.0	88.4	90.1	88.9	89.5	96.7	95.5	96.1	94.4	99.7	97.0
04	MASONRY	99.5	93.0	95.4	97.7	95.7	96.4	103.1	77.9	87.4	95.1	100.4	98.4	102.3	89.7	94.5	127.0	98.9	109.5
05	METALS	99.5	121.2	106.2	102.1	120.5	107.8	97.7	117.6	103.8	99.7	115.9	104.7	97.4	120.0	104.4	97.4	120.6	104.5
06	WOOD, PLASTICS & COMPOSITES	98.5	88.1	92.6	100.7	86.2	92.6	87.6	83.2	85.1	86.4	88.4	87.5	88.9	88.1	88.5	76.7	95.4	87.2
07	THERMAL & MOISTURE PROTECTION	101.8	112.1	106.0	102.2	95.6	99.5	101.5	90.6	97.0	102.0	85.3	95.2	100.0	104.9	102.0	102.6	100.7	101.8
08	OPENINGS	91.3	98.9	93.1	95.0	91.9	94.3	92.0	89.3	91.4	95.6	88.2	93.9	94.2	92.8	93.9	97.9	104.7	99.5
0920	Plaster & Gypsum Board	96.8	87.5	90.5	98.0	85.6	89.6	89.6	82.5	84.8	88.0	87.8	87.9	92.7	87.5	89.2	89.2	95.2	93.2
0950, 0980	Ceilings & Acoustic Treatment	76.9	87.5	83.9	91.1	85.6	87.5	82.9	82.5	82.6	82.0	87.8	85.8	80.7	87.5	85.2	94.2	95.2	94.8
0960	Flooring	96.8	91.7	95.4	94.6	107.0	98.2	92.3	97.5	93.8	89.9	51.3	78.9	95.9	87.7	93.5	90.4	105.3	94.7
0970, 0990	Wall Finishes & Painting/Coating	97.3	106.4	102.8	98.5	106.4	103.2	98.5	109.2	105.0	98.5	64.6	78.0	105.9	94.3	98.9	100.2	109.2	105.6
09	FINISHES	92.0	90.8	91.3	93.9	93.1	93.4	89.4	85.8	87.4	88.9	80.0	84.0	91.3	88.7	89.9	93.2	97.5	95.6
COVERS	DIVS. 10 - 14, 25, 28, 41, 43, 44, 46	100.0	100.3	100.1	100.0	101.0	100.2	100.0	93.1	98.6	100.0	62.4	92.4	100.0	97.9	99.6	100.0	102.2	100.4
21, 22, 23	FIRE SUPPRESSION, PLUMBING & HVAC	100.1	108.8	103.7	100.2	99.2	99.8	95.5	85.9	91.6	95.5	100.0	97.3	95.5	92.0	94.1	95.2	91.1	93.5
26, 27, 3370	ELECTRICAL, COMMUNICATIONS & UTIL.	99.6	95.5	97.5	99.2	97.7	98.4	91.9	112.2	102.6	93.6	143.3	119.9	89.5	90.8	90.2	92.0	112.2	102.7
MF2014	WEIGHTED AVERAGE	97.1	102.1	99.3	98.6	99.2	98.9	95.3	92.8	94.2	95.3	101.4	97.9	95.8	95.8	95.8	97.3	101.7	99.2

DIVISION		PENNSYLVANIA																	
		WASHINGTON			WELLSBORO			WESTCHESTER			WILKES-BARRE			WILLIAMSPORT			YORK		
		153			169			193			186 - 187			177			173 - 174		
		MAT.	INST.	TOTAL	MAT.	INST.	TOTAL	MAT.	INST.	TOTAL	MAT.	INST.	TOTAL	MAT.	INST.	TOTAL	MAT.	INST.	TOTAL
015433	CONTRACTOR EQUIPMENT		109.6	109.6		112.9	112.9		98.8	98.8		112.9	112.9		112.9	112.9		112.1	112.1
0241, 31 - 34	SITE & INFRASTRUCTURE, DEMOLITION	96.6	102.9	101.1	95.9	102.0	100.3	101.6	97.8	98.9	86.5	104.7	99.5	87.9	103.4	98.9	85.2	103.5	98.2
0310	Concrete Forming & Accessories	84.8	96.2	94.6	85.8	86.7	86.6	89.8	128.2	122.9	89.9	88.9	89.1	92.5	61.0	65.3	86.9	88.1	87.9
0320	Concrete Reinforcing	92.6	109.3	101.1	93.0	106.2	99.7	89.3	113.6	101.7	93.7	107.1	100.6	92.5	64.6	78.3	93.2	106.2	99.8
0330	Cast-in-Place Concrete	97.1	96.4	96.8	91.2	88.9	90.3	94.1	126.4	107.4	81.7	93.0	86.4	76.4	74.2	75.5	83.0	98.1	89.2
03	CONCRETE	94.9	99.8	97.3	97.8	92.2	95.1	96.9	124.5	110.5	87.2	95.1	91.1	84.7	68.2	76.6	94.5	96.3	95.4
04	MASONRY	109.4	100.5	103.9	103.6	89.7	94.9	104.7	124.9	117.3	110.6	95.3	101.0	93.9	93.7	93.8	103.6	90.1	95.2
05	METALS	97.3	120.9	104.6	97.8	114.2	102.9	99.2	117.2	104.8	97.9	121.0	105.0	97.5	100.4	98.4	99.1	121.4	105.9
06	WOOD, PLASTICS & COMPOSITES	84.9	95.4	90.8	84.9	88.1	86.7	87.1	130.6	111.5	88.9	86.6	87.6	85.2	51.0	66.1	84.1	86.7	85.5
07	THERMAL & MOISTURE PROTECTION	102.7	101.1	102.1	102.5	89.6	97.2	101.7	129.9	113.3	101.8	106.4	103.7	99.4	102.4	100.6	99.0	110.4	103.7
08	OPENINGS	97.9	104.7	99.5	95.1	93.0	94.6	86.9	126.1	96.0	92.2	98.8	93.8	94.2	53.5	84.7	94.1	94.0	94.1
0920	Plaster & Gypsum Board	92.4	95.2	94.3	87.9	87.5	87.6	87.0	131.4	117.0	88.8	86.0	86.9	93.5	49.3	63.6	94.3	86.0	88.7
0950, 0980	Ceilings & Acoustic Treatment	94.2	95.2	94.8	82.9	87.5	85.9	84.5	131.4	115.3	85.4	86.0	85.8	84.0	49.3	61.2	83.1	86.0	85.0
0960	Flooring	93.8	105.3	97.1	89.1	50.8	78.2	95.7	139.2	108.1	90.7	94.6	91.8	94.8	49.7	81.9	96.4	91.7	95.1
0970, 0990	Wall Finishes & Painting/Coating	100.2	109.2	105.6	98.5	106.4	103.2	98.6	147.8	128.3	98.5	106.4	103.2	105.9	106.4	106.2	105.9	89.6	96.1
09	FINISHES	94.6	97.8	96.4	89.5	82.2	85.5	92.1	131.0	113.6	89.9	90.8	90.4	91.1	62.3	75.2	90.8	88.1	89.3
COVERS	DIVS. 10 - 14, 25, 28, 41, 43, 44, 46	100.0	102.2	100.4	100.0	97.9	99.6	100.0	117.3	103.5	100.0	100.8	100.2	100.0	94.6	98.9	100.0	97.6	99.5
21, 22, 23	FIRE SUPPRESSION, PLUMBING & HVAC	95.2	97.2	96.0	95.5	91.1	93.7	95.2	125.8	107.6	95.5	99.0	96.9	95.5	93.1	94.5	100.2	92.4	97.1
26, 27, 3370	ELECTRICAL, COMMUNICATIONS & UTIL.	94.3	112.2	103.8	92.7	86.5	89.5	92.9	114.9	104.5	93.6	91.7	92.6	90.0	74.6	81.8	91.0	88.3	89.5
MF2014	WEIGHTED AVERAGE	97.0	103.2	99.7	96.3	92.6	94.7	95.9	121.5	107.1	95.1	98.7	96.6	93.8	83.0	89.1	96.7	95.9	96.3

DIVISION		PUERTO RICO			RHODE ISLAND						SOUTH CAROLINA								
		SAN JUAN			NEWPORT			PROVIDENCE			AIKEN			BEAUFORT			CHARLESTON		
		009			028			029			298			299			294		
		MAT.	INST.	TOTAL	MAT.	INST.	TOTAL	MAT.	INST.	TOTAL	MAT.	INST.	TOTAL	MAT.	INST.	TOTAL	MAT.	INST.	TOTAL
015433	CONTRACTOR EQUIPMENT		90.5	90.5		102.4	102.4		102.4	102.4		101.3	101.3		101.3	101.3		101.3	101.3
0241, 31 - 34	SITE & INFRASTRUCTURE, DEMOLITION	133.8	91.2	103.5	89.2	105.0	100.4	91.4	105.0	101.1	118.9	87.4	96.5	114.3	85.6	93.9	99.7	86.3	90.2
0310	Concrete Forming & Accessories	92.4	17.8	28.1	101.5	122.4	119.5	99.8	122.4	119.3	97.5	67.4	71.5	96.4	37.9	46.0	95.4	63.0	67.5
0320	Concrete Reinforcing	188.3	12.7	98.9	106.4	149.2	128.2	102.3	149.2	126.2	93.7	65.2	79.2	92.8	25.8	58.7	92.7	59.3	75.7
0330	Cast-in-Place Concrete	103.8	30.4	73.7	79.5	124.1	97.8	95.5	124.1	107.2	79.2	69.8	75.4	79.2	47.2	66.1	92.9	49.5	75.1
03	CONCRETE	108.7	22.2	66.2	93.2	127.5	110.0	100.4	127.5	113.7	102.7	68.9	86.1	99.9	40.9	70.9	94.4	58.9	77.0
04	MASONRY	90.1	16.9	44.5	95.4	132.5	118.5	101.3	132.5	120.7	79.6	61.4	68.3	93.9	33.3	56.1	95.1	40.9	61.4
05	METALS	118.4	34.6	92.6	97.4	125.6	106.1	103.1	125.6	110.0	101.8	83.8	96.3	101.8	68.2	91.5	103.8	80.4	96.6
06	WOOD, PLASTICS & COMPOSITES	94.4	17.5	51.3	99.0	120.9	111.3	100.1	120.9	111.7	97.0	68.6	81.1	95.3	37.8	63.1	94.0	68.4	79.7
07	THERMAL & MOISTURE PROTECTION	129.1	21.6	85.0	100.6	121.5	109.2	100.5	121.5	109.1	102.7	66.4	87.8	102.4	38.8	76.3	101.6	47.3	79.3
08	OPENINGS	152.0	15.4	120.2	102.2	128.2	108.3	107.3	128.2	112.2	99.0	64.8	91.1	99.0	36.7	84.5	102.9	63.3	93.7
0920	Plaster & Gypsum Board	159.8	14.9	61.9	92.9	120.9	111.9	94.5	120.9	112.4	104.0	67.4	79.3	107.4	35.7	59.0	109.0	67.3	80.8
0950, 0980	Ceilings & Acoustic Treatment	225.7	14.9	87.2	93.2	120.9	111.4	90.1	120.9	110.4	86.4	67.4	73.9	89.8	35.7	54.3	89.8	67.3	75.0
0960	Flooring	224.3	16.6	164.9	97.7	141.5	110.2	98.0	141.5	110.4	105.4	68.4	94.8	106.8	51.2	90.9	106.5	58.2	92.7
0970, 0990	Wall Finishes & Painting/Coating	213.5	16.8	94.8	99.9	129.2	117.6	97.7	129.2	116.7	109.5	70.3	85.8	109.5	34.7	64.3	109.5	66.9	83.8
09	FINISHES	210.6	17.6	104.2	96.6	126.9	113.3	93.7	126.9	112.0	98.3	68.0	81.6	99.5	39.8	66.6	97.8	63.5	78.9
COVERS	DIVS. 10 - 14, 25, 28, 41, 43, 44, 46	100.0	17.5	83.3	100.0	108.6	101.7	100.0	108.6	101.7	100.0	71.8	94.3	100.0	70.4	94.0	100.0	69.0	93.7
21, 22, 23	FIRE SUPPRESSION, PLUMBING & HVAC	103.3	14.0	67.3	100.1	110.6	104.3	99.9	110.6	104.2	95.7	63.1	82.5	95.7	36.5	71.8	100.5	53.7	81.6
26, 27, 3370	ELECTRICAL, COMMUNICATIONS & UTIL.	125.9	13.2	66.3	100.0	99.0	99.5	99.6	99.0	99.2	96.9	65.9	80.5	100.7	33.6	65.2	99.0	88.8	93.6
MF2014	WEIGHTED AVERAGE	122.7	24.4	79.8	98.4	117.5	106.7	100.6	117.5	108.0	98.6	69.1	85.7	99.3	44.9	75.6	99.9	65.2	84.8

DIVISION		SOUTH CAROLINA															SOUTH DAKOTA		
		COLUMBIA			FLORENCE			GREENVILLE			ROCK HILL			SPARTANBURG			ABERDEEN		
		290 - 292			295			296			297			293			574		
		MAT.	INST.	TOTAL	MAT.	INST.	TOTAL	MAT.	INST.	TOTAL	MAT.	INST.	TOTAL	MAT.	INST.	TOTAL	MAT.	INST.	TOTAL
015433	CONTRACTOR EQUIPMENT		101.3	101.3		101.3	101.3		101.3	101.3		101.3	101.3		101.3	101.3		98.2	98.2
0241, 31 - 34	SITE & INFRASTRUCTURE, DEMOLITION	99.8	86.3	90.2	108.8	86.3	92.8	104.1	85.9	91.2	101.7	85.0	89.8	103.9	86.0	91.1	99.2	93.7	95.3
0310	Concrete Forming & Accessories	94.4	45.1	51.9	83.2	45.3	50.5	95.0	45.1	52.0	93.2	38.7	46.2	98.0	45.3	52.5	94.8	36.6	44.6
0320	Concrete Reinforcing	95.8	59.1	77.1	92.3	59.3	75.5	92.2	44.5	67.9	93.0	44.1	68.1	92.2	57.0	74.3	96.1	38.1	66.6
0330	Cast-in-Place Concrete	95.8	50.1	77.0	79.2	49.4	66.9	79.2	49.3	66.9	79.2	43.9	64.7	79.2	49.4	66.9	105.6	43.5	80.1
03	CONCRETE	96.2	51.1	74.1	94.2	51.0	73.0	93.0	48.2	71.0	90.9	43.3	67.5	93.2	50.6	72.3	102.8	40.7	72.3
04	MASONRY	91.5	37.6	57.9	79.8	40.9	55.6	77.4	40.9	54.7	101.6	34.6	59.8	79.8	40.9	55.6	115.9	55.4	78.2
05	METALS	100.9	79.6	94.3	102.6	80.0	95.6	102.6	74.5	93.9	101.8	71.2	92.4	102.6	79.1	95.4	95.9	63.6	86.0
06	WOOD, PLASTICS & COMPOSITES	98.0	44.8	68.2	79.7	44.8	60.2	93.7	44.8	66.4	92.0	38.9	62.3	98.0	44.8	68.2	99.6	36.0	64.0
07	THERMAL & MOISTURE PROTECTION	97.2	43.3	75.1	101.9	44.8	78.5	101.8	44.8	78.4	101.6	39.7	76.3	101.8	44.8	78.5	99.2	47.7	78.1
08	OPENINGS	104.6	50.5	92.0	99.1	50.5	87.8	99.0	47.0	86.9	99.0	39.4	85.2	99.0	50.0	87.6	99.1	36.7	84.6
0920	Plaster & Gypsum Board	102.3	42.9	62.2	97.1	42.9	60.5	102.6	42.9	62.3	101.9	36.8	57.9	105.5	42.9	63.2	103.7	34.3	56.8
0950, 0980	Ceilings & Acoustic Treatment	91.3	42.9	59.5	87.3	42.9	58.1	86.4	42.9	57.8	86.4	36.8	53.8	86.4	42.9	57.8	93.2	34.3	54.5
0960	Flooring	100.5	43.6	84.2	98.4	43.6	82.7	104.3	57.2	90.8	103.3	44.4	86.5	105.6	57.2	91.8	108.4	50.3	91.8
0970, 0990	Wall Finishes & Painting/Coating	106.2	66.9	82.5	109.5	66.9	83.8	109.5	66.9	83.8	109.5	39.2	67.0	109.5	66.9	83.8	102.6	37.1	63.1
09	FINISHES	95.9	46.6	68.7	94.4	47.1	68.3	96.3	49.3	70.4	95.7	39.4	64.6	97.1	49.3	70.8	100.3	38.8	66.4
COVERS	DIVS. 10 - 14, 25, 28, 41, 43, 44, 46	100.0	66.3	93.2	100.0	66.3	93.2	100.0	66.3	93.2	100.0	64.6	92.9	100.0	66.3	93.2	100.0	40.5	88.0
21, 22, 23	FIRE SUPPRESSION, PLUMBING & HVAC	100.0	53.0	81.1	100.5	53.1	81.4	100.5	53.0	81.3	95.7	44.2	74.9	100.5	53.1	81.3	100.1	39.8	75.8
26, 27, 3370	ELECTRICAL, COMMUNICATIONS & UTIL.	99.0	59.7	78.3	96.8	59.7	77.2	99.1	57.4	77.0	99.1	56.6	76.6	99.1	57.4	77.1	101.7	51.1	75.0
MF2014	WEIGHTED AVERAGE	99.3	56.5	80.6	98.2	56.9	80.2	98.3	55.8	79.8	97.8	50.3	77.1	98.6	56.7	80.3	100.5	49.6	78.3

DIVISION		SOUTH DAKOTA																	
		MITCHELL			MOBRIDGE			PIERRE			RAPID CITY			SIOUX FALLS			WATERTOWN		
		573			576			575			577			570 - 571			572		
		MAT.	INST.	TOTAL	MAT.	INST.	TOTAL	MAT.	INST.	TOTAL	MAT.	INST.	TOTAL	MAT.	INST.	TOTAL	MAT.	INST.	TOTAL
015433	CONTRACTOR EQUIPMENT		98.2	98.2		98.2	98.2		98.2	98.2		98.2	98.2		99.2	99.2		98.2	98.2
0241, 31 - 34	SITE & INFRASTRUCTURE, DEMOLITION	96.0	93.7	94.3	95.9	93.7	94.3	100.5	93.7	95.7	97.7	93.8	94.9	94.3	95.4	95.0	95.8	93.7	94.3
0310	Concrete Forming & Accessories	94.0	37.0	44.8	84.9	36.8	43.4	97.7	38.4	46.6	102.2	36.7	45.7	98.8	40.5	48.5	81.6	36.6	42.8
0320	Concrete Reinforcing	95.5	46.3	70.5	98.0	38.1	67.5	99.0	71.1	84.8	89.9	71.3	80.4	98.0	71.2	84.3	92.9	38.2	65.1
0330	Cast-in-Place Concrete	102.6	42.3	77.8	102.6	43.5	78.3	98.5	42.2	75.4	101.8	42.8	77.6	92.3	42.9	72.0	102.6	45.9	79.3
03	CONCRETE	100.5	41.8	71.7	100.3	40.7	71.0	99.0	47.3	73.6	99.8	46.7	73.7	95.9	48.4	72.6	99.2	41.5	70.9
04	MASONRY	103.9	53.3	72.4	113.4	55.4	77.2	117.4	53.5	77.6	113.5	56.4	77.9	105.1	53.6	73.0	141.1	57.4	88.9
05	METALS	95.0	63.8	85.4	95.0	63.6	85.3	98.4	79.1	92.4	97.8	79.7	92.2	98.2	79.3	92.4	95.0	64.1	85.5
06	WOOD, PLASTICS & COMPOSITES	98.5	36.5	63.8	87.6	36.2	58.8	104.5	36.8	66.6	103.5	33.3	64.2	96.8	39.3	64.6	83.8	36.0	57.0
07	THERMAL & MOISTURE PROTECTION	98.9	45.6	77.1	99.0	47.8	78.0	101.8	45.9	78.9	99.6	47.8	78.3	101.9	48.2	79.9	98.8	48.3	78.1
08	OPENINGS	97.7	37.3	83.6	100.4	36.3	85.5	105.7	46.7	92.0	103.1	44.8	89.6	106.5	48.1	92.9	97.7	36.5	83.4
0920	Plaster & Gypsum Board	102.1	34.8	56.7	96.2	34.6	54.5	100.1	35.1	56.2	103.2	31.6	54.8	90.7	37.7	54.9	93.9	34.3	53.6
0950, 0980	Ceilings & Acoustic Treatment	89.9	34.8	53.7	93.2	34.6	54.7	93.3	35.1	55.0	95.7	31.6	53.6	90.6	37.7	55.8	89.9	34.3	53.3
0960	Flooring	107.9	50.3	91.4	103.5	50.3	88.3	107.4	35.8	86.9	107.6	78.6	99.3	103.8	74.9	95.5	102.1	50.3	87.3
0970, 0990	Wall Finishes & Painting/Coating	102.6	40.6	65.2	102.6	41.7	65.8	105.7	44.6	68.8	102.6	44.6	67.6	101.7	44.6	67.2	102.6	37.1	63.1
09	FINISHES	99.1	39.5	66.2	97.7	39.4	65.6	100.8	37.3	65.8	100.5	43.9	69.3	97.1	46.5	69.2	96.2	38.8	64.6
COVERS	DIVS. 10 - 14, 25, 28, 41, 43, 44, 46	100.0	37.6	87.4	100.0	40.5	88.0	100.0	75.8	95.1	100.0	75.7	95.1	100.0	76.2	95.2	100.0	40.5	88.0
21, 22, 23	FIRE SUPPRESSION, PLUMBING & HVAC	95.3	39.4	72.8	95.3	39.9	73.0	100.0	64.5	85.7	100.1	65.0	85.9	100.0	38.6	75.2	95.3	39.9	73.0
26, 27, 3370	ELECTRICAL, COMMUNICATIONS & UTIL.	99.9	44.0	70.3	101.7	44.8	71.7	104.8	51.8	76.8	98.0	51.8	73.6	101.3	75.1	87.5	99.0	44.8	70.4
MF2014	WEIGHTED AVERAGE	97.8	48.4	76.3	98.5	48.8	76.8	101.7	58.4	82.8	100.4	59.5	82.6	100.0	57.7	81.5	99.0	49.1	77.2

DIVISION		TENNESSEE																	
		CHATTANOOGA			COLUMBIA			COOKEVILLE			JACKSON			JOHNSON CITY			KNOXVILLE		
		373 - 374			384			385			383			376			377 - 379		
		MAT.	INST.	TOTAL	MAT.	INST.	TOTAL	MAT.	INST.	TOTAL	MAT.	INST.	TOTAL	MAT.	INST.	TOTAL	MAT.	INST.	TOTAL
015433	CONTRACTOR EQUIPMENT		103.8	103.8		98.2	98.2		98.2	98.2		104.5	104.5		97.7	97.7		97.7	97.7
0241, 31 - 34	SITE & INFRASTRUCTURE, DEMOLITION	104.5	97.9	99.8	89.1	86.9	87.5	94.5	86.4	88.8	97.7	96.8	97.1	111.1	86.6	93.7	90.7	87.3	88.3
0310	Concrete Forming & Accessories	97.1	56.4	62.0	82.1	62.0	64.8	82.2	35.4	41.8	89.0	44.9	51.0	83.7	38.9	45.1	95.6	61.0	65.7
0320	Concrete Reinforcing	91.2	67.4	79.1	87.4	61.2	74.1	87.4	61.2	74.1	87.4	62.0	74.4	91.7	60.3	75.7	91.2	62.3	76.5
0330	Cast-in-Place Concrete	101.3	62.7	85.5	94.2	51.6	76.7	106.7	43.5	80.7	104.2	49.3	81.6	81.6	60.2	72.8	95.2	65.7	83.1
03	CONCRETE	96.0	62.3	79.4	96.7	59.8	78.6	106.9	45.2	76.6	97.7	51.6	75.1	103.3	52.3	78.3	93.4	64.4	79.1
04	MASONRY	109.7	51.7	73.5	127.7	54.6	82.1	121.9	43.8	73.2	127.9	47.2	77.6	125.7	44.8	75.3	86.6	55.3	67.1
05	METALS	98.7	89.2	95.8	96.4	86.0	93.2	96.5	85.9	93.2	98.8	85.9	94.8	95.9	85.2	92.6	99.3	87.0	95.5
06	WOOD, PLASTICS & COMPOSITES	104.0	57.3	77.8	70.1	64.4	66.9	70.3	33.7	49.8	85.4	45.4	63.0	76.9	36.9	54.5	91.0	60.6	73.9
07	THERMAL & MOISTURE PROTECTION	99.4	59.9	83.2	93.6	59.2	79.5	94.0	45.7	74.2	96.0	54.1	78.8	94.5	52.6	77.3	92.5	61.6	79.8
08	OPENINGS	99.7	57.6	89.9	92.6	57.9	84.6	92.7	42.2	80.9	100.3	51.6	89.0	96.2	45.1	84.3	93.3	58.6	85.2
0920	Plaster & Gypsum Board	82.8	56.5	65.0	87.7	63.7	71.5	87.7	32.1	50.1	89.3	44.1	58.8	102.4	35.5	57.2	109.0	59.8	75.8
0950, 0980	Ceilings & Acoustic Treatment	96.6	56.5	70.2	77.0	63.7	68.3	77.0	32.1	47.5	86.2	44.1	58.5	93.0	35.5	55.2	93.8	59.8	71.5
0960	Flooring	99.1	58.9	87.7	89.3	20.5	69.7	89.4	58.6	80.6	87.1	40.8	73.9	96.7	40.6	80.7	101.3	55.8	88.3
0970, 0990	Wall Finishes & Painting/Coating	102.5	70.7	83.3	94.0	33.4	57.4	94.0	36.5	59.3	95.8	49.6	67.9	99.6	42.1	64.9	99.6	81.2	88.5
09	FINISHES	97.2	58.2	75.7	89.8	52.0	69.0	90.3	38.4	61.7	89.7	43.7	64.4	101.9	38.1	66.7	94.5	62.0	76.6
COVERS	DIVS. 10 - 14, 25, 28, 41, 43, 44, 46	100.0	41.8	88.3	100.0	47.9	89.5	100.0	41.0	88.1	100.0	43.0	88.5	100.0	75.5	95.1	100.0	81.7	96.3
21, 22, 23	FIRE SUPPRESSION, PLUMBING & HVAC	100.2	62.6	85.0	97.4	78.1	89.6	97.4	72.6	87.4	100.1	68.0	87.1	99.9	59.1	83.4	99.9	67.9	87.0
26, 27, 3370	ELECTRICAL, COMMUNICATIONS & UTIL.	102.3	69.2	84.8	94.2	57.4	74.7	95.9	61.6	77.8	101.0	58.5	78.5	92.5	46.3	68.1	98.1	56.0	75.8
MF2014	WEIGHTED AVERAGE	100.0	66.2	85.2	96.9	66.1	83.5	98.1	59.1	81.1	99.9	61.2	83.0	99.8	56.5	80.9	96.6	66.9	83.7

DIVISION		TENNESSEE									TEXAS								
		MCKENZIE			MEMPHIS			NASHVILLE			ABILENE			AMARILLO			AUSTIN		
		382			375,380 - 381			370 - 372			795 - 796			790 - 791			786 - 787		
		MAT.	INST.	TOTAL	MAT.	INST.	TOTAL	MAT.	INST.	TOTAL	MAT.	INST.	TOTAL	MAT.	INST.	TOTAL	MAT.	INST.	TOTAL
015433	CONTRACTOR EQUIPMENT		98.2	98.2		102.8	102.8		103.7	103.7		89.9	89.9		89.9	89.9		89.4	89.4
0241, 31 - 34	SITE & INFRASTRUCTURE, DEMOLITION	94.3	86.5	88.7	94.1	93.7	93.8	98.9	97.5	97.9	99.4	88.7	91.8	97.3	88.6	91.1	102.4	88.4	92.4
0310	Concrete Forming & Accessories	90.0	38.1	45.2	94.3	64.9	69.0	98.9	65.8	70.3	99.0	65.5	70.1	97.9	57.1	62.7	98.6	59.1	64.5
0320	Concrete Reinforcing	87.5	61.9	74.5	99.7	68.7	83.9	96.0	67.7	81.6	92.9	52.5	72.3	99.0	51.3	74.7	96.5	48.5	72.1
0330	Cast-in-Place Concrete	104.4	57.3	85.1	94.1	61.3	80.6	90.3	65.9	80.3	96.6	63.7	83.1	94.1	63.7	81.6	93.4	66.3	82.3
03	CONCRETE	105.5	51.3	78.9	94.1	65.8	80.2	92.4	67.6	80.2	95.0	63.0	79.3	96.7	59.0	78.2	95.7	60.2	78.3
04	MASONRY	126.2	46.2	76.3	99.8	62.7	76.7	93.8	59.0	72.1	104.2	63.6	78.9	109.7	63.6	81.0	107.6	56.5	75.7
05	METALS	96.5	86.1	93.3	100.1	90.6	97.2	100.4	89.9	97.2	105.3	69.8	94.4	101.6	69.2	91.7	100.5	66.3	90.0
06	WOOD, PLASTICS & COMPOSITES	79.1	36.3	55.1	95.4	67.0	79.5	101.6	66.6	82.0	101.1	68.5	82.8	100.8	57.1	76.4	94.4	59.9	75.0
07	THERMAL & MOISTURE PROTECTION	94.0	48.7	75.4	95.5	63.4	82.4	95.7	61.5	81.7	100.0	67.7	86.7	97.9	65.3	84.5	104.2	64.8	88.1
08	OPENINGS	92.7	44.1	81.4	100.1	66.0	92.2	98.5	65.9	90.9	97.1	63.3	89.2	101.1	56.8	90.8	104.3	55.6	92.9
0920	Plaster & Gypsum Board	90.5	34.8	52.8	95.6	66.3	75.8	95.8	66.0	75.6	84.0	68.0	73.2	92.1	56.3	67.9	86.7	59.2	68.1
0950, 0980	Ceilings & Acoustic Treatment	77.0	34.8	49.2	93.9	66.3	75.8	95.5	66.0	76.1	88.9	68.0	75.2	98.0	56.3	70.6	87.9	59.2	69.0
0960	Flooring	92.1	39.3	77.0	99.4	36.1	81.3	97.7	64.2	88.2	110.9	76.5	101.1	106.5	76.5	97.9	105.7	64.1	93.8
0970, 0990	Wall Finishes & Painting/Coating	94.0	49.6	67.2	97.0	55.7	72.1	100.5	72.0	83.3	106.6	56.1	76.1	100.8	56.1	73.8	104.3	48.8	70.8
09	FINISHES	91.3	38.2	62.1	97.3	58.5	75.9	100.2	66.0	81.4	93.7	67.1	79.0	96.8	60.3	76.7	95.6	58.7	75.2
COVERS	DIVS. 10 - 14, 25, 28, 41, 43, 44, 46	100.0	27.4	85.3	100.0	82.4	96.4	100.0	82.9	96.5	100.0	82.2	96.4	100.0	68.6	93.7	100.0	80.5	96.1
21, 22, 23	FIRE SUPPRESSION, PLUMBING & HVAC	97.4	68.3	85.6	100.0	73.9	89.5	100.0	84.0	93.5	100.3	49.0	79.6	100.0	54.9	81.8	100.1	60.1	84.0
26, 27, 3370	ELECTRICAL, COMMUNICATIONS & UTIL.	95.6	62.9	78.3	101.7	65.3	82.4	97.6	63.1	79.4	99.3	48.7	72.5	101.1	64.1	81.6	96.8	63.9	79.4
MF2014	WEIGHTED AVERAGE	98.2	59.2	81.2	99.0	71.3	86.9	98.4	74.1	87.8	99.7	62.2	83.3	100.2	63.2	84.1	100.0	63.6	84.1

DIVISION		TEXAS																	
		BEAUMONT			BROWNWOOD			BRYAN			CHILDRESS			CORPUS CHRISTI			DALLAS		
		776 - 777			768			778			792			783 - 784			752 - 753		
		MAT.	INST.	TOTAL	MAT.	INST.	TOTAL	MAT.	INST.	TOTAL	MAT.	INST.	TOTAL	MAT.	INST.	TOTAL	MAT.	INST.	TOTAL
015433	CONTRACTOR EQUIPMENT		91.5	91.5		89.9	89.9		91.5	91.5		89.9	89.9		96.6	96.6		99.1	99.1
0241, 31 - 34	SITE & INFRASTRUCTURE, DEMOLITION	90.0	89.9	89.9	105.2	89.6	94.1	81.8	90.5	88.0	109.7	88.0	94.3	140.1	84.4	100.5	104.3	89.8	94.0
0310	Concrete Forming & Accessories	101.8	61.9	67.4	99.0	60.5	65.8	81.0	67.1	69.0	97.3	65.5	69.8	100.2	58.1	63.9	101.2	65.8	70.7
0320	Concrete Reinforcing	95.9	65.2	80.3	92.9	52.1	72.1	98.3	48.6	73.0	93.0	52.1	72.2	88.0	48.5	67.9	100.2	52.5	75.9
0330	Cast-in-Place Concrete	89.5	64.7	79.3	101.4	63.6	85.9	71.7	65.4	69.1	99.1	63.7	84.5	109.8	64.4	91.2	98.0	64.6	84.3
03	CONCRETE	96.3	64.2	80.5	101.2	60.7	81.3	80.9	63.7	72.5	104.0	62.9	83.8	100.1	59.9	80.4	99.0	64.1	81.9
04	MASONRY	106.8	63.9	80.0	141.8	56.6	88.7	146.3	60.4	92.7	108.5	56.6	76.2	91.0	56.6	69.6	102.4	56.7	73.9
05	METALS	95.2	75.5	89.1	99.9	69.1	90.5	94.8	70.1	87.2	102.7	69.2	92.4	94.7	77.9	89.5	100.4	80.0	94.1
06	WOOD, PLASTICS & COMPOSITES	106.1	61.8	81.3	99.5	61.9	78.4	73.5	68.5	70.7	100.4	68.5	82.6	116.0	58.4	83.7	104.1	68.6	84.2
07	THERMAL & MOISTURE PROTECTION	101.9	66.7	87.5	98.2	65.2	84.7	93.9	66.3	82.5	100.5	64.8	85.9	104.5	66.2	88.8	93.0	66.4	82.1
08	OPENINGS	94.0	62.3	86.7	94.7	59.7	86.6	97.2	61.6	88.9	94.2	63.3	87.1	109.1	54.7	96.5	99.5	63.5	91.1
0920	Plaster & Gypsum Board	102.3	61.2	74.5	87.8	61.2	69.8	90.3	68.0	75.2	83.6	68.0	73.1	98.4	57.5	70.7	96.7	68.0	77.3
0950, 0980	Ceilings & Acoustic Treatment	98.2	61.2	73.9	80.6	61.2	67.8	91.2	68.0	76.0	87.3	68.0	74.6	92.7	57.5	69.6	97.8	68.0	78.2
0960	Flooring	112.4	73.0	101.1	95.3	64.1	86.3	85.8	64.1	79.6	109.1	64.1	96.2	118.7	64.1	103.1	102.8	64.1	91.7
0970, 0990	Wall Finishes & Painting/Coating	95.1	59.4	73.6	104.2	56.1	75.1	92.5	62.2	74.2	106.6	56.1	76.1	120.1	56.5	81.7	107.0	56.1	76.3
09	FINISHES	93.7	63.4	77.0	87.7	60.7	72.8	82.3	66.2	73.4	94.0	64.6	77.8	105.9	58.7	79.9	100.0	64.7	80.5
COVERS	DIVS. 10 - 14, 25, 28, 41, 43, 44, 46	100.0	83.4	96.7	100.0	81.4	96.3	100.0	82.7	96.5	100.0	82.2	96.4	100.0	82.6	96.5	100.0	82.5	96.5
21, 22, 23	FIRE SUPPRESSION, PLUMBING & HVAC	100.2	64.1	85.6	95.4	50.0	77.1	95.4	66.5	83.8	95.5	54.9	79.1	100.2	58.9	83.5	100.0	62.4	84.9
26, 27, 3370	ELECTRICAL, COMMUNICATIONS & UTIL.	95.2	70.6	82.2	93.4	42.8	66.7	93.4	67.4	79.7	99.2	64.1	80.7	92.8	60.0	75.5	95.2	64.6	79.0
MF2014	WEIGHTED AVERAGE	97.4	68.6	84.9	98.9	59.4	81.7	94.9	68.2	83.3	99.3	64.3	84.0	100.9	63.5	84.6	99.5	67.4	85.5

DIVISION		TEXAS																	
		DEL RIO			DENTON			EASTLAND			EL PASO			FORT WORTH			GALVESTON		
		788			762			764			798 - 799,885			760 - 761			775		
		MAT.	INST.	TOTAL	MAT.	INST.	TOTAL	MAT.	INST.	TOTAL	MAT.	INST.	TOTAL	MAT.	INST.	TOTAL	MAT.	INST.	TOTAL
015433	CONTRACTOR EQUIPMENT		89.4	89.4		95.8	95.8		89.9	89.9		89.9	89.9		89.9	89.9		100.5	100.5
0241, 31 - 34	SITE & INFRASTRUCTURE, DEMOLITION	121.8	88.4	98.0	105.5	81.3	88.3	108.0	87.6	93.5	101.7	87.6	91.7	102.1	88.7	92.6	107.2	88.4	93.8
0310	Concrete Forming & Accessories	96.5	57.8	63.1	106.2	65.3	70.9	99.8	65.3	70.0	97.8	63.4	68.1	98.0	65.6	70.1	89.8	64.6	68.1
0320	Concrete Reinforcing	88.7	48.4	68.2	94.3	52.1	72.8	93.1	50.9	71.6	99.2	52.0	75.2	98.4	52.4	75.0	97.8	62.4	79.8
0330	Cast-in-Place Concrete	118.2	63.2	95.6	78.5	64.6	72.8	107.2	63.6	89.3	88.4	63.7	78.3	95.0	63.7	82.1	95.2	66.4	83.4
03	CONCRETE	121.3	58.6	90.5	80.0	63.9	72.1	105.9	62.6	84.6	94.0	61.9	78.3	97.0	63.1	80.4	97.5	66.4	82.2
04	MASONRY	106.9	56.5	75.5	152.1	56.7	92.6	106.8	56.6	75.5	99.0	57.7	73.3	104.7	56.6	74.7	102.5	60.4	76.3
05	METALS	94.4	66.2	85.7	99.5	80.6	93.7	99.7	68.5	90.1	98.5	67.4	89.0	101.5	69.5	91.7	96.3	88.3	93.9
06	WOOD, PLASTICS & COMPOSITES	96.5	58.3	75.1	111.3	68.6	87.4	106.1	68.5	85.1	91.3	65.7	76.9	97.3	68.5	81.2	89.4	64.7	75.6
07	THERMAL & MOISTURE PROTECTION	101.3	65.3	86.5	96.0	66.7	84.0	98.6	65.9	85.1	99.8	64.4	85.3	96.0	65.9	83.6	93.0	67.1	82.4
08	OPENINGS	101.8	54.7	90.8	112.2	63.4	100.8	70.4	63.0	68.7	93.3	58.1	85.1	99.4	63.4	91.0	101.6	63.2	92.7
0920	Plaster & Gypsum Board	94.7	57.5	69.6	92.1	68.0	75.8	87.8	68.0	74.4	95.9	65.1	75.1	91.8	68.0	75.7	96.8	64.0	74.7
0950, 0980	Ceilings & Acoustic Treatment	89.1	57.5	68.3	84.7	68.0	73.8	80.6	68.0	72.3	91.4	65.1	74.1	90.5	68.0	75.7	94.6	64.0	74.5
0960	Flooring	100.7	64.1	90.2	90.4	64.1	82.9	121.4	64.1	105.0	110.4	64.1	97.1	117.9	64.1	102.5	100.1	64.1	89.8
0970, 0990	Wall Finishes & Painting/Coating	106.4	48.8	71.6	115.6	56.1	79.7	105.8	56.1	75.8	106.2	53.7	74.5	103.2	56.1	74.7	103.4	60.0	77.2
09	FINISHES	98.2	57.7	75.9	85.8	64.7	74.2	95.8	64.6	78.6	97.3	63.0	78.4	98.4	64.6	79.8	91.4	63.7	76.2
COVERS	DIVS. 10 - 14, 25, 28, 41, 43, 44, 46	100.0	80.5	96.1	100.0	82.5	96.5	100.0	82.2	96.4	100.0	81.0	96.2	100.0	82.3	96.4	100.0	84.5	96.9
21, 22, 23	FIRE SUPPRESSION, PLUMBING & HVAC	95.4	57.3	80.0	95.4	42.6	74.1	95.4	49.9	77.1	100.0	50.3	80.0	100.0	58.4	83.2	95.4	66.7	83.8
26, 27, 3370	ELECTRICAL, COMMUNICATIONS & UTIL.	94.9	68.8	81.1	95.9	64.6	79.3	93.3	64.5	78.1	99.1	57.4	77.0	94.7	64.5	78.8	95.0	67.9	80.7
MF2014	WEIGHTED AVERAGE	100.8	63.2	84.4	98.9	62.5	83.0	95.9	63.2	81.6	98.0	61.7	82.2	99.3	65.3	84.5	97.0	70.0	85.2

DIVISION		TEXAS																	
		GIDDINGS			GREENVILLE			HOUSTON			HUNTSVILLE			LAREDO			LONGVIEW		
		789			754			770 - 772			773			780			756		
		MAT.	INST.	TOTAL	MAT.	INST.	TOTAL	MAT.	INST.	TOTAL	MAT.	INST.	TOTAL	MAT.	INST.	TOTAL	MAT.	INST.	TOTAL
015433	CONTRACTOR EQUIPMENT		89.4	89.4		96.7	96.7		100.4	100.4		91.5	91.5		89.4	89.4		91.4	91.4
0241, 31 - 34	SITE & INFRASTRUCTURE, DEMOLITION	107.8	88.4	94.0	97.9	85.4	89.0	106.0	88.2	93.4	96.6	89.9	91.8	102.1	88.4	92.3	95.8	92.3	93.3
0310	Concrete Forming & Accessories	94.1	57.9	62.9	92.3	65.3	69.0	91.8	64.6	68.4	87.8	61.6	65.2	96.6	58.0	63.3	88.4	64.9	68.1
0320	Concrete Reinforcing	89.2	45.9	67.2	100.7	52.1	76.0	97.5	61.4	79.1	98.5	50.6	74.1	88.7	48.8	68.4	99.6	52.0	75.4
0330	Cast-in-Place Concrete	100.2	63.2	85.0	89.3	63.9	78.9	92.3	66.4	81.7	98.6	64.6	84.6	84.6	63.3	75.8	103.8	62.8	86.9
03	CONCRETE	97.6	58.2	78.2	91.6	63.6	77.8	95.2	66.2	80.9	103.9	61.3	83.0	92.5	58.8	75.9	107.5	62.3	85.3
04	MASONRY	115.7	56.5	78.8	160.0	56.6	95.6	102.3	66.5	80.0	145.1	58.9	91.4	100.2	63.5	77.3	155.7	56.6	93.9
05	METALS	93.9	65.3	85.1	97.7	79.3	92.1	99.2	88.2	95.8	94.7	69.2	86.8	97.0	66.9	87.8	91.1	67.8	83.9
06	WOOD, PLASTICS & COMPOSITES	95.6	58.3	74.7	93.5	68.6	79.5	91.7	64.7	76.6	82.0	61.8	70.7	96.5	58.3	75.1	87.2	68.4	76.7
07	THERMAL & MOISTURE PROTECTION	102.0	64.5	86.6	93.0	66.0	81.9	92.6	67.8	82.4	94.9	65.9	83.0	100.2	66.5	86.4	94.5	65.2	82.5
08	OPENINGS	100.8	54.0	89.9	97.7	63.4	89.7	104.3	62.8	94.6	97.2	58.5	88.2	101.6	54.7	90.7	87.8	63.3	82.1
0920	Plaster & Gypsum Board	93.7	57.5	69.2	90.3	68.0	75.3	99.1	64.0	75.4	94.2	61.2	71.9	95.8	57.5	69.9	89.1	68.0	74.9
0950, 0980	Ceilings & Acoustic Treatment	89.1	57.5	68.3	93.6	68.0	76.8	99.6	64.0	76.2	91.2	61.2	71.5	93.3	57.5	69.8	90.3	68.0	75.7
0960	Flooring	101.1	64.1	90.5	98.6	64.1	88.7	101.2	64.1	90.6	89.4	64.1	82.1	100.5	64.1	90.1	103.5	64.1	92.2
0970, 0990	Wall Finishes & Painting/Coating	106.4	48.8	71.6	107.0	56.1	76.3	103.4	62.2	78.5	92.5	59.4	72.5	106.4	53.2	74.3	96.8	56.1	72.2
09	FINISHES	97.1	57.7	75.4	96.6	64.6	79.0	98.9	64.0	79.7	84.7	61.6	72.0	97.7	58.2	75.9	101.7	64.5	81.2
COVERS	DIVS. 10 - 14, 25, 28, 41, 43, 44, 46	100.0	80.8	96.1	100.0	82.3	96.4	100.0	84.5	96.9	100.0	81.0	96.2	100.0	80.5	96.1	100.0	82.0	96.4
21, 22, 23	FIRE SUPPRESSION, PLUMBING & HVAC	95.4	64.6	83.0	95.3	49.0	76.6	100.1	66.7	86.6	95.4	65.9	83.5	100.2	64.1	85.6	95.3	33.6	70.4
26, 27, 3370	ELECTRICAL, COMMUNICATIONS & UTIL.	91.8	63.9	77.1	92.0	64.6	77.5	96.9	67.5	81.4	93.4	68.0	80.0	95.1	60.9	77.0	92.2	55.6	72.9
MF2014	WEIGHTED AVERAGE	97.7	64.0	83.0	98.8	64.0	83.6	99.4	70.5	86.8	98.0	66.7	84.4	98.3	64.5	83.6	98.5	58.7	81.2

	DIVISION	TEXAS																	
		LUBBOCK			LUFKIN			MCALLEN			MCKINNEY			MIDLAND			ODESSA		
		793 - 794			759			785			750			797			797		
		MAT.	INST.	TOTAL	MAT.	INST.	TOTAL	MAT.	INST.	TOTAL	MAT.	INST.	TOTAL	MAT.	INST.	TOTAL	MAT.	INST.	TOTAL
015433	CONTRACTOR EQUIPMENT		98.5	98.5		91.4	91.4		96.8	96.8		96.7	96.7		98.5	98.5		89.9	89.9
0241, 31 - 34	SITE & INFRASTRUCTURE, DEMOLITION	123.8	86.1	97.0	91.2	94.0	93.1	144.1	84.4	101.7	94.6	85.3	88.0	126.7	86.1	97.8	99.7	88.0	91.4
0310	Concrete Forming & Accessories	97.9	57.2	62.8	91.7	61.7	65.9	101.2	57.5	63.5	91.3	65.3	68.8	101.9	65.4	70.5	98.9	65.4	70.0
0320	Concrete Reinforcing	94.1	52.5	72.9	101.3	64.9	82.8	88.2	48.4	67.9	100.7	50.9	75.3	95.1	52.1	73.2	92.9	52.1	72.1
0330	Cast-in-Place Concrete	96.8	64.7	83.6	92.9	64.0	81.0	119.3	64.1	96.6	83.8	63.8	75.6	102.9	64.6	87.2	96.6	63.6	83.1
03	CONCRETE	93.7	60.4	77.4	99.8	63.6	82.0	107.9	59.5	84.1	87.0	63.3	75.3	98.2	64.0	81.4	95.0	62.8	79.2
04	MASONRY	103.6	64.0	78.9	119.4	58.8	81.6	106.8	56.4	75.4	172.2	56.6	100.2	121.5	56.7	81.1	104.2	56.6	74.6
05	METALS	109.0	81.3	100.5	97.9	72.4	90.0	94.3	77.5	89.2	97.6	78.8	91.8	107.2	80.7	99.0	104.6	69.1	93.7
06	WOOD, PLASTICS & COMPOSITES	101.1	57.3	76.5	95.3	61.8	76.5	114.6	58.4	83.1	92.3	68.6	79.0	106.0	68.6	85.1	101.1	68.5	82.8
07	THERMAL & MOISTURE PROTECTION	89.3	66.2	79.8	94.3	65.6	82.5	104.8	59.8	86.3	92.8	66.0	81.8	89.5	65.4	79.7	100.0	65.9	86.0
08	OPENINGS	108.5	57.2	96.6	66.7	63.0	65.9	105.7	54.7	93.9	97.7	63.0	89.6	107.5	63.4	97.2	97.1	63.3	89.2
0920	Plaster & Gypsum Board	84.4	56.3	65.4	88.0	61.2	69.9	99.4	57.5	71.1	90.0	68.0	75.1	85.9	68.0	73.8	84.0	68.0	73.2
0950, 0980	Ceilings & Acoustic Treatment	90.6	56.3	68.1	84.4	61.2	69.1	93.3	57.5	69.8	93.6	68.0	76.8	88.1	68.0	74.9	88.9	68.0	75.2
0960	Flooring	104.4	64.1	92.8	138.2	64.1	117.0	118.1	63.8	102.6	98.2	64.1	88.4	105.7	64.1	93.8	110.9	64.1	97.5
0970, 0990	Wall Finishes & Painting/Coating	119.0	56.1	81.0	96.8	56.1	72.2	120.1	48.8	77.1	107.0	56.1	76.3	119.0	56.1	81.0	106.6	56.1	76.1
09	FINISHES	96.4	58.0	75.2	110.2	61.2	83.2	106.4	57.8	79.6	96.2	64.6	78.8	96.8	64.7	79.1	93.7	64.6	77.7
COVERS	DIVS. 10 - 14, 25, 28, 41, 43, 44, 46	100.0	81.3	96.2	100.0	80.9	96.1	100.0	80.7	96.1	100.0	82.3	96.4	100.0	81.2	96.2	100.0	80.9	96.1
21, 22, 23	FIRE SUPPRESSION, PLUMBING & HVAC	99.8	50.7	80.0	95.3	64.4	82.8	95.4	58.7	80.6	95.3	45.4	75.2	95.0	49.1	76.5	100.3	49.1	79.6
26, 27, 3370	ELECTRICAL, COMMUNICATIONS & UTIL.	97.9	64.2	80.1	93.5	70.6	81.4	92.6	36.3	62.8	92.1	64.6	77.5	97.9	64.2	80.1	99.4	64.1	80.7
MF2014	WEIGHTED AVERAGE	101.6	63.6	85.0	95.6	67.8	83.5	101.0	59.8	83.0	98.7	63.1	83.2	101.5	64.2	85.2	99.6	63.1	83.7

	DIVISION	TEXAS																	
		PALESTINE			SAN ANGELO			SAN ANTONIO			TEMPLE			TEXARKANA			TYLER		
		758			769			781 - 782			765			755			757		
		MAT.	INST.	TOTAL	MAT.	INST.	TOTAL	MAT.	INST.	TOTAL	MAT.	INST.	TOTAL	MAT.	INST.	TOTAL	MAT.	INST.	TOTAL
015433	CONTRACTOR EQUIPMENT		91.4	91.4		89.9	89.9		91.9	91.9		89.9	89.9		91.4	91.4		91.4	91.4
0241, 31 - 34	SITE & INFRASTRUCTURE, DEMOLITION	96.6	92.6	93.8	101.5	89.7	93.1	100.9	92.0	94.5	89.8	89.1	89.3	85.9	92.0	90.2	95.1	92.4	93.2
0310	Concrete Forming & Accessories	82.7	65.4	67.8	99.3	58.7	64.3	95.0	58.1	63.2	102.8	57.8	64.0	99.0	57.4	63.1	93.6	65.4	69.3
0320	Concrete Reinforcing	98.8	52.1	75.1	92.7	52.1	72.1	93.6	49.8	71.3	92.9	49.1	70.6	98.7	52.3	75.1	99.6	52.4	75.6
0330	Cast-in-Place Concrete	84.9	63.0	75.9	95.7	64.8	83.0	85.6	65.2	77.2	78.5	63.1	72.2	85.5	62.7	76.2	101.9	63.0	85.9
03	CONCRETE	101.6	62.6	82.4	96.7	60.3	78.8	91.7	59.7	76.0	82.9	58.7	71.0	92.8	59.0	76.2	107.1	62.7	85.3
04	MASONRY	113.8	56.6	78.2	138.0	58.7	88.5	96.8	65.3	77.2	151.0	56.4	92.1	176.5	56.6	101.8	166.0	56.6	97.8
05	METALS	97.6	68.3	88.6	100.1	69.1	90.6	96.9	68.6	88.2	99.8	66.5	89.6	91.0	67.8	83.9	97.5	68.4	88.5
06	WOOD, PLASTICS & COMPOSITES	85.3	68.4	75.8	99.8	58.3	76.5	94.7	57.4	73.8	108.9	58.3	80.6	99.7	58.2	76.5	97.0	68.4	81.0
07	THERMAL & MOISTURE PROTECTION	94.8	65.3	82.7	98.0	66.0	84.9	96.4	68.2	84.8	97.7	64.8	84.2	94.1	64.2	81.9	94.6	65.3	82.6
08	OPENINGS	66.7	63.3	65.9	94.7	57.8	86.1	102.7	54.4	91.5	67.1	57.0	64.7	87.7	57.8	80.8	66.6	63.4	65.9
0920	Plaster & Gypsum Board	85.2	68.0	73.6	87.8	57.5	67.3	98.0	56.5	70.0	87.8	57.5	67.3	93.4	57.5	69.1	88.0	68.0	74.5
0950, 0980	Ceilings & Acoustic Treatment	84.4	68.0	73.6	80.6	57.5	65.4	102.7	56.5	72.3	80.6	57.5	65.4	90.3	57.5	68.7	84.4	68.0	73.6
0960	Flooring	129.9	64.1	111.1	95.3	64.1	86.4	104.1	64.1	92.6	123.2	64.1	106.3	111.1	64.1	97.6	140.1	64.1	118.4
0970, 0990	Wall Finishes & Painting/Coating	96.8	56.1	72.2	104.2	56.1	75.1	106.3	53.2	74.2	105.8	48.8	71.4	96.8	56.1	72.2	96.8	56.1	72.2
09	FINISHES	108.0	64.5	84.0	87.5	59.1	71.8	103.4	58.1	78.4	95.1	57.7	74.5	104.0	58.5	78.9	111.1	64.5	85.4
COVERS	DIVS. 10 - 14, 25, 28, 41, 43, 44, 46	100.0	82.0	96.4	100.0	81.0	96.2	100.0	81.2	96.2	100.0	81.1	96.2	100.0	80.9	96.1	100.0	82.0	96.4
21, 22, 23	FIRE SUPPRESSION, PLUMBING & HVAC	95.3	59.6	80.9	95.4	50.1	77.1	100.0	63.2	85.1	95.4	54.7	79.0	95.3	37.0	71.8	95.3	62.4	82.0
26, 27, 3370	ELECTRICAL, COMMUNICATIONS & UTIL.	89.8	55.6	71.7	97.4	44.8	69.6	96.4	60.9	77.6	94.5	59.8	76.2	93.4	60.1	75.8	92.2	55.5	72.8
MF2014	WEIGHTED AVERAGE	95.0	64.4	81.7	98.5	59.6	81.5	98.5	65.1	84.0	94.7	61.7	80.3	98.0	58.4	80.7	98.5	65.0	83.9

	DIVISION	TEXAS															UTAH		
		VICTORIA			WACO			WAXAHACKIE			WHARTON			WICHITA FALLS			LOGAN		
		779			766 - 767			751			774			763			843		
		MAT.	INST.	TOTAL	MAT.	INST.	TOTAL	MAT.	INST.	TOTAL	MAT.	INST.	TOTAL	MAT.	INST.	TOTAL	MAT.	INST.	TOTAL
015433	CONTRACTOR EQUIPMENT		99.3	99.3		89.9	89.9		96.7	96.7		100.5	100.5		89.9	89.9		97.5	97.5
0241, 31 - 34	SITE & INFRASTRUCTURE, DEMOLITION	111.8	85.7	93.2	98.6	88.7	91.6	96.0	85.4	88.4	117.1	87.8	96.2	99.3	88.7	91.8	96.0	95.7	95.8
0310	Concrete Forming & Accessories	90.0	59.0	63.3	101.2	65.5	70.4	91.3	65.5	69.0	84.9	61.7	64.9	101.2	65.5	70.4	104.7	58.1	64.5
0320	Concrete Reinforcing	93.9	48.5	70.8	92.5	48.5	70.1	100.7	50.9	75.4	97.7	50.6	73.7	92.5	51.3	71.5	99.6	81.3	90.3
0330	Cast-in-Place Concrete	106.9	65.6	89.9	84.9	66.8	77.4	88.4	63.9	78.3	109.8	65.5	91.6	90.7	63.7	79.6	87.3	73.4	81.6
03	CONCRETE	104.9	60.9	83.3	89.7	63.3	76.7	90.6	63.4	77.2	109.1	62.5	86.2	92.4	62.8	77.9	106.0	68.3	87.5
04	MASONRY	120.0	58.9	81.9	105.2	56.6	74.9	160.7	56.6	95.8	103.6	58.9	75.8	105.7	63.6	79.5	105.8	63.2	79.3
05	METALS	94.9	80.7	90.5	102.2	67.8	91.6	97.7	79.0	92.0	96.3	81.6	91.8	102.2	69.8	92.2	102.2	79.0	95.0
06	WOOD, PLASTICS & COMPOSITES	92.5	58.4	73.4	107.1	68.5	85.5	92.3	68.6	79.0	83.1	62.0	71.3	107.1	68.5	85.5	85.3	54.9	68.3
07	THERMAL & MOISTURE PROTECTION	96.8	66.9	84.6	98.5	65.8	85.1	92.9	66.0	81.9	93.4	66.6	82.4	98.5	67.7	85.9	97.4	67.9	85.3
08	OPENINGS	101.4	56.1	90.8	77.7	62.4	74.1	97.7	63.0	89.6	101.6	58.6	91.6	77.7	63.3	74.4	94.3	56.5	85.5
0920	Plaster & Gypsum Board	93.8	57.5	69.3	88.2	68.0	74.6	90.4	68.0	75.3	92.5	61.2	71.3	88.2	68.0	74.6	78.9	53.5	61.7
0950, 0980	Ceilings & Acoustic Treatment	95.4	57.5	70.5	82.2	68.0	72.9	95.3	68.0	77.4	94.6	61.2	72.6	82.2	68.0	72.9	97.2	53.5	68.5
0960	Flooring	99.5	64.1	89.4	122.4	64.1	105.8	98.2	64.1	88.4	97.7	64.1	88.1	123.4	76.5	110.0	101.2	47.7	85.9
0970, 0990	Wall Finishes & Painting/Coating	103.6	59.4	76.9	105.8	48.8	71.4	107.0	56.1	76.3	103.4	59.4	76.8	108.5	56.1	76.8	103.3	60.7	77.5
09	FINISHES	89.4	59.6	73.0	95.8	63.8	78.1	96.8	64.6	79.1	90.8	61.7	74.8	96.3	67.1	80.2	96.2	55.3	73.6
COVERS	DIVS. 10 - 14, 25, 28, 41, 43, 44, 46	100.0	80.8	96.1	100.0	82.3	96.4	100.0	82.3	96.4	100.0	81.3	96.2	100.0	69.8	93.9	100.0	85.1	97.0
21, 22, 23	FIRE SUPPRESSION, PLUMBING & HVAC	95.4	65.9	83.5	100.2	54.7	81.8	95.3	58.5	80.5	95.4	65.0	83.1	100.2	54.2	81.6	99.9	69.5	87.6
26, 27, 3370	ELECTRICAL, COMMUNICATIONS & UTIL.	99.8	57.6	77.5	97.6	63.6	79.6	92.1	64.6	77.5	98.6	68.0	82.4	99.3	60.1	78.6	97.5	73.0	84.6
MF2014	WEIGHTED AVERAGE	99.0	65.6	84.4	96.5	64.1	82.4	98.6	65.9	84.4	98.9	67.7	85.3	97.1	64.4	82.8	99.8	70.1	86.9

DIVISION		UTAH												VERMONT					
		OGDEN			PRICE			PROVO			SALT LAKE CITY			BELLOWS FALLS			BENNINGTON		
		842,844			845			846 - 847			840 - 841			051			052		
		MAT.	INST.	TOTAL	MAT.	INST.	TOTAL	MAT.	INST.	TOTAL	MAT.	INST.	TOTAL	MAT.	INST.	TOTAL	MAT.	INST.	TOTAL
015433	CONTRACTOR EQUIPMENT		97.5	97.5		96.6	96.6		96.6	96.6		97.5	97.5		100.9	100.9		100.9	100.9
0241, 31 - 34	SITE & INFRASTRUCTURE, DEMOLITION	84.9	95.7	92.6	93.3	93.8	93.7	92.3	94.1	93.6	84.5	95.6	92.4	88.2	100.5	96.9	87.5	100.5	96.7
0310	Concrete Forming & Accessories	104.7	58.1	64.5	107.1	48.3	56.4	106.3	58.0	64.7	107.1	58.0	64.8	98.3	85.5	87.3	95.9	106.6	105.2
0320	Concrete Reinforcing	99.2	81.3	90.1	106.8	81.1	93.7	107.7	81.3	94.3	101.5	81.3	91.2	83.2	85.7	84.5	83.2	85.7	84.5
0330	Cast-in-Place Concrete	88.7	73.4	82.4	87.4	59.5	75.9	87.4	73.3	81.6	96.9	73.3	87.2	87.0	110.7	96.7	87.0	110.7	96.7
03	CONCRETE	95.9	68.3	82.3	107.3	59.0	83.6	105.9	68.2	87.4	114.8	68.2	91.9	91.0	94.2	92.6	90.8	103.6	97.1
04	MASONRY	99.7	63.2	77.0	110.9	61.3	80.0	111.1	63.2	81.2	113.1	63.2	82.0	103.6	95.2	98.4	112.4	95.2	101.6
05	METALS	102.7	79.0	95.4	99.4	76.6	92.4	100.3	78.9	93.7	106.2	78.9	97.8	94.8	88.3	92.8	94.7	88.2	92.7
06	WOOD, PLASTICS & COMPOSITES	85.3	54.9	68.3	88.6	44.3	63.8	86.8	54.9	68.9	87.2	54.9	69.1	103.7	83.5	92.4	100.6	112.4	107.2
07	THERMAL & MOISTURE PROTECTION	96.4	67.9	84.7	99.0	59.9	82.9	99.0	67.9	86.2	103.4	67.9	88.9	97.0	82.7	91.1	97.0	81.1	90.4
08	OPENINGS	94.3	56.5	85.5	98.2	48.9	86.7	98.2	56.5	88.5	96.2	56.5	86.9	105.0	86.5	100.7	105.0	102.3	104.3
0920	Plaster & Gypsum Board	78.9	53.5	61.7	81.7	42.5	55.2	79.2	53.5	61.8	90.5	53.5	65.5	106.0	82.4	90.1	104.2	112.2	109.6
0950, 0980	Ceilings & Acoustic Treatment	97.2	53.5	68.5	97.2	42.5	61.3	97.2	53.5	68.5	91.5	53.5	66.5	91.5	82.4	85.6	91.5	112.2	105.1
0960	Flooring	99.0	47.7	84.3	102.3	35.5	83.2	102.0	47.7	86.5	103.2	47.7	87.3	100.6	104.6	101.8	99.7	104.6	101.1
0970, 0990	Wall Finishes & Painting/Coating	103.3	60.7	77.5	103.3	37.5	63.6	103.3	62.6	78.7	106.7	62.6	80.1	97.4	105.8	102.5	97.4	105.8	102.5
09	FINISHES	94.3	55.3	72.8	97.3	43.5	67.7	96.8	55.5	74.0	96.6	55.5	74.0	95.9	90.7	93.0	95.4	107.8	102.2
COVERS	DIVS. 10 - 14, 25, 28, 41, 43, 44, 46	100.0	85.1	97.0	100.0	45.8	89.0	100.0	85.0	97.0	100.0	85.0	97.0	100.0	97.6	99.5	100.0	100.7	100.1
21, 22, 23	FIRE SUPPRESSION, PLUMBING & HVAC	99.9	69.5	87.6	97.5	67.7	85.5	99.9	69.4	87.6	100.1	69.4	87.7	95.3	96.4	95.7	95.3	96.4	95.7
26, 27, 3370	ELECTRICAL, COMMUNICATIONS & UTIL.	97.9	73.0	84.7	103.5	73.0	87.4	98.4	59.8	78.0	100.5	59.8	79.0	101.5	89.3	95.1	101.5	61.2	80.2
MF2014	WEIGHTED AVERAGE	98.1	70.1	85.9	100.2	64.6	84.7	100.3	68.2	86.3	102.2	68.3	87.4	97.2	93.0	95.3	97.5	93.5	95.8

DIVISION		VERMONT																	
		BRATTLEBORO			BURLINGTON			GUILDHALL			MONTPELIER			RUTLAND			ST. JOHNSBURY		
		053			054			059			056			057			058		
		MAT.	INST.	TOTAL	MAT.	INST.	TOTAL	MAT.	INST.	TOTAL	MAT.	INST.	TOTAL	MAT.	INST.	TOTAL	MAT.	INST.	TOTAL
015433	CONTRACTOR EQUIPMENT		100.9	100.9		100.9	100.9		100.9	100.9		100.9	100.9		100.9	100.9		100.9	100.9
0241, 31 - 34	SITE & INFRASTRUCTURE, DEMOLITION	88.9	100.5	97.1	93.0	100.4	98.2	87.2	99.0	95.6	91.5	100.1	97.6	91.7	100.1	97.7	87.3	99.0	95.6
0310	Concrete Forming & Accessories	98.6	85.4	87.2	98.0	87.6	89.0	96.1	78.3	80.7	97.9	84.1	86.0	98.9	87.6	89.1	94.4	78.3	80.5
0320	Concrete Reinforcing	82.3	85.7	84.0	101.1	85.6	93.2	83.9	85.6	84.8	94.1	85.6	89.8	103.9	85.6	94.6	82.3	85.6	84.0
0330	Cast-in-Place Concrete	89.7	110.7	98.3	100.9	109.6	104.5	84.3	101.0	91.2	100.9	109.6	104.5	85.2	109.6	95.2	84.3	101.0	91.2
03	CONCRETE	93.0	94.1	93.6	101.2	94.7	98.0	88.6	87.6	88.1	100.1	93.1	96.7	94.3	94.7	94.5	88.2	87.6	87.9
04	MASONRY	111.5	95.2	101.3	115.7	93.7	102.0	111.8	78.2	90.8	109.6	93.7	99.7	93.9	93.7	93.8	138.8	78.2	101.0
05	METALS	94.7	88.2	92.7	103.5	87.5	98.6	94.8	87.3	92.5	99.6	87.4	95.9	100.6	87.4	96.5	94.8	87.3	92.5
06	WOOD, PLASTICS & COMPOSITES	104.0	83.5	92.5	99.9	88.2	93.3	100.0	83.5	90.8	97.7	83.5	89.7	104.2	88.2	95.2	94.8	83.5	88.5
07	THERMAL & MOISTURE PROTECTION	97.1	78.0	89.3	103.8	80.4	94.2	96.8	70.7	86.1	103.3	79.9	93.7	97.2	80.4	90.3	96.7	70.7	86.1
08	OPENINGS	105.0	86.5	100.7	108.4	85.0	103.0	105.0	82.4	99.7	105.8	82.4	100.3	108.4	85.0	102.9	105.0	82.4	99.7
0920	Plaster & Gypsum Board	106.0	82.4	90.1	109.7	87.2	94.5	112.4	82.4	92.2	108.9	82.4	91.0	106.6	87.2	93.5	113.8	82.4	92.6
0950, 0980	Ceilings & Acoustic Treatment	91.5	82.4	85.6	101.0	87.2	91.9	91.5	82.4	85.6	95.1	82.4	86.8	95.9	87.2	90.2	91.5	82.4	85.6
0960	Flooring	100.7	104.6	101.8	102.4	104.6	103.1	103.6	104.6	103.9	103.2	104.6	103.6	100.6	104.6	101.8	106.9	104.6	106.3
0970, 0990	Wall Finishes & Painting/Coating	97.4	105.8	102.5	102.9	85.2	92.2	97.4	85.2	90.0	101.7	85.2	91.7	97.4	85.2	90.0	97.4	85.2	90.0
09	FINISHES	96.0	90.7	93.1	99.7	90.8	94.8	97.5	84.2	90.2	97.9	88.0	92.5	97.1	90.8	93.6	98.7	84.2	90.7
COVERS	DIVS. 10 - 14, 25, 28, 41, 43, 44, 46	100.0	97.6	99.5	100.0	97.5	99.5	100.0	91.8	98.3	100.0	97.0	99.4	100.0	97.5	99.5	100.0	91.8	98.3
21, 22, 23	FIRE SUPPRESSION, PLUMBING & HVAC	95.3	96.4	95.7	99.9	71.8	88.6	95.3	64.0	82.7	95.1	71.8	85.7	100.1	71.8	88.7	95.3	64.0	82.7
26, 27, 3370	ELECTRICAL, COMMUNICATIONS & UTIL.	101.5	89.3	95.1	102.2	61.2	80.5	101.5	61.2	80.2	99.7	61.2	79.3	101.6	61.2	80.2	101.5	61.2	80.2
MF2014	WEIGHTED AVERAGE	97.8	92.8	95.6	102.4	83.6	94.2	97.4	77.8	88.8	99.5	82.8	92.2	99.7	83.6	92.7	98.6	77.8	89.6

DIVISION		VERMONT			VIRGINIA														
		WHITE RIVER JCT.			ALEXANDRIA			ARLINGTON			BRISTOL			CHARLOTTESVILLE			CULPEPER		
		050			223			222			242			229			227		
		MAT.	INST.	TOTAL	MAT.	INST.	TOTAL	MAT.	INST.	TOTAL	MAT.	INST.	TOTAL	MAT.	INST.	TOTAL	MAT.	INST.	TOTAL
015433	CONTRACTOR EQUIPMENT		100.9	100.9		103.0	103.0		101.7	101.7		101.7	101.7		106.1	106.1		101.7	101.7
0241, 31 - 34	SITE & INFRASTRUCTURE, DEMOLITION	91.5	99.2	97.0	114.3	89.8	96.9	124.5	87.7	98.3	108.8	86.1	92.6	113.5	88.1	95.4	111.8	87.6	94.6
0310	Concrete Forming & Accessories	93.1	78.9	80.8	92.2	72.8	75.5	91.2	72.0	74.7	87.1	42.5	48.6	85.4	48.7	53.7	82.5	71.4	72.9
0320	Concrete Reinforcing	83.2	85.6	84.4	83.0	87.0	85.1	93.7	84.4	88.9	93.7	68.4	80.8	93.1	72.8	82.7	93.7	84.2	88.9
0330	Cast-in-Place Concrete	89.7	101.9	94.7	105.1	81.8	95.6	102.3	80.3	93.3	101.8	48.5	79.9	105.9	55.5	85.2	104.8	69.2	90.2
03	CONCRETE	94.7	88.2	91.5	102.0	79.6	91.0	106.7	78.3	92.8	102.9	51.2	77.5	103.5	57.2	80.8	100.6	74.1	87.6
04	MASONRY	124.5	79.7	96.6	91.3	72.8	79.8	106.3	70.7	84.1	95.4	51.4	68.0	120.3	53.2	78.5	107.5	70.7	84.6
05	METALS	94.8	87.3	92.5	105.7	97.0	103.0	104.3	97.2	102.1	103.1	85.7	97.8	103.4	90.2	99.3	103.5	95.3	101.0
06	WOOD, PLASTICS & COMPOSITES	97.4	83.5	89.6	95.0	71.0	81.6	91.5	71.0	80.0	83.7	40.0	59.2	82.2	45.6	61.7	80.8	71.0	75.3
07	THERMAL & MOISTURE PROTECTION	97.1	71.4	86.6	102.7	81.5	94.0	104.7	80.6	94.8	104.2	60.1	86.1	103.8	67.7	89.0	104.0	78.1	93.4
08	OPENINGS	105.0	82.4	99.7	100.9	75.2	94.9	99.0	75.2	93.5	102.0	46.5	89.1	100.2	52.0	89.0	100.5	75.2	94.6
0920	Plaster & Gypsum Board	102.8	82.4	89.0	107.4	69.9	82.1	104.2	69.9	81.0	99.7	38.0	58.0	99.7	43.0	61.4	99.9	69.9	79.6
0950, 0980	Ceilings & Acoustic Treatment	91.5	82.4	85.6	91.4	69.9	77.3	89.8	69.9	76.7	88.9	38.0	55.4	88.9	43.0	58.8	89.8	69.9	76.7
0960	Flooring	98.6	104.6	100.4	106.0	82.9	99.4	104.5	81.9	98.0	101.0	66.5	91.1	99.7	66.5	90.2	99.7	81.9	94.6
0970, 0990	Wall Finishes & Painting/Coating	97.4	85.2	90.0	124.0	80.0	97.5	124.0	80.0	97.5	109.6	54.3	76.2	109.6	76.1	89.3	124.0	76.1	95.1
09	FINISHES	95.2	84.6	89.3	100.1	74.6	86.0	100.0	73.8	85.6	96.5	47.2	69.3	96.2	53.5	72.6	96.9	73.2	83.9
COVERS	DIVS. 10 - 14, 25, 28, 41, 43, 44, 46	100.0	92.3	98.4	100.0	88.6	97.7	100.0	87.6	97.5	100.0	67.8	93.5	100.0	79.3	95.8	100.0	87.6	97.5
21, 22, 23	FIRE SUPPRESSION, PLUMBING & HVAC	95.3	64.8	83.0	100.3	86.9	94.9	100.3	85.4	94.3	95.6	50.5	77.4	95.6	69.3	85.0	95.6	72.9	86.4
26, 27, 3370	ELECTRICAL, COMMUNICATIONS & UTIL.	101.5	61.2	80.2	97.1	102.4	99.9	94.7	99.2	97.1	96.8	35.4	64.3	96.7	71.9	83.6	99.3	99.2	99.2
MF2014	WEIGHTED AVERAGE	98.5	78.3	89.7	101.1	85.5	94.3	101.9	84.0	94.1	99.4	54.9	80.0	100.5	67.1	86.0	99.9	80.5	91.4

DIVISION		VIRGINIA																	
		FAIRFAX			FARMVILLE			FREDERICKSBURG			GRUNDY			HARRISONBURG			LYNCHBURG		
		220 - 221			239			224 - 225			246			228			245		
		MAT.	INST.	TOTAL	MAT.	INST.	TOTAL	MAT.	INST.	TOTAL	MAT.	INST.	TOTAL	MAT.	INST.	TOTAL	MAT.	INST.	TOTAL
015433	CONTRACTOR EQUIPMENT		101.7	101.7		106.1	106.1		101.7	101.7		101.7	101.7		101.7	101.7		101.7	101.7
0241, 31 - 34	SITE & INFRASTRUCTURE, DEMOLITION	123.1	87.8	98.0	109.2	87.6	93.8	111.4	87.7	94.5	106.6	84.8	91.1	120.1	86.3	96.1	107.5	86.3	92.4
0310	Concrete Forming & Accessories	85.4	72.0	73.9	98.0	43.7	51.2	85.4	69.4	71.6	90.3	36.4	43.8	81.4	42.6	47.9	87.1	60.3	64.0
0320	Concrete Reinforcing	93.7	87.0	90.3	90.9	49.9	70.0	94.4	87.0	90.6	92.4	50.4	71.0	93.7	64.9	79.0	93.1	68.5	80.5
0330	Cast-in-Place Concrete	102.3	80.6	93.4	103.1	53.5	82.7	103.9	69.3	89.7	101.8	47.5	79.5	102.3	58.0	84.1	101.8	57.7	83.7
03	CONCRETE	106.3	78.8	92.8	101.2	49.9	76.0	100.3	73.8	87.3	101.6	44.4	73.5	104.1	53.8	79.4	101.5	62.3	82.3
04	MASONRY	106.2	71.4	84.5	103.5	44.9	66.9	106.6	70.7	84.2	96.5	47.3	65.8	104.2	50.3	70.6	112.0	53.2	75.4
05	METALS	103.6	97.0	101.5	101.0	77.1	93.6	103.5	96.2	101.3	103.1	69.9	92.9	103.4	83.9	97.4	103.3	86.3	98.1
06	WOOD, PLASTICS & COMPOSITES	83.7	71.0	76.6	96.4	44.1	67.1	83.7	68.1	75.0	86.7	35.1	57.8	79.8	39.0	56.9	83.7	62.6	71.9
07	THERMAL & MOISTURE PROTECTION	104.6	73.7	91.9	104.3	49.3	81.8	104.0	78.5	93.6	104.2	45.3	80.0	104.4	63.9	87.8	104.0	63.9	87.6
08	OPENINGS	99.0	75.2	93.5	100.7	43.0	87.3	100.2	73.6	94.0	102.0	34.4	86.3	100.5	49.4	88.6	100.5	58.8	90.8
0920	Plaster & Gypsum Board	99.9	69.9	79.6	108.6	41.5	63.2	99.9	66.9	77.6	99.7	32.9	54.5	99.7	36.9	57.3	99.7	61.3	73.7
0950, 0980	Ceilings & Acoustic Treatment	89.8	69.9	76.7	84.8	41.5	56.3	89.8	66.9	74.8	88.9	32.9	52.1	88.9	36.9	54.8	88.9	61.3	70.8
0960	Flooring	101.4	81.9	95.8	106.5	60.3	93.3	101.4	81.9	95.8	102.3	33.7	82.6	99.4	74.0	92.1	101.0	66.5	91.1
0970, 0990	Wall Finishes & Painting/Coating	124.0	80.0	97.5	112.8	38.7	68.1	124.0	76.1	95.1	109.6	35.1	64.6	124.0	54.3	81.9	109.6	54.3	76.2
09	FINISHES	98.5	74.0	85.0	98.0	46.0	69.4	97.4	71.5	83.1	96.7	35.1	62.8	97.3	47.8	70.1	96.3	60.2	76.4
COVERS	DIVS. 10 - 14, 25, 28, 41, 43, 44, 46	100.0	87.8	97.5	100.0	45.7	89.0	100.0	83.4	96.6	100.0	43.7	88.6	100.0	67.6	93.5	100.0	70.9	94.1
21, 22, 23	FIRE SUPPRESSION, PLUMBING & HVAC	95.6	85.7	91.6	95.5	43.4	74.5	95.6	85.1	91.3	95.6	62.1	82.1	95.6	66.9	84.0	95.6	69.0	84.8
26, 27, 3370	ELECTRICAL, COMMUNICATIONS & UTIL.	97.9	99.2	98.6	90.1	48.2	68.0	94.9	99.2	97.2	96.8	44.6	69.2	96.9	96.4	96.7	97.9	52.7	74.0
MF2014	WEIGHTED AVERAGE	100.7	84.0	93.4	98.7	52.4	78.5	99.5	82.7	92.1	99.3	52.5	78.9	100.2	67.2	85.8	99.9	65.6	85.0

DIVISION		VIRGINIA																	
		NEWPORT NEWS			NORFOLK			PETERSBURG			PORTSMOUTH			PULASKI			RICHMOND		
		236			233 - 235			238			237			243			230 - 232		
		MAT.	INST.	TOTAL	MAT.	INST.	TOTAL	MAT.	INST.	TOTAL	MAT.	INST.	TOTAL	MAT.	INST.	TOTAL	MAT.	INST.	TOTAL
015433	CONTRACTOR EQUIPMENT		106.1	106.1		106.7	106.7		106.1	106.1		106.0	106.0		101.7	101.7		106.1	106.1
0241, 31 - 34	SITE & INFRASTRUCTURE, DEMOLITION	108.2	88.8	94.4	108.9	89.8	95.3	111.7	89.0	95.6	106.7	88.6	93.8	105.9	85.5	91.4	103.6	89.0	93.2
0310	Concrete Forming & Accessories	97.3	64.1	68.7	97.9	64.3	68.9	90.6	57.2	61.8	86.4	53.1	57.7	90.3	38.7	45.8	97.6	57.2	62.8
0320	Concrete Reinforcing	90.6	72.0	81.2	94.7	72.1	83.2	90.3	73.0	81.5	90.3	72.0	81.0	92.4	64.6	78.3	97.5	73.0	85.0
0330	Cast-in-Place Concrete	100.2	66.1	86.2	105.2	66.1	89.1	106.3	53.7	84.7	99.2	66.0	85.6	101.8	48.2	79.8	98.1	53.7	79.8
03	CONCRETE	98.4	67.6	83.3	101.5	67.7	84.9	103.6	60.4	82.4	97.1	62.6	80.2	101.6	48.7	75.6	98.5	60.4	79.8
04	MASONRY	97.8	53.9	70.5	103.8	53.9	72.7	111.9	52.6	75.0	103.8	53.9	72.7	91.5	48.2	64.5	102.5	52.6	71.4
05	METALS	103.3	89.8	99.2	103.1	89.9	99.1	101.1	90.7	97.9	102.2	89.1	98.2	103.2	82.5	96.8	107.1	90.7	102.0
06	WOOD, PLASTICS & COMPOSITES	95.3	66.2	79.0	93.2	66.2	78.1	86.4	58.6	70.8	82.9	51.5	65.3	86.7	36.8	58.7	95.0	58.6	74.6
07	THERMAL & MOISTURE PROTECTION	104.3	68.4	89.6	100.9	68.4	87.5	104.3	68.3	89.5	104.3	66.8	88.9	104.2	51.1	82.4	102.2	68.3	88.3
08	OPENINGS	101.1	62.2	92.1	100.4	62.2	91.5	100.4	59.0	90.8	101.2	54.1	90.2	102.0	42.2	88.1	104.8	59.0	94.1
0920	Plaster & Gypsum Board	109.4	64.3	78.9	105.5	64.3	77.6	101.3	56.4	71.0	101.6	49.1	66.1	99.7	34.7	55.7	104.2	56.4	71.9
0950, 0980	Ceilings & Acoustic Treatment	88.2	64.3	72.5	88.9	64.3	72.7	85.7	56.4	66.4	88.2	49.1	62.5	88.9	34.7	53.3	90.6	56.4	68.1
0960	Flooring	106.5	66.5	95.1	105.6	66.5	94.4	102.5	71.2	93.6	99.5	66.5	90.0	102.3	66.5	92.0	103.4	71.2	94.2
0970, 0990	Wall Finishes & Painting/Coating	112.8	42.0	70.1	113.0	76.1	90.7	112.8	76.1	90.6	112.8	76.1	90.6	109.6	54.3	76.2	110.0	76.1	89.5
09	FINISHES	98.7	62.2	78.6	98.2	66.0	80.5	96.3	61.2	76.9	95.6	56.7	74.2	96.7	43.7	67.5	98.5	61.2	77.9
COVERS	DIVS. 10 - 14, 25, 28, 41, 43, 44, 46	100.0	82.2	96.4	100.0	82.2	96.4	100.0	79.6	95.9	100.0	72.0	94.3	100.0	44.0	88.7	100.0	79.6	95.9
21, 22, 23	FIRE SUPPRESSION, PLUMBING & HVAC	100.3	63.6	85.5	100.0	64.3	85.6	95.5	67.5	84.2	100.3	64.3	85.8	95.6	62.2	82.1	99.9	67.5	86.8
26, 27, 3370	ELECTRICAL, COMMUNICATIONS & UTIL.	92.8	63.5	77.3	95.5	58.3	75.8	93.0	71.9	81.8	91.2	58.3	73.8	96.8	54.4	74.4	96.8	71.9	83.6
MF2014	WEIGHTED AVERAGE	100.0	68.2	86.1	100.5	68.2	86.4	99.4	68.7	86.0	99.4	65.3	84.5	99.0	57.3	80.9	101.3	68.7	87.1

DIVISION		VIRGINIA									WASHINGTON								
		ROANOKE			STAUNTON			WINCHESTER			CLARKSTON			EVERETT			OLYMPIA		
		240 - 241			244			226			994			982			985		
		MAT.	INST.	TOTAL	MAT.	INST.	TOTAL	MAT.	INST.	TOTAL	MAT.	INST.	TOTAL	MAT.	INST.	TOTAL	MAT.	INST.	TOTAL
015433	CONTRACTOR EQUIPMENT		101.7	101.7		106.1	106.1		101.7	101.7		91.2	91.2		101.6	101.6		101.6	101.6
0241, 31 - 34	SITE & INFRASTRUCTURE, DEMOLITION	106.4	86.3	92.1	109.9	87.7	94.1	118.7	87.6	96.6	99.2	90.1	92.7	91.5	108.4	103.5	93.4	108.4	104.1
0310	Concrete Forming & Accessories	96.8	60.5	65.5	89.9	50.0	55.5	83.8	69.5	71.4	112.7	67.4	73.6	113.4	100.4	102.2	99.0	100.3	100.1
0320	Concrete Reinforcing	93.4	68.5	80.7	93.1	48.9	70.6	93.1	86.4	89.7	112.0	87.2	99.4	104.2	98.8	101.5	110.8	98.7	104.6
0330	Cast-in-Place Concrete	115.7	57.7	91.9	105.9	51.5	83.6	102.3	69.3	88.7	96.3	81.9	90.4	103.5	106.2	104.6	98.9	106.2	101.9
03	CONCRETE	105.9	62.4	84.5	102.9	52.0	77.9	103.5	73.7	88.9	107.2	76.3	92.0	98.7	101.6	100.1	98.9	101.6	100.2
04	MASONRY	97.7	53.2	70.0	106.9	51.4	72.3	101.7	59.7	75.5	102.3	81.5	89.3	110.6	100.8	104.5	103.0	99.1	100.6
05	METALS	105.5	86.5	99.7	103.4	80.5	96.3	103.5	94.7	100.8	88.3	81.6	86.2	101.8	92.4	98.9	101.2	92.1	98.4
06	WOOD, PLASTICS & COMPOSITES	95.9	62.6	77.3	86.7	49.9	66.1	82.2	68.1	74.3	113.2	63.8	85.5	111.7	100.0	105.2	91.1	100.0	96.1
07	THERMAL & MOISTURE PROTECTION	103.8	66.8	88.7	103.8	52.0	82.6	104.5	75.6	92.7	143.8	77.2	116.5	103.7	101.7	102.9	103.5	96.8	100.7
08	OPENINGS	100.9	58.8	91.1	100.5	48.1	88.3	102.1	73.7	95.5	119.0	68.7	107.3	101.3	99.8	101.0	106.9	99.9	105.3
0920	Plaster & Gypsum Board	107.4	61.3	76.2	99.7	47.5	64.4	99.9	66.9	77.6	145.7	62.5	89.5	109.4	99.9	103.0	101.3	99.9	100.4
0950, 0980	Ceilings & Acoustic Treatment	91.4	61.3	71.6	88.9	47.5	61.7	89.8	66.9	74.8	97.6	62.5	74.5	99.8	99.9	99.9	98.1	99.9	99.3
0960	Flooring	106.0	66.5	94.7	101.9	39.5	84.0	100.7	81.9	95.4	91.6	47.1	78.9	113.8	91.6	107.4	103.8	91.7	100.3
0970, 0990	Wall Finishes & Painting/Coating	109.6	54.3	76.2	109.6	34.5	64.3	124.0	87.2	101.8	100.0	58.5	75.0	95.4	91.5	93.0	99.0	91.5	94.5
09	FINISHES	99.0	61.1	78.1	96.6	46.6	69.0	97.9	72.7	84.0	110.3	61.3	83.3	104.7	97.6	100.8	98.9	97.6	98.2
COVERS	DIVS. 10 - 14, 25, 28, 41, 43, 44, 46	100.0	70.9	94.1	100.0	70.9	94.1	100.0	87.3	97.4	100.0	75.4	95.0	100.0	99.0	99.8	100.0	100.5	100.1
21, 22, 23	FIRE SUPPRESSION, PLUMBING & HVAC	100.3	64.8	86.0	95.6	58.9	80.8	95.6	87.7	92.4	95.6	83.4	90.7	100.1	99.1	99.7	100.0	99.1	99.7
26, 27, 3370	ELECTRICAL, COMMUNICATIONS & UTIL.	96.8	54.4	74.4	95.6	74.9	84.7	95.3	99.2	97.4	92.4	94.8	93.7	104.9	98.5	101.5	104.4	99.1	101.6
MF2014	WEIGHTED AVERAGE	101.5	65.1	85.6	99.7	61.8	83.2	100.0	82.2	92.2	101.5	80.2	92.2	101.6	99.6	100.7	101.2	99.4	100.4

	DIVISION	WASHINGTON																	
		RICHLAND			SEATTLE			SPOKANE			TACOMA			VANCOUVER			WENATCHEE		
		993			980 - 981,987			990 - 992			983 - 984			986			988		
		MAT.	INST.	TOTAL	MAT.	INST.	TOTAL	MAT.	INST.	TOTAL	MAT.	INST.	TOTAL	MAT.	INST.	TOTAL	MAT.	INST.	TOTAL
015433	CONTRACTOR EQUIPMENT		91.2	91.2		101.6	101.6		91.2	91.2		101.6	101.6		96.7	96.7		101.6	101.6
0241, 31 - 34	SITE & INFRASTRUCTURE, DEMOLITION	101.8	91.0	94.1	95.1	106.7	103.4	101.1	91.0	93.9	94.5	108.4	104.4	104.2	96.1	98.4	103.2	107.1	106.0
0310	Concrete Forming & Accessories	112.8	78.5	83.2	108.3	100.9	102.0	118.2	78.1	83.6	104.2	100.4	100.9	104.9	90.5	92.5	105.8	77.6	81.5
0320	Concrete Reinforcing	107.4	87.4	97.2	105.1	98.9	101.9	108.2	87.3	97.6	103.1	98.8	100.9	103.9	98.4	101.1	103.9	87.7	95.6
0330	Cast-in-Place Concrete	96.5	84.4	91.5	100.9	106.4	103.2	100.3	84.2	93.7	106.4	106.2	106.3	118.8	98.4	110.4	108.6	77.3	95.7
03	CONCRETE	106.8	82.2	94.7	99.6	102.1	100.8	109.1	82.0	95.8	100.4	101.6	101.0	110.7	94.6	102.8	108.4	79.6	94.2
04	MASONRY	104.3	84.3	91.8	108.2	100.9	103.6	104.9	84.3	92.0	106.2	100.9	102.9	106.7	95.6	99.8	108.6	91.4	97.9
05	METALS	88.7	84.9	87.5	102.5	94.2	99.9	90.9	84.5	89.0	103.6	92.3	100.1	101.0	91.8	98.2	101.0	86.2	96.4
06	WOOD, PLASTICS & COMPOSITES	113.4	76.1	92.5	104.9	100.0	102.2	122.4	76.1	96.5	101.0	100.0	100.4	92.9	90.2	91.4	102.5	75.8	87.5
07	THERMAL & MOISTURE PROTECTION	145.0	78.8	117.8	102.1	101.8	102.0	141.4	79.5	116.0	103.5	98.5	101.4	104.0	91.6	98.9	104.1	81.5	94.8
08	OPENINGS	121.4	71.3	109.7	100.1	99.9	100.0	121.9	71.8	110.3	102.0	99.9	101.5	98.3	92.8	97.0	101.5	71.0	94.4
0920	Plaster & Gypsum Board	145.7	75.2	98.1	103.6	99.9	101.1	137.7	75.2	95.5	107.8	99.9	102.5	105.7	90.1	95.1	110.2	74.9	86.4
0950, 0980	Ceilings & Acoustic Treatment	104.2	75.2	85.1	110.2	99.9	103.5	99.5	75.2	83.5	103.1	99.9	101.0	100.2	90.1	93.5	95.4	74.9	82.0
0960	Flooring	92.0	80.1	88.6	108.3	106.8	107.8	91.2	85.3	89.5	107.2	91.7	102.8	113.5	102.2	110.2	109.9	62.6	96.4
0970, 0990	Wall Finishes & Painting/Coating	100.0	67.6	80.4	96.3	91.5	93.4	100.2	70.9	82.5	95.4	91.5	93.0	97.6	70.8	81.4	95.4	72.6	81.6
09	FINISHES	112.0	76.7	92.6	105.0	100.6	102.6	109.8	78.1	92.3	103.4	97.6	100.2	101.5	90.6	95.5	103.8	73.5	87.1
COVERS	DIVS. 10 - 14, 25, 28, 41, 43, 44, 46	100.0	85.4	97.0	100.0	100.5	100.1	100.0	85.3	97.0	100.0	100.5	100.1	100.0	60.7	92.1	100.0	77.2	95.4
21, 22, 23	FIRE SUPPRESSION, PLUMBING & HVAC	100.5	107.6	103.3	100.0	115.1	106.1	100.4	84.3	93.9	100.2	99.2	99.8	100.3	87.0	94.9	95.4	86.7	91.9
26, 27, 3370	ELECTRICAL, COMMUNICATIONS & UTIL.	89.7	94.8	92.4	102.6	107.0	104.9	88.0	79.0	83.3	104.7	99.1	101.7	109.7	101.0	105.1	105.5	98.4	101.8
MF2014	WEIGHTED AVERAGE	103.1	89.3	97.1	101.4	104.7	102.8	103.4	82.3	94.2	101.8	99.7	100.9	102.7	92.1	98.1	101.6	86.6	95.0

	DIVISION	WASHINGTON			WEST VIRGINIA														
		YAKIMA			BECKLEY			BLUEFIELD			BUCKHANNON			CHARLESTON			CLARKSBURG		
		989			258 - 259			247 - 248			262			250 - 253			263 - 264		
		MAT.	INST.	TOTAL	MAT.	INST.	TOTAL	MAT.	INST.	TOTAL	MAT.	INST.	TOTAL	MAT.	INST.	TOTAL	MAT.	INST.	TOTAL
015433	CONTRACTOR EQUIPMENT		101.6	101.6		101.7	101.7		101.7	101.7		101.7	101.7		101.7	101.7		101.7	101.7
0241, 31 - 34	SITE & INFRASTRUCTURE, DEMOLITION	96.8	107.8	104.7	100.5	89.9	92.9	100.6	89.9	92.9	106.8	89.9	94.8	101.5	90.7	93.8	107.4	89.9	95.0
0310	Concrete Forming & Accessories	104.6	95.8	97.0	83.5	93.6	92.2	87.0	93.6	92.7	86.4	94.3	93.2	96.7	94.1	94.4	83.8	94.2	92.8
0320	Concrete Reinforcing	103.5	87.4	95.3	92.7	89.0	90.8	92.0	84.1	88.0	92.6	84.2	88.3	101.0	89.1	94.9	92.6	84.1	88.3
0330	Cast-in-Place Concrete	113.6	83.9	101.4	98.1	103.0	100.1	99.6	102.9	101.0	99.3	101.8	100.3	96.8	103.0	99.3	108.8	100.0	105.2
03	CONCRETE	105.3	89.9	97.7	97.6	96.4	97.0	97.4	95.5	96.5	100.4	95.4	98.0	99.1	96.7	97.9	104.4	94.8	99.7
04	MASONRY	99.8	83.1	89.4	98.9	98.9	98.9	93.3	98.9	96.8	104.5	98.9	101.0	95.8	97.9	97.1	108.4	98.9	102.4
05	METALS	101.6	86.3	96.9	100.9	98.1	100.0	103.4	96.3	101.2	103.6	97.1	101.6	99.6	98.9	99.4	103.6	96.9	101.5
06	WOOD, PLASTICS & COMPOSITES	101.4	100.0	100.6	83.5	93.4	89.1	85.6	93.4	90.0	84.9	93.4	89.7	98.4	93.4	95.6	81.5	93.4	88.2
07	THERMAL & MOISTURE PROTECTION	103.6	82.7	95.0	102.9	90.6	97.9	104.0	90.6	98.5	104.2	91.7	99.1	99.4	90.4	95.7	104.1	90.6	98.6
08	OPENINGS	101.5	81.0	96.7	101.3	85.8	97.7	102.5	84.6	98.3	102.5	84.6	98.4	103.6	85.8	99.4	102.5	85.1	98.5
0920	Plaster & Gypsum Board	107.4	99.9	102.4	97.0	93.0	94.3	98.9	93.0	94.9	99.3	93.0	95.1	100.3	93.0	95.4	97.2	93.0	94.4
0950, 0980	Ceilings & Acoustic Treatment	97.3	99.9	99.0	83.9	93.0	89.9	87.3	93.0	91.1	88.9	93.0	91.6	92.2	93.0	92.7	88.9	93.0	91.6
0960	Flooring	108.1	58.0	93.8	102.4	118.0	106.8	98.9	118.0	104.3	98.6	118.0	104.1	106.4	118.0	109.7	97.6	118.0	103.4
0970, 0990	Wall Finishes & Painting/Coating	95.4	70.9	80.6	107.4	91.5	97.8	109.6	94.6	100.6	109.6	94.3	100.3	106.9	95.9	100.2	109.6	94.3	100.3
09	FINISHES	102.5	86.6	93.7	94.7	98.6	96.9	94.8	98.9	97.1	95.7	98.9	97.5	99.1	98.8	99.0	95.0	98.9	97.2
COVERS	DIVS. 10 - 14, 25, 28, 41, 43, 44, 46	100.0	97.6	99.5	100.0	56.9	91.3	100.0	56.9	91.3	100.0	101.5	100.3	100.0	101.2	100.2	100.0	101.5	100.3
21, 22, 23	FIRE SUPPRESSION, PLUMBING & HVAC	100.2	107.2	103.0	95.5	93.5	94.7	95.6	92.8	94.4	95.6	95.1	95.4	100.0	93.4	97.4	95.6	95.0	95.3
26, 27, 3370	ELECTRICAL, COMMUNICATIONS & UTIL.	107.8	94.9	101.0	93.9	91.1	92.4	95.7	91.1	93.3	97.1	96.1	96.6	99.1	91.1	94.8	97.1	96.1	96.6
MF2014	WEIGHTED AVERAGE	102.0	94.1	98.6	97.9	93.3	95.9	98.3	92.9	96.0	99.6	95.5	97.8	99.9	94.8	97.6	100.1	95.4	98.1

	DIVISION	WEST VIRGINIA																	
		GASSAWAY			HUNTINGTON			LEWISBURG			MARTINSBURG			MORGANTOWN			PARKERSBURG		
		266			255 - 257			249			254			265			261		
		MAT.	INST.	TOTAL	MAT.	INST.	TOTAL	MAT.	INST.	TOTAL	MAT.	INST.	TOTAL	MAT.	INST.	TOTAL	MAT.	INST.	TOTAL
015433	CONTRACTOR EQUIPMENT		101.7	101.7		101.7	101.7		101.7	101.7		101.7	101.7		101.7	101.7		101.7	101.7
0241, 31 - 34	SITE & INFRASTRUCTURE, DEMOLITION	104.2	89.9	94.0	105.1	90.9	95.0	116.3	89.8	97.5	104.2	90.2	94.3	101.6	90.5	93.7	109.8	90.6	96.2
0310	Concrete Forming & Accessories	85.8	92.2	91.4	95.3	95.9	95.8	84.2	93.5	92.2	83.6	82.8	82.9	84.2	93.9	92.6	88.6	90.3	90.1
0320	Concrete Reinforcing	92.6	84.0	88.3	94.1	92.9	93.5	92.6	83.9	88.2	92.7	80.6	86.6	92.6	84.1	88.3	92.0	83.9	87.9
0330	Cast-in-Place Concrete	104.0	103.5	103.7	106.9	102.5	105.1	99.7	102.9	101.0	102.8	96.7	100.3	99.3	101.6	100.2	101.5	96.4	99.4
03	CONCRETE	100.8	95.1	98.0	102.8	98.0	100.4	107.1	95.4	101.4	101.2	87.9	94.7	97.2	95.2	96.2	102.5	91.8	97.2
04	MASONRY	109.2	95.9	100.9	96.4	100.8	99.1	95.9	95.9	95.9	99.7	89.5	93.3	126.5	98.9	109.3	82.9	91.7	88.4
05	METALS	103.5	96.8	101.4	103.3	100.4	102.4	103.5	96.0	101.2	101.3	94.2	99.1	103.6	96.6	101.5	104.2	96.6	101.9
06	WOOD, PLASTICS & COMPOSITES	84.0	90.7	87.8	94.7	95.3	95.1	81.8	93.4	88.3	83.5	81.6	82.5	81.8	93.4	88.3	85.0	89.1	87.3
07	THERMAL & MOISTURE PROTECTION	104.0	90.6	98.5	103.0	90.8	98.0	104.8	89.8	98.7	103.1	79.3	93.4	104.0	90.6	98.5	103.9	89.0	97.8
08	OPENINGS	100.7	83.2	96.6	100.5	87.6	97.5	102.5	84.6	98.4	103.3	73.4	96.3	103.8	85.1	99.4	101.4	82.3	96.9
0920	Plaster & Gypsum Board	98.6	90.2	92.9	104.8	95.0	98.1	97.2	93.0	94.4	97.6	80.9	86.3	97.2	93.0	94.4	99.7	88.5	92.1
0950, 0980	Ceilings & Acoustic Treatment	88.9	90.2	89.8	86.4	95.0	92.0	88.9	93.0	91.6	86.4	80.9	82.8	88.9	93.0	91.6	88.9	88.5	88.7
0960	Flooring	98.4	118.0	104.0	109.8	121.7	113.2	97.7	118.0	103.5	102.4	118.0	106.8	97.7	118.0	103.5	101.5	118.0	106.2
0970, 0990	Wall Finishes & Painting/Coating	109.6	95.9	101.3	107.4	94.6	99.7	109.6	75.4	88.9	107.4	41.4	67.5	109.6	94.3	100.3	109.6	94.6	100.6
09	FINISHES	95.2	97.5	96.5	98.5	100.8	99.8	96.1	96.8	96.5	95.6	83.1	88.7	94.6	98.9	97.0	96.6	95.9	96.2
COVERS	DIVS. 10 - 14, 25, 28, 41, 43, 44, 46	100.0	101.2	100.2	100.0	101.8	100.4	100.0	56.9	91.3	100.0	97.7	99.5	100.0	65.8	93.1	100.0	100.5	100.1
21, 22, 23	FIRE SUPPRESSION, PLUMBING & HVAC	95.6	93.9	94.9	100.3	90.5	96.4	95.6	93.5	94.7	95.5	86.0	91.7	95.6	95.0	95.3	100.3	88.3	95.4
26, 27, 3370	ELECTRICAL, COMMUNICATIONS & UTIL.	97.1	91.1	93.9	97.8	95.3	96.5	93.0	91.1	92.0	100.1	81.8	90.4	97.3	96.1	96.7	97.2	92.7	94.8
MF2014	WEIGHTED AVERAGE	99.5	93.9	97.1	100.7	95.7	98.5	99.8	92.4	96.6	99.3	86.4	93.7	100.2	94.4	97.6	100.1	91.8	96.5

	DIVISION	WEST VIRGINIA									WISCONSIN								
		PETERSBURG			ROMNEY			WHEELING			BELOIT			EAU CLAIRE			GREEN BAY		
		268			267			260			535			547			541 - 543		
		MAT.	INST.	TOTAL	MAT.	INST.	TOTAL	MAT.	INST.	TOTAL	MAT.	INST.	TOTAL	MAT.	INST.	TOTAL	MAT.	INST.	TOTAL
015433	CONTRACTOR EQUIPMENT		101.7	101.7		101.7	101.7		101.7	101.7		102.3	102.3		100.4	100.4		98.2	98.2
0241, 31 - 34	SITE & INFRASTRUCTURE, DEMOLITION	100.9	90.5	93.5	103.6	90.5	94.3	110.6	90.4	96.2	94.7	108.3	104.3	96.1	102.9	101.0	99.7	98.9	99.2
0310	Concrete Forming & Accessories	87.4	91.6	91.0	83.4	91.8	90.6	90.3	93.2	92.8	100.2	100.3	100.3	97.1	99.4	99.1	105.5	98.7	99.7
0320	Concrete Reinforcing	92.0	82.5	87.2	92.6	81.0	86.7	91.4	84.0	87.7	92.3	116.0	104.3	90.4	99.2	94.9	88.7	92.4	90.6
0330	Cast-in-Place Concrete	99.3	99.1	99.2	104.0	99.1	102.0	101.5	100.1	100.9	100.6	100.9	100.7	104.0	99.2	102.0	107.6	99.7	104.3
03	CONCRETE	97.3	93.0	95.2	100.6	92.8	96.8	102.5	94.4	98.5	98.8	103.5	101.1	99.7	99.4	99.6	102.8	98.0	100.5
04	MASONRY	99.5	98.9	99.1	96.2	98.9	97.9	107.6	93.9	99.1	100.6	104.3	102.9	94.7	102.1	99.3	126.7	101.4	110.9
05	METALS	103.7	95.8	101.2	103.7	95.5	101.2	104.4	96.8	102.1	93.1	107.7	97.6	95.0	101.4	97.0	97.4	98.2	97.6
06	WOOD, PLASTICS & COMPOSITES	85.9	90.7	88.6	80.9	90.7	86.4	86.7	93.4	90.5	99.9	99.2	99.5	105.1	99.3	101.9	110.0	99.3	104.0
07	THERMAL & MOISTURE PROTECTION	104.0	86.0	96.6	104.1	83.4	95.6	104.3	90.0	98.4	102.1	96.8	99.9	102.4	86.7	95.9	104.3	86.0	96.8
08	OPENINGS	103.8	82.8	98.9	103.7	82.4	98.8	102.2	85.1	98.2	102.5	108.8	103.9	103.7	96.2	102.0	100.1	96.0	99.2
0920	Plaster & Gypsum Board	99.3	90.2	93.2	96.8	90.2	92.4	99.7	93.0	95.2	94.5	99.6	97.9	102.9	99.6	100.6	98.7	99.6	99.3
0950, 0980	Ceilings & Acoustic Treatment	88.9	90.2	89.8	88.9	90.2	89.8	88.9	93.0	91.6	88.1	99.6	95.6	90.4	99.6	96.4	81.9	99.6	93.5
0960	Flooring	99.4	118.0	104.7	97.5	118.0	103.4	102.3	118.0	106.8	99.8	118.1	105.1	91.4	114.1	97.9	110.2	114.1	111.3
0970, 0990	Wall Finishes & Painting/Coating	109.6	94.3	100.3	109.6	94.3	100.3	109.6	94.3	100.3	98.3	97.2	97.7	94.5	86.4	89.6	105.4	83.2	92.0
09	FINISHES	95.4	97.3	96.4	94.7	97.3	96.1	96.9	98.2	97.6	97.4	103.1	100.5	94.6	101.2	98.3	99.0	100.7	99.9
COVERS	DIVS. 10 - 14, 25, 28, 41, 43, 44, 46	100.0	65.5	93.0	100.0	101.2	100.2	100.0	95.4	99.1	100.0	96.6	99.3	100.0	96.8	99.3	100.0	96.5	99.3
21, 22, 23	FIRE SUPPRESSION, PLUMBING & HVAC	95.6	93.7	94.8	95.6	93.4	94.7	100.3	93.6	97.6	100.1	98.6	99.5	100.1	87.6	95.0	100.3	85.1	94.2
26, 27, 3370	ELECTRICAL, COMMUNICATIONS & UTIL.	100.5	81.8	90.6	99.8	81.8	90.3	94.4	96.1	95.3	101.0	86.7	93.4	103.8	84.9	93.8	98.2	82.0	89.6
MF2014	WEIGHTED AVERAGE	99.3	91.3	95.8	99.4	92.2	96.3	101.2	94.3	98.2	98.9	100.7	99.7	99.3	95.3	97.5	101.2	93.4	97.8

	DIVISION	WISCONSIN																	
		KENOSHA			LA CROSSE			LANCASTER			MADISON			MILWAUKEE			NEW RICHMOND		
		531			546			538			537			530,532			540		
		MAT.	INST.	TOTAL	MAT.	INST.	TOTAL	MAT.	INST.	TOTAL	MAT.	INST.	TOTAL	MAT.	INST.	TOTAL	MAT.	INST.	TOTAL
015433	CONTRACTOR EQUIPMENT		100.2	100.2		100.4	100.4		102.3	102.3		102.3	102.3		90.5	90.5		100.8	100.8
0241, 31 - 34	SITE & INFRASTRUCTURE, DEMOLITION	100.3	105.2	103.8	90.1	102.8	99.1	93.9	108.2	104.1	92.4	108.3	103.7	94.6	98.5	97.4	94.4	103.3	100.8
0310	Concrete Forming & Accessories	108.0	100.5	101.6	84.1	99.1	97.1	99.5	97.9	98.2	102.0	100.1	100.4	104.1	118.0	116.1	92.4	100.9	99.7
0320	Concrete Reinforcing	92.1	97.9	95.0	90.1	91.5	90.9	93.5	91.3	92.3	97.9	91.6	94.7	95.7	98.2	97.0	87.8	99.0	93.5
0330	Cast-in-Place Concrete	109.4	107.3	108.5	93.5	95.9	94.5	100.0	100.9	100.4	91.1	100.8	95.1	99.8	109.6	103.8	108.3	82.5	97.7
03	CONCRETE	103.5	102.4	103.0	90.8	96.8	93.7	98.5	97.8	98.1	95.7	98.8	97.2	99.7	110.5	105.0	97.5	94.3	95.9
04	MASONRY	97.9	114.1	108.0	93.8	105.9	101.4	100.7	104.3	102.9	101.5	104.3	103.2	104.9	120.6	114.7	121.6	102.3	109.6
05	METALS	93.9	101.2	96.2	94.9	98.2	95.9	90.6	95.0	92.0	91.9	97.2	93.6	97.5	94.8	96.7	95.1	99.7	96.5
06	WOOD, PLASTICS & COMPOSITES	105.0	96.4	100.2	90.1	99.3	95.3	99.1	99.2	99.2	96.7	99.2	98.1	104.7	118.5	112.5	94.0	102.5	98.7
07	THERMAL & MOISTURE PROTECTION	102.2	108.7	104.9	101.8	90.2	97.1	101.9	82.5	93.9	102.8	105.1	103.8	100.8	114.0	106.2	103.6	92.2	98.9
08	OPENINGS	96.9	101.1	97.9	103.7	85.0	99.4	97.9	85.8	95.1	107.9	100.5	106.2	101.0	113.2	103.8	89.1	97.5	91.1
0920	Plaster & Gypsum Board	85.3	96.7	93.0	97.7	99.6	99.0	93.3	99.6	97.5	99.6	99.6	99.6	101.0	119.3	113.4	87.7	103.0	98.1
0950, 0980	Ceilings & Acoustic Treatment	88.1	96.7	93.7	89.5	99.6	96.1	83.1	99.6	93.9	94.0	99.6	97.6	95.9	119.3	111.3	58.5	103.0	87.8
0960	Flooring	118.8	110.4	116.4	85.6	114.1	93.7	99.5	110.5	102.7	103.2	110.5	105.3	98.2	114.0	102.7	103.2	114.1	106.3
0970, 0990	Wall Finishes & Painting/Coating	108.3	115.5	112.6	94.5	77.9	84.5	98.3	66.4	79.1	102.4	97.2	99.3	94.4	122.9	111.6	108.3	86.4	95.1
09	FINISHES	102.5	103.6	103.1	91.6	100.3	96.4	96.0	96.5	96.3	98.6	101.9	100.4	100.9	118.6	110.7	90.0	101.3	96.2
COVERS	DIVS. 10 - 14, 25, 28, 41, 43, 44, 46	100.0	101.1	100.2	100.0	96.7	99.3	100.0	49.7	89.8	100.0	99.7	99.9	100.0	104.6	100.9	100.0	96.1	99.2
21, 22, 23	FIRE SUPPRESSION, PLUMBING & HVAC	100.3	97.0	98.9	100.1	87.5	95.0	95.3	87.3	92.1	100.0	95.6	98.2	100.0	103.5	101.4	94.8	86.9	91.6
26, 27, 3370	ELECTRICAL, COMMUNICATIONS & UTIL.	101.5	98.7	100.1	104.2	85.1	94.1	100.8	85.1	92.5	102.1	92.1	96.8	100.8	99.5	100.1	101.9	85.1	93.0
MF2014	WEIGHTED AVERAGE	99.5	102.2	100.7	97.8	94.5	96.3	96.7	92.5	94.8	99.1	99.0	99.1	100.0	107.2	103.1	96.8	94.6	95.9

	DIVISION	WISCONSIN																	
		OSHKOSH			PORTAGE			RACINE			RHINELANDER			SUPERIOR			WAUSAU		
		549			539			534			545			548			544		
		MAT.	INST.	TOTAL	MAT.	INST.	TOTAL	MAT.	INST.	TOTAL	MAT.	INST.	TOTAL	MAT.	INST.	TOTAL	MAT.	INST.	TOTAL
015433	CONTRACTOR EQUIPMENT		98.2	98.2		102.3	102.3		102.3	102.3		98.2	98.2		100.8	100.8		98.2	98.2
0241, 31 - 34	SITE & INFRASTRUCTURE, DEMOLITION	91.5	98.9	96.8	85.0	108.0	101.4	94.5	109.0	104.8	103.3	98.8	100.1	91.3	103.3	99.9	87.4	98.9	95.6
0310	Concrete Forming & Accessories	88.6	98.1	96.8	91.7	98.9	97.9	100.5	100.5	100.5	86.2	97.8	96.2	90.5	91.9	91.7	88.1	98.7	97.2
0320	Concrete Reinforcing	88.8	92.3	90.6	93.5	91.4	92.5	92.3	97.9	95.1	89.0	90.9	90.0	87.8	91.4	89.6	89.0	91.3	90.2
0330	Cast-in-Place Concrete	99.7	99.4	99.6	85.9	101.8	92.4	98.7	107.0	102.1	113.1	99.3	107.4	102.0	97.9	100.3	93.0	99.6	95.7
03	CONCRETE	92.4	97.6	94.9	86.4	98.5	92.4	97.9	102.3	100.1	104.7	97.2	101.0	91.8	94.2	93.0	87.3	97.8	92.5
04	MASONRY	107.9	101.4	103.8	99.6	104.3	102.5	100.6	114.0	109.0	125.3	101.4	110.4	120.8	104.5	110.6	107.4	101.4	103.6
05	METALS	95.4	96.7	95.8	91.3	96.0	92.7	94.8	101.2	96.8	95.3	96.7	95.7	96.1	98.2	96.8	95.1	97.7	95.9
06	WOOD, PLASTICS & COMPOSITES	89.8	99.3	95.1	89.2	99.2	94.8	100.2	96.4	98.1	87.3	99.3	94.0	92.3	90.1	91.0	89.2	99.3	94.9
07	THERMAL & MOISTURE PROTECTION	103.4	85.7	96.1	101.4	92.6	97.7	102.1	108.3	104.7	104.2	83.5	95.7	103.3	92.2	98.7	103.3	84.4	95.5
08	OPENINGS	95.9	95.5	95.8	98.1	92.3	96.8	102.5	101.1	102.2	96.0	87.1	93.9	88.6	91.1	89.2	96.1	95.7	96.0
0920	Plaster & Gypsum Board	86.2	99.6	95.2	86.4	99.6	95.3	94.5	96.7	96.0	86.2	99.6	95.2	87.6	90.2	89.4	86.2	99.6	95.2
0950, 0980	Ceilings & Acoustic Treatment	81.9	99.6	93.5	85.6	99.6	94.8	88.1	96.7	93.7	81.9	99.6	93.5	59.3	90.2	79.6	81.9	99.6	93.5
0960	Flooring	101.6	114.1	105.2	96.2	118.1	102.5	99.8	110.4	102.8	100.9	114.1	104.7	104.4	122.4	109.6	101.5	114.1	105.1
0970, 0990	Wall Finishes & Painting/Coating	102.2	101.4	101.7	98.3	66.4	79.1	98.3	115.5	108.7	102.2	60.4	76.9	96.0	102.8	100.1	102.2	83.2	90.7
09	FINISHES	94.0	100.9	97.8	94.1	98.6	96.6	97.4	103.6	100.8	94.8	97.0	96.0	89.3	98.6	94.4	93.7	100.7	97.5
COVERS	DIVS. 10 - 14, 25, 28, 41, 43, 44, 46	100.0	64.6	92.8	100.0	63.4	92.6	100.0	101.1	100.2	100.0	55.0	90.9	100.0	94.1	98.8	100.0	96.5	99.3
21, 22, 23	FIRE SUPPRESSION, PLUMBING & HVAC	95.6	85.0	91.3	95.3	95.3	95.3	100.1	97.0	98.8	95.6	86.7	92.0	94.8	89.4	92.6	95.6	86.4	91.9
26, 27, 3370	ELECTRICAL, COMMUNICATIONS & UTIL.	102.5	80.4	90.9	104.8	92.1	98.1	100.8	98.6	99.6	101.8	81.8	91.3	107.1	99.0	102.8	103.7	81.8	92.1
MF2014	WEIGHTED AVERAGE	96.9	91.9	94.7	95.3	96.6	95.9	99.1	102.4	100.5	99.3	91.2	95.8	96.6	96.3	96.5	96.2	93.5	95.0

City Cost Indexes

DIVISION		WYOMING																	
		CASPER			CHEYENNE			NEWCASTLE			RAWLINS			RIVERTON			ROCK SPRINGS		
		826			820			827			823			825			829 - 831		
		MAT.	INST.	TOTAL	MAT.	INST.	TOTAL	MAT.	INST.	TOTAL	MAT.	INST.	TOTAL	MAT.	INST.	TOTAL	MAT.	INST.	TOTAL
015433	**CONTRACTOR EQUIPMENT**		98.2	98.2		98.2	98.2		98.2	98.2		98.2	98.2		98.2	98.2		98.2	98.2
0241, 31 - 34	**SITE & INFRASTRUCTURE, DEMOLITION**	98.6	94.4	95.6	92.5	94.4	93.8	85.1	94.0	91.5	98.4	94.0	95.3	92.1	94.0	93.5	89.2	94.0	92.6
0310	Concrete Forming & Accessories	102.5	50.1	57.3	104.4	53.5	60.5	94.8	62.2	66.7	99.2	62.2	67.3	93.7	62.1	66.5	101.5	62.1	67.5
0320	Concrete Reinforcing	105.7	70.6	87.8	98.0	70.6	84.1	105.5	70.8	87.9	105.2	70.8	87.7	106.1	70.6	88.0	106.1	70.6	88.0
0330	Cast-in-Place Concrete	102.8	72.0	90.2	95.0	72.1	85.6	95.9	61.6	81.8	96.0	61.6	81.9	96.0	60.0	81.2	95.9	59.9	81.1
03	**CONCRETE**	101.7	62.1	82.3	98.6	63.7	81.4	99.0	63.9	81.8	111.8	63.9	88.3	106.7	63.3	85.4	99.5	63.2	81.7
04	**MASONRY**	108.2	51.9	73.1	103.7	51.9	71.4	100.3	54.8	71.9	100.3	54.8	71.9	100.3	50.5	69.3	160.4	50.4	91.8
05	**METALS**	99.3	72.1	90.9	101.7	72.1	92.6	97.8	72.0	89.9	97.9	72.0	89.9	98.0	71.6	89.8	98.7	71.1	90.2
06	**WOOD, PLASTICS & COMPOSITES**	97.2	49.7	70.6	96.5	54.3	72.9	86.4	66.9	75.5	90.4	66.9	77.2	85.3	66.9	75.0	95.4	66.9	79.4
07	**THERMAL & MOISTURE PROTECTION**	105.9	55.1	85.1	100.1	55.6	81.8	101.8	55.8	82.9	103.2	55.8	83.7	102.7	59.9	85.1	101.9	59.9	84.7
08	**OPENINGS**	108.0	53.3	95.3	108.3	55.8	96.1	112.6	61.7	100.8	112.3	61.7	100.5	112.5	61.7	100.7	113.0	61.3	101.0
0920	Plaster & Gypsum Board	97.4	48.1	64.1	89.3	52.9	64.7	85.2	65.8	72.1	85.5	65.8	72.2	85.2	65.8	72.1	97.3	65.8	76.0
0950, 0980	Ceilings & Acoustic Treatment	99.2	48.1	65.6	91.0	52.9	66.0	92.9	65.8	75.1	92.9	65.8	75.1	92.9	65.8	75.1	92.9	65.8	75.1
0960	Flooring	112.0	56.4	96.1	109.9	56.4	94.6	103.5	56.4	90.0	106.2	56.4	92.0	103.0	56.4	89.7	108.4	56.4	93.5
0970, 0990	Wall Finishes & Painting/Coating	104.7	43.1	67.5	111.2	43.1	70.0	107.1	55.4	75.9	107.1	55.4	75.9	107.1	55.4	75.9	107.1	55.4	75.9
09	**FINISHES**	103.2	49.6	73.6	100.0	52.3	73.7	95.0	60.8	76.1	97.2	60.8	77.1	95.7	60.8	76.5	98.1	60.8	77.5
COVERS	**DIVS. 10 - 14, 25, 28, 41, 43, 44, 46**	100.0	87.6	97.5	100.0	88.2	97.6	100.0	89.1	97.8	100.0	89.1	97.8	100.0	81.4	96.2	100.0	80.8	96.1
21, 22, 23	**FIRE SUPPRESSION, PLUMBING & HVAC**	99.9	64.1	85.4	99.9	64.1	85.4	97.4	63.4	83.7	97.4	63.4	83.7	97.4	63.4	83.7	99.8	63.3	85.1
26, 27, 3370	**ELECTRICAL, COMMUNICATIONS & UTIL.**	103.3	64.1	82.6	105.8	68.6	86.1	104.5	69.0	85.7	104.5	69.0	85.7	104.5	62.9	82.5	102.4	62.9	81.5
MF2014	**WEIGHTED AVERAGE**	102.0	63.9	85.4	101.5	65.2	85.7	99.9	66.9	85.5	101.8	66.9	86.6	101.0	65.4	85.5	103.7	65.3	87.0

DIVISION		WYOMING												CANADA					
		SHERIDAN			WHEATLAND			WORLAND			YELLOWSTONE NAT'L PA			BARRIE, ONTARIO			BATHURST, NEW BRUNSWICK		
		828			822			824			821								
		MAT.	INST.	TOTAL	MAT.	INST.	TOTAL	MAT.	INST.	TOTAL	MAT.	INST.	TOTAL	MAT.	INST.	TOTAL	MAT.	INST.	TOTAL
015433	**CONTRACTOR EQUIPMENT**		98.2	98.2		98.2	98.2		98.2	98.2		98.2	98.2		102.1	102.1		102.2	102.2
0241, 31 - 34	**SITE & INFRASTRUCTURE, DEMOLITION**	92.4	94.4	93.8	89.6	94.0	92.8	87.1	94.0	92.0	87.2	94.2	92.2	118.5	100.4	105.6	103.4	96.6	98.6
0310	Concrete Forming & Accessories	102.2	50.6	57.7	96.8	52.7	58.8	96.9	62.0	66.8	96.9	63.1	67.8	124.4	89.4	94.2	104.4	63.7	69.3
0320	Concrete Reinforcing	106.1	70.9	88.2	105.5	70.8	87.8	106.1	70.6	88.0	107.9	70.6	88.9	173.2	83.8	127.7	137.0	56.5	96.0
0330	Cast-in-Place Concrete	99.2	72.0	88.1	100.2	61.5	84.3	95.9	59.9	81.1	95.9	61.6	81.8	165.6	89.0	134.2	131.6	61.9	103.0
03	**CONCRETE**	106.9	62.4	85.0	103.7	59.7	82.1	99.2	63.2	81.6	99.5	64.3	82.2	150.3	88.4	119.9	124.6	62.4	94.1
04	**MASONRY**	100.6	56.3	73.0	100.7	54.7	72.0	100.3	50.4	69.2	100.4	53.4	71.1	171.5	98.0	125.7	160.8	64.9	101.1
05	**METALS**	101.6	72.4	92.6	97.8	71.7	89.8	98.0	71.5	89.8	98.6	71.5	90.3	110.2	90.2	104.0	109.0	71.5	97.5
06	**WOOD, PLASTICS & COMPOSITES**	97.9	50.4	71.3	88.3	54.3	69.2	88.3	66.9	76.3	88.3	66.9	76.3	119.8	88.2	102.1	101.8	64.0	80.7
07	**THERMAL & MOISTURE PROTECTION**	102.9	56.3	83.8	102.1	55.8	83.1	101.9	57.5	83.7	101.4	58.8	83.9	111.8	89.1	102.5	106.0	62.2	88.1
08	**OPENINGS**	113.2	52.7	99.1	111.1	55.8	98.3	112.8	61.7	100.9	105.6	61.7	95.4	93.9	86.5	92.2	87.9	55.9	80.5
0920	Plaster & Gypsum Board	109.4	48.8	68.5	85.2	52.8	63.3	85.2	65.8	72.1	85.4	65.8	72.1	151.1	87.7	108.2	137.6	62.7	87.0
0950, 0980	Ceilings & Acoustic Treatment	95.7	48.8	64.9	92.9	52.8	66.6	92.9	65.8	75.1	93.7	65.8	75.4	91.5	87.7	89.0	106.4	62.7	77.7
0960	Flooring	107.5	56.4	92.9	105.0	56.4	91.1	105.0	56.4	91.1	105.0	56.4	91.1	127.2	98.9	119.1	107.9	47.2	90.6
0970, 0990	Wall Finishes & Painting/Coating	109.4	43.1	69.4	107.1	34.4	63.3	107.1	55.4	75.9	107.1	55.4	75.9	116.2	89.9	100.3	117.6	51.6	77.8
09	**FINISHES**	102.8	50.0	73.7	95.8	51.0	71.1	95.5	60.8	76.3	95.7	61.5	76.8	115.0	91.5	102.1	111.3	59.7	82.8
COVERS	**DIVS. 10 - 14, 25, 28, 41, 43, 44, 46**	100.0	87.7	97.5	100.0	80.3	96.0	100.0	81.4	96.2	100.0	82.4	96.4	139.2	74.4	126.1	131.1	65.3	117.8
21, 22, 23	**FIRE SUPPRESSION, PLUMBING & HVAC**	97.4	64.1	83.9	97.4	63.3	83.6	97.4	63.3	83.6	97.4	64.9	84.3	101.8	99.3	100.8	102.0	69.8	89.0
26, 27, 3370	**ELECTRICAL, COMMUNICATIONS & UTIL.**	107.4	63.0	83.9	104.5	68.6	85.5	104.5	62.9	82.5	103.2	62.9	81.9	116.8	90.5	102.9	112.7	61.8	85.8
MF2014	**WEIGHTED AVERAGE**	102.6	64.3	85.9	100.4	64.3	84.7	100.0	65.3	84.9	99.3	66.3	84.9	117.7	93.0	106.9	111.5	67.3	92.2

DIVISION		CANADA																	
		BRANDON, MANITOBA			BRANTFORD, ONTARIO			BRIDGEWATER, NOVA SCOTIA			CALGARY, ALBERTA			CAP-DE-LA-MADELEINE, QUEBEC			CHARLESBOURG, QUEBEC		
		MAT.	INST.	TOTAL	MAT.	INST.	TOTAL	MAT.	INST.	TOTAL	MAT.	INST.	TOTAL	MAT.	INST.	TOTAL	MAT.	INST.	TOTAL
015433	**CONTRACTOR EQUIPMENT**		104.5	104.5		102.1	102.1		101.9	101.9		107.1	107.1		102.8	102.8		102.8	102.8
0241, 31 - 34	**SITE & INFRASTRUCTURE, DEMOLITION**	133.0	99.3	109.1	118.0	100.7	105.7	102.6	98.2	99.5	124.0	105.4	110.7	98.4	99.9	99.4	98.4	99.9	99.4
0310	Concrete Forming & Accessories	146.5	72.5	82.7	124.9	96.9	100.7	97.1	71.6	75.1	124.4	97.4	101.1	130.7	88.0	93.8	130.7	88.0	93.8
0320	Concrete Reinforcing	185.9	53.2	118.4	165.4	82.5	123.2	140.6	46.7	92.8	135.6	71.9	103.2	140.6	78.8	109.2	140.6	78.8	109.2
0330	Cast-in-Place Concrete	133.4	76.7	110.1	153.6	110.3	135.8	157.8	72.2	122.7	189.0	107.3	155.5	124.0	97.8	113.2	124.0	97.8	113.2
03	**CONCRETE**	145.6	71.0	109.0	141.8	98.8	120.7	139.4	67.9	104.3	159.1	96.1	128.2	124.3	89.8	107.4	124.3	89.8	107.4
04	**MASONRY**	233.5	67.6	130.1	168.0	102.6	127.2	163.4	71.9	106.4	196.3	91.4	130.9	163.9	87.1	116.0	163.9	87.1	116.0
05	**METALS**	125.6	76.8	110.6	109.1	90.9	103.5	108.2	74.2	97.7	137.0	89.3	122.3	107.4	87.6	101.3	107.4	87.6	101.3
06	**WOOD, PLASTICS & COMPOSITES**	154.2	73.5	109.0	123.7	95.7	108.1	92.9	71.1	80.7	102.3	97.2	99.4	135.3	88.0	108.8	135.3	88.0	108.8
07	**THERMAL & MOISTURE PROTECTION**	135.0	71.8	109.1	113.8	94.7	106.0	108.6	69.9	92.8	117.8	94.6	108.3	107.4	89.9	100.3	107.4	89.9	100.3
08	**OPENINGS**	105.5	64.8	96.1	92.4	92.7	92.5	86.1	64.9	81.1	89.9	86.2	89.1	93.6	81.1	90.7	93.6	81.1	90.7
0920	Plaster & Gypsum Board	119.3	72.2	87.4	130.1	95.5	106.7	130.7	70.0	89.7	139.4	96.6	110.5	157.0	87.4	109.9	157.0	87.4	109.9
0950, 0980	Ceilings & Acoustic Treatment	109.6	72.2	85.0	96.5	95.5	95.8	96.5	70.0	79.1	138.8	96.6	111.0	96.5	87.4	90.5	96.5	87.4	90.5
0960	Flooring	138.6	70.2	119.1	120.5	98.9	114.3	102.7	67.2	92.5	122.3	92.6	113.8	120.5	99.5	114.5	120.5	99.5	114.5
0970, 0990	Wall Finishes & Painting/Coating	126.5	58.7	85.6	118.3	98.6	106.4	118.3	64.1	85.6	127.4	112.4	118.3	118.3	91.7	102.2	118.3	91.7	102.2
09	**FINISHES**	122.5	71.0	94.1	112.4	97.9	104.4	106.6	70.3	86.6	125.5	98.7	110.7	115.3	90.8	101.8	115.3	90.8	101.8
COVERS	**DIVS. 10 - 14, 25, 28, 41, 43, 44, 46**	131.1	67.6	118.3	131.1	76.6	120.1	131.1	66.6	118.1	131.1	96.9	124.2	131.1	85.2	121.9	131.1	85.2	121.9
21, 22, 23	**FIRE SUPPRESSION, PLUMBING & HVAC**	102.1	84.1	94.9	102.0	102.3	102.1	102.0	84.4	94.9	101.2	92.9	97.8	102.3	91.0	97.8	102.3	91.0	97.8
26, 27, 3370	**ELECTRICAL, COMMUNICATIONS & UTIL.**	118.6	69.6	92.7	111.5	89.9	100.1	117.1	64.4	89.2	113.5	101.3	107.1	111.2	72.6	90.8	111.2	72.6	90.8
MF2014	**WEIGHTED AVERAGE**	125.2	75.8	103.7	114.9	96.9	107.1	112.9	74.6	96.2	123.5	95.7	111.4	112.4	87.6	101.6	112.4	87.6	101.6

	DIVISION	CANADA																	
		CHARLOTTETOWN, PRINCE EDWARD ISLAND			CHICOUTIMI, QUEBEC			CORNER BROOK, NEWFOUNDLAND			CORNWALL, ONTARIO			DALHOUSIE, NEW BRUNSWICK			DARTMOUTH, NOVA SCOTIA		
		MAT.	INST.	TOTAL	MAT.	INST.	TOTAL	MAT.	INST.	TOTAL	MAT.	INST.	TOTAL	MAT.	INST.	TOTAL	MAT.	INST.	TOTAL
015433	**CONTRACTOR EQUIPMENT**		101.8	101.8		102.8	102.8		103.3	103.3		102.1	102.1		102.2	102.2		101.9	101.9
0241, 31 - 34	**SITE & INFRASTRUCTURE, DEMOLITION**	121.3	95.7	103.1	100.7	99.1	99.6	137.7	97.1	108.9	116.1	100.2	104.8	100.9	96.6	97.8	124.8	98.2	105.9
0310	Concrete Forming & Accessories	110.4	58.2	65.4	133.2	93.1	98.6	121.8	61.7	69.9	122.6	89.6	94.1	103.5	63.9	69.3	111.8	71.6	77.2
0320	Concrete Reinforcing	141.5	46.8	93.3	103.3	95.9	99.5	170.3	48.6	108.4	165.4	82.2	123.1	142.7	56.6	98.9	178.2	46.7	111.3
0330	Cast-in-Place Concrete	166.8	61.2	123.4	124.0	99.8	114.0	155.3	70.6	120.5	138.2	100.6	122.7	131.8	62.0	103.2	149.8	72.2	118.0
03	**CONCRETE**	147.1	57.9	103.3	114.6	95.9	105.4	182.2	63.2	123.7	134.3	92.2	113.6	132.5	62.5	98.1	162.1	67.9	115.8
04	**MASONRY**	178.4	62.6	106.3	162.7	95.0	120.5	229.3	64.1	126.4	166.8	94.0	121.4	160.2	64.9	100.8	244.8	71.9	137.1
05	**METALS**	133.9	67.6	113.5	108.8	92.2	103.7	126.0	72.9	109.7	109.1	89.6	103.1	104.4	71.8	94.3	125.9	74.2	110.0
06	**WOOD, PLASTICS & COMPOSITES**	97.8	57.8	75.4	137.4	93.6	112.9	128.3	60.8	90.5	122.1	89.0	103.6	100.2	64.0	80.0	116.1	71.1	90.9
07	**THERMAL & MOISTURE PROTECTION**	121.3	60.4	96.3	106.5	96.4	102.3	139.8	61.8	107.8	113.6	89.3	103.6	110.8	62.2	90.9	136.8	69.9	109.4
08	**OPENINGS**	88.2	50.0	79.3	92.2	80.6	89.5	112.4	57.1	99.5	93.6	86.2	91.9	89.1	55.9	81.4	95.3	64.9	88.2
0920	Plaster & Gypsum Board	131.6	56.3	80.7	159.4	93.2	114.6	149.0	59.3	88.4	189.4	88.6	121.3	141.2	62.7	88.2	144.6	70.0	94.2
0950, 0980	Ceilings & Acoustic Treatment	120.2	56.3	78.2	105.6	93.2	97.4	110.5	59.3	76.9	99.0	88.6	92.2	98.9	62.7	75.1	117.2	70.0	86.2
0960	Flooring	111.7	62.9	97.8	123.6	99.5	116.8	121.0	57.2	102.8	120.5	97.5	113.9	106.5	72.1	96.7	116.1	67.2	102.1
0970, 0990	Wall Finishes & Painting/Coating	125.6	42.9	75.7	117.6	106.4	110.8	126.4	61.8	87.4	118.3	91.9	102.4	120.7	51.6	79.0	126.4	64.1	88.8
09	**FINISHES**	117.1	57.8	84.4	118.4	96.5	106.4	122.0	60.9	88.4	121.0	91.6	104.8	110.6	64.5	85.2	120.1	70.3	92.6
COVERS	**DIVS. 10 - 14, 25, 28, 41, 43, 44, 46**	131.1	64.8	117.7	131.1	86.5	122.1	131.1	65.6	117.9	131.1	74.0	119.6	131.1	65.3	117.8	131.1	66.6	118.1
21, 22, 23	**FIRE SUPPRESSION, PLUMBING & HVAC**	102.3	63.8	86.8	102.0	92.3	98.1	102.1	71.8	89.9	102.3	100.1	101.4	102.0	69.8	89.0	102.1	84.4	95.0
26, 27, 3370	**ELECTRICAL, COMMUNICATIONS & UTIL.**	110.6	52.3	79.8	109.7	86.8	97.6	116.0	58.3	85.5	112.4	90.9	101.0	114.5	58.4	84.8	120.7	64.4	90.9
MF2014	**WEIGHTED AVERAGE**	120.0	62.8	95.1	111.4	92.9	103.3	129.6	67.6	102.6	115.0	93.3	105.5	111.9	67.5	92.5	126.1	74.6	103.7

	DIVISION	CANADA																	
		EDMONTON, ALBERTA			FORT MCMURRAY, ALBERTA			FREDERICTON, NEW BRUNSWICK			GATINEAU, QUEBEC			GRANBY, QUEBEC			HALIFAX, NOVA SCOTIA		
		MAT.	INST.	TOTAL	MAT.	INST.	TOTAL	MAT.	INST.	TOTAL	MAT.	INST.	TOTAL	MAT.	INST.	TOTAL	MAT.	INST.	TOTAL
015433	**CONTRACTOR EQUIPMENT**		107.1	107.1		104.4	104.4		102.2	102.2		102.8	102.8		102.8	102.8		101.9	101.9
0241, 31 - 34	**SITE & INFRASTRUCTURE, DEMOLITION**	144.1	105.4	116.6	123.1	101.3	107.6	105.0	96.6	99.1	98.2	99.8	99.4	98.7	99.8	99.5	109.2	98.3	101.5
0310	Concrete Forming & Accessories	123.2	97.4	100.9	122.8	92.3	96.5	120.3	64.2	71.9	130.7	87.8	93.7	130.7	87.8	93.7	108.4	78.8	82.9
0320	Concrete Reinforcing	134.2	71.9	102.5	152.9	71.8	111.7	132.8	56.7	94.1	148.7	78.8	113.2	148.7	78.8	113.2	144.3	62.7	102.7
0330	Cast-in-Place Concrete	183.9	107.3	152.4	203.9	105.1	163.3	128.3	62.0	101.1	122.3	97.8	112.2	126.4	97.7	114.6	149.4	81.9	121.7
03	**CONCRETE**	156.4	96.1	126.8	163.6	93.1	129.0	126.6	62.7	95.2	124.8	89.8	107.6	126.7	89.7	108.5	136.4	77.3	107.4
04	**MASONRY**	187.2	91.4	127.5	207.9	89.1	133.8	182.6	66.5	110.2	163.7	87.1	116.0	164.1	87.1	116.1	185.9	84.7	122.8
05	**METALS**	136.9	89.3	122.3	135.4	88.9	121.1	129.1	72.3	111.6	107.4	87.4	101.2	107.4	87.4	101.2	134.9	81.5	118.4
06	**WOOD, PLASTICS & COMPOSITES**	105.2	97.2	100.7	117.6	91.8	103.2	116.7	64.0	87.2	135.3	88.0	108.8	135.3	88.0	108.8	97.1	77.8	86.3
07	**THERMAL & MOISTURE PROTECTION**	122.7	94.6	111.2	121.9	91.9	109.6	116.0	63.2	94.4	107.4	89.9	100.3	107.4	88.3	99.6	119.0	79.4	102.8
08	**OPENINGS**	89.2	86.2	88.5	93.6	83.3	91.2	89.3	54.8	81.2	93.6	76.3	89.6	93.6	76.3	89.6	93.5	72.0	88.5
0920	Plaster & Gypsum Board	134.3	96.6	108.8	128.2	91.1	103.1	137.4	62.7	86.9	126.6	87.4	100.1	129.0	87.4	100.8	127.0	77.0	93.2
0950, 0980	Ceilings & Acoustic Treatment	139.5	96.6	111.3	104.9	91.1	95.8	116.1	62.7	81.0	96.5	87.4	90.5	96.5	87.4	90.5	118.7	77.0	91.3
0960	Flooring	122.5	92.6	114.0	120.5	92.6	112.5	116.8	75.3	104.9	120.5	99.5	114.5	120.5	99.5	114.5	109.2	88.3	103.3
0970, 0990	Wall Finishes & Painting/Coating	122.4	112.4	116.3	118.4	95.5	104.5	123.1	65.8	88.5	118.3	91.7	102.2	118.3	91.7	102.2	124.9	87.4	102.2
09	**FINISHES**	126.6	98.7	111.2	115.3	93.0	103.0	117.7	66.7	89.6	111.2	90.8	100.0	111.5	90.8	100.1	115.6	81.8	97.0
COVERS	**DIVS. 10 - 14, 25, 28, 41, 43, 44, 46**	131.1	96.9	124.2	131.1	95.2	123.9	131.1	65.3	117.8	131.1	85.2	121.9	131.1	85.2	121.9	131.1	68.5	118.5
21, 22, 23	**FIRE SUPPRESSION, PLUMBING & HVAC**	101.1	92.9	97.8	102.4	99.0	101.0	102.3	79.3	93.0	102.3	91.0	97.8	102.0	91.0	97.6	100.6	77.6	91.3
26, 27, 3370	**ELECTRICAL, COMMUNICATIONS & UTIL.**	110.6	101.3	105.7	105.8	85.4	95.1	116.1	77.1	95.5	111.2	72.6	90.8	111.8	72.6	91.1	117.0	81.2	98.1
MF2014	**WEIGHTED AVERAGE**	123.1	95.7	111.2	123.6	92.8	110.2	117.4	72.5	97.9	112.1	87.4	101.3	112.3	87.4	101.4	119.6	80.9	102.7

	DIVISION	CANADA																	
		HAMILTON, ONTARIO			HULL, QUEBEC			JOLIETTE, QUEBEC			KAMLOOPS, BRITISH COLUMBIA			KINGSTON, ONTARIO			KITCHENER, ONTARIO		
		MAT.	INST.	TOTAL	MAT.	INST.	TOTAL	MAT.	INST.	TOTAL	MAT.	INST.	TOTAL	MAT.	INST.	TOTAL	MAT.	INST.	TOTAL
015433	**CONTRACTOR EQUIPMENT**		108.9	108.9		102.8	102.8		102.8	102.8		106.1	106.1		104.4	104.4		104.3	104.3
0241, 31 - 34	**SITE & INFRASTRUCTURE, DEMOLITION**	115.5	112.4	113.3	98.2	99.8	99.4	98.8	99.9	99.6	120.5	103.3	108.3	116.1	104.0	107.5	99.6	104.8	103.3
0310	Concrete Forming & Accessories	120.9	94.7	98.3	130.7	87.8	93.7	130.7	88.0	93.8	123.4	90.8	95.3	122.8	89.6	94.2	113.0	87.2	90.8
0320	Concrete Reinforcing	146.4	92.0	118.7	148.7	78.8	113.2	140.6	78.8	109.2	110.3	76.8	93.2	165.4	82.2	123.1	100.8	91.9	96.3
0330	Cast-in-Place Concrete	143.5	101.0	126.1	122.3	97.8	112.2	127.4	97.8	115.2	110.3	101.6	106.7	138.2	100.6	122.7	129.0	93.6	114.4
03	**CONCRETE**	132.8	96.3	114.8	124.8	89.8	107.6	125.9	89.8	108.2	132.9	92.0	112.8	136.2	92.2	114.6	110.8	90.4	100.8
04	**MASONRY**	183.0	101.2	132.0	163.7	87.1	116.0	164.1	87.1	116.1	170.7	95.0	123.5	173.5	94.1	124.0	150.8	97.6	117.6
05	**METALS**	123.4	92.5	113.9	107.4	87.4	101.2	107.4	87.6	101.3	109.8	87.5	102.9	110.8	89.5	104.3	117.4	92.2	109.6
06	**WOOD, PLASTICS & COMPOSITES**	100.4	93.9	96.8	135.3	88.0	108.8	135.3	88.0	108.8	105.2	89.4	96.4	122.1	89.2	103.7	109.7	85.5	96.1
07	**THERMAL & MOISTURE PROTECTION**	118.2	95.7	109.0	107.4	89.9	100.2	107.4	89.9	100.2	123.3	86.8	108.3	113.6	90.4	104.1	108.6	92.8	102.1
08	**OPENINGS**	91.3	91.6	91.4	93.6	76.3	89.6	93.6	81.1	90.7	90.2	85.7	89.2	93.6	85.9	91.8	84.7	85.3	84.8
0920	Plaster & Gypsum Board	149.5	93.6	111.7	126.6	87.4	100.1	157.0	87.4	109.9	113.6	88.5	96.6	192.8	88.7	122.4	120.2	84.9	96.3
0950, 0980	Ceilings & Acoustic Treatment	122.1	93.6	103.4	96.5	87.4	90.5	96.5	87.4	90.5	96.5	88.5	91.2	112.4	88.7	96.8	98.6	84.9	89.6
0960	Flooring	123.1	102.3	117.2	120.5	99.5	114.5	120.5	99.5	114.5	119.6	55.6	101.3	120.5	97.5	113.9	110.5	102.3	108.1
0970, 0990	Wall Finishes & Painting/Coating	121.6	102.6	110.2	118.3	91.7	102.2	118.3	91.7	102.2	118.3	84.0	97.6	118.3	84.9	98.1	115.3	92.6	101.6
09	**FINISHES**	122.0	97.0	108.2	111.2	90.8	100.0	115.3	90.8	101.8	111.7	84.2	96.6	124.4	90.9	105.9	106.4	90.2	97.5
COVERS	**DIVS. 10 - 14, 25, 28, 41, 43, 44, 46**	131.1	98.8	124.6	131.1	85.2	121.9	131.1	85.2	121.9	131.1	95.6	124.0	131.1	74.0	119.6	131.1	97.0	124.2
21, 22, 23	**FIRE SUPPRESSION, PLUMBING & HVAC**	101.3	89.2	96.4	102.0	91.0	97.6	102.0	91.0	97.6	102.0	94.4	98.9	102.3	100.3	101.5	100.4	87.7	95.3
26, 27, 3370	**ELECTRICAL, COMMUNICATIONS & UTIL.**	108.3	100.5	104.2	112.9	72.6	91.6	111.8	72.6	91.1	115.1	81.9	97.6	112.4	89.6	100.3	111.5	98.0	104.4
MF2014	**WEIGHTED AVERAGE**	117.0	96.7	108.2	112.2	87.4	101.4	112.5	87.6	101.7	114.5	90.6	104.1	116.0	93.4	106.2	109.6	92.9	102.3

	DIVISION	CANADA																	
		LAVAL, QUEBEC			LETHBRIDGE, ALBERTA			LLOYDMINSTER, ALBERTA			LONDON, ONTARIO			MEDICINE HAT, ALBERTA			MONCTON, NEW BRUNSWICK		
		MAT.	INST.	TOTAL	MAT.	INST.	TOTAL	MAT.	INST.	TOTAL	MAT.	INST.	TOTAL	MAT.	INST.	TOTAL	MAT.	INST.	TOTAL
015433	CONTRACTOR EQUIPMENT		102.8	102.8		104.4	104.4		104.4	104.4		104.4	104.4		104.4	104.4		102.2	102.2
0241, 31 - 34	SITE & INFRASTRUCTURE, DEMOLITION	98.7	99.8	99.5	115.6	101.9	105.9	115.4	101.3	105.4	113.1	104.9	107.3	114.2	101.4	105.1	102.8	96.8	98.5
0310	Concrete Forming & Accessories	130.8	87.8	93.8	124.1	92.4	96.8	122.3	82.6	88.1	119.1	88.5	92.7	124.1	82.6	88.3	104.4	64.5	69.9
0320	Concrete Reinforcing	148.7	78.8	113.2	152.9	71.8	111.7	152.9	71.8	111.7	132.8	91.9	112.0	152.9	71.8	111.7	137.0	61.5	98.6
0330	Cast-in-Place Concrete	126.4	97.8	114.6	153.0	105.1	133.3	141.9	101.4	125.3	147.4	98.6	127.4	141.9	101.4	125.3	126.8	78.7	107.0
03	CONCRETE	126.7	89.8	108.6	139.5	93.2	116.7	134.1	87.5	111.3	132.4	92.7	112.9	134.3	87.5	111.3	122.3	69.7	96.5
04	MASONRY	164.0	87.1	116.1	181.7	89.1	124.0	163.4	82.5	113.0	182.6	98.7	130.3	163.4	82.5	113.0	160.5	65.0	101.0
05	METALS	107.4	87.4	101.3	128.8	88.9	116.5	110.0	88.8	103.4	124.9	91.8	114.7	110.0	88.7	103.4	109.0	80.4	100.2
06	WOOD, PLASTICS & COMPOSITES	135.5	88.0	108.9	121.3	91.8	104.8	117.6	82.0	97.7	108.2	86.6	96.1	121.3	82.0	99.3	101.8	64.0	80.7
07	THERMAL & MOISTURE PROTECTION	107.9	89.9	100.5	119.1	91.9	108.0	116.0	87.5	104.3	119.1	93.5	108.6	122.3	87.5	108.0	110.2	65.3	91.8
08	OPENINGS	93.6	76.3	89.6	93.6	83.3	91.2	93.6	77.9	89.9	86.4	86.6	86.4	93.6	77.9	89.9	87.9	61.0	81.6
0920	Plaster & Gypsum Board	129.3	87.4	100.9	119.1	91.1	100.2	114.4	81.0	91.8	152.4	86.1	107.6	117.2	81.0	92.8	137.6	62.7	87.0
0950, 0980	Ceilings & Acoustic Treatment	96.5	87.4	90.5	104.1	91.1	95.5	96.5	81.0	86.3	121.1	86.1	98.1	96.5	81.0	86.3	106.4	62.7	77.7
0960	Flooring	120.5	99.5	114.5	120.5	92.6	112.5	120.5	92.6	112.5	117.2	102.3	113.0	120.5	92.6	112.5	107.9	72.1	97.7
0970, 0990	Wall Finishes & Painting/Coating	118.3	91.7	102.2	118.2	104.2	109.8	118.4	81.2	96.0	122.0	99.4	108.3	118.2	81.2	95.9	117.6	51.6	77.8
09	FINISHES	111.5	90.8	100.1	112.9	94.0	102.5	110.7	84.0	96.0	120.5	91.9	104.7	110.9	84.0	96.1	111.3	64.5	85.5
COVERS	DIVS. 10 - 14, 25, 28, 41, 43, 44, 46	131.1	85.2	121.9	131.1	95.2	123.9	131.1	91.9	123.2	131.1	97.5	124.3	131.1	91.9	123.2	131.1	65.3	117.8
21, 22, 23	FIRE SUPPRESSION, PLUMBING & HVAC	100.4	91.0	96.6	102.2	95.6	99.6	102.3	95.6	99.6	101.4	86.6	95.4	102.0	92.3	98.1	102.0	70.1	89.1
26, 27, 3370	ELECTRICAL, COMMUNICATIONS & UTIL.	112.9	72.6	91.6	107.1	85.4	95.7	104.9	85.4	94.6	107.0	97.5	102.0	104.9	85.4	94.6	117.2	61.8	88.0
MF2014	WEIGHTED AVERAGE	112.1	87.4	101.3	118.4	92.2	107.0	113.4	88.9	102.8	116.4	93.3	106.3	113.6	88.2	102.5	111.8	70.1	93.6

	DIVISION	CANADA																	
		MONTREAL, QUEBEC			MOOSE JAW, SASKATCHEWAN			NEW GLASGOW, NOVA SCOTIA			NEWCASTLE, NEW BRUNSWICK			NORTH BAY, ONTARIO			OSHAWA, ONTARIO		
		MAT.	INST.	TOTAL	MAT.	INST.	TOTAL	MAT.	INST.	TOTAL	MAT.	INST.	TOTAL	MAT.	INST.	TOTAL	MAT.	INST.	TOTAL
015433	CONTRACTOR EQUIPMENT		104.5	104.5		100.6	100.6		101.9	101.9		102.2	102.2		102.1	102.1		104.3	104.3
0241, 31 - 34	SITE & INFRASTRUCTURE, DEMOLITION	112.1	99.4	103.1	115.5	95.9	101.6	118.5	98.2	104.1	103.4	96.6	98.6	133.9	99.8	109.7	110.7	103.9	105.9
0310	Concrete Forming & Accessories	124.3	93.4	97.7	107.1	60.5	66.9	111.7	71.6	77.1	104.4	63.9	69.5	147.4	87.0	95.3	118.4	89.1	93.1
0320	Concrete Reinforcing	134.0	96.0	114.7	107.9	61.9	84.5	170.3	46.7	107.4	137.0	56.6	96.1	201.7	81.8	140.7	159.6	84.4	121.3
0330	Cast-in-Place Concrete	161.7	101.2	136.8	137.7	71.1	110.4	149.8	72.2	118.0	131.6	62.0	103.0	144.3	87.1	120.8	149.1	87.9	124.0
03	CONCRETE	139.7	96.5	118.4	117.9	65.1	92.0	160.9	67.9	115.2	124.6	62.5	94.1	163.4	86.3	125.5	135.6	88.0	112.2
04	MASONRY	167.9	95.1	122.5	162.0	64.4	101.1	228.9	71.9	131.1	160.8	64.9	101.1	235.3	89.9	144.7	153.8	94.4	116.8
05	METALS	128.0	92.7	117.1	106.3	73.9	96.3	123.6	74.2	108.4	109.0	71.8	97.5	124.6	89.2	113.7	108.6	91.4	103.3
06	WOOD, PLASTICS & COMPOSITES	114.4	93.9	102.9	102.7	59.1	78.3	116.1	71.1	90.9	101.8	64.0	80.7	157.4	87.6	118.3	116.7	88.2	100.7
07	THERMAL & MOISTURE PROTECTION	116.5	96.9	108.5	106.8	64.1	89.3	136.8	69.9	109.4	110.2	62.2	90.5	143.5	85.6	119.8	109.4	87.4	100.4
08	OPENINGS	90.1	82.4	88.3	89.3	56.8	81.7	95.3	64.9	88.2	87.9	55.9	80.5	104.2	83.7	99.4	90.4	87.9	89.8
0920	Plaster & Gypsum Board	135.1	93.2	106.8	110.7	57.8	74.9	142.7	70.0	93.6	137.6	62.7	87.0	140.3	87.1	104.4	124.7	87.7	99.7
0950, 0980	Ceilings & Acoustic Treatment	126.1	93.2	104.4	96.5	57.8	71.0	109.6	70.0	83.6	106.4	62.7	77.7	109.6	87.1	94.8	95.3	87.7	90.3
0960	Flooring	122.0	99.5	115.6	110.6	63.3	97.1	116.1	67.2	102.1	107.9	72.1	97.7	138.6	97.5	126.9	113.7	104.9	111.2
0970, 0990	Wall Finishes & Painting/Coating	120.8	106.4	112.1	118.3	66.8	87.2	126.4	64.1	88.8	117.6	51.6	77.8	126.4	91.2	105.1	115.3	106.7	110.1
09	FINISHES	119.9	96.7	107.1	107.1	61.4	81.9	118.2	70.3	91.8	111.3	64.5	85.5	125.1	89.8	105.7	107.9	93.9	100.2
COVERS	DIVS. 10 - 14, 25, 28, 41, 43, 44, 46	131.1	87.1	122.3	131.1	64.6	117.7	131.1	66.6	118.1	131.1	65.3	117.8	131.1	72.7	119.3	131.1	97.2	124.3
21, 22, 23	FIRE SUPPRESSION, PLUMBING & HVAC	101.2	92.4	97.7	102.3	76.1	91.8	102.1	84.4	95.0	102.0	69.8	89.0	102.1	98.1	100.5	100.4	100.7	100.5
26, 27, 3370	ELECTRICAL, COMMUNICATIONS & UTIL.	112.6	86.8	99.0	113.9	62.8	86.9	116.3	64.4	88.9	112.0	61.8	85.5	116.4	90.8	102.9	112.6	90.5	100.9
MF2014	WEIGHTED AVERAGE	117.8	93.2	107.1	110.7	69.5	92.7	124.1	74.6	102.5	111.6	67.9	92.5	127.3	91.1	111.5	112.3	94.3	104.4

	DIVISION	CANADA																	
		OTTAWA, ONTARIO			OWEN SOUND, ONTARIO			PETERBOROUGH, ONTARIO			PORTAGE LA PRAIRIE, MANITOBA			PRINCE ALBERT, SASKATCHEWAN			PRINCE GEORGE, BRITISH COLUMBIA		
		MAT.	INST.	TOTAL	MAT.	INST.	TOTAL	MAT.	INST.	TOTAL	MAT.	INST.	TOTAL	MAT.	INST.	TOTAL	MAT.	INST.	TOTAL
015433	CONTRACTOR EQUIPMENT		104.3	104.3		102.1	102.1		102.1	102.1		104.5	104.5		100.6	100.6		106.1	106.1
0241, 31 - 34	SITE & INFRASTRUCTURE, DEMOLITION	108.2	104.6	105.7	118.5	100.3	105.5	118.0	100.2	105.3	116.6	99.3	104.3	110.9	96.1	100.3	124.0	103.3	109.3
0310	Concrete Forming & Accessories	121.2	91.4	95.5	124.4	85.5	90.9	124.9	88.1	93.1	124.3	72.0	79.2	107.1	60.3	66.8	112.7	85.5	89.3
0320	Concrete Reinforcing	138.9	91.9	115.0	173.2	83.8	127.7	165.4	82.3	123.1	152.9	53.2	102.2	112.7	61.8	86.8	110.3	76.8	93.2
0330	Cast-in-Place Concrete	145.6	99.9	126.8	165.6	83.1	131.7	153.6	88.7	127.0	141.9	76.1	114.9	124.8	71.0	102.7	138.2	101.6	123.2
03	CONCRETE	132.6	94.4	113.8	150.3	84.6	118.1	141.8	87.4	115.1	127.8	70.6	99.7	112.6	65.0	89.2	145.4	89.7	118.0
04	MASONRY	168.0	98.8	124.9	171.5	95.6	124.2	168.0	96.7	123.5	166.5	66.6	104.2	161.1	64.4	100.8	172.9	95.0	124.3
05	METALS	127.9	91.7	116.8	110.2	90.0	104.0	109.1	89.7	103.1	110.0	76.7	99.7	106.3	73.7	96.3	109.8	87.6	102.9
06	WOOD, PLASTICS & COMPOSITES	109.4	90.6	98.9	119.8	84.3	99.9	123.7	86.3	102.8	121.1	73.5	94.4	102.7	59.1	78.3	105.2	82.1	92.3
07	THERMAL & MOISTURE PROTECTION	124.4	93.8	111.8	111.8	86.4	101.4	113.8	91.2	104.5	107.3	71.3	92.5	106.7	63.0	88.8	117.2	86.1	104.4
08	OPENINGS	93.9	88.7	92.7	93.9	83.1	91.4	92.4	85.5	90.8	93.6	64.8	86.9	88.2	56.8	80.9	90.2	81.8	88.2
0920	Plaster & Gypsum Board	168.3	90.2	115.5	151.1	83.7	105.6	130.1	85.8	100.1	114.2	72.2	85.8	110.7	57.8	74.9	113.6	81.0	91.6
0950, 0980	Ceilings & Acoustic Treatment	125.6	90.2	102.4	91.5	83.7	86.4	96.5	85.8	89.4	96.5	72.2	80.5	96.5	57.8	71.0	96.5	81.0	86.3
0960	Flooring	112.3	97.5	108.1	127.2	98.9	119.1	120.5	97.5	113.9	120.5	70.2	106.1	110.6	63.3	97.1	116.0	76.2	104.6
0970, 0990	Wall Finishes & Painting/Coating	123.2	94.2	105.7	116.2	89.9	100.3	118.3	93.5	103.3	118.4	58.7	82.4	118.3	57.0	81.3	118.3	84.0	97.6
09	FINISHES	122.6	92.9	106.2	115.0	88.6	100.5	112.4	90.5	100.3	110.5	70.8	88.6	107.1	60.3	81.3	110.6	83.4	95.6
COVERS	DIVS. 10 - 14, 25, 28, 41, 43, 44, 46	131.1	95.7	124.0	139.2	73.2	125.9	131.1	74.2	119.6	131.1	67.2	118.2	131.1	64.6	117.7	131.1	94.8	123.8
21, 22, 23	FIRE SUPPRESSION, PLUMBING & HVAC	101.4	87.2	95.7	101.8	98.1	100.3	102.0	101.7	101.9	102.0	83.6	94.5	102.3	68.6	88.7	102.0	94.4	98.9
26, 27, 3370	ELECTRICAL, COMMUNICATIONS & UTIL.	107.3	98.7	102.8	118.2	89.6	103.1	111.5	90.4	100.4	113.6	60.1	85.3	113.9	62.8	86.9	111.7	81.9	96.0
MF2014	WEIGHTED AVERAGE	117.3	94.0	107.2	117.8	91.2	106.2	114.9	93.1	105.4	113.4	74.2	96.3	109.9	67.7	91.5	115.4	89.9	104.3

		CANADA																	
	DIVISION	QUEBEC, QUEBEC			RED DEER, ALBERTA			REGINA, SASKATCHEWAN			RIMOUSKI, QUEBEC			ROUYN-NORANDA, QUEBEC			SAINT HYACINTHE, QUEBEC		
		MAT.	INST.	TOTAL	MAT.	INST.	TOTAL	MAT.	INST.	TOTAL	MAT.	INST.	TOTAL	MAT.	INST.	TOTAL	MAT.	INST.	TOTAL
015433	CONTRACTOR EQUIPMENT		104.9	104.9		104.4	104.4		100.6	100.6		102.8	102.8		102.8	102.8		102.8	102.8
0241, 31 - 34	SITE & INFRASTRUCTURE, DEMOLITION	110.8	99.5	102.8	114.2	101.4	105.1	124.4	98.0	105.6	98.7	99.1	99.0	98.2	99.8	99.4	98.7	99.8	99.5
0310	Concrete Forming & Accessories	125.6	93.7	98.1	139.6	82.6	90.4	117.6	90.9	94.6	130.7	93.1	98.3	130.7	87.8	93.7	130.7	87.8	93.7
0320	Concrete Reinforcing	138.3	96.0	116.8	152.9	71.8	111.7	138.0	84.8	111.0	106.0	95.9	100.9	148.7	78.8	113.2	148.7	78.8	113.2
0330	Cast-in-Place Concrete	141.8	101.6	125.2	141.9	101.4	125.3	171.5	95.7	140.4	128.6	99.8	116.8	122.3	97.8	112.2	126.4	97.8	114.6
03	CONCRETE	131.0	96.7	114.1	135.3	87.5	111.8	148.9	91.4	120.6	121.1	95.9	108.7	124.8	89.8	107.6	126.7	89.8	108.5
04	MASONRY	163.0	95.1	120.7	163.4	82.5	113.0	183.5	96.7	129.4	163.6	95.0	120.9	163.7	87.1	116.0	164.0	87.1	116.1
05	METALS	128.0	93.0	117.2	110.0	88.7	103.4	129.3	86.4	116.1	106.9	92.2	102.4	107.4	87.4	101.2	107.4	87.4	101.2
06	WOOD, PLASTICS & COMPOSITES	115.8	93.9	103.6	121.3	82.0	99.3	101.9	91.9	96.3	135.3	93.6	112.0	135.3	88.0	108.8	135.3	88.0	108.8
07	THERMAL & MOISTURE PROTECTION	118.4	97.0	109.6	132.2	87.5	113.9	123.2	86.0	107.9	107.4	96.4	102.9	107.4	89.9	100.2	107.7	89.9	100.4
08	OPENINGS	93.0	90.1	92.3	93.6	77.9	89.9	91.9	80.7	89.3	93.2	80.6	90.2	93.6	76.3	89.6	93.6	76.3	89.6
0920	Plaster & Gypsum Board	142.1	93.2	109.0	117.2	81.0	92.8	130.5	91.6	104.2	156.8	93.2	113.8	126.4	87.4	100.0	128.7	87.4	100.8
0950, 0980	Ceilings & Acoustic Treatment	126.1	93.2	104.4	96.5	81.0	86.3	125.4	91.6	103.2	95.7	93.2	94.0	95.7	87.4	90.2	95.7	87.4	90.2
0960	Flooring	125.8	99.5	118.3	123.0	92.6	114.3	120.0	65.5	104.4	120.5	99.5	114.5	120.5	99.5	114.5	120.5	99.5	114.5
0970, 0990	Wall Finishes & Painting/Coating	127.8	106.4	114.9	118.2	81.2	95.9	123.9	92.9	105.2	118.3	106.4	111.1	118.3	91.7	102.2	118.3	91.7	102.2
09	FINISHES	122.2	96.8	108.2	111.7	84.0	96.4	122.7	87.2	103.1	115.1	96.5	104.9	111.0	90.8	99.9	111.3	90.8	100.0
COVERS	DIVS. 10 - 14, 25, 28, 41, 43, 44, 46	131.1	87.3	122.3	131.1	91.9	123.2	131.1	72.4	119.3	131.1	86.5	122.1	131.1	85.2	121.9	131.1	85.2	121.9
21, 22, 23	FIRE SUPPRESSION, PLUMBING & HVAC	101.2	92.4	97.7	102.0	92.3	98.1	100.7	84.8	94.3	102.0	92.3	98.1	102.0	91.0	97.6	98.8	91.0	95.7
26, 27, 3370	ELECTRICAL, COMMUNICATIONS & UTIL.	111.4	86.8	98.4	104.9	85.4	94.6	114.4	92.4	102.8	111.8	86.8	98.6	111.8	72.6	91.1	112.5	72.6	91.4
MF2014	WEIGHTED AVERAGE	117.1	93.6	106.8	114.1	88.2	102.8	120.7	89.0	106.9	111.8	92.9	103.6	112.0	87.4	101.3	111.6	87.4	101.1

		CANADA																	
	DIVISION	SAINT JOHN, NEW BRUNSWICK			SARNIA, ONTARIO			SASKATOON, SASKATCHEWAN			SAULT STE MARIE, ONTARIO			SHERBROOKE, QUEBEC			SOREL, QUEBEC		
		MAT.	INST.	TOTAL	MAT.	INST.	TOTAL	MAT.	INST.	TOTAL	MAT.	INST.	TOTAL	MAT.	INST.	TOTAL	MAT.	INST.	TOTAL
015433	CONTRACTOR EQUIPMENT		102.2	102.2		102.1	102.1		100.6	100.6		102.1	102.1		102.8	102.8		102.8	102.8
0241, 31 - 34	SITE & INFRASTRUCTURE, DEMOLITION	103.5	98.0	99.6	116.5	100.3	105.0	111.7	98.0	102.0	106.7	99.8	101.8	98.7	99.8	99.5	98.8	99.9	99.6
0310	Concrete Forming & Accessories	124.6	67.0	74.9	123.6	95.3	99.1	107.4	90.9	93.2	112.8	87.9	91.3	130.7	87.8	93.7	130.7	88.0	93.8
0320	Concrete Reinforcing	137.0	62.0	98.8	117.4	83.6	100.2	114.5	84.8	99.4	106.1	82.1	93.9	148.7	78.8	113.2	140.6	78.8	109.2
0330	Cast-in-Place Concrete	129.7	80.1	109.3	141.7	102.1	125.5	135.4	95.7	119.1	127.4	87.2	110.9	126.4	97.8	114.6	127.4	97.8	115.2
03	CONCRETE	125.2	71.4	98.8	128.6	95.4	112.3	119.6	91.4	105.7	113.1	86.8	100.2	126.7	89.8	108.5	125.9	89.8	108.2
04	MASONRY	180.8	73.7	114.0	179.1	99.5	129.5	168.8	96.7	123.8	164.5	94.4	120.8	164.1	87.1	116.1	164.1	87.1	116.1
05	METALS	108.9	81.7	100.5	109.1	90.1	103.2	104.5	86.4	99.0	108.3	90.2	102.7	107.4	87.4	101.2	107.4	87.6	101.3
06	WOOD, PLASTICS & COMPOSITES	125.7	65.4	91.9	122.6	94.7	107.0	100.5	91.9	95.7	109.6	88.3	97.6	135.3	88.0	108.8	135.3	88.0	108.8
07	THERMAL & MOISTURE PROTECTION	110.5	70.9	94.3	113.9	94.7	106.0	107.1	86.0	98.4	112.7	88.4	102.7	107.4	89.9	100.3	107.4	89.9	100.2
08	OPENINGS	87.8	60.6	81.5	94.7	89.4	93.4	88.9	80.7	87.0	86.5	85.6	86.3	93.6	76.3	89.6	93.6	81.1	90.7
0920	Plaster & Gypsum Board	152.7	64.2	92.9	150.6	94.4	112.6	123.5	91.6	101.9	119.6	87.8	98.1	128.7	87.4	100.8	156.8	87.4	109.9
0950, 0980	Ceilings & Acoustic Treatment	111.5	64.2	80.4	100.7	94.4	96.6	117.4	91.6	100.4	96.5	87.8	90.8	95.7	87.4	90.2	95.7	87.4	90.2
0960	Flooring	119.5	72.1	105.9	120.5	106.6	116.5	111.0	65.5	98.0	113.4	100.8	109.8	120.5	99.5	114.5	120.5	99.5	114.5
0970, 0990	Wall Finishes & Painting/Coating	117.6	80.2	95.0	118.3	105.7	110.7	120.7	92.9	103.9	118.3	97.9	106.0	118.3	91.7	102.2	118.3	91.7	102.2
09	FINISHES	117.9	69.0	91.0	116.1	98.9	106.6	114.3	87.2	99.3	108.1	91.5	99.0	111.3	90.8	100.0	115.1	90.8	101.7
COVERS	DIVS. 10 - 14, 25, 28, 41, 43, 44, 46	131.1	66.3	118.0	131.1	75.7	119.9	131.1	74.1	119.6	131.1	96.4	124.1	131.1	85.2	121.9	131.1	85.2	121.9
21, 22, 23	FIRE SUPPRESSION, PLUMBING & HVAC	102.0	79.8	93.0	102.0	107.8	104.3	100.5	84.8	94.2	102.0	94.8	99.1	102.3	91.0	97.8	102.0	91.0	97.6
26, 27, 3370	ELECTRICAL, COMMUNICATIONS & UTIL.	120.3	88.2	103.3	114.2	92.7	102.8	116.2	92.4	103.6	113.0	90.8	101.3	111.8	72.6	91.1	111.8	72.6	91.1
MF2014	WEIGHTED AVERAGE	114.0	77.9	98.3	114.8	97.5	107.3	111.1	89.1	101.5	110.3	92.1	102.3	112.4	87.4	101.5	112.5	87.6	101.7

		CANADA																	
	DIVISION	ST CATHARINES, ONTARIO			ST JEROME, QUEBEC			ST JOHNS, NEWFOUNDLAND			SUDBURY, ONTARIO			SUMMERSIDE, PRINCE EDWARD ISLAND			SYDNEY, NOVA SCOTIA		
		MAT.	INST.	TOTAL	MAT.	INST.	TOTAL	MAT.	INST.	TOTAL	MAT.	INST.	TOTAL	MAT.	INST.	TOTAL	MAT.	INST.	TOTAL
015433	CONTRACTOR EQUIPMENT		102.1	102.1		102.8	102.8		103.3	103.3		102.1	102.1		101.8	101.8		101.9	101.9
0241, 31 - 34	SITE & INFRASTRUCTURE, DEMOLITION	100.0	101.4	101.0	98.2	99.8	99.4	123.4	99.9	106.7	100.1	100.9	100.7	128.4	95.7	105.2	114.3	98.2	102.8
0310	Concrete Forming & Accessories	110.9	94.2	96.5	130.7	87.8	93.7	123.9	84.8	90.1	107.0	89.0	91.5	112.0	58.3	65.7	111.7	71.6	77.1
0320	Concrete Reinforcing	101.7	92.0	96.7	148.7	78.8	113.2	158.9	75.5	116.4	102.5	91.3	96.8	168.1	46.8	106.4	170.3	46.7	107.4
0330	Cast-in-Place Concrete	123.2	100.6	113.9	122.3	97.8	112.2	167.5	96.4	138.3	124.2	97.0	113.0	142.3	61.2	109.0	115.6	72.2	97.8
03	CONCRETE	108.0	95.9	102.1	124.8	89.8	107.6	152.6	87.2	120.5	108.4	92.2	100.4	170.1	57.9	115.0	144.7	67.9	106.9
04	MASONRY	150.3	100.7	119.4	163.7	87.1	116.0	187.7	92.4	128.3	150.4	96.6	116.8	228.1	62.6	125.0	226.2	71.9	130.1
05	METALS	107.7	92.2	102.9	107.4	87.4	101.2	136.3	84.3	120.3	107.7	91.5	102.7	123.6	67.6	106.4	123.6	74.2	108.4
06	WOOD, PLASTICS & COMPOSITES	107.4	93.9	99.8	135.3	88.0	108.8	114.7	82.8	96.8	103.6	88.3	95.0	116.4	57.8	83.6	116.1	71.1	90.9
07	THERMAL & MOISTURE PROTECTION	108.6	96.5	103.6	107.4	89.9	100.2	127.1	91.4	112.5	108.0	91.7	101.3	136.1	61.1	105.4	136.8	69.9	109.4
08	OPENINGS	84.1	90.2	85.5	93.6	76.3	89.6	93.7	74.5	89.2	84.8	85.6	85.0	108.0	50.0	94.5	95.3	64.9	88.2
0920	Plaster & Gypsum Board	108.5	93.6	98.4	126.4	87.4	100.0	143.9	82.1	102.1	112.9	87.8	95.9	143.1	56.3	84.4	142.7	70.0	93.6
0950, 0980	Ceilings & Acoustic Treatment	95.3	93.6	94.1	95.7	87.4	90.2	120.6	82.1	95.3	90.3	87.8	88.6	109.6	56.3	74.6	109.6	70.0	83.6
0960	Flooring	109.3	98.9	106.3	120.5	99.5	114.5	116.0	59.3	99.8	107.3	100.8	105.5	116.2	62.9	100.9	116.1	67.2	102.1
0970, 0990	Wall Finishes & Painting/Coating	115.3	116.0	115.7	118.3	91.7	102.2	120.4	95.9	105.6	115.3	93.6	102.2	126.4	42.9	76.0	126.4	64.1	88.8
09	FINISHES	103.8	97.8	100.4	111.0	90.8	99.9	122.0	81.4	99.6	102.7	91.6	96.5	119.2	57.8	85.4	118.2	70.3	91.8
COVERS	DIVS. 10 - 14, 25, 28, 41, 43, 44, 46	131.1	75.2	119.8	131.1	85.2	121.9	131.1	72.6	119.3	131.1	97.1	124.3	131.1	64.8	117.7	131.1	66.6	118.1
21, 22, 23	FIRE SUPPRESSION, PLUMBING & HVAC	100.4	87.8	95.3	102.0	91.0	97.6	100.8	82.1	93.3	101.9	87.8	96.2	102.1	63.8	86.7	102.1	84.4	95.0
26, 27, 3370	ELECTRICAL, COMMUNICATIONS & UTIL.	112.9	98.8	105.5	112.4	72.6	91.3	114.4	83.2	97.9	110.3	99.2	104.4	115.4	52.3	82.0	116.3	64.4	88.9
MF2014	WEIGHTED AVERAGE	107.6	94.5	101.9	112.1	87.4	101.3	122.7	85.2	106.4	107.7	93.0	101.3	126.7	62.8	98.8	122.1	74.6	101.4

	DIVISION	CANADA																	
		THUNDER BAY, ONTARIO			TIMMINS, ONTARIO			TORONTO, ONTARIO			TROIS RIVIERES, QUEBEC			TRURO, NOVA SCOTIA			VANCOUVER, BRITISH COLUMBIA		
		MAT.	INST.	TOTAL	MAT.	INST.	TOTAL	MAT.	INST.	TOTAL	MAT.	INST.	TOTAL	MAT.	INST.	TOTAL	MAT.	INST.	TOTAL
015433	CONTRACTOR EQUIPMENT		102.1	102.1		102.1	102.1		104.4	104.4		102.8	102.8		101.9	101.9		112.7	112.7
0241, 31 - 34	SITE & INFRASTRUCTURE, DEMOLITION	104.8	101.3	102.3	118.0	99.8	105.1	133.8	105.6	113.7	114.8	99.9	104.2	102.8	98.2	99.5	118.2	107.6	110.6
0310	Concrete Forming & Accessories	118.3	92.2	95.8	124.9	87.0	92.2	124.3	100.8	104.0	155.5	88.0	97.3	97.1	71.6	75.1	121.1	91.0	95.1
0320	Concrete Reinforcing	90.9	90.8	90.9	165.4	81.8	122.9	146.0	94.4	119.7	170.3	78.8	123.8	140.6	46.7	92.8	134.1	79.2	106.2
0330	Cast-in-Place Concrete	135.6	99.7	120.8	153.6	87.1	126.3	136.8	111.8	126.5	119.7	97.8	110.7	159.5	72.2	123.7	143.5	97.9	124.8
03	CONCRETE	115.8	94.5	105.3	141.8	86.3	114.5	129.7	103.1	116.7	147.4	89.8	119.1	140.2	67.9	104.7	137.1	91.4	114.6
04	MASONRY	151.0	100.6	119.6	168.0	89.9	119.3	182.0	110.9	137.7	230.7	87.1	141.2	163.6	71.9	106.5	166.7	91.4	119.7
05	METALS	107.5	91.1	102.5	109.1	89.2	103.0	132.5	93.9	120.7	122.7	87.6	111.9	108.2	74.2	97.7	139.3	90.3	124.2
06	WOOD, PLASTICS & COMPOSITES	116.7	90.9	102.2	123.7	87.6	103.5	115.1	98.9	106.0	173.0	88.0	125.4	92.9	71.1	80.7	104.8	91.5	97.4
07	THERMAL & MOISTURE PROTECTION	108.8	93.8	102.7	113.8	85.6	102.2	123.9	102.8	115.2	135.2	89.9	116.6	108.6	69.9	92.8	125.2	86.9	109.5
08	OPENINGS	83.3	88.0	84.4	92.4	83.7	90.4	90.5	96.8	92.0	105.5	81.1	99.8	86.1	64.9	81.1	88.9	86.9	88.4
0920	Plaster & Gypsum Board	139.8	90.5	106.5	130.1	87.1	101.1	133.0	98.8	109.9	173.4	87.4	115.3	130.7	70.0	89.7	129.7	90.6	103.3
0950, 0980	Ceilings & Acoustic Treatment	90.3	90.5	90.4	96.5	87.1	90.3	114.9	98.8	104.3	108.8	87.4	94.7	96.5	70.0	79.1	132.2	90.6	104.8
0960	Flooring	113.7	105.9	111.4	120.5	97.5	113.9	117.0	108.4	114.6	138.6	99.5	127.5	102.7	67.2	92.5	124.6	96.0	116.4
0970, 0990	Wall Finishes & Painting/Coating	115.3	94.6	102.8	118.3	91.2	101.9	120.4	106.7	112.1	126.4	91.7	105.4	118.3	64.1	85.6	121.7	95.6	106.0
09	FINISHES	108.5	95.1	101.1	112.4	89.8	99.9	116.9	103.0	109.2	128.7	90.8	107.8	106.6	70.3	86.6	123.8	92.7	106.6
COVERS	DIVS. 10 - 14, 25, 28, 41, 43, 44, 46	131.1	75.2	119.8	131.1	72.7	119.3	131.1	101.2	125.1	131.1	85.2	121.9	131.1	66.6	118.1	131.1	94.9	123.8
21, 22, 23	FIRE SUPPRESSION, PLUMBING & HVAC	100.4	88.0	95.4	102.0	98.1	100.4	101.0	97.3	99.5	102.1	91.0	97.7	102.0	84.4	94.9	101.2	80.5	92.8
26, 27, 3370	ELECTRICAL, COMMUNICATIONS & UTIL.	111.5	97.0	103.8	113.0	90.8	101.3	108.8	101.0	104.7	116.6	72.6	93.4	111.5	64.4	86.6	109.0	83.3	95.4
MF2014	WEIGHTED AVERAGE	108.8	93.4	102.1	115.1	91.1	104.6	118.3	101.3	110.9	124.8	87.6	108.6	112.5	74.6	96.0	119.5	89.1	106.2

	DIVISION	CANADA																	
		VICTORIA, BRITISH COLUMBIA			WHITEHORSE, YUKON			WINDSOR, ONTARIO			WINNIPEG, MANITOBA			YARMOUTH, NOVA SCOTIA			YELLOWKNIFE, NWT		
		MAT.	INST.	TOTAL	MAT.	INST.	TOTAL	MAT.	INST.	TOTAL	MAT.	INST.	TOTAL	MAT.	INST.	TOTAL	MAT.	INST.	TOTAL
015433	CONTRACTOR EQUIPMENT		109.3	109.3		102.1	102.1		102.1	102.1		106.5	106.5		101.9	101.9		101.9	101.9
0241, 31 - 34	SITE & INFRASTRUCTURE, DEMOLITION	123.0	107.1	111.7	142.3	96.7	109.9	95.9	101.3	99.7	118.2	100.9	105.9	118.3	98.2	104.0	154.1	101.0	116.4
0310	Concrete Forming & Accessories	112.7	90.4	93.5	129.4	59.6	69.2	118.3	90.5	94.3	128.5	69.3	77.4	111.7	71.6	77.1	132.7	80.2	87.4
0320	Concrete Reinforcing	112.0	79.1	95.2	160.8	60.0	109.5	99.6	91.9	95.7	132.9	57.9	94.7	170.3	46.7	107.4	151.4	62.7	106.2
0330	Cast-in-Place Concrete	140.3	97.2	122.6	209.3	70.3	152.2	126.2	101.9	116.2	163.9	75.0	127.4	148.3	72.2	117.0	211.2	88.1	160.6
03	CONCRETE	150.1	90.7	120.9	182.4	64.1	124.3	109.6	94.7	102.3	144.4	69.7	107.7	160.1	67.9	114.8	182.0	80.0	131.9
04	MASONRY	171.7	91.3	121.6	260.3	62.8	137.2	150.5	99.9	119.0	176.6	70.8	110.7	228.8	71.9	131.0	260.3	74.8	144.7
05	METALS	107.0	86.3	100.6	135.3	73.6	116.3	107.6	91.7	102.7	141.7	75.0	121.2	123.6	74.2	108.4	137.9	78.7	119.7
06	WOOD, PLASTICS & COMPOSITES	104.4	91.4	97.1	117.6	58.3	84.4	116.7	89.0	101.2	113.6	70.2	89.3	116.1	71.1	90.9	124.0	81.9	100.4
07	THERMAL & MOISTURE PROTECTION	116.9	86.7	104.5	139.1	62.0	107.5	108.6	93.9	102.6	120.8	71.8	100.7	136.8	69.9	109.4	151.5	78.4	121.5
08	OPENINGS	90.5	83.2	88.8	107.6	55.5	95.4	83.1	87.5	84.1	85.5	62.6	80.2	95.3	64.9	88.2	102.2	68.9	94.4
0920	Plaster & Gypsum Board	115.8	90.6	98.8	150.5	56.8	87.2	125.7	88.6	100.6	129.5	68.6	88.3	142.7	70.0	93.6	163.7	81.1	107.9
0950, 0980	Ceilings & Acoustic Treatment	98.1	90.6	93.2	144.6	56.8	86.9	90.3	88.6	89.2	136.2	68.6	91.8	109.6	70.0	83.6	138.9	81.1	100.9
0960	Flooring	116.4	76.2	104.9	141.0	61.3	118.2	113.7	103.1	110.6	117.0	75.6	105.2	116.1	67.2	102.1	135.5	92.5	123.2
0970, 0990	Wall Finishes & Painting/Coating	120.7	95.6	105.5	143.1	56.0	90.5	115.3	95.5	103.3	124.5	56.5	83.4	126.4	64.1	88.8	135.3	80.3	102.1
09	FINISHES	112.1	89.3	99.5	140.9	59.3	95.9	106.3	93.4	99.2	122.3	69.5	93.2	118.2	70.3	91.8	137.5	82.0	106.9
COVERS	DIVS. 10 - 14, 25, 28, 41, 43, 44, 46	131.1	71.5	119.1	131.1	63.9	117.6	131.1	74.7	119.7	131.1	68.7	118.5	131.1	66.6	118.1	131.1	67.1	118.2
21, 22, 23	FIRE SUPPRESSION, PLUMBING & HVAC	102.0	80.5	93.3	102.5	74.8	91.3	100.4	88.0	95.4	101.1	66.9	87.3	102.1	84.4	95.0	102.7	94.0	99.2
26, 27, 3370	ELECTRICAL, COMMUNICATIONS & UTIL.	113.1	82.7	97.0	133.5	62.2	95.8	116.0	99.0	107.0	113.3	68.0	89.4	116.3	64.4	88.9	130.5	85.2	106.5
MF2014	WEIGHTED AVERAGE	115.7	87.2	103.3	135.3	68.5	106.2	108.2	93.4	101.7	121.0	71.8	99.5	124.0	74.6	102.5	135.3	84.2	113.0

Location Factors

Costs shown in RSMeans cost data publications are based on national averages for materials and installation. To adjust these costs to a specific location, simply multiply the base cost by the factor and divide by 100 for that city. The data is arranged alphabetically by state and postal zip code numbers. For a city not listed, use the factor for a nearby city with similar economic characteristics.

STATE/ZIP	CITY	MAT.	INST.	TOTAL
ALABAMA				
350-352	Birmingham	100.6	76.7	90.2
354	Tuscaloosa	99.8	61.2	83.0
355	Jasper	100.2	57.5	81.6
356	Decatur	99.8	57.9	81.5
357-358	Huntsville	99.8	72.1	87.8
359	Gadsden	99.9	60.0	82.5
360-361	Montgomery	99.9	57.1	81.2
362	Anniston	98.6	61.8	82.6
363	Dothan	99.0	49.4	77.4
364	Evergreen	98.6	50.7	77.7
365-366	Mobile	99.9	65.4	84.9
367	Selma	98.7	49.9	77.4
368	Phenix City	99.5	54.6	79.9
369	Butler	98.9	49.4	77.3
ALASKA				
995-996	Anchorage	121.1	115.6	118.7
997	Fairbanks	121.0	116.4	119.0
998	Juneau	121.3	115.6	118.8
999	Ketchikan	132.0	115.6	124.8
ARIZONA				
850,853	Phoenix	99.2	73.8	88.1
851,852	Mesa/Tempe	98.6	72.8	87.4
855	Globe	98.8	70.0	86.3
856-857	Tucson	97.4	72.4	86.5
859	Show Low	98.9	72.5	87.4
860	Flagstaff	101.3	71.8	88.4
863	Prescott	99.2	68.9	86.0
864	Kingman	97.6	72.0	86.4
865	Chambers	97.6	71.5	86.2
ARKANSAS				
716	Pine Bluff	99.4	63.3	83.6
717	Camden	97.1	50.1	76.6
718	Texarkana	98.2	49.2	76.8
719	Hot Springs	96.4	51.8	76.9
720-722	Little Rock	98.5	64.0	83.4
723	West Memphis	96.1	62.4	81.4
724	Jonesboro	96.7	59.1	80.3
725	Batesville	94.6	54.6	77.1
726	Harrison	95.9	50.4	76.0
727	Fayetteville	93.4	52.0	75.4
728	Russellville	94.6	53.7	76.8
729	Fort Smith	97.2	61.6	81.7
CALIFORNIA				
900-902	Los Angeles	99.2	117.5	107.2
903-905	Inglewood	94.8	114.1	103.2
906-908	Long Beach	96.2	114.1	104.0
910-912	Pasadena	96.4	114.6	104.3
913-916	Van Nuys	99.1	114.6	105.9
917-918	Alhambra	98.3	114.8	105.5
919-921	San Diego	100.5	109.6	104.5
922	Palm Springs	97.2	112.0	103.6
923-924	San Bernardino	95.1	111.3	102.1
925	Riverside	99.4	114.6	106.0
926-927	Santa Ana	96.9	111.8	103.4
928	Anaheim	99.4	114.9	106.2
930	Oxnard	99.6	114.4	106.0
931	Santa Barbara	98.9	115.1	106.0
932-933	Bakersfield	100.7	113.6	106.3
934	San Luis Obispo	99.4	112.0	104.9
935	Mojave	96.8	109.8	102.5
936-938	Fresno	100.1	115.3	106.7
939	Salinas	100.2	121.6	109.5
940-941	San Francisco	105.2	145.4	122.7
942,956-958	Sacramento	100.5	119.3	108.7
943	Palo Alto	98.9	134.5	114.4
944	San Mateo	100.9	134.6	115.6
945	Vallejo	100.1	127.2	111.9
946	Oakland	103.1	134.3	116.7
947	Berkeley	102.7	135.1	116.8
948	Richmond	101.8	133.1	115.4
949	San Rafael	103.5	136.7	118.0
950	Santa Cruz	104.5	121.8	112.0

STATE/ZIP	CITY	MAT.	INST.	TOTAL
CALIFORNIA (CONT'D)				
951	San Jose	102.5	136.6	117.4
952	Stockton	100.9	117.2	108.0
953	Modesto	100.8	115.7	107.3
954	Santa Rosa	100.8	134.7	115.6
955	Eureka	102.1	116.4	108.4
959	Marysville	101.3	115.9	107.7
960	Redding	107.8	117.7	112.1
961	Susanville	106.8	118.1	111.7
COLORADO				
800-802	Denver	100.8	81.7	92.5
803	Boulder	97.8	81.3	90.6
804	Golden	99.8	78.9	90.7
805	Fort Collins	101.3	79.0	91.6
806	Greeley	98.9	75.5	88.7
807	Fort Morgan	98.3	79.0	89.9
808-809	Colorado Springs	100.4	84.0	93.2
810	Pueblo	101.1	80.3	92.0
811	Alamosa	102.2	73.7	89.8
812	Salida	101.9	75.3	90.3
813	Durango	102.7	75.7	90.9
814	Montrose	101.4	75.4	90.1
815	Grand Junction	104.8	75.3	91.9
816	Glenwood Springs	102.3	78.1	91.8
CONNECTICUT				
060	New Britain	100.1	120.6	109.1
061	Hartford	101.4	120.8	109.8
062	Willimantic	100.8	120.2	109.2
063	New London	97.1	120.7	107.4
064	Meriden	98.9	120.7	108.4
065	New Haven	101.6	120.7	109.9
066	Bridgeport	101.3	120.8	109.8
067	Waterbury	100.9	120.7	109.5
068	Norwalk	100.8	128.7	113.0
069	Stamford	100.9	128.7	113.0
D.C.				
200-205	Washington	101.1	91.9	97.1
DELAWARE				
197	Newark	98.8	109.2	103.3
198	Wilmington	99.3	109.2	103.6
199	Dover	99.3	109.2	103.6
FLORIDA				
320,322	Jacksonville	98.8	66.4	84.7
321	Daytona Beach	99.0	72.9	87.6
323	Tallahassee	100.8	56.9	81.7
324	Panama City	100.2	57.8	81.7
325	Pensacola	102.7	61.2	84.6
326,344	Gainesville	100.3	66.7	85.7
327-328,347	Orlando	100.5	70.0	87.2
329	Melbourne	101.7	76.3	90.6
330-332,340	Miami	99.8	73.1	88.2
333	Fort Lauderdale	98.6	72.6	87.2
334,349	West Palm Beach	97.4	71.1	85.9
335-336,346	Tampa	100.1	78.4	90.6
337	St. Petersburg	102.6	63.3	85.5
338	Lakeland	99.4	77.8	90.0
339,341	Fort Myers	98.6	72.6	87.3
342	Sarasota	100.8	74.3	89.2
GEORGIA				
300-303,399	Atlanta	97.6	74.4	87.5
304	Statesboro	97.3	55.8	79.2
305	Gainesville	96.0	61.6	81.0
306	Athens	95.5	63.9	81.7
307	Dalton	97.2	59.0	80.5
308-309	Augusta	96.3	63.6	82.1
310-312	Macon	97.2	64.7	83.0
313-314	Savannah	99.2	60.9	82.5
315	Waycross	98.3	59.4	81.4
316	Valdosta	98.5	61.4	82.3
317,398	Albany	98.5	61.3	82.3
318-319	Columbus	98.4	65.4	84.0

Location Factors

STATE/ZIP	CITY	MAT.	INST.	TOTAL
HAWAII				
967	Hilo	114.8	120.2	117.2
968	Honolulu	119.5	120.2	119.8
STATES & POSS.				
969	Guam	136.2	58.0	102.1
IDAHO				
832	Pocatello	101.1	77.4	90.8
833	Twin Falls	102.2	67.3	87.0
834	Idaho Falls	99.8	74.7	88.9
835	Lewiston	108.7	84.9	98.3
836-837	Boise	100.1	78.6	90.7
838	Coeur d'Alene	107.9	82.2	96.7
ILLINOIS				
600-603	North Suburban	99.3	135.1	114.9
604	Joliet	99.2	138.4	116.3
605	South Suburban	99.3	135.1	114.9
606-608	Chicago	99.8	139.7	117.2
609	Kankakee	95.6	131.7	111.3
610-611	Rockford	99.5	124.8	110.5
612	Rock Island	97.2	100.5	98.6
613	La Salle	98.4	124.4	109.7
614	Galesburg	98.2	106.7	101.9
615-616	Peoria	100.5	109.3	104.4
617	Bloomington	97.5	109.4	102.7
618-619	Champaign	101.0	107.6	103.9
620-622	East St. Louis	96.4	108.7	101.8
623	Quincy	98.1	101.0	99.3
624	Effingham	97.3	106.0	101.1
625	Decatur	99.2	104.9	101.7
626-627	Springfield	100.0	106.5	102.8
628	Centralia	95.0	108.5	100.9
629	Carbondale	94.7	107.9	100.5
INDIANA				
460	Anderson	97.0	83.8	91.3
461-462	Indianapolis	99.1	84.8	92.9
463-464	Gary	98.3	110.4	103.6
465-466	South Bend	98.0	84.4	92.1
467-468	Fort Wayne	97.6	79.1	89.5
469	Kokomo	94.8	81.5	89.0
470	Lawrenceburg	94.0	77.7	86.9
471	New Albany	95.1	75.4	86.5
472	Columbus	97.2	80.9	90.1
473	Muncie	97.9	82.3	91.1
474	Bloomington	99.3	81.8	91.6
475	Washington	95.8	81.8	89.7
476-477	Evansville	97.1	86.9	92.6
478	Terre Haute	97.9	86.5	92.9
479	Lafayette	96.9	82.7	90.7
IOWA				
500-503,509	Des Moines	100.5	85.2	93.8
504	Mason City	97.6	65.8	83.8
505	Fort Dodge	97.9	58.2	80.6
506-507	Waterloo	99.5	71.8	87.4
508	Creston	98.2	77.3	89.1
510-511	Sioux City	99.7	72.6	87.9
512	Sibley	98.3	56.4	80.0
513	Spencer	99.9	56.6	81.0
514	Carroll	97.0	71.2	85.8
515	Council Bluffs	100.7	78.1	90.9
516	Shenandoah	97.6	71.2	86.1
520	Dubuque	99.1	80.2	90.9
521	Decorah	98.0	64.2	83.3
522-524	Cedar Rapids	100.0	83.2	92.7
525	Ottumwa	97.9	73.7	87.3
526	Burlington	97.2	75.3	87.6
527-528	Davenport	99.0	93.1	96.4
KANSAS				
660-662	Kansas City	99.5	94.8	97.5
664-666	Topeka	99.5	66.0	84.9
667	Fort Scott	97.9	73.0	87.0
668	Emporia	97.9	69.2	85.4
669	Belleville	99.6	64.9	84.4
670-672	Wichita	100.7	66.0	85.6
673	Independence	100.9	69.2	87.1
674	Salina	101.5	67.6	86.7
675	Hutchinson	96.5	65.0	82.8
676	Hays	100.5	65.4	85.2
677	Colby	101.1	65.0	85.4

STATE/ZIP	CITY	MAT.	INST.	TOTAL
KANSAS (CONT'D)				
678	Dodge City	102.7	66.0	86.7
679	Liberal	100.4	65.5	85.2
KENTUCKY				
400-402	Louisville	96.6	83.3	90.8
403-405	Lexington	96.6	81.4	90.0
406	Frankfort	98.1	80.7	90.5
407-409	Corbin	94.0	71.8	84.3
410	Covington	95.9	91.4	93.9
411-412	Ashland	94.6	99.3	96.6
413-414	Campton	95.8	78.6	88.3
415-416	Pikeville	96.9	88.6	93.3
417-418	Hazard	95.2	78.7	88.0
420	Paducah	93.9	84.5	89.8
421-422	Bowling Green	96.2	83.9	90.8
423	Owensboro	96.3	88.9	93.1
424	Henderson	93.7	86.1	90.4
425-426	Somerset	93.2	79.9	87.4
427	Elizabethtown	92.8	82.0	88.1
LOUISIANA				
700-701	New Orleans	101.4	69.5	87.5
703	Thibodaux	98.2	65.4	83.9
704	Hammond	95.8	54.5	77.8
705	Lafayette	97.9	63.0	82.7
706	Lake Charles	98.1	65.0	83.7
707-708	Baton Rouge	99.1	63.7	83.7
710-711	Shreveport	101.0	59.6	83.0
712	Monroe	99.3	56.5	80.6
713-714	Alexandria	99.4	57.2	81.0
MAINE				
039	Kittery	95.2	84.7	90.6
040-041	Portland	102.0	90.1	96.8
042	Lewiston	99.6	90.1	95.5
043	Augusta	102.1	79.5	92.3
044	Bangor	99.0	89.5	94.9
045	Bath	97.5	81.9	90.7
046	Machias	97.1	85.5	92.0
047	Houlton	97.2	85.6	92.1
048	Rockland	96.3	85.5	91.6
049	Waterville	97.5	79.5	89.7
MARYLAND				
206	Waldorf	97.1	81.7	90.4
207-208	College Park	97.2	83.4	91.2
209	Silver Spring	96.5	83.2	90.7
210-212	Baltimore	100.8	81.9	92.6
214	Annapolis	101.5	82.0	93.0
215	Cumberland	96.8	81.8	90.2
216	Easton	98.3	68.6	85.3
217	Hagerstown	97.6	83.3	91.4
218	Salisbury	98.7	62.1	82.7
219	Elkton	95.9	81.9	89.8
MASSACHUSETTS				
010-011	Springfield	100.2	112.6	105.6
012	Pittsfield	99.7	111.3	104.8
013	Greenfield	97.7	112.3	104.1
014	Fitchburg	96.5	126.5	109.6
015-016	Worcester	100.2	126.4	111.6
017	Framingham	95.7	135.1	112.9
018	Lowell	99.5	134.5	114.8
019	Lawrence	100.3	134.6	115.3
020-022, 024	Boston	101.6	139.5	118.1
023	Brockton	99.9	128.1	112.2
025	Buzzards Bay	94.4	128.1	109.1
026	Hyannis	97.0	128.1	110.6
027	New Bedford	99.1	128.3	111.8
MICHIGAN				
480,483	Royal Oak	94.9	106.9	100.1
481	Ann Arbor	97.1	107.5	101.7
482	Detroit	97.8	109.6	102.9
484-485	Flint	96.8	96.0	96.4
486	Saginaw	96.4	91.8	94.4
487	Bay City	96.6	91.7	94.5
488-489	Lansing	99.3	94.7	97.3
490	Battle Creek	98.3	87.6	93.7
491	Kalamazoo	98.7	85.7	93.0
492	Jackson	96.5	95.1	95.9
493,495	Grand Rapids	99.7	85.1	93.3
494	Muskegon	97.2	84.8	91.8

STATE/ZIP	CITY	MAT.	INST.	TOTAL
MICHIGAN (CONT'D)				
496	Traverse City	96.3	80.1	89.3
497	Gaylord	97.2	83.6	91.3
498-499	Iron mountain	99.3	86.2	93.6
MINNESOTA				
550-551	Saint Paul	101.4	117.9	108.6
553-555	Minneapolis	100.4	120.0	108.9
556-558	Duluth	101.3	107.3	103.9
559	Rochester	99.9	104.7	102.0
560	Mankato	97.2	101.3	99.0
561	Windom	95.9	92.3	94.3
562	Willmar	95.5	101.3	98.0
563	St. Cloud	96.9	116.0	105.2
564	Brainerd	96.9	100.5	98.5
565	Detroit Lakes	98.9	95.1	97.3
566	Bemidji	98.2	98.2	98.2
567	Thief River Falls	97.8	92.4	95.4
MISSISSIPPI				
386	Clarksdale	97.3	52.6	77.8
387	Greenville	101.0	61.7	83.9
388	Tupelo	98.8	54.9	79.7
389	Greenwood	98.4	52.2	78.3
390-392	Jackson	100.7	62.2	83.9
393	Meridian	97.6	64.1	83.0
394	Laurel	98.9	55.8	80.1
395	Biloxi	99.2	58.1	81.3
396	Mccomb	97.4	51.7	77.5
397	Columbus	98.8	53.9	79.2
MISSOURI				
630-631	St. Louis	98.7	107.8	102.7
633	Bowling Green	97.3	94.4	96.0
634	Hannibal	96.2	92.4	94.6
635	Kirksville	99.5	89.9	95.3
636	Flat River	98.1	95.0	96.8
637	Cape Girardeau	98.2	92.4	95.7
638	Sikeston	96.7	92.2	94.8
639	Poplar Bluff	96.2	92.0	94.4
640-641	Kansas City	101.4	105.7	103.3
644-645	St. Joseph	100.6	93.5	97.5
646	Chillicothe	97.6	94.5	96.3
647	Harrisonville	97.1	101.5	99.1
648	Joplin	99.5	81.7	91.8
650-651	Jefferson City	99.9	93.7	97.2
652	Columbia	100.2	93.9	97.5
653	Sedalia	100.2	97.3	98.9
654-655	Rolla	97.9	94.6	96.5
656-658	Springfield	101.7	83.7	93.8
MONTANA				
590-591	Billings	102.7	77.1	91.5
592	Wolf Point	102.3	74.2	90.0
593	Miles City	100.1	74.0	88.7
594	Great Falls	104.1	76.2	91.9
595	Havre	101.2	73.1	89.0
596	Helena	102.4	73.8	89.9
597	Butte	102.6	77.3	91.6
598	Missoula	99.9	75.5	89.3
599	Kalispell	99.0	75.3	88.7
NEBRASKA				
680-681	Omaha	100.2	78.6	90.8
683-685	Lincoln	100.5	74.9	89.4
686	Columbus	98.4	77.1	89.1
687	Norfolk	100.0	76.7	89.8
688	Grand Island	100.0	77.1	90.0
689	Hastings	99.4	79.8	90.9
690	Mccook	98.7	71.1	86.7
691	North Platte	99.0	78.3	90.0
692	Valentine	101.2	72.5	88.7
693	Alliance	101.0	70.5	87.7
NEVADA				
889-891	Las Vegas	100.9	108.7	104.3
893	Ely	99.4	102.2	100.6
894-895	Reno	99.6	93.3	96.8
897	Carson City	99.7	93.5	97.0
898	Elko	98.3	84.9	92.4
NEW HAMPSHIRE				
030	Nashua	100.5	93.8	97.6
031	Manchester	101.4	93.8	98.1

STATE/ZIP	CITY	MAT.	INST.	TOTAL
NEW HAMPSHIRE (CONT'D)				
032-033	Concord	101.1	92.5	97.4
034	Keene	97.2	73.3	86.7
035	Littleton	97.1	76.3	88.0
036	Charleston	96.7	69.5	84.8
037	Claremont	95.9	69.5	84.4
038	Portsmouth	97.9	94.3	96.4
NEW JERSEY				
070-071	Newark	101.2	128.4	113.0
072	Elizabeth	98.8	127.6	111.3
073	Jersey City	97.9	127.9	111.0
074-075	Paterson	99.6	127.8	111.9
076	Hackensack	97.6	128.1	110.9
077	Long Branch	97.1	126.3	109.8
078	Dover	97.9	127.8	110.9
079	Summit	97.9	127.6	110.8
080,083	Vineland	97.5	124.7	109.4
081	Camden	99.5	124.8	110.5
082,084	Atlantic City	98.2	124.8	109.8
085-086	Trenton	100.9	125.1	111.5
087	Point Pleasant	99.2	125.9	110.8
088-089	New Brunswick	99.7	127.6	111.9
NEW MEXICO				
870-872	Albuquerque	99.6	72.6	87.8
873	Gallup	99.5	72.6	87.8
874	Farmington	100.1	72.6	88.1
875	Santa Fe	100.8	72.6	88.5
877	Las Vegas	98.0	72.6	86.9
878	Socorro	97.6	72.6	86.7
879	Truth/Consequences	97.5	69.4	85.2
880	Las Cruces	96.1	69.4	84.4
881	Clovis	98.3	72.5	87.1
882	Roswell	100.0	72.6	88.0
883	Carrizozo	100.3	72.6	88.2
884	Tucumcari	99.0	72.5	87.4
NEW YORK				
100-102	New York	103.4	168.7	131.8
103	Staten Island	99.2	168.4	129.4
104	Bronx	97.7	166.1	127.5
105	Mount Vernon	97.5	141.4	116.6
106	White Plains	97.6	141.4	116.7
107	Yonkers	102.1	143.7	120.2
108	New Rochelle	97.9	141.3	116.8
109	Suffern	97.6	129.0	111.3
110	Queens	100.9	166.8	129.6
111	Long Island City	102.4	166.8	130.5
112	Brooklyn	102.8	168.5	131.4
113	Flushing	102.5	166.8	130.5
114	Jamaica	100.8	166.8	129.6
115,117,118	Hicksville	100.9	153.3	123.7
116	Far Rockaway	102.6	166.8	130.6
119	Riverhead	101.6	148.2	121.9
120-122	Albany	100.1	104.7	102.1
123	Schenectady	99.3	104.7	101.7
124	Kingston	102.6	118.5	109.6
125-126	Poughkeepsie	101.8	136.7	117.0
127	Monticello	101.1	115.1	107.2
128	Glens Falls	94.0	98.6	96.0
129	Plattsburgh	99.1	96.7	98.0
130-132	Syracuse	98.9	98.0	98.5
133-135	Utica	97.1	97.1	97.1
136	Watertown	98.5	96.3	97.5
137-139	Binghamton	98.5	101.4	99.8
140-142	Buffalo	100.3	106.2	102.9
143	Niagara Falls	97.8	106.4	101.5
144-146	Rochester	101.1	99.1	100.2
147	Jamestown	96.8	92.5	94.9
148-149	Elmira	96.7	97.4	97.0
NORTH CAROLINA				
270,272-274	Greensboro	101.3	55.7	81.4
271	Winston-Salem	101.0	56.5	81.6
275-276	Raleigh	100.5	54.1	80.3
277	Durham	102.9	55.3	82.1
278	Rocky Mount	98.5	52.3	78.3
279	Elizabeth City	99.3	51.6	78.5
280	Gastonia	101.6	55.1	81.3
281-282	Charlotte	101.9	56.8	82.2
283	Fayetteville	104.5	58.1	84.2
284	Wilmington	100.3	55.9	81.0
285	Kinston	98.4	52.7	78.5

Location Factors

STATE/ZIP	CITY	MAT.	INST.	TOTAL
NORTH CAROLINA (CONT'D)				
286	Hickory	98.8	53.9	79.2
287-288	Asheville	100.8	55.3	80.9
289	Murphy	99.6	48.3	77.2
NORTH DAKOTA				
580-581	Fargo	102.7	67.6	87.4
582	Grand Forks	102.7	51.6	80.4
583	Devils Lake	102.0	58.6	83.1
584	Jamestown	102.1	45.2	77.3
585	Bismarck	103.1	66.2	87.0
586	Dickinson	102.8	60.9	84.6
587	Minot	102.7	71.3	89.0
588	Williston	101.2	60.9	83.7
OHIO				
430-432	Columbus	98.6	89.2	94.5
433	Marion	95.1	85.3	90.8
434-436	Toledo	98.7	98.1	98.4
437-438	Zanesville	95.6	86.5	91.6
439	Steubenville	96.8	94.9	96.0
440	Lorain	98.9	93.0	96.3
441	Cleveland	99.0	100.3	99.6
442-443	Akron	99.9	94.2	97.4
444-445	Youngstown	99.2	89.0	94.7
446-447	Canton	99.3	85.6	93.3
448-449	Mansfield	96.7	87.3	92.6
450	Hamilton	97.9	85.1	92.3
451-452	Cincinnati	98.3	84.5	92.2
453-454	Dayton	98.0	84.4	92.1
455	Springfield	98.0	84.7	92.2
456	Chillicothe	96.8	93.0	95.1
457	Athens	99.6	80.4	91.2
458	Lima	100.0	86.8	94.3
OKLAHOMA				
730-731	Oklahoma City	100.3	65.7	85.2
734	Ardmore	98.3	61.9	82.4
735	Lawton	100.7	62.8	84.2
736	Clinton	99.6	63.1	83.7
737	Enid	100.2	60.5	82.9
738	Woodward	98.3	63.4	83.1
739	Guymon	99.4	59.7	82.1
740-741	Tulsa	98.1	61.6	82.2
743	Miami	94.8	67.5	82.9
744	Muskogee	97.4	55.7	79.2
745	Mcalester	94.5	60.4	79.7
746	Ponca City	95.1	62.1	80.7
747	Durant	95.1	63.5	81.3
748	Shawnee	96.5	61.8	81.4
749	Poteau	94.3	61.4	80.0
OREGON				
970-972	Portland	98.8	100.4	99.5
973	Salem	100.5	99.4	100.0
974	Eugene	98.6	99.3	98.9
975	Medford	100.0	97.1	98.8
976	Klamath Falls	99.8	97.1	98.6
977	Bend	98.9	99.4	99.1
978	Pendleton	94.8	101.4	97.7
979	Vale	92.9	98.1	95.2
PENNSYLVANIA				
150-152	Pittsburgh	99.9	104.6	102.0
153	Washington	97.0	103.2	99.7
154	Uniontown	97.3	101.7	99.2
155	Bedford	98.2	91.3	95.2
156	Greensburg	98.2	100.9	99.4
157	Indiana	97.1	99.3	98.0
158	Dubois	98.5	96.8	97.8
159	Johnstown	98.1	97.2	97.7
160	Butler	92.5	102.0	96.6
161	New Castle	92.5	97.9	94.8
162	Kittanning	92.9	103.3	97.4
163	Oil City	92.4	96.9	94.4
164-165	Erie	94.5	95.8	95.1
166	Altoona	94.6	91.9	93.4
167	Bradford	95.7	96.4	96.0
168	State College	95.3	92.8	94.2
169	Wellsboro	96.3	92.6	94.7
170-171	Harrisburg	99.8	95.6	98.0
172	Chambersburg	96.2	88.0	92.6
173-174	York	96.7	95.9	96.3
175-176	Lancaster	95.1	89.0	92.4

STATE/ZIP	CITY	MAT.	INST.	TOTAL
PENNSYLVANIA (CONT'D)				
177	Williamsport	93.8	83.0	89.1
178	Sunbury	95.8	95.8	95.8
179	Pottsville	94.9	98.5	96.5
180	Lehigh Valley	95.8	114.0	103.7
181	Allentown	97.9	108.6	102.5
182	Hazleton	95.4	98.3	96.6
183	Stroudsburg	95.3	101.4	97.9
184-185	Scranton	98.6	99.2	98.9
186-187	Wilkes-Barre	95.1	98.7	96.6
188	Montrose	94.7	96.9	95.7
189	Doylestown	94.9	120.6	106.1
190-191	Philadelphia	99.7	133.6	114.5
193	Westchester	95.9	121.5	107.1
194	Norristown	95.0	129.5	110.0
195-196	Reading	97.1	102.1	99.3
PUERTO RICO				
009	San Juan	122.7	24.4	79.8
RHODE ISLAND				
028	Newport	98.4	117.5	106.7
029	Providence	100.6	117.5	108.0
SOUTH CAROLINA				
290-292	Columbia	99.3	56.5	80.6
293	Spartanburg	98.6	56.7	80.3
294	Charleston	99.9	65.2	84.8
295	Florence	98.2	56.9	80.2
296	Greenville	98.3	55.8	79.8
297	Rock Hill	97.8	50.3	77.1
298	Aiken	98.6	69.1	85.7
299	Beaufort	99.3	44.9	75.6
SOUTH DAKOTA				
570-571	Sioux Falls	100.0	57.7	81.5
572	Watertown	99.0	49.1	77.2
573	Mitchell	97.8	48.4	76.3
574	Aberdeen	100.5	49.6	78.3
575	Pierre	101.7	58.4	82.8
576	Mobridge	98.5	48.8	76.8
577	Rapid City	100.4	59.5	82.6
TENNESSEE				
370-372	Nashville	98.4	74.1	87.8
373-374	Chattanooga	100.0	66.2	85.2
375,380-381	Memphis	99.0	71.3	86.9
376	Johnson City	99.8	56.5	80.9
377-379	Knoxville	96.6	66.9	83.7
382	Mckenzie	98.2	59.2	81.2
383	Jackson	99.9	61.2	83.0
384	Columbia	96.9	66.1	83.5
385	Cookeville	98.1	59.1	81.1
TEXAS				
750	Mckinney	98.7	63.1	83.2
751	Waxahackie	98.6	65.9	84.4
752-753	Dallas	99.5	67.4	85.5
754	Greenville	98.8	64.0	83.6
755	Texarkana	98.0	58.4	80.7
756	Longview	98.5	58.7	81.2
757	Tyler	98.5	65.0	83.9
758	Palestine	95.0	64.4	81.7
759	Lufkin	95.6	67.8	83.5
760-761	Fort Worth	99.3	65.3	84.5
762	Denton	98.9	62.5	83.0
763	Wichita Falls	97.1	64.4	82.8
764	Eastland	95.9	63.2	81.6
765	Temple	94.7	61.7	80.3
766-767	Waco	96.5	64.1	82.4
768	Brownwood	98.9	59.4	81.7
769	San Angelo	98.5	59.6	81.5
770-772	Houston	99.4	70.5	86.8
773	Huntsville	98.0	66.7	84.4
774	Wharton	98.9	67.7	85.3
775	Galveston	97.0	70.0	85.2
776-777	Beaumont	97.4	68.6	84.9
778	Bryan	94.9	68.2	83.3
779	Victoria	99.0	65.6	84.4
780	Laredo	98.3	64.5	83.6
781-782	San Antonio	98.5	65.1	84.0
783-784	Corpus Christi	100.9	63.5	84.6
785	Mc Allen	101.0	59.8	83.0
786-787	Austin	100.0	63.6	84.1

STATE/ZIP	CITY	MAT.	INST.	TOTAL
TEXAS (CONT'D)				
788	Del Rio	100.8	63.2	84.4
789	Giddings	97.7	64.0	83.0
790-791	Amarillo	100.2	63.2	84.1
792	Childress	99.3	64.3	84.0
793-794	Lubbock	101.6	63.6	85.0
795-796	Abilene	99.7	62.2	83.3
797	Midland	101.5	64.2	85.2
798-799,885	El Paso	98.0	61.7	82.2
UTAH				
840-841	Salt Lake City	102.2	68.3	87.4
842,844	Ogden	98.1	70.1	85.9
843	Logan	99.8	70.1	86.9
845	Price	100.2	64.6	84.7
846-847	Provo	100.3	68.2	86.3
VERMONT				
050	White River Jct.	98.5	78.3	89.7
051	Bellows Falls	97.2	93.0	95.3
052	Bennington	97.5	93.5	95.8
053	Brattleboro	97.8	92.8	95.6
054	Burlington	102.4	83.6	94.2
056	Montpelier	99.5	82.8	92.2
057	Rutland	99.7	83.6	92.7
058	St. Johnsbury	98.6	77.8	89.6
059	Guildhall	97.4	77.8	88.8
VIRGINIA				
220-221	Fairfax	100.7	84.0	93.4
222	Arlington	101.9	84.0	94.1
223	Alexandria	101.1	85.5	94.3
224-225	Fredericksburg	99.5	82.7	92.1
226	Winchester	100.0	82.2	92.2
227	Culpeper	99.9	80.5	91.4
228	Harrisonburg	100.2	67.2	85.8
229	Charlottesville	100.5	67.1	86.0
230-232	Richmond	101.3	68.7	87.1
233-235	Norfolk	100.5	68.2	86.4
236	Newport News	100.0	68.2	86.1
237	Portsmouth	99.4	65.3	84.5
238	Petersburg	99.4	68.7	86.0
239	Farmville	98.7	52.4	78.5
240-241	Roanoke	101.5	65.1	85.6
242	Bristol	99.4	54.9	80.0
243	Pulaski	99.0	57.3	80.9
244	Staunton	99.7	61.8	83.2
245	Lynchburg	99.9	65.6	85.0
246	Grundy	99.3	52.5	78.9
WASHINGTON				
980-981,987	Seattle	101.4	104.7	102.8
982	Everett	101.6	99.6	100.7
983-984	Tacoma	101.8	99.7	100.9
985	Olympia	101.2	99.4	100.4
986	Vancouver	102.7	92.1	98.1
988	Wenatchee	101.6	86.6	95.0
989	Yakima	102.0	94.1	98.6
990-992	Spokane	103.4	82.3	94.2
993	Richland	103.1	89.3	97.1
994	Clarkston	101.5	80.2	92.2
WEST VIRGINIA				
247-248	Bluefield	98.3	92.9	96.0
249	Lewisburg	99.8	92.4	96.6
250-253	Charleston	99.9	94.8	97.6
254	Martinsburg	99.3	86.4	93.7
255-257	Huntington	100.7	95.7	98.5
258-259	Beckley	97.9	93.3	95.9
260	Wheeling	101.2	94.3	98.2
261	Parkersburg	100.1	91.8	96.5
262	Buckhannon	99.6	95.5	97.8
263-264	Clarksburg	100.1	95.4	98.1
265	Morgantown	100.2	94.4	97.6
266	Gassaway	99.5	93.9	97.1
267	Romney	99.4	92.2	96.3
268	Petersburg	99.3	91.3	95.8
WISCONSIN				
530,532	Milwaukee	100.0	107.2	103.1
531	Kenosha	99.5	102.2	100.7
534	Racine	99.1	102.4	100.5
535	Beloit	98.9	100.7	99.7
537	Madison	99.1	99.0	99.1

STATE/ZIP	CITY	MAT.	INST.	TOTAL
WISCONSIN (CONT'D)				
538	Lancaster	96.7	92.5	94.8
539	Portage	95.3	96.6	95.9
540	New Richmond	96.8	94.6	95.9
541-543	Green Bay	101.2	93.4	97.8
544	Wausau	96.2	93.5	95.0
545	Rhinelander	99.3	91.2	95.8
546	La Crosse	97.8	94.5	96.3
547	Eau Claire	99.3	95.3	97.5
548	Superior	96.6	96.3	96.5
549	Oshkosh	96.9	91.9	94.7
WYOMING				
820	Cheyenne	101.5	65.2	85.7
821	Yellowstone Nat'l Park	99.3	66.3	84.9
822	Wheatland	100.4	64.3	84.7
823	Rawlins	101.8	66.9	86.6
824	Worland	100.0	65.3	84.9
825	Riverton	101.0	65.4	85.5
826	Casper	102.0	63.9	85.4
827	Newcastle	99.9	66.9	85.5
828	Sheridan	102.6	64.3	85.9
829-831	Rock Springs	103.7	65.3	87.0
CANADIAN FACTORS (reflect Canadian currency)				
ALBERTA				
	Calgary	123.5	95.7	111.4
	Edmonton	123.1	95.7	111.2
	Fort McMurray	123.6	92.8	110.2
	Lethbridge	118.4	92.2	107.0
	Lloydminster	113.4	88.9	102.8
	Medicine Hat	113.6	88.2	102.5
	Red Deer	114.1	88.2	102.8
BRITISH COLUMBIA				
	Kamloops	114.5	90.6	104.1
	Prince George	115.4	89.9	104.3
	Vancouver	119.5	89.1	106.2
	Victoria	115.7	87.2	103.3
MANITOBA				
	Brandon	125.2	75.8	103.7
	Portage la Prairie	113.4	74.2	96.3
	Winnipeg	121.0	71.8	99.5
NEW BRUNSWICK				
	Bathurst	111.5	67.3	92.2
	Dalhousie	111.9	67.5	92.5
	Fredericton	117.4	72.5	97.9
	Moncton	111.8	70.1	93.6
	Newcastle	111.6	67.9	92.5
	St. John	114.0	77.9	98.3
NEWFOUNDLAND				
	Corner Brook	129.6	67.6	102.6
	St Johns	122.7	85.2	106.4
NORTHWEST TERRITORIES				
	Yellowknife	135.3	84.2	113.0
NOVA SCOTIA				
	Bridgewater	112.9	74.6	96.2
	Dartmouth	126.1	74.6	103.7
	Halifax	119.6	80.9	102.7
	New Glasgow	124.1	74.6	102.5
	Sydney	122.1	74.6	101.4
	Truro	112.5	74.6	96.0
	Yarmouth	124.0	74.6	102.5
ONTARIO				
	Barrie	117.7	93.0	106.9
	Brantford	114.9	96.9	107.1
	Cornwall	115.0	93.3	105.5
	Hamilton	117.0	96.7	108.2
	Kingston	116.0	93.4	106.2
	Kitchener	109.6	92.9	102.3
	London	116.4	93.3	106.3
	North Bay	127.3	91.1	111.5
	Oshawa	112.3	94.3	104.4
	Ottawa	117.3	94.0	107.2
	Owen Sound	117.8	91.2	106.2
	Peterborough	114.9	93.1	105.4
	Sarnia	114.8	97.5	107.3
	Sault Ste Marie	110.3	92.1	102.3

STATE/ZIP	CITY	MAT.	INST.	TOTAL
ONTARIO (CONT'D)				
	St. Catharines	107.6	94.5	101.9
	Sudbury	107.7	93.0	101.3
	Thunder Bay	108.8	93.4	102.1
	Timmins	115.1	91.1	104.6
	Toronto	118.3	101.3	110.9
	Windsor	108.2	93.4	101.7
PRINCE EDWARD ISLAND				
	Charlottetown	120.0	62.8	95.1
	Summerside	126.7	62.8	98.8
QUEBEC				
	Cap-de-la-Madeleine	112.4	87.6	101.6
	Charlesbourg	112.4	87.6	101.6
	Chicoutimi	111.4	92.9	103.3
	Gatineau	112.1	87.4	101.3
	Granby	112.3	87.4	101.4
	Hull	112.2	87.4	101.4
	Joliette	112.5	87.6	101.7
	Laval	112.1	87.4	101.3
	Montreal	117.8	93.2	107.1
	Quebec	117.1	93.6	106.8
	Rimouski	111.8	92.9	103.6
	Rouyn-Noranda	112.0	87.4	101.3
	Saint Hyacinthe	111.6	87.4	101.1
	Sherbrooke	112.4	87.4	101.5
	Sorel	112.5	87.6	101.7
	St Jerome	112.1	87.4	101.3
	Trois Rivieres	124.8	87.6	108.6
SASKATCHEWAN				
	Moose Jaw	110.7	69.5	92.7
	Prince Albert	109.9	67.7	91.5
	Regina	120.7	89.0	106.9
	Saskatoon	111.1	89.1	101.5
YUKON				
	Whitehorse	135.3	68.5	106.2

R011105-05 Tips for Accurate Estimating

1. Use pre-printed or columnar forms for orderly sequence of dimensions and locations and for recording telephone quotations.
2. Use only the front side of each paper or form except for certain pre-printed summary forms.
3. Be consistent in listing dimensions: For example, length x width x height. This helps in rechecking to ensure that, the total length of partitions is appropriate for the building area.
4. Use printed (rather than measured) dimensions where given.
5. Add up multiple printed dimensions for a single entry where possible.
6. Measure all other dimensions carefully.
7. Use each set of dimensions to calculate multiple related quantities.
8. Convert foot and inch measurements to decimal feet when listing. Memorize decimal equivalents to .01 parts of a foot (1/8″ equals approximately .01′).
9. Do not "round off" quantities until the final summary.
10. Mark drawings with different colors as items are taken off.
11. Keep similar items together, different items separate.
12. Identify location and drawing numbers to aid in future checking for completeness.
13. Measure or list everything on the drawings or mentioned in the specifications.
14. It may be necessary to list items not called for to make the job complete.
15. Be alert for: Notes on plans such as N.T.S. (not to scale); changes in scale throughout the drawings; reduced size drawings; discrepancies between the specifications and the drawings.
16. Develop a consistent pattern of performing an estimate. For example:
    a. Start the quantity takeoff at the lower floor and move to the next higher floor.
    b. Proceed from the main section of the building to the wings.
    c. Proceed from south to north or vice versa, clockwise or counterclockwise.
    d. Take off floor plan quantities first, elevations next, then detail drawings.
17. List all gross dimensions that can be either used again for different quantities, or used as a rough check of other quantities for verification (exterior perimeter, gross floor area, individual floor areas, etc.).
18. Utilize design symmetry or repetition (repetitive floors, repetitive wings, symmetrical design around a center line, similar room layouts, etc.). Note: Extreme caution is needed here so as not to omit or duplicate an area.
19. Do not convert units until the final total is obtained. For instance, when estimating concrete work, keep all units to the nearest cubic foot, then summarize and convert to cubic yards.
20. When figuring alternatives, it is best to total all items involved in the basic system, then total all items involved in the alternates. Therefore you work with positive numbers in all cases. When adds and deducts are used, it is often confusing whether to add or subtract a portion of an item; especially on a complicated or involved alternate.

R011105-50 Metric Conversion Factors

Description: This table is primarily for converting customary U.S. units in the left hand column to SI metric units in the right hand column. In addition, conversion factors for some commonly encountered Canadian and non-SI metric units are included.

	If You Know		Multiply By		To Find
Length	Inches	x	25.4[a]	=	Millimeters
	Feet	x	0.3048[a]	=	Meters
	Yards	x	0.9144[a]	=	Meters
	Miles (statute)	x	1.609	=	Kilometers
Area	Square inches	x	645.2	=	Square millimeters
	Square feet	x	0.0929	=	Square meters
	Square yards	x	0.8361	=	Square meters
Volume (Capacity)	Cubic inches	x	16,387	=	Cubic millimeters
	Cubic feet	x	0.02832	=	Cubic meters
	Cubic yards	x	0.7646	=	Cubic meters
	Gallons (U.S. liquids)[b]	x	0.003785	=	Cubic meters[c]
	Gallons (Canadian liquid)[b]	x	0.004546	=	Cubic meters[c]
	Ounces (U.S. liquid)[b]	x	29.57	=	Milliliters[c, d]
	Quarts (U.S. liquid)[b]	x	0.9464	=	Liters[c, d]
	Gallons (U.S. liquid)[b]	x	3.785	=	Liters[c, d]
Force	Kilograms force[d]	x	9.807	=	Newtons
	Pounds force	x	4.448	=	Newtons
	Pounds force	x	0.4536	=	Kilograms force[d]
	Kips	x	4448	=	Newtons
	Kips	x	453.6	=	Kilograms force[d]
Pressure, Stress, Strength (Force per unit area)	Kilograms force per square centimeter[d]	x	0.09807	=	Megapascals
	Pounds force per square inch (psi)	x	0.006895	=	Megapascals
	Kips per square inch	x	6.895	=	Megapascals
	Pounds force per square inch (psi)	x	0.07031	=	Kilograms force per square centimeter[d]
	Pounds force per square foot	x	47.88	=	Pascals
	Pounds force per square foot	x	4.882	=	Kilograms force per square meter[d]
Bending Moment Or Torque	Inch-pounds force	x	0.01152	=	Meter-kilograms force[d]
	Inch-pounds force	x	0.1130	=	Newton-meters
	Foot-pounds force	x	0.1383	=	Meter-kilograms force[d]
	Foot-pounds force	x	1.356	=	Newton-meters
	Meter-kilograms force[d]	x	9.807	=	Newton-meters
Mass	Ounces (avoirdupois)	x	28.35	=	Grams
	Pounds (avoirdupois)	x	0.4536	=	Kilograms
	Tons (metric)	x	1000	=	Kilograms
	Tons, short (2000 pounds)	x	907.2	=	Kilograms
	Tons, short (2000 pounds)	x	0.9072	=	Megagrams[e]
Mass per Unit Volume	Pounds mass per cubic foot	x	16.02	=	Kilograms per cubic meter
	Pounds mass per cubic yard	x	0.5933	=	Kilograms per cubic meter
	Pounds mass per gallon (U.S. liquid)[b]	x	119.8	=	Kilograms per cubic meter
	Pounds mass per gallon (Canadian liquid)[b]	x	99.78	=	Kilograms per cubic meter
Temperature	Degrees Fahrenheit		(F-32)/1.8	=	Degrees Celsius
	Degrees Fahrenheit		(F+459.67)/1.8	=	Degrees Kelvin
	Degrees Celsius		C+273.15	=	Degrees Kelvin

[a]The factor given is exact
[b]One U.S. gallon = 0.8327 Canadian gallon
[c]1 liter = 1000 milliliters = 1000 cubic centimeters
1 cubic decimeter = 0.001 cubic meter
[d]Metric but not SI unit
[e]Called "tonne" in England and "metric ton" in other metric countries

R011105-60 Weights and Measures

Measures of Length
1 Mile = 1760 Yards = 5280 Feet
1 Yard = 3 Feet = 36 inches
1 Foot = 12 Inches
1 Mil = 0.001 Inch
1 Fathom = 2 Yards = 6 Feet
1 Rod = 5.5 Yards = 16.5 Feet
1 Hand = 4 Inches
1 Span = 9 Inches
1 Micro-inch = One Millionth Inch or 0.000001 Inch
1 Micron = One Millionth Meter + 0.00003937 Inch

Surveyor's Measure
1 Mile = 8 Furlongs = 80 Chains
1 Furlong = 10 Chains = 220 Yards
1 Chain = 4 Rods = 22 Yards = 66 Feet = 100 Links
1 Link = 7.92 Inches

Square Measure
1 Square Mile = 640 Acres = 6400 Square Chains
1 Acre = 10 Square Chains = 4840 Square Yards = 43,560 Sq. Ft.
1 Square Chain = 16 Square Rods = 484 Square Yards = 4356 Sq. Ft.
1 Square Rod = 30.25 Square Yards = 272.25 Square Feet = 625 Square Lines
1 Square Yard = 9 Square Feet
1 Square Foot = 144 Square Inches
An Acre equals a Square 208.7 Feet per Side

Cubic Measure
1 Cubic Yard = 27 Cubic Feet
1 Cubic Foot = 1728 Cubic Inches
1 Cord of Wood = 4 x 4 x 8 Feet = 128 Cubic Feet
1 Perch of Masonry = 16½ x 1½ x 1 Foot = 24.75 Cubic Feet

Avoirdupois or Commercial Weight
1 Gross or Long Ton = 2240 Pounds
1 Net or Short ton = 2000 Pounds
1 Pound = 16 Ounces = 7000 Grains
1 Ounce = 16 Drachms = 437.5 Grains
1 Stone = 14 Pounds

Shipping Measure
For Measuring Internal Capacity of a Vessel:
1 Register Ton = 100 Cubic Feet

For Measurement of Cargo:
Approximately 40 Cubic Feet of Merchandise is considered a Shipping Ton, unless that bulk would weigh more than 2000 Pounds, in which case Freight Charge may be based upon weight.

40 Cubic Feet = 32.143 U.S. Bushels = 31.16 Imp. Bushels

Liquid Measure
1 Imperial Gallon = 1.2009 U.S. Gallon = 277.42 Cu. In.
1 Cubic Foot = 7.48 U.S. Gallons

R011110-10 Architectural Fees

Tabulated below are typical percentage fees by project size, for good professional architectural service. Fees may vary from those listed depending upon degree of design difficulty and economic conditions in any particular area.

Rates can be interpolated horizontally and vertically. Various portions of the same project requiring different rates should be adjusted proportionately. For alterations, add 50% to the fee for the first $500,000 of project cost and add 25% to the fee for project cost over $500,000.

Architectural fees tabulated below include Structural, Mechanical and Electrical Engineering Fees. They do not include the fees for special consultants such as kitchen planning, security, acoustical, interior design, etc

Civil Engineering fees are included in the Architectural fee for project sites requiring minimal design such as city sites. However, separate Civil Engineering fees must be added when utility connections require design, drainage calculations are needed, stepped foundations are required, or provisions are required to protect adjacent wetlands.

	Total Project Size in Thousands of Dollars						
Building Types	100	250	500	1,000	5,000	10,000	50,000
Factories, garages, warehouses, repetitive housing	9.0%	8.0%	7.0%	6.2%	5.3%	4.9%	4.5%
Apartments, banks, schools, libraries, offices, municipal buildings	12.2	12.3	9.2	8.0	7.0	6.6	6.2
Churches, hospitals, homes, laboratories, museums, research	15.0	13.6	12.7	11.9	9.5	8.8	8.0
Memorials, monumental work, decorative furnishings	—	16.0	14.5	13.1	10.0	9.0	8.3

R011110-30 Engineering Fees

Typical **Structural Engineering Fees** based on type of construction and total project size. These fees are included in Architectural Fees.

Type of Construction	Total Project Size (in thousands of dollars)			
	$500	$500-$1,000	$1,000-$5,000	Over $5000
Industrial buildings, factories & warehouses	Technical payroll times 2.0 to 2.5	1.60%	1.25%	1.00%
Hotels, apartments, offices, dormitories, hospitals, public buildings, food stores		2.00%	1.70%	1.20%
Museums, banks, churches and cathedrals		2.00%	1.75%	1.25%
Thin shells, prestressed concrete, earthquake resistive		2.00%	1.75%	1.50%
Parking ramps, auditoriums, stadiums, convention halls, hangars & boiler houses		2.50%	2.00%	1.75%
Special buildings, major alterations, underpinning & future expansion	▼	Add to above 0.5%	Add to above 0.5%	Add to above 0.5%

For complex reinforced concrete or unusually complicated structures, add 20% to 50%.

Typical **Mechanical and Electrical Engineering Fees** are based on the size of the subcontract. The fee structure for both are shown below. These fees are included in Architectural Fees.

Type of Construction	Subcontract Size							
	$25,000	$50,000	$100,000	$225,000	$350,000	$500,000	$750,000	$1,000,000
Simple structures	6.4%	5.7%	4.8%	4.5%	4.4%	4.3%	4.2%	4.1%
Intermediate structures	8.0	7.3	6.5	5.6	5.1	5.0	4.9	4.8
Complex structures	10.1	9.0	9.0	8.0	7.5	7.5	7.0	7.0

For renovations, add 15% to 25% to applicable fee.

R012157-20 Construction Time Requirements

Table lists average construction time in months for different types, sizes, and values of building projects. Design time runs 25% to 40% of construction time.

Building Type	Size S.F.	Project Value	Construction Duration
Industrial/Warehouse	100,000	$8,000,000	14 Months
	500,000	$32,000,000	19 Months
	1,000,000	$75,000,000	21 Months
Offices/Retail	50,000	$7,000,000	15 Months
	250,000	$28,000,000	23 months
	500,000	$58,000,000	34 months
Institutional/Hospitals/Laboratory	200,000	$45,000,000	31 months
	500,000	$110,000,000	52 months
	750,000	$160,000,000	55 months
	1,000,000	$210,000,000	60 months

R012909-80 Sales Tax by State

State sales tax on materials is tabulated below (5 states have no sales tax). Many states allow local jurisdictions, such as a county or city, to levy additional sales tax.

Some projects may be sales tax exempt, particularly those constructed with public funds.

State	Tax (%)	State	Tax (%)	State	Tax (%)	State	Tax (%)
Alabama	4	Illinois	6.25	Montana	0	Rhode Island	7
Alaska	0	Indiana	7	Nebraska	5.5	South Carolina	6
Arizona	5.6	Iowa	6	Nevada	6.85	South Dakota	4
Arkansas	6	Kansas	6.3	New Hampshire	0	Tennessee	7
California	7.5	Kentucky	6	New Jersey	7	Texas	6.25
Colorado	2.9	Louisiana	4	New Mexico	5.13	Utah	4.7
Connecticut	6.35	Maine	5	New York	4	Vermont	6
Delaware	0	Maryland	6	North Carolina	4.75	Virginia	5
District of Columbia	6	Massachusetts	6.25	North Dakota	5	Washington	6.5
Florida	6	Michigan	6	Ohio	5.5	West Virginia	6
Georgia	4	Minnesota	6.88	Oklahoma	4.5	Wisconsin	5
Hawaii	4	Mississippi	7	Oregon	0	Wyoming	4
Idaho	6	Missouri	4.23	Pennsylvania	6	Average	5.04 %

Sales Tax by Province (Canada)

GST - a value-added tax, which the government imposes on most goods and services provided in or imported into Canada. PST - a retail sales tax, which three of the provinces impose on the price of most goods and some services. QST - a value-added tax, similar to the federal GST, which Quebec imposes. HST - Five provinces have combined their retail sales tax with the federal GST into one harmonized tax.

Province	PST (%)	QST (%)	GST(%)	HST(%)
Alberta	0	0	5	0
British Columbia	7	0	5	0
Manitoba	8	0	5	0
New Brunswick	0	0	0	13
Newfoundland	0	0	0	13
Northwest Territories	0	0	5	0
Nova Scotia	0	0	0	15
Ontario	0	0	0	13
Prince Edward Island	0	0	0	14
Quebec	0	9.975	5	0
Saskatchewan	5	0	5	0
Yukon	0	0	5	0

R012909-85 Unemployment Taxes and Social Security Taxes

State unemployment tax rates vary not only from state to state, but also with the experience rating of the contractor. The federal unemployment tax rate is 6.0% of the first $7,000 of wages. This is reduced by a credit of up to 5.4% for timely payment to the state. The minimum federal unemployment tax is 0.6% after all credits.

Social security (FICA) for 2015 is estimated at time of publication to be 7.65% of wages up to $117,000.

R012909-90 Overtime

One way to improve the completion date of a project or eliminate negative float from a schedule is to compress activity duration times. This can be achieved by increasing the crew size or working overtime with the proposed crew.

To determine the costs of working overtime to compress activity duration times, consider the following examples. Below is an overtime efficiency and cost chart based on a five, six, or seven day week with an eight through twelve hour day. Payroll percentage increases for time and one half and double time are shown for the various working days.

Days per Week	Hours per Day	Production Efficiency					Payroll Cost Factors	
		1st Week	2nd Week	3rd Week	4th Week	Average 4 Weeks	@ 1-1/2 Times	@ 2 Times
5	8	100%	100%	100%	100%	100 %	1.000	1.000
	9	100	100	95	90	96	1.056	1.111
	10	100	95	90	85	93	1.100	1.200
	11	95	90	75	65	81	1.136	1.273
	12	90	85	70	60	76	1.167	1.333
6	8	100	100	95	90	96	1.083	1.167
	9	100	95	90	85	93	1.130	1.259
	10	95	90	85	80	88	1.167	1.333
	11	95	85	70	65	79	1.197	1.394
	12	90	80	65	60	74	1.222	1.444
7	8	100	95	85	75	89	1.143	1.286
	9	95	90	80	70	84	1.183	1.365
	10	90	85	75	65	79	1.214	1.429
	11	85	80	65	60	73	1.240	1.481
	12	85	75	60	55	69	1.262	1.524

R013113-40 Builder's Risk Insurance

Builder's risk insurance is insurance on a building during construction. Premiums are paid by the owner or the contractor. Blasting, collapse and underground insurance would raise total insurance costs above those listed. Floater policy for materials delivered to the job runs $.75 to $1.25 per $100 value. Contractor equipment insurance runs $.50 to $1.50 per $100 value. Insurance for miscellaneous tools to $1,500 value runs from $3.00 to $7.50 per $100 value.

Tabulated below are New England Builder's Risk insurance rates in dollars per $100 value for $1,000 deductible. For $25,000 deductible, rates can be reduced 13% to 34%. On contracts over $1,000,000, rates may be lower than those tabulated. Policies are written annually for the total completed value in place. For "all risk" insurance (excluding flood, earthquake and certain other perils) add $.025 to total rates below.

Coverage	Frame Construction (Class 1)		Brick Construction (Class 4)		Fire Resistive (Class 6)	
	Range	Average	Range	Average	Range	Average
Fire Insurance	$.350 to $ .850	$.600	$.158 to $.189	$.174	$.052 to $.080	$.070
Extended Coverage	.115 to .200	.158	.080 to .105	.101	.081 to .105	.100
Vandalism	.012 to .016	.014	.008 to .011	.011	.008 to .011	.010
Total Annual Rate	$.477 to $1.066	$.772	$.246 to $.305	$.286	$.141 to $.196	$.180

R013113-50 General Contractor's Overhead

There are two distinct types of overhead on a construction project: Project overhead and main office overhead. Project overhead includes those costs at a construction site not directly associated with the installation of construction materials. Examples of project overhead costs include the following:

1. Superintendent
2. Construction office and storage trailers
3. Temporary sanitary facilities
4. Temporary utilities
5. Security fencing
6. Photographs
7. Cleanup
8. Performance and payment bonds

The above project overhead items are also referred to as general requirements and therefore are estimated in Division 1. Division 1 is the first division listed in the CSI MasterFormat but it is usually the last division estimated. The sum of the costs in Divisions 1 through 49 is referred to as the sum of the direct costs.

All construction projects also include indirect costs. The primary components of indirect costs are the contractor's main office overhead and profit. The amount of the main office overhead expense varies depending on the following:

1. Owner's compensation
2. Project managers' and estimators' wages
3. Clerical support wages
4. Office rent and utilities
5. Corporate legal and accounting costs
6. Advertising
7. Automobile expenses
8. Association dues
9. Travel and entertainment expenses

These costs are usually calculated as a percentage of annual sales volume. This percentage can range from 35% for a small contractor doing less than $500,000 to 5% for a large contractor with sales in excess of $100 million.

R013113-60 Workers' Compensation Insurance Rates by Trade

The table below tabulates the national averages for workers' compensation insurance rates by trade and type of building. The average "Insurance Rate" is multiplied by the "% of Building Cost" for each trade. This produces the "Workers' Compensation" cost by % of total labor cost, to be added for each trade by building type to determine the weighted average workers' compensation rate for the building types analyzed.

Trade	Insurance Rate (% Labor Cost)		% of Building Cost			Workers' Compensation		
	Range	Average	Office Bldgs.	Schools & Apts.	Mfg.	Office Bldgs.	Schools & Apts.	Mfg.
Excavation, Grading, etc.	3.4 % to 21.2%	9.7%	4.8%	4.9%	4.5%	0.47%	0.48%	0.44%
Piles & Foundations	6.1 to 28.3	14.3	7.1	5.2	8.7	1.02	0.74	1.24
Concrete	4.2 to 35.8	12.9	5.0	14.8	3.7	0.65	1.91	0.48
Masonry	4.9 to 36.4	13.7	6.9	7.5	1.9	0.95	1.03	0.26
Structural Steel	6.1 to 101.1	31.7	10.7	3.9	17.6	3.39	1.24	5.58
Miscellaneous & Ornamental Metals	4.3 to 31.5	12.2	2.8	4.0	3.6	0.34	0.49	0.44
Carpentry & Millwork	5.0 to 36.8	14.9	3.7	4.0	0.5	0.55	0.60	0.07
Metal or Composition Siding	6.1 to 69.2	18.4	2.3	0.3	4.3	0.42	0.06	0.79
Roofing	6.1 to 100.3	31.7	2.3	2.6	3.1	0.73	0.82	0.98
Doors & Hardware	4.1 to 36.8	11.3	0.9	1.4	0.4	0.10	0.16	0.05
Sash & Glazing	4.8 to 30.7	13.3	3.5	4.0	1.0	0.47	0.53	0.13
Lath & Plaster	3.1 to 36.2	10.9	3.3	6.9	0.8	0.36	0.75	0.09
Tile, Marble & Floors	3 to 24.8	8.9	2.6	3.0	0.5	0.23	0.27	0.04
Acoustical Ceilings	3.8 to 33.8	8.4	2.4	0.2	0.3	0.20	0.02	0.03
Painting	4.5 to 36.1	11.7	1.5	1.6	1.6	0.18	0.19	0.19
Interior Partitions	5.0 to 36.8	14.9	3.9	4.3	4.4	0.58	0.64	0.66
Miscellaneous Items	2.5 to 132.3	13.4	5.2	3.7	9.7	0.70	0.50	1.30
Elevators	1.3 to 12.8	5.3	2.1	1.1	2.2	0.11	0.06	0.12
Sprinklers	2.9 to 18.5	7.3	0.5	—	2.0	0.04	—	0.15
Plumbing	2.2 to 19.3	7.0	4.9	7.2	5.2	0.34	0.50	0.36
Heat., Vent., Air Conditioning	3.8 to 18.1	8.8	13.5	11.0	12.9	1.19	0.97	1.14
Electrical	2.4 to 13.5	5.8	10.1	8.4	11.1	0.59	0.49	0.64
Total	1.3 % to 132.3%	—	100.0%	100.0%	100.0%	13.61%	12.45%	15.18%
		Overall Weighted Average	13.75%					

Workers' Compensation Insurance Rates by States

The table below lists the weighted average Workers' Compensation base rate for each state with a factor comparing this with the national average of 13.5%.

State	Weighted Average	Factor	State	Weighted Average	Factor	State	Weighted Average	Factor
Alabama	20.6%	153	Kentucky	12.5%	93	North Dakota	8.1%	60
Alaska	12.5	93	Louisiana	19.6	145	Ohio	8.1	60
Arizona	13.8	102	Maine	11.4	84	Oklahoma	10.3	76
Arkansas	7.7	57	Maryland	13.5	100	Oregon	11.3	84
California	27.1	201	Massachusetts	11.5	85	Pennsylvania	18.3	136
Colorado	7.3	54	Michigan	13.5	100	Rhode Island	14.7	109
Connecticut	25.9	192	Minnesota	22.8	169	South Carolina	16.9	125
Delaware	14.0	104	Mississippi	14.0	104	South Dakota	14.1	104
District of Columbia	10.4	77	Missouri	13.9	103	Tennessee	12.2	90
Florida	10.8	80	Montana	8.3	61	Texas	9.3	69
Georgia	29.5	219	Nebraska	17.0	126	Utah	8.8	65
Hawaii	8.7	64	Nevada	9.2	68	Vermont	13.3	99
Idaho	10.5	78	New Hampshire	18.8	139	Virginia	8.7	64
Illinois	26.9	199	New Jersey	15.7	116	Washington	9.5	70
Indiana	5.0	37	New Mexico	15.3	113	West Virginia	7.4	55
Iowa	15.5	115	New York	19.0	141	Wisconsin	13.2	98
Kansas	9.2	68	North Carolina	17.3	128	Wyoming	6.5	48
			Weighted Average for U.S. is	13.7% of payroll = 100%				

Rates in the following table are the base or manual costs per $100 of payroll for Workers' Compensation in each state. Rates are usually applied to straight time wages only and not to premium time wages and bonuses.

The weighted average skilled worker rate for 35 trades is 13.5%. For bidding purposes, apply the full value of Workers' Compensation directly to total labor costs, or if labor is 38%, materials 42% and overhead and profit 20% of total cost, carry 38/80 x 13.5% =6.4% of cost (before overhead and profit) into overhead. Rates vary not only from state to state but also with the experience rating of the contractor.

Rates are the most current available at the time of publication.

R013113-80 Performance Bond

This table shows the cost of a performance bond for a construction job scheduled to be completed in 12 months. Add 1% of the premium cost per month for jobs requiring more than 12 months to complete. The rates are "standard" rates offered to contractors that the bonding company considers financially sound and capable of doing the work. Preferred rates are offered by some bonding companies based upon financial strength of the contractor. Actual rates vary from contractor to contractor and from bonding company to bonding company. Contractors should prequalify through a bonding agency before submitting a bid on a contract that requires a bond.

Contract Amount	Building Construction Class B Projects	Highways & Bridges	
		Class A New Construction	Class A-1 Highway Resurfacing
First $ 100,000 bid	$25.00 per M	$15.00 per M	$9.40 per M
Next 400,000 bid	$ 2,500 plus $15.00 per M	$ 1,500 plus $10.00 per M	$ 940 plus $7.20 per M
Next 2,000,000 bid	8,500 plus 10.00 per M	5,500 plus 7.00 per M	3,820 plus 5.00 per M
Next 2,500,000 bid	28,500 plus 7.50 per M	19,500 plus 5.50 per M	15,820 plus 4.50 per M
Next 2,500,000 bid	47,250 plus 7.00 per M	33,250 plus 5.00 per M	28,320 plus 4.50 per M
Over 7,500,000 bid	64,750 plus 6.00 per M	45,750 plus 4.50 per M	39,570 plus 4.00 per M

General Requirements R0151 Temporary Utilities

R015113-65 Temporary Power Equipment

Cost data for the temporary equipment was developed utilizing the following information.

1) Re-usable material-services, transformers, equipment and cords are based on new purchase and prorated to three projects.
2) PVC feeder includes trench and backfill.
3) Connections include disconnects and fuses.
4) Labor units include an allowance for removal.
5) No utility company charges or fees are included.
6) Concrete pads or vaults are not included.
7) Utility company conduits not included.

R015423-10 Steel Tubular Scaffolding

On new construction, tubular scaffolding is efficient up to 60′ high or five stories. Above this it is usually better to use a hung scaffolding if construction permits. Swing scaffolding operations may interfere with tenants. In this case, the tubular is more practical at all heights.

In repairing or cleaning the front of an existing building the cost of tubular scaffolding per S.F. of building front increases as the height increases above the first tier. The first tier cost is relatively high due to leveling and alignment.

The minimum efficient crew for erecting and dismantling is three workers. They can set up and remove 18 frame sections per day up to 5 stories high. For 6 to 12 stories high, a crew of four is most efficient. Use two or more on top and two on the bottom for handing up or hoisting. They can also set up and remove 18 frame sections per day. At 7' horizontal spacing, this will run about 800 S.F. per day of erecting and dismantling. Time for placing and removing planks must be added to the above. A crew of three can place and remove 72 planks per day up to 5 stories. For over 5 stories, a crew of four can place and remove 80 planks per day.

The table below shows the number of pieces required to erect tubular steel scaffolding for 1000 S.F. of building frontage. This area is made up of a scaffolding system that is 12 frames (11 bays) long by 2 frames high.

For jobs under twenty-five frames, add 50% to rental cost. Rental rates will be lower for jobs over three months duration. Large quantities for long periods can reduce rental rates by 20%.

Description of Component	Number of Pieces for 1000 S.F. of Building Front	Unit
5′ Wide Standard Frame, 6′-4″ High	24	Ea.
Leveling Jack & Plate	24	
Cross Brace	44	
Side Arm Bracket, 21″	12	
Guardrail Post	12	
Guardrail, 7′ section	22	
Stairway Section	2	
Stairway Starter Bar	1	
Stairway Inside Handrail	2	
Stairway Outside Handrail	2	
Walk-Thru Frame Guardrail	2	▼

Scaffolding is often used as falsework over 15′ high during construction of cast-in-place concrete beams and slabs. Two foot wide scaffolding is generally used for heavy beam construction. The span between frames depends upon the load to be carried with a maximum span of 5′.

Heavy duty shoring frames with a capacity of 10,000#/leg can be spaced up to 10′ O.C. depending upon form support design and loading.

Scaffolding used as horizontal shoring requires less than half the material required with conventional shoring.

On new construction, erection is done by carpenters.

Rolling towers supporting horizontal shores can reduce labor and speed the job. For maintenance work, catwalks with spans up to 70′ can be supported by the rolling towers.

R015423-20 Pump Staging

Pump staging is generally not available for rent. The table below shows the number of pieces required to erect pump staging for 2400 S.F. of building frontage. This area is made up of a pump jack system that is 3 poles (2 bays) wide by 2 poles high.

Item	Number of Pieces for 2400 S.F. of Building Front	Unit
Aluminum pole section, 24′ long	6	Ea.
Aluminum splice joint, 6′ long	3	
Aluminum foldable brace	3	
Aluminum pump jack	3	
Aluminum support for workbench/back safety rail	3	
Aluminum scaffold plank/workbench, 14″ wide x 24′ long	4	
Safety net, 22′ long	2	
Aluminum plank end safety rail	2	▼

The cost in place for this 2400 S.F. will depend on how many uses are realized during the life of the equipment.

R015433-10 Contractor Equipment

Rental Rates shown elsewhere in the book pertain to late model high quality machines in excellent working condition, rented from equipment dealers. Rental rates from contractors may be substantially lower than the rental rates from equipment dealers depending upon economic conditions; for older, less productive machines, reduce rates by a maximum of 15%. Any overtime must be added to the base rates. For shift work, rates are lower. Usual rule of thumb is 150% of one shift rate for two shifts; 200% for three shifts.

For periods of less than one week, operated equipment is usually more economical to rent than renting bare equipment and hiring an operator.

Costs to move equipment to a job site (mobilization) or from a job site (demobilization) are not included in rental rates, nor in any Equipment costs on any Unit Price line items or crew listings. These costs can be found elsewhere. If a piece of equipment is already at a job site, it is not appropriate to utilize mob/demob costs in an estimate again.

Rental rates vary throughout the country with larger cities generally having lower rates. Lease plans for new equipment are available for periods in excess of six months with a percentage of payments applying toward purchase.

Rental rates can also be treated as reimbursement costs for contractor-owned equipment. Owned equipment costs include depreciation, loan payments, interest, taxes, insurance, storage, and major repairs.

Monthly rental rates vary from 2% to 5% of the cost of the equipment depending on the anticipated life of the equipment and its wearing parts. Weekly rates are about 1/3 the monthly rates and daily rental rates about 1/3 the weekly rate.

The hourly operating costs for each piece of equipment include costs to the user such as fuel, oil, lubrication, normal expendables for the equipment, and a percentage of mechanic's wages chargeable to maintenance. The hourly operating costs listed do not include the operator's wages.

The daily cost for equipment used in the standard crews is figured by dividing the weekly rate by five, then adding eight times the hourly operating cost to give the total daily equipment cost, not including the operator. This figure is in the right hand column of the Equipment listings under Equipment Cost/Day.

Pile Driving rates shown for pile hammer and extractor do not include leads, crane, boiler or compressor. Vibratory pile driving requires an added field specialist during set-up and pile driving operation for the electric model. The hydraulic model requires a field specialist for set-up only. Up to 125 reuses of sheet piling are possible using vibratory drivers. For normal conditions, crane capacity for hammer type and size are as follows.

Crane Capacity	Hammer Type and Size		
	Air or Steam	Diesel	Vibratory
25 ton	to 8,750 ft.-lb.		70 H.P.
40 ton	15,000 ft.-lb.	to 32,000 ft.-lb.	170 H.P.
60 ton	25,000 ft.-lb.		300 H.P.
100 ton		112,000 ft.-lb.	

Cranes should be specified for the job by size, building and site characteristics, availability, performance characteristics, and duration of time required.

Backhoes & Shovels rent for about the same as equivalent size cranes but maintenance and operating expense is higher. Crane operators rate must be adjusted for high boom heights. Average adjustments: for 150′ boom add 2% per hour; over 185′, add 4% per hour; over 210′, add 6% per hour; over 250′, add 8% per hour and over 295′, add 12% per hour.

Tower Cranes of the climbing or static type have jibs from 50′ to 200′ and capacities at maximum reach range from 4,000 to 14,000 pounds. Lifting capacities increase up to maximum load as the hook radius decreases.

Typical rental rates, based on purchase price are about 2% to 3% per month.

Erection and dismantling runs between 500 and 2000 labor hours. Climbing operation takes 10 labor hours per 20′ climb. Crane dead time is about 5 hours per 40′ climb. If crane is bolted to side of the building add cost of ties and extra mast sections. Climbing cranes have from 80′ to 180′ of mast while static cranes have 80′ to 800′ of mast.

Truck Cranes can be converted to tower cranes by using tower attachments. Mast heights over 400′ have been used.

A single 100′ high material **Hoist and Tower** can be erected and dismantled in about 400 labor hours; a double 100′ high hoist and tower in about 600 labor hours. Erection times for additional heights are 3 and 4 labor hours per vertical foot respectively up to 150′, and 4 to 5 labor hours per vertical foot over 150′ high. A 40′ high portable Buck hoist takes about 160 labor hours to erect and dismantle. Additional heights take 2 labor hours per vertical foot to 80′ and 3 labor hours per vertical foot for the next 100′. Most material hoists do not meet local code requirements for carrying personnel.

A 150′ high **Personnel Hoist** requires about 500 to 800 labor hours to erect and dismantle. Budget erection time at 5 labor hours per vertical foot for all trades. Local code requirements or labor scarcity requiring overtime can add up to 50% to any of the above erection costs.

Earthmoving Equipment: The selection of earthmoving equipment depends upon the type and quantity of material, moisture content, haul distance, haul road, time available, and equipment available. Short haul cut and fill operations may require dozers only, while another operation may require excavators, a fleet of trucks, and spreading and compaction equipment. Stockpiled material and granular material are easily excavated with front end loaders. Scrapers are most economically used with hauls between 300′ and 1-1/2 miles if adequate haul roads can be maintained. Shovels are often used for blasted rock and any material where a vertical face of 8′ or more can be excavated. Special conditions may dictate the use of draglines, clamshells, or backhoes. Spreading and compaction equipment must be matched to the soil characteristics, the compaction required and the rate the fill is being supplied.

R015433-15 Heavy Lifting

Hydraulic Climbing Jacks

The use of hydraulic heavy lift systems is an alternative to conventional type crane equipment. The lifting, lowering, pushing, or pulling mechanism is a hydraulic climbing jack moving on a square steel jackrod from 1-5/8″ to 4″ square, or a steel cable. The jackrod or cable can be vertical or horizontal, stationary or movable, depending on the individual application. When the jackrod is stationary, the climbing jack will climb the rod and push or pull the load along with itself. When the climbing jack is stationary, the jackrod is movable with the load attached to the end and the climbing jack will lift or lower the jackrod with the attached load. The heavy lift system is normally operated by a single control lever located at the hydraulic pump.

The system is flexible in that one or more climbing jacks can be applied wherever a load support point is required, and the rate of lift synchronized.

Economic benefits have been demonstrated on projects such as: erection of ground assembled roofs and floors, complete bridge spans, girders and trusses, towers, chimney liners and steel vessels, storage tanks, and heavy machinery. Other uses are raising and lowering offshore work platforms, caissons, tunnel sections and pipelines.

R015436-50 Mobilization

Costs to move rented construction equipment to a job site from an equipment dealer's or contractor's yard (mobilization), or to move the equipment off the job site (demobilization), are not included in the rental or operating rates, nor in the equipment cost on a unit price line or in a crew listing. These costs can be found consolidated in the Mobilization section of the data and elsewhere in particular site work sections. If a piece of equipment is already on the job site, it is not appropriate to include mob/demob costs in a new estimate that requires use of that equipment. The following table identifies approximate sizes of rented construction equipment that would be hauled on a towed trailer. Because this listing is not all-encompassing, the user can infer as to what size trailer might be required for a piece of equipment not listed.

3-ton Trailer	20-ton Trailer	40-ton Trailer	50-ton Trailer
20 H.P. Excavator	110 H.P. Excavator	200 H.P. Excavator	270 H.P. Excavator
50 H.P. Skid Steer	165 H.P. Dozer	300 H.P. Dozer	Small Crawler Crane
35 H.P. Roller	150 H.P. Roller	400 H.P. Scraper	500 H.P. Scraper
40 H.P. Trencher	Backhoe	450 H.P. Art. Dump Truck	500 H.P. Art. Dump Truck

R024119-10 Demolition Defined

Whole Building Demolition - Demolition of the whole building with no concern for any particular building element, component, or material type being demolished. This type of demolition is accomplished with large pieces of construction equipment that break up the structure, load it into trucks and haul it to a disposal site, but disposal or dump fees are not included. Demolition of below-grade foundation elements, such as footings, foundation walls, grade beams, slabs on grade, etc., is not included. Certain mechanical equipment containing flammable liquids or ozone-depleting refrigerants, electric lighting elements, communication equipment components, and other building elements may contain hazardous waste, and must be removed, either selectively or carefully, as hazardous waste before the building can be demolished.

Foundation Demolition - Demolition of below-grade foundation footings, foundation walls, grade beams, and slabs on grade. This type of demolition is accomplished by hand or pneumatic hand tools, and does not include saw cutting, or handling, loading, hauling, or disposal of the debris.

Gutting - Removal of building interior finishes and electrical/mechanical systems down to the load-bearing and sub-floor elements of the rough building frame, with no concern for any particular building element, component, or material type being demolished. This type of demolition is accomplished by hand or pneumatic hand tools, and includes loading into trucks, but not hauling, disposal or dump fees, scaffolding, or shoring. Certain mechanical equipment containing flammable liquids or ozone-depleting refrigerants, electric lighting elements, communication equipment components, and other building elements may contain hazardous waste, and must be removed, either selectively or carefully, as hazardous waste, before the building is gutted.

Selective Demolition - Demolition of a selected building element, component, or finish, with some concern for surrounding or adjacent elements, components, or finishes (see the first Subdivision (s) at the beginning of appropriate Divisions). This type of demolition is accomplished by hand or pneumatic hand tools, and does not include handling, loading, storing, hauling, or disposal of the debris, scaffolding, or shoring. "Gutting" methods may be used in order to save time, but damage that is caused to surrounding or adjacent elements, components, or finishes may have to be repaired at a later time.

Careful Removal - Removal of a piece of service equipment, building element or component, or material type, with great concern for both the removed item and surrounding or adjacent elements, components or finishes. The purpose of careful removal may be to protect the removed item for later re-use, preserve a higher salvage value of the removed item, or replace an item while taking care to protect surrounding or adjacent elements, components, connections, or finishes from cosmetic and/or structural damage. An approximation of the time required to perform this type of removal is 1/3 to 1/2 the time it would take to install a new item of like kind (see Reference Number R220105-10). This type of removal is accomplished by hand or pneumatic hand tools, and does not include loading, hauling, or storing the removed item, scaffolding, shoring, or lifting equipment.

Cutout Demolition - Demolition of a small quantity of floor, wall, roof, or other assembly, with concern for the appearance and structural integrity of the surrounding materials. This type of demolition is accomplished by hand or pneumatic hand tools, and does not include saw cutting, handling, loading, hauling, or disposal of debris, scaffolding, or shoring.

Rubbish Handling - Work activities that involve handling, loading or hauling of debris. Generally, the cost of rubbish handling must be added to the cost of all types of demolition, with the exception of whole building demolition.

Minor Site Demolition - Demolition of site elements outside the footprint of a building. This type of demolition is accomplished by hand or pneumatic hand tools, or with larger pieces of construction equipment, and may include loading a removed item onto a truck (check the Crew for equipment used). It does not include saw cutting, hauling or disposal of debris, and, sometimes, handling or loading.

R024119-20 Dumpsters

Dumpster rental costs on construction sites are presented in two ways:

The cost per week rental includes the delivery of the dumpster; its pulling or emptying once per week, and its final removal. The assumption is made that the dumpster contractor could choose to empty a dumpster by simply bringing in an empty unit and removing the full one. These costs also include the disposal of the materials in the dumpster.

The alternate pricing can be used when actual planned conditions are not approximated by the weekly numbers. For example, these lines can be used when a dumpster is needed for 4 weeks and will need to be emptied 2 or 3 times per week. Conversely the alternate pricing lines can be used when a dumpster will be rented for several weeks or months but needs to be emptied only a few times over this period.

R024119-30 Rubbish Handling Chutes

To correctly estimate the cost of rubbish handling chute systems, the individual components must be priced separately. First choose the size of the system; a 30-inch diameter chute is quite common, but the sizes range from 18 to 36 inches in diameter. The 30-inch chute comes in a standard weight and two thinner weights. The thinner weight chutes are sometimes chosen for cost savings, but they are more easily damaged.

There are several types of major chute pieces that make up the chute system. The first component to consider is the top chute section (top intake hopper) where the material is dropped into the chute at the highest point. After determining the top chute, the intermediate chute pieces called the regular chute sections are priced. Next, the number of chute control door sections (intermediate intake hoppers) must be determined. In the more complex systems, a chute control door section is provided at each floor level. The last major component to consider is bolt down frames; these are usually provided at every other floor level.

There are a number of accessories to consider for safe operation and control. There are covers for the top chute and the chute door sections. The top chute can have a trough that allows for better loading of the chute. For the safest operation, a chute warning light system can be added that will warn the other chute intake locations not to load while another is being used. There are dust control devices that spray a water mist to keep down the dust as the debris is loaded into a dumpster. There are special breakaway cords that are used to prevent damage to the chute if the dumpster is removed without disconnecting from the chute. There are chute liners that can be installed to protect the chute structure from physical damage from rough abrasive materials. Warning signs can be posted at each floor level that is provided with a chute control door section.

In summary, a complete rubbish handling chute system will include one top section, several intermediate regular sections, several intermediate control door (intake hopper) sections and bolt down frames at every other floor level starting with the top floor. If so desired, the system can also include covers and a light warning system for a safer operation. The bottom of the chute should always be above the Dumpster and should be tied off with a breakaway cord to the Dumpster.

Existing Conditions — R0265 Underground Storage Tank Removal

R026510-20 Underground Storage Tank Removal

Underground storage tank removal can be divided into two categories: non-leaking and leaking. Prior to removing an underground storage tank, tests should be made, with the proper authorities present, to determine whether a tank has been leaking or the surrounding soil has been contaminated.

To safely remove liquid underground storage tanks:

1. Excavate to the top of the tank.
2. Disconnect all piping.
3. Open all tank vents and access ports.
4. Remove all liquids and/or sludge.
5. Purge the tank with an inert gas.
6. Provide access to the inside of the tank and clean out the interior using proper personal protective equipment (PPE).
7. Excavate soil surrounding the tank using proper PPE for on-site personnel.
8. Pull and properly dispose of the tank.
9. Clean up the site of all contaminated material.
10. Install new tanks or close the excavation.

Existing Conditions — R0282 Asbestos Remediation

R028213-20 Asbestos Removal Process

Asbestos removal is accomplished by a specialty contractor who understands the federal and state regulations regarding the handling and disposal of the material. The process of asbestos removal is divided into many individual steps. An accurate estimate can be calculated only after all the steps have been priced.

The steps are generally as follows:

1. Obtain an asbestos abatement plan from an industrial hygienist.
2. Monitor the air quality in and around the removal area and along the path of travel between the removal area and transport area. This establishes the background contamination.
3. Construct a two part decontamination chamber at entrance to removal area.
4. Install a HEPA filter to create a negative pressure in the removal area.
5. Install wall, floor and ceiling protection as required by the plan, usually 2 layers of fireproof 6 mil polyethylene.
6. Industrial hygienist visually inspects work area to verify compliance with plan.
7. Provide temporary supports for conduit and piping affected by the removal process.
8. Proceed with asbestos removal and bagging process. Monitor air quality as described in Step #2. Discontinue operations when contaminate levels exceed applicable standards.
9. Document the legal disposal of materials in accordance with EPA standards.
10. Thoroughly clean removal area including all ledges, crevices and surfaces.
11. Post abatement inspection by industrial hygienist to verify plan compliance.
12. Provide a certificate from a licensed industrial hygienist attesting that contaminate levels are within acceptable standards before returning area to regular use.

R028319-60 Lead Paint Remediation Methods

Lead paint remediation can be accomplished by the following methods.

1. Abrasive blast
2. Chemical stripping
3. Power tool cleaning with vacuum collection system
4. Encapsulation
5. Remove and replace
6. Enclosure

Each of these methods has strengths and weakness depending on the specific circumstances of the project. The following is an overview of each method.

1. **Abrasive blasting** is usually accomplished with sand or recyclable metallic blast. Before work can begin, the area must be contained to ensure the blast material with lead does not escape to the atmosphere. The use of vacuum blast greatly reduces the containment requirements. Lead abatement equipment that may be associated with this work includes a negative air machine. In addition, it is necessary to have an industrial hygienist monitor the project on a continual basis. When the work is complete, the spent blast sand with lead must be disposed of as a hazardous material. If metallic shot was used, the lead is separated from the shot and disposed of as hazardous material. Worker protection includes disposable clothing and respiratory protection.
2. **Chemical stripping** requires strong chemicals be applied to the surface to remove the lead paint. Before the work can begin, the area under/adjacent to the work area must be covered to catch the chemical and removed lead. After the chemical is applied to the painted surface it is usually covered with paper. The chemical is left in place for the specified period, then the paper with lead paint is pulled or scraped off. The process may require several chemical applications. The paper with chemicals and lead paint adhered to it, plus the containment and loose scrapings collected by a HEPA (high efficiency particulate air filter) vac, must be disposed of as a hazardous material. The chemical stripping process usually requires a neutralizing agent and several wash downs after the paint is removed. Worker protection includes a neoprene or other compatible protective clothing and respiratory protection with face shield. An industrial hygienist is required intermittently during the process.
3. **Power tool cleaning** is accomplished using shrouded needle blasting guns. The shrouding with different end configurations is held up against the surface to be cleaned. The area is blasted with hardened needles and the shroud captures the lead with a HEPA vac and deposits it in a holding tank. An industrial hygienist monitors the project, protective clothing and a respirator is required until air samples prove otherwise. When the work is complete the lead must be disposed of as a hazardous material.
4. **Encapsulation** is a method that leaves the well bonded lead paint in place after the peeling paint has been removed. Before the work can begin, the area under/adjacent to the work must be covered to catch the scrapings. The scraped surface is then washed with a detergent and rinsed. The prepared surface is covered with approximately 10 mils of paint. A reinforcing fabric can also be embedded in the paint covering. The scraped paint and containment must be disposed of as a hazardous material. Workers must wear protective clothing and respirators.
5. **Remove and replace** is an effective way to remove lead paint from windows, gypsum walls and concrete masonry surfaces. The painted materials are removed and new materials are installed. Workers should wear a respirator and tyvek suit. The demolished materials must be disposed of as hazardous waste if it fails the TCLP (toxicity characteristic leachate process) test.
6. **Enclosure** is the process that permanently seals lead painted materials in place. This process has many applications such as covering lead painted drywall with new drywall, covering exterior construction with tyvek paper then residing, or covering lead painted structural members with aluminum or plastic. The seams on all enclosing materials must be securely sealed. An industrial hygienist monitors the project, and protective clothing and a respirator is required until air samples prove otherwise.

All the processes require clearance monitoring and wipe testing as required by the hygienist.

R031113-10 Wall Form Materials

Aluminum Forms

Approximate weight is 3 lbs. per S.F.C.A. Standard widths are available from 4″ to 36″ with 36″ most common. Standard lengths of 2′, 4′, 6′ to 8′ are available. Forms are lightweight and fewer ties are needed with the wider widths. The form face is either smooth or textured.

Metal Framed Plywood Forms

Manufacturers claim over 75 reuses of plywood and over 300 reuses of steel frames. Many specials such as corners, fillers, pilasters, etc. are available. Monthly rental is generally about 15% of purchase price for first month and 9% per month thereafter with 90% of rental applied to purchase for the first month and decreasing percentages thereafter. Aluminum framed forms cost 25% to 30% more than steel framed.

After the first month, extra days may be prorated from the monthly charge. Rental rates do not include ties, accessories, cleaning, loss of hardware or freight in and out. Approximate weight is 5 lbs. per S.F. for steel; 3 lbs. per S.F. for aluminum.

Forms can be rented with option to buy.

Plywood Forms, Job Fabricated

There are two types of plywood used for concrete forms.

1. Exterior plyform which is completely waterproof. This is face oiled to facilitate stripping. Ten reuses can be expected with this type with 25 reuses possible.
2. An overlaid type consists of a resin fiber fused to exterior plyform. No oiling is required except to facilitate cleaning. This is available in both high density (HDO) and medium density overlaid (MDO). Using HDO, 50 reuses can be expected with 200 possible.

Plyform is available in 5/8″ and 3/4″ thickness. High density overlaid is available in 3/8″, 1/2″, 5/8″ and 3/4″ thickness.

5/8″ thick is sufficient for most building forms, while 3/4″ is best on heavy construction.

Plywood Forms, Modular, Prefabricated

There are many plywood forming systems without frames. Most of these are manufactured from 1-1/8″ (HDO) plywood and have some hardware attached. These are used principally for foundation walls 8′ or less high. With care and maintenance, 100 reuses can be attained with decreasing quality of surface finish.

Steel Forms

Approximate weight is 6-1/2 lbs. per S.F.C.A. including accessories. Standard widths are available from 2″ to 24″, with 24″ most common. Standard lengths are from 2′ to 8′, with 4′ the most common. Forms are easily ganged into modular units.

Forms are usually leased for 15% of the purchase price per month prorated daily over 30 days.

Rental may be applied to sale price, and usually rental forms are bought. With careful handling and cleaning 200 to 400 reuses are possible.

Straight wall gang forms up to 12′ x 20′ or 8′ x 30′ can be fabricated. These crane handled forms usually lease for approx. 9% per month.

Individual job analysis is available from the manufacturer at no charge.

R031113-30 Slipforms

The slipform method of forming may be used for forming circular silo and multi-celled storage bin type structures over 30′ high, and building core shear walls over eight stories high. The shear walls, usually enclose elevator shafts, stairwells, mechanical spaces, and toilet rooms. Reuse of the form on duplicate structures will reduce the height necessary and spread the cost of building the form. Slipform systems can be used to cast chimneys, towers, piers, dams, underground shafts or other structures capable of being extruded.

Slipforms are usually 4′ high and are raised semi-continuously by jacks climbing on rods which are embedded in the concrete. The jacks are powered by a hydraulic, pneumatic, or electric source and are available in 3, 6, and 22 ton capacities. Interior work decks and exterior scaffolds must be provided for placing inserts, embedded items, reinforcing steel, and concrete. Scaffolds below the form for finishers may be required. The interior work decks are often used as roof slab forms on silos and bin work. Form raising rates will range from 6″ to 20″ per hour for silos; 6″ to 30″ per hour for buildings; and 6″ to 48″ per hour for shaft work.

Reinforcing bars and stressing strands are usually hoisted by crane or gin pole, and the concrete material can be hoisted by crane, winch-powered skip, or pumps. The slipform system is operated on a continuous 24-hour day when a monolithic structure is desired. For least cost, the system is operated only during normal working hours.

Placing concrete will range from 0.5 to 1.5 labor-hours per C.Y. Bucks, blockouts, keyways, weldplates, etc. are extra.

R031113-40 Forms for Reinforced Concrete

Design Economy

Avoid many sizes in proportioning beams and columns.

From story to story avoid changing column dimensions. Gain strength by adding steel or using a richer mix. If a change in size of column is necessary, vary one dimension only to minimize form alterations. Keep beams and columns the same width.

From floor to floor in a multi-story building vary beam depth, not width, as that will leave slab panel form unchanged. It is cheaper to vary the strength of a beam from floor to floor by means of steel area than by 2″ changes in either width or depth.

Cost Factors

Material includes the cost of lumber, cost of rent for metal pans or forms if used, nails, form ties, form oil, bolts and accessories.

Labor includes the cost of carpenters to make up, erect, remove and repair, plus common labor to clean and move. Having carpenters remove forms minimizes repairs.

Improper alignment and condition of forms will increase finishing cost. When forms are heavily oiled, concrete surfaces must be neutralized before finishing. Special curing compounds will cause spillages to spall off in first frost. Gang forming methods will reduce costs on large projects.

Materials Used

Boards are seldom used unless their architectural finish is required. Generally, steel, fiberglass and plywood are used for contact surfaces. Labor on plywood is 10% less than with boards. The plywood is backed up with 2 × 4′s at 12″ to 32″ O.C. Walers are generally 2 - 2 × 4′s. Column forms are held together with steel yokes or bands. Shoring is with adjustable shoring or scaffolding for high ceilings.

Reuse

Floor and column forms can be reused four or possibly five times without excessive repair. Remember to allow for 10% waste on each reuse.

When modular sized wall forms are made, up to twenty uses can be expected with exterior plyform.

When forms are reused, the cost to erect, strip, clean and move will not be affected. 10% replacement of lumber should be included and about one hour of carpenter time for repairs on each reuse per 100 S.F.

The reuse cost for certain accessory items normally rented on a monthly basis will be lower than the cost for the first use.

After fifth use, new material required plus time needed for repair prevent form cost from dropping further and it may go up. Much depends on care in stripping, the number of special bays, changes in beam or column sizes and other factors.

Costs for multiple use of formwork may be developed as follows:

2 Uses

$$\frac{\text{(1st Use + Reuse)}}{2} = \text{avg. cost/2 uses}$$

3 Uses

$$\frac{\text{(1st Use + 2 Reuse)}}{3} = \text{avg. cost/3 uses}$$

4 Uses

$$\frac{\text{(1st use + 3 Reuse)}}{4} = \text{avg. cost/4 uses}$$

R031113-60 Formwork Labor-Hours

Item	Unit	Hours Required			Total Hours	Multiple Use		
		Fabricate	Erect & Strip	Clean & Move	1 Use	2 Use	3 Use	4 Use
Beam and Girder, interior beams, 12″ wide	100 S.F.	6.4	8.3	1.3	16.0	13.3	12.4	12.0
Hung from steel beams		5.8	7.7	1.3	14.8	12.4	11.6	11.2
Beam sides only, 36″ high		5.8	7.2	1.3	14.3	11.9	11.1	10.7
Beam bottoms only, 24″ wide		6.6	13.0	1.3	20.9	18.1	17.2	16.7
Box out for openings		9.9	10.0	1.1	21.0	16.6	15.1	14.3
Buttress forms, to 8′ high		6.0	6.5	1.2	13.7	11.2	10.4	10.0
Centering, steel, 3/4″ rib lath			1.0		1.0			
3/8″ rib lath or slab form			0.9		0.9			
Chamfer strip or keyway	100 L.F.		1.5		1.5	1.5	1.5	1.5
Columns, fiber tube 8″ diameter			20.6		20.6			
12″			21.3		21.3			
16″			22.9		22.9			
20″			23.7		23.7			
24″			24.6		24.6			
30″			25.6		25.6			
Columns, round steel, 12″ diameter			22.0		22.0	22.0	22.0	22.0
16″			25.6		25.6	25.6	25.6	25.6
20″			30.5		30.5	30.5	30.5	30.5
24″			37.7		37.7	37.7	37.7	37.7
Columns, plywood 8″ x 8″	100 S.F.	7.0	11.0	1.2	19.2	16.2	15.2	14.7
12″ x 12″		6.0	10.5	1.2	17.7	15.2	14.4	14.0
16″ x 16″		5.9	10.0	1.2	17.1	14.7	13.8	13.4
24″ x 24″		5.8	9.8	1.2	16.8	14.4	13.6	13.2
Columns, steel framed plywood 8″ x 8″			10.0	1.0	11.0	11.0	11.0	11.0
12″ x 12″			9.3	1.0	10.3	10.3	10.3	10.3
16″ x 16″			8.5	1.0	9.5	9.5	9.5	9.5
24″ x 24″			7.8	1.0	8.8	8.8	8.8	8.8
Drop head forms, plywood		9.0	12.5	1.5	23.0	19.0	17.7	17.0
Coping forms		8.5	15.0	1.5	25.0	21.3	20.0	19.4
Culvert, box			14.5	4.3	18.8	18.8	18.8	18.8
Curb forms, 6″ to 12″ high, on grade		5.0	8.5	1.2	14.7	12.7	12.1	11.7
On elevated slabs		6.0	10.8	1.2	18.0	15.5	14.7	14.3
Edge forms to 6″ high, on grade	100 L.F.	2.0	3.5	0.6	6.1	5.6	5.4	5.3
7″ to 12″ high	100 S.F.	2.5	5.0	1.0	8.5	7.8	7.5	7.4
Equipment foundations		10.0	18.0	2.0	30.0	25.5	24.0	23.3
Flat slabs, including drops		3.5	6.0	1.2	10.7	9.5	9.0	8.8
Hung from steel		3.0	5.5	1.2	9.7	8.7	8.4	8.2
Closed deck for domes		3.0	5.8	1.2	10.0	9.0	8.7	8.5
Open deck for pans		2.2	5.3	1.0	8.5	7.9	7.7	7.6
Footings, continuous, 12″ high		3.5	3.5	1.5	8.5	7.3	6.8	6.6
Spread, 12″ high		4.7	4.2	1.6	10.5	8.7	8.0	7.7
Pile caps, square or rectangular		4.5	5.0	1.5	11.0	9.3	8.7	8.4
Grade beams, 24″ deep		2.5	5.3	1.2	9.0	8.3	8.0	7.9
Lintel or Sill forms		8.0	17.0	2.0	27.0	23.5	22.3	21.8
Spandrel beams, 12″ wide		9.0	11.2	1.3	21.5	17.5	16.2	15.5
Stairs			25.0	4.0	29.0	29.0	29.0	29.0
Trench forms in floor		4.5	14.0	1.5	20.0	18.3	17.7	17.4
Walls, Plywood, at grade, to 8′ high		5.0	6.5	1.5	13.0	11.0	9.7	9.5
8′ to 16′		7.5	8.0	1.5	17.0	13.8	12.7	12.1
16′ to 20′		9.0	10.0	1.5	20.5	16.5	15.2	14.5
Foundation walls, to 8′ high		4.5	6.5	1.0	12.0	10.3	9.7	9.4
8′ to 16′ high		5.5	7.5	1.0	14.0	11.8	11.0	10.6
Retaining wall to 12′ high, battered		6.0	8.5	1.5	16.0	13.5	12.7	12.3
Radial walls to 12′ high, smooth		8.0	9.5	2.0	19.5	16.0	14.8	14.3
2′ chords		7.0	8.0	1.5	16.5	13.5	12.5	12.0
Prefabricated modular, to 8′ high		—	4.3	1.0	5.3	5.3	5.3	5.3
Steel, to 8′ high		—	6.8	1.2	8.0	8.0	8.0	8.0
8′ to 16′ high		—	9.1	1.5	10.6	10.3	10.2	10.2
Steel framed plywood to 8′ high		—	6.8	1.2	8.0	7.5	7.3	7.2
8′ to 16′ high		—	9.3	1.2	10.5	9.5	9.2	9.0

R032110-10 Reinforcing Steel Weights and Measures

Bar Designation No.**	Nominal Weight Lb./Ft.	U.S. Customary Units Nominal Dimensions* Diameter in.	Cross Sectional Area, in.^{2}	Perimeter in.	SI Units Nominal Weight kg/m	Nominal Dimensions* Diameter mm	Cross Sectional Area, cm^2	Perimeter mm
3	.376	.375	.11	1.178	.560	9.52	.71	29.9
4	.668	.500	.20	1.571	.994	12.70	1.29	39.9
5	1.043	.625	.31	1.963	1.552	15.88	2.00	49.9
6	1.502	.750	.44	2.356	2.235	19.05	2.84	59.8
7	2.044	.875	.60	2.749	3.042	22.22	3.87	69.8
8	2.670	1.000	.79	3.142	3.973	25.40	5.10	79.8
9	3.400	1.128	1.00	3.544	5.059	28.65	6.45	90.0
10	4.303	1.270	1.27	3.990	6.403	32.26	8.19	101.4
11	5.313	1.410	1.56	4.430	7.906	35.81	10.06	112.5
14	7.650	1.693	2.25	5.320	11.384	43.00	14.52	135.1
18	13.600	2.257	4.00	7.090	20.238	57.33	25.81	180.1

* The nominal dimensions of a deformed bar are equivalent to those of a plain round bar having the same weight per foot as the deformed bar.
** Bar numbers are based on the number of eighths of an inch included in the nominal diameter of the bars.

R032110-20 Metric Rebar Specification - ASTM A615-81

Grade 300 (300 MPa* = 43,560 psi; +8.7% vs. Grade 40)
Grade 400 (400 MPa* = 58,000 psi; –3.4% vs. Grade 60)

Bar No.	Diameter mm	Area mm^2	Equivalent in.^{2}	Comparison with U.S. Customary Bars
10M	11.3	100	.16	Between #3 & #4
15M	16.0	200	.31	#5 (.31 in.^{2})
20M	19.5	300	.47	#6 (.44 in.^{2})
25M	25.2	500	.78	#8 (.79 in.^{2})
30M	29.9	700	1.09	#9 (1.00 in.^{2})
35M	35.7	1000	1.55	#11 (1.56 in.^{2})
45M	43.7	1500	2.33	#14 (2.25 in.^{2})
55M	56.4	2500	3.88	#18 (4.00 in.^{2})

* MPa = megapascals

R032110-40 Weight of Steel Reinforcing Per Square Foot of Wall (PSF)

Reinforced weights: The table below suggests the weights per square foot for reinforcing steel in walls. Weights are approximate and will be the same for all grades of steel bars. For bars in two directions, add weights for each size and spacing.

C/C Spacing in Inches	Bar Size								
	#3 Wt. (PSF)	#4 Wt. (PSF)	#5 Wt. (PSF)	#6 Wt. (PSF)	#7 Wt. (PSF)	#8 Wt. (PSF)	#9 Wt. (PSF)	#10 Wt. (PSF)	#11 Wt. (PSF)
2″	2.26	4.01	6.26	9.01	12.27				
3″	1.50	2.67	4.17	6.01	8.18	10.68	13.60	17.21	21.25
4″	1.13	2.01	3.13	4.51	6.13	8.10	10.20	12.91	15.94
5″	.90	1.60	2.50	3.60	4.91	6.41	8.16	10.33	12.75
6″	.752	1.34	2.09	3.00	4.09	5.34	6.80	8.61	10.63
8″	.564	1.00	1.57	2.25	3.07	4.01	5.10	6.46	7.97
10″	.451	.802	1.25	1.80	2.45	3.20	4.08	5.16	6.38
12″	.376	.668	1.04	1.50	2.04	2.67	3.40	4.30	5.31
18″	.251	.445	.695	1.00	1.32	1.78	2.27	2.86	3.54
24″	.188	.334	.522	.751	1.02	1.34	1.70	2.15	2.66
30″	.150	.267	.417	.600	.817	1.07	1.36	1.72	2.13
36″	.125	.223	.348	.501	.681	.890	1.13	1.43	1.77
42″	.107	.191	.298	.429	.584	.753	.97	1.17	1.52
48″	.094	.167	.261	.376	.511	.668	.85	1.08	1.33

R032110-50 Minimum Wall Reinforcement Weight (PSF)

This table lists the approximate minimum wall reinforcement weights per S.F. according to the specification of .12% of gross area for vertical bars and .20% of gross area for horizontal bars.

Location	Wall Thickness	Bar Size	Horizontal Steel Spacing C/C	Sq. In. Req'd per S.F.	Total Wt. per S.F.	Bar Size	Vertical Steel Spacing C/C	Sq. In. Req'd per S.F.	Total Wt. per S.F.	Horizontal & Vertical Steel Total Weight per S.F.
Both Faces	10″	#4	18″	.24	.89#	#3	18″	.14	.50#	1.39#
	12″	#4	16″	.29	1.00	#3	16″	.17	.60	1.60
	14″	#4	14″	.34	1.14	#3	13″	.20	.69	1.84
	16″	#4	12″	.38	1.34	#3	11″	.23	.82	2.16
	18″	#5	17″	.43	1.47	#4	18″	.26	.89	2.36
One Face	6″	#3	9″	.15	.50	#3	18″	.09	.25	.75
	8″	#4	12″	.19	.67	#3	11″	.12	.41	1.08
	10″	#5	15″	.24	.83	#4	16″	.14	.50	1.34

R032110-70 Bend, Place, and Tie Reinforcing

Placing and tying by rodmen for footings and slabs runs from nine hrs. per ton for heavy bars to fifteen hrs. per ton for light bars. For beams, columns, and walls, production runs from eight hrs. per ton for heavy bars to twenty hrs. per ton for light bars. Overall average for typical reinforced concrete buildings is about fourteen hrs. per ton. These production figures include the time for placing of accessories and usual inserts, but not their material cost (allow 15% of the cost of delivered bent rods). Equipment handling is necessary for the larger-sized bars so that installation costs for the very heavy bars will not decrease proportionately.

Installation costs for splicing reinforcing bars include allowance for equipment to hold the bars in place while splicing as well as necessary scaffolding for iron workers.

R032110-80 Shop-Fabricated Reinforcing Steel

The material prices for reinforcing, shown in the unit cost sections of the book, are for 50 tons or more of shop-fabricated reinforcing steel and include:

1. Mill base price of reinforcing steel
2. Mill grade/size/length extras
3. Mill delivery to the fabrication shop
4. Shop storage and handling
5. Shop drafting/detailing
6. Shop shearing and bending
7. Shop listing
8. Shop delivery to the job site

Both material and installation costs can be considerably higher for small jobs consisting primarily of smaller bars, while material costs may be slightly lower for larger jobs.

R032205-30 Common Stock Styles of Welded Wire Fabric

This table provides some of the basic specifications, sizes, and weights of welded wire fabric used for reinforcing concrete.

	New Designation	Old Designation		Steel Area per Foot				Approximate Weight per 100 S.F.	
	Spacing — Cross Sectional Area (in.) — (Sq. in. 100)	Spacing — Wire Gauge (in.) — (AS & W)		Longitudinal		Transverse			
				in.	cm	in.	cm	lbs	kg
Rolls	6 x 6 — W1.4 x W1.4	6 x 6 — 10 x 10		.028	.071	.028	.071	21	9.53
	6 x 6 — W2.0 x W2.0	6 x 6 — 8 x 8	1	.040	.102	.040	.102	29	13.15
	6 x 6 — W2.9 x W2.9	6 x 6 — 6 x 6		.058	.147	.058	.147	42	19.05
	6 x 6 — W4.0 x W4.0	6 x 6 — 4 x 4		.080	.203	.080	.203	58	26.91
	4 x 4 — W1.4 x W1.4	4 x 4 — 10 x 10		.042	.107	.042	.107	31	14.06
	4 x 4 — W2.0 x W2.0	4 x 4 — 8 x 8	1	.060	.152	.060	.152	43	19.50
	4 x 4 — W2.9 x W2.9	4 x 4 — 6 x 6		.087	.227	.087	.227	62	28.12
	4 x 4 — W4.0 x W4.0	4 x 4 — 4 x 4		.120	.305	.120	.305	85	38.56
Sheets	6 x 6 — W2.9 x W2.9	6 x 6 — 6 x 6		.058	.147	.058	.147	42	19.05
	6 x 6 — W4.0 x W4.0	6 x 6 — 4 x 4		.080	.203	.080	.203	58	26.31
	6 x 6 — W5.5 x W5.5	6 x 6 — 2 x 2	2	.110	.279	.110	.279	80	36.29
	4 x 4 — W1.4 x W1.4	4 x 4 — 4 x 4		.120	.305	.120	.305	85	38.56

NOTES: 1. Exact W—number size for 8 gauge is W2.1
2. Exact W—number size for 2 gauge is W5.4

The above table was compiled with the following excerpts from the WRI Manual of Standard Practices, 7th Edition, Copyright 2006. Reproduced with permission of the Wire Reinforcement Institute, Inc.:

1. Chapter 3, page 7, Table 1 Common Styles of Metric Wire Reinforcement (WWR) With Equivalent US Customary Units
2. Chapter 6, page 19, Table 5 Customary Units
3. Chapter 6, Page 23, Table 7 Customary Units (in.) Welded Plain Wire Reinforcement
4. Chapter 6, Page 25 Table 8 Wire Size Comparison
5. Chapter 9, Page 30, Table 9 Weight of Longitudinal Wires Weight (Mass) Estimating Tables
6. Chapter 9, Page 31, Table 9M Weight of Longitudinal Wires Weight (Mass) Estimating Tables
7. Chapter 9, Page 32, Table 10 Weight of Transverse Wires Based on 62″ lengths of transverse wire (60″ width plus 1″ overhand each side)
8. Chapter 9, Page 33, Table 10M Weight of Transverse Wires

R033053-50 Industrial Chimneys

Foundation requirements in C.Y. of concrete for various sized chimneys.

Size Chimney	2 Ton Soil	3 Ton Soil	Size Chimney	2 Ton Soil	3 Ton Soil	Size Chimney	2 Ton Soil	3 Ton Soil
75′ x 3′-0″	13 C.Y.	11 C.Y.	160′ x 6′-6″	86 C.Y.	76 C.Y.	300′ x 10′-0″	325 C.Y.	245 C.Y.
85′ x 5′-6″	19	16	175′ x 7′-0″	108	95	350′ x 12′-0″	422	320
100′ x 5′-0″	24	20	200′ x 6′-0″	125	105	400′ x 14′-0″	520	400
125′ x 5′-6″	43	36	250′ x 8′-0″	230	175	500′ x 18′-0″	725	575

R033105-10 Proportionate Quantities

The tables below show both quantities per S.F. of floor areas as well as form and reinforcing quantities per C.Y. Unusual structural requirements would increase the ratios below. High strength reinforcing would reduce the steel weights. Figures are for 3000 psi concrete and 60,000 psi reinforcing unless specified otherwise.

Type of Construction	Live Load	Span	Per S.F. of Floor Area				Per C.Y. of Concrete		
			Concrete	Forms	Reinf.	Pans	Forms	Reinf.	Pans
Flat Plate	50 psf	15 Ft.	.46 C.F.	1.06 S.F.	1.71lb.		62 S.F.	101 lb.	
		20	.63	1.02	2.40		44	104	
		25	.79	1.02	3.03		35	104	
	100	15	.46	1.04	2.14		61	126	
		20	.71	1.02	2.72		39	104	
		25	.83	1.01	3.47		33	113	
Flat Plate (waffle construction) 20″ domes	50	20	.43	1.00	2.10	.84 S.F.	63	135	53 S.F.
		25	.52	1.00	2.90	.89	52	150	46
		30	.64	1.00	3.70	.87	42	155	37
	100	20	.51	1.00	2.30	.84	53	125	45
		25	.64	1.00	3.20	.83	42	135	35
		30	.76	1.00	4.40	.81	36	160	29
Waffle Construction 30″ domes	50	25	.69	1.06	1.83	.68	42	72	40
		30	.74	1.06	2.39	.69	39	87	39
		35	.86	1.05	2.71	.69	33	85	39
		40	.78	1.00	4.80	.68	35	165	40
Flat Slab (two way with drop panels)	50	20	.62	1.03	2.34		45	102	
		25	.77	1.03	2.99		36	105	
		30	.95	1.03	4.09		29	116	
	100	20	.64	1.03	2.83		43	119	
		25	.79	1.03	3.88		35	133	
		30	.96	1.03	4.66		29	131	
	200	20	.73	1.03	3.03		38	112	
		25	.86	1.03	4.23		32	133	
		30	1.06	1.03	5.30		26	135	
One Way Joists 20″ Pans	50	15	.36	1.04	1.40	.93	78	105	70
		20	.42	1.05	1.80	.94	67	120	60
		25	.47	1.05	2.60	.94	60	150	54
	100	15	.38	1.07	1.90	.93	77	140	66
		20	.44	1.08	2.40	.94	67	150	58
		25	.52	1.07	3.50	.94	55	185	49
One Way Joists 8″ x 16″ filler blocks	50	15	.34	1.06	1.80	.81 Ea.	84	145	64 Ea.
		20	.40	1.08	2.20	.82	73	145	55
		25	.46	1.07	3.20	.83	63	190	49
	100	15	.39	1.07	1.90	.81	74	130	56
		20	.46	1.09	2.80	.82	64	160	48
		25	.53	1.10	3.60lb.	.83	56	190	42
One Way Beam & Slab	50	15	.42	1.30	1.73		84	111	
		20	.51	1.28	2.61		68	138	
		25	.64	1.25	2.78		53	117	
	100	15	.42	1.30	1.90		84	122	
		20	.54	1.35	2.69		68	154	
		25	.69	1.37	3.93		54	145	
	200	15	.44	1.31	2.24		80	137	
		20	.58	1.40	3.30		65	163	
		25	.69	1.42	4.89		53	183	
Two Way Beam & Slab	100	15	.47	1.20	2.26		69	130	
		20	.63	1.29	3.06		55	131	
		25	.83	1.33	3.79		43	123	
	200	15	.49	1.25	2.70		41	149	
		20	.66	1.32	4.04		54	165	
		25	.88	1.32	6.08		41	187	

R033105-10 Proportionate Quantities (cont.)

4000 psi Concrete and 60,000 psi Reinforcing—Form and Reinforcing Quantities per C.Y.					
Item	Size	Forms	Reinforcing	Minimum	Maximum
Columns (square tied)	10″ x 10″	130 S.F.C.A.	#5 to #11	220 lbs.	875 lbs.
	12″ x 12″	108	#6 to #14	200	955
	14″ x 14″	92	#7 to #14	190	900
	16″ x 16″	81	#6 to #14	187	1082
	18″ x 18″	72	#6 to #14	170	906
	20″ x 20″	65	#7 to #18	150	1080
	22″ x 22″	59	#8 to #18	153	902
	24″ x 24″	54	#8 to #18	164	884
	26″ x 26″	50	#9 to #18	169	994
	28″ x 28″	46	#9 to #18	147	864
	30″ x 30″	43	#10 to #18	146	983
	32″ x 32″	40	#10 to #18	175	866
	34″ x 34″	38	#10 to #18	157	772
	36″ x 36″	36	#10 to #18	175	852
	38″ x 38″	34	#10 to #18	158	765
	40″ x 40″	32 ▼	#10 to #18	143	692

Item	Size	Form	Spiral	Reinforcing	Minimum	Maximum
Columns (spirally reinforced)	12″ diameter	34.5 L.F.	190 lbs.	#4 to #11	165 lbs.	1505 lb.
		34.5	190	#14 & #18	—	1100
	14″	25	170	#4 to #11	150	970
		25	170	#14 & #18	800	1000
	16″	19	160	#4 to #11	160	950
		19	160	#14 & #18	605	1080
	18″	15	150	#4 to #11	160	915
		15	150	#14 & #18	480	1075
	20″	12	130	#4 to #11	155	865
		12	130	#14 & #18	385	1020
	22″	10	125	#4 to #11	165	775
		10	125	#14 & #18	320	995
	24″	9	120	#4 to #11	195	800
		9	120	#14 & #18	290	1150
	26″	7.3	100	#4 to #11	200	729
		7.3	100	#14 & #18	235	1035
	28″	6.3	95	#4 to #11	175	700
		6.3	95	#14 & #18	200	1075
	30″	5.5	90	#4 to #11	180	670
		5.5	90	#14 & #18	175	1015
	32″	4.8	85	#4 to #11	185	615
		4.8	85	#14 & #18	155	955
	34″	4.3	80	#4 to #11	180	600
		4.3	80	#14 & #18	170	855
	36″	3.8	75	#4 to #11	165	570
		3.8	75	#14 & #18	155	865
	40″	3.0	70	#4 to #11	165	500
		3.0	70	#14 & #18	145	765

R033105-10 Proportionate Quantities (cont.)

3000 psi Concrete and 60,000 psi Reinforcing—Form and Reinforcing Quantities per C.Y.						
Item	**Type**	**Loading**	**Height**	**C.Y./L.F.**	**Forms/C.Y.**	**Reinf./C.Y.**
Retaining Walls	Cantilever	Level Backfill	4 Ft.	0.2 C.Y.	49 S.F.	35 lbs.
			8	0.5	42	45
			12	0.8	35	70
			16	1.1	32	85
			20	1.6	28	105
		Highway Surcharge	4	0.3	41	35
			8	0.5	36	55
			12	0.8	33	90
			16	1.2	30	120
			20	1.7	27	155
		Railroad Surcharge	4	0.4	28	45
			8	0.8	25	65
			12	1.3	22	90
			16	1.9	20	100
			20	2.6	18	120
	Gravity, with Vertical Face	Level Backfill	4	0.4	37	None
			7	0.6	27	↓
			10	1.2	20	↓
		Sloping Backfill	4	0.3	31	↓
			7	0.8	21	↓
			10	1.6	15	↓

Item	Span	Live Load in Kips per Linear Foot							
		Under 1 Kip		2 to 3 Kips		4 to 5 Kips		6 to 7 Kips	
		Forms	**Reinf.**	**Forms**	**Reinf.**	**Forms**	**Reinf.**	**Forms**	**Reinf.**
Beams	10 Ft.	—	—	90 S.F.	170 #	85 S.F.	175 #	75 S.F.	185 #
	16	130 S.F.	165 #	85	180	75	180	65	225
	20	110	170	75	185	62	200	51	200
	26	90	170	65	215	62	215	—	—
	30	85	175	60	200	—	—	—	—

Item	Size	Type	Forms per C.Y.	Reinforcing per C.Y.
Spread Footings	Under 1 C.Y.	1,000 psf soil	24 S.F.	44 lbs.
		5,000	24	42
		10,000	24	52
	1 C.Y. to 5 C.Y.	1,000	14	49
		5,000	14	50
		10,000	14	50
	Over 5 C.Y.	1,000	9	54
		5,000	9	52
		10,000	9	56
Pile Caps (30 Ton Concrete Piles)	Under 5 C.Y.	shallow caps	20	65
		medium	20	50
		deep	20	40
	5 C.Y. to 10 C.Y.	shallow	14	55
		medium	15	45
		deep	15	40
	10 C.Y. to 20 C.Y.	shallow	11	60
		medium	11	45
		deep	12	35
	Over 20 C.Y.	shallow	9	60
		medium	9	45
		deep	10	40

R033105-10 Proportionate Quantities (cont.)

3000 psi Concrete and 60,000 psi Reinforcing — Form and Reinforcing Quantities per C.Y.						
Item	Size	Pile Spacing	50 T Pile	100 T Pile	50 T Pile	100 T Pile
Pile Caps (Steel H Piles)	Under 5 C.Y.	24″ O.C.	24 S.F.	24 S.F.	75 lbs.	90 lbs.
		30″	25	25	80	100
		36″	24	24	80	110
	5 C.Y. to 10 C.Y.	24″	15	15	80	110
		30″	15	15	85	110
		36″	15	15	75	90
	Over 10 C.Y.	24″	13	13	85	90
		30″	11	11	85	95
		36″	10	10	85	90

Item		8″ Thick		10″ Thick		12″ Thick		15″ Thick	
	Height	Forms	Reinf.	Forms	Reinf.	Forms	Reinf.	Forms	Reinf.
Basement Walls	7 Ft.	81 S.F.	44 lbs.	65 S.F.	45 lbs.	54 S.F.	44 lbs.	41 S.F.	43 lbs.
	8	↓	44	↓	45	↓	44	↓	43
	9	↓	46	↓	45	↓	44	↓	43
	10	↓	57	↓	45	↓	44	↓	43
	12	↓	83	↓	50	↓	52	↓	43
	14	↓	116	↓	65	↓	64	↓	51
	16	↓		↓	86	↓	90	↓	65
	18	↓		↓		↓	106	↓	70

R033105-20 Materials for One C.Y. of Concrete

This is an approximate method of figuring quantities of cement, sand and coarse aggregate for a field mix with waste allowance included.

With crushed gravel as coarse aggregate, to determine barrels of cement required, divide 10 by total mix; that is, for 1:2:4 mix, 10 divided by 7 = 1-3/7 barrels.

If the coarse aggregate is crushed stone, use 10-1/2 instead of 10 as given for gravel.

To determine tons of sand required, multiply barrels of cement by parts of sand and then by 0.2; that is, for the 1:2:4 mix, as above, 1-3/7 x 2 x .2 = .57 tons.

Tons of crushed gravel are in the same ratio to tons of sand as parts in the mix, or 4/2 x .57 = 1.14 tons.

1 bag cement = 94#
4 bags = 1 barrel

1 C.Y. sand or crushed gravel = 2700#
1 ton sand or crushed gravel = 20 C.F.

1 C.Y. crushed stone = 2575#
1 ton crushed stone = 21 C.F.

Average carload of cement is 692 bags; of sand or gravel is 56 tons.

Do not stack stored cement over 10 bags high.

R033105-30 Metric Equivalents of Cement Content for Concrete Mixes

94 Pound Bags per Cubic Yard	Kilograms per Cubic Meter	94 Pound Bags per Cubic Yard	Kilograms per Cubic Meter
1.0	55.77	7.0	390.4
1.5	83.65	7.5	418.3
2.0	111.5	8.0	446.2
2.5	139.4	8.5	474.0
3.0	167.3	9.0	501.9
3.5	195.2	9.5	529.8
4.0	223.1	10.0	557.7
4.5	251.0	10.5	585.6
5.0	278.8	11.0	613.5
5.5	306.7	11.5	641.3
6.0	334.6	12.0	669.2
6.5	362.5	12.5	697.1

a. If you know the cement content in pounds per cubic yard, multiply by .5933 to obtain kilograms per cubic meter.

b. If you know the cement content in 94 pound bags per cubic yard, multiply by 55.77 to obtain kilograms per cubic meter.

R033105-40 Metric Equivalents of Common Concrete Strengths (to convert other psi values to megapascals, multiply by 0.006895)

U.S. Values psi	SI Value Megapascals	Non-SI Metric Value kgf/cm²*
2000	14	140
2500	17	175
3000	21	210
3500	24	245
4000	28	280
4500	31	315
5000	34	350
6000	41	420
7000	48	490
8000	55	560
9000	62	630
10,000	69	705

* kilograms force per square centimeter

R033105-50 Quantities of Cement, Sand and Stone for One C.Y. of Concrete per Various Mixes

This table can be used to determine the quantities of the ingredients for smaller quantities of site mixed concrete.

	Mix = 1:1:1-3/4			Mix = 1:2:2.25			Mix = 1:2.25:3			Mix = 1:3:4		
Concrete (C.Y.)	Cement (sacks)	Sand (C.Y.)	Stone (C.Y.)	Cement (sacks)	Sand (C.Y.)	Stone (C.Y.)	Cement (sacks)	Sand (C.Y.)	Stone (C.Y.)	Cement (sacks)	Sand (C.Y.)	Stone (C.Y.)
1	10	.37	.63	7.75	.56	.65	6.25	.52	.70	5	.56	.74
2	20	.74	1.26	15.50	1.12	1.30	12.50	1.04	1.40	10	1.12	1.48
3	30	1.11	1.89	23.25	1.68	1.95	18.75	1.56	2.10	15	1.68	2.22
4	40	1.48	2.52	31.00	2.24	2.60	25.00	2.08	2.80	20	2.24	2.96
5	50	1.85	3.15	38.75	2.80	3.25	31.25	2.60	3.50	25	2.80	3.70
6	60	2.22	3.78	46.50	3.36	3.90	37.50	3.12	4.20	30	3.36	4.44
7	70	2.59	4.41	54.25	3.92	4.55	43.75	3.64	4.90	35	3.92	5.18
8	80	2.96	5.04	62.00	4.48	5.20	50.00	4.16	5.60	40	4.48	5.92
9	90	3.33	5.67	69.75	5.04	5.85	56.25	4.68	6.30	45	5.04	6.66
10	100	3.70	6.30	77.50	5.60	6.50	62.50	5.20	7.00	50	5.60	7.40
11	110	4.07	6.93	85.25	6.16	7.15	68.75	5.72	7.70	55	6.16	8.14
12	120	4.44	7.56	93.00	6.72	7.80	75.00	6.24	8.40	60	6.72	8.88
13	130	4.82	8.20	100.76	7.28	8.46	81.26	6.76	9.10	65	7.28	9.62
14	140	5.18	8.82	108.50	7.84	9.10	87.50	7.28	9.80	70	7.84	10.36
15	150	5.56	9.46	116.26	8.40	9.76	93.76	7.80	10.50	75	8.40	11.10
16	160	5.92	10.08	124.00	8.96	10.40	100.00	8.32	11.20	80	8.96	11.84
17	170	6.30	10.72	131.76	9.52	11.06	106.26	8.84	11.90	85	9.52	12.58
18	180	6.66	11.34	139.50	10.08	11.70	112.50	9.36	12.60	90	10.08	13.32
19	190	7.04	11.98	147.26	10.64	12.36	118.76	9.84	13.30	95	10.64	14.06
20	200	7.40	12.60	155.00	11.20	13.00	125.00	10.40	14.00	100	11.20	14.80
21	210	7.77	13.23	162.75	11.76	13.65	131.25	10.92	14.70	105	11.76	15.54
22	220	8.14	13.86	170.05	12.32	14.30	137.50	11.44	15.40	110	12.32	16.28
23	230	8.51	14.49	178.25	12.88	14.95	143.75	11.96	16.10	115	12.88	17.02
24	240	8.88	15.12	186.00	13.44	15.60	150.00	12.48	16.80	120	13.44	17.76
25	250	9.25	15.75	193.75	14.00	16.25	156.25	13.00	17.50	125	14.00	18.50
26	260	9.64	16.40	201.52	14.56	16.92	162.52	13.52	18.20	130	14.56	19.24
27	270	10.00	17.00	209.26	15.12	17.56	168.76	14.04	18.90	135	15.02	20.00
28	280	10.36	17.64	217.00	15.68	18.20	175.00	14.56	19.60	140	15.68	20.72
29	290	10.74	18.28	224.76	16.24	18.86	181.26	15.08	20.30	145	16.24	21.46

R033105-65 Field-Mix Concrete

Presently most building jobs are built with ready-mixed concrete except at isolated locations and some larger jobs requiring over 10,000 C.Y. where land is readily available for setting up a temporary batch plant.

The most economical mix is a controlled mix using local aggregate proportioned by trial to give the required strength with the least cost of material.

R033105-70 Placing Ready-Mixed Concrete

For ground pours allow for 5% waste when figuring quantities.

Prices in the front of the book assume normal deliveries. If deliveries are made before 8 A.M. or after 5 P.M. or on Saturday afternoons add 30%. Negotiated discounts for large volumes are not included in prices in front of book.

For the lower floors without truck access, concrete may be wheeled in rubber-tired buggies, conveyer handled, crane handled or pumped. Pumping is economical if there is top steel. Conveyers are more efficient for thick slabs.

At higher floors the rubber-tired buggies may be hoisted by a hoisting tower and wheeled to location. Placement by a conveyer is limited to three floors and is best for high-volume pours. Pumped concrete is best when building has no crane access. Concrete may be pumped directly as high as thirty-six stories using special pumping techniques. Normal maximum height is about fifteen stories.

Best pumping aggregate is screened and graded bank gravel rather than crushed stone.

Pumping downward is more difficult than pumping upward. Horizontal distance from pump to pour may increase preparation time prior to pour. Placing by cranes, either mobile, climbing or tower types, continues as the most efficient method for high-rise concrete buildings.

R033105-85 Lift Slabs

The cost advantage of the lift slab method is due to placing all concrete, reinforcing steel, inserts and electrical conduit at ground level and in reduction of formwork. Minimum economical project size is about 30,000 S.F. Slabs may be tilted for parking garage ramps.

It is now used in all types of buildings and has gone up to 22 stories high in apartment buildings. Current trend is to use post-tensioned flat plate slabs with spans from 22′ to 35′. Cylindrical void forms are used when deep slabs are required. One pound of prestressing steel is about equal to seven pounds of conventional reinforcing.

To be considered cured for stressing and lifting, a slab must have attained 75% of design strength. Seven days are usually sufficient with four to five days possible if high early strength cement is used. Slabs can be stacked using two coats of a non-bonding agent to insure that slabs do not stick to each other. Lifting is done by companies specializing in this work. Lift rate is 5′ to 15′ per hour with an average of 10′ per hour. Total areas up to 33,000 S.F. have been lifted at one time. 24 to 36 jacking columns are common. Most economical bay sizes are 24′ to 28′ with four to fourteen stories most efficient. Continuous design reduces reinforcing steel cost. Use of post-tensioned slabs allows larger bay sizes.

R033543-10 Polished Concrete Floors

A polished concrete floor has a glossy mirror-like appearance and is created by grinding the concrete floor with finer and finer diamond grits, similar to sanding wood, until the desired level of reflective clarity and sheen are achieved. The technical term for this type of polished concrete is bonded abrasive polished concrete. The basic piece of equipment used in the polishing process is a walk-behind planetary grinder for working large floor areas. This grinder drives diamond-impregnated abrasive discs, which progress from coarse- to fine-grit discs.

The process begins with the use of very coarse diamond segments or discs bonded in a metallic matrix. These segments are coarse enough to allow the removal of pits, blemishes, stains, and light coatings from the floor surface in preparation for final smoothing. The condition of the original concrete surface will dictate the grit coarsness of the initial grinding step which will generally end up being a three- to four-step process using ever finer grits. The purpose of this initial grinding step is to remove surface coatings and blemishes and to cut down into the cream for very fine aggregate exposure, or deeper into the fine aggregate layer just below the cream layer, or even deeper into the coarse aggregate layer. These initial grinding steps will progress up to the 100/120 grit. If wet grinding is done, a waste slurry is produced that must be removed between grit changes and disposed of properly. If dry grinding is done, a high performance vacuum will pick up the dust during grinding and collect it in bags which must be disposed of properly.

The process continues with honing the floor in a series of steps that progress from 100-grit to 400-grit diamond abrasive discs embedded in a plastic or resin matrix. At some point during, or just prior to, the honing step, one or two coats of stain or dye can be sprayed onto the surface to give color to the concrete, and two coats of densifier/hardener must be applied to the floor surface and allowed to dry. This sprayed-on densifier/hardener will penetrate about 1/8″ into the concrete to make the surface harder, denser and more abrasion-resistant.

The process ends with polishing the floor surface in a series of steps that progress from resin-impregnated 800-grit (medium polish) to 1500-grit (high polish) to 3000-grit (very high polish), depending on the desired level of reflective clarity and sheen.

The Concrete Polishing Association of America (CPAA) has defined the flooring options available when processing concrete to a desired finish. The first category is aggregate exposure, the grinding of a concrete surface with bonded abrasives, in as many abrasive grits necessary, to achieve one of the following classes:

A. Cream – very little surface cut depth; little aggregate exposure
B. Fine aggregate (salt and pepper) – surface cut depth of 1/16″; fine aggregate exposure with little or no medium aggregate exposure at random locations
C. Medium aggregate – surface cut depth of 1/8″; medium aggregate exposure with little or no large aggregate exposure at random locations
D. Large aggregate – surface cut depth of 1/4″; large aggregate exposure with little or no fine aggregate exposure at random locations

The second CPAA defined category is reflective clarity and sheen, the polishing of a concrete surface with the minimum number of bonded abrasives as indicated to achieve one of the following levels:

1. Ground – flat appearance with none to very slight diffused reflection; none to very low reflective sheen; using a minimum total of 4 grit levels up 100-grit
2. Honed – matte appearance with or without slight diffused reflection; low to medium reflective sheen; using a minimum total of 5 grit levels up to 400-grit
3. Semi-polished – objects being reflected are not quite sharp and crisp but can be easily identified; medium to high reflective sheen; using a minimum total of 6 grit levels up to 800-grit
4. Highly-polished – objects being reflected are sharp and crisp as would be seen in a mirror-like reflection; high to highest reflective sheen; using a minimum total of up to 8 grit levels up to 1500-grit or 3000-grit

The CPAA defines reflective clarity as the degree of sharpness and crispness of the reflection of overhead objects when viewed 5′ above and perpendicular to the floor surface; and reflective sheen as the degree of gloss reflected from a surface when viewed at least 20′ from and at an angle to the floor surface. These terms are relatively subjective. The final outcome depends on the internal makeup and surface condition of the original concrete floor, the experience of the floor polishing crew, and expectations of the owner. Before the grinding, honing, and polishing work commences on the main floor area, it might be beneficial to do a mock-up panel in the same floor but in an out of the way place to demonstrate the sequence of steps with increasingly fine abrasive grits and to demonstrate the final reflective clarity and reflective sheen. This mock-up panel will be within the area of, and part of, the final work.

R034105-30 Prestressed Precast Concrete Structural Units

Type	Location	Depth	Span in Ft.		Live Load Lb. per S.F.
Double Tee (8' to 10')	Floor	28" to 34"	60 to 80		50 to 80
	Roof	12" to 24"	30 to 50		40
	Wall	Width 8'	Up to 55' high		Wind
Multiple Tee (8')	Roof	8" to 12"	15 to 40		40
	Floor	8" to 12"	15 to 30		100
Plank or	Roof or Floor		Roof	Floor	40 for Roof 100 for Floor
		4"	13	12	
		6"	22	18	
		8"	26	25	
		10"	33	29	
		12"	42	32	
Single Tee (8' to 10')	Roof	28" 32" 36" 48"	40 80 100 120		40
AASHTO Girder	Bridges	Type 4 5 6	100 110 125		Highway
Box Beam (4')	Bridges	15" 27" 33"	40 to 100		Highway

The majority of precast projects today utilize double tees rather than single tees because of speed and ease of installation. As a result casting beds at manufacturing plants are normally formed for double tees. Single tee projects will therefore require an initial set up charge to be spread over the individual single tee costs.

For floors, a 2" to 3" topping is field cast over the shapes. For roofs, insulating concrete or rigid insulation is placed over the shapes.

Member lengths up to 40' are standard haul, 40' to 60' require special permits and lengths over 60' must be escorted. Over width and/or over length can add up to 100% on hauling costs.

Large heavy members may require two cranes for lifting which would increase erection costs by about 45%. An eight man crew can install 12 to 20 double tees, or 45 to 70 quad tees or planks per day.

Grouting of connections must also be included.

Several system buildings utilizing precast members are available. Heights can go up to 22 stories for apartment buildings. Optimum design ratio is 3 S.F. of surface to 1 S.F. of floor area.

R034136-90 Prestressed Concrete, Post-Tensioned

In post-tensioned concrete the steel tendons are tensioned after the concrete has reached about 3/4 of its ultimate strength. The cableways are grouted after tensioning to provide bond between the steel and concrete. If bond is to be prevented, the tendons are coated with a corrosion-preventative grease and wrapped with waterproofed paper or plastic. Bonded tendons are usually used when ultimate strength (beams & girders) are controlling factors.

High strength concrete is used to fully utilize the steel, thereby reducing the size and weight of the member. A plasticizing agent may be added to reduce water content. Maximum size aggregate ranges from 1/2″ to 1-1/2″ depending on the spacing of the tendons.

The types of steel commonly used are bars and strands. Job conditions determine which is best suited. Bars are best for vertical prestresses since they are easy to support. The trend is for steel manufacturers to supply a finished package, cut to length, which reduces field preparation to a minimum.

Bars vary from 3/4″ to 1-3/8″ diameter. Table below gives time in labor-hours per tendon for placing, tensioning and grouting (if required) a 75′ beam. Tendons used in buildings are not usually grouted; tendons for bridges usually are grouted. For strands the table indicates the labor-hours per pound for typical prestressed units 100′ long. Simple span beams usually require one-end stressing regardless of lengths. Continuous beams are usually stressed from two ends. Long slabs are poured from the center outward and stressed in 75′ increments after the initial 150′ center pour.

Labor Hours per Tendon and per Pound of Prestressed Steel

Length	100′ Beam		75′ Beam		100′ Slab	
Type Steel	**Strand**		**Bars**		**Strand**	
Diameter	**0.5″**		**3/4″**	**1-3/8″**	**0.5″**	**0.6″**
Number	**4**	**12**	**1**	**1**	**1**	**1**
Force in Kips	**100**	**300**	**42**	**143**	**25**	**35**
Preparation & Placing Cables	3.6	7.4	0.9	2.9	0.9	1.1
Stressing Cables	2.0	2.4	0.8	1.6	0.5	0.5
Grouting, if required	2.5	3.0	0.6	1.3		
Total Labor Hours	8.1	12.8	2.3	5.8	1.4	1.6
Prestressing Steel Weights (Lbs.)	215	640	115	380	53	74
Labor-hours per Lb. Bonded	0.038	0.020	0.020	0.015		
Non-bonded					0.026	0.022

Flat slab construction — 4000 psi concrete with span-to-depth ratio between 36 and 44. Two way post-tensioned steel averages 1.0 lb. per S.F. for 24′ to 28′ bays (usually strand) and additional reinforcing steel averages .5 lb. per S.F.

Pan and joist construction — 4000 psi concrete with span-to-depth ratio 28 to 30. Post-tensioned steel averages .8 lb. per S.F. and reinforcing steel about 1.0 lb. per S.F. Placing and stressing averages 40 hours per ton of total material.

Beam construction — 4000 to 5000 psi concrete. Steel weights vary greatly.

Labor cost per pound goes down as the size and length of the tendon increase. The primary economic consideration is the cost per kip for the member.

Post-tensioning becomes feasible for beams and girders over 30′ long; for continuous two-way slabs over 20′ clear; also in transferring upper building loads over longer spans at lower levels. Post-tension suppliers will provide engineering services at no cost to the user. Substantial economies are possible by using post-tensioned lift slabs.

R034513-10 Precast Concrete Wall Panels

Panels are either solid or insulated with plain, colored or textured finishes. Transportation is an important cost factor. Prices shown in the unit cost section of the book are based on delivery within 50 miles of a plant including fabricators' overhead and profit. Engineering data is available from fabricators to assist with construction details. Usual minimum job size for economical use of panels is about 5000 S.F. Small jobs can double the prices shown. For large, highly repetitive jobs, deduct up to 15% from the prices shown.

2″ thick panels cost about the same as 3″ thick panels, and maximum panel size is less. For building panels faced with granite, marble or stone, add the material prices from those unit cost sections to the plain panel price shown. There is a growing trend toward aggregate facings and broken rib finish rather than plain gray concrete panels.

No allowance has been made in the unit cost section for supporting steel framework. On one story buildings, panels may rest on grade beams and require only wind bracing and fasteners. On multi-story buildings panels can span from column to column and floor to floor. Plastic-designed steel-framed structures may have large deflections which slow down erection and raise costs.

Large panels are more economical than small panels on a S.F. basis. When figuring areas include all protrusions, returns, etc. Overhangs can triple erection costs. Panels over 45′ have been produced. Larger flat units should be prestressed. Vacuum lifting of smooth finish panels eliminates inserts and can speed erection.

Concrete R0347 Site-Cast Concrete

R034713-20 Tilt Up Concrete Panels

The advantage of tilt up construction is in the low cost of forms and the placing of concrete and reinforcing. Panels up to 75′ high and 5-1/2″ thick have been tilted using strongbacks. Tilt up has been used for one to five story buildings and is well-suited for warehouses, stores, offices, schools and residences.

The panels are cast in forms on the floor slab. Most jobs use 5-1/2″ thick solid reinforced concrete panels. Sandwich panels with a layer of insulating materials are also used. Where dampness is a factor, lightweight aggregate is used. Optimum panel size is 300 to 500 S.F.

Slabs are usually poured with 3000 psi concrete which permits tilting seven days after pouring. Slabs may be stacked on top of each other and are separated from each other by either two coats of bond breaker or a film of polyethylene. Use of high early-strength cement allows tilting two days after a pour. Tilting up is done with a roller outrigger crane with a capacity of at least 1-1/2 times the weight of the panel at the required reach. Exterior precast columns can be set at the same time as the panels; interior precast columns can be set first and the panels clipped directly to them. The use of cast-in-place concrete columns is diminishing due to shrinkage problems. Structural steel columns are sometimes used if crane rails are planned. Panels can be clipped to the columns or lowered between the flanges. Steel channels with anchors may be used as edge forms for the slab. When the panels are lifted the channels form an integral steel column to take structural loads. Roof loads can be carried directly by the panels for wall heights to 14′.

Requirements of local building codes may be a limiting factor and should be checked. Building floor slabs should be poured first and should be a minimum of 5″ thick with 100% compaction of soil or 6″ thick with less than 100% compaction.

Setting times as fast as nine minutes per panel have been observed, but a safer expectation would be four panels per hour with a crane and a four-man setting crew. If crane erects from inside building, some provision must be made to get crane out after walls are erected. Good yarding procedure is important to minimize delays. Equalizing three-point lifting beams and self-releasing pick-up hooks speed erection. If panels must be carried to their final location, setting time per panel will be increased and erection costs may approach the erection cost range of architectural precast wall panels. Placing panels into slots formed in continuous footers will speed erection.

Reinforcing should be with #5 bars with vertical bars on the bottom. If surface is to be sandblasted, stainless steel chairs should be used to prevent rust staining.

Use of a broom finish is popular since the unavoidable surface blemishes are concealed.

Precast columns run from three to five times the C.Y. price of the panels only.

Concrete R0352 Lightweight Concrete Roof Insulation

R035216-10 Lightweight Concrete

Lightweight aggregate concrete is usually purchased ready mixed, but it can also be field mixed.

Vermiculite or Perlite comes in bags of 4 C.F. under various trade names. Weight is about 8 lbs. per C.F. For insulating roof fill use 1:6 mix. For structural deck use 1:4 mix over gypsum boards, steeltex, steel centering, etc., supported by closely spaced joists or bulb trees. For structural slabs use 1:3:2 vermiculite sand concrete over steeltex, metal lath, steel centering, etc., on joists spaced 2′-0″ O.C. for maximum L.L. of 80 P.S.F. Use same mix for slab base fill over steel flooring or regular reinforced concrete slab when tile, terrazzo or other finish is to be laid over.

For slabs on grade use 1:3:2 mix when tile, etc., finish is to be laid over. If radiant heating units are installed use a 1:6 mix for a base. After coils are in place, cover with a regular granolithic finish (mix 1:3:2) to a minimum depth of 1-1/2″ over top of units.

Reinforce all slabs with 6 × 6 or 10 × 10 welded wire mesh.

R040130-10 Cleaning Face Brick

On smooth brick a person can clean 70 S.F. an hour; on rough brick 50 S.F. per hour. Use one gallon muriatic acid to 20 gallons of water for 1000 S.F. Do not use acid solution until wall is at least seven days old, but a mild soap solution may be used after two days.

Time has been allowed for cleanup in brick prices.

R040513-10 Cement Mortar (material only)

Type N - 1:1:6 mix by volume. Use everywhere above grade except as noted below. - 1:3 mix using conventional masonry cement which saves handling two separate bagged materials.

Type M - 1:1/4:3 mix by volume, or 1 part cement, 1/4 (10% by wt.) lime, 3 parts sand. Use for heavy loads and where earthquakes or hurricanes may occur. Also for reinforced brick, sewers, manholes and everywhere below grade.

Mix Proportions by Volume and Compressive Strength of Mortar

Where Used	Allowable Proportions by Volume					Compressive Strength @ 28 days
	Mortar Type	Portland Cement	Masonry Cement	Hydrated Lime	Masonry Sand	
Plain Masonry	M	1	1	—	6	2500 psi
		1	—	1/4	3	
	S	1/2	1	—	4	1800 psi
		1	—	1/4 to 1/2	4	
	N	—	1	—	3	750 psi
		1	—	1/2 to 1-1/4	6	
	O	—	1	—	3	350 psi
		1	—	1-1/4 to 2-1/2	9	
	K	1	—	2-1/2 to 4	12	75 psi
Reinforced Masonry	PM	1	1	—	6	2500 psi
	PL	1	—	1/4 to 1/2	4	2500 psi

Note: The total aggregate should be between 2.25 to 3 times the sum of the cement and lime used.

The labor cost to mix the mortar is included in the productivity and labor cost of unit price lines in unit cost sections for brickwork, blockwork and stonework.

The material cost of mixed mortar is included in the material cost of those same unit price lines and includes the cost of renting and operating a 10 C.F. mixer at the rate of 200 C.F. per day.

There are two types of mortar color used. One type is the inert additive type with about 100 lbs. per M brick as the typical quantity required. These colors are also available in smaller-batch-sized bags (1 lb. to 15 lb.) which can be placed directly into the mixer without measuring. The other type is premixed and replaces the masonry cement. Dark green color has the highest cost.

R040519-50 Masonry Reinforcing

Horizontal joint reinforcing helps prevent wall cracks where wall movement may occur and in many locations is required by code. Horizontal joint reinforcing is generally not considered to be structural reinforcing and an unreinforced wall may still contain joint reinforcing.

Reinforcing strips come in 10′ and 12′ lengths and in truss and ladder shapes, with and without drips. Field labor runs between 2.7 to 5.3 hours per 1000 L.F. for wall thicknesses up to 12″.

The wire meets ASTM A82 for cold drawn steel wire and the typical size is 9 ga. sides and ties with 3/16″ diameter also available. Typical finish is mill galvanized with zinc coating at .10 oz. per S.F. Class I (.40 oz. per S.F.) and Class III (.80 oz per S.F.) are also available, as is hot dipped galvanizing at 1.50 oz. per S.F.

R042110-10 Economy in Bricklaying

Have adequate supervision. Be sure bricklayers are always supplied with materials so there is no waiting. Place best bricklayers at corners and openings.

Use only screened sand for mortar. Otherwise, labor time will be wasted picking out pebbles. Use seamless metal tubs for mortar as they do not leak or catch the trowel. Locate stack and mortar for easy wheeling.

Have brick delivered for stacking. This makes for faster handling, reduces chipping and breakage, and requires less storage space. Many dealers will deliver select common in 2′ × 3′ × 4′ pallets or face brick packaged. This affords quick handling with a crane or forklift and easy tonging in units of ten, which reduces waste.

Use wider bricks for one wythe wall construction. Keep scaffolding away from wall to allow mortar to fall clear and not stain wall.

On large jobs develop specialized crews for each type of masonry unit.

Consider designing for prefabricated panel construction on high rise projects.

Avoid excessive corners or openings. Each opening adds about 50% to labor cost for area of opening.

Bolting stone panels and using window frames as stops reduces labor costs and speeds up erection.

R042110-20 Common and Face Brick

Common building brick manufactured according to ASTM C62 and facing brick manufactured according to ASTM C216 are the two standard bricks available for general building use.

Building brick is made in three grades; SW, where high resistance to damage caused by cyclic freezing is required; MW, where moderate resistance to cyclic freezing is needed; and NW, where little resistance to cyclic freezing is needed. Facing brick is made in only the two grades SW and MW. Additionally, facing brick is available in three types; FBS, for general use; FBX, for general use where a higher degree of precision and lower permissible variation in size than FBS is needed; and FBA, for general use to produce characteristic architectural effects resulting from non-uniformity in size and texture of the units.

In figuring the material cost of brickwork, an allowance of 25% mortar waste and 3% brick breakage was included. If bricks are delivered palletized with 280 to 300 per pallet, or packaged, allow only 1-1/2% for breakage. Packaged or palletized delivery is practical when a job is big enough to have a crane or other equipment available to handle a package of brick. This is so on all industrial work but not always true on small commercial buildings.

The use of buff and gray face is increasing, and there is a continuing trend to the Norman, Roman, Jumbo and SCR brick.

Common red clay brick for backup is not used that often. Concrete block is the most usual backup material with occasional use of sand lime or cement brick. Building brick is commonly used in solid walls for strength and as a fire stop.

Brick panels built on the ground and then crane erected to the upper floors have proven to be economical. This allows the work to be done under cover and without scaffolding.

R042110-50 Brick, Block & Mortar Quantities

Running Bond					
Number of Brick per S.F. of Wall - Single Wythe with 3/8″ Joints				C.F. of Mortar per M Bricks, Waste Included	
Type Brick	Nominal Size (incl. mortar) L H W	Modular Coursing	Number of Brick per S.F.	3/8″ Joint	1/2″ Joint
Standard	8 x 2-2/3 x 4	3C=8″	6.75	8.1	10.3
Economy	8 x 4 x 4	1C=4″	4.50	9.1	11.6
Engineer	8 x 3-1/5 x 4	5C=16″	5.63	8.5	10.8
Fire	9 x 2-1/2 x 4-1/2	2C=5″	6.40	550 # Fireclay	—
Jumbo	12 x 4 x 6 or 8	1C=4″	3.00	22.5	29.2
Norman	12 x 2-2/3 x 4	3C=8″	4.50	11.2	14.3
Norwegian	12 x 3-1/5 x 4	5C=16″	3.75	11.7	14.9
Roman	12 x 2 x 4	2C=4″	6.00	10.7	13.7
SCR	12 x 2-2/3 x 6	3C=8″	4.50	21.8	28.0
Utility	12 x 4 x 4	1C=4″	3.00	12.3	15.7

For Other Bonds Standard Size Add to S.F. Quantities in Table to Left		
Bond Type	Description	Factor
Common	full header every fifth course	+20%
	full header every sixth course	+16.7%
English	full header every second course	+50%
Flemish	alternate headers every course	+33.3%
	every sixth course	+5.6%
Header = W x H exposed		+100%
Rowlock = H x W exposed		+100%
Rowlock stretcher = L x W exposed		+33.3%
Soldier = H x L exposed		—
Sailor = W x L exposed		-33.3%

Concrete Blocks Nominal Size	Approximate Weight per S.F.		Blocks per 100 S.F.	Mortar per M block, waste included	
	Standard	Lightweight		Partitions	Back up
2″ x 8″ x 16″	20 PSF	15 PSF	113	27 C.F.	36 C.F.
4″	30	20		41	51
6″	42	30		56	66
8″	55	38		72	82
10″	70	47		87	97
12″	85	55		102	112

Brick & Mortar Quantities
©Brick Industry Association. 2009 Feb. Technical Notes on Brick Construction 10:
Dimensioning and Estimating Brick Masonry. Reston (VA): BIA. Table 1 Modular Brick Sizes and Table 4 Quantity Estimates for Brick Masonry.

R042210-20 Concrete Block

The material cost of special block such as corner, jamb and head block can be figured at the same price as ordinary block of same size. Labor on specials is about the same as equal-sized regular block.

Bond beam and 16″ high lintel blocks are more expensive than regular units of equal size. Lintel blocks are 8″ long and either 8″ or 16″ high.

Use of motorized mortar spreader box will speed construction of continuous walls.

Hollow non-load-bearing units are made according to ASTM C129 and hollow load-bearing units according to ASTM C90.

R050516-30 Coating Structural Steel

On field-welded jobs, the shop-applied primer coat is necessarily omitted. All painting must be done in the field and usually consists of red oxide rust inhibitive paint or an aluminum paint. The table below shows paint coverage and daily production for field painting.

See Division 05 05 13.50 for hot-dipped galvanizing and Division 09 97 13.23 for field-applied cold galvanizing and other paints and protective coatings.

See Division 05 01 10.51 for steel surface preparation treatments such as wire brushing, pressure washing and sand blasting.

Type Construction	Surface Area per Ton	Coat	One Gallon Covers		In 8 Hrs. Person Covers		Average per Ton Spray	
			Brush	Spray	Brush	Spray	Gallons	Labor-hours
Light Structural	300 S.F. to 500 S.F.	1st	500 S.F.	455 S.F.	640 S.F.	2000 S.F.	0.9 gals.	1.6 L.H.
		2nd	450	410	800	2400	1.0	1.3
		3rd	450	410	960	3200	1.0	1.0
Medium	150 S.F. to 300 S.F.	All	400	365	1600	3200	0.6	0.6
Heavy Structural	50 S.F. to 150 S.F.	1st	400	365	1920	4000	0.2	0.2
		2nd	400	365	2000	4000	0.2	0.2
		3rd	400	365	2000	4000	0.2	0.2
Weighted Average	225 S.F.	All	400	365	1350	3000	0.6	0.6

R050521-20 Welded Structural Steel

Usual weight reductions with welded design run 10% to 20% compared with bolted or riveted connections. This amounts to about the same total cost compared with bolted structures since field welding is more expensive than bolts. For normal spans of 18′ to 24′ figure 6 to 7 connections per ton.

Trusses — For welded trusses add 4% to weight of main members for connections. Up to 15% less steel can be expected in a welded truss compared to one that is shop bolted. Cost of erection is the same whether shop bolted or welded.

General — Typical electrodes for structural steel welding are E6010, E6011, E60T and E70T. Typical buildings vary between 2# to 8# of weld rod per ton of steel. Buildings utilizing continuous design require about three times as much welding as conventional welded structures. In estimating field erection by welding, it is best to use the average linear feet of weld per ton to arrive at the welding cost per ton. The type, size and position of the weld will have a direct bearing on the cost per linear foot. A typical field welder will deposit 1.8# to 2# of weld rod per hour manually. Using semiautomatic methods can increase production by as much as 50% to 75%.

R050523-10 High Strength Bolts

Common bolts (A307) are usually used in secondary connections (see Division 05 05 23.10).

High strength bolts (A325 and A490) are usually specified for primary connections such as column splices, beam and girder connections to columns, column bracing, connections for supports of operating equipment or of other live loads which produce impact or reversal of stress, and in structures carrying cranes of over 5-ton capacity.

Allow 20 field bolts per ton of steel for a 6 story office building, apartment house or light industrial building. For 6 to 12 stories allow 25 bolts per ton, and above 12 stories, 30 bolts per ton. On power stations, 20 to 25 bolts per ton are needed.

R051223-10 Structural Steel

The bare material prices for structural steel, shown in the unit cost sections of the book, are for 100 tons of shop-fabricated structural steel and include:

1. Mill base price of structural steel
2. Mill scrap/grade/size/length extras
3. Mill delivery to a metals service center (warehouse)
4. Service center storage and handling
5. Service center delivery to a fabrication shop
6. Shop storage and handling
7. Shop drafting/detailing
8. Shop fabrication
9. Shop coat of primer paint
10. Shop listing
11. Shop delivery to the job site

In unit cost sections of the book that contain items for field fabrication of steel components, the bare material cost of steel includes:

1. Mill base price of structural steel
2. Mill scrap/grade/size/length extras
3. Mill delivery to a metals service center (warehouse)
4. Service center storage and handling
5. Service center delivery to the job site

R051223-20 Steel Estimating Quantities

One estimate on erection is that a crane can handle 35 to 60 pieces per day. Say the average is 45. With usual sizes of beams, girders, and columns, this would amount to about 20 tons per day. The type of connection greatly affects the speed of erection. Moment connections for continuous design slow down production and increase erection costs.

Short open web bar joists can be set at the rate of 75 to 80 per day, with 50 per day being the average for setting long span joists.

After main members are calculated, add the following for usual allowances: base plates 2% to 3%; column splices 4% to 5%; and miscellaneous details 4% to 5%, for a total of 10% to 13% in addition to main members.

The ratio of column to beam tonnage varies depending on type of steels used, typical spans, story heights and live loads.

It is more economical to keep the column size constant and to vary the strength of the column by using high strength steels. This also saves floor space. Buildings have recently gone as high as ten stories with 8″ high strength columns. For light columns under W8X31 lb. sections, concrete filled steel columns are economical.

High strength steels may be used in columns and beams to save floor space and to meet head room requirements. High strength steels in some sizes sometimes require long lead times.

Round, square and rectangular columns, both plain and concrete filled, are readily available and save floor area, but are higher in cost per pound than rolled columns. For high unbraced columns, tube columns may be less expensive.

Below are average minimum figures for the weights of the structural steel frame for different types of buildings using A36 steel, rolled shapes and simple joints. For economy in domes, rise to span ratio = .13. Open web joist framing systems will reduce weights by 10% to 40%. Composite design can reduce steel weight by up to 25% but additional concrete floor slab thickness may be required. Continuous design can reduce the weights up to 20%. There are many building codes with different live load requirements and different structural requirements, such as hurricane and earthquake loadings which can alter the figures.

Structural Steel Weights per S.F. of Floor Area

Type of Building	No. of Stories	Avg. Spans	L.L. #/S.F.	Lbs. Per S.F.	Type of Building	No. of Stories	Avg. Spans	L.L. #/S.F.	Lbs. Per S.F.
		20′x20′		8	Apartments	2-8	20′x20′	40	8
Steel Frame Mfg.	1	30′x30′	40	13		9-25			14
		40′x40′		18		to 10			10
Parking garage	4	Various	80	8.5	Office	20	Various	80	18
Domes	1	200′	30	10		30			26
(Schwedler)*		300′		15		over 50			35

R051223-25 Common Structural Steel Specifications

ASTM A992 (formerly A36, then A572 Grade 50) is the all-purpose carbon grade steel widely used in building and bridge construction.

The other high-strength steels listed below may each have certain advantages over ASTM A992 stuctural carbon steel, depending on the application. They have proven to be economical choices where, due to lighter members, the reduction of dead load and the associated savings in shipping cost can be significant.

ASTM A588 atmospheric weathering, high-strength low-alloy steels can be used in the bare (uncoated) condition, where exposure to normal atmosphere causes a tightly adherant oxide to form on the surface protecting the steel from further oxidation. ASTM A242 corrosion-resistant, high-strength low-alloy steels have enhanced atmospheric corrosion resistance of at least two times that of carbon structural steels with copper, or four times that of carbon structural steels without copper. The reduction or elimination of maintenance resulting from the use of these steels often offsets their higher initial cost.

Steel Type	ASTM Designation	Minimum Yield Stress in KSI	Shapes Available
Carbon	A36	36	All structural shape groups, and plates & bars up thru 8″ thick
	A529	50	Structural shape group 1, and plates & bars up thru 2″ thick
High-Strength Low-Alloy Quenched & Self-Tempered	A913	50 60 65 70	All structural shape groups
High-Strength Low-Alloy Columbium-Vanadium	A572	42	All structural shape groups, and plates & bars up thru 6″ thick
		50	All structural shape groups, and plates & bars up thru 4″ thick
		55	Structural shape groups 1 & 2, and plates & bars up thru 2″ thick
		60	Structural shape groups 1 & 2, and plates & bars up thru 1-1/4″ thick
		65	Structural shape group 1, and plates & bars up thru 1-1/4″ thick
High-Strength Low-Alloy Columbium-Vanadium	A992	50	All structural shape groups
Weathering High-Strength Low-Alloy	A242	42	Structural shape groups 4 & 5, and plates & bars over 1-1/2″ up thru 4″ thick
		46	Structural shape group 3, and plates & bars over 3/4″ up thru 1-1/2″ thick
		50	Structural shape groups 1 & 2, and plates & bars up thru 3/4″ thick
Weathering High-Strength Low-Alloy	A588	42	Plates & bars over 5″ up thru 8″ thick
		46	Plates & bars over 4″ up thru 5″ thick
		50	All structural shape groups, and plates & bars up thru 4″ thick
Quenched and Tempered Low-Alloy	A852	70	Plates & bars up thru 4″ thick
Quenched and Tempered Alloy	A514	90	Plates & bars over 2-1/2″ up thru 6″ thick
		100	Plates & bars up thru 2-1/2″ thick

R051223-30 High Strength Steels

The mill price of high strength steels may be higher than A992 carbon steel but their proper use can achieve overall savings through total reduced weights. For columns with L/r over 100, A992 steel is best; under 100, high strength steels are economical. For heavy columns, high strength steels are economical when cover plates are eliminated. There is no economy using high strength steels for clip angles or supports or for beams where deflection governs. Thinner members are more economical than thick.

The per ton erection and fabricating costs of the high strength steels will be higher than for A992 since the same number of pieces, but less weight, will be installed.

R051223-35 Common Steel Sections

The upper portion of this table shows the name, shape, common designation and basic characteristics of commonly used steel sections. The lower portion explains how to read the designations used for the above illustrated common sections.

Shape & Designation	Name & Characteristics	Shape & Designation	Name & Characteristics
W	W Shape Parallel flange surfaces	MC	Miscellaneous Channel Infrequently rolled by some producers
S	American Standard Beam (I Beam) Sloped inner flange	L	Angle Equal or unequal legs, constant thickness
M	Miscellaneous Beams Cannot be classified as W, HP or S; infrequently rolled by some producers	T	Structural Tee Cut from W, M or S on center of web
C	American Standard Channel Sloped inner flange	HP	Bearing Pile Parallel flanges and equal flange and web thickness

Common drawing designations follow:

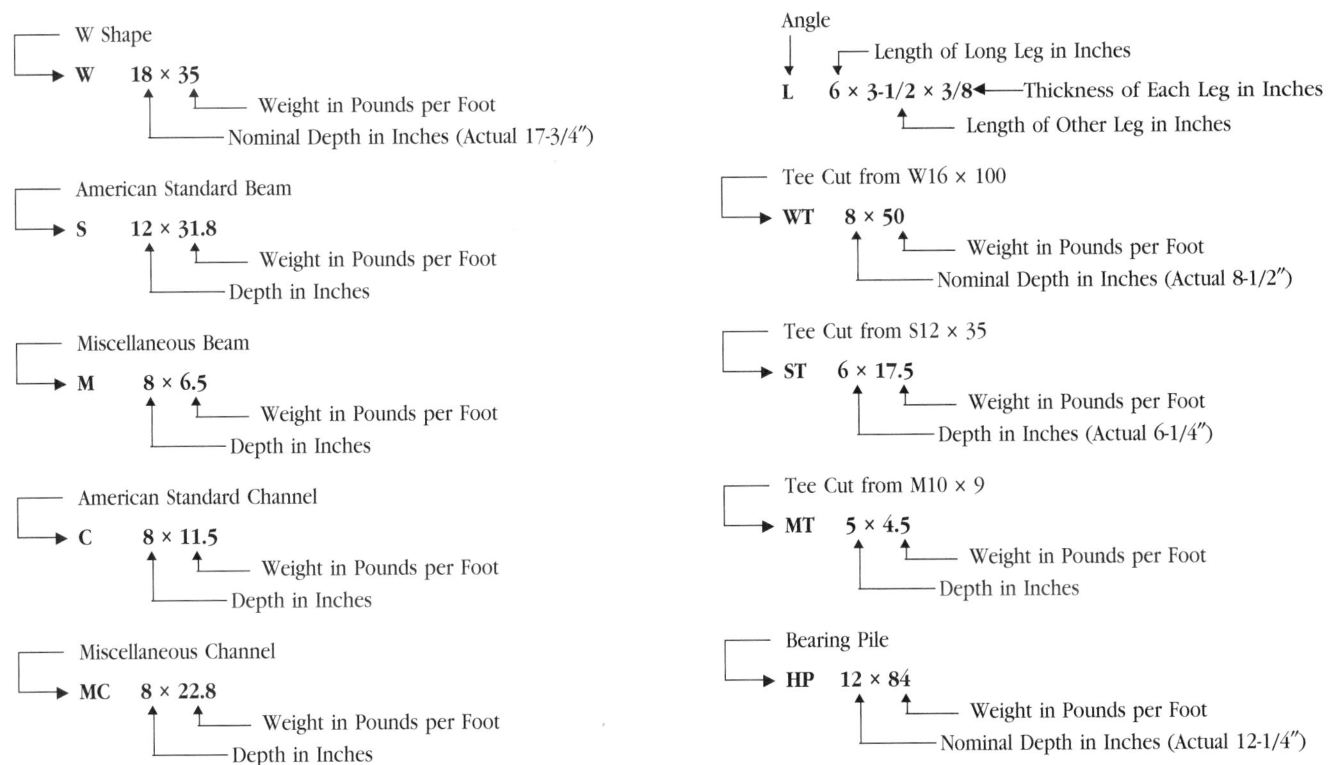

R051223-45 Installation Time for Structural Steel Building Components

The following tables show the expected average installation times for various structural steel shapes. Table A presents installation times for columns, Table B for beams, Table C for light framing and bolts, and Table D for structural steel for various project types.

Table A

Description	Labor-Hours	Unit
Columns		
Steel, Concrete Filled		
3-1/2″ Diameter	.933	Ea.
6-5/8″ Diameter	1.120	Ea.
Steel Pipe		
3″ Diameter	.933	Ea.
8″ Diameter	1.120	Ea.
12″ Diameter	1.244	Ea.
Structural Tubing		
4″ x 4″	.966	Ea.
8″ x 8″	1.120	Ea.
12″ x 8″	1.167	Ea.
W Shape 2 Tier		
W8 x 31	.052	L.F.
W8 x 67	.057	L.F.
W10 x 45	.054	L.F.
W10 x 112	.058	L.F.
W12 x 50	.054	L.F.
W12 x 190	.061	L.F.
W14 x 74	.057	L.F.
W14 x 176	.061	L.F.

Table B

Description	Labor-Hours	Unit	Labor-Hours	Unit
Beams, W Shape				
W6 x 9	.949	Ea.	.093	L.F.
W10 x 22	1.037	Ea.	.085	L.F.
W12 x 26	1.037	Ea.	.064	L.F.
W14 x 34	1.333	Ea.	.069	L.F.
W16 x 31	1.333	Ea.	.062	L.F.
W18 x 50	2.162	Ea.	.088	L.F.
W21 x 62	2.222	Ea.	.077	L.F.
W24 x 76	2.353	Ea.	.072	L.F.
W27 x 94	2.581	Ea.	.067	L.F.
W30 x 108	2.857	Ea.	.067	L.F.
W33 x 130	3.200	Ea.	.071	L.F.
W36 x 300	3.810	Ea.	.077	L.F.

Table C

Description	Labor-Hours	Unit
Light Framing		
Angles 4″ and Larger	.055	lbs.
Less than 4″	.091	lbs.
Channels 8″ and Larger	.048	lbs.
Less than 8″	.072	lbs.
Cross Bracing Angles	.055	lbs.
Rods	.034	lbs.
Hanging Lintels	.069	lbs.
High Strength Bolts in Place		
3/4″ Bolts	.070	Ea.
7/8″ Bolts	.076	Ea.

Table D

Description	Labor-Hours	Unit	Labor-Hours	Unit
Apartments, Nursing Homes, etc.				
1-2 Stories	4.211	Piece	7.767	Ton
3-6 Stories	4.444	Piece	7.921	Ton
7-15 Stories	4.923	Piece	9.014	Ton
Over 15 Stories	5.333	Piece	9.209	Ton
Offices, Hospitals, etc.				
1-2 Stories	4.211	Piece	7.767	Ton
3-6 Stories	4.741	Piece	8.889	Ton
7-15 Stories	4.923	Piece	9.014	Ton
Over 15 Stories	5.120	Piece	9.209	Ton
Industrial Buildings				
1 Story	3.478	Piece	6.202	Ton

R051223-50 Subpurlins

Bulb tee subpurlins are structural members designed to support and reinforce a variety of roof deck systems such as precast cement fiber roof deck tiles, monolithic roof deck systems, and gypsum or lightweight concrete over formboard. Other uses include interstitial service ceiling systems, wall panel systems, and joist anchoring in bond beams. See Unit Price section for pricing on a square foot basis at 32-5/8″ O.C. Maximum span is based on a 3-span condition with a total allowable vertical load of 40 psf.

R051223-80 Dimensions and Weights of Sheet Steel

Gauge No.	Approximate Thickness				Weight		
	Inches (in fractions)	Inches (in decimal parts)		Millimeters			
	Wrought Iron	Wrought Iron	Steel	Steel	per S.F. in Ounces	per S.F. in Lbs.	per Square Meter in Kg.
0000000	1/2″	.5	.4782	12.146	320	20.000	97.650
000000	15/32″	.46875	.4484	11.389	300	18.750	91.550
00000	7/16″	.4375	.4185	10.630	280	17.500	85.440
0000	13/32″	.40625	.3886	9.870	260	16.250	79.330
000	3/8″	.375	.3587	9.111	240	15.000	73.240
00	11/32″	.34375	.3288	8.352	220	13.750	67.130
0	5/16″	.3125	.2989	7.592	200	12.500	61.030
1	9/32″	.28125	.2690	6.833	180	11.250	54.930
2	17/64″	.265625	.2541	6.454	170	10.625	51.880
3	1/4″	.25	.2391	6.073	160	10.000	48.820
4	15/64″	.234375	.2242	5.695	150	9.375	45.770
5	7/32″	.21875	.2092	5.314	140	8.750	42.720
6	13/64″	.203125	.1943	4.935	130	8.125	39.670
7	3/16″	.1875	.1793	4.554	120	7.500	36.320
8	11/64″	.171875	.1644	4.176	110	6.875	33.570
9	5/32″	.15625	.1495	3.797	100	6.250	30.520
10	9/64″	.140625	.1345	3.416	90	5.625	27.460
11	1/8″	.125	.1196	3.038	80	5.000	24.410
12	7/64″	.109375	.1046	2.657	70	4.375	21.360
13	3/32″	.09375	.0897	2.278	60	3.750	18.310
14	5/64″	.078125	.0747	1.897	50	3.125	15.260
15	9/128″	.0713125	.0673	1.709	45	2.813	13.730
16	1/16″	.0625	.0598	1.519	40	2.500	12.210
17	9/160″	.05625	.0538	1.367	36	2.250	10.990
18	1/20″	.05	.0478	1.214	32	2.000	9.765
19	7/160″	.04375	.0418	1.062	28	1.750	8.544
20	3/80″	.0375	.0359	.912	24	1.500	7.324
21	11/320″	.034375	.0329	.836	22	1.375	6.713
22	1/32″	.03125	.0299	.759	20	1.250	6.103
23	9/320″	.028125	.0269	.683	18	1.125	5.490
24	1/40″	.025	.0239	.607	16	1.000	4.882
25	7/320″	.021875	.0209	.531	14	.875	4.272
26	3/160″	.01875	.0179	.455	12	.750	3.662
27	11/640″	.0171875	.0164	.417	11	.688	3.357
28	1/64″	.015625	.0149	.378	10	.625	3.052

R053100-10 Decking Descriptions

General - All Deck Products

Steel deck is made by cold forming structural grade sheet steel into a repeating pattern of parallel ribs. The strength and stiffness of the panels are the result of the ribs and the material properties of the steel. Deck lengths can be varied to suit job conditions, but because of shipping considerations, are usually less than 40 feet. Standard deck width varies with the product used but full sheets are usually 12″, 18″, 24″, 30″, or 36″. Deck is typically furnished in a standard width with the ends cut square. Any cutting for width, such as at openings or for angular fit, is done at the job site.

Deck is typically attached to the building frame with arc puddle welds, self-drilling screws, or powder or pneumatically driven pins. Sheet to sheet fastening is done with screws, button punching (crimping), or welds.

Composite Floor Deck

After installation and adequate fastening, floor deck serves several purposes. It (a) acts as a working platform, (b) stabilizes the frame, (c) serves as a concrete form for the slab, and (d) reinforces the slab to carry the design loads applied during the life of the building. Composite decks are distinguished by the presence of shear connector devices as part of the deck. These devices are designed to mechanically lock the concrete and deck together so that the concrete and the deck work together to carry subsequent floor loads. These shear connector devices can be rolled-in embossments, lugs, holes, or wires welded to the panels. The deck profile can also be used to interlock concrete and steel.

Composite deck finishes are either galvanized (zinc coated) or phosphatized/painted. Galvanized deck has a zinc coating on both the top and bottom surfaces. The phosphatized/painted deck has a bare (phosphatized) top surface that will come into contact with the concrete. This bare top surface can be expected to develop rust before the concrete is placed. The bottom side of the deck has a primer coat of paint.

Composite floor deck is normally installed so the panel ends do not overlap on the supporting beams. Shear lugs or panel profile shape often prevent a tight metal to metal fit if the panel ends overlap; the air gap caused by overlapping will prevent proper fusion with the structural steel supports when the panel end laps are shear stud welded.

Adequate end bearing of the deck must be obtained as shown on the drawings. If bearing is actually less in the field than shown on the drawings, further investigation is required.

Roof Deck

Roof deck is not designed to act compositely with other materials. Roof deck acts alone in transferring horizontal and vertical loads into the building frame. Roof deck rib openings are usually narrower than floor deck rib openings. This provides adequate support of rigid thermal insulation board.

Roof deck is typically installed to endlap approximately 2″ over supports. However, it can be butted (or lapped more than 2″) to solve field fit problems. Since designers frequently use the installed deck system as part of the horizontal bracing system (the deck as a diaphragm), any fastening substitution or change should be approved by the designer. Continuous perimeter support of the deck is necessary to limit edge deflection in the finished roof and may be required for diaphragm shear transfer.

Standard roof deck finishes are galvanized or primer painted. The standard factory applied paint for roof deck is a primer paint and is not intended to weather for extended periods of time. Field painting or touching up of abrasions and deterioration of the primer coat or other protective finishes is the responsibility of the contractor.

Cellular Deck

Cellular deck is made by attaching a bottom steel sheet to a roof deck or composite floor deck panel. Cellular deck can be used in the same manner as floor deck. Electrical, telephone, and data wires are easily run through the chase created between the deck panel and the bottom sheet.

When used as part of the electrical distribution system, the cellular deck must be installed so that the ribs line up and create a smooth cell transition at abutting ends. The joint that occurs at butting cell ends must be taped or otherwise sealed to prevent wet concrete from seeping into the cell. Cell interiors must be free of welding burrs, or other sharp intrusions, to prevent damage to wires.

When used as a roof deck, the bottom flat plate is usually left exposed to view. Care must be maintained during erection to keep good alignment and prevent damage.

Cellular deck is sometimes used with the flat plate on the top side to provide a flat working surface. Installation of the deck for this purpose requires special methods for attachment to the frame because the flat plate, now on the top, can prevent direct access to the deck material that is bearing on the structural steel. It may be advisable to treat the flat top surface to prevent slipping.

Cellular deck is always furnished galvanized or painted over galvanized.

Form Deck

Form deck can be any floor or roof deck product used as a concrete form. Connections to the frame are by the same methods used to anchor floor and roof deck. Welding washers are recommended when welding deck that is less than 20 gauge thickness.

Form deck is furnished galvanized, prime painted, or uncoated. Galvanized deck must be used for those roof deck systems where form deck is used to carry a lightweight insulating concrete fill.

Wood, Plastics & Comp. R0611 Wood Framing

R061110-30 Lumber Product Material Prices

The price of forest products fluctuates widely from location to location and from season to season depending upon economic conditions. The bare material prices in the unit cost sections of the book show the National Average material prices in effect Jan. 1 of this book year. It must be noted that lumber prices in general may change significantly during the year.

Availability of certain items depends upon geographic location and must be checked prior to firm-price bidding.

Wood, Plastics & Comp. R0616 Sheathing

R061636-20 Plywood

There are two types of plywood used in construction: interior, which is moisture-resistant but not waterproofed, and exterior, which is waterproofed.

The grade of the exterior surface of the plywood sheets is designated by the first letter: A, for smooth surface with patches allowed; B, for solid surface with patches and plugs allowed; C, which may be surface plugged or may have knot holes up to 1″ wide; and D, which is used only for interior type plywood and may have knot holes up to 2-1/2″ wide. "Structural Grade" is specifically designed for engineered applications such as box beams. All CC & DD grades have roof and floor spans marked on them.

Underlayment-grade plywood runs from 1/4″ to 1-1/4″ thick. Thicknesses 5/8″ and over have optional tongue and groove joints which eliminate the need for blocking the edges. Underlayment 19/32″ and over may be referred to as Sturd-i-Floor.

The price of plywood can fluctuate widely due to geographic and economic conditions.

Typical uses for various plywood grades are as follows:

AA-AD Interior — cupboards, shelving, paneling, furniture

BB Plyform — concrete form plywood

CDX — wall and roof sheathing

Structural — box beams, girders, stressed skin panels

AA-AC Exterior — fences, signs, siding, soffits, etc.

Underlayment — base for resilient floor coverings

Overlaid HDO — high density for concrete forms & highway signs

Overlaid MDO — medium density for painting, siding, soffits & signs

303 Siding — exterior siding, textured, striated, embossed, etc.

Thermal & Moist. Protec. R0731 Shingles & Shakes

R073126-20 Roof Slate

16″, 18″ and 20″ are standard lengths, and slate usually comes in random widths. For standard 3/16″ thickness use 1-1/2″ copper nails. Allow for 3% breakage.

Thermal & Moist. Protec. R0751 Built-Up Bituminous Roofing

R075113-20 Built-Up Roofing

Asphalt is available in kegs of 100 lbs. each; coal tar pitch in 560 lb. kegs. Prepared roofing felts are available in a wide range of sizes, weights and characteristics. However, the most commonly used are #15 (432 S.F. per roll, 13 lbs. per square) and #30 (216 S.F. per roll, 27 lbs. per square).

Inter-ply bitumen varies from 24 lbs. per sq. (asphalt) to 30 lbs. per sq. (coal tar) per ply, MF4@ 25%. Flood coat bitumen also varies from 60 lbs. per sq. (asphalt) to 75 lbs. per sq. (coal tar), MF4@ 25%. Expendable equipment (mops, brooms, screeds, etc.) runs about 16% of the bitumen cost. For new, inexperienced crews this factor may be much higher.

Rigid insulation board is typically applied in two layers. The first is mechanically attached to nailable decks or spot or solid mopped to non-nailable decks; the second layer is then spot or solid mopped to the first layer. Membrane application follows the insulation, except in protected membrane roofs, where the membrane goes down first and the insulation on top, followed with ballast (stone or concrete pavers). Insulation and related labor costs are NOT included in prices for built-up roofing.

R075213-30 Modified Bitumen Roofing

The cost of modified bitumen roofing is highly dependent on the type of installation that is planned. Installation is based on the type of modifier used in the bitumen. The two most popular modifiers are atactic polypropylene (APP) and styrene butadiene styrene (SBS). The modifiers are added to heated bitumen during the manufacturing process to change its characteristics. A polyethylene, polyester or fiberglass reinforcing sheet is then sandwiched between layers of this bitumen. When completed, the result is a pre-assembled, built-up roof that has increased elasticity and weatherablility. Some manufacturers include a surfacing material such as ceramic or mineral granules, metal particles or sand.

The preferred method of adhering SBS-modified bitumen roofing to the substrate is with hot-mopped asphalt (much the same as built-up roofing). This installation method requires a tar kettle/pot to heat the asphalt, as well as the labor, tools and equipment necessary to distribute and spread the hot asphalt.

The alternative method for applying APP and SBS modified bitumen is as follows. A skilled installer uses a torch to melt a small pool of bitumen off the membrane. This pool must form across the entire roll for proper adhesion. The installer must unroll the roofing at a pace slow enough to melt the bitumen, but fast enough to prevent damage to the rest of the membrane.

Modified bitumen roofing provides the advantages of both built-up and single-ply roofing. Labor costs are reduced over those of built-up roofing because only a single ply is necessary. The elasticity of single-ply roofing is attained with the reinforcing sheet and polymer modifiers. Modifieds have some self-healing characteristics and because of their multi-layer construction, they offer the reliability and safety of built-up roofing.

R078413-30 Firestopping

Firestopping is the sealing of structural, mechanical, electrical and other penetrations through fire-rated assemblies. The basic components of firestop systems are safing insulation and firestop sealant on both sides of wall penetrations and the top side of floor penetrations.

Pipe penetrations are assumed to be through concrete, grout, or joint compound and can be sleeved or unsleeved. Costs for the penetrations and sleeves are not included. An annular space of 1″ is assumed. Escutcheons are not included.

Metallic pipe is assumed to be copper, aluminum, cast iron or similar metallic material. Insulated metallic pipe is assumed to be covered with a thermal insulating jacket of varying thickness and materials.

Non-metallic pipe is assumed to be PVC, CPVC, FR Polypropylene or similar plastic piping material. Intumescent firestop sealant or wrap strips are included. Collars on both sides of wall penetrations and a sheet metal plate on the underside of floor penetrations are included.

Ductwork is assumed to be sheet metal, stainless steel or similar metallic material. Duct penetrations are assumed to be through concrete, grout or joint compound. Costs for penetrations and sleeves are not included. An annular space of 1/2″ is assumed.

Multi-trade openings include costs for sheet metal forms, firestop mortar, wrap strips, collars and sealants as necessary.

Structural penetrations joints are assumed to be 1/2″ or less. CMU walls are assumed to be within 1-1/2″ of metal deck. Drywall walls are assumed to be tight to the underside of metal decking.

Metal panel, glass or curtain wall systems include a spandrel area of 5′ filled with mineral wool foil-faced insulation. Fasteners and stiffeners are included.

R081313-20 Steel Door Selection Guide

Standard steel doors are classified into four levels, as recommended by the Steel Door Institute in the chart below. Each of the four levels offers a range of construction models and designs, to meet architectural requirements for preference and appearance, including full flush, seamless, and stile & rail. Recommended minimum gauge requirements are also included.

For complete standard steel door construction specifications and available sizes, refer to the Steel Door Institute Technical Data Series, ANSI A250.8-98 (SDI-100), and ANSI A250.4-94 Test Procedure and Acceptance Criteria for Physical Endurance of Steel Door and Hardware Reinforcements.

Level		Model	Construction	For Full Flush or Seamless		
				Min. Gauge	Thickness (in)	Thickness (mm)
I	Standard Duty	1 2	Full Flush Seamless	20	0.032	0.8
II	Heavy Duty	1 2	Full Flush Seamless	18	0.042	1.0
III	Extra Heavy Duty	1 2 3	Full Flush Seamless *Stile & Rail	16	0.053	1.3
IV	Maximum Duty	1 2	Full Flush Seamless	14	0.067	1.6

*Stiles & rails are 16 gauge; flush panels, when specified, are 18 gauge

Openings — R0851 Metal Windows

R085123-10 Steel Sash

Ironworker crew will erect 25 S.F. or 1.3 sash unit per hour, whichever is less.

Mechanic will point 30 L.F. per hour.

Painter will paint 90 S.F. per coat per hour.

Glazier production depends on light size.

Allow 1 lb. special steel sash putty per 16″ x 20″ light.

Openings — R0852 Wood Windows

R085216-10 Window Estimates

To ensure a complete window estimate, be sure to include the material and labor costs for each window, as well as the material and labor costs for an interior wood trim set.

R087110-10 Hardware Finishes

This table describes hardware finishes used throughout the industry. It also shows the base metal and the respective symbols in the three predominate systems of identification. Many of these are used in pricing descriptions in Division Eight.

US"	BMHA*	CDN^	Base	Description
US P	600	CP	Steel	Primed for Painting
US 1B	601	C1B	Steel	Bright Black Japanned
US 2C	602	C2C	Steel	Zinc Plated
US 2G	603	C2G	Steel	Zinc Plated
US 3	605	C3	Brass	Bright Brass, Clear Coated
US 4	606	C4	Brass	Satin Brass, Clear Coated
US 5	609	C5	Brass	Satin Brass, Blackened, Satin Relieved, Clear Coated
US 7	610	C7	Brass	Satin Brass, Blackened, Bright Relived, Clear Coated
US 9	611	C9	Bronze	Bright Bronze, Clear Coated
US 10	612	C10	Bronze	Satin Bronze, Clear Coated
US 10A	641	C10A	Steel	Antiqued Bronze, Oiled and Lacquered
US 10B	613	C10B	Bronze	Antiqued Bronze, Oiled
US 11	616	C11	Bronze	Satin Bronze, Blackened, Satin Relieved, Clear Coated
US 14	618	C14	Brass/Bronze	Bright Nickel Plated, Clear Coated
US 15	619	C15	Brass/Bronze	Satin Nickel, Clear Coated
US 15A	620	C15A	Brass/Bronze	Satin Nickel Plated, Blackened, Satin Relieved, Clear Coated
US 17A	621	C17A	Brass/Bronze	Nickel Plated, Blackened, Relieved, Clear Coated
US 19	622	C19	Brass/Bronze	Flat Black Coated
US 20	623	C20	Brass/Bronze	Statuary Bronze, Light
US 20A	624	C20A	Brass/Bronze	Statuary Bronze, Dark
US 26	625	C26	Brass/Bronze	Bright Chromium
US 26D	626	C26D	Brass/Bronze	Satin Chromium
US 20	627	C27	Aluminum	Satin Aluminum Clear
US 28	628	C28	Aluminum	Anodized Dull Aluminum
US 32	629	C32	Stainless Steel	Bright Stainless Steel
US 32D	630	C32D	Stainless Steel	Stainless Steel
US 3	632	C3	Steel	Bright Brass Plated, Clear Coated
US 4	633	C4	Steel	Satin Brass, Clear Coated
US 7	636	C7	Steel	Satin Brass Plated, Blackened, Bright Relieved, Clear Coated
US 9	637	C9	Steel	Bright Bronze Plated, Clear Coated
US 5	638	C5	Steel	Satin Brass Plated, Blackened, Bright Relieved, Clear Coated
US 10	639	C10	Steel	Satin Bronze Plated, Clear Coated
US 10B	640	C10B	Steel	Antique Bronze, Oiled
US 10A	641	C10A	Steel	Antiqued Bronze, Oiled and Lacquered
US 11	643	C11	Steel	Satin Bronze Plated, Blackened, Bright Relieved, Clear Coated
US 14	645	C14	Steel	Bright Nickel Plated, Clear Coated
US 15	646	C15	Steel	Satin Nickel Plated, Clear Coated
US 15A	647	C15A	Steel	Nickel Plated, Blackened, Bright Relieved, Clear Coated
US 17A	648	C17A	Steel	Nickel Plated, Blackened, Relieved, Clear Coated
US 20	649	C20	Steel	Statuary Bronze, Light
US 20A	650	C20A	Steel	Statuary Bronze, Dark
US 26	651	C26	Steel	Bright Chromium Plated
US 26D	652	C26D	Steel	Satin Chromium Plated

* - BMHA Builders Hardware Manufacturing Association
" - US Equivalent
^ - Canadian Equivalent
Japanning is imitating Asian lacquer work

R088110-10 Glazing Productivity

Some glass sizes are estimated by the "united inch" (height + width). The table below shows the number of lights glazed in an eight-hour period by the crew size indicated, for glass up to 1/4" thick. Square or nearly square lights are more economical on a S.F. basis. Long slender lights will have a high S.F. installation cost. For insulated glass reduce production by 33%. For 1/2" float glass reduce production by 50%. Production time for glazing with two glaziers per day averages: 1/4" float glass 120 S.F.; 1/2" float glass 55 S.F.; 1/2" insulated glass 95 S.F.; 3/4" insulated glass 75 S.F.

Glazing Method	United Inches per Light							
	40"	**60"**	**80"**	**100"**	**135"**	**165"**	**200"**	**240"**
Number of Men in Crew	**1**	**1**	**1**	**1**	**2**	**3**	**3**	**4**
Industrial sash, putty	60	45	24	15	18	—	—	—
With stops, putty bed	50	36	21	12	16	8	4	3
Wood stops, rubber	40	27	15	9	11	6	3	2
Metal stops, rubber	30	24	14	9	9	6	3	2
Structural glass	10	7	4	3	—	—	—	—
Corrugated glass	12	9	7	4	4	4	3	—
Storefronts	16	15	13	11	7	6	4	4
Skylights, putty glass	60	36	21	12	16	—	—	—
Thiokol set	15	15	11	9	9	6	3	2
Vinyl set, snap on	18	18	13	12	12	7	5	4
Maximum area per light	2.8 S.F.	6.3 S.F.	11.1 S.F.	17.4 S.F.	31.6 S.F.	47 S.F.	69 S.F.	100 S.F.

R092000-50 Lath, Plaster and Gypsum Board

Gypsum board lath is available in 3/8″ thick x 16″ wide x 4′ long sheets as a base material for multi-layer plaster applications. It is also available as a base for either multi-layer or veneer plaster applications in 1/2″ and 5/8″ thick–4′ wide x 8′, 10′ or 12′ long sheets. Fasteners are screws or blued ring shank nails for wood framing and screws for metal framing.

Metal lath is available in diamond mesh pattern with flat or self-furring profiles. Paper backing is available for applications where excessive plaster waste needs to be avoided. A slotted mesh ribbed lath should be used in areas where the span between structural supports is greater than normal. Most metal lath comes in 27″ x 96″ sheets. Diamond mesh weighs 1.75, 2.5 or 3.4 pounds per square yard, slotted mesh lath weighs 2.75 or 3.4 pounds per square yard. Metal lath can be nailed, screwed or tied in place.

Many **accessories** are available. Corner beads, flat reinforcing strips, casing beads, control and expansion joints, furring brackets and channels are some examples. Note that accessories are not included in plaster or stucco line items.

Plaster is defined as a material or combination of materials that when mixed with a suitable amount of water, forms a plastic mass or paste. When applied to a surface, the paste adheres to it and subsequently hardens, preserving in a rigid state the form or texture imposed during the period of elasticity.

Gypsum plaster is made from ground calcined gypsum. It is mixed with aggregates and water for use as a base coat plaster.

Vermiculite plaster is a fire-retardant plaster covering used on steel beams, concrete slabs and other heavy construction materials. Vermiculite is a group name for certain clay minerals, hydrous silicates or aluminum, magnesium and iron that have been expanded by heat.

Perlite plaster is a plaster using perlite as an aggregate instead of sand. Perlite is a volcanic glass that has been expanded by heat.

Gauging plaster is a mix of gypsum plaster and lime putty that when applied produces a quick drying finish coat.

Veneer plaster is a one or two component gypsum plaster used as a thin finish coat over special gypsum board.

Keenes cement is a white cementitious material manufactured from gypsum that has been burned at a high temperature and ground to a fine powder. Alum is added to accelerate the set. The resulting plaster is hard and strong and accepts and maintains a high polish, hence it is used as a finishing plaster.

Stucco is a Portland cement based plaster used primarily as an exterior finish.

Plaster is used on both interior and exterior surfaces. Generally it is applied in multiple-coat systems. A three-coat system uses the terms scratch, brown and finish to identify each coat. A two-coat system uses base and finish to describe each coat. Each type of plaster and application system has attributes that are chosen by the designer to best fit the intended use.

Gypsum Plaster Quantities for 100 S.Y.	2 Coat, 5/8″ Thick		3 Coat, 3/4″ Thick		
	Base	Finish	Scratch	Brown	Finish
	1:3 Mix	2:1 Mix	1:2 Mix	1:3 Mix	2:1 Mix
Gypsum plaster	1,300 lb.		1,350 lb.	650 lb.	
Sand	1.75 C.Y.		1.85 C.Y.	1.35 C.Y.	
Finish hydrated lime		340 lb.			340 lb.
Gauging plaster		170 lb.			170 lb.

Vermiculite or Perlite Plaster Quantities for 100 S.Y.	2 Coat, 5/8″ Thick		3 Coat, 3/4″ Thick		
	Base	Finish	Scratch	Brown	Finish
Gypsum plaster	1,250 lb.		1,450 lb.	800 lb.	
Vermiculite or perlite	7.8 bags		8.0 bags	3.3 bags	
Finish hydrated lime		340 lb.			340 lb.
Gauging plaster		170 lb.			170 lb.

Stucco–Three-Coat System Quantities for 100 S.Y.	On Wood Frame	On Masonry
Portland cement	29 bags	21 bags
Sand	2.6 C.Y.	2.0 C.Y.
Hydrated lime	180 lb.	120 lb.

R092910-10 Levels of Gypsum Drywall Finish

In the past, contract documents often used phrases such as "industry standard" and "workmanlike finish" to specify the expected quality of gypsum board wall and ceiling installations. The vagueness of these descriptions led to unacceptable work and disputes.

In order to resolve this problem, four major trade associations concerned with the manufacture, erection, finish, and decoration of gypsum board wall and ceiling systems have developed an industry-wide *Recommended Levels of Gypsum Board Finish.*

The finish of gypsum board walls and ceilings for specific final decoration is dependent on a number of factors. A primary consideration is the location of the surface and the degree of decorative treatment desired. Painted and unpainted surfaces in warehouses and other areas where appearance is normally not critical may simply require the taping of wallboard joints and 'spotting' of fastener heads. Blemish-free, smooth, monolithic surfaces often intended for painted and decorated walls and ceilings in habitated structures, ranging from single-family dwellings through monumental buildings, require additional finishing prior to the application of the final decoration.

Other factors to be considered in determining the level of finish of the gypsum board surface are (1) the type of angle of surface illumination (both natural and artificial lighting), and (2) the paint and method of application or the type and finish of wallcovering specified as the final decoration. Critical lighting conditions, gloss paints, and thin wall coverings require a higher level of gypsum board finish than do heavily textured surfaces which are subsequently painted or surfaces which are to be decorated with heavy grade wall coverings.

The following descriptions were developed jointly by the Association of the Wall and Ceiling Industries-International (AWCI), Ceiling & Interior Systems Construction Association (CISCA), Gypsum Association (GA), and Painting and Decorating Contractors of America (PDCA) as a guide.

Level 0: Used in temporary construction or wherever the final decoration has not been determined. Unfinished. No taping, finishing or corner beads are required. Also could be used where non-predecorated panels will be used in demountable-type partitions that are to be painted as a final finish.

Level 1: Frequently used in plenum areas above ceilings, in attics, in areas where the assembly would generally be concealed, or in building service corridors and other areas not normally open to public view. Some degree of sound and smoke control is provided; in some geographic areas, this level is referred to as "fire-taping," although this level of finish does not typically meet fire-resistant assembly requirements. Where a fire resistance rating is required for the gypsum board assembly, details of construction should be in accordance with reports of fire tests of assemblies that have met the requirements of the fire rating acceptable.

All joints and interior angles shall have tape embedded in joint compound. Accessories are optional at specifier discretion in corridors and other areas with pedestrian traffic. Tape and fastener heads need not be covered with joint compound. Surface shall be free of excess joint compound. Tool marks and ridges are acceptable.

Level 2: It may be specified for standard gypsum board surfaces in garages, warehouse storage, or other similar areas where surface appearance is not of primary importance.

All joints and interior angles shall have tape embedded in joint compound and shall be immediately wiped with a joint knife or trowel, leaving a thin coating of joint compound over all joints and interior angles. Fastener heads and accessories shall be covered with a coat of joint compound. Surface shall be free of excess joint compound. Tool marks and ridges are acceptable.

Level 3: Typically used in areas that are to receive heavy texture (spray or hand applied) finishes before final painting, or where commercial-grade (heavy duty) wall coverings are to be applied as the final decoration. This level of finish should not be used where smooth painted surfaces or where lighter weight wall coverings are specified. The prepared surface shall be coated with a drywall primer prior to the application of final finishes.

All joints and interior angles shall have tape embedded in joint compound and shall be immediately wiped with a joint knife or trowel, leaving a thin coating of joint compound over all joints and interior angles. One additional coat of joint compound shall be applied over all joints and interior angles. Fastener heads and accessories shall be covered with two separate coats of joint compound. All joint compounds shall be smooth and free of tool marks and ridges. The prepared surface shall be covered with a drywall primer prior to the application of the final decoration.

Level 4: This level should be used where residential grade (light duty) wall coverings, flat paints, or light textures are to be applied. The prepared surface shall be coated with a drywall primer prior to the application of final finishes. Release agents for wall coverings are specifically formulated to minimize damage if coverings are subsequently removed.

The weight, texture, and sheen level of the wall covering material selected should be taken into consideration when specifying wall coverings over this level of drywall treatment. Joints and fasteners must be sufficiently concealed if the wall covering material is lightweight, contains limited pattern, has a glossy finish, or has any combination of these features. In critical lighting areas, flat paints applied over light textures tend to reduce joint photographing. Gloss, semi-gloss, and enamel paints are not recommended over this level of finish.

All joints and interior angles shall have tape embedded in joint compound and shall be immediately wiped with a joint knife or trowel, leaving a thin coating of joint compound over all joints and interior angles. In addition, two separate coats of joint compound shall be applied over all flat joints and one separate coat of joint compound applied over interior angles. Fastener heads and accessories shall be covered with three separate coats of joint compound. All joint compounds shall be smooth and free of tool marks and ridges. The prepared surface shall be covered with a drywall primer like Sheetrock® First Coat prior to the application of the final decoration.

Level 5: The highest quality finish is the most effective method to provide a uniform surface and minimize the possibility of joint photographing and of fasteners showing through the final decoration. This level of finish is required where gloss, semi-gloss, or enamel is specified; when flat joints are specified over an untextured surface; or where critical lighting conditions occur. The prepared surface shall be coated with a drywall primer prior to the application of final decoration.

All joints and interior angles shall have tape embedded in joint compound and be immediately wiped with a joint knife or trowel, leaving a thin coating of joint compound over all joints and interior angles. Two separate coats of joint compound shall be applied over all flat joints and one separate coat of joint compound applied over interior angles. Fastener heads and accessories shall be covered with three separate coats of joint compound.

A thin skim coat of joint compound shall be trowel applied to the entire surface. Excess compound is immediately troweled off, leaving a film or skim coating of compound completely covering the paper. As an alternative to a skim coat, a material manufactured especially for this purpose may be applied such as Sheetrock® Tuff-Hide primer surfacer. The surface must be smooth and free of tool marks and ridges. The prepared surface shall be covered with a drywall primer prior to the application of the final decoration.

R096613-10 Terrazzo Floor

The table below lists quantities required for 100 S.F. of 5/8″ terrazzo topping, either bonded or not bonded.

Description	Bonded to Concrete 1-1/8″ Bed, 1:4 Mix	Not Bonded 2-1/8″ Bed and 1/4″ Sand
Portland cement, 94 lb. Bag	6 bags	8 bags
Sand	10 C.F.	20 C.F.
Divider strips, 4′ squares	50 L.F.	50 L.F.
Terrazzo fill, 50 lb. Bag	12 bags	12 bags
15 Lb. tarred felt		1 C.S.F.
Mesh 2 x 2 #14 galvanized		1 C.S.F.
Crew J-3	0.77 days	0.87 days

2′ × 2′ panels require 1.00 L.F. divider strip per S.F.
3′ × 3′ panels require 0.67 L.F. divider strip per S.F.
4′ × 4′ panels require 0.50 L.F. divider strip per S.F.
5′ × 5′ panels require 0.40 L.F. divider strip per S.F.
6′ × 6′ panels require 0.33 L.F. divider strip per S.F.

R097223-10 Wall Covering

The table below lists the quantities required for 100 S.F. of wall covering.

Description	Medium-Priced Paper	Expensive Paper
Paper	1.6 dbl. rolls	1.6 dbl. rolls
Wall sizing	0.25 gallon	0.25 gallon
Vinyl wall paste	0.6 gallon	0.6 gallon
Apply sizing	0.3 hour	0.3 hour
Apply paper	1.2 hours	1.5 hours

Most wallpapers now come in double rolls only.
To remove old paper, allow 1.3 hours per 100 S.F.

R099100-10 Painting Estimating Techniques

Proper estimating methodology is needed to obtain an accurate painting estimate. There is no known reliable shortcut or square foot method. The following steps should be followed:

- List all surfaces to be painted, with an accurate quantity (area) of each. Items having similar surface condition, finish, application method and accessibility may be grouped together.
- List all the tasks required for each surface to be painted, including surface preparation, masking, and protection of adjacent surfaces. Surface preparation may include minor repairs, washing, sanding and puttying.
- Select the proper Means line for each task. Review and consider all adjustments to labor and materials for type of paint and location of work. Apply the height adjustment carefully. For instance, when applying the adjustment for work over 8' high to a wall that is 12' high, apply the adjustment only to the area between 8' and 12' high, and not to the entire wall.

When applying more than one percent (%) adjustment, apply each to the base cost of the data, rather than applying one percentage adjustment on top of the other.

When estimating the cost of painting walls and ceilings remember to add the brushwork for all cut-ins at inside corners and around windows and doors as a LF measure. One linear foot of cut-in with brush equals one square foot of painting.

All items for spray painting include the labor for roll-back.

Deduct for openings greater than 100 SF, or openings that extend from floor to ceiling and are greater than 5' wide. Do not deduct small openings.

The cost of brushes, rollers, ladders and spray equipment are considered to be part of a painting contractor's overhead, and should not be added to the estimate. The cost of rented equipment such as scaffolding and swing staging should be added to the estimate.

R099100-20 Painting

Item	Coat	One Gallon Covers			In 8 Hours a Laborer Covers			Labor-Hours per 100 S.F.		
		Brush	Roller	Spray	Brush	Roller	Spray	Brush	Roller	Spray
Paint wood siding	prime	250 S.F.	225 S.F.	290 S.F.	1150 S.F.	1300 S.F.	2275 S.F.	.695	.615	.351
	others	270	250	290	1300	1625	2600	.615	.492	.307
Paint exterior trim	prime	400	—	—	650	—	—	1.230	—	—
	1st	475	—	—	800	—	—	1.000	—	—
	2nd	520	—	—	975	—	—	.820	—	—
Paint shingle siding	prime	270	255	300	650	975	1950	1.230	.820	.410
	others	360	340	380	800	1150	2275	1.000	.695	.351
Stain shingle siding	1st	180	170	200	750	1125	2250	1.068	.711	.355
	2nd	270	250	290	900	1325	2600	.888	.603	.307
Paint brick masonry	prime	180	135	160	750	800	1800	1.066	1.000	.444
	1st	270	225	290	815	975	2275	.981	.820	.351
	2nd	340	305	360	815	1150	2925	.981	.695	.273
Paint interior plaster or drywall	prime	400	380	495	1150	2000	3250	.695	.400	.246
	others	450	425	495	1300	2300	4000	.615	.347	.200
Paint interior doors and windows	prime	400	—	—	650	—	—	1.230	—	—
	1st	425	—	—	800	—	—	1.000	—	—
	2nd	450	—	—	975	—	—	.820	—	—

Special Construction R1311 Swimming Pools

R131113-20 Swimming Pools

Pool prices given per square foot of surface area include pool structure, filter and chlorination equipment, pumps, related piping, ladders/steps, maintenance kit, skimmer and vacuum system. Decks and electrical service to equipment are not included.

Residential in-ground pool construction can be divided into two categories: vinyl lined and gunite. Vinyl lined pool walls are constructed of different materials including wood, concrete, plastic or metal. The bottom is often graded with sand over which the vinyl liner is installed. Vermiculite or soil cement bottoms may be substituted for an added cost.

Gunite pool construction is used both in residential and municipal installations. These structures are steel reinforced for strength and finished with a white cement limestone plaster.

Municipal pools will have a higher cost because plumbing codes require more expensive materials, chlorination equipment and higher filtration rates.

Municipal pools greater than 1,800 S.F. require gutter systems to control waves. This gutter may be formed into the concrete wall. Often a vinyl/stainless steel gutter or gutter/wall system is specified, which will raise the pool cost.

Competition pools usually require tile bottoms and sides with contrasting lane striping, which will also raise the pool cost.

R133113-10 Air Supported Structures

Air supported structures are made from fabrics that can be classified into two groups: temporary and permanent. Temporary fabrics include nylon, woven polyethylene, vinyl film, and vinyl coated dacron. These have lifespans that range from five to fifteen plus years. The cost per square foot includes a fabric shell, tension cables, primary and back-up inflation systems and doors. The lower cost structures are used for construction shelters, bulk storage and pond covers. The more expensive are used for recreational structures and warehouses.

Permanent fabrics are teflon coated fiberglass. The life of this structure is twenty plus years. The high cost limits its application to architectural designed structures which call for a clear span covered area, such as stadiums and convention centers. Both temporary and permanent structures are available in translucent fabrics which eliminates the need for daytime lighting.

Areas to be covered vary from 10,000 S.F. to any area up to 1000 foot wide by any length. Height restrictions range from a maximum of 1/2 of width to a minimum of 1/6 of the width. Erection of even the largest of the temporary structures requires no more than a week.

Centrifugal fans provide the inflation necessary to support the structure during application of live loads. Airlocks are usually used at large entrances to prevent loss of static pressure. Some manufacturers employ propeller fans which generate sufficient airflow (30,000 CFM) to eliminate the need for airlocks. These fans may also be automatically controlled to resist high wind conditions, regulate humidity (air changes), and provide cooling and heat.

Insulation can be provided with the addition of a second or even third interior liner, creating a dead air space with an "R" value of four to nine. Some structures allow for the liner to be collapsed into the outer shell to enable the internal heat to melt accumulated snow. For cooling or air conditioning, the exterior face of the liner can be aluminized to reflect the sun's heat.

R133113-90 Seismic Bracing

Sometimes referred to as anti-sway bracing, this support system is required in earthquake areas. The individual components must be assembled to make a required system.

Example

C-Clamp 3/8″ rod	2 ea.
Rod, continuous thread 3/8″	10 L.F.
Field, weld 1″	2 ea.

Additionally, height factors must be taken into account. Add the following percentages to labor for elevated installations:

15′ to 20′ high	10%
21′ to 25′ high	20%
26′ to 30′ high	30%
31′ to 35′ high	40%
36′ to 40′ high	50%
41′ to 50′ high	60%

Special Construction R1334 Fabricated Engineered Structures

R133419-10 Pre-Engineered Steel Buildings

These buildings are manufactured by many companies and normally erected by franchised dealers throughout the U.S. The four basic types are: Rigid Frames, Truss type, Post and Beam and the Sloped Beam type. Most popular roof slope is low pitch of 1″ in 12″. The minimum economical area of these buildings is about 3000 S.F. of floor area. Bay sizes are usually 20′ to 24′ but can go as high as 30′ with heavier girts and purlins. Eave heights are usually 12′ to 24′ with 18′ to 20′ most typical.

Material prices shown in the Unit Price section are bare costs for the building shell only and do not include floors, foundations, anchor bolts, interior finishes or utilities. Costs assume at least three bays of 24′ each, a 1″ in 12″ roof slope, and they are based on 30 psf roof load and 20 psf wind load (wind load is a function of wind speed, building height, and terrain characteristics; this should be determined by a Registered Structural Engineer) and no unusual requirements. Costs include the structural frame, 26 ga. non-insulated colored corrugated or ribbed roofing and siding panels, fasteners, closures, trim and flashing but no allowance for insulation, doors, windows, skylights, gutters or downspouts. Very large projects would generally cost less for materials than the prices shown. For roof panel substitutions and wall panel substitutions, see appropriate Unit Price sections.

Conditions at the site, weather, shape and size of the building, and labor availability will affect the erection cost of the building.

R133423-30 Dome Structures

Steel — The four types are Lamella, Schwedler, Arch and Geodesic. For maximum economy, rise should be about 15 to 20% of diameter. Most common diameters are in the 200′ to 300′ range. Lamella domes weigh about 5 P.S.F. of floor area less than Schwedler domes. Schwedler dome weight in lbs. per S.F. approaches .046 times the diameter. Domes below 125′ diameter weigh .07 times diameter and the cost per ton of steel is higher. See R051223-20 for estimating weight.

Wood — Small domes are of sawn lumber, larger ones are laminated. In larger sizes, triaxial and triangular cost about the same; radial domes cost more. Radial domes are economical in the 60′ to 70′ diameter range. Most economical range of all types is 80′ to 200′ diameters. Diameters can run over 400′. All costs are quoted above the foundation. Prices include 2″ decking and a tension tie ring in place.

Plywood — Stock prefab geodesic domes are available with diameters from 24′ to 60′.

Fiberglass — Aluminum framed translucent sandwich panels with spans from 5′ to 45′ are commercially available.

Aluminum — Stressed skin aluminum panels form geodesic domes with spans ranging from 82′ to 232′. An aluminum space truss, triangulated or nontriangulated, with aluminum or clear acrylic closure panels can be used for clear spans of 40′ to 415′.

R142000-10 Freight Elevators

Capacities run from 2,000 lbs. to over 100,000 lbs. with 3,000 lbs. to 10,000 lbs. most common. Travel speeds are generally lower and control less intricate than on passenger elevators. Costs in the Unit Price sections are for hydraulic and geared elevators.

R142000-20 Elevator Selective Costs See R142000-40 for cost development.

	Passenger		Freight		Hospital	
A. Base Unit	Hydraulic	Electric	Hydraulic	Electric	Hydraulic	Electric
Capacity	1,500 lb.	2,000 lb.	2,000 lb.	4,000 lb.	4,000 lb.	4,000 lb.
Speed	100 F.P.M.	200 F.P.M.	100 F.P.M.	200 F.P.M.	100 F.P.M.	200 F.P.M.
#Stops/Travel Ft.	2/12	4/40	2/20	4/40	2/20	4/40
Push Button Oper.	Yes	Yes	Yes	Yes	Yes	Yes
Telephone Box & Wire	"	"	"	"	"	"
Emergency Lighting	"	"	No	No	"	"
Cab	Plastic Lam. Walls	Plastic Lam. Walls	Painted Steel	Painted Steel	Plastic Lam. Walls	Plastic Lam. Walls
Cove Lighting	Yes	Yes	No	No	Yes	Yes
Floor	V.C.T.	V.C.T.	Wood w/Safety Treads	Wood w/Safety Treads	V.C.T.	V.C.T.
Doors, & Speedside Slide	Yes	Yes	Yes	Yes	Yes	Yes
Gates, Manual	No	No	No	No	No	No
Signals, Lighted Buttons	Car and Hall	Car and Hall	Car and Hall	Car and Hall	Car and Hall	Car and Hall
O.H. Geared Machine	N.A.	Yes	N.A.	Yes	N.A.	Yes
Variable Voltage Contr.	"	"	N.A.	"	"	"
Emergency Alarm	Yes	"	Yes	"	Yes	"
Class "A" Loading	N.A.	N.A.	"	"	N.A.	N.A.

R142000-30 Passenger Elevators

Electric elevators are used generally but hydraulic elevators can be used for lifts up to 70′ and where large capacities are required. Hydraulic speeds are limited to 200 F.P.M. but cars are self leveling at the stops. On low rises, hydraulic installation runs about 15% less than standard electric types but on higher rises this installation cost advantage is reduced. Maintenance of hydraulic elevators is about the same as electric type but underground portion is not included in the maintenance contract.

In electric elevators there are several control systems available, the choice of which will be based upon elevator use, size, speed and cost criteria. The two types of drives are geared for low speeds and gearless for 450 F.P.M. and over.

The tables on the preceding pages illustrate typical installed costs of the various types of elevators available.

R142000-40 Elevator Cost Development

To price a new car or truck from the factory, you must start with the manufacturer's basic model, then add or exchange optional equipment and features. The same is true for pricing elevators.

Example:

Requirement: One-passenger elevator, five-story hydraulic, 2,500 lb. capacity, 12′ floor to floor, speed 150 F.P.M., emergency power switching and maintenance contract.

Description	Adjustment
A. Base Elevator: Hydraulic Passenger, 1500 lb. Capacity, 100 fpm, 2 Stops, Standard Finish	1 Ea.
B. Capacity Adjustment (2,500 lb.)	1 Ea.
C. Excess Travel Adjustment: 48′ Total Travel (4 x 12′) minus 12′ Base Unit Travel =	36 V.L.F.
D. Stops Adjustment: 5 Total Stops minus 2 Stops (Base Unit) =	3 Stops
E. Speed Adjustment (150 F.P.M.)	1 Ea.
F. Options:	
1. Intercom Service	1 Ea.
2. Emergency Power Switching, Automatic	1 Ea.
3. Stainless Steel Entrance Doors	5 Ea.
4. Maintenance Contract (12 Months)	1 Ea.
5. Position Indicator for main floor level (none indicated in Base Unit)	1 Ea.

R143210-20 Moving Ramps and Walks

These are a specialized form of conveyor 3′ to 6′ wide with capacities of 3,600 to 18,000 persons per hour. Maximum speed is 140 F.P.M. and normal incline is 0° to 15°.

Local codes will determine the maximum angle. Outdoor units would require additional weather protection.

R220102-20 Labor Adjustment Factors

Labor Adjustment Factors are provided for Divisions 21, 22, and 23 to assist the mechanical estimator account for the various complexities and special conditions of any particular project. While a single percentage has been entered on each line of Division 22 01 02.20, it should be understood that these are just suggested midpoints of ranges of values commonly used by mechanical estimators. They may be increased or decreased depending on the severity of the special conditions.

The group for "existing occupied buildings", has been the subject of requests for explanation. Actually there are two stages to this group: buildings that are existing and "finished" but unoccupied, and those that also are occupied. Buildings that are "finished" may result in higher labor costs due to the workers having to be more careful not to damage finished walls, ceilings, floors, etc. and may necessitate special protective coverings and barriers. Also corridor bends and doorways may not accommodate long pieces of pipe or larger pieces of equipment. Work above an already hung ceiling can be very time consuming. The addition of occupants may force the work to be done on premium time (nights and/or weekends), eliminate the possible use of some preferred tools such as pneumatic drivers, powder charged drivers etc. The estimator should evaluate the access to the work area and just how the work is going to be accomplished to arrive at an increase in labor costs over "normal" new construction productivity.

R220105-10 Demolition (Selective vs. Removal for Replacement)

Demolition can be divided into two basic categories.

One type of demolition involves the removal of material with no concern for its replacement. The labor-hours to estimate this work are found under "Selective Demolition" in the Fire Protection, Plumbing and HVAC Divisions. It is selective in that individual items or all the material installed as a system or trade grouping such as plumbing or heating systems are removed. This may be accomplished by the easiest way possible, such as sawing, torch cutting, or sledge hammer as well as simple unbolting.

The second type of demolition is the removal of some item for repair or replacement. This removal may involve careful draining, opening of unions, disconnecting and tagging of electrical connections, capping of pipes/ducts to prevent entry of debris or leakage of the material contained as well as transport of the item away from its in-place location to a truck/dumpster. An approximation of the time required to accomplish this type of demolition is to use half of the time indicated as necessary to install a new unit. For example; installation of a new pump might be listed as requiring 6 labor-hours so if we had to estimate the removal of the old pump we would allow an additional 3 hours for a total of 9 hours. That is, the complete replacement of a defective pump with a new pump would be estimated to take 9 labor-hours.

R221113-50 Pipe Material Considerations

1. Malleable fittings should be used for gas service.
2. Malleable fittings are used where there are stresses/strains due to expansion and vibration.
3. Cast fittings may be broken as an aid to disassembling of heating lines frozen by long use, temperature and minerals.
4. Cast iron pipe is extensively used for underground and submerged service.
5. Type M (light wall) copper tubing is available in hard temper only and is used for nonpressure and less severe applications than K and L.
6. Type L (medium wall) copper tubing, available hard or soft for interior service.
7. Type K (heavy wall) copper tubing, available in hard or soft temper for use where conditions are severe. For underground and interior service.
8. Hard drawn tubing requires fewer hangers or supports but should not be bent. Silver brazed fittings are recommended, however soft solder is normally used.
9. Type DMV (very light wall) copper tubing designed for drainage, waste and vent plus other non-critical pressure services.

Domestic/Imported Pipe and Fittings Cost

The prices shown in this publication for steel/cast iron pipe and steel, cast iron, malleable iron fittings are based on domestic production sold at the normal trade discounts. The above listed items of foreign manufacture may be available at prices of 1/3 to 1/2 those shown. Some imported items after minor machining or finishing operations are being sold as domestic to further complicate the system.

Caution: Most pipe prices in this book also include a coupling and pipe hangers which for the larger sizes can add significantly to the per foot cost and should be taken into account when comparing "book cost" with quoted supplier's cost.

Heating, Ventilating & A.C. R2356 Solar Energy Heating Equipment

R235616-60 Solar Heating (Space and Hot Water)

Collectors should face as close to due South as possible, however, variations of up to 20 degrees on either side of true South are acceptable. Local climate and collector type may influence the choice between east or west deviations. Obviously they should be located so they are not shaded from the sun's rays. Incline collectors at a slope of latitude minus 5 degrees for domestic hot water and latitude plus 15 degrees for space heating.

Flat plate collectors consist of a number of components as follows: Insulation to reduce heat loss through the bottom and sides of the collector. The enclosure which contains all the components in this assembly is usually weatherproof and prevents dust, wind and water from coming in contact with the absorber plate. The cover plate usually consists of one or more layers of a variety of glass or plastic and reduces the reradiation by creating an air space which traps the heat between the cover and the absorber plates.

The absorber plate must have a good thermal bond with the fluid passages. The absorber plate is usually metallic and treated with a surface coating which improves absorptivity. Black or dark paints or selective coatings are used for this purpose, and the design of this passage and plate combination helps determine a solar system's effectiveness.

Heat transfer fluid passage tubes are attached above and below or integral with an absorber plate for the purpose of transferring thermal energy from the absorber plate to a heat transfer medium. The heat exchanger is a device for transferring thermal energy from one fluid to another.

Piping and storage tanks should be well insulated to minimize heat losses.

Size domestic water heating storage tanks to hold 20 gallons of water per user, minimum, plus 10 gallons per dishwasher or washing machine. For domestic water heating an optimum collector size is approximately 3/4 square foot of area per gallon of water storage. For space heating of residences and small commercial applications the collector is commonly sized between 30% and 50% of the internal floor area. For space heating of large commercial applications, collector areas less than 30% of the internal floor area can still provide significant heat reductions.

A supplementary heat source is recommended for Northern states for December through February.

The solar energy transmission per square foot of collector surface varies greatly with the material used. Initial cost, heat transmittance and useful life are obviously interrelated.

Heating, Ventilating & A.C. R2360 Central Cooling Equipment

R236000-20 Air Conditioning Requirements

BTUs per hour per S.F. of floor area and S.F. per ton of air conditioning.

Type of Building	BTU/Hr per S.F.	S.F. per Ton	Type of Building	BTU/Hr per S.F.	S.F. per Ton	Type of Building	BTU/Hr per S.F.	S.F. per Ton
Apartments, Individual	26	450	Dormitory, Rooms	40	300	Libraries	50	240
Corridors	22	550	Corridors	30	400	Low Rise Office, Exterior	38	320
Auditoriums & Theaters	40	300/18*	Dress Shops	43	280	Interior	33	360
Banks	50	240	Drug Stores	80	150	Medical Centers	28	425
Barber Shops	48	250	Factories	40	300	Motels	28	425
Bars & Taverns	133	90	High Rise Office—Ext. Rms.	46	263	Office (small suite)	43	280
Beauty Parlors	66	180	Interior Rooms	37	325	Post Office, Individual Office	42	285
Bowling Alleys	68	175	Hospitals, Core	43	280	Central Area	46	260
Churches	36	330/20*	Perimeter	46	260	Residences	20	600
Cocktail Lounges	68	175	Hotel, Guest Rooms	44	275	Restaurants	60	200
Computer Rooms	141	85	Corridors	30	400	Schools & Colleges	46	260
Dental Offices	52	230	Public Spaces	55	220	Shoe Stores	55	220
Dept. Stores, Basement	34	350	Industrial Plants, Offices	38	320	Shop'g. Ctrs., Supermarkets	34	350
Main Floor	40	300	General Offices	34	350	Retail Stores	48	250
Upper Floor	30	400	Plant Areas	40	300	Specialty	60	200

*Persons per ton
12,000 BTU = 1 ton of air conditioning

R260519-92 Minimum Copper and Aluminum Wire Size Allowed for Various Types of Insulation

Minimum Wire Sizes									
	Copper		Aluminum			Copper		Aluminum	
Amperes	THW THWN or XHHW	THHN XHHW *	THW XHHW	THHN XHHW *	Amperes	THW THWN or XHHW	THHN XHHW *	THW XHHW	THHN XHHW *
15A	#14	#14	#12	#12	195	3/0	2/0	250kcmil	4/0
20	#12	#12	#10	#10	200	3/0	3/0	250kcmil	4/0
25	#10	#10	#10	#10	205	4/0	3/0	250kcmil	4/0
30	#10	#10	# 8	# 8	225	4/0	3/0	300kcmil	250kcmil
40	# 8	# 8	# 8	# 8	230	4/0	4/0	300kcmil	250kcmil
45	# 8	# 8	# 6	# 8	250	250kcmil	4/0	350kcmil	300kcmil
50	# 8	# 8	# 6	# 6	255	250kcmil	4/0	400kcmil	300kcmil
55	# 6	# 8	# 4	# 6	260	300kcmil	4/0	400kcmil	350kcmil
60	# 6	# 6	# 4	# 6	270	300kcmil	250kcmil	400kcmil	350kcmil
65	# 6	# 6	# 4	# 4	280	300kcmil	250kcmil	500kcmil	350kcmil
75	# 4	# 6	# 3	# 4	285	300kcmil	250kcmil	500kcmil	400kcmil
85	# 4	# 4	# 2	# 3	290	350kcmil	250kcmil	500kcmil	400kcmil
90	# 3	# 4	# 2	# 2	305	350kcmil	300kcmil	500kcmil	400kcmil
95	# 3	# 4	# 1	# 2	310	350kcmil	300kcmil	500kcmil	500kcmil
100	# 3	# 3	# 1	# 2	320	400kcmil	300kcmil	600kcmil	500kcmil
110	# 2	# 3	1/0	# 1	335	400kcmil	350kcmil	600kcmil	500kcmil
115	# 2	# 2	1/0	# 1	340	500kcmil	350kcmil	600kcmil	500kcmil
120	# 1	# 2	1/0	1/0	350	500kcmil	350kcmil	700kcmil	500kcmil
130	# 1	# 2	2/0	1/0	375	500kcmil	400kcmil	700kcmil	600kcmil
135	1/0	# 1	2/0	1/0	380	500kcmil	400kcmil	750kcmil	600kcmil
150	1/0	# 1	3/0	2/0	385	600kcmil	500kcmil	750kcmil	600kcmil
155	2/0	1/0	3/0	3/0	420	600kcmil	500kcmil		700kcmil
170	2/0	1/0	4/0	3/0	430		500kcmil		750kcmil
175	2/0	2/0	4/0	3/0	435		600kcmil		750kcmil
180	3/0	2/0	4/0	4/0	475		600kcmil		

*Dry Locations Only

Notes:

1. Size #14 to 4/0 is in AWG units (American Wire Gauge).
2. Size 250 to 750 is in kcmil units (Thousand Circular Mils).
3. Use next higher ampere value if exact value is not listed in table.
4. For loads that operate continuously increase ampere value by 25% to obtain proper wire size.
5. Refer to Table R260519-91 for the maximum circuit length for the various size wires.
6. Table R260519-92 has been written for estimating purpose only, based on ambient temperature of 30°C (86° F); for ambient temperature other than 30°C (86° F), ampacity correction factors will be applied.

R260533-22 Conductors in Conduit

Table below lists maximum number of conductors for various sized conduit using THW, TW or THWN insulations.

Copper Wire Size	Conduit Size																										
	½″			¾″			1″			1-¼″			1-½″			2″			2-½″			3″		3-½″		4″	
	TW	THW	THWN	TW	THW	THWN	TW	THW	THWN	TW	THW	THWN	TW	THW	THWN	TW	THW	THWN	TW	THW	THWN	THW	THWN	THW	THWN	THW	THWN
#14	9	6	13	15	10	24	25	16	39	44	29	69	60	40	94	99	65	154	142	93		143		192			
#12	7	4	10	12	8	18	19	13	29	35	24	51	47	32	70	78	53	114	111	76	164	117		157			
#10	5	4	6	9	6	11	15	11	18	26	19	32	36	26	44	60	43	73	85	61	104	95	160	127		163	
#8	2	1	3	4	3	5	7	5	9	12	10	16	17	13	22	28	22	36	40	32	51	49	79	66	106	85	136
#6		1	1		2	4		4	6		7	11		10	15		16	26		23	37	36	57	48	76	62	98
#4		1	1		1	2		3	4		5	7		7	9		12	16		17	22	27	35	36	47	47	60
#3		1	1		1	1		2	3		4	6		6	8		10	13		15	19	23	29	31	39	40	51
#2		1	1		1	1		2	3		4	5		5	7		9	11		13	16	20	25	27	33	34	43
#1					1	1		1	1		3	3		4	5		6	8		9	12	14	18	19	25	25	32
1/0					1	1		1	1		2	3		3	4		5	7		8	10	12	15	16	21	21	27
2/0					1	1		1	1		1	2		3	3		5	6		7	8	10	13	14	17	18	22
3/0					1	1		1	1		1	1		2	3		4	5		6	7	9	11	12	14	15	18
4/0						1		1	1		1	1		1	2		3	4		5	6	7	9	10	12	13	15
250 kcmil								1	1		1	1		1	1		2	3		4	4	6	7	8	10	10	12
300								1	1		1	1		1	1		2	3		3	4	5	6	7	8	9	11
350									1		1	1		1	1		1	2		3	3	4	5	6	7	8	9
400											1	1		1	1		1	1		2	3	4	5	5	6	7	8
500											1	1		1	1		1	1		1	2	3	4	4	5	6	7
600												1		1	1		1	1		1	1	3	3	4	4	5	5
700														1	1		1	1		1	1	2	3	3	4	4	5
750 ▼														1	1		1	1		1	1	2	2	3	3	4	4

R312316-40 Excavating

The selection of equipment used for structural excavation and bulk excavation or for grading is determined by the following factors.

1. Quantity of material
2. Type of material
3. Depth or height of cut
4. Length of haul
5. Condition of haul road
6. Accessibility of site
7. Moisture content and dewatering requirements
8. Availability of excavating and hauling equipment

Some additional costs must be allowed for hand trimming the sides and bottom of concrete pours and other excavation below the general excavation.

When planning excavation and fill, the following should also be considered.

1. Swell factor
2. Compaction factor
3. Moisture content
4. Density requirements

A typical example for scheduling and estimating the cost of excavation of a 15′ deep basement on a dry site when the material must be hauled off the site is outlined below.

Assumptions:

1. Swell factor, 18%
2. No mobilization or demobilization
3. Allowance included for idle time and moving on job
4. No dewatering, sheeting, or bracing
5. No truck spotter or hand trimming

Number of B.C.Y. per truck = 1.5 C.Y. bucket × 8 passes = 12 loose C.Y.

$$= 12 \times \frac{100}{118} = 10.2 \text{ B.C.Y. per truck}$$

Truck Haul Cycle:

Load truck, 8 passes	=	4 minutes
Haul distance, 1 mile	=	9 minutes
Dump time	=	2 minutes
Return, 1 mile	=	7 minutes
Spot under machine	=	1 minute
		23 minute cycle

Fleet Haul Production per day in B.C.Y.

$$4 \text{ trucks} \times \frac{50 \text{ min. hour}}{23 \text{ min. haul cycle}} \times 8 \text{ hrs.} \times 10.2 \text{ B.C.Y.}$$

$$= 4 \times 2.2 \times 8 \times 10.2 = 718 \text{ B.C.Y./day}$$

Add the mobilization and demobilization costs to the total excavation costs. When equipment is rented for more than three days, there is often no mobilization charge by the equipment dealer. On larger jobs outside of urban areas, scrapers can move earth economically provided a dump site or fill area and adequate haul roads are available. Excavation within sheeting bracing or cofferdam bracing is usually done with a clamshell and production is low, since the clamshell may have to be guided by hand between the bracing. When excavating or filling an area enclosed with a wellpoint system, add 10% to 15% to the cost to allow for restricted access. When estimating earth excavation quantities for structures, allow work space outside the building footprint for construction of the foundation and a slope of 1:1 unless sheeting is used.

R312316-45 Excavating Equipment

The table below lists THEORETICAL hourly production in C.Y./hr. bank measure for some typical excavation equipment. Figures assume 50 minute hours, 83% job efficiency, 100% operator efficiency, 90° swing and properly sized hauling units, which must be modified for adverse digging and loading conditions. Actual production costs in the front of the book average about 50% of the theoretical values listed here.

Equipment	Soil Type	B.C.Y. Weight	% Swell	1 C.Y.	1-1/2 C.Y.	2 C.Y.	2-1/2 C.Y.	3 C.Y.	3-1/2 C.Y.	4 C.Y.
Hydraulic Excavator "Backhoe" 15′ Deep Cut	Moist loam, sandy clay	3400 lb.	40%	165	195	200	275	330	385	440
	Sand and gravel	3100	18	140	170	225	240	285	330	380
	Common earth	2800	30	150	180	230	250	300	350	400
	Clay, hard, dense	3000	33	120	140	190	200	240	260	320
Power Shovel Optimum Cut (Ft.)	Moist loam, sandy clay	3400	40	170	245	295	335	385	435	475
				(6.0)	(7.0)	(7.8)	(8.4)	(8.8)	(9.1)	(9.4)
	Sand and gravel	3100	18	165	225	275	325	375	420	460
				(6.0)	(7.0)	(7.8)	(8.4)	(8.8)	(9.1)	(9.4)
	Common earth	2800	30	145	200	250	295	335	375	425
				(7.8)	(9.2)	(10.2)	(11.2)	(12.1)	(13.0)	(13.8)
	Clay, hard, dense	3000	33	120	175	220	255	300	335	375
				(9.0)	(10.7)	(12.2)	(13.3)	(14.2)	(15.1)	(16.0)
Drag Line Optimum Cut (Ft.)	Moist loam, sandy clay	3400	40	130	180	220	250	290	325	385
				(6.6)	(7.4)	(8.0)	(8.5)	(9.0)	(9.5)	(10.0)
	Sand and gravel	3100	18	130	175	210	245	280	315	375
				(6.6)	(7.4)	(8.0)	(8.5)	(9.0)	(9.5)	(10.0)
	Common earth	2800	30	110	160	190	220	250	280	310
				(8.0)	(9.0)	(9.9)	(10.5)	(11.0)	(11.5)	(12.0)
	Clay, hard, dense	3000	33	90	130	160	190	225	250	280
				(9.3)	(10.7)	(11.8)	(12.3)	(12.8)	(13.3)	(12.0)
				Wheel Loaders				Track Loaders		
				3 C.Y.	4 C.Y.	6 C.Y.	8 C.Y.	2-1/4 C.Y.	3 C.Y.	4 C.Y.
Loading Tractors	Moist loam, sandy clay	3400	40	260	340	510	690	135	180	250
	Sand and gravel	3100	18	245	320	480	650	130	170	235
	Common earth	2800	30	230	300	460	620	120	155	220
	Clay, hard, dense	3000	33	200	270	415	560	110	145	200
	Rock, well-blasted	4000	50	180	245	380	520	100	130	180

R312319-90 Wellpoints

A single stage wellpoint system is usually limited to dewatering an average 15′ depth below normal ground water level. Multi-stage systems are employed for greater depth with the pumping equipment installed only at the lowest header level. Ejectors with unlimited lift capacity can be economical when two or more stages of wellpoints can be replaced or when horizontal clearance is restricted, such as in deep trenches or tunneling projects, and where low water flows are expected. Wellpoints are usually spaced on 2-1/2′ to 10′ centers along a header pipe. Wellpoint spacing, header size, and pump size are all determined by the expected flow as dictated by soil conditions.

In almost all soils encountered in wellpoint dewatering, the wellpoints may be jetted into place. Cemented soils and stiff clays may require sand wicks about 12″ in diameter around each wellpoint to increase efficiency and eliminate weeping into the excavation. These sand wicks require 1/2 to 3 C.Y. of washed filter sand and are installed by using a 12″ diameter steel casing and hole puncher jetted into the ground 2′ deeper than the wellpoint. Rock may require predrilled holes.

Labor required for the complete installation and removal of a single stage wellpoint system is in the range of 3/4 to 2 labor-hours per linear foot of header, depending upon jetting conditions, wellpoint spacing, etc.

Continuous pumping is necessary except in some free draining soil where temporary flooding is permissible (as in trenches which are backfilled after each day's work). Good practice requires provision of a stand-by pump during the continuous pumping operation.

Systems for continuous trenching below the water table should be installed three to four times the length of expected daily progress to ensure uninterrupted digging, and header pipe size should not be changed during the job.

For pervious free draining soils, deep wells in place of wellpoints may be economical because of lower installation and maintenance costs. Daily production ranges between two to three wells per day, for 25′ to 40′ depths, to one well per day for depths over 50′.

Detailed analysis and estimating for any dewatering problem is available at no cost from wellpoint manufacturers. Major firms will quote "sufficient equipment" quotes or their affiliates offer lump sum proposals to cover complete dewatering responsibility.

Description for 200′ System with 8″ Header		Quantities
Equipment & Material	Wellpoints 25′ long, 2″ diameter @ 5′ O.C.	40 Each
	Header pipe, 8″ diameter	200 L.F.
	Discharge pipe, 8″ diameter	100 L.F.
	8″ valves	3 Each
	Combination jetting & wellpoint pump (standby)	1 Each
	Wellpoint pump, 8″ diameter	1 Each
	Transportation to and from site	1 Day
	Fuel for 30 days x 60 gal./day	1800 Gallons
	Lubricants for 30 days x 16 lbs./day	480 Lbs.
	Sand for points	40 C.Y.
Labor	Technician to supervise installation	1 Week
	Labor for installation and removal of system	300 Labor-hours
	4 Operators straight time 40 hrs./wk. for 4.33 wks.	693 Hrs.
	4 Operators overtime 2 hrs./wk. for 4.33 wks.	35 Hrs.

R312323-30 Compacting Backfill

Compaction of fill in embankments, around structures, in trenches, and under slabs is important to control settlement. Factors affecting compaction are:

1. Soil gradation
2. Moisture content
3. Equipment used
4. Depth of fill per lift
5. Density required

Production Rate:

$$\frac{1.75' \text{ plate width} \times 50 \text{ F.P.M.} \times 50 \text{ min./hr.} \times .67' \text{ lift}}{27 \text{ C.F. per C.Y.}} = 108.5 \text{ C.Y./hr.}$$

Production Rate for 4 Passes:

$$\frac{108.5 \text{ C.Y.}}{4 \text{ passes}} = 27.125 \text{ C.Y./hr.} \times 8 \text{ hrs.} = 217 \text{ C.Y./day}$$

Example:

Compact granular fill around a building foundation using a 21″ wide x 24″ vibratory plate in 8″ lifts. Operator moves at 50 F.P.M. working a 50 minute hour to develop 95% Modified Proctor Density with 4 passes.

R314116-40 Wood Sheet Piling

Wood sheet piling may be used for depths to 20′ where there is no ground water. If moderate ground water is encountered Tongue & Groove sheeting will help to keep it out. When considerable ground water is present, steel sheeting must be used.

For estimating purposes on trench excavation, sizes are as follows:

Depth	Sheeting	Wales	Braces	B.F. per S.F.
To 8′	3 x 12′s	6 x 8′s, 2 line	6 x 8′s, @ 10′	4.0 @ 8′
8′ x 12′	3 x 12′s	10 x 10′s, 2 line	10 x 10′s, @ 9′	5.0 average
12′ to 20′	3 x 12′s	12 x 12′s, 3 line	12 x 12′s, @ 8′	7.0 average

Sheeting to be toed in at least 2′ depending upon soil conditions. A five person crew with an air compressor and sheeting driver can drive and brace 440 SF/day at 8′ deep, 360 SF/day at 12′ deep, and 320 SF/day at 16′ deep. For normal soils, piling can be pulled in 1/3 the time to install. Pulling difficulty increases with the time in the ground. Production can be increased by high pressure jetting.

R314116-45 Steel Sheet Piling

Limiting weights are 22 to 38#/S.F. of wall surface with 27#/S.F. average for usual types and sizes. (Weights of piles themselves are from 30.7#/L.F. to 57#/L.F. but they are 15″ to 21″ wide.) Lightweight sections 12″ to 28″ wide from 3 ga. to 12 ga. thick are also available for shallow excavations. Piles may be driven two at a time with an impact or vibratory hammer (use vibratory to pull) hung from a crane without leads. A reasonable estimate of the life of steel sheet piling is 10 uses with up to 125 uses possible if a vibratory hammer is used. Used piling costs from 50% to 80% of new piling depending on location and market conditions. Sheet piling and H piles can be rented for about 30% of the delivered mill price for the first month and 5% per month thereafter. Allow 1 labor-hour per pile for cleaning and trimming after driving. These costs increase with depth and hydrostatic head. Vibratory drivers are faster in wet granular soils and are excellent for pile extraction. Pulling difficulty increases with the time in the ground and may cost more than driving. It is often economical to abandon the sheet piling, especially if it can be used as the outer wall form. Allow about 1/3 additional length or more for toeing into ground. Add bracing, waler and strut costs. Waler costs can equal the cost per ton of sheeting.

Earthwork R3145 Vibroflotation & Densification

R314513-90 Vibroflotation and Vibro Replacement Soil Compaction

Vibroflotation is a proprietary system of compacting sandy soils in place to increase relative density to about 70%. Typical bearing capacities attained will be 6000 psf for saturated sand and 12,000 psf for dry sand. Usual range is 4000 to 8000 psf capacity. Costs in the front of the book are for a vertical foot of compacted cylinder 6′ to 10′ in diameter.

Vibro replacement is a proprietary system of improving cohesive soils in place to increase bearing capacity. Most silts and clays above or below the water table can be strengthened by installation of stone columns.

The process consists of radial displacement of the soil by vibration. The created hole is then backfilled in stages with coarse granular fill which is thoroughly compacted and displaced into the surrounding soil in the form of a column.

The total project cost would depend on the number and depth of the compacted cylinders. The installing company guarantees relative soil density of the sand cylinders after compaction and the bearing capacity of the soil after the replacement process. Detailed estimating information is available from the installer at no cost.

R316326-60 Caissons

The three principal types of cassions are:

(1) **Belled Caissons,** which except for shallow depths and poor soil conditions, are generally recommended. They provide more bearing than shaft area. Because of its conical shape, no horizontal reinforcement of the bell is required.

(2) **Straight Shaft Caissons** are used where relatively light loads are to be supported by caissons that rest on high value bearing strata. While the shaft is larger in diameter than for belled types this is more than offset by the saving in time and labor.

(3) **Keyed Caissons** are used when extremely heavy loads are to be carried. A keyed or socketed caisson transfers its load into rock by a combination of end-bearing and shear reinforcing of the shaft. The most economical shaft often consists of a steel casing, a steel wide flange core and concrete. Allowable compressive stresses of .225 f′c for concrete, 16,000 psi for the wide flange core, and 9,000 psi for the steel casing are commonly used. The usual range of shaft diameter is 18″ to 84″. The number of sizes specified for any one project should be limited due to the problems of casing and auger storage. When hand work is to be performed, shaft diameters should not be less than 32″. When inspection of borings is required a minimum shaft diameter of 30″ is recommended. Concrete caissons are intended to be poured against earth excavation so permanent forms which add to cost should not be used if the excavation is clean and the earth sufficiently impervious to prevent excessive loss of concrete.

Soil Conditions for Belling		
Good	**Requires Handwork**	**Not Recommended**
Clay	Hard Shale	Silt
Sandy Clay	Limestone	Sand
Silty Clay	Sandstone	Gravel
Clayey Silt	Weathered Mica	Igneous Rock
Hard-pan		
Soft Shale		
Decomposed Rock		

R329219-50 Seeding

The type of grass is determined by light, shade and moisture content of soil plus intended use. Fertilizer should be disked 4″ before seeding. For steep slopes disk five tons of mulch and lay two tons of hay or straw on surface per acre after seeding. Surface mulch can be staked, lightly disked or tar emulsion sprayed. Material for mulch can be wood chips, peat moss, partially rotted hay or straw, wood fibers and sprayed emulsions. Hemp seed blankets with fertilizer are also available. For spring seeding, watering is necessary. Late fall seeding may have to be reseeded in the spring. Hydraulic seeding, power mulching, and aerial seeding can be used on large areas.

R331113-80 Piping Designations

There are several systems currently in use to describe pipe and fittings. The following paragraphs will help to identify and clarify classifications of piping systems used for water distribution.

Piping may be classified by schedule. Piping schedules include 5S, 10S, 10, 20, 30, Standard, 40, 60, Extra Strong, 80, 100, 120, 140, 160 and Double Extra Strong. These schedules are dependent upon the pipe wall thickness. The wall thickness of a particular schedule may vary with pipe size.

Ductile iron pipe for water distribution is classified by Pressure Classes such as Class 150, 200, 250, 300 and 350. These classes are actually the rated water working pressure of the pipe in pounds per square inch (psi). The pipe in these pressure classes is designed to withstand the rated water working pressure plus a surge allowance of 100 psi.

The American Water Works Association (AWWA) provides standards for various types of **plastic pipe.** C-900 is the specification for polyvinyl chloride (PVC) piping used for water distribution in sizes ranging from 4″ through 12″. C-901 is the specification for polyethylene (PE) pressure pipe, tubing and fittings used for water distribution in sizes ranging from 1/2″ through 3″. C-905 is the specification for PVC piping sizes 14″ and greater.

PVC pressure-rated pipe is identified using the standard dimensional ratio (SDR) method. This method is defined by the American Society for Testing and Materials (ASTM) Standard D 2241. This pipe is available in SDR numbers 64, 41, 32.5, 26, 21, 17, and 13.5. Pipe with an SDR of 64 will have the thinnest wall while pipe with an SDR of 13.5 will have the thickest wall. When the pressure rating (PR) of a pipe is given in psi, it is based on a line supplying water at 73 degrees F.

The National Sanitation Foundation (NSF) seal of approval is applied to products that can be used with potable water. These products have been tested to ANSI/NSF Standard 14.

Valves and strainers are classified by American National Standards Institute (ANSI) Classes. These Classes are 125, 150, 200, 250, 300, 400, 600, 900, 1500 and 2500. Within each class there is an operating pressure range dependent upon temperature. Design parameters should be compared to the appropriate material dependent, pressure-temperature rating chart for accurate valve selection.

Transportation R3472 Railway Construction

R347216-10 Single Track R.R. Siding

The costs for a single track RR siding in the Unit Price Section include the components shown in the table below.

Description of Component	Qty. per L.F. of Track	Unit
Ballast, 1-1/2″ crushed stone	.667	C.Y.
6″ x 8″ x 8′-6″ Treated timber ties, 22″ O.C.	.545	Ea.
Tie plates, 2 per tie	1.091	Ea.
Track rail	2.000	L.F.
Spikes, 6″, 4 per tie	2.182	Ea.
Splice bars w/ bolts, lock washers & nuts, @ 33′ O.C.	.061	Pair
Crew B-14 @ 57 L.F./Day	.018	Day

R347216-20 Single Track, Steel Ties, Concrete Bed

The costs for a R.R. siding with steel ties and a concrete bed in the Unit Price section include the components shown in the table below.

Description of Component	Qty. per L.F. of Track	Unit
Concrete bed, 9′ wide, 10″ thick	.278	C.Y.
Ties, W6x16 x 6′-6″ long, @ 30″ O.C.	.400	Ea.
Tie plates, 4 per tie	1.600	Ea.
Track rail	2.000	L.F.
Tie plate bolts, 1″, 8 per tie	3.200	Ea.
Splice bars w/bolts, lock washers & nuts, @ 33′ O.C.	.061	Pair
Crew B-14 @ 22 L.F./Day	.045	Day

Change Orders

Change Order Considerations

A change order is a written document, usually prepared by the design professional, and signed by the owner, the architect/engineer, and the contractor. A change order states the agreement of the parties to: an addition, deletion, or revision in the work; an adjustment in the contract sum, if any; or an adjustment in the contract time, if any. Change orders, or "extras" in the construction process occur after execution of the construction contract and impact architects/ engineers, contractors, and owners.

Change orders that are properly recognized and managed can ensure orderly, professional, and profitable progress for all who are involved in the project. There are many causes for change orders and change order requests. In all cases, change orders or change order requests should be addressed promptly and in a precise and prescribed manner. The following paragraphs include information regarding change order pricing and procedures.

The Causes of Change Orders

Reasons for issuing change orders include:

- Unforeseen field conditions that require a change in the work
- Correction of design discrepancies, errors, or omissions in the contract documents
- Owner-requested changes, either by design criteria, scope of work, or project objectives
- Completion date changes for reasons unrelated to the construction process
- Changes in building code interpretations, or other public authority requirements that require a change in the work
- Changes in availability of existing or new materials and products

Procedures

Properly written contract documents must include the correct change order procedures for all parties—owners, design professionals and contractors—to follow in order to avoid costly delays and litigation.

Being "in the right" is not always a sufficient or acceptable defense. The contract provisions requiring notification and documentation must be adhered to within a defined or reasonable time frame.

The appropriate method of handling change orders is by a written proposal and acceptance by all parties involved. Prior to starting work on a project, all parties should identify their authorized agents who may sign and accept change orders, as well as any limits placed on their authority.

Time may be a critical factor when the need for a change arises. For such cases, the contractor might be directed to proceed on a "time and materials" basis, rather than wait for all paperwork to be processed—a delay that could impede progress. In this situation, the contractor must still follow the prescribed change order procedures including, but not limited to, notification and documentation.

Lack of documentation can be very costly, especially if legal judgments are to be made, and if certain field personnel are no longer available. For time and material change orders, the contractor should keep accurate daily records of all labor and material allocated to the change.

Owners or awarding authorities who do considerable and continual building construction (such as the federal government) realize the inevitability of change orders for numerous reasons, both predictable and unpredictable. As a result, the federal government, the American Institute of Architects (AIA), the Engineers Joint Contract Documents Committee (EJCDC) and other contractor, legal, and technical organizations have developed standards and procedures to be followed by all parties to achieve contract continuance and timely completion, while being financially fair to all concerned.

Pricing Change Orders

When pricing change orders, regardless of their cause, the most significant factor is when the change occurs. The need for a change may be perceived in the field or requested by the architect/engineer *before* any of the actual installation has begun, or may evolve or appear *during* construction when the item of work in question is partially installed. In the latter cases, the original sequence of construction is disrupted, along with all contiguous and supporting systems. Change orders cause the greatest impact when they occur *after* the installation has been completed and must be uncovered, or even replaced. Post-completion changes may be caused by necessary design changes, product failure, or changes in the owner's requirements that are not discovered until the building or the systems begin to function.

Specified procedures of notification and record keeping must be adhered to and enforced regardless of the stage of construction: *before, during,* or *after* installation. Some bidding documents anticipate change orders by requiring that unit prices including overhead and profit percentages—for additional as well as deductible changes—be listed. Generally these unit prices do not fully take into account the ripple effect, or impact on other trades, and should be used for general guidance only.

When pricing change orders, it is important to classify the time frame in which the change occurs. There are two basic time frames for change orders: *pre-installation change orders,* which occur before the start of construction, and *post-installation change orders,* which involve reworking after the original installation. Change orders that occur between these stages may be priced according to the extent of work completed using a combination of techniques developed for pricing *pre-* and *post-installation* changes.

Factors To Consider When Pricing Change Orders

As an estimator begins to prepare a change order, the following questions should be reviewed to determine their impact on the final price.

General

- *Is the change order work* pre-installation *or* post-installation?

  Change order work costs vary according to how much of the installation has been completed. Once workers have the project scoped in their minds, even though they have not started, it can be difficult to refocus. Consequently they may spend more than the normal amount of time understanding the change. Also, modifications to work in place, such as trimming or refitting, usually take more time than was initially estimated. The greater the amount of work in place, the more reluctant workers are to change it. Psychologically they may resent the change and as a result the rework takes longer than normal. Post-installation change order estimates must include demolition of existing work as required to accomplish the change. If the work is performed at a later time, additional obstacles, such as building finishes, may be present which must be protected. Regardless of whether the change occurs pre-installation or post-installation, attempt to isolate the identifiable factors and price them separately. For example, add shipping costs that may be required pre-installation or any demolition required post-installation. Then analyze the potential impact on productivity of psychological and/or learning curve factors and adjust the output rates accordingly. One approach is to break down the typical workday into segments and quantify the impact on each segment.

Change Order Installation Efficiency

The labor-hours expressed (for new construction) are based on average installation time, using an efficiency level. For change order situations, adjustments to this efficiency level should reflect the daily labor-hour allocation for that particular occurrence.

- *Will the change substantially delay the original completion date?*

  A significant change in the project may cause the original completion date to be extended. The extended schedule may subject the contractor to new wage rates dictated by relevant labor contracts. Project supervision and other project overhead must also be extended beyond the original completion date. The schedule extension may also put installation into a new weather season. For example, underground piping scheduled for October installation was delayed until January. As a result, frost penetrated the trench area, thereby changing the degree of difficulty of the task. Changes and delays may have a ripple effect throughout the project. This effect must be analyzed and negotiated with the owner.

- *What is the net effect of a deduct change order?*

  In most cases, change orders resulting in a deduction or credit reflect only bare costs. The contractor may retain the overhead and profit based on the original bid.

Materials

- *Will you have to pay more or less for the new material, required by the change order, than you paid for the original purchase?*

  The same material prices or discounts will usually apply to materials purchased for change orders as new construction. In some instances, however, the contractor may forfeit the advantages of competitive pricing for change orders. Consider the following example:

  A contractor purchased over $20,000 worth of fan coil units for an installation, and obtained the maximum discount. Some time later it was determined the project required an additional matching unit. The contractor has to purchase this unit from the original supplier to ensure a match. The supplier at this time may not discount the unit because of the small quantity, and the fact that he is no longer in a competitive situation. The impact of quantity on purchase can add between 0% and 25% to material prices and/or subcontractor quotes.

- *If materials have been ordered or delivered to the job site, will they be subject to a cancellation charge or restocking fee?*

  Check with the supplier to determine if ordered materials are subject to a cancellation charge. Delivered materials not used as result of a change order may be subject to a restocking fee if returned to the supplier. Common restocking charges run between 20% and 40%. Also, delivery charges to return the goods to the supplier must be added.

Labor

- *How efficient is the existing crew at the actual installation?*

  Is the same crew that performed the initial work going to do the change order? Possibly the change consists of the installation of a unit identical to one already installed; therefore, the change should take less time. Be sure to consider this potential productivity increase and modify the productivity rates accordingly.

- *If the crew size is increased, what impact will that have on supervision requirements?*

  Under most bargaining agreements or management practices, there is a point at which a working foreman is replaced by a nonworking foreman. This replacement increases project overhead by adding a nonproductive worker. If additional workers are added to accelerate the project or to perform changes while maintaining the schedule, be sure to add additional supervision time if warranted. Calculate the hours involved and the additional cost directly if possible.

- *What are the other impacts of increased crew size?*

  The larger the crew, the greater the potential for productivity to decrease. Some of the factors that cause this productivity loss are: overcrowding (producing restrictive conditions in the working space), and possibly a shortage of any special tools and equipment required. Such factors affect not only the crew working on the elements directly involved in the change order, but other crews whose movement may also be hampered. As the crew increases, check its basic composition for changes by the addition or deletion of apprentices or nonworking foreman, and quantify the potential effects of equipment shortages or other logistical factors.

- *As new crews, unfamiliar with the project, are brought onto the site, how long will it take them to become oriented to the project requirements?*

  The orientation time for a new crew to become 100% effective varies with the site and type of project. Orientation is easiest at a new construction site, and most difficult at existing, very restrictive renovation sites. The type of work also affects orientation time. When all elements of the work are exposed, such as concrete or masonry work, orientation is decreased. When the work is concealed or less visible, such as existing electrical systems, orientation takes longer. Usually orientation can be accomplished in one day or less. Costs for added orientation should be itemized and added to the total estimated cost.

- *How much actual production can be gained by working overtime?*

  Short term overtime can be used effectively to accomplish more work in a day. However, as overtime is scheduled to run beyond several weeks, studies have shown marked decreases in output. The following chart shows the effect of long term overtime on worker efficiency. If the anticipated change requires extended overtime to keep the job on schedule, these factors can be used as a guide to predict the impact on time and cost. Add project overhead, particularly supervision, that may also be incurred.

Days per Week	Hours per Day	Production Efficiency: 1 Week	2 Weeks	3 Weeks	4 Weeks	Average 4 Weeks	Payroll Cost Factors: @ 1-1/2 Times	@ 2 Times
5	8	100%	100%	100%	100%	100%	100%	100%
	9	100	100	95	90	96.25	105.6	111.1
	10	100	95	90	85	91.25	110.0	120.0
	11	95	90	75	65	81.25	113.6	127.3
	12	90	85	70	60	76.25	116.7	133.3
6	8	100	100	95	90	96.25	108.3	116.7
	9	100	95	90	85	92.50	113.0	125.9
	10	95	90	85	80	87.50	116.7	133.3
	11	95	85	70	65	78.75	119.7	139.4
	12	90	80	65	60	73.75	122.2	144.4
7	8	100	95	85	75	88.75	114.3	128.6
	9	95	90	80	70	83.75	118.3	136.5
	10	90	85	75	65	78.75	121.4	142.9
	11	85	80	65	60	72.50	124.0	148.1
	12	85	75	60	55	68.75	126.2	152.4

Effects of Overtime

Caution: Under many labor agreements, Sundays and holidays are paid at a higher premium than the normal overtime rate.

The use of long-term overtime is counterproductive on almost any construction job; that is, the longer the period of overtime, the lower the actual production rate. Numerous studies have been conducted, and while they have resulted in slightly different numbers, all reach the same conclusion. The figure above tabulates the effects of overtime work on efficiency.

As illustrated, there can be a difference between the *actual* payroll cost per hour and the *effective* cost per hour for overtime work. This is due to the reduced production efficiency with the increase in weekly hours beyond 40. This difference between actual and effective cost results from overtime work over a prolonged period. Short-term overtime work does not result in as great a reduction in efficiency and, in such cases, effective cost may not vary significantly from the actual payroll cost. As the total hours per week are increased on a regular basis, more time is lost due to fatigue, lowered morale, and an increased accident rate.

As an example, assume a project where workers are working 6 days a week, 10 hours per day. From the figure above (based on productivity studies), the average effective productive hours over a 4-week period are:

$$0.875 \times 60 = 52.5$$

Depending upon the locale and day of week, overtime hours may be paid at time and a half or double time. For time and a half, the overall (average) *actual* payroll cost (including regular and overtime hours) is determined as follows:

$$\frac{40 \text{ reg. hrs.} + (20 \text{ overtime hrs.} \times 1.5)}{60 \text{ hrs.}} = 1.167$$

Based on 60 hours, the payroll cost per hour will be 116.7% of the normal rate at 40 hours per week. However, because the effective production (efficiency) for 60 hours is reduced to the equivalent of 52.5 hours, the effective cost of overtime is calculated as follows:

For time and a half:

$$\frac{40 \text{ reg. hrs.} + (20 \text{ overtime hrs.} \times 1.5)}{52.5 \text{ hrs.}} = 1.33$$

Installed cost will be 133% of the normal rate (for labor).

Thus, when figuring overtime, the actual cost per unit of work will be higher than the apparent overtime payroll dollar increase, due to the reduced productivity of the longer workweek. These efficiency calculations are true only for those cost factors determined by hours worked. Costs that are applied weekly or monthly, such as equipment rentals, will not be similarly affected.

Equipment

- *What equipment is required to complete the change order?*

  Change orders may require extending the rental period of equipment already on the job site, or the addition of special equipment brought in to accomplish the change work. In either case, the additional rental charges and operator labor charges must be added.

Summary

The preceding considerations and others you deem appropriate should be analyzed and applied to a change order estimate. The impact of each should be quantified and listed on the estimate to form an audit trail.

Change orders that are properly identified, documented, and managed help to ensure the orderly, professional and profitable progress of the work. They also minimize potential claims or disputes at the end of the project.

Estimating Tips

- The cost figures in this section were derived from approximately 11,000 projects contained in the RSMeans database of completed construction projects. They include the contractor's overhead and profit, but do not generally include architectural fees or land costs. The figures have been adjusted to January of the current year. New projects are added to our files each year, and outdated projects are discarded. For this reason, certain costs may not show a uniform annual progression. In no case are all subdivisions of a project listed.
- These projects were located throughout the U.S. and reflect a tremendous variation in square foot (S.F.) and cubic foot (C.F.) costs. This is due to differences, not only in labor and material costs, but also in individual owners' requirements. For instance, a bank in a large city would have different features than one in a rural area. This is true of all the different types of buildings analyzed. Therefore, caution should be exercised when using these square foot costs. For example, for courthouses, costs in the database are local courthouse costs and will not apply to the larger, more elaborate federal courthouses. As a general rule, the projects in the 1/4 column do not include any site work or equipment, while the projects in the 3/4 column may include both equipment and site work. The median figures do not generally include site work.
- None of the figures "go with" any others. All individual cost items were computed and tabulated separately. Thus, the sum of the median figures for plumbing, HVAC, and electrical will not normally total up to the total mechanical and electrical costs arrived at by separate analysis and tabulation of the projects.
- Each building was analyzed as to total and component costs and percentages. The figures were arranged in ascending order with the results tabulated as shown. The 1/4 column shows that 25% of the projects had lower costs and 75% had higher. The 3/4 column shows that 75% of the projects had lower costs and 25% had higher. The median column shows that 50% of the projects had lower costs and 50% had higher.
- There are two times when square foot costs are useful. The first is in the conceptual stage when no details are available. Then, square foot costs make a useful starting point. The second is after the bids are in and the costs can be worked back into their appropriate categories for information purposes. As soon as details become available in the project design, the square foot approach should be discontinued and the project priced as to its particular components. When more precision is required, or for estimating the replacement cost of specific buildings, the current edition of *RSMeans Square Foot Costs* should be used.
- In using the figures in this section, it is recommended that the median column be used for preliminary figures if no additional information is available. The median figures, when multiplied by the total city construction cost index figures (see City Cost Indexes) and then multiplied by the project size modifier at the end of this section, should present a fairly accurate base figure, which would then have to be adjusted in view of the estimator's experience, local economic conditions, code requirements, and the owner's particular requirements. There is no need to factor the percentage figures, as these should remain constant from city to city. All tabulations mentioning air conditioning had at least partial air conditioning.
- The editors of this book would greatly appreciate receiving cost figures on one or more of your recent projects, which would then be included in the averages for next year. All cost figures received will be kept confidential, except that they will be averaged with other similar projects to arrive at square foot cost figures for next year's book. See the last page of the book for details and the discount available for submitting one or more of your projects. ■

	50 17 00 | S.F. Costs		UNIT	UNIT COSTS 1/4	UNIT COSTS MEDIAN	UNIT COSTS 3/4	% OF TOTAL 1/4	% OF TOTAL MEDIAN	% OF TOTAL 3/4	
01	0010	**APARTMENTS Low Rise (1 to 3 story)**	S.F.	74.50	94	125				01
	0020	Total project cost	C.F.	6.70	8.85	10.95				
	0100	Site work	S.F.	5.45	8.70	15.30	6.05%	10.55%	13.95%	
	0500	Masonry		1.47	3.62	5.95	1.54%	3.92%	6.50%	
	1500	Finishes		7.90	10.85	13.45	9.05%	10.75%	12.85%	
	1800	Equipment		2.44	3.70	5.50	2.71%	3.99%	5.95%	
	2720	Plumbing		5.80	7.45	9.50	6.65%	8.95%	10.05%	
	2770	Heating, ventilating, air conditioning		3.70	4.56	6.70	4.20%	5.60%	7.60%	
	2900	Electrical		4.33	5.75	7.80	5.20%	6.65%	8.35%	
	3100	Total: Mechanical & Electrical		15.40	19.95	24.50	16.05%	18.20%	23%	
	9000	Per apartment unit, total cost	Apt.	69,500	106,000	156,500				
	9500	Total: Mechanical & Electrical	"	13,100	20,700	27,000				
02	0010	**APARTMENTS Mid Rise (4 to 7 story)**	S.F.	99	119	147				02
	0020	Total project costs	C.F.	7.70	10.65	14.55				
	0100	Site work	S.F.	3.95	8	15.45	5.25%	6.70%	9.20%	
	0500	Masonry		4.13	9.05	12.40	5.10%	7.25%	10.50%	
	1500	Finishes		12.95	17.35	20.50	10.70%	13.50%	17.70%	
	1800	Equipment		2.77	4.30	5.65	2.54%	3.47%	4.31%	
	2500	Conveying equipment		2.24	2.76	3.33	2.05%	2.27%	2.69%	
	2720	Plumbing		5.80	9.30	9.85	5.70%	7.20%	8.95%	
	2900	Electrical		6.50	8.85	10.75	6.35%	7.20%	8.95%	
	3100	Total: Mechanical & Electrical		21	26	31.50	18.25%	21%	23%	
	9000	Per apartment unit, total cost	Apt.	112,000	132,000	218,500				
	9500	Total: Mechanical & Electrical	"	21,100	24,400	25,600				
03	0010	**APARTMENTS High Rise (8 to 24 story)**	S.F.	112	129	155				03
	0020	Total project costs	C.F.	10.90	12.65	14.95				
	0100	Site work	S.F.	4.07	6.60	9.20	2.58%	4.84%	6.15%	
	0500	Masonry		6.50	11.80	14.65	4.74%	9.65%	11.05%	
	1500	Finishes		12.45	15.55	18.35	9.75%	11.80%	13.70%	
	1800	Equipment		3.61	4.44	5.85	2.78%	3.49%	4.35%	
	2500	Conveying equipment		2.55	3.87	5.25	2.23%	2.78%	3.37%	
	2720	Plumbing		7.15	9.75	11.95	6.80%	7.20%	10.45%	
	2900	Electrical		7.70	9.75	13.15	6.45%	7.65%	8.80%	
	3100	Total: Mechanical & Electrical		23	29.50	35.50	17.95%	22.50%	24.50%	
	9000	Per apartment unit, total cost	Apt.	116,500	128,500	178,000				
	9500	Total: Mechanical & Electrical	"	23,200	28,800	30,500				
04	0010	**AUDITORIUMS**	S.F.	117	161	234				04
	0020	Total project costs	C.F.	7.30	10.15	14.55				
	2720	Plumbing	S.F.	6.90	10.20	12.20	5.85%	7.20%	8.70%	
	2900	Electrical		9.40	13.55	22.50	6.85%	9.50%	11.40%	
	3100	Total: Mechanical & Electrical		62.50	83	101	24.50%	27.50%	31%	
05	0010	**AUTOMOTIVE SALES**	S.F.	86.50	119	146				05
	0020	Total project costs	C.F.	5.70	6.80	8.85				
	2720	Plumbing	S.F.	3.93	6.85	7.45	2.89%	6.05%	6.50%	
	2770	Heating, ventilating, air conditioning		6.05	9.25	10	4.61%	10%	10.35%	
	2900	Electrical		6.95	10.90	16	7.25%	8.80%	12.15%	
	3100	Total: Mechanical & Electrical		19.35	31	37.50	17.30%	20.50%	22%	
06	0010	**BANKS**	S.F.	170	211	268				06
	0020	Total project costs	C.F.	12.10	16.45	21.50				
	0100	Site work	S.F.	19.40	30.50	43	7.90%	12.95%	17%	
	0500	Masonry		7.70	16.50	29.50	3.36%	6.90%	10.05%	
	1500	Finishes		15.15	23	28.50	5.85%	8.65%	11.70%	
	1800	Equipment		6.35	14	28.50	1.34%	5.55%	10.50%	
	2720	Plumbing		5.30	7.55	11.05	2.82%	3.90%	4.93%	
	2770	Heating, ventilating, air conditioning		10.05	13.45	17.90	4.86%	7.15%	8.50%	
	2900	Electrical		16.05	21.50	28	8.25%	10.20%	12.20%	
	3100	Total: Mechanical & Electrical		38.50	51.50	62	16.55%	19.45%	24%	
	3500	See also division 11 22 00								

50 17 | Square Foot Costs

	50 17 00 | S.F. Costs		UNIT	UNIT COSTS 1/4	UNIT COSTS MEDIAN	UNIT COSTS 3/4	% OF TOTAL 1/4	% OF TOTAL MEDIAN	% OF TOTAL 3/4	
13	0010	**CHURCHES**	S.F.	114	145	190				13
	0020	Total project costs	C.F.	7.05	8.90	11.75				
	1800	Equipment	S.F.	1.22	3.19	6.80	.83%	2.04%	4.30%	
	2720	Plumbing		4.43	6.20	9.15	3.51%	4.96%	6.25%	
	2770	Heating, ventilating, air conditioning		10.35	13.50	19.15	7.50%	10%	12%	
	2900	Electrical		9.60	13.20	18	7.30%	8.80%	10.95%	
	3100	Total: Mechanical & Electrical	↓	29.50	39.50	53	18.30%	22%	25%	
	3500	See also division 11 91 00								
15	0010	**CLUBS, COUNTRY**	S.F.	122	147	185				15
	0020	Total project costs	C.F.	9.85	12	16.55				
	2720	Plumbing	S.F.	7.40	10.95	25	5.60%	7.90%	10%	
	2900	Electrical		9.60	13.15	17.15	7%	8.95%	11%	
	3100	Total: Mechanical & Electrical	↓	51	64	67.50	19%	26.50%	29.50%	
17	0010	**CLUBS, SOCIAL Fraternal**	S.F.	103	141	188				17
	0020	Total project costs	C.F.	6.10	9.25	11				
	2720	Plumbing	S.F.	6.15	7.65	11.55	5.60%	6.90%	8.55%	
	2770	Heating, ventilating, air conditioning		7.15	10.70	13.75	8.20%	9.25%	14.40%	
	2900	Electrical		7.15	12.05	13.75	5.95%	9.30%	10.55%	
	3100	Total: Mechanical & Electrical	↓	38	41	52	21%	23%	23.50%	
18	0010	**CLUBS, Y.M.C.A.**	S.F.	134	167	252				18
	0020	Total project costs	C.F.	5.65	9.45	14.05				
	2720	Plumbing	S.F.	7.75	15.40	17.25	5.65%	7.60%	10.85%	
	2900	Electrical		10.05	13	22.50	6.45%	7.95%	8.90%	
	3100	Total: Mechanical & Electrical	↓	41	46.50	76	20.50%	21.50%	28.50%	
19	0010	**COLLEGES Classrooms & Administration**	S.F.	118	167	216				19
	0020	Total project costs	C.F.	7.10	12.50	20				
	0500	Masonry	S.F.	9.30	17.15	21	5.10%	8.05%	10.50%	
	2720	Plumbing		5.30	12.90	24	5.10%	6.60%	8.95%	
	2900	Electrical		10.05	15.85	22	7.70%	9.85%	12%	
	3100	Total: Mechanical & Electrical	↓	42.50	58.50	77.50	23%	28%	31.50%	
21	0010	**COLLEGES Science, Engineering, Laboratories**	S.F.	230	269	310				21
	0020	Total project costs	C.F.	13.20	19.25	22				
	1800	Equipment	S.F.	6.25	29	31.50	2%	6.45%	12.65%	
	2900	Electrical		18.95	27	41.50	7.10%	9.40%	12.10%	
	3100	Total: Mechanical & Electrical	↓	70.50	83.50	129	28.50%	31.50%	41%	
	3500	See also division 11 53 00								
23	0010	**COLLEGES Student Unions**	S.F.	147	199	241				23
	0020	Total project costs	C.F.	8.20	10.70	13.25				
	3100	Total: Mechanical & Electrical	S.F.	55	59.50	70.50	23.50%	26%	29%	
25	0010	**COMMUNITY CENTERS**	S.F.	122	150	203				25
	0020	Total project costs	C.F.	7.90	11.30	14.65				
	1800	Equipment	S.F.	2.45	4.80	7.90	1.47%	3.01%	5.30%	
	2720	Plumbing		5.75	10.05	13.70	4.85%	7%	8.95%	
	2770	Heating, ventilating, air conditioning		8.65	13.35	19.15	6.80%	10.35%	12.90%	
	2900	Electrical		10.25	13.55	19.90	7.15%	8.90%	10.40%	
	3100	Total: Mechanical & Electrical	↓	34	43	61.50	18.80%	23%	30%	
28	0010	**COURT HOUSES**	S.F.	175	206	284				28
	0020	Total project costs	C.F.	13.40	16	20				
	2720	Plumbing	S.F.	8.30	11.60	13.15	5.95%	7.45%	8.20%	
	2900	Electrical		18.55	21	30.50	9.05%	10.65%	12.15%	
	3100	Total: Mechanical & Electrical	↓	52	68	74.50	22.50%	26.50%	30%	
30	0010	**DEPARTMENT STORES**	S.F.	64.50	87.50	110				30
	0020	Total project costs	C.F.	3.46	4.48	6.10				
	2720	Plumbing	S.F.	2.01	2.54	3.85	1.82%	4.21%	5.90%	
	2770	Heating, ventilating, air conditioning	↓	4.88	9.05	13.65	8.20%	9.10%	14.80%	

	50 17 00 | S.F. Costs		UNIT	UNIT COSTS 1/4	UNIT COSTS MEDIAN	UNIT COSTS 3/4	% OF TOTAL 1/4	% OF TOTAL MEDIAN	% OF TOTAL 3/4	
30	2900	Electrical	S.F.	7.40	10.15	12	9.05%	12.15%	14.95%	30
	3100	Total: Mechanical & Electrical		13.05	16.65	29	13.20%	21.50%	50%	
31	0010	**DORMITORIES Low Rise (1 to 3 story)**	S.F.	122	170	212				31
	0020	Total project costs	C.F.	6.90	11.20	16.75				
	2720	Plumbing	S.F.	7.35	9.85	12.45	8.05%	9%	9.65%	
	2770	Heating, ventilating, air conditioning		7.80	9.35	12.45	4.61%	8.05%	10%	
	2900	Electrical		8.10	12.35	16.90	6.40%	8.65%	9.50%	
	3100	Total: Mechanical & Electrical		42	45	70.50	22%	25%	27%	
	9000	Per bed, total cost	Bed	52,000	57,500	123,500				
32	0010	**DORMITORIES Mid Rise (4 to 8 story)**	S.F.	150	196	242				32
	0020	Total project costs	C.F.	16.55	18.20	22				
	2900	Electrical	S.F.	15.95	18.15	24.50	8.20%	10.20%	11.95%	
	3100	Total: Mechanical & Electrical	″	44.50	89	90.50	25%	30.50%	35.50%	
	9000	Per bed, total cost	Bed	21,400	48,800	282,500				
34	0010	**FACTORIES**	S.F.	57	85	131				34
	0020	Total project costs	C.F.	3.65	5.45	9.05				
	0100	Site work	S.F.	6.50	11.85	18.75	6.95%	11.45%	17.95%	
	2720	Plumbing		3.07	5.70	9.45	3.73%	6.05%	8.10%	
	2770	Heating, ventilating, air conditioning		5.95	8.55	11.55	5.25%	8.45%	11.35%	
	2900	Electrical		7.05	11.20	17.05	8.10%	10.50%	14.20%	
	3100	Total: Mechanical & Electrical		20	32.50	41	21%	28.50%	35.50%	
36	0010	**FIRE STATIONS**	S.F.	113	156	213				36
	0020	Total project costs	C.F.	6.60	9.05	12.05				
	0500	Masonry	S.F.	15.40	30	38	7.50%	10.95%	15.55%	
	1140	Roofing		3.68	9.95	11.35	1.90%	4.94%	5.05%	
	1580	Painting		2.86	4.27	4.37	1.37%	1.57%	2.07%	
	1800	Equipment		1.40	2.69	4.97	.62%	1.63%	3.42%	
	2720	Plumbing		6.30	10.05	14.25	5.85%	7.35%	9.45%	
	2770	Heating, ventilating, air conditioning		6.25	10.15	15.65	5.15%	7.40%	9.40%	
	2900	Electrical		8.15	14.35	19.35	6.90%	8.60%	10.60%	
	3100	Total: Mechanical & Electrical		41.50	53	60	18.40%	23%	27%	
37	0010	**FRATERNITY HOUSES & Sorority Houses**	S.F.	113	145	199				37
	0020	Total project costs	C.F.	11.25	11.70	14.10				
	2720	Plumbing	S.F.	8.55	9.75	17.90	6.80%	8%	10.85%	
	2900	Electrical		7.45	16.10	19.70	6.60%	9.90%	10.65%	
38	0010	**FUNERAL HOMES**	S.F.	119	162	295				38
	0020	Total project costs	C.F.	12.15	13.55	26				
	2900	Electrical	S.F.	5.25	9.65	10.55	3.58%	4.44%	5.95%	
39	0010	**GARAGES, COMMERCIAL (Service)**	S.F.	67.50	104	144				39
	0020	Total project costs	C.F.	4.43	6.55	9.50				
	1800	Equipment	S.F.	3.80	8.55	13.30	2.21%	4.62%	6.80%	
	2720	Plumbing		4.67	7.20	13.10	5.45%	7.85%	10.65%	
	2730	Heating & ventilating		4.60	7.95	10.15	5.25%	6.85%	8.20%	
	2900	Electrical		6.40	9.75	14.10	7.15%	9.25%	10.85%	
	3100	Total: Mechanical & Electrical		12.10	27	40	12.35%	17.40%	26%	
40	0010	**GARAGES, MUNICIPAL (Repair)**	S.F.	105	135	192				40
	0020	Total project costs	C.F.	6.20	7.85	13.50				
	0500	Masonry	S.F.	5.20	18.20	28	4.03%	9.15%	12.50%	
	2720	Plumbing		4.44	8.55	16.05	3.59%	6.70%	7.95%	
	2730	Heating & ventilating		7.60	11	21	6.15%	7.45%	13.50%	
	2900	Electrical		7.65	12.35	22.50	6.90%	9.40%	13%	
	3100	Total: Mechanical & Electrical		34.50	56.50	68	21.50%	25.50%	35.50%	
41	0010	**GARAGES, PARKING**	S.F.	39.50	56	96.50				41
	0020	Total project costs	C.F.	3.62	4.92	7.15				

	50 17 00 | S.F. Costs		UNIT	UNIT COSTS 1/4	UNIT COSTS MEDIAN	UNIT COSTS 3/4	% OF TOTAL 1/4	% OF TOTAL MEDIAN	% OF TOTAL 3/4
41	2720	Plumbing	S.F.	.66	1.69	2.61	1.72%	2.70%	3.85%
	2900	Electrical		2.12	2.80	4.08	4.52%	5.40%	6.35%
	3100	Total: Mechanical & Electrical		4.11	6.05	7.50	7%	8.90%	11.05%
	3200								
	9000	Per car, total cost	Car	16,300	20,400	26,000			
43	0010	**GYMNASIUMS**	S.F.	108	143	196			
	0020	Total project costs	C.F.	5.35	7.25	8.90			
	1800	Equipment	S.F.	2.55	4.78	8.60	1.76%	3.26%	6.70%
	2720	Plumbing		5.90	7.90	10.40	4.65%	6.40%	7.75%
	2770	Heating, ventilating, air conditioning		6.40	11.10	22.50	5.15%	9.05%	11.10%
	2900	Electrical		8.50	11.25	14.95	6.75%	8.50%	10.30%
	3100	Total: Mechanical & Electrical		29	40.50	49.50	19.75%	23.50%	29%
	3500	See also division 11 66 00							
46	0010	**HOSPITALS**	S.F.	206	258	355			
	0020	Total project costs	C.F.	15.60	19.40	28			
	1800	Equipment	S.F.	5.20	10.05	17.30	.80%	2.53%	4.80%
	2720	Plumbing		17.70	25	32	7.60%	9.10%	10.85%
	2770	Heating, ventilating, air conditioning		26	33.50	46.50	7.80%	12.95%	16.65%
	2900	Electrical		22.50	30.50	45.50	10%	11.75%	14.10%
	3100	Total: Mechanical & Electrical		66.50	92.50	137	28%	33.50%	37%
	9000	Per bed or person, total cost	Bed	238,000	328,500	378,500			
	9900	See also division 11 71 00							
48	0010	**HOUSING For the Elderly**	S.F.	101	128	157			
	0020	Total project costs	C.F.	7.20	10	12.80			
	0100	Site work	S.F.	7.05	10.95	16.05	5.05%	7.90%	12.10%
	0500	Masonry		1.91	11.55	16.85	1.30%	6.05%	11%
	1800	Equipment		2.45	3.37	5.35	1.88%	3.23%	4.43%
	2510	Conveying systems		2.29	3.31	4.49	1.78%	2.20%	2.81%
	2720	Plumbing		7.50	9.60	12.10	8.15%	9.55%	10.50%
	2730	Heating, ventilating, air conditioning		3.86	5.45	8.15	3.30%	5.60%	7.25%
	2900	Electrical		7.55	10.25	13.10	7.30%	8.50%	10.25%
	3100	Total: Mechanical & Electrical		26	32	41	18.10%	22.50%	29%
	9000	Per rental unit, total cost	Unit	94,000	110,000	122,500			
	9500	Total: Mechanical & Electrical	"	21,000	24,100	28,100			
50	0010	**HOUSING Public (Low Rise)**	S.F.	85	118	154			
	0020	Total project costs	C.F.	6.75	9.45	11.75			
	0100	Site work	S.F.	10.85	15.60	25.50	8.35%	11.75%	16.50%
	1800	Equipment		2.31	3.77	5.75	2.26%	3.03%	4.24%
	2720	Plumbing		6.15	8.10	10.25	7.15%	9.05%	11.60%
	2730	Heating, ventilating, air conditioning		3.08	6	6.55	4.26%	6.05%	6.45%
	2900	Electrical		5.15	7.65	10.65	5.10%	6.55%	8.25%
	3100	Total: Mechanical & Electrical		24.50	31.50	35	14.50%	17.55%	26.50%
	9000	Per apartment, total cost	Apt.	93,500	106,500	133,500			
	9500	Total: Mechanical & Electrical	"	19,900	24,600	27,200			
51	0010	**ICE SKATING RINKS**	S.F.	72.50	170	187			
	0020	Total project costs	C.F.	5.35	5.45	6.30			
	2720	Plumbing	S.F.	2.72	5.10	5.20	3.12%	3.23%	5.65%
	2900	Electrical		7.80	11.95	12.65	6.30%	10.15%	15.05%
52	0010	**JAILS**	S.F.	189	286	370			
	0020	Total project costs	C.F.	19.95	28	32.50			
	1800	Equipment	S.F.	8.65	25.50	43.50	2.80%	6.95%	9.80%
	2720	Plumbing		21.50	28.50	37.50	7%	8.90%	13.35%
	2770	Heating, ventilating, air conditioning		19.95	26.50	51.50	7.50%	9.45%	17.75%
	2900	Electrical		23.50	31.50	40	9.40%	11.70%	15.25%
	3100	Total: Mechanical & Electrical		62	110	131	28%	31%	36%
53	0010	**LIBRARIES**	S.F.	143	190	248			
	0020	Total project costs	C.F.	9.55	12	15.30			

50 17 00 | S.F. Costs

		50 17 00 | S.F. Costs	UNIT	UNIT COSTS			% OF TOTAL			
				1/4	MEDIAN	3/4	1/4	MEDIAN	3/4	
53	0500	Masonry	S.F.	9.50	19.35	32.50	5.60%	7.15%	10.95%	53
	1800	Equipment		1.91	5.15	7.75	.28%	1.39%	4.07%	
	2720	Plumbing		5.15	7.30	10.20	3.38%	4.60%	5.70%	
	2770	Heating, ventilating, air conditioning		11.45	19.35	23.50	7.80%	10.95%	12.80%	
	2900	Electrical		14.40	19	25.50	8.40%	10.55%	12%	
	3100	Total: Mechanical & Electrical		47	57	67	21%	24%	28%	
54	0010	**LIVING, ASSISTED**	S.F.	129	154	180				54
	0020	Total project costs	C.F.	10.90	12.70	14.45				
	0500	Masonry	S.F.	3.75	4.53	5.55	2.36%	3.16%	3.86%	
	1800	Equipment		2.87	3.42	4.38	2.12%	2.45%	2.87%	
	2720	Plumbing		10.85	14.50	15.05	6.05%	8.15%	10.60%	
	2770	Heating, ventilating, air conditioning		12.85	13.45	14.75	7.95%	9.35%	9.70%	
	2900	Electrical		12.70	14.25	16.50	9%	9.95%	10.65%	
	3100	Total: Mechanical & Electrical		35	41	48	24%	28.50%	31.50%	
55	0010	**MEDICAL CLINICS**	S.F.	132	162	207				55
	0020	Total project costs	C.F.	9.65	12.50	16.60				
	1800	Equipment	S.F.	1.68	7.50	11.65	1.05%	2.94%	6.35%	
	2720	Plumbing		8.75	12.30	16.50	6.15%	8.40%	10.10%	
	2770	Heating, ventilating, air conditioning		10.45	13.70	20	6.65%	8.85%	11.35%	
	2900	Electrical		11.30	16.10	21	8.10%	10%	12.20%	
	3100	Total: Mechanical & Electrical		36	49	67.50	22%	27%	33.50%	
	3500	See also division 11 71 00								
57	0010	**MEDICAL OFFICES**	S.F.	124	154	190				57
	0020	Total project costs	C.F.	9.25	12.50	16.95				
	1800	Equipment	S.F.	4.09	8.10	11.35	.66%	4.87%	6.50%	
	2720	Plumbing		6.85	10.55	14.25	5.60%	6.80%	8.50%	
	2770	Heating, ventilating, air conditioning		8.25	11.95	15.75	6.10%	8%	9.70%	
	2900	Electrical		9.90	14.50	20.50	7.50%	9.80%	11.70%	
	3100	Total: Mechanical & Electrical		27	39.50	56	19.30%	22.50%	27.50%	
59	0010	**MOTELS**	S.F.	78	113	148				59
	0020	Total project costs	C.F.	6.95	9.30	15.25				
	2720	Plumbing	S.F.	7.90	10.10	12.05	9.45%	10.60%	12.55%	
	2770	Heating, ventilating, air conditioning		4.83	7.20	12.90	5.60%	5.60%	10%	
	2900	Electrical		7.40	9.35	11.65	7.45%	9.05%	10.45%	
	3100	Total: Mechanical & Electrical		22.50	31.50	54	18.50%	24%	25.50%	
	5000									
	9000	Per rental unit, total cost	Unit	39,800	75,500	82,000				
	9500	Total: Mechanical & Electrical	"	7,750	11,700	13,600				
60	0010	**NURSING HOMES**	S.F.	123	158	197				60
	0020	Total project costs	C.F.	9.65	12.05	16.45				
	1800	Equipment	S.F.	3.34	5.10	8.55	2%	3.62%	4.99%	
	2720	Plumbing		10.50	15.90	19.20	8.75%	10.10%	12.70%	
	2770	Heating, ventilating, air conditioning		11.05	16.80	22.50	9.70%	11.45%	11.80%	
	2900	Electrical		12.15	15.20	21	9.40%	10.60%	12.50%	
	3100	Total: Mechanical & Electrical		29	40.50	68	26%	28%	30%	
	9000	Per bed or person, total cost	Bed	54,500	68,000	88,000				
61	0010	**OFFICES Low Rise (1 to 4 story)**	S.F.	103	135	175				61
	0020	Total project costs	C.F.	7.35	10.15	13.40				
	0100	Site work	S.F.	8.30	14.45	21.50	5.95%	9.70%	13.55%	
	0500	Masonry		4	8.05	14.70	2.61%	5.40%	8.45%	
	1800	Equipment		.93	2.15	5.85	.57%	1.50%	3.42%	
	2720	Plumbing		3.69	5.70	8.35	3.66%	4.50%	6.10%	
	2770	Heating, ventilating, air conditioning		8.15	11.40	16.65	7.20%	10.30%	11.70%	
	2900	Electrical		8.45	12.10	17.20	7.45%	9.65%	11.40%	
	3100	Total: Mechanical & Electrical		23.50	32.50	48.50	18.20%	22%	27%	
62	0010	**OFFICES Mid Rise (5 to 10 story)**	S.F.	109	132	180				62
	0020	Total project costs	C.F.	7.75	9.90	14				

50 17 00 | S.F. Costs

	50 17 00 | S.F. Costs		UNIT	UNIT COSTS 1/4	UNIT COSTS MEDIAN	UNIT COSTS 3/4	% OF TOTAL 1/4	% OF TOTAL MEDIAN	% OF TOTAL 3/4	
62	2720	Plumbing	S.F.	3.30	5.10	7.35	2.83%	3.74%	4.50%	62
	2770	Heating, ventilating, air conditioning		8.30	11.85	18.95	7.65%	9.40%	11%	
	2900	Electrical		8.10	10.40	15.70	6.35%	7.80%	10%	
	3100	Total: Mechanical & Electrical		21	27	51.50	18.95%	21%	27.50%	
63	0010	OFFICES High Rise (11 to 20 story)	S.F.	134	169	208				63
	0020	Total project costs	C.F.	9.40	11.75	16.85				
	2900	Electrical	S.F.	8.15	9.95	14.80	5.80%	7.85%	10.50%	
	3100	Total: Mechanical & Electrical		26.50	35.50	59.50	16.90%	23.50%	34%	
64	0010	POLICE STATIONS	S.F.	161	212	270				64
	0020	Total project costs	C.F.	12.85	15.70	21.50				
	0500	Masonry	S.F.	16.10	26.50	33.50	7.80%	9.10%	11.35%	
	1800	Equipment		2.31	11.25	17.80	.98%	3.35%	6.70%	
	2720	Plumbing		9	17.95	22.50	5.65%	6.90%	10.75%	
	2770	Heating, ventilating, air conditioning		13.20	18.70	26.50	5.85%	10.55%	11.70%	
	2900	Electrical		17.60	25.50	33	9.80%	11.70%	14.50%	
	3100	Total: Mechanical & Electrical		67	70.50	94	28.50%	31.50%	32.50%	
65	0010	POST OFFICES	S.F.	127	157	200				65
	0020	Total project costs	C.F.	7.65	9.70	11				
	2720	Plumbing	S.F.	5.15	7.10	8.95	4.24%	5.30%	5.60%	
	2770	Heating, ventilating, air conditioning		8	11.05	12.30	6.65%	7.15%	9.35%	
	2900	Electrical		10.50	14.75	17.50	7.25%	9%	11%	
	3100	Total: Mechanical & Electrical		30.50	39.50	45	16.25%	18.80%	22%	
66	0010	POWER PLANTS	S.F.	880	1,175	2,150				66
	0020	Total project costs	C.F.	24.50	53	113				
	2900	Electrical	S.F.	62.50	132	197	9.30%	12.75%	21.50%	
	8100	Total: Mechanical & Electrical		155	505	1,125	32.50%	32.50%	52.50%	
67	0010	RELIGIOUS EDUCATION	S.F.	103	137	170				67
	0020	Total project costs	C.F.	5.70	8.15	10.20				
	2720	Plumbing	S.F.	4.28	6.05	8.60	4.40%	5.30%	7.10%	
	2770	Heating, ventilating, air conditioning		10.80	12.25	17.30	10.05%	11.45%	12.35%	
	2900	Electrical		8.15	11.65	17.10	7.70%	9.05%	10.35%	
	3100	Total: Mechanical & Electrical		35.50	45	53	22%	23.50%	26%	
69	0010	RESEARCH Laboratories & Facilities	S.F.	157	222	325				69
	0020	Total project costs	C.F.	11.40	22.50	27				
	1800	Equipment	S.F.	6.80	13.40	32.50	1.22%	4.80%	9.35%	
	2720	Plumbing		11.20	19.70	31.50	6.15%	8.30%	10.80%	
	2770	Heating, ventilating, air conditioning		13.80	46.50	55	7.25%	16.50%	17.50%	
	2900	Electrical		18.45	29.50	48.50	9.45%	11.15%	14.60%	
	3100	Total: Mechanical & Electrical		59	104	147	29.50%	36%	41%	
70	0010	RESTAURANTS	S.F.	149	192	249				70
	0020	Total project costs	C.F.	12.50	16.40	21.50				
	1800	Equipment	S.F.	8.80	23.50	35.50	6.10%	13%	15.65%	
	2720	Plumbing		11.80	14.30	18.75	6.10%	8.15%	9%	
	2770	Heating, ventilating, air conditioning		14.95	21	25	9.20%	12%	12.40%	
	2900	Electrical		15.70	19.40	25	8.35%	10.55%	11.55%	
	3100	Total: Mechanical & Electrical		48.50	52	67.50	21%	25%	29.50%	
	9000	Per seat unit, total cost	Seat	5,450	7,275	8,600				
	9500	Total: Mechanical & Electrical	″	1,375	1,825	2,150				
72	0010	RETAIL STORES	S.F.	69.50	93.50	124				72
	0020	Total project costs	C.F.	4.71	6.70	9.35				
	2720	Plumbing	S.F.	2.52	4.20	7.15	3.26%	4.60%	6.80%	
	2770	Heating, ventilating, air conditioning		5.45	7.45	11.20	6.75%	8.75%	10.15%	
	2900	Electrical		6.25	8.55	12.35	7.25%	9.90%	11.60%	
	3100	Total: Mechanical & Electrical		16.65	21.50	28.50	17.05%	21%	23.50%	
74	0010	SCHOOLS Elementary	S.F.	114	141	172				74
	0020	Total project costs	C.F.	7.40	9.50	12.25				
	0500	Masonry	S.F.	8	17.30	25.50	4.89%	10.50%	14%	
	1800	Equipment		2.63	4.96	9.40	1.83%	3.13%	4.61%	

	50 17 00 | S.F. Costs		UNIT	UNIT COSTS			% OF TOTAL			
				1/4	MEDIAN	3/4	1/4	MEDIAN	3/4	
74	2720	Plumbing	S.F.	6.50	9.20	12.30	5.70%	7.15%	9.35%	74
	2730	Heating, ventilating, air conditioning		9.80	15.55	22.50	8.15%	10.80%	14.90%	
	2900	Electrical		10.70	14.25	18.20	8.45%	10.05%	11.85%	
	3100	Total: Mechanical & Electrical	↓	39	47.50	60	25%	27.50%	30%	
	9000	Per pupil, total cost	Ea.	13,000	19,300	43,300				
	9500	Total: Mechanical & Electrical	"	3,675	4,650	11,700				
76	0010	**SCHOOLS Junior High & Middle**	S.F.	117	145	177				76
	0020	Total project costs	C.F.	7.40	9.60	10.75				
	0500	Masonry	S.F.	12.55	18.80	23	7.55%	11.10%	14.30%	
	1800	Equipment		3.20	6	8.80	1.80%	3.03%	4.80%	
	2720	Plumbing		6.80	8.40	10.40	5.30%	6.80%	7.25%	
	2770	Heating, ventilating, air conditioning		13.60	16.55	29	8.90%	11.55%	14.20%	
	2900	Electrical		11.90	14.70	18.60	8.05%	9.55%	10.60%	
	3100	Total: Mechanical & Electrical	↓	37.50	48	59.50	23%	25.50%	29.50%	
	9000	Per pupil, total cost	Ea.	14,800	19,400	26,100				
78	0010	**SCHOOLS Senior High**	S.F.	123	150	188				78
	0020	Total project costs	C.F.	7.35	10.30	17.40				
	1800	Equipment	S.F.	3.24	7.55	10.45	1.88%	2.67%	4.30%	
	2720	Plumbing		6.80	10.25	18.70	5.60%	6.90%	8.30%	
	2770	Heating, ventilating, air conditioning		11.75	15.95	24.50	8.95%	11.60%	15%	
	2900	Electrical		12.30	16.40	24	8.70%	10.35%	12.50%	
	3100	Total: Mechanical & Electrical	↓	40.50	48	77.50	24%	26.50%	28.50%	
	9000	Per pupil, total cost	Ea.	11,500	23,300	29,200				
80	0010	**SCHOOLS Vocational**	S.F.	98.50	143	177				80
	0020	Total project costs	C.F.	6.10	8.80	12.15				
	0500	Masonry	S.F.	4.33	14.25	22	3.20%	4.61%	10.95%	
	1800	Equipment		3.08	7.65	10.60	1.24%	3.10%	4.26%	
	2720	Plumbing		6.30	9.35	13.80	5.40%	6.90%	8.55%	
	2770	Heating, ventilating, air conditioning		8.80	16.40	27.50	8.60%	11.90%	14.65%	
	2900	Electrical		10.25	14	19.15	8.45%	11%	13.20%	
	3100	Total: Mechanical & Electrical	↓	38.50	67.50	86.50	27.50%	29.50%	31%	
	9000	Per pupil, total cost	Ea.	13,700	36,700	54,500				
83	0010	**SPORTS ARENAS**	S.F.	86	115	177				83
	0020	Total project costs	C.F.	4.67	8.35	10.80				
	2720	Plumbing	S.F.	4.31	7.55	15.95	4.35%	6.35%	9.40%	
	2770	Heating, ventilating, air conditioning		10.75	12.70	17.60	8.80%	10.20%	13.55%	
	2900	Electrical	↓	7.95	12.15	15.70	7.65%	9.75%	11.90%	
85	0010	**SUPERMARKETS**	S.F.	79.50	92	108				85
	0020	Total project costs	C.F.	4.42	5.35	8.10				
	2720	Plumbing	S.F.	4.44	5.60	6.50	5.40%	6%	7.45%	
	2770	Heating, ventilating, air conditioning		6.55	8.65	10.55	8.60%	8.65%	9.60%	
	2900	Electrical		9.30	11.35	13.50	10.40%	12.45%	13.60%	
	3100	Total: Mechanical & Electrical	↓	25.50	27	36.50	23.50%	27.50%	28.50%	
86	0010	**SWIMMING POOLS**	S.F.	129	300	460				86
	0020	Total project costs	C.F.	10.30	12.85	14				
	2720	Plumbing	S.F.	11.90	13.60	18.25	4.80%	9.70%	20.50%	
	2900	Electrical		9.75	15.65	34.50	6.05%	6.95%	7.60%	
	3100	Total: Mechanical & Electrical	↓	61	80.50	106	11.15%	14.10%	23.50%	
87	0010	**TELEPHONE EXCHANGES**	S.F.	160	243	315				87
	0020	Total project costs	C.F.	10.65	17.05	23.50				
	2720	Plumbing	S.F.	7.20	11.15	16.30	4.52%	5.80%	6.90%	
	2770	Heating, ventilating, air conditioning		16.75	33.50	42	11.80%	16.05%	18.40%	
	2900	Electrical		17.40	27.50	49	10.90%	14%	17.85%	
	3100	Total: Mechanical & Electrical	↓	51.50	97.50	138	29.50%	33.50%	44.50%	
91	0010	**THEATERS**	S.F.	102	138	203				91
	0020	Total project costs	C.F.	4.96	7.35	10.80				

		50 17 00 | S.F. Costs	UNIT	UNIT COSTS 1/4	UNIT COSTS MEDIAN	UNIT COSTS 3/4	% OF TOTAL 1/4	% OF TOTAL MEDIAN	% OF TOTAL 3/4	
91	2720	Plumbing	S.F.	3.34	3.88	15.85	2.92%	4.70%	6.80%	91
	2770	Heating, ventilating, air conditioning		10.45	12.65	15.65	8%	12.25%	13.40%	
	2900	Electrical		9.40	12.70	26	8.05%	9.35%	12.25%	
	3100	Total: Mechanical & Electrical	↓	24	32.50	38.50	21.50%	25.50%	27.50%	
94	0010	**TOWN HALLS City Halls & Municipal Buildings**	S.F.	112	152	199				94
	0020	Total project costs	C.F.	8.10	12.60	18.45				
	2720	Plumbing	S.F.	5	9.35	16.45	4.31%	5.95%	7.95%	
	2770	Heating, ventilating, air conditioning		9.05	17.95	26.50	7.05%	9.05%	13.45%	
	2900	Electrical		11.40	15.70	21.50	8.05%	9.45%	11.65%	
	3100	Total: Mechanical & Electrical	↓	39.50	50	76.50	22%	26.50%	31%	
97	0010	**WAREHOUSES & Storage Buildings**	S.F.	44.50	63.50	95				97
	0020	Total project costs	C.F.	2.34	3.67	6.05				
	0100	Site work	S.F.	4.62	9.15	13.80	6.05%	12.95%	19.55%	
	0500	Masonry		2.54	6.35	13.70	3.60%	7.35%	12%	
	1800	Equipment		.68	1.55	8.70	.72%	1.69%	5.55%	
	2720	Plumbing		1.49	2.68	5	2.90%	4.80%	6.55%	
	2730	Heating, ventilating, air conditioning		1.70	4.80	6.45	2.41%	5%	8.90%	
	2900	Electrical		2.65	4.98	8.20	5.15%	7.20%	10.05%	
	3100	Total: Mechanical & Electrical	↓	7.40	11.35	22.50	13.30%	18.90%	26%	
99	0010	**WAREHOUSE & OFFICES Combination**	S.F.	55	73.50	101				99
	0020	Total project costs	C.F.	2.82	4.09	6.05				
	1800	Equipment	S.F.	.95	1.84	2.74	.42%	1.19%	2.40%	
	2720	Plumbing		2.13	3.70	5.50	3.74%	4.76%	6.30%	
	2770	Heating, ventilating, air conditioning		3.36	5.25	7.35	5%	5.65%	10.05%	
	2900	Electrical		3.75	5.55	8.65	5.85%	8%	10%	
	3100	Total: Mechanical & Electrical	↓	10.45	15.85	25	14.55%	19.95%	24.50%	

Square Foot Project Size Modifier

One factor that affects the S.F. cost of a particular building is the size. In general, for buildings built to the same specifications in the same locality, the larger building will have the lower S.F. cost. This is due mainly to the decreasing contribution of the exterior walls, plus the economy of scale usually achievable in larger buildings. The area conversion scale shown below will give a factor to convert costs for the typical size building to an adjusted cost for the particular project.

The square foot base size table lists the median costs, most typical project size in our accumulated data, and the range in size of the projects.

The size factor for your project is determined by dividing your project area in S.F. by the typical project size for the particular building type. With this factor, enter the area conversion scale at the appropriate size factor and determine the appropriate cost multiplier for your building size.

Example: Determine the cost per S.F. for a 100,000 S.F. Mid-rise apartment building.

$$\frac{\text{Proposed building area} = 100{,}000 \text{ S.F.}}{\text{Typical size from below} = 50{,}000 \text{ S.F.}} = 2.00$$

Enter area conversion scale at 2.0, intersect curve, read horizontally the appropriate cost multiplier of .94. Size adjusted cost becomes .94 × \$119.00 = \$112.00 based on national average costs.

Note: For size factors less than .50, the cost multiplier is 1.1
For size factors greater than 3.5, the cost multiplier is .90

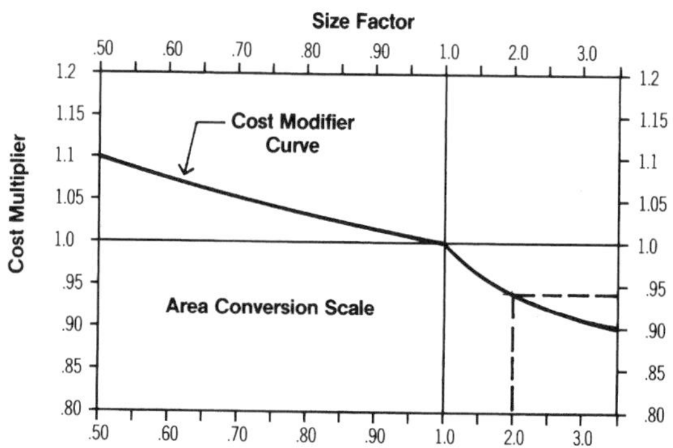

Square Foot Base Size							
Building Type	Median Cost per S.F.	Typical Size Gross S.F.	Typical Range Gross S.F.	Building Type	Median Cost per S.F.	Typical Size Gross S.F.	Typical Range Gross S.F.
Apartments, Low Rise	\$ 94.00	21,000	9,700 - 37,200	Jails	\$ 286.00	40,000	5,500 - 145,000
Apartments, Mid Rise	119.00	50,000	32,000 - 100,000	Libraries	190.00	12,000	7,000 - 31,000
Apartments, High Rise	129.00	145,000	95,000 - 600,000	Living, Assisted	154.00	32,300	23,500 - 50,300
Auditoriums	161.00	25,000	7,600 - 39,000	Medical Clinics	162.00	7,200	4,200 - 15,700
Auto Sales	119.00	20,000	10,800 - 28,600	Medical Offices	154.00	6,000	4,000 - 15,000
Banks	211.00	4,200	2,500 - 7,500	Motels	113.00	40,000	15,800 - 120,000
Churches	145.00	17,000	2,000 - 42,000	Nursing Homes	158.00	23,000	15,000 - 37,000
Clubs, Country	147.00	6,500	4,500 - 15,000	Offices, Low Rise	135.00	20,000	5,000 - 80,000
Clubs, Social	141.00	10,000	6,000 - 13,500	Offices, Mid Rise	132.00	120,000	20,000 - 300,000
Clubs, YMCA	167.00	28,300	12,800 - 39,400	Offices, High Rise	169.00	260,000	120,000 - 800,000
Colleges (Class)	167.00	50,000	15,000 - 150,000	Police Stations	212.00	10,500	4,000 - 19,000
Colleges (Science Lab)	269.00	45,600	16,600 - 80,000	Post Offices	157.00	12,400	6,800 - 30,000
College (Student Union)	199.00	33,400	16,000 - 85,000	Power Plants	1175.00	7,500	1,000 - 20,000
Community Center	150.00	9,400	5,300 - 16,700	Religious Education	137.00	9,000	6,000 - 12,000
Court Houses	206.00	32,400	17,800 - 106,000	Research	222.00	19,000	6,300 - 45,000
Dept. Stores	87.50	90,000	44,000 - 122,000	Restaurants	192.00	4,400	2,800 - 6,000
Dormitories, Low Rise	170.00	25,000	10,000 - 95,000	Retail Stores	93.50	7,200	4,000 - 17,600
Dormitories, Mid Rise	196.00	85,000	20,000 - 200,000	Schools, Elementary	141.00	41,000	24,500 - 55,000
Factories	85.00	26,400	12,900 - 50,000	Schools, Jr. High	145.00	92,000	52,000 - 119,000
Fire Stations	156.00	5,800	4,000 - 8,700	Schools, Sr. High	150.00	101,000	50,500 - 175,000
Fraternity Houses	145.00	12,500	8,200 - 14,800	Schools, Vocational	143.00	37,000	20,500 - 82,000
Funeral Homes	162.00	10,000	4,000 - 20,000	Sports Arenas	115.00	15,000	5,000 - 40,000
Garages, Commercial	104.00	9,300	5,000 - 13,600	Supermarkets	92.00	44,000	12,000 - 60,000
Garages, Municipal	135.00	8,300	4,500 - 12,600	Swimming Pools	300.00	20,000	10,000 - 32,000
Garages, Parking	56.00	163,000	76,400 - 225,300	Telephone Exchange	243.00	4,500	1,200 - 10,600
Gymnasiums	143.00	19,200	11,600 - 41,000	Theaters	138.00	10,500	8,800 - 17,500
Hospitals	258.00	55,000	27,200 - 125,000	Town Halls	152.00	10,800	4,800 - 23,400
House (Elderly)	128.00	37,000	21,000 - 66,000	Warehouses	63.50	25,000	8,000 - 72,000
Housing (Public)	118.00	36,000	14,400 - 74,400	Warehouse & Office	73.50	25,000	8,000 - 72,000
Ice Rinks	170.00	29,000	27,200 - 33,600				

Abbreviations

A	Area Square Feet; Ampere
AAFES	Army and Air Force Exchange Service
ABS	Acrylonitrile Butadiene Stryrene; Asbestos Bonded Steel
A.C., AC	Alternating Current; Air-Conditioning; Asbestos Cement; Plywood Grade A & C
ACI	American Concrete Institute
ACR	Air Conditioning Refrigeration
ADA	Americans with Disabilities Act
AD	Plywood, Grade A & D
Addit.	Additional
Adj.	Adjustable
af	Audio-frequency
AFUE	Annual Fuel Utilization Efficiency
AGA	American Gas Association
Agg.	Aggregate
A.H., Ah	Ampere Hours
A hr.	Ampere-hour
A.H.U., AHU	Air Handling Unit
A.I.A.	American Institute of Architects
AIC	Ampere Interrupting Capacity
Allow.	Allowance
alt., alt	Alternate
Alum.	Aluminum
a.m.	Ante Meridiem
Amp.	Ampere
Anod.	Anodized
ANSI	American National Standards Institute
APA	American Plywood Association
Approx.	Approximate
Apt.	Apartment
Asb.	Asbestos
A.S.B.C.	American Standard Building Code
Asbe.	Asbestos Worker
ASCE.	American Society of Civil Engineers
A.S.H.R.A.E.	American Society of Heating, Refrig. & AC Engineers
ASME	American Society of Mechanical Engineers
ASTM	American Society for Testing and Materials
Attchmt.	Attachment
Avg., Ave.	Average
AWG	American Wire Gauge
AWWA	American Water Works Assoc.
Bbl.	Barrel
B&B, BB	Grade B and Better; Balled & Burlapped
B&S	Bell and Spigot
B.&W.	Black and White
b.c.c.	Body-centered Cubic
B.C.Y.	Bank Cubic Yards
BE	Bevel End
B.F.	Board Feet
Bg. cem.	Bag of Cement
BHP	Boiler Horsepower; Brake Horsepower
B.I.	Black Iron
bidir.	bidirectional
Bit., Bitum.	Bituminous
Bit., Conc.	Bituminous Concrete
Bk.	Backed
Bkrs.	Breakers
Bldg., bldg	Building
Blk.	Block
Bm.	Beam
Boil.	Boilermaker
bpm	Blows per Minute
BR	Bedroom
Brg.	Bearing
Brhe.	Bricklayer Helper
Bric.	Bricklayer
Brk., brk	Brick
brkt	Bracket
Brng.	Bearing
Brs.	Brass
Brz.	Bronze
Bsn.	Basin
Btr.	Better
Btu	British Thermal Unit
BTUH	BTU per Hour
Bu.	bushels
BUR	Built-up Roofing
BX	Interlocked Armored Cable
°C	degree centegrade
c	Conductivity, Copper Sweat
C	Hundred; Centigrade
C/C	Center to Center, Cedar on Cedar
C-C	Center to Center
Cab	Cabinet
Cair.	Air Tool Laborer
Cal.	caliper
Calc	Calculated
Cap.	Capacity
Carp.	Carpenter
C.B.	Circuit Breaker
C.C.A.	Chromate Copper Arsenate
C.C.F.	Hundred Cubic Feet
cd	Candela
cd/sf	Candela per Square Foot
CD	Grade of Plywood Face & Back
CDX	Plywood, Grade C & D, exterior glue
Cefi.	Cement Finisher
Cem.	Cement
CF	Hundred Feet
C.F.	Cubic Feet
CFM	Cubic Feet per Minute
CFRP	Carbon Fiber Reinforced Plastic
c.g.	Center of Gravity
CHW	Chilled Water; Commercial Hot Water
C.I., CI	Cast Iron
C.I.P., CIP	Cast in Place
Circ.	Circuit
C.L.	Carload Lot
CL	Chain Link
Clab.	Common Laborer
Clam	Common maintenance laborer
C.L.F.	Hundred Linear Feet
CLF	Current Limiting Fuse
CLP	Cross Linked Polyethylene
cm	Centimeter
CMP	Corr. Metal Pipe
CMU	Concrete Masonry Unit
CN	Change Notice
Col.	Column
CO_2	Carbon Dioxide
Comb.	Combination
comm.	Commercial, Communication
Compr.	Compressor
Conc.	Concrete
Cont., cont	Continuous; Continued, Container
Corr.	Corrugated
Cos	Cosine
Cot	Cotangent
Cov.	Cover
C/P	Cedar on Paneling
CPA	Control Point Adjustment
Cplg.	Coupling
CPM	Critical Path Method
CPVC	Chlorinated Polyvinyl Chloride
C.Pr.	Hundred Pair
CRC	Cold Rolled Channel
Creos.	Creosote
Crpt.	Carpet & Linoleum Layer
CRT	Cathode-ray Tube
CS	Carbon Steel, Constant Shear Bar Joist
Csc	Cosecant
C.S.F.	Hundred Square Feet
CSI	Construction Specifications Institute
CT	Current Transformer
CTS	Copper Tube Size
Cu	Copper, Cubic
Cu. Ft.	Cubic Foot
cw	Continuous Wave
C.W.	Cool White; Cold Water
Cwt.	100 Pounds
C.W.X.	Cool White Deluxe
C.Y.	Cubic Yard (27 cubic feet)
C.Y./Hr.	Cubic Yard per Hour
Cyl.	Cylinder
d	Penny (nail size)
D	Deep; Depth; Discharge
Dis., Disch.	Discharge
Db	Decibel
Dbl.	Double
DC	Direct Current
DDC	Direct Digital Control
Demob.	Demobilization
d.f.t.	Dry Film Thickness
d.f.u.	Drainage Fixture Units
D.H.	Double Hung
DHW	Domestic Hot Water
DI	Ductile Iron
Diag.	Diagonal
Diam., Dia	Diameter
Distrib.	Distribution
Div.	Division
Dk.	Deck
D.L.	Dead Load; Diesel
DLH	Deep Long Span Bar Joist
dlx	Deluxe
Do.	Ditto
DOP	Dioctyl Phthalate Penetration Test (Air Filters)
Dp., dp	Depth
D.P.S.T.	Double Pole, Single Throw
Dr.	Drive
DR	Dimension Ratio
Drink.	Drinking
D.S.	Double Strength
D.S.A.	Double Strength A Grade
D.S.B.	Double Strength B Grade
Dty.	Duty
DWV	Drain Waste Vent
DX	Deluxe White, Direct Expansion
dyn	Dyne
e	Eccentricity
E	Equipment Only; East; emissivity
Ea.	Each
EB	Encased Burial
Econ.	Economy
E.C.Y	Embankment Cubic Yards
EDP	Electronic Data Processing
EIFS	Exterior Insulation Finish System
E.D.R.	Equiv. Direct Radiation
Eq.	Equation
EL	elevation
Elec.	Electrician; Electrical
Elev.	Elevator; Elevating
EMT	Electrical Metallic Conduit; Thin Wall Conduit
Eng.	Engine, Engineered
EPDM	Ethylene Propylene Diene Monomer
EPS	Expanded Polystyrene
Eqhv.	Equip. Oper., Heavy
Eqlt.	Equip. Oper., Light
Eqmd.	Equip. Oper., Medium
Eqmm.	Equip. Oper., Master Mechanic
Eqol.	Equip. Oper., Oilers
Equip.	Equipment
ERW	Electric Resistance Welded

Abbreviations

E.S.	Energy Saver	H	High Henry	Lath.	Lather
Est.	Estimated	HC	High Capacity	Lav.	Lavatory
esu	Electrostatic Units	H.D., HD	Heavy Duty; High Density	lb.; #	Pound
E.W.	Each Way	H.D.O.	High Density Overlaid	L.B., LB	Load Bearing; L Conduit Body
EWT	Entering Water Temperature	HDPE	High density polyethelene plastic	L. & E.	Labor & Equipment
Excav.	Excavation	Hdr.	Header	lb./hr.	Pounds per Hour
excl	Excluding	Hdwe.	Hardware	lb./L.F.	Pounds per Linear Foot
Exp., exp	Expansion, Exposure	H.I.D., HID	High Intensity Discharge	lbf/sq.in.	Pound-force per Square Inch
Ext., ext	Exterior; Extension	Help.	Helper Average	L.C.L.	Less than Carload Lot
Extru.	Extrusion	HEPA	High Efficiency Particulate Air Filter	L.C.Y.	Loose Cubic Yard
f.	Fiber stress	Hg	Mercury	Ld.	Load
F	Fahrenheit; Female; Fill	HIC	High Interrupting Capacity	LE	Lead Equivalent
Fab., fab	Fabricated; fabric	HM	Hollow Metal	LED	Light Emitting Diode
FBGS	Fiberglass	HMWPE	high molecular weight polyethylene	L.F.	Linear Foot
F.C.	Footcandles	HO	High Output	L.F. Nose	Linear Foot of Stair Nosing
f.c.c.	Face-centered Cubic	Horiz.	Horizontal	L.F. Rsr	Linear Foot of Stair Riser
f'c.	Compressive Stress in Concrete; Extreme Compressive Stress	H.P., HP	Horsepower; High Pressure	Lg.	Long; Length; Large
F.E.	Front End	H.P.F.	High Power Factor	L & H	Light and Heat
FEP	Fluorinated Ethylene Propylene (Teflon)	Hr.	Hour	LH	Long Span Bar Joist
F.G.	Flat Grain	Hrs./Day	Hours per Day	L.H.	Labor Hours
F.H.A.	Federal Housing Administration	HSC	High Short Circuit	L.L., LL	Live Load
Fig.	Figure	Ht.	Height	L.L.D.	Lamp Lumen Depreciation
Fin.	Finished	Htg.	Heating	lm	Lumen
FIPS	Female Iron Pipe Size	Htrs.	Heaters	lm/sf	Lumen per Square Foot
Fixt.	Fixture	HVAC	Heating, Ventilation & Air-Conditioning	lm/W	Lumen per Watt
FJP	Finger jointed and primed	Hvy.	Heavy	LOA	Length Over All
Fl. Oz.	Fluid Ounces	HW	Hot Water	log	Logarithm
Flr.	Floor	Hyd.; Hydr.	Hydraulic	L-O-L	Lateralolet
FM	Frequency Modulation; Factory Mutual	Hz	Hertz (cycles)	long.	longitude
Fmg.	Framing	I.	Moment of Inertia	L.P., LP	Liquefied Petroleum; Low Pressure
FM/UL	Factory Mutual/Underwriters Labs	IBC	International Building Code	L.P.F.	Low Power Factor
Fdn.	Foundation	I.C.	Interrupting Capacity	LR	Long Radius
FNPT	Female National Pipe Thread	ID	Inside Diameter	L.S.	Lump Sum
Fori.	Foreman, Inside	I.D.	Inside Dimension; Identification	Lt.	Light
Foro.	Foreman, Outside	I.F.	Inside Frosted	Lt. Ga.	Light Gauge
Fount.	Fountain	I.M.C.	Intermediate Metal Conduit	L.T.L.	Less than Truckload Lot
fpm	Feet per Minute	In.	Inch	Lt. Wt.	Lightweight
FPT	Female Pipe Thread	Incan.	Incandescent	L.V.	Low Voltage
Fr	Frame	Incl.	Included; Including	M	Thousand; Material; Male; Light Wall Copper Tubing
F.R.	Fire Rating	Int.	Interior	M^2CA	Meters Squared Contact Area
FRK	Foil Reinforced Kraft	Inst.	Installation	m/hr.; M.H.	Man-hour
FSK	Foil/scrim/kraft	Insul., insul	Insulation/Insulated	mA	Milliampere
FRP	Fiberglass Reinforced Plastic	I.P.	Iron Pipe	Mach.	Machine
FS	Forged Steel	I.P.S., IPS	Iron Pipe Size	Mag. Str.	Magnetic Starter
FSC	Cast Body; Cast Switch Box	IPT	Iron Pipe Threaded	Maint.	Maintenance
Ft., ft	Foot; Feet	I.W.	Indirect Waste	Marb.	Marble Setter
Ftng.	Fitting	J	Joule	Mat; Mat'l.	Material
Ftg.	Footing	J.I.C.	Joint Industrial Council	Max.	Maximum
Ft lb.	Foot Pound	K	Thousand; Thousand Pounds; Heavy Wall Copper Tubing, Kelvin	MBF	Thousand Board Feet
Furn.	Furniture	K.A.H.	Thousand Amp. Hours	MBH	Thousand BTU's per hr.
FVNR	Full Voltage Non-Reversing	kcmil	Thousand Circular Mils	MC	Metal Clad Cable
FVR	Full Voltage Reversing	KD	Knock Down	MCC	Motor Control Center
FXM	Female by Male	K.D.A.T.	Kiln Dried After Treatment	M.C.F.	Thousand Cubic Feet
Fy.	Minimum Yield Stress of Steel	kg	Kilogram	MCFM	Thousand Cubic Feet per Minute
g	Gram	kG	Kilogauss	M.C.M.	Thousand Circular Mils
G	Gauss	kgf	Kilogram Force	MCP	Motor Circuit Protector
Ga.	Gauge	kHz	Kilohertz	MD	Medium Duty
Gal., gal.	Gallon	Kip	1000 Pounds	MDF	Medium-density fibreboard
gpm, GPM	Gallon per Minute	KJ	Kiljoule	M.D.O.	Medium Density Overlaid
Galv., galv	Galvanized	K.L.	Effective Length Factor	Med.	Medium
GC/MS	Gas Chromatograph/Mass Spectrometer	K.L.F.	Kips per Linear Foot	MF	Thousand Feet
Gen.	General	Km	Kilometer	M.F.B.M.	Thousand Feet Board Measure
GFI	Ground Fault Interrupter	KO	Knock Out	Mfg.	Manufacturing
GFRC	Glass Fiber Reinforced Concrete	K.S.F.	Kips per Square Foot	Mfrs.	Manufacturers
Glaz.	Glazier	K.S.I.	Kips per Square Inch	mg	Milligram
GPD	Gallons per Day	kV	Kilovolt	MGD	Million Gallons per Day
gpf	Gallon per flush	kVA	Kilovolt Ampere	MGPH	Thousand Gallons per Hour
GPH	Gallons per Hour	kVAR	Kilovar (Reactance)	MH, M.H.	Manhole; Metal Halide; Man-Hour
GPM	Gallons per Minute	KW	Kilowatt	MHz	Megahertz
GR	Grade	KWh	Kilowatt-hour	Mi.	Mile
Gran.	Granular	L	Labor Only; Length; Long; Medium Wall Copper Tubing	MI	Malleable Iron; Mineral Insulated
Grnd.	Ground	Lab.	Labor	MIPS	Male Iron Pipe Size
GVW	Gross Vehicle Weight	lat	Latitude	mj	Mechanical Joint
GWB	Gypsum wall board			m	Meter
				mm	Millimeter
				Mill.	Millwright
				Min., min.	Minimum, minute

Abbreviations

Misc.	Miscellaneous
ml	Milliliter, Mainline
M.L.F.	Thousand Linear Feet
Mo.	Month
Mobil.	Mobilization
Mog.	Mogul Base
MPH	Miles per Hour
MPT	Male Pipe Thread
MRGWB	Moisture Resistant Gypsum Wallboard
MRT	Mile Round Trip
ms	Millisecond
M.S.F.	Thousand Square Feet
Mstz.	Mosaic & Terrazzo Worker
M.S.Y.	Thousand Square Yards
Mtd., mtd., mtd	Mounted
Mthe.	Mosaic & Terrazzo Helper
Mtng.	Mounting
Mult.	Multi; Multiply
M.V.A.	Million Volt Amperes
M.V.A.R.	Million Volt Amperes Reactance
MV	Megavolt
MW	Megawatt
MXM	Male by Male
MYD	Thousand Yards
N	Natural; North
nA	Nanoampere
NA	Not Available; Not Applicable
N.B.C.	National Building Code
NC	Normally Closed
NEMA	National Electrical Manufacturers Assoc.
NEHB	Bolted Circuit Breaker to 600V.
NFPA	National Fire Protection Association
NLB	Non-Load-Bearing
NM	Non-Metallic Cable
nm	Nanometer
No.	Number
NO	Normally Open
N.O.C.	Not Otherwise Classified
Nose.	Nosing
NPT	National Pipe Thread
NQOD	Combination Plug-on/Bolt on Circuit Breaker to 240V.
N.R.C., NRC	Noise Reduction Coefficient/ Nuclear Regulator Commission
N.R.S.	Non Rising Stem
ns	Nanosecond
nW	Nanowatt
OB	Opposing Blade
OC	On Center
OD	Outside Diameter
O.D.	Outside Dimension
ODS	Overhead Distribution System
O.G.	Ogee
O.H.	Overhead
O&P	Overhead and Profit
Oper.	Operator
Opng.	Opening
Orna.	Ornamental
OSB	Oriented Strand Board
OS&Y	Outside Screw and Yoke
OSHA	Occupational Safety and Health Act
Ovhd.	Overhead
OWG	Oil, Water or Gas
Oz.	Ounce
P.	Pole; Applied Load; Projection
p.	Page
Pape.	Paperhanger
P.A.P.R.	Powered Air Purifying Respirator
PAR	Parabolic Reflector
P.B., PB	Push Button
Pc., Pcs.	Piece, Pieces
P.C.	Portland Cement; Power Connector
P.C.F.	Pounds per Cubic Foot
PCM	Phase Contrast Microscopy
PDCA	Painting and Decorating Contractors of America
P.E., PE	Professional Engineer; Porcelain Enamel; Polyethylene; Plain End
P.E.C.I.	Porcelain Enamel on Cast Iron
Perf.	Perforated
PEX	Cross linked polyethylene
Ph.	Phase
P.I.	Pressure Injected
Pile.	Pile Driver
Pkg.	Package
Pl.	Plate
Plah.	Plasterer Helper
Plas.	Plasterer
plf	Pounds Per Linear Foot
Pluh.	Plumbers Helper
Plum.	Plumber
Ply.	Plywood
p.m.	Post Meridiem
Pntd.	Painted
Pord.	Painter, Ordinary
pp	Pages
PP, PPL	Polypropylene
P.P.M.	Parts per Million
Pr.	Pair
P.E.S.B.	Pre-engineered Steel Building
Prefab.	Prefabricated
Prefin.	Prefinished
Prop.	Propelled
PSF, psf	Pounds per Square Foot
PSI, psi	Pounds per Square Inch
PSIG	Pounds per Square Inch Gauge
PSP	Plastic Sewer Pipe
Pspr.	Painter, Spray
Psst.	Painter, Structural Steel
P.T.	Potential Transformer
P. & T.	Pressure & Temperature
Ptd.	Painted
Ptns.	Partitions
Pu	Ultimate Load
PVC	Polyvinyl Chloride
Pvmt.	Pavement
PRV	Pressure Relief Valve
Pwr.	Power
Q	Quantity Heat Flow
Qt.	Quart
Quan., Qty.	Quantity
Q.C.	Quick Coupling
r	Radius of Gyration
R	Resistance
R.C.P.	Reinforced Concrete Pipe
Rect.	Rectangle
recpt.	receptacle
Reg.	Regular
Reinf.	Reinforced
Req'd.	Required
Res.	Resistant
Resi.	Residential
RF	Radio Frequency
RFID	Radio-frequency identification
Rgh.	Rough
RGS	Rigid Galvanized Steel
RHW	Rubber, Heat & Water Resistant; Residential Hot Water
rms	Root Mean Square
Rnd.	Round
Rodm.	Rodman
Rofc.	Roofer, Composition
Rofp.	Roofer, Precast
Rohe.	Roofer Helpers (Composition)
Rots.	Roofer, Tile & Slate
R.O.W.	Right of Way
RPM	Revolutions per Minute
R.S.	Rapid Start
Rsr	Riser
RT	Round Trip
S.	Suction; Single Entrance; South
SBS	Styrene Butadiere Styrene
SC	Screw Cover
SCFM	Standard Cubic Feet per Minute
Scaf.	Scaffold
Sch., Sched.	Schedule
S.C.R.	Modular Brick
S.D.	Sound Deadening
SDR	Standard Dimension Ratio
S.E.	Surfaced Edge
Sel.	Select
SER, SEU	Service Entrance Cable
S.F.	Square Foot
S.F.C.A.	Square Foot Contact Area
S.F. Flr.	Square Foot of Floor
S.F.G.	Square Foot of Ground
S.F. Hor.	Square Foot Horizontal
SFR	Square Feet of Radiation
S.F. Shlf.	Square Foot of Shelf
S4S	Surface 4 Sides
Shee.	Sheet Metal Worker
Sin.	Sine
Skwk.	Skilled Worker
SL	Saran Lined
S.L.	Slimline
Sldr.	Solder
SLH	Super Long Span Bar Joist
S.N.	Solid Neutral
SO	Stranded with oil resistant inside insulation
S-O-L	Socketolet
sp	Standpipe
S.P.	Static Pressure; Single Pole; Self-Propelled
Spri.	Sprinkler Installer
spwg	Static Pressure Water Gauge
S.P.D.T.	Single Pole, Double Throw
SPF	Spruce Pine Fir; Sprayed Polyurethane Foam
S.P.S.T.	Single Pole, Single Throw
SPT	Standard Pipe Thread
Sq.	Square; 100 Square Feet
Sq. Hd.	Square Head
Sq. In.	Square Inch
S.S.	Single Strength; Stainless Steel
S.S.B.	Single Strength B Grade
sst, ss	Stainless Steel
Sswk.	Structural Steel Worker
Sswl.	Structural Steel Welder
St.; Stl.	Steel
STC	Sound Transmission Coefficient
Std.	Standard
Stg.	Staging
STK	Select Tight Knot
STP	Standard Temperature & Pressure
Stpi.	Steamfitter, Pipefitter
Str.	Strength; Starter; Straight
Strd.	Stranded
Struct.	Structural
Sty.	Story
Subj.	Subject
Subs.	Subcontractors
Surf.	Surface
Sw.	Switch
Swbd.	Switchboard
S.Y.	Square Yard
Syn.	Synthetic
S.Y.P.	Southern Yellow Pine
Sys.	System
t.	Thickness
T	Temperature; Ton
Tan	Tangent
T.C.	Terra Cotta
T & C	Threaded and Coupled
T.D.	Temperature Difference
Tdd	Telecommunications Device for the Deaf
T.E.M.	Transmission Electron Microscopy
temp	Temperature, Tempered, Temporary
TFFN	Nylon Jacketed Wire

Abbreviations

TFE	Tetrafluoroethylene (Teflon)	U.L., UL	Underwriters Laboratory	w/	With
T. & G.	Tongue & Groove; Tar & Gravel	Uld.	unloading	W.C., WC	Water Column; Water Closet
Th., Thk.	Thick	Unfin.	Unfinished	W.F.	Wide Flange
Thn.	Thin	UPS	Uninterruptible Power Supply	W.G.	Water Gauge
Thrded	Threaded	URD	Underground Residential Distribution	Wldg.	Welding
Tilf.	Tile Layer, Floor	US	United States	W. Mile	Wire Mile
Tilh.	Tile Layer, Helper	USGBC	U.S. Green Building Council	W-O-L	Weldolet
THHN	Nylon Jacketed Wire	USP	United States Primed	W.R.	Water Resistant
THW.	Insulated Strand Wire	UTMCD	Uniform Traffic Manual For Control Devices	Wrck.	Wrecker
THWN	Nylon Jacketed Wire	UTP	Unshielded Twisted Pair	W.S.P.	Water, Steam, Petroleum
T.L., TL	Truckload	V	Volt	WT., Wt.	Weight
T.M.	Track Mounted	VA	Volt Amperes	WWF	Welded Wire Fabric
Tot.	Total	VAT	Vinyl Asbestos Tile	XFER	Transfer
T-O-L	Threadolet	V.C.T.	Vinyl Composition Tile	XFMR	Transformer
tmpd	Tempered	VAV	Variable Air Volume	XHD	Extra Heavy Duty
TPO	Thermoplastic Polyolefin	VC	Veneer Core	XHHW, XLPE	Cross-Linked Polyethylene Wire Insulation
T.S.	Trigger Start	VDC	Volts Direct Current	XLP	Cross-linked Polyethylene
Tr.	Trade	Vent.	Ventilation	Xport	Transport
Transf.	Transformer	Vert.	Vertical	Y	Wye
Trhv.	Truck Driver, Heavy	V.F.	Vinyl Faced	yd	Yard
Trlr	Trailer	V.G.	Vertical Grain	yr	Year
Trlt.	Truck Driver, Light	VHF	Very High Frequency	Δ	Delta
TTY	Teletypewriter	VHO	Very High Output	%	Percent
TV	Television	Vib.	Vibrating	~	Approximately
T.W.	Thermoplastic Water Resistant Wire	VLF	Vertical Linear Foot	Ø	Phase; diameter
UCI	Uniform Construction Index	VOC	Volitile Organic Compound	@	At
UF	Underground Feeder	Vol.	Volume	#	Pound; Number
UGND	Underground Feeder	VRP	Vinyl Reinforced Polyester	<	Less Than
UHF	Ultra High Frequency	W	Wire; Watt; Wide; West	>	Greater Than
U.I.	United Inch			Z	zone

A

B

C

E

F

G

M

N

O

P

Q

R

S

T

U

V

W

X

Y

Z

Notes

Notes

Notes

Notes

Notes

Notes

Division Notes

	CREW	DAILY OUTPUT	LABOR-HOURS	UNIT	BARE COSTS MAT.	BARE COSTS LABOR	BARE COSTS EQUIP.	BARE COSTS TOTAL	TOTAL INCL O&P

Division Notes

		CREW	DAILY OUTPUT	LABOR-HOURS	UNIT	BARE COSTS MAT.	BARE COSTS LABOR	BARE COSTS EQUIP.	BARE COSTS TOTAL	TOTAL INCL O&P

Division Notes

		CREW	DAILY OUTPUT	LABOR-HOURS	UNIT	BARE COSTS MAT.	BARE COSTS LABOR	BARE COSTS EQUIP.	BARE COSTS TOTAL	TOTAL INCL O&P

Division Notes

	CREW	DAILY OUTPUT	LABOR-HOURS	UNIT	BARE COSTS				TOTAL INCL O&P
					MAT.	LABOR	EQUIP.	TOTAL	

Division Notes

	CREW	DAILY OUTPUT	LABOR-HOURS	UNIT	BARE COSTS				TOTAL INCL O&P
					MAT.	LABOR	EQUIP.	TOTAL	

Division Notes

I	CREW	DAILY OUTPUT	LABOR-HOURS	UNIT	BARE COSTS MAT.	BARE COSTS LABOR	BARE COSTS EQUIP.	BARE COSTS TOTAL	TOTAL INCL O&P

RSMeans—
a tradition of excellence in construction cost information and services since 1942

For more information visit the RSMeans website at **www.rsmeans.com**

Unit prices according to the latest MasterFormat!

Book Selection Guide

The following table provides definitive information on the content of each cost data publication. The number of lines of data provided in each unit price or assemblies division, as well as the number of crews, is listed for each book. The presence of other elements such as reference tables, square foot models, equipment rental costs, historical cost indexes, and city cost indexes, is also indicated. You can use the table to help select the RSMeans book that has the quantity and type of information you most need in your work.

Unit Cost Divisions	Building Construction	Mechanical	Electrical	Commercial Renovation	Square Foot	Site Work Landsc.	Green Building	Interior	Concrete Masonry	Open Shop	Heavy Construction	Light Commercial	Facilities Construction	Plumbing	Residential
1	573	393	410	506		509	191	312	456	572	501	246	1042	403	178
2	777	279	85	733		993	178	399	213	776	734	481	1221	286	257
3	1690	340	230	1085		1471	987	355	2032	1689	1685	481	1788	316	388
4	957	21	0	922		720	179	612	1155	925	612	529	1172	0	441
5	1893	158	155	1090		841	1799	1095	720	1893	1037	979	1909	204	721
6	2452	18	18	2110		110	589	1528	281	2448	123	2140	2124	22	2660
7	1584	215	128	1623		581	757	532	524	1581	26	1317	1685	227	1037
8	2082	81	45	2609		265	1123	1753	105	2077	0	2193	2879	0	1526
9	1971	72	26	1767		313	449	2056	390	1911	15	1659	2220	54	1435
10	1026	17	10	642		215	27	842	158	1026	29	511	1118	235	224
11	1087	208	165	546		124	54	932	28	1071	0	230	1107	169	110
12	544	0	2	319		217	137	1654	14	532	0	352	1704	23	310
13	719	140	137	242		343	111	248	66	695	244	83	746	94	79
14	273	36	0	221		0	0	257	0	273	0	12	293	16	6
21	90	0	16	37		0	0	250	0	87	0	68	426	431	220
22	1157	7544	150	1190		1568	1067	838	20	1121	1677	857	7440	9347	712
23	1177	6971	580	928		158	902	776	38	1101	111	878	5187	1918	469
26	1354	454	10081	1021		793	611	1136	55	1349	562	1298	10030	399	617
27	72	0	297	34		13	0	71	0	72	39	52	279	0	22
28	96	58	136	71		0	21	78	0	99	0	40	151	44	25
31	1498	735	610	803		3217	289	7	1208	1443	3256	600	1560	660	611
32	822	53	8	886		4426	356	405	291	793	1841	420	1701	165	468
33	530	1079	538	252		2178	41	0	237	522	2159	128	1699	1286	154
34	107	0	47	4		190	0	0	31	62	193	0	128	0	0
35	18	0	0	0		327	0	0	0	18	442	0	84	0	0
41	60	0	0	32		7	0	22	0	60	30	0	67	14	0
44	75	79	0	0		0	0	0	0	0	0	0	75	75	0
46	23	16	0	0		274	261	0	0	23	264	0	33	33	0
48	12	0	25	0		0	25	0	0	12	17	12	12	0	12
Totals	24719	18967	13899	19673		19853	10154	16158	8022	24231	15597	15566	49880	16421	12682

Assem Div	Building Construction	Mechanical	Elec-trical	Commercial Renovation	Square Foot	Site Work Landscape	Assemblies	Green Building	Interior	Concrete Masonry	Heavy Construction	Light Commercial	Facilities Construction	Plumbing	Asm Div	Residential
A		15	0	188	150	577	598	0	0	536	571	154	24	0	1	376
B		0	0	848	2498	0	5658	56	329	1975	368	2089	174	0	2	211
C		0	0	647	926	0	1304	0	1629	146	0	816	250	0	3	588
D		1067	941	712	1859	72	2538	330	825	0	0	1345	1105	1088	4	851
E		0	0	86	260	0	300	0	5	0	0	257	5	0	5	392
F		0	0	0	114	0	114	0	0	0	0	114	3	0	6	357
G		527	447	318	249	3365	729	0	0	535	1350	205	293	677	7	307
															8	760
															9	80
															10	0
															11	0
															12	0
Totals		1609	1388	2799	6056	4014	11241	386	2788	3192	2289	4980	1854	1765		3922

Reference Section	Building Construction Costs	Mechanical	Electrical	Commercial Renovation	Square Foot	Site Work Landscape	Assem.	Green Building	Interior	Concrete Masonry	Open Shop	Heavy Construction	Light Commercial	Facilities Construction	Plumbing	Resi.
Reference Tables	yes	yes	yes	yes	no	yes	yes	yes	yes	yes	yes	yes	yes	yes	yes	yes
Models					111			25					50			28
Crews	571	571	571	550		571		571	571	571	548	571	548	550	571	548
Equipment Rental Costs	yes	yes	yes	yes		yes		yes	yes	yes	yes	yes	yes	yes	yes	yes
Historical Cost Indexes	yes	yes	yes	yes	yes	yes	yes	yes	yes	yes	yes	yes	yes	yes	yes	no
City Cost Indexes	yes	yes	yes	yes	yes	yes	yes	yes	yes	yes	yes	yes	yes	yes	yes	yes

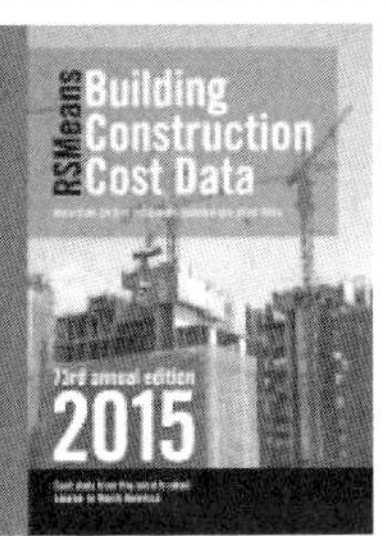

RSMeans Building Construction Cost Data 2015

Offers you unchallenged unit price reliability in an easy-to-use format. Whether used for verifying complete, finished estimates or for periodic checks, it supplies more cost facts better and faster than any comparable source. More than 24,700 unit prices have been updated for 2015. The City Cost Indexes and Location Factors cover more than 930 areas, for indexing to any project location in North America. Order and get *RSMeans Quarterly Update Service* FREE.

$206.95 | Available Sept. 2014 | Catalog no. 60015

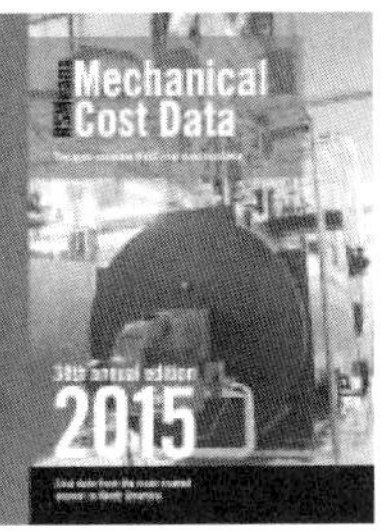

RSMeans Mechanical Cost Data 2015

Total unit and systems price guidance for mechanical construction . . . materials, parts, fittings, and complete labor cost information. Includes prices for piping, heating, air conditioning, ventilation, and all related construction.

Plus new 2015 unit costs for:

- Thousands of installed HVAC/controls, sub-assemblies and assemblies
- "On-site" Location Factors for more than 930 cities and towns in the U.S. and Canada
- Crews, labor, and equipment

$203.95 | Available Oct. 2014 | Catalog no. 60025

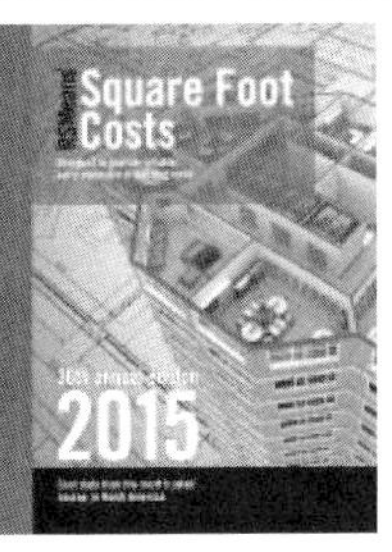

RSMeans Square Foot Costs 2015

Accurate and Easy-to-Use

- **Updated price information** based on nationwide figures from suppliers, estimators, labor experts, and contractors.
- Green building models
- Realistic graphics, offering true-to-life illustrations of building projects
- Extensive information on using square foot cost data, including sample estimates and alternate pricing methods

$218.95 | Available Oct. 2014 | Catalog no. 60055

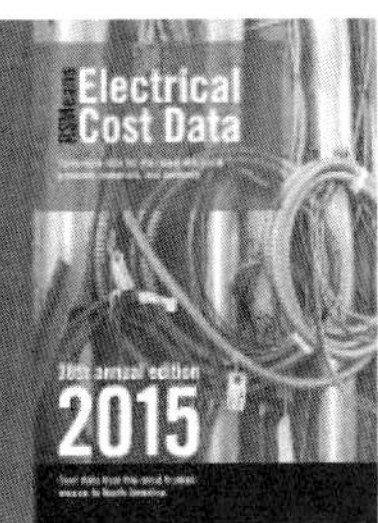

RSMeans Electrical Cost Data 2015

Pricing information for every part of electrical cost planning. More than 13,800 unit and systems costs with design tables; clear specifications and drawings; engineering guides; illustrated estimating procedures; complete labor-hour and materials costs for better scheduling and procurement; and the latest electrical products and construction methods.

- A variety of special electrical systems, including cathodic protection
- Costs for maintenance, demolition, HVAC/mechanical, specialties, equipment, and more

$208.95 | Available Oct. 2014 | Catalog no. 60035

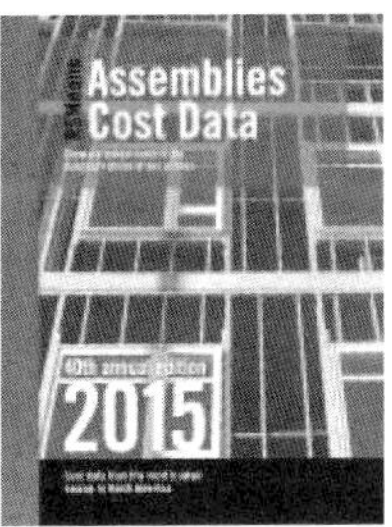

RSMeans Assemblies Cost Data 2015

RSMeans Assemblies Cost Data takes the guesswork out of preliminary or conceptual estimates. Now you don't have to try to calculate the assembled cost by working up individual component costs. We've done all the work for you.

Presents detailed illustrations, descriptions, specifications, and costs for every conceivable building assembly—over 350 types in all—arranged in the easy-to-use UNIFORMAT II system. Each illustrated "assembled" cost includes a complete grouping of materials and associated installation costs, including the installing contractor's overhead and profit.

$334.95 | Available Sept. 2014 | Catalog no. 60065

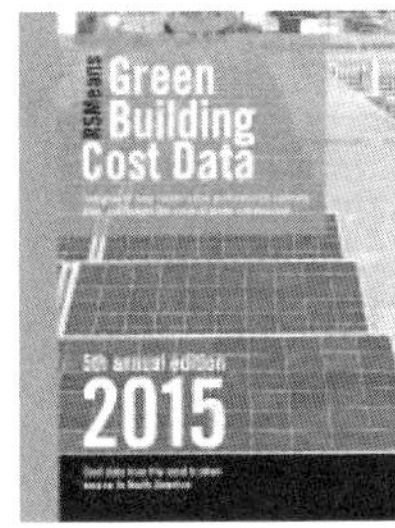

RSMeans Green Building Cost Data 2015

Estimate, plan, and budget the costs of green building for both new commercial construction and renovation work with this fifth edition of *RSMeans Green Building Cost Data*. More than 10,000 unit costs for a wide array of green building products plus assemblies costs. Easily identified cross references to LEED and Green Globes building rating systems criteria.

$170.95 | Available Nov. 2014 | Catalog no. 60555

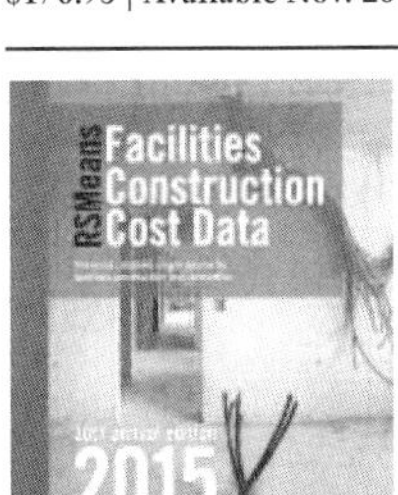

RSMeans Facilities Construction Cost Data 2015

For the maintenance and construction of commercial, industrial, municipal, and institutional properties. Costs are shown for new and remodeling construction and are broken down into materials, labor, equipment, and overhead and profit. Special emphasis is given to sections on mechanical, electrical, furnishings, site work, building maintenance, finish work, and demolition.

More than 49,800 unit costs, plus assemblies costs and a comprehensive Reference Section are included.

$524.95 | Available Nov. 2014 | Catalog no. 60205

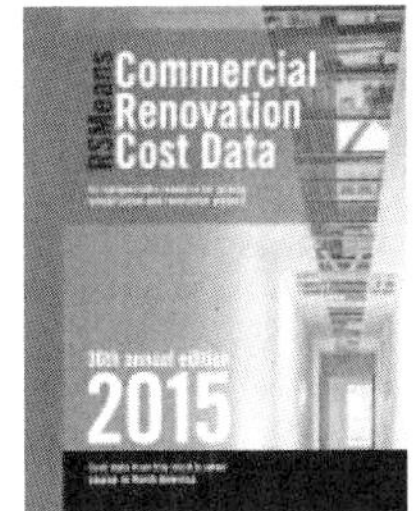

RSMeans Commercial Renovation Cost Data 2015

Commercial/Multi-family Residential

Use this valuable tool to estimate commercial and multi-family residential renovation and remodeling.

Includes: Updated costs for hundreds of unique methods, materials, and conditions that only come up in repair and remodeling, PLUS:

- Unit costs for more than 19,600 construction components
- Installed costs for more than 2,700 assemblies
- More than 930 "on-site" localization factors for the U.S. and Canada

$166.95 | Available Oct. 2014 | Catalog no. 60045

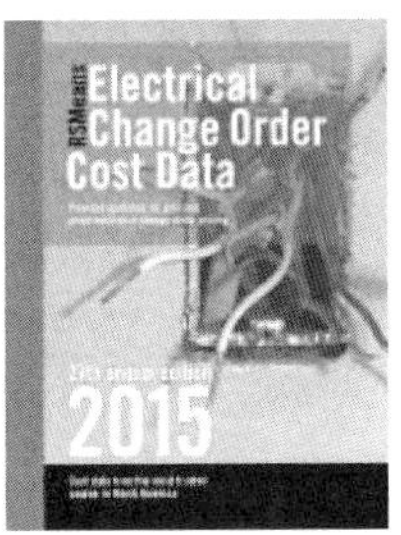

RSMeans Electrical Change Order Cost Data 2015

RSMeans Electrical Change Order Cost Data provides you with electrical unit prices exclusively for pricing change orders—based on the recent, direct experience of contractors and suppliers. Analyze and check your own change order estimates against the experience others have had doing the same work. It also covers productivity analysis, and change order cost justifications. With useful information for calculating the effects of change orders and dealing with their administration.

$199.95 | Available Dec. 2014 | Catalog no. 60235

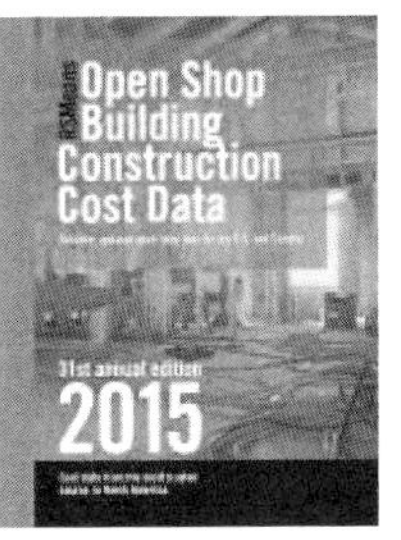

RSMeans Open Shop Building Construction Cost Data 2015

The latest costs for accurate budgeting and estimating of new commercial and residential construction . . . renovation work . . . change orders . . . cost engineering.

RSMeans Open Shop "BCCD" will assist you to:

- Develop benchmark prices for change orders.
- Plug gaps in preliminary estimates and budgets.
- Estimate complex projects.
- Substantiate invoices on contracts.
- Price ADA-related renovations.

$175.95 | Available Dec. 2014 | Catalog no. 60155

For more information visit the RSMeans website at **www.rsmeans.com**

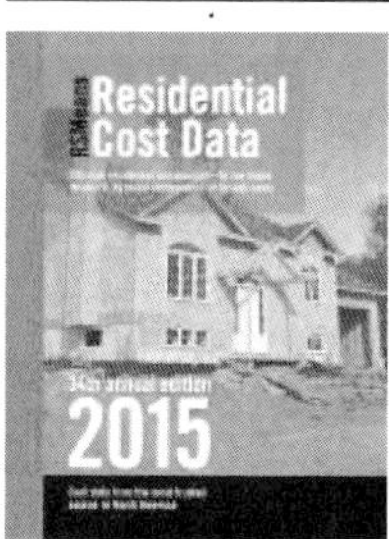

RSMeans Residential Cost Data 2015

Contains square foot costs for 28 basic home models with the look of today, plus hundreds of custom additions and modifications you can quote right off the page. Includes more than 3,900 costs for 89 residential systems. Complete with blank estimating forms, sample estimates, and step-by-step instructions.

Contains line items for cultured stone and brick, PVC trim, lumber, and TPO roofing.

$148.95 | Available Oct. 2014 | Catalog no. 60175

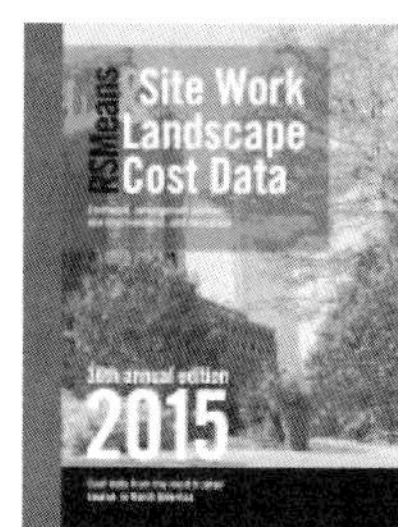

RSMeans Site Work & Landscape Cost Data 2015

Includes unit and assemblies costs for earthwork, sewerage, piped utilities, site improvements, drainage, paving, trees and shrubs, street openings/repairs, underground tanks, and more. Contains more than 60 types of assemblies costs for accurate conceptual estimates.

Includes:

- Estimating for infrastructure improvements
- Environmentally-oriented construction
- ADA-mandated handicapped access
- Hazardous waste line items

$199.95 | Available Dec. 2014 | Catalog no. 60285

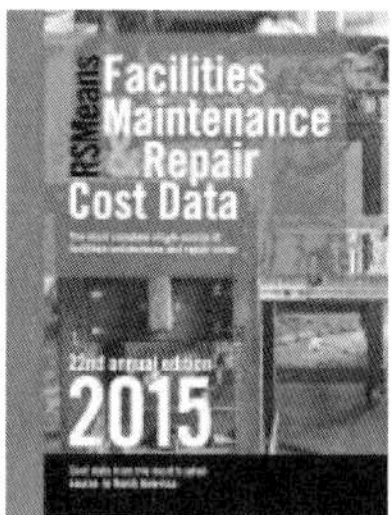

RSMeans Facilities Maintenance & Repair Cost Data 2015

RSMeans Facilities Maintenance & Repair Cost Data gives you a complete system to manage and plan your facility repair and maintenance costs and budget efficiently. Guidelines for auditing a facility and developing an annual maintenance plan. Budgeting is included, along with reference tables on cost and management, and information on frequency and productivity of maintenance operations.

The only nationally recognized source of maintenance and repair costs. Developed in cooperation with the Civil Engineering Research Laboratory (CERL) of the Army Corps of Engineers.

$461.95 | Available Nov. 2014 | Catalog no. 60305

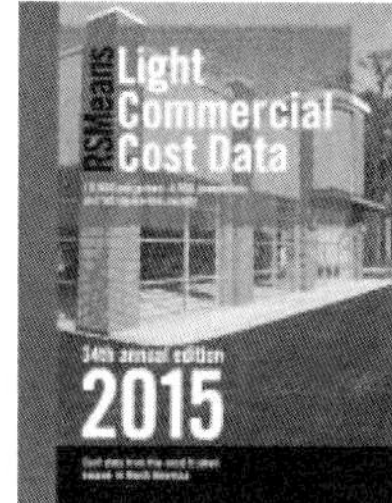

RSMeans Light Commercial Cost Data 2015

Specifically addresses the light commercial market, which is a specialized niche in the construction industry. Aids you, the owner/designer/contractor, in preparing all types of estimates—from budgets to detailed bids. Includes new advances in methods and materials.

Assemblies Section allows you to evaluate alternatives in the early stages of design/planning.

More than 15,500 unit costs ensure that you have the prices you need, when you need them.

$153.95 | Available Nov. 2014 | Catalog no. 60185

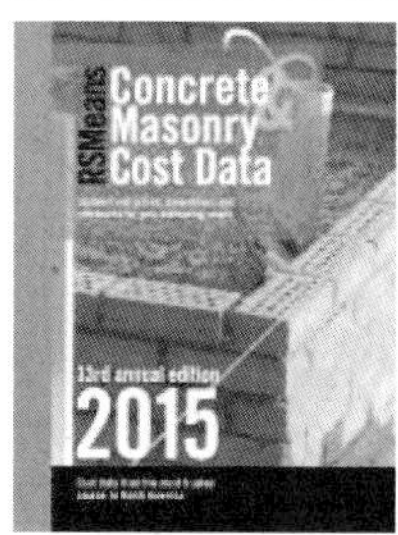

RSMeans Concrete & Masonry Cost Data 2015

Provides you with cost facts for virtually all concrete/masonry estimating needs, from complicated formwork to various sizes and face finishes of brick and block—all in great detail. The comprehensive Unit Price Section contains more than 8,000 selected entries. Also contains an Assemblies [Cost] Section, and a detailed Reference Section that supplements the cost data.

$188.95 | Available Dec. 2014 | Catalog no. 60115

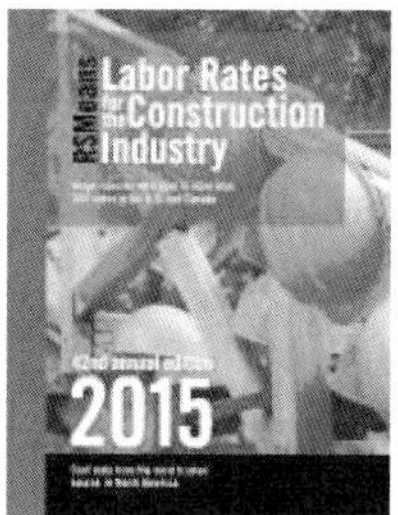

RSMeans Labor Rates for the Construction Industry 2015

Complete information for estimating labor costs, making comparisons, and negotiating wage rates by trade for more than 300 U.S. and Canadian cities. With 46 construction trades in each city, and historical wage rates included for comparison. Each city chart lists the county and is alphabetically arranged with handy visual flip tabs for quick reference.

$450.95 | Available Dec. 2014 | Catalog no. 60125

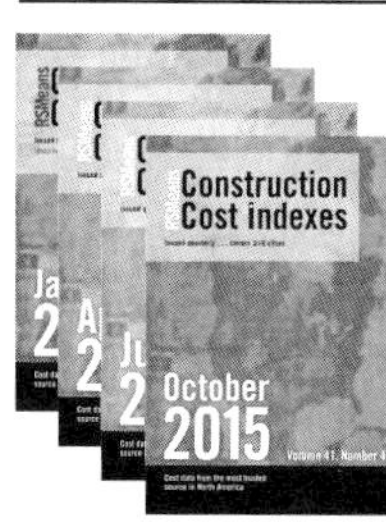

RSMeans Construction Cost Indexes 2015

What materials and labor costs will change unexpectedly this year? By how much?

- Breakdowns for 318 major cities
- National averages for 30 key cities
- Expanded five major city indexes
- Historical construction cost indexes

$359.95 per year (subscription) | Catalog no. 50145
$94.95 individual quarters | Catalog no. 60145 A,B,C,D

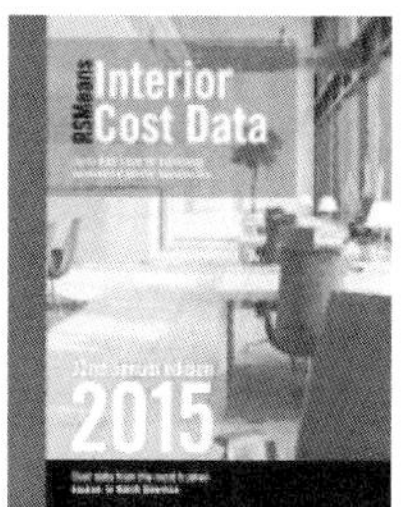

RSMeans Interior Cost Data 2015

Provides you with prices and guidance needed to make accurate interior work estimates. Contains costs on materials, equipment, hardware, custom installations, furnishings, and labor for new and remodel commercial and industrial interior construction, including updated information on office furnishings, and reference information.

$208.95 | Available Nov. 2014 | Catalog no. 60095

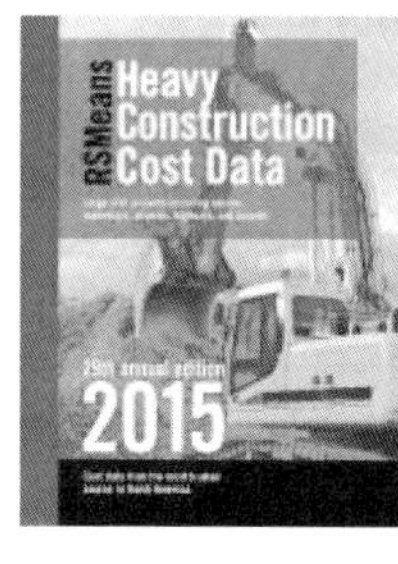

RSMeans Heavy Construction Cost Data 2015

A comprehensive guide to heavy construction costs. Includes costs for highly specialized projects such as tunnels, dams, highways, airports, and waterways. Information on labor rates, equipment, and materials costs is included. Features unit price costs, systems costs, and numerous reference tables for costs and design.

$204.95 | Available Dec. 2014 | Catalog no. 60165

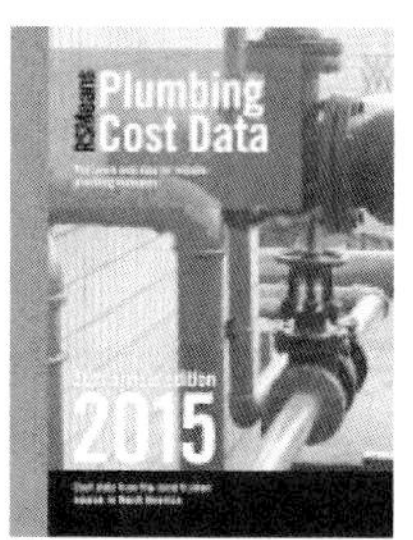

RSMeans Plumbing Cost Data 2015

Comprehensive unit prices and assemblies for plumbing, irrigation systems, commercial and residential fire protection, point-of-use water heaters, and the latest approved materials. This publication and its companion, *RSMeans Mechanical Cost Data*, provide full-range cost estimating coverage for all the mechanical trades.

Contains updated costs for potable water, radiant heat systems, and high efficiency fixtures.

$206.95 | Available Oct. 2014 | Catalog no. 60215

RSMeans Online and RSMeans CostWorks CD Comparison Matrix

Both the RSMeans CostWorks CD package and the RSMeans Online estimating tool provide the same comprehensive, up-to-date, and localized cost information to improve planning, estimating, and budgeting with RSMeans—the industry's most-trusted source of construction cost data.

In addition to key differences noted in the comparison matrix below, both products let you do the following:

- Quickly locate costs in the searchable RSMeans database.
- Instantly adjust the data to your specific location for accurate costs anywhere in the U.S. and Canada.
- Create reliable, updated cost estimates in minutes.
- Manage, share, and export your estimates to MS Excel® with ease.
- Access a wide range of reference materials, including crew rates and national average labor rates.

Both the CD and Online options are available in a convenient one-year subscription!

Use the handy comparison matrix below to help you decide the best product for your cost estimating needs!

Why us?

For more than 70 years, RSMeans has provided the quality cost data construction professionals everywhere depend on to estimate with confidence, save time, improve decision-making, and increase profits.

Improve planning and decision-making

Back your estimates with complete, accurate, and up-to-date cost data for informed business decisions.

- Verify construction costs from third parties.
- Check validity of subcontractor proposals.
- Evaluate material and assembly alternatives.

Increase profits

Use RSMeans to estimate projects quickly and accurately, so you can gain an edge over your competition.

- Create accurate and competitive bids.
- Minimize the risk of cost overruns.
- Reduce variability.
- Gain control over costs.

The following matrix summarizes the key comparison points between RSMeans' two electronic cost data products.

	Question	CostWorks CD	RSMeans Online.com
1	How is the data presented?	• The data is organized by book title and sold at comparable book cost. • Several packaged bundles of five titles are available at a cost savings. • Available in a convenient one-year subscription	• The data is organized by book title and sold at comparable book cost. • Several packaged bundles of five titles are available at a cost savings. • RSMeans Online also offers a Complete Library title with all online data at a cost savings. • Available in a convenient one-year subscription
2	Where is the data stored?	With a typical install, the data is read off of your hard drive or off of the CD.	All data is stored on our secure server.
3	How are updates delivered?	Updates are available on a quarterly basis for download from the Costworks HomePort website.	Data is updated automatically on a quarterly basis, and users are notified when updates occur.
4	Where are my projects and estimates stored?	Upon installation, a "My CostWorks Projects" folder is created within the "My Documents" directory. All projects are defaulted to save here unless otherwise specified.	Estimates are stored on our secure server.
5	Does my IT team need to assist me in installing and running the program?	Yes, depending on the level of your permission. See #11 for system requirements.	While the product is entirely Web-based, you may need IT support with permissions for certain tasks. See #11 for system requirements.
6	Can users collaborate on projects/estimates?	Yes, you can copy project folders between computers.	You can create shared folders that can be viewed, copied, and edited by anyone within your account workgroup.

Comparison matrix continued

7	Can I create and use my own data with the RSMeans database?	Custom lines cannot be added to the CD database. With the Estimator tool, however, you can create and add custom cost lines to any estimate. You can also save custom lines as an Estimator template for use in creating new estimates.	You can create • a custom line within an estimate • a custom cost book consisting of custom unit (not assembly) cost lines, no more than once per quarter Note that custom data is searchable using the Include My Custom Data feature.
8	Can I enter my own trade/labor rates and markups?	Although you cannot adjust the trade labor rates, you can add custom markups for each report template using the Estimator tool.	For each quarterly update, you can create one custom unit cost book. You can edit the labor rates by individual trade and the crew equipment costs, and also edit the associated material and equipment markups.
9	What are my reporting options?	The Estimator tool lets you create editable custom reports and report templates. Reports can also be exported to MS Excel®.	You can generate pre-formatted estimate summaries and cost lists in both PDF and Excel formats.
10	How do I find specific cost data?	You can search by CSI line number and by keyword within a division.	You can search by CSI line number and by keyword within a division. There are several additional search options, including keyword and advanced search using special characters.
11	What are the system requirements?	• Operating system – Windows XP or later • PC requirements - 256 MB RAM (recommended), 1024 x 768 High Color Monitor, 800MHz Computer (recommended), 24X CD-ROM, 250 MB Hard Disk Space	• Resolution – Recommended resolution is 1024 x 768 or higher • Compatibility - Internet Explorer version 8.x, 9.x and Mozilla Firefox version 10 and higher • iPad - iOS 5 and higher

2015 RSMeans Seminar Schedule

Note: call for exact dates and details.

Location	Dates
Seattle, WA	January and August
Dallas/Ft. Worth, TX	January
Austin, TX	February
Anchorage, AK	March and September
Las Vegas, NV	March and October
Washington, DC	April and September
Phoenix, AZ	April
Kansas City, MO	April
Toronto	May
Denver, CO	May
San Francisco, CA	June
Bethesda, MD	June
Columbus, GA	June
El Segundo, CA	August

Location	Dates
Jacksonville FL	September
Dallas, TX	September
Charleston, SC	October
Houston, TX	October
Atlanta, GA	November
Baltimore, MD	November
Orlando, FL	November
San Diego, CA	December
San Antonio, TX	December
Raleigh, NC	December

1-800-334-3509, Press 1

Professional Development

For more information visit the RSMeans website at **www.rsmeans.com**

eLearning Training Sessions

Learn how to use *RSMeans Online®* or *RSMeans CostWorks®* CD from the convenience of your home or office. Our eLearning training sessions let you join a training conference call and share the instructors' desktops, so you can view the presentation and step-by-step instructions on your own computer screen. The live webinars are held from 9 a.m. to 4 p.m. eastern standard time, with a one-hour break for lunch.

For these sessions, you must have a computer with high speed Internet access and a compatible Web browser. Learn more at www.rsmeansonline.com or call for a schedule: 1-800-334-3509 and press 1. Webinars are generally held on selected Wednesdays each month.

RSMeans Online®
$299 per person

RSMeans CostWorks® CD
$299 per person

RSMeans Online™ Training

Construction estimating is vital to the decision-making process at each state of every project. RSMeansOnline works the way you do. It's systematic, flexible and intuitive. In this one day class you will see how you can estimate any phase of any project faster and better.

Some of what you'll learn:

- Customizing RSMeansOnline
- Making the most of RSMeans "Circle Reference" numbers
- How to integrate your cost data
- Generate reports, exporting estimates to MS Excel, sharing, collaborating and more

Also available: RSMeans Online® training webinar

Maintenance & Repair Estimating for Facilities

This *two-day* course teaches attendees how to plan, budget, and estimate the cost of ongoing and preventive maintenance and repair for existing buildings and grounds.

Some of what you'll learn:

- The most financially favorable maintenance, repair, and replacement scheduling and estimating
- Auditing and value engineering facilities
- Preventive planning and facilities upgrading
- Determining both in-house and contract-out service costs
- Annual, asset-protecting M&R plan

Who should attend: facility managers, maintenance supervisors, buildings and grounds superintendents, plant managers, planners, estimators, and others involved in facilities planning and budgeting.

Facilities Construction Estimating

In this *two-day* course, professionals working in facilities management can get help with their daily challenges to establish budgets for all phases of a project.

Some of what you'll learn:

- Determining the full scope of a project
- Identifying the scope of risks and opportunities
- Creative solutions to estimating issues
- Organizing estimates for presentation and discussion
- Special techniques for repair/remodel and maintenance projects
- Negotiating project change orders

Who should attend: facility managers, engineers, contractors, facility tradespeople, planners, and project managers.

Practical Project Management for Construction Professionals

In this *two-day* course, acquire the essential knowledge and develop the skills to effectively and efficiently execute the day-to-day responsibilities of the construction project manager.

Covers:

- General conditions of the construction contract
- Contract modifications: change orders and construction change directives
- Negotiations with subcontractors and vendors
- Effective writing: notification and communications
- Dispute resolution: claims and liens

Who should attend: architects, engineers, owner's representatives, project managers.

Construction Cost Estimating: Concepts and Practice

This *one-day* introductory course to improve estimating skills and effectiveness starts with the details of interpreting bid documents, ending with the summary of the estimate and bid submission.

Some of what you'll learn:

- Using the plans and specifications for creating estimates
- The takeoff process—deriving all tasks with correct quantities
- Developing pricing, using various sources; how subcontractor pricing fits in
- Summarizing the estimate to arrive at the final number
- Formulas for area and cubic measure, adding waste and adjusting productivity to specific projects
- Evaluating subcontractors' proposals and prices
- Adding Insurance and bonds
- Understanding how labor costs are calculated
- Submitting bids and proposals

Who should attend: project managers, architects, engineers, owner's representatives, contractors, and anyone who's responsible for budgeting or estimating construction projects.

Mechanical & Electrical Estimating

This *two-day* course teaches attendees how to prepare more accurate and complete mechanical/electrical estimates, avoiding the pitfalls of omission and double-counting, while understanding the composition and rationale within the RSMeans mechanical/electrical database.

Some of what you'll learn:

- The unique way mechanical and electrical systems are interrelated
- M&E estimates—conceptual, planning, budgeting, and bidding stages
- Order of magnitude, square foot, assemblies, and unit price estimating
- Comparative cost analysis of equipment and design alternatives

Who should attend: architects, engineers, facilities managers, mechanical and electrical contractors, and others who need a highly reliable method for developing, understanding, and evaluating mechanical and electrical contracts.

Unit Price Estimating

This interactive *two-day* seminar teaches attendees how to interpret project information and process it into final, detailed estimates with the greatest accuracy level.

The most important credential an estimator can take to the job is the ability to visualize construction and estimate accurately.

Some of what you'll learn:

- Interpreting the design in terms of cost
- The most detailed, time-tested methodology for accurate pricing
- Key cost drivers—material, labor, equipment, staging, and subcontracts
- Understanding direct and indirect costs for accurate job cost accounting and change order management

Who should attend: corporate and government estimators and purchasers, architects, engineers, and others who need to produce accurate project estimates.

RSMeans CostWorks® CD Training

This one-day course helps users become more familiar with the functionality of *RSMeans CostWorks* program. Each menu, icon, screen, and function found in the program is explained in depth. Time is devoted to hands-on estimating exercises.

Some of what you'll learn:

- Searching the database using all navigation methods
- Exporting RSMeans data to your preferred spreadsheet format
- Viewing crews, assembly components, and much more
- Automatically regionalizing the database

This training session requires you to bring a laptop computer to class.

When you register for this course you will receive an outline for your laptop requirements.

Also offering web training for RSMeans CostWorks CD!

Facilities Estimating Using RSMeans CostWorks® CD

Combines hands-on skill building with best estimating practices and real-life problems. Brings you up-to-date with key concepts, and provides tips, pointers, and guidelines to save time and avoid cost oversights and errors.

Some of what you'll learn:

- Estimating process concepts
- Customizing and adapting RSMeans cost data
- Establishing scope of work to account for all known variables
- Budget estimating: when, why, and how
- Site visits: what to look for—what you can't afford to overlook
- How to estimate repair and remodeling variables

This training session requires you to bring a laptop computer to class.

Who should attend: facility managers, architects, engineers, contractors, facility tradespeople, planners, project managers and anyone involved with JOC, SABRE, or IDIQ.

Conceptual Estimating Using RSMeans CostWorks® CD

This *two-day* class uses the leading industry data and a powerful software package to develop highly accurate conceptual estimates for your construction projects. All attendees must bring a laptop computer loaded with the current year *Square Foot Costs* and the *Assemblies Cost Data* CostWorks titles.

Some of what you'll learn:

- Introduction to conceptual estimating
- Types of conceptual estimates
- Helpful hints
- Order of magnitude estimating
- Square foot estimating
- Assemblies estimating

Who should attend: architects, engineers, contractors, construction estimators, owner's representatives, and anyone looking for an electronic method for performing square foot estimating.

Assessing Scope of Work for Facility Construction Estimating

This *two-day* practical training program addresses the vital importance of understanding the SCOPE of projects in order to produce accurate cost estimates for facility repair and remodeling.

Some of what you'll learn:

- Discussions of site visits, plans/specs, record drawings of facilities, and site-specific lists
- Review of CSI divisions, including means, methods, materials, and the challenges of scoping each topic
- Exercises in SCOPE identification and SCOPE writing for accurate estimating of projects
- Hands-on exercises that require SCOPE, take-off, and pricing

Who should attend: corporate and government estimators, planners, facility managers, and others who need to produce accurate project estimates.

Unit Price Estimating Using RSMeans CostWorks® CD

Step-by-step instruction and practice problems to identify and track key cost drivers—material, labor, equipment, staging, and subcontractors—for each specific task. Learn the most detailed, time-tested methodology for accurately "pricing" these variables, their impact on each other and on total cost.

Some of what you'll learn:

- Unit price cost estimating
- Order of magnitude, square foot, and assemblies estimating
- Quantity takeoff
- Direct and indirect construction costs
- Development of contractor's bill rates
- How to use *RSMeans Building Construction Cost Data*

This training session requires you to bring a laptop computer to class.

Who should attend: architects, engineers, corporate and government estimators, facility managers, and government procurement staff.

Registration Information

Register early . . . Save up to $100! Register 30 days before the start date of a seminar and save $100 off your total fee. *Note: This discount can be applied only once per order. It cannot be applied to team discount registrations or any other special offer.*

How to register Register by phone today! The RSMeans toll-free number for making reservations is **1-800-334-3509, Press 1.**

Two-day seminar registration fee - $935. One-day *RSMeans CostWorks*® training registration fee - $375. To register by mail, complete the registration form and return, with your full fee, to: RSMeans Seminars, 700 Longwater Drive, Norwell, MA 02061.

Government pricing All federal government employees save off the regular seminar price. Other promotional discounts cannot be combined with the government discount.

Team discount program for two to four seminar registrations. Call for pricing: 1-800-334-3509, Press 1

Multiple course discounts When signing up for two or more courses, call for pricing.

Refund policy Cancellations will be accepted up to ten business days prior to the seminar start. There are no refunds for cancellations received later than ten working days prior to the first day of the seminar. A $150 processing fee will be applied for all cancellations. Written notice of cancellation is required. Substitutions can be made at any time before the session starts. **No-shows are subject to the full seminar fee.**

AACE approved courses Many seminars described and offered here have been approved for 14 hours (1.4 recertification credits) of credit by the AACE International Certification Board toward meeting the continuing education requirements for recertification as a Certified Cost Engineer/ Certified Cost Consultant.

AIA Continuing Education We are registered with the AIA Continuing Education System (AIA/CES) and are committed to developing quality learning activities in accordance with the CES criteria. Many seminars meet the AIA/CES criteria for Quality Level 2. AIA members may receive (14) learning units (LUs) for each two-day RSMeans course.

Daily course schedule The first day of each seminar session begins at 8:30 a.m. and ends at 4:30 p.m. The second day begins at 8:00 a.m. and ends at 4:00 p.m. Participants are urged to bring a hand-held calculator, since many actual problems will be worked out in each session.

Continental breakfast Your registration includes the cost of a continental breakfast, and a morning and afternoon refreshment break. These informal segments allow you to discuss topics of mutual interest with other seminar attendees. (You are free to make your own lunch and dinner arrangements.)

Hotel/transportation arrangements RSMeans arranges to hold a block of rooms at most host hotels. To take advantage of special group rates when making your reservation, be sure to mention that you are attending the RSMeans seminar. You are, of course, free to stay at the lodging place of your choice. **(Hotel reservations and transportation arrangements should be made directly by seminar attendees.)**

Important Class sizes are limited, so please register as soon as possible.

Note: *Pricing subject to change.*

Registration Form

ADDS-1000

Call 1-800-334-3509, Press 1 to register or FAX this form 1-800-632-6732. Visit our website: www.rsmeans.com

Please register the following people for the RSMeans construction seminars as shown here. We understand that we must make our own hotel reservations if overnight stays are necessary.

☐ Full payment of $____________________ enclosed.

☐ Bill me.

Please print name of registrant(s).

(To appear on certificate of completion)

____

____

____

____

P.O. #: ______________________________

GOVERNMENT AGENCIES MUST SUPPLY PURCHASE ORDER NUMBER OR TRAINING FORM.

Firm name ______________________________

Address ______________________________

City/State/Zip ______________________________

Telephone no. ____________________ Fax no. ____________________

E-mail address ______________________________

Charge registration(s) to: ☐ MasterCard ☐ VISA ☐ American Express

Account no. ____________________ Exp. date ____________________

Cardholder's signature ______________________________

Seminar name ______________________________

Seminar City ______________________________

Please mail check to: RSMeans Seminars, 700 Longwater Drive, Norwell, MA 02061 USA

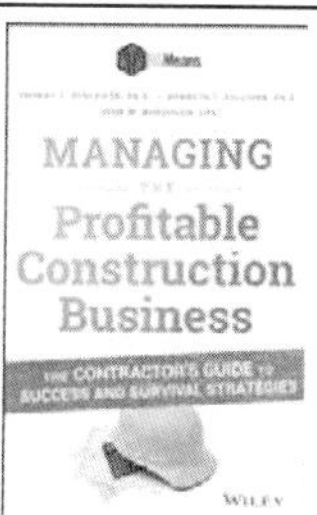

Managing the Profitable Construction Business: The Contractor's Guide to Success and Survival Strategies, 2nd Ed.

by Thomas C. Scheifler, Ph.D.

Learn from a team of construction business veterans led by Thomas C. Schleifer, who is commonly referred to as a construction business "turnaround" expert due to the number of construction companies he has rescued from financial distress. His financial acumen, combined with his practical, hands-on experience, has made him a sought-after private consultant.

$60.00 | 288 pages, hardover | Catalog no. 67370

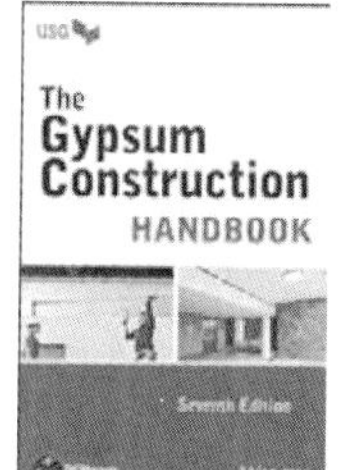

The Gypsum Construction Handbook, 7th Edition

by USG

The tried-and-true Gypsum Construction Handbook is a systematic guide to selecting and using gypsum drywall, veneer plaster, tile backers, ceilings, and conventional plaster building materials. A widely respected training text for aspiring architects and engineers, the book provides detailed product information and efficient installation methodology.

The 7th edition features updates in gypsum products, including ultralight panels, glass-mat panels, paper-faced plastic bead, and ultralightweight joint compound, and modern specialty acoustical and ceiling product guidelines. This comprehensive reference also incorporates the latest in sustainable products.

$40.00 | 576 pages, softcover | Catalog no. 67357B

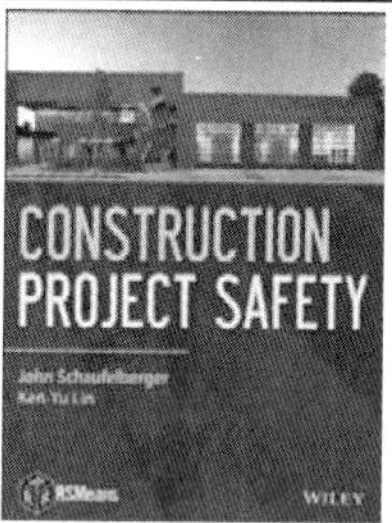

Construction Project Safety

by John Schaufelberger, Ken-Yu Lin

This essential introduction to construction safety for construction management personnel takes a project-based approach to present potential hazards in construction and their mitigation or prevention.

Beginning with an introduction to Accident Prevention Programs and OSHA compliance requirements, the book integrates safety instruction into the building process by following a building project from site construction through interior finish. Reinforcing this applied approach are photographs, drawings, contract documentation, and an online 3D BIM model to help visualize the on-site scenarios.

$85.00 | 320 pages, hardover | Catalog no. 67368

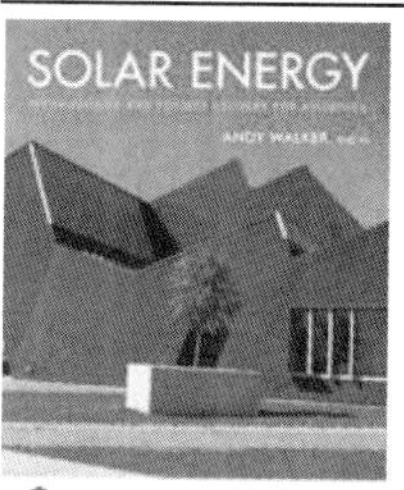

Solar Energy: Technologies & Project Delivery for Buildings

by Andy Walker

An authoritative reference on the design of solar energy systems in building projects, with applications, operating principles, and simple tools for the construction, engineering, and design professionals, the book simplifies the solar design and engineering process and provides sample documentation for the complete design of a solar energy system for buildings.

$85.00 | 320 pages, hardover | Catalog no. 67365

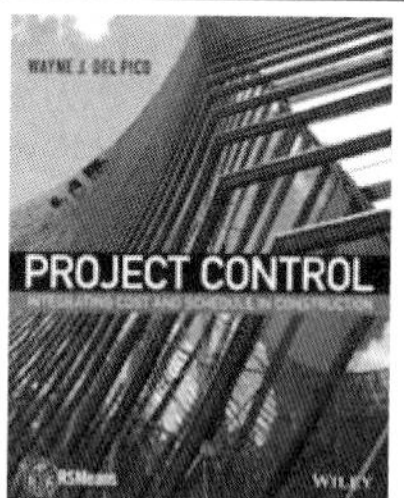

Project Control: Integrating Cost & Schedule in Construction

by Wayne J. Del Pico

Written by a seasoned professional in the field, *Project Control: Integrating Cost and Schedule in Construction* fills a void in the area of project control as applied in the construction industry today. It demonstrates how productivity models for an individual project are created, monitored, and controlled, and how corrective actions are implemented as deviations from the baseline occur.

$65.00 | 240 pages, softcover | Catalog no. 67366

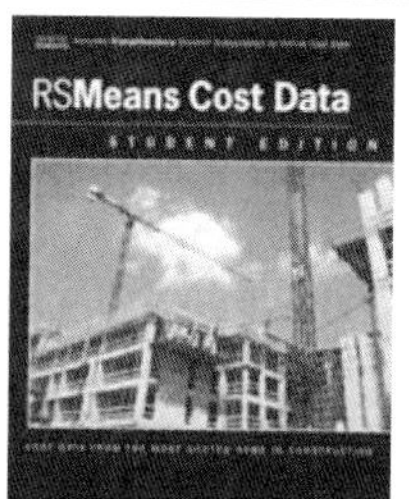

RSMeans Cost Data, Student Edition

Provides a thorough introduction to cost estimating in a print and online package. Clear explanations and example-driven. The ideal reference for students and new professionals. Features include:

- Commercial and residential construction cost data in print and online formats
- Complete how-to guidance on the essentials of cost estimating
- A supplemental website with plans, problem sets, and a full sample estimate

$110.00 | 512 pages, softcover | Catalog no. 67363

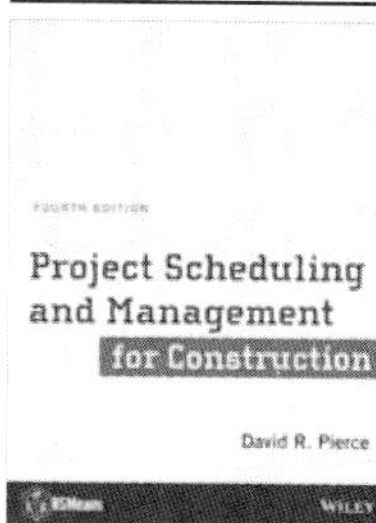

Project Scheduling and Management for Construction, 4th Edition

by David R. Pierce

First published in 1988 by RSMeans, the new edition of *Project Scheduling and Management for Construction* has been substantially revised for both professionals and students enrolled in construction management and civil engineering programs. While retaining its emphasis on developing practical, professional-level scheduling skills, the new edition is a relatable, real world case study.

$95.00 | 272 pages, softcover | Catalog no. 67367

How to Estimate with Means Data & CostWorks, 4th Edition

by RSMeans and Saleh A. Mubarak, Ph.D.

This step-by-step guide takes you through all the major construction items with extensive coverage of site work, concrete and masonry, wood and metal framing, doors and windows, and other divisions. The only construction cost estimating handbook that uses the most popular source of construction data, RSMeans, this indispensible guide features access to the instructional version of *CostWorks* in electronic form, enabling you to practice techniques to solve real-world estimating problems.

$75.00 | 320 pages, softcover | Includes CostWorks CD | Catalog no. 67324C

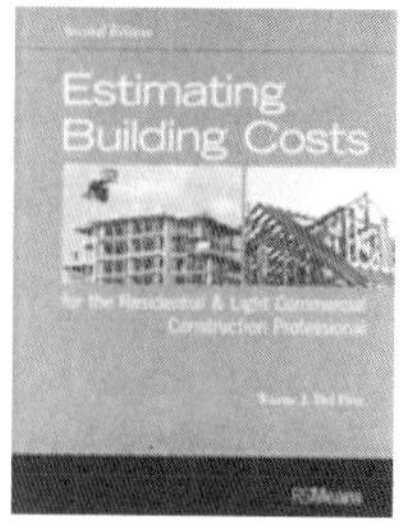

Estimating Building Costs, 2nd Edition

For the Residential & Light Commercial Construction Professional

by Wayne J. Del Pico

Explains, in detail, how to put together a reliable estimate that can be used not only for budgeting, but also for developing a schedule, managing a project, dealing with contingencies, and, ultimately, making a profit.

Completely revised and updated to reflect the CSI MasterFormat 2010 system.

$75.00 | 528 pages, softcover | Catalog no. 67343A

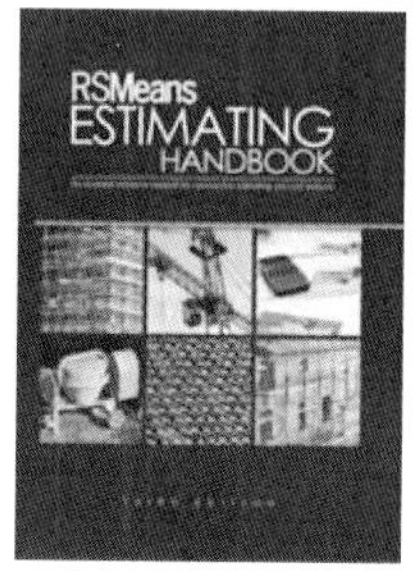

RSMeans Estimating Handbook, 3rd Edition

Widely used in the industry for tasks ranging from routine estimates to special cost analysis projects. This handbook will help construction professionals:

- Evaluate architectural plans and specifications
- Prepare accurate quantity takeoffs
- Compare design alternatives and costs
- Perform value engineering
- Double-check estimates and quotes
- Estimate change orders

$110.00 | Over 976 pages, hardcover | Catalog No. 67276B

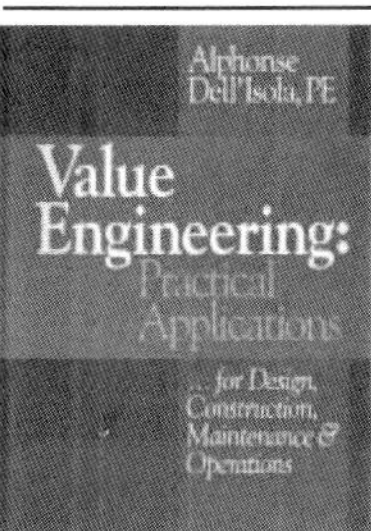

Value Engineering: Practical Applications

For Design, Construction, Maintenance & Operations

by Alphonse Dell'Isola, PE

A tool for immediate application—for engineers, architects, facility managers, owners, and contractors. Includes making the case for VE—the management briefing; integrating VE into planning, budgeting, and design; conducting life cycle costing; using VE methodology in design review and consultant selection; and case studies.

$90.00 | Over 427 pages, illustrated, softcover | Catalog no. 67319A

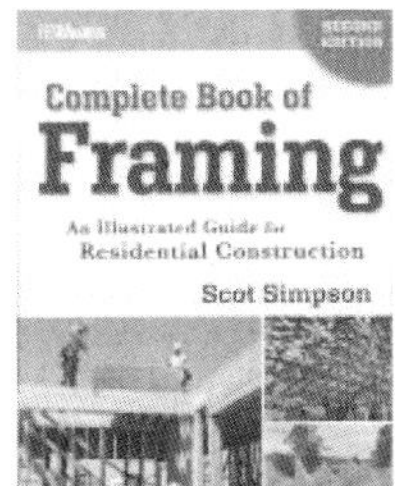

Complete Book of Framing, 2nd Edition

by Scot Simpson

This updated, easy-to-learn guide to rough carpentry and framing is written by an expert with more than thirty years of framing experience. Starting with the basics, this book begins with types of lumber, nails, and what tools are needed, followed by detailed, fully illustrated steps for framing each building element. Framer-Friendly Tips throughout the book show how to get a task done right—and more easily.

$29.95 | 368 pages, softcover | Catalog no. 67353A

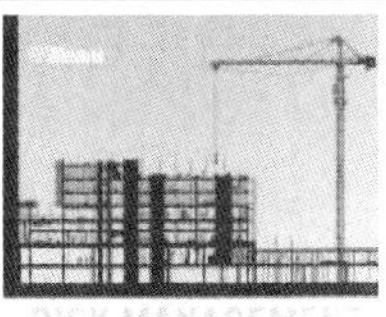

Risk Management for Design and Construction

Introduces risk as a central pillar of project management and shows how a project manager can prepare to deal with uncertainty. Includes:

- Integrated cost and schedule risk analysis
- An introduction to a ready-to-use system of analyzing a project's risks and tools to proactively manage risks
- A methodology that was developed and used by the Washington State Department of Transportation
- Case studies and examples on the proper application of principles
- Combining value analysis with risk analysis

$125.00 | Over 288 pages, softcover | Catalog no. 67359

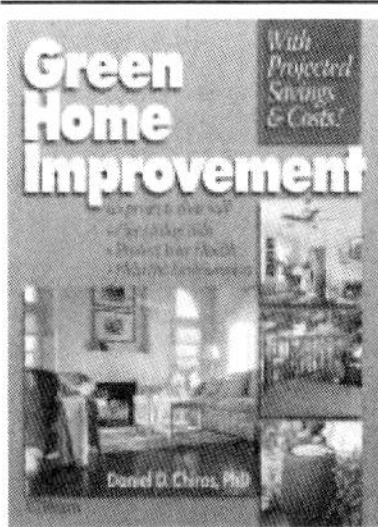

Green Home Improvement

by Daniel D. Chiras, PhD

With energy costs rising and environmental awareness increasing, people are looking to make their homes greener. This book, with 65 projects and actual costs and projected savings, helps homeowners prioritize their green improvements.

Projects range from simple water savers that cost only a few dollars, to bigger-ticket items such as HVAC systems. With color photos and cost estimates, each project compares options and describes the work involved, the benefits, and the savings.

$34.95 | 320 pages, illustrated, softcover | Catalog no. 67355

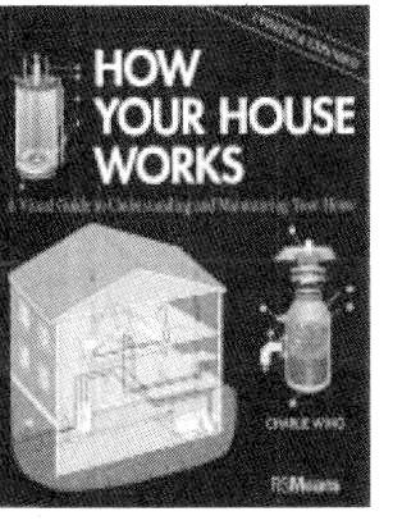

How Your House Works, 2nd Edition

by Charlie Wing

Knowledge of your home's systems helps you control repair and construction costs and makes sure the correct elements are being installed or replaced. This book uncovers the mysteries behind just about every major appliance and building element in your house. See-through, cross-section drawings in full color show you exactly how these things should be put together and how they function, including what to check if they don't work. It just might save you having to call in a professional.

$22.95 | 208 pages, softcover | Catalog no. 67351A

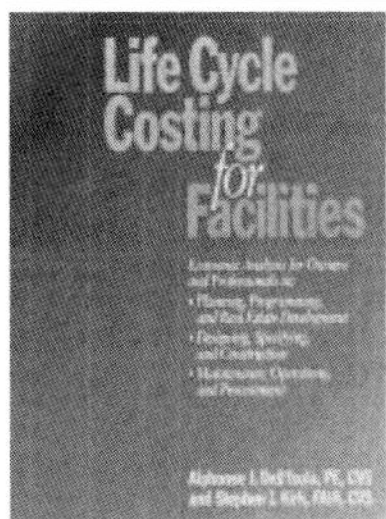

Life Cycle Costing for Facilities

by Alphonse Dell'Isola and Dr. Stephen Kirk

Guidance for achieving higher quality design and construction projects at lower costs! Cost-cutting efforts often sacrifice quality to yield the cheapest product. Life cycle costing enables building designers and owners to achieve both. The authors of this book show how LCC can work for a variety of projects—from roads to HVAC upgrades to different types of buildings.

$110.00 | 396 pages, hardcover | Catalog no. 67341

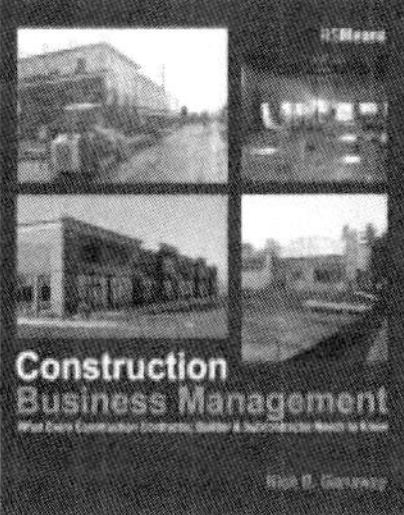

Construction Business Management

by Nick Ganaway

Only 43% of construction firms stay in business after four years. Make sure your company thrives with valuable guidance from a pro with 25 years of success as a commercial contractor. Find out what it takes to build all aspects of a business that is profitable, enjoyable, and enduring. With a bonus chapter on retail construction.

$60.00 | 201 pages, softcover | Catalog no. 67352

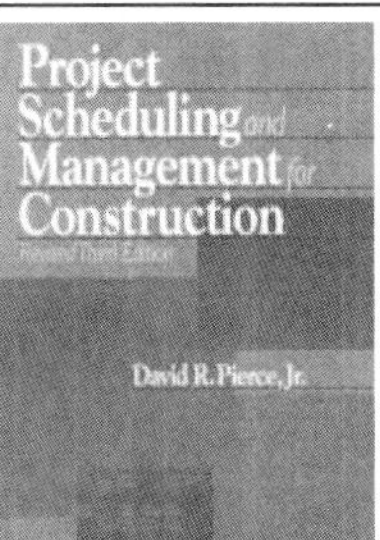

Project Scheduling and Management for Construction, 3rd Edition

by David R. Pierce, Jr.

A comprehensive, yet easy-to-follow guide to construction project scheduling and control—from vital project management principles through the latest scheduling, tracking, and controlling techniques. The author is a leading authority on scheduling, with years of field and teaching experience at leading academic institutions. Spend a few hours with this book and come away with a solid understanding of this essential management topic.

$64.95 | Over 286 pages, illustrated, hardcover | Catalog no. 67247B

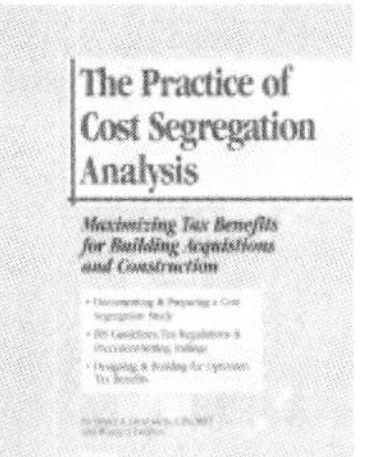

The Practice of Cost Segregation Analysis

by Bruce A. Desrosiers and Wayne J. Del Pico

This expert guide walks you through the practice of cost segregation analysis, which enables property owners to defer taxes and benefit from "accelerated cost recovery" through depreciation deductions on assets that are properly identified and classified.

With a glossary of terms, sample cost segregation estimates for various building types, key information resources, and updates via a dedicated website, this book is a critical resource for anyone involved in cost segregation analysis.

$105.00 | Over 240 pages | Catalog no. 67345

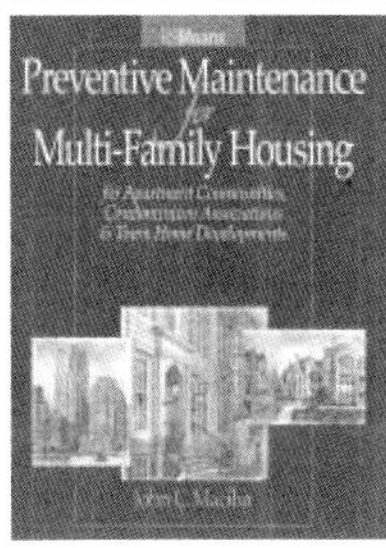

Preventive Maintenance for Multi-Family Housing

by John C. Maciha

Prepared by one of the nation's leading experts on multi-family housing.

This complete PM system for apartment and condominium communities features expert guidance, checklists for buildings and grounds maintenance tasks and their frequencies, and a dedicated website featuring customizable electronic forms. A must-have for anyone involved with multi-family housing maintenance and upkeep.

$95.00 | 290 pages | Catalog no. 67346

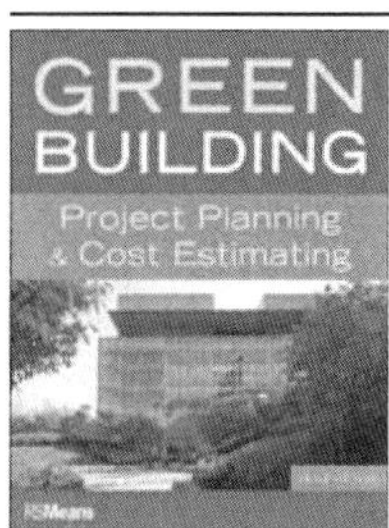

Green Building: Project Planning & Cost Estimating, 3rd Edition

Since the widely read first edition of this book, green building has gone from a growing trend to a major force in design and construction.

This new edition has been updated with the latest in green building technologies, design concepts, standards, and costs. Full-color with all new case studies—plus a new chapter on commercial real estate.

$110.00 | 480 pages, softcover | Catalog no. 67338B

RSMeans Illustrated Construction Dictionary Unabridged 4th Edition, with CD-ROM

Long regarded as the industry's finest, *Means Illustrated Construction Dictionary* is now even better. With nearly 20,000 terms and more than 1,400 illustrations and photos, it is the clear choice for the most comprehensive and current information. The companion CD-ROM that comes with this new edition adds extra features, such as larger graphics and expanded definitions.

$110.00 | Over 880 pages, illust., hardcover | Catalog no. 67292B

Construction Supervision

Inspires excellence with proven tactics and techniques applied by thousands of construction supervisors. Leadership guidelines carve out a practical blueprint for motivating work performance and increasing productivity through effective communication. Features:

- A unique focus on field supervision and crew management
- Coverage of supervision from the foreman to the superintendent level
- An overview of technical skills whose mastery will build confidence and success for the supervisor
- A detailed view of "soft" management and communication skills

$95.00 | 464 pages, softcover | Catalog no. 67358

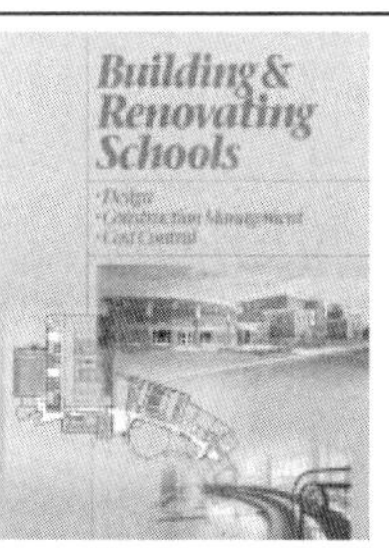

Building & Renovating Schools

This all-inclusive guide covers every step of the school construction process—from initial planning, needs assessment, and design, right through moving into the new facility. A must-have resource for anyone concerned with new school construction or renovation. With square foot cost models for elementary, middle, and high school facilities, and real-life case studies of recently completed school projects.

The contributors to this book—architects, construction project managers, contractors, and estimators who specialize in school construction—provide start-to-finish, expert guidance on the process.

$110.00 | Over 412 pages, hardcover | Catalog no. 67342

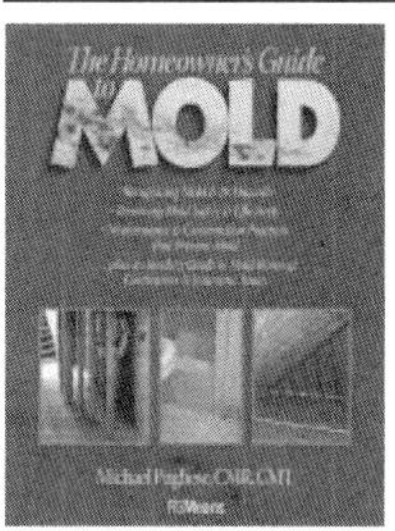

The Homeowner's Guide to Mold

By Michael Pugliese

Mold, whether caused by leaks, humidity or flooding, is a real health and financial issue—for homeowners and contractors. This full-color book explains:

- Construction and maintenance practices to prevent mold
- How to inspect for and remove mold
- Mold remediation procedures and costs
- What to do after a flood
- How to deal with insurance companies

$21.95 | 144 pages, softcover | Catalog no. 67344

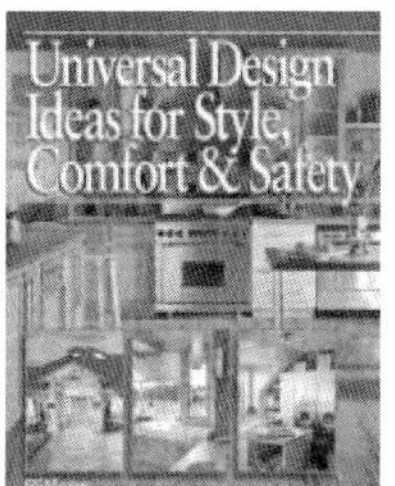

Universal Design Ideas for Style, Comfort & Safety

by RSMeans and Lexicon Consulting, Inc.

Incorporating universal design when building or remodeling helps people of any age and physical ability more fully and safely enjoy their living spaces. This book shows how universal design can be artfully blended into the most attractive homes. It discusses specialized products like adjustable countertops and chair lifts, as well as simple ways to enhance a home's safety and comfort. With color photos and expert guidance, every area of the home is covered. Includes budget estimates that give an idea how much projects will cost.

$21.95 | 160 pages, illustrated, softcover | Catalog no. 67354

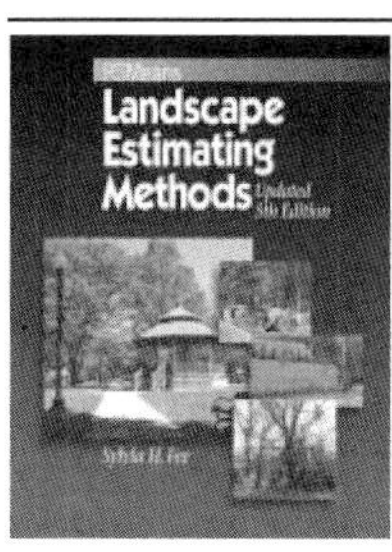

Landscape Estimating Methods, 5th Edition

Answers questions about preparing competitive landscape construction estimates, with up-to-date cost estimates and the new MasterFormat classification system. Expanded and revised to address the latest materials and methods, including new coverage on approaches to green building. Includes:

- Step-by-step explanation of the estimating process
- Sample forms and worksheets that save time and prevent errors

$75.00 | 336 pages, softcover | Catalog no. 67295C

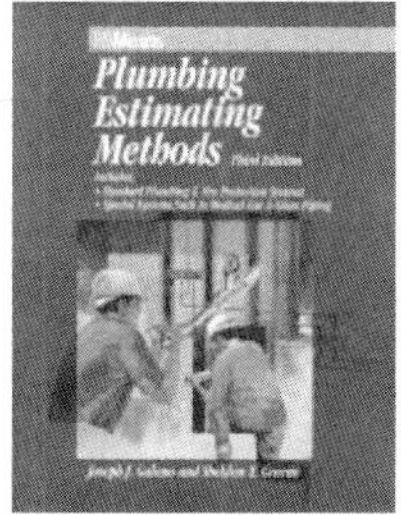

Plumbing Estimating Methods, 3rd Edition

by Joseph J. Galeno and Sheldon T. Greene

Updated and revised! This practical guide walks you through a plumbing estimate, from basic materials and installation methods through change order analysis. *Plumbing Estimating Methods* covers residential, commercial, industrial, and medical systems, and features sample takeoff and estimate forms and detailed illustrations of systems and components.

$65.00 | 381+ pages, softcover | Catalog no. 67283B

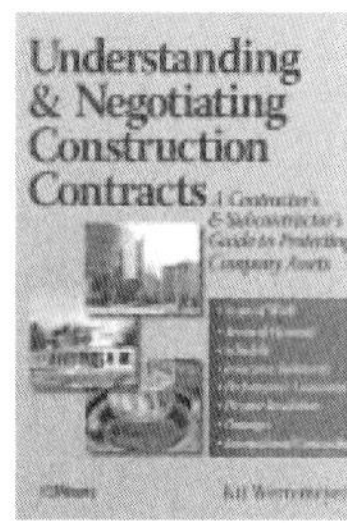

Understanding & Negotiating Construction Contracts

by Kit Werremeyer

Take advantage of the author's 30 years' experience in small-to-large (including international) construction projects. Learn how to identify, understand, and evaluate high risk terms and conditions typically found in all construction contracts—then, negotiate to lower or eliminate the risk, improve terms of payment, and reduce exposure to claims and disputes. The author avoids "legalese" and gives real-life examples from actual projects.

$80.00 | 320 pages, softcover | Catalog no. 67350

2015 Order Form

ORDER TOLL FREE 1-800-334-3509
OR FAX 1-800-632-6732

For more information visit the RSMeans website at **www.rsmeans.com**

Qty.	Book no.	COST ESTIMATING BOOKS	Unit Price	Total
	60065	Assemblies Cost Data 2015	$334.95	
	60015	Building Construction Cost Data 2015	206.95	
	60045	Commercial Renovaion Cost Data 2015	166.95	
	60115	Concrete & Masonry Cost Data 2015	188.95	
	50145	Construction Cost Indexes 2015 (subscription)	359.95	
	60145A	Construction Cost Index–January 2015	94.95	
	60145B	Construction Cost Index–April 2015	94.95	
	60145C	Construction Cost Index–July 2015	94.95	
	60145D	Construction Cost Index–October 2015	94.95	
	60345	Contr. Pricing Guide: Resid. R & R Costs 2015	39.95	
	60235	Electrical Change Order Cost Data 2015	199.95	
	60035	Electrical Cost Data 2015	208.95	
	60205	Facilities Construction Cost Data 2015	524.95	
	60305	Facilities Maintenance & Repair Cost Data 2015	461.95	
	60555	Green Building Cost Data 2015	170.95	
	60165	Heavy Construction Cost Data 2015	204.95	
	60095	Interior Cost Data 2015	208.95	
	60125	Labor Rates for the Const. Industry 2015	450.95	
	60185	Light Commercial Cost Data 2015	153.95	
	60025	Mechanical Cost Data 2015	203.95	
	60155	Open Shop Building Const. Cost Data 2015	175.95	
	60215	Plumbing Cost Data 2015	206.95	
	60175	Residential Cost Data 2015	148.95	
	60285	Site Work & Landscape Cost Data 2015	199.95	
	60055	Square Foot Costs 2015	218.95	
	62014	Yardsticks for Costing (2014)	189.95	
	62015	Yardsticks for Costing (2015)	201.95	
		REFERENCE BOOKS		
	67329	Bldrs Essentials: Best Bus. Practices for Bldrs	45.00	
	67307	Bldrs Essentials: Plan Reading & Takeoff	50.00	
	67342	Building & Renovating Schools	110.00	
	67261A	The Building Prof. Guide to Contract Documents	75.00	
	67353A	Complete Book of Framing, 2nd Ed.	29.95	
	67146	Concrete Repair & Maintenance Illustrated	75.00	
	67352	Construction Business Management	60.00	
	67364	Construction Law	115.00	
	67358	Construction Supervision	95.00	
	67363	Cost Data Student Edition	110.00	
	67314	Cost Planning & Est. for Facil. Maint.	95.00	
	67230B	Electrical Estimating Methods, 3rd Ed.	75.00	
	67343A	Est. Bldg. Costs for Resi. & Lt. Comm., 2nd Ed.	75.00	
	67276B	Estimating Handbook, 3rd Ed.	110.00	
	67318	Facilities Operations & Engineering Reference	115.00	
	67338B	Green Building: Proj. Planning & Cost Est., 3rd Ed.	110.00	
	67355	Green Home Improvement	34.95	
	67357B	The Gypsum Construction Handbook	40.00	
	67148	Heavy Construction Handbook	110.00	
	67308E	Home Improvement Costs–Int. Projects, 9th Ed.	24.95	
	67309E	Home Improvement Costs–Ext. Projects, 9th Ed.	24.95	
	67344	Homeowner's Guide to Mold	21.95	
	67324C	How to Est.w/Means Data & CostWorks, 4th Ed.	75.00	

Qty.	Book no.	REFERENCE BOOKS (Cont.)	Unit Price	Total
	67351A	How Your House Works, 2nd Ed.	22.95	
	67282A	Illustrated Const. Dictionary, Condensed, 2nd Ed.	65.00	
	67292B	Illustrated Const. Dictionary, w/CD-ROM, 4th Ed.	110.00	
	67362	Illustrated Const. Dictionary, Student Ed.	60.00	
	67348	Job Order Contracting	130.00	
	67295C	Landscape Estimating Methods, 5th Ed.	75.00	
	67341	Life Cycle Costing for Facilities	110.00	
	67370	Managing the Profitable Construction Business	60.00	
	67294B	Mechanical Estimating Methods, 4th Ed.	75.00	
	67345	Practice of Cost Segregation Analysis	105.00	
	67366	Project Control: Integrating Cost & Sched.	95.00	
	67346	Preventive Maint. for Multi-Family Housing	95.00	
	67367	Project Sched. and Manag. for Constr., 4th Ed.	95.00	
	67322B	Resi. & Light Commercial Const. Stds., 3rd Ed.	65.00	
	67359	Risk Management for Design and Construction	125.00	
	67365	Solar Energy: Technology & Project Delivery	85.00	
	67145B	Sq. Ft. & Assem. Estimating Methods, 3rd Ed.	75.00	
	67321	Total Productive Facilities Management	85.00	
	67350	Understanding and Negotiating Const. Contracts	80.00	
	67303B	Unit Price Estimating Methods, 4th Ed.	75.00	
	67354	Universal Design	21.95	
	67319A	Value Engineering: Practical Applications	90.00	

MA residents add 6.25% state sales tax	
Shipping & Handling**	
Total (U.S. Funds)*	

Prices are subject to change and are for U.S. delivery only. *Canadian customers may call for current prices. **Shipping & handling charges: Add 8% of total order for check and credit card payments. Add 8% of total order for invoiced orders.

Send order to: ADDV-1000

Name (please print) ______________________

Company ______________________

☐ Company

☐ Home Address ______________________

____

City/State/Zip ______________________

Phone # ______________________ P.O. # ______________________

(Must accompany all orders being billed)

Mail to: RSMeans, 700 Longwater Drive, Norwell, MA 02061

RSMeans Project Cost Report

By filling out this report, your project data will contribute to the database that supports the RSMeans® Project Cost Square Foot Data. When you fill out this form, RSMeans will provide a $50 discount off one of the RSMeans products advertised in the preceding pages. Please complete the form including all items where you have cost data, and all the items marked (☑).

$50.00 Discount per product for each report you submit

Project Description (NEW construction only, please)

☑ Building Use (Office School . . .) ____________________

☑ Address (City, State) ____________________

____

☑ Total Building Area (SF) ____________________

☑ Ground Floor (SF) ____________________

☑ Frame (Wood, Steel . . .) ____________________

☑ Exterior Wall (Brick, Tilt-up . . .) ____________________

☑ Basement: (checkone) ☐ Full ☐ Partial ☐ None

☑ Number of Stories ____________________

☑ Floor-to-Floor Height ____________________

☑ Volume (C.F.) ____________________

% Air Conditioned ____________ Tons ____________

Total Project Cost $ ____________

Owner ____________________

Architect ____________________

General Contractor ____________________

☑ Bid Date ____________________

☑ Typical Bay Size ____________________

☑ Occupant Capacity ____________________

☑ Labor Force: _____ % Union _____ % Non-Union

☑ Project Description (Circle one number in each line.)

1. Economy 2. Average 3. Custom 4. Luxury

1. Square 2. Rectangular 3. Irregular 4. Very Irregular

Comments ____________________

____

A	☑	General Conditions	$
B	☑	Site Work	$
C	☑	Concrete	$
D	☑	Masonry	$
E	☑	Metals	$
F	☑	Wood & Plastics	$
G	☑	Thermal & Moisture Protection	$
GR		Roofing & Flashing	$
H	☑	Doors and Windows	$
J	☑	Finishes	$
JP		Painting & Wall Covering	$

K	☑	Specialties	$
L	☑	Equipment	$
M	☑	Furnishings	$
N	☑	Special Construction	$
P	☑	Conveying Systems	$
Q	☑	Mechanical	$
QP		Plumbing	$
QB		HVAC	$
R	☑	Electrical	$
S	☑	Mech./Elec. Combined	$

Please specify the RSMeans product you wish to receive.
Complete the address information.

Product Name ____________________

Product Number ____________________

Your Name ____________________

Title ____________________

Company ____________________

☐ Company

☐ Home Street Address ____________________

City, State, Zip ____________________

Email Address ____________________

Return by mail or fax 888-492-6770.

Method of Payment:

Credit Card # ____________________

Expiration Date ____________________

Check ____________________

Purchase Order ____________________

Phone Number ____________________
(if you would prefer a sales call)

RSMeans
Square Foot Costs Department
700 Longwater Drive
Norwell, MA 02061
www.RSMeans.com